Radio Communication Handbook

EIGHTH EDITION

Editors
Mike Dennison, G3XDV
Chris Lorek, G4HCL

ISBN 190508608-3

9 781905 086085

Radio Society of Great Britain

Published by the Radio Society of Great Britain, Lambda House, Cranborne Rd, Potters Bar, Herts EN6 3JE.
Tel 0870 904 7373. Web www.rsgb.org

First published 2005.

© Radio Society of Great Britain, 2005. All rights reserved. No part of this publication may be reproduced, stored in a retrieval system, or transmitted, in any form or by any means, electronic, mechanical, photo-copying, recording or otherwise, without the prior written permission or the Radio Society of Great Britain.

Cover design: Jodie Escott, M3TPQ

Production: Mark Allgar, M1MPA

Design and layout: Mike Dennison, G3XDV, Emdee Publishing

Printed in Great Britain by Nuffield Press Ltd of Abingdon, England

Companion CD printed by DBMasters of Faversham, England (www.dbmasters.co.uk)

The opinions expressed in this book are those of the author(s) and are not necessarily those of the Radio Society of Great Britain. Whilst the information presented is believed to be correct, the publishers and their agents cannot accept responsibility for consequences arising from any inaccuracies or omissions.

ISBN 1-905086-08-3   EAN 9781-905086-08-5

For updates, web links and time-sensitive information, see **www.rsgb.org/books/extra/handbook.htm**

Acknowledgements

The principal contributors to this book were:

Chapter 1:	Principles	Alan Betts, G0HIQ
Chapter 2:	Passive components	'Phosphor'
		Stuart Swain, G0FYX
Chapter 3:	Semiconductors and valves	Alan Betts, G0HIQ
		Fred Ruddell, GI4MWA
Chapter 4:	Building blocks 1: Oscillators	Peter Goodson, G4PCF
Chapter 5:	Building blocks 2: Amplifiers, mixers etc	Peter Goodson, G4PCF
Chapter 6:	HF receivers	Hans Summers, G0UPL
Chapter 7:	HF transmitters and transceivers	Hans Summers, G0UPL
		Peter Hart, G3SJX
Chapter 8:	PIC-A-STAR: A software transmitter and receiver	Peter Rhodes, G3XJP
Chapter 9:	VHF/UHF receivers, transmitters and transceivers	Andy Barter, G8ATD
		Chris Lorek, G4HCL
Chapter 10:	LF: The 136kHz band	Jim Moritz, M0BMU
Chapter 11:	Practical microwave receivers and transmitters	Andy Barter, G8ATD
Chapter 12:	Propagation	Martin Harrison, G3USF
		Gwyn Williams, G4FKH
		Alan Melia, G3NYK
Chapter 13:	Antenna basics and construction	Peter Dodd, G3LDO
Chapter 14:	Transmission lines	Peter Dodd, G3LDO
Chapter 15:	Practical HF antennas	Peter Dodd, G3LDO
Chapter 16:	Practical VHF/UHF antennas	Peter Swallow, G8EZE
Chapter 17:	Practical microwave antennas	Andy Barter, G8ATD
Chapter 18:	Morse code	Dave Lawley, G4BUO
Chapter 19:	Data communications	Murray Greenman, ZL1BPU
		Chris Lorek, G4HCL
Chapter 20:	Imaging techniques	Brian Kelly, GW6BWX
Chapter 21:	Satellites and space	John Heath, G7HIA
Chapter 22:	Computers in the shack	Mike Dennison, G3XDV
Chapter 23:	Electromagnetic compatibility	Robin Page-Jones, G3JWI
Chapter 24:	Power supplies	'Phosphor'
Chapter 25:	Measurement and test equipment	Clive Smith, GM4FZH
Chapter 26:	Construction and workshop practice	Terry Kirk, G3OMK
		David Mackenzie, GM4HJQ

Thanks also go to Paul Hubbard of SpaceMatters, David Bowman G0MRF, the contributors to previous editions of this book, the authors of the published *RadCom* articles which provided source material, the *RadCom* editorial staff and the members of the RSGB LF Group.

Contents

Note: Many chapters have references to the *RSGB Bulletin*, *Radio Communication* or *RadCom*. These are historic names of the RSGB members' monthly journal. The magazines are available on a series of CD-ROMs from: RSGB, Lambda House, Cranborne Road, Potters Bar, Herts, EN6 3JE (www.rsgb.org)

Preface

This, 8th, edition of the *Radio Communication Handbook* has been extensively revised and updated. Some chapters continue to incorporate the tried and tested designs from previous editions, others have been completely re-written, whilst the rest are a combination of old and new.

The Building Blocks chapter has been split into two: the first bringing together the information on oscillators from various parts of the previous edition; the second covering everything else. Antenna basics, and transmission lines, whether for HF or VHF/UHF now have their own chapters, with practical antenna designs having their own frequency-related chapters. Also, the old Microwave section has been split into radios and antennas. The valves chapter is now combined with semiconductors.

The use of several new authors has led to many chapters being completely re-written. This includes those covering antennas, 136kHz, VHF/UHF, microwaves, Morse and digital communications.

The main editorial change between the 7th edition and this one is to recognise that computers play a major part in most radio amateurs' workshops and shacks. The world wide web is a much more stable publishing environment nowadays, so web addresses are frequently quoted in the References sections as sources of further information. A new chapter describes the software and computer-based information resources that are now available to assist the constructor, and several chapters include the use of computer programs in the design and analysis of circuits and antennas.

One new chapter brings together all twenty parts of 'Pic-a-Star', an innovative *RadCom* series describing a transmitter and receiver, controlled by software and using modern construction techniques.

The only area that has been dropped from this edition is Operating. If you need information on this topic, please note that the *RSGB Amateur Radio Operating Manual*, which was completely revised in 2004, covers all aspects of amateur radio operating in detail.

Once again, the *Radio Communication Handbook* combines the knowledge and experience of many experts. The book deserves a place on the shelf of everyone who designs or builds radio equipment for any frequency from LF to microwaves, or who just wants to understand more about what goes on inside their radio equipment.

Finally, I am indebted to Technical Editor Eur Ing Chris Lorek, B Sc (Hons), C Eng, G4HCL, for his valuable contributions, comments and proof-reading.

Mike Dennison, G3XDV

1 Principles

A good understanding of the basic principles and physics of matter, electronics and radio communication is essential if the self-training implicit in amateur radio is to be realised. These principles are not particularly difficult and a good grasp will allow the reader to understand the following material rather than simply accepting that it is true but not really knowing why. This will, in turn, make more aspects of the hobby both attractive and enjoyable.

STRUCTURE OF MATTER

All matter is made up of atoms and molecules. A molecule is the smallest quantity of a substance that can exist and still display the physical and chemical properties of that substance. There is a very great number of different sorts of molecule. Each molecule is, in turn, made up of a number of atoms. There are about 102 different types of atom which are the basic elements of matter. Two atoms of hydrogen will bond with one atom of oxygen to form a molecule of water for example. The chemical symbol is H_2O. The H stands for hydrogen and the subscript 2 indicates that two atoms are required; the O denotes the oxygen atom.

A more complex substance is H_2SO_4. Two hydrogen atoms, one sulphur atom and four oxygen atoms form a molecule of sulphuric acid, a rather nasty and corrosive substance used in lead-acid batteries.

Atoms are so small that they cannot be seen even under the most powerful optical microscopes. They can, however, be visualised using electronic (not electron) microscopes such as the scanning tunnelling microscope (STM) and the atomic force microscope (AFM). **Fig 1.1** shows an AFM representation of the surface of a near-perfect crystal of graphite with the carbon atoms in a hexagonal lattice.

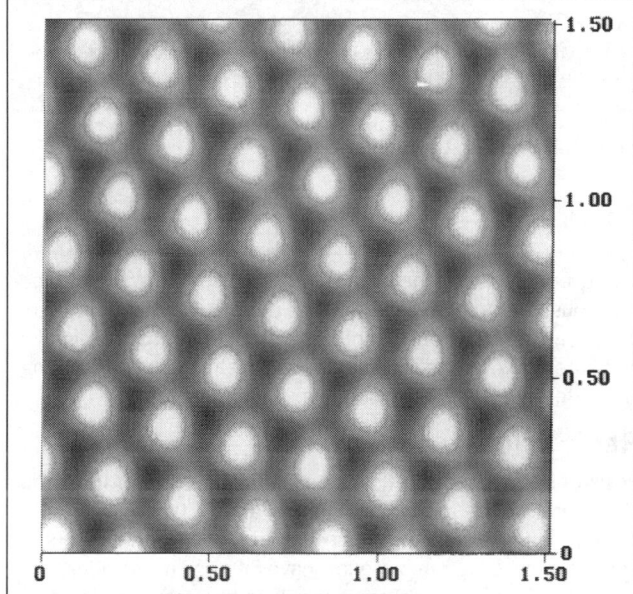

Fig 1.1: Image of the atoms in a piece of high-purity graphite (distances in nanometres). The magnification is approximately 45 million times. Note that no optical microscope can produce more than about 1000 times magnification

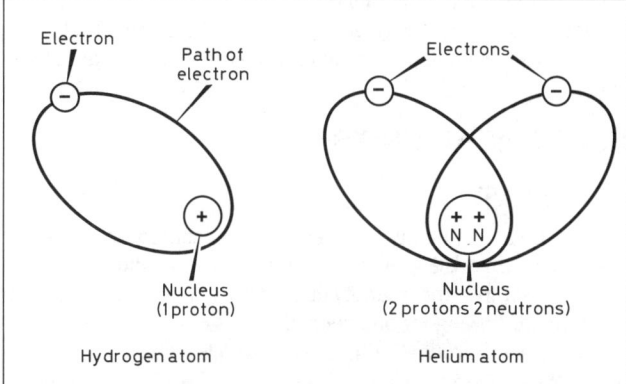

Fig 1.2: Structure of hydrogen and helium atoms

Atoms are themselves made up of yet smaller particles; the electron, the proton and the neutron, long believed to be the smallest things that could exist. Modern atomic physics has shown that this is not so and that not only are there smaller particles but that these particles, energy and waves are, in certain scenarios, indistinguishable from each other. Fortunately we need only concern ourselves with particles down to the electron level but there are effects, such as in the tunnel diode, where the electron seems to 'disappear' and 'reappear' on the far side of a barrier.

The core of an atom comprises one or more protons and may include a number of neutrons. The electrons orbit the core, or 'nucleus' as it is called, rather like the planets orbit the sun. Electrons have an electrical charge that we now know to be a negative charge. Protons have an equal positive charge. The neutrons are not charged.

A hydrogen atom has a single proton and a single orbiting electron. A helium atom has a nucleus of two protons and two neutrons with two orbiting electrons; this is shown in **Fig 1.2**. An atom is electrically neutral; the positive charge on the nucleus is balanced by the negative charge on the electrons. The magnitude of the charge is tiny; it would require 6,250,000,000,000,000,000 (6.25×10^{18}) electrons to produce a charge of 1 coulomb, that is 1A flowing for 1s.

Conductors and Insulators

The ease with which the electrons in a substance can be detached from their parent atoms varies from substance to substance. In some substances there is a continual movement of electrons in a random manner from one atom to another and the application of an electrical force or potential difference (for example from a battery) to the two ends of a piece of wire made of such a substance will cause a drift of electrons along the wire called an electric current; electrical conduction is then said to take place. It should be noted that if an electron enters the wire from the battery at one end it will be a different electron which immediately leaves the other end of the wire.

To visualise this, consider a long tube such as a scaffold pole filled with snooker balls. As soon as another ball is pushed in one end, one falls out the other but the progress of any particular ball is much slower. The actual progress of an individual electron along a wire, the drift velocity, is such that it could take some minutes to move even a millimetre.

The flow of current is from a point of positive charge to negative. Historically, the decision of what represented 'positive' was arbitrary and it turns out that, by this convention, electrons have a negative charge and the movement of electrons is in the opposite direction to conventional current flow.

Materials that exhibit this property of electrical conduction are called conductors. All metals belong to this class. Materials that do not conduct electricity are called insulators, and **Table 1.1** shows a few examples of commonly used conductors and insulators.

ELECTRICAL UNITS

Charge (q)

Charge is the quantity of electricity measured in units of coulombs. **Table 1.2** gives the units and their symbol.

One coulomb is the quantity of electricity given by a current of one ampere flowing for one second.

Charge q = current (A) x time (s), normally written as: q=I x t

Current Flow, the Ampere (A)

The Ampere, usually called Amp, is a fundamental (or base) unit in the SI (System International) system of units. It is actually defined in terms of the magnetic force on two parallel conductors each carrying 1A.

Energy (J)

Energy is the ability to do work and is measured in joules. One joule is the energy required to move a force of one Newton through a distance of one metre. As an example, in lifting a 1kg bag of sugar 1m from the floor to a table, the work done or energy transferred is 9·81 joules.

Power (W)

Power is simply the rate at which work is done or energy is transferred and is measured in watts, W.

$$\text{Power} = \frac{\text{energy transferred}}{\text{time taken}}$$

For example, if the bag of sugar was lifted in two seconds, the power would be 9·81/2 or approximately 5W.

Electromotive Force, EMF (V)

A battery is a source of energy. If one coulomb of charge in a battery contains one joule of energy, the battery has an electromotive force (emf) of one volt. If one coulomb of charge flows in, for example, a bulb, and 12 joules of energy are transferred into heat and light, the EMF is 12 volts.

This leads to the definition of the volt as the number of joules of energy per coulomb of electricity.

Conductors	Insulators
Silver	Mica
Copper	Quartz
Gold	Glass
Aluminium	Ceramics
Brass	Ebonite
Steel	Plastics
Mercury	Air and other gasses
Carbon	Oil
Solutions of salts or acids in water	Pure water

Table 1.1: Examples of conducting and insulating materials

Quantity	Symbol	Unit	Abbreviation
Charge	q	coulomb	C
Current	I	ampere (amp)	A
Voltage*	E or V	volt	V
Time	t	second	s or sec
Resistance	R	ohm	Ω
Capacitance	C	farad	F
Inductance	L	henry	H
Mutual inductance	M	henry	H
Power	P	watt	W
Frequency	f	hertz	Hz
Wavelength	l	metre	m

* 'Voltage' includes 'electromotive force' and 'potential difference'.

Since the above units are sometimes much too large (eg the farad) and sometimes too small, a series of multiples and sub-multiples are used:

Unit	Symbol	Multiple
Microamp	μA	1 millionth (10^{-6}) amp
Milliamp	mA	1 thousandth (10^{-3}) amp
Microvolt	μV	10^{-6}V
Millivolt	mV	10^{-3}V
Kilovolt	kV	10^{3}V
Picofarad	pF	10^{-12}F
Nanofarad	nF	10^{-9}F
Microfarad	μF	10^{-6}F
Femtosecond	fs	10^{-15}s
Picosecond	ps	10^{-12}s
Microsecond	μs	10^{-6}s
Millisecond	ms	10^{-3}s
Microwatt	μW	10^{-6}W
Milliwatt	mW	10^{-3}W
Kilowatt	kW	10^{3}W
Gigahertz	GHz	10^{9}Hz
Megahertz	MHz	10^{6}Hz
Kilohertz	kHz	10^{3}Hz
Centimetre	cm	10^{-2}m
Kilometre	km	10^{3}m

Note: The sub-multiples abbreviate to lower case letters. All multiples or sub-multiples are in factors of a thousand except for the centimetre.

Table 1.2: Units and symbols

$$1 \text{ volt} = \frac{1 \text{ joule}}{1 \text{ coulomb}}$$

In everyday usage the voltage is thought of as a unit of electrical push or force, a higher voltage denotes a greater force trying to cause a flow of current. Potential difference (PD) is also measured in volts and, in everyday use is almost the same thing. The differences are discussed later in the chapter.

Resistance

Resistance restricts the flow of charge, the current. In forcing electrons through a conductor, some energy is lost as heat. A longer, thinner conductor will have a greater loss, that is a higher resistance. Different materials have differing resistivities, that is, a wire of the same dimensions will have different resistances depending on the material. The conductors in the list in Table 1.1 are in conductivity (inverse of resistivity) order.

Materials such a Nichrome, Manganin and Eureka are alloys with a deliberately high resistivity and are used in power resis-

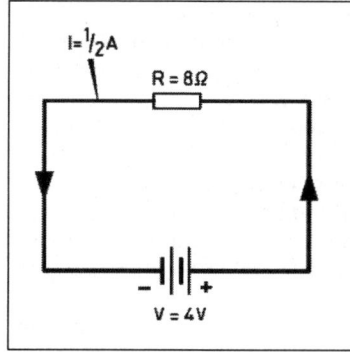

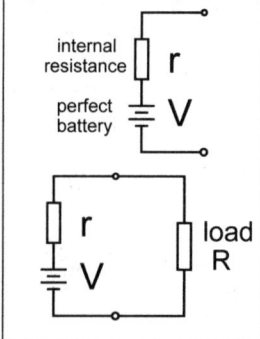

Fig 1.3: Application of Ohm's Law   **Fig 1.4: A real battery**

tors and wire-wound variable resistors. Tungsten has a relatively high resistance but its key property is a high melting point and relative strength when white hot. It is used to make the filament of incandescent light bulbs.

Specific resistance: This is simply the resistance of a standard size piece of the subject material, normally a 1 metre cube and is quoted in units of ohm metre, Ωm and has the symbol ρ, the Greek letter rho. Its purpose is to compare the resistivity of different materials.

The unit of resistance is the ohm, symbol Ω, the Greek upper case letter omega. It is defined as the ratio of the applied EMF and the resulting current.

$$\text{Resistance R ohms} = \frac{\text{applied EMF}}{\text{current flowing}}$$

Ohm's Law

Ohm's Law is simply a restatement of the definition of resistance. In words it reads:

In a circuit at constant temperature the current flowing is directly proportional to the applied voltage and inversely proportional to the resistance. The reference to temperature is important. In a practical test, the energy will be converted to heat and most materials change their resistance as the temperature changes.

In algebraic form, where V is the applied voltage or potential difference, I is the current flowing and R is the resistance:

$$V = I \times R \qquad I = \frac{V}{R} \quad \text{and} \quad R = \frac{V}{I}$$

Example.

Consider the circuit shown in **Fig 1.3** which consists of a 4V battery and a resistance R of 8Ω. What is the magnitude of the current in the circuit?

Here V = 4V and R = 8Ω. Let I be the current flowing in amperes. Then from Ohm's Law:

$$I = \frac{V}{R} = \frac{4}{8} = \frac{1}{2} = 0 \cdot 5A$$

It should be noted that in all calculations based on Ohm's Law care must be taken to ensure that V, I and R are in consistent units, ie in amperes, volts and ohms respectively, if errors in the result are to be avoided. In reality , typical currents may be in mA or μA and resistances in kΩ or MΩ.

Conductance

Conductance is simply the inverse or reciprocal of resistance. Many years ago it was measured in a unit called the mho; today the correct unit is the siemen. A resistance of 10 ohms is the same as a conductance of 0·1 siemen.

EMF, PD and Source Resistance

Sources of electrical energy, such as batteries, hold a limited amount of energy and there is also a limit to the rate at which this energy can be drawn, that is the power is limited. A battery or power supply can be considered as a perfect source (which can supply any desired current) together with a series resistance. The series resistance will limit the total current and will also result in a drop in the voltage or potential difference at the terminals. This is shown in **Fig 1.4**.

The voltage source V has a voltage equal to the open circuit terminal voltage and is the electromotive force (EMF) of the device. On load, that is when a current is being drawn, the potential difference at the terminals will drop according to the current drawn. The drop may be calculated as the voltage across the internal resistor 'r' using the formula:

Voltage drop = current drawn (I) x rΩ and the terminal voltage will be

$$V_{Terminal} = V_{Supply} - I \times r$$

Maximum Power Transfer

It is interesting to consider what is the maximum power that can be drawn from a particular source. As the load resistance decreases, the current drawn increases but the potential difference across the load decreases. Since the power is the product of the load current and the terminal voltage, there will be a maximum point and attempts to draw more power will be thwarted by the drop in terminal voltage.

Fig 1.5 shows the power in the load as the load resistance is varied from 0 to 10Ω, connected to a source of EMF, 10V and internal resistance 2Ω.

Maximum power transfer occurs when the load resistance is the same as the source resistance. For DC and power circuits this is never done since the efficiency drops to 50% but in RF and low level signal handling, maximum signal power transfer may be a key requirement.

Sources of Electricity

When two dissimilar metals are immersed in certain chemical solutions, or electrolytes, an electromotive force (EMF or voltage) is created by chemical action within the cell so that if these pieces of metal are joined externally, there will be a continuous flow of electric current. This device is called a simple cell and such a cell, comprising copper and zinc rods immersed in diluted sulphuric acid, is shown in **Fig 1.6(a)**. The flow of current is from the copper to the zinc plate in the external circuit; ie the copper forms the positive (+) terminal of the cell and the zinc forms the negative (-) terminal.

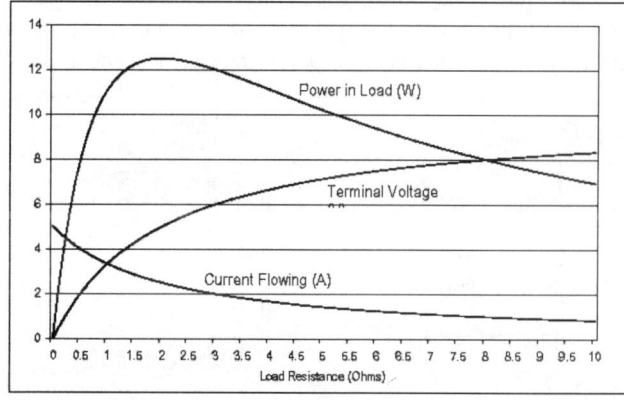

Fig 1.5: Power dissipated in the load

In a simple cell of this type hydrogen forms on the copper electrode, and this gas film has the effect of increasing the internal resistance of the cell and also setting up within it a counter or polarising EMF which rapidly reduces the effective EMF of the cell as a whole.

This polarisation effect is overcome in practical cells by the introduction of chemical agents (depolarisers) surrounding the anode for the purpose of removing the hydrogen by oxidation as soon as it is formed.

Primary Cells

Practical cells in which electricity is produced in this way by direct chemical action are called primary cells; a common example is the Leclanché cell, the construction of which is shown diagrammatically in **Fig 1.6(b)**. The zinc case is the negative electrode and a carbon rod is the positive electrode. The black paste surrounding the carbon rod may contain powdered carbon, manganese dioxide, zinc chloride, ammonium chloride and water, the manganese dioxide acting as depolariser by combining with hydrogen formed at the anode to produce manganese oxide and water. The remainder of the cell is filled with a white paste which may contain plaster of Paris, flour, zinc chloride, ammonium chloride and water. The cell is sealed with a soft plastic insulator except for a small vent which allows accumulated gas to escape. It is often surrounded by a further tin-plate outer case. This is said to be 'leakproof' but is not so over a prolonged period. Therefore cells should always be removed from equipment when they are exhausted or when the latter is going to be laid up or unused for some time.

The EMF developed by a single dry cell is about 1·5V and cells may be connected in series (positive terminal to negative terminal and so on) until a battery of cells, usually referred to simply as a battery, of the desired voltage is obtained. The symbol used to denote a cell in a circuit diagram is shown in **Fig 1.6(c)**. The long thin stroke represents the positive terminal and the short thick stroke the negative terminal. Several cells joined in series to form a battery are shown; for higher voltages it becomes impracticable to draw all the individual cells involved and it is sufficient to indicate merely the first and last cells with a dotted line between them with perhaps a note added to state the actual voltage. The amount of current which can be derived from a dry cell depends on its size and the life required, and may range from a few milliamperes to an amp or two.

Secondary Cells

In primary cells some of the various chemicals are used up in producing the electrical energy - a relatively expensive and wasteful process. The maximum current available also is limited. Another type of cell called a secondary cell or accumulator offers the advantage of being able to provide a higher current and is capable of being charged by feeding electrical energy into the cell to be stored chemically, and be drawn out or discharged later as electrical energy again. This process of charging and discharging the cell is capable of repetition for a large number of cycles depending on the chemistry of cell.

A common type of secondary cell is the lead-acid cell such as that used in vehicle batteries. Vehicle batteries are of limited use for amateur radio for two reasons. Firstly they are liable to leak acid if tipped and give off hydrogen (explosive in confined spaces) and secondly, they are designed to float charge and start vehicle engines with very high current surges rather than undergo deep discharge.

Sealed or 'maintenance free' types are available but they must still be used the correct way up or leakage will occur as gas is generated and the pressure increases.

Deep cycle batteries (sometimes called leisure batteries for caravans) are available that are designed to be fully charge/discharge cycled but are considerably more expensive. These batteries are often available at amateur rallies. It is necessary to check which type they are and their origin. Those removed from alarm systems or uninterruptible power supplies (UPS) are likely to have been changed at the five-year maintenance review. They may well have a couple of years' service left and perhaps considerably more. It is a risk but can be a cheap way of obtaining otherwise expensive batteries.

Nickel-cadmium batteries (NiCd) are now ubiquitous and Nickel-metal-hydride (NiMh) and common. These don't have quite the energy capacity per unit volume (energy density) but are much easier to handle. Lithium-ion batteries are still relatively expensive and offer a better energy density. All these types can be damaged by misuse, especially overcharging or the use of the wrong type of charger. As always, check with the manufacturer.

The Power Supplies chapter discusses batteries in more detail.

Mechanical Generators

Mechanical energy may be converted into electrical energy by moving a coil of wire in a magnetic field. Direct-current or alternating-current generators are available in all sizes but the commonest types likely to be met in amateur radio work are:

(a) Petrol-driven AC generators of up to 1 or 2kW output such as are used for supplying portable equipment; and

(b) Small motor generators, sometimes called dynamotors or rotary converters, which furnish up to about 100W of power and comprise a combined low-voltage DC electric motor and a high voltage AC or DC generator. These have two origins; ex military devices providing high voltage DC for use in valve transmitter/receiver equipments and those supplied for use in mobile caravans to provide mains voltage AC from the 12V DC battery. The latter function is now normally achieved by wholly electronic means but the waveform can be a compromise.

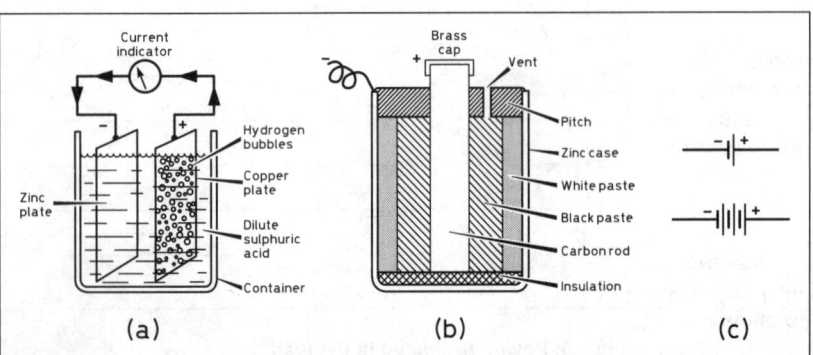

(a) (b) (c)

Fig 1.6: The electric cell. A number of cells connected in series is called a battery. (a) A simple electric cell consisting of copper and zinc electrodes immersed in dilute sulphuric acid. (b) Sectional drawing showing construction of a dry cell. (c) Symbol used to represent single cells and batteries in circuit diagram

ELECTRICAL POWER

When a current flows through a resistor, eg in an electric fire, the resistor gets hot and electrical energy is turned into heat. The actual rise in temperature depends on the amount of power dissipated in the resistor and its size and shape. In most circuits the power dissipated is insignificant but it is a factor the designer must consider, both in fitting a resistor of adequate dissipation and the effect of the heat generated on nearby devices.

The unit of electrical power is the watt (W). The amount of power dissipated in a resistor is equal to the product of the potential difference across the resistor and the current flowing in it. Thus:

Power (watts) = Voltage (volts) x Current (amperes)

$$W = V \times I$$

Ohm's Law states: $V = I \times R$ and $I = V/R$

Substituting V or I in the formula for power gives two further formulas:

$$W = I^2 R$$

and

$$W = \frac{V^2}{R}$$

These formulas are useful for finding, for example, the power input to a transmitter or the power dissipated in various resistors in an amplifier so that suitably rated resistors can be selected. To take a practical case, consider again the circuit of Fig 1.3. The power dissipated in the resistor may be calculated as follows:

Here V = 4V and R = 8Ω. so the correct formula to use is

$$W = \frac{V^2}{R}$$

Inserting the numbers gives

$$W = \frac{4^2}{8} = \frac{16}{8} = 2W$$

It must be stressed again that the beginner should always see that all values are expressed in terms of volts, amperes and ohms in this type of calculation. The careless use of megohms or milliamperes, for example, may lead to an answer several orders too large or too small.

RESISTORS

Resistors used in Radio Equipment

As already mentioned, a resistor through which a current is flowing may get hot. It follows therefore that in a piece of radio equipment the resistors of various types and sizes that are needed must be capable of dissipating the power as required without overheating.

Generally speaking, radio resistors can be divided roughly into

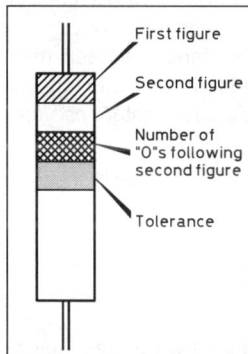

Fig 1.7: Standard resistance value markings

Colour	Value (numbers)	Value (multiplier)	Value (tolerance)
Black	0	1	-
Brown	1	10	1%
Red	2	100	2%
Orange	3	1000 (103)	-
Yellow	4	104	-
Green	5	105	-
Blue	6	106	-
Violet	7	-	-
Grey	8	-	-
White	9	-	-
Silver	-	0.01	10%
Gold	-	0.1	5%
No Colour	-	-	20%

Note: A pink band may be used to denote a 'high-stability' resistor. Sometimes an extra band is added to give two figures before the multiplier.

Table 1.3: Resistor colour code

two classes, (a) low power up to 3W and (b) above 3W. The low-power resistors are usually made of carbon or metal film and may be obtained in a wide range of resistance values from about 10Ω to 10MΩ and in power ratings of 0·1W to 3W. For higher powers, resistors are usually wire-wound on ceramic formers and the very fine wire is protected by a vitreous enamel coating. Typical resistors are shown in the Passive Components chapter.

Resistors, particularly the small carbon types, are usually colour-coded to indicate the value of the resistance in ohms and sometimes also the tolerance or accuracy of the resistance. The standard colour code is shown in **Table 1.3**.

The colours are applied as bands at one end of the resistor as shown in **Fig 1.7** As an example, what would be the value of a resistor with the following colour bands: yellow, violet, orange, silver?

The yellow first band signifies that the first figure is 4, the violet second band signifies that the second figure is 7, while the orange third band signifies that there are three zeros to follow; the silver fourth band indicates a tolerance of ±10%. The value of the resistor is therefore 47,000Ω ±10% (47kΩ ±10%).

So far only fixed resistors have been mentioned. Variable resistors, sometimes called potentiometers or volume controls, are also used. The latter are usually panel-mounted by means of a threaded bush through which a quarter inch or 6mm diameter spindle protrudes and to which the control knob is fitted. Low-power high-value variable resistances use a carbon resistance element, and high-power lower-resistance types (up to 10,000Ω) use a wire-wound element. These are not colour coded but the value is printed on the body either directly, eg '4k7', or as '472' meaning '47' followed by two noughts.

Resistors in Series and Parallel

Resistors may be joined in series or parallel to obtain a specific value of resistance. Some caution may be called for since the tolerance will affect the actual value and close tolerance resistors may be needed if the value is critical such as a divider used in a measuring circuit.

When in series, resistors are connected as shown in **Fig 1.8(a)** and the total resistance is equal to the sum of the separate resistances. The parallel connection is shown in **Fig 1.8(b)**, and with this arrangement the reciprocal of the total resistance is equal to the sum of the reciprocals of the separate resistances.

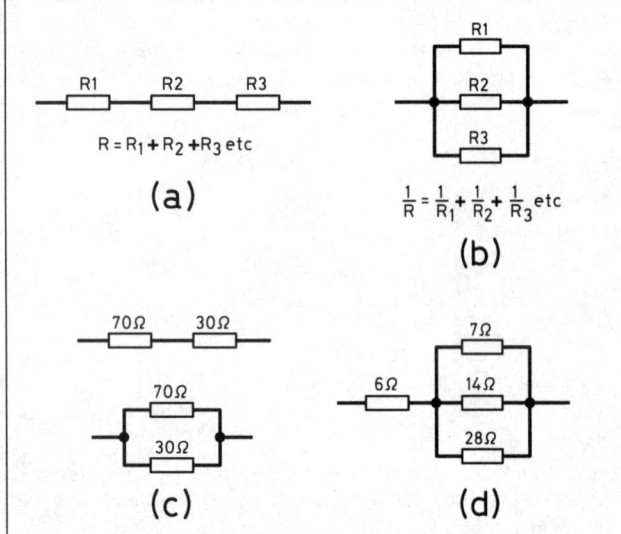

Fig 1.8: Resistors in various combinations: (a) series, (b) parallel, (c) series and parallel, (d) series-parallel. The calculation of the resultant resistances in (c) and (d) is explained in the text

Series connection:
$R_{total} = R1 + R2 + R3 + $ etc

Parallel connection

$$\frac{1}{R_{total}} = \frac{1}{R1} + \frac{1}{R2} + \frac{1}{R3} \quad \text{etc}$$

If only two resistors are in parallel, an easier calculation is possible:

$$\frac{1}{R_{total}} = \frac{1}{R1} + \frac{1}{R2} = \frac{R2}{R1 \times R2} + \frac{R1}{R1 \times R2} = \frac{R1 + R2}{R1 \times R2}$$

Inverting this and writing in 'standard form' (omitting multiplication signs) gives

$$R_{total} = \frac{R1\,R2}{R1 + R2}$$

This is a useful formula since the value of two resistors in parallel can be calculated easily.

It is also useful to remember that for equal resistors in parallel the formula simplifies to

$$R_{total} = \frac{R}{n}$$

where R is the value of one resistor and n is the number of resistors.

Example 1.

Calculate the resistance of a 30Ω and a 70Ω resistor connected first in series and then in parallel.

In series connection:
R = 30 + 70 = 100Ω
In parallel connection, using the simpler formula for two resistors in parallel:

$$R_{total} = \frac{R1\,R2}{R1 + R2} = \frac{30 \times 70}{30 + 70} = \frac{2100}{100} = 21\Omega$$

These two calculations are illustrated in **Fig 1.8(c)**.

Example 2.

Three resistors of 7Ω, 14Ω and 28Ω are connected in parallel. If another resistor of 6Ω is connected in series with this combination, what is the total resistance of the circuit?

The circuit, shown in **Fig 1.8(d)**, has both a series and parallel configuration. Taking the three resistors in parallel first, these are equivalent to a single resistance of R ohms given by:

$$\frac{1}{R_{total}} = \frac{1}{R1} + \frac{1}{R2} + \frac{1}{R3} = \frac{1}{7} + \frac{1}{14} + \frac{1}{28}$$

$$= \frac{4}{28} + \frac{2}{28} + \frac{1}{28} = \frac{7}{28}$$

Inverting (don't forget to do this!)

$$R_{total} = \frac{28}{7} = 4\Omega$$

This parallel combination is in series with the 6Ω resistor, giving a total resistance of 10Ω for the whole circuit.

Note: Often the maths is the most awkward issue and numerical mistakes are easy to make. A calculator helps but whether done with pencil and paper or by calculator, it is essential to estimate an answer first so mistakes can be recognised.

Let us consider the parallel calculation. The answer will be less than the lowest value. The 14Ω and 28Ω resistors will form a resistor less than 14Ω but greater than 7Ω since two 14Ω parallel resistors will give 7Ω and one of the actual resistors is well above 14Ω.

We now have this 'pair' in parallel with the 7Ω resistor. By the same logic, the answer will be less than 7Ω but greater than 3·5Ω. The calculated answer of 4Ω meets this criterion so stands a good chance of being correct. Results outside the 3·5-7Ω range must be wrong.

This may seem a bit long winded. In words it is. In practice, with a little bit of experience it takes moments.

Capacitors and Capacitance

Capacitors have the property of being able to store a charge of electricity. They consist of two parallel conducting plates or strips separated by an insulating medium called a dielectric. When a capacitor is charged there is a potential difference between its plates.

The capacitance of a capacitor is defined in terms of the amount of charge it can hold for a given potential difference. The capacitance 'C' is given by:

$$C = \frac{q}{V}$$

in units of Farads, F. (q in coulombs and V in volts)

A larger capacitor can hold a greater charge for a given voltage. Or to put it another way, a smaller capacitor will need a greater voltage if it is to store the same charge as a large capacitor.

The Farad is too large a unit for practical use, and typical values range from a few picofarads (pf) to some thousands of microfarads (µF). Unusually the convention is to write 20,000µF rather than 20mF which is, technically, correct notation.

The area of the two plates and the distance between them determines the capacitance. The material of the dielectric also has an effect; materials with a high dielectric constant can considerably enhance the capacitance.

The capacitance can be calculated from the formula

$$C = \frac{\varepsilon_0\,\varepsilon_r\,A}{d}$$

where:

C is in Farads
ε_0 is a natural constant, the permitivity of free space

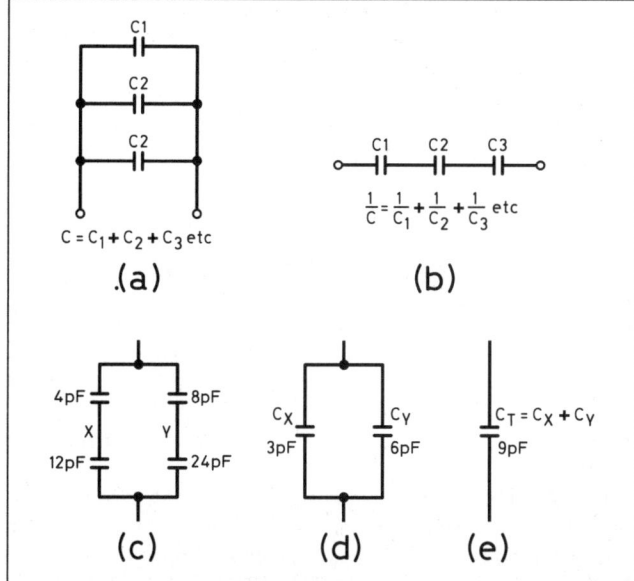

Fig 1.9: Capacitors in various combinations: (a) parallel, (b) series, (c) series-parallel. The calculation of the resultant capacitance of the combination shown in (c) required first the evaluation of each series arm X and Y as shown in (d). The single equivalent capacitance of the combination is shown in (e)

ε_r is the relative permitivity (or dielectric constant K)

A is the area of the plates in m² and

d is their separation in m.

Example:

A tuning capacitor has six fixed plates and five movable plates meshed between them, with a gap of 1mm. Fully meshed at maximum capacitance the area of overlap is 8cm². The dielectric is air, $\varepsilon_r=1$, and ε_0 is 8·85 x 10⁻¹² F/m. What is the capacitance?

Firstly, it will be numerically easier to re-arrange the formula so the separation may be given in centimetres and the area in cm² and the answer in picofarads. The formula becomes

$$C = \frac{0 \cdot 0885\,A}{d} = \frac{0 \cdot 0885 \times 8 \times 10}{0.1} = 71\text{pF}$$

The factor 10 above comes from the fact that there are 10 'surfaces' to consider in the meshed capacitor.

Such a capacitor will probably be acceptable for receiving and transmitting up to about 10-20W of power, depending in part on the matching and voltages involved. Higher powers and voltages will require considerably more separation between the plates, reducing the capacitance.

The factor by which the dielectric increases the capacitance compared with air is called the dielectric constant ε_r (or relative permitivity K) of the material. Physicists tend to use the symbol ε_r and engineers the symbol K. You may meet both, depending on which texts you are reading.

Typical values of K are: air 1, paper 2·5, glass 5, mica 7. Certain ceramics have much higher values of K of 10,000 or more. If the dielectric is a vacuum, as in the case of the inter-electrode capacitance of a valve, the same value of K as for air may be assumed. (Strictly, K = 1 for a vacuum and is very slightly higher for air.) The voltage at which a capacitor breaks down depends on the spacing between the plates and the type of dielectric used. Capacitors are often labelled with the maximum working voltage which they are designed to withstand and this figure should not be exceeded.

Capacitors Used in Radio Equipment

The values of capacitors used in radio equipment extend from below 1pF to 100,000µF. They are described in detail in the Passive Components chapter.

Capacitors in Series and Parallel

Capacitors can be connected in series or parallel, as shown in **Fig 1.9**, either to obtain some special capacitance value using a standard range of capacitors, or perhaps in the case of series connection to obtain a capacitor capable of withstanding a greater voltage without breakdown than is provided by a single capacitor. When capacitors are connected in parallel, as in **Fig 1.9(a)**, the total capacitance of the combination is equal to the sum of the separate capacitances. When capacitors are connected in series, as in **Fig 1.9(b)**, the reciprocal of the equivalent capacitance is equal to the sum of the reciprocals of the separate capacitances.

If C is the total capacitance these formulas can be written as follows:

Parallel connection

C = C1 + C2 + C3 etc

Series connection

$$\frac{1}{C_{total}} = \frac{1}{C1} + \frac{1}{C2} + \frac{1}{C3} \quad \text{etc}$$

Similar to the formula for resistors in parallel, a useful equivalent formula for two capacitors in series is

$$C_{total} = \frac{C1\,C2}{C1 + C2}$$

The use of these formulas is illustrated by the following *example:*

Two capacitors of 4pF and 12pF are connected in series; two others of 8pF and 24pF are also connected in series. What is the equivalent capacitance if these series combinations are joined in parallel?

The circuit is shown in **Fig 1.9(c)**. Using the formula for two capacitors in series, the two series arms X and Y can be reduced to single equivalent capacitances C_x and C_y as shown in **Fig 1.9(d)**.

$$C_X = \frac{4 \times 12}{4 + 12} = \frac{48}{16} = 3\text{pF}$$

$$C_Y = \frac{8 \times 24}{8 + 24} = \frac{96}{32} = 6\text{pF}$$

These two capacitances are in parallel and may be added to give the total effective capacitance represented by the single capacitor C_T in **Fig 1.9(e)**.

C = C_X + C_Y = 3 + 6 = 9pF

The total equivalent capacitance of the four capacitors connected as described is therefore 9pF.

This is not just an academic exercise or something to do for an examination. Consider the circuit in **Fig 1.10**. C is a variable capacitor covering the range 5-100pF, a 20:1 range. It will be seen later that, in a tuned circuit, this will give a tuning range of about 4·5:1 which may well be rather greater than required and

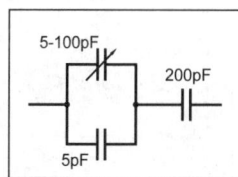

Fig 1.10: Padding a variable capacitor to change its range

cramping the desired range over a relatively small part of the 180 degree rotation normally available. A 5pF capacitor C1 is connected in parallel with C, giving a capacitance range for the pair of 10-105pF by addition of the capacitor values. This is now in series with a 200pF capacitor, so the capacitance range becomes 9·5-69pF, a ratio of 7·3 and a tuning range of 2·7:1.

Many variable capacitors have a small parallel trimmer capacitor included in their construction. Often that is used to set the highest frequency of the tuning range and a series capacitor or a variable inductor (see later in this chapter) used to set the lower end.

MAGNETISM

Permanent Magnets

A magnet will attract pieces of iron towards it by exerting a magnetic force upon them. The field of this magnetic force can be demonstrated by sprinkling iron filings on a piece of thin card-

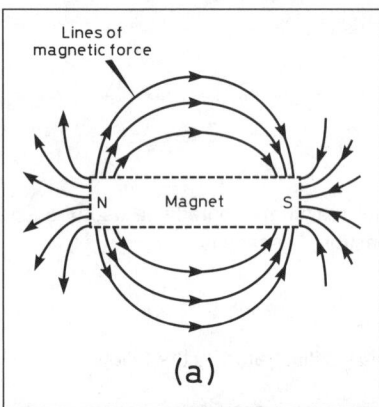

(a)

Fig 1.11: magnetic field produced by a bar magnet

Fig 1.12: Iron filings mapping out the magnetic field of (top) a bar magnet, and (bottom) a solenoid carrying current

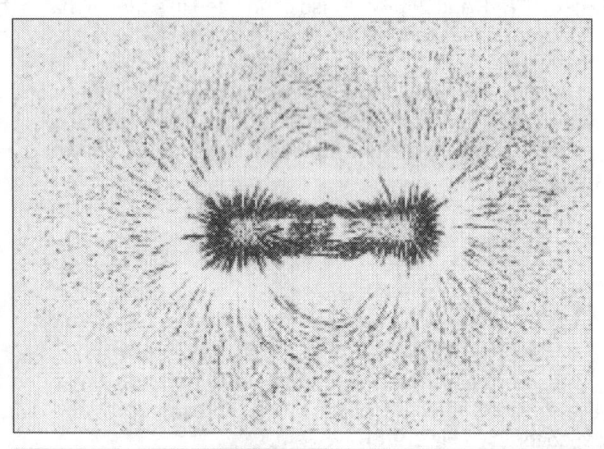

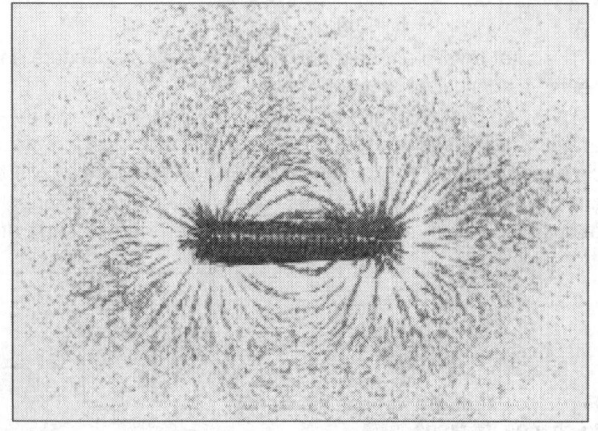

board under which is placed a bar magnet. The iron filings will map out the magnetic field as sketched in **Fig 1.11** and the photograph **Fig 1.12**. It will be seen that the field is most intense near the ends of the magnet, the centres of intensity being called the poles, and lines of force spread out on either side and continue through the material of the magnet from one end to the other.

If such a magnet is suspended so that it can swing freely in a horizontal plane it will always come to rest pointing in one particular direction, namely towards the Earth's magnetic poles, the Earth itself acting as a magnet. A compass needle is simply a bar of magnetised steel. One end of the magnet (N) is called the north pole, which is an abbreviation of 'north-seeking pole' and the other end (S) a south pole or south-seeking pole. It is an accepted convention that magnetic force acts in the direction from N to S as indicated by the arrows on the lines of force in Fig 1.11.

If two magnets are arranged so that the north pole of one is near the south pole of another, there will be a force of attraction between them, whereas if similar poles are opposite one another, the magnets will repel: see **Fig 1.13**.

Permanent magnets are made from certain kinds of iron, nickel and cobalt alloys and certain ceramics, the hard ferrites (see Passive Components chapter) and retain their magnetism more or less indefinitely. They have many uses in radio equipment, such as loudspeakers, headphones, some microphones, cathode-ray tube focusing arrangements and magnetron oscillators.

Other types of iron and nickel alloys and some ceramics (the soft ferrites), eg soft iron, are not capable of retaining magnetism, and cannot be used for making permanent magnets. They are effective in transmitting magnetic force and are used as cores in electromagnets and transformers. These materials concentrate the magnetic field by means of a property called permeability. The permeability is, essentially, the ratio of the magnetic field with a core to that without it.

Electromagnets

A current of electricity flowing through a straight wire exhibits a magnetic field, the lines of force of which are in a plane perpendicular to the wire and concentric with the wire. If a piece of cardboard is sprinkled with iron filings, as shown in **Fig 1.14**, they will arrange themselves in rings round the wire, thus illustrating the magnetic field associated with the flow of current in the wire. Observation of a small compass needle placed near the wire would indicate that for a current flow in the direction illustrated the magnetic force acts clockwise round the wire. A reversal of current would reverse the direction of the magnetic field.

The corkscrew rule enables the direction of the magnetic field round a wire to be found. Imagine a right-handed corkscrew being driven into the wire so that it progresses in the direction of current flow; the direction of the magnetic field around the wire will then be in the direction of rotation of the corkscrew.

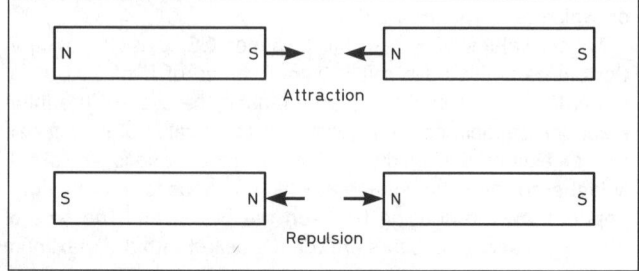

Fig 1.13: Attraction and repulsion between bar magnets

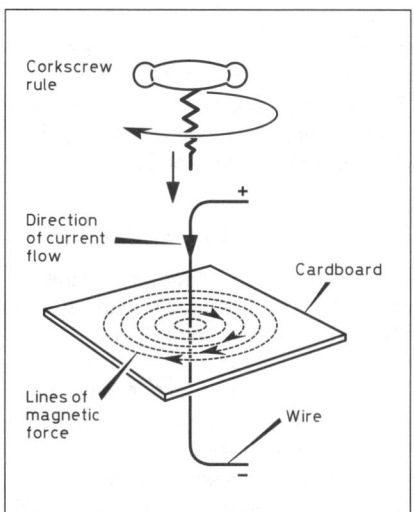

Fig 1.14: Magnetic field produced by current flowing in a straight wire

The magnetic field surrounding a single wire is relatively weak. Forming the wire into a coil will combine the field of each turn producing a stronger field. A much greater increase in the magnetic field can then be achieved by inserting a piece of soft iron, called a core, inside the coil.

Fig 1.15 and the bottom photograph in Fig 1.12 show the magnetic field produced by a coil or solenoid as it is often called. It will be seen that it is very similar to that of a bar magnet also shown in Fig 1.12. A north pole is produced at one end of the coil and a south pole at the other. Reversal of the current will reverse the polarity of the electromagnet. The polarity can be deduced from the S rule, which states that the pole that faces an observer looking at the end of the coil is a south pole if the current is flowing in a clockwise direction; see Fig 1.15. The current is the conventional current, not the direction of flow of electrons.

The strength of a magnetic field produced by a current is directly proportional to the current, a fact made use of in moving coil meters (see Test equipment chapter). It also depends on the number of turns of wire, the area of the coil, and the permeability of the core.

Interaction of Magnetic Fields

Just as permanent magnets can attract or repel, so can electromagnets. If one of the devices, a coil for example, is free to move, then a current will cause the coil to move with a force or at a rate related to the magnitude of the current. The moving coil meter relies on this effect, balancing the force caused by the current against a return spring so that the movement or deflection indicates the magnitude of the current.

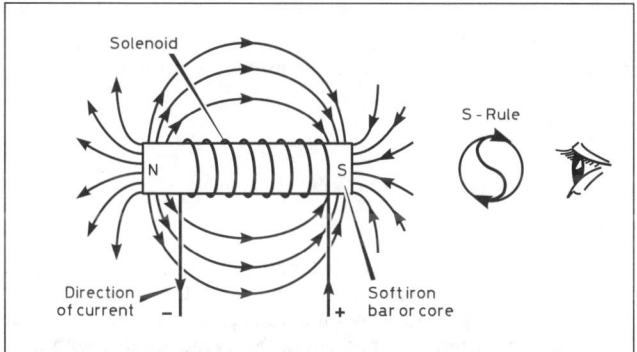

Fig 1.15: The S rule for determining the polarity of an electromagnet

Electromagnetic Induction

If a bar magnet is plunged into a coil as indicated in **Fig 1.16(a)**, the moving-coil microammeter connected across the coil will show a deflection. The explanation of this phenomenon, known as electromagnetic induction, is that the movement of the magnet's lines of force past the turns of the coil causes an electromotive force to be induced in the coil which in turn causes a current to flow through the meter. The magnitude of the effect depends on the strength and rate of movement of the magnet and the size of the coil. Withdrawal of the magnet causes a reversal of the current. No current flows unless the lines of force are moving relative to the coil. The same effect is obtained if a coil of wire is arranged to move relative to a fixed magnetic field. Dynamos and generators depend for their operation on the principle of electromagnetic induction.

Consider a pair of coils of wire are arranged as shown in **Fig 1.16(b)**. When the switch K is open there is no magnetic field from the coil P linking the turns of the coil S, and the current through S is zero. Closing K will cause a magnetic field to build up due to the current in the coil P. This field, while it is building up, will induce an EMF in coil S and cause a current to flow through the meter for a short time until the field due to P has reached a steady value, when the current through S falls to zero again. The effect is only momentary while the current P is changing.

The fact that a changing current in one circuit can induce a voltage in another circuit is the principle underlying the operation of transformers.

Self-inductance

Above we considered the effect of a change of current in coil P inducing a voltage is coil S. In fact the changing field also induces an EMF in coil P even though it is the current in coil P that is causing the effect. The induced EMF is of a polarity such that it tends to oppose the original change in current.

This needs some care in understanding. On closing the switch K the EMF induced in P will tend to oppose the build up of current, that is it will oppose the voltage from the battery and the current will build up more slowly as a result. However if K is opened the current will now fall and the EMF induced in P will be of opposite polarity, trying to keep the current flowing. In reality the current falls more slowly.

This effect of the induced EMF due to the change in current is known as *inductance*, usually denoted by the letter L. If the current is changing at the rate of one amp per second (1A/s), and the induced EMF is one volt, the coil has an inductance of one henry (1H). A 2H coil will have 2V induced for the same changing current. Since the induced voltage is in the coil containing the changing current this inductance is properly called *self-inductance*.

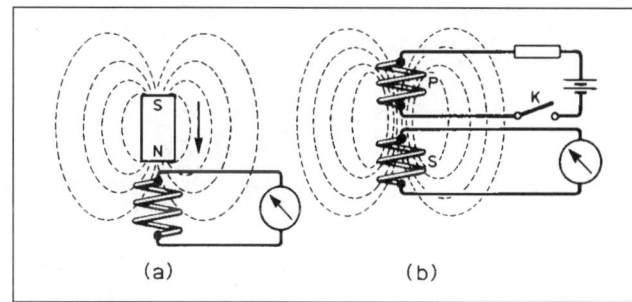

Fig 1.16: Electromagnetic induction. (a) Relative movement of a magnet and a coil causes a voltage to be induced in the coil. (b) When the current in one of a pair of coupled coils changes in value, current is induced in the second coil

$$\text{Inductance } L = \frac{V}{\partial I / \partial t}$$

where δI is a small change in current and δt is the time for that change.

That is

$$\frac{\partial I}{\partial t}$$

is the rate of change of current in A/s

The inductance values used in radio equipment may be only a very small fraction of a henry, the units millihenry (mH) and microhenry (µH) meaning one thousandth and one millionth of a henry respectively are commonly used.

The inductance of a coil depends on the number of turns, on the area of the coil and the permeability of the core material on which the coil is wound. The inductance of a coil of a certain physical size and number of turns can be calculated to a fair degree of accuracy from formulas or they can be derived from coil charts.

Mutual Inductance

A changing current in one circuit can induce a voltage in a second circuit: as in **Fig 1.16(b)**. The strength of the voltage induced in the second circuit depends on the closeness or tightness of the magnetic coupling between the circuits; for example, if both coils are wound together on an iron core practically all the magnetic flux from the first circuit will link with the turns of the second circuit. Such coils would be said to be tightly coupled whereas if the coils were both air-cored and spaced some distance apart they would be loosely coupled.

The mutual inductance between two coils is also measured in henrys, and two coils are said to have a mutual inductance of one henry if, when the current in the primary coil changes at a rate of one ampere per second, the voltage across the secondary is one volt. Mutual inductance is often denoted in formulas by the symbol M.

Inductors in Series and Parallel

Provided that there is no mutual coupling between inductors when they are connected in series, the total inductance obtained is equal to the sum of the separate inductances. When they are in parallel the reciprocal of the total inductance is equal to the sum of the reciprocals of the separate inductances.

If L is the total inductance (no mutual coupling) the relationships are as follows:

Series connection

$L_{total} = L1 + L2 + L3$ etc

Parallel connection

$$\frac{1}{L_{total}} = \frac{1}{L1} + \frac{1}{L2} + \frac{1}{L3} \quad \text{etc}$$

These two formulas are of the same format as the formula for resistors and the special case of only two resistors is also true for inductors. In reality paralleled inductors are uncommon but may be found in older radio receivers that are permeability tuned. That is where the inductance of a tuned circuit (see later) is varied rather than the more common case of variable capacitors.

CR Circuits and Time Constants

Fig 1.17(a) shows a circuit in which a capacitor C can either be charged from a battery of EMF V, or discharged through a resistor R, according to whether the switch S is in position a or b.

If the switch is thrown from b to a at time ta, current will start to flow into the capacitor with an initial value V/R. As the capacitor charges the potential difference across the capacitor increases, leaving less PD across the resistor, and the current through the circuit therefore falls away, as shown in the charging portion of **Fig 1.17(b)**. When fully charged to the voltage V the current will have dropped to zero.

At time tb, the switch is thrown back to b, the capacitor will discharge through the resistor R, the current being in the opposite direction to the charging current, starting at a value -V/R and dying away to zero. As the capacitor discharges, the PD across its plates falls to zero as shown in the discharge portion of Fig 1.17(b).

The voltage at any point during the charge cycle is given by the formula

$$V = V_b \left(1 - e^{-\frac{t}{CR}} \right)$$

Where V_b is the battery voltage and t is the time, in seconds, after the switch is thrown. C and R are given in Farads and Ohms respectively and e is the base of natural logarithms.

This is known as an exponential formula and the curve is one of a family of exponential curves.

As the capacitor approaches fully charged, or discharged, the current is very low. In theory an infinite time is needed to fully complete charging. For practical purposes the circuit will have charged to 63% (or 1/e) in the time given by CR and this time is known as the *time constant* of the circuit. On discharge the voltage will have fallen to 37% of the initial voltage (1-1/e) after one time constant. Over the next time constant it will have fallen to 37% of its new starting voltage, or 14% of the original voltage. Five time constants will see the voltage down to 0·7%.

Time constant τ = CR seconds

As an example, the time constant of a capacitance of 0·01µF (10^{-8}F) and a resistance of 47kΩ (4·7 x 10^4Ω) is:

$$\tau = 10^{-8} \times 4 \cdot 7 \times 10^4$$
$$= 4 \cdot 7 \times 10^{-4}\text{s} = 0 \cdot 47\text{ms}$$

High voltage power supplies should have a bleeder resistor across the smoothing capacitor to ensure lethal voltages are removed before anyone can remove the lid after switching off,

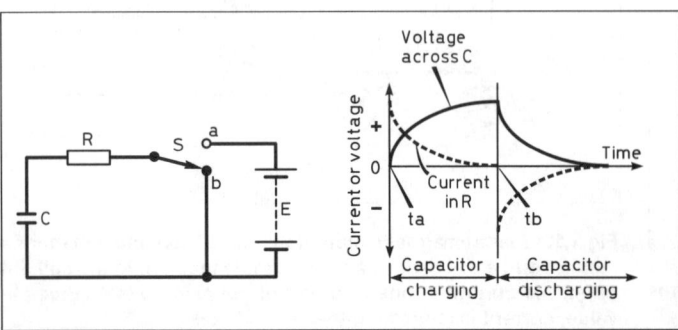

Fig 1.17: In (a) a capacitor C can be charged or discharged through the resistor R by operating the switch S. The curves of (b) show how the voltage across the capacitor and the current into and out of the capacitor vary with time as the capacitor is charged and discharged. The curve for the rise and fall of current in an LR circuit is similar to the voltage curve for the CR circuit

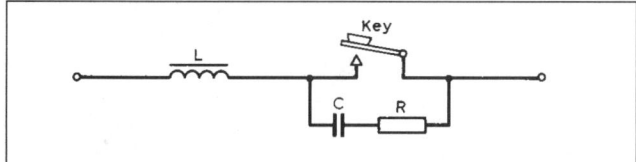

Fig 1.18: Typical key-click filter. L serves to prevent a rise of current. C, charging through R, serves to continue flow of current momentarily when key contacts open. Typical values: L = 0.01 to 0.1H, C = 0.01 to 0.1μF, R = 10 to 100

thinking it is safe! The time constant here may be a few tens of seconds.

In audio detector circuits, the capacitor following the detector diode is chosen so that, along with the load resistance, the time constant is rather longer than the period of the intermediate frequency to give good smoothing or filtering. However it also needs to be rather shorter than the period of the highest audio frequency, or unwanted attenuation of the higher audio notes will occur.

A similar constraint applies in AGC (automatic gain control) circuits where the receiver must be responsive to variation caused by RF signal fading without affecting the signal modulation. Digital and pulse circuits may rely on fast transient waveforms. Here very short time constants are required, unwanted or stray capacitance must be avoided.

LR Circuits

Inductors oppose the change, rise or fall, of current. The greater the inductance, the greater the opposition to change. In a circuit containing resistance and inductance the current will not rise immediately to a value given by V/R if a PD V is applied but will rise at a rate depending on the L/R ratio. LR circuits also have a time constant, the formula is:

τ = L/R   where L is in henries and R in ohms

Fig 1.17 showed the rise in capacitor voltage as it charged through a resistor; the curve of the rise in current in an LR circuit is identical. In one time constant, the current will have risen to 63%, (1 - 1/e) of its final value, or to decay to 37% of its initial value.

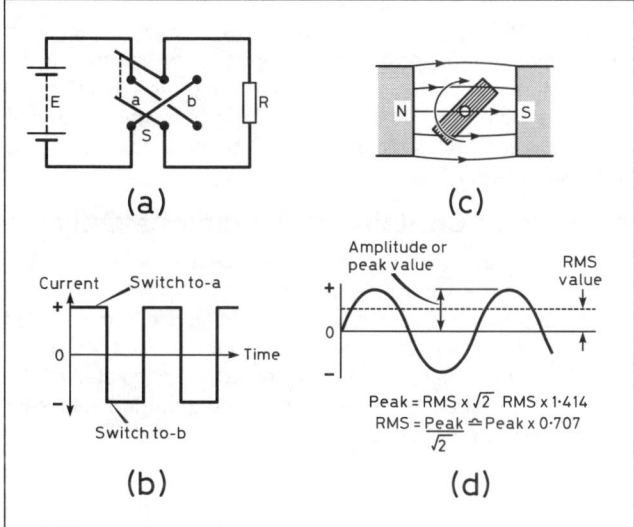

Fig 1.19: Alternating current. A simple circuit with a current-reversing switch shown at (a) produces a square-wave current through the resistor R as shown in (b). When a coil is rotated in a magnetic field as in (c) the voltage induced in the coil has a sinusoidal waveform (d)

An example of the use of an inductor to slow the rise and fall of current is the Morse key filter shown in **Fig 1.18**. Slowing the rise of current as the key is depressed is simple enough but when the key is raised the circuit is cut and a high voltage can be developed which will cause sparking at the key contacts. The C and R across the key provide a path for the current to flow momentarily as it falls to zero. The reason for the use of this filter is discussed in the Morse chapter.

ALTERNATING CURRENT

So far we have concerned ourselves with uni-directional current flow or direct current (DC). Audio and radio circuits rely heavily on currents (and voltages) that change their polarity continuously; alternating currents (AC).

Fig 1.19(a) shows a battery and reversing switch. Continually operating the switch would cause the current in the resistor to flow in alternate directions as shown by the waveform in **Fig 1.19(b)**. The waveform is known as a square wave.

Alternating current normally has a smoother waveform shown in **Fig 1.19(d)** and can be produced by an electrical generator shown in **Fig 1.19(c)**. Consider for a moment a single wire in the coil in Fig 1.19(c) and remember that if a wire moves through a magnetic field an EMF is induced in the wire proportional to the relative velocity of the wire.

At the bottom the wire is travelling horizontally and not cutting through the field; no EMF is induced. After 90 degrees of rotation the wire is travelling up through the field at maximum velocity and the induced EMF is at a maximum. At the top the EMF has fallen to zero and now reverses polarity as the wire descends, falling to zero again at the bottom where the cycle begins again. The vertical component of the velocity follows a sinusoidal pattern, as does the EMF; it is a sine wave. The precise shape can be plotted using mathematical 'sine' tables used in geometry.

Specifying an AC Waveform

For AC there are two parameters that must be quoted to define the current or voltage. Like DC we must give the magnitude or *amplitude* as it is called and we must also say how fast the cycles occur.

If t is the time for one cycle (in seconds) then 1/t will give the number of cycles occurring in a second; this is known as the frequency.

t = 1/f and f=1/t

The unit of frequency is cycles per second and is given the name Hertz, abbreviation Hz.

Example: the UK mains has a frequency of 50Hz, what is the periodic time?

f =50Hz so the time for 1 cycle t =1/f = 1/50 second or 0·02s.

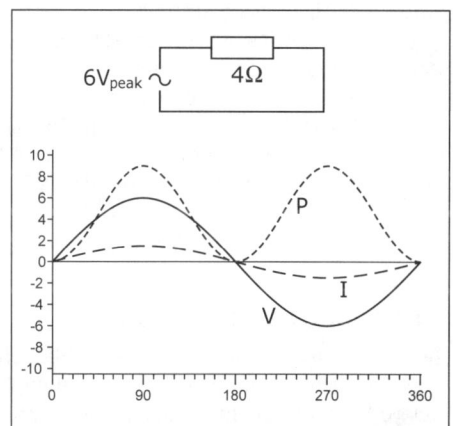

Fig 1.20: The power dissipated in a resistor with a sine wave of current

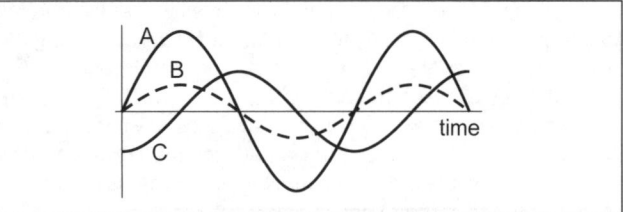

Fig 1.21: Relative phase

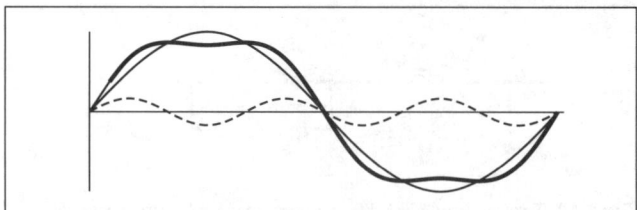

Fig 1.22: Harmonics of a sine wave

RMS Values

Specifying the amplitude is more interesting. It would seem sensible to quote the maximum amplitude and allow the reader to find the amplitude at other parts of the cycle by looking up the values in sine tables. In actual fact the RMS (root mean square) value is used but what does that mean?

Consider passing an AC current through a resistor. The power dissipated in the resistor will vary over the cycle as shown in **Fig 1.20**.

The peak current is 6V/4Ω = 1·5A and peak power is 6V x 1·5A = 9W. The power curve is symmetrical about 4·5W and the average power over the cycle is indeed 4·5W. The DC voltage that would produce 4·5W in a 4Ω resistor is just over 4V (4·24V). This is the RMS value of the 6V peak waveform. It is that value of DC voltage (or current) that would produce the same heating effect in a resistor.

The RMS value is related to the peak value by

$$RMS = \frac{Peak}{\sqrt{2}}$$

or

$$Peak = RMS \times \sqrt{2}$$

It may help to remember that

$$\sqrt{2} = 1 \cdot 4142$$

and

$$\frac{1}{\sqrt{2}} = 0 \cdot 707$$

Example: suppose your mains supply is 240V RMS, what is the peak value?

$$Peak = RMS \times \sqrt{2} \quad = 240 \times \sqrt{2} \quad = 340V$$

Other Frequencies

Audible frequencies range from around 50-100Hz up to 20kHz for a young child, rather less for adults. Some animals can hear up to around 40-50kHz. Dogs will respond to an ultrasonic whistle (above human hearing) and bats use high pitch sounds around 50kHz for echo location.

Radio frequencies legally start at 9kHz, and 20kHz is used for worldwide maritime communication. Below 30kHz is known as the VLF band (very low frequency; LF is 30-300kHz; MF (medium frequencies) is 300-3000kHz. The HF, high frequency, band is 3 to 30MHz, although amateurs often refer to the 1·8MHz amateur band as being part of HF. VHF is 30-300MHz and UHF (ultra-high frequencies) is 300-3000MHz. Above that is the SHF band 3-30GHz, super-high frequencies. Also, by common usage, the 'microwave band' is regarded as being above 1GHz.

Phase and Harmonics

Two waveforms that are of the same frequency are *in phase* if they both start at the same point in time. If one waveform is delayed with respect to the other then they are not in phase and the phase difference is usually expressed as a proportion of a complete cycle of 360°. This is shown in **Fig 1. 21**.

Waveforms A and B are in phase but are of different amplitudes.

Waveform C is not in phase with A, it lags A by 1/4 cycle or 90°, or A leads C by 90°. It lags because the 'start' of the cycle is to the right on the time axis of the graph, that is, it occurs later in time. The 'start' is conventionally regarded as the zero point, going positive.

If two waveforms are of different frequencies then their phase relationships are continuously changing. However, if the higher frequency is an exact multiple of the lower, the phase relationship is again constant and the pattern repeats for every cycle of the lower frequency waveform. These multiple frequencies are known as harmonics of the lower 'fundamental' frequency. The second harmonic is exactly twice the frequency and the third, three times the frequency.

Fig 1.22 shows a fundamental and a third harmonic at 1/6 amplitude. The heavy line is the sum of the fundamental and the harmonic. It shows a phenomenon known as harmonic distortion.

If a sine wave is distorted, perhaps by over-driving a loudspeaker or by imperfections (often deliberate) in electronic circuits, then the distortion can be regarded as the original sine wave plus the right amplitudes and phases of various harmonics required to produce the actual distorted waveform. The procedure to determine the required harmonics is known as Fourier analysis.

It is important to appreciate that in distorting an otherwise clean sine wave, the harmonics really are created; new frequencies are now present that were not there prior to the distortion. If a distorted sine wave is observed using a spectrum analyser, the harmonic frequencies can be seen at the relative amplitudes. Moreover, if they can be filtered out, the distortion will have been removed. This will prove a useful technique later in the discussions on radio receivers, intermediate frequencies and AGC (automatic gain control).

AC Circuit Containing Resistance Only

Ohm's Law is true for a resistor at every point in a cycle of alternating current or voltage. The current will flow backwards and forwards through the resistor under the influence of the applied voltage and will be in phase with it as shown in **Fig 1.23(a)**.

The power dissipated in the resistor can be calculated directly from the small power formulas, provided that RMS values for voltage and current are used.

$$P = VI \qquad P = \frac{V^2}{R} \qquad P = I^2R$$

If peak values of voltage and current are used, these formulas become:

$$P = \frac{V_P}{\sqrt{2}} \times \frac{I_P}{\sqrt{2}} \quad = \frac{V_P I_P}{2} \qquad P = \frac{V_P^2}{2R}$$

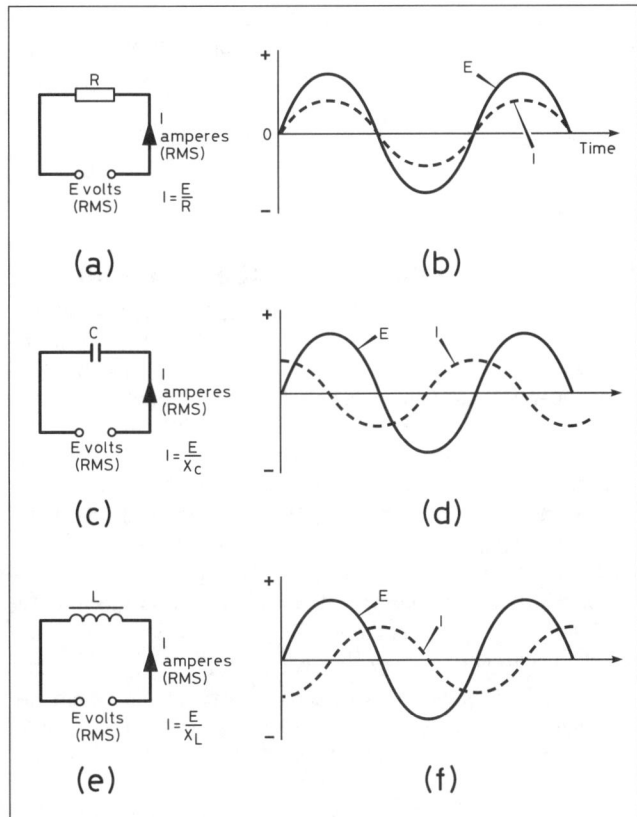

Fig 1.23: Voltage and current relationships in AC circuits comprising (a) resistance only, (c) capacitance only and (e) inductance only

AC Circuit Containing Capacitance Only; Reactance of a Capacitor

If an alternating current is applied to a capacitor the capacitor will charge up. The PD at its terminals will depend on the amount of charge gained over the previous part cycle and the capacitance of the capacitor. With an alternating current, the capacitor will charge up first with one polarity, then the other. **Fig 1.23(c)** shows the circuit and **Fig 1.23(d)** shows the voltage (E) and current (I) waveforms. Inspection of **Fig 1.23(d)** shows that the voltage waveform is 90 degrees lagging on the current. This is explained by remembering that the voltage builds up as more charge flows into the capacitor. When the current is a maximum, the voltage is increasing at its maximum rate. When the current has fallen to zero the voltage is (momentarily) constant at its peak value.

There is no such thing as the resistance of a capacitor. Consider **Fig 1.23(d)** again. At time zero the voltage is zero yet a current is flowing, a quarter cycle (90°) later the voltage is at a maximum yet the current is zero. The resistance appears to vary between zero and infinity! Ohm's Law does not apply.

What we can do is to consider the value of V_{rms}/I_{rms}. It is not a resistance but it is similar to resistance because it does relate the voltage to the current. The quantity is called *reactance* (to denote to 90° phase shift between V and I) and is measured in ohms, Ω.

Remember that for the same charge, a larger capacitor will have a lower potential difference between the plates. This suggests a larger capacitor will have a lower reactance. Also, as the frequency rises, the periodic time falls and the charge, which is current x time, also falls. The reactance is lower as the frequency rises, it varies with frequency. The proof of this requires inte-

gral calculus which is beyond the scope of the book but the formula for the reactance of a capacitor is

$$X_C = \frac{1}{2\pi fC}$$

where f is the frequency in Hertz and C the capacitance in Farads. You may also meet this formula written using the symbol ω instead of $2\pi f$. The lower case Greek letter omega, ω, gives the frequency in radians per second - remember there are 2π radians in 360° and ω is called the angular frequency.

Example:
A capacitor of 500pF is used in an antenna matching unit and is found to have 400V across it when the transmitter is set to 7MHz. Calculate the reactance and current flowing.

$$X_C = \frac{1}{2\pi fC} = \frac{1}{2\pi \times 7 \times 10^6 \times 500 \times 10^{-12}}$$

$$X_C = \frac{1}{7\pi \times 10^{-3}} = \frac{1000}{7\pi} = 45 \cdot 5\Omega$$

With 400V applied the current is

$$I = \frac{V}{X_C} = \frac{400}{45 \cdot 5} = 8 \cdot 8A$$

AC Circuit Containing Inductance Only; Reactance of a Coil

The opposition of an inductance to alternating current flow is called the inductive reactance of the coil: **Fig 1.23(e)** shows the circuit and the current and voltage waveforms. If a potential difference is applied to a coil, the current builds up slowly, how slowly depending on the inductance. Similarly, if the PD is reduced, the current in the coil tries to keep flowing. With AC the current waveform lags the voltage waveform by 90°. As the frequency increases the current in the coil is being expected to change more rapidly and the opposition is greater, the magnitude of the current is less and the reactance higher.

This leads to the formula for the reactance of a coil as:

$$X_L = 2\pi fL$$

As before X_L is in ohms, L in henries and f in Hertz.

Example. A coil has an inductance of 5μH, calculate it reactance at 7MHz.

$$X_L = 2\pi fL = 2\pi \times 7 \times 10^6$$
$$= 70\pi = 220\Omega$$

AC Circuits with R and C or R and L

Consider **Fig 1.24**, a resistor and capacitor in series. The same current flows through both R and C and is shown by the dashed waveform. The voltage across R will be in phase with the current, shown solid, and the voltage across C will lag by 90°, shown dotted. The supply voltage is the sum of the volt-

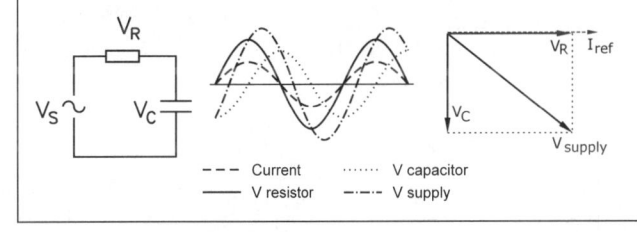

Fig 1.24: Voltage and current in a circuit containing a resistor and capacitor

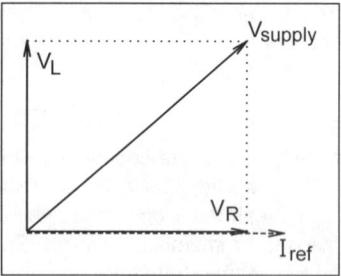

Fig 1.25: Phasor diagram for R and L

Fig 1.26: L and C in series

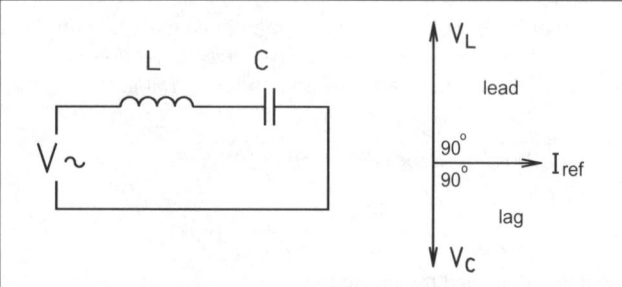

ages across R and C. Since the voltages across R and C are not in phase, they cannot simply be added together; they must be added graphically, shown by the dash-dot line. For example when V_R is at a maximum, V_C is zero, so the sum should be coincident with V_R.

The waveform diagram is messy to say the least, it has been shown once to illustrate the situation. Fortunately there is a simpler way, the phasor diagram shown to the right in Fig 1.24. By convention the current, which is common to both components, is drawn horizontally to the right. The vector or phasor representing the voltage across the resistor is drawn parallel to it and of a length representing the magnitude of the voltage across R. V_C lags by 90°, shown as a phasor downwards, again of the correct length to represent the magnitude of the voltage across C. The vector sum, the resultant, gives the supply voltage. The length of the arrow represents the actual voltage and the angle will give its phase.

Since V_R and V_C are at right angles we can use Pythagoras' Theorem to obtain a formula for the voltage.

$$V_{supply} = \sqrt{V_R{}^2 + V_C{}^2}$$

By a similar argument we can also obtain:

$$Z = \sqrt{R^2 + X_C{}^2}$$

where Z is the impedance of the CR circuit, R is the resistance and X_C the reactance.

Impedance (symbol Z) is the term used to denote the 'resistance' of a circuit containing both resistance and reactance and is also measured in ohms. The convention is that *resistance* means V and I are in phase; *reactance* means 90° (capacitive or inductive) and *impedance* means somewhere between, or indeterminate.

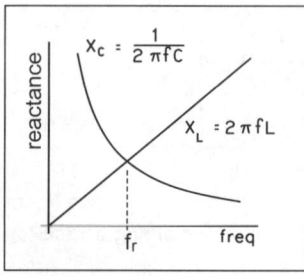

Fig 1.27: Reactance of L and C

Fig 1.25 shows the phasor diagram for R and L. Since the coil voltage now leads the current, it is drawn upwards but the geometry and the formula are the same.

$$V_{supply} = \sqrt{V_R{}^2 + V_L{}^2}$$

$$Z = \sqrt{R^2 + X_L{}^2}$$

Capacitance and Inductance, Resonance

In the series circuit containing C and L the current is still the common factor and the voltage across C will lag the current by 90° whilst the voltage across the inductor will lead by 90°. Consequently these two voltages will be 180° out of phase; that is, in anti-phase. The two voltages will tend to cancel rather than add. **Fig 1.26** shows the circuit and the phasor diagram. As drawn, the voltage across the capacitor is less than the voltage across the inductor, indicating the inductor has the greater reactance. In general the total reactance is given by:

$X = X_L - X_C$.

However the reactance varies with frequency as shown in **Fig 1.27**. If the reactances of C and L are equal, the voltages V_C and V_L will also be equal and will cancel exactly. The voltage across the circuit will be zero. This occurs at a frequency where the curves intersect in Fig 1.27, it is called the resonant frequency.

At resonance the total reactance $X = X_L - X_C$ will be zero and if $X_L = X_C$ then:

$$2\pi fL = \frac{1}{2\pi fC}$$

Rearranging this equation gives:

$$f = \frac{1}{2\pi\sqrt{LC}} \quad or \quad C = \frac{1}{4\pi^2 f^2 L} \quad or \quad L = \frac{1}{4\pi^2 f^2 C}$$

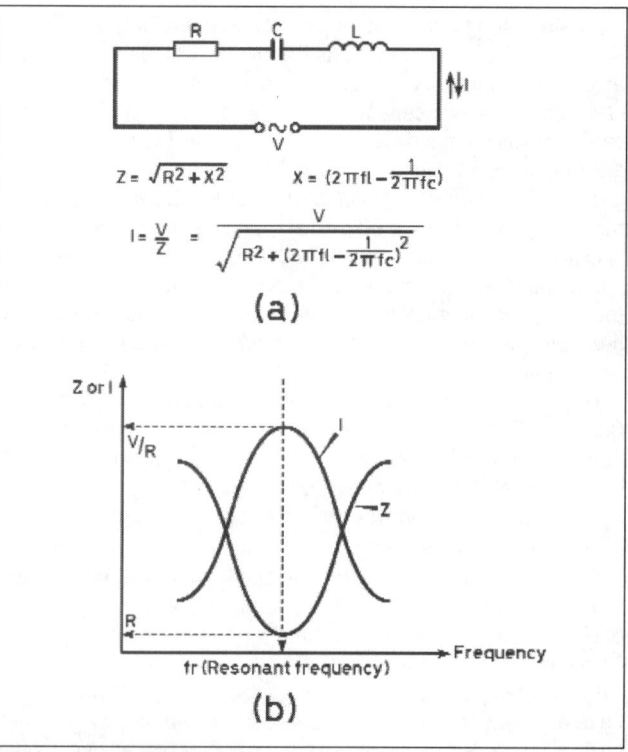

Fig 1.28: The series-resonant circuit. The curves shown at (b) indicate how the impedance and the current vary with frequency in the type of circuit shown at (a)

The Radio Communication Handbook

Fig 1.28 shows a circuit containing a capacitor, inductor and resistor. This is representative of real circuits, even if the resistance is merely that of the coil. This may be higher than expected due to 'skin effect'; a phenomenon explained later in the chapter. The overall impedance of the circuit is high when away from resonance but it falls, towards 'R' the value of resistance, as resonance is approached. The series resonant circuit is sometimes called an acceptor circuit because it accepts current at resonance. The current is also shown in Fig 1.28.

Example 1:

A 50µH inductor and a 500pF capacitor are connected in series; what is the resonant frequency?

The formula is:

$$f = \frac{1}{2\pi\sqrt{LC}}$$

So inserting the values:

$$f = \frac{1}{2\pi\sqrt{50\times10^{-6}\times500\times10^{-12}}} = \frac{1}{2\pi\sqrt{5\times10^{-5}\times5\times10^{-10}}}$$

$$f = \frac{1}{2\pi\sqrt{25\times10^{-15}}} = \frac{1}{2\pi\sqrt{2\cdot5\times10^{-14}}}$$

$$f = \frac{1}{2\pi\times1.58\times10^{-7}} = \frac{10^7}{3\cdot16\pi} = 1\text{MHz}$$

Note: The calculation has been set out deliberately to illustrate issues in handling powers and square roots. In line 1 the numbers were put in standard form, that is a number between 1 and 10, with the appropriate power of 10. This then resulted in 25 x 10⁻¹⁵ inside the square root sign. The root of 25 is easy but the root of an odd power of 10 is not. Consequently the sum has been changed to 2·5 10⁻¹⁴. It so happens that this is also then in standard form but the real purpose was to obtain an even power of ten, the square root of which is found by halving the 'power' or exponent.

There is another method which may be easier to calculate; namely to use the formula for f² but remembering to take the square root at the end.

The formula is:

$$f^2 = \frac{1}{4\pi^2 LC}$$

An advantage is that π² is 9·87 which is close enough to 10, remembering the error is much less than the tolerance on the components.

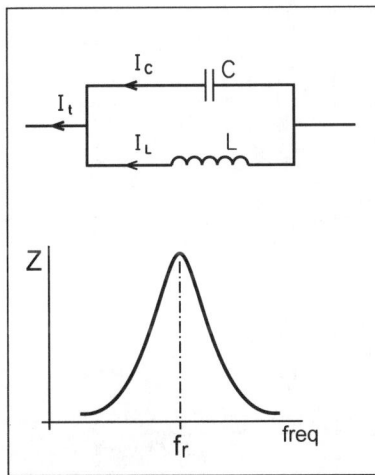

Fig 1.29: Parallel resonance

Example 2:

A vertical antenna has a series inductance of 20µH and capacitance of 100pF. What value of loading coil is required to resonate at 1·8MHz.

The formula is

$$L = \frac{1}{4\pi^2 f^2 C} = \frac{1}{40\times\left(1\cdot8\times10^6\right)^2\times10^{-10}}$$

$$L = \frac{1}{40\times3\cdot24\times10^{12}\times10^{-10}} = \frac{1}{12960}$$

which is 77µH.

The antenna already has 20µH of inductance, so the loading coil needs to add another 57µH. A coil some 10cm diameter of 30 turns spread over 10cm could be suitable.

Parallel Resonance

The L and C can also be connected in parallel as shown in **Fig 1.29**. Resistor 'r' is the internal resistance of the coil which should normally be as low as reasonably possible. The voltage is common to both components and the current in the capacitor will lead by 90° while that in the coil lags by 90°.

This time the two currents will tend to cancel out, leaving only a small supply current. The impedance of the circuit will increase dramatically at resonance as shown in the graph in Fig.1.29. This circuit is sometimes called a rejector circuit because it rejects current at resonance; the opposite of the series resonant circuit.

Magnification Factor, Q

Consider again the series resonant circuit in Fig 1.28. At resonance the overall impedance falls to the resistance R and the current is comparatively large. The voltage across the coil is still given by $V = I \times X_L$ and similarly for the capacitor $V = I \times X_C$. Thus the voltages across L and C are very much greater than the voltage across the circuit as a whole which is simply $V = I \times R$. Remember X_C and X_L are very much larger than R.

The magnification factor is the ratio of the voltage across the coil (or capacitor) to that across the resistor.

The formula are:

$$Q = \frac{X_L}{R} \quad \text{or} \quad \frac{X_C}{R}$$

that is:

$$Q = \frac{2\pi fL}{R} \quad \text{or} \quad \frac{1}{2\pi fCR}$$

This can also be written as:

$$Q = \frac{\omega L}{R} \quad \text{or} \quad \frac{1}{\omega CR}$$

where ω=2πf.

Resonance Curves and Selectivity

The curves in Figs 1.28 and 1.29 showed how tuned circuits were more responsive at their resonant frequency than neighbouring frequencies. These resonance curves show that a tuned circuit can be used to select a particular frequency from a multitude of frequencies that might be present. An obvious example is in selecting the wanted radio signal from the thousands that are transmitted.

The sharpness with which a tuned circuit can select the wanted frequency and reject nearby ones is also determined by the

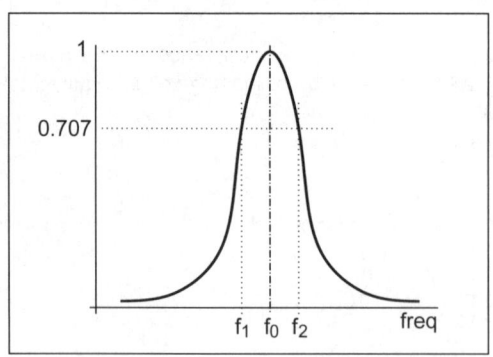

Fig 1.30: Bandwidth of a tuned circuit

Relative bandwidth	Percentage output (voltage)	Loss (dB)
f/3Q	95	0·2
f/2Q	90	0·46
f/Q	70	1·55
2f/Q	45	3·5
4f/Q	24	6·2
8f/Q	12	9·2

Table 1.4: Selectivity of tuned circuits

Q factor. **Fig 1.30** shows a resonance curve with the centre frequency and the frequencies each side at which the voltage has fallen to $1/\sqrt{2}$ or 0·707 of its peak value, that is the *half power bandwidth*. The magnification or Q factor is also given by:

$$Q = \frac{\text{resonant frequency}}{\text{bandwidth}}$$

where the resonant frequency is f_0 and the bandwidth $f_2 - f_1$ in Fig 1.20.

Example:

A radio receiver is required to cover the long and medium wave broadcast bands, from 148·5kHz to 1606·5kHz. Each station occupies 6kHz of bandwidth. What Q factors are required at each end of the tuning range.

For long wave the Q factor is 148·5/6 ≈ 25. At the top of the medium wave broadcast band the required Q factor is 1606·6/6 ≈ 268. It is not realistic to achieve a Q factor of 268 with one tuned circuit.

Fig 1.31 shows several resonance curves offering different selectivities. Curve A is for a single tuned circuit with a resonant frequency of 465kHz and a Q factor of 100. The width of the curve will be 4·65kHz at 70·7% of the full height. An unwanted signal 4kHz off-tune at 461kHz will still be present with half its voltage amplitude or one quarter of the power.

The dashed curve, B, has a Q of 300; the higher Q giving greater selectivity. In a radio receiver it is common to have several stages of tuning and amplification, and each stage can have its own tuned circuit, all contributing to the overall selectivity. Curve C shows the effect of three cascaded tuned circuits, each

with a Q of 100. In practice this is better than attempting to get a Q of 300 because, as will be seen later chapters, the usual requirement is for a response curve with a reasonably broad top but steeper sides. For a single tuned circuit the response, as a proportion of the bandwidth, is shown in **Table 1.4**. The values are valid for series circuits and parallel circuits where the Q-factor is greater than 10, as is normally the case.

Dynamic Resistance

Fig 1.32 shows a parallel tuned circuit with the resistance of the coil separately identified. This resistance means that the currents I_L and I_C are not exactly equal and do not fully cancel. The impedance of the circuit at resonance is high, but not infinite. This impedance is purely resistive, that is V and I are in phase, and is known as the dynamic impedance or resistance.

To find its value, the series combination of R and L must be transformed into its parallel equivalent.

$$R_P = \frac{R_s^2 + X_s^2}{R_s} \quad \text{and} \quad X_P = \frac{R_s^2 + X_s^2}{X_s}$$

The proof of these transformations is outside the scope of this book.

If the resistance is considerably lower than the reactance of L; normal in radio circuits, then this simplifies to

$$R_P = \frac{X_s^2}{R_s} \quad \text{and} \quad X_P = X_s$$

Remembering that $X_s = 2\pi f L$ and that, at resonance, $2\pi f L = 1/2\pi f C$, then the dynamic resistance, normally written R_D can be written as

$$R_D = \frac{L}{Cr}$$

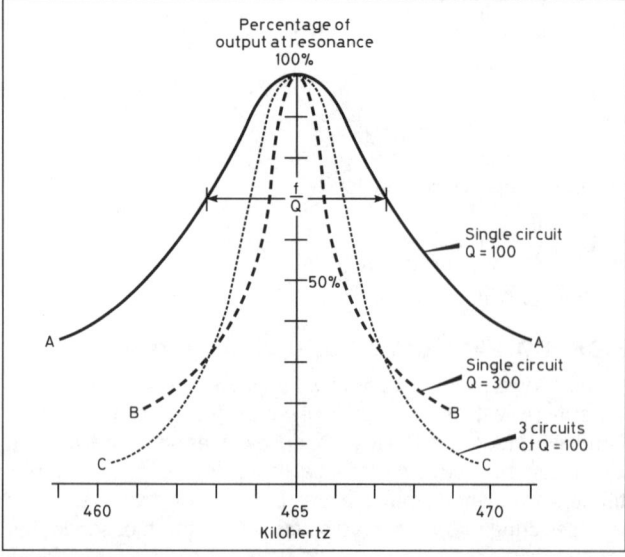

Fig 1.31: Selectivity curves of single and cascaded tuned circuits

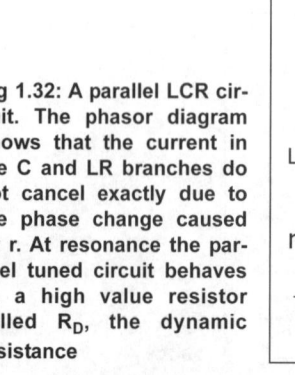

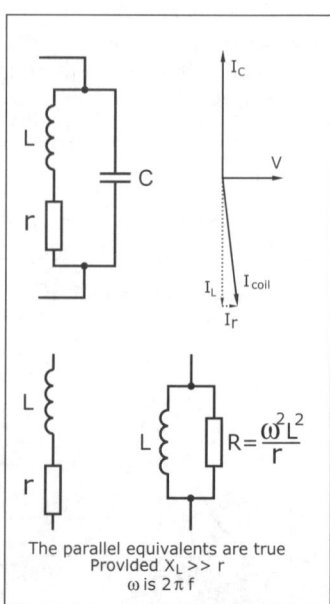

Fig 1.32: A parallel LCR circuit. The phasor diagram shows that the current in the C and LR branches do not cancel exactly due to the phase change caused by r. At resonance the parallel tuned circuit behaves as a high value resistor called R_D, the dynamic resistance

The parallel equivalents are true
Provided $X_L \gg r$
ω is 2πf

$$R = \frac{\omega^2 L^2}{r}$$

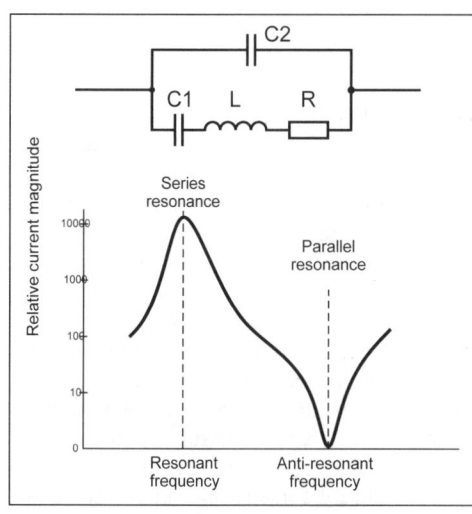

Fig 1.33: Typical variation of current through a quartz crystal with frequency. The resonant and anti-resonant frequencies are normally separated by less than 0·1%; a few hundred hertz for a 7MHz crystal

where L and C are the coil and capacitor values and r is the series resistance of the coil.

Remembering also that Q=2πfL/r, allows another substitution in the formula to get:

Q=2πfR$_D$.

The significance of this is to note that a high R$_D$ implies a high Q.

L/C Ratio

From the formula for R$_D$ above, it can be seen that as well as keeping r as low as practicable, the values of L and C will influence R$_D$. The value of L x C is fixed (from the formula for the resonant frequency) but we can vary the ratio L/C as required with the choice determined by practical considerations.

A high L/C ratio is normal in HF receivers, allowing a high dynamic resistance and a high gain in the amplifier circuits. Stray capacitances limit the minimum value for C. In variable frequency oscillators (VFOs) on the other hand a lower ratio is normal so the capacitor used swamps any changes in the parameters of the active devices, minimising frequency instability.

External components in the rest of the circuit may appear in parallel with the tuned circuit, affecting the actual R$_D$ and reducing overall Q factor. The loading of power output stages has a significant effect and the L/C ratio must then be a compromise between efficiency and harmonic suppression.

Skin Effect

Skin effect is a phenomenon that affects the resistance of a conductor as the frequency rises. The magnetic field round the conductor also exists inside the conductor but the magnetic field at the surface is slightly weaker than at the centre. Consequently the inductance of the centre of the conductor is slightly higher than at the outer surface. The difference is small but at higher frequencies is sufficient that the current flows increasingly close to the surface. Since the centre is not now used, the cross-section of the conductor is effectively reduced and its resistance rises.

The skin depth is the depth at which the current has fallen to 1/e of its value at the surface. (e is the base of natural logarithms). For copper this works out as:

$$\text{skin depth } \delta = \frac{66 \cdot 2}{\sqrt{f}} \quad mm$$

At 1MHz this is 0·066 mm. 26SWG wire, which might be used on a moderately fine coil, has a diameter of 0·46mm, (0·23mm radius) so current is flowing in approximately half the copper cross-section and the resistance will be roughly doubled.

At UHF, 430MHz, however, a self-supporting coil of 16SWG coil (wire radius 0·8mm) the skin depth is 0·003mm and only the surface will carry a current, resulting in a considerable increase in resistance. If possible it will be better to use an even larger wire diameter which has a larger surface or use silver plated wire because the resistivity of silver is lower.

Quartz Crystals

A quartz crystal is a very thin slice of quartz cut from a naturally occurring crystal. Quartz exhibits the piezo-electric effect, which is a mechanical-electrical effect. If the crystal is subjected to a mechanical stress, a voltage is developed between opposite faces. Similarly, if a voltage is applied then the crystal changes shape slightly.

When an AC signal is applied to a crystal at the correct frequency, its mechanical resonance produces an electrical resonance. The resonant frequency depends on the size of the crystal slice. The electrical connections are mde by depositing a thin film of gold or silver on the two faces and connecting two very thin leads.

Below 1MHz, the crystal is usually in the form of a bar rather than a thin slice. At 20kHz the bar is about 70mm long. Up to 22MHz the crystal can operate on its fundamental mode; above that a harmonic or overtone resonance is used. This is close to a multiple of the fundamental resonance but the term 'overtone' is used since the frequency is close to odd multiples of the fundamental but not exact.

Fig 1.33 shows the equivalent circuit of a crystal and the two resonances possible with L and either C1 or C2. The two frequencies are within about 0·1% of each other. The crystal is supplied and calibrated for one particular resonance, series or parallel and should be used in the designed mode.

The key advantage of using a crystal is that its Q factor is very much higher than can be achieved with a real LC circuit. Qs for crystals are typically around 50,000 but can reach 1,000,000. Care must be taken to ensure circuit resistances do not degrade this. There are more details in the oscillators chapter.

Coupled Circuits

Pairs of coupled tuned circuits are often used in transmitters and receivers. The coupling is by transformer action (see later in this chapter), and the two tuned circuits interact as shown in **Fig 1.34**.

When the coupling is loose, that is the coils are separated, the overall response is as shown in curve 1. As the coils are moved closer, the coupling and the output increase until curve 2 is reached which shows *critical coupling*. Closer coupling results in curve 3 which is *over-coupling*. Critical coupling is the closest

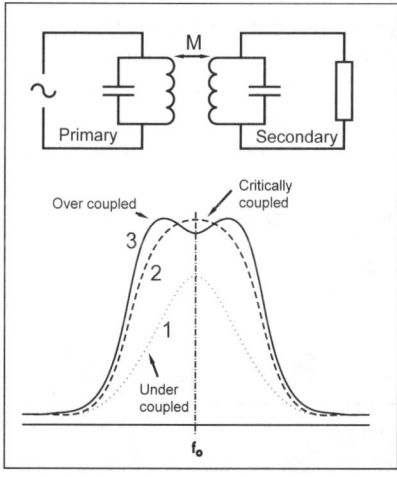

Fig 1.34: Inductively coupled tuned circuits. The mutual inductance M and the coupling increase as the coils are moved together. The response varies according to the degree of coupling with close coupling giving a broad response

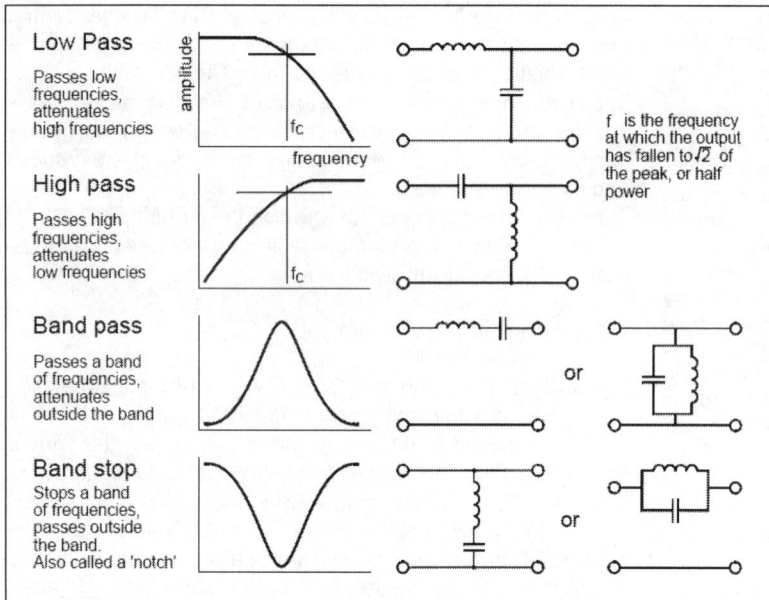

Low Pass

Passes low frequencies, attenuates high frequencies

High pass

Passes high frequencies, attenuates low frequencies

Band pass

Passes a band of frequencies, attenuates outside the band

Band stop

Stops a band of frequencies, passes outside the band. Also called a 'notch'

f is the frequency at which the output has fallen to $\sqrt{2}$ of the peak, or half power

Fig 1.35: The circuits and frequency response of low pass, high pass, band pass and band stop (or notch) filters

coupling before a dip appears in the middle of the response curve. Often some over-coupling is desirable because it causes a broader 'peak' with steep sides which rapidly attenuate signals outside the desired frequency range.

Two tuned circuits are often mounted in a screening can, the coils generally being wound the necessary distance apart on the same former to give the required coupling. The coupling is then said to be fixed.

Some alternative arrangements for coupling tuned circuits are shown in the Building Blocks chapter.

Filters

Filters may be 'passive', that is an array of capacitors and inductors or 'active' where an amplifier and, usually, capacitors and resistors are used. Mostly active filters are used at lower frequencies, typically audio passive LC filters at RF. In both cases the aim is to pass some frequencies and block or attenuate others.

Fig 1.35 shows the four basic configurations; low pass, high pass, band pass and band stop or notch; together with the circuit of a typical passive filter, and the circuits for passive filters. The Building Blocks chapter discusses the topic further.

Transformers

Mutual inductance was introduced in Fig 1.16(b); a changing current in one coil inducing an EMF in another. This is the basis of the transformer. In radio frequency transformers the degree of coupling, the mutual inductance, was one of the design features. In power transformers, the windings, the primary and secondary, are tightly coupled by being wound on a bobbin sharing the same laminated iron core to share and maximise the magnetic flux (the name for magnetic 'current'). The size of core used in the transformer depends on the amount of power to be handled.

A transformer can provide DC isolation between the two coils and also vary the current and voltage by varying the relative number of turns. **Fig 1.36** shows a transformer with n_P turns on the primary and n_S turns on the secondary. Since the voltage induced in each coil will depend on the number of turns, we can conclude that:

$$V_S = V_P \times \frac{n_S}{n_P}$$

With a load on the secondary (the output winding) a secondary current. I_S will flow. Recognising that, neglecting losses, the power into the primary must equal the power out of the secondary, we can also conclude that

$$I_P = I_S \times \frac{n_S}{n_P}$$

It is simple to say that if power is drawn from the secondary, the primary current must increase to provide that power, but that does not really provide an explanation. When current flows in the secondary the magnetic field due to the secondary current will weaken the overall field, resulting in a reduced back EMF in the primary. This allows more primary current to flow, restoring the field. The primary current does not fall to zero with no secondary load, the residual current, known as the *magnetising current*, is that which would flow through the inductance of the primary. Normally this current and any small losses in the transformer can be neglected but it should be noted that the windings do rise in temperature during use and an overload could cause breakdown of the internal insulation or start a small fire before the fuse blows.

Since the current and voltages can be changed, it follows that the impedances, given by $Z_P = V_P/I_P$ and $Z_S = V_S/I_S$ will also change. If, for a step-down transformer, the secondary voltage is halved, the secondary current will be double the primary current. The impedance has reduced to a quarter of the primary value.

$$Z_P = Z_S \left(\frac{n_P}{n_S} \right)^2$$

The transformation of impedance is a valuable property and transformers are widely used for impedance matching. Several examples will be seen in the transmitters chapters.

Auto-transformers

If DC isolation between the primary and secondary is not required, an auto-transformer can be used, which has a single winding. It is tapped at an appropriate point so that, for a step-down transformer, the whole winding forms the primary but only a few of those turns form the secondary. An example is a mains transformer with 1000 turns, tapped at 478 turns to give a 110V output from the 230V supply. The advantage of the auto-transformer is partly simplified construction but mainly space and weight saving.

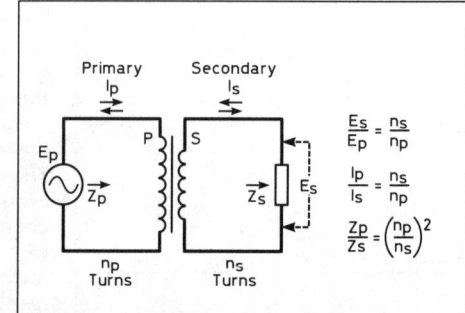

$$\frac{E_S}{E_P} = \frac{n_S}{n_P}$$

$$\frac{I_P}{I_S} = \frac{n_S}{n_P}$$

$$\frac{Z_P}{Z_S} = \left(\frac{n_P}{n_S}\right)^2$$

Fig 1.36: The low-frequency transformer

The Radio Communication Handbook

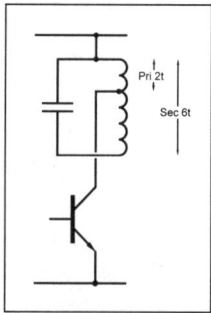

Fig 1.37: An autotransformer in a parallel tuned circuit. Transformer tapping will reduce the loading of the transistor on the tuned circuit R$_D$

Fig 1.37 shows a step-up transformer in a tuned circuit in the collector of a transistor. The loading effect of the transistor will be considerably reduced, thereby preserving the Q of the unloaded tuned circuit.

Screening

When two circuits are near one another, unwanted coupling may exist between them due to stray capacitance between them, or due to stray magnetic coupling.

Placing an earthed screen of good conductivity between the two circuits, as shown in **Fig 1.38(b)**, can eliminate stray capacitance coupling. There is then only stray capacitance from each circuit to earth and no direct capacitance between them. A useful practical rule is to position screens so that the two circuits are not visible from one another.

Stray magnetic coupling can occur between coils and wires due to the magnetic field of one coil or wire intersecting the other. At radio frequency, coils can be inductively screened (as well as capacitively) by placing them in closed boxes or cans made from material of high conductivity such as copper, brass or aluminium. Eddy currents are induced in the can, setting up

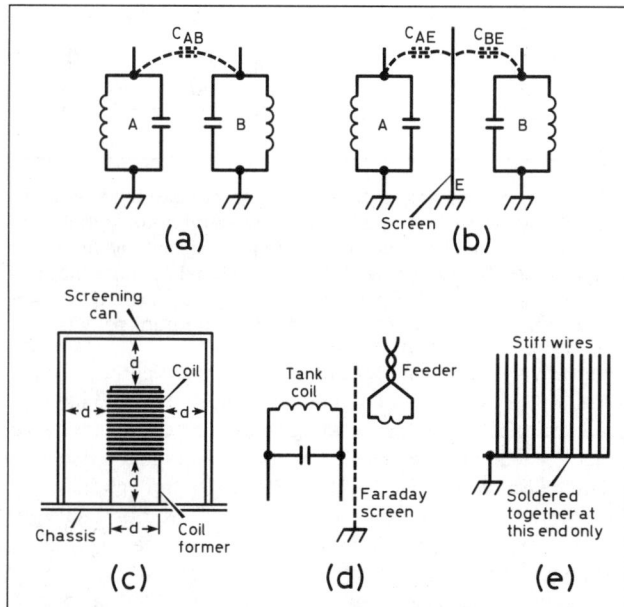

Fig 1.38: (a) Stray capacitance coupling C$_{AB}$ between two circuits A and B. The introduction of an earthed screen E in (b) eliminates direct capacitance coupling, there being now only stray capacitance to earth from each circuit C$_{AE}$ and C$_{BE}$. A screening can (c) should be of such dimensions that it is nowhere nearer to the coil it contains than a distance equal to the diameter of the coil d. A Faraday screen between two circuits (d) allows magnetic coupling between them but eliminates stray capacitance coupling. The Faraday screen is made of wires as shown at (e)

a field which opposes and practically cancels the field due to the coil beyond the confines of the can.

If a screening can is too close to a coil the performance of the coil, ie its Q and also its inductance, will be considerably reduced. A useful working rule is to ensure the can in no closer to the coil than its diameter, see **Fig 1.38(c)**.

At low frequencies eddy current screening is not so effective and it may be necessary to enclose the coil or transformer in a box of high-permeability magnetic material such as Mumetal in order to obtain satisfactory magnetic screening. Such measures are not often required but a sensitive component such as a microphone transformer may be enclosed in such a screen in order to make it immune from hum pick-up.

It is sometimes desirable to have pure inductive coupling between two circuits with no stray capacitance coupling. In this case a Faraday screen can be employed between the two coils in question, as shown in Fig 1.27(d). This arrangement is sometimes used between an antenna and a receiver input circuit or between a transmitter tank circuit and an antenna. The Faraday screen is made of stiff wires (**Fig 1.38e**) connected together at one end only, rather like a comb. The 'open' end may be held by non-conductive material. The screen is transparent to magnetic fields because there is no continuous conducting surface in which eddy currents can flow. However, because the screen is connected to earth it acts very effectively as an electrostatic screen, eliminating stray capacitance coupling between the circuits.

SEMICONDUCTORS

Early semiconductors were based on germanium but this has now been replaced by silicon, which forms the basis of most transistors and diodes. Specialist devices may be formed of the compound gallium arsenide.

A simplified picture of the silicon atom is shown in **Fig 1.39**. Silicon has an atomic number of 14 indicating it has 14 electrons and 14 protons. The electrons are arranged in shells that must be filled before starting the next outermost layer. It has four electrons in its outer shell, which are available for chemical bonding with neighbouring atoms, forming a crystal lattice. A single near perfect crystal can be grown from a molten pool of pure material. As a pure crystal, the four outer electrons are all committed to bonding and none are available to support the flow of current. Pure crystaline silicon is an insulator. To obtain the semiconductor, carefully controlled impurities are added. This is known as doping.

If a small quantity of an element with five outer electrons, for example phosphorus, antimony or arsenic, is added at around 1 part in 10^7, then some of the atoms in the lattice will now appear to have an extra electron. Four will form bonds but the fifth is unattached. Although it belongs to its parent nucleus, it is relatively free to move about and support an electric current. Since this material has mobile electrons it is called an n-type semiconductor.

If an element with three outer electrons is added to the crystal, eg indium, boron or aluminium, these three electrons form bonds but a hole is left in the fourth place. An electron that is part of a nearby bond may move to complete the vacant bond.

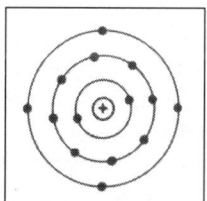

Fig 1.39: The silicon atom

Fig 1.40: Signals on balanced and unbalanced lines

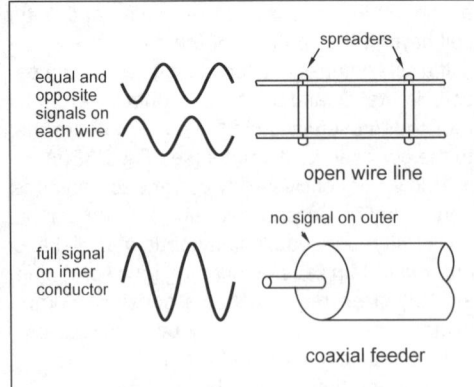

equal and opposite signals on each wire

spreaders

open wire line

full signal on inner conductor

no signal on outer

coaxial feeder

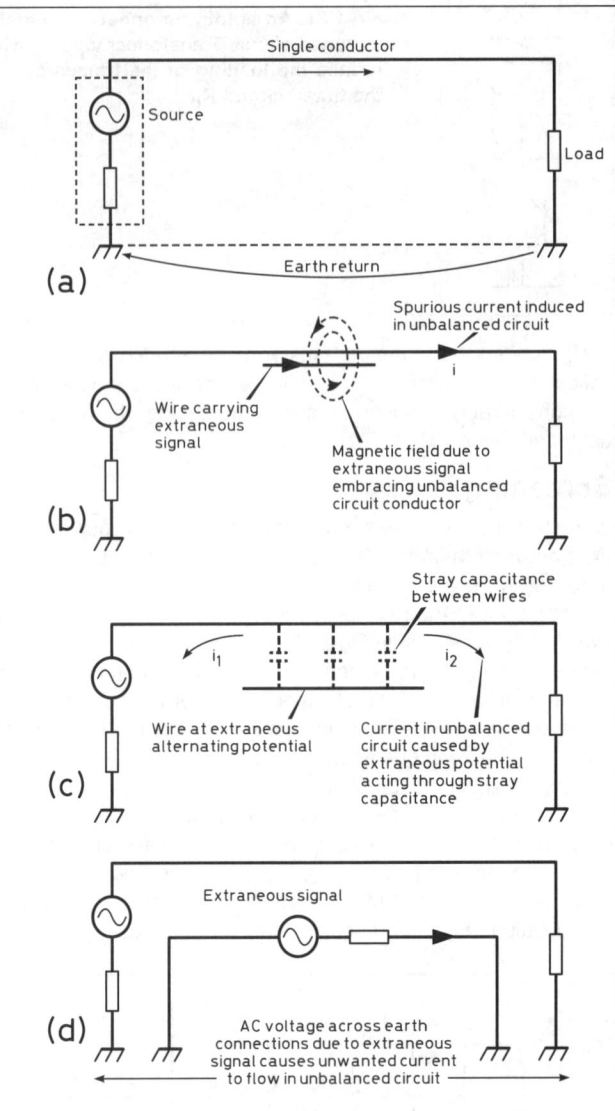

(a)

Single conductor

Source

Load

Earth return

(b)

Spurious current induced in unbalanced circuit

Wire carrying extraneous signal

Magnetic field due to extraneous signal embracing unbalanced circuit conductor

i

(c)

Stray capacitance between wires

i_1

i_2

Wire at extraneous alternating potential

Current in unbalanced circuit caused by extraneous potential acting through stray capacitance

(d)

Extraneous signal

AC voltage across earth connections due to extraneous signal causes unwanted current to flow in unbalanced circuit

Fig 1.41: The unbalanced circuit. (a) The basic unbalanced circuit showing earth return path. (b) How extraneous signals and noise can be induced in an unbalanced circuit by magnetic induction. (c) Showing how extraneous signals can be induced by stray capacitance coupling. (d) Showing how extraneous signals can be induced due to a common earth return path

The new hole being filled, in turn, by another nearby electron. It is easier to consider the hole as being the mobile entity than to visualise several separate electrons each making single jumps. This material is a p-type semiconductor.

A *diode* is formed from a p-n junction and will only allow current to flow in one direction. This is used in rectification, changing AC into unidirectional half-cycles which can then be smoothed to conventional DC using a large capacitor.

A transistor uses a three layer semiconductor device formed either as n-p-n or, perhaps less common, as p-n-p sandwich. The transistor may act as an amplifier, increasing the amplitude or power of weak signals, or it can act as a switch in a control circuit. These uses, and more detailed discussion of the operation of transistors and diodes, are contained in this book.

BALANCED AND UNBALANCED CIRCUITS

Fig 1.40 shows a balanced and unbalanced wire or line. The balanced line (top) has equal and opposite signals on the two conductors, the voltages and currents are of equal magnitude but opposite in direction. For AC, this would mean a 180 degree phase difference between the two signals. The input and output from the line is taken between the two conductors. If each conductor had a 2V AC signal, the potential difference between the two conductors would be 4V.

The unbalanced line (bottom) has a single conductor and an 'earth return' which is at 0V. In the example above, the full 4V signal would be on the single (centre) conductor. Most electrical circuits are unbalanced. In a car the 'live' side is at +12V from the battery and the chassis of the car is often used as the earth return path.

In line telegraph use it was common for there to be a single wire, often alongside railway tracks and the earth literally was used as the return path to complete the circuit.

Any wire carrying a changing current has a tendency to radiate part of the signal as electromagnetic waves or energy. This is the principle of the antenna (aerial). The efficiency with which radiation occurs is a function of the length of the wire as a proportion of the wavelength of the signal. This is covered later in the chapters on HF and VHF/UHF antennas. A radiator can equally pick up any stray electromagnetic signals which will then get added to the wanted signal in the wire.

Unbalanced circuits are particularly prone to this effect. **Fig 1.41** shows an unbalanced wire with magnetic (mutual inductance) coupling and capacitive coupling to another nearby conductor. Circuits of this type are very commonly used in radio equipment and are perfectly satisfactory provided leads are kept short and are spaced well away from other leads. It is, however, prone to the pick-up of extraneous noise and signals from neighbouring circuits by three means: inductive pick-up, capacitive pick-up and through a common earth return path.

Inductive pick-up, **Fig 1.41(b)**, can take place due to transformer action between the unbalanced circuit wire and another nearby wire carrying an alternating current; a common example is hum pick-up in audio circuits due to the AC mains wiring.

Capacitive pick-up, **Fig 1.41(c)**, takes place through the stray capacitance between the unbalanced circuit lead and a neighbouring wire. Such pick-up can usually be eliminated by introducing an earthed metal screen around the connecting wire.

If the unbalanced circuit has an earth return path that is *common* to another circuit, **Fig 1.41(d)**, unwanted signals or noise may be injected by small voltages appearing between the two earth return points of the unbalanced circuit. Interference of this type can be minimised by using a low-resistance chassis and avoiding common earth paths as far as possible.

A balanced circuit is shown in **Fig 1.42**. As many signal sources, and often loads as well, are inherently unbalanced (ie one side is earthed) it is usual to use transformers to connect a source of signal to a remote load via a balanced circuit. In the balanced circuit, separate wires are used to conduct current to

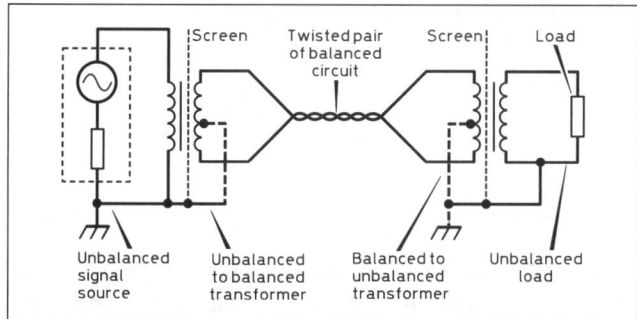

Fig 1.42: The balanced circuit

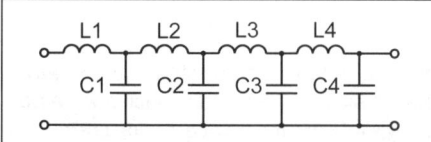

Fig 1.43: A long feeder can be represented by series inductors and parallel capacitors

and back from the load; no current passes through a chassis or earth return path.

The circuit is said to be 'balanced' because the impedances from each of the pair of connecting wires to earth are equal. It is usual to use twisted wire between the two transformers as shown in Fig 1.42. For a high degree of balance, and therefore immunity to extraneous noise and signals, transformers with an earthed screen between primary and secondary windings are used. In some cases the centre taps of the balanced sides of the transformers are earthed as shown dotted in Fig 1.42.

The balanced circuit overcomes the three disadvantages of the unbalanced circuit. Inductive and capacitive pick-up are eliminated since equal and opposite currents are induced in each of the two wires of the balanced circuit and these cancel out. The same applies to interfering currents in the common earth connection in the case where the centre taps of the windings are earthed.

The argument also applies to radiation from a balanced circuit. The two conductors will tend to radiate equal and opposite signals which will cancel out. Some care is required however because if conducting objects are close to the feeder, comparable in distance to the separation of the two conductors, then the layout may not be symmetrical and some radiation due to the imbalance may occur.

Looking again at Fig 1.40, the balanced feeder must be symmetrically run, away from walls and other conductors, but the unbalanced feeder has an earthed conducting sleeve, typically an outer braid or metal tape to screen the centre conductor and minimise radiation outside the cable. This is known as coaxial cable. It may be run close to walls and conductors but the manufacturers advice on minimum bending radius should be heeded.

FEEDERS

The feeder is the length of cable from the transmitter/receiver to the antenna. It must not radiate, using the properties of balance or screening outlined above and it must be as loss free as possible. Consider the circuit shown in **Fig 1.43.** Each short length of feeder is represented by its inductance, and the capacitance between the conductors is also shown. Assume now a battery is connected to the input. The first length of cable will charge up to the battery voltage but the rate of rise will be limited by the inductance of L1 and the need for C1 to charge. This will occur progressively down the cable, drawing some current from the battery all the while.

Clearly with a short cable, this will take no time at all. However an infinitely long cable will be drawing current from the battery for quite some time. The ratio of the battery voltage to the current drawn will depend on the values of L and C and is the characteristic impedance of the cable.

It is given by:

$$Z_0 = \sqrt{\frac{L}{C}}$$

where L and C are the values per unit length of the feeder.

This argument and the formula assume the series resistance of the feeder is negligible, as is the leakage resistance of the insulation. This is acceptable from the point of view of Z_0 but the series resistance is responsible, in part, for feeder losses; some of the power is lost as heat. The other facet of the loss is the dielectric loss in the insulation, that is the heating effect of the RF on the plastic, easiest viewed as the heating of the plastic in a microwave oven. The loss is frequency dependant and, to a simple approximation, rises as the square root of the frequency.

A very long length of feeder will look like a resistance of Z_0. So will a shorter length terminated in an actual resistor of value Z_0. Signals travelling down the feeder will be totally absorbed in a load of value Z_0 (as if it were yet more feeder) but if the feeder is terminated in a value other than Z_0, some of the energy will be absorbed and some will be reflected back to the source. This has a number of effects which are discussed in the chapters on antennas. Properly terminating the feeder in its characteristic impedance (actually a pure resistance) is termed correct matching. It is also, of course, necessary to use a balanced load on a balanced line (twin feeder) and an unbalanced load on an unbalanced line such as coaxial cable.

THE ELECTROMAGNETIC SPECTRUM

Radio frequencies are regarded by the International Telecommunication Union to comprise frequencies from 9kHz to 400GHz but the upper limit rises occasionally due to advances in technology. This, however is only a small part of the electromagnetic spectrum shown in **Fig 1.44.**

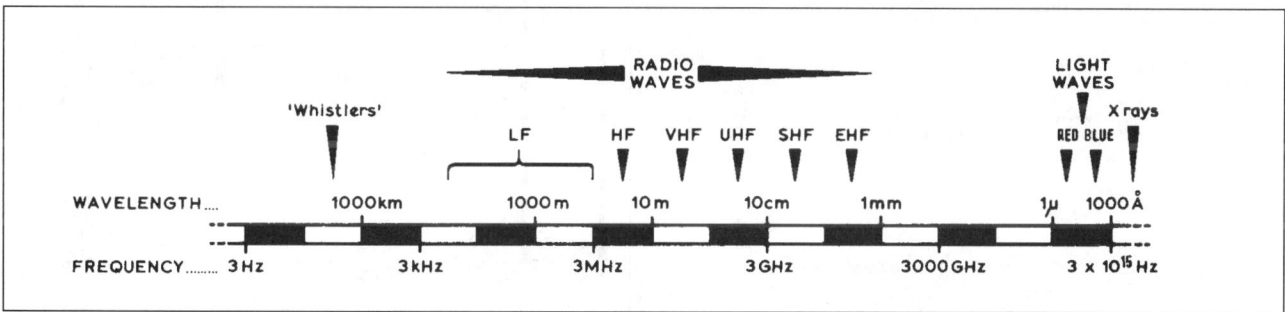

Fig 1.44: The electromagnetic spectrum

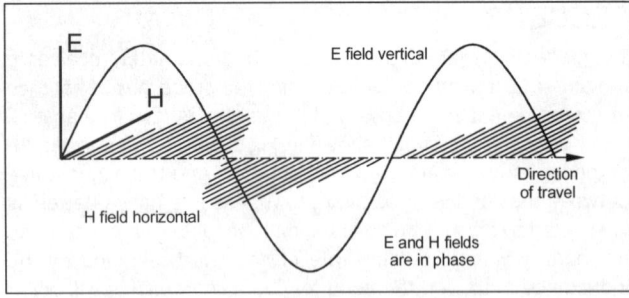

Fig 1.45: An electromagnetic wave. By convention the polarisation is taken from the electric field, so this is a vertically polarised wave

The whole spectrum comprises radio waves, heat, light, ultraviolet, gamma rays, and X-rays. The various forms of electromagnetic radiation are all in the form of oscillatory waves, and differ from each other only in frequency and wavelength. They all travel through space with the same speed, approximately 3×10^8 metres per second. This is equivalent to about 186,000 miles per second or once round the world in about one-seventh of a second.

As might be surmised from the name, an electromagnetic wave consists of an oscillating electric field and a magnetic field. It can exist in a vacuum and does not need a medium in which to travel. **Fig 1.45** shows one and a half cycles of an e-m wave. The electric (E) and magnetic (H) fields are at right angles and both are at right angles to the direction of propagation. E and H are in phase and their magnitudes are related by the formula:

$$\frac{E}{H} = \sqrt{\frac{\mu_0}{\varepsilon_0}} = 120\pi \ \Omega$$

where μ_0 is the permeability of free space and ε_0 is the permitivity of free space, both natural physical constants. The quantity E/H is known as the impedance of free space and may be likened to the characteristic impedance of a feeder.

Frequency and Wavelength

The distance travelled by a wave in the time taken to complete one cycle of oscillation is called the *wavelength*. It follows that wavelength, frequency and velocity of propagation are related by the formula

Velocity = Frequency x Wavelength, or c = fλ

where c is the velocity of propagation, f is the frequency (Hz), and λ is the wavelength in metres.

Example:

What are the frequencies corresponding to wavelengths of (i) 150m, (ii) 2m and (iii) 75cm?

From the formula c = fλ, the frequencies are given by:

150m

$$f = \frac{c}{\lambda} = \frac{3 \times 10^8}{150} = \frac{300 \times 10^6}{150} = 2 \cdot 0 \text{MHz}$$

2m:

$$f = \frac{c}{\lambda} = \frac{3 \times 10^8}{2} = \frac{300 \times 10^6}{2} = 150 \text{MHz}$$

75cm:

$$f = \frac{c}{\lambda} = \frac{3 \times 10^8}{0 \cdot 75} = \frac{300 \times 10^6}{0 \cdot 75} = 400 \text{MHz}$$

It is important to remember to work in metres and hertz in these formula. However, it will also be noticed that if the frequency is expressed in MHz throughout and λ in metres, a simplified formula is:

$$f = \frac{300}{\lambda} \text{ MHz}$$

or

$$\lambda = \frac{300}{f} \text{ metres}$$

Fig 1.46 shows how the radio spectrum may be divided up into various bands of frequencies, the properties of each making them suitable for specific purposes. Amateur transmission is permitted on certain frequency bands in the LF, MF, HF, VHF and UHF, SHF/microwave ranges.

ANTENNAS

An antenna (or aerial) is used to launch electromagnetic waves into space or conversely to pick up energy from such a wave travelling through space. Any wire carrying an alternating current will radiate electromagnetic waves and conversely an electromagnetic wave will induce a voltage in a length of wire. The issue in antenna design is to radiate as much transmitter power as possible in the required direction or, in the case of a receiver, to pick up as strong a signal as possible, very often in the presence of local interference.

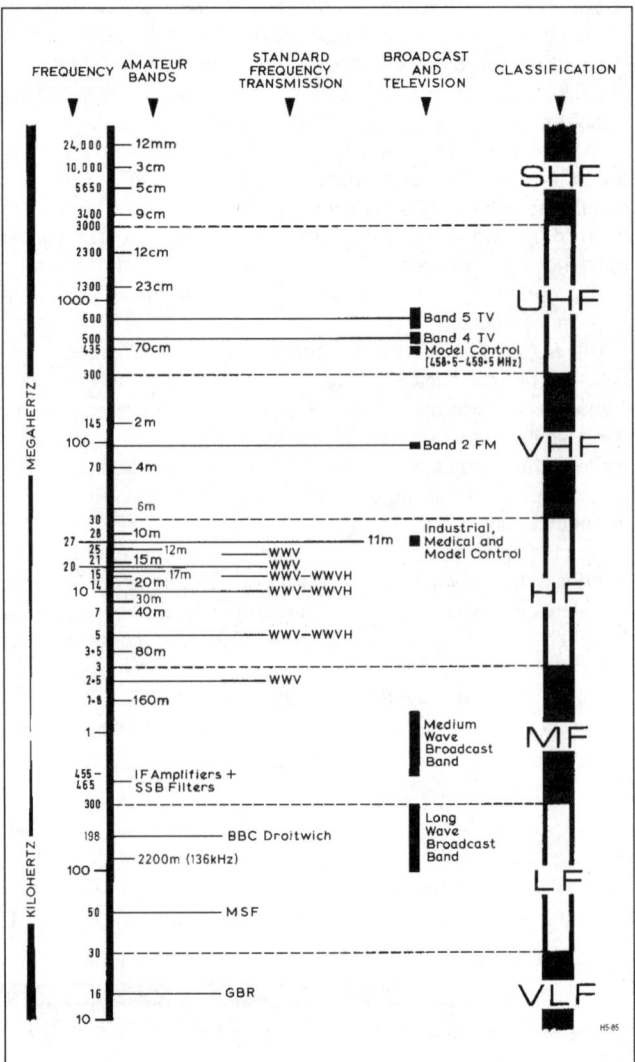

Fig 1.46: Amateur bands in relation to other services

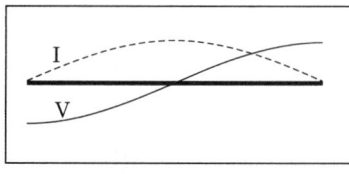

Fig 1.47: Current and voltage distribution on a half-wave dipole

Isotropic Radiator

An isotropic radiator is one that radiates equally in all, three-dimensional, directions. In practical terms, it does not exist but it is easy to define as a concept and can be used as a reference.

The power flux density p from such an antenna, at a distance r is given by:

$$p = \frac{P}{4\pi r^2} \quad W/m^2$$

where P is the power fed to the antenna.

This leads to the electric field strength E, recalling the formula for the impedance of free space above:

$$p = \frac{E^2}{R} = \frac{E^2}{120\pi} = \frac{P}{4\pi r^2}$$

rearranging gives

$$E = \frac{\sqrt{30P}}{r} = \frac{5\cdot5\sqrt{P}}{r}$$

Note that this formula is for an isotropic radiator.

Radiation Resistance

The radiation resistance of the antenna can be regarded as:

$$R_r = \frac{\text{total power radiated}}{\left(\text{RMS current at antenna input}\right)^2}$$

This will vary from one antenna type to another. Electrically short antennas, much less than a wavelength, have very low radiation resistances leading to considerable inefficiency if this approaches the resistance of the various conductors. This resistance is not a physical resistor but the antenna absorbs power from the feeder as if it were a resistor of that value.

Antenna Gain

Antennas achieve gain by focussing the radiated energy in a particular direction. This leads to the concept of Effective Radiated Power (ERP) which is the product of the power fed to the antenna multiplied by the antenna gain. The figure can be regarded as equivalent to the power required to be fed to an antenna without gain to produce the same field strength (in the wanted direction) as the actual antenna.

It should be noted that this gain also applies on receive because antennas are passive, reciprocal devices that work identically on transmit or receive.

Effective Aperture

Consider a signal of field strength E arriving at an antenna and inducing an EMF. This EMF will have a source resistance of R_r, the radiation resistance and will transfer maximum power to a load of the same value. This power can be viewed as the area required to capture sufficient power from the incident radio wave.

An antenna of greater gain will have a larger effective aperture; the relationship is:

$$A_{eff} = \frac{\lambda^2}{4\pi} \times G$$

where λ is the wavelength of the signal and G is the antenna gain in linear units.

This aperture will be much larger that the physical appearance of a high gain antenna and is an indication of the space required to allow the antenna to operate correctly without loss of gain.

A Dipole

The current and voltage distribution on a dipole are shown in **Fig 1.47**. The current is a half sine wave. By integrating the field set up by each element of current over the length of the dipole, it can be shown that the dipole has a gain of 1·64 or 2·16dB and a radiation resistance (or feed impedance) of 73 ohms (see below for a discussion of dB, decibels).

The dipole is often used as the practical reference antenna for antenna gain measurements. It must then be remembered that the gain of an antenna quoted with reference to a dipole is 1·64 times or 2·16dB less than if the gain is quoted with reference to isotropic.

Effective Radiated Power

The effective radiated power (ERP) of an antenna is simply the product of the actual power to the antenna and the gain of the antenna. Again it is necessary to know if this is referenced to isotropic (EIRP) or a dipole (EDRP). ERP is normally quoted with reference to a dipole. This figure indicates the power that would need to be fed to an actual dipole in order to produce the field strength, in the intended direction, that the gain antenna produces. The total radiated power is still that supplied to the antenna, the enhancement in the intended direction is only as a result of focussing the power in that direction. A side benefit, but an important one, is that the power in other directions is much reduced.

The field from such an antenna is given by:

$$E = \frac{7\cdot01\sqrt{ERP}}{d} \quad V/m$$

where d is the distance from the antenna (shown as r above for the isotropic radiator).

The different coefficient of 7·01 rather than 5·5 accounts for the fact that the ERP is reference to a dipole, which already has some gain over isotropic.

Feeding an Antenna

The concept of maximum power transfer applies to an antenna, but there is a additional complication if the antenna is not a resonant length. Fig 1.47 showed the current and voltage distribution on the dipole. The dipole is a half wavelength long and the current can be viewed as millions of electrons 'sloshing' from one end to the other rather like water in a bath.

As a half wavelength (or a multiple) the antenna is resonant and can be regarded as an LC resonant circuit with a resistor, the radiation resistance R_r, to account for the power radiated. Below resonance the antenna will appear partially inductive, and capacitive above resonance.

This means that the antenna will no longer appear as a good match to the feeder and some power will be reflected back down the feeder. This, in turn, will result in the feeder being a poor match to the transmitter resulting in less than maximum power being transferred. To recover this situation, the opposite type of reactance must be inserted to offset that presented to the transmitter. The antenna/feeder/inserted reactance now represents a pure resistance to the transmitter. This function is usually performed by an antenna tuning (or matching) unit and is covered more fully in the chapter on HF antennas.

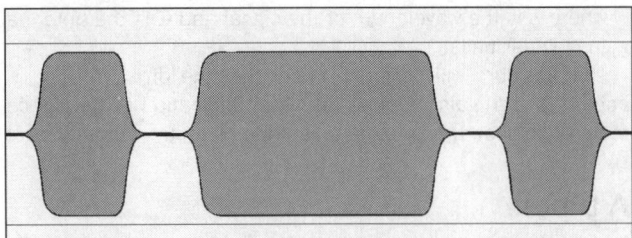

Fig 1.48: The Morse letter R

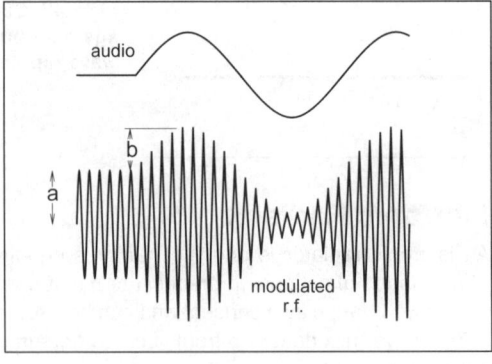

Fig 1.49: Amplitude modulation

MODULATION

A simple sine wave radio signal conveys no information except that it is there and perhaps the direction of its origin. In order to convey information we must change the signal in some agreed way, a process called modulation.

The simplest form of modulation is to turn the signal on and off and this was all that early pioneers of radio could achieve. This led to the development of the Morse code, which remains very effective at sending text messages and is still practiced by many amateurs although Morse has given way to other techniques in commercial applications. A Morse signal is shown in **Fig 1.48**. It will be noticed that the edges of the Morse are rounded off; the reason will be explained shortly in the section below. It is called CW or carrier wave because the carrier (the sine wave signal) is all that is transmitted. However we now know that is not quite correct.

Amplitude Modulation (AM)

In amplitude modulation, the amplitude of the carrier is varied according to the instantaneous amplitude of the audio modulating signal as shown in **Fig 1.49**. Only the carrier is transmitted in the absence of modulation, the audio (or other) signal increasing and decreasing the transmitted amplitude. The depth of modulation, m, is given by the quantity B/A or B/A x 100%.

It is desirable to achieve a reasonably high level of modulation but the depth must never exceed 100%. To do so results in distortion of the audio signal, the production of audio harmonics and a transmitted signal that is wider than intended and likely to splatter over adjacent frequencies causing interference.

Sidebands

The process of modulation leads to the production of sidebands. These are additional transmitted frequencies each side of the carrier, separated from the carrier by the audio frequency. This is first shown mathematically and then the result used to demonstrate the effect in a more descriptive manner.

A single carrier has the form:

$$A \sin \omega_c t$$

where A is the amplitude and ω_c is $2\pi f_c$, the frequency of the signal. When amplitude modulated the formula becomes:

$$A(1 + m \sin \omega_a t) \sin \omega_c t$$

where m is the depth of modulation and ω_a the audio frequency used to modulate the carrier ω_c.

Expanding this gives:

$$A \sin \omega_c t + Am \sin \omega_a t \times \sin \omega_c t$$

The second term is of the form sinA sinB which can be represented as
½(sin(A-B)+sin(A+B)), so our signal becomes:

$$A \sin \omega_c t + \frac{Am}{2}(\sin(\omega_c t - \omega_a t) + \sin(\omega_c t + \omega_a t))$$

Inspection will show that these three terms are the original carrier plus two new signals, one at a frequency f_c - f_a and the other at f_c+f_a and both at an amplitude dependant on the depth of modulation but with a maximum amplitude of half the carrier. This is shown in **Fig 1.50**.

It still might not be all that clear just why the sidebands need to exist. Consider the effect of just the carrier and one of the sidebands but as if they were both high pitch audio notes close together in frequency. A beat note would be heard. Add the other sideband, giving the same beat note. The overall effect is the carrier varying in amplitude.

The bandwidth needed for the transmission depends on the audio or modulating frequency. The two sidebands are equidistant from the carrier so the total bandwidth is twice the highest audio frequency.

Modulating the carrier also increases the transmitted power. The formula above gave the amplitude (voltage) of the signal. If the power in the carrier is P_c then the total transmitted power is:

$$P_c + 2 \times P_C \left(\frac{m}{2}\right)^2 = P_c \left(1 + \frac{m^2}{2}\right)$$

For 100% modulation, the total transmitted power is 150% of the carrier power, that is the carrier plus 25% in each sideband.

It can now be shown why the keying of the Morse signal was rounded. Morse can be regarded as amplitude modulation with only two values, 0 or 100% modulation. Simplistically, there would be sidebands separated from the carrier dependant on the keying speed. That is not the whole story however.

The rise and fall of the transmitted envelope greatly affect the overall bandwidth. **Fig 1.51** shows the Morse keying with fast and slow rises and falls of the keying and thus the transmitted RF envelope. The fast rise can be regarded as a small part of a

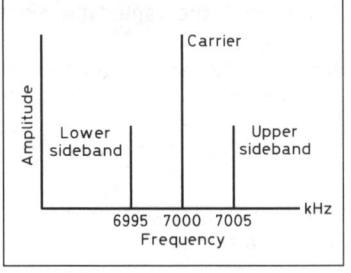

Fig 1.50: Sideband spectrum of a 7MHz carrier which is amplitude modulated by a 5kHz tone

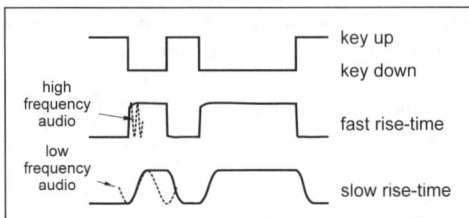

Fig 1.51: Fast and slow rise times on a CW keying signal

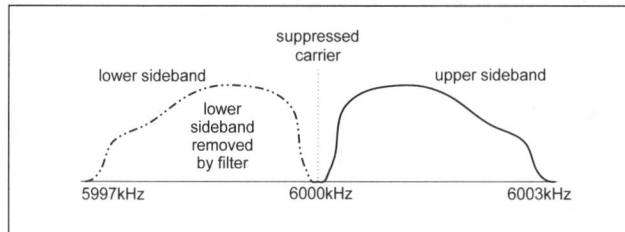

Fig 1.52: SSB signal from a balanced modulator, with one sideband filtered out

high frequency audio signal; a slower rise time being part of a lower frequency audio signal. The sidebands at the moment of keying will be those applicable to the 'audio'. The bandwidth will be much wider than that assumed from the Morse speed alone. A well designed keying circuit is inaudible outside the tuning range over which the Morse tone can be heard. Bad keying causes 'key clicks' that can be heard several kHz each side of the Morse signal proper. It should be avoided.

Single Sideband (SSB)

Fig 1.52 shows an AM signal where the audio is speech containing a range of frequencies from just above zero to 3kHz. The upper sideband is a copy of the audio, albeit shifted up in frequency to just above the carrier, and the lower sideband is a mirror image just below the carrier. In fact all the audio information is contained in the sidebands; as stated at the very beginning, the carrier contains no information. We can, therefore, dispense with the carrier and either one of the sidebands. In the figure the lower sideband is shown as filtered out and the carrier has been suppressed.

The modulation and detection techniques needed to do this are shown in the chapters on receivers and transmitters. SSB transmission requires only half the bandwidth and represents a considerable saving in power, as evidenced by the formula for the power in the various components of the AM signal above.

It will be recalled that the frequency of the audio signal was retained in the transmitted signal by the frequency separation between the carrier and the sideband. If the carrier and the other sideband are removed the reference point for determining the audio frequency is lost. It is 're-inserted' at the receiver by the act of tuning in the signal. The receiver must be accurately tuned or all the received frequencies will be offset by the same amount, namely the error in the tuning. This is why the pitch of the audio of an SSB signal varies as the receiver is tuned.

The bandwidth of an SSB signal is simply the audio bandwidth, half that of the AM transmission. The power varies continuously from zero, when unmodulated, to its peak value depending on the audio volume and the designed power of the transmitter.

Frequency Modulation (FM)

In FM, the instantaneous frequency of the carrier is varied according to the instantaneous amplitude of the carrier as shown in **Fig 1.53**. Unlike AM there is no natural limit to the amount by which the frequency may be deviated from the nominal centre frequency. The peak deviation is a system design parameter chosen as a compromise between audio quality and occupied bandwidth. The ability of an FM detector or discriminator to combat received noise increases as the cube of the deviation. However, the total received noise increases linearly with the bandwidth so the overall effect is a square law. For instance, doubling the deviation and bandwidth improves the recovered audio signal to noise ratio by a factor of four (6dB).

For FM the deviation ratio, corresponding to the depth of modulation of AM, is defined as:

$$\text{Deviation ratio} = \frac{\text{actual deviation}}{\text{peak deviation}}$$

The modulation index relates the peak deviation to the maximum audio frequency:

$$\text{Modulation index} = \frac{\text{peak deviation}}{\text{audio frequency}}$$

A narrow band FM system is one where the modulation index is about or less than unity. Amateur FM systems are narrow band (NBFM). At 2m, with 12·5kHz channels, the max audio frequency is around 2·8kHz and the peak deviation 2·5kHz; a mod index of 0·9. At 70cm, with 25kHz-spaced channels, the calculation is 5/3·5, a mod index of 1·4.

VHF broadcast stations have an audio range up to 15kHz and a peak deviation of 75kHz, giving a mod index of 5. This is a wide band FM system (WBFM). TV sound has a deviation of 50kHz but good quality TV sound systems now use the NICAM digital sound channel which also offers stereo reproduction. Simplistically, the mod index will give a guide to the audio quality to be expected.

The bandwidth of an FM transmission is considerably wider than twice the peak deviation, as might be expected. The modulation gives rise to sidebands but even for a single audio modulating tone, there may be several sidebands. The mathematical derivation is much more complex and the 'sin' function requires Bessel functions to calculate. The formula is:

$$A \sin(\omega_c t + m \sin \omega_a t)$$

where m is the modulation index.

This gives a transmitted spectrum shown in **Fig 1.54** which has, theoretically, an infinite number of sidebands of decreasing amplitude. The rule of thumb often applied, known as Carson's rule, is that the necessary bandwidth is given by:

Bandwidth = 2(peak deviation + max audio frequency)

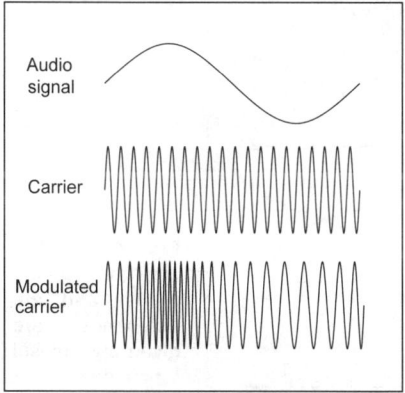

Fig 1.53: Frequency modulation

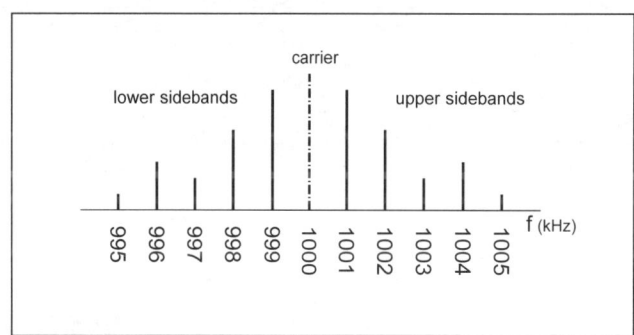

Fig 1.54: The spectrum of an FM signal

For a typical 2m system this works out as:

Bandwidth = 2(2·5 + 2·8) = 10·6kHz

The number of relevant sidebands depends on the modulation index, according to the graph in **Fig 1.55**. This shows the carrier and the first three sidebands. For those with access to *Microsoft Excel®*, the function BesselJ, available in the maths and statistics pack, will give the values. This pack is provided on the CD but frequently not loaded. The help menu gives guidance on how to load the pack if it is not loaded.

It is interesting to note that the amplitude of the carrier also varies with the modulation index, and goes to zero at an index of 2·4. This provides a very powerful method to set the deviation. Set the transmitter up with an audio signal generator at full amplitude at 1·042kHz (2·5/2·4). Listen to the carrier on a CW narrow bandwidth receiver and adjust the deviation on the transmitter for minimum carrier; this can also be done on a spectrum analyser if one is available. The deviation is now correctly set provided signals of greater amplitude are not fed into the microphone socket. Ideally the transmitter will have a microphone gain control, followed by a limiter and then followed by a deviation control. Thus excessive amplitude signals will not reach the deviation control and the modulator.

Phase Modulation (PM)

This is a subtle variant of frequency modulation. In FM it is the instantaneous frequency of the transmitter that is determined by the instantaneous modulating input voltage. Clearly, as the frequency varies, the phase will differ from what it would have been had the carrier been unmodulated. There are phase changes but these are as a consequence of the frequency modulation. In phase modulation, it is the phase that is directly controlled by the input audio voltage and the frequency varies as a consequence.

Frequency modulation is usually achieved by a variable capacitance diode in the oscillator circuit. This has the side effect of introducing another source of frequency drift or instability. An RC circuit situated after the oscillator can achieve phase modulation by using a variable capacitance diode for C. This way the oscillator is untouched and more stable. Many transceivers do in fact use phase modulation and pre-condition the audio so the transmitted signal appears frequency modulated.

Many data transmissions use phase modulation, the digital signal directly changing the carrier phase.

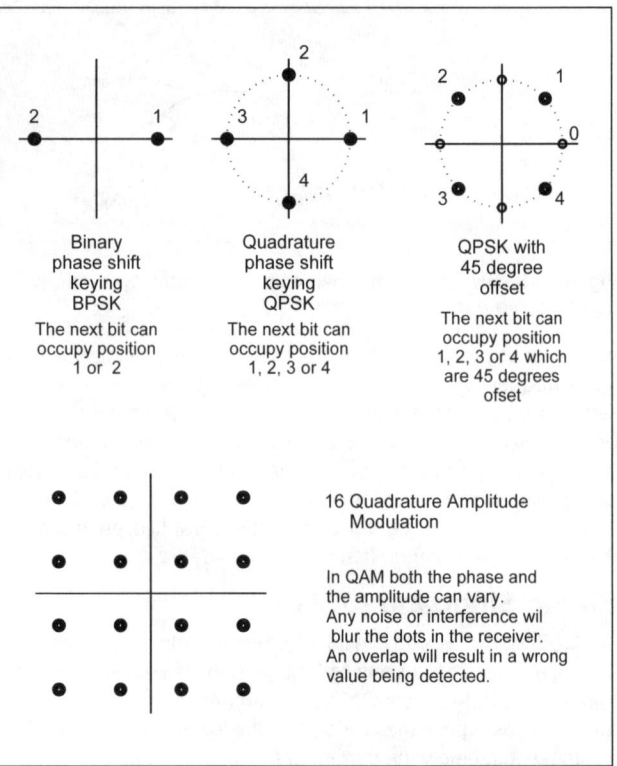

Fig 1.56: Digital modulation schemes often vary the phase of the signal to represent the next transmitted symbol. 16 symbols can transmit four binary bits at once, but use amplitude as well as phase changes

Digital Modulation

The techniques of digital modulation are little different, except the modulating input will normally comprise a data signal having one of two discrete values, normally denoted '0' and '1' but in practise two different voltages. In CMOS digital logic this may be 0V and 10V, or for RS232 signals from a computer ±6V.

Amplitude shift keying, that is two different amplitudes, either of the carrier or of an audio modulating signal is uncommon, partly because nearly all interference mechanisms are amplitude related and better immunity can be achieved by other methods.

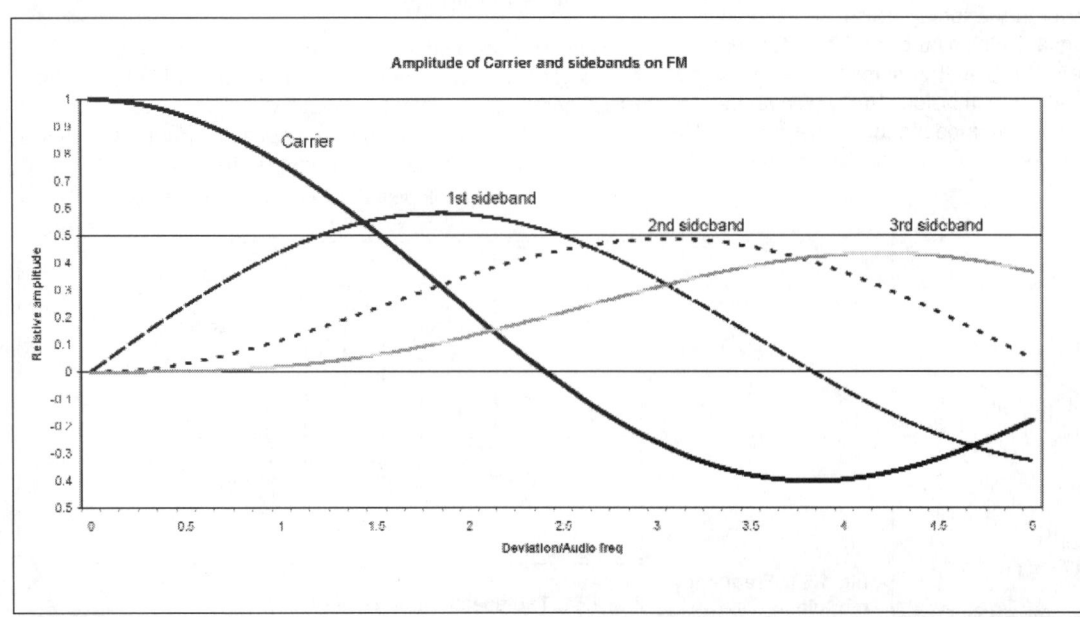

Fig 1.55: The amplitude of the carrier and the sidebands are given by Bessel functions

Frequency shift keying is popular, the carrier having two frequencies separated by 170Hz for narrow shift or 450 and 750Hz wide shift typically used commercially. The occupied bandwidth is again greater than the frequency shift since each of the two 'carriers' will have sidebands depending on the keying speed and method. The mode can be generated either by direct keying of the modulator or feeding audio tones (1275 and 1455Hz for narrow shift amateur use) into an SSB transmitter.

Phase shift modulation is used in many amateur modes. These modes are discussed in more detail in the data communications chapter.

Two phases, 0° and 180° may be used. This is binary phase shift keying, BPSK. Another method, known as QPSK, quadrature phase shift keying has 4 states or phases; 0°, 90°, 180° and 270°. This allows two binary digits (each with two states 0 and 1, giving four in total) to be combined into a single character or symbol to be sent. By this means the symbol rate (the number of different symbols per second) is lower, half in this case, than the baud rate, defined as the number of binary bits per second.

Each symbol may be advanced in phase by 45° in addition to the phase change required by the data so that, irrespective of the data, there is always some phase change. This is needed to ensure the receiver keeps time with the transmitter. If this did not happen and the data happened to be a string of 20 zeros with no phase changes, there is a risk that the receiver would lose timing and output either 19, 20 or 21 zeros. It is then out of synchronisation with the transmitter and the output data signal may well be meaningless.

Higher level modulation schemes can combine even more binary digits into a symbol, allowing even higher data rates. Quadrature Amplitude Modulation (QAM) is used professionally but not often in amateur circles. This utilises both amplitude and phase changes to encode the data. Whilst higher data rates are achievable, the difference between each symbol gets progressively less and less as each symbol is used to contain more and more binary digits. **Fig 1.56** shows these systems diagrammatically. In isolation this is all very well but the effects of noise and interference get progressively worse as the complexity of the system increases.

Amateurs have tended to stick to relatively straightforward modulation schemes but have utilised then in novel and clever ways to combat interference or send signals with very low bandwidth, finding a slot in an otherwise crowded band. The data communications chapter has the details.

Intermodulation

Mixing and modulation often use non-linear devices to perform the required function. Non-linear means that the output is not just a larger (or smaller) but otherwise identical copy of the input. Consider the equation y=4x. The output (y) is simply 4 times the input (x). Plotting the function on graph paper would give a straight line - linear.

Now consider y=x². Plotting this would give a curve, not a straight line; it is non-linear.

If the input was a sine wave, for example $\sin \omega_c t$, then the output would be of the form:

$$\sin^2 \omega_c t \quad \text{which is} \quad \tfrac{1}{2}(1 - \cos 2\omega_c t)$$

The output will contain a DC component and a sine wave of twice the frequency. This is a perfect frequency doubler.

Similarly if two signals are applied, $\sin \omega_a$ and ω_b, the output will be:

$$(\sin \omega_a + \sin \omega_b)^2$$

which is

$$\sin^2 \omega_a + 2\sin \omega_a \sin \omega_b + \sin^2 \omega_b$$

We already know the two sin² terms are frequency doubling and we also know (from the amplitude modulation section) that sinA x sinB gives us sin (A+B) and sin (A-B), that is the sum and difference frequencies. We now have a mixer.

Unfortunately no device is quite that simple, and amplifiers are not perfectly linear. The errors are usually small enough to neglect but when badly designed, or overdriven, the non-linearities will produce both harmonics of the input frequencies and various sums and differences of those frequencies.

For two sine wave inputs A and B, the output contains frequencies m x A ± n x B where m and n are any integers from 0 upwards. These are called intermodulation products (IMPs).

The frequencies 2A, 3A; 2B, 3B etc are far removed from the original frequencies and are usually easily filtered out. However, the frequencies m x A - n x B where m and n differ by 1, are close to the original frequencies.

This can happen in an SSB power amplifier where two (or more) audio tones are mixed up to RF and then amplified. The unwanted intermodulation products may well be inside the passband of the amplifier or close to it such that they are not well attenuated by the filter. This, then represents distortion of the signal, or, worse, interference to adjacent channels. **Fig 1.57** illustrates this.

The value of m+n is known as the 'order' of the intermodulation product. It will be noticed that it is only the odd orders that result in frequencies close to the wanted signals. It is also the

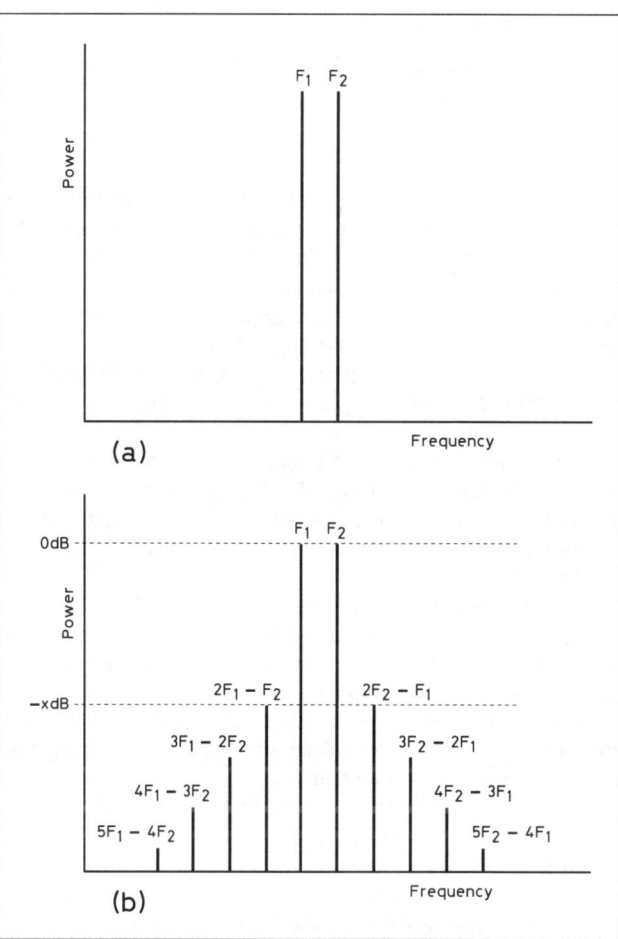

Fig 1.57: Spectrum of two frequencies (a) before and (b) after passing through a non-linear device

case that the lower order products (ie 3rd) will have a greater amplitude than higher order products. The power in these relative to the fundamentals, xdB in **Fig 1.57(b)**, should be at least 30dB down and preferably more.

Intermodulation can also occur in other systems. In receivers it can occur in the 'front-end' if this operating at a non-linear level (due for example by being overloaded). For instance, in a crowded band such as 7MHz with high-powered broadcast signals close to (and even in!) the amateur band, intermodulation can occur between them. It should be noted that these signals may be outside the bandwidth of the later stages of the radio receiver but inside the band covered by the RF front end. The IMPs occur in the overloaded front end, producing extra signals which may well be close to, or on top of, the wanted signal.

This degrades the receiver performance if the products are within the pass-band of the receiver. The use of high-gain RF amplifiers in receivers can be a major source of this problem and may be countered by reducing the gain by using an input attenuator. Alternatively, only just enough gain should be used to overcome the noise introduced by the mixer. For frequencies up to 14MHz, it is not necessary to have any pre-mixer gain provided the mixer is a good low-noise design (see the chapter on HF receivers).

Intermodulation can also occur in unlikely places such as badly constructed or corroded joints in metalwork. This is the so-called rusty-bolt effect and can result in intermodulation products which are widely separated from the original transmitter frequency and which cause interference with neighbours' equipment (see the chapter on electromagnetic compatibility).

DECIBELS

Consider the numbers:

$$100 = 10^2$$
$$1000 = 10^3$$
$$10000 = 10^4$$

The logarithm of a number is defined as the power to which the base of the logarithm (in this case 10) must be raised to equal the number. So the logarithm (log for short) of 1000 is 3, because 10^3 is 1000. This is written as Log_{10} (1000) = 3. However it is usually understood that base 10 is used so the expression is often simplified to Log1000 = 3.

It should also be noted that 10^3 (1,000) x 10^4 (10,000) is 10,000,000 or 10^7. In multiplying the two numbers the powers have been added.

It is possible to consider the Logs of numbers other than powers of 10. For example, the Log of 2 is 0·3010 and the Log of 50 is 1·6990. If these two Logs are added, we get 0·3010+1·6990 =2. We know the Log of 100 is 2, so the anti-Log of 2 is 100. We also know that 2 x 50 is 100.

Today this seems an incredibly awkward way of working out that 2 x 50=100. However, computers and calculators are relatively recent devices and Logs were a simple way of multiplying rather more cumbersome numbers.

The aim of the above is really to set out a principle. Multiplying and dividing can be achieved by adding and subtracting the Logs of the numbers concerned.

Consider the route from a microphone to the distant loudspeaker. A microphone amplifier provides some gain, the modulator may provide a gain or a loss, the RF power amplifier will provide gain, the feeder to the antenna will cause a loss and the antenna may well have some gain. The radio path will have losses associated with distance, trees and building clutter, hills and weather conditions. The receive antenna will provide gain, the feeder a loss and then the many stages in the receiver will have

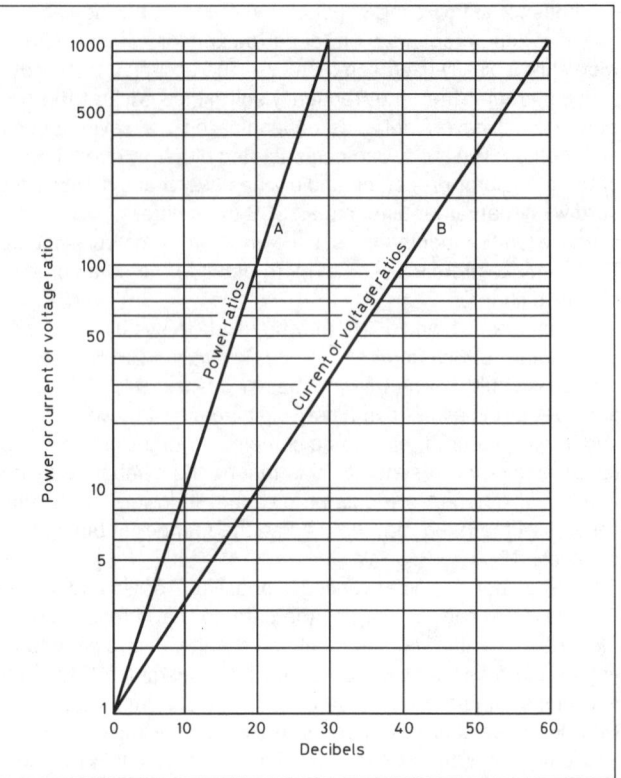

Fig 1.58: Graph relating decibels to power or current or voltage ratio

multiplier	construction	dB construction	dB value
2	-	-	3
4	2 x 2	3+3	6
8	2 x 2 x 2	3+3+3	9
16	2 x 2 x 2 x 2	3+3+3+3	12
10	-	-	10
100	10 x 10	10+10	20
20	2 x 10	3+10	13
40	2 x 2 x 10	3+3+10	16
5	10 ÷ 2	10-3	7
50	10 x 10 ÷ 2	10+10-3	17
25	10 x 10 ÷ 2 ÷ 2	10+10-3-3	14
1·26	-	-	1
1·58	-	-	2

Table 1.5: Construction of dB values

their own gains and losses. These gains and losses will not be simple round numbers and they all need multiplying or dividing. It is normal in radio and line transmission engineering to express all these gains and losses in terms of units proportional to the logarithm of the gain (positive) or loss (negative). In this way gains and losses in amplifiers, networks and transmission paths can be added and subtracted instead of having to be multiplied together.

The unit is known as the decibel or dB. It is defined as:

$$\text{Gain (or loss)} = 10 Log_{10}\left(\frac{P_{out}}{P_{in}}\right)$$

It is always a power ratio, not a current or voltage ratio. If current or voltage ratios were used, a voltage step-up transformer would appear to have a gain, which is not the case. The power out of a transformer is never greater than the power in.

However, provided and only provided, the currents or voltages are being measured in the same resistance (so a voltage increase does signify a power increase) then it is possible to use the definition:

$$\text{Gain (or loss)} = 20\text{Log}_{10}\left(\frac{V_{out}}{V_{in}}\right) \quad \text{or} \quad 20\text{Log}_{10}\frac{I_{out}}{I_{in}}$$

The change to 20x follows from the formula P=V²/R or I²R. If a number is squared (ie multiplied by itself) then the Log will be doubled (added to itself).

Line A in **Fig 1.58** can be used for determining the number of decibels corresponding to a given power ratio and vice versa. Similarly line B, shows the relationship between decibels and voltage or current ratios.

The use of Log tables (or the Log function on a calculator) can often be avoided by remembering two values and understanding how decibels apply in practice.

It was stated above that Log2 = 0·3010, so a doubling of power is 3·01dB, equally a halving is -3·01dB. The ·01 is usually omitted so doubling is simply assumed to be 3dB. Quadrupling is double double and, like Logs, decibels are added, giving 6dB. The other one to remember is 10 times is 10dB.

Table 1.5 can be constructed from just those two basic values. It can also be remembered that, with the same resistance (or impedance), the ratios in column 1 applied to voltage or current will double the dB value.

REFERENCES

[1] *Tables of Physical and Chemical Constants*, Kaye and Laby, 14th edn, Longman, London, 1972, pp102-105
[2] *Electronic and Radio Engineering*, Terman, 4th edn, McGraw Hill
[3] *Advance! The Full Licence Manual*, Alan Betts, 1st edn, RSGB
[4] *The RSGB Guide to EMC,* Robin Page-Jones, 2nd edn, RSGB
[5] *Four Figure mathematical tables*, Castle, Macmillan

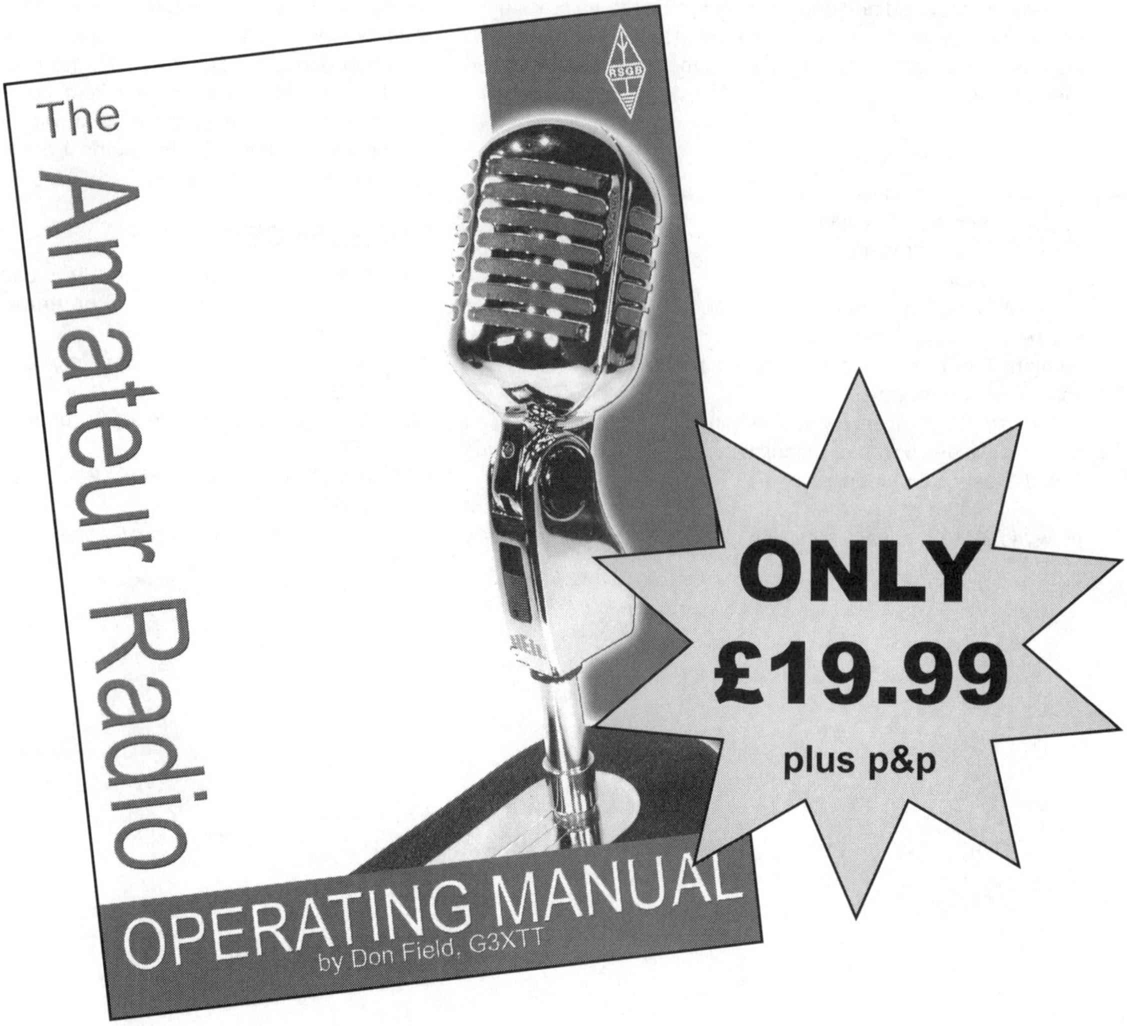

The
Amateur Radio
OPERATING MANUAL
by Don Field, G3XTT

ONLY £19.99 plus p&p

RSGB Radio Amateur Operating Manual
Don Field, G3XTT

This 6th edition of the RSGB Amateur Radio Operating Manual has been completely updated and redesigned this edition reflects the huge changes in hobby in recent years.

The impact of licensing changes and the ubiquity of PCs and the Internet are just some of the challenges in the hobby in the 21st Century. To deal with these, RSGB Amateur Radio Operating Manual has a completely new look at the content and approach. For example, some of the traditional demarcations between HF and VHF and between the various operating modes have been overturned. New and comprehensive chapters can be found on topics such as PCs in the Shack and Operating Modes. There is also a huge amount of new material included, for example, the 136kHz and 5MHz allocations, new data modes and the WSJT software suite, APRS and VoIP. Much of the book has been heavily updated and there is a complete rewrite of the chapter on Satellites and Space communications.

If you are interested in amateur radio the RSGB Amateur Radio Operating Manual is the book you should not be without. This book provides a comprehensive guide to operating across the amateur radio spectrum. Packed with information and tips this book has long been a standard reference work found on the bookshelf of radio amateurs. The RSGB Amateur Radio Operating Manual is a valuable addition to your bookshelf and the must have book for everyone interested in amateur radio.

RSGB Members Price: £16.99 plus p&p

E&OE

Radio Society of Great Britain
Lambda House, Cranborne Road, Potters Bar, Herts. EN6 3JE
Tel. 0870 904 7373 Fax. 0870 904 7374

ORDER 24 HOURS A DAY ON OUR WEBSITE
www.rsgb.org/shop

2 Passive Components

RELIABILITY

While most passive components are very reliable, reliability is just as important for the amateur as for the professional. To us, a breakdown may mean a frustrating search with inadequate instruments for the defective component. This can be especially trying when you have just built a published circuit, and it does not work. Reliable components do tend to be expensive when bought new, but may sometimes be bought at rallies if you know what to look for. The aim of this chapter is to give some guidance.

If you test components before using, you save yourself from this troubleshooting.

Components should be de-rated as much as possible. In 1889 a Swedish chemist, Arrhenius found an exponential law connecting chemical reaction speed with temperature. As most failures are caused by high temperature, reducing the temperature will greatly reduce failure rate. Some components, particularly electrolytic capacitors, do not take kindly to being frozen, so take the above advice with proper caution. Similarly, operating at a voltage or current below the maker's maximum rating favours long life.

WIRE

Copper is the conductor most often used for wire, being of low resistivity, easily worked and cheaper than silver. PVC or PTFE covered wire is most often used in equipment and for leads. Kynar covering is favoured for wire-wrap layouts. **Table 2.1** gives the characteristic of some coverings. Silicone rubber is cheaper than PTFE and can be used in high temperature locations but has not the strength of PTFE.

For inductors, the oleo enamel, cotton, regenerated cellulose and silk of yesteryear have given way to polyester-imide or similar plastics. These have better durability and come with a variety of trade names. They are capable of much higher temperature operation, and resist abrasion and chemicals to a greater degree. Either trade names or maximum temperature in degrees C are used to describe them. They are more difficult to strip, especially fine or 'Litz' wires. Commercially, molten sodium hydroxide (caustic soda) is used, leaving bright easily soldered copper, but there is a hazard, not only is the material harmful to the skin but, if wet, is liable to spit when heated. Paint stripper is a good substitute, but is liable to 'wick' with Litz wire. Be sure

Material	Temperature guide (maximum °C)	Remarks
PVC	70-80	Toxic fume hazard
Cross-linked PVC	105	Toxic fume hazard
Cross-linked polyolefine	150	
PTFE (Tefzel)	200	Difficult to strip
Kynar	130	Toxic fume hazard. ε_r=7*
Silicone rubber	150	Mechanically weak
Glassfibre composite	180-250	Replaces asbestos

If temperature range is critical, check with manufacturer's data.
* ε_r is the dielectric constant.

Table 2.1: Equipment and sleeving insulation

Metal	Resistivity at 20°C ($\mu\Omega$-cm)	Temperature coefficient of resistivity per °C from 20°C (ppm/°C)	Notes
Annealed silver	1.58	4000	
Annealed pure copper	1.72	3930	
Aluminium	2.8	4000	(1)
Brass	9	1600	(1)
Soft iron	11	5500	
Cast iron	70	2000	
Stainless steel	70	1200	(1)
Tungsten	6.2	5000	
Eureka	50	40	(1, 2)
Manganin	43	±10	(3)
Nichrome	110	100	(1)

Notes:
(1) Depends on composition and purity. Values quoted are only approximate.
(2) Copper 60%, nickel 40% approx. Also known as 'Constantan', 'Ferry' and 'Advance'. High thermal EMF against copper.
(3) Copper 84%, manganese 12%, nickel 4%. Low-temperature coefficient only attained after annealing at 135°C in an inert atmosphere and a low thermal EMF against copper. Cannot be soft-soldered.

Table 2.2: Resistivity and temperature coefficient for commonly used metals

to clean it off.

There are also polyurethane coatings, usually pink or, for identification purposes, green, purple or yellow. These can be used up to 130°C, and are self-fluxing, if a sufficiently hot iron is used. The fumes are irritating and not very good for you, especially if you are asthmatic.

Dual coatings of polyester-imide with a lower melting point plastic layer outside are made. When a coil has been wound with these, it is heated by passage of current and the outer layers fuse, forming a solid bond. Dual coatings are also found where the outer coating is made of a more expensive material than the inner.

Oleo enamel and early polyester-imide cracks with age and kinking so old wire should be viewed with suspicion. The newer synthetic coatings can resist abrasion to a remarkable degree, but kinked wire, even of this type should not be used for inductors or transformers.

Wire size can be expressed as the diameter of the conductor in millimetres, Standard Wire Gauge (SWG) or American Wire Gauge (AWG). The sizes and recommended current carrying capacity are given in the general data chapter at the end of this book. Precise current carrying capacity is difficult to quote as it depends upon the allowable temperature rise and the nature of the cooling. **Tables 2.1 and 2.2** give data and a guide to usage.

Radio frequency currents only penetrate to a limited depth; for copper this is $6.62/\sqrt{(F_{HZ})}$ cm. For this reason, coils for the higher frequencies are sometimes silver plated; silver has a marginally better conductivity than copper (see Table 2.2) and is less liable to corrosion, though it will tarnish if not protected. Generally the cost is not justified. For high currents, tubing may

be used for reduced cost, as the current flows on the outside.

From some 10kHz up to about 1 MHz, a special stranded wire called 'Litz' gives lower radio frequency resistance due to the skin effect mentioned above. The strands are insulated from each other and are woven so that each strand takes its turn in the centre and outside of the wire. Modern Litz wire has plastic insulation, which has to be removed by one of the methods already described. Self-fluxing covering is also made. Older Litz wire has oleo enamel and silk insulation, needing careful heating in a meths flame for removal. There has been considerable dispute over the effect of failing to solder each strand; it is the writer's experience that it is essential to solder each strand, at least if the coil is to be used at the lower range of Litz effectiveness. Litz loses its usefulness at the high end of the range because of the inter strand capacitance making it seem to be a solid wire. This may explain why some authorities maintained that it was unnecessary to ensure that each strand was soldered - at the top end of the range it may not make much difference.

Data on radio frequency cables will be found in the general data chapter.

For some purposes high resistivity is required, with a low temperature coefficient of resistivity. Table 2.2 gives values of resistivity and temperature coefficient of commonly used materials. For alloys, the values depend very much on the exact composition and treatment.

FIXED RESISTORS

These are probably the components used most widely and are now among the most reliable if correctly selected for the purpose. As the name implies, they resist the passage of electric current - in fact the unit 'Ohm' is really the Volt per Amp.

Carbon composition resistors in the form of rods with radial wire ends were once very common. There is no excuse for using them now, as they are bulky, and suffer from a large negative temperature coefficient, irreversible and erratic change of value with age or heat due to soldering. Due to the granular nature of the carbon and binder, more than thermal noise is generated when current flows, another reason not to use them!

Tubes of carbon composition in various diameters with metallised ends are made for high wattage dissipation. These may bear the name 'Morganite' and are excellent for RF dummy loads (Fig. 2.1).

Carbon films deposited by 'cracking' a hydrocarbon on to ceramic formers are still used. To adjust the value, a helix is cut into the film, making the component slightly inductive. Protection is either by varnish, conformal epoxy or ceramic tube. Tolerance down to 5% is advisable, as the temperature coefficient (tempco) of between -100 and -900 parts per million per degree Celsius (ppm/°C) makes the value of closer tolerance doubtful.

In the 1930s, Dubilier made metal film resistors similar in appearance to carbon composition ones of that era, but they never became popular. In the 1950s, metal oxide film resistors

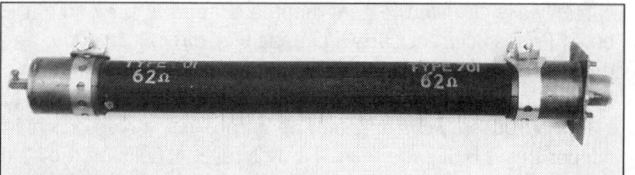

Fig 2.1: An amateur-made 62-ohm dummy load with UHF coaxial connector. It is made from a 50 watt resistor but is suitable for 200W for short periods

Fig 2.2: Surface mounting components. The small dark components on the left labelled '473', '153', '333' and '243' are resistors, the larger grey one on the right is a capacitor and the light-coloured one labelled 'Z1' (bottom left) is a transistor. The scale at the bottom is in millimetres

were introduced, with tempco of 300ppm/°C, soon to be replaced for lower wattage by the forgotten metal film types. These were more stable, less noisy and more reliable than carbon film, now only little more expensive. They use the same helical groove technique as carbon film, and little difficulty should be experienced due to this up to say 50 MHz. The tempco (temperature coefficient) is of the order of ±15ppm/°C, allowing tolerance of 1% or better. Protection is either by conformal or moulded epoxy coating.

Surface mounting has now become universal in the commercial world, and amateurs have to live with it. Fig. 2.2 shows a typical board layout. More about this later in this chapter, and in the construction chapter.

Resistor values are quoted from a series that allows for the tolerance. The series is named according to the number of members in a decade. For example, E12 (10%) has 12 values between 1 and 10 (including 1 but not 10). In this case, each will be the twelfth root of ten times the one below, rounded off to two significant figures (Table 2.3) E192 allows for 1% tolerance, the figures now being rounded off to three significant figures. Unfortunately E12 is not a subset of E192, and some unfamiliar numbers will be met if E12 resistors are required from E192 stock. Values can either be colour coded (see the general data chapter) or for surface mounted ones, marked by two significant digits followed by the number of noughts. This is called the 'exponent system'. For example, 473 indicates 47k, but 4.7 Ohms would be 4R7. Metal film resistors can also be made into integrated packages either as dual in line, single in line or surface mount. By using these, space is saved on printed circuit boards.

For higher powers than metal oxide will permit, wire wound resistors are used. The cheaper ones are wound on a fibre substrate, protected by a rectangular ceramic tube or, rarely, moulded epoxy. More reliable (and expensive) ones are wound on a ceramic substrate covered by vitreous or silicone enamel. For even greater heat dissipation, it may be encased in an aluminium body designed to be bolted to the chassis or other heat sink. Vitreous enamelled wound resistors can be made with a portion of the winding left free from enamel so that one or more taps can be fitted. Wire wound resistors, even if so-called 'non-inductively' wound, do have more reactance than is desirable for RF use. However Ayrton Perry wound resistors in gas filled glass bulbs are suitable for use in the HF bands, but hard to come by.

Unless an aluminium-clad resistor is used for high power, care should be taken to ensure that the heat generated does not damage the surroundings, particularly if the resistor is mounted on a printed circuit board. The construction chapter gives practi-

E3 ±40%	E6 ±20%	E12 ±10%	E24 ±5%
1.0	1.0	1.0	1.0
-	-	-	1.1
-	-	1.2	1.2
-	-	-	1.3
-	1.5	1.5	1.5
-	-	-	1.6
-	-	1.8	1.8
-	-	-	2.0
2.2	2.2	2.2	2.2
-	-	-	2.4
-	-	2.7	2.7
-	-	-	3.0
-	3.3	3.3	3.3
-	-	-	3.6
-	-	3.9	3.9
-	-	-	4.3
4.7	4.7	4.7	4.7
-	-	-	5.1
-	-	5.6	5.6
-	-	-	6.2
-	6.8	6.8	6.8
-	-	-	7.5
-	-	8.2	8.2
-	-	-	9.1

The values shown above are multiplied by the appropriate power of 10 to cover the range.

Table 2.3: 'E' range of preferred values, with approximate tolerance values from next number in range

cal advice about this. Marking is by printed value, perhaps including wattage and tolerance as well.

Fig. 2.3 shows photographs of many of the fixed resistors that have just been described.

VARIABLE RESISTORS

Often a resistor has to be made variable, either for control or precise adjustment purposes; the latter is usually called a trimmer. Generally variable resistors are made with a moveable taping point, and the arrangement is called a potentiometer or 'pot.'

for short, the name coming from a laboratory instrument for use in measuring voltage. If only a variable component is required, it is advisable to connect one end of the track to the slider.

Tracks are made from carbon composition, cermet or conductive plastic, or they may be wire wound. Circular tracks with either multi-turn or single turn sliders are made, both for pots and trimmers, in varying degrees of accuracy and stability. The HiFi world likes linear tracks for control purposes, as do some rigs available to amateurs. Wire wound types have the disadvantage of limited resolution and higher price, but have the advantage of higher wattage and good stability. It is not always possible to tell the type from the external appearance as the photograph of **Fig. 2.4** shows.

Single turn rotary pots for volume control and many other uses usually turn over a range of 250°-300°, with a log, anti-log or linear law connecting resistance with rotation. Ganged units and pots with switches are common. Precision components generally have a linear law (some for special purposes may have another law) and turn some 300°. For greater accuracy, multi-turn pots are used, some in conjunction with turn counting dials. There are multi turn linear trimmers using a screw mechanism to move the slider over the linear track. Available values are frequently only in the E3 range, to 20% tolerance, the value being printed on the component, possibly in the exponent system.

NON-LINEAR RESISTORS

All the resistors so far described obey Ohm's Law closely, that is to say that the current through them is proportional to the applied voltage (provided that the physical conditions remain constant, particularly temperature). Resistors, except metal film and wire wound, do change their value slightly with applied voltage, but the effect can generally be ignored. For some purposes, however, specialist types of resistor have been developed.

Thermistors are sintered oxides or sulphides of various metals that have a large temperature coefficient of resistance, and are deliberately allowed to get hot. Both positive (PTC) and negative (NTC) tempco thermistors are made. Of all the forms of construction, uninsulated rods or discs with metallised ends are the most interesting to amateurs. Bead thermistors have special uses, such as stabilising RC oscillators or sensing temperature, one use being in re-chargeable battery packs.

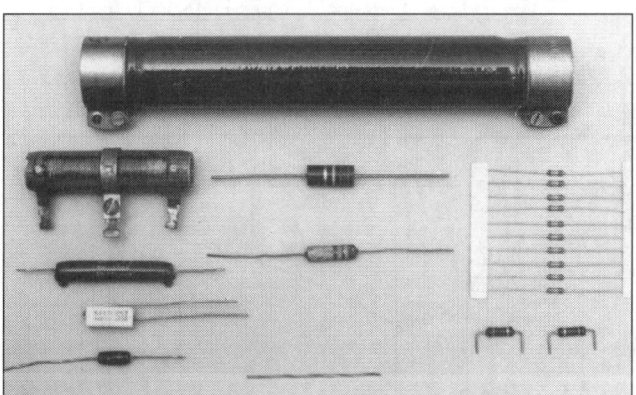

Fig 2.3: Fixed resistors. Top: 50W wire wound, vitreous-enamel covered. Left, top to bottom: 10W wire wound with adjustable tapping, 12W vitreous enamelled, 4W cement coated and 2.5W vitreous enamelled. Centre: 2W carbon composition, 1W carbon film. Right, top: ten ¼W metal film resistors in a 'bandolier' of adhesive tape, two carbon film resistors with preformed leads. Bottom: striped marker is 50mm long

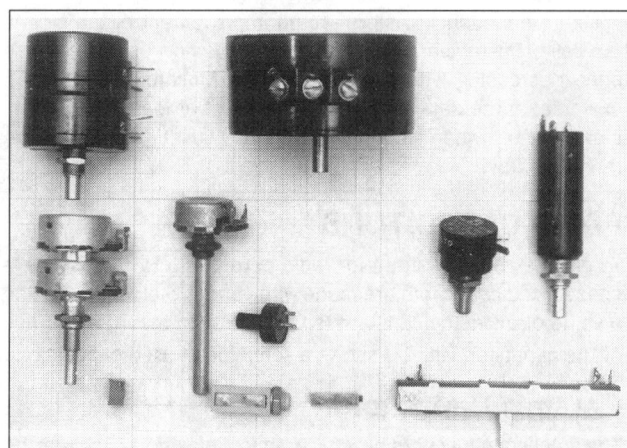

Fig 2.4: Variable resistors. Top, left to right: double-gang 2W wire wound, 5W wire wound. Middle, left to right: two independent variable resistors with coaxial shafts, standard carbon track, single-turn preset, standard wire wound and 10-turn variable resistor. Bottom, left to right: 10-turn PCB vertical mounting preset, 10-turn panel and PCB horizontal mounting preset. Right: a slider variable resistor

PTC thermistors are used for current limitation. The thermistor may either operate in air or be incorporated into the device to be protected. Self-generated heat due to excess current or rise of ambient temperature will raise the resistance, and prevent further rise of current They should not be used in constant current circuits because this can result in ever increasing power dissipation and therefore temperature (thermal runaway), so destroying the device.

Unlike metals, PTC thermistors do not exhibit a positive tempco over the whole range of temperature likely to be encountered. **Fig. 2.5(a)** shows this behaviour for a particular thermistor.

One type of PTC 'thermistor', though not usually classed as such, is the tungsten filament lamp. Tungsten has a large tempco of about 5000ppm/°C (Table 2.2) and lamps operate with a temperature rise of some 2500°C. An increase to about 12 times the cold resistance can be expected if the lamp is lit to full brilliance. These lamps are relatively non-inductive and can be used as RF loads if the change of resistance is acceptable.

NTC thermistors find a use for limiting in-rush currents with capacitor input rectifier circuits. They start with a high resistance, which decreases as they heat up. (**Fig. 2.5(b)**). Remember that they retain the heat during a short break, and the in-rush current will not be limited after the break. When self-heated, the thermistor's temperature may rise considerably and the mounting arrangements must allow for this.

PTC bead thermistors are the ones used to stabilise RC oscillators. Also, they are used in temperature compensated crystal oscillators where self-heating is arranged to be negligible and the response to temperature is needed to offset the crystal's tempco.

Non-ohmic resistors in which the voltage coefficient is large and positive are called voltage dependant resistors (VDRs) or varistors. VDRs having no rectification properties, ie symmetrical VDRs are known by various trade names (eg Metal Oxide Varistors). Doped zinc oxide is the basis, having better properties than the silicon carbide (Atmite, Metrosil or Thyrite) formerly used, as its change of resistance is more marked (**Fig. 2.5(c)**). VDRs are used as over-voltage protectors; above a certain voltage the current rises sharply. As the self-capacitance is of the order of nanofarads, they cannot be used for RF. Alternative over-voltage protectors are specially made zener diodes or gas discharge tubes.

Light dependent resistors are an improvement of the selenium cell of yesteryear. Cadmium sulphide is now used, its resistance decreasing with illumination. The Mullard ORP12 is an example. In the dark it has a resistance of at least 1MΩ, dropping to 400Ω when illuminated with 100 Lux. Photo-diodes are more useful!

FIXED CAPACITORS

To get the values wanted for radio purposes into a reasonable space, fixed capacitors are made with various dielectrics, having a value of dielectric constant (ε_r) according to the intended use of the capacitor. **Fig. 2.6** shows a selection of fixed capacitors.

Ceramic Capacitors

The smallest capacitors have a ceramic dielectric, which may be one of three classes. Class 1 has a low ε_r and therefore a larger size than the other two for a given capacitance. The loss factor for Class 1 is very low, comparable with silvered mica, and the value stays very nearly constant with applied voltage and life.

The designations 'COG' and 'NPO' refer to some members of this class of ceramic capacitor, which should be used where stability of value and low loss factor are of greater importance than

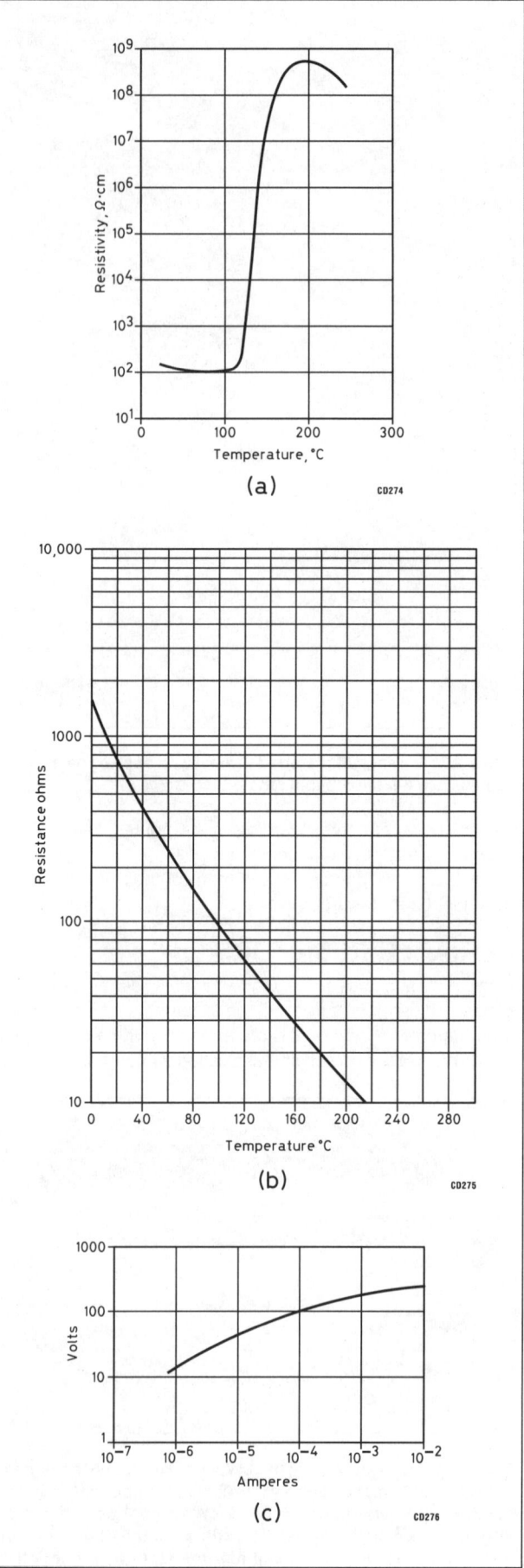

Fig 2.5: Characteristics of thermistors

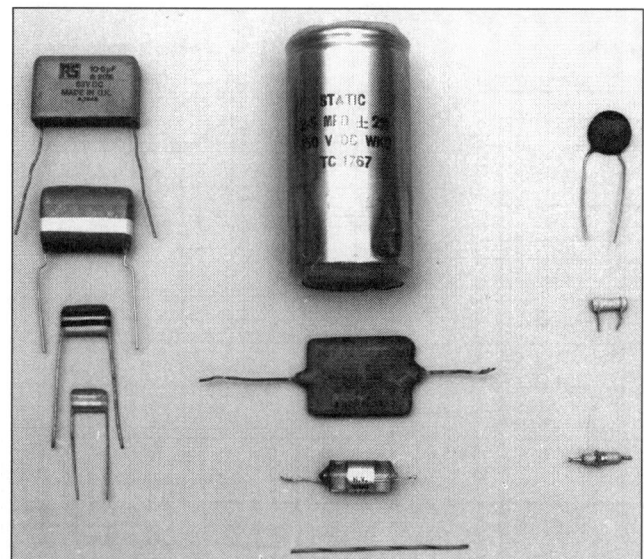

Fig 2.6: Fixed capacitors. Left column (top to bottom): 10µF, 63V polycarbonate dielectric, 2.7µF 250V, 0.1 F, 400V, 0.033µF, 250V all polyester dielectric. Centre column from top: 2.5µF 150V ±2% polyester, 240pF ±1% silver mica, 1nF ±2% polystyrene (all close tolerance). Right column from top: 4.7nF disc ceramic, 220pF tubular ceramic, 3nF feedthrough. Striped marker is 50mm long

small size. Other low loss ceramic capacitors are available with either N*** or P*** identification, where *** is the intentional negative or positive tempco in ppm/°C. These are for compensation to allow for the opposite temperature coefficient of other components in the circuit.

Class 2 comprises medium and high ε_r capacitors. They both have a value depending on voltage, temperature and age, these effects increasing with ε_r. It is possible to restore the capacitance lost by ageing by heating above the Curie temperature, this being obtained from the manufacturer - not really a technique to be done at home!

'X7R' refers to a medium ε_r material, Z5U, 2F4 and Y5V to higher ε_r ceramic. **Fig. 2.7** shows the performance of COG, X7R and Y5V types. This class should not be used where stability is paramount, but they are very suitable for RF bypassing, lead through capacitors being made where low inductance is required. Class 1 and 2 capacitors are made in surface mount, either marked in the exponent system, or unmarked.

Class 3 have a barrier layer dielectric, giving small size, poor loss factor and wide tolerance. Z5U dielectric has made this class obsolete, but some may be found as radial leaded discs in surplus equipment.

Large high voltage tubular or disc capacitors are made for transmitter use which, if they can be found, are particularly useful for coupling the valves in a linear amplifier to the matching network.

Table 2.4 summarises ceramic capacitor properties and recommended usage. The value colour coding is to be found in the general data chapter and **Table 2.5** gives the dielectric codes used.

Mica Capacitors

These are larger than similar ceramic capacitors for a given value and have a tempco of between +35 and +75ppm/°C, the variation being caused by them being natural rather than synthetic. Today, silver electrodes are plated directly on to the mica sheets and the units encapsulated for protection.

Function	Type	Advantages
AF/IF coupling	Paper, polyester, polycarbonate	High voltage, cheap
RF coupling	X7R ceramic	Small, cheap but lossy
	Polystyrene	Very low loss, low leakage but bulky and not for high temperature
	COG ceramic or silver mica	Close tolerance
	Stacked mica	For use in power amplifiers
RF decoupling	X7R or Z5U ceramic disc or feedthrough	Very low inductance
Tuned circuits	Polystyrene	Close tolerance, low loss, negative temperature coefficient (150ppm)
	Silver mica	Close tolerance, low loss, positive temperature coefficient (+50ppm)
	COG ceramic	Close tolerance, low loss
	Class 1 ceramic	Various temperature coefficients available, more lossy than COG

Table 2.4: Use of fixed capacitors

Class 1 - r < 500
COG, NPO temperature coefficient ±30ppm/°C
N*** Temperature coefficient: ***ppm/°C
P*** Temperature coefficient: +***ppm/°C

Class 2 - r > 500

EIA coding

Working temperature range				Capacitance change over range	
Lower		Upper			
Letter	Temp	Figure	Temp	Letter	Change
Z	+10°C	2	+45°C	R	±15%
Y	30°C	4	+65°C	S	±22%
X	55°C	5	+85°C	T	+22 to 33%
		6	+105°C	U	+22 to 56%
		7	+125°C	V	+22 to 82%
		8	+150°C		

X7R and Z5U are commonly met. X7R was formerly denoted W5R.

CECC 32100 coding

Code	Capacitance change over range		Temperature range (°C)				
			55	55	40	25	+10
			+125	+85	+85	+85	+85
	At 0V DC	At rated voltage	Final code figure				
			1	2	3	4	6
2B*	±10%	+10 to 15%		*	*	*	
2C*	±20%	+20 to 30%	*	*	*		
2D*	+20 to 30%	+20 to 40%				*	
2E*	+22 to 56%	+22 to 70%		*	*	*	*
2F*	+30 to 80%	+30 to 90%		*	*	*	*
2R*	±15%		*				
2X*	±15%	+15 to 25%	*				

Reference temperature +20°C.

Example: X7R (EIA code) would be 2R1 in CECC 32100 code, tolerance ±15% over 55 to +125°C.

Class 3 barrier layer
Not coded but tolerance approximately +50% to 25% over temperature range of 40 to +85°C. Refer to maker for exact details.

Table 2.5: Ceramic capacitor dielectric codes

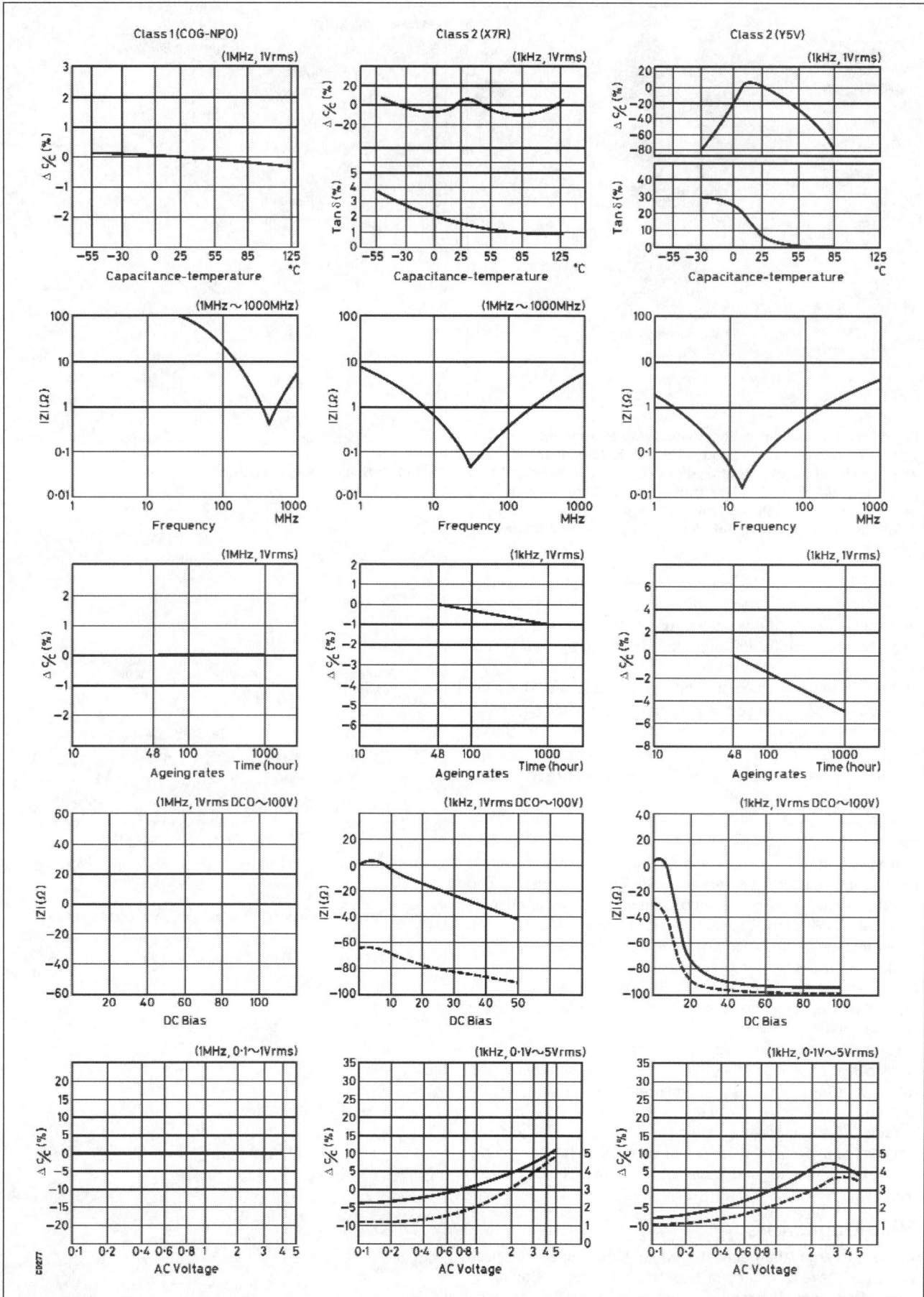

Fig 2.7: Ceramic capacitor networks

Because they generally remain stable over life and have low loss, they are popular for tuned circuits and filters. Occasionally however, mica capacitors will jump in value and this effect is unpredictable. High voltage and current stacked foil types were used for coupling in transmitters, but are not easy to get now. The value and voltage rating is usually marked on the body of mica capacitors.

Glass Capacitors

These are used for rare applications where they have to work up to 200°C. Because of this, they do not appear in most distributors catalogues, but may be found in surplus equipment.

Paper Dielectric Capacitors

At one time paper dielectric capacitors were much used, but now they may only be found in some high voltage smoothing capacitors and interference suppression capacitors. Paper capacitors are large and expensive for a given value, but are very reliable, especially when de-rated. There can be trouble if inferior paper is used, with voids in it. The capacitors tend to break down at the voids, where the electric stress is concentrated. Both foil and metallised electrodes are used and the units can be somewhat inductive if not non-inductively wound. A resistor can be included if the unit is to be used for spark suppression.

Plastic Film Capacitors

These have replaced paper capacitors, and there are two distinct types of plastic used, polar and non-polar.

Polar plastics include polycarbonate, polyester and cellulose acetate. These are characterised by a moderate loss, which can increase at frequencies where the polar molecules resonate. They also suffer from dielectric absorption, meaning that after a complete discharge some charge reappears later. This is of some importance in DC applications. The insulation resistance is very high, making the capacitors suitable for coupling the lower frequencies of AC across a potential difference (within the voltage rating of course).

Commonly, values range from about a nanofarad to several microfarads at voltages from 30V up to kilovolts, thus replacing paper. Tolerances are not usually important as these capacitors are not recommended for tuned circuits, but as they may be used in RC oscillators or filters, 5% or better can be bought at increased cost. Polycarbonate is the most stable of this group and cellulose the worst and cheapest.

Polyethylene terephthalate (PETP) is used for polyester film and is also known as Mylar (in USA) and domestically as Terylene or Melinex. PETP exhibits piezo-electric properties, as can be demonstrated by connecting a PETP capacitor to a high impedance voltmeter and squeezing the capacitor. PETP capacitors therefore behave as rather poor microphones, which could introduce noise when used in mobile applications.

Non-polar plastics suitable for capacitors are polystyrene, polypropylene and polytetrafluoroethylene (PTFE). These exhibit very low loss, independent of frequency, but tend to be bulky. Polystyrene capacitors in particular have a tempco around -150 ppm/°C which enables them to offset a similar positive tempco of ferrites cores used for tuned circuits. Unfortunately polystyrene cannot be used above 70°C and care must be taken when soldering polystyrene capacitors particularly if you use lead free solder. The end connected to the outer foil may be marked with a colour and this should be made the more nearly grounded electrode where possible.

Polypropylene capacitors can be used up to 85°C and are recommended for pulse applications. They are also used at 50Hz

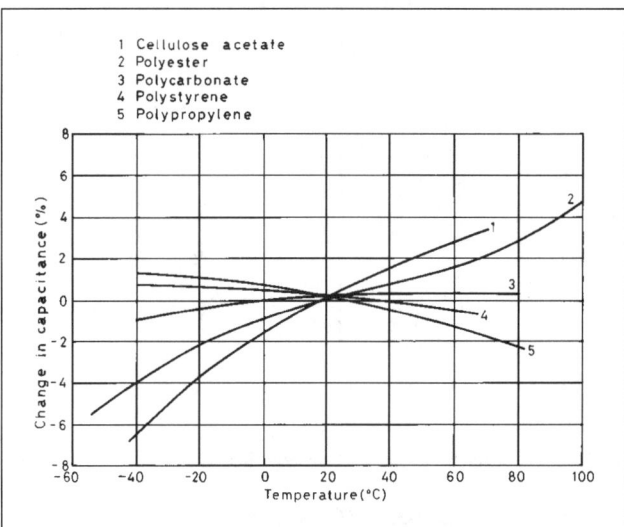

Fig 2.8: Temperature coefficients of plastic capacitors

for power factor correction and motor starting. Again, take care when soldering.

Encapsulation of plastic foil capacitors is either by dipping or moulding in epoxy or it may be omitted altogether for cheap components where environmental protection is not considered necessary. The value is marked by colour code or printing along with tolerance and voltage rating. Small values are mostly tubular axial leaded, but single ended tubular and boxed are commonly available. You can get surface mount ones.

Fig. 2.8 shows the effect of temperature on five different types.

PTFE fixed capacitors are not generally available, but variable ones are, see later in this chapter. Small lengths of PTFE or, if unobtainable polystyrene coaxial cable, make very good low loss high voltage capacitors for use as tuning elements in multiband HF antenna systems.

Cutting to an exact value is easy, if you know the type, as the capacitance per metre is quoted for cables in common use. The length must be much shorter than the wavelength for which the cut cable is to be used, to avoid trouble due to the inductance.

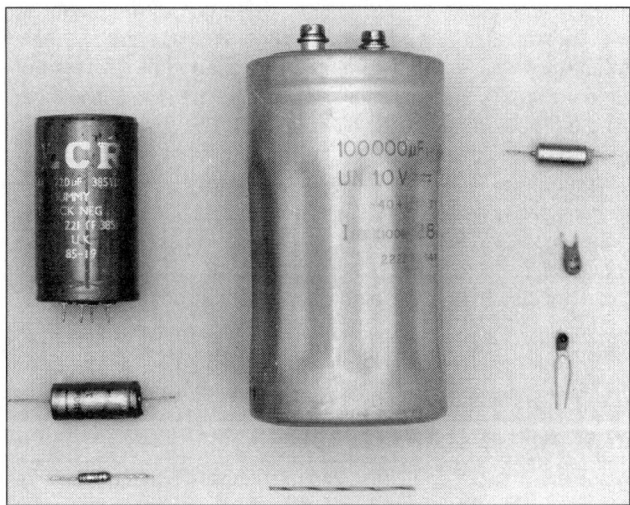

Fig 2.9: Electrolytic capacitors. Left, from top: 220μF 385V, 150μF 63V, 2.5μF 15V, electrodes. Centre: 100,000μF, 10V, all with aluminium. Right, from top: 68μF 15V, 22μF 25V and 22μF 6.3V, all tantalum types. Striped marker is 50mm long

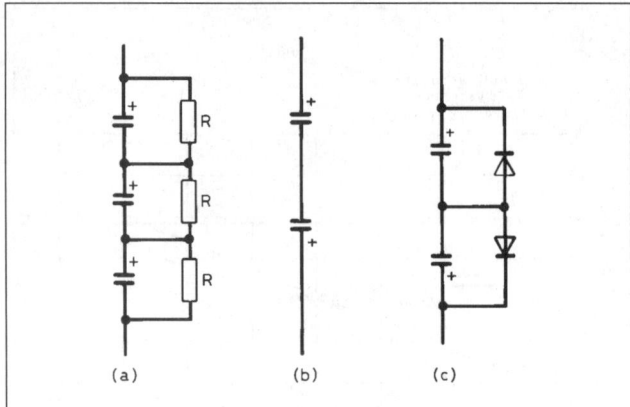

Fig 2.10: Circuits for electrolytic capacitors

Electrolytic Capacitors

This type is much used for values bigger than about 1µF. The dielectric is a thin film of either aluminium, tantalum, or niobium oxide on the positive plate.

An electrolyte (liquid or solid) is used to connect to the negative plate (**Fig. 2.9**). For this reason the electrolytic will be polarised, unless constructed with two oxide coated plates. This is not common in amateur usage, but can be used for motor starting on AC, and in the output or cross-over circuits of some HiFi amplifiers. Alternatively back-to-back capacitors can be used (see **Fig. 2.10**).

Both tantalum and niobium oxide capacitors can stand a reverse voltage up to 0.3 volts. Aluminium capacitors cannot withstand any reverse voltage. The oxide film is formed during manufacture, and deteriorated if the capacitor is not used. Capacitors, unused for a long time, should be reformed by applying an increasing voltage until the rated voltage is reached, and leakage current falls.

Electrolytics are rated for voltage and ripple current at a specified maximum temperature, and it is here that de-rating is most advisable. At full rating some electrolytics have only 1000 hours advertised life. There is always an appreciable leakage current, worst with liquid aluminium ones.

Tantalum and niobium oxide capacitors should not be allowed to incur large in-rush currents. As the capacitor may fail, add a small resistance in series. Niobium oxide capacitors fail to an open circuit; all others go to a short circuit. For all types the maximum ripple current should not be exceeded to prevent rise of temperature. This is particularly important when the capacitor is used after a rectifier (reservoir capacitor). There is more on this in the power supplies chapter.

Although capacitive reactance decreases with frequency, electrolytics are unsuitable for radio frequencies. The innards may have considerable inductance and the equivalent series resistance increases with frequency. To avoid this, a small ceramic capacitor should be paralleled with the electrolytic, 0.1µF to 1nF being common. For switch-mode converters, special considerations apply. Again see the power supplies chapter for more information.

VARIABLE CAPACITORS

These are most often used in conjunction with inductors to form tuned circuits, but there is another use, with resistors in RC oscillators - see the oscillators chapter. Small values (say less than 100pF) with a small capacitance swing are used as trimmers to adjust circuit capacitance to the required value (**Figs. 2.11 (a) and (b)**).

Capacitors with vacuum as the dielectric are used to withstand large RF voltages and currents, for example in valve power amplifiers. The capacitance is adjusted through bellows, by linear motion. The range of adjustment usually allows one amateur band to be covered at a time. For the power allowed, the expense of new vacuum capacitors is not justified.

Air spaced variable capacitors are widely used both for tuning and trimming. For precise tuning of a variable frequency oscillator (VFO) the construction must be such that the capacitance does not vary appreciably with temperature or vibration. This implies low loss, high grade insulating supports for the plates and good mechanical design. These are not factors that make for cheapness.

For valve power amplifiers, the plate spacing must be wide enough to withstand both the peak RF voltage, including the effects of any mismatch, and any superimposed DC. Rounded plate edges minimise the risk of flash over, and application of DC should be avoided by use of choke-capacity coupling. However, extreme stability is not required owing to the low Q of the circuit in which they are used.

As the moving plates are generally connected to the frame, it may be necessary to insulate this from the chassis in some applications. Ceramic shafts have been used to avoid this. It is possible to gang two or more sections, which need not have the

Fig 2.11(a): Variable capacitors. Left: 500pF maximum, four-gang air-spaced receiver capacitor. Right: 125pF max transmitter capacitor with wide spacing. Bottom, left to right: 500pF max twin-gang receiver capacitor. Solid dielectric receiver capacitor. Single-gang 75pF max receiver capacitor. Twin-gang 75pF max receiver capacitor

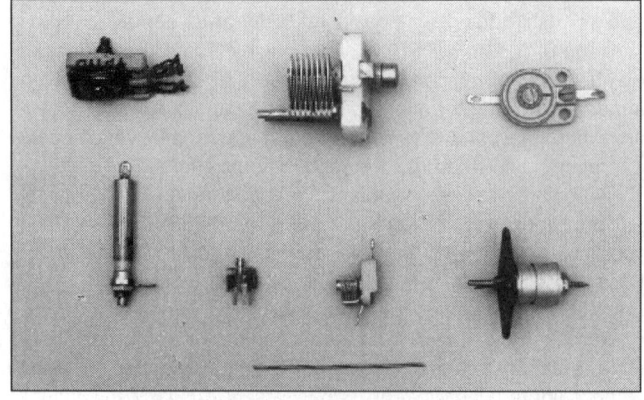

Fig 2.11(b): Trimmer capacitors. Top, from left: 50pF max 'postage stamp', 70pF max air dielectric, 50pF max flat ceramic. Bottom, from left: 27pF max tubular ceramic, 30pF max plastic film, 12pF max miniature air dielectric, 30pF max 'beehive' air dielectric

same value, but it is usual for the moving vanes to be commoned, with the exception mentioned above. The plates need not be semi-circular, but can be shaped so as to give a wanted law to capacitance versus rotation angle. In early superhet receivers specially shaped vanes solved the oscillator-tracking problem.

For small equipment, plastic film dielectric variable capacitors are much used but do not have the stability for a VFO. Trimmers may have air or film dielectric, the latter being a low loss non-polar plastic or even mica. Both parallel plate and tubular types are used for precise applications, and mica compression ones now only occasionally where stability is of less importance than cost.

Junction diodes can be used as voltage variable capacitors - they are described in the semiconductors chapter.

INDUCTORS

Inductors are used for resonant circuits and filters, energy storage and presenting an impedance to AC whilst allowing the passage of DC, or transforming voltage, current or impedance. Sometimes an inductor performs more than one of these functions simultaneously, but it is easier to describe the wide range of inductors by these categories.

Tuned Circuits

Either air or magnetic cored coils are used - in the latter case different cores are selected for optimum performance at the operating frequency and power level. The criterion for 'goodness' for

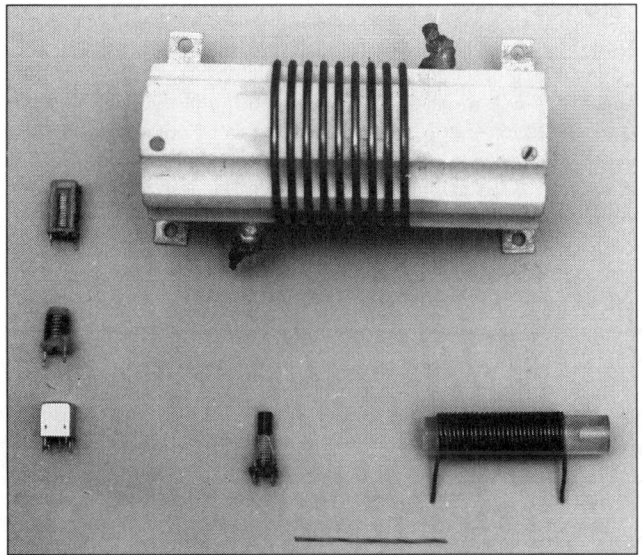

Fig 2.12(a): Air-cored inductors. Top: high-power transmitting on ceramic former. Left: VHF inductors moulded in polythene, the bottom one in a shielding can. Centre: small air-cored HF inductor (could have an iron dust core). Right: HF inductor wound on polystyrene rod. The striped marker is 50mm long

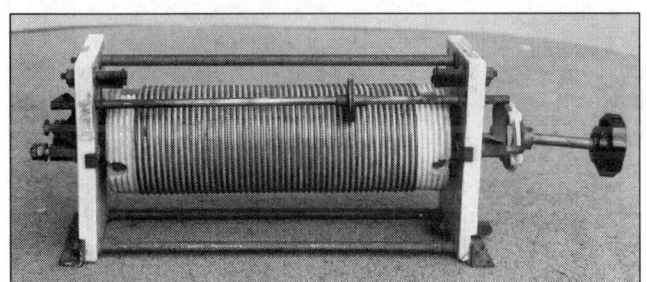

Fig 2.12(b): 'Roller-coaster' variable inductor

Fig 2.13: A variometer made by ON6ND for use with his 136kHz transmitter. The inductance is varied by altering the mutual coupling between the fixed and variable coils

a tuning coil is Q, the ratio of reactance to resistance at the operating frequency. With air-cored coils, the ratio of diameter to length should be greater than 1 for maximum Q, but Q falls off slowly as the length increases. It is often convenient to use a long thin coil rather than a short fat one, accepting the slight loss of Q.

Calculation of the inductance of air-cored coils is difficult and usually done by the use of charts or computer programs. However there is an approximate formula {1} which may be used:

$$L(\mu H) = \frac{r^2N^2}{25.4\,(9r + 10l)}$$

where r is the radius, l is the length of the coil, both in millimetres and N the number of turns. If you use inches, omit the 25.4 in the denominator. The formula is correct to 1% provided that $l > 0.8r$, ie the coil is not too short.

Figs 2.12 (a) and (b) show some different air-cored coils including a much sought after 'roller coaster' used for antenna and power amplifier tuning networks. **Fig 2.13** shows another type of variable inductor.

Magnetic cores may be either ferrite or iron (carbonyl powder for RF, iron alloy for AF). Screw cores can be used for adjustment in nominally air-cored coils or in pot cores (**Fig 2.14**). For closed magnetic circuits, a factor called A_L is quoted for extreme posi-

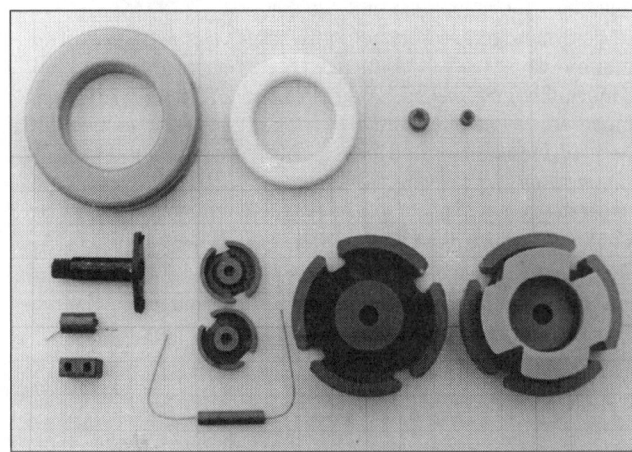

Fig 2.14: RF magnetic materials. Top, left to right: large, medium and small ferrite rings, and a ferrite bead with a single hole. Below, left top: tuneable RF coil former; middle: six-hole ferrite bead with winding as an RF choke; bottom: ferrite RF transformer or balun former. Centre: small pot cores and an RF choke former with leads moulded into the ends. Right: large pot core with former

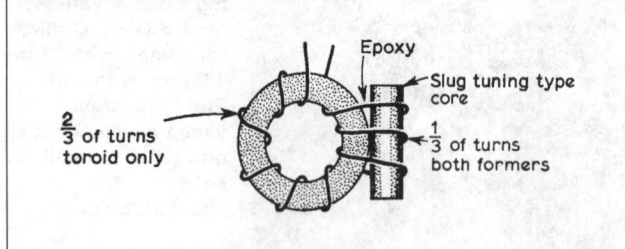

Fig 2.15: Tuneable toroid technique. About 10% variation in inductance can be achieved

tions of the core. A_L is the inductance in nanohenries that a one turn coil would have, or alternatively millihenries per thousand turns. For a coil of N turns the inductance will be N^2A_L nanohenries. Checking by grid dip oscillator is difficult because of the closed nature of the magnetic circuit.

Fig 2.15 shows a technique for adjusting the inductance where it is not important to preserve a completely closed magnetic circuit. Trimming by removing turns is more easily done with the aid of a small crochet hook. Also shown in Fig. 2.14 are toroidal cores.

There are two common types of ferrite, manganese-zinc (Mn/Zn) and nickel-zinc (Ni/Zn). Mn/Zn has lower resistivity than Ni/Zn and lower losses due to hysteresis. The latter is caused by the magnetic flux lagging behind the magnetic field, so offering a resistive component to the alternating current in any coil wound round the material. As the applied frequency is increased, eddy currents and possibly dimensional resonances play a greater part and the higher resistivity Ni/Zn has to be used. Unfortunately, Ni/Zn has greater hysteresis loss. Both types of ferrite saturate in the region of 400mT, this being a disadvantage in high power uses (saturation is when increasing magnetic field fails to produce the same increase in magnetic flux as it did at lower levels). More will be said about this in non-tuned circuit applications.

There are many proprietary ferrites on the market and reference must be made to the catalogues for details. If you have a piece of unmarked ferrite, one practical test is to apply the prods of an ohmmeter to it. Mn/Zn ferrite will show some resistance on a 100k range, but Ni/Zn will not. Further tests with a winding could give you some idea of its usefulness.

Microwave ferrites such yttrium iron garnet have uses for the amateur who operates on the Gigahertz bands.

Inevitably, coils will possess resistance and distributed capacitance as well as inductance. The latter confuses the measurement of inductance, and hence both confuses and limits the tuning range with a variable capacitor. When winding a tuning coil, it may be necessary to use a coating for environmental protection. All 'dopes' have an ε_r greater than one and will increase the self-capacitance. Polystyrene with an ε_r of only 2.5, dissolved in toluene (a hazardous vapour) or cellulose thinners is recommended. Polythene or PTFE would be attractive, but there is no known solvent for them.

Energy Storage Inductors

The type of core will depend on the operating frequency. At mains supply frequency, silicon iron or amorphous iron are the only choices To minimise eddy current loss caused by the low resistivity of iron, either thin laminations or grain orientated silicon strip (GOSS - the grains of which lie along the length of the strip in which the magnetic flux lies), is used. If magnetic saturation would be a problem, an air gap is left in the magnetic circuit. The calculation of core size, gap, turns and wire size is out-

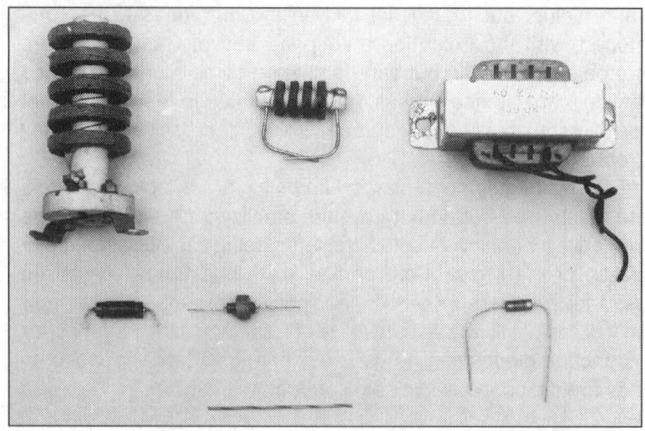

Fig 2.16: RF and AF chokes. Top, left to right: 1mH large, 1mH small RF chokes and 15H AF type. Bottom, left to right: 10µH single layer, 300µH 'pile' wound, 22µH encapsulated. Striped marker is 50mm long

side the scope of this book. Advice on winding is given later, under the heading 'Transformers'.

At higher frequencies such as for switch mode converter use, ferrites and amorphous alloys are widely used in the form of E shaped half cores, in pairs. As with silicon iron, an air gap is required to prevent saturation. The core manufacturers' data books give the information needed for design

Chokes

While energy storage inductors are most often called chokes, the term was originally applied to inductors used to permit the passage of DC, while opposing AC, and it still does. The cores used are as for energy storage for the lower frequencies, but at higher frequencies, where ferrites are too lossy, carbonyl powdered iron is used. At VHF and above, air-cored formers are used, surface mounted ones being available.

There is a problem with self-resonance in RF chokes. It is impossible to avoid capacitance, and at some frequency it will resonate with the inductance to give a very high impedance - ideal - but at higher frequencies the so-called inductor will behave as a capacitor whose impedance decreases with frequency. The situation is complicated by the fact that the self-capacitance is distributed and there may be multiple parallel and series resonances. For feeding HT to valve anodes, the choke should not have series resonances in any of the bands for which it is designed. Where available, a wave winder should be used to wind different numbers of turns in separate coils which, being on the same former, are in series. A less efficient, but simpler alternative, is to sectionalise a single layer solenoid winding. **Fig. 2.16** shows a variety of chokes.

Transformers

The Principles chapter described mutual inductance, and this property enables transformers to be made. The magnetic flux from one coil is allowed to link one or more other coils, where it produces an EMF. Two types of transformer are met, tightly coupled and loosely coupled. With tight coupling the object is to make as much as possible of the flux from the first (primary) coil link the other (secondary). In this case the ratio of the applied EMF to induced EMF is that of the turns ratio secondary to primary. Formulas for calculating the required turns are given in the general data chapter.

Leakage inductance caused by incomplete flux linkage is minimised by interleaving the primary and secondary windings, how-

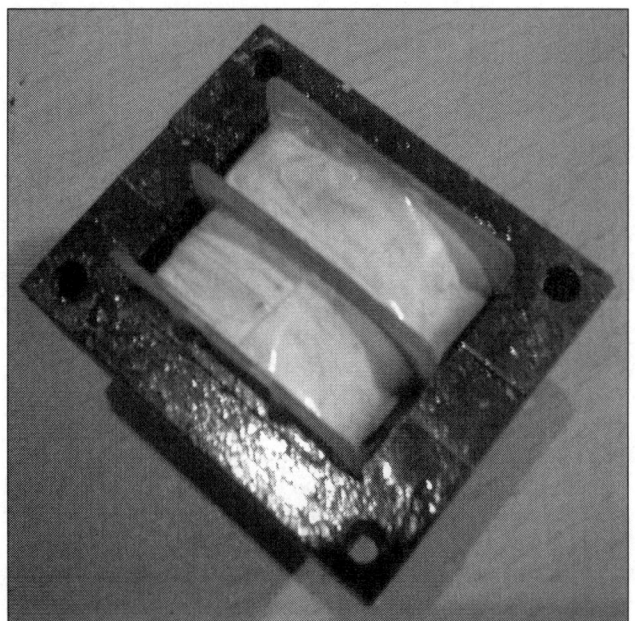

Fig 2.17: A transformer with the primary and secondary wound separately with an insulated divider

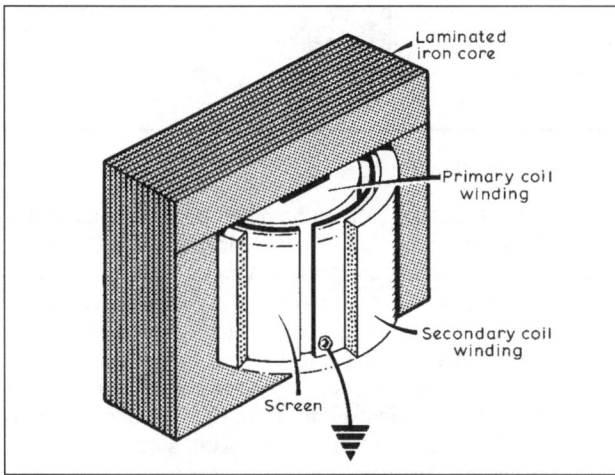

Fig 2.18: A Faraday screen is an earthed sheet of copper foil between the primary and secondary windings. For clarity, a gap is shown here between the ends, but in reality they must overlap while being insulated from one another to avoid creating a shorted turn. As copper is non-magnetic, mutual induction is not affected. The screen prevents mains-borne interference from reaching the secondary winding by capacitive coupling. Some transformers may have another screen (shorted this time) outside the magnetic circuit to prevent stray fields

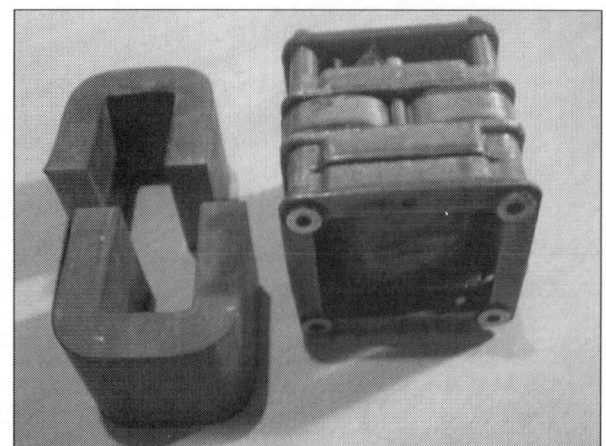

Fig 2.19: Examples of the use of 'C' cores and toroids

ever this does increase the self self-capacitance. This may not matter at 50Hz. Some modern transformers wind the primary and secondary separately, with an insulating divider instead of on top of each other (**Fig 2.17**). This provides greater safety at the cost of poorer regulation (the decrease of output voltage with current). An alternative is to use two concentric bobbins. Many small commercial transformers suffer from poor regulation due to this effect.

If capacitance between primary and secondary is to be minimised, a Faraday screen is inserted - see **Fig 2.18** for the way to do this. It is essential that the screen should not cause a shorted turn.

Insulation, if required, is best made with Kapton or Nomex tape, which will withstand as high a temperature as the wire.

Toroidal or 'C' cores (**Fig 2.19**) are used to reduce the external magnetic field and in Variacstm (see later and **Fig 2.20**). Beware of making a shorted turn with the fixing arrangements. As toroidal transformers are worked near the maximum permissible flux density, a large in-rush current may be noticed when switching on. If the supply is switched on as the voltage crosses zero, doubling of the current will occur as the voltage and current establish the correct phase relationship; it could even saturate the core momentarily.

This is particularly noticeable with the Variac. **Fig 2.21** shows a comparison between a switch mode transformer, which runs at 500kHz, and a standard 50Hz transformer. Where the stray

Fig 2.20: A variac

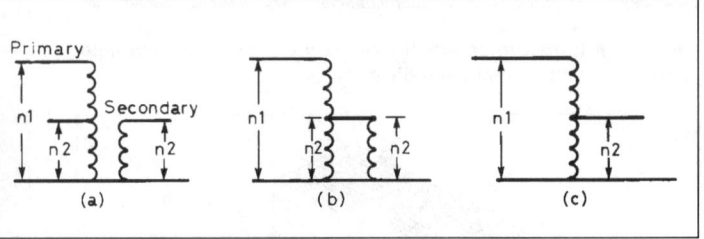

Fig 2.21: The two transformers on the left are 125W devices for 500kHz operation and weigh 40g each, while the one on the right is a 12W transformer for mains frequency (50Hz) and weighs in at 420g. The standard graph paper background has 1cm squares and 1mm subdivisions

Fig 2.22: Copper foil under the 'Danger' label surrounds the winding and the outer limbs of the core to reduce radiation

Fig 2.23: The auto-transformer

magnetic field has to be reduced, a shorted turn of copper foil outside the whole magnetic circuit is used - see **Fig 2.22** .

Practical note: When winding on E and I or T and U laminations, remember that part of the bobbin's cheeks will later be covered by the core, so do not bring the leads out in this region! If the wires are very thin, skeining them before bringing them out makes a sounder job.

If isolation is not required between input and output, an autotransformer can be used. This will increase the power handling capacity of a particular core. The primary and secondary are continuous - the transformation ratio being still the ratio of the number of turns, as in **Fig 2.23** . Wire gauge is chosen to suit the current. Since the secondary is part of the primary (or vice versa), more use is made of the winding window. The tapping point can be made variable to allow adjustment of the output voltage, the device then being known as a Variac.

In loosely coupled transformers, if the windings are tuned by capacitors, the degree of coupling controls the bandwidth of the combination (**Fig 2.24**). Such transformers are widely used in intermediate frequency stages and wideband couplers. Again, ferrite, iron dust or brass slugs may be used for trimming in the appropriate frequency range. Loose coupling is used on mains in ballasts for sodium lamps and in microwave ovens, whose magnetron needs a constant current supply. Beware of trying to use a microwave oven to supply a valve linear amplifier, apart from loose coupling, the inner of the secondary is either earthed or not well insulated.

MATERIALS

Earlier in this chapter conductors were described - here some of the insulators used in amateur radio will be considered first. These will be grouped into three categories: inorganic, polar and non-polar plastics. Mica (mainly impure aluminium silicate), glass and ceramics are the most likely to be used (asbestos is potentially dangerous). Vitreous quartz is one of the best insulators, but is not generally available. It has nearly zero temperature coefficient of expansion, and would make a good former for VFO inductors. Mica is not generally used except in capacitors. Glass has a tendency to collect a film of moisture, but the leakage is hardly likely to be troublesome. If it is, the glass should be coated with silicone varnish. Porcelain antenna insulators do not suffer from this trouble. Ceramic coil formers leave little to be desired but are becoming increasingly more difficult to obtain.

Non-polar plastics like polystyrene, polypropylene and PTFE are all very good. Polypropylene rope avoids the use of antenna insulators (except where very high voltages are encountered as when using a Marconi antenna on the 136kHz band) which is particularly useful in portable operation.

Polar plastics such as phenol-formaldehyde (Tufnol and Bakelite), Nylon, Perspex and PVC have higher losses that are frequency sensitive. The test using a microwave oven is not the best, as the loss at 2450MHz may not be the same as at the required frequency. Tests involving burning are not advised. Because of possible loss, they are not recommended to withstand high RF voltages. **Fig 2.25** shows two coils, one on polythene and the other on Tufnol. The one on the polythene former has a Q of 300, but the Tufnol one only 130 at 5MHz. Some protective varnishes are polar, being made for 50Hz. Only polystyrene dope previously described should be used where RF loss is important.

Permanent magnets may be a special ferrous alloy, ceramic rare earth alloy or neodymium-boron-iron. Ferrous alloys such as Alnico and Ticonal were much used in DC machines, headphones and loudspeakers. The others now replace them.

Fig 2.24: Bandwidth of coupled circuits

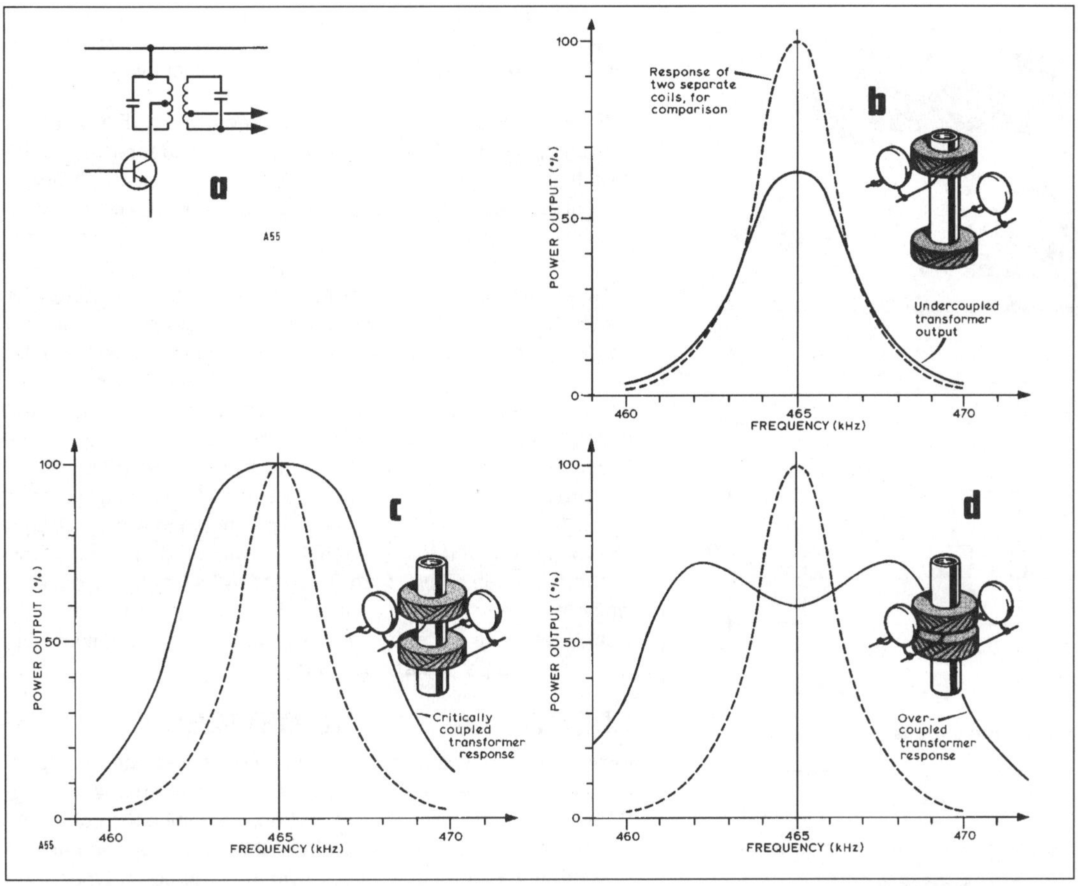

Ceramic magnets are cheap and withstand being left without a keeper better than ferrous alloys. The other two store more magnetic energy, are very good on open magnetic circuits but are expensive.

RELAYS

The 'Post Office' types 3000 and 600 are still available in amateur circles (**Fig 2.26**). However they are not sealed, and many sealed relays are now available. Newer relays need less operating current, some being able to operate directly from logic ICs. Contacts are made from different materials according to the intended use, such as very high or low currents.

The abbreviations 'NO', 'NC' and 'CO' apply to the contacts and stand for 'normally open', 'normally closed' and 'change over', the first two referring to the unenergised state. Some manufacturers use 'Form A' for NO, 'Form B' for NC and 'Form C' for CO. Multiple contacts are indicated by a number in front of the abbreviation.

Latching relays are stable in either position and only require a set or reset momentary energisation through separate coils or a pulse of the appropriate polarity on a single coil. These relays use permanent magnet assistance, meaning that correct polarity must be observed.

It is possible to get relays for operation on AC without the need for a rectifier, the range including 240V, 50 Hz. Part of the core has a short circuited turn of copper around its face - this holds the relay closed while the current through the coil falls to zero twice per cycle.

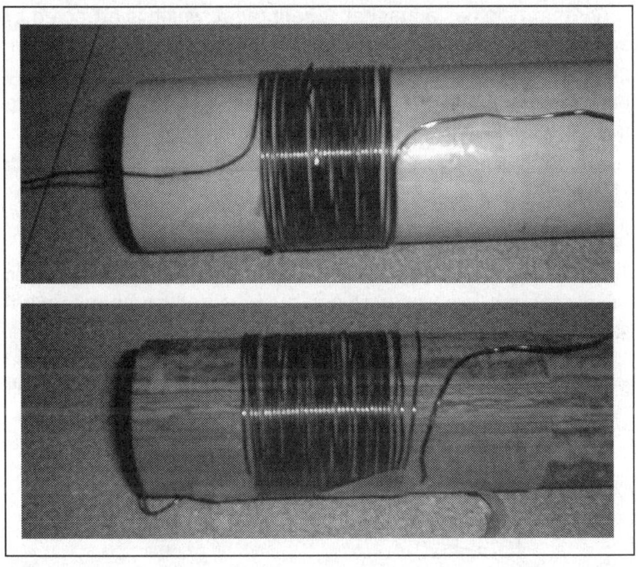

Fig 2.25: Similar coils wound on (a) polythene and (b) Tufnol have different Q values

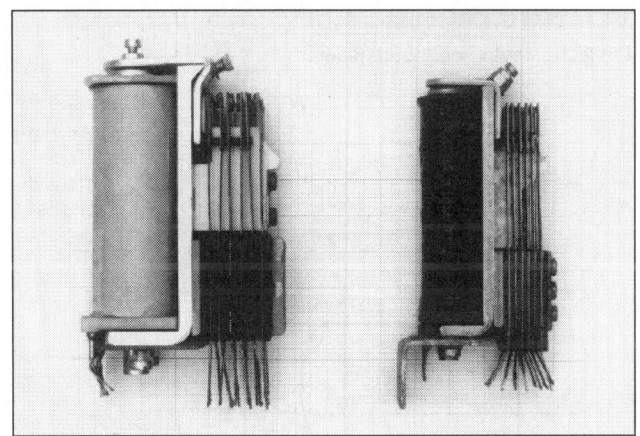

Fig 2.26: Post Office relays. Left type 3000, right type 600

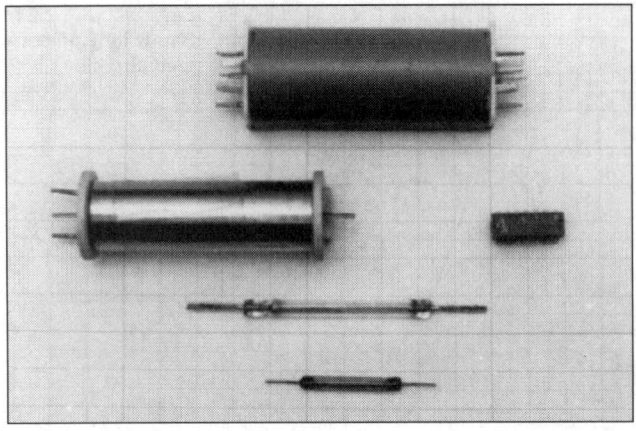

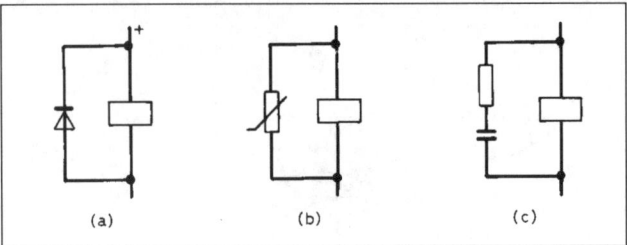

Fig 2.28: Relay coil suppression circuits

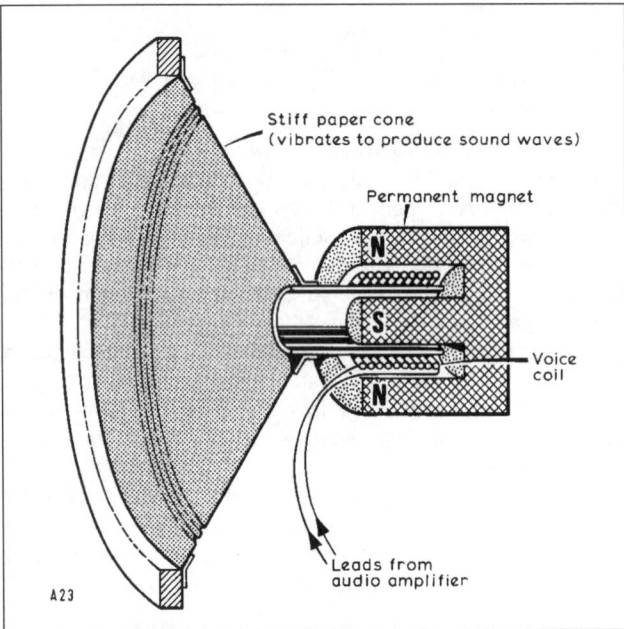

Stiff paper cone
(vibrates to produce sound waves)

Permanent magnet

N
S
N

Voice coil

Leads from audio amplifier

A23

Fig 2.29: Moving-coil loudspeaker

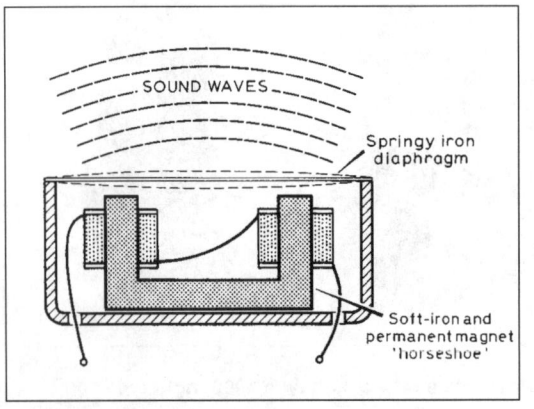

SOUND WAVES

Springy iron diaphragm

Soft-iron and permanent magnet 'horseshoe'

Fig 2.30: Moving iron head-phone

Fig 2.27: Reed relays. Top: four reeds in one shielded coil. Middle left: reed in open coil. Right: miniature reed relay in 14-pin DIL case. Below: large and small reed elements

If RF currents are to be switched, reed relays (**Fig. 2.27**) or coaxial relays which maintain a stated impedance along the switched path should be used. Reed relays, either with dry or mercury wetted contacts, are among the fastest operating types, and should be considered if a relay is needed in a break-in circuit.

As a relay is an inductive device, when the coil current is turned off, the back EMF tries to maintain, possibly creating a spark. The back EMF may damage the operating transistor or IC, so it must be suppressed (**Fig. 2.28**). A diode across the coil, connected so that it does not conduct during energisation will conduct when the back EMF is created, but in so doing it slows down the release of the relay. Varistors or a series RC combination are also effective. If rapid release is required, a high voltage transistor must be used with an RC combination which will limit the voltage created to less than the maximum of this transistor.

Another undesirable feature of relays is contact bounce, which may introduce false signals in a digital system. Mercury wetted reed relays are one way over this problem, but some will only work in a particular orientation.

ELECTRO-ACOUSTIC DEVICES
Moving coil loudspeakers (and headphones) use the force generated when a magnetic flux acts on a current carrying coil (**Fig. 2.29**). The force moves the coil and cone to which the coil is fastened. Sound is then radiated from the cone. Some means of preventing the entry of dust is provided in the better types.

Sounders
The piezo-electric effect is used to move a diaphragm by applying a voltage. The frequency response has a profound resonance so the device is often used as a sounder at this frequency.

Headphones
Small versions of the moving coil and piezo-electric loudspeakers are used as headphones. The in-ear headphone is a moving coil element, with a neodymium-boron-iron magnet to reduce the size. Deaf aid in-ear headphones use the piezo-electric effect, with a better response than the sounder has.

Another type has a magnetic diaphragm, which is attracted by a permanent magnet with the audio frequency field superimposed on it (**Fig. 2.30**) The permanent field is necessary because magnetic attraction is proportional to the square of the flux (and current). Without the permanent flux, second harmonic production would take place. The over-riding steady flux prevents this.

Microphones
The moving coil loudspeaker in miniature forms the dynamic microphone. It is possible to use the same unit both as a loudspeaker and microphone, often in hand-held equipment.

The piezo-electric element shown in **Fig 2.31** is reversible and forms the basis of the crystal microphone. The equivalent circuit

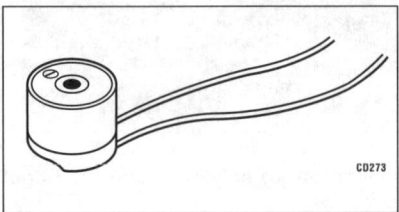

CD273

Fig 2.31: Piezoelectric loudspeaker

The Radio Communication Handbook

Fig 2.32: Microphone inserts. Left: carbon mic. Right: dynamic mic

Fig 2.33: Electret microphone

is a small capacitance in series with a small resistance. This causes a loss of low frequencies if the load has too low a resistance. FETs make a very suitable load. Moisture should be avoided, as some crystals are prone to damage.

It is possible to use the diaphragm headphone as a microphone since the effect is reversible, and some ex-service capsules are made this way.

Probably the carbon microphone is the oldest type still in telephone use; one type is shown in **Fig 2.32**. Sound waves alternatively compress and relieve the pressure on the carbon granules, altering the current when energised by DC. It amplifies, but is basically noisy and subject to even more unpleasant blasting noises if overloaded acoustically. It is rarely used nowadays in amateur radio. The impedance is some hundreds of ohms.

The above types can be made (ambient) noise cancelling by allowing noise to be incident on both sides of the diaphragm (or element) with little effect. Close speaking will influence the nearer side more, improving the voice to ambient noise ratio.

The so-called condenser microphone has received a new lease of life under the name of electret microphone. The con-

denser microphone had a conductive diaphragm stretched in front of another electrode and a large DC potential difference maintained, through a resistive load, across the gap. Variations in the position of the diaphragm by sound waves caused a variation of capacitance, and so an AF current flows through the load resistance. As the variation in capacitance is small, a large load resistance is required. A valve amplifier was included in the assembly to avoid loss of signal in leads and possible unwanted pick up. In the electret microphone, the need for a high polarising voltage is obviated by using an electret as the rear electrode and an FET in place of the valve. An electret is the electrostatic counterpart of a permanent magnet, and maintains a constant potential across the electrodes without requiring any power. The FET provides impedance transformation to feed low impedance leads. A point worth remembering it that the field from the electret attracts dust! Two small ones are shown in **Fig 2.33**.

QUARTZ CRYSTALS

Quartz is a mineral, silicon dioxide and the major constituent of sand. In its crystalline form it exhibits piezo-electric properties. The quartz crystal is hexagonal in section, and has hexagonal points. The useful piezo-electric effect takes place only at certain angles to the main axis from point to point. If a slice is cut at one of the correct angles, it can act like a bell and ring if struck. Ringing is an oscillation and with small slices this will take place in the MHz region. Mechanical ringing is accompanied by electrical ringing, and with proper amplifier connection this can be sustained (see the chapter on oscillators).

Charge is applied to the quartz element from metal plates sprung on to it, or better by silver electrodes plated on.

The type of deformation that the charge produces, stretching, shearing or bending, depends on the angle of cut and mode of mechanical oscillation to be excited. Various different cuts have different temperature coefficients, DT having a parabolic curve, AT a third order curve and SC virtually no tempco. More detail of this can be found in the oscillators chapter.

Naturally occurring quartz always had defects in the crystal, which made for unpredictability, but now crystals are made by dissolving sand in water, and crystallising it out. Very high pressure and temperature are used.

The Q of the crystal is much higher than can be obtained with LC circuits. As an example, a 7MHz crystal may have a Q of 10,000. The equivalent circuit is in **Fig. 2.34** where C_s, L_s and R_s represent the properties of the crystal and C_p the shunt capacitance of the holder. The measured values of these parameters are strange when compared with the LC circuit, as **Table 2.6** reveals. Note the very large inductance of the crystal; also a Q of 300 for the LC circuit is good.

A quartz crystal is not a primary standard of frequency; it has to be compared with another. Not only this, all crystals age, even if not excited. Switching on and over excitation add to ageing, which is usually towards a higher frequency.

Crystals can normally be cut, ground or etched for the kilohertz region up to about 30MHz, above this they are too fragile to be of practical value. An overtone crystal can be used, which

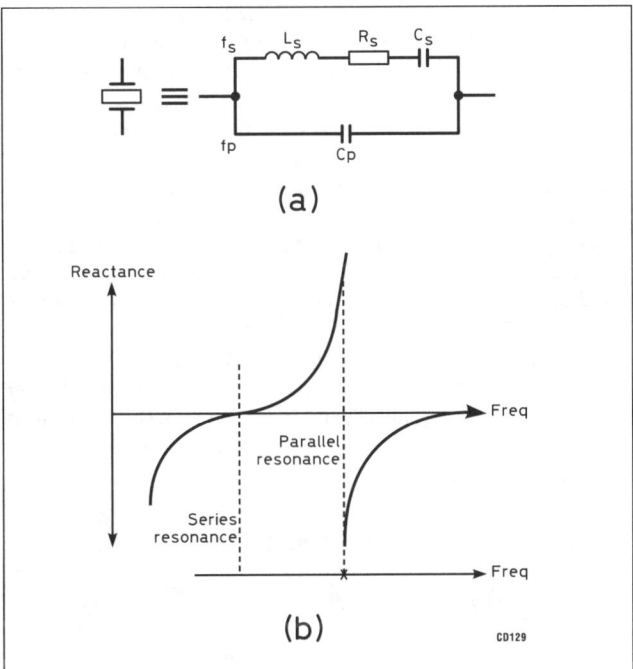

Fig 2.34: (a) Equivalent circuit of a piezoelectric crystal as used in RF oscillators or in filter circuits. Typical values for a 7MHz crystal are given in Table 2.6. Cp represents the capacitance of the electrodes and the holder. (b) Typical variation of reactance with frequency. The series and parallel reactances are normally separated by about 0.01% of the frequencies themselves (eg about a kilohertz for a 7MHz crystal)

Parameter	7MHz crystal	7MHz LC
Ls	42.5mH	12.9 H
Cs	0.0122pF	40pF
Rs	19	0.19
Q	98,000	300

Table 2.6: Parameters of a crystal compared with an LC circuit

is excited at a frequency that is nearly equal to an odd number times the fundamental. It will still try to oscillate at the fundamental, but the external circuit must be made to prevent this.

Since crystals act as tuned circuits, they can also be used as filters; piezo-ceramics being much used. More information can be found in the chapters on receivers and transmitters. The use of cheap, readily available computer or colour TV crystals in filters has been well documented [3]. [4].

Mechanical Design

Quartz crystals must be protected from the environment and are mounted in holders; these are usually evacuated. There are several types and sizes (**Fig 2.35**) but basically they are either in metal or glass. If in metal, the cheapest is solder sealed, next is resistance welded and the most expensive cold welded.

Ceramic Resonators

Certain ceramics, mainly based on titanium and/or zirconium oxides, have similar properties to quartz and are used as oscil-

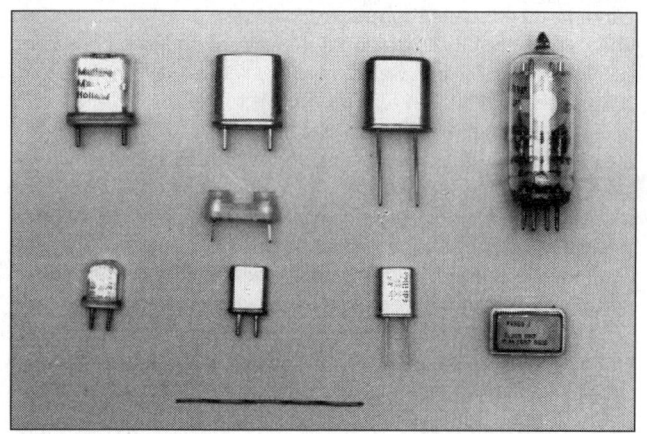

Fig 2.35: Quartz crystals in different holders. Top, from left: HC27U, HC6U, HC47U (wire ended) and B7G. Holder for HC6U between the rows. Bottom, from left: HC29U, HC25U, HC18U (wire ended) and a packaged crystal oscillator. The striped bar is 50mm long

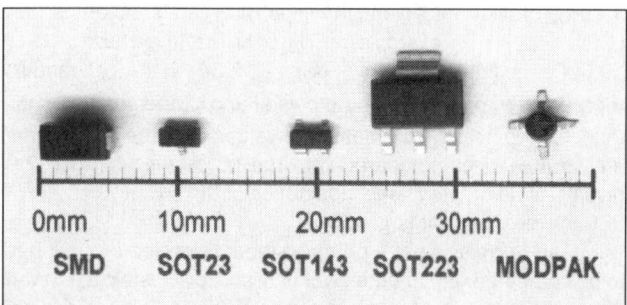

Fig 2.36: Discrete SMT packages

lators or filters. They have a lower Q, a lower maximum frequency and a higher temperature coefficient, but these are offset by a lower price. They are also only available for a limited range of frequencies.

There is another type of ceramic resonator that works at microwave frequencies and is widely used in satellite TV receivers.

SURFACE MOUNT TECHNOLOGY

Although most people have seen surface mount technology (SMT), relatively few without a professional involvement in electronics have used it in construction. SMT dominates commercial technology where its high-density capability and automated PCB manufacture compatibility is of great value. The arrival of SMT combined with the increasing dominance of monolithic IC parts over discrete components means that amateurs will have to move with the technology or rapidly lose the ability to build and modify equipment. SMT has an image of being difficult but this is not necessarily the case and it is quite practical for amateurs to use the technology. An incentive is that traditional leaded components are rapidly disappearing with many recently common parts no longer being manufactured and new parts not being offered in through-hole format. The cost of SMT parts in small quantities has fallen dramatically and through-hole construction will become increasingly expensive in comparison when it is possible at all.

The variety of available SM components is enormous and growing rapidly. While the fundamental component types are familiar, the packaging, benefits and limitations of SM components are significantly different to their through-hole equivalents and a brief review is worthwhile (**Fig 2.36**).

Surface Mount Resistors

Typical sizes of resistors continue to shrink due to market pressure for ever-higher densities and lower costs. The surface-mount parts are often referred to by their nominal case sizes such as '1206' which is approximately 12 hundredths of an inch long by 6 hundredths of an inch wide. **Fig 2.37**.

Table 2.7 lists the commoner sizes but the largest volume is currently in the 0603 size. The 0805 is probably slightly better for the amateur as smaller parts tend not to bear any markings. One standard technique with surface mount is to use the space between the two pads under a component to run a surface track under the component and save a wire link. Below 0805 size this technique becomes demanding of amateur PCB processes. Tolerances of resistors have also improved with standard parts being 1% tolerance and 100ppm stability. Such has been the improvement in manufacturing processes looser tolerance parts are generally not available. The 0201 size is in restricted commercial use but is of limited relevance to the

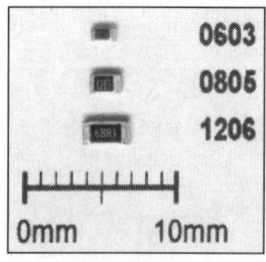

Fig 2.37: SMT resistors

Style	Size (mm)		Rating
	L	W	(mW)
1206	3.2	1.60	250
0805	2.0	1.25	125
0603	1.6	0.80	100
0402	1.0	0.50	63
0201	0.6	0.30	50

Table 2.7: SMT resistors

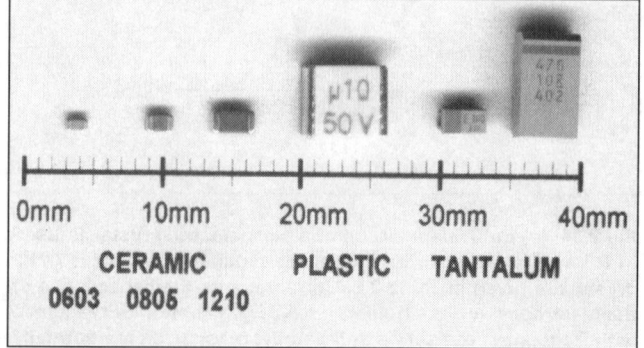

Fig 2.38: SMT capacitors

Dielectric type	Relative permittivity	Tolerance (%)	Typical temperature stability (%)
COG, NPO	low	±5	±0.5
X7R	medium	±10	±10
Z5U	high	+80/ 20	60/+20
Y5V	high	+80/ 20	80/+20

Table 2.8: Ceramic capacitors

amateur. The power ratings are for comparison purposes only as in practice they are partly dependent on the PCB design and other factors.

Surface Mount Capacitors

Capacitors are available in similar sizes to resistors but due to physical constraints it is not possible to settle on a single case size for all requirements. The dominant type is the monolithic multi-layer ceramic type with tantalum dielectric as the next most common. Progress has been made with ceramic capacitors to decrease size and to increase maximum values to over 10µF, displacing electrolytic parts in some applications. Plastic film parts are also available but suffer from a number of problems. See **Fig. 2.38**.

Ceramic capacitors

When selecting these, choices must be made on tolerance, stability and voltage rating which will result in a number of size alternatives. The voltage rating is important since larger values may have quite low ratings. The choice of dielectrics is wide but some of the commoner choices are given in **Table 2.8** with some typical properties. The structure of these parts is a multi-layer monolithically fabricated structure.

Although the parasitic inductances of these surface-mount devices are superior to leaded equivalents there are special parts available for microwave frequencies that have better-defined parasitic properties and high-frequency Q, albeit at a cost penalty. The monolithic multi-layer capacitors are more fragile than they may appear at first and unfortunately seldom give any external sign of failure.

Mechanical shock can damage the parts quite easily and at least one supplier warns that the parts should not be dropped onto a hard surface. When hand soldering, avoid any contact between the iron and the unplated body of the component. The damage will often take the form of shattering of some of the internal plates resulting in a significant capacitance error. Commercially an ultrasonic microscope or x-ray investigation can identify the problem.

The parts can suffer permanent value changes due to soldering and the values also change slightly with the DC voltage applied to them. The higher-permittivity dielectrics are particularly subject to these effects.

Tantalum and niobium oxide capacitors

These capacitors are available in many variants and should be chosen carefully for the particular application. In addition to the standard types there are parts that provide lower effective series resistance (ESR), usually for PSU filter applications, and parts that provide either surge resistance or built-in fuses. It is important to consider safety, as an electrolytic failure on a power rail will often result in debris being blown off the PCB and even a small capacitor can cause eye damage. This means using appropriate voltage ratings, and using parts that will tolerate maximum current surges and have appropriate power dissipation and temperature ratings.

Plastic dielectric capacitors

Although available, these have only made a limited impact due to the problem that the dielectric has to withstand direct solder temperatures and many manufacturers simply will not approve them for their solder processes. Parts typically use polyester and care must be taken to avoid overheating them while soldering. Even with care the parts may change value significantly due to heating and the resultant mechanical distortion.

Aluminium electrolytic capacitors

These are available in SMT, often as slight variations of a leaded package to allow surface soldering. Although they have been displaced to an extent by tantalum parts they still have a role to play.

Surface Mount Inductors and Ferrites

A large range of inductive devices exists, based on either traditional wound structures or layered ferrite structures (**Fig. 2.39**). The advantage of smaller size in many SMT parts may be offset by reduced performance, often with lower Q and poorer stability in comparison to large leaded parts. The wound parts come in a number of styles and an additional option is often whether the part is to be shielded, ie the magnetic field is mostly self-contained as in a toroid. The current rating needs to be carefully watched as the parts will saturate and lose performance long before any other effects are noticed. The ferrite parts are available in multi-layer structures, a little like a monolithic multi-layer capacitor and can be very useful for RF suppression. A vast range of devices exist that integrate capacitors into the package to make various filters usually for EMC purposes, but monolithic low-pass filters are available for VHF and UHF transmitter harmonic suppression.

Surface Mount Connectors

Surface mount connectors are now relatively common for both multi-way interconnects and for coaxial connections. There is little to say specifically about them since they are essentially similar to the equivalent through-hole parts. The small size and the reliance on solder connections for mechanical strength means that they should be treated with care as they can readily be torn

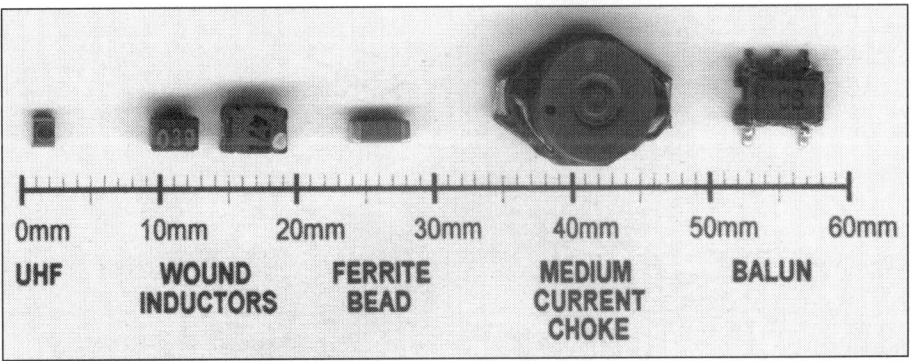

Fig 2.39: SMT inductors

Alpha chr	9	0	1	Numeral 2	3	4	5	6	7
A	0.10p	1.0p	10p	100p	1.0n	10n	100n	1.0	10
B	0.11p	1.1p	11p	110p	1.1n	11n	110n	1.1	11
C	0.12p	1.2p	12p	120p	1.2n	12n	120n	1.2	12
D	0.13p	1.3p	13p	130p	1.3n	13n	130n	1.3	13
E	0.15p	1.5p	15p	150p	1.5n	15n	150n	1.5	15
F	0.16p	1.6p	16p	160p	1.6n	16n	160n	1.6	16
G	0.18p	1.8p	18p	180p	1.8n	18n	180n	1.8	18
H	0.20p	2.0p	20p	200p	2.0n	20n	200n	2.0	20
J	0.22p	2.2p	22p	220p	2.2n	22n	220n	2.2	22
K	0.24p	2.4p	24p	240p	2.4n	24n	240n	2.4	24
L	0.27p	2.7p	27p	270p	2.7n	27n	270n	2.7	27
M	0.30p	3.0p	30p	300p	3.0n	30n	300n	3.0	30
N	0.33p	3.3p	33p	330p	3.3n	33n	330n	3.3	33
P	0.36p	3.6p	36p	360p	3.6n	36n	360n	3.6	36
Q	0.39p	3.9p	39p	390p	3.9n	39n	390n	3.9	39
R	0.43p	4.3p	43p	430p	4.3n	43n	430n	4.3	43
S	0.47p	4.7p	47p	470p	4.7n	47n	470n	4.7	47
T	0.51p	5.1p	51p	510p	5.1n	51n	510n	5.1	51
U	0.56p	5.6p	56p	560p	5.6n	56n	560n	5.6	56
V	0.62p	6.2p	62p	620p	6.2n	62n	620n	6.2	62
W	0.68p	6.8p	68p	680p	6.8n	68n	680n	6.8	68
X	0.75p	7.5p	75p	750p	7.5n	75n	750n	7.5	75
Y	0.82p	8.2p	82p	820p	8.2n	82n	820n	8.2	82
Z	0.91p	9.1p	91p	910p	9.1n	91n	910n	9.1	91
a	0.25p	2.5p	25p	250p	2.5n	25n	250n	2.5	25
b	0.35p	3.5p	35p	350p	3.5n	35n	350n	3.5	35
d	0.40p	4.0p	40p	400p	4.0n	40n	400n	4.0	40
e	0.45p	4.5p	45p	450p	4.5n	45n	450n	4.5	45
f	0.50p	5.0p	50p	500p	5.0n	50n	500n	5.0	50
m	0.60p	6.0p	60p	600p	6.0n	60n	600n	6.0	60
n	0.70p	7.0p	70p	700p	7.0n	70n	700n	7.0	70
t	0.80p	8.0p	80p	800p	8.0n	80n	800n	8.0	80
y	0.90p	9.0p	90p	900p	9.0n	90n	900n	9.0	90

The letter may be preceded by a manufacturer's mark such as a letter or symbol. Typically, a part by Kemet Electronics may be marked as in Fig 17.45.

Table 2.9: EIA-198 capacitor marking system, showing the capacitance (pF, nF, μF) for various identifiers

from a PCB. An additional problem is that due to their small size and low mating forces the connectors are very vulnerable to pollution on the contacts arising from flux or dirt and it is essential that they are kept very clean if problems are to be avoided.

Although Pressfit connectors are through-hole components they are included here as they are common on surface-mount boards. These connectors typically have square pins that are inserted into slightly undersize round-plated through holes so that the pins cut into the copper hole plating, resulting in a cold weld requiring no soldering. Considerable force may be required for large multi-way connectors but the construction is attractive to PCA assemblers as it can eliminate an additional solder process.

This technique is starting to appear on miniature coaxial PCB-mounting sockets. If unsoldered connectors are seen on a PCA it should not be automatically assumed that some mistake has been made! These parts can be used by the amateur - they should be treated as ordinary through-hole parts and hand soldered.

SMT Component Markings

Resistors
The small size of these parts means that often no markings are printed onto parts smaller than 0805. If they are marked, resistors are printed with their value using a three or four digit number, for example:

$$393 = 39 \times 10^3 = 39k\Omega$$
$$1212 = 121 \times 10^2 = 12.1k\Omega$$
$$180 = 18 \times 10^0 = 18.0\Omega$$
$$3R3 = 3.3 = 3.3\Omega$$

Capacitors
Unfortunately capacitors are seldom marked in this manner and if marked normally use a code system (EIA-198) of a letter followed by a number to indicate value. Access to a capacitance measurement device is extremely useful since the majority is unmarked. The EIA-198 capacitor marking system is shown in **Table 2.9**.

Inductors
These are usually marked with their value in microhenrys using a similar system to resistors, for example:

$$3u3 = 3.3\mu H$$
$$333 = 33mH$$

REFERENCES
[1] 'Simple inductance formulae for radio coils. Proc.IEE, Vol. 16, Oct 1928, p 1398
[2] *Shortwave Wireless Communications*, Ladner and Stoner, 4th edition, Chapman & Hall, London 1946 pp 308-339. This is an old reference and it does not show modern crystal designs, but does explain the 'cuts' very well.
[3] G3JIR, *Radio Communication* 1976 p896, 1972, pp28, 122 and 687, RSGB
[4] G3OUR, *Radio Communication* 1980, p1294, 1982 p863.

The chapter author is grateful for the assistance given by the AVX Corporation of the USA on the subject of Niobium Oxide capacitors.

3 Semiconductors and Valves

The development of semiconductor technology has had a profound impact on daily life by facilitating unprecedented progress in all areas of electronic engineering and telecommunications. Since the first transistors were made in the late 1940s, the number and variety of semiconductor devices have increased rapidly with the application of advanced technology and new materials. The devices now available range from the humble silicon rectifier diode used in power supplies to specialised GaAsFETs (gallium arsenide field effect transistors) used for low-noise microwave amplification. Digital ULSI ICs (ultra large-scale integrated circuits), involving the fabrication of millions of tiny transistors on a single chip of silicon, are used for the low-power microprocessor and memory functions which have made powerful desktop personal computers a reality.

Within amateur radio, there are now virtually no items of electronic equipment that cannot be based entirely on semiconductor, or 'solid-state', engineering. However, thermionic valves may still be found in some high-power linear amplifiers used for transmission (see the Building Blocks chapter).

SILICON

Although the first transistors were actually made using the semiconductor germanium (atomic symbol Ge), this has now been almost entirely replaced by silicon (Si). The silicon atom, shown pictorially in **Fig 3.1**, has a central nucleus consisting of 14 protons, which carry a positive electrical charge, and also 14 neutrons. The neutrons have no electrical charge, but they do possess the same mass (atomic weight) as the protons. Arranged around the nucleus are 14 electrons. The electrons, which are much lighter particles, carry a negative charge. This balances the positive charge of the protons, making the atom electrically neutral and therefore stable.

The 14 electrons are arranged within three groups, or shells. The innermost shell contains two electrons, the middle shell eight, and the outer shell four. The shells are separated by for-

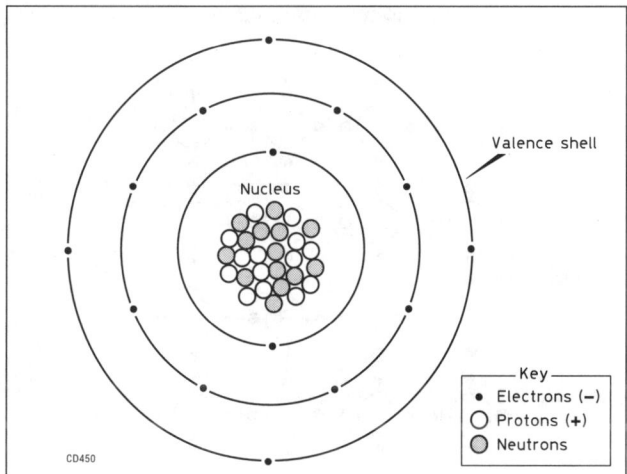

Key
- ● Electrons (−)
- ○ Protons (+)
- ◉ Neutrons

CD450

Fig 3.1: The silicon atom, which has an atomic mass of 28.086. 26% of the Earth's crust is composed of silicon, which occurs naturally in the form of silicates (oxides of silicon), eg quartz. Bulk silicon is steel grey in colour, opaque and has a shiny surface

Fig 3.2: Band gaps of conductors (a), a typical insulator (b) and silicon (c)

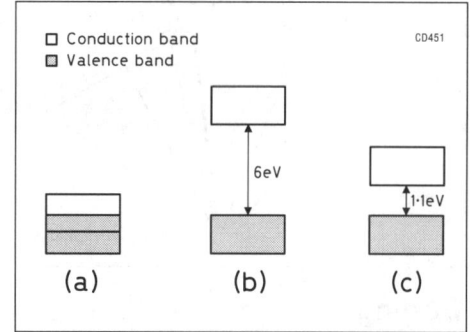

bidden regions, known as energy band gaps, into which individual electrons cannot normally travel. However, at temperatures above absolute zero (0 degrees Kelvin or minus 273 degrees Celsius) the electrons will move around within their respective shells due to thermal excitation. The speed, and therefore the energy, of the excited electrons increases as temperature rises.

It is the outermost shell, or valence band, which is of importance when considering whether silicon should be classified as an electrical insulator or a conductor. In conductors (see **Fig 3.2(a)**), such as the metals copper, silver and aluminium, the outermost (valence) electrons are free to move from one atom to another, thus making it possible for the material to sustain electron (current) flow. The region within which this exchange of electrons can occur is known, not surprisingly, as the conduction band, and in a conductor it effectively overlaps the valence band. In insulating materials (eg most plastics, glass and ceramics), however, there is a forbidden region (bandgap) which separates the valence band from the conduction band (**Fig 3.2(b)**). The width of the bandgap is measured in electron volts (eV), 1eV being the energy imparted to an electron as it passes through a potential of 1V. The magnitude of the bandgap serves as an indication of how good the insulator is and a typical insulator will have a bandgap of around 6eV. The only way to force an insulator to conduct an electrical current is to subject it to a very high potential difference, ie many thousands of volts. Under such extreme conditions, the potential may succeed in imparting sufficient energy to the valence electrons to cause them to jump into the conduction band. When this happens, and current flow is instigated, the insulator is said to have broken down. In practice, the breakdown of an insulator normally results in its destruction. It may seem odd that such a high potential is necessary to force the insulator into conduction when the bandgap amounts to only a few volts. However, even a very thin slice of insulating material will have a width of many thousand atoms, and a potential equal to the band gap must exist across each of these atoms before current flow can take place.

Fig 3.2(c) shows the relationship between the valence and conduction bands for intrinsic (very pure) silicon. As can be seen, there is a bandgap of around 1.1eV at room temperature. This means that intrinsic silicon will not under normal circumstances serve as an electrical conductor, and it is best described as a narrow bandgap insulator. The relevance of the term 'semiconductor', which might imply a state somewhere between conduction and insulation, or alternatively the intriguing possibility of being able to move from one to the other, is partly explained by the methods that have been developed to modify the electrical behaviour of materials like silicon.

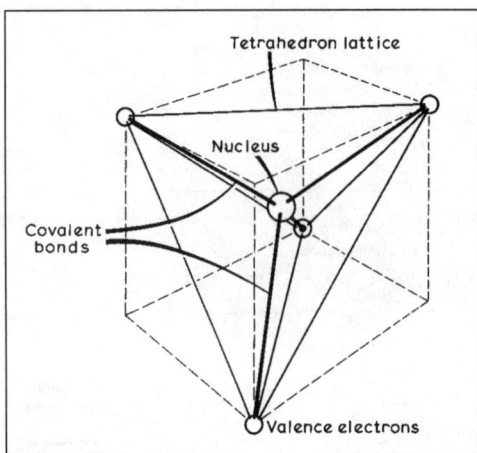

Fig 3.3: A three-dimensional view of the crystal lattice filter

Doping

The atoms within a piece of silicon form themselves into a criss-cross structure known as a tetrahedral crystal lattice - this is illustrated three-dimensionally by **Fig 3.3**. The lattice is held together by a phenomenon called covalent bonding, where each atom shares its valence electrons with those of its four nearest neighbours by establishing electron pairs. For greater clarity, **Fig 3.4** provides a simplified, two-dimensional, representation of the crystal lattice.

It is possible to alter slightly the crystal lattice structure of silicon, and through doing so modify its conductivity, by adding small numbers of atoms of other substances. These substances, termed dopants (also rather misleadingly referred to as 'impurities'), fall into two distinct categories:

- Group III - these are substances that have just three valence electrons: one less than silicon. Boron (atomic symbol B) is the most commonly used. Material doped in this way is called P-type.

- Group V - substances with five valence electrons: one more than silicon. Examples are phosphorus (P) and arsenic (As). This produces N-type material.

The level of doping concentration is chosen to suit the requirements of particular semiconductor devices, but in many cases the quantities involved are amazingly small. There may be only around one dopant atom for every 10,000,000 atoms of silicon (1 in 10^7). Silicon modified in this way retains its normal structure because the Group III and Group V atoms are able to fit themselves into the crystal lattice.

In the case of N-type silicon **(Fig 3.5(a)**, each dopant atom has one spare electron that cannot partake in covalent bonding.

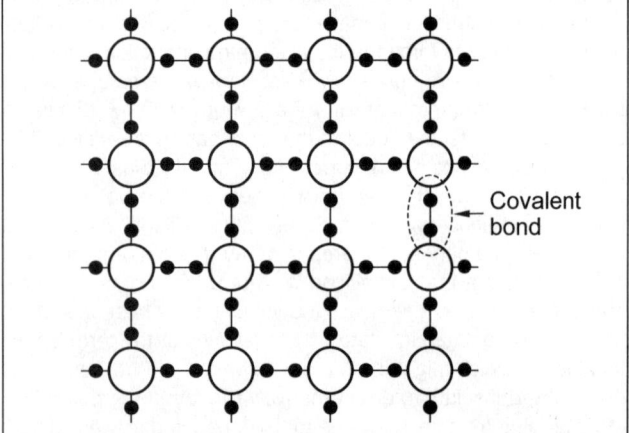

Fig 3.4: Covalent bonding in the crystal lattice filter

These 'untethered' electrons, are free to move into the conduction band and can therefore act as current carriers. The dopant atom is known as a donor, because it has 'donated' an electron to the conduction band. In consequence, N-type silicon has a far higher conductivity than intrinsic silicon. The conductivity of silicon is also enhanced by P-type doping, although the reason for this is less obvious. **Fig 3.5(b)** shows that when a Group III atom fits itself into the crystal lattice an additional electron is 'accepted' by the dopant atom (acceptor), creating a positively charged vacant electron position in the valence band known as a hole. Holes are able to facilitate electron (current) flow because they exert a considerable force of attraction for any free electrons. Semiconductors which utilise both free electrons and holes for current flow are known as bipolar devices. Electrons, however, are more mobile than holes and so semiconductor devices designed to operate at the highest frequencies will usually rely on electrons as their main current carriers.

THE PN JUNCTION DIODE

The PN junction diode is the simplest bipolar device. It serves a vital role both in its own right (in rectification and switching applications), and as an important building block in more complex devices, such as transistors. **Fig 3.6** shows that the junction consists of a sandwich of P- and N-type silicon. This representation suggests that a diode can be made by simply bonding together small blocks of P- and N-type material. Although this is now technically possible, such a technique is not in widespread use.

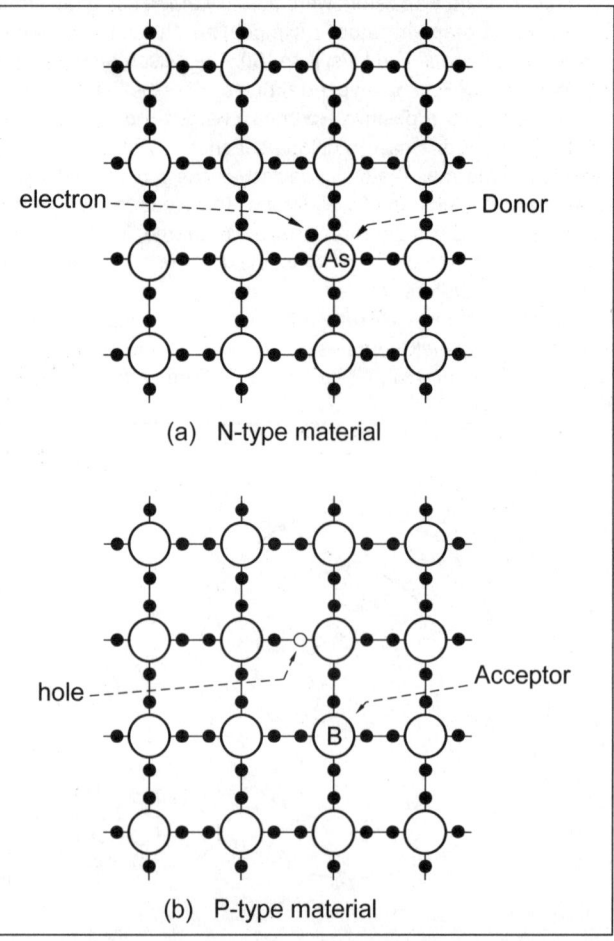

(a) N-type material

(b) P-type material

Fig 3.5: (a) The introduction of a dopant atom having five valence electrons (arsenic) into the crystal lattice. (b) Doping with an atom having three valence electrons (boron) creates a hole

Fig 3.6: The PN junction and diode symbol

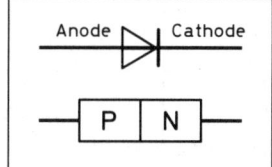

The process generally used to fabricate junction diodes is shown in **Fig 3.7**. This is known as the planar process, and variations of this technique are employed in the manufacture of many other semiconductor devices. The starting point is a thin slice of N-type silicon which is called the substrate (see **Fig 3.7(a)**). An insulating layer of silicon dioxide (SiO_2) is thermally grown on the top surface of the substrate in order to protect the silicon underneath. The next step (**Fig 3.7(b)**) is to selectively etch a small hole, referred to as a window, into the oxide layer - often using dilute hydrofluoric acid. A P-type region is then formed by introducing boron dopant atoms into the silicon substrate through the oxide window, using either a diffusion or ion implantation process (**Fig 3.7(c)**). This means, of course, that a region of the N-type substrate has been converted to P-type material. However, the process does not involve removing any of the N-type dopant, as it is simply necessary to introduce a higher concentration of P-type atoms than there are N-type already present. Ohmic contacts (ie contacts in which the current flows equally in either direction) are then formed to both the substrate and window (**Fig 3.7(d)**). In practice, large numbers of identical diodes will be fabricated on a single slice (wafer) of silicon which is then cut into 'chips', each containing a single diode. Finally, metal leads are bonded onto the ohmic contacts before the diode is encapsulated in epoxy resin, plastic or glass.

The diode's operation is largely determined by what happens at the junction between the P-type and N-type silicon. In the absence of any external electric field or potential difference, the large carrier concentration gradients at the junction cause carrier diffusion. Electrons from the N-type silicon diffuse into the P side, and holes from the P-type silicon diffuse into the N-side. As electrons diffuse from the N-side, uncompensated positive donor ions are left behind near the junction, and as holes leave the P-side, uncompensated negative acceptor ions remain. As **Fig 3.8(a)** shows, a negative space charge forms near the P-side of the junction (represented by '-' symbols), and a positive space charge forms near the N-side (represented by '+' symbols). Therefore this diffusion of carriers results in an electrostatic potential difference across the junction within an area known as the depletion region. The magnitude of this built-in potential depends on the doping concentration, but for a typical silicon diode it will be around 0.7V. It is important to realise, however, that this potential cannot be measured by connecting a voltmeter externally across the diode because there is no net flow of current through the device.

If a battery is connected to the diode as in **Fig 3.8(b)**, the external applied voltage will increase the electrostatic potential across the depletion region, causing it to widen (because the polarity of the external voltage is the same as that of the built-in potential). Under these conditions practically no current will flow through the diode, which is said to be reverse biased. However, if the external bias is raised above a certain voltage the diode may 'break down', resulting in reverse current flow.

Reversing the polarity of the battery (**Fig 3.8(c)**) causes the external applied voltage to reduce the electrostatic potential across the depletion region (by opposing the built-in potential), therefore narrowing the depletion region. Under this condition, known as forward bias, current is able to flow through the diode by diffusion of current carriers (holes and electrons) across the depletion region.

The diode's electrical characteristics are summarised graphically in **Fig 3.9**, where a linear scale is used for both current and voltage. In Region A, where only a small forward bias voltage is applied (less than about 0·7V), the current flow is dominated by carrier diffusion, and in fact rises exponentially as the forward bias voltage is increased. Region B commences at a point often termed the 'knee' of the forward bias curve, and here series resistance (both internal and external) becomes the major factor in determining current flow. There remains, however, a small voltage drop which, in the case of an ordinary silicon diode, is around 0·7V .

Under reverse bias (region C) only a very small leakage current flows, far less than 1µA for a silicon diode at room temperature. This current is typically due to generation of electron-hole pairs in the reverse-biased depletion region, and does not vary with reverse voltage. In Region D, where the reverse bias is increased above the diode's breakdown voltage, significant reverse current begins to flow. This is often due to a mechanism

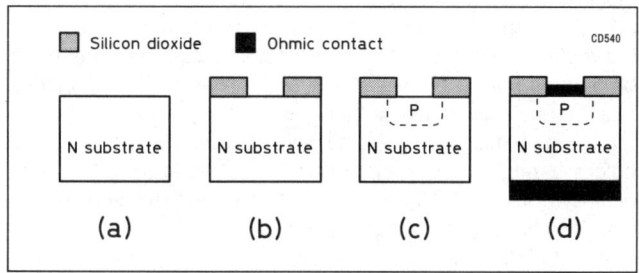

Fig 3.7: Fabrication of a PN junction diode

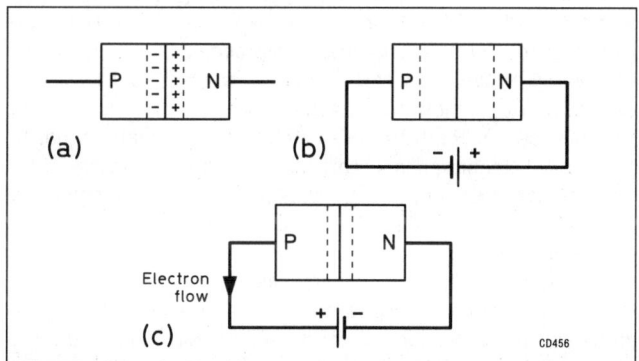

Fig 3.8: Behaviour of the junction diode under zero bias (a), reverse bias (b) and forward bias (c)

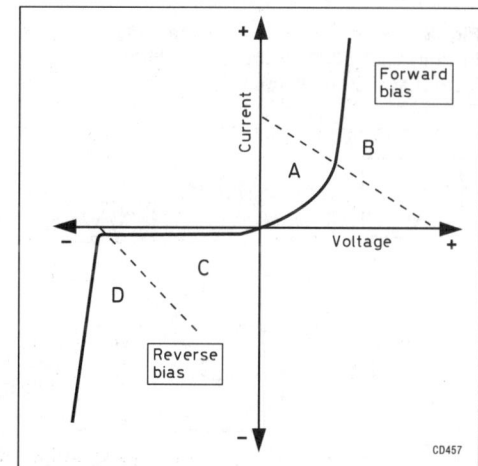

Fig 3.9: Characteristic curve of a PN junction diode

Fig 3.10: The junction diode used as a half-wave rectifier

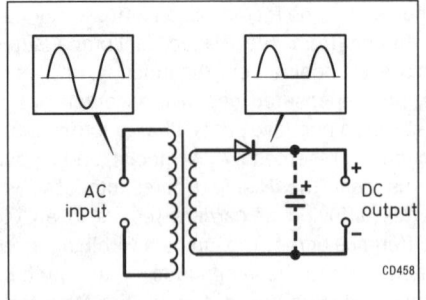

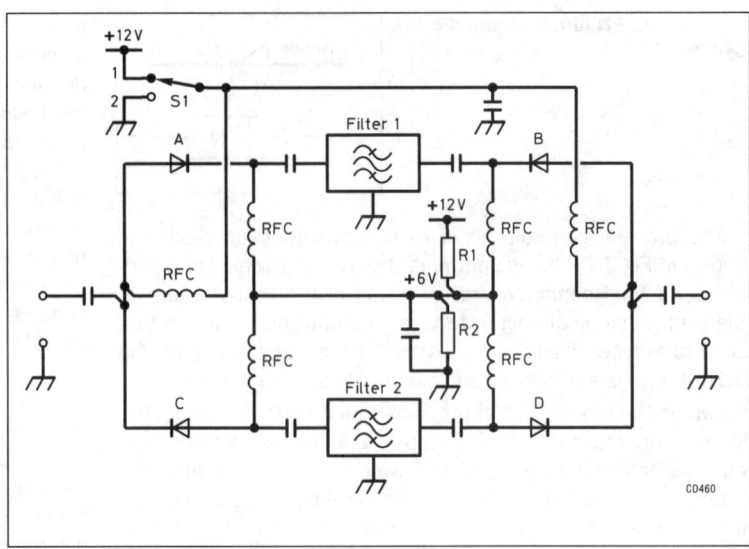

Fig 3.13: Diodes are frequently used as switches in the signal circuitry of transceivers. The setting of S1 determines which of the two filters is selected. R1 and R2 will both have a value of around 2.2kΩ (this determines the forward current of the switching diodes). For HF applications the RFCs are 100µH

of avalanche multiplication of electron-hole pairs in the high electric field in the depletion region. Although this breakdown process is not inherently destructive, the maximum current must be limited by an external circuit to avoid excessive heating of the diode.

Diodes have a very wide range of applications in radio equipment, and there are many types available with characteristics tailored to suit their intended use. **Fig 3.10** shows a diode used as a rectifier of alternating current in a simple power supply. The triangular part of the diode symbol represents the connection to the P-type material and is referred to as the anode. The single line is the N-type connection, or cathode (Fig 3.6). Small diodes will have a ring painted at one end of their body to indicate the cathode connection (see the lower diode in **Fig 3.11**). The rectifying action allows current to flow during positive half-cycles of the input waveform, while preventing flow during negative half-cycles. This is known as half-wave rectification because only 50% of the input cycle contributes to the DC output. **Fig 3.12** shows how a more efficient power supply may be produced using a bridge rectifier, which utilises four diodes. Diodes A and B conduct during positive half-cycles, while diodes C and D conduct during negative cycles. Notice that C

Fig 3.11: Rectifiers. From the left, clockwise: 35A, 100PIV bridge, 1.5A 50PIV bridge, 6A 100PIV bridge, 3A 100PIV diode. Note that the small squares are 1mm across

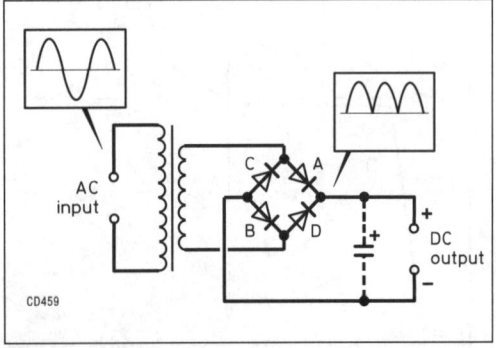

Fig 3.12: A full-wave, or bridge, rectifier using four unijunction diodes

and D are arranged so as to effectively reverse the connections to the transformer's secondary winding during negative cycles, thus enabling the negative half-cycles to contribute to the 'positive' output. Because this arrangement is used extensively in equipment power supplies, bridge rectifier packages containing four interconnected diodes are readily available. When designing power supplies it is necessary to take account of the voltages and currents that the diodes will be subjected to. All rectifiers have a specified maximum forward current at a given case temperature. This is because the diode's forward voltage drop gives rise to power dissipation within the device - for instance, if the voltage across a rectifier diode is 0·8V at a current of 5A, the diode will dissipate 4W (0·8 x 5 = 4). This power will be converted to heat, thus raising the diode's temperature. Consequently, the rectifier diodes of high-current power supplies may need to be mounted on heatsinks. In power supplies especially, diodes must not be allowed to conduct in the reverse direction. For this reason, the maximum reverse voltage that can be tolerated before breakdown is likely to occur must be known. The term peak inverse voltage (PIV) is used to describe this characteristic (see the chapter on power supplies).

Miniature low-current diodes, often referred to as small-signal types, are very useful as switching elements in transceiver circuitry. **Fig 3.13** shows an arrangement of four diodes used to select one of two band-pass filters, depending on the setting of switch S1. When S1 is set to position 1, diodes A and B are forward biased, thus bringing filter 1 into circuit. Diodes C and D, however, are reverse biased, which takes filter 2 out of circuit. Setting S1 at position 2 reverses the situation. Notice how the potential divider R1/R2 is used to develop a voltage equal to half that of the supply rail. This makes it easy to arrange for a forward bias of 6V, or a reverse bias of the same magnitude, to appear across the diodes. Providing that the peak level of signals at the filter terminations does not reach 6V under any circumstances, the diodes will behave as almost perfect switches. One of the main advantages of diode switching is that signal paths can be kept short, as the leads (or PCB tracks) forming connections to the front-panel switches carry only a DC potential which is isolated from the signal circuitry using the RF chokes and decoupling capacitors shown.

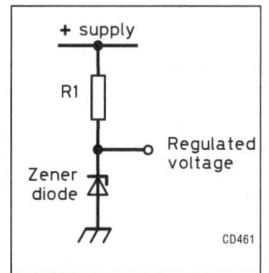

Fig 3.14: A zener diode used as a simple voltage regulator

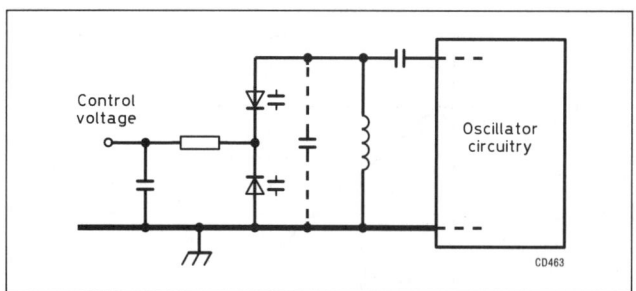

Fig 3.16: Two varactor diodes used to tune a voltage-controlled oscillator. The capacitor drawn with dotted lines represents the additional component which may be added to the tank circuit in order to modify the LC ratio and tuning range

Zener Diodes

These diodes make use of the reverse breakdown characteristic discussed previously. The voltage at which a diode begins to conduct when reverse biased depends on the doping concentration. As the doping level is increased, the breakdown voltage drops. This fact can be exploited during the manufacture of the diodes, enabling the manufacturer to specify the breakdown voltage for a given component. Zener diodes with breakdown voltages in the range 2·7V to over 150V are available and can be used to provide reference voltages for power supplies and bias generators.

Fig 3.14 shows how a zener diode can be used in conjunction with a resistor to provide voltage regulation (note the use of a slightly different circuit symbol for the zener diode). When power is initially applied, the zener diode will start to conduct as the input voltage is higher than the diode's reverse breakdown value. However, as the diode begins to pass current, an increasingly high potential difference will appear across resistor R1. This potential will tend to rise until the voltage across R1 becomes equal to the difference between the input voltage and the zener's breakdown voltage. The net result is that the output voltage will be forced to settle at a level close to the diode's reverse breakdown potential. The value of R1 is chosen so as to limit the zener current to a safe value (the maximum allowable power dissipation for small zener diodes is around 400mW), while ensuring that the maximum current to be drawn from the regulated supply will not increase the voltage across R1 to a level greater than the difference between the zener voltage and the minimum expected input voltage (see the chapter on power supplies).

Varactor Diodes

The varactor, or variable capacitance, diode makes use of the fact that a reverse biased PN junction behaves like a parallel-plate capacitor, where the depletion region acts as the spacing between the two capacitor plates. As the reverse bias is increased, the depletion region becomes wider. This produces the same effect as moving the plates of a capacitor further apart - the capacitance is reduced (see the fundamentals chapter). The varactor can therefore be used as a voltage-controlled vari-

able capacitor, as demonstrated by the graph in Fig 3.15. The capacitance is governed by the diode's junction area (ie the area of the capacitor plates), and also by the width of the depletion region for a given value of reverse bias (which is a function of the doping concentration). Varactors are available covering a wide range of capacitance spreads, from around 0·5-10pF up to 20-400pF. The voltages at which the stated maximum and minimum capacitances are obtained will be quoted in the manufacturer's literature, but they normally fall in the range 2-20V. The maximum reverse bias voltage should not be exceeded as this could result in breakdown.

Varactors are commonly used to achieve voltage control of oscillator frequency in frequency synthesisers. Fig 3.16 shows a typical arrangement where two varactors are connected 'back to back' and form part of the tank circuit of a voltage-controlled oscillator (VCO). The use of two diodes prevents the alternating RF voltage appearing across the tuned circuit from driving the varactors into forward conduction, which is most likely to happen when the control voltage is low. Because the varactor capacitances appear in series, the maximum capacitance swing is half that obtainable when using a single diode. Three-lead packages containing dual diodes internally connected in this way are readily available. It is also possible to obtain multiple-diode packages containing two or three matched diodes but with separate connections. These are used to produce voltage-controlled versions of two-gang or three-gang variable capacitors.

When used as a capacitive circuit element, the varactor's Q may be significantly lower than that of a conventional capacitor. This factor must be taken into account when designing high-performance frequency synthesisers (see the chapter on oscillators).

PIN Diodes

A PIN diode, Fig 3.17, is a device that operates as a variable resistor at RF and even into microwave frequencies. Its resistance is determined only by its DC excitation. It can also be used to switch quite large RF signals using smaller levels of DC excitation.

The PIN diode chip consists of a chip of pure (intrinsic or I-type) silicon with a layer of P-type silicon on one side and a layer of N-type on the other. The thickness of the I-region (W) is only a little smaller than the thickness of the original wafer from which the chip was cut.

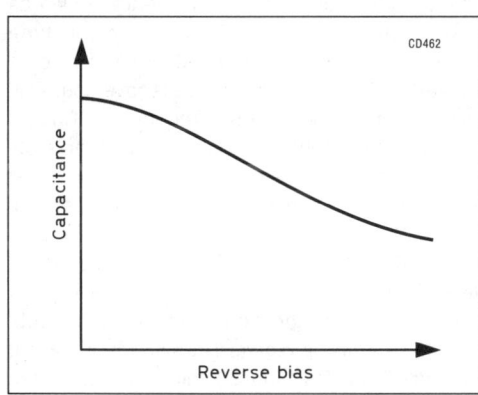

Fig 3.15: Relationship between capacitance and reverse bias of a varactor (variable capacitance diode

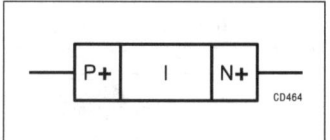

Fig 3.17: The PIN diode

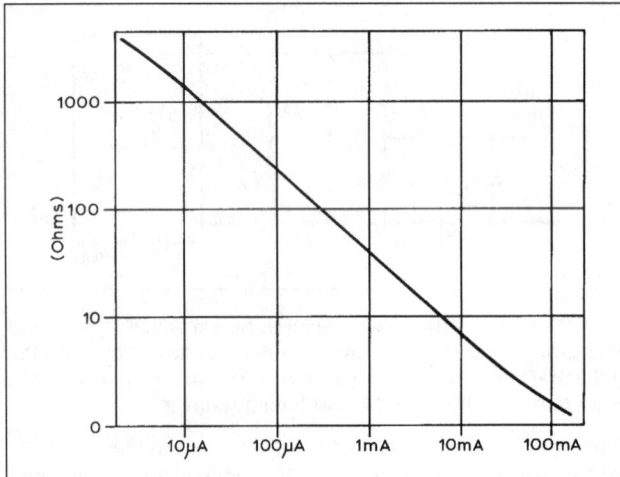

Fig 3.18: Relationship between forward current and resistance for a PIN dode

Forward-biased PIN diodes

When forward-biased electrons and holes are injected into the I-region from the N- and P-regions respectively, these charges don't recombine immediately but a finite charge remains stored in the I-layer.

The quantity of stored charge, Q, depends on the recombination time (t, usually called the carrier lifetime) and the forward current I as:

$$Q = It$$

The resistance of the I-layer is proportional to the square of its thickness (W) and inversely proportional to Q, and is given by:

$$R = \frac{W^2}{(N+p)Q}$$

Combining these two equations, we get:

$$R = \frac{W^2}{(N+p)It}$$

Thus the resistance is inversely proportional to the DC excitation I. For a typical PIN diode, R varies from 0·1Ω at 1A to 10kΩ at 1µA. This is shown in **Fig 3.18**.

This resistance-current relationship is valid over a wide frequency range, the limits at low resistance being set by the parasitic inductance of the leads, and at high resistance by the junction capacitance. The latter is small owing to the thickness of the I-layer which also ensures that there is always sufficient charge so that the layer presents a constant resistance over the RF cycle.

Reverse-biased PIN diodes

At high RF, a reverse-biased PIN diode behaves as a capacitor independent of the reverse bias with a value of:

$$C = εA/W \quad \text{farads}$$

where ε is the permittivity of silicon. Note that although the dielectric constant is quite high, the capacitance of the diode is small because of the small area of the diode and the thickness of the intrinsic layer. Therefore as a switch the isolation in the

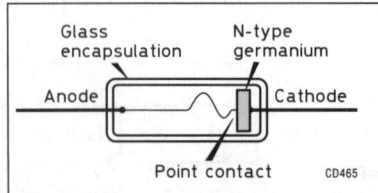

Fig 3.19: The germanium point-contact diode

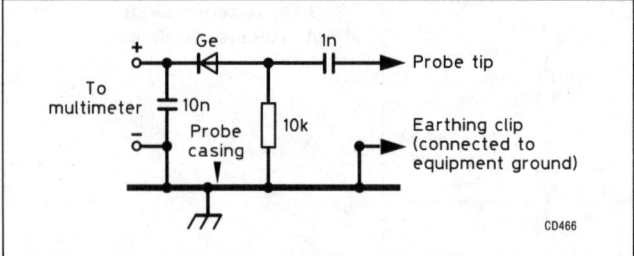

Fig 3.20: A simple RF probe using a germanium (Ge) point-contact diode

reverse bias state is good but gets slightly worse as the frequency increases. This is not as serious as it seems since nearly all PIN diode switches at VHF are operated in 50Ω circuits and the reactance is high compared to 50Ω.

Germanium Point-contact Diodes

Ironically, one of the earliest semiconductor devices to find widespread use in telecommunications actually pre-dates the thermionic valve. The first broadcast receivers employed a form of envelope detector (RF rectifier) known colloquially as a cat's whisker. This consisted of a spring made from a metal such as bronze or brass (the 'whisker'), the pointed end of which was delicately bought into contact with the surface of a crystal having semiconducting properties, such as galena, zincite or carborundum.

The germanium point-contact diode is a modern equivalent of the cat's whisker and consists of a fine tungsten spring which is held in contact with the surface of an N-type germanium crystal (**Fig 3.19**). During manufacture, a minute region of P-type material is formed at the point where the spring touches the crystal. The point contact therefore functions as a PN diode. In most respects the performance of this device is markedly inferior to that of the silicon PN junction diode. The current flow under reverse bias is much higher - typically 5mA, the highest obtainable PIV is only about 70V and the maximum forward current is limited by the delicate nature of the point contact.

Nevertheless, this device has a number of saving graces. The forward voltage drop is considerably lower than that of a silicon junction diode - typically 0·2V - and the reverse capacitance is also very small. There is also an improved version known as the gold-bonded diode, where the tungsten spring is replaced by one made of gold. **Fig 3.20** shows a simple multimeter probe which is used to rectify low RF voltages. The peak value can then be read with the meter switched to a normal DC range. The low forward voltage drop of the point-contact diode leads to more accurate readings.

The Schottky Barrier Diode

Ordinary PN junction diodes suffer from a deficiency known as charge storage, which has the effect of increasing the time taken for a diode to switch from forward conduction to reverse cut-off when the polarity of the applied voltage is reversed. This reduces the efficiency of the diode at high frequencies. Charge storage occurs because holes, which are less mobile than electrons, require a finite time to migrate back from the N-doped cathode material as the depletion region widens under the influence of reverse bias. The fact that in most diodes the P-type anode is more heavily doped than the N-type cathode tends to exacerbate matters.

The hot-carrier, or Schottky barrier, diode overcomes the problem of charge storage by utilising electrons as its main current carriers. It is constructed (**Fig 3.21**) in a similar fashion to the

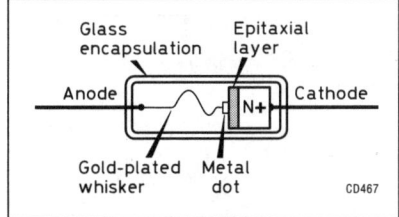

(above) **Fig 3.21: Hot-carrier (Schottky) diode**

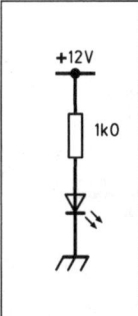

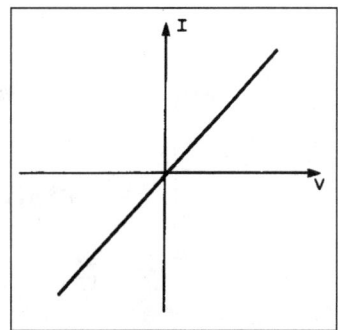

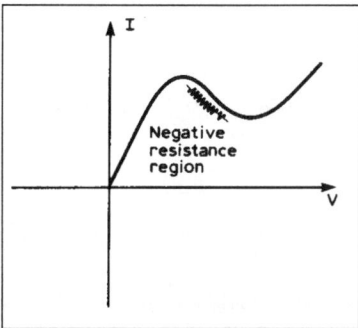

Fig 3.22: An LED emits light when forward biased. The series resistor limits the forward current to a safe value

Fig 3.23: Voltage-current through a resistor

Fig 3.24: Voltage-current through a Gunn diode

germanium point-contact type, but there are a few important differences. The semiconductor used is N-type silicon which is modified by growing a layer (the epitaxial region) of more lightly doped material onto the substrate during manufacture. The device is characterised by its high switching speed and low capacitance. It is also considerably more rugged than the germanium point-contact diode and generates less noise.

Hot-carrier diodes are used in high-performance mixers of the switching, or commutating, type capable of operating into the microwave region.

Light-emitting diodes (LEDs)

The LED consists of a PN junction formed from a compound semiconductor material such as gallium arsenide (GaAs) or gallium phosphide (GaP). As gallium has a valency of three, and arsenic and phosphorus five, these materials are often referred to as Group III-V semiconductors.

When electrons recombine with holes across the energy gap of a semiconductor, as happens around the depletion layer of a forward-biased PN junction, particles of light energy known as photons are released. The energy, and therefore wavelength, of the photons is determined by the semiconductor band gap. Pure gallium arsenide has a band gap of about 1·43eV, which produces photons with a wavelength of 880 nanometres (nm). This lies at the infrared end of the spectrum. Adding aluminium to the gallium arsenide has the effect of increasing the band gap to 1·96eV which shortens the light wavelength to 633nm. This lies in the red part of the visible spectrum. Red LEDs can also be made by adding phosphorus to gallium arsenide. Green LEDs (wavelength 560nm) are normally made from gallium phosphide.

The LEDs used as front-panel indicators consist of a PN junction encapsulated within translucent plastic. At a current of 10mA they will generate a useful amount of light without overheating. The forward voltage drop at this current is about 1·8V. **Fig 3.22** shows a LED operating from a 12V power rail (note the two arrows representing rays of light which differentiates the LED circuit symbol from that of a normal diode). The series resistor determines the forward current and so its inclusion is mandatory. LEDs are also used in more complex indicators, such the numeric (seven-segment) displays employed in frequency counters (see the chapter on test equipment).

The Gunn diode

The Gunn diode, named after its inventor, B J Gunn, comprises little more than a block of N-type gallium arsenide. It is not, in fact, a diode in the normally accepted sense because there is no P-N junction and consequently no rectifying action. It is properly called the Gunn effect device. However, 'diode' has become accepted by common usage and merely explains that it has two connections.

It consists of a slice of low-resistivity N-type gallium arsenide on which is grown a thin epitaxial layer, the active part, of high-resistivity gallium arsenide with a further thicker layer of low-resistivity gallium arsenide on top of that. Since the active layer is very thin, a low voltage across it will produce a high electric field strength. The electrons in gallium arsenide can be in one of two conduction bands. In one they have a much higher mobility than in the other and they are initially in this band. As the electric field increases, more and more are scattered into the lower mobility band and the average velocity decreases. The field at which this happens is called the threshold field and is 320V/mm. Since current is proportional to electron velocity and voltage is proportional to electric field, the device has a region of negative resistance. This odd concept is explained by the definition of resistance as the slope of the voltage-current graph and a pure resistor has a linear relationship (**Fig 3.23**). On the other hand, the Gunn diode has a roughly reversed 'S'-shaped curve (**Fig 3.24**) and the negative resistance region is shown by the hatching. The current through the device takes the form of a steady DC with superimposed pulses (**Fig 3.25**) and their fre-

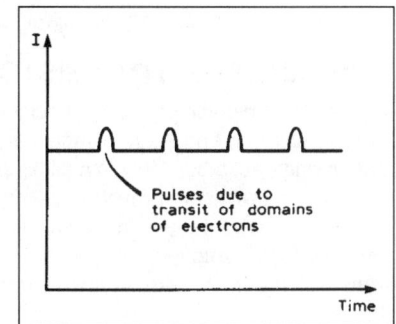

Fig 3.25: Current in Gunn diode

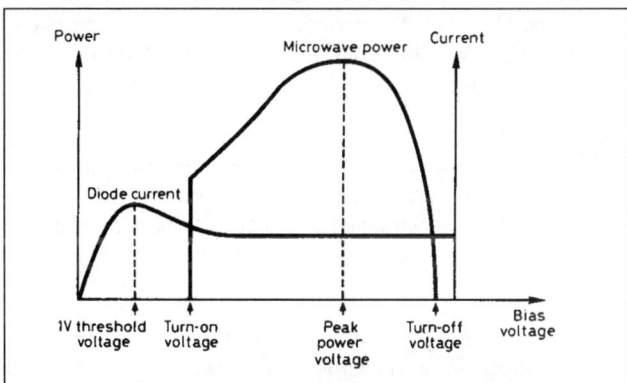

Fig 3.26: Characteristic shape of a Gunn oscillator's bias power curve

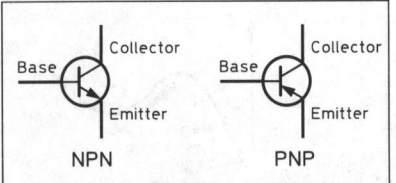

Fig 3.27: Circuit symbols for the bipolar transistor

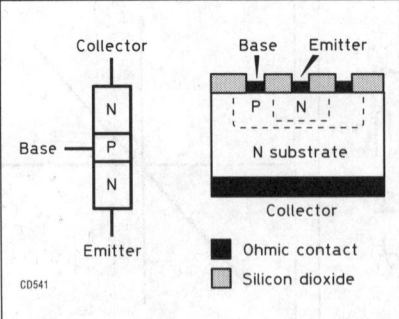

Fig 3.28: Construction of an NPN transistor

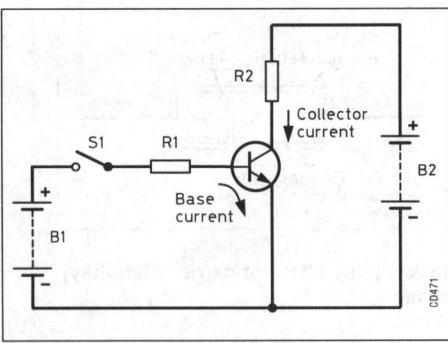

Fig 3.29: Applying forward bias to the base-emitter junction of a bipolar transistor causes a much larger current to flow between the collector and emitter. The arrows indicate conventional current flow, which is opposite in direction to electron flow

quency is determined by the thickness of the active epitaxial layer. As each pulse reaches the anode, a further pulse is generated and a new domain starts from the cathode. Thus the rate of pulse formation depends on the transit time of these domains through the epitaxial layer. In a 10GHz Gunn diode, the layer is about 10μm thick, and a voltage of somewhat greater than 3·5V gives the high field state and microwave pulses are generated. The current through the device shows a peak before the threshold voltage followed by a plateau. See **Fig 3.26**. The power output reaches a peak at between 7·0 and 9·0V for 10GHz devices.

The Gunn diode is inherently a wide-band device so it is operated in a high-Q cavity (tuned circuit) and this determines the exact frequency. It may be tuned over a narrow range by altering the cavity with a metallic screw, a dielectric (PTFE or Nylon) screw or by loading the cavity with a varactor. With a 10GHz device, the whole of the 10GHz amateur band can be covered with reasonable efficiency. Gunn diodes are not suitable for narrow-band operation since they are of low stability and have relatively wide noise sidebands. The noise generated has two components, thermal noise and a low frequency 'flicker' noise. Analysis of the former shows that it is inversely proportional to the loaded Q of the cavity and that FM noise close to the carrier is directly proportional to the oscillator's voltage pushing, ie to the variation of frequency caused by small variations of voltage. Clearly, the oscillator should be operated where this is a minimum and that is often near the maximum safe bias.

THE BIPOLAR TRANSISTOR

The very first bipolar transistor, a point-contact type, was made by John Bardeen and Walter Brattain at Bell Laboratories in the USA during December 1947. A much-improved version, the bipolar junction transistor, arrived in 1950 following work done by another member of the same team, William Schockley. In recognition of their pioneering work in developing the first practical transistor, the three were awarded the Nobel Prize for physics in 1956.

The bipolar transistor is a three-layer device which exists in two forms, NPN and PNP. The circuit symbols for both types are shown in **Fig 3.27**. Note that the only difference between the symbols is the direction of the arrow drawn at the emitter connection.

Fig 3.28 shows the three-layer sandwich of an NPN transistor and also how this structure may be realised in a practical device. The emitter region is the most heavily doped.

In **Fig 3.29** the NPN transistor has been connected into a simple circuit to allow its operation to be described. The PN junction between the base and emitter forms a diode which is forward biased by battery B1 when S1 is closed. Resistor R1 has been included to control the level of current that will inevitably flow, and R2 provides a collector load. A voltmeter connected between the base and emitter will indicate the normal forward voltage drop of approximately 0·6V typical for a silicon PN junction.

The collector-base junction also forms the equivalent of a PN diode, but one that is reverse biased by battery B2. This suggests that no current will flow between the collector and the emitter. This is indeed true for the case where S1 is open, and no current is flowing through the base-emitter junction. However, when S1 is closed, current does flow between the collector and emitter. Due to the forward biasing of the base-emitter junction, a large number of electrons will be injected into the P-type base from the N-type emitter. Crucially, the width of the base is made less than the diffusion length of electrons, and so most of these injected carriers will reach the reverse biased collector-base junction. The electric field at this junction is such that these electrons will be swept across the depletion region into the collector, thus causing significant current to flow between the emitter and collector.

However, not all the electrons injected into the base will reach the collector, as some will recombine with holes in the base. Also, the base-emitter junction forward bias causes holes to be injected from the P-type base to the N-type emitter, and electrons and holes will recombine in the base-emitter depletion region. These are the three main processes which give rise to the small base current. The collector current will be significantly larger than the base current, and the ratio between the two (known as the transistor's ß) is related to the doping concentration of the emitter divided by the doping concentration of the base. The PNP transistor operates in a similar fashion except that the polarities of the applied voltages, and also the roles of electrons and holes, are reversed. The graph at **Fig 3.30** shows

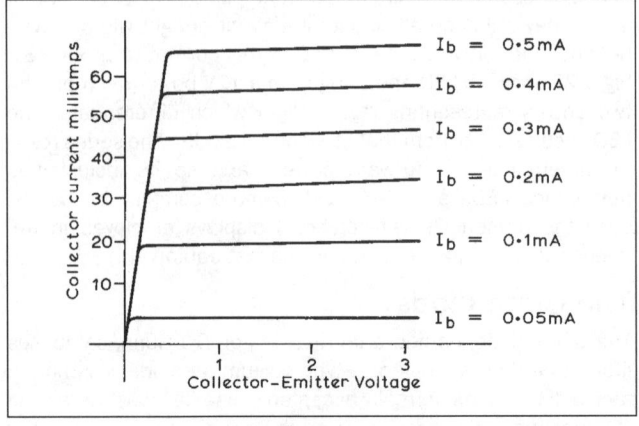

Fig 3.30: The relationship between base current and collector current for a bipolar transistor

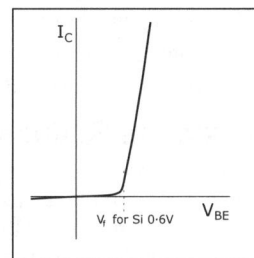

Fig 3.31: The relationship between base-emitter voltage and collector current for a bipolar transistor

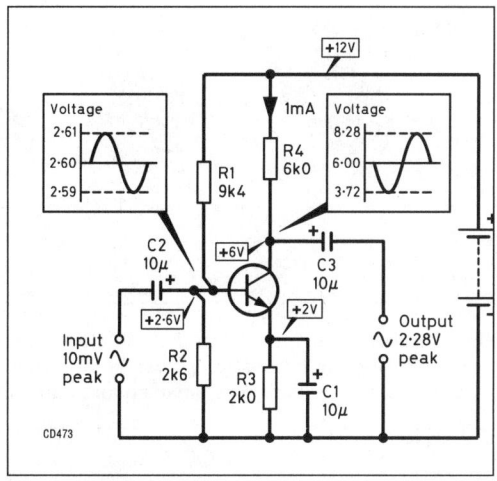

Fig 3.33: A practical common-emitter amplifier

the relationship between base current (I_B) and collector current (I_C) for a typical bipolar transistor. The point to note is that the collector current is very much determined by the base current ($I_C = \text{ß}I_B$) and the collector-emitter voltage, V_{CE}, has comparatively little effect. The relationship between base-emitter voltage, V_{BE}, and collector current I_C is shown in **Fig 3.31**. The graph should be compared with the diode characteristics previously shown in Fig 3.9. The graph is almost identical but the current axis is ß times greater due to the current gain of the transistor. It should be noted that the graph of transistor base current against applied voltage I_B/V_{BE} is simply the graph of the base-emitter 'diode'.

The circuit shown in Fig 3.29 therefore provides the basis for an amplifier. Small variations in base current will result in much larger variations in collector current. The DC current gain (ß or h_{FE}), of a typical bipolar transistor at a collector current of 1mA will be between 50 and 500, ie the change in base current required to cause a 1mA change in collector current lies in the range 2 to 20µA. The value of ß is temperature dependent, and so is the base-emitter voltage drop (V_{BE}), which will fall by approximately 2mV for every 1°C increase in ambient temperature.

A Transistor Amplifier

If the battery (B1) used to supply base current to the transistor in Fig 3.29 is replaced with a signal generator set to give a sinewave output, the collector current will vary in sympathy with the input waveform as shown in **Fig 3.32**. A similar, although inverted, curve could be obtained by plotting the transistor's collector voltage.

Two problems are immediately apparent. First, because the base-emitter junction only conducts during positive half-cycles of the input waveform, the negative half-cycles do not appear at the output. Second, the barrier height of the base-emitter junction results in the base current falling exponentially as the amplitude of the input waveform drops below +0·6V. This causes significant distortion of the positive half-cycles, and it is clear that if the amplitude of the input waveform were to be significantly reduced then the collector current would hardly rise at all.

The circuit can be made far more useful by adding bias. **Fig 3.33** shows the circuit of a practical common-emitter amplifier

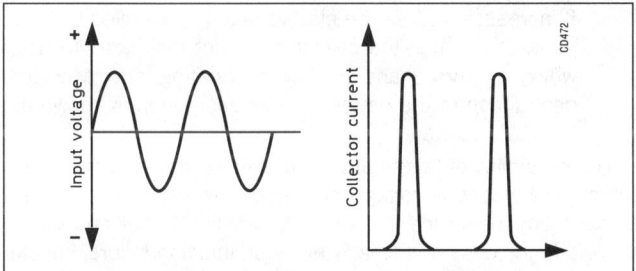

Fig 3.32: The circuit shown in Fig 3.29 would not make a very good amplifier

operating in Class A. The term 'common emitter' indicates that the emitter connection is common to both the input and output circuits. Resistors R1 and R2 form a potential divider which establishes a positive bias voltage at the base of the transistor. The values of R1 and R2 are chosen so that they will pass a current at least 10 times greater than that flowing into the base. This ensures that the bias voltage will not alter as a result of variations in the base current. R3 is added in order to stabilise the bias point, and its value has been calculated so that a potential of 2V will appear across it when, in the absence of an input signal, the desired standing collector current of 1mA (0·001A) flows (0·001A x 2000 = 2V). For this reason, the ratio between the values of R1 and R2 has been chosen to establish a voltage of 2·6V at the base of the transistor. This allows for the expected forward voltage drop of around 0·6V due to the barrier height of the base-emitter junction. Should the collector current attempt to rise, for instance because of an increase in ambient temperature causing V_{BE} to fall, the voltage drop across R3 will increase, thus reducing the base-emitter voltage and preventing the collector current rising. Capacitor C1 provides a low-impedance path for alternating currents so that the signal is unaffected by R3, and coupling capacitors C2 and C3 prevent DC potentials appearing at the input and output. The value of the collector load resistor (R4) is chosen so that in the absence of a signal the collector voltage will be roughly 6V - ie half that of the supply rail. Where a transistor is used as an RF or IF amplifier, the collector load resistor may be replaced by a parallel tuned circuit (see the building blocks chapters).

An alternating input signal will now modulate the base voltage, causing it to rise slightly during the positive half-cycles, and fall during negative half-cycles. The base current will be similarly modulated and this, in turn, will cause far larger variations in the collector current. Assuming the transistor has a ß of 100, the voltage amplification obtained can gauged as follows:

In order to calculate the effect of a given input voltage, it is necessary to develop a value for the base resistance. This can be approximated for a low-frequency amplifier using the formula:

$$\text{Base resistance} = \frac{26 \times \beta}{\text{Emitter current in mA}}$$

The emitter current is the sum of the base and collector currents but, as the base current is so much smaller than the collector current, it is acceptable to use just the collector current in rough calculations:

$$\text{Base resistance} = \frac{26 \times 100}{1} = 2600\Omega$$

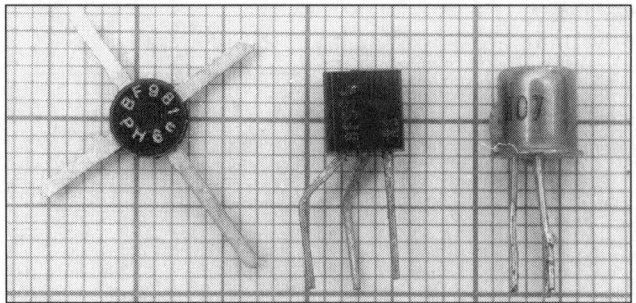

Fig 3.34: Power devices. From the left, clockwise: NPN bipolar power Darlington transistor, power audio amplifier IC, NPN bipolar power transistor. The small squares are 1mm across

Fig 3.35: Power devices. From the left, clockwise: NPN bipolar power Darlington transistor, power audio amplifier IC, NPN bipolar power transistor. The small squares are 1mm across

Using Ohm's Law, and assuming that the peak amplitude of the input signal is 10mV (0·01V), the change in base current will be:

$$\frac{0\cdot01}{2600} = 3\cdot8\mu A$$

This will produce a change in collector current 100 times larger, that is 380µA.

The change in output voltage (again using Ohm's Law) is:

$$380\mu A \times 6k\Omega \ (R4) = 2\cdot28V$$
$$(3\cdot8 \times 10\text{-}4 \ \ 6 \ 103 = 2\cdot28)$$

The voltage gain obtained is therefore:

$$\frac{2\cdot28}{0\cdot01} = 228 \quad \text{or } 47dB$$

Note that the output voltage developed at the junction between R4 and the collector is phase reversed with respect to the input. As operating frequency is increased the amplifier's gain will start to fall. There are a number of factors which cause this, but one of the most significant is charge storage in the base region. For this reason, junction transistors designed to operate at high frequencies will be fabricated with the narrowest possible base width. A guide to the maximum frequency at which a transistor can be operated is given by the parameter f_T. This is the frequency at which the ß falls to unity (1). Most general-purpose junction transistors will have an f_T of around 150MHz, but specialised devices intended for use at UHF and microwave frequencies will have an f_T of 5GHz or even higher. An approximation of a transistor's current gain at frequencies below f_T can be obtained by:

$$ß = f_T \div \text{operating frequency}$$

For example, a device with an f_T of 250MHz will probably exhibit a current gain of around 10 at a frequency of 25MHz.

Maximum Ratings

Transistors may be damaged by the application of excessive voltages, or if made to pass currents that exceed the maximum values recommended in the manufacturers' data sheets. Some, or all, of the following parameters may need to be considered when selecting a transistor for a particular application:

V_{CEO} The maximum voltage that can be applied between the collector and emitter with the base open-circuit (hence the 'O'). In practice the maximum value is dictated by the reverse breakdown voltage of the collector-base junction. In the case of some transistors, for instance many of those intended for use as RF power amplifiers, this rating may seem impracticably low, being little or no higher than the intended supply voltage. However, in practical amplifier circuits, the base will be connected to the emitter via a low-value resistance or coupling coil winding. Under these conditions the collector-base breakdown voltage will be raised considerably (see below).

V_{CBO} The maximum voltage that can be applied between the collector and base with the emitter open-circuit. This provides a better indication of the collector-base reverse breakdown voltage. An RF power transistor with a V_{CEO} of 18V may well have a V_{CBO} rating of 48V. Special high-voltage transistors are manufactured for use in the EHT (extra high tension) generators of television and computer displays which can operate at collector voltages in excess of 1kV.

V_{EBO} The maximum voltage that can be applied between the emitter and base with the collector open-circuit. In the case of an NPN transistor, the emitter will be held at a positive potential with respect to the base. Therefore, it is the reverse breakdown voltage of the emitter-base junction that is being measured. A rating of around 5V can be expected.

I_C The maximum continuous collector current. For a small-signal transistor this is usually limited to around 150mA, but a rugged power transistor may have a rating as high as 30A.

P_D The maximum total power dissipation for the device. This figure is largely meaningless unless stated for a particular case temperature. The more power a transistor dissipates, the hotter it gets. Excessive heating will eventually lead to destruction, and so the power rating is only valid within the safe temperature limits quoted as part of the P_D rating. A reasonable case temperature for manufacturers to use in specifying the power rating is 50°C. It is unfortunate that as a bipolar transistor gets hotter, its V_{BE} drops and its ß increases. Unless the bias voltage is controlled to compensate for this, the collector current may start to rise, which in turn leads to further heating and eventual destruction of the device. This phenomenon is known as thermal runaway.

The possibility of failure due to the destruction of a transistor junction by excessive voltage, current or heating is best avoided by operating the device well within its safe limits at all times (see below). In the case of a small transistor, junction failure, should it occur, is normally absolute, and therefore renders the device useless. Power transistors, however, have a more complex construction. Rather than attempting to increase the junction area

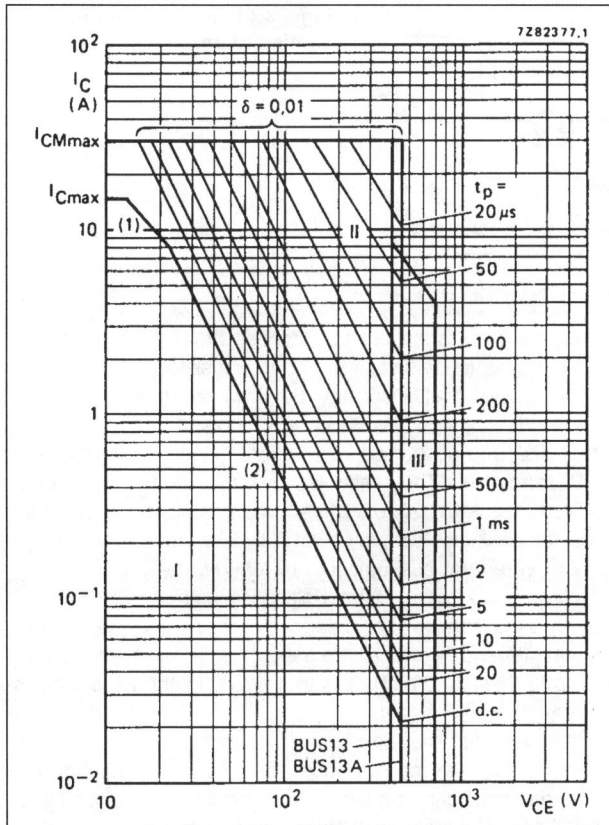

Fig 3.36: The safe operating (SOAR) curves for a BUS13A power bipolar transistor. When any power transistor is used as a pass device in a regulated PSU, it is important to ensure that the applied voltage and current are inside the area on the graph marked 'I - region of permissible DC operation'. Failure to do so may result in the device having a very short life because of secondary breakdown. This is due to the formation of hot spots in the transistor's junction. Note that power FETs do not suffer from this limitation. Key to regions: I - Region of permissible DC operation. II - Permissible extension for repetitive pulse operation. III - Area of permissible operation during turn-on in single-transistor converters, provided $R_{BE} \leq 100\Omega$ and $t_p \leq 0.6\mu s$. (Reproduction courtesy Philips Semiconductors)

of an individual transistor in order to make it more rugged, power transistors normally consist of large numbers of smaller transistors fabricated on a single chip of silicon. These are arranged to operate in parallel, with low-value resistances introduced in series with the emitters to ensure that current is shared equally between the individual transistors. A large RF power transistor may contain as many as 1000 separate transistors. In such a device, the failure of a small number of the individual transistors may not unduly affect its performance. This possibility must sometimes be taken into account when testing circuitry which contains power transistors.

Safe operating area - SOAR

All bipolar transistors can fail if they are over-run and power bipolars are particularly susceptible since, if over-run by excessive power dissipation, hot spots will develop in the transistor's junction, leading to total destruction. To prevent this, manufacturers issue SOAR data, usually in the form of a graph or series of graphs plotted on log-log graph paper with current along one axis and voltage along the other. A typical example is shown in **Fig 3.36**.

Most amateur use will be for analogue operation and should be confined to area I in the diagram (or, of course, to the corresponding area in the diagram of the transistor being considered). For pulse operation, it is possible to stray out into area II.

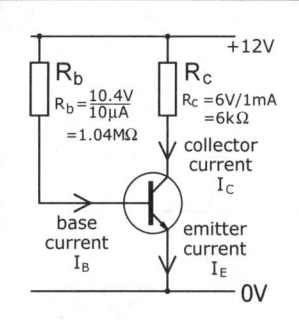
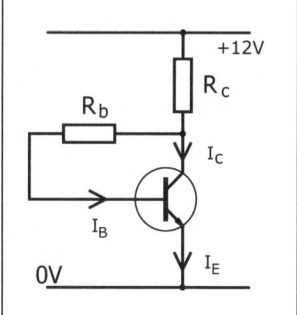

Fig 3.37: Simple bias circuit offering no bias stability

Fig 3.38: Improved bias circuit with some bias stability

How far depends on the height of the current pulses and the duty cycle. Generally speaking, staying well within area I will lead to a long life.

All bipolar transistors have this SOAR but, in the case of small devices, it is not often quoted by the manufacturers and, in any case, is most unlikely to be exceeded.

Other Bias Circuits and Classes of Bias

Fig 3.33 showed a transistor amplifier biased to bring the operating point of the transistor, that is the standing or quiescent voltages and currents, to a mid-range value allowing a signal to vary those currents and voltages both up and down to provide a maximum available range or swing. It is important that these values are predictable and tolerant of the variation in characteristics, particularly gain, from one transistor to another, albeit of the same type and part number.

The circuit in **Fig 3.37** would provide biasing. The base current required is calculated as $1mA/\beta$ (assuming $I_c = 1mA$) and the resistor R_b chosen accordingly. ß has been assumed to be 100. However the value of ß is not well controlled and may well vary over a 4:1 range. In this circuit, if ß actually was 200, the collector current would double, with 12V across R_c and almost none across the transistor. Clearly unsatisfactory, the design is not protected against the vagaries of the transistor ß.

If the top of R_b was moved from the supply rail to the collector of the transistor, as shown in **Fig 3.38**, then an increase in collector current (due to ß being higher than expected) would cause a fall in collector voltage and a consequent drop in the voltage across R_b. This would reduce the base current into the transistor, offsetting the rise in collector current. The circuit is reason-

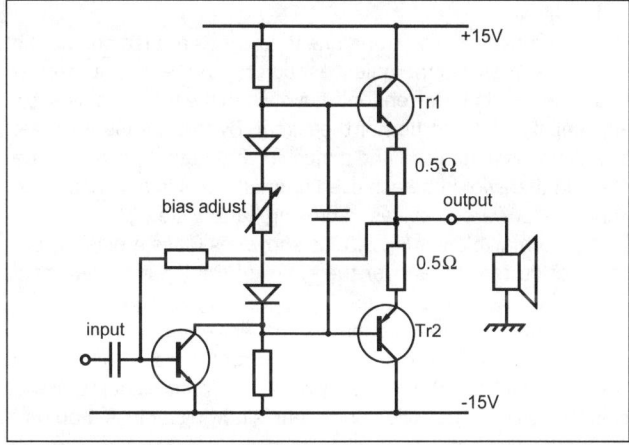

Fig 3.39: A push-pull amplifier using two transistors biased in Class B

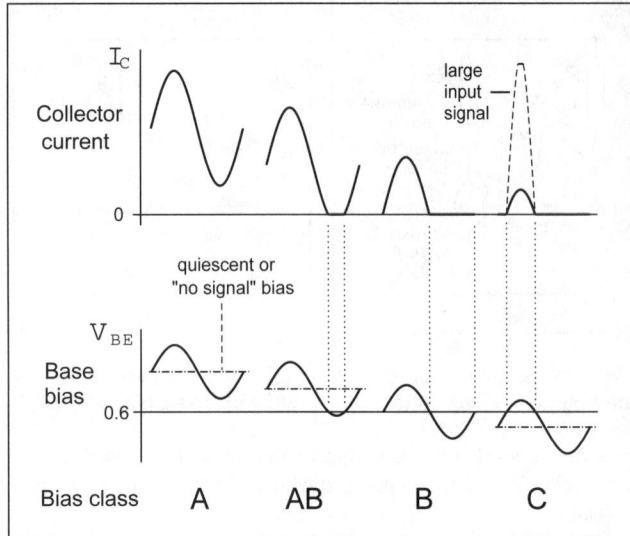

Fig 3.40: Classes of bias relate to the proportion of the signal waveform for which collector current flows

ably stable against variations in gain and is sometimes used although R_b does feed some of the output signal back to the base, which, being inverted or in anti-phase, does reduce the signal gain.

The method of bias stability in Fig 3.33 is better. If the collector current is higher than calculated, perhaps because ß is greater, the potential difference across R3 will rise. However the base voltage is held constant by R1 and R2, so the base-emitter voltage V_{BE} will fall. Reference to Fig 3.31 will show that a reduction in V_{BE} has a marked effect in reducing I_c quickly offsetting the assumed rise.

Bias Classes

The choice of base and collector voltages is normally such that an input signal is amplified without the distortion shown in Fig 3.32. However there are occasions where this is not intended. Take, for example, the circuit in **Fig 3.39**. The two output transistors may be biased just into conduction, drawing a modest quiescent current. If the signal voltage at the collector of the input transistor rises, then the V_{BE} on transistor 1 will tend to increase and transistor 1 will conduct, amplifying the signal. The V_{BE} on transistor 2 will reduce and it will not conduct. Conversely, if the voltage at the collector of the input transistor falls then transistor 1 will not conduct and transistor 2 will amplify the signal.

The purpose of this arrangement, know as a push-pull amplifier, is that it allows the quiescent current in the output transistors to be very low. Current only flows when the transistor is actually amplifying its portion of the signal. By this means the heat dissipation is minimised and a much higher output power can be realised than would be achieved normally. This is also desirable in battery-powered devices to prolong battery life.

The 'normal' bias of Fig 3.33 is known as Class A bias, where collector current flows over the entire of the input signal waveform. The push-pull amplifier has Class B bias, where collector current flows over half the input signal waveform. Class AB bias denotes biasing such that collector current flows for more than half but less than the whole cycle of the input signal. Class C bias tells us that the collector current is flowing for less than half the input waveform. This is illustrated in **Fig 3.40**.

The distortion of class B bias is avoided simply by the two transistors sharing the task. In reality there is a small overlap

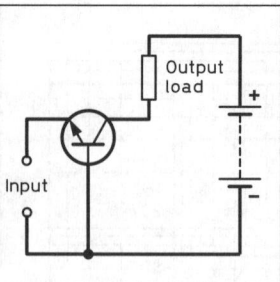

Fig 3.41: The common-base configuration

between the two transistors to minimise any cross-over distortion where conduction swaps from one transistor to the other. It could be argued that, in reality, the transistors are biased in class AB, but close to class B. Some schools of thought use the notation AB1 and AB2 to denote closer to class A and closer to class B respectively. The design issue is distortion versus standing current in the transistor. The standing current having implications for heat dissipation/power handling and, especially in battery powered devices, current drain and battery life.

No such option is available for class C bias and distortion will occur. Class C can only be used at radio frequencies where tuned circuits and filters can be used. It is, however highly efficient, up to 66% or 2/3. That is to say 2/3 of the DC supply is transformed into RF signal.

It will be recalled that any form of waveform or harmonic distortion introduces harmonic frequencies into the signal. Filtering and tuned circuits can remove the harmonics, thereby removing the distortion. This enables a transistor to produce even higher output powers, but, as will be seen in later chapters, the technique is not useable with any form of amplitude modulation of the input signal although a class C RF power amplifier stage can itself be amplitude modulated using collector or anode modulation. This is a technique where the RF input drive is unmodulated and the audio modulating signal is superimposed on the DC supply to the RF power amplifier transistor.

Transistor Configurations

The transistor has been used so far with its emitter connected to the 0V rail and the input to the base. This is not the only way of connecting and using a transistor. The important things to remember are that the transistor is a current driven device, which responds to base current and the potential difference V_{BE}. The output is a variation in collector current, which is normally translated into a voltage by passing that current through a collector resistor. It must also be noted that $I_E=I_B+I_C$ and that $I_c>>I_B$.

The common-emitter configuration, already seen in Fig 3.33, is usually preferred because it provides both current and voltage gain. The input impedance of a common-emitter amplifier using a small, general-purpose transistor will be around 2kΩ at audio frequencies but, assuming that the emitter resistor is bypassed with a capacitor, this will drop to approximately 100Ω at a frequency of a few megahertz. The output impedance will be in the region of 10kΩ. The chapter on Building Blocks provides more information on the design of common-emitter amplifiers.

Common-base configuration

The common-base configuration is shown in **Fig 3.41**. The input and output coupling and bias circuitry has been omitted for the sake of clarity. As the emitter current is the sum of the collector and base currents, the current gain will be very slightly less than unity. There is, however, considerable voltage gain. The input impedance is very low, typically between 10 and 20Ω, assuming a collector current of 2mA. The output impedance is much high-

The Radio Communication Handbook

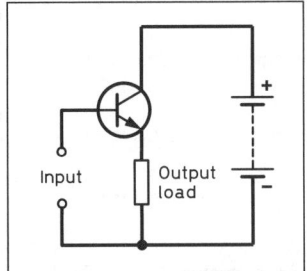

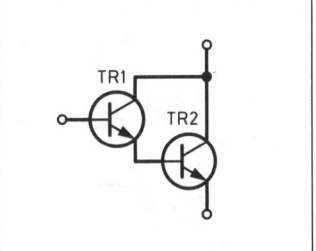

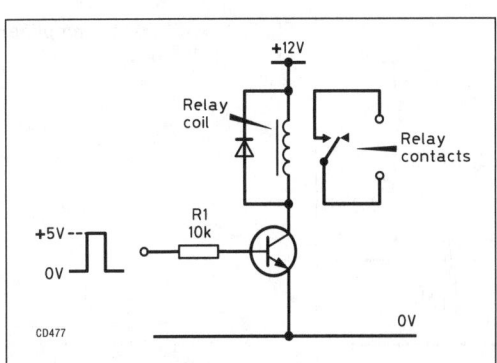

Fig 3.44: A transistor used to switch a relay

Fig 3.42: The common-collector, or emitter-follower, configuration

Fig 3.43: A Darlington pair is the equivalent of a single transistor with extremely high current gain

er, perhaps 1MΩ at audio frequencies. The output voltage will be in phase with the input. The common-base configuration will sometimes be used to provide amplification where the transistor must operate at a frequency close to its f_T.

Common-collector configuration

The common-collector configuration is normally referred to as the emitter follower, shown in **Fig 3.42**. This circuit provides a current gain equal to the transistor's ß, but the voltage gain is very slightly less than unity. The input impedance is much higher than for the common-emitter configuration and may be approximated by multiplying the transistor's ß by the value of the emitter load impedance. The output impedance, which is much lower, is normally calculated with reference to the impedance of the circuitry which drives the follower, and is approximated by:

$$\text{Output impedance} = \frac{Z_{out}}{\beta}$$

where Z_{out} is the output impedance of the circuit driving the emitter follower.

For example, if the transistor used has a ß of 100 at the frequency of operation and the emitter follower is driven by circuitry with an output impedance (Z_0) of 2kΩ, the output impedance of the follower is roughly 20Ω. As with the common-base circuit, the output voltage is in phase with the input. Emitter followers are used extensively as 'buffers' in order to obtain impedance transformation, and isolation, between stages (see the later chapters).

The Darlington pair

Fig 3.43 shows how two transistors may be connected to produce the equivalent of a single transistor with extremely high current gain (ß). A current flowing into the base of TR1 will cause a much larger current to flow into the base of TR2. TR2 then provides further current gain. Not surprisingly, the overall current gain of the Darlington pair is calculated by multiplying the ß of TR1 by the ß of TR2. Therefore, if each transistor has a ß of 100, the resultant current gain is 10,000.

As an alternative to physically connecting two separate devices, it is possible to obtain 'Darlington transistors'. These contain a pair of transistors, plus the appropriate interconnections, fabricated onto a single chip.

The transistor as a switch

There is often a requirement in electronic equipment to activate relays, solenoids, electric motors and indicator devices etc using control signals generated by circuitry that cannot directly power the device which must be turned on or off. The transistor in **Fig 3.44** solves such a problem by providing an interface between the source of a control voltage (+5V in this example) and a 12V relay coil. If the control voltage is absent, only a minute leakage current flows between the collector and emitter of the switching

transistor and so the relay is not energised. When the control voltage is applied, the transistor draws base current through R1 and this results in a larger current flow between its collector and emitter. The relay coil is designed to be connected directly across a potential of 12V and so the resistance between the collector and emitter of the transistor must be reduced to the lowest possible value. The desired effect is therefore the same as that which might otherwise be obtained by closing a pair of switch contacts wired in series with the relay coil.

Assuming that R1 allows sufficient base current to flow, the transistor will be switched on to the fullest extent (a state referred to as saturation). Under these conditions, the voltage at the collector of the transistor will drop to only a few tenths of a volt, thus allowing a potential almost equal to that of the supply rail (12V) to appear across the relay coil, which is therefore properly energised. Under these conditions the transistor will not dissipate much power because, although it is passing considerable current, there is hardly any resistance between the emitter and collector. Assuming that the relay coil draws 30mA when energised from a 12V supply, and that the ß of the transistor is 150 at a collector current of this value, the base current will be:

$$\frac{30 \times 10^{-3}}{150} = 200 \mu A$$

Using Ohm's Law, this suggests that the current limiting resistor R1 should have a value of around 25kΩ. In practice, however, a lower value of 10kΩ would probably be chosen in order to make absolutely sure that the transistor is driven into saturation. The diode connected across the relay coil, which is normally reverse biased, protects the transistor from high voltages by absorbing the coil's back EMF on switching off.

Constant-current generator

Fig 3.45 shows a circuit that will sink a fixed, predetermined current into a load of varying resistance. A constant-current battery charger is an example of a practical application which might use such a circuit. Also, certain low-distortion amplifiers will employ constant-current generators, rather than resistors, to act as collector loads. The base of the PNP transistor is held at a potential

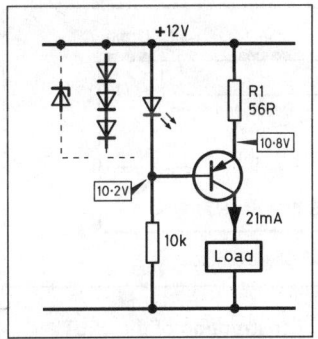

Fig 3.45: The constant-current generator

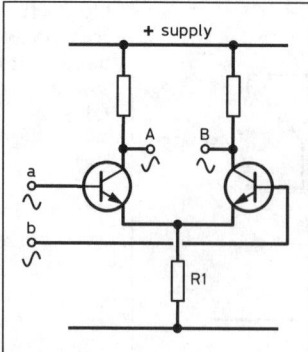

Fig 3.46: A differential amplifier or long-tailed pair

of 10·2V by the forward voltage drop of the LED (12 - 1·8 = 10·2V). Note that although LEDs are normally employed as indicators, they are also sometimes used as reference voltage generators. Allowing for the base-emitter voltage drop of the transistor (approximately 0·6V), the emitter voltage is 10·8V. This means that the potential difference across R1 will be held at 1·2V. Using Ohm's Law, the current flowing through R1 is:

$$\frac{1\cdot 2}{56} = 0\cdot 021A \quad or \quad 21mA$$

The emitter current, and also the collector current, will therefore be 21mA. Should the load attempt to draw a higher current, the voltage across R1 will try and rise. As the base is held at a constant voltage, it is the base-emitter voltage that must drop, which in turn prevents the transistor passing more current.

Also shown in Fig 3.45 are two other ways of generating a reasonably constant reference voltage. A series combination of three forward-biased silicon diodes will provide a voltage drop similar to that obtained from the LED (3 x 0·6 = 1.8V). Alternatively, a zener diode could be used, although as the lowest voltage zener commonly available is 2·7V, the value of R1 must be recalculated to take account of the higher potential difference.

The long-tailed pair

The long-tailed pair, or differential amplifier, employs two identical transistors which share a common emitter resistor. Rather than amplifying the voltage applied to a single input, this circuit provides an output which is proportional to the difference between the voltages presented to its two inputs, labelled 'a' and 'b' in **Fig 3.46**.

Providing that the transistors are well matched, variations in V_{BE} and   will cause identical changes in the potential at the two outputs, A and B. Therefore, if both A and B are used, it is the voltage difference between them that constitutes the wanted output. The long-tailed pair is very useful as an amplifier of DC potentials, an application where input and output coupling capacitors cannot be used. The circuit can be improved by replacing the emitter resistor (R1) with a constant-current generator, often referred to in this context as a current source. The differential amplifier is used extensively as an input stage in operational amplifiers.

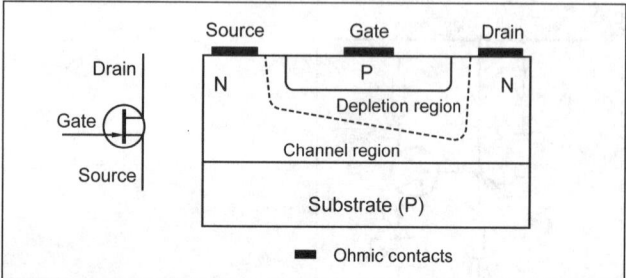

Fig 3.47: The circuit symbol and construction of the JFET

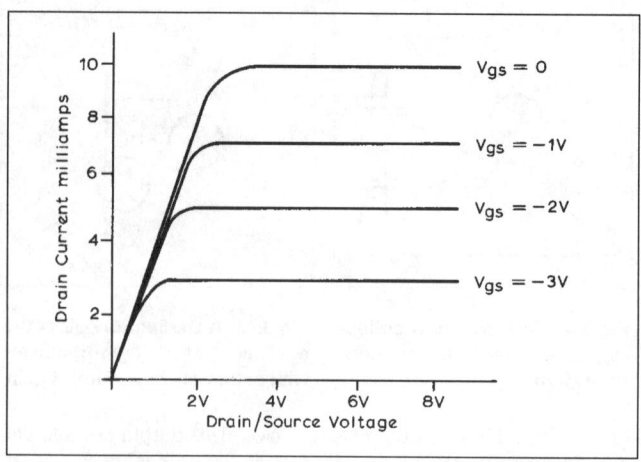

Fig 3.48: The relationship between gate voltage and drain current for a JFET

JUNCTION FIELD EFFECT TRANSISTOR

Like its bipolar counterpart, the junction field effect transistor (JFET) is a three-terminal device which can be used to provide amplification over a wide range of frequencies. **Fig 3.47** shows the circuit symbol and construction of an N-channel JFET. This device differs from the bipolar transistor in that current flow between its drain and source connections is controlled by a voltage applied to the gate. The current flowing to the gate terminal is very small, as it is the reverse bias leakage current of a PN junction. Thus the input (gate) impedance of a JFET is very high, and the concept of current gain, as applied to bipolar devices, is meaningless. The JFET may be referred to as a unipolar device, since the current flow is by majority carriers only, ie electrons in an N-channel transistor. In such a device holes (which have lower mobility than electrons) have no part in the current transport process, thus offering the possibility of very good high-frequency performance.

If both the gate and source are at ground potential (V_{GS}=0), and a positive voltage is applied to the drain (V_{DS}), electrons will flow from source to drain through the N-type channel region (known as the drain-source current, I_{DS}). If V_{DS} is small, the gate junction depletion region width will remain practically independent of V_{DS}, and the channel will act as a resistor. As V_{DS} is increased, the reverse bias of the gate junction near the drain increases, and the average cross-sectional area for current flow is reduced, thus increasing the channel resistance. Eventually V_{DS} is large enough to cause the gate depletion region to expand to fill the channel at the drain end, thus separating the source from the drain. This condition is known as pinch-off. It is important to note that current (I_{DS}) continues to flow across this depletion region, since carriers are injected into it from the channel and accelerated across it by the electric field present. If V_{DS} is increased beyond pinch-off, the edge of the depletion region will move along the channel towards the source. However the voltage drop along the channel between the source and the edge of the depletion region will remain fixed, and thus the drain-source current will remain essentially unchanged. This current saturation is evident from the JFET current-voltage characteristics shown in **Fig 3.48** (V_{GS}=0).

A negative voltage applied to the P-type gate (V_{GS}) establishes a reverse biased depletion region intruding into the N-type channel (see Fig 3.47). Thus for small values of of V_{DS} the channel will again act as a resistor, however its resistance will be larger than when V_{GS}=0 because the cross-sectional area of the channel

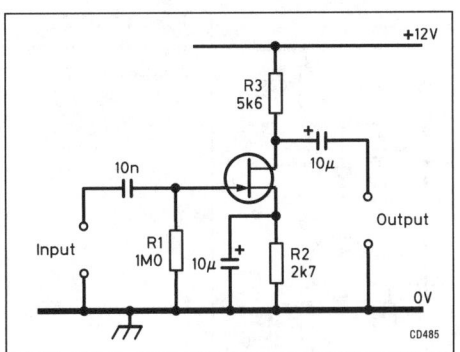

Fig 3.49: A JFET amplifier

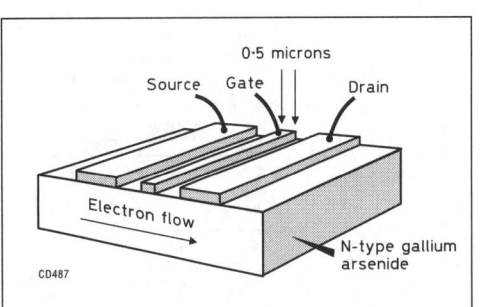

Fig 3.51: Construction of a GaAsFET

has decreased due to the wider depletion region. As V_{DS} is increased the gate depletion region again expands to fill the channel at the drain end, causing pinch-off and current saturation. However, the applied gate voltage reduces the drain voltage required for the onset of pinch-off, thus also reducing the saturation current. This is evident from Fig 3.48 for V_{GS} = -1V, -2V and -3V. Furthermore, Fig 3.48 also suggests that a varying signal voltage applied to the gate will cause proportional variations in drain current. The gain, or transconductance (G_m or Y_{fs}), of a field effect device is expressed in siemens (see Chapter 1). A small general-purpose JFET will have a G_m of around 5 milli-siemens.

The circuit of a small-signal amplifier using a JFET is shown in **Fig 3.49**. The potential difference across R2 provides bias by establishing a positive voltage at the source, which has the same effect as making the gate negative with respect to the source. R1 serves to tie the gate at ground potential (0V), and in practice its value will determine the amplifier's input impedance at audio frequencies. The inherently high input impedance of the JFET amplifier is essentially a result of the source-gate junction being reverse biased. If the gate were to be made positive with respect to the source (a condition normally to be avoided), gate current would indeed flow, thus destroying the field effect. The value of R3, the drain load resistor (R_L), dictates the voltage gain obtained for a particular device transconductance (assumed to be 4 millisiemens in this case) as follows:

$$\text{Voltage gain} = G_m \times R_L$$
$$= 4 \times 10^{-3} \times 5 \cdot 6 \times 10^3$$
$$= 4 \times 5 \cdot 6$$
$$= 22 \cdot 4 \quad \text{or} \quad 27 \text{dB}$$

The voltage gain obtainable from a common-source JFET amplifier is therefore around 20dB, or a factor of 10, lower than that provided by the equivalent common-emitter bipolar amplifier. Also, the characteristics of general-purpose JFETs are subject to considerable variation, or 'spread', a fact that may cause problems in selecting the correct value of bias resistor for a particular device. However, the JFET does offer the advantage of high input impedance, and this is exploited in the design of sta-

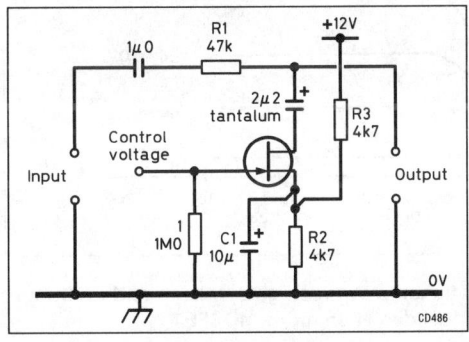

Fig 3.50: The JFET may be used as a voltage-controlled attenuator

ble, variable frequency oscillators. JFETs are also employed in certain types of RF amplifier and switching mixer.

The JFET is often used as a voltage-controlled variable resistor in signal gates and attenuators. The channel of the JFET in **Fig 3.50** forms a potential divider working in conjunction with R1. Here R2 and R3 develop a bias voltage which is sufficient to ensure pinch-off. This means that the JFET exhibits a very high resistance between the source and drain and so, providing that the following stage also has a high input impedance, say at least five times greater than the value of R1, the signal will suffer practically no attenuation. Conversely, if a positive voltage is applied to the gate sufficient to overcome the effect of the bias, the channel resistance will drop to the lowest possible value - typically 400Ω for a small-signal JFET. The signal will now be attenuated by a factor nearly equal to the ratio between R1 and the channel resistance - 118, or 41dB (note that C1 serves to bypass R2 at signal frequencies). The circuit is not limited to operation at these two extremes, however, and it is possible to achieve the effect of a variable resistor by adjusting the gate voltage to achieve intermediate values of channel resistance.

GaAsFETs

Although field effect devices are generally fabricated from silicon, it is also possible to use gallium arsenide. GaAsFETs (gallium arsenide field effect transistors) are N-channel field effect transistors designed to exploit the higher electron mobility provided by gallium arsenide (GaAs). The gate terminal differs from that of the standard silicon JFET in that it is made from gold, which is bonded to the top surface of the GaAs channel region (see **Fig 3.51**). The gate is therefore a Schottky barrier junction, as used in the hot-carrier diode. Good high-frequency performance is achieved by minimising the electron transit time between the source and drain. This is achieved by reducing the source drain spacing to around 5 microns and making the gate from a strip of gold only 0·5 microns wide (note that this critical measurement is normally referred to as the gate length because it is the dimension running parallel to the electron flow).

The very small gate is particularly delicate, and it is therefore essential to operate GaAsFETs with sufficient negative bias to ensure that the gate source junction never becomes forward biased. Protection against static discharge and supply line transients is also important.

GaAsFETs are found in very-low-noise receive preamplifiers operating at UHF and microwave frequencies up to around 20GHz. They can also be used in power amplifiers for microwave transmitters.

MOSFETs

The MOSFET (metal oxide field effect transistor), also known as the IGFET (insulated gate field effect transistor), is a very important device, with applications ranging from low-noise preamplification at microwave frequencies to high-power amplifiers in HF and VHF transmitters. Ultra large scale integrated circuits

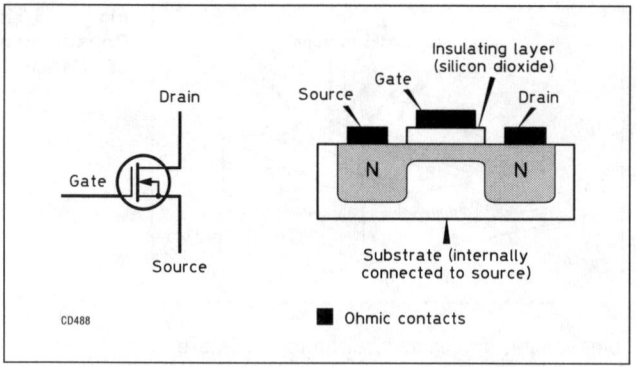

Fig 3.52: Circuit symbol and method of construction of a deple-tion-mode N-channel MOSFET (metal oxide semiconductor field effect transistor). The circuit symbol for the P-channel type is the same, except that the direction of the arrow is reversed

(ULSICs), including microprocessors and memories, also make extensive use of MOS transistors.

The MOSFET differs from the JFET in having an insulating layer, normally composed of silicon dioxide (SiO_2), interposed between the gate electrode and the silicon channel. The ability to readily grow a high quality insulating material on silicon is a key reason for the dominance of silicon device technology. The source and drain are formed by diffusion into the silicon sub-strate. This insulation prevents current flowing into, or out of, the gate, which makes the MOSFET easier to bias and guarantees an extremely high input resistance. The insulating layer acts as a dielectric, with the gate electrode forming one plate of a capacitor. Gate capacitance depends on the area of the gate, and its general effect is to lower the impedance seen at the gate as frequency rises. The main disadvantage of this structure is that the very thin insulating layer can be punctured by high volt-ages appearing on the gate. Therefore, in order to protect these devices against destruction by static discharges, internal zener diodes are normally incorporated. Unfortunately, the protection provided by the zener diodes is not absolute, and all MOS devices should therefore be handled with care.

MOSFETs have either N-type or P-type channel regions, con-duction being provided by electrons in N-channel devices, and holes in P-channel devices. However, as electrons have greater mobility than holes, the N-channel device is often favoured because it promises better high-frequency performance.

The current-voltage characteristics of MOSFETs are analagous in form to JFETs. For low values of drain-source voltage (V_{DS}) the channel acts as a resistor, with the source-drain current (I_{DS}) pro-portional to V_{DS}. As V_{DS} increases, pinch-off eventually occurs, and beyond this point I_{DS} remains essentially constant. The gate-source voltage (V_{GS}) controls current flow in the channel by caus-

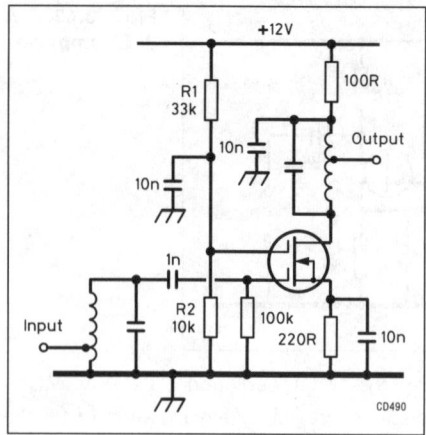

Fig 3.54: An RF amplifier using a dual-gate MOSFET

ing carrier accumulation, carrier depletion or carrier inversion at the silicon channel surface. In accumulation, V_{GS} causes an enhanced concentration of majority carriers (eg electrons in N-type silicon), and in depletion the majority carrier concentration is reduced. In inversion the applied gate voltage is sufficient to cause the number of minority carriers at the silicon surface (ie holes in N-type silicon) to exceed the number of majority carri-ers. This means that, for example, an N-type silicon surface becomes effectively P-type. The value of V_{GS} required to invert the silicon surface is known as the threshold voltage.

There are basically two types of N-channel MOSFETs. If at $V_{GS}=0$ the channel resistance is very high, and a positive gate threshold voltage is required to form the N-channel (thus turning the transistor on), then the device is an enhancement-mode (normally off) MOSFET. If an N-channel exists at $V_{GS}=0$, and a negative gate threshold voltage is required to invert the channel surface in order to turn the transistor off, then the device is a depletion-mode (normally on) MOSFET. Similarly there are both enhancement and depletion-mode P-channel devices. Adding further to the variety of MOSFETs available, there are also dual-gate types.

Fig 3.52 shows the circuit symbol and construction for a sin-gle-gate MOSFET, and Fig 3.53 features the dual-gate equiva-lent. Dual-gate MOSFETs perform well as RF and IF amplifiers in receivers. They contribute little noise and provide good dynamic range. Transconductance is also higher than that offered by the JFET, typically between 7 and 15 millisiemens for a general-pur-pose device.

Fig 3.54 shows the circuit of an RF amplifier using a dual-gate MOSFET. The signal is presented to gate 1, and bias is applied separately to gate 2 by the potential divider comprising R1 and R2. Selectivity is provided by the tuned circuits at the input and

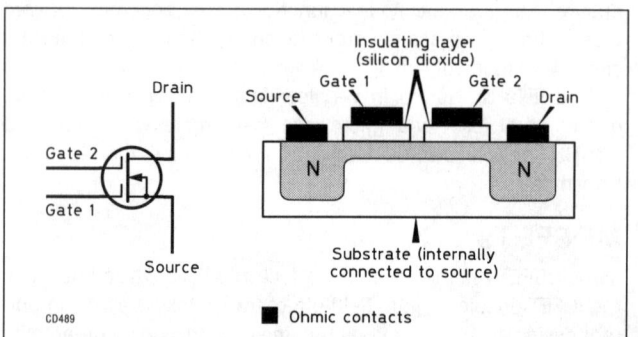

Fig 3.53: A depletion-mode dual-gate MOSFET

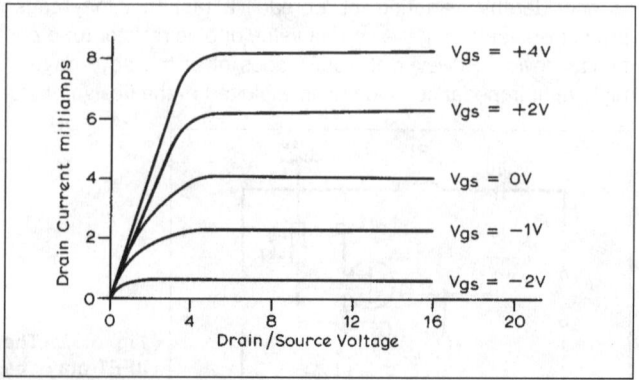

Fig 3.55: The relationship between gate voltage and drain cur-rent for an N-channel depletion mode MOSFET

The Radio Communication Handbook

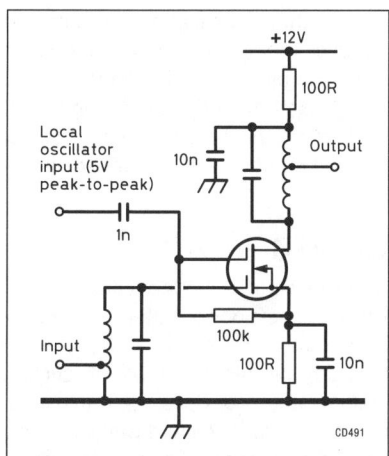

Fig 3.56: A mixer circuit based on a dual-gate MOSFET

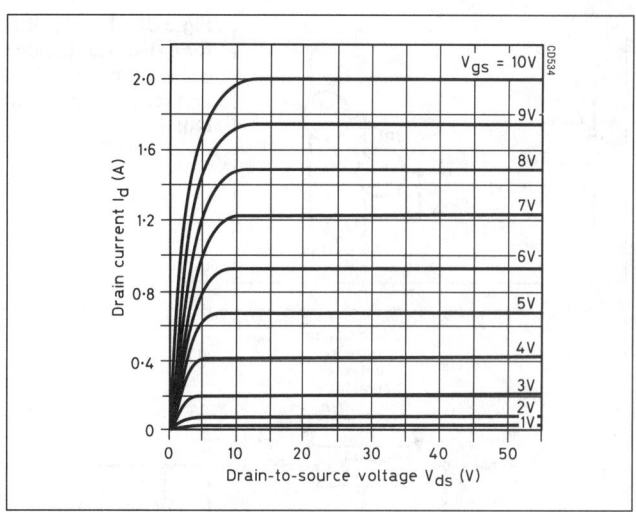

Fig 3.58: Relationship between gate voltage and drain current for a VMOS power transistor

output. Care must be taken in the layout of such circuits to prevent instability and oscillation. A useful feature of the dual-gate amplifier is the ability to control its gain by varying the level of the gate 2 bias voltage. This is particularly useful in IF amplifiers, where the AGC voltage is often applied to gate 2. The characteristic curve at **Fig 3.55** shows the effect on drain current of making the bias voltage either negative or positive with respect to the source.

Dual-gate MOSFETs may also be used as mixers. In **Fig 3.56**, the signal is applied to gate 1 and the local oscillator (LO) drive to gate 2. There is a useful degree of isolation between the two gates, and this helps reduce the level of oscillator voltage fed back to the mixer input. For best performance, the LO voltage must be sufficient to turn the MOSFET completely off and on, so that the mixer operates in switching mode. This requires an LO drive of around 5V peak to peak. However, as the gate impedance is high, very little power is required.

VMOS Transistors

As already discussed, MOSFETs exhibit very low gate leakage current, which means that they do not require complex input drive circuitry compared with bipolar devices. In addition, unipolar MOSFETs have a faster switching speed than bipolar transistors. These features make the MOSFET an attractive candidate for power device applications. The VMOS™ (vertical metal oxide semiconductor), also known as the power MOSFET, is constructed in such a way that current flows vertically between the drain, which forms the bottom of the device, to a source terminal at the top (see **Fig 3.57**). The gate occupies either a V- or U-shaped groove etched into the upper surface. VMOS devices feature a four-layer sandwich comprising N+, P, N- and N+ material and

operate in the enhancement mode. The vertical construction produces a rugged device capable of passing considerable drain current and offering a very high switching speed. These qualities are exploited in power control circuits and transmitter output stages. **Fig 3.58** shows the characteristic curves of a typical VMOS transistor. Note that the drain current is controlled almost entirely by the gate voltage, irrespective of drain voltage. Also, above a certain value of gate voltage, the relationship between gate voltage and drain current is highly linear. Power MOSFETs fabricated in the form of large numbers of parallel-connected VMOS transistors are termed HEXFETs™.

Although the resistance of the insulated gate is for all intents and purposes infinite, the large gate area leads to high capacitance. A VMOS transistor intended for RF and high-speed switching use will have a gate capacitance of around 50pF, whereas devices made primarily for audio applications have gate capacitances as high as 1nF. A useful feature of these devices is that the relationship between gate voltage and drain current has a negative temperature coefficient of approximately 0·7% per degree Celsius. This means that as the transistor gets hotter, its drain current will tend to fall, thus preventing the thermal runaway which can destroy bipolar power transistors.

Fig 3.59 shows the circuit of a simple HF linear amplifier using a single VMOS transistor. Forward bias is provided by the poten-

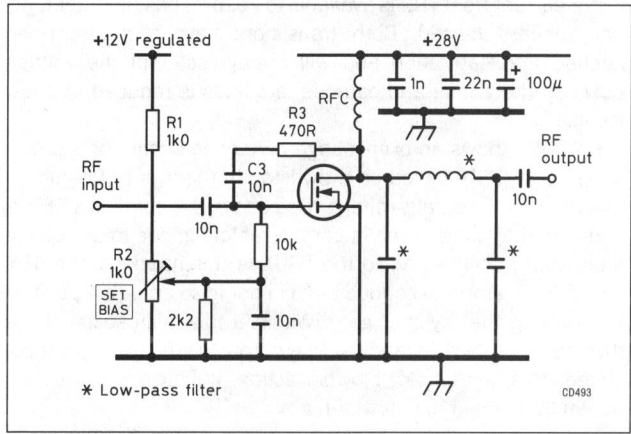

Fig 3.59: A linear amplifier utilising a VMOS power transistor. Assuming a 50-ohm output load, the RFC value is chosen so that it has an inductive reactance (X_L) of approximately 400Ω at the operating requency

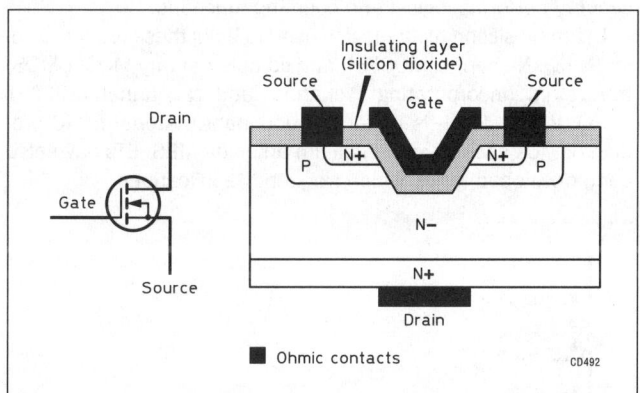

Fig 3.57: Circuit symbol and method of construction of a VMOS transistor

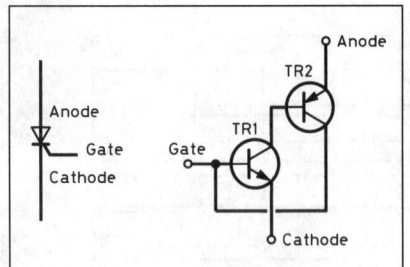

Fig 3.60: The symbol for a thyristor (silicon controlled rectfier) and its equivalent circuit

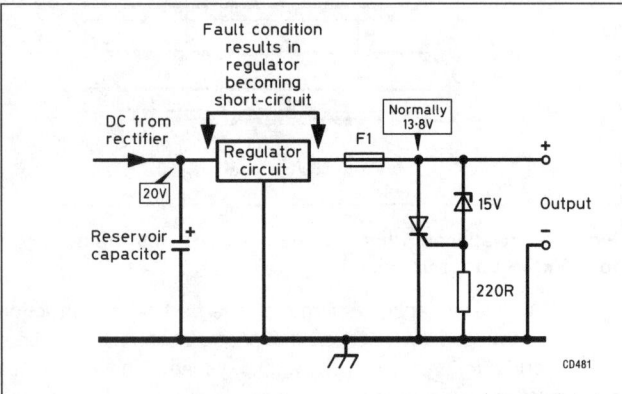

Fig 3.61: Crowbar over-voltage protection implemented wth a thyristor

tial divider R1, R2 so that the amplifier operates in Class AB. In this circuit R3 and C3 provide a small amount of negative feedback to help prevent instability. More complex push-pull amplifiers operating from supplies of around 50V can provide RF outputs in excess of 100W.

THYRISTORS

The thyristor, or silicon controlled rectifier (SCR), is a four-layer PNPN device which has applications in power control and power supply protection systems. The thyristor symbol and its equivalent circuit is shown in **Fig 3.60**. The equivalent circuit consists of two interconnected high-voltage transistors, one NPN and the other PNP. Current flow between the anode and cathode is initiated by applying a positive pulse to the gate terminal, which causes TR1 to conduct. TR2 will now also be switched on because its base is forward biased via TR1's collector. TR1 continues to conduct after the end of the trigger pulse because collector current from TR2 is available to keep its base-emitter junction forward biased. Both transistors have therefore been latched into saturation and will remain so until the voltage between the anode and cathode terminals is reduced to a low value.

Fig 3.61 shows an over-voltage protection circuit for a power supply unit (PSU) based on a thyristor. In the event of regulator failure, the PSU output voltage rises above the nominal 13·8V, a situation that could result in considerable damage to any equipment that is connected to the PSU. As this happens, the 15V zener diode starts to conduct, and in doing so applies a positive potential to the thyristor gate. Within a few microseconds the thyristor is latched on, and the PSU output is effectively short-circuited. This shorting, or crowbar action, will blow fuse F1 and, hopefully, prevent any further harm.

The thyristor will only conduct in one direction, but there is a related device, called the triac (see **Fig 3.62** for symbol), which effectively consists of two parallel thyristors connected anode to cathode. The triac will therefore switch currents in either direction and is used extensively in AC power control circuits, such as the ubiquitous lamp dimmer. In these applications a trigger circuit varies the proportion of each mains cycle for which the triac conducts, thus controlling the average power supplied to a load. Having been latched on at a predetermined point during the AC cycle, the triac will switch off at the next zero crossing point of the waveform, the process being repeated for each following half cycle.

INTEGRATED CIRCUITS

Having developed the techniques used in the fabrication of individual semiconductor devices, the next obvious step for the electronics industry was to work towards the manufacture of complete integrated circuits (ICs) on single chips of silicon. Integrated circuits contain both active devices (eg transistors) and passive devices (eg resistors) formed on and within a single semiconductor substrate, and interconnected by a metallisation pattern.

ICs offer significant advantages over discrete device circuits, principally the ability to densely pack enormous numbers of devices on a single silicon chip, thus achieving previously unattainable functionality at low processing cost per chip. Another advantage of integrated, or 'monolithic', construction is that because all the components are fabricated under exactly the same conditions, the operational characteristics of the transistors and diodes, and also the values of resistors, are inherently well matched. The first rudimentary hybrid IC was made in 1958 by Jack Kilby of Texas Instruments, just eight years after the birth of the bipolar junction transistor. Since then, advances in technology have resulted in a dramatic reduction in the minimum device dimension which may be achieved, and today MOSFET gate lengths of only 50nm (5 x 10⁻⁶ cm) are possible. This phenomenal rate of progress is set to continue, with 10nm gate lengths expected within the next decade.

The earliest ICs could only contain less than 50 components, but today it is possible to mass-produce ICs containing billions of components on a single chip. For example, a 32-bit microprocessor chip may contain over 42 million components, and a 1Gbit dynamic random access memory (DRAM) chip may contain over 2 billion components.

ICs fall into two broad categories - analogue and digital. Analogue ICs contain circuitry which responds to finite changes in the magnitude of voltages and currents. The most obvious example of an analogue function is amplification. Indeed, virtually all analogue ICs, no matter what their specific purpose may be, contain an amplifier. Conversely, digital ICs respond to only two voltage levels, or states. Transistors within the IC are normally switched either fully on, or fully off. The two states will typically represent the ones and zeros of binary numbers, and the circuitry performs logical and counting functions.

The main silicon technologies used to build these ICs are bipolar (NPN), N-channel MOSFET, and complementary MOS (CMOS) transistors (incorporating P-channel and N-channel MOSFET pairs). While silicon is by far the dominant material for IC production, ICs based around gallium arsenide MESFETs have also been developed for very high frequency applications.

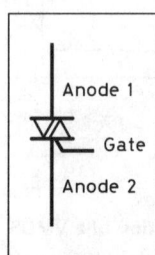

Fig 3.62: The triac or AC thyristor

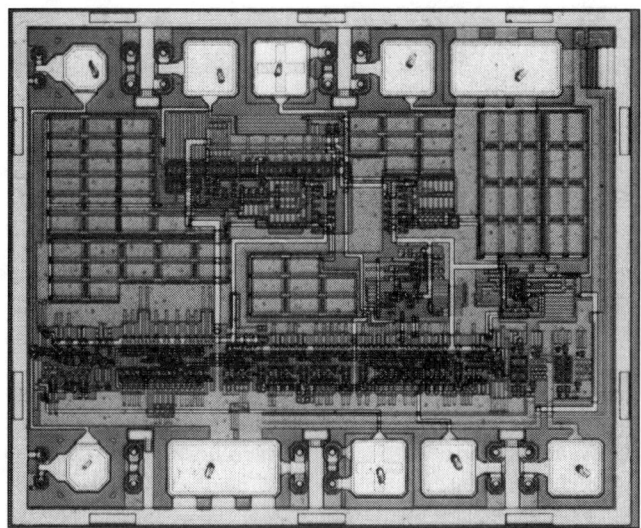

Fig 3.63: Photograph of the SP8714 integrated circuit chip showing pads for bonding wires (GEC-Plessey Semiconductors)

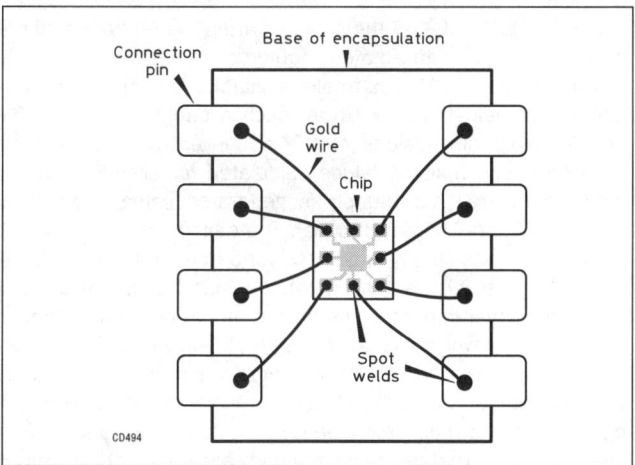

Fig 3.64: Internal construction of an integrated circuit

ICs are produced using a variety of layer growth/deposition, photolithography, etching and doping processes, all carried out under dust-free cleanroom conditions. All chemicals and gases used during processing are purified to remove particulates and ionic contamination. The starting point is a semiconductor wafer, up to 300mm in diameter, which will contain many individual chips. The patterns of conducting tracks and semiconductor junctions are defined by high resolution photolithography and etching. Doping is carried out by diffusion or ion implantation. These processes are repeated a number of times in order to fabricate different patterned layers. Strict process control at each stage minimises yield losses, and ensures that the completed wafer contains a large number of correctly functioning circuits. The wafer is then cut into individual chips and automatically tested to ensure compliance with the design specification. Each selected chip is then fixed to the base of its encapsulation, and very fine gold wires are spot welded between pads located around the chip's

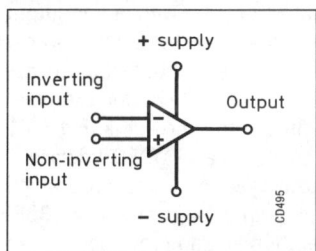

Fig 3.65: Circuit symbol for an operational amplifier (op-amp)

periphery and the metal pins which serve as external connections - see **Figs 3.63 and 3.64**. Finally, the top of the encapsulation is bonded to the base, forming a protective seal. Most general-purpose ICs are encapsulated in plastic, but some expensive devices that must operate reliably at high temperatures are housed in ceramic packages.

LINEAR INTEGRATED CIRCUITS

Operational Amplifiers

The operational amplifier (op-amp) is a basic building-block IC that can be used in a very wide range of applications requiring low-distortion amplification and buffering. Modern op-amps feature high input impedance, an open-loop gain of at least 90dB (which means that in the absence of gain-reducing negative feedback, the change in output voltage will be at least 30,000 times greater than the change in input voltage), extremely low distortion, and low output impedance. Often two, or even four, separate op-amps will be provided in a single encapsulation. The first operational amplifiers were developed for use in analogue computers and were so named because, with suitable feedback, they can perform mathematical 'operations' such as adding, subtracting, logging, antilogging, differentiating and integrating voltages.

A typical op-amp contains around 20 transistors, a few diodes and perhaps a dozen resistors. The first stage is normally a long-tailed pair and provides two input connections, designated inverting and non-inverting (see the circuit symbol at **Fig 3.65**). The input transistors may be bipolar types, but JFETs or even MOSFETs are also used in some designs in order to obtain very high input impedance. Most op-amps feature a push-pull output stage operating in Class AB which is invariably provided with protection circuitry to guard against short-circuits. The minimum value of output load when operating at maximum supply voltage is normally around 2kΩ, but op-amps capable of driving 500Ω loads are available. Between the input and output circuits there will be one or two stages of voltage amplification. Constant-current generators are used extensively in place of collector load resistors, and also to stabilise the emitter (or source) current of the input long-tailed pair.

In order to obtain an output voltage which is in phase with the input, the non-inverting amplifier circuit shown at **Fig 3.66** is used. Resistors R1 and R2 form a potential divider which feeds a proportion of the output voltage back to the inverting input. In most cases the open-loop gain of an op-amp can be considered infinite. Making this assumption simplifies the calculation of the closed-loop gain obtained in the presence of the negative feedback provided by R1 and R2. For example, if R1 has a value of 9k and R2 is 1k, the voltage gain will be:

$$\frac{R_1 + R_2}{R_2} = \frac{9000 + 1000}{1000} = 10 \text{ or } 20dB$$

If R2 is omitted, and R1 replaced by a direct connection between the output and the inverting input, the op-amp will function as a unity gain buffer.

Fig 3.63 shows an inverting amplifier. In this case the phase of the output voltage will be opposite to that of the input. The closed-loop gain is calculated by simply dividing the value of R1 by R2. Therefore, assum-

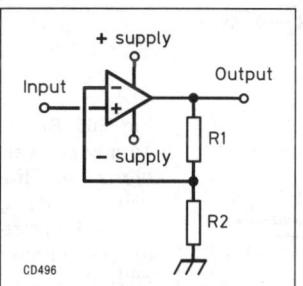

Fig 3.66: A non-inverting amplifier

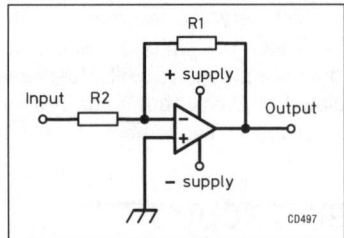

Fig 3.67: An inverting amplifier

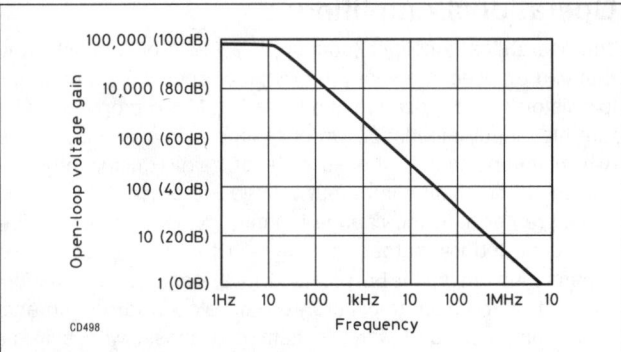

Fig 3.68: The relationship between open loop gain and frequency for a typical internally compensated op-amp

ing the values are the same as used for the non-inverting amplifier:

$$\frac{R_1}{R_2} = \frac{9000}{1000} = 9 \text{ or } 19\text{dB}$$

At low frequencies, the operation of the negative feedback networks shown in **Figs 3.66 and 3.67** is predictable in that the proportion of the output voltage they feed back to the input will be out of phase with the input voltage at the point where the two voltages are combined. However, as the frequency rises, the time taken for signals to travel through the op-amp becomes a significant factor. This delay will introduce a phase shift that increases with rising frequency. Therefore, above a certain frequency - to be precise, where the delay contributes a phase shift of more than 135 degrees - the negative feedback network actually becomes a positive feedback network, and the op-amp will be turned into an oscillator. However, if steps are taken to reduce the open-loop gain of the op-amp to below unity (ie less than 1) at this frequency, oscillation cannot occur. For this reason, most modern op-amps are provided with an internal capacitor which is connected so that it functions as a simple low-pass filter. This measure serves to reduce the open-loop gain of the amplifier by a factor of 6dB per octave (ie for each doubling of frequency the voltage gain drops by a factor of two), and ensures that it falls to unity at a frequency below that at which oscillation might otherwise occur. Op-amps containing such a capacitor are

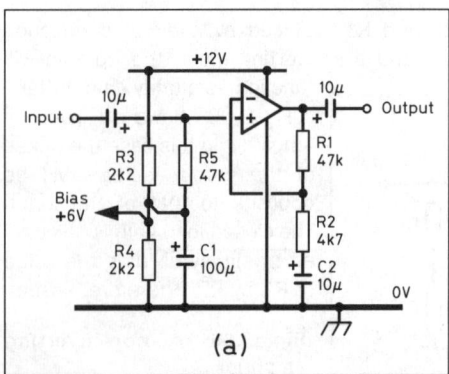

(a)

Fig 3.65: Single supply rail versions of the non-inverting (a) and inverting (b) amplifier circuits

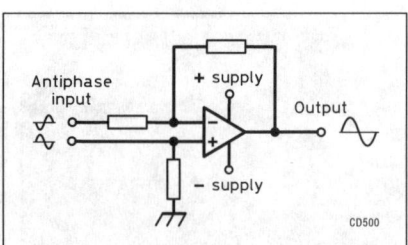

Fig 3.70: A differential amplifier

designated as being 'internally compensated'.

Compensated op-amps have the advantage of being absolutely stable in normal use. The disadvantage is that the open-loop gain is considerably reduced at high frequencies, as shown by the graph at **Fig 3.68**. This means that general-purpose, internally compensated op-amps are limited to use at frequencies below about 1MHz and they will typically be employed as audio frequency preamplifiers, and in audio filters. There are, however, special high-frequency types available, usually featuring external compensation. An externally compensated op-amp has no internal capacitor, but connections are provided so that the user may add an 'outboard' capacitor of the optimum value for a particular application, thus maximising the gain at high frequencies.

In Figs 3.66 and 3.67 the op-amps are powered from dual-rail supplies. However, in amateur equipment only a single supply rail of around +12V is normally available. Op-amps are quite capable of being operated from such a supply, and **Fig 3.69** shows single-rail versions of the non-inverting and inverting amplifiers with resistor values calculated for slightly different gains. A mid-rail bias supply is generated using a potential divider (R3 and R4 in each case). The decoupling capacitor C1 enables the bias supply to be used for a number of op-amps (in simpler circuits it is often acceptable to bias the op-amp using a potential divider connected directly to the non-inverting input of the op-amp, in which case C1 is omitted). The value of R5 is normally made the same as R1 in order to minimise the op-amp input current, although this is less important in the case of JFET op-amps due to their exceptionally high input resistance. C2 is incorporated to reduce the gain to unity at DC so that the output voltage will settle to the half-rail potential provided by the bias generator in the absence of signals.

Fig 3.70 shows a differential amplifier where the signal drives both inputs in antiphase. An advantage of this balanced arrangement is that interference, including mains hum or ripple, tends to impress voltages of the same phase at each input and so will be eliminated by cancellation. This ability to reject in-phase, or common-mode, signals when operating differentially is known as the common-mode rejection ratio (CMRR). Even an inexpensive op-amp will provide a CMRR of around 90dB, which means that an in-phase signal would have to be 30,000 times greater in magnitude than a differential signal in order to generate the same output voltage.

Further information on op-amps and their circuits is given in the Building Blocks chapter.

Audio Power Amplifiers

The audio power IC is basically just an op-amp with larger output transistors. Devices giving power outputs in the range 250mW to 40W are available, the bigger types being housed in encapsulations featuring metal mounting tabs (TO220 for example) that enable the IC to be bolted directly to a heatsink. **Fig 3.71** shows a 1W audio output stage based on an LM380N device.

The LM380N has internal negative feedback resistors which provide a fixed closed loop voltage gain of approximately 30dB. A bias network for single-rail operation is also provided, and so

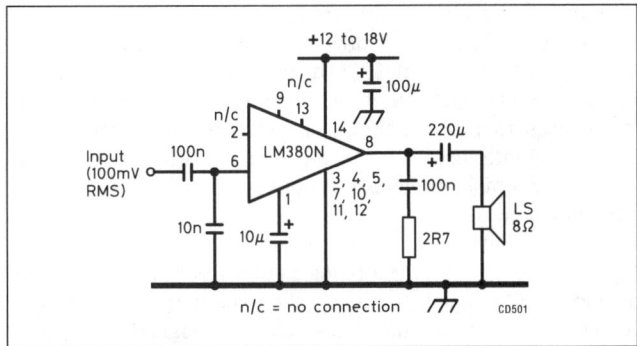

Fig 3.71: An audio output stage using the LM380N IC. The series combination of the 2.7Ω resistor and the 100nF capacitor at the output form a Zobel network. This improves stability by compensating for the inductive reactance of the loudspeaker voice coil at high frequencies

very few external components are required. Not all audio power ICs incorporate the negative feedback and single-rail bias networks 'on-chip', however, and so some, or all, of these components may have to be added externally.

Voltage Regulators

These devices incorporate a voltage reference generator, error amplifier and series pass transistor. Output short-circuit protection and thermal shut-down circuitry are also normally provided. There are two main types of regulator IC - those that generate a fixed output voltage, and also variable types which enable a potentiometer, or a combination of fixed resistors, to be connected externally in order to set the output voltage as required. Devices capable of delivering maximum currents of between 100mA and 5A are readily available, and fixed types offering a wide variety of both negative and positive output voltages may be obtained. **Fig 3.72** shows a simple mains power supply unit (PSU), based on a type 7812 regulator, providing +12V at 1A maximum.

Switched-mode power regulator ICs, which dissipate far less power, are also available. Further information on regulator ICs is given in the chapter on power supplies.

RF Building Blocks

Although it is now possible to fabricate an entire broadcast receiver on a single chip, this level of integration is rarely possible in amateur and professional communications equipment. To achieve the level of performance demanded, and also provide a high degree of operational flexibility, it is invariably necessary to consider each section of a receiver or transceiver separately, and then apply the appropriate technology to achieve the design goals. In order to facilitate this approach, ICs have been developed which perform specific circuit functions such as mixing, RF amplification and IF amplification.

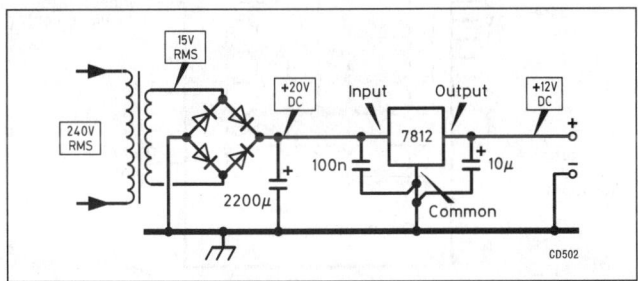

Fig 3.72: A simple 12V power supply unit using a fixed voltage IC regulator type 7812

An example of an IC mixer based on bipolar transistors is the Philips/Signetics NE602AN (featuring very low current consumption as demanded in battery-operated equipment). A useful wide-band RF amplifier is the NE5209D, and there is a similar dual version (ie containing two RF amplifiers within the same encapsulation), the NE5200D. There are also devices offering a higher level of integration - often termed sub-system ICs - which provide more than one block function. For more information on how to use such ICs, see chapters on receivers and transmitters. Amplification at UHF and microwave frequencies calls for special techniques and components. A wide range of devices, known generically as MMICs (microwave monolithic integrated circuits) are available, see the microwaves chapter.

DIGITAL INTEGRATED CIRCUITS

Logic Families

In digital engineering, the term 'logic' generally refers to a class of circuits that perform relatively straightforward gating, latching and counting functions. Historically, logic ICs were developed as a replacement for computer circuitry based on large numbers of individual transistors, and, before the development of the bipolar transistor, valves were employed. Today, very complex ICs are available, such as the microprocessor, which contain most of the circuitry required for a complete computer fabricated on to a single chip. It would be wrong to assume, however, that logic ICs are now obsolescent because there are still a great many low-level functions, many of them associated with microprocessor systems, where they are useful. In amateur radio, there are also applications which do not require the processing capability of a digital computer, but nevertheless depend upon logic - the electronic Morse keyer and certain transmit/receive changeover arrangements, for example.

By far the most successful logic family is TTL (transistor, transistor logic) although CMOS devices have been replacing them for some while. Originally developed in the 'sixties, these circuits have been continuously developed in order to provide more complex functions, increase speed of operation and reduce power consumption. Standard (type 7400) TTL requires a 5V power rail stabilised to within ±250mV. Logic level 0 is defined as a voltage between zero and 800mV, whereas logic 1 is defined as 2·4V or higher. **Fig 3.73** shows the circuit of a TTL NAND gate ('NAND' is an abbreviation for 'NOT AND' and refers to the gate's function, which is to produce an output of logic 0 when both inputs are at logic 1. Really this is an AND function followed by an inverter or

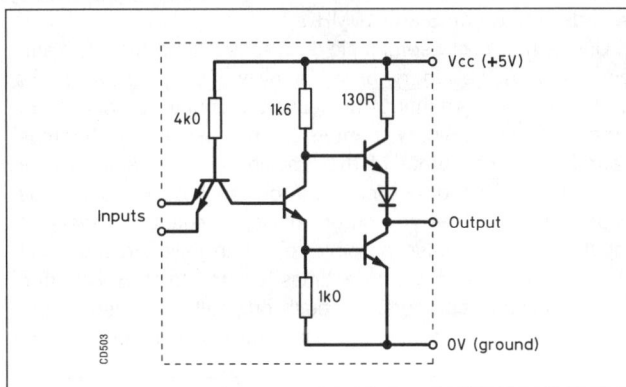

Fig 3.73: The internal circuit of a standard TTL two-input NAND gate. 74LS (low-power Schottky) TTL logic uses higher-value resistors in order to reduce power consumption, and the circuitry is augmented with clamping diodes to increase the switching speed of the transistors

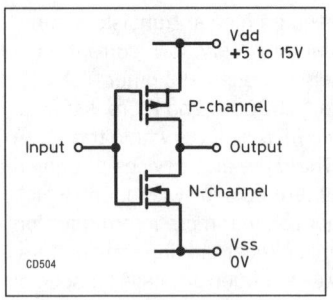

Fig 3.74: Simplified circuit of a CMOS inverter as used in the CD4000 logic family

NOT function. Note the use of a dual-emitter transistor at the input which functions in the same way as two separate transistors connected in parallel.

One of the first improvements made to TTL was the incorporation of Schottky clamping diodes in order to reduce the turn-off time of the transistors - this enhancement produced the 74S series. Latterly, a low-power consumption version of Schottky TTL, known as 74LS, has become very popular. The latest versions of TTL are designed around CMOS (complementary metal oxide silicon) transistors. The 74HC (high-speed CMOS) series is preferred for general use but the 74HCT type must be employed where it is necessary to use a mixture of CMOS and 74LS circuits to implement a design. A 74HC/HCT counter will operate at frequencies up to 25MHz. The 'complementary' in 'CMOS' refers to the use of gates employing a mixture of N-channel and P-channel MOS transistors.

Fig 3.74 shows the simplified circuit of a single CMOS inverter. If the input of the gate is held at a potential close to zero volts (logic 0), the P-channel MOSFET will be turned on, reducing its channel resistance to approximately 400Ω, and the N-channel MOSFET is turned off. This establishes a potential very close to the positive supply rail (V_{dd} - logic 1) at the gate's output. An input of logic 1 will have the opposite effect, the N-channel MOSFET being turned on and the P-channel MOSFET turned off. As one of the transistors is always turned off, and therefore has a channel resistance of about 10,000MΩ, virtually no current flows through the gate under static conditions. However, during transitions from one logic state to another, both transistors will momentarily be turned on at the same time, thus causing a measurable current to flow. The average current consumption will tend to increase as the switching frequency is raised because the gates spend a larger proportion of time in transition between logic levels. The popular CD4000 logic family is based entirely on CMOS technology. These devices can operate with supply voltages from 5 to 15V, and at maximum switching speeds of between 3 and 10MHz.

One of the most useful logic devices is the counter. The simplest form is the binary, or 'divide-by-two' stage shown at **Fig 3.75**. For every two input transitions the counter produces one output transition. Binary counters can be chained together (cascaded) with the output of the first counter connected to the input of the second, and so on. If four such counters are cascaded, there will be one output pulse, or count, for every 16 input pulses. It is also possible to obtain logic circuits which divide by 10 - these are sometimes referred to as BCD (binary coded decimal) counters. Although originally intended to perform the arithmetic function of binary division in computers, the

counter can also be used as a frequency divider. For instance, if a single binary counter is driven by a series of pulses which repeat at a frequency of 100kHz, the output frequency will be 50kHz. Counters can therefore be used to generate a range of frequencies that are sub-multiples of a reference input. The crystal calibrator uses this technique, and frequency synthesisers employ a more complex form of counter known as the programmable divider.

Counters that are required to operate at frequencies above 100MHz use a special type of logic known as ECL (emitter coupled logic). ECL counters achieve their speed by restricting the voltage swing between the levels defined as logic 0 and logic 1 to around 1V. The bipolar transistors used in ECL are therefore never driven into saturation (turned fully on), as this would reduce their switching speed due to charge storage effects. ECL logic is used in frequency synthesisers operating in the VHF to microwave range, and also in frequency counters to perform initial division, or prescaling, of the frequency to be measured.

Memories

There are many applications in radio where it is necessary to store binary data relating to the function of a piece of equipment, or as part of a computer program. These include the spot frequency memory of a synthesised transceiver, for instance, or the memory within a Morse keyer which is used to repeat previously stored messages.

Fig 3.76 provides a diagrammatic overview of the memory IC. Internally, the memory consists of a matrix formed by a number of rows and columns. At each intersection of the matrix is a storage cell which can hold a single binary number (ie either a zero or a one). Access to a particular cell is provided by the memory's address pins. A suitable combination of logic levels, constituting a binary number, presented to the address pins will instruct the IC to connect the addressed cell to the data pin. If the read/write control is set to read, the logic level stored in the cell will appear at the data pin. Conversely, if the memory is set to write, whatever logic level exists on the data pin will be stored in the cell, thus overwriting the previous value. Some memory ICs contain eight separate matrixes, each having their own data pin. This enables a complete binary word (byte) to be stored at each address.

The memory described above is known as a RAM (random access memory) and it is characterised by the fact that data can be retrieved (read) and also stored (written) to individual locations. There are two main types of RAM - SRAM (static RAM) in which data is latched within each memory cell for as long as the power supply remains connected, and DRAM (dynamic RAM)

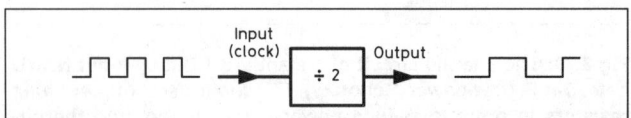

Fig 3.75: A binary counter may be used as a frequency divider

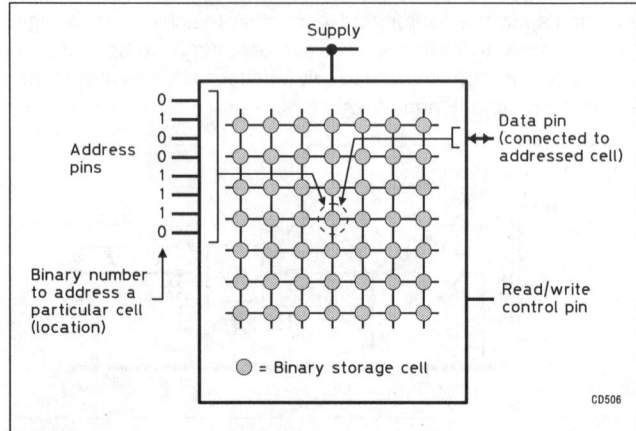

Fig 3.76: Representation of an IC memory

which uses the charge, or the absence of a charge, on a capacitor to store logic levels. The capacitors within a DRAM cannot hold their charge for more than a few milliseconds, and so a process known as refreshing must be carried out by controlling circuitry in order to maintain the stored levels. The method of creating addresses for the memory locations within a DRAM is somewhat complex, in that the rows and columns of the matrix are dealt with separately. DRAM memory chips are used extensively in desktop computers and related equipment because the simple nature of the capacitor memory cell means that a large number of cells can easily be fabricated on a single chip (there are now DRAMs capable of storing 16 million binary digits, or bits). SRAMs fabricated with CMOS transistors are useful for storing data that must be retained while equipment is turned off. The low quiescent current consumption of these devices makes it practicable to power the memory from a small battery located within the equipment, thus providing an uninterrupted source of power.

The ROM (read-only memory) has data permanently written into it, and so there is no read/write pin. ROMs are used to store computer programs and other data that does not need to be changed. The ROM will retain its data indefinitely, irrespective of whether it is connected to a power supply. A special form of ROM known as the EPROM (eraseable programmable ROM) may be written to using a special programmer. Data may later be erased by exposing the chip to ultra-violet light for a prescribed length of time. For this reason, EPROMs have a small quartz window located above the chip which is normally concealed beneath a UV-opaque protective sticker. The EEPROM (electrically eraseable programmable ROM) is similar to the EPROM, but may be erased without using UV light. The PROM (programmable ROM) has memory cells consisting of fusible links. Assuming that logic 0 is represented by the presence of a link, logic 1 may be programmed into a location by feeding a current into a special programming pin which is sufficient to fuse the link at the addressed location. However, once a PROM has been written to in this way, the cells programmed to logic 1 can never be altered.

Analogue-to-digital converters

Analogue-to-digital conversion involves measuring the magnitude of a voltage or current and then generating a numeric value to represent the result. The digital multimeter works in this way, providing an output in decimal format which is presented directly to the human operator via an optical display. The analogue to digital (A/D) converters used in signal processing differ in two important respects. Firstly, the numeric value is generated in binary form so that the result may be manipulated, or 'processed' using digital circuitry. Secondly, in order to 'measure' a signal, as opposed to, say, the voltage of a battery, it is necessary to make many successive conversions so that amplitude changes occurring over time may be captured. In order to digitise speech, for instance, the instantaneous amplitude of the waveform must be ascertained at least 6000 times per second. Each measurement, known as a sample, must then be converted into a separate binary number. The accuracy of the digital representation depends on the number of bits (binary digits) in the numbers - eight digits will give 255 discrete values, whereas 16 bits provides 65,535.

Maximum sampling frequency (ie speed of conversion) and the number of bits used to represent the output are therefore the major parameters to consider when choosing an A/D converter IC. The fastest 8-bit converters available, known as flash types, can operate at sampling frequencies of up to 20MHz, and are used to digitise television signals. 16-bit converters are unfortunately much slower, with maximum sampling rates of

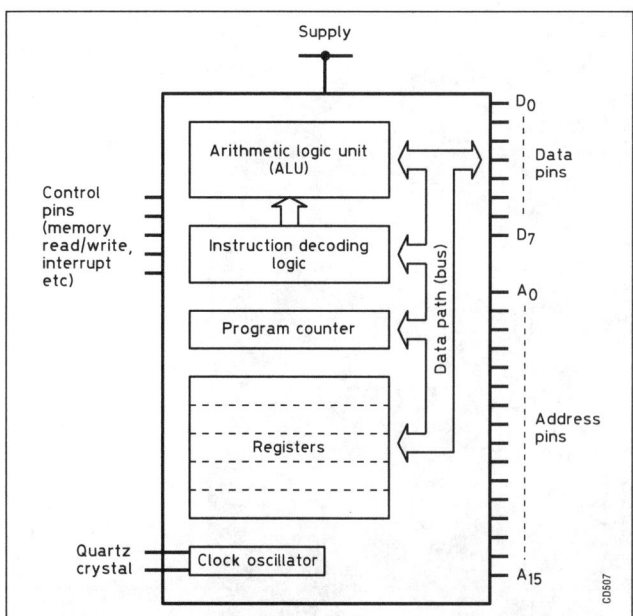

Fig 3.77: An 8-bit microprocessor

around 100kHz. It is also possible to obtain converters offering intermediate levels of precision, such as 10 and 12 bits.

Having processed a signal digitally, it is often desirable to convert it back into an analogue form - speech and Morse are obvious examples. There are a variety of techniques which can be used to perform digital-to-analogue conversion, and ICs are available which implement these.

Microprocessors

The microprocessor is different from other digital ICs in that it has no preordained global function. It is, however, capable of performing a variety of relatively straightforward tasks, such as adding two binary numbers together. These tasks are known as instructions, and collectively they constitute the microprocessor's instruction set. In order to make the microprocessor do something useful, it is necessary to list a series of instructions (write a program) and store these as binary codes in a memory IC connected directly to the microprocessor. The microprocessor has both data and address pins (see **Fig 3.77**), and when power is first applied it will generate a pre-determined start address and look for an instruction in the memory location accessed by this. The first instruction in the program will be located at the starting address. Having completed this initial instruction, the microprocessor will fetch the next one, and so on. In order to keep track of the program sequence, and also provide temporary storage for intermediate results of calculations, the microprocessor has a number of internal counters and registers (a register is simply a small amount of memory). The manipulation of binary numbers in order to perform arithmetic calculations is carried out in the arithmetic logic unit (ALU). A clock oscillator, normally crystal controlled, controls the timing of the program-driven events. The microprocessor has a number of control pins, including an interrupt input. This allows normal program execution to be suspended while the microprocessor responds to an external event, such as a keyboard entry.

Microprocessors exist in a bewildering variety of forms. Some deal with data eight bits at a time, which means that if a number is greater than 255 it must be processed using a number of separate instructions. 16-bit and 32-bit and more recently 64-bit microprocessors are therefore generally faster. A special class of microprocessor known as the microcontroller is designed specif-

Fig 3.78: Surface mounting components - the light coloured one labelled 'Z1' (bottom left) is a transistor. The scale at the bottom is in millimetres

ically to be built into equipment other than computers. As a result, microcontrollers tend to be more self-sufficient than general microprocessors, and will often be provided with internal ROM, RAM and possibly an A/D converter. Microcontrollers are used extensively in transceivers in order to provide an interface between the frequency synthesiser's programmable divider, the tuning controls - including the memory keypad - and the frequency display.

Following the development of the first microprocessors in the 1970s, manufacturers began to compete with each other by offering devices with increasingly large and ever-more-complex instruction sets. The late 1990s saw something of a backlash against this trend, with the emergence of the RISC (reduced instruction set microprocessor). The rationale behind the RISC architecture is that simpler instructions can be carried out more quickly, and by a processor using a smaller number of transistors.

Computers developed using the 86 series followed briefly by the 168,286 and 386 devices. The 586 series was renamed the Pentium™ and following a ruling that a number could not be trademarked the P1, P2, P3, P4 series had a variety of names for a number of foundries.

Amateurs today make frequent use of the 'PIC' series of microprocessors and these are discussed in more detail elsewhere in the handbook.

DSP (digital signal processing) ICs are special microprocessors which have their instruction sets and internal circuitry optimised for fast execution of the mathematical functions associated with signal processing - in particular the implementation of digital filtering.

SURFACE MOUNT DEVICES

As the semiconductor industry has developed, all packaging has been reduced in size and manufacturing techniques have changed so that devices are mounted on the copper-side surface of the printed circuit board. These SMDs (surface mounting devices) are now the norm and we as amateurs have to live with them. An example is shown in **Fig 3.78** with a section of steel rule with a millimetre scale to show just how small they are. Further details on SMDs and how to handle them are given in the Passive Components, and Construction and Workshop Practice chapters.

ELECTRONIC TUBES AND VALVES

Modern electronic tubes and valves have attained a high degree of reliability and are still available in many forms for a number of the more specialist applications. They have been superseded for virtually all low-power purposes by semiconductors and many of the high power amateur applications. Their use tends to be confined to the highest powers at HF, VHF, UHF and above.

Fundamentals

An electronic valve comprises a number of electrodes in an evacuated glass or ceramic envelope. The current in the valve is simply a large number of electrons emitted from a heated electrode, the cathode, and collected by the anode, which is maintained at a high positive potential. Other electrodes control the characteristics of the device.

Emission

In most types of valve the emission of electrons is produced by heating the cathode, either directly by passing a current through it, or indirectly by using an insulated heater in close proximity. The construction and surface coating of the cathode and the temperature to which it is heated govern the quantity of electrons emitted. This is known as thermionic emission.

Emission may also be produced when electrons impinge on to a surface at a sufficient velocity. For example, electrons emitted from a hot cathode may be accelerated to an anode by the latter's positive potential. If the velocity is high enough, other electrons will be released from the anode. This is known as secondary emission.

The emission of electrons from metals or coated surfaces heated to a certain temperature is a characteristic property of that metal or coated surface. The value of the thermionic emission may be calculated from Richardson's formula:

$$I_S = AT^2 e^{-\frac{b}{T}}$$

where I_S is the emission current in amperes per square centimetre of cathode surface; A and b are constants depending on the material of the emitting surface and T is absolute temperature in Kelvin (K).

Space charge

The thermionic electrons given off by the cathode form a cloud or space charge round the cathode. It tends to repel further electrons leaving the cathode due to its negative charge repelling the negatively charged electrons.

Electron flow

If the anode, see **Fig 3.79** is raised to a positive potential, the electron flow or current will increase to a point where the space charge is completely neutralised and the total emission from the cathode reaches the anode. The flow can be further increased by raising the cathode temperature. **Fig 3.80** shows the rise in anode current up to the saturation point for different heater temperatures.

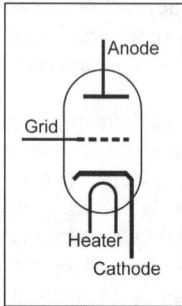

Fig 3.79: A triode valve

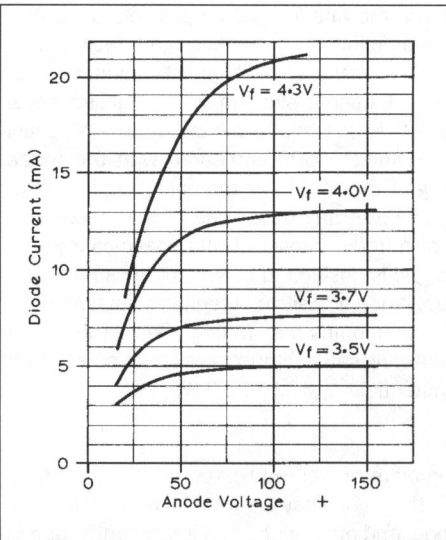

Fig 3.80: Diode saturation curve showing the effect of different filament voltages and hence temperatures

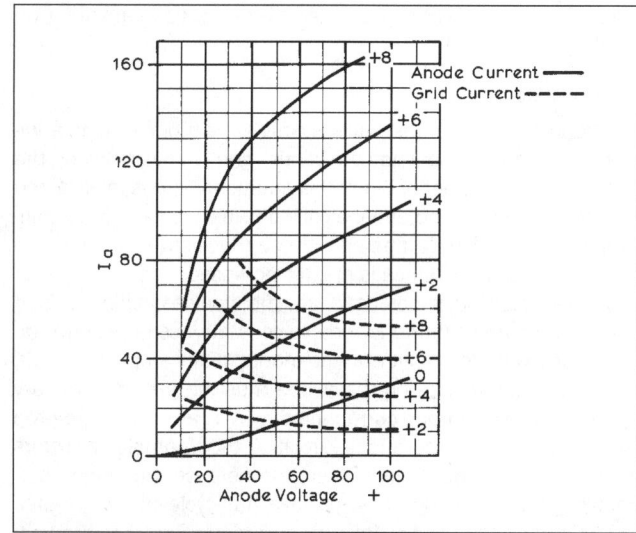

Fig 3.82: Positive-region triode characteristic curves showing how enhanced values of positive grid voltage increase anode current flow

As the electrons traverse the space between one electrode and another they may collide with gas molecules (because no vacuum can be perfect) and such collisions will impede their transit. For this reason the residual gas left inside the evacuated envelope must be minimal. An electronic tube that has been adequately evacuated is termed hard.

However, if a significant amount of gas is present, the collisions between electrons and gas molecules will cause it to ionise. The resultant blue glow between the electrodes indicates that the tube is soft. This blue glow should not be confused with a blue haze which may occur on the inside of the envelope external to the electrode structure: this is caused by bombardment of the glass, and in fact indicates that the tube is very hard.

Diodes

As described so far, we have a diode, a two-electrode valve. Like its semiconductor counterpart it will only permit current flow in one direction, from anode to cathode. The anode cannot emit electrons because it is not heated.

Triodes

By introducing a grid between the cathode and anode of an electronic tube the electron flow may be controlled. This flow may be varied in accordance with the voltage applied to the grid, the

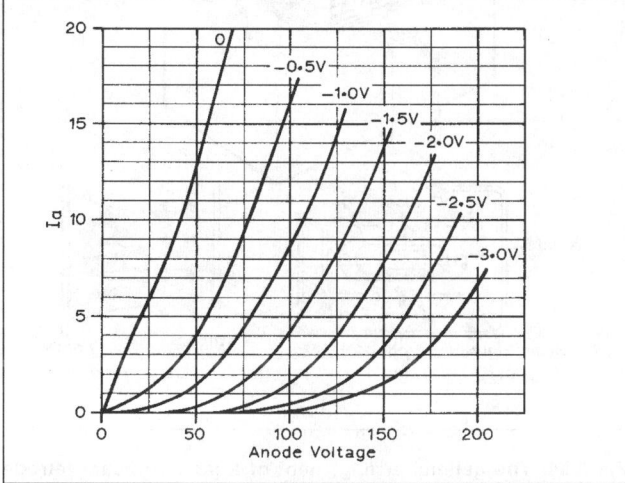

Fig 3.81: Negative-region characteristic curves of a triode showing the reduction in anode current which occurs with increase of negative grid voltage

required value being determined by the geometry of the grid and the desired valve characteristics. **Fig 3.81** shows the relationship between anode current (I_a) and voltage (V_a) for various negative bias voltages on the grid.

Normally the grid will operate at a negative potential, limiting the electron flow to the anode. However positive potentials can enhance the anode current but also lead to grid current and possibly more grid dissipation than the designer intended, reducing reliability. Some valves, typically high power devices, are designed to permit grid dissipation and should be used if grid current is expected. **Fig 3.82** shows both the increasing anode current and the grid current for different positive values of grid potential.

The grid voltage (grid bias) for a small general-purpose valve may be obtained by one of several methods:
1. A separate battery - historically common but no longer used.
2. A resistor connected between the cathode and the chassis (earth) so that when current flows the voltage drop across it renders the cathode more positive with respect to the chassis (earth), and the grid circuit return becomes negative with respect to cathode.
3. A resistor connected between the grid and the chassis (earth). When the grid is so driven that appreciable current flows (as in an RF driver, amplifier or multiplier), the grid resistor furnishes a potential difference between grid and chassis (earth), and with cathode connected to chassis a corresponding negative voltage occurs at the grid. A combination of grid resistor and cathode resistor is good practice and provides protection against failure of drive, which a grid resistor alone would not give.

As power amplifiers, triodes have the virtue of simplicity, especially when used in the grounded-grid mode (see later). At VHF and UHF the electrons can have a very short transit time if the valve is built on the planar principle. In this, the cathode, grid and anode are all flat and it is possible to have only a very short distance between them. Also the connection to each electrode has a very low impedance. Both factors promise good operation at HF/VHF/UHF. From this design, it is clear that earthed-grid operation is best when the grid with its disc-shaped connector acts as a screen between the input (cathode) and the output (anode) circuits. (See the section on 'Disc seal valves' later in

this chapter. The 2C39A illustrated there is a good amplifier up to at least 2.3GHz.)

Tetrodes

A tetrode ('four-electrode') valve is basically a triode with an additional grid mounted outside the control grid. That is between the first grid and the anode. When this additional grid is maintained at a steady positive potential a considerable increase in amplification factor occurs compared with the triode state; at the same time the valve impedance is greatly increased.

The reason for this increased amplification lies in the fact that the anode current in the tetrode valve is far less dependent on the anode voltage than it is in the triode. In any amplifier circuit, of course, the voltage on the anode must be expected to vary since the varying anode current produces a varying voltage-drop across the load in the anode circuit. A triode amplifier suffers from the disadvantage that when, for instance, the anode current begins to rise due to a positive half-cycle of grid voltage swing, the anode voltage falls (by an amount equal to the voltage developed across the load) and the effect of the reduction in anode voltage is to diminish the amount by which the anode current would otherwise increase. Conversely, when the grid voltage swings negatively, the anode current falls and the anode voltage rises. Because of this increased anode voltage the anode current is not so low as it would have been if it were independent of anode voltage. This means that the full amplification of the triode cannot be achieved. The introduction of the screen grid, however, almost entirely eliminates the effect of the anode voltage on the anode current, and the amplification obtainable is thus much greater.

A screen functions best when its voltage is below the mean value of the anode voltage. Most of the electrons from the cathode are thereby accelerated towards the anode, but some of them are unavoidably caught by the screen. The resulting screen current serves no useful purpose, and if it becomes excessive it may cause overheating of the screen.

If in low-voltage applications the anode voltage swings down to the screen voltage or lower, the anode current falls rapidly while that of the screen rises due to secondary emission from the anode to the screen. It should be noted that the total cathode current is equal to the sum of the screen and anode currents.

The I_a/V_a characteristics of the tetrode are shown in **Fig 3.83**. It will be noticed that there is a kink in the curve where an increase in anode voltage results in a decrease in anode current. This occurs where the anode voltage is lower than the screen grid but high enough that electrons from the cathode are accelerated fast enough to knock electrons off the anode, which are captured by the higher potential screen grid. As soon as the anode potential is above that of the screen grid, the effect ceases.

Another important effect of introducing the screen grid (G2) is that it considerably reduces the capacitive coupling between the input (control) grid and the anode, making possible the use of stable, high-gain RF amplification. To utilise this facility additional shields are added to the grid (electrically connected) so that the input connection cannot 'see' the anode or its supports. With such a structure it is possible to reduce the unit's capacitance by a factor of almost 1000 compared with the triode. Adequate decoupling of the screen at the operating frequency by the use of a suitable external bypass capacitor is essential.

In another type of tetrode, known as the space-charge grid tetrode, the second grid is positioned between the usual control grid and the cathode. When a positive potential is applied to this space-charge grid, it overcomes the limiting effect of the negative space charge, allowing satisfactory operation to be achieved at very low anode potentials, typically 12-24V.

Pentodes

To overcome the problem presented by secondary emission in the pure tetrode, a third grid may be introduced between the screen and the anode, and maintained at a low potential or connected to the cathode. Anode secondary emission is overcome and much larger swings of the anode voltage may be realised. This third grid is known as the suppressor grid (G3). Other methods which achieve the same effect are:

- Increasing the space between screen grid and anode;
- Fitting small fins to the inside surface of the anode; or
- Fitting suppressor plates to the cathode to produce what is known as the kinkless tetrode, which is the basis of the beam tetrode suppression system.

In some special types of pentode where it is necessary for application reasons to provide two control grids, the No 3 (sup-

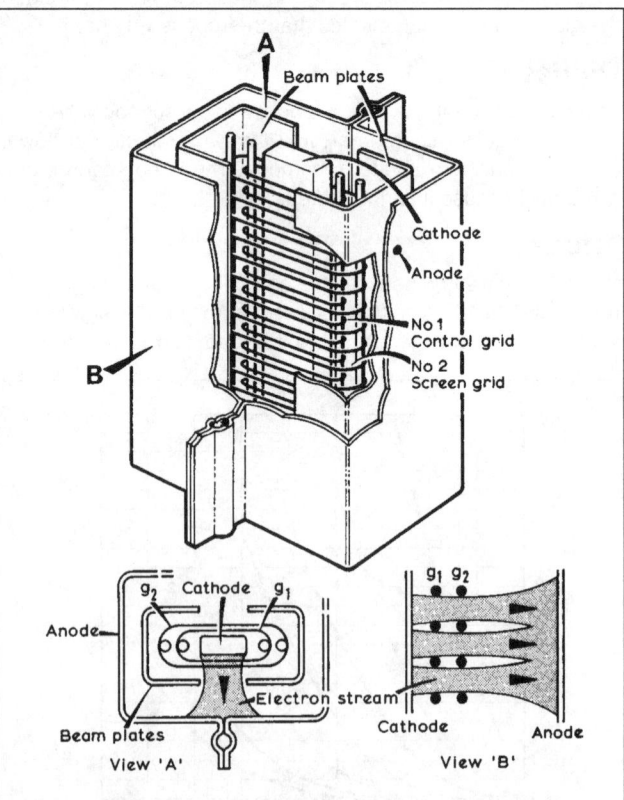

Fig 3.84: The general arrangement of a modern beam tetrode showing the aligned grid winding and the position of the beam forming plates. View 'A': looking vertically into a beam tetrode. View 'B': showing how the aligned electrode structure focuses electrons from cathode to anode

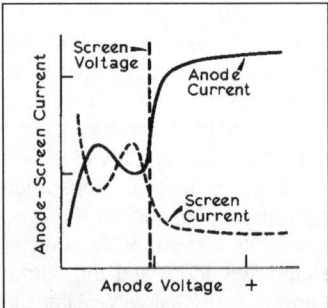

Fig 3.83: Characteristic curves of pure tetrode (often termed screened grid) showing the considerable secondary emission occurring when no suppression is used

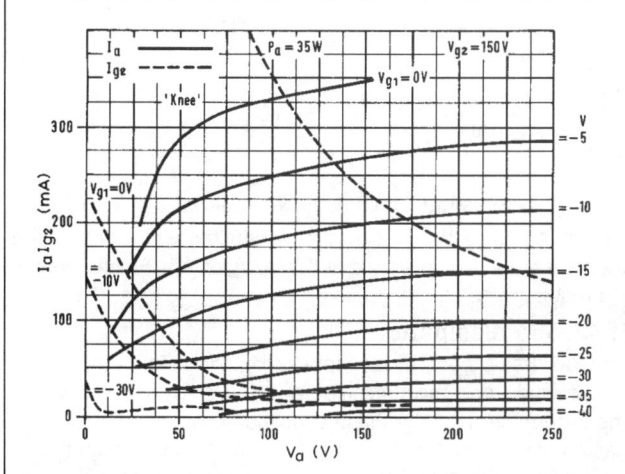

Fig 3.85: Characteristic curves of a beam tetrode. Anode secondary emission is practically eliminated by the shape and position of the suppressor plates

pressor) grid is used as the second and lower-sensitivity control for gating, modulation or mixing purposes. Units of this type need to have a relatively high screen grid (G2) rating to allow for the condition when the anode current is cut off by the suppressor grid (G3).

Beam Tetrodes

A beam tetrode employs principles not found in other types of valve: the electron stream from the cathode is focused (beamed) towards the anode. The control grid and the screen grid are made with the same winding pitch and they are assembled so that the turns in each grid are in optical alignment: see **Fig 3.84**. The effect of the grid and screen turns being in line is to reduce the screen current compared with a non-beam construction. For example, in a pentode of ordinary construction the screen current is about 20% of the anode current, whereas in a beam valve the figure is 5-10%.

The pair of plates for suppressing secondary emission referred to above is bent round so as to shield the anode from any electrons coming from the regions exposed to the influence of the grid support wires at points where the focusing of the electrons is imperfect. These plates are known as beam-confining or beam-forming plates.

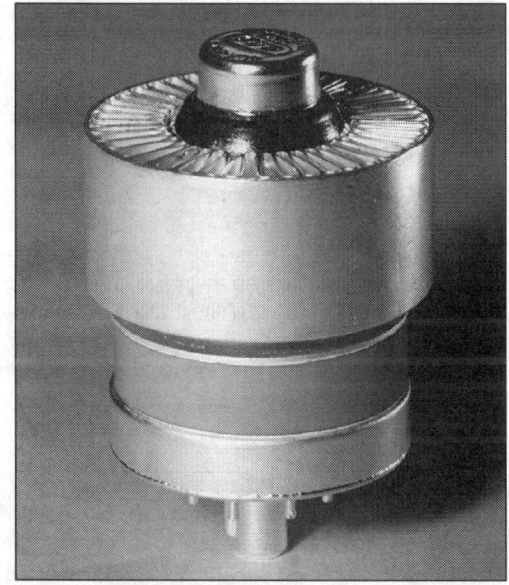

Fig 3.86: 4CX250B showing its air-cooled vanes

Beam valves were originally developed for use as audio-frequency output valves, but the principle has been applied to many types of RF tetrodes, both for receiving and transmitting. Their superiority over pentodes for AF output is due to the fact that the distortion is caused mainly by the second harmonic and only very slightly by the third harmonic, which is the converse of the result obtained with a pentode. Two such valves used in push-pull give a relatively large output with small harmonic distortion because the second harmonic tends to cancel out with push-pull connection.

Fig 3.85 shows the characteristic curves of a beam tetrode. By careful positioning of the beam plates a relatively sharp 'knee' can be produced in the anode current/anode voltage characteristic, at a lower voltage than in the case of a pentode, thus allowing a larger anode voltage swing and greater power output to be achieved. This is a particularly valuable feature where an RF beam tetrode is to be used at relatively low anode voltages.

Valve Construction and Characteristics

Cathodes

Although several types of cathode are used in modern valves, the differences are only in the method of producing thermionic emission. The earliest type is the bright emitter in which a pure tungsten wire is heated to a temperature in the region of 2500-2600K. At such a temperature emission of 4-40mA per watt of heating power may be obtained. These are not normally found in amateur applications.

Oxide-coated cathodes are the most common type of thermionic emitter found in both directly and indirectly heated valves. In this type, the emissive material is usually some form of nickel ribbon, tube or thimble coated with a mixture of barium and strontium carbonate, often with a small percentage of calcium. During manufacture, the coating is reduced to its metallic form and the products of decomposition removed during the evacuation process. The active ingredient is the barium, which provides much greater emission than thoriated tungsten at lower heating powers. Typically, 50-150mA per watt is obtained at temperatures of 950-1050K.

An indirectly heated cathode is a metal tube, sleeve or thimble shape, having a coating of emissive material on the outer surface. The cathode is heated by radiation from a metal filament, called the heater, which is mounted inside the cathode. The heater is electrically insulated from the cathode. The heater is normally made of tungsten or molybdenum-tungsten alloy. The heater or filament voltage should be accurately measured at the valve base and adjusted to the correct value as specified by the makers. This needs great care as it must be done with the stage operating at its rated power.

The life of valves with oxide-coated cathodes is generally good provided the ratings are not exceeded. Occasionally there is some apparent reduction in anode current due to the formation of a resistive layer between the oxide coating and the base metal, which operates as a bias resistor.

Anodes

In most valves the anode takes the form of an open-ended cylinder or box surrounding the other electrodes, and is intended to collect as many as possible of the electrons emitted from the cathode; some electrons will of course be intercepted by the grids interposed between the cathode and the anode.

The material used for the anode of the small general-purpose type of valve is normally bright nickel or some form of metal, coated black to increase its thermal capacity. Power dissipated in the anode is radiated through the glass envelope, a process which is assisted when adequate circulation of air is provided

around the glass surface. In some cases a significant improvement in heat radiation is obtained by attaching to the valve envelope a close-fitting finned metal radiator which is bolted to the equipment chassis so that this functions as a worthwhile heatsink.

Higher-power valves with external anodes are cooled directly by forced air (**Fig 3.86**), by liquid, or by conduction to a heatsink. Forced-air cooling requires a blower, preferably of the turbine type (rather than fan), capable of providing a substantial quantity of air at a pressure high enough to force it through the cooler attached to the anode.

Liquid cooling calls for a suitable cooler jacket to be fitted to the anode; this method is generally confined to large power valves. If water is used as the coolant, care must be taken to ensure that no significant leakage occurs through the water by reason of the high voltage used on the anode. A radiator is then used to cool the water.

In certain UHF disc seal valves a different form of conduction cooling is used, the anode seal being directly attached to an external tuned-line circuit that doubles as the heatsink radiator. Needless to say, it must be suitably isolated electrically from the chassis.

Whatever the type of valve and whatever method is used to cool it, the limiting temperatures quoted by the makers, such as bulb or seal temperatures, should never be exceeded. Underrunning the device in terms of its dissipation will generally greatly extend its life.

Grids

Mechanically, the grid electrode takes many forms, dictated largely by power and the frequency of operation. In small general-purpose valves the grids are usually in the form of a helix (molybdenum or other suitable alloy wire) with two side support rods (copper or nickel) and a cross-section varying from circular to flat rectangular, dependent on cathode shape.

In some UHF valves the grid consists of a single winding of wire or mesh attached to a flat frame fixed directly to a disc seal.

Characteristics

Technical data available from valve and tube manufacturers includes static characteristics and information about typical operating performances obtainable under recommended conditions. Adherence to these recommendations - indeed, to use the valve at lower than the quoted values - will increase life and reliability, which can otherwise easily be jeopardised. In particular, cathodes should always be operated within their rated power recommendations. The following terms customarily occur in manufacturers' data:

Mutual conductance (slope, g_m, transconductance)

This is the ratio of change of anode current to the change of grid voltage at a constant anode voltage. This factor is usually expressed in milliamperes per volt (or sometimes mili-siemens).

Amplification factor (μ)

This is the ratio of change of anode voltage to change of grid voltage for a constant anode current. In the case of triodes classification is customarily in three groups, low μ, where the amplification factor is less than 10, medium μ (10-50) and high μ (greater than 50).

Impedance (r_a, AC resistance, slope resistance)

When the anode voltage is changed while grid voltage remains constant, the anode current will change, an Ohm's Law effect. Consequently, impedance is measured in ohms. The relationship between these three primary characteristics is given by:

$$\text{Impedance }(\Omega) = \frac{\text{Amplification factor}}{\text{Mutual conductance}} \times 1000$$

or

$$r_a = \frac{\mu}{g_m} \quad k\Omega$$

where g_m is the mutual conductance in mA/V.

It will be noted that the mutual conductance and the impedance are equal to the slopes of the Ia/Vg and Ia/Va characteristics respectively.

Electrode dissipation

The conversion from anode input power to useful output power will depend upon the tube type and the operating conditions. The difference between these two values, known as the anode dissipation, is radiated as heat. If maximum dissipation is exceeded overheating will cause the release of occluded gas, which will poison the cathode and seriously reduce cathode emission. The input power to be handled by any valve or tube will, in the limiting case, depend on the class of operation. Typical output efficiencies expressed as percentages of the input power are:

Class A	33%
Class AB1	60-65%
Class AB2	60-65%
Class B	65%
Class C	75-80%

Considering a valve with a 10W anode dissipation, the above efficiencies would give outputs as follows (assuming there are no other limiting factors such as peak cathode current):

Class A	5W
Class AB1 and AB2	15-18.5W
Class B	18.5W
Class C	30-40W

Hum

When a cathode is heated by AC, the current generates a magnetic field which can modulate the electron stream, and a modulating voltage is injected into the control grid through the interelectrode capacitance and leakages: additionally there can be emission from the heater in an indirectly heated valve (see below).

When operating directly heated valves such as transmitting valves with thoriated tungsten filaments, the filament supply should be connected to earth by a centre tap or a centre-tapped resistor connected across the filament supply (a hum-bucking resistor). The hum is usually expressed as an equivalent voltage (in microvolts) applied to the control grid. Valve hum should not be confused with hum generated in other circuit components.

Valve Applications

Amplifiers

When an impedance is connected in series with the anode of a valve and the voltage on the grid is varied, the resulting change of anode current will cause a voltage change across the impedance. The curves in **Fig 3.87** illustrate the classifications of valve amplifier operating conditions, showing anode current/grid voltage characteristics and the anode current variations caused by varying the grid voltage.

Class A: The mean anode current is set to the middle of the straight portion of the characteristic curve. If the input signal is allowed either to extend into the curved lower region or to approach zero grid voltage, distortion will occur because grid current is caused to flow by the grid contact potential (usually 0.7-1.0V). Under Class A conditions anode current should show

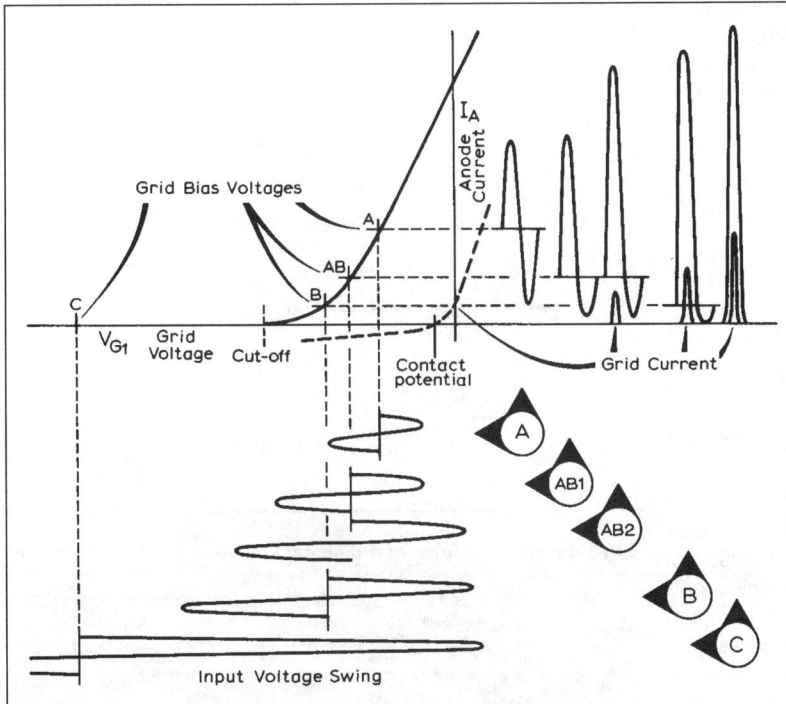

Fig 3.87: The valve as an amplifier: the five classes of operation

no movement with respect to the signal impressed on the grid. The amplifier is said to be linear.

Class AB1: The amount of distortion produced by a non-linear amplifier may be expressed in terms of the harmonics generated by it. When a sine wave is applied to an amplifier the output will contain the fundamental component but, if the valve is allowed to operate on the curved lower portion of its characteristic, ie running into grid current, harmonics will be produced as well. Harmonic components are expressed as a percentage of the fundamental. Cancellation of even harmonics may be secured by connecting valves in push-pull, a method which has the further virtue of providing more power than a single valve can give, and was once widely found in audio amplifiers and modulators.

Class AB2: If the signal input is increased beyond that used in the Class AB1 condition peaks reaching into the positive region will cause appreciable grid current to be drawn and the power output to be further increased. In both the Class AB1 and AB2 conditions the anode current will vary from the zero signal mean level to a higher value determined by the peak input signal.

Class B: This mode, an extension of Class AB2, uses a push-pull pair of valves with bias set near to the cut-off voltage. For zero signal input the anode current of the push-pull pair is low, but rises to high values when the signal is applied. Because grid current is considerable an appreciable input power from the drive source is required. Moreover, the large variations of anode current necessitate the use of a well-regulated power (HT) supply.

Class C: This condition includes RF power amplifiers and frequency multipliers where high efficiency is required without linearity, as in CW, AM and NBFM transmitters. Bias voltage applied to the grid is at least twice, sometimes three times, the cut-off voltage, and is further increased for pulse operation. The input signal must be large compared with the other classes of operation outlined above, and no anode current flows until the drive exceeds the cut-off voltage. This could be for as little as 120° in the full 360° cycle, and is known as the conduction angle. Still smaller conduction angles increase the efficiency fur-

ther, but more drive power is then needed. Pulse operation is simply 'super Class C'; very high bias is applied to the grid and a very small angle of conduction used.

Grid driving power

An important consideration in the design of Class B or Class C RF power amplifiers is the provision of adequate driving power. The driving power dissipated in the grid-cathode circuit and in the resistance of the bias circuit is normally quoted in valve manufacturers' data.

These figures frequently do not include the power lost in the valveholder and in components and wiring, or the valve losses due to electron transit-time phenomena, internal lead impedances and other factors. Where an overall figure is quoted, it is given as driver power output.

If this overall figure is not quoted, it can be taken that at frequencies up to about 30MHz the figure given should be multiplied by two, but at higher frequencies electron transit-time losses increase so rapidly that it is often necessary to use a driver stage capable of supplying 3-10 times the driving power shown in the published data.

The driving power available for a Class C amplifier or frequency multiplier should be sufficient to permit saturation of the driven valve, ie a substantial increase or decrease in driving power should produce no appreciable change in the output of the driven stage. This is particularly important when the driven stage is anode-modulated.

Passive grid

In linear amplifiers the driver stage must work into an adequate load, and the use of the passive grid arrangement is to be recommended. A relatively low resistance (typically 1kΩ) is applied between grid and cathode with a resonant grid circuit where appropriate. This arrangement helps to secure stable operation but should not be used as a cure for amplifier instability.

Grounded cathode

Most valves are used with the cathode connected to chassis or earth, or where a cathode-bias resistor is employed it is shunted with a capacitor of low reactance at the lowest signal frequency used so that the cathode is effectively earthed. In modulated amplifiers two capacitors, one for RF and the other for AF, must be used.

Grounded grid

Although a triode must be neutralised to avoid instability when it is used as an RF amplifier, this is not always essential if an RF type of tetrode or pentode is employed. However, above about 100MHz a triode gives better performance than a tetrode or pentode, providing that the inherent instability can be overcome. One way of achieving this is to earth the grid instead of the cathode so that the grid acts as an RF screen between cathode and anode, the input being applied to the cathode. The capacitance tending to make the circuit unstable is then that between cathode and anode, which is much smaller than the grid-to-anode capacitance.

The input impedance of a grounded-grid stage is normally low, of the order of 100Ω, and therefore appreciable grid input power is required. Since the input circuit is common to the anode-cathode circuit, much of this power is, however, transferred directly to the output circuit, ie not all of the driving power is lost.

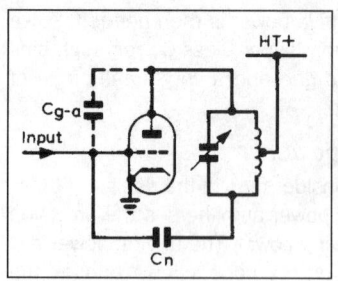

Fig 3.88: Neutralising a grounded-cathode triode amplifier. The circuit is equally suitable for a tetrode or a pentode

Grounded anode

For some purposes it is desirable to apply the input to the grid and to connect the load in the cathode circuit, the anode being decoupled to chassis or earth through a low-reactance capacitor. Such circuits were employed in cathode followers and infinite-impedance detectors.

Neutralising amplifiers

Instability in RF amplifiers results from feedback from the anode to the grid through the grid-to-anode capacitance and is minimised by using a tetrode or pentode. At high frequencies, particularly if the grid and/or anode circuit has high dynamic impedance, this capacitance may still be too large for complete stability. A solution is to employ a circuit in which there is feedback in opposite phase from the anode circuit to the grid so that the effect of this capacitance is balanced out. The circuit is then said to be neutralised.

A typical arrangement is shown in **Fig 3.88**. Here the anode coil is centre-tapped in order to produce a voltage at the 'free' end which is equal and opposite in phase to that at the anode end. If the free end is connected to the grid by a capacitor (Cn) having a value equal to that of the valve grid-to-anode capacitance (Cg-a) shown dotted, any current flowing through Cg-a will be exactly balanced by that through Cn. This is an idealised case because the anode tuned circuit is loaded with the valve anode impedance at one end but not at the other; also the power factor of Cn will not necessarily be equal to that of Cg-a.

The importance of accurate neutralisation in transmitter power amplifier circuits cannot be overstressed, and will be achieved if the layout avoids multiple earth connections and inductive leads; copper strip is generally preferable to wire for valve socket cathode connections.

Disc Seal Valves

In the disc seal triode (**Figs 3.89 and 3.90**), characteristically of high mutual conductance, the electrode spacing is minimal. The 'top hat' cathode contains the insulated heater, one side of which is connected to the cathode and the other brought out coaxially through the cathode sleeve connection. The fine-wire grid stretched across a frame emerges through the envelope by an annular connection. Because the clearance between grid and cathode is very small, the cathode surface is shaved during construction to provide a plane surface. The anode also emerges via an external disc for coaxial connection. On larger disc seal valves the anode may form part of the valve envelope.

Disc seal valves are available for power dissipations of a few watts to 100W with forced air cooling and outputs in the frequency range 500-6000MHz. It should be noted that maximum power and frequency are not available simultaneously.

Disc seal valves, although intended for coaxial circuits, may be effectively employed with slab-type circuits. Important points to be observed are (a) only one electrode may be rigidly fixed, in order to avoid fracture of the seals (which is more likely to occur in the glass envelope types); and (b) except in the case of forced-air anode cooling the anode is cooled by conduction into its asso-

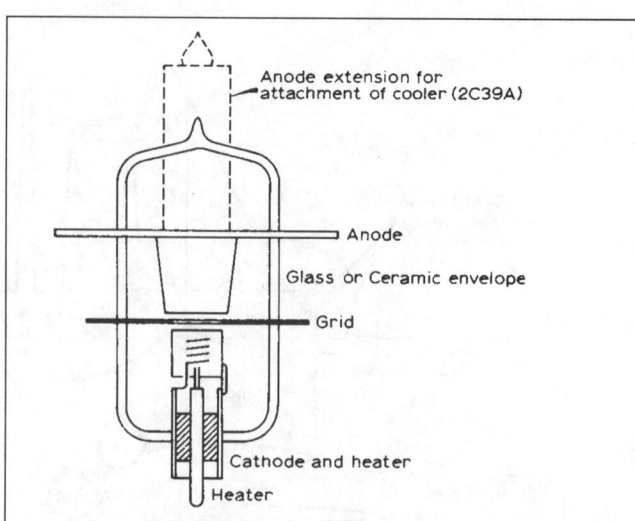

Fig 3.89: General form of a disc seal valve

Fig 3.90: The 2C39 disc seal triode, showing glass version (left and ceramic version (right). The anode cooling fins on the ceramic version are removable, clamped in place with Allen bolts

ciated circuit. With a shunt-fed circuit thin mica insulation will function both as a capacitance and a good transmitter of heat.

Certain sub-miniature metal ceramic envelope types such as the 7077, although of the generic disc seal form, require special sockets if they are used in conventional circuits. Many of them give significant output at the lower SHF bands.

CATHODE-RAY TUBES

A cathode-ray tube contains an electron gun, a deflection system and a phosphor-coated screen for the display. The electron gun, which is a heated cathode, is followed by a grid consisting of a hole in a plate exerting control on the electron flow according to the potential applied to it, followed in turn by an accelerating anode or anodes.

The simplest form of triode gun is shown at **Fig 3.91**. The beam is focused by the field between the grid and the first anode. In tubes where fine line spots and good linearity are essential (eg in measurement oscilloscopes), the gun is often extended by the addition of a number of anodes to form a lens system. Beam focusing may be by either electrostatic or electromagnetic means.

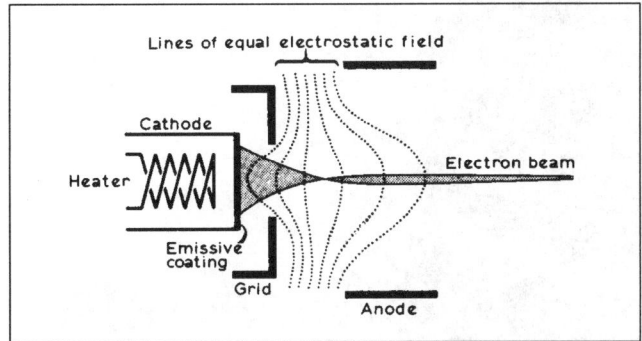

Fig 3.91: Diagram of the electron gun used in cathode-ray tubes, travelling wave tubes and klystrons

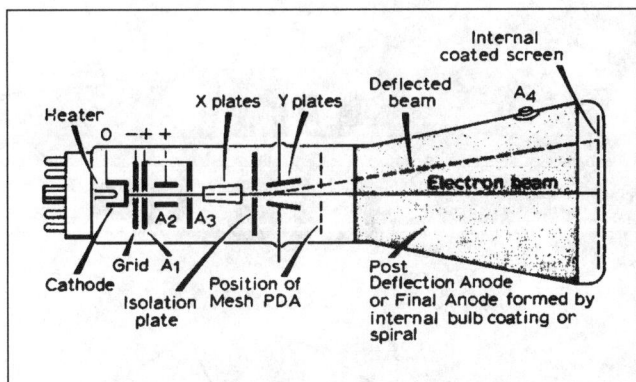

Fig 3.92: Diagrammatic arrangement of a cathode-ray tube with electrostatic focusing and deflection

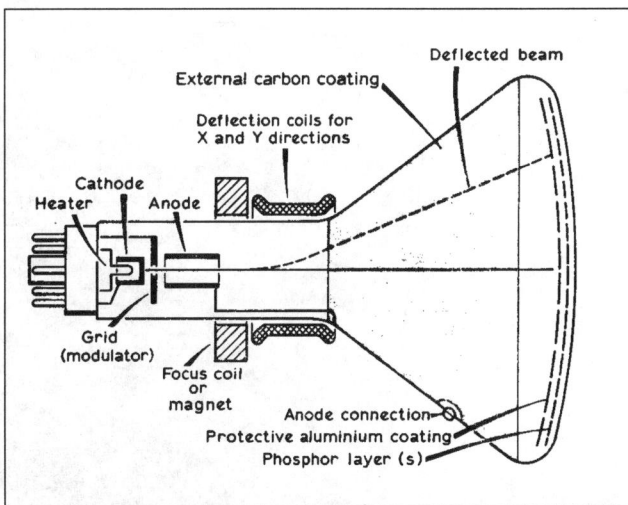

Fig 3.93: Diagrammatic arrangement of a cathode-ray tube having magnetic focusing and deflection

Oscilloscope Tubes

Electrostatic focusing is used in oscilloscope tubes, the deflection system consisting of pairs of plates to deflect the beam from its natural centre position, depending on the relative potentials applied. Interaction between the two pairs of deflection plates is prevented by placing an isolation plate between them (Fig 3.91).

After its deflection the beam is influenced by a further accelerating electrode known as the post-deflection accelerator (PDA) which may take the form of a wide band of conducting material on the inside of the cone-shaped part of the bulb, or of a close-pitch spiral of conducting material connected to the final anode. For some purposes when it is important to maintain display size constant irrespective of the final anode voltage, a mesh post-deflection accelerator is fitted close to the deflecting plates.

If a double beam is required, the electron flow is split into two and there are two sets of deflection plates. Alternatively, two complete systems are enclosed in one tube. Although a common X-deflection plate may be fitted, the advantage of two complete systems lies in providing complete alignment of the timing (horizontal) deflection. By setting two systems at an angle to one another adequate overlap of each of the displays is provided.

Radar and Picture Tubes

In radar and television tubes magnetic focusing and deflection are common (**Fig 3.93**), the deflection angle being vastly greater than with an oscilloscope tube. Such tubes employ a simple triode or tetrode electron gun with an anode potential of 20kV or more.

Screen Phosphors

Many types of phosphor are used for coating the screens of cathode-ray tubes, their characteristics varying according to the application. In all of them the light output is determined by the final anode voltage used, but where this exceeds about 4kV the phosphor is protected against screen burn by a thin backing layer of evaporated aluminium. Oscilloscope tubes require a phosphor with a wide optical band to give a bright display for direct viewing. This phosphor, yellow-green in colour, extends into the blue region to enable direct photographs to be taken from the display.

CR Tube Power Supplies

Unlike the valve, the cathode ray tube's current requirements are low, the beam current being only a few tens of microamperes. For oscilloscope work the deflector plates need to be at earth potential, cathode and other electrodes consequently being at a high negative potential to earth.

In magnetically focused tubes the cathode may be at earth potential and the final anode many kilovolts above. Power supplies for either type of tube need to be of very high impedance for safety reasons, and the short-circuit current should not exceed about 0.5mA.

Further Details

Klystrons, Magnetrons and travelling wave tubes are covered in earlier versions of this handbook along with technical details of valve circuit design and operation.

ONLY £16.99

plus p&p

COMMAND

Computers, Microcontrollers and DSP for the Radio Amateur
Andy Talbot, G4JNT

This book is for the radio amateur and home experimenter who wants to use modern digital and computer technology to make more use of his/her radio equipment. It is aimed at those who want to add additional hardware and try new signalling and communications methods, and to build hardware and systems for controlling existing equipment. The home computer is first covered, writing programmes and software, with much detail on how to interface the PC to external hardware via its various ports. Particular emphasis is placed on what can be done with the older computer, now very cheaply available. Software techniques for detecting signals in noise and for automatic beacon monitoring are described.

Then come Microcontrollers, covered in depth, particularly the PIC family of devices. From the basics of writing the first PIC programme and programming the device, many different types of hardware are described, such as A/D converters and relays. This includes simple arithmetic and coding issues for security and remote control. Finally, basic Digital Signal Processing is covered, with aspects such as digital filtering, time/frequency transformations and very narrow bandwidth working being described. How to start using DSP techniques at home is explained; evaluation modules and DSP routes using some simple additional hardware and a PC are covered. Programming in the Windows operating system, with particular emphasis on using the soundcard for DSP purposes, is then introduced.

Written for the experienced amateur, this book is aimed at the experimenter and home constructor who wants to get involved in the subject, and to understand how to take it further.

Size: 240 x 173 mm, 232 pages, ISBN: 1-872309-94-1
RSGB Members Price: £14.44 plus p&p

E&OE

Radio Society of Great Britain
Lambda House, Cranborne Road, Potters Bar, Herts. EN6 3JE
Tel. 0870 904 7373 Fax. 0870 904 7374

ORDER 24 HOURS A DAY ON OUR WEBSITE
www.rsgb.org/shop

RSGB SHOP

Building Blocks 1: Oscillators

4

Oscillators are fundamental to radio and are used in almost all types of equipment. This edition of the *Handbook* has collected together much of the material on oscillators that was spread over several chapters in previous editions.

The purpose of an oscillator is to generate an output at a specific frequency. For most applications, an oscillator would ideally generate a pure sine wave. If a spectrum of voltage against frequency were to be plotted, it would consist of a single line at the required frequency. **Fig 4.1(a)** shows the spectrum of the 'ideal' oscillator. Spectral plots like this (from a spectrum analyser) are much used in professional electronics but are rarely available to the amateur. A more practical approach for the amateur is to tune across the frequency band of interest with a reasonably good receiver, preferably in single sideband (SSB) mode. The ideal oscillator would then produce a single clean beat note, of amplitude related to the amplitude of the oscillations and, of course, the coupling into the receiver.

No oscillator produces this ideal output, and there are always harmonics, noise and often sidebands in addition to the wanted output frequency. A more realistic plot is shown in **Fig 4.1(b)**. Tuning a receiver across this would show where the problems are. The noise floor (the background level of noise which is equal at all frequencies) is apparent, as is the '$1/f$' noise, which is an increase in noise close to the centre (wanted, or carrier) frequency. The harmonics also show these effects, as do the side-

bands and spurious oscillations which often occur in synthesisers because of practical limitations in loop design.

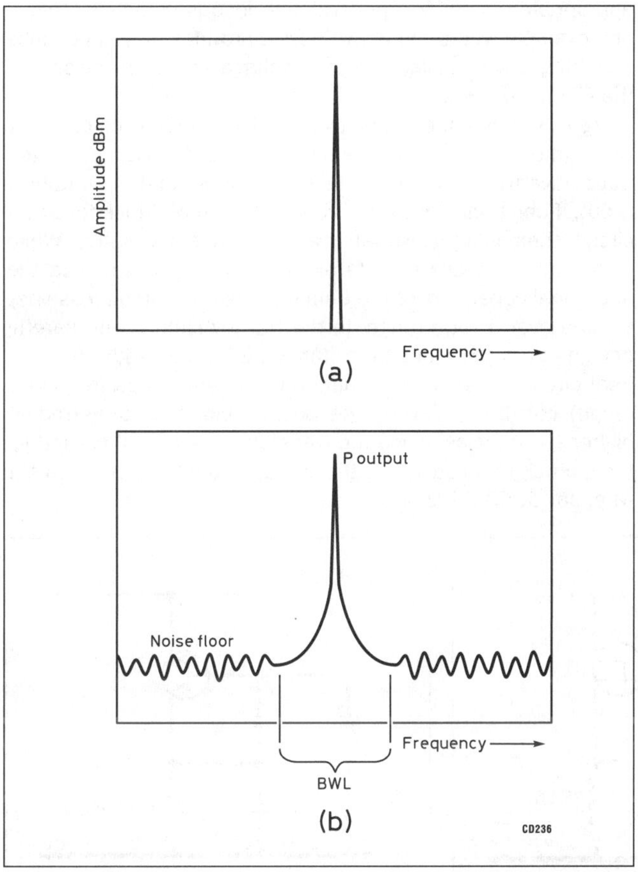

Noise and sidebands are always undesirable, and can be very difficult to remove once generated, so one purpose of this chapter is to indicate ways of minimising them. Harmonics are less of a problem. They are a long way in frequency terms from the wanted signal, and can often be simply filtered out. In some cases, they may even be wanted, since it is often convenient to take an output frequency from an oscillator at a harmonic instead of the fundamental. Examples of this are shown below for crystal oscillators.

Oscillators may be classed as variable frequency oscillators (VFOs), crystal oscillators (COs) (including the class of variable crystal and ceramic resonator oscillators (VXOs)), phase-locked loop synthesisers (PLLs), which include VFOs as part of their system and, more recently, direct digital synthesisers (DDSs). Any or all of these may be used in a particular piece of equipment, although recent trends commercially are to omit any form of 'free-running' (ie not synthesised) VFO. The name is retained, however, and applied to the sum of the oscillators in the equipment. Since these are digitally controlled, it is possible to have two (or more) virtual 'VFOs' in a synthesised rig, where probably only a single variable oscillator exists physically. It can be retuned in milliseconds to any alternative frequency by the PLL control circuit.

All oscillators must obey certain basic design rules, and examination of preferred designs will show that these have been obeyed by the designer, whether deliberately or empirically. The first essential for an oscillator is an amplifying element. This will be an active device such as a bipolar transistor (sometimes called a 'BJT' for bipolar junction transistor), a junction field-effect transistor (JFET), a metal-oxide semiconductor FET (MOS-FET) or a gallium arsenide FET (GaAsFET). There are lesser-used devices such as gallium arsenide bipolars, and gallium arsenide HEMTs (high electron mobility transistors), a variant of the GaAsFET with lower noise.

Operational amplifiers are rarely used at RF, since very few are capable of high-frequency operation. Some integrated circuits do contain oscillator circuits, but they are very often variants of standard discrete circuits. Valves, now long obsolete in the context of oscillator design, will be neglected in this text, but it is worth observing that the fundamental circuit configurations owe their origins to valve designs from the early part of the 20th century. The Colpitts and Hartley oscillators were both first published in 1915. The Colpitts oscillator is named after its developer, Edwin Henry Colpitts (1872-1949) and the Hartley oscillator was developed by Ralph Hartley (1889-1970). For those interested in valve design, reference should be made to earlier editions of this handbook, and to references [1] and [2]. A more recent and readily available source is Section 4 of reference [3].

The basic requirements for an oscillator are:

1. There must be gain over the whole frequency range required of the oscillator.

2. There must be a feedback path such that the product of the forward gain and the feedback attenuation still leaves a net loop gain greater than 1.

3. The feedback must occur in such a way that it is in phase with the input to the gain stage.

Fig 4.1: (a) Spectrum of 'ideal' oscillator. (b) Spectrum of local oscillator showing noise

4. There must be some form of resonant circuit to control the oscillator frequency. This is most often an inductor and capacitor (L-C). More precise frequency control can be achieved using electromechanical structures such as crystals, ceramic resonators or surface acoustic wave (SAW) resonators. This latter category is only really feasible in specialist applications where the high cost can be justified.

The resonant circuit may not be the one apparent because, as in many non-intended oscillations, a feedback around many paths including the parasitic components in the circuit can control the oscillations, which can of course be at more than one frequency. At low frequencies, a resistor-capacitor (RC) or inductor-resistor (LR) network may be used for frequency control. These cannot be said to be truly resonant themselves but, with active gain stages, resonant peaks which are sometimes of very high Q can be achieved.

A potential successor to the crystal and ceramic resonator is the electrochemically etched mechanical resonator, usually in the same silicon chip as the oscillator circuit. These devices are still in the development stage, but may well come into common use soon.

An oscillator therefore needs gain, feedback and a resonant circuit. However, there are more criteria that can be applied:

5. For stable oscillations, the resonant circuit should have high Q. This tends to rule out RC and LR oscillators for high frequencies, although they are often used for audio applications. More information on low-frequency oscillators can be found in audio texts [4].

6. For stable oscillations, the active components should also show a minimum of loading on the resonant circuit, ie the loaded Q should be as high as possible. Although an oscillator could be said to be the ultimate in Q-multipliers, high inherent Q gives better stability to the final product.

7. The choice of active device is critical in achieving low-noise operation. This will be covered extensively below, since in almost all cases low-noise oscillations are required, and the difference between 'noisy' and 'quiet' oscillators may be only in the choice of a transistor and its matching circuit.

8. The output stage of an oscillator can be very important. It must drive its load adequately, without changing the loading of the oscillator and thereby changing the oscillator frequency (an effect known as load pulling).

9. The power supply to an oscillator should be very carefully considered. It can affect stability, noise performance and output amplitude. Separate supply decoupling and regulation arrangements are usually essential in good oscillator design.

VARIABLE-FREQUENCY OSCILLATORS (VFOS)

A VFO is a type of oscillator in which the oscillation frequency is adjustable by the operator. Normally this is done by the tuning control on the front panel of the equipment. This control operates through a variable capacitor or variable inductor to control the oscillation frequency. Most VFOs consist of a single active device with a tuned circuit and a feedback network to sustain the oscillation. If the active device is an inverting stage, then this is sometimes referred to as a 180° oscillator, since the active device produces a 180° phase shift and the tuned circuit produces a further 180° shift, giving the 360° total shift needed for oscillation. Some circuits do not invert in the active device, but use it as a voltage-follower. In this case, the gain of the active device may not be used, or may only be used as a buffer. However, the real gain of the circuit comes about through the impedance transformation occurring in the follower, so that a high transformer ratio at the tuned circuit, produced by the Q, is buffered down to a lower impedance for the feedback network.

Two such examples are shown in **Figs 4.2 and 4.3**. Some comments on these diagrams will show the design compromises in action. Fig 4.2 uses a tapped inductor and is known as the Hartley oscillator. The diagram shows the basic circuit and does not include DC bias components. In the Hartley oscillator of Fig 4.2, the tuned circuit is designed for high Q. The active device is an FET, chosen for good gain at the intended frequency of oscillation. A dual-gate MOSFET could be used or, especially at UHF and microwave frequencies, a GaAsFET. The high input impedance of the FET puts little loading onto the tuned circuit, while the FET source provides feedback at low impedance. The feedback voltage is transformed up to a higher impedance and appropriate phase, so the oscillation loop is complete. This is one example where the active device provides only impedance matching to the oscillator loop; no voltage gain or phase shift in the FET is involved.

Fig 4.3 shows the capacitively tapped version of basically the same circuit. This is known as the Colpitts oscillator. Capacitive transformers are less easy to understand intuitively but, if one thinks of the whole as a high-impedance resonant circuit, then a capacitive tap is a reasonable alternative. While in principle the two circuits are very similar, in practice the additional capacitors of the Colpitts make the circuit less easy to tune over a wide range. In the Hartley circuit, with careful design and a low-input-capacitance FET, a very wide range of total circuit capacitance variation (and hence wide frequency range) can be achieved. The Colpitts circuit is preferred at higher frequencies, since the transformer action of the inductor is much reduced where the 'coil' is in the form of a straight wire, say at 500MHz or above.

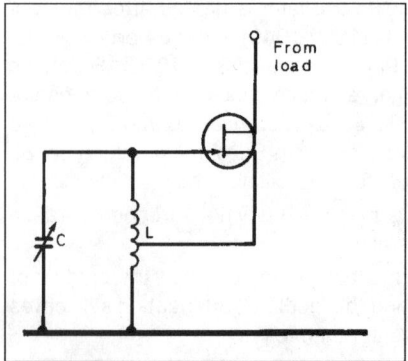

Fig 4.2: Hartley oscillator

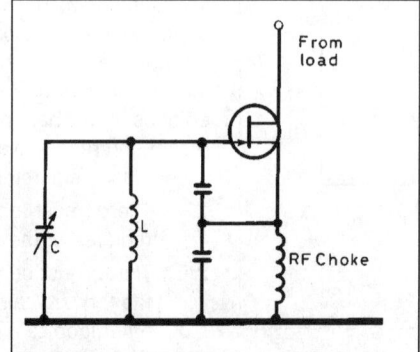

Fig 4.3: Colpitts oscillator

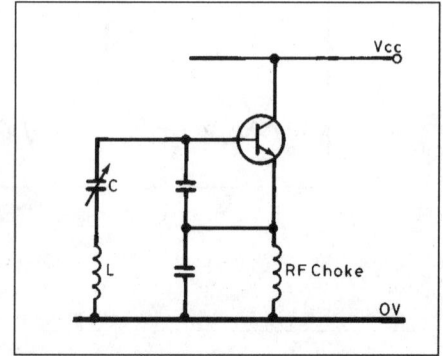

Fig 4.4: Clapp oscillator

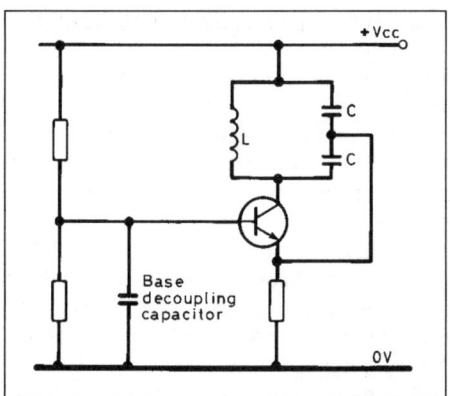

Fig 4.5: 180 degree phase-shift oscillator

Sometimes described as a variant of the Colpitts circuit is the Clapp oscillator: **Fig 4.4**. This shows a series-resonant circuit, which is more easily matched into the bipolar transistor shown. This circuit is especially suitable to UHF and low microwave work, where it is capable of operating very close to the cut-off frequency (Ft) of the transistor.

Variants of both the Hartley and Colpitts circuits have been derived for feedback around the active device including a 180° phase shift. An example is shown in **Fig 4.5** and this is again more suited to a bipolar transistor.

In the technical press, there has been much discussion on whether the best choice for the active device in an oscillator is a bipolar transistor or an FET. There are many factors involved in this choice, but some simple rules can be derived. First, since the loaded Q must be maintained as high as possible, an FET is attractive for its high input impedance. A bipolar transistor could be used as an emitter follower with an undecoupled emitter resistor but a noise analysis of the circuit would put all of that resistance in the noise path. As an example, **Figs 4.6 and 4.7** compare noise sources in oscillator (or amplifier) input circuits. The FET, provided it is operated well within its frequency range, has a high input impedance. Any input capacitance is absorbed into the tuned circuit fairly directly, since it looks just like a capacitor with at most an ohm or two of input series resistance and perhaps 1 or 2nH of series inductance from the bondwire on the chip. The output impedance at the source, ie in a follower, is $1/g_m$, (plus a small ohmic resistance term) where g_m is the mutual conductance at the operating frequency. As a gain stage, the output impedance of most FETs is very high, so the gain is determined by gm and the load impedance. An exception to this

is the GaAs MESFET, where the output impedance is typically a few hundred ohms, so there is a serious limit to the gain available per stage. Some silicon junction FETs (JFETs) have low g_m, so it is worth choosing the device carefully for the frequency used. GaAsFETs tend to have much higher g_m, especially at high frequencies, and so are recommended for microwave work, with the proviso that they are rather prone to $1/f$ noise. This will be described below.

Turning now to the bipolar device, it can be seen that the input impedance is inherently low. At low frequencies, the input impedance is approximately h_{fe} (the AC current gain) multiplied by the emitter resistance ($R_e + r_e$). This term is comprised of R_e, the $1/g_m$ term as in the FET, and r_e, the ohmic series resistance.

This latter can be several ohms, while the former is determined by the current through the transistor; at room temperature, $R_e = 26/I_e$, where I_e is the emitter current in milliamps. Thus at, say, 10mA, the total emitter resistance of a transistor may be an ohm or two, and the input impedance say 50 to 100Ω. This will severely degrade the Q of most parallel-tuned circuits. There is no advantage in using a smaller emitter current, since this leads to lower F_t and hence lower gain; as the frequency approaches F_t, the AC current gain is degraded to unity by definition at F_t. For completeness, the bipolar noise sources are included. In this respect, the best silicon bipolar transistors compare roughly equally with the best silicon FETs. GaAs devices tend to be better again, especially above 1GHz, but the effect of $1/f$ noise has to be considered, so that in an oscillator (but not an amplifier) a silicon device has many advantages right into the microwave region. GaAs devices only predominate because most of them are designed for amplifier service; if sub-0.5 micron geometry, silicon discrete FETs became commercially available, they could become the mainstay of oscillators to 10GHz and beyond.

So what is this $1/f$ noise, and how does noise affect an oscillator? Noise is more familiarly the province of the low-noise amplifier builders, but the principles are the same for oscillators. Where noise comes into the receiver context is that if the signal to be generated is the local oscillator (LO) in a receiver, then most or all of the local oscillator noise is modulated onto the wanted signal at the intermediate frequency (IF). There are several mechanisms for this, including the straightforward modulation of the wanted signal, and the intrusion in the IF of reciprocal mixing (see the chapter on HF receivers), where a strong

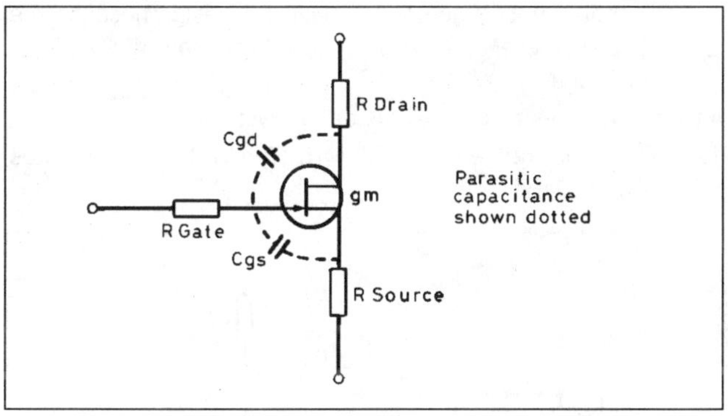

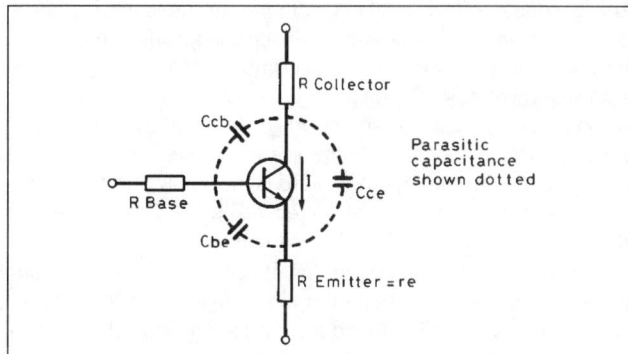

Fig 4.6: FET input characteristics. Input resistance is very high. Input capacitance is approximately equal to C_{gs} + (voltage gain) C_{gd}. Input equivalent noise resistance is approximately equal to R_{GATE} + R_{SOURCE} + $1/g_m$. Typically R_{GATE} is less than 10Ω and R_{SOURCE} is approximately equal to R_{DRAIN} and less than 100Ω. Power devices may be less than 1Ω

Fig 4.7: Bipolar input characteristics. Input resistance is approximately R_{BASE} + h_{fe} x (r_e + R_e) where R_e = $1/g_m$ and g_m = (q/kT) x I. Input capacitance is very approximately equal to C_{be} + V_{gain} x C_{cb} but transistor effects will usually increase this dynamically. Input equivalent noise resistance (R_{IN}) is equal to R_{BASE} + (R_e + r_e)/2. Typically, for a small device, R_{BASE} is about 100Ω, R_e is 26Ω at 1mA, and r_e is 3Ω. Therefore R_{IN} is approximately 129Ω

but unwanted signal, which is close to the wanted one within the front-end pass-band, mixes with LO noise. A low-noise oscillator is therefore a major contributor to a low-noise receiver. On transmit, the effects may not be so obvious to the operator, but the transmission of noisy sidebands at potentially high powers is inconvenient to other band users, and may in extreme cases cause transmissions outside the band. In practice, an oscillator with low enough noise for reception is unlikely to be a problem on transmit.

All devices, active and passive, generate noise when not at absolute zero temperature. Inductors and capacitors only do so through their non-ideal resistive terms, so can be neglected in all practical cases. Resistors and transistors (FET and bipolar) are the real sources of noise in the circuit. The noise power generated by a resistor is:

$$\text{Noise power} = kTB$$

where k is Boltzmann's constant, a fundamental constant in the laws of physics; T is the absolute temperature (room temperature is usually approximated to 300K, ie 27°C); and B is the measurement bandwidth.

This equation is sufficiently accurate for most modern resistors like metal film and modern carbon film resistors. Note that some older types, notably carbon-composition types, do generate additional noise. Active devices are less so, and an equivalent noise resistance (R_{IN} in Fig 4.6 and 4.7) can be derived or measured for any active device which approximates the device to a resistor.

This may have a value close to the metallic resistance around the circuit, or it may be greater. For many applications, this resistance can be used to estimate the noise contribution of the whole circuit. However, there are other noise sources in the device, most noticeably in the oscillator context those due to $1/f$. Fig 4.1(b) shows $1/f$ noise diagrammatically. Noise sources of this type have been studied for many years in many different types of device. Classically, this noise source increases as frequency is reduced. It was thought that there had to be a turnover somewhere in frequency, but this need not be the case.

If we take a DC power supply, it has a voltage at the output, but it has not always been so. There was a time before the supply was switched on, so the 'noise' is infinite, ie $1/f$ holds good at least in qualitative terms. This gives rise to the concept of noise within a finite bandwidth, usually 1Hz for specification purposes. Practical measurements are made in a sensible bandwidth, say a few kilohertz, and then scaled appropriately to 1Hz. The goodness of an oscillator is therefore measured in noise power per hertz of bandwidth at a specified offset frequency from the carrier. A good crystal oscillator at 10MHz would show a noise level of -130dBc (decibels below carrier) at 10kHz offset. A top-class professional source might be -150dBc. There are several possible causes of $1/f$ noise and, when examining the output of an oscillator closely, there is invariably a region close to the carrier where the noise increases as $1/f$. This is clearly a modulation effect, but where does it come from?

The answer is that most (but not all) oscillators are very non-linear circuits. This is inherent in the design. Since it is necessary to have a net gain around the oscillation loop (which is the gain of gain stage(s) divided by the loss of feedback path), then the signal in the oscillator must grow. Suppose the net gain is 2. Then the signal after one pass round the loop is twice what it was, after a further pass four times and so on. After many passes, the amplitude should be 'infinite'. There clearly must be a practical limit. This is usually provided by voltage limitations due to power supply rails, or gain reduction due to overdrive and saturation of the gain stage(s). Occasionally, in a badly designed

circuit, component breakdown can occur where breakdown would not be present under DC conditions. This can be disastrous as in a burn-out or, more sinister, as in base-emitter breakdown of bipolar transistors, which will lead to permanent reduction of transistor gain and eventual device failure in service.

So, almost all practical oscillators are highly non-linear and have sources of noise, be they wide-band (white noise) or $1/f$ (pink noise). The non-linearity thus modulates the noise onto the output frequency. In contrast, a properly designed amplifier is not non-linear, has very little modulation process, and is therefore only subject to additive noise.

The exceptions to this are the very few oscillators which have a soft limitation in the output amplitudes, ie they are inherently sine wave producers. This requires very fast-acting AGC circuitry, and is very rarely used because it adds much cost and complication. Most practical oscillators are inherently square-wave generators but a tuned circuit in the output will normally remove most harmonic energy so that the output looks like a sine wave on an oscilloscope. A spectrum analyser normally gives the game away, with harmonics clearly visible. The DDS devices are an exception to all this for reasons explained below.

Generally, a silicon device will have very much lower levels of $1/f$ noise than a gallium arsenide device. This is because silicon is a very homogeneous material, of very high purity, and in modern processes with very little surface contamination or surface states. Gallium arsenide is a heterogeneous material which is subject to surface states. The $1/f$ knee is the frequency below which the noise levels associated with the surface states start to increase, typically at 6dB/octave of frequency. In, say, 1980 this frequency would have been 100MHz or more for most gallium arsenide processes. In 1990 the figure was as low as 5MHz in some processes. Silicon typically has a $1/f$ knee of less than 100Hz, sometimes less than 10Hz, so $1/f$ noise is unimportant for most practical cases. It means that there is a possibility of increased close-to-carrier noise in GaAs-based oscillators. Whether this makes the oscillator better or worse than a silicon design depends on the exact choice of devices and circuits. More information on these topics can be found in [5].

PRACTICAL VFOS

A VFO design should start with a set of clear objectives, such as

- The required frequency tuning range.
- The exact means of tuning.
- The requirement for frequency stability.
- The RF power output required from the oscillator generally should not influence the oscillator design because the RF power level can easily be raised using a buffer / amplifier stage.

A typical VFO design objective is as follows:

1. Frequency range 5.0 to 5.5 MHz. The VFO therefore tunes over 500kHz

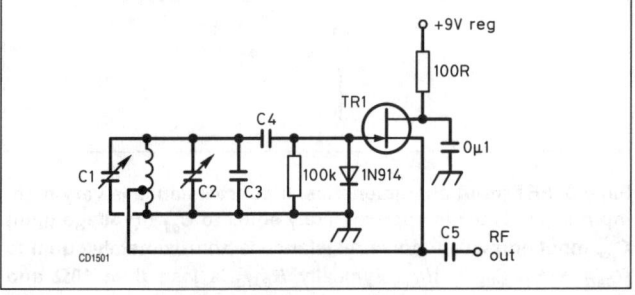

Fig 4.8: Hartley VFO (*W1FB's QRP Notebook*)

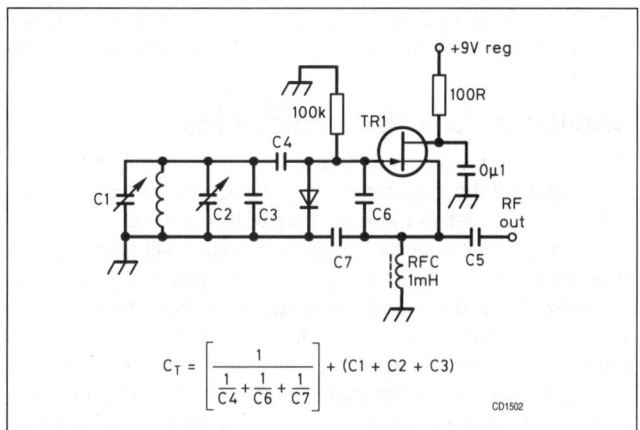

$$C_T = \left[\frac{1}{\frac{1}{C4} + \frac{1}{C6} + \frac{1}{C7}} \right] + (C1 + C2 + C3)$$

Fig 4.9: Colpitts VFO (*W1FB's QRP Notebook*)

2. The VFO is required to be tuned by a variable capacitor.

3. The VFO is required to tune an SSB / CW transceiver. Therefore this requires that the frequency should not drift by more than 100Hz in an hour, and should always be repeatable to within 200Hz.

4. RF Power output needs to be 10mW.

The third requirement is defined by the fact that an SSB receiver needs to be tuned to within about 50Hz of the correct frequency. This puts a difficult requirement on VFO stability because small changes in temperature caused by, for example switching on a heater in the room, can cause considerably more change than 50Hz.

Figs 4.8 to 4.10 show a range of practical VFO circuits, which include the DC bias components. In each circuit, C1 is the main tuning capacitor and C2 provides bandspread. The circuit in Fig 4.8 is a Hartley oscillator. The feedback tap should attach at about 25% of the coil from the earthy end; actually, the lower down the better, consistent with easy starting over the whole tuning range. A bipolar transistor could be used in this circuit with bias modification, but it will tend to reduce circuit Q and hence be less stable than a FET. The FET should be chosen for low noise, and JFETs are usually preferred. However this circuit will also work well with modern dual-gate MOSFETs. Its second gate should be connected to a bias point at about 4V. Fig 4.9 shows the Colpitts version of the circuit. The additional input capacitance restricts the available tuning range but this is rarely significant. Again, a dual-gate MOSFET version is possible. A voltage-tuned version of the Hartley is shown in Fig 4.10. This is actually a VCO (see later) whose tuning voltage is derived from a potentiometer, so to the operator it operates like a VFO.

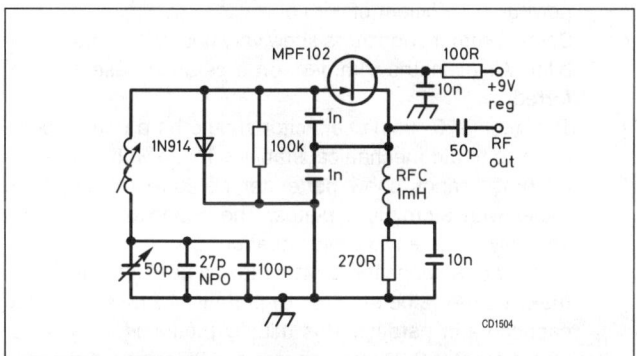

Fig 4.11: A JFET series-tuned Clapp VFO for 1.8MHz (*ARRL Electronic Data Book*)

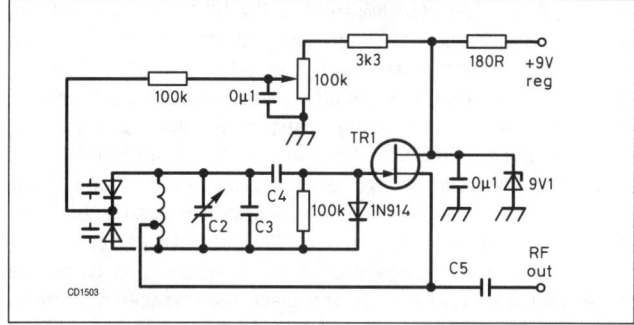

Fig 4.10: Varactor-tuned Hartley VFO. Suggested values for reactances are: X_{C1} 1200Ω; X_{C2} 3500Ω; X_{C3}, X_{C4}, X_{C5} 880Ω; X_{C6}, X_{C7} 90Ω; X_{L1} 50Ω (*W1FB's QRP Notebook*)

Many other variants on these basic circuits have been published over the years; some examples are shown in **Figs 4.11**, a Clapp oscillator and **4.12**, a Colpitts oscillator using a dual-gate MOSFET as the active device. In Fig 4.12, the biasing of the dual-gate MOSFET in oscillator service is typical; G1 is at source DC voltage level and G2 at 25% of the drain voltage.

As well as the Colpitts. Clapp and Hartley oscillator, other designs described in this chapter are the Franklin and Vackar oscillators.

Fig 4.13 illustrates a Franklin oscillator. This uses two active devices which in this example are GaAs MESFETs. As mentioned above, the MESFET tends to have relatively low gain per stage, so there is an advantage in using two gain stages in series to provide positive gain over a very wide frequency range. The devices are shown running with self-bias on the gates. Ideally, they should be run close to I_{dss}, where the gain is usually greatest. The feedback path is arranged to give 360° phase shift, ie two device inversions. The feedback is then directed into the tuned circuit.

The output impedance of the MESFET is relatively low, but it is prevented from damping the resonant circuit by very light cou-

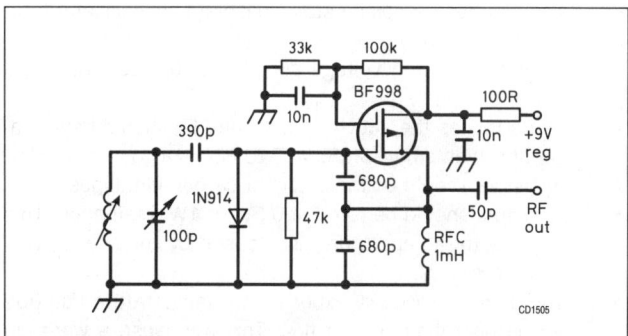

Fig 4.12: A dual-gate MOSFET in a common-drain Colpitts circuit (*ARRL Electronic Data Book*)

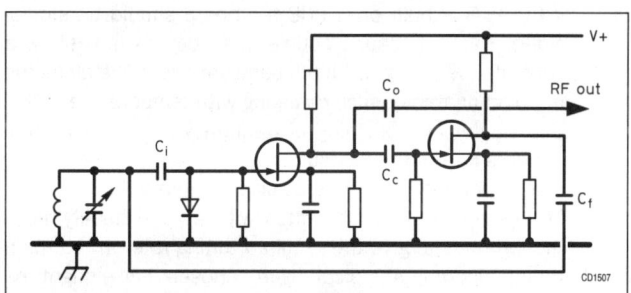

Fig 4.13: The Franklin oscillator

pling; C_i and C_f typically are 1-2pF in the HF region, smaller still at higher frequencies. The high gain ensures oscillation even in this lightly coupled state, so that very high circuit Q is maintained, and the oscillation is stable and relatively noise-free. A particular feature of this configuration is the insensitivity of the circuit to supply variation; but the supply should still be well decoupled. A disadvantage of the circuit is the total delay through two stages, which limits the upper frequency, but with gallium arsenide devices this can still yield 5GHz oscillators.

GaAsFETs are relatively expensive, so no specific types are suggested. Surplus 'red spot' and 'black spot' devices have been successfully tried in this circuit. A useful source of GaAsFETs is old satellite television LNBs.

DESIGNING / BUILDING VFOs

The following are recommendations for building variable frequency oscillators:

- For best stability, operate the VFO within the range 5 - 10MHz.
- Build the VFO in a sturdy box, eg a die-cast box. This will provide mechanical stability.
- Screen the oscillator to prevent the transfer of energy to/from other electronic circuits nearby. The screening box mentioned above will achieve this, but note that a box will not provide screening against magnetic fields.
- The VFO should be positioned away from circuits which generate high magnetic fields eg transformers.
- Protect the circuit from radiant and convected heat sources. Placing the VFO in a box will achieve this, but minimise the number of heat sources inside this box. For example, voltage regulators should be placed outside the box.
- Avoid draughts across the tuned circuit.
- Operate the VFO at a low power level. This will minimise the heating effect of RF currents in the oscillator components.
- Provide adequate decoupling.
- Use a buffer / amplifier stage to isolate the oscillator from the load.
- Provide regulated voltage supplies to the oscillator and buffer amplifier.
- Use a FET as the active device. The FET should have reasonably high gm (> 5mS). A J310 FET is a good example.
- Resistors should be metal film or carbon film types.
- Resistors should be rated at 0.5W or 1W dissipation. This reduces the temperature rise caused by currents flowing in the resistors.
- The active device will experience a temperature rise due to the operating current flow. This will cause a warm-up drift in the oscillator frequency. For this reason a conventional leaded package should be used, not a surface mounted package, and a small heat sink should be fitted.
- If the VFO is built on a PCB the board should be single-sided. This is because double-sided boards may show a capacitance change with temperature due to the dielectric material characteristics changing with temperature.
- Use single-point earthing of frequency-determining components in the tuned circuit.
- All components should be as clean as possible.
- The oscillator should be fitted with a good quality slow-motion drive to give the required tuning rate on the main tuning control. A typical figure chosen here might be 5kHz/turn of the tuning control for SSB mode.
- The receiver/transceiver operating frequency is most eas-

ily indicated using a digital counter/display. This should have a resolution of 0.1kHz or better, to allow easy re-setting to a certain frequency.

Oscillator Tuning Components.

The oscillation frequency of an oscillator is primarily determined by the value of the inductor and capacitor(s) in the tuned circuit. Undesired changes in the reactance of these components (due to temperature change or some other factor) will have a direct effect on the oscillator frequency. It can be easily shown mathematically that if the inductor reactance changes by +100ppm then the oscillator frequency will move by -50ppm. A similar result is obtained if the capacitor reactance changes by +100ppm. This shows that there is a direct relationship between component value and oscillator frequency. For example, consider a 5MHz oscillator which employs a silvered mica capacitor with a temperature coefficient of +100ppm/°C. The capacitor would cause the oscillator to drift by -250Hz/°C. This is significant drift, and if the VFO was used in an SSB receiver, would require frequent retuning as the room temperature changed.

All components in the tuned circuit chosen should therefore be as stable as possible, and this is normally achieved by making the right choice of the type of component for each capacitor and inductor. A single poor quality component in an otherwise good oscillator design may ruin oscillator stability.

Capacitors

The following is a description of various capacitor types and their suitability for oscillator use.

- X7R dielectric ceramic capacitors must be avoided. These have low Q and are microphonic. They also exhibit capacitance change with applied voltage.
- COG dielectric ceramic capacitors are significantly better than X7R. They have higher Q and better temperature stability. These are available in surface-mounted (leadless) form and these are recommended. This is because they can be fitted directly onto a PCB and are then mechanically very stable.
- Silvered Mica. These are excellent for oscillators. They have high Q, and are temperature stable. However, the quantities used in the electronics industry are declining and they are becoming difficult to obtain. Typical temperature coefficient is -20 to +100 ppm/°C
- Polystyrene. These have a high Q at lower radio frequencies and are temperature stable. These are also becoming difficult to obtain. Typical temperature coefficient is typically -150 ppm/°C
- Porcelain capacitors. These are specialised capacitors, which are designed for use in high power microwave amplifiers. These offer very high Q and have typical temperature coefficient of +90 ppm/°C.
- Some trimmer capacitors show very poor temperature stability. Air dielectric trimmers on a ceramic base are preferred.
- The main VFO tuning capacitor should be a double bearing type to aid mechanical stability. Those with silver-plated brass vanes show better temperature stability than those with aluminium plates. The capacitor should be securely mounted to avoid vibration.
- If the Q of a capacitor is rather low, a useful technique is make up the value required by putting several small value capacitors in parallel. This usually produces a capacitor with a greater Q. It also reduces the RF current flowing in each capacitor, which reduces the heating effect of that current.

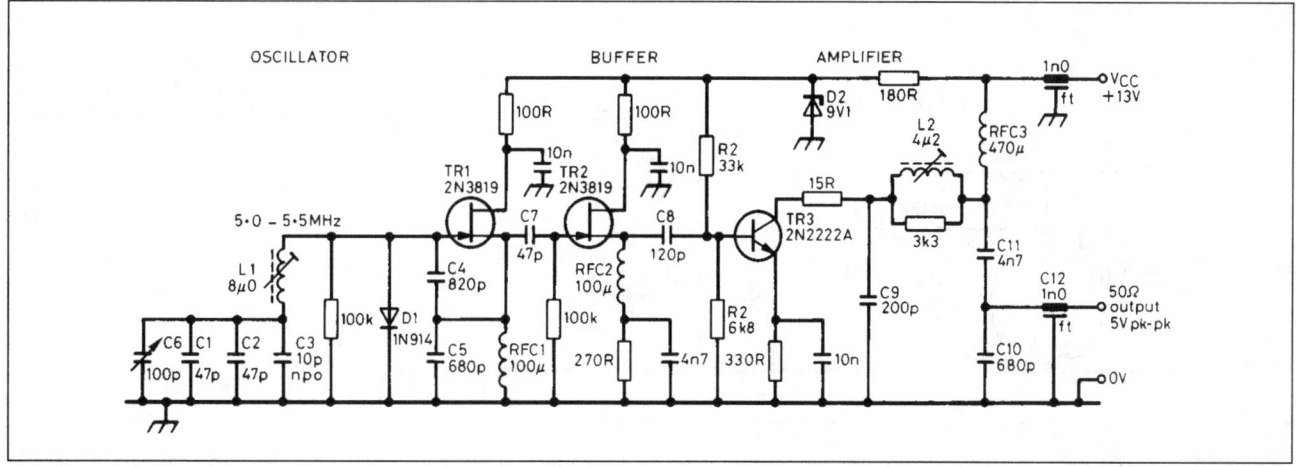

Fig 4.14: Clapp VFO with amplifiers for 5.0-5.5MHz. Reactance values are L1 = 265Ω, L2 = 140Ω, C1 = 690Ω, C2 = 690Ω, C3 = 2275Ω, C4 = 33Ω, C5 = 48Ω, C6 = 303Ω min, C7 = 690Ω, C8 = 227Ω, C9 = 152Ω, C10 = 48Ω, C11 = 4.5Ω, C12 = 23Ω, RFC1 = 4400Ω, RFC2 = 4400Ω

Inductors

If the guidelines (above) about capacitors are followed, the Q of an HF oscillator coil will usually be less than that of the capacitors. Therefore, the coil is the limiting factor on oscillator Q.

- The oscillator Q can usually be increased by raising the Q of the coil, and this can be done by winding the coil from Litz wire.
- If the oscillator is operated on VHF, coil Q is usually high (typically 400) and the tuned circuit Q may be limited by the capacitors.
- Ceramic or fused silica formers are preferred for coil formers.
- Plastic formers should avoided unless there is no other alternative.
- If the coil is fitted with an adjustable screw slug the slug should be of a low permeability type. The number of turns on the coil should be adjusted so that the final adjustment position of the slug is just into one end of the coil.
- Toko coils are sometimes used in projects which must be easily reproducible. These are slug-tuned coils, and do not provide the best stability, but are commercially available as a ready-made part.
- If the coil is wound on a toroidal core the type of core material should be chosen carefully because some core materials have a high temperature coefficient of permeability. The resulting oscillator would show very high temperature drift. Powdered iron cores are more stable than ferrite.

Many of the specialised parts mentioned above may be obtained at radio rallies.

TWO PRACTICAL VFOS

Gouriet-Clapp Oscillator

The series-tuned Colpitts oscillator, often referred to as the Clapp or Gouriet-Clapp (devised by G C Gouriet of the BBC), has been a favourite with designers for many years. The series-tuned oscillator enables a higher value of inductance to be used than would normally be required for a parallel-tuned design, resulting in claims of improved stability. The circuit of a practical VFO, **Fig 4.14**, has component values selected for a nominal operating frequency of 5MHz, but reactance values are tabulated for frequency-determining components to allow calculation of opti-

mum values for operation on other frequencies in the range 1.8 to around 10MHz. Simply substitute the desired VFO frequency into the reactance formula and the required L and C values can be deduced. Round off the calculated value to the nearest preferred value.

L1 should be wound on a ceramic low-loss former and the inductance is made variable by fitting a low-permeability, powdered-iron core. The series capacitors C1, C2 and C3 are the most critical components in the circuit; they carry high RF currents and the use of three capacitors effectively decreases the current through each one. A single capacitor in this location may cause frequency-jumping brought about by dielectric stress as a result of heat generated by the RF current. C3 is selected to counteract the drift of the circuit. In most circuits the tuning capacitor C6 will be located in series with the inductor and in parallel with C1-C3; a trimmer capacitor may also be placed in parallel with the main tuning capacitor. By placing the tuning capacitor in parallel with C5, a smaller tuning range is possible. This may be desirable for a 40m VFO where the frequency range required is only 200kHz.

The active device is a JFET and another JFET, TR2, acts as a buffer amplifier. This is lightly coupled to the source of TR1. Both transistors operate from a zener-stabilised 9V power supply. TR3 is a voltage amplifier operating in Class A and uses a bipolar device such as the 2N2222A. Bias for Class A operation is established by the combination of the emitter resistor and the two base bias resistors which are also connected to the stabilised 9V supply. The collector impedance of TR3 is transformed to 50Ω using a pi-network which also acts as a low-pass filter to attenuate harmonics. The loaded Q of the output network is reduced to approximately 4 by adding a parallel resistor across the inductor, and this broadens the bandwidth of the amplifier. For optimum output the network must be tuned either by adjusting L2 or by altering the value of C9. The 15Ω series resistor is included to aid stability.

The VFO described in Fig 4.14 is capable of driving a solid-state CW transmitter, and it is ideally suited to provide the local oscillator drive for a diode ring mixer. It also has sufficient output to drive a valve amplifier chain in a hybrid design.

Vackar Oscillator

A notable example of the Vackar oscillator was a design by P G Martin, G3PDM (**Fig 4.15**) which is described more fully in [6]. This design has adjustable temperature-compensation and

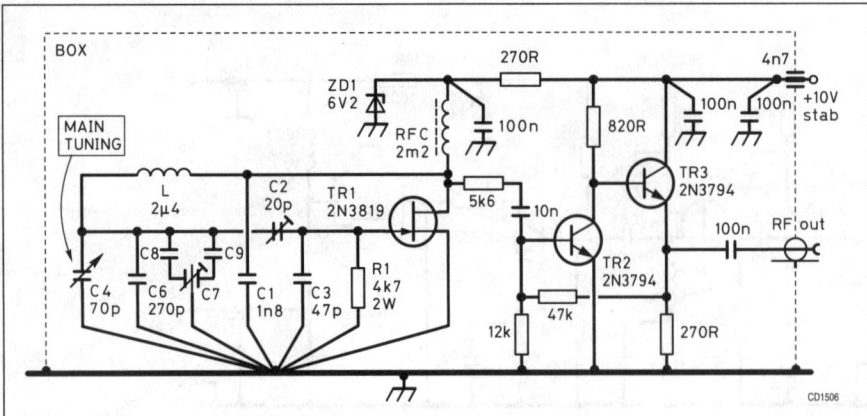

Fig 4.15: This Vackar oscillator covering 5.88-6.93MHz and two-stage buffering is based on a G3PDM design. A temperature compensation scheme consisting of C7, C8 and C9 is described in the text

employed most of the measures described earlier to produce a stable oscillator. The design used a silver-plated copper wire for the tank coil (wound on a ceramic former), and included a two-transistor buffer stage.

The temperature compensation components are the 100 + 100pF differential air-spaced preset capacitor C7, the 100pF ceramic C8 with negative temperature coefficient and the 100pF ceramic C9 with positive temperature coefficient. Turning C7 effects a continuous variation of temperature coefficient of the combination between negative and positive with little variation of the total capacitance of about 67pF. Differential capacitors are hard to find but can be constructed by ganging two singles with semicircular plates, offset by 180 degrees. Adjustment for, and confirmation of, zero temperature coefficient of the whole oscillator is a time-consuming business. An updated version of this circuit would probably use a dual-gate MOSFET, which should offer marginally better performance. The buffer stages could use more easily obtainable devices such as the 2N3904.

VFOs are generally not used commercially because every VFO needs setting-up. This is costly and it is cheaper to use a synthesiser system, even if the synthesiser is fairly complex. However, for the amateur, building and adjusting a VFO for an SSB receiver/transceiver is a viable option to obtain variable frequency operation. It requires a significant amount of time and effort to build the VFO and adjust the temperature compensation.

EVALUATION OF A NEWLY-BUILT VFO

It is essential that a VFO is properly tested for frequency stability before being used, particularly if it is to be used for transmitting. The process of testing the VFO is described below.

The oscillator should be located in the chassis in which it will finally be used and the output connected to a dummy load. The normal supply voltages supplied to the oscillator and buffer should be applied and the total current consumption checked. This should be no more than 30mA.

Check that the circuit is oscillating by looking at the output on an oscilloscope, checking the frequency on a frequency counter, or listening on a receiver tuned to the expected frequency of oscillation. You may have to tune around on the receiver slightly because the oscillator probably won't be on the exact frequency you expected.

Adjust the oscillator to the required frequency range. This is normally done by adjusting the value of the inductor with the tuning capacitor set to mid-travel. The inductor is adjusted to set the oscillation frequency to the middle of the required tuning range. The padding capacitors are then adjusted to set the upper and lower frequency range required.

Frequency measurements should then be done to check the temperature stability. Temperature drift can be adjusted by exchanging fixed capacitors for ones having the same capacitance value, but different temperature coefficients (see the chapter on passive components).

VOLTAGE-CONTROLLED OSCILLATORS

These are a class of oscillator which are similar to a VFO. However a tuning capacitor or tuning inductor is not used and the oscillation frequency is controlled by an applied voltage. This is achieved by applying the control voltage to a component (or components) in the oscillator tuned circuit whose reactance varies with the applied voltage. The usual element used for this is a tuning diode. This is a junction diode which is operated in reverse bias, and exhibits a capacitance which varies with the applied bias voltage. Diodes are available which are designed specifically for operation as a tuning diode. These are optimised for wide capacitance variation with applied voltage, and good Q. Other means of electronic tuning are possible in some circuits, but the use of tuning diodes is most common. The important parameter of the VCO is its tuning rate in MHz/volt and a typical design figure might be 1MHz/volt. VCO design is not without problems, particularly if the design needs a high tuning rate.

A VCO is normally used in a closed-loop control system. The system is designed to control the frequency of oscillation by means of the DC input control voltage. The requirements for short term stability (ie phase noise) are similar to VFOs. Therefore, most of the design rules presented above are equally applicable for VCOs as for VFOs. There is one exception to this: Because of the closed-loop nature of the VCO control system, the design rules for long-term stability of the oscillator no longer apply or can be significantly relaxed.

VCOs can be built for virtually any frequency range from LF to microwave.

CRYSTAL OSCILLATORS

A crystal oscillator is a type of oscillator which is designed to operate on one frequency only. The frequency of oscillation is determined by a quartz crystal and the crystal must be manufactured specifically for the frequency required. Quartz shows the property of the Piezo-electric effect and can be made to oscillate at radio frequencies. The main virtue of quartz is that it has a very low temperature coefficient so the resulting oscillator will be very temperature stable. It also has a very high Q. The types of crystal are similar to the ones used in quartz crystal filters.

Crystal oscillators exhibit the following characteristics.

- Good temperature stability.
- The high Q gives good oscillator phase noise.

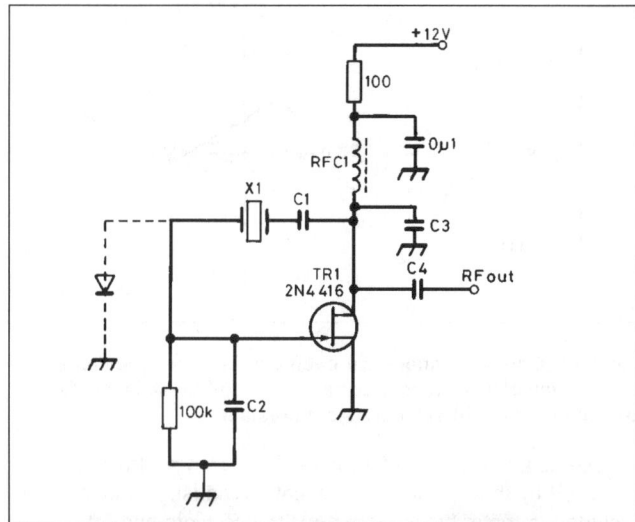

Fig 4.16: Pierce oscillator. Reactance values are: X_{C1}, X_{C3} 230Ω; X_{C2}, X_{C4} 450Ω. (*W1FB's QRP Notebook*)

They are used where an accurate frequency source is needed. Examples of this are in electronic instruments, frequency synthesisers and RF signal generators.

For a description of the equivalent electrical circuit of a crystal see the next building blocks chapter.

Crystal oscillators employ one of two modes of oscillation:

- Fundamental mode, which is normally used up to about 24MHz.
- Overtone mode, which is normally used above about 24MHz.

There really are no user serviceable parts inside a crystal, but it is interesting to open up an old one carefully to examine the construction. The crystal plate is held in fairly delicate metal springs to avoid shocks to the crystal. The higher-frequency crystals are very delicate indeed.

In the last few years, crystals have become available which are intended to be operated in fundamental mode up to about 250MHz. They are manufactured using ion beam milling techniques to produce local areas of thinning in otherwise relatively robust crystal blanks. This maintains some strength while achieving previously unattainable frequencies. These are intended for professional applications and are generally too expensive for the amateur, and not easily available.

Crystals intended for fundamental-mode operation are generally specified at their anti-resonant or 'parallel' frequency with a specific 'load' capacity, most often between 12 and 30pF. When a crystal is operated in fundamental mode, it is possible to tune, or trim, the oscillation frequency slightly by adjustment of a reac-

tance in the oscillator circuit. This is usually done with a variable capacitor and is done to adjust for the initial tolerance of the crystal frequency and can be done subsequently to adjust the frequency as the crystal ages.

When a crystal is operated in the overtone mode, it is in a resonant, series mode. Overtones available are always odd integers and the normal overtones used are 3rd and 5th. Overtone operation allows an oscillator to operate up to about 130MHz (5th overtone). A trimmer capacitor can be used to effect a slight adjustment of the frequency but the ability to adjust the frequency is much less than with the fundamental mode of operation.

Crystal Oscillator Circuits

The operation of a crystal oscillator is basically the same as the principles of any oscillator (VFO or VCO) described so far in this chapter. The best option for the amateur is to follow one of the circuits described here. Some crystal oscillator circuits do not need to use inductors, so the Hartley circuit is less common than the Colpitts type of circuit. Most circuits use a single active device.

For fundamental-mode crystal oscillators, any small HF transistor is adequate, even a BC109. For overtone oscillators any RF type is recommended, eg NPN transistor types 2N2222A (metal can), BF494 or 2N3904 (plastic); N-channel plastic JFETs include J310 and MPF102 and 2N3819. Virtually any dual-gate MOSFETs can also be used.

Excessive feedback should be avoided, primarily because it tends to reduce stability. Excessive oscillation levels at the crystal can lead to damage, but this is very unlikely in circuits operating on a supply of 12V or less. Sufficient feedback is needed to guarantee starting of the oscillator. Some experimentation may be needed, especially if the crystal is of poor quality. Pressure-mounted crystals and those in epoxy-sealed aluminium cans, eg colour-burst crystals from early colour television sets, are often poor.

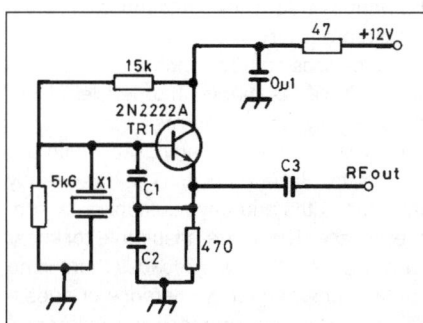

Fig 4.18: Colpitts oscillator. Reactance values are: X_{C1}, X_{C2}, X_{C3} 450Ω.. (*W1FB's QRP Notebook*)

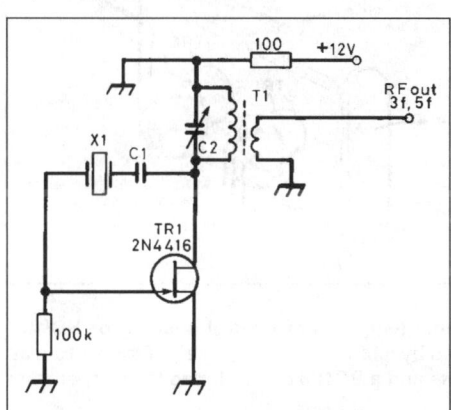

Fig 4.17: Overtone oscillator. Reactance values are: X_{C1} 22Ω., X_{C2} 150Ω. at resonance. (*W1FB's QRP Notebook*)

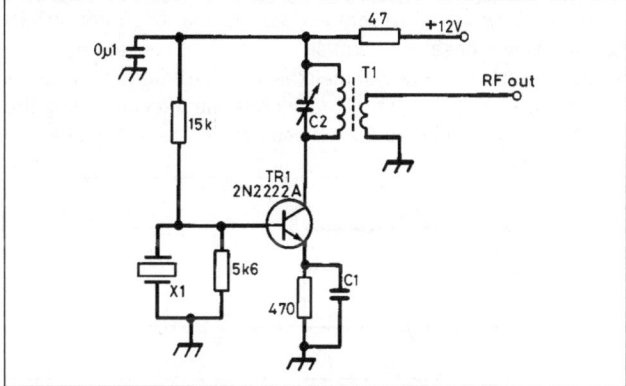

Fig 4.19: The tuning of the LC collector circuit determines the mode of oscillation: fundamental or overtone

This section describes a range of fixed-frequency crystal oscillators with practical circuit details. Crystal oscillator circuits are not hard to build, and most reasonable designs work fairly well since the crystal itself provides extremely high Q, and in such a way that it is difficult to degrade the Q without deliberately setting out to do so.

Figs 4.16 to 4.19 show four basic crystal oscillator circuits. Fig 4.16 is a Pierce oscillator using a FET as the active device. The circuit operates as a fundamental oscillator and the crystal is connected in the feedback path from drain to gate. The only frequency setting element is the crystal. For this to be the case, the RFC must tune with the circuit parasitic capacitances to a frequency lower than the crystal. Most small JFETs will work satisfactorily in this circuit. Note that the FET source is grounded, which gives maximum gain.

Fig 4.17 shows an overtone version of Fig 4.16. In this, a tuned circuit is inserted in the drain of the FET. The tuned circuit is tuned to the overtone frequency required and hence the FET only has gain at that frequency. This causes the circuit to operate only on the wanted overtone.

Fig 4.18 shows a Colpitts fundamental oscillator circuit, which uses a bipolar transistor. The chief advantage of this configuration is that the crystal is grounded at one end. This facilitates switching between crystals, which is much less easily achieved in the Pierce circuits.

Fig 4.19 shows an overtone oscillator using a bipolar transistor. One side of the crystal is earthed, which makes switching crystals easier. However, switching crystals in an overtone oscillators is not recommended, since the stray reactance associated with the switching can lead to loss of stability or moding of the crystal. This causes some crystals to oscillate at an unintended overtone, or on another frequency altogether. If switching of overtone crystals is required, it is better to build a separate oscillator for each crystal and then switch oscillators.

Oscillators can also be built using an integrated circuit as the active device. **Fig 4.20** shows an oscillator using a standard CMOS gate package. Almost any of the CMOS logic families will work in this circuit. The primary frequency limitation is the crystal. This oscillator will generate a very square output waveform, and with the faster CMOS devices can be a rich source of harmonics to be used, for example, as marker frequencies. Note that a square wave contains only odd harmonics.

Self-contained crystal oscillators can be purchased which are intended for use as clocks for digital circuits. They frequently contain the crystal plate, the gates and any resistors on a substrate in a hermetic metal case. These are unsuitable for most amateur receiver or transmitter applications because there is no way of tuning out manufacturing frequency tolerance or subsequent ageing. Also, the phase noise performance is usually poor.

Where an oscillator with very high temperature stability is required, a temperature-compensated crystal oscillator (TCXO) can be used. These are normally supplied in a sealed package and they were developed to maintain a precise frequency over a wide temperature range for long periods without consuming the stand-by power required for a crystal oven. These are suitable

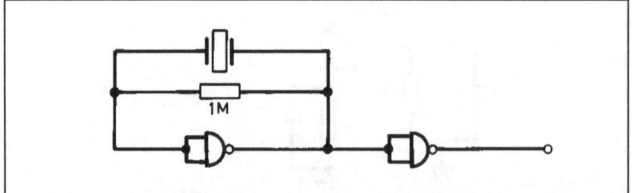

Fig 4.20: CMOS oscillator circuit. With modern 'HC' CMOS, this will oscillate to over 20MHz with a good square-wave output

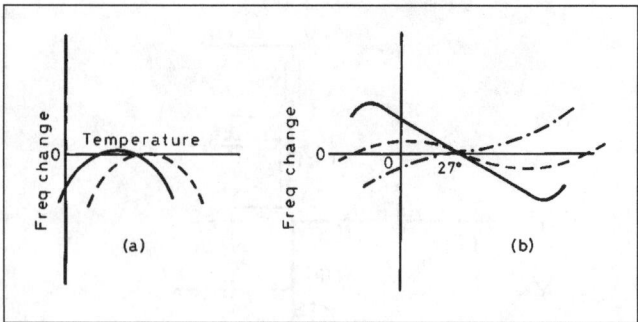

Fig 4.21: Crystal temperature coefficients. Frequency/temperature curves of 'zero-temperature coefficient' crystals. (a) Typical BT-cut crystals. (b) Typical AT-cut crystals

for use as a frequency reference in frequency synthesisers and are used by the mobile communication industry. They are now included in top-of-the-line amateur transceivers and offered as an accessory in the mid-price field.

A TCXO is pre-aged by temperature cycling well above and below its normal operating range. During the last cycle, the frequency-vs-temperature characteristic is computer logged; the computer then calculates a set of temperature compensation values, which are stored in an on-board ROM. A good TCXO might have a tolerance of 1ppm over 0 - 60°C and age no more than 1ppm in the first year after manufacture, less thereafter. There is normally a provision to make small adjustments to the output frequency so that it can be adjusted from time to time.

Buying Crystals

Crystals must be bought for the frequency and mode that you require. Many common frequencies can be bought ready-made (eg exact frequencies like 5MHz or 10MHz). If you need to get a crystal made you will need to specify (as a minimum):

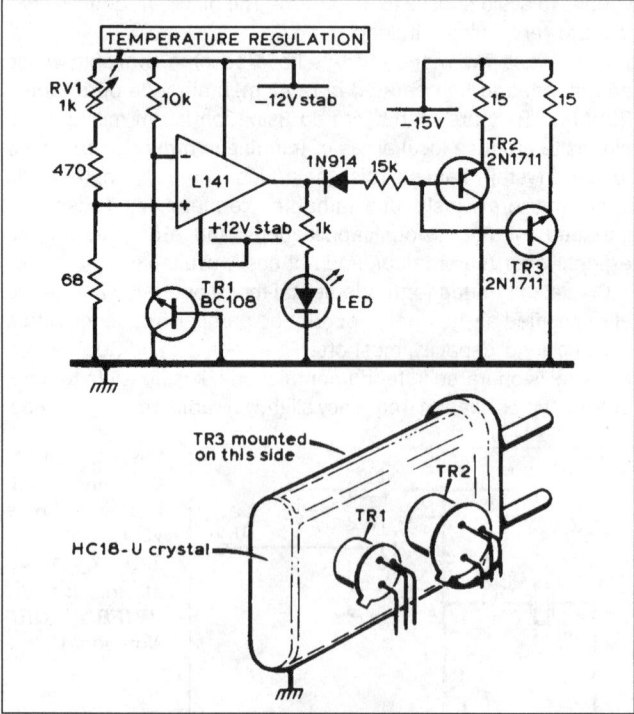

Fig 4.22: Proportional temperature control system for HC/16U crystals as proposed by I6MCF and using a pair of transistors as the heating elements and a BC108 transistor as the temperature sensor

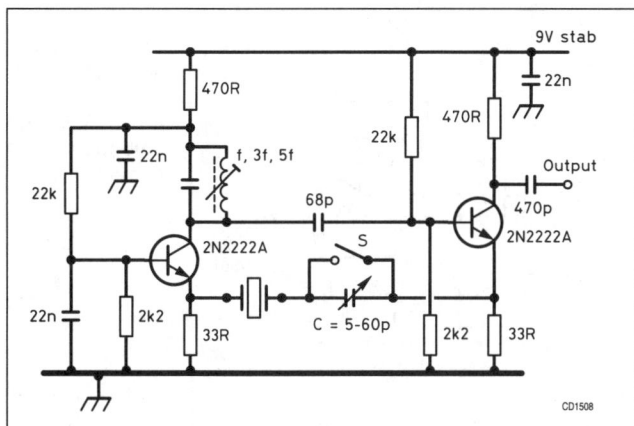

Fig 4.23: A crystal test oscillator using the Butler circuit. With S closed, the resonant frequency is generated. With S open and C set to the specified load capacitance, the anti-resonant frequency can be measured

- Frequency
- Mode
- Loading
- Operating temperature

Crystals are normally available in HC6/U, HC18/U, HC25/U or similar holders. If crystals are bought at a rally, much older, larger packages may be found.

Crystals made for commercial use are similar in specification to amateur devices; they may be additionally aged or, more commonly, ground slightly more precisely to the nominal frequency. For most purposes this is not significant unless multiplication into the microwave region is required.

Occasionally, crystals are specified at a temperature other than room temperature (see **Fig 4.21**). Ideally, the crystal should have a zero temperature coefficient around the operating temperature, so that temperature variations in the equipment or surroundings have little effect. In some cases, the crystal is placed in a crystal oven at 70°C. It is important to note, however, that the best result will only be obtained if the crystal is cut for the oven temperature. Although any temperature stabilisation is useful, a room-temperature-cut crystal may have little advantage in an oven. A scheme for thermal stabilisation of crystals is shown in **Fig 4.22**.

Fig 4.23 shows a Butler oscillator which is a favourite for testing oscillator crystals. With the switch S closed, the circuit will oscillate at the crystal's resonant frequency; the output amplitude is then measured. Next, a 100Ω variable composition resistor is substituted for the crystal and adjusted to produce the same output. Its final resistance value is equal to the crystal resistance, R_s in **Fig 4.24**.

VARIABLE CRYSTAL (VXOS) AND CERAMIC-RESONATOR OSCILLATORS

A VXO is a type of crystal oscillator which is designed to have a much greater frequency adjustment range than a standard crystal oscillator. A typical use for a VXO is to obtain crystal control on an HF amateur band. The frequency can be adjust-

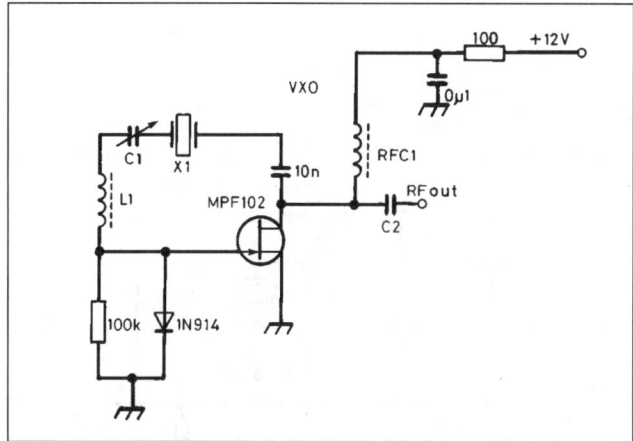

Fig 4.25: A Pierce VXO for fundamental-mode crystals. The diode limits the output amplitude variation over the tuning range. Reactances at the crystal frequency are: C1 = 110Ω, C2 = 450Ω, L = 1.3Ω, RFC = 12.5kΩ

ed by several kilohertz to avoid causing interference to other band users.

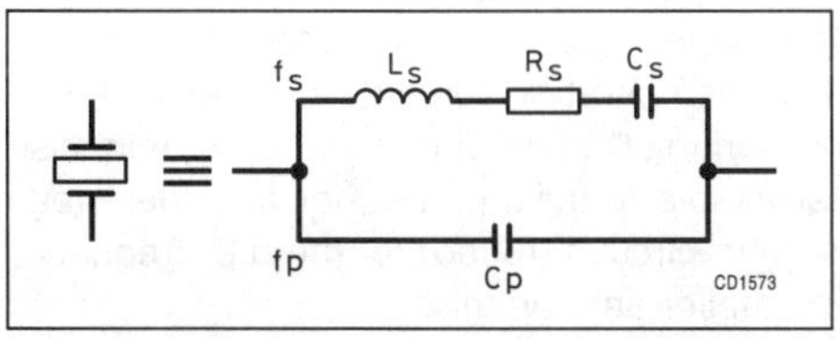

VXOs are used frequently in portable equipment where the virtues of simplicity and stability are particularly useful. Fundamental mode oscillators should be used because these can be pulled in frequency much more than overtone oscillators. The principle can be understood if one replaces the piezo-electric crystal by the electrical equivalent circuit shown in Fig 4.24. If a variable inductor is placed in series with that equivalent circuit, the resonant frequency of the combination can be 'pulled' somewhat below that of the crystal alone; for convenience, tuning is usually carried out with a variable capacitor in series with this coil which is over-dimensioned for that purpose.

When building VXOs the precautions described for VFO design should be used, but VXOs are less sensitive to external influences since the oscillation frequency is controlled primarily by the crystal. The two problems of VXOs are the very limited tuning range, typically of 0.1%, and the tendency to instability if this range is exceeded. A good VXO should be very close to crystal stability, with the added ability to change frequency just a little. Of course, multiplication to higher frequencies is possible, and several of the older types of commercial equipment successfully used VXOs on the 2m band.

Typical VXO circuits are shown in **Figs 4.25 and 4.26**. Fig 4.25 is a Pierce oscillator using an FET as the active device. Stray capacitance in the circuit should be minimised. Ideally, the VXO

Fig 4.24: Crystal equivalent circuit

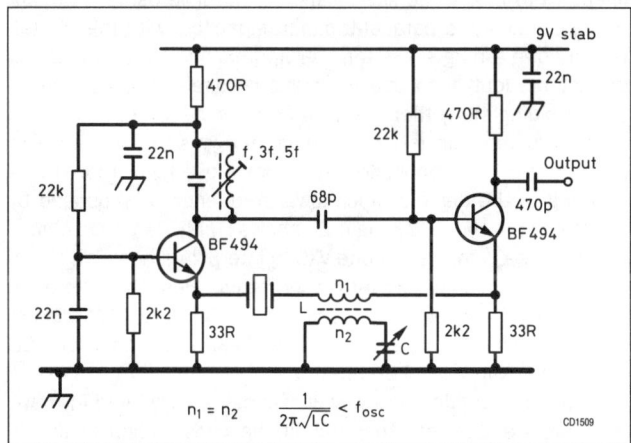

Fig 4.26: A Butler VXO for fundamental or overtone crystals

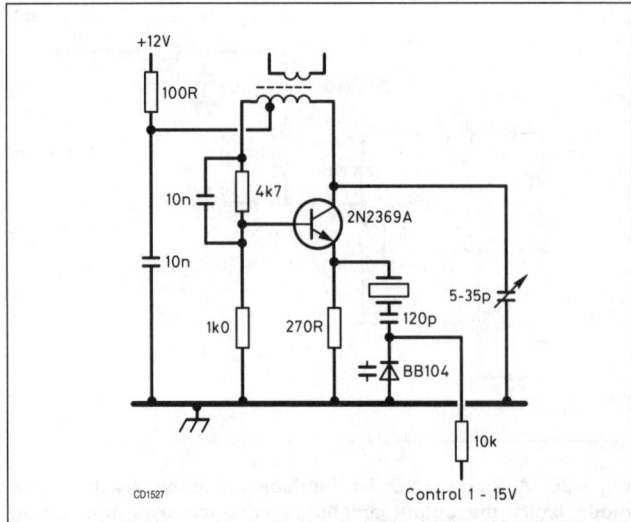

Fig 4.27: Voltage-controlled 21.4 and 20MHz oscillators used by G3MEV in his 'heterodyne' 1.4MHz VXO system

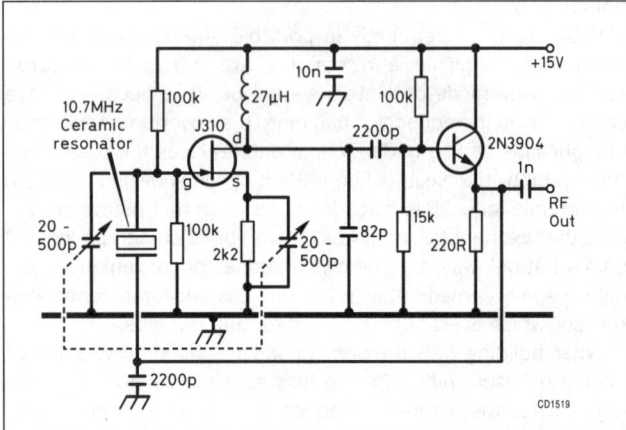

Fig 4.28: K2BLA's variable ceramic-resonator oscillator covers a 2% frequency range

should be able to shift from below to just slightly above the nominal crystal frequency. Fig 4.26 shows a two-transistor Butler oscillator.

Crystal switching in VXOs should be avoided and it is better to switch oscillators. Non-linear operation of capacitance against frequency should be expected, so the tuning scale will be non-linear. Note that some linearisation is possible using combinations of series and parallel trimmer capacitors with the crystal. Instead of a tuning scale, a better approach is to fit an LCD frequency readout or to use the station digital frequency meter, thus avoiding altogether the need for a tuning scale.

One way to achieve a wider tuning range is to use two VXOs ganged to shift in opposite directions at two high frequencies differing by the desired much lower frequency, as proposed by G3MEV in reference [7]. Fig 4.27 shows G3MEV's VXO circuit. Another idea is the use of one VXO to interpolate between closely-spaced multi-channel synthesiser frequencies; The difference frequencies between a VXO tuning 24.10125 to 24.11125MHz (only 10kHz in 24MHz) and the 40 synthesised channels of a CB transmitter will cover 3.5 - 3.9MHz.

Ceramic resonators are now inexpensively available in many standard frequencies. They permit the construction of stable variable-frequency oscillators with a wider pulling range than VXOs. See reference [8].

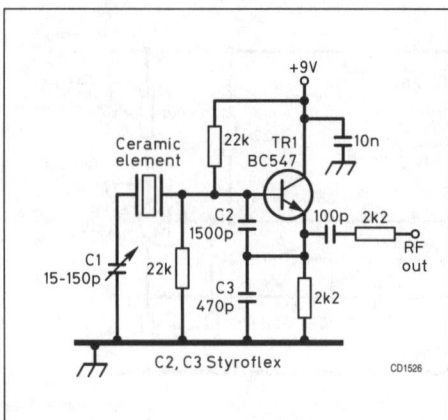

Fig 4.29: LA8AK's oscillator can tune 3.5-3.6MHz with a 3.58MHz ceramic resonator

Ceramic resonators, like quartz crystals, are piezo-electric devices; they have a worse temperature coefficient than crystals and a lower Q, but for most HF applications it is adequate and the lower Q permits pulling over a wider frequency range. The oscillator circuits are similar to those for VXOs. **Figs 4.28 and 4.29** are examples.

USING CRYSTAL CONTROL ON VHF

To obtain a crystal controlled source on VHF an overtone oscillator could be used. Crystals for fundamental use are, however, much cheaper than those intended for overtone use. Also if the source is intended for use in an FM transmitter, it will be easier to achieve the frequency deviation required if a fundamental oscillator is used. It is necessary to multiply the frequency and this can be achieved using frequency multiplier stages.

A frequency multiplier is a non-linear circuit which has its output tuned to a multiple of its input frequency, eg x2, x3 or x4. These multipliers are known as doublers, triplers and quadruplers, respectively. VHF/UHF bipolar transistors or FETs can perform adequately as multipliers. Fig 4.30(a) illustrates a typical multiplier.

If a higher multiplication factor is required, it is better to use two or more multipliers, eg for x9 multiplication use two tripler stages. The multiplier may be operated with a small amount of forward bias dependent on the transistor forward transfer characteristic but the stage must be driven hard enough (with RF) for it to function correctly. The bias is adjusted for maximum efficiency (output) after the multiplier has been tuned to the desired output frequency.

The multiplier can be single-ended, as in **Fig 4.30(a)**, or with two FETs configured as a push-push doubler as in **(b)**, or two transistors as a push-pull tripler as in **(c)**. The efficiency of **(b)** and **(c)** is typically higher than (a). The push-push doubler discriminates against odd-order multiples and the push-pull tripler discriminates against even-order multiples, so additional attenuation of unwanted outputs is much improved over the single-ended multiplier. R1 in (b) and (c) is adjusted for optimum dynamic balance between TR1 and TR2, ie maximum multiplier output.

It is always good practice to start with a reasonably high oscillator frequency in a receiver LO multiplier chain, thus requiring fewer multiplier stages. This is because there is the distinct possibility that one of the unwanted multiples will reach the mixer and cause a spurious response (or responses) within the receiver tuning range. The magnitude of any unwanted injection frequencies depends on the operating conditions of the final multiplier, including the working Q of its output circuit. To minimise this problem it is advisable to include a buffer amplifier stage with input and output circuits tuned to the LO frequency between the final multiplier and the mixer.

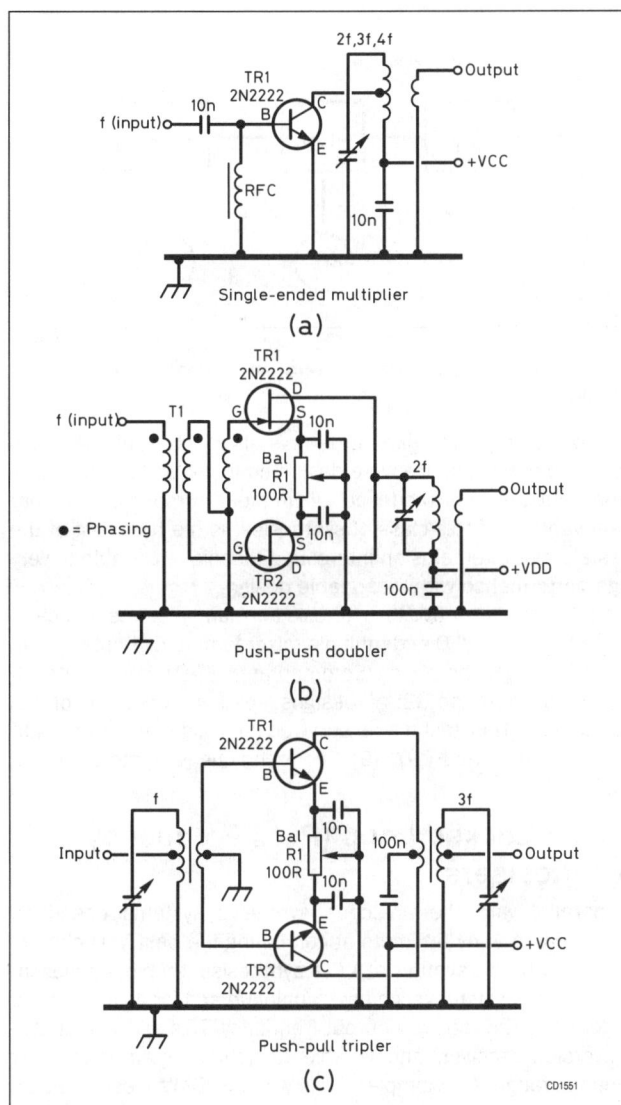

Fig 4.30: Frequency multipliers. (a) Single-ended. [b] Push-push doubler. (c) Push-pull tripler

These circuits may consist of a single high-Q tuned circuit called a high-Q break or two loosely coupled circuits. Helical filters provide an ideal solution to this problem. Two filters may be cascaded to give greater attenuation.

Adequate filtering is essential in transmitter frequency multipliers, particularly between the final multiplier and the first RF amplifier, to attenuate all unwanted multiplier output frequencies from the oscillator to the final multiplier. In order to establish the crystal frequency it is useful to draw a chart such as **Fig 4.31**. This shows the combination of frequencies and multiplication factors which could be used to obtain an output on 118MHz.

It is important to understand that any frequency change in the final multiplier output is the frequency change (or drift) of the oscillator multiplied by the multiplication factor. As an example, if the frequency change in a 6MHz crystal oscillator was 50Hz and the multiplication factor is 24, then at 144MHz the change will be 1.2kHz. This change is not too serious for NBFM but would not be acceptable for SSB equipment.

BUFFER STAGE

The output of an oscillator is not normally connected directly to the load. This is because changes in load impedance will affect the oscillator frequency. Changes in load impedance may be caused by:
- Changes in temperature
- Switching from receive to transmit
- Changes in supply voltage

A buffer amplifier must always be used to isolate the oscillator from the load. A buffer should be used with VFOs, VCOs and crystal oscillators. It is an RF amplifier which is designed for high reverse isolation. This greatly reduces the effect of load changes on the oscillator frequency.

The following needs to be known when designing a buffer amplifier:
- The expected load impedance.
- The gain required.
- The output level required.

Some of the oscillator circuits in this chapter show a buffer amplifier. Examples are:
- Fig 4.15. In this circuit the buffer is formed by TR2 and TR3.
- **Fig 4.32**. In this circuit the buffer is formed by TR2 which operates as an emitter follower.
- **Fig 4.33** shows a two-stage buffer amplifier which is designed to drive a 50-ohm load.

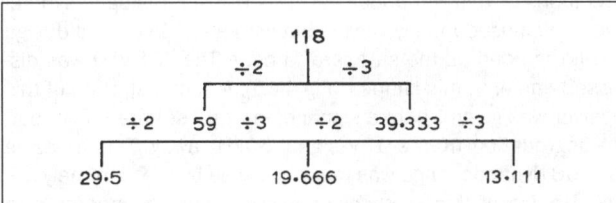

Fig 4.31: Type of chart for determining the fundamental crystal frequency required in an oscillator multiplier chain

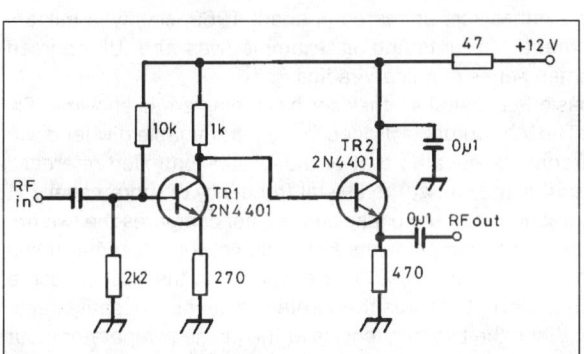

Fig 4.32: Direct-coupled buffer circuit. (*W1FB's QRP Notebook*)

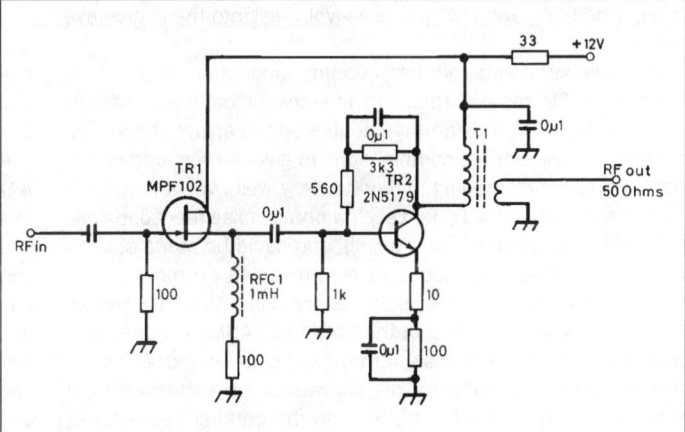

Fig 4.33: Transformer output buffer circuit. (*W1FB's QRP Notebook*)

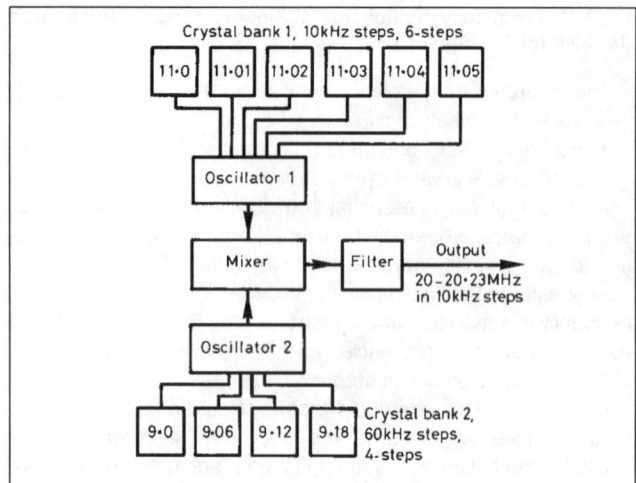

Fig 4.34: Simplified crystal bank sythesiser

Amplifiers employing feedback should not be used, on their own, as a buffer because these have poor reverse isolation. Buffers should be built using high quality components and should be solidly built like a VFO.

FREQUENCY SYNTHESIS

The description 'frequency synthesiser' is generally applied to a circuit in which the output frequency is non-harmonically related to the synthesiser reference frequency. Therefore, this excludes circuits which use a crystal oscillator and frequency multiplier stage to generate the output frequency. This section contains a description of the operation of analogue synthesis, PLL frequency synthesis and Direct Digital synthesis (DDS). Early forms of synthesis are generally no longer used commercially, but are described here because they are still of relevance to the radio amateur.

Analogue Synthesis Systems

Analogue (or Direct) synthesis was the first form of synthesis. It involves the production of an output frequency, by the addition of several (sometimes many) oscillators, usually by a combination of mixing, multiplying, re-mixing etc with many stages of filtering. This was difficult and expensive to set up really well. Examples were primarily limited to specialist, often military, synthesisers in the 'fifties. The output spectrum could be made very clean, especially in respect of close-to-carrier phase noise, but only by significant amounts of filtering. Direct synthesisers are still used in specialist areas, because they can give very fast frequency hopping and low spurious levels well into the microwave region.

A simple version did go into quantity production, in the early US-market CB radios. This was the crystal bank synthesiser shown in **Fig 4.34**. This arrangement used two arrays of crystals, selected in appropriate combinations to give the required coverage and channel spacing. It worked very well over a restricted frequency range, could offer very low phase noise (because crystal oscillators are inherently 'quiet'), and could be made acceptably (for the time) compact and low power. An example on the UK market in the 1970s, was the Belcom Liner 2, which was an SSB transceiver operating in the 2m (144-146MHz) band. This used crystal bank synthesis with a front-panel-tuned VXO on a further crystal oscillator for the conversion up and down from the effective 'tuneable IF' at 30MHz to the working frequency of 145MHz. The need for the VXO illustrates the weakness of this type of synthesiser; close channel spacing is essentially prohib-

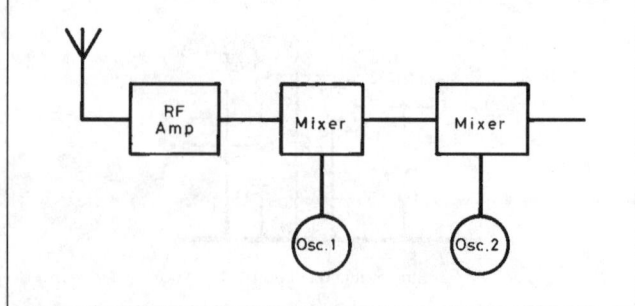

Fig 4.35: Multiple oscillators/mixers, an alternative to frequency synthesis

ited by the very large number of crystals which would be needed. Triple bank versions were described in the professional literature, but were not produced for amateur purposes. The other disadvantage of this class of synthesiser is the high cost of the crystals, but it remains an interesting technique capable of very high performance with reasonable design.

A popular alternative to synthesis for many years used a carefully engineered VFO and multiple mixer format. Examples are in the KW range, the Yaesu/Sommerkamp FT-DX series and, in homebrew form, the G2DAF designs (see earlier editions of this *Handbook*). This technique survived in solid-state form, with among others, the FT101 [9]. The technique is illustrated in **Fig 4.35** [10].

Phase-Locked Loop (PLL) Frequency Synthesisers

In parallel with the analogue synthesis systems described above, came early attempts at combining the best VFO characteristics with PLL synthesis. A PLL synthesiser employs a closed-loop control system, which uses analogue and/or digital circuits to generate the required output frequency. This allows a single-conversion receiver, and is able to achieve greater receiver dynamic range. An example of this was G3PDM's classic phase-locked oscillator, as used in his receiver and described in earlier editions of this *Handbook* [6]. This was an analogue synthesiser which used mainly valves, but with a FET VFO. This design is still regarded as the standard to beat. The FET VFO was discussed earlier in this chapter (Fig 4.15). A demonstration of this receiver was given where a signal below 1μV suffered no apparent degradation from a 10V signal 50kHz away. This amazing >140dB dynamic range was made possible by the extremely low phase noise of this synthesiser, and the superb linearity of a beam-deflection mixer valve.

Most modern PLLs contain digital circuits and the PLL output frequency is normally driven by digital control inputs. The first digital synthesisers appeared in about 1969, chiefly in military equipment. An outstanding example was the UK-sourced Clansman series of military radios.

A basic PLL digital synthesiser block diagram is shown in **Fig 4.36**. The VCO output is divided in a programmable divider down to a frequency equal to that of the crystal-controlled reference source; this may be at the crystal frequency or more usually at some fraction of it. The phase comparator compares the two frequencies, and then produces a DC voltage which is proportional to the difference in phase. This is applied to the VCO in such a way as to drive it towards the wanted frequency (negative feedback). When the two frequencies at the phase comparator input are identical, the synthesiser is said to be in-lock. Ideally, the phase comparator is actually a phase and frequency comparator, so that it can bring the signals into lock from well away. The

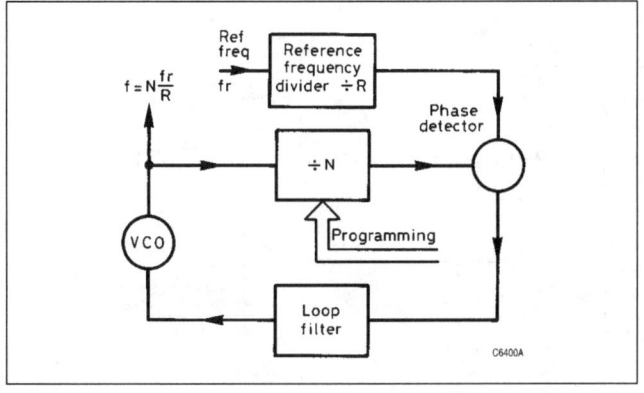

Fig 4.36: Basic PLL synthesiser

'DC' term referred to is a varying voltage determined by the frequency and phase errors. The bandwidth available to this voltage is said to be the loop bandwidth and it determines the speed at which lock can be achieved.

The programmable divider is a frequency divider using digital counters. Fully programmable dividers of appropriate frequency range were not immediately available to the early PLL designers. Two solutions were proposed for this:

• The dual-modulus pre-scaler.
• A variant of the crystal mix scheme.

The dual-modulus counter (**Fig 4.37** uses a dual-modulus pre-scaler and is a particularly clever use of circuit tricks to achieve the objective of high-speed variable ratio division. Dividers were designed which could be switched very rapidly between two different division ratios, for example 10 and 11. If the divider is allowed to divide by 11 until the first programmable counter (the A-counter) reaches a preset count (say, A) and then divides by 10 until the second counter reaches M, the division ratio is:

(11 x A) + 10 x (M - A) or (10M + A)

The advantage in the system is that the fully programmable counters M and A need only respond, in this example, to one-tenth of the input frequency, and can therefore use lower cost, slower logic. The disadvantage is that there is a minimum available count of A x (11), where A is the largest count possible in the A-counter. This is rarely a problem in practice.

Fig 4.37 shows the complete loop consisting of the VCO, the dual-modulus counter, a main counter, a reference oscillator and divider chain, a phase detector and a loop amplifier/ filter.

This type of synthesiser is relatively easy to design when a synthesiser is needed with large step sizes. It has been described

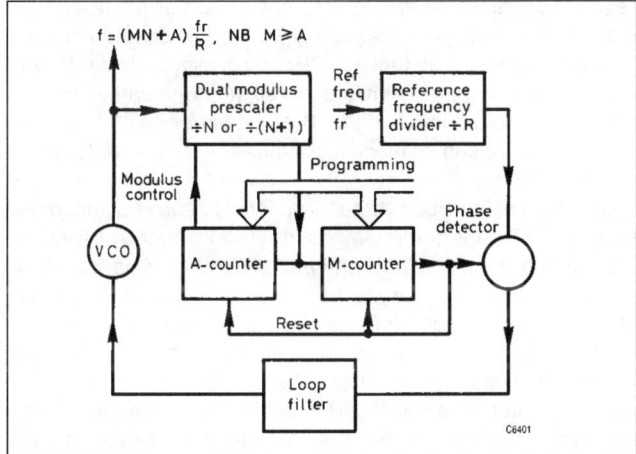

Fig 4.37: Dual modulus prescaler

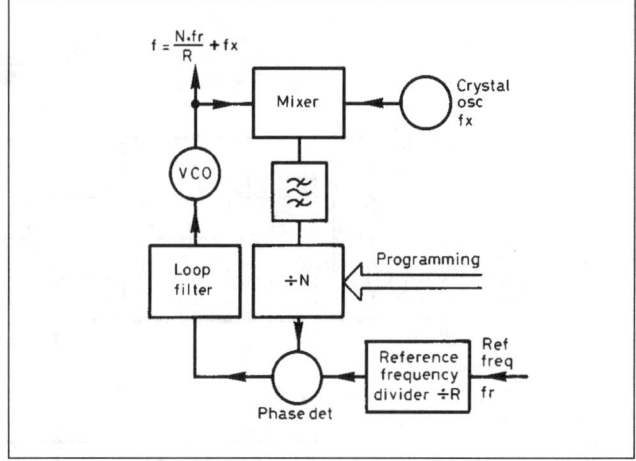

Fig 4.38: Mixer-loop PLL

many times in the professional and amateur radio fields [11]. An example of this is a synthesiser for an FM receiver, which is required to step in 25kHz or 12.5kHz steps. This is also used in TV receivers, satellite receiver and Band II FM tuners. Most commercially available PLL synthesiser chips contain the digital parts of the PLL. That is the main divider, the reference divider and the phase detector, and the counters are normally programmed by serial input. Most modern PLL chips will operate up to VHF without the need for a dual-modulus pre-scaler. They also provide two (or more) phase detectors to cover the far-from-lock and locked-in cases.

The second method, that of crystal mixing with a digital synthesiser, is probably the most widely used technique in amateur equipment, where wide operating frequency range is not needed. It offers advantages of simplicity and low component count, especially when the digital functions of the system can be contained in a single chip; see **Fig 4.38**. Unlike the dual modulus scheme, no pre-scaler is needed, but a high-frequency mixer is necessary, usually with two crystals, one for mixing down and one for the reference.

The PLL is a closed-loop system, which employs negative feedback. Like any feedback system, the loop has a finite speed at which it can operate and this is defined by the closed-loop bandwidth. There is a (usually) wider bandwidth called the lock-in bandwidth, which is the range of starting frequencies from which lock can be achieved.

The closed loop bandwidth is under control of the designer and is set by the open-loop parameters. These are:

• VCO gain
• Divider ratios
• Phase detector gain
• Loop filter amplitude/frequency and phase/frequency responses

The control loop is able to control and hence reduce the effects of VCO noise within the closed-loop bandwidth, this includes long term drift in the VCO. Outside the loop bandwidth, VCO noise is unaffected by the loop and the closed-loop bandwidth is normally set to about one tenth of the reference frequency. The closed loop is a second-order system and so it is possible for the loop to exhibit loop instability. The designer controls the loop stability primarily by controlling the response of the loop filter. The lock-up time is inherently related to the closed-loop bandwidth. For an FM-only rig, with say 12.5kHz minimum frequency step, lock-up time can be very quick.

For SSB receiver and transceivers, there are strong commercial pressures to replace the VFO with a full digital synthesis sys-

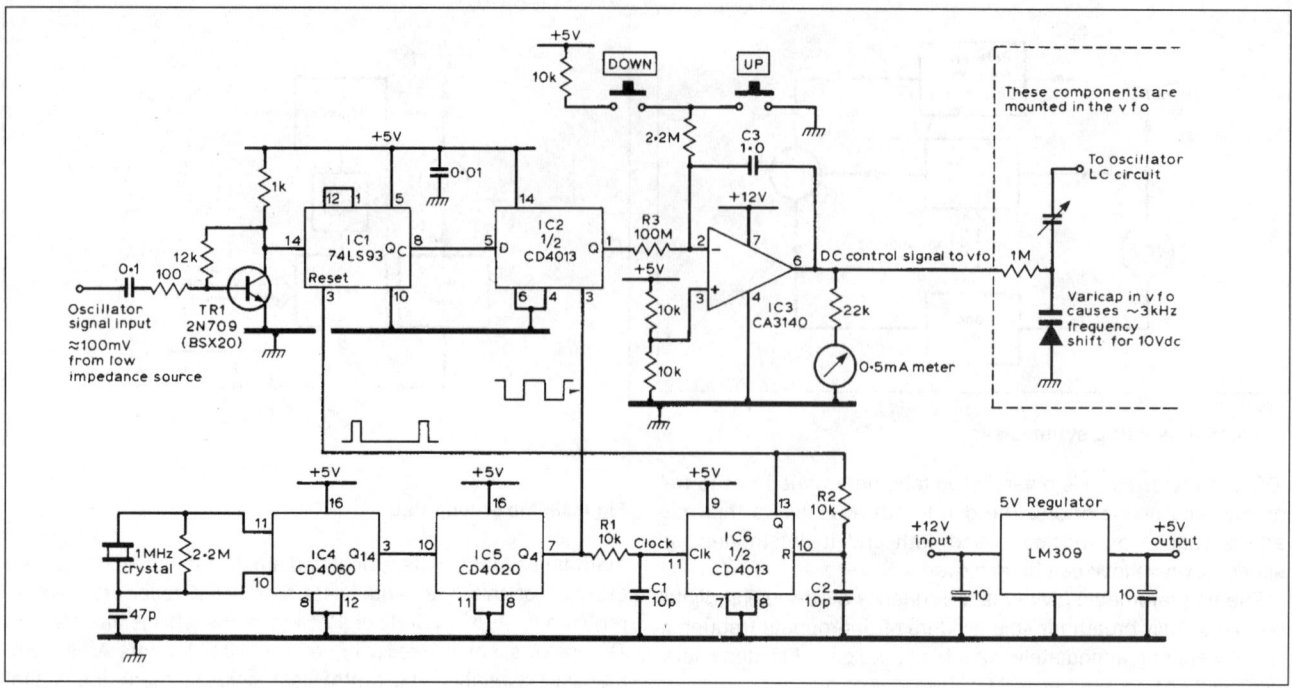

Fig 4.39: Circuit details of the CMOS form of the 'huff-and-puff' VFO stabilising system - a version of PA0KSB's system as described in *Ham Radio*

tem because this avoids many of the problems and costs of building and setting up a good VFO. However, trying to reproduce the action of a good VFO in digital synthesiser form is difficult. A PLL synthesiser can only tune in steps and so to reproduce the VFO action for SSB requires very small steps, typically 20Hz. This cannot be done in a single loop because the loop bandwidth would have to be very narrow, and the loop would be very slow. Professional systems overcome this by complex, multiple-loop synthesis, which can be expensive to design. They can also be expensive to manufacture and are sometimes bulky.

The operating frequency of a digital synthesiser is normally controlled by pulses from a shaft encoder system. This allows control of the frequency by a tuning knob in the same way as a VFO. Frequently, the tuning rate of the knob is programmable so that the tuning rate (in kHz per turn) is adjustable for different modes of signal. AM or FM modes can use faster tuning rates than SSB. The equipment uses a digital frequency display and this normally has resolution of 100Hz or 10Hz. It should be noted that this does not mean that the frequency is accurate to 100Hz or 10Hz. (Resolution doesn't equal accuracy). The frequency accuracy of the synthesiser is determined entirely by the accuracy of the reference oscillator.

PLL synthesiser design is a very demanding and essentially analogue task. The VCO is particularly difficult to design because very small unwanted voltages induced onto the tuning voltage will generate noise or spurious outputs from the synthesiser. Nevertheless, truly excellent performance can be achieved by the professional engineer in factory-built equipment. Properly designed, PLL synthesis works very well indeed, and most commercial amateur equipment uses some form of PLL synthesiser. However, the problems of providing narrow frequency increments, with the demand for complete HF band coverage (at least on receive) have lead to rather compromised synthesiser designs. This has been seen in particular in reviews of otherwise excellent radios limited in performance by synthesiser noise and spurious outputs causing reciprocal mixing.

It is possible for the very dedicated and well-equipped amateur to design a PLL system for SSB use, particularly as he does

not cost his time into the project. It should be said that a spectrum analyser, or at least a continuously tuneable receiver covering the frequency of interest and a harmonic or two, is essential for any synthesiser work. Loop stability and noise should be checked at several frequencies in the synthesiser tuning range.

Amateur equipment practice has tended to take a different route. A synthesiser is employed which has a minimum step size of typically 1kHz. Increments between the 1kHz steps are achieved either with a separate control knob, on older equipment, or by a digital-to-analogue converter (DAC) operated by the last digit of the frequency setting control. The analogue voltage tunes a VXO in the rig, which may either be the conversion crystal in the synthesiser or the reference crystal itself. Care must be taken to ensure that the pulling range is accurate or the steps will show a jump in one direction or the other at the 1kHz increments. Actually, many rigs do show this if observed carefully; the reason is that until recently the DAC was a fairly simple affair consisting of a resistor array and switching transistors. More recently, commercial DACs have been used, but this is not a complete solution, since the VXO is unlikely to be linear to the required degree, so some step non-linearity is inevitable. At least if the end points are not seriously wrong, this should not be a problem. Some rigs do tune in 10Hz steps, especially on HF; this is an extension of the technique to 100 steps instead of just 10. Again, there is the issue of the 1kHz crossover points; but with 10Hz steps this can be made less noticeable on a well-designed and adjusted rig.

One variant on PLL synthesis which appeared some years ago, and which is particularly suitable to amateur construction, is the so-called huff-and-puff VFO [12]. This could be described as a frequency-locked loop and consists (Fig 4.39) of a VFO, and a digital locking circuit. This has the same frequency reference chain and phase-sensitive detector, but the divider chain is largely omitted. The VFO is tuned using a variable capacitor, and the loop locks the VFO frequency to the nearest multiple of the reference frequency possible. The reference is typically below 10Hz, so that analogue feel is retained in the tuning. To the operator, this behaves like a

PLL. The loop can take a significant time to lock up, so various provisions to speed this up can be employed. The loop must not jitter between different multiples of the reference, hence the need for very high initial stability in the VFO. As a commercial proposition, this scheme is unattractive compared to a true PLL, but as an amateur approach it can overcome the VFO tuning problem in an elegant way.

Direct Digital Synthesis (DDS)

Direct Digital Synthesis is a method of frequency synthesis which differs from the methods described so far. A DDS is virtually all-digital and the only analogue part is a digital-to-analogue convertor (DAC) at the DDS output. In fact many modern DDS chips now contain two DACs, supplying two separate outputs. A DDS does not contain a negative feedback closed-loop system like a PLL. The DDS offers the possibility of generating an output which has very small frequency steps but, unlike a PLL, very rapid tuning to new frequency. Compared to a digital synthesis PLL type system, a DDS is relatively easy to design because the complex, critical electronics are contained within the DDS chip. Construction is relatively easy because it is built up as a conventional logic PCB and there are none of the problems of setting-up which occur with a conventional VFO. The output frequency stability (both short-term and long-term) is determined by the characteristics of the reference oscillator.

For the amateur, the best option for the reference oscillator is to use a quartz crystal oscillator and it is worthwhile using a high quality oscillator circuit to obtain low phase noise and good frequency stability. The alternative is to purchase a ready-made high- performance crystal oscillator module.

Direct digital synthesis has been frequently mentioned in *RadCom*, and is a feature of current amateur equipment [13, 14]. The basic direct digital synthesiser consists of an arrangement to generate the output frequency directly from the clock and the input data. The simplest conception is shown in **Fig 4.40**. This consists of a digital accumulator, a ROM containing, in digital form, the pattern of a sine wave, and a digital-to-analogue converter. Dealing with the accumulator first, this is simply an adder with a store at each bit. It adds the input data word to that in the store. The input data word only changes when the required frequency is to be changed. In the simplest case the length of accumulator is the clock frequency divided by the channel spacing, although it is usually calculated the other way round, ie if a 5kHz channel spacing up to 150MHz is needed, and since at least two clock pulses are required per output cycle, a clock frequency of at least 300MHz will be required. More conveniently, a 16-bit accumulator gives 65,536 steps. If the step size is to be 5kHz, then a clock frequency to the accumulator of 65,536 x 5kHz is needed, ie 327.68MHz. Such a DDS would produce any frequency in the range covered, ie 5kHz to over 100MHz in a single range without any tuned circuits.

Other frequency increments are available; any multiple of 5kHz by selection of input data, and others by choice of clock

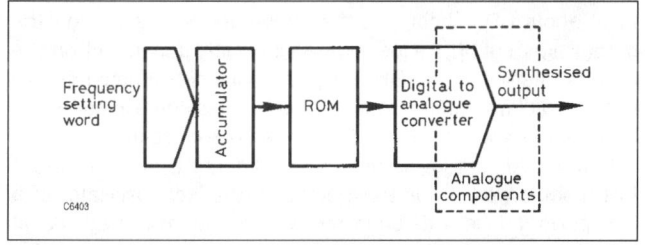

Fig 4.40: DDS concept

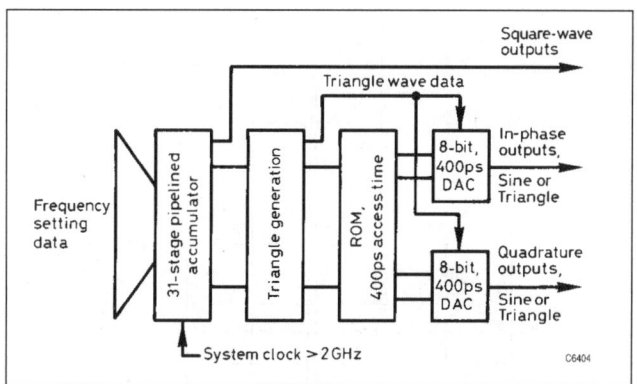

Fig 4.41: More complex DDS

frequency, eg 6.25kHz requires a clock at 204.8MHz with a two-channel (2 x 3.125kHz) program word. Of course, frequency multiplication or mixing can be used to take the output to higher frequencies. Programming is very easy, DDS devices have either a parallel-input data format or are designed to be driven from a serial input.

A more advanced device is shown in **Fig 4.41** [15]. This has on-chip DACs and facilities for square, triangle and sine outputs. Two DACs are used, because both phase and quadrature output signals are available in true and complement form. The most significant bit (MSB) from the accumulator feeds the square-wave output buffer direct. In parallel, the next seven bits from the accumulator, which digitally represent a sawtooth waveform, feed a set of XOR gates, under control of the MSB, so that a triangle output is generated, in digital form, at this point in the circuit. Actually, two triangles in quadrature are generated. These can be digitally steered to the output DACs or can be used to address a ROM containing data for sine and cosine waves; only 90° is needed, since all four quadrants can be generated from one. Finally, if selected, the digital sine/cosine is fed to the DACs for conversion to analogue form.

This device was designed to operate at up to 500MHz output frequency with 1Hz steps, so the full range is 1Hz to 500MHz, again in a single 'range' with no means of or requirement for tuning. The clock frequency is necessarily very high; the quadrature requirement adds a further factor of two, so nominal clock frequency is 2^{31} Hz, ie 2.147483648GHz. This must be supplied with crystal-controlled stability, although since phase noise is effectively divided down in the synthesiser and the frequency is fixed, it can be generated by a simple multiplier from a crystal source.

Output spectra are shown in **Figs 4.42 and 4.43**. Fig 4.42 shows a clean output at exactly one quarter of the clock frequency. Fig 4.43 shows a frequency not integrally related to the clock frequency. Here the spurious sidebands have come up to a level about 48dB below the carrier. This is the fundamental limitation of the DDS technique, and where most development work is going on. The limit comes from the finite word size and accuracy of the DAC, and incidentally the ROM, although this could have been made bigger fairly easily.

Fast DACs are difficult to make accurately, for two reasons. The first is technological. IC processes in general have a limit to a component matching accuracy of about 0.1%, ie 9 or 10 bits accuracy in a careful design. Other techniques, such as laser trimming of the resistors, can be used on some processes. However, this tends to use older, slower processes, and especially requires large resistors for trimming which slow the DAC settling time. The second problem is simply the requirements on fast settling: the DAC is required to get to the final

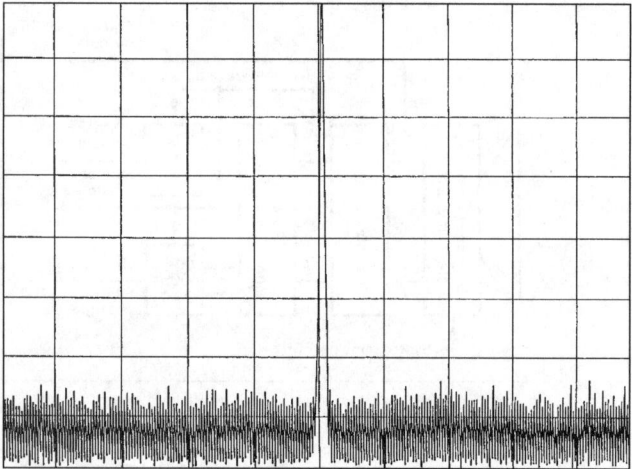

Fig 4.42: DDS spectrum at 250MHz with a 1GHz clock. 10MHz/div, X axis; 10dB/div, Y axis

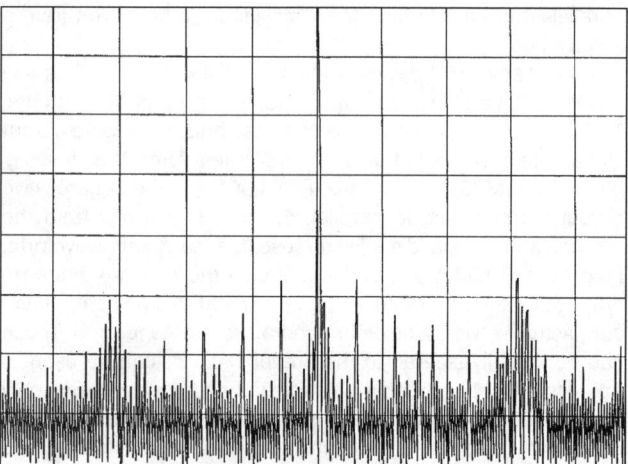

Fig 4.43: DDS spectrum at 225MHz with a 1GHz clock. 10MHz/div, X axis; 10dB/div, Y axis

value quickly and this will in general happen with the smallest number of bits. An 8-bit system as in this example was chosen to fit the process capabilities and the device requirements, but this leads to a limitation in the high level of spurious signals present.

Basically, although some frequencies are very clean, in the worst case the spurious level is 6*N* dB below the carrier, where *N* is the number of effective DAC bits. For an 8-bit system, this gives -48dB. Oversampling, ie running the clock at more than twice the output frequency, gives some improvement (at 6dB per octave) by improvement of the DAC resolution, but only up to a limit of the DACs accuracy. Typically this is about 9 bits or -54dB. In some applications, this may not be serious, since filtering can remove all but the close-in spurs. Phase-locked translation loops into the microwave region also act as filters of relatively narrow bandwidth, while retaining the 1Hz step capability.

In the amateur rigs using DDS, the synthesisers operate at relatively low frequencies and are raised to the working frequency by PLL techniques. This is complicated, but makes fine frequency increments available without the compromise of PLL design. The devices used are CMOS types, with DAC accuracies of 8 and 10 bits. This gives spurious signals of theoretically -60dB referred to the carrier, which is adequate with filtering from the PLL.

Direct Digital Synthesiser for Radio Projects

This project was written for *RadCom* by Andy Talbot, G4JNT [16]. In this article, a DDS chip is included in a small stand-alone module controlled by straightforward text-based messages from a PC. The module has been designed to be used as a drop-in component in larger projects such as receivers or transceivers.

An onboard PIC microcontroller allows control of the DDS chip from a standard PC via the serial port, and a straightforward command syntax has been developed so that standard software commands, written in any language, can be used to set the operating parameters. This PIC can easily be reprogrammed to suit any user's requirements in a stand-alone project and enough spare Input / Output lines are provided to allow for this.

This article is intended to provide an overview of the DDS module as a component for larger projects and, therefore, only limited details are included.

DDS module design

The module is based around an Analog Devices AD9850 DDS chip, full details of which are available from Analog Devices [18]. The device will accept a clock signal up to 120MHz, although a suitable source is not provided within the module. This is best left to individual constructors.

The DDS can generate an output up to approximately one third of the clock frequency with a resolution of over 4 billion so, for a 120MHz clock, frequencies from DC to 40MHz can be produced in steps of approximately 28 millihertz (mHz), and so the actual frequency of the clock is unimportant, provided of course it is known with reasonable accuracy.

The RF output is at a level of 1V p-p (0.35V RMS) and the output impedance is 100 ohms.

The supplied PIC controller translates text based messages received from the serial port into command codes for the AD9850, and (optionally) stores these in non-volatile RAM for immediate setting at switch-on. Spare memory in the PIC allows user information, such as the exact clock frequency, to be readout on request, allowing common software to be written that can drive individual modules, each having different clock frequencies.

Another option available in the firmware is a 'times-four' output, where the output from the module is designed to drive a quadrature frequency generator which performs the final frequency division.

A single hex-digit module address is included as part of the command syntax, to allow multiple modules to be driven in a multi-drop arrangement from a common controller with one COM port.

Output spurii

Spurious output signals from a DDS are complex in their nature and are not harmonically related - the device data sheet [18] gives more details.

All spurii from the design here are at levels of -60dB or better. This figure is obtained from the manufacturer's specification and is affected by the design of the output filter. One useful facet about DDS circuitry is that the spurious levels below the output filter cut-off frequency are inherently dependent on the circuits internal to the DDS chip, and are not affected by any poor circuit layout; this aspect is often trouble-some in other synthesiser designs, as filtering cannot remove close-in products.

60dBc spurious levels may be considered a bit excessive if this module were to be used alone as the local oscillator of a high-performance wide-band receiver, but there are additional techniques such as Phase Locked Loops that can clean up this signal.

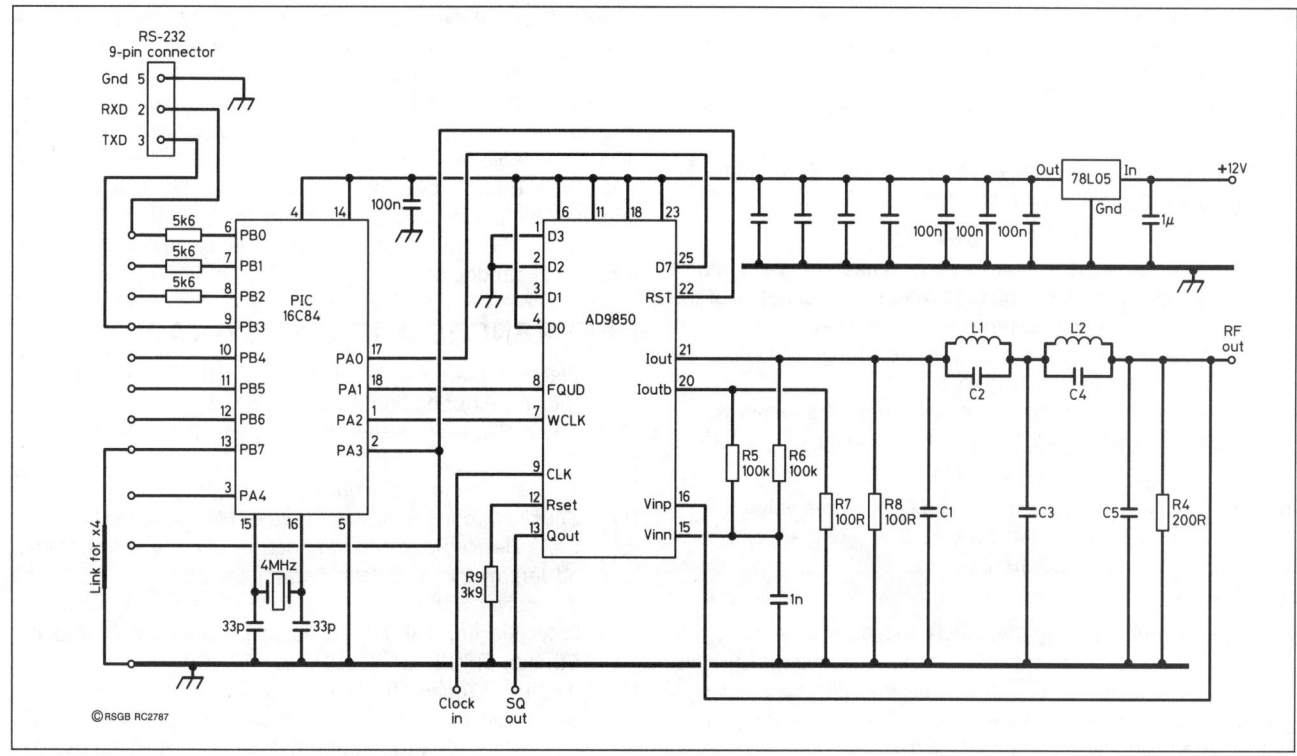

Fig 4.44: Circuit diagram of the AD9850 DDS module. Filter components and some decoupling components are dependent upon clock frequency

Construction

The circuit diagram is shown in **Fig 4..44**, from which it can be seen that the module actually has very few components, most of them being for decoupling and output filtering. Apart from the integrated circuits, all other components are either of the surface-mount 1206 or 0805 style; for the output filter, wire-ended components are used. The PIC controller is mounted in a socket for easy re-programming.

Fig 4.45: A prototype version of the G4JNT DDS board

The photograph **Fig 4.45** shows a prototype version of the DDS board, together with a high-stability 94MHz source on the PCB top right and an output buffer amplifier bottom left.

Values of components for the output filter depend on the clock frequency chosen, as do the values of decoupling capacitors. With a clock input that can range from a few kHz up to 120MHz, the optimum values of these can vary over a wide range. Guidelines for selecting appropriate values are given in the constructional and programming data available from [17].

Software control

Commands are sent using ASCII / Hex characters over a bi-directional RS-232 link with no handshaking. Parameters are 19,200 baud, 8 data bits, no parity, 1 stop bit. A simple terminal programme such as HYPERTERM (Windows®) or PROCOMM (DOS) can be used to command the frequency source.

Set this to 19200 N 8 1, full-duplex, no flow control and all start up, and modem commands set to null. Alternatively, custom software can be written to drive the COM port with the commands.

The first character sent is a board address which precedes all commands. This is a single Hex character sent as ASCII 0 - F and potentially allows up to 16 modules to be driven from the same COM port.

The next character is a command which may have hex data following it.

Q   followed by eight hex digits for the frequency command word terminated by a carriage return [CR]

P   followed by two hex digits for phase word and [CR]

U   writes the data sent above to the AD9850 DDS chip

W   as for U, and also stores all data in the PIC's non-volatile EEPROM memory for switch-on next time

Y   followed by one Hex digit, changes the board address and stores in EEPROM. No [CR] needed

K   followed by 10 hex digits and [CR]. User data, not used for driving the DDS. (In practice, read as decimal number for user data, typically clock frequency)

R   read back current data values - not necessarily those in EEPROM

The 32 bit or 8 hexadecimal character, value N (required for frequency-setting), can be derived from:

$$N = F_{out} / F_{clock} \times 2^{32}$$

Phase can be set to any one of 32 values in increments of 11.25 degrees. These form the five highest significant bits of the phase word Pxx. The lowest three bits are ignored.

Data is sent back from the DDS in text strings which can be read directly by application software.

An example of a command to set the output frequency is:

5Q03D70A3D [CR]   Board address 5, set frequency word N = hex 03D70A3D

5U   Programme the DDS to this value (with a 120MHz clock, this gives an output at 1.8MHz).

Applications

Any experiments or testing that needs an agile frequency source is a candidate. Just about any programming language that includes commands to drive the serial port can be employed, and does not even have to be PC-based. The only requirement is that you can actually write suitable software!

One application written to demonstrate the functionality of the module is for generating a narrow-band Multi-Tone Hellschreiber signal for LF use. SMT Hell transmits visible text as an image, and can be received on a frequency / time plot, commonly known as a spectrogram or waterfall display, using public domain commonly-available audio analysis software. The software generates SMT-Hell signals by directly commanding the DDS module (in real time) to set the frequencies that make up the vertical elements of each character sent. The horizontal components are made up by appropriately setting software delays. Transmissions as narrow as 2.5Hz bandwidth have been sent on 137kHz and successfully decoded as visible letters even where the signal is completely inaudible below the noise.

The ability to set the output signal phase to one of 32 values means that slow Phase Shift Keying is also possible by direct command. An additional command code (T) allows frequency and phase updates to be synchronised to an external trigger input on port B1 such as that from a GPS receiver. The DDS chip is updated within 3µs of the trigger signal rising edge and will allow, for example, precisely-timed low data rate signalling experiments.

By using the DDS output to drive the reference input to a conventional Phase Locked Loop synthesiser, the best of both worlds becomes possible. The high frequency capability of PLL synthesisers, up to many GHz, can be coupled with the tiny step size of the DDS. For example, a PLL operating with an output at 2.4GHz could be made with a step size of 0.55Hz.

DDS frequency resolution is considerably better than the stability of most crystals will allow. In fact, the actual crystal frequency can be measured, stored in the user data area of memory and then used in subsequent high-accuracy output frequency calculations.

DDS chips are evolving rapidly, to operate at higher frequencies and with more bits, but at reasonable cost. This trend will continue, and will make DDS more and more attractive for amateur applications. Web sites are the best source of up-to-date information on DDS chips. At the time of writing, the fastest device available from Analog Devices is the AD5898 which is a 10bit DDS. This can by driven by serial or parallel data input and operates with a clock frequency up to 1000MHz. For other examples of the use of DDS techniques, see [19] and later chapters of this handbook.

REFERENCES AND BIBLIOGRAPHY

[1]   *Electronic and Radio Engineering*, F E Terman, McGraw-Hill, any older edition, eg the fourth, 1955.

[2]   *Radio Receiver Design*, K R Sturley, Chapman and Hall, 1953.

[3]   *Amateur Radio Techniques*, 7th edn, Pat Hawker, RSGB.

[4]   *Special Circuits Ready-Reference*, J Markus, McGraw-Hill.

[5]   *GaAs Devices and their Impact on Circuits and Systems*, ed Jeremy Everard, Peter Peregrinus.

[6]   'A receiver with noise immunity and frequency synthesis', Peter Martin, G3PDM, *Radio Communication Handbook*, 5th edn, RSGB, 1976, pp10.104-10.108.

[7]   'Technical Topics', *Radio Communication* May 1991, p31.

[8]   See 'Using ceramic resonators in oscillators', Ian Braithwaite, G4COL, *RadCom*, February 1994, pp38-39.

[9]   *Maintenance Service Manual*, FT-101 series, Yaesu Musen Co Ltd.

[10]   'An easy-to-set-up amateur band synthesiser', Ian Keyser, G3ROO, *Radio Communication* December 1993, pp33-36.

[11]   The Plessey Company, part of GEC, have, when they were still making telecommunications ICs, published several applications books describing PLL synthesis; any of these are well worth acquiring, although they are only obtainable second-hand. Examples are *Radio Communications Handbook*, 1977; *Professional Radio Communications*, 1979; *Radio Telecoms IC Handbook*, 1987; *Professional Data Book*, 1991; and *Frequency Dividers And Synthesisers IC Handbook*.

[12]   Klaas Spaargaren, PAOKSB, in the ARRL publication *QEX* (Feb 1996, pp19-23), Chas F Fletcher, G3DXZ in *RadCom* December 1997, as well as corrections, improvements, additions and comments in 'Technical Topics' in *RadCom*, Jul, Aug, Sep 1996, Feb, Dec 1997, and Feb, Apr 1998.

[13]   'Technical Topics', Pat Hawker, G3VA, *Radio Communication* Dec 1988, pp957-958.

[14]   'Direct digital synthesis, what is it and how can I use it?', P H Saul, *Radio Communication* Dec 1990, pp44-46.

[15]   Plessey Semiconductors Data Sheet *SP2002*.

[16]   'Direct Digital Synthesis for Radio Projects'. By Andy Talbot G4JNT. RadCom Nov 2000.

[17]   The Printed Circuit Board is available (as a Gerber file,) and the PIC programme are available at the handbook web site.

[18]   Analog Devices web site. http://www.analog.com

[19]   'CDG 2000 HF Transceiver'. *RadCom* Jun 2002 to Dec 2002.

5 Building Blocks 2: Amplifiers, Mixers etc

Many of the building blocks described in this chapter can, to a certain extent, be plugged together to make up any circuit function required. Some building blocks are standard circuits using discrete semiconductors, and some are made up from application-specific integrated circuits. The performance of building blocks can be described by the following parameters.

- **Gain** describes the available AC power gain of the circuit from input to output. The source and load impedance may also need to be specified for the gain specification to be useful. Gain is normally expressed in dB.

- **Isolation** (normally between the ports of a mixer) refers to the input applied to one port affecting whatever is connected to another port. In the simplest of receivers that uses a mixer without much isolation, peaking the antenna tuning circuit connected to the signal input may 'pull' the frequency of an unbuffered free-running oscillator connected to the local oscillator (LO) port. Conversely, the LO signal may unintentionally be transmitted from the antenna. Isolation is normally expressed in dB.

- **Noise** is generated in all circuits. It is quantified as a 'noise figure', expressed in decibels over the noise generated by a resistor of the same value as the impedance of the mixer port at the prevailing temperature, eg 50Ω at 27°C. For example, if front-end mixer noise is significant as compared to the smallest signal to be processed, the signal-to-noise ratio will suffer. The same noise level or 'noise floor', in terms of dBm, is shown in **Fig 5.1**. Below 10MHz, however, atmospheric noise will override receiver noise and most mixers will be adequate in this respect.

- **Overload** occurs if an input signal exceeds the level at which the output is proportional to it. The overloading input may be at a frequency other than the desired one. Overload is normally expressed in dBm.

- **Compression** is the gain reduction which occurs when the signal input magnitude exceeds the maximum the circuit can handle in a linear manner. The input level for 1dB of gain reduction is often specified. Note the bending of the 'fundamental component' line in **Fig 5.1**. The signal input level should be kept below the one causing this bending.

- **Intermodulation** products are the result of non-linearity in the circuit, and are generated when the input contains two or more signals. Intermodulation performance is often specified as the power level in dBm of the 'third-order intercept point'. This is the fictitious intersection on a signal input power versus output power plot, **Fig 5.1**, of the extended (dashed) fundamental (wanted) line and the third-order intermodulation line. Note that the third-order line rises three times steeper than the fundamental line.

- **Harmonic distortion** is a result of non-linearity in the circuit. The result of harmonic distortion is to produce, at the output, harmonics of the input fundamental frequency. Even-order harmonics (ie 2nd, 4th, 6th etc) can be virtually eliminated by the use of push-pull circuits.

- **Operating impedance**. This defines the input and output impedances of the circuit, or the source and load impedances for which the circuit is designed. Circuits which operate at low impedance (typically 50 ohms) can be designed to have wide bandwidth. They normally operate at low voltage, high current. Circuits which operate at high impedance (typically 1000 ohms) tend to have reduced bandwidth. They tend to operate at a higher voltage, but with lower current consumption than low-impedance circuits.

- **Bandwidth** defines the range of frequencies over which the circuit is designed to function. A wideband circuit design is one which typically covers several octaves and would be used, for example, for a circuit required to operate over 1.8 to 52MHz. Wideband design is not always necessary for amateur use, and is of no value for those who only operate on one amateur band.

- **Power consumption** is the power which the circuit draws from the DC power supply. The power drawn is given by the product of the DC supply voltage multiplied by the operating current. Power consumption is most significant when the equipment is operated from batteries, because high power consumption will reduce battery life.

Building blocks may be classed as passive or active. Passive circuits require no power supply and therefore provide no gain. In fact they may have a significant insertion loss. Examples of passive circuits are diode mixers, diode detectors, LC (inductor + capacitor) filters, quartz filters and attenuators. Examples of active circuits are amplifiers, active mixers, oscillators and active filters.

MIXERS

Terms and Specifications

A mixer is a three-port device, of which two are inputs and the third is the output. The output voltage is the mathematical product of the two input voltages. In the frequency domain, this can be shown to generate the sum and difference of the two input frequencies. The following circuits are better known by the specific

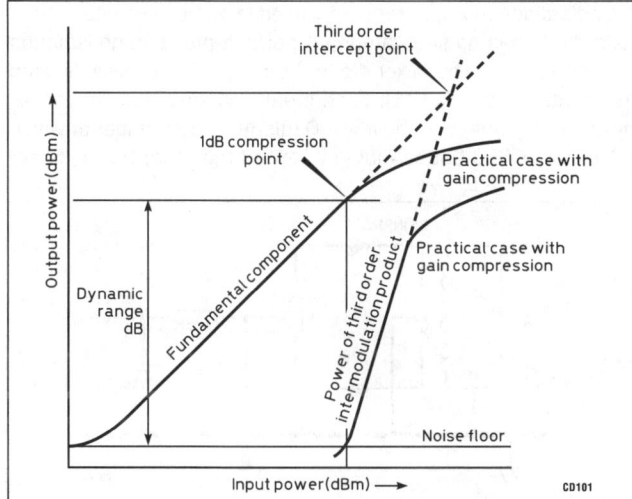

Fig 5.1: Noise floor, 1dB compression point, dynamic range and 3rd-order intercept indicated on a mixer output vs input plot (*GEC-Plessey Professional Products IC Handbook*)

function that they perform, but they are all examples of circuits which operate as a mixer, or multiplier.

- Front-end mixer
- Detector
- Product detector
- Synchronous detector
- Demodulator
- Modulator
- Phase detector
- Quadrature FM detector
- FM Stereo decoder
- TV Colour demodulator
- Analogue multipliers
- Multiplexer
- Sampling gate

These all use the principle of multiplying the two input signals. However, there are differences in the performance of the circuits. For example, a mixer required to operate as the front-end mixer in a high-performance receiver would be designed for good high dynamic range, whereas a mixer operating as a detector in the back-end of a receiver only needs very limited dynamic range. This is because the level of signal into the demodulator is constant (due to the receiver AGC system).

It should be noted that there is a distinction between mixers (or multipliers) in this context and mixers which are used in audio systems. The mixers used in audio systems operate by adding the input voltages, and the function is therefore mathematically different.

Of the two inputs to a mixer, one, f_s, contains the intelligence, the second, f_o, is specially generated to shift that intelligence to (any positive value of) $\pm f_s$, $\pm f_o$, of which only one is the desired output. In addition, the mixer output also contains both input frequencies, their harmonics, and the sum and difference frequencies of any two of all those. If any one of these many

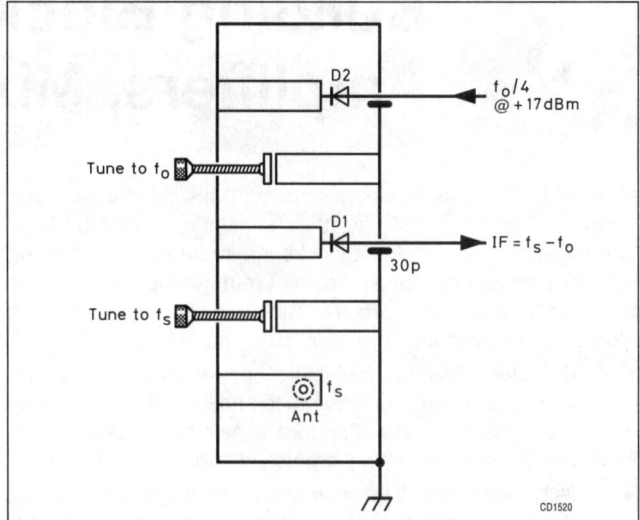

Fig 5.2: Single-diode mixer for 1296MHz. An interdigital filter provides isolation between ports. D1 is the mixer and D2 is the last multiplier in the local oscillator chain.

unwanted mixer products almost coincides with the wanted signal at some spot on a receiver dial, there will be an audible beat note or birdie at that frequency. Similarly, in a transmitter, a spurious output may result.

In a general-coverage receiver and multi-band transmitter design, the likelihood of spurious responses occuring somewhere in the tuning range is very high, but this problem can be reduced by the use of balanced mixers. When a signal is applied to a balanced input port of a mixer, the signal, its even harmonics and their mixing products will not appear at the output. If both input ports are balanced, this applies to both f_o and f_s, and the device is called a double-balanced mixer. **Table 5.1** shows how balancing reduces the number of mixing products. Note that products such as $2f_o \pm f_s$ are known as third-order products, $3f_o \pm 3f_s$ as sixth-order and so on. Lower-order products are generally stronger and therefore more bothersome than higher-order products.

Practical Mixer Circuits

Passive mixers

A single-diode mixer is frequently used in microwave equipment. As a diode has no separate input ports, it provides no isolation between inputs. The mixer diode D1 in **Fig 5.2**, however, is used in an interdigital filter [1]. Each input frequency readily passes by its high-Q 'finger' to the lower-Q (diode-loaded) finger to which D1 is connected but cannot get beyond the other high-Q finger

Unbalanced mixer

	f_o	$2f_o$	$3f_o$	$4f_o$	$5f_o$
fs	$f_o \pm f_s$	$2f_o \pm f_s$	$3f_o \pm f_s$	$4f_o \pm f_s$	$5f_o \pm f_s$
2fs	$2f_s \pm f_o$	$2f_o \pm 2f_s$	$3f_o \pm 2f_s$	$4f_o \pm 2f_s$	$5f_o \pm 2f_s$
3fs	$3f_s \pm f_o$	$3f_s \pm 2f_o$	$3f_o \pm 3f_s$	$4f_o \pm 3f_s$	$5f_o \pm 3f_s$
4fs	$4f_s \pm f_o$	$4f_s \pm 2f_o$	$4f_s \pm 3f_o$	$4f_o \pm 4f_s$	$5f_o \pm 4f_s$
5fs	$5f_s \pm f_o$	$5f_s \pm 2f_o$	$5f_s \pm 3f_o$	$5f_s \pm 4f_o$	$5f_o \pm 5f_s$

Balanced mixer - half the number of mixer products

	f_o	$2f_o$	$3f_o$	$4f_o$	$5f_o$
fs	$f_o \pm f_s$	$2f_o \pm f_s$	$3f_o \pm f_s$	$4f_o \pm f_s$	$5f_o \pm f_s$
3fs	$3f_s \pm f_o$	$3f_s \pm 2f_o$	$3f_o \pm 3f_s$	$4f_o \pm 3f_s$	$5f_o \pm 3f_s$
5fs	$5f_s \pm f_o$	$5f_s \pm 2f_o$	$5f_s \pm 3f_o$	$5f_s \pm 4f_o$	$5f_o \pm 5f_s$

Double-balanced mixer - one quarter the number of mixer products

	f_o	$2f_o$	$3f_o$	$4f_o$	$5f_o$
fs	$f_o \pm f_s$	-	$3f_o \pm f_s$	-	$5f_o \pm f_s$
3fs	$3f_s \pm f_o$	-	$3f_o \pm 3f_s$	-	$5f_o \pm 3f_s$
5fs	$5f_s \pm f_o$	-	$5f_s \pm 3f_o$	-	$5f_o \pm 5f_s$

f_o is the local oscillator. Note that a product such as $2f_o$ +-f_s is known as a third-order product, $3f_s$+-$3f_o$ as a sixth-order product and so on

Table 5.1: Mixing products in single, balanced and double-balanced mixers

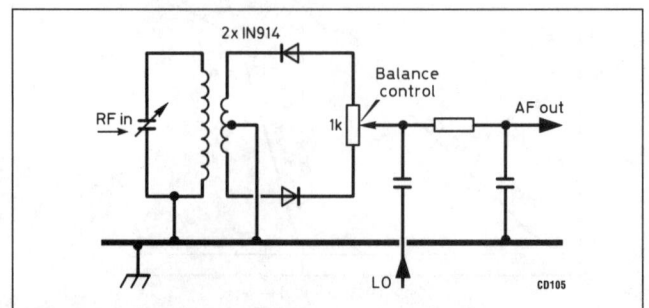

Fig 5.3: In this two-diode balanced mixer for direct conversion receivers, balancing helps to keep LO drive from reaching the antenna

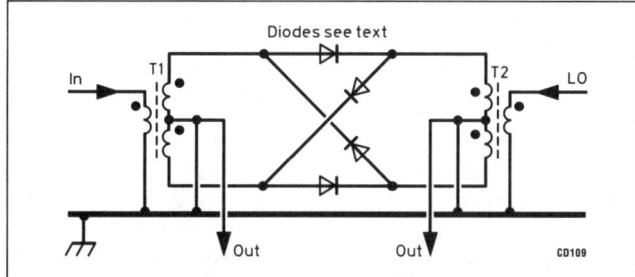

Fig 5.4: Diode ring mixers are capable of very high performance at all the usual signal levels but several sometimes costly precautions have to be taken to realise that potential; see text. Further improvement of the balance can be obtained by the use of push-pull feed to the primaries of T1 and T2 and feeding the output from between the T1 and T2 centre-taps into a balanced load

which is tuned to the other input frequency. In this design f_s = 1296MHz and f_o = 1268MHz. The 30pF feedthrough capacitor at the mixer output terminal acts as a short to earth for both input frequencies, but is part of a tuned circuit (not shown) at the output frequency of 1296 - 1268 = 28MHz. Diode D2 is not part of the mixer, but is the last link in a local oscillator chain. D2 operates as a quadrupler from 317 to 1268MHz. To get a good noise figure, the mixer D1 is a relatively expensive microwave diode. In the multiplier spot, a cheaper diode suffices.

Fig 5.3 is a single-balanced version of the diode mixer. It is popular for direct-conversion receivers where the balance reduces the amount of local oscillator radiation from the receiving antenna. The RC output filter reduces local oscillator RF reaching the following AF amplifier.

Fig 5.4 shows a diode ring double-balanced mixer. This design has been used for many decades and its performance can be second to none. Diode mixers are low-impedance devices, typically 50 ohms. This means that the local oscillator must deliver power, and typically 5mW (7dBm) is required for best dynamic range. The low impedances mean that the circuit is essentially a wideband design.

Diode mixers can be built by the amateur, and for very little cost. The four diodes should be matched for both forward and reverse resistance. Inexpensive silicon diodes like the 1N914 can

be used, but Schottky barrier types such as the BA481 (for UHF) and BAT85 (for lower frequencies and large-signal applications) are capable of higher performance. The bandwidth of the mixer is determined largely by the bandwidth of the transformers. For wide-band operation, these are normally wound on small ferrite toroids. For HF and below, the Amidon FT50-43 is suitable, and 15 trifilar turns of 0.2mm diameter enamelled copper wire are typical. The dots indicate the same end of each wire.

Complete diode mixers can be bought ready-made and are now available quite cheaply. These are usually manufactured as double-balanced mixers and one of the cheapest is the model SBL-1. They contain four matched diodes and two transformers and are available in a sealed metal package. The advantage of these is that their mixing performance is specified by the manufacturer. When studying mixer data sheets, it can be seen that the main difference between mixer models is the intermodulation performance. The models which have the best intermodulation performance also require much greater local oscillator input power. Mixers can be purchased which have local oscillator requirements from about 0dBm up to about +27dBm. Commercial diodes mixers are typically specified to operate over 5 - 500MHz, but should still offer reasonable performance beyond these frequencies.

Passive mixers have a conversion loss of about 7dB. If such a mixer is used as a receiver front-end mixer, this loss will contribute significantly to the noise figure of the receiver. It is recommended that the mixer is followed immediately by an IF amplifier. If the receiver is being used above about 25MHz, a low noise-figure receiver is important, and this IF amplifier must, therefore, be a low-noise type.

The performance of passive diode mixers is normally specified with the three ports all terminated with 50 ohms. Failure to this on any port may increase third-order intermodulation. On the LO port, as it is difficult to predict the output impedance of an oscillator circuit over a wide frequency range, it is best to generate more LO power than that required by the mixer and insert a resistive attenuator, 3-6dB being common.

Reactive termination of the output port can increase conversion loss, spurious responses and third-order IMD. Most mixers work into a filter to select the desired frequency. At that frequency, the filter may have a purely resistive impedance of the proper value but at the frequencies it is designed to reject it certainly does not. If there is more desired output than necessary, an attenuator is indicated; if not, the filtering may have to be

Table 5.2: Comparison of semiconductor performance in mixers

Device	Advantages	Disadvantages
Bipolar transistor	Low noise figure	High intermodulation
	High gain	Easy overload
	Low DC power	Subject to burn-out
Diode	Low noise figure	High LO drive
	High power handling	Interface to IF
	High burn-out level	Conversion loss
JFET	Low noise figure	Optimum conversion gain
	Conversion gain	not at optimum square-law
	Excellent square-law	response level
	characteristic	High LO power
	Excellent overload	
	High burn-out level	
Dual-gate MOSFET	Low IM distortion	High noise figure
	AGC	Poor burn-out level
	Square-law	Unstable
	characteristic	

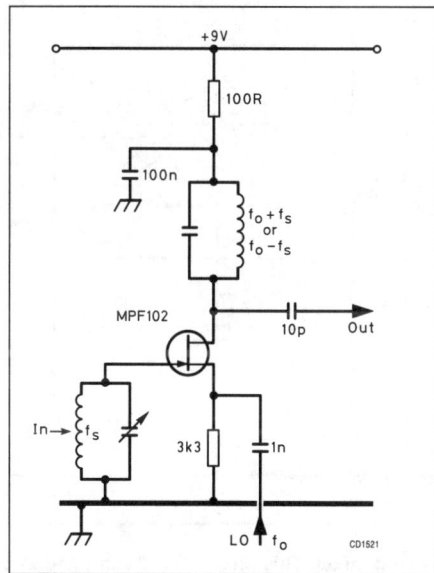

Fig 5.5: A single JFET makes a simple inexpensive mixer. The LO drives a low-impedance port, requiring it to supply some power

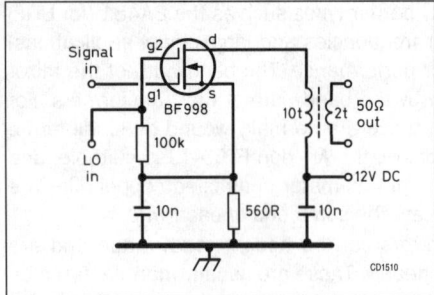

Fig 5.6: The two gates of a dual-gate MOSFET present high impedances to both the signal input and the local oscillator. A self-biassing configuration is shown here

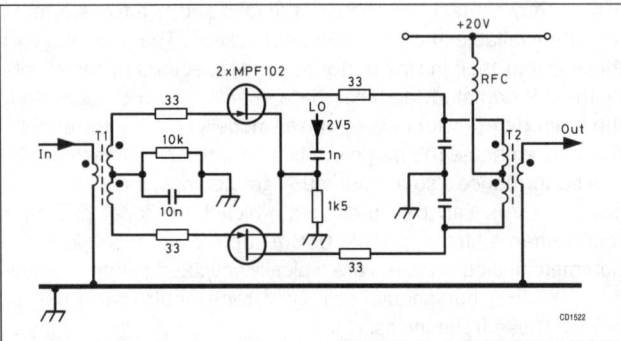

Fig 5.7: Two JFETs in a balanced mixer are capable of the performance required from all but the most expensive communications receivers

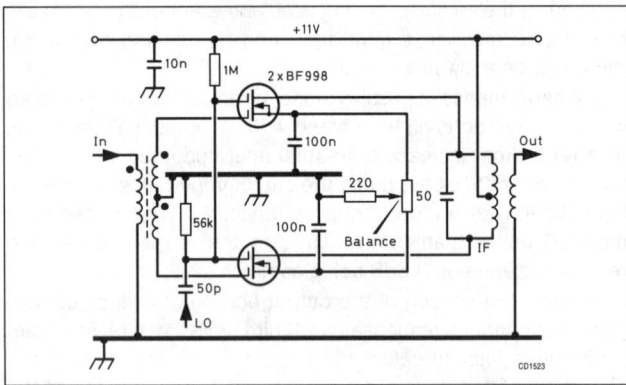

Fig 5.8: Two dual-gate MOSFETs in this balanced mixer require low LO power. At the lower frequencies, where noise is not a major consideration, they perform as well as JFETs

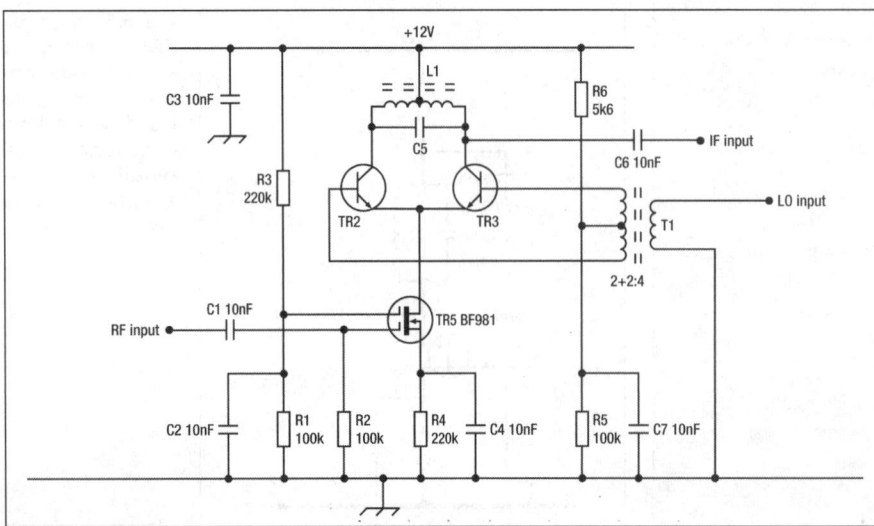

Fig 5.9: A single-balanced mixer. This circuit has been used successfully

done after the following amplifier, which may be the one making up for the conversion loss. A cascode amplifier is sometimes used as it can have a good noise figure and, over a wide frequency range, a predictable resistive input impedance.

Active mixers

It is normal to use active devices like bipolar transistors, junction FETs, single and dual-gate MOSFETs, or their valve equivalents. FETs are particularly good as mixers because they have a strong second-order response in their characteristic curve, and it is the second-order response which gives the mixing action in active mixers.(see Table 5.2). An active mixer normally has a useful amount of conversion gain.

Fig 5.5 shows a junction FET in what is probably the simplest active mixer circuit. It is a non-balanced mixer but does provide some isolation between ports. It also provides low noise, high conversion gain and a reasonable dynamic range. The latter can be improved by the use of an FET, such as the J310, biased for high source current, say 20mA, depending on the device. JFETs require careful adjustment of bias (source resistor) and local oscillator input level for best performance.

The signal is applied to the high-impedance gate; hence, loading on the resonant input circuit is minimised and gain is high. The local oscillator, however, feeds into the low-impedance source; this implies that power is required from the local oscillator, which may need buffering. One could reverse the inputs, thereby reducing the local oscillator power requirements, but that would reduce the gain. The output is taken from the drain with a tuned circuit selecting the sum or difference frequency.

Fig 5.6 shows a dual-gate MOSFET in a mixer circuit in which both inputs feed into high-impedance ports, but the dynamic range of these devices is somewhat limited. In this circuit, the G2 voltage equals that of the source. The local oscillator injection must be large, 1-3V p-p. Alternatively, G2 could be biased at approximately 25% of the supply voltage.

Figs 5.7 and 5.8 are single-balanced versions of JFET and dual-gate MOSFET mixers.

The circuit of Fig 5.9 shows a single-balanced mixer, which is balanced with respect to the RF input. The circuit consists of a Dual-gate MOSFET (TR1), which amplifies the RF input and drives an RF current into the bipolar transistors TR1 and TR2. These two transistors achieve the mixing action and need to be reasonably well matched (a matched pair is ideal). The two bipolar transistors can be virtually any low to medium f_T devices. They should not have a very high f_T because they may tend to be unstable at VHF or UHF. For the MOSFET, virtually any Dual-gate FET can be used. The output inductor L1 was bifilar wound on a ferrite core and is resonant with C5 at the intermediate frequency. The local oscillator input transformer T1 was trifilar wound on a ferrite two-hole bead.

Bipolar ICs can also be configured as single-balanced or double-balanced mixers for use in the VHF range and below. ICs developed for battery-powered instruments such as portable and cellular telephones have very low power consumption and are used to advantage in QRP amateur applications. Often, a single IC contains not only a mixer but other functions such as a local oscillator and RF or IF amplifiers.

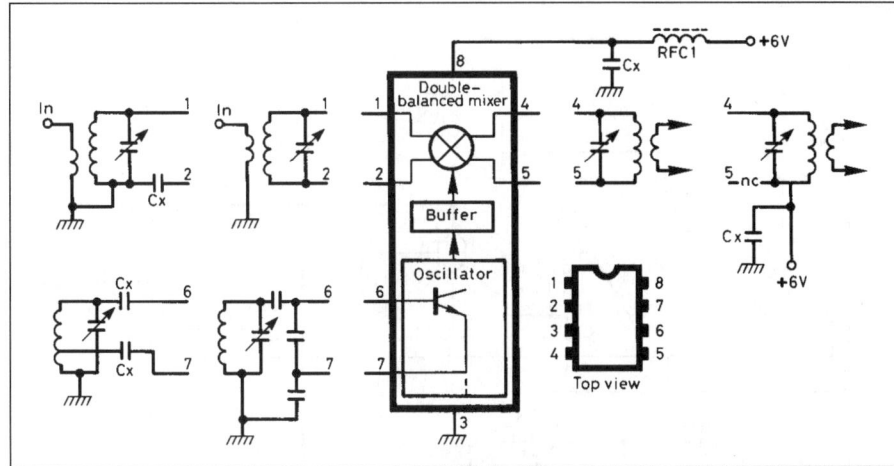

Fig 5.10: The NE602N double-balanced-mixer/oscillator is shown here in block diagram form with some options for external circuitry. This IC combines good performance with low DC requirements and a reasonable price. The option on the right for pins 1 and 2 should preferably use a split-stator capacitor. The option on the left for pins 6 and 7 is for a Hartley oscillator, the one on the right for a Colpitts

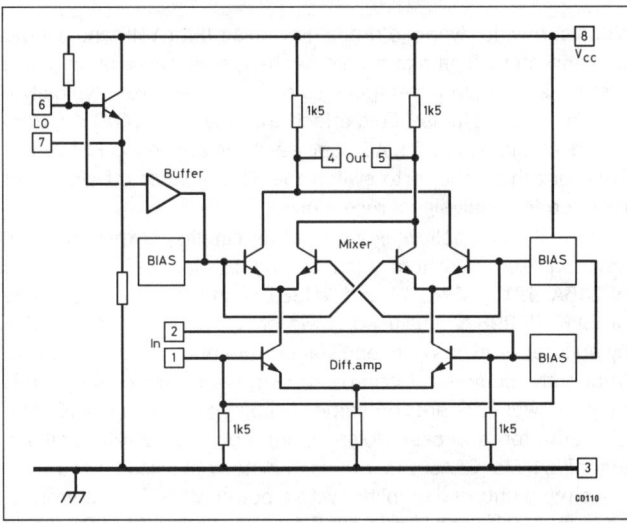

Fig 5.11: How the NE602N monolithic mixer/oscillator works is described in the text with reference to this equivalent circuit (*Signetics RF Communications Handbook*)

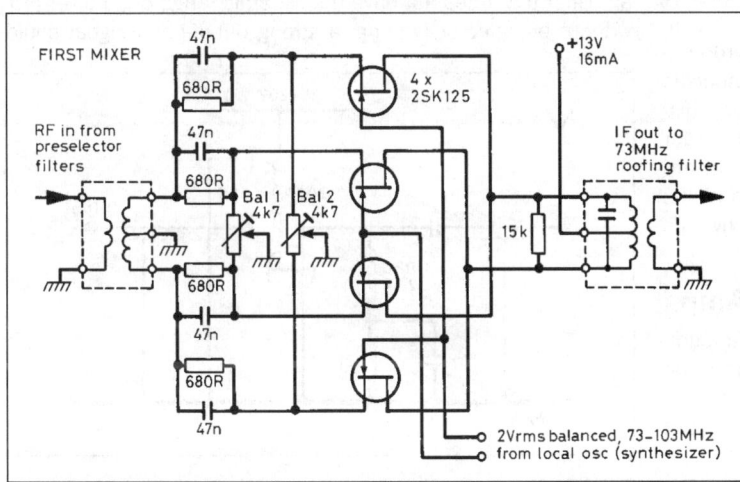

Fig 5.12: This four-JFET double-balanced mixer is as good as any seen in amateur receivers but does not have the problems of conversion loss, termination and high LO drive associated with diode ring mixers

Monolithic technology relieves the home constructor of the task of matching components and adjusting bias, while greatly simplifying layout. Also, the results are more predictable, if not always up to the best obtainable with discrete components.

The NE602N mixer/local oscillator IC may serve as an example. Fig 5.10 shows a block diagram with several possible input, output and local oscillator options. Signal input and output can be either balanced or single-ended. Balanced inputs enhance the suppression of unwanted mixing product and reduce LO leakage into the signal input.

Hartley and Colpitts connections are shown for the local oscillator. In the latter, a crystal can be substituted for the parallel-tuned LC circuit, or an external LO signal can be fed into pin 6. All external connections should be blocked for DC by capacitors as all biasing is done within the IC.

Fig 5.11, the equivalent circuit of the IC and the following description of how it works were taken from reference [2]. The IC contains a Gilbert cell (also called a transistor tree), an oscillator/buffer and a temperature-compensated bias network. The Gilbert cell is a differential amplifier (pins 1 and 2) which drives a balanced switching cell. The differential input stage provides gain and determines the noise figure and the signal-handling performance.

Performance of this IC is a compromise. Its conversion gain is typically 18dB at 45MHz. The noise figure of 5dB is good enough to dispense with an RF amplifier in receivers for HF and below. The large-signal handling capacity is not outstanding, as evidenced by a third-order intercept point of only -15dBm (typically +5dBm referred to output because of the conversion gain). This restricts the attainable dynamic range to 80dB, well below the 100dB and above attainable with the preceding discrete-component mixers, but this must be seen in the light of this IC's low power consumption (2.4mA at 6V) and its reasonable price. The on-chip oscillator is good up to about 200MHz, the actual upper limit depending on the Q of the tuned circuit.

The following double-balanced mixers can be used where low noise and good dynamic range must be combined with maximum suppression of unwanted outputs, eg in continuous-coverage receivers and multiband transmitters. Two different approaches are presented.

Fig 5.12 shows a double-balanced mixer using four JFETs. This is capable of high performance and is used in the Yaesu FT-1000 HF transceiver. The circuit offers some conversion gain and the local oscillator needs to supply little power as it feeds into the high-impedance gates.

Fig 5.13 shows a double-balanced mixer, which was derived from the single-balanced mixer in Fig 5.9. The circuit is designed on the principle of the Gilbert Cell. In this design however, the lower pair of transistors in the Gilbert Cell have been replaced with two dual-gate MOSFETs. The two MOSFETs do not operate as mixers but as amplifiers. In the prototype, matching was done by selection, finding two devices (from a batch of BF981s) that would operate at the same DC conditions when fitted into the circuit. However, pairs of MOSFETs in a single SMD package are available, eg

Fig 5.13: A double-balanced version of the mixer shown in Fig 5.9

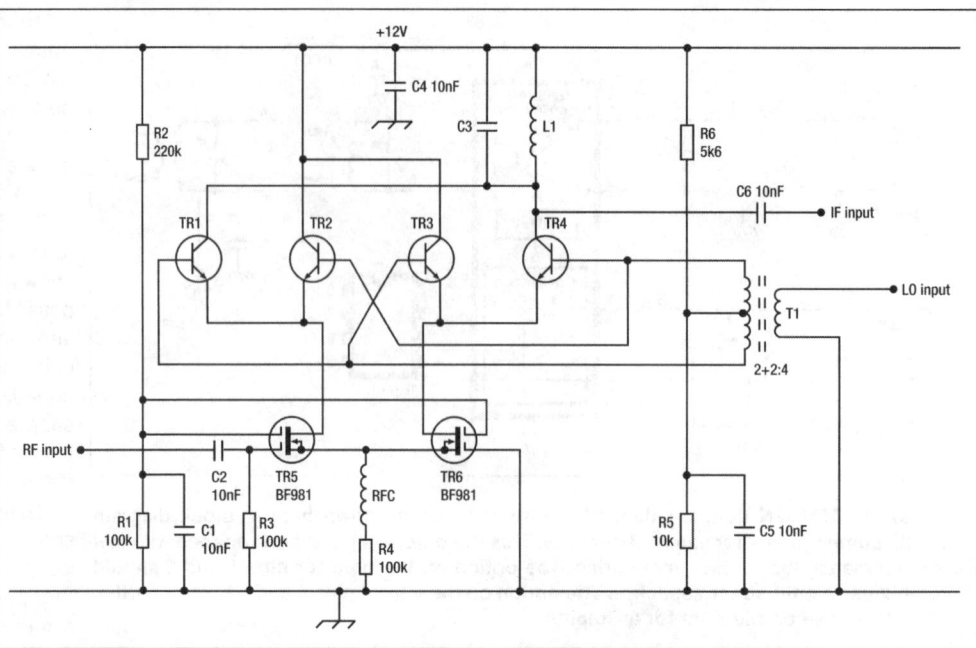

Philips BF1102, and the use of these devices is recommended. The top four bipolar transistors achieve the mixing action in the same way as in the Gilbert Cell. The four transistors (TR1 to TR4) were contained within a 3046 integrated circuit so they are very well matched. C3 and L1 were selected to be resonant at the intermediate frequency. T1 is similar to transformer T1 in Fig 5.9. The circuit gives extremely good balance, good noise figure, reasonable intermodulation performance, and a useful amount of gain.

The 3rd order intercept is about 25dB higher than the NE602, and the circuit is suitable as a front-end mixer in a medium performance HF receiver and would not require an RF amplifier. The circuit operates with a total supply current of about 20mA.

RADIO-FREQUENCY AMPLIFIERS

Amplifiers are an essential part of radio frequency electronics, and are used anywhere it is necessary to raise the power of a signal. Examples of this are described below.

- An RF amplifier is used on the antenna input of a VHF or UHF receiver. The input signals here may be extremely small, less then 1µV.

- In a receiver, an IF amplifier is used to provide most of the overall gain.

- A transmitter contains a power amplifier to raise the power of the signal up to the level required for transmission. The output power may be many hundreds of watts.

The following section considers amplification of signals from just above audio frequency up to UHF. This includes examples of amplifiers for receiver input stages through to amplifiers which will transmit the normal power limit for UK amateurs (400W PEP), especially those which might be considered for home construction projects. Integrated circuits are available which require few extra components but they are often application-specific and so the performance is fixed. The advantage of using discrete components is that the performance of the circuit is under control of the designer. Therefore, included in this section are small-signal amplifiers with discrete semiconductors and ICs and power amplifiers with discrete semiconductors, hybrids (semi-integrated modules) and valves.

Low-level Discrete-semiconductor RF Amps

Discrete semiconductors are indicated when the function requires no more gain than that which can be realised in one stage, generally between 6 and 20dB, or when the required noise level or dynamic range cannot be met by an IC. In receiver input stages, ie ahead of the first mixer, both conditions may apply.

Modern discrete devices for use in RF amplifiers are extremely cheap, and the cost is therefore not a deciding factor in the decision on which device to use. Modern design of the first stage in a receiver demands filters rejecting strong out-of-band

signals, a wide dynamic range, low-noise IF at VHF and above, and only enough gain to overcome the greater noise of an active first mixer or the conversion loss of a passive mixer, typically a gain of 10dB. Bipolar, field-effect and dual-gate MOSFET transistors can provide that. AGC can easily be applied to FET amplifiers, but the trend is to switch the RF amplifier off when not required for weak-signal reception.

In most of the following low-level RF circuits inexpensive general-purpose transistors such as the bipolar 2N2222A, FETs BF245A, J310, MPF102 and 2N3819, and the SMD dual-gate MOSFET BF998 can be used. However, substitution of one device by another or even by its equivalent from another manufacturer frequently requires a bias adjustment. For exceptional dynamic range, power FETs are sometimes used at currents up to 100mA. Similarly, for the best noise figure above 100MHz, gallium-arsenide FETs (GaAsFETs) are used instead of silicon types.

A simple untuned amplifier with a power MOSFET in a grounded-gate circuit is shown in **Fig 5.14**. The Philips BLF221 would be suitable. This circuit is used in HF antenna distribution amplifiers, ie where several receivers tuned to different frequencies must work off one antenna. Beware, however, of operating any amplifier that does not have the dynamic range of a power FET without preselector filtering; a strong out-of-band signal could

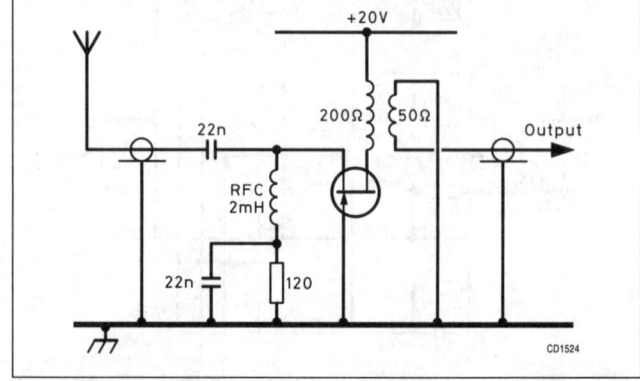

Fig 5.14: 10dB broad-band RF amplifier using a power FET in the common-gate mode. It will handle 0.5 to 40MHz signals from 0.3 V to almost 3V p-p with a noise figure of 2.5dB. The drain current is 40mA

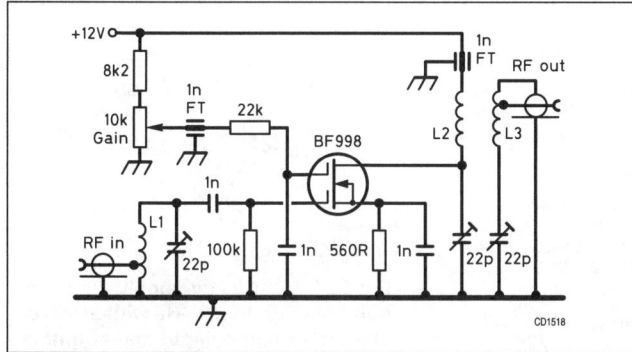

Fig 5.15: A 144MHz preamp using a BF998 SMD dual-gate MOS-FET. L1: 5t 0.3mm tinned copper tapped 1t from earthy end, 10mm long, 6mm dia. L2, L3: 6t 1mm tinned copper 18mm long, 8mm dia. L3 tapped 1t from earthy end. L2, L3 mounted parallel, 18mm between centres

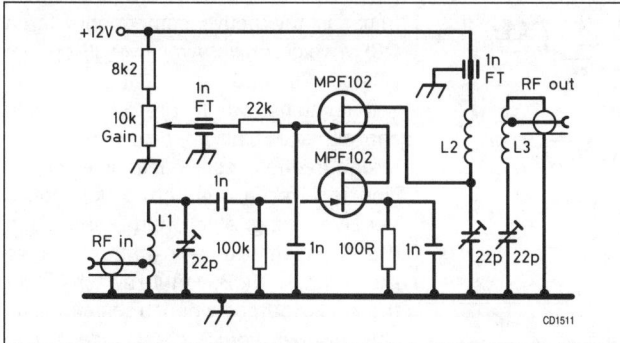

Fig 5.16: Two JFETs in series (cascode) can be substituted for a dual-gate MOSFET. As a rule of thumb, the upper gate should be biased at one-half of the supply voltage. AGC can be applied to that gate, going downward from one-half supply voltage

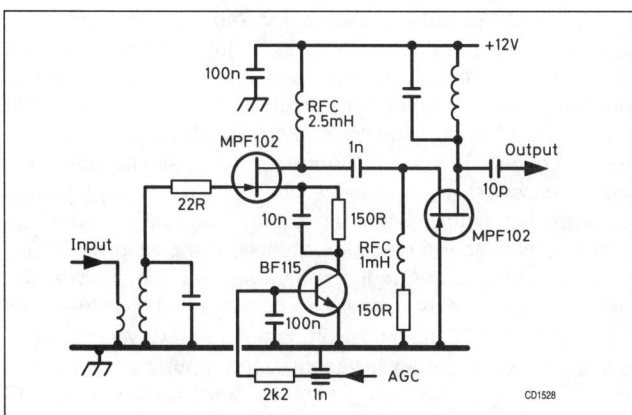

Fig 5.17: A cascode amplifier with two JFETs which are in series for RF but in parallel for DC. The bipolar transistor controls the gain in response to the applied AGC voltage

overload it and block your receiver. Another use for this grounded-gate amplifier is as a buffer stage between a diode mixer and crystal filter. The amplifier provides the correct load for the mixer and can also be designed to provide the correct termination impedance for a crystal filter as well.

The VHF amplifier in **Fig 5.15** uses a dual-gate MOSFET. The MOSFET is now normally available in a SMD package. If a MOSFET is not convenient, two JFETs in the circuit of **Fig 5.16** may be substituted. In either case, the best noise figure requires adjustment of the input tuning with the help of a noise generator

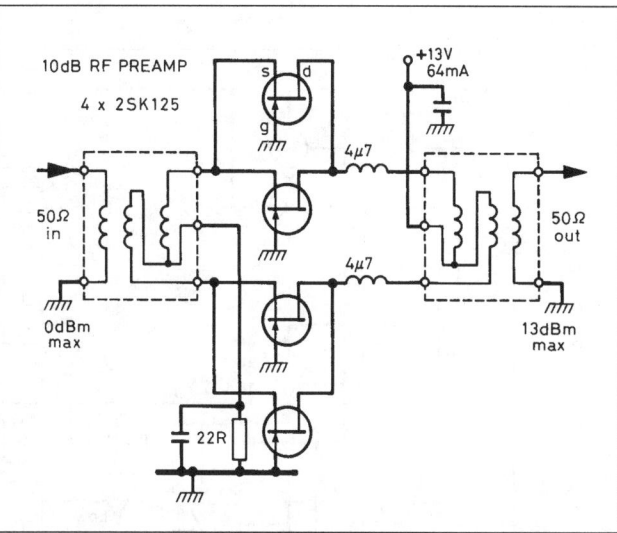

Fig 5.18: Four grounded-gate FETs in push-pull parallel running high drain currents give this RF stage of the Yaesu FT-1000 HF transceiver its excellent dynamic range, suppression of second-order intermodulation, and proper termination of preceding and following bandpass filters

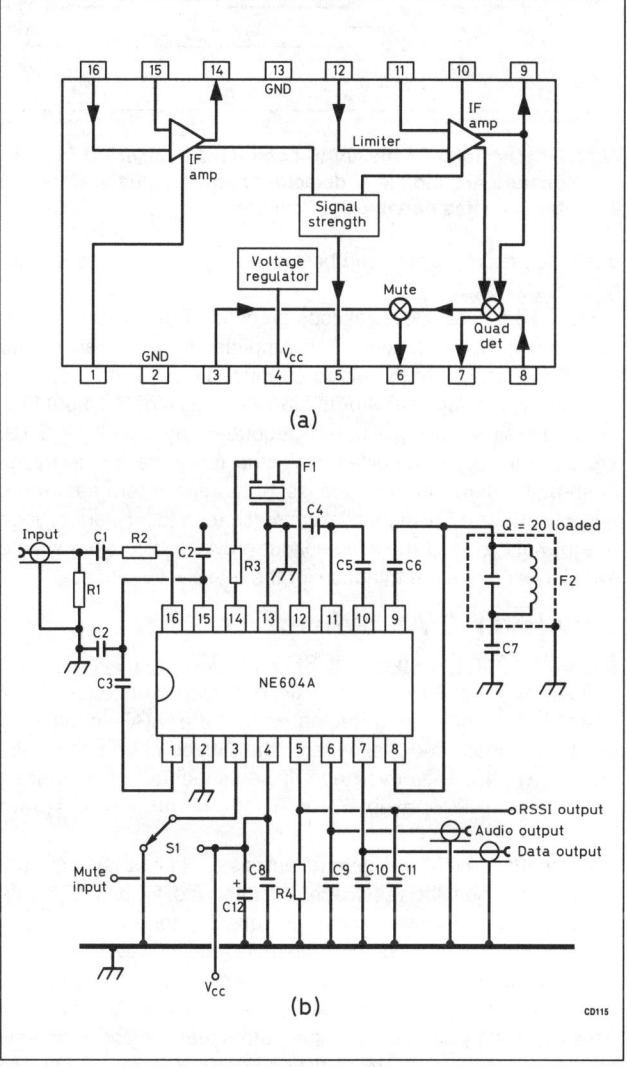

Fig 5.19: The NE604A FM receiver IFIC. (a) Internal block diagram. (b) External connections

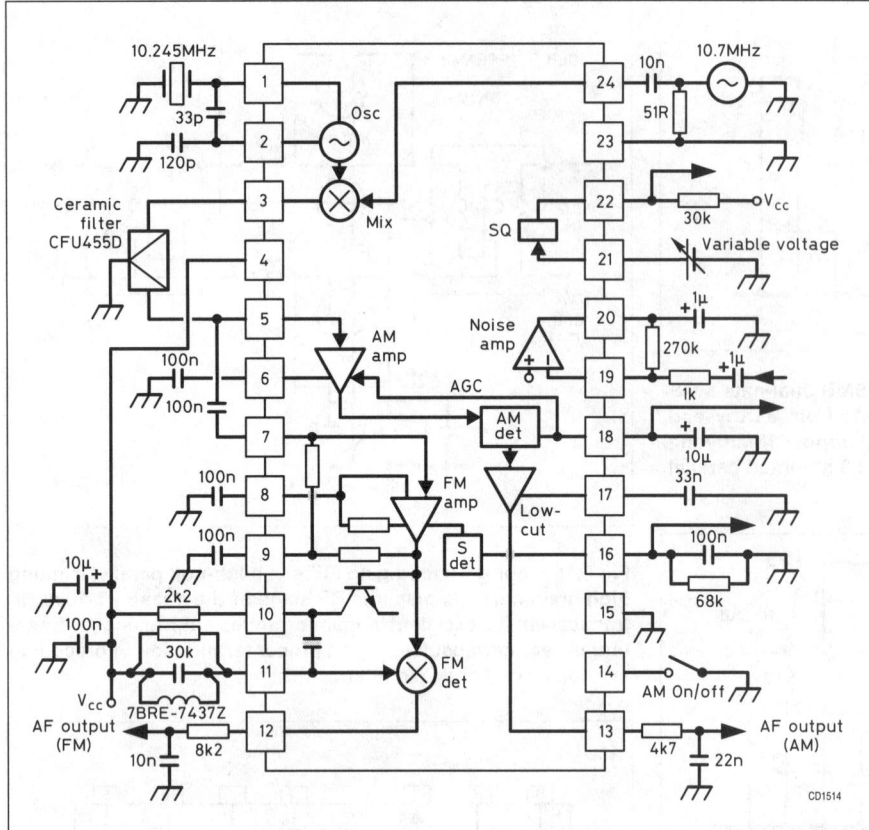

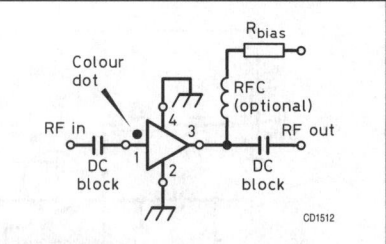

Fig 5.21: MMICs can supply high wide-band gain up to 2.3GHz with a reasonable noise figure. Input and output are 50 ohms and no tuned circuit is

Fig 5.20: The Toko TK10930V IC comprises a mixer/LO from a 10.7MHz 1st IF to 455kHz and separate AM and FM IF-detector channels, plus auxiliary circuitry. Low drain, from a 3V battery, invites portable applications

because maximum gain and best noise figure do not necessarily coincide.

Fig 5.17 represents a cascode amplifier. The tuned input circuit is only lightly loaded. The amplifier is very stable, and smoothly changes from gain to attenuation as the AGC voltage causes the bipolar transistor to reduce the gain of the input FET.

Four FETs in a push-pull parallel grounded-gate circuit, **Fig 5.18**, are used in the RF amplifiers of some top-grade HF receivers. Push-pull operation reduces second-order intermodulation. Several smaller FETs in parallel approach the wide dynamic range of a power FET, and the source input provides, through a simple transformer, proper termination for the preselector filters.

Low-level IC Amplifiers

Monolithic integrated-circuit RF amplifiers come in a great variety, sometimes combined on one chip with other functions such as a mixer, detector, oscillator or AGC amplifier. A linear IC, though more expensive than its components in discrete form, has the advantage of well-specified performance in the proven PCB layouts given in the manufacturer's data sheet.

Some ICs are labelled 'general purpose', while others are optimised for a specific application. If intended for a VHF hand-portable FM transceiver, low power consumption might be a key feature, but linearity would be unimportant. However, an IC for fixed-station SSB use would be selected for good linearity, but the current consumption would be unimportant. Any potential user would do well to consult the data sheets or books on several alternative ICs before settling on any one. New devices are being introduced frequently and the price of older ones, often perfectly adequate for the intended function, is then reduced.

The following elaborates on a few popular devices, but only a fraction of the data sheet information can be accommodated here.

The Philips-Signetics NE604AN contains all the active components for a NBFM voice or data receiver IF system [2] - see **Fig 5.19**.

It contains two IF amplifiers, which can either be cascaded or used at different frequencies in a double-conversion receiver, a quadrature detector, a 'received signal-strength indicator' (RSSI) circuit with a logarithmic range greater than 90dB, a mute switch (to cut the audio output when transmitting in simplex operation) and a voltage regulator to assure constant operation on battery voltages between 4.5V (at 2.5mA) and 8V (at 4mA). The IC is packaged in a 16-pin DIP. The operating temperature range of the NE604AN is 0 to 70°C.

The Toko TK10930V IC is an AM-FM 2nd IF-detector system in a 24-pin DIP. In it, the 10.7MHz 1st IF input is converted to a 455kHz 2nd IF and processed in separate AM and FM amplifier and detector channels to simultaneously provide audio output in each. AVC, RSSI and squelch circuitry is included. DC requirements are only 3V @ 4mA (FM only) or 7mA (both channels on). Applications are in portable scanners, airband and marine receivers but, unfortunately, there is no provision for SSB reception. Internal and external components are shown in **Fig 5.20**.

Agilent and Mini-Circuits make several types of MMIC (microwave monolithic integrated circuit) which provide wide-band gain from DC up to the 2.3GHz band with very few external components [3, 4]. They have a 50 ohms input, can be cascaded without interstage tuning, have noise figures as low as 3dB at 1GHz and are capable of 10-20mW of output into 50Ω. Packaging is for direct soldering to PCB tracks. The only external components required are a choke and series resistor in the 12VDC supply lead and DC blocking capacitors at their input and output as shown in **Fig 5.21**. As new models appear frequently, no listing is given here.

The Philips NE592N is an inexpensive video amplifier IC with sufficient bandwidth to provide high gain throughout the HF region. Its differential input (and output) enabled G4COL to use it, without a transformer, in a very sensitive antenna bridge in which both the RF source and the unknown impedance are earthed [5] The circuit is shown in **Fig 5.22**. For video response down to DC this IC requires both positive and negative supply voltages; for AC-only, ie RF, a single supply suffices, as explained for operational amplifiers later in this chapter.

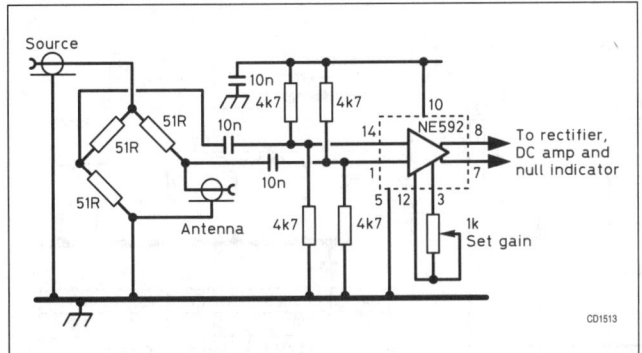

Fig 5.22. An inexpensive video amplifier IC, NE592N, can serve as a wide-band HF amplifier, as it does in G4COL's sensitive HF antenna bridge

The NE602 monolithic mixer oscillator IC can also be used as an amplifier. The circuit is shown in **Fig 5.23**. The circuit has about 20dB of gain and the 3rd order linearity is the same as the NE602 used as a mixer. The circuit is most suitable for use in the limiting stage of a FM receiver.

RF Solid-state Power Amplifiers

Representative professional designs of high-power amplifiers are given and explained in later chapters.

The reader will notice the many measures required to protect the often very costly transistors under fault conditions, some of which may occur during everyday operation. That does not encourage experimentation with those designs but amateurs have discovered that devices either not designed for RF service or widely available on the surplus market cost much less and can be made to perform in homebrew amplifiers. One might be tempted to use VHF transistors salvaged from retired AM PMR (VHF or UHF) transmitters, in HF amplifiers. This is attractive because of the very high power gain of VHF transistors used a long way below their design frequency. However, stability problems must be expected.

Switching MOSFETs at HF

As 'real' HF power transistors remain expensive, amateurs have found ways to use inexpensive MOSFETs intended for audio, digital switching, switch-mode power supplies or ultrasonic power applications at ever-increasing frequencies.

MOSFETs, as compared with bipolar power transistors, facilitate experimentation for several reasons. Their inherent reduction of drain current with increasing temperature contrasts with bipolar power transistors where 'thermal run-away' is an ever-present danger. MOSFET gates require only driving voltage, not power; though developing that voltage across the high gate-to-source capacitance requires some ingenuity, it greatly simplifies biasing. Where driving power is available, a swamping resistor across the gate-to-source capacitance helps, especially if that capacitance is made part of a pi-filter dimensioned to match the swamping resistor to the desired input impedance, eg 50Ω. A swamping resistor also reduces the danger of oscillation caused by the considerable drain-gate capacitance of a MOSFET. The most important limitation is that MOSFETs require a high supply voltage for efficient operation, 24-50V being most common; on 13.8V, output is limited.

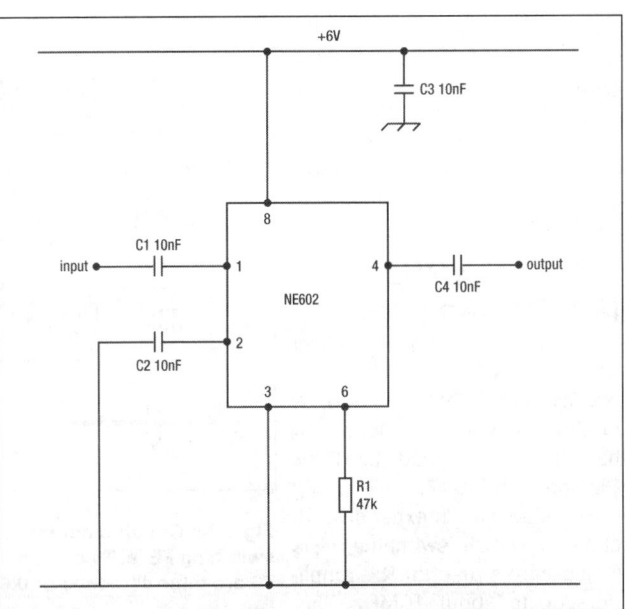

Fig 5.23: NE602 mixer IC used as an RF or IF amplifier. Pins 7 and 5 on the IC are not connected

Here follow some such designs. Note that most use push-pull circuits. This suppresses even harmonics and thereby reduces the amount of output filtering required. With any wide-band amplifier, a proper harmonic filter for the band concerned must be used; without it even the third or fifth harmonic is capable of harmful interference. See **Table 5.3**.

Parasitic oscillations have also dogged experimenters. They can destroy transistors before they are noticed. A spectrum analyser is the professional way to find them but with patience a continuous coverage HF/VHF receiver can also be used to search for them. It is useful to key or modulate the amplifier to create a 'worst case' situation during the search. If found, a ferrite bead in each gate lead and/or a resistor and capacitor in series between each source and drain are the usual remedies. W1FB found negative feedback from an extra one-turn winding on the output transformer to the gate(s) helpful. VHF-style board

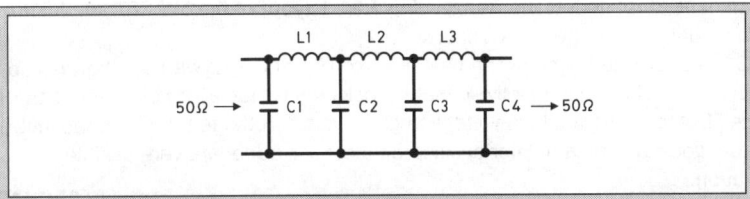

Band (MHz)	C1, C4 (pF)	C2, C3 (pF)	L1, L3 (H)	L2 (turns)	Wire (H)	(turns)	Cores (Cu enam)	
3.5	560	1200	2.46	21	2.89	23	0.4mm, 26AWG	T-50-2
7	470	820	1.4	17	1.56	18	0.5mm, 24AWG	T-50-2
10.1	220	470	0.96	15	1.13	17	0.5mm, 24AWG	T-50-6
14	110	300	0.6	11	0.65	14	0.5mm, 24AWG	T-50-6
18	100	250	0.52	11	0.65	13	0.6mm, 22AWG	T-50-6
21	110	240	0.48	11	0.56	12	0.6mm, 22AWG	T-50-6
24	120	270	0.54	12	0.63	13	0.6mm, 22AWG	T-50-6
28	56	150	0.3	8	0.38	10	0.8mm, 20AWG	T-50-6

Notes. Various cut-off frequencies and ripple factors used to achieve preferred-value capacitors. Coil turns may be spread or compressed with an insulated tool to peak output. Cores are Amidon toroids. Capacitors are silver mica or polystyrene, 100V or more.

Table 5.3: Low pass output filters for HF amplifiers up to 25W

layout and bypassing, where the back of the PCB serves as a ground plane, are of prime importance [6].

The first example is aimed at constructors without great experience; those following represent more difficult projects.

A 1.8 - 10.1MHz 5W MOSFET Amplifier

A push-pull broad-band amplifier using DMOS FETs and providing 5W CW or 6W PEP SSB with 0.1W of drive and a 13.8VDC supply has been described by Drew Diamond, VK3XU [7].

He used two inexpensive N-channel DMOS switching FETs which make useful RF amplifiers up to about 10MHz. This amplifier had a two-tone IMD of better than -30dBc (-35dBc typical) and, with the output filter shown in **Fig 5.24**, all harmonics were better than -50dBc. The amplifier should survive open or shorted output with full drive and remain stable at any load SWR.

The drain-to-drain impedance of the push-pull FETs is 2 x 24 = 48Ω so that no elaborate impedance transformation is needed to match into 50Ω. T3 serves as a balun transformer. T2 is a balanced choke to supply DC to the FETs. Negative RF feedback is provided by R3 and R4, stabilising the amplifier and helping to keep the frequency response constant throughout the range. The heatsink of the bias zener

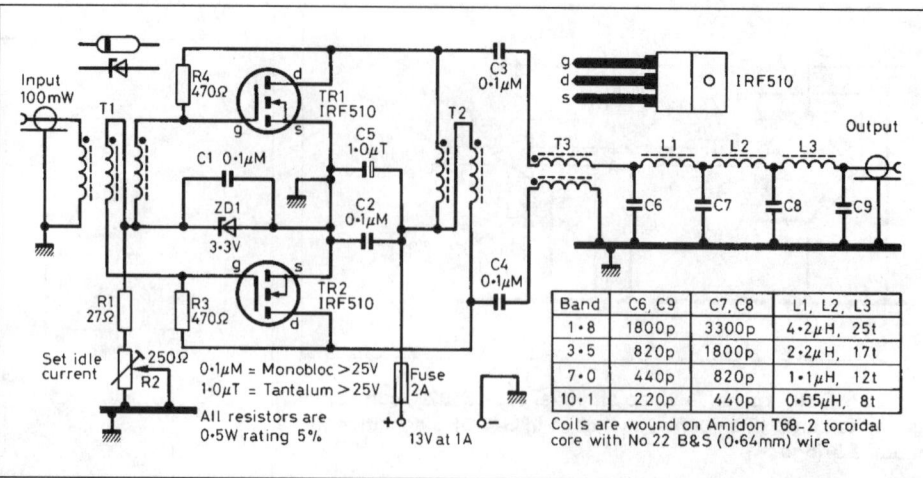

Band	C6, C9	C7, C8	L1, L2, L3
1·8	1800p	3300p	4·2μH, 25t
3·5	820p	1800p	2·2μH, 17t
7·0	440p	820p	1·1μH, 12t
10·1	220p	440p	0·55μH, 8t

Coils are wound on Amidon T68-2 toroidal core with No 22 B&S (0·64mm) wire

Fig 5.24: Circuit diagram of the 1.8-10.1MHz amplifier providing 5W CW or 6W PEP SSB using switching FETs. T1 comprises three 11t loops of 0.5mm enam wire on Amidon FT50-43 core. T2, T3 are three 11t loops of 0.64mm enam wire on Amidon FT50-43. * Indicates start of winding

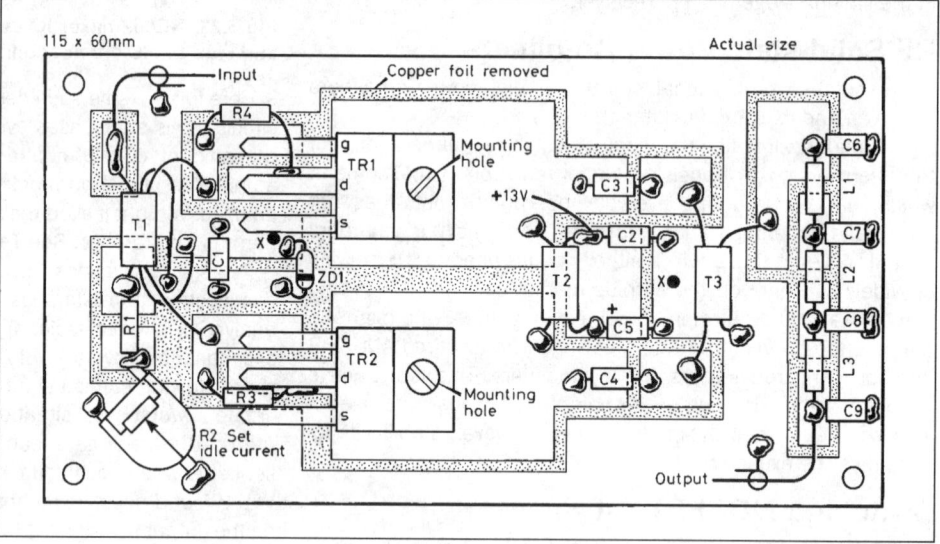

Fig 5.25: Layout of the 5W FET amplifier on double-sided PCB

ZD1 is positioned against the heatsinks of TR1 and TR2 with a small blob of petroleum jelly so that it tracks the temperature of the FETs, causing the bias voltage to go down when the temperature goes up. The amplifier enclosure must have adequate ventilation.

The complete amplifier, with an output filter for one band, can be built on a double-sided 115 x 60mm PCB. For stability, the unetched 'ground plane' should be connected to the etched-side common/earth in at least two places marked 'X' in **Fig 5.25**. Drill 1mm holes, push wires through and solder top and bottom.

If multiband operation is required, the filter for the highest band should be accommodated on the amplifier board and kept in the circuit on all bands. Lower-frequency filters can then be mounted on an additional board. Polystyrene or silvered-mica capacitors should be used in the filters. Hard-to-get values can be made up of several smaller ones in parallel.

When setting up, with R2 at minimum resistance the desired no-signal current is 200-300mA. With 100mW drive and a 50Ω (dummy) load, the supply current should be about 1A. After several minutes of operation at this level and with suitable heatsinks, the latter should not be uncomfortably hot when touched lightly. While 100mW drive should suffice on the lower

bands, up to 300mW may be needed at 10.1MHz, which is about the limit. Overdriving will cause flat-topping. With larger heatsinks and higher supply voltage, more output would be possible.

A MOSFET 1.8-7MHz 25W Power Amplifier

VK3XU later described a more powerful version of his 5W amplifier using Motorola MTP4N08 MOSFETs (80V at 4A) made for switch-mode power supplies [8]. The IRF510, having the same pin-out and voltage/current ratings and lower input and output capacitances, should give better performance above 7MHz; This amplifier is shown in **Fig 5.26**.

VK3XU lists performance with MTP4N08 devices as: frequency range 1.8-7MHz (with reduced output at 14MHz); output power nominally 25W, typically 30W PEP or CW with 1W drive; input SWR less than 1.2:1; two-tone IMD about -35dBc; harmonic output, depending on low-pass output filter, -50dBc; No output protection is required; the amp will withstand any output SWR, including short- and open-circuit at full drive; DC supply is 25V at 2A (no regulation required) or 13.8V at reduced output. Proper output filters must be used for each band, eg those filters shown in Table 5.3.

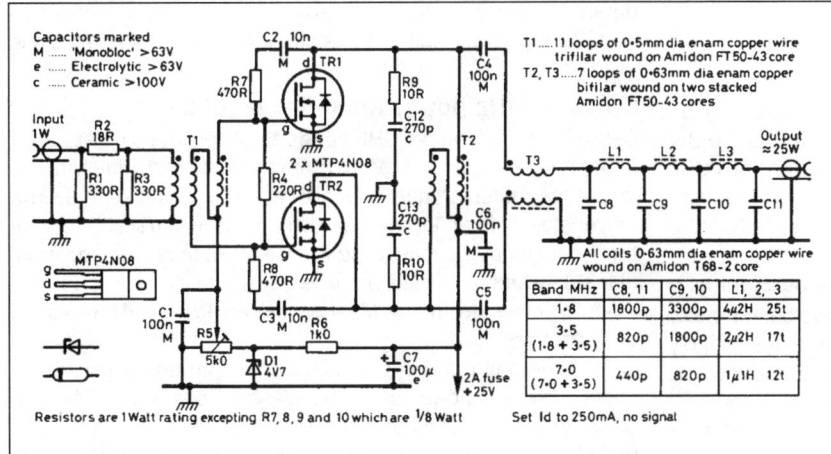

Fig 5.26: Circuit of the 1.8-7MHz amplifier providing 25W PEP of SSB using FETs intended for switching power supplies

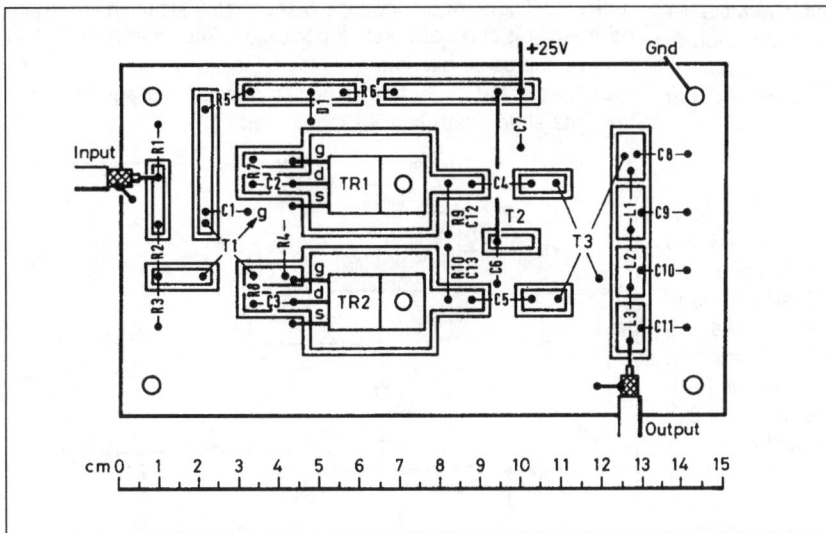

Fig 5.27: Layout of the 25W FET amplifier on double-sided PCB

The construction, **Fig 5.27**, is very much like that of the 5W version. The MOSFETs must each be fitted with an adequate heatsink. If this amplifier were to be operated at its rated 50W input and 25W output, each transistor has to dissipate 12.5W; To hold the FET tab temperature to 125°C at an in-cabinet ambient of 50°, ie a rise of 75°, heatsinks rated less than 75/12.5 = 6°/W would be required; this would be correct in duplex FM or RTTY service but when using Morse, AMTOR or SSB, the average dissipation would be roughly half that and 10°/W heatsinks would do. Pushing the amplifier to its high-frequency limits, however, would reduce efficiency and require larger heatsinks. If in doubt, use a small blower. Do remember that the tabs of these FETs are connected to the drains, so they are 'hot' both for DC and RF.

The large capacitor in the bias supply allows time for the antenna change-over relay to close before forward bias reaches the FET gates; arrangements should be made, eg with a diode in the PTT line (not shown) to 'kill' that bias as soon as the PTT switch is released, ie before the antenna relay opens. These measures assure that there is no output from the amplifier unless the antenna relay is connecting the load, thus preventing damage to the relay or the FETs.

100W Multiband HF Linear

David Bowman, G0MRF, built and described [9] in great detail the design, construction and testing of his amplifier, complete with power supply, T/R switching and output filtering. With a pair of inexpensive 2SK413 MOSFETs, it permits an output of 100W from topband through 14MHz and somewhat less up to 21MHz. The RF part is shown in **Fig 5.28**.

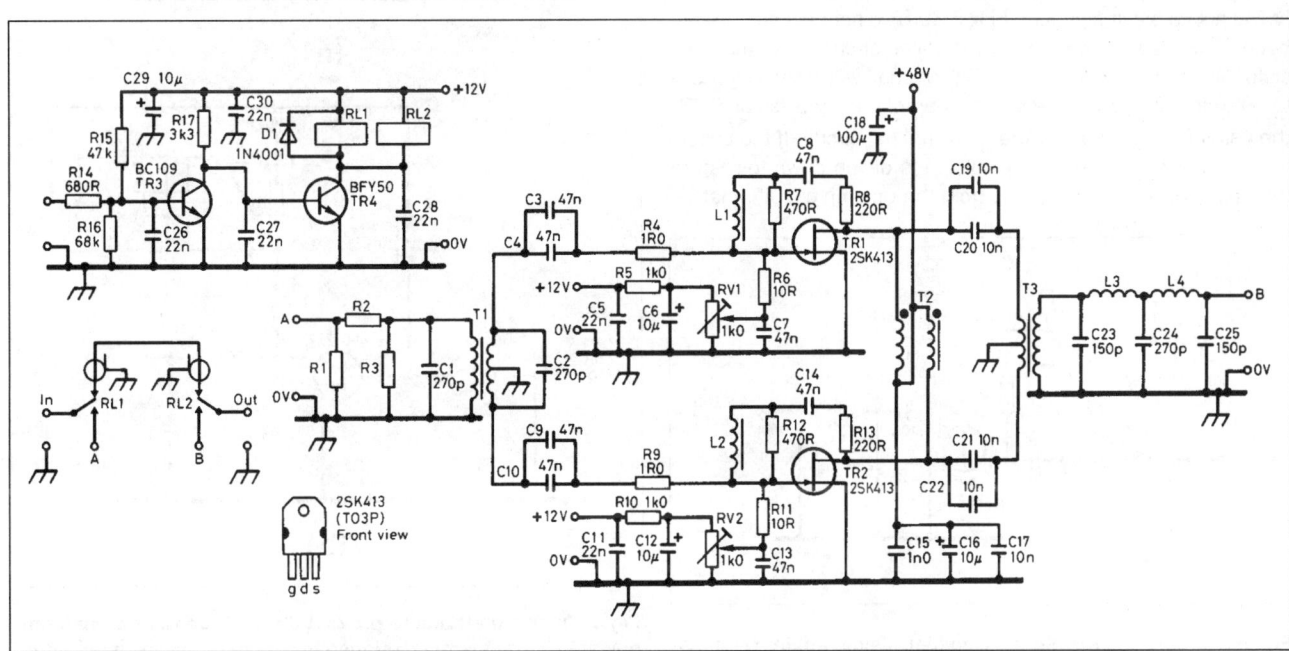

Fig 5.28: G0MRF's 100W HF linear amplifier uses inexpensive MOSFETs

A Linear 50MHz Amplifier

PAOKLS found that useful 50MHz output could be obtained from a single IRF610 MOSFET in the circuit of **Fig 5.29** [10]. With a 50V supply, the forward bias for Class AB operation was adjusted for an idling current of 50mA. A two-tone input of 0.2-0.3W then produced an output of 16W at a drain current of 0.5A. Third-order distortion was 24dB below either tone, not brilliant but adequate. Voice quality was reported to be good. On CW, the output exceeded 20W.

Best output loading of $(V_d)^2/2P_o = 78\Omega$ was obtained with another pi-network; while this provides only single-band matching, it reduces harmonics. The drain choke was made roughly resonant with the IRF610's drain-to-source capacitance of 53pF, so that the pi-coupler only has to match near-pure resistances.

As in a single-ended amplifier there is no cancellation of even harmonics and the second harmonic of 50MHz is in the 88-108MHz FM broadcast band, additional low-pass filtering is mandatory.

Set up for 28MHz, this amplifier produced 30-40W CW; at this power level, with an insufficiently large heatsink, the MOSFET got very hot but it did not fail due to the inherent reduction of drain current mentioned above.

The difference between 28 and 50MHz output shows that 50MHz is about the upper limit for the IRF610.

VHF Linear Amplifiers

For medium-power VHF linear amplifiers there is another cost-saving option. Because transistors intended for Class C (AM, FM or data) cost a fraction of their linear counterparts, it is tempting and feasible to use them for linear applications. If simple diode biasing alone is used, however, the transistors, such as the MRF227, go into thermal runaway. Two solutions to this problem were suggested in the ARRL lab for further experimentation [11]. Both methods preserve the advantages of a case earthed for DC.

The current-limiting technique of **Fig 5.30(a)** should work with any device, but is not recommended where current drain or power dissipation are important considerations. The approach is simply to use a power supply which is current limited, eg by a LM317 regulator, and to forward bias the transistor to operate in Class B. Forward bias is set by the values of RB1 and RB2.

The active biasing circuit of **Fig 5.30(b)** is not new but has not been much used by amateurs. The collector current of the transistor is sampled as it flows through a small-value sensing resistor R_s; the voltage drop across it is amplified by a factor R_f/R_i, the gain of the op-amp differential amplifier circuit. If the collector current goes up, the base voltage is driven down to restore the balance. The gain required from the op-amp circuit must be

determined empirically. While experimenting, the use of a current-limiting power supply is recommended to avoid destroying transistors.

50-1296MHz power amplifier modules

Designing a multi-stage UHF power amplifier with discrete components is no trivial task. More often than not, individually tuned interstage matching networks are required. Duplicating even a proven design in an amateur workshop has its pitfalls. One solution, though not the least expensive, is the use of a sealed modular sub-assembly. These are offered by several semiconductor manufacturers, including Mitsubishi (see **Table 5.4**), Motorola and Toshiba.

For each band, there is a choice of output power levels, frequency ranges, power gain and class of operation (linear in Class AB for all modes of transmission or non-linear in Class C for FM or data modes only). All require a 13.8V DC supply, sometimes with separate external decoupling for each built-in stage, and external heatsinking. In general, their 50-ohm input provides a satisfactory match for the preceding circuitry, and a pi-tank (with linear inductor on the higher bands) is used for antenna matching and harmonic suppression. They are designed to survive and be stable under any load, including open- and short-circuit. As each model has its peculiarities, it is essential to follow data sheet instructions to the letter.

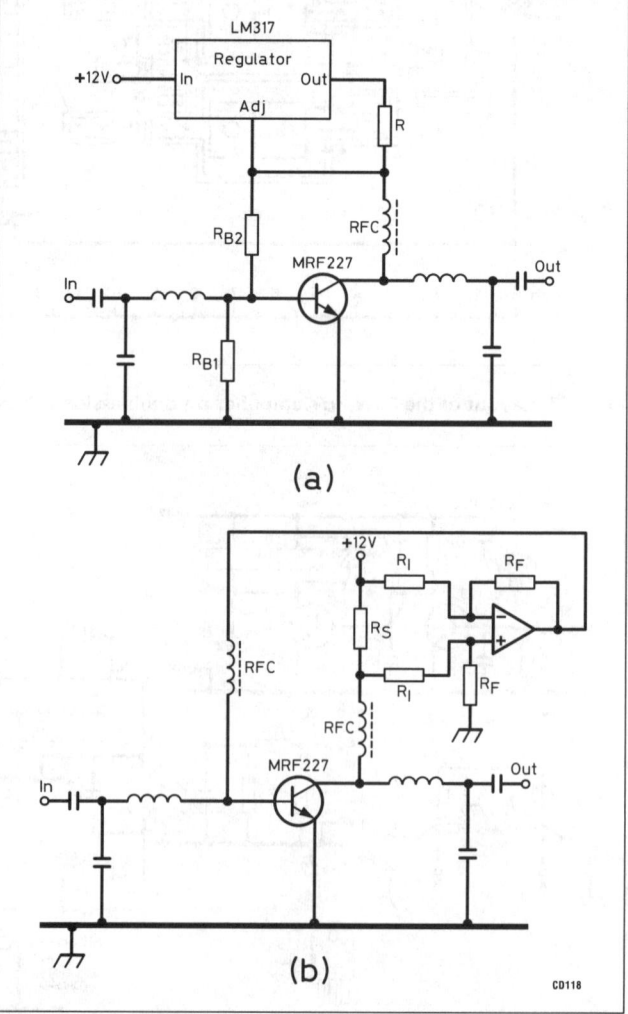

(a)

(b)

Fig 5.30: Two methods to prevent thermal runaway of medium-power VHF transistors designed for Class C only in linear amplifiers. (a) A supply current regulator. (b) An active biasing circuit (*QST*)

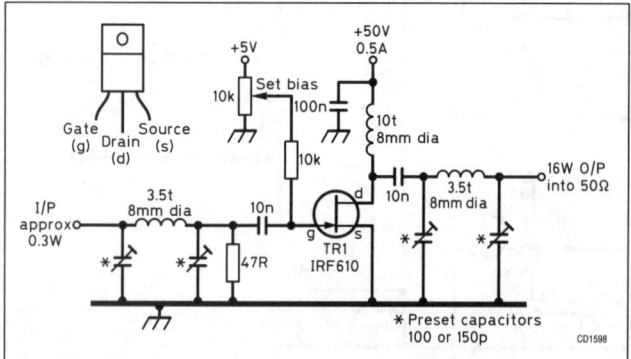

Fig 5.29: This 50MHz linear amplifier using an inexpensive switching MOSFET will deliver 16W output with less than 0.3W of drive

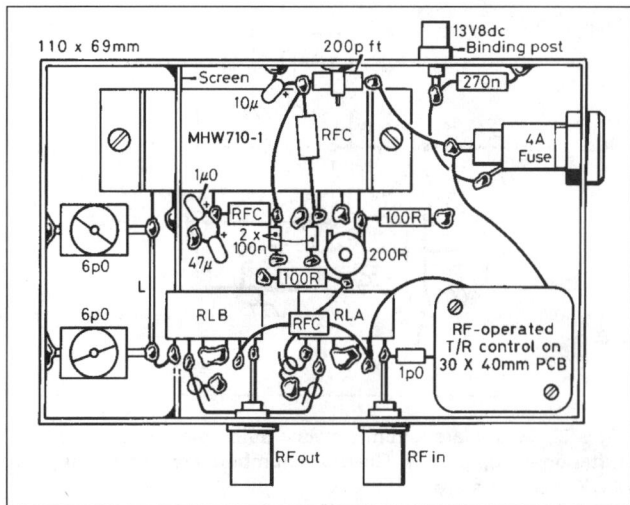

Fig 5.32: Construction of the PA0GMS 15W 435MHz mobile booster. The RF-operated T/R control is a sub-assembly on a PCB

Part No.	Po (W)	Pin (mW)	Band (MHz)	Mode
M57735	19	200	50	lin
M57706L	8	200	145	-
M57719L	14	200	145	-
M57732L	7	20	145	lin
M57741UL	28	200	145	-
M57796L	7	300	145	lin
M67748L	7	20	145	lin
M67781L	40	300	145	-
M57704M	13	200	435	-
M57714M	7	100	435	-
M57716	17	200	435	lin
M57729	30	300	435	-
M57745	33	300	435	lin
M57788M	40	300	435	-
M57797MA	7	100	435	lin
M67709	13	10	435	lin
M67728	60	10	435	lin
M67749M	7	20	435	lin
M57762	18	1000	1296	lin

Table 5.4: Some Mitsubishi RF power modules. All operate on 12V mobile systems, but those rated over 25W cannot be supplied from a 5A (cigar lighter) circuit and generally require forced air cooling of the heat sink. Suitable for SSB only if marked 'lin' (Extracted from a 1998 Mainline Electronics pamphlet)

A 435MHz 15W mobile booster

PA0GMS built this 70cm power amplifier with RF VOX to boost the FM output from his hand-held transceiver to a respectable 15W for mobile use [12]. The circuit is shown in **Fig 5.31**. Its top half represents the amplifier proper, consisting of (right to left) the

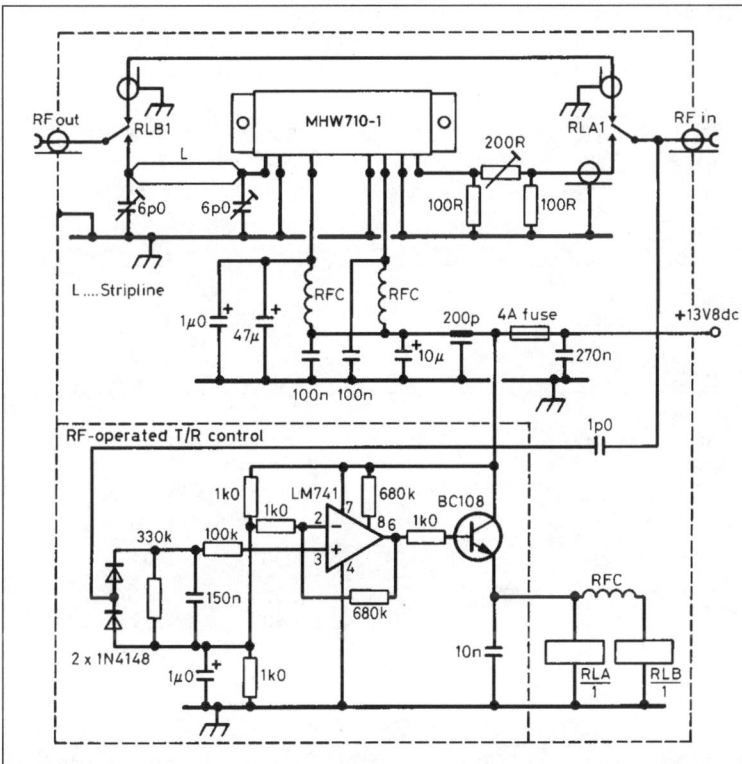

Fig 5.31: The PA0GMS 15W 435MHz mobile booster. The original MHW710-1 power module is obsolete, but model M57704M (different pin-out) is suitable

input T/R relay, input attenuator to limit input power to what the module requires for rated output, the power module with its power lead decoupling chokes (ferrite 6 x 3mm beads with three turns of 0.5mm diameter enamelled copper wire) and bypass capacitors (those 0.1µF and larger are tantalum beads with their negative lead earthed; lower values are disc ceramics, except for the 200pF which is a feedthrough type). The output tank (two air trimmers and a 25mm long straight piece of 2mm silvered wire) feeds into the output T/R relay. Both relays are National model RH12 shielded miniature relays not designed for RF switching but adequate for UHF at this power level. The choke between the relay coils is to stop RF feedback through the relays.

The relays are activated by what is sometimes misnamed a VOX (voice-operated control) but is in fact an RF-operated control. Refer to the diagram at the bottom of Fig 5.31. A small fraction of any RF drive applied to the amplifier input is rectified and applied to the high-gain op-amp circuit here used as a comparator. Without RF input, the op-amp output terminal is at near-earth potential, the NPN transistor is cut off, and the relays do not make, leaving the amplifier out of the circuit as is required for receiving. When the operator has pressed his PTT (push-to-talk) switch, RF appears at the booster input, the op-amp output goes positive, the transistor conducts, and the relays make, routing the RF circuit through the booster.

This kind of circuit is widely used for bypassing receiving amplifiers and/or inserting power boosters in the antenna cable when transmitting; at VHF or UHF, this is often done to make up for losses in a long cable by placing these amplifiers at masthead or in the loft just below. Though not critical, the 1pF capacitor may have to be increased for lower frequencies and lower power levels, and reduced for higher frequencies and higher input power. The NPN transistor must be capable of passing the coil currents of the relays. While not an issue with 0.5W in and 15W out, at much higher power levels proper relay sequencing will be required.

No PCB is used for the amplifier proper. It is built into a 110 x 70 x 30mm tin-plate box, **Fig 5.32**. The module is bolted through the bottom of the box to a 120 x 70mm heatsink which must dissipate up to

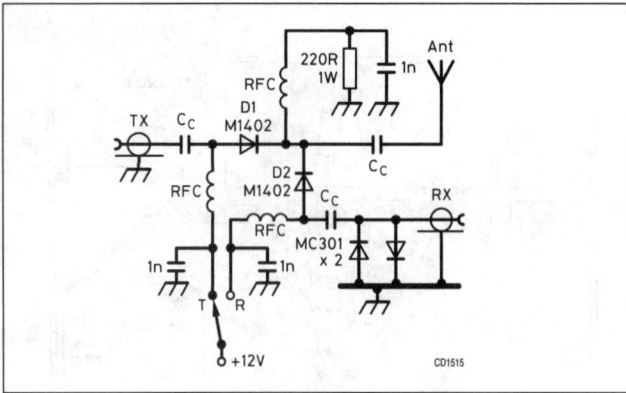

Fig 5.33: Solid-state antenna switch using pin diodes for transmitter power up to 25W. The diode numbers are Mitsubishi models. Typical data are:

Frequency (MHz)	Isolation Tx to Rx (dB)	Insertion loss (dB)
29	40	0.3
50	39	0.3
144	39	0.3
220	38	0.4
440	36	0.5

30W. Heat transfer compound is used between the module and the box and also between the box and the heatsink; both must be as flat as possible. Care is required when tightening the bolts as the ceramic substrate of the module is brittle. The other components are soldered in directly. The RF-operated relay control is a sub-assembly mounted on a 40 x 30mm PCB.

Adjustments are simple. Set the input attenuator to maximum resistance and the two air trimmers to half-mesh. Apply 0.6 or 0.7W input from the hand-held and verify that both relays make. A few watts of output should be generated at this stage. Adjust the two air trimmers to peak the output, then reduce the input attenuation until the output reaches 15W with a supply current of about 3A. This completes the adjustment. Harmonics were found to be -60dBc and intermodulation -35dBc.

Pin Diode T/R Switching

If the same antenna is to be used both for transmitting and receiving, change-over switching is required. This normally is accomplished by means of an electro-mechanical relay. If, however, the receiver must function as quickly as possible after the end of each transmission, as in high-speed packet operation, the release plus settling time of a relay, several milliseconds, restricts the data throughput and solid-state switching is indicated.

This is done by using pin diodes as switches. When such a diode is 'biased on' by (typically) 50mA DC in its pass direction, its RF resistance is almost as low as that of closed relay contacts; when biased by a voltage in the opposite direction, it acts like a pair of open relay contacts. To separate bias and RF circuits, bias is applied through RF chokes; capacitors pass only the RF. **Fig 5.33** shows the basic circuit. The resistor limits the bias current and provides the 'off bias' voltage. The back-to-back diodes limit the transmitter power reaching the receiver under fault condition. The insertion loss and isolation (the fraction of transmitter power getting through to the receiver) achievable with discrete components is tabulated.

At UHF and above, less-effective chokes and stray capacitances limit performance. For the 432 and 1296MHz bands, there are hybrid modules available containing the whole circuit optimised for one band.

Valve Power Amplifiers

Warning - all valve power amplifiers use lethal voltages. Never reach into one before making sure that the HT is off and HT filter capacitors discharged; do not rely on bleeder resistors - they can fail open-circuit without you knowing it. If there is mains voltage in the amplifier enclosure, be sure that the plug is pulled before working on it.

Virtually all current factory-made HF amplifiers up to the normal power limit (400W = 26dBW PEP output in the UK) are solid state. The same is true for VHF and UHF equipment up to about 100W. For home construction, however, the criteria are different. Power transistors are expensive and can be instantly

Type	Base	Heater (V)	Pa (A)	Va (W)	Vg2 (V)	Fmax full (V)	Po max rating (MHz)	Socket (W)	
QQV02-6*	B9A (Fig 5.48)	6.3 12.6	0.8 0.4	2 3	275	200	500	5	B9A
QQV03-10* 6360	B9A (Fig 5.48)	6.3 12.6	0.8 0.4	2 5	300	200	225	12.5	B9A
QQV03-20A* 6252	B7A (Fig 5.72)	6.3 12.6	1.3 0.65	2 10	600	250	200 600	48 20	B7A
QQV06-40A* 5894	B7A (Fig 5.72)	6.3 12.6	1.8 0.9	2 20	750	250	200 475	90 60	B7A
PL519/40KG6A 40KD6A	B9D (Fig 5.73) 9RJ (Fig 5.73)	42	0.3	35	2500	275	21	100	B9D
EL519/6KG6A 6KD6A	B9D (Fig 5.73) 9RJ (Fig 5.73)	6.3	2.0	35	2500	275	21	100	B9D
813	5BA	10	5.0	125	2500	750	30	250	5BA
4X150A QV1-150A	B8F Special	6.0	2.6	150	1250	200	165 500	195 140	2m SK600A 70cm SK620A
2C39A 7289 3CX100A5	Disc seal	6.3	1.05	100	1000	-	2500	40 17	500MHz special 2500MHz
4CX250B QE61-250	B8F Special	6.0	2.6	250	2000	400	500	300 (AB1) 390 (C)	2m SK600A 70cm SK620A

Table 5.5: Ratings of some commonly used PA valves

Test frequency	3.7MHz
Anode voltage	710V
Anode current (no signal)	20mA
Anode current (max signal)	162mA*
Control-grid bias	5V set for minimum cross-over distortion on 'scope
Average RF current into 70	1.15A
PEP RF output	185W*
PEP RF input	20W* (estimated)
Anode load resistance	2k (estimated)
Valve inter-electrode capacitances	
Anode to all other electrodes	22pF measured cold
Anode to control grid	2.5pF from data sheet
Control grid to cathode	20pF measured cold

At this level, the solder of the anode caps would melt even with fan cooling! It is reassuring to know that overdriving the amplifier will not distort the output, ie cause splatter, or instantly destroy the valve but it obviously is not for routine operation.

Table 5.6: Two-tone measurements by G4DTC on his PL519 grounded-grid amplifier with fan cooling

destroyed, even by minor abuse; some components, such as a transformer for a 50V/20A power supply, are not easy to find at a reasonable price. By contrast, valves do tolerate some abuse and those who consider the search for inexpensive components an integral part of their hobby find rich pickings at rallies and surplus sales. **Table 5.5** gives rough operating conditions (PEP or CW) for some valves still popular with amateur constructors.

All high-power amplifiers require an RF-tight enclosure and RF filtering of supply and control leads coming out of them. Doing otherwise invites unnecessary interference with any of the electronic devices in your and your neighbours' homes.

Technical alternatives will now be considered and several economical ways of implementing them are given. All are single-stage linear amplifiers intended to boost the output power of a transceiver.

Choosing valves

The first choice concerns the valves themselves; one to four of these must be capable of delivering the desired power gain and output power at the highest intended operating frequency at reasonable filament and anode voltage and, where required, control-grid bias and screen-grid voltage. The availability and cost of the valve(s), valve socket(s), power supply transformer(s) and anode tank capacitor will frequently restrict the choice.

Some valves of interest to the economy-minded constructor of HF amplifiers were not designed for radio transmitters at all but to drive deflection yokes and high-voltage transformers at the horizontal sweep frequency in CTV receivers; accordingly, the manufacturers' data sheets are not much help to the designer of RF circuits. However, amateurs have published the results of their experiments over the years. For the popular (originally Philips) pentode PL519, which can produce 100W PEP at HF and reduced output up to 50MHz, some information is given in **Tables 5.5 and 5.6**. With fan cooling, long valve life can be expected.

Another HF favourite is the (originally GE) 'beam power' tetrode type 813. This valve, first made more than 50 years ago, is now available new from Eastern European and Chinese manufacturers at a reasonable price but there are good buys available in used American JAN-813s. These can be tested by comparison with an 813 which is known to be good in the type of amplifier and at the approximate frequency and ratings of intended usage. This valve is rated at 100W anode dissipation;

Tuned grid, Class AB1

Advantages
(a) Low driving power.
(b) As there is no grid current, the load on the driver stage is constant.
(c) There is no problem of grid bias supply regulation.
(d) Good linearity and low distortion.

Disadvantages
(a) Requires tuned grid input circuit and associated switching or plug-in coils for multiband operation.
(b) Amplifier must be neutralised.*
(c) Lower efficiency than Class AB2 operation.

Tuned grid, Class AB2

Advantages
(a) Less driving power than passive grid or cathode-driven operation.
(b) Higher efficiency than Class AB1.
(c) Greater power output.

Disadvantages
(a) Requires tuned grid input circuit.
(b) Amplifier must be neutralised*.
(c) Because of wide changes in input impedance due to grid current flow, there is a varying load on the driver stage.
(d) Bias supply must be very 'stiff' (have good regulation).
(e) Varying load on driver stage may cause envelope distortion with possibility of increased harmonic output and difficult with TVI.

Passive grid

Advantages
(a) No tuned grid circuit.
(b) Due to relatively low value of passive grid resistor, high level of grid damping makes neutralising unnecessary.
(c) Constant load on driver stage.
(d) Compact layout and simplicity of tuning.
(e) Clean signal with low distortion level.
(f) Simple circuitry and construction lending itself readily to compact layout without feedback troubles.

Disadvantages
(a) Requires higher driving power than tuned grid operation.

Cathode driven

Advantages
(a) No tuned grid circuit.
(b) No neutralising (except possibly on 28MHz).
(c) Good linearity due to inherent negative feedback.
(d) A small proportion of the driving power appears in the anode circuit as 'feedthrough' power.

Disadvantages
(a) High driving power - greater than the other methods.
(b) Isolation of the heater circuit with ferrite chokes or special low-capacitance-wound heater transformer.
(c) Wide variation in input impedance throughout the driving cycle causing peak limiting and distortion of the envelope at the driver.
(d) The necessity for a high-C tuned cathode circuit to stabilise the load impedance as seen by the driver stage and overcome the disadvantage of (c).

Neutralisation is built into the QQV-series of double tetrodes.

Table 5.7: Advantages and disadvantages of tuned-grid, passive-grid and grounded-grid PA valves

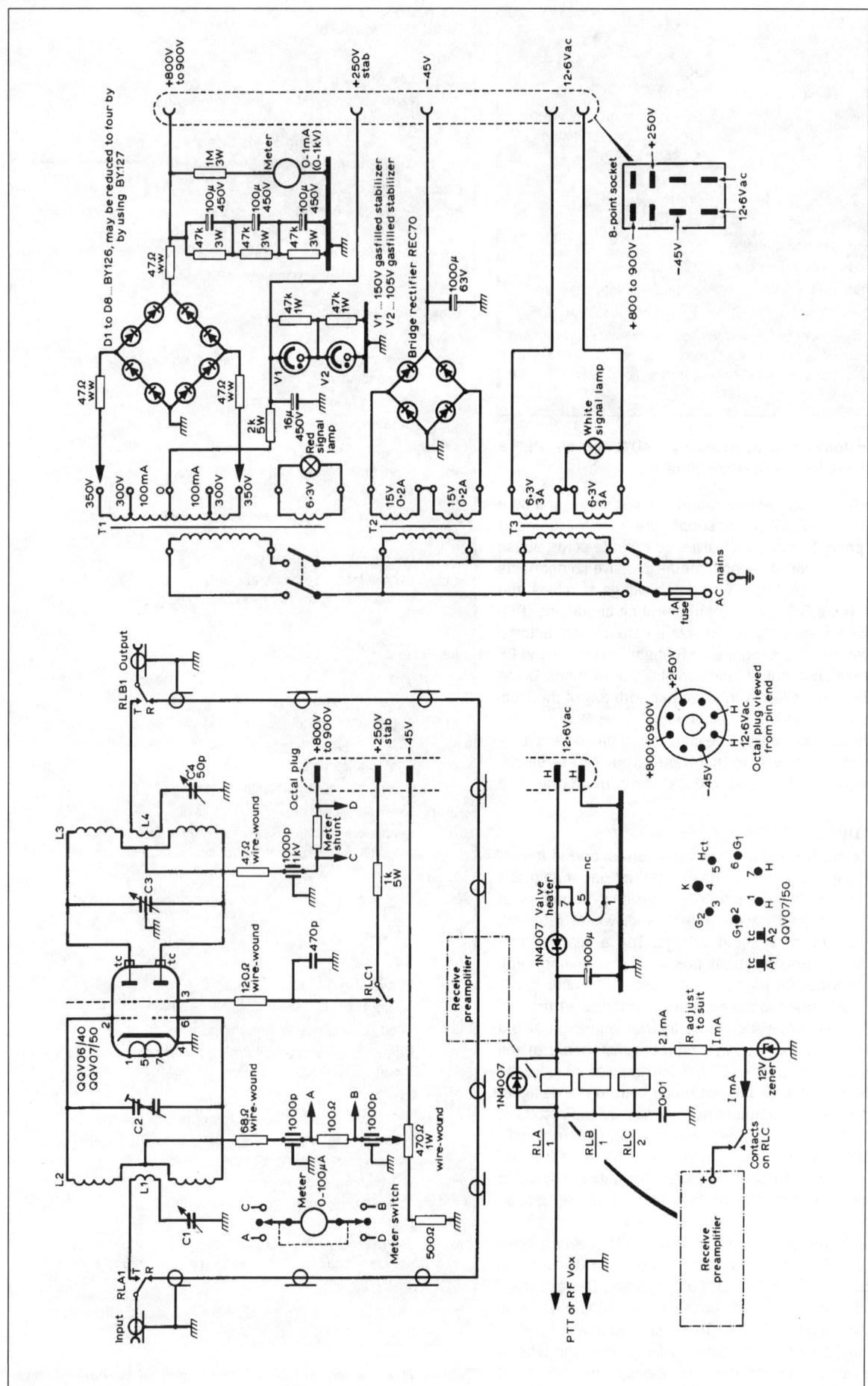

Fig 5.34: A tuned-grid double-tetrode linear VHF amplifier and its power supply. 50, 70 and 144MHz versions differ in tuned circuit values only

with 2500V on the anode and forced-air cooling a single valve can provide an output of 400W PEP, but it is better practice to use two valves in push-pull or parallel to produce this output without such cooling. One disadvantage is the requirement for a hefty filament transformer; each 813 requires 10V at 5A, AC or DC. A detailed description of a 400W HF amplifier using two 813s appeared in reference [13].

For VHF and UHF, two series of valves stand out. One comprises the double-tetrodes (originally Philips) QQV02/6, QQV03/10, QQV03/20A and QQV06/40A; the other is a family of (originally Eimac) 'external-anode' valves including the beam-power tetrodes 4X150A and 4CX250B. Though expensive when new, these valves can be found 'good used' as a result of the military and avionics practice of replacing valves on a 'time expired' rather than 'when worn out' basis. Test them as recommended for the 813 above.

On 144MHz and below, a QQV06/40A in Class AB1 can produce up to 100W PEP on a plate voltage of 1kV. In 432MHz operation, the efficiency will be lower, but 60W can be obtained with reduced input; fan cooling will increase the life expectancy of valves run near their dissipation limits.

External-anode valves can produce 'maximum legal' power on VHF and UHF but they always require forced-air cooling and the construction of the amplifiers is mechanically demanding. Convertible surplus equipment with them is not often found but several proven designs are detailed in the *RSGB VHF/UHF Handbook*.

For high power in the 23 and 13cm bands, the disc-seal triode 2C39A is universally used. Amplifier designs appear in the *RSGB Microwave Handbook*.

Valve configuration

There are three different ways to drive an amplifier valve: tuned grid, passive grid and grounded grid; the advantages and disadvantages of each are shown in **Table 5.7**. One disadvantage listed for tuned-grid amplifiers is the requirement for neutralisation; this may need explanation. The anode-to-control-grid capacitance within a valve feeds part of the RF output back to the control grid. The impedance of that capacitance decreases with increasing frequency, until at some frequency the stage will oscillate.

In tetrodes and pentodes, the RF-earthed screen grid reduces the anode-to-control-grid capacitance, so that most are stable on the lower HF bands but, depending on valve characteristics and external circuitry, oscillation will become a problem at some frequency at which the valve otherwise works well.

Neutralisation

One remedy is neutralisation, which is the intentional application of feedback from the anode to the grid, outside the valve

and in opposite phase to the internal feedback, so that the two feedback voltages will cancel. Neutralisation was universally required in tuned-grid triode amplifiers but recent trends have been to avoid the need for neutralisation. Generally, neutralisation is not required in amplifiers using passive-grid tetrodes or pentodes and grounded-grid triodes, tetrodes or pentodes, or external-anode tetrodes with flat screen grids and hence extremely small anode-to-control-grid capacitance. Exceptions are tuned-grid push-pull amplifiers with the QQV-series double-tetrodes which have the neutralisation capacitors built-in.

A Double-tetrode Linear VHF Amplifier

The amplifier shown in **Fig 5.34** [14] is popular for 50, 70 and 144MHz; differences are in tuned circuit values only. With 1.5-3W PEP drive, 70-90W PEP output can be expected, depending on frequency. For CW and SSB, convection cooling is sufficient. For FM or data, the input power should be held to 100W or a little more with fan cooling. The anodes should not be allowed to glow red. The circuit exemplifies several principles with wider applicability. Typical values for this particular amplifier are given below in (*italics*).

- It is a push-pull amplifier, which cancels out even harmonics; this helps against RFI, eg from a 50MHz transmitter into the FM broadcast band.

- It also makes neutralisation easier; the neutralisation achieved by capacitors from each control grid to the anode of the other tetrode (here within the valve envelope) is frequency independent as long as both grid and anode tuned circuits are of balanced construction and well shielded from each other.

- Control-grid bias (*-30V*) is provided from a fairly low-impedance source set for a small (*35mA*) 'standing' (ie zero signal) current between 10 and 20% of the peak signal current (*250mA*); the latter must be established by careful adjustment of the RF input. This is different from Class C amplifiers (in CW, AM and FM transmitters) with large (tens of kilohms) grid resistors which make the stage tolerant of a wide range of drive levels.

- The screen-grid voltage is stabilised; this also differs from the practice in Class C amplifiers to supply the screen through a dropping resistor from the anode voltage; that could cause non-linear amplification and excessive screen-grid dissipation in a Class AB1 linear amplifier. Where very expensive valves are used, it is wise to arrange automatic removal of the screen-grid voltage in case of failure of the anode voltage; a valve with screen-grid but no anode voltage will not survive long. This can be com-

Band	C1	L1	L2	C2	C3	L3	L4
144MHz	30pF	1½t, 8mm ID, 2mm dia insulated	2 + 2t, 8mm ID, 2mm dia, length 28mm, gap 8mm	20 + 20pF	12 + 12pF	2 + 2t, 16mm ID, 3mm dia Cu tubing, length 25mm, gap 8mm	1t 16mm ID, 2mm dia well-insulated
70MHz	50pF	2t 9.5mm ID, 0.9mm dia insulated	8 + 8t, 9.5mm ID, 1.2mm dia	20 + 20pF	20 + 20pF	8 + 8t 25mm ID, 1.2mm dia	1t 25mm ID, 1.2mm dia well-insulated
50MHz	50pF	2t 15mm ID, 0.8mm dia insulated 7cm to C1 twisted	3 + 3t 15mm ID, 0.8mm dia	50 + 50pF	30 + 30pF	6 + 6t 15mm ID, 1mm dia	2t 15mm ID, 0.8mm dia well-insulated, 10cm to C4 twisted

C2 and C3 are split stator. C4 is 50pF

Table 5.8: Tuned circuits for the double-tetrode linear VHF amplifier

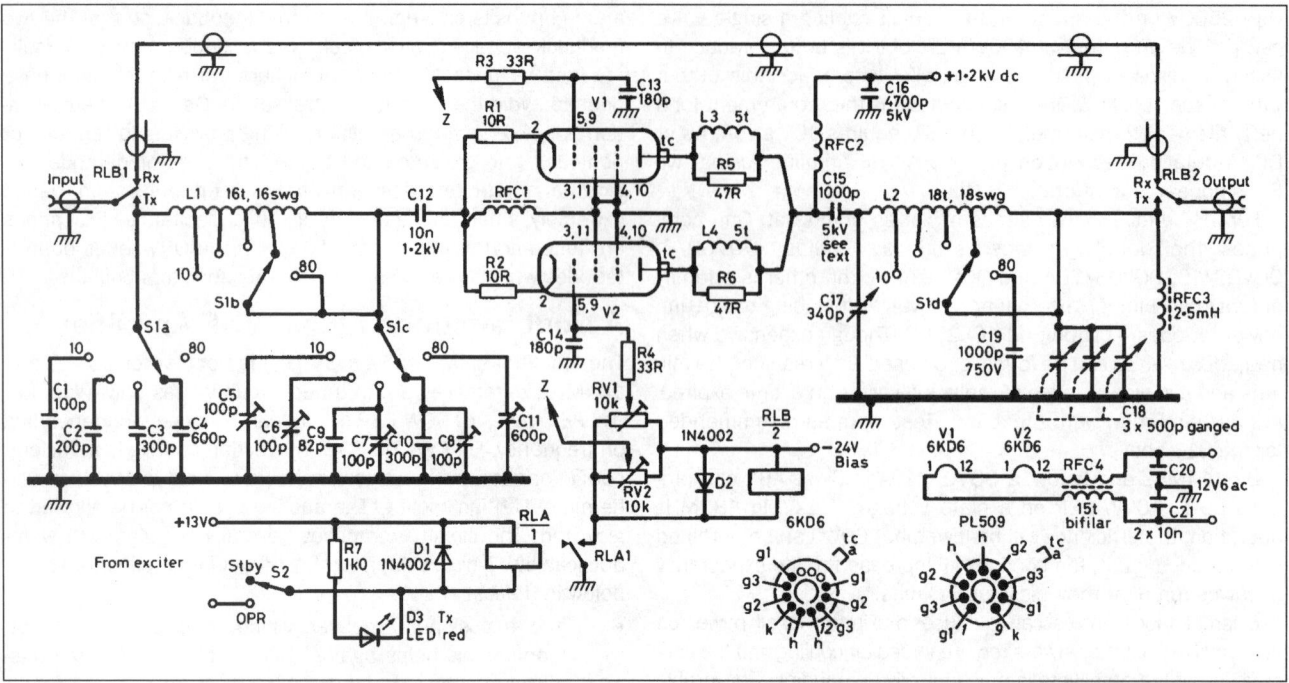

Fig 5.35: A grounded-grid 200W linear HF amplifier built by G3TSO, mostly from junk-box parts. No screen grid supply is required

bined with the function of relay RLC, which removes screen-grid voltage during receive periods, thereby eliminating not only unnecessary heat generation and valve wear but also the noise which an 'idling' PA sometimes generates in the receiver.

- Tuning and coupling of both the grid and plate tanks of a linear amplifier should be for maximum RF output from a low but constant drive signal. This again differs from

L1	16t, 1.6mm Cu-tinned, 25mm dia, 51mm long, taps at 3, 4, 6, 10t
L2	18t, 1.2mm Cu-tinned, 38mm dia, first 4t treble spaced, remainder double spaced. Taps at 3, 4, 6, 9t
L3, 4	5t 1.2mm Cu-enam wound on 47 1W carbon resistor
RFC1	40t 0.56mm Cu-enam on ferrite rod or toroid (μ = 800)
RFC2	40t 0.56mm Cu-enam on 12mm dia ceramic or PTFE former
RFC3	2.5mH rated 500mA
RFC4	15 bifilar turns, 1.2mm Cu-enam on ferrite rod 76mm 9.5mm dia
C1-4, 9, 10, 12-14	Silvered mica
C5-8	Compression trimmers
C15	Two 500p ceramic TV EHT capacitors in parallel
C16	4.7n, 3kV or more ceramic, transmitting type or several smaller discs in parallel
C18	3 x 500p ganged air-spaced variable (ex valved AM receiver)
C19	1n 750V silver mica, transmitting type or several smaller units in parallel
R1, 2	10R 3W carbon (beware of carbon film resistors, which may not be non-inductive)
RV1, 2	10k wirewound or linear-taper cermet potentiometer
Coaxial cable	RG58

Table 5.9: Components for the G3TSO grounded-grid 200W linear HF amplifier

Class C amplifiers, which are tuned for maximum 'dip' of the DC anode current. This amplifier circuit has been used on 50, 70 and 144MHz. **Table 5.8** gives tuned-circuit data; there is no logical progression to the coils from band to band because the data were taken from the projects of three different amateurs. Using a 'dip' oscillator, ie before applying power to the amplifier, all tuned circuits should, by stretching or squeezing the air-wound coils, be pre-adjusted for resonance in the target band, so that there is plenty of capacitor travel either side of resonance.

- Note the cable from the power supply to the amplifier: the power supply has an eight-pin chassis socket and the amplifier an eight-pin chassis plug; the cable has a plug on the power supply end and a socket on its amplifier end. Disconnecting this cable at either end never exposes dangerous voltages.

Grounded-grid 200W Linear HF Amplifier

The amplifier shown in **Fig 5.35** was designed by G3TSO to boost the output of HF transceivers in the 5 to 25W class by about 10dB [15]. It uses a pair of TV sweep valves in parallel; originally equipped with the American 6KD6, the European EL519 or PL519 are equivalent and the pin numbers in the diagram are for them. The choice depends only on the availability and price of the valves, sockets and filament supply; the 6KD6 and EL519 have a 6.3V, 2A filament, while the PL509 requires 42V, 0.3A.

The two filaments can be connected in series, as shown in the diagram, or in parallel. **Table 5.9** provides details of some other components.

A few of the design features are:

- It may seem that the pi-input filter, L1 and C1-C11, is superfluous; it is if the driving stage is a valve with a resonant anode circuit. The cathodes do not, however, present to a wide-band semiconductor driver a load which is constant over the RF cycle, causing distortion of the signal in the driver stage. This is avoided by inclusion of a tuner which does double duty as an impedance transformer

from 50Ω. The pi-filter is sufficiently flat to cover each band without retuning, but not to include any of the WARC bands.

- The valves, with all three grids earthed, ie connected as zero-bias triodes and without RF drive, would remain within their dissipation rating. With separately adjustable bias to each control grid, however, it is possible not only to reduce the no-signal anode current to the minimum necessary for linearity (20mA each valve) but also, by equalising the two idling currents, assure that the two valves will share the load when driven. This is particularly important when valves with unequal wear or of different pedigree are used together. It makes the purchase of 'matched pairs' unnecessary.

- When valves are operated in parallel, the wires strapping like electrodes together, along with valve and socket capacitances, will resonate at some VHF or UHF frequency. If the valves have sufficient gain at that frequency, the stage may burst into parasitic oscillation, especially at the modulation peaks when power gain is greatest. Even in a single-valve or push-pull stage this can happen as HF tuning capacitors are short-circuits at VHF or UHF. In this amplifier, several precautions are taken. Both pins of each screen and suppressor grid are earthed to chassis by the shortest possible straps; the control grids are by-passed to earth through C13 and C14 in a like manner; some valve sockets have a metal rim with four or more solder lugs for that purpose. Carbon stopper resistors R5 and R6 are used to lower the Q of parasitic resonances. Mere resistors would dissipate an intolerable fraction of the output, especially on the highest operating frequency where the valve's anode-to-earth capacitances are a large part of the total tank capacitance and carry high circulating currents; small coils L3 and L4 are therefore wound on the stoppers. These coils are dimensioned to have negligible impedance at all operating frequencies but not to short the stopper resistor at a potential parasitic frequency. The resistor with coil is called a parasitic suppressor and considerable experimentation is often required; it is sometimes possible to get away without them.

- The insulation between valves' cathodes and filaments was never made to support RF voltages. The cathodes of grounded-grid amplifiers, however, must be 'hot' for RF; the filaments, therefore, should be RF-insulated from earth; this is the purpose of RFC4. The bypass capacitors C20 and C21 prevent RF getting into the power supply. The filament wiring, including that on RFC4, must carry the filament current (4A for two parallel EL519 filaments) without undue voltage drop. Figure-8 hi-fi speaker cable is suitable.

- The anode tuning capacitor C17 is a 'difficult' component. Not only must it be rated for more than twice the off-load DC plate voltage (3.5mm spacing), but its maximum capacitance should be sufficient for a loaded Q of 12 at 3.5MHz, while its minimum capacitance should be low lest the Q gets too high on 29MHz. If an extra switch section S1e is available, C17 can be 100pF maximum with fixed capacitors of 100pF and 250pF respectively being switched across it on 7 and 3.5MHz. These fixed capacitors must be rated not only for high RF voltages (3.5kV) but also for the high circulating currents in a high-Q tuned circuit. Several low-capacitance silvered-mica capacitors in parallel will carry more current than one high-capacitance unit of the same model. High-voltage, high-current

'transmitting type' ceramic capacitors are available. Beware of Second World War surplus moulded mica capacitors; after more than 50 years, many have become useless.

- The anode choke requires special attention. It must have sufficient inductance to isolate the anode at RF from its power supply. Even at the lowest operating frequency (3.5MHz) and its self-resonance(s) (the inductance resonating with the distributed capacitance within the coil) must not fall within 20% or so of any operating frequency lest circulating current destroys the choke. Self-resonances can be revealed by shorting the choke and coupling a dip meter to it. The traditional approach was to divide the choke winding into several unequal series-connected segments, thereby moving their individual self-resonance frequencies into the VHF region; one problem is that surrounding metallic objects cause in situ resonances to differ from those measured on the bench. Another approach, adopted in RFC2, is to use one single-layer winding with a self-inductance not much greater than the whole pi-filter inductor (L2). Its self-resonance is well above the highest operating frequency (29MHz) but one has to accept that at 3.5MHz a substantial fraction of the tank current flows through RFC2 rather than through L2. At full output, the choke will get very hot; that is why it should be wound on a ceramic or PTFE rod or tube; also, the wire connections should be mechanically well made so that melting solder will not cause failure. The bypass capacitor (C16) must be rated for the RF current through RFC2 and also for more than the DC supply voltage.

- The bandswitch in the pi-output filter (S1d) must withstand high RF voltages across open contacts and high circulating currents through closed contacts; above the 100W level, ceramic wafers of more-than-receiver size are required.

- The 'safety choke' (RFC3) is there to 'kill' the HT supply (and keep lethal voltages off the antenna!) in case the capacitor (C15) should fail short-circuit. RFC3 must be wound of sufficiently thick wire to carry the HT supply's short-circuit current long enough to blow its fuse.

- At HF, coaxial relays are unnecessary luxuries; power switching relays with 250VAC/5A contacts are adequate. In some relays with removable plastic covers all connections are brought out on one side with fairly long wires running from the moving contacts to the terminals. If so, remove the cover and those long wires and solder the (flexible) centre conductors of input and output coaxial cables directly to the fixed ends of the moving contact blades. In this amplifier, one relay, RLB, switches both input and output. This requires care in dressing RF wiring to avoid RF feedback from output to input. A relay with three changeover contacts side by side is helpful; the outer contact sets are used for switching, the spare centre set's three contacts are earthed as a shield. The foolproof way is to use two relays.

Passive-grid 400W Linear HF Amplifier

PAOFRI's Frinear-400 is shown in **Fig 5.36** [16]. It has several interesting features.

- Being a passive-grid amplifier, most of the input power is dissipated in a hefty carbon resistor. The voltage across it is applied to the control-grids of the valves and, considering the low value of the resistor (50 or 68Ω), one might expect this arrangement to be frequency-independent.

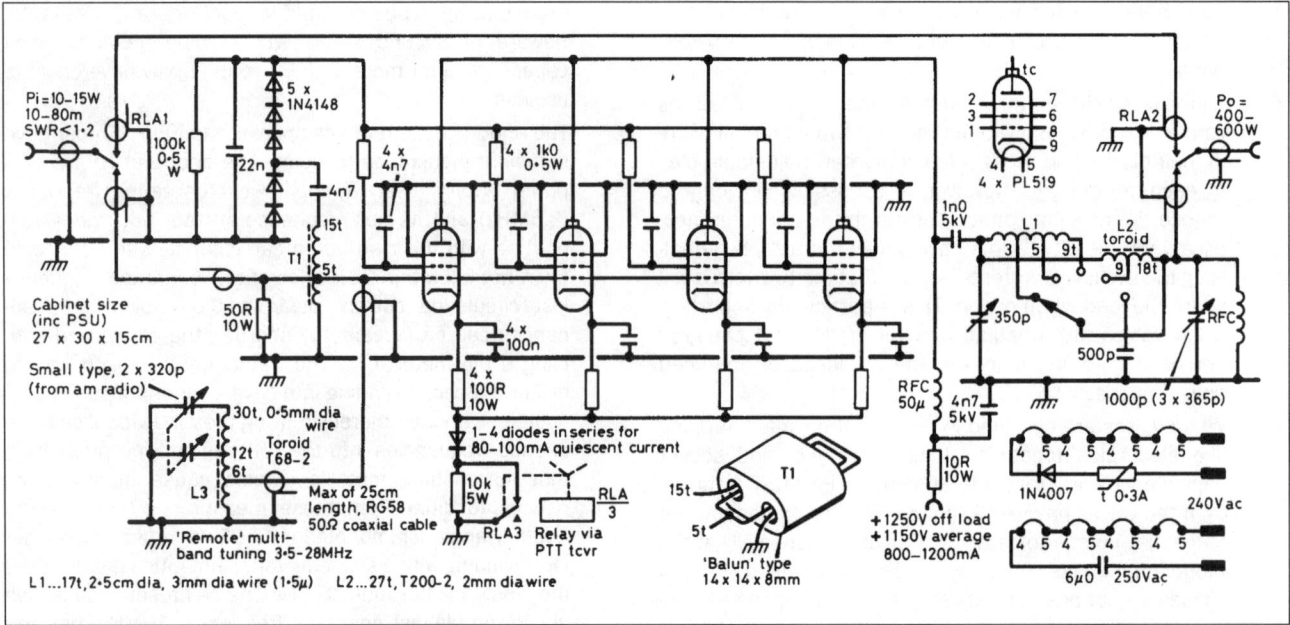

Fig 5.36: A passive-grid 400W linear HF amplifier, the PA0FRI Frinear. Screen-grid voltage is derived from the RF input and no grid bias supply is required

However, the capacitances of the four grids, sockets and associated wiring add up to about 100pF which is only 55Ω at 29MHz! This capacitance must be tuned out if what is adequate drive on 3.5MHz is to produce full output on the higher-frequency bands. PA0FRI does this with a dual-resonant circuit (L3 and ganged tuning capacitors) similar to the well-known E-Z-Match antenna tuner; it covers 3.5-29MHz without switching.

- The screen grids in this amplifier are neither at a fixed high voltage nor at earth potential but at a voltage which is proportional to the RF drive. To that end, the RF input is transformed up 3:1 in T1, rectified in a voltage doubler and applied to the four bypassed screen grids through individual resistors. This method is consistent with good linearity.

- Control-grid bias is not taken from a mains-derived negative supply voltage but the desired effect, reducing the standing current to 20-25mA per valve, is obtained by raising the cathodes above earth potential. The bias voltage is developed by passing each cathode current through an individual 100Ω resistor and the combined currents through as many forward-biased rectifier diodes as are required to achieve a total standing current of 80-100mA. The individual cathode resistors help in equalising the currents in the four valves. During non-transmit periods the third contact set (RLA3) on the antenna changeover relay opens and inserts a large (10kΩ) resistor into the combined cathode current, which is reduced to a very low value.

- The pi-filter coil for 3.5 and 7MHz is wound on a powdered-iron toroid which is much smaller than the usual air-core coil. This is not often seen in high-powered amplifiers due to the fear that the large circulating current might saturate the core and spoil the intermodulation performance but no distortion was discernible in a two-tone test [17].

- In Fig 5.36, the 42V filaments of the four valves and a capacitor are shown series connected to the 230V mains. This 0.3A chain is the way these valves were intended to be used in TV sets and it does save a filament trans-

former, but this method is not recommended for experimental apparatus such as a home construction project. Besides, a 6μF 250VAC capacitor is neither small nor inexpensive, and generally not available from component suppliers. Also, with lethal mains voltage in the amplifier chassis, the mains plug must be pulled every time access to the chassis is required and after the change or adjustment is made there is the waiting for filaments to heat up before applying HT again. It is much safer and more convenient to operate the filaments in parallel on a 42V transformer (3 x 12.6 + 5V will do), or to use EL519 valves in parallel, series-parallel or series on 6.3, 12.6 or 25.2V respectively.

DC AND AF AMPLIFIERS

This section contains information on the analogue processing of signals from DC (ie zero frequency) up to 5kHz. This includes audio amplifiers for receivers and transmitters for frequencies generally between 300 and 3000Hz as well as auxiliary circuitry, which may go down to zero frequency.

Operational Amplifiers (op-amps)

An op-amp is an amplifier which serves to drive an external network of passive components, some of which function as a feedback loop. They are normally used in IC form, and many different types are available. The name operational amplifier comes from their original use in analogue computers where they were used to perform such mathematical operations as adding, subtracting, differentiating and integrating. The response of the circuit (for example the gain versus frequency response) is determined entirely by the components used in the feedback network around the op-amp. This holds true provided the op-amp has sufficient open-loop gain and bandwidth.

Why use IC op-amps rather than discrete components? Op-amps greatly facilitate circuit design. Having established that a certain op-amp is adequate for the intended function, the designer can be confident that the circuit will reproducibly respond as calculated without having to worry about differences between individual transistors or changes of load impedance or

supply voltages. Furthermore, most op-amps will survive accidental short-circuits of output or either input to earth or supply voltage(s). General-purpose op-amps are often cheaper than the discrete components they replace and most are available from more than one maker. Manufacturers have done a commendable job of standardising the pin-outs of various IC packages. Many devices are available as one, two or four in one DIL package; this is useful to reduce PCB size and cost but does not facilitate experimenting.

Many books have been written about the use of op-amps; some of these can be recommended to amateur circuit designers as most applications require only basic mathematics; several titles are found in the Maplin catalogue. In the following, no attempt is made to even summarise these books, but some basics are explained and a few typical circuits are included for familiarisation.

The symbol of an op-amp is shown in **Fig 5.37**. There are two input terminals, an inverting or (-) input and a non-inverting or (+) input, both with respect to the single output, on which the voltage is measured against earth or 'common'. The response of an op-amp goes down to DC and an output swing both positive and negative with respect to earth may be required; therefore a dual power supply is used, which for most op-amps is nominally ±15V. Where only AC (including audio) signals are being processed, a single supply suffices and blocking capacitors are used to permit the meaningless DC output to be referred to a potential halfway between the single supply and earth established by a resistive voltage divider. In application diagrams, the power supply connections are often omitted as they are taken for granted.

To understand the use of an op-amp and to judge the adequacy of a given type for a specific application, it is useful to define an ideal op-amp and then compare it with the specifications of real ones. The ideal op-amp by itself, open loop, ie without external components, has the following properties:

- Infinite gain, ie the voltage between (+) and (-) inputs is zero.
- Output is zero when input is zero, ie zero offset.
- Infinite input resistance, ie no current flows in the input terminals.
- Zero output resistance, ie unlimited output current can be drawn.
- Infinite bandwidth, ie from zero up to any frequency.

No real op-amp is ideal but there are types in which one or two of the specifications are optimised in comparison with general-purpose types, sometimes at the expense of others and at a much higher price. The most important of these specifications will now be defined; the figures given will be those of the most popular and least expensive of general-purpose op-amps, the µA741C. Comparative specifications for a large variety of types are given in the Maplin catalogue.

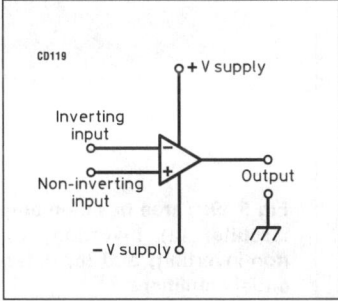

Fig 5.37: Symbol of an operational amplifier (op-amp)

Maximum ratings

Values which the IC is guaranteed to withstand without failure include:

- Supply voltage: ±18V.
- Internal power dissipation: 0.5W.
- Voltage on either input: not exceeding applied positive and negative supply voltages
- Output short-circuit: indefinite.

Static electrical characteristics

These are measured at DC (V_s = ±15V, T = 25°C).

- Input offset voltage (V_{oi}): the DC voltage which must be applied to one input terminal to give a zero output. Ideally zero. 6mV max.
- Input bias current (I_b): the average of the bias currents flowing into the two input terminals. Ideally zero. 500nA max.
- Input offset current (I_{os}): the difference between the two input currents when the output is zero. Ideally zero. 200nA max.
- Input voltage range (V_{cm}): the common-mode input, ie the voltage of both input terminals to power supply common. Ideally unlimited. ±12V min.
- Common-mode rejection ratio (CMRR): the ratio of common-mode voltage to differential voltage to have the same effect on output. Ideally infinite. 70dB min.
- Input resistance (Z_i): the resistance 'looking into' either input while the other input is connected to power supply common. Ideally infinite. 300kΩ min.
- Output resistance (Z_o): the resistance looking into the output terminal. Ideally zero. 75Ω typ.
- Short-circuit current (I_{sc}): the maximum output current the amplifier can deliver. 25mA typ.
- Output voltage swing (±Vo): the peak output voltage the amplifier can deliver without clipping or saturation into a nominal load. ±10V min (R_L = 2kΩ).
- Open-loop voltage gain (A_{OL}): the change in voltage between input terminals divided into the change of output voltage caused, without external feedback. 200,000 typ, 25,000 min. (V_o = ±10V, R_L = 2kΩ).
- Supply current (quiescent, ie excluding I_o) drawn from the power supply: 2.8mA max.

Dynamic electrical characteristics

- Slew rate is the fastest voltage change of which the output is capable. Ideally infinite. 0.5V/ms typ. (R_L 3 2kΩ).
- Gain-bandwidth product: the product of small-signal open-loop gain and the frequency (in Hz) at which that gain is measured. Ideally infinite. 1MHz typ.

Understanding the gain-bandwidth product concept is basic to the use of op-amps at other than zero frequency. If an amplifier circuit is to be unconditionally stable, ie not given to self-oscillation, the phase shift between inverting input and output must be kept below 180° at all frequencies where the amplifier has gain. As a capacitive load can add up to 90°, the phase shift within most op-amps is kept, by internal frequency compensation, to 90°, which coincides with a gain roll-off of 6dB/octave or 20dB/decade. In **Fig 5.38**, note that the open-loop voltage gain from zero up to 6Hz is 200,000 or 10^6dB. From 6Hz, the long slope is at -6dB/octave until it crosses the unity gain (0dB) line at 1MHz. At any point along that line the product of gain and frequency is the same: 10^6. If one now applies external feedback to achieve a signal voltage or closed-loop gain of, say, 100 times,

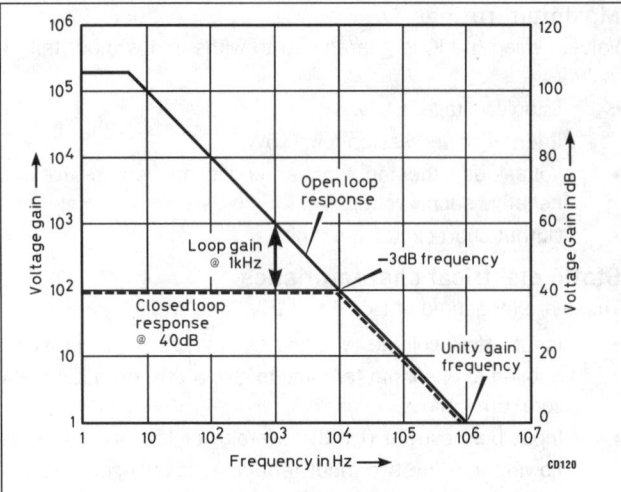

Fig 5.38: Open-loop, closed-loop and loop gain of an op-amp, internally compensated for 6dB/octave roll-off

the -3dB point of the resulting audio amplifier would be at $10^6/100 = 10\text{kHz}$ and from one decade further down, ie 1kHz to DC, the closed-loop gain would be flat within 1%. At 3kHz, ie the top of the communication-quality audio range, this amplifier would be only 1dB down and this would just make a satisfactory 40dB voice amplifier. If one wished to make a 60dB amplifier for the same frequency range, an op-amp with a gain-bandwidth product of at least $10^7 = 10\text{MHz}$ should be used or one without internal frequency compensation.

Op-amp types

The specifications given above apply to the general-purpose bipolar op-amp μA741C. There are special-purpose op-amps for a variety of applications, only a few are mentioned here.

Fast op-amps:

The OP-37GP has high unity gain bandwidth (63MHz), high slew rate (13.5V/μs) and low noise, (3nV/√Hz) which is important for low-level audio.

The AD797 has a gain bandwidth product of 110MHz. It has a slew rate of 20V/μs and has an extremely low noise. The input noise voltage is 0.9nV/√Hz. This device is fairly expensive but has sufficient bandwidth that it can be used as an IF amplifier at 455kHz. The low input noise voltage means that it will give a good signal/noise when driven from a low impedance source (eg 50 ohms). The AD797 also has very good DC characteristics. The input offset voltage is typically 25μV. if this device is to be used, the manufacturers data sheet should be consulted [18]. This shows how the op-amp should be connected and decoupled to ensure stability.

FET-input op-amps:

The TL071-C has the very low input current characteristic of FET gates ($I_{os} \leq 5\text{pA}$) necessary if very-high feedback resistances (up to 10MΩ) must be used, but it will work on a single battery of 4.5V, eg in a low-level microphone amplifier, for which it has the necessary low noise. The input impedance is in the teraohm ($10^{12}\Omega$) range.

Power op-amps:

The LM383 is intended for audio operation with a closed-loop gain of 40dB and can deliver up to 7W to a 4Ω speaker on a single 20V supply, 4W on a car battery, and can dissipate up to 15W if mounted on an adequate heatsink.

Basic op-amp circuits

Fig 5.39 shows three basic configurations: inverting amplifier (a), non-inverting amplifier (b) and differential amplifier (c).

The operation of these amplifiers is best explained in terms of the ideal op-amp defined above. In (a), if V_o is finite and the op-amp gain is infinite, the voltage between the (-) and (+) input terminals must approach zero, regardless of what V_i is; with the (+) input earthed, the (-) input is also at earth potential, which is called virtual earth. The signal current Is driven by the input voltage V_i through the input resistor R_i will, according to Ohm's Law, be V_i/R_i. The current into the ideal op-amp's input being zero, the signal current has nowhere to go but through the feedback resistor R_f to the output, where the output voltage V_o must be $-I_sR_f$; hence:

$$V_o = - V_i \frac{R_f}{R_i} \quad \text{in which} \quad - \frac{R_f}{R_i} = A_{cl}, \text{ the closed-loop gain}$$

The signal source 'sees' a load of R_i; it should be chosen to suitably terminate that signal source, consistent with a feedback resistor which should, for general purpose bipolar op-amps, be between 10kΩ and 100kΩ; with FET-input op-amps, feedback resistors up to several megohms can be used.

The junction marked 'S' is called the summing point; several input resistors from different signal sources can be connected to the summing point and each would independently send its signal current into the feedback resistor, across which the algebraic sum of all signal currents would produce an output voltage. A summing amplifier is known in the hi-fi world as a mixing amplifier; in amateur radio, it is used to add, not mix, the output of two audio oscillators together for two-tone testing of SSB transmitters.

For the non-inverting amplifiers, **Fig 5.39(b)**, the output voltage and closed-loop gain derivations are similar, with the (-) input assuming Vi:

$$V_o = V_i \frac{R_i + R_f}{R_i} \quad \text{in which} \quad \frac{R_i + R_f}{R_i} = A_{cl}, \text{ the closed loop gain}$$

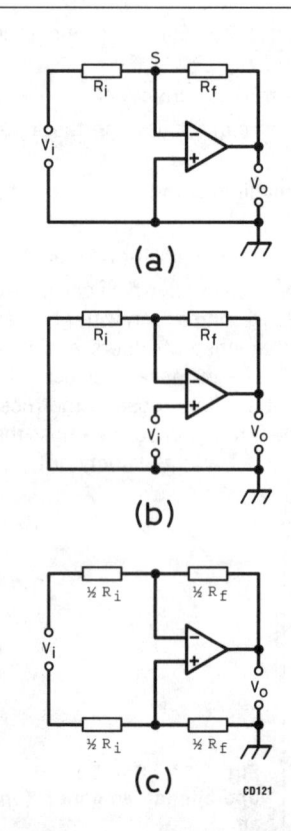

Fig 5.39: Three basic op-amp circuits: (a) inverting; (b) non-inverting; and (c) differential amplifiers

The signal source 'sees' the very high input impedance of the bare op-amp input. In the extreme case where $R_i = \infty$, ie left out, and $R_f = 0$, $V_o = V_i$ and the circuit is a unity-gain voltage follower, which is frequently used as an impedance transformer having an extremely high input impedance and a near-zero output impedance. Non-inverting amplifiers cannot be used as summing amplifiers.

A differential amplifier is shown in **Fig 5.39(c)**. The differential input, ie V_i, is amplified according to the formulas given above for the inverting amplifier. The common-mode input, ie the average of the input voltages measured against earth, is rejected to the extent that the ratio of the resistors connected to the (+) input equals the ratio of the resistors connected to the (-) input. A differential amplifier is useful in 'bringing to earth' and amplifying the voltage across a current sampling resistor of which neither end is at earth potential; this is a common requirement in current-regulated power supplies; see also Fig 5.30.

Power supplies for op-amps

Where the DC output voltage of an op-amp is significant and required to assume earth potential for some inputs, dual power supplies are required. Where a negative voltage cannot be easily obtained from an existing mains supply, eg in mobile equipment, a voltage mirror can be used. It is an IC which converts a positive supply voltage into an almost equal negative voltage, eg type Si7661CJ. Though characterised when supplied with ±15V, most op-amps will work off a wide range of supply voltages, typically ±5V to ±18V dual supplies or a single supply of 10V to 36V, including 13.8V. One must be aware, however, that the output of most types cannot swing closer to either supply bus than 1-3V.

In audio applications, circuits can be adapted to work off a single supply bus. In **Fig 5.40**, the AC-and-DC form of a typical inverting amplifier with dual power supplies (a) is compared with the AC-only form (b) shown operating off a single power supply.

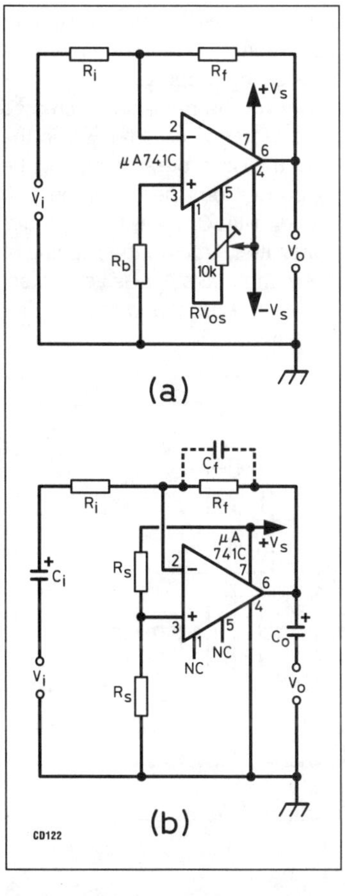

Fig 5.40: DC to audio amplifier (a) requires dual power supplies; it has input bias current and voltage offset compensation. The audio-only amplifier (b) needs only a single power supply

The DC input voltage and current errors are only of the order of millivolts but if an amplifier is programmed for high gain, these errors are amplified as much as the signal and spoil the accuracy of the output.

In **Fig 5.40(a)**, two measures reduce DC errors. The input error created by the bias current into the (-) input is compensated for by the bias current into the (+) input flowing through R_b, which is made equal to the parallel combination of R_i and R_f. The remaining input error, $I_{os} \cdot R_b + V_{os}$, is trimmed to zero with RV_{os}, which is connected to a pair of amplifier pins intended for that purpose. Adjustment of RV_{os} for $V_o = 0$ must be done with the V_i terminals shorted.

Note that some other amplifier designs use different offset arrangements such as connecting the slider of the offset potentiometer to $+V_s$. Consult the data sheet or catalogue.

In audio-only applications, DC op-amp offsets are meaningless as input and output are blocked by capacitors C_i and C_o in **Fig 5.40(b)**. Here, however, provisions must be made to keep the DC level of the inputs and output about half-way between the single supply voltage and earth. This is accomplished by connecting the (+) input to a voltage divider consisting of the two resistors R_s. Note that the DC closed loop gain is unity as C_i blocks R_i.

In the audio amplifier of **Fig 5.40(b)**, the high and low frequency responses can be rolled off very easily. To cut low-frequency response below, say 300Hz, C_i is dimensioned so that at 300Hz, $XC_i = R_i$. To cut high-frequency response above, say 3000Hz, a capacitor C_f is placed across Rf; its size is such that at 3000Hz, $XC_f = R_f$.

For more sophisticated frequency shaping, see the section on active filters.

A PEP-reading module for RF power meters

The inertia of moving-coil meters is such that they cannot follow speech at a syllabic rate and even if they could, the human eye would be too slow to follow. The usual SWR/power meter found in most amateur stations, calibrated on CW, is a poor PEP indicator.

GW4NAH designed an inexpensive circuit [19], **Fig 5.41**, on a PCB small enough to fit into the power meter, which will rise to a peak and hold it there long enough for the meter movement, and the operator, to follow. The resistance of RV1 + RV2 takes the place of the meter movement in an existing power meter; the voltage across it is fed via R1 and C1 to the (+) input of op-amp IC1a. Its output charges C3 via D2 and R6 with a rise time of

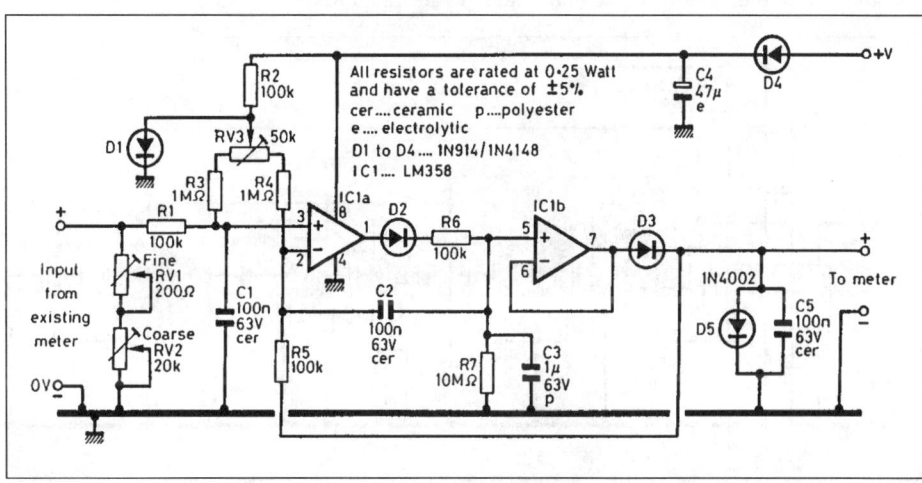

Fig 5.41: A PEP-reading module for RF power meters using a dual op-amp (GW4NAH)

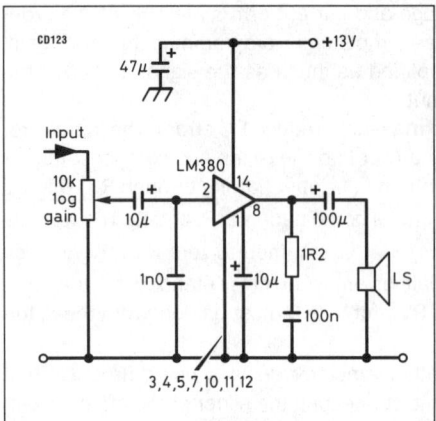

Fig 5.42: This simple receiver audio IC provides up to 40dB voltage gain and over a watt of low-distortion output power to drive an 8-ohm loudspeaker

0.1s, but C3 can discharge only through R7 with a decay time constant of 10s. The voltage on C3 is buffered by the unity-gain voltage follower IC1b and is fed via D3 to the output terminals to which the original meter movement is now connected.

The input-to-output gain of the circuit is exactly unity by virtue of R5/R1 = 1. C2 creates a small phase advance in the feedback loop to prevent overshoot on rapid transients. The LM358 dual op-amp was chosen because, unlike most, it will work down to zero DC output on a single supply. The small voltage across D1 is used to balance out voltage and current offsets in the op-amps, for which this IC has no built-in provisions, via R3, R4 and RV3. D4 protects against supply reversals and C4 is the power supply bypass capacitor. D5 and C5 protect the meter movement from overload and RF respectively.

Receiver Audio

The audio signal obtained from the demodulator of a radio receiver generally requires filtering and voltage amplification; if a loudspeaker is to be the 'output transducer' (as distinct from headphones or an analogue-to-digital converter for computer processing) some power amplification is also required. For audio filtering, see the section on filters. For voltage amplification, the op-amp is the active component of choice for reasons explained in the section on them. There is also a great variety of ICs which contain not only the op-amp but also its gain-setting resistors and other receiver functions such as demodulator, AGC generator and a power stage dimensioned to drive a loudspeaker. A fraction of a watt is sufficient for a speaker in a quiet shack but for mobile operation in a noisy vehicle several watts are useful. Design, then, comes down to the selection from a catalogue or the junk box of the right IC, ie one that offers the desired output

from the available input signal at an affordable price and on available power supply voltages.

The data sheet of the IC selected will provide the necessary details of external components and layout. **Fig 5.42** is an example of voltage and power amplification in an inexpensive 14-pin DIL IC; the popular LM380 provides more than 1W into an 8Ω speaker on a single 13.8V supply. Even the gain-setting resistors are built-in.

Note that many ICs which were popular as audio amplifiers are now being withdrawn by the manufacturers. The newer devices which replace them are often switched-mode amplifiers. These offer advantages for the user because they are more efficient than linear amplifiers and hence require less heatsinking. They are not recommended however, for use in a receiver because of the significant amount of radio-frequency energy generated by the amplifier.

Transmitter Audio

With the virtual disappearance from the amateur scene of high-level amplitude modulation, the audio processing in amateur as in commercial and military transmitters consists of low-level voltage amplification, compression, limiting and filtering, all tasks for op-amp circuitry as described in the section on them. The increase in the consumer usage of radio transmitters in cellular telephones and private mobile radios has given incentive to IC manufacturers to integrate ever more of the required circuitry onto one chip.

First, an explanation of a few terms used in audio processing: clipper, compressor, VOGAD and expander.

The readability of speech largely resides in the faithful reproduction of consonant sounds; the vowels add little to the readability but much to the volume. Turning up the microphone gain does enhance the consonant sounds and thereby the readability, but the vowel sounds would then overload the transmitter, distort the audio and cause RF splatter. One remedy is to linearly amplify speech up to the amplitude limit which the transmitter can process without undue distortion and remove amplitude peaks exceeding that limit. This is called clipping.

If a waveform is distorted, however, harmonics are created and as harmonics of the lower voice frequencies fall within the 300-2700Hz speech range where they cannot subsequently be filtered out, too much clipping causes audible distortion and reduces rather than enhances readability. One way to avoid this, at least in a single-sideband transmitter, is to do the clipping at a higher frequency, eg the IF where the SSB signal is generated; the harmonics then fall far outside the passband required for speech and can be readily filtered out.

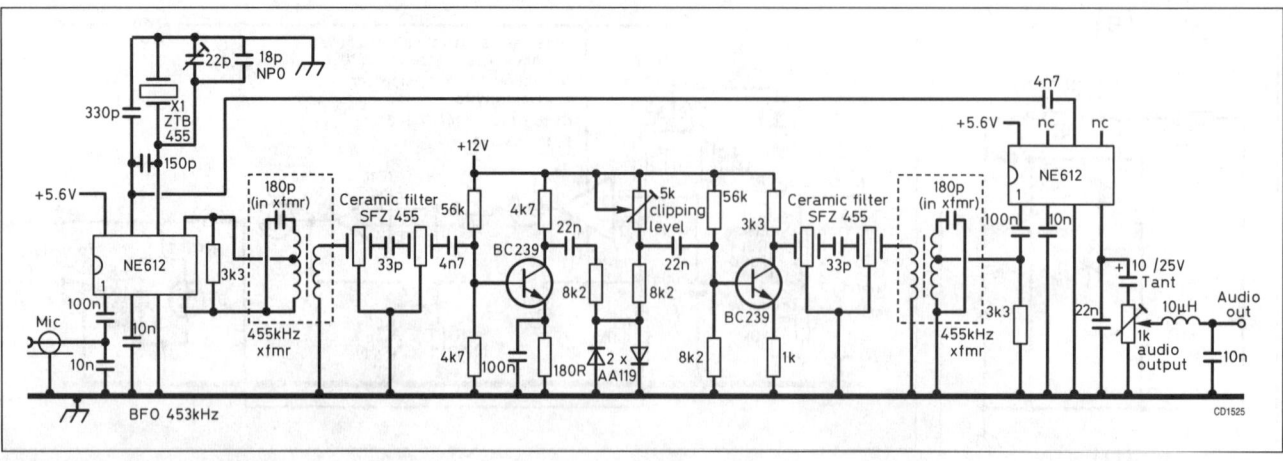

Fig 5.43: The DF4ZS RF speech clipper combines high compression with low distortion

VOGAD stands for 'voice-operated gain adjustment device'. It automatically adjusts the gain of the microphone amplifier so that the speech level into the transmitter, averaged over several syllables, is almost independent of the voice level into the microphone. It is widely used, eg in hand-held FM transceivers.

A compressor is an amplifier of which the output is proportional to the logarithm of the input. Its purpose is to reduce the dynamic range of the modulation, in speech terms the difference between shouting and whispering into the microphone, and thereby improve the signal to noise ratio at the receiving end. If best fidelity is desired, the original contrast can be restored in the receiver by means of an expander, ie an exponential amplifier; this is desirable for hi-fi music but for speech it is seldom necessary.

A compander is a compressor and an expander in one unit, nowadays one IC. It could be used in a transceiver, with the compressor in the transmit and the expander in the receive chain.

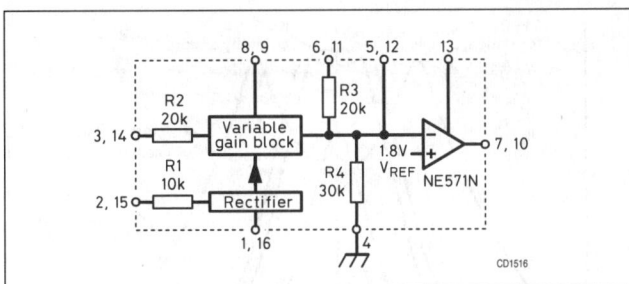

Fig 5.44: Block diagram of the Philips NE571N two-channel compander. Basic input-to-output characteristics are as follows:

Compressor input level or expander output level (dBm)	Compressor output level or expander input level (dBm)
+20	+10
0	0
20	10
40	20
60	30
80	40

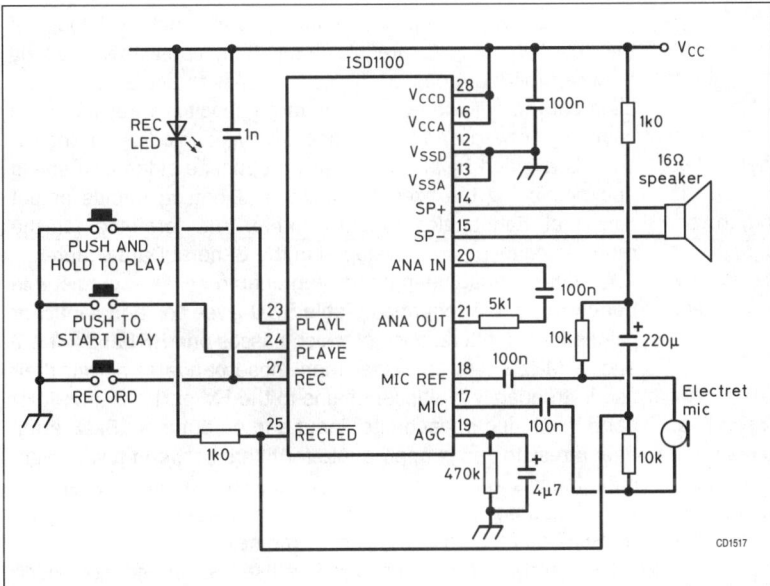

Fig 5.45: A record-playback IC, with its associated circuitry, will record 10-20 seconds of speech, store it in non-volatile memory, and play it back at the push of a button

An RF speech clipper

As explained above, clipping is better done at a radio frequency than at audio. Analogue speech processors in the better pre-DSP SSB transceivers clip at the intermediate frequency at which the SSB signal is generated.

Rigs without a speech processor will benefit from the stand-alone unit designed by DF4ZS [20]. In **Fig 5.43** the left NE612 IC (a cheaper version of the NE602 described above) mixes the microphone audio with a built-in 453kHz BFO to yield 453kHz + audio, a range of 453.3 to 455.7kHz. The following filter removes all other mixing products. The signal is then amplified, clipped by the back-to-back diodes, amplified again, passed through another filter to remove the harmonics generated by the clipping process, and reconverted to audio in the right-hand NE612. A simple LC output filter removes non-audio mixing products.

A compander IC

The Philips NE571N Compander IC contains two identical channels, each of which can be externally connected as a compressor or as an expander. Referring to **Fig 5.44**, this can be explained (in a very simplified way) as follows:

To expand, the input signal to the device is also fed into the rectifier which controls a 'variable gain block' (VGB); if the input is high, the current gain of the VGB is also high. For example, if the input goes up 6dB, the VGB gain increases by 6dB as well and the current into the summing point of the op-amp, and hence the output, go up 12dB, R3 being used as the fixed feedback resistor.

To compress, the output is fed into the rectifier and the VGB is connected in the feedback path of the op-amp with R3 being connected as the fixed input resistor. Now, if both the output and the VGB gain are to go up 6dB, the input must rise 12dB.

Having two channels enables application as a stereo compressor or expander, or, in a transceiver, one channel can compress the transmitted audio while the other expands the receiver output.

A voice record-playback device

Contesters used to get sore throats from endlessly repeated 'CQs'. Repeated voice messages can now be sent with an IC which can record into non-volatile erasable analogue memory a message of 10-20 seconds in length from a microphone, and play it back with excellent fidelity through a loudspeaker or into the microphone socket of a transmitter. Playback can be repeated as often as desired and then instantly cleared, ready for a new message.

The US company Information Storage Devices' ISD1100-series ICs sample incoming audio at a 6.4kHz rate, which permits an audio bandwidth up to 2.7kHz. **Fig 5.45** shows a simple application diagram.

Velleman, a Belgian manufacturer, makes a kit, including a PCB, on which to assemble the IC and the required passive components and switches. ON5DI showed how to interface this assembly with a transceiver and its microphone [21].

FILTERS AND LC COUPLERS

Filters are circuits designed to pass signals of some frequencies and to reject or stop signals of others. Amateurs use filters ranging in operation from audio to microwaves. Applications include:
• Preselector filters which keep strong out-of-band signals from overloading a receiver.

- IF (intermediate frequency) filters which provide adjacent-channel selectivity in superheterodyne receivers.

- Audio filters which remove bass and treble, which are not essential for speech communication, from a microphone's output. This minimises the bandwidth taken up by a transmitted signal.

- A transmitter output is filtered, using a low-pass filter, to prevent harmonics being radiated.

- Mains filters, which are low-pass filters, used to prevent mains-borne noise from entering equipment.

Filters are classified by their main frequency characteristics. High-pass filters pass frequencies above their cut-off frequency and stop signals below that frequency. In low-pass filters the reverse happens. Band-pass filters pass the frequencies between two cut-off frequencies and stop those below the lower and above the upper cut-off frequency. Band-reject (or band-stop) filters stop between two cut-off frequencies and pass all others. Peak filters and notch filters are extremely sharp band-pass and band-reject filters respectively which provide the frequency characteristics that their names imply.

A coupler is a unit that matches a signal source to a load having an impedance which is not optimum for that source. An example is the matching of a transmitter's transistor power amplifier requiring a 2-ohm load to a 50-ohm antenna. Frequently, impedance matching and filtering is required at the same spot, as it is in this example, where harmonics must be removed from the output before they reach the antenna. There is a choice then, either to do the matching in one unit, eg a wideband transformer with a 1:√(50/2) = 1:5 turns ratio and the filtering in another, ie a 'standard' filter with 50Ω input and output, or to design a special filter-type LC circuit with a 2Ω input and 50Ω output impedance. In multiband HF transceivers, transformers good for all bands and separate 50Ω/50Ω filters for each band would be most practical. For UHF, however, and for the high impedances in RF valve anode circuitry, there are no satisfactory wide-band transformers; the use of LC circuits is required.

Ideal Filters and the Properties of Real Ones

Ideal filters would let all signals in their intended pass-band through unimpeded, ie have zero insertion loss, suppress completely all frequencies in their stop-band, ie provide infinite attenuation, and have sharp transitions from one to the other at their cut-off frequencies (Fig 5.46). Unfortunately such filters do not exist. In practice, the cut-off frequency, that is the transition point between pass-band and stop-band, is generally defined as the frequency where the response is -3dB (down to 70.7% in voltage) with respect to the response in the pass-band; in very sharp filters, such as crystal filters, the -6dB (half-voltage) points are frequently considered the cut-off frequencies. There are several practical approximations of ideal filters but each of these optimises one characteristic at the expense of others.

LC Filters

If two resonant circuits are coupled together, a band-pass filter can be made. The degree of coupling between the two resonant circuits, both of which are tuned to the centre frequency, determines the shape of the filter curve (Fig 5.47). Undercoupled, critically coupled and overcoupled two-resonator filters all have their applications.

Four methods of achieving the coupling are shown in Fig 5.48. The result is always the same and the choice is mainly one of convenience. If the signal source and load are not close together, eg on different PCBs, placing one resonant circuit with each

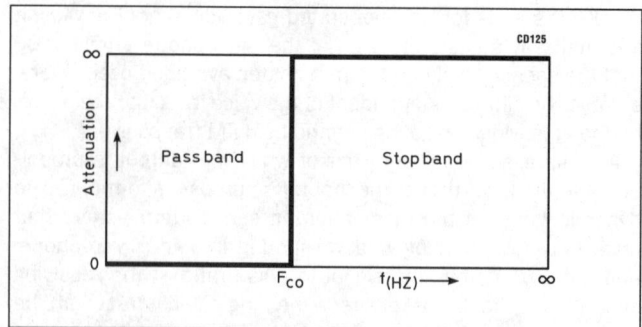

Fig 5.46: Attenuation vs frequency plot of an ideal low-pass filter. No attenuation (insertion loss) in the pass-band, infinite attenuation in the stop-band, and a sharp transition at the cut-off frequency

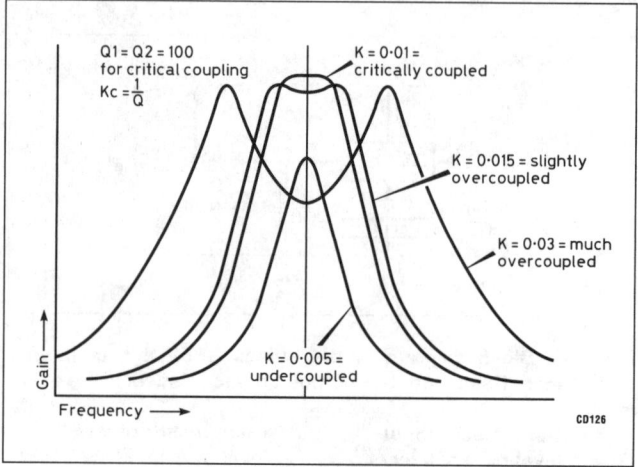

Fig 5.47: Frequency response of a filter consisting of two resonant circuits tuned to the same frequency as a function of the coupling between them (after Terman). Under-coupling provides one sharp peak, critical coupling gives a flat top, mild overcoupling widens the top with an acceptably small dip and gross overcoupling results in peaks on two widely spaced frequencies. All these degrees of coupling have their applications

and using link coupling is recommended to avoid earth loops. If the resonant circuits are close together, capacitive coupling between the 'hot ends' of the coils is easy to use and the coupling can be adjusted with a trimmer capacitor. Stray inductive coupling between adjacent capacitively coupled resonant circuits is avoided by placing them on opposite sides of a shield and/or placing the axes of the coils (if not on toroids or pot cores) at right-angles to one-another. The formulae for the required coupling are developed in the General Data chapter.

All filters must be properly terminated to give predictable bandwidth and attenuation. **Table 5.10** gives coil and capacitor values for five filters, each of which passes one HF band. (The 7 and 14MHz filters are wider than those bands to permit their use in frequency multiplier chains to the FM part of the 29MHz band.) Included in each filter input and output is a 15kΩ termination resistor, as is appropriate for filters between a very high-impedance source and a similar load, eg the anode of a pentode valve and the grid of the next stage, or the drain of one dual-gate MOSFET amplifier and the gate of the next.

Frequently, a source or load has itself an impedance much lower than 15kΩ, eg a 50Ω antenna. These 50Ω then become the termination and must be transformed to 'look like' 15kΩ by tapping down on the coil. As the circulating current in high-Q resonant circuits is many times larger than the source or load cur-

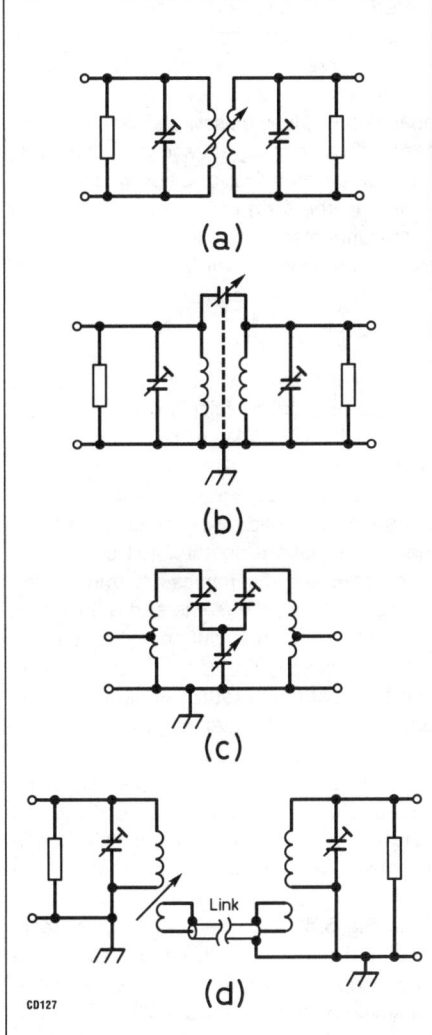

Fig 5.48: Four ways of arranging the coupling between two resonant circuits. Arrows mark the coupling adjustment. (a) Direct inductive coupling; the coils are side-by-side or end-to-end and coupling is adjusted by varying the distance between them. (b) 'Top' capacitor coupling; if the coils are not wound on toroids or pot cores they should be shielded from each other or installed at right-angles to avoid uncontrolled inductive coupling. (c) Common capacitor coupling; if the source and/or load have an impedance lower than the proper termination, they can be 'tapped down' on the coil (regardless of coupling method). (d) Link coupling is employed when the two resonant circuits are physically separated

rent and if the magnetic flux in a coil is the same in all its turns, true in coils wound on powdered iron or ferrite toroids or pot cores, the auto-transformer formula may be used to determine where the antenna tap should be: at $\sqrt{(50/15,000)} \approx 6\%$ up from the earthy end of the coil. In coils without such cores, the flux in the end turns is less than in the centre ones, so the tap must be experimentally located higher up the coil. Impedances lower than 50Ω can be accommodated by placing that source or load in series with the resonant circuit rather than across all or part of it.

DC operating voltages to source and load devices are often fed though the filter coils. If properly bypassed, this does not affect filter operation. To avoid confusion, DC connections and bypass capacitors are not shown in the filter circuitry in this chapter.

Several more sophisticated LC filter designs are frequently used. All can be configured in high-pass, low-pass, band-pass and band-reject form.

Butterworth filters have the flattest response in the pass-band. Chebyshev filters have a steeper roll-off to the stop-band but exhibit ripples in the pass-band, their number depending on the number of filter sections. Elliptic filters have an even steeper roll-off, but have ripples in the stop-band (zeros) as well as in the pass-band (poles): see **Fig 5.49**. Chebyshev and elliptical filters have too much overshoot, particular near their cut-off frequencies, for use where pulse distortion must be kept down, eg in RTTY filters.

The calculation of component values for these three types would be a tedious task but for filter tables normalised for a cut-off frequency of 1Hz and termination resistance of 1Ω (or 1MHz and 50Ω where indicated). These can be easily scaled to the desired frequency and termination resistance. See the General Data chapter for more information.

M-derived and constant-k filters are older designs with less-well-defined characteristics but amateurs use them because component values are more easily calculated 'long-hand'. The diagrams and formulae to calculate component values are contained in the General Data chapter.

From audio frequency up to, say, 100MHz, filter inductors are mostly wound on powdered-iron pot cores or toroids of a material and size suitable for the frequency and power, and capacitors ranging from polystyrene types at audio, to mica and ceramic types at RF, with voltage and current ratings commensurate with

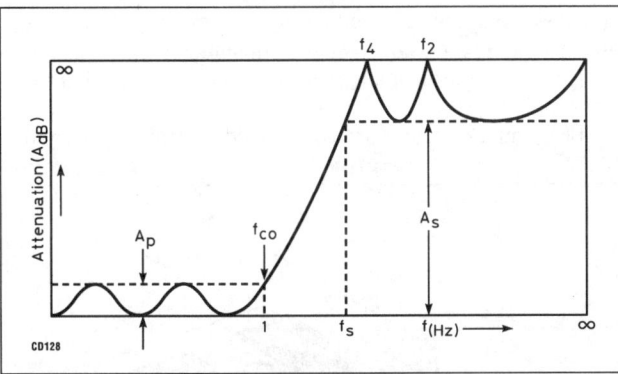

Fig 5.49: Attenuation vs frequency plot of a two-section elliptic low-pass filter. A is the attenuation (dB); A_p is the maximum attenuation in the pass-band or ripple; f_4 is the first attenuation peak; f_2 is the second attenuation peak with two-section filter; f_{co} is the frequency where the attenuation first exceeds that in the pass-band; A_s is the minimum attenuation in the stop-band; and f_s is the frequency where minimum stop-band attenuation is first reached

Lowest frequency (MHz)	Centre frequency (MHz)	Highest frequency (MHz)	Coupling (pF)	Parallel capacitance (pF)	L (µH)	Winding details (formers ¾in long, 3/8in dia)
3.5	3.65	3.8	6	78	24	60t 32SWG close-wound
7	7.25	7.5	3	47	10	40t 28SWG close-wound
14	14.5	15	1.5	24	5	27t 24SWG close-wound
21	21.225	21.45	1	52	1	12t 20SWG spaced to ¾in
28	29	30	0.6	10 primary	3	21t 24SWG spaced to ¾in
				30 secondary	1	12t 20SWG spaced to ¾in

The use of tuning slugs in the coils is not recommended. Capacitors can be air, ceramic or mica compression trimmers. Adjust each coupler to cover frequency range shown.

Table 5.10: Band-pass filters for five HF bands and 15kΩ input and output terminations as shown in Fig 5.49(b)

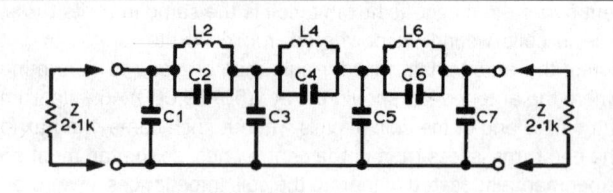

Fig 5.50: A three-section elliptic low-pass filter with 3kHz cut-off. C1 = 37.3nF, C2 = 3.87nF, C3 = 51.9nF, C4 = 19.1nF, C5 = 46.4nF, C6 = 13.5nF, C7 = 29.9nF, mica, polyester or polystyrene. L2 = 168mH, L4 = 125mH, L6 = 130mH are wound on ferrite pot cores

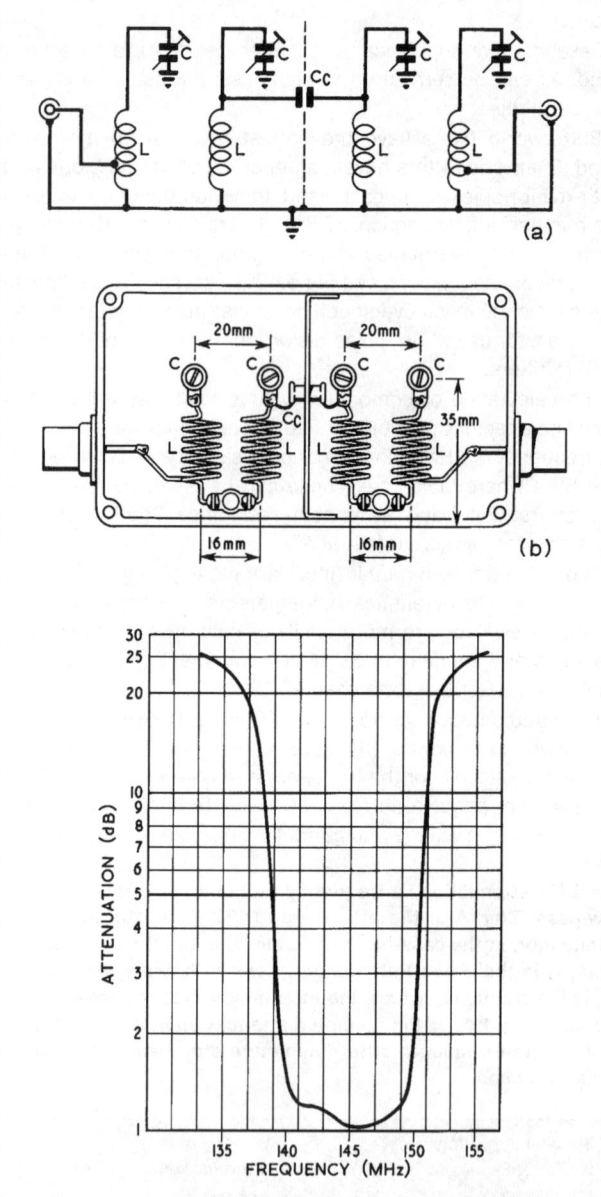

Fig 5.51: A four-section band-pass filter for 145MHz. (a) Direct inductive coupling between sections 1-2 and 3-4, top capacitor (0.5pF) coupling between sections 2-3; both input and output are tapped down on the coils for 50 ohm terminations. (b) The filter fits into an 11 x 6 x 3cm die-cast box; coils are 3/8in (9.5mm) inside dia, 6½ turns bare 18SWG (1.22mm dia) spaced 1 wire dia; taps 1t from earthy end; C are 1-6pF piston ceramic trimmers for receiving and QRP transmitting; for higher power there is room in the box for air trimmers. (c) Performance curve

the highest to be expected, even under fault conditions.

Fig 5.50 shows an audio filter to provide selectivity in a direct-conversion receiver. It is a three-section elliptic low-pass filter with a cut-off frequency of 3kHz, suitable for voice reception.

To make any odd capacitance values of, say, 1% accuracy, start with the next lower standard value (no great accuracy required), measure it precisely (ie to better than 1%), and add what is missing from the desired value in the form of one or more smaller capacitors which are then connected in parallel with the first one. The smaller capacitors, having only a small fraction of the total value, need not be more accurate than, say, ±5%.

Filtering at VHF and Above

At HF and below, it may be assumed that filters will perform as designed if assembled from components which are known to have the required accuracy, either because they were bought to tight specifications or were selected or adjusted with precise test equipment. It was further assumed that capacitor leads had negligible self-inductance and that coils had negligible capacitance.

At VHF and above these assumptions do not hold true. Though filter theory remains the same, the mechanics are quite different. Even then, the results are less predictable and adjustment will be required after assembly to tune out the stray capacitances and inductances. A sweep-generator and oscilloscope provide the most practical adjustment method. A variable oscillator with frequency counter and a voltmeter with RF probe, plus a good deal of patience, can also do the job.

At VHF, self-supporting coils and mica, ceramic or air-dielectric trimmer capacitors give adequate results for most applications. For in-band duplex operation on one antenna, as is common in repeaters, bulky and expensive very-high-Q cavity resonators are required, however.

The band-pass filter of **Fig 5.51** includes four parallel resonant circuits [14]. Direct inductive coupling is used between the first two and the last two; capacitive coupling is used between the centre two, where a shield prevents stray coupling. The input and output connections are tapped down on their respective coils to transform the 50Ω source and load into the proper terminations. This filter can reduce harmonics and other out-of-band spurious emissions when transmitting and suppress strong out-of-band incoming signals which could overload the receiver.

At UHF and SHF, filters are constructed as stripline or coaxial

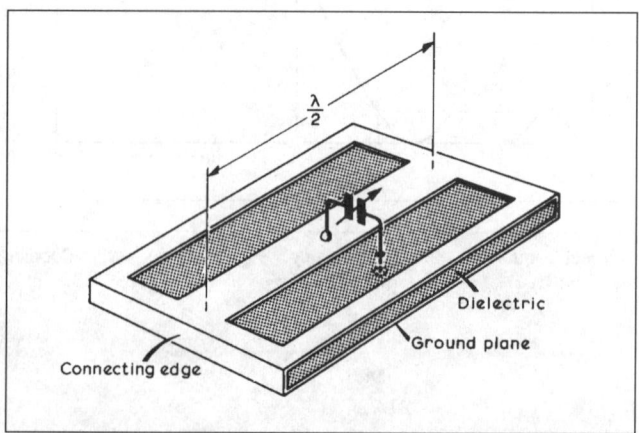

Fig 5.52; Basic microstrip quarter-wave (top) and half-wave resonators. The electrical length of the strips is made shorter than their nominal length so that trimmers at the voltage maxima can be used to tune to resonance; the mechanical length of the strips is shorter than the electrical length by the velocity factor arising from the dielectric constant of the PCB material

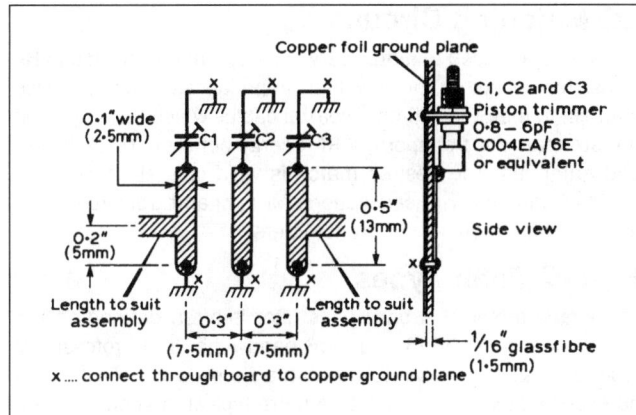

Fig 5.53: A 1.3GHz microstrip filter consisting of three quarter-wave resonators. Coupling is by the stray capacitance between trimmer stators. The input and output lines of 50 ohm microstrip are tapped down on the input and output resonators

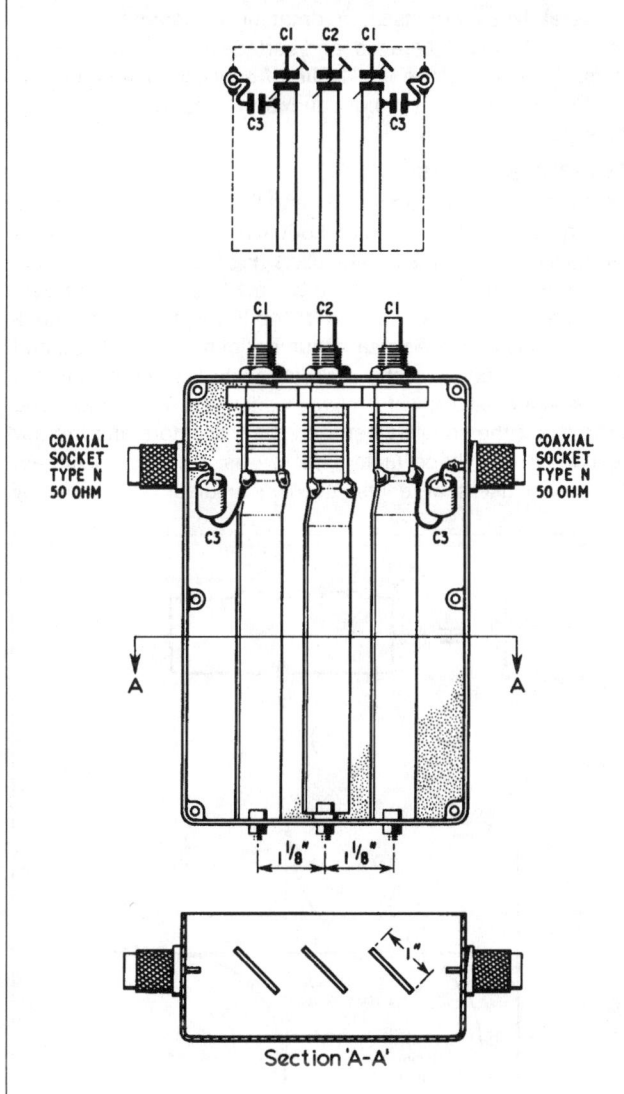

Fig 5.54: The circuit and layout for a 100W 145MHz slab-line filter. The strips are 1 x 1/16in (25 x 1.5mm) sheet copper, offset at 45° to avoid overcoupling. Input and output lines are 6½in (165mm) long; the centre resonator is slightly shorter to allow for the greater length of C2 and a rib in the cast box. C1 = 50pF, C2 = 60pF, C3 = 4.4pF

transmission line sections with air-dielectric trimmer capacitors. One type of stripline, sometimes called microstrip, consists of carefully dimensioned copper tracks on one side of high-grade (glass-filled or PTFE, preferably the latter) printed circuit board 'above' a ground plane formed by the foil on the other side of that PCB. The principle is illustrated in **Fig 5.52** and calculations are given in the General Data chapter. Many amateurs use PCB strip lines wherever very high Q is not mandatory, as they can be fabricated with the PCB-making skills and tools used for many other home-construction projects. Frequently, in fact, such filters are an integral part of the PCB on which the other components are assembled.

Fig 5.53 exemplifies a PCB band-pass filter for the 1.3GHz band. It consists of three resonators, each of which is tuned by a piston-type trimmer. It is essential that these trimmers have a low-impedance connection to earth (the foil on the reverse side of the PCB). With 1-5pF trimmers, the tuning range is 1.1-1.5GHz. The insertion loss is claimed to be less than 1dB. The input and output lines, having a characteristic impedance of 50Ω, may be of any convenient length. This filter is not intended for high-power transmitters.

For higher power, the resonators can be sheet copper striplines, sometimes called slab lines, with air as the dielectric. **Fig 5.54** gives dimensions for band-pass filters for the 144MHz band [14]. The connections between the copper fingers and the die-cast box are at current maxima and must have the lowest possible RF resistance. In the prototype, the ends of the strips were brazed into the widened screwdriver slots in cheesehead

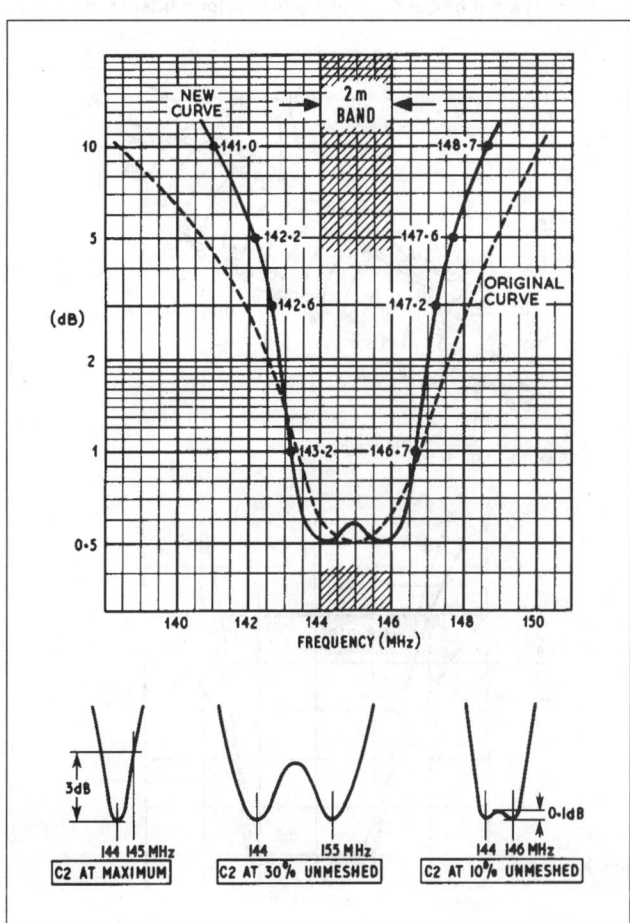

Fig 5.55: Performance curves of the 145MHz slab-line filter as affected by the setting of C2. Note that the insertion loss is only 0.6dB. Also, compare the 10dB bandwidth of this filter with that of the four-section LC filter: 7.7MHz for this filter when tuned to 'new curve', 13MHz for the four-section filter with small, lower-Q coils

brass bolts. The 'hot' ends of the strips are soldered directly to the stator posts of the tuning capacitors. After the input and output resonators are tuned for maximum power throughput at the desired frequency with the centre capacitor fully meshed, the latter is then adjusted to get the desired coupling and thereby pass-band shape (Fig 5.55).

For the highest Q at UHF and SHF, and with it the minimum insertion loss and greatest out-of-band attenuation, coaxial cavity resonators are used. Their construction requires specialised equipment and skills, as brass parts must be machined, brazed together and silver plated.

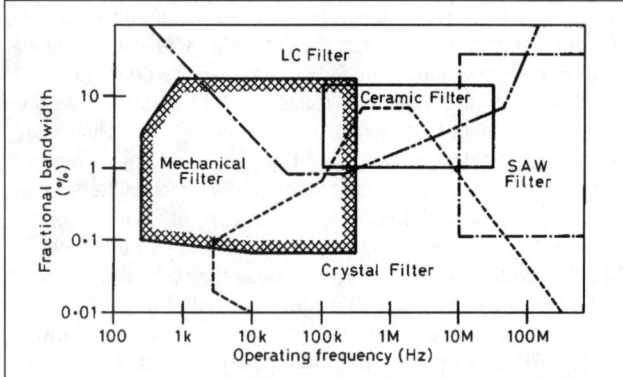

Fig 5.56: Typical frequency ranges for various filter techniques

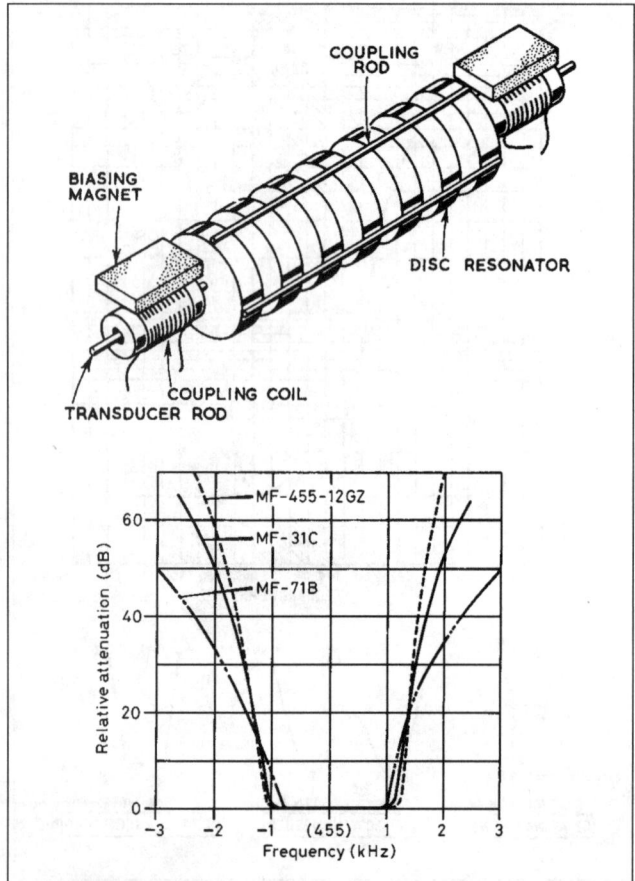

Fig 5.57: The Collins mechanical filter. An IF signal is converted into mechanical vibrations in a magnetostrictive transducer, is then passed, by coupling rods, along a series of mechanical resonators to the output transducer which reconverts into an electrical signal. Below are response curves of three grades of Goyo miniature mechanical filters

LC Matching Circuits

LC circuits are used to match very low impedances, such as VHF transistor collectors, or very high impedances, such as valve anodes, to the 50 or 75-ohm coaxial cables which have become the standard for transporting RF energy between 'black boxes' and antennas. The desired match is valid only at or near the design frequency. The calculations for L, pi and L-pi circuits are given in the chapter on HF transmitters.

High-Q Filter Types

The shape factor of a band-pass filter is often defined as the ratio between its -60dB and -6dB bandwidth. In a professional receiver, a single-sideband filter with a shape factor of 1.8 could be expected, while 2.0 might be more typical in good amateur equipment. It is possible to make LC filters with such performance, but it would have to be at a very low intermediate frequency (10-20kHz), have many sections, and be prohibitively bulky, costly and complicated. The limited Q of practically realisable inductors, say 300 for the best, is the main reason. Hence the search for resonators of higher Q.

Several types are used in amateur equipment, including mechanical, crystal, ceramic, and surface acoustic wave (SAW) filters. Each is effective in a limited frequency range and fractional bandwidth (the ratio of bandwidth to centre frequency in percent). See Fig 5.56.

Mechanical filters

There are very effective SSB, CW and RTTY filters for intermediate frequencies between 60 and 600kHz based on the mechanical resonances of small metal discs (Fig 5.57). The filter comprises three types of component: two magnetostrictive or piezoelectric transducers which convert the IF signals into mechanical vibrations and vice versa, a number of resonator discs, and coupling rods between those disks. Each disc represents the mechanical equivalent of a high-Q series resonant circuit, and the rods set the coupling between the resonators and thereby the bandwidth. Shape factors as low as 2 can be achieved. Mechanical filters were first used in amateur equipment by

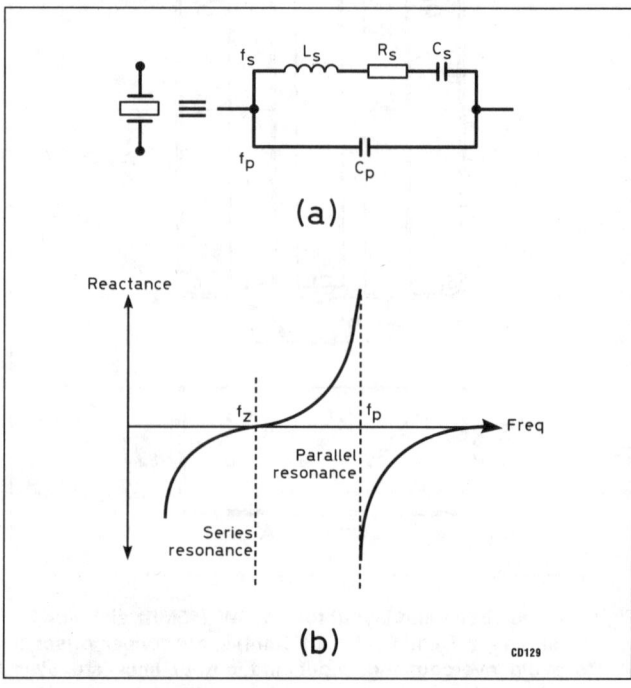

Fig 5.58: The crystal equivalent circuit (a), and its reactance vs frequency plot (b)

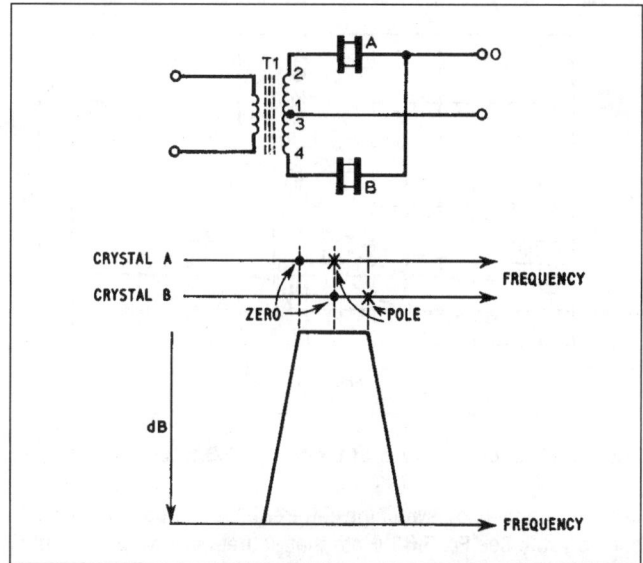

Fig 5.59: The basic single-section half-lattice filter diagram (top) and its idealised frequency plot (bottom). Note the poles and zeros of the two crystals in relation to the cut-off frequencies; if placed correctly, the frequency response is symmetrical. The bifilar transformer is required to provide balanced inputs

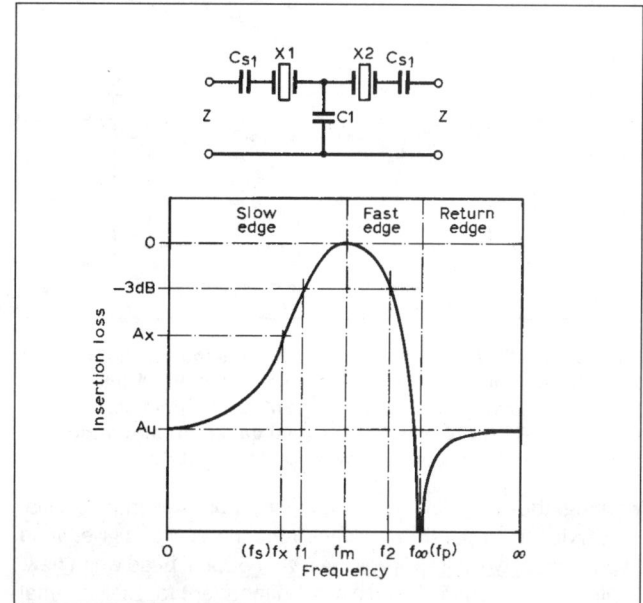

Fig 5.60: The basic two-pole ladder filter diagram (top) and its generalised frequency plot (bottom). Note that the crystals are identical but that the resulting frequency response is asymmetric

Collins Radio (USA). They still are offered as options in expensive amateur transceivers but with the advent of digital signal processing (DSP) in IF systems, typically between 10 and 20kHz, similar performance can be obtained at lower cost.

Crystal filters

Piezo-electric crystals, cut from man-made quartz bars, are resonators with extremely high Q, tens of thousands being common. Their electrical equivalent, **Fig 5.58(a)**, shows a very large inductance L_s, an extremely small capacitance C_s and a small loss resistance R_s, in a series-resonant circuit. It is shunted by the parallel capacitance C_p, which is the real capacitance of the crystal electrodes, holder, socket, wiring and any external load capacitor one may wish to connect across the crystal. At frequencies above the resonance of the series branch, the net impedance of that branch is inductive. This inductance is in parallel resonance with C_p at a frequency slightly above the series resonance: **Fig 5.58(b)**. The essentially mechanical series resonance, the zero, may be considered user-immovable but the parallel resonance or pole can be pulled down closer to the series resonant frequency by increasing the load capacitance, eg with a trimmer.

Though their equivalent circuit would suggest that crystals are linear devices, this is not strictly true. Crystal filters can therefore cause intermodulation, especially if driven hard, eg by a

strong signal just outside the filter's pass-band. Sometimes, interchanging input and output solves the problem. Very little is known about the causes of crystal non-linearity and the subject is rarely mentioned in published specifications.

The two most common configurations are the half-lattice filter, **Fig 5.59**, and the ladder filter, **Fig 5.60**.

Half-lattice filter curves are symmetrical about the centre frequency, an advantage in receivers, but they require crystals differing in series resonant frequency by somewhat more than half the desired pass-band width and an RF transformer for each two filter sections. Two to four sections, four to eight crystals, are required in a good HF SSB receiver IF filter. Model XF-9B in **Table 5.11** is a high-performance filter which seems to be of this type. Half-lattice filters make good home construction projects only for the well-equipped and experienced.

Ladder filters require no transformers and use crystals of only one frequency but they have asymmetrical pass-band curves; this creates no problem in SSB generators but is less desirable in a receiver with upper and lower sideband selection. With only five crystals a good HF SSB generator filter can be made. Model XF-9A in Table 5.11 seems to be of this type.

Ladder crystal filters using inexpensive consumer (3.58MHz telephone and 3.579, 4.433 or 8.867MHz TV colour-burst) crystals have been successfully made by many amateurs. A word of

Filter type	XF-9A	XF-9B	XF-9C	XF-9D	XF-9E	XF-9M
Application	SSB TX	SSB TX/RX	AM	AM	NBFM	CW
Number of poles	5	8	8	8	8	4
6dB bandwidth (kHz)	2.5	2.4	3.75	5.0	12.0	0.5
Passband ripple (dB)	<1	<2	<2	<2	<2	<1
Insertion loss (dB)	<3	<3.5	<3.5	<3.5	<3	<5
Termination	500 /30pF	500 /30pF	500 /30pF	500 /30pF	1200 /30pF	500 /30pF
Shape factor	1.7 (6-50dB)	1.8 (6-60dB)	1.8 (6-60dB)	1.8 (6-60dB)	1.8 (6-60dB)	2.5 (6-40dB)
		2.2 (6-80dB)	2.2 (6-80dB)	2.2 (6-80dB)	2.2 (6-80dB)	4.4 (6-60dB)
Ultimate attenuation (dB)	>45	>100	>100	>100	>90	>90

Table 5.11: KVG 9MHz crystal filters for SSB, AM, FM and CW bandwidths

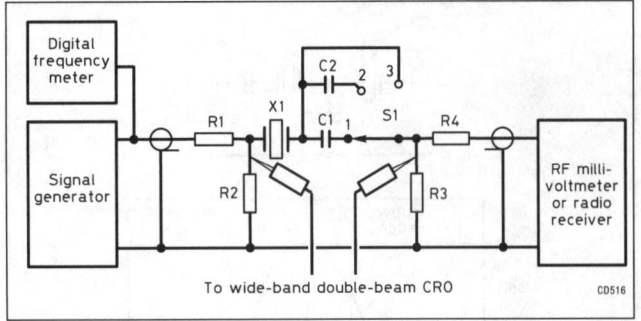

Fig 5.61: G3JIR's crystal test circuit. At resonance, both series (switch position 3) and parallel (switch positions 1 and 2), the two 'scope traces are in-phase. For 9MHz filter crystals, R1 = R4 = 1k , R2 = R3 = 220 ohms. The signal generator and frequency meter should have a frequency resolution of 10Hz

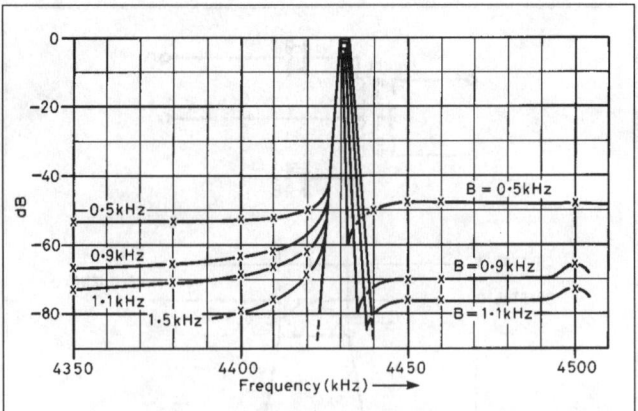

Fig 5.63: Response curves of the the Y27YO filter

warning: these crystals were made for parallel-resonant oscillator service; their parallel-resonant frequencies with a given load capacitance, typically 20pF, were within 50ppm or so when new. Their series-resonant frequencies, unimportant for their original purpose but paramount for filter application, can differ by much more. It therefore is very useful to have many more crystals than needed so that a matched set can be selected.

To design a filter properly around available crystals one must know their characteristics. They can be measured with the test circuit of **Fig 5.61**, where the three positions of S1 yield three equations for the three unknowns L_s, C_s and C_p. R_s can be measured after establishing series-resonance with S1 in position 3 by substituting non-inductive resistors for the crystal without touching the frequency or output level of the signal generator. A resistor which gives the same reading on the output meter as the crystal is equal to R_s. Note that the signal generator must be capable of setting and holding a frequency within 10Hz or so. All relevant information and PC programs to simplify the calculations are given in [22].

If a sweep generator and 'scope are used to adjust or verify a completed filter, the sweep speed must be kept very low: several seconds per sweep. Traditionally this has been done by using 'scopes with long-persistence CRTs. Sampling 'scopes, of course, can take the place of long-persistence screens.

The home-made filter of **Fig 5.62**, described by Y27YO, uses six 4.433MHz PAL colour TV crystals [23]. The filter bandwidth can be changed by switching different load capacitors across each crystal. See **Fig 5.63**. Note that at the narrower bandwidths the ultimate attenuation is reduced because of the greater load capacitors shunting each crystal. Each crystal with its load capacitors and switch wafer should be in a separate shielding compartment.

The input impedance of any crystal filter just above and below its pass-band is far from constant or non-reactive; therefore it is unsuitable as a termination for a preceding diode-ring mixer, which must have a fixed purely resistive load to achieve the desired rejection of unwanted mixing products. The inclusion in the Y27YO filter of a common-gate FET buffer solves this problem.

Virtually all available crystals above about 24MHz are overtone crystals, which means that they have been processed to have high Q at the third or fifth mechanical harmonic of their fundamental frequency. Such crystals can be used in filters. The marking on overtone crystals is their series resonant frequency, which is the one that is important in filters.

48MHz is a common microprocessor clock frequency and third-overtone crystals for it are widely available and relatively inexpensive. 48MHz also is a suitable first IF for dual- or triple-conversion HF receivers, which then require a roofing filter at that frequency. PAOSE reported on the 48MHz SSB filter design shown in **Fig 5.64**. Prototype data are shown in **Table 5.12** - these were measured with a 50Ω source and 150Ω//7pF as shown in Fig 5.64.

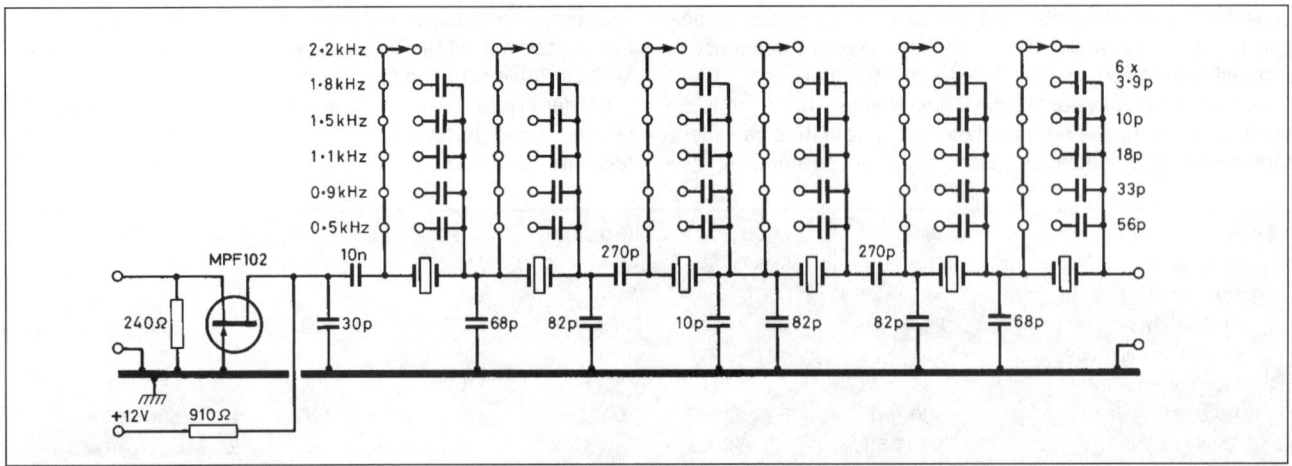

Fig 5.62: Y27YO described this six-section 4.433MHz ladder filter with switch-selectable bandwidth in *Funkamateur* 1/85. The FET buffer is included to present a stable, non-reactive load to the preceding diode mixer and the proper source impedance to the filter. While it uses inexpensive components, including PAL TV colour-burst crystals, careful shielding between sections is required and construction and test demand skill and proper instrumentation

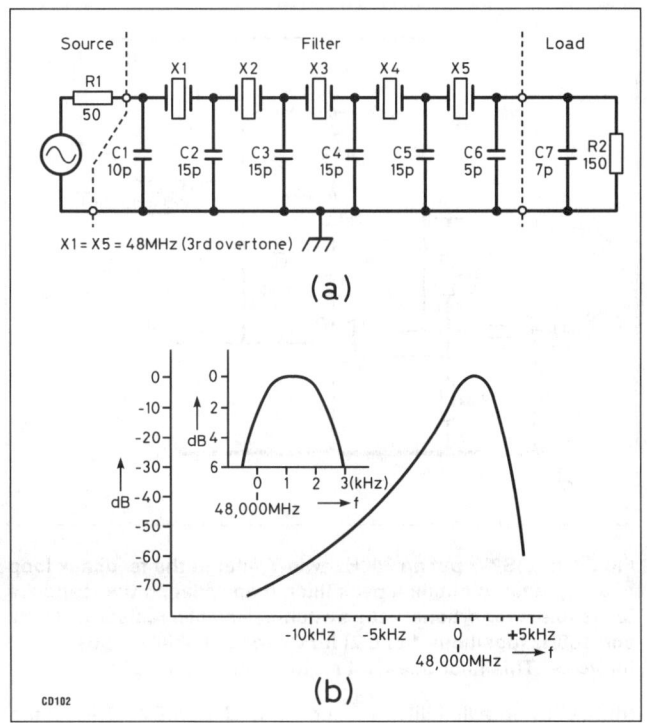

Fig 5.64: Overtone crystals, always marked with their series-resonant frequency, are used in this 48MHz ladder filter. This design by J Wieberdink from the Dutch magazine *Radio Bulletin* 10/83 makes a good roofing filter for an HF SSB/CW receiver, but its top is too narrow to pass AM or NBFM and its shallow skirt slope on the low side requires that it be backed up by another filter at a second or third IF

Centre frequency (f0)	48.0012MHz
3dB bandwidth	2.6kHz
6dB bandwidth	3kHz
40dB bandwidth	9kHz
60dB bandwidth	15.1kHz
Spurious responses	< 70dB
Pass-band ripple	<0.2dB
Insertion loss	2.1dB

Table 5.12: Performance of the 48MHz crystal filter

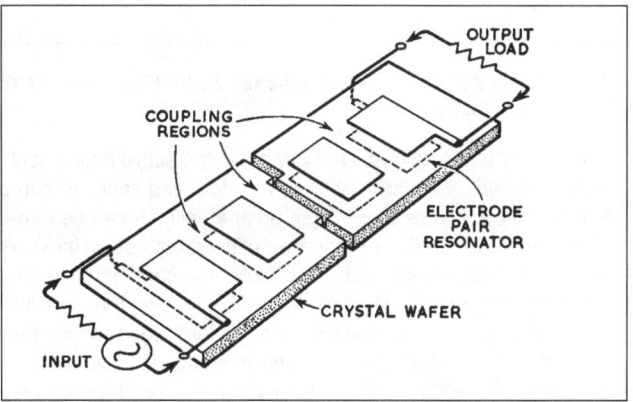

Fig 5.65: A monolithic crystal band-pass filter consists of several resonators, ie pairs of electrodes, on a single quartz plate. The coupling between the resonators is essentially mechanical through the quartz. Monolithic filters are mounted in three or four-pin hermetically sealed crystal holders. At HF, they are less expensive though less effective than the best discrete-component crystal filters, but they do make excellent VHF roofing filters

Adequate shielding of the whole filter and between its sections is essential at this high frequency.

Monolithic crystal filters

Several pairs of electrodes can be plated onto a single quartz blank as shown in **Fig 5.65**. This results in a multi-section filter with the coupling between elements being mechanical through the quartz. Monolithic crystal filters in the 10-100MHz range are used as IF filters where the pass-band must be relatively wide, ie in AM and FM receivers and as roofing filters in multimode receivers.

Ceramic filters

Synthetic (ceramic) piezo-electric resonators are being made into band-pass filters in the range of 400kHz to 10.7MHz. Monolithic, ladder and half-lattice crystal filters all have their ceramic equivalents. Cheaper, ceramic resonators have lower Q than quartz crystals and their resonant frequencies have a wider tolerance and are more temperature dependent. While ceramic band-pass filters with good shape factors are made for bandwidths commensurate with AM, FM and SSB, they require more sections to achieve them, hence their insertion loss is greater. Care must be taken that BFOs used with them have sufficient frequency adjustment range to accommodate the centre frequency tolerance of a ceramic filter, eg 455±2kHz at 25°C. Input and output matching transformers are included in some ceramic filter modules. **Fig 5.66** shows a Murata ladder filter and a Toko monolithic filter.

Active filters

When designing LC audio filters, it is soon discovered that the inductors of values one would wish to use are bulkier, more expensive and of lower Q than the capacitors. Moreover, this lower Q requires the use of more sections, hence more insertion loss and

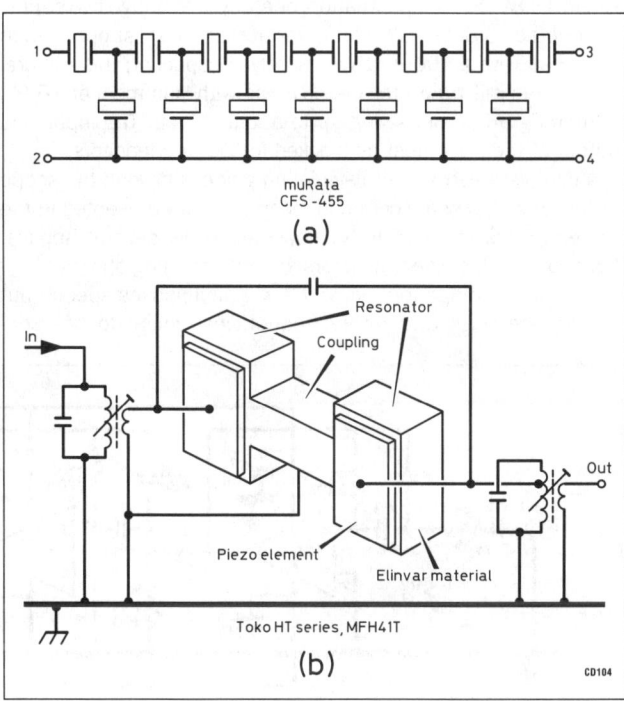

Fig 5.66: Ceramic band-pass filters are made in several configurations resembling those of crystal filters. While cheaper than the latter, they require more sections for a given filter performance, hence have greater insertion loss. Centre-frequency tolerances are 0.5% typical. Shown here are a ladder filter (a) and a monolithic filter (b)

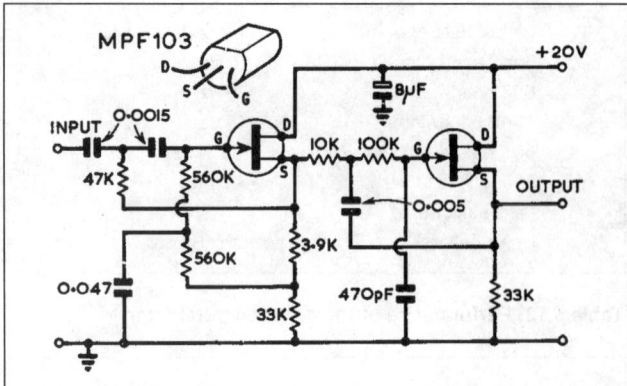

Fig 5.67: The ZL2APC active band-pass audio filter using FETs and a single DC supply

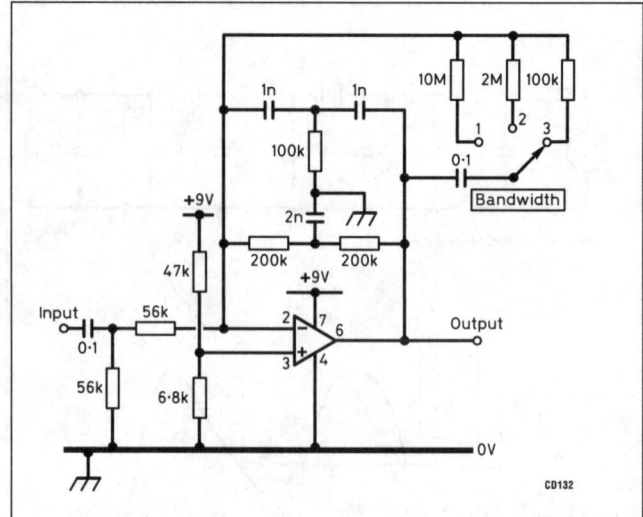

Fig 5.68: G3SZW put an 800Hz twin-T filter in the feedback loop of an op-amp to obtain a peak filter, then widened the response to usable proportions with switch-selectable resistors: to 60 and 180Hz (positions 1 and 2) for CW or 300-3500Hz (position 3) for voice. This filter does not require dual DC supplies

even greater bulk and cost. One way out is the active filter, a technique using an amplifier to activate resistors and capacitors in a circuit which emulates an LC filter. Such amplifiers can be either single transistors or IC operational amplifiers, both being inexpensive, small, miserly with their DC supply, and capable of turning insertion loss into gain. Most active audio filters in amateur applications use two, three or four two-pole sections, each section having an insertion voltage gain between one and two. Filter component (R and C) accuracies of better than 5% are generally adequate, polystyrene capacitors being preferred.

The advantage of op-amps over single transistors is that the parameters of the former do not appear in the transfer (ie output vs input) function of the filter, thereby simplifying the calculations.

Note that most IC op-amps are designed to have a frequency response down to DC. To allow both positive and negative outputs, they require both positive and negative supply voltages. In AC-only applications, however, this can be circumvented. In a single 13.8V DC supply situation, one way is to bridge two series-connected 6.8kΩ 1W resistors across that supply, each bypassed with a 100µF/16V electrolytic capacitor; this will create a three-rail supply for the op-amps with 'common' at +6.9V, permitting an output swing up to about 8V p-p. The input and output of the filter must be blocked for DC by capacitors.

A complete active filter calculation guide is beyond the scope of this book but some common techniques are presented in the General Data chapter. Here, however, follow several applications, one with single transistors and others using op-amps.

A discrete-component active filter, which passes speech but rejects the bass and treble frequencies which do not con-

tribute to intelligibility, is shown in **Fig 5.67**. This filter, designed by ZL2APC, might be used to provide selectivity for phone reception with a direct-conversion receiver (though it must be pointed out that active filters do not have sufficient dynamic range to do justice to those very best DC receivers, which can detect microvolt signals in the presence of tens of millivolts of QRM on frequencies which the audio filter is required to reject).

A twin-T filter used in the feedback loop of an op-amp is shown in **Fig 5.68**. A twin-T filter basically is a notch filter which rejects one single frequency and passes all others. With the R and C values shown, that frequency is about 800Hz. Used in the feedback loop of an op-amp, as in this design by G3SZW, the assembly becomes a peak filter which passes only that one frequency, too sharp even for CW. G3SZW broadened the response by shunting switch-selectable resistors across the twin-T; to 60Hz with 10MΩ, 180Hz with 2MΩ, and from 300Hz to 3500Hz, for phone, with 100kΩ.

Twin-T filters require close matching of resistors and capacitors. That would, in this example, be best accomplished by using four identical 1nF capacitors and four identical 200kΩ resistors, using two of each in parallel for the earthed legs.

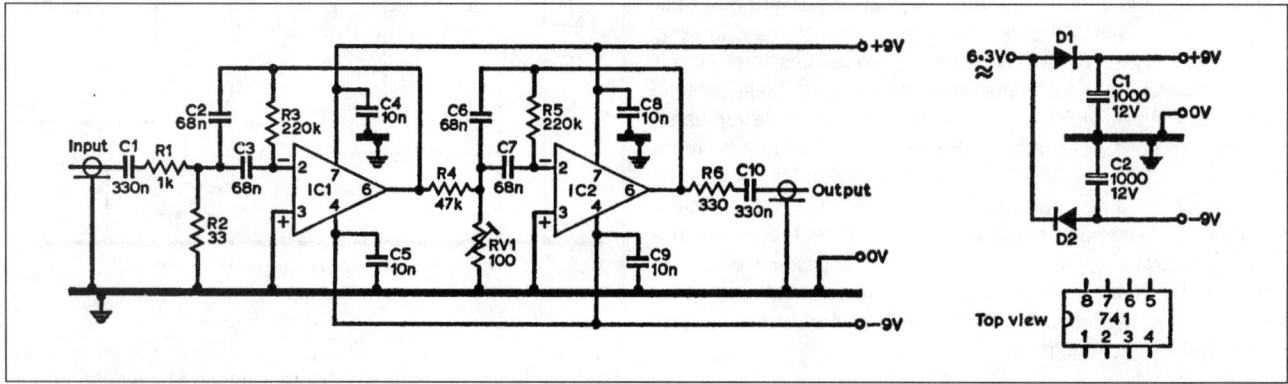

Fig 5.69(a): This CW band-pass filter using two multiple-feedback stages comes from LA2IJ and LA4HK. The centre frequency of the second stage can be equal to or offset from the first. If the two frequencies coincide, the overall bandwidth is 50Hz at 6dB (640Hz at 50dB), if staggered 200Hz (1550Hz). The dual power supply is derived from 6.3VAC, available in most valved receivers or transmitters

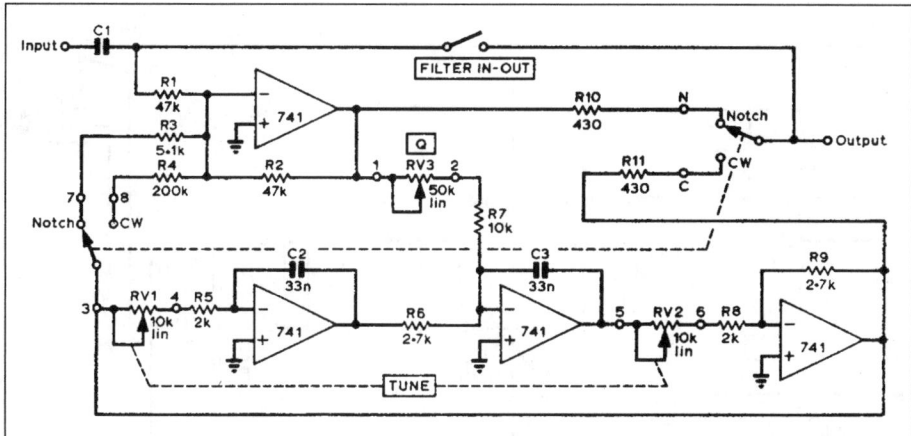

Fig 5.70: A universal filter scheme described by DJ6HP provides variable-Q band-pass filtering for CW or a tunable notch for 'phone. Dual DC supplies between ±9V and ±15V are required

This circuit also demonstrates another technique for the use of op-amps on a single supply, here 9V. The DC level of both inputs is set by the voltage divider to which the (+) input is connected:

$$[6.8/(6.8 + 47)] \times 9 = 1.14V$$

The op-amp's DC output level is set by its DC input voltage and inverting gain:

$$1.14 + [(200 + 200)/(56 + 56)] \times 1.14 = 5.2V$$

roughly half-way between +9V and earth. The capacitor in series with the bandwidth switch is to prevent the lowest bandwidth resistor from upsetting the DC levels. The input and output blocking capacitors have the same purpose with respect to any DC paths through the signal source or load.

A CW filter with two 741 or 301A-type op-amps in a multiple-feedback, band-pass configuration was described by LA2IJ and LA4HK [24], and is shown in **Fig 5.69**. It would be a worthwhile accessory with a modern transceiver lacking a narrow crystal filter.

R9	10M, 0.25W	C13, 15	22μ, 25V
R22	330R, 0.25W	C14	1μ tantalum
RV1	4k7 preset	C16, 18	10μ
C1, 11	0.47μ	IC1	S3529
C2	0.01μ	IC2	S3528
C3, 6-8	680p	IC3	LM386
C4, 5, 9, 10	0.1μ	IC4	78L05
C12, 17	100μ, 16V	IC5	7660
S1	3-pole changeover toggle switch		
S2	Four BCD thumbwheel switches or two 40-pos CB channel switches		
X	3.58MHz crystal		
Two 18-pin DIL sockets			
Two 8-pin DIL sockets			
Red LED			
12V power input socket			
Audio input/output sockets/jacks according to choice			
Phones jack			
All resistors except R9 and R22 are 47k, 0.25W.			

Table 5.13: Components list for the BARTG R5 switched capacitor filter

The first stage is fixed-tuned to about 880Hz, depending partly on the value of R2. The corresponding resistor in the second stage is variable and with it the resonant frequency can be adjusted to match that of the first stage, or to a slight offset for a double-humped band-pass characteristic. In the first state the filter has a pass-band width of only 50Hz at -6dB (about 640Hz at -50dB). When off-tuned the effective pass-band can be widened to about 200Hz (1550Hz).

Note that a bandwidth as narrow as 50Hz is of value only if the frequency stability of both the transmitter and receiver is such that the beat note does not drift out of the pass-band during a transmission, a stability seldom achieved with free-running home-built VFOs. Few analogue filters can compete with the ears of a skilled operator when it comes to digging out a weak wanted signal from among much stronger QRM.

A scheme to provide second-order CW band-pass filtering or a tuneable notch for voice reception was described by DJ6HP [25]. It is shown in **Fig 5.70**, and is based on the three-op-amp so-called state variable or universal active filter. The addition of a fourth op-amp, connected as a summing amplifier, provides the notch facility.

The resonant Q can be set between 1 and 5 with a single variable resistor and the centre frequency can be tuned between 450 and 2700Hz using two ganged variable resistors.

Switched-capacitor filters

The switched-capacitor filter is based on the digital processing of analogue signals, ie a hybrid between analogue and digital signal processing. It depends heavily on integrated circuits for its implementation. It pays to study the manufacturers' data sheets of devices under consideration before making a choice.

While switched-capacitor filters had been introduced to amateurs before, eg the Motorola MC14413/4 (superseded by MC145414) by W1JF [26] and the National Semiconductor MF10 by AI2T [27], it was an article by WB4TLM/KB4KVE [28] featuring the AMI S3528/9 ICs in the AFtronics SuperSCAF that led to the BARTG R5 design constructed and described by G3ISD [29] shown in **Tables 5.13 and 5.14** and in **Fig 5.71**.

WB4TLM/KB4KVE describe the operation of switched-capacitor filters as follows: "The SCF works by storing discrete samples of an analogue signal as a charge on a capacitor. This charge is transferred from one capacitor to another down a chain of capacitors forming the filter. The sampling and transfer operations take place at regular intervals under control of a precise frequency source or clock. Filtering is achieved by combining the charges on the different capacitors in specific ratios and by feeding charges back to the prior stages in the capacitor chain. In this way, filters of much higher performance and complexity may be synthesised than is practical with analogue filters".

The BARTG R5 filter includes a seven-pole high-pass and a seven-pole low-pass SCAF of which the cut-off frequencies can be chosen by selecting one of 40 positions of two switches according to **Table 5.15**. If used as an audio filter behind a receiver, CW might be listened to with the filter switches set to

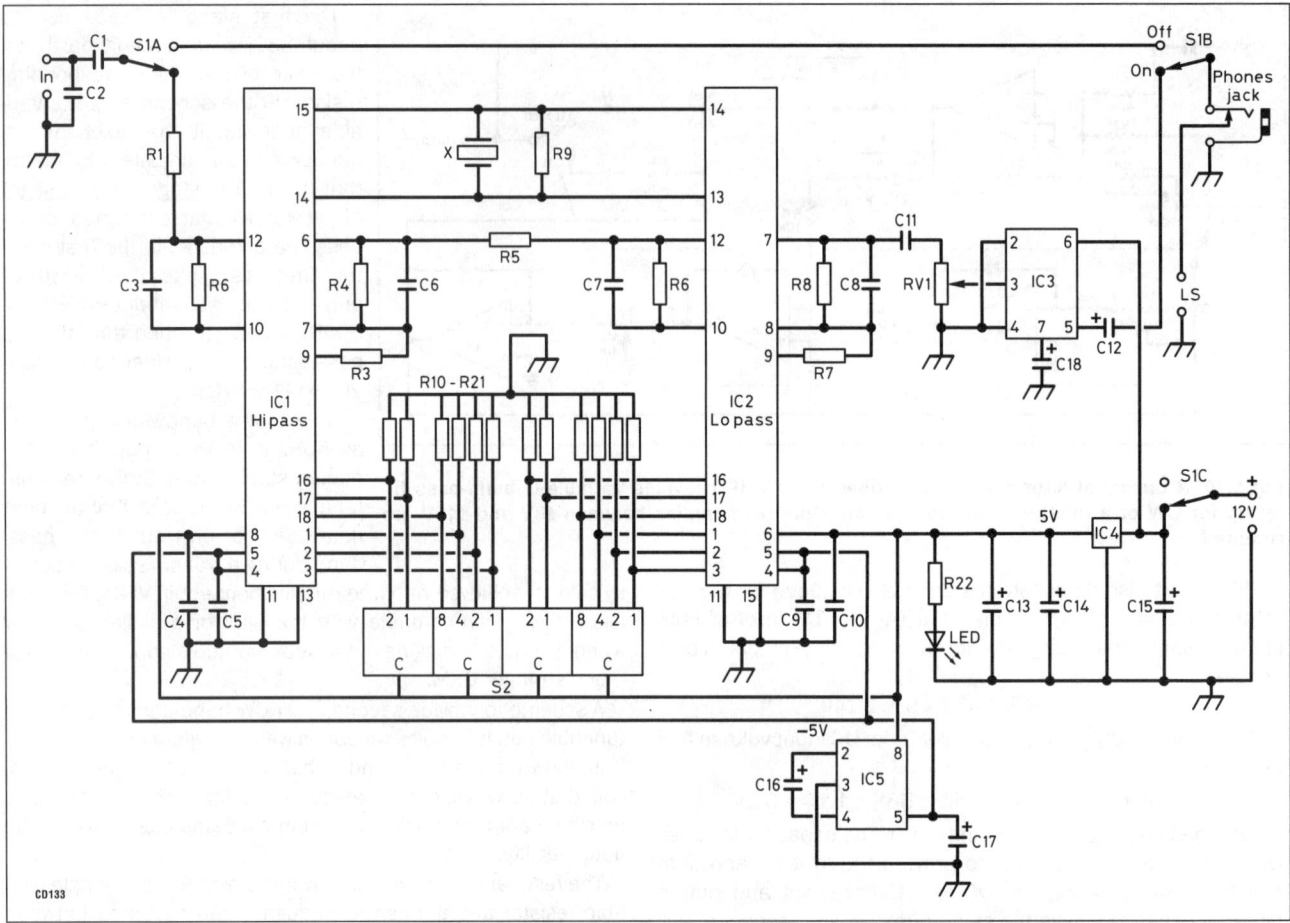

Fig 5.71: The schematic of the BARTG R5 switched capacitor filter. 40 different cut-off frequencies can be selected independently for the high-pass and low-pass filter to configure the optimum pass-band for CW, speech or data (*Datacom*)

H7 and L9 for a pass-band of 635-904Hz, while for voice reception H3 and L27, 273-2711Hz would be appropriate. Other selections would be useful for RTTY tones. In addition to the filters, the board contains ICs for audio amplification to speaker level and ±5VDC from the 12VDC power input, as well as the necessary passive components.

Fig 5.72 shows the circuit of an active low-pass filter which has a Butterworth (maximally flat) type of response and is a fourth-order design. Outside the passband, the attenuation of a fourth-order filter increases at 24dB/octave. The filter was designed using an active filter design reference book [30].

The values of R and C in Fig 5.72 were set to R = 10kΩ and C = 5n6. This gives a cut-off frequency (where the amplitude vs frequency response curve is 3dB down) of 2.8kHz. The 5n6 capacitors should be polystyrene types because this gives the best audio quality. The filter was designed to improve the overall selectivity of a receiver which has poor IF selectivity, but could also be used in a transmitter to limit the bandwidth of the radiated signal. The op-amps were contained within a 5532 type dual op-amp package, and pin numbers are given for this device. The resistor Rb is a bias resistor, and is required only if the source does not have a DC path to supply bias current for input pin 2 of the 5532. The power supply used was ±12V.

VOLTAGE REGULATORS

The voltage regulator is an important building block. Ideally, it provide a constant output DC voltage, which is independent of the input voltage, load current and temperature. If the voltage regulator is used to regulate the supply to a VFO (for example)

Switch position	High-pass	Low-pass	Switch position	High-pass	Low-pass
00	40	44	21	1892	2081
01	91	100	22	1985	2183
02	182	200	23	2086	2295
03	273	300	24	2198	2418
04	363	399	25	2260	2486
05	455	500	26	2392	2632
06	546	601	27	2465	2711
07	635	699	28	2543	2797
08	726	799	29	2625	2887
09	822	904	30	2712	2983
10	914	1005	31	2805	3086
11	1005	1105	32	2905	3196
12	1099	1209	33	3013	3314
13	1179	1297	34	3129	3442
14	1271	1398	35	5423*	5965*
15	1355	1491	36	3254	3579
16	1453	1598	37	3389	3728
17	1535	1688	38	5811*	6392*
18	1627	1790	39	3537	3891
19	1731	1904			
20	1808	1989		* Note these frequencies. Table taken from *Datacom*.	

Table 5.14: The BARTG R5 filter. The cut-off frequencies are selected by each of the forty high-pass and low-pass switch positions

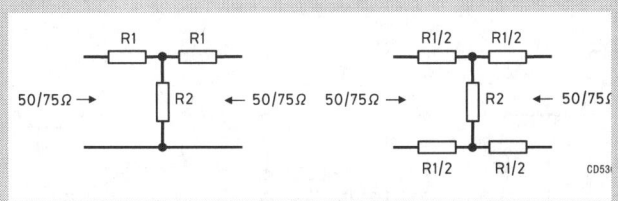

T-pad

Attenuation (dB)	50Ω		75Ω	
	R1	R2	R1	R2
1	2.9	433	4.3	647
2	5.7	215	8.6	323
3	8.5	142	12.8	213
4	11.3	105	17.0	157
5	14.0	82	21.0	123.4
6	16.6	67	25.0	100
7	19.0	56	28.7	83.8
8	21.5	47	32.3	71
9	23.8	41	35.7	61
10	26.0	35	39.0	52.7
11	28.0	30.6	42.0	45.9
12	30.0	26.8	45.0	40.2
13	31.7	23.5	47.6	35.3
14	33.3	20.8	50.0	31.2
15	35.0	18.4	52.4	25.0
20	41.0	10.0	61.4	15.2
25	44.7	5.6	67.0	8.5
30	47.0	3.2	70.4	4.8
35	48.2	1.8	72.4	2.7
40	49.0	1.0	73.6	1.5

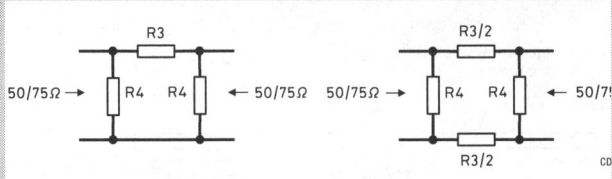

π-pad

Attenuation (dB)	50Ω		75Ω	
	R3	R4	R3	R4
1	5.8	870	8.6	1305
2	11.6	436	17.4	654
3	17.6	292	26.4	439
4	23.8	221	35.8	331
5	30.4	179	45.6	268
6	37.3	151	56.0	226
7	44.8	131	67.2	196
8	52.3	116	79.3	174
9	61.6	105	92.4	158
10	70.7	96	107	144
11	81.6	89	123	134
12	93.2	84	140	125
13	106	78.3	159	118
14	120	74.9	181	112
15	136	71.6	204	107
20	248	61	371	91.5
25	443	56	666	83.9
30	790	53.2	1186	79.7
35	1406	51.8	2108	77.7
40	2500	51	3750	76.5

* Note these frequencies.   Table taken from *Datacom*.

Table 5.17. Resistor values for 50Ω and 75Ω T- and pi-attenuators

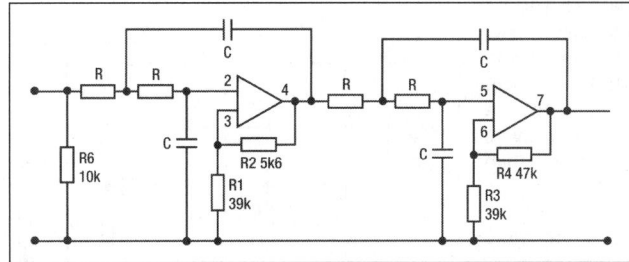

Fig 5.72: Audio low-pass filter, fourth-order Butterworth design

then it will not be affected by variations in supply voltage. For sensitive circuits like the VFO, this is a significant benefit. This allows circuits to be designed to operate at one voltage, which eases the design.

As well as the DC parameters of regulators, most circuits also provide a significant isolation of high frequencies. This isolation is provided partly by the decoupling capacitors which accompany the regulator, and partly by the electronic isolation of the regulator itself. IC type regulators are now so cheap (typically 50p for an LM317) that it is economical to use one for each stage of a complex circuit. For example, in a frequency synthesiser separate regulators could be used for:

- The VCO
- The VCO buffer/amplifier stage
- The reference oscillator
- The digital divider circuits

More on voltage regulators can be found in the power supplies chapter.

ATTENUATORS

Attenuators are resistor networks which reduce the signal level in a line while maintaining its characteristic impedance. Table 5.15 gives the resistance values to make up 75Ω and 50Ω unbalanced RF T- and pi-attenuators; the choice between the T- and pi- configurations comes down to the availability of resistors close to the intended values, which can also be made up of two or more higher values in parallel; the end result is the same. Attenuators are also discussed in the chapter on test equipment. Here are some of the applications:

Receiver Overload

Sensitive receivers often suffer overload (blocking, cross-modulation) from strong out-of-band unwanted signals which are too close to the wanted signal to be adequately rejected by preselector filters. This condition can often be relieved by an attenuator in the antenna input. In modern receivers, one reduction of sensitivity is usually provided for by a switch which removes the RF amplifier stage from the circuit; and another reduction by switching in an attenuator, usually 20dB.

There is an additional advantage of an attenuator. However carefully an antenna may be matched to a receiver at the wanted frequency, it is likely to be grossly mismatched at the interfering frequency, leaving the receiver's preselector filter poorly terminated and less able to do its job. A 10dB attenuator keeps the nominally 50Ω termination of the preselector filter between 41 and 61Ω under all conditions of antenna mismatch.

An S-meter as a Field Strength Meter

When plotting antenna patterns, the station receiver is often used as a field strength indicator. As the calibration of S-meters is notoriously inconsistent, it is better to always use the same S-meter reading, say S9, and to adjust the signal source or the

Fig 5.73: 50-ohm attenuator bank with a range of 0-41dB in 1dB steps. At RF, it is unwise to try to get more than 20dB per step because of the effect of stray capacitances. (*ARRL Handbook*)

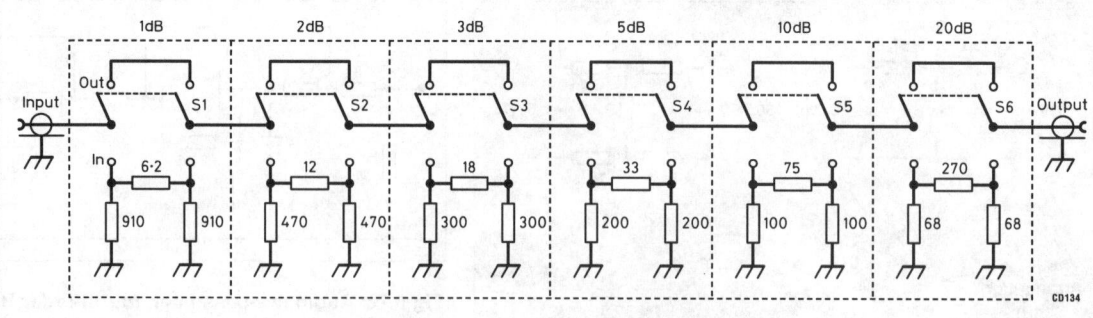

receiver sensitivity to get that reading at a pattern null. All higher field strengths are then reduced to S9 by means of a calibrated switchable attenuator.

A suitable attenuator, with a range of 0 - 41dB in 1dB steps, is shown in **Fig 5.73**. Such an attenuator bank can be constructed from 5% carbon composition or film resistors of standard values and DPDT slide switches, eg RS 337-986. The unit is assembled in a tin-plate box; shields between individual attenuators reduce capacitive coupling and can be made of any material that is easy to cut, shape and solder, ie tin plate, PCB material or copper gauze. At VHF, the unit is still useful but stray capacitances and the self-inductance of resistors are bound to reduce accuracy. Verification of the accuracy of each of the six sections can be done at DC, using a volt or two from a battery or PSU and a DVM; do not forget to terminate the output with a 50Ω resistor.

Driving VHF Transverters

RF attenuators may sometimes be used at higher powers, eg where an HF transmitter without a low-level RF output is to drive, but not to overdrive, a VHF transverter. Care must be taken to ensure that the resistors can handle the power applied to them.

ANALOGUE-DIGITAL INTERFACES

Most real-world phenomena are analogue in nature, meaning they are continuously variable rather than only in discrete steps; examples are a person's height above ground going up a smooth wheel-chair ramp, the shaft angle of a tuning capacitor, the temperature of a heatsink and the wave shape of someone's voice. However, some are digital in nature, meaning that they come in whole multiples of a smallest quantity called least-significant bit (LSB); examples are a person's height going up a staircase (LSB is one step), telephone numbers (one cannot dial in between two numbers) and money (LSB is the penny and any sum is a whole multiple thereof).

Frequently, it is useful to convert from analogue to digital and vice versa in an analogue-to-digital converter (ADC) or a digital-to-analogue converter (DAC). If anti-slip grooves with a pitch of, say, 1cm are cut across the wheelchair ramp, it has, in fact, become a staircase with minuscule steps (LSBs) which one can count to determine how far up one is, with any position between two successive steps being considered trivial. Without going into details of design, a number of conversions commonly employed in amateur radio equipment will now be explained.

Shaft Encoders and Stepper Motors

Amateurs expect to twist a knob, an analogue motion, when they want to change frequency. Traditionally, that motion turned the shaft of a variable capacitor or screwed a core into or out of a coil. Now that frequencies in many radios are generated by direct digital synthesis, a process more compatible with key-

board entry and up/down switches, amateurs not only still like to twist knobs, they also like them to feel like the variable capacitors of yesteryear; equipment manufacturers comply but what the knob actually does drive is an optical device called an incremental shaft encoder: **Fig 5.74** [31].

The encoder is often a disc divided into sectors which are alternately transparent and opaque. A light source is positioned at one side of the disc and a light detector at the other. As the tuning knob is spun and the disc rotates, the output from the detector goes on or off when a transparent or an opaque disc sector is in the light path. Thus, the spinning encoder produces a stream of pulses which, when counted, indicate the change of angular position of the shaft. A second light source and detector pair, at an angle to the main pair, indicates the direction of rotation. A third pair sometimes is used to sense the one-per-revolution marker seen at the right on the disc shown.

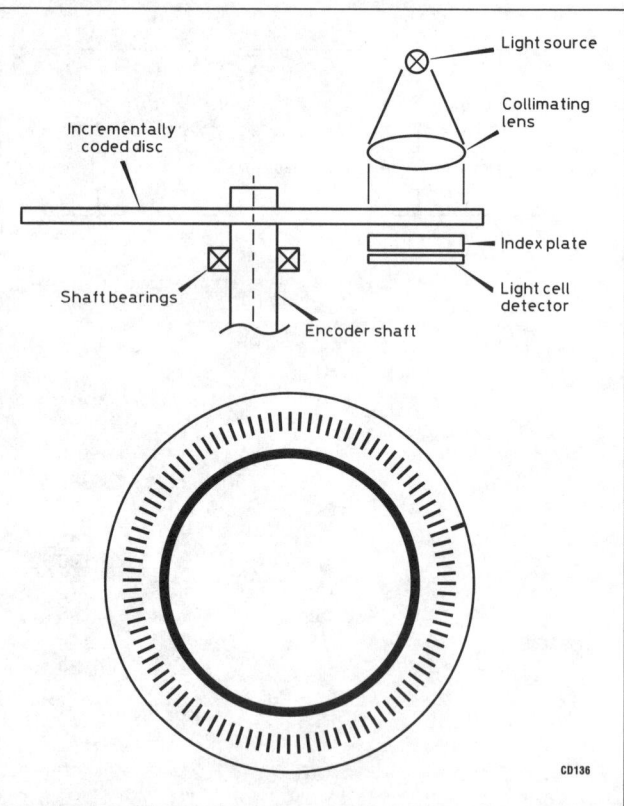

Fig 5.74: The incremental shaft encoder. In some radios with digital frequency synthesis, the tuning knob turns the shaft which turns the disc which alternatingly places transparent and opaque sectors in the light path; each pulse from the light detector increases or decreases the frequency by one step. (Taken from *Analog-Digital Conversion Handbook*, 3rd edn, Prentice-Hall, 1986, p444, by permission. Copyright Analog Devices Inc, Norwood, MA, USA)

Available encoder resolutions (the number of opaque and transparent sectors per disc) range from 100 to 65,000. The SSB/CW tuning rate of one typical radio was found to be 2kHz per knob rotation in 10Hz steps; this means an encoder resolution of 200. The ear cannot detect a 10Hz change of pitch at 300Hz and above, so the tuning feels completely smooth, though its output is a digital signal in which each pulse is translated into a 10Hz frequency increment or decrement.

Stepper motors do the opposite of shaft encoders; they turn a shaft in response to digital pulses, a step at a time. Typical motors have steps of 7.5 or 1.8 degrees; these values can be halved by modifying the pulse sequence. Pulses vary from 12V at 0.1A to 36V at 3.5A per phase (most motors have four phases) and are applied through driver ICs for small motors or driver boards with ICs plus power stages for big ones. The drivers are connected to a control board which in turn may be software-programmed by a computer, eg via an RS232 link (a standard serial data link). In amateur radio they are used on variable capacitors and inductors in microprocessor-controlled automatic antenna tuning units and on satellite-tracking antenna azimuth and elevation rotators.

Digital Panel (volt) Meters (DPMs)

Analogue methods for DC voltage measurements, say with a resolution of a millivolt on a 13.8V power supply, are possible but cumbersome. With a digital voltmeter they are simple and comparatively inexpensive. Digital panel meters are covered in the Test Equipment chapter.

Digitising Speech

While speech can be stored on magnetic tape, and forwarded at speeds a few times faster than natural, each operation would add a bit of noise and distortion and devices with moving parts, ie tape handling mechanisms, are just not suitable for many environments. Digital speech has changed all that. Once digitised, speech can be stored, filtered, compressed and expanded by computer, and forwarded at the maximum speed of which the transmission medium is capable, all without distortion; it can then be returned to the analogue world where and when it is to be listened to.

If speech were to be digitised in the traditional way, ie sampled more than twice each cycle on the highest speech fre-

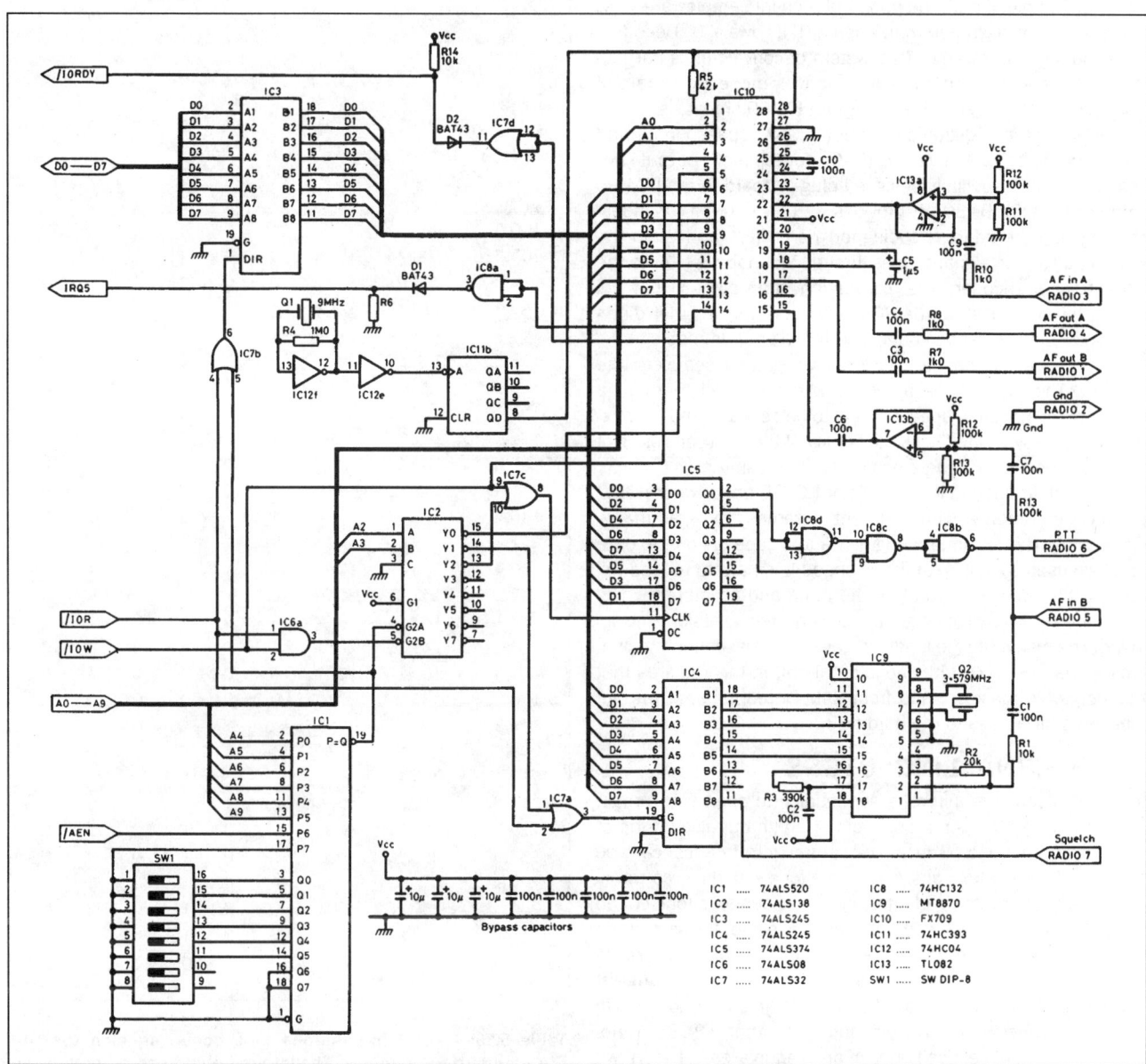

Fig 5.75: Speech encoder/decoder designed by DG3RBU and DL8MBT for voice mailboxes on UHF FM repeaters

quency, eg at 7kHz for speech up to 3kHz, and 10-bit resolution of the amplitude of each sample were required for reasonable fidelity, a bit rate of 70kbit/s would result, about 20 times the 3.5kHz bandwidth considered necessary for the SSB transmission of analogue speech. Also, almost a megabyte of memory would be required to store each minute of digitised speech. Differential and adaptive techniques are used to reduce these requirements.

DG3RBU and DL8MBT developed hardware and software for the digital storage and analogue retransmission of spoken messages on normal UHF FM voice repeaters and the digital forwarding of such messages between repeaters via the packet network. They described their differential analogue-digital conversion as follows.

"The conversion is by continuously variable slope decoding (CVSD), a form of delta modulation; in this process it is not the instantaneous value of an analogue voltage that is being sampled and digitised but its instantaneous slope at the moment of sampling. Binary '1' represents an increasing voltage, '0' a decreasing voltage.

"Encoding the slope of the increase or decrease depends on the prior sample. If both the prior and present samples are '1', a steeper slope is assumed than when a '1' follows a '0'. Decoding does the same in reverse. This system of conversion is particularly suitable for speech; even at a relatively modest data rate of 16kbit/s it yields good voice quality. An FX709 chip is used.

"The analogue-to-digital converter (ADC) for speech input and digital-to-analogue converter (DAC) for speech output are assembled on a specially designed plug-in board for an IBM PC computer, **Fig 5.75** [32]. It provides all the required functions, starting with the address coding for the PC (IC1 and 2). In the FX709 (IC10), the signal passes through a bandpass filter to the one-bit serial encoder; after conversion to an eight-bit parallel format the data pass to the PC bus via IC3; in the other direction, voice signals pass through a software-programmable audio filter; registers for pause and level recognition complete the module. The FX709 has a loopback mode which permits a received and encoded speech signal to be decoded and retransmitted, an easy way to check the fidelity of the loop consisting of the radio receiver-encoder-decoder-radio transmitter.

"The built-in quartz clock oscillator (IC12f) and divider (IC11) permit experiments with different externally programmable clock rates. The maximum length of a text depends on the data rate. We used 32kbit/s, at which speed it is hard to tell the difference between the sound on the input and an output which has gone through digitising, storage, and reconversion to analogue. The maximum file length is 150s. The reason for this time limit is that the FX709 has no internal buffer; this requires that the whole file must be read from RAM in real time, ie without interrupts for access to the hard disk."

TONE SIGNALLING: CTCSS

If an FM receiver requires, for its squelch to open and remain open, not only a carrier of sufficient strength but also a tone of a specific frequency, much co-channel interference can be avoided. If a repeater transmits such a tone with its voice transmissions but not with its identification, a continuous-tone coded squelch system (CTCSS) equipped receiver set for the same tone would hear all that repeater's voice transmissions but not its idents. Its squelch would not be opened either, eg during lift conditions, by a repeater on the same channel but located in another service area and sending another CTCSS tone. Conversely, a mobile, positioned in an overlap area between two repeaters on the same channel but which have their receivers set for different CTCSS tones, could use one repeater without

opening the other by sending the appropriate tone. Similar advantages can be had where several groups of stations, each with a different CTCSS tone, share a common frequency. With all transceivers within one group set for that group's CTCSS tone, conversations within the group would not open the squelch of stations of other groups monitoring the same frequency.

The Electronic Industries Association (EIA) has defined 38 CTCSS standard sub-audible tone frequencies. They are all between 67 and 250.3Hz, ie within the range of human hearing but outside the 300-3000Hz audio pass-band of most communications equipment and their level is set at only 10% of maximum deviation for that channel; hence sub-audible.

The left column of **Table 5.16** gives the frequency list. CTCSS encoders (tone generators) and also decoders (tone detectors)

Nominal frequency (Hz)		FX365 frequency (Hz)	Δfo (%)	D0	D1	D2	D3	D4	D5
67.0	A	67.05	+0.07	1	1	1	1	1	1
71.9	B	71.90	0.0	1	1	1	1	1	0
74.4		74.35	0.07	0	1	1	1	1	1
77.0	C	76.96	0.05	1	1	1	1	0	0
79.7		79.77	+0.09	1	0	1	1	1	1
82.5	D	82.59	+0.10	0	1	1	1	1	0
85.4		85.38	0.02	0	0	1	1	1	1
88.5	E	88.61	+0.13	0	1	1	1	0	0
91.5		91.58	+0.09	1	1	0	1	1	1
94.8	F	94.76	0.04	1	0	1	1	1	0
97.4		97.29	0.11	0	1	0	1	1	1
100.0		99.96	0.04	1	0	1	1	0	0
103.5	G	103.43	0.07	0	0	1	1	1	0
107.2		107.15	0.05	0	0	1	1	0	0
110.9	H	110.77	0.12	1	1	0	1	1	0
114.8		114.64	0.14	1	1	0	1	0	0
118.8	J	118.80	0.0	0	1	0	1	1	0
123.0		122.80	0.17	0	1	0	1	0	0
127.3		127.08	0.17	1	0	0	1	1	0
131.8		131.67	0.10	1	0	0	1	0	0
136.5		136.61	+0.08	0	0	0	1	1	0
141.3		141.32	+0.02	0	0	0	1	0	0
146.2		146.37	+0.12	1	1	1	0	1	0
151.4		151.09	0.20	1	1	1	0	0	0
156.7		156.88	+0.11	0	1	1	0	1	0
162.2		162.31	+0.07	0	1	1	0	0	0
167.9		168.14	+0.14	1	0	1	0	1	0
173.9		173.48	0.19	1	0	1	0	0	0
179.9		180.15	+0.14	0	0	1	0	1	0
186.2		186.29	+0.05	0	0	1	0	0	0
192.8		192.86	+0.03	1	1	0	0	1	0
203.5		203.65	+0.07	1	1	0	0	0	0
210.7		210.17	0.25	0	1	0	0	1	0
218.1		218.58	+0.22	0	1	0	0	0	0
225.7		226.12	+0.18	1	0	0	0	1	0
233.6		234.19	+0.25	1	0	0	0	0	0
241.8		241.08	0.30	0	0	0	0	1	0
250.3		250.28	0.01	0	0	0	0	0	0
No tone		No tone	-	0	0	0	0	1	1

Taken from *Consumer Microelectronics Ltd IC Data Book*, 1st edn

Table 5.16: CTCSS (continuous tone coded squelch system) EIA-standard frequencies. The letters behind some frequencies refer to the tones used by UK voice repeaters. The eight right-hand columns refer to the FX365 LSI chip (see text)

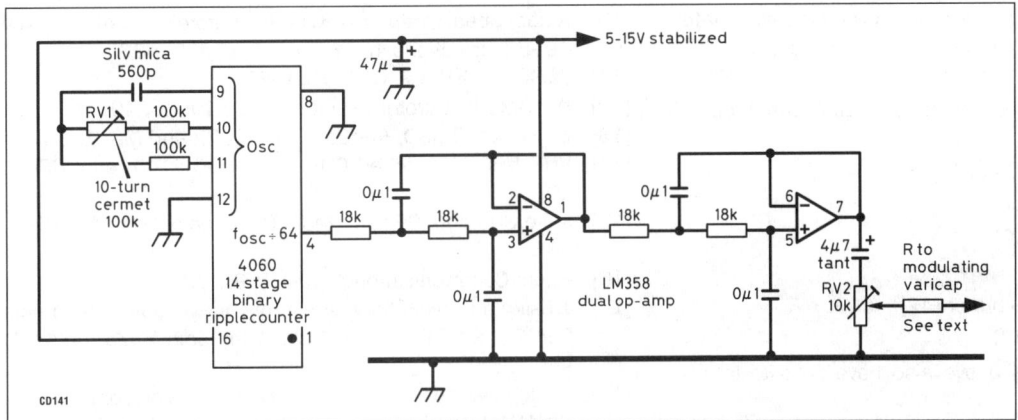

Fig 5.76: The Tuppenny simple CTCSS encoder (G0CBM) consists of an RC oscillator and ÷64 divider in one CMOS IC, together with a two-stage active LP filter built around a dual op-amp (G8HLE) (*Kent Repeater Newsletter*)

are offered as options for many earlier mobile and hand-held VHF and UHF FM transceivers. Most current transceivers have CTCSS encode built-in as standard.

Commercial standards require encoder frequencies to be within 0.1% of the nominal tone frequency under all operating conditions, attainable only with crystal control; most amateur repeaters are more tolerant, however, and RC-oscillators have been used successfully. The encoder tone output must be a clean sine wave lest its harmonic content above 300Hz becomes audible; this requires good filtering.

G0CBM's very simple Tuppenny CTCSS tone generator was designed for retrofitting in a surplus PMR transmitter to access local repeaters: **Fig 5.76** [33].

It is built around the inexpensive CMOS oscillator-frequency divider IC 4060. The parts connected to pins 9, 10 and 11 are the frequency-determining components of the RC oscillator; they have a tolerance of 5% but parts with the lowest possible temperature coefficient should be chosen to obtain adequate frequency stability, especially under mobile operation. RV1 allows frequency adjustment over the range 4288-7603Hz, which, after dividing by 64, yields at pin 4 any CTCSS frequency between 67 and 119Hz; this includes tones A-J assigned to UK repeaters.

A two-stage active low-pass filter was designed by G8HLE to get sufficient suppression of any harmonics above 300Hz of all

tones. The LM358 dual op-amp was chosen because it is small, cheap and works on a low, single supply voltage. The -6dB frequency was chosen at 88Hz, ie the higher tones to be passed fall outside the pass-band. As the tone amplitude is far greater than required, this attenuation is no disadvantage, but each tone tuned in with RV1 requires a different setting of RV2 to get the same deviation. The output resistor R depends on the modulator circuitry in the transmitter. It should be dimensioned to get the proper CTCSS deviation at the highest tone frequency to be used with RV2 set near maximum.

Encoders/decoders are more complicated and the LSI CMOS device used, eg CML FX365, is expensive: **Fig 5.77** [34]. It contains not only the encoder and decoder proper, but also a high-pass filter which prevents any received CTCSS tone becoming audible, a low-pass filter to suppress harmonics of CTCSS tones and a crystal-controlled reference oscillator from which the tones are derived; see columns 2 and 3 of Table 5.16. Tone selection is according to the six right-hand columns of that table, either by microprocessor or by hard-wired switches. Another feature is transmit phase reversal upon release of the PTT switch; this shortens the squelch tail at the receiver.

SOFTWARE BUILDING BLOCKS

The PIC [35] is a microcontroller which is being used more and more in amateur radio projects. A PIC is an economical way of

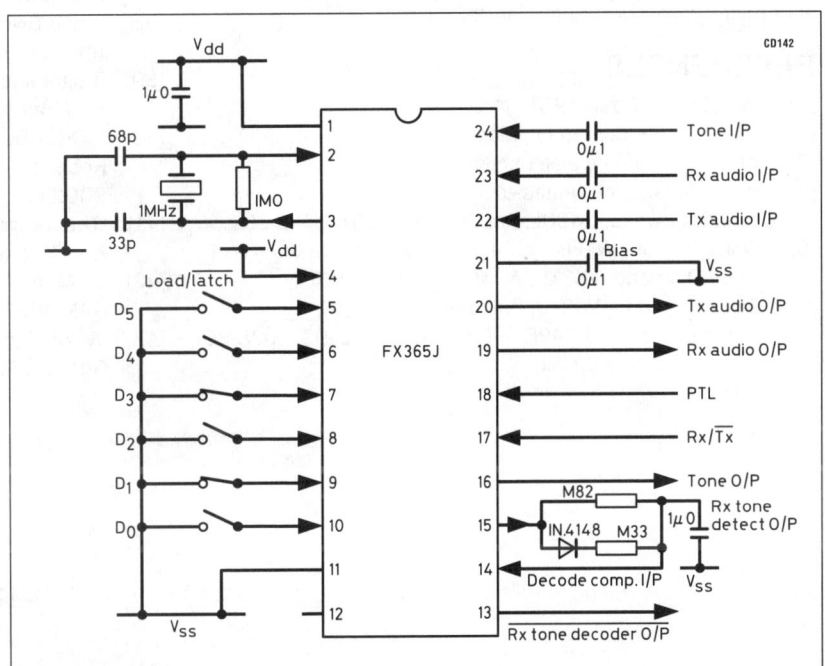

Fig 5.77: CTCSS encoder/decoder using a CML CMOS LSI device. The chip also contains 300Hz-cut-off HP and LP filters to separate tones and speech and a crystal reference oscillator. Tone selection can be by microprocessor or hard-wired switches (*Consumer Microcircuits IC Data Book*)

providing software control of functions in home-built amateur radio equipment. Examples of the use of PIC devices are:

- ATU control [36]
- DDS controller [37] (see also the oscillators chapter of this Handbook)
- Frequency-dependant switch [38]
- Bug key [39]
- Transceiver control [40]
- Keyer [41]
- Morse code speed calibrator [42]
- Morse code reader [43]

Note that the references above also have links to internet sites.

PIC Code

In some of the above projects, the PIC can be purchased ready programmed and this is an easy solution to reproduce a published design. However, there are strong reasons for the constructor to write his or her own code:

- The constructor may wish to modify existing code to either modify an existing function or add extra requirements to an existing design.
- The constructor may wish to write all of the code from scratch.
- The existing design may be suitable, but further functions will be added in the future. For example, this may be caused by a change in amateur bands used, or a change of mode.

The ideal environment in which to develop PIC software is the MPLAB integrated package, which is available from the manufacturers of the PIC [41]. This runs on a PC, but the PC doesn't need to be sophisticated. An old, unused PC may be adequate. For the purpose of developing software, it is recommended that PIC devices with on-board EEPROM are used.

Practical PICs

Later in this book is a chapter dealing in detail with a software transceiver based on a PIC. Many of the parts of this radio may be used on their own, or as building blocks in other conventional or microprocessor-based projects.

REFERENCES

[1]   W2CQH, QST, Jan 1974
[2]   Signetics RF Communications Handbook
[3]   http://www.home.agilent.com
[4]   http://www.mini-circuits.com
[5]   Ian Braithwaite, G4COL, RadCom, Jul 1997, pp38-39
[6]   Doug DeMaw, W1FB, QST, Apr 1989, pp30-33
[7]   Drew Diamond, VK3XU, Amateur Radio, 10/88
[8]   Drew Diamond, VK3XU, Amateur Radio, 1/91
[9]   David Bowman, G0MRF, in RadCom, Feb 1993, pp28-30, & Mar 1993, pp28-29
[10] Klaas Spaargaren, PA0KLS, in 'Eurotek', RadCom, Nov 1998 (corrections in RadCom, Jan 1999, p20
[11] Zack Lau, KH6CP, QST, Oct 1987
[12] PA0GMS in 'Eurotek', RadCom, Dec 1992, p49
[13] E J Hatch, G3ISD, RadCom, May 1982 and Mar 1984
[14] VHF/UHF Manual, ed G R Jessop, G6JP, 4th edn, RSGB, 1983
[15] Mike Grierson, G3TSO, Radio Communication, Mar 1990, p35
[16] Radio Communication August 1992, p39
[17] 'Design a toroidal tank circuit for your vacuum tube amplifier', Robert E Bloom, W6YUY, Ham Radio Magazine (USA) Aug 1985
[18] Analog Devices web site, http://www.analog.com
[19] GW4NAH, Radio Communication, Jan 1989, p48
[20] DF4ZS in 'Eurotek', RadCom, Oct 1998
[21] ON5DI in 'Eurotek', RadCom, Dec 1998
[22] 'Computer-aided ladder crystal filter design', J A Hardcastle, G3JIR, Radio Communication, May 1983
[23] H R Langer, Y27YO, Funkamateur, Jan 1985; Radio Communication, Jun 1985, p452
[24] LA2IJ and LA4HK, Amator Radio, Nov 1974
[25] DJ6HP, cq-DL, Feb 1974
[26] W1JF, QST, Nov 1982, Jul 1984 and Jan 1985
[27] AI2T, CQ, Jan 1986
[28] WB4TLM and KB4KVE, QST, Apr 1986
[29] E J Hatch, G3ISD, Datacom (British Amateur Radio Teledata Group), Autumn 1989 and Jan 1990
[30] Active Filter Cookbook, Don Lancaster. Howard W. Sams and Co. ISBN 0-672-21168-8
[31] Analogue-Digital Conversion Handbook, 3rd edition, Prentice-Hall, 1986, p215
[32] RadCom, May 1992, p62
[33] Kent Repeater Newsletter, Jan 1993
[34] IC Data Book, 1st edition, Consumer Microcircuits Ltd, p2.40
[35] http://www.microchip.com
[36] 'picATUne - the intelligent ATU', Peter Rhodes, G3XJP, RadCom, Sep 2000 to Jan 2001
[37] 'Use of PICs in DDS Design' in 'Technical Topics' column, Pat Hawker G3VA. RadCom, Jan 2001
[38] 'Pic-A-Switch A Frequency Dependent Switch', Peter Rhodes, G3XJP, RadCom, Sep - Dec 2001
[39] 'Bugambic: Son of Superbug', Chas Fletcher, G3DXZ, RadCom, Apr 2002
[40] 'CDG2000 HF Transceiver', Colin Horrabin, G3SBI, Dave Roberts, G3KBB, and George Fare, G3OGQ, RadCom, Jun 2002 to Dec 2002
[41] 'The programmer and the keyer', Ed Chicken, G3BIK, RadCom, Nov 2004
[42] 'A Morse Code Speed Calibrator'. Jonathan Gudgeon, G4MDU, RadCom, Aug 2004
[43] 'A Talking Morse Code Reader', Jonathan Gudgeon, G4MDU, RadCom, Jun 2001

6 HF Receivers

Amateur HF operation, whether for two-way contacts or for listening to amateur transmissions, imposes stringent requirements on the receiver. The need is for a receiver that enables an experienced operator to find and hold extremely weak signals on frequency bands often crowded with much stronger signals from local stations or from the high-power broadcast stations using adjacent bands. The wanted signals may be fading repeatedly to below the external noise level, which limits the maximum usable sensitivity of HF receivers, and which will be much higher than in the VHF and UHF spectrum.

Although the receivers now used by most amateurs form part of complex, factory-built HF transceivers, the operator should understand the design parameters that determine how well or how badly they will perform in practice, and appreciate which design features contribute to basic performance as HF communications receivers, as opposed to those which may make them more user-friendly but which do not directly affect the reception of weak signals. This also applies to dedicated receivers that are factory-built, such as the one shown in **Fig 6.1**.

Ideally, an HF receiver should be able to provide good intelligibility from signals which may easily differ in voltage delivered from the antenna by up to 10,000 times and occasionally by up to one million times (120dB) - from less than 1µV from a weak signal to nearly 1V from a near-neighbour. To tune and listen to SSB or to a stable CW transmission while using a narrow-band filter, the receiver needs to have a frequency stability of within a few hertz over periods of 15 minutes or so, representing a stability of better than one part in a million. It should be capable of being tuned with great precision, either continuously or in increments of at most a few hertz.

A top-quality receiver may be required to receive transmissions on all frequencies from 1.8MHz to 30MHz (or even 50MHz) to provide 'general coverage' or only on the bands allotted to amateurs. Such a receiver may be suitable for a number of different modes of transmission - SSB, CW, AM, NBFM, data (RTTY/packet) etc - with each mode imposing different requirements in selectivity, stability and demodulation (decoding). Such a receiver would inevitably be complex and costly to buy or build.

On the other hand, a more specialised receiver covering only a limited number of bands and modes such as CW-only or CW/SSB-only, and depending for performance rather more on the skill of the operator, can be relatively simple to build at low cost.

As with other branches of electronics, the practical implementation of high-performance communications receivers has undergone a number of radical changes since their initial development in the mid-1930s, some resulting from the improved stability needed for SSB reception and others aimed at reducing costs by substituting electronic techniques in place of mechanical precision.

However, it needs to be emphasised that, in most cases, progress in one direction has tended to result in the introduction of new problems or the enhancement of others: "What we call progress is the exchange of one nuisance for another nuisance" (Havelock Ellis) or "Change is certain; progress is not" (A J P Taylor). As late as 1981, an Australian amateur was moved to write: "Solid-state technology affords commercial manufacturers cheap, large-scale production but for amateur radio receivers and transceivers of practical simplicity, valves remain incomparably superior for one-off, home-built projects." The availability of linear integrated circuits capable of forming the heart of communications receivers combined with the increasing scarcity and hence cost of special valve types has tended to reverse this statement. It is still possible to build reasonably effective HF receivers, particularly those for limited frequency coverage, on the kitchen table with the minimum of test equipment.

Furthermore, since many newcomers will eventually acquire a factory-built transceiver but require a low-cost, stand-alone HF receiver in the interim period, the need can be met either by building a relatively simple receiver, or by acquiring, and if necessary modifying, one of the older valve-type receivers that were built in very large numbers for military communications during the second world war, or those marketed for amateur operation in the years before the virtually universal adoption of the transceiver.

Even where an amateur has no intention of building or servicing his or her own receiver, it is important that he or she should have a good understanding of the basic principles and limitations that govern the performance of all HF communications receivers.

BASIC REQUIREMENTS

The main requirements for a good HF receiver are:

- Sufficiently high sensitivity, coupled with a wide dynamic range and good linearity to allow it to cope with both the very weak and very strong signals that will appear together at the input; it should be able to do this with the minimum impairment of the signal-to-noise ratio by receiver noise, cross-modulation, blocking, intermodulation, reciprocal mixing, hum etc.

- Good selectivity to allow the selection of the required signal from among other (possibly much stronger) signals on adjacent or near-adjacent frequencies. The selectivity characteristics should 'match' the mode of transmission, so that interference susceptibility and noise bandwidth should be as close as possible to the intelligence bandwidth of the signal.

- Maximum freedom from spurious responses - that is to say signals which appear to the user to be transmitting on specific frequencies when in fact this is not the case. Such spurious responses include those arising from image

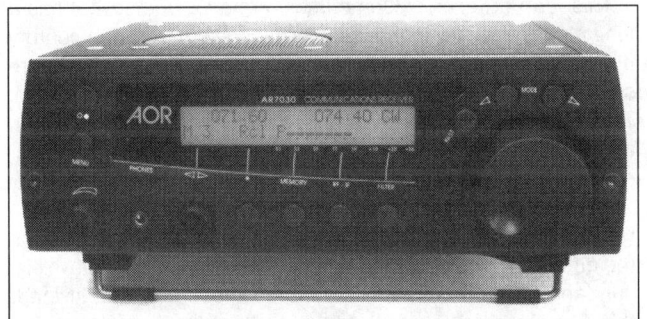

Fig 6.1: The AOR 7030 is a sophisticated receiver covering the frequency range 0-32MHz

responses, breakthrough of signals and harmonics of the receiver's internal oscillators.

- A high order of stability, in particular the absence of short-term frequency drift or jumping.

- Good read-out and calibration of the frequency to which the set is tuned, coupled with the ability to reset the receiver accurately and quickly to a given frequency or station.

- Means of receiving SSB and CW, normally requiring a stable beat frequency oscillator preferably in conjunction with product detection.

- Sufficient amplification to allow the reception of signals of under 1μV input; this implies a minimum voltage gain of about one million times (120dB), preferably with effective automatic gain control (AGC) to hold the audio output steady over a very wide range of input signals.

- Sturdy construction with good-quality components and with consideration given to problems of access for servicing when the inevitable occasional fault occurs.

A number of other refinements are also desirable: for example it is normal practice to provide a headphone socket on all communications receivers; it is useful to have ready provision for receiver 'muting' by an externally applied voltage to allow voice-operated, push-to-talk or CW break-in operation; an S-meter to provide immediate indication of relative signal strengths; a power take-off socket to facilitate the use of accessories; an IF signal take-off socket to allow use of external special demodulators for NBFM, FSK, DSBSC, data etc.

In recent years, significant progress has continued to be made in meeting these requirements - although we are still some way short of being able to provide them over the entire signal range of 120dB at the ideal few hertz stability. The introduction of more and more semiconductor devices into receivers has brought a number of very useful advantages, but has also paradoxically made it more difficult to achieve the highly desirable wide dynamic range. Professional users now require frequency read-out and long-term stability of an extremely high order (better than 1Hz stability is needed for some applications) and this has led to the use of frequency synthesised local oscillators and digital read-out systems; although these are effective for the purposes which led to their adoption, they are not necessarily the correct approach for amateur receivers since, unless very great care is taken, a complex frequency synthesiser not only adds significantly to the cost but may actually result in a degradation of other even more desirable characteristics.

So long as continuous tuning systems with calibrated dials were used, the mechanical aspects of a receiver remained very important; it is perhaps no accident that one of the outstanding early receivers (HRO) was largely designed by someone whose early training was that of a mechanical engineer.

It should be recognised that receivers which fall far short of ideal performance by modern standards may nevertheless still provide entirely usable results, and can often be modified to take advantage of recent techniques. Despite all the progress made in recent decades, receiver designs dating from the 'thirties and early 'forties are still capable of being put to good use, provided that the original electrical and mechanical design was sound. Similarly, the constructor may find that a simple, straightforward and low-cost receiver can give good results even when its specification is well below that now possible. It is ironical that almost all the design trends of the past 30 years have, until quite recently, impaired rather than improved the performance of receivers in the presence of strong signals!

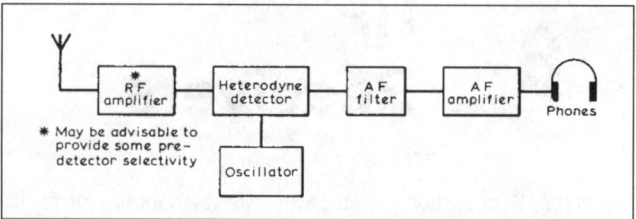

Fig 6.2: Outline of a simple direct-conversion receiver in which high selectivity can be achieved by means of audio filters

BASIC TYPES OF RECEIVERS

Amateur HF receivers fall into one of two main categories:

(a) 'straight' regenerative and direct-conversion receivers in which the incoming signal is converted directly into audio by means of a demodulator working at the signal frequency;

(b) single- and multiple-conversion superhet receivers in which the incoming signal is first converted to one or more intermediate frequencies before being demodulated. Each type of receiver has basic advantages and disadvantages.

Regenerative Detector ('Straight' or TRF) Receivers

At one time valve receivers based on a regenerative (reaction) detector, plus one or more stages of AF amplification (ie 0-V-1, 0-V-2 etc), and sometimes one or more stages of RF amplification at signal frequency (1-V-1 etc) were widely used by amateurs. High gain can be achieved in a correctly adjusted regenerative detector when set to a degree of positive feedback just beyond that at which oscillation begins; this makes a regenerative receiver capable of receiving weak CW and SSB signals. However, this form of detector is non-linear and cannot cope well in situations where the weak signal is at all close to a strong signal; it is also inefficient as an AM detector since the gain is much reduced when the positive feedback (regeneration) is reduced below the oscillation threshold. Since the detector is non-linear, it is usually impossible to provide adequate selectivity by means of audio filters.

Simple Direct-conversion Receivers

A modified form of 'straight' receiver which can provide good results, even under modern conditions, becomes possible by using a linear detector which is in effect simply a frequency converter, in conjunction with a stable local oscillator set to the signal frequency (or spaced only the audio beat away from it). Provided that this stage has good linearity in respect of the signal path, it becomes possible to provide almost any desired degree of selectivity by means of audio filters (Fig 6.2).

This form of receiver (sometimes termed a homodyne) has a long history but only in the last few decades of the 20th century did it become widely used for amateur operation since it is more suited (in its simplest form) to CW and SSB reception than AM. The direct-conversion receiver may be likened to a superhet with an IF of 0kHz or alternatively to a straight receiver with a linear rather than a regenerative detector.

In a superhet receiver the incoming signal is mixed with a local oscillator signal and the intermediate frequency represents the difference between the two frequencies; thus as the two signals approach one another the IF becomes lower and lower. If this process is continued until the oscillator is at the same frequency as the incoming signal, then the output will be at audio (baseband) frequency; in effect one is using a frequency chang-

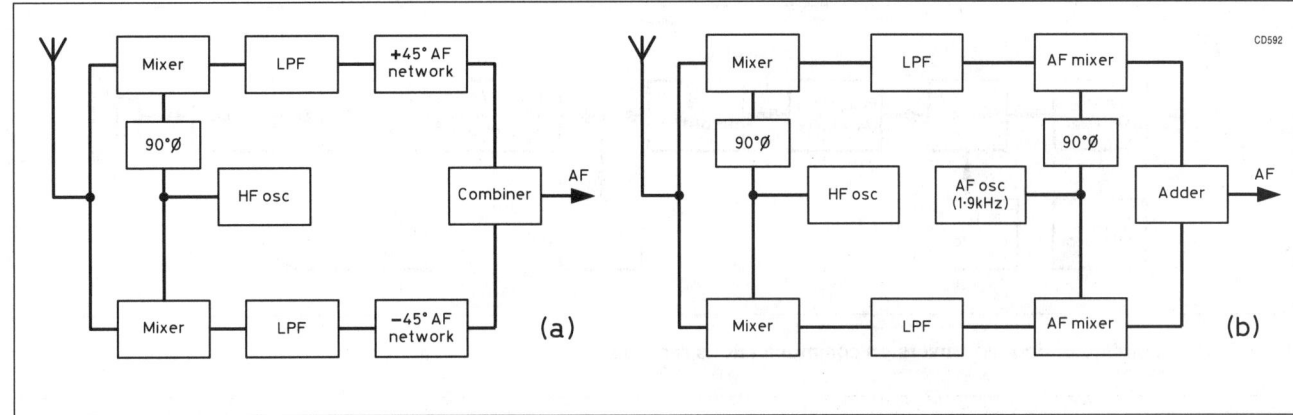

Fig 6.3: Block outline of two-phase ('autophasing') form of direct-conversion receiver. (b) Block outline of 'third method' (Weaver or Barber) SSB direct-conversion receiver

er or translator to demodulate the signal. Because high gain cannot be achieved in a linear detector, it is necessary to provide very high AF amplification. Direct-conversion receivers can be designed to receive weak signals with good selectivity but in this form do not provide true single-sideband reception (see later); another problem often found in practice is that very strong broadcast signals (eg on 7MHz) drive the detector into non-linearity and are then demodulated directly and not affected by any setting of the local oscillator.

A crystal-controlled converter can be used in front of a direct-conversion receiver, so forming a superhet with variable IF only. Alternatively a frequency converter with a variably tuned local oscillator providing output at a fixed IF may be used in front of a direct-conversion receiver (regenerative or linear demodulator) fixed tuned to the IF output. Such a receiver is sometimes referred to as a supergainer receiver.

Two-phase and 'Third-method' Direct-Conversion Receivers

An inherent disadvantage of the simple direct-conversion receiver is that it responds equally to signals on both sides of its local oscillator frequency, and cannot reject what is termed the audio image no matter how good the audio filter characteristics; this is a serious disadvantage since it means that the selectivity of the receiver can only be made half as good as the theoretically ideal bandwidth. This problem can be overcome, though at the cost of additional complexity, by phasing techniques similar to those associated with SSB generation. Two main approaches are possible: see **Fig 6.3**.

Fig 6.3(a) shows the use of broad-band AF 45 degree phase-shift networks in an 'outphasing' system, and with care can result in the reduction of one sideband to the extent of 30-40dB. Another possibility is the polyphase SSB demodulator which does not require such critical component values as conventional SSB phase-shift networks.

Fig 6.3(b) shows the 'third method' (sometimes called the Weaver or Barber system) which requires the use of additional balanced mixers working at AF but eliminates the need for accurate AF phase-shift networks. The 'third method' system, particularly in its AC-coupled form [1] provides the basis for high-performance receivers at relatively low cost, although suitable designs for amateur operation are rare. Two-phase direct-conversion receivers based on two diode-ring mixers in quadrature (90° phase difference) are capable of the high performance of a good superhet.

HF Superhet Receivers

The vast majority of receivers are based on the superhet principle. By changing the incoming signals to a fixed frequency (which may be lower or higher than the incoming signals) it becomes possible to build a high-gain amplifier of controlled selectivity to a degree which would not be possible over a wide spread of signal frequencies. The main practical disadvantage with this system is that the frequency conversion process involves unwanted products which give rise to spurious responses, and much of the design process has to be concentrated on minimising the extent of these spurious responses in practical situations.

A single-conversion superhet is a receiver in which the incoming signal is converted to its intermediate frequency, amplified and then demodulated at this second frequency. Virtually all domestic AM broadcast receivers use this principle, with an IF of about 455-470kHz, and a similar arrangement but with refinements was found in many communications receivers. However, for reasons that will be made clear later in this chapter, some receivers convert the incoming signal successively to several different frequencies; these may all be fixed IFs: for example the first IF might be 9MHz and the second 455kHz and possibly a third at 35kHz. Or the first IF may consist of a whole spectrum of frequencies so that the first IF is variable when tuning a given

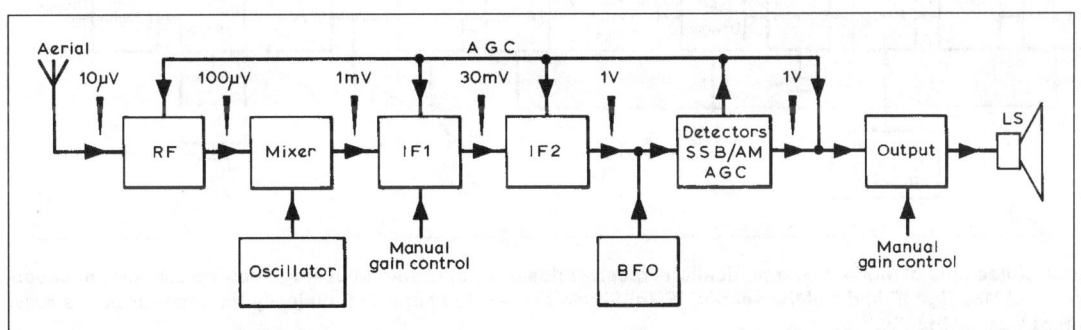

Fig 6.4: Block outline of representative single-conversion

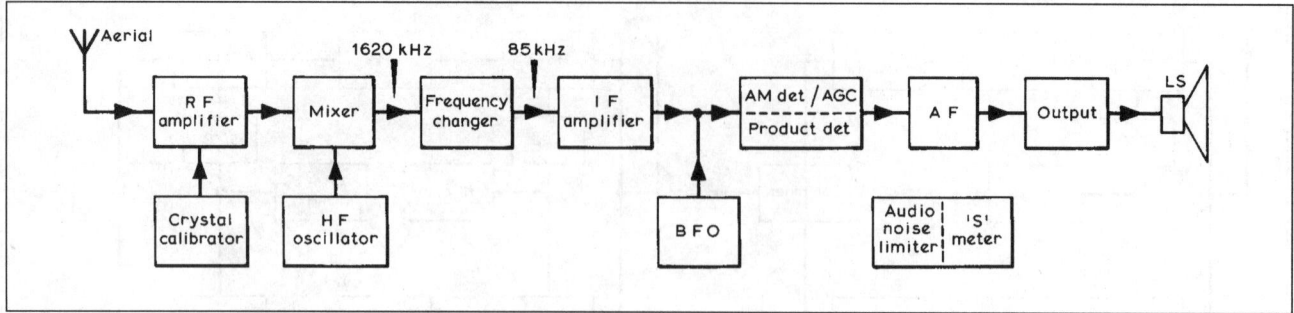

Fig 6.5: Block outline of double-conversion communications receiver with both IFs fixed

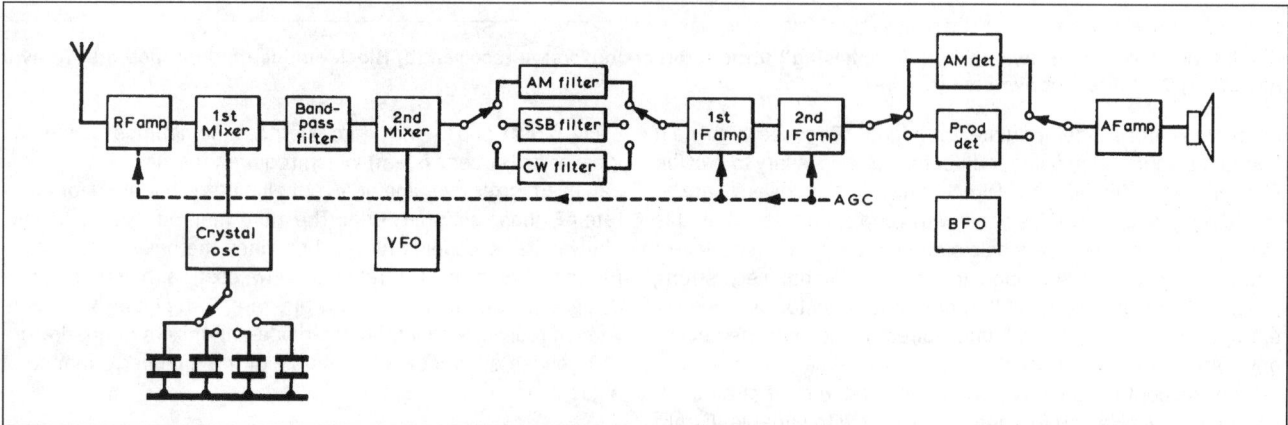

Fig 6.6: Block diagram of a double-conversion receiver with crystal-controlled first oscillator - typical of many late 20th-century designs

band, with a subsequent second conversion to a fixed IF. There are in fact many receivers using double or even triple conversion, and a few with even more conversions, though unless care is taken each conversion makes the receiver susceptible to more spurious responses. The block diagram of a typical single-conversion receiver is shown in **Fig 6.4**. **Fig 6.5** illustrates a double-conversion receiver with fixed IFs, while **Fig 6.6** is representative of a receiver using a variable first IF in conjunction with a crystal-controlled first local oscillator (HFO).

Many modern factory-built receivers up-convert the signal fre-

quency to a first IF at VHF as this makes it more convenient to use a frequency synthesiser as the first HF oscillator: **Fig 6.7(a)**. As the degree of selectivity provided in a receiver increases, it reaches the stage where the receiver becomes a single-side-band receiver, although this does not mean that only SSB signals can be received. In fact the first application of this principle was the single-signal receiver for CW reception where the selectivity is sufficient to reduce the strength of the audio image (resulting from beating the IF signal with the BFO) to an insignif-icant value, thus virtually at one stroke halving the apparent

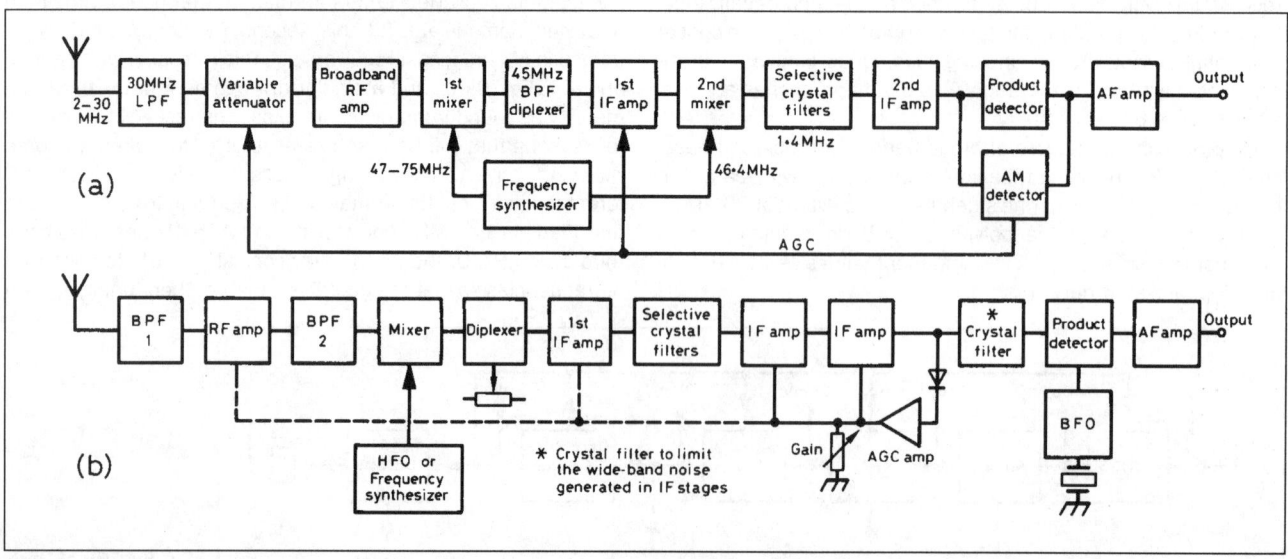

Fig 6.7: Representative architectures of modern communications receiver designs. (a) General-coverage double-conversion super-het with up-conversion to 45MHz first IF and 1.4MHz second IF. (b) Single-conversion superhet, typically for amateur bands only, with an IF in the region of 9 or 10MHz

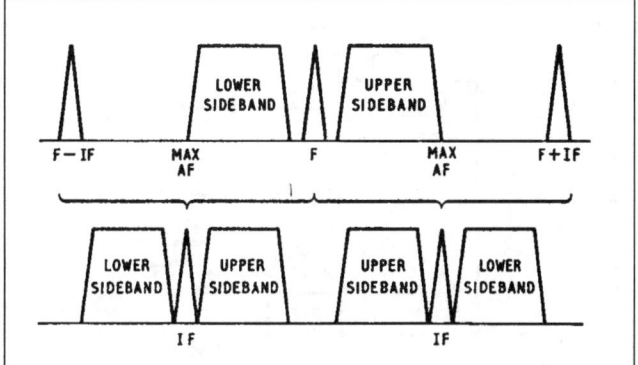

Fig 6.8: A local oscillator frequency lower than the signal frequency (ie f - IF) keeps the upper and lower sidebands of the intermediate frequency signals in their original positions. However, when the local oscillator is placed higher in frequency than the signal frequency (f + IF), the positions of the sidebands are transposed. By incorporating two oscillators, one above and the other below the input signal, sideband selection is facilitated (this is generally carried out at the final IF by switching the BFO or carrier insertion oscillator)

number of CW stations operating on the band (previously each CW signal was heard on each side of the zero beat). Similarly, double-sideband AM signals can be received on a set having a carefully controlled pass-band as though they were SSB, with the possibility of receiving either sideband should there be interference on the other. This degree of selectivity can be achieved with good IF filters or alternatively the demodulator can itself be designed to reject one or other of the sidebands, by using phasing techniques similar to those sometimes used to generate SSB signals and for two-phase direct-conversion receivers. But most receivers rely on the use of crystal or mechanical filters to provide the necessary degree of sideband selectivity, and then use heterodyne oscillators placed either side of the nominal IF to select upper or lower sidebands.

It is important to note that whenever frequency conversion is accomplished by beating with the incoming signal an oscillator lower in frequency than the signal frequency, the sidebands retain their original position relative to the carrier frequency, but when conversion is by means of an oscillator placed higher in frequency than the carrier, the sidebands are inverted. That is to say, an upper sideband becomes a lower sideband and vice versa (see **Fig 6.8**).

DESIGN TRENDS

After the 'straight' receiver, because of its relatively poor performance and lack of selectivity on AM phone signals, had fallen into disfavour in the mid-1930s, came the era of the superhet communications receiver. Most early models were single conversion designs based on an IF of 455-470kHz, with two or three IF stages, a multi-electrode triode-hexode or pentagrid mixer, sometimes but not always with a separate oscillator valve. This approach made at least one RF amplifying stage essential in order to raise the level of the incoming signal before it was applied to the relatively noisy mixer; two stages were to be preferred since this meant they could be operated in less critical conditions and provided the additional pre-mixer RF selectivity needed to reduce 'image' response on 14MHz and above. Usually a band-switched LC HF oscillator was gang-tuned so as to track with two or three signal-frequency tuned circuits, calling for fairly critical and expensive tuning and alignment systems. These receivers were often designed basically to provide full coverage on the HF band (and often also the MF band), some-

times with a second tuning control to provide electrical bandspread on amateur bands, or with provision (as on the HRO) optionally to limit coverage to amateur bands only. Selectivity depended on the use of good-quality IF transformers (sometimes with a tertiary tuned circuit) in conjunction with a single-crystal IF filter which could easily be adjusted for varying degrees of selectivity and which included a phasing control for nulling out interfering carriers.

In later years, to overcome the problem of image response with only one RF stage, there was a trend towards double- or triple-conversion receivers with a first IF of 1.6MHz or above, a second IF about 470kHz and (sometimes) a third IF about 50kHz.

With a final IF of 50kHz it was possible to provide good single-signal selectivity without the use of a crystal filter.

The need for higher stability than is usually possible with a band-switched HF oscillator and the attraction of a similar degree of band-spreading on all bands has led to the widespread adoption of an alternative form of multi-conversion superhet; in effect this provides a series of integral crystal-controlled converters in front of a superhet receiver (single or double conversion) covering only a single frequency range (for example 5000-5500kHz) This arrangement provides a fixed tuning span (in this example 500kHz) for each crystal in the HF oscillator. Since a separate crystal is needed for each band segment, most receivers of this type are designed for amateur bands only (though often with provision for the reception of a standard frequency transmission, for example on 10MHz); more recently some designs have eliminated the need for separate crystals by means of frequency synthesis, and in such cases it is economically possible to provide general coverage. The selectivity in these receivers is usually determined by a band-pass crystal filter, mechanical filter or multi-pole ceramic filter, a separate filter is being used for SSB, CW and AM reception (although for economic reasons sets may be fitted with only one filter, usually intended for SSB reception). In this system the basic 'superhet' section forms in effect a variable IF amplifier.

In practice the variable IF type of receiver provides significantly enhanced stability and lower tuning rates on the higher frequency bands, compared with receivers using fixed IF, though it is considerably more difficult to prevent breakthrough of strong signals within the variable IF range, and to avoid altogether the appearance of 'birdies' from internal oscillators. With careful design a high standard of performance can be achieved; the use of multiple conversion (with the selective filter further from the antenna input stage) makes the system less suitable for semiconductors than for valves, particularly where broadband circuits are employed in the front-end and in the variable IF stage.

There is now a trend back to the use of fixed IF receivers, either with single conversion or occasionally with double conversion (provided that in this case an effective roofing filter is used at the first IF). A roofing filter is a selective filter intended to reduce the number of strong signals passing down an IF chain without necessarily being of such high grade or as narrow bandwidth as the main selective filter. To overcome the problem of image reception a much higher first IF is used; for amateur band receivers this is often 9MHz since effective SSB and CW filters at this frequency are available. This reduces (though does not eliminate) the need for pre-mixer selectivity; while the use of low-noise mixers makes it possible to reduce or eliminate RF amplification. To overcome frequency stability problems inherent in a single-conversion approach, it is possible to obtain better stability with FET oscillators than was usually possible with valves; another approach is to use mixer-VFO systems (essentially a

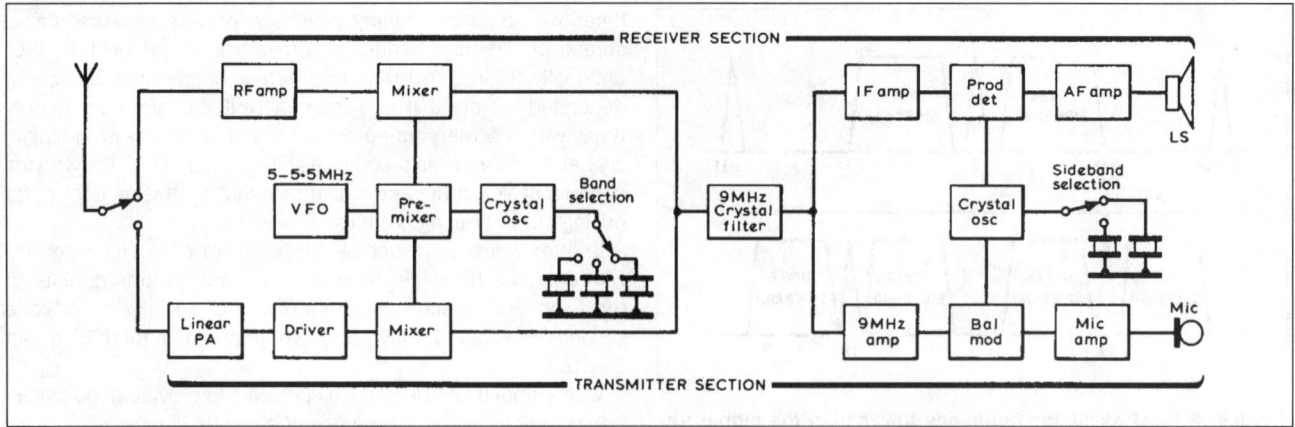

Fig 6.9: Block diagram of a typical modern SSB transceiver in which the receiver is a single-conversion superhet with 9MHz IF in conjunction with the pre-mixer form of partial frequency synthesis

simple form of frequency synthesis) and such systems can provide identical tuning rates on all bands, though care has to be taken to reduce to a minimum spurious injection frequencies resulting from the mixing process.

To achieve the maximum possible dynamic range, particular attention has to be given to the mixer stage, and it is an advantage to make this a balanced, or double-balanced (see later) arrangement using either double-triodes, Schottky (hot-carrier) diodes or FETs (particularly power FETs).

A further significant reduction of spurious responses may prove possible by abandoning the superhet in favour of high-performance direct-conversion receivers (such as the Weaver or 'third-method' SSB direct-conversion arrangement); however, such designs are still only at an early stage of development.

Most modern receivers are built in the form of compact transceivers functioning both as receiver and transmitter, and with some stages common to both functions (**Fig 6.9**). Modern transceivers use semiconductor devices throughout. Dual-gate FET devices are generally found in the signal path of the receiver. Most transceivers have a common SSB filter for receive and transmit; this may be a mechanical or crystal filter at about 455kHz but current models more often use crystal filters at about 3180, 5200, 9000kHz or 10.7MHz, since the use of a higher frequency reduces the total number of frequency conversions necessary.

One of the fundamental benefits of a transceiver is that it provides common tuning of the receiver and transmitter so that both are always 'netted' to the same frequency. It remains, however, an operational advantage to be able to tune the receiver a few kilohertz around the transmit frequency and vice versa, and provision for this incremental tuning is often incorporated; alternatively many transceivers offer two oscillators s that the transmit and receive frequencies may be separated when required.

The most critical aspect of modern receivers is the signal-handling capabilities of the early (front-end) stages. Various circuit techniques are available to enhance such characteristics: for example the use of balanced (push-pull) rather than single-ended signal frequency amplifiers; the use of balanced or double-balanced mixer stages; the provision of manual or AGC-actuated antenna-input attenuators; and careful attention to the question of gain distribution.

An important advantage of modern techniques such as linear integrated circuits and wide-band fixed-tuned filters rather than tuneable resonant circuits is that they make it possible to build satisfactory receivers without the time-consuming and constructional complexity formerly associated with high-performance receivers. Nevertheless a multiband receiver must still be

regarded as a project requiring considerable skill and patience.

The widespread adoption of frequency synthesisers as the local HF oscillator has led to a basic change in the design of most factory-built receivers, although low-cost synthesisers may not be the best approach for home-construction. Such synthesisers cannot readily (except under microprocessor control) be 'ganged' to band-switched signal-frequency tuned circuits; additionally, mechanically-ganged variable tuning of band-switched signal-frequency and local oscillator tuned circuits as found in older communications receivers would today be a relatively high-cost technique.

These considerations have led to widespread adoption in factory-built receivers and transceivers of 'single-span' up-conversion multiple-conversion superhets with a first IF in the VHF range, up to about 90MHz, followed by further conversions to lower IFs at which the main selectivity filter(s) are located.

In such designs, pre-selection before the first mixer or preamplifier (often arranged to be optionally switched out of circuit) may simply take the form of a low-pass filter (cut off at 30MHz) or a single wide-band filter covering the entire HF band. Higher-performance receivers usually fit a series of sub-octave bandpass filters, with electronic switching (preferably with PIN diodes).

With fixed filtering, even of the sub-octave type, very strong HF broadcast transmissions will be present at the mixer(s) and throughout the 'front-end' up to the main selectivity filter(s). To enable weak signals to be received free of intermodulation products, this places stringent requirements on the linearity of the front-end. The use of relatively noisy low-cost PLL frequency synthesisers also raises the problem of 'reciprocal mixing' (see later). For home-construction of high-performance receivers, the earlier design approaches are still attractive, including the 'old-fashioned' concept of achieving good pre-mixer selectivity with high-Q tuned circuits using variable capacitors rather than electronic tuning diodes. Diode switching rather than mechanical switching can also significantly degrade the intermodulation performance of receivers. Further, it should be noted that reed relay switching can often introduce sufficient series resistance to seriously degrade the Q of tuned circuits unless reeds especially selected for their RF properties are used.

DIGITAL TECHNIQUES

The availability of general-purpose, low-cost digital integrated circuit devices made a significant impact on the design of communication receivers although, until the later introduction of digital signal processing, their application was primarily for operator convenience and their use for stable, low-cost frequency syn-

thesisers rather than their use in the signal path.

By incorporating a digital frequency counter or by operation directly from a frequency synthesiser, it is now normal practice to display the frequency to which the receiver (or transceiver) is tuned directly on matrices of light-emitting diodes or liquid crystal displays. This requires that the display is offset by the IF from the actual output of the frequency synthesiser or free-running local oscillator. Such displays have virtually replaced the use of calibrated tuning dials.

Frequency synthesisers are commonly 'tuned' by a rotary shaft-encoded switch which can have the 'feel' of mechanical capacitor tuning of a VFO, but this may be supplemented by pushbuttons which enable the wanted frequency to be punched in. Common practice is for the frequency change per knob revolution to be governed by the rate at which the knob is spun, to speed up large changes of frequency. With a synthesiser the frequency may change in steps of 100Hz, 10Hz or even 1Hz, and it is desirable that this should be free of clicks and should appear to be almost instantaneous. Many of the factory-built equipments incorporate digital memory chips which can be programmed with frequencies to which it is desired to return to frequently. It should be noted that a phase-locked oscillator has an inherent jitter that appears as phase and amplitude noise that can give rise to reciprocal mixing and an apparent raising of the noise floor of the receiver. Digital direct frequency synthesis (DDS) can reduce phase noise and is being increasingly used.

Digital techniques may be used to stabilise an existing free-running VFO by continuously 'sampling' the frequency over predetermined timing periods and then applying a DC correction to a varactor forming part of the VFO tuned circuit. The timing periods can be derived from a stable crystal oscillator and the technique is capable of holding a reasonably good VFO to within a few hertz. Here again some care is needed to prevent the digital pulses, with harmonics extending into the VHF range, from affecting reception.

Microprocessor control of user interfaces can include driving the tuning display, memory management (including BFO and pass-band tuning), frequency and channel scanning; data bus to RS-232 conversion. As stressed by Dr Ulrich Rohde, DJ2LR, several key points need to be observed:

- Keypad and tuning-knob scanning must not generate any switching noises that can reach the signal path.

- All possible combination of functions such as frequency steps, operating modes, BFO frequency offset and pass-band tuning should be freely and independently programmable and storable as one data string in memory.

- It is desirable that multilevel menus should be provided for easy use and display of all functions. This includes not only modes (USB, LSB, CW etc) but also AGC attack and decay times. Such parameters should be freely accessible and independently selectable.

It should be understood that most of the above microprocessor functions are for user convenience and do not add to the

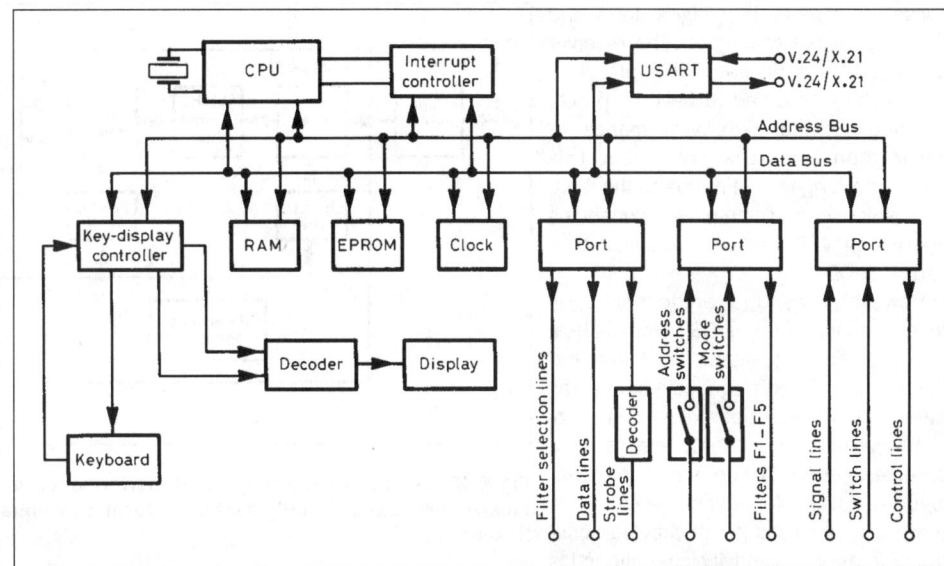

Fig 6.10: Architecture of the elaborate internal computer system found in a modern professional fully-synthesised receiver (DJ2LR)

basic performance (apart from frequency stability) of a receiver; and unless care is exercised may in practice degrade performance. **Fig 6.10** shows an example of the computer control architecture of a modern receiver.

Digital Signal Processing

Digital techniques have been used in HF receivers for several decades, notably in the form of frequency synthesisers, digital frequency displays, frequency memories, microprocessor controlled mode-switching, variable tuning rates etc. By the mid-1980s, professional designers were engaged in further extensions in the use of digital rather than analogue approaches in the main signal path of HF receivers.

Digital signal processing (DSP) is seen as opening the way to the 'all-digital receiver' in which the incoming signals would be

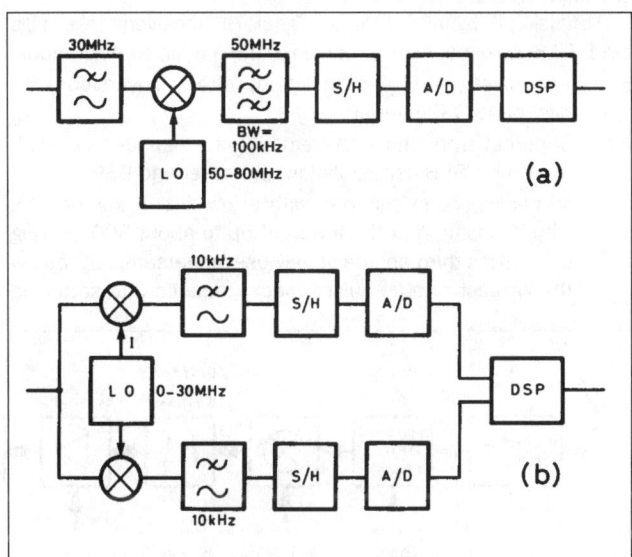

Fig 6.11: Basic arrangements for hybrid analogue/digital receivers incorporating digital signal processing. (a) Single-channel, single-conversion superhet (in practice a double-conversion analogue front-end is more likely to be employed to bring the IF signal down to 1.6 or 3MHz). (b) Dual-channel homodyne approach: S/H, sample and hold; A/D, analogue/digital converter; DSP, digital signal processing

converted directly into digital form and then processed throughout the receiver as a digital bitstream. In time, this approach is expected to become practical and provide high performance at lower than analogue costs, but this remains a long-term aim due to the limited resolution and sampling frequencies of existing VLSI analogue-to-digital converters (ADC).

However, hybrid analogue/digital receivers incorporating DSP, with digitisation of incoming signals at audio frequency (baseband frequency) or at a relatively low IF have been developed, providing performance characteristics comparable to good (but not very good) analogue receivers. An advantage of DSP is that such processing can provide accurately shaped selectivity characteristics at a lower cost than a full complement of analogue (crystal) band-pass filters as fitted in high-grade, multimode professional receivers.

Potentially, digital technology offers a lower component count, easier factory assembly, higher reliability and the lower costs that can ensue from using the standard digital devices developed for use in computers etc. The radio amateur, however, should not expect an overall performance significantly better than is possible in a well-designed receiver based on analogue technology in the signal path; furthermore, unless the designer exercises great care, receivers using sophisticated DSP may suffer in comparison with an all-analogue signal path.

Digital technology does, however, open new possibilities not only in filter characteristics but also in such areas as AGC, noise cancellation etc.

Professional hybrid analogue-digital HF receivers (see **Figs 6.11-6.13**) have been implemented in three main configurations:

- Two-phase (SSB demodulation) direct-conversion with baseband DSP filtering.
- Superhet front-end with 'zero-IF' (super-gainer type) with two-phase SSB demodulation and baseband DSP.
- Multi-conversion superhet with a low final IF (LF or MF), with digitisation at the low IF of up to about 500kHz (**Fig 6.14**). This third approach may use 'undersampling' below the Nyquist rate (Nyquist's theory requires that sampling

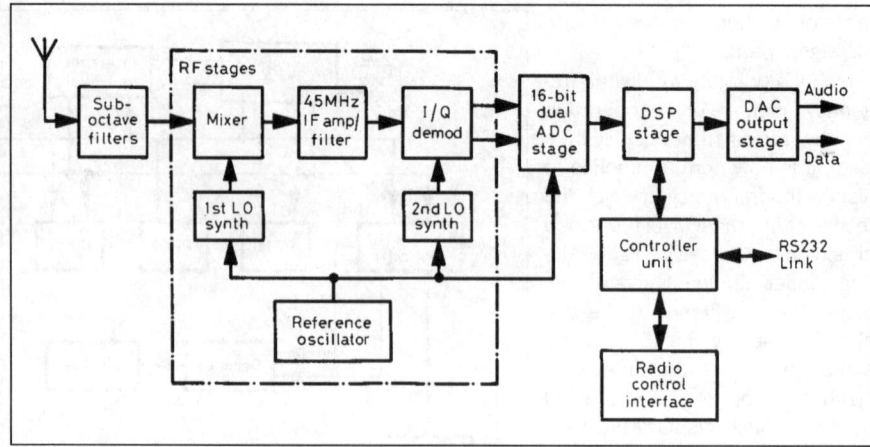

Fig 6.12: Outline of prototype high-performance analogue / digital communications receiver with baseband digitisation following two-phase I/Q demodulation (Roke Manor Research Ltd)

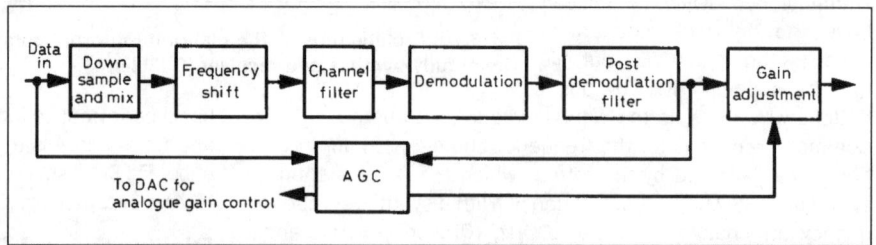

Fig 6.13: Software structure for the STC receiver

should be at least twice the maximum frequency of the sampled waveform) with the aliasing products eliminated by means of relatively low-grade analogue filtering (eg ceramic filter). For example, a sampling frequency of 96kHz may be used to digitise signals at an IF of 456kHz.

Amateur-radio transceivers have been marketed with DSP filtering as an optional extra, in this case at baseband frequencies behind a conventional SSB front-end with a product detector after analogue SSB filters.

DSP at other than audio frequency has the problem that the dynamic range is adversely affected by the need that much of the receiver gain must be ahead of the analogue to digital converter (ADC), and therefore must be under automatic gain control. With effective AGC the dynamic range may be specified as better than 120dB but it should be appreciated that the instantaneous dynamic range will be much lower than this, since current ADCs cannot cope with large dynamic ranges. It should be appreciated that a large instantaneous dynamic range is highly

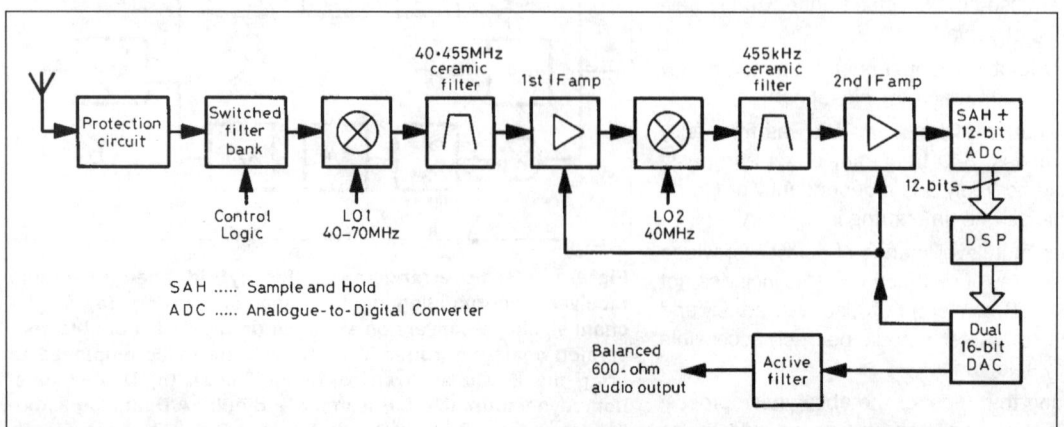

Fig 6.14: Block diagram of the STC marine HF band analogue / digital receiver with sub-Nyquist sampling of 455kHz IF

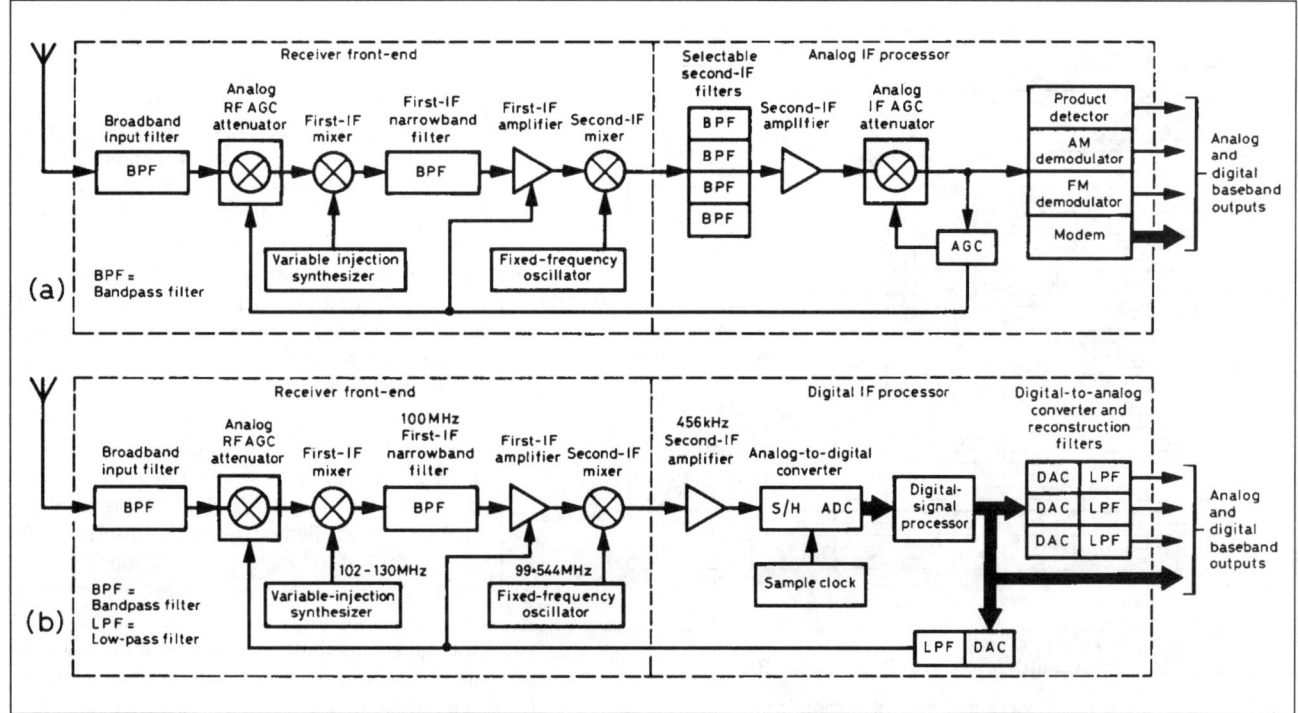

Fig 6.15: (a) Outline of typical professional analogue communications receiver with several band-pass filters and demodulation circuits for the different modes. (b) Use of a software-reconfigurable digital signal processor at the 455kHz second IF potentially reduces cost by eliminating the multiple band-pass filters etc

desirable when trying to copy a weak narrow-band signal immediately alongside a very strong signal. The linearity and resolution of the ADC is thus a key requirement for a high-performance receiver. It should be realised that if a 1 microvolt input signal is to be digitised such that a 20dB SINAD ratio (Signal plus Noise plus Distortion to Noise plus Distortion) is required, the digitisation of the wanted signal needs to be to 4 bits. Handling a 0.25V signal off-tune then demands an ADC with a resolution of 24 bits per sample if the ADC is to have a linear transfer characteristic (ie equal voltage steps per bit).

There is no doubt, however, that DSP can provide excellent filter characteristics programmed to match the required bandwidth of different modes. For example, an SSB filter's attenuation at a 500Hz offset from the upper band edge can be over 60dB without any adjustments required and is insensitive to external component and environmental changes, and without the phase delays that can affect performance of data transmissions.

The same DSP filter can be programmed to provide a series of band-pass filters (eg 500Hz, 1kHz, 2.7kHz, 6kHz etc), each having shape factors superior to a complete set of high-cost crystal filters. (**Fig 6.15**) In practice, many of the advantages of DSP filtering may be lost due to limitations in the analogue front-end and in the ADC as noted above. It seems certain that DSP will come to play an increasingly important role in HF receivers and many other aspects of amateur radio.

RECEIVER SPECIFICATION

The performance of a communications receiver is normally specified by manufacturers or given in equipment reviews, or stated in the various constructional articles. It is important to understand what these specifications mean and how they relate to practical requirements in order to know what to look for in a good receiver. It will be necessary to study any specifications with some caution, since a manufacturer or designer will usual-

ly wish to present his receiver in the most favourable light, and either omit unfavourable characteristics or specify them in obscure terms. The specifications do not tell the whole story: the operational 'feel' may be as important as the electrical performance; the 'touch' of the tuning control, the absence of mechanical backlash or other irregularities, the convenient placing of controls, the positive or uncertain action of the band-change switch and so on will all be vitally important.

Furthermore, there is a big difference between receiver measurements made under laboratory conditions, with only locally generated signals applied to the input, and the actual conditions under which it will be used, with literally hundreds of amateur, commercial and broadcasting signals being delivered by the antenna in the presence of electrical interference and possibly including one or more 'blockbusting' signals from a nearby transmitter.

Sensitivity

Weak signals clearly need to be amplified more than strong ones in order to provide a satisfactory output to the loudspeaker, headphones or data modem. However, there are limits to this process set by the noise generated within the receiver and the external noise picked up by the antenna. What is important to the operator is the signal-to-noise ratio (SNR) of the output signal and how this compares with the SNR of the signal delivered by the antenna. Ideally these would be the same, in which case the noise factor (NF) would be unity.

Noise generated within the receiver is most important when it arises in those parts of the receiver where the incoming signal is still weak, ie in the early (front-end) stages; this noise is, within limits, under the control of the designer.

External noise includes atmospheric noise, which is dependent on both frequency, time of day and ionospheric conditions, and also on local man-made electromagnetic noise, which will usually be more significant in urban locations than in rural sites

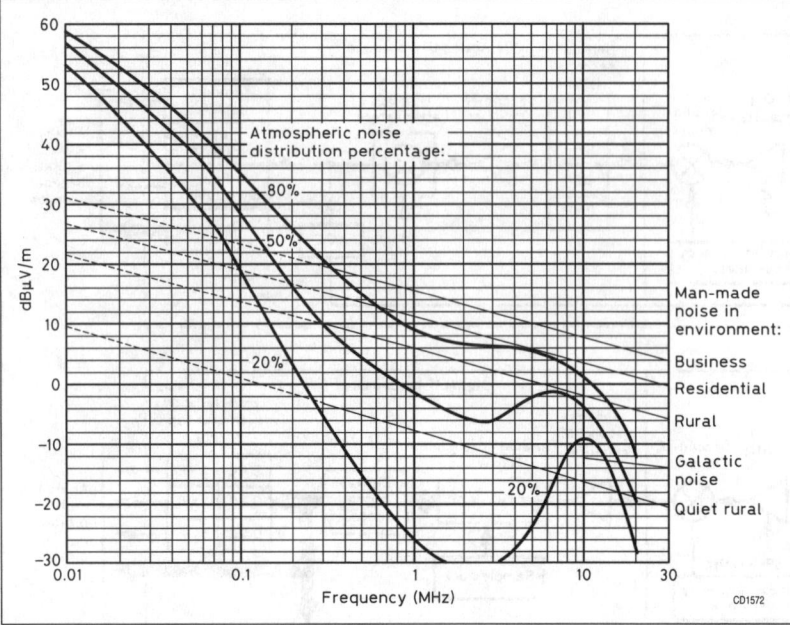

Fig 6.16: Noise field strength levels in 2.7kHz bandwidth

due to the multiplicity of electrical and electronic appliances. See **Fig 6.16**.

Noise power is usually regarded as distributed evenly over the receiver's bandwidth so that the effective noise level is reduced if the receiver is operated at the minimum bandwidth appropriate for the mode in use (eg about 2.1 or 2.7kHz for SSB, about 300Hz for CW).

Noise Figure and Noise Factor

Noise figure defines the maximum sensitivity of a receiver without regard to its pass-band (bandwidth) or its input impedance, and will be determined in the front-end of the receiver. It is defined as the ratio:

Noise figure = SNRin/SNRout

where the SNR is noise power ratio. When expressed logarithmically:

Noise factor = $10 \log_{10}$ (SNRin/SNRout) dB

The minimum equivalent input noise for a receiver at room temperature (290K) is -174dBm/Hz. This is given by:

P = kTB

where P is the power in watts, k is Boltzmann's constant (1.38 x 10^{-23} J/K) and B is the bandwidth in Hertz. The noise floor of the receiver is then given by:

Noise floor = $-174 + NF + 10 \log_{10}B$ dBm

where NF is the noise factor in decibels, dBm is the level in decibels relative to 1 milliwatt and, for an input resistance of 50Ω, this means that -107dBm is 1 microvolt, and B is the receiver noise bandwidth in Hertz. For a CW or SSB receiver, this is approximately the 6dB bandwidth of the narrowest filter: for FM the 6dB bandwidth of the narrowest IF filter can be used as a reasonable approximation. However, the SNR out of an FM detector displays a non-linear relationship to the input SNR, depending upon the detector and IF amplifier type. So a receiver with a 3kHz bandwidth and a 10dB noise figure will have an equivalent input noise floor of -130dBm. That is to say, if the receiver was perfect, it would behave as if the input noise was at -130dBm. The relationship between sensitivity and the various ways of describing noise is shown in **Fig 6.17**.

Because of galactic, atmospheric and man-made noise always present on HF there is little need for a receiver noise fac-

tor of less than 15-17dB on bands up to about 18-20MHz, or less than about 10dB up to 30MHz, even in quiet sites. It may, however, be an advantage if the first stages (preamplifier or mixer or post-mixer) have a lower noise figure since this will permit good reception with an electrically short antenna or allow the use of a narrow-band filter, which attenuates the signal power even if providing an impedance step-up, to be used between antenna and the mixer stage (ie improved pre-mixer selectivity). It should be noted that excessive sensitivity is likely to impair the strong-signal handling capabilities of the receiver.

Signal-to-noise ratio (or, as measured in practice, more accurately signal-plus-noise to noise ratio) gives the minimum antenna input voltage to the receiver needed to give a stated output SNR with a specified noise bandwidth. The input voltage for a given input power depends on the input impedance of the receiver; for modern sets this is invariably 50 or 75Ω but for older (though still useful) receivers this may be 400Ω and such sets may thus appear wrongly to be less sensitive unless the impedance is taken into account. The signal delivered by the antenna from a weak incoming signal may well be of the order of -130dBm, representing 0.14µV across a matched 50Ω line. It should be noted that where sensitivity is defined in terms of SNR, this depends not only on input impedance but also on the output SNR which, while usually 10dB, may occasionally be 6dB; it will also depend as noted above on the noise bandwidth of the receiver.

It is useful to note also the minimum discernible signal (MDS) which is defined as where the signal is equal to the noise voltage, and thus 10dB less than the minimum signal for a SNR

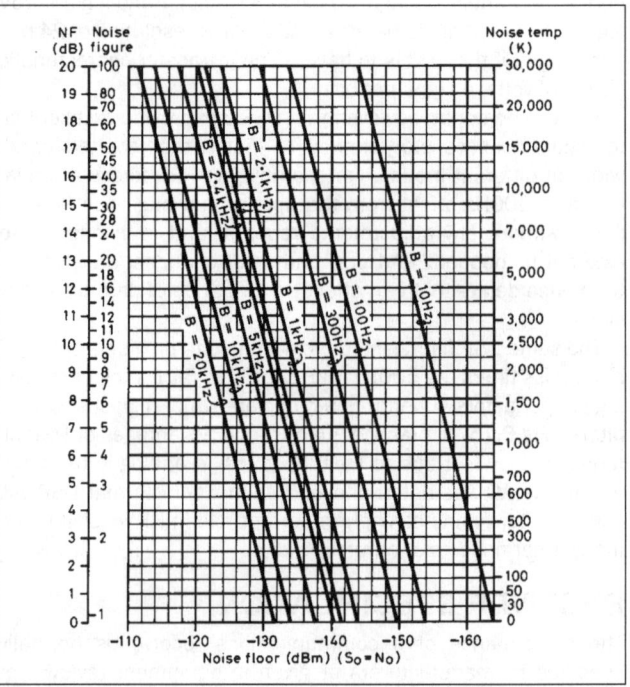

Fig 6.17: Relationship between the sensitivity of a receiver as defined in terms of noise factor (dB), noise figure or noise temperature and the noise floor (noise = signal) in dBm for various receiver bandwidths. Note how the minimum detectable signal reduces with narrower bandwidths for the same noise factor

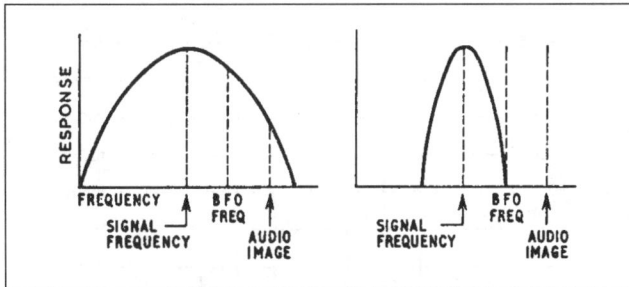

Fig 6.18: How a really selective receiver provides 'single-signal' reception of CW. The broad selectivity of the response curve on the left is unable to provide substantial rejection of the audio 'image' frequency whereas with the more selective curve the audio image is inaudible and CW signals are received only on one side of zero-beat

(NOT in this case (S + N)/R) output of 10dB, although normally given in dBm. The MDS is thus the noise floor of the receiver and represents the power required from a signal generator to produce a 3dB (S+N)/N output, ie where the input signal equals the background noise. MDS (dBm) = -174dBm + 10 log BW + NF where NF is in decibels.

The noise floor of HF receivers is conventionally given in -dBm in a 3kHz noise bandwidth and in practice may range from about -120dBm to about -145dBm. Alternatively, receiver noise in a 3kHz bandwidth at 50Ω impedance may be represented by an equivalent signal expressed in dBmV EMF, which will equal the noise factor less 26dB.

With modern forms of mixers, it is possible to achieve a noise factor better than 10dB so that no preamplification is required to achieve maximum usable sensitivity on HF up to about 21MHz; however, low-gain amplification may still be advisable in order to minimise radiation of the local oscillator output from the antenna and to permit the use of narrow-band or resonant input filtering. Oscillator radiation is limited for amateur equipment to a maximum level of -57dBm conducted to the antenna for emissions below 1GHz, and to -47dBm above 1GHz, by the European Radiocommunications Committee Recommendation on Spurious Emissions.

The effect of cascading stages is given by the well-known equation due to Friis:

$$F_t = F_1 + [(F_2 - 1)/g_1]+ [(F_3 - 1)/g_1g_2] + \ldots$$

where (all in power ratios and NOT decibels) F_t is the overall

noise figure, F_2, F_3 are the individual stage noise figures and g_1, g_2 are the individual stage gains. Obviously, the series can be extended to cover as many stages as necessary.

For design work, this can usefully be incorporated into a spreadsheet approach for rapid evaluation of changes; however, the values must be converted from decibels to absolute power ratios.

Selectivity

The ability of a receiver to separate stations on closely adjacent frequencies is determined by its selectivity. The limit to usable selectivity is governed by the bandwidth of the type of signal which is being received.

For high-fidelity reception of a double-sideband AM signal the response of a receiver would need ideally to extend some 15kHz either side of the carrier frequency, equivalent to 30kHz bandwidth; any reduction of bandwidth would cause some loss of the information being transmitted. In practice, for average MF broadcast reception the figure is reduced to about 9kHz or even less; for communications-quality speech in a double-sideband system we require a bandwidth of about 6kHz; for single-sideband speech about half this figure or 3kHz is adequate, and filters with a nose bandwidth of 2.7 or 2.1kHz are used, providing, in the case of a 2.7kHz filter, audio frequencies from 300 to 3000Hz with little loss of intelligibility. For CW, at manual keying speeds, the minimum theoretically possible bandwidth will reduce with speed from about 100Hz to about 10Hz for very slow Morse. **Fig 6.18** shows how 'single signal' reception is achieved with a narrow filter. Excessively narrow filters with good shape factors make searching difficult - and many operators like to have some idea of signals within a few hundred hertz of the wanted signal. Ideally, again, we would like to receive just the right bandwidth, with the response of the receiver then dropping right off as shown in **Fig 6.19**, to keep the noise bandwidth to a minimum. Although modern filters can approach this response quite closely, in practice the response will not drop away as sharply over as many decibels as the ideal.

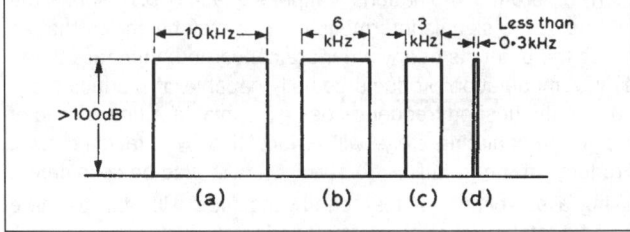

Fig 6.19: The ideal characteristics of the overall band-pass of a receiver are affected by the type of signals to be received. (a) This would be suitable for normal broadcast reception (DSB signals) permitting AF response to 5kHz. (b) Suitable for AM phone (AF to 3kHz). (c) For SSB the band-pass can be halved without affecting the AF response (in the example shown this would be about 300 to 3300Hz). (d) Extremely narrow channels (under 100Hz) are occupied by manually keyed CW signals but some allowance must usually be made for receiver or transmitter drift and a 300Hz bandwidth is typical - by selection of the carrier insertion oscillator frequency any desired AF beat note can be produced

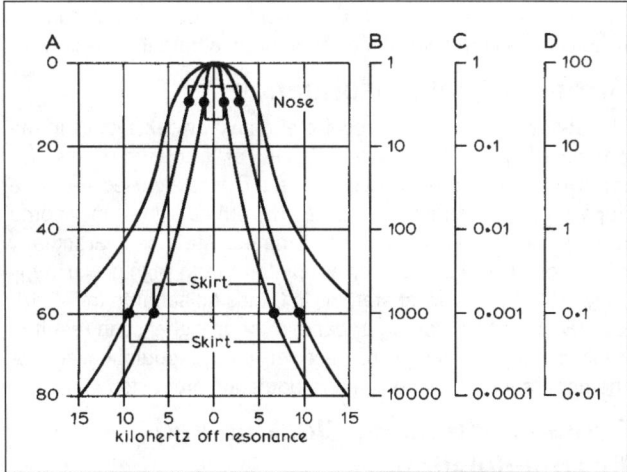

Fig 6.20: The ideal vertical sides of Fig 6.19 cannot be achieved in practice. The curves shown here are typical. These three curves represent the overall selectivity of receivers varying from the 'just adequate' broadcast curve of a superhet receiver having about four tuned IF circuits on 470kHz to those of a moderately good communications receiver. A, B, C and D indicate four different scales often used to indicate similar results. A is a scale based on the attenuation in decibels from maximum response; B represents the relative signal outputs for a constant output; C is the output voltage compared with that at maximum response; D is the response expressed as a percentage

To compare the selectivity of different receivers, or the same receiver for different modes, a series of curves of the type shown in **Fig 6.20** may be used. There are two ways in which these curves should be considered: first the bandwidth at the nose, representing the bandwidth over which a signal will be received with little loss of strength; the other figure - in practice every bit as important - is the bandwidth over which a powerful signal is still audible, termed the skirt bandwidth.

The nose bandwidth is usually measured for a reduction of not more than 6dB, the skirt bandwidth for a reduction of one thousand times on its strength when correctly tuned in, that is 60dB down. These two figures can then be related by what is termed the shape factor, representing the bandwidth at the skirt divided by the bandwidth at the nose. The idealised curves of Fig 6.19, which have the same bandwidth regardless of signal strength, would represent a shape factor of 1; such a receiver cannot be designed at the present state of the art, although it can be approached by digital filters and some SSB filters; the narrower CW filters, although much sharper at the nose, tend to broaden out to about the same bandwidth as the SSB filter and thus have a rather worse shape factor, although this may not be a handicap. Typically a high-grade modern receiver might have an SSB shape factor of 1.2 to 2 with a skirt bandwidth of less than 5kHz. Note that, although 6 and 60dB are conventionally used for nose and skirt, some radios are specified at 3 and 30dB, or even 6 and 30dB - it is important to be aware of what figures are used when comparing one filter or receiver with another.

It should be noted that such specifications are determined when applying only one signal to the input of the receiver; unfortunately in practice this does not mean that a receiver will be unaffected by very strong signals operating many kilohertz away from the required signal and outside the IF pass-band; this important point will be considered later in this chapter.

It therefore needs to be stressed that the effective selectivity cannot be considered solely in terms of static characteristics determined when just one test signal is applied to the input, but rather in the real-life situation of hundreds of signals present at the input: in other words it is the dynamic selectivity which largely determines the operational value of an amateur HF receiver.

Strong-signal Performance

The ability of a receiver to receive effectively weak signals in the presence of much stronger signals at frequencies not far removed from the desired signal is today recognised as more important than extreme sensitivity. The strong signals may come from local or medium-distance amateur stations (particularly during contest operation) or at any time from high-power (typically 500kW) broadcast stations in bands adjacent to the 7, 10, 14, 18 and 21MHz amateur bands. Strong signals can result in desensitising (blocking) the receiver, cross-modulation and/or the generation of spurious intermodulation products.

Cross-modulation, Blocking and Intermodulation

Even with a receiver that is highly selective to one signal down to the -60dB level, there remains the problem of coping with numbers of extremely strong signals. When an unwanted signal is transmitting on a frequency that is well outside the IF passband of the receiver, it may unfortunately still affect reception as a result of cross-modulation, blocking or intermodulation.

When any active device such as a transistor or valve is operated with an input signal that is large enough to drive the device into a non-linear part of its transfer characteristic (ie so that some parts of the input waveform are distorted and amplified to

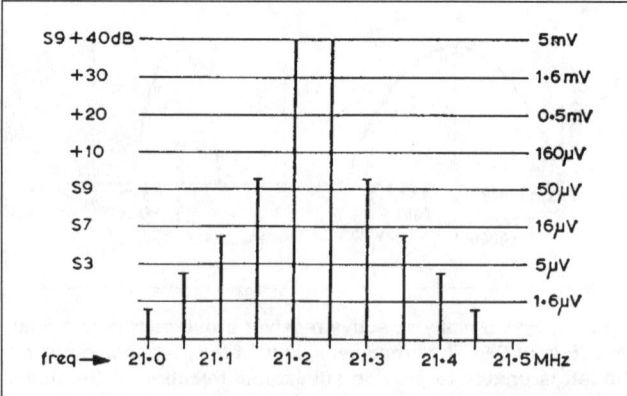

Fig 6.21: Intermodulation products. This diagram shows the effect of two very strong signals on 21,200kHz and 21,250kHz reaching the front-end of a typical modern transceiver and producing spurious signals at 50kHz intervals. Note the S9 signals produced as the third-order products f1 + (f1 - f2) and f2 - (f1 - f2). Three very strong signals will produce far more spurious signals, and so on

different degrees) the device acts as a 'modulator', impressing on the wanted signal the modulation of the strong signal by the normal process of mixing. When a very strong signal reaches a receiver, the broad selectivity of the signal-frequency tuned circuits (and often any tuned circuits, including IF circuits, prior to the main selective filter) means it will be amplified, along with the wanted signal, until one or more stages are likely to be driven into a non-linear condition. It should be noted that the strong signal may be many kilohertz away from the wanted signal, but once this cross-modulation has occurred there is no means of separating the wanted and unwanted modulation. A strong CW carrier can reduce the amplification of the wanted signal by a similar process which is in this case called desensitisation, or in extreme cases blocking.

In these processes there need be no special frequency relationship between wanted and unwanted signals. However, a further condition arises when there are specific frequency relationships between wanted and unwanted signals, or between two strong unwanted signals; a process called intermodulation: see **Fig 6.21**.

Intermodulation is closely allied to the normal mixing process but with strong signals providing the equivalent of local oscillators, unwanted intermodulation products (IPs) can result from many different combinations of input signal. Fig 6.18 shows the effects of two signals intermodulating; it may be shown that as far as the products of any order are concerned, there are $2n(n - 1)$ intermodulation products actually capable of producing signals on the desired frequency. Using a sub-octave filter ahead of the intermodulating stage will reduce this by a factor of two. Products of the from $f_1 + f_2 - f_3 = f_{tune}$ must also be considered. Using a sub-octave or less bandwidth filter will lead to there being a total number of intermodulation products:

$$T = n(n - 1) + 0.5n(n - 1)(n - 2)$$

where T is the total number of intermodulation products and n is the total number of input signals.

Because of the potentially large number of products, the effect is to raise the effective noise floor of the receiver. This occurs in a manner analogous to white light being created from a mixture of all colours (frequencies): similarly, noise is produced from a mixture of a large number of frequencies of random amplitude and phase distribution.

It should be appreciated that cross-modulation, blocking and intermodulation products can all result from the presence of

extremely strong unwanted signals applied to any stage having insufficient linearity over the full required dynamic range. The solution to this problem is either to reduce the strength of unwanted signals applied to the stage, or alternatively to improve the dynamic range of the stage.

Clearly the more amplifying stages there are in a receiver before the circuits or filters which determine its final selectivity, the greater are the chances that one or more may be overloaded unless particular care is paid to the gain distribution in the receiver (see later). From this it follows that multiple-conversion receivers are more prone to these problems than single-conversion or direct-conversion receivers.

It is often difficult for an amateur to assess accurately the performance of a receiver in this respect; undoubtedly many multi-conversion superhets fall far below the desired performance. Even high-grade receivers are likely to be affected by S9 + 60dB signals 50kHz or more away from the wanted signal.

The susceptibility of receivers to these forms of interference depends on a number of factors: notably the type of active devices used in the front-end of the receiver; the phase noise performance of the various oscillators, the pattern of gain distribution through the receiver and how this is modified by the action of AGC. On receivers suffering from this problem (often most apparent on 7MHz where weak amateur stations may be sought alongside extremely strong broadcast stations), considerable improvement often results from the inclusion of an attenuator in the antenna feeder, since this will reduce the unwanted signals to an extent where they may not cause spurii before reducing the wanted signal to the level where it cannot be copied.

If sufficient dynamic range could be achieved in all stages before the final selectivity filter, then selectivity could be determined at any point in the receiver (for example at first, second or third IF or even at AF).

In practice, some early semiconductor receivers were unable to cope with undesired signals more than about 20-30dB stronger than a required signal of moderate strength, though their 'static' selectivity was extremely good. The overall effect of limited dynamic range depends upon the actual situation and band on which the receiver is used; it is of most importance on 7MHz or where there is another amateur within a few hundred yards. Even where there are no local amateurs, an analysis of commercial HF signals has shown that typically, out of a total of some 3800 signals logged between 3 and 29MHz at strengths more than 10dB above atmospheric noise, 154 were between 60-70dBµV, 72 between 70-80dBµV, 36 between 80-90dBµV and 34 between 90-100dBµV. If weak signals are to be received satisfactorily the receiver needs to be able to cope with signals some 100dB stronger (ie up to about S9 + 50dB). This underlines the importance of achieving front-end stages of extremely wide dynamic range or alternatively providing sufficient pre-mixer selectivity to cut down the strength of all unwanted signals reaching the mixer (preferably to well under 100mV). It has been shown that even the use of a small antenna such as a 3/4 G5RV at 8m (25ft) can lead to a substantial intermodulation problem with receivers having only a 0dBm third-order intercept point.[2]

The effects arise from non-linearity in the receiver and can be specified in terms of the receiver's dynamic range, although it is important to note that the dynamic range needs to be specified separately for blocking, cross-modulation, intermodulation and reciprocal mixing (see later). The most useful and most critical specification is given by the third-order intercept point or as the spurious-free dynamic range.

Strong-signal performance and a wide dynamic range have become widely recognised as important and highly desirable

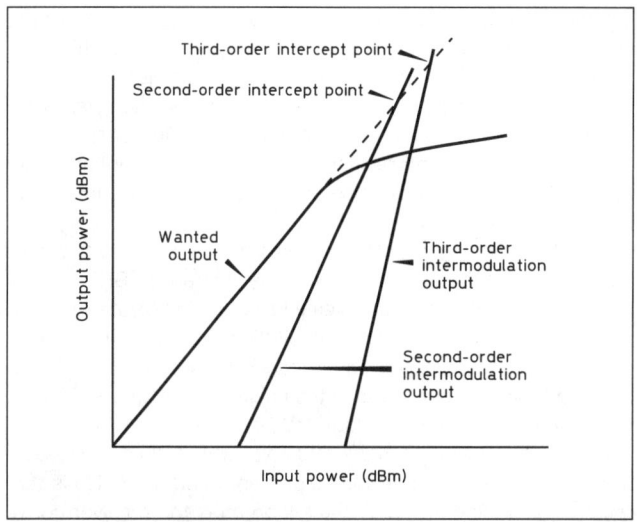

Fig 6.22: Graphical representation of receiver dynamic range performance

characteristics of receiver performance, although it can be argued that, at least for the Amateur Service (particularly CW), accepted methods of laboratory measurement may not provide a realistic specification. It is conventional practice to make dynamic range measurements using two signal generators with frequencies spaced 20kHz apart. However, where receivers incorporate relatively low-cost frequency synthesisers, the spacing between the two signals may have to be increased to 50 or even 100kHz in order to overcome the effects of synthesiser phase noise (made even more significant by the trend to up-conversion to a VHF intermediate frequency because VHF oscillators are generally worse than HF ones in this respect). It would be more useful if instantaneous dynamic range measurements were made with frequencies spaced 2 or 5kHz apart. It should also be remembered that in practice, the receiver will be required to cope not with just two locally generated signals but with dozens of strong broadcast carriers which will reach the vulnerable mixer stage unless filtered out (or at least reduced) by pre-mixer selectivity. Multiple carriers produce a multitude of IMPs which may resemble noise, thus raising the noise floor and decreasing the sensitivity of the receiver.

In specifying receiver dynamic range use is made of the concept of intercept points, although it is important to realise that these are purely graphic presentations based on two-signal laboratory measurements and do not represent directly the signal-handling properties of a receiver. Third-order intermodulation products are measured in the laboratory with the aid of a spectrum analyser and two high-performance signal generators. The intercept point is found with the aid of a graph with axes representing output power in dBm plotted against input power in dBm, on which the wanted signal and the N-order intermodulation product outputs are extended to the points where the IMPs intercept the wanted signal plot. This will occur at signal inputs well above those that can in practice be handled by the receiver (Fig 6.22). Note that there are two intercept points which may be quoted - the input intercept point, and the output intercept point. These differ by the gain of the device, it not always being specified as to which is being used. Generally, commercial considerations lead to whichever is the biggest being the one quoted.

With both axes logarithmic, the second-order IMPs will have twice, and the third-order IMPs approximately three times, the slope of a wanted signal. For third-order IMPs, N = 3 and the IMP power is approximately proportional to the cube of the input powers. In other words, for every 1dB increase in the input pow-

ers, the IP3 output power increases by approximately 3dB. When plotted graphically there will thus be a point at which the IP3 output crosses and overtakes the plot of a wanted signal.

This shows that with very strong off-frequency signals, IMPs would completely block out the wanted signal. But for specification purposes we are more concerned to know the point at which the IMPs rise above the noise floor and can just be heard as unwanted interference.

A useful guide to the signal-handling performance of a receiver is given by the spurious-free dynamic range (SFDR). The upper end is defined by the signal level applied to the receiver input which produces third-order IMD products equal to the noise floor of the receiver under test. The lower end is defined as the minimum input signal at the noise floor of the receiver, ie the minimum discernible signal (MDS).

The SFDR is given by 2/3 x (IP3 - No) where SFDR is in decibels, IP3 is the third-order intercept point in dBm and No is the receiver noise floor in dBm. In the laboratory the receiver noise floor will be that of the internally generated noise and this will usually be significantly below the noise floor with the receiver connected to an antenna. The full SFDR may thus not be usable in practice, and it may be more useful to know the maximum input signal level, Pi(max) that produces third-order IMD products equal to the receiver noise floor:

$$Pi(max) = 1/3 \ (2IP + No)$$

where Pi(max) is the maximum input signal in dBm, IP is the third-order intercept point in dBm and No the receiver noise floor in dBm.

SFDR is sometimes termed the two-tone dynamic range, but note that receiver dynamic range may be defined differently, for example as the blocking dynamic range (BDR) representing the difference between the 1dB compression point and the noise floor. With double-balanced mixers, a rule of thumb puts the third-order intercept point some 10 to 15dB above the 1dB compression point but, since the spurious-free IMD range is restricted to where the IMPs rise above the noise floor, the blocking dynamic range will be significantly greater than the SFDR. For example, an extremely-high-performance front-end with a third-order input intercept point of +33dBm, a 1dB input compression point of +14.3dBm and a minimum discernible signal of -133.4dBm would have a spurious-free dynamic range of 111dB but a blocking dynamic range of 147.7dB. In practice most receivers would have performance specifications considerably below this example. However, as will be shown later, for these levels of performance to be of any use, the reciprocal mixing performance must also be considered. In this respect, the phase noise limited dynamic range (PNDR) is as important as the SFDR, and the two should be approximately equal.

In practice, the noise levels at the receiver input are quite likely to be high enough to provide a bottom end limit to sensitivity, and certain unpublished measurements suggest that a SFDR of about 95 to 100dB, used in conjunction with a switched 10 or 20dB antenna attenuator is able to provide a more than adequate performance. The use of an antenna attenuator may well be considered as an admission of defeat: it should NOT be considered as a substitute for anything less than 95dB of SFDR.

Merely because crystal and mechanical filters are passive devices, they should not be considered as non-contributors to the intermodulation performance of the receiver. Crystal filters are notorious for their IM contribution: the higher the frequency, the worse the performance. One cause for this is the mechanical stress in the crystal for a given voltage across it increases with frequency because of the decrease in crystal thickness, and when the limit of Hook's law is approached, intermodulation occurs. Another factor is concerned with finishing of the crystal, where the final lapping can have marked effects. It is worth while turning a filter round, as it may well perform far better one way round than the other. Mechanical filters of the older variety are also quite poor, although modern filters have performances comparable to crystal filters - input intermodulation intercept points of +16 to +30dBm being achieved.

Generally speaking, cross-modulation is not a common effect in modern receivers: the effects of intermodulation and reciprocal mixing are more prevalent. On FM, of course, cross-modulation as such does not occur.

It needs to be stressed that in order to achieve state-of-the-art meaningful dynamic performance of a receiver, extreme care must be taken not only with the vulnerable mixer stages but throughout the receiver, up to and including the final selectivity filters. On the other hand, budget-conscious constructors/purchasers can take heart from the fact that receivers with far less stringent specifications may still prove entirely adequate for many forms of amateur activity not involving the reception of very weak signals under the most hostile conditions such as international HF contests, or, for example, 7MHz night-time operation in the presence of strong broadcast carriers.

Automatic Gain Control (AGC)

The fading characteristics of HF signals and the absence of a carrier wave with SSB make the provision of effective AGC an important characteristic of a communications receiver, although it should be stressed that no AGC may be preferable to a poor AGC system that can degrade overall performance of the receiver. Unfortunately, unwanted dynamic effects of an AGC system cannot be deduced from the usual form of receiver specification which usually provides only limited information on the operation of the AGC circuits, indicating only the change of audio output for a specified change of RF input. For example, the specification may state that there will be a 3dB rise in audio output for an RF signal input change from 1mV to 50mV. This represents a high standard of control, provided that the sensitivity of the receiver is not similarly reduced by strong off-tune signals, and that the control acts smoothly throughout its range without introducing intermodulation distortion etc.

Basically, AGC is applied to a receiver to maintain the level of the wanted signal output at a more or less constant value, while ensuring that none of the stages is overloaded, with consequent production of IMPs etc. The control voltages, derived usually at the end of the IF amplifying stages, are applied to a number of stages in the signal path while usually ensuring that the IF gain is reduced first, and the RF/first mixer later, in order to preserve the SNR.

When AGC is applied to an amplifier, this shifts the operating point and may affect both its dynamic range and the production of intermodulation distortion. One partial solution is the use of an AGC-controlled RF attenuator(s) ahead of the first stage(s).

All AGC systems are designed with an inherent delay based on resistance-capacitance time-constants in their response to changes in the incoming signal; too-rapid response would result in the receiver following the audio envelope or impulsive noise peaks. The delays are specified as the attack time, ie the time taken for the AGC to act, and the decay time for which it continues to act in the absence or fade of the wanted signal (Fig 6.23).

For AM signals, now only rarely encountered in amateur radio, the envelope detector may be used directly to generate the AGC voltage. In this case the attack and delay times need to be fast enough to allow the AGC system to respond to fading, but slow

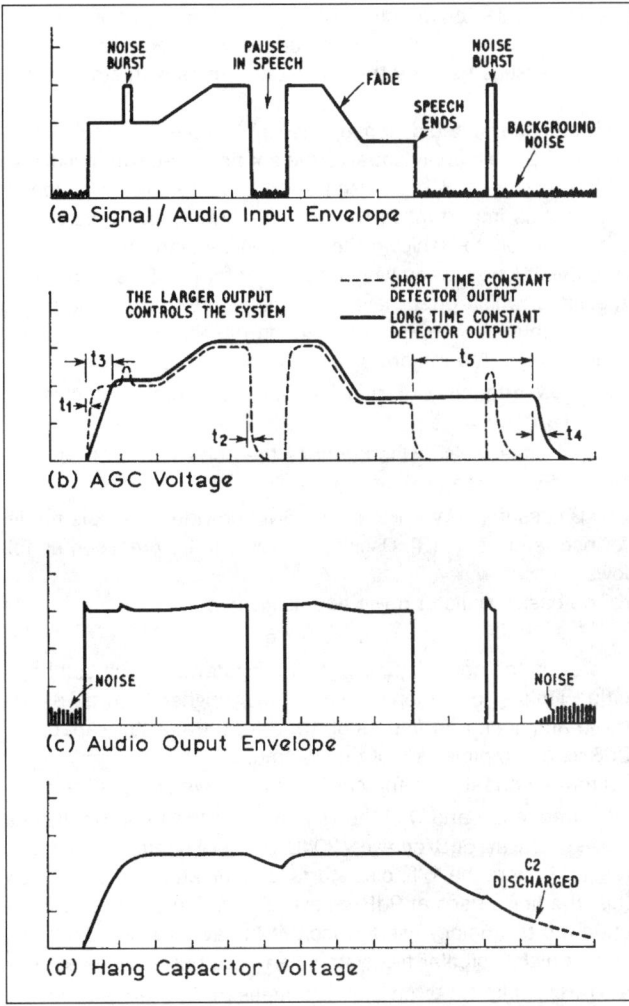

(a) Signal / Audio Input Envelope

(b) AGC Voltage

SHORT TIME CONSTANT DETECTOR OUTPUT
LONG TIME CONSTANT DETECTOR OUTPUT

(c) Audio Output Envelope

(d) Hang Capacitor Voltage

Fig 6.23: The behaviour of dual time-constant AGC circuits under various operating conditions, where t_1 is the fast detector rise time, t_2 the fast detector decay time, t_3 the slow detector rise time, t_4 the slow detector fall time, and t_5 the hang time

enough not to respond to noise pulses or the modulation of the carrier. Typically time-constants are about 0.1 to 0.2s.

For amateur SSB signals there is no carrier level and the AGC voltage must be derived from the peak signal level. This is sometimes done by using the AF signal from a product detector but it is more satisfactory to incorporate a dedicated envelope detector to which a portion of the final IF signal is fed. This avoids the effects of the AGC signal disappearing at low audio beat notes. For SSB, a fast attack time is needed but the release (decay) time needs to be slow in order not to respond to the brief pauses of a speech transmission. Such a system is termed hang AGC and can be used also for CW reception. Typically the attack time needs to be less than 20ms and the release time some 200 to 1000ms; with a receiver intended also for AM reception it is useful to have a shorter release time (say 25ms) available for fast AGC. Hang AGC systems are often based on two time-constants to allow the system to be relatively unaffected by noise pulses while retaining fast attack and hang characteristics.

For both SSB and CW reception, it is important that the BFO or carrier insertion oscillator should not affect the AGC system and for many years this problem was sidestepped by turning the AGC system off during CW reception. Even with modern AGC systems, it may be useful to be able to turn the system off, particularly where final narrow-band CW selectivity depends on post-demodulation filtering, with the result that the AGC system will

react to signals within the bandwidth of the final IF system which may be about 3kHz unless narrow-band IF crystal filters are fitted.

The distribution of gain control is an important, but often neglected function. Obviously, gain could be controlled by an antenna attenuator, but the result would be that the signal-to-noise ratio was that obtained for the weakest input signal. Ideally, an increase of input signal of 20dB would provide an increase in SINAD of 20dB, but there is usually an ultimate receiver SINAD limit of 40 to 50dB. A test of SINAD improvement ratio is usefully but rarely done: a signal is fed into the receiver to provide a 20dB SINAD ratio, and is increased in 10dB steps. A reasonable receiver will show about 28 to 29dB SINAD for the first step, and no less than 37dB SINAD for the second.

Oscillator Noise and Reciprocal Mixing

A single-conversion superhet requires a variable HF oscillator which should not only be stable but should provide an extremely 'pure' signal with the minimum of noise sidebands. Any variation of the oscillator output in terms of frequency drift or sudden 'jumps' will cause the receiver to detune from the incoming signal. The need for a spectrally pure output is less readily grasped, yet it is this feature which often represents a practical limitation on the performance of modern receivers. This is due to noise sidebands or jitter in the oscillator output. Noise voltages, well known in amplifiers, occur also in oscillators, producing output voltages spread over a wide frequency band and rising rapidly immediately adjacent to the wanted oscillator output.

The noise jitter and sidebands immediately adjacent to the oscillator frequency are particularly important. When a large interfering signal reaches the mixer on an immediately adjacent channel to the wanted signal, this signal will mix with the tiny noise sidebands of the oscillator (the sidebands represent in effect a spread of oscillator frequencies) and so may produce output in the IF pass-band of the receiver: this effect is termed reciprocal mixing. Such a receiver will appear noisy, and the effect is usually confused with a high noise factor.

At HF the 'noise' output of an oscillator falls away very sharply either side of the oscillator frequency, yet it is now recognised that this noise may be sufficient to limit the performance of a receiver. Particular care is necessary with some forms of frequency synthesisers, including those based on the phase-locking of a free-running oscillator, since these can often produce significantly more noise sidebands and jitter than that from a free-running oscillator alone.

Synthesisers involving a number of mixing processes may easily have a noise spectrum 40 to 50dB higher than a basic LC oscillator. A tightly controlled phase-locked oscillator with variable divider might be some 20 to 30dB higher than an LC oscillator, but possibly less than this where the VCO is inherently very stable and needs only infrequent 'correction'.

For LC and crystal oscillators, FETs are often recommended. However, suitable bipolar transistors operated in common base (where the a cut-off frequency is highest) with a 27 ohm unbypassed emitter resistor have been shown to be capable of equal results. The use of fairly high supply voltages and currents (increasing the tank circuit power) allows high SNRs to be obtained, albeit at the price of power consumption and possible thermal effects. In VCOs, the use of two varactor diodes back-to-back to avoid conduction at any point of the RF cycle is also recommended. Although 'hyperabrupt' diodes offer the greatest tuning range, they are often of lower Q than the graded junction types, and it is in any case best to have the lowest possible sensitivity in megahertz per volt that can be used.

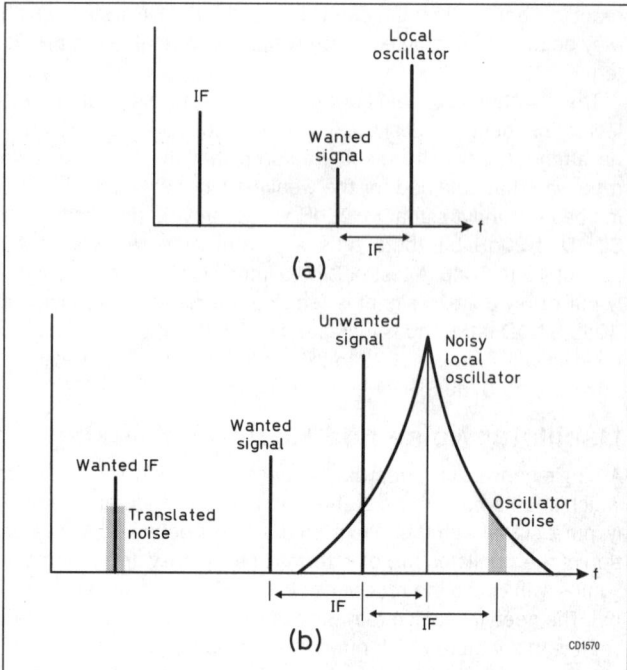

Fig 6.24: Superhet mixing. (a) Local oscillator separated by the IF from the wanted signal. (b) Local oscillator noise sideband separated by the IF from a strong unwanted signal

Fig 6.24 shows in simplified form the basic mechanism of reciprocal mixing. Because of the noise sidebands (or 'skirts') of the local oscillator, some part of the strong unwanted signal is translated into the receiver's IF pass-band and thus reduces the SNR of a weak wanted signal; further (not shown) a small fragment of the oscillator noise also spreads out to the IF, enters the IF channel and reduces the SNR. In practice the situation is even more complex since the very strong unwanted carrier will itself have noise sidebands which spread across the frequency of the wanted signal and degrade SNR no matter how pure the output of the local oscillator.

Fig 6.25 indicates the practical effect of reciprocal mixing on high-performance receivers; in the case of receiver 'A' (typical of many high-performance receivers) it is seen that the dynamic selectivity is degraded to the extent that the SNR of a very weak signal will be reduced by strong unwanted signals (1 to 10mV or more) up to 20kHz or more off-tune, even assuming that the front-end linearity is such that there is no cross-modulation, blocking or intermodulation. Reciprocal mixing thus tends to be

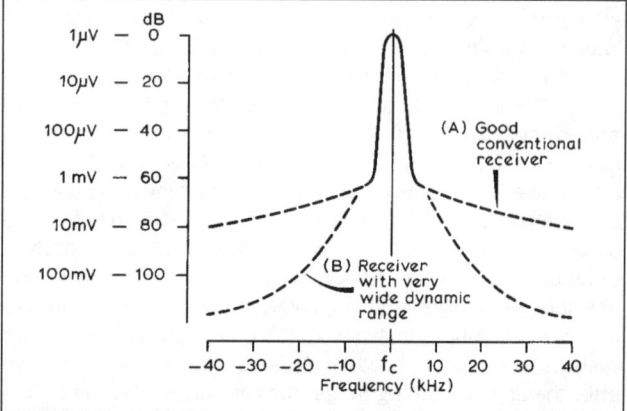

Fig 6.25: Reciprocal mixing due to oscillator noise can modify the overall selectivity curve of an otherwise very good receiver

the limiting factor affecting very weak station performance in real situations, although intermodulation or cross-modulation characteristics become the dominant factors with stronger signals.

The phase limited dynamic range of the receiver should therefore be approximately equal to the spurious-free dynamic range of the receiver: it is of no use having only one them very good.

It is thus important in the highest-performance receivers to pay attention to achieving low noise-sidebands in oscillators, and this is one reason why the simpler forms of frequency synthesisers, which often have appreciable jitter and noise in the output, must be viewed with caution despite the high stability they achieve. The major noise mechanisms are:

(a) low frequency (1/f) noise which predominates close to carrier;

(b) thermal noise which defines the noise floor at frequency offsets greater than f/Q.

D B Leeson (ex-W6QHS, now W6NL) provided a simple model for oscillator noise [3]. Oscillator noise can be predicted as follows.

The basic oscillator noise floor is given by:

$$-174 + 3 + f \text{ dBm}$$

where f is the noise figure of the active device under self-oscillating limiting conditions. This is much higher than as a low-noise amplifier, and figures of 10dB for a transistor capable of 3dB as an amplifier are not uncommon.

From an offset from the carrier of f_0 / Q (where f_0 is the oscillator frequency, and Q is the working Q of the tank circuit), the noise rises at 6dB/octave (20dB/decade), until the point is reached where the 1/f noise starts to dominate. At offsets below this, the noise rises at 9dB/octave. This 1/f knee is dependent upon the technology used: silicon FETs have very low 1/f noise knees: high-f_T bipolar transistors can have noise knees as high as 1kHz, while GaAs devices are measured in terms of megahertz.

A practical case is that of a transistor oscillator at 70MHz, running at 0dBm, with a 10dB noise figure, and a tank circuit Q of 70. From the above equations, the noise in a 1Hz bandwidth would -174 + 3 + 10 or -161dBm at 1MHz offset. At 100kHz, it would be -140dBm/Hz, at 10kHz, -120dBm/Hz, and at 1kHz -100dBm/Hz. The total amount of noise in the receiver IF bandwidth would be increased by the bandwidth: a 3kHz IF bandwidth receives 34dB more noise than a 1Hz bandwidth, so from a strong signal 10kHz away, 120 - 34dB or -86dB noise relative to wanted signal would be received. A filter with a 90dB stopband would be compromised by this level of noise.

It can be seen that the selectivity of a receiver can be determined not by the IF filters, but by the phase noise of the local oscillator.

Since the shot noise of an oscillator spreads across the IF channel, care is needed with low-noise mixers to limit the amount of this injection-source noise that enters the IF amplifier. Balanced and double-balanced mixers provide up to about 30dB rejection of oscillator noise. Note that this noise only affects the sensitivity of the receiver; it is not related to reciprocal mixing. Another technique is to use a rejector trap (tuned circuit) resonant at the IF between oscillator and mixer.

Optimum oscillator performance calls for the use of a high unloaded tank-circuit Q. For switching-mode mixers and product detectors the optimum oscillator output waveform would be a square wave but this refinement is comparatively rare in practice.

Much more information on oscillators can be found in the first Building Blocks chapter.

Frequency Stability

The ability of a receiver to remain tuned to a wanted frequency without drift depends upon the electrical and mechanical stability of the internal oscillators. The primary cause of instability in an oscillator is a change of temperature, usually as the result of internally generated heat. With valves, even in the best designs, there will usually be steady frequency variation of any oscillator using inductors and capacitors during a period of perhaps 15 minutes or more after first switching on; transistor and FET oscillators reach thermal stability in a few seconds, provided that they are not then affected by other local sources of heat.

After undergoing a number of heat cycles, some components do not return precisely to their original values, making it difficult to maintain accurate calibration over a long period. Some receivers include a crystal-controlled oscillator of high stability providing marker signals (for example every 100kHz) so that receiver calibration can be checked and brought into adjustment.

Receiver drift can be specified in terms of maximum drift in Hertz over a specified time, usually quoting a separate figure to cover the warming-up period. For example a high-stability receiver might specify drift as "not worse than 200Hz in any five-hour period at constant ambient temperature and constant mains voltage after one hour warm-up".

Mechanical instability, which may appear as a shift of frequency when the receiver is subjected to mechanical shock or vibration, cannot easily be defined in the form of a performance specification. Sturdy construction on a mechanically rigid chassis can help. The need for high stability for SSB reception has led to much greater use of crystal-controlled oscillators and various forms of frequency synthesis.

Spurious Responses

A most important test of any receiver is the extent to which it receives signals when it is tuned to frequencies on which they are not really present, so adding to interference problems and misleading the operator. Every known type of receiver suffers from various forms of 'phantom' signals but some are much worse than others. Unfortunately, the mixing process is inherently prone to the generation of unwanted 'products' and receiver design is concerned with minimising their effect rather than their complete elimination.

Spurii may take the form of:

(a) external signals heard on frequencies other than their true frequency;

(b) external signals which cannot be tuned out but are heard regardless of the setting of the tuning knob;

(c) carriers heard within the tuning range of the receiver but stemming not from external signals but from the receiver's own oscillators ('birdies').

Cases (a) and (b) are external spurious responses, while case (c) is an internal spurious.

In any superhet receiver tuneable signals may be created in the set whenever the interfering station or one of its harmonics (often produced within the receiver) differs from the intermediate frequency by a frequency equal to the local oscillator or one of its harmonics. This is reflected in the general expression:

$$mf_u \pm nf_0 = f_i$$

where m, n are any integers, including 0, f_u is the frequency of the unwanted signal, f_0 is the frequency of the local oscillator, and f_i is the intermediate frequency.

An important case occurs when m and n are 1, giving $f_u \pm f_0 = f_i$. This implies that f_u will either be on the frequency to which the set is correctly tuned (f_s) or differs from it by twice the IF (2 x f_i), producing the so-called 'image' frequency. This is either $f_u = f_s + 2f_i$ (for cases where the local oscillator is higher in frequency than the wanted signal), or $f_u = f_s - 2f_i$ (where the local oscillator is lower in frequency than the wanted signal).

For an example, take a receiver tuned to about 14,200kHz with an IF of 470kHz and the oscillator high (ie about 14,670kHz). Such a receiver may, because of 'image', receive a station operating on 14,200 + (2 x 470) = 15,140kHz. Since 15,140kHz is within the 19m broadcast band, there is thus every likelihood that as the set is tuned around 14,200kHz strong broadcast signals will be received.

To reduce such undesirable effects, pre-mixer selectivity must be provided in the form of more RF tuned circuits or RF bandpass filters, or by increasing the Q of such circuits, or alternatively by increasing the frequency difference between the wanted and unwanted image signals. This frequency separation can be increased by increasing the factor $2f_i$, in other words by raising the intermediate frequency, and so allowing the broadly tuned circuits at signal frequency to have more effect in reducing signals on the unwanted image frequency before they reach the mixer.

It should also be noted from the general formula $mf_u \pm nf_0$ that 'image' is only one (though usually the most important) of many possible frequency combinations that can cause unwanted signals to appear at the intermediate frequency, even on a single-conversion receiver. The problem is greatly increased when more than one frequency conversion is employed.

Even with good pre-mixer selectivity it is still possible for a number of strong signals to reach the mixer, drive this or an RF stage into non-linearity and then produce a series of intermodulation products as spurious signals within an amateur band (see Fig 6.21).

The harmonics of the HF oscillator(s) may beat against incoming signals and produce output in the IF pass-band; strong signals may generate harmonics in the receiver stages and these can be received as spurii. Most such forms of spurii can be reduced by increasing pre-mixer selectivity to decrease the number of strong signals reaching the mixer, by increasing the linearity of the early stages of the receiver, or by reducing the amplitude of signals within these stages by the use of an antenna attenuator.

The case of internal spurious responses is similar. Here signals at $mf_1 + nf_2$ equalling the first IF or, more rarely, the tune or image frequency or the second IF can cause problems. Where a third or further IF is used, the oscillators there have also to be added to the equations. Usually, careful screening and filtering and particular attention to earth loops are required to minimise these problems in multi-conversion general-coverage receivers. In receivers for the amateur bands, careful choice of IF is necessary to minimise problems, although not even then can such difficulties be entirely eliminated. A typical 'built in' problem of this type is the use of a 'backwards tuning' 5 - 5.5MHz VFO, producing a spur from its 4th harmonic on 21.2MHz.

Very strong signals on or near the IF may break directly into the IF amplifier and then appear as untuneable interference. This form of interference (although it then becomes tuneable) is particularly serious with the variable IF type of multiple-conversion receiver since there is almost certain to be a number of very strong signals operating over the segment of the HF spectrum chosen to provide the variable IF. Direct breakthrough may occur if the screening within the receiver is insufficient or if sig-

nals can leak in through the early stages due to lack of pre-mixer selectivity. For single-conversion and double-conversion with fixed first IF, it is common practice to include a resonant 'trap' (tuned to the IF) to reject incoming signals on this frequency. The multiple-conversion superhet contains more internal oscillators, and harmonics of the second and third (and occasionally the BFO) can be troublesome. For amateur-bands-only receivers every effort should be made to choose intermediate and oscillator frequencies that avoid as far as possible the effects of oscillator harmonics (birdies).

Ideally one would like an IF rejection (compared with the wanted signal) of almost 120dB but most amateurs would be well satisfied with 80-100dB of protection; in practice many receivers with variable first IF do not provide more than about 40-60dB protection.

Because of the great difficulty in eliminating spurious responses in double- and triple-conversion receivers, the modern designer tends to think more in terms of single conversion with high IF (eg 9MHz). Potentially the direct-conversion receiver is even more attractive, though it needs to be fairly complex to eliminate the audio-image response. It must also have sufficient linearity or pre-detector selectivity to reduce any envelope detection of very strong signals which may otherwise break through into the audio channel, regardless of the setting of the heterodyne oscillator. The direct-conversion receiver can also suffer from spurii resulting from harmonics of the signal or oscillator and this needs to be reduced by RF selectivity. A difficulty that can arise is that close in phase noise on the local oscillator being transferred by leakage to the mixer input, and then treated as an input signal, being demodulated, resulting in the noise appearing at AF.

Choice of IF

Choice of the intermediate frequency or frequencies is a most important consideration in the design of any superhet receiver. The lower the frequency, the easier it is to obtain high gain and good selectivity and also to avoid unwanted leakage of signals round the selective filter. On the other hand, the higher the IF, the greater will be the frequency difference between the wanted signal and the 'image' response, so making it simpler to obtain good protection against image reception of unwanted signals and also reducing the 'pulling' of the local oscillator frequency. These considerations are basically opposed, and the IF of a single-conversion receiver is thus a matter of compromise; however, in recent years it has become easier to obtain good selectivity with higher-frequency band-pass crystal filters and it is no longer any problem to obtain high gain at high frequencies. The very early superhet receivers used an IF of about 100kHz; then for many years 455-470kHz was the usual choice - many modern designs use between about 3 and 9MHz, and SSB IF filters are now available to 40MHz. A good rule of thumb is that the IF should be no less than 5%, and preferably 10% of the signal frequency. Where the IF is higher than the signal frequency the action of the mixer is to raise the frequency of the incoming signal, and this process is now often termed up-conversion (a term formerly reserved for a special form of parametric mixer). Up-conversion, in conjunction with a low-pass filter at the input, is an effective means of reducing IF breakthrough as well as image response.

A superhet receiver, whether single- or multi-conversion, must have its first IF outside its tuning range. For general-coverage HF receivers tuning between, say, 1.5 and 30MHz, this limits the choice to below 1.5MHz or above 30MHz. To reduce image response without having to increase pre-mixer RF selectivity (which can involve costly gang-tuned circuits) professional

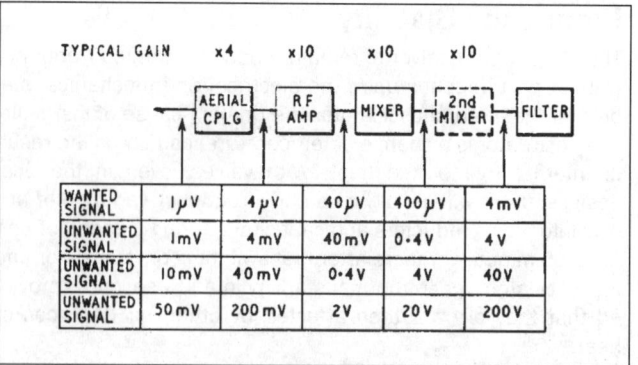

Fig 6.26: This diagram shows how unwanted signals are built up in high-gain front-ends to levels at which cross-modulation, blocking and intermodulation are virtually bound to occur

designers are increasingly using a first IF well above 30MHz. This trend is being encouraged by the availability of VHF crystal filters suitable for use either directly as an SSB filter or more often with relaxed specification as a roofing filter.

The use of a very high first IF, however, tends to make the design of the local oscillator more critical (unless the Wadley triple-conversion drift-free technique is used - when the intermodulation problems are more difficult because of the number of stages prior to the selectivity determining filters). For amateur-bands-only receivers the range of choice for the first IF is much wider, and 3.395MHz and 9MHz are typical.

A number of receivers have adopted 9MHz IF with 5.0-5.5MHz local oscillator: this enables 4MHz-3.5MHz and 14.0-14.5MHz to be received without any band switching in the VFO; other bands are received using crystal-controlled converters with outputs at 3.5 or 14MHz. With frequency synthesisers, up-conversion is commonly found, with the first IF between 40-70MHz.

Gain Distribution

In many receivers of conventional design, it has been the practice to distribute the gain throughout the receiver in such a manner as to optimise signal-to-noise ratio and to minimise spurious responses. So long as relatively noisy mixer stages were used it was essential to amplify the signal considerably before it reached the mixer. This means that any strong unwanted signals, even when many kilohertz from the wanted signal, pass through the early unselective amplifiers and are built up to levels where they cause cross-modulation within the mixer: see **Fig 6.26**. Today it is recognised that it is more satisfactory if pre-mixer gain can be kept low to prevent this happening. Older multi-grid valve mixers had an equivalent noise resistance as high as $200,000\Omega$ (representing some 4-5mV of noise referred to the grid). Later types such as the ECH81 and 6BA7 reduced this to an ENR of about $60,000\Omega$ (about 2.25mV of noise) while the ENR of pentode and triode mixers was lower still (although these may not be as satisfactory for mixers in other respects).

The noise contribution of semiconductor mixers is also low - for example an FET mixer may have a noise factor as low as 3dB - so that generally the designer need no longer worry unduly about the requirement for pre-mixer amplification to overcome noise problems. Nevertheless a signal frequency stage may still be useful in helping to overcome image reception by providing a convenient and efficient method of coupling together signal-frequency tuned circuits, and when correctly controlled by AGC it becomes an automatic large-signal attenuator. Pre-mixer selectivity limits the number of strong signals reaching the mixer. The use of double-balanced mixers is generally advisable to reduce the spurious response possibilities.

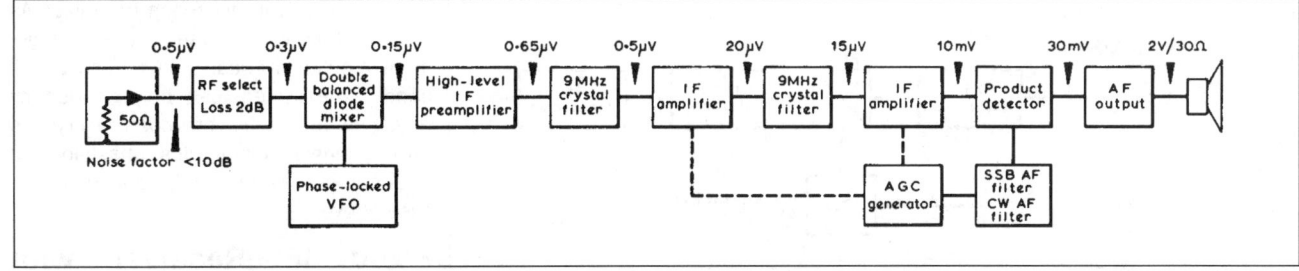

Fig 6.27: Gain distribution in a high-performance semiconductor single-conversion receiver built by G3URX using seven integrated circuits, 27 transistors and 14 diodes. 1µV signals can be received 5kHz off-tune from a 60mV signal and the limiting factor for weak signals is the noise sidebands of the local oscillator, although the phase-locked VFO gives lower noise and spurious responses than the more usual pre-mixer VFO system

Fig 6.27 shows a typical gain distribution as found in a modern design in which the signal applied to the first mixer is much lower than was the case with older (valved) receivers.

The significant conversion losses of modern diode and FET ring mixers (6-10dB), the losses of input band-pass filtering and the need for correct impedance termination means that in the highest-performance receivers it is desirable to include low-gain, low-noise, high dynamic range pre- and post-mixer amplifiers and a diplexer ahead of the (main or roofing) crystal filter to achieve constant input impedance over a broad band of frequencies.

If the diplexer is a simple resistive network this will introduce a further loss of some 6dB. Stage-by-stage gain, noise figure, third-order intercept, 1dB compression point and 1dB desensitisation point performance of the N6NWP high-dynamic-range MF/HF front-end of a single-conversion (9MHz IF) receiver [4](QST February 1993, and see later) is shown in **Fig 6.28**. More complex diplexers may be used to divert the local oscillator feedthrough signal from entering the IF strip.

For double- and multiple-conversion receivers, the gain distribution and performance characteristics of all the stages preceding the selective filter need to be considered. In general the power loss in any band-pass filter will decrease as the bandwidth increases: for example a 75MHz, 25kHz-bandwidth filter might have a power loss of 1dB whereas a 75MHz, 7kHz-bandwidth filter might have a power loss of 3.5dB.

FREQUENCY STABILITY OF RECEIVERS

The resolution of SSB speech and the reception of a CW signal requires that a receiver can be tuned to, and remain within, about 25-30Hz of the frequency of the incoming signal. At 29MHz this represents a tolerance of only about one part in a million. For amateur operation the main requirement is that this degree of stability should be maintained over periods of up to about 30min. Long-term stability is less important for amateurs than short-term stability; it will also be most convenient if a receiver reaches this degree of stability within a fairly short time of switching on.

It is extremely difficult to achieve or even approach this order of stability with a free-running, band-switched variably tuned oscillator working on the fundamental injection frequency, although with care a well-designed FET oscillator can come fairly close. This has led (in the same way as for transmitting VFO units) to various frequency-synthesis techniques in which the stability of a free-running oscillator is enhanced by the use of crystals. The following are among the techniques used:

Multi-conversion Receiver with Crystal-controlled First Oscillator and Variable First IF

This very popular technique has a tuneable receiver section covering only one fixed frequency band; for example 5.0-5.5MHz. The oscillator can be carefully designed and temperature compensated over one band without the problems arising from the uncertain action of wavechange switches, and can be separated from the mixer stage by means of an isolating or buffer stage. For the front-end section a separate crystal is needed for each tuning range (the 28MHz band may require four or more crystals to provide full coverage of 28.0-29.7MHz).

Partial Synthesis

The arrangement of (1) becomes increasingly costly to implement as the frequency coverage of the variable IF section is reduced below about 500kHz, or is required to provide general coverage throughout the HF band. Beyond a certain number of crystals it becomes more economical (and offers potentially higher stability) if the separate crystals are replaced by a single

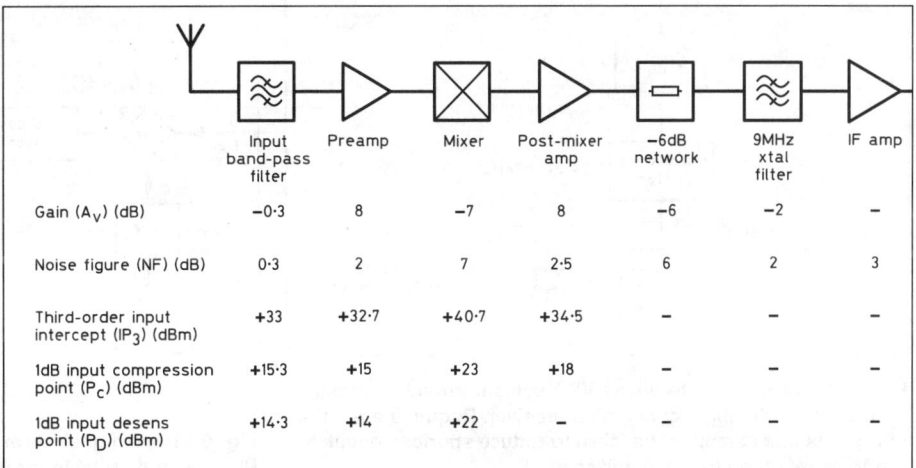

Fig 6.28: Stage by stage gain and characteristsics of the N6NWP front-end of single-conversion 14MHz HF receiver with 9MHz IF

	Input band-pass filter	Preamp	Mixer	Post-mixer amp	−6dB network	9MHz xtal filter	IF amp
Gain (A_V) (dB)	−0·3	8	−7	8	−6	−2	−
Noise figure (NF) (dB)	0·3	2	7	2·5	6	2	3
Third-order input intercept (IP_3) (dBm)	+33	+32·7	+40·7	+34·5	−	−	−
1dB input compression point (P_C) (dBm)	+15·3	+15	+23	+18	−	−	−
1dB input desens point (P_D) (dBm)	+14·3	+14	+22	−	−	−	−

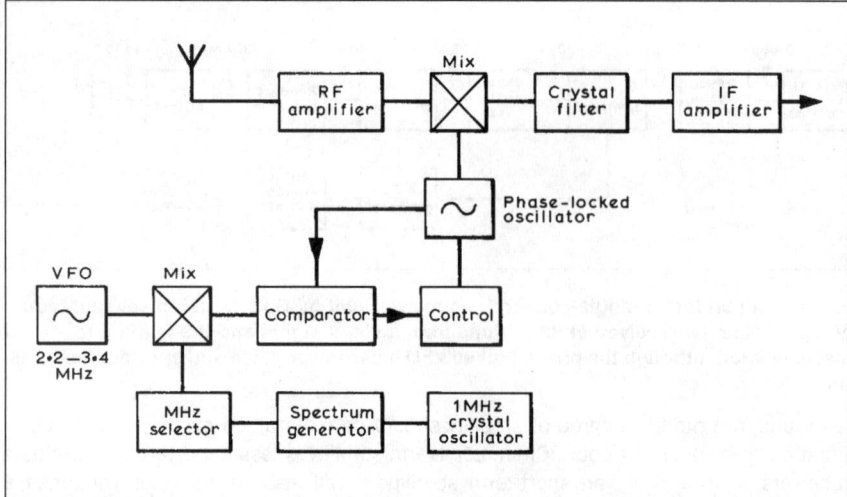

Fig 6.29: Practical frequency synthesis using fixed-range VFO with Megahertz signals derived from a single 1MHz crystal

high-stability crystal (eg 1MHz) from which the various band-setting frequencies are derived (**Fig 6.29**). This may be done, for example, by digital techniques or by providing a spectrum of harmonics to one of which a free-running oscillator is phase-locked. Note that with this system the tuning within any band still depends on the VFO and for this reason is termed partial synthesis.

Single-conversion Receiver with Heterodyne (Pre-mixer) VFO

In this system of partial synthesis, the receiver may be a single-conversion superhet (or dual-conversion with fixed IFs) (**Fig 6.30**). The variable HF injection frequency is obtained using a heterodyne-type VFO, in which the output of a crystal-controlled oscillator is mixed with that of a single-range VFO, and the output is then filtered and used as the injection frequency. The overall stability will be much the same as for (1) but the system allows the selective filter to be placed immediately after the first mixer stage. However, to reduce spurious responses the unwanted mixer products of the heterodyne-VFO must be reduced to a

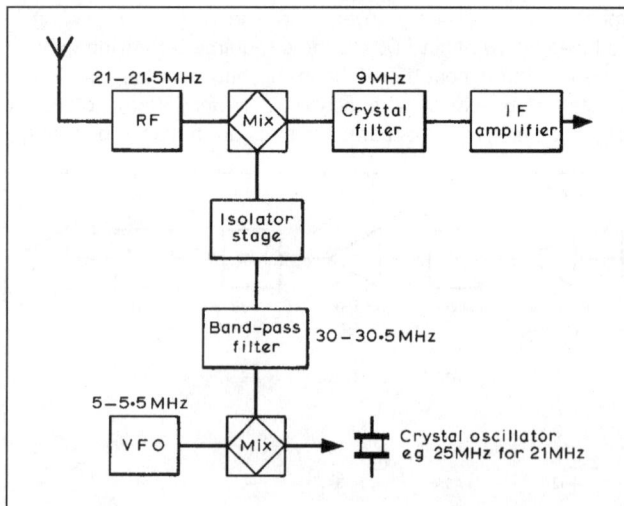

Fig 6.30: Pre-mixer heterodyne VFO system provides constant tuning rate with single-conversion receiver. Requires a number of crystals and care must be taken to reduce spurious oscillator products reaching the main mixer

very low level and not reach the mixer. As with the tuneable IF system, this arrangement results in equal tuning rates on all bands. The system can be extended by replacing the series of separate crystals with a single crystal plus phase-locking arrangement as in the Partial Synthesis case above.

Fixed IF Receiver with Partial Frequency Synthesis

Fig 6.31 outlines an ingenious frequency synthesiser (due to Plessey) incorporating an interpolating LC oscillator and suitable for use with single- or multiple-conversion receivers having fixed intermediate frequencies. The output of the VFO is passed through a variable-ratio divider and then added to the reference frequency by means of a mixer. The sum of the two frequencies applied to the mixer is selected by means of a band-pass filter and provides one input to a phase comparator; the other input to this phase comparator is obtained from the output of the voltage-controlled oscillator after it has also been divided in the same ratio as the interpolating frequency. The phase comparator can then be used to phase lock the VCO to the frequency $mf_{ref} + f_{VFO}$ where m represents the variable-ratio division. If for example f_{ref} is 1MHz, f_{VFO} covers a tuning range of 1-2MHz and the variable ratio dividers are set to 16, then the VCO output can be controlled over 17,000-18,000kHz; if the ratio divider is changed to 20 then the tuning range becomes 21,000-22,000kHz and so on. The use of two relatively simple variable ratio divider chains thus makes it possible to provide output over the full HF range, with the VFO at a low frequency (eg 1-2MHz).

VXO Local Oscillator

For reception over only small segments of a band or bands, a variable-crystal oscillator (VXO) can be used to provide high stability for mobile or portable receivers. As explained in the chapter on oscillators, the frequency of a crystal can be 'pulled' over a small percentage of its nominal frequency without significant loss of stability. The system is attractive for small transceivers. Oscillators based on ceramic resonators can be 'pulled' over significantly greater frequency ranges.

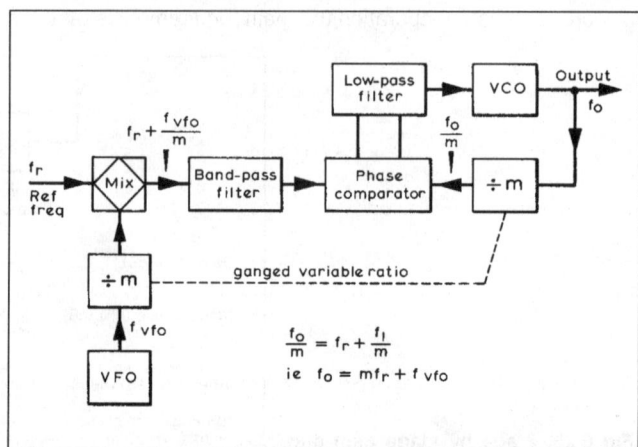

Fig 6.31: A digital form of partial synthesis developed by Plessey and suitable for use in single-conversion receivers

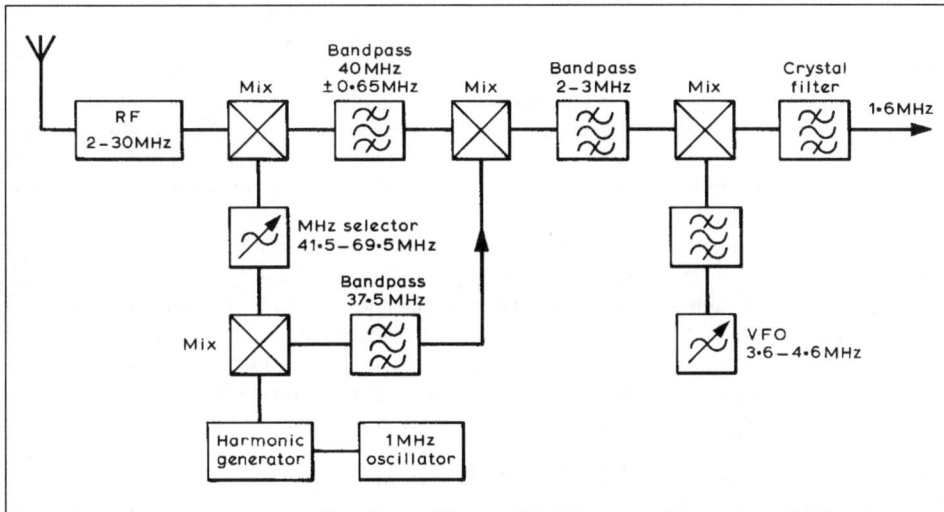

Fig 6.32. The Wadley drift-cancelling loop system as used on some early Racal HF receivers but requiring a considerable number of mixing processes and effective VHF band-pass filters

Drift-cancelling Wadley Loop

A stable form of front-end for use with variable IF-type receivers is the multiple-conversion Wadley loop which was pioneered in the Racal RA17 receiver (**Fig 6.32**). By means of an ingenious triple-mixing arrangement a variable oscillator tuning 40.5 to 69.5MHz and a 1MHz crystal oscillator provides continuous tuning over the range 0.5 to 30MHz as a series of 1MHz segments. Any drift of the variable VHF oscillator is automatically corrected. Although the system has been used successfully in home-constructed receivers, it is essential to use a good VHF band-pass filter (eg 40MHz ± 0.65MHz) and extremely good screening if spurious responses are to be minimised.

PLL frequency synthesisers and Direct Digital Synthesis (DDS).

These systems produce an output waveform which is locked precisely to a crystal reference frequency. The frequency stability is therefore equal to that of the crystal oscillator. Both these techniques are described in detail in the chapter on oscillators.

ACTIVE DEVICES FOR RECEIVERS

The radio amateur is today faced with a wide and sometimes puzzling choice of active devices around which to design a receiver: valves, bipolar transistors, field-effect transistors including single- and dual-gate MOSFETs and junction FETs; special diodes such as Schottky (hot-carrier) diodes, and an increasing number of integrated circuits, many designed specifically for receiver applications.

Each possesses advantages and disadvantages when applied to high-performance receivers, and most recent designs tend to draw freely from among these different devices.

Valves

In general the valve is bulky, requires additional wiring and power supplies for heaters, generates heat, and is subject to ageing in the form of a gradual change of characteristics throughout its useful life. On the other hand it is not easily damaged by high-voltage transients; was manufactured in a wide variety of types for specific purposes to fairly close tolerances; and is capable of handling small signals with good linearity (and special types can cope well with large signals). However, the future lies with the semiconductor.

Bipolar Transistors

These devices can provide very good noise performance with high gain, are simple to wire and need only low-voltage supplies, consuming very little power and so generating very little heat (except power types needed to form the audio output stage). On the other hand, they are low-impedance, current-operated devices, making the interstage matching more critical and tending to impose increased loading on the tuned circuits; they have feedback capacitances that may require neutralising; they are sensitive to heat, changing characteristics with changing temperature; and they can be damaged by large input voltages or transients. Their main drawback in the signal path of a receiver is the difficulty of achieving wide dynamic range and satisfactory AGC characteristics. On the other hand, bipolar transistors are suitable for most AF applications. The bipolar transistors developed for CATV (wire distribution of TV) such as the BFW17, BFW17A, 2N5109 etc, used with a heatsink, can form excellent RF stages or mixers, especially where feedback is used to enhance linearity. Not all transistors are equal in linearity, and those developed for CATV show advantages.

Field-effect Transistors

These devices can offer significant advantages over bipolar devices for the low-level signal path. Their high input impedance makes accurate matching less important; their near square-law characteristics make them comparable with variable-mu valves in reducing susceptibility to cross-modulation; at the same time, second-order intermodulation and responses to signals at 'half the IF' removed necessitate careful consideration of front-end filtering.

They can readily be controlled by AGC systems, although attempts to obtain too much gain reduction can lead to severe distortion. The dual-gate form of device is particularly useful for small-signal applications, and forms an important device for modern receivers. They tend, however, to be limited in signal-handling capabilities. Special types of high-current field-effect transistors have been developed capable of providing extremely wide dynamic range (up to 140dB) in the front-ends of receivers. A problem with FET devices is the wide spread of characteristics between different devices bearing the same type number, and this may make individual adjustment of the bias levels of FET stages desirable. The good signal-handling capabilities of MOSFET mixers can be lost by incorrect signal or local oscillator levels.

Integrated Circuits

Special-purpose integrated circuits use large numbers of bipolar transistors in configurations designed to overcome many of the problems of circuits based on discrete devices. Because of the extremely high gain that can be achieved within a single IC they also offer the home-constructor simplification of design and construction. High-performance receivers can be designed around a few special-purpose linear ICs, or one or two consumer-type ICs may alternatively form the 'heart' of a useful communications receiver.

It should be recognised, however, that their RF signal-handling capabilities are less than can be achieved with special-purpose discrete devices and their temperature sensitivity and heat generation (due to the large number of active devices in close proximity) usually make them unsuitable for oscillator applications. They also have a spread of characteristics which may make it desirable to select devices from a batch for critical applications. The electronics industry has tended towards digital techniques where possible, and generally to application specific integrated circuits (ASICs) for both analogue and digital applications. In general, such devices as operational amplifiers and some audio amplifiers will continue to be available, but many of the small-scale integration (SSI) devices that were the mainstay of the amateur throughout the 1970s and 1980s have now disappeared. Of the circuits manufactured for radio applications, many are aimed at markets such as cellular radio, and have become so dedicated that their use (and availability in small quantities) for amateur radio is doubtful.

Integrated-circuit precautions

As with all semiconductor devices it is necessary to take precautions with integrated circuits, although if handled correctly high reliability may be expected.

Recommended precautions include:

- Do not use excessive soldering heat and ensure that the tip is at earth potential
- Take precautions against static discharge - eg don't walk across a synthetic fibre carpet on a cold day and then touch a device. It is advisable if handling CMOS devices to use a static bleed wriststrap, connected to earth.
- Check and recheck all connections several times before applying any voltages.
- Keep integrated circuits away from strong RF fields.
- Keep supply voltages within ±10% of those specified for the device from well-smoothed supplies.

Integrated circuit amplifiers can provide very high gains (eg up to 80dB or so) within a single device having input and output leads separated by only a small distance: this means that careful layout is needed to avoid instability, and some devices may require the use of a shield between input and output circuits. Some devices have earth leads arranged so that a shield can be connected across the underside of the device.

Earth returns are important in high-gain devices: some have input and output earth returns brought out separately in order to minimise unwanted coupling due to common earth return impedances, but this is not true of all devices.

Normally IC amplifiers are not intended to require neutralisation to achieve stability; unwanted oscillation can usually be traced to unsatisfactory layout or circuit arrangements. VHF parasitics may generally be eliminated by fitting a 10 ohm resistor in series with either the input or output lead, close to the IC.

With high-gain amplifiers, particular importance attaches to the decoupling of the voltage feeds. At the low voltages involved, values of series decoupling resistors must generally be kept low so that the inclusion of low-impedance bypass capacitors is usually essential. Since high-Q RF chokes may be a cause of RF oscillation it may be advisable to thread ferrite beads over one lead of any RF choke to reduce the Q.

As with bipolar transistors, IC devices (if based on bipolar transistors) have relatively low input and output impedances so that correct matching is necessary between stages. The use of a FET source follower stage may be a useful alternative to step-down transformers for matching.

Maximum and minimum operating temperatures should be observed. Many linear devices are available at significantly lower cost in limited temperature ranges which are usually more than adequate for operation under normal domestic conditions. Because of the relatively high temperature sensitivity and noise (especially LF noise) of bipolar-type integrated circuits, they are not generally suitable as free-running oscillators in high-performance receivers.

For the very highest grade receivers, discrete components and devices are still required in the front-end since currently available integrated circuits do not have comparable dynamic range. The IC makes possible extremely compact receivers; in practice miniaturisation is now limited - at least for general-purpose receivers - by the need to provide easy-to-use controls for the non-miniaturised operator.

SELECTIVE FILTERS

The selectivity characteristics of any receiver are determined by filters: these filters may be at signal frequency (as in a straight receiver); intermediate frequency; or audio frequency (as in a direct-conversion receiver). Filters at signal frequency or IF are usually of band-pass characteristics; those at AF may be either band-pass or low-pass. With a very high first IF (112-150MHz) low-pass filters may be used at RF.

A number of different types of filters are in common use: LC (inductor-capacitor) filters as in a conventional IF transformer or tuned circuit; crystal filters; mechanical filters; ceramic filters; RC (resistor-capacitor) active filters (usually only at AF but feasible also at IF).

Roofing Filters

If the main selective filter is placed early in the signal path (eg immediately following a diode mixer or low-gain, low-noise, post-mixer amplifier) where the signal voltage is very low, subsequent amplification (of the order of 100dB or more) will introduce considerable broad-band noise unless further narrow-band filtering is provided (which may be AF filtering). Another answer is to use an initial roofing filter with the final, more-selective filter(s) further down the signal path.

Crystal Filters

The selectivity of a tuned circuit is governed by its frequency and by its Q (ratio of reactance to resistance). There are practical limits to the Q obtainable in coils and IF transformers. In 1929, Dr J Robinson, a British scientist, introduced the quartz crystal resonator into radio receivers. The advantages of such a device for communications receivers were appreciated by James Lamb of the American Radio Relay League and he made popular the IF crystal filter for amateur operators.

For this application a quartz crystal may be considered as a resonant circuit with a Q of from 10,000 to 100,000 compared with about 300 for a very-high-grade coil and capacitor tuned circuit. From earlier chapters, it will be noted that the electrical equivalent of a crystal is not a simple series- or parallel-tuned circuit, but a combination of the two: it has (a) a fixed series resonant frequency (f_s) and (b) a parallel resonant frequency (f_p). The frequency fp is determined partly by the capacitance of the crystal holder and by any added parallel capacitance and can be varied over a small range.

The crystal offers low impedance to signals at its series resonant frequency; a very high impedance to signals at its parallel resonant frequency, and a moderately high impedance to signals on other frequencies, tending to decrease as the frequency increases due to the parallel capacitance.

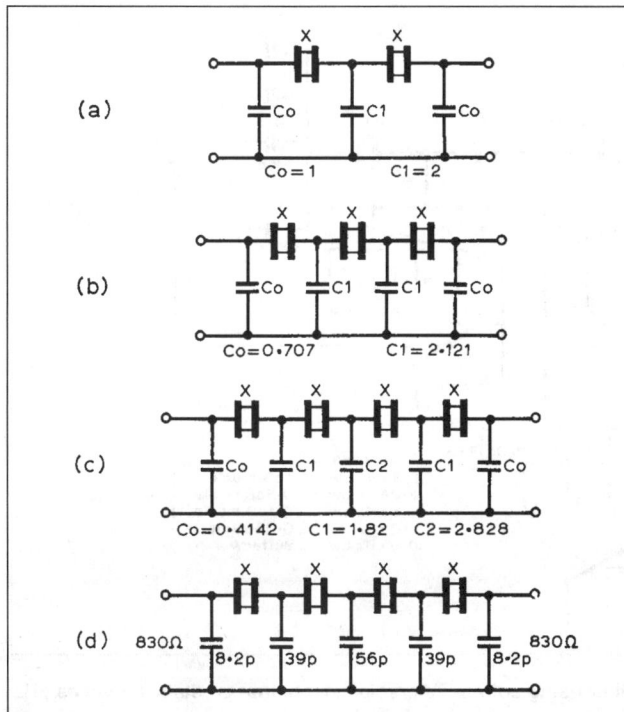

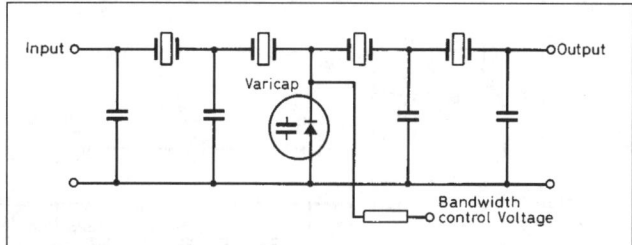

Fig 6.34: Simple technique for varying the bandwidth of an 8MHz crystal ladder filter from about 12.8kHz down to 1.1kHz

Fig 6.33: Crystal ladder filters, investigated by F6BQP, can provide effective SSB and CW band-pass filters. All crystals (X) are of the same resonant frequency and preferably between 8 and 10MHz for SSB units. To calculate values for the capacitors multiply the coefficients given above by $1/(2\pi fR)$ where f is frequency of crystal in hertz (MHz by 10^6), R is input and output termination impedance and 2π is roughly 6.8. (a) Two-crystal unit with relatively poor shape factor. (b) Three-crystal filter can give good results. (c) Four-crystal unit capable of excellent results. (d) Practical realisation of four-crystal unit using 8314kHz crystals, 10% preferred-value capacitors and termination impedance of 820Ω. Note that for crystals between 8 and 10MHz the termination impedance should be between about 800 and 1000Ω for SSB. At lower crystal frequencies use higher design impedances to obtain sufficient bandwidth. For CW filters use lower impedance and/or lower frequency crystals

Ladder crystal filters

Single or multiple crystal half-lattice filters were used in the valve era. The modern filter of choice is the ladder filter, which can provide excellent SSB filters at frequencies between about 4-11MHz. This form of crystal filter uses a number of crystals of the same (or nearly the same) frequency and so avoids the need for accurate crystal etching or selection. Further, provided it is correctly terminated, it does not require the use of transformers or inductors. Plated crystals such as the HC6U or 10XJ types are more likely to form good SSB filters, although virtually any type of crystal may be used for CW filters.

A number of practical design approaches have been described by J Pochet, F6BQP, [5] and in a series of articles [6]. Fig 6.33 outlines the F6BQP approach. By designing for lower

termination impedances and/or lower frequency crystals excellent CW filters can be formed. A feature of the ladder design is the very high ultimate out-of-band rejection that can be achieved (75-95dB) in three- or four-section filters. For SSB filters at about 8MHz a suitable design impedance would be about 800Ω with a typical 'nose' band-pass of 2.0-2.1kHz.

Intercept points of SSB and roofing filters can be over +50dBm with low insertion losses, and over +45dBm for narrowband CW filters.

The ladder configuration is particularly attractive for homebuilt receivers since they can be based on readily available, low-cost 4.43MHz PAL colour-subcarrier crystals produced for use in domestic colour-TV receivers to P129 or P128 specifications. In NTSC countries, including North America and Japan, 3.58MHz TV crystals can be used, although this places the IF within an amateur band and would be unsuitable for single-conversion superhet receivers. Low-cost crystals at twice the PAL sub-carrier frequency (ie 8.86MHz) are also suitable.

Virtually any combination of crystals and capacitors produces a filter of some sort, but published equations or guidance should be followed to achieve optimum filter shapes and desired bandwidth. If this is done, ladder filters can readily be assembled from a handful of nominally identical crystals (ideally selected with some small offsets of up to about 50 or 100Hz) plus a few capacitors, yet providing SSB or CW filters with good ultimate rejection, plus reasonably low insertion loss and passband ripple. Ladder filters have intercept points significantly above those of most economy-grade, lattice-type filters.

Figs 6.34 and 6.35 show typical ladder filter designs based on 4.43MHz crystals.

A valuable feature of the ladder configuration is that it lends itself to variable selectivity by changing the value of some, or preferably all, of the capacitors. The bandwidth can be varied over a restricted but useful range simply by making the middle capacitor variable, using a mechanically variable capacitor or electronic tuning diode (which may take the form of a 1W zener diode). Fig 6.34 shows an 8MHz filter which has a 'nose' bandwidth that can be varied from about 2.8kHz down to 1.1kHz.

Fig 6.35 illustrates a 4.43MHz nine-crystal filter built by R Howgego, G4DTC, for AM/SSB/CW/RTTY reception; the 3dB points can be varied from 4.35kHz down to 600Hz. In development, he noted that the bandwidth is determined entirely by the 'vertical' capacitors. If, however, these are reduced below about

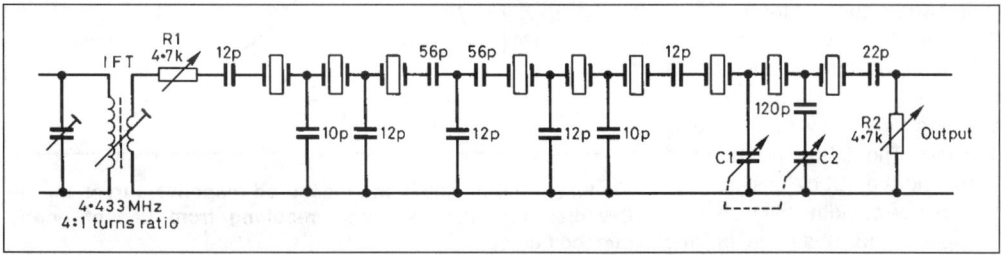

Fig 6.35: Variable-selectivity ladder filter using low-cost PAL colour-TV crystals for AM/SSB/CW/RTTY reception

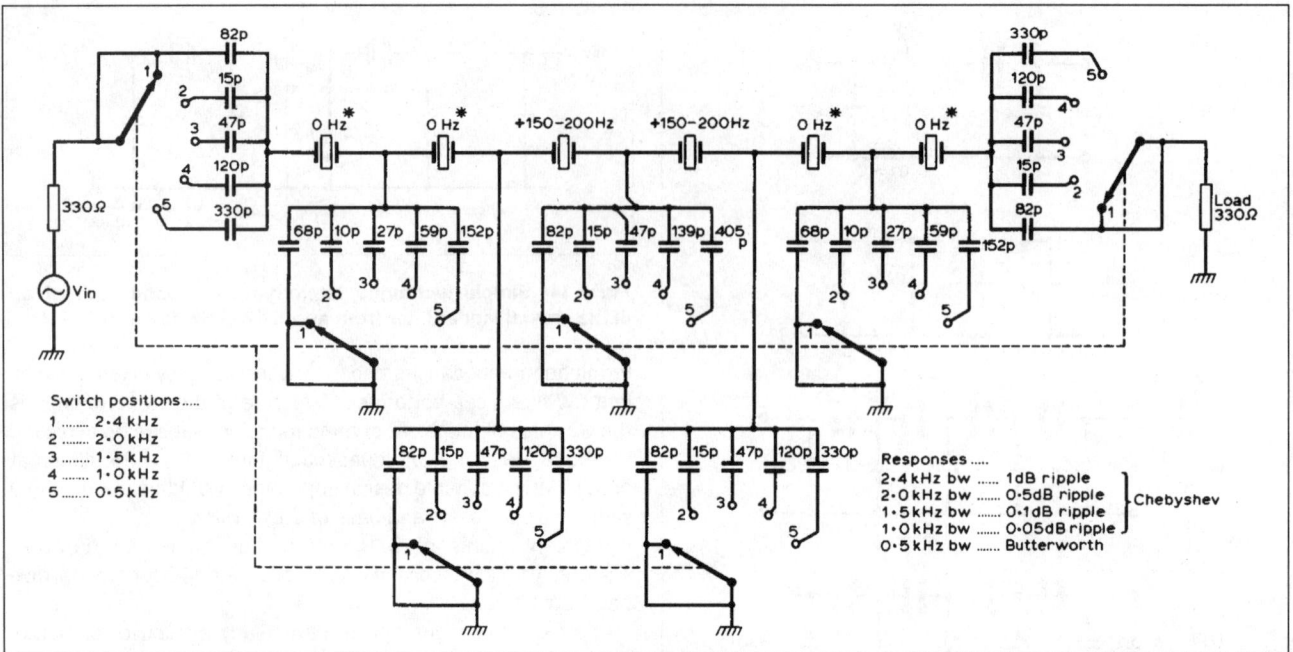

Fig 6.36: G3UUR's design for a switched variable-bandwidth ladder filter using colour-TV crystals. Note that crystals shown as 0Hz offset can be in practice ±0.5Hz without having too detrimental an effect on the pass-band ripple

10pF, the bandwidth begins to narrow rather than widen. The maximum bandwidth that could be achieved was about 4.5kHz. This could be widened by placing resistors (1kΩ to 10kΩ) across the capacitors but this increases insertion loss. Terminating impedances affect the pass-band ripple, not the bandwidth.

The filter of Fig 6.35 gives continuously variable selectivity yet is relatively easy to construct. It is basically a six-pole roofing filter followed by a variable three-pole filter. It was based on low-cost Philips HC18-U type crystals and these were found to be all within a range of 80Hz. Crystals in the large case style (eg HC6-U) tend to be about 200Hz lower. C1, C2 is a 60 + 142pF miniature tuning capacitor as found in many portable broadcast receivers. The integral trimmers are set for maximum bandwidth when capacitor plates are fully unmeshed. Set R1 for best compromise between minimum bandwidth and insertion loss, 1.2kΩ nominal.

The following specification should be achievable. C1, C2 plates unmeshed: 3dB points at 4437.25kHz and 4432.90kHz, bandwidth 4.35kHz. C1, C2 plates half-meshed: 3dB points at 4434.0kHz and 4432.90kHz, bandwidth 1.10kHz. C1, C2 plates meshed: 3dB points at 4433.5kHz and 4432.90kHz, bandwidth 600Hz. Insertion loss in pass-band: maximum (R1 2.5kΩ, R2 1.2kΩ) 10dB, minimum (R1 0kΩ , R2 1.2kΩ) 6dB. Pass-band ripple 1-3dB (dependent on R1). Ripple reduces with bandwidth. Stop-band attenuation better than 60dB. -20dB bandwidth typically 1kHz wider than the -3dB bandwidth.

The filter shown in **Fig 6.36**, designed by D Gordon-Smith, G3UUR, provides six different bandwidths suitable for both SSB and CW operation, switching the value of all capacitors, and ideally preceded by a roofing filter. In order to reduce the number of switched components, the terminating resistors remain constant and the ripple merely decreases with bandwidth. This also reduces the variation in insertion loss. The 2.4kHz position has a 1dB ripple Chebyshev response, and the 500Hz position represents a Butterworth response. A 5:1 bandwidth change is possible if the maximum tolerable ripple is 1dB. This range is fixed

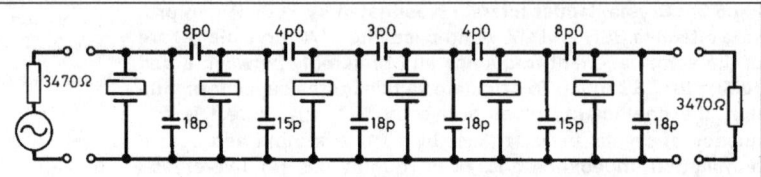

Fig 6.37: Shunt-type crystal ladder CW filter using six 3.58MHz NTSC crystals designed for 3470Ω terminations. Crystals should usually be matched to within 100Hz. (TV crystals are often only specified as within 300Hz but are usually within 200Hz.)

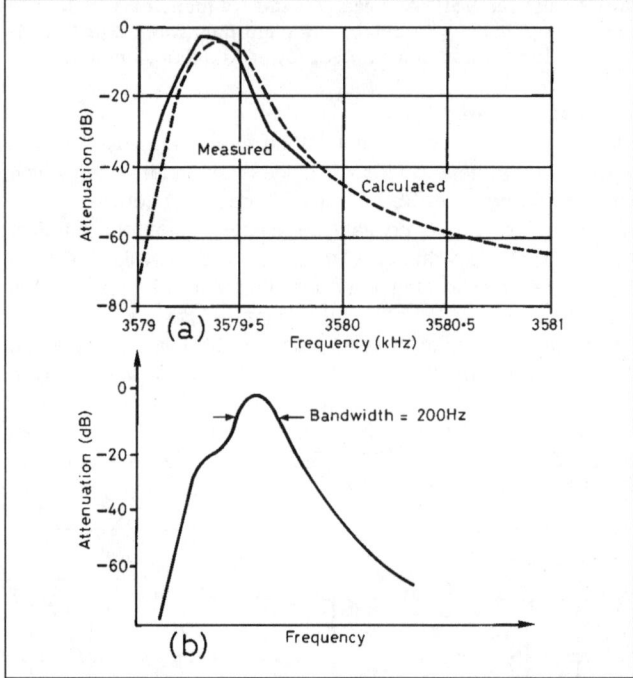

Fig 6.38: (a) Calculated and measured response curves for the CW filter. (b) Filter response resulting from use of poorly matched crystals

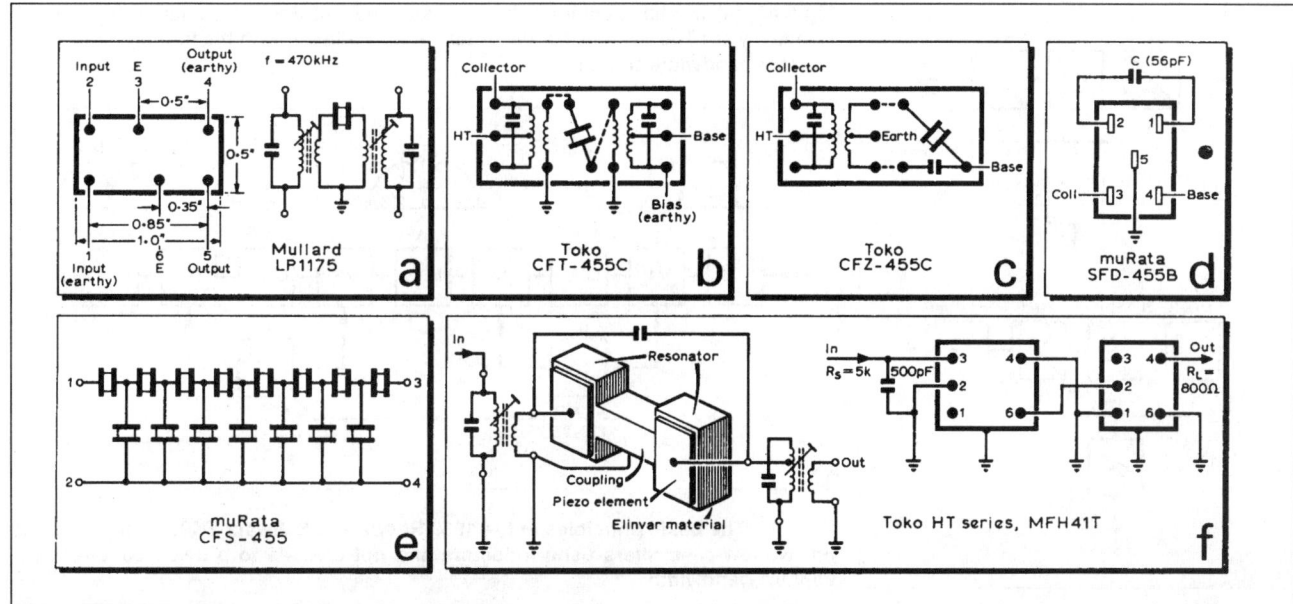

Fig 6.39: Representative types of ceramic filters

by design constraints: the ratio of 1dB to 0dB ripple response terminating resistance is approximately 5:1 for the same bandwidth, and therefore the same terminating resistance satisfies bandwidths that have a ratio of about 5:1. The main disadvantage is that the pass-band moves low in frequency as it is narrowed. This could be compensated for by moving the carrier crystals down in sympathy with the filter centre frequency; a total shift of 1kHz or less would probably be adequate, and could be corrected at the VFO by the RIT shift control.

A different approach to ladder filters is to put the crystals in shunt with the signal rather than in series. The filter then has its steeper slope on the low-frequency side instead of the high-frequency side, and this may be preferable for narrow-bandwidth filters (eg CW filters), particularly when using relatively low-Q plated crystals in HC-18 holders. **Fig 6.37** shows a shunt-type crystal filter designed by John Pivnichy, N2DCH. The filter response is shown in **Fig 6.38**.

Ceramic Filters

Piezoelectric effects are not confined to quartz crystals; in recent years increasing use has been made of certain ceramics, such as lead zirconate titanate (PZT). Small discs of PZT, which resonate in the radial dimension, can form economical selective filters in much the same way as quartz, though with considerably lower Q.

Ceramic IF transfilters are a convenient means of providing the low impedances needed for bipolar transistor circuits. The simplest ceramic filters use just one resonator but numbers of resonators can be coupled together to form filters of required bandwidth and shape factor. While quite good nose selectivity is achieved with simple ceramic filters, multiple resonators are required if good shape factors are to be achieved. Some filters are of 'hybrid' form using combinations of inductors and ceramic resonators.

Examples of ceramic filters include the Philips LP1175 in which a hybrid unit provides the degree of selectivity associated with much larger conventional IF transformers; a somewhat similar arrangement is used in the smaller Toko filters such as the CFT455C which has a bandwidth (to -6dB) of 6kHz. A more complex 15-element filter is the Murata CFS-455A with a bandwidth of 3kHz at -6dB, 7.5kHz at -70dB and insertion loss 9dB, with

input and output impedances of 2kΩ and centre frequency of 455kHz. In general ceramic filters are available from 50kHz to about 10.7MHz centre frequencies. **Fig 6.39** provides more details.

Ceramic filters tend to be more economical than crystal or mechanical filters but have lower temperature stability and may have greater pass-band attenuation.

Mechanical and Miscellaneous Filters

Very effective SSB and CW filters at intermediate frequencies from about 60 to 600kHz depended on the mechanical resonances of a series of small elements usually in the form of discs. The mechanical filter consisted of three basic elements: two magneto-striction transducers which convert the IF signals into mechanical vibrations and vice versa; a series of metal discs mechanically resonated to the required frequency; and disc coupling rods. Each disc represents a high-Q series resonant circuit and the bandwidth of the filter is determined by the coupling rods. 6-60dB shape factors can be as low as about 1.2, with low pass-band attenuation. The limitation of mechanical filters to frequencies of about 500kHz or below has led to their virtual disappearance from the amateur markets and they are now found only in older models despite their excellent performance below 500kHz.

Other forms of mechanical filters have been developed which include ceramic piezoelectric transducers with mechanical coupling: they thus represent a combination of ceramic and mechanical techniques. These filters may consist of an H-shaped form of construction; such filters include a range manufactured by the Toko company of Japan. Generally the performance of such filters is below that of the disc resonator type, but can still be useful.

Surface acoustic wave (SAW) filters are available for possible band-pass filter applications where discrete-element filters have previously been used, including IF filters. Filters in the 80 to 150MHz region are manufactured for cellular telephone applications, and those designed for the American AMPS and IS136 standards may be narrow enough (<30kHz) to be useful as roofing filters. It should be noted that the mechanism of signal propagation in the SAW filter is such that they are extremely resistant to intermodulation difficulties.

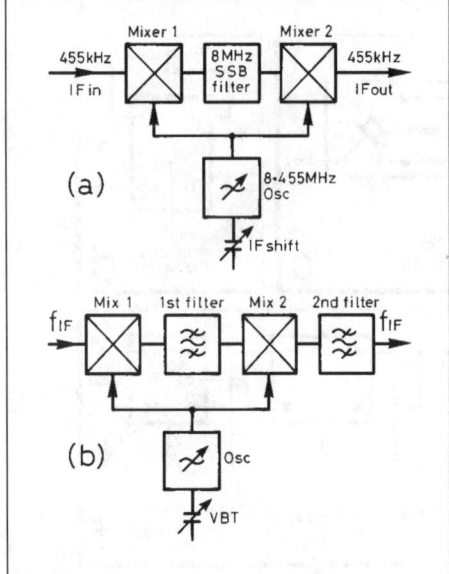

(a)

(b)

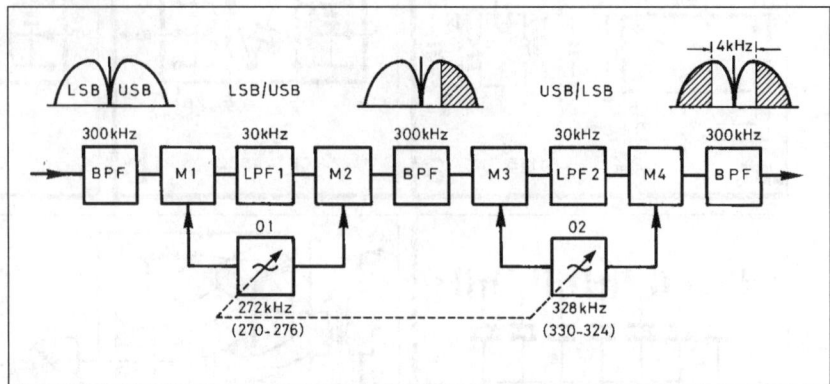

Fig 6.40: (a) IF shift positions the IF pass-band but does not change the overall selectivity. (b) The use of a second filter variably aligned with the first filter provides variable-bandwidth tuning

Fig 6.41: The basic principles of the 1969 Rhode and Schwarz EK07-80 filter, based on two low-pass filters usng inductors and not crystals to provide continuously variable bandwidth

Variable IF Pass-Bands

The use of two mixers both controlled from the same variable oscillator can provide pass-band tuning (IF shift) and variable bandwidth tuning (VBT), techniques outlined in **Fig 6.40**. IF shift positions the IF pass-band of the signal passing through the SSB filter but does not change the overall selectivity. With VBT two filters are used, with the second filter variably aligned with the first filter and thus providing variable bandwidth tuning. A more ambitious version of VBT requiring four mixers and two ganged oscillators tuning in opposite directions has been used by Rohde & Schwarz in a professional receiver developed in the late 'sixties.

In this form, two high-grade low-pass filters at 30kHz, using inductors rather than crystals, were arranged so that they provided a band-pass filter acting on both upper and lower sidebands: **Fig 6.41**.

As shown in **Fig 6.42** this filter had an excellent shape factor, giving a bandwidth continuously adjustable from ±6kHz down to ±150Hz with substantially similar slope right down to -70dB at all settings, without introducing the non-linearities and limited dynamic range inherent in crystal filters.

A more recent, and potentially more economical, approach to variable selectivity is now possible with a single pre-programmed digital signal processing (DSP) filter; such filtering has been imple-

mented at audio (baseband) frequencies and at IF up to about 455kHz. As described earlier, the limiting factor tends to be the dynamic range and resolution of the analogue-to-digital converter. Such filters can also provide multiple notches in the pass-band that automatically suppress unwanted carriers.

CIRCUITRY

Receiver Protection

Receivers, particularly where they are to be used alongside a medium- or high-powered transmitter, need to be protected from high transient or other voltages induced by the local transmitter or by build-up of static voltages on the antenna. Semiconductors used in the first stage of a receiver are particularly vulnerable and invariably require protection. The simplest form of protection is the use of two diodes in back-to-back configuration. Such a combination passes signals less than the potential hill of the diodes (about 0.3V for germanium diodes, about 0.6V for silicon diodes) but provides virtually a short-circuit for higher-voltage signals. This system is usually effective but has the disadvantage that it introduces non-linear devices into the signal path and may occasionally be the cause of cross- and inter-modulation.

The MOSFET devices are particularly vulnerable to static puncture and some types include built-in zener diodes to protect the 'gates' of the main structure. Since these have limited rating it may still be advisable to support them with external diodes or small gas-filled transient suppressors.

Input Circuits and RF Amplifiers

It has already been noted that with low-noise mixers it is now possible to dispense with high-gain RF amplification. Amplifiers at the signal frequency may, however, still be advisable to provide: pre-mixer selectivity; an AGC-controlled stage which is in effect a controlled attenuator on strong signals; to counter the effects of conversion loss in diode and FET-array mixer stages.

In practice semiconductor RF stages are often based on junction FETs as shown in **Fig 6.43** or dual-gate MOSFETs, or alternatively integrated circuits in which large numbers of bipolar transistors are used in configurations designed to increase their signal-handling capabilities.

Tuned circuits between the antenna and the first stage (mixer or RF amplifier) have two main functions: to provide high atten-

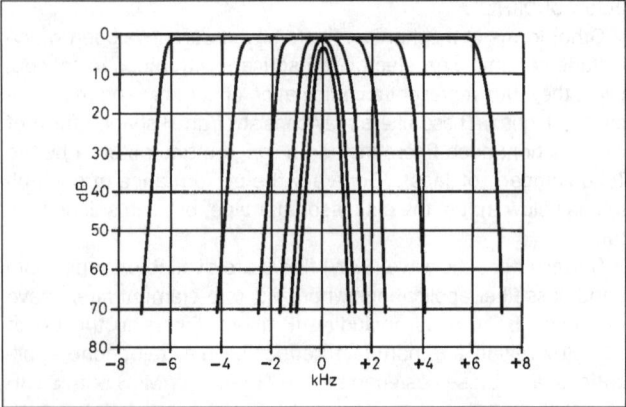

Fig 6.42: Selectivity curves of the EK07-80 filter at bandwidths of ±0.15, ±0.30, ±0.75, ±1.5, ±3.0, ±6.0kHz. Note the similar slope at all settings down to -70dB

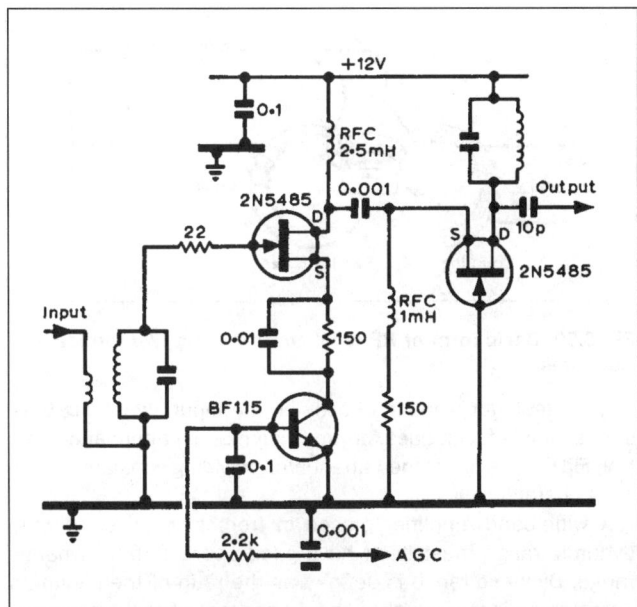

Fig 6.43: Cascode RF amplifier using two JFETs with transistor as AGC

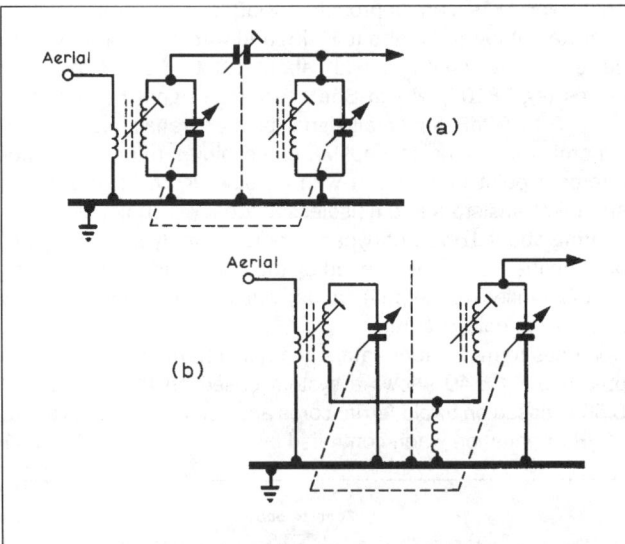

Fig 6.44: Typical RF input circuits used to enhance RF selectivity and capable of providing more than 40dB attenuation of unwanted signals 10% off tune

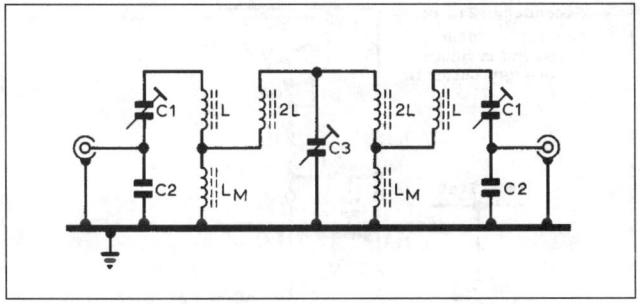

Fig 6.45. Cohn minimum-loss band-pass filter suitable for providing additional RF selectivity to existing receivers. Values for 14MHz: L 2.95µH, 2L 5.9µH, Lm 0.27µH, C1 22pF with 25pF trimmer, C2 340pF, C3 10-60pF (about 34pF nominal). Values for 3.5MHz: L 8µH, 2L 16µH, Lm 2.4µH, C1 150pF + 33pF + 5-25pF trimmer, C2 1nF, C3 150pF + 10-60pF trimmer

uation at the image frequency; to reduce as far as possible the amplitude of all signals outside the IF pass-band.

Coupled tuned circuits

Most amateur receivers still require good pre-mixer selectivity; this can be achieved by using a number of tuned circuits coupled through low-gain amplifiers, or alternatively by tuneable or fixed band-pass filters that attenuate all signals outside the amateur bands (**Fig 6.44**). The most commonly used input arrangement consists of two tuned circuits with screening between them and either top-coupled through a small-value fixed capacitor, or bottom-coupled through a small common inductance.

In order that the coupling is maintained constant over the tuning range, the coupling element should be the same sort as the fixed element, as in **Fig 6.44(b)**. Critical coupling is achieved when the coupling coefficient, k, multiplied by the working Q is equal to 1. For **Fig 6.44(a)**, k is given by:

$$k = Cc/\sqrt{C_1C_2}$$

where C_1, C_2 are the tuning capacitors, and Cc is the top coupling capacitor. In Fig 6.44(b):

$$k = Lc/\sqrt{L_1L_2}$$

where L_1, L_2 are the tuning inductors, and Lc is the coupling inductor.

Critical coupling can be hard to achieve, and for amateur use, it is frequently the case that slight undercoupling is preferable to overcoupling.

Slightly more complex but capable of rather better results is the minimum-loss Cohn filter; this is capable of reducing signals 10% off-tune by as much as 60dB provided that an insertion loss of about 4dB is acceptable. This compares with about 50dB (and rather less insertion loss) for an undercoupled pair of tuned circuits. The Cohn filter is perhaps more suited for use as a fixed band-pass filter which can be used in front of receivers having inadequate RF selectivity. **Fig 6.45** shows suitable values for 3.5 and 14MHz filters.

RF amplifiers and attenuators

Broad-band and untuned RF stages are convenient in construction but can be recommended only when the devices used in the front-end of the receiver have wide dynamic range. An example shown in **Fig 6.46** is a power FET designed specifically for this application and operated in the earthed-gate mode suitable for use on incoming low-impedance coaxial feeders.

Unless the front-end of the receiver is capable of coping with the full range of signals likely to be received, it may be useful to

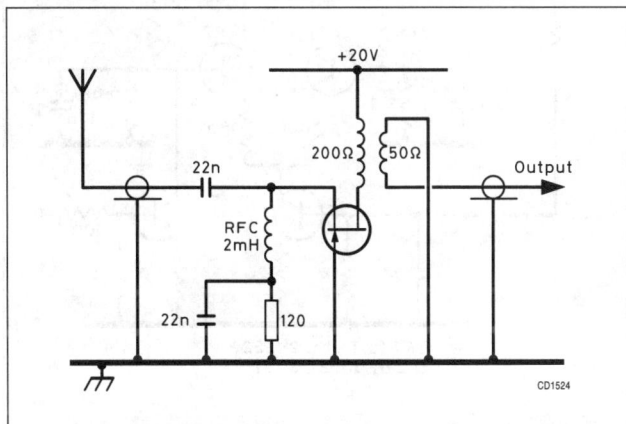

Fig 6.46: Broad-band RF amplifier using power FET and capable of handling signals to almost 3V p-p, 0.5 to 40MHz with 2.5dB noise figure and 140dB dynamic range. Drain current 40mA. Voltage gain 10dB. A suitable device would be a 2N5435 FET

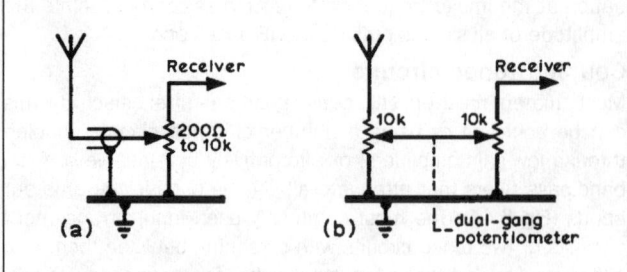

Fig 6.47: Simple attenuators for use in front of a receiver of restricted dynamic range. (a) No attempt is made to maintain constant impedance. (b) Represents less change in impedance

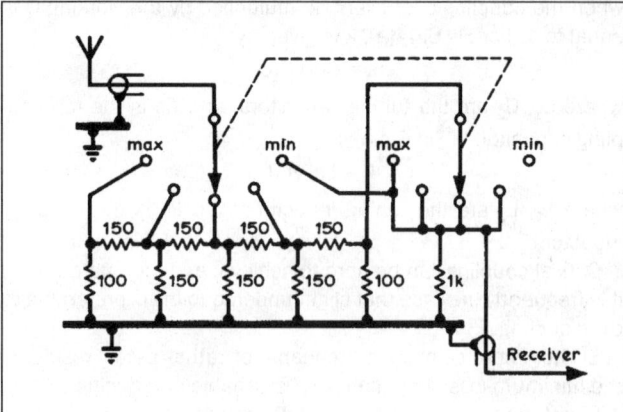

Fig 6.48: Switched antenna attenuator for incorporation in a receiver

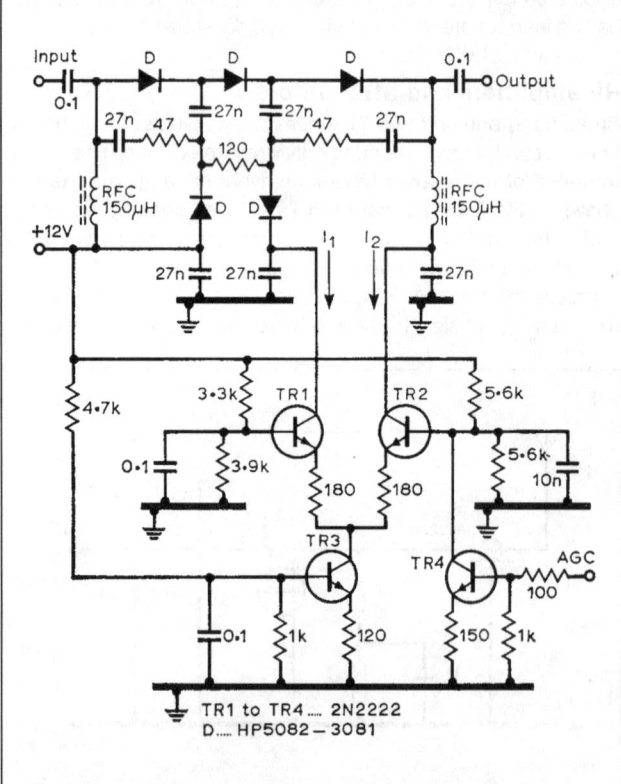

Fig 6.49: Five PIN diodes in a double-T arrangement form an AGC-controlled attenuator. The sum of the transistor collector currents is maintained constant to keep input and output impedances constant

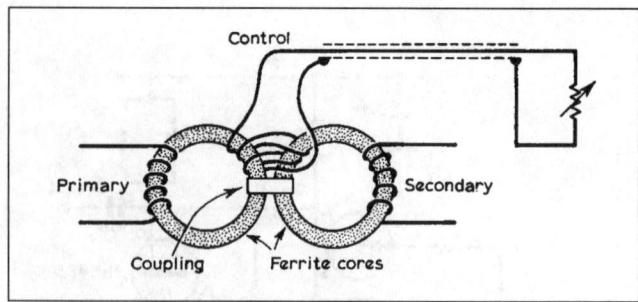

Fig 6.50: Basic form of RF level control using two toroidal ferrite cores

fit an attenuator working directly on the input signal. Fig 6.47 shows simple techniques for providing manual attenuation control; Fig 6.48 is a switched attenuator providing constant impedance characteristics.

A wide-band amplifier placed in front of a mixer of wide dynamic range must itself have good spurious free dynamic range. Dynamic range is defined as the ratio of the minimum detectable signal (equal to the noise floor of the receiver, ie 3dB signal-plus-noise-to-noise ratio) to that signal level which leads to the intermodulation products being equal o the noise floor (ie 3dB signal-plus-noise-to-noise ratio). Amplifiers based on power FETs can approach 140dB dynamic range when operated at low gain (about 10dB) and with about 40mA drain current. This compares with about 90 to 100dB for good valves (eg E810F), 80 to 85dB for small-signal FET devices; and 70 to 90dB for small-signal bipolar transistors. A high-dynamic-range bipolar amplifier can exhibit a third-order input intercept point of +33dBm with a noise figure of 3dB, using suitable transistors and noiseless feedback techniques, representing about 160dB of dynamic range. The dynamic range of an amplifier can be increased by the operation of two devices in a balanced (push-pull) mode, but at a penalty of a 3dB increase in noise figure.

Various forms of attenuators controlled from the AGC line are possible. Fig 6.49 shows a system based on PIN diodes; Fig 6.50 is based on toroid ferrite cores and can provide up to about 45dB attenuation when controlled by a potentiometer. Fig 6.51

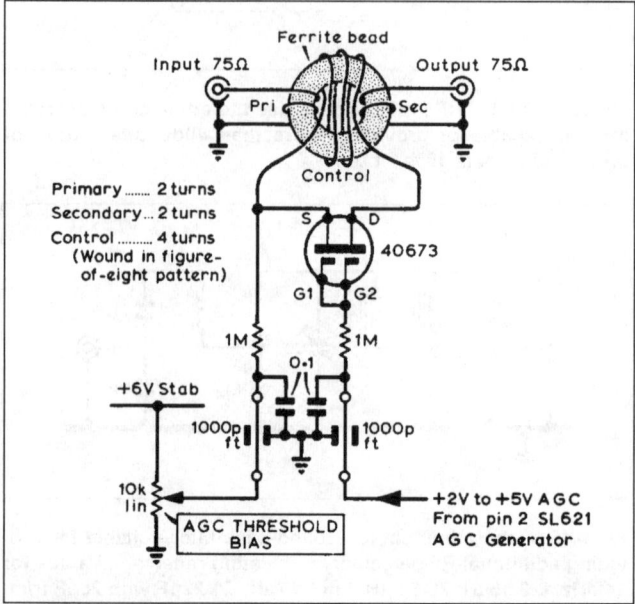

Fig 6.51: Automatic antenna attenuator based on the technique shown in Fig 6.50

shows an adaptation with MOSFET control element for use on AGC lines, although the range is limited to about 20dB.

The tuned circuits used in front-ends may be based on toroid cores since these can be used without screening with little risk of oscillation due to mutual coupling. However, the available Q cannot always compete with that of air-cored coils of thick wire, or of bunched conductor wire and pot cores at the lower frequencies.

It is important to check filters and tuned circuits for non-linearity in iron or ferrite materials. Intermodulation can be caused by the cores where the flux level rises above the point at which saturation effects begin to occur. This is not usually a problem with dust iron cores, however.

Fig 6.46 shows an amplifier with a dynamic range approaching 140dB and suitable for use in front of, or immediately behind, a double-balanced Schottky diode mixer. It should be appreciated, however, that power FETs are relatively expensive devices, although there are some lower-cost devices such as the J310.

Since the optimum dynamic range of an amplifier is usually achieved when the device is operated at a specific working point (ie bias potential) it may be an advantage to design the stage for fixed (low) gain with front-end gain controlled by means of an antenna attenuator (manual or AGC-controlled). Attenuation of signals ahead of a stage subject to cross-modulation is often beneficial since 1dB of attenuation reduces cross-modulation by 2dB.

For semiconductor stages using FETs and bipolar transistors the grounded-gate and grounded-base configuration is to be preferred. Special types of bipolar transistors (such as the BF314, BF324 developed for FM radio tuners) provide a dynamic range comparable with many junction FETs, a noise figure of about 4dB and a gain of about 15dB with a collector current of about 5mA. Generally, the higher the input power of the device, the greater is likely to be its signal-handling capabilities: overlay and multi-emitter RF power transistors or those developed for CATV applications can have very good intermodulation characteristics. For general-purpose, small-signal, un-neutralised amplifiers the zener-protected dual-gate MOSFET is probably the best and most versatile of the low-cost discrete semiconductor devices, with its inherent cascode configuration. If gate 2 is based initially at about 30 to 40% of the drain voltage, gain can be reduced (manually or by AGC action) by lowering this gate 2 voltage, with the advantage that this then increases the signal-handling capacity at this stage. This type of device can be used effectively for RF, mixer, IF, product detector, AF and oscillator applications in HF receivers.

Mixers

Much attention has been given in recent years to improving mixer performance in order to make superhet designs less subject to spurious responses and to improve their ability to handle weak signals in the presence of strong unwanted signals. In particular there has been increasing use of low-noise balanced and double-balanced mixers, sometimes constructed in wide-band form. Ideally mixers are simply multipliers, as are product detectors.

Mixers operate either in the form of switching mixers (the normal arrangement with diode mixers) or in what are termed continuous non-linear (CNL) modes. Generally, switching mixers can provide better performance than CNL modes but require more oscillator injection, preferably in near-square-wave form. The concept of 'linearity' in mixers may seem a contradiction in terms, since in order to introduce frequency conversion the device must behave in a highly non-linear fashion in so far as the oscillator/signal

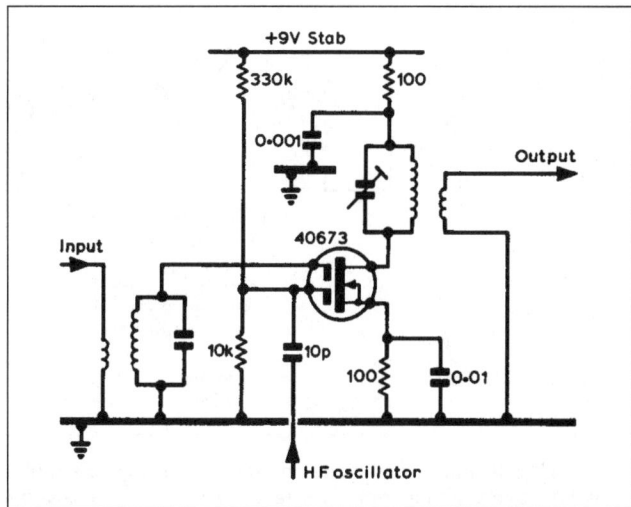

Fig 6.52. Typical dual-gate MOSFET mixer - one of the best 'simple' semiconductor mixers providing gain and requiring only low oscillator injection, but of rather limited dynamic range

mixing process is concerned, and the term 'linearity' refers only to the signal path.

Because of their near square-law characteristics field-effect devices make successful mixers provided that care is taken on the oscillator drive level and the operating point (ie bias resistor); preferably both these should be adjusted to suit the individual device used. Even so, they are not recommended for high-performance receivers except as a balanced mixer.

Optimum performance of a mixer requires correct levels of the injected local oscillator signal and operation of the device at the correct working point. This is particularly important for FET devices. Some switching-mode mixers require appreciable oscillator power.

Junction FETs used as mixers can be operated in three ways:
(a)   RF signal applied to gate, oscillator signal to source;
(b)   RF signal to source, oscillator signal to gate; and
(c)   RF and oscillator signals applied to gate.

Approach (a) provides high conversion gain but requires high oscillator power and may result in oscillator pulling; (b) gives

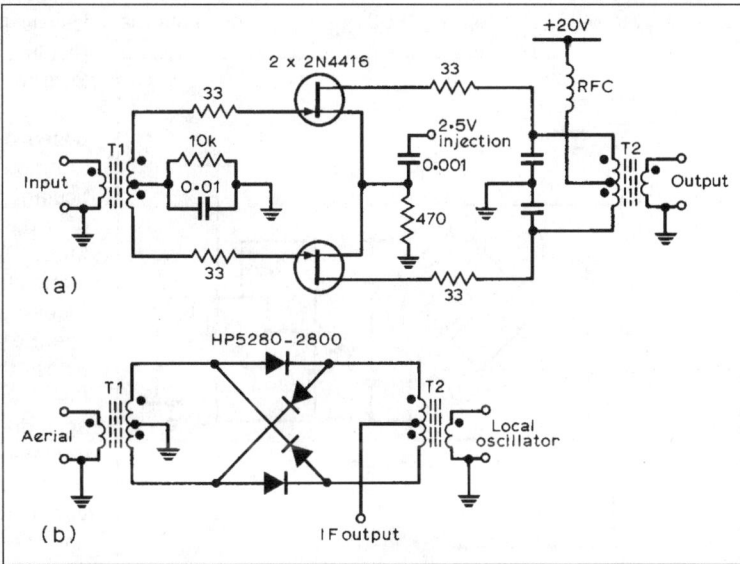

Fig 6.53. Low-noise mixers of wide dynamic range. (a) Balanced FET mixer (preferably used with devices taking fairly high current). (b) Diode ring mixer using Schottky (hot-carrier) diodes

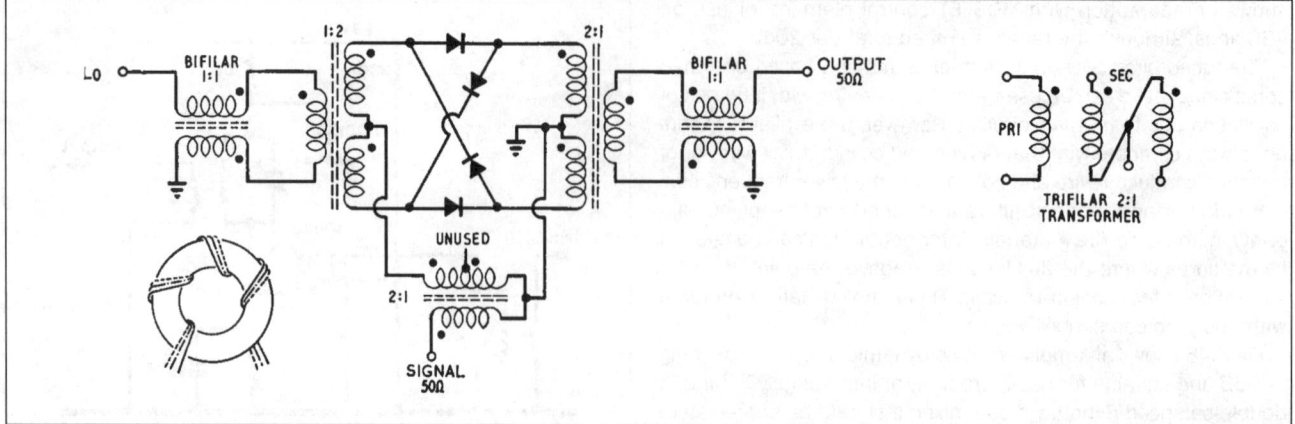

Fig 6.54. Double-balanced diode ring mixer showing how additional bifilar-wound transformers can be added to improve balance, with details of the transformers. The three strands of wire should be twisted together before winding; each winding consists of 12 to 20 turns (depending on frequency range) of No 32 enamelled wire. Injection signal should be 0.8 to 3V across 50 ohms (4-12mW). The toroid material is dependent on the frequency range

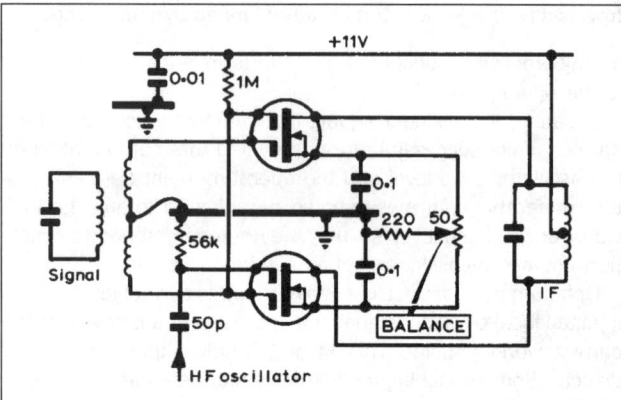

Fig 6.55: Balanced mixer using dual-gate MOSFETs

good freedom from oscillator pulling and requires low oscillator power, but provides significantly lower gain; (c) gives fairly high gain with low oscillator power and may often be the optimum choice. For all FET mixers careful attention must be paid to operating point and local oscillator drive level. For most applications the dual-gate MOSFET mixer (**Fig 6.52**) probably represents the

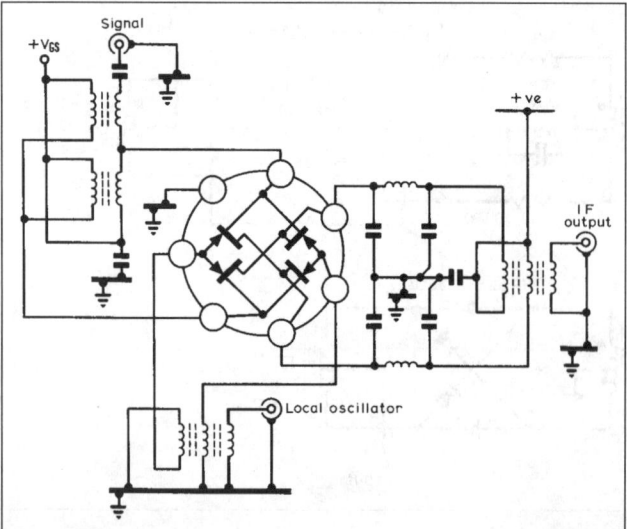

Fig 6.56: Double-balanced active FET mixer of wide dynamic range using JFET quad of power-type FETs

best of the 'simple' single-device arrangements. However, care needs to be taken in that coupling of signals from gate 2 to gate 1 can lead to undesirable radiation and in some cases, difficult-to-find spurious responses.

During the 1980s and '90s, much attention was given to improving the input intercept point of receivers in order to cope better with the range of strong signals that reach low-noise broad-band mixers with a minimum of pre-mixer selectivity. A number of high-performance, double-balanced mixers were been developed which available in package form for home-construction.

It has been noted that a double-balanced mixer offers advantages over single-device and single-balanced mixers in reducing the number of IMPs, and also the oscillator radiation from the antenna and the oscillator noise entering the IF channel.

Double-balanced mixers using four hot-carrier diodes (eg the ones shown in **Fig 6.53 and 6.54**) when driven at a suitable level and with correct output impedance can provide third-order intercept points (IP3) of up to about +40dBm, representing a useful margin over that required for high-performance amateur operation (about +20-25dBm), a figure that is more likely to be achieved with packaged DBMs at reasonable cost. Conversion loss is likely to be at least 6dB, and optimum performance requires a high-level of oscillator injection (up to about +20-30dBm) with a near-square waveform.

An active form of double-balanced ring modulator can be based on four symmetrical medium-power FETs. Such mixers are more sensitive than diodes to changes in termination, particularly reactive components. A less costly approach is the use of a push-pull arrangement with N-junction FETs or dual-gate MOSFETs (which provide slightly higher conversion gain) Although the gates of MOSFETs do not consume power, they do require an appreciable voltage swing in order for the mixer to operate correctly, owing to the considerable signal voltage across the FET switch and the signal current through it unless the mixer is configured so that these signal currents and voltages do not appear across the FETs. An ultra-low-distortion HF switched FET mixer was devised by Eric Kushnik [7] in which an IP3 of +25dBm has been measured with -3dBm local oscillator power. Some of the 74HCT series of CMOS transmission gate ICs have been successfully used in this application. FET mixers are shown in **Figs 6.55 and 6.56**.

Both active and passive FET switching-mode mixers can provide a wide dynamic range; the active mixer provides some conversion gain, the passive arrangement in which the devices act

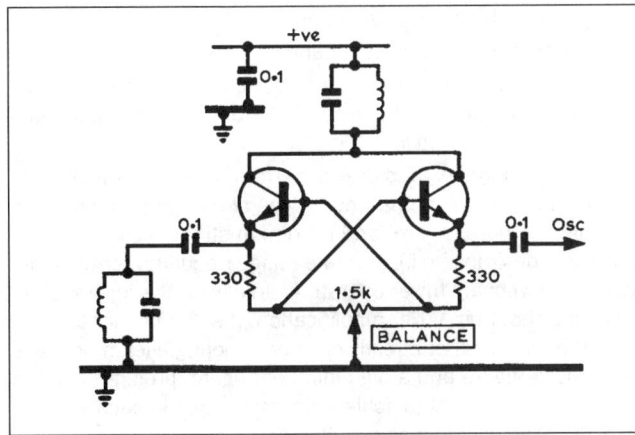

Fig 6.57: Use of cross-coupled transistors to form a double-balanced mixer without special balanced input transformers

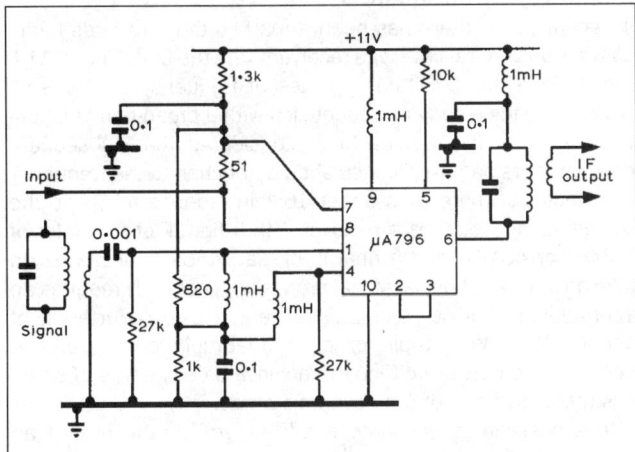

Fig 6.58: Double-balanced IC mixer circuit for use with µA796 or MC1596G etc devices

Characteristic	Single-ended	Single-balanced	Double-balanced
Bandwidth	Several decades	Decade	Decade
Relative intermodulation density	1	0.5	0.25
Interport isolation	Little	10-20dB	>30dB
Relative oscillator power	0dB	+3dB	+6dB

Table 6.1: Basic mixer arrangements

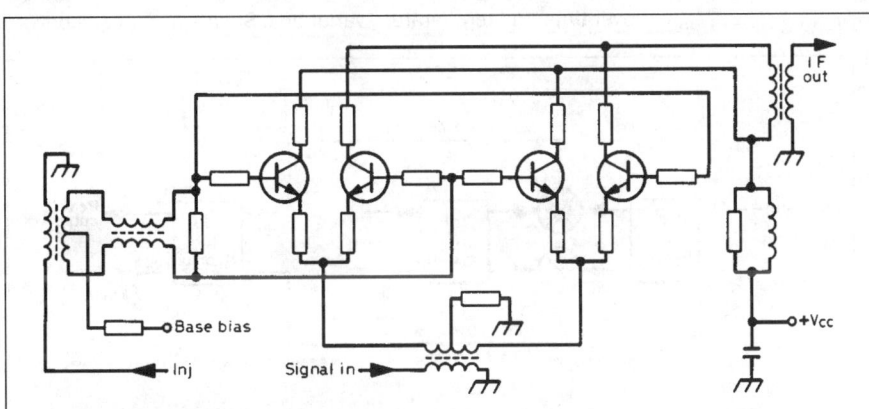

Fig 6.59: A double-balanced active mixer using bipolar transistors in a denegrated version of the balanced transconductance mixer (*Radio Receivers* - Gosling (Ed))

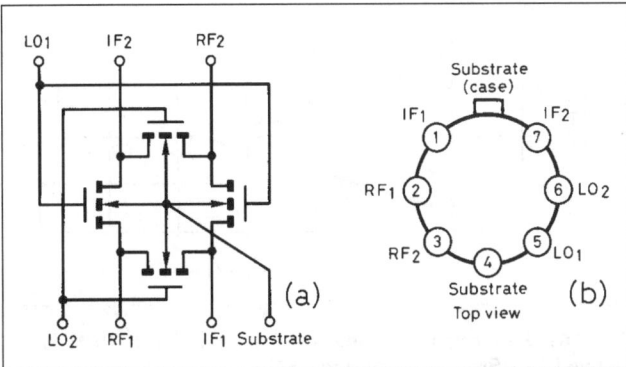

Fig 6.60: Siliconix Si8901 ring demodulator/balanced mixer. (a) Functional block diagram. (b) Pin configuration with Si8901A in TO-78 and Si8901Y as surface-mounted So14 configuration

Device	Advantages	Disadvantages
Bipolar transistor	Low noise figure High gain Low DC power	High intermodulation Easily overloaded Subject to burnout
Diode	Low noise figure High power handling High burn-out level	High LO drive Interface to IF Conversion loss
JFET	Low noise figure Conversion gain Excellent in performance Square-law characteristic Excellent overload High burn-out level	Optimum conversion gain not possible at optimum square law response High LO power
Dual-gate MOSFET	Low IM distortion AGC Square-law characteristic	High noise figure Poor burn-out level Unstable

Table 6.2: Device comparisons

basically only as switches results, as with diode mixers, in some conversion loss and must be followed by a low-noise IF amplifier, preferably with a diplexer arrangement that presents a constant impedance over a wide frequency range.

High-level ring-type mixers can also be based on the use of medium-power bipolar transistors in cross-coupled 'tree' arrangements (**Fig 6.57**). With such an arrangement there need be no fundamental requirement for push-pull drive or balanced input/output transformers, although their use minimises local oscillator conducted emissions. This approach is used at low-level in the Motorola MC1596 and now-obsolete Plessey SL640 IC packages, and at high-level in the obsolete Plessey SL6440. The SL6440 was specifically developed as a high-level, low-noise mixer and is capable of +30dBm intercept point, +15dBm 1dB compression point with a conversion 'gain' of -1dB. A circuit for the µA796 and similar ICs is shown in **Fig 6.58**.

Fig 6.59 shows the basic arrangement of a similar class of mixer, comprising a pair of transconductance mixers with emitter resistors added for IM improvement. Resistors in the base and collector leads add loss of ultra-high frequencies to suppress parasitic oscillations caused by resonances formed by circuit and

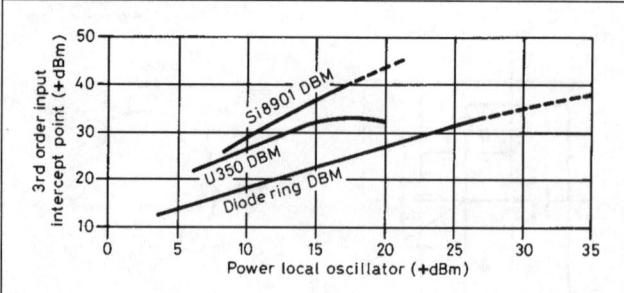

Fig 6.61: Performance comparison between Si8901 DBM, U350 active FET DBM and diode ring DBM

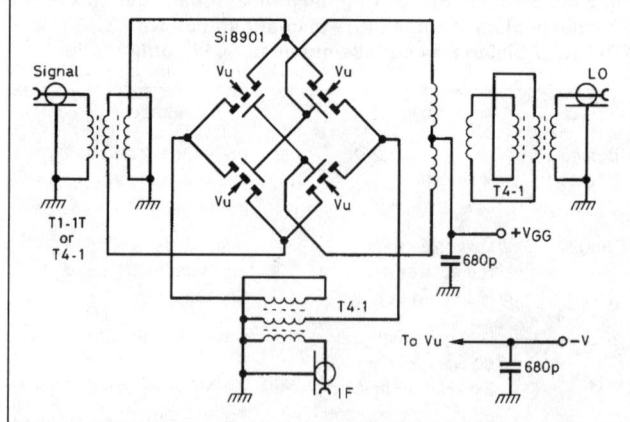

Fig 6.62: Prototype commutation double-balanced mixer as described in the Siliconix applications notes

transistor capacitances together with the leakage reactance of the associated transformers. Injection is via a balancing transformer to the bases of the bipolar transistors which are overdriven, resulting in signal switching action. Collector supply voltage is applied to the output transformer centre-tap through a parallel resistor-inductor to further suppress oscillation. With medium-power transistors, this active mixer can give a 3dB gain, 9dB noise figure and +25dBm input intercept over the 2 to 30MHz band as an up-converter to the 100MHz range. This device can be used with or without preamplication and with little pre-mixer selectivity in high-performance home-built receivers.

Common to both diode and switching-mode FET mixers is their square-law characteristics, an important factor in maintaining low distortion during mixing. Equally important for high dynamic range is the ability to withstand overload that can be a major cause of distortion. Some designs with passive ring mixers use paralleled diodes to provide greater current handling; the penalty attached to this approach is the need for a very large increase in local-oscillator power.

A form of high-performance mixer using monolithic quad-ring double-diffused MOSFETs developed by Ed Oxner, KB6QJ [8] used a resonant-gate drive transformer to provide sufficient switching voltage without a corresponding increase in switching power, an effective technique but not easily implemented with broadband mixers.

Switching mixers, if they are to achieve a high third-order intercept point, require a drive that must:

- approach ideal square wave;
- ensure a 50% duty cycle; and
- have sufficient amplitude to switch the devices fully 'on' and 'off', and, in the case of FET devices, offer minimum resistance when 'on'.

Further, to maintain good overall performance in terms of minimum conversion loss, maximum dynamic range (ie taking into account the noise figure) and maximum strong signal performance, it is desirable to incorporate image-frequency termination. With any switching mixer operating directly on the incoming RF signals without pre-mixer amplification, the IF amplifier that follows the mixer, either directly or after a roofing filter, must have a low noise figure and a high intercept figure, preferably with a diplexer arrangement to achieve constant input impedance over a broad band of frequencies.

Details of Siliconix mixers and their performance are shown in **Figs 6.60 to 6.62**.

Image-rejection mixer

In recent years, there has been a marked trend towards multi-conversion general-coverage receivers with the first IF in the VHF range since this facilitates the use of frequency synthesisers covering a single span of frequencies with a broad-band up-conversion mixer. It has long been considered that subsequent down-conversion mixer stages should not change the frequency by a factor of more than about 10:1 in order to minimise the 'image' response. Thus a receiver with a first IF of the order of 70MHz or 45MHz and a final IF of, say, 455kHz or 50kHz (to take advantage of digital signal processing) normally requires an intermediate IF of, say, 9 or 10.7MHz and possibly a further IF of about 1MHz. With triple or even quadruple conversions, it becomes increasingly difficult to achieve a design free of spurious responses and of wide dynamic range.

It is possible to eliminate mid-IF stages by the use of an image-rejecting two-phase mixer akin to the form of demodulator used in two-phase direct conversion receivers; this is best implemented using two double-balanced diode-ring mixers or the equivalent.

Fig 6.63 outlines the basic arrangement of an image-rejection mixer. Image rejection mixers offer slightly better intermodulation performance, all other considerations being equal, because the input signal level is reduced by 3dB. This reduces the absolute power of the 3rd order intermodulation products by 9dB, and summing the products increases this level by a maximum of 3dB. However, there is a 3dB greater noise figure, and so the absolute improvement in dynamic range disappears. Image rejection mixers are generally limited to about 30dB rejection: with careful trimming, possibly 40dB may be attained over time and temperature variations. Such a mixer can convert

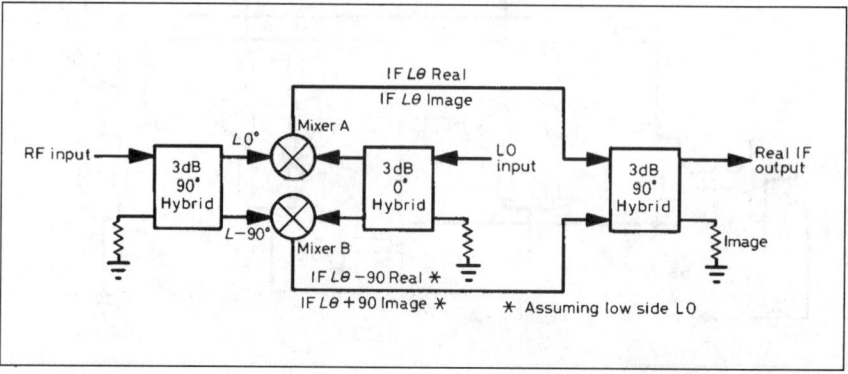

Fig 6.63: Image rejecting mixer from *Microwave Solid-state Circuit Design* by Bahl and Bhartia (1988)

The Radio Communication Handbook

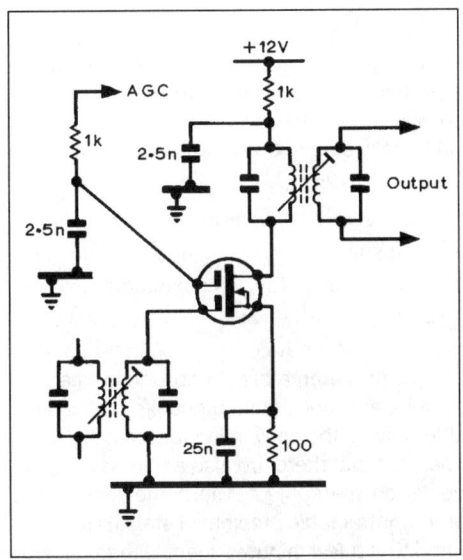

Fig 6.64: Typical automatic gain-controlled IF amplifier a dual-gate MOSFET

a VHF signal directly to, say, 50kHz while maintaining image response at a low level.

More about mixers can be found in the one of the Building Blocks chapters earlier in this book.

The HF oscillator(s)

The frequency to which a superhet or direct-conversion receiver responds is governed not by the input signal frequency circuits but by the output of the local oscillator. Any frequency variations or drift of the oscillator are reflected in variation of the received signal; for SSB reception variations of more than about 50Hz will render the signal unintelligible unless the set is retuned.

Much more on this topic can be found in the chapter Building Blocks: Oscillators.

IF Amplifiers

The IF remains the heart of a superhet receiver, for it is in this section that virtually all of the voltage gain of the signal and the selectivity response are achieved. Whereas with older superhets (which had significant front-end gain) the IF gain was of the order of 70-80dB, today it is often over 100dB.

Where the output from the mixer is low (possibly less than 1µV) it is essential that the first stage of the IF section should have low-noise characteristics and yet not be easily overloaded. Although it is desirable that the crystal filter (or roofing filter) should be placed immediately after the mixer, the very low output of diode and passive FET mixers may require that a stage of amplification takes place before the signal suffers the insertion loss of the filter.

Similarly it is important that where the signal passes through the sideband filter at very low levels the subsequent IF amplifier must have good noise characteristics. Further, for optimum CW reception, it will often be necessary to ensure that the noise bandwidth of the IF amplifier after the filter is kept narrow. The noise bandwidth of the entire amplifier should be little more than that of the filter. This can be achieved by including a further narrow-band filter (for example a single-crystal filter with phasing control) later in the receiver, or alternatively by further frequency conversion to a low IF.

To achieve a flat AGC characteristic it may be desirable for all IF amplifiers to be controlled by the AGC loop, and it is important that amplifier distortion should be low throughout the dynamic range of the control loop.

The dual-gate MOSFET (**Fig 6.64**) with reverse AGC on gate 1 and partial forward AGC on gate 2 has excellent cross-modula-

tion properties but the control range is limited to about 35dB per stage. Integrated circuits with high-performance gain-controlled stages are available.

Where a high-grade SSB or CW filter is incorporated it is vital to ensure that signals cannot 'leak' around the filter due to stray coupling; good screening and careful layout are needed.

In multiple-conversion receivers, it is possible to provide continuously variable selectivity by arranging to vary slightly the frequency of a later conversion oscillator so that the band-pass of the two IF channels overlap to differing degrees. For optimum results this requires that the shape factor of both sections of the IF channel should be good, so that the edges are sharply defined.

With double-tuned IF transformers, gain will be maximum when the product kQ is unity (where k is the coupling between the windings). IF transformers designed for this condition are said to be critically coupled; when the coupling is increased beyond this point (over-coupled) maximum gain occurs at two points equally spaced about the resonant frequency with a slight reduction of gain at exact resonance: this condition may be used in broadcast receivers to increase bandwidth for good-quality reception. If the coupling factor is lowered (under-coupled) the stage gain falls but the response curve is sharpened, and this may be useful in communications receivers.

FETs as Small-signal Amplifiers

Field effect transistors make good small-signal RF or IF amplifiers and offer advantages when used properly. They can provide low-noise, rugged amplifiers once wired in circuits having an easy, low-resistance path between their gate(s) and earth, though they are vulnerable to electrostatic discharge when out of circuit, unless protected by an internal diode(s). However, despite being, like thermionic valves, voltage-controlled devices, they should not be used as replacements for valves in similar circuits.

FETs have very high slope of up to 30mA/V (much higher than most valves) but also much greater drain-to-gate capacitance than the inter-electrode capacitances of triode valves. This is a sure prescription for self-oscillation if connected in a typical pentode-type amplifier circuit.

However, the FET can be used with a very low output load impedance and thus can be used as a stable low-gain device, yet providing excellent stage gain because of the voltage step-up that can be readily achieved with a resonant input transformer.

For example, the 21MHz preamplifier shown in **Fig 6.65** using the 2N3819 FET (10mA/V) with a 330Ω resistor as load has a device gain of only three, but the input tuned circuit can provide a voltage gain of about seven, resulting in a stable voltage gain of about 21, with the FET's very high input impedance presenting only light loading of the input transformer. This pro-

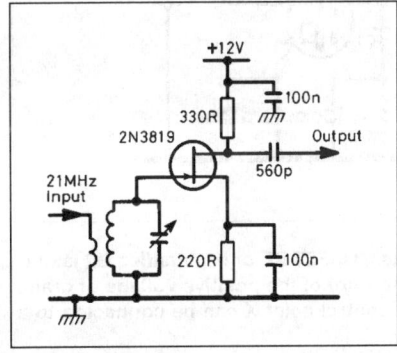

Fig 6.65: Stable FET preamp with low-impedance output load and with the main part of the gain coming from the step-up input transformer

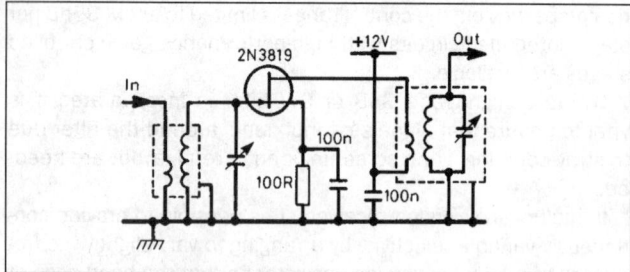

Fig 6.66: Stable FET IF amplifier using two bipolar transistor-type IF transformers in reverse configuration

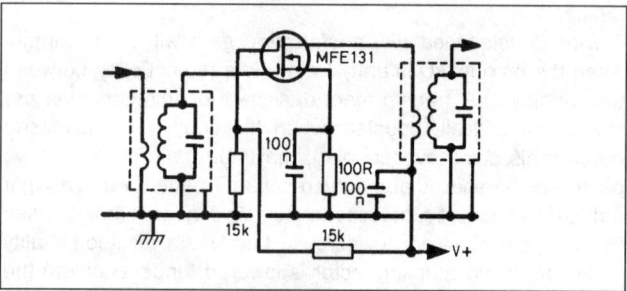

Fig 6.67: Dual-gate FET IF amplifier

vides an unconditionally stable stage gain of over 20dB. **Figs 6.66, 6.67 and 6.68** show typical FET and dual-gate FET small-signal amplifiers.

The most critical amplifier in a modern, high-performance receiver is usually the post-mixer amplifier, ie the IF amplifier that follows the mixer, either directly or after a crystal filter. It needs to be of low noise in order to cope with the conversion loss of a ring mixer, and if preceding the crystal filter will have to cope with large off-frequency signals, requiring a high intercept characteristic.

For the highest-performance receivers, push-pull bipolar transistors with a noise figure of about 2dB and a third-order intercept point equal or better than that of the mixer are required.

Possibly the simplest post-mixer amplifier for high-performance receivers is a power-FET common-gate stage, such as that shown earlier. With a 2N5435 FET this can provide a 2dB noise figure with a 50-ohm system gain of 9dB and an output third-order intercept point of +30dBm when biased at V_{DD} of 12-15V at 50mA.

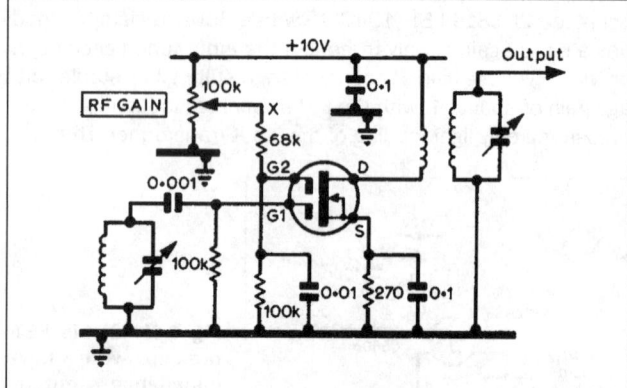

Fig 6.68: Typical dual-gate MOSFET RF or IF amplifier. G2 is normally biased to about one-third of the positive voltage of drain. In place of manual gain, control point X can be connected to a positive AGC line

Amplifier Stability Criteria

For any amplifying device, a consideration of its stability may be achieved by the application of a stability criterion. The Linville stability criterion C provides an equation for C such that when C < 1, the circuit is unconditionally stable. C is given by:

$$C = (y_{rs} \cdot y_{fs})/(2g_{is}\, g_{os} - R(y_{rs}y_{fs})$$

where y_{rs} is the magnitude of reverse transmission; y_{fs} is the magnitude of forward transmission; g_{is} is the input conductance; and g_{os} is the output conductance. $R(y_{rs}y_{fs})$ means the real part of the product of $y_{rs}y_{fs}$.

Where the characteristics of a device are given in terms of 's' or other parameters (eg 'h' parameters), a conversion can be made. The Motorola RF transistor data handbooks are particularly recommended to provide the conversion formulas between the various sets of parameters: there are also a number of standard student textbooks on the subject. The home-constructor, faced with an apparently intractable problem of stability in an RF amplifier, may well find that a few minutes spent with the device data sheet and a calculator can provide not only the explanation of the instability but also a cure. Because of this, and being based on engineering, rather than 'green fingered' principles, it is far more likely to be more repeatable and stable in the long term than a guesswork approach.

Demodulation

For many years, the standard form of demodulation for communications receivers, as for broadcast receivers, was the envelope detector using diodes. Envelope detection is a non-linear process (part mixing, part rectification) and is inefficient at very low signal levels. On weak signals this form of detector distorts or may even lose the intelligence signals altogether. On the other hand, synchronous or product detection preserves the signal-to-noise ratio, enabling post-detector signal processing and audio filters to be used effectively (**Fig 6.69**).

Synchronous detection is essentially a frequency conversion process and the circuits used are similar to those used in mixer stages. The IF or RF signal is heterodyned by a carrier at the same frequency as the original carrier frequency and so reverts back to the original audio modulation frequencies (or is shifted from these frequencies by any difference between the inserted carrier and the original carrier as in CW where such a shift is used to provide an audio output between about 500 and 1000Hz). Typical product detector circuits are shown in **Fig 6.70**.

It should be noted that a carrier is needed for both envelope and product detection: the carrier may be radiated along with the sidebands, as in AM, or locally generated and inserted in the receiver (either at RF or IF - usually at IF in superhets, at RF in direct-conversion receivers).

Synchronous or product detection has been widely adopted for SSB and CW reception in amateur receivers; the injected carrier frequency is derived from the beat frequency oscillator,

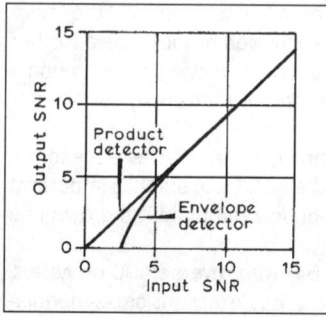

Fig 6.69: Synchronous (product) detection maintaining the SNR of signals down to the lowest levels whereas the efficiency of envelope detection falls off rapidly at low SNR, although as efficient on strong signals

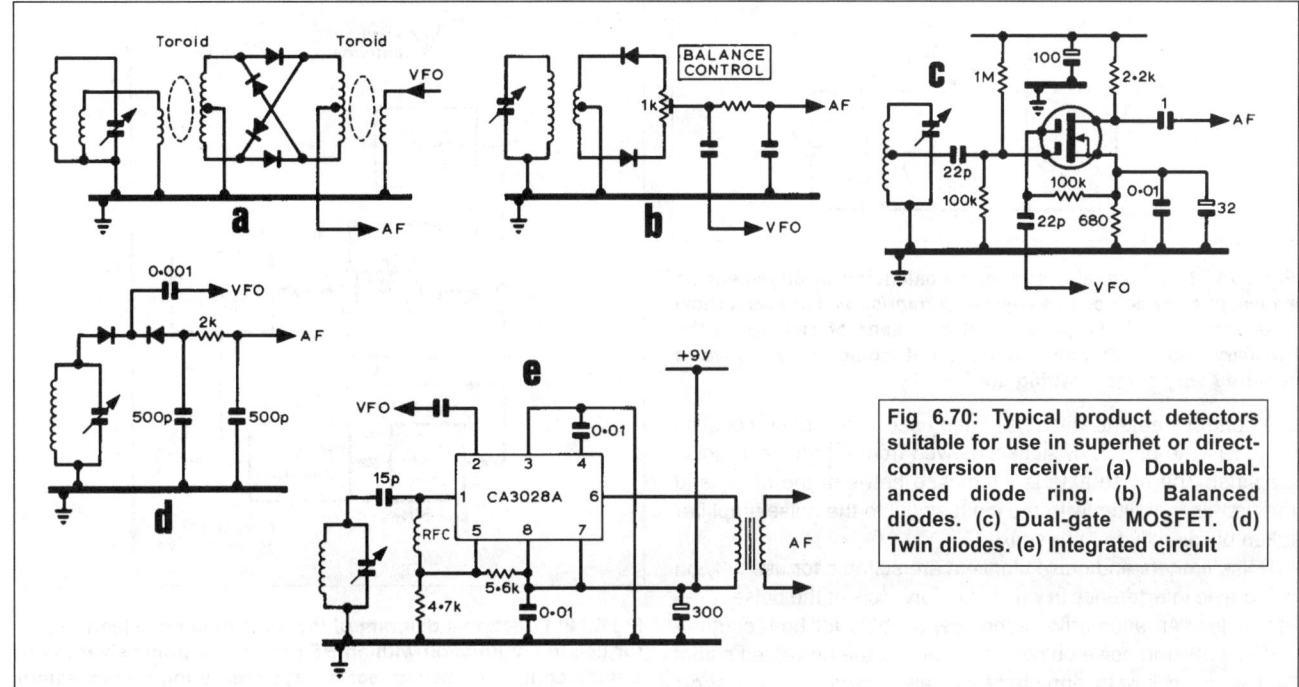

Fig 6.70: Typical product detectors suitable for use in superhet or direct-conversion receiver. (a) Double-balanced diode ring. (b) Balanced diodes. (c) Dual-gate MOSFET. (d) Twin diodes. (e) Integrated circuit

which is either LC or crystal controlled. By using two crystals it is possible to provide selectable upper or lower sideband reception.

The use of synchronous detection can be extended further to cover AM, DSBSC, NBFM and RTTY but for these modes the injected carrier really needs to be identical to the original carrier, not only in frequency but also in phase: that is to say the local oscillator needs to be in phase coherence with the original carrier (an alternative technique is to provide a strong local carrier that virtually eliminates the original AM carrier - this is termed exalted carrier detection).

Phase coherence cannot be achieved between two oscillators unless some effective form of synchronisation is used. The simplest form of synchronisation is to feed a little of the original carrier into a local oscillator, so forcing a phase lock on a free-running oscillator; such a technique was used in the synchrodyne receiver. The more usual technique is to have a phase-lock loop. At one time such a system involved a large number of components and would have been regarded as too complex for most purposes; today, however, complete phase-lock loop detectors are available in the form of a single integrated circuit, both for AM and NBFM applications.

Apart from the phase-lock loop approach a number of alternative forms of synchronous multi-mode detectors have been developed. One interesting technique which synthesises a local phase coherent carrier from the incoming signal is the reciprocating detector.

Noise Limiters, Null-steerers and Blankers

The HF spectrum, particularly above 15MHz or so, is susceptible to man-made electrical impulse interference stemming from electric motors and appliances, car ignition systems, thyristor light controls, high-voltage power lines and many other causes. Static and locally generated interference from appliances, TV receivers etc can be a serious problem below about 5MHz, and may be maximum at LF/MF.

These interference signals are usually in the form of high-amplitude, short-duration pulses covering a wide spectrum of frequencies. In many urban and residential areas this man-made interference sets a limit to the usable sensitivity of

receivers and may spoil the reception of even strong amateur signals.

Because the interference pulses, though of high amplitude, are often of extremely short duration, a considerable improvement can be obtained by 'slicing' off all parts of the audio signal which are significantly greater than the desired signal. This can be done by simple AF limiters such as back-to-back diodes. For AM reception more elegant noise limiters develop fast-acting biasing pulses to reduce momentarily the receiver gain during noise peaks. The ear is much less disturbed by 'holes of silence' than by peaks of noise. Many limiters of this type have been fitted in the past to AM-type receivers.

Unfortunately, since the noise pulses contain high-frequency transients, highly selective IF filters will distort and broaden out the pulses. To overcome this problem, noise blankers have been developed which derive the blanking bias potentials from noise pulses which have not passed through the receiver's selective filters. In some cases a parallel broadly tuned receiver is used, but more often the noise signals are taken from a point early in the receiver. For example, the output from the mixer goes to two channels: the signal channel which includes a blanking control element which can rapidly reduce gain when activated; and a wide-band noise channel to detect the noise pulse and initiate the gain reduction of the signal channel. To be most effective it is necessary for the gain reduction to take place virtually at the instant that the interference pulse begins. In practice, because of the time constants involved, it is difficult to do this unless the signal channel incorporates a time delay to ensure that the gain reduction can take place simultaneously with or even just before the noise pulse. One form of time delay which has been described in the literature utilises a PAL-type glass ultrasonic television delay line to delay signals by 64ms. It is, however, difficult to eliminate completely transients imposed on the incoming signal.

One possible approach, which has been investigated at the University College, Swansea, is to think in terms of receivers using synchronous demodulation at low level so that a substantial part of the selectivity, but not all of it, is obtained after demodulation. This allows noise blankers to operate at a fairly low level on AF signals.

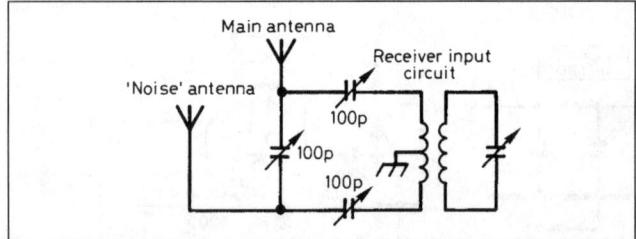

Fig 6.71: The original Jones noise-balancing arrangement as shown in early editions of *The Radio Handbook*. Local electrical interference could be phased out by means of pick-up on the auxiliary 'noise' antenna. Although it could be effective it required very careful setting up

A control element which has been used successfully consists of a FET gate pulsed by signals derived from a wide-band noise amplifier. The noise gate is interposed between the mixer and the first crystal filter, with the input signal to the noise amplifier taken off directly from the mixer.

Noise limiters and noise blankers are suitable for use only on pulse-type interference in which the duty cycle of the pulse is relatively low. An alternative technique, suitable for both continuous signals and noise pulses, is to null out the unwanted signal by balancing it with anti-phase signals picked up on a short 'noise antenna'. This has led to the revival of the 'thirties Jones noise-balancing technique in which local interference could be phased out by means of pick-up on an auxiliary noise antenna. This system (**Fig 6.71**) was capable of reducing a specific unwanted source of interference by tenf od decibels while reducing the wanted signal by only about 3-6dB, but required critical adjustment of the controls.

In the early 1980s John K Webb, W1ETC, developed a more sophisticated method of phasing out interference using coiled lengths of coaxial cable as delay lines to provide the necessary phase shifts: **Fig 6.72**. This included a compact null-steerer located alongside the receiver/transceiver, capable of generating deep nulls against a single source of interference, resembling the nulls of an efficient MW ferrite-rod antenna. The two controls adjusted phase and amplitude of the signals from the auxiliary noise antenna. W1ETC summarised results [9] with such a unit as follows:

1. The available null depth in signals propagated over short paths of up to 20 miles is large and stable, limited only by how finely the controls are adjusted.

2. Nulls on signals arriving over short skywave paths of up to a few hundred miles are in the order of 30dB, provided there is a single mode of propagation and one direction of arrival. Such nulls are usually stable.

3. Signals propagated over paths of 10 to 100 miles may arrive as a mixture of ground-wave and skywave. A single null is thus ineffective.

4. Signals propagated by skywave over long distances frequently involve several paths, each having a different path length so that a single null has little effect on what is usually the 'wanted' signal.

5. Broad-band radiated noise can be nulled as deeply as any radio signal. This seems to be a more effective counter to noise than blanking or limiting techniques

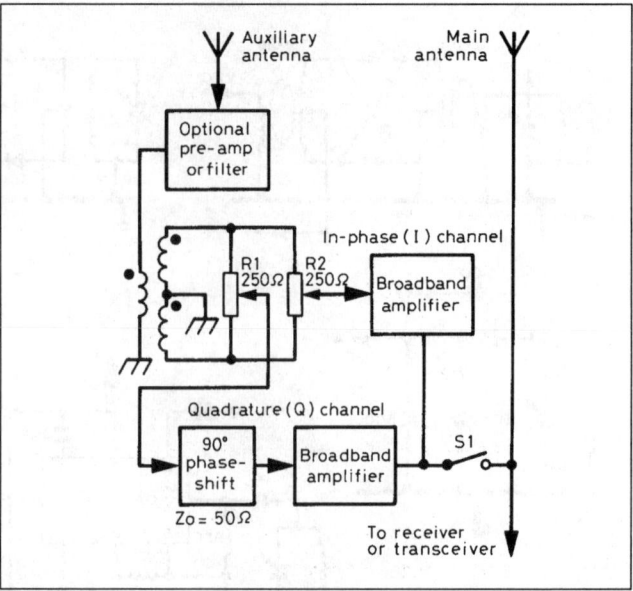

Fig 6.72: Functional diagram of the electronic null steering unit for use in conjunction with an HF receiver or transceiver (S1 is a relay contact to disconnect the system during transmission) as described by John Webb, W1ETC, of the Mitre Corporation in 1982

with local electrical noise deeply nulled, and with little effect on wanted long-distance signals.

To meet result (5) the interference has to be directly radiated to the receiver antennas and not enter the receiver in a less-directional manner (for example re-radiation from mains cabling etc). Various methods of implementing null-steering have been described: **Figs 6.73-6.75** show a design by Lloyd Butler, VK5BR, utilising the phase shifts of off-tune resonant circuits.

AF stages

The AF output from an envelope or product detector of a superhet receiver is usually of the order of 0.5 to 1V, and many receivers incorporate relatively simple one- or two-stage audio amplifiers, typically using an IC device and providing about 2W output. On the other hand the direct-conversion receiver may require a high-gain audio section capable of dealing with signals of less than 1µV.

Provided that all stages of the receiver up to and including the product detector are substantially linear, many forms of post-demodulation signal processing are possible: for example band-

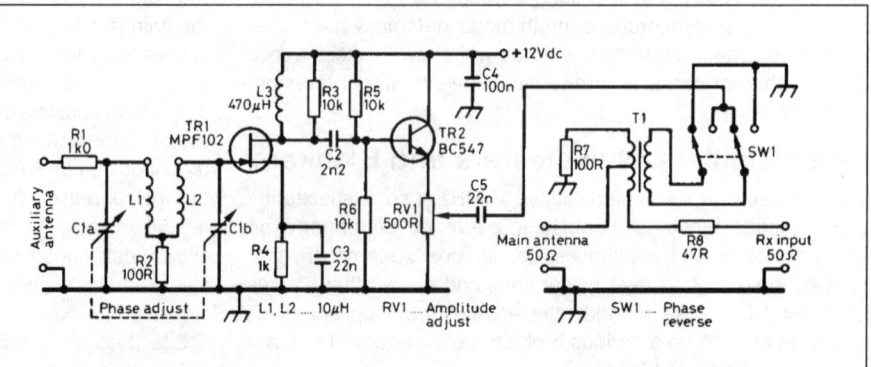

Fig 6.73: VK5BR's Mk 2 interference-cancelling circuit as described in the January 1993 issue of Amateur Radio. As shown this covers roughly 3.5 to 7MHz. C1 ganged 15-250pF variable capacitor or similar. L1, L2 miniature 10µH RF chokes. T1 11 turns quadfilar wound on Amidon FT-50-75m toroidal core

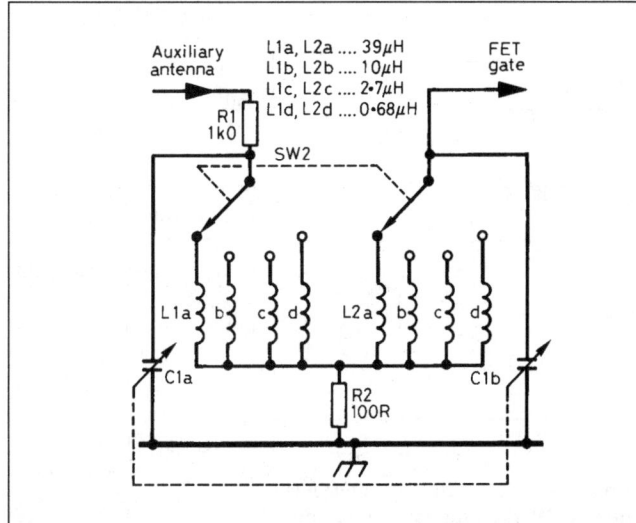

Fig 6.74: Bandswitching modification to provide 1.8 to 30MHz in four ranges

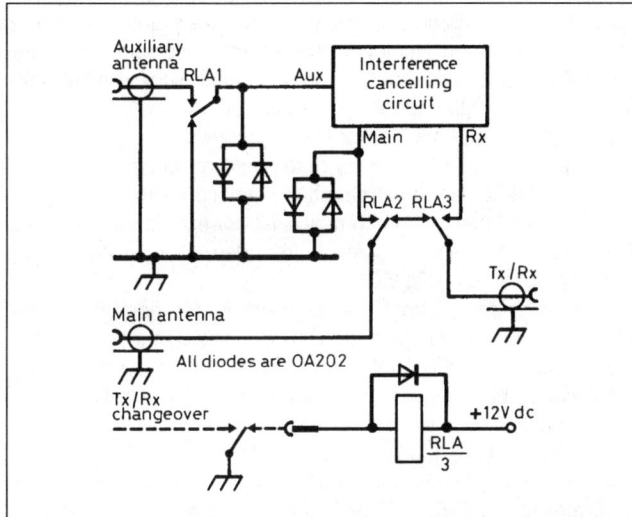

Fig 6.75: Transmit-receive switching with protection diodes for use with the VK5BR interference-cancelling circuit with a transceiver

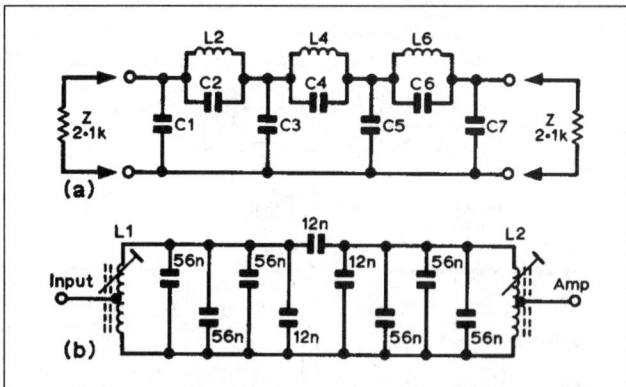

Fig 6.76: (a) Phone and (b) CW AF filters suitable for use in direct-conversion or other receivers requiring very sharply defined AF responses. The CW filter is tuned to about 875Hz. Values for (a) can be made from preferred values as follows: C1 37.26nF (33,000 + 2200 + 1800 + 220pF); C2 3.871nF (3300 + 560pF); C3 51.87nF (47000 + 4700 + 150pF); C4 19.06nF (18,000 + 1000pF); C5 46.41nF (39,000 + 6800 + 560pF); C6 13.53nF (12,000 + 1500pF); C7 29.85nF (27,000 + 2700 + 150pF). All capacitors mica or polyester or styroflex types. L2 168.2mH (540 turns), L4 124.5mH (460 turns); L6 129.5mH (470 turns) using P30/19 3H1 pot cores and 0.25mm enam wire. Design values based on 2000-ohm impedance

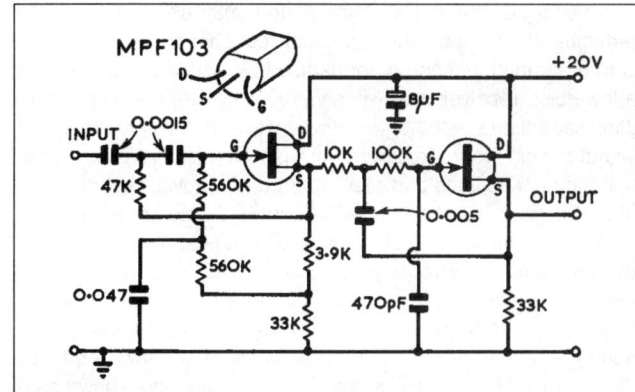

Fig 6.77: Active band-pass AF filter for amateur telephony. The -6dB points are about 380 and 3200Hz, -18dB about 160 and 6000Hz

pass or narrow-band filtering to optimise signal-to-noise ratio of the desired signal, audio compression or expansion; the removal of audio peaks, AF noise blanking, or (for CW) the removal by gating of background noise. Audio phasing techniques may be used to convert a DSB receiver into an SSB receiver (as in two-phase or third-method SSB demodulation) or to insert nulls into the audio pass-band for the removal of heterodynes. Then again, in modern designs the AGC and S-meter circuits are usually operated from a low-level AF stage rather than the IF-derived techniques used in AM-type receivers.

It should be appreciated that linear low-distortion demodulation and AF stages are necessary if full advantage is to be taken of such signal processing, since strong intermodulation products can easily be produced in these stages. Thus, despite the restricted AF bandwidth of speech and CW communications, the intermodulation distortion characteristics of the entire audio section should preferably be designed to high-fidelity audio standards. Very sophisticated forms of audio filtering, notching and noise reduction are now marketed in the form of add-on digital-signal-processing (DSP) units, or may be incorporated into a transceiver.

Audio filters may be passive, using inductors and capacitors, or active, usually with resistors and capacitors in conjunction with op-amps or FETs (Figs 6.76-6.79). Many different circuits have been published covering AF filters of variable bandwidth, tuneable centre frequencies and for the insertion of notches. The full theoretical advantage of a narrow-band AF filter for CW reception may not always be achieved in operational use: this is because the human ear can itself provide a 'filter' bandwidth of about 50Hz with a remarkably large dynamic range and the ability to tune from 200 to 1000Hz without introducing 'ringing'.

MODIFICATIONS TO RECEIVERS

While the number of amateurs who build their own receivers from scratch is today in a minority, some newcomers or those with limited budgets buy relatively low-cost models or older second-hand receivers and then set about improving the performance. Old, but basically well-designed and mechanically satisfactory, valved receivers can form the basis of excellent receivers, often rather better than is possible by modifying some more recent low-cost receivers. The main drawback of the older receivers is their long warm-up period, making it difficult to

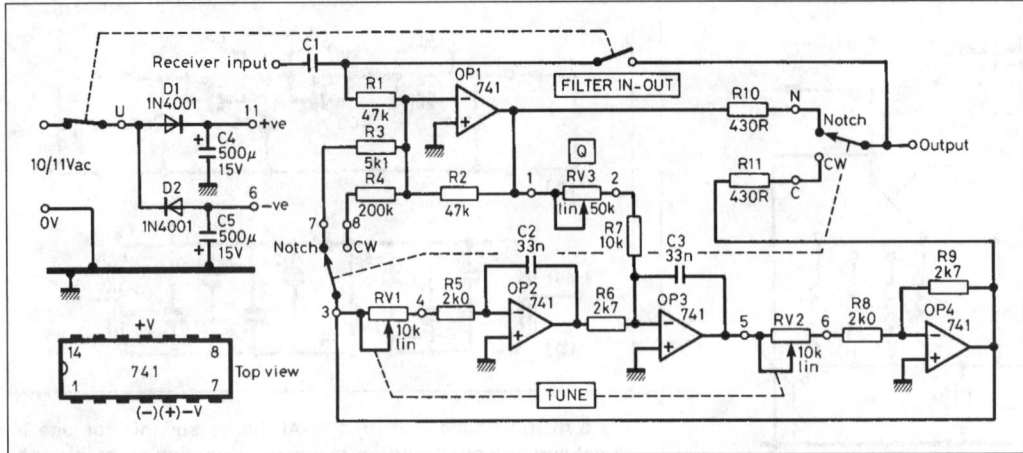

Fig 6.78: Versatile active analogue AF filter for speech or CW reception as described originally by DJ6HP in 1974 and which continues to represent an effective design. It provides a CW filter tuneable over about 450 to 2700Hz with the Q (bandwidth) variable over a range of about 5:1. For speech the filter can be switched to a notch mode. Although modern digital audio filters could provide more precisely shaped tuneable filtering, this analogue filter has received many endorsements over the years

receive SSB signals satisfactorily until the receiver has been switched on for perhaps 15 or 20 minutes.

Some of the older models using relatively noisy mixer stages may be improved on 14, 21 and 28MHz by the addition of an external preamplifier and such a unit may also be useful in reducing image and other spurious responses. However, high-gain preamplifiers should not be used indiscriminately since on a low-noise receiver they will seriously degrade the signal-handling capabilities without providing a worthwhile improvement of signal-to-noise ratio. Receivers having low noise but poor signal-handling capabilities can more often be improved by the fitting of a switched, adjustable or AGC-controlled antenna attenuator. Such an attenuator is likely to prove of most use on 7MHz where the presence of extremely strong broadcast signals will often result in severe cross-modulation and intermodulation.

A receiver deficient in selectivity can often be improved by adding a second frequency changer followed by a low-frequency (50 to 100kHz) IF amplifier (a technique sometimes known as a Q5-er); or by adding a crystal or mechanical filter, or by fitting a Q-multiplier. CW reception can be improved by the use of narrow-band audio filters, although the degree of improvement may not always be as much as might be expected theoretically because of the ability of an experienced operator to provide a high degree of discrimination.

Older receivers having only envelope detection may be improved for SSB and CW operation by the fitting of a product detector and possibly adding a good mechanical or crystal band-pass filter. NBFM reception may be included by adding an FM discriminator.

Many older receivers use single rather than band-pass crystal filters (and the excellence of the single crystal plus phasing control for CW reception should not be underestimated) and these often provide a degree of nose selectivity too sharp for satisfactory AM or SSB phone reception: speech may sound 'woolly' and virtually unintelligible due to the loss of high- and low-frequency components. However, because the response curve of such filters is by no means vertical, the addition of a high degree of tone correction (about 6dB/octave) can do much to restore intelligibility and the combination then provides an effective selectivity filter for SSB reception. The tone correction circuit shown in **Fig 6.80** is suitable for high-impedance circuits and can be adapted by using higher C and lower R for low-impedance circuits.

The addition of an antenna matching unit between receiver and antenna can improve reception significantly in those cases where appreciable mismatch may exist (for example when using long-wire antennas with receivers intended for use with a 50 or 70Ω dipole feeder) (**Fig 6.81**).

A common fault with older receivers is deterioration of the Yaxley-type wave-change switch and/or the connection to the rotor spindle of the variable tuning capacitors; such faults may often cause bad frequency instability and poor reset performance. Improvement is often possible by the careful use of modern switch-cleaning lubricants and aerosols.

A simple accessory for older receivers (or those modern receivers not already incorporating one) is a crystal calibrator providing 'marker' signals derived from a 100kHz or 1MHz crystal. While a simple 100kHz oscillator will usually provide harmonics throughout the HF range, the availability of integrated circuit dividers makes it practicable to provide markers which are not direct harmonics of the crystal. For example 10kHz or 25kHz or even 1kHz markers can be provided using TTL decade divider logic or divide-by-two devices.

A receiver deficient in HF oscillator stability on the higher frequency bands may still form the basis of a good tuneable IF strip when used on a low frequency band in conjunction with a crystal-controlled converter. Again, when the basic problem is oscillator drift due to heat, this can sometimes be reduced by fitting silicon power diodes in place of a hot-running rectifier valve or by

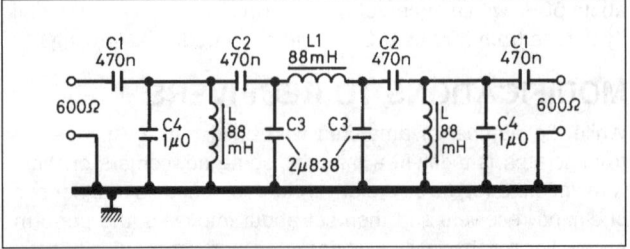

Fig 6.79: Passive AF filter by DJ1ZB using standard 88mH toroids and with a centre frequency of about 420Hz and bandwidth about 80Hz. Note that this design is for 600-ohm input/output impedance [*Sprat* No 58]

Fig 6.80: The single-crystal filter can be used effectively for phone reception by incorporating AF tone correction to remove the 'wooliness' of the heavily top-cut speech. A simple network such as the above provides top lift that restores intelligibility when used with the response curve of a typical single-crystal filter

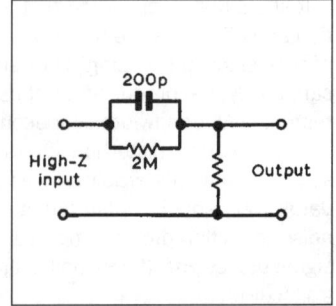

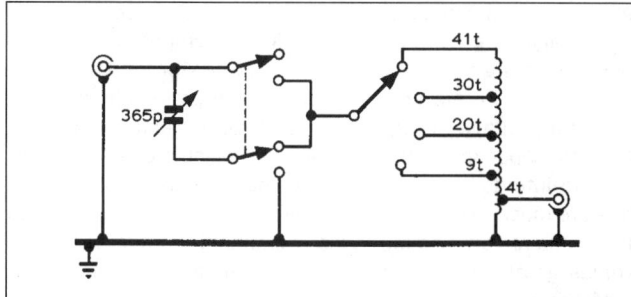

Fig 6.81: Typical antenna tuning and matching unit to cover 0.55 to 30MHz

adding temperature compensation to the HFO. A more drastic modification is to replace an existing valve HFO with an internal or external FET VFO. Excessive tuning rate can sometimes be overcome by fitting an additional or improved slow-motion drive. A receiver with a good VFO can be modified for really high-stability performance (better than about 20Hz) by means of external 'huff and puff' digital stabilisation using crystal-derived timing periods. See the chapter on oscillators in this book and [10, 11].

In brief, the excellent mechanical and some of the electrical characteristics of the large and solidly built receivers, such as the AR88, HRO, SUPER-PRO and some Eddystone models which featured single conversion with two tuned RF stages, are seldom equalled in modern 'cost-effective' designs. It may prove well worth spending time and trouble to up-grade these vintage models into receivers which can be excellent even by modern standards. Post-war Collins and Racal RA17L valved receivers remain highly regarded HF receivers.

The following summary indicates some common faults with older models and ways in which these can be overcome.

Poor sensitivity

Due to atmospheric noise this usually only degrades performance on 21 and 28MHz and then only on older valve models. Sensitivity can be improved by the addition of a preamplifier, but gain should not be more than is necessary to overcome receiver noise. Note that the sensitivity of a receiver may have been impaired by poor alignment, or by mismatched antennas, or due to the ageing of valves.

Image response

This can be reduced by additional pre-mixer selectivity, often most conveniently by means of a low-gain preamplifier with two or more tuned circuits. It is also possible to use a pre-tuned filter such as the Cohn minimum-loss filter for particular bands.

Stability

This is a direct function of the oscillators within the receiver. Excessive drift and frequency 'jumping' may be due to a faulty valve or band-change switch, or to incorrect adjustment of any temperature-compensation adjustments. Drift can sometimes be reduced by reducing the amount of heating of the oscillator coil by fitting heat screens, or by the addition of temperature compensation. But often with older receivers it will be found difficult to achieve sufficient stability on the higher frequency bands. In such cases considerably greater stability may be achieved by using the receiver as a variable IF system on one of the lower frequency bands, with the addition of one or more crystal-controlled converters for the higher frequency bands. It is worth noting that all oscillators (not only the first 'HFO') may be the cause of instability (eg second or third frequency conversion oscillator or even the beat frequency oscillator).

Tuning rate

The tuning rate of some older but still good receivers tends to be too fast for easy tuning of SSB and CW signals. Often this problem can be overcome by the fitting of an additional slow-motion drive on the main tuning control.

Selectivity

It is possible to improve the selectivity of a receiver by fitting an external low IF section, or by fitting a (better) crystal filter, or a Q-multiplier. Many SSB receivers make little provision for narrow-band CW reception and in such cases it may be possible to include a single-crystal filter with phasing control in one of the later IF stages, or to add a Q-multiplier or audio filter.

Blocking and intermodulation

Performance of many semiconductor (and some valve) receivers can be improved by the addition of even a simple antenna attenuator for use on 7MHz in the presence of extremely strong broadcast signals.

BUILDING RECEIVERS

For many years the percentage of home-built HF receivers in use on the amateur bands has been very low and increasingly has been confined to specialised sectors of the hobby such as compact, low-power (QRP) portable operation, often on one specific band and for a single mode of operation (often CW only) or as introductory receivers for those to whom the rather daunting cost of factory-built receivers or transceivers can represent a deterrence to HF operation.

There remain, however, valid reasons to encourage home-construction, not only for those with limited budgets but also as an ideal form of 'hands-on' learning process. A major advantage of building or modifying older receivers is that the constructor can then be confident of maintaining it in good trim. It is a substantial advantage for the amateur, particularly if located a long way away from the suppliers, to use equipment that he or she feels capable of keeping in good condition, and carrying out his or her own repairs when necessary. Increasingly factory-built equipment for the amateur does not lend itself to home-servicing.

Nor should it be forgotten that for those with the necessary practical experience, the availability of complex integrated circuits developed for consumer electronics and ceramic resonators that do not require skilled alignment has eased the construction of both simple and high-performance receivers. The factory designer must usually cater for all possible modes and bands, usually said to contain all the 'bells and whistles', most of which are not used in practice, whereas the constructor can build a no-compromise receiver to suit precisely his or her own particular interests. The amateur can still provide himself with a station receiver or transceiver, or a portable receiver or transceiver that can bear comparison with the best available factory models, and in doing so prove that the communications receiver is not a 'black box' or 'consumer appliance' of which the technology remains a largely unknown quantity.

It has traditionally been a feature of the hobby of amateur radio that the enthusiast strives to understand the technology; the home-construction and home-maintenance of equipment are not assets that should be surrendered lightly despite the undoubted attractions of factory-designed equipment.

Simple receivers, converters and single-band QRP transceivers can be built in an evening or two, but an advanced receiver - or receiver section of a transceiver - may take several months of work and adjustment.

Although the homodyne-type direct conversion receiver and the 'zero-IF' form in which the output from a first mixer is fed directly into a product detector make possible low-cost HF

receivers which are simple to build, care must be taken if optimum performance is to be achieved.

Because the necessary high overall gain (about 100dB) is achieved virtually entirely in the AF amplifier, these stages are very sensitive to hum pick-up at the 50Hz mains supply frequency and its harmonics. The AF stages should be well decoupled with the mixer and first AF stage in close proximity to enable the connections between them to be as short as possible; power supplies need to be well smoothed with care taken not to introduce 'earth loops' and/or direct pick-up from the magnetic field surrounding mains transformers and inductors. AF hum can usu-

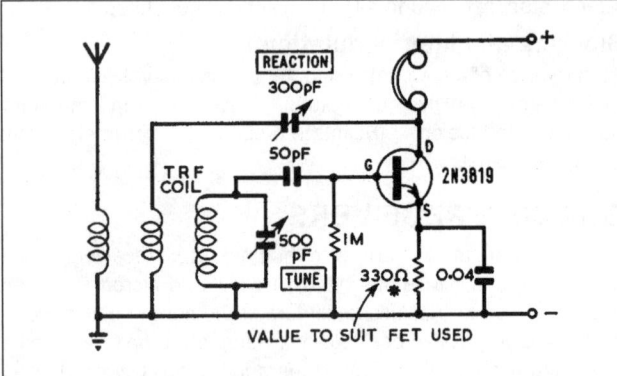

Fig 6.82: A single-FET receiver

ally be avoided altogether with battery-operated receivers.

The high AF gain may also introduce 'microphony' with components acting as microphones. A loudspeaker tends to make components vibrate and headphones are preferable, with the added advantage of requiring less AF output and overall gain. Ceramic capacitors, which exhibit piezo-electric characteristics, tend to introduce microphony; moulded polycarbonate capacitors are much to be preferred. Similarly, ferrite toroid cores used for AF filters can also introduce microphony. For inductors with values greater than 0.1mH, screened air-cored inductors are preferable.

Direct-conversion receivers of all types, including those with regenerative detectors, tend to suffer from local oscillator radiation from the receiving antenna and can cause interference in the locality, unless RF leakage through even a double-balanced mixer/demodulator is reduced, eg by the use of a broad-band isolator. A low-gain resonant or broad-band RF stage reduces oscillator radiation.

Microphony may also arise in the signal-frequency components, with the mixer acting as a phase detector, reflecting the interaction between the LO radiation or leakage and the incoming signal in a high-Q tuned circuit. Remember that the LO voltage is very much stronger than the incoming signal and it may be difficult to eliminate RF microphony altogether without effective screening and isolation.

Oscillator radiation can also result in RF hum that appears as 50Hz hum in the headphones. This arises from the oscillator signal becoming hum modulated in the mains wiring/rectifiers and then being re-radiated back to the receiver antenna. The cure is to stop the LO signal from radiating; all connections to the receiver, such as power supply and headphones, should be RF grounded to the receiver case using decoupling capacitors.

Designs for Home Construction

The simplest types of HF receiver suitable for home construction 'on the kitchen table' without the need for high-grade measuring equipment etc are undoubtedly the various forms of 'straight' (direct-conversion) receivers using either a regenerative

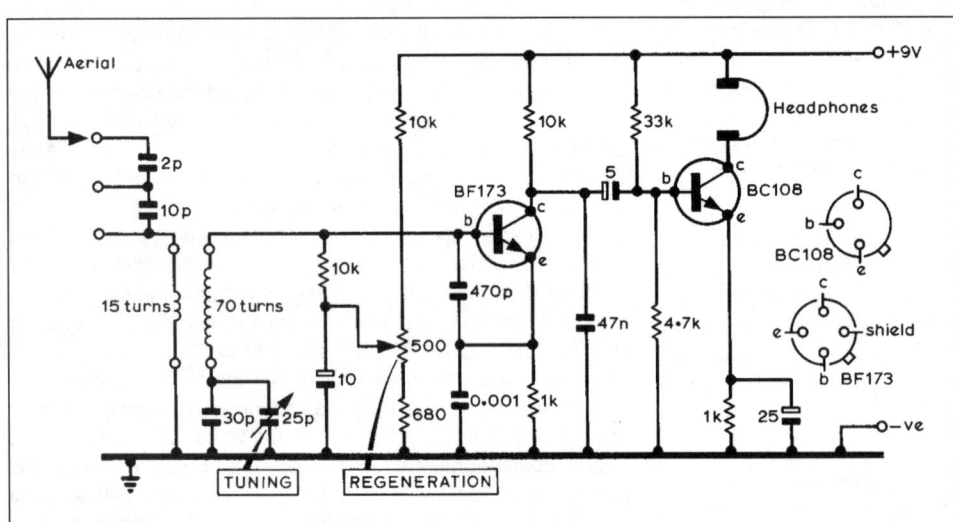

Fig 6.83: A simple 'straight' receiver intended for 3-5MHz SSB/CW reception and using Clapp-type oscillator to improve stability

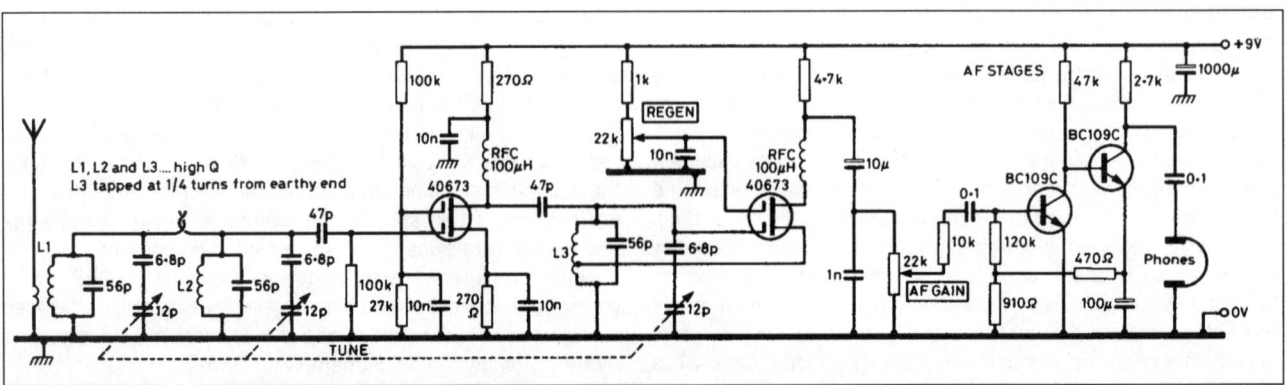

Fig 6.84: Solid-state 14MHz TRF 'straight' receiver originally described by F9GY in *Radio-REF* in the 1970s and intended primarily as a monoband CW receiver

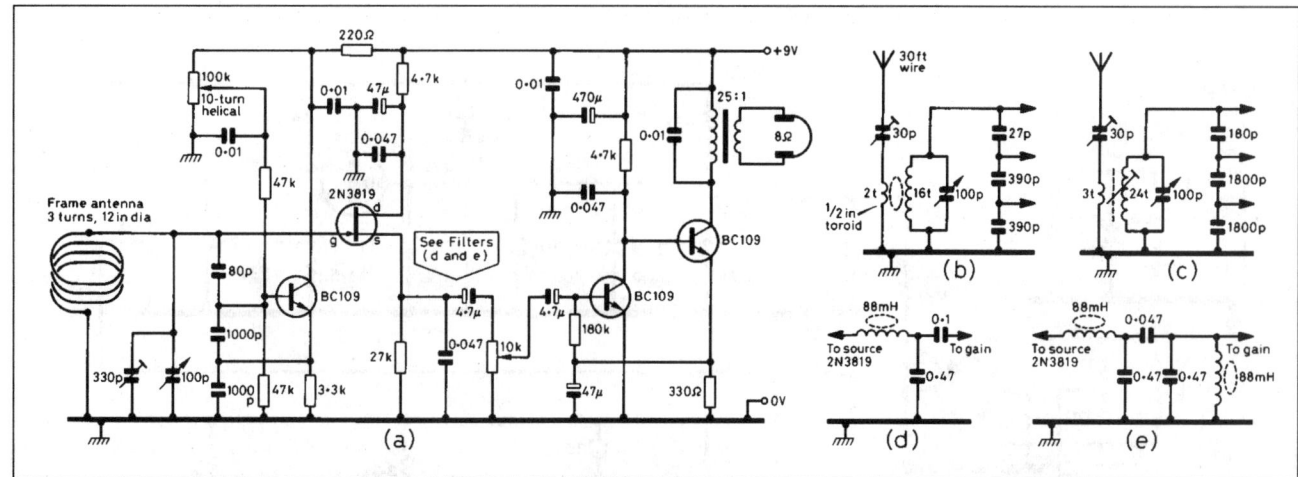

Fig 6.85: GI3XZM's solid-state regenerative 'blooper' receivers. (a) 3.5MHz version with frame antenna mounted about 12in above chassis with miniature coaxial-cable 'download'. (b) Input circuit for 9-16MHz version. (c) Conventional input circuit for 3.5MHz receiver using wire antenna (coil 26SWG close-wound on 0.5in slug-tuned former). (d) Audio filter that replaces the 4.7µF capacitor shown in (a). (e) CW filter 14MHz Mk2 version

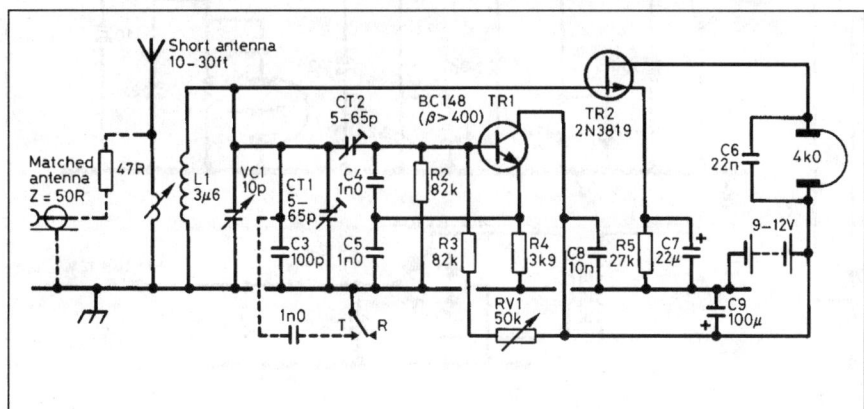

Fig 6.86: G3RJT's 'two-transistor communications receiver' or 'active crystal-set receiver', based on the design approach of GI3XZM but using a higher-gain drain-bend detector that permits the omission of the two-transistor AF amplifier provided that high-impedance headphones of good sensitivity are used

detector, or Q-multiplier plus source-follower detector, or the now more popular homodyne form of direct conversion, preferably using a balanced or double-balanced product detector. In practice, to avoid the complications of band-switching, most of the simple models tend to be either single-band receivers or may still use the once-popular 'plug-in' coils.

Figs 6.82-6.87 show a representative selection of circuit diagrams for simple receivers, suitable for use directly with high-impedance headphones, or with output transformers permitting the use of modern low-impedance headphones.

For simple receivers where a very high dynamic range is not sought, construction can be simplified by the use of SA602 type IC devices which contain a double-balanced mixer, oscillator and isolator stages. This device, originally developed for VHF portable radiophones, has been widely adopted by amateurs for use as frequency converters, complete front-ends for direct-conversion receivers, and as product detectors etc. **Fig 6.88** outlines the SA602 and some ways it can be used, while **Fig 6.89** shows its use as the front-end of a 28MHz direct-conversion receiver which could be adapted for lower HF bands.

The SA602 is equally suitable for use as a crystal-controlled frequency converter to provide extra bands in front of an existing receiver or for the 'super-gainer' form of simple superhet or as the mixer/oscillator stages of a conventional superhet, possibly using a second SA602 as a product-detector.

Receivers based on standard IC devices have particular application for 'listening' or as the receiver section of compact low-power transceivers.

Fig 6.90 shows how a Motorola MC3362 IC can form the complete front-end of a single-band superhet including IF and detector stages as part of a compact transceiver. The chip pinout is in **Fig 6.91**.

Super-linear Front-ends

The front-end of a superhet or direct-conversion receiver comprises all circuitry preceding the main selectivity filter. For a superhet this includes the passive preselector, the RF amplifier(s), the mixer(s) and heterodyne oscillator(s), the diplexer between mixer and post-mixer amplifier, the roofing filter, and any IF stages up to an including the main (crystal) filter. For a direct-conversion receiver, the front-end comprises all stages up to the selective audio-filter(s), including the product detector (which for a high-performance receiver may be of the two-phase, audio-image-rejecting type).

For any receiver, superhet or direct-conversion, in which digital signal processing at IF or audio baseband is used to determine the selectivity, the A/D converter and digital filtering must be considered in determining the front-end performance in terms of linearity and dynamic range.

In designing a receiver for the highest possible front-end performance, attention must be paid to all of the circuitry involved, to the gain distribution, to the noise characteristics, to both the strong-signal handling characteristics and to the intermodulation intercept points. The ability to hold and copy an extremely weak signal, barely above the atmospheric noise level, adjacent to a strong local signal or close to signals from super-high-power

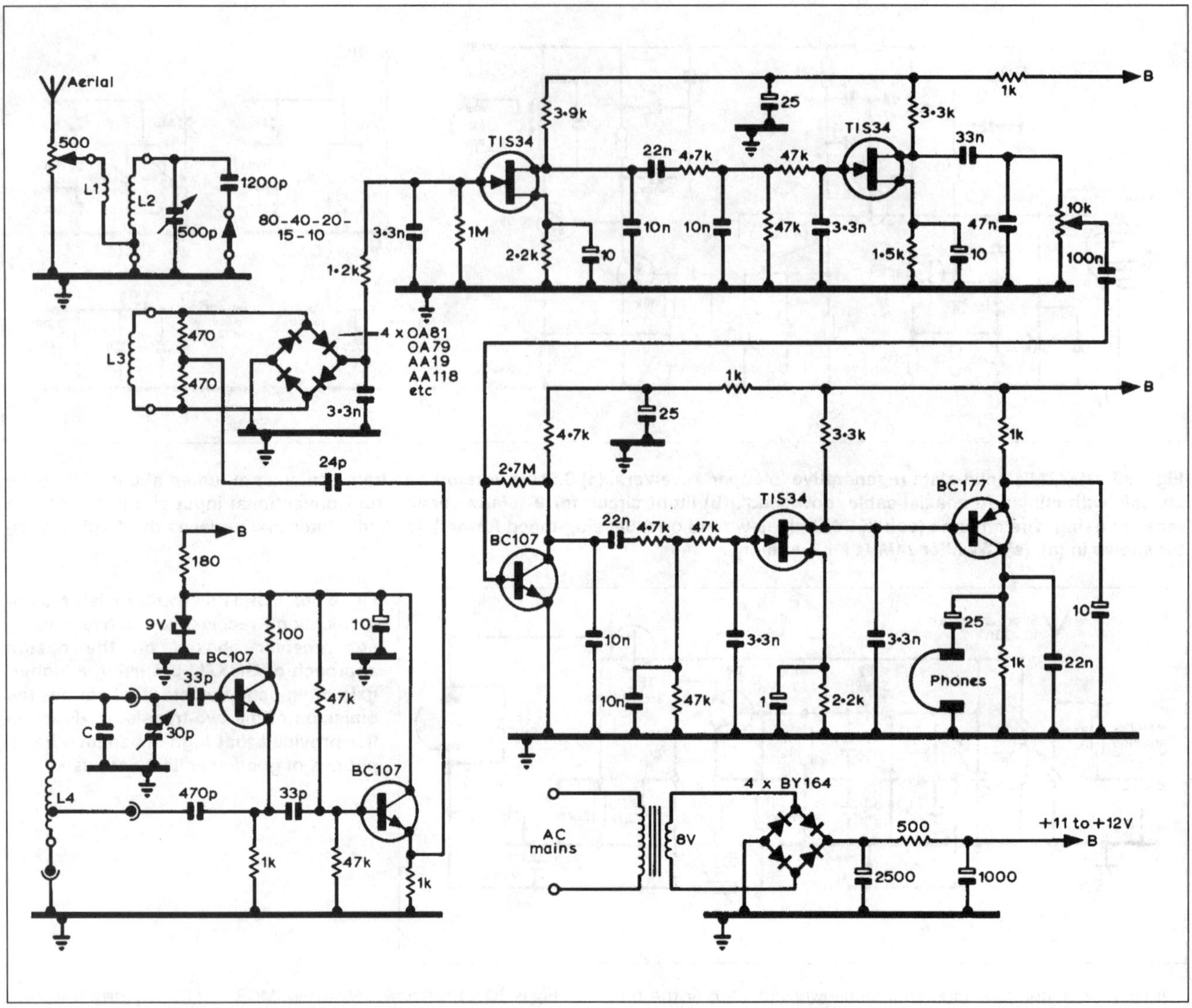

Fig 6.87: A multiband direct-conversion receiver using diode ring demodulator and plug-in coils for oscillator section

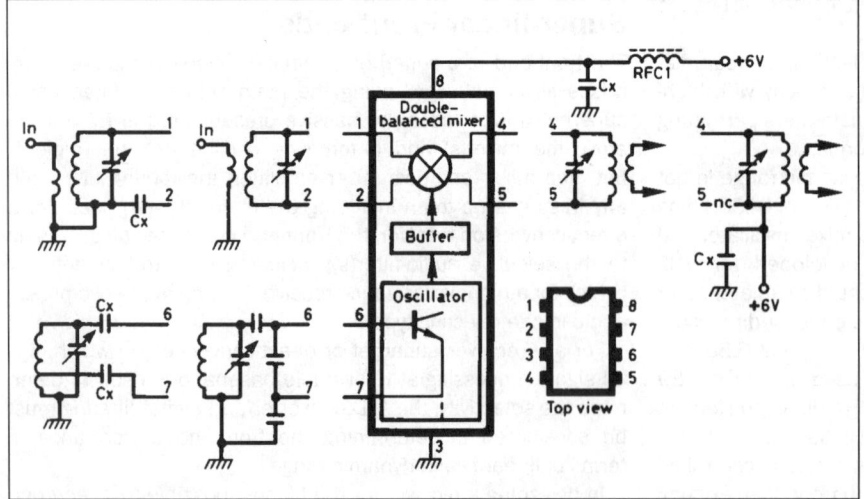

Fig 6.88: Typical configurations of the SA602. Balanced circuits are to be preferred but may be more difficult to implement. Cx blocking capacitor 0.001 to 0.1µF depending on frequency. RFC1 (ferrite beads or RF choke) recommended at higher frequencies. Supply voltage should not exceed 6V (2.5mA). Noise figure about 5dB. Mixer gain 20dB. Third-order intercept 15dBm (do not use an RF preamplifier stage). Input and output impedances are both 2 x 1.5kΩ.

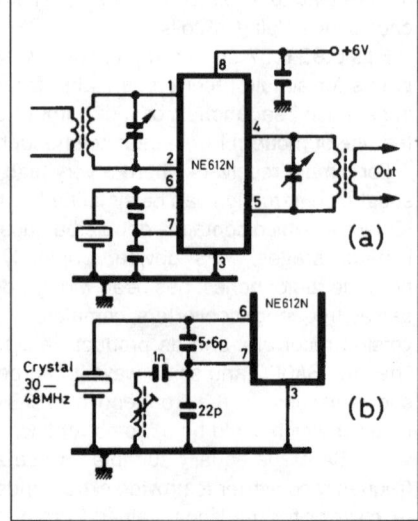

Fig 6.89: Use of SA602 as a crystal-controlled converter. The 5dB noise figure is low enough for optimum sensitivity up to about 50MHz without an RF amplifier

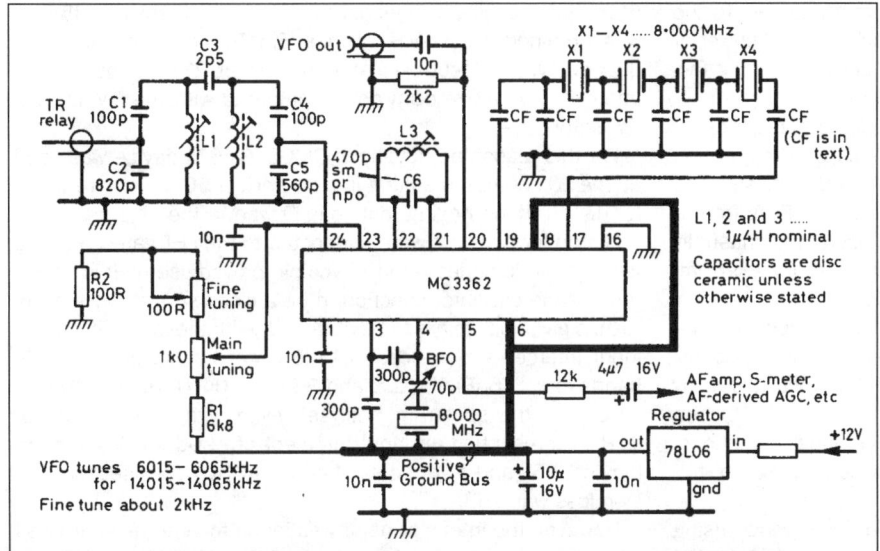
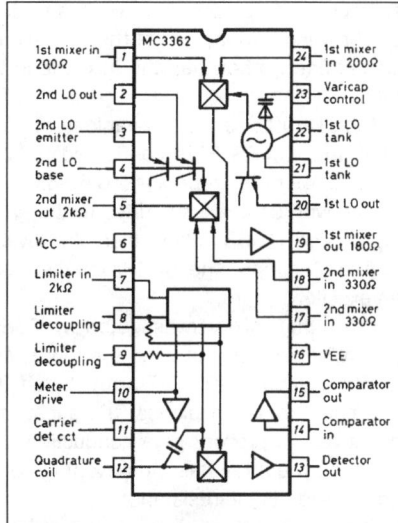

Fig 6.90: K9AY uses the MC3362 as the complete front-end of the 14MHz superhet receiver

Fig 6.91: The MC3362 chip showing pinout and basic functions

broadcast signals places heavy demands on the active and passive components available within amateur budgets, including any ferrite-cored transformers and the crystal filters.

The limiting factors in the design of high-performance, solid-state HF receivers remain the spurious-free dynamic range (SFDR) of the mixer-stage, the noise and stability of the associated oscillator and strong-signal performance of the filters. Jacob Makhinson, N6NWP [12] believes, by applying known design principles, radio amateurs can construct a high-performance front-end which combines a very high intercept point with excellent sensitivity. Used with a low-noise local oscillator, a front-end based on a DMOS FET quad device as a double-balanced switching mixer and low-noise square-wave injection, obtained by means of a dual flip-flop followed by a simple diplexer network at an IF of 9MHz, can achieve a wide dynamic range even when a suitable push-pull RF low-noise amplifier is used ahead of the mixer. N6NWP stresses that: "A receiver incorporating such a front-end can provide strong-signal performance that rivals or exceeds that of most commercial equipment available to the amateur."

Investigations of the N6NWP front-end (**Fig 6.92**) have found that it is possible to achieve extremely high third-order intercept points (+50dBm on 1.8, 3.5 and 7MHz, +45dBm on 14MHz

Fig 6.93: G3SBI's modified N6NWP-type mixer test assembly using the SD5000 FET array and 74AC74N to provide square-wave injection from a high-quality signal generator source at twice the required frequency

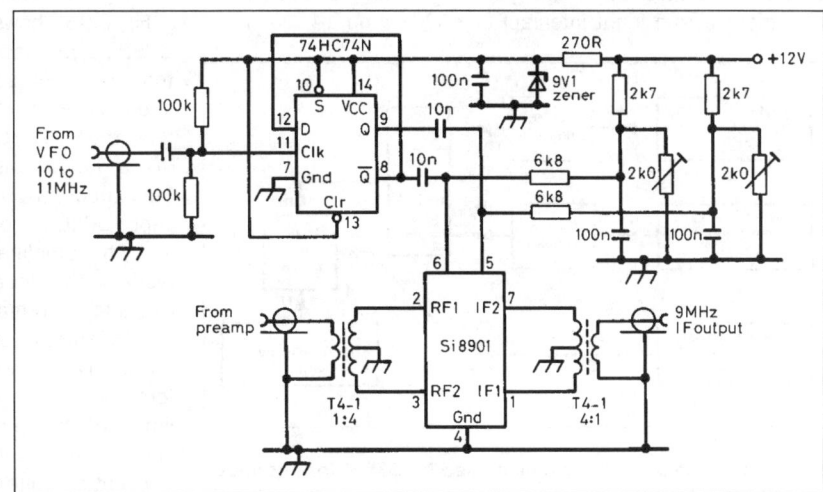

Fig 6.92: The basic high-dynamic-range MF/HF receiver front-end mixer circuitry as developed by N6BWP

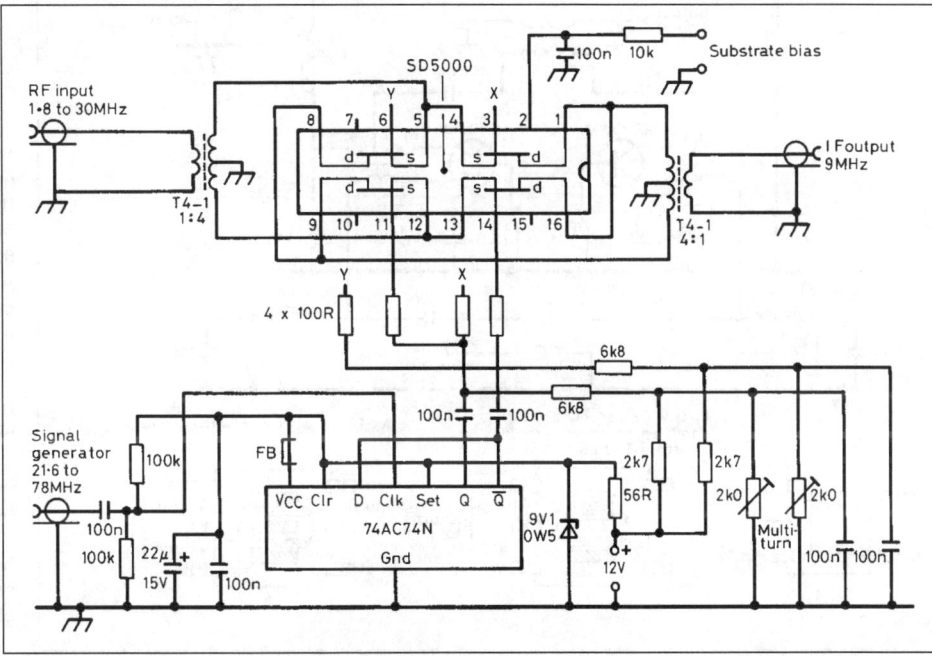

subject to some spread in devices in different batches) to the extent where other parts of the circuitry, such as the diplexer or crystal filter rather than the mixer, tend to become limiting factors.

The SD8901 device (available from Calogic Corporation) contains four DMOS FETs configured for use as a commutation (switching) mixer (see Figs 6.60-6.62). N6NWP utilises this device with square-wave drive to the gates of the FETs from a high-speed CMOS D-type bistable device operating unusually from a 9V supply. The 14MHz intercept point is about +39dBm, an excellent figure.

The initial mixer investigated by G3SBI was based on the N6NWP approach but used the more widely available Siliconix SD5000 quad DMOS FET array (batch 9042). Since this array has gate-protection diodes, the substrate needs to be biased negatively to prevent gate conduction under some conditions; however, the array has the advantage of close matching of the drain-to-source on-resistance.

The test board using the circuit of **Fig 6.93** was made using earth-plane construction, with all transformers and ICs fitted into turned-pin DIL construction, so that they could be changed easily. With the test set-up of **Fig 6.94**, it initially proved possible to achieve a true input intercept of +42dBm on 14MHz using

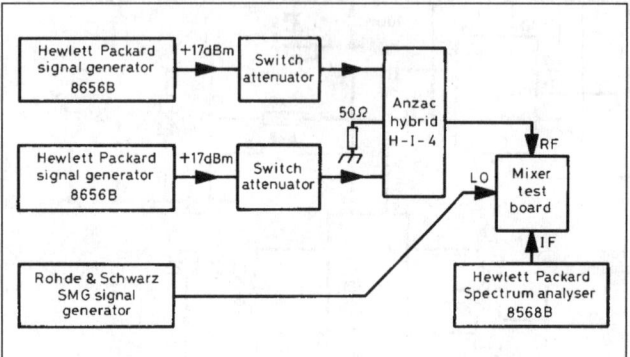

Fig 6.94: The test instrumentation used by G3SBI for intermodulation tests on the mixer

5MHz local oscillator injection. An input intercept of +45dBm was obtained on 3.5MHz with a 5.5MHz local oscillator frequency. With a local oscillator running at 23MHz the 14MHz intercept was a few dBm down compared with the 5MHz local oscillator.

For this reason an advanced CMOS 74AC74 device was used as the LO squarer, resulting in near-perfect 50-50 square waveform. To reduce ringing, only one D-type in the chip was used and stopper resistors were connected to the FET gates; a single ferrite bead in series with the Vcc pin proved useful. It is important for the oscillator injection to be a clean square wave if the results given above are to be obtained. With these modifications input intercepts of at least +42dBm were achieved on all HF bands and +46dBm on 1.8 and 3.5MHz. On 7MHz no IMD was visible on the spectrum analyser, even with a bandwidth of 10Hz, representing an input intercept of +50dBm. A substrate bias of -7.5V and a gate bias of about +4.5V were used; conversion loss was 7dB.

However, the intercept point was found to degrade sharply as soon as the input signal exceeds +7dBm (0.5V); the situation can be recovered by dropping the gate bias voltage, but it is then no longer possible to achieve intercept points above +45dBm.

Fig 6.95 shows a 9MHz post-mixer amplifier developed by G3SBI, again based on the N6NWP approach but with changes that provide improved performance in terms of gain and output intercept point, with a noise figure of 0.5dB. Whereas N6NWP used the MRF586 device, G3SBI used the MRF580A device, giving a lower noise figure at a collector current of 60mA. Measured performance showed a gain of 8.8dB, output intercept +56dBm, noise figure 0.5dB. However, a crystal filter driven by the amplifier would present a complex impedance, particularly on the slope and near the stop-band, and would seriously degrade performance of the amplifier.

G3SBI also investigated the performance of quadrature hybrid 9MHz crystal-filter combinations, and he found that the performance of budget-priced crystal lattice-type filters is a serious limitation (home-made ladder filters appear to have higher intercept points and lower insertion loss although the shape factor may not be quite so good). The problem with budget-priced lat-

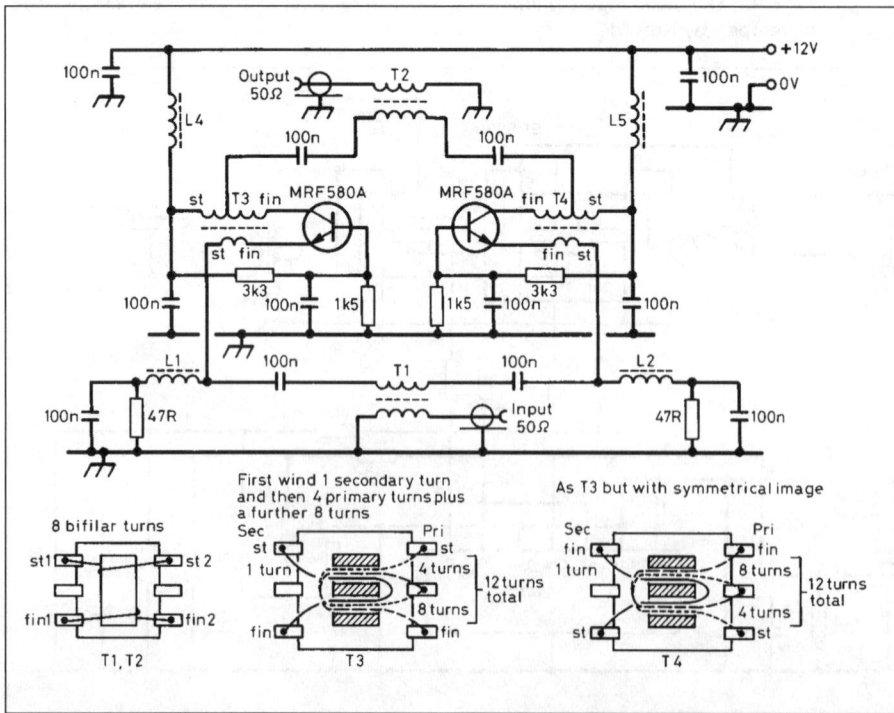

Fig 6.95: G3SBI's modified post-mixer test amplifier adapted from the N6NWP design but using the MRF580A devices. All resistors 0.25W metal-film RS Components. All 0.1μF capacitors monolithic ceramic RS Components. L4, L5 4t of 0.315mm dia bicelflux wire on RS Components ferrite bead. L1, L2 5t 0.315mm dia bicelflux wire (RS Components). T1-T4 use 40swg bicelflux wire. Take two glassfibre Cambion 14-pin DIP component headers, cut each into two parts and bend the tags 90° outwards. Stick a piece of double-sided tape onto the header and mount the balun cores on this. Wind the transformers as shown above. The amplifier is constructed with earth-plane layout

tice filters can be reduced by eliminating the post-mixer amplifier with the mixer going immediately to a quadrature hybrid network 2.4kHz-bandwidth filter, followed by a low-noise amplifier. The 2.4kHz filter is then used as a roofing filter. Although this is not an ideal arrangement, it can result in an overall noise figure of about 13.5dB (5dB noise figure due to the filter and amplifier, another 7dB from mixer loss, and a further 2.5dB loss due to the antenna input band-pass filter). This would be adequate sensitivity on 7MHz without a pre-mixer amplifier.

The N6NWP-type mixer followed basically accepted practice in commutation (switching mixers) achieving a +50dBm input intercept point on 7MHz when used with a precise square-wave drive, but the performance fell off on bands lower and higher in frequency than 7MHz. Results could be improved on the lower frequency bands by altering the capacitive balance of the RF input and above, but this had no significant effect on 14MHz and above.

Fig 6.96 illustrates a conventional commutation ring mixer. If A is 'on', FETs F1 and F3 are 'on' and the direction of the RF signal across transformer T2 is given by the 'F' arrows. A deficiency of this arrangement is that as the RF input signal level increases, it has a significant effect on the true gate-to-source voltage needed to switch the FET 'on' or keep it switched 'off'. Larger local oscillator amplitudes are then required, but linearity problems may still exist because of the difference in the FET 'on' resistance between negative and positive RF signal states.

The new mixer developed by G3SBI (intellectual title held by SERC) is shown in **Fig 6.97** which illustrates why it has been given the name 'H-mode'. Operation is as follows: Inputs A and B are complementary square-wave inputs derived from the sine-wave local oscillator at twice the required frequency. If A is 'on' then FETs F1 and F3 are 'on' and the direction of the RF signal across T1 is given by the 'E' arrows. When B is 'on', FETs F2 and F4 are 'on' and the direction of the RF signal across T1 reverses (arrows 'F'). This is still the action of a commutation mixer, but now the source of each FET switch is grounded, so that the RF signal switched by the FET cannot modulate the gate source voltage.

In this configuration the transformers are important: T1 is a Mini-Circuits type T4-1; T2 is two Mini-Circuits T4-1 transformers with their primaries connected in parallel. The parallel-connected transformers give good balance and perform well.

A practical test circuit of the H-mode mixer is shown in **Fig 6.98**. It was constructed on an earth plane board with all transformers and ICs mounted in turned-pin DIL sockets. The printed circuit tracks connecting T1 to T2 and from T2 to the SD5000 are kept short and of 0.015in width to minimise capacitance to ground. The local oscillator is divided by two in frequency and squared by a 74AC74 advanced CMOS bistable similar to that used in the N6NWP-type mixer. However, the bistable is run from +10V instead of +9V and a cut-down RS Components ferrite bead is inserted over the ground pin of the 74AC74 to clean up the square wave.

The preferred method of setting the gate-bias potentiometers with the aid of professional-standard test equipment is as follows. One potentiometer is set to the desired bias voltage for a

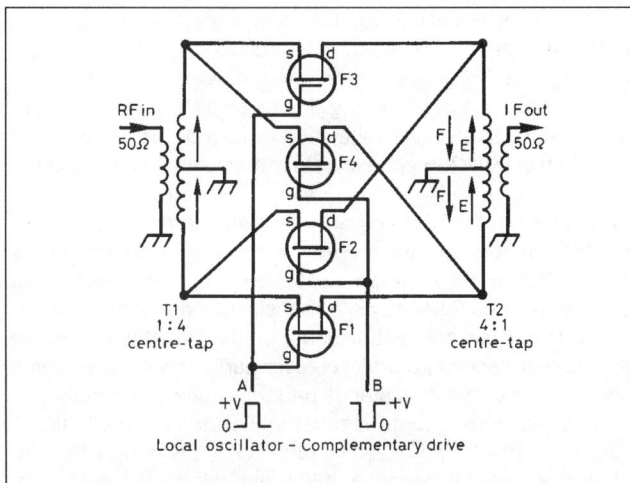

Fig 6.96: Conventional commutation ring mixer

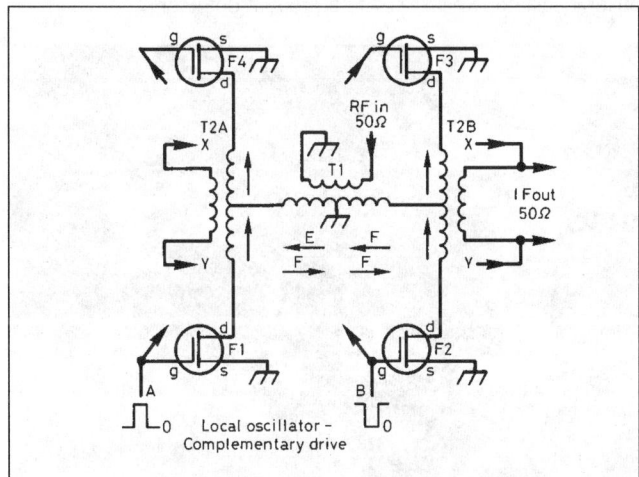

Fig 6.97: G3SBI 'H-mode' mixer

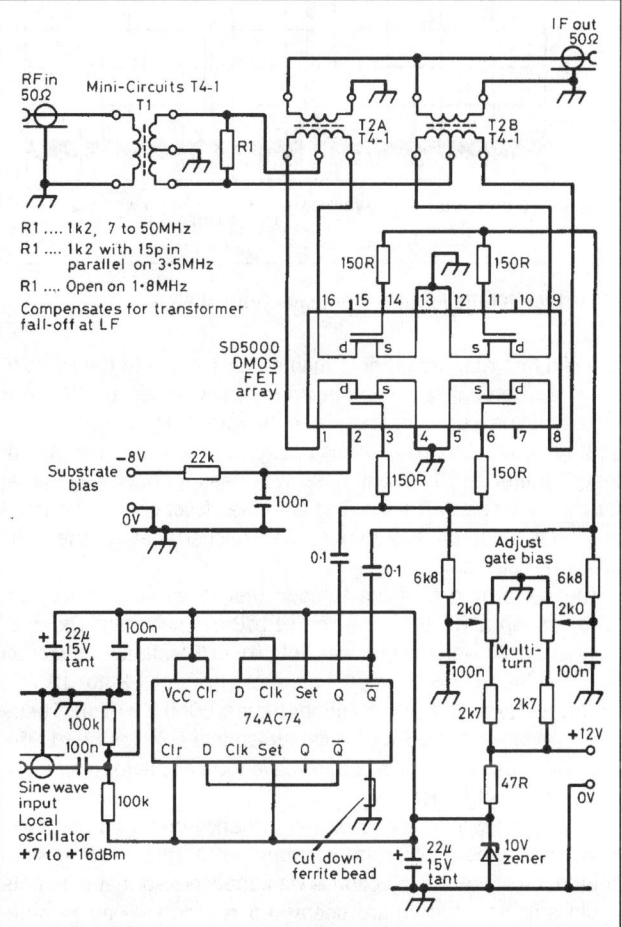

Fig 6.98: Test assembly for 'H-mode' mixer

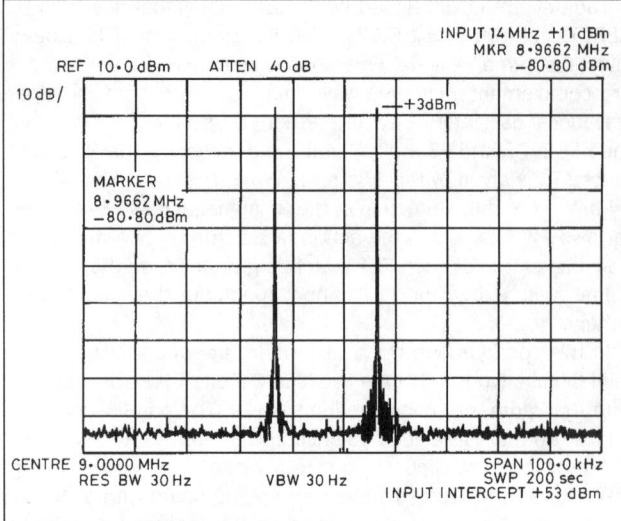

Fig 6.99: 14MHz input intermodulation spectrum

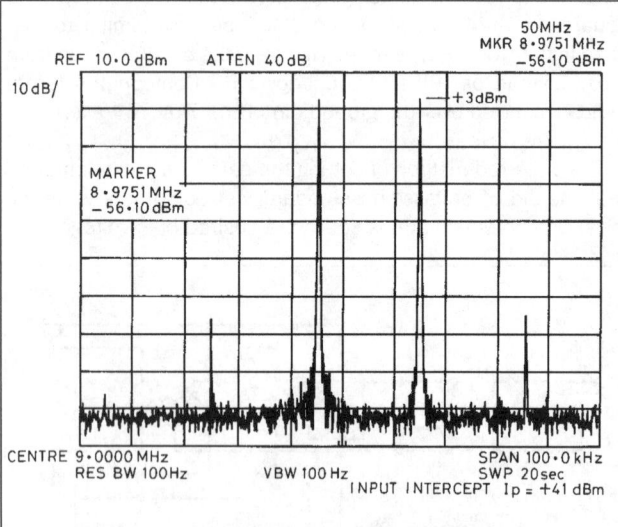

Fig 6.100: 50MHz input, output spectrum at 9MHz

specific test run, the other is then set by looking at the RF-to-IF path feedthrough on the spectrum analyser at 14MHz and adjusting the potentiometer for minimum IF feedthrough. The setting is quite sharp and ensures good mixer balance. An RF test signal of 11dBm (0.8V RMS) was used for each test signal for the two-tone IMD tests. The gate bias level chosen enabled an input level of +12dBm to be reached before the IMD increased sharply.

The performance of the H-mode test mixer was as follows. With an input RF test level of +11dBm (spaced at 2kHz or 20kHz) the conversion loss was 8dB; RF to IF isolation -68dB; LO to IF isolation -66dB. Input intercept points: 1.8 to 18MHz +53dBm; 21 to 28MHz +47dB, or better; 50MHz +41dB. These results were achieved with a gate-to-source DC bias of +1.95V and -8V substrate bias, a square-wave local oscillator amplitude of 9V and IF at 9MHz.

It seems likely that a good performance could be achieved with an H-mode mixer transformer-driven from a sine-wave source, provided the injection is via capacitors so that bias pots could still be used. There seems no reason why an H-mode mixer should not be used in an up-conversion arrangement rather than for a 9MHz IF.

RECEIVER PROJECTS

The Yearling

This easy-to-build receiver design (**Fig 6.101**) by Paul Lovell, G3YMP, was first published in *D-i-Y Radio* and later in *Radio Communication*. The original design was for the 20m band but it has been extended to cover 80m as well. The receiver is powered from either a 9V PP3 battery or mains adapter and can be built either on the printed circuit board (PCB) supplied with the kit or (with the help of an experienced constructor) on a prototype board. The building instructions here are based on the PCB and kit. Either headphones or a loudspeaker can be used.

Circuit design

Direct conversion was the first option considered. This type of receiver has the merit of simplicity but, unless carefully designed and constructed, can suffer from problems such as strong breakthrough from broadcast stations on nearby frequencies. Also, a stable VFO at 14MHz, while certainly not impossible, is not something to be undertaken lightly by a beginner. The superhet was then considered. This overcame some of the problems associated with direct conversion but created others such as the need for a rather expensive IF filter. So the result was a happy compromise, which in effect is a direct-conversion receiver preceded by a frequency converter. This means that the VFO runs at about 5MHz instead of 14MHz, so stability is much better.

The circuit is given in **Fig 6.102**. Incoming signals at 14MHz or 3.5MHz are selected by the tuned circuit L1/C1/D1 or filter FL1. Note that the ANT TUNE control is not needed on 80m as FL1 provides the necessary filtering. Tuning on 14MHz is carried out by RV2 which adjusts the voltage on varicap D1. This is one half of diode type KV1236 - note carefully the polarity of this component.

A Philips SA602 (IC1) converts the signal to the range 5.0 to 5.5MHz (approx) by means of its internal oscillator. This has a crystal (X1) working at about 8.9MHz. In fact any crystal between 8.8 and 9.0MHz will be satisfactory but a frequency of 8.95MHz will give greatest accuracy on the dial. It will be noted that D1 is in fact forward biased over part of its voltage range. However, the circuit as it stands performs quite adequately.

The signal output from the mixer in IC1 then passes to the IF filter formed by C5 and L2. The tuned circuit is damped by the rather low output impedance of IC1, and this gives a nice compromise between selectivity and insertion loss. The balanced output of the tuned circuit is applied to IC2, another SA602 mixer/oscillator which acts as a product detector.

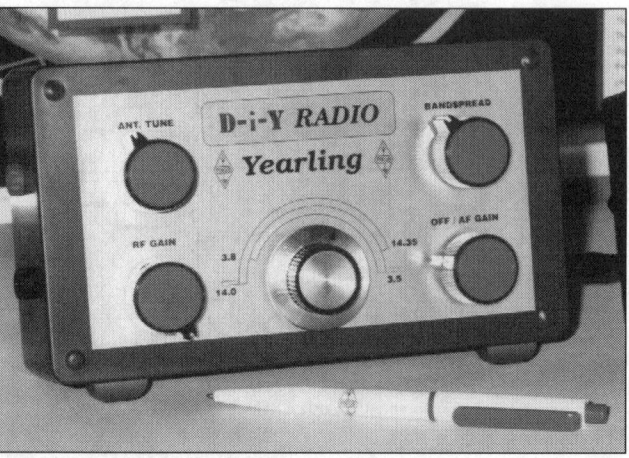

Fig 6.101: The Yearling receiver

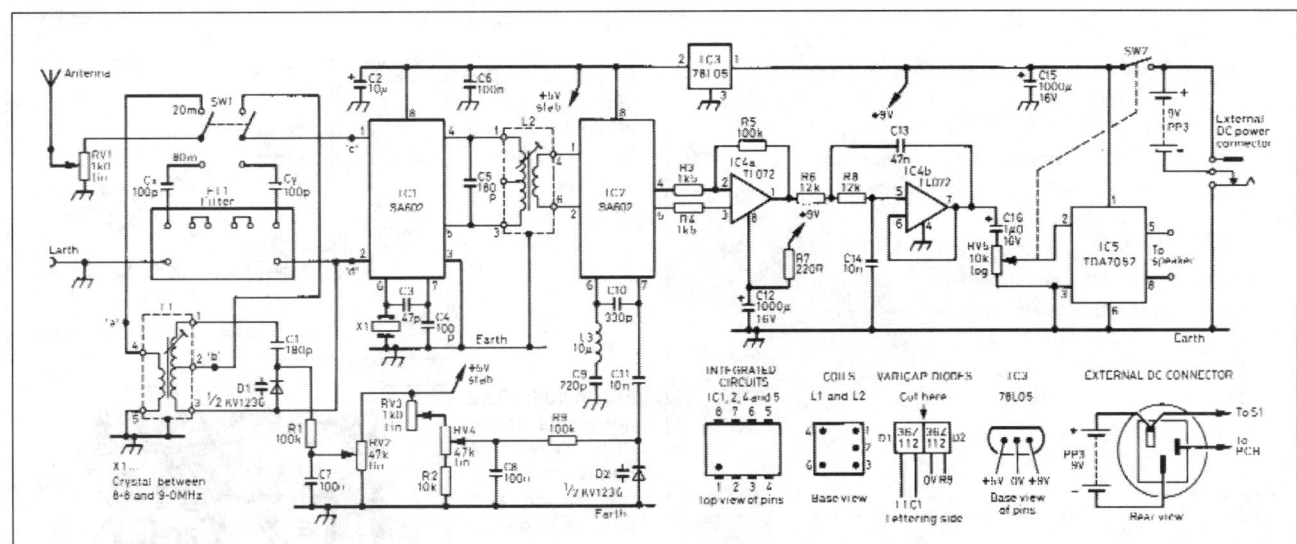

Fig 6.102: Circuit diagram of the Yearling

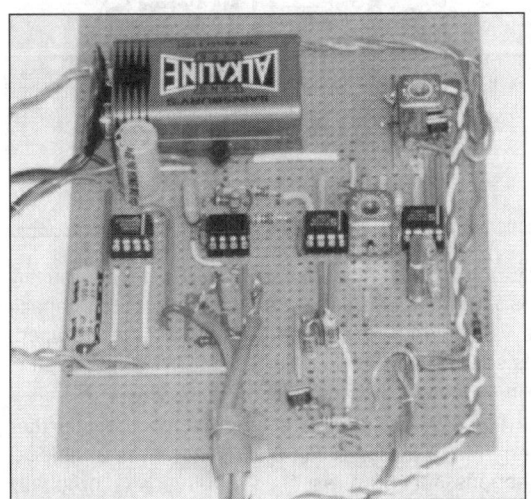

Fig 6.103: The Yearling may also be built on a proto-type board

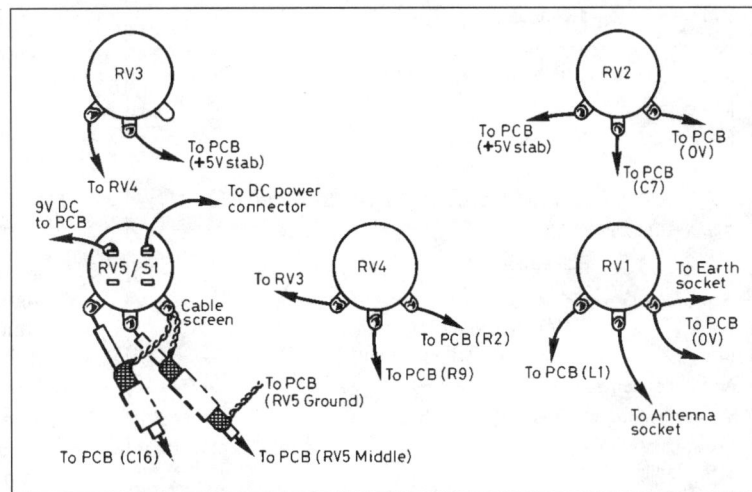

Fig 6.104: Rear view of the variable resistors. Check the connections carefully to make sure that the wires fit the correct holes on the board

The main VFO uses the oscillator section of IC2, which covers a range of approximately 5.0-5.5MHz. Assuming the use of a 9MHz crystal for X1, the 20m band will track within the range 5.00-5.35MHz and the 80m band from 5.2-5.5MHz. Note that the LF ends of the respective bands will be at opposite ends of the dial, since 20m makes of the sum of the receiver's two oscillator frequencies, and 80m uses the difference between them.

Main tuning is carried out by RV4, and RV5 provides the bandspread control. Tuning is by means of the voltage on varicap D2 which, in association with C9, C10 and L3, determines the frequency of oscillation. The varicap is a dual type - cut in two with a sharp knife. Voltage regulator IC3 in the supply lines to the early stages makes stability surprising good. The audio output from IC2 is amplified by IC4a and filtered by low-pass filter IC4b, before being further amplified to speaker level by IC5.

Construction

Many Yearlings have been built from kits without problems, while a number of constructors have successfully used a prototype board instead of the PCB. The prototype was built using just such a method as illustrated in **Fig 103**. No special precautions are needed but, as with any radio, neat wiring makes the tracing of faults a much easier process.

Fig 6.104 shows the connections to the gain and tuning controls. Screened cable should be used for the leads to the volume control but stranded bell wire should be satisfactory elsewhere. Incidentally, there was no problem with using IC sockets for all the 8-pin devices. The coils are colour coded, with L1 having a pink core and L2 a yellow one.

It is rather easy to wire the varicaps incorrectly but, if using the PCB, the lettering on D1 should be next to coil L1 and the lettering on D2 should be facing resistor R7. **Fig 6.105** shows the bandchange switch and the 80m filter which is glued to the side of the case. Holes for the five controls are 10.5mm diameter, and the speaker and power connectors have 6.3mm and 11mm holes respectively. The antenna and earth sockets need 8mm holes.

Setting it up

Connect a 9V battery and a reasonable antenna, and, on switching on, some stations - or at least some whistles - should be heard. It is suggested that a start is made with the 20m band, and the adjustments made before fitting the controls and sockets to the case.

A signal generator is useful, of course, but not essential to get the receiver working. The following steps should be followed, where the Yearling should burst into life.

Capacitors

All rated at 16V or more

C1, 5	180p polystyrene, 5% or better tolerance
C2	10µ electrolytic
C3	47p polystyrene, 5% or better tolerance
C4, Cx, Cy	100p polystyrene, 5% or better tolerance
C6-8	100n ceramic
C9	220p polypropylene, 2% or better tolerance
C10	330p polypropylene, 2% or better tolerance
C11, 14	10n ceramic
C12, 15	1000µ electrolytic
C13	47n, 5% polyester
C16	1µ electrolytic

Resistors

All 0.25W 5%

R1, 5, 9	100k
R2	10k
R3, 4	1k5
R6, 8	12k
R7	220R
RV1, 3	1k linear
RV2, 4	47k linear
RV5	10k log with switch (SW2)

Inductors

L1	Toko KANK3335R
L2	Toko KANK3334R
L3	10µ, 5% tolerance (eg Toko 283AS-100)

Semiconductors

IC1, 2	Philips SA602
IC3	78L05 5V 100mA regulator
IC4	TL072 dual op-amp
IC5	Philips TDA7052 audio amp

Additional items

Varicap diode, Toko KV1236 (cut into two sections)

Crystal, between 8.8 and 9.0MHz. An 8.86MHz type is available from JAB, Maplin, etc

Wavechange switch, DPDT changeover type

8-pin sockets for IC1, 2, 4 and 5

4mm antenna (red) and earth (black) sockets

3.5mm chassis-mounting speaker socket

DC power socket for external power supply (if required)

4 knobs, approx 25mm diameter with pointer

Tuning knob with pointer, eg 37mm PK3 type

Printed circuit board or prototype board

Plastic case, approx 17 x 11 x 6cm, eg Tandy 270-224.

Speaker between 8 and 32 ohm impedance (or headphones)

Kits of components and PCB are available from JAB Electronic Components, The Industrial Estate, 1180 Aldridge Road, Great Barr, Birmingham B44 8PE. E-Mail jabdog@blueyonder.co.uk

Table 6.3: Components list for the Yearling receiver

1. Set the core of L2 to mid-position.
2. Set RV1, RV2 and RV4 to mid-position and rotate the core of L1 until you hear a peak of noise. Now adjust L2 for maximum noise.
3. Tune carefully with the main tuning control RV4 until you hear amateur signals. Adjustment of the bandspread may be needed to clarify the speech.
4. Switch off the receiver and fit the controls and sockets to the case.

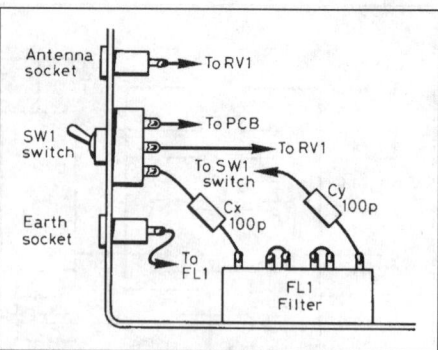

Fig 6.105: Internal view of the yearling case. The 80m filter is attached to the base with glue

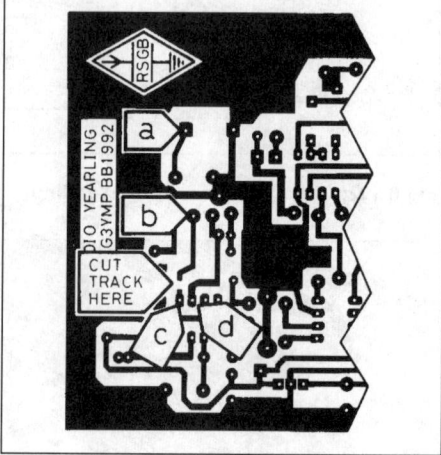

Fig 6.106: The underside of the PCB. Wires are connected from the switch and filter as shown

5. Finally, adjust the tuning knob so that the pointer roughly agrees with the dial. Due to the spread of varicap capacitance values, you may find the tuning a little cramped. This is easily fixed by adding a resistor (try 22kΩ to start with) in series with RV3 or adjusting the value of R2.

6. Check the 80m band - this should work without further adjustments to the coils. **Fig 6.106** shows the additional connections for 80m as the Yearling was originally designed for 20m only.

7. Finally, fix the PCB inside the case (double-sided sticky tape works well).

Experimental Direct Conversion Polyphase Receiver

This experimental design by Hans Summers, G0UPL, combines new and old techniques to produce a simple but high performance direct conversion receiver. As discussed above, the most obvious disadvantage of direct conversion is that both sidebands are detected. This can be solved via the use of audio phasing networks such that the unwanted sideband is mathematically cancelled by summation of correctly phased signals. Conventionally this required the use of two mixers, fed by phase shifted RF inputs.

A recent commutative mixer design by Dan Tayloe, N7VE, produces four audio outputs at 90-degree phase angles using a very simple circuit. Despite its simplicity the detector boasts impressive performance as shown in **Table 6.4**.

The circuit is not really a mixer producing both sum and difference frequencies: it might more accurately be described as a switching integrator. It possesses a very useful bandpass filter characteristic, tracking the local oscillator frequency with a Q of typically 3000.

The quadruple phased audio output of the Tayloe Detector is ideally suited to drive a circuit of much older heritage: the pas-

Conversion Loss:	1dB
Minimum Discernable Signal (MDS):	-136dBm
Two Tone Dynamic Range (2TDR):	111dB
Third Order Intercept (IP3):	+30dBm
Bandpass RF filter characteristic:	Q ~3000

Table 6.4: N7VE's performance figures for his commutative mixer

BAND MHz TYPE	T1 - T2 Inductance	T1 - T2 [pF]	C1 - C3 [pf]	C2
1.8 - 2.0	3333	45µH	150	12
3.5 - 3.8	3333	45µH	39	3.3
7.0 - 7.1	3334	5.5µH	100	8.2
10.1 - 10.15	3334	5.5µH	47	6.8
14.0 - 14.35	3334	5.5µH	22	3.3
18.07 - 18.17	3335	1.2µH	68	6.8
21.0 - 21.17	3335	1.2µH	47	4.7
24.89 - 24.99	3335	1.2µH	33	3.3
28.0 - 29.7	3335	1.2µH	22	3.3

Table 6.5: Details of using the receiver on all HF amateur bands

sive polyphase network. This consists of a resistor-capacitor network with resonant frequencies (poles) tuned such that they occur at evenly spaced intervals across the audio band of interest. The effect is a quite precise 90-degree phase shift throughout the audio band. In the current experimental receiver, the polyphase network connection is such that the unwanted sideband is mathematically completely cancelled inside the network, resulting in a single sideband audio output that is then amplified and filtered in a conventional way.

Fig 6.107 shows the block diagram of the experimental receiver design.

Input filter and detector

The high performance characteristics of the Tayloe Detector make a preceeding RF amplifier unnecessary. In this receiver (**Fig 6.108**), the RF signal is filtered by a simple bandpass filter consisting of two TOKO KANK3333 canned transformers. Circuit values are shown for 80m, but the circuit can readily be adapted to any HF amateur band, see **Table 6..5** from [13].

The input to the Tayloe Detector is biased to mid-rail (2.5V) in order to obtain maximum dynamic range. The FST3253 IC is a dual 1-4 way analogue multiplexer designed for memory bus switching applications and possessing high bandwidth and low ON-resistance. A local oscillator signal at four times the reception frequency is required to accurately generate the necessary switching signals: a synchronous binary counter type 74HC163 accomplishes this easily. The four audio outputs are buffered by low noise NE5534 op-amps configured for a gain of 33dB. The gain of three of the op-amps is made adjustable by mult-turn 1kΩ preset potentiometers to allow the amplitude of each of the four paths to be matched precisely.

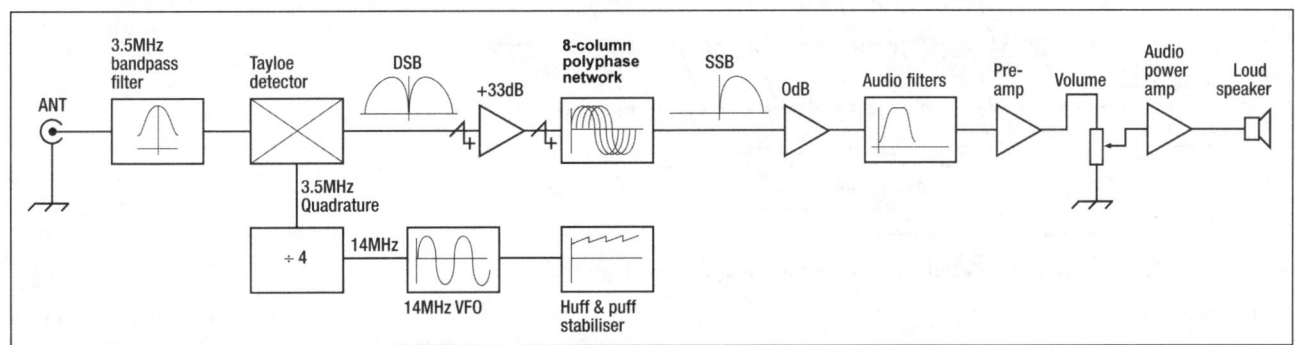

Fig 6.107: Block diagram of experimental Direct Conversion Polyphase receiver

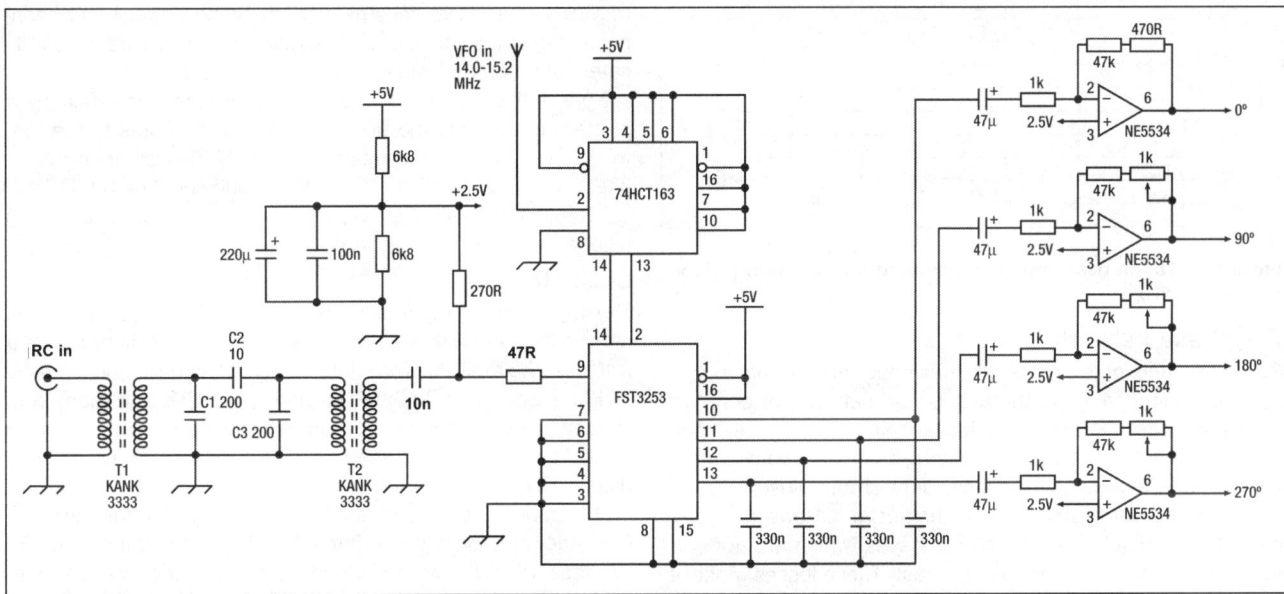

Fig 6.108: Front-end: RF filter and Tayloe Detector. Op-amps pin 4 is grounded and Pin 7 is +12V

Fig 6.109: Prototype of G0UPL's experimental direct conversion polyphase receiver

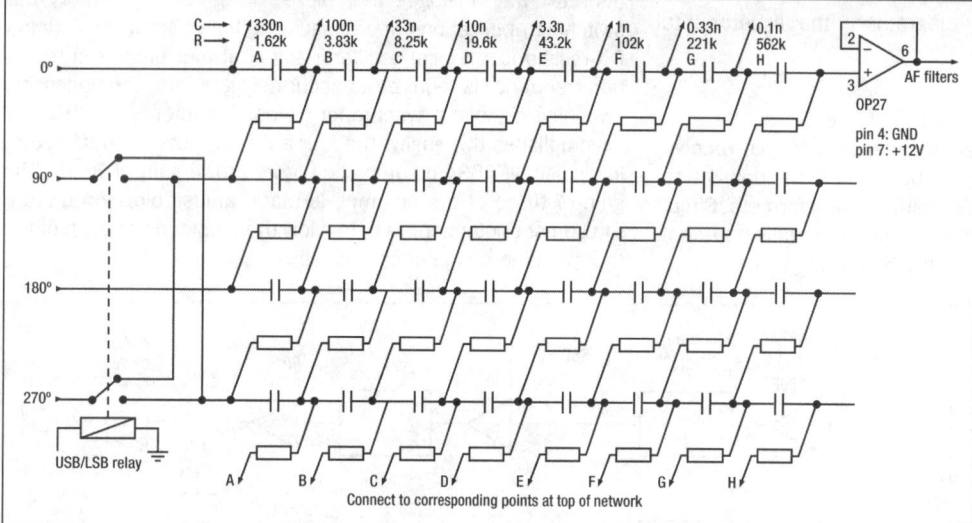

Fig 6.110: Polyphase network. Resistors and capacitors should be high tolerance types

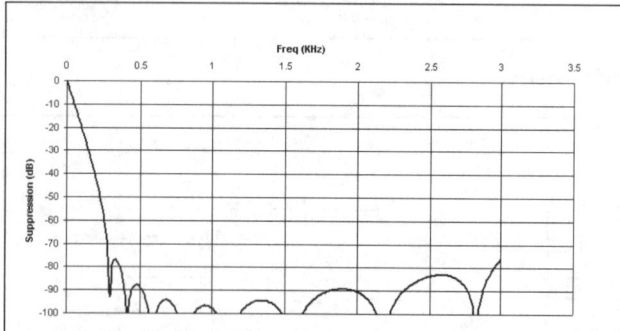

Fig 6.111: Theoretical opposite sideband suppression (dB) vs frequency (kHz)

Polyphase network

Polyphase networks are usually designed using either a constant capacitance value throughout the network, or constant resistance. Since capacitors are available in less-closely specified tolerances compared to resistors, a constant value capacitance made it possible to choose sets of four closely matching capacitors for each column in the network. However such a network can result in considerable losses, which are not constant across the audio band of interest. These losses must be compensated by higher amplification elsewhere in the signal chain, which generally degrades noise performance and

dynamic range of the receiver.

This component values for this design were calculated by reference to an excellent article by Tetsuo Yoshida JA1KO [14]. Tetsuo's unique design process increases the resistance value from column to column in such a way as to produce a lossless passive polyphase network. This counterintuitive result has been verified by measurement.

An 8-column polyphase network was designed (**Fig 6.110**). The theoretical opposite sideband suppression of this network (**Fig 6.111**) assumes precise component values. In practice, this is impossible to achieve and the real world performance will be degraded somewhat compared to the ideal curve. This degradation can be mitigated as far as possible by the use of high accuracy (0.1%) resistors, together with 'padding' capacitors by connecting smaller capacitors in parallel until measurement indicates four matched capacitors for each column.

Note that selection of Upper or Lower sideband is simply a matter of swapping the 90- and 270-degree inputs to the network. This could be accomplished by a DPDT relay or switch; or alternatively for single-band use the circuit could be hard-wired.

A single unity-gain high impedance low-noise op-amp (OP27) follows the polyphase network.

Oscillator

The Tayloe switching detector requires a VFO at four times the receive frequency. One way to obtain a stable VFO is by use of a Huff & Puff stabiliser. This simple circuit was developed initially by Klaas Spargaren PA0KSB in the early 1970s and many subsequent modifications have appeared in *RadCom's* 'Technical Topics' column and elsewhere (see also the chapter on Building Blocks: Oscillators).

The stabiliser compares the VFO to a stable crystal reference frequency and locks the VFO to this reference in small frequency steps. It might be described as a 'frequency locked loop' rather than a 'phase locked loop'. The locking process is a slow loop, and lacks the complication of phase locked loops and fre-

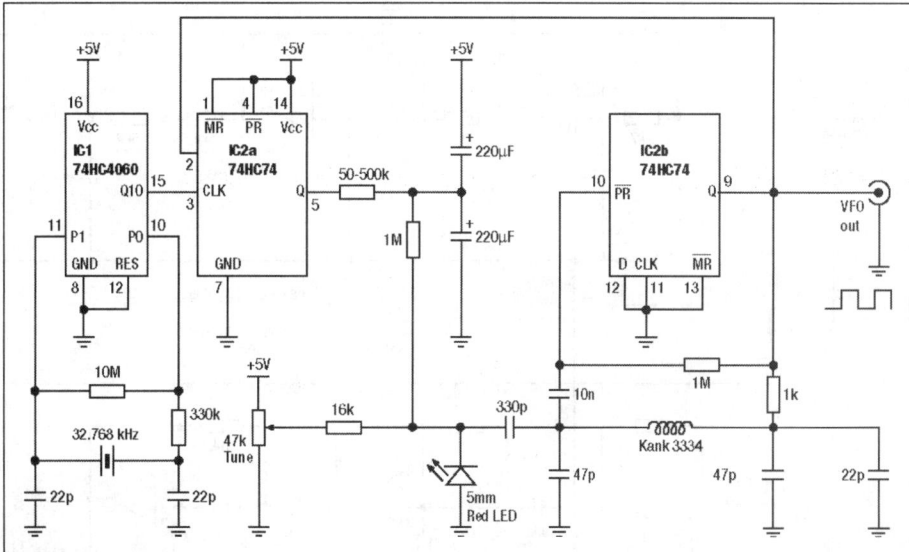

Fig 6.112: Two-chip simple Huff & Puff stabilised VFO

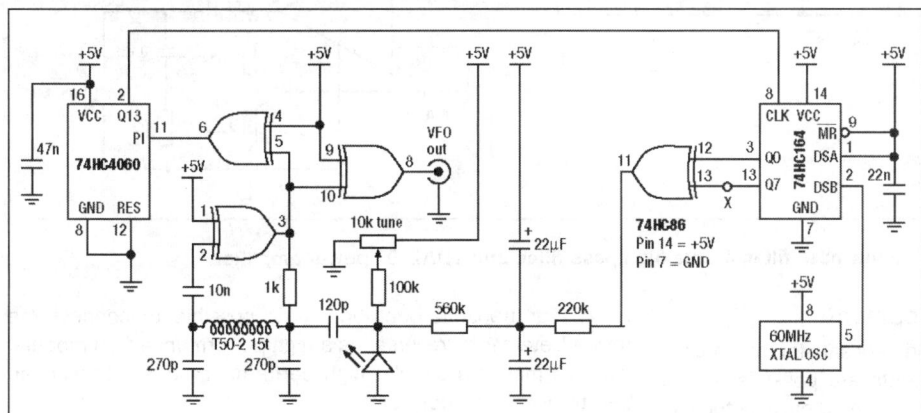

Fig 6.113: Three-chip 'fast' Huff & Puff stabilised VFO, offering improved performance

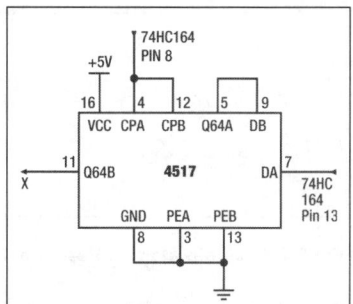

Fig 6.114: For higher frequencies, a 4517 CMOS IC may be added to improve the Huff & Puff stabiliser even further

quency synthesisers, whilst easily achieving a low phase noise output.

Two simple, minimalist designs (**Figs 6.112 and 6.113**) developed recently by G0UPL make it even simpler to build a Huff & Puff stabilised VFO. These designs combine the VFO and stabiliser, resulting in a stable output frequency at TTL-logic levels, perfectly suited for driving the Tayloe Detector.

Fig 6.112 shows an implementation of PA0KSB's original stabiliser configuration. This design uses one half of a 74HC74 D-type flip flop uniquely forced to behave as a simple inverter gate, and pressed into service as an oscillator. The crystal reference frequency is generated from a cheap 32.768kHz watch crystal. The frequency step size of the resulting VFO is determined by the division ratio of the 32.768kHz reference frequency, eg 32Hz. Remember that in the Tayloe Detector (Fig 6.107) the VFO is divided by four, which will also divide the tuning step by the same factor.

A later development was the 'fast' stabiliser by Peter Lawton, G7IXH, which was described in an article in *QEX* magazine [15]. He used a shift register as a n-stage delay line and compared the input and output of the delay line using an exclusive OR-gate (XOR). The effect is a statistical averaging of the output control signal. The 'fast' method makes it possible to stabilise a worse VFO compared to the standard method, or stabilise a comparable VFO with much less frequency ripple. The frequency step-size is given by:

$$\text{Step} = 10^6 \times \text{VFO}^2 / (z \times M \times \text{xtal})$$

where VFO is the VFO frequency in MHz, z is the number of

stages of delay, xtal is the crystal reference frequency in MHz, and $M = 2^n$ where n is the number of divide-by-2 stages in the VFO divider.

The minimalist design in Fig 6.113 uses only three ICs to implement G7IXH's 'fast' Huff & Puff method. One XOR gate is used as the VFO. The shift 74HC164 register effectively provides a 7-stage delay line.

To increase this further, a 4517 CMOS IC (128-stage shift register) could be cascaded in series with the 74HC164 to provide a 135-stage delay line (**Fig 6.114**). Note that the 4517 (part numbers HEF4517, CD4517 etc) is a member of the original CMOS 4000-series and was not produced in later, higher speed families such as the 74HC-series. Therefore it must be connected AFTER the 74HC164 so that the '164 is responsible for detecting the fast edges of the 60MHz reference oscillator. This 'fast' design is recommended for higher frequency VFOs such as might be used to build this experimental receiver for higher HF bands.

Audio stages

The remainder of the experimental receiver is relatively conventional and non-critical. Low-noise NE5534 op-amps are used to construct high-pass, low-pass filters to restrict the SSB bandwidth to 300Hz - 2.8kHz.

A switchable narrow filter at 800Hz could be added for CW reception. A standard TDA2002 audio power amplifier, produces sufficient output power to drive a 4-ohm loudspeaker for comfortable 'arm-chair' copy. An example audio section is shown in **Fig 6.115**.

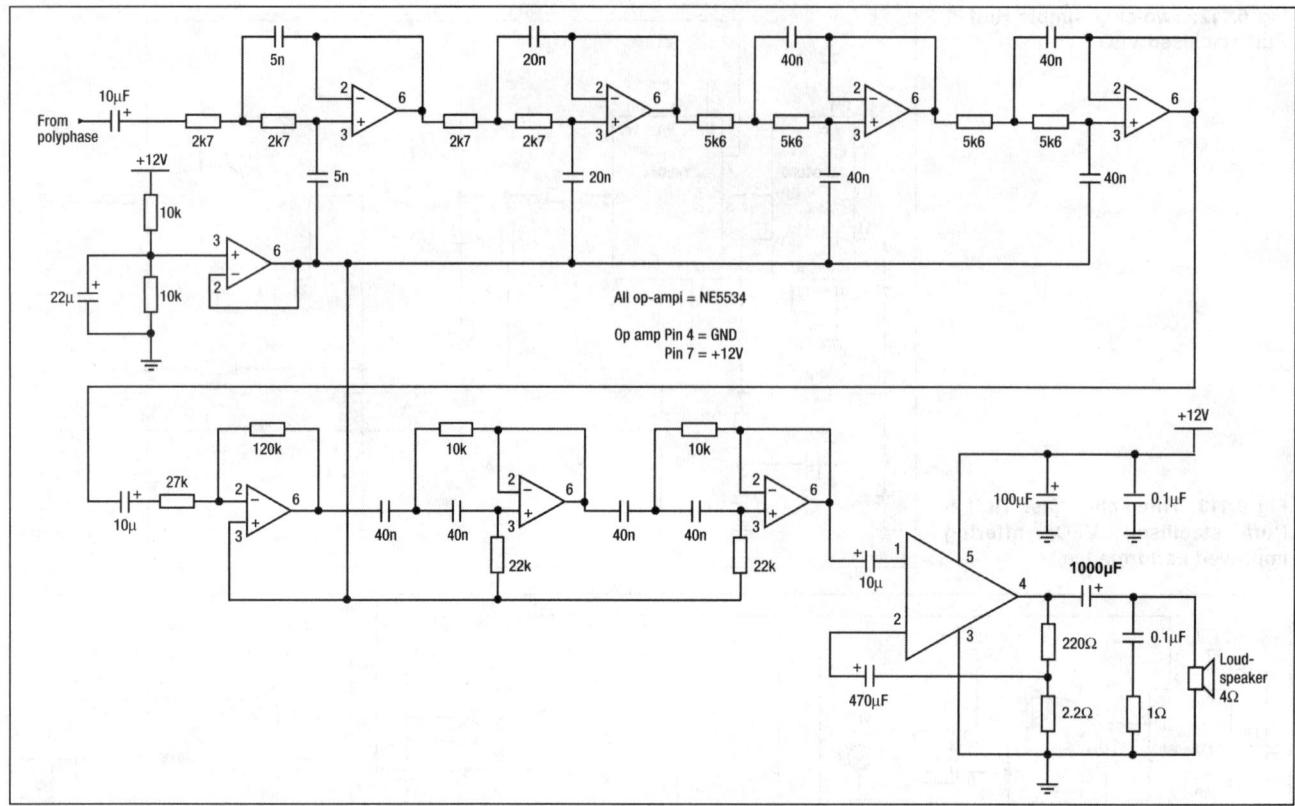

Fig 6.115: Audio stages: Pre-amp, 8-pole low pass filter, 4-pole high pass filter, and TDA2002 power amplifier.

Conclusions and further development

The receiver described has been found very satisfactory in use. Further developments and improvements are possible. Other designers of similar receivers have implemented parts of the circuit slightly differently. Many of these modifications would make the circuit more complex, but the following points might suggest avenues for further experiment:

1.   The switching order of the divide-by-4 circuit in the Tayloe detector can be altered to a 'gray code' sequence such that only one of the 2-bit outputs changes state at each clock pulse. This is said to produce lower switching noise, though on 80m the atmospheric noise probably swamps any such effects anyway.

2.   A clock-squarer circuit can be employed to generate a precise 50% duty cycle from a VFO at two times the reception frequency; this obviously imposes less stringent requirements on the VFO, which is harder to construct for higher frequencies.

3.   The VFO could be replaced by one generated by Direct Digital Synthesis or other precise oscillator methods (see the Building Blocks: Oscillators chapter).

4.   It is possible to use the other half of the FST3253 switch in parallel to halve the ON-resistance; alternatively a double-balanced Tayloe detector may be constructed using the second switch and a phase-splitting transformer at the input. This would make the detector more immune to noisy VFO signals such as might be produced by digital methods, eg DDS.

5.   Many designers combine 0, 180 and 90, 270 degree outputs of the Tayloe detector prior to feeding the polyphase network. This can reduce certain common-mode noise sources.

6.   Instead of using just one output from the polyphase network, all four can be combined thereby averaging errors and improving the signal-to-noise ratio.

7.   For transmit operation, it is possible to connect the Tayloe Detector in 'reverse' as a high-performance SSB modulator. This makes a simple, high performance direct conversion SSB transceiver a possibility.

REFERENCES

[1]   *The Radio & Electronic Engineer*, Vol 43, No 3, Mar 1973, pp209-215

[2]   'Phase noise, intermodulation and dynamic range', P E Chadwick, *RF Expo Proceedings*, Anaheim, Jan 1986

[3]   'A simple model of a feedback oscillator noise spectrum', D B Leeson, *Proc IEEE*, Vol 54, Feb 1966

[4]   N6NWP high-dynamic-range MF/HF front-end of a single-conversion receiver, *QST*, Feb 1993, ARRL

[5]   'Technical Topics', *Radio Communication*,Sep 1976 and Wireless World July 1977

[6]   'Some experiments with HF ladder crystal filters', J A Hardcastle, G3JIR, *Radio Communication*, Dec 1976, Jan, Feb and Sep 1977

[7]   *RF Design*, Sep 1992

[8]   Siliconix Application Note AN85-2

[9]   *QST*, Oct 1982, ARRL

[10] 'The "fast" digital oscillator stabiliser', Pete Lawton, G7IXH, *QEX*, Nov/Dec 1998, p17ff, ARRL

[11] 'Frequency stabilization of L-C oscillators', Klaas Spaargaern, PA0KSB, *QEX*, Feb 1996, pp19-23, ARRL

[12] *QST*, Feb 1993, ARRL

[13] *The Filter Handbook*, by Stefen Niewiadomski and presented on the G-QRP club website http://www.gqrp.com)

[14] *QEX* magazine, Nov 1995, ARRL

[15] *QEX* magazine, Nov 1998, ARRL

7 HF Transmitters and Transceivers

The purpose of a transmitter is to generate RF energy which may be keyed or modulated and thus employed to convey intelligence to one or more receiving stations. This chapter deals with the design of that part of the transmitter which produces the RF signal, while keying, data modulation etc are described separately in other chapters.

Transmitters operating on frequencies between 1.7 and 30MHz only are discussed here; methods of generating frequencies higher than 30MHz are contained in other chapters. Where the frequency-determining oscillators of a combined transmitter and receiver are common to both functions, the equipment is referred to as a transceiver; the design of such equipment operating in the HF spectrum is also included in this chapter.

One of the most important requirements of any transmitter is that the desired frequency of transmission shall be maintained within fine limits to prevent interference with other stations and to ensure that the operator remains within the allowed frequency allocation.

Spurious frequency radiation capable of causing interference with other services, including television and radio broadcasting, must also be avoided. These problems are considered in the chapter on electromagnetic compatibility.

The simplest form of transmitter is a single-stage, self-excited oscillator coupled directly to an antenna system: **Fig 7.1(a)**. Such an elementary arrangement has, however, three serious limitations:

- The limited power which is available with adequate frequency stability;
- The possibility of spurious (unwanted) radiation;
- The difficulty of securing satisfactory modulation or keying characteristics.

In order to overcome these difficulties, the oscillator must be called upon to supply only a minimum of power to the following stages. Normally an amplifier is used to provide a constant load on the oscillator: **Fig 7.1(b)**. This will prevent phase shifts caused by variations in the load from adversely affecting the oscillator frequency.

A FET source follower with its characteristic high input impedance makes an ideal buffer after a VFO. A two-stage buffer amplifier is often used, **Fig 7.1(c)**, incorporating a source follower followed by a Class A amplifier.

Transmitters for use on more than one frequency band often employ two or more oscillators; these signals are mixed together to produce a frequency equal to the sum or difference of the original two frequencies, the unwanted frequencies being removed using a suitable filter. This heterodyne transmitter, **Fig 7.1(d)**, is very similar in operation to the superhet receiver. More recently, frequency synthesisers have become popular for the generation of RF energy, **Fig 7.1(e)**, and have almost entirely replaced conventional oscillators in commercial amateur radio equipment. Synthesisers are ideally suited for microprocessor control and multi-frequency coverage (see the chapter on the POC-controlled transceiver). Oscillators are covered in detail in the earlier chapter, Building Blocks: Oscillators.

INTERSTAGE COUPLING

Correct impedance matching between a stage and its load is important if the design power output and efficiency for the stage are to be achieved. In order to keep dissipation in the stage within acceptable limits, the load impedance is often higher than that needed for maximum power transfer. Thus a transmitter designed to drive a 50Ω load usually has a much lower source impedance than this. The output power available from a stage can be calculated approximately from the formula:

$$P_{out} = \frac{V_{cc}^2}{2_{ZL}}$$

where ZL is the load impedance in ohms.

Determination of the base input impedance of a stage is difficult without sophisticated test equipment, but if the output of a stage is greater than 2W its input impedance is usually less than 10Ω and may be as low as 1 to 2Ω. For this reason some kinds of LC matching do not lend themselves to this application. With the precise input impedance of a stage unknown, adjustable LC networks often lend themselves best to matching a wide range of impedances. A deliberate mismatch may be introduced by the designer to control power distribution and aid stability.

In the interest of stability it is common practice to use low-Q networks between stages in a solid-state transmitter, but the penalty is poor selectivity and little attenuation of harmonic or spurious energy. Most solid-state amplifiers use loaded Qs of 5 or less compared to Qs of 10 to 15 found in valve circuitry. All calculations must take into account the input and output capacitance of the solid-state devices, which must be included in the

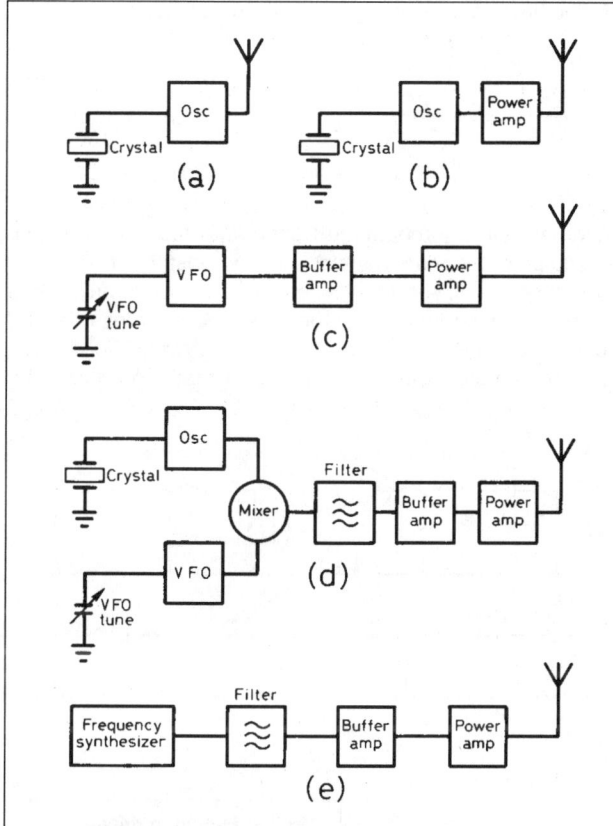

Fig 7.1: Block diagrams of basic transmitter types

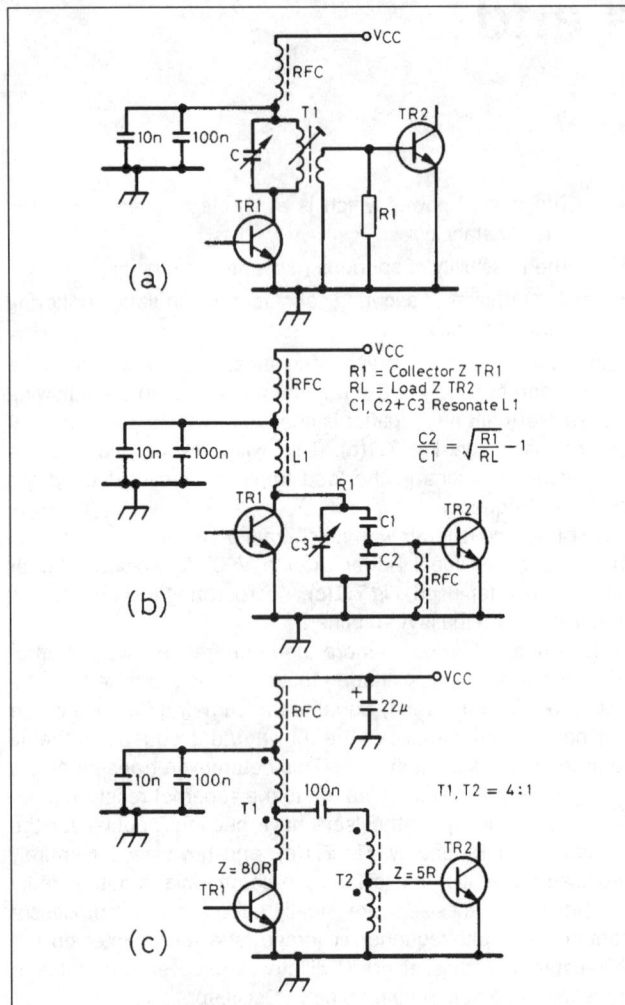

Fig 7.2: Interstage coupling. (a) Transformer coupling; (b) capacitive divider coupling; (c) broad-band transformer matching

network calculations. The best source of information on input and output capacitance of power transistors is the manufacturers' data sheets. Impedances vary considerably with frequency and power level, producing a complex set of curves, but input and output capacitance values are independent of these parameters.

Transformer coupling has always been popular with solid-state devices: **Fig 7.2(a)**. Toroidal transformers are the most efficient and satisfactory up to 30MHz, the tapping and turns ratio of T1 being arranged to match the collector impedance of TR1 to the base impedance of TR2. In some circumstances only one turn may be required on the secondary winding. A low value of resistor is used to slug this secondary winding to aid stability.

An alternative to inductive coupling is to use a capacitive divider network which resonates L1 as well as providing a suitable impedance tap: **Fig 7.2(b)**. Suitable RF chokes wound on high-permeability (μ = 800) ferrite beads, adequately decoupled, are added to aid stability. This circuit is less

prone to VHF parasitics than the one in **Fig 7.2(a)**, especially if C2 has a relatively high value of capacitance.

When the impedance values to be matched are such that specific-ratio broad-band transformers can be used, typically 4:1 and 16:1, one or more fixed-ratio transformers may be cascaded as shown in **Fig 7.2(c)**. The system exhibits a lack of selectivity, but has the advantage of offering broad-band characteristics. Dots are normally drawn on to transformers to indicate the phasing direction of the windings (all dots start at the same end).

Simple LC networks provide practical solutions to matching impedances between stages in transmitters. It is assumed that normally high output impedances will be matched to lower-value input impedances. However, if the reverse is required, networks can be simply used in reverse to effect the required transformation.

Networks 1 and 2, **Figs 7.3(a) and 7.3b)**, are variations on the L-match and may used fixed or with adjustable inductors and capacitors. Network 3, **Fig 7.3(c)**, is used by many designers because it is capable of matching a wide range of impedances. It is a low-pass T-network and offers harmonic attenuation to a degree determined by the transformation ratio and the total network Q. For stages feeding an antenna, additional harmonic suppression will normally be required. The value of Q may vary from 4 to 20 and represents a compromise between bandwidth and attenuation.

Network calculations are shown in **Table 7.1**.

Conventional broad-band transformers are very useful in the construction of solid-state transmitters. They are essentially devices for the transformation of impedances relative to the ratio of the transformer and are not specific impedance devices, eg a solid-state PA with a nominal 50Ω output impedance may employ a 3:1 ratio broad-band output transformer. The impedance ratio is given by the square of the turns ratio and will be 9:1. The impedance seen by the collector is thus:

$$Z_C = \frac{50}{9} = 5.55\Omega$$

The power delivered to the load is then determined by:

$$P_{out} = \frac{V_{CC}^2}{2Z_C}$$

Another type of broad-band transformer found in transmitting equipment is the transmission-line transformer. This acts as a conventional transformer at the lower frequencies but, as the frequency increases, the core becomes less 'visible' and the transmission-line properties take over. The calculations are complex, but it is well known that a quarter-wave-

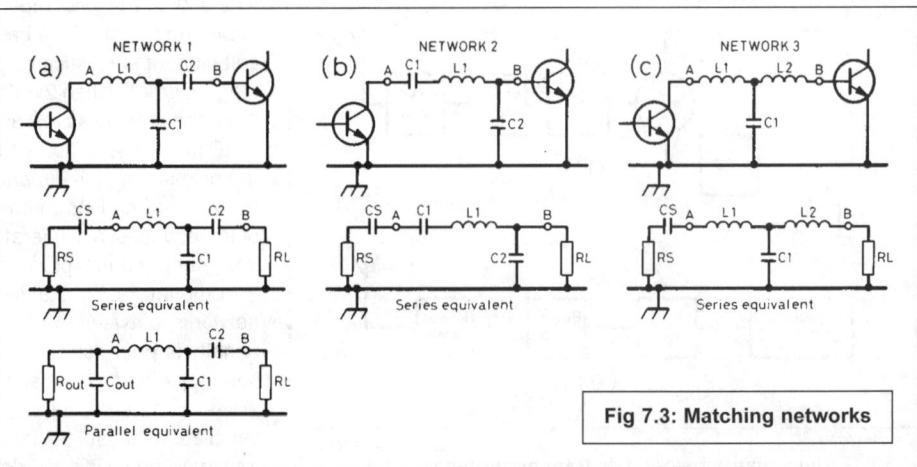

Fig 7.3: Matching networks

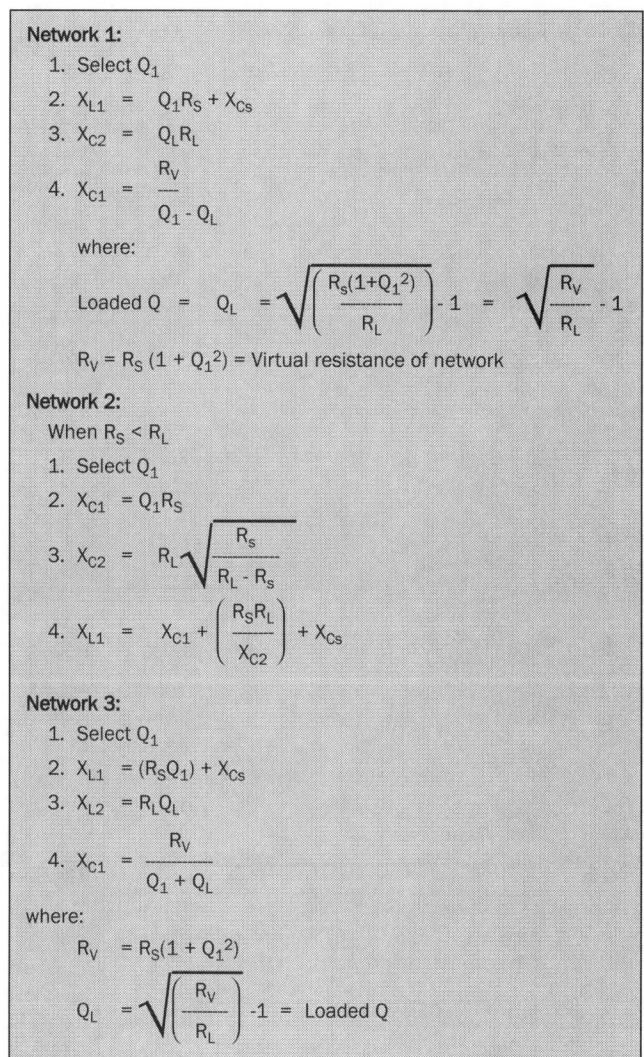

Network 1:

1. Select Q_1
2. $X_{L1} = Q_1 R_S + X_{Cs}$
3. $X_{C2} = Q_L R_L$
4. $X_{C1} = \dfrac{R_V}{Q_1 - Q_L}$

where:

Loaded Q $= Q_L = \sqrt{\left(\dfrac{R_s(1+Q_1^2)}{R_L}\right) - 1} = \sqrt{\dfrac{R_V}{R_L} - 1}$

$R_V = R_S(1 + Q_1^2)$ = Virtual resistance of network

Network 2:

When $R_S < R_L$

1. Select Q_1
2. $X_{C1} = Q_1 R_S$
3. $X_{C2} = R_L \sqrt{\dfrac{R_s}{R_L - R_s}}$
4. $X_{L1} = X_{C1} + \left(\dfrac{R_S R_L}{X_{C2}}\right) + X_{Cs}$

Network 3:

1. Select Q_1
2. $X_{L1} = (R_S Q_1) + X_{Cs}$
3. $X_{L2} = R_L Q_L$
4. $X_{C1} = \dfrac{R_V}{Q_1 + Q_L}$

where:

$R_V = R_S(1 + Q_1^2)$

$Q_L = \sqrt{\left(\dfrac{R_V}{R_L}\right) - 1}$ = Loaded Q

Table 7.1: Calculations for the three networks shown in Fig 7.3

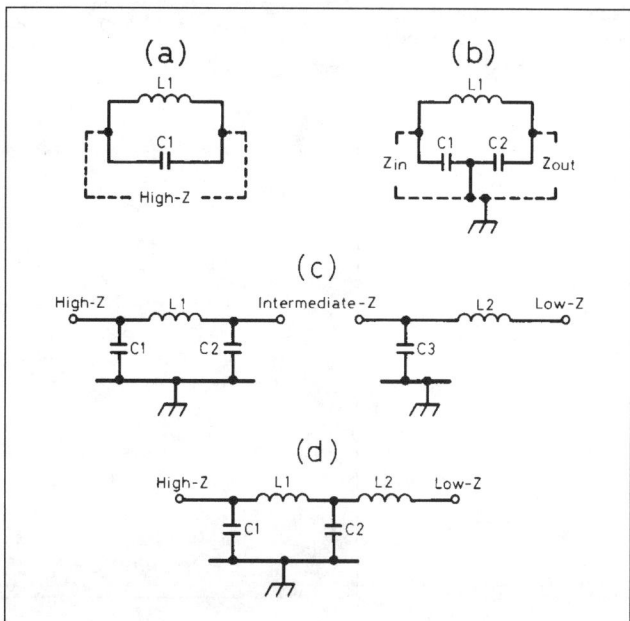

Fig 7.4: Pi-network and derivation. (a) Parallel-tuned tank circuit; (b) parallel-tuned tank with capacitive tap; (c) pi-network and L-network; (d) pi-L network

length of line exhibits impedance-transformation properties. When a quarter-wave line of impedance Z_0 is terminated with a resistance R_1, a resistance R_2 is seen at the other end of the line.

$Z_0^2 = R_1 R_2$

Transmission-line transformers are often constructed from twisted pairs of wires wound onto a ferrite toroidal core having a initial permeability (μ) of at least 800.

This ensures relatively high values of inductance with a small number of turns. It should be borne in mind that with high-power solid-state transmitters some of the impedances to be transformed are very low, often amounting to only a few ohms.

ANODE TANK CIRCUITS

Whilst solid state devices have replaced valves in most amateur equipment, high power valve linear amplifiers are likely to be in use for some considerable time.

Pi Network

The pi-tank (**Fig 7.4**) has been the most popular matching network since the introduction of multiband transmitters in the 1950s. It is easily bandswitched and owes its name to its resemblance to the Greek letter π.

The anode tank circuit is required to meet the following conditions:

1   The anode circuit of the valve must be presented with the proper resistance in relation to its operating conditions to ensure efficient generation of power.
2.   This power must be transferred to the output without appreciable loss.
3   The circuit Q must be sufficient to ensure good flywheel action in order to achieve a close approximation of a sinusoidal RF output voltage. This is especially important in Class C amplifiers where the drive from the valve is in the form of a series of pulses of RF energy.

Tank circuit Q

In order to quantify the ability of a tank circuit to store RF energy (essential for flywheel action), a quality factor Q is defined. Q is the ratio of energy stored to energy lost in the circuit.

$$Q = 2\pi \dfrac{W_S}{W_L} = \dfrac{X}{R}$$

where W_S is the energy stored in the tank circuit; W_L is the energy lost to heat and the load; X is the reactance of either the inductor or capacitor in the tank circuit; R is the series resistance.

Since both circulating current and Q are inversely proportional to R, then the circulating current is proportional to Q. By Ohm's Law, the voltage across the tank circuit components must also be proportional to Q.

When the circuit has no load the only resistance contributing to R are the losses in the tank circuit. The unloaded Q_U is given by:

$$Q_U = \dfrac{X}{R_{loss}}$$

where X is the reactance in circuit and R_{loss} is the sum of the resistance losses in the circuit.

To the tank circuit, a load acts in the same way as circuit losses. Both consume energy but only the circuit losses produce

Table 7.2: Pi-network values for selected anode loads (QL = 12)

	MHz	1500	2000	2500	3000	3500	4000	5000	6000	8000
	1.8	708	531	424	354	303	264	229	206	177
	3.5	364	273	218	182	156	136	118	106	91
C1	7.0	182	136	109	91	78	68	59	53	46
(pF)	14.0	91	68	55	46	39	34	30	27	23
	21.0	61	46	36	30	26	23	20	18	15
	28.0	46	34	27	23	20	17	15	13	11
	1.8	3413	2829	2415	2092	1828	1600	1489	1431	1392
	3.5	1755	1455	1242	1076	940	823	766	736	716
C2	7.0	877	728	621	538	470	411	383	368	358
(pF)	14.0	439	364	310	269	235	206	192	184	179
	21.0	293	243	207	179	157	137	128	123	119
	28.0	279	182	155	135	117	103	96	92	90
	1.8	12.81	16.60	20.46	24.21	27.90	31.50	36.09	39.96	46.30
	3.5	6.59	8.57	10.52	12.45	14.35	16.23	18.56	20.55	23.81
L1	7.0	3.29	4.29	5.26	6.22	7.18	8.12	9.28	10.26	11.90
(µH)	14.0	1.64	2.14	2.63	3.11	3.59	4.06	4.64	5.14	5.95
	21.0	1.10	1.43	1.75	2.07	2.39	2.71	3.09	3.43	3.97
	28.0	0.82	1.07	1.32	1.56	1.79	2.03	2.32	2.57	2.98

Table 7.2: Pi-L network values for selected anode loads (Q_L = 12)

	MHz	1500	2000	2500	3000	3500	4000	5000	6000	8000
	1.8	784	591	474	397	338	297	238	200	152
	3.5	403	304	244	204	174	153	123	103	78
C1	7.0	188	142	114	94	81	71	57	48	36
(pF)	14.0	93	70	56	47	40	35	29	24	18
	21.0	62	47	38	32	27	23	19	16	12
	28.0	48	36	29	24	21	18	15	13	9
	1.8	2621	2355	2168	2026	1939	1841	1696	1612	1453
	3.5	1348	1211	1115	1042	997	947	872	829	747
C2	7.0	596	534	493	468	444	418	387	368	337
(pF)	14.0	292	264	240	222	215	204	186	172	165
	21.0	191	173	158	146	136	137	125	117	104
	28.0	152	135	127	115	106	107	95	87	86
	1.8	14.047	17.933	21.730	25.466	29.155	32.805	40.011	47.118	61.119
	3.5	7.117	9.086	11.010	12.903	14.772	16.621	20.272	23.873	30.967
L1	7.0	3.900	4.978	6.030	7.070	8.094	9.107	11.108	13.081	16.968
(µH)	14.0	1.984	2.533	3.069	3.597	4.118	4.633	5.651	6.655	8.632
	21.0	1.327	1.694	2.053	2.406	2.755	3.099	3.780	4.452	5.775
	28.0	0.959	1.224	1.483	1.738	1.989	2.238	2.730	3.215	4.171
	1.8	8.917	*The value of L2 remains constant for all values of anode impedance.*							
	3.5	4.518								
L2	7.0	2.476								
(µH)	14.0	1.259								
	21.0	0.843								
	28.0	0.609								

heat energy. When energy is coupled from the tank circuit to the load, the loaded Q (Q_L) is given by:

$$Q_L = \frac{X}{R_{load} + R_{loss}}$$

It follows that if the circuit losses are kept to a minimum the loaded Q value will rise.

Tank efficiency can be calculated from:

$$\text{Tank efficiency (\%)} = \left(1 - \frac{Q_L}{Q_U} \right) \times 100$$

where Q_U is the unloaded Q and Q_L is the loaded Q.

Typically the unloaded Q for a pi-tank circuit will be between 100 and 300 while a value of 12 is accepted as a good compromise for the loaded Q. In order to assist in the design of anode tank circuits for different frequencies, inductance and capacitance values for a pi-network with a loaded Q of 12 are provided in **Table 7.2** for different values of anode load impedance.

Pi-L Network

The pi-L network, **Fig 7.4(d)**, is a combination of the pi-network and the L-network. The pi-network transforms the load resistance to an intermediate impedance, typically several hundred ohms, and the L-network then transforms this intermediate impedance to the output impedance of 50Ω. The output capacitor of the pi-network is in parallel with the input capacitor of the L-network and is combined into one capacitor equal to the sum of the two individual values.

The major advantage of the pi-L network over a pi-network is considerably greater harmonic suppression, making it particu-

larly suitable for high-power linear amplifier applications. A table of values for a pi-L network having a loaded Q of 12 for different values of anode load impedance is given in **Table 7.3**. Both Table 7.2 and Table 7.3 assume that source and load impedances are purely resistive; the values will have to be modified slightly to compensate for any reactance present in the circuits to be matched. Under certain circumstances matching may be compromised by high values of external capacitance, in which case a less-than-ideal value of Q may have to be accepted.

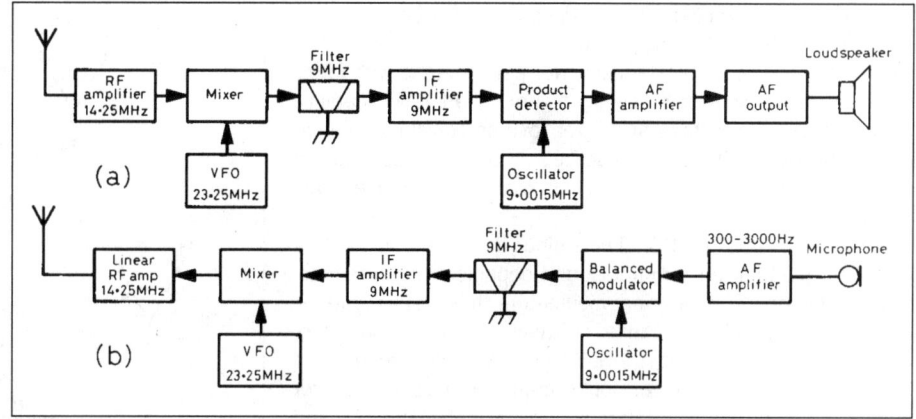

Fig 7.5: (a) Typical single-conversion superhet receiver. (b) Single conversion SSB transmitter

VOICE MODULATION TECHNIQUES

A description of the theory of Amplitude Modulation (AM), Single Sideband (SSB), and Frequency and Phase Modulation (FM and PM) can be found in the Principles chapter.

Although currently common for broadcasting, AM is rarely used on the amateur bands. The use of FM and PM is confined to the bands above 30MHz (with the exception of a little activity in the upper part of the 28MHz band), so they are covered in the chapter on VHF/UHF transmitters and receivers. This chapter will deal with the practical aspects of SSB modulation.

An SSB transmitter has a similar architecture to an SSB receiver as shown in **Fig 7.5**. This makes it very suitable for combining into a transceiver, see later.

BALANCED MODULATORS

Balanced modulators are essentially the same as balanced mixers, balanced demodulators and product detectors; they are tailored to suit different circuit applications especially with respect to the frequencies in use. The balanced modulator is a circuit which mixes or combines a low-frequency (audio) signal with a higher frequency (RF) signal in order to obtain the sum-and-difference frequencies (sidebands); the original RF frequency is considerably attenuated by the anti-phase or balancing action of the circuit.

A *singly* balanced modulator is designed to balance out only one of the input frequencies, either f_1 or f_2, normally the higher frequency. In a *doubly* balanced modulator, both f_1 and f_2 are balanced out, leaving the sum and difference frequencies $f_1 + f_2$ and $f_1 - f_2$. In addition, intermodulation products (IMD) will appear in the form of spurious signals caused by the interaction and mixing of the various signals and their harmonics.

Balanced modulators come in many different forms, employing a wide variety of devices from a simple pair of diodes to complex ICs. In their simplest form they are an adaptation of the bridge circuit, but it should be noted that diodes connected in a modulator circuit are connected differently to those in a bridge rectifier. Simple diode balanced modulators can provide high

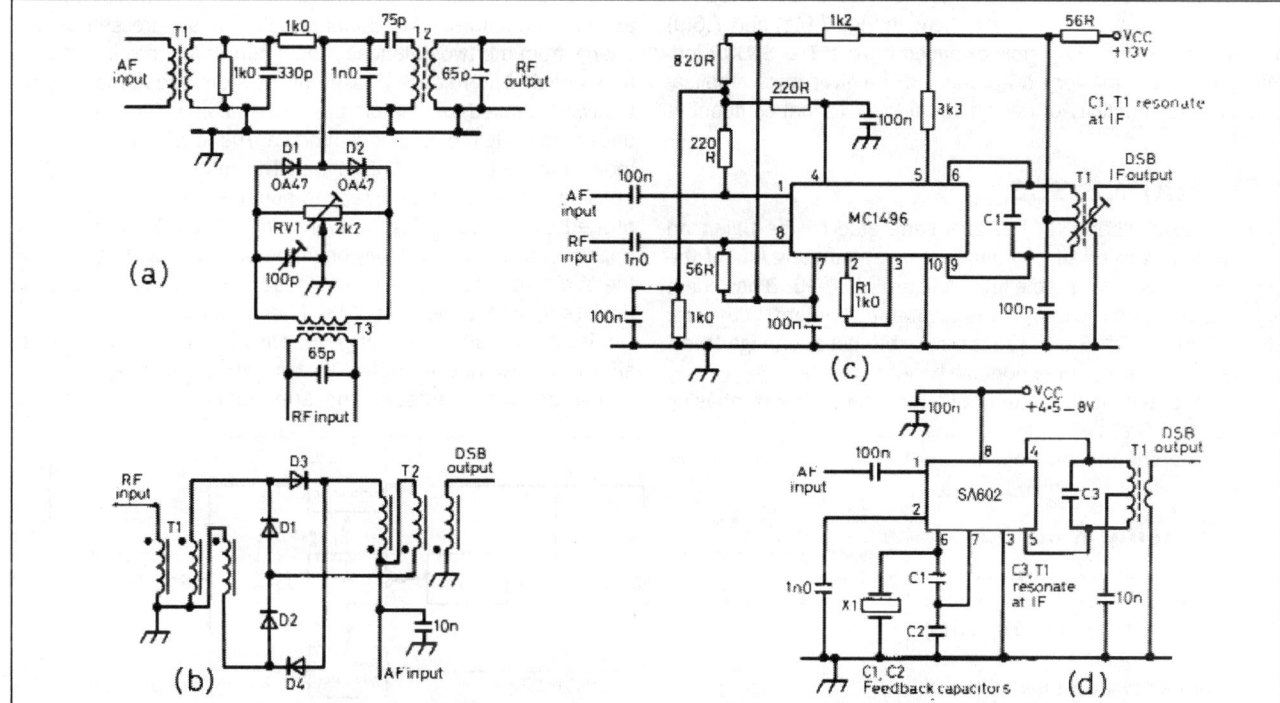

Fig 7.6: Balanced modulators. (a) Singly balanced diode modulator. (b) Doubly balanced diode ring modulator. (c) MC1496 doubly balanced modulator. (d) SA602 doubly balanced modulator with internal oscillator

performance at low cost. Early designs used point-contact ger-manium diodes while more recent designs use hot-carrier diodes (HCD). The HCD offers superior performance with lower noise, higher conversion efficiency, higher square law capability, higher breakdown voltage and lower reverse current combined with a lower capacitance. In practice, almost any diode can be used in a balanced modulator circuit, including the ubiquitous 1N914.

In the early days of SSB, simple diode balanced modulators were very popular, easy to adjust and capable of good results. Doubly balanced diode ring modulators have subsequently proved very popular because of their higher performance. However, they incur at least a 6dB signal loss while requiring a high level of oscillator drive. Doubly balanced modulators offer greater isolation between inputs as well as between input and output ports when compared to singly balanced types.

The introduction of integrated circuits resulted in a multitude of ICs suitable for use in balanced modulator and mixer applica-tions. These include the popular Philips SA602 which combines an input amplifier, local oscillator and double balanced modula-tor in a single package, and the MC1496 double balanced mixer IC from Motorola [1]. The majority of IC mixers are based upon a doubly balanced transistor tree circuit, using six or more tran-sistors on one IC. The major difference between different types of IC lies in the location of resistors which may be either internal or external to the IC.

IC mixers offer conversion gain, lower oscillator drive require-ments and high levels of balance, but IMD performance can be inferior to that of diode ring modulators. Some devices, such as the Analog Devices AD831, permit control of the bias current. This allows the designer freedom to improve IMD performance at the expense of power consumption.

Fig 7.6 illustrates a range of practical balanced modulator cir-cuits. The shunt-type diode modulator, Fig 7.6(a), was common in early valve SSB transmitters, offering a superior balance to that achievable with conventional valve circuitry. Fig 7.6(b) shows a simple diode ring balanced modulator capable of very high performance - devices such as the MD108 and SBL1 are derivations of this design. The designs in Figs 7.6(c) and 7.6(d) illustrate the use of IC doubly balanced mixers. The SA602, pri-marily designed for very-low-power VHF receiver mixer applica-tions, offers simplicity of design and a low external component count.

GENERATING SSB

The double sideband (DSB) signal generated by the balanced modulator has to be turned into SSB by attenuating one of the sidebands to an acceptable level. A figure of 30-35dB has come to be regarded as the minimum acceptable standard. With care, suppression of 50dB or more is attainable but such high levels of attenuation are of questionable benefit.

The unwanted sideband may be attenuated either by phasing or filtering. The two methods are totally different in conception, and will be discussed in detail.

The Phasing Method

The phasing method of SSB genera-tion can be simply explained with the aid of vector diagrams. Fig 7.7(a) shows two carriers, A and B, of the same frequency and phase, one of which is modulated in a balanced modulator by an audio tone to pro-duce contra-rotating sidebands A1

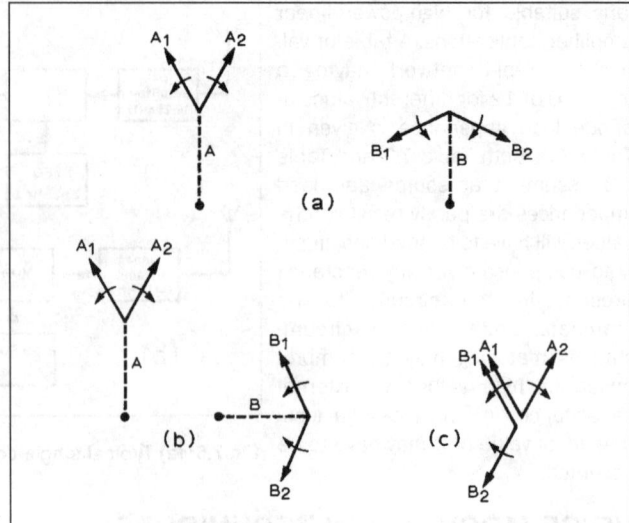

Fig 7.7: Phasing system vectors

and A2, and the other modulated by a 90° phase-shifted ver-sion of the same audio tone. This produces sidebands B1 and B2 which have a 90° phase relationship with their A counter-parts. The carrier vector is shown dotted since the carrier is absent from the output of the balanced modulators. Fig 7.7(b) shows the vector relationship if the carrier B is shifted in phase by 90° and Fig 7.7(c) shows the addition of these two signals. It is evident that sidebands A2 and B2 are in antiphase and there-fore cancel whereas A1 and B1 are in phase and are additive. The result is that a single sideband is produced by this process.

A block diagram of a phasing-type transmitter is shown in Fig 7.8 from which will be seen that the output of an RF oscillator is fed into a network in which it is split into two separate compo-nents, equal in amplitude but differing in phase by 90°

Similarly, the output of an audio amplifier is split into two components of equal amplitude and 90° phase difference. One RF and one AF component are combined in each of two bal-anced modulators. The double-sideband, suppressed-carrier energy from the two balanced modulators is fed into a common tank circuit. The relative phases of the sidebands produced by the two balanced modulators are such that one sideband is bal-anced out while the other is reinforced. The resultant in the com-mon tank circuit is an SSB signal. The main advantages of a phasing exciter are that sideband suppression may be accom-plished at the operating frequency and that selection of the upper or lower sideband may be made by reversing the phase of the audio input to one of the balanced modulators. These facili-ties are denied to the user of the filter system.

If it were possible to arrange for absolute precision of phase shift in the RF and AF networks, and absolute equality in the amplitude of the outputs, the attenuation of the unwanted

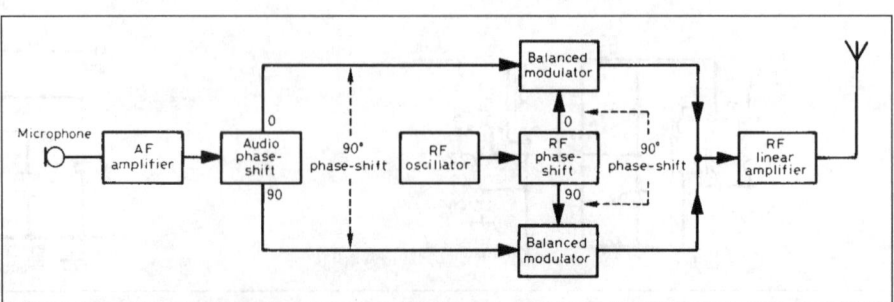

Fig 7.8: Phasing-type exciter

sideband would be infinite. In practice, perfection is impossible to achieve, and some degradation of performance is inevitable. Assuming that there is no error in the amplitude adjustment, a phase error of 1° in either the AF or the RF network will reduce the suppression to 40dB, while an error of 2° will produce 35dB, and 3.5° will result in 30dB suppression. If, on the other hand, phase adjustment is exact, a difference of amplitude between the two audio channels will similarly reduce the suppression. A difference between the two voltages of 1% would give 45dB, and 4% 35dB approximately. These figures are not given to discourage the intending constructor, but to stress the need for high precision workmanship and adjustment if a satisfactory phasing-type SSB transmitter is to be produced.

The early amateur phasing transmitters were designed for fundamental-frequency operation, driven directly from an existing VFO tuning the 80m band, and used a low-Q phase-shift network. This low-Q circuit has the ability to maintain the required 90° phase shift over a small frequency range, and this made the network suitable for use at the operating frequency in single-band exciters designed to cover only a portion of the chosen band. The RF phase-shift network is incapable of maintaining the required accuracy of phase shift for operation over ranges of 200kHz or more, and the available sideband suppression deteriorates to a point at which the exciter is virtually radiating a double sideband signal.

For amateur band operation a sideband suppression of 30-35dB and a carrier suppression of 50dB should be considered the minimum acceptable standard. Any operating method that is fundamentally incapable of maintaining this standard should not be used on the amateur bands. For this reason, the fundamental type of phasing unit is not recommended. For acceptable results, the RF phase shift must be operated at a fixed frequency outside the amateur bands. The SSB output from the balanced modulator is then heterodyned to the required bands by means of an external VFO.

Audio phase-shift network

Achieving the audio phase shift necessary for SSB generation in a phasing exciter traditionally required the use of high-tolerance components, often necessitating the use of a commercially made phase-shift network. Such devices can prove more costly than the crystal filter required for a filter-type exciter.

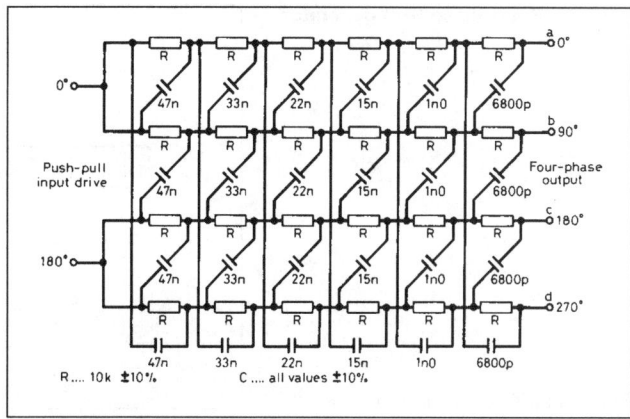

Fig 7.9: Gingell polyphase network

An alternative method of devising the required 90° phase shift using off-the-shelf values was devised by M J Gingell and is referred to as the polyphase network (**Fig 7.9**). Standard 10% tolerance resistors and capacitors are used in the construction of a six-pole network capable of providing four outputs of equal amplitude, all lagging one another by 90°. The network is designed to phase shift audio signals between 300Hz and 3000Hz, and it is therefore necessary to limit the bandwidth of the audio input using a filter or clipper circuit which can be either active or passive. Audio input to the polyphase network is derived using a simple phase splitter to provide the two phase inputs required by the network. Resistors used in the network are of one common value and Mylar audio-grade capacitors are suitable for the capacitive elements.

RF phase-shift network

Traditionally the most satisfactory way to produce a 90° RF phase shift was to employ a low-Q network comprising of two loosely coupled tuned circuits which exhibits a combination of inductance, resistance and capacitance. **Fig 7.10(a)** illustrates such a network in the anode circuit of a valve amplifier. The primary coils are inductively coupled while the link couplings are connected in series. When both circuits are tuned to resonance there will be exactly 90° phase shift between them. Difficulties occur when the frequency is changed, and the network has to be retuned, restricting the bandwidth to no more than 200kHz.

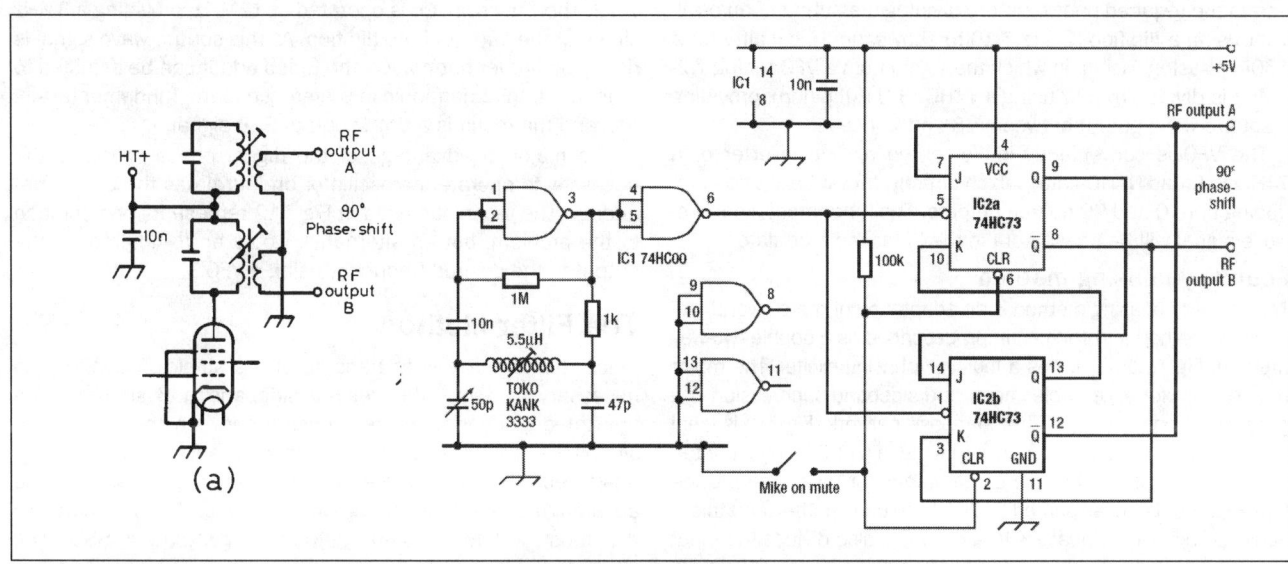

Fig 7.10: Methods of obtaining RF phase shift. (a) Traditional method of obtaining 90° phase shift using loosely coupled tuned circuits. (b) Active RF phase shifter 7.2MHz VFO providing 1.8MHz output with 90° shifts

Fig 7.11: Four-phase SSB generator

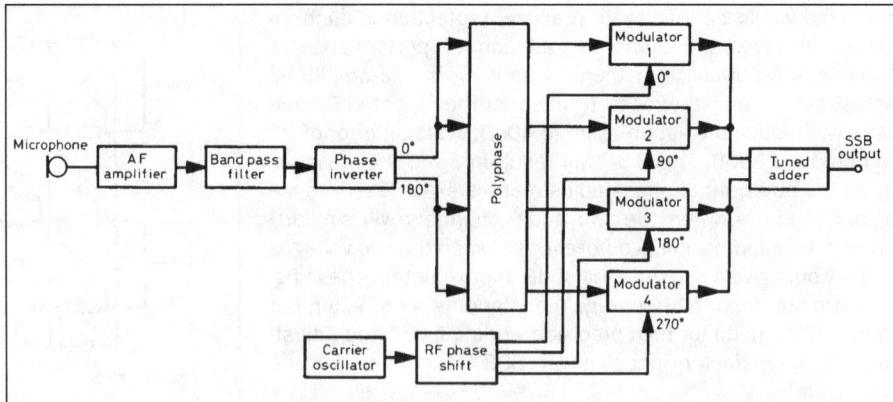

Fig 7.12: 9MHz phasing exciter

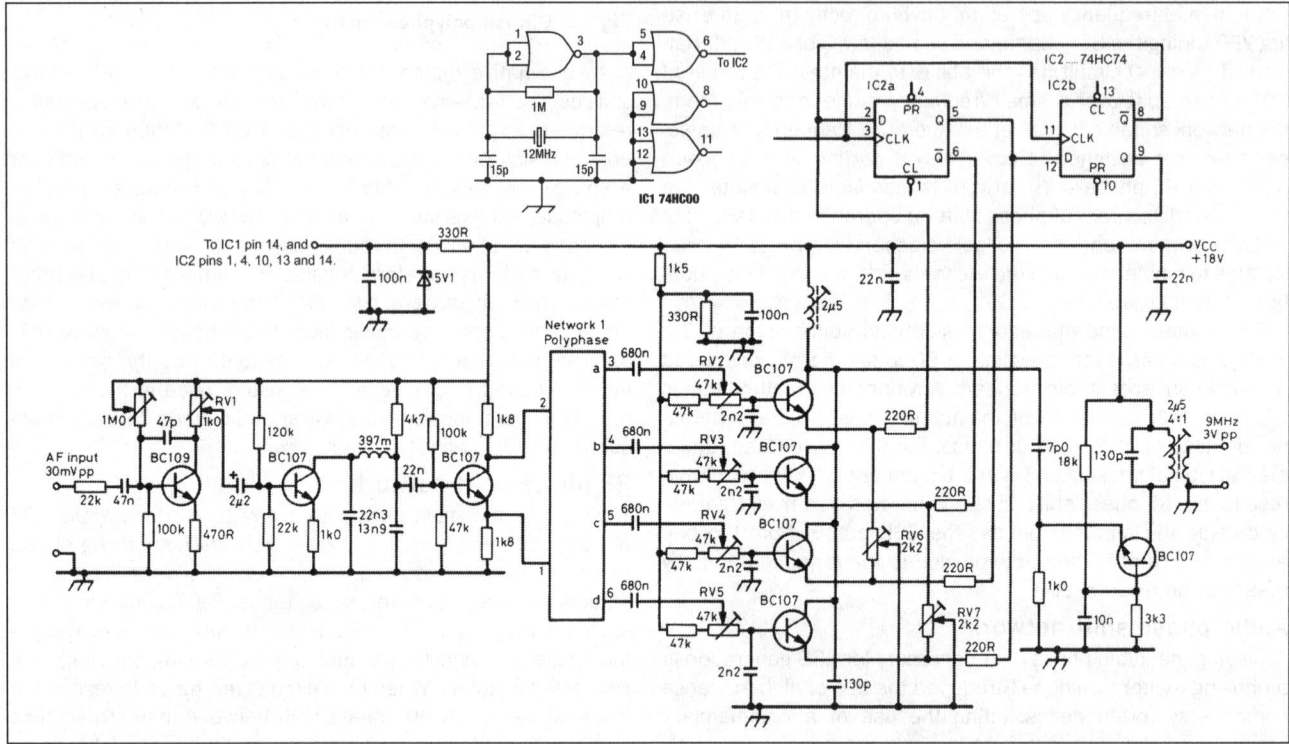

With the advent of digital ICs, it became relatively easy to obtain the required phase shift by dividing the output of an oscillator using a flip-flop IC. **Fig 7.10(b)** shows the RF circuitry for a 160m phasing exciter, in which the signal from a VFO tuning 7.2-8MHz is divided by four using a 74HC73 (J-K flip-flop), providing a square-wave output between 1.8 and 2MHz.

The VFO is conveniently implemented by one inverter of a 74HC00 (quad NAND-gate), which directly drives the flip-flop producing both 0 and 90 degree outputs. The fundamental square-wave signal will be phased out in the balanced modulator.

Four-way phasing method

The four-way phasing method is an adaptation of the conventional phasing method and can be simply described as a double two-way method. **Fig 7.11** illustrates a four-way phasing exciter. The major requirement for acceptable carrier and sideband suppression is a good audio phase shifter. The polyphase network (Fig 7.9) is ideal and provides the required four output signals at 90° phase intervals. The RF output from the carrier generator must also provide four RF outputs phase-shifted by 90° from one another, and this is achieved by using a dual J-K flip-flop which also divides the input frequency by a factor of four. The phase-shifted AF and RF signals are fed to four modulators, the outputs of which are summed in a tuned adder, resulting in a SSB output signal.

A practical 9MHz four-way SSB generator is illustrated in **Fig 7.12**. The TTL oscillator is operated at 12MHz, providing a 3MHz signal at the output of the flip-flop. As this square wave signal is rich in odd-order harmonics, the tuned adder can be adjusted to tune to the third harmonic in preference to the fundamental signal, and the result is a 9MHz output SSB signal.

One major disadvantage of the digital phase shifter is the necessity to operate the oscillator on four times the output frequency. The technique used in Fig 7.12 represents one solution to the problem, but an alternative would be to heterodyne the output to the desired frequency using a VFO.

The Filter Method

Since the objective is to transmit only a single sideband, it is necessary to select the desired sideband and suppress the unwanted sideband. The relationship between the carrier and sidebands is shown in the Principles chapter. Removing the unwanted sideband by the use of a selective filter has the advantage of simplicity and good stability, and this is therefore the most widely adopted method of generating SSB. The unwanted sideband suppression is determined by the attenuation of the sideband filter, while the stability of this suppression is determined by the stability of the elements used in construct-

ing the filter. High stability can be achieved by using materials that have a very low temperature coefficient. Commonly used materials are quartz, ceramic and metallic plates.

The filter method, because of its proven long-term stability, has become the most popular method used by amateurs. At present three types of selective sideband filters are in common use:

- High-frequency crystal filter
- Low-frequency mechanical filter
- Low-frequency ceramic filter

Crystal filters

The crystal filter is the most widely used type of filter found in SSB transmitters. In a transceiver, one common filter can be used for both transmit and receive functions. Generation of a SSB signal in the 9MHz range permits single-frequency conversion techniques to be employed to cover the entire HF spectrum, and for this reason 9MHz and 10.7MHz filters have virtually dominated the market. A number of other frequencies have been employed for filters, including 5.2MHz, 3.18MHz and 1.6MHz, the latter mainly for commercial applications.

The principle of operation of a crystal or quartz filter is based upon the piezo-electric effect. When the crystal is excited by an alternating electric current, it mechanically resonates at a frequency dependent upon its physical shape, size and thickness. A crystal will easily pass current at its natural resonant frequency but attenuates signals either side of this frequency (see the chapter on Passive Components). By cascading a number of crystals having the same, or very closely related, resonant frequencies it is possible to construct a filter having a high degree of attenuation either side of a band of wanted frequencies, typically 40-60dB with a six-pole filter, 60-80dB using a eight-pole filter, and 80-100dB with a 10-pole filter. The characteristic bandwidth of a SSB filter is selected to pass a communications-quality audio spectrum of typically 300-3000Hz. For SSB transmission the best-sounding results can be achieved using a 3kHz wide filter, but for receiver applications a slightly narrower filter is preferable and 2.4kHz has become the accepted compromise. SSB transmissions made using narrower filters have a very restricted audio sound when received. Filter bandwidths are normally quoted at the 6dB and 60dB attenuation levels, and the ratio of the two quoted bandwidths is referred to as the

shape factor, 1:1 being the ideal but not realistically achievable. Anything better than 2.5:1 is regarded as acceptable.

The purpose of the crystal filter is to attenuate the unwanted sideband but it has a secondary function: to attenuate further the already-suppressed carrier. A balanced modulator seldom attenuates the carrier by more than 40dB, which equates to 1mW of carrier from a 100W SSB transmitter. A further 20dB of carrier suppression is available from the sideband filter, making a carrier suppression of 60dB possible.

The passband of a crystal filter may be symmetrical in shape, **Fig 7.13(a)**, or asymmetric as in **Fig 7.13(b)**. Home-constructed filters are invariably asymmetric to some degree whereas commercially made filters will be designed to fall into one of the two categories. Assuming that we wish to generate a LSB signal with a carrier frequency of 9MHz, a SSB filter will be required to pass the frequency range 8.9975MHz to 9.0MHz, whereas if we wished to change to USB, the filter would be required to pass the frequency range 9.0MHz to 9.0025MHz. At first sight it would appear that two filters are required. In commercial equipment, the use of two filters is common practice, in which case the filters will most probably be asymmetric and annotated with the sideband that they are designed to generate and the intended carrier frequency. In amateur radio equipment a much cheaper technique is adopted: it is easier to use a symmetrical filter with a centre frequency of 9.000MHz and move the carrier frequen-

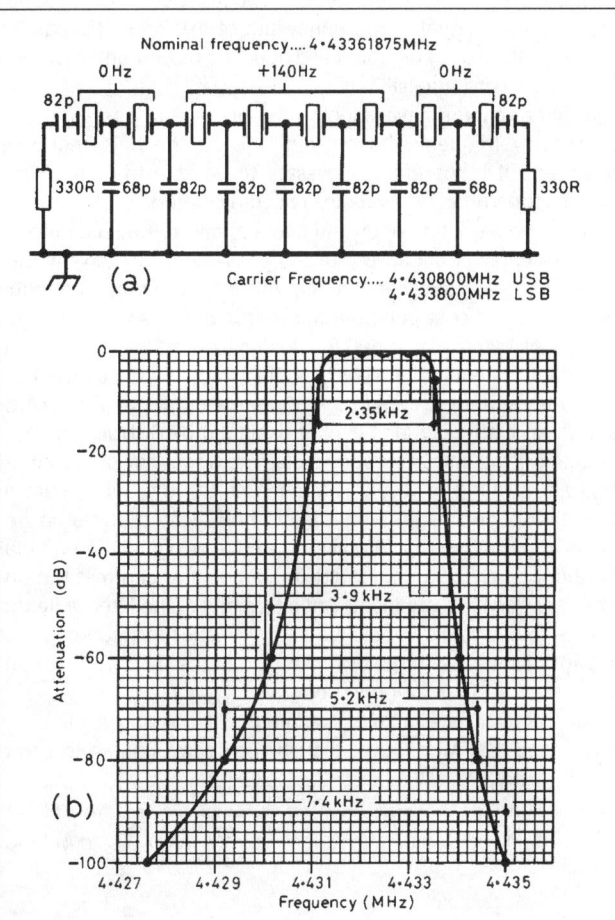

Fig 7.14: (a) Ladder crystal filter using 4.43MHz TV crystals. (b) Ladder crystal filter using P129 specification colour-TV 4.43MHz crystals. It provides a performance comparable with the ladder filter in the Atlas 180 and 215 transceivers. Insertion loss 4-5dB, shape factor (6/60dB) 1.66. Note that the rate of attenuation on the low-frequency side of the response is as good as an eight-crystal lattice design; on the HF side it is better

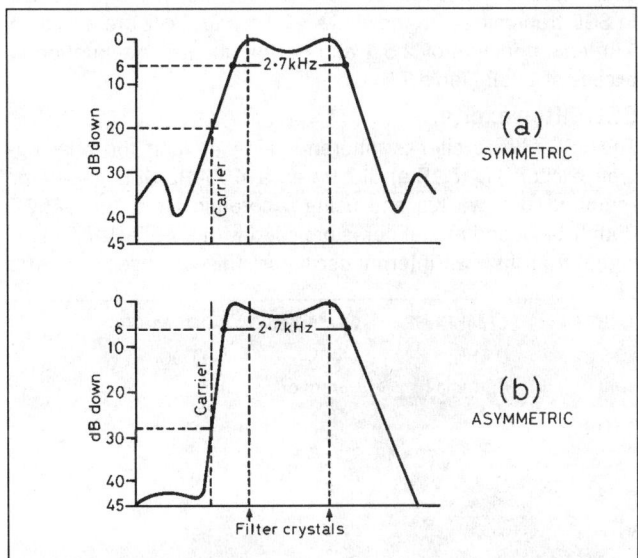

Fig 7.13: Response curves of two- and four-pole symmetric and asymmetric crystal filters

Filter type	XF-9A	XF-9B	XF-9C	XF-9D	XF-9E	XF-9M
Application	SSB TX	SSB TX/RX	AM	AM	FM	CW
Number of poles	5	8	8	8	8	4
6dB bandwidth (kHz)	2.5	2.4	3.75	5.0	12.0	0.5
Passband ripple (dB)	<1	<2	<2	<2	<2	<1
Insertion loss (dB)	<3	<3.5	<3.5	<3.5	<3	<5
Termination	500 /30pF	500 /30pF	500 /30pF	500 /30pF	1200 /30pF	500 /30pF
Shape factor	1.7 (6-50dB)	1.8 (6-60dB)	1.8 (6-60dB)	1.8 (6-60dB)	1.8 (6-60dB)	2.5 (6-40dB)
		2.2 (6-80dB)	2.2 (6-80dB)	2.2 (6-80dB)	2.2 (6-80dB)	4.4 (6-60dB)
Ultimate attenuation (dB)	>45	>100	>100	>100	>90	>90

Table 7.4: KVG 9MHz crystal filters for SSB, AM, FM and CW applications

cy from one side of the filter to the other in order to change sidebands. Typically 8.9985MHz for USB and 9.0015MHz for LSB are used; note the filter frequency is above the carrier frequency to give USB and below the carrier frequency to give LSB. Asymmetric filters are invariably marked with the carrier frequency and have the advantage of higher attenuation of the unwanted carrier and sideband, due to the steeper characteristic of the filter on the carrier frequency side. This is the primary reason why they are used in commercial applications where they are required to meet a higher specification. One minor disadvantage of the symmetrical filter and switched carrier frequency method is that changing sideband causes a shift in frequency approximately equal to the bandwidth of the filter. This can be compensated for by an equal and opposite movement of the frequency-conversion oscillator.

Radio amateurs have adopted the practice of operating LSB on the low-frequency bands and USB on the high-frequency ones, and it is therefore necessary to be able to switch sidebands if operation on all bands is contemplated.

Home construction of crystal filters is only to be recommended if a supply of cheap crystals is available. Clock crystals for microprocessor applications are manufactured in enormous quantities and cost pence rather than pounds. Another source of crystals is those intended for TV colour-burst; the UK and continental frequency of 4.43MHz is the more suitable as the USA colour-burst frequency is in the 80m amateur band. Fig 7.14(a) shows an eight-pole ladder filter designed by G3UUR and constructed using colour-burst crystals. The frequency of individual crystals is found to vary by as much as 200Hz, but by careful selection of crystals it is possible to find those that are on frequency and those that are slightly above or below the nominal frequency. For optimum results it is recommended that the on-frequency crystals are located at either end of the filter while the centre four crystals should be slightly higher in frequency. Fig 7.14(b) shows the frequency response of the G3UUR filter constructed from TV colour-burst crystals.

Commercially available crystal filters are primarily confined to 9MHz and 10.7MHz types, though the sources seem to come and go. Numerous filters are available new, and surplus filters can often be found at bargain prices. Table 7.4 lists the KVG range of filters - alternative filters are listed in Table 7.5.

Mechanical filters

The mechanical filter was developed by the Collins Radio Company for low-frequency applications in the range 60-500kHz. The F455 FA-21 was designed specifically for the amateur radio market with a nominal centre frequency of 455kHz, a 6dB bandwidth of 2.1kHz and a 60dB bandwidth of 5.3kHz, providing a shape factor of just over 2:1.

The mechanical filter is made up from a number of metal discs joined by coupling rods. The discs are excited by magnetostrictive transducers employing polarised biasing magnets and must not be used in circuits where DC is present. The input and output transducers are identical and are balanced to ground so the filter can be used in both directions. Mechanical filters have always been expensive but provide exceptional performance, and examples of the Collins filters can often be found in surplus equipment. Japanese Kokusai filters were available for a number of years and offered a more cost-effective alternative. Mechanical filters are now seldom used in new equipment.

Ceramic filters

Ceramic filters have been developed for broadcast radio applications; they are cheap, small in size and available in a wide range of frequencies and bandwidths. Narrow-bandwidth ceramic filters with a nominal centre frequency of 455kHz have been manufactured for use in SSB receiver IF applications and provide a level of performance making them ideally suited for use in SSB transmitters. Bandwidths of 2.4 and 3kHz are available with shape factors of 2.5:1 and having ultimate attenuation in excess of 90dB (Table 7.5).

SSB filter exciter

The SSB filter exciter is inherently simpler than the phasing-type exciter. Fig 7.15(a) illustrates a 455kHz SSB generator requiring just two ICs and using a ceramic filter. The SA602 doubly balanced modulator is provided with a 455kHz RF input signal from its own internal oscillator. This can use a 455kHz

Filter type	90H2.4B	10M02DS	CFS455J	CFJ455K5	CFJ455K14	QC1246AX
Centre frequency (MHz)	9.0000	10.7000	0.455	0.455	0.455	9.0000
Number of poles	8	8	Ceramic	Ceramic	Ceramic	8
6dB bandwidth (kHz)	2.4	2.2	3	2.4	2.2	2.5
60dB bandwidth (kHz)	4.3	5	9 (80dB)	4.5	4.5	4.3
Insertion loss (dB)	<3.5	<4	8	6	6	<3
Termination	500 /30pF	600 /20pF	2k	2k	2k	500 30pF
Ultimate attenuation (dB)	>100	-	>60	>60	60	>90
Manufacturer/UK supplier	IQD	Cirkit	Cirkit	Bonex	Bonex	SEI

Table 7.5: Popular SSB filters

The Radio Communication Handbook

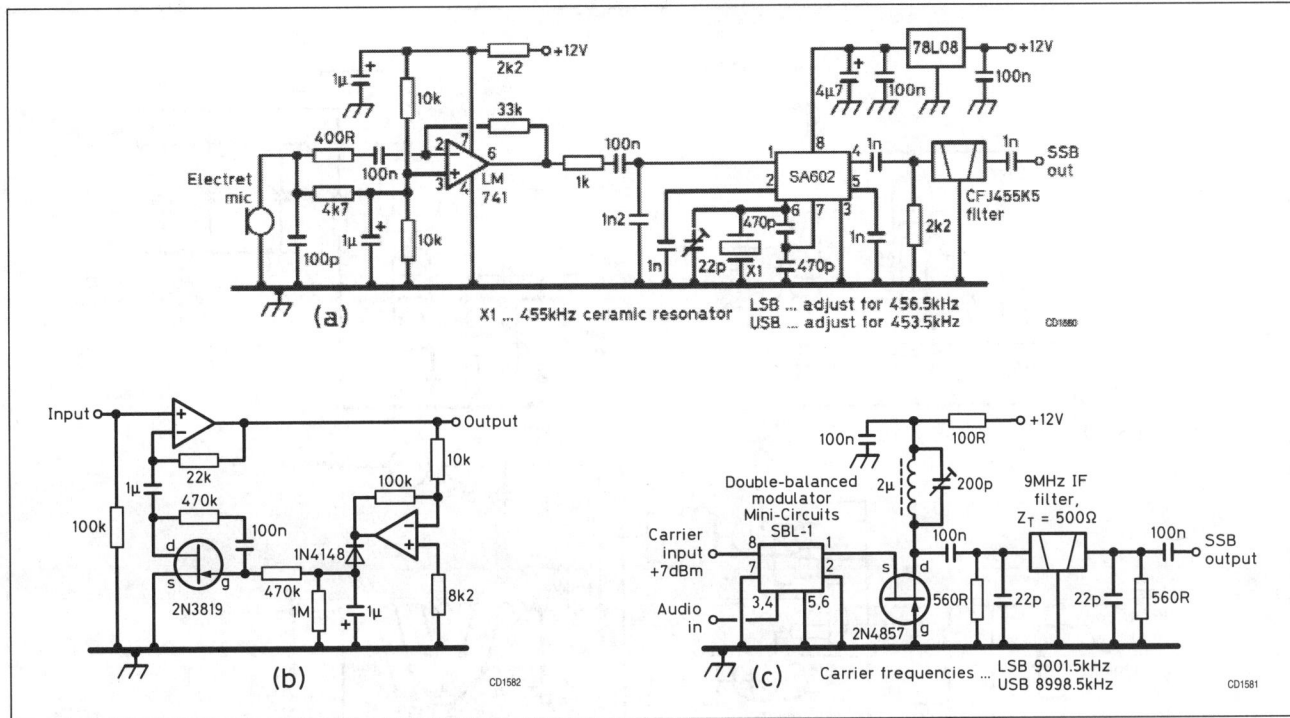

Fig 7.15: (a) 455kHz SSB generator. (b) Example of an audio AGC circuit. (c) 9MHz SSB generator

crystal or the cheaper ceramic resonator, and the latter can be pulled in frequency to generate either a LSB or USB carrier. A DSB signal at the output of the balanced modulator is fed directly to the ceramic filter for removal of the unwanted sideband, resulting in a 455kHz SSB output signal. This can be heterodyned directly to the lower-frequency amateur bands or via a second higher IF to the HF bands. The circuit performance can be improved by the addition of an audio AGC circuit, such as the one shown in **Fig 7.15(b)**. **Fig 7.15(c)** illustrates a design for a 9MHz SSB generator. Microphone audio would need to pass through a suitable audio amplifier before being applied to the balanced modulator at pins 3 and 4. Output from the 9MHz SSB exciter can be heterodyned directly to any of the LF and HF amateur bands.

The only adjustment required in either circuit of **Fig 7.16(a) and (c)** is to adjust the oscillator to the correct frequency in relation to the filter; this frequency is normally located 20dB down the filter response curve. Quite often the oscillator can simply be adjusted for the best audio response at the receiver.

POWER AMPLIFIERS

The RF power amplifier is normally considered to be that part of a transmitter which provides RF energy to the antenna. It may be a single valve or transistor, or a composite design embodying numerous devices to take low-level signals to the final output power level. RF amplifiers are also discussed in detail in the Building Blocks chapter.

Push-pull valve amplifiers, Fig 7.16(a), were popular until the 1950s and offered a number of advantages over single-ended output stages as in **Fig 7.16(b)**. The inherent balance obtained when two similar valves having almost identical characteristics are operated in push-pull results in improved stability, while even-order harmonics are phased out in the common tank circuit. One major shortcoming of the push-pull valve amplifier is that switching of the output tank circuit for operation on more than one frequency band is exceedingly difficult due to the high RF voltages present.

During the 1960s, push-pull amplifiers were superseded by the single-ended output stage, often comprising two valves in parallel, Fig 7.16(c), coupled to the antenna via a pi-output network and low-impedance coaxial cable. The changes in design were brought about by two factors. First, the rapid expansion in TV broadcasting introduced problems of harmonically related TVI. This demanded greater attenuation of odd-order harmonics than was possible with the conventional tank circuit used in push-pull amplifiers. Second, there was a trend towards the development of smaller, self-contained transmitters capable of multiband operation, ultimately culminating in the transceiver concept which has almost totally replaced the separate transmitter and receiver in amateur radio stations.

The pi-output tank circuit differs from the conventional parallel-tuned tank circuit in that it uses two variable capacitors in series: Figs 7.4(a) and 7.4(b). The junction of the two capacitors is grounded and the network is isolated from DC using a high-voltage blocking capacitor. The input capacitor has a low value, while the output capacitor has a considerably higher value. This provides high-to-low impedance transformation across the network, enabling the relatively high anode impedance of the power amplifier to be matched to an output impedance of typically 75 or 50Ω. The pi-tank circuit performs the function of a low-pass filter and provides better attenuation of harmonics than a link-coupled tank circuit. The low-impedance output is easily band-switched over the entire HF spectrum by shorting out turns, and facilitates direct connection to a dipole-type antenna, an external low-pass filter for greater reduction of harmonics or an antenna matching unit for connection to a variety of antennas.

The Class C non-linear power amplifiers commonly used in CW and AM transmitters produced high levels of harmonics, making compatibility with VHF TV transmissions exceedingly difficult. Fortunately the introduction of amateur SSB occurred at much the same time as the rapid expansion in television operating hours, and the introduction of 'linear' Class AB amplifiers, essential for the amplification of SSB signals, greatly reduced the problems of TVI with TV frequencies that were often exact multi-

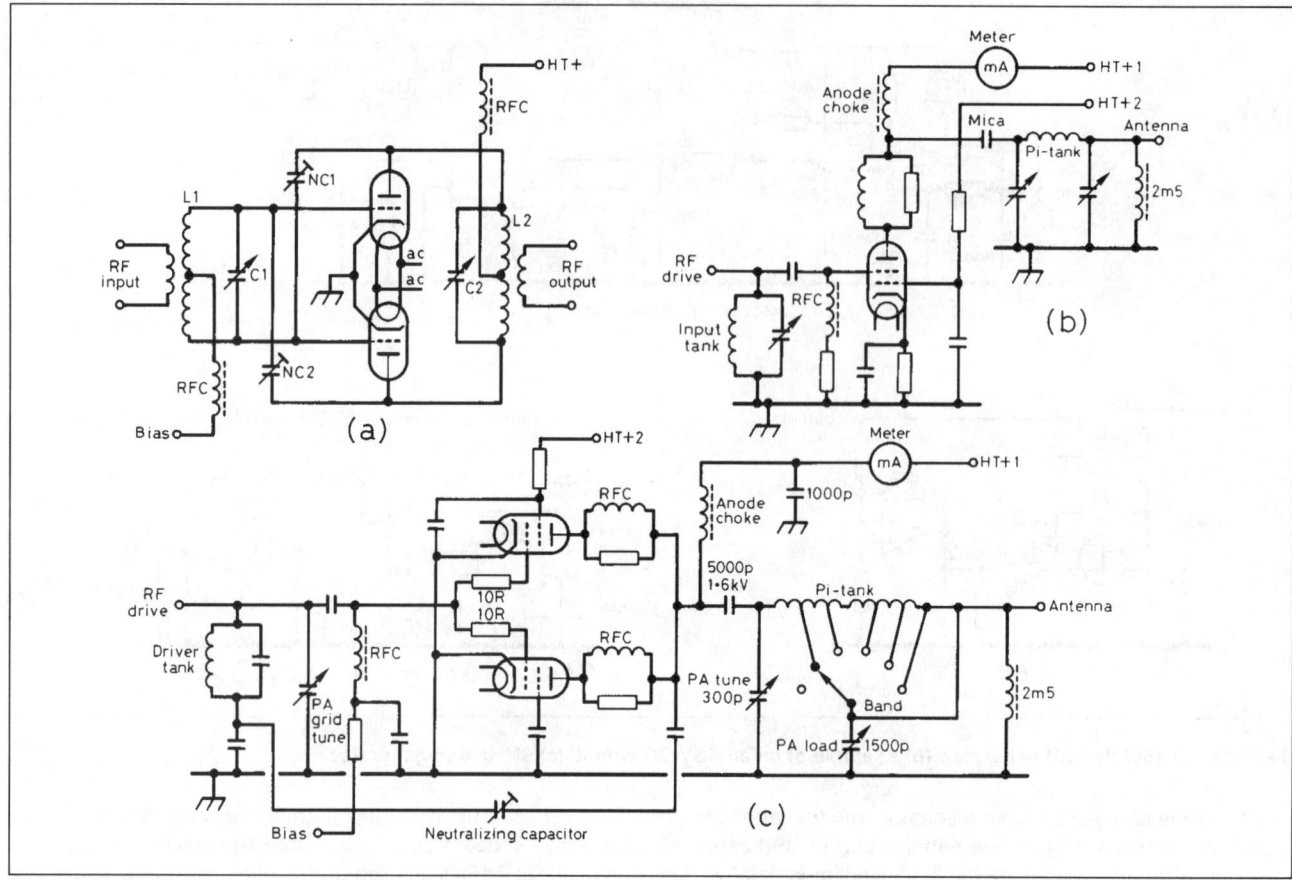

Fig 7.16: Valve power amplifiers. (a) Push-pull amplifier (b) single-ended Class-C amplifier; (c) parallel-pair RF amplifier

ples of the 14, 21 and 28MHz amateur bands. The parallel-pair valve power amplifier became the standard output stage for amateur band transmitters and is still in common use in many amateur radio transceivers and external linear amplifiers.

Solid-state power amplifiers first appeared in the late 1970s and were capable of up to 100W RF output. Initially the reliability was poor but improved rapidly, and within 10 years virtually all commercially made amateur radio transmitters were equipped with a 100W solid-state PA unit. Transistors for high-power RF amplification are specially designed for the purpose and differ considerably internally from low-frequency switching transistors. Initial attempts by amateurs to use devices not specifically designed for RF amplification resulted in a mixture of success and failure, probably giving rise to initial claims that solid-state power amplifiers were less reliable than their valve counterparts.

Due to the very low output impedances of solid-state bipolar devices, it is almost impossible to match the RF output from a solid-state power amplifier to a resonant tank circuit with its characteristic high impedance. As a result, low-Q matching circuits (Fig 7.2 and 7.3) are used to transform the very low (typically 1 or 2Ω) output impedance encountered at the collector of an RF power amplifier to the now-standard output impedance of 50Ω.

As the matching circuits are not resonant, they are broadband in nature, making it possible to operate the amplifier over a wide range of frequencies with no need for any form of tuning. Unfortunately, the broad-band amplifier also amplifies harmonics and other unwanted products, making it essential to use a low-pass filter immediately after the amplifier in order to achieve an adequate level of spectral purity. It is also essential that the amplifier itself should not contribute to the production of

unwanted products. For this reason solid-state amplifier designs have reverted to the balanced, push-pull mode of operation, with its inherent suppression of even-order harmonics. As band switching and tuning of the amplifier is neither possible or necessary, construction of the amplifier is relatively simple. Typically, HF broad-band amplifiers will provide an output over the entire HF spectrum from 1.8 to 30MHz.

For powers in excess of 100W, valves are still popular and are likely to remain in use for some time to come on the grounds of cost, simplicity and superior linearity.

Power amplifiers can be categorised into two basic types, valve and solid-state, and then further into sub-groups based upon the output power. Low-power amplifiers can be regarded as 10W or less, including QRP (usually regarded as less than 5W output power), medium power up to 100W, and high power in excess of 100W.

The class of operation of a power amplifier is largely determined by its function. Class C amplifiers are commonly used for CW, AM and FM transmissions because of their high efficiency. Class C is also a pre-requisite for successful high-level modulation in an AM transmitter. Due to the non-linear operation of Class C amplifiers, the harmonic content is high and must be adequately filtered to minimise interference.

Single sideband transmission demands linear amplification if distortion is to be avoided, and amplifiers may be operated in either Class A or B. Class A operation is inefficient and normally confined to driver stages, the high standing current necessary to achieve this class of operation usually being unacceptable in amplifiers of any appreciable power rating. Most linear amplifiers designed for SSB are operated in between Class A and B, in what are known as Classes AB1 and AB2, in order to achieve a compromise between efficiency and linearity.

Solid-state Versus Valve Amplifiers

The standard RF power output for the majority of commercially produced amateur radio transmitters/transceivers is 100W, and at this power level solid-state amplifiers offer the following:

- Compact design.
- Simpler power supplies requiring only one voltage (normally 13.8V) for the entire equipment.
- Broad-band, no tune-up operation, permitting ease of operation.
- Long life with no gradual deterioration due to loss of emission.
- Ease of manufacture and reduced cost.

There are of course some disadvantages with solid-state amplifiers and it is for this reason that valves have not disappeared entirely. However, valve amplifiers have considerably more complex power supply requirements and great care must be taken with the high voltages involved if home construction is contemplated. Although solid state amplifiers only require relatively low voltages, they require very high currents for the generation of any appreciable power, placing demands upon the devices and their associated power supplies. Heatsinking and voltage stabilisation become very important.

While the construction of solid-state amplifiers up to 1kW is feasible using modern devices operated from a high-current 50V power supply, valve designs offer a simpler and more cost-effective alternative. However, the development of VMOS devices with much higher input and output impedances is beginning to bridge the gap. Quite possibly within a few years VMOS devices will offer a cheaper alternative to the valve power amplifier at the kilowatt level.

Impedance Matching

All types of power amplifier have an internal impedance made up from a combination of the internal resistance, which dissipates power in the form of heat, and reactance. As both source and load impedances are fixed values and not liable to change, it is necessary to employ some form of impedance transformation or matching in order to obtain the maximum efficiency from an amplifier. The power may be expressed as:

$$P_{input} = P_{output} + P_{dissipated}$$

where P_{input} is the DC input power to the stage; P_{output} is the RF power delivered to the load and $P_{dissipated}$ is the power absorbed in the source resistance and dissipated as heat.

$$\text{Efficiency} = \frac{P_{output}}{P_{input}} \times 100\%$$

When the source impedance is equal to the load impedance, the current through either will be equal as they are in series, with the result that 50% of the power will be dissipated by the source and 50% will be supplied to the load. The object of a power amplifier is to provide maximum power to the load. Design of a power amplifier must also take into account the maximum dissipation of the output device as specified by the manufacturer. An optimum load resistance is selected to ensure maximum output from a power amplifier while not exceeding the amplifying device's power dissipation. Efficiency increases as the load resistance to source resistance ratio increases and vice versa. The optimum load resistance is determined by the device's current transfer characteristics and for a solid-state device is given by:

$$R_L = \frac{V_{cc}^2}{2P_{out}}$$

Valves have more complex current transfer characteristics which differ for different classes of operation; the optimum load resistance is proportional to the ratio of the DC anode voltage to the DC anode current divided by a constant which varies from 1.3 in Class A to approximately 2 in Class C.

$$R_L = \frac{V_a}{KI_a}$$

The output from a RF power amplifier is usually connected to an antenna system of different impedance, so a matching network must be employed. Two methods are commonly used: pi-tank circuit matching for valve circuits and transformer matching for solid-state amplifiers. The variable nature of a pi-tank circuit permits matching over a wide range of impedances, whereas the fixed nature of a matching transformer is dependent upon a nominal load impedance of typically 50Ω, and it is therefore almost essential to employ some form of antenna matching unit between the output and the final load impedance. Matching networks serve to equalise load and source resistances while providing inductance and capacitance to cancel any reactance.

Valve Power Amplifiers

Valve power amplifiers are commonly found in the output stages of older amateur transmitters. Usually two valves will be operated in parallel, providing twice the output power possible with a single valve. The output stage may be preceded by a valve or solid-state driver stage.

The valve amplifier is capable of high gain when operated in the tuned input, tuned output configuration often referred to as TPTG (tuned plate tuned grid). Two 6146 valves are capable of producing in excess of 100W RF output with as little as 500mV of RF drive signal at the input. Operation of two valves in parallel increases the inter-electrode capacitances by a factor of two and ultimately affects the upper frequency operating limit. **Fig 7.16(c)** illustrates a typical output stage found in amateur transceivers.

Neutralisation

The anode-to-grid capacitance of a valve provides a path for RF signals to feed back energy from the anode to the grid. If this is sufficiently high, oscillation will occur. This can be overcome by feeding back a similar level of signal, but of opposite phase, thus cancelling the internal feedback, and this process is referred to as neutralisation.

Once set, the neutralisation should not require further adjustment unless the internal capacitance of the valve changes or the valve has to be replaced.

While there are numerous ways of achieving neutralisation, the most common method still in use is series-capacitance neutralisation. A low-value ceramic variable capacitor is connected from the anode circuit to the earthy end of the grid input circuit as in Fig 7.16(c).

The simplest way to adjust the neutralising capacitor is to observe the anode current when the PA output tuning capacitor is adjusted through resonance. The current should reduce gradually to a minimum value and then rise smoothly again to its previous value. Any asymmetry of the current either side of the resonant point will indicate that the PA is not correctly neutralised. Neutralising capacitors should be adjusted with non-metallic trimming tools as metal tools will interfere with the adjustment.

Parasitic oscillation

It is not uncommon for a power amplifier to oscillate at some frequency other than one in the operating range of the amplifier. This can usually be detected by erratic tuning characteristics

and a reduction in efficiency. The parasitics are often caused by the resonance of the connecting leads in the amplifier circuit with the circuit capacitance. To overcome problems at the design stage it is common to place low-value resistors in series with the grid, and low-value RF chokes, often wound on a resistor body, in series with the anode circuit. Ferrite beads may also be strategically placed in the circuit to damp out any tendency to oscillate at VHF.

HIGH-POWER AMPLIFICATION

Output powers in excess of 100W are invariably achieved using an add-on linear amplifier. In view of the high drive power available, the amplifier can be operated at considerably lower gain, with the advantages of improved stability and no requirement for neutralisation. Input circuits are usually passive, with valves operated in either passive-grid or grounded-grid modes. The grounded-grid amplifier has a cathode impedance ideally suited to matching the pi-output circuit of a valve exciter. Pi-input networks are usually employed to provide the optimum 50-ohm match for use with solid-state exciters. Passive grid amplifiers often employ a grid resistor of 200 to 300Ω, which is suitable for connecting to a valve exciter but will require a matching network such as a 4:1 auto-transformer for connection to a solid-state exciter.

Output matching

There are only two output circuits in common use in valve power amplifiers - the pi-output network is by far the most common and suitable values for a range of anode impedances are provided in Table 7.2. More recently, and especially for applications in high-power linear amplifiers, an adaptation of the circuit has appeared called the pi-L output network, and here the conventional pi-network has been combined with an L-network to provide a matching network with a considerable improvement in attenuation of unwanted products. The simple addition of one extra inductor to the circuit provides a considerable improvement in performance. Suitable values for a pi-L network are given in Table 7.3 for a range of anode impedances.

Valve amplifiers employing pi-output networks require a suitable RF choke to isolate the anode of the power amplifier from the high voltage power supply. This choke must be capable of carrying the anode current as well as the high anode voltages likely to be encountered in such an amplifier. The anode choke must not have any resonances within the operating range of the amplifier or it will overheat with quite spectacular results. Chokes are often wound in sections to reduce the capacitance between turns and may employ sections in varying diameters. Ready-made chokes are available for high-power operation from Barker and Williamson RF Engineering Ltd in the UK.

Solid-state Power Amplifiers

Solid-state power amplifiers are invariably designed for 50Ω input and output impedances, and they employ broad-band transformers to effect the correct matching to the devices. While single-ended amplifiers, **Fig 7.17(a)**, are often shown in test circuits, their use is not recommended for the following reasons:

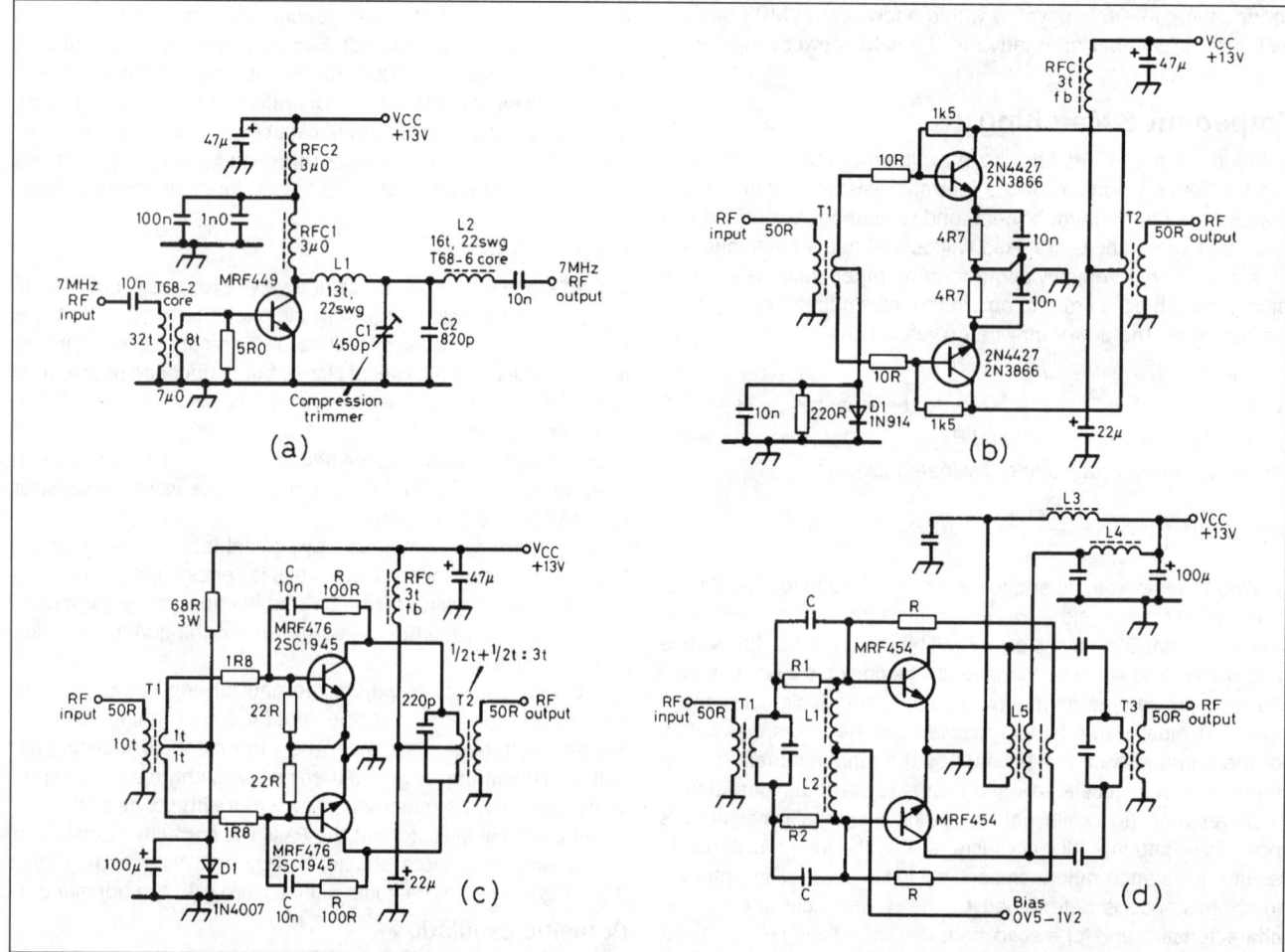

Fig 7.17: Solid-state power amplifiers. (a) Single-ended Class C amplifier (15W); (b) low-power amplifier driver (c) low-power push-pull amplifier (10-25W); (d) medium-power push-pull amplifier (100W)

(a) high levels of second harmonic are present and are difficult to attenuate using standard low-pass filters;

(b) multiband operation is more difficult to achieve due to the more complex filter requirements.

The broad-band nature of a solid-state amplifier with no requirement for bandswitching enables push-pull designs to be used, taking advantage of their improved balance and natural suppression of even-order harmonics. The gain of a solid-state broad-band amplifier is considerably lower than that of a valve power amplifier (typically 10dB) and may necessitate cascading a number of stages in order to achieve the desired power level. This is easily achieved using common input and output impedances.

The reduced gain has the advantage of aiding stability, but the gain rises rapidly with a reduction in frequency and demands some form of frequency-compensated gain reduction. The latter is achieved using negative feedback with a series combination of R and C: **Figs 7.17(b) and (c)**. Good decoupling of the supply down to audio frequencies is essential. The use of VHF power transistors is not recommended in the HF spectrum as instability can result even when high levels of negative feedback are employed.

Solid-state amplifiers can be operated in Class C for use in CW and FM transmitters, but their use for AM transmission should be treated very carefully for the following reasons:

- It is almost impossible to achieve symmetrical amplitude modulation of a solid-state PA.
- Device ratings must be capable of sustaining double the collector voltage and current on modulation peaks, ie four times the power of the carrier. Additional safety margins must be included to allow for high RF voltages generated by a mismatched load. Invariably suitable devices are not available for other than low-power operation.

For AM transmission it is recommended that the signal be generated at a low level and then amplified using a linear amplifier. Again, allowance should be made for the continuous carrier and the power on modulation peaks.

Output filters

Solid-state amplifiers must not be operated into an antenna without some form of harmonic filtering. The most common design is the pi-section filter, comprising typically of a double pi-section (five-element) and in some cases a triple pi-section (seven-element) filter. Common designs are based upon the Butterworth and Chebyshev filters and derivations of them. The purpose of the low-pass filter is to pass all frequencies below the cut off frequency (f0), normally located just above the upper band edge, while providing a high level of attenuation to all frequencies above the cut-off frequency. Different filter designs provide differing attenuation characteristics versus frequency, and it is desirable to achieve a high level of attenuation by at least three times the cut-off frequency in order to attenuate the third harmonic.

The second harmonic, which should be considerably lower in value due to the balancing action of the PA, will be further attenuated by the filter which should have achieved approximately 50% of its ultimate attenuation.

Elliptic filters are designed to have tuned notches which can provide higher levels of attenuation at selected frequencies such as 2f

and 3f. Filters are discussed in detail in the chapters on Building Blocks and General Data.

The desire to achieve high levels of signal purity may tempt constructors to place additional low-pass filters between cascaded broad-band amplifiers, but this practice will almost certainly result in spurious VHF oscillations. These occur when the input circuit resonates at the same frequency as the output circuit, ie when the filter acts as a short-circuit between the amplifier input and output circuits. Low-pass filters should only be employed at the output end of an amplifier chain. If it is essential to add an external amplifier to an exciter which already incorporates a low-pass filter, it is important either to use a resistive matching pad between the exciter and the amplifier, or modify the output low-pass filter to ensure that it has a different characteristic to the input filter. If a capacitive input filter is used in the exciter, an inductive input filter should be employed at the output of the linear amplifier.

The resulting parasitics caused by the misuse of filters may not be apparent without the use of a spectrum analyser, and the only noticeable affect may be a rough-sounding signal and warm low-pass filters.

Amplifier matching

Broad-band transformers used in solid-state amplifiers consist of a small number of turns wound on a stacked high-permeability ferrite core. The secondary winding may be wound through the primary winding which may be constructed from either brass tube or copper braid. The grade of ferrite is very important and will normally have an initial permeability of at least 800 (Fairite 43 grade).

Too low a permeability will result in poor low-frequency performance and low efficiency. Some designs use conventional centre-tapped transformers for input and output matching (Fig 7.17(b) and (c)); these transformers carry the full DC bias and PA currents.

Other designs, **Fig 7.17(d)**, include phasing transformers to supply the collector current while the output transformer is blocked to DC by series capacitors. The latter arrangement provides an improvement in IMD performance of several decibels but is often omitted in commercial amateur radio equipment on grounds of cost.

Amplifier protection

The unreliability of early solid-state PAs was largely due to a lack of suitable protection circuitry. ALC (automatic level control) has been used for controlling the output of valve amplifiers for many years - by sampling the grid current in a valve PA it is possible to provide a bias that can be fed back to the exciter to reduce the drive level. A similar system is used for solid-state amplifiers but

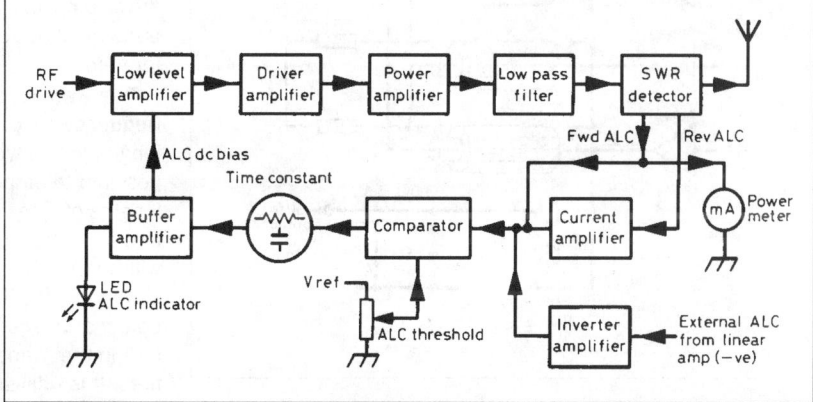

Fig 7.18: ALC system incorporating forward and reverse protection

it is usually derived by sampling some of the RF output present in a SWR bridge circuit.

ALC is very similar to the AGC system found in a receiver. One disadvantage of sampling the RF output signal is that excessive ALC levels will cause severe clipping of the RF signal with an associated degradation of the IMD performance of the amplifier. For optimum performance the ALC system should only just be operating.

One of the major differences between solid-state and valve amplifying devices is that the maximum voltage ratings of solid-state devices are low and cannot be exceeded without disastrous consequences.

Reverse ALC is provided to overcome this problem and works in parallel with the conventional or forward ALC system. High RF voltages appearing at the PA collectors are attributable to operating into mismatched loads which can conveniently be detected using a SWR bridge.

The reverse or reflected voltage can be sampled using the SWR bridge and amplified and used to reduce the exciter drive by a much greater level than that used with the forward ALC. The output power is cut back to a level which then prevents the high RF voltages being generated and so protects the solid-state devices. Some RF output devices are fitted internally with zener diodes to prevent the maximum collector voltage being exceeded. **Fig 7.18** illustrates a typical ALC protection system.

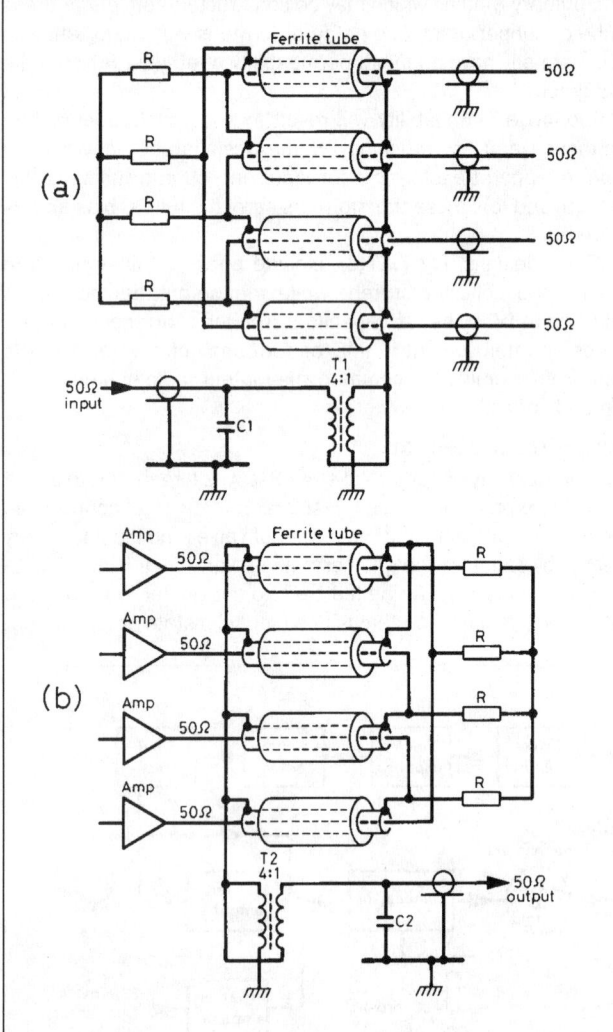

Fig 7.19: Combining multiple power amplifiers. (a) Four port power divider. (b) Four-port output combiner

Heatsinking for solid-state amplifiers

The importance of heatsinking for solid-state amplifiers cannot be overstressed - on no account should any power amplifier be allowed to operate without a heatsink, even for a short period of time.

Low-power amplifiers can often be mounted directly to an aluminium chassis if there is adequate metalwork to dissipate the heat, but a purpose-made heatsink should be included for power levels in excess of 10W. For SSB operation the heatsinking requirements are less stringent than for CW or data operation due to the much lower average power dissipation. Nevertheless, an adequate safety margin should be provided to allow for long periods of key-down operation.

Solid-state power amplifiers are of the order of 50% efficient, and therefore a 100W output amplifier will also have to dissipate 100W of heat. It can be seen that for very-high-power operation heatsinking becomes a major problem due to the compact nature of solid-state amplifiers. The use of copper spreaders is advisable for powers above 200W. The amplifier is bolted directly to a sheet of copper at least 6mm (0.25in) thick, which in turn is bolted directly to the aluminium heatsink. The use of air blowers to circulate air across the heatsink should also be considered.

The actual mounting of power devices requires considerable care; the surface should ideally be milled to a flatness of ±0.012mm (0.5 thousandths of an inch). For this reason, die-cast boxes must not be used for high-power amplifiers as the conductivity is poor and the flatness is nowhere near to being acceptable. Heatsinking compounds are also essential for aiding the conduction of heat rapidly away from the device. Motorola Application Note AN1041/D [1] provides guidance in the mounting of power devices in the 200-600W range.

Power dividers and combiners

It is possible to increase the power output of a solid-state amplifier by combining the outputs of a number of smaller amplifiers. For instance it is practical to combine the outputs of four 100W amplifiers to produce 400W, or two 300W amplifiers to produce 600W.

Initially the drive signal is split using a power divider and then fed to a number of amplifiers that are effectively operated in parallel. The outputs of the amplifiers are summed together in a power combiner that is virtually the reverse of the divider circuit. **Fig 7.19** illustrates a four-way divider and combiner.

The purpose of the power divider is to divide the input power into four equal sources, providing an amount of isolation between each. The outputs are designed for 50Ω impedance, which sets the common input impedance to 12.5. A 4:1 step-down transformer provides a match to the 50Ω output of the driver amplifier. The phase shift between the input and output ports must be zero and this is achieved by using 1:1 balun transformers.

These are loaded with ferrite tubes to provide the desired low-frequency response without resort to increasing the physical length. In this type of transformer, the currents cancel, making it possible to employ high-permeability ferrite and relatively short lengths of transmission line. In an ideally balanced situation, no power will be dissipated in the magnetic cores and the line loss will be low. The minimum inductance of the input transformer should be 16μH at 2MHz - a lower value will degrade the isolation characteristics between the output ports and this is important in the event of a change in input VSWR to one of the amplifiers. It is unlikely that the splitter will be subjected to an open- or short-circuit load at the amplifier input, due to the base-frequency compensation networks in the amplifier modules. The

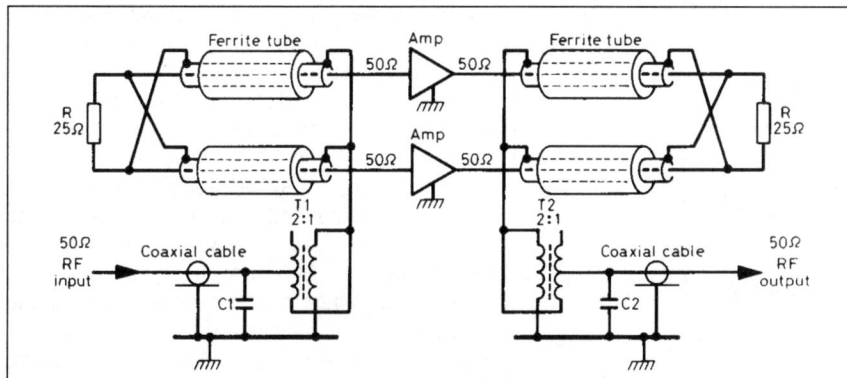

Fig 7.20: Two-port divider and combiner

purpose of the balancing resistors R is to dissipate any excess power if the VSWR rises. The value of R is determined by the number of 50Ω sources assumed unbalanced at any one time. Except for a two-port divider, the resistor values can be calculated for an odd or even number of ports as:

$$R = \left(\frac{R_L - R_{IP}}{n + 1} \right) n$$

where R_L is the impedance of output ports (50Ω); R_{IP} is the impedance of input port (12.5Ω); n is the number of correctly terminated output ports.

Although the resistor values are not critical for the input divider, the same formula applies to the output combiner where mismatches have a larger effect on the total power output and linearity.

The power divider employs ferrite sleeves having a μ of 2500 and uses 1.2in lengths of RG-196 coaxial cable; the inductance is approximately 10μH. The input transformer is wound on a 63-grade ferrite toroid with RG-188 miniature coaxial cable. Seven turns are wound bifilar and the ends are connected inner to outer, outer to inner at both ends to form a 4:1 transformer.

The output combiner is a reverse of the input divider and performs in a similar manner. The ballast resistors R must be capable of dissipating large amounts of heat in the event that one of the sources becomes disabled, and for this reason they must be mounted on a heatsink and be non-inductive. For one source disabled in a four-port system, the heat dissipated will be approximately 15% of the total output power, and if phase differences occur between the sources, it may rise substantially. The resistors are not essential to the operation of the divider/combiner; their function is to provide a reduced output level in the event of an individual amplifier failure. If they are not included, failure of any one amplifier will result in zero output from the amplifier combination. The output transformer must be capable of carrying the combined output power and must have

sufficient cross-sectional area. The output transformer is likely to run very warm during operation. High-frequency compensating capacitors C1 and C2 may be fitted to equalise the gain distribution of the amplifier but they are not always necessary.

A two-port divider combiner is illustrated in **Fig 7.20**; operation is principally the same as in the four-port case but the input and output transformers are tapped to provide a 2:1 ratio. Detailed constructional notes of two- and four-port dividers and combiners are given in Motorola Application Notes AN749 and AN758 respectively [1]

TRANSCEIVERS

Separate transmitter and receiver combinations housed in one cabinet may be referred to as a 'transceiver'. This is not strictly correct as a transceiver is a combined transmitter and receiver where specific parts of the circuit are common to both functions. Specifically, the oscillators and frequency-determining components are common and effectively synchronise both the transmitter and receiver to exactly the same frequency. This synchronisation is a pre-requisite for SSB operation and transceivers owe their existence to the development of SSB transmission. While many early attempts at SSB generation used the phasing method, the similarity of the filter-type SSB generator circuit to a superhet SSB receiver circuit (Fig 7.5) makes interconnection of the two circuits an obvious development (**Fig 7.21**). True transceive operation is possible by simply using common oscillators, but it is also advantageous to use a common SSB filter in the IF amplifier stages, which provides similar audio characteristics on both transmit and receive as well as providing a considerable saving in cost.

Initially low-frequency SSB generation necessitated double-conversion designs, often employing a tuneable second IF and a crystal-controlled oscillator for frequency conversion to the desired amateur bands. This technique was superseded by single-conversion designs using a high-frequency IF in the order of 9MHz, with a heterodyne-type local oscillator consisting of a

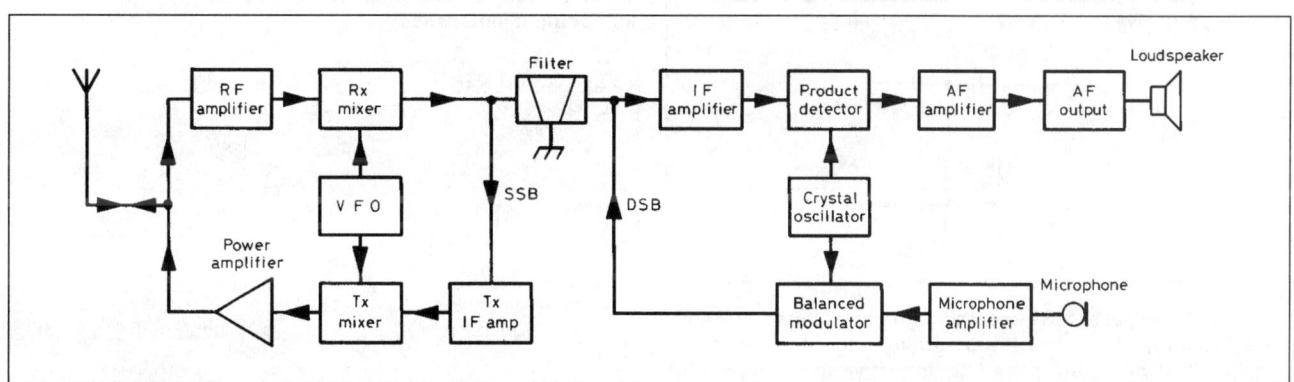

Fig 7.21: SSB transceiver block diagram

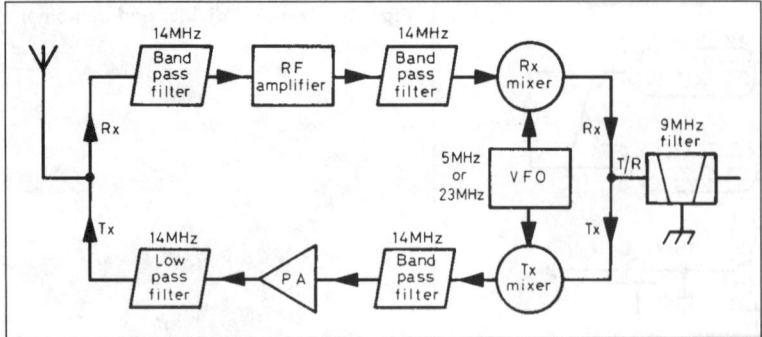

Fig 7.22: HF transceiver RF filtering

medium-frequency VFO and a range of high-frequency crystal oscillators. With the advent of the phase-locked loop (PLL) synthesiser and the trend towards wide-band equipment, modern transceivers typically up-convert the HF spectrum to a first IF in the region of 40-70MHz. This is then mixed with a synthesised local oscillator and converted down to a working IF in the order of 9MHz. Often a third IF in the order of 455kHz will be employed, giving a total of three frequency conversions.

In order to simplify the construction of equipment, designers have attempted to combine as many parts of the transceiver circuit as possible. Front-end filtering can be cumbersome and requires a number of filters for successful operation. **Fig 7.22** illustrates the typical filtering requirements in a 14MHz transceiver. Traditionally, the receiver band-pass filter and even the transmit band-pass filter would have employed a variable tuning capacitor, often referred to as a preselector, to provide optimum selectivity and sensitivity when correctly peaked. To arrange for

a number of filters to tune and track with one another requires careful design and considerable care in alignment, especially in multiband equipment. The introduction of the solid-state PA with its wide-band characteristics has lead to the development of wide-band filters possessing a flat response across an entire amateur band. By necessity these filters are of low Q and consequently must have more sections or elements if they are to exhibit any degree of out-of-band selectivity. By employing low-Q, multi-section, band-pass filters it is possible to eliminate one filter between the receiver RF amplifier and the receive mixer. The gain of the RF amplifier should be kept as low as possible and in most cases can be eliminated entirely for use below 21MHz. By providing the band-pass filters with low input and output impedances, typically 50Ω, switching of the filters can simplified to the extent that one filter can be used in both the transmit and receive paths, thus reducing the band-pass filter requirement to one per band.

Fig 7.23(a) illustrates a typical band-pass filter configuration for amateur band use, and suitable component values are listed in **Table 7.6**.

A transmit low-pass filter is essential for attenuation of all unwanted products and harmonics amplified by the broad-band amplifier chain. Suitable values for a typical Chebyshev filter, **Fig 7.23(b)**, suitable for use at the output of a power amplifier chain providing up to 100W output, are given in **Table 7.7**. One disadvantage of the Chebyshev filter design is its slow roll-off and hence limited second harmonic suppression. The elliptic filter, **Fig 7.23(d)**, has a greatly improved characteristic. The addition of two capacitors placed in parallel with the two filter inductors results in the circuits L1, C4 and L2, C5 being resonant at approximately two and three times the input frequency respec-

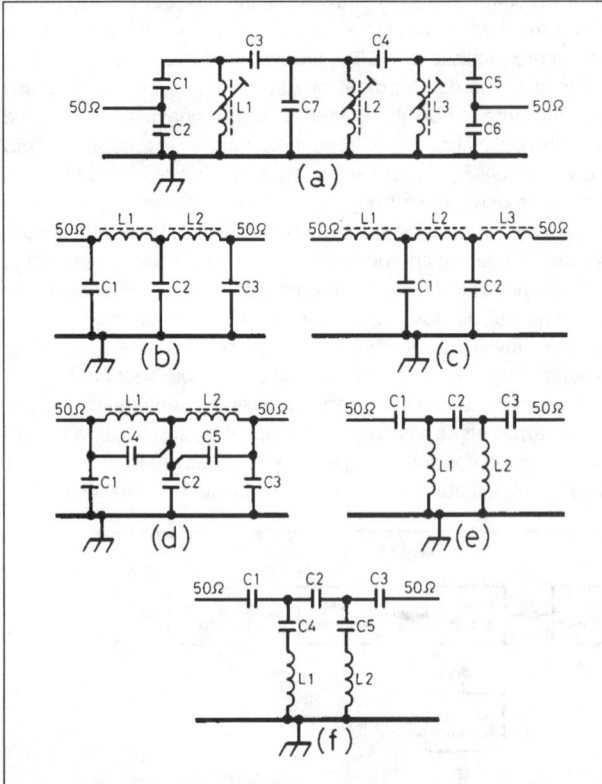

Fig 7.23: Band-pass, low-pass and high-pass filters. (a) Band-pass filter; (b) Chebyshev low-pass filter capacitive input/output; (c) Chebyshev low-pass filter inductive input/output; (d) elliptic function low-pass filter; (e) Chebyshev high-pass filter; (f) elliptic high-pass filter

Band (m)	L1, L3 (H)	L2 (H)	L-type	C1, C5 (pF)	C2, C6 (pF)	C3, C4 (pF)	C7 (pF)
160	8.0	8.0	27t KANK3335R	1800	2700	180	750
80	5.8	5.8	KANK3334R	390	1800	47	270
40	2.8	2.8	KXNK4173AO	220	1000	10	150
30	1.3	1.3	KANK3335R	220	1000	10	180
20	1.2	1.2	KANK3335R	120	560	4.7	100
17	0.29	-	Toko S18 Blue	220	750	8.2	-
15	0.29	-	Toko S18 Blue	180	560	6.8	-
12	0.29	-	Toko S18 Blue	100	560	2.7	-
10	0.29	-	Toko S18 Blue	82	390	2.7	-

All capacitors are polystyrene except those less than 10pF which are ceramic. L2 and C7 are not used on the 10-17m bands.

Table 7.6: Band-pass filter of Fig 7.23(a). 50 ohms nominal input/output impedance

Band (m)	L1, L2	Core	C1, C3 (pF)	C2 (pF)
160	31t/24swg	T50-2	1200	2500
80	22t/20swg	T50-2	820	1500
40	18t/20swg	T50-6	360	680
30/20	12t/20swg	T50-6	220	360
17/15	10t/20swg	T50-6	100	220
12/10	9t/20swg	T50-6	75	160

All capacitors are silver mica or polystyrene - for 100W use 300VDC wkg; for <50W use 63VDC wkg. Cores are Amidon: T50-2 Red or T50-6 Yellow.

Table 7.7. Chebyshev low-pass filter of Fig 7.23(b)

tively to provide peaks of attenuation at the second and third harmonic frequencies. Where possible, the use of elliptic filters is recommended - typical values for amateur band use are given in **Table 7.8**.

For an external solid-state amplifier being driven by an exciter that already contains a capacitive-input low-pass filter such as that in Fig 7.23(b), the inductive input design, **Fig 7.23(c)**, may be necessary at the final amplifier output. The combination of the two different types of filter should eliminate parasitic oscillations which will almost certainly occur if two filters with similar characteristics are employed. Values for the inductive input filter are given in **Table 7.9** for use at powers up to 300W.

The development of HF synthesisers has led to the development of HF transceivers providing general-coverage facilities and requiring yet further changes in the design of band-pass filters for transceiver front-ends. There is a finite limit to the bandwidth that can be achieved using conventional parallel-tuned circuit filters. For general-coverage operation from 1 to 30MHz, approximately 30-40 filters would be required, and this is obviously not a practical proposition. While returning to the mechanically tuned filter might reduce the total number of filters required, the complexity of electronic band-changing would be formidable. By combining the characteristics of both low-pass and high-pass filters, **Figs 7.23(e) and (f)**, the simple action of cascading the two such filters will result in a band-pass filter, **Fig 7.24(a)**, having a bandwidth equal to the difference between the two filter cut-off frequencies. The limiting bandwidth will be one octave, ie the highest frequency is double the lowest frequency. Practically, filter bandwidth is restricted to slightly less than one octave. Typical filters in a general-coverage HF transceiver might cover the following bands:

(a)   1.5-2.5MHz
(b)   2.3-4.0MHz
(c)   3.9-7.5MHz
(d)   7.4-14.5MHz
(e)   14.0-26.0MHz
(f)   20.0-32.0MHz

It can be seen that six filters will permit operation on all the HF amateur bands as well as providing general coverage of all the in-between frequencies. Hybrid low/high-pass filters invariably use fixed-value components and require no alignment, thus simplifying construction. The transmit low-pass filter may also be left in circuit on receive in order to enhance the high-frequency rejection; it has no effect on low-frequency signals. The use of separate high-pass filters in the receiver input circuit prior to the band-pass filter serves to eliminate low-frequency broadcast signals. Typically, a high-pass filter of the multi-pole elliptic type, **Fig 7.24(b)**, having a cut-off of 1.7MHz, is fitted to most commercial amateur band equipment.

TRANSVERTERS

Transverters are transmit/receive converters that permit equipment to be operated on frequencies not covered by that equipment. Traditionally HF equipment was transverted to the VHF/UHF bands but, with the increase in availability of 144MHz SSB equipment, down-conversion to the HF bands has become popular. **Fig 7.25** shows the schematic of a typical 144/14MHz transverter providing HF operation with a VHF transceiver.

A transverter takes the output from a transmitter, attenuated to an acceptable level, heterodynes it with a crystal-controlled oscillator to the desired frequency and then amplifies it to the

Band (m)	L1	L2	Core	C1 (pF)	C2 (pF)	C3 (pF)	C4 (pF)	C5 (pF)
160	28t/22SWG	25t/22SWG	T68-2	1200	2200	1000	180	470
80	22t/22SWG	20t/22SWG	T50-2	680	1200	560	90	250
40	18t/20SWG	16t/20SWG	T50-6	390	680	330	33	100
30/20	12t/20SWG	11t/20SWG	T50-6	180	330	150	27	75
17/15	10t/20SWG	9t/20SWG	T50-6	120	220	100	12	33
12/10	8t/20SWG	7t/20SWG	T50-6	82	150	68	12	39
Capacitors 300VDC wkg silver mica up to 200W. All cores Amidon.								

Table 7.8: Elliptic low-pass filter of Fig 7.23(d)

required level. The receive signal is converted by the same process in reverse to provide transceive capabilities on the new frequency band. The techniques employed in transverters are the same as those used in comparable frequency transmitting and receiving equipment. Where possible it is desirable to provide low-level RF output signals for transverting rather than having to attenuate the high-level output from a transmitter with its associated heatsinking requirements.

Occasionally transverters may be employed from HF to HF in order to include one of the 'WARC bands' on an older transceiver, or to provide 160m band facilities where they have been omitted. In some cases it may prove simpler to add an additional frequency band to existing equipment in preference to using a transverter. Transverters have also been built to convert an HF transceiver to operate on the 136kHz band.

Band (m)	L1, L3	L2	Core	C1, C2 (pF)
160	8.1μH/24t	11.4μH	T106-2	1700
80	4.1μH/17t	5.8μH/21t	T106-2	860
40	2.3μH/13t	3.2μH/15t	T106-2	470
30/20	1.18μH/10t	1.65μH/12t	T106-6	240
17/15	0.79μH/8t	1.11μH/10t	T106-6	160
12/10	0.57μH/7t	0.8μH/8t	T106-6	120
This filter is for use with a high-power external amplifier, when capacitive input is fitted to exciter. Capacitors silver mica: 350VDC wkg up to 200W; 750VDC wkg above 300W. Use heaviest possible wire gauge for inductors. All cores Amidon.				

Table 7.9: Chebyshev low-pass filter inductive input of Fig 7.23(c)

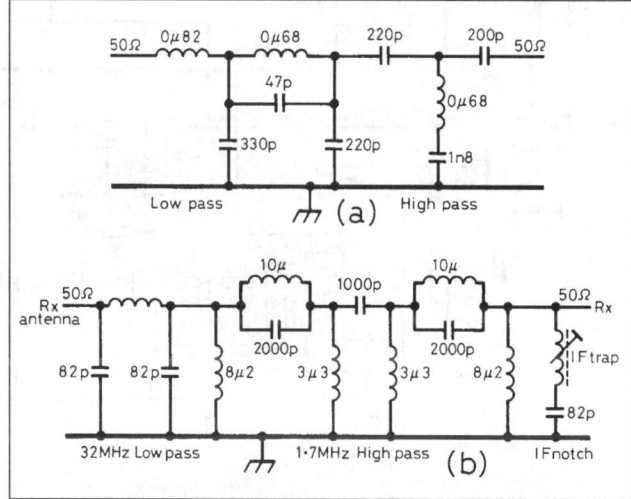

Fig 7.24: Multifunction filters. (a) Band-pass filter using high/low pass filters; (b) composite receiver input filter

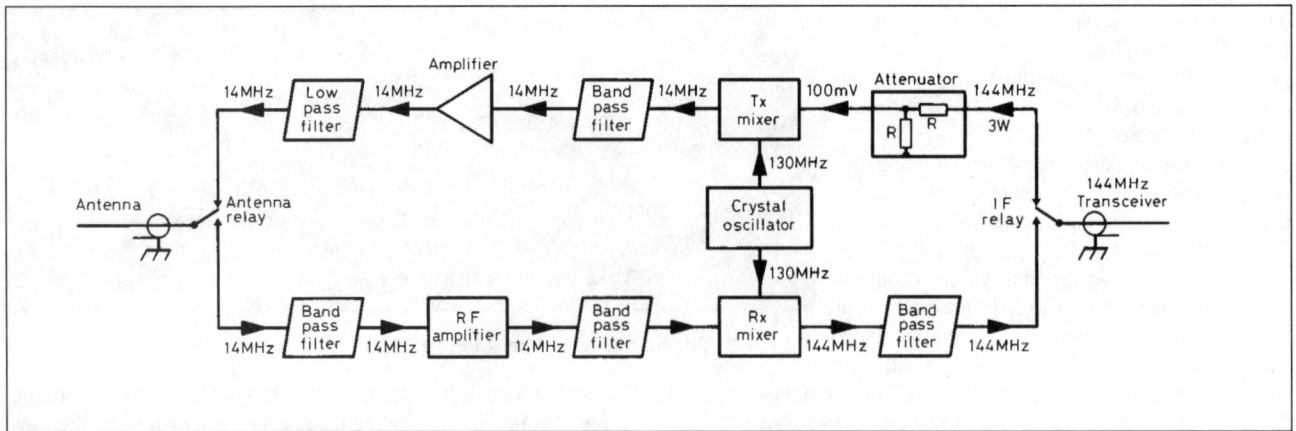

Fig 7.25: 144MHz to 14MHz transverter

Fig 7.26: The G4ENA transceiver

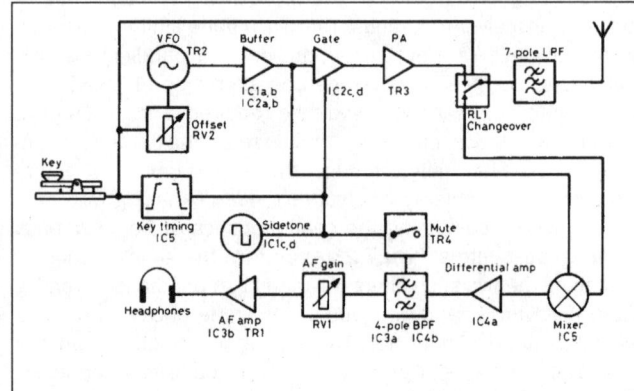

Fig 7.27: Block diagram of the G4ENA transceiver

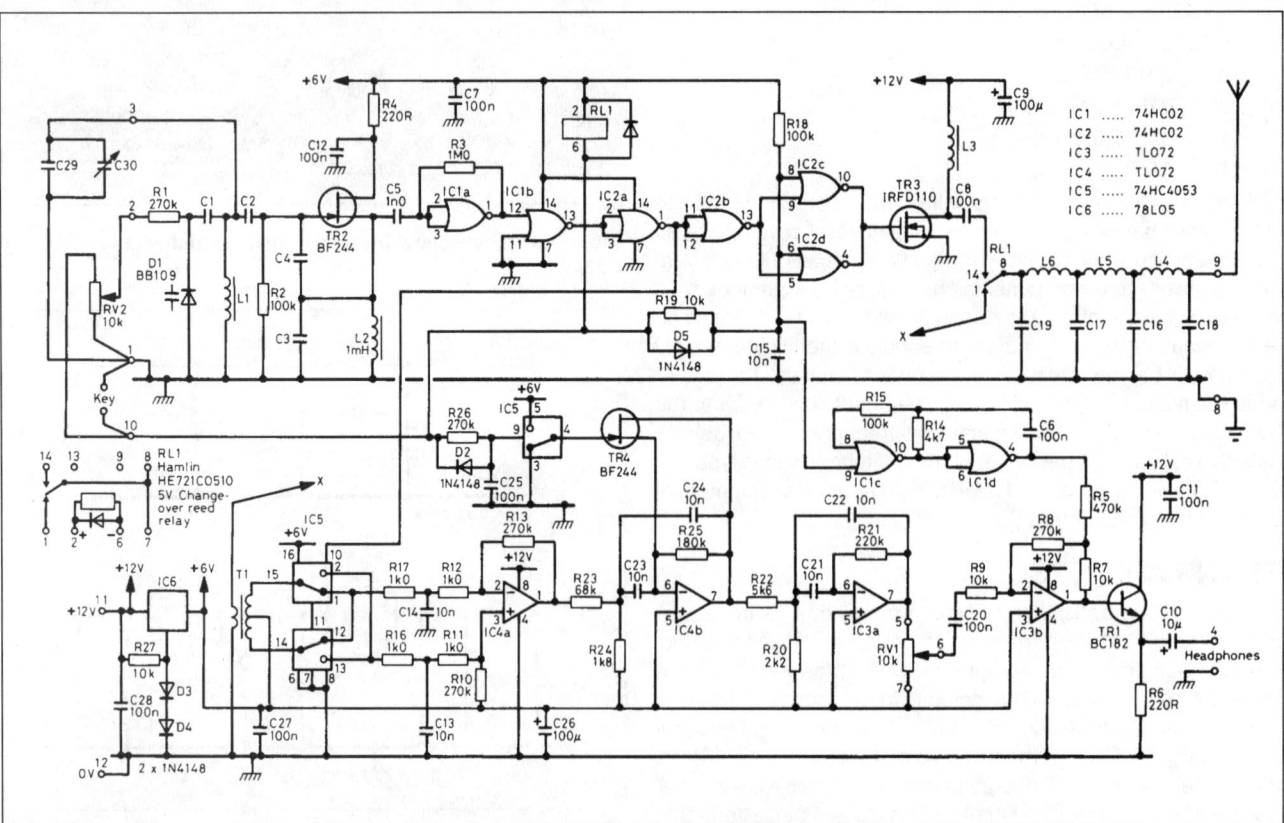

Fig 7.28: The switching PA stage requires a seven-pole filter, as shown in the circuit diagram above

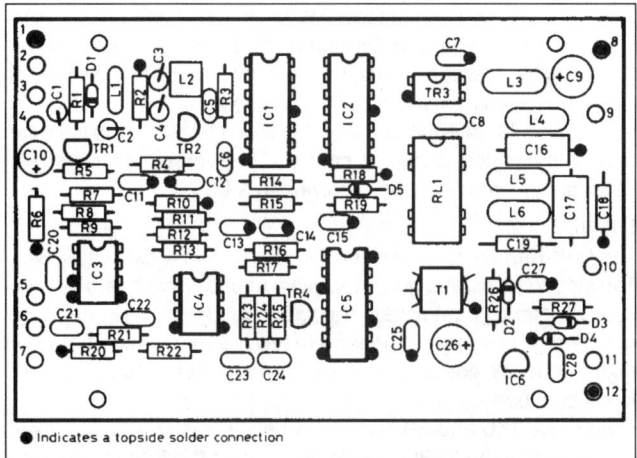

● Indicates a topside solder connection

Fig 7.29: A neat component layout results in a compact unit suitable for portable operation (see also Fig 7.30)

Fig 7.30: Inside view of the G4ENA transceiver

PRACTICAL TRANSMITTER DESIGNS

QRP + QSK - A Novel Transceiver with Full Break-in

This design by Peter Asquith, G4ENA, originally appeared in *Radio Communication* [2].

Introduction

The late 20th century saw significant advances in semiconductor development. One such area was that of digital devices. Their speed steadily improved to the point which permitted them to be used in low band transceiver designs. One attractive feature of these components is their relatively low cost.

The QSK QRP Transceiver (**Figs 7.26 and 7.27**) employs several digital components which, together with simple analogue circuits, provide a small, high-performance and low-cost rig. Many features have been incorporated in the design to make construction and operation simple.

One novel feature of this transceiver is the switching PA stage. The output transistor is a tiny IRFD110 power MOSFET. This device has a very low 'ON' state resistance which means that very little power is dissipated in the package, hence no additional heatsinking is required. However, using this concept does mean that good harmonic filtering must be used.

The HC-type logic devices used in the rig are suitable for operation on the 160m and 80m bands. The transmitter efficiency on 40m is poor and could cause overheating problems. Future advances in component design should raise the top operating frequency limit.

Circuit description

VFO

The circuit diagram of the transceiver is shown in **Fig 7.28**. TR2 is used in a Colpitts configuration to provide the oscillator for both receive and transmit.

The varicap diode D1 is switched via RV2/R1 by the key to offset the receive signal by up to 2kHz, such that when transmitting, the output will appear in the passband of modern transceivers operating in the USB mode. C1 controls the RIT range and C29/30 the band coverage. IC1a, IC1b and IC2a buffer the VFO. It is important that the mark/space ratio of the square wave at IC1b is about 50:50. Small variations will affect output power.

Transmitter

When the key is operated, RL1 will switch and, after a short delay provided by R19/C15, IC2c and IC2d will gate the buffered VFO to the output FET, TR3. TR3 operates in switch mode and is therefore very efficient. The seven-pole low-pass filter after the changeover relay removes unwanted harmonics, which are better than -40dB relative to the output.

Receiver

The VFO signal is taken from IC2a to control two changeover analogue switches in IC5, so forming a commutating mixer and providing direct conversion to audio of the incoming stations. IC4a is a low-noise, high-gain, differential amplifier whose output feeds the four-pole CW filter, IC4a and IC3a, before driving the volume attenuator, RV1. IC3b amplifies and TR1 buffers the audio to drive headphones or a small speaker. When the key is down TR4 mutes the receiver, and audio oscillator IC1a and b injects a sidetone into the audio output stage, IC3b. The value of R5 sets the sidetone level.

Constructional notes

The component layout is shown in **Figs 7.29 and 7.30**, and the Components List is in **Table 7.10**.

1. Check that all top-side solder connections are made.
2. Wind turns onto toroids tightly and fix to PCB with a spot of glue.
3. Do not use IC sockets. Observe anti-static handling precautions for all ICs and FETs.
4. All VFO components should be earthed close to VFO.
5. Fit a 1A fuse in the supply line.
6. Component suppliers: Bonex, Cirkit, Farnell Electronic Components, JAB Electronic Components etc.

Test and calibration

Before TR3 is fitted, the transceiver must be fully operational and calibrated. Prior to switch-on, undertake a full visual inspection for unsoldered joints and solder splashes. Proceed as follows.

1. Connect external components C29/30, RV1/2, headphones and power supply.
2. Switch on power supply. Current is approx 50mA.
3. Check +6V supply, terminal pin 7. Voltage is approx 6.3V.
4. Select values for C29 to bring oscillator frequency to CW portion of band (1.81-1.86MHz/3.50-3.58MHz). Coverage should be set to fall inside the band limits of 1.810/3.500MHz.
5. Connect an antenna or signal generator to terminal pin 9 and monitor the received signal on headphones. Tuning through the signal will test the response of the CW filter which will peak at about 500Hz.
6. Connect key and check operation of sidetone and antenna changeover relay. Sidetone level can be changed by selecting value of R5.

RESISTORS

R1, 8, 10, 13, 26	270k	R14	4k7
R2, 15, 18	100k	R20	2k2
R3	1M	R21	220k
R4, 6	220R	R22	5k6
R5	470k	R23	68k
R7, 9, 19, 27	10k	R24	1k8
R11, 12, 16, 17	1k	R25	180k

All resistors 0.25W 2%

RV1, 2	10k lin

CAPACITORS

Ref	Type	Pitch	Value (80m)	Value (160m)
C1	Ceramic plate 9	2.54	4p7	15p
C2	Polystyrene	-	47p	100p
C3, 4	Polystyrene	-	220p	470p
C5	Ceramic monolithic	2.54	1n	1n
C6, 7, 8, 11, 12, 20, 25, 27, 28	Ceramic monolithic	2.54	100n	100n
C9, 26	Aluminium radial 16V	2.5	100	100µ
C10	Aluminium radial 16V	2.0	10µ	10µ
C13, 14, 15, 21, 22, 23, 24	Ceramic monolithic	2.54	10n 10%	10n 10%
C16, 17	Polystyrene	-	1n5	2n7
C18, 19	Polystyrene	-	470p	1n
C29*	Polystyrene	-	470p	820p
C30*	Air-spaced VFO	-	25p	75p

Select on test component

INDUCTORS

Ref	Type	80m	160m
L1	T37-2 (Amidon)	31t 27SWG (0.4mm)	41t 30SWG (0.315mm)
L2	7BS (Toko)	1mH	1mH
L3	T37-2	2.2µH 23t 27SWG	4.5µH 33t 30SWG
L4, 6	T37-2	2.9µH 26t 27SWG	5.45µH 36t 30SWG
L5	T37-2	4.0µH 31t 27SWG	6.9µH 41t 30SWG
T1	Balun	2t primary, 5+5t secondary, 0.2mm 36SWG (28-43002402)	

SEMICONDUCTORS

D1	BB109	IC1, 2	74HC02*
D2, 3, 4, 5	1N4148	IC3, 4	TL072*
TR1	BC182 (not 'L')	IC5	74HC4053*
TR2, 4	BF244	IC6	78L05
TR3	IRFD110*		

Static-sensitive devices

MISCELLANEOUS

RL1	5V change-over reed relay, Hamlin, HE721CO510; PED/Electrol, 17708131551-RA30441051
PCB	Boards and parts are available from JAB Electronic Components [3]

Table 7.10: G4ENA transceiver components list

7. Monitor output of IC2c and IC2d (TR3 gate drive) and check correct operation. A logic low should be present with key-up, and on key-down the VFO frequency will appear. This point can be monitored with an oscilloscope or by listening on a receiver with a short antenna connected to IC2c or IC2d.

8. When all checks are complete fit TR3 (important! - static-sensitive device) and connect the transceiver through a power meter to a dummy load. On key-down the output

power should be at least 5W for +12V supply, rising to 8W for 13.8V supply. Note: should the oscillator stop when the key is pressed it will instantly destroy TR3. Switch off power when selecting VFO components.

9. Connect antenna and call CQ. When a station replies note the position of the RIT control. The average receive offset should be used when replying to a CQ call.

On air

The QSK (full break-in) concept of the rig is very exciting in use. The side tone is not a pure sine wave and is easily heard if there is an interfering beat note of the same frequency. One important note is to remember to tune the receiver into a station from the high-frequency side so that when replying your signal falls within his passband.

Both the 160 and 80m versions have proved very successful on-air. During the 1990 Low Power Contest the 80m model was operated into a half-wave dipole and powered from a small nicad battery pack. This simple arrangement produced the highest 80m single band score!

Its small size and high efficiency makes this rig ideal for portable operation. A 600mAh battery will give several hours of QRP pleasure - no problem hiding away a complete station in the holiday suitcase!

A QRP Transceiver for 1.8MHz

This design by S E Hunt, G3TXQ, originally appeared in *Radio Communication* [4].

Introduction

This transceiver was developed as part of a 1.8MHz portable station, the other components being a QRP ATU, a battery-pack and a 200ft kite-supported antenna. It would be a good constructional project for the new licensee or for anyone whose station lacks 1.8MHz coverage. The 2W output level may seem a little low, but it results in low battery drain and is adequate to give many 1.8MHz contacts.

The designer makes no claim for circuit originality. Much of the design was adapted from other published circuitry; however, he does claim that the design is repeatable - six transceivers have been built to this circuit and have worked first time. Repeatability is achieved by extensive use of negative feedback; this leads to lower gain-per-stage (and therefore the need for more stages) but makes performance largely independent of transistor parameter variations.

Circuit description

The transceiver circuit (**Fig 7.31**) comprises a direct-conversion receiver together with a double-sideband (DSB) transmitter. This approach results in much simpler equipment than a superhet design, and is capable of surprisingly good performance, particularly if care is taken over the mixer circuitry.

During reception, signals are routed through the band-pass filter (L1, L2 and C25-C31) to a double-balanced mixer, M1, where they are translated down to baseband. It is vital for the mixer to be terminated properly over a wide range of frequencies, and this is achieved by a diplexer comprising R34, RFC2 and C32-R34. Unwanted RF products from the mixer, rejected by RFC2, pass through C32 to the 47-ohm terminating resistor R34. The wanted audio products pass through RFC2 and C34 to a common-base amplifier stage which is biased such that it presents a 50-ohm load impedance. The supply rail for this stage comes via an emitter-follower, TR5, which has a long time-constant (4s) RC circuit across its base. This helps to prevent any hum on the 12V rail reaching TR6 and being amplified by IC3.

The voltage gain of the common-base stage (about x20) is controlled by R37 which also determines the source resistance

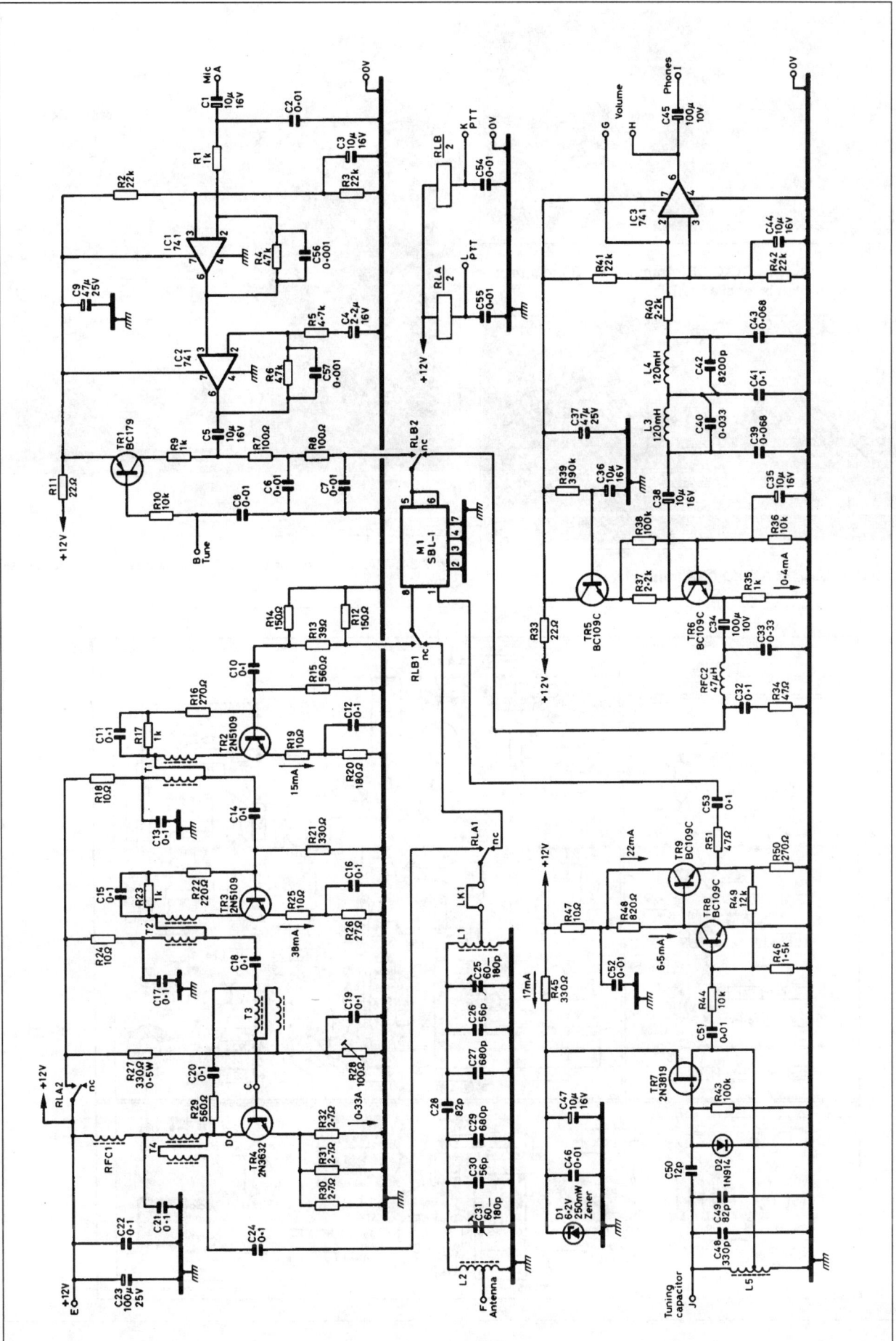

Fig 7.31: Circuit of the G3TXQ transceiver

Fig 7.32: Front view of G3TXQ QRP transceiver. On this proto-type the TUNE switch is labelled TEST

Fig 7.33: Rear view of G3TXQ QRP transceiver

Fig 7.34: The top view of the G3TXQ low power 1.8MHz trans-ceiver

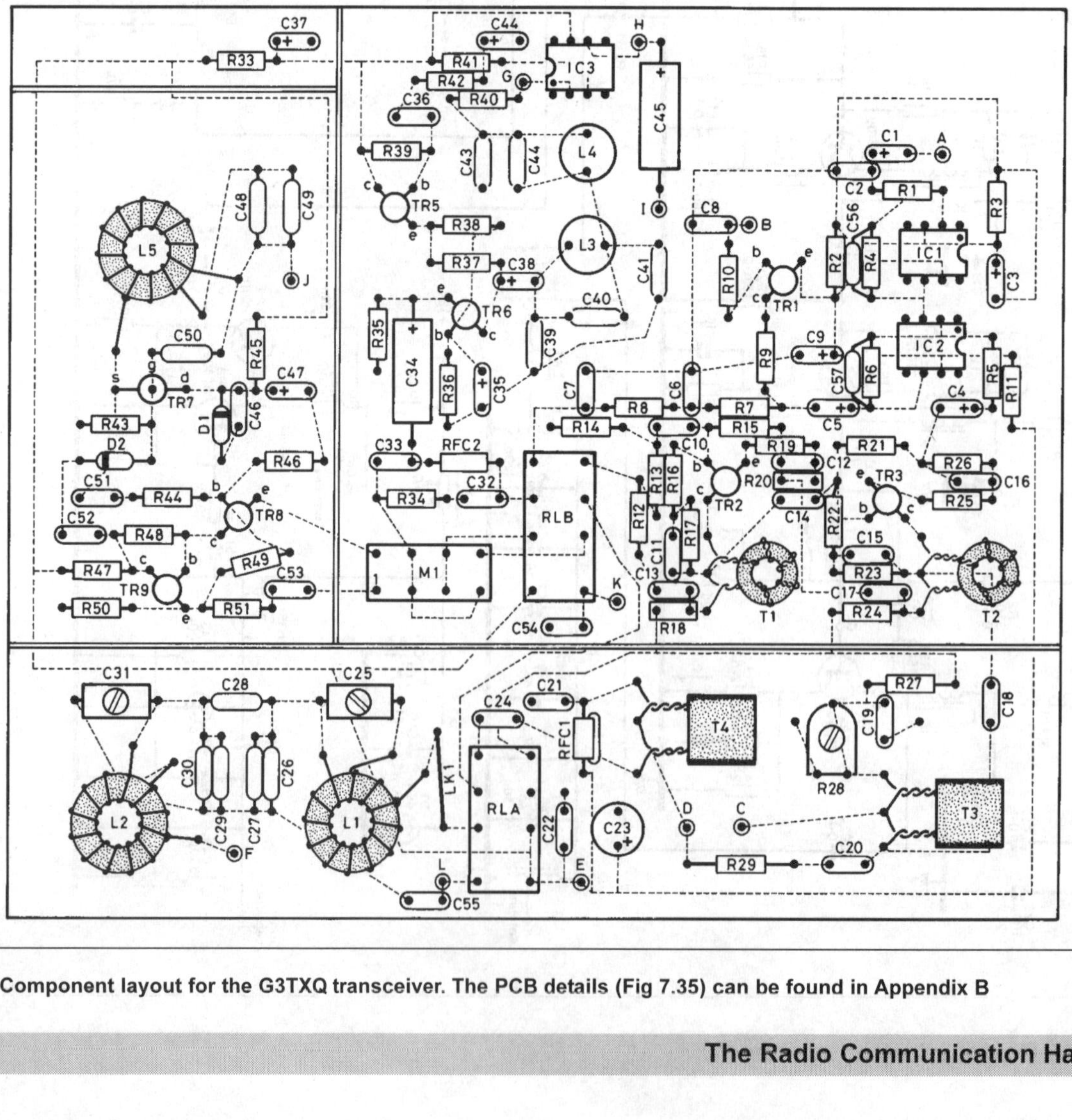

Fig 7.36: Component layout for the G3TXQ transceiver. The PCB details (Fig 7.35) can be found in Appendix B

for the following low-pass filter (L3, L4 and C39-C43). This filter is a Chebyshev design and it determines the overall selectivity of the receiver. The filter is followed by a single 741 op-amp stage which give adequate gain for headphone listening; however, an LM380 audio output stage can easily be added if you require loudspeaker operation.

On transmit, audio signals from the microphone are amplified in IC1 and IC2, and routed to the double balanced mixer where they are heterodyned up into the 1.8MHz band as a double-side-band, suppressed-carrier signal. Capacitors C56 and C57 cause some high-frequency roll-off of the audio signal and thereby restrict the transmitted bandwidth. A 6dB attenuator (R12-R14) provides a good 50Ω termination for the mixer. The DSB signal is amplified by two broad-band feedback amplifiers, TR2 and TR3, each having a gain of 15dB. TR3 is biased to a higher standing current to keep distortion products low.

The PA stage is a single-ended design by VE5FP [5]. The inclusion of unbypassed emitter resistors R30-R32 establishes the gain of the PA and also helps to prevent thermal runaway by stabilising the bias point. Additional RF negative feedback is provided by the shunt feedback resistor R29. The designer chose to run the PA at a moderately high standing current (330mA) in order to reduce distortion products, thinking that at some stage he might use the transceiver as a 'driver' for a 10-15W linear amplifier.

The PA output (about 2W PEP) is routed through the band-pass filter to the antenna. The designer used a 2N3632 transistor in the PA because he happened to have one in the junk-box; the slightly less expensive 2N3375 would probably perform just all well. VE5FP used a 2N5590 transistor but this would need different mounting arrangements.

At the heart of the transceiver is a Hartley VFO comprising TR7 and associated components. The supply of this stage is stabilised at 6.2V by zener diode D1 and decoupled by C46 and

C47. It is important for best stability that the Type 6 core material is used for L5 as this has the lowest temperature coefficient. Output from the VFO is taken from the low impedance tap on L5. The VFO buffer is a feedback amplifier comprising TR8 and TR9. The input impedance of this buffer is well-defined by R44 and presents little loading of the VFO. Its gain is set by the ratio R49/R44 and R51 has been included to define the source resistance of this stage at approximately 50Ω.

Changeover between transmit and receive is accomplished by two DPDT relays which are energised when the PTT lines are grounded. A CW signal for tuning purposes can be generated by grounding the TUNE pin - this switches on TR1, which in turn unbalances the mixer, allowing carrier to leak through to the driver and PA stages.

Construction

The transceiver, pictured in **Figs 7.32-7.34**, is constructed on a single 6 by 5in PCB. The artwork, component layout and wiring diagram are shown in **Figs 7.35 (see Appendix B), 7.36** and **7.37** respectively. The PCB is double-sided - the top (component) surface being a continuous ground plane of unetched copper. The Components List is shown in **Table 7.11**.

Without the facility to plate-through holes, some care needs to be taken that components are grounded correctly. Where a component lead is not grounded, a small area of copper must be removed from the ground plane, using a spot-cutter or a small twist drill.

Where a component lead needs to be grounded, the copper should not be removed and the lead should be soldered to the ground plane as well as to the pad on the underside. This is easy to achieve with axial-lead components (resistors, diodes etc) but can be difficult with radial-lead components. In most cases the PCB layout overcomes this by tracking radial leads to ground via nearby resistor leads. A careful look at the circuit diagram as each component is loaded soon shows what is needed.

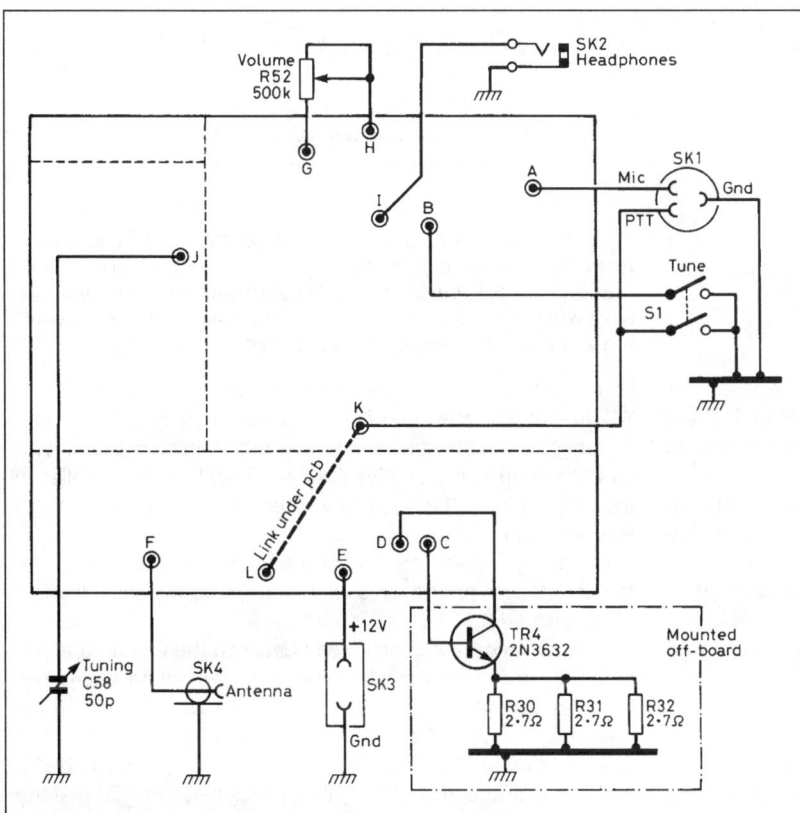

Fig 7.37: Wiring diagram for the G3TXQ transceiver

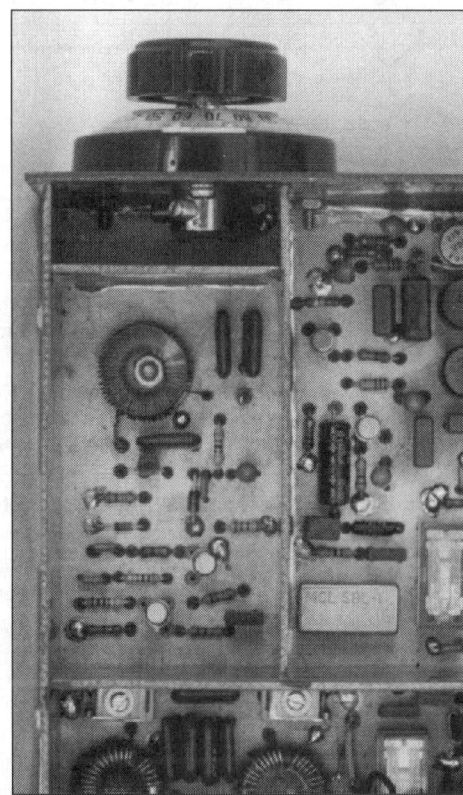

Fig 7.38: Detail of top view with C58 removed to show mounting arrangements

R1, 9, 17, 23, 35	1k		C1, 3, 5, 35, 36, 38, 44, 47	10µ 16V tant bead		R52	500k log pot
R2, 3, 41, 42	22k		C2, 6, 7, 8, 46, 51,			L1, 2	37t on T68-2 core tapped at 7t from ground
R4, 6	47k		52, 54, 55	0.01µ ceramic		L3, 4	120mH (eg Cirkit 34-12402)
R5	4k7		C4	2µ2 16V tant bead		L5	57t on T68-6 core tapped at 14t from ground
R7, 8	100R		C9, 37	47µ 25V tant bead		RFC1	2t on small ferrite bead
R10, 36, 44	10k		C10-22, 24,			RFC2	47µH choke
R11, 33	22R		32, 53	0.1µ ceramic		T1, 2	10t twisted wire on 10mm OD ferrite toroidal core
R12, 14	150R		C23	100µ 25V elect			Al = 1µH/t (eg SEI type MM622). See Fig 7.44.
R13	39R		C25, 31	60-180p trimmer		T3, 4	4t twisted wire on two 2-hole ferrite cores. Al = 4 H/t
R15, 29	560R			(Cirkit 06-18006)			(eg Mullard FX2754). See Fig 7.45.
R16, 50	270R		C26, 30	56p silver mica		TR1	BC179
R18, 19, 24,			C27, 29	680p silver mica		TR2, 3	2N5109 or 2N3866
25, 47	10R		C28	82p silver mica		TR4	2N3632 (see text)
R20	180R		C33	0.33µ		TR5, 6, 8, 9	BC109C
R21, 45	330R		C34, 45	100µ 10V elect		TR7	2N3819
R22	220R		C39, 43	0.068µ		D1	6.2V 250mW zener
R26	27R		C40	0.033µ		D2	1N914
R27	330R 0.5W		C41	0.1µ polystyrene		IC1, 2, 3	741 op-amp
R28	100R preset		C42	8200p silver mica		M1	Mini-circuits SBL-1 double-balanced mixer
R30, 31, 32	2R7		C48	330p silver mica		RLA, B	DPDT 12V relay (eg RS Electromail 346-845)
R34, 51	47R		C49	82p silver mica		SK1	Microphone socket
R37, 40	2k2		C50	12p silver mica		SK2	Headphone socket
R38, 43	100k		C56, 57	0.001µ ceramic		SK3	DC power socket (eg Maplin YX34M)
R39	390k		C58	50p air-spaced		SK4	Antenna socket
R46	1k5			variable, SLC law		S1	DPDT toggle switch
R48	820R			(Maplin FF45Y)			Slow-motion drive for C58 (eg Maplin RX40T)
R49	12k						Heatsink approx 1.5 by 2in
							Knob for R52

Table 7.11: 1.8MHz QRP transceiver components list

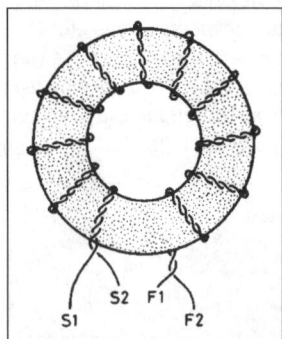

Fig 7.39: Winding details of T1 and T2. Connect S2 and F1 to form the centre tap. Note that the two wires are twisted together before winding. S1, F1: start and finish respectively of winding 1. S2, F2: start and finish respectively of winding 2. Core: 10mm OD ferrite toroid

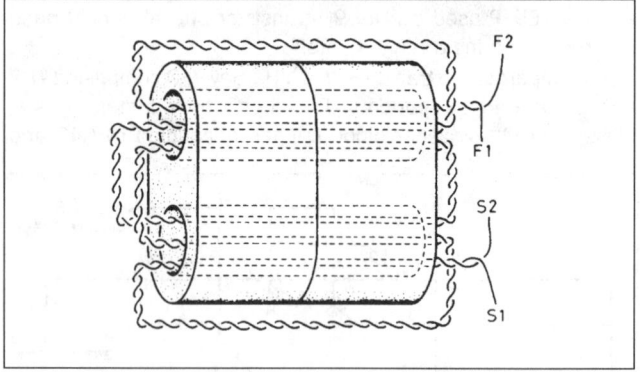

Fig 7.40: Winding details of T3 and T4. Connect S2 and F1 to form the centre tap. Note that the two wires are twisted together before winding. S1, F1: start and finish respectively of winding 1. S2, F2: start and finish respectively of winding 2. Core: two 2-hole cores stacked end-to-end

Remember to put in a wire link between pins L and K, and in position LK1. Screened cable was used for connecting pins G and H to the volume control - connect the outer to pin H.

There are no PCB pads for C56 and C57, so these capacitors should be soldered directly across R4 and R6 respectively. TR4 must be adequately heatsinked as it dissipates almost 4W even under no-drive conditions. TR4 was bolted through the rear panel to a 1.5 by 2.5in finned heatsink. Resistors R30-R32 are soldered directly between the emitter of TR4 and the ground plane.

It is important that the VFO coil L5 is mechanically stable. Ensure that it is wound tightly and fixed rigidly to the PCB; the coil was 'sandwiched' between two Perspex discs and bolted through the discs to the PCB (Fig 7.38). Also, be sure to use rigid heavy-gauge wire for connecting to C58.

The designer used a 6:1 vernier slow-motion drive which, with the limited tuning range of 100kHz, provides acceptable bandspread; the 0-100 vernier scale (0 = 1.900MHz, 100 = 2.000MHz) gives a surprisingly accurate read-out of frequency, the worst-case error being 1kHz across the tuning range.

The broad-band transformers, T1-T4, are wound by twisting together two lengths of 22SWG enamelled copper wire. The twisted pair is then either wound on a ferrite toroidal core (T1

and T2), or wound through ferrite double-holed cores (T3 and T4). Identify the start and finish of each winding using an ohmmeter - connect the start of one wire to the finish of the other to form the centre tap (see Figs 7.39 and 7.40 for more details). All transformers and the band-pass filter coils were secured to the PCB with adhesive.

The designer fabricated all of the transceiver, other than the top and bottom panels, by soldering together double-sided PCB materials.

It is vital to have a good screen between the PA and the VFO, otherwise the transmitter will frequency modulate badly. Two-inch high screens were used around the PA and VFO area, and a screen was included at the front of the VFO compartment on which to mount C58. If you use lower screens you may need to put a lid over the VFO; cut a tightly fitting piece of PCB material and bolt it in position to four nuts soldered into the corners of the VFO compartment.

Alignment

Check the PCB thoroughly for correct placement of components and absence of solder bridges.

Turn the volume control fully counter-clockwise, the TUNE switch to the off position and R28 fully counter-clockwise. Connect the transceiver to a 12V supply and switch on. Check that the current drawn from the supply is about 50mA.

Check the frequency of the VFO either by using a frequency counter connected to the source of TR7, or by monitoring the VFO on another receiver. With C58 set to mid-position, the frequency should be about 1.95MHz; if it is very different, you can adjust L5 slightly by spreading or squeezing together the turns. Alternatively, major adjustments can be made by substituting alternative values for C49. Check that the range of the VFO is about 1.9 to 2.0MHz.

Plug in a pair of headphones and slowly advance the volume control; you should hear receiver noise (a hissing sound). If you have a signal generator, set it to 1.95MHz and connect it to the antenna socket; if not, you will have to connect the transceiver to an antenna and make the next adjustment using an off-air received signal. Tune to a signal at 1.95MHz and alternately adjust C25 and C31 for a peak in its level.

Connect the transceiver to a 50Ω power meter or through an SWR bridge to a 50Ω load. Plug in a low-impedance microphone and operate the PTT switch. Note the current drawn from the supply - it should be about 200mA. Slowly turn R28 clockwise and note that the supply current increases; adjust R28 until the supply current has increased by 330mA. Release the PTT switch and operate the TUNE switch; the power meter should indicate between 1 and 2W.

At this stage, final adjustments can be made to C25 and C31. Swing the VFO from end to end of its range and note the variation in output power. The desired response is a slight peak in power at either end of the VFO range with a slip dip at mid-range. It should be possible to achieve by successive adjustments to C25 and C31.

For those of you lucky enough to have access to a spectrum analyser and tracking generator, LK1 was included to allow isolation of the band-pass filter.

If you have any problems, refer to **Tables 7.12 to 7.14** which show typical AC and DC voltages around the circuit. If necessary, you can tailor the gain of IC2 to suit the sensitivity of your microphone by changing the value of R5.

Final thoughts

In retrospect it would have been useful to have included the low-pass filter (L3, L4, C39-C43) in the transmit audio path in order to further restrict the bandwidth. Normally the roll-off achieved by C57 and C56 combined with the low output power means that you are unlikely to cause problems for adjacent contacts. However, when using a 200ft vertical antenna during portable operation, the transceiver puts out a potent signal and a reduction in bandwidth would then be more 'neighbourly'.

A CW facility could be added fairly easily using the TUNE pin as a keying point. You would need to add RIT (receive independent tune) facilities - probably by placing a varactor diode between TR7 source and ground. You might also consider changing to a band-pass audio filter rather than a low-pass audio filter in the receiver.

The transceiver can be adapted for other bands by changing the VFO components and the band-pass filter components - all other circuitry is broad-band. You will need to worry more about VFO stability as you increase frequency, and you may find the gain of the buffer falls - you can overcome this by decreasing the value of R44. The noise figure of the receiver is adequate for operation on the lower frequency bands but on 14MHz and above a preamplifier will probably be needed. Those who enjoy experimentation might try changing the VFO to a VXO, adding a preamplifier to the receiver, and seeing if operation on 50MHz is possible!

	Emitter	Base	Collector	Note
TR1	12.2	11.6	11.8	Tune switch operated
TR2	2.85	3.6	12	Transmit
TR3	1.4	2.15	11.6	Transmit
TR4	0.3	1	12.2	Transmit
TR5	11.2	11.8	12.2	
TR6	0.4	1	10.3	
TR8	0	0.65	6.75	
TR9	6	6.75	12	

Table 7.12: Bipolar transistor DC voltages (with 12.2V supply)

	Source	Gate	Drain
TR7	0	0	6.2

Table 7.13: FET DC voltages (with 12.2V supply)

Circuit node	AC voltage	Notes
TR7 source	2.6V p/p	1.8MHz RF
TR9 emitter	2.6V p/p	1.8MHz RF
Mic input	4mV p/p	Transmit audio
IC1 pin 6	200mV p/p	Transmit audio
IC2 pin 6	2.2V p/p	Transmit DSB RF
TR2 base	200mV p/p	Transmit DSB RF
TR4 collector	15V p/p	Transmit DSB RF
Ant (50)	30V p/p	Transmit DSB RF

Table 7:14: AC voltages

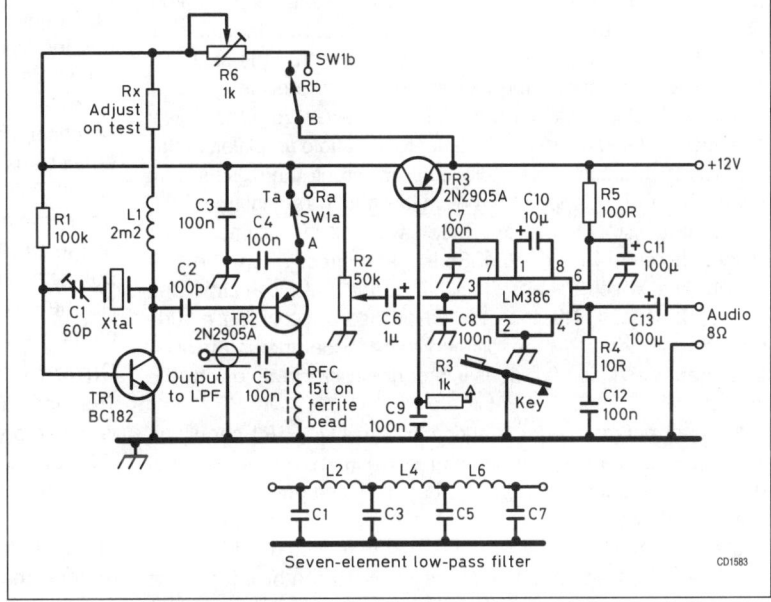

Fig 7.41: FOXX2 circuit diagram

Band (MHz)	C1, 7 (pF)	C3, 5 (pF)	L2, 6 (turns)	L4 (turns)	Core	Wire (SWG)
3.5	470	1200	25	27	T37-2	28
7.0	270	680	19	21	T37-6	26
10.1	270	560	19	20	T37-6	26
14.0	180	390	16	17	T37-6	24

Table 7.15: FOXX2 low-pass filter values

Finally, it has been interesting to note that, despite theory, with careful tuning it is quite possible to resolve DSB signals on the direct-conversion receiver.

The FOXX2 Transceiver

The FOXX2 is based on a design by George Burt, GM3OXX, and modified by Rev George Dobbs, G3RJV. This article originally appeared in *SPRAT* [6] and shows just how simple a CW transceiver can be.

Introduction

The FOXX transceiver is an elegant little circuit which uses the same transistor for the transmit power amplifier and the receive mixer. It is capable of transceiver operation on several bands and generates around 1W of RF power out. The original FOXX circuit has been revised with a few design changes.

Circuit design

The circuit diagram is shown in **Fig 7.41**. TR1 is a VXO (variable crystal oscillator) stage. The feedback loop formed by the crystal and the trimmer capacitor (C1) tunes the circuit to the desired frequency. C1 provides a small amount of frequency shift. The output is coupled to a power amplifier stage. This stage is unusual in that a PNP transistor is used with the emitter connected to the positive supply and the output taken from the collector load, which goes to ground. The output of the transmitter may be adjusted by a resistor (Rx - a few hundred ohms) to around 1W. TR2 should be fitted with a clip-on heatsink. TR3, another PNP transistor, allows the transmitter to be keyed with respect to ground. TR3 and TR2 are both 2N2905A PNP switching transistors.

The low-pass filter is a seven-element circuit based on the circuit and constants described by W3NQN. The transmit-receive function is performed by a double-pole, double-throw switch, SW1A and SW1B. The receive position has two functions. It bypasses the keying transistor, TR3, to ensure that the oscillator TR1 remains on during the receive position to provide the local oscillator. It also switches the supply line away from the power amplifier TR3 and connects the latter to the audio amplifier. In this position TR3 functions as a diode mixer, mixing the signals from the antenna which appear at the emitter and the signal from TR1.

The audio amplifier is an LM386 working in maximum gain mode. The supply for the LM386 is taken directly from the 12V supply line which means it is on during both transmit and receive functions. This has the advantage of providing a rudimentary sidetone to monitor the keying. 'Sidetone' is an overstatement because all it does is produce clicks in time with the keying.

A preset potentiometer is added in series with TR1 supply on receive. This is a very simple form of RIT (receiver incremental tuning). If the supply voltage to TR1 is reduced enough, it shifts the frequency of the oscillations. Assuming the value of Rx to be in the order of a few hundred ohms (just to reduce the drive from TR1 a little on transmit), a 1kΩ preset at R6 can be set to shift the frequency by around 700-800Hz, giving a comfortable offset for CW reception.

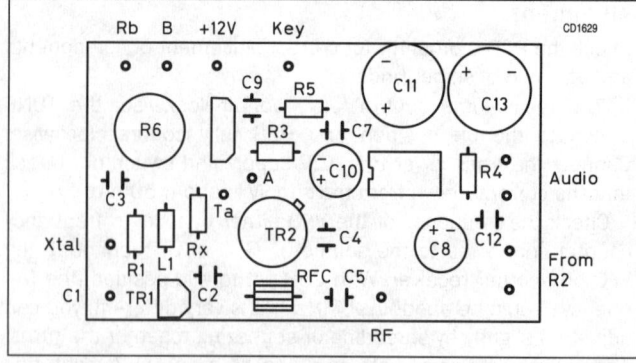

Fig 7.43: FOXX2 component layout (See Appendix B for PCB)

R1	100k		C3-5, 7-9, 12	100n
R2	50k		C6	1µ
R3	1k		C10	10µ
R4	10R		C11, 13	100µ
R5	100R		L1	2.2mH
R6	1k		RFC	15t on ferrite bead
Rx	See text		TR1	BC182
C1	60p trimmer		TR2, 3	2N2905A
C2	100p		IC1	LM386

Table 7.16. FOXX2 component list

Values for the seven-element low pass filter are shown in **Table 7.15**. The wire gauge is not critical but wind the coil so as to comfortably fill the core over about three-quarters of its full circumference.

The PCB layout is shown in **Fig 7.42** (in Appendix B), the component layout is in **Fig 7.43**, and the Components List is in **Table 7.16**.

The Epiphyte-2

This section is based on articles by Derry Spittle, VE7QK [7] and Rev George Dobbs, G3RJV [8].

Introduction

The Epiphyte is a remarkable little transceiver which has introduced many QRP operators to the pleasure of building their own SSB equipment. It was designed by Derry Spittle, VE7QK, who needed reliable radio communication when journeying into wilderness areas in British Columbia. Beyond the range of VHF repeaters, simple battery-operated HF equipment offers the only practical means of communication. The objective was to build a small transceiver capable of providing effective voice communication with the British Columbia Public Service Net on 3729kHz from anywhere in the province. The Epiphyte began as a project of the QRP Club of British Columbia and the design was first published in 1994 by the G-QRP Club and the QRP Club of Northern California. The Epiphyte-2 now includes many modifications and suggestions since made by their members.

Circuit description

The block diagram of the EP-2 is shown in **Fig 7.44**. The circuit is based around a pair of SA602 double-balanced mixers (IC2, IC3), and a MuRata miniature 455kHz ceramic SSB filter (F1). IC1 switches the LF and HF oscillators, permitting the same mixers to be used for both transmitting and receiving. On transmit, the modulated signal passes though a band-pass filter to remove the image, before being applied to the driver (IC5). Transmit/receive switching is accomplished with a DPDT relay (K1) and IC1. When receiving, B+ is disconnected from the

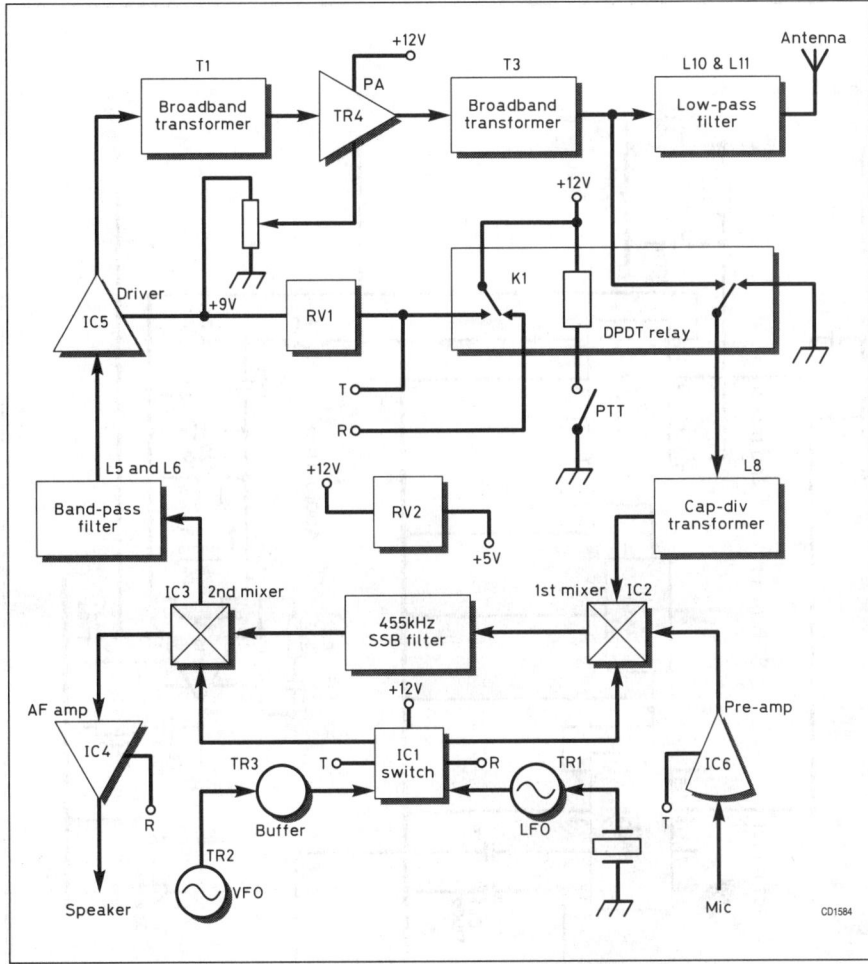

Fig 7.44: Block diagram of Epiphyte-2

microphone preamplifier (IC6) and the driver (IC5), and forward bias is removed from the PA (TR4). When transmitting, B+ is removed from the audio amplifier (IC6) and the RF input to the first mixer is disconnected from the low-pass filter and grounded. At the same time, the relay provides switching voltages to IC1. The antenna remains connected to the low-pass filter and PA at all times.

The circuit is shown in **Figs 7.45 and 7.46. Table 7.17** shows the Components List.

The VFO (TR2) is a varactor-tuned Vackar circuit using a Toko 3.3µH variable coil. This is buffered by an emitter follower (TR3). The LF oscillator (TR1) uses a 455kHz ceramic resonator adjusted to 452.8kHz. An MPF102 FET is used for both oscillators.

The RF bandpass filter was modelled in a series-tuned configuration to give a reasonably flat response over 200kHz and a sharp roll-off on the high-frequency side for rejection of the image frequency. It is designed for an input impedance of 1500Ω (to match the SA602 mixer) and to terminate in a 100Ω resistive load to ensure stability in the driver. It uses a pair of Toko 4.7µH coils and the fixed capacitors are standard values.

The driver stage (IC5) is a CA3020A differential amplifier. Operating from a 9V supply, this stage has a power gain of 60dB and is capable of 500mW output. The broad-band transformer (T1) has a bifilar-wound primary link-coupled to a 47Ω resistive load (R23) at the gate of TR4. The power amplifier (TR4) is an IRF510 MOSFET with an RF output of 5W PEP. A broad-band transformer (T3) matches the output to a conventional 50Ω low-pass filter.

The receiver RF input from the low-pass filter is a capacitance-divider matching circuit to the first mixer and is tuned to the cen-tre of the phone band. The AF amplifier (IC4) is a LM386 with a balanced input. The receiver is quite sensitive and with a dipole or inverted-V antenna at 25ft it provides adequate speaker volume from all but the weakest signals. Band noise is usually the limiting factor.

R19 provides the polarising voltage for an electret microphone (two-terminal type) and should be omitted if a dynamic microphone is used. The speech amplifier (IC6) is an LM741. The value of R20 should match the impedance of the microphone and the stage gain may be set by adjusting the value of R17.

The component layout is shown in **Fig 7.47** and the PCB layout is in **Fig 7.48** (in Appendix B). A ground loop present in earlier versions (eg **Fig 7.49**) has now been eliminated.

Assembly

Assembly is fairly straightforward. Ensure that the Toko coils (L3, 5, 6, 8), SSB filter (F1), ceramic resonator (X1) and trimmer capacitor (C1) fit the PCB and enlarge the holes if necessary. Install the CA3020A (IC5) first as it is easier to align the 12 pins without the other components in place. The tab is over pin 12. Be sure to solder in the two bare jumper wires on top of the PCB before installing the socket for IC1. Some fairly large value poly-styrene capacitors are specified and their physical size should be ascertained before ordering if they are to fit comfort-ably on the board. Alternatively, NPO/COG ceramic capacitors may be substituted. Cut off the centre pin (drain) before mounting the IRF510 (TR4) and heatsink with a 4-40 machine screw, nut and star washers. Output from TR4 is taken from the tab. Remove or cut-off unused terminals from the relay socket. Finally, don't bother soldering the three unconnected pins on the Toko coils to the ground plane - you may need to remove them one day!

Alignment and testing

This must be carried out with the single-sided PCB fastened to a ground plane with four metal stand-offs.

1. Remove both metering jumpers. Install the relay (K1) and PTT switch if not built into the microphone. Remove all socketed ICs and connect to a 12-14V FUSED supply. Verify that RV2 is delivering 5V and that RV1 is delivering 9V on transmit. With an RF probe, check that both oscillators are functioning. Install IC1 and verify that the VFO is switching between transmit and receive.

2. Set the LFO to 453kHz with the trimmer (C10). Monitor the ninth harmonic at 4.075MHz on a communications receiver. It may be necessary to change the value of the padder (C11).

3. Set R24 to mid position and adjust L3 until the frequency at the 'test point' (Con9) measures approximately 4.2MHz. To avoid having the ferrite core protrude above the case of L3, screw it completely into the coil and tune it upwards (anti-clockwise). The core will remain firmly in position without further 'fixing'. R24 limits the overall frequency

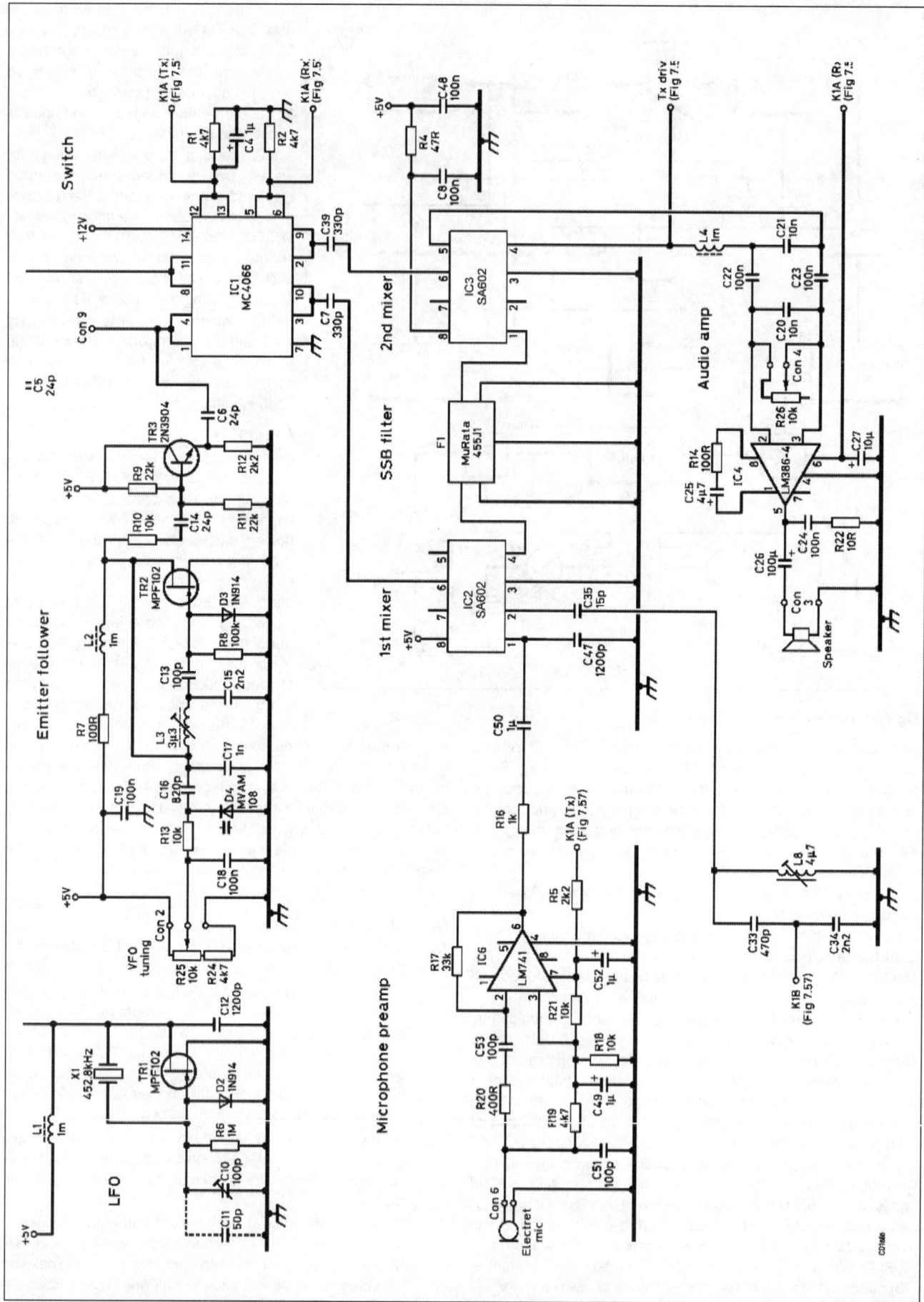

Fig 7.45: Circuit of Epipyhte-2 receiver and SSB generator

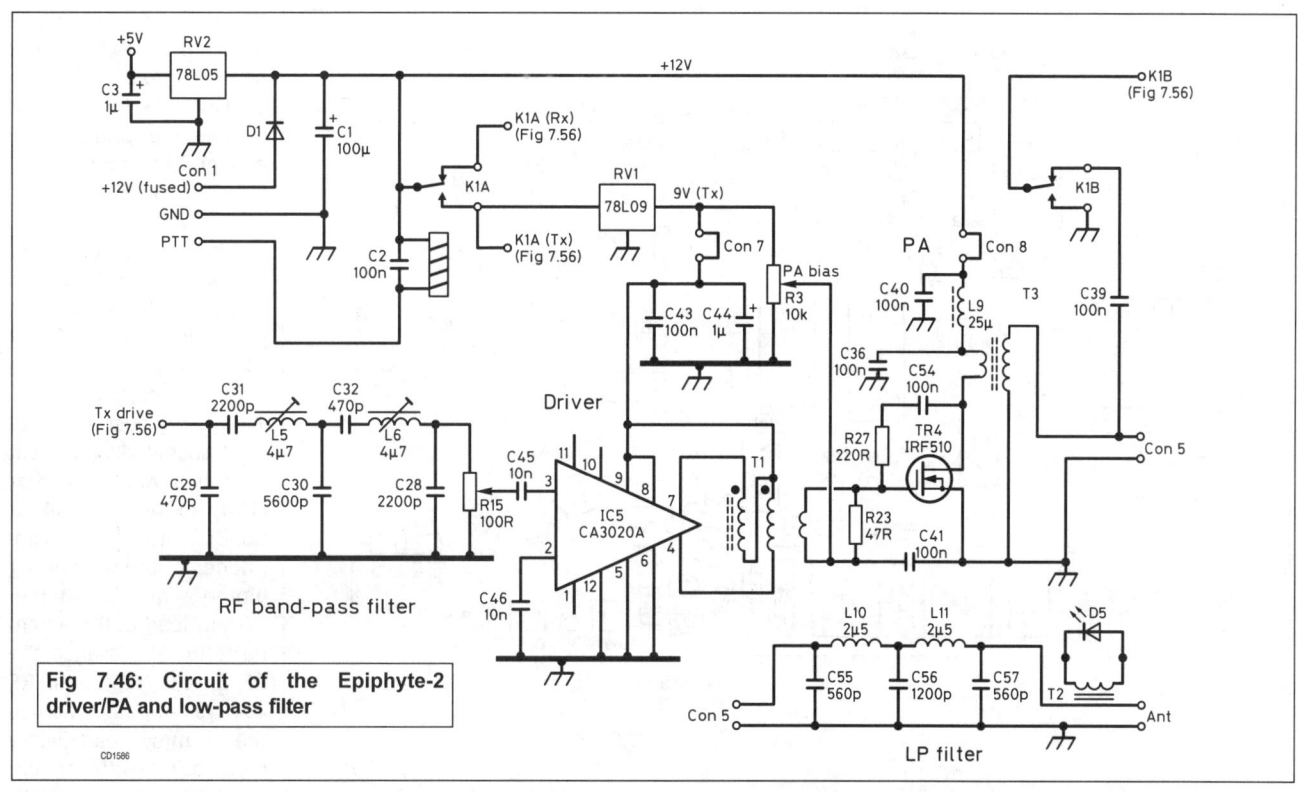

Fig 7.46: Circuit of the Epiphyte-2 driver/PA and low-pass filter

CD1586

Table 7.17: Epiphyte-2 component list

R1, 2, 19, 24	4k7		C27	10μ tantalum
R3	10k lin (multi-turn trim pot)		C29, 32, 33	470p axial polystyrene
R4, 23	47R		C30	5600p axial polystyrene
R5, 12	2k2		C35	15p NPO
R6	1M		C55, 57	560p axial polystyrene
R7, 14	100R		C56	1200p axial polystyrene
R8	100k		L1, 2, 4	1mH choke
R9, 11	22k		L3	3.3μH (Toko BTKANS9445)
R10, 13, 18, 21	10k		L5, 6, 8	4.7μH (Toko 154AN-T1005)
R15	100R lin (multi-turn trim pot)		L9	RFC (7t on Amidon FB-43-801)
R16	1k		L10, 11	2.5μH (25t on Amidon T-32-2)
R17	33k		T1	4:1 broad-band transformer (5t bifilar on FB-43-801 with 3t link)
R20	400R			
R22	10R		T2	2t on Amidon FB-43-2401
R25	10k (precision 10 turn pot)		T3	binocular broadband transformer (2t pri, 5t sec on Amidon BM-43-202)
R26	10k log			
R27	270R		IC1	MC14066
RV1	78L09		IC2, 3	SA602
RV2	78L05		IC4	LM386-4
C1, 26	100μ electrolytic		IC5	CA3020A
C3, 4, 44, 49, 52	1μ tantalum		IC6	LM741
			D1	polarity protection diode, eg 1N4001
C50	1μ non-polar ceramic		D2, 3	1N914
C5, 6, 14	24p NPO		D5	LED
C7, 9	330p ceramic		D4	MVAM108
C2, 8, 18, 19, 22-24, 36, 39-43, 48, 53, 56	100n monolithic ceramic		TR1, 2	MPF102
			TR3	2N3904
			TR4	IRF510
C15, 28, 31, 34	2200p axial polystyrene		F1	455kHz SSB filter (MuRata 455J1)
C10	100p trimmer		K1	Miniature DPDT relay
C11	50p NPO		X1	455kHz ceramic resonator
C12, 47	1200p ceramic		Two 3-pin, four 2-pin polarised Molex terminals	
C13, 51	100p NPO		Two metering jumpers and terminals	
C32	470p NPO		One 1-pin test point	
C16	820p axial polystyrene		Heatsink for TR4	
C17	1n axial polystyrene		Four 8-pin, one 14-pin, one 16-pin IC sockets	
C20, 21, 45, 46	10n		Four ¼in metal stand-offs	
C25	4μ7		Printed circuit boards and a kit of parts for the project are available from JAB Electronic Components [3]	

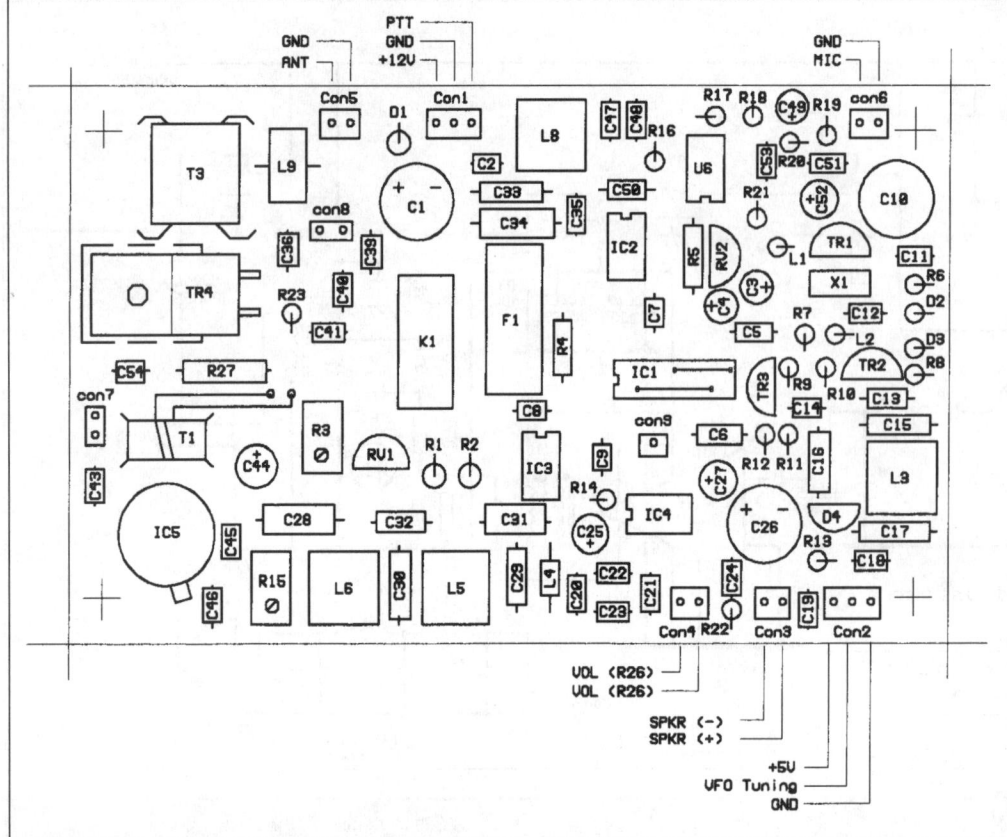

Fig 7.47: Epiphyte-2 component layout

5. The RF voltage at pin 6 on IC2 and IC3 should read 100 to 150mV RMS. If necessary, change the value of C5 and/or C6 to adjust.

6. Set the RF drive control (R15) to minimum. The transmit idling current in the driver (IC5) should measure 25mA +10% at Con 7. If not then TR1 or IC5 has probably been installed incorrectly.

7. Adjust R3 to set the transmit idling current in the power amplifier (TR4) to read 50mA at Con8. Install IC6, microphone, both metering jumpers and a 50-ohm dummy load at the antenna terminal. Advance the RF driver (R15) until RF voltage appears across the dummy load while modulating with a constant tone (whistle!). Adjust the band-pass filter (L5, L6) to maximise, but do not 'stagger tune' them. Continue increasing the drive until it peaks at around 16V RMS with modulation. The driver current should rise to 60 or 70mA and the PA current to around 600mA with a continuous heavy tone. The 'average' current with normal speech modulation will be considerably less. Monitoring your own signal with phones on a receiver will disclose any serious problem before testing on the air with a local amateur.

coverage and, at the same time, ensures that the varactor diode is always biased positive. The value of R24 may be changed to alter the bandspread.

4. Install IC4 and connect the antenna, speaker and volume control (R26). The leads to R26 should be kept as short as possible. Test the receiver and adjust L8.

The GQ-40 CW Transceiver

This section describes the GQ-40 (**Figs 7.50 and 7.51**) a high-performance, single-band CW transceiver by GW8ELR which first appeared in *SPRAT* [9].

The main Components List is shown in **Table 7.18**, and com-

Fig 7.49: An assembled Epiphyte-2 which uses a earlier version of the PCB

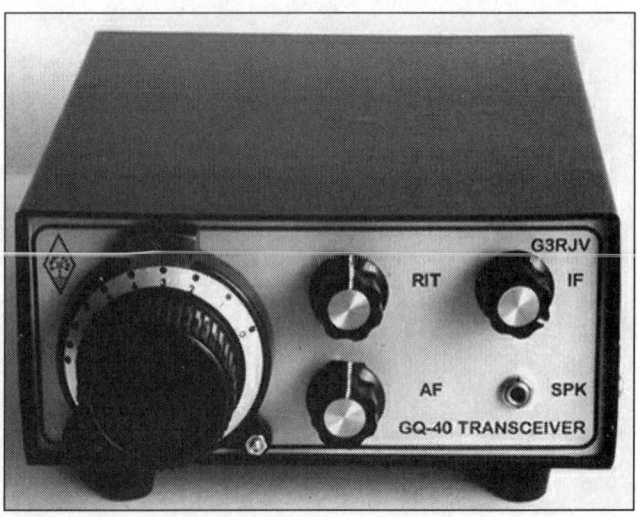

Fig 7.50: GQ-40 exterior view

R1, 41	47k
R2, 4	180R
R3, 5	10R
R6	560R
R7, 9, 24, 51, 52, 65	4k7
R8, 19, 27, 45, 46, 50, 53-56	10k
R15,16	68k
R17, 20, 23, 25, 26, 28	1k
R18	220R
R21	5R6
R22	220k
R29, 42, 44, 47-49, 57, 58, 67	100k
R31	3k
R32	27R
R33, 35, 36	22k
R34	2k2
R10, 38-40	not used
R59	390R
R63	680R
R64	12R
RFC2-5	1mH
VR1	1k preset
VR2, 3	10k preset
RV1, 2	10k log
RV3	22k log switched
RV4	100k lin
RV5, 6	22k lin
C14, 20, 28	100p
C8, 12, 23-27, 29, 34, 35, 40, 42, 45, 55-60, 62, 66, 68,79, 80, 83, 84, 86-91, 100, 101, 110	100n
C9-11, 43, 61, 65, 71, 72	1n
C13, 15, 16, 18, 19, 21, 31	180p
C17	220p
C13a,15a, 9a, 21a	22p
C14a, 20a	4p7
C16a, 18a	8p2
C17a	18p
C22, 32, 38, 39, 49, 53, 63, 64, 69, 70, 73, 78, 97, 111	10n
C30, 48	47p
C33	1µ
C37, 67	10µ
C41	4n7
C44	47µ
C46, 47, 81	100µ
C50	150p
C51	560p
C52	68p
C82	4µ7
C85	22µ tantalum
C92	6p8
TC1, 2	60p (brown)
TC3	6-10p (blue)
T1-3, 6, 7	37KX830 (matt black)
T4, 5, 10	BLN43002402 (two hole balun core)
T8, 9	56-61001101 (matt black)
X1-8	4.4336MHz
RFC1	100µH (Toko 7BS)
RFC2-5	1mH (Toko 7BS)
RFC6,7	15µH (Toko 8RBSH)
RFC8	10µH (Toko 7BS)
D1, 2, 4, 9, 10, 11	1N4148
D3, 8	BA243
D5-7	1N4007
ZD1	BZY88C4V7
ZD3 [varicap D12]	BB105
TR1, 2, 5, 16	J310
TR3, 12	2N3904
TR4	2N2222
TR6	PN2222/MPS3392
TR7	MFE201
TR8	VN66AFD
TR9, 10	IRF510
TR11	2N3906
TR13, 14	BD140
TR15	BS170
TR17	2N3866
IC1	HPF-505X/SBL-1
IC2	MC1350P
IC3, 9	SA612/602
IC4	LF351
IC5	LM380
IC6	4093
IC7, 8	78L08
IC10	78L05
IC11	78L06

Table 7.18: GQ transceiver parts list - all bands

Fig 7.51: GQ40 interior view

R61	not fitted
R37	820R
C1, 4, 6	100p
C2, 7	1n
C3, 5	3p9
C74, 77	47p + 220p polystyrene 63V
C75, 76	680p polystyrene
C93	47p + 47p + 33p
C94	470p polystyrene
C95, 96	1800p polystyrene
C99	82p
C102-109	not fitted - install bypass link
L1-3	Toko KANK 3334 (yellow)
L4	40t 28SWG on T68-6 or T50-6 (yellow)
L5, 7	21t 26SWG on T37-6 (yellow)
L6	24t 26SWG on T37-6 (yellow)
RFC9	1mH
TC4	not fitted
VFO frequency 2566-2666kHz for 7000-7100kHz coverage	

Table 7.19: GQ transceiver parts list - additional parts for 40m version

ponent values are provided for 7, 10, and 14MHz versions (**Tables 7.19, 7.20 and 21**).

General

The receiver is a conventional superhet design at an IF of 4.4MHz, with a high-dynamic-range, double-balanced, mixer, a six-pole 500Hz crystal IF ladder filter and high-power audio stage. The transmitter is of the mixer type, utilising the receive mixer in a bi-directional mode. The driver and PA are MOSFETs of the inexpensive commercial switching type;

adjustable gate bias allows user selection of operating class. The PA is run in push-pull to give a low harmonic content and when in Class AB1 an output of 7W is typical. Power output in normal operation is fully variable via a power control (drive) potentiometer.

Full QSK is achieved by electronic timing of the antenna pin diodes, positive supply lines and IF gain control. Conventional rectifier diodes 1N4007 are used for the antenna changeover

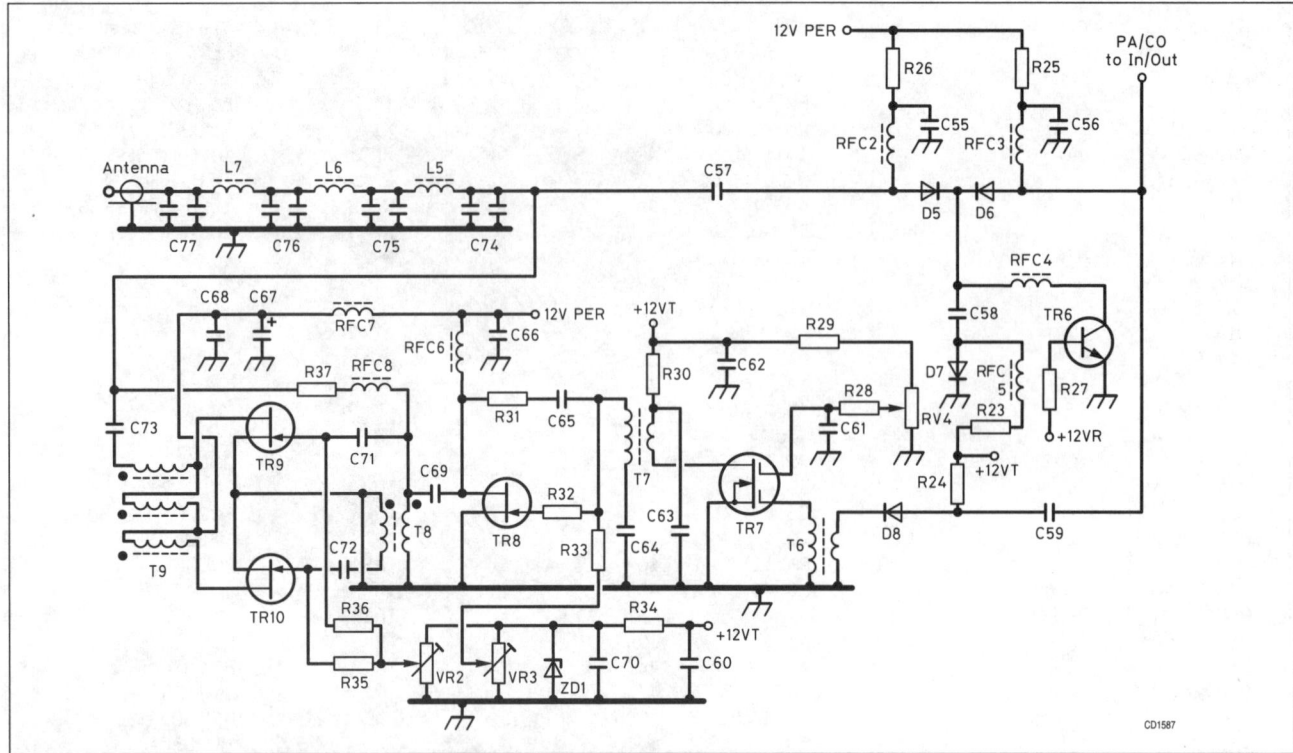

Fig 7.52: GQ-40 receiver LPF, diode change-over, transmit driver and pre-driver

R61	not fitted
R37	820R
C1, 6	47p
C2, 7	470p
C5	fit wire link
C74, 77	270p polystyrene
C75, 76	560p polystyrene
C93	33p + 39p NP0
C94	220p + 220p polystyrene
C95	180p polystyrene
C96	220p + 220p polystyrene
C99	82p
C102-109	not fitted - install bypass link
C104	47p
TC4	not fitted
L1	not fitted
L2,3	Toko KANK 3334 (yellow)
L4	32t 28SWG on T68-6 or T50-6 (yellow)
L5,7	19t 26SWG on T37-6 (yellow)
L6	20t 26SWG on T37-6 (yellow)
RFC9	1mH
VFO frequency 5666-5766kHz for 10,100-10,200kHz coverage.	
Adjust VC1S series cap to bandspread	

Table 7.20: GQ transceiver parts list - additional parts for 30m version

R61	100R
R37	4k7
C1, 6	120p
C2, 7	1n
C5	4p7
C3	fit wire link
C4	not used
C74, 77	180p polystyrene
C75, 76	390p polystyrene
C93	33p + 39p NP0
C94	220p + 220p polystyrene
C95	180p polystyrene
C96	220p + 220p polystyrene
C98, 102	10n
C99	82p
C103	not used
C104	47p
TC4	18p fixed
C105	56p
C106	100n
C107	1p8
C108	56p
C109	10n
X9	24.00MHz
L1	not fitted
L2, 3	Toko KANK 3335 (pink)
L4	32T 28SWG on T68-6 or T50-6 (yellow)
L5, 7	16T 26SWG on T37-6 (yellow)
L6	17T 26SWG on T37-6 (yellow)
L8, 9	1.2µH Toko KANK 3335 (pink)
RFC9	1mH
IC9	SA612/602
VFO frequency 5566-5466kHz (LO = 24 VFO) for 14000-14100kHz	
coverage	

Table 7.21: GQ transceiver parts list - additional parts for 20m version

T1	K37X830	3t IC1		13t TR2	32SWG
T2	-	15t TR2	10t C13		32SWG
T3	-	6t C21	24t IC2		32SWG
T4	BLN 43002402	2t TR5	Ct D3	6t 12VT	32SWG
T5	-	2t TR1	Ct C11	6t 12VP	32SWG
T6	K37X830	4t D8		15t TR7	32SWG
T7	-	14t TR7	3t R31		32SWG
T8	59-61001101	12t PA		12t TR8	bifilar 28SWG
T9	-	12T			trifilar 28SWG
T10	BLN 43002402	6t R66		2t R3	32SWG

Table 7.22: GQ transceiver transformer winding details

system as these are inexpensive and have a similar doping profile to more expensive PIN diodes. Insertion loss is less than 0.1dB at 10MHz with an IP3 of +50dBm when biased at 5mA minimum.

Frequency control is by a Colpitts VFO with a high-quality variable capacitor. Due to the high frequency required for the local oscillator on 14MHz, the VFO is pre-mixed with a crystal oscillator.

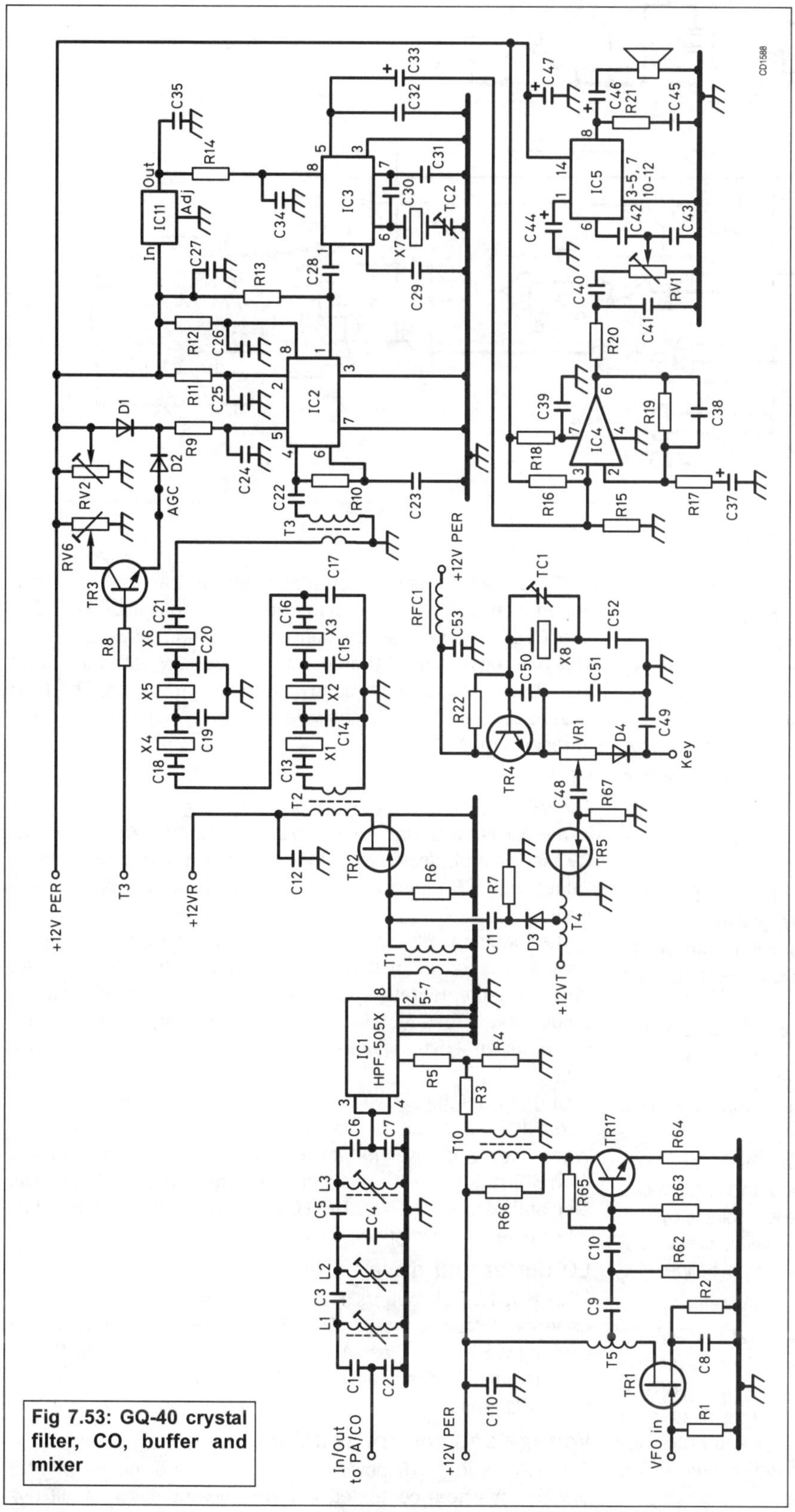

Receiver front-end

Signals from the antenna are routed through the LPF L5-7/C74-77 to the diode changeover system (**Fig 7.52**). D5/6 are biased from the permanent 12V line when TR6 is switched on by the receive 12V line. Bias is regulated by R26/25 with RFC2-4 and C55, 56 keeping RF from entering the DC supply. D7 prevents signal loss during receive through T6. Signals now enter the three-pole input filter formed by C1-7/L1-3 (**Fig 7.53**). The input filter is a low-loss Butterworth design with a bandwidth of 300kHz. Capacitive dividers C1/2 and C6/7 match the characteristic filter impedance to 50Ω.

The filter output feeds IC1, a double-balanced hot-carrier mixer, type SBL-1/HPF-505 etc. LO drive for the mixer is provided from a J310 buffer amplifier, and to ensure a correct termination a 50-ohm T pad is used. The mixer is terminated by a simple active termination which uses a step-up transformer to a J310 FET amplifier with a gate resistor. Whilst this is not as good as the more common diplexer arrangement, it requires no setting up.

IF and crystal filter

The filter (Fig 7.53) is a six-pole ladder design with a centre frequency of 4.433MHz. The unit uses high-quality, high-volume crystals designed for TV colour burst timing. The filter has a 3dB bandwidth of 500Hz and a Gaussian shape by use of Butterworth design constants. T2, 3 match the input and output impedance of the IF amplifier and product detector. IC2, a MC1350P, provides up to 60dB gain at the IF frequency.

During transmit the amplifier is muted by applying bias to the AGC control, pin 5. This bias voltage is switched by TR3 controlled from the T3 line. A front-panel IF gain control is fed to the AGC control input via D1 and external AGC may also be applied via D2.

The exclusion of an AGC system was a deliberate design policy. Many of the audio-derived AGC systems found in current QRP transceiver designs are far from satisfactory and the inclusion of AGC is often counter-productive to the reception of weak CW signals. If you must add an AGC system, add a good one.

The amplified IF frequency is converted to audio frequency in IC3, an SA602 balanced mixer. LO (BFO) injection for the conversion is provided by an on-chip crystal oscillator. The BFO frequency is adjusted by TC2. The SA602 requires a 6V power supply and this is provided from IC11, a three-terminal regulator.

AF preamp and amplifier

A low-noise LM351 preamplifier, IC4, with some frequency tailoring drives an LM380 audio output amplifier (Fig 7.53). The output is capable of about 1W of audio, and has a Zobell network to ensure stability.

Fig 7.53: GQ-40 crystal filter, CO, buffer and mixer

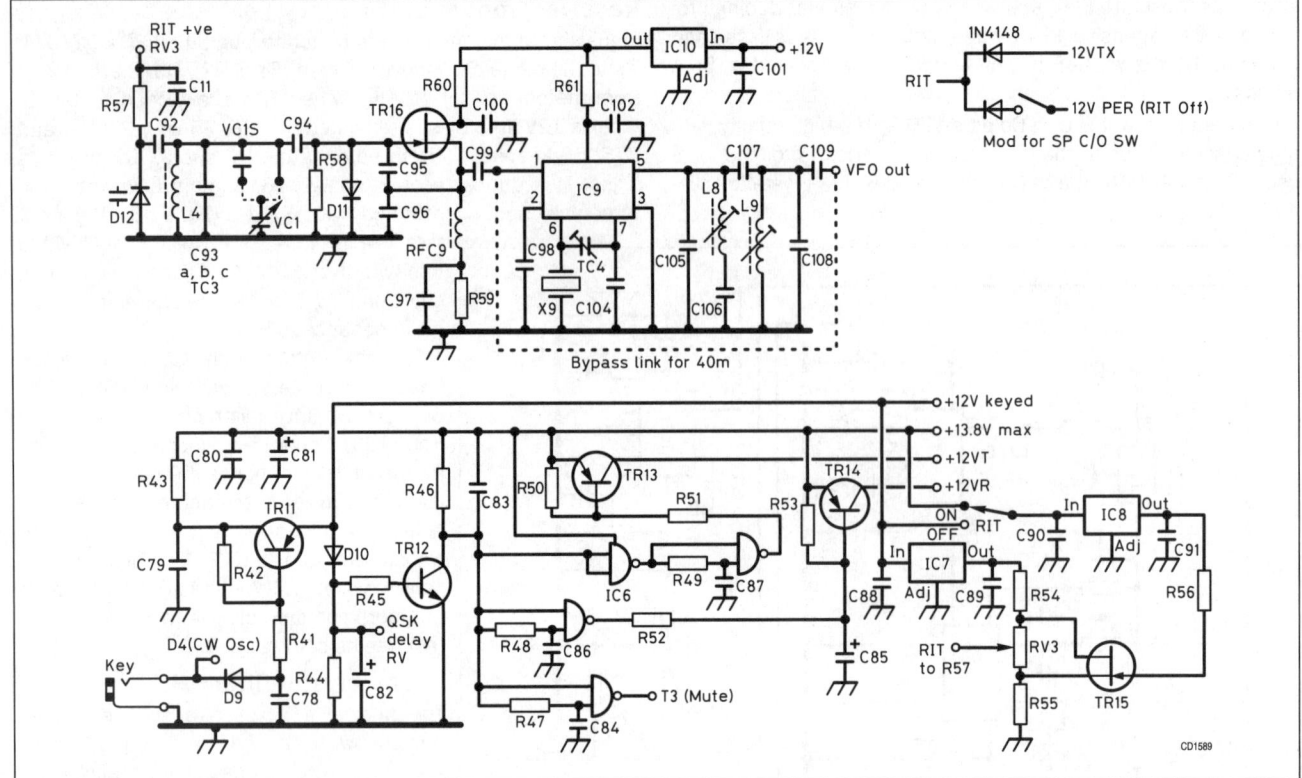

Fig 7.54: VFO, pre-mix and voltage control circuits

Transmit carrier oscillator, buffer and mixer

TR4 is the transmit oscillator (Fig 7.53) and is crystal controlled at 4.433MHz. TC1 allows the frequency to be offset to match the receive BFO audio note. The oscillator is keyed directly at the keying rate via the key line. TR5 is a J310 buffer amplifier. This feeds the mixer via D3 which is biased on during transmit, via the 12VT line. The output is coupled to the mixer transformer T1 via C11, a DC blocking capacitor. The transmit signal is mixed in IC1 to the output frequency and routed back through the band-pass filter to the pre-driver, TR7.

Transmit pre-driver and driver

When in transmit, D8 is biased on by the 12VT line, coupling the transmit signal to TR7 by T6 (see Fig 7.62). D7 is also biased on to provide an RF ground between D5/6 to improve the input/output receive bypass isolation. C58 provides a DC block to prevent D5/6 being switched on.

The pre-driver is a dual-gate MOSFET; gate 2 is used as a control element by varying its voltage by RV4, a panel-mounted potentiometer. This stage is transformer coupled to the next by T7, a broad-band matching transformer. C64 on the secondary provides an RF ground but provides DC isolation.

The driver, TR8, a VN66 FET, is biased as a Class A amplifier. Feedback is applied to the driver to ensure stability; this is controlled by R3. Bias adjustment is by VR3. The bias voltage is stabilised by ZD1 and supplied from the 12VT line. As the amplifier has no bias during the receive period, TR8 is connected to the 12V permanent line.

Power amplifier

TR9, 10 are run as a push-pull amplifier (Fig 7.52) to maintain stability. R37/RFC8 provide feedback. Push-pull drive for the amplifier is provided from a phase-splitter transformer T8. The amplifier is provided with bias to allow the operating class to be varied. The best balance between output power and efficiency will be in Class B. VR2 is the bias control potentiometer and is adjusted for the required quiescent current.

T9, a trifilar transformer matches the transistors' output impedance to the load presented by the output filter. LPF L5-7 and C74-77 form a seven-element Chebyshev low-pass filter. This has a low SWR at the operating frequency, with the cut-off just above the maximum bandwidth upper frequency. The filter is used bi-directionally to cascade with the bandpass filter when in receive mode.

VFO and pre-mix

The VFO is a standard Colpitts oscillator (**Fig 7.54**). The voltage to the oscillator is stabilised by a three-terminal regulator at 5V. C93 is the main fixed capacitance. This is made up from three values with NP0 dielectric, and a small 10pF trimmer for calibration.

The inductance is wound on type 6 material which is a good compromise between thermal stability and the winding size. Receiver incremental tuning is provided by a variable capacitance diode. To maintain a reasonable voltage swing on the diode, its effective capacitance is reduced by series capacitor C92.

Output from the oscillator is coupled by C99, either to the pre-mixer, IC9, or direct to the main mixer buffer amplifier TR1. When the transceiver model requires a LO frequency greater than 6MHz, IC9 pre-mixer is employed. This is a SA602A used as a crystal oscillator/mixer. The LO pre-mixer output is filtered by a two-pole bandpass filter, C105-109/L8-L9.

LO buffer and drive amplifier

TR1, a J310 FET, buffers either the pre-mix or VFO frequency dependent on the model (Fig 7.53). Its output is transformer coupled via T5 to a Class A driver, TR17. This is a 2N3866, and the amplifier is designed as a 50-ohm line driver feeding the mixer IC1 T-pad.

Voltage and control switching

The 12VT and 12VR power supply lines are switched by PNP power transistors controlled from a time-and-delay circuit (Fig

7.54). The timing switch IC6 is a quad NAND Schmitt trigger package, gate timing being adjusted with small ceramic capacitors C84-87. Timing is initiated by TR11 keyer which directly keys the CW oscillator and the delay generator TR12. TR12 will hold on, dependent on the charge across C82. The charge is adjustable via RV5 which is mounted on the rear panel.

RIT voltage is supplied through a divider network controlled by RV3, a front-panel potentiometer. Voltage to the divider is stabilised at 8V by a three-terminal regulator IC7. During transmit, or when the RIT is switched off, gate voltage is applied to TR15 from IC8. This turns on the FET to short RV3 and centre the voltage equivalent to the mid-position of RV3.

Test and alignment
The minimum requirements for test and alignment are a high-impedance multimeter, power meter, 50-ohm dummy load and a general-coverage receiver. The power meter can be a simple peak diode detector across the dummy load used in conjunction with the multimeter. If available, a signal generator or in-band test oscillator are very useful, as is a frequency counter and oscilloscope.

Before starting, carry out a resistance check across the 13.8V input. This should be about 200Ω. If all is OK, connect the multimeter in series with the DC+ lead on its highest amperage range, and switch on. With no signal present and minimum volume, expect the current to be 100-200mA. If the current is appreciably more, switch off and check for faults. Once this test is complete, remove the meter and connect normally to the supply.

Turn up the volume control and check for some background noise.

Next test and set up the VFO. Check the VFO output frequency using a counter or a general-coverage receiver with a short antenna near the VFO. Adjust TC3 for the LF band edge. If you are unable to correct the frequency with TC3, then spread or compress the turns on L4. If still outside the range, remove or add turns to L1, or add capacitance using 5mm NP0 discs at C93a, C93b, C93c. Next unmesh VC1 and note the HF frequency. If the coverage is more than required, reconnect VC1 through a fixed capacitor to the VC1 pin. VC1 is now in series with the capacitor and its effective capacitance swing will be reduced dependent on the value. Check the HF and LF limits again and readjust TC3 as necessary. If necessary alter the fixed value and repeat the checks. Once you are happy with the coverage, cement the turns on L4, then cement it to the board using Balsa or plastic adhesive.

Once the VFO is up and running correctly, any temperature drift may be corrected. The capacitance of C93 can be made up with three capacitors using the A and B pads. If required, it should be possible to correct long-term drift by mixing dielectric types.

If a VFO crystal mixer is installed, check for the correct output frequency. If an oscilloscope is available, check the output at R5 or use a general-coverage receiver. Peak the output by adjusting L8/L9.

Attach an antenna to the input pin and adjust the cores of L1-3 for maximum received signal strength - do not use a screwdriver for adjustment as the core is very brittle. If you do not have the correct tool, an old plastic knitting needle with the end filed is ideal.

Tune to a test signal and adjust the BFO frequency using TC2. Adjust for the best filter response and opposite sideband suppression.

Test the transmitter next. Connect a 50-ohm dummy load and a power meter or diode detector and multimeter to the antenna output. Fit a temporary heatsink to TR8-10; a clip-on TO220 type

is ideal (the tab is part of the drain so be careful that during the checks it is not grounded). Set VR2/3 to the earthy end of their travel. (If you have connected RV4, this must also be at the earth end of its range.) Set VR1 wiper to maximum (the TR5 end of its travel).

Connect a multimeter on its 'amps' range in the 13.8V supply line, and key a dash length by grounding the KEY pin. If all is well, the transceiver should change to transmit and hold for a short time. With a multimeter on its DC voltage range, check at R27 that the 12VR line switches correctly. Re-check that 4.7V appears across ZD1 on transmit.

Current consumption will be around 200mA providing there is no drive from TR7. If the consumption is excessive, remove D8 to isolate the driver and PA, and re-check. Possible faults include reversed diodes, ICs or transistors, incorrect values on decoupling or bias resistors, or shorted or incorrectly installed transformers. Once any errors have been corrected, reconnect D8.

Adjust TC1 to obtain a 700Hz beat note from the sidetone. If the audio level is excessive, check that the mute voltage to the IF amp of 8V plus appears at the junction of R9/D1.

Attach a multimeter on its current range in series with the DC supply line. Key the transmitter and set TR8 quiescent current by adjusting VR3 so that the meter current rises by 200mA (ie TR8 standing current is 200mA). Now adjust VR2 so that the current increases by a further 100mA (this sets TR9/10 quiescent to 100mA).

The adjustments should be smooth with no power output on the meter. Jumps on the current or the power meter would indicate instability at some level of drive. In this case, remove D8 to isolate the drive and re-check. If there is no improvement, check that T9 is correctly wired. Also check the antenna isolation: confirm that TR6 is switching and that D7 is biased on by the 12VT line.

Finally, attach RV4 and adjust it to give 2-3V on the gate of TR7. Key the transmitter and check for power output. This completes the tests.

GQ transceiver PCB
The PCB, a complete kit of parts and a manual containing the latest revisions can be obtained from Hands Electronics, Tegryn, Llanfyrnach, Dyfed SA35 0BL. Tel 01239 698427.

MEDIUM AND HIGH-POWER SOLID-STATE HF LINEAR AMPLIFIERS
The construction of a solid-state linear amplifier now represents a realistic alternative to valve designs even up to the normal maximum UK power level of 400W. The construction of solid-state amplifiers has often been discouraged in the amateur press with claims that such projects can only be made with the use of expensive test equipment, and in particular, a spectrum analyser. The 'simple spectrum analyser' described by Roger Blackwell, G4PMK, in *Radio Communication*, November 1989 (also described at http://www.qsl.net/g3pho/ssa.html) is ideal for the purpose, and fully justifies the small amount of money necessary for its construction.

Number	Power out (W)	Supply voltage (V)
AN762	140	12-14
EB63	140	12-14
EB27A	300	28
AN758	300-1200	50
EB104	600	50

Table 7.23: Motorola RF amplifiers

Motorola Applications Notes AN762, AN758 and EB104, available from Motorola in the UK [1] provide the designs for solid-state linear amplifiers with outputs in the range 140W to over 1kW. The application notes (AN) and engineering bulletins (EB) describe the construction and operation of suitable amplifiers using Motorola devices, and include printed circuit foil information, making construction relatively easy. It is interesting to note that the PA units fitted to virtually all of the currently available commercial amateur radio equipment are based upon these designs by Motorola, with only a few individual differences to the original design.

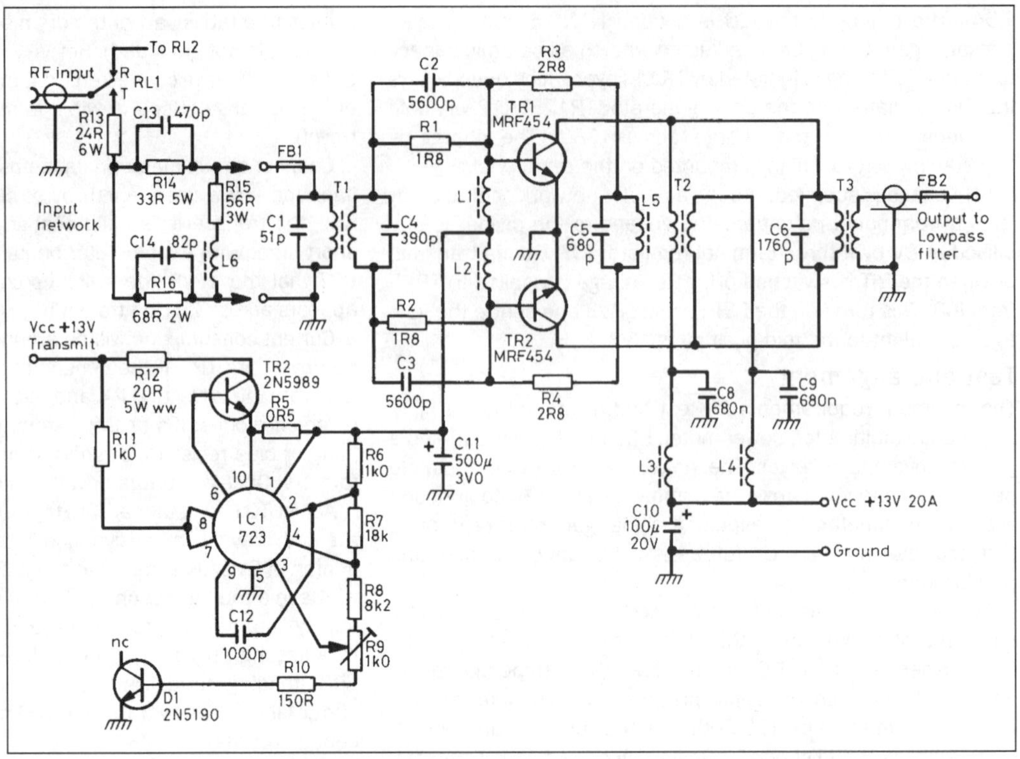

With the availability of suitable application notes it is perhaps surprising that few amateurs seem to have embarked upon such a project. High prices for solid-state power devices, combined with a lack of faith in their reliability, may well account for the reluctance to construct solid-state linear amplifiers. Suitable components have been readily available in the USA for some time: Communications Concepts Inc of Xenia, Ohio [10] have offered a complete range of kits and components for the Motorola designs, and many of these are now available in the UK from Mainline Electronics, PO Box 235, Leicester.

The kits include the PCB, solid-state devices, and all the components and the various ferrite transformers already wound, so that all the constructor has to do is solder them together. A large heatsink is necessary and is not supplied.

Home construction of PCBs for an amplifier is possible from the foil patterns available in the application notes, but it should be borne in mind that the thickness of copper on much of the laminate available to amateurs is unknown and inadequate for the high current requirements of a linear amplifier. A ready-made PCB is therefore a very sensible purchase and the kit of parts as supplied by CCI represents a very-cost-effective way of building any of the Motorola designs.

Choosing an Amplifier

Perhaps the most useful amplifier for the radio amateur is described in AN762. It operates from a 13V supply and is capable of providing up to 160W output with only 5W of drive. This design is the basis for the majority of commercial 100W PA units and lends itself to both mobile and fixed station operation using readily available power supplies. **Table 7.23** illustrates a number of alternative designs.

AN758 describes the construction of a single 300W output amplifier operating from a 50V supply and further describes a method of using power combiners to sum the outputs of a number of similar units to provide power outputs of 600 and 1200W respectively. Two such units would be ideal for a full-power linear amplifier for UK use, giving an output comparable to the FL2100

Fig 7.55: The HF SSB 140-300W linear amplifier circuit

Fig 7.57: The HF linear amplifier

type of commercial valve amplifier. It is recommended that any prospective constructor reads the relevant applications note before embarking upon the purchase and construction of such an amplifier.

After constructing a number of low-power transceivers it was decided to commence the construction of the AN762 amplifier. The EB63 design is similar, but uses a slightly simpler bias circuit.

AN762 describes three amplifier variations: 100W, 140W and 180W using MRF453, MRF454 and MRF421 devices respectively, and the middle-of-the road MRF454 140W variant was chosen (**Fig 7.55**).

Fig 7.58: Circuit board of the linear amplifier

Fig 7.59: Low-pass filter and ALC circuits

Constructing the Amplifier

Construction of the amplifier is very straightforward, especially to anyone who has already built up a solid-state amplifier. The PCB drawings (**Fig 7.56**, see Appendix B) and layouts are very good and a copy of the relevant application note was included with the kit. **Figs 7.57, 7.58 and 7.59** show how the amplifier was constructed. A Components List is is **Table 7.24**.

Fixing the transformers to the PCB posed a slight problem: on previous amplifiers that the author has built they were soldered directly to the PCB, but not this one. Approximately 20 turret tags (not supplied) are required: they are riveted to the board and soldered on both sides. A modification is required if you wish to switch the PA bias supply to control the T/R switching: the bias supply must be brought out to a separate terminal rather than being connected to the main supply line. This is achieved by fitting a stand-off insulator to the PCB at a convenient point; the bias supply components are then soldered directly to this stand-off rather than directly to the PCB.

A number of small ceramic chip capacitors have to be soldered directly to the track on the underside of the PCB; it should be borne in mind when doing this that the clearance between the PCB and heatsink is slightly less than 1/8in and the capacitors must not touch the heatsink.

It is advisable to mount the PCB to the heatsink before attempting to make any solder connections to it, as it is necessary to mark the mounting holes accurately, ideally drilling and tapping them either 6BA or 3mm. The power transistor mounting is very critical in order to avoid stress on the ceramic casing of the devices. The devices mount directly onto the heatsink and should ideally be fitted by drilling and tapping it. The PCB is raised above the heatsink on stand-offs made from either 6BA or 3mm nuts, so that the tabs on the transistors are flush with the PCB - they must not be bent up or down. When the PCB and transistors have been mounted to the heatsink correctly, they may be removed for the board to be assembled. The transistors should not be soldered in at this stage. The nuts to be used as the stand-offs can be soldered to the PCB, if required, to make refitting the latter to the heatsink a little simpler, but ensure the alignment of the spacers is concentric with the holes.

Assembly should commence with the addition of the turret tags and stand-offs. Then the ceramic chip capacitors should be added under the PCB. D1, which is really a transistor, is also mounted under the PCB; only the emitter and base are connected - the collector lead is cut off and left floating. This transistor is mounted on a mica washer and forms a central stand-off when the PCB is finally screwed down to the heatsink. The mounting screw passes through the device which must be carefully aligned with the hole in the PCB. A number of holes on the

HF LINEAR AMPLIFIER						
C1	51p chip		(2673021801) on	C15, 18	330p	RL2	OM1 type
C2, 3	5600p chip	L5	16SWG wire / 1t through T2	C17	27p	RL3-14	2A SPCO PCB mtg, 6V coil
C4	390p chip	L6	0.82µH (T50-6)	C19	75p		
C5	680p chip	T1	2 x Fairite beads	C20, 28	150p	D1, 2	0A91 or 0A47
C6 (C7)	1760p (2 x 470p chips plus 820p silver mica in parallel)		0.375in x 0.2in 0.4in, 3:1 turns	C21	120p		
		T2	6t 18SWG ferrite 57-9322 toroid	C22, 27	12p		SWITCHING AND ALC
C8, 9	0.68µ chip	T3	2 x 57-3238 ferrite	C23	220p	R1, 17, 20	10k
C10	100µ 20V		cores (7d grade) 4:1	C25	100p	R2	4k7
C11	500µ 3V		turns	C26	82p	R3-7	1k
C12	1000p disc	FB1, 2	Fairite 26-	C29	39p	R8, 9	33k
C13	470p silver mica	43006301		C30	68p	R10, 13, 21, 22	1M
C14	82p silver mica		cores	C31, 32	10n ceramic	R11	3k3
R1, 2	2 x 3.6R in parallel	RL1	OUD type	C33	10p trimmer	R12, 14, 18	47k
R3, 4	2 x 5.6R in parallel	TR1, 2	MRF454	C34	220p silver mica	R15	47k trimpot
R5	0.5R	TR3	2N5989	Note: C1-C30 silver mica 350VDC		R16	390R
R6	1k	D1	2N5190	L1	28t 22SWG T68-2	R19	220k
R7	18k	IC1	723 regulator	L2	25t 22SWG T68-2	R23	8k2
R8	8k2			L3	22t 22SWG T50-2	TR1	BC212
R9	1k trimpot		LOW-PASS FILTER	L4	20t 22SWG T50-2	TR2	BC640
R10	150R	C1	1200p	L5	18t 20SWG T50-6	IC1	LM3900
R11	1k	C2, 16	180p	L6	16t 20SWG T50-6	C1, 14-26	100n
R12	20R 5W WW	C3	2200p	L7	12t 20SWG T50-6	C2-5, 7, 9-11, 13	10n
R13	24R carbon 6W*	C4	470p	L8	11t 20SWG T50-6	C6	0.22µ
R14	33R carbon 5W*	C5	1000p	L9	10t 20SWG T50-6	C8	10 x 16V
R15	56R carbon 3W*	C6, 13	680p	L10	9t 20SWG T50-6	C12	1µ 16V
R16	68R carbon 2W*	C7	90p	L11	8t 20SWG T50-6	D1-3	LEDs
* Make up from several higher values in parallel.		C9	250p	L12	7t 20SWG T50-6	D4, 5, 7-18	1N914
		C10	560p	T1	18t bifilar T50-43 pri: 1t	D6	10V zener
L1, 2	VK200 19/4B choke (6-hole ferrite beads)	C11	390p	R1	68R	S1	1-pole 6-way
		C12, 24	33p	R2, 3	22k trimpot	S2, 3	SPCO
L3, 4	Fairite beads x 2	C14	100p	R4	1k8	Meter	500µA or similar
				L13	1mH RFC		

Table 7.24: HF linear amplifier components list

PCB are plated through and connect the upper and lower ground planes together, and it is a good idea to solder through each of these holes. The upper-side components can be mounted starting with the resistors and capacitors, and finally the transformers can be soldered directly to the turret tags. Soldering should be to a high standard as some of the junctions will be carrying up to 10A or more.

When the board is complete it should be checked at least twice for errors and any long leads removed from the underside to ensure clearance from the heatsink. Mount the PCB to the heatsink and tighten it down. Now mount the power transistors which should fit flush with the upper surface of the PCB. Tighten them down, ensuring that there is no stress on the ceramic cases. If any of the connections need to be slightly trimmed to fit, cut them with metal cutters. Ensure the collector tab is in the correct place. Once the transistors fit correctly, they can be removed again and very lightly tinned. The PCB should also be lightly tinned. The transistors can now be refitted and tightened down. Now they can be soldered in but, once in, they are very difficult to remove, so take great care at this stage. The amplifier board is now complete. It will need to be removed once more to allow the application of silicon heatsink compound to the devices in contact with the heatsink. Always unscrew the transistors first and then the PCB; refitting is a reverse of this process.

Construction of the amplifier takes very little time but requires considerable care to avoid damage to the output devices; the metal work may take a little longer. A large heatsink is required for 140W and an even larger one for 300W. It is recommended that the higher-power amplifiers are mounted onto a sheet of 6mm (0.25in) copper which is in turn bolted to the main heatsink. (Note: on no account should the transistors be mounted onto a diecast box due to surface imperfections and poor thermal conductivity.) Blowing may be necessary at the higher powers or if a less-than-adequate heatsink is used.

Setting Up and Testing the Amplifier

There is only one adjustment on the amplifier, making setting up relatively simple. Before connecting any power supplies, check and re-check the board for any possible errors. The first job is to test the bias supply. This must always be done before connecting the collector supply to the amplifier as a fault here could destroy the devices instantly. With +13V connected to the bias supply only, it should be possible to vary the base bias from approximately 0.5V to 0.9V. Set it to the lowest setting, ie 0.5V. Disconnect the bias supply from the 13V line. When conducting any tests on the amplifier always ensure that it is correctly terminated in a 50-ohm resistive dummy load. Apply +13V to the amplifier and observe the collector current on a suitable meter. It should not exceed a few milliamps - if it does something is wrong, so stop and check everything. Assuming that your amplifier only draws 3 to 4mA, connect the bias supply to the +13V supply and observe an increase in current, partly caused by the bias supply itself and also by the increased standing current in the output devices. The current can be checked individually in each of the output devices by unsoldering the wire links L3 and L4 on the PCB. Set the bias to 100mA per device by adjusting R9. The current should be approximately the same in each device: if it is not it could indicate a fault in either device or the bias circuitry to it. Increase the standing current to ensure that it rises smoothly before returning it to 100mA per device. Once the total standing current is set to 100 + 100 = 200mA, the amplifier is ready for operation.

With a power meter in series with the dummy load, apply a drive signal to the input, steadily increasing the level. The output should increase smoothly to a maximum of about 160W. It will go to 200W but will exceed the device specification. Observing the output on a spectrum analyser should reveal the primary signal, together with its second, third and higher harmonics. Check that there are no other outputs. Removal of the input signal should cause the disappearance of the other signals displayed. It is helpful during initial setting up to monitor the current drawn by the amplifier. At full output, efficiency should be in the order of 50%, perhaps lowering slightly at the upper and lower frequency limits and increasing a little somewhere in the 20MHz range. The maximum current likely to be drawn by the 140W amplifier is in the order of 24A.

Putting the Amplifier to Use

Building and setting up the amplifier is undoubtedly a simple operation, and may lead one into a false sense of security. Before the amplifier can be used it must have a low-pass filter added to the output to remove the harmonics generated. For single-band operation only one filter would be required, but for operation on the HF amateur bands a range of filters is required with typically six switched filters covering the range 2 to 30MHz. For most applications a five-pole Chebyshev filter will provide all the rejection required, but the majority of commercial designs now use the elliptic type of filter providing peaks of rejection centred around the second and third harmonics. Such filters can be tuned to maximise the rejection at specific frequencies. An elliptic function filter was decided upon as it only requires two extra components over and above the standard Chebyshev design and setting up is not critical.

The construction of a suitable low-pass filter (**Fig 7.60**) may take the form of the inductors and capacitors mounted around a suitable wafer switch, or they may be mounted on a PCB and switched in and out of circuit using small low-profile relays. This makes lead lengths shorter and minimises stray paths across the filter. Unused filters may be grounded easily using relays permitting only one filter path to be open at a time. The relays need only to be able to carry the output current; they are not required to switch it. 2A contacts are suitable in the 100-140W range. Amidon cores ensure the duplication of suitable inductors while silver mica capacitors should be used to tune the filters. The voltage working of the capacitors should be scaled to suit the power level being used; ideally, 350V working should be used in the 100-150W range and, for powers in the region of 400-600W, 750V working capacitors should be used, the latter being available from CCI.

The antenna change-over relay may be situated at either end of the low-pass filter. If it is intended to use the filter on receive then the relay will be placed between the amplifier and the filter, but if the amplifier is an add-on unit then it may not be necessary to use the filter on receive and the relay may be located at the output end of the filter. The filter performance is enhanced if it is mounted in a screened box with all DC leads suitably decoupled.

SWR Protection and ALC

One of the major shortcomings of early solid-state amplifiers was PA failure resulting from such abuse as overdriving, short-circuited output, open-circuited output and other situations causing a high SWR. A high SWR destroys transistors either by exceeding the collector-base breakdown voltage for the device or through overcurrent and dissipation.

ALC (automatic level control) serves two functions in a modern-day transmitter: it controls the output power to prevent overdriving and distortion and can be combined with a SWR detector to reduce the power if a high SWR is detected, which reduces the voltages that can appear across the output device and so protects it.

A conventional SWR detector provides indication of power (forward) and SWR (reflected power) which can be amplified and compared with a reference. If the forward power exceeds the preset reference an ALC voltage is fed back to the exciter to reduce the drive and hence hold the power at the preset level. A high SWR will produce a signal that is amplified more than the forward signal and will reach the reference level more quickly, again causing a reduction in the drive level.

The circuit shown in **Fig 7.61** has been designed to work in conjunction with the ALC system installed in the G3TSO modular transceiver [11] and produces a positive-going output voltage. The LM3900 IC used to generate the ALC voltage contains two unused current-sensing op-amps which have been used as a meter buffer with a sample-and-hold circuit providing a power meter with almost a peak reading capability. In practice it reads about 85% of the peak power compared to the 25% measured on a typical SWR meter.

T/R Control

There are many ways of controlling the T/R function of an amplifier. It was decided to make this one operate from the PTT line but unfortunately direct connection resulted in a hang-up when relays

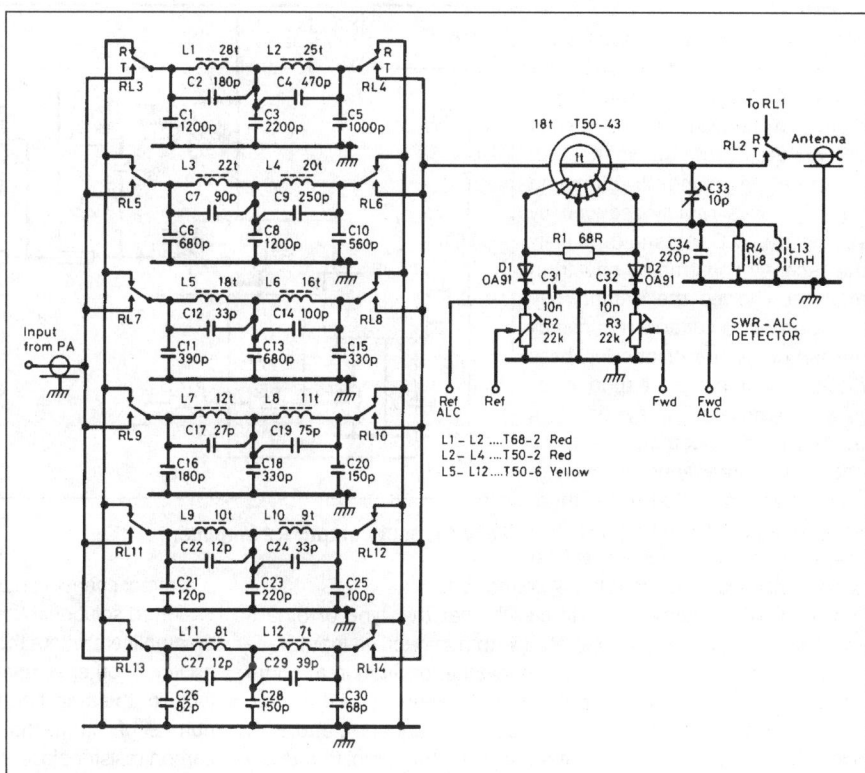

Fig 7.60: The amplifier is followed by six switched filters and a detector for SWR and ALC circuitry

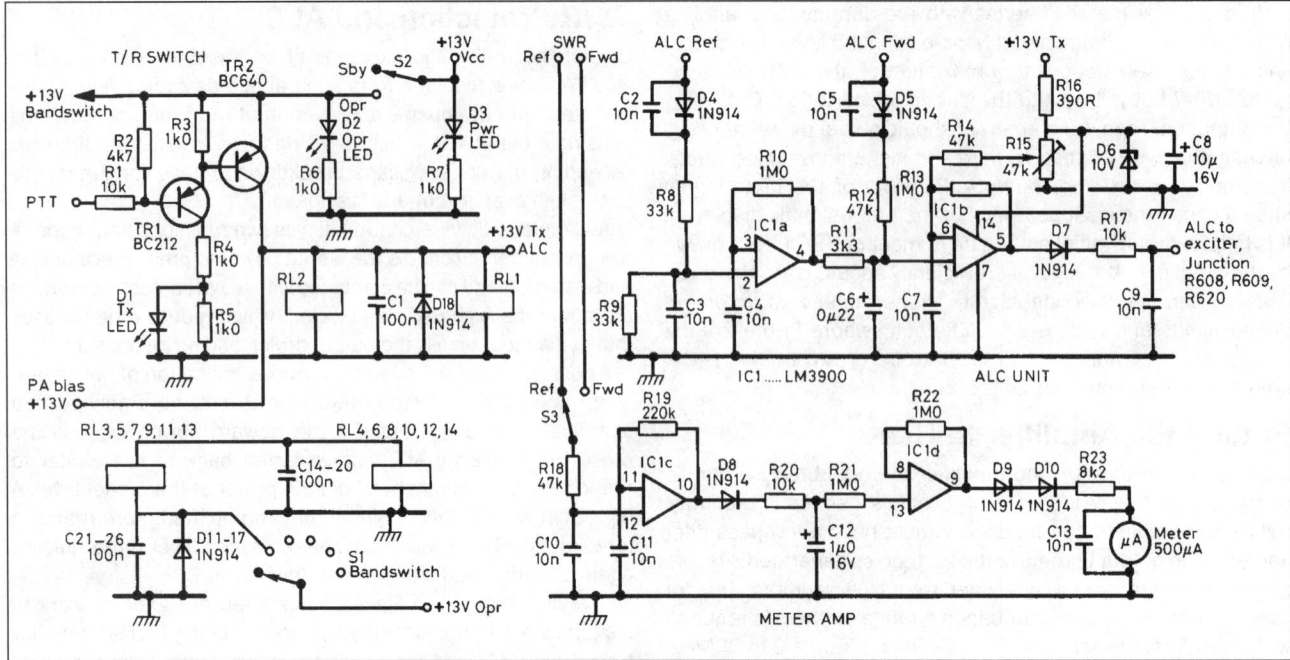

Fig 7.61: Transmit/receive switching, ALC and SWR circuits

in the main transceiver remained activated after the PTT line was released. The buffer circuit comprising TR4 and TR5 simply switch the input and output relays from receive to transmit and provide a PA bias supply on transmit.

Interfacing Amplifier to Exciter

The G3TSO modular transceiver was used as the drive source for the AN762 solid-state amplifier, a mere 5W of drive producing a solid 140W output from the linear. A little more drive and 200W came out. This was rapidly reduced by setting the ALC threshold. Initially the recently constructed spectrum analyser showed the primary signal, harmonics suitably reduced by the action of the elliptic low-pass filter, but alas a response at 26MHz and not many decibels down on the fundamental. A quick check with the general-coverage transceiver revealed that there really was something there, while a finger on the 80m low-pass filter showed quite a lot of heat being generated.

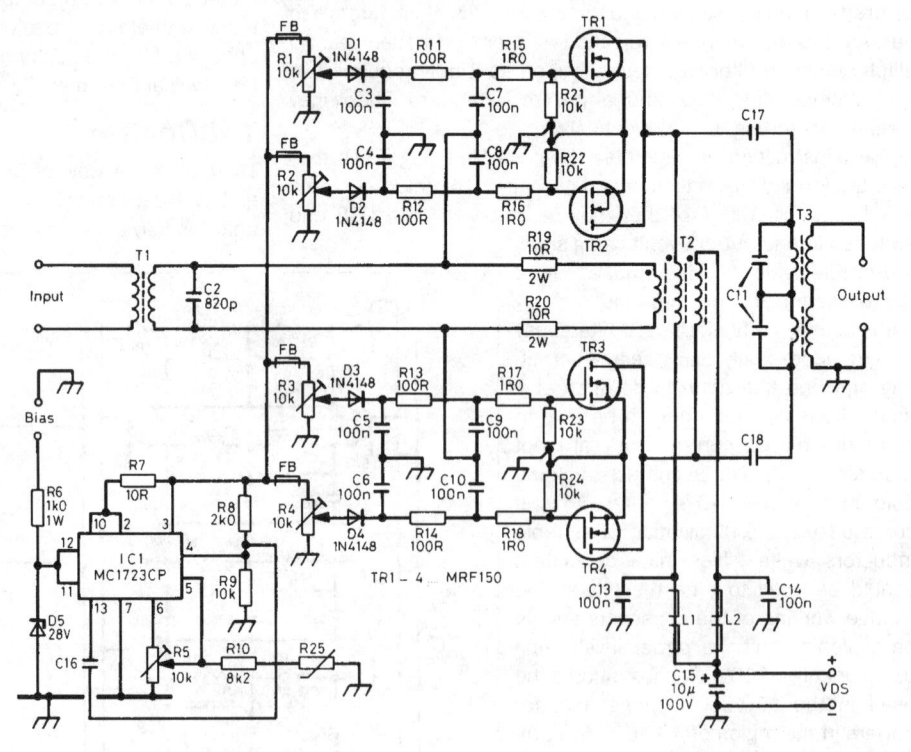

Fig 7.62: 600W output RF amplifier

Investigations revealed quite clearly that this type of broad-band amplifier cannot be operated with a capacitive-input, low-pass filter at either end without it going into oscillation at some frequency, usually well above the cut-off frequency of the filter. The filter input impedance decreases with frequency, and with two such filters located at either end of the amplifier there comes a point where the input circuit and output circuit resonate at the same frequency and a spurious oscillation occurs. Removal of either filter solves the problem.

A direct connection between the exciter and linear amplifier is the preferred solution, but is not always practicable in the case of add-on amplifiers where there is already a low-pass filter installed in the exciter. Another solution is to provide a resistive termination at the input of the amplifier; this is far simpler to effect and is used in a number of commercial designs. The network used comprises a 50Ω carbon resistor placed across the input of the amplifier which effectively reduces the input impedance to 25W, so a 30W resistor is placed in series with the drive source to present a near-50Ω impedance to the exciter. Power from the exciter will be absorbed in these

The Radio Communication Handbook

Fig 7.64: 600W amplifier component layout

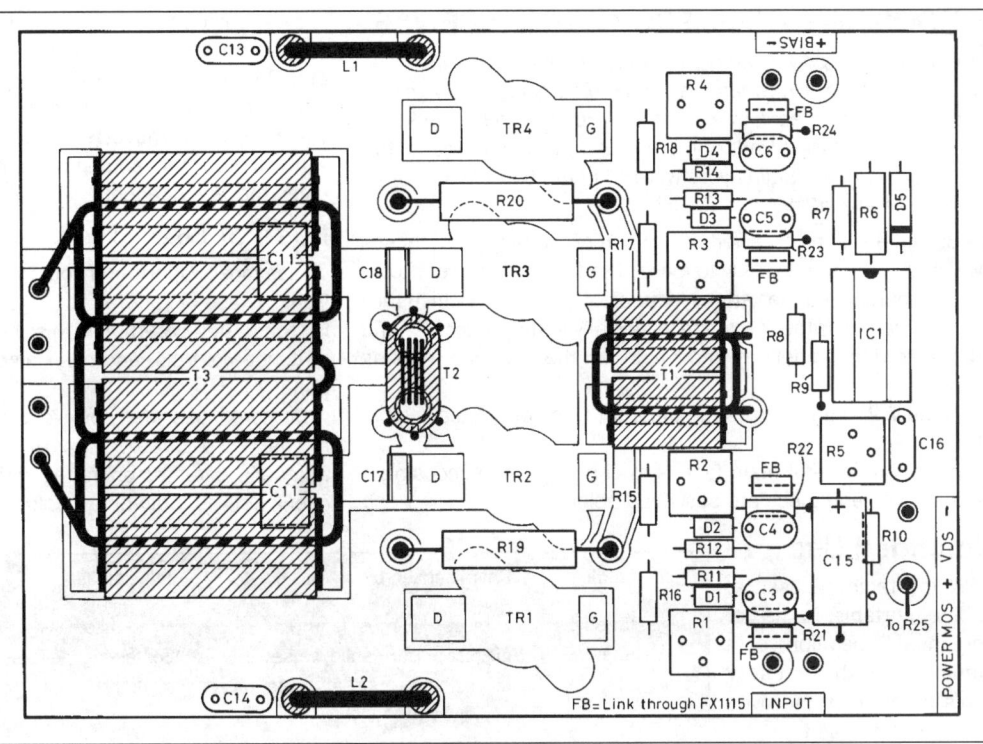

resistors which must be made up from a number of lower-wattage resistors in parallel, ie five 150Ω resistors make a 30Ω resistor with five times the power rating. In addition it was found necessary to add a ferrite bead to the input and output leads to the amplifier to effect a complete cure to the parasitic problem which was at its worst on 21MHz, the parasitics occurring above 40MHz. An alternative solution would be to use an inductive-input low-pass filter (Fig 7.23(c) and Table 7.9).

Conclusion

The construction of a solid-state high-power amplifier is very simple, especially as the parts are obtainable in kit form in the UK and via international mail order from the USA. The use of a spectrum analyser, no matter how simple, greatly eases the setting up of such an amplifier and ensures peace of mind when operating it.

Amplifiers are available for a number of different power levels, and can be combined to provide higher power levels. The use of 13V supplies practically limits powers to about 180W and below, while 28 or 50V facilitates higher-power operation without the need for stringent PSU regulation. The full-power solid-state linear is now possible at a price showing a considerable saving on the cost of a commercially made unit.

There is a great similarity between all the Motorola designs: the 300W amplifier described in AN758 is virtually the same as the AN762 amplifier, the fundamental difference being the 28V DC supply voltage. The 600W MOSFET amplifier (**Figs 7.62, 7.63** in Appendix B, **7.64 and 7.65**, plus **Table 7.25**) described in EB104 represents a real alternative to the valve linear amplifier and operates from a 50V supply, but considerably more attention must be paid to the dissipation of heat.

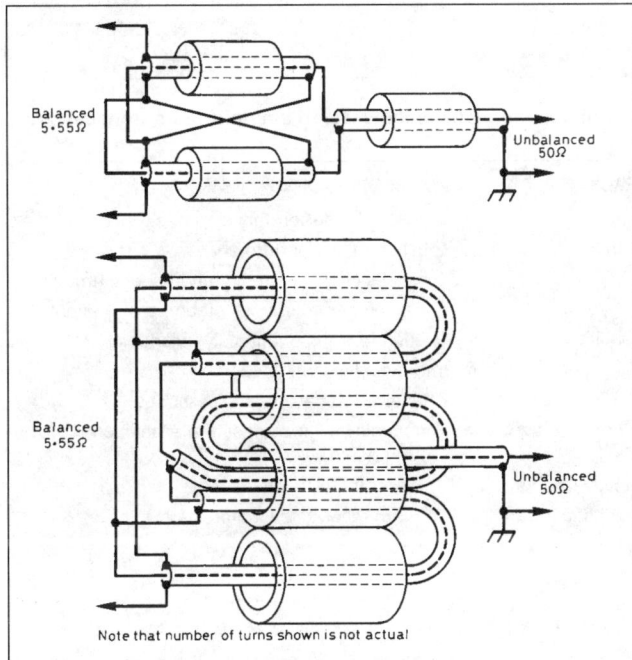

Fig 7.65: Winding the transformers

R1-R5	10k trimpot
R6	1k, 1W
R7	10R
R8	2k
R9, 21-24	10k
R10	8k2
R11-14	100R
R15-18	1R
R19, 20	10R, 2W carbon
R25	Thermistor, 10k (25°C), 2k5 (75°C)
All resistors ½W carbon or metal film unless otherwise noted.	
C1	Not used
C2	820p ceramic chip
C3-6, 13, 14	100n ceramic
C7-10	100n ceramic chip
C11	1200p each, 680p mica in parallel with an Arco 469 variable or three or more small mica capacitors in parallel
C12	Not used
C15	10µ, 100V elec
C16	1000p ceramic
C17, 18	Two 100n, 100V ceramic each (ATC 200/823 or equivalent)
D1-4	1N4148
D5	28V zener, 1N5362 or equivalent
L1, 2	Two Fair-Rite 2673021801 ferrite beads each or equivalent, 4 H
T1	9:1 ratio (3t:1t)
T2	2µH on balun core (1t line)
T3	See Fig 7.64
T1-T3 can be obtained from Mainline Electronics.	
TR1-4	MRF150
IC1	MC1723CP

Table 7.25: Components list for 600W output RF amplifier

CHOOSING A COMMERCIAL TRANSCEIVER

Apart from low power radios, the majority of transceivers used in the shacks of UK amateurs are commercial, usually manufactured in the far east. Many of these are reviewed in the RSGB members' magazine RadCom, by Peter Hart, G3SJX. The following is his guide to finding your way round the bewildering range of choices available.

Buying an HF transceiver can be one of the bigger purchases which the amateur is likely to make. The large number of radios on the market and the various features and functions which they provide can be rather daunting, especially to the newcomer and the following guidelines may help in the decision making process.

Table 7.26 summarises the different types of radio available for HF use.

There are several main factors to consider when choosing a transceiver, and these are described below.

Intended Use

Will the radio be used at home, portable, in the car or on a DXpedition, or a multiple of these uses? This determines the overall size and weight.

Is the main use for general operating or for competitive activities such as contesting and DXing? A higher performance radio is desirable for contesting whereas a lower performance and cheaper radio is entirely satisfactory for more casual operating.

Is the main use on a specific mode such as SSB, CW or Data modes? Some radios have more features suited to different modes, such as full break-in and built-in keyers for CW and extensive PC/audio interfacing and dedicated data mode selection and filters for RTTY, PSK and other specialist modes.

Features and Functions

Most modern radios are very well equipped with all the features you are likely to need for most general purpose operation. See Table 7.27.

Filtering functions and noise reduction capabilities are better implemented on the higher and top end models. The larger base station radios provide more extensive interfacing capabilities to multiple and receive-only antennas, linears, PC and audio lines etc. Radios at home are more likely to be used in combination with a linear, PC control and sound card for data, CW and voice keyers, transverters etc.

If using a small radio at home make sure that it has the interfaces that are needed.

Ease of Use

This is most important if you are to obtain maximum enjoyment from the use of your radio. Large well-spaced control knobs and buttons, dedicated controls rather than menu or context switched controls, and clear displays all make for easy operation.

Compromises are, however, inevitable on the smaller sized radios. Ease of use is particularly important in minimising fatigue in extended operating periods such as in contests. Try to check out the radio on air before buying.

Type of radio	Characterised by	Principle use	Typical models
Base stations mains and 12V	Largest size, most features, highest performance, versatile, highest cost, may need external mains PSU	Home use, DXing, contesting, general use, all modes	FT1000MP IC-756PROIII TS-950D Ten-Tec Orion
Mid size 12V	Mid size, needs external mains PSU, many features, mid/high performance, mid cost	General all round use principally at home.	FT-847 IC-737 TS-2000 K2/100
Small size 12V	Small size, multifunction knobs, mid performance, mid/low cost, external PSU for mains use	General all round use at home, portable, transportable, lightweight DXpeditions	FT-897 IC-725 TS-480
Mobile	Dash mounting or detachable panel, 12V operated, multifunction knobs, mid cost, mid performance	Mobile, transportable	FT-857 IC-706 TS-50S
Battery operated portables	Small size, lightweight, low power, fewer features, internal or external batteries	Hand portable, take anywhere	FT-817 IC-703

Table 7.26: Comparison of the categories of commercially manufactured amateur radio transceivers

ALL RADIOS	HIGHER END RADIOS	TOP CLASS RADIOS
Multimode	Selectable IF filters	Excellent RF performance
Twin VFOs	Variable bandwidth	Excellent audio performance
Memories	Notch filters	Top class channel filtering and filtering armoury
Clarifier	AF Filters	Ability to operate in multi station environment
Switchable front end	Dual watch capability	Uncompromised dual receiver capability
Variable Power	CW keyer	Interfacing to QSK linears
Interfacing	Auto ATU	Extensive interfacing to accessories
Switchable AGC	50MHz and higher coverage	Ergonomic use in contest and expedition environments
Many software features	More comprehensive displays	Upgradeable software
Noise blankers	More user customisation	Use with receive only antennas
Computer interface	DSP filters and noise reduction	
	Improved data capability	
	More features	
	Spectrum display	

Table 7.27: Features provided on HF Transceivers

Table 7.28:
Key receive
performance
parameters

RECEIVE PERFORMANCE	
ON-CHANNEL	Sensitivity
	Distortion / AF quality
	Bandwidth
	AGC
OFF-CHANNEL	Stopband selectivity
	Spurious responses
	Non-linearity / intermodulation
	Phase noise

Table 7.29:
Key transmit
performance
parameters

TRANSMIT PERFORMANCE	
ON-CHANNEL	Power output
	Distortion / AF quality
	Bandwidth
	Keying characteristics
OFF-CHANNEL	Spurious outputs
	Sideband splatter/clicks
	Non-linearity
	Phase noise
	Wideband noise

TRANSCEIVER		Recip. Mix @10kHz	SFDR @5kHz	SFDR @50kHz
Ten-Tec	ORION	95dB	93dB	93dB
Elecraft	K2/100	94dB	91dB	95dB
Ten-Tec	CORSAIR	98dB	90dB	90dB
Ten-Tec	OMNI-VI	96dB	88dB	88dB
Icom	IC-7800	100dB	88dB	111dB
Kenwood	TS-950	102dB	83dB	102dB
Yaesu	FT-1000MP	94dB	82dB	97dB
JRC	JST-245	81dB	80dB	92dB
Icom	IC-737	104dB	80dB	102dB
Yaesu	FT-747	85dB	80dB	97dB
Icom	IC-725	83dB	79dB	95dB
Yaesu	FT-990	93dB	78dB	97dB
Kenwood	TS-930	89dB	77dB	95dB
Icom	IC-746	90dB	77dB	98dB
Icom	IC-707	93dB	77dB	91dB
Kenwood	TS-940	98dB	76dB	94dB
Drake	R8E	89dB	76dB	92dB
Icom	IC-736/8	94dB	76dB	100dB
Kenwood	TS-850	100dB	75dB	98dB
Kenwood	TS-480	89dB	75dB	98dB
Ten-Tec	JUPITER	78dB	75dB	89dB
Icom	IC-751A	105dB	75dB	104dB
Alinco	DX-70TH	78dB	75dB	92dB
Icom	IC-756PROIII	93dB	74dB	105dB
Icom	IC-756	91dB	74dB	89dB
Icom	IC-756PROII	92dB	73dB	98dB
Icom	IC-756PRO	92dB	73dB	92dB
Icom	IC-7400	93dB	73dB	97dB
Icom	IC-703	84dB	73dB	91dB
Yaesu	FT-1000MP mk5	95dB	73dB	95dB
Icom	IC-775DSP	100dB	72dB	93dB
Yaesu	FT-920	97dB	72dB	96dB
Kenwood	TS-50	95dB	69dB	96dB
Yaesu	FT-900	91dB	69dB	97dB
Yaesu	FT-890	91dB	69dB	95dB
Kenwood	TS-2000	90dB	68dB	95dB
Yaesu	FT-817	85dB	68dB	90dB
Yaesu	FT-100	87dB	68dB	90dB
Yaesu	FT-847	90dB	67dB	94dB
Yaesu	FT-897	82dB	65dB	92dB
Yaesu	FT-857	84dB	65dB	93dB
Yaesu	FT-1000	91dB	65dB	96dB
Kenwood	TS-430	82dB	62dB	96dB
Icom	IC-729	100dB	62dB	101dB
Kenwood	TS-870	97dB	61dB	94dB
Yaesu	FT-767	73dB	59dB	95dB
Icom	IC-781	103dB	57dB	103dB
Yaesu	FT-757	85dB	55dB	88dB

Performance

The most important performance parameters for both the receiver and the transmitter are shown in **Tables 7.28 and 7.29**. A high performance receiver is most beneficial in contesting and DXing where wanted signals can be weak, bands crowded and unwanted signals strong. A 100dB dynamic range in SSB bandwidths represents a target for the very best receivers, 95dB is very good, 85-95dB is typical for mid price radios and 80-85dB for budget priced radios.

Third order intermodulation and reciprocal mixing are the principal performance measuring parameters of significance for dynamic range. Achieving a respectable dynamic range at frequency offsets greater than 20kHz from the receive frequency is achieved by most commercial amateur receivers these days, but closer in, particularly inside the roofing filter bandwidth, the performance of most receivers degrades sharply. There are very few radios which achieve even 80dB dynamic range at 5kHz offset, and cost is not necessarily an indicator of best performance at these close spacings. **Table 7.30** summarises the performance of receivers measured for *RadCom* equipment reviews over the last 20 years in respect of their dynamic range due to third order intermodulation at 5kHz and 50kHz offset and reciprocal mixing at 10kHz offset.

Cost

Cost is generally related to performance and features. Second hand purchases can be a good buy as radios do not normally wear out and significant savings can be made. Some useful guidance on buying second hand is contained in the RSGB Publication *The Rig Guide*, edited by Steve White., G3ZVW

Frequency Coverage

All HF transceivers cover the bands 1.8 to 30MHz but many also provide coverage of the VHF and UHF bands. A growing number provide coverage of 50MHz and a lesser number cover 144, 432 and even 1296MHz. If your interests cover VHF and UHF as well as the HF bands then a multiband radio may be of particular interest. Many such radios allow simultaneous reception and transmission (full duplex) between main band groupings eg between HF, 144 and 432MHz with frequency tracking making them particularly suitable for satellite working.

RF Power Output

Most HF transceivers provide a transmit output power of 100W unless they are intended for use on internal batteries in which case the power is significantly less. A few provide a higher power of 200W and even 400W and this extra power can be an advantage. However where a linear amplifier is also used the extra power is unnecessary, and indeed unwanted where there is a danger of overdrive of the linear amplifier.

REFERENCES

[1]   Motorola UK, 69 Fairfax House, Buckingham Road, Aylesbury, HP20 2NF.

[2]   'QRP + QSK - a novel transceiver with full break-in', Peter Asquith, G4ENA, *Radio Communication*, May 1992

[3]   Kit and component supplier JAB Electronic Components, PO Box 5774, Great Barr, Birmingham B44 8PJ. Tel: 0121 682 7045. Fax 0121 681 1329. Web http://www.jab-dog.com/. E-mail jabdog@blueyonder.co.uk

Fig 7.30: Dynamic range receive performance of HF transceivers

[4]   'A QRP transceiver for 1.8MHz', S E Hunt, G3TXQ, *Radio Communication*, Sep 1987

[5]   'Wideband linear amplifier', J A Koehler, VE5FP, *Ham Radio*, Jan 1976

[6]   'The FOXX 2 - an old favourite revisited', George Dobbs, G3RJV and George Burt, GM3OXX, *SPRAT*, Summer 1997

[7]   'Epiphytes for the Third World', George Dobbs, G3RJV, *Radio Communication*, Jul 1997

[8]   'The EP-2 portable 75m SSB transceiver', Derry Spittle, VE7QK, *SPRAT*, Winter 1995/96

[9]   'The GQ-40 (GQ-20) CW transceiver', Sheldon Hands, GW8ELR, *SPRAT*, Summer 1985

[10] Communications Concepts Inc, 508 Millstone Drive, Xenia, Ohio 45385, USA. Tel (513) 426 8600.

[11] 'A modular multiband transceiver', Mike Grierson, *Radio Communication*, Oct/Nov 1988

PIC-A-STAR
a Software Transmitter And Receiver

Unusually, this chapter is devoted to a single project. It is, however, a project that can be dipped into by those wishing to learn about modern design and construction techniques. It is by no means necessary to build the entire project as many of the techniques and modules can be used elsewhere. Pic-a-STAR may simply be read as a tutorial for anyone embarking on electronics construction in the 21st century.

PIC-A-STAR was originally published as a 20-part series in RadCom [1] and was written by Peter Rhodes, BSc, G3XJP [2]. It is reproduced here almost in its entirety.

This is a detailed construction project aimed at those of modest experience who would like to enhance both their craft and technology skills.

At the outset -like me - it may well be that you don't have the skills or knowledge to build this project. By the end, you will have. That is, as I see it, the whole idea.

By design, this is a project without end. From my perspective, it is the basis for years of happy building to come -and is my first

investment in a new core transceiver platform in some 25 years. A glance at **Fig 8.1** tells you why I needed a new one.

From your perspective, it is a source of ideas for improving an existing transceiver - not least, upgrading the back-end with a powerful Digital Signal Processing (DSP) capability. There are also some craft techniques for handling small-size high-function components. So, there is something in this for all, with an eye on the self-education requirement of their licence.

SUMMARY

The heart of PIC-A-STAR is the DSP module. This provides both the back-end receiver functionality, as well as SSB/CW generation on transmit. The bottom line is absolutely superb audio quality on both transmit and receive. If you want to test the former, come on the home-brew net frequency (3.727MHz) any day around lunchtime where you will find at least one STAR in operation most days. If you want to test the latter then you will just have to make one.

Being implemented by software, it provides the opportunity to address both absolute performance as well as the delights of

Fig 8.1: Early integration testing. Bottom left is the author's Third Method transceiver (borrowed front-end and PA), top right is Pic 'N' Mix DDS still on its original breadboard (injection and controls) - and in the middle is the new DSP module. Note that Pic 'N' Mix provides all the transceiver controls, leading to a clean and compact front panel

- SSB and CW detection and generation
- a bank of high-performance Rx filters
- impulse noise blanking
- non-coherent noise reduction
- auto-notch heterodyne removal
- variable AGC decay time
- synthetic stereo reception
- adjustable RF clipping on transmit
- very fast VOX and QSK operation
- the flexibility to change!

Table 8.1: A brief outline of the features of PIC-A-STAR

operational convenience - at zero incremental cost. This is precisely the basis for future developments, but the fundamental functionality together with some bells and even the odd whistle has been in daily use here for about nine months at the time of writing. This is the project on offer - but by the time you get there it will have moved on.

PIC-A-STAR is explicitly designed to be upgraded over the web, so there will no incremental DSP enhancement costs.

An outline of the main features of the project is shown in **Table 1**.

POSITIONING DSP

You might reasonably expect the author of a DSP project to have some serious knowledge in the field. So would I! Actually, in many ways, it is important to get this published before I acquire more than enough to be merely dangerous.

If, like me, you are at least in your late 50s, it is unlikely that DSP theory featured even in a formal engineering education. And if, equally like me, you have never worked in the engineering profession then you could reasonably start from the position that DSP is some kind of black magic which you could never understand in a life-time of trying. You might well be correct in this assumption because some of the theory is indeed very heavy.

But my personal discovery was that you don't need to understand DSP at other than a superficial level to be able to build it at home and to use it.

From a position of not being able to spell DSP, it took me two weeks to get my first DSP receiver working. The attraction is that everything since then has been incremental and I have not been off-air for a single day.

Design mistakes - and there have been many - have cost me my time but never any money - which is about perfect for a hobby. So this lends itself to a learn-as-you-go approach. In other words, unlike conversational French, you don't have to learn a lot before you can even get started.

SKILLS AND FACILITIES

A requirement of all my projects is that they can be built on the kitchen table with no access to professional facilities. Otherwise, it would not be amateur radio.

This one is no exception - though I have had to acquire new skills and hone them to the point of repeatability in order to build some of the hardware. This is all part of the adventure, part of the fun.

A simple (and in-expensive) technique for making precision PCBs will be covered - which includes mounting a 48-pin chip with a mere 0.5mm interval between pins. And you get to practice on a really easy one of 128 pins by 0.8mm first. If the prospect of this puts you off, I really can't help. If it sparks a 'can-do' spirit of adventure then we are in business.

INSPIRATION

THREE THINGS made this project possible. In the order in which I found them:-.

- *The Scientist and Engineer's Guide to Digital Signal Processing*, by Steven W Smith. This book is a little gem. If you flick through quickly, you will see copious examples and illustrations. What you do not see are lots of equations and impenetrable notation. I need just one quote:-"[this book] . . . is written for those who want to use DSP as a tool, not a new career." My kind of book!

- The Analog Devices website [3]. This contains a wealth of both theoretical and practical information - and specifically the electronic version of the above book. Most valuable to me were lots of DSP code examples for the ADSP-218x processors. The first incarnation of STAR was built by six of us on the ADSP-2181EZLITE evaluation board which had become somewhat of a standard over the years. Then over one fateful weekend when this project was 'finished', its price went from $90 to $275 - which spurred the chal-

lenge to home-brew a compatible and reproducible DSP board.

- DSP-10, a 2m DSP transceiver project published by *QST* in September - November 1999 - and reviewed in *RadCom*, Feb 2000. Although featured for VHF/UHF applications, the DSP core is totally universal. This project was designed by Bob Larkin, W7PUA [4], and I am indebted to Bob not only for the inspiration for this project, but for a significant amount of advice and help - including some code written specifically for PIC-A-STAR. Above all, Bob showed it can be done and whenever I get into problems, his material is the first place I look for clarification and understanding.

INTEGRATING PIC-A-STAR

The DSP module - designed to combine with the Tx/Rx RF stages of your choice - operates at a final IF of 15kHz as shown in **Fig 8.2**. This is a high enough frequency to make it immune from image responses, yet low enough to be affordable. And it is not a DSP audio add-on - which, coming after the product detector, will always struggle.

RF STAGES

Your HF IF can be derived from any reasonable transceiver front-end. My third method transceiver [5] and G3TSO's modular transceiver [6] have both been tested as representative - and there are lots of them out there. CDG2000 looks like a powerful approach

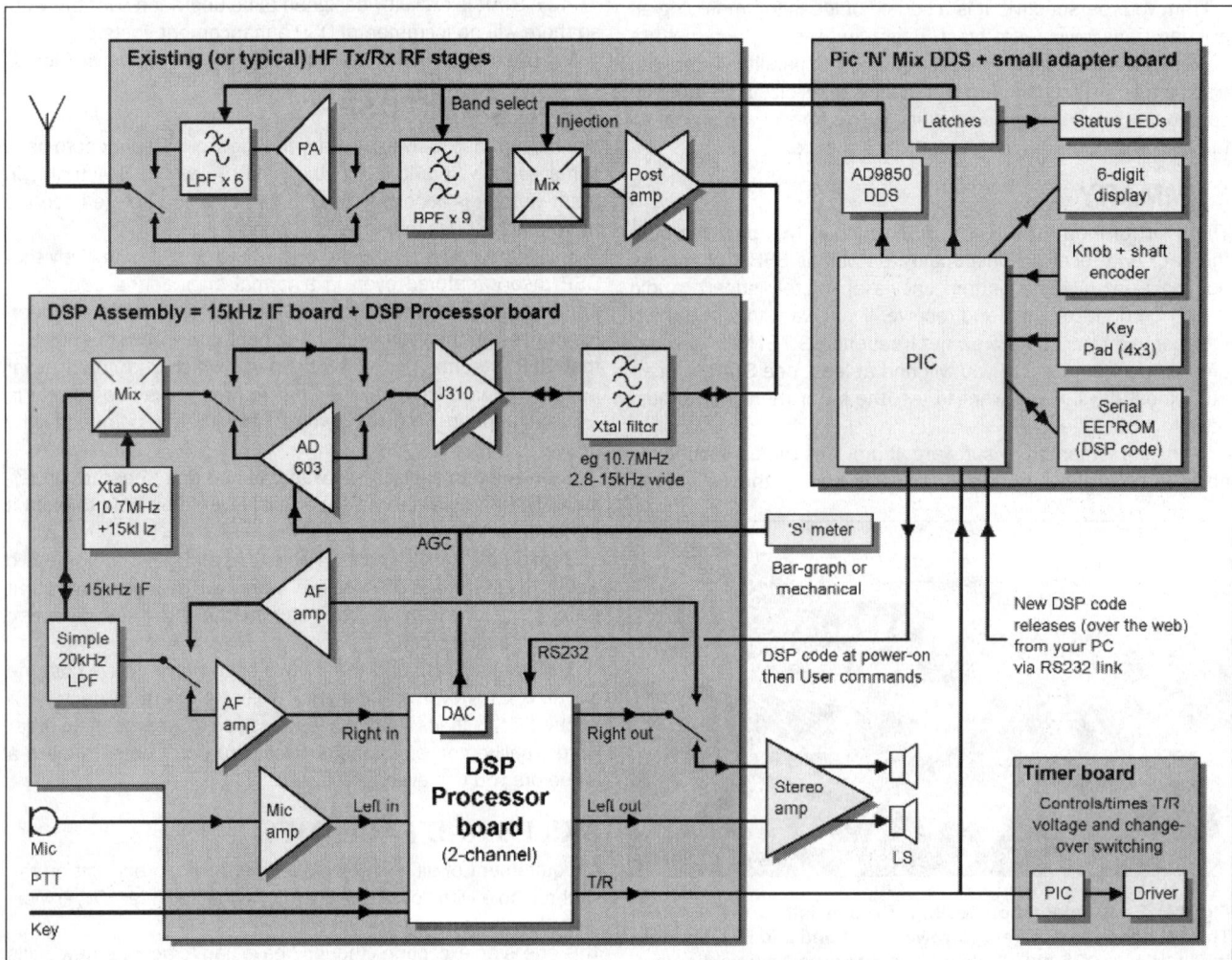

Fig 8.2: A typical transceiver incorporating PIC-A-STAR at a final IF of 15kHz. See text for a discussion of the major hardware elements

and its front-end could well be my next increment. The choice will substantially determine the overall receiver strong-signal handling capability - but not the intrinsic benefits of downstream DSP.

IF STAGES

In principle (and possibly in practice), you could modify an existing IF board so that its product detector produced a 15kHz output instead of straight audio but I don't recommend it in the long run. To cut a long story short, this design includes an IF board built for the job.

Details of this follow later, but it needs only a modest roofing filter (at any HF IF of your choosing) since all the serious filtering is implemented in DSP.

CONVERSION INJECTION

You won't be surprised to see Pic 'N' Mix [7] used as the injection source to mix from RF to your chosen HF IF. This is not mandatory, but my records show 281 of them out there, so it is a non-trivial population.

COMMAND AND CONTROL

You need the ability to command the DSP for all the functions normally associated with front-panel controls. You may be somewhat surprised to see Pic 'N' Mix used for this purpose as well.

A small adapter board is used to fit a more versatile and powerful PIC - which not only controls all the original DDS capability, but extends the existing keypad, display and tuning knob to control all the transceiver features.

Although highly recommended, use of Pic 'N'Mix is not mandatory. As an alternative, you can use your PC to load and control the transceiver - and a BASIC utility is provided to achieve this.

PIC-A-STAR DSP

Illustrated in **Fig 8.3** is the functionality implemented in software. You will have seen not dissimilar block diagrams implemented in analogue hardware - but not at this price and not inside a 28mm² chip!

Actually, there is an intrinsic overhead, namely that the analogue signals need converting to digital form before processing and back to analogue after. This is the purpose of the CODEC (encoder/decoder) referenced on several inputs and outputs - and is implemented on a separate chip.

The greatest appeal of the software approach, not least to the amateur, is the flexibility to change the line-up at a touch of the keyboard, so to speak. This allows easy experimentation (or overt tinkering, if you prefer), since at any time you can abandon the change and go back to the previous version. There are other subtleties.

For example, you will find five 15kHz oscillators scattered around the diagram. In fact their frequencies change depending on mode ie USB/LSB/CW. In DSP software terms the sinusoidal oscillator is simply a subroutine. To invoke it, all you need do is tell it what frequency/phase you want on any given occasion - and it is done.

Another example is the 'delay' in the receiver front-end image-cancelling I/Q mixer. In one path, there is a 90° phase shift, in

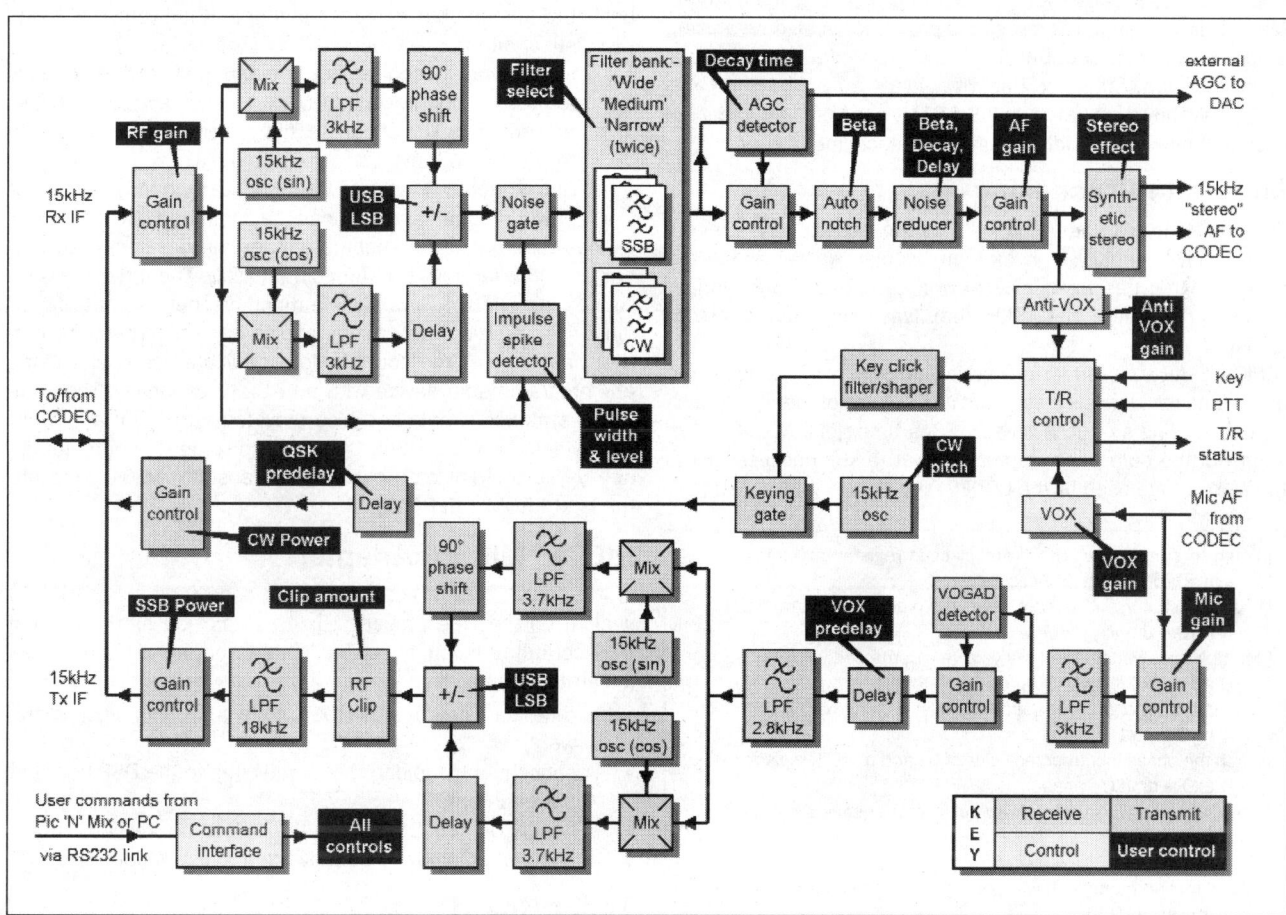

Fig 8.3: Software block diagram of PIC-A-STAR DSP functionality. Not shown are simple on / off switches associated with VOX, noise blanker, auto-notch, noise reducer and the RF clipper. The filter bank also has an off (ie bypass) switch to give a net maximum bandwidth of some 3kHz

the other a delay. The latter arises because it takes real elapsed time to produce the phase shift, so an equal amount of time has to be 'wasted' in the other channel to maintain that phase relationship.

Time is of the Essence

The basic understanding you need in order to grasp how DSP works is to note that time is the critical commodity. Every functional box in **Fig 8.3** takes time to execute. So does every individual instruction that goes to make up that functionality.

This would be of little concern were it not for our old friend Nyquist. He stated that, in order to process a signal faithfully, you must sample it at (at least) twice the rate of the highest frequency present.

For example, the incoming receive signal is around 15kHz, and so needs to be sampled at 30kHz or more. In fact, 48kHz is used to provide a useful margin.

The consequence of this is that, having grabbed one sample, you have no more than 20.83μs (by simple arithmetic) to do all the processing required before you have to get back to handle the next one.

So just how much processing can be achieved in 20 millionths of a second? The ADSP-2181 processor in this design executes an instruction in 30 nanoseconds. The simplistic answer is therefore 666 instructions-worth. But this is far from the whole story. During one processor cycle it can, for example, fetch two 16-bit numbers, multiply them to give a 32-bit product and add the result to a 40-bit accumulator. This MAC (Multiply & Accumulate) instruction is the essence of filter implementation and is critical because you need to loop around it many times. Meanwhile, in the background, the processor is also organising data samples in and out of the CODEC as well as handling any serial communications port activity.

Fig 8.4 shows a snatch of PIC-A-STAR code, so you can visualise just how much radio you get from each line of code.

Multi-rate Processing

There is a more structural solution to the issue of buying some time - which, equally, derives from Nyquist. Namely, once the receive signal has been mixed down to audio, you no longer need to process it at the 48kHz rate. Twice the audio frequency is fast enough.

PIC-A-STAR runs audio processing at 8kHz - by grouping the audio functions into six blocks and running one of them - but each in turn - during six successive 20.83μs time-slots. At the end of each slot the data is again processed at 48kHz because that is the sample rate used by the CODEC for outbound signals also.

```
{ Fetch Rx sample via CODEC and place in register mx0 ...}
        mx0=dm(Rx_in_buffer);
{ ... and fetch current RF gain value and place in register my0. }
        my0 = dm(RF_gain);
{ Multiply the two together to give a gain-controlled value ...}
        mr = mx0 * my0 (SU);
{ ... and keep the gain-controlled signal in register my0. }
        my0 = mr1;
{ Fetch the phase incremented value of LO and place in register ax0 ...}
        ax0 = dm(LO_phase);
{ Pass the phase value to sin to get instantaneous sinusoid amplitude ...}
        call sin;
{ ... and mix (ie multiply) it with the signal in register my0 }
        mr=ar*my0(SS);
```

Fig 8.4: Some early lines of code for the receiver. Yes, the last line truly is a mixer (otherwise known as a product detector)

So you can see that, all the way along the line, Nyquist is satisfied -and so am I because there is plenty of time for some exotic as well as the more mundane processing.

SOFTWARE PACKAGING

The desire to provide choice and flexibility, but above all upgradability (if that is an English word), leads to some complication in describing the various modules. The context will become clearer once the hardware functionality has been covered. Suffice it to say at this stage that, from an operator's perspective, the system is totally transparent, ie you just switch it on, wait about 20 seconds (as if for the valves to warm up) and then use it. The software comes in the following modules:

DSP Boot Utility

This code resides in PROM on the DSP board. At power-on time, besides running some basic hardware checks, it manages the on-board serial port to load the target DSP code. This utility was written by Bob Larkin, W7PUA, for PIC-A-STAR based on the original AD code.

DSP Transmit/Receive Code

This runs on the DSP board and provides the core functionality as in Fig 8.3. It needs to be loaded at power-on time, a process which takes some 20 seconds. Subsequent to loading it, you also need to be able to command it.

DSP Loader

This is a QBASIC utility which runs on your PC. It is written in very basic BASIC to enable you to adapt it or port it if you wish. It has two distinct alternative functions:
- to load and subsequently command the DSP code directly to the DSP board, via a COM port and a 9.6kB serial link.
- to load a new (or, of course, first) release of the DSP code to the PicAdapter board (see next) in Pic 'N' Mix. Subsequently, Pic 'N' Mix automatically loads the code at power-on time -and provides the command user interface.

These alternatives are not mutually exclusive. For early testing and use, the former gets you going quickly. The latter frees up your PC and, in my view, gives a much cleaner user interface - albeit with a little practice. The choice is yours. (There is a further option here. You could build a dedicated controller using any programmable device with an RS232 capability. The command syntax is simple and also provided - and is in any event self-evident from the QBASIC code. With some loss of maintainability, you could also burn the entire transmit/receive code into the boot PROM / EPROM.)

PIC 'N' MIX PicAdapter

Written in MicroChip Assembler, this code runs on a 16F870 (which replaces the present 16x84) to provide all the original DDS control functionality of Pic 'N' Mix and, in addition, it now integrates the ability to:
- download new release DSP code from your PC (via the web);
- subsequently upload that same code to the DSP board at power-on time;
- command the DSP using the self-same keypad, tuning knob and display as already fitted to Pic 'N' Mix.

Timer Board

Also in MicroChip Assembler, this code runs on a 16F627. It provides the sequencing and timing of receive/transmit transitions - both ways - to make them as clean and fast as possible. This

board is designed to be general and will find uses on other transceiver projects.

Bargraph S-Meter

This is both optional for PIC-A-STAR and equally of general application. Also running on a 16F627, it controls a 12-LED bargraph on the Status board. It was built at all because 10 LEDs, as provided by most control chips, are not enough - and in any event, the PIC provides a lower-cost solution.

To Summarise

The programmed chips are a PROM, a 16F870 and two 16F627s. These provide the base infrastructure. The target DSP code - where most of the future enhancements will occur - is loaded from your PC using the QBASIC utility. No further hardware (eg a programmer) is needed.

SOFTWARE DISTRIBUTION

All the software itemised above will be available for your personal use at no charge. However, this does not mean it comes entirely free. The 'price' is that you need to send me an e-mail note requesting the software - giving an explicit undertaking that it is for your personal use and for the purpose of self-education.

Not least, this allows me to maintain a list of 'customers' to advise when updates become available; as mentioned previously, this is, by intent, a project without end. By software, I mean at this stage the loadable object code.

If, however, you want me to use my resources to program chips for you then I can supply them ready-programmed at a few pounds per chip - plus return postage (see [2] for author's address). It is worth pointing out in this context that you could build a programmer yourself for about £10.

Resistors 1/8-1/4W, 5-10%

R1	47R
R2, R48, R49, R76, R77	100R
R3	220R
R4	470R
R5, R6	560R
R7, R50, R64-R71	1k
R10, R43, R52, R72	2k2
R11	2k7
R12, R34, R37	22k
R13-R15	3k3
R16-R19	4k7
R20	5k6
R21-R32, R45-R47, R53-R55, R63, R73, R78	10k
R33, R51, R59-R61	47k
R35, R36	100k
R38	120k
R56-R58	10R
R62	56k
R74	1k x 4 SIL network (5 pin)
R75	1k x 8 SIL network (9 pin)
RV1, RV2, RV6	100k horizontal preset
RV3, RV4	10k horizontal preset
RV5	47k horizontal preset

The following resistors are all SMD 1206 size, 5%

R8, R9, R93	1k
R39, R40	22k
R41, R42, R97	220k
R44	3k3
R79-R89	10k
R90	330
R91, R92	2k2
R94-R96, R100-R103	4k7
R98, R99	100k
R104,R105	47k

Capacitors, 16V rating

C1-C191n solder-in feedthrough (C11 and C12 may be replaced with a stereo jack socket)	
C21	10p ceramic plate
C20, C22, C23, C26, C69	220µ axial
C24	0µ radial
C25, C75, C76, C84	33µ radial
C27	47µ axial
C28	2µ2 radial
C29	4µ7 radial
C30	33µ radial
C31-C34	220n ceramic
C35, C36	1n ceramic
C37-C44, C59, C70, C71	10n ceramic
C45	22n ceramic

C46-C58, C60-C65, C67, C68 C74, C77-C83, C85-C89	100n ceramic
C72, C73	15p ceramic plate
C90, C91	22p ceramic plate
C92, C93	270p NPO ceramic
C94	47n ceramic
C95, C96	220n ceramic
C97	220p ceramic plate
C120, C121	18p ceramic plate
C122, C131	4µ7 radial
C123-C125, C128, C132	10µ radial
C126, C127, C129, C130, C133	1µ radial
C66, C98-C119	100n SMD 1206 size
VC1-3	100p polyethylene trimmer (1)

Inductors

RFC1-RFC3, RFC6, RFC7	1mH axial choke
RFC4, 5, RFC8-RFC10	100µH axial choke
RFC12	330µH axial choke
L1	4.85µH 38t 30SWG on T50-2 (2), (3)
T1	6t:3t 30SWG on FT37-43 (2)
T2	10t:3t 30SWG on FT37-43
T3	20t:4t 24SWG on T50-2 (3)
FB	1 turn through small ferrite bead

Semiconductors

IC1, IC13, IC15, IC17-IC19	78L05 regulator, 5V
IC2	AD603AQ (DIL)
IC3, IC4	FST3125M (SMD)
IC5	SBL-1 (7)
IC6, IC7, IC22	TL072 (DIL)
IC8	TLC7524CD (SMD)
IC9	PIC 16F870-ISP (DIL)
IC10	24LC256-IP (DIL)
IC11, IC28	PIC 16F627-04P(DIL)
IC12	ULN2803A (DIL)
IC14	7810 regulator, 10V
IC16	4094 (DIL)
IC20	74LVX125 (SMD)
IC21	ST232N (DIL)
IC23	74HC14 (SMD)
IC24	LE33CZ regulator, 3V3
IC25	AD1885JST
IC26	27C512 512KB PROM, programmed
IC27	ADSP-2181 KS130 or KS160
TR1, TR2	J310
TR3-TR5, TR12	2N3904
TR6	2N3906
TR7, TR8	BC517
TR9-TR11	VP0300LS P-ch MOSFET
D1-D6, D11, D15	1N4148
D7-D10	BA244 (8)
D12-D14, D34	1N4007 or similar

D16-D20, D35	3mm LED, red
D21	3mm LED, red/green tricolour
D22-D33	1.8mm LED, colours for S-meter
ZD1	4V7 200mW Zener diode

Miscellaneous

FL1	Crystal roofing filter (4)
X1	wire-ended BFO crystal (5)
X2	4MHz wire-ended crystal
X3	16.67MHz, probably custom
X4	24.576MHz low profile, Rapid Electronics
S-meter	1mA, optional
S1	8-pole on/off PCB DIL switch
S2, S3	skeleton push-to-make switch
All screened RF leads	RG174
Other screened leads	thin microphone lead
Spacers and 3mm nuts/bolts	4 sets (6)
PL1	2 strips of 9-way SIL header ie male/male
SK2-SK8	SIL socket strip
PL2-PL8	mating SIL plug strip
Turned pin socket IC6, 7, 10, 22	8 pin
Turned pin socket IC16, 21	16 pin
Turned pin socket IC11, 12, 28	18 pin
Turned pin socket used as spacer	18 pin
Turned pin socket IC9	28 pin by 0.3in
Turned pin socket IC26	8 pin by 0.6in

NOTES

(1) These three trimmers may all be replaced with fixed capacitors after adjustment.

(2) Assumes a 2.2k-ohm impedance.

(3) Assumes FL1 and X1 around 10.7MHz.

(4) Any HF centre frequency. Filter width between 2.7kHz and 15kHz. Around 3 - 6kHz is ideal.

(5) Centre frequency of FL1 plus 15kHz.

(6) Spacing specified to give clearance between FL1 and the DSP board.

(7) The SBL-1 mixer may be replaced by a stronger mixer and / or one using discrete components at your discretion. It is specified here as a well-established datum and works well.

(8) These diodes can be replaced with 1N4148 if you can't obtain RF switching diodes.

SUPPLIERS

Most of the components were procured from Farnell or Rapid Electronics. The exceptions are the toroids which I bought from Mainline - and the crystal filter, the specification of which is loose enough that it should be possible to find a surplus one. The two critical chips, namely IC25 and IC27, are manufactured by Analog Devices.

Table 8.2: Components list for all options, ie Timer board, DSP boards and assembly, IF board, PicAdapter board, Status board

PCB MANUFACTURE

IT HAS BECOME a tradition with my projects that each has been produced by a different one-off PCB production technique - in an attempt to dispel any unwarranted mystique and even some phobia. Frankly, I want to advance my own craft skills with every project. I used iron-on laser film to speed-up the development cycle and as the only realistic approach to making the DSP board - and found the results to be excellent.

Process Overview

In outline, the process is to photocopy the published artwork onto the film; then transfer the toner from the film to the board using heat and pressure from a clothes iron. This toner then acts as a superb resist while etching the copper in the normal way.

This is not entirely a precision engineering process and there are some experiential skills. Cleanliness is everything! Although incredibly fine lines and small spacings reproduce well in my experience, ironically there is sometimes difficulty with large black areas. But equally, these are the easiest to touch up with an indelible pen before etching - and, at times, you absolutely will need to.

The finished and populated board is illustrated in **Fig 8.5**.

Resources

You need access to a black-and-white laser photocopier. Almost all modern machines use this technology. It may well be that the copier in the corner shop will not be up to the job and you should probably consider the small cost of taking it to a professional copy shop as money well spent.

Some sheets of laser film (sold nowadays by several suppliers for this purpose) are required; a block of flat scrap wood larger than the PCB; and a domestic clothes iron constitute the tools.

The latter should preferably not be a steam-iron. If it is, ensure it is fully-drained, since water/steam and this process do not mix. Also, the steam holes in the sole-plate are unhelpful. Avoid if possible the more modern easy-iron technology which has fine ridges on the sole-plate. Check that the sole-plate is flat. Some have a slight curvature. If faced with these problems, it is best to use several (say, three) intervening layers of clean paper to provide a more evenly distributed heat source. Experiment!

The PCB material (all double-sided) can start out badly discoloured, but must not be mechanically damaged, ie no scratches. 1oz (or more) copper is better, but I used merely 0.5oz on GRP for all my boards.

The Process

1. Firstly, test-copy the artwork onto plain paper in order to check for copier quality and acceptable scaling error. Use the maximum contrast consistent with retaining a clean white background.

2. Copy the artwork onto the film. When viewed with the toner (matt) side down, you want to end up looking at the tracking with the correct orientation ie as if viewing the finished board. For most published artwork this requires the extra step of firstly copying it to a transparency, flipping it over and then copying that to the iron on film. In order to avoid this extra step - with some inevitable degradation - the PCB artwork in this project will be printed pre-flipped so to speak -and therefore should be copied directly to the film. The film itself is not "sided'.

3. Cut the PCB to size (or, preferably, somewhat over-size for now).

Remove all burrs and sharp edges. With cold water, wet a soap-impregnated wire-wool pad and use it to polish the copper - with increasingly light strokes - until immaculate; do not touch

Fig 8.5: The completed CODEC board, illustrating the quality of PCB production available (0.2mm wide tracks at 0.5mm intervals) - using domestic kitchen resources. The CODEC chip is 7mm square

the surface thereafter. Polish both sides and then wash off all traces of soap residue with a clean paint-brush and cold water - and dry with kitchen paper.

4. Place the PCB on the scrap wood and clean it with some kitchen paper (uncoloured) moistened in acetone, isopropyl alcohol or cellulose thinners.

5. Heat the iron to about 140°C (cotton setting) and leave it for a few minutes to attain an even temperature across the sole-plate. At this temperature it should just scorch plain 80gsm copier paper.

6. Cut out the artwork to no larger than the PCB and register it toner side to the board.

7. With at least one sheet of clean paper interposed, lower the iron vertically onto the middle of the board and let it rest there for some five seconds. This will establish the registration of the artwork to the copper.

8. If the board is bigger than the iron, lift the iron off and relocate it every few seconds. Under no circumstances use an 'ironing' motion. Simply raise and lower it vertically - and frequently - until all the board has seen the iron and some applied pressure for about 20 seconds. For pressure, the weight of the iron plus about as much again is near enough and is not critical. Too little pressure and the toner will not transfer. Too much and the toner will migrate to widen the lines (ie smear) and reduce its depth. The former is correctable, the latter is absolutely not.

9. Inspect the result. You may see any areas which have not transferred as still retaining a somewhat glossy appearance. Repeat selectively as necessary. Pay particular attention to the edges.

10. Allow the board to cool naturally back to room temperature.

11. Carefully peel back the film from each corner and note that the toner has transferred. If any critical areas have not taken, the artwork will still be registered and you can selectively repeat.

12. Touch up any blemishes or areas of visibly thin toner with an indelible pen.

13. Now spray-mask the opposite side of the board and etch as normal.

14. Before removing the etch resist, centre-pop and/or drill the holes. They are easier to see at this stage. Clean off the etch resist with cellulose thinners and gently re-polish the board.

The Radio Communication Handbook

Fig 8.6: Timer board circuit diagram. This provides timed transitions between transmit and receive - in both directions. S1 allows you to set up the timing for your installation - and covers the range from slow relay-based RF T / R switching through to solid-state QSK

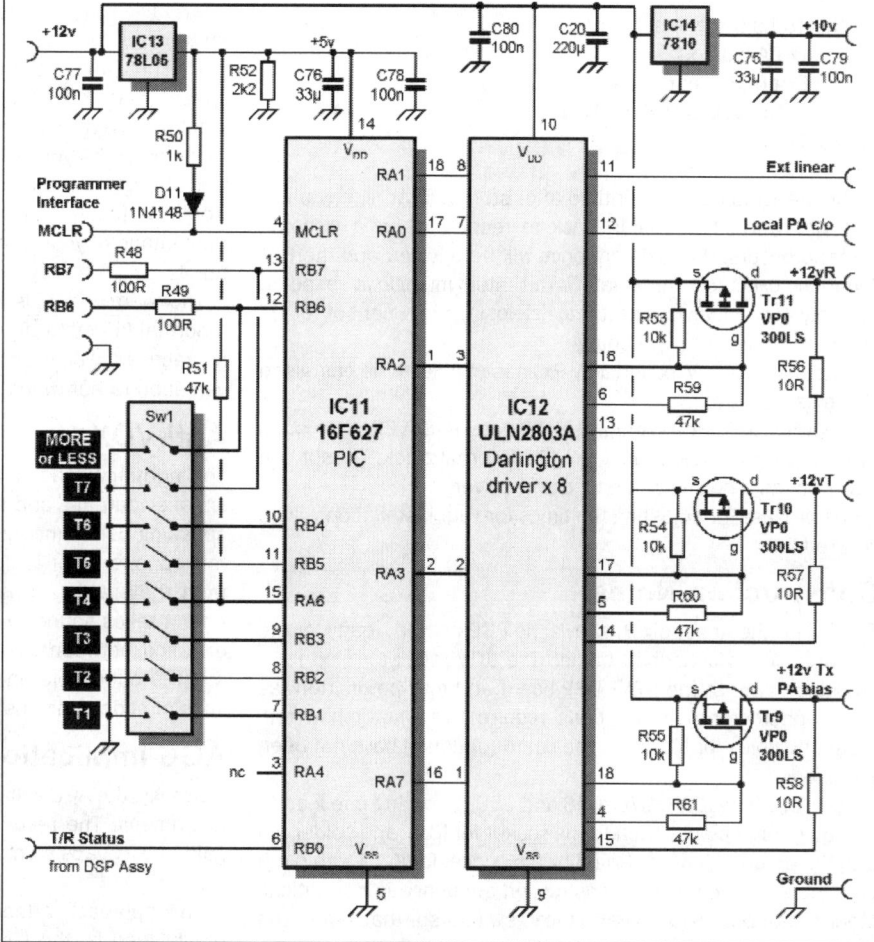

Fig 8.7: The Timer board, which will fit in a small corner in most transceivers and get rid of those DC switching relays

15. At this stage I lightly spray the board with SK10, which is both a protective lacquer and a flux. Available from Rapid Electronics, this makes for clean soldering and prevents contamination of the copper.

SUPERVOX

I have always operated VOX and, indeed, for many years did not bother to fit any PTT capability at all. Perhaps it would help the psychology if that switch on the microphone were known as "Lift To Listen'. The system has two elements, namely a timed solid-state switch for the various DC T/R lines and any relays, and an intelligent VOX system, implemented within DSP. But note that the Timer Board has been designed as a flexible and stand-alone general solution to manage the T/R switching in any transceiver.

Prerequisites

To be effective, any VOX system needs both T/R transitions to be free from clicks and thumps - electrical and mechanical. The first step to achieving this is to leave the maximum amount of circuitry powered up on both transmit and receive. Certainly all DC switching should be solid-state (hence the Timer Board), but the RF changeover can be more of an issue, especially at higher power levels. I use the circuit published in the 1988 *ARRL Handbook*, but hope to do better before the end of this series. Even if you use relays there are still significant benefits, though personally I just hate those acoustic rattling noises.

Timer Board

The T/R timer board circuit is shown in **Fig 8.6** and manages both the R to T and the T to R transitions. A completed board is in **Fig 8.7**.

The benefit of this approach is that the two transition sequences and timing can be - as indeed they should be - different. This cannot be achieved with the typical window comparator approach.

This board has one significant input, namely T/R Status from the DSP Assembly. This is a +5V logic signal (or floating) when on receive - and grounded to 0V to switch to transmit.

The PIC is a 16F627 which has the benefit of not needing an external crystal if timing accuracy requirements are modest; as a result the two crystal pins can be used for digital I/O purposes.

There are five timed outputs - which have been arbitrarily named for their most obvious general use. The 'External Linear' and 'Local PA c/o' lines are grounded on transmit, can sink 500mA -and are thus suitable for relay or solid-state switch control. The other three lines are at +12V when active and are explicitly grounded otherwise. They can each source/sink 500mA. They behave as follows.

Receive to Transmit (R/T)

The sequence for this transition follows, each step being followed by a timed delay:

External linear to Tx	T1
Local PA c/o to Tx	T2
12V Rx off	T3
12V Tx on	T4
Tx PA bias on	

There then follows a re-triggerable hang time, T5. This whole transition is not interruptible, see below.

Transmit to Receive (T/R)

For this transition the sequence is:

Tx PA bias off
External linear to Rx
Local PA c/o to Rx   T6 12V
Tx off   T7
12V Rx on

This transition is interruptible after step 1. That is, if you are part way through dropping back to receive when a transmit demand occurs, the T/R sequence will be aborted and the R/T sequence executed immediately. This interrupt logic is based on the view that it is always better to risk losing a moment of reception than to risk 'hot' switching.

Note that the 12V Tx and 12V Rx lines can never be energised at the same time.

Following a T/R transition, the PIC goes to SLEEP; that is, all dynamic activity ceases including its internal clock. Thus it can never act as a noise source to your receiver.

The process for adjusting the times for your installation will be covered later.

Construction Notes

Fig 8.8 (in the Appendix B) shows the PCB artwork ready for the iron-on process described earlier. The 10V regulator chip IC14 provides power to the STAR DSP board and may be omitted (as in the photograph) if you don't require this unswitched rail. Mounting holes for IC14 and the board (optional) have not been specified.

Start by fitting IC13, C76, C78 and C80, soldering one lead to the top ground plane. Then fit the socket for IC11 and solder pin 5 to the ground plane, followed by the socket for IC12 with pin 9 grounded. The remaining construction sequence is not critical. Mount the otherwise symmetrical switch so that, with the switches set away from the PIC, they are open circuit.

SUPER VOX IN DSP

Conventional VOX

When VOX detects the beginning of your speech, it initiates the R to T transition - which is going to take at least 3ms to complete. Further, if you have a T/R relay, the design must ensure the relay has settled in the transmit position before letting the RF through, typically adding a further 20 - 30ms. The result is that the leading edge of your speech is clipped off. Not by much in a good design, but often noticeable.

To disguise this effect, a VOX hang time is incorporated which is set to drop back to receive if you pause for breath, which at least minimises the number of truncated words. The other workaround you often hear from VOX operators is that they do not answer a direct question with - Yes". They tend to say "um, yes", probably subconsciously in order to avoid it coming over as merely "-esss".

How much better it would be if your transceiver started the R to T transition in anticipation, ie just before you started to speak! Sounds fanciful? In effect, this is what Super VOX does. And by the way, it applies equally to QSK CW operation.

Super VOX

The idea is to trigger the R to T transition immediately on detection of your voice, but then delay the 'voice' in DSP for the time it takes for the transition to complete. Thus the leading edge can never be clipped off.

Critically, this means in turn that you need no hang-time, since there is now no desire to minimise the number of transitions. Of course your delayed voice is still coming 'out of the antenna' for a few milliseconds after you stopped talking so you need to stay on transmit for that time - but absolutely no longer.

The net effect is that at a normal conversational speaking speed, you drop back onto receive not between breaths and sentences, but between every word - and often enough, between syllables, and if your T/R transitions are fast enough, you can listen through. Equally, someone listening to your transmission would be totally unaware that you were spending a significant percentage of your over on receive - in short, but very frequent, bursts.

The overall effect is very close in sensation to full duplex as in a normal (and therefore interruptible) conversation and, if widely practised, would do much to turn many a contact into a conversation rather than a series of speeches.

Anti-VOX

This normally works by comparing the microphone input with the speaker output - and if the same, concludes that it is not you speaking. STAR incorporates a further refinement in that the microphone input is compared with the output that did come from the speaker 4ms earlier. Why 4ms? Because this is the time it takes sound to travel 4ft in air, an assumed reasonable distance between the speaker and microphone. The improvement is noticeable and is worth having because the few extra lines of code don't cost anything.

AGC Implications

Normally, the AGC voltage decays to nothing shortly after you go to transmit. The result is that the receiver comes back on full gain in VOX gaps - which is not very comfortable in an 'S9' contact.

The approach adopted by STAR is to retain the AGC level established by the last 2s period of continuous receive - and apply that level during the gaps. It is important to ignore AGC levels established during the gaps for this purpose, so 2s was chosen as an arbitrary interval which is clearly longer than a casual pause. If, at any time, somebody other than you starts speaking, the normal AGC attack takes care of any adjustment in a few milliseconds.

So this is, if you like, extended-hang AGC where the 'hang' is extended over periods of transmission.

DSP BOARD

The board is based on (and is not incompatible with) the Analog Devices 2181 EZLITE board. That is, aspects of that board which are not used either by STAR or by W7PUA's DSP-10 have been omitted; the physical construction is completely different, and a current production and superior CODEC chip has been used. But conversely, the EZLITE board can be (and has been) used in this application and, if you already have your hands on one, e-mail me [2] for further details. Signal names as defined by Analog Devices are used throughout.

Mother Board

The mother board with her two daughters is shown in **Fig 8.9**. This form of construction was adopted to spread the risk during board manufacture and to allow upgrade of either the CODEC or Processor chips later.

Each board has its own regulator chips to spread the heat dissipation and to maintain modularity.

The CODEC daughter converts analogue signals to/from digital/analogue form for the benefit of the Processor. The digital signals are passed back and forth using a 12.288MHz industry

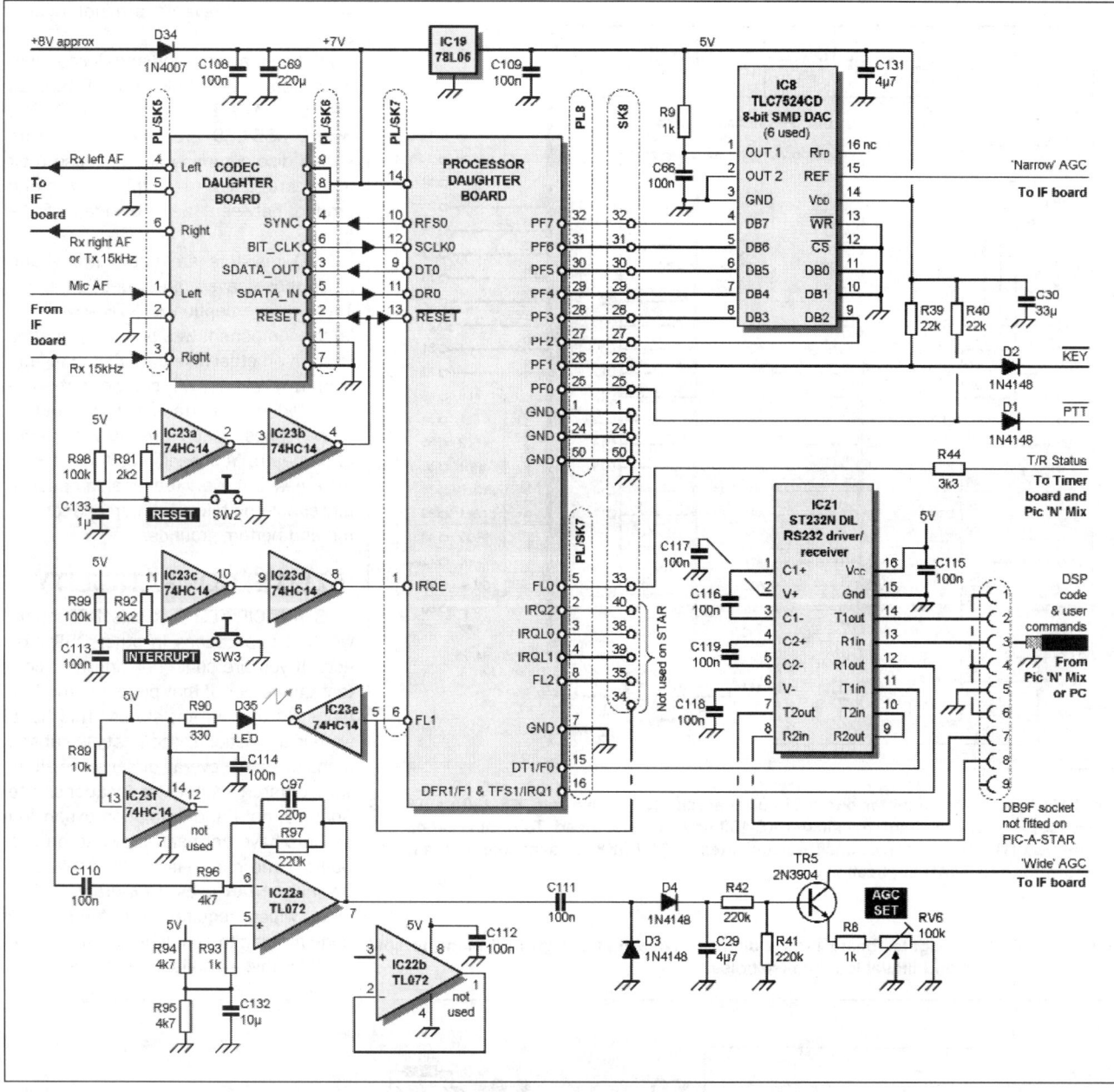

Fig 8.9: DSP mother / daughters relationship - and mother board circuit diagram

standard AC '97 serial bus -which multiplexes data in, data out and commands.

The Processor daughter does the DSP processing (no surprises there) -but has other control inputs / outputs as well.

Unlike the EZLITE board, the Mother Board carries IC8 and IC22/TR5 for generating AGC voltages for use on the IF Board accepts inputs from KEY and PTT lines -and generates the controlling system T/R line as a function of mode and control parameters eg VOX/QSK operation -thus customising it from the general to this particular transceiver application.

IC21 controls RS232 communications from a host - either your PC or the PIC in Pic 'N' Mix - and is used to upload the operational DSP code. It also accepts user commands to control the entire transceiver. IC23 buffers manual resets and interrupts , and drives an LED to show status.

Processor Board

This comprises the processor chip -and some memory used only at power-on (or Reset) time to boot load the real operational

code. See **Fig 8.10**. For further detail, see the ADSP-2181 data sheet [3]. Being mostly track, the board is very quick and easy to build.

CODEC Board

This is a standard (albeit minimal) implementation of the AD1885JST CODEC chip. See **Fig 8.11**. For further detail, consult the data sheet [3]. The CODEC uses a 3V3 digital rail, but 5V on the analogue side; IC20 translates the 5V logic signals from the Processor to this 3V3 level. Outbound 3V3 lines to the Processor are already within its logic 1/0 definition range.

COMPONENT SPECIFICATION

SMD components have been specified here where space, cost, or performance considerations requires them - but not otherwise. 1206-size devices are used and these are no more difficult to handle than conventional leaded components.

Specifically, SMD electrolytic capacitors are not used because these are expensive - and the small space savings

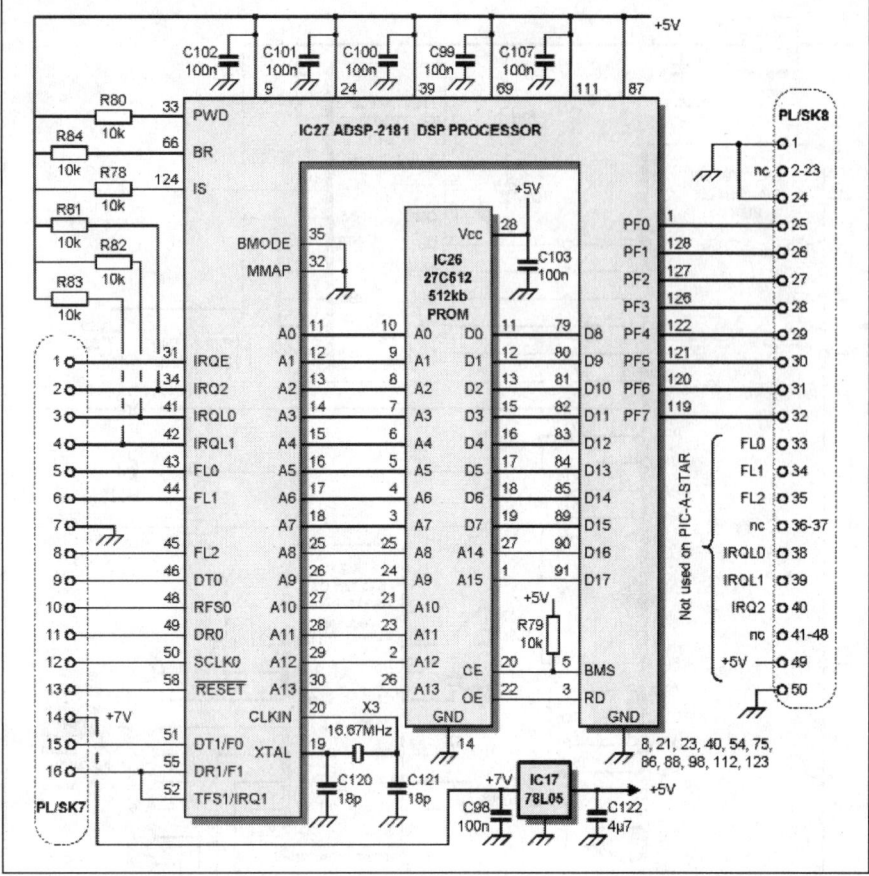

Fig 8.10: Processor daughter board circuit diagram. IC27 may be a KS-130 or KS-160 processor but, in any event, the slower KS-130 device is assumed. To retain compatibility with EZLITE, the unused or unconnected lines on PL / SK8 are available on the mother board for non-STAR applications

Fig 8.11: CODEC daughter board circuit diagram. Note that the ground plane is split between analogue and digital to minimise noise.

which are achievable are not necessary.

All the small coupling and decoupling capacitors are 100nF and, in general, they are SMD. However, on the CODEC board, C85-C89 are specified as wire ended disc ceramic units because their leads are used to couple power and ground between the two sides of the board.

SMD resistors are used throughout, since these save a great deal of space. The single exception is R78 - where a larger component was positively needed to span an otherwise unbridgeable gap.

In any event, all components are mounted on the track surface, but in some cases, leads are also soldered underneath. You need not take this to extremes, but every reasonable opportunity should be taken to interconnect the top and bottom grounds.

'EZLITE' COMPATIBILITY

IT IS ANTICIPATED that this DSP board will find applications in other DSP projects. If you are contemplating this, contact the author of that project in the first place for the current status. This hardware is a functionally compatible subset. It has the same overall dimensions albeit with different connector locations. The address, data and emulation expansion sockets have not been provided on this board - and nor, realistically, could they be. PIC-A-STAR uses a different CODEC chip, which requires a different DSP code module to handle it. A source code shell for this is available on request.

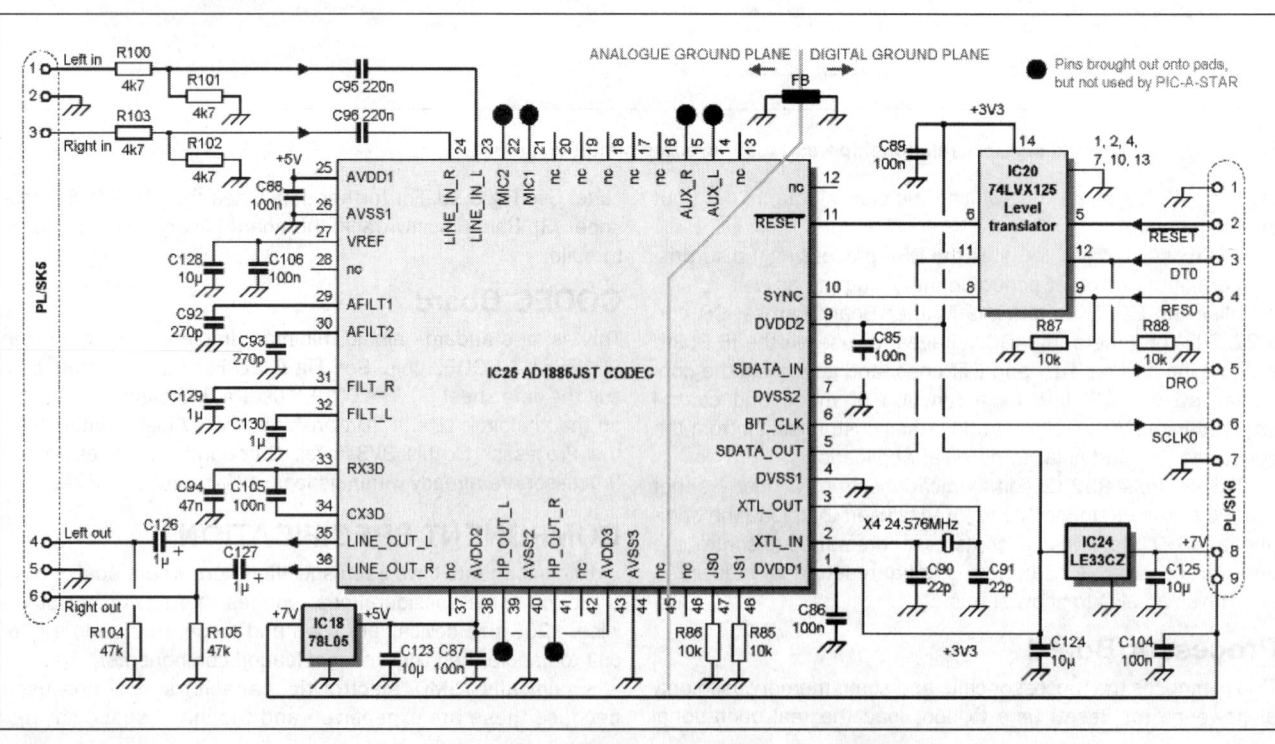

HARDWARE TEST

As the DSP board is progressively completed, it is highly desirable to test it in stand-alone mode before moving on. This process also proves the interface to your PC - which will be needed operationally later.

Prerequisites

The first requirement is that you are running *QBASIC* under Windows on your PC.

On older machines it is a standard application; later it was provided on the archive disc and on Windows ME it is not provided at all - but does run. In any event, it is an absolute prerequisite. The PC itself is totally uncritical.

You need to make up a lead from your computer's serial port - but only two of the lines are used. These are pin 3, the signal - and pin 5, the ground.

These connect (temporarily) to the mother board at 'DSP code and user commands' as per **Fig 8.12**. Ensure the ground lead is indeed grounded.

Set Up

On your PC, establish a new directory. The software assumes C:\STAR but you can edit the software for any other location.

In that directory, place the files testxx.xjp (where xx is the current version number of the test program) and XJPload.bas which is the utility used to load all STAR DSP software, not least this test program.

Open *QBASIC* and from there, open XJPload.bas. To run the test program, just follow the on-screen instructions!

Processor Test

This requires the Mother board with Processor daughter - but not necessarily the CODEC. The test process starts with D35 flashing at 1Hz. Once you start to load the test program, the LED will be permanently lit.

Once the program has loaded, the LED will be off if the CODEC was successfully initialised, on if it was not (particularly if it is not yet even fitted). In either event, if you press the Interrupt switch, S3, the LED will toggle on and off. This verifies that the

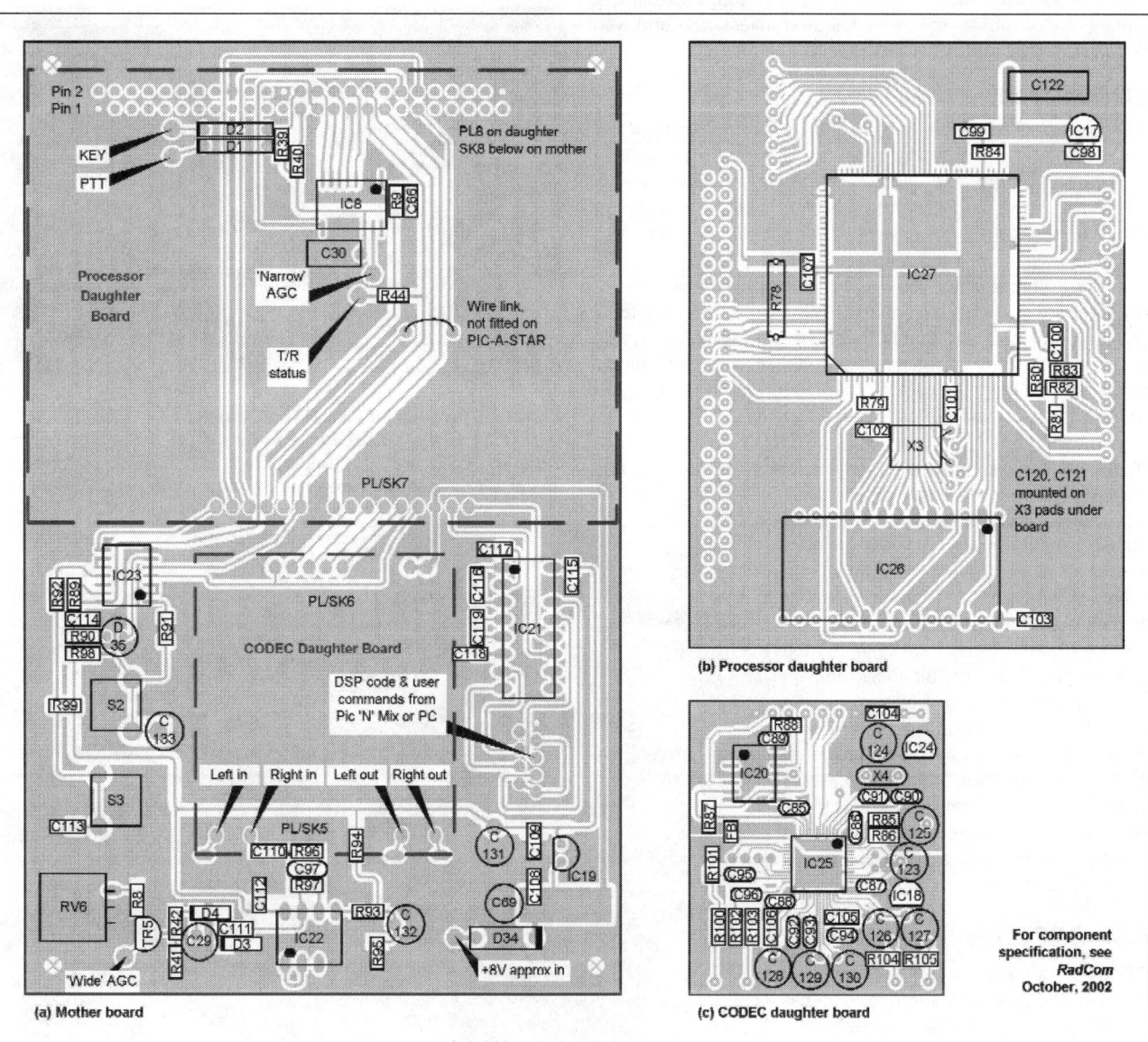

(a) Mother board

(b) Processor daughter board

(c) CODEC daughter board

Fig 8.12: DSP boards component location diagram. SMD capacitors are shown as rectangles, disc ceramics with 'rounded' ends. The Processor daughter should be rotated clockwise through a right-angle to visualise the fit on the mother board. A significant number of pins on PL8 and SK8 are not used by PIC-A-STAR, but were included for compatibility with Analog Devices EZLITE board. These locations need not be populated for STAR

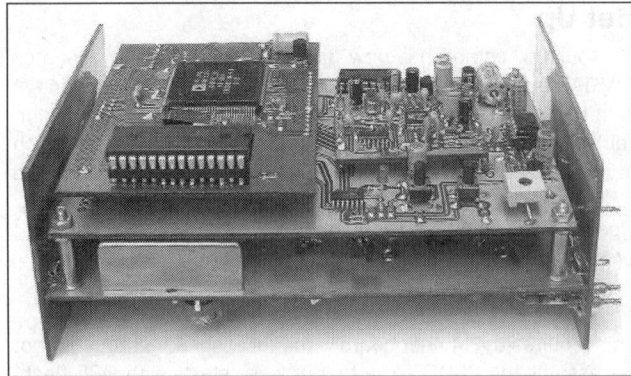

Fig 8.13: The DSP assembly. That is, the DSP mother board with CODEC and Processor daughter boards - mounted back-to-back with the IF board in its enclosure. The top, bottom and side screening panels are not fitted until after final test

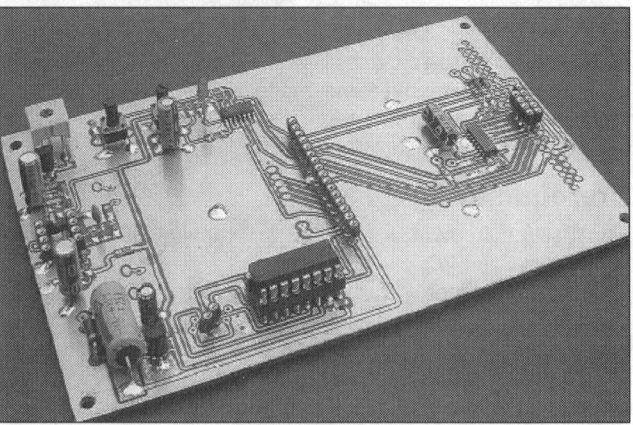

Fig 8.14: The DSP mother board, ready for daughter board fitting and test. Note IC8 and its associated components are located under the Processor daughter board. In fact, neither they nor IC22 need be fitted for stand-alone testing

code has loaded and that the processor is running and is in (or indeed, under) control. This also establishes your capability of loading any code over the serial link and unless and until you can achieve this, no further progress can be made.

CODEC Test

Once the CODEC daughter has been fitted and the previous test successfully repeated, power down and connect a patch lead from the CODEC left and right outputs to a stereo amplifier.

Power on again and reload the test program. A damp finger placed on the CODEC left or right inputs should now produce a corresponding hum on the respective output. Should you prefer something more exciting, you could connect up a microphone or any standard line-level stereo input. This is a test of a full loopback on both channels. That is, the input is being digitised, sent to the processor where a minimal operation occurs in the digital domain before it comes back to the CODEC, where it is converted back to analogue form and thence to your ears.

Fig 8.15: The finished Processor daughter board. Note that the crystal X3 is fitted after bending its leads - to reduce height. C120 and C121 are fitted under the board. IC26, when fitted in its socket defines the overall height of the complete DSP board

Thus, when this test works, you have completely proved the CODEC and the vast majority of the processor functionality - and the interface between them.

If, however, it should fail, yet the processor successfully loaded the test program in the first place, the problem almost certainly lies on the CODEC board itself - or the link between it and the processor.

A 12.288MHz clock train is generated by the CODEC on the Bit Clock line. In response, the processor provides a 48kHz clock on the Sync line. If these are both present and you can see data pulses on the Data in / out lines, then the problem is probably on the analogue side of the CODEC. But, if you rigorously checked the board in the first place, there can't be a problem, can there?

DSP CODE DEVELOPMENT

If you want to develop your own code, you will need the tools. The author's code was developed using Analog Devices' older DOS-based development tools - which used to be supplied with its EZLITE board. These are available from its FTP site [8]. Nowadays it supplies its *VisualDSP++* environment which has the merit of a 'C'

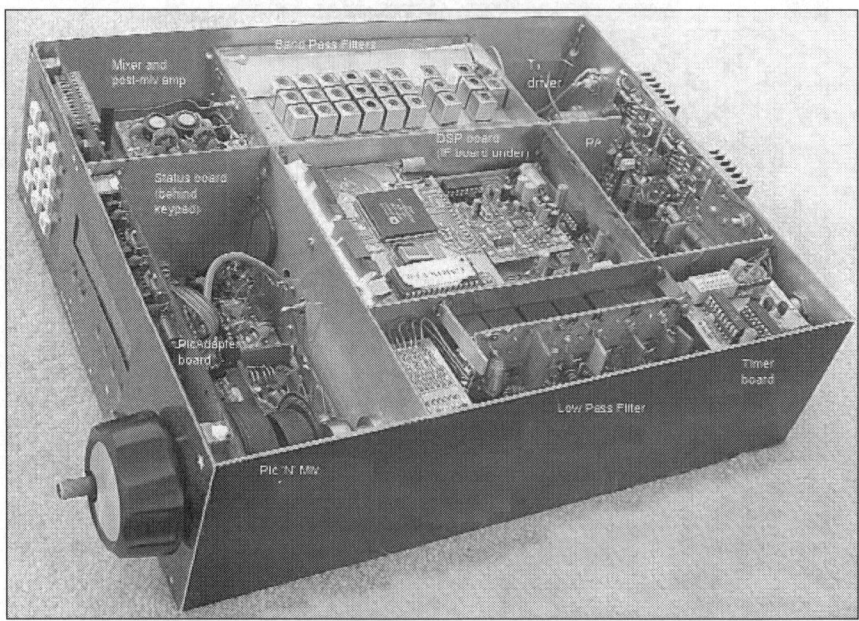

Fig 8.16: G3XJP's STAR built in a PCB enclosure - shown with all compartment coverplates removed. The overall dimensions of the case are 310mm deep by 240mm wide by 85mm high. This generous size allows good in situ access to all the boards

Fig 8.17: The view from underneath, traditionally somewhat less beautiful - so shown smaller

compiler. As an evaluation package, it also has a program memory limit but, at the time of writing, STAR would only use about half this limit. This world can change very quickly, so visit the AD site for the latest information.

DSP BOARD CONSTRUCTION

Although targeted specifically at the STAR DSP board, the technique for mounting the chips is totally general. There are no special tools required to mount these .difficult. chips - except a positive attitude. The author has heard much moaning about how these chips spell the end of home-brew construction - but it turns out the opposite is true. You can lay these chips down with a minimum of histrionics, and the following process - which is completely repeatable - came from AA7QU.

Tools

Firstly, the soldering iron. I used an Antex CS series iron (17W) with a 0.1mm tip, filed back from a mere point to a small chisel. Any bit about 1-2mm is fine. The other ingredients are:

- laser film, Farnell 895-945;
- some common solder;
- desolder braid, 2.7mm or less;
- a flux pen, Farnell 891-186;
- jam (home-brew, of course) or toothpaste.

The latter is for holding the CODEC chip in place long enough to tack its legs down. In fact, any water soluble non-setting stick is fine.

Construction Sequence

Make all three PCBs first as per **Fig 8.20** (in Appendix B) using the iron-on process previously described.

The daughter boards are double-sided but, by design, only just. Under all circumstances, treat these as two-pass single sided boards. Any attempt to etch both sides in one pass is simply taking unnecessary risks. Do the complex topside first. If you want to use the artwork for the second side, drill all the holes, register the artwork with pins through those holes and then iron it on. But much easier, just sketch the trivial track and ground-plane in with an indelible pen, joining up the dots. When etching either side, merely spray mask the other.

When you have fully etched a board, absolutely check every track for continuity or shorts, either inter-track or to ground. If you get an open-circuit track the likelihood is that it will merely not work till you find the problem. If you have shorted tracks, however, the likelihood is that you will cook a chip and never find the problem.

If you have not used SMD Rs and Cs before, just tack one end down crudely, while holding it in position with a vertical screwdriver. Then solder the other end properly - and then revisit the first end.

Mother Board

Build this first, less the daughter board sockets. This board is completely unetched on the reverse (ground-plane) side. The only points to watch are the sockets for IC21 and IC22. Cut all their pins back to the shoulder except the grounded ones, which are soldered both sides. Check that all the obviously grounded areas on the board are indeed continuous and if not, add links through to the ground-plane side.

For the external connections, I simply countersunk the holes on the ground side, soldered stub wires to the pads on the track side - and then applied epoxy resin on the ground side to fabricate instant feed-through insulators.

Processor Board

Fit the inter-side links first. Then check the integrity of the tracking.

IC26 socket comes next. Cut back the pins which solder only to the top track; note that, exceptionally, pins 14 and 28 are soldered both sides.

Next the processor chip. Although it has more pins than the CODEC, it is somewhat easier to mount, since the pin spacing is greater and the chip is quite heavy so it is less inclined to skid around. The target time to mount this 128-pin PQFP chip is no more than 15 minutes - or you are doing something wrong!

Line the chip to the pads. Please check the orientation as you only have a 25% chance if you leave it to luck. The good news is

Fig 8.18: Processor chip before . . .

. . . and after removing excess solder. The target time to mount this 128-pin chip and clean up is 15 minutes

Fig 8.19: The CODEC chip before . . .

. . . and after desoldering. Don't panic, it works!

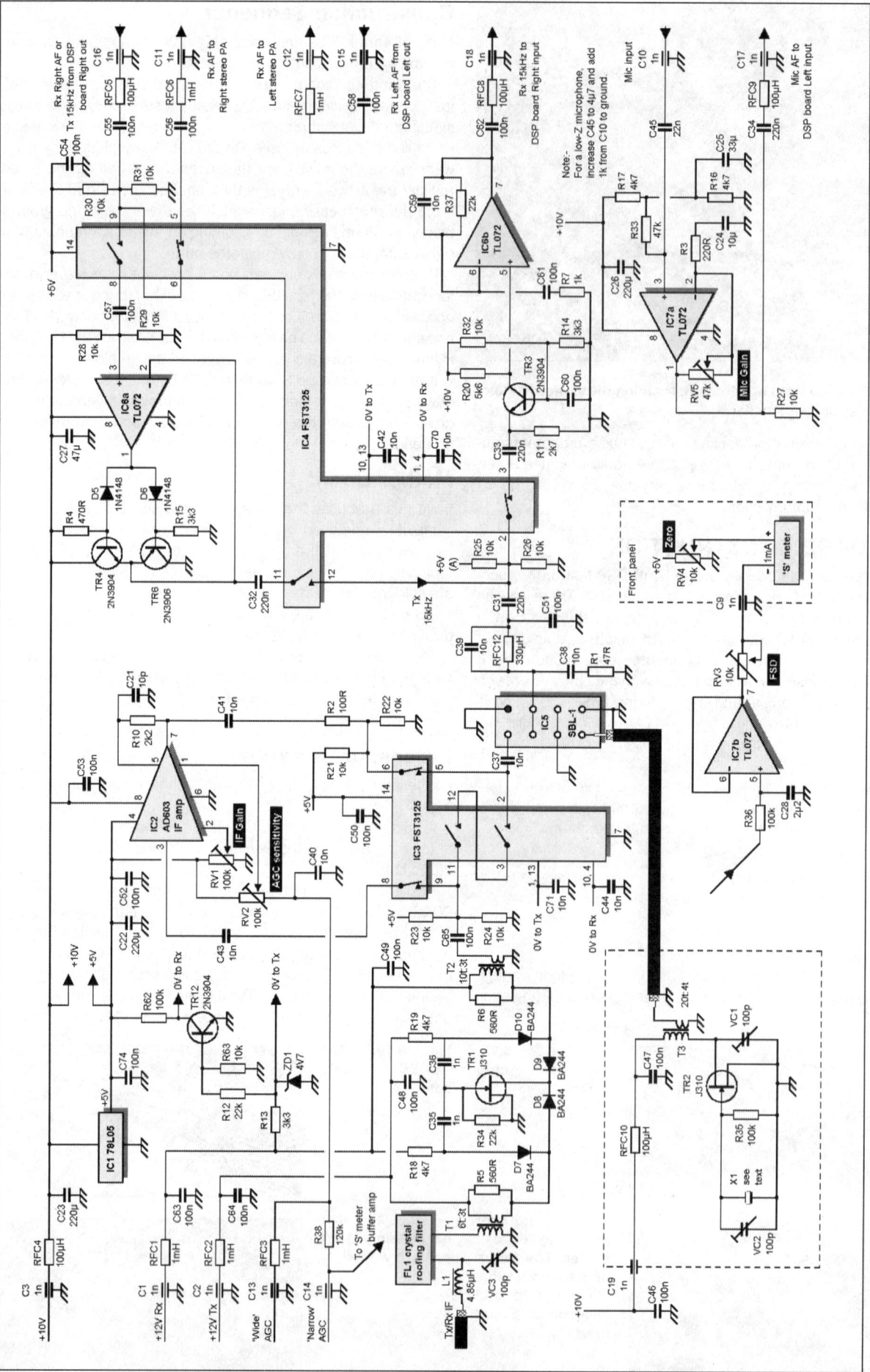

Fig 8.21: IF board circuit diagram. Takes an HF IF feed from a typical bi-directional mixer and post-mix amplifier, and translates it to/from 15kHz. This board mounts back to back with the DSP board. All components are mounted on the track side except for the crystal filter, FL1, and the SBL-1. The roofing filter is your choice and X1 must correspond. The switches in the receive path are shown as closed for illustration purposes only

that the correct quad-pack chip location on the board is totally unambiguous. Get someone else to hold it down while you roughly tack down a few legs in the middle of each side. It sounds cruel, but trust me, it feels no pain.

Running the iron and solder along each side at the point where the pins meet the track, run in a fillet of solder paying (almost) no attention to bridging the pins or the tracks. The only requirement at this stage is that every pin is indeed soldered to its track.

Three minutes elapsed.

Saturate some desolder braid with flux. Rest some fresh braid - over the top of the chip - on the bridged pins. Lightly apply the iron to the braid and, when you see the solder appear on the braid, withdraw. Then repeat as needed. Lay the braid on any bridged tracks - and repeat until all surplus solder has been removed. Do not draw the braid across the tracks, only along them.

Eight minutes elapsed.

Using a continuity meter, preferably with a 'beep' - and fabricating some probes from sewing needles - check that all bridges have indeed been removed. Finally, wash off any surplus flux under tepid water, and air dry.

Job done, seven seconds per pin.

Note that C120 and C121 mount on the pads of X3 on the underside of the board. Use SIL plug strip for both PL7 and PL8 - but use only the minimum population needed for the latter. Ensure the smaller diameter end of the plugs mates with the sockets.

For the sockets on the mother board, cut back the pins - except the grounded ones which solder both sides.

Fit the connectors dry to both the mother and daughter to ensure alignment - and then solder them to their respective boards.

The partial assembly may now be tested. Apply 8-10V power to the mother board and check the voltage rails before and then after fitting IC21 and IC22. Then plug in the Processor daughter and, after power up, D35 should flash at about 1Hz. Pressing the Reset button, S2 should cause a momentary hesitation before the flash resumes. Now run the test program, as described earlier.

CODEC Board

Having established that the digital and analogue ground-planes are mutually isolated, fit a wire link via a ferrite bead (FB) to join them. Then mount the CODEC chip as described for the processor, but in this case, use a *very* small amount of jam to hold the chip in register at first.

Then fit IC20 and C85-C89 and check integrity of ground and power. The other components should be mounted working outward from the chip, leaving the electrolytics till last. Finally, after rigorous checking and probing of every pin and every track (it must be right first time), I mounted the daughter to the mother using short lengths of component lead. In the case of the left and right inputs and outputs, their leads pass right through the mother board.

With this approach, should you ever want to remove the daughter subsequently, cut each wire first and then desolder both ends.

The complete board may now be tested by again running the test program, details of which were given earlier. The pleasure and pride of success at this stage is indescribable! .

Fig 8.22: The IF board

IF BOARD OVERVIEW

The block diagram was shown in Fig 8.2 at the start of this chapter. The board has a bi-directional IF port - which is then translated to / from 15kHz where the DSP takes over both on transmit and receive.

The IF frequency can be at any HF frequency of your choosing, typically in the range 5 to 12MHz. The determinant is the availability of the crystal filter FL1.

Pic 'N' Mix [7] allows you to change the IF frequency injection offset in a matter of seconds, so there are no issues there.

IF CIRCUIT DESCRIPTION

Referring to **Fig 8.21**, the 50-ohm IF is matched to FL1 by L1/VC3. This is a standard L-match and should be modified - applying the textbook L-match equations - for your filter's frequency and impedance. VC3 is adjusted for maximum output in the first place, but thereafter for minimum passband ripple. The turns ratio of T1 also needs to be established for your filter impedance. The values given assume a 10.7MHz filter with 2200Ω impedance.

TR1 is the ubiquitous bi-directional J310 IF amplifier. It offers modest and quiet gain, and stable load and much convenience.

IC34 and IC4 provide fast (and silent) T/R signal switching - with high isolation and only a few ohms on-resistance. Resistive divider networks are used throughout to bias the signal paths to mid-rail.

On Receive

IC3 routes the signal to IC2, the AD603 IF amplifier. This is a quiet device with good AGC characteristics. As used here, it has a gain range from 0 to 40dB - with a linear dB response to a linear control voltage (an *increase* in control voltage produces an *increase* in gain). It is not the most inexpensive device available, but if you have been brought up on IF amplifiers that emulate snakes, you will appreciate the difference.

This 40dB AGC range is combined with a further 45dB in DSP to give 85dB in total - more than plenty by most standards.

The process for setting RV1 and RV2 follows later. The two AGC control signals ('wide' and 'narrow') are generated on the DSP board and summed at the junction of R38 and RFC3. The 'wide' AGC voltage is generated by detection over the full FL1 bandwidth - and is there only for emergency gain back-off in the presence of a very strong signal outside the DSP filter bandwidth. Normally the 'narrow' control voltage dominates - and it is

Fig 8.23: DSP assembly illustration and recognition drawing, not to scale. The IF board and two end-plates are seam-soldered to form an H-section. The height of the endplates is typically 6cm as a minimum, but can be up to the full height of the Tx / Rx enclosure. The DSP board is bolted to the back of the IF board. Note that critically, the external connections are brought out at different 'levels' depending on which side of which board they connect to - and at different ends depending on the destination. Two further sides and a top and bottom (not shown) complete the screening but are not added until after final commissioning

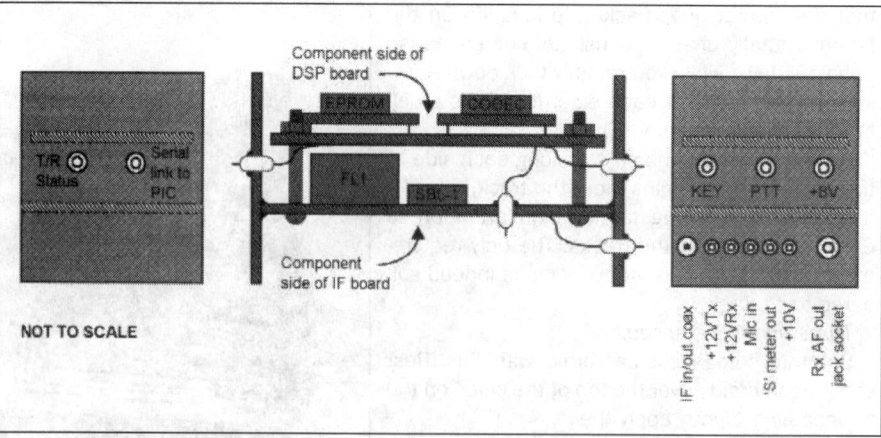

also routed via the buffer, IC7b, to drive the S-meter. This latter is shown as a 1mA movement - but later a bar-graph alternative will be offered - in which case RV3 sets the zero point and RV4 is not fitted at all.

The output from IC2 is routed via IC3 to the SBL-1 mixer, IC5. You may ultimately wish to fit a stronger device here - depending on the width of your roofing filter and your operating needs. The mixer injection port is fed from a basic crystal oscillator - and this could also be 'beefed-up' if required.

C38 and R1 terminate the sum (HF) mixer product - whereas RFC12, C39 and C51 pass the wanted 15kHz difference component.

TR3 is a low-noise, modest-gain amplifier which feeds IC6b. This latter has modest gain at low frequencies with the response rolled off rapidly by heavy negative feedback provided by C59.

On Transmit

IC7a provides modest shaping of the microphone audio and significant gain to get the level up to that required by the CODEC on the DSP board.

You should alter the input arrangements of IC7a to suit your microphone impedance - and C45 in particular for a good mid-range audio response with your voice. Some tailoring options may later be added in DSP as well.

The output of IC7a is routed unconditionally to one input of the DSP since it needs to monitor the mic input continuously for VOX purposes.

The transmit signal next appear as a 15kHz SSB or CW signal from the DSP which is routed via IC4 to the buffer, IC6a. This, in turn, drives the complementary pair, TR4 and TR6, which are there to deliver power into the low-impedance load presented by the SBL-1.

On transmit, the AD603 is out of circuit and IC3 routes the signal directly to the J310, TR1, and from there to the filter FL1, and thence out to your transmit IF strip.

T/R Switching Control

The J310 is switched by the 12V Rx and 12V Tx lines, the inactive one being taken to near ground. All other T/R switching is managed by IC3 and IC4. The switching voltages ae derived from the 12V Rx line only, with TR12 acting as a simple inverter. This approach is designed to prevent you from being on transmit and receive at the same time in the event of the loss of either the 12V Tx or 12V Rx supplies.

BUILDING THE IF BOARD

The IF board comprises a traditional PCB with two end-plates soldered on to form an H-section, as shown in **Fig 8.23**. The DSP board is subsequently mounted on the IF board as illustrated.

This form of construction is not strictly necessary. You could build the IF board and DSP board into two separate enclosures, but this approach was chosen because these two boards are highly interconnected.

The IF PCB dimensions are determined by the size of (and are just larger than) the DSP board - resulting in generous spacing between the functional blocks. The surplus board area has been allocated around the crystal filter and the crystal oscillator; the former so that any reasonably-sized filter may be fitted, the latter to give room for a more sophisticated oscillator if desired.

The PCB is assembled by soldering most of the components to the track side. This approach makes signal tracing easier and minimises the amount of hole-drilling. SMD components were not specified here because they are not needed, but most of the components are in fact mounted SMD-style.

Mask, etch and drill the PCB using the iron-on laser film technique covered earlier. On the ground-plane side, countersink the ungrounded holes associated with FL1 and the SBL-1. Both the coax lead to the SBL-1 and C37 are soldered directly to the SBL-1 pins - as opposed to PCB track - so drill generous clearance holes for these pins. All other holes are grounded both sides of the board and are not countersunk.

End-plate Dimensions

The width of the end-plates is that of the IF board. The task now is to determine their height - which is principally (but not entirely) determined by that of your crystal filter.

Fit FL1 and then, using spacers somewhat longer than the height of this filter, crudely trial-mount the DSP board as in Fig 8.23.

The height of the end-plates is now that of this assembly plus at least 20mm for the IF board components. The approximate sum is 24mm for the DSP board, plus 20mm for the IF board components, plus 2mm for the PCB thickness, plus the height of your chosen crystal filter plus 3mm margin. The latter two measurements also sum to give you the length of the four mounting spacers. Be generous.

End-plate Fitting

The end-plates are fitted before mounting the components, because this makes the board easier to build and handle without contaminating it with finger marks.

Mark the target position of both boards on the inside of the end-plates and then, looking at **Fig 8.24**, lay off the position of the feedthrough capacitors from the IF board. Drill holes for these now, but leave fitting the capacitors until later.

Clean both sides of the IF board and end-plates immaculately, and apply a light coat of spray flux / lacquer to both sides.

Now seam-solder the end-plates to the IF board with a large

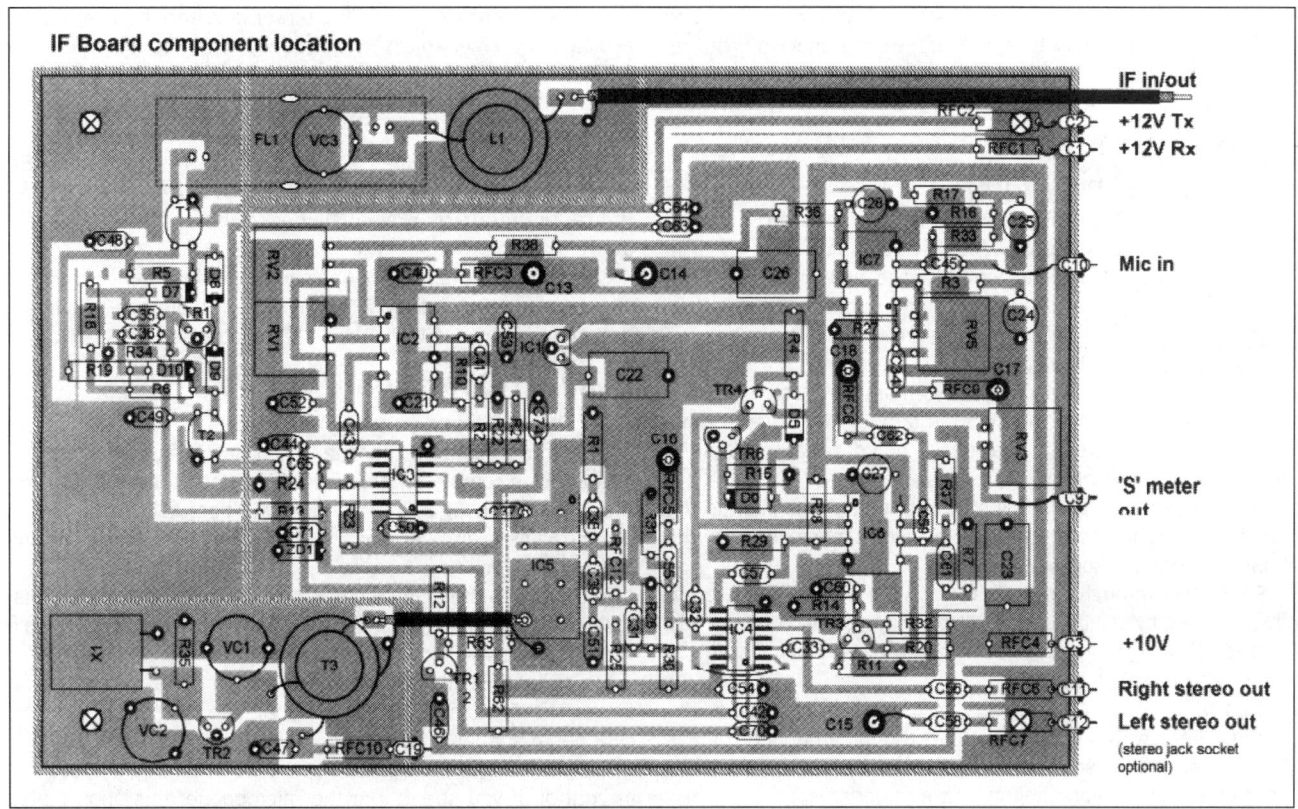

IF Board component location

IF in/out
+12V Tx
+12V Rx

Mic in

'S' meter out

+10V

Right stereo out
Left stereo out
(stereo jack socket optional)

Fig 8.24: IF board PCB layout on double-sided board (see Appendix B for the PCB artwork). The reverse side is completely unetched to form a continuous ground plane and screen. All components with the exception of the crystal filter, FL1, and the SBL-1 mixer, IC5, are mounted on the track side. You may need to customise the tracking to suit your crystal filter. The 'holes' are shown on the component layout only to define the tracks should you be producing the PCB by some manual method. Only components which feed through to the back of the board require actual drilled holes and these are as defined on the tracking template. The tracking template image is mirrored (ie flipped left-to-right) for direct copying to iron-on laser film. The basic drilling size is 0.7mm - with holes for mounting, feedthroughs etc drilled larger to suit. Some internal screening partitions are made from PCB material or brass shim stock. Those

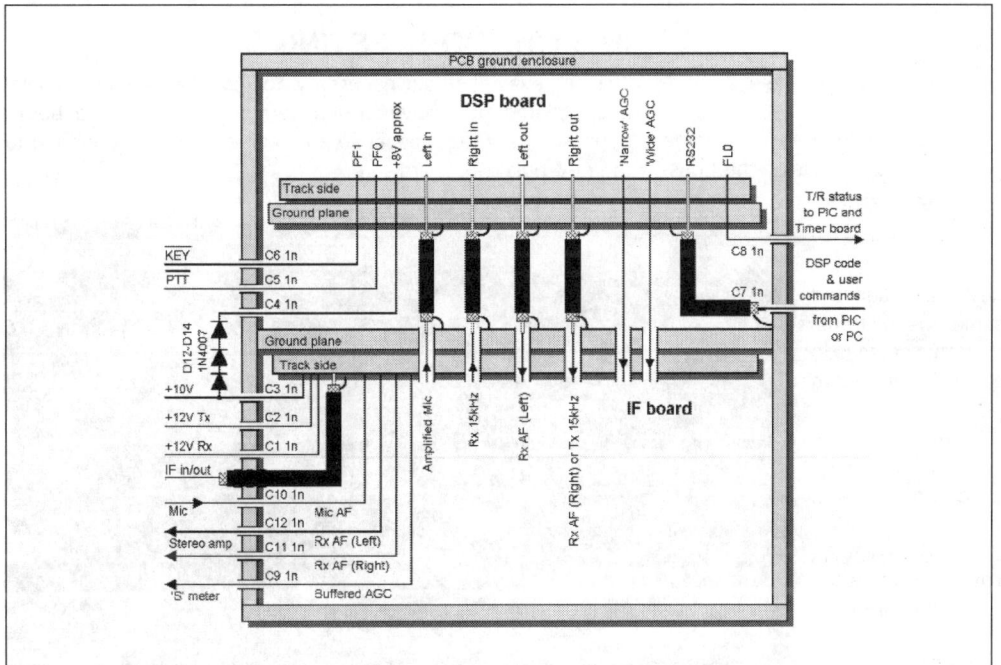

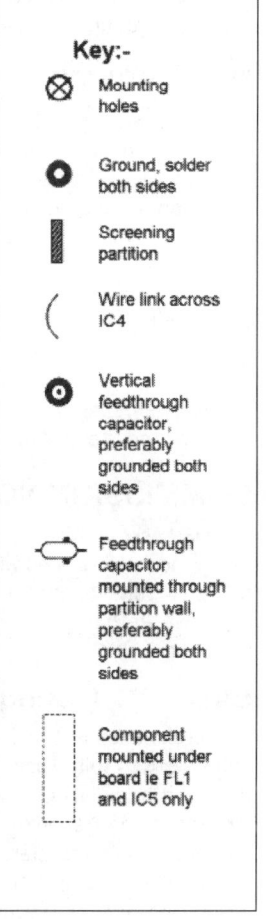

Key:-

⊗ Mounting holes

◉ Ground, solder both sides

▨ Screening partition

(Wire link across IC4

◉ Vertical feedthrough capacitor, preferably grounded both sides

⊸ Feedthrough capacitor mounted through partition wall, preferably grounded both sides

▯ Component mounted under board ie FL1 and IC5 only

Fig 8.25: DSP sub-assembly. The IF board is bolted back-to-back with the DSP board using nuts, bolts and spacers. Both their ground planes form a screen to isolate the two halves of the box. Feedthrough capacitors are used to route between the two halves of the box - and to the rest of the transceiver. C11 and C12 may be replaced with a stereo jack socket

iron. If you mount both at the same time, you will be able to check on the geometry by eye. Progressively checking that all remains true and working both sides of the IF board, use small single tacks first of all, then multiple tacks and finally form neat fillets.

Component Mounting

Refer to Fig 8.24 and **Fig 8.25**. Tin all the pads except those under the FST3125s. Mount both FST3125 chips. Align the chip and tack down two opposite corners to the larger pads provided. For the remaining pins, offer the iron and solder to the track just short of the pin - and the solder will spread along the board and wet the pins by capillary action.

Fit the wire link across IC4. Cutting their ungrounded leads so that they sit just above the board, mount all the other components - except the preset capacitors. A pair of tweezers is useful for handling the smaller components.

To surface-mount the DIL ICs, cut off all the pins back to the shoulder except any grounded pins which pass through the board. Do not use sockets.

Fit the feedthrough capacitors ,which are typically made off to the IF board, by using a series RFC as a flying lead. Mount RFC1 and RFC2 at right-angles to each other in the vertical plane to minimise mutual coupling.

Trim and solder all the grounded leads on the back of the board. Check with a continuity meter that all grounded track is in fact grounded. Also perform all the usual basic tests such as checking isolation and integrity of the power rails.

Mask off the preset resistors and give both sides of the board a final and generous coat of spray lacquer. Finally, fit the preset capacitors, definitely unlacquered.

DSP ASSEMBLY – ASSEMBLY

This process starts when the IF board and DSP board are fully built, and the latter has been tested using the test program. The required DC supplies come from the Timer Board (or some equivalent arrangement).

Make off all the leads between the boards as shown in Fig 8.25. With the two boards at right-angles (but preferably less), trim their lengths and make off the other ends to their respective feedthroughs. Ground the braids to the adjacent groundplane.

Mate the two boards, and in the process, perhaps trim some excess lead lengths. Fit diodes D12 to D14 – outside the housing – to drop the 10V rail to a nominal 8V.

COMMISSIONING

The DSP Assembly is first proved in isolation and then crudely integrated with some existing transceiver for verification. The idea at this stage is to demonstrate hardware functionality, not system performance.

Basic DC Testing

As a preliminary, set RV1, RV2, RV5 to mid-travel and RV3, RV4 fully clockwise.

On the end-plates, connect up 10V, +12V Tx (grounded on receive), +12V Rx (+12V on receive) – and the stereo outputs, typically to some domestic amplifier.

For the first few seconds after power-on, a voltmeter on the S-meter feedthrough should show definite activity on a 5V range. Check that the T/R Status line is near +5V.

On the IF board, check all the power rails and then get the X1 oscillator working. Adjust its frequency to the centre frequency of FL1 + 15kHz.

Loading Test Software

Connect the serial cable from your PC's COM port to the DSP Assembly. Also, a microphone (both audio and PTT). Load the test program as previously described and re-verify operation.

Speaking into the microphone should produce audio from one stereo channel. Adjust RV5 for maximum undistorted output – but, in any event, no more than 2V peak-to-peak on C17.

Loading Operational Software

Reset the DSP board (ie press and release S2) and load in the operational software as per the loader onscreen instructions.

Loading is complete when you are looking at user controls on the screen as in **Fig 8.26** – and the DSP board LED is out. If the LED remains – or reverts to – flashing, this indicates a comms failure during loading.

At this stage the DSP Rx should be operational. To verify this, feed a sniff of RF at your IF frequency into the IF in/out coax. Just tack a few inches of wire to the coax inner and put it near some suitable signal source eg the DDS or a GDO. As you tune across the IF, you should hear the beat note, and the LED on the DSP board should light in the presence of signal.

Turn the RF gain up and down on the PC to verify that you are in control. If you speak into the microphone, this should also light the LED. Grounding the PTT line should mute the Rx – and the T/R Status line should go to near 0V.

On the PC, switch to CW. Grounding the KEY line should then produce sidetone.

The T/R Status line may now be connected to the Timer board and its operation verified. Under no circumstances be tempted to connect T/R Status to some external PTT line, say on your transceiver.

Getting to this point is a major milestone. But if any of the preceding fails, stop and correct the problem before going further.

INTEGRATION TESTING

This stage is not strictly necessary. You could wait until you have a completed transceiver. But I commend this as the better approach - not least because any problems will be confined to the new-build DSP hardware.

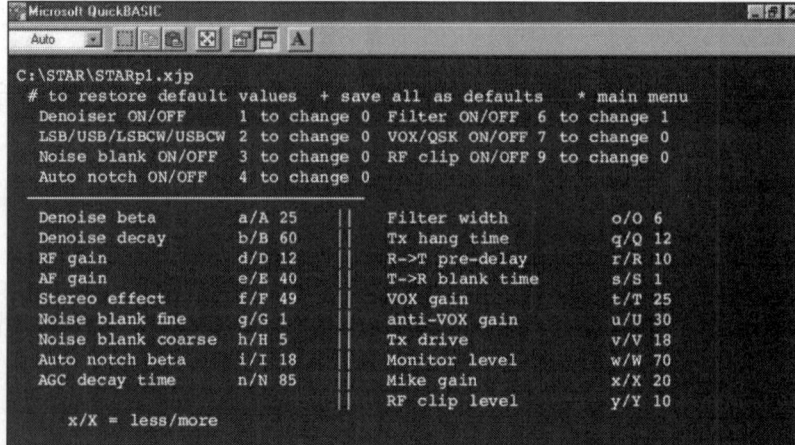

Fig 8.26: The PC screen running under QBASIC. Illustrated are the STAR parameters for the SSB mode. For development and proving purposes only this user interface is designed to be rather more functional than beautiful

Receiver

Connect a short fat ground strap from your transceiver to the DSP Assembly. Locate a suitable bi-directional 50-ohm point on your transceiver; after the mixer, after any post-mix amplifier, after any pad is best – but the 50-ohm IF port on your mixer will suffice for test purposes. Patch in the IF in/out coax via a series 100nF instead of your existing IF strip.

Arrange to be able to switch your transceiver between transmit and receive. Turn the AF gain down on your transceiver – and any other Rx gain controls to maximum; and Tx gain controls to minimum.

Power up on an low frequency band and then load the operational software as previously. Inject signal frequency plus FL1 centre frequency into your transceiver mixer – and you should hear resolved LSB signals from both speakers. On a quiet frequency (LED is out), peak VC3 for maximum band noise. Then peak RV1.

Connect a CRO (DC, 1V/cm) to the S-meter output. This should show about 4V on weak signals and progressively less as AGC action occurs. Find a signal giving about 2.5V and adjust RV2 until it is slightly less. While listening on a noisy band, adjust RV6 until the AGC loop is clearly unstable and hunting – and then back it off until it is smooth. That completes a crude setting up of the AGC system, enough to verify that the hardware is working.

At this stage, with the DSP Assembly unscreened, there may be evidence of white noise on the higher bands.

Transmitter Integration Test

With your Tx drive level well down, set Tx Drive on the PC to 10. With your transceiver connected to a dummy load, and preferably monitoring on another receiver, set up to observe the Tx output on a CRO for flat topping etc.

Put your transceiver on transmit. When you ground the DSP PTT line, this will put the DSP assembly onto transmit as well. The mic gain on the PC should be increased as far as possible – but only so long as there is no evidence of any clipping, compression or distortion.

If all is well, bring up the drive on your Tx to its normal setting. Then increase the Tx Drive on the PC, ensuring the output remains clean – up to your normal power level.

Now would be a good time to screen the crystal oscillator and add the other screens on top of the IF board. The fully-screened enclosure is best left until the very end.

PC CONTROL OPTION

This is a timely opportunity to outline the behaviour of the PC control panel. Fig 8.26 shows the screen of the loader after the DSP code and the controllable parameter values have been downloaded to the DSP assembly – at 9.6kB.

Adjustment and use of the various DSP features themselves follows later. Here, we are concerned only with the mechanics. Simply key the appropriate number to change a switch state, the upper-case letter to increase a parameter value – and the lower-case letter to decrease it.

Syntax

The BASIC control software has been optimised for simplicity. That is, only the most basic syntax has been used – and if you have ever written any software in any language (very nearly, English will suffice) – then you will have no trouble following it or editing it. Equally if you want to build some controller other than Pic 'N' Mix [7] – either in dedicated hardware or on your PC – then this acts as a model. If you have the background to undertake this, then equally you will have no issues following the code.

Control Parameters

These are held in a separate file, param01.xjp. It is the controller's responsibility to handle parameter values and to constrain them to be within maximum and minimum values, and in any event, within an 8-bit byte. The value 255 is assigned to any parameter that does not apply in a given mode.

Following any user change, the new parameter value is sent to the DSP as three bytes. The first is always a tilde '~', the second is unique and identifies the parameter, and the third is the new value. Nothing could be simpler.

Frequency Control

One of the virtues of controlling the whole transceiver from Pic 'N' Mix [7] is that it can handle the injection offset needed when switching between SSB and CW and transmit and receive. Obviously, the PC has no intrinsic ability to do this, so you need to make other arrangements – eg operate your Tx/Rx split when on CW.

PIC A TIME

In the STAR environment, the sole purpose of the Timer board (see earlier) is to provide click- and spike-free R/T and T/R transitions. Hang times for VOX or QSK operation are controlled by the DSP.

All the switching times may be independently set between 1ms (very fast) and 63ms (incredibly slow). In general, it is best to start with the times set to incredibly slow, and then reduce them progressively until there are any signs of switching spikes on the transmitted output – or clicks on reverting to receive. Having said that, if your transceiver has inherently noisy switching, a click on reverting to receive is inevitable. The DSP code has a feature for blanking any such click – but it is obviously best avoided by design.

Adjustment Process

The process for altering the timing is as follows, starting with all the switches OFF, ie away from the adjacent PIC:

1 Set the More/Less switch as required to increase or decrease the time delay. (More is towards the PIC).

2 Set the switch(es) for the time(s) you want to alter to ON, ie towards the PIC.

3 Key the T/R Status line down and up once for each required millisecond of change.

The altered time(s) will be implemented immediately – but not stored. When all the required changes have been made, put all the switches to ON, key the T/R Status line down/up one final time – and all the new times will be stored and retained. As evidence of success, this particular R/T/R transition sequence will not occur. Conversely, to abort all changes since power-on simply miss out this stage completely, power off and wait 20 seconds before powering on again.

Finally, set all the switches to OFF. Note that this process can be used to change several (but not all) of the time delays simultaneously though you may wish to avoid this practice unless gross changes are required. Note also that the PIC cannot be programmed via the programmer interface if the switches are ON. .

USER INTERFACE

The rationale behind the PIC-A-STAR User Interface (UI) has had significant impact on the front-panel layout and ultimately on the entire transceiver enclosure.

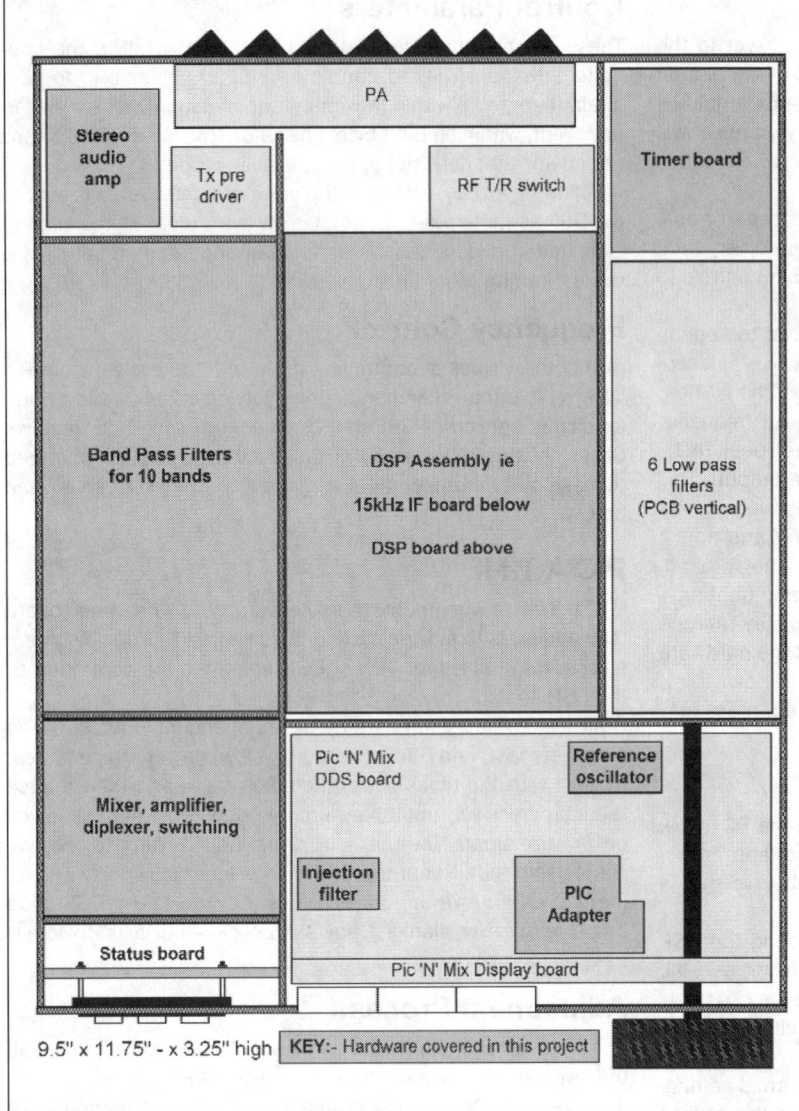

Fig 8.27: A possible transceiver enclosure illustrated at halfscale. The author's is fabricated from 2mm doublesided fibreglass PCB stock. This gives excellent access from both sides

Until recently, the author had followed a philosophy of fixing the front panel controls with those anybody could reasonably need to drive any transceiver, on the grounds that these controls were essentially independent of the inner workings. That approach bought me 25 years development of four fundamentally different transceivers - all using the same housing and front panel.

But times they are a-changing! I have come to appreciate that the opposite approach is more appropriate for a transceiver with a significant software content. **Fig 8.27** shows, not least, the consequences to the overall dimensions.

A significant amount of effort has gone into achieving an effective UI. The challenge is to avoid the extremes. On the one hand, it is easy to end up with a system whose complexity exceeds human intellectual capacity. We have all listened in on those amazing menu comparison contacts which usually end in "I know I am supposed to hit 'Return'. How hard?". On the other hand, personal preferences do vary and you shouldn't be prevented by the designer from adjusting a parameter merely for the sake of a simpler UI.

The clue to the best approach came very early.

No Pain – No AF Gain

During the early evolutionary development, there was, of necessity, a period of several weeks when I had absolutely no adjustable controls whatsoever. To increase the AF Gain, for example, I had to edit the DSP code, re-assemble it and download the whole suite – including this one changed parameter. As you can imagine, I did not bother very often. Actually, it encouraged me to focus on improving the functionality so that the built-in control systems would take care of the 'variables' without undue manual intervention.

The background thought here is that as a design evolves – perhaps over several years – the number and purpose of the controls can swing wildly. And as technology evolves, so do the opportunities. Who would have thought I would need an Auto-notch on/off switch even 15 years ago? And when I designed Pic 'N' Mix, I can assure you the thought that it had the intrinsic flexibility to control the entire transceiver never crossed my mind.

So this time I have taken the minimalist approach - starting with the observation that all controls can be classified under two generic categories, namely "switches" and "amounts". Thus PIC-A-STAR has two (and exactly *only* two) corresponding physical controls, namely: a knob which alters 'amounts', and a keypad which handles the 'switches' – as well as specifying which 'amount' the knob is connected to.

By 'amounts' I mean, for example, amount of AF Gain, amount of RF Clipping and, indeed, amount of Frequency. By 'switches', I probably better mean 'choices' eg '80m' instead of '15m' and 'Autonotch on' as opposed to 'Autonotch off'.

Having settled the mechanical format, then at any time I can have more or less as many 'free' knobs as I like - by simply assigning them in the software. And at the same time, saving much cash on real pots, knobs, switches – and that real nightmare, the consequential system cabling.

Now that my STAR is in daily operational use – besides changing bands and frequency – the biggest strain on the UI has been turning the VOX off when the fast jets go over - and turning the transmitter power up and down to suit conditions. All the other controls have to be set up correctly, but most are essentially set-and-forget.

FRONT PANEL

The template used to make the front panel is shown in **Fig 8.28**. This is designed to last a lifetime in the sense that I can allocate any function to any switch - including a cluster of related controls as a simple sequential menu list. And thereafter, any range of values to the menu items – and so on.

The worst-case change issue is that one day I may need some new legends on my keypad overlay. I have never been an advocate of beautiful homemade radios versus functional homemade radios (given a finite life-time, you have to choose) – but one non-trivial benefit of this approach is that the front panel is less than A4 (and US Letter) in size – so a new one can be print-

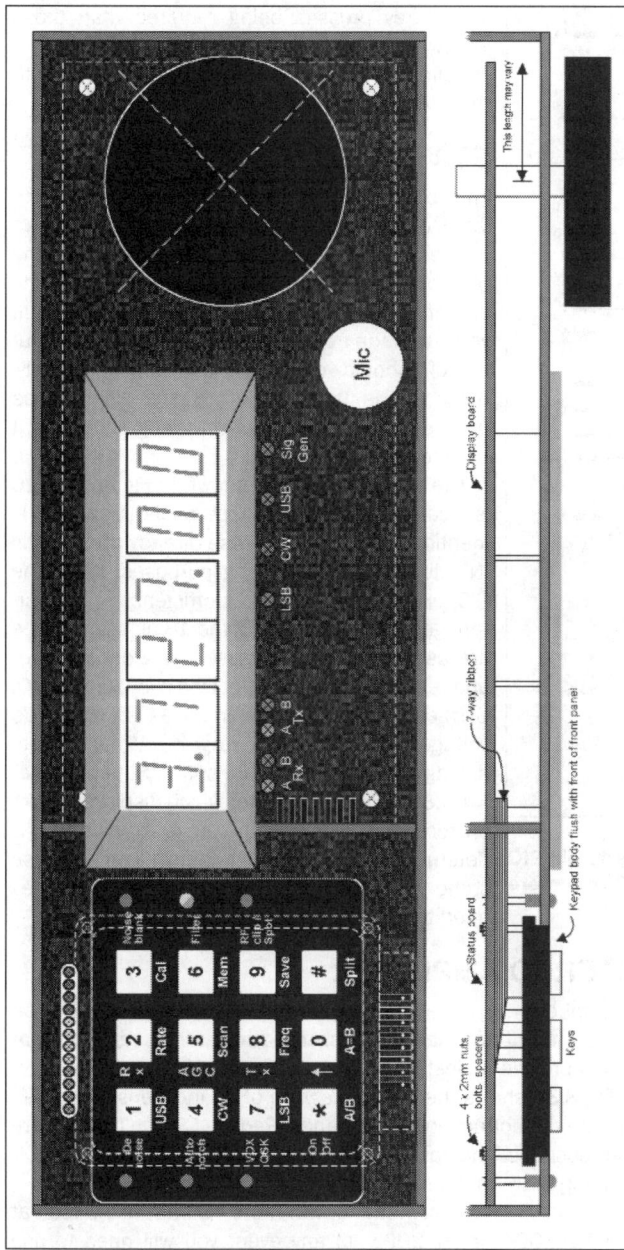

Fig 8.28: The author's STAR front panel layout, to scale. In this case a bargraph S-meter has been used. The nine most frequently used DSP control groups are assigned to the 1–9 numeric keys

ed off on photo-paper and stuck on anytime. Upholstery or foam backed carpet adhesive is the answer to your next question.

A bar-graph S-meter is shown, but is not mandatory. A small edge-wise movement could be accommodated above the keypad, but a 'real' one would need an increase in the front-panel width to accommodate. The bar-graph LEDs, six status LEDs and the keypad all mount on the Status board.

PICADAPTOR AND STATUS BOARDS

Next the circuits of the PicAdapter and Status boards. Constructional detail follows later. You don't need these boards to commission STAR initially, since you can load and control the DSP software from your PC.

Thereafter, in terms of constructional sequence, you need the PicAdapter first, which then allows DSP code download from your PC and subsequent upload to the DSP assembly. Thereafter, you need the Status board to complete the user interface.

RS-232 Connections

First the wires! The required cabling at any one time is one of the following:

- From PC to DSP – early test.
- From PC to PicAdapter – load new code.
- From PicAdapter to DSP – normal use.

Fig 8.30 shows a simple implementation. The lead with the female connector for mating with the PC serial cable should be fitted for occasional use – if at all. Certainly the lead should not be routed via the RF section of the transceiver to the rear panel , to avoid any potential EMC coupling. It could be kept in a drawer and got out when needed.

The link to the PC is needed only to load new releases of DSP code, whereas the link to the DSP assembly is used continuously. Normally, the lead(s) plug into the PicAdapter board, but for loading and controlling the DSP assembly directly from the PC, a trivial connector with TX wired to RX can be used for pass-through operation instead.

PICADAPTER BOARD

This plugs into the original PIC socket on the Pic 'N' Mix DDS board. The circuit diagram is shown in **Fig 8.31**.

For compatibility reasons, the PIC, IC9, uses essentially the original Pic 'N' Mix code to provide all the original Pic 'N' Mix functionality. However for STAR purposes, it has four incremental tasks:

1. To download new release DSP code together with control parameter values from your PC – and retain these in IC10, a serial EEPROM – for subsequent normal use.

Fig 8.29: PicAdapter board in situ in Pic 'N' Mix

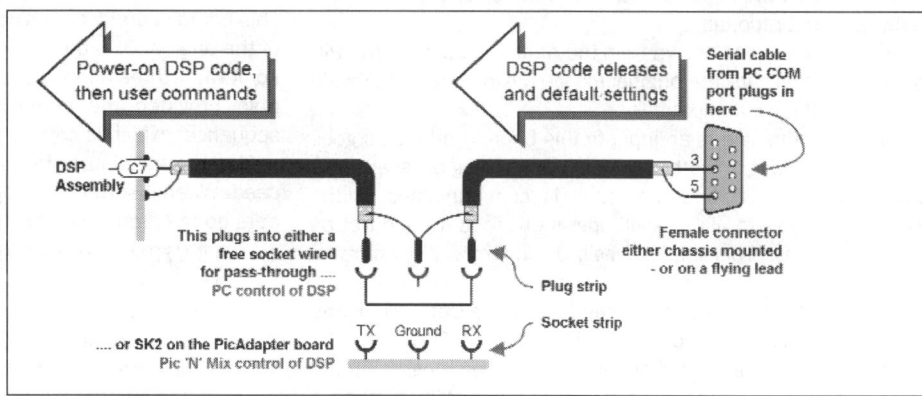

Fig 8.30: RS-232 connections. This can be made up as one composite loom – or as two separate leads

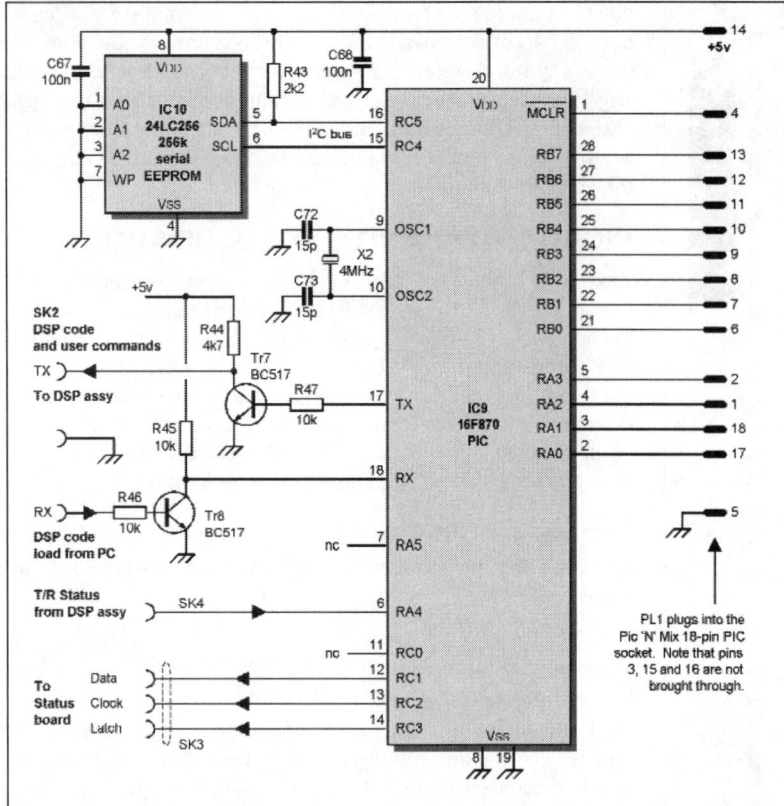

Fig 8.31: PicAdapter board circuit diagram. This board plugs into the 18-pin PIC socket on the original Pic 'N' Mix DDS board and upgrades the PIC to a more recent and versatile PIC, the 16F870

2. At power-on time, to read the DSP code and parameters from the EEPROM – and load them to the DSP board.

3. During normal operational use, to communicate any changes you make to the control parameters (eg AF Gain) – to the DSP board. The changed values are also retained in EEPROM and thereby survive power-down.

4. To drive data out to the Status board to control the state of the LEDs.

To these ends, minimal RS-232 links from your PC at 1.2kb/s and to the DSP Assembly at 9.6Kb/s are controlled by the PIC. The PIC also controls an I2C link to the serial EEPROM.

The slower downlink speed from the PC is used to give the EEPROM time to write each byte of the DSP code and control settings. However reading from the EEPROM is a much faster process, hence the higher baud value – which is the mode normally used operationally.

All the connections shown from the right-hand side of the PIC in Fig 8.31 duplicate the original Pic 'N' Mix pinout and provide much-valued code compatibility.

The T/R status line is an input to this board (and it also goes to the Timer board). Whether STAR is transmitting or receiving is determined by DSP and the result is communicated to the PicAdapter solely to allow 'split' operation. Thus it need not be fitted at first test. This line is at logic '1' on receive, '0' on transmit.

The only other pins worthy of mention are the Data, Clock and Latch pins, which simply drive the Status board LEDs.

Because this board is self-contained, it can be programmed – as an assembly – in the 18-pin socket of a PIC programmer.

For test purposes this board, once programmed, should provide full normal Pic 'N' Mix DDS operation, albeit with slightly longer key presses being required than those with which Pic'N' Mix users will be familiar.

Split operation should finally be verified with the T/R Status line connected.

STATUS BOARD

This board carries the bar-graph S-meter, the latch/driver for the status LEDs and the passive connections to – and the mechanical mounting of – the keypad. You have the option of not fitting any of these functional elements should it suit you – and the PCB is laid out so that you can 'cut bits off' should you wish. Equally and conversely, these elements were designed explicitly to be used standalone in totally differing situations if required.

The circuit diagram is shown in **Fig 8.32**. IC16 is a conventional serial-in, parallel-out driver. It is identical in function to those already fitted to Pic 'N' Mix for band-switching purposes. IC28, the PIC, exemplifies the pinout efficiency of the current generation of PICs. Of the 18 pins, only three are assigned to 'overheads' – ie ground, power and reset, the other 15 all being available for I/O. Of these, one is programmed as an analogue voltage input (the AGC voltage) – 12 as outputs driving LEDs – and two are spare. So far, that is.

IC28 is a mere voltmeter which displays S-units on receive and relative power on transmit. R73 determines its sensitivity – and I can visualise some non-STAR applications where you may need to tune its value.

STEREO AMPLIFIER

It is not easy to find a good stereo amplifier which works well on a 12V rail. If you have a scrap car radio with speakers, that would provide an instant solution.

Fig 8.33 shows the circuit diagram of an inexpensive amplifier from the bottom end of the range. Reduce C7 and C8 for more top response. The component layout is in **Fig 8.34** and the PCB is in **Fig 8.35** (in Appendix B).

Should you need something with more output, take a look at the TDA2004 or TDA2005. In any event you will need to use decent speakers to get the full benefit of PIC-ASTAR's audio quality. Several builders have found it difficult to move away from 20W per channel into 12in speakers - including me.

PICADAPTER BOARD

This board is pretty tight since it needs to fit within the envelope of the original DDS board. So, as you can see from **Fig 8.36**, it is somewhat three-dimensional. The spacing is eminently achievable provided the components are loaded in the correct sequence - which is critical.

Check meticulously for continuity and isolation as you proceed. When inserting the sockets / plugs, ensure the pin shoulders do not ground on the opposite side.

1. Fit C73 underneath (ie on the groundplane side).

2. Fit IC9 socket and ground pins 8 and 19 on the groundplane side.

3. Fit R43, underneath. 4. Cut a 9-way SIL header strip. Noting the larger diameter end fits to the PCB, cut off that end of pin 3 and insert to make PL1 pins 1-9, ie the inner strip. Solder all track-side pins - and pin 5 to ground underneath.

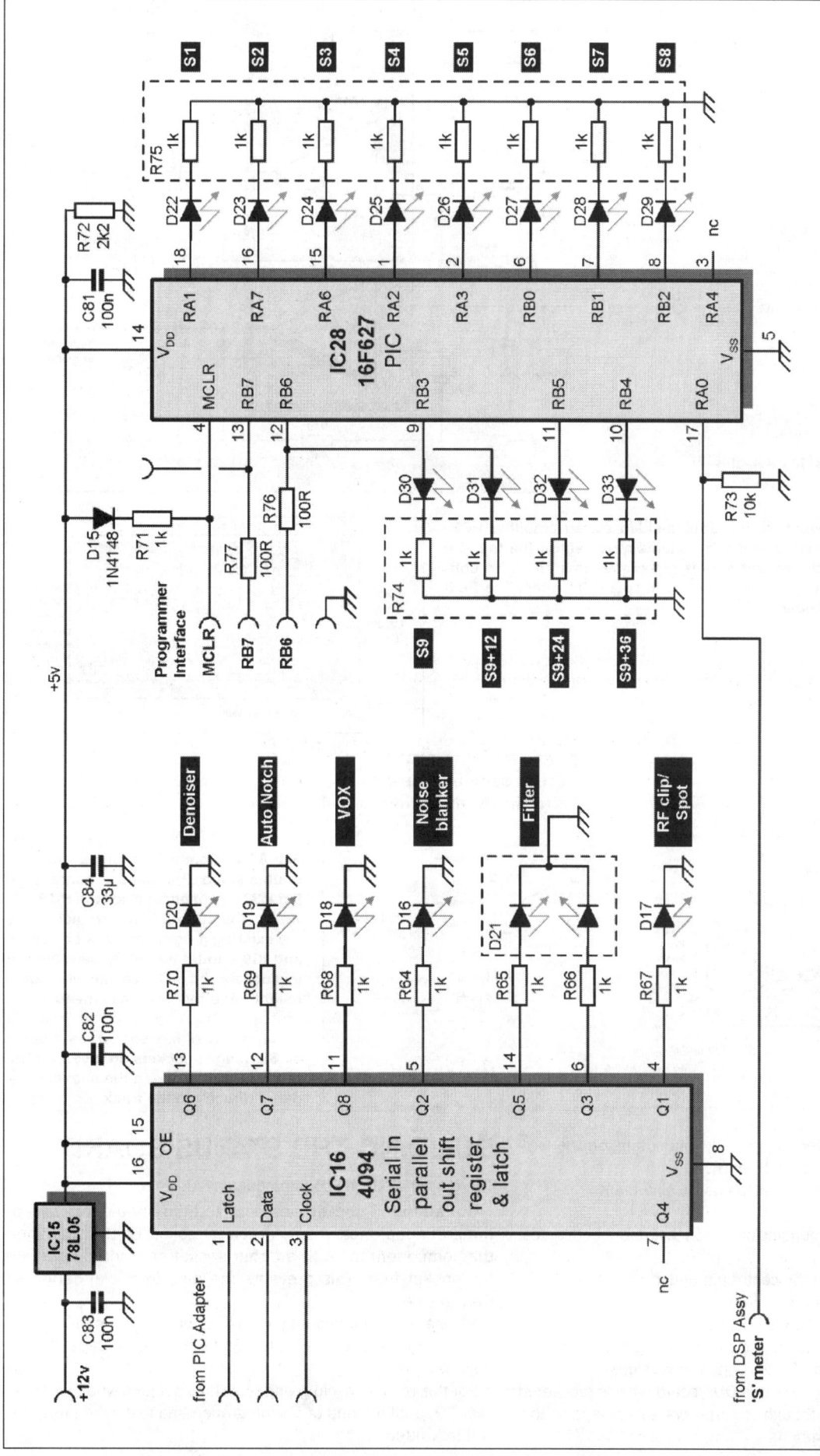

Fig 8.32: Status board circuit diagram. The status LEDs and driver are functionally unrelated to the S-meter LEDs and driver, but they are co-located around the keypad. Not shown here are some merely passive tracks which are used to make off the 7-way ribbon cable to the keypad. The connector shown on RB7 is for future development only

Fig 8.33: A suitable 1W + 1W stereo audio amplifier. Stereo balance and gain are controlled in DSP

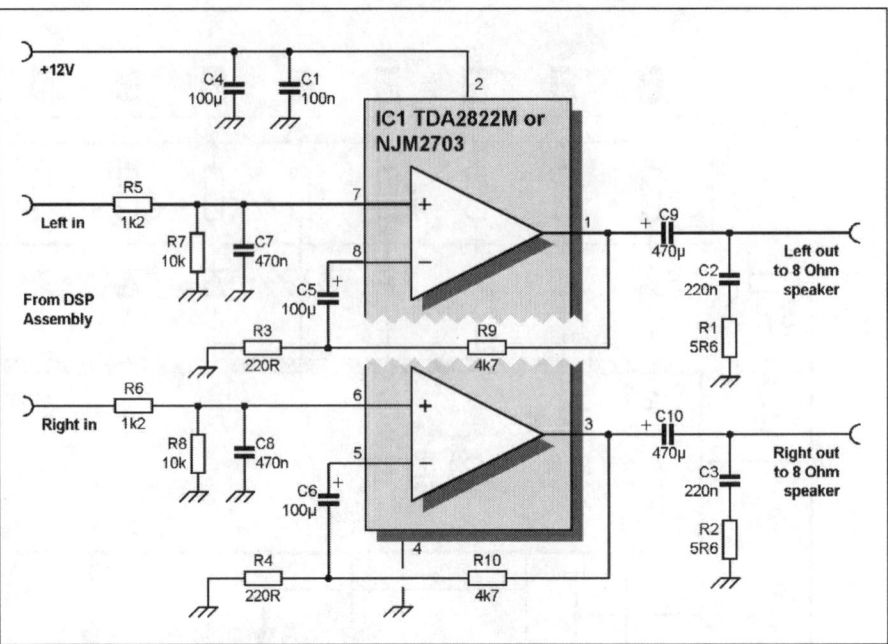

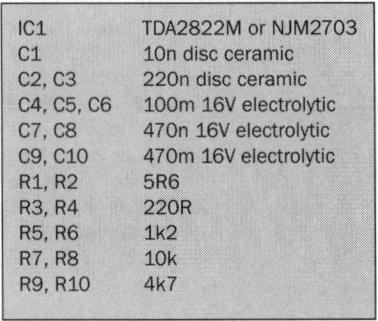

IC1	TDA2822M or NJM2703
C1	10n disc ceramic
C2, C3	220n disc ceramic
C4, C5, C6	100m 16V electrolytic
C7, C8	470n 16V electrolytic
C9, C10	470m 16V electrolytic
R1, R2	5R6
R3, R4	220R
R5, R6	1k2
R7, R8	10k
R9, R10	4k7

Table 8.3: Components list for the stereo amplifier

Fig 8.34: Component layout for the double-sided stereo amplifier PCB. This is a 'conventional' board with components mounted on the top, the track underneath. The component side is completely unetched. No connectors are specified, but the relevant pads have a 0.1in pitch. The PCB artwork (Fig 8.35) is in Appendix B

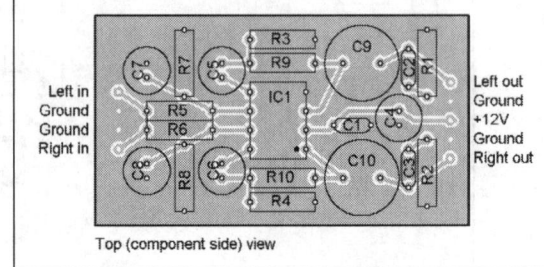

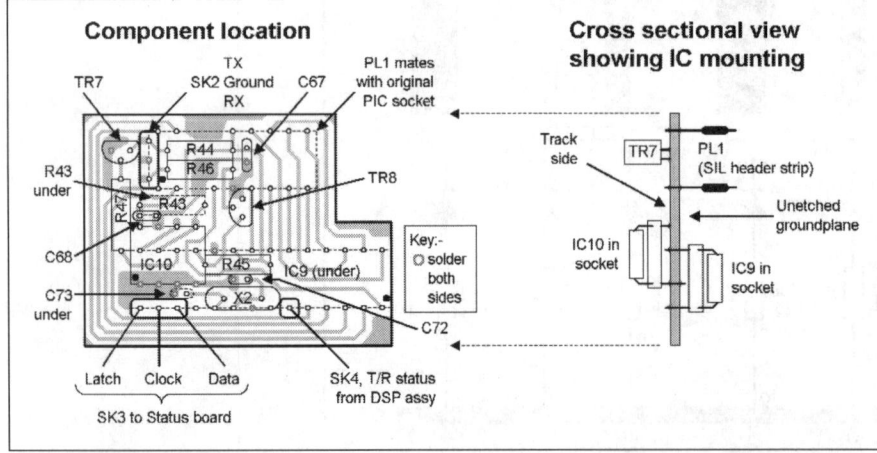

Fig 8.36: Component layout for the double-sided PicAdapter board (see Fig 8.37 in Appendix B for the PCB artwork). The cut-out is to give access to the existing programming socket. IC10 and IC9 should definitely be mounted in sockets. To achieve the clearance height required you may need to fit first an extra 18-pin socket into the existing Pic 'N' Mix socket -as a spacer. SIL plugs / sockets are used for the remaining leads - with the sockets soldered directly to the track

5. Repeat for the outer strip of PL1 - but cutting off the larger diameter end of pins 15 and 16.
Use a spare 18-pin socket to ensure alignment.
6. Fit TR8, C72, R44.
7. Fit IC10 socket, grounding pins 1, 2, 3, 4 and 7 both sides.
8. Fit C68.
9. Fit SK2, grounding the centre pin underneath.
10. Fit TR7, grounding the emitter underneath.
11. Fit R46, R45, C67, R47 and last, X2.
12. If there is any chance of the board fouling the Pic 'N' Mix Display board, chamfer the copper both sides.

The original 4MHz crystal on the DDS board may be recovered and reused elsewhere, though you may want to postpone this until the PicAdapter is working.

BUILDING THE STATUS BOARD

There are no special constructional issues here.

Cut all the IC socket pins back to their shoulders - except the grounded ones. A small trick for soldering the socket on the component side; fit another socket or a scrap chip into the socket first. This prevents the pins from wandering as they get hot.

Please note that should you wish to program this PIC in situ, you may not be able do so with D31 fitted, since it loads the programmer.

For this reason, avoid giving reports to stations who are "12dB over S9" until the end of full integration and test -when this LED is finally fitted.

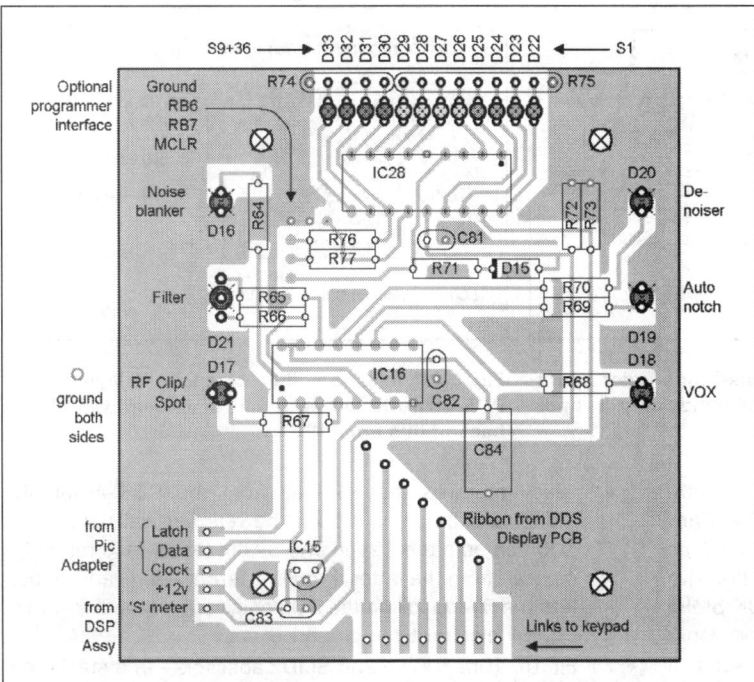

Fig 8.38: Component layout for the double-sided Status board PCB with the ground-plane side unetched (see Fig 8.39 in Appendix B for the PCB artwork). The picture is of the track side, viewed from the rear of the front panel. The width of this board (3in) accommodates that of the RF front-end which sits behind the Status board. All components except the LEDs are surface-mounted on the track side. This allows access to both the components and to the LED pads. The latter is needed to adjust the LED lead lengths for flush-fit to the front panel. The four mounting holes are for the keypad which is mounted using short spacers. The 7-way ribbon cable from the Pic 'N' Mix Display board is routed between this board and the keypad and made off to pads / tracks provided on this board and thence via short wire links to the keypad itself. IC16 and IC28 are mounted in sockets with all pins - except their respective ground pin - cut back for surface mounting. No connectors are specified. The relevant pads have a 0.1in pitch

PIC 'N' MIX FUNCTIONALITY

DDS Key Sequences

The keypad sequences are summarised in **Table 8.4** These include the increments for STAR operation.

Sequences shown with a leading 9 save the current frequency in the respective location if the 2-key sequence is preceded by the 9 (Save) key.

Pic 'N' Mix Code Changes

The following apply to code shipped explicitly for use with PIC-A-STAR:

- The opportunity has been taken to de-bounce the keypad more vigorously. The simple consequence of this is that key presses now need to be somewhat more deliberate.

- CW offset calibration (previously 34) is no longer required, since the CW offset is now managed in DSP. The procedure for calibrating the reference clock and SSB IF offsets is unchanged. However, the reference clock frequency calibration may now be loaded directly from your PC. For IF offset calibration you will find it useful to switch the DSP filters off . so that you can hear right down to zero-beat. A CRO connected to the receiver audio output is also invaluable for seeing the exact zero-beat point. When both SSB offsets are correct, there should be a difference of precisely 2.7kHz between them.

- CW is now received either upper or lower sideband . so you have the choice of tuning direction. Either in the same direction as SSB signals on the same band - or always tuning CW in the same direction whatever the band. The frequency readout always shows your transmitted frequency (not least for legal reasons) and your receive frequency is displaced from this by the CW offset . which is your chosen and preferred beat note.

- If you are in CW mode and you key 44 again, this will toggle reverse CW. This switches both receive sideband and injection frequency to give the same beat-note, but 'from the other side'. It can be useful in clearing QRM and for checking you are properly netted since if not, the beat note will change.

---

* - toggle between VFOs
0 - VFO A = VFO B
- Split operation
11 - select USB
44 - select CW . again for reverse CW
77 - select LSB
22 - toggle rate tuning mode on / off
26 - toggle software flywheel on / off
9 31 - calibrate / save USB offset
9 37 - calibrate / save LSB offset
9 33 - calibrate / save reference clock
41 - go up to nearest kHz point
48 - toggle ¡®spare¡¯ latch output
47 - go down to nearest kHz point
55 - monitor guard channel
56 - scan between memory frequencies
58 - scan frequency range (wobbulator)
5* - scan both VFOs
9 6x - where x is 0 - 9.Go/save to memory - may be used for 5MHz frequencies
70 - display tuning rate as bargraph
72 - spare user on / off switch
73 - go to SLEEP
74 - toggle display auto-dim
78 - display LSD as 10Hz / 100Hz
79 - toggle x6 ref clock for AD9851
81 - high side injection (default)
87 - low side injection
83 - signal generator mode (no offset)
88- direct keypad frequency entry
9 10, 12, 15, 17, 20, 30, 40, 60, 80,
16 - go/save to respective band
990- save as power-on frequency
999 - reboot DDS software

Table 8.4: DDS key sequences (see text)

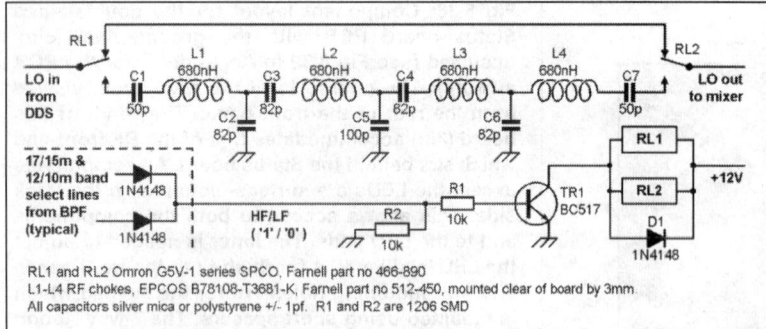

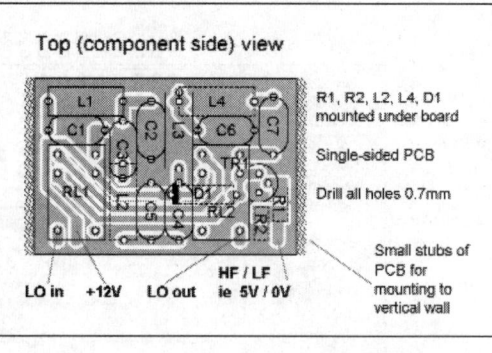

Fig 8.40: 26-40MHz injection filter to strip both unwanted high-order spurs and those at the IF from the DDS output on the higher HF bands only. It is not needed (ie switched out) on lower bands

Fig 8.41: Injection filter component layout. The PCB artwork can be found in Appendix B at the back of this book

- Pic 'N' Mix now has 5MHz (60m) capability. The latch output previously marked "15MHz WWV" is now active if any 5MHz frequency is selected and . exceptionally . if you go to 5MHz, the default sideband is now USB. The first five memory locations (60.64) are loaded with the UK 5MHz channels. These are the correct frequencies for upper sideband operation. If you don't want this feature, you can re-program these allocations with any other frequencies (and re-enter the 5MHz ones yourself any time).

- You may now fit an AD9851 DDS chip which has the ability to multiply the reference clock frequency by six. 79 toggles this feature. See later for more detail.

- There is a latched output bit labelled 'spare'. This may be toggled by keying 48. It was designed so that you can switch any device - a pre-amp, attenuator, transverter etc - from the keypad. This switch has been used to configure the STAR mixer and post-mixer amplifier to be described later.

- The output bit labelled 'broadband' 72 is now a spare uncommitted toggle switch.

- QSK Split operation has been improved. Previously it was limited to about 20WPM.

- The frequency display is dimmed after about three minutes of user inactivity - to reduce heat dissipation and to prolong LED life. 74 toggles this feature on/off.

- All frequencies from 0.29MHz now activate the nearest band select line.

- RIT and XIT operation may be selected instead of Split. The # and . keys and the Rx/Tx A/B LEDs then change meaning. This is still under development. Detail follows.

- Part-way through any DDS key sequence, the # key now aborts it.

PIC 'N' MIX HARDWARE

Since first designing Pic 'N' Mix five years ago, some detailed improvements have evolved. What will never change is the requirement for meticulous (albeit textbook) screening, decoupling and filtering practice . if the DDS spur level is to be contained. This cannot be overstated and is the starting point for what follows.

The following modifications produce incremental reductions in DDS spurs and are easy enough to do to make them all worthwhile. If you are building STAR, I countenance you not to implement any of them until you have everything else working . and until you have truly attended to the meticulous bits just mentioned.

- Fit a separate 7805 regulator for the DDS chip carrier assembly. There is a simple track cut under the DDS board

which removes the +5V rail from the DDS Assembly. Mount a separate 7805 on the rear vertical panel with 100n on both the 12V in and 5V out leads - and run a flying lead from the latter through a ferrite bead - and solder it to the 5V top foil on the DDS chip carrier together with a 100¥iF electrolytic to ground.

- Fit 1n, 10n, 100n 1206 SMD capacitors - in a stack - on at least two corners of the DDS assembly from the +5V foil to ground. In other words, whatever it takes to ensure that the +5V rail and ground-plane are at the same AC potential - AF to VHF.

- Add a filter in the LO feed to the mixer. **Figs 8.40 and 8.41** show a suitable arrangement for high-side injection with any IF between 8MHz and 11MHz. This filter offers at least 30dB attenuation at the IF and a similar figure at 50MHz - and rising. If all else is right, this produces dramatic results. It does not need any exceptional screening if mounted within the already-screened volume of Pic 'N' Mix. It is shown switched in by band-select lines, diode-ORed. You should wire in diodes for any bands for which your LO falls in the range 26 . 40MHz. However, for evaluation purposes, it would be prudent to control it with a simple toggle switch to +5V / 0V in the first instance.

- You may simply substitute an AD9851 on the DDS Assembly in lieu of an AD9850 under all circumstances, ie it is completely hardware and software compatible. The AD9851 allows higher reference clock frequencies . which may be useful if you have a relatively high IF. This gets you away from the 1/3 reference clock zone. In addition, it offers a 6x multiplier option for the reference clock. So, for example, you could clock it at a mere 30MHz yet have the effective benefit of a 180MHz clock. There are small spur and significant phase noise performance disadvantages for which see the AD9851 data sheet - but convenience issues may predominate for you. A facility for specifying a 6x clock is now built into STAR software. You may instead want to try clocking your existing AD9850 much faster.

Fig 8.42 shows a reference oscillator which will operate on either the 5th or 7th overtone of a crystal in the 22 - 25MHz range . simply by tuning VC1. Typically, the 5th overtone is used with an AD9850 and the 7th overtone with an AD9851. In this latter case, the x6 feature would not be invoked. Also included in Fig 8.42 is an arrangement for stabilising the crystal temperature, the detailed design of which is due to G3NHR. It works so well, I have since stuck (literally) a similar arrangement on my IF board translation oscillator. The thermistor, TH1, and the TIP122 tab are secured to opposite faces of the crystal can. Heat-

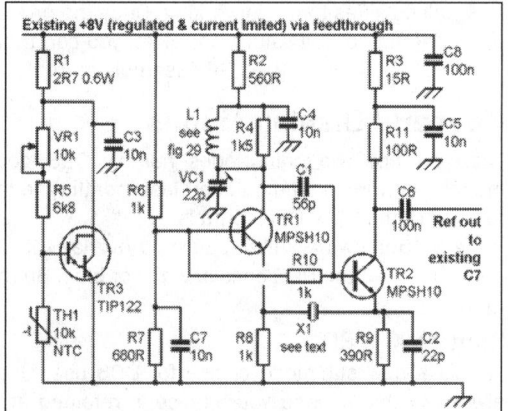

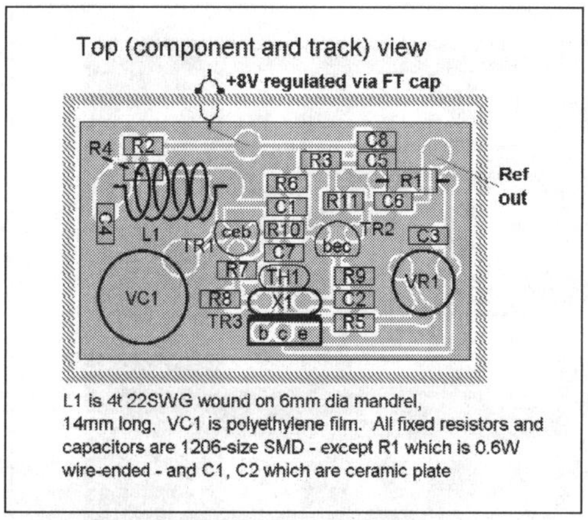

Existing +8V (regulated & current limited) via feedthrough
R1 2R7 0.6W
R2 560R
R3 15R
C8 100n
VR1 10k
C3 10n
L1 see fig 29
R4 1k5
C4 10n
R11 100R
C5 10n
R5 6k8
R6 1k
VC1 22p
C1 56p
C6 100n
Ref out to existing C7
TR1 MPSH10
R10 1k
TR2 MPSH10
TR3 TIP122
TH1 10k NTC
R7 680R
C7 10n
R8 1k
X1 see text
R9 390R
C2 22p

Fig 8.42: Butler oscillator for either 5th or 7th overtone operation . with crystal temperature control

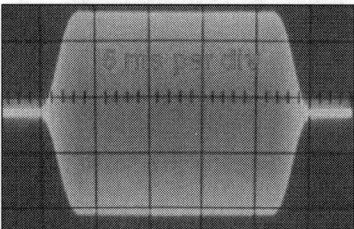

Top (component and track) view
+8V regulated via FT cap
Ref out
L1 is 4t 22SWG wound on 6mm dia mandrel, 14mm long. VC1 is polyethylene film. All fixed resistors and capacitors are 1206-size SMD - except R1 which is 0.6W wire-ended - and C1, C2 which are ceramic plate

Fig 8.43: Reference oscillator component layout. The PCB art-work can be found in Appendix B

shrink sleeving is highly recommended . or failing that, super-glue. For smaller crystals, cut off the excess TIP122 tab to reduce height. Set VR1 to maximum resistance and then adjust for 100mV drop across R1. Repeat every minute for five minutes and the result should be a crystal thermally stable at 35 degrees C.

In use, the frequency will change rapidly for the first five minutes after switch-on . but stabilise thereafter. This is not an on-off oven. This is proportional control.

The small PCB, shown in **Fig 8.43**, is designed as a drop-in replacement . after simply removing the components from the original Pic 'N' Mix oscillator.

- Fit transformer-coupled output from the DDS chip. This gives 6dB more LO output and further spur reduction. The core for this transformer is the EPCOS B62152A4X1 available from ElectroValue [9]. (You will need four more of them for the mixer later). The primary is three bifilar turns 32SWG and the secondary is 12 turns wound over the top. The core is mounted in lieu of the 100 and 200 ohm resistors on the DDS carrier. Connect the primary instead to pins 20 and 21 of the DDS chip, grounding the centre-tap. Ground one side of the secondary and take the other via a 100n blocking capacitor to pin 19 on the 28-pin carrier. If you are not confident your mixer will stand the extra 6dB, fit a pad on its LO port. (The STAR mixer . yet to be described . is fine.)

To reduce DDS spurs to a highly acceptable level (ie virtually none) you may or may not need any or all of these changes.

As with all flexible designs, there are intelligent user choices to be made. Such is amateur radio!

FACILITIES

PIC-A STAR has more useful facilities than most homebrew designs - and getting a feel for their value may well determine if this is the project for you.

Those who don't have these facilities refer to them as 'bells and whistles'. Those of us who do just grin - and keep ringing them bells and blowing them whistles! Always assuming they are underpinned by a rock-solid base receiver performance, that is.

SSB/CW Mode Management

Switching between SSB and CW - and transmit and receive for that matter - are non-trivial design problems if the result is to be user-friendly. So some background discussion is helpful in understanding what follows. See also ref [10]. You may also care to compare critically the behaviour of commercial transceivers. If I can find anything friendlier than STAR, I will simply change the design until it isn't. I have, as I write, already invested in the architectural infrastructure to make all this possible.

Fig 8.44: The STAR CW waveform from G4HMC, photographed off-air on 80m from the author's STAR

CW operation

In days of old, especially with 'separates', it was easy, reliable - but a bit tortuous. You would zero-beat an incoming CQ on your receiver, then zero-beat your transmitter - and finally move your receiver off to get a comfortable beat note.

With modern filters there is a snag. You can't hear anywhere near down to zero beat, so this process produces totally unacceptable errors.

However, it does establish two critical principles, namely: both stations must transmit on the same frequency - and both must operate 'split' if they are to hear a beat note. CW is inherently a 'split' mode.

With a multi-mode transceiver, you must either explicitly operate 'split' for CW - or the design must take care of it transparently.

PIC-A-STAR is in the latter category. If you are in SSB mode and hear a CW station you want to work, when you switch to CW mode the received pitch will not alter and you will not have to retune. And vice versa if starting out from CW mode.

While on the topic of CW, take a look at the photograph of STAR's transmitted waveform. You won't find better.

SSB operation

When you change sideband, neither your indicated nor actual frequency should alter. This is common currency nowadays. There are some, but not many, occasions when this matters - given that we don't usually operate on the 'wrong' sideband. If you operate via OSCAR or into a transverter or on 60m, it is critical.

This point also reinforces a principle which should be obvious - namely that if a given feature is critical to a minority interest, provided it is not imposed on the majority - then there is no good excuse for not providing it.

Pic 'N' Mix also gives you the option to switch between high- and low-side injection. Since this implies a sideband inversion, the software puts in an equal and opposite change of sideband - and you end up on the same net frequency.

But be aware that many band-pass filters are optimised for injection from a preferred side. The STAR front-end is optimised for use with high-side injection; the optional filter in the LO line also assumes high-side injection on the higher bands.

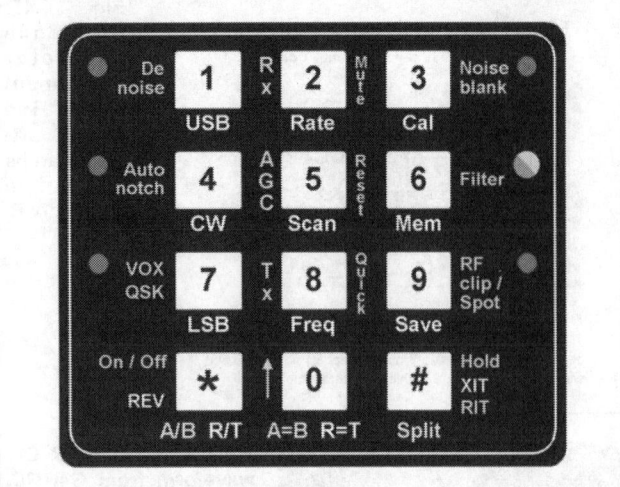

Fig 8.45: Keypad allocations and corresponding status LEDs. This is to scale and may be used as a keypad overlay

PTT and key behaviour

PIC-A-STAR uses the simple conventions that, on CW, the microphone audio is ignored – and on SSB the key is ignored.

If on CW and you merely key, you will produce only sidetone. This is for CW practice since, to transmit in earnest, you either need to switch QSK on – or hold down the PTT line for non-QSK operation.

If on SSB, you must either switch on VOX – or hold down the PTT line – before anything will happen.

AT POWER-ON TIME

The display shows your chosen startup frequency, but flashing. You then have four structurally-different options:

- Upload DSP code from Pic 'N' Mix to the DSP Assembly. This is normal every-day operational use. The Status board LEDs flash strangely so you know something is happening.
- Run the DDS without loading DSP code. This is mainly a diagnostic mode.
- Enter a DDS reference clock frequency from your PC. A useful utility.
- Download a new (or your first) DSP software release – via the Internet to your PC – and thence to Pic 'N' Mix. This latter process takes several minutes. During this period the incoming bytes are counted on the display – albeit faster than the eye can follow – much like money on a petrol pump. Unlike petrol it gives you a warm feeling you are getting good value – and the fact that it is counting at

all signifies that it is working. Once the new code and the default control values are all in, you can then proceed to upload them to the DSP Assembly thereafter.

Keypad / Display Modes

There are now two main modes, namely 'DDS mode' and 'DSP mode'. The former was outlined last month. The latter lets you tune all of the STAR DSP controls.

Fig 8.45 shows a suitable keypad overlay with the DSP legends in yellow; the DDS ones in white or blue. But before we get to that . . .

Split or XIT/RIT

This is a new sub-mode choice for DDS use. One of these is always enabled – and your choice is retained at power-down. Both give you the potential to transmit and receive on different frequencies. See also [11] for a general discussion.

Split mode is unchanged from Pic 'N' Mix – but with enhancements. It operates on the two independent VFOs, 'A' and 'B'.

Conversely, XIT/RIT operates on either one of the VFOs – and that choice determines the initial Tx and Rx frequencies. But, thereafter, the Tx and/or the Rx frequencies can be independently changed – and retained throughout an XIT/RIT session. Meanwhile, the 'other' VFO remains uncontaminated – and available.

What are the differences between Split and XIT/RIT? Split can be cross-band and/or cross-mode and 'rests' on your Rx frequency when off. Conversely, the XIT/RIT tuning range is the current band, current mode. It 'rests' on your Tx frequency when off – thus providing an RIT on/off capability – which Split doesn't give you.

Split and XIT/RIT use

Besides pure transceive when switched off, both modes have the following options:

- Tune only the receive frequency while your transmit frequency remains fixed (ie RIT). If you call CQ and a station answers off-frequency, this option (in either mode) is the answer. However, in a net with one station off-frequency, XIT/RIT mode is better since you can turn XIT/RIT off when the offending station is not transmitting.
- Tune only the transmit frequency while continuing to monitor your receive channel (ie XIT). To call a DX station who is operating split, use in either mode to tune your Tx quickly to the specified DX listening frequency – while not missing a word on your receive channel.
- Tune the transmit frequency while monitoring what is about to be your transmit channel (ie REV XIT). Use this to check for a quiet spot before calling.

You can do any of the above – independently – in any order using merely one key-press to define your choice. On receive,

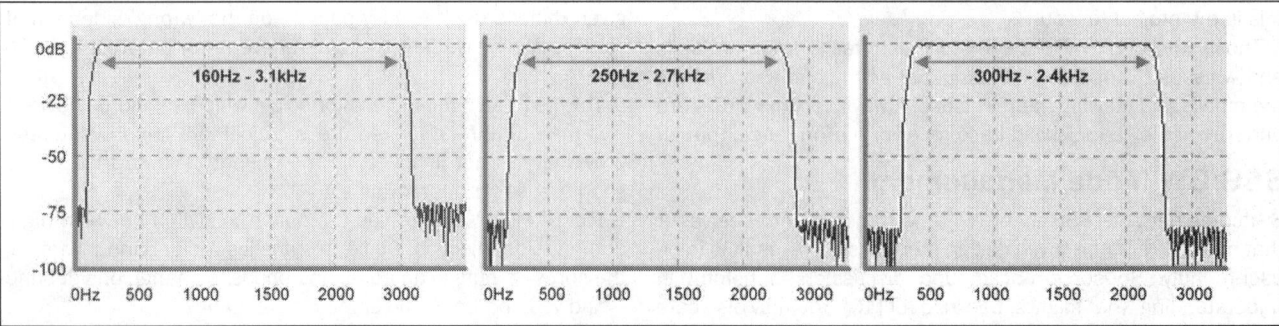

Fig 8.46: Plot of wide, medium and narrow Rx SSB filters. Other filters in the DSP receiver give a further theoretical 30dB of ultimate stop band. The pass-band ripple in all cases is less than 0.2dB. Because STAR is not a mere audio add-on, these widths are actually achievable – and usable

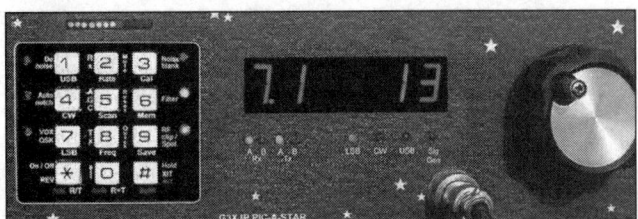

Fig 8.47: STAR display in DSP mode

1*	Denoiser ON/OFF	
1.1	Denoise beta	
1.2	Denoise decay	
2*	Mute Rx and suspend VOX	both
2.1	AF gain	both
2.2	RF gain (per band)	both
2.3	Stereo effect (ie amount)	both
2.4	Stereo balance	both
3*	Noise blank ON/OFF	
3.1	Noise blank threshold	
4*	Auto notch ON/OFF	
4.1	Auto notch beta	
4.2	CW tones, 1 or 2	CW only
4.3	CW offset frequency	CW only
4.4	Sidetone frequency	CW only
5*	Reset downloaded values	both
5.1	AGC hang time	
6*	Filter ON/OFF	
6.1	Filter width (1 – 6)	
6.2	Filter depth	
7*	VOX/QSK ON/OFF	
7.1	VOX/QSK hang time	
7.2	Rx – Tx pre-delay	
7.3	Tx – Rx blank time	
7.4	VOX gain	SSB only
7.5	Anti-VOX gain	SSB only
8*	Quick Switch ON/OFF	both
8.1	Tx drive level (per band)	
8.2	Monitor level	
8.3	Mic gain	SSB only
8.4	Tx Top boost	SSB only
8.5	Tx Bass boost	SSB only
9*	RF clip/Spot ON/OFF	
9.1	RF clip ON/OFF	SSB only
9.1	Spot level	CW only
#	Hold DSP mode ON/OFF	both

Table 8.5: DSP menu structure

the frequency displayed will be that which you are changing – and on transmit, always your Tx frequency.

Further utility options let you swap the two VFO frequencies – or initialise them as the same. Likewise for the XIT/RIT frequencies.

Initialisation is done for you in XIT/RIT mode should you change VFO – or band – or frequency by more than 2.5kHz while on pure transceive – on the grounds that any difference must then be irrelevant.

DDS/DSP mode switching

As supplied, the DDS mode is permanently engaged and you would be unaware that there is any other. This is deliberate in order that you can check out the DDS functionality after first commissioning – without distraction.

Once you are happy here, the DSP mode becomes available following the first time you download DSP code from your PC to Pic 'N' Mix. The remainder of this discussion assumes that this has happened.

In normal use, STAR 'rests' in DDS mode and displays frequency. The key to switching to DSP mode is the duration of the first key press. A quick press on a key activates DSP mode; whereas a longer press invokes the 'business-as-usual' DDS function.

The resultant displays are quite different, so it will only take you a few minutes to get the 'feel' ingrained. The DSP functionality is given this priority because most DSP functions are needed quickly in real operating conditions. For example, turning the auto-notch on when that tuner starts up is a more immediate issue than, say, changing bands. Certainly a 'quick press' need not be tentative, merely not overtly sustained. Once the mode has been determined by the duration of the first key press, the duration of subsequent key presses is unimportant.

The very deliberate exception is the bottom row of keys – which act to give vital DDS functions – and which therefore have no DSP functionality as the first key-press.

Once you are in DSP mode if you neither press a key nor alter a value for about three seconds, PIC-A-STAR will revert to DDS mode. You can prevent this by pressing the # key – which toggles holding the display in DSP mode. This is invaluable when setting up the DSP control settings.

USING THE DSP MENU

The DSP controls are grouped to form a menu, as shown in **Table 8.5**. This menu detail will change over time, but not the intrinsic structure. Project without end, right?

Menu Groupings

The menu is ordered with the more commonly-used controls near the 'top' of each menu group – and rarely used ones near the 'bottom'. In practice, some of these latter items can be regarded as presets.

There are no sub-menus, so the system is inherently limited to 99 controls in nine groups times two modes .

The menu is intrinsically SSB/CW modal in the sense that if you are in SSB mode then you simply can't get at controls which are unique to CW – and vice versa. These controls are annotated 'SSB only' or 'CW only'. Some controls share one common value for both modes and are annotated 'both'. Two controls, namely 2.2 and 8.1, have different values per band.

Conversely, all other menu items that are not peculiar to mode have different settings stored for SSB and CW. For example, different AGC time constants, Denoiser settings and so on can be set up, varied – and retained independently for SSB and CW. This applies also to the on/off switch settings.

All this is designed to foster a 'set-and-forget' philosophy.

How to PIC from the Menu

Immediately after you (quickly) press a 1-9 key, the display will switch to DSP mode. For example, a dab on the 7 key will show '7.1 13' where 7.1 denotes the first control in that menu group, namely VOX/QSK hang time – and 13 is its present value. See also **Fig 8.47**.

If you want to move on to the second control in the same group, press the 7 key again – and so on. If you want to move back up through the group, press the 0. If you want to move to a completely different menu group, simply press the corresponding key.

Changing Values

After you have selected a control, if you want to change its value, turn the knob – clockwise for more, anticlockwise for less.

There are maximum and minimum values for each control – and the rate of change is proportional to the range.

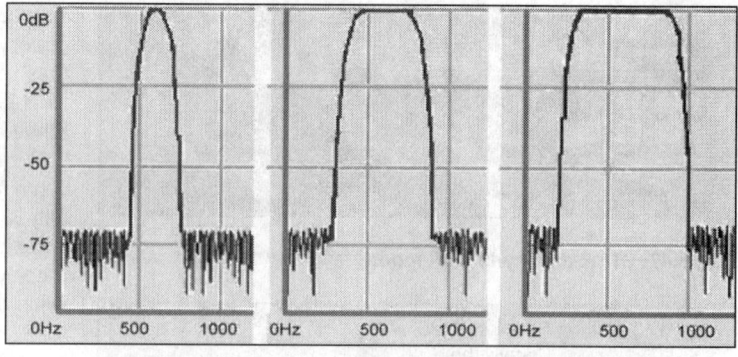

Fig 8.48: Plot of the Rx CW filters. This is the bank centred on 600Hz – and those on different centre frequencies are otherwise similar. Their widths are approximately 200Hz, 500Hz and 750Hz but this absolutely depends on where you measure them. This filter shape gives less ringing than a 'brick wall' type. Other filters in the receive path give a further theoretical 30dB of ultimate stop band

ON/OFF Switches

To switch a DSP feature on/off, press the corresponding key (1, 4, 7, 3, 6, 9) followed by *. The adjacent LED will change to provide visible status thereafter.

For example 7*. will switch VOX on/off if in SSB – and QSK on/off if in CW mode.

In fact, irrespective of which menu item in a group you are addressing, the . key will toggle the associated switch and you will revert immediately to DDS mode.

Mute

2* near-mutes the receiver and suspends VOX operation. This is the 'panic' button for unexpected interruptions eg when the phone rings. Any subsequent key press or knob turn restores your pre-panic state.

Quick Switch Option

8* toggles the quick switch facility. When engaged, any one of the 1, 4, 7, 3, 9 keys – when pressed – simply toggles its respective DSP switch.

Because you are thereby not presented with the values for those menu groups, you would not want to use this option until those groups are set up. Conversely, once the values are tuned and you have gained familiarity, this could be the mode of choice.

Resetting Control Values

5* resets the control values (both SSB and CW) to those that you last downloaded from the PC – with the exception of RF Gain and Tx Drive – the latest per-band values of which are retained. All the DSP values are remembered across a power-down, but not switch settings – which initialise to off but with the DSP filter on – and in SSB mode.

DSP FEATURES DESCRIPTION

A few words are in order for some of the more esoteric features you may not have met before. In roughly menu order:

Denoiser

This is rather more a comfort feature than a performance one. It acts to reduce background white noise – and when working well, is not unlike squelch on FM. It is especially effective on CW and very useful if just monitoring a quiet (albeit noisy) channel.

The 'right' combination of settings is somewhat subjective and can occasionally vary from one signal to another – and certainly by mode. It is best with the RF gain turned up and with longer AGC hang times. Experiment!

Both the Denoiser and Autonotch (see later) are essentially as implemented in DSP-10 by Bob Larkin. See also [12-15] for the pioneering work and the theory.

RF Gain

This comes right at the front of the DSP receiver chain and is used to set the SNR for different conditions. It should normally be turned well up so that AGC action produces constant audio output – and the best possible SNR. This also contributes to clean VOX operation.

Stereo Effect

This gives body and presence to signals and warrants a decent stereo audio amplifier and speakers. Some folks report an increase in readability on weak signals. Personally, I just love it! For me, it completely transforms the listening experience.

Stereo balance

Values above 100 decrease the right channel output; those below 100 decrease the left channel output. Another scratchy pot saved.

Noise Blanker

As opposed to the Denoiser which acts on white noise, this acts on impulse interference – eg ignition noise, thermostats, electric fences and the like.

Auto Notch

This removes an interfering heterodyne – and, in many circumstances, several. It works best on pure tones and especially lower-pitched ones. It has exactly one use in CW mode, namely for monitoring key clicks ie what's left after removing the tone. One very popular commercial transceiver shows up here every time.

Auto notch is applied after the filter bank and is outside the DSP AGC loop – to avoid strong-signal overload of the DSP.

CW Offset

This is your preferred beat note and may be pre-set (when you download from your PC) to 5, 6, 7, 8 or 900Hz. The centre frequency of the CW filters is changed to match.

If you are interfacing with other STAR display in DSP mode than Pic 'N' Mix, you will need to adjust your Tx/Rx mixer injection frequency by mode. So, for the record, the following are the exact DSP IFs (in kHz) used by PIC-A-STAR:

LSB	Tx = Rx = 16.35
LSB CW	Tx = 16.35, Rx = Tx + CW offset
USB	Tx = Rx = 13.65
USB CW	Tx = 13.65, Rx = Tx − CW offset

There is also a Reverse CW option and if set:

LSB CW	Tx = 16.35, Rx = Tx − CW offset
USB CW	Tx = 13.65, Rx = Tx + CW offset

Sidetone Frequency

Not to be confused with CW Offset, this is the tone you hear when sending CW. The QSK experts tell me it can be useful to

have this at a different pitch from an inbound signal – so you can intuitively tell the difference between you and the station being worked when using fast break-in.

To enhance this effect, the sidetone comes from one speaker only, the inbound signal from both. The pitch may be excessively varied between 10Hz and 2.54kHz in 10Hz increments – this extended range being useful should you need it to double as an instant audio signal generator.

CW Tones 1 or 2

This allows two-tone testing. The two tones are 700Hz and 600Hz. If you have not used a two-tone test for linearity checking before, be aware that the duty cycle is very high – so use only short or pulsed bursts.

AGC Hang Time

This can be set anywhere between very short and very long. I understand that most people usually only change this per QSO for CW work – and so you can vary the CW setting without altering the SSB setting. A value of 0 turns DSP AGC off.

While actually changing frequency, the hang-time is set to short to avoid annoying hangs after tuning across large signals.

Filter Width

This allows you to set the receiver filter width by turning the knob (or on/off using 6.). There are six filters currently provided, three each for SSB and CW (though any one can be used in either mode) – see Figs 8.46 and 8.48. The status LED is tri-colour and corresponds to wide, medium, narrow – or off. Turning the filter off is a useful way of checking if a signal has stopped transmitting or has just slipped out of the pass-band since, when you switch the filter back on again, it reverts to the previous filter width.

Filter Depth

This concept was inspired by yet another conversation with Bill Carver, W7AAZ. In traditional analogue terms, it allows a controlled leak past the filter. For CW operation it is in many ways more useful than controlling the filter width. In use, having tuned a wanted signal to the centre of the pass-band, you simply increase the filter depth (ie the stopband rejection) until the QRM is reduced to any level with which you feel comfortable. Putting it another way, you come up to periscope depth to find a target – centre it up – and then go down again so the nearby destroyers can't get you.

VOX and QSK

The hang time is the duration spent on transmit after you have stopped speaking (or keying) before PIC-ASTAR reverts to receive. If you have a relay-free T/R system, then there is no need to set other than a very low value here. With relays, any greater setting will minimise the number of rattling occasions. It is adjusted to allow the trailing edge of your transmission to pass before switching to receive.

The Rx-Tx pre-delay is the time your signal is delayed by DSP to allow for relay settling when switching to transmit. It is adjusted so that the leading edge of a transmission is not truncated – just.

The Tx-Rx blank time is the duration of DSP receiver blanking immediately after reverting to receive. It should be set to the minimal value possible, consistent with no objectionable click coming from the receiver after the transition. With a full STAR configuration, this is simply zero.

The above three parameters are separately set for SSB and CW. For SSB only, VOX and anti-VOX gains may also be set. See later for further discussion.

Tx Drive

This is the power-setting control. When on transmit, the S-meter reading corresponds to Tx drive level – and is therefore modal.

Monitor Level

On SSB this control sets the level at which you monitor your own voice – after all VOX processing and filtering. If, for example, you turn the Rx-Tx pre-delay up high, you will – somewhat disconcertingly – hear your very delayed voice. And you will hear leading or trailing edge truncation if the timing is not set up properly. For operational use, however, the level should be kept low to avoid any feedback; or worse, confusion of the VOX software.

On CW, this control sets the sidetone level; you do not hear a delayed signal – since this would play havoc with your sending.

Microphone Gain

In conjunction with the Mic Gain preset on the IF board (RV5), this should be set to provide adequate input to the software VOGAD. The latter will hold the audio amplitude at a substantially constant level even in moments of excitement.

Bass and Treble Boost

These act independently to tailor the transmitted audio profile.

RF Clipping

I have always found this the most effective form of SSB processing – as opposed to audio compression. It increases the average power while holding the peak power steady. In mechanical engineering terms, it increases your transmitted signal's power-to-weight ratio. So it also increases the strain on your power supply, linear and ATU.

Use it only sparingly and when necessary (and not because it is there), bearing in mind that any form of processing – by definition – introduces distortion.

CW Spot Level

A 'spot' tone (equal in frequency to your CW offset) may be injected into the receiver output. As you net onto an incoming CW signal you will hear it beat with the 'spot' tone – and when they are on the same frequency, you are indeed netted.

This control alters the minimum amplitude. However, the amplitude is also increased automatically with the strength of the incoming signal. This is done because it is easier to beat two notes of similar amplitude – especially when the 'spot' tone amplitude tracks the inbound keying.

Fig 8.49: The 'Magic Roundabout' of G3TIE

THE FRONT END

How Good?

When it comes to receiver front-ends, there are those who would scale the highest IP3 mountains, and those who explore the depths of classic simplicity and minimalism.

Each pursuit is valid and fascinating in its own right. I am typically to be found sitting on a fence, aware that better performance is always achievable – but unclear if it is of real operational value. Aware also that the law of diminishing returns cuts in exponentially when it comes to cost and complexity.

Conversely, faced with the delightful quality emanating from the STAR DSP, it would have been remiss not to provide it with a proportionate frontend. "How good?" is, as ever, the question.

I much enjoyed 'HF Receiver Dynamic Range: How Much Do We Need?' [16], not least because it was written in practical and tangible terms (see also [17] for a summary).

I conclude from this paper:

•   Phase noise performance is critical especially when there are many unwanted strong signals in the pass-band.

•   The strong-signal dynamic range (DR) requirement is not horrendous – if you are prepared to shift the DR up and down to suit conditions. That is, sometimes you need good sensitivity; sometimes you need good strong-signal performance. Rarely, in practice, can you use both. So I don't think I can use all the dynamic range offered by, for example, the CDG2000 design [18].

This argument assumes absolutely that you are not blessed with a specific point-source problem such as a strong nearby transmitter. Normally propagated signals are assumed to apply. In real life – in Europe – the ability to handle 40m at night is the pragmatic test.

Pic 'N' Mix has intrinsically superb phase noise performance. The trick is to avoid degrading it (eg by adding a crude PLL) to get round the issue of DDS spurs.

The strategy of moving the DR up and down is equally compatible with the need to extract the maximum possible range from a finite number of bits in down-stream DSP.

STAR achieves this with the ability to reconfigure the receiver gain distribution dynamically.

THE MAGIC ROUNDABOUT

With early STAR, I used a mere SBL-1 mixer both with a bi-directional 2N3866 post-mix amplifier – and with a bi-directional J310. They both 'work'.

But, in 1998, Colin Horrabin, G3SBI, showed that there are no technical excuses for not using an H-mode mixer [19] – and Giancarlo Moda, I7SWX, showed that, with a fast bus switch, it can be truly affordable – except – in my view – for those very expensive transformers.

However, I can't turn away the opportunity of an H-mode mixer with home-brew transformers since, although this degrades the intercept somewhat, there is still plenty in hand. And no great cost.

In fact, there is so much in hand that I contemplated the heresy of fitting an RF amplifier before the mixer. Or should I settle for less sensitivity and instead fit an IF amplifier after the mixer? I wanted both options; and

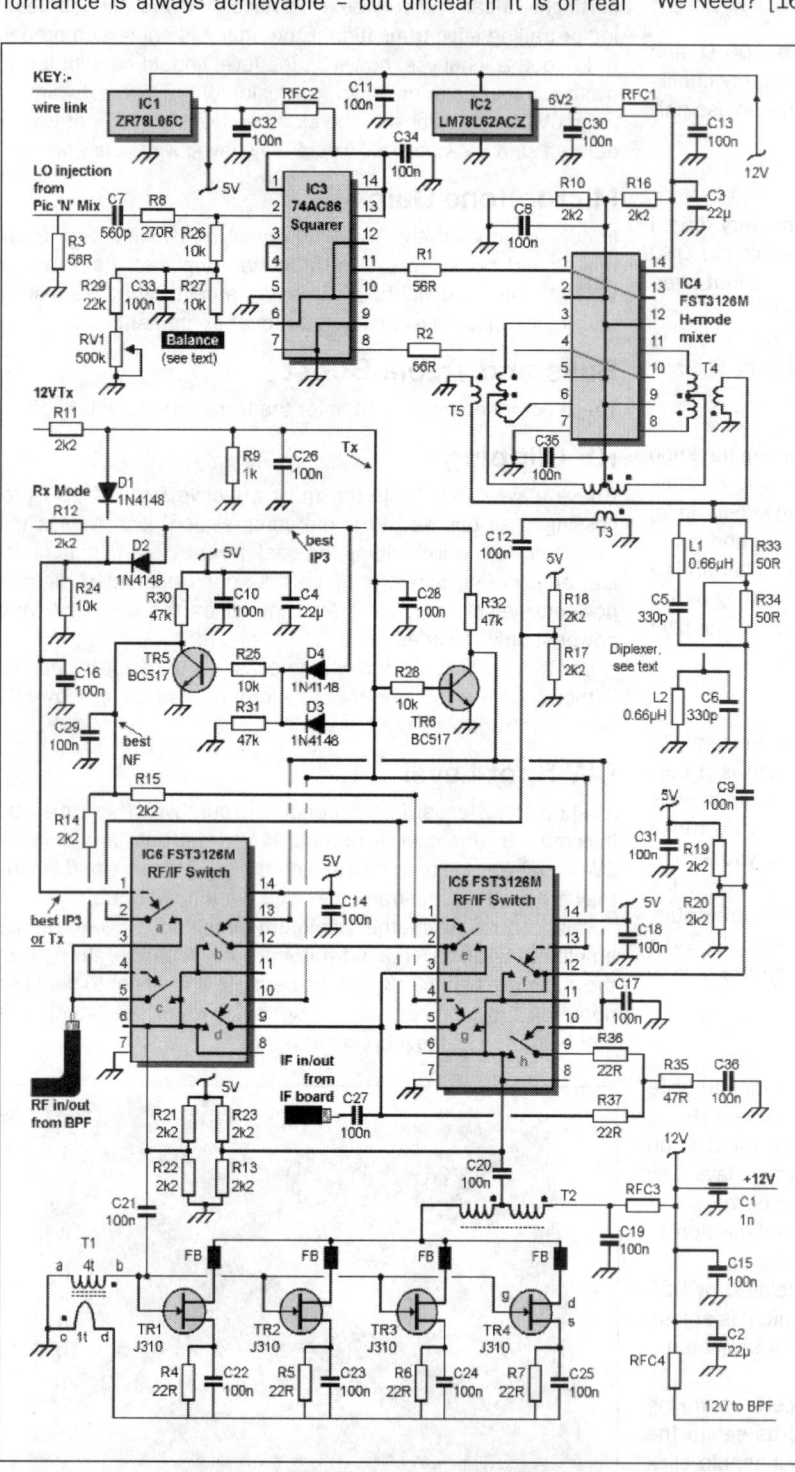

Fig 8.50: Dynamically configurable H-mode mixer – and quad J310 amplifier, better known as the 'Magic Roundabout'. The 'Rx Mode' line is set to +5V ie logic '1' for best intercept (IP3) – or to logic '0' for best noise figure (NF). 12VTx – when taken to +12V – selects the transmit configuration. The mixer requires fundamental frequency injection between -10 and +10dBm

Resistors 1206 SMD	
R1-R3	56R
R4-R7, R36, R37	22R
R35	47R
R33, R34	each 2 off 100R in parallel to give 50R
R8	270R
R9	1k
R10-R23	2k2
R24-R28	10k
R29	22k
R30-R32	47k
RV1	500k multi-turn preset pot
Capacitors	
C1	1n feedthrough
C2-C4	22μ radial electrolytic
C5, C6	330p ceramic plate (see next month)
C7	560p ceramic plate
C8-C36	100n SMD 1206
Semiconductors	
D1-D4	1N4148 or similar
IC1	5V regulator, 150mA eg ZR78L05C
IC2	6V2 regulator, LM78L62ACZ
IC3	74AC86 SMD
IC4-IC6	FST3126M (IC4 could be FST3125M)
TR1-TR4	J310
TR5, 6	BC517 Darlington
Inductors	
L1, L2	0.66μH on T25-2 (see next month)
RFC1-RFC4	6t thin enam wire on type 43 FB
FB	small ferrite bead on J310 drain lead
T1-T6	wound with 32SWG self-fluxing copper wire on EPCOS (was Siemens) B62152A4X1 binocular core (available from ElectroValue)
T1, T6	details follow next month
T2	4 bifilar turns (wire wound at 5tpi)
T3-T5	4 trifilar turns (wire wound at 5tpi)

Table 8.6: Front end component list

after doing the performance arithmetic, I was convinced I needed both under different operational circumstances. Typically, on 10m I want the sensitivity; on 40m I want the higher intercept.

So, pragmatically, I decided to make it configurable – so that I indeed have both options and can compare and contrast them under differing real-life conditions – at the touch of a button.

Thus I can switch between 'best NF' and 'best IP3' modes to suit the prevailing conditions – with the key related benefit that I can have enough RF gain for DDS spurs to be below the band noise.

This approach takes care of T/R switching also – and the whole concept has become known in STAR circles as the 'Magic Roundabout'.

The circuit diagram is shown in **Fig 8.50** and the switching arrangements are summarised in **Fig 8.51**. The switch references are common to both.

Circuit Description

The LO injection is squared up and made symmetrical by IC3. Critically, this removes the even harmonics. RV1 is best adjusted for minimum DDS spurii on 10m. In 'best NF' mode these should be very hard to find. IC4 is a conventional fundamental-injection H-mode mixer. See also [19].

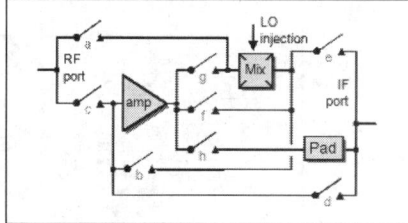

Fig 8.51: Modified squarer by I7SWX. Please note that neither the changed nor the incremental parts are included in the component list.

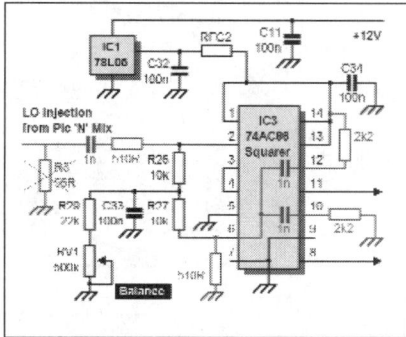

Fig 8.52: Magic Roundabout block diagram. For 'best NF', switches c, g and e are closed. For 'best IP3', switches a, b and h are closed and the pad improves the intercept. For 'transmit', switches d, f and a are closed

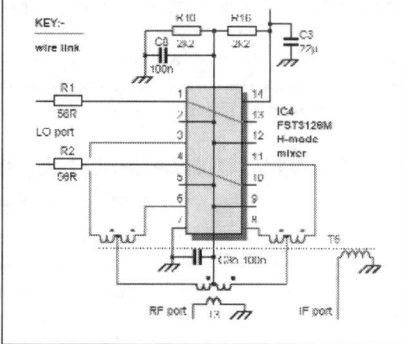

Fig 8.53: Two transformer H-mode mixer by I7SWX. Details of transformer T6 (which replaces T4 and T5) will be provided next month. It is wound on the same ferrite as the other transformers

TR1-TR4 comprise a quad J310 amplifier used either as an RF or IF amplifier. I first saw the feedback arrangement in Introduction to RF Design by Wes Hayward, W7ZOI; and the use of multiple FETs (to raise the intercept) in an IF amplifier design by Bill Carver, W7AAZ.

With an 8dB pad (R35 – R37) in 'best IP3' mode only, the system gain is essentially constant in either receive mode. On transmit, the J310s are always used as an IF amplifier – irrespective of the receive mode. IC5 and 6 with TR5 and TR6 control the roundabout switching.

I7SWX Improvements

Fig 8.50 uses the original squarer and mixer from [19] – but in private correspondence with Giancarlo in early 2003, he suggested two improvements. You may wish to incorporate either or both. I certainly have and commend them.

Fig 8.52 shows changes to the squarer which improve the symmetry and 'squareness' of the switching waveform. Also, IC3 now does not require (nor gets hot in the absence of) LO drive.

Some people have reported unstable lumps of RF energy apparently emanating from the original squarer – and with this modification I have seen/heard no further evidence of them.

Giancarlo omitted the balance arrangements; but I prefer to retain them for the STAR application. The LO injection level requirement is between 0dBm and +10dBm.

Fig 8.53 shows Giancarlo's two-transformer mixer. This is to be preferred in principle, since when it comes to improving the mixer intercept, the only really good ferrite transformer is an eliminated one.

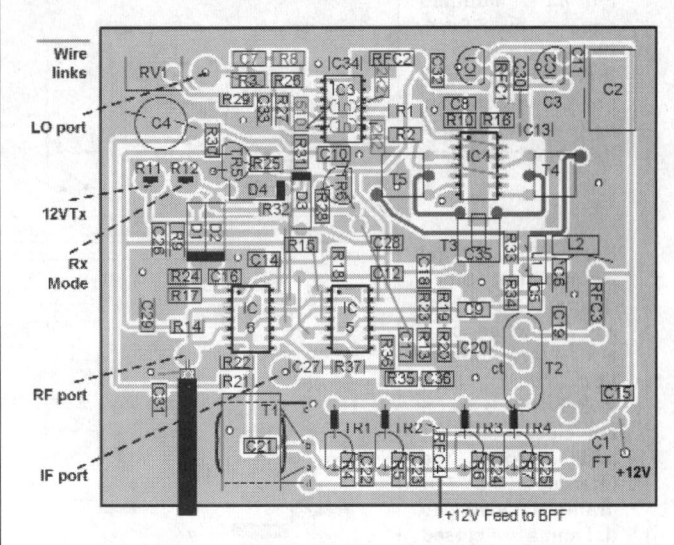

Fig 8.54: 'Magic Roundabout' component layout (see Fig 8.55 in Appendix B for the PCB artwork). This board is designed to mate mechanically with the BPF assembly described next month. Components shown in blue are for the improved squarer. Track shown in blue may optionally be removed (see text) for the two-transformer H-mode mixer

MAGIC ROUNDABOUT CONSTRUCTION

The component layout is illustrated in **Fig 8.54**, and the PCB artwork is in **Fig 8.55** (in Appendix B). It is again made using the iron-on process. The board is double-sided with the underside unetched except for small pads to mount R11 and R12 - and to make off the Rx Mode and 12V Tx lines.

Thus the underside of the board provides a ground-plane as well as screening. R11 and R12 are mounted in the thickness of the PCB as 'feedthrough resistors'.

There are several wire links on the board, for which I apologise. These derive from the need to make this board as small as possible to minimise the risk of pick-up and radiation.

The RF Port requires a DC blocking capacitor. For STAR, this is fitted on the BPF board. Do not omit it!

Track Options

The board is shown tracked for all options. If you are definitely fitting the two-transformer H-mode mixer, you should remove all the tracking between T4 and T5 and T3 - leaving only a small pad to connect a wire from T6 to the IF port diplexer. To 'remove' the track before etching, simply fill and join up the ground with an indelible pen. Alternatively, after etching, you can cut the tracks feeding the diplexer - and then bridge all the unwanted track to ground. The latter is easy and reversible so gives you more options for experimentation.

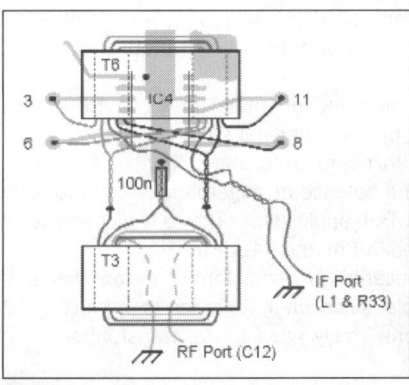

Fig 8.56: Details of T6/T3 connections to each other and to IC4. This is not strictly to scale, but does show the correct relative positioning of the components. T6 and T3 will end up separated by about 6mm in practice

Squarer Options

These are shown in dark grey on Fig 8.54 with the component values taken from Fig 8.52.

This is a classic example of where there is absolutely no way this board could be made commercially, since there is no suitable track layout. For one-off purposes, however, the two 1n wire-ended capacitors fit beautifully - as shown - across the top of IC3. If you want to retain the original mixer configuration, simply replace them with wire-links; omit the dark-outlined components; and revert to the values in Fig 8.50.

Construction Sequence

The holes in the grounded areas of the board are for links through to the ground-plane. These should be fitted first and soldered both sides.

Then mount all the SMD components; then the discrete devices with the exception of TR1; then the wire links (ideally using thin self-fluxing wire) and finally the transformers.

Note the gap between TR2 and TR3 to give space for the feedthrough to the BPF board - via RFC4. Slip a small ferrite bead over each J310 drain lead before soldering.

IF Port Diplexer

L1/2, C5/6, R33/34 form a diplexer on the mixer IF port. L1, L2, C5 and C6 should each have a reactance of about 50 ohms at the IF - using the nearest preferred capacitor value. The values given are for 10.7MHz. L1 and L2 in this case are each 14 turns of 32SWG wire wound on a T25-2 core spread over about 2/3 of the circumference.

This derives from an A_L value of 34µH per 100 turns for this core. Before fitting, connect each coil in parallel with its resonating capacitor, pass a single turn through the toroid and loosely couple it to a GDO; and dip it at your IF.

Mixer Transformers

T3, T4 and T5 are illustrated as MCL transformers and the tracking is appropriate should you want to use these.

If using the home-brew 3-transformer mixer, the EPCOS ferrites mount vertically as shown for T2.

Mixer balance is critically dependent on the transformers all being the same. To this end, make up enough trifilar wire in one length for T3-T5. Rather than winding one end of the wire through the core continuously, wind alternate ends.

If building the two-transformer version, then see Fig 8.53 for the circuit diagram and **Fig 8.56** for the mechanical result. This may indeed turn out to be the definitive test of your understanding of 'the phasing dot convention' but if you follow the steps below you can build it by rote.

T6 is wound with five parallel strands of self-fluxing wire - untwisted. Cut five lengths of wire to some 20cm long. Solder one end of all of them together to retain them. Then trim them all to exactly the same length and solder the other ends together.

Now wind four turns on the core under modest and continuous tension - passing alternate ends through the core.

Cut off the surplus wire equally and initially to approximately 5cm. Tin all 10 ends. Using a continuity meter, locate one pair and twist them to form the IF port feed. That was easy!

Now locate two more 'pairs' and cross-connect a start/end (and twist them together) to form the centre-tap. Repeat for the two pairs.

Locate T6 on top of IC4 and trim the four leads that are made off to the track to the same length. Those going to pins 6 and 8

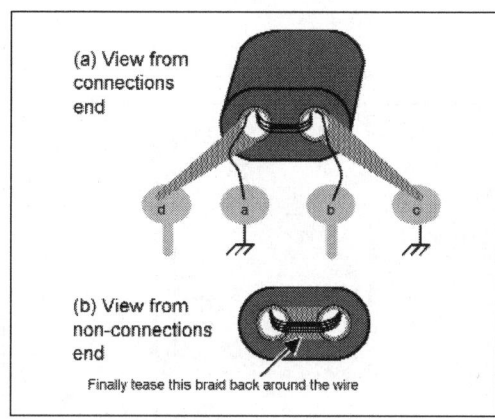

Fig 8.57: Transformer T1 detail. The pad designations correspond to those of Fig 8.50

pads define that length. Tin and solder them to their pads as per Fig 8.56.

Then make off the IF port leads to the end of the diplexer.

Now wind the trifilar transformer T3. Solder the centre-tap to the track first, then the RF port feed. Finally, trim off the flying leads to T6 equally, as short as reasonably possible, and solder them up.

J310 Transformers

T2 is a conventional bifilar transformer with no complications. The input transformer T1 needs a little explanation. See **Fig 8.57**.

The T1 feedback turn is made from a length of miniature coax braid. Form the U-shaped primary turn - with plenty of excess lead length. Pierce and spread the braid to fill the tubular holes in the core - from both ends of the core. Make off the braid leads to the PCB track.

Wind the four turns - inside the braid - out the back of the core, across and back through the braid on the other side - and solder to the track. As far as possible, tease the braid back round the turns on the non-lead end (as W4ZCB says, so nobody can see how it was done).

TR1 may now be fitted.

Incremental Testing

Depending on your personal style and confidence, you may want to perform the functional testing of this board progressively. By lifting one end of some RF chokes you can selectively enable parts of the circuit. 100n wire ended capacitors should be used to couple RF in and out of each circuit element under test.

You can test the J310 amplifier is indeed amplifying by placing it in the down-lead of some receiver. Equally, the squarer and mixer - with suitable injection - can be tested as a crude converter.

System Interface

You could control the Roundabout with a simple switch and a status LED. Arguably, you could do worse than driving it from the SSB select lines, ie 'best NF' on the USB bands, 'best IP3' on the LSB bands. But 60m is exceptional and complete user choice is desirable until you see how it works for you. It depends not least on what antennas you have. I control it from the 'spare' output on Pic 'N' Mix - with a 1k5 resistor in series with a tell-tale LED across the line. On STAR, this line is toggled by keypad sequence 48.

STAR PERFORMANCE

These performance measurements were made by Harold, W4ZCB, using professional test equipment. They were corroborated by the author using test equipment borrowed from I7SWX - and his own. Thanks go to both because these are very important numbers.

All measurements relate to the complete STAR line-up - ie including the band-pass filters which follow next month. These have, by design, decreasing insertion loss with increasing frequency. So, to get the complete picture, you need to consider the performance on each band. Four representative bands are shown, the rest being somewhere 'in between'. Since the design features a 'best IP3' or 'best NF' mode, that adds a further dimension.

The mixer is Giancarl's two-transformer topology driven by the modified 74AC86 squarer. MDS was measured in a 3kHz bandwidth; IP3 at 20kHz tone spacing.

'Best IP3' mode			'Best NF' mode		
Band (m)	MDS (dBm)	IP3 (dBm)	Band (m)	MDS (dBm)	IP3 (dBm)
80	-123	+33	80	-127	+30
40	-122	+35	40	-123	+30
20	-124	+31	20	-127	+28
10	-127	+28	10	-130	+25

AGC RANGE

AGC holds the audio output constant within 1dB for a 100dB change of signal. You can place this range anywhere on the amplitude scale by adjusting the RF Gain, but from -95dBm to +5dBm would be typical.

OBSERVATIONS

Note that excess sensitivity is not provided on the lower bands where it could never be used. Instead, it is traded for superior strong-signal performance.

For comparison with commercial transceivers see [20].

Finishing Off

When all is working well, trim RV1 to minimise DDS spurs on a high band, eg 10m, in 'best NF' mode. When connected to a dummy load, you may still be able to hear some. But then, when connected to even a modest antenna, although the band may be essentially closed, the ambient band noise should mask them to the point that you will need to try very hard to find them.

Finally, enclose and screen the whole board and re-trim RV1.

10-BAND BANDPASS FILTER

The author felt a need to improve on the Third Method front-end (and add 60m) and could find nothing suitable in the literature that met the requirements. This development is of general interest and application.

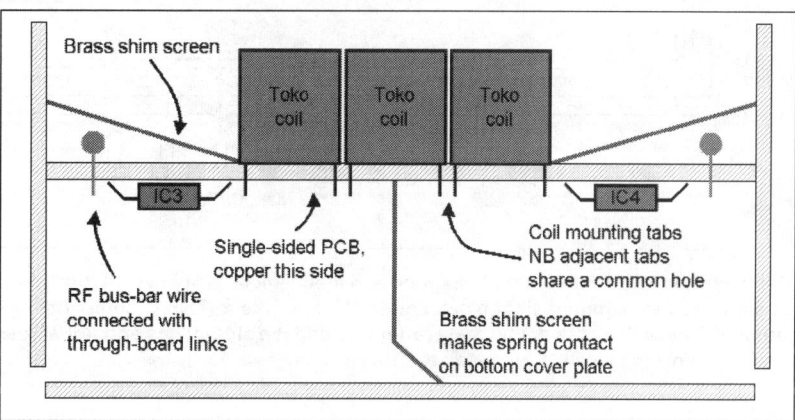

Fig 8.58: Cross-section of bandpass filter enclosure. The main board is single-sided with critical incremental screening on the top provided by brass shim. The same material is used to form a central spine shield running up the middle of the filters in the lower compartment. It is bent over to make spring contact with a bottom cover plate - made from PCB stock

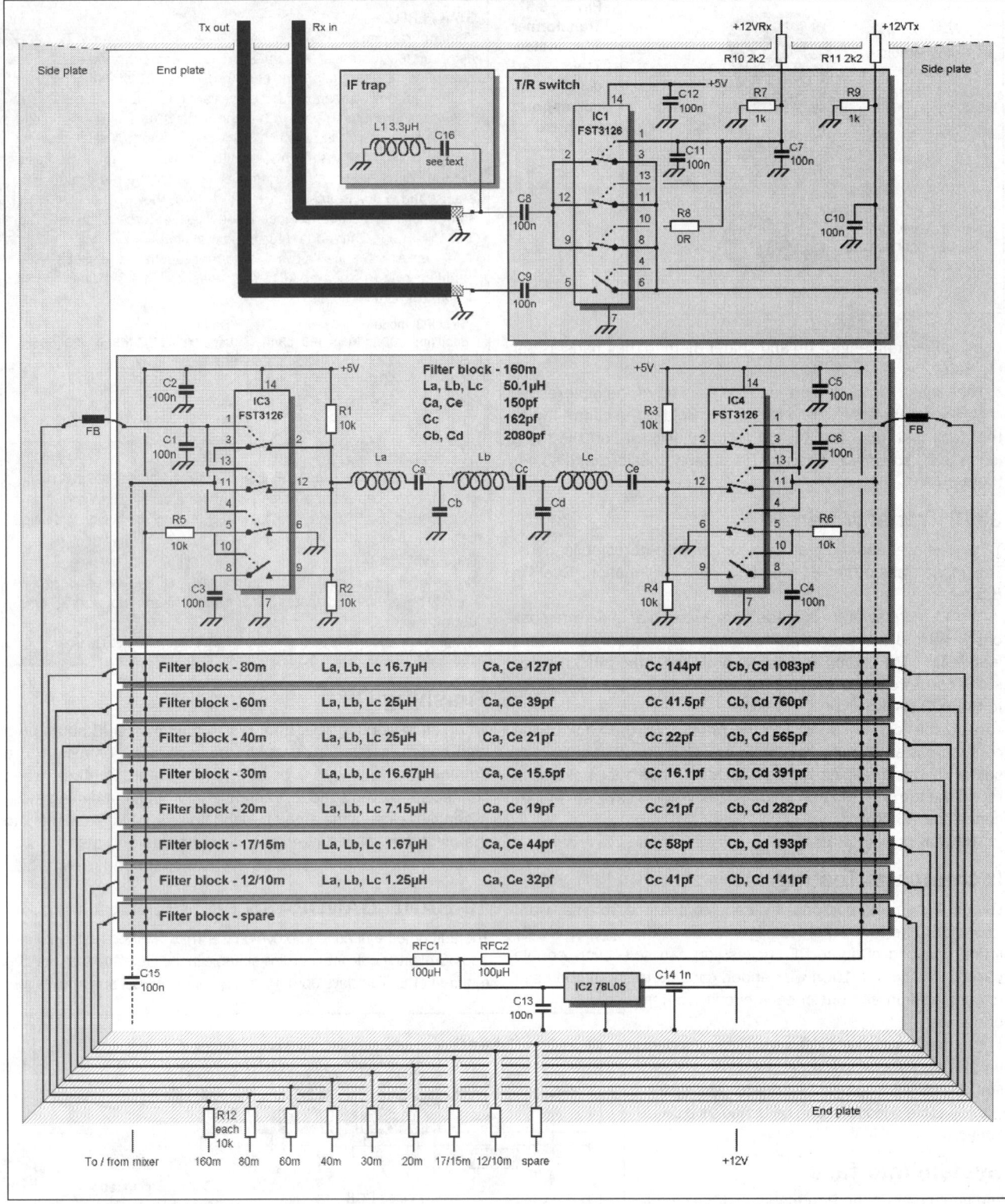

Fig 8.59: Bandpass filters circuit diagram and mechanical overview. All the filters have identical circuit diagrams. A band-select line is set to +5V to engage a filter block and to 0V to isolate it. This is compatible with direct interfacing to the Pic 'N' Mix band-select outputs. These lines are deliberately routed around the side-plates and not across the filter. For illustration, the 160m filter is shown 'on' and the T/R switch is 'on receive'

Design Aims

Like everyone else, I think I want narrow filters with no insertion loss, superb IP3 performance - and acceptable cost. I definitely want a finished size that does not impact on the overall dimensions of my transceiver. For a fact, you can't simultaneously optimise all these parameters, so this design - like all others - is a careful compromise.

The prime function of any BPF is to reject the image frequency adequately - and it is achieved by this design (with a little help from an ATU and/or your LPF on the highest bands), provided your IF is 9MHz or higher - and you use high-side injection.

You could use different filter topologies. The mechanical construction is for three inductors (see **Fig 8.58**). That is the only

FOR THE OVERALL BPF ASSEMBLY		PER BPF FILTER BLOCK	
Resistors 1206 SMD		**Resistors, 1206 SMD**	
R7, R9	1k	R1-R6, R12	10k
R8	OR link (or wire!)	**Capacitors**	
R10, R11	2k2	C1-C6	100n 1206 SMD
Capacitors		Filter capacitors Polystyrene or silver mica. For values, see Fig 8.59.	
C7-C13	100n 1206 SMD	**Inductors for filters (all Toko coils, 3 off)**	
C14	1n feedthrough	160m	154ANS-T1017Z
C15	100n disc ceramic	80m	154ANS-T1012Z
C16	see text	60m	154ANS-T1014Z
Inductors		40m	154ANS-T1014Z
RFC1, RFC2	100µH axial choke	30m	154ANS-T1012Z but see text
L1	Toko BTKANS-9445HM 3µ3	20m	154ANS-T1007Z
Integrated circuits		17m/15m	TKAN-9448HM
IC1	FST3126M	12m/10m	BTKANS-9450HM
IC2	78L05 regulator	**Integrated circuits**	
		IC3, IC4	FST3126M
		Miscellaneous	
		Small ferrite bead, 2 off	

Table 8.7: Band-pass filter components list

practical constraint. The filter capacitors are soldered directly to the coil terminals and there is plenty of room for lots of them in different configurations.

Performance requirements
Since, on the higher bands, Noise Figure is everything, I need low insertion loss above all else. Say 1.5dB. The consequence is wider filters which have the significant benefit of spanning more than one band.

The other great benefit - and integral to the whole front-end design strategy - is that I need the highest possible signal level going into the mixer so that any DDS spurii are below this level on the higher bands.

As you move down in frequency, Noise Figure becomes less and less important and greater insertion loss is a positive benefit, adding directly to the mixer intercept. It also helps to keep the power output flat on transmit if it tends to drop off at the higher frequencies.

Cost considerations
On the cost front, diode switching is the cheapest - but unacceptable for strong-signal performance. Good relays (and you would not want to use bad ones) are very expensive given the quantity involved. I settled on integrated bus switches and use the FST3126. This is a close relative of the FST3125 as typically used in the H-mode mixer, the difference being that the switch control logic is inverted and is compatible with direct drive from the Pic 'N' Mix band-switching latches.

As I have configured them, the ON insertion loss is less than I can measure, at well under 0.5dB; the OFF isolation is better than 90dB; and IP3 is better than 40dB. They are very inexpensive.

I settled for Toko coils, because they are readily available and also inexpensive - but I used the larger cup-core inductors where possible, ie on all bands up to and including 20m.

As a gross alternative, you could use fixed toroidal inductors and trimmer capacitors if the Toko coil Q or IP3 performance are issues for you.

Circuit Description
The filters (see Fig 42) are all 0.01dB Chebyshev designs. They are switched by identical switches at each end (IC3 and IC4). For each switch, two sections are paralleled to reduce 'on' insertion loss; one section inverts the control logic and one section grounds the filter when 'off'. For the T/R switch (IC1), three sections are paralleled on receive and one is used on transmit.

L1 and C16 form a series-tuned IF trap and C16 should be chosen to resonate with L1 at your chosen IF frequency.

A spare filter position is provided for experimental purposes, eg different topologies, different frequencies.

The filter capacitors' exact theoretical values are given in **Fig 8.59**, and the nearer you can get to them, the better. I obtained a large bag of assorted polystyrene capacitors and arrived at the values to within 1pF (as measured on my DVM) with never more than two in parallel - by measuring the actual values within the tolerance range.

For example, the 40m and 20m blocks require six capacitors of near 20pF. The exact values were found from a small selection of 20pF and 22pF 5% capacitors.

Performance summary
Up to 20m, the filters give >100dB image rejection. By 12m/10m, this has fallen to some 50dB, so you could benefit from some incremental filtering provided by low-pass filters, an ATU, or even a beam.

Insertion loss is around 5dB on the lower bands, falling to 3dB on 20m, 1.6dB on 17m, 1dB on 15m, 1.5dB on 12m and a delightful 1dB at 28MHz rising to a mere 1.3dB at 29.7MHz.

30m is exceptional because of the proximity to my IF. Here a 5dB inband insertion loss rises to 18dB at 10.7MHz and 55dB at 9MHz. Depending on your mixer balance and trap tuning, this could be marginal with a 10.7MHz IF.

Filter Construction
The layout is shown in **Fig 8.60**. Three filters are illustrated, the remainder being constructionally identical. More space has been allocated to the 160m and 80m filters to accommodate larger filter capacitors. The components are all mounted on the track-side except for the Toko coils, the RF in/out bus-bars (22SWG tinned wire) and C15. The bus-bars are connected to the IC switches with throughboard wire links. Note that the middle Toko coil is rotated by 180 degrees relative to the outer two. Unused coil pins are cut back so they do not appear on the opposite side.

The PCB artwork is shown in **Fig 8.61** (in Appendix B). This assembles into an H-section brick (for the want of a better term). The filter board is single-sided - and SRBP if you want to save on drill bits; the side and end-plates are doublesided. Ensure opposite sides of all the double-sided boards are intimately connected. The outside faces of both end-plates have simple oval pads

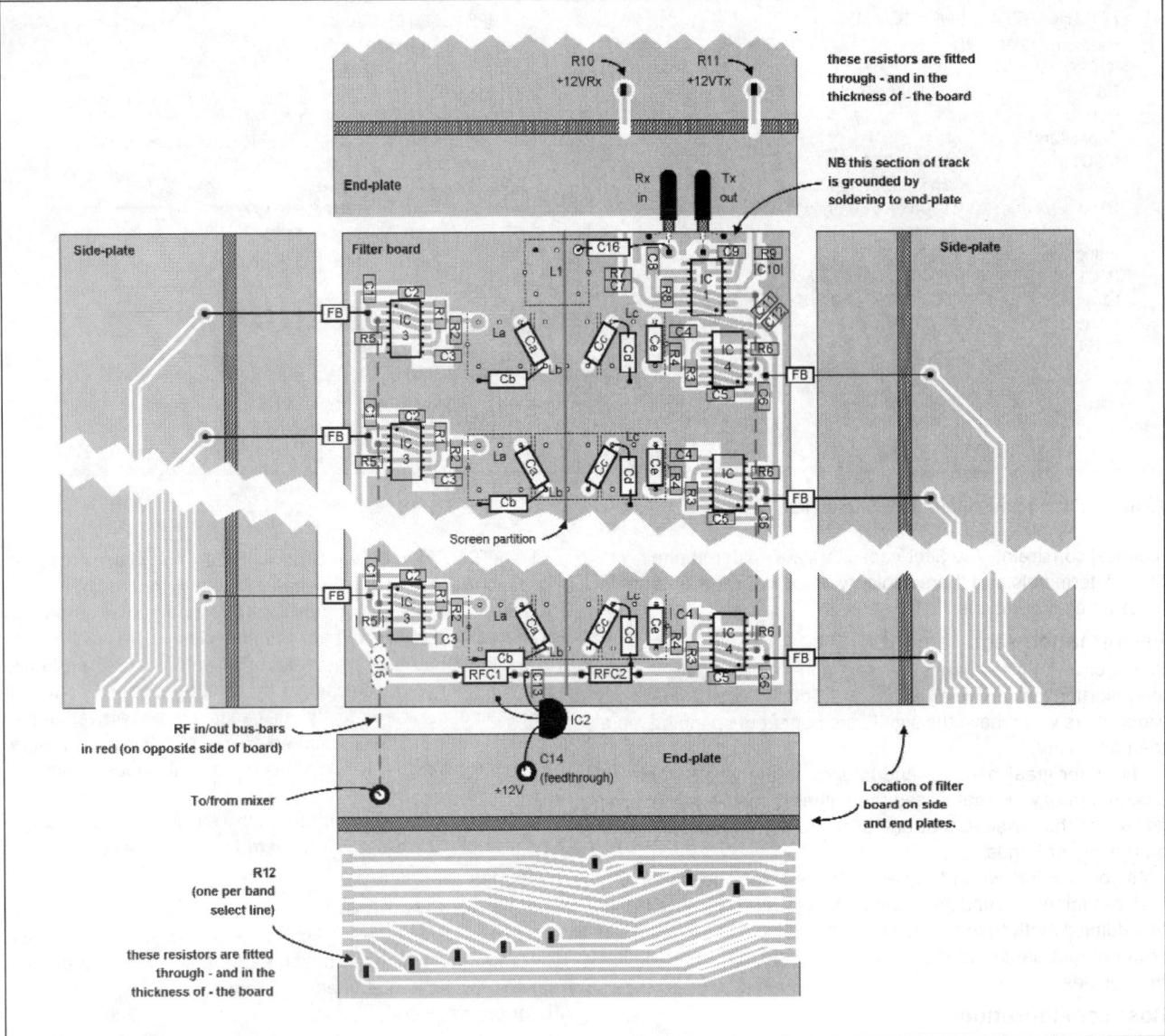

Fig 8.60: Component layout for band-pass filters

to make off the feedthrough resistors to the band-select lines at one end - and the 12VTx and Rx lines at the other. The artwork for this is not provided since some elementary removal of the masking spray after drilling and prior to etching will achieve the desired and non-critical result.

Filter Adjustment

In the first instance, put the entire filter assembly in series with the antenna of an existing receiver and check that all the filter switches work and the coils peak. There are sophisticated ways of tuning these filters, but if all coils are peaked mid-band and then the two end coils are peaked at the two band edges, you will not be far out. A refinement of this is to put Pic 'N' Mix in wobbulator mode across the segment of interest and, while on low power CW, transmit into a dummy load and tweak for a flat pass-band.

Acknowledgements

These filters were designed by Harold, W4ZCB, to whom I am much indebted. He in turn credits ELSIE, a filter design package by Jim Tonne, WB6BLD - which did all the sums. I have subsequently become a fan of this software.

Harold also independently measured the switching perform-ance.

DIY DSP FILTERS

There is not much prescriptive that can be said here. The trick is to search the web for a program that will generate coefficients for FIR filters. From time to time these are available in the pub-lic domain. These coefficients then need converting to a format suitable for loading.

Experimental Methods in RF Design, by Hayward, Campbell and Larkin is also a useful source of understanding - pages 10.13 to 10.19 in particular. This tells you how to do it - and pro-vides the software on a CD to let you.

I suggest the best way to prove the process in the first place is to see if you can build the existing medium-width SSB filter (FL5), plug it in and prove to yourself that yours is no different.

To this end, all the Rx filters are packaged individually and dis-cretely. The specification for the existing FL5 is:

Sample rate = 8kHz

Remez equiripple - but not mandatory

Order = 198, ie 199 coefficients

Summary: modularity is everything

From the outset, this project has been modular by design. Both in the sense that several people have substituted different blocks of hardware and software in their STARs and, conversely, have used STAR blocks in other applications.

In summary, and in chronological order, the project and its description comprises:

- A discussion of STAR hardware integration possibilities.
- A discussion of software options and flexibility.
- A generalised process for producing precision one-off PCBs.
- A T/R changeover timer that would suit any transceiver.
- Details of a DSP processor assembly that could be the basis for any DSP project, with modular daughter boards to allow future upgrades.
- A completely repeatable generalised process for mounting SMD ICs with lots of closely-spaced (eg 0.5mm) pins.
- A bi-directional IF strip that could be readily adapted for use in any home-brew design.
- A PC-based loader and controller that could be adapted for any Analog Devices 218x DSP project.
- One of a number of possible physical implementations. A glance at **Fig 8.62** shows you that every Betabuilder exercised completely different options here. There are no two the same.
- An adapter to replace an 18-pin PIC to give greater I/O and more processing capability generally, and a bargraph S-meter.
- A spur reduction filter for any DDS, and a stable reference oscillator.
- A useful DSP shopping list, at the least. Check out the competition!
- A general purpose bi-directional mixer/amplifier with configurable topology. Of the strong and silent type.
- A universal front-end that could drop into almost any existing HF transceiver.

Table 8.8: This chapter has dealt with an entire transceiver, but in such a way that many of the modules and techniques can be used in other projects

Passband = 320Hz - 2284Hz

Lower stop = 177Hz

Upper stop = 2400Hz

Once you have replicated this and proved the process, you can rapidly produce filters for any specialised application, eg RTTY.

CONCLUSION

A summary of the features of this project are shown in **Table 8.8**. You don't have to be building a STAR to find something of interest here. Equally and oppositely, you don't have to build it all to benefit from STAR DSP.

If you just want to build an error-free transceiver design that works beautifully, now is the moment. All the fruits of this development are available to you at no charge, provided only that they are for your personal use.

To obtain all the software - including the source code - follow the process given earlier. For all the PCB artwork, any enhancements and all ongoing support, simply join the Interest Group (**see box**).

You don't need to understand DSP. At least not to use my STAR code and get your transceiver going. Thereafter, it is entirely both up to you - and down to you.

STAR Build Sequence

Most receiver designs start from the antenna and logically follow the signal flow to the loudspeaker. This project has taken (more or less) the opposite approach. This is to encourage you to build the trickiest bits first.

I think it is a better strategy anyway, since once you have some sort of noise coming out of the speakers, you can work back towards the antenna, using the completed elements to test the

PIC-A-STAR web interest group

http://uk.groups.yahoo.com/group/picastar/

new build. I commend it to you.

Once you have the STAR receiver working, the few incremental components to get the transmitter going can be taken from any of the many HF designs. You need a driver, PA and low-pass filters - commensurate with the amount of power you want to run.

The Linearity Conundrum

Why does a STAR sound so good both on receive and transmit? There can only ever be one answer. It is because it is linear.

Because STAR is an IF processor it has a built-in head start over any AF add-on. The latter may indeed be better than nothing, but if the DSP has neither control of the AGC nor of the detection process then the damage has already been done - so to speak - before there is any chance to benefit from subsequent DSP.

Thereafter, most of the 'star quality' derives from the inherent nature of digital (as opposed to linear analogue) signal processing.

With analogue processing, the quality is determined ultimately by device linearity. Any non-linearity - and all devices have some - results in intermodulation distortion products which can fall within the pass-band. And to a varying extent they grate on the human ear - which is particularly sensitive to their presence.

With digital processing, the nature of any distortion is quite different. It arises from rounding errors, quantisation errors and lack of arithmetic precision generally.

As long as the system is not grossly overloaded (in which case it would fall apart in a big way) these errors do not result in discrete in-band IMD products.

Rather, they result in a general noise floor - of trivial amplitude. And in any practical HF radio communication system, this noise is indistinguishable from and is buried well below the band

noise. There are simply no conventional IMD products to hear.

So the answer to the conundrum is this - and it may not be instinctive. Digital signal processing is inherently more linear than linear analogue processing. This is the technical rationale for the PIC-A-STAR project.

Fig 8.62: The Constellation Beta. Starry-eyed and legless after their much-acclaimed performance in the accordion band competition. They are (from left to right): Top row Alan, G3TIE; Harry, G3NHR; Les, GW3PEX and Peter, G3XJS. Bottom row: David, G4HMC; Eddie, G0SEY; Peter, G3XJP; Bill, W7AAZ, and Harold, W4ZCB, with Harold's STAR

ACKNOWLEDGEMENTS

There are lots of acknowledgements, since the PIC-A-STAR project was and continues to be a truly collaborative and international effort. They are in summary and in no particular order:

The Original Beta Team

The photograph above (Fig 8.62) shows the original team that built, evaluated, tested and continuously suggested improvements - for both the hardware and software of PIC-A-STAR.

Infrastructure and Utilities

My thanks to David Tait for his latest and greatest TOPIC (PIC programming software); Jim Tonne, WB6BLD, for ELSIE (filter design software); Analog Devices for their DSP utilities and code fragments.

Inspiration and Actual Help

Lee, G3SEW, for much useful discussion on the UI in the early days; Gian, I7SWX, for use of his test equipment and mixer design.

Bill, W7AAZ, for many, many ideas and encouragement; Harold, W4ZCB, for the design of the front-end filters, for much performance evaluation - and his unwavering enthusiasm.

Keith, G3OHN, Paul, G0OER, Mike, G3XYG, Jim, G3ZQC, Michel, ON4MJ, and John, G6AK, for much building and testing and the benefit of their diverse skills and wide-ranging experience.

Fran for the proof reading. George Brown, M5ACN, the Technical Editor of RadCom, for steering all this into print

And last, but by no means least, our thanks to Bob Larkin, W7PUA, for sharing his original DSP-10 work, the Digital Signal Processing chapters in Experimental Methods in RF Design, the adaptation of the STAR boot code, his advice and suggestions - and the ultimate inspiration for the whole STAR project.

REFERENCES

[1] RadCom, Aug 2002 - Mar 2004, RSGB
[2] Peter Rhodes, G3XJP, Danvers House, Wigmore, Herefordshire HR6 9UF. E-mail: G3XJP@qsl.net
[3] Analog Devices. http://www.analog.com
[4] Bob Larkin, W7PUA. http://www.proaxis.com/~boblark/dsp10.htm
[5] 'HF Third Method SSB Transceiver', Peter Rhodes, G3XJP, RadCom, Jun -Oct 1996
[6] 'Multiband HF Transceiver', Mike Grierson, G3TSO, RadCom, Oct - Nov 1988
[7] 'Pic 'N' Mix Dgital Injection System', Peter Rhodes, G3XJP, RadCom, Jan - May 1999
[8] Analog Devices FTP download ftp://ftp.analog.com/pub/dsp/21xx/218x/ez-kit-lite/ (You need the two files entitled disk1.zip and disk2.zip)
[9] ElectroValue Ltd., Unit 5, Beta Way, Thorpe Industrial Park, Egham, Surrey TW20 8RE, England. UK. Tel: +44 (0) 1784 433604. Fax: +44 (0) 1784 433605. E-mail: sales@electrovalue.co.uk. Web: http://www.electrovalue.co.uk/
[10] 'In Practice', Ian White, G3SEK, RadCom, Feb 2003
[11] 'HF', Don Field, G3XTT, RadCom, Jun 2003
[12] 'A DSP-Based Audio Signal Processor', Johan Forrer, KC7WW, QEX, Sep 1996
[13] 'Low-Cost Digital Signal Processing for the Radio Amateur', D Hershberger, QST, Sep 1992
[14] 'DSP – An Intuitive Approach', D Hershberger, QST, Feb 1996
[15] 'Using the LMS Algorithms for QRM and QRN Reduction', S E Reyer and D Hershberger, QEX, Sept 1992
[16] 'HF Receiver Dynamic Range: How Much Do We Need?', Peter Chadwick, G3RZP, QEX, May/Jun 2002
[17] 'Technical Topics', RadCom, Feb 2003
[18] 'The CDG2000 HF Transceiver', Colin Horrabin, G3SBI, Dave Roberts, G8KBB, and George Fare, G3OGQ, RadCom Jun – Dec 2002
[19] 'Technical Topics', RadCom, Sep 1998
[20] 'Technical Topics', RadCom, Dec 2002 - Table 1, the IP3 column in particular

9 VHF/UHF Receivers, Transmitters and Transceivers

The purpose of this chapter is to give the reader an insight into what is available to the radio amateur on the VHF and UHF bands in the UK. There is some theory about the choice of the equipment to use but the emphasis is on the practical aspects of choosing equipment and enhancing it with preamplifiers, and power amplifiers to give you a station that can be used to its maximum effect. Chris Lorek, G4HCL, has written a section on choosing a commercial rig or a PMR rig for conversion to the amateur bands. There are references in the bibliography to help you locate equipment and component suppliers.

GETTING THE BEST OUT OF YOUR VHF/UHF STATION

The following paragraphs comprise an abridged version of an overview of the VHF/UHF bands by David Butler, G4ASR, published in the February 1999 edition of RadCom.

One of the great attractions of operating on the VHF/UHF bands is that there are so many different aspects of the hobby that can be utilised at these frequencies. Interested in voice communications? You can use the VHF bands for both local and international contacts. Perhaps your interest lies in digital communications. Well you can join the growing band of enthusiasts that use packet radio (AX25) to access mailboxes or the DX Cluster.

A further aspect of this technology is the automatic packet reporting system (APRS) that allows real-time tracking of mobile (or fixed) stations. Image communication such as slow scan television (SSTV) is also popular, especially now that most of the processing is achieved using a computer and sound card. And don't forget Morse! This 'digital' mode is still very much used on the VHF bands by the DX community. Once you get hooked on working DX you'll then discover exotic propagation modes such as trans-equatorial propagation (TEP), Sporadic-E (Sp-E), Aurora and meteor scatter (MS). And it doesn't have to be two-way terrestrial contacts.

You can also make use of amateur satellites or even bounce your VHF/UHF signals off the moon (earth moon earth EME) to make world-wide contacts. You can operate from home, in the car or go out back-packing from the hill tops. Other activities include low power or high power, rag chews or contesting. The VHF/UHF bands really do have something for everyone.

Prime Mover

The one piece of equipment that determines exactly what facilities you can ultimately use on the VHF bands is the station transceiver. This will either be a single-mode or multi-mode base station, mobile unit or portable hand-held radio. Most single-mode transceivers available today are mobile units (often pressed into service for home use) and portable hand-helds. These are designed to operate exclusively on FM and are very popular, as they can be used for short-range telephony (either direct or via a repeater) and for data communications such as packet radio. Single-mode FM transceivers can be obtained from amateur radio retailers, but that's not the only source of this type of equipment.

Commercial operators regularly upgrade their private mobile radio (PMR) equipment, and this can be obtained from traders who specialise in electronic surplus. It will get you operational on

the VHF/UHF bands very quickly and at a price that will suit most pockets. Indeed, for many fixed station applications, I would recommend that you use dedicated PMR equipment as it does possess many advantages. It is designed for use by a wide range of operators in varying environments. Because of this the equipment is normally of rugged construction. Drop it and it will probably keep working. The majority of PMR equipment has to be built to a high technical performance and reliability. Spectral purity of the transmitted signal is very good and the equipment is designed to run 24 hours a day without a break. Some amateur band allocations are very close to the commercial PMR bands. By looking around you should find equipment suitable for the 50MHz, 70MHz, 144MHz and 430MHz bands. Most equipment is relatively easy to modify and in some instances may not need any modification at all. However, before you hand over your money there are a few points to note. Is the equipment working on a frequency range close to an amateur band? What transmission mode does it use? Is it AM or FM? What is the bandwidth of the IF filters? Is it 50kHz, 25kHz or 12.5kHz? The latter two are preferable, whereas the 50kHz bandwidth would indicate that the equipment is many years old and may not be suitable for use on the VHF bands today.

The VHF/UHF Bands

The three UK VHF amateur bands are 50MHz, 70MHz and 144MHz, and the only UHF band is 430MHz. Profiles of these bands, their different propagation characteristics and the band-plans can be found in the *Amateur Radio Operating Manual*, 6th edition, by Don Field, G3XTT, and available from the RSGB.

Long Distance

As I've just mentioned, the use of FM equipment is for short-range communication links. If you want to broaden your horizons and contact stations much further away then you'll need to procure a multi-mode rig which in addition to FM includes CW and SSB transmission modes. Unfortunately you won't be able to find surplus PMR equipment that can be pressed into service as a multi-mode rig, so it really is a case of digging deep into your pockets and buying a suitable transceiver.

If you already possess a multi-mode HF transceiver. you may wish to consider the use of a transverter. A transverter is a

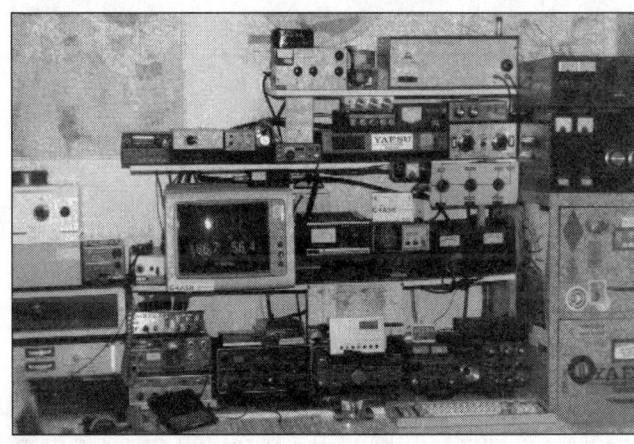

Fig 9.1: The VHF station of David Butler, G4ASR

transmitting converter, a receiving converter and a local oscillator source all combined into one unit. It connects to the antenna socket of an existing transceiver that provides the driving signal, typically at 28MHz. The transverter then mixes the IF drive from the transceiver with its own local oscillator to produce an output on the VHF/UHF band of your choice. On receive a similar process takes place, the VHF signals being down-converted to provide an output signal in the 28MHz band. In practice transverters are available for all VHF/UHF bands and for a variety of IF drive frequencies. Although the majority will be at 28MHz you'll also find models that will accept drive at 144MHz. So if you already have a transceiver on this VHF/UHF band you should have no problem finding a transverter that will allow you to operate on the 50MHz band. The advantage of using a transverter is that it allows all the functions and performance of the driving transceiver to be used on the VHF/UHF band of your choice; more on this topic later.

Optimisation

Now it's time to take a look at what a VHF/UHF station comprises and how you can make simple improvements. **Fig 9.1** shows G4ASR's station. No matter what VHF/UHF band or transmission mode you wish to use, the basic system will always be the same. It's a transceiver feeding an antenna via a length of coaxial cable.

So why do some stations consistently perform better than others? One of the most important factors is the site on which the VHF/UHF station is located. Ideally, a hill-top location is best, but good results can be obtained in low lying areas that are clear of local obstructions. Results depend very much on the band used, obstructions having considerably less effect at 50MHz than at 144MHz or 70cm. We can't all live at 250m above sea level with a clear take-off, so you need to pay special attention to the most significant item in your station. That of course is the antenna; **Fig 9.2** shows a typical VHF/UHF antenna array.

Convention dictates that FM operation, for both telephony and digital communications, an antenna with vertical polarisation is required. If you want to make local contacts then you'll probably need omni-directional coverage. For packet radio you will require a similar vertical antenna, although you might consider using a small 4 or 5-element beam a fixed link.

For serious VHF DX work, using CW or SSB, a horizontally polarised directional Yagi is recommended. There are many types of beam antennas available, some very good and some, well, not so good. But the difference between the poorest designs to that of the very best may only amount to 4dB or so.

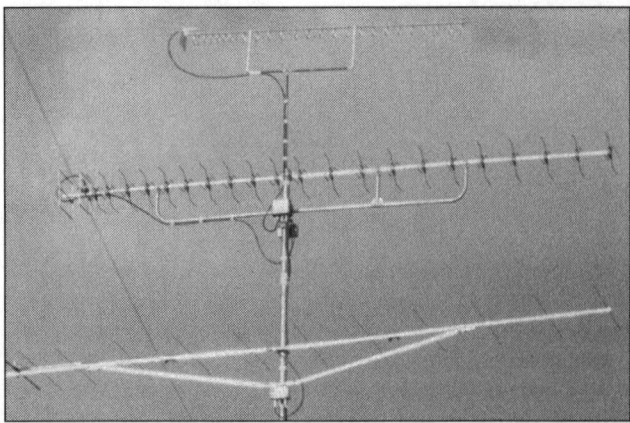

Fig 9.2: VHF/UHF antenna system of Richard Girling, G4FCD (circa 1991)

The point here is that if you are only interested in working occasional DX when the band is open, what 'real' difference do a few decibels make when propagation conditions can vary by many tens of dB? So, unless you really want to eke out the very last vestige of antenna gain, the most important criterion is not ultimate gain but build quality. After all, a long boom antenna is no good if it folds in half during the winter gales. Similarly, the longer the antenna boom the sharper the directivity of the array becomes. The possibility of missing stations away from the main antenna lobe becomes increasingly likely. So you might consider trading off some gain for an increase in beamwidth. Taking all these factors into account you might find that a pair of stacked 9-element Yagis (on the 144MHz band) will provide a more practical solution than using a single 18-element Yagi.

Much more information can be found in the chapter on VHF/UHF antennas.

Siting and Cabling

The siting of an antenna is just as important as the type of antenna used. A ground-plane antenna located on a chimney top, clear of any obstructions, may give better results than a beam antenna located in a loft space. Unless you have restrictions imposed at your QTH, the best location for a VHF/UHF antenna is always outside in an uncluttered location. If possible, mount it on a suitable pole, elevating it above the roof and away from nearby television aerials.

The coaxial feeder connecting the antenna to the transceiver should have a low loss at the frequency in use and this is especially important on the VHF/UHF bands. A poor quality cable will lose valuable transmit and receive signal power, so be prepared to spend more money on the main feeder than on the antenna. It really will be an investment.

Finally, make sure that the connectors you use are of the highest quality. Although the use of N-type plugs and sockets is recommended, they are not essential, especially on the lower VHF bands.

Background Noise

Having paid attention to the antenna and feeder, it's now time to look at the receiver. The background sky noise arriving at the antenna effectively limits the maximum receiver sensitivity required for normal communications. On the lower VHF bands of 50MHz and 70MHz man-made noise often exceeds the background noise by 10dB or so. Consequently, receiver noise figures as high as 12dB and 10dB respectively are quite adequate for these bands. At 144MHz and 70cm however, the sky noise is much less and a receiver noise figure of around 2.5dB will be quite adequate for most types of terrestrial communication. Unfortunately you probably won't find out the overall noise figure of your commercially made transceiver because it is rarely given. Normally the specification is given in terms of so many μV for a signal to noise ratio of so many dB. For example, one 144MHz transceiver quotes "better than 0.5μV for 11dB s/n", making the most favourable assumptions, this translates to a noise figure of 11dB.

Now you can see how little some manufacturers are really offering the VHF/UHF enthusiast. Much effort seems to be exerted in producing rigs with 100 memories, air-band receive facilities, computer control and displays that say "Hello", when what is really required is a VHF/UHF transceiver with a low noise figure, a dynamic range in excess of 100dB, switchable filters, IF shift, notch filtering, adjustable noise blankers and full CW break-in. All these features can be found on a modern HF radio, which brings us nicely back to the original suggestion of using a VHF/UHF transverter with an HF transceiver. You really do get the best system performance by adopting this technique.

Pre-amps

Another way of overcoming the basic lack of sensitivity is to use an external pre-amplifier and, if this is mounted at the antenna, it will also eliminate the effect of feeder loss in the receive direction. Unfortunately, the receive sensitivity is only improved if the pre-amplifier has sufficient gain, but this extra gain also decreases the strong-signal handling capability of the receiver. Therefore the use of a pre-amplifier may show overload effects on some signals that originally didn't cause any problems. Try to use a pre-amplifier that has adjustable gain, so that you can adjust it to suit your receiver. Typically, a gain of between 6-15dB will be sufficient for most needs.

Summary

The biggest improvements to your VHF/UHF station always come first. Changes to the antenna system, coaxial feeder, making the receiver more sensitive and increasing your transmit power will easily improve your system performance. After that it becomes a little bit more difficult. The rewards are still available but each improvement will be less significant.

DESIGN THEORY

This section will explore some of the theory and practice of designing receiving and transmitting equipment for the 50, 70, 144 and 432MHz amateur bands.

Receivers

Standards for VHF/UHF receivers are strongly based on the performance expected from HF receivers, in particular the ability of the receiver to detect, without any deterioration in performance, a weak signal in the presence of one or more unwanted strong signals present at the same time.

Above 50MHz background noise is much lower, so with a good receiver it is possible to realise a performance superior in terms of sensitivity and signal-to-noise ratio. An HF signal of a few microvolts is often down in the noise but at VHF and UHF, communication between stations can be achieved with signal levels as low as a few nanovolts ($1nV = 1V \times 10^{-9}$). On the HF bands a limit is imposed by both man-made and natural interference, beyond which any attempt to recover signals is fruitless. In VHF/UHF signal reception, there is no appreciable atmospheric noise with the exception of that caused by lightning discharges or from electrically charged rain drops. The limiting factor, when the receiver (and antenna) is in a good location, is extraterrestrial noise but the receiver can be designed to respond to signals only slightly above this level.

Definition of noise

Broadly, noise is unwanted signal of a more or less random nature within the pass-band of the receiver. It may be natural or man-made. Examples of natural noise are the radiation from the Sun or, as described earlier, that from electrical storms and charged rain drops. These can only be avoided by excluding the Sun or the electrical storms from the 'field of view' of the antenna. Also, there is the inescapable noise generated in a resistor at any temperature above absolute zero, and shot noise produced in semiconductors, caused by the random generation and recombination of electron-hole pairs in their operations.

Examples of man-made noise are the radiation from switches and thermostats when they break current, and the radiation from computers caused by their processing pulses with very fast rise and fall times.

In the design of VHF/UHF receivers only the inescapable natural noise needs to be considered. Resistors introduce thermal noise, due to the random motion of charge carriers that produce

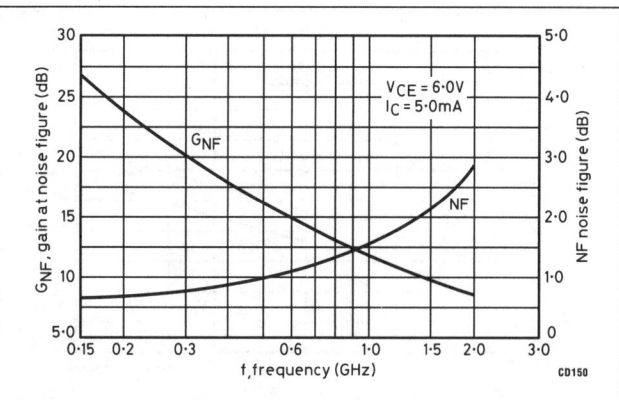

Fig 9.3: Gain at noise figure and noise figure versus frequency

random voltages and currents in the resistive element. There is unfortunately no resistor that will not produce these random products unless the receiver is operated at a temperature at absolute zero (OK). However, resistor noise generation can be minimised, particularly in the receiver front-end, by the correct choice of resistor. Metal film types are recommended. Thermal noise is also known as Johnson or white noise.

Shot noise in semiconductors is due to charge carriers of a particle-like nature having fluctuations at any one instance of time when direct current is flowing through the device. The random fluctuations cause random instantaneous current changes. Shot noise is also known as Schottky noise.

Noise factor and noise figure

The noise factor is the ratio of the input signal-to-noise ratio to the output signal-to-noise ratio. The noise figure is the noise factor expressed in decibels and is used as a figure of merit for VHF and UHF circuits:

$$f = \frac{\text{Input S}/\text{N}}{\text{Output S}/\text{N}}$$

$$NF = 10\log_{10} f$$

It is measured as the noise power present at the receiver output assuming a conventional S/N ratio of 1 at the input. An ideal noiseless receiver does not produce any noise in any stage. Thus the equation becomes 1/1 or a noise factor of 1 or 0dB. The noise factor of a practical receiver that will generate noise in any stage, particularly the front-end, is the factor by which the receiver falls short of perfection.

Amateur communication receiver manufacturers usually rate the noise characteristics with respect to the signal input at the antenna socket. It is commonly expressed as: (signal + noise) / noise, or "signal-to-noise ratio".

The sensitivity is usually expressed as the voltage in microvolts at the antenna terminal required for a (signal + noise) / noise ratio of 10dB. Sensitivity can also be specified as the minimum discernible signal or noise floor of the receiver.

An important point to remember in VHF/UHF receivers is that the optimum noise figure of an RF amplifier does not necessarily coincide with the highest maximum usable gain from that stage.

Figs 9.3 and 9.4 illustrate this feature. The transistor is capable of operation up to 2.0GHz. As Fig 9.4 shows, however, the gain and noise figure falls but the actual noise figure increases with increasing frequency. This characteristic is also shown in Fig 9.4 but here the gain at noise figure and the actual noise figure are plotted for 500MHz and 1GHz against variations in collector current. Note that the maximum gain occurs with good

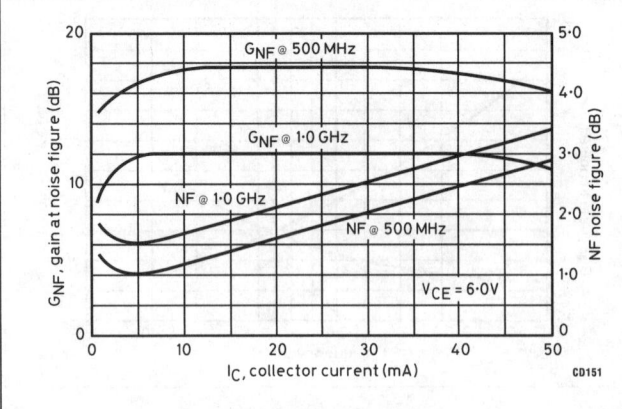

Fig 9.4: Gain at noise figure and noise figure versus collector current

input matching but minimum noise does not. For example, a GaAsFET preamplifier may have an input VSWR as high as 10:1 when tuned for the lowest noise figure.

The definition of noise figure and degradation of receiver performance due to noise implies that the front-end stages, namely the RF amplifier and mixer, must use active devices, either bipolar or field effect types, with a low inherent noise figure. The noise figure quoted for the transistor illustrated in Figs 9.3 and 9.4 applies only when the device is connected to a 50Ω source. As can be seen the noise figure increases by almost two times between 1 and 2GHz. This figure can be reduced by mismatching to the source at 1GHz and above.

Modern design theory and practice now employ S-parameters (scatter parameters) to obtain maximum performance from an RF amplifier and mixer while maintaining the noise figure at low levels. However, S-parameters require an advanced knowledge of design which includes the use of the Smith chart and the availability of some sophisticated equipment such as a network analyser. Manufacturers' data on RF devices includes tables of S-parameters. This method of design of amplifiers and mixers is outside the scope of this handbook.

Intermodulation

Intermodulation occurs when two or more signals combine to produce additional (spurious) signals that were not originally present at the receiver input and possibly causing interference to a weak wanted signal. The receiver front-end is handling many incoming signals of different strengths but only these signals passing through the selective (IF) filters will eventually be detected. The RF circuits can have a bandwidth of several megahertz but the selective IF filters reduce the bandwidth to that required for adequate resolution of signals, dependent on the method of modulation of the wanted carrier. At low signal levels the front-end will have optimum linearity, ie there is no unwanted mixing between signals. However, as stated previously, very strong signals will cause the front-end to go into its non-linear region of operation and then these signals will mix together and produce new signals which can appear in the IF pass-band. Second-order intermodulation products (IPs) are caused by two signals mixing, viz f_1 and f_2, and generating new frequencies which appear as $(f_1 + f_2)$ and $(f_1 - f_2)$ and the second harmonics of each signal ($2f_1$ and $2f_2$), generating the second-order IPs. However, if f_1 and f_2 are close spaced, their second-order IPs will be well spaced and can be easily filtered out by the selective IF filters.

However, if the f_1 and f_2 signals are increased in strength then another set of IPs is generated. These are third-order intermodulation products, due to the fact that mixing occurs between

three signals. The three signals can be independent but the same products can be generated by f_1 and f_2 by themselves. These frequencies f_1 and f_2 can add or subtract to produce the following third-order IPs:

Third harmonics:	$(f_1+f_1+f_1)$ and $(f_2+f_2+f_2)$
Sum products:	$(f_1+f_1+f_2)$ and $(f_2+f_2+f_1)$
Difference products:	$(f_1+f_1-f_2)$ and $(f_2+f_2-f_1)$

It is clear that if f_1 and f_2 are equally spaced above and below the wanted frequency, interference will be severe. When $f_1 + f_2$ are close to the wanted frequency the third-order sum products will appear in the third harmonic area of this frequency and will be attenuated by the selective filters in the receiver. However, when the difference products containing a minus sign are close to f_1 and f_2 and are generated by the receiver, the filters, however selective, will not remove these spurious signals. These products could cause unwanted interference to a wanted weak signal. When third-order intermodulation products are generated in the receiver, they will increase in level by 3dB for every 1dB increase in the levels of f_1 and f_2. Thus the appearance of intermodulation products above the receiver noise floor is quite usual. When further levels of f_1 and f_2 occur higher odd-order intermodulation products are generated, eg fifth, seventh etc, which can interfere with a weak wanted signal. These higher-order products will appear even more quickly than third-order products but require stronger f_1 and f_2 signals, e.g. fifth-order products will be generated five times as fast as f_1 and f_2 as the level of f_1 and f_2 is increased. Significant intermodulation products can only result from f_1 and f_2 when their strength is high. If either the f_1 or f_2 signal disappears, leaving only one signal, the intermodulation product will disappear. Optimising linearity in the receiver front-end will minimise generation of these unwanted intermodulation products and hence interference to wanted signals.

Gain compression

This occurs when a strong incoming signal appearing at the antenna socket causes one (or more) stages in the front-end to drive into the non-linear region of its output characteristic. As an example, when an amplifier stage is operating in its linear region, an increase by 3dB in signal level at its input will cause by linear transfer, a 3dB increase in signal level at the output. However, a further increase in input signal level could cause non-linear transfer and limit the output level increase to 1dB. A very strong signal could drive the stage into extreme non-linearity, making the stage degenerative (gain less than 1) and desensitising the receiver. Background noise will decrease in level together with all other signals, including wanted weak signals. As with intermodulation, optimising front-end linearity will minimise receiver desensitisation by strong signals.

Reciprocal mixing

This phenomenon occurs when the receiver local oscillator is allowed to produce excessive sideband noise on its carrier and a strong off-channel RF carrier mixes with this noise to produce the IF. Reciprocal mixing causes an increase in receiver noise level when a strong carrier appears, the opposite effect of gain compression. The receiver selectivity is not necessarily defined by front-end RF filters. The 'cleanest' local oscillators are LC (VFOs) and crystal-controlled oscillators. Some of the `noisiest' oscillators are found in receivers employing a synthesised system, for example a phase-locked-loop synthesised oscillators. Some earlier receivers suffered from reciprocal mixing effects, ie generation of an unwanted spurious signal in the IF pass-band, for example, noise on the voltage-controlled oscillator con-

trol line leading to FM noise sidebands. However, modern receivers now employ 'quiet' synthesised local oscillators that minimise reciprocal mixing.

Dynamic range

The main problems of front-end overload are gain compression, intermodulation and reciprocal mixing. Each phenomenon has its own characteristic and level at which strong unwanted signal(s) cause degradation in receiving wanted signals. Just one strong signal causes gain compression or reciprocal mixing, whereas two are required to cause intermodulation products. The gain of the front-end, ie the RF stage and the mixer, should be kept as low as possible, consistent with good sensitivity and signal/noise performance. Gain compression and intermodulation are caused when either or both stages are driven beyond their linear transfer range. The front-end should be designed so it cannot be overloaded by even the strongest amateur band signals. Intermodulation products will not be a problem if they are restricted to the level of the background noise and gain compression and reciprocal mixing effects are not a problem if they do not significantly change the system noise level.

The lowest end of the dynamic range will be designed for the lowest power audible signal and, conversely, the highest end will be designed for the unwanted signal of the highest power level, ie signals without any overload effects degrading the front-end performance. This principle is called spurious-free dynamic range but the range will change according to the differing power levels of unwanted signals.

Receiver Front-End Stages

Thus the requirements for front-end stage design in a VHF/UHF receiver are:

- Low noise figure
- Large dynamic range
- Power gain consistent with good sensitivity and signal-to-noise ratio

The noise figure and dynamic range requirements have already been described in detail. However, 'power gain' must be brought into the equation to complete the design philosophy. 'Power gain' needs some explanation because sheer power gain is not sufficient in itself or even desirable. In a multistage receiver with, say, eight stages of gain, input noise originating in the first stage, normally an RF amplifier, will be amplified by the eight gain stages, that in the second by seven and so on. If the effective noise voltages are denoted by V_1, V_2, V_3, ... V_8 and the stage gains by G_1, G_2, G_3,... G_8, the total noise present at the receiver detector will be:

$$V_1(G_1 G_2 G_3. .. G_8) + V_2(G_2 G_3. .. G_8) + V_3(G_3 G_4 . ..G_8) \text{ and so on.}$$

If the voltage gain of the RF amplifier (G_1) is high, for example 20dB (10 times) or more, the important noise contribution is due to G_1. Provided the remaining gain stages are correctly designed and provide evenly distributed gain, the overall noise contribution from them will be very small.

Additional noise generated by any stage that is degenerative or is actually oscillating will degrade the overall receiver noise performance, and might actually cause receiver desensitisation and consequent poor weak-signal performance.

However, at the output of G_3 (eg the first IF stage) an amplified signal will be large compared to the noise contributed and G_4 ... G_8 should not degrade noise performance. The function of the RF stage is to provide just sufficient gain to overcome the noise contribution of the mixer stage. The mixer is by definition a non-linear device and normally contributes more noise than

any other stage no matter how well designed. The RF stage therefore considerably improves the receiver signal-to-noise ratio, improving weak-signal performance. If the RF stage gain is too high then the problems of strong signals, ie intermodulation, gain compression and reciprocal mixing products, will, as described, degrade receiver performance.

A system of gain control on the RF amplifier may appear to be the answer. However, the use of AGC must be carefully considered, otherwise the weak-signal performance will be degraded if the onset of AGC is not delayed. Reducing the amplifier gain, using an AGC current for bipolar transistors or a voltage for field effect transistors, can degrade the RF stage linearity. The noise factor of a transistor amplifier is also dependent on emitter (source) current. A large variation from the manufacturer's data given for noise factor V I_E (I_S) will also degrade weak-signal performance. Circuit layout and correct shielding for VHF/UHF RF amplifiers is of paramount importance for stable operation. Instability and spurious oscillation can be produced via RF feedback through the amplifier transistor. Even the latest designs of transistor, both bipolar and field effect types, can give rise to these effects. Regeneration or actual oscillation can be prevented by neutralising the internal feedback, using an external circuit from output to input of the amplifier. This will feed back an equal amount of out-of-phase signal and, providing there is no other feedback path, the amplifier will be stable.

Circuit noise

Noise due to devices other than transistors is produced solely by the resistive component; inductive or capacitive reactances do not produce noise. Inductors of any form have negligible resistance at VHF and UHF but the leakage resistance of capacitors and insulators is important. It is imperative to choose high-quality capacitors such as silver mica, ceramic, polycarbonate, or polystyrene types with negligible leakage current and high-Q properties. Tantalum capacitors should be used where it is necessary to decouple at LF, as well as VHF and UHF. Attention should be given to the use of low-noise resistors. Carbon or metal film types (of adequate power dissipation) must be used. The old-style carbon composition resistors are very good noise generators. 25dB noise difference between film and composition resistors has been observed when a direct current was passed through the two types, both being the same value.

Circuit noise due to regeneration has already been discussed. Common causes of regeneration are:

(a) Insufficient decoupling of voltage supplies (at each stage) and particularly emitter (source) and collector (drain) circuits. Suitable capacitor values and types are as follows:

10MHz	47nF (0.047µF) ceramic disc or plate
10-20MHz	22nF (0.022µF) ceramic disc or plate
20-30MHz	10nF (0.01µF} ceramic disc or plate
30-100MHz	4.7nF (0.0047µF) ceramic disc or plate
100-200MHz	2.2nF (0.0022µF) ceramic disc or plate
200-500MHz	1nF (0.001µF) ceramic disc or plate
500-1000MHz	220pF (0.00022µF) ceramic disc or plate

Leadless capacitors should be used above about 400MHz.

(b) Closely sited input and output circuits.

Every attempt should be made to build the amplifier(s) in a straight line and where possible, to use shielded coils, particularly for 6m, 4m and 2m equipment.

(c) Insufficient or wrongly placed screening between input and output circuits.

In transistor amplifiers where a screen is required, it should be mounted close fitting across the transistor with input and output circuits (normally base/gate and collector/drain) on

Fig 9.5: Two conventional superheterodyne configurations: (a) Self-contained single super-het with tuneable oscillator. (b) Self-contained double superhet or converter in front of a single superhet. Either or both oscillators may be tuneable

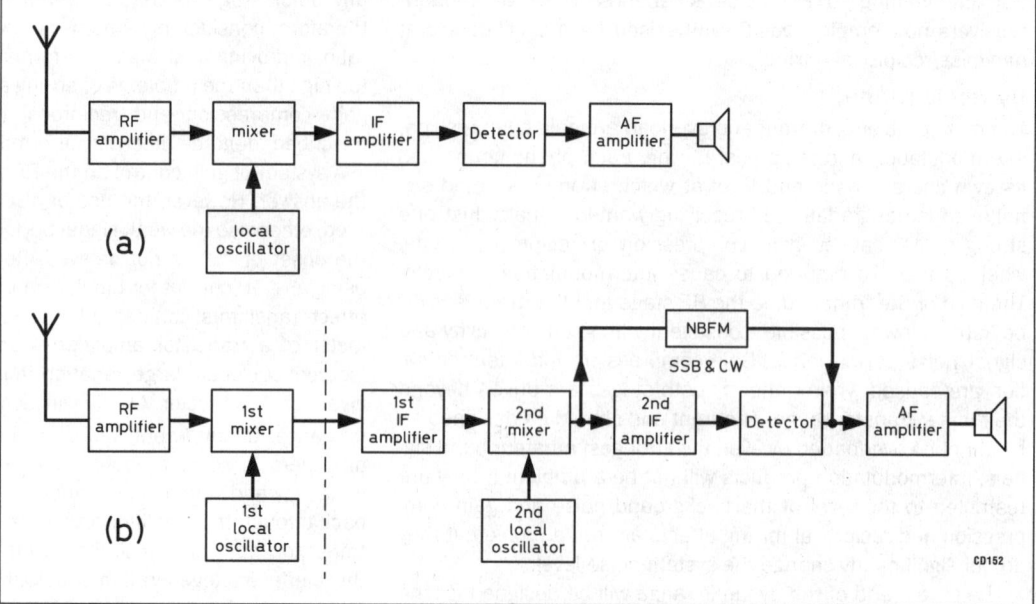

opposite edges of the screen. The screen can conveniently be made of double-sided copper laminate and soldered to the main PCB.

(d) Circulating IF currents in the PCB due to multipoint grounding.

Decoupling capacitors for each stage should be grounded at a single point very close to the emitter (source). It is preferable to use a double-sided PCB using one side as the ground plane.

Effect of bandwidth on noise

If the noise factor of a receiver is measured with a noise generator it is independent of receiver bandwidth. Generator noise has the same characteristics as circuit noise so, for instance, if the bandwidth doubled, the overall noise is doubled. For the reception of a signal of finite bandwidth however, the optimum signal-to-noise ratio is obtained when the bandwidth of the receiver is only just sufficient to accommodate the signal. Any further increase in bandwidth results in increased noise. The signal-to-noise ratio at the receiver detector therefore depends on the power per unit bandwidth of the transmitted signal.

As an example, a receiver may generate 0.25µV of noise for each 2.5kHz of bandwidth. Assuming an SSB transmitter radiates a sideband signal of 2.5kHz bandwidth and produces 2.5µV of signal at the receiver detector, the signal-to-noise ratio is therefore 10, provided the receiver overall bandwidth is 2.5kHz. If the transmission bandwidth is reduced to 1.25kHz for a CW signal and the radiated power is unchanged, the receiver input will remain at 0.25µV but if the bandwidth is also reduced to 1.25kHz, the receiver detects only 0.125µV of noise and the signal-to-noise ratio will increase to 20.

Using receiver bandwidths that exceed transmission bandwidths is therefore undesirable when optimum signal-to-noise ratio is the prime factor. Transmitters with poor frequency stability will either require the receiver to be retuned or the use of wider bandwidth, resulting in a degraded signal-to-noise ratio. Fortunately, well designed transmitters with PLL synthesised oscillators or crystal-controlled oscillators are now employed in the majority of the VHF/UHF bands.

Choice of Receiver Configuration

Receivers using other than superheterodyne techniques are rare on VHF or UHF. Modern superheterodyne receivers may have one, two or three frequency changes before the final IF, each

with its own oscillator which may be tuneable (by the receiver tuning control) or of fixed frequency. Receivers may have a variety of configurations; two are illustrated here in **Fig 9.5**. **Fig 9.5(a)** shows a conventional single superheterodyne for use on the HF bands. The local oscillator will be partially synthesised, ie use a pre-mixer driven by an HF crystal-controlled oscillator and a LF VFO to produce the local oscillator for the main mixer. **Fig 9.5(b)** shows a double superheterodyne. This can be a purpose-built receiver (or as illustrated in Fig 9.5a)) to which is added (to the left of the dotted line) a VHF/UHF converter. The first local oscillator is crystal controlled and tuning is accomplished using the HF receiver (second) oscillator.

The main disadvantage of this method of using an HF receiver as a 'tuneable IF' amplifier preceded by a converter is again the problem of overloading the first amplifier and second mixer with strong signals. This will result in intermodulation and reciprocal mixing products, if not gain compression, particularly if the converter gain is high, say, 20 to 30dB (x10 to x30).

A superior arrangement is to build a tuneable IF amplifier containing all the refinements of a normal HF receiver, including an NBFM IF amplifier and detector, and restrict the tuning range to a few megahertz to cover the VHF/UHF ranges of the converters. The HF receiver gain in front of the second mixer must be low. The first IF amplifier can be omitted but the pre-mixer selectivity should be retained. The converter gain (from VHF/UHF to first IF) should also be low: 10 to 14dB (x3 to x4). This will result in a VHF/UHF receiver with a very good noise factor and dynamic range, and demodulation of NBFM in addition to CW and SSB signals.

Choice of the first IF

In any superheterodyne receiver it is possible for two incoming frequencies to mix with the local oscillator to give the IF; these are the desired signal and the image frequency. A few figures should make the position clear. It will be assumed that the receiver is to cover the 144 to 146MHz band, and that the first IF is to be 4 to 6MHz. The crystal oscillator frequency must differ from that of the signal by this range of frequencies as the band is tuned and could therefore be 144 - 4MHz = 140MHz or, alternatively, 144 + 4MHz. Assuming that the lower of the two crystal frequencies is used, a signal on 136MHz would also produce a difference of 4MHz and unless the RF and mixer stages are selective enough to discriminate against such a signal, it will

be heard along with the desired signal on 144MHz. From the foregoing, it will be appreciated that the image frequency is always removed from the signal frequency by twice the IF and is on the same side as the local oscillator.

It should be noted that even if no actual signal is present at the image frequency, there will be some contributed noise which will be added to that already present on the desired signal. It is usual to set the RF and mixer tuned circuits to the centre of the band in use so that on the 144MHz band they should be at least 2MHz wide in order to respond to signals anywhere in the band. This bandwidth only represents approximately 1.4% of the mid-band frequency and it is not surprising that appreciable response will be obtained over the image frequency range of 134 to 136MHz unless additional RF filtering is employed. Naturally the higher the first IF, the greater the separation between desired and image frequencies. However, an IF as low as 4 to 6MHz is feasible, provided some attempt is made to restrict the bandwidth of the converter by, for example, employing two inductively-coupled tuned circuits between the RF and mixer stages, thus providing a band-pass effect.

The choice of the first IF is also conditioned by other factors. Firstly, it is desirable that no harmonic of the oscillator in the main receiver should fall in the VHF band in use and secondly, there should be no breakthrough from stations operating on the frequency or band of frequencies selected for the first IF.

Many HF receiver oscillators produce quite strong harmonics in the VHF bands and, although these are high-order harmonics and are therefore tuned through quickly, they can be distracting when searching for signals in the band in question. The problem only exists when the converter oscillator is crystal controlled, as freedom from harmonic interference is then required over a band equal in width to the VHF band to be covered. This also applies of course to IF breakthrough.

As it is practically impossible to find a band some hundreds of kilohertz wide which is unoccupied by at least some strong signals, it is necessary to take steps to ensure that the main receiver does not respond to them when an antenna is not connected. Frequencies in the range 20 to 30MHz are often chosen, since fewer strong signals are normally found there than on the lower frequencies but this state of affairs may well be reversed during periods of high sunspot activity.

With the greatly increased use of general-coverage receivers covering 100kHz to 30MHz, the best part of the spectrum for 6m, 4m and 2m is from 28 to 30MHz. Full coverage of the 70cm band will require the receiver to be tuned from 10 to 30MHz. IF breakthrough is minimised and frequency calibration is simple.

Tuned Circuits

Tuning is readily achieved at HF by lumped circuits, ie those in which the inductor and capacitor are substantially discrete components. At VHF the two components are never wholly separate, the capacitance between the turns of the inductor often being a significant part of the total circuit capacitance. The self inductance of the plates of the capacitor is similarly important. Often the capacitance required is equal to, or less than, the necessary minimum capacitance associated with the wiring and active devices, in which case no physical component identifiable as 'the capacitor' is present and the circuit is said to be tuned by the 'stray' circuit capacitance.

As the required frequency of a tuned circuit increases, obviously the physical sizes of the inductor and capacitor become smaller until they can no longer be manipulated with conventional tools. For amateur purposes the limits of physical coils

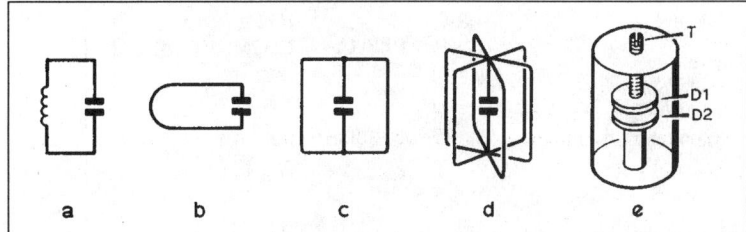

Fig 9.6: Progressive development of tuned circuits from a coil to a cavity as the frequency is increased

and capacitors occur in the lower UHF bands: lumped circuits are often used in the 432MHz band but are rare in the 1.3GHz band.

Distributed circuits

Fig 9.6 illustrates how progressively lower inductances are used to tune a fixed capacitor to higher frequencies. In **Fig 9.6(b)** the 'coil' is reduced to a single hairpin loop, this configuration being commonly used at 432MHz.

Two loops can be connected to the same capacitor as in **Fig 9.6(c)**. This halves the inductance and can be very convenient for filters.

Fig 9.6(d) represents a multiplication of this structure and in **Fig 9.6(e)** there are in effect an infinite number of loops in parallel, ie a cylinder closed at both ends with a central rod in series with the capacitor. If the diameter of the structure is greater than its height it is termed a rhumbatron, otherwise it is a coaxial cavity.

The simple hairpin, shown at Fig 9.6(b), is a very convenient form of construction: it can be made of wide strip rather than wire and is especially suitable for push-pull circuits. It may be tuned by parallel capacitance at the open end, or by a series capacitance at the closed end.

In a modification of the hairpin loop, the loop can be produced from good-quality double-sided printed circuit board and such an arrangement is known as microstripline. The loop is formed on one side; the ground-plane side of the PCB, through the dielectric, makes the stripline. When the PCB is very thin, the result is called microstrip and that is used in many commercial receivers.

Bandpass circuits

Tuning of antenna and RF circuits to maintain selectivity in the front-end of a VHF/UHF receiver cannot be undertaken with normal ganged tuning capacitors, not only due to the effects of strong capacitance coupling between circuits, which become prominent in these frequencies, but also due to the difficulty of procuring small-swing (say 20pF) multi-ganged capacitors. The varicap diode can be used to replace mechanical capacitors, but can degrade the receiver performance when strong signals are present by rectifying these signals and introducing intermodulation products into the mixer, thus seriously reducing the receiver's dynamic range.

Fortunately modern construction techniques have enabled coil manufacturers to introduce a band-pass circuit in a very small screened unit, namely the helical filter.

The helical filter in simple terms is a coil within a shield. However, a more accurate description is a shielded, resonant section of helically wound transmission line, having relatively high characteristic impedance. The electrical length is approximately 94% of an axial quarter-wavelength. One lead of the winding is connected to the shield; the other end is open-circuit. The Q_u of the resonator is dictated by the size of the shield, which can be round or square. Q_u is made higher by silver plat-

Parameter	HRW (231MT-10001A)	HRQ (232MT-1001A)
Centre frequency (MHz)	435	435
Bandwidth at 3dB (MHz)	12 min	11
Attenuation (dB)	20 min at ±30MHz	25 min at ±15MHz
Max ripple (dB)	1.5	2
Max insertion loss (dB)	2.5	4
Impedance (Ω)	50	50

Table 9.1: Electrical characteristics of Toko HRW and HRQ helical resonators

Fig 9.7: (left) Toko resonators. (left) HRW and (right) HRQ

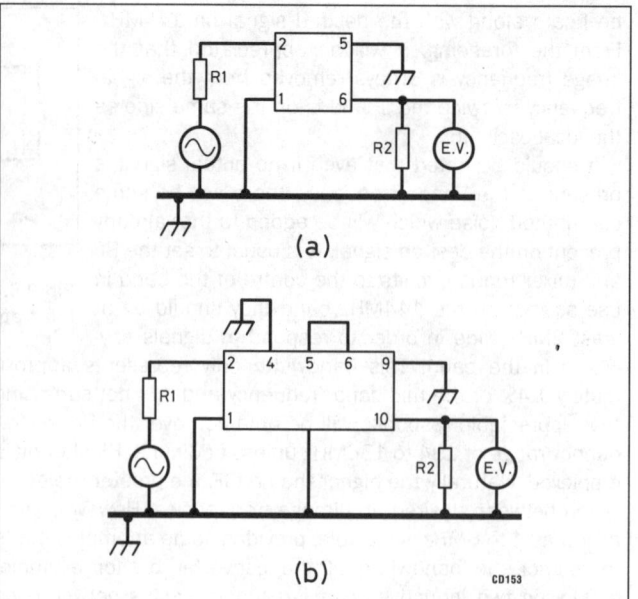

Fig 9.8: Test circuit for (a) Toko HRW and (b) Toko HRQ filters. The case lugs must be grounded. R1 = R2 = 50Ω

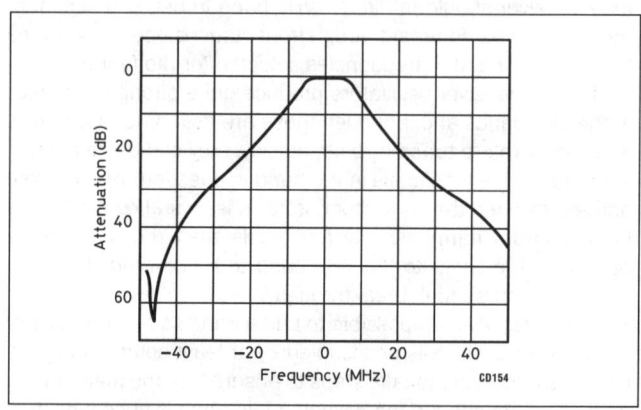

Fig 9.9: The Toko HRW frequency response

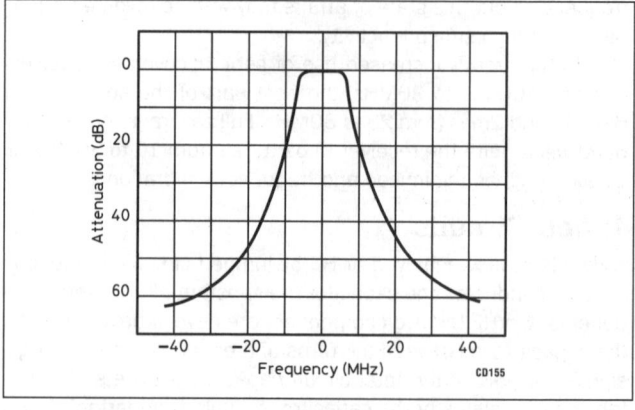

Fig 9.10: The Toko HRQ frequency response

ing the shield. Resonance can be adjusted over a small range by opening or closing the turns of the helix. The adjustment is limited over the small frequency range to prevent degradation of Q_u. Modern miniature helical resonators can be obtained in a shield only 5mm square, but the minimum resonant frequency is normally 350MHz. Maximum F_r can be 1.5GHz. Large-size resonators (10mm square) will resonate down to 130MHz.

The band-pass filter is obtained by combining two to four resonators in one unit with slots cut in each resonator screen of defined shape to couple the resonators. This forms a high-selectivity tuned circuit with minimum in-band insertion loss and maximum out-of-band attenuation.

Helical filters can be cascaded to increase out-of-band attenuation. As an example, a quadruple filter with a centre frequency of 435MHz might have a 3dB bandwidth of 11MHz and 25dB attenuation at 15MHz; with a ripple factor of 2dB and insertion loss of 4dB (see Table 9.1). Fig 9.7 shows pictures of the helical filters tested. The test circuit is shown in Fig 9.8, and Figs 9.9 and 9.10 show the test results.

Tuning is accomplished by brass screws in the top of the screen, one for each helix. The nominal input/output impedance is 50Ω formed by placing a tap on the helix. This impedance is ideal for matching to antennas and to RF amplifiers and mixers designed using S-parameters. Thus the helical filter replaces conventional tuned circuits in the receiver front-end, resulting in a considerable improvement in selectivity. One note of caution; the screening can lugs must be soldered perfectly to the PCB, otherwise the out-of-band attenuation characteristics of the filters will be degraded. These filters can be used for 2m, 70cm and 23cm receiver front-ends.

Preamplifiers

As available devices improve and new circuit designs are published, it will become apparent that a receiver that may have been considered a first-rate design when built is no longer as good as may be desired. Specifically, a receiver using early types of transistor may not be as sensitive as required, although the local oscillator may perform satisfactorily. The sensitivity of such a receiver can be improved without radical redesign by means of an additional separate RF amplifier, usually referred to as a pre-amplifier. Such an amplifier should have the lowest possible noise figure and just sufficient gain to ensure that the overall performance is satisfactory.

Fig 9.11 shows the improvement to be expected from a pre-amplifier given its gain and the noise factor of the preamplifier alone and the main receiver.

An example will suffice to show the application. An existing

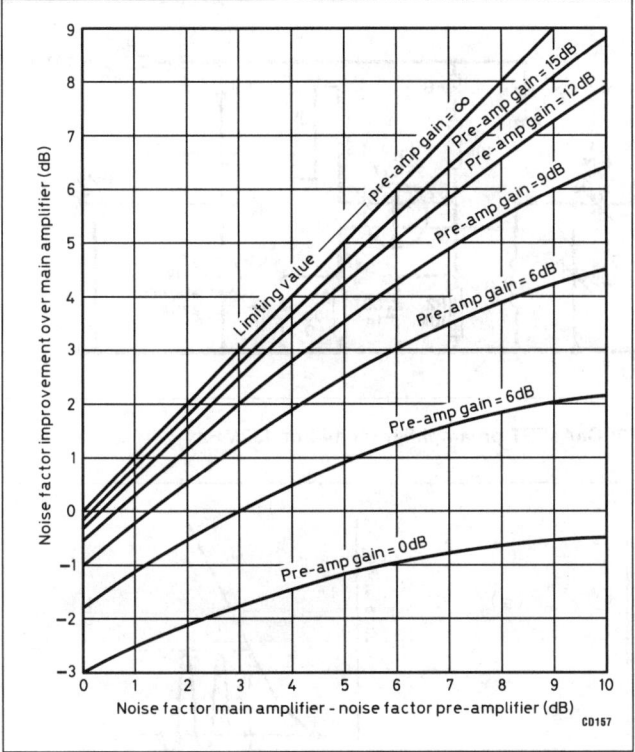

Fig 9.11: Receiver noise figures

145MHz receiver has a measured noise figure of 6dB and is connected to its antenna via a feeder with 3dB loss. It is desired to fit a preamplifier at the masthead; what is the performance required of the preamplifier?

Suppose a BF180 transistor is available: this has a specified maximum noise figure of 2.5dB at 200MHz and will be slightly better than this at 145MHz. The main receiver and feeder can be treated as having an overall noise figure of 3 + 6 = 9dB. From Fig 9.11, if the preamplifier has a gain of 10dB, the overall noise figure will be better than 4.1dB. Increasing the gain of the preamplifier to 15dB will only reduce the overall noise figure to 3.6dB and may lead to difficulty due to the effect of varying temperatures on critical adjustments. The addition of so much gain in front of an existing receiver is also very likely to give rise to intermodulation products from strong local signals. If it is desired to operate under such conditions, it is essential that provision is made for disconnecting the preamplifier when a local station is transmitting.

RF Amplifier Design

The advent of the field effect transistor (FET) eased the design of VHF RF amplifiers in two ways. The relatively high input resistance at the gate permits reasonably high-Q tuned circuits providing protection against strong out-of-band signals such as from broadcast or vehicle mobile stations. Also the drain current is exactly proportional to the square of the gate voltage; this form of non-linearity gives rise to harmonics (and the FET is a very efficient frequency doubler) but a very low level of intermodulation.

The use of a square-law RF stage is not, however, as straightforward as it first appears. Second-order products are still present, such as the sum of two strong signals from transmitters outside the band, typically Band II broadcasts, mixing to generate the unwanted signals on the 144MHz band. It follows, of course, that if two FET amplifiers are operated in cascade a band-pass filter is required between them to ensure that the

distortion products generated in the first stage are not passed to the next stage where they will be re-mixed with the wanted signal.

A development of the FET was the metal oxide semiconductor FET (MOSFET), the gate is insulated by a very thin layer of silica. The gate therefore draws no current and a high input resistance is possible, limited only by the losses in the gate capacitance. These devices may be damaged by static charges and must be protected against antenna pick-up during electrical storms and also by RF from the transmitter feeding through an antenna changeover relay with an excessively high contact capacitance. MOSFETs now have protective diodes incorporated in the device which limit the input voltage to a safe level and these devices are thus much more rugged.

In the dual-gate MOSFET the drain current is controlled by two gates, resulting in various useful circuit improvements. If it is desired to apply gain control to a bipolar transistor stage the control voltage is applied to the same electrode as the signal but the result can be a reduction in signal handling capacity, showing up as intermodulation and/or blocking. The dual-gate FET avoids this problem and automatic or manual gain control can be applied to gate 2 without reducing the signal-handling capability at gate 1. **Fig 9.12** shows how such an RF stage for a converter is arranged. When a strong local station causes intermodulation at the mixer or an early stage of the main receiver, the RF gain can be reduced until interference-free reception is again possible.

The GaAsFET is similar to the MOSFET but is based on gallium arsenide rather than silicon. Gallium arsenide has larger electron mobility than silicon and therefore has a better performance at UHF. GaAsFETs were originally designed for use in television receiver tuners and are entirely suitable for 144MHz and 432MHz preamplifiers or as a replacement for an existing RF amplifier. These devices are available from major semiconductor manufacturers and include the 3SK97, 3SK112, C3000 and the CF739 (Siemens).

In a correctly designed amplifier the GaAsFET will provide excellent performance at these frequencies. The quoted noise figure for the 3SK97 is 1dB at 900MHz. Silicon diodes are mounted in the GaAsFET chip to protect the gates against ESD (electrostatic discharge) breakdown.

The noise figure obtainable is related to the manufacturer's data sheet. Biasing is the same as for silicon MOSFETs, ensur-

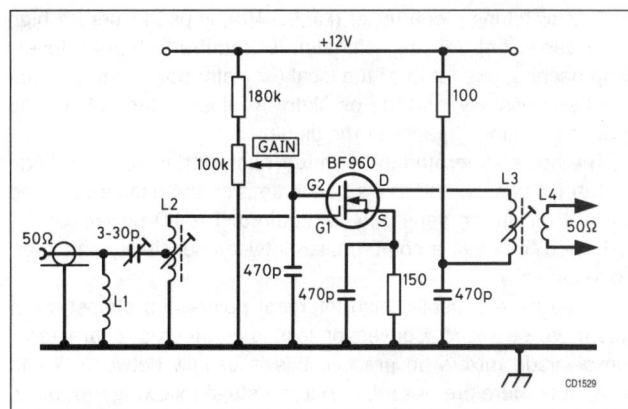

Fig 9.12: Dual-gate MOSFET RF amplifier with gain control. L1: 3 turns 1.0mm enamelled wire, 6.0mm ID, 8.0mm long. L2: 2 turns 1.0mm tinned copper wire on 8.0mm former 10mm long, tapped at 1½ turns and tuned with dust core. L3: 6 turns 1.0mm enamelled wire on 8mm former tuned with dust core and coupled to L4. L4: 2 turns 1.0mm enamelled wire on same former, close spaced to capacitor end of L3

ing, however, that the gates are never positively biased with respect to the channel.

The circuit for a GaAsFET preamplifier [1] that incorporates a self-biasing arrangement is shown in **Fig 9.13**. Construction of L1 and T1 for 144MHz and 432MHz is given below.

L1   *144MHz:* 6 turns 2.0mm tinned copper wire 6mm ID 13mm long

   432MHz: copper line 15mm wide 57mm long spaced 4mm above ground plane

T1   *144MHz:* 12 turns bifilar wound 0.5mm enamelled copper wire

   432MHz: 5 turns 0.5mm enamelled copper wire centre tapped as a 4:1 transformer on an Amidon T-20-12 toroid Core

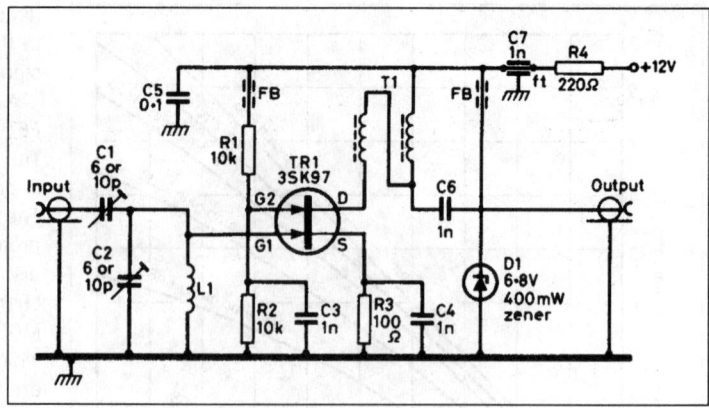

Fig 9.13: GaAsFET preamplifier for 144 or 432MHz

Double-sided copper laminate is used for mounting most of the components with a vertical screen of tinplate forming a screen between input and output circuits. Care must be taken against static when mounting the GaAsFET in the circuit. After careful checks for constructional errors the current should be checked before alignment. This should be between 25-30mA.

Power gain alignment will be in the order of 26dB at 144MHz and 23dB at 432MHz. Any tendency towards instability can be cured by fitting a ferrite head over the drain lead close to the FET.

An attenuator must be used between the preamplifier and input to an existing receiver or converter to prevent degradation of the strong-signal performance.

Mixers

The most common types of mixers in use today are designed around diodes, bipolar and field effect transistors; integrated circuits are now available for use in mixer applications.

Diode mixers

A diode mixer operates non-linearly, either around the bottom bend of its characteristic curve where the current through the diode is proportional to the square of the applied voltage - see **Fig 9.14(a)**, or by the switching action between forward and reverse conduction as shown in **Fig 9.14(b)**. For optimum working conditions in (a), it is usually necessary to apply DC forward bias to the diode of typically 100 to 200mV; this will vary with the type of diode.

The switching diode mixer (Fig 9.14(b)) is used where a high overload level (strong signals) is required. Signal levels approaching one tenth of the local oscillator power can be handled successfully, and the oscillator level is limited only by the power handling capacity of the diode.

The noise generated in the mixer rises with increasing diode current, however, and this sets the limit on the usable overload level if maximum sensitivity is required. The LO power can be adjusted to select a compromise between sensitivity and overload capacity.

Diode mixers display high intercept points and almost all of them are balanced. Conversion loss is an inherent characteristic of diode mixers. In practice this is usually between 3 and 6dB. It is therefore essential that the stage following the mixer, eg an IF amplifier, has the lowest possible noise figure.

Balanced-diode mixers

Noise component transfer from the LO and the mixer to the post-mixer stage can be reduced to a low level by using a balanced two-diode design employing modern low-noise diodes.

However with modern diodes it is not usually necessary to provide adjustment of balance at the LO frequency for best

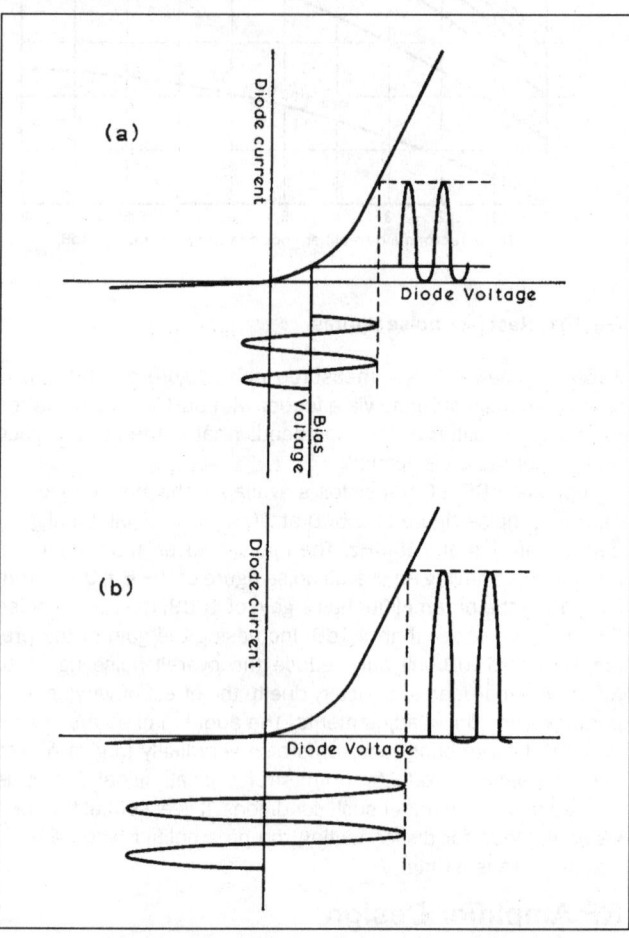

Fig 9.14: Working conditions of diodes mixers. In (a) forward bias is required to provide an optimum point. In (b) the local oscillator is higher and no bias is required

noise performance. It is always necessary to include a low-pass filter between the mixer and the antenna to reduce LO radiation from the antenna.

Conversion loss is minimised with as little injection as 1mW (0dBm), but IMD (intermodulation) product rejection is improved by increasing the LO drive level to +7 to +10dBm (5mW to 10mW).

Diodes suitable for VHF and UHF receiver mixers are the hot-carrier, high-speed silicon switching types and Schottky barrier diodes. The latter are the only types used in mixers designed specifically to have a high overload capability. A Schottky barrier diode such as the HP 2800 is probably the best choice for mix-

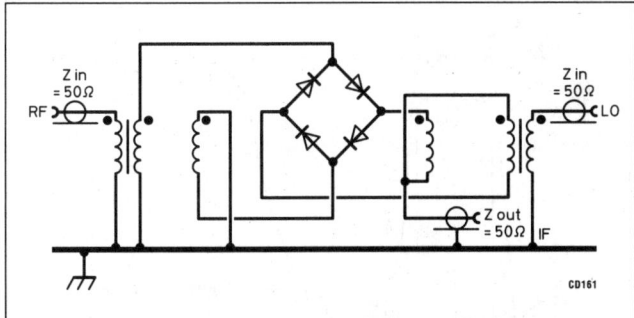

Fig 9.15: Typical diode ring mixer circuit (*ARRL Handbook*)

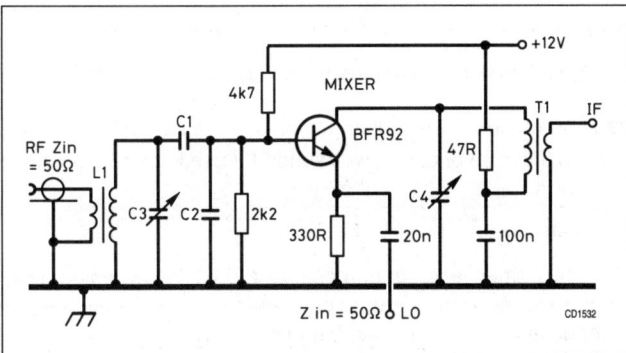

Fig 9.16: Bipolar mixer circuit (*ARRL Handbook*)

ers operating at these frequencies. The diode ring mixer as shown in **Fig 9.15** is capable of very high performance at VHF and UHF at all the usual signal levels. The transformer cores are normally a low-permeability ferrite with a Q value of typically 125. Toroids are not the best type to use but multi-hole ferrite beads are ideal and make transformer construction simple. For optimum balance the diodes themselves must be dynamically balanced, and for this reason it is easier to purchase a ready-built diode ring mixer, which is usually built into a screened and potted assembly.

Bipolar Transistor Mixers

A bipolar transistor can be used as a mixer at VHF and will provide a reasonable performance. From the circuit shown in **Fig 9.16** it will be seen that RF input is made to the base and LO injection is made to the emitter. The injection is made at low impedance - the LO coupling capacitor C5 has a low reactance at the IF. Bipolar transistor mixers require a fairly low value of LO input power. This reduces the output power requirement of the LO. A typical level is -10dBm. Conversion gain is moderate and correct choice of the transistor will minimise the noise figure.

However, the intrinsic exponential forward-transfer characteristic of the bipolar transistor severely limits the large-signal handling capabilities. Blocking and IMD products become evident even with moderate signal input levels to the mixer. The gain of the preceding RF stage (if used) must be kept low, eg 6dB, to minimise these unwanted products.

Junction field effect transistor mixers

Junction FETs can provide very good performance as mixers at VHF. A typical JFET mixer circuit is shown in **Fig 9.17**. The input impedance is high but the conversion gain is only 25% of the gain of the same device used as a VHF amplifier. Bias is critical, and is normally chosen so the gate-source voltage is 50% of the pinch-off voltage of the device. The LO voltage is injected into the source, the level being chosen to avoid the pinch-off region, but not sufficient to cause the source-gate diode to be driven into conduction. Normally the LO peak-to-peak amplitude should be kept a little below the pinch-off voltage.

The JFET has moderate output impedance, typically 10kΩ. This eases impedance matching to the following IF filter, which is usually performed via a step-down transformer as shown in Fig 9.17. The JFET does not have an ideal square-law input characteristic due to the effect of bulk resistance associated with the source. However, the generation of unwanted IMD products under strong signal conditions is much lower than the bipolar mixer, although the noise figure is similar to that of the latter.

Dual-gate MOSFET mixers

MOSFET mixers can provide a superior performance compared with both bipolar and junction field transistors. They have excellent characteristics including a low noise figure, almost perfect square-law forward transfer, together with high input and output impedances. Conversion gain can be high and at the same time there is very low generation of IMD products at the IF output under large signal input levels. Overall performance is extremely good at both VHF and UHF.

Fig 9.18 shows the dual-gate MOSFET in a mixer circuit. The signal input is applied to gate 1 as for an RF amplifier, while the LO is applied to gate 2, the IF output at the drain is controlled by the input levels. Optimum conversion gain is obtained with about 5V peak-to-peak of LO. This mixer has the advantage of having inherent isolation between the signal and LO gate inputs. Next to the ring diode mixer the dual-gate MOSFET mixer has the highest 'overload' level, at the same time giving conversion gain instead of loss.

Balanced mixers, whether they use bipolar transistors, junction FETs or MOSFETs, will give a much improved performance where low noise and the largest dynamic range must be achieved at the same time as maximum suppression of

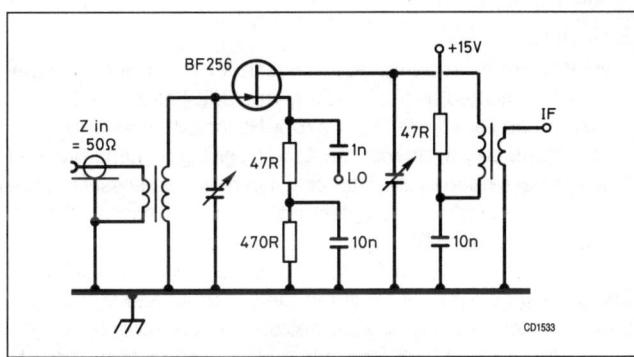

Fig 9.17: A JFET mixer (*ARRL Handbook*)

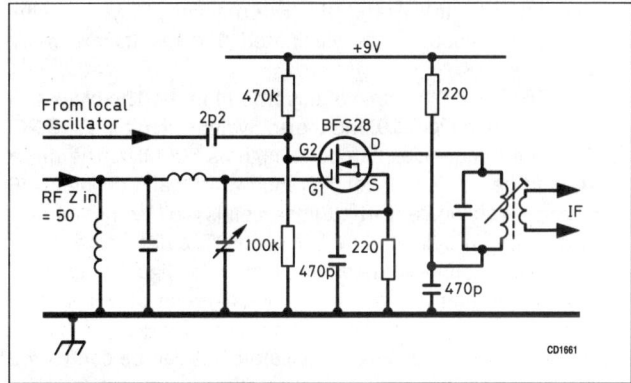

Fig 9.18: A dual-gate MOSFET mixer

The Radio Communication Handbook

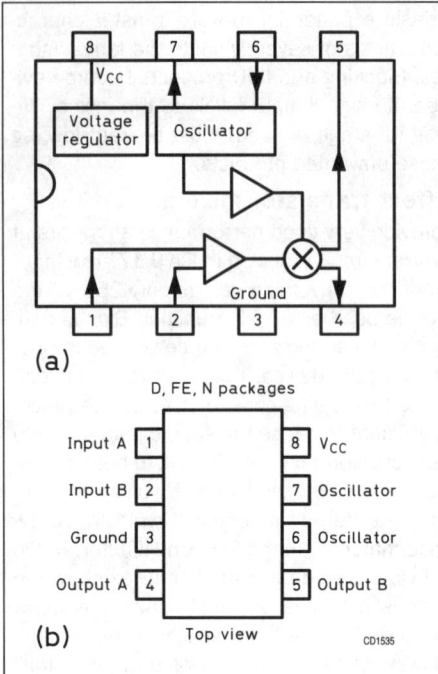

Fig 9.19: NE612 double-bal-anced mixer and oscillator. (a) Block dia-gram. (b) Pin configuration (Signetics)

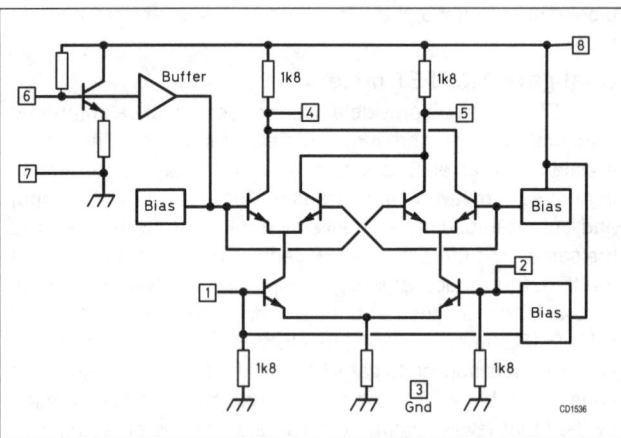

Fig 9.20: Equivalent of NE612 double-balanced mixer and oscil-lator (Signetics)

unwanted mixer outputs (typically IMD products and LO feedthrough) either to the antenna or to the IF amplifier.

Integrated circuit mixers

Using IC types can obviate the problems of matching transis-tors and components in balanced mixers. These are available in the form of a monolithic bipolar double-balanced mixer intended for use at VHF and UHF. External circuit layout is sim-plified, bias adjustment is eliminated and results are more predictable.

The NE612 is an example of this type of mixer. The block dia-gram is shown in **Fig 9.19** and the equivalent circuit in **Fig 9.20**. Input signal frequencies can be as high as 500MHz. The mixer is a Gilbert cell multiplier configuration which can provide 14dB or more conversion gain. The Gilbert cell is a differential ampli-fier that drives a balanced switching cell. The differential input stage provides gain and determines the noise figure and signal handling performance. The mixer noise figure at 50MHz is typi-cally < 6dB.

The NE612 contains a local oscillator. This can be configured as a CCO or a VFO. The oscillator can also be reconfigured as a buffer amplifier for an external oscillator (the latter is to be pre-

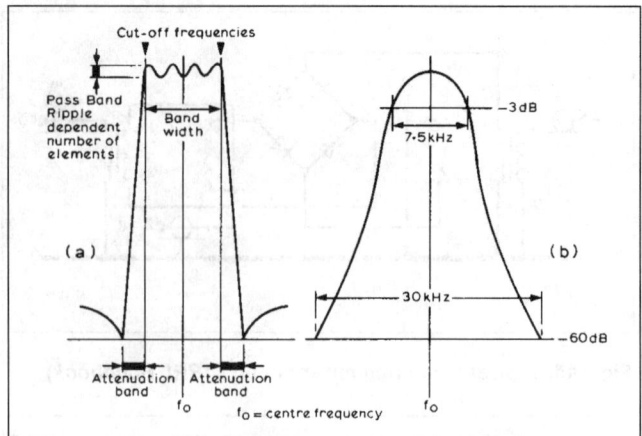

Fig 9.21: Block filter characteristics (a) SSB, (b) NBFM

ferred). The low power consumption, typically 2.4mA at V_{cc} = 6V, makes this type of device well suited for use in battery operated receivers.

IF Filters

The performance of a purpose-built double-conversion VHF/ UHF receiver with a 'tuneable' first local oscillator can be enhanced by using a crystal filter between the first mixer and IF amplifier. Commercial crystal filters are now readily available, not only for the well-known IFs of 10.7 and 21.4MHz, but also for 45 and 75MHz. Until recently, crystal filters above about 25MHz were only available with third-overtone-mode crystal elements. Now 45 and 75MHz filters are available with fundamental crys-tal elements. Recommended filters for VHF/UHF bands are:

6m and 4m	10.7MHz
2m	21.4MHz
70cm	45.0MHz

These filters are available in two, four or six-pole versions; a six-pole filter will give the best attenuation to out-of-pass-band signals. Good selectivity must be achieved before the first IF amplifier and the crystal filter will give superior selectivity com-pared to an LC filter. Crystal filters have low input and output impedances, the actual impedance being dependent on the number of poles. These impedances vary from 3kΩ/3pF at 10.7MHz to 500Ω/1pF at 75MHz and are given in the maker's literature.

The preferred matching mechanism to the first mixer is the pi-network as illustrated in the mixer circuits. This network has an impedance inversion property. The input impedance of the crys-tal filter appears as an open-circuit to out-of-pass-band signals while the load presented to the drain (of a MOSFET mixer) appears as a near-short-circuit. This will enhance the mixer IM product attenuation.

Bandwidth

Crystal filters must be correctly chosen for the type of modula-tion to be detected by the receiver. The -3dB bandwidth for SSB filters can be as little as 2.1kHz (500Hz for CW receivers).

The -60dB bandwidth can be 4.2kHz, giving a shape factor of 2:1. The pass-band of the filter characteristic can possess ripple and will not introduce pre-detector audio modulation distortion. However, filters for NBFM receivers must have a -3dB bandwidth of ±3.75kHz for 12.5kHz channel spacing and ±7.5kHz for 25kHz channel spacing for linear 'detection' of NBFM transmit-ters using maximum frequency deviation of the modulator.

Pass-band ripple must be minimal (not more than 1dB) to avoid modulation distortion. The -60dB bandwidth for a typical

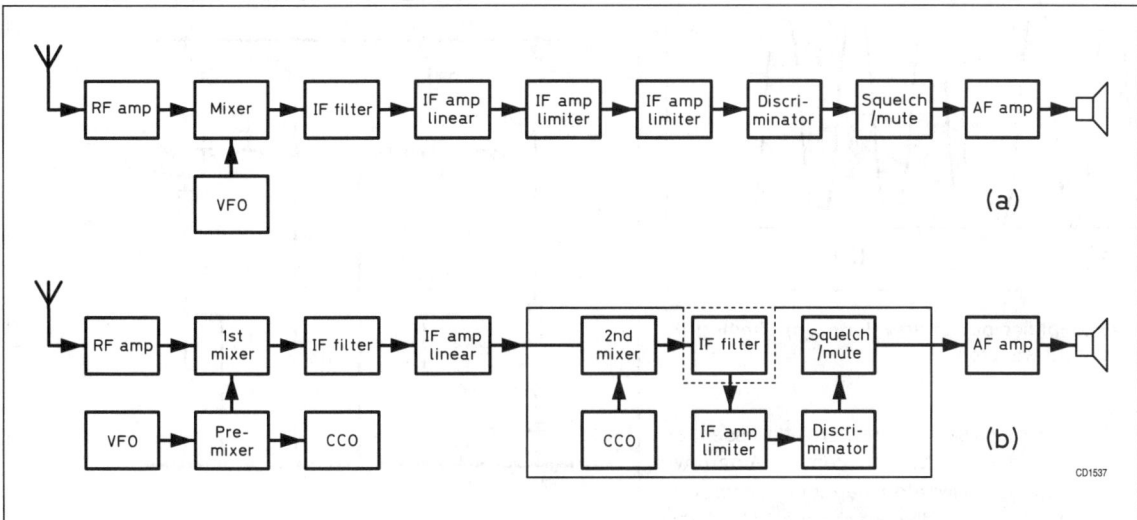

Fig 9.22: Comparison of the essential stages of receivers for (a) NBFM with discrete circuits and (b) NBFM with integrated circuits

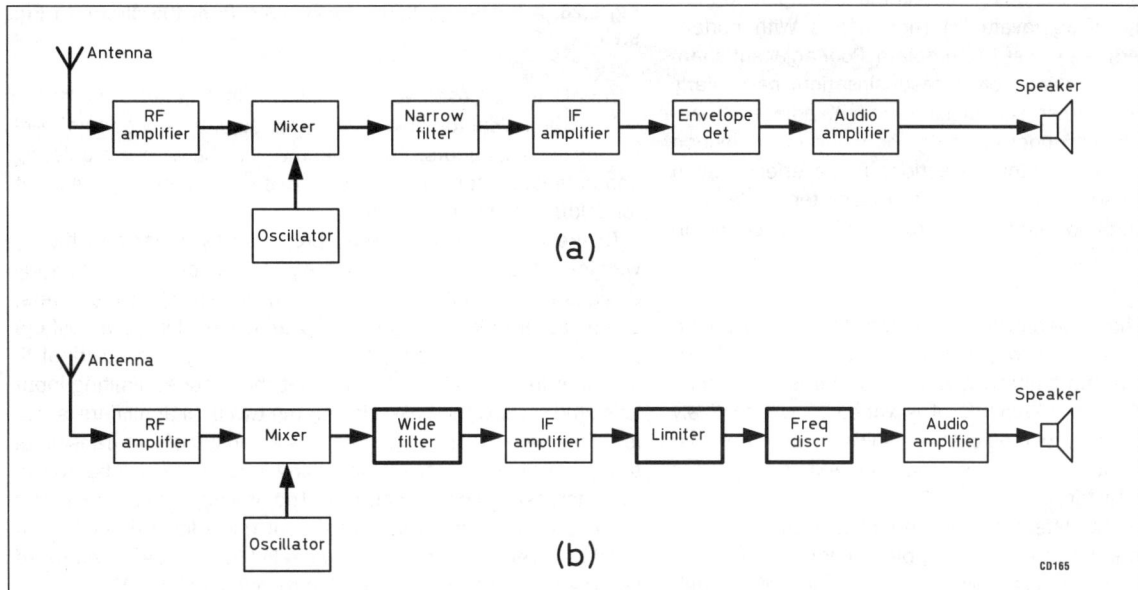

Fig 9.23: Block diagrams of (a) an AM and (b) an FM receiver. Dark borders outline the sections that are different from the AM set (*ARRL Handbook*)

six-pole NBFM crystal filter can be 30kHz for 12.5kHz channel spacing and 60kHz for 25kHz channel spacing, giving a shape factor of 4:1 All crystal filters have some insertion loss and it is usual to build a post-filter IF amplifier in the receiver to negate this loss. **Fig 9.21** illustrates the difference in SSB and NBFM IF filter characteristics.

RECEPTION OF FM SIGNALS

There are two principal features in receivers designed to receive FM signals, namely limiting rather than linear amplifiers precede the detector and the latter is designed to convert IF variations into AF signals of varying amplitude, dependent on the degree of frequency variation in the transmitter carrier.

The FM Receiver

The block diagrams of an FM receiver and AM/SSB receiver are shown in **Figs 9.22 and 9.23**. The principal difference between the receivers are the IF filter bandwidths (see above) and the IF amplifier gains required before the detector.

It is necessary to provide sufficient gain between the antenna and detector of an FM receiver to ensure receiver quieting; ie optimum signal-to-noise ratio with the weakest signal. Usually this is less than 0.35µV PD or -116dBm (into 50 ohms).

Thus it is necessary to use the double superheterodyne principle to achieve the required gain, usually greater than 1 million or 120dB, whilst ensuring optimum stability independent of the input frequency. Other receiver stages, particularly the RF amplifier, mixer, oscillator and audio stages, can be identical to those employed in AM/SSB/CW receivers.

In a multimode receiver designed for reception and detection of all principal methods of modulation, the difference in signal-to-noise ratio and effect of interference is very noticeable between FM and AM/SSB signals. The limiter and detector (discriminator) for FM signals reduce interference effects, usually impulse noise, to a very low level, thus achieving a high signal-to-noise ratio. However, it is necessary to align the detector correctly and in use tune the receiver accurately to achieve noise suppression.

An unusual effect peculiar only to FM receivers, and known as capture effect, occurs when a strong signal appears exactly on the frequency to which the receiver is tuned. If this strong signal has a carrier amplitude more than two to three times that of the wanted signal, the strong signal will be detected. This effect can be a problem in mobile operation, particularly in a geographical area between two repeater outputs on the same frequency.

Weak-signal reception in AM and FM receivers can be degraded by a much stronger carrier on or near the frequency of the weak carrier.

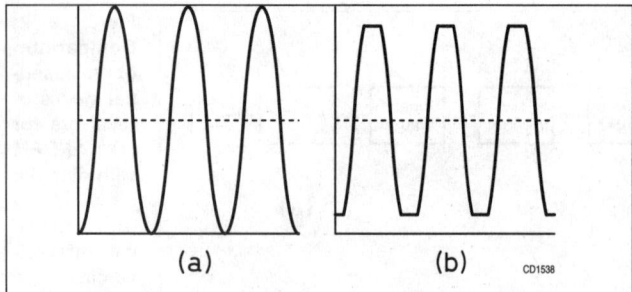

Fig 9.24: (a) Linear amplifier output waveform. (b) Limiting IF amplifier 'clipped' output waveform

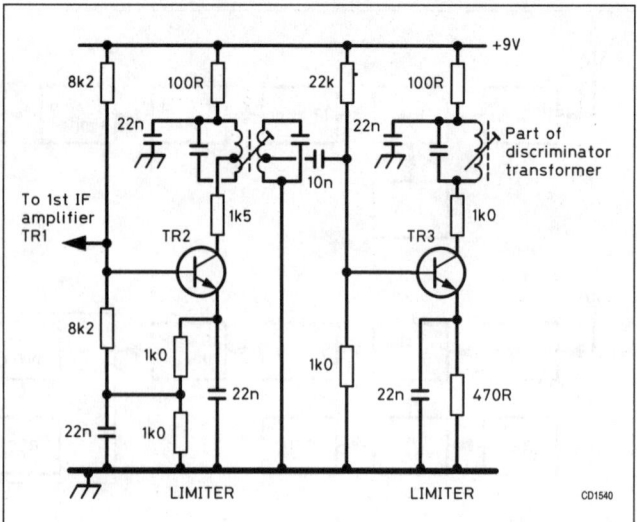

Fig 9.26: A two-stage limiter developed from the circuit of Fig 9.25

Selectivity

As already stated in the previous section on filters, it is essential to choose the IF filters designed for NBFM reception. A narrow-bandwidth filter will introduce unwanted harmonic distortion. Too wide a bandwidth will degrade adjacent channel selectivity. However, transmitters exceeding the recommended limit on frequency deviation, will aggravate distortion effects. With modern NBFM transmitters this is less of a problem. Poor adjacent channel selectivity can cause receiver desensitisation, particularly when strong signals are present on either or both adjacent channels. A transmitter with poor adjacent channel power rejection can also degrade weak-signal reception. These effects again cause receiver desensitisation but the transmitter modulation will not appear on the wanted signal (cross-modulation cannot in theory occur in FM systems).

Limiters

Limiting IF amplifiers are specifically designed to introduce gain compression into their forward-transfer characteristics. If an amplifier is driven into limiting, its output signal level remains unchanged as the input signal level is varied. This effectively removes any sudden amplitude change, which is important as it is necessary to remove any impulse noise and AM on the carrier prior to an FM detector.

Fig 9.24 shows the difference between a linear and a limiting IF amplifier output waveform. The clipping action removes the AM component. The overall amplifier gain must be high enough to ensure the limiting stages are limiting even with weak signals or with large changes of signal level IF to the receiver input. With an IF input of typically 5.0μV (equivalent to 0.25μV RF input to a receiver front-end with a conversion gain of 26dB) to the IF amplifier input, a minimum of three stages are required to raise the level of the signal for limiting action to commence. As the IF

carrier level increases above 5.0μV the limiting action starts.

Now the signal-to-noise ratio improves until at a certain level the noise disappears. This is known as the receiver quieting characteristic referred to earlier in this section, usually the input for 20dB signal-to-noise ratio.

Discrete limiting amplifier(s) preceded by linear amplifier(s) with interstage IF transformers are still to be found in some early designs but are not now employed in modern NBFM receivers. Examples are shown in **Figs 9.25 and 9.26**. Linear amplifiers precede the limiting amplifiers. The base bias on the final IF amplifier in Fig 9.25 is varied to set the required limiting input level. In Fig 9.26 the base bias on the two final amplifiers is varied. This sets the limiting knee characteristic of the transistors to a point at which, for an increasing input, there will be no further increase in collector current. The amplifiers saturate, giving the required limiting and consequent noise level reduction for good receiver quieting. The circuit in Fig 9.26 gives an improved limiting performance compared to the circuit in Fig 9.25.

Some FM receiver manufacturers incorporated an IC limiting amplifier containing two or more stages, as they gave superior limiting action compared to discrete designs such as the MC1590 and the CA3028A. These ICs became obsolete with the introduction of devices containing six or more differential DC coupled IF amplifiers with improved and consistent limiting characteristics, operating at a relatively low IF, typically 455kHz. This system removed the requirement for IF transformers between each stage, but caused layout problems. However, it considerably simplified alignment. These ICs are called NBFM IF subsystems, and will be described after the section on FM demodulators.

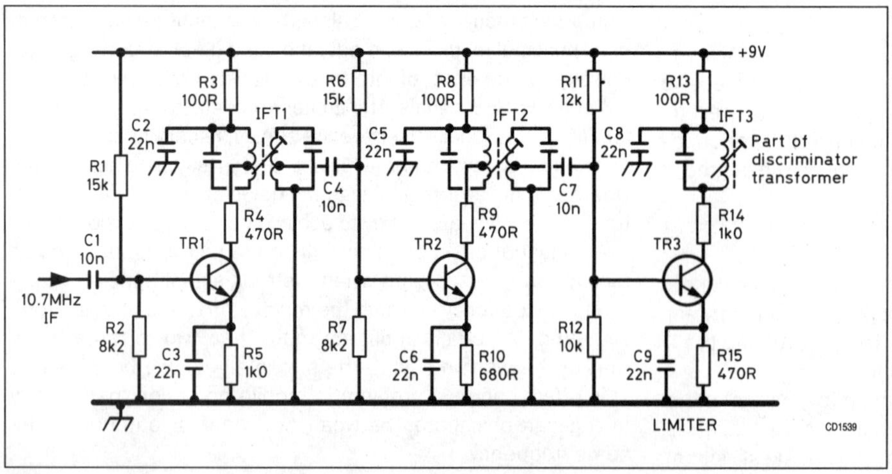

FM demodulators

The FM detector, or more correctly the FM discriminator, was evolved to be able to respond only to changes in frequency as received from an FM transmitter and not to amplitude variations as received from an AM transmitter when the carriers are modulated. The degree of frequency change (frequency

Fig 9.25: A three-stage IF amplifier and limiter. Transistors are BF194 or similar

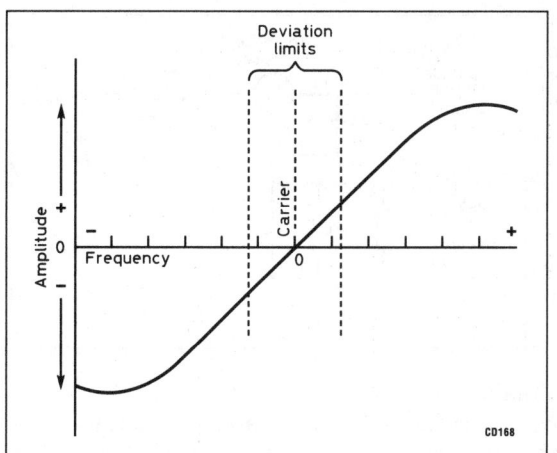

Fig 9.27: The characteristic of an FM discriminator

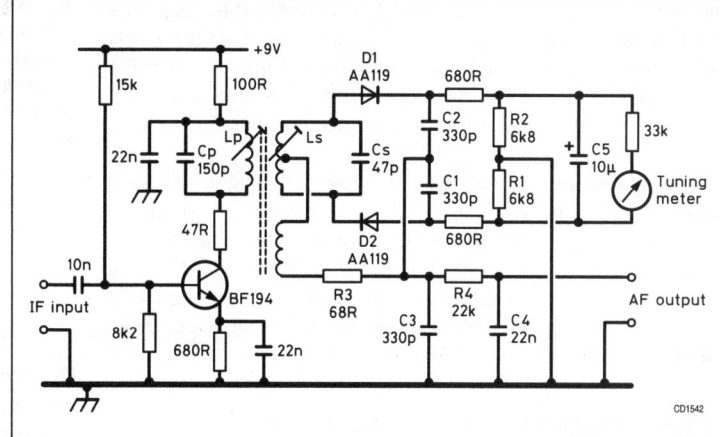

Fig 3.29: A practical ratio detector circuit

deviation) corresponds to the amplitude of the audio signal applied to the transmitter modulator.

For example, if the transmitted carrier is deviated by ±2.5kHz by the modulator, the received carrier will be changing in frequency by ±2.5kHz.

This frequency 'swing' is detected by an FM discriminator. **Fig 9.27** shows clearly the 'S' curve characteristic of the discriminator. Provided the swing is in the linear (straight) portion of the curve, ie about ±6kHz, very little distortion will be present in the recovered audio signal.

Maximum frequency deviation in amateur FM equipment is limited to ±5kHz compared to ±75kHz for Band 2 broadcast transmitters and receivers. This explains why FM voice and data communications systems are normally known as narrow-band FM systems.

A practical discriminator, known as the Foster-Seeley discriminator, after its inventors, is shown in **Fig 9.28**. T1 is the discriminator transformer. Voltage, due to the IF carrier, is developed across the primary of T1. Primary-to-secondary inductive coupling induces a current in the secondary 90° out of phase with the primary current. The IF carrier is also coupled to the secondary centre tap via a coupling capacitor.

The voltages on the secondary are combined in such a way that these lead and lag the primary voltage by equal amounts (degrees) when an unmodulated carrier is present. The resultant rectified voltages are of equal and opposite polarity.

When the received carrier is deviated the phase is changed between primary and secondary, resulting in an increased output level on one side and a decreased output on the other side. These voltage-level differences, after rectification by D1 and D2, represent recovered audio. This discriminator responds to carrier amplitude variations (AM) unless driven hard by the limiting IF amplifier stages and, for this reason, fell from favour.

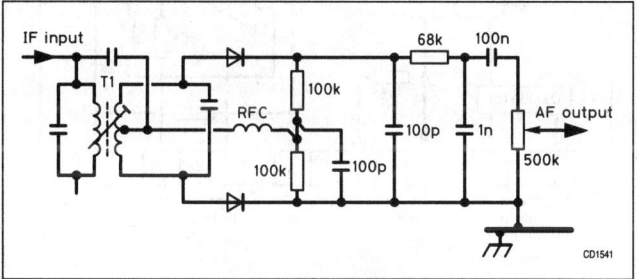

Fig 9.28: The Foster-Seeley discriminator

Ratio Detectors

This type of FM detector was developed from the frequency discriminator and became very popular in broadcast FM receivers - it has also been widely used in amateur receivers. The ratio detector is far less susceptible to carrier amplitude variations, hence less limiting is required before it. The detector is shown in **Fig 9.29**. T1 is a ratio detector transformer comprising of a primary, centre-tapped secondary and a tertiary winding, tightly coupled to the primary.

As its name implies, it functions by dividing the rectified DC voltages from D1 and D2 appearing across R1 and R2 into a ratio equal to the amplitude ratio present on each side of transformer secondary (L_s). The DC voltage sum appears across the electrolytic capacitor C5. This has sufficient capacitance to maintain this voltage at a constant level during fluctuations in carrier levels, caused for example by AM or noise signals being present on the carrier. Hence the ratio detector has its own

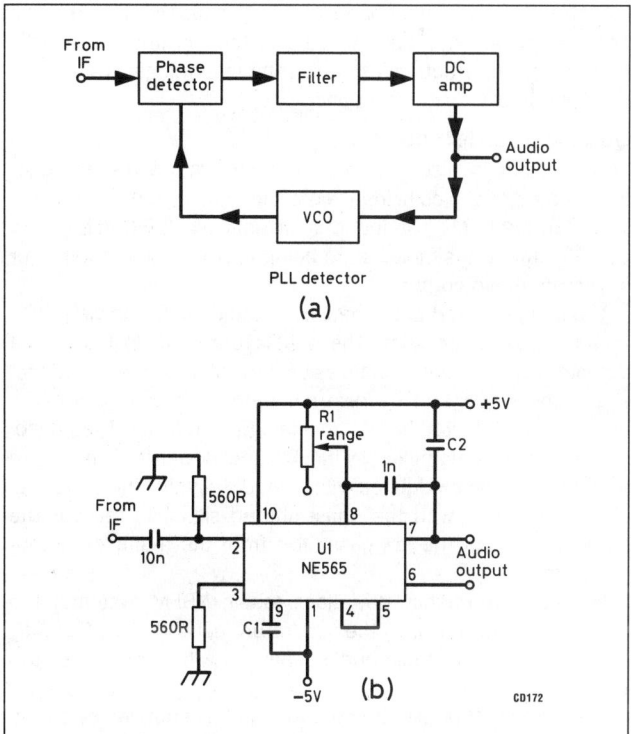

Fig 9.30: (a) Block diagram of a PLL demodulator. (b) Complete PLL circuit

inherent limiting characteristic. With a detector that responds only to ratios, the strength of the IF carrier can vary considerably without causing the output level to change. Therefore only FM can be detected and not AM. The carrier level should not fall below a level to cause D1 and D2 to become partially or fully non-conductive.

When the carrier is deviated by FM the audio signal is recovered from the tertiary winding L1, The IF carrier is filtered out by R3 and C3. The diode load resistors are lower in value than for the discriminator. It will be noted the diodes are connected in series rather than series-opposing as in the discriminator. This makes the ratio detector 6dB less sensitive. The diodes such as the gold-bonded AA 119 should preferably be matched dynamically.

Phase-Lock-Loop Detectors (PLL)

With the advent of the single-chip PLL it has been possible to design a reliable NBFM detector without tuned circuits and, therefore, without the necessity for alignment.

An example of an IC PLL detector is shown in **Fig 9.30**. The block diagram is shown in **Fig 9.30(a)** and the circuit including external components in **Fig 9.30(b)**.

Referring to Fig 9.30(a), the VCO oscillates close to the carrier frequency, in most cases 455kHz. The phase detector produces an error voltage when the VCO frequency and the carrier frequency are not identical. This error voltage is a DC voltage that is amplified after filtering, and 'corrects' the VCO frequency. When the carrier is deviated by audio modulation the frequency change is sensed by the phase detector and the resultant error voltage corrects the VCO frequency, causing it to remain locked to the carrier frequency, The system bandwidth is controlled by the loop filter. As the error voltage corresponds exactly to the frequency deviation, the PLL circuit functions as a precise FM detector.

It has a high sensitivity, requiring typically a 1mV carrier level for the PLL circuit to function correctly. Referring to Fig 9.30(b), R1 and C1 set the VCO frequency close to the carrier frequency. C2 controls the loop filter bandwidth that in turn controls the PLL capture range. The capture range is the maximum deviation from the carrier frequency to which the loop will gain and maintain lock on the carrier frequency.

Quadrature detectors

This detector is also known as the quadrature discriminator (sometimes as the coincidence detector). The symbolic circuit is shown in **Fig 9.31**. It is found in virtually all NBFM subsystem ICs. Alignment is simple, there being only one coil to align for maximum audio output.

The IF input is fed to the detector, both directly and via a 90° (quadrature) phase shift. The quadrature input is fed to the detector using an appropriate value capacitor (C2) and a phase-shift network C1 and L which resonates at the IF centre frequency. The detector itself is an analogue multiplier. The phase-shifted signal is multiplied by the IF signal deviation from centre frequency (when modulated with an audio signal). The direct signal is multiplied with the phase-shifted signal to recover the audio signal, via the low-pass filter from the multiplex output spectrum.

For small frequency deviations (as in NBFM systems) the phase shift, controlled by the quadrature network, is sufficiently linear to give acceptable audio quality, ie with very little distortion.

The working Q of L in the network can be controlled by shunting it with a resistor (R). The lower the resistor value, the better the linearity, as this will increase the peak deviation capability of the detector. However, the audio recovery level is reduced.

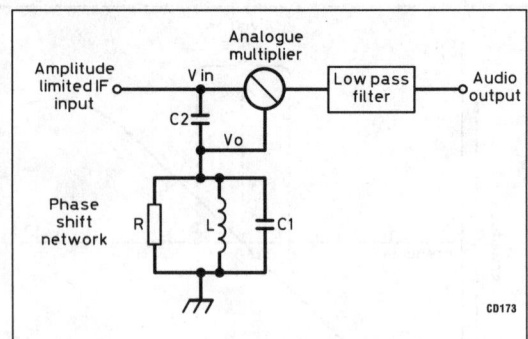

Fig 9.31: Symbolic circuit of a quadrature detector

De-emphasis

It is normal practice to insert a de-emphasis network, usually a resistor and capacitor combination, in the post-detector section of the receiver, irrespective of the type of detector employed, to attenuate noise and audio frequencies above 3kHz by 6dB/octave from 3kHz to 1kHz.

NBFM IF subsystems

These are used in modern dual-conversion receivers. The IC integrates the limiting amplifiers with a second mixer and oscillator for converting to a low IF, typically 455kHz, a quadrature detector, an active filter for driving a squelch circuit and a post-detector AF preamplifier in a single monolithic block. This results in considerable space saving, particularly for hand-held and mobile equipment. Power consumption is low and setting up and alignment is simple. Early examples of this type of IC are the Motorola MC3357 and the MC3359; these ICs can be found in many professional and amateur NBFM receivers.

More recently improved performance versions have appeared, eg the MC3361 and MC3362. The MC3361 is shown in block diagram form in **Fig 9.32** and in a practical circuit in **Fig 9.33**.

The mixer is balanced to reduce spurious radiation. It converts the first IF input signal to the second IF of 455kHz. After passing through an external band-pass filter, normally a multi-pole ceramic type, the IF signal is fed to the five-stage limiting amplifier and then to the quadrature detector where the audio signal is recovered. The 10pF on-chip capacitor produces the 90°

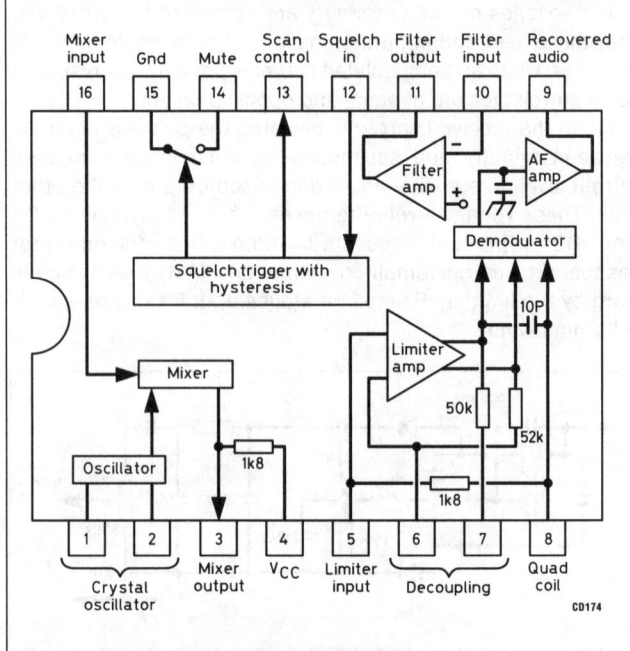

Fig 9.32: Symbolic circuit of a quadrature detector

Fig 9.33: Practical circuit using the MC3361 (Motorola)

phase shift between one output port of the final limiting amplifier and one input port of the quadrature detector. The quadrature coil is shunted by parallel resistor R2. This controls the linearity of the detector swing from centre frequency, hence the harmonic distortion in the recovered AF output which is buffered by the internal amplifier. R3 and C7 form the de-emphasis network. R4, R5 and C9 apply attenuated output to the filter amplifier. C10 and R6 peak the filter response to approximately 10kHz, ie above the normal audio pass-band of 300Hz to 3kHz.

In the absence of a carrier, only the noise signal is amplified this is detected by D1, which conducts, causing the DC voltage at the junction of' R9/R10 to fall. This level change is applied to the squelch switch input. The internal switch that stops noise output appearing across RV1 grounds the mute output pin. R10 and C12 filter the DC output from D1 anode. R7 and R8, with RV2, the squelch control, set the initial biasing of D1 and hence the squelch threshold level. When a carrier is detected and audio signals are recovered, D1 will not conduct and the squelch mute switch will remain open, removing the muting of the audio signal across RVI. The scan control at pin 13 can be used in conjunction with a digital PLL tuning system for 'locking' onto channels where the receiver has scanning facilities.

Fig 9.33 shows the external components required to complete the circuit of a practical NBFM receiver IF system. An IF block filter, as described earlier, must be inserted between the 1st mixer output and the IC input (2nd mixer) to provide adequate adjacent selectivity. The oscillator is an internally biased Colpitts type. The circuit shows a 10.245MHz fundamental-mode crystal oscillator for converting the 1st IF of 10.7MHz into the 2nd IF of 455kHz. However, a 20.945MHz crystal can be used for a 21.4MHz IF input and a 44.454MHz third overtone crystal for a 45MHz IF input. C1 and C2 form the crystal load capacitance.

The doubly balanced mixer has a high input impedance of about 3kΩ. This characteristic enables crystal filters to be matched easily to the mixer input. Similarly the output impedance is fixed by the internal 1.8kΩ resistor for correct matching to the 455kHz ceramic bandpass filter (FL1). The filter -3dB bandwidth should be 7.5kHz for 12.5kHz and 15kHz for 25kHz channel spacing respectively. Ultimate adjacent-channel selectivity is a function of the filter stop-band attenuation. A

Fig 9.34: Typical application for the MC3371 at 10.7MHz with a parallel LC discriminator (Motorola)

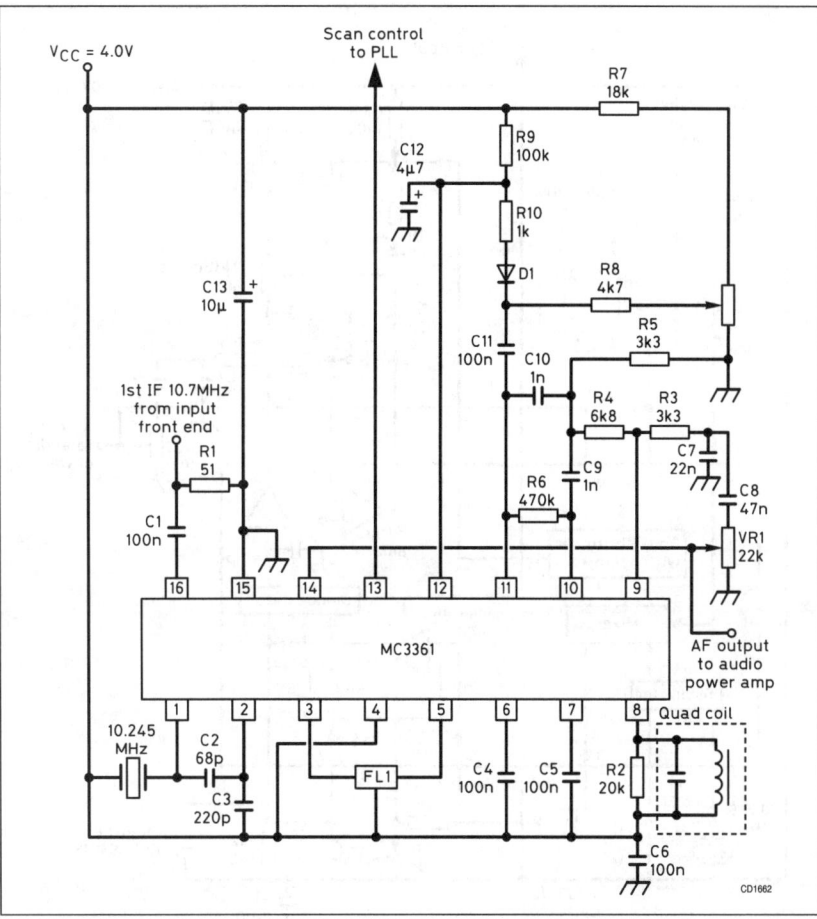

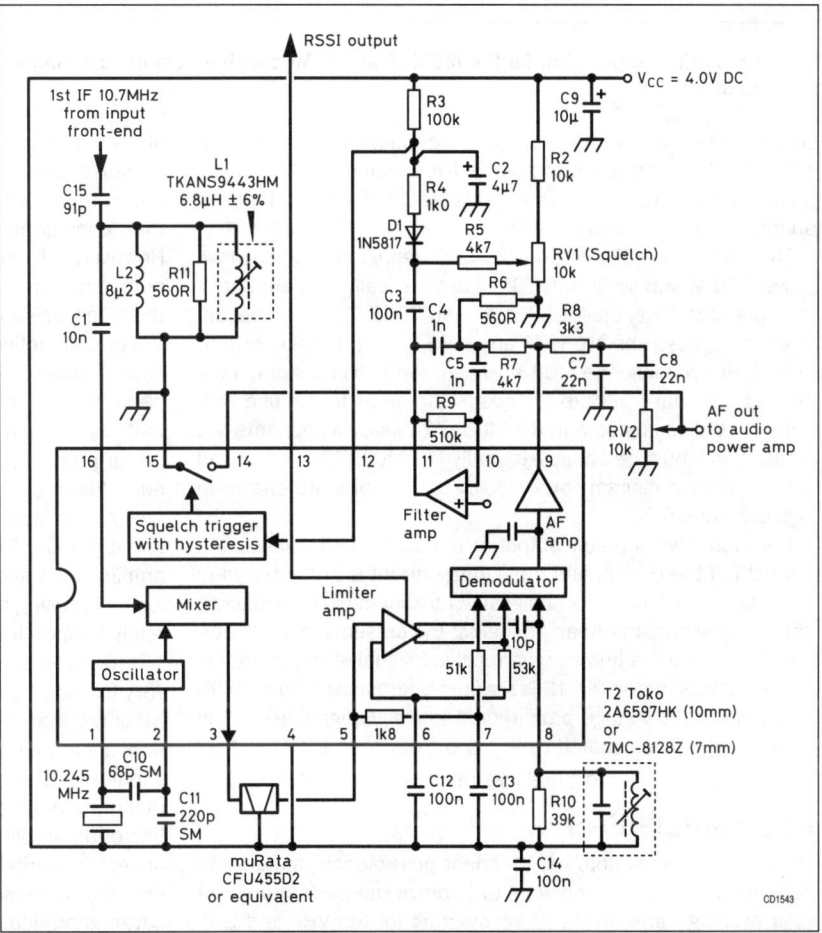

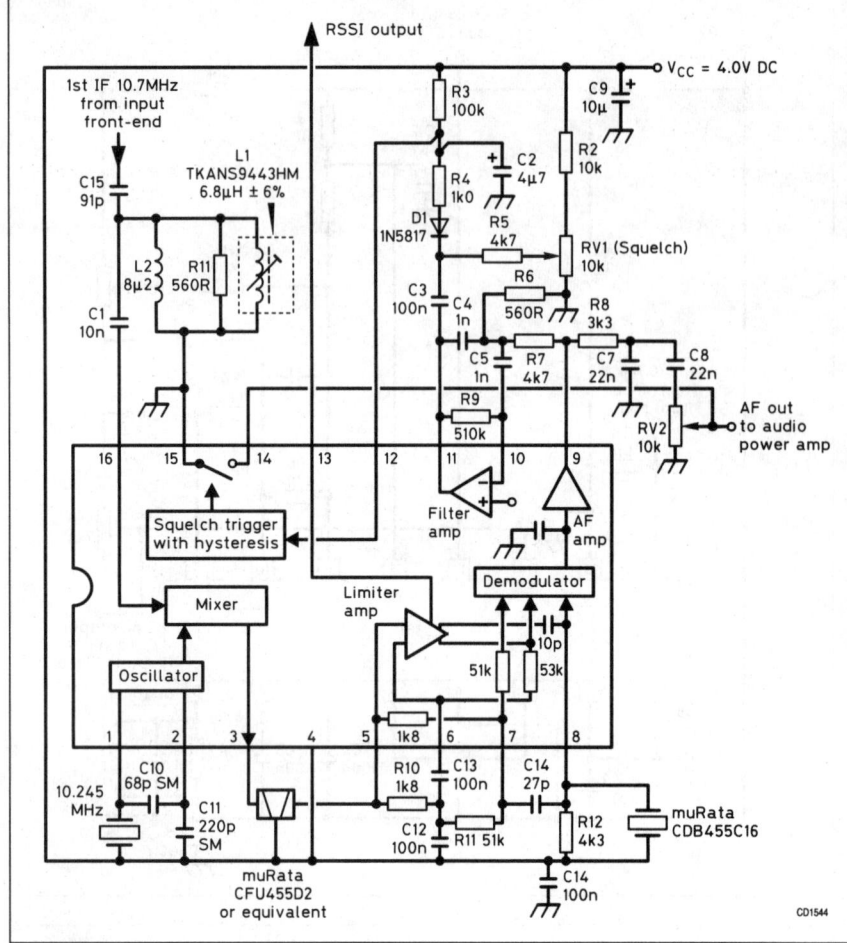

Fig 9.35: Typical application for the MC3372 at 10.7MHz with a ceramic discriminator (Motorola)

bands. The examples included in this chapter have been designed by amateurs to meet the specific needs of operation on the VHF and UHF bands. The three designs for the lower bands are from Dragoslav Dodricic, YU1AW [2] and show that bipolar transistors rather than FET transistors still have their place. The design for 70cm comes from Ole Nykjaer, OZ2OE [3] uses a modern E-PHEMT transistor.

Preamplifiers for 6m, 4m & 2m

A new type of low noise preamplifier is described here, which is recommended for its exceptional noise and inter-modulation characteristics not only for normal DX operation, but also for operation under difficult conditions when there are a significant number of powerful local stations, for example during competitions. The amplifiers are designed to have low noise, unconditional stability and exceptional linearity, thanks to the use of special ultra-linear, bipolar, low noise transistors designed for TV signals amplifiers. Since they are widely used, they are readily available at low cost. Construction is extremely simple with a small number of components, very simple adjustments and a high repeatability. This has been achieved using extensive computer non-linear and statistical optimisation. Designs are available for all amateur bands from 6m to 23cm [2]; watch that web site for latest updates to these designs.

For three decades, MOSFET or GaAsFET transistors have been used almost exclusively in preamplifier designs. The reason for this is their superior noise performance and amplification. What we inevitably encounter when using GaAsFETs is a stability problem due to their conditional stability on VHF and UHF frequencies [4 - 7]. However, with an increasing number of stations using greater output powers, especially during competitions, the majority of these low noise preamplifiers that are successfully used for DX, MS or EME activity, become overloaded. This is manifested by a large number of inter-modulation products that contaminate the band, this is attributed to other stations and especially those that use powerful amplifiers.

The problem, of course, could be in non-linear power amplifiers due to excess input power causing saturation. This generates a high level of inter-modulation products. However, in practice it is more frequently due to the receiver's excessively high amplification and insufficient linearity, ie its input stage is overloaded causing it to generate products that look as though they really exist on the band.

In order to understand how to cure this problem, it is necessary to know how, where and under what conditions it occurs. It turns out that the source of this problem is very high amplification, a feature that the majority of amateurs praise the most and should be praised the least or even avoided. Technically it is much more difficult to achieve the two other important properties of an amplifier: low noise factor and strong signal performance, ie linearity. These are the most important properties for an amplifier to those for whom decibels are not just numbers that cover ignorance.

six-pole filter will provide sufficient attenuation. Measured at ±100kHz the attenuation should be 40dB minimum: this is adequate for NBFM receivers. The filter is matched at the IF input by an internal 1.8kΩ resistor.

The Motorola MC3371 and MC3372 represent the two low-power NBFM sub-system ICs. They are basically very similar to their predecessors, the MC3361 and MC3362. The principal difference between the MC3371 an MC3372 is in the limiter and quadrature circuits. The MC3371 has internal components connecting the final limiter to the quadrature detector for use with parallel LC discriminator. In the MC3372 these components are omitted and must be added externally The MC3372 can be used with a ceramic discriminator. Application circuits are shown in **Figs 9.34 and 9.35**.

Both ICs have a meter output at pin 13 which indicates the strength of the IF level and the output current is proportional to the logarithm of the IF input signal. A maximum of 60µA is available to drive an S-meter and to detect the presence of an IF carrier. This feature is known as a received signal strength indicator (RSSI) or S-meter. Pin 13 is resistively terminated (to ground) to provide a DC voltage proportional to the IF signal level. The resistor value is estimated by V_{cc} - 1.0V/60µA, so for V_{cc} = 4.0V a 50kΩ resistor will provide a maximum swing of 3.0V.

PREAMPLIFIERS

Modern semiconductors have made it possible for amateurs to build high quality preamplifiers to improve the performance of their existing transceivers or transverters for the VHF and UHF

How do we determine which amplifier is of good quality? In order to resolve this dilemma a measure of an amplifier's quality has been introduced which encompasses all of an amplifier's three characteristics: noise factor, amplification and the output level of a signal for a determined level of non-linear distortions. This measure of quality is called the dynamic range of an amplifier and represents a range in which the level of a signal on an amplifier's input can be changed, while the output signal degradation stays within defined limits. The lower limit of this range is determined by the minimum allowable signal/noise ratio of the output signal and it is directly determined by the amplifier's noise factor, and the upper limit is the allowable level of non-linear distortion.

The lower limit of a dynamic range is the level of the input signal that gives a previously determined minimal signal/noise ratio (S/N) at the output. If the lower limit value is a S/N = 0 (incoming signal and following noise are equal) and if the upper limit of this range is limited by the maximum output signal voltage at which the amplifier, due to non-linear distortion, generates products equal to the level of noise on the output of the amplifier, then this is the so-called SFDR (spurious free dynamic range) or a dynamic range free from distortion, ie products of inter-modular distortions or IMD.

Since the third order inter-modulation distortion (IMD3) is dependant on the cube of the input signal, that is with each increase or decrease of the input signal by 1dB, the third order inter-modulation products increase or decrease by 3dB. It is therefore possible to calculate the maximum output level for different values of relations between products and the signal that is being used, or the value of IMD3 products, at different output signal levels. Using an attenuator enables us to also check whether an amplifier is overloaded, ie to recognize whether an audible signal on our receiver really exists on the band or whether it is simply the 'imagination' of our overloaded receiver. This enables us to dispose of overload and IMD3.

Since the level of products rise faster than the basic signal, by increasing the input signal we reach the point at which third order inter-modulation products, IMD3, reach the level of a useful signal at the output and that point is known as IP3 (Intercept Point). When the IP3 value is quoted it is necessary to state if it is referenced to the input or output of the amplifier. These values naturally differ by the value of the amplifier's amplification. Occasionally, it is stated as the TOI (third order intercept). This point is often taken as a measure of an amplifier's linearity and is highly convenient when comparing different amplifiers. Knowing the value of an amplifier's IP3 enables us to precisely

calculate the value of IMD3 products at some arbitrarily chosen output or input signal level.

If excessive amplification is used, for example in a multi stage amplifier, a danger exists where aerial noise and the noise of the first amplifier are amplified to such an extent that they exceed the limit of linear operation of the last transistor, at which point the amplifier is saturated with the noise itself without any signal.

The conclusion is clear: An amplifier is worth as much as its dynamic range value, rather than how great its amplification is!

Therefore, if we want to construct an amplifier with the maximum amount of SFDR we have to fulfil the following conditions:

• the noise factor is as low as possible
• the IP3 is as high as possible
• it has acceptable amplification

On the one hand, amplification should be as large as possible, to prevent second degree influence on noise factor, and on the other hand it should be as small as possible so that the IP3 input is as high as possible, ie so that the amplifier should withstand the highest possible input signals without distortion. Compromise is essential and it usually ranges between 13-20dB amplification, depending on which parameter is more important for us.

If we want a low noise amplifier with a high dynamic range, then the choice of a corresponding transistor is extremely important. It is necessary to choose the type of transistor that besides low noise and sufficient amplification on the given frequency fulfils the condition of good linearity, that is high IP3 along with unconditional stability. Hitherto, MOSFET and GaAsFET transistors did not fulfil this condition in a satisfactory manner. Specially built transistors for ultra linear working, primarily for CATV do fulfil these criteria. For that reason, Siemens BFP196 bipolar transistors in SMD packaging were chosen. The Philips transistor BFG540/X corresponds closely to the Siemens device, it requires only slightly different base bias resistors. This Philips transistor should be used on 1296MHz because it gives several decibels greater amplification. It should be stressed that BFG540 without /X could be used, but the layout of pins is different, - it is not pin-to-pin compatible with the BFP196 - therefore the printed circuit board has to be changed, which is not recommended.

Since we are talking about a broadband transistor whose Z_{nf} and S11 values are relatively close to 50Ω, the input circuit has been chosen to optimally match the transistor with regards to noise, while at the same time it provides some selectivity at the input. By varying the circuit values a compromise is found which provides the highest selectivity with minimal degradation of the noise factor. On lower bands where the noise factor is not as

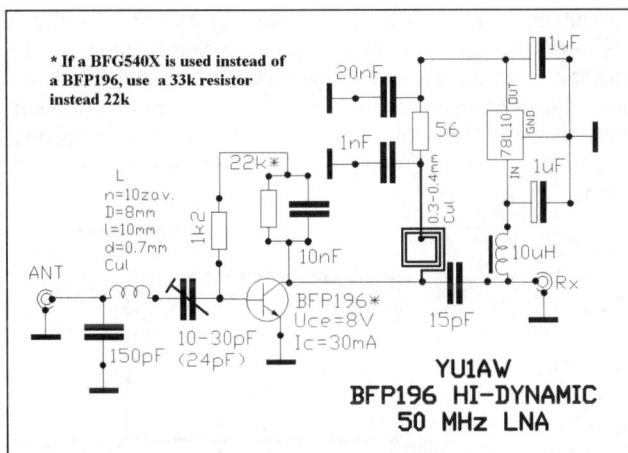

Fig 9.39: Circuit diagram for the 6m low noise amplifier

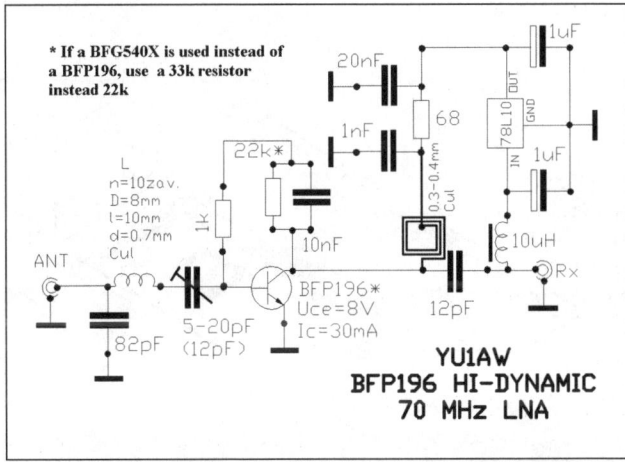

Fig 9.40: Circuit diagram for the 4m low noise preamplifer

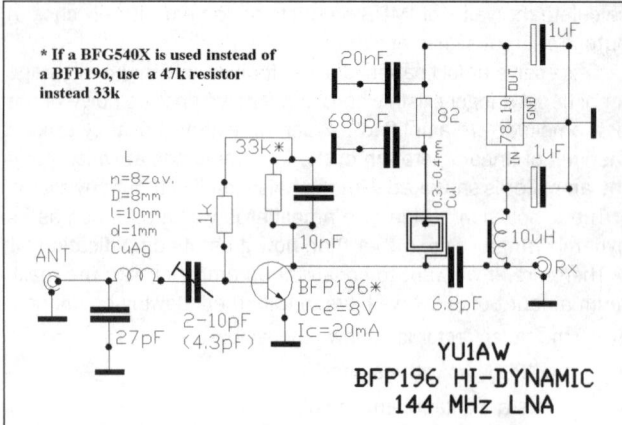

Fig 9.41: Circuit diagram for the 2m low noise amplifier

important, the compromise was in favour of selectivity which is more important than noise on these bands. The operating point of the transistor was also chosen as a compromise between minimal noise and maximum IP3. The output circuit is relatively broadband and it is implemented using a printed inductor to reduce coupling with the input and to provide high repeatability. In order to maintain optimal output matching that gives minimal IMD, any matching by trimmer capacitors or by variable inductances is forbidden on the output. In order to achieve unconditional stability, minimal IMD, optimal amplification and minimal noise, negative feedback is applied which cannot be changed arbitrarily.

The printed circuit board is made with the dimensions shown in the relevant figure (**Fig 9.36** for 6m, **Fig 9.37** for 4m and **Fig 9.38** for 2m - all located in the Appendix B). Double sided board, type G10 or FR4 is suitable. The bottom copper surface is an unetched ground plane. SMD components are the 1206 type and the ground connections are made using through plated holes or with wire links through the holes soldered on both sides. The parallel resistor and capacitor in the base bias circuit are soldered on top of each other and not next to each other. The transistor collector is connected to the wider track.

The trimmer used is either of the air or PTFE foil type, although a ceramic one can also be used if it has a suitable capacity range. It is especially important for the higher band amplifiers that the trimmer capacitor has a low enough minimum capacity.

The coil is wound, as shown in the relevant circuit diagram (**Fig 9.39** for 6m, **Fig 9.40** for 4m and **Fig 9.41** for 2m), with silver plated copper wire, thickness 'd' and 'n' turns with a body diam-

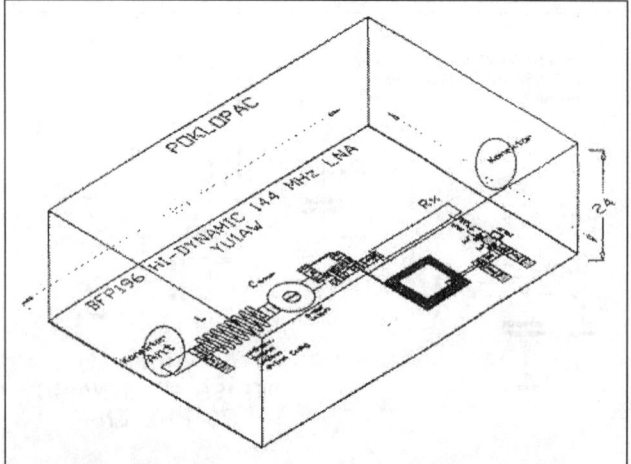

Fig 9.42: Mechanical layout of the low-noise amplifier

eter 'D'. The coil is to be expanded to length 'l'. When the coil is fitted it has to be positioned so that the bottom is approximately 3mm above the printed circuit board.

The box for the amplifier is made so that the printed board is the bottom side of the box, as can be seen in **Fig 9.42**. The easiest way to do this is to solder 25-30mm wide copper or brass strip, 0.3mm thick, around the edges of the printed board. Connections are mounted onto the box created, and a lid is made out of the same kind of sheet metal.

Once everything is carefully soldered, check for any possible mistakes such as short circuits, then connect the DC supply voltage and measure the collector current and voltage. If everything is correct and properly connected and the transistor is functioning properly, the values should be close to the ones given in the circuit diagram. If the differences are within 10%, everything is OK. If the differences are greater, check the supply voltage and then reduce the value of the base bias resistor, which is in parallel with the capacitor. Make it lower to raise the collector current and vice-versa. Do not change the value of the other resistor in the base bias circuit. If the collector voltage is not correct at the correct value of collector current, adjust the value of the resistor in the supply line. Such corrections are extremely rare and are necessary only if the particular transistor used has different characteristics from the common characteristics for that transistor type.

When both collector voltage and current are within the expected range, connect an aerial to the amplifier input and the receiver to the output and adjust the trimmer capacitor for maximum received signal using a weaker station. This completes the final adjustment; the performance will be very close to the predicted values. With higher band amplifiers, especially 23cm, there is a slight difference in matching for maximum amplification and for minimum noise. The amplifier should be set to maximum amplification and then adjusted to a slighter lower frequency, ie slightly raise the trimmer capacity, until the amplification falls by 1-2dB.

Any further changes or modifications except the ones stated above are absolutely not recommended, because the amplifier is optimised so that it immediately reaches the required characteristics. Any modification would prevent that and would produce much worse results than the expected.

The amplifier should always be mounted on the aerial pole and connected to the aerial with the shortest possible cable, using coaxial relays to switch the aerial from receive to transmit. Its power supply should be fed through the coaxial cable that connects it to the receiver, using the adapter shown in **Fig 9.43**.

As expected, very good noise characteristics have been proved in practical use, which mainly satisfy every requirement for serious DX work. Only for EME work at 432 and 1296MHz you might try using lower noise value amplifiers, ie the GaAsFET amplifiers [4 - 7], but in all other cases the amplifier satisfies even the most rigorous noise requirements. These amplifiers have shown exceptional linearity with IP3 values far exceeding 30dBm on all bands, except at 1296MHz where it is 3-4dB lower.

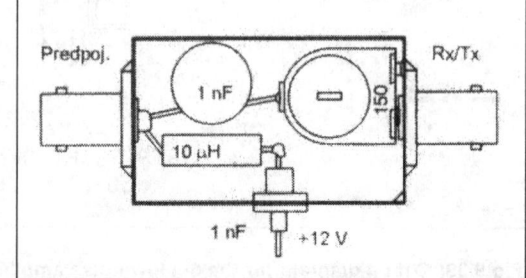

Fig 9.43: Power feed for the pre-amplifiers

As a comparison, **Figs 9.44 and 9.45** show the performance of this amplifier and a common amplifier which uses an MGF1302 GaAsFET. Both amplifier inputs have three signals of 7.1mV (-30dBm) to simulate three strong stations on the band. The graphs show what would be heard with an ideal receiver without its own IMD. With a real receiver, because of its possible IMD, things would look even worse! Before you accuse someone of band wasting check with an input attenuator whether your receiver might actually be creating IMD due to a strong input signal!

The amplifier using the BFP196 is superior to the one using the MGF1302. The difference in the IMD products appearing on the output of an ideal receiver was over 30dB! Of course, in both cases the amplifiers had approximately the same amplification.

The component layouts for the amplifiers are shown in **Figs 9.46 - 9.48** (in Appendix B) and the predicted performance is shown in **Figs 9.49 - 9.56**. The values shown have been simulated on a computer. Also, in real life the values have been proven on a sufficient number of built and measured amplifiers that they do not differ more than usual for this type of construction. Strict adherence to the guidelines given here will produce amplifiers with performance very close to those shown.

The final results achieved with these amplifiers in real life conditions largely depend on the IMD characteristics of the receiver used. If it has weaker characteristics, then the results may even be worse in respect to IMD because when signals, amplified in the preamplifier, reach the input of a bad receiver they cause

overload and IMD and the results are poor. That is why the minimum necessary amplification is recommended between this amplifier and the first mixer in the receiver or the transverter in order to preserve as much of the dynamic range of the whole receiving system as possible.

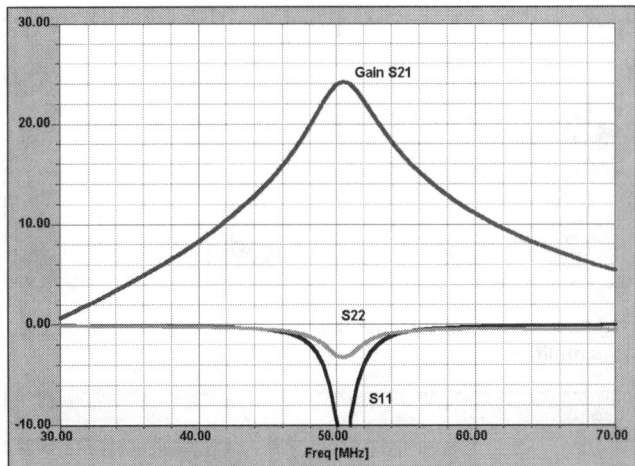

Fig 9.49: Amplification, input and output adjustment for the 6m low noise preamplifier

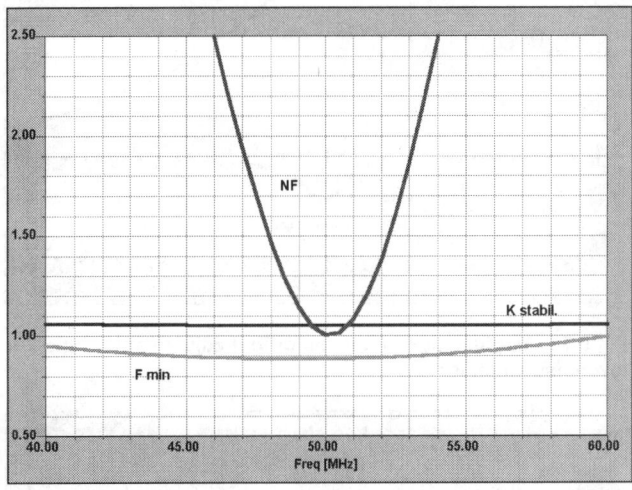

Fig 9.50: Noise figure, minimum noise and stability of the 6m low noise preamplifier

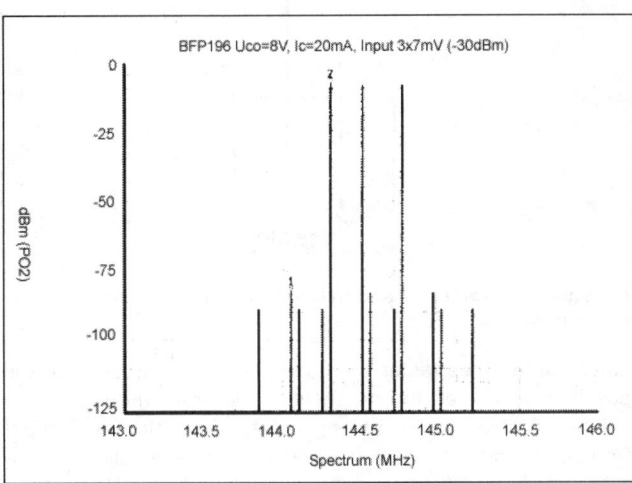

Fig 9.44: IMD products from a BFP196 preamplifier

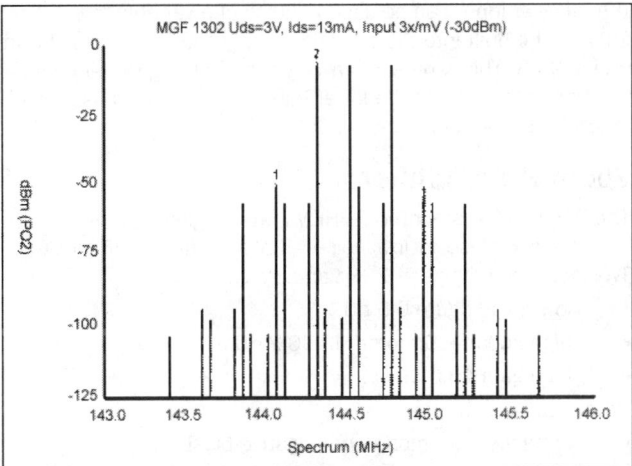

Fig 9.45: IMD products from a MGF1302 preamplifier

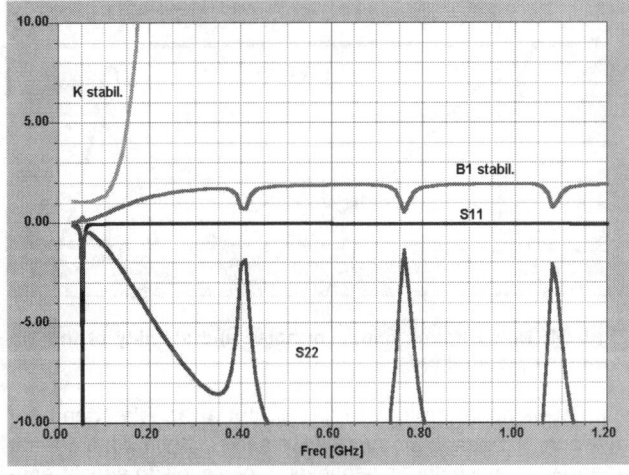

Fig 9.51: Stability factor and adjustment for the 6m low noise preamplifier

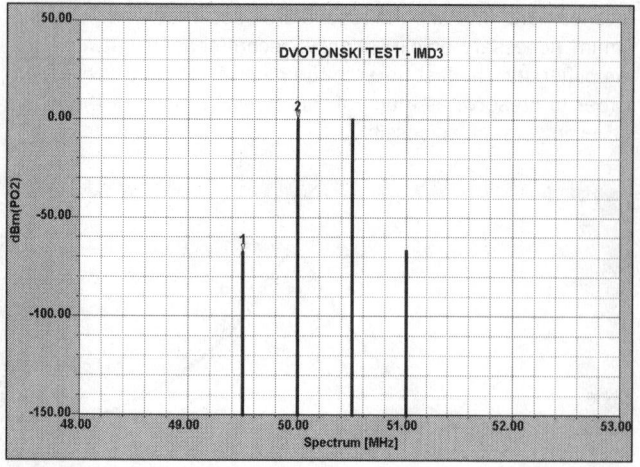

Fig 9.52: Two tone test and IMD products for the 6m low noise preamplifier

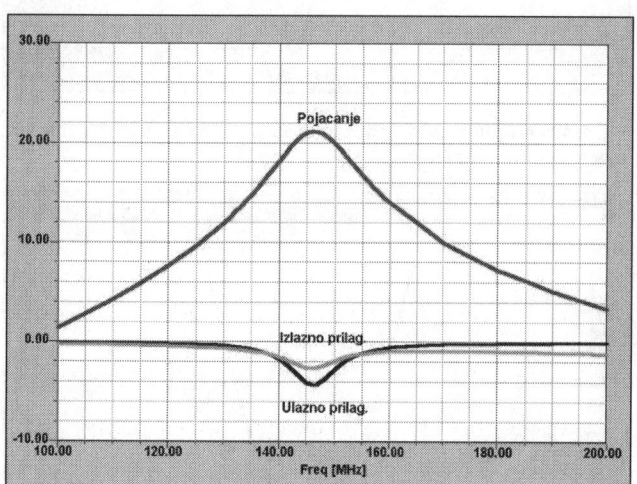

Fig 9.53: Amplification, input and output adjustment for the 2m low noise amplifier

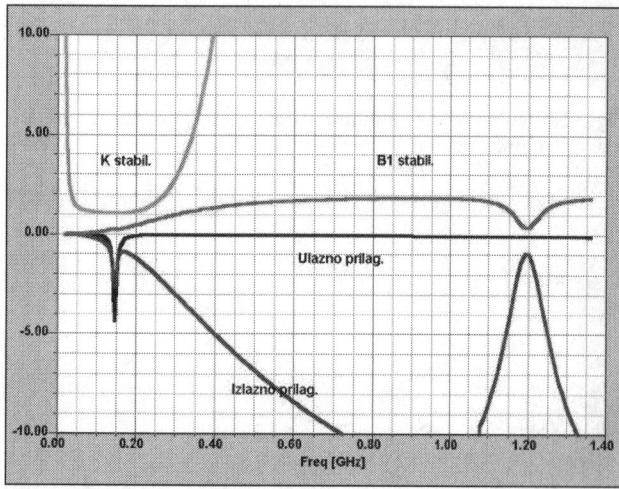

Fig 9.54: Noise figure, minimum noise and stability of the 2m low noise preamplifier

If IMD is apparent in the receiver, put a variable attenuator between the amplifier and the receiver, define the lowest attenuation at which it disappears, replace it with a fix attenuation of the same value and work with that in circuit. This method is highly efficient because the IMD products are atten-

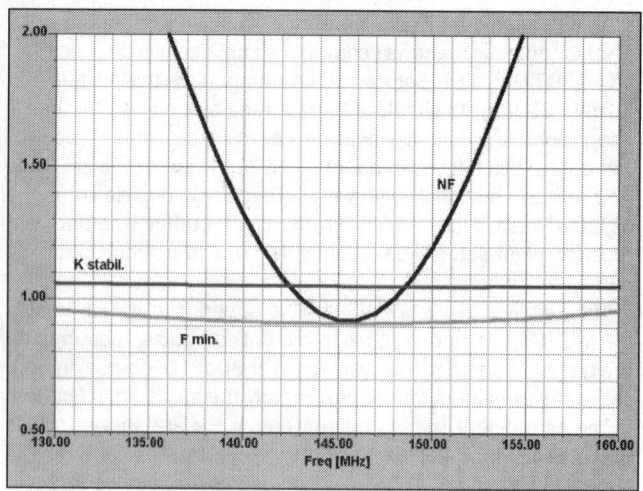

Fig 9.55: Stability factor and adjustment for the 2m low noise preamplifier

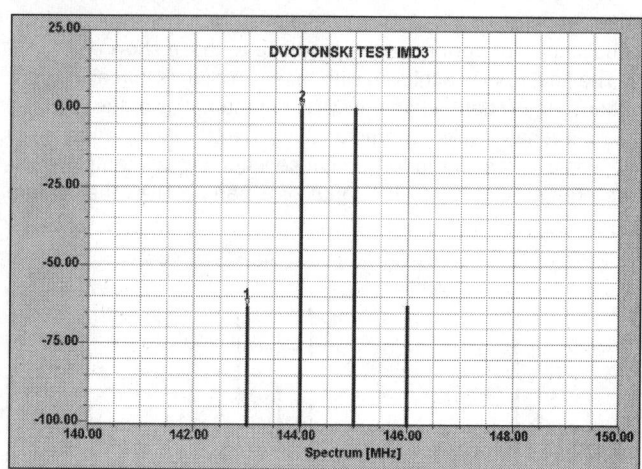

Fig 9.56: Two tone test and IMD products for the 2m low noise preamplifier

uated three times faster than the wanted signal, so that it is possible to weaken the products to the level where they are not heard whilst the useful signal has very little attenuation! Don't be afraid that you will not hear the desired signal, there is too much amplification as soon as IMD appears - feel free to lower it!

A miniature 100-500Ω trimmer potentiometer connected to the receiver input can be used in place of a variable attenuator. This can be built into the amplifier supply adaptor box as shown on Fig 9.43. This represents a very practical and rather elegant solution at least on of the lower bands. You can also use a variable 20dB attenuator used in CATV.

70cm Preamplifier

The ATF-54143 is a new E-PHEMT from Agilent (formerly HP), developed for use in low noise amplifiers from VHF to 6GHz. Typical specifications at 2GHz and 3V, 60mA:
- low noise figure 0.5 dB
- high linearity 3rd order IP 36.2 dBm
- high gain 16.6 dB

Other useful features:
- enhancement mode - no negative bias!
- easy matching for best noise figure
- low cost

Fig 9.57: Pin connections and package markings for an Agilent AFT-54143 E-PHEMT

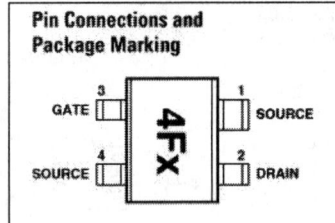

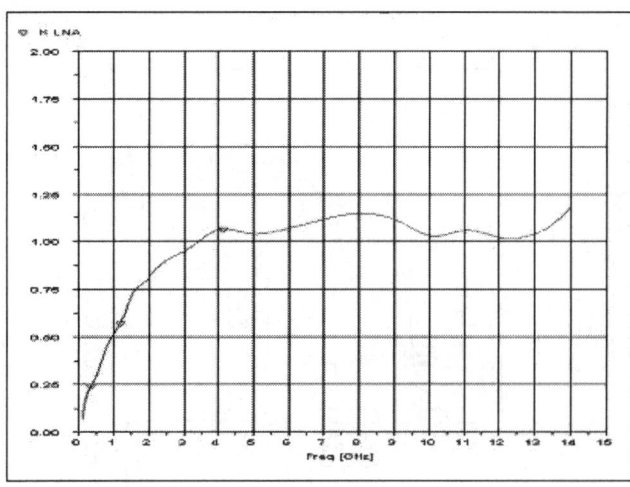

Fig 9.59: Stability factory versus frequency for an Agilent AFT-54143 E-PHEMT

Fig 9.58: Illustration of stability in a transistor amplifier. Feedback can occur inside or outside the device

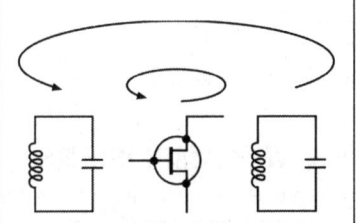

Not so useful features:

- high gain (>30 dB at VHF), can give instability
- small size 1.2 x 2mm - you need a steady hand

The package and pin connections are shown in **Fig 9.57**. For more information look up the datasheet on Agilent's homepage [8], select Products and RF & Microwave and find ATF-54143.

An amplifier must be stable, not only around the working frequency, but also on frequencies far away; **Fig 9.58** illustrates this point. Stability can be expressed in terms of k-factor. The k-factor is calculated from the transistor S-parameters. With k>1 the circuit is unconditionally stable - so whatever you put on the input and output terminal, it will not oscillate. Using S-parameters for ATF-54143 and with a little help from the 'ARRL Radio Designer' circuit simulator, we get a plot of k-factor from 0.1 to 14GHz (**Fig 9.59**). As you can see, the transistor is potentially unstable below 3.5GHz - this is a wild animal!

To increase stability, some attenuation is put directly on the drain terminal. This can be in the form of a ferrite bead or a resistor. Resistors on the gate terminal will also improve stability, and must be placed after the UHF decoupling to avoid any impact on noise figure. Another stabilising technique is using just a little inductance in the source (1-2mm of source lead).

The designer used a computer program 'ARRL Radio Designer' for designing the amplifier. It ran under Windows and was based on the professional Super-Compact circuit simulator program - at a fraction of the cost. (Unfortunately ARRL has stopped selling this program so get it 'surplus' if you can). The design procedure is this:

- load S-parameters (from internet)
- run stability analysis and then increase stability with resistors and/or source feed-back as required
- next calculate input and output matching for optimum noise (a special feature of 'ARRL Radio Designer')

The program can now calculate gain, noise figure, stability, input VSWR and a lot of other parameters for the finished amplifier. After that, all that is needed is to build the thing!

The input circuit transforms the 50Ω antenna into the optimum generator impedance for the transistor. Looking at a Smith Chart we have to go from the centre of the chart to a point towards the periphery. The more transformation, the more complex (and lossy) the input circuit. Unfortunately FET transistors - at low UHF frequencies - have rather high impedances for optimum noise match, making it difficult to achieve the very low noise figure the device itself is capable of.

A great advantage of ATF-54143 is an optimum noise match close to the centre of the Smith Chart. To give some low fre-

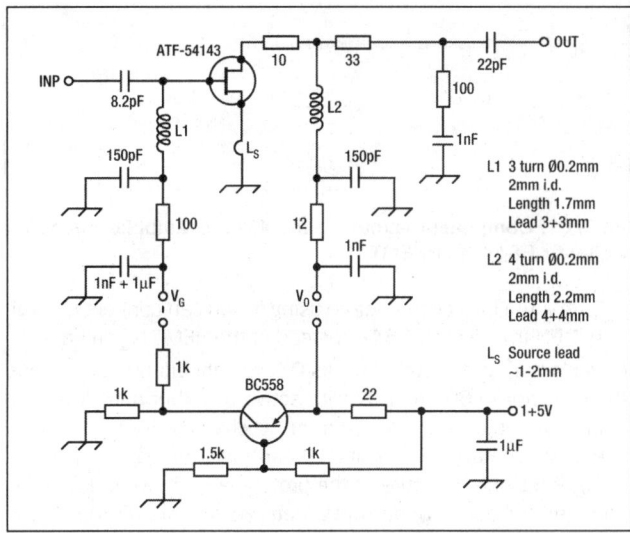

Fig 9.60: Circuit of the 432MHz preamplifier using an Agilent AFT-54143 PHEMT. L1 is 3 turns of 0.2mm wire, 2mm ID, length 1.7mm with a lead length of 3 +4mm. L2 is 4 turns of 0.3mm wire, 2mm ID, length 2.2mm with a lead length of 4 + 4mm

quency cut-off the transformation is carried out by a series capacitor and a parallel inductor.

Similar to the input matching, the output match transforms the 50Ω output impedance into the load impedance for maximum gain (or max. output power depending on what you want). Again transformation is done by a parallel inductor and a series capacitor.

The ATF-54143 works with a drain voltage of 2 to 4V. Maximum drain current is 120mA. The noise figure is almost independent of drain current and (as you might expect) large signal handling improves with drain current. The designer selected 3V and 60mA as bias. (but with 10Ω drain resistor I end up with 0.6V less voltage).

Being an enhancement mode device the transistor needs a positive voltage on the gate in order to draw drain current - typically + 0.59V on the gate to give 60mA drain current. If the gate is put to ground potential the current will drop to a few microamperes of leakage current. This is more like biasing a bipolar transistor! Therefore we have the possibility of making a simple resistor bias as well as building an active bias. The circuit is shown in **Fig 9.60**.

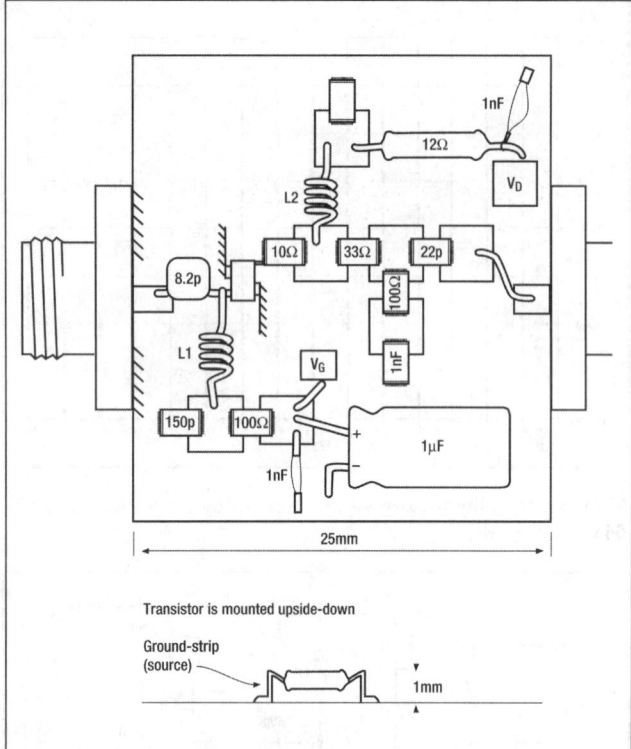

Fig 9.61: Component layout for the 432MHz amplifier using an Agilent AFT-54143 PHEMT

To test the bias circuit before using it, you can connect an ordinary NPN transistor (BC547) instead of the HEMT (V_g on base, V_d on collector). If everything works OK, you should be able to bias the collector to 60mA at 3V with active bias (but only 30mA 4V using passive bias due to different DC characteristic)

Prototypes have been built on breadboard, the layout is shown in **Fig 9.61** and a picture of the prototype is shown in **Fig 9.62**. The material used is piece of 0.8mm epoxy board (FR4). Islands were cut out, about 2 x 2mm, for mounting the components. The SMA connectors are soldered to copper 'ground'. SMD components were used around the UHF signal and decoupling path.

Simulation shows the 432MHz amplifier to be more stable with some source self inductance. This is achieved by mounting the transistor upside down on two small ground strips each 1mm high by 4mm long. These strips add about 0.5nH to the source. The gate (and drain) is in free air, so there is no loss due to the epoxy PCB.

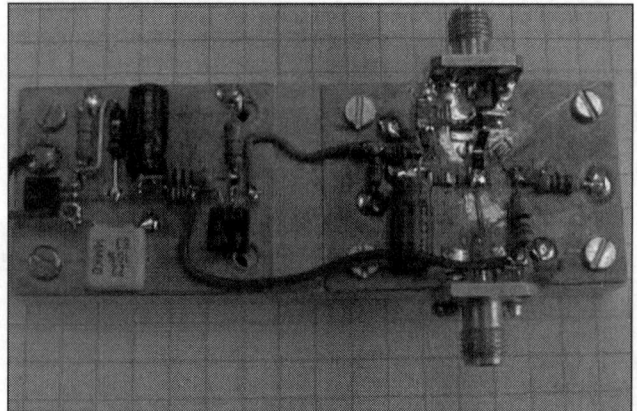

Fig 9.62: The prototype 432MHz preamplifier using an Agilent AFT-54143 PHEMT

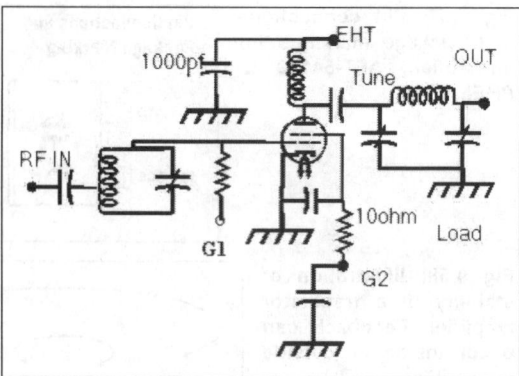

Fig 9.63: Circuit diagram of the 6 metre power amplifier

POWER AMPLIFIERS

There is a simple decision to be made when you are thinking of making a power amplifier: should it use valves or should it use semiconductors?

There is no doubt that you can get more power for less money with valves, the down side is that high power valve amplifiers require very high voltages, 1- 2kV, that can be very dangerous if you don't take the correct precautions. There are many designs for valve power amplifiers for all of the VHF and UHF bands, in this chapter two more novel designs are included. The 6m power amplifier was designed by Geoffrey Brown, G4ICD [9] and the 70cm power amplifier is from *VHF Communications Magazine* 4/1998 [10]. Semiconductor amplifiers fall into two types, those that use hybrid modules and those that use discrete semiconductor devices. Hybrid modules are easier to use because they require very few external components to get them working. Unfortunately the most common modules used by radio amateurs, from Mitsubishi, have been discontinued. They are still available form some suppliers and on the second hand market and there are other modules available that can be used. The designs for transverters shown later in this chapter have optional amplifiers using hybrid modules. The transistor amplifiers by Dragoslav Dobricic, YU1AW [2], for 6m and 2m are a good example of discrete semiconductors being used, and give details of several suitable devices.

If you don't want to embark on a big constructional project you can get more power on the VHF and UHF bands by modifying a commercial amplifier. Modification of the ex PMR A200 amplifier is shown as an example of this possibility.

6m Amplifier

This 6m amplifier can be built in under a day and will provide 250+ watts out for about a couple of watts in. The circuit diagram is shown in **Fig 9.63** and the parts list is shown in **Table 9.2**.

Anode Section:	
Anode Tuning Capacitor	Jackson C804 25pf (wide-spaced)
Loading Capacitor	Jackson C804 150pf
Anode Isolating Capacitor	1000pf 20kV 'door knob'
Anode Coil	Connects between Tune and Loading capacitors, 5t 12SWG, 1.375in dia
EHT RFC	36t 22SWG enamelled wire on 5/8in dia PTFE rod
Grid Section:	
Grid Tuning	Jackson C804 50pf
Grid Tuning Coil	6t 14SWG 1/2 in diameter
C1 input capacitor	Connects between input connector and Grid Tuning Coil, 1000pf mica

Table 9.2: Parts list for 6m power amplifier

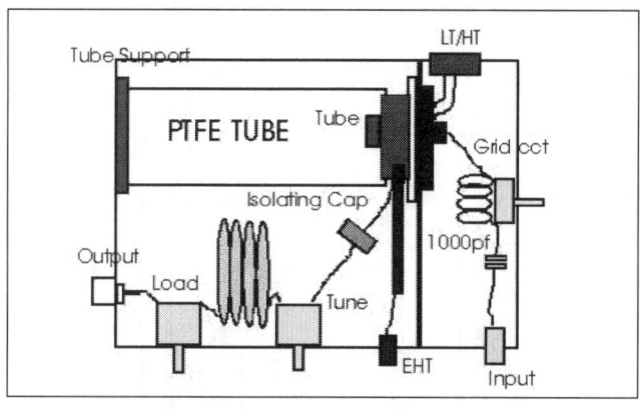

Fig 9.64: Component layout for the 6m power amplifier

Fig 9.66: The A200 amplifier. (Picture supplied by the Pye Museum [24])

Fig 9.65: The 6m power amplifier

Fig 9.67: The identification plate on the A200 amplifier (Picture supplied by the Pye Museum [24])

You will need a diecast box, fan, SK600 or SK610 surplus socket for the valve, a 4CX250B valve, a couple of tuning capacitors, some PTFE sheet to make the anode cooling chimney, and a multiple connector for the various voltages, plus a high voltage connector such as a PET 100 or TNC for the EHT of 1kV to 1.5kV.

Remember that any valve type amplifier has to operate with high voltages in order to work. The typical voltages applied to the 4CX250B series of valves is as follows:

- EHT (anode) is between 1 and 2kV. For this design 1.8kV at 250mA is ideal
- Screen grid + which MUST be regulated by zeners or stabilising valves, this should be 300 volts with a current capacity of 50mA
- A bias supply for the G1, this should be variable to -75 volts
- A heater voltage is also required which is 6.3 volts at a couple of amps
- A relay voltage is also required for the bias circuit and the antenna changeover relays

This amplifier was built in a die-cast box. The box measures 9 x 5 x 5 inches and a plate is fitted across the right hand end about 2.5 inches in. This plate (made of aluminium) has the SK600/SK610 socket fitted on it but offset towards the back (see **Fig 9.64**). The tuning capacitor and loading capacitor are fitted towards the front of the box along with the EHT choke, isolating capacitor and tuning coil. Cooling is via the anode compartment, so no manufacturer's chimney is used. A chimney made of PTFE sheet bonded together with silicon rubber is fabricated to fit onto the 4CX250B and run to the left hand end of the box. A small printed circuit board is used to support the chimney, it has a hole cut in it for the exhaust and a brass shim

soldered to support the chimney. The blower is fitted to the lid of the box.

Anode and loading capacitors are fitted in the front panel of the box, as can be seen in **Figs 9.63 and 9.65**. A band of brass, or an electrolytic capacitor mounting clamp is utilised to connect the isolating capacitor and the EHT choke onto the 4CX250B. RF out is fitted on the right hand end of the box. The grid circuit is straightforward with the tuned circuit being fitted behind the SK600/610 socket. A power connector is fitted to the rear wall of the grid compartment.

4m Amplifier

The Pye (later Philips) A200 was designed as a boot mounting linear amplifier to give more output power for their range of mobile radios. They are still available on the second hand market but many radio amateurs do not realise the potential of these units to add a useful amount of extra power to a 4m station, they can also be used on 2m and 6m. The A200 is built to last in a heavy weatherproof case with automatic RF sensing for transmit/receive switching, so it unlikely that you will buy one that does not work.

It is easy to spot the A200 by the chunky black case shown in **Fig 9.66**. There are three connections at one end, these are RF input, RF output and a thick DC power lead. The DC power lead is actually heavy duty mains cable with brown being the positive 13.8V supply, blue is negative and green/yellow is for switching. Do not connect this lead to a mains supply, this is a sure fire way to destroy your new acquisition. Also be careful not to confuse a VR200 24V to 12V converter for and A200, it has a similar case but two DC cables coming out of the side.

Two types of A200 were manufactured; early models had a TNC connector for the RF input and an N-Type RF output socket. Later models had a flying lead for the RF input and an SO239 RF output socket. Both models are very similar inside.

To decide if the unit is suitable for 4m, look at the identification plate on the side, see **Fig 9.67**. They are marked "Cat No.

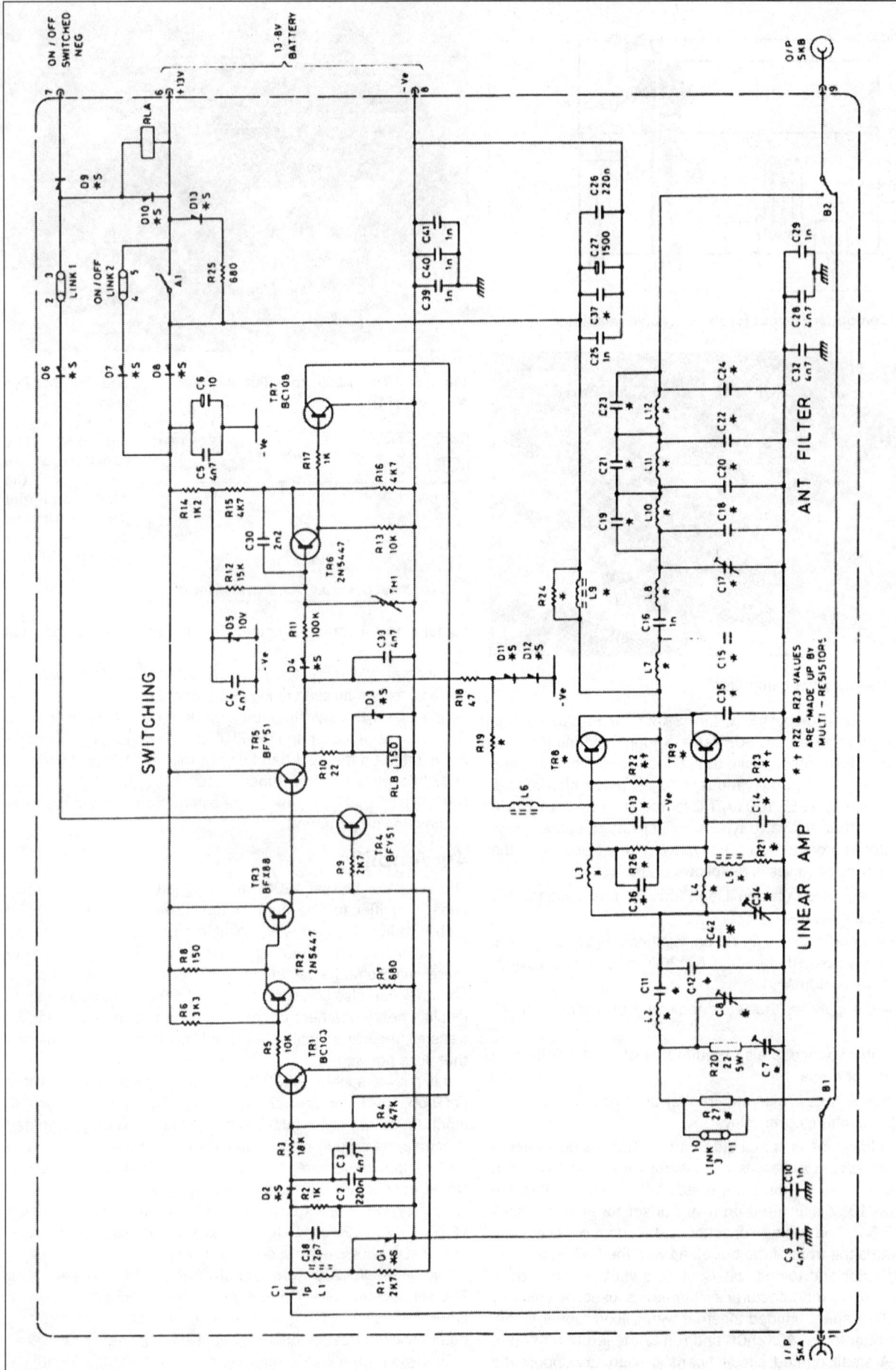

Fig 9.68: Circuit of the A200 amplifier. Reproduced with the permission of Pye Telecommunications

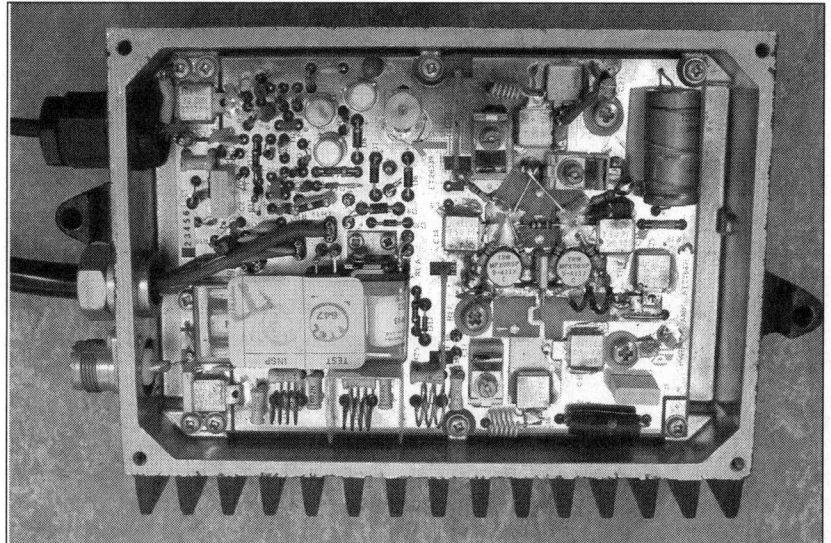

Fig 9.69: Internal view of the A200 amplifier. (Picture supplied by the Pye Museum [24])

A200" but the aligned frequency is often blank. Fortunately the 'Code' should be marked, something like "01 E0", this will tell you the frequency range:

E0:	68 - 88MHz
M1:	105 - 108MHz
B0:	132 - 156MHz
A0:	148 - 174MHz

The E0 model is suitable for 4m and the A0 or B0 models will work on 2m. The E0 model can be modified to work on 6m, this is described in [11].

Fig 9.68 shows the circuit diagram of the A200. A pair of MPX085P or BLW60 transistors are used in the output stage with bias derived using a wire wound resistor and two forward biased diodes. Printed circuit inductors are used for input and output circuits, tuned with compression trimmers. There is a three-stage low pass filter in the output. The RF sensing circuit switches the amplifier into circuit if DC power is applied to the A200.

The amplifier is well protected including a thermal cut out to shut down the unit if the output transistors are overheating. **Fig 9.69** shows an internal picture of the amplifier and **Fig 9.70** shows the component layout.

As an initial check, ensure that links between 2 and 3 plus 4 and 5 are fitted. This will ensure that the RF sensing is enabled. This switches power to the amplifier via relay A and the RF path through the amplifiers via relay B when RF is sensed on the input. If you want to use direct switching, remove these two links and switch the green/yellow wire to 0V to enable the amplifier.

The amplifier requires about 10W drive to produce 60 - 70W output and will draw 10 - 15A from a 13.8V supply. To align for 4m the following steps should be used:

• Set C7 to minimum, this reduces the input drive to the amplifier

• With 2 - 15W input power, check that the relays operate

• Tune C8 and C17 to achieve maximum output power. It may be necessary to repeat adjustment of these two capacitors to achieve optimum tuning.

• If an SWR bridge is available insert it between your transmitter and the A200 and tune C8 for minimum SWR. This should coincide with maximum power output.

Because the amplifier is linear, it will operate on AM, FM or SSB.

For AM operation C7 should be set for a maximum output of 25W with no modulation to prevent over driving the amplifier.

For SSB operation either direct switching should be used or the 'hang time' of the RF sensing circuit should be increased to prevent

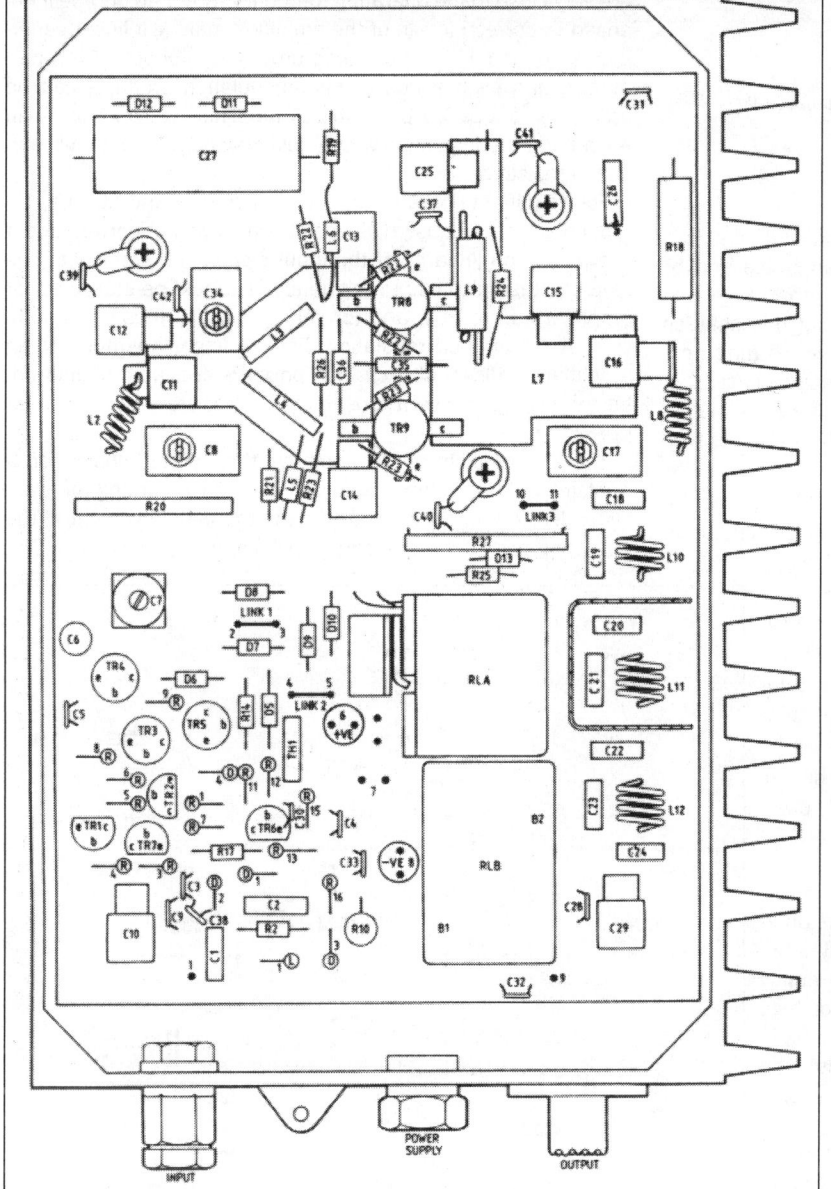

Fig 9.70: Component layout of the A200 amplifier. Reproduced with the permission of Pye Telecomm

chatter. Fitting a 0.68µF across C2 and C3 will give a 'hang time' of approximately 0.75 seconds. It is also necessary to increase the sensitivity of the circuit by fitting a 4.7pF capacitor in parallel with C1. C7 should be adjusted to reduce the maximum power by about 10% from the maximum, to prevent overdriving the amplifier. This will still mean that you get 45-50W PEP output with the third order IMD products at least 28db down.

6m and 2m Transistor Amplifiers

Several transistorised VHF power amplifiers are described here; examples are given for the 50 and 144MHz bands. Two types of amplifiers will be considered, amplifiers with one or two transistors. Assembly instructions are given as well as explanations of some general issues with regard to the construction of VHF power amplifiers.

The first step is to choose a transistor, this ultimately dictates what will be achieved. The quality of an amplifier is defined by several parameters that define its quality. An amplifier can be deemed to be a high quality amplifier if it has the following characteristics:

- High efficiency level
- High power amplification
- Good suppression of harmonics and low level of IMD products
- Good electric and thermal stability
- Simple design and adaptability

Some of these are contradictory and often have to be solved by various compromises, while others are compatible and by creating the first we automatically create the other. High efficiency is the most important factor, maximising this ensures that the majority of the other parameters approach their optimum. Beside an arbitrarily chosen working point or class in which the transistor will operate, the matching as well as the circuit losses influences efficiency. For highest efficiency it is necessary to:

- Strictly respect AC and DC current and voltage recommendations by the manufacturer for the class of operation
- Match the transistor to the output conditions recommended by the manufacturer for maximum efficiency. Also losses in the matching circuits should be minimal
- Keep the temperature of the transistors under the maximum allowed temperature, this is achieved by adequate cooling

By fulfilling these conditions a high degree of efficiency is obtained, as well as highest output power, relatively low operating temperatures and slight harmonic and IMD distortions.

The second condition, high power amplification, is achieved by matching the input to ensure stable operation. Losses in the matching circuits directly influence the power amplification. Another extremely important condition that has to be fulfilled for high amplification is extremely good grounding of the emitter or source in FET transistors. Leads should be extremely short and therefore of extremely low resistance!

Stability of the amplifier depends on the transistor itself, but also in a large measure on other factors such as: matching, mechanical construction and separation of electric paths in the input and output circuits. This will eliminate unwanted feedback between input and output. The most important issue for the stability of any amplifier concerns the low frequency performance where the transistors have enormous gain. It is important to make sure that the base and collector 'see' several dozen ohms. If a collector or base 'sees' a short circuit via a large capacitor or open circuit via a high inductance at low frequencies, an amplifier oscillates very easily.

The third condition, good suppression of higher harmonics and low level of generated inter-modular products (IMD) also depends on many factors. We will list just a few:

- choice of the optimal value of efficiency (Q) of output circuit
- linearity of a transistor's chosen operating point
- output matching
- small output resistance of the base bias circuit
- value of excitation power

By correct design, the unwanted secondary products generated by the amplifier could be reduced to a minimum. Output filtering can reduce these even further. Filtering cannot reduce some of the odd order inter-modulation products because they are very close to the operating frequency. They can only be influenced by correct design of the amplifier; making it linear generates a very small level of these unwanted products. Even harmonics and even series of inter-modulation products can be efficiently decreased by push-pull amplifier design. This is an advantage in comparison with amplifiers operating with two parallel transistors.

For amplifiers operating in linear classes (A and AB) the correct choice of the base bias circuit is also very important and it provides good linearity of the input signal. The stability of collector voltage also has an influence, it should be stabilised for amplifiers up to 20V, and for voltages above these it is sufficient if it has good regulation. For the linear operation of an amplifier, the level of excitation power is crucial. It should not be too high, because it 'pushes' the transistor into non-linear operation.

Thermal stability is provided by the correct choice of a heatsink and more importantly, by correctly mounting the transistor. In order to compensate for the thermal movement of the operating point it is necessary to provide thermal feedback that

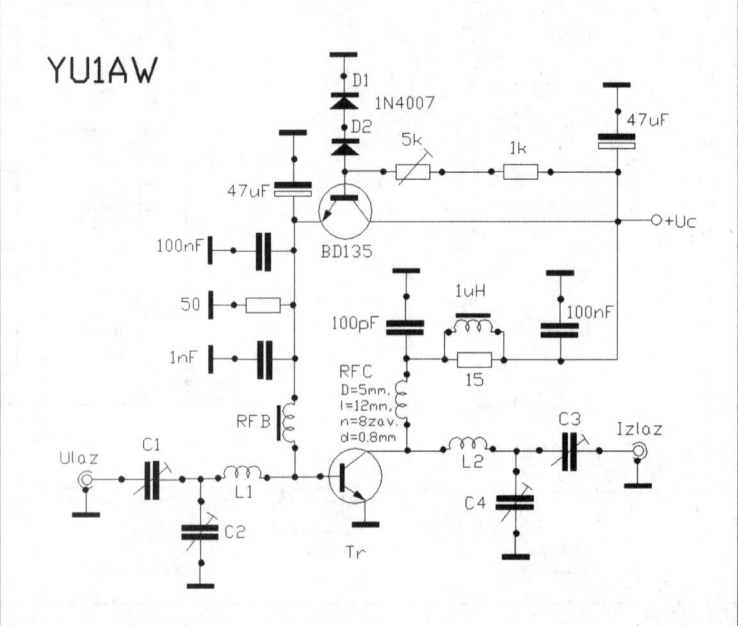

Fig 9.71: Circuit diagram of the one transistor power amplifier

Transistor	Freq MHz	C1 pF	C2 pF	C3 pF	C4 pF	L1 Ind nH	D mm	L mm	n turns	d mm	L2 Ind nH	D mm	L mm	n turns	d mm
BLW76	50	36	200	21	46	44	6.5	6	3	1.2	157	12	10	4	1.5
BLX15	50	56	363	20	37	23	6	4	2	1.2	195	15	12	4	1.5
MRF317	50	40	210	24	57	47	10	10	2.6	2	140	15	15	3.7	2
2xBLW84	145	12	33	5	3.3	15	6	3	1.5	1.2	113	10	10	4	1.5
2xMRF245	145	21	50	21	33	7.5	7	3	1	1.5	14	10	4	1	2
2xMRF247	145	21	50	21	33	7.5	7	3	1	1.5	14	10	4	1	2
MRF317	145	6.7	18	13	46	48	10	10	2.7	2	22	15	5	1	2
MRF247	145	18	110	18	80	8	7	3	1	1.5	13	10	4	1	2
MRF239	145	13	67	10	31	14	8	4	1.3	1.5	29	12	5	1.5	2

Table 9.3: Components required for various transistors

Transistor	Frequency MHz	V_C Volts	I_{CO} mA	I_{Cmax} A	Pdrive W	Pout W	Efficiency %
BLX15	50	50	50	6.5	15	150	65
BLW76	50	28	50	8	6	90	60
MRF317	50/145	28	10	6.5	12	100	60
MRF238	145	13.8	20	4	3	30	55
2xMRF245	145	13.5	2x30	2x18	35	200	50
2xBLW84	145	28	2x30	2x2	6	60	55

will, by monitoring the temperature of the tran-

Table 9.4: Transistor parameters

sistor, change the base bias to maintain the DC operating point stability.

All these conditions have to be fulfilled if an amplifier can be described as a high quality amplifier; they are only discussed briefly here. They are elaborated in literature [12] and [2].

One Transistor Power Amplifiers

The circuit of the one transistor power amplifier is shown in **Fig 9.71**. A PCB layout is shown in **Fig 9.72** and the component layout in **Fig 9.73** - these are both in Appendix B. The PCB may have to be varied slightly depending in the actual RF transistor chosen. **Table 9.3** shows the component values for various transistors and **Table 9.4** shows the expected performance. A circuit consisting of C1, C2 and L1 performs input matching. The Q of this circuit is about 10 - 15, which is the optimal value in this case. Base bias is fed via an RF choke, which is a VK 200 ferrite bead with six holes. The output circuit that matches the output of the transistor to 50 ohms is performed by L2, C4 and C3. All capacitors in the matching circuits are ARCO, ceramic-mica trimmers of corresponding capacity. The values of the matching circuit elements are shown in Table 9.3. The number printed on the trimmer gives the range of certain trimmers (**Fig 9.74**). Types of trimmers and the corresponding range of capacity are given in **Table 9.5**. The voltage values, currents, power and efficiency for each separate type of transistor are illustrated in Table 9.4. The collector supply is fed via an RF choke which instead of a ferrite is open wound. The diameter of this choke is 5mm, it is 12mm long, and has 8 turns of 0.8mm wire. After the choke, a fairly

Fig 9.74: An ARCO trimmer

small value capacitor decouples VHF frequencies to the ground. The 100nF capacitor decouples low frequencies via a 15Ω resistor to ground. The choke in parallel with the resistor is a VK 200 ferrite or

Type of ARCO trimmer	Range
404	7pF - 60pF
423	7pF - 156pF
426	37pF - 250pF
462	5pF - 80pF

Table 9.5: Types of ARC trimmer

a similar one that can carry the high currents that flow through the transistor.

A BD135 transistor connected as an emitter follower with extremely low output resistance carries out base bias, this ensures a stable operating point. Thermal monitoring and compensation of the operation point is done by diode D1, positioned so that it has physical and thermal contact with the transistor (not with the heatsink!). This ensures stable operation over a large temperature range. The other diode, D2, monitors the ambient temperature. Both diodes are of the 1N4007 type or similar. Transistor BD135 is mounted on the same heatsink as the RF transistor. Adjusting the operating point is performed by 5kΩ trimmer potentiometers. Apart from the electrolytic (or tantalum) capacitors all the others are disc ceramic or of some similar VHF quality.

Two Transistor Amplifiers

As can be seen in the circuit diagram shown in **Fig 9.75** there are two identical amplifiers connected by wideband baluns, they operate in a symmetrical anti-phase or push-pull connection. A PCB layout is shown in **Fig 9.76** and the component layout in **Fig 9.77** - both are in Appendix B. The PCB may have to be varied slightly depending in the actual RF transistor chosen. Table 9.3 shows the component values for various transistors and Table 9.4 shows the expected performance. Circuits for supplying the base and collector of the transistors are the same as the one transistor amplifiers. Even the input and output matching circuits are similar. The only novelties are two pieces of 50Ω coaxial cable 95mm long that transform the asymmetric input and output of the amplifier onto the symmetric connection of two

transistors. Teflon cable several millimetres thick, RG-142, type or similar should be used. The coaxial cable can be wound in a coil, or even better on some small ferrite toroid or onto a larger bead. Ferrite with two small holes can be used for this purpose. Using two such ferrites, two windings could be wound. If you do not have these, you could thread two small ferrite beads with one hole each onto the cable so that there is one bead on each end, but it is even better if you thread several beads and place them along the cable length. This will improve the symmetry of the transformer, especially at low frequencies, which can be significant for a stable amplifier.

A self-resonant balun has been found to give an

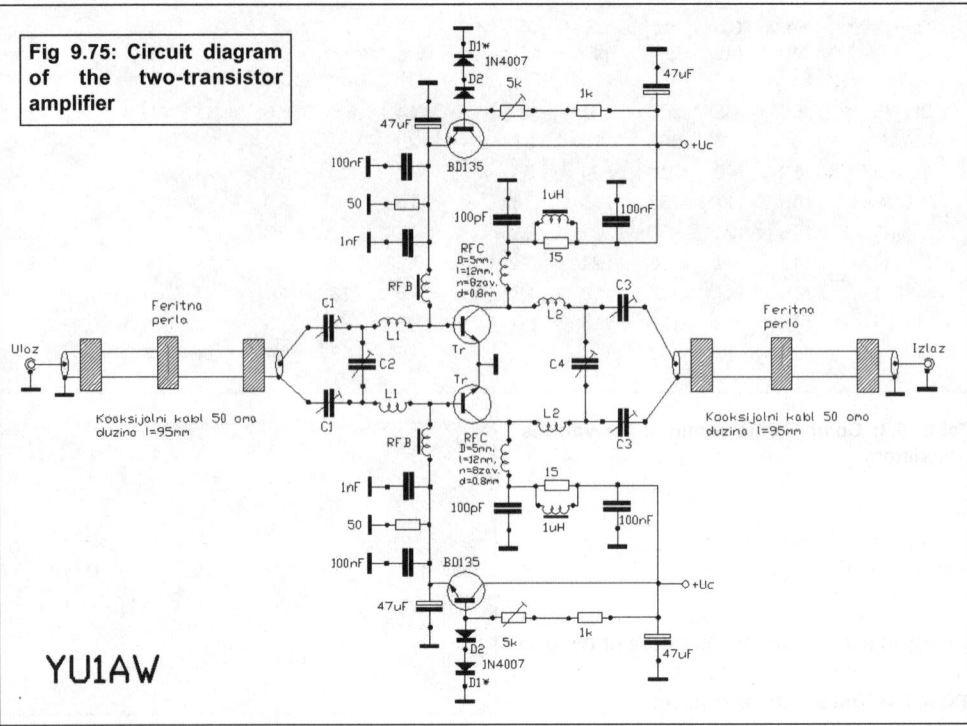

Fig 9.75: Circuit diagram of the two-transistor amplifier

improvement over the wideband balun made from 95mm of coax cable. A 144MHz resonant balun can be made from 620mm of 5mm diameter PTFE coax cable, RG142 or similar. Close wind into 6 turns with 30mm inner diameter and about 30mm long with no space between turns. For a 50MHz balun, the coax cable length is 1770mm, coil inner diameter is 50mm, and coil length is around 55mm and 11 turns.

It is very important that the entire amplifier has to be totally symmetrical with regards to the mechanical layout of components and electrical parameters (values of elements, currents and voltages, etc), so that more power, amplification, efficiency and suppression of even harmonics can be achieved. While adjusting, it is very important to maintain the same capacitances of C1 and C3 in both amplifiers.

Mechanical Construction

The whole amplifier has to be built on a relatively small piece of a single layer FR4 printed circuit board (Fig 9.76 in Appendix B). Emitter leads are soldered as short as possible onto the ground of the board. The transistor should be mounted onto a large heatsink using thermal paste as illustrated in **Fig 9.78**. The surface of the board should be small to enable the fins of the cooler to be as close to the transistor as possible. To provide better cooling it is important to make sure that the hole drilled for mounting the transistor onto the heatsink is big enough to allow a screw to pass through it. The transistors should be mounted with their entire surface on to the heatsink. The surface of the heatsink where the transistors are mounted has to be smooth. Input and output connectors can be fixed onto the board but also onto the heatsink or eventually the box and connected with coaxial cable to the board with its grounds soldered on both ends. The coils, L1 and L2, in the matching circuits should be mounted so that their axes are at a right angle, to decrease interactive coupling.

Once everything is connected, as illustrated in Fig 9.77, check once again to ensure that there are no mistakes and short circuits to the ground, and adjust the potentiometers for maximum resistance. Connect the supply to one transistor and adjust the collector current to the value from Table 9.4. The same proce-

dure should be carried out with the second transistor. Even more important than the exact value of quiescent current is that they are identical in both transistors! Then connect both transistors to the supply and connect a 50Ω dummy load to the output via a wattmeter or SWR meter. Supply minimal excitation and by measuring the output power alternately adjust trimmers until maximum output power is achieved. Repeat the adjustment several times, gradually increasing the excitation power. Finally, with full excitation, which should not exceed the permitted output power, adjust all trimmers to the highest output power. At the same time the transistor's current should be measured so as not to exceed the maximum allowed value. If input trimmers, C1 or C2, need to be at maximum or minimum capacity during adjusting, it is necessary to change the length of the cable between the exciter and the amplifier itself. The optimal length of the cable should be determined experimentally to obtain adjustment with approximately the same values of C1 and C2. This experimentally determined cable should always be used when operating the amplifier. A change of exciter could occasionally require a new length of cable to be determined. In

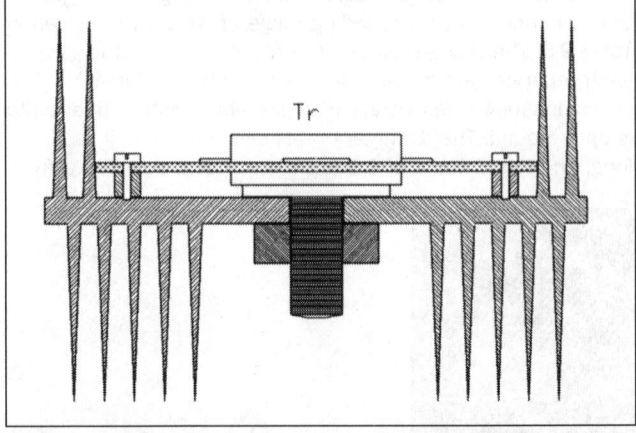

Fig 9.78: Mounting the power transistor on a heatsink for the transistor power amplifiers

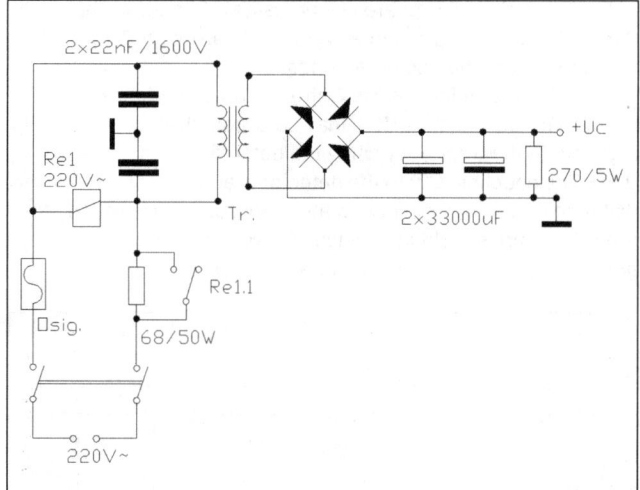

Fig 9.79: Circuit diagram of a power supply for the transistor power amplifiers

push-pull amplifiers it is extremely important to perform adjustment so that the corresponding trimmers on each transistor are adjusted simultaneously to ensure that they have approximately the same capacity during adjustment. By maintaining symmetry during adjustments, extremely dangerous situations are avoided which can cause the amplifier to self-oscillate.

For transistors with a power supply higher than 20V, a non-stabilized supply can be used, but it is constructed so that it has very good voltage regulation as illustrated in **Fig 9.79**. It should be well specified with good quality electrolytic capacitors. The transformer should be slightly over specified. To avoid blown fuses caused by the charging current of electrolytic capacitors, it is necessary to build in a delayed switching device. This is performed simply by a 220V relay connected as shown in the circuit. At the moment when power is switched on, the transformer is connected to a power supply via a resistor that limits high charging currents. When the capacitors are charged and transients in the transformer settle down, the current through the resistor decreases, the voltage on the primary increases and the relay that bridges the resistor with its contacts is switched on. It is also possible to use relays that switch on via some electronic timer after a couple of dozen seconds. Although this appears to be a more elegant solution, it is a far worse solution, for two reasons: first, in the case of a very short interruption of supply voltage the timer has not been reset, voltage is switched on without delay; and second: in the case of a fault that causes high current consumption, when the relay would not be switched on, the entire system would protect itself, whereas the timer would switch on the relay and subsequently provide full power.

Transistors with 12-18V collector voltage must be supplied with stabilized voltage. In this case it is vital to build in a thyristor device for over voltage protection. This device must connect the voltage via a thyristor to the ground and thereby induce a forced blown fuse if the voltage exceeds some preset value; this

prevents the RF transistor blowing up. Any type of relay protection is not recommended in such cases as it simply cannot react quickly enough and by the time the relay is switched, the transistor is already blown up!

70cm Amplifier

The 70cm linear power amplifier described here can be built in little more than a weekend. It delivers the power required for satellite working, small antennas and short cables or larger antennas and longer cables. Three readily available 2C39 disc-seal triodes are used in parallel delivering 300W output for an input drive of 15W.

As can be seen in **Fig 9.80**, the three triodes operate in a grounded grid circuit with the cathodes being driven in parallel. The amplifier requires only two supply voltages for reliable operation: the anode and the filament voltages. The anode voltage may be between 1.3 and 1.5kV and the filament between 5.8 and 6.0VAC (at 3A).

With 1.3kV on the anode, the anode current can be driven up to 400mA giving an RF output power of some 300W for 15W drive power. It is quite possible that, if good tubes are used, the output power will be even more but they should not be overdriven. A good axial air-blower should be used for the anode cooling.

The RF circuits were computed with the aid of a computer program. Special attention was given to optimise the half-wave anode line so that with the given impedance the lowest possible loaded Q was obtained. Through a careful selection of the parameters, a loaded Q of 39 was achieved which, for this application, is the lowest possible value. This ensures that the anode tank circuit and the amplifier work with the maximum efficiency.

The variable capacitors, shown in **Fig 9.81**, should have the following calculated values for optimum operation:

C1	4.3pF
C2	5.3pF
C3	1.4pF
C4	5.3pF

The dimensions for the cathode (L_K) and the anode (L_A) line resonators as well as the coupling and tuning plates (C1 to C4) are shown in Fig 9.81.

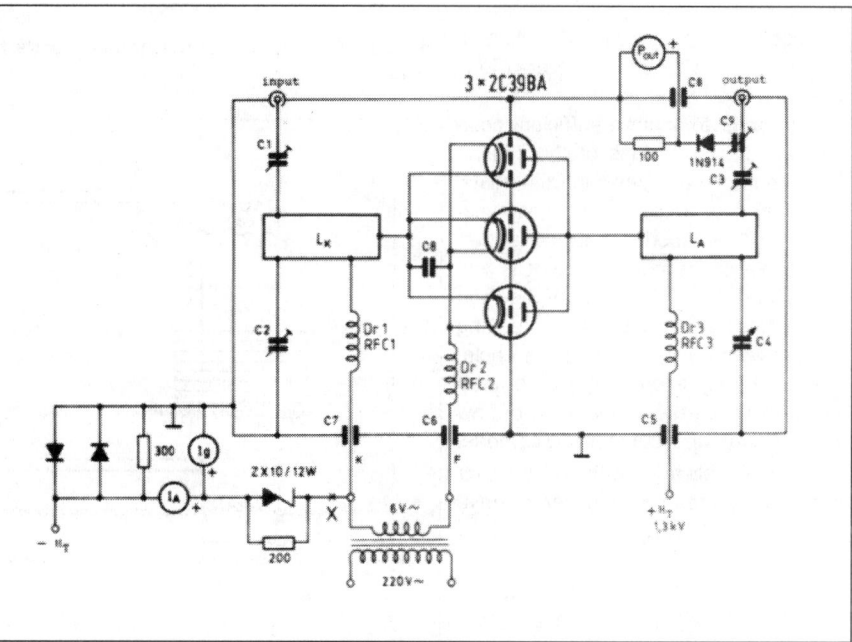

Fig 9.80: Circuit diagram of the 300W 12dB gain, 2C39A power amplifier for the 70cm band

The construction is very simple as can be seen from **Figs 9.82 to 9.84**. A few special parts of the circuit should be explained.

In order that the anode resonator (L_A) can be properly connected to the valves, the latter should be modified in the following manner:

- The cooling-fins are taken off and tapped, 4mm. The strip line can then be held tightly between the cooling-fin body and the tube's anode. A 25mm ceramic pillar supports the other end of the strip line

- Strip contact fingers are used to make the grid-ring contact. The cathode contact, on the other hand, can be fashioned from 10mm outer diameter copper tube of 0.5mm wall thickness. This tube is 12mm long and slit longitudinally down to the middle. The slotted half is then press-fitted over the cathode contact and the other end soldered to the cathode strip line L_K. The remote end of the strip line is secured to a PTFE or ceramic pillar.

The strip lines for the anode and cathode resonators are cut from 1 to 1.5mm stock and silvered, if at all possible.

The amplifier is built into a housing made from 1mm thick brass plate, see Figs 9.82, 9.83 and 9.84. The sides are soldered together.

Tuning capacitors C2 and C4 are made from 0.5mm thick brass plate (Fig 9.81) and hinged and rotated using nylon fishing line. A piece of insulating material - PTFE or polystyrene - is positioned as a stop to prevent direct contact with the opposite electrode. A couple of thick knots tied in the fishing line serve the same purpose.

These tubes require a lot of cooling air if they are to work reliably over a long period. The air blast must also be powerful in order to achieve sufficient cooling over all the surface of the cooling fins. The forced air comes in from above via C4 and cools both the anode resonator and the anode itself and is then vented out of the anode area. It is recommended that a couple of not too small holes be provided in the screening wall between anode and cathode enclosures (Fig 9.82) in order to allow a weak flow of air from the mainstream to flow over the cathode resonator and cathode.

The HT supply as with the drive power is connected to the amplifier by BNC panel sockets. An N socket is used for the RF output.

The valve heaters are connected in parallel. Between the inner heater contact and the cathode lead of every tube, a 1nF disc ceramic (C8) is fitted using the shortest possible connections.

The RF chokes (RFCs) are wound using a 6 to 8mm shaft with 0.8 to 1mm diameter copper wire. They are 6 to 7 turn coils, supported from their soldered ends.

Tuning the amplifier is very straightforward, simply tune for maximum output power. This may be accomplished with the aid of a UHF SWR meter or by using the detector circuit shown in Fig 9.80. The coupling (C9) to the detector is adjusted by varying the distance of the silicon diode to the N socket centre pin. The first tuning attempt should take place with very low input drive power and then gradually increase it to maximum.

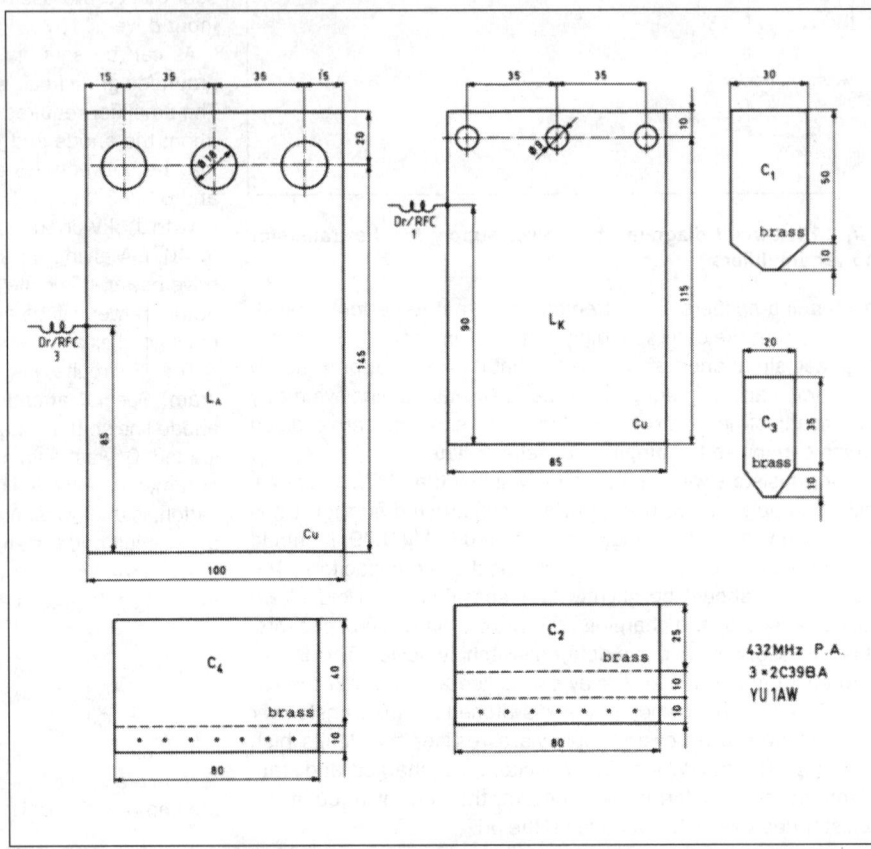

Fig 9.81: Dimensions of the housing parts for the 2C39A power amplifier for the 70cm band

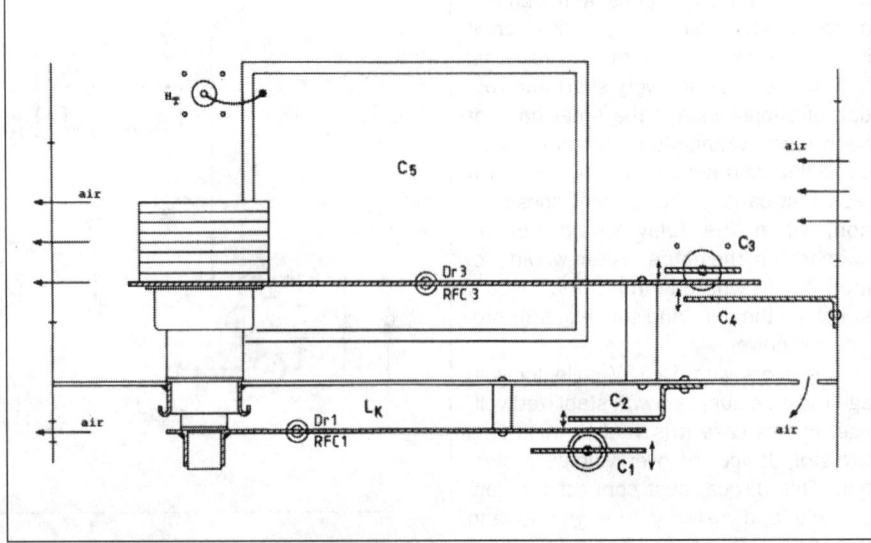

Fig 9.82: Side view of the 2C39A power amplifier for the 70cm band

The Radio Communication Handbook

When the amplifier is in tune the following conditions should exist:

Anode voltage	1300V
Grid voltage	-10 to -12V
Filament voltage	5.8 to 5.9V
Filament current	3A
Quiescent anode current	120mA (40mA per valve)
Maximum anode current	400mA (130mA per valve)
Maximum grid current	100mA (32mA per valve)
Output power	280 to 300W
Power dissipation	210W (70W per valve)
DC Input power	520W
Efficiency	60%
Gain	13dB

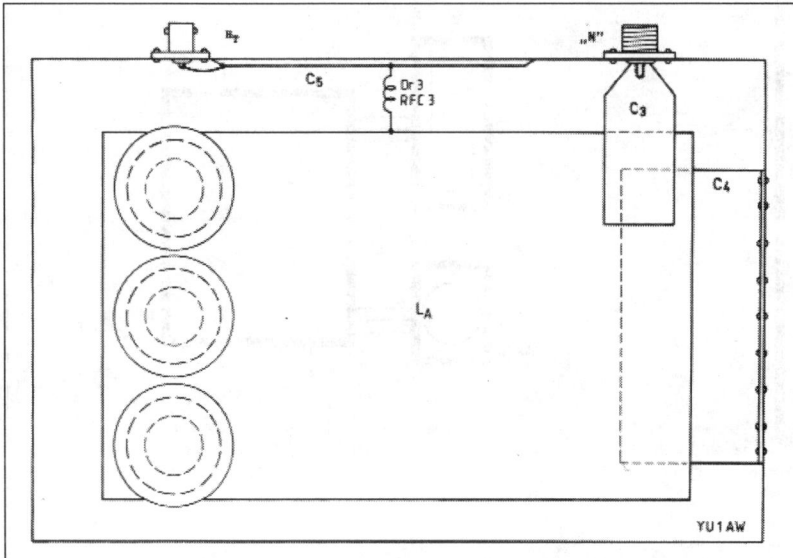

Fig 9.83: Bottom view of the 2C39A amplifier for the 70cm band

Fig 9.84: View of the anode enclosure of the 2C39A power amplifier for the 70cm band

It has been found that the anode voltage can remain on during transmit breaks and receive periods. If noise interference can be heard in the receiver, a 10kΩ resistor can be included in the circuit at the point marked X. This resistor must, of course, be short-circuited during transmit. Any type of available relay will do this job.

RECEIVERS

With the large number of commercial receivers available for the VHF and UHF bands, not many amateurs build their own. This design by Andy Talbot, G4JNT, from *RadCom* shows a new type of receiver [13].

The converter was designed with the primary aim of using it for the IF stage on microwave transverters. A linear receiver was needed with no AGC, but with a calibrated gain control to make accurate relative measurements of microwave beacons using a PC soundcard-based system for the actual level and signal-to-noise ratio measurements. A straightforward gain calibration could then be used to convert these into absolute readings, making this a useful piece of test equipment for propagation studies.

There is nothing inherently narrowband in the design - filtering limits the RF bandwidth to around 8MHz to eliminate strong signals from broadcast and PMR and the audio bandwidth is kept to about 20kHz, wide enough for the normal maximum soundcard sampling rate of 44100Hz. Any subsequent audio filtering for listening purposes is performed by the software or in separate audio processing circuitry.

The circuit diagram is shown in **Fig 9.85**. In the RF path two MMICs, a MAR-6 and a MAR-3, amplify the RF; there is a two stage bandpass filter between them with l0MHz bandwidth. The output feeds into two SRA-1 type DBMs via a resistive splitter, with the quadrature local oscillator (LO) signal generated using a MiniCircuits PSCQ-2-160 90° power splitter. This device guarantees less than 3° phase error over 100 to 160MHz; as 144MHz is near the middle of the range, we can expect better performance here.

The local oscillator is an AD9851 DDS, currently clocked at 100MHz, generating 16 to 16.67MHz followed by a x9 RF multiplier. The DDS source is not described here, but the module in a basic form is described in reference [14]. The active stages in the multiplier consist of MAR-6 MMICs configured as a pair of cascaded tuned x3 stages with a final MAR-6 as amplifier/limiter, this combination forming probably the simplest tuned RF multiplier possible! There are a couple of CW spurii generated by the DDS, but once you know where they are they can be ignored. All filtering is designed to allow the LO to tune over 144 to 150MHz to cover more than the normal 2MHz narrowband segments on the microwave bands, and allow for odd LO frequencies. The multiplier output level is +l0dBm drive to the quadrature hybrid.

By using the internal x6 option in the AD9851 DDS chip the LO could be driven from a 10MHz frequency reference, producing a clock of 60MHz, but this has not been tried.

The mixer outputs drive a pair of identical NE5532 op-amps with a voltage gain approach-

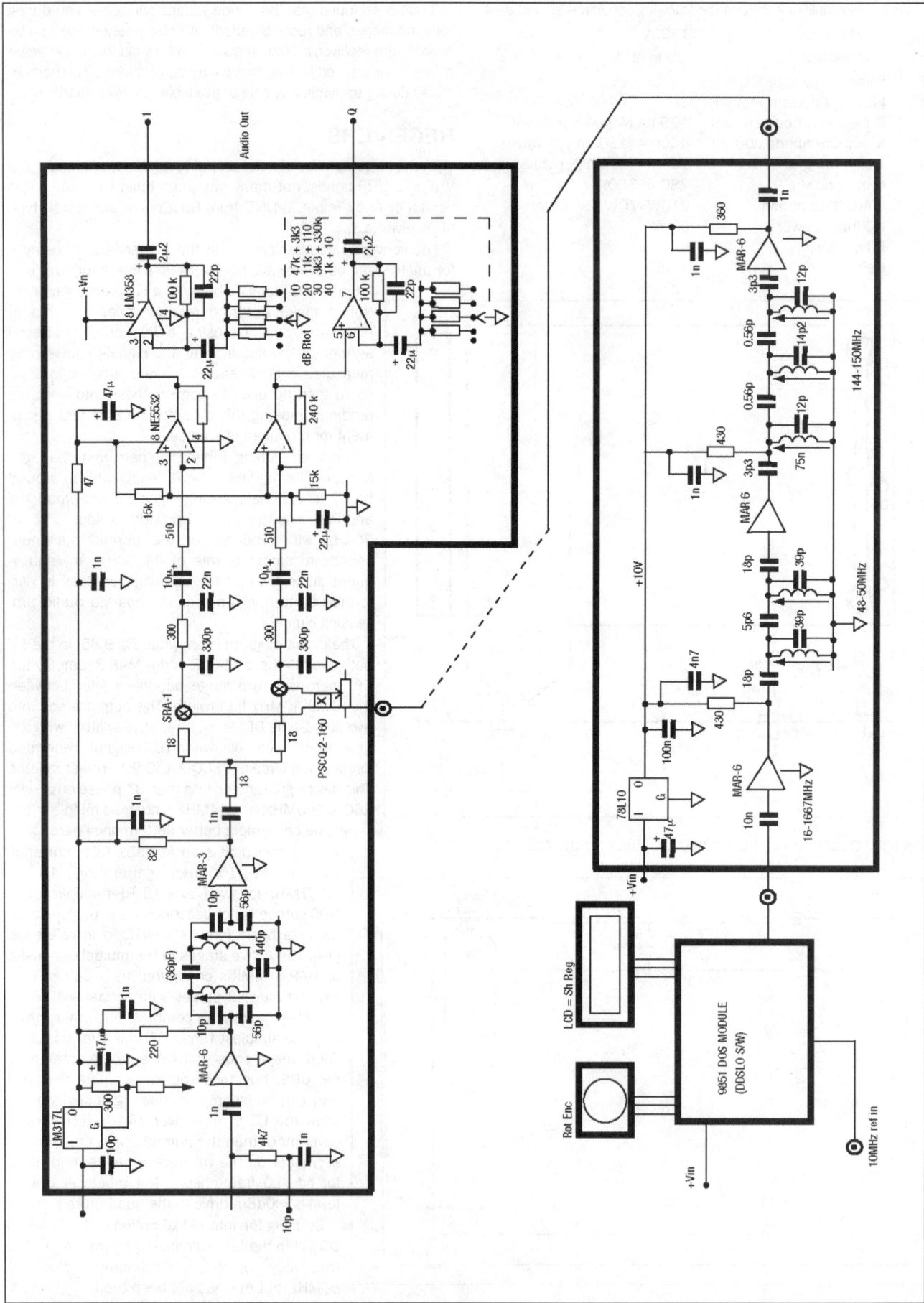

Fig 9.85: Circuit diagram of the 144MHz direct conversion receiver

ing 300 (the exact value is a bit uncertain due to the internal impedance of the mixer IF port). No clever matching is used, just the mixer feeding the inverting input, giving 800Ω input resistance at audio, and low-pass filtering to get rid of RF leakage. The I/Q outputs feed another pair of op-amps with precisely switchable gain from 0 to 40dB in 10dB steps. Audio bandwidth is not especially tailored, but rolls off gently from about 20kHz to allow for 44100Hz sampling rate in a soundcard.

The total system gain and dynamic range is based on 16-bit digitisation, and is sufficient at maximum (+40dB) to place its own thermal noise least 10dB above the quantisation noise pedestal. Strong signals and extra RF gain in transverters are catered for by backing off the audio gain. For signals too strong even for this (80db S/N in 20kHz) an external (calibrated), an RF attenuator can be added.

No attempt was made to put this on a proper PCB. The converter and audio stages were built birds-nest style on a piece of un-etched copper clad PCB as can be seen in the photograph. Plenty of decoupling and short direct wires ensure stable performance. As there is a lot of gain - particularly at audio - the whole unit was built into a tinplate box for screening

Using parallel and series 1% resistors for the switchable gain stage, no special trimming or adjustment was necessary, the traces looked well enough matched on an oscilloscope and, as the aim was only 20 - 25dB sideband rejection to make opposite sideband noise insignificant, tweaking was not necessary. 3° phase error will give 25dB rejection, assuming the amplitude is correct, which is about equivalent to 5% amplitude imbalance. All power rails are regulated and well-filtered for operation from a portable 12V supply.

The LO multiplier was made by cutting a 50Ω microstrip line into a double-sided PCB. To make a 50Ω line quickly without etching, score two lines 2.8mm apart through the copper on the top face of the PCB for the full width; use a Stanley knife or similar, making sure you penetrate the copper fully. A 2.8mm width on normal 1.6mm-thick fibreglass PBC gives about 50Ω characteristic impedance. Then, score two more lines about 1mm from

each of these. Using a hot soldering iron, use this to soften the adhesive and with a pair of tweezers, lift up and remove the two 1mm wide strips, which will give a single 50Ω line surrounded by a copper ground-plane. Drill a number of 0.8 to 1mm holes through the top ground plane to the underside and fit wire links to give a solid RF ground structure. Wire links are best fitted close to where grounding and decoupling components are connected.

Cut the 50Ω line into segments with gaps for the MMICs, DC blocking capacitors and filters. Other connections around the filters are made up bird's-nest style. When completed and aligned, coils can be held in place with glue (a hot glue gun is a useful accessory to have around).

For the stand alone unit for use as a receiver in the field, a simple quadrature network and loudspeaker amplifier can be added to make a complete receiver. A high/low pass pair of all-pass networks will give 15dB sideband rejection over 400Hz to 2kHz, which is good enough for listening to beacon signals on hill tops. Alternatively, look at [15] for phasing-type SSB networks to give an improved SSB performance.

A meter driven from the audio level via a precision rectifier circuit can be added to allow quite precise signal strength measurements to be made in conjunction with the calibrated attenuator. Alternatively, take at look at the Software-Defined Radio software [16] from I2PHD, for another solution

The DDS module as described in [14] has new PIC software, along with a rotary encoder and LCD display to give a user friendly interface. For anyone who has the original DDS board, G4JNT can supply PIC software for this modification. However, the AD9850 and AD9851 chips are in short supply now - they have been replaced in most cases by larger, faster, new devices in a different package. G4JNT has also developed a rotary encoder / display for the AD9852 DDS which gives a better route for a local oscillator as it can generate up to 100MHz. He can be reached at [17].

Alternatively, emulate the venerable IC-202 transceiver and build a VCXO to supply the signal to the multiplier. Or use a VFO/mixer, or a synthesiser - the choice is yours!

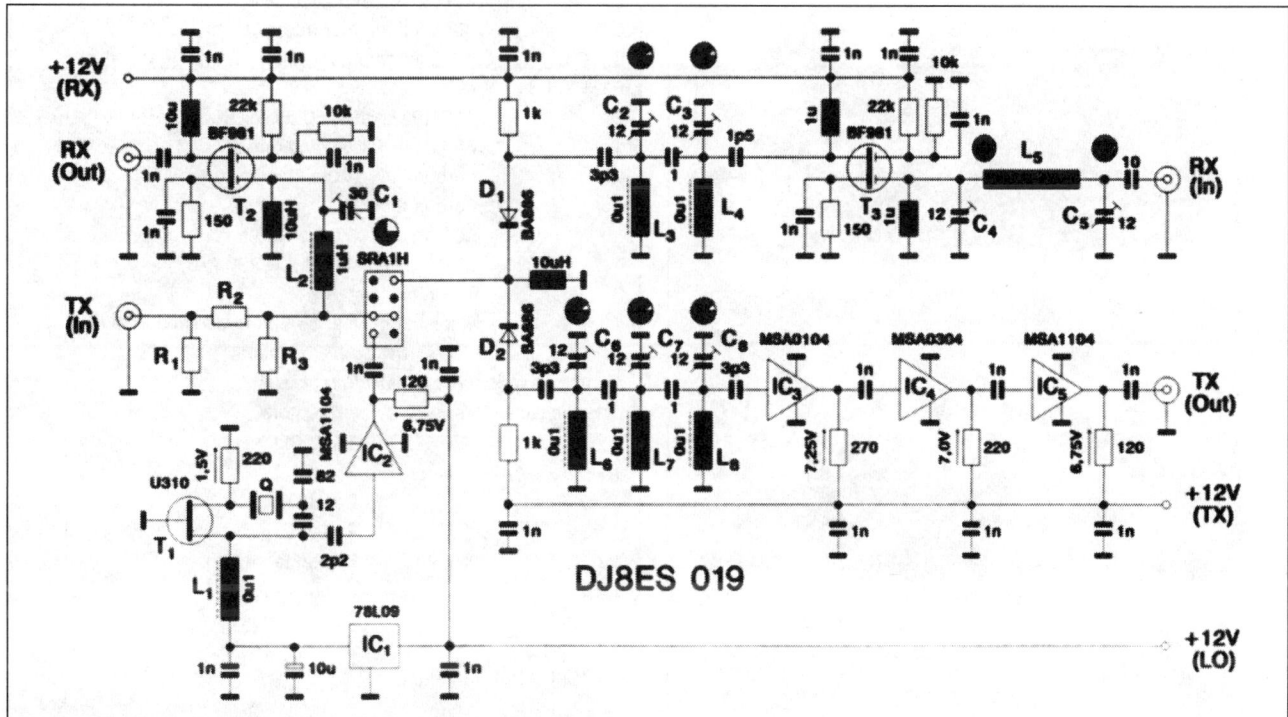

Fig 9.86: Circuit diagram of the 2m transverter

Pin	dB	R1 in	R2 in	R3 in
1mW	0	–	0	51
2mW	3	300	18	300
5mW	7	120	47	120
10mW	10	100	68	100
20mW	13	82	100	82
50mW	17	68	180	68
100mW	20	62	240	62

Table 9.6: Resistance values for attenuator used in 6m, 2m, and 70cm transverters.

TRANSVERTERS

If you already have an HF band transceiver, one of the easiest ways of getting onto the VHF and UHF bands is to use a transverter. This takes the output of your transceiver, usually the 28 - 30MHz band and converts it to the chosen VHF or UHF band and converts received signals on the VHF or UHF band so that they are received on your transceiver. The transceiver output and input is commonly called a tuneable IF. The advantage of this approach is that all of the facilities of your HF transceiver are available on the VHF or UHF band.

This section contains transverter designs for all of the VHF and UHF bands. The first set of designs cover 6m, 2m and 70cm using similar circuits. The 4m transverter design was used as a club project by the Andover Radio Amateurs Club [18].

2m Transverter

In 1990, Wilhem Schüerings, DK4TJ and Wolfgang Schneider, DJ8ES, presented a paper at the 35th VHF Congress in Weinheim on a universal transverter concept [19] and [20]. The following design is the resulting 28/144MHz transverter [21]. It should be possible for the transverter to directly feed a standard power amplifier, the design of a suitable amplifier using a hybrid amplifier module is shown.

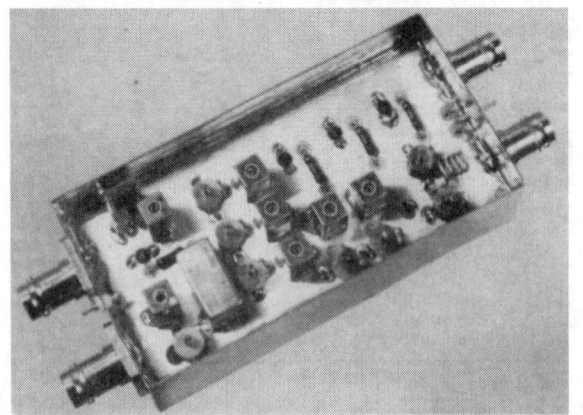

Fig 9.87: The completed 2m transverter

Transverters for the 2m band are always of interest, in an attempt to match the current state of art in amateur radio technology this transverter was developed using modern components to convert the 144 - 146MHz range into the 10m band. Concepts such as high-level signal strength and oscillator signal spectral purity have taken on increasing significance. It is also important that the equipment can be reproduced easily. The transverter described below represents a circuit that corresponds to today's requirements.

Fig 9.86 shows the circuit diagram of the 28/144MHz transverter. The local oscillator uses a tried and tested U310 crystal oscillator at 116MHz.

This signal is amplified by the next stage, an MSA1104 MMIC, giving an output level of 50mW. The SRA1H high-level ring mixer requires a local oscillator level of +17dBm (50mW) and can be used at up to 500MHz.

A pi attenuator, consisting of R1 to R3 is used to the control the output from the driving transmitter (IF). For a 'clean' signal (intermodulation products <50 dB), the ring mixer must be driven by a maximum of 1mW (0dBm) at the IF port. **Table 9.6** shows the resistance values needed for the attenuator for various IF input power levels. The attenuator uses standard value resistors. The attenuator also acts as a 50 ohm termination for the ring mixer.

The converted receiver signal is fed from the mixer by L2 and C1 to a high impedance amplifier using a BF981 low-noise tran-

IC1	TA78L09F SMD voltage regulator
IC2, IC5	MSA1104 (Avantek)
IC3	MSA0104 (Avantek)
IC4	MSA0304 (Avantek)
T1	U310 (Siliconix)
T2, T3	BF981 (Siliconix)
D1, D2	BA886 PIN diode (SMD)
L1, L3, L4, L6, L7, L8	Neosid BV5061 0.1µH blue/brown coil
L2	Neosid BV5048 1µH yellow/grey coil
L5	4.5 turns, 1mm gold plated copper wire
C1	30pF foil trimmer (red), 7.5mm grid (Valvo)
C2, C3	12pF foil trimmer (yellow), 7.5mm grid (Valvo)
R1, R2, R3	Attenuator, see Table
Q	116MHz crystal, HC18U or HC25U
1x	SRA1H high-level ring mixer
2 x	Carbon film: 120Ω, 0.5 W
1 x	Carbon film: 220Ω, 0.5 W
1 x	Carbon film: 270Ω , 0.5 W
4 x	BNC flanged socket (UG-290 A/U)
3x	Teflon bushing
1 x	Tinplate housing: 55.5mm x 111mm x 30mm
9x	Copper rivets (1.5mm dia.)

All other components in SMD format

Ceramic capacitors		Resistors	
3 x 1pF		1 x 150Ω	
1 x 1.5pF		2 x 220Ω	
1 x 2.2pF		2 x 1kΩ	
4 x 3.3pF		2x 10kΩ	
1 x 10pF		2 x 22kΩ	
1 x 12pF		**Inductors**	
1 x 82pF		2 x 1µH choke	
17 x 1nF		3 x 10µH choke	
1 x 10µF / 20V Tantalum			

Table 9.7: Parts list for 2m transverter.

The Radio Communication Handbook

sistor to give the required intermediate frequency amplification.

The 2m received signal is fed to the gate of the BF981 RF amplifier through a pi filter from the 50Ω aerial input. The RF amplifier is followed by a two-pole filter. The received signal is switched to the ring mixer by the +12V receiver supply voltage through the PIN diode, D1 (BA886).

In transmit mode, diode D2 is activated. The 2m signal from the ring mixer first passes through a three-pole filter. The signal is then amplified by three MMIC amplifiers (IC3, IC4, IC5). The combination of MSA0104, MSA0304 and MSA1104 guarantees an output level of 50mW (+17dBm). The transverter can be used with any power amplifier but additional harmonic filtering is recommended.

The 28/144MHz transverter is assembled on a double sided PCB measuring 54mm x 108mm. The board can be mounted in standard tinplate housing of 55.5mm x 111mm x 30mm, **Fig 9.87** show pictures of the completed transverter. The board should either be made as a through-plated PCB, or copper rivets used to make the earth connections for the coils and ring mixer. The parts list is shown in **Table 9.7** and component layouts for both sides of the PCB are shown in **Figs 9.88 and 9.89** (see Appendix B). Suitable holes for the crystal, trimmers and Neosid coils, etc are drilled on the earth side of the boards (fully coated side) using a 2.5mm drill. Holes that are not used for earth through connections should be countersunk using an 8mm drill. Suitable slots are to be sawn out in the printed circuit board for the BNC connectors. When the board has been soldered to the sides of the housing, the actual assembly can be undertaken. The boards are fitted into the housing so that the connector pins of the RF connectors are level with the surface of the PCB (cut off projecting Teflon collars with a knife first). When the mechanically large components (filter coils, trimmers, crystal and ring mixer) have been fitted it must be possible to fit the housing cover without any obstruction.

When the equipment is used for the first time, the following test equipment should be available: multimeter, frequency counter, diode probe, wattmeter and received signal (eg beacon). First the crystal oscillator is tuned by adjusting L1, the power consumption should be approximately 65mA. In transmit modes the only adjustment required is to tune the three-pole filter, the power consumption should be approximately 130mA. In receive mode the input filter should be tuned using a weak signal eg a beacon, then the 28MHz filter, C1 / L2, should be tuned.

A power amplifier can be added to the 2m transverter to increase the output power to 20W. The amplifier uses a Mitsubishi hybrid module, these are still available from some suppliers and can be found in surplus equipment.

Fig 9.90 shows the relatively simple circuit for the 144MHz power amplifier. The core of the circuit is a Mitsubishi M57727 hybrid module (IC1). This module operates at a working voltage of 12 volts. With exactly 27dB amplification, the transverter signal is raised to an output voltage of 20W. The output power to input power ratio is shown in **Fig 9.91**. The current consumption of the module is directly proportional to this ratio.

Such amplifier modules are constructed using thick film technology. This module is designed for the 144 - 148MHz frequency range, and the amplification is achieved in two stages. **Fig 9.92** shows what a typical module looks like from the inside. The 50Ω input and output matching circuits are clearly visible.

A low-pass filter (**Fig 9.93**) on the output provides the harmonic suppression required. Amazing suppression is obtained

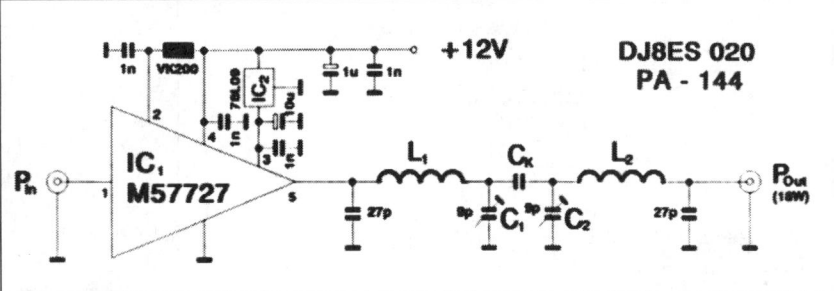

Fig 9.90: Circuit of the 2m power amplifier to be used with the 2m transverter

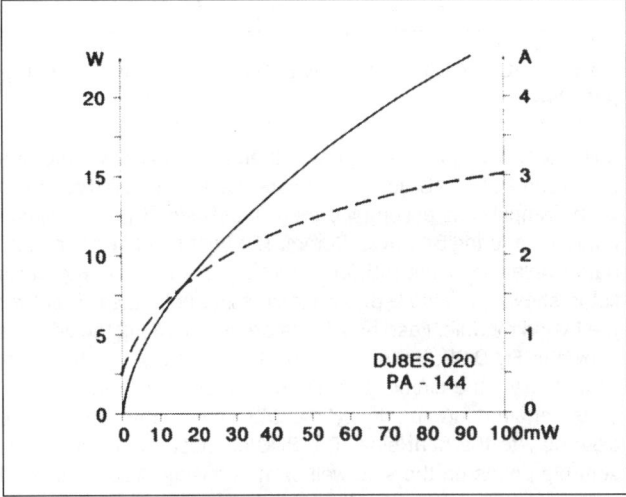

Fig 9.91: Power transfer characteristics of the Mitsubishi M57727 hybrid module

Fig 9.92: An internal view of a hybrid module. This is the Toshiba S-AU4 which is a 70cm amplifier

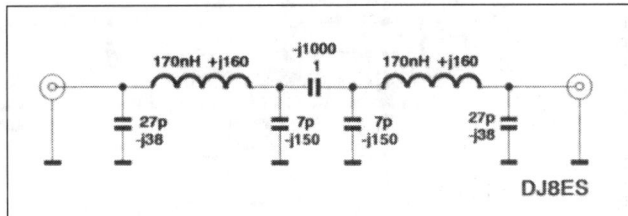

Fig 9.93: Circuit of the low pass filter used in the 2m power amplifier

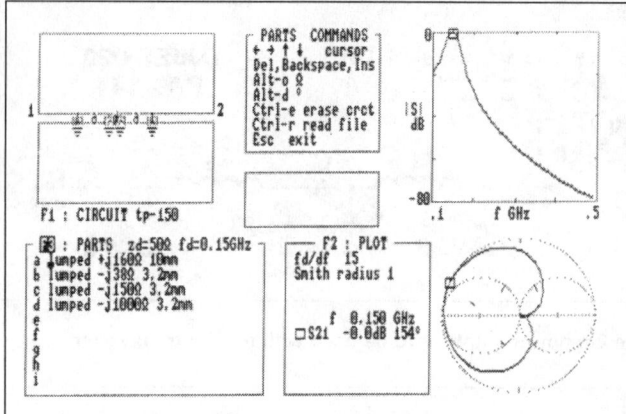

Fig 9.94: PUFF CAD software output used to design the 2m low pass filter

IC1	M57727 (Mitsubishi)
IC2	TA78L09F voltage regulator (SMD)
L1, L2, C_K	see text
C1, C2	9pF trimmer with soldering lug
1 x VK200 VHF broad-band choke	
1 x 1nF feed-through capacitor, solderable	
2 x BNC flanged bush (UG-290 A/U)	
1 x Tinplate housing 55.5 x 111 x 30mm	
All other components in SMD format:	
1 x lµF/20V tantalum	
1 x 10µF/20V tantalum	
2 x 27pF, ATC chip	
3 x 1nF, ceramic capacitor	

Table 9.8: Parts list for 2m hybrid amplifier.

using only two pi filters wired together. **Fig 9.94** shows the output of the PUFF CAD design software used to design the filter.

The amplifier is assembled on a double-sided printed circuit board measuring 54mm x 108mm. The board can fit into a standard tinplate housing (55.5mm x 111 mm x 30mm), the parts list is shown in **Table 9.8**. A suitable sized hole is sawn out for the hybrid module. Fastening holes are drilled along the edge as shown in **Fig 9.95** (in Appendix B). Good earth connections are essential for the circuit to operate correctly. The through contacts required are made by the M3 screws that secure the assembly to the heat sink. The BNC connectors are placed at suitable points on the side wall of the housing. Also positioned in the side wall is the feed-through capacitor for the power supply. The components are not inserted until the board has been soldered to the sides of the housing. The board should be fitted at the edges of the sides, so that the amplifier module will lie flat on the heat sink.

The two coils (L1, L2) and the coupling capacitor, C_K, are hand made. The coils are 8.5 turns of silvered copper wire with a

diameter of 1mm. The wire is wound around a 6mm mandrel (eg a 6mm drill shank) and soldered on with a 1mm clearance from the PCB. The coupling capacitor, C_K, is made from a 1cm long piece of coaxial cable (RG174), the length is chosen to give the required 1pF capacitance. A standard chip capacitor cannot be used here, due to the relatively high power level. A thin copper plate is soldered between the two pi filters for screening, see **Fig 9.95** (in Appendix B). Finally, the remaining components are added. The module is screwed directly onto the heatsink using two M4 screws after applying heat conducting paste.

A power meter and a multimeter are required for putting the equipment into operation. The quiescent current should be approximately 400mA, which rises to around 2.5A under full drive with an input power of 60mW. This gives an output power of the order of 18W.

Only the low-pass filter (C1, C2) requires tuning in the hybrid amplifier. The trimmers are normally screwed about half way in when the unit is correctly tuned. In order protect the hybrid module, carry out this tuning procedure with only a low drive power level (max 10mW).

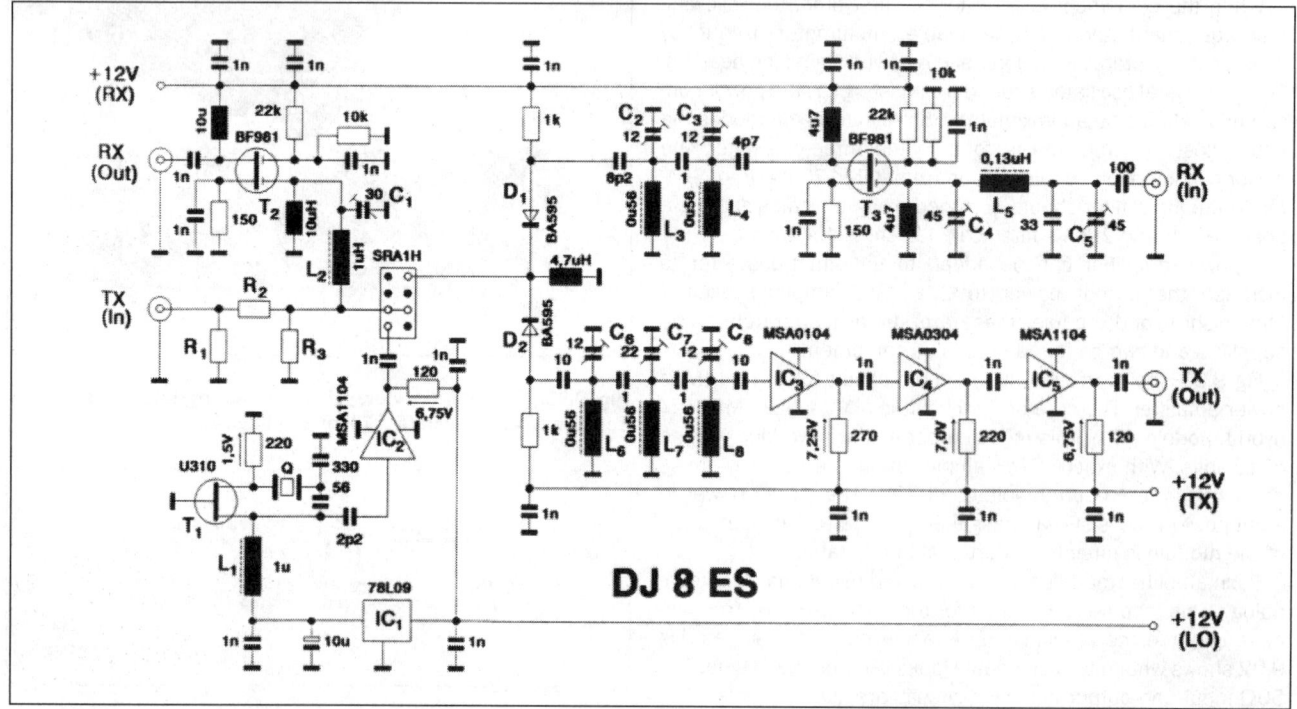

Fig 9.96: Circuit diagram of the 6m transverter

IC1	TA78L diameter 9F voltage regulator (SMD)	2 x 120Ω / 0.5W Carbon film	
IC2, IC5	MSA1104 (Avantek)	1 x 220Ω / 0.5W Carbon film	
IC3	MSA0104 (Avantek)	1 x 270Ω / 0.5 W Carbon layer	
IC4	MSA0304 (Avantek)	4 x BNC flanged connector (UG-290 A/U)	
T1	U310 (Siliconix)	3 x Teflon bushing	
T2, T3	BF981 (Siemens)	1 x Tinplate housing 55.5 x 111 x 30mm	
DI, D2	BA595 PIN diode (SMD)	9 x 1.5mm dia. Copper rivets	
L1, L2	BV5048 Neosid coil,1 µH, yellow/grey	*All other components are SMD format:*	
L3, L4	BV5036 Neosid coil, 0.58µH, orange/blue	**Ceramic capacitors**	**Resistors etc**
L5	BV5063 Neosid coil, 0.58µH, blue-orange	3 x 1pF	1 x 150Ω
L6, L7, L8	BV5063 Neosid coil, L8 0.58µH, orange/blue	1 x 2.2pF	2 x 220Ω
C1	30pF foil trimmer (red) 7.5 mm grid (Valvo)	1 x 4.7pF	2 x 1kΩ
C2, C3	l2pF foil trimmer (yellow) 7.5 mm grid (Valvo)	1 x 8.2pF	2 x 10kΩ
C4, C5	45pF foil trimmer (violet) 7.5 mm grid (Valvo)	2 x 10nF	2 x 22kΩ
C6, C8	12pF foil trimmer (yellow) 7.5 mm grid (Valvo)	1x 33pF	1x 10µF / 20VTantalum
C7	22pF foil trimmer (green) 7.5 mm grid (Valvo)	1 x 56pF	3x Choke, 4.7µH
Q	22MHz crystal, HC18U or HC25U	1 x 330pF	2x Choke, 10µH
1 x SRA1H ring mixer		17 x 1nF	

Table 9.9: Parts list for 6m transverter

6m Transverter

A transverter for the 6m band can be produced based on the 28/144MHz transverter described above [22]. All that is required is modification of the oscillator and the filter.

Fig 9.96 shows the complete circuit for the 28/50MHz transverter. The circuit can be assembled using the printed circuit board used for the 2m transverter. The pi filter at the input of the receiver needs to be altered; **Fig 9.97** (in Appendix B) shows details of the modification. All the coils and some of the capacitors have different values for the lower frequency range, **Table 9.9** show the parts list. To make it easier to produce the 6m version of the transverter, the layout of the printed circuit board with the appropriate components for the 50MHz version is illustrated in **Fig 9.98** and **Fig 9.99** (both in Appendix B).

To increase the power output an M57735 hybrid module is used in a separate amplifier stage for the 6m band, the circuit diagram is shown in **Fig 9.100**. The M57735 module was developed for use around 50MHz and is still available from some suppliers or in surplus radio equipment. About 10W can be expected at the output of the PA from the 50mW output from the transverter.

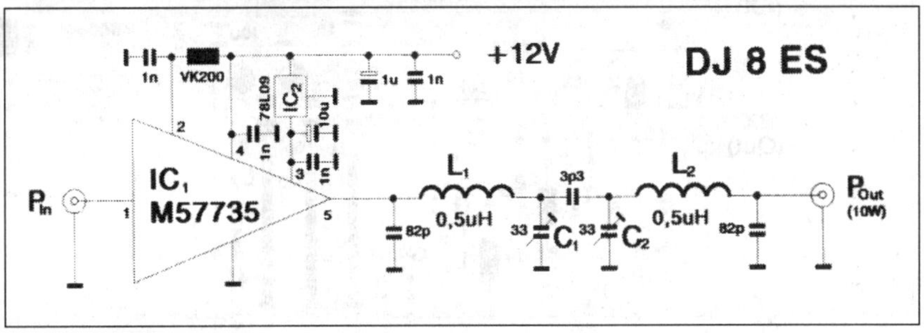

Fig 9.100: Circuit of the 6m power amplifier to be used with the 6m converter

The low-pass filter provides the harmonic filtration required. Only components of appropriate quality (eg air-core coils and air-spaced trimmers) should be used here. The 50MHz amplifier can be assembled on the printed circuit board used for the 2m version using the same construction techniques. The parts list is shown in **Table 9.10**.

IC1	M57735 (Mitsubishi)
IC2	TA78L09F voltage regulator (SMD)
L1, L2	0.5µH air-core coil
C1, C2	33pF trimmer with soldering lugs
1 x VK200 VHF wide-band choke	
1 x 1nF feed through capacitor, solderable	
2 x BNC flanged bush (UG-290 A/U)	
1x Tinplate housing 55.5 x 111 x 30mm	
1x 1µF / 20V Tantalum	
1x 10µF / 20V Tantalum	
1x 3.3pF, ATC chip	
2x 82pF, ATC chip	
3x 1nF, ceramic capacitor	

Table 9.10: Parts list for 6m hybrid amplifier

70cm Transverter

The following design for a 28/432 MHz transverter [23] is similar to the 28/144MHz transverter described above. It uses two boards; the oscillator and the transverter **Fig 9.101** shows a pic-

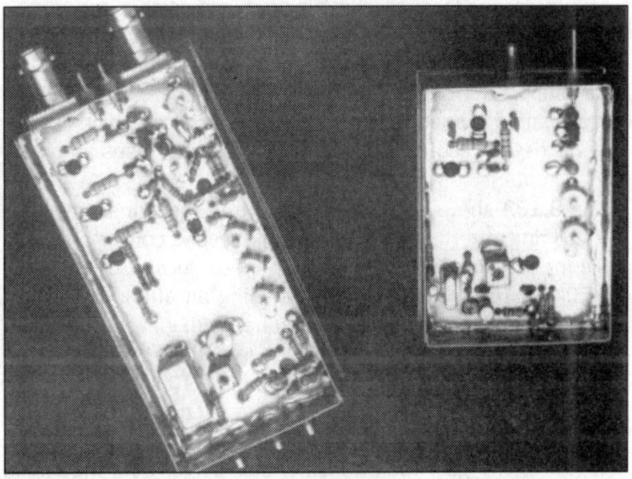

Fig 9.101: The completed local oscillator and transverter units of the 70cm transverter

Fig 9.102: Circuit of the local oscillator used in the 70cm transverter

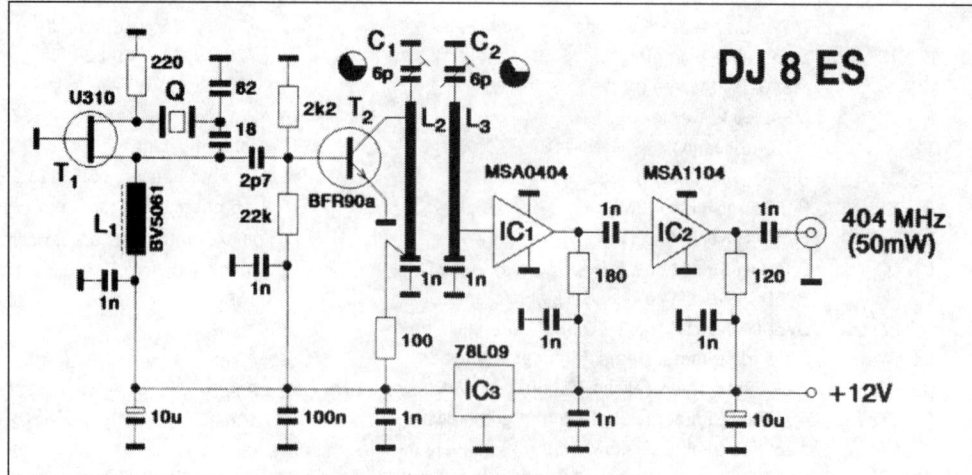

Fig 9.103: Circuit diagram of the 70cm transverter

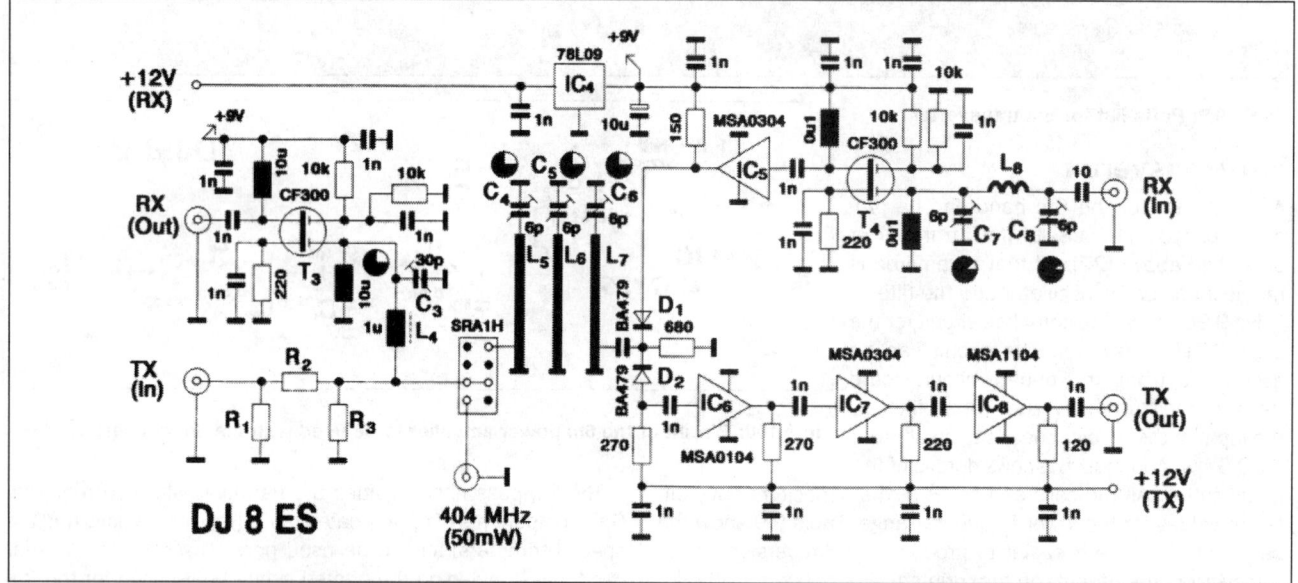

ture of the completed units. It should be possible for the transverter to directly feed a standard power amplifier.

Using wide-band amplifier ICs and a ring mixer makes the circuit very flexible, by just changing the filters and the crystal oscillator, the tuning range can be changed to suit the requirements.

Fig 9.102 shows the circuit diagram of the local oscillator, it uses a U310 as the crystal oscillator at 101MHz. The 404MHz required for the local oscillator is produced using a quadrupler. A printed circuit 2-pole filter provides the necessary filtering. Two wide-band integrated amplifiers, MSA0404 (IC1) and MSA1104 (IC2) supply the desired output of 50mW. The correct level of amplification is important, only the amplification, which is actually necessary, should be used. Any excess increases the spurious outputs.

Fig 9.103 shows the circuit diagram of the transverter. The SRA1H ring mixer used in the transmit/receive converter is suitable for use up to 500MHz, and requires a local oscillator level of 50mW. The mixer is controlled using an attenuator, which should provide an intermediate frequency (IF) level of no more than 1mW at the ring mixer. The attenuator must be designed on the basis of the IF output available. Table 9.10 shows the resistor values required for the attenuator in relation to the IF power level. All the values are based on the standard values from the E12 to E24 ranges. The attenuator also serves as a wide-band 50Ω termination for the ring mixer (SRA1H). The received signal is matched at high impedance to the CF300 (T3) using L4 and

C3. This low-noise transistor stage provides the necessary intermediate frequency amplification.

The 70cm received signal is passed to the gate of the CF300 (T4) through a pi filter (aerial impedance 50 ohms) that is followed by an MSA0304 amplifier. When the receive +12V power supply is connected, the PIN diode D1 (BA479) is biased on and the signal passed through.

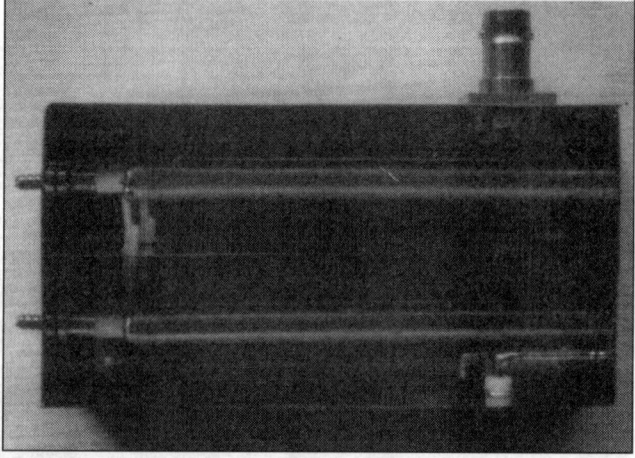

Fig 9.104: 70cm bandpass filter that can be used with the 70cm transverter

The printed circuit 3-pole 70cm filter is used for both receive and transmit. The transmit signal initially passes through the filter and diode D2 which is biased on. The subsequent amplifier uses integrated wide-band amplifiers (IC6, IC7, IC8). The combination of MSA0104, MSA0304 and MSA1104 provides an output of 50mW (+17dBm) with 40dB of amplification.

In practical operation, such transverters are used with the same driving unit; an additional filter for harmonics and spurious transmissions is recommended. **Fig 9.104** shows a possible two-pole bandpass filter. It can be assembled as an air core construction using a standard tinplate housing measuring 55.5 x 111 x 30mm.

The 28/432 MHz transverter is divided into two independent assemblies: the local oscillator and the transmit/receive converter. The double sided printed circuit boards measure 54mm x 72mm for the local oscillator and a 54mm x 108mm for the transmit/receive converter. The parts list for the local oscillator and transverter are show in **Table 9.11 and 9.12** respectively. The component layout for the local oscillator is shown in **Figs 9.105 and 9.106**, and for the transverter in **Figs 9.107 and 9.108** (all four of these are in Appendix B). The PCB are mounted in standard tinplate housings, suitable holes are drilled for the stripline transistors and the wide-band amplifiers; these components are mounted level with the surface of the board. The holes for the crystal, trimmers and Neosid coils, etc are drilled on the earth side of the boards (fully coated side) using a 2.5mm drill. Suitable slots are to be sawn out in the printed circuit board for the SMC or BNC connectors. The same applies to the 1nF capacitors at the source connection of the amplifier transistors, T3 and T4. When the individual boards have been soldered to the sides of the housing, the actual assembly can be undertaken. The boards are fitted into the housing so that the connector pins of the RF connectors are level with the surface of the PCB (cut off projecting Teflon collars with a knife first). When the mechanically large components (filter coils, trimmers, crystal and ring mixer) have been fitted it must be possible to fit the housing cover without any obstruction.

When the equipment is used for the first time, the following test equipment should be available: Multimeter, Frequency counter, Wattmeter and Received signal (eg beacon). The assemblies switched on one after another.

Firstly, the crystal oscillator is set to its operating frequency of 101MHz by adjusting coil, L1. The onset of oscillation results in a slight increase in the collector current of T2 (monitoring voltage drop across 100 ohm resistor). A frequency counter is loosely coupled and the oscillator frequency measured.

The two-pole filter after the quadrupler T2 (BFR90a) filters out the 404MHz frequency required. To adjust this, the two trimmers, C1 and C2 are adjusted one at a time for maximum output. The local oscillator should supply an output of at least 50mW. The current consumption for an operating voltage of +12V is about 120mA.

The transmit branch of the transmit/receive converter is put into operation first. Only the three-pole filter (C4, C5, C6) has to be adjusted. A current of approximately 130mA should be measured for an operating voltage of +12V. This is an indication that the amplifier stages are operating satisfactorily. If the input attenuator is selected as described in Table 9.6, an output greater than 50mW can be expected. Possible spurious outputs (oscillator, image frequency, etc.) are suppressed by better than 50dB.

The receiver can be calibrated using a strong received signal (eg a beacon). Because the same filter is used as in the transmit branch, the beacon signal should be audible immediately. Another filter is used at the intermediate frequency (28MHz)

IC1	MSA0404 (Avantek)
IC2	MSA1104 (Avantek)
IC3	78L09 voltage regulator
T1	U310 (Siliconix)
T2	BFR90a (Valvo)
L1	Neosid BV5061 0.1µH blue/brown coil
L2, L3	λ/4 stripline, etched
C1, C2	6pF foil trimmer (grey), 7.5mm grid (Valvo)
Q	101MHz crystal, HC18U or HC25U
1 x	Carbon film: 180Ω, 0.5 W
1 x	Carbon film: 120Ω, 0.5 W
1 x	SMC or BNC flanged socket (UG-290 A/U)
1 x	Teflon bushing
1 x	Tinplate housing: 55.5mm x 74mm x 30mm
2 x	1nF trapezoid capacitor
2 x	10µF 20V tantalum capacitor

Ceramic Capacitors (2.5mm grid)	Resistors (1/8W, 10mm)
1 x 2.7pF	1 x 100
1 x 18nF	1 x 220
1 x 82pF	1 x 2.2k
6 x 1nF	1 x 22k
1 x 100nF	

SMD Capacitor (model 1206 or 0805)	
2 x 1nF	

Table 9.11: Parts list for 70cm transverter local oscillator

IC4	78L09 voltage regulator
IC6	MSA0104 (Avantek)
IC5, IC7	MSA0304 (Avantek)
IC8	MSA1104 (Avantek)
T3, T4	CF300 (Telefunken)
DI, D2	PIN diode BA479
L4	BV5048 Neosid coil, 1µH, yellow/grey
L5, L6, L7	λ/4 stripline, etched
L8	1.5 turns, 1mm CuAg wire
1x	SRA1H high-level ring mixer
C3	30pF foil trimmer (red) 7.5mm grid (Valvo)
C4, C5, C6	6pF foil trimmer (grey) 7.5mm grid (Valvo)
C7, C8	6pF foil trimmer (grey) 7.5mm grid (Valvo)
R1, R2, R3	Attenuator, see Table
1 x Carbon film: 120Ω , 0.5W	
1 x Carbon film: 150Ω, 0.5W	
1 x Carbon film: 220Ω, 0.5W	
I x Carbon film: 270Ω, 5W	
5 x SMC sockets (some of which may be BNC flanged: UG-290 A/U) (see photo of specimen assembly)	
2 x Teflon bushing	
1 x Tinplate housing 55.5 x 111 x 30mm	
4 x 1nF trapezoid capacitor	
2 x 0.1 µH choke, I0mm grid, axial	
2 x 10µH choke, I0mm grid, axial	
1 x 10µF 20V tantalum	

Resistors (1/8W, 10mm)	Ceramic Capacitors (2.5mm grid)
2 x 220Ω	1 x 10pF
1 x 270Ω	12 x 1nF
1 x 680Ω	SMD Capacitor (model 1206 or 0805)
4 x 10kΩ	6 x1n

Table 9.12: Parts list for 70cm transverter

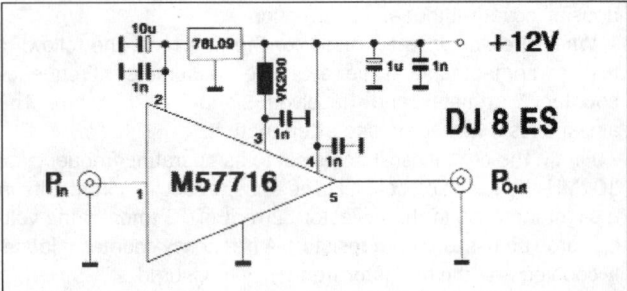

Fig 9.109: Circuit diagram of the 70cm power amplifier to be used with the 70cm transverter

after the mixer. The trimmer C3 should be adjusted to give maximum signal output. Optimising the signal-to-noise ratio using the pi filter, C7, C8 and L8 completes the calibration. The current consumption of the receive converter is very low, only 50mA. The noise factor is approximately 2dB with a conversion gain of approximately 30dB.

The designer uses the transverter described in association with an external preamplifier and power amplifier. Modern hybrid modules are just the thing for power amplifier stages. The output signal can be increased from 50mW to 10 - 20W in one go using such components. **Fig 9.109** shows the circuit for such a module using a Mitsubishi M55716. A 2C39 valve PA can be fully driven using this 10W output.

4m Transverter

This transverter was designed as a club project for the Andover Radio Amateur Club [18]. Various schemes such as transverting from two meters were considered but after some discussion, it was decided to transvert from 27MHz or optionally 28MHz. Using 27MHz as the drive source was particularly attractive for a number of reasons. Many members own or could cheaply obtain a 27MHz CB rig which would mean that the completed transverter could be permanently connected to a dedicated drive source without 'tying-up' and restricting the use of equipment used regularly on other bands. This would make it more practical to establish a club frequency that could be monitored whenever one is in the shack. It was felt this would help to build up band activity. Although CB radios have 10kHz channel spacing whilst the 4m band uses 25kHz, the two spacings coincide on a number of frequencies including all the calling frequencies (70.45 FM, 70.26 All Mode and 70.2MHz SSB) and two of the most commonly used simplex FM working frequencies 70.35 and 70.40MHz.

The design of this transverter was not intended to push the frontiers of technology, but to provide a simple, repeatable design based on readily available components many of which could be found in the 'junk box' thus keeping down costs. The circuit diagram for the transverter is shown in **Figs 9.110 and 9.111**.

The receive converter uses Dual Gate MOSFET RF and mixer stages which provide good gain, noise performance and stability. L1 is resonated by the series combination of C1 and C2 to provide the first stage of input filtering, while the ratio C2:C1 provides matching from the 50 ohm input to the higher impedance of Q1 Gate 1. The RF and mixer stages are band-pass coupled (L2 and L3) to improve rejection of unwanted, out of band signals.

The local oscillator uses a third overtone crystal and is shared between the receive and transmit mixers. The choice of crystal frequency depends on the drive source to be used. For 28MHz the crystal frequency is 42MHz, for a CB 27/81 driver a frequency of 42.49875MHz is required while for the newer PR 27/94 CB Rigs a frequency of 43.0850MHz is required. The zener diode ZD1 is an option which may be required to improve stability if the transverter is to be used for SSB or CW operation (Initial tests indicate that it is not necessary so long as a regulated supply is used).

The transmit mixer uses a proprietary doubly balanced mixer to provide additional suppression of the oscillator and driver frequencies and their products. There are a number of different mixer units that will operate satisfactorily, but if substituting a different device beware of its pin connections as two different pin layouts are in common use and only the right one will work! Also, some types are not as shown in **Fig 9.112** but have pins 2, 5, 6 and 7 bonded directly to the case. This is not a problem. Note that pin 1 is identified by a different coloured bead that may be lighter or darker than its neighbours.

These doubly balanced mixers require about +7dbm of local oscillator injection and 0dBm (1mW) of Drive (IF). The output of the oscillator is loosely coupled to L6 providing a band-pass arrangement to reduce unwanted harmonics. C19 and C20 in series resonate L6 while their ratios provide an impedance transformation to match the oscillator output to the mixer (U1) at 50 ohms.

The input from the driving 'rig' is first coupled through C42 to a voltage doubling detector D3/D4 which acts as an RF sensor driving Q7 to operate the change over relays RL1 and RL2. Note

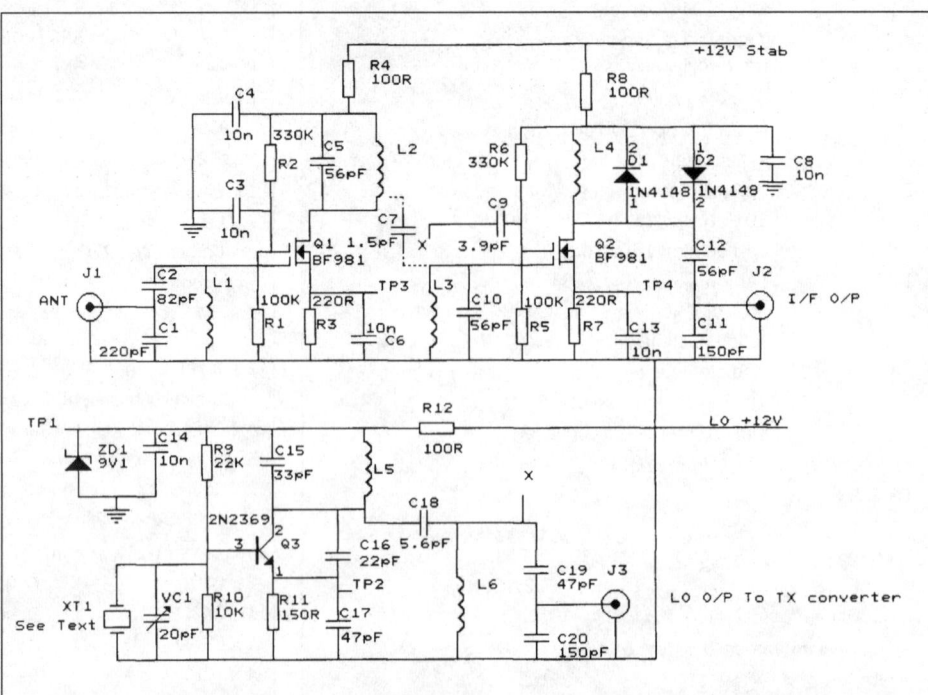

Fig 9.110: Receive converter circuit diagram for the 4m transverter

The Radio Communication Handbook

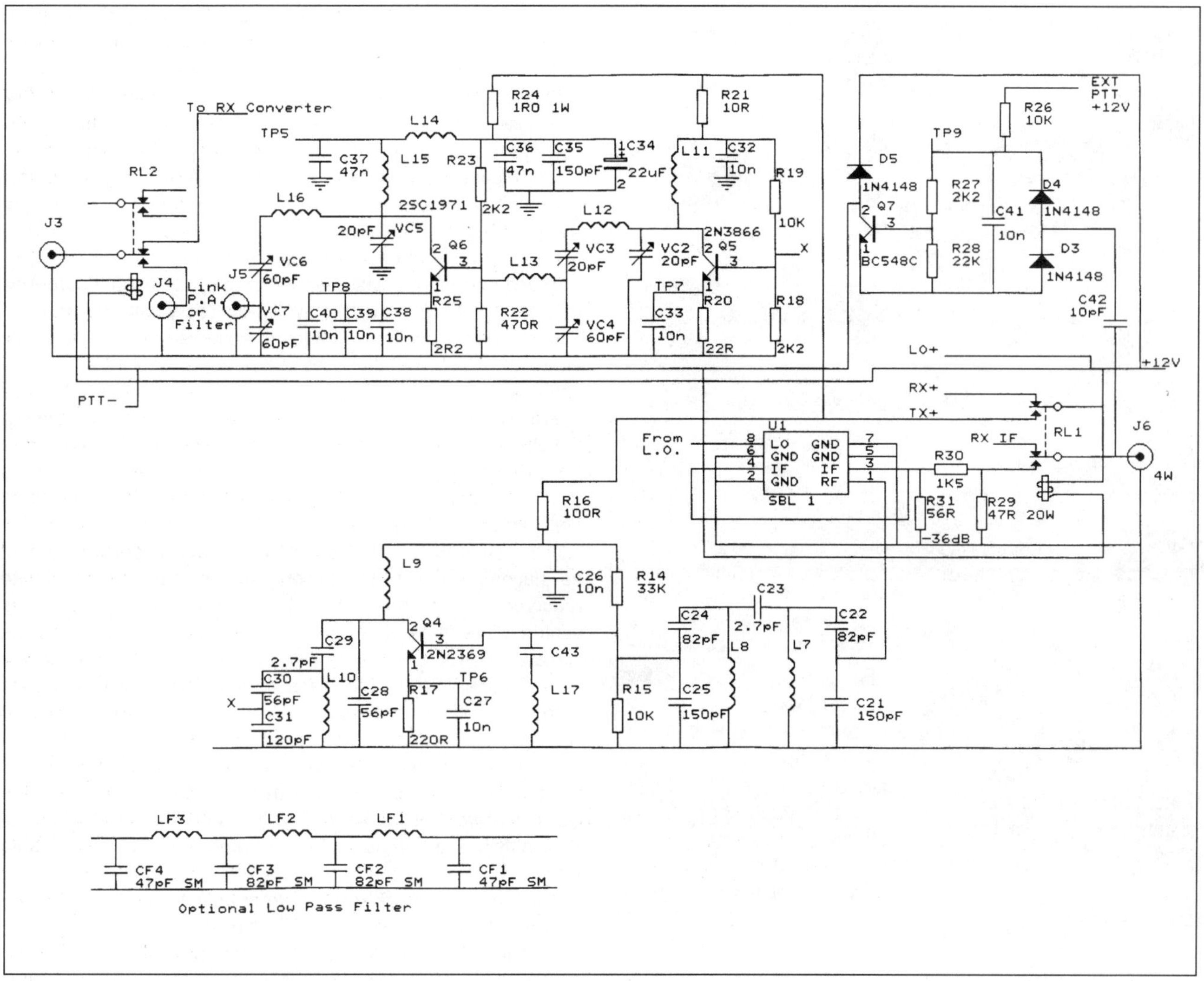

Fig 9.111: Transmitter converter circuit diagram for the 4m transverter

that the choice of device for Q7 was based largely on the need for a high current gain (H_{fe}), the specified BC548C having an H_{fe} min. of 420. Once a sufficient level of RF is present at the input, Q7 turns ON and relays RL1 (and RL2) close routing the RF through an attenuator, R29,R30 and R31 to reduce its level to about 0dbm (1mW). Please note that RF switching is generally acceptable for FM but can be problematic for SSB operation so direct switching is recommended for this mode. There are two options for external switching: a positive voltage on Transmit applied to the EXT PTT + input or a contact closure to ground applied to EXT PTT LO. Note that the latter should be from volt-age free contacts or possibly through a series diode.

Assuming that a standard CB rig is being used with an output of 4 watts, the attenuation required is -36dB. The values shown for these resistors produce about 36dB of attenuation. If you wish to use a different input level, you will need to change these values, **Table 9.13** shows some common values. Bear in mind that R29 will dissipate the bulk of the power output from the driving rig and should be rated accordingly. This design uses a TO220 style non-inductive power resistor for R29, and this is bolted to the front panel and hence chassis of the transverter to dissipate the heat. Although rated at 20W, a maximum drive level of 10 watts should not be exceeded and if you intend to have long overs using FM, you should keep to a maximum input level of 4 watts or the case will get mightily hot!

The mixer is band pass coupled (L7/L8) to a pre driver stage Q4 which is in turn band-pass coupled (L9/L10) to the driver Q5. Note that space has been provided for a trap (L17/C43) from the base of Q4 to ground. This trap which is expected to be necessary

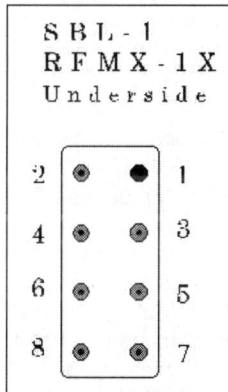

Fig 9.112: Pin connections for the mixer U1 used in the 4m transverter

Input Level	Required Attenuation	R30
1W	30dB	820
4W	36dB	1.5K
10W	40dB	2.7K

Table 9.13: Attenuator values for 4m transverter

C3, 4, 6, 8, 13, 14 ,26, 27, 32, 33, 38, 36, 37, 39, 40, 41, 44	0.1µF Multi layer ceramic		R9, 28		22k
			R14		33k
			R1, 5		100k
			R2, 6		330k
C7	1pF/1.5pF		Q3, 4		2N2369
C23, 29	2.7pF		Q1, 2		BF981
C9	3.3/3.9pF		Q5		2N3866
C18	5.6pF		Q6		2SC1971
C42	10pF		Q7		BC548C
C16	22pF		U1		DB Mixer SBL-1
C15	33pF		D1, 2, 3, 4, 5		1N4148
C17, 19	47pF		ZD1		Zener 9V1
C5, 10, 12, 28, 30	56pF		RL1, 2		DPCO Relays
C2, 22, 24	82pF		L1, 2, 3, 7, 8, 9, 10		Toko style MC119 3.5T
C31	120pF		L4		Toko style MC119 15.5T without can
C11, 20, 21, 25, 35	150pF				
C1	220pF				
C34	22 F 25V		L5, 6		Toko style MC119 9.5T without can
VC1, 2,3,5	Variable 22pF				
VC4,6,7	Var 5-60pF		10mm screening cans		
R24,25	1R0 2W		L11, 12, 13, 14, 15, 16		Airspaced coils
R21	10R				
R20	22R		XT1		Crystal, see text
R29	47R 20W		Enclosure AB10		Maplin LF11M
R31	56R		Heatsink		Farnell 170-071
R4, 8, 12, 16	100R		Sockets BNC		
R11	150R		Misc. Nuts bolts wire		
R3, 7, 17	220R		M3 x 6mm PH Screw		
R22	470R		M3 x 10mm Spacer		
R30	1k5		M3 x 10mm Screw		
R18, 23, 27	2k2		Nuts + washer + spw		
R10, 15, 19, 26	10k		Feet		

Table 9.14: Parts list for 4m transverter

only if a 144 or 145MHz IF is used should be resonated at the LO frequency (74/75MHz). With a 28MHz IF and no trap, the local oscillator output level was measured at about 47dB below full output power which was felt to be acceptable. As in the receive part of the transverter, split capacitance is used (C21/C22, C24/C25 and C30/C31) to effect impedance matching.

The Driver and PA devices were chosen on the basis of price and availability. The PA is characterised for FM (Class C) operation (5W at 150MHz) but it was felt that there was a good chance that it could be made to operate satisfactorily in Class B and that would allow the transverter to be used for SSB (and AM) if required.

L1,2,3,7,8,9,10	Cirkit	35-11934		
L4	Cirkit	35-13415		
L5,6	Cirkit	35-13492		
L11	5t	22SWG	6mm dia close wound	
L12	6t	22SWG	6mm dia	- ditto -
L13	1t	22SWG	6mm dia	- ditto -
L14	7t	22SWG	6mm dia	- ditto -
L15	7t	22SWG	6mm dia	- ditto -
L16	6t	22SWG	6mm dia	- ditto -
L17	2m IF version only.			
LF1,2,3	6t	22SWG	6mm dia close wound	

Table 9.15: Coil winding data for 4m transverter.

The first prototype appeared to provide about 4W and this power level was considered entirely adequate for the local activity that this project was designed to stimulate.

Subsequent improvements to the layout yielded the full 5W. For those wanting higher power, the layout provides for connection of an external power amplifier within the RF switching provided. Please note, however, that relays with a higher contact current rating may be needed to perform the DC switching of such a PA as their absolute maximum limit is 2A DC and in practice this should not be closely approached.

Space on the PCB has been left free for additional output filtering which it was thought may be necessary to reduce the harmonic output from the -35dB level observed on the prototype. Although not absolutely necessary at the 5 watt level, it would nonetheless represent good practice and tests suggest that this filter reduces harmonics to about 60dB down on the wanted signal at full power. Tests on a sample filter suggest a loss of about 0.85dB reducing the output from 5W to 4W. In real terms this is insignificant. If a PA is to be fitted a filter is essential and should be fitted to its output.

PCB assembly commences with the smallest components first as placement of the larger components makes it difficult to reach and inspect the smaller ones. Start with the pins, whose positions are marked on the PCB component layout overlay as circles, the PCB is shown in **Figs 9.113 and 9.114** in Appendix B. Push them firmly into place and this is most easily done before any other components have been fitted. The ridges on the pins hold them in place until soldering which need not begin until almost all the components are fitted. Leave the fitting of the following components until much later: MOSFETS Q1 and Q2. The power resistor R29 is not fitted until the unit is assembled in its box, the parts list for the transverter is shown in **Table 9.14**.

When fitting components, pre-form all leads so that the component will sit as close to the PCB as possible except for the self supporting coils which should sit 2mm above the board. Some capacitors such as C3, 7, 8, 9 must be pre-formed to a slightly wider pitch (than 0.1in) as conductors pass between their pins and more clearance is required. Don't cut the leads until you have fitted the component and bent its leads over to about 45 degrees from the board. Leave 1.5-2mm protruding. This will hold them in place until soldering which should not start until all the components except the MOSFETS, Inductors and R29 have been fitted. Note that the PCB layout shows VC5 as a large capacitor, in fact a small (green) one should be fitted in this position. The circuit diagram and parts list are correct. Note that Q5, the driver transistor should be mounted on a TO5 transipad.

After fitting the small components, carry out a careful inspection to confirm that everything is correctly placed before soldering. Carefully inspect your soldering to ensure that all the joints are good. It is helpful to remove all surplus flux using a PCB cleaning solvent and a stiff brush (such as a half-inch paint brush with its bristles cut short) before inspection if possible as surplus flux has been known to mask a bad joint. When all the small components have been fitted it is time to fit the pre-wound inductors. These should be fitted one at a time and one pin 'tack soldered' merely to hold them in place. Note that L17 and C43 are only required if a 2m IF is used.

Next, the self supporting coils should be wound using a mandrel such as a drill bit with the correct diameter (6mm). All coils are specified for close (no) spacing between turns, details for winding the coils are shown in **Table 9.15**. While each coil is still on its mandrel, carefully scrape off the insulation enamel at the ends, then remove it from the mandrel and pre-tin the ends. Next, pre-form the leads to fit the PCB in the appointed space

The Radio Communication Handbook

observing the orientation as shown on the component layout. Fit the coils so that they lie 2mm above the PCB and bend the leads over at the rear of the PCB by about 45 degrees to hold them in place as you did with the other components. Be sure to observe the correct orientation of the axis of the self supporting coils. The sense (clockwise/anticlockwise) in which they are wound doesn't matter. When all inductors are fitted, inspect the PCB again for correct placement then solder them. With the pre-wound coils in particular, solder the unsoldered pin first to avoid them dropping out of the PCB!

Next fit the wire link LK1 and, if you have opted to fit the low pass filter, place a link from the transmitter output to the filter input which is nearby. If you are not fitting the filter, run a miniature co-axial link from the PA stage to the change-over relay RL2 using additional pins where CF4 would have been fitted.

Next, fit and solder the MOSFETS Q1 and Q2 keeping the leads as short as possible and taking sensible anti-static precautions (see below). First pre-form the leads downwards 2mm from the body so that they will pass through the holes on the PCB.

Fit 'tails' of 22SWG Tinned copper wire about 1in (25mm) long to the pins adjacent to the IF socket and 50mm long to the ANT and GND Pins adjacent to RL2 by making one turn tightly around the pin then soldering. Bend these tails up at right angles to the PCB. They will be used to connect to the IF and antenna sockets when the PCB is fitted in its box.

Finally, make a trial assembly of the PCB into its case, using the spacers provided and fit the IF BNC socket. Leave the other BNC socket off at this stage as it will prevent removal of the PCB from the box once fitted. Next, fit R29 in place, bolt it firmly in position and 'tack solder' its leads on the top of the PCB. Remember that it will need to be un-bolted from the box when the PCB is removed. Now remove the PCB from the box and trim and solder R29's leads on the underside of the PCB. It is recommended that initial testing is carried out before final assembly into the box.

To test the transverter you will need a multi-meter with an input impedance of 20kΩ/V or better and either an oscilloscope with an input sensitivity of from 10mV per division (bandwidth immaterial) or a millivolt meter. A frequency counter, a general coverage receiver and a 70MHz signal source are also helpful, but not essential. In addition to the basic equipment mentioned, you will need two very simple and useful pieces of test equipment, an RF 'sniffer' and a diode probe to turn your oscilloscope or millivolt meter into a wideband RF level indicator.

Before final assembly into the box, connect a red 14/0.2mm lead to one of the pins at either end of LK1 and a black lead to any convenient point on the ground plane. Carefully apply 13.8 volts DC from a current limited supply. If possible limit to 100mA to avoid damage if there are any serious errors in the construction. If you do not have a current limit on your supply or if it cannot be set to 1 amp or less, connect the supply through a resistor of 47 to 100 Ohms.

With a voltmeter having an input impedance of 20kΩ/V or more, set to a range which extends to at least 15 volts, check the following points:

Local Oscillator supply.	TP1	12V
Local Oscillator Emitter.	TP2	1.8V
Q1 Source	TP3	0.55V
Q2 Source.	TP4	0.8V

With a diode sniffer, tune L5 for maximum deflection, then tune L6. There will be slightly less deflection for L6 but the meter should move significantly. These two circuits interact so repeat the process. Set VC1 so that it is about half engaged and re-tune

L5 for the correct frequency. If you have a frequency counter, couple it loosely to L6 using a single or two turn loop and re-tune L5 for the correct crystal frequency. If you do not have a counter don't worry as the oscillator is unlikely to be far out and can be trimmed to frequency using an off air signal later. For fine trimming adjust VC1.

Now increase the current limit on your power supply to 500mA, or reduce the series resistor to about 22 ohms and connect the EXT PTT input (R26) to the +ve supply. You should hear the relays click. Now check the following voltages:

TX Supply	TP5	13V
Q4 Emitter	TP6	1.8V
Q5 Emitter	TP7	0.9V
Q6 Emitter	TP8	0.3V

Satisfactory results to these tests give us some confidence that resistors and semiconductors are correctly placed and there are no disastrous short circuits between tracks!

As taking the PCB out of the box to correct errors is very tedious, it was found best to run through the tune-up procedure with the PCB out of the box first and carry out a final re-tune later after fitting the fully tested PCB into the box.

First, however, fit the 10mm spacers in the box and screw the PCB in place temporarily, Next, fit R29 in place and bolt it firmly to the box before soldering it in place. Now un-bolt R29 and remove the PCB for initial set-up and test. Fit 'tails' of 22SWG tinned copper wire to the IF and Antenna pins and adjacent ground pins then connect the loose BNC sockets to these tails with the ground 'tails' soldered to the large solder tags supplied with the sockets.

Having already tuned the oscillator, the front end, mixer and IF stages (L1, L2 and L3 and L4) can be tuned by connecting a receiver (or transceiver) to the IF socket, and using a strong local signal.

Before tuning the transmit section, you should adjust the output of the driving transceiver to a level compatible with the attenuator R29, R30 and R31. If you are using a CB transceiver, set it to the low power (0.4W) power setting.

The input level to the transmit mixer is just 0dBm (1mw). If you are using an HF transceiver which might be capable of 100 watts or more, test the power level into a dummy load first. Once you have got the level about right, a simple way to check it is by measuring the voltage at TP9. For 4W input it should be about 15-20V and for 0.4W about 6-8V. Remember however that if you apply much too much power you could damage the transverter and you definitely will produce a poor quality signal.

With enough power applied to operate the relays and switch the transverter to transmit, work through the tuned circuits in the transmit path using the diode probe and an oscilloscope, millivolt meter or a 50µA meter movement. Start at the junction of L7, C22 and C23 and tune L7 for maximum. Next move on to the base of Q4 and tune L8 for Maximum. Repeat the process for the base of Q5, tuning L9 and L10 for maximum.

After this, sufficient power should be present in the self supporting coils to give a reading on the RF sniffer. Hold its loop near the circuit to be tuned and tune for 'maximum smoke'. At this stage, a dummy load should be fitted to the output. If you have a power meter, put this in the output circuit. To tune the driver and PA stages a number of variable capacitors need to be tuned and some will interact with others so you should expect to adjust each several times to achieve maximum output. Next, check that the output is at the correct frequency within the 4m band ie between 70.000 and 70.500MHz. This can be done using a frequency counter and a wavemeter, taking a wide sweep on the later to ensure there are no measurable spurious

products. Also, check that the output falls to zero when the drive is removed while the transverter is held in transmit mode using the PTT. If it does not, there is self oscillation. Under no circumstances transmit until this problem is solved!

Assuming all is well, you are ready to go on the air for a test.

BIBLIOGRAPHY

Books

Amateur Radio Operating Manual, 6th edition, Don Field, G3XTT, RSGB.

The VHF/UHF Handbook, Dick Biddulph, M0CGN, RSGB

Guide to VHF/UHF, Ian Poole, G3YWX, RSGB

VHF Contesting Handbook, RSGB VHF Contest Committee, RSGB (Members only)

UHF/Microwave Projects Manual, Volumes 1 & 2, ARRL

UHF Compendium, Parts I, II, III and IV, editor K Weiner, DJ9HO, available from UKW Berichte. E-mail: info@ukwberichte.com or from K M Publications (E-mail: andy@vhfcomm.co.uk)

Magazines

CQ-TV, BATC Publications, Fern House, Church Road, Harby, Notts, NG23 7ED. E-mail: publications@batc.org.uk. http://www.batc.org.uk

Dubus magazine, available in the UK from G4PMK (QTHR). E-mail: dubus@marsport.demon.co.uk

RadCom, RSGB. This is the RSGB members' monthly journal; it has a monthly VHF operating column and a bi-monthly microwave column. It also carries occasional VHF/UHF and microwave articles. http://www.rsgb.org

VHF Communications magazine, K M Publications, 63 Ringwood Road, Luton, Beds, LU2 7BG. E-mail: andy@vhfcomm.co.uk. http://www.vhfcomm.co.uk

Component or Kit suppliers

Farnell Electronic Components Limited, Canal Road, Leeds, LS12 2TU, Tel: 0870 1200 200, Fax: 0870 1200 296. http://www.farnellinone.co.uk

Maplin Electronics, P.O. Box 777, Wombwell, S73 0ZR, Tel: 0870 429 6000, Fax: 0870 429 6001. http://www.maplin.co.uk

Mini-Circuits Europe, Dale House, Wharf Road, Frimley Green, Camberley, Surrey, GU16 6LF, Tel: 01252-832600. Fax: 01252-837010. http://www.minicircuits.com

Walters & Stanton, 22 Main Road, Hockley, Essex, SS5 4QS, Tel 0800 737388. http://www.wsplc.com

Quasar Electronics Ltd, PO Box 6935, Bishops Stortford, CM23 4WP, Tel: 0870 246 1826. http://www.quasarelectronics.com

Sycom, PO Box 148, Leatherhead, Surrey, KT22 9YW, Tel: 01372 372587. http://www.sycomcomp.co.uk

REFERENCES

[1]   'Gallium arsenide FETs for 144 and 432MHz' John Regnault, G4SWX, *Radio Communication*, Apr 1984

[2]   These designs are shown on the web site: www.qsl.net/yu1aw/. The author, Dragoslav Dobricic, can be contacted on: dobricic@eunet.yu

[3]   This design can be found on the homepage of the author, Ole Nkkjaer, OZ2OE: http://www.qsl.net/oz2oe

[4]   'Low noise aerial amplifier for 144 MHz', *Radioamater*, Dragoslav Dobricic, YU1AW, 10/1998, pp 12-14 (part I) and 11/1998, pp 12-15 (part II). Also: 'Low noise aerial amplifier for 144 MHz', KKE Lecture text, Dec 1998

Choosing Commercial Equipment for the VHF/UHF Bands

When choosing equipment, everyone will naturally have a budget in mind. But within this budget, it's often totally useless in asking "what's the best transceiver to buy?". This is because each person will have his or her own particular interests and operating restrictions, or indeed lack of them. For those just starting out on the hobby, the best advice would be to go for a general coverage multimode receiver initially, possibly a second-hand model so that its resale value, for when eventually 'trading up', isn't compromised too much. Once licensed, or on the verge of getting one, a transceiver would be a better bet, but in this case it is useful first to decide where your interests primarily lie and then your operating possibilities.

If you're very restricted in terms of the size and height of antennas you can erect at home, yet want to work long distances, then aim to get a transceiver that can also be used in portable and mobile situations, such as from a local elevated site using either a mobile antenna or a temporary portable mast-mounted antenna. If however you're only interested in operating through your local 2m or 70cm repeater, a simple single-band or dual-band FM transceiver will suffice. But beware, once the 'bug' bites, you may quickly want to move on to bigger and better things.

One of the best improvements you can make to your station is to get a better antenna system; it's little use having a £5000 transceiver operating into a compact mobile whip mounted in the loft or on a window ledge when you've a garden or rooftop at your disposal. You'd usually be a lot better spending somewhat less on the transceiver and rather more on improving the antenna system, provided of course that you can physically get something more substantial erected without falling foul of planning laws or the neighbours!

Take some time to sit back and think what you're interested in and where and who you'd like to contact. If you enjoy a challenge, then either low-power 'QRP' operation, possibly with a home-built transceiver (but again with the very best antenna system you can manage) is one way to go. Another could be the challenge of gaining operating awards and competing against other amateurs around the world, in which case you'll need the utmost in transceiver performance and transmit power level, again not forgetting a good antenna system. If however you're keen to use the hobby more as a social outlet and technical discussion scenario, typically having 'raghews' with local amateurs and exchanging information, then an appropriate transceiver system that's easy to operate from a comfortable position, without scores of tiny knobs and buttons on the front panel, would be more appropriate.

If you'd like to be able to have a wide choice of operating bands, modes and facilities, there are a growing number of reasonably priced multimode transceivers offering HF, VHF and UHF all combined in a compact portable 'carry around' unit which can also be used at home and in the car. Once you get more experienced and you possibly find your interests lie in a more defined aspect of the hobby, a transceiver and antenna system dedicated to this can be obtained, with your earlier transceiver being used ether as a 'trade-in' or as a secondary transceiver for other modes and bands.

Whichever transceiver you go for, bear in mind one final but very important point. Even if you're not considering integrating computer control into your station right now, do make sure that whichever transceiver you eventually choose for your base station or even portable / mobile station has this facility built in. There will eventually come a day when you'll want to at least try this, and there is a plethora of very powerful freeware and shareware programs available for download or on CD/DVD. These offer everything from simple transceiver frequency control right up to complete station management including automatic log keeping (taking the frequency, mode etc automatically from your transceiver), QSL card printing (again with time, date, callsign etc. entered for you), DX Cluster handling and automatic QSY to the frequency of that rare station on your 'wanted' list, automatic antenna rotor control to the identified station, satellite tracking and automatic Doppler shift correction, even down to automatic generation of award sheets for you after a contest.

[5]   'Low Noise aerial amplifier for 144MHz', Dragoslav Dobricic, YU1AW, *CQ ZRS,* Dec 1999, pp 26-31

[6]   'Low noise aerial amplifier for 432MHz', Dragoslav Dobricic, YU1AW, Radioamater, 1/2001 and 2/2001

[7]   'Low noise aerial amplifier for 432MHz', Dragoslav Dobricic, YU1AW, *CQ ZRS,* 6/2000, pp 27-31

[8]   The Agilent homepage: http://www.home.agilent.com. The datasheet for the ATF-54143 is: http://cp.literature.agilent.com/litweb/pdf/5989-1922EN.pdf

[9]   http://www.btinternet.com/~geoffrey.brown3/4CX250B.html

[10   '432MHz Linear PA using 3 x 2C39BA', Dragoslav Dobricic, YU1AW, *VHF Communications Magazine,* 4/1988, pp 233 - 237

[11] *Surplus 2-Way Radio Conversion Handbook,* Chris Lorek, pp 85 - 93

[12] 'Transistorised power amplifier for 144MHz', Dragoslav Dobricic, YU1AW, *Radioamater,* 2/1988, pp 34-37, *Radioamater,* 3/1988, pp 66-68

[13] '144MHz direct-conversion receiver with I/Q outputs for use with software defined radio', Andy Talbot, G4JNT, *RadCom,* Nov 2004, pp 102 - 103

[14] 'AD9850 DDS Module', *RadCom,* Nov 2000

[15] *RadCom* series on SSB phasing net works, Feb to June 2004

[16] http://www.sdradio.org

[17] Update firmware available from the author, Andy Talbot, G4JNT: actalbot@southsurf.com

[18] The transverter is described on the 70MHz organisation web site: http://www.70MHz.org/transvert.htm. The 4m transverter design is on the web site: http://myweb.tiscali.co.uk/g4nns/ARACTVT.html

[19] 'Transverters for the 70, 23 or 13cm tuning areas', 1992 Weinheim Congress proceedings, Wolfgang Schneider, DJ8ES

[20] '28/432MHz Transverter instructions, tips and improvements', 1993 Weinheim Congress proceedings, Wolfgang Schneider, DJ8ES

[21] '28/144MHz Transverter', Wolfgang Schneider, DJ8ES, *VHF Communications Magazine,* 4/1993, pp 221 - 226

[22] '28/50MHz Transverter', Wolfgang Schneider, DJ8ES, *VHF Communications Magazine,* 2/1994, pp 107 - 111

[23] '28/432MHz Transverter', Wolfgang Schneider, DJ8ES, *VHF Communications Magazine,* 2/1994, pp 98 - 106

[24] Pye museum, run by G8EPR, www.qsl.net/g8mgk/pye/Pye.htm

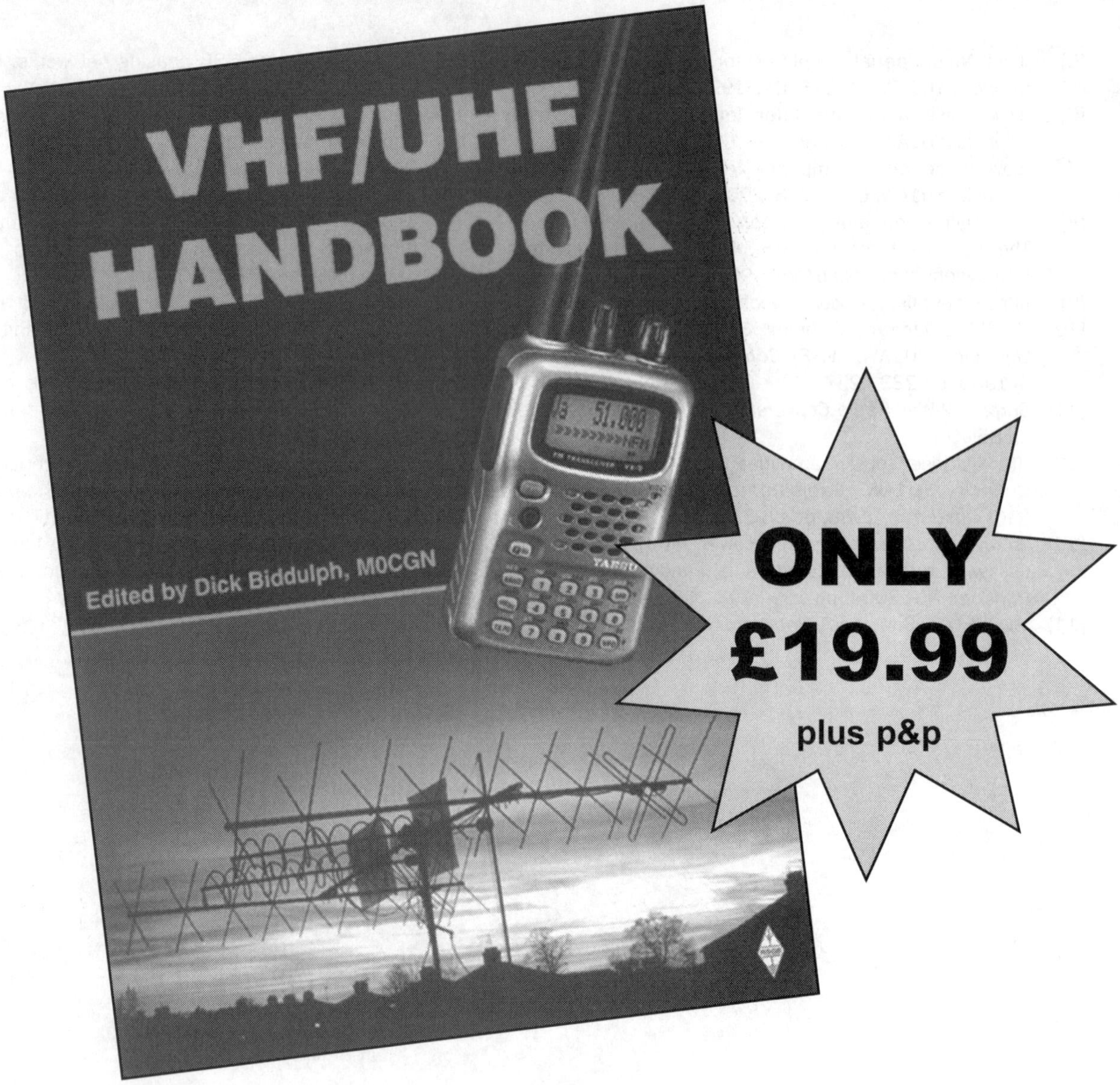

VHF/UHF HANDBOOK

Edited by Dick Biddulph, M0CGN

ONLY £19.99 plus p&p

The VHF/UHF Handbook
Edited by Dick Biddulph, M0CGN

This guide to the theory and practice of amateur radio reception and transmission on the VHF and UHF bands gives the reader the background to such essential topics as antennas, EMC, propagation, receivers and transmitters, together with constructional details of many items of equipment. As most amateurs today use commercial transceivers, the emphasis is on accessories and add-ons which are relatively simple to build.

Specialised modes such as data and television are also covered, making this handbook one of the most complete guides around for VHF/UHF operators.

RSGB, paperback, hundreds of illustrations and photos.
Size: 272 x 199mm; 320 pages; ISBN: 1-872309-42-9
RSGB Members Price: £16.99 plus p&p

E&OE

Radio Society of Great Britain
Lambda House, Cranborne Road, Potters Bar, Herts. EN6 3JE
Tel. 0870 904 7373 Fax. 0870 904 7374

ORDER 24 HOURS A DAY ON OUR WEBSITE
www.rsgb.org/shop

10 LF: The 136kHz Band

The 136kHz band, 135.7kHz – 137.8kHz, introduced in January 1998, is one of the most recent additions to the amateur radio spectrum, and is unique in being in the LF frequency range (Low Frequency, defined as 30kHz – 300kHz). As the only true LF amateur band, it has several unique characteristics. LF propagation is different from any other band, being mostly ground wave or D-layer. Due to the narrow bandwidth available (a total of only 2.1kHz for the whole band), and the high noise levels on the band, several specialized operating techniques have been developed specifically for 136kHz [1, 2]

LF RECEIVING

The majority of LF stations currently use commercially available receivers. Many amateur HF receivers and transceivers include general coverage that extends to 136kHz, and in many cases these can be successfully pressed into service. However, unlike HF reception where reasonable results are often achieved simply by connecting a 'random' wire antenna to the receiver input, successful LF reception is a bit more difficult, for a number of reasons. First, there is usually a very large mismatch between the impedance of a wire antenna at LF and the typically 50Ω receiver input impedance, which leads to a large reduction in signal level. This is often exacerbated by degraded receiver sensitivity at LF, particularly in amateur-type equipment. Secondly, amateurs with their relatively tiny radiated signals share the LF spectrum with vastly more powerful broadcast and utility signals (**Fig 10.1**). Unless effective filtering is provided, this results in severe problems with overloading at the receiver front end. Fortunately, very satisfactory results can usually be achieved by using quite simple antenna matching, preamplifier, and preselector arrangements, as will be seen later in this section.

Receiver Requirements

These depend on the type of operation that is envisaged. Adequate sensitivity is obviously required; the internal receiver noise level should be well below the natural band noise at all

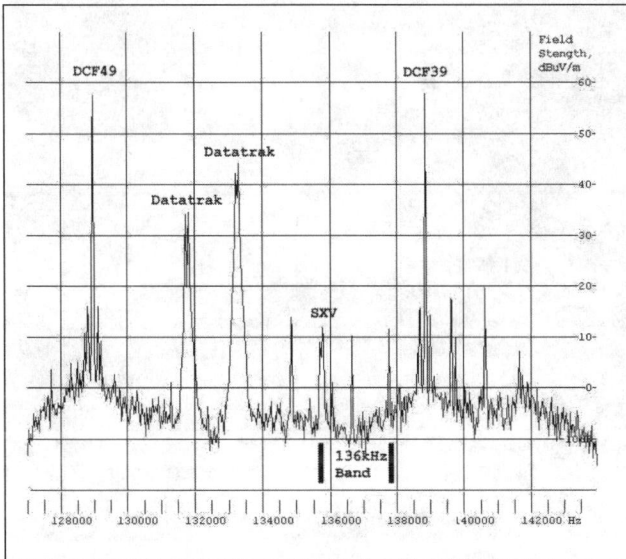

Fig 10.1: LF spectrum in the vicinity of 136kHz

times. As a guide, similar figures to those for HF receivers (a few tenths of a microvolt in a CW bandwidth) will suffice. If a large, transmitting-type antenna is used, the signal level will be high enough to allow a receiver with considerably poorer sensitivity to be used. Small loop and whip receiving antennas usually require a dedicated high gain, low noise preamplifier.

The most widely used operating mode on 136kHz is CW. Because the 136kHz band is only 2.1kHz wide, a CW filter is almost essential. A 500Hz bandwidth is adequate, but 250Hz or narrower bandwidths can be used to advantage. Due to the strong utility signals just outside the amateur band, good filter shape factor is important since the utility signals can be 60dB or more above the level of readable amateur signals, with a frequency separation of less than 1kHz. For specialised extremely narrow-bandwidth modes such as QRSS (see the chapter on Morse), selectivity is also provided at audio frequencies using DSP techniques in a personal computer, but good basic receiver selectivity is still desirable to prevent strong out-of-band signals entering the audio stages.

Frequency stability requirements depend on the operating mode. For CW operation, maintaining frequency within 100Hz during a contact is not usually a problem. Narrow band modes such as QRSS require better stability, and frequency setting accuracy. For the popular QRSS3 operating speed (widely used for DX contacts around Europe), an initial setting accuracy, and drift of perhaps 10-20Hz per hour is acceptable if somewhat irritating. This level of stability can often be achieved by older receivers with mechanically tuned VFOs, provided the receiver is allowed to reach thermal equilibrium, and some means of calibrating the receiver frequency is available. These difficulties are eliminated in fully synthesised receivers, which generally exhibit a setting accuracy within 1 or 2 Hertz and drift a fraction of a Hertz over an extended period. This degree of frequency stability is adequate for the vast majority of LF applications, including the reception of inter-continental beacon signals using QRSS30, QRSS60 and QRSS120 speeds, with bandwidths of as little as 0.01Hz. For some specialised narrow band operating modes, a higher order of stability is required. This is achieved by using high stability synthesiser reference frequency sources, such as TCXOs, OCXOs, and even atomic clock or GPS derived frequency standards.

Amateur Receivers and Transceivers at LF

Many amateur HF receivers and transceivers can tune to 136kHz, and since they are already available in many shacks, probably a majority of LF amateur stations use receivers of this type. All modern equipment is fully synthesised, so frequency accuracy and stability are good. Older receivers using multiple crystal oscillators or an interpolating VFO have relatively poor stability. Modern crystal or mechanical CW filters have excellent shape factors, giving good rejection of strong adjacent signals. In some receivers, multiple filters can be cascaded, giving further enhanced selectivity.

Since LF reception is generally included as an afterthought, manufacturers rarely specify sensitivity of amateur receivers at 136kHz. Unfortunately it can often be poor. There is little relation between the HF performance, cost, or sophistication of a particular model, and the sensitivity at LF. Therefore it may well be that

Fig 10.2: The Kenwood TS-850 is the most popular amateur bands transceiver used to receive on 136kHz

older, cheaper models perform better at LF than their newer successors. Few laboratory-quality sensitivity measurements are available for the LF sensitivity of amateur transceivers and receivers, but the following lists some models which have been used successfully as LF receivers. Classified as "good" are the Kenwood TS-850 (probably the most popular HF rig with LF operators - **Fig 10.2**), TS-440 and Yaesu FT-990 transceivers. Receivers include the AOR 7030, Icom R-75, JRC NRD345, NRD525 and NRD545, and Yaesu FRG100. Classified as "adequate" are the Icom IC-706, IC-718, IC-756Pro, IC-761, IC-765, and IC-781, Kenwood TS-140, TS-870 and Yaesu FT-817 and FT-1000MP. Equipment classified as "adequate" requires either a large antenna and/or an external preamplifier to achieve enough sensitivity. The IC-718 is fairly typical in this respect, requiring 1μV to achieve 10dB SNR in a 250Hz bandwidth, a figure about 20dB poorer than it achieves in the HF bands.

The reason for poor sensitivity lies within the receiver front-end design. The inter-stage coupling components, in particular the first mixer input transformer, are optimised for operation at HF, and often have high losses at LF, reducing the signal level. Internally generated synthesiser noise may also be higher at LF. The front end filter used at LF is normally a simple low-pass filter with a cut-off frequency of 1 – 2MHz, often including an attenuator pad to reduce overloading due to medium wave broadcast signals; this further reduces sensitivity, without eliminating the broadcast signals. Some LF operators have improved receiver performance substantially by replacing the mixer input transformer with one having extended low frequency response [3]; this component must be carefully designed if receiver HF performance is to be maintained. A simpler and more common approach is to use an external preamplifier, and provide additional signal frequency selectivity, as described later in this section.

Commercial Equipment for LF

Many professional communications receivers, made by such firms as Racal, Plessey, Harris, Collins, Eddystone, Rohde & Schwarz, include coverage of the LF spectrum; surplus prices are often competitive with the amateur-type equipment discussed above. Ex-professional equipment is usually fully specified at LF frequencies, so sensitivity and dynamic range are usually good at 136kHz. Fully synthesised professional receivers often have precision reference oscillators with excellent stability; they also usually have inputs for an external frequency reference. These features are not often found on amateur-type equipment, making them attractive if the more specialised LF communications modes are to be explored. A drawback is that affordable examples are usually fairly old, so servicing may be required. Also, they have a rather basic feel, with few of the 'bells and whistles' operator facilities found on modern amateur rigs. The Racal RA1792 (**Fig 10.3**) has been popular with UK amateurs on LF. The older RA1772 also performs well.

Fig 10.3: Racal RA1792 (top), and RA1772 perform well on LF

A number of vintage receivers, including the HRO, Marconi CR100, AR88LF, cover the 136kHz range. Also, valve-era equipment designed for marine service often includes LF coverage. A few amateurs have used vintage receivers for 136kHz operation. The antenna input circuit of this type of equipment is generally designed to be operated at LF using an un-tuned wire antenna, so usually gives good sensitivity at 136kHz without requiring additional antenna tuners or preamplifiers. The major disadvantage of most vintage receivers is that their single-pole crystal filters have poor skirt selectivity compared to modern IF filters. This results in strong utility signals several kilohertz from the receive frequency reaching the IF and detector stages of the receiver, causing blocking and heterodyne whistles which swamp the weak amateur signals. Unmodified vintage receivers are therefore usually poor performers at 136kHz, although for the experimentally minded, the addition of a modern IF filter and product detector could result in an effective LF CW receiver.

Selective level meters (SLMs), also called selective measuring sets or selective voltmeters, are instruments designed for measuring signal levels in the now-obsolete frequency division multiplex telephone systems; consequently, they are sometimes available surplus at low cost. SLMs are designed for precision measurement of signals down to sub-microvolt levels; their fre-

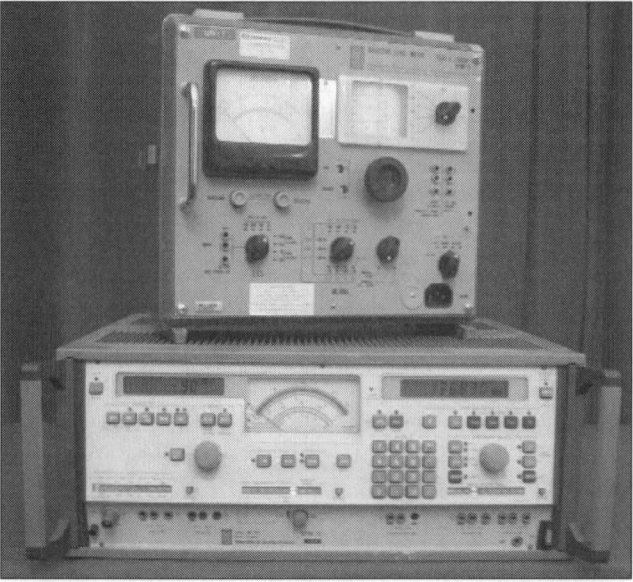

Fig 10.4: SPM-19 (Bottom), and portable SPM-3 (top) selective level meters can be used as LF receivers

Fig 10.5: Receive antenna tuning circuits for 136kHz: (a) simple series inductor; (b) series capacitor and (c) parallel capacitor

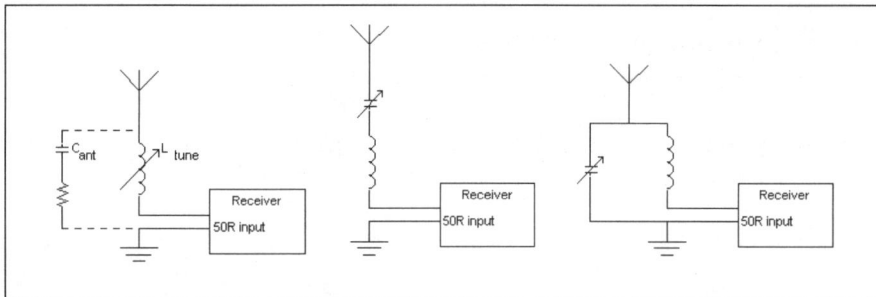

quency range extends from a few kilohertz into the MF or HF range, so can make effective LF receivers. Well-known manufacturers are Hewlett-Packard (HP3586) and the German companies Wandel and Golterman (the SPM- "selektiver pegelmesser" series; **Fig 10.4**) and Siemens.

SLMs are not purposely designed as receivers and do not have many normal receiver features, such as AGC and selectable operating modes, or sometimes even an audio output. Filter bandwidths are designed for telephony systems and are not always suited for LF operating modes. Normally the CW audio pitch is fixed at around 2kHz, so they are not well suited to CW operating, although this presents no obstacle for sound card operating modes. The area where SLMs excel is in LF signal measurement; they have been used by a number of amateurs for LF field strength measurements (see LF Measurements and Instrumentation section). They are often available with a tracking level generator, which is very useful for measurements on filters, or bridge-type impedance measurements.

Receive Antenna Tuning

The impedance of a typical long-wire antenna at LF can be modelled as a series resistor and capacitor. Taking the example in the Transmitting Antennas section of a typical long-wire antenna, the capacitance might be 287pF, in series with 40Ω. Assuming receiver input impedance of 50Ω, the SWR at the feed point of the antenna is about 8200 to 1! This mis-match results in an unacceptable signal loss of about 32dB. Most of the loss is caused by the capacitive reactance; signal levels can be greatly improved by resonating the antenna at 137kHz with a series inductance. In the example above, the antenna with resonating inductance form a tuned circuit with Q around 40 and bandwidth of only a few kilohertz, which is very effective in filtering out powerful broadcast band signals. The practical effect of resonating the antenna is dramatic. Normally with a long-wire antenna connected directly to the receiver, the only signals heard in the 136kHz range are numerous intermodulation products. With the antenna tuned, these disappear and the band noise is audible above the noise floor of reasonably sensitive receivers.

Fig 10.5 shows typical circuits used to tune wire antennas for reception. **Fig 10.5(a)** is a simple series inductor; the value required is:

$$L_{tune} = \left(\frac{1}{2\pi f \sqrt{C_{ant}}} \right)$$

with L_{tune} in henries, C_{ant} in farads and f in hertz.

A useful rule of thumb is that the antenna capacitance C_{ant} will be roughly 6pF for each metre of wire, typically L_{tune} of a few millihenries will be required. Because of the high Q the inductance must be adjustable; this can be done using the same techniques as for transmitting antennas (see later), or a slug-tuned coil can be used for receive-only.

It is often more convenient to use a fixed inductor, and adjust to resonance using a variable capacitor, as shown in **Fig 10.5(b)**. This can be a broadcast-type variable, with both sections paralleled to give about 1000pF maximum. A higher tuning inductance is required to make up for the overall reduction in capacitance. The shunt-tuned circuit of **Fig 10.5(c)** has one side of the tuning capacitor grounded. The impedance match will not be quite as good, although normally perfectly adequate. Capacitor tuning is not suitable for antennas carrying transmit signals because of the very high voltages encountered.

LF Preamps

To overcome reduced LF sensitivity, many amateur-type receivers require a preamplifier. Because of the strong broadcast signals in the LF and MF frequency ranges, it is important that adequate selectivity is provided at the signal frequency. To obtain good signal to noise ratio, it is also necessary to pay attention to impedance matching between antenna and preamplifier.

A useful and well tried design from G3YXM is shown in **Fig 10.6**, and is available as a kit with PCB from G0MRF [4]. It incorporates a double-tuned input filter which provides a bandwidth of around 3kHz centred on the amateur band. The preamp is designed for 50Ω input impedance, so antenna matching as described in the previous section will normally be required.

The LF antenna tuner/preamp circuit of **Fig 10.7** combines the antenna matching, filtering and preamplifier functions. It is quite flexible and can be used with a wide range of long-wire and loop antenna elements. It has been used successfully with an IC-718 transceiver, which has fairly poor sensitivity at 136kHz.

The preamp is a compound follower, with a high-impedance JFET input, and a bipolar output to drive a low impedance load

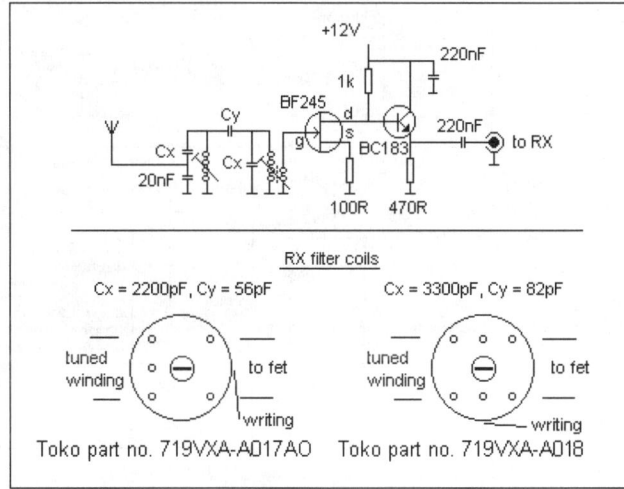

Fig 10.6: G3YXM's 136kHz preamplifier

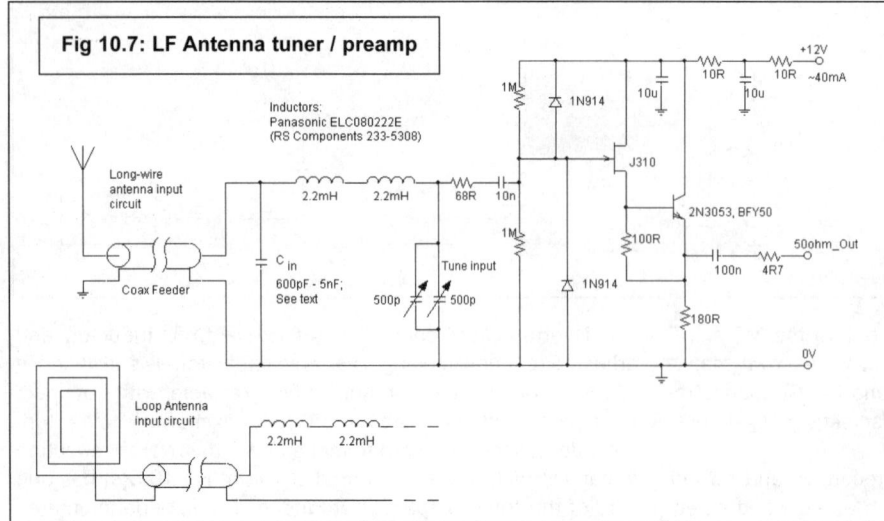

Fig 10.7: LF Antenna tuner / preamp

with low distortion. The gain of the follower is about unity, but the high Q, peaked low-pass filter input circuit provides voltage gain, and also gives substantial attenuation of unwanted broadcast signals at higher frequencies. The gain of the circuit depends on the type of antenna element used, of the order of 10dB with a long wire element and 30 – 40dB with a loop element. The 2.2mH inductors are the type wound on small ferrite bobbins with radial leads, and have a Q around 80 at 136kHz; other types of inductor with similar or greater Q could also be used. For wire antennas, C_{in} should be in the range 600pF – 5000pF, with large values giving a reduced signal level with longer wires, and smaller values suiting short wire antennas. The antenna can be fed with coaxial cable, in which case the distributed capacitance of the coax (about 100pF/m for 50Ω cable) makes up part or all of C_{in}. This allows the receiving antenna to be located remote from the shack, which is often useful in reducing interference pick-up.

This circuit has given good results with wire antennas ranging from a 5m vertical whip to a 55m long wire. For loop antennas, C_{in} is omitted. Satisfactory sensitivity was obtained using a 1m² loop with 10 turns of 1mm² insulated wire, and also with larger single-turn loops with area around 10m² (see the Receiving Antennas section of this chapter). Loop antennas can also be fed via quite long lengths of coax cable.

Converters

For HF receivers without coverage of 136kHz, or those that have very poor performance at LF, a converter can offer excellent 136kHz reception. A number of LF/VLF – HF converters have been manufactured in the past (Datong VLF converter, Palomar VLF-S, VLF-A), although these have mainly been aimed at broadcast reception, and may not give good results in the more demanding circumstances of the 136kHz band. A number of "low frequency adaptors" have been manufactured for professional HF receivers (e.g. Racal RA37, RA137).

The following converter was developed as an 'easy build' design – no coil winding is required, all inductors used are standard value ferrite-cored chokes. No alignment is required, apart from peaking the antenna tuning capacitor in use, and optionally trimming the conversion oscillator to the nominal frequency. A 137kHz input frequency is mixed with a 4MHz crystal oscillator to produce an output frequency of 4.137MHz, which is suitable for many modern receivers with general coverage at HF. The performance is excellent when used with an HF receiver equipped with a good CW filter. The complete circuit diagram is shown in **Fig 10.8**, and component details are shown in **Table 10.1.**.

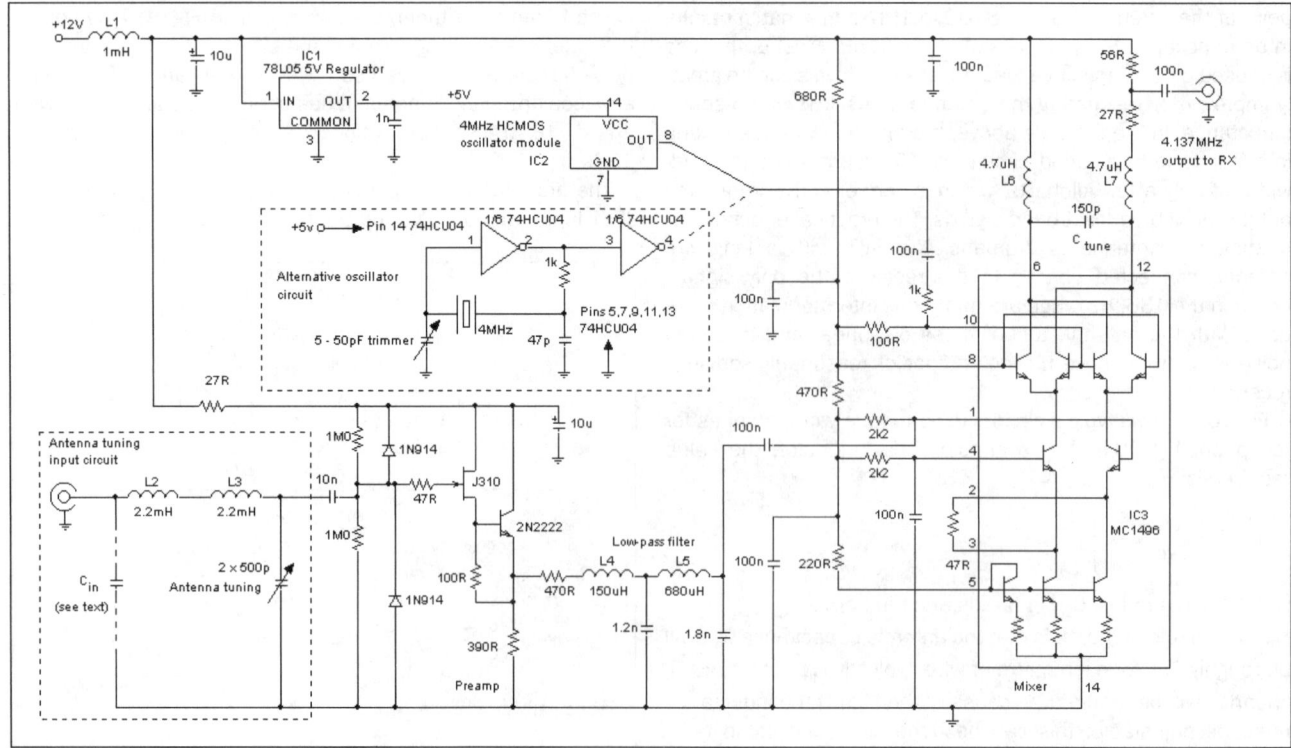

Fig 10.8: 'Easy-build' LF – HF converter

L1	1mH Panasonic ELC08D102E (233-5291)
L2, L3	2.2mH Panasonic ELC08D222E (233-5308)
L4	150µH axial choke Epcos BC type (191-0627)
L5	680µH axial choke Epcos BC type (191-0699)
IC2	14 pin DIL HCMOS crystal oscillator module, 4MHz or 10MHz(see text)
IC3	MC1496 - 14 pin DIP
L6, L7	4MHz oscillator - 4.7µH axial choke Epcos BC type (191-0447)
	10MHz oscillator - 1.8µH axial choke Epcos BC type (191-0419)
C_{in}	600pF - 5000pF, depending on antenna - see text
C_{tune}	150pF ceramic (4MHz oscillator), 68pF ceramic (10MHz oscillator)
The numbers in brackets are RS Components catalogue references [5]	

Table 10.1: Components details for the 'Easy-build' converter

The input tuning circuit is a peaked low-pass filter, identical to that used in the tuner/preamplifier circuit of Fig 10.7 above, and can be adapted for long or short wire antennas, or large loop elements in the same way. The tuned input network provides front-end selectivity, and voltage gain due to the impedance step-up. The voltage follower input stage drives a low-pass filter, which provides additional image rejection and prevents breakthrough of strong signals at the IF frequency.

The low-pass filter output feeds a MC1496 double-balanced mixer IC; a balanced mixer is desirable, since it cancels out the 4MHz oscillator signal which would otherwise be present at the mixer output, and would be likely to overload the following receiver front-end stages. The MC1496 is biased to operate at

relatively high current (about 12mA total), which improves strong signal handling capability. The mixer output is tuned broadly to the 4.136MHz output frequency, and since the Q is only around 4, no tuning adjustment is required. A resistive pad at the converter output reduces interaction with the input preselector of the receiver.

The 4MHz conversion oscillator used in the prototype was a DIP clock oscillator module, generating a logic-compatible 5V p-p square wave output which, attenuated by the potential divider formed by the 1kΩ and 100Ω resistors, is ideal to drive the mixer. The oscillator modules are normally within a few hundred hertz of the nominal frequency. An alternative oscillator circuit using a 74HCU04 un-buffered HCMOS inverter IC with a separate crystal allows trimming to the nominal frequency. If a different IF output frequency is desired, the oscillator frequency can be changed, and the mixer output tuning components L6, L7, C_{tune} scaled appropriately. A suitable value for inductors L6, L7 is (20µH/f), where f is the oscillator frequency in megahertz. The required value of C_{tune} is that required to resonate the total inductance (L6+L7) at the output frequency, minus about 5pF stray capacitance. One prototype of this circuit used a 10MHz oscillator, giving an output at 10.137MHz, falling within the 30m amateur band; this is convenient for an amateur bands only receiver. L6 and L7 in this case were 1.8µH, with C_{tune} 68pF. Oscillator frequencies between about 2MHz and 10MHz are satisfactory; at lower frequencies, image rejection of signals in the medium wave range is likely to be a problem, whilst higher frequencies will lead to increased oscillator drift. Clock oscillator modules are available at several frequencies within this range; the alternative oscillator circuit also worked well with several crystals between 2 and 12MHz.

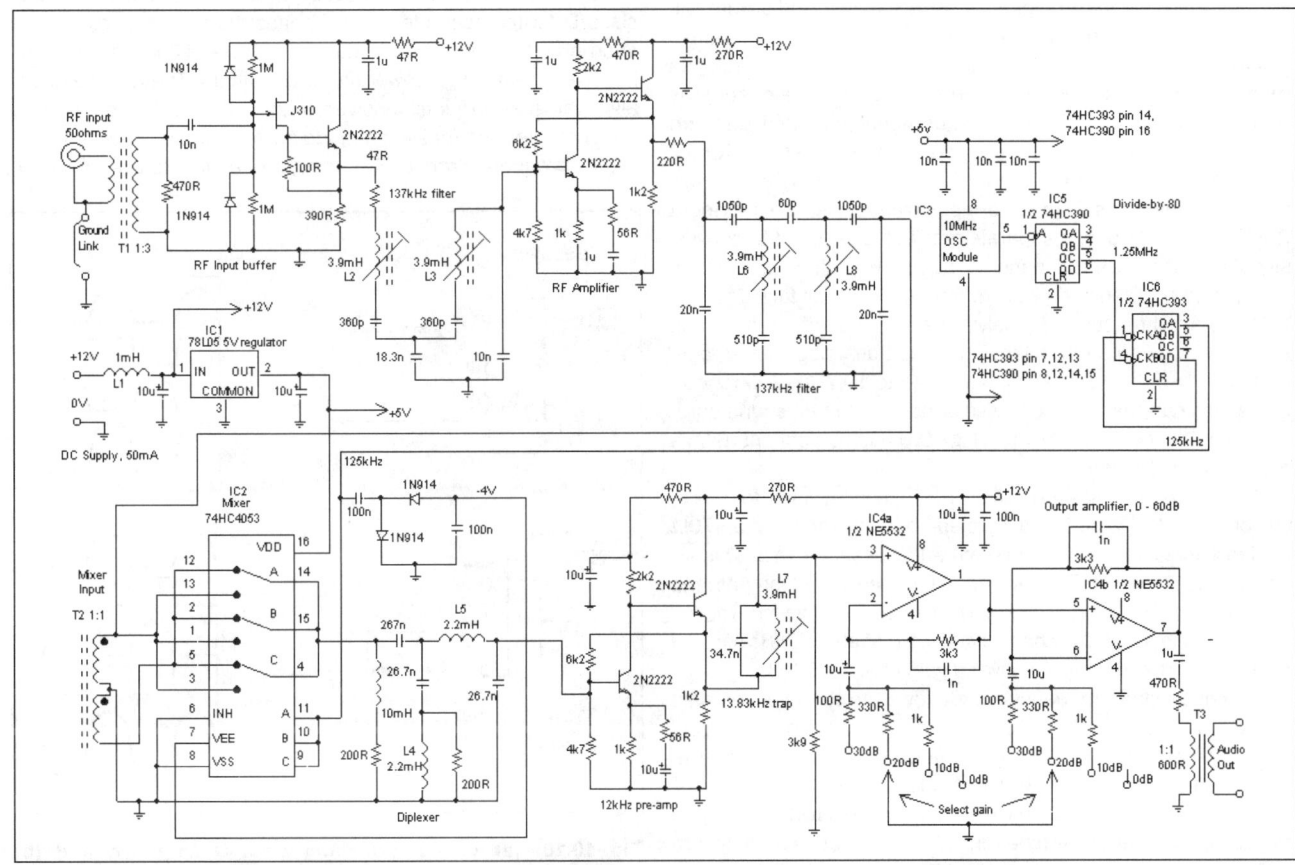

Fig 10.9: Soundcard receiver converter

L1	1mH Panasonic ELC08D102E (233-5291)
L2, L3, L6, L7, L8	120 turns on AL = 250nH/t RM7 adjustable pot core (eg 228-236)
L4, L5	2.2mH Panasonic ELC08D222E (233-5308)
T1	primary 25 turns, secondary 75 turns on RM6 pot core, AL = 2000nH/t (eg 231-8735)
T2	2 x 50 turns bifilar wound on RM6 pot core, AL = 2000nH/t (eg 231-8735)
T3	1:1, 600-ohm audio line transformer
The numbers in brackets are RS Components catalogue references [5]	

Table 10.2: Component details for the soundcard receiver converter

The prototype converters were either built on ground-plane prototyping board, or 'dead bug style' on un-etched PCB board. Setting up involves checking the frequency calibration; tune the receiver to exactly 4MHz, and use the RIT control to centre the oscillator carrier in the receiver passband (or trim the oscillator frequency, if this is adjustable). Having selected an appropriate value of C_{in} (see notes on the LF antenna tuner/preamplifier in the previous section), set the receiver frequency to 4.137MHz, and adjust the antenna tuning capacitor to peak the band noise. The antenna tuning adjustment is very sharp, and will require tweaking when tuning across the band. The frequency drift of the prototypes was of the order of 1 or 2 Hertz over a period of hours, which is adequate for all except the most extreme narrow band modes.

A 136kHz Converter for Sound Card 'Software Receivers'

Most of the specialised modes used on LF utilise a PC sound card as an analogue to digital converter, taking audio from the receiver output, and performing digital signal processing algorithms on the digitised audio. All the functions of an analogue receiver can also be performed in the digital domain, such as frequency conversion, filtering and demodulation, and software is available that can perform these 'Software Defined Radio' functions on the PC. However, the sound card, being essentially an audio device, is limited to input frequencies below about 20kHz. So for LF use, a separate receiver is required to convert signals in the 136kHz band to audio. But since all that is required is frequency down-conversion, the other facilities provided by a full-functioned receiver are not needed, and a much simplified circuit can be used. The circuit described below converts the 135.7 – 137.8kHz input range to 10.7 – 12.8kHz output, which can then be processed directly by the PC sound card. The complete circuit is shown in **Fig 10.9** and component details are shown in **Table 10.2.**

The input buffer uses a similar FET compound follower to the circuit of Fig 10.8. A broadband 1:3 step-up transformer and a 470Ω load resistor is used to provide some voltage gain with a 50Ω input impedance. The circuit was intended for use with a loop antenna but a tuned input circuit as in Fig 10.7 can be used instead. The following double-tuned filter has a bandwidth of about 3kHz centred on 137kHz. The RF amplifier gives a gain of about 24dB.

A second signal frequency filter follows the RF amplifier, also with a bandwidth of 3kHz; this is designed to have a deep rejection notch at about 113kHz, giving good rejection of image frequency signals (125 – 113kHz = 12kHz). The signal then passes to a switching mixer based on the 74HC4053 analogue switch IC; the three switches in this device are paralleled to reduce overall 'on' resistance. The 125kHz local oscillator signal is obtained by dividing the output of a 10MHz HCMOS crystal

oscillator module by 80 using the 74HC390 and 74HC393 counter ICs. A negative supply voltage at the V_{EE} pin of the CMOS switch slightly improves performance, about –4V at low current is obtained by rectifying the local oscillator signal.

The mixer output uses a diplexer to maintain the mixer input impedance at a well defined 200Ω; it also reduces the level of local oscillator signal reaching the 12kHz IF preamplifier, which is similar to the RF amplifier. A 13.83kHz trap is included to attenuate the strong IF signal resulting from the DCF39 transmitter carrier on 138.83kHz, which is otherwise only attenuated by about 10dB by the input filters. Two cascaded amplifier stages using a NE5532 dual op-amp increase the 12kHz output signal by a gain selectable from 0 – 60dB, driving the sound card microphone or line input.

The converter has been used with two pieces of software, DL4YHF's "Spectrum Lab" [6] and I2PHD's "SDRadio" [7], both of which may be downloaded free from the authors' web sites. Spectrum lab contains a comprehensive set of user-programmable features that may be used to down-convert and filter the sound card input signal; the output can be displayed as a scrolling spectrogram for 'visual' modes such as QRSS, or to generate an audio output from the PC speakers for aural signal reception. SDradio is specifically for aural reception, and includes a spectrum display, and capability to adjust tuning and gain from the PC screen. Both of these programs have on-line help files to aid setting up.

136KHZ TRANSMITTERS

LF transmitters are usually required to produce between 100W and a few kilowatts of output. One approach to LF transmitter design is to use HF circuit techniques, with appropriate scaling of components for a lower operating frequency. One such design, using MOSFET devices in a 'linear' output stage, is described in [4]. However, most LF operators are currently using class D switching-mode output stages; these can achieve very good efficiency, which considerably simplifies cooling problems associated with high-power linear amplifiers. These circuits are also well suited to inexpensive power MOSFETs and other components intended for switch-mode power supplies operating in a similar frequency range. Switching mode circuits are, however,

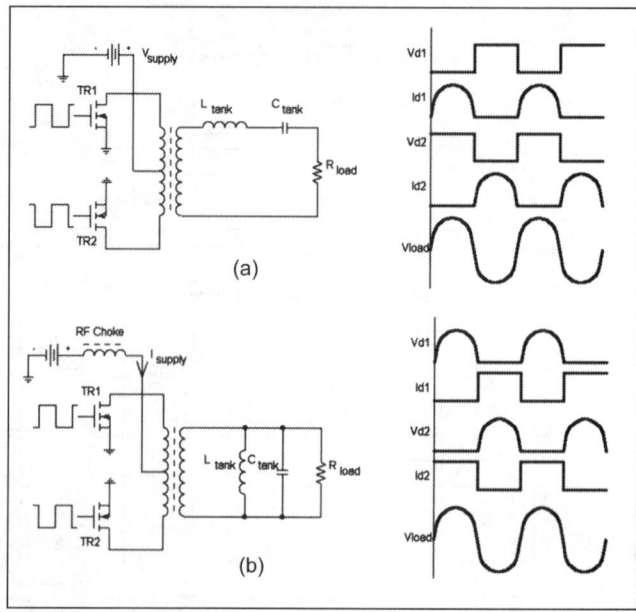

Fig 10.10: (a) Voltage-switching class-D amplifier, and (b) Current-switching class-D amplifier with MOSFET drain waveforms

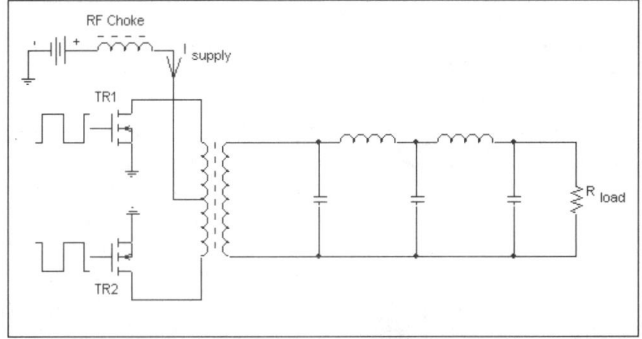

Fig 10.11: Practical form of Class D output stage

more difficult to key or modulate satisfactorily. Fortunately, most LF operation uses simple on-off keying of the transmitter, or frequency-shift keying.

Class D Transmitters

Class-D amplifiers fall into two distinct types, voltage-switching, **Fig 10.10(a)**, or current-switching, **Fig 10.10(b)**. In each case, the load is connected to the output stage via a resonant tank circuit. In the voltage-switching type the switching MOSFETs develop a square-wave voltage at the input side of the series tank circuit, however the tank circuit ensures the current flowing in the load is almost a pure sine wave.

The current-switching type has a parallel-tuned tank circuit; the supply to the output devices is a constant current which is applied to the tank circuit in alternate directions depending on which MOSFET is switched on. The resulting square wave current applied to the tank circuit again results in an almost sinusoidal voltage across the load.

Since a constant-current DC supply is not very practical, a constant voltage supply is used with a series RF choke. Provided the impedance of the choke is much greater than the load resistance, the supply current is almost constant. The major advantage of class D is that the MOSFETs are either fully 'on', in which case the only power loss is due to the MOSFET 'on' resistance, $r_{DS(on)}$, or fully off, with essentially zero power dissipation. In practice, there are additional losses, but these are small compared to linear amplifiers, and efficiency can exceed 90%.

In many amateur circuits, the tank circuit is replaced by cascaded low-Q pi-networks, **Fig 10.11**. This circuit is 'quasi-parallel resonant'; it provides a resistive load at the output frequency, but a low shunt impedance at the harmonics. The low Q leads to the voltage waveform being an imperfect sine wave, but has the advantages that smaller inductors and capacitors are required, tolerances are less critical, and better rejection of higher harmonics is provided by the multiple filter sections.

The voltage and current waveforms of a real-world class D output stage using this circuit are shown in **Fig 10.12**. Compared to the idealised waveforms, some high frequency 'ringing' is visible. This is due to stray capacitance and inductance which inevitably exists in the circuit, and is undesirable since it causes increased losses, as well as the potential for generating high–order harmonics. It is therefore important to minimise stray reactance, two important causes of which are the parasitic capacitance of the MOSFETs themselves, and the leakage inductance of the output transformer. Adding damping RC 'snubber' networks can also usefully reduce the level of ringing.

As with other types of amplifier, the output power of class D amplifiers is defined by the supply voltage V_{cc}, output trans-

former turns ration n and load impedance R_L. For the voltage switching amplifier:

$$P_L = \frac{8n^2 V_{cc}^2}{\pi^2 R_L}$$

While for the current-switching class D:

$$P_L = \frac{\pi^2 n^2 V_{cc}^2}{8R_L}$$

These formulas assume losses in the circuit are negligible; in practice, some losses do occur but since they are small, the results given by are reasonably accurate.

Class D PA design example

The design process for a class D transmitter output stage is best illustrated by an example. The following design is for a LF transmitter with about 200W output, using a current-switching class D circuit. This is a modest power level for 136kHz, but the principles discussed have been applied equally well to designs with 1kW or more output using this circuit configuration, which is probably the most popular in use at present. The complete circuit is shown in **Fig 10.13**.

The first design decision is what DC supply voltage to use, since the power supply is normally the most expensive and bulky part. In this case 13.8V was selected; it can use the standard DC supply found in many amateur shacks. The DC input power required will be about 10% greater than the RF output due to losses, so the expected supply current will be 220W ÷ 13.8V = 16A, a level that most 13.8V supplies can readily deliver. For higher power designs, 40 – 60V is often a good compromise, since the problem of large DC and RF currents is then reduced. It is perfectly possible to use an 'off line' directly rectified AC mains supply [8], with no bulky mains transformer – although design for electrical safety is absolutely essential in this case. Inexpensive switching MOSFETs are available suitable for any of these supply voltages.

In the ideal push-pull current switching circuit, the peak MOSFET drain-source voltage will theoretically be π times the DC supply voltage, in practice about 4 times the 13.8VDC supply is likely. MOSFETs TR2 and TR3 should be selected so that only a few percent of the DC input power will be dissipated in their 'on' resistance, $r_{DS(on)}$. This condition also ensures the MOSFET will have adequate drain current rating. STW60NE10 devices were available with BV_{DS} of 100V, and a typical $r_{DS(on)}$ of 0.016Ω, lead-

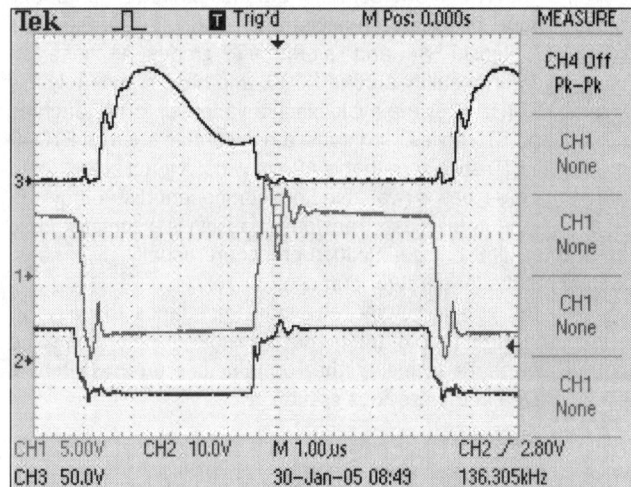

Fig 10.12: Class D waveforms. Upper trace, drain voltage; middle trace, drain current; lower trace, gate drive voltage

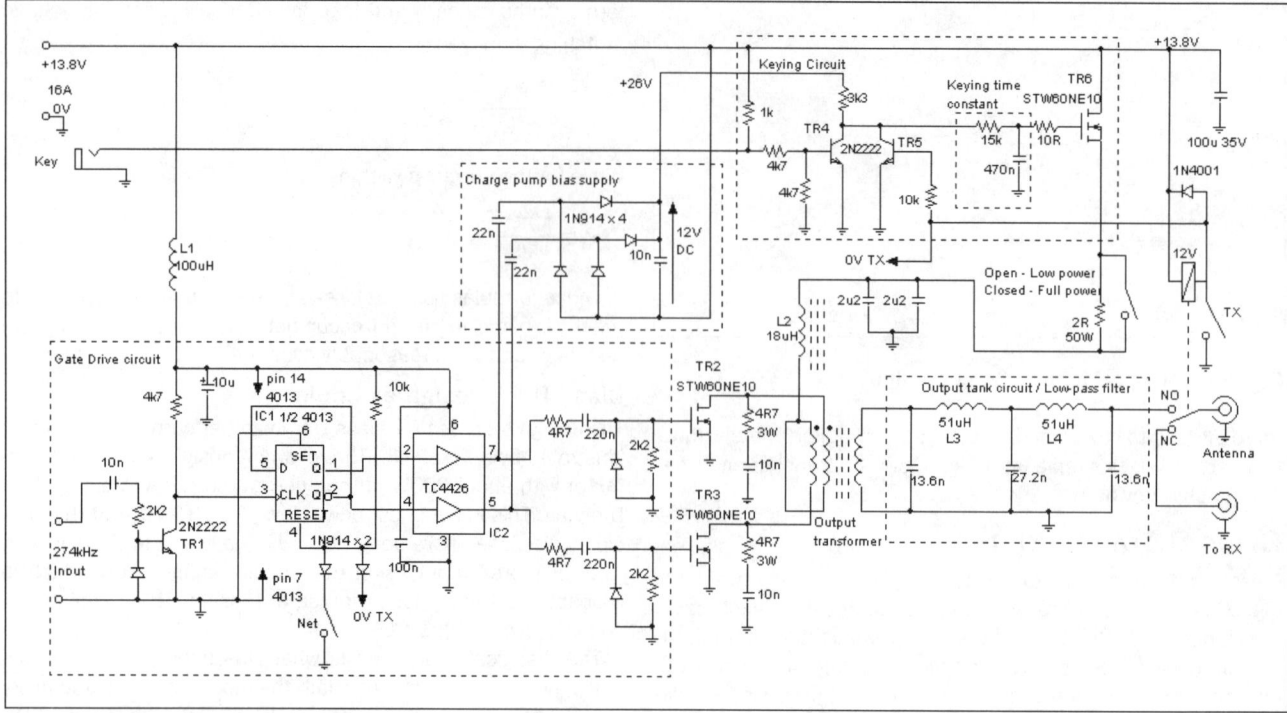

Fig 10.13: 200W Class D transmitter circuit

ing to about 4W dissipation due to the 'on' resistance and 16A supply current (I^2 x $r_{DS(on)}$). Additional dissipation occurs during the transient period where the device is switching 'on' or 'off'. This can be determined from measurement of circuit waveforms, but can be assumed to be similar to that due to $r_{DS(on)}$. In normal operation therefore, each MOSFET will only dissipate a few watts; however with a severe mismatch, power dissipation can be much higher, especially without DC supply current limiting. For a robust design, the MOSFETs and their heatsink should be able to dissipate of the order of 50% of the total DC input power, at least during a short overload period. The STW60NE10 devices have a TO-247 package and can dissipate 90W each at a case temperature of 100°C, which is adequate.

The output transformer is the most important part of the design. It is normally wound on a core using the same ferrite grades that are used for switch-mode power supplies. These may be large toroidal cores, pot cores, or 'E' cores with plastic bobbins. Several manufacturers produce suitable materials; these include Ferroxcube (Philips) 3C8, 3C85, 3C90, Siemens N27, N87, Neosid F44, and Fair-Rite #77 grades. All these ferrites have permeability around 2000, and have reasonably low loss at 137kHz. They are available in a variety of forms, such as EE, EC, and ETD styles, and sizes; a designation such as ETD49 means an ETD style core that is 49mm wide. A good selection of different core types is available from component distributors [5, 9] at reasonably low cost. Transformer design is a complex topic in its own right, but a simplified procedure usually gives satisfactory results for amateur purposes, as follows.

Given the supply voltage and load impedance (usually 50 ohms), the turns ratio of the output transformer determines the output power. Rearranging the formula for current-switching class D given in the previous section gives:

$$n = \sqrt{\frac{8P_L R_L}{\pi^2 V_{cc}^2}}$$

For V_{cc} = 13.8V, R_L = 50Ω, and P_L = 220Ω, this gives n = 1:6.8.

Next, a suitable sized core is chosen. As a guide, using the types of ferrite listed above, and ETD34 core is suitable for powers up to 250W, an ETD44 core for 500W, and an ETD49 for up to 1kW. Similar sizes in different styles have similar power handling. If in doubt, use a bigger core!

The number of turns N in the secondary winding can then be determined. N must be large enough to keep the peak magnetic flux B_{peak} to a value well below the saturation level at the expected output voltage level, V_{RMS}:

$$B_{peak} = \frac{V_{RMS}}{4.44fNA_e}, \quad V_{RMS} = \sqrt{P_L R_L}$$

Where A_e is the effective area of the core in m². The number of turns must also be large enough so that the inductance of the winding has a large reactance compared to the load impedance: A value of X_L about 5 – 10 times the load impedance is desirable. An ETD34 core and bobbin of 3C85 ferrite material was available , which according to the manufacturer's data has an A_e of 97.1mm² (97.1 x 10⁻⁶ m²), and A_L of 2500nH/T². A suitable maximum value of B_{peak} for power-grade ferrite materials is around 0.15 tesla. For 220W output, V_{RMS} is 105V A few trials using the formulas resulted in N = 14 turns B_{peak} = 0.127T and L = 490μH, X_L = 422Ω, which meets the criteria given. Two turn primary windings result in a turns ratio of 1:7, close enough in practice to the 1:6.8 design value. The primary windings were 4 x 2 turns, quadrifilar wound of 1mm enamelled copper wire, using 2 windings in parallel for each half of the primary winding. The secondary of 14 turns of 0.8mm enamelled copper was wound on top of the primaries, and insulated from them with polyester tape.

The DC feed choke L2 must be capable of handling the full DC supply current without saturation and also have a high reactance at 137kHz compared to the load impedance at the transformer primary, which is (50Ω ÷ 7²), about 1Ω. A reactance of 10Ω or greater is adequate, requiring at least 12μH. A high Q is not required, since only a small RF current flows in the choke.

The 18µH choke used a Micrometals T-106-26 iron dust core. Iron dust cores of similar types to this can often be salvaged from defunct PC switch-mode PSUs. The winding used 2 x 17 turns in parallel of 1mm² enamelled copper wire. An air-cored inductor would also be feasible, if more bulky.

The output filter consists of two identical cascaded pi-sections. The filter should provide a resistive load at the 137kHz output frequency, but a low capacitive reactance at harmonics. This can be achieved by designing the pi-sections as low-Q matching networks, with equal source and load resistances. This yields a circuit with two equal capacitors. The standard pi-section design formulas can be used, modified for $R_{in} = R_{out} = R$:

$$X_C = \frac{R}{Q}, \quad X_L = \frac{2QR}{Q^2+1}$$
$$C = \frac{1}{2\pi f X_C}, \quad L = \frac{X_L}{2\pi f}$$

Most designers select Q between 0.5 and 1. Metallized polypropylene capacitors are a good choice, since they have low losses at 137kHz, and are available with large values and high voltage ratings. The DC voltage rating should be several times larger than the RMS RF voltage present; the main limitation is the heating effect of the RF current causing internal heating of the capacitor. Several 6.8nF, 1kV polypropylene capacitors were available, so C = 2 x 6.8nF = 13.6nF was used. This has reactance of 85.4Ω at 137kHz, forcing Q = 0.585, and giving X_L = 43.6Ω, L = 50.6µH. The inductors L3 and L4 must have low loss at 137kHz to avoid excessive heating. Micrometals T130-2 iron dust cores were used, wound with 68 turns of 0.7mm enamelled wire. It is a good idea to check the capacitance and inductance of the filter components using an LCR meter or bridge. However, the main effect of small errors is only to slightly alter the output power from the circuit, without greatly affecting the efficiency.

The drive signal applied to the class D output stage is a 50% duty cycle square wave. The 137kHz gate drive signal is obtained from a 274kHz input using a D-type flip-flop in a divide-by-2 configuration, guaranteeing an accurate 50% duty cycle. When the circuit is switched to receive, the flip-flop is disabled by pulling the reset input high, preventing a 137kHz signal leaking to the receiver input and causing interference. For netting the transmitter, the 'net' switch enables the flip-flop. The MOSFETs require zero or negative gate voltage to switch the transistor fully off, and +10 to +15 volts to bias them fully on. The MOSFET gates behave essentially as capacitors, requiring transient charging and discharging currents as the drive voltage switches on and off, but drawing no current while the gate voltage remains stable. In order to achieve fast MOSFET switching, a TC4426 gate driver IC is used. These driver ICs accept a TTL-compatible logic level input signal, and are designed to produce peak output currents of 1 amp or more which charge and discharge the MOSFET gate in a fraction of a microsecond. A disadvantage of using a flip-flop to generate the drive signal is that if the input signal is lost, one MOSFET will remain switched on and act as a virtual short across the supply. To avoid this, the gate driver is capacitively coupled to the MOSFETs; the shunt diodes perform a DC restoration function, making the full positive peak voltage available to drive the gate. If drive is lost, the gates discharge through the 2.2kΩ resistors, switching both MOSFETs off. The 4.7Ω resistors in series with the gate drivers help to reduce ringing.

Each MOSFET has a series RC damping network from drain to source, reducing high frequency 'ringing' superimposed on the drain waveform. The component values are best determined experimentally, since they depend on the individual circuit. A good starting point is to make the capacitor about five times

larger than the MOSFET output capacitance. A resistance between 2Ω and 20Ω is usually effective. Effectiveness is best checked by examining the MOSFET drain waveforms with an oscilloscope, and compromising between minimising high-frequency ringing and excessive power dissipation in the resistors. Larger capacitors and smaller resistors normally result in reduced ringing, but increased dissipation. These components should be appropriate for high frequency use; in this circuit, 4.7Ω, 3W metal film resistors, and 10nF, 250V polypropylene capacitors were satisfactory.

The transmitter is keyed using series MOSFET TR6. The MOSFET should have low $r_{DS(on)}$ to minimise loss when switched on. A third STW60NE10 was used, although since the maximum voltage applied to this device is the 13.8V DC supply, a lower voltage device could be used instead. During the rise and fall of the keying waveform, dissipation in the keying MOSFET peaks at about 25% of the maximum DC input, 55W in this case. However, when the MOSFET is fully on, it dissipates only a few watts due to $r_{DS(on)}$, and when fully off dissipation is practically zero, so the average power dissipated is small, under 10W in this circuit, provided it is not keyed very rapidly. In order to bias this MOSFET fully on, a voltage around 10V higher than the 13.8V DC rail must be applied to the gate. Only a few milliamps bias are required; a small auxiliary DC supply could be used in a mains-powered transmitter, but in this case the bias voltage was obtained by rectifying the capacitively-coupled gate drive waveform using a charge-pump circuit. The bias is controlled by the key input via TR4, and the keying waveform is shaped by the RC time constant to give around 10ms rise and fall times. TR5 maintains the keying 'off' when the circuit is switched to receive. The 15V zener across the MOSFET gate and source limits the gate voltage to prevent damage. The output power of a class D transmitter can be varied either by changing the supply voltage, or by having multiple taps on the output winding; both these techniques are used in the G0MRF and G3YXM designs described later. In this design, a resistor can be switched in series with the DC supply, reducing the supply voltage to the output stage to around 4V, and RF output to 18W, for tuning-up purposes. The 2Ω wirewound resistor dissipates nearly 40W in low power mode, so greatly reduces efficiency, but does make it very difficult to damage the PA due to its inherent current limiting, useful when using a battery supply or when initially testing the circuit.

Construction of this type of transmitter is reasonably non-critical. The low-power parts of the transmitter can be assembled using 'veroboard' or similar, but the gate driver IC must have a 0.1µF ceramic decoupling capacitor directly across the supply pins due to the large transient currents present. Also for this reason, the gate leads, and the ground return from the MOSFET sources should be kept very short (<30mm). The MOSFETs, output transformer, keying circuit and DC feed carry heavy currents, so connections should be as short as possible and use thick wire (at least 2.5mm²). The RC damping components should be mounted directly across the MOSFET drain and source pins, and the connections to the output transformer kept short. The circuit described above was assembled on an aluminium plate about 160 x 200 x 3mm, which provided ample heatsinking for the three MOSFETs when air was allowed to circulate freely. In an enclosed box, a fan would probably be desirable.

Testing a class D circuit should start by checking operation of the gate-drive circuit, ensuring that complementary 12Vp-p square waves are present at the MOSFET gates at the correct frequency. A dummy load is almost obligatory for testing LF transmitters (see section on LF measurements). If possible, apply a reduced DC supply voltage to the output stage (but not

Fig 10.14: G3YXM one kilowatt transmitter

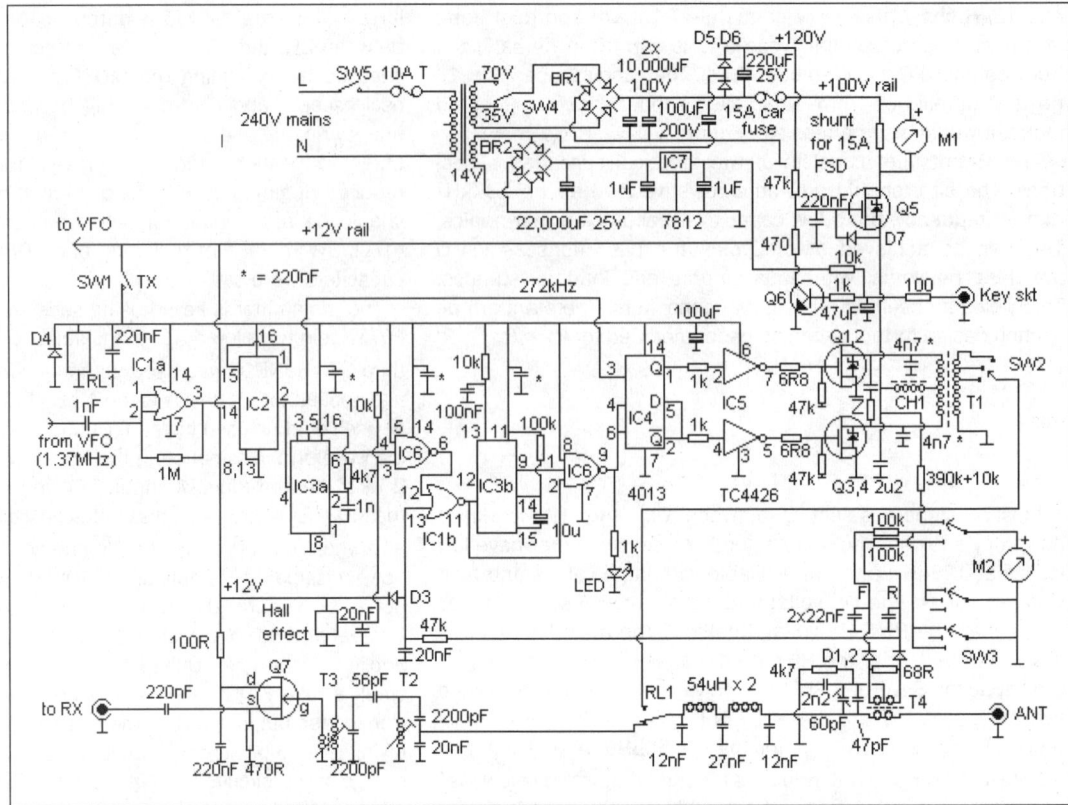

to the gate drive circuit!), or use a series resistor to reduce the supply voltage, as included in this circuit. An oscilloscope is the ideal tool to check the correct waveforms are present. A useful check is the efficiency; the ratio of RF output power to DC input power should be well over 80% if the circuit is working correctly.

G3YXM 136kHz 1kW transmitter

G3YXM set out to design a transmitter that is reasonably small, produces around 1kW RF output, and will withstand antenna mis-

match and other mishaps. The description here is an abridged version of the original article available via the internet [10].

The circuit is shown in **Fig 10.14**, and the main components are listed in **Table 10.3**.

An input signal at 1.36MHz from the VFO (**Fig 10.15**), or from a crystal oscillator and further divider, is divided by ten, the output at IC4 being a symmetrical square wave, driving the output MOSFETs via gate driver IC6. Each MOSFET shown is actually two devices in parallel. The output transformer ratio is set by switch S2. Higher output is obtained with more turns selected. Across the primary of the transformer the Zobel network marked 'Z' (22Ω and 4n7 in series) reduces ringing. The output is fed to the antenna via a low-pass filter. The cut-off frequency is quite high at about 220kHz as virtually no second harmonic is produced. The SWR bridge consists of T4 and associated components. It is a bifilar winding of 2x18 turns which forms the centre-tapped secondary and the coax inner passing through the toroid core forms the single-turn primary. The protection circuit which cuts the drive for about a second is triggered by high SWR via IC1B, or over-current signal from the Hall-effect device, which is triggered by the magnetic field of CH1, which is made from a 50mm piece of ferrite antenna rod wound with 20 turns of 1.5mm enameled copper wire. The Hall-effect detector is placed

IC1	HEF4001
IC2	HEF4017
IC3	HEF4538
IC4	HEF4013
IC5	TC4426
IC6	HEF4023
IC7	7812
Q1, 2, 3, 4	IRFP450
Q5	IRFP260
D1, D2	1N4936
D3, D4, D4, D6	1N4006
Hall effect device	OHN3040U (Farnell 405-656)
BR1, BR2	35A, 600V (Farnell 234-151)
R (for Zobel network "Z")	22Ω, 25W (Farnell 345-090)
Mains transformer	2 x 35V, 530W
T1	Primary 2 x 8 turns, secondary 20 turns, tapped at 12 and 16 turns
CH1	20 turns on 50mm length of antenna-type ferrite rod
T2, T3	Toko 719VXA-A017AO (Bonex)
T4	Primary 1 turn, secondary 2 x 18 turns bifilar
Output filter inductors	54µH. 65 turns 1mm enamelled wire on Micrometals T200-2 toroid core

Table 10.3: Component details for the G3YXM 1kW transmitter

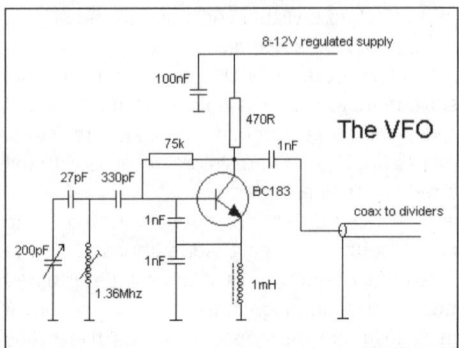

Fig 10.15: 1.36MHz VFO for the G3YXM 136kHz transmitter

near one end of CH1 and the spacing adjusted to trip at about 20 amps. The receive pre-amp uses coupled tuned circuits giving a band-pass response over the 135 to 138kHz range A single JFET (Q7) makes up for the filter loss.

The mains transformer used in the power supply has two series 35V windings. The DC voltage is either 50V from the centre tap or 100V from both windings. An auxiliary 12V winding was added by winding 30 turns of 16SWG wire through the toroid. At full output the HT will drop to about 80V. The keying circuit uses a series MOSFET with shaping of rise and fall times to prevent key-clicks. To turn this MOSFET fully on, its gate must be at least 5V positive of its source which is close to the main supply voltage. Diodes D5 and 6 are supplied via a high voltage capacitor from the 12V winding to produce an extra 20V bias for this purpose.

The low-level circuitry was built on strip-board, taking care to keep the tracks short and earth unused inputs. The TC4426 chip IC5 is capable of driving 1.5A into the gate capacitances of the MOSFETs and the decoupling capacitors must be fitted close to the chip with short leads. The 6R8 series resistors are mounted on the gate pins of the MOSFETs, the resistor leads forming the connections to the strip board. It is probably best to use one resistor for each gate. The strip board should be grounded to the earth plane as near as possible to the MOSFETs, which should have the source leads soldered to the ground plane. The two 4n7 capacitors should be connected directly across the MOSFETs. Output transformer T1 should be constructed from two-core 'figure of eight' speaker cable wound eight times through the ferrite toroid, connected as a centre-tapped primary by connecting one end of one winding to the opposite end of the other. The secondary is wound over it with 20 turns of thin wire tapped at 12 and 16 turns. The Zobel network should be wired from drain to drain with short wires.

Get the PSU, VFO and CMOS stages working first. Check with a scope that you have complementary 12V square waves on the gates, the waveform will be slightly rounded off due to the gate capacitance. Connect the transmitter to a 50 ohm load and, having selected the first tap on SW2, apply 50V (SW4 in low position) with a resistor in place of the fuse to limit the current. The MOSFETs should draw no current without drive. Press the key and the output stage should draw a few amps and produce a few watts into the dummy load. If the shut-down LED comes on, either the load is mis-matched, the SWR bridge is connected backwards or the 60pF capacitor needs adjustment. If all seems well, remove the current limiting resistor and increase the power by selecting taps, key the rig in short bursts and check for overheating of MOSFETs and cores. When you are happy that the transmitter is working OK, load it up to 15A PA current and slowly move the Hall device nearer to the end of the ferrite rod (CH1) until the protection circuit trips. Move it just a tiny bit further away and fix in position with silicone rubber. The receive preamplifier filter inductors can be aligned using a signal near 137kHz; the tuning is very sharp.

Transmitter Drive Sources

Several different types of frequency source are in use for 136kHz transmitters. As in the case of receivers, frequency stability requirements vary depending on operating mode. For straightforward CW operation, a simple VFO is quite adequate, the low output frequency tends to mean low drift also. Higher stability is needed for the extreme narrow band modes; frequency synthesisers of various forms are usually used. Crystal oscillators are sometimes used, usually in conjunction with a frequency divider, but the crystal frequency can usually only be pulled a few tens of hertz at the LF output frequency, and being able to change frequency is very desirable even in this narrow band.

VFO and divider

VFOs for 136kHz usually operate at higher frequencies, and are divided using digital counters to the LF range. It is easier to find suitable components for higher frequency VFOs; high stability inductors suitable for LF are difficult to produce. An example of this type of VFO is used in the G3YXM transmitter described above, where the VFO operates around 1.37MHz and is divided by 10. This is convenient for use with a frequency counter.

Ceramic resonators have been widely used in HF VFOs because of their good frequency stability, whilst the frequency can be 'pulled' over a much wider range than crystal oscillators. Ceramic resonators are not available for 137kHz, but it is possible to use a 3.58MHz resonator in conjunction with a divide-by-26 counter to produce a VFO covering the whole 136kHz band. The complete circuit is shown in **Fig 10.16**.

The oscillator is pulled over the range 3.5282 – 3.5828 MHz using a 60pF series tuning capacitor and 18pF padding capacitor. The frequency tolerances of the ceramic resonators are fairly loose, so these capacitor values may need to be altered to suit individual resonators. The 1V pk-pk VFO output is buffered and then converted to a logic-level signal by an amplifier using three cascaded CMOS inverters. The signal frequency is divided by 13 by the 74HC161 programmable counter, and the resulting 274kHz signal is further divided by two by the 74HC74 flip-flop to give an accurate 50% duty cycle square wave at 137kHz. In class-D transmitter designs that use a divide-by-two stage to obtain a square wave drive signal, this flip-flop can be omitted, and the 274kHz signal used directly. The resonator used has a negative temperature coefficient; the prototype circuit exhibited drift over a range of about 3Hz during a 24 hour period in an indoor environment.

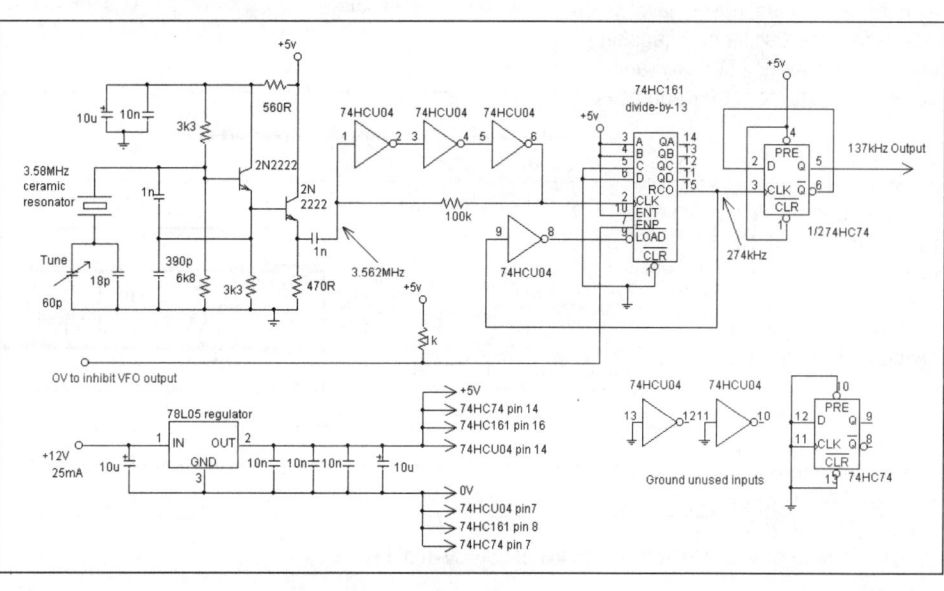

Fig 10.16: 137kHz ceramic resonator VFO

Crystal mixer

Several amateurs have used a so-called 'crystal mixer' scheme. Two crystal oscillators are operated at frequencies separated by about 137kHz (or sometimes 274kHz), and mixed together. A low-pass filter at the mixer output selects the difference frequency component. Because the crystals operate at relatively high frequency, their 'pulling' range is relatively large, and by making one or both oscillators a VXO, the whole band may be covered. A version of this system is used in the G0MRF 300W transmitter [11]. In **Fig 10.17**, the crystal oscillators are both tuned by varicap diodes, and the oscillators and mixer are implemented using a CMOS quad NAND gate IC. Another implementation by DJ1ZB is shown in **Fig 10.18**, using discrete components. This type of frequency source is less stable than a simple crystal oscillator, since drift is due to differential changes in frequency between the two oscillator frequencies, but typically the output frequency is maintained within a few hertz over a considerable period, which is adequate for many applications.

Synthesiser

Several types of frequency synthesiser have been used as LF transmitter drive sources. A number of DDS (Direct digital synthesis – see the chapter on oscillators) synthesisers have been produced in kit form [12, 13, 14]. These are well suited to extreme narrow-band LF application, since they are capable of high tuning resolution, often tuning in steps of small fractions of 1Hz. Some older synthesised signal generators are available as surplus quite cheaply, especially ones that do not cover the UHF range. These often have precision reference oscillators, capable of very high frequency stability, and are obviously useful for other purposes in the shack.

Another synthesised signal source that has been widely used for 136kHz operation is an HF transceiver, whose output is digitally divided down to the LF range. A typical example of this scheme due to G3KAU is shown in **Fig 10.19**. Output from an HF rig at 13.6MHz is applied

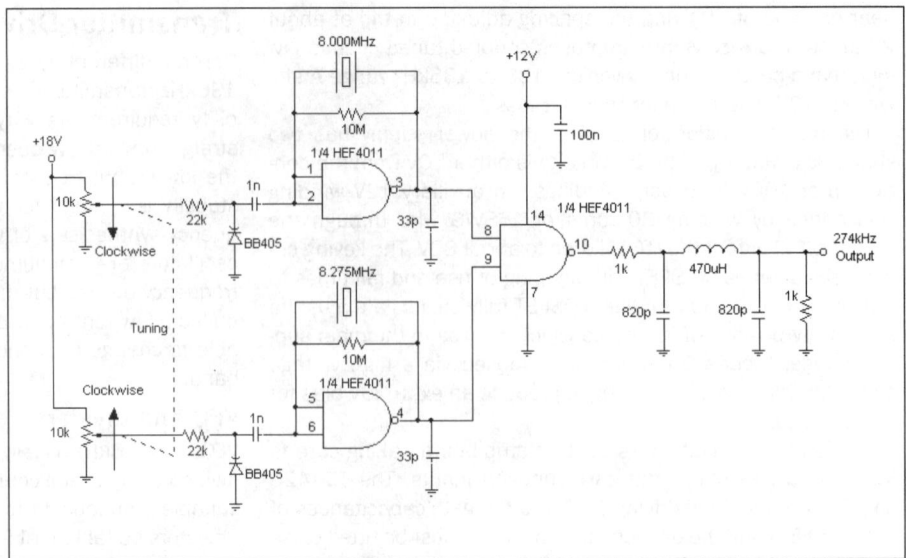

Fig. 9.17: 'Crystal mixer' VFO from G0MRF transmitter

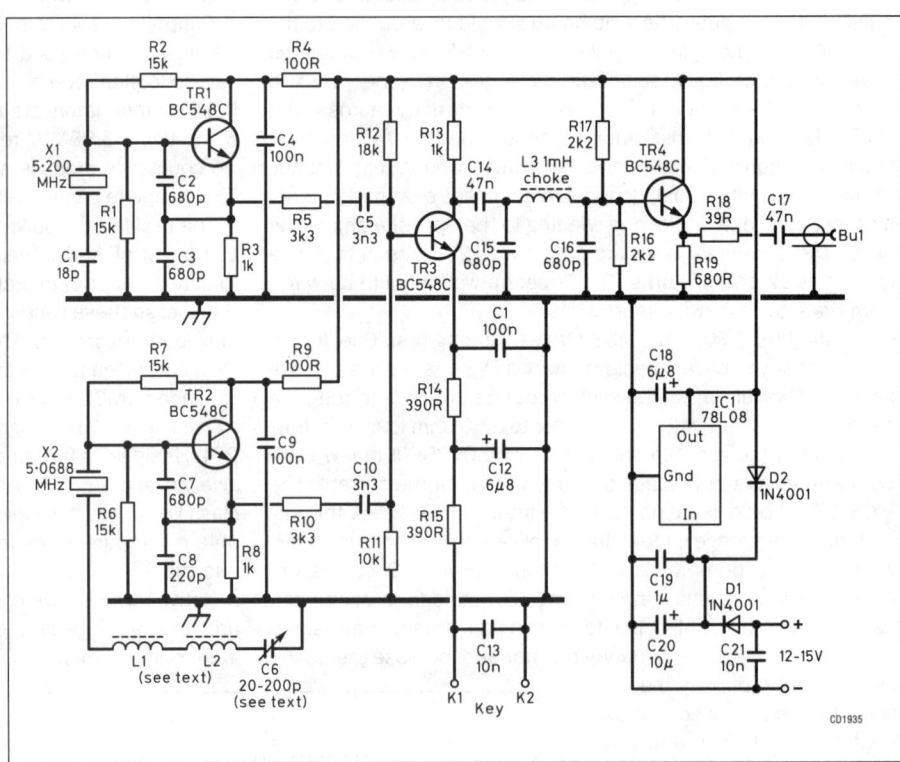

Fig. 9.18: DJ1ZB 'crystal mixer' VFO

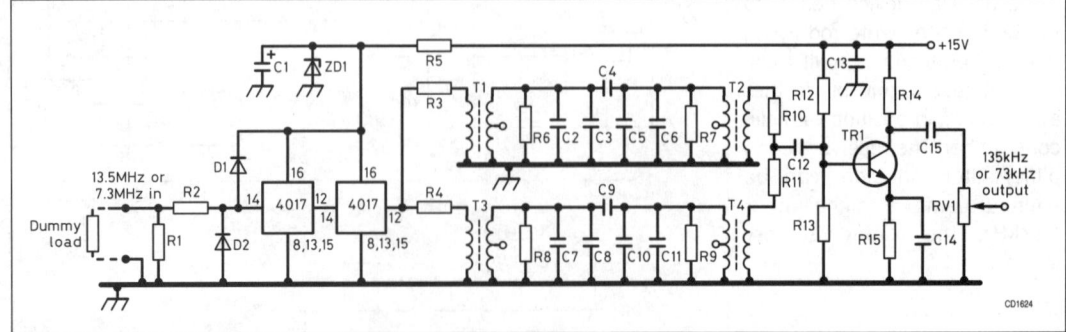

Fig 10.19: G3KAU divide-by-100 circuit

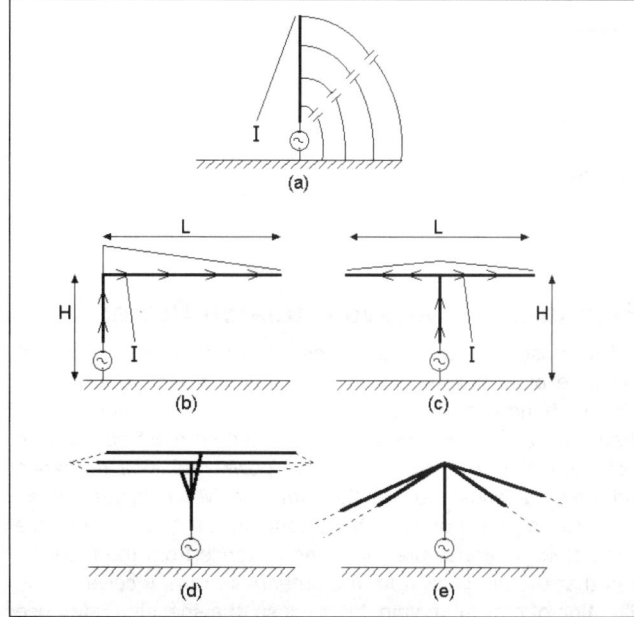

Fig. 9.20: Vertical antenna configurations

to two cascaded decade counters, giving output at 136kHz. This design also includes filtering to produce a sinusoidal output waveform for a linear transmitter. Note that most modern HF rigs inhibit transmission outside the amateur bands at frequencies such as 13.6MHz. However, in most cases only simple modification is required to enable transmission at out-of-band frequencies; contact the manufacturers for information.

TRANSMITTING ANTENNAS AND MATCHING

The underlying principles of antennas are of course the same at 136kHz as at any other frequency, but the distinguishing feature of amateur LF antennas is that, with an operating wavelength of around 2200m, they are almost invariably no more than a few percent of a wavelength long. This means they are always operated far below their self-resonant frequency, and require a large amount of inductive and/or capacitive loading in order to present a resistive load to the transmitter. 'Electrically small' antennas of this type are very inefficient, typically 0.1% efficiency would be a good figure for a back-garden amateur antenna. However, in spite of this, most of the long-distance operation on LF is done using quite ordinary wire antennas of similar dimensions to those used for HF operation. Electrically small antennas fall into two types, vertical or loop. In the United Kingdom and Europe, the vast majority of LF amateurs have used vertical antennas, however LF loops have been popular with operators in North America, for reasons that will be discussed in the section on loop antennas.

Vertical Antennas

The vertical or Marconi antenna is the most widely used LF transmitting antenna. It consists of a vertical monopole element a small fraction of a wavelength long, driven against a ground plane, **Fig 10.20(a)**. The voltage on the element is nearly equal at all points, while the current is a maximum at the feed point, tapering to zero at the end, as the current flows to the ground plane through the distributed capacitance of the antenna element. A figure of merit for a vertical antenna is the effective height, H_{eff}. This is the height of a notional 'ideal' vertical antenna element with a uniform current all the way along its length,

that would generate the same radiated field as the real antenna when fed with the same current. Because of the non-uniform current distribution, the effective height of a real antenna is always less than the physical height. The effective height can be increased by adding top loading to the basic vertical element, as is done with the T and inverted-L shown in **Figs 10.20(b) and (c)**. These can often be existing HF dipole or long wire antennas. A large proportion of the antenna current flows to ground through the distributed capacitance of the top loading wire, increasing the current flowing in the upper part of the vertical section. In either the T or inverted-L, the distributed capacitance of the vertical section, C_V, and the horizontal section C_H is approximately:

$$C_V = \frac{24H}{Log_{10}\left(\frac{1.15H}{d}\right)}, \quad C_H = \frac{24L}{Log_{10}\left(\frac{4H}{d}\right)}$$

Where C_V, C_H are in picofarads, H is the height of the vertical section and L the length of the horizontal section in metres, and d is the wire diameter in metres. For plastic covered wire, d can be taken as the overall diameter of the insulation. A handy approximation for C_V is 6pF per metre of height, and C_H 5pF per metre of length. To maximise the amount of top loading in a limited amount of space, multiple top-loading wires are sometimes used, as in the "flat-top" T antenna of **Fig 10.20(d)**. Due to proximity effects, multiple parallel wires have less capacitance than a single wire of the same total length. As a guide, two 1mm wires spaced 100mm apart will have about 39% greater capacitance than a single wire over the same span, spacing 1m apart will increase the capacitance by 68% compared to the single wire.

The effective height of the top-loaded vertical depends on the physical height H, and the relative values of C_V and C_H:

$$H_{eff} = H \frac{\frac{C_H}{C_V} + \frac{1}{2}}{\frac{C_H}{C_V} + 1}$$

With no top loading, ie with $C_H = 0$, $H_{eff} = 1/2H$. With a very large top loading capacitance, $C_H \gg C_V$ and H_{eff} is nearly equal to H. Therefore, adding a large amount of top loading can nearly double the effective height of the basic vertical element. The actual shape of the top loading is not very important; the main objective is to maximise its capacitance.

Another form of vertical antenna often used at LF is the umbrella, **Fig 10.20(e)**. In this case, the top loading consists of sloping wires, with the advantage that only a single tall support is required. With only two top loading wires, the umbrella becomes an inverted-V, or with one wire just a sloping long wire. The drawback is that the current flowing in the sloping wires will have a 'downwards' component, partly cancelling the 'upwards' current of the vertical section. Bringing the ends of the loading wires close to ground level is therefore likely to reduce the effective height of the antenna. Many combinations of length and angle of slope are possible, but provided the lower ends of the loading wires are at least half the height of the central vertical element, the overall effect of the top loading will be beneficial. The formulas for the T and inverted L antennas can still be used to calculate the approximate effective height of the umbrella, with the modification that H is now the average height of the sloping wires, rather than the highest vertical point of the antenna, and L is the horizontal length of the sloping wires.

Although the horizontal parts of the antenna wire are often much longer than the vertical section, little horizontally polarised radiation is generated. This is because, when the

Fig 10.21: (a) Cancellation of horizontally-polarised radiation from vertical. (b) Radiation pattern of short vertical antenna

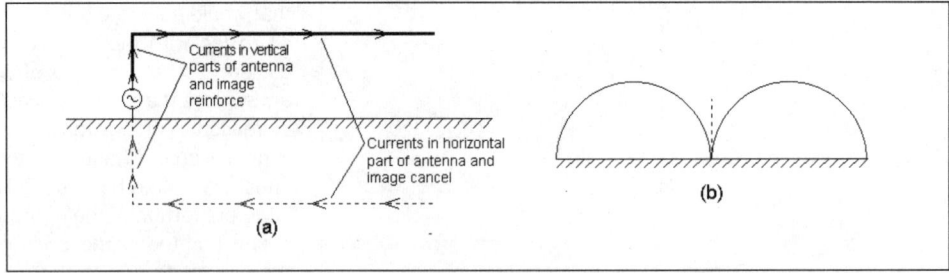

height of the horizontal section is a tiny fraction of the wavelength, the effect of the horizontally-flowing current components are almost completely cancelled out by the 'image' currents reflected in the ground plane, **Fig 10.21(a)**. Thus, these antennas are still classified as verticals, and the radiation produced is almost entirely vertically polarised. The radiation pattern of any electrically short vertical antenna is virtually the same. It is omnidirectional in the azimuth plane, and has field strength proportional to the cosine of elevation, giving rise to maximum radiation towards the horizon, and a null vertically upwards, **Fig 10.21(b)**.

As an example, consider a typical antenna that might be used for LF operation, a 40m long horizontal wire 10m above ground, made from 2mm diameter wire. Using the formulas given above, $C_V = 64pF$ and $C_H = 223pF$. The antenna capacitance C_A is the sum of C_V and C_H, 287pF. H_{eff} becomes 8.9m, as expected somewhat less than the physical height.

The impedance of this and other electrically short vertical antennas can be represented by a series combination of two resistances, R_{RAD} and R_{loss}, and C_A. R_{RAD} is the radiation resistance, which represents the conversion of transmitter power into radiated electromagnetic waves. R_{RAD} is related to the effective height H_{eff}, and the wavelength of the radiated signal λ, in metres, by the formula:

$$R_{RAD} = 160\pi^2 \frac{H_{eff}^2}{\lambda^2}$$

For 137kHz, this becomes:

$$R_{RAD} = 0.000329 \times H_{eff}^2$$

So R_{rad} is typically very small; for the example antenna with 8.9m effective height, R_{rad} is only 0.026Ω. The power radiated from the antenna as electromagnetic waves is $I^2 R_{rad}$, so quite large antenna currents are required for appreciable power to be radiated. The other resistive component of the impedance is the loss resistance, R_{loss}, which represents the power losses in the antenna and its surroundings. These include the resistance of the antenna wires and the ground system, and power dissipated due to dielectric losses in the ground under the antenna, and in objects near the antenna, such as trees and buildings. Power dissipated in R_{loss} is converted to heat, and is therefore wasted.

The same antenna current flowing in R_{rad} also has to flow through the loss resistance, resulting in a power $I^2 R_{loss}$ being wasted, so the ratio R_{rad}/R_{loss} is a measure of the antenna efficiency. There is no reliable way of determining R_{loss} by theoretical calculation; measured values for typical amateur antennas vary from perhaps 15Ω to 200Ω, with larger antennas having lower loss resistance; for our example antenna an optimistic figure would be 40Ω. The efficiency of this antenna is therefore 0.026Ω/40Ω = 0.00065, or only 0.065%; a minuscule figure compared to even mediocre antennas on the higher frequency amateur bands.

Estimating Effective Radiated Power

Unlike most other amateur bands, the UK amateur licence (at the time of writing) specifies a maximum power level for the 136kHz band of "1W (0dBW) Effective Radiated Power", rather than the transmitter power limit specified on most bands. One definition of ERP is the amount of power fed to a reference antenna that would produce the same received field strength as the actual antenna does, with the conditions that both reference and actual antennas are the same distance from the receiver, and that the direction from the antennas to the receiver is the direction of maximum gain. In this case, the specified reference antenna is a loss-free, half-wave dipole in free space. At 136kHz, the reference dipole is obviously a theoretical abstraction, however it is easy to calculate what field strength E (in volts per metre) it would produce if it really existed:

$$E = 7\frac{\sqrt{P}}{d}$$

where P is the power fed to the ideal dipole and d is the distance in metres. For example, at a distance of 10km from the dipole, with 1W feeding it, the field strength would be 700µV/m. Therefore, any transmitter and antenna combination producing a received field strength of 700µV/m at 10km distance is radiating 1W ERP. The definitive way of determining ERP is to measure the field strength at a known distance from the station, but is by no means a simple measurement to make (see LF Measurements section). A simpler way to estimate ERP uses another definition:

ERP = (Radiated power) x (directional gain of antenna with respect to a dipole)

The radiated power is $I^2 R_{rad}$, which can be estimated by measuring the antenna current and calculating the radiation resistance, using the formulas given above. The directional gain of all electrically short vertical antennas is close to 2.62dB, or 1.83 as a power ratio, because they have the same radiation pattern irrespective of their shape or size. Estimating ERP therefore involves the following steps:

- Using the formulas given previously, calculate the radiation resistance R_{rad} using the measured dimensions of the antenna.
- Measure the antenna current (see LF Measurements section for details of suitable RF ammeter designs).
- Calculate the ERP:

$$P_{ERP} = 1.83 I^2 R_{rad}$$

Using the previous example of a 10m high, 40m long wire antenna with radiation resistance of 0.026Ω, it can be seen that an antenna current of 4.6A is required to obtain 1W ERP. The power wasted in the 40Ω loss resistance of the antenna in producing this current is 846W, emphasising the need for large transmitter powers at LF! It is usually found that the true ERP is actually somewhat lower than the estimated value, making the estimate conservative for ensuring the station's compliance with the licence conditions.

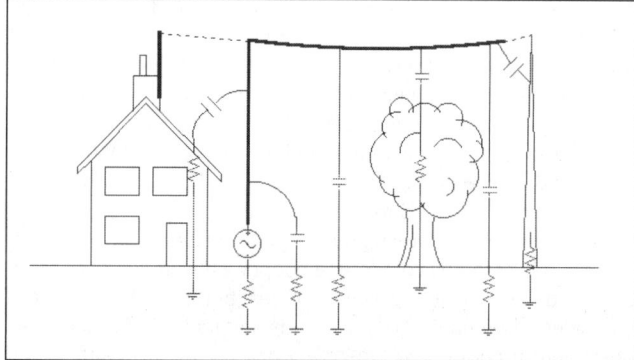

Fig 10.22: Loss in environment around antenna

Practical LF Antenna Considerations

"As much wire as possible as high as possible" is a good guiding principle for LF transmitting antennas. As seen in the previous section, the radiated power is proportional to the radiation resistance of the antenna, and the radiation resistance is proportional to the square of the effective height, H_{eff}. Therefore, for a given value of antenna current, and other things being equal, the amount of power radiated is proportional to the square of the height of the antenna. The most important dimension of an LF antenna is therefore always its height. The effective height can also be increased by maximising the capacitance of the top loading section, so the second most important dimension is the length of wire making up the top loading section of the antenna; making this as long as possible will make the effective height of the antenna as close as possible to the physical height. However, as noted in the discussion of umbrella antennas and sloping wires, it is the average height of the top loading that is important, so having long loading wires that have a large sag, or droop close to the ground is counterproductive.

After achieving the maximum possible effective height, the next most important goal is to minimise antenna losses. The vertical antenna can be thought of as a capacitor, one plate of the capacitor being the antenna wire, and the other plate being the ground system. Losses occur due to the resistance of the 'plates', and also due to the 'dielectric', made up of the air in between, the ground underneath, and the other objects surrounding the antenna, such as buildings and trees (**Fig 10.22**). The major source of losses depends on the shape and size of the antenna. For large commercial antennas, most of the loss occurs in the resistance of the ground system, but for much smaller amateur antennas, experiments have shown that most of the loss occurs in the "dielectric". Part of the electric field of the antenna penetrates the ground, which, with its high moisture content, has high losses at 137kHz. The electric field also induces RF currents to flow in building materials and the wood and leaves of trees and plants, leading to further loss.

Maximising the height of the antenna obviously keeps the antenna wires as far as possible from the ground and other lossy materials. Also, maximising the size of the top loading will reduce losses; a higher antenna capacitance will lead to a reduced voltage (see later). This results in a reduced electric field intensity, reducing dielectric losses, which are proportional to the square of the field strength. The capacitance between the antenna wires and poorly-conducting objects such as trees and buildings should be kept to a minimum by keeping the antenna, including the downlead, as far from them as possible. If the antenna is supported by a metal mast, the current capacitively coupled to the mast and flowing to ground will act as a parasitic antenna element in phase opposition to the rest of the antenna,

so it is desirable to maximise the clearance between mast and antenna, supporting the antenna wires some metres clear using insulating halyards, or perhaps replacing the top part of the mast with fibreglass. This problem will be reduced if the mast is well insulated from ground, but if it is not possible to adequately insulate the mast, ensure it has a good earth connection in order to minimise power loss due to the circulating RF current.

Antenna Voltage and Safety

An important practical consideration at LF is the antenna voltage. The RF voltage V_{ant} is almost the same at all points on the antenna, and is approximately equal to the antenna current multiplied by the reactance of the antenna capacitance:

$$V_{ant} = I_{ant} \times \frac{1}{2\pi f C_A}$$

At 137kHz, and with C in picofarads, this formula reduces to:

$$V_{ant} = 1.16 \times 10^6 \times \frac{I_{ant}}{C_A}$$

For our previous 10m high, 40m long wire example, C_{ant} = 287pF, I_{ant} = 4.6A, V_{ant} becomes 18,600V! Very large voltages are typical, particularly when using high powers with small antennas. Special attention must therefore be paid to the insulation of LF antennas. A particular problem experienced by many LF operators is corona discharge. Corona occurs where the electric field around the antenna is most intense; typically at the ends of wires where they attach to insulators, at sharp bends, where loose ends of wire project, or where the antenna passes near another object that is at ground potential. Corona manifests as a continuous, diffuse electrical discharge that produces a hissing sound; it is often very hard to see, even in the dark it may appear only as a faint glow. However, it absorbs substantial RF power, and so generates a lot of heat, which can ignite plastic insulators, support ropes and antenna tuning components, leading to the collapse of the antenna. Several instances have occurred where plastic insulators, ignited by corona discharge, have dripped burning plastic on to the ground beneath. If plastic insulators are used, ensure that they are in a position where burning debris cannot fall on a building or people. If insulators must be positioned over a building, glass or ceramic ones are safer. Even glass or ceramic insulators are eroded and cracked by corona over a period of time. The downlead of the antenna is equally prone to corona, and this represents a particular hazard if the antenna wire is brought directly into the shack; a number of minor fires have resulted from corona discharges igniting nearby woodwork. Measures to prevent corona include the following:

- Make the capacitance of the antenna top loading as large as possible to minimise the antenna voltage.

- Use 'corona rings' (**Fig 10.23**) at ends of antenna wires, or at sharp corners, to reduce the field gradient. These can be made from loops of stiff wire, 100mm or more in diameter. Dress the ends of the wire so there are no sharp projecting points.

Fig 10.23: Insulator fitted with 'corona ring'

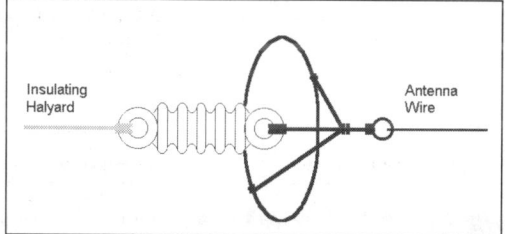

- Use insulating rope halyards, rather than conducting wires, to support insulators; this will reduce the voltage gradient across the insulator.
- Keep antenna wires well clear of buildings and trees, and other antennas. Locate the antenna download and antenna tuner away from the house or shack.

It will be seen that these are mostly the same guidelines as those given for reducing antenna losses, so taking precautions against corona also help to improve efficiency.

High antenna voltages are obviously also hazardous to people. Coming too close to an LF antenna wire whilst transmitting at high power can result in a severe RF burn, even without actually coming into direct contact. Be sure that it is not possible for a person to come within a few metres of any part of the antenna while operating. It is also possible for large metal objects near the antenna, such as ladders, garden furniture or other antennas, to have sufficient RF voltage induced on them to cause burns to a person touching them. This can be prevented by earthing all such objects.

LF Ground Systems

A ground system is required for any vertical antenna. In professional LF and VLF antenna systems, the ground system is normally the dominant source of losses, and earth mats containing many kilometres of buried wire are used. But for amateur LF antennas, it is usually easy to produce a ground system that has a negligible contribution to overall antenna loss, since other losses in the antenna system are much higher. The most widely used ground system consists of a number of ground rods connected to a common point close to the antenna tuner. This type of ground works well over much of the UK, where soil conductivity is quite high. As a very rough guide, a single 1m long ground rod will have a resistance of the order of 20 ohms; where several rods are used, spaced a few metres from each other, this figure is roughly divided by the number of rods. The losses in other parts of the antenna system are normally tens of ohms or more, so a point of diminishing returns is quickly reached where the ground system resistance is only a few ohms, and further improvements to the ground system yield little reduction in overall loss resistance. If a large number of ground rods are used, it is found that relatively little RF current flows in the rods that are

further away from the feed point. This appears to be because the distributed inductance of the longer connecting wire has a large impedance compared to the rest of the ground system. This may not be true where the ground conductivity is very low. In this situation, a ground system distributed over a wider area could be expected to give a useful improvement, although little practical data is available.

Ground rods intended for domestic mains earthing are ideal; these are designed to be rigid enough to be hammered into the ground. It is also possible to use copper water pipe in very soft ground, or inserted into pre-made holes, but this material quickly buckles when hammered. The rods should be as long as possible, and in contact with permanently damp soil. Systems of buried radial wires as used for HF verticals have also proved effective.

Matching Vertical Antennas

In principle, the same matching networks used for HF antenna matching, such as the pi- or T networks, could be used to match an LF vertical antenna. But in practice it is found that the component values, particularly for capacitors, are impracticably large, and due to the high antenna voltage requires very high ratings. The two most popular LF antenna matching circuits are shown in Fig 10.24. In Fig 10.24(a), a series loading coil has an inductive reactance that cancels out the capacitance, C_A, of the antenna.

The resistive component of the impedance (practically equal to the loss resistance) is then matched to 50Ω, or other value of transmitter output impedance, using a ferrite-cored transformer. The capacitance of back garden amateur antennas, typically hundreds of picofarads, corresponds to a loading inductance of a few millihenries. For most antennas, R_{loss} is between perhaps 15Ω and 200Ω, requiring transformer turns ratios between about 1:2 and 2:1 to match to 50Ω. One design of matching transformer, satisfactory at power levels up to 1kW, uses an ETD49 transformer core in 3C90 ferrite material wound with 32 turns of 1.5mm enamelled copper wire, tapped every two turns.

The 50Ω transmitter output is connected at the 16 turn tapping point, and the "cold" end of the loading coil connected to the tap that gives optimum matching. Because the antenna reactance is much larger than the resistance, the loading inductor must be capable of fine adjustment to obtain resonance accurately. Coarse adjustment is usually achieved by a series of taps on the loading coil.

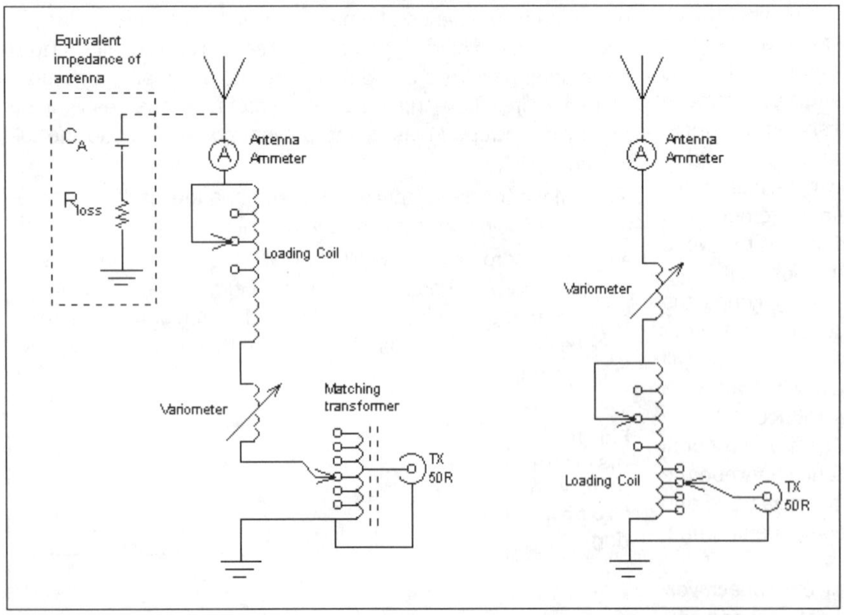

For fine tuning, the inductance is made variable over a narrow range using a variometer, described below. This matching arrangement is very straightforward to use, since the adjustment of antenna resonance and resistance loading are almost completely independent.

Another popular matching network uses the loading coil as a matching transformer as shown in Fig 10.24(b). The low potential end of the coil is equipped with closely-spaced taps, so the loading coil also performs the function of the matching transformer.

Although this is physically simpler than Fig 10.24(a), the electrical behaviour of this circuit is more complicated. The primary and secondary of the transformer are not

Fig 10.24: (a) LF antenna tuner. (b) Alternative antenna tuner circuit

The Radio Communication Handbook

tightly coupled, so the transformer impedance ratio will not closely correspond to the turns ratio and the adjustment of the antenna to resonance and the selection of the impedance-matching tap will be somewhat interdependent. However, it is not difficult to find a suitable tapping point by trial and error, and this will not then often need to be changed.

Loading coil construction

The loading coil is the most critical component in the LF antenna tuning system. It must have low losses (ie a high Q factor) if it is not to substantially reduce antenna efficiency and the same time must withstand large RF voltages and currents. With high power transmitters, the loading coil may need to dissipate hundreds of watts. It is also necessary to make the inductance easily adjustable. To meet these requirements, the most practical form of loading coil is a single layer solenoid of large dimensions. The former for the coil may be large diameter plastic tubing; many other roughly cylindrical plastic objects such as buckets and bins have been used to good effect. A visit to a garden or DIY centre may yield useful materials (**Fig 10.25**). Wooden coil forms have been tried, but these result in disappointingly high losses, probably due to the poor dielectric properties of wood. Enamelled or plastic insulated wire are usually used for winding. If available, Litz wire has considerably lower loss. Litz wire is composed of many fine strands of enamelled copper wire twisted together in the form of a rope. The purpose of this construction is to reduce the 'skin effect', where at high frequencies current flow is confined to the surface of a conductor, increasing its

resistance. The strands if litz wire weave in and out of the bundle, forcing current to flow throughout the thickness of the wire. This reduces the RF resistance typically by a factor of 2 or 3 at 137kHz.

The best form of loading coil usually depends on the materials available. Data on some practical loading coils (see **Fig. 10.26**) is given in **Table 10.4**. Coils 1 and 2 are close-wound with enamelled copper wire on PVC drain pipe formers. This gives a compact coil for a given inductance, but a relatively low Q. Coil 3 is wound with teflon-insulated, stranded core equipment wire. Teflon is an excellent dielectric, and withstands high temperatures and voltages, so produces a robust, higher rated coil with lower losses. This type of wire is quite expensive, but surplus bargains sometimes appear. Coils 4 and 5 are wound with Litz wire, the 'Rolls-Royce' of loading coil materials, but hard to obtain and relatively very expensive. Coil 6 is wound with wire obtained by stripping the outer sheathing from inexpensive 2.5mm^2 'twin and earth' cable available for domestic mains wiring.

The resistance of the loading coil results in the loss of a proportion of the transmitter power in the coil. The percentage loss of power is given by:

$$\% \, \mathrm{Loss} = \frac{R_{coil}}{R_{loss} + R_{coil}} \times 100\%$$

or, in decibels, $\mathrm{Loss(dB)} = 10 \mathrm{Log}_{10}\left(\frac{R_{coil}}{R_{loss} + R_{coil}}\right)$

Fig 10.25: Typical Loading coils

Fig 10.26: Some of the loading coils of Table 10.4

	Former	Winding length	Wire	Turns	L, mH	Series R @137kHz	Q @ 137kHz
1	110mm dia PVC tube	280mm	0.9mm Enamelled	281	2.75	9.4	250
2	156mm dia PVC tube	190mm	0.9mm enamelled	190	3.50	15	220
3	200mm dia PVC tube.	435mm	2mm dia (1.25mm dia conductor) Teflon insulated	225	3.75	10.3	330
4	395mm dia Polythene bucket	300mm	2.7 dia 729 strand Litz	109	3.89	3.0	1100
5	485mm dia ribbed "sectional manhole", polypropylene	200mm (multi-layer)	4.1mm dia Polythene insulated 729 strand Litz	79	3.83	5.1	650
6	ditto	440mm	3.8mm dia PVC insulated	95	3.37	9.6	300

Table 10.4: Practical loading coil data

Fig 10.27: Forms of vari-ometer

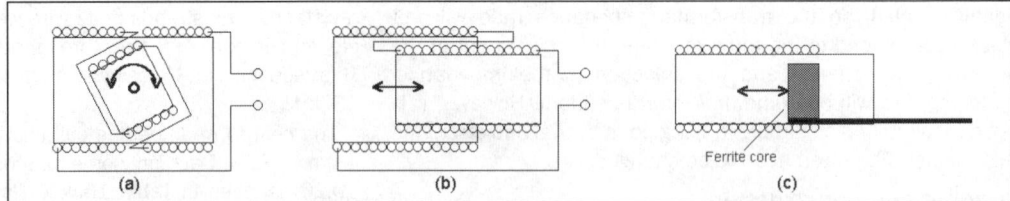

(a)   (b)   (c)

Where R_{coil} is the coil resistance, and R_{loss} is the loss resistance of the antenna. Taking the inverted L antenna discussed above as an example, with R_{loss} of 40Ω, using coil 2 in Table 10.4 with Q = 220 will result in 27% of the transmitter power being dissipated in the coil, while if coil 4 with a much higher Q of 1100 is used, only 6% of the power is dissipated in the coil. The loss in radiated power in decibels is 1.4dB for coil 2 and 0.25dB for coil 4. In either case, this loss amounts to only a fraction of an S-point at the receiving station, so the effect on overall system performance is minimal. What is more significant is the power-handling capability of the coil. Physically small coils such as 1 and 2 are suitable for transmitter power levels of up to a few hundred watts. Larger coils with higher Q dissipate a smaller proportion of the transmitter power, and also have greater surface area for cooling. Coils 3, 4 and 5 are suitable for kilowatt power levels, as is coil 6, which in spite of being relatively low Q has a very large surface area.

It is wise to make the loading coil inductance somewhat higher than that calculated to resonate the antenna, and provide several taps to accommodate changes in antenna capacitance. The winding should be attached to the former in several places, otherwise there is a tendency for wire to spill off when the former expands and contracts with changes in temperature. Loading coils are often located out-of-doors, and require protection from the elements by some sort of housing. The housing may itself cause considerable loss, especially if made of materials with high RF loss such as wood. A good protective housing is a large plastic dustbin. It is advisable to lift the housing at least a metre clear of the ground, to reduce dielectric losses in the ground underneath.

Variometers

A variometer is essentially a variable inductor, which is invaluable for fine adjustment of the loading inductance. The most common type of variometer is shown in **Fig 10.27 (a)**, and consists of two coils, one rotating inside the other. When the coils are aligned so their magnetic fields add together, the mutual inductance between the coils adds to the self inductance of each coil, resulting in maximum overall inductance. Rotating the inner coil by 180° results in the fields opposing, and the mutual inductance subtracting, giving minimum inductance. An adjustment range of about 2:1 in inductance is possible. Another form of variometer is shown in **Fig 10.27(b)**, in this case the two coils "telescope" together to vary the mutual inductance.

A very simple arrangement is shown in **Fig 10.27(c)**; this uses a ferrite core for permeability tuning of an air-cored coil. This may simply be a coil of a few hundred microhenries inductance wound on a horizontal tube former, with the ferrite core, glued to a plastic handle, free to slide inside the tube. It is important to use a short, thick piece of ferrite rather than a long rod, since in the latter case the flux density can easily become very high, leading to saturation and severe heating of the core. A large switch-mode power supply transformer core works well.

Loop Antennas

A large, single-turn loop is the 'alternative' LF transmitting antenna. While the LF vertical is a high voltage, relatively low

current device, the loop features relatively low voltage, and high current. Like smaller receiving loop antennas, the transmitting loop has a figure-of-eight radiation pattern in the azimuth plane, with deep nulls at right angles to the plane of the loop. The impedance of the loop can be modelled as the loop inductance in series with a radiation resistance R_{rad} and a loss resistance R_{loss}. The value of R_{rad} depends on the area of the loop A in square metres :

$$R_{rad} = 2 \times 320\pi^4 \frac{N^2 A^2}{\lambda^4}$$

where N is the number of turns, normally one, and λ is the wavelength in metres. The factor of 2 is included in the formula due to the effect of the ground plane underneath the loop, which doubles R_{rad} due to the 'image' antenna reflected in the ground plane. R_{rad} for back-garden sized loop antennas, which typically have areas of hundreds of square metres, is normally below a milli-ohm. This requires loss resistance R_{loss} to be below a couple of ohms to achieve efficiency comparable with vertical antennas. The dominant source of losses for loop antennas is the AC resistance of the loop conductor. To achieve reasonable efficiency, very thick wire, or multiple parallel wires are required; several builders have resorted to using the braid of UR67 coax, or even copper water pipe!

Matching a loop antenna normally uses one of the circuits shown in **Fig 10.28**. A step-down transformer is used in **Fig 10.28(a)** to match the loop R_{loss} to the 50Ω transmitter output, and a series capacitance to resonate the loop inductance L_{ant}. L_{ant} in henrys is given approximately by the formula:

$$L_{ant} = 2 \times 10^{-7} P \cdot Log_e \left(\frac{3440A}{dP} \right)$$

Where P is the overall length of the loop perimeter (m), A is the loop area (m²), and d is the conductor diameter (mm). C_{tune} is therefore:

$$C_{tune} = \left(\frac{1}{2\pi f \sqrt{L_{ant}}} \right)^2$$

C_{tune} is often divided into two series capacitors as shown, to make the loop voltages approximately balanced with respect to ground. The required transformer turns ratio is $\sqrt{(R_{load}/R_{loss})}$. An alternative matching scheme uses a capacitive matching network, **Fig 10.28(b)**.

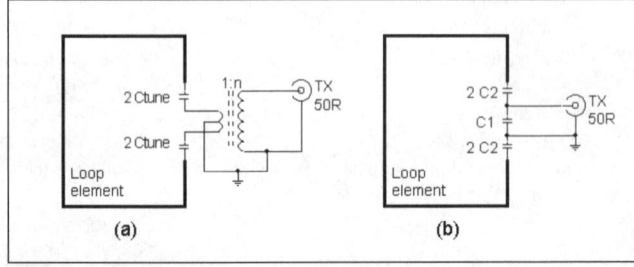

Fig 10.28: Transmitting loop matching circuits

The values of C1 and C2 are:

$$C_1 = \frac{\sqrt{\dfrac{R_{load} - R_{loss}}{R_{loss}}}}{2\pi f R_{load}}, \quad C_2 = \frac{1}{2\pi f \left(2\pi f L_{ant} - \sqrt{R_{loss}(R_{load} - R_{loss})}\right)}$$

As with vertical LF antennas, achieving good performance from loop antennas depends mostly on size. The radiation resistance is proportional to the square of the loop area, so every attempt should be made to make the loop as long and as high as possible. The main advantage of transmitting loops is that the loop voltages are much lower than for the vertical, resulting in lower dielectric losses in objects around the antenna. This makes loops a good choice for wooded surroundings, where many trees close to the antenna would lead to very poor efficiency with a vertical. This seems to be a common situation in North America, where several loop antennas have been constructed using branches of tall trees to support the loop element. In 'open field' sites, loops are usually less efficient than similarly sized verticals. Loops also do not rely on a low resistance ground connection, so may be an improvement where there is very dry or rocky soil. A disadvantage is that stronger antenna supports are required to support the thick loop conductor. A further drawback is the directional pattern of the antenna; the radiated signal will be reduced in some directions due to the nulls in the radiation pattern and options for changing the orientation of a large loop are usually limited.

Receiving Antennas and Interference Reduction

The 136kHz band is subject to high levels of noise, both naturally occurring and man-made. The major source of naturally occurring noise is thunderstorms, which give rise to the characteristic crackling lightning static (QRN) heard at most times on the band. When QRN is low, the audible band noise floor in the UK is usually dominated by the rhythmic chattering, "galloping horses", sound produced by Loran C. This is a hyperbolic radio navigation system which uses pulsed transmissions, occupying a band from 90–110kHz. Although radiation outside this band is suppressed, due to the very high powers used there is sufficient leakage of power in the 136kHz range for the Loran sidebands to be the dominant noise source under quiet band conditions.

A more serious problem for many amateur stations is locally generated, man-made noise. This has many sources, most associated with mains electrical wiring. Many devices use switch-mode power supplies, operating at switching frequencies in the LF range. Equipment with rectifiers or triac-based phase-control circuits can generate significant levels of harmonics of the mains frequency throughout the LF range. Digital systems can also generate wide-band noise in the LF spectrum, a particular problem if the source is a computer within the shack.

A good first stage in isolating a local noise source is to switch off all mains-operated equipment. This usually requires unplugging the equipment completely, since often the same or sometimes greater noise level can be generated when switched to 'standby'. The most certain method is to switch off the house mains supply at the main switch, whilst using a battery-operated receiver to monitor the noise level. If the offending device is on the premises, the noise can simply be eliminated by switching it off while operating on 136kHz. Unfortunately, LF noise can propagate considerable distances along the mains wiring, so often the LF operator has no control over the noise source.

Few operator's locations are entirely free of mains noise, however in most cases it is possible to reduce the problem to a tolerable level. Often it is most effective to move the antenna further from the noise source. This is not usually practical when the transmitting antenna is used for reception, since it normally occupies all available space, but properly designed, physically small loop and whip antennas are capable of sensitivity limited only by external band noise, and can easily be moved around to find the point where the noise level is lowest. Many LF operators therefore use separate transmit and receive antennas. Where the noise originates from a localised point, noise-canceling schemes are sometimes very effective. Even if attempting to reduce the noise level is not effective, it is often found that the noise level varies a lot during the day, and is sometimes low enough for LF operation.

Receiving Loop Antennas

Much has been said about the 'noise reducing' properties of receiving loops. Loop antennas respond essentially to the magnetic field component of an electromagnetic wave, and so reject noise that exists as a local electric field. Unfortunately, most local noise sources at LF involve common-mode noise currents flowing through mains wiring, giving rise to magnetic fields which the loop will pick up. Using an LF loop antenna inside or near a building almost always yields very poor results. However, these fields rapidly decrease in strength as the antenna is moved away from the offending wiring. Often moving the receive antenna by only a few metres results in substantial reduction in noise. It is therefore most important to experiment with different

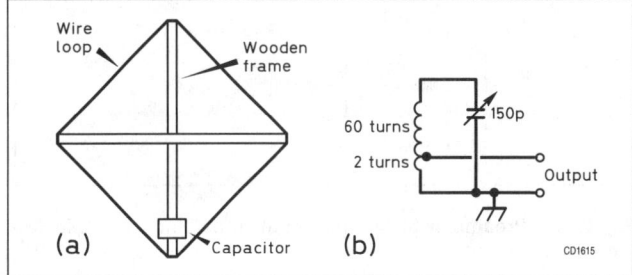

Fig 10.29: LF tuned loop antenna

Fig 10.30: Example of a well-constructed LF loop by PA0SE

positions for receiving loops, often a quiet spot can be found even where high noise levels exist all around. All LF loop antennas have a figure-of-eight directional pattern, with nulls at right angles to the plane of the loop, and maximum sensitivity along the plane of the loop. The directional null of a loop is often very effective in eliminating distant noise sources such as Loran. Also, the loop null can sometimes be used to suppress local noise.

A traditional tuned loop antenna is shown in **Fig 10.29**. It consists of about 60 turns of wire wound onto a wooden cross-shaped frame, and is typically about 1m², tuned by a variable capacitor. A practical realisation of this type of antenna is shown in **Fig 10.30**. The receiver input is fed via a low impedance tap two turns from the grounded end of the winding.

The output of the loop is small, and a low-noise preamplifier will normally be required, such as the one shown in **Fig 10.31**. The Q of the loop is typically 100 or more, so re-tuning will be required even within the narrow 136kHz band. This selectivity is very useful in reducing intermodulation due to strong out-of-band signals. A number of amateurs have used much larger

tuned loops for reception, which achieve higher output signal levels and so can dispense with the preamplifier, at the expense of being more bulky.

An alternative is the "lazy loop" of **Fig 10.32**. This uses a large single-turn loop, the area of which is around 10 - 20m². The shape is not at all important, and it can be normal insulated wire slung from bushes or fence posts, etc, hence the name! The loop can be fed through a coax feeder, allowing it to be positioned remote from the shack to reduce noise, whilst the tuning components are easily accessible at the receiver end of the feeder. This tuning arrangement is not optimum from the point of view of minimising losses, but due to the large loop area the signal-to-noise ratio is more than adequate. It is also possible to use somewhat smaller, multi-turn loops. Again, a low-noise preamp, such as Fig 10.31 will be required. This type of loop element also gives good results with the LF Antenna tuner/preamp circuit of Fig 10.7.

The need to frequently re-tune a high Q loop is something of a drawback; the Q can be reduced by adding resistive loading, but unfortunately this also reduces the signal level available to the receiver. An alternative approach to increasing loop bandwidth is to combine the tuned circuit formed by the loop with other capacitors and inductors to form a bandpass filter. This results in a wider bandwidth, while at the same time improving rejection of out-of-band signals. Two such bandpass loops are shown in **Fig 10.33(a) and (b)**. The 1m² loop uses a loop consisting of 10 turns of 1mm² PVC insulated stranded wire, divided into two windings of five turns each. The windings are taped together in a bundle and supported on a cross-shaped wooden frame. The loop inductance is resonated by a total capacitance of approximately 3.6nF, and coupled through the 55nF capacitor to a second, series-tuned circuit of 640pF/2.12mH. Output is taken from the series tuned circuit to a preamplifier. The two coupled tuned circuits give a flat pass-band about 4kHz wide centred on 137kHz. The response of the antenna is –40dB down at 115kHz and 160kHz, giving good rejection of LF broadcast stations and Loran at 100kHz. The preamplifier has an input impedance of 50Ω, and uses the circuit shown in Fig 10.31. The 2.12mH inductor is wound on an RM6/250n adjustable pot core; the capacitors should be low-loss types such as silver-mica, polystyrene or polypropylene. To align this antenna, inject a signal at 137kHz between ground and the 10kΩ resistor, and temporarily disconnect one of the loop windings. Monitor the preamp output on a receiver, and tune the 2.12mH pot

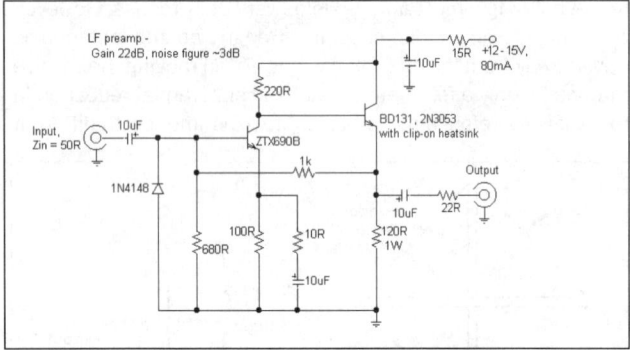

Fig 10.31: Preamp with 50-ohm input impedance suitable for loop antennas

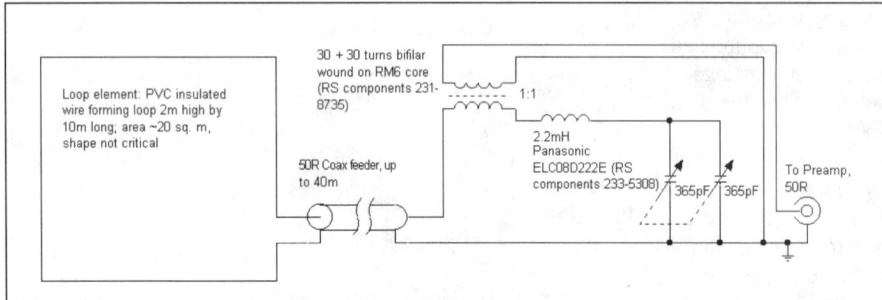

Fig 10.32 'Lazy loop' and tuning arrangement

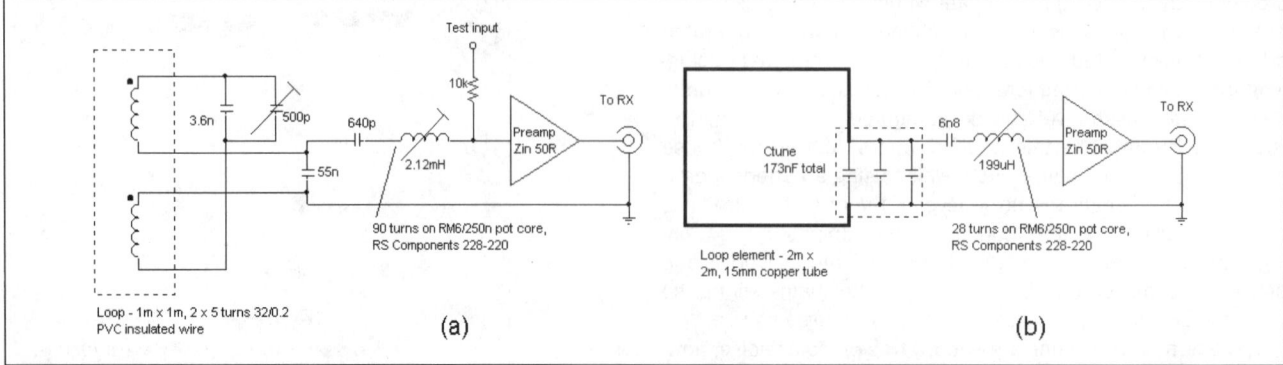

Fig 10.33: (a) 1m² Bandpass loop. (b) 2m x 2m Bandpass loop

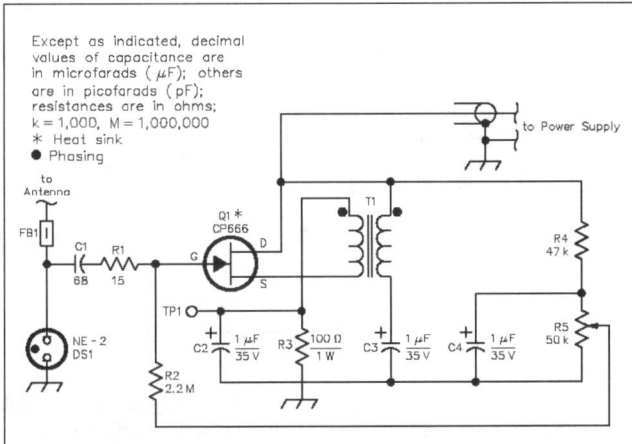

Fig 10.34: Amplifier for the AMRAD active antenna [ARRL]

core for a minimum signal. Then reconnect the loop, and tune the 500pF trimmer for maximum signal.

The 2m x 2m loop uses a single-turn element made from copper water pipe; this is easier to weatherproof compared to a multi-turn winding, and self-supporting. The loop element has 7.9μH inductance, and is tuned by a total of approximately 173nF, made up from smaller-value polypropylene capacitors connected in parallel.

This feeds the same 50Ω preamp through a series tuned circuit made up of 6.8nF/199μH. The bandwidth of this larger loop is about 26kHz at the 3dB points.

The tuning is less critical than for the smaller loop, and it is sufficient to adjust the loop tuning capacitance and the series-tuned circuit separately for resonance at 137kHz, before connecting them together.

Both these loop designs are capable of providing ample signal to noise ratio even under quiet band conditions. The noise level at the preamp output is around 1μV in 300Hz bandwidth, so a receiver with reasonably good LF sensitivity is required; some amateur-type receivers with poor LF sensitivity may require additional preamplification.

Fig 10.35: The AMRAD active antenna in place

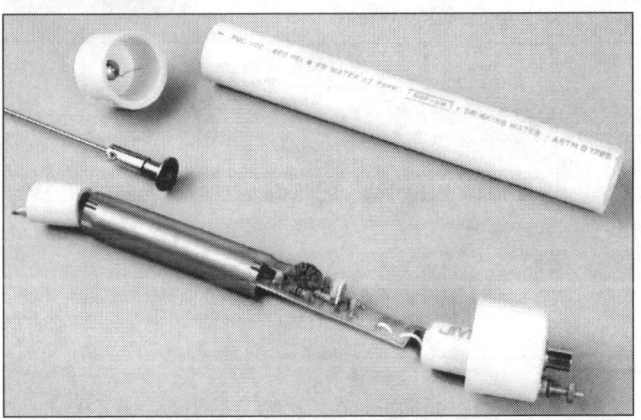

Fig 10.36: The amplifier can be built into the base of the antenna

Active Whip Antennas

Quite short wire or whip antennas provide adequate signal to noise ratio at LF when matched to the receiver input by a high impedance buffer amplifier. The response of the resulting Active Whip antenna, or E-field antenna is broad band, and can extend from the VLF range to the VHF range, depending on the amplifier used.

The preamplifier is located at the base of the antenna, and its DC supply is usually fed up the coax. The whip element need only be 1 – 2m long, and the small size makes the antenna easy to site in an electrically quiet location. As with the loop antennas, it is important to experiment with the antenna location to find a site that has a low noise level. Since all signals over a wide frequency range are present within the buffer amplifier, good dynamic range is important.

An antenna well suited to LF reception has been developed by the American experimenter's group AMRAD. The circuit is shown in **Fig 10.34**, with photos in **Figs 10.35 and 10.36**, but this should not be regarded as full construction information. A detailed article by Frank Gentges, K0BRA, describes the design, construction and performance of the antenna together with the power supply design [15].

Noise Cancelling

By combining the signals received from two separate antennas, with suitable adjustment of amplitude and phase, it is sometimes possible to cancel out local noise sources. The success of this depends on the noise source being localized so that one of the antennas picks up the same noise at a substantially greater level than the other antenna.

The basic circuit of a noise canceller built by G3GRO is shown in **Fig 10.37**. The signal from a long wire antenna feeds an adjustable phase shift network, with phase adjusted by RV1 over a range of nearly 180 degrees. S2 is provided to allow a fixed zero degrees phase shift, since the adjustment range of RV1 does not quite reach this.

Changeover switch S1 allows a 180 degree phase shift to be added. The phase shifted output is simply combined in parallel with the input from a loop antenna via a variable series resistor RV2, which allows adjustment of the nulling signal amplitude. The circuit is based on the premise that the signal from the long wire is generally a much larger amplitude than that from the loop, so adjustable attenuation allows the amplitudes of both signals to be made the same, and so cancel out. This could also be achieved with other types of antennas by using a preamplifier to increase the gain sufficiently. In use, the amplitude and phase controls are iteratively adjusted to achieve nulling of the noise signal.

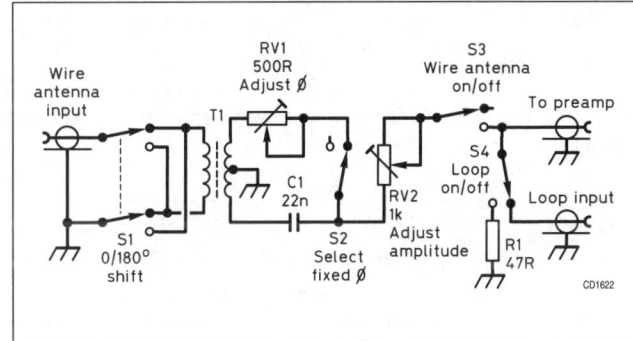

Fig 10.37: G3GRO noise canceller. RV1 and RV2 are linear cermet or carbon. T1 is 3 x 18 turns trifilar wound on a ferrite toroid (13.25mm dia. 3C85 ferrite, Fair-Rite FT-50-43 or similar)

Fig 10.38: Two type of RF ammeters

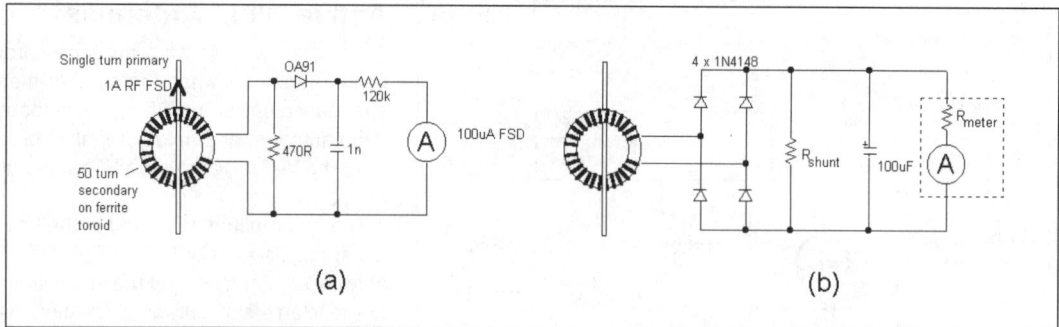

(a)

(b)

LF MEASUREMENTS & INSTRUMENTATION

RF Ammeters

As was seen in the section on antennas, the most useful way of estimating the effective radiated power, or determining the efficiency of an LF station requires measurement of the antenna current. Most LF stations have some form of RF ammeter so that antenna current can be measured. The traditional RF ammeter uses a thermojunction; these are difficult to obtain, and easily damaged by overload. Fortunately, it is easy to make a rectifier-based RF ammeter that is much more robust and adaptable.

The RF ammeter of **Fig 10.38** has a full-scale deflection of 1A, and uses a ferrite toroid as an RF current transformer. The single-turn primary is the current-carrying wire threaded through the toroid, which is wound with a 50 turn secondary. The secondary current, which is thus 1/50 times the primary current, or 20mA maximum, develops a voltage of 9.4V RMS across the 470Ω load resistor. This voltage is measured by a simple diode voltmeter; the peak voltage across the smoothing capacitor is 1.414 x V_{RMS}, or 13.3V, less around 0.5V diode forward voltage, giving approximately 12.8V DC which the 120kΩ series resistor sets as the full scale deflection of the 100µA meter. The current range can be increased by proportionally reducing the value of the load resistor; it is desirable to maintain the voltage across the load at around 10V in order to maintain the linearity of the diode voltmeter. Note that with higher maximum currents, the power dissipation in the load resistor can reach a few watts, so the resistor should be appropriately rated.

A slightly different ammeter circuit is shown in **Fig 10.38(b)**. In this circuit, the output current from the secondary of the transformer is fed directly to a bridge rectifier, whose mean output is measured by a DC milliameter. If I_{RF} is the RMS RF current, and the secondary has N turns, the mean DC current at the rectifier output is 0.90 x I_{RF}/N. The meter movement requires a shunt resistor to read the desired full-scale value. For example, if the current transformer has a 50 turn secondary, and 6A RMS full scale is required:

$$I_{mean} = 0.90 \times \frac{I_{RF}}{N} = 0.90 \times \frac{6}{50} = 0.108A$$

A 1mA meter with 75Ω resistance was used, so the required shunt resistor R_{shunt} was:

$$R_{shunt} = R_{meter} \times \frac{I_{meter}}{(I_{mean} - I_{meter})} = 75 \times \frac{1mA}{(108mA - 1mA)} = 0.70\Omega$$

which was made up of larger value resistors in parallel. This circuit has good linearity down to low currents because the diodes are fed from the high impedance of the transformer secondary winding. The voltages in this circuit are essentially just the forward voltage drop of the rectifier diodes, resulting in somewhat less power dissipation than the previous circuit, and also leading to less error due to the shunting effect of the transformer inductance.

A high permeability ferrite core is required for either of these circuits, so that the impedance of the secondary winding is much larger than the load impedance presented by the load/meter circuit. Toroids with a permeability of 5000 or above are ideal; with a 50 turn secondary, these typically yield inductances of several millihenries. A split ferrite core can be used to allow the ammeter to be put into the circuit without disconnecting the wire. If the RF ammeter is to be used at a high voltage point, such as at the feed point of the antenna wire, a screening metal case enclosing the transformer and meter circuit is advisable, to prevent the possibility of stray currents flowing in the meter circuit due to capacitive coupling, and causing errors.

The Scopematch Tuning Aid

Tuning LF antennas is critical; the bandwidth of the antenna is usually only a few hundred hertz. This tuning aid displays the voltage and current waveforms on a dual-trace oscilloscope. This makes it a simple matter to see if the antenna is adjusted to resonance, since the waveforms will be in phase when this occurs. The resistance at the transmitter output can then easily be determined from the ratio of voltage to current. The circuit diagram and construction of the scopematch are shown in **Fig 10.39(a) and (b)**. Any high permeability ferrite core of about 18mm diameter works for the transformers, both of which have

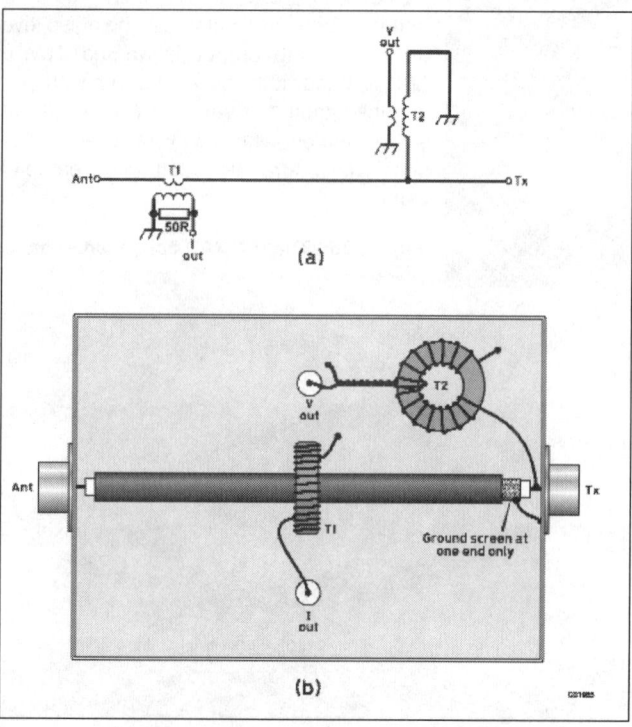

Fig 10.39: Scopematch tuning aid

a single turn primary winding and 50 turn secondary. Adequate insulation is required between primary and secondary to withstand the applied RF voltage. The 1:50 transformer and 50Ω load give a scale factor of 1V = 1A at the 'I' output, the other 50:1 transformer gives a scale factor of 1V = 50V at the 'V' output. Thus when the load resistance is 50Ω, the amplitudes at the V and I outputs will be equal.

In use, transmitter power is applied to the antenna tuner, and the loading coil adjusted until V and I waveforms are in phase. If the amplitude of V is greater than I, loading is adjusted to reduce the resistance, or to increase the resistance if V is less than I.

LF Dummy Loads

Conventional dummy loads intended for the HF and higher bands will of course work perfectly well at LF. However, higher powers are often required for LF. Fortunately, many types of relatively cheap, high power resistors will give good results at 137kHz.

Wirewound resistors with values from a few tens to a few hundred ohms usually have manageable inductance at LF. For example, a 50Ω load made up of two 100Ω, 150W (Arcol HS150 metal clad wirewound) resistors in parallel was found to have an impedance of (50.5 + j6.2)Ω at 137kHz. This amounts to an SWR of 1.13:1, which would be adequate for many applications. Connecting 2 x 1.5nF, 1kV polypropylene capacitors in parallel with the resistors to tune out the inductive component of the impedance resulted in SWR of 1.02:1. More recently, 'power metal film' resistors have become available at reasonable cost in ratings of tens to hundreds of watts (eg Vishay RCH50 series rated at 50W, available from RS components [5]). These have very low inductance and give good performance as dummy loads into the HF range and above. Like the metal-clad wirewound resistors, these are designed to be bolted to a heatsink, and suitable series-parallel combinations can be used to build a high power dummy load.

Several amateurs have pressed various types of heating element into service as high power LF loads, including such things as toasters and electric fan heaters. These often have resistance in the vicinity of 50Ω, with elements of wire or strip in a zigzag shape that has a reasonably low inductance. It is advisable to run such a load at considerably less than its mains rated power, since the resistance rises considerably when at its normal operating temperature.

Field Strength Measurement

As pointed out in the section on antennas, the definitive way of determining the effective radiated power of an LF station is to measure the field strength at a known distance from the antenna. The relation between field strength E (volts/metre) at a distance d metres, and ERP is given by the formula:

$$P_{ERP} = \frac{E^2 d^2}{50}$$

For this relationship to be valid, the distance d must be in the far field region of the antenna. In the far field region, the field strength falls away in inverse proportion to the distance from the antenna. The near field region is closer to the antenna, where the field strength decreases more rapidly with distance, and the formula above does not apply. There is no definite dividing line between near and far field, it depends on the operating wavelength, the nature and the dimensions of the antenna, and the required accuracy, but for small antennas at 136kHz, a safe minimum distance is about 1km. At much greater distances, the formula also becomes invalid, due to the effects of ground loss on the propagating wave close to the ground, and ionospheric reflections (see the chapter on propagation). For amateur stations, the signal is also likely to be too weak to accurately measure at large distances. A practical maximum distance is usually a few tens of kilometres. Field strength measurement is normally a two man job, with portable equipment required!

Two pieces of equipment are required to measure field strength, a calibrated receiver and a calibrated 'measuring' antenna. The calibrated receiver must be capable of accurately measuring signal levels down to a few microvolts, and have sufficient selectivity to reject unwanted adjacent signals; the ideal amateur equipment for this purpose is the selective level meter (see section on receivers). Calibrated antennas have a specified antenna factor (AF), which is the number of decibels which must be added to the signal voltage measured at their terminals to obtain the field strength. Quite good accuracy at LF can be obtained using a simple single turn loop antenna. Such loops have a low feed point impedance, so the received signal level is little affected by the load impedance. The output voltage of an N-turn loop with area A square metres at a frequency f hertz is given by:

$$V = 2.1 \times 10^{-8} \times fNAE$$

From this, the antenna factor of a single turn loop at 137kHz is:

$$AF(dB) = 20Log_{10}\left(\frac{1}{2.1 \times 10^{-8} \times 137 \times 10^{3} \times A}\right) = 20Log_{10}\left(\frac{350}{A}\right)$$

A square or circular loop made of tubing is usually used, with an area between 0.5m² and 1m². As an example, suppose a signal level of 7.5dBμV (ie 7.5 decibels above 1μV, or 2.4μV; selective level meters usually give a decibel–scaled reading) is measured at a distance of 5km from the transmitting antenna, using a 1m² loop. From the formula above, AF is 51dB, so the field strength is 58.5dBμV/m, or 840μV/m. Using the ERP formula gives P_{ERP} = 350mW.

A more compact alternative to the loop is a tuned ferrite rod antenna, however this requires calibration with a known field strength to determine the antenna factor. A field strength measuring system, including ferrite rod antenna, measuring receiver, and calibration set-up has been described by PAOSE [16].

Field strength measurements are prone to errors caused by environmental factors. The measured field strength is particularly affected by conducting objects giving rise to parasitic antenna effects. Such parasitic antennas can be large steel-framed structures such as buildings and road bridges, overhead power and telephone wires, even such things as fence wires and shallow buried cables. Such factors are difficult to entirely avoid, so several field strength measurements should be made at different locations over as wide an area as possible. Locations giving widely different values of ERP can then be rejected; it will be found that a few decibels of variation still exists between different measurement sites, so the ERP should be taken as an average of several measuring sites [17].

REFERENCES

[1] *LF Today*, Mike Dennison, G3XDV, RSGB, 2004
[2] *Amateur Radio Operating Manual*, Don Field, G3XTT, RSGB, 2004
[3] http://www.qsl.net/i2phd/TS950/change.html
[4] David Bowman, G0MRF web site, http://g0mrf.com
[5] RS Components. http://rswww.com
[6] http://www.qsl.net/dl4yhf/spectra1.html
[7] http://www.weaksignals.com]
[8] http://www.wireless.org.uk/jnt.htm

[9]   Farnell InOne. http://www.farnell.com

[10] http://www.wireless.org.uk/136rig.htm

[11] 'A Class-D Transmitter for 136kHz', David Bowman, G0MRF, *RadCom* Jan/Feb 2003. Also reproduced in [1]

[12] 'Build a PIC controlled DDS VFO, 0 to 6MHz' by Johan Bodin, SM6LKM. http://home.swipnet.se/~w-41522/minidds/minidds.html

[13] 'A software based DDS for 137kHz'. http://wireless.org.uk/swdds.htm

[14] ZL1BPU's DDS/exciter. http://www.qsl.net/zl1bpu/micro/EXCITER/

[15] *QST*, Sep 2002, ARRL. Alternatively download at http://www.arrl.org/tis/info/pdf/0109031.pdf

[16] http://www.wireless.org.uk/pa0se.htm

[17] 'Experimental investigation of very small low frequency transmitting antennas', J R Moritz, IEE 9th International Conference on HF Radio Systems and Techniques, Jun 2003, IEE Conference Publication no 493 pp 51 - 56

11 Practical Microwave Receivers and Transmitters

Many of the techniques for generating and receiving microwave frequencies were investigated and developed more than 70 years ago, in the 1930s. Microwave usage was given added impetus by the development of radar and the advent of the Second World War. Before 1940, the definition of the higher parts of the radio frequency spectrum [1, 2] read like this:

- 30 to 300Mc/s   Very high frequencies (VH/F)
- 300 to 3000Mc/s   Decimetre waves (dc/W)
- 3000 to 30.000Mc/s   Centimetre, waves (cm/W)

Radio frequencies above 30,000Mc/s (now 30GHz) apparently did not exist! Various definitions have appeared in the intervening years. These have included terms such as super-high frequencies (SHF) and extra-high frequencies (EHF).

In the course of time, the unit of frequency cycles per second (c/s), its decimal multiples, kilocycles per second (kc/s) and megacycles per second (Mc/s), have been replaced by the unit hertz (Hz), its decimal multiples kilohertz = 10^3Hz (kHz), megahertz = 10^6Hz (MHz), gigahertz = 10^9Hz (GHz) and terahertz = 10^{12}Hz (THz).

Today, the term microwave has come to mean all radio frequencies above 1000MHz (1GHz). The division between radio frequencies and other electromagnetic frequencies, such as infra-red, visible (light) frequencies, ultra-violet and X-rays, is still not well defined since many of the techniques overlap, just as they do in the transition between HF and VHF or UHF and microwaves. There has been keen interest in amateur communications using infra-red and visible light, over the last few years, so a section on it has been included in this chapter. To a large extent the divisions are artificial insofar as the electromagnetic spectrum is a frequency continuum, although there are several good reasons for these divisions.

Around 1GHz (30cm wavelength) the lumped circuit techniques used at lower frequencies are replaced by distributed circuit techniques such as resonators and microstrip. Conventional components, such as resistors and capacitors, become a significant fraction of a wavelength in size so surface-mount devices (SMDs) are used which are very small and leadless. These require special techniques for constructors; these are described later.

Conventional valves (vacuum tubes) and silicon bipolar solid state devices are usable beyond 1GHz - perhaps to about 3.5GHz - and, as frequencies increase, these devices are replaced by special valves such as klystrons, magnetrons and travelling-wave tubes. The first semiconductor devices to be used at the higher microwave frequencies were varactor diodes, PIN diodes and Gunn diodes. Because of the massive development in semiconductors for use by commercial telecommunications systems there are now a wide range of transistors available to radio amateurs including gallium arsenide field effect transistors (GaAsFETs), metal epitaxial semiconductor field effect transistors (MESFETs), pseudomorphic high electron mobility transistors (pHEMTs) and many more.

The other device that has revolutionised microwave designs is the microwave monolithic integrated circuit (MMIC); these are usually designed to work in matched 50-ohm networks, making them easy to use and (usually) free from instability problems. They are used as building blocks, in a similar way that the oper-

ational amplifier (OP-amp) is used in DC and audio designs, but are extremely wide band and often require bandpass filtering to obtain the desired results.

Many of the more exotic semiconductor devices are now appearing on the surplus market making them more acceptable to the amateur's pocket. These can be used to build new amateur equipment or whole surplus units can be modified to work on amateur bands. Quite a few articles have appeared, describing the modification of commercial equipment, so there is an example of such an article later in this chapter.

Other advantages of operation in the microwave spectrum are compact, high-gain antennas and available bandwidth. None of these advantages are attainable in an amateur station operating on the lower frequency amateur bands. High gain antennas are impossibly large below VHF, and the levels of spectral pollution from man-made and natural noise are such that low noise receivers, needed to handle weak signals, cannot now be effectively used, even at VHF. Communication on the many available microwave bands over distances of hundreds of kilometres is now quite common (sometimes over thousands of kilometres, given favourable tropospheric propagation conditions or the use of amateur satellites or moonbounce). This destroys the perception that microwaves are useful only over limited line-of-sight paths! A good place to find details of the latest records and operating conditions is *Dubus Magazine*; all of the main amateur bands are reported with details of activity using various types of propagation.

There is an increasing amount of commercially produced equipment available from amateur radio retailers for all of the microwave bands and plenty of designs for the constructor. It is true that attaining really high transmitter power output above about 3 or 4GHz is still difficult and expensive for most amateurs. Many successful amateur operators settle for comparatively low power output, ranging from perhaps 50 to 100W in the lower-frequency bands, to milliwatts in the 'centimetre' bands, or even microwatts in the 'millimetre' bands. This is compensated for by using very high antenna gain and, as already mentioned, receivers with very low noise figures.

Since the first essential requirement of microwave construction is easy availability of designs and components, many leading microwave amateurs have launched small-quantity component sources or have designed and can supply either kits of parts for home construction or ready-made equipment to these designs. Most microwave equipment now uses printed circuit boards (PCBs) and surface-mount components. This avoids the use of the precision engineering usually associated with older, waveguide based designs and means that construction of microwave equipment is not restricted to the amateur who has his own mechanical workshop. Conventional tools can be used together with some fairly simple test equipment to construct some very sophisticated equipment that produces excellent results.

There are many examples of designs that can be purchased in kit form or ready built; suppliers are listed in the Bibliography at the end of this chapter. The most widely used designs in the UK come from Michael Kuhne (DB6NT) who has equipment for all bands up to 241GHz; most have been described in the pages of *Dubus Magazine* or the *Dubus Technik* publications.

Free and easy access to practical information is important to the microwave amateur enthusiast. The question most often asked by amateurs new to microwaves is "Where do I get reliable information and (possibly) help?" Many microwave designs have appeared in the amateur press, in published books or magazines and in the various national amateur radio societies' journals. Some of the more prolific or rewarding titles are given in the bibliography. In addition to these sources, obtaining up-to-date designs, component information and design tools is extremely easy using the internet. The number of suppliers of microwave components has mushroomed with the expansion of the mobile phone networks, so has the sophistication of the design tools available.

Many suppliers of design tools have student or 'Lite' versions of their software free to download from their websites. These generally have reduced functionality compared with the full versions of the software, which may cost several thousand pounds, but are more than adequate for most amateur use. A selection of website addresses for the more significant, current sites is given at the web page associated with this book (the address is given at the end of this chapter in the bibliography). Please note that website addresses (URLs) change quite frequently, if the links supplied do not lead to the expected web pages a quick search with one of the popular search engines, using the relevant key words, should find the new location.

The range of current amateur microwave allocations offers scope to try out all of the modes and techniques available to amateurs. All amateurs are encouraged to try out some of these which will help retain our allocations. The lowest microwave frequency amateur allocation, the so called 23cm band, (1240MHz to 1325MHz in the UK), can be regarded as the transition point from 'conventional' radio techniques and components to the 'special' microwave techniques and components to be reviewed here.

In the space of a single chapter it will only be possible to give a flavour of some of the practical techniques involved, by outlining a few representative designs for most of the bands currently used by amateurs. If you need more detail, there are plenty of pointers to other sources of information shown in the bibliography.

The microwave bands support a wide range of activities such as:

- All narrow band modes
- Amateur TV, including wide band colour transmission
- Moon bounce (EME)
- Amateur satellite operation
- Meteor scatter

Since a significant amount of amateur microwave interest centres on the use of narrow band modes to achieve long distance, weak signal communication, the majority of the designs outlined here will concentrate on such equipment. More details of components and techniques (including wide band modes) are available in other publications [3-5].

In some instances construction and alignment procedures are described in some detail, again to illustrate the techniques used by amateurs in the absence of elaborate or costly test equipment, such as microwave noise sources, power meters, frequency counters or spectrum analysers. Most of the designs described are capable of being home constructed without elaborate workshop facilities (most can be constructed using hand tools, a generous helping of patience and some basic knowledge and skills!) and aligned with quite ordinary test equipment such as matched loads, directional couplers, attenuators, detectors, multimeters and calibrated absorption wavemeters.

AMATEUR MICROWAVE ALLOCATIONS

Most countries in the world have amateur microwave allocations extending far into the millimetre wave region, ie above 30GHz. Many of these allocations are both 'common' and 'shared Secondary', ie they are similar in frequency in many countries but are shared with professional (in this case 'Primary') users who take precedence. Amateur usage must, therefore, be such that interference to Primary users is avoided and amateurs must be prepared to accept interference from the Primary services, especially in those parts of the spectrum designated as Industrial, Scientific and Medical (ISM) bands. The UK Amateur Service allocations are summarised in **Table 11.1** and the UK Amateur Satellite Service allocations are shown in **Table 11.2**.

All the familiar transmission modes are allowed under the terms of the amateur licence: in contrast to the lower frequency bands, most of the microwave bands are sufficiently wide to support such modes as full-definition fast-scan TV (FSTV) or very high speed data transmissions as well as the more conventional amateur narrow-band modes, such as CW, NBFM and SSB.

Many of the bands are so wide (even though they may be Secondary allocations) that it may be impracticable for amateurs to produce equipment, particularly receivers that cover a whole allocation without deterioration of performance over some part of the band. Most amateur operators do possess a high-performance multimode receiver (or transceiver) as part of their station equipment and this will frequently form the 'tuneable IF' for a microwave receiver or transverter: commonly used intermediate frequencies are 144-146MHz or 432-434MHz. either of which are spaced far enough away from the signal frequency to simplify the design of good image and local oscillator carrier sideband noise rejection filters. An intermediate fre-

Allocation (MHz)	Amateur Status	Preferred (alternative) narrow band segment
1,240 - 1,325	Secondary	1,296 - 1,298
2,310 - 2,450	Secondary	2,320 - 2,322
3,400 - 3,475	Secondary	3,400 - 3,402
		(3,456 - 3,458)
5,650 - 5,680	Secondary	5,668 - 5,670
5,755 - 5,765	Secondary	5,760 - 5,762
5,820 - 5,850	Secondary	
10,000 - 10,125	Secondary	All modes
10,225 - 10,475	Secondary	10,368 - 10,370
		(10,450 - 10-452)
10,475 - 10,500	Secondary	Space only
24,000 - 24,050	Primary	24,048 - 24,050
24,150 - 24,250	Secondary	24,192 - 24,194
47,000 - 47,200	Primary	47,088 centre of activity
75,500 - 76,000	Primary	75,976 centre of activity
		(until December 2006)
76,000 - 77,500	Secondary	
77,500 - 78,000	Primary	77,500 - 77,502
		(after January 2007)
78,000 - 81,000	Secondary	
122,250 - 123,000	Secondary	
134,000 - 136,000	Primary	
136,000 - 141,000	Secondary	
142,000 - 144,000	Primary (until December 2006)	
241,000 - 248,000	Secondary	
248,000 - 250,000	Primary	

Table 11.1: UK Amateur Service Allocations at Spring 2005

Allocation (MHz)	Amateur Status	Comments
1,260 - 1,270	Secondary	ETS only
2,400 - 2,450	Secondary	ETS/STE. *(Note 1)*
5,650 - 5,568	Secondary	ETS only
5,830 - 5,850	Secondary	STE only
10,475 - 10,500	Secondary	ETS/STE
24,000 - 24,050	Primary	ETS/STE
47,000 - 47,200	Primary	ETS/STE
75,500 - 76,000	Primary	ETS/STE
76,000 - 77,500	Secondary	ETS/STE
77,500 - 78,000	Primary	ETS/STE
79,000 - 81,000	Secondary	ETS/STE
122,250 - 123,000	Secondary	ETS/STE
134,000 - 136,000	Primary	ETS/STE
136,000 - 141,000	Secondary	ETS/STE
142,000 - 144,000	Primary (until Dec 2006)	ETS/STE
241,000 - 248,000	Secondary	ETS/STE
248,000 - 250,000	Primary	ETS/STE

ETS = Earth to space, STE = Space to earth,
ISM = Industrial, scientific and medical applications
Note 1: Users must accept interference from ISM users

Table 11.2: UK Amateur Satellite Service Allocations at Spring 2005

Table 11.3: Some harmonic relationships for the microwave bands

Starting frequency	Multiplication	Output frequency
144MHz	x3	432MHz
	x9	1296MHz
	x16	2304MHz
	x24	3456MHz
	x46	5760MHz
	x72	10,368MHz
	x108	24,192MHz
432MHz	x3	1296MHz
	x8	3456MHz
	x24	10,368MHz
	x56	24,192MHz
1152MHz	+144*	1296MHz
	x2	2304MHz
	x3	3456MHz
	x5	5760MHz
	x9	10,368MHz
	x21	24,192MHz

* Note: additive mixing, not multiplication

quency of 1296-1298MHz is often used for the millimetre bands, ie 24GHz and higher.

There are 'preferred' sub-bands in virtually all of the amateur allocations where the majority of narrow band (especially weak signal DX) operation takes place. Typically 2MHz wide sub bands, often harmonically related to 144MHz as shown in **Table 11.3**, were originally adopted for this purpose.

Some of these harmonic relationships are no longer universally available or usable because the lower microwave bands are rapidly filling up with Primary user applications. Indeed, the position is changing particularly rapidly at the time of publication and the reader should refer to the latest ITU/IARU band plans (see RSGB web site) to get up-to-date information on current amateur usage, even though the current narrow band segments are indicated in Table 11.1.

MODERN MICROWAVE COMPONENTS AND CONSTRUCTION TECHNIQUES

Static Precautions

Some types of microwave components, for example Schottky diodes (mixers and detectors), microwave bipolar transistors and GaAsFETs can be damaged or destroyed by static charges induced by handling, and thus certain precautions should be taken to minimise the risk of damage.

Such sensitive devices are delivered in foil lined, sealed envelopes in conductive (carbon filled) foam plastic or wrapped in metal foil. The first precaution to be taken is to leave the device in its wrapping until actually used. The second precaution is to ensure that the device is always the last component to be soldered in place in the circuit. Once in circuit the risk is minimised since other components associated with the device will usually provide a 'leakage' path of low impedance to earth that will give protection against static build up.

Before handling such devices the constructor should be aware of the usual sources of static. Walking across nylon or polyester carpets and the wearing of clothes made from the same materials

are potent sources of static, especially under cold dry conditions. The body may carry static to a potential of several thousand volts although much lower leakage potentials existing on improperly earthed mains voltage soldering irons are still sufficient to cause damage. Some precautions are listed below:

- Avoid walking across synthetic fibre carpets immediately before handling sensitive devices.
- Avoid wearing clothes of similar materials.
- Ensure that the soldering iron is properly earthed whilst it is connected to its power supply. This is a common sense precaution in any case. Preferably use a low voltage soldering iron.
- Use a pair of crocodile clips and a flexible jumper wire to connect the body of the soldering iron to the earth plane of the equipment into which the device is being soldered.
- If the component lead configuration allows (and the usual flat pack will), place a small metal washer over the device before removing it from its packing in such a position that all leads are shorted together before and during handling. Alternatively, it might be possible to use a small piece of aluminium foil to perform the same function, removing the foil once the device has been soldered in place.
- A useful precaution that will minimise the risk of heat damage, rather than static damage, is to ensure that the surfaces to be soldered are very clean and preferably pre tinned.
- Immediately before handling the device, touch the earth plane of the equipment and the protective foil to ensure that both are essentially at the same potential.
- Place the device in position, handling as little as possible.
- Disconnect the soldering iron from its power supply and quickly solder the device in place. It may be necessary to repeat some of the operations if the soldering iron has little heat capacity.

Finally, when assembling items of equipment to form a complete operating system, for instance when installing a masthead pre-amplifier and associated transmit/receive switching, it is important to keep leads carrying supply voltages to the sensitive devices well away from other leads carrying appreciable RF levels

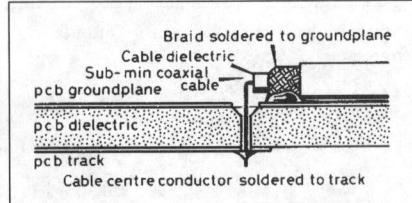

Fig 11.1: Correct coaxial cable connection technique

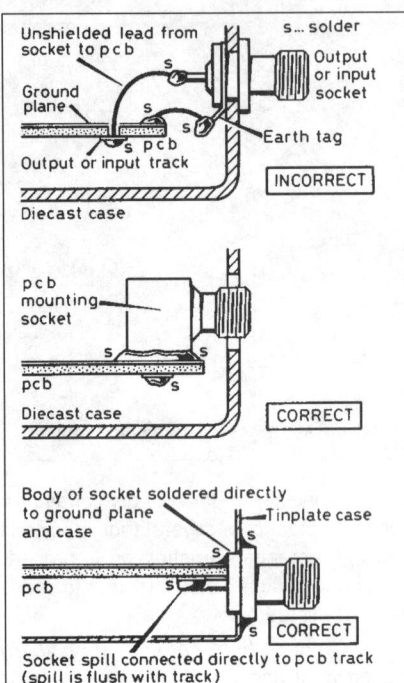

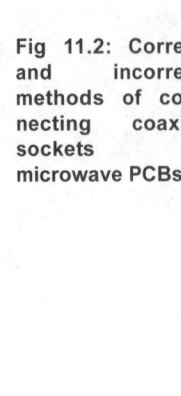

Fig 11.2: Correct and incorrect methods of connecting coaxial sockets to microwave PCBs

or those leads which might carry voltage transients arising from inductive (relay) switching. Such supply lines should be well screened and decoupled in any case, but physical separation can minimise pick up, thus making the task of decoupling easier.

PCB Materials

A printed circuit board (PCB) in a microwave design is not like the one you find in HF equipment. The printed tracks are an integral part of the circuit not just there to interconnect the components. The tracks are microstrip transmission lines, used to form matching circuits, tuned circuits and filters. The design process takes into account the base material of the board used, the thickness of the copper deposit and the dimensions of the track etched. It is therefore important to use the material specified in the design that you are using otherwise the circuit may not perform as expected. Conventional Epoxy/glass PCB board is usable, with care, up to about 3GHz. Most designs use special PCB materials such as Rogers RO 4003 or RT/duroid 5870.

If possible try to use PCBs that have been professionally produced using the correct PCB material and good artwork. If that is not possible you can produce your own PCBs using conventional etching techniques. Microwave PCBs always have an earth plane on one side and etched tracks on the other side; the earth connections from one side to the other are important and are formed by plating through the appropriate holes on professionally manufactured boards. For home made PCBs the best technique is to use small rivets to make these connections, they are fitted and soldered in place forming a good, low inductance, interconnection.

Earthing and Interconnections

Earthing is a very important topic for the microwave constructor. As mentioned above, the earth connections from one side of a PCB to the other must be as good as possible to reduce the effects of stray inductance. The same is true for all other earth connections; they must be as short and solid as possible. It is important to house finished PCBs properly in order to screen the circuitry from stray pick-up and provide a good earth. Small die-cast boxes can be used but they are quite expensive and difficult to use. Piper Communications stock tinplate boxes of various sizes that are an acceptable substitute for the die-cast boxes. These are widely used in Europe for housing such PCBs and are much less expensive. They consist of two L-shaped side pieces, and top and bottom lids. It is intended that the PCB be put into the box, joining the edges of the ground plane to the sides of the box.

To interconnect circuits it is necessary to use RF connectors. N-type connectors are too large and BNC connectors can be unreliable. UHF sockets *must not* be used (they are absolutely useless at UHF, despite their name, and significantly mismatched even at 144MHz) and the only really reliable types are SMA, SMB or SMC, all of which are expensive. You might like to consider taking the outputs away by directly connecting miniature 50Ω coaxial cable as shown in **Fig 11.1**. Do not take the output away as shown in **Fig 11.2**, this is disas-

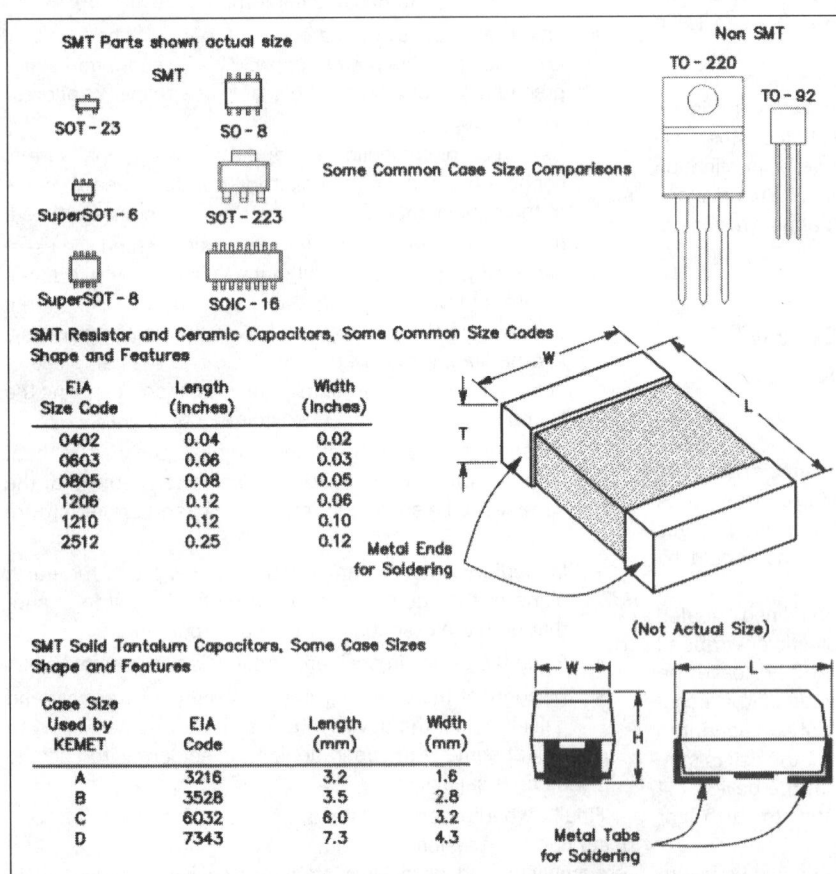

Fig 11.3: Size comparison of some surface mount devices and their dimensions

Product Family	Min Freq (GHz)	Max Freq. (GHz)	Pout (dBm)	Gain (dB)	Noise Figure (dB)	Isolation (dB)	Bias current (mA)	@ Vdd (V)
MGA-641	1.2	10	12	12	7.5	35	50	10
MGA-725	0.1	6	17	14	1.4	23	20	3
MGA-815	0.5	6	14	12	2.8	24	42	3
MGA-825	0.5	6	17	13	2.2	22	84	3
MGA-835	0.5	6	22	22	6	32	142	3
MGA-855	0.5	8	9	18	1.6	41	15	3
MGA-865	0.8	8	6	20	2	46	16	5

Table 11.4: Data for Agilent MMICs. (Copyright Agilent Technologies. Reproduced with permission)

Model, Mini Circuits	Equivalent MAR/MAV	Equivalent Avantek	Alphanumeric	Dot Colour
MAR-1	MAV-1	MSA0185	A01	Brown
MAR-2	MAV-2	MSA0285	A02	Red
MAR-3	MAV-3	MSA0385	A03	Orange
MAR-4	MAV-4	MSA0485	A04	Yellow
MAR-6		MSA0685	A06	White
		MSA0735		
MAR-7			A07	Violet
MAR-8		MSA0885	A08	Blue
		MSA0835		
MAV-1	MAR-1	MSA0104	1	
MAV-2	MAR-2	MSA0204	2	
MAV-3	MAR-3	MSA0304	3	
MAV-4	MAR-4	MSA0404	4	
		MSA0504	5	
		MSA0604	6	
		MSA0704	7	
		MSA0804	8	
MAV-11		MSAO 1 104	A	
ERA-1			E1	
ERA-2			E2	
ERA-3			E3	
ERA-4			E4	
ERA-5			E5	
ERA-6			E6	

Model	Gain typical dB at Frequency (GHz)							Max power out 1dB comp. 1GHz	Noise figure	IP3 dBm
	0.1	0.5	1.0	2.0	3.0	4.0	6.0			
MAR-1	18.5	17.5	15.5					+1.5dBm	5.5	+14.0
MAR-2	12.5	12.3	12.0	11.0				+4.5dBm	6.6	+17.0
MAR-3	12.5	12.2	12.0	11.5				+10dBm	6.0	+23.0
MAR-4	8.3	8.2	8.0					+12.5dBm	6.5	+25.5
MAR-6	20.0	18.5	16.0	11.0				+2.0dBm	3.0	+14.5
MAR-7	13.5	13.1	12.5	11.0				+5.5dBm	5.0	+19.0
MAR-8	32.5	28.0	22.5					+12.5DbM	3.3	+27.0
MAV-11	12.7	12.0	10.5					+17.5dBm	3.6	+30.0
ERA-1				11.6	11.2		10.5	+13dBm (2GHz)	7.0	+26.0
ERA-2	16.0			14.9	13.9		11.8	+14dBm (2GHz)	6.0	+27.0
ERA-3	22.2			20.2	18.2			+11dBm (2GHz)	4.5	+23.0
ERA-4	13.8		14.0	13.9	13.9	13.4		+19.1dBm	5.2	+36.0
ERA-5	20.4		20.0	19.0	17.6	15.8		+19.6dBm	4.0	+36.0
ERA-6	11.1		11.1	11.8	11.5	11.3		+18.5dBm	8.4	+36.5

Application	Model	Application	Model
High frequency gain	ERA-1, usable to 10GHz	Stable high gain	MAR-1, ERA-3
Low noise amplifier	MAR-6, MAR-8, MAV-11	Medium output	MAV-11, MAR-3, MAR-4
Medium noise	ERA-3, ERA-5	High output	MAV-11, ERA-4, ERA-5
High dynamic range	MAV-11	Multiplier	ERA-3 (clean harmonics)

Table 11.5: MMIC data, an extract from data on the Mini Circuits web site

trous as it will almost certainly cause mismatch, stray inductive losses and may detune the output lines so that they will not resonate properly.

Surface Mount Components

Surface mount components are ideally suited for microwave construction because there are no leads to introduce extra inductance in series with the actual component being used. This means that circuit performance can be reproduced more easily. It does introduce a new construction challenge for the newcomer to microwaves. Dealing with tiny components and soldering them in place can seem a daunting task but after some practice it is easy.

Fig 11.3 shows some of the more common SMD component sizes, obviously some special tools are needed to cope with these small components. The essential tools are:

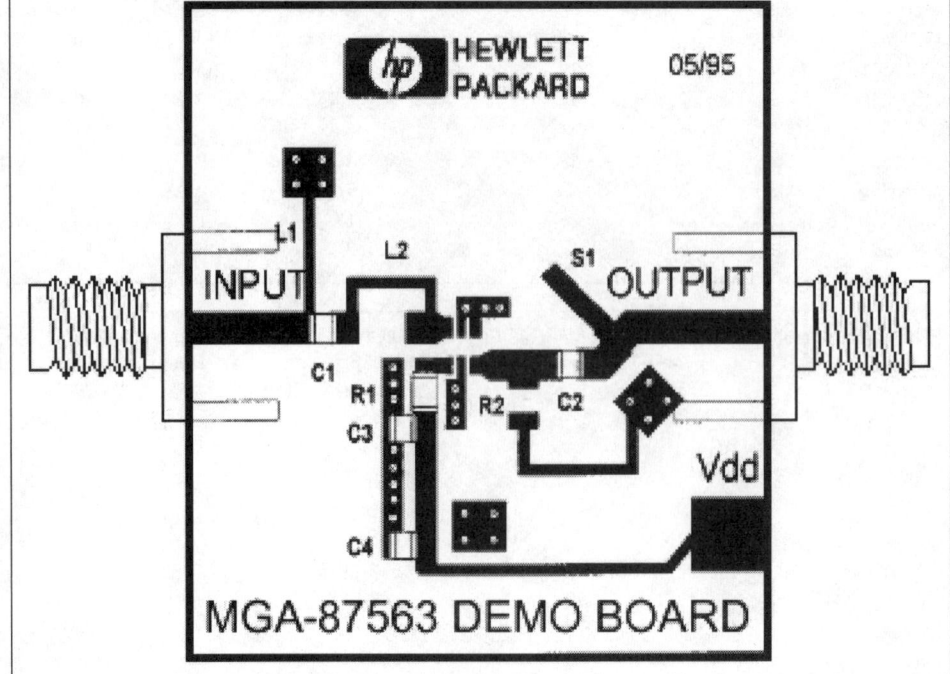

Fig 11.4: Agilent PCB for 2,400MHz LNA. For more details see [6]

- An illuminated magnifying glass. For instance, a bench-mounted magnifier with a five inch glass.
- A low power temperature-controlled soldering iron. This should have its tip earthed to reduce problems with static. A fine conical tip is best.
- Thin flux cored solder, preferably 26SWG (0.5mm). Larger diameter solder is difficult to use because it floods the solder pads and tends to create short circuits between solder pads.
- De-soldering braid, for use when too much solder has been applied.
- A flux pen to apply a small amount of flux before components are mounted.
- A good pair of non-magnetic tweezers.
- A PCB frame or some other method to hold your printed circuit board whilst soldering. If you don't hold the PCB down you will run out of hands to hold everything else!

Before you start to mount components, the PCB should be lightly tinned, just enough solder for the solder to flow onto the component but not too much otherwise short-circuits will be a problem. As with wired components the assembly sequence should start with the low value parts such as resistors and capacitors.

Position the component in position and apply heat just long enough for the solder to flow; you will need to hold the component in place otherwise it will move as the solder flows and may well land up on the end of your soldering iron rather than on the PCB!

Multi-leaded devices, like ICs, should be tacked in place by soldering leads at opposite corners and then flowing solder to all the other legs. Some of the latest ICs have an earthing pad underneath which makes life very difficult. The only successful technique that I have heard of is to mount such components first by heating the complete PCB over an electric cooker to flow the solder under the IC.

Monolithic Microwave Integrated Circuit (MMIC) Amplifiers

MMICs are now widely used in amateur radio designs and are available from several manufacturers including Mini-Circuits and Agilent (Hewlett Packard) at a price that makes them very attractive for many applications.

Keeping track of the devices that are available can be a problem; two useful sources of information are the Agilent (Hewlett Packard) website [6] and the Mini-Kits website [7]. Some useful information from these suppliers is reproduced in **Tables 11.4 and 11.5**.

The Agilent website has a wide range of application notes to show how to use their MMICs. They also supply evaluation boards so that the complete circuit can be built and tested. One such application note is for a 2,400MHz LNA, designed to provide an optimum noise match from 2,400 through 2,500MHz, making it useful for applications that operate in the 2,400 to

Frequency (MHz)	Gain (dB)	Noise Figure (dB)
1,700	10.4	2.60
1,800	13.8	2.57
1,900	11.3	2.45
2,000	11.9	2.38
2,100	13.3	2.06
2,200	12.8	2.02
2,300	12.9	2.12
2,400	11.5	2.05
2,500	11.5	2.14
2,600	10.5	2.25
2,700	10.9	2.29
2,800	10.3	2.33
2,900	9.8	2.35
3,000	9.6	3.42

Table 11.6: 2,400MHz LNA noise figure and gain with Vd = 3V

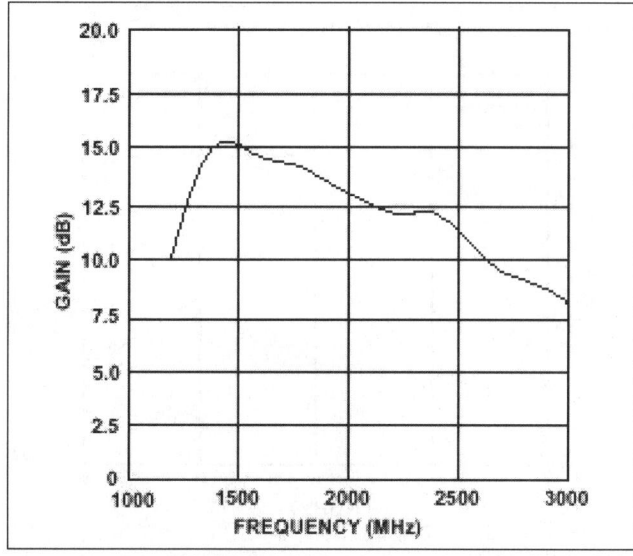

Fig 11.5: 2,400MHz LNA associated gain at maximum noise figure. With Vd = 5V

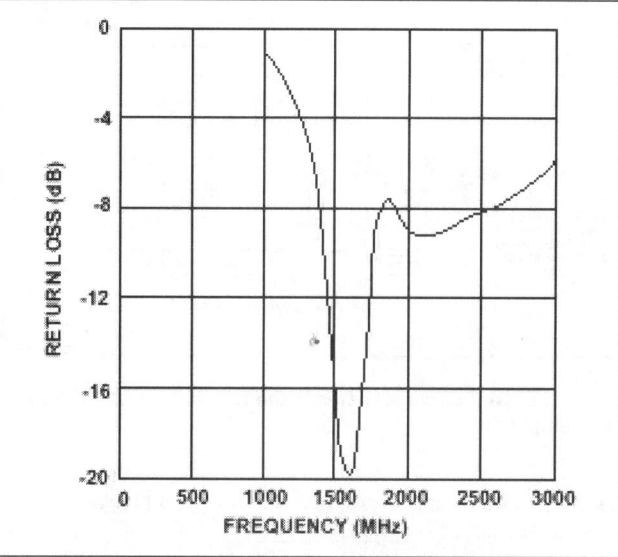

Fig 11.7: 2,400MHz LNA input return loss. With Vd = 5V

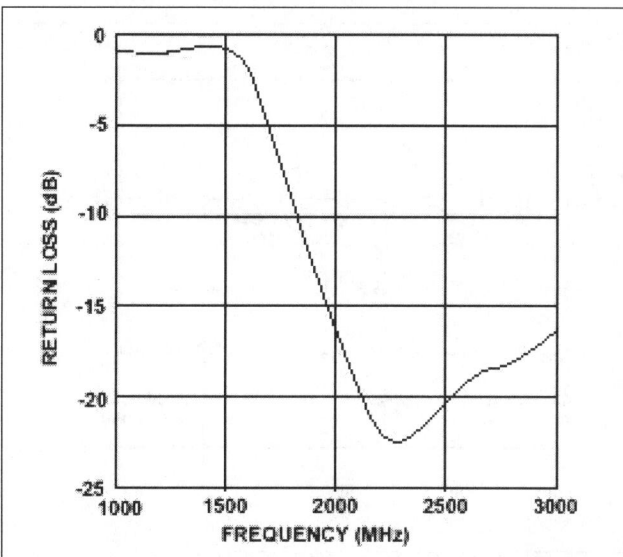

Fig 11.6: 2,400MHz LNA output return loss. With Vd = 5V

2,483MHz ISM band. The component labels appearing in the following paragraphs refer to positions shown in **Fig 11.4**. The input match consists of a shunt inductor at L1 and a series inductor at L2. Both of these inductors use the tracks as originally etched on the circuit board without modification. The output is matched with a simple shunt open circuited stub (S1) on the output 50-ohm microstripline. 22pF capacitors were used for both the input (C1) and output (C2) blocking capacitors.

A 16-ohm chip resistor placed at R1 and decoupled by a 100pF capacitor at C3 provides a proper termination for the device power terminal. An additional bypass capacitor (100 to 1000pF) placed further down the power supply line at location C4 may be required to further decouple the supply terminal, especially if this stage is to be cascaded with an additional one. Proper decoupling of device VCC terminals of cascaded amplifier stages is required if stable operation is to be obtained. If desired, a 50-ohm resistor placed at R2 will provide low frequency loading of the device. This termination reduces low frequency gain and enhances low frequency stability. The MGA-87563 has three ground leads, all of which need to be well grounded for proper RF performance. This can be especially critical at 2.4GHz where common lead inductance can significantly decrease gain.

The performance of the LNA as measured on the HP 8970 Noise Figure Meter is shown in **Table 6**. At 2.4GHz, the loss of the FR-4/G-10 epoxy glass material can add several tenths of a dB to noise figure and lower gain by double the amount. The swept plots, **Figs 5 – 7** were taken on a scalar analyser and show the performance of the amplifier.

Another simple design that is usable up to 5.7GHz, using MMIC amplifiers as gain blocks, is shown here. Two designs are presented, a two and a three stage amplifier using similar circuitry [8]. The first amplifier is shown in **Fig 11.8**. It uses two MSA0835 devices, in cascade, with a gain that varies from 27dB at 2GHz down to 7dB at 6GHz. Output power is only a few milliwatts at the upper end of the range but the amplifier is not intended to form the output stage of a transmitter. This function is better provided by a

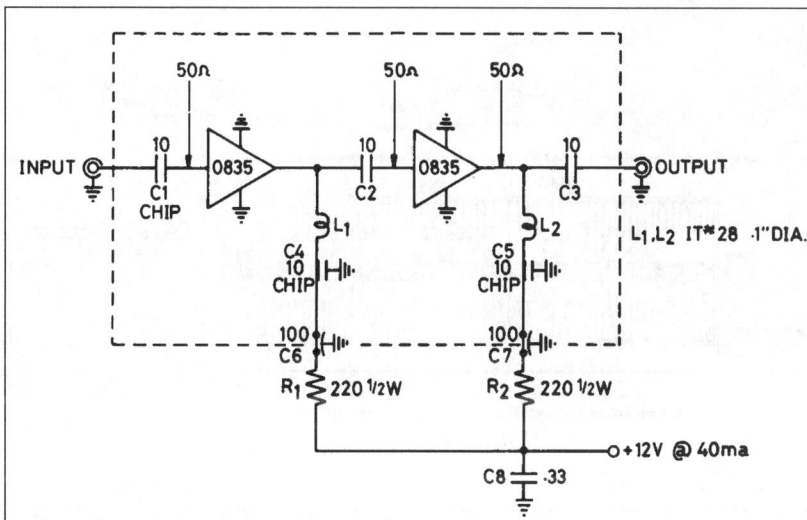

Fig 11.8: Circuit of the two stage MSA0835 wide band amplifier

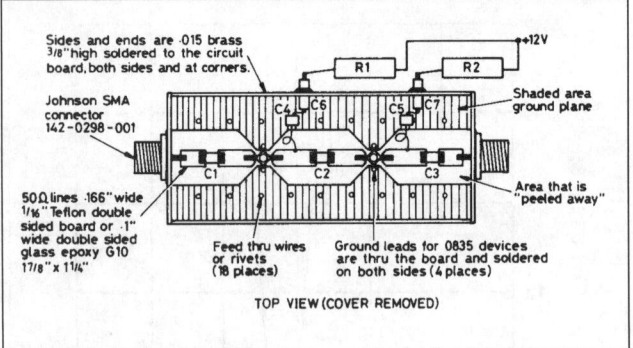

Fig 11.9: Layout of components in the two stage MSA0835 wide band amplifier

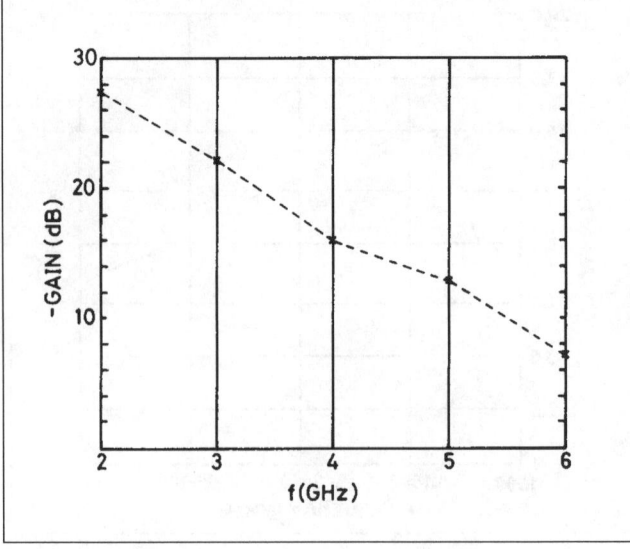

Fig 11.10: The frequency response of the two stage MSA0835 wide band amplifier

GaAsFET power amplifier above 3GHz. and by a bipolar amplifier below 3GHz.

Construction of the two stage amplifier is straightforward using a simple microstrip line track of 50-ohm impedance, on either glass fibre or PTFE double sided board, as shown in **Fig 11.9**. PTFE board material is preferred if operation above 3GHz is required.

Small IOpF ceramic chip capacitors prevent the bias supply to the MSA0835 devices being shorted out by the source and load. Another capacitor prevents the collector supply to the first stage shorting out the second stage base bias. Simple resistor current limiting from the amplifier 12V supply provides bias. Single-turn chokes in the collector leads of the MMICs prevent the bias resistors shunting the output signal to ground. The frequency response of the two stage amplifier is shown in **Fig 11.10**.

The three stage amplifier shown in **Fig 11.11** offers more gain than the two stage amplifier right across the frequency range. Construction is similar to the latter but uses a slightly longer board as shown in **Fig 11.12**. Its frequency response is shown in **Fig 11.13**.

More output can be achieved in this design if a Siemens CGY40 MMIC is used in the output stage in preference to the MSA0835. Gain will be slightly less and the maximum frequency of operation will fall to 3.4GHz. With this modification the output power at 3.4GHz can be as high as 50mW. This power level should be satisfactory for short links or longer line-of-sight paths.

These amplifiers have very high gain and are only conditionally stable when used between good 50-ohm load and source impedances. In practice, the insertion of a 3 or 6dB attenuator at the input and output of the amplifier should ensure stable operation, although in many cases the attenuators will not be required.

Replacing the MSA0835 MMIC with the lower gain 0735 device results in unconditional stability and much less gain variation across the frequency range. The penalty for

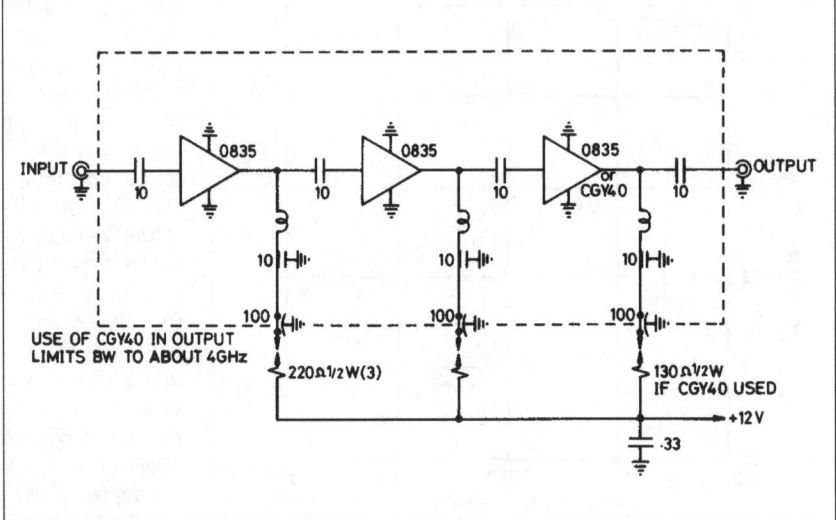

Fig 11.11: Circuit of the three stage MSA0835 wide band amplifier

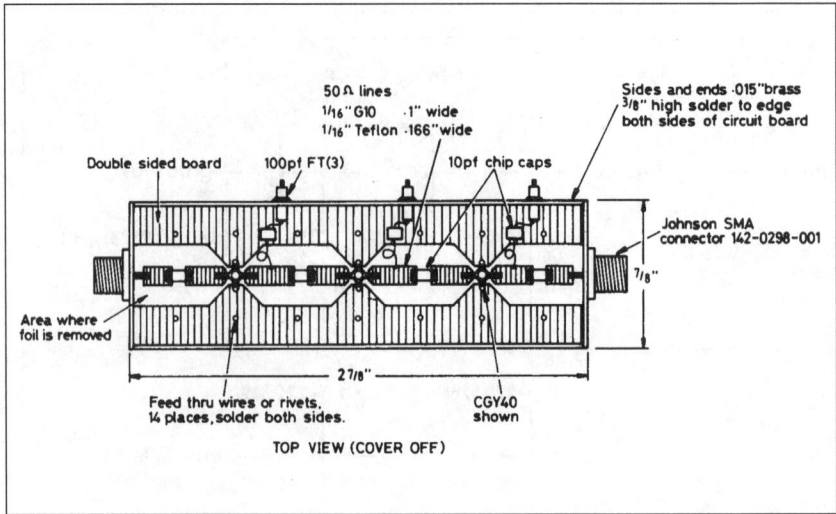

Fig 11.12: Layout of the components in the three stage MSA0835 wide band amplifier

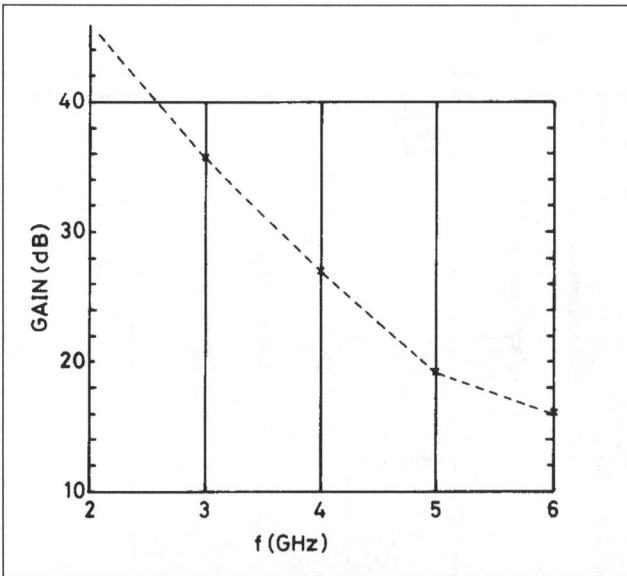

Fig 11.13: The frequency response of the three stage MSA0835 wide band amplifier

this is lower gain and output power. About 32dB gain at 2GHz, falling to 22dB at 3.4GHz can be expected with this design.

It cannot be stressed enough how important the grounding is around the emitter leads of the MMIC devices on this board. In the absence of a plated through-hole board, through-board wire links must be provided where shown, and especially under the emitter lead connections to the top ground plane of the board. Soldering copper foil along the edges of the board is just not sufficient to ensure good ground integrity and the amplifier will self oscillate if links are not used.

MICROWAVE LOCAL OSCILLATOR SOURCES

Early experimenters on the amateur microwave bands used wide band modes, so frequency accuracy was not that important. Many designs in the 1960s used free running oscillators at the operating frequency using Gunn diodes. Then improvements were made and the free running oscillators were locked to a known stable source. The quest for communications using narrow band modes, such as SSB, necessitated a different approach. A stable crystal controlled oscillator, followed by a multiplier chain to produce the required local oscillator frequency became commonplace. The local oscillator is mixed with the output of a commercial transceiver to produce the required sig-

nal on the amateur band to be used (see **Fig 11.14**). The local oscillator and mixer are usually combined into a single unit – a transverter. This still remains the technique of choice used by serious microwave operators. Any 144MHz transceiver can be used but the IC202 is still often used because of its clean and stable output. The main design criteria for a good local oscillator source for microwave use are:

- Good short term frequency stability. Short term frequency variations may be caused by such things as the type of crystal used, the type of oscillator circuit, stability of supply voltages and temperature changes. Small changes in the frequency of the crystal oscillator are multiplied, eg if a local oscillator is used to generate an output on the 10GHz band, a 106.5MHz crystal frequency will be multiplied by 96 to give a local oscillator frequency of 10,244MHz. Thus a change of 25Hz in crystal frequency will change the frequency at 10GHz by about 2.5kHz.

- Good long term frequency stability. Long term frequency variations may be caused by ageing of the crystal used. Also crystals suffer from a hysteresis effect that causes them to operate on a slightly different frequency each time they are started.

- Good signal purity. This is governed by the design of the oscillator. It is important to keep phase noise and spurious outputs of the crystal oscillator to a minimum because the multiplier chain magnifies these.

Other techniques, such as Phase Locked Loops (PLL) and Direct Digital Synthesisers (DDS) are used to generate the local oscillator signal, but these can suffer from poor signal purity.

A High Quality Microwave Source for 1.0 to 1.3GHz

This high quality microwave source, with output in the range 1.0 to 1.3GHz, was originally designed by G4DDK [9] to provide two +10dBm (10mW) outputs at 1152MHz for use in a 1296MHz transverter with a 144MHz IF. It is known as the G4DDK-001 microwave source. Several modified versions have appeared since the original design, but this one remains the 'standard'. Later work showed it was possible to use the same board, with a suitable crystal, anywhere in the range 1000 to 1400MHz with only minor changes in component values and output spectral purity. Versions of the board have since been produced to provide outputs between 700 and 1500MHz. At these two extremes it has been found necessary to change the length of some of the tuned microstrip lines in order to maintain resonance. PCBs are available from the Microwave Component Service [10].

The circuit of the oscillator source is shown in **Fig 11.15** and the component values in **Table 11.7**. The crystal oscillator section uses the Butler design. In this design the oscillator operates with a crystal frequency between 90-110MHz, although by changing the value of C3 the circuit can be made to operate reliably between about 84-120MHz. Operation outside this range may require changes in the value of other components. A 9V integrated circuit regulator (78L09) stabilises the supply to the oscillator and limiter stages.

The third stage is a times two multiplier. Input, in the frequency range 250-330MHz, is taken from the high imped-

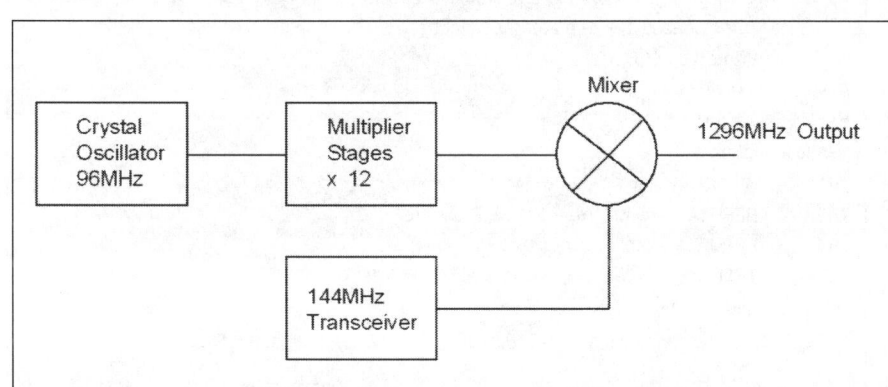

Fig 11.14: Local oscillator mixed with transceiver to give microwave output

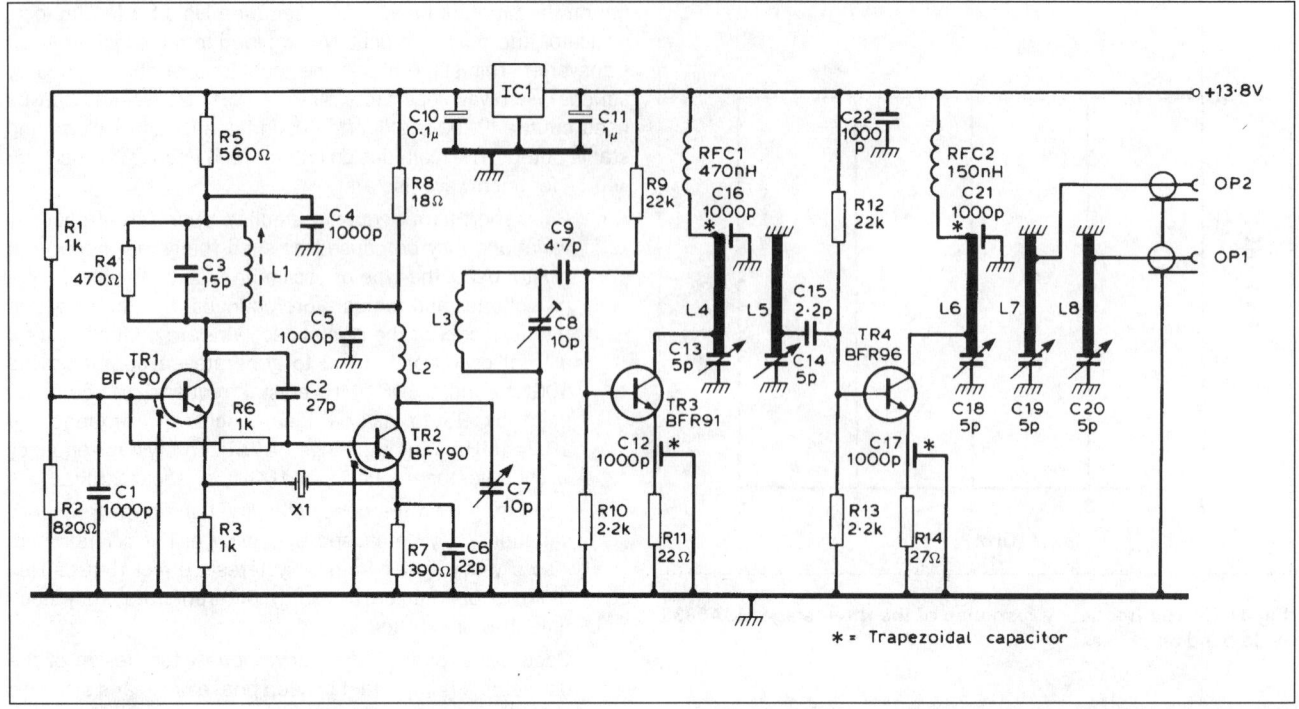

Fig 11.15; Schematic diagram of the microwave source G4DDK-001

ance end of the tuned circuit formed by L3 and C8 via C9. The coupled microstripline tuned circuits L4/C13 and L5/C14 tune the output of TR3 to the range 500-660MHz.

A high impedance tap on L5 couples the doubled signal to the base of the final multiplier stage, TR4. The output of this stage is tuned by three coupled tuned microstriplines; L6/C18, L7/C19 and L8/C20.

The tuning range of this filter is highly dependent upon the type of trimmer capacitors chosen for C18, 19 and 20. By using the specified capacitors, the filter will tune from 980 to 1400MHz, encompassing the range of second harmonics from the previous stage (1000 to 1320MHz).

Three stages of filtering are used to achieve a very clean spectrum at output I. A slightly less clean output is available at output 2, since this output is taken from the second stage of the filter. Even so, the output here is more than adequate for use in a receive converter with output 1 being reserved for the transmit converter. Each output is at a level of +10dBm (10mW), but if only one output is required then this should be taken from output 1 and the track to output 2 cut where it leaves L7. A single output of +13dBm (20mW) should be available in this configuration. **Fig 11.16(a) and (b)** are analyser plots of the outputs of the board.

PCB artwork for the UHF source is shown full size in **Fig 11.17** (in Appendix

Resistors

R1, 3, 6	1k0	R9,12	22k0
R2	820R	R10, 13	2k2
R4	470R	R11	22R
R5	560R	R14	27R
R7	390R		0.25W miniature carbon film or metal film
R8	18R		

Capacitors

C1,4,5,22	1000pF high-K ceramic plate, e.g. Philips 629 series
C12, 16, 17, 21	1nF trapezoidal capacitor from Microwave Component Service or Piper
C10	0.1μF tantalum bead, 16V working
C11	1μF tantalum bead, 16V working
C3	15pF low-K ceramic plate, e.g. Philips 632 Series
C7, 8	5mm trimmer, 10pF maximum
C13,14	5mm trimmer, 5pF maximum
C18.19.20	5mm trimmer, 5pF maximum. Must be able to reach 0.9pF minimum, e.g. SKY (green) or Murata TZ03 (black)

Inductors

L1	TOKO S18 5.5t (green) with aluminium core
L2, 3	3t of 22swg tinned or enamelled copper wire, 3mm ID. Turns spaced 1 wire diameter, height of coil 5mm above board.
L4-8	Printed on PCB
RFC1	470nH
RFC2	150nH

Semiconductors

TR1, 2	BFY90, available from Bonex, Piper etc.
TR3	BFR91A, available from Bonex, Piper etc
TR4	BFR96, available from Bonex, Piper etc
IC1	78L08, Piper, STC Components etc

Miscellaneous

X1	5th overtone crystal in HC18/U case. Frequency of crystal = Fout/12

Table 11.7: Component list for G4DDK-001 oscillator/multiplier

The Radio Communication Handbook

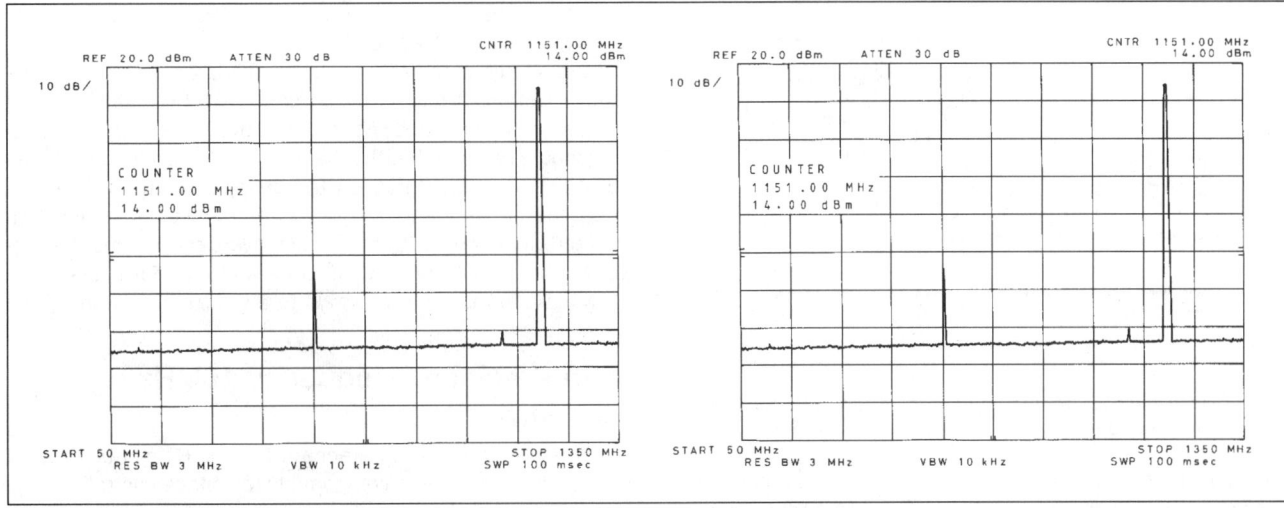

Fig 11.16: A spectrum analyser plot of the output from the microwave source G4DDK-001. (a) Output 1, output 2 terminated in 50Ω. (b) Output 2, output 1 terminated in 50Ω. If output 2 is not required, cut the track near the line. Output 1 should then look like (a), but be at +13dBm

B). **Fig 11.18** is a photograph of a finished unit. Construction is generally straightforward, with the trapezoidal capacitors being carefully soldered into the slots in the board as shown in the component overlay diagram **Fig 11.19**. Capacitors C18, 19 and 20 must be miniature 5mm diameter types such as SKY, Murata or Oxley. The use of larger 7mm trimmers such as the popular Dau or Philips types will inevitably lead to tuning problems. The circuit was designed to take the small types. This also applies to C7, 8, 13 and 14, where overcoupling and consequently tuning problems can also occur.

Alignment is straightforward. Connect DC power and check that the current drawn is no more than about 150mA. If significantly more, check for short-circuits or wrongly placed components.

When all is well, proceed with the alignment. Place an absorption wavemeter pick-up coil close to L1 and tune to the crystal frequency. A strong reading should be indicated on the meter. Peak the indication by turning the core of L1. Turn the oscillator on and off to check that it re-starts satisfactorily. If it doesn't, turn the core of L1 about a quarter turn and try again. Too close coupling of the wavemeter (ie too much power absorption) may also inhibit easy re-starting, so use the loosest coupling possible to give adequate indication.

Set an analogue (moving coil) multimeter to the 2.5V range (or nearest equivalent) and measure the voltage across R11. This should be no more than a few hundred millivolts. Peak the reading by tuning C7 and C8. Confirm that the frequency selected is three times the crystal frequency by placing the coil of the wavemeter close to L3. The reason for using an analogue meter is that digital meters do not measure in real time and therefore tend to show what you have just done, rather than what you are doing, whilst adjusting the

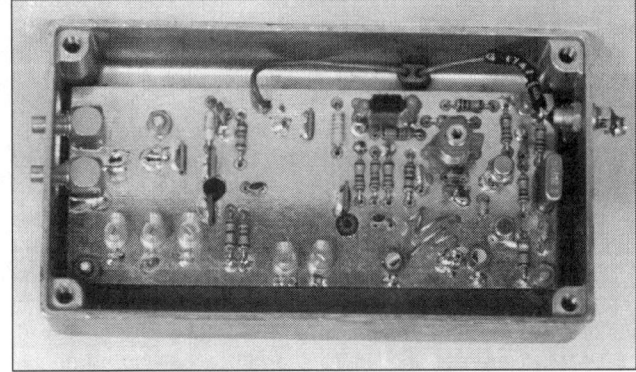

Fig 11.18: Photograph of the G4DDK-001 microwave source in the recommended size diecast box. The picture shows an early version of the source. L1 and L2 were laid out slightly differently in later versions

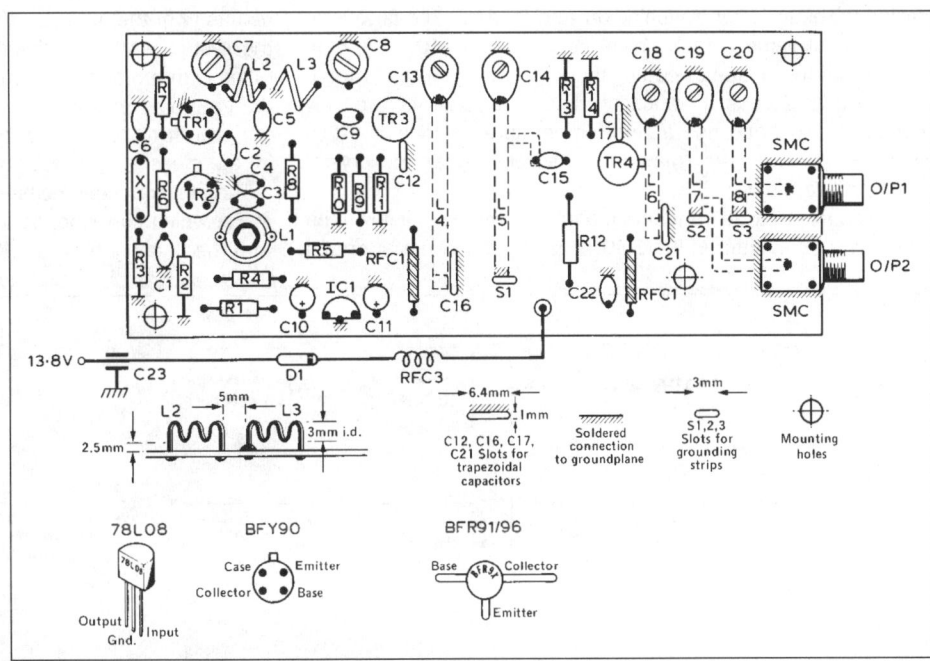

Fig 11.19: Component layout diagram of the G4DDK-001 microwave source

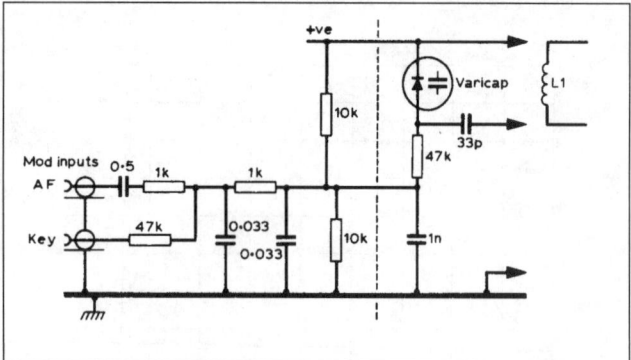

Fig 11.20: A simple FM/PSK modulator circuit for the G4DDK-001 microwave source

trimmers: indeed, if a digital meter is used, it is possible to miss the increase in measured voltage which occurs as each circuit is brought to resonance - the tuning is sharp!

Transfer the meter leads across R14 and peak the reading by tuning C13 and C14. Again check that the correct harmonic (twice the preceding stage) has been selected by using the wavemeter.

Finally, connect a low power wattmeter (+10 to +20dBm full scale, ie 10 to 100mW) to the output and tune C18, 19 and 20 for a maximum reading.

By using the wavemeter, confirm that the correct harmonic (now the output frequency and twice the preceding stage) has been selected. It may now be necessary to go back and slightly re-peak the trimmers for an absolute maximum reading at the final output frequency.

Final setting of the frequency of the crystal oscillator can now be done by either using a known high accuracy frequency counter or by connecting the source as the local oscillator of your 1296MHz converter and listening for a beacon whose frequency is known. L1 can then be adjusted to bring the signal onto the correct receiver dial calibration.

If difficulty is experienced in pulling the frequency to that marked on the crystal, it is very likely you have a non standard crystal. Pulling the frequency too far can result in the oscillator failing to restart after switching off and then on. The cure is to put a small value ceramic plate capacitor, say 10 - 33pF, in series with the crystal by cutting the PCB track near the latter. If you have to do this modification, keep the leads of the new capacitor short and use a zero temperature coefficient (NPO) capacitor, or frequency may drift unacceptably as the crystal oscillator warms up.

The source may be frequency modulated, although if it is to be used as a transmitter in the 1.3GHz band, greater deviation will

be required than when multiplying to 10GHz. The modulator circuit shown in **Fig 11.20** can be used.

If the source is to be used as a 1.3GHz beacon source, the required 800Hz frequency shift can easily be obtained using this same circuit. Note, though, that the 'sense' of the deviation should be such that the marker signal (carrier only) should be at the nominated beacon frequency and 'space' keys the beacon low in frequency by 800Hz, returning to 'mark' for each character element. With the circuit given, this means the keying voltage should be low for mark and high for space (conventional), ie mark corresponds to an earth condition on the keying lead.

High Precision Frequency 10MHz Standard

A high stability frequency standard for 10MHz can be created using only three system components, a voltage controlled crystal oscillator (VCXO), a counter controlled by the signal from the GPS satellite system and a D/A converter for fine control of the VXCO. The short term and long term frequency stability that can be obtained by this simple means far exceed the requirements for practical amateur radio operations. The 10MHz standard can thus be used as the basis of a highly accurate method of generating local oscillator signals for microwave use.

The most commonly used method for precision time comparisons nowadays makes use of the satellites of the Global Positioning System (GPS). The GPS satellites carry atomic clocks of the highest accuracy, the operation of which is carefully monitored by the ground stations. A stable quartz oscillator regulated with the aid of the GPS ensures that its maximum frequency deviation always remains better than 1×10^{-11}. This is a precision of 0.0001Hz in 10MHz! Or, for the microwave amateur, 1Hz in 100GHz.

The frequency control of a 10MHz oscillator using GPS, shown in **Fig 11.21**, was designed by Wolfgang Schneider, DJ8ES, and Frank-Peter Richter, DL5HAT, and published in *VHF Communications* [11]. It uses an HP10544A VXCO (**Fig 11.22**), these are now available on the surplus market and often found for sale on Ebay. In practice an accuracy of approximately 4×10^{-10} can be achieved or, in other words, 4Hz in 10GHz. This value results from the inaccuracies of the counting process built into the system. In all frequency counters, the last bit should be taken with a pinch of salt. Depending on the phase position of the gate time to the counting signal, an error occurs of ±1 bit (phase error ±100ns). For a gate time of 1s, that would be 1Hz for the measuring frequency 10MHz ($\pm 1 \times 10^{-7}$).

The first practical measurements were based on a gate time of 8s, which corresponds to a resolution of 0.125Hz. Together with the phase jitter of the GPS signal, there should have been

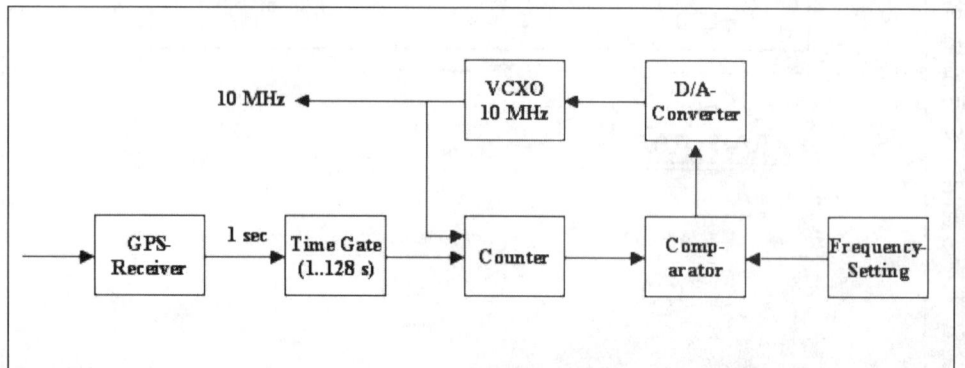

Fig 11.21: Block diagram of frequency control via GPS

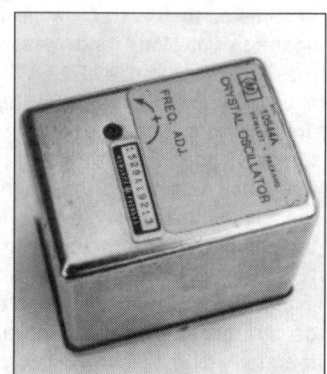

Fig 11.22: HP10544A VXCO

uniform distribution and thus a levelling off of the reading over a relatively long period of time (max. 64 measurements). This turned out to be wishful thinking. On investigation, it was established that the oscillator frequency varied very slowly around the required value of 10MHz. The absolute frequency here was 10.0MHz ±0.0305Hz.

If the gate time is increased to 128s, in theory the reading improves to at least ±0.0078125Hz. However, the influence of the GPS phase jitter is now reduced. This results in an effective usable precision for the 10MHz signal of approximately 4×10^{-10} or 4Hz at 10GHz.

The control stage (**Fig 11.23**) operates like a frequency counter with an additional numerical comparator. The 10MHz output of the HP10544A oscillator is counted. The gate time of the counter is generated from the 1pps signal of the GPS receiver with a 74LS393. For control operation a gate time of 128 seconds is used, and 8s is used in the comparison mode for the OCXO.

The 74HC590 8-bit counter can be used, with a gate time of 8s, to measure the input frequency 10MHz ±16Hz. The minimum resolution here is 0.125Hz. In control operation (gate time 128s), this is improved by a factor of 16, resulting in the system determined precision of 0.78×10^{-9}, based on the 10MHz frequency oscillator.

The frequency of the HP oscillator can be finely adjusted using a tuning voltage of ±5V. This is done by the digital to analogue converter (AD 1851). It has a resolution of 16 bits for a control voltage range of ±3V. This gives a setting range for the OCXO of approximately ±0.5Hz.

The AT89C52 micro-controller controls all the functions including the D/A converter and the status in the LC display. The software in the micro-controller performs two tasks. Firstly, it enables a rough comparison operation to be carried out, and secondly it will continuously carry out the final fine adjustment using the GPS signal.

If pin 4 of K13 is open, then when the voltage is applied to the control circuit board the LC display shows "Warming Up". In order to eliminate any artificial jitter in the GPS signal, a mean value is formed and displayed from 64 readings from the 74HC590 counter. A change in the oscillator frequency using its frequency adjustment control will thus not display any effect for some time. So after a change of the control we must just wait for approximately 64 x 8s until the next adjustment takes place. If a value of ≥ ±0.250Hz is attained, we can switch over to the basic control by earthing Pin 4 of K13 and selecting the long gate time of 128s by means of a bridge between pins 15 and 16 at K14. The first message on the display is "Warming Up", the first value is displayed after approximately 15 minutes, it is not the deviation in Hertz but the value that is written in the AD 1851 D/A converter. This value can reach a maximum of ±32767, which means approximately ±3V. The software assesses the output of the counter and calculates the value for the AD 1851 D/A converter. From the present value of the counter and the mean of the last 64 counter results, the figure is determined which is to be added to or subtracted from the current D/A converter value. This can be seen on the display.

Theoretically, with the gate time of 128s, and with a mean value formed over 64 readings, the precision of 4×10^{-10} Hz (ie 4Hz in 10GHz) is achieved after 4.5 hours. However, it has been demonstrated in practice that this value has already been reached after approximately two hours.

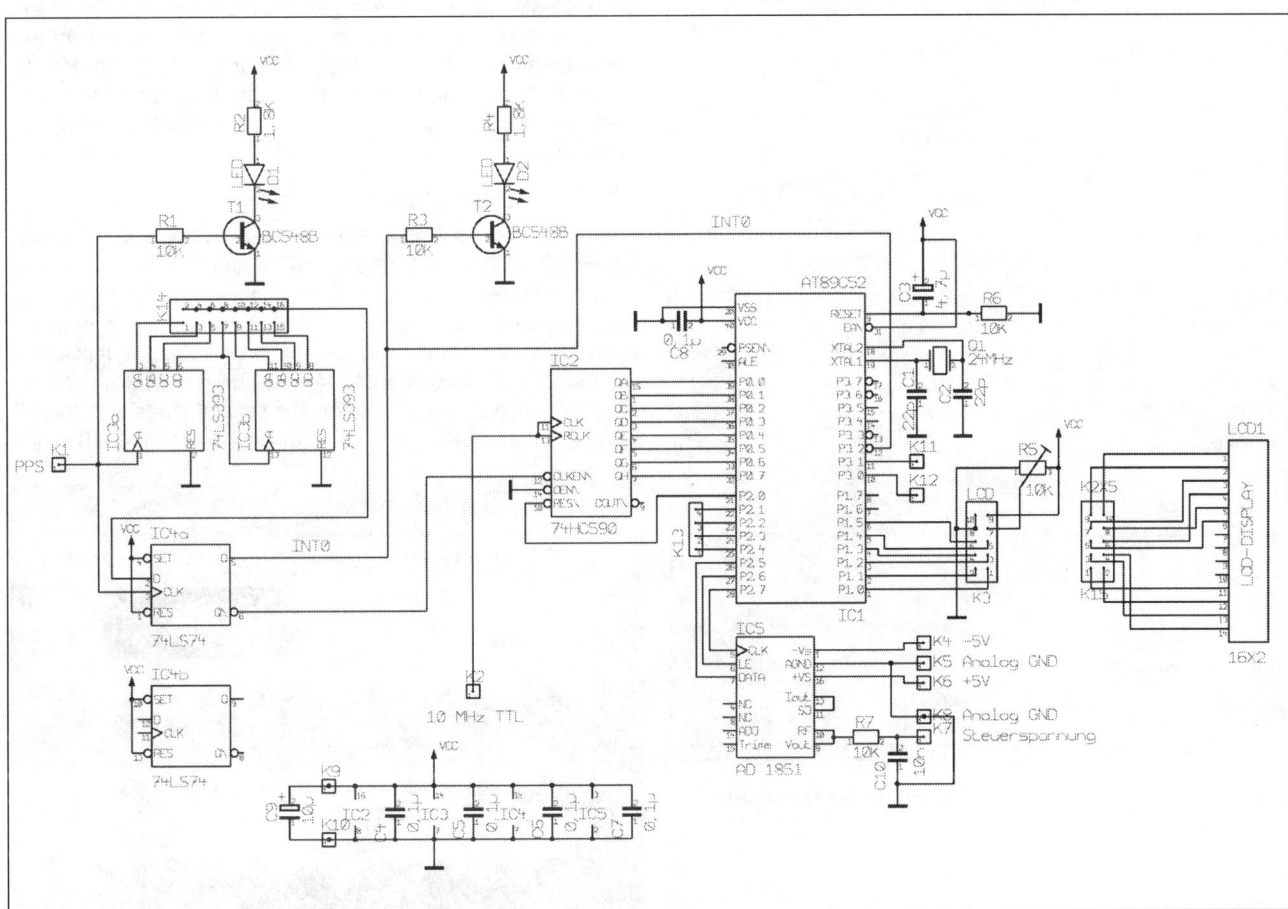

Fig 11.23: Circuit diagram of GPS control stage of high precision frequency standard for 10MHz

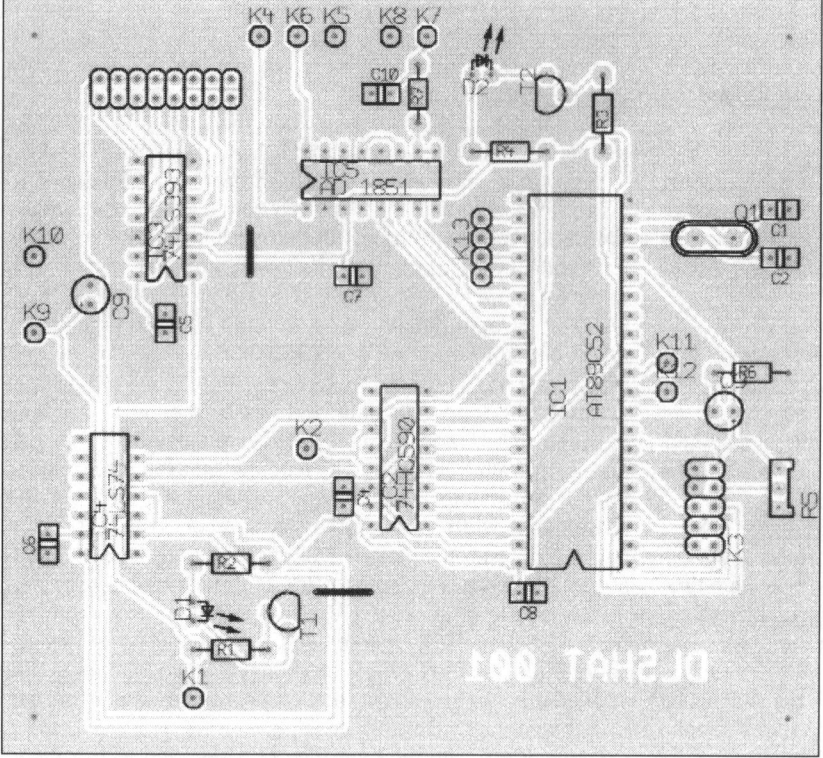

Fig 11.25. Component layout for GPS control stage of high precision frequency standard for 10MHz

Miscellaneous components	
1 x micro-controller	AT89C52
1 x A/D converter	AD1851
1 x TTL-IC	74LS74
1 x TTL-IC	74LS393
1 x TTL-IC	74HC590
2 x Transistor	BC848B
2 x LED	green, low current
1 x crystal	24 MHz
1 x potentiometer	10k
1 x socket terminal strip, 10-pin	
1 x plug strip, 10-pin	
1 x stud strip, 14-pin	
1 x jumper	
1 x circuit board	DL5HAT 001
Resistors	
2 x	1.8k
4 x	10k
Ceramic capacitors	
5 x	0.1µF
2 x	22pF
1 x	10nF
Tantalum capacitors	
1 x	4.7µF/25V
1 x	10µF/25V

Table 11.8: Component list for GPS controller for 10MHz frequency standard.

Fig 11.26: Picture of the prototype high precision frequency standard for 10MHz controlled by GPS

The frequency controller circuit is assembled on 100mm x 100mm double sided PCB (**Fig 11.24** in Appendix B), the component layout can be see in **Fig 11.25** and the component list is shown in **Table 11.8**.

In the author's prototype (**Fig 11.26**) the HP10544A oscillator was used to drive a buffer stage designed by DJ8ES [12]. This gives TTL outputs at 1, 5 and 10MHz and 3 separate 10MHz sine wave outputs. The 1pps signal was generated from a GPS receiver manufactured by Garmin (GPS 25-LVS receiver board). The control assembly output supplies the control voltage for the HP oscillator. The tuning voltage, ±5V, *must* be separately generated in the frequency controller power supply. The following power supplies are required:

- +24V   HP oscillator
- +5V   GPS receiver
- +5V   control assembly
- ±5V   control voltage

Long-term observations of the 10MHz frequency standard over approximately four weeks confirmed the design criteria.

RECEIVE PREAMPLIFIERS

Receive preamplifiers are an important part of the microwave station. They are often used as masthead preamplifiers to overcome feeder loss that can be quite considerable for the higher frequency bands. There are many suitable semiconductors available to produce very low noise preamplifiers for all the amateur bands up to 10GHz. Above that, the devices are available but they are fairly expensive. Two designs are shown here to illustrate the technology available.

13cm Preamplifier

This design is by Rainer Bertelsmeier, DJ9BV, and originally appeared in *Dubus Magazine*. A preamp equipped with a PHEMT provides a top-notch performance in noise figure and gain as well as unconditional stability for the 13cm amateur band. The noise figure is 0.35dB at a gain of 15dB. It utilises the C band PHEMT, NEC NE42484A and provides a facility for an optional second stage on board. The second stage with the HP GaAs MMIC MGA86576 can boost the gain to about 40dB in one

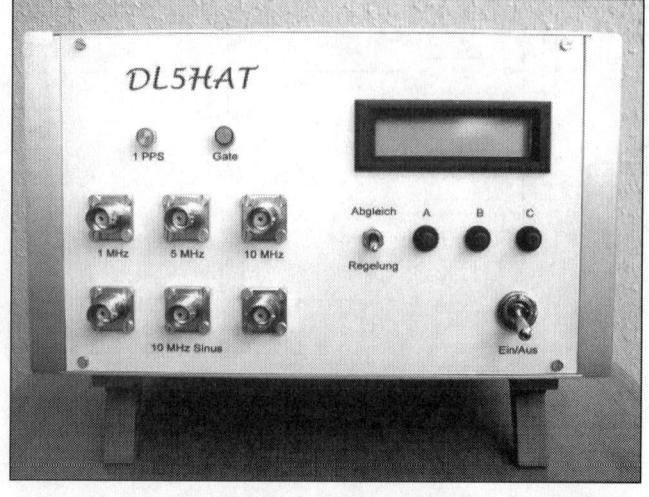

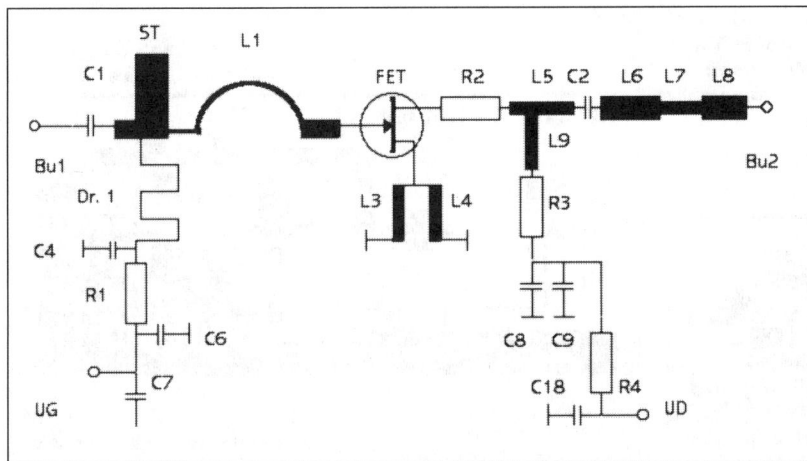

Fig 11.27: Circuit diagram of single stage 13cm PHEMT preamplifier

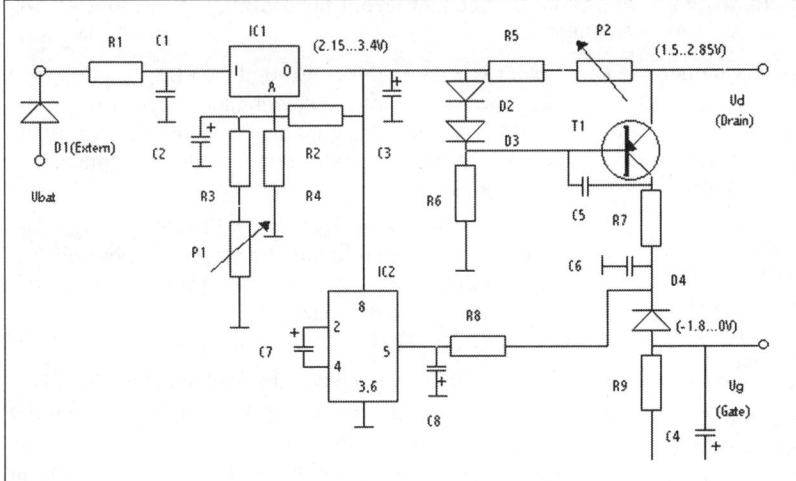

Fig 11.28: Circuit diagram of bias circuit for 13cm PHEMT preamplifier

Table 11.9: Components for 13cm PHEMT preamp

Capacitors	
C1	4.7pF Chip-C 50mil (500 DHA 4R7 JG)
C2, 3, 6	100pF SMD-C, size 0805
C4	5.6pF SMD-C, size 0805
C7, 8, 19	1000pF SMD, size 0805
C9, 18, 20	10nF SMD-C, size 0805
C10, 12, 17	0.1µF SMD-C, size 1206
C11, 14, 15	10µF SMD-Electro, size 1210
C13	1µF SMD-Electro, size 1206
C16	1000pF Feedthrough
Resistors	
R1, 3	470 SMD-R size 1206
R2, 14	390 SMD-R size 1206
R4, 5	100 SMD-R size 1206
R12	6.8kO SMD-R size 1206
P1	1000 SMD-Pot, Murata 4310
Miscellaneous	
Dr.1	Printed λ/4
L1	Wire loop, 0.5mm gold plated copper wire 18mm long, 1mm above board
L2	Wire loop, 0.5mm gold plated copper wire 8mm long, on PCB
D1	IN4007
FET	NE42484A, NEC
MMIC	MGA-86576, HP
T1	PNP eg BC807, BC856, BC857, BC858, BC859, SOT-23
IC1	LTC1044SN8
Bu1, 2	N small flange or SMA
PCB	Taconix TLX, 35 x 72mm, 0.79mm er = 2.55
Box	Tinplate 35 x 74 x 30mm

enclosure. The preamplifier is rather broadband and usable from 2300 to 2450MHz.

The construction of this LNA follows the proven design of the 23cm HEMT LNA [13] by using a wire loop with an open stub as an input circuit (**Fig 11.27**). The FET's grounded source requires a bias circuit to provide the negative voltage for the gate. A special active bias circuit (**Fig 11.28**) is integrated into the RF board that provides regulation of voltage and current for the FET. The component list is shown in **Table 11.9**.

Stub ST and inductance L1 provide a match for optimum source impedance for minimum noise figure. L1 is as a dielectric transmission line above a PTFE board and has somewhat lower loss than a microstripline. L3 and L4 provide inductive feedback to increase the stability factor and input return loss. R1, R2, L9 and R3 increase the stability factor. The system of C2/L5/C6/L7 and L8 is specially designed to match the output of the single stage version to 50 ohms and to allow easy insertion of the GaAs MMIC for the two stage version.

In the two stage version (**Fig 11.29**) it provides the appropriate input and output match to the MMIC. This solution was found by doing some hours of design work with the software design package *Microwave Harmonica*. It allows

the two versions to have the same PCB. C4 provides a short on 2.3GHz, because it is in series resonance at this frequency. On all frequencies outside the operating band the gate structure is terminated by R1. Dr1 is a printed λ/4 choke to decouple the gate bias supply.

The two stage version utilises a HP GaAs MMIC, MGA86576 in the second stage. It provides about 2dB noise figure and 24dB gain. Input is matched by a wire loop for optimum noise figure. Output is terminated by a resistor R5 and a short transmission line L10. Together with L7/L8 and C3 a good output return loss

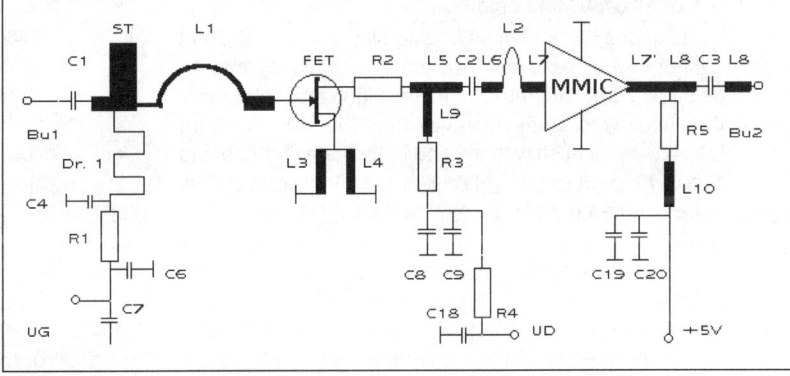

Fig 11.29: Circuit diagram of two stage 13cm PHEMT preamplifier

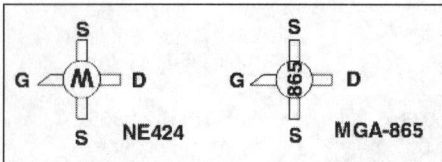

Fig 11.31: Top view of FET and MMIC used in 13cm preamplifier

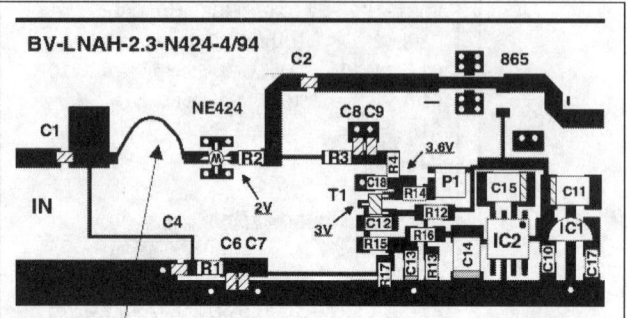

Fig 11.32: Component layout for single stage 13cm PHEMT pre-amplifier

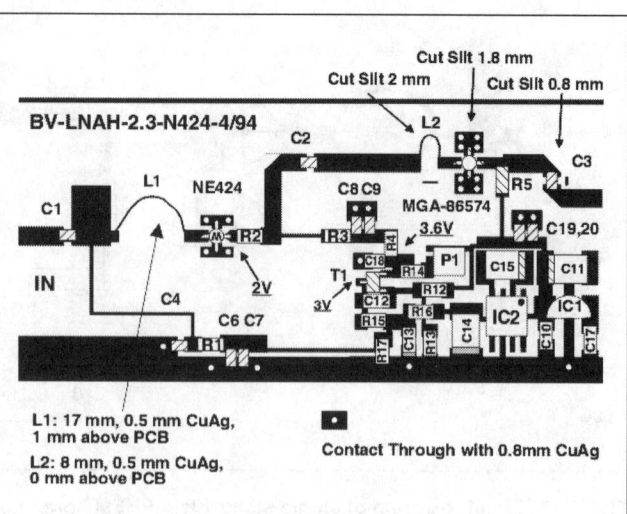

L1: 17 mm, 0.5 mm CuAg, 1 mm above PCB

L2: 8 mm, 0.5 mm CuAg, 0 mm above PCB

Contact Through with 0.8mm CuAg

Fig 11.33: Component layout for two stage 13cm PHEMT pre-amplifier

is measured. The source pads have to provide a very low inductance path to the ground plane, to preserve the MMIC's inherent unconditional stability.

To achieve unconditional stability four ground connections are needed on each source. Appropriate source pads are provided on the PCB. Simulation indicates a minimum K factor of 1.2 in this arrangement on a 0.79mm thick substrate. A thicker substrate is prohibitive. The MMIC typically adds 0.07dB to the noise figure of the first stage. This is somewhat difficult to measure, because most converters will exhibit gain compression, when the noise power of the source, amplified by more than 40dB, will enter the converter.

The construction uses microstripline techniques on glass PTFE substrate Taconix TLX with 0.79mm thickness with the PCB having dimensions of 34 x 72mm. An active bias circuit, which provides constant voltage and current, is integrated into the PCB (Fig 30 in Appendix B). Fig 31 shows a top view of FETs and Fig 11.32 shows the component layout. The construction process is:

- Prepare tinned box (solder side walls).
- Prepare PCB to fit into box.
- Prepare holes for N connectors. Note, Input and output connector are asymmetrical. Use PCB to do the markings.
- Drill holes for through connections (0.9mm diameter) and use 0.8mm gold plated copper wire at the positions indicated.
- Solder all resistors onto PCB.
- Solder all capacitors onto PCB.
- For L1, cut a 17mm length of gold plated copper 0.5mm diameter wire. Bend down the ends at 1mm length to 45 degrees. Form wire into a half circle loop and solder into the circuit with 1mm clearance from the PCB. The wire loop has to be flush with the end of the gate stripline and should be soldered a right angles to it. The wire loop has to be oriented flat and parallel to the PCB.
- Verify the open circuit function of bias circuit. Adjust P1 to 45 ohms. Solder a 100-ohm test resistor from the drain terminal on the PCB to ground. Apply +12V to IC1 and measure +5V at output of IC1, -5V at IC2 Pin5, -2.5V> at collector of T1, +3.6V at emitter of T1, +3V at base of T1, -2.5V at R17 and +2.0V across the 100 ohms. If OK, remove 100-ohm test resistor.

- Solder PHEMT onto PCB. Ground the PCB, your body and the power supply of the soldering iron. Never touch the PHEMT on the gate, only on the source or the drain, when applying it to the PCB and solder fast (much less than 5 seconds).
- Solder N Connectors into sides of the box.
- Solder the finished PCB into the box, solder both sides of the PCB at the sides of the box and solder centre pins of the connectors to the microstriplines.
- Solder feed-through capacitors into box.
- Connect D1 between feed-through capacitor and PCB.
- Connect 12V and adjust P1 for 16mA drain current (measure 160mV across R4 on RF Board). Voltages should be around +2.0V at the drain terminal, -0.4V at the gate, +3.6V at emitter of T1.
- Connect LNA to a noise figure meter, if you have one, and adjust input wire loop, adjust the clearance to PCB as well as drain current by adjusting P1 for minimum noise figure. Even without tuning, the noise figure should be within 0.1dB of minimum because of the limited tuning range of the wire loop.
- Glue conducting foam inside the top cover and slip into the top of the box.

To add the MMIC Amplifier, refer to Fig 11.33 for Construction.
- Prepare PCB by cutting slits into the microstriplines around the MGA865. These are a 2mm slit for L2, a 1.8mm slit for the MMIC and a 0.8mm slit for C3.
- For L2 cut an 8mm length of gold plated copper 0.5mm diameter wire. Form wire into a half circle loop and solder wire loop into the circuit. The wire loop has to lie flat on the PCB, flush with the end of the gate stripline and should be soldered in a right angle to it.
- Follow other instructions given above

Measurements were taken using an HP8510 network analyser and HP8970B/HP346A noise figure analyser, transferred to a PC and plotted. Figs 11.34 and 11.35 show the measurement results for gain and noise figure for the one stage and two stage version respectively. Using a special PHEMT, NEC NE42484A optimised for C Band, a typical noise figure of 0.35dB at a gain of 15dB can be measured on 2.32GHz. An optional second stage on the same PCB using the GaAs MMIC MGA86576 from HP will boost the gain from 15db

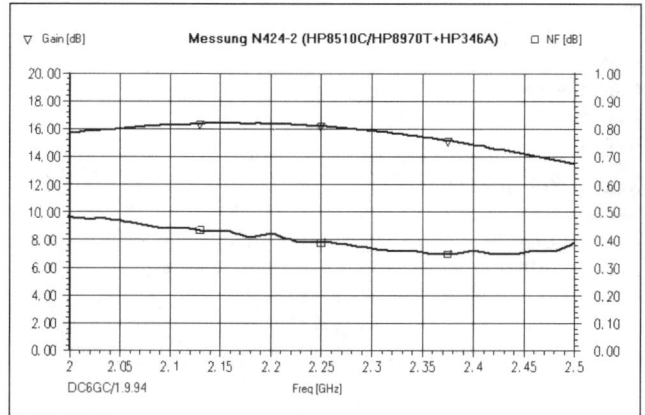

Fig 11.34: Noise figure and gain measurements for single stage 13cm preamplifier

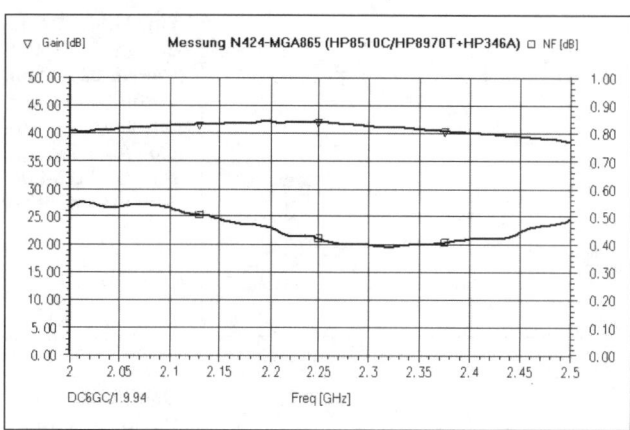

Fig 11.35: Noise figure and gain measurements for two stage 13cm preamplifier

to 41dB. The two stage version measures with a noise figure of 0.45dB. This version can be used for satellite operation. For EME, where lowest noise figure is at premium, a cascade of two identical one stage LNAs may be more appropriate. Both versions are rather broadband. They can cover the various portions of the 13cms amateur allocation from 2300 to 2450MHz without re-tuning. The real surprise is the performance of the C band PHEMT NE424. It performs better than several other HEMTs (FHX35, FHX06, NE324, and NE326) tried in this circuit, and it measures 0.15dB better than its published noise figure. In fact the *Microwave Harmonica* simulation predicts a 0.5dB noise figure based on the data sheet value. The lower noise figure measured seems to be due to a special bias current and the lower value of gamma at approximately 0.75 which is due to the gate length of 0.35μm. This provides optimum properties for application in 2 - 4GHz LNAs. Stability is excellent. This has been achieved by a carefully controlled combination of inductive source feedback, resistive loading in the drain and non resonant DC feed structures for drain and gate. A broadband sweep from 0.2 to 20GHz showed a stability factor K of not less than 1.2 and the B1 measure was always greater than zero. These two properties indicate unconditional stability. At the operating frequency of 2.3GHz, stability factor is about 1.6. The two-stage version with the MGA865 measures with K>>4 at all frequencies.

The preamplifier provides quantum leap towards the perfect noiseless preamplifier. It uses a low cost and rugged C band PHEMT instead of relying on expensive X band HEMTs. An

improvement of about 0.2dB in noise figure has been achieved in comparison to the no tune HEMT preamplifier described in [14]. This improvement provides roughly 1.5dB more S/N in EME or satellite operation but is not noticeable in terrestrial links. However, the new preamplifier has to be tuned. This requires a noise figure meter for alignment. For those who like a no tune device, the HEMT preamplifier in [14] provides adequate performance with a typical NF of 0.55dB.

3cm Preamplifier

Much of the equipment used by amateurs in the UK comes from designs by G3WDG. This preamplifier design, G3WDG-004, is one of the full range of designs for 10GHz. Two versions of the -004 amplifier are available, one with SMA input and one with WG16 waveguide input. All prototypes averaged 1dB NF or slightly under. Where the noise figure was fractionally above 1dB there was some evidence of degradation during construction, possibly by static. The circuit for either amplifier is identical and is shown in **Fig 11.36**. A short kit is available from the Microwave Components Service [10], it contains both the HEMT and a detailed construction sheets. The advantage of the waveguide input version is that a probe inputs signal directly from the waveguide to the HEMT. The SMA version will almost certainly require a least one more SMA connector and a short length of semi-rigid 'cable' to the amplifier input. Thus the noise figure will be degraded by a few tenths of a dB, it is vital to win that little loss back if you are searching for the ultimate performance, for instance for moonbounce!

The layout details of the board of either version are given in **Fig 11.37** to give you some idea of the work involved. There is also a universal regulated power supply that is used on many of the microwave designs available, the circuit diagram is shown in

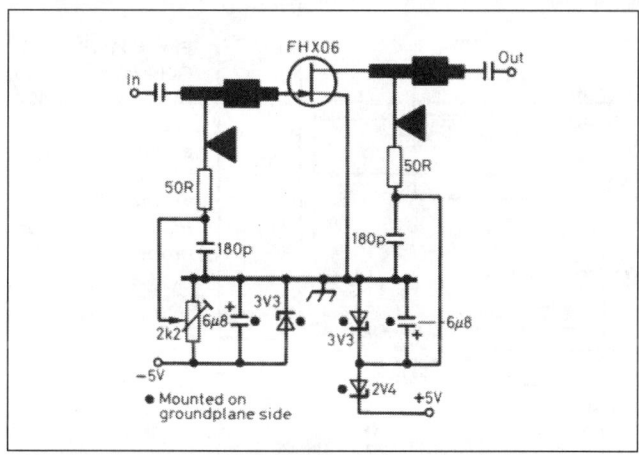

Fig 11.36: Circuit diagram of the G3WDG-004 3cm HEMT preamplifier

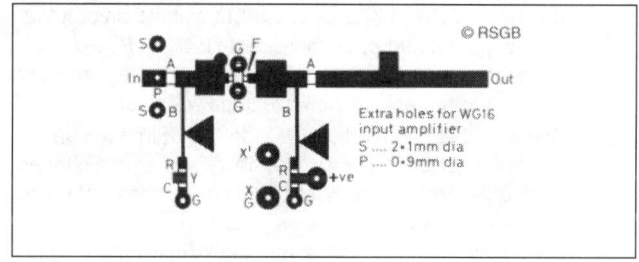

Fig 11.37: PCB and component layout of the G3WDG-004 HEMT preamplifier. Key to components: A- 2.2pF ATC capacitor, B - bias wires, C - 180pF cip capacitors, F - HEMT, G - grounding vero pins, R - 47O chip resistors, X and Y positions of connection to bias pot (X is -ve input)

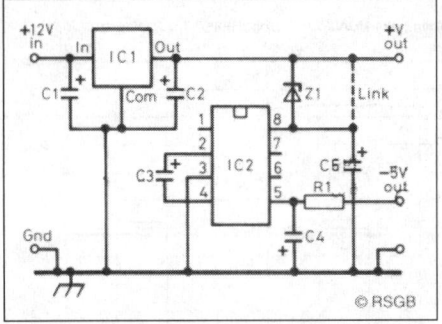

Fig 11.38: Basic circuit of the G4FRE-023 dual output power regulator module. As used on the G3WDG-004 3cm preamplifier

R1	2k20
C1, 2	1µF SMD tantalum
C3, 4	22µF SMD tantalum
C5	10µF SMD tantalum
Z1	Not fitted - replace with a link
IC1	78L05
IC2	ICL7660SCPA

Table 11.10: Component list for dual output power regulator used on G3WDG-004 HEMT pre-amplifier

Fig 11.39: (a) Preparing a tin plate box for the G3WDG-004 HEMT pre-amplifier (waveguide version). (b) Waveguide 16 dimensions and drilling sizes

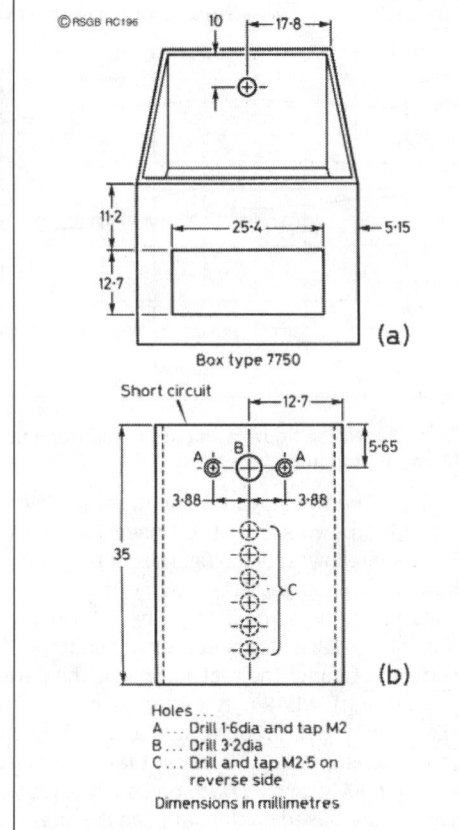

Fig 11.38 . Details of the components for this LNA are given in Table 11.10. The preamplifier gives a gain of about 10dB with a noise figure of 1dB or a little less. The additional holes shown on the layout are needed for the waveguide 16 (WG16) version of the amplifier which is more difficult to construct than the SMA input version. This is because more 'mechanical' work is needed for this version in order to gain those few valuable tenths of a dB noise figure where ultimate performance is needed. Fig 11.39(a) shows the engineering work needed to adapt a Type 7750 tin plate box (Piper Communications) to take the short length of WG16 (input), the PCB, and output SMA connector. Construction should follow the order suggested by G3WDG:

- Connect 12V and adjust P1 for 16mA drain current (measure 160mV across R4 on RF Board). Voltages should be around +2.0V at the drain terminal, -0.4V at the gate, +3.6<N>V at emitter of T1.Solder the tin plate box together in the usual fashion. Then cut out the hole needed for the WG16 by drilling a lot of small holes inside the outline of the large hole and then join them and file out to be a tight fit on the WG16.

- Connect 12V and adjust P1 for 16mA drain current (measure 160mV across R4 on RF Board). Voltages should be around +2.0V at the drain terminal, -0.4V at the gate, +3.6V at emitter of T1. Cut the piece of WG16 to length, square off the ends and drill/tap all the holes shown in Fig 11.39(b) .

- Connect 12V and adjust P1 for 16mA drain current (measure 160mV across R4 on RF Board). Voltages should be around +2.0V at the drain terminal, -0.4V at the gate, +3.6V at emitter of T1. Cut a short circuit plate (thin copper sheet or double sided FR4/GlO PCB material can be used) slightly larger than the waveguide and solder it to the end of the waveguide where shown.

- Connect 12V and adjust P1 for 16mA drain current (measure 160mV across R4 on RF Board). Voltages should be around +2.0V at the drain terminal, -0.4V at the gate, +3.6V at emitter of T1. File off any excess, especially where the PCB abuts the WG.

Now refer to Fig 11.40 and prepare the PCB by very carefully cutting away the copper around the hole P with a Vero cutter, to about 3mm diameter. Trim the PCB to be a neat fit into the box, be sure to leave enough on the input end of the board to allow 2 x 4mm screws to fit inside the box! These screws serve to clamp the PCB to the waveguide (holes S on the PCB to holes A on the WG). Cut off part of the input line, leaving 1mm to the left of the centre line of hole P. Fit and solder all the grounding pins. Fit the Veropin probe and solder it in place in hole P. Cut and fit the PTFE washer, referring to Fig 11.40(b), on to the pin on the ground plane side of the PCB, pushing it firmly down to the PCB, but taking care not to damage the pin or board.

Fit the piece of WG into the box, through the hole, offer up the PCB, check alignment and loosely fit the 4mm screws. If the board does not align with the probe pin in the centre of the WG probe hole, Fig 11.40(b), then file the hole in the box as required for proper alignment. Mark the centre of the hole for the SMA output connector (the 17.8mm dimension shown in Fig 11.39(a) may need to be altered!). Remove the screws and dismantle PCB and WG, drill the holes in the box for the SMA connector and open the centre clearance hole to 3.3mm. Refit the WG, PCB and 4mm

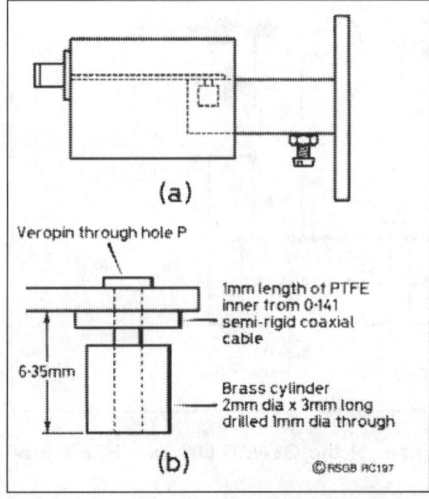

Fig 11.40. (a) General side view of G3WDG-004 WG16 HEMT pre-amplifier assembly. (b) Detail of waveguide to PCB transition probe

screws. Then cut the spill of the SMA connector to 1.5mm and fit it to the box. Fit the PCB into place so that the output track touches the spill, and solder it to the spill. Solder three sides of the WG to the tinplate box, but not to the side to which the PCB mates. Remove the SMA fixing screws, unsolder the spill and remove the connector. Remove the 4mm screws and the PCB. Carefully solder the last face of the wave-guide, but do not melt the solder on the WG shorting plate or you'll have to start again! After soldering, make sure the mating surface of both the waveguide and the PCB are clean, bright and free from oxide and flux. Fit the brass collar, as shown in Fig 11.40((b), to the probe pin, the length shown is critical to 0.05mm. Complete the WG assembly by sol-dering a WG16 flange to the input

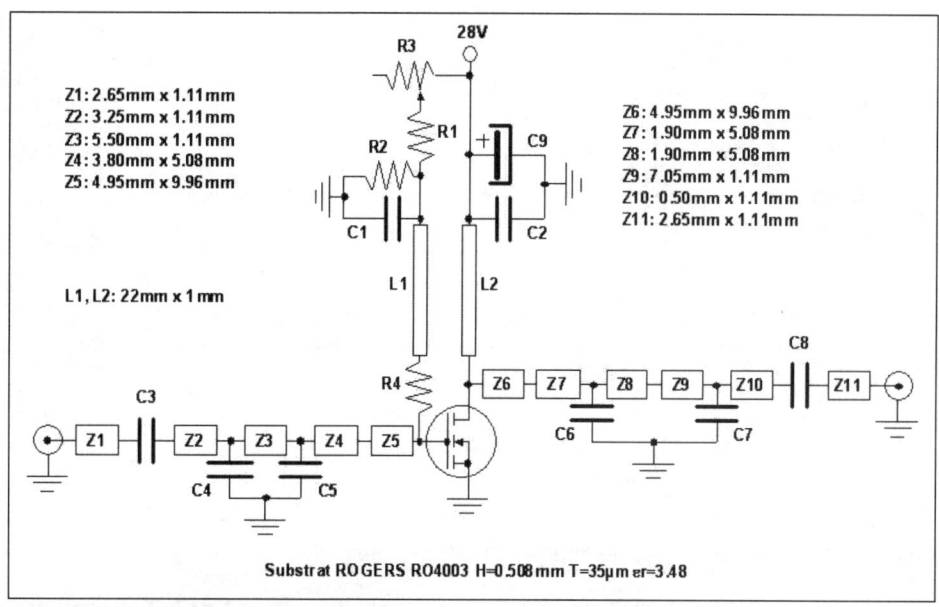

Fig 11.41: Circuit diagram for MRF284 23cm power amplifier designed by F4CIB

end, again being careful not to melt the solder on the rest of the assembly. Apply a small amount of conductive epoxy to the wave-guide where the PCB mates, especially around the probe hole, but not so much that a short circuit occurs when the PCB and WG are squeezed together! Reassemble the PCB and SMA connector. Solder the output track to the SMA spill and tighten the 4mm screws. Check there is not a short between the WG and input probe caused by 'squeezing' of excess epoxy. Cure the epoxy by heating at about 150 degrees C for an hour or so. After curing, recheck for absence of short circuit on the input probe - if there is a short, remove the PCB and try again! This may damage the PCB, so do it with care and try to get it right the first time! Finally, solder all around the ground plane / box junction. Finish off the amplifi-er by mounting all the components as for the SMA version. Incidentally, the holes C on the WG16 are used for matching screws to literally "screw the last tenth of a dB" out of the amplifi-er!

TRANSMIT POWER AMPLIFIERS

Generating any reasonable amount of power on the microwave bands was the speciality of the valve. On the lower bands the trusty 2C39A was well used in anything up to eight valve designs. On the higher bands the travelling wave tube was used but this needed special powers supplies. These are being replaced with semiconductor devices, with their compact design and more manageable power requirements. These can also be masthead mounted.

On 23cm, the Mitsubishi M57762 hybrid amplifier has been used for many years but this is now going out of production and other semiconductors are replacing them, an example shown here. As the frequency increases it becomes more difficult to find, or afford, semiconductors for power ampli-fiers. Fortunately the gain from the anten-nas used comes to the rescue reducing the input power needed to achieve the required radiated power. On the higher bands spe-cial techniques are still required, such a direct bonding to semiconductor sub-strates, and example of these techniques is shown here.

LDMOSFET Power Amplifier for 23cm

This amplifier design was researched by Franck Rousseau, F4CIB, [15] and published in *Dubus Magazine*. Conversion of Motorola devices on 23cm and 13cm was the subject of articles by Maurice Niquel, F5EFD, [16] and [17] and Jean-Pierre Lecarpentier, F1ANH, [18]. These devices provide a really good opportunity to replace expensive power bricks like the M57762, and the 2C39 cavity amplifiers. The transistor characteristics, extracted from product datasheet [19] are really interesting for ham applications:

- Broadband (1GHz-2.6GHz), class A & AB specified

Specified Two–Tone Performance @ 2000 MHz, 26V
- Output Power = 30 Watts PEP
- Power Gain = 9dB
- Efficiency = 30%
- Intermodulation Distortion = –29dBc

Typical Single–Tone Performance at 2000MHz, 26V
- Output Power = 30 Watts CW
- Power Gain = 9.5dB
- Efficiency = 45%
- Capable of Handling 10:1 VSWR, @ 26V DC, 2000MHz, 30 Watts CW Output Power

The circuit diagram (**Fig 11.41**) is an adaptation from the orig-inal demonstration board with some component values changed. The PCB layout is shown in **Fig 11.42**.

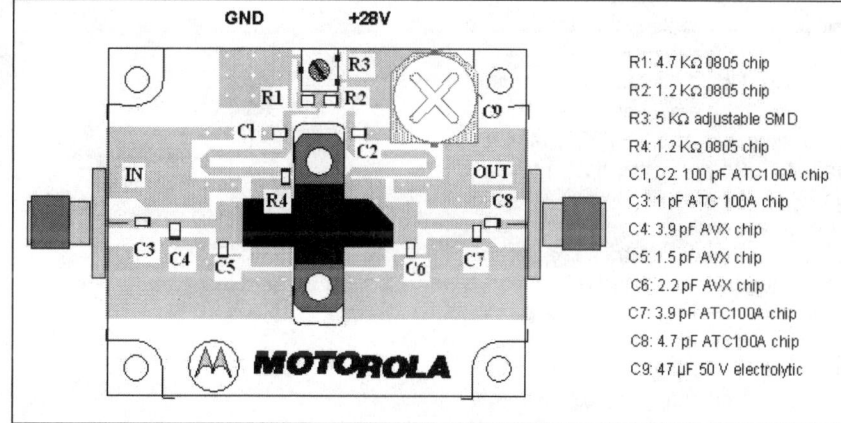

R1: 4.7 KΩ 0805 chip
R2: 1.2 KΩ 0805 chip
R3: 5 KΩ adjustable SMD
R4: 1.2 KΩ 0805 chip
C1, C2: 100 pF ATC100A chip
C3: 1 pF ATC 100A chip
C4: 3.9 pF AVX chip
C5: 1.5 pF AVX chip
C6: 2.2 pF AVX chip
C7: 3.9 pF ATC100A chip
C8: 4.7 pF ATC100A chip
C9: 47 µF 50 V electrolytic

Fig 11.42: PCB layout for MRF284 23cm power amplifier

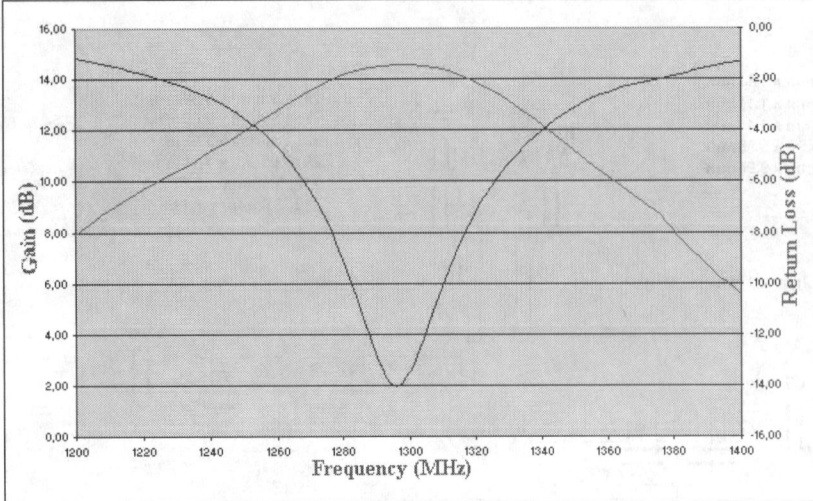

Fig 11.44: Photograph of the completed MRF284 23cm power amplifier

Fig 11.43: Gain and return loss for MRF284 23cm power amplifier

All measurements have been done on professional calibrated equipment:

- Network Analyser HP8753: Forward Gain S21 and Input Return Loss S11.
- Power Meter HP438A: RF Output Power.
- DC Power Supply HP6632: DC Sources and Measures.

A single DC supply is required, so the first step is to set bias current with the help of the adjustable resistor. Set bias to 200mA and forget it! Input matching was achieved with a 3.9pF capacitor close to the DC blocking capacitor (C3) and a 1.5pF capacitor on the 30-ohm line. The position of the 3.9pF capacitor is the most critical, the 1.5pF capacitor along 30-ohm line helps to centre the tuned frequency at 1296MHz. Once the input circuit is set up like this, the gain is around 9-10dB, and the Input Return Loss (IRL) is -13 to 14dB. The output network should be a 2.2pF capacitor soldered at the centre of the 30-ohm line and a 3.9pF as close as possible to the DC blocking capacitor. This will increase the gain to 14dB, while IRL remains at -14dB (**Fig 11.43**). The 1dB compressed power, P1dB, (Output power level for a gain drop of 1dB) has been measured at 45.5dBm (35W) in accordance with device specifications. For the following conditions:

- VCC = 28 V
- Input Power = 31.6dBm
- Output power measured at 45.2dBm
- DC current 2.41A.

Computing the Power Added Efficiency (PAE) gives 47% which exceeds specifications, but that's not a surprise as the device is able to work up to 2 GHz. **Fig 11.44** shows a picture of the completed amplifier.

As expected, the MRF282, MRF286 and MRF284 work without any problems on 1296MHz. Output power is twice that of a power brick, and gain should be a good alternative to these expensive power modules. The trick is to find a good and cheap source of these transistors; as a clue they are used in cellular base stations so they should be available on the second hand market. The design does need a 2.5A 28V DC power supply, a suitable design is shown in [20]

GaAsFET Amplifier for 3cm

The power amplifier for 3cm described here was developed by Peter Vogl, DL1RQ, [21] with a view to ease of copying and reliable long term operation. Numerous examples of the 1W power amplifier have been running for some years (some of them installed on masts) and none has yet given rise to any problems. This circuit was expanded using a TIM 0910-4 from Toshiba to give 5W output, the design for the 5W amplifier is shown in [21] and [3]. In spite of all efforts at reproducibility, the cumulative total of the small tolerances in the component values and the assembly can eventually lead to significant individual deviations (-3 dB is normal) in the amplification and output power. But there is some comfort in the fact that, with patience, experience and good measurement facilities, a power amplifier can be trimmed to the rated values with a fine calibration using the "small disc method" (see below).

The circuit diagram of the 1W amplifier, shown in **Fig 11.45**, is similar to DL1RQ's two stage 5.7GHz power amplifier, published in [22] and [23], including the additional voltage inverter for the negative gate voltage. Good experiences with the reliable FSX52WF transistors (drive) and FLC103WG transistors (high level stage) from Fitjitsu led to trials at 10GHz, which were immediately successful, although the FLC103WG is only specified for use up to 8GHz by the manufacturer. 0805 model 1pF SMD capacitors were used as high frequency coupling elements

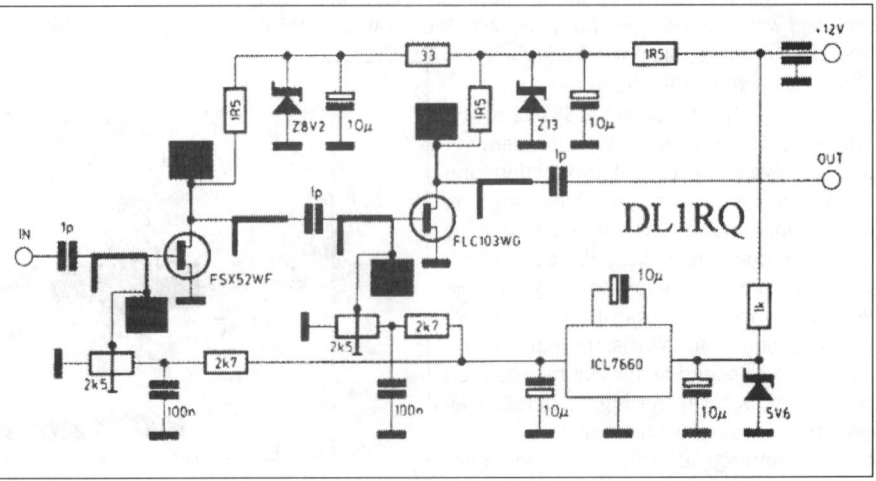

Fig 11.45: Circuit diagram of the 1W GaAsFET amplifier for 3cm

(2.0 mm x 1.25 mm). Research carried out only recently as part of a specialist project [24] showed that the SMD capacitors used by the author, with a series inductance of L_{series} = 0.66 ±0.01nH, differ from those components available by normal mail order, with L_{series} = 0.72 ±0.02nH. Unfortunately, DL1RQ was not in a position to identify the individual manufacturer. In any case, a rough calculation quickly shows that the series resonance frequency of these SMD capacitors lies around 6.0GHz. So, at 10GHz the coupling "capacitors" should be considered more as

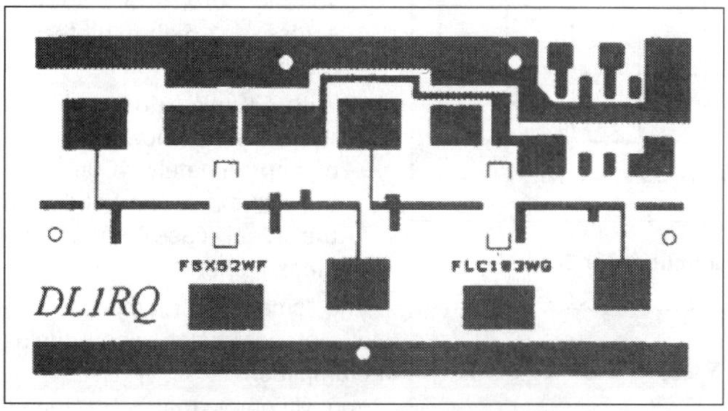

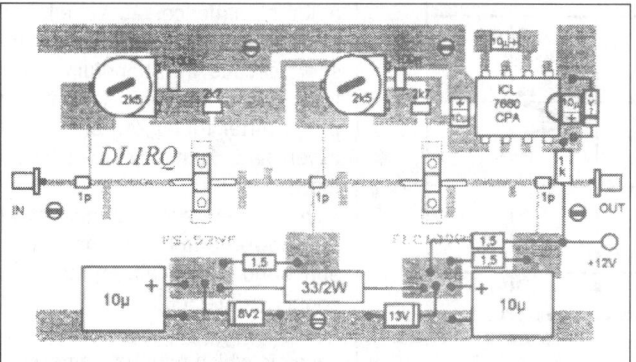

Fig 11.46: PCB layout for the 1W GaAsFET amplifier for 3cm

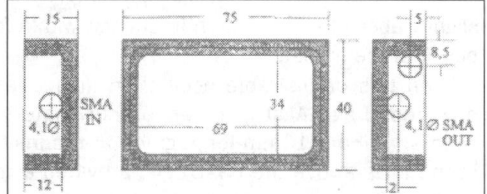

Fig 11.47: Component layout for the 1W GaAsFET amplifier for 3cm

Fig 11.48: Dimensions of a milled housing for the 1W GaAsFET amplifier for 3cm

DC disconnecting components with inductive behaviour. Naturally, this inductance affects the matching by the striplines (at the cost of narrowing the band!).

The power supply was deliberately made simple. Stabilisation was provided through 1.3W zener diodes, which provide protection against over voltage and reverse polarity. The 1.5-ohm/0.25W axial carbon film resistors act as both isolation resistances and safety resistances. A circuit for protection if the negative power supply failed was dispensed with following an involuntary 24 hour test without any negative supply voltage, which did not damage the semiconductors.

The assembly and board layout of an amplifier for microwaves are determined by two essential requirements:

• The high frequency transition from the earth surface of the board to the source flange of the transistor must be as close to ideal and as smooth as possible

• The extraction of the transistor's lost heat must be as close to ideal as possible

A board with a sandwich construction has proved to be a way of being able to fulfil both requirements. The PCB layout (Fig 11.46) is etched onto an RT/Ditroid D-5870 board measuring 68.5mm x 34mm x 0.25mm. After pre-tinning the earth surface, the board is soldered onto a 1.0mm thick copper plate under high pressure. Next, two oval grooves 0.75mm deep are milled, using a 2.5mm diameter bore groove milling cutter, for the source flanges of the transistors. When the five 2.1mm diameter holes have been made for the board to be screwed into the housing and for the contacts to be connected up, and when the tracks have been tin plated (or silver plated), the board is assembled as far as the two 10µF capacitors (on the drain side) and the transistors (Fig 11.47). In order to guarantee good heat transfer between the copper plate and the milled aluminium housing (Fig 11.48), some heat conducting paste is smeared over the aluminium base in the vicinity of the transistors. To ensure good connection between the high frequency section and the earth in the input and output areas, silver conducting lacquer can be smeared there (very sparingly, of course). The partly assembled board is now fitted into the suitably prepared aluminium housing and screwed down by five M2 brass screws. When the connections to the feed-through capacitor have been completed, it is possible to check the DC function. For this purpose, the two trimmers are pre set to a gate voltage of about -1.5V. The trickiest stage in the procedure is the soldering of the GaAsFET into the milled grooves. To this end, the aluminium housing is first heated, with the board inside, to precisely 150 degrees C. Each milled groove is then pre tinned, using low temperature solder with a melting temperature of 140 degrees C.

Excess tin is then removed using a de-soldering pump. The transistors are next placed in the grooves; all the relevant safety measures known must be taken. Normally, the tin binds very well with the gold plated flanged base, something that can easily be tested by a visual check of the flanged holes. Naturally, this soldering process should be carried out as rapidly as possible. The housing is then immediately placed on a cold copper block or a large cooling body, so that the temperature quickly falls. Drain and gate connections are soldered onto the striplines; all the relevant safety measures must be taken. The two 10µF capacitors on the drain side are fitted and the SMA flanged bushes are screwed on. The power amplifier is ready for tuning (Fig 11.49).

Fig 11.49: Photograph of the 1W GaAsFET amplifier for 3cm

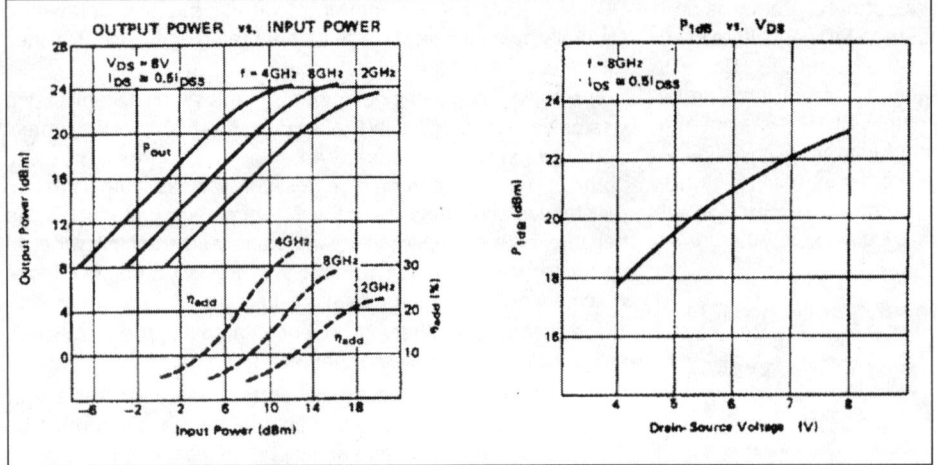

Fig 11.50: Data for the Fujitsu FSX53WF used in the 1W GaAsFET amplifier for 3cm

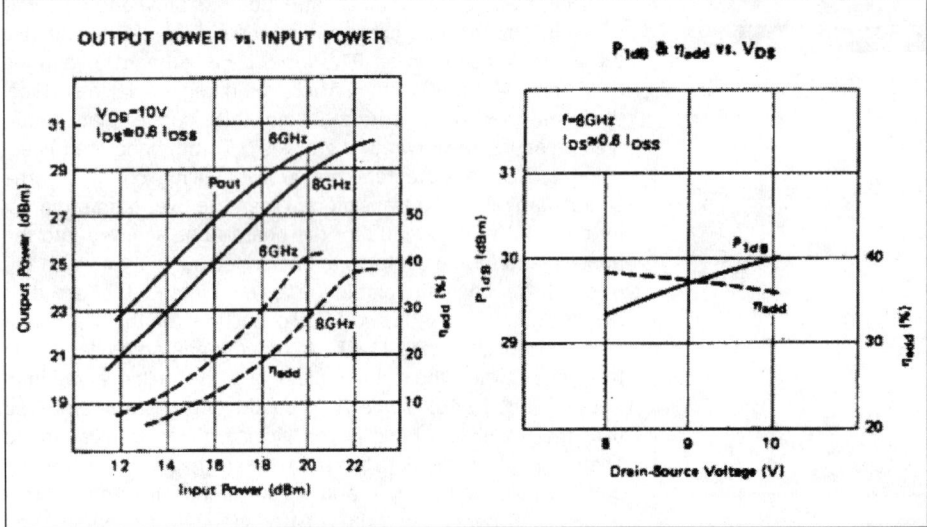

Fig 11.51: Data for the Fujitsu FSX53WF used in the 1W GaAsFET amplifier for 3cm

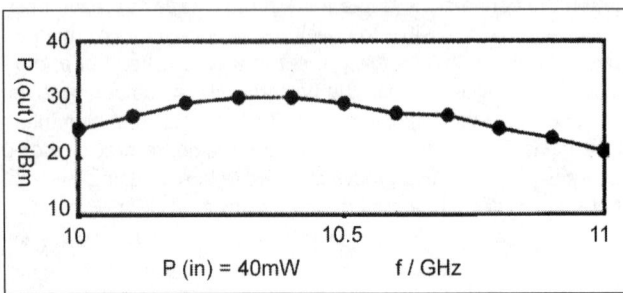

Fig 11.52: Power bandwidth of the 1W GaAsFET amplifier for 3cm

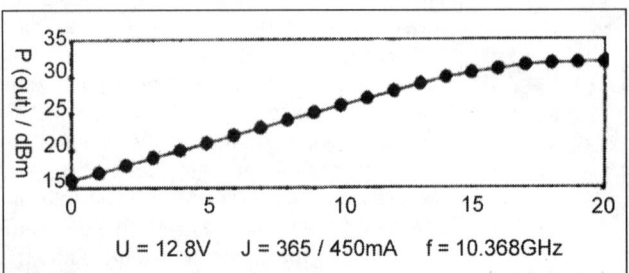

Fig 11.53: Input matching of the 1W GaAsFET amplifier for 3cm

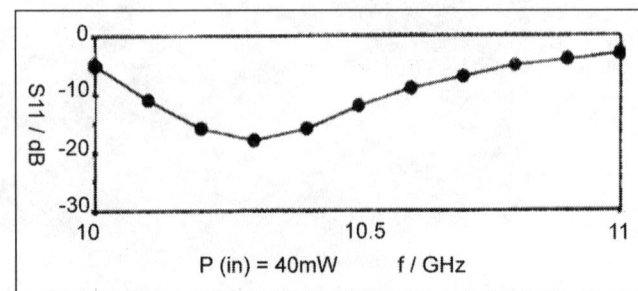

Fig 11.54: Linearity of the 1W GaAsFET amplifier for 3cm.

First, the no-signal currents are set as follows:

- For the FSX52WF at approximately 70mA, this corresponds to a voltage drop of 105mV, across a 1.5-ohm protective resistor. For the FLC103WG, at approximately 240mA, it corresponds to a voltage drop of 360mV, across a 1.5-ohm protective resistor.

- With 30mW drive at the desired frequency, an output of approximately 400mW (in the worst case) and of 1W (in the ideal case) should be measurable.

The "small disc method" is normally of assistance when tuning the amplifier.

You will need small discs, measuring about 2 - 4mm^2, a few toothpicks to press down and push and a lot of patience. Above all, the greatest care in watching out for short circuits will (hopefully) soon lead you to achieve full output power. After tuning, an aluminium cover plate 1mm thick can be fitted.

In the overwhelming majority of the power amplifiers measured, almost no influence from the cover could be detected.

Of course, there were just a few cases in which minimal self excitation were detected when the cover was fitted. This is caused by astonishingly stable housing resonance, slightly above the calibration frequency, with a few milliwatts of power at the output.

Even this undesirable oscillation disappeared with a low powered drive. A strip of absorbent material about 5mm wide and about 10mm long, glued to the inside of the cover in the area above the FSX52WF, provided a reliable remedy here.

A comparison of the output data from the semiconductors (**Figs 11.50 and 11.51**) with the readings from a typical power amplifier (**Figs 11.52 - 54**) makes clear how successful the project is in practice.

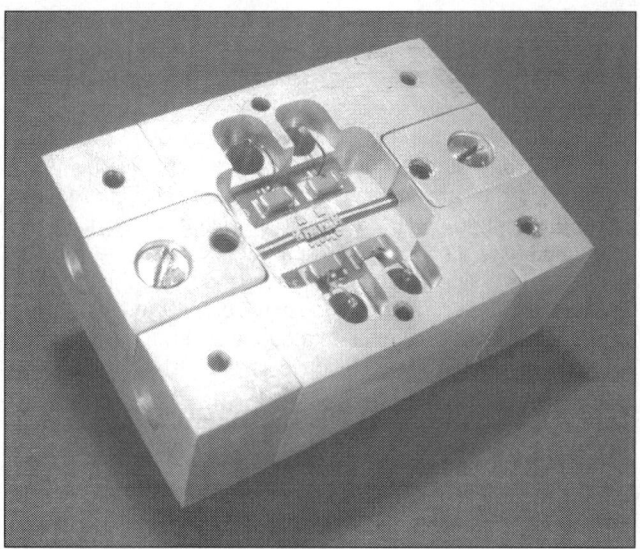

Fig 11.55: Picture of the completed 76GHz amplifier

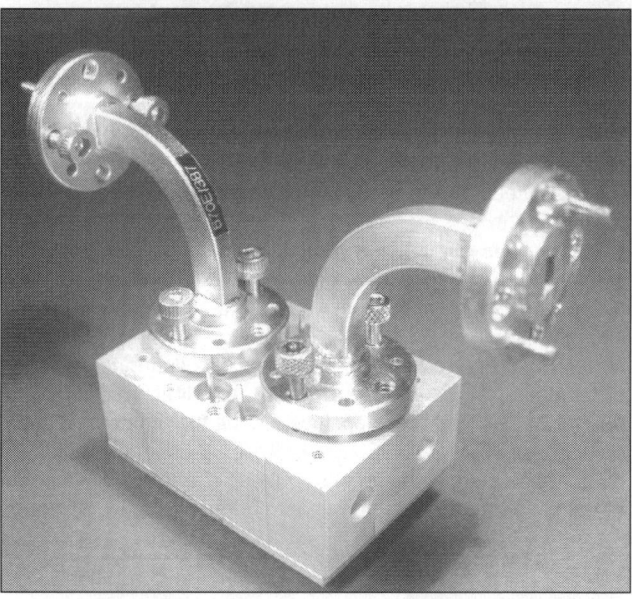

Fig 11.56: Picture of the completed 76GHz amplifier with WR-12 waveguide connections fitted.

Fig 11.57: The 1AF-MPA7710 chip, used in the 76GHz amplifier, magnified about 120 times

76GHz Amplifier

This design [25] is part of a series of articles by Sigurd Werner, DL9MFV, published in *VHF Communications* magazine. It describes an amplifier that uses two MMIC's (IAF-MPA7710) connected in series, and originates from development work by the Fraunhofer Institute for Applied Solid-State Physics in Freiburg. The gain of the amplifier is 24dB at 76,088MHz.

The first difficulty lies in the procurement (selection would be something of an exaggeration) of suitable MMIC's for this frequency. Siemens manufacture a two-stage GaAs amplifier chip (T602B-MPA-2) for use in its car collision radar, which amplifies small signals by approximately 9.5dB [27]. One genuine alternative to this is an MMIC that has been developed at the Fraunhofer Institute. This is another two-stage amplifier chip (1.5mm x 1mm x 0.635mm), with the designation IAF-MPA7710. It has the following characteristics:

- Frequency range: 73 - 80GHz
- Gain: >11dB
- Output >14dBm at 1dB compression, and with a power consumption of approximately 800mW [28] and [29].

Since gate 1 is earthed, only two positive voltages of approximately 1.5V (for Gate 2) and approximately 4V each for all drain connections are required. In order to attain a usable level of amplification, two MMIC's are wired in series, without any additional matching circuit between the two chips.

The housing with dimensions of 27.5mm x 39.8mm x 13.5mm was milled from brass and subsequently gold-plated (**Fig 11.55**). The two chips were mounted in a 0.5mm hollow in the 4.1mm machined cavity. This balances out the difference in height between the MMIC's (0.635mm!) and the connection substrates (0.254mm). Strict attention was paid to ensure that the distance between the two MMIC's and their distance from the substrates were as small as possible (approximately 60 or 75µm). The connection substrates that link the chips with the two WR-12 waveguides (amplifier input and output) are made from aluminium nitride (approximately 1.1mm wide). They were sawn out of an existing ceramic, a substrate designed for other purposes by R&S. Since the chips have co-planar RF connections, the substrates likewise have a short co-planar section (0.5mm), which then goes into a 50-ohm stripline (8mm). This line projects approximately λ/8 into the waveguide. This construction technique is described in more detail in [30]. The ground plane of the ceramic projecting into the waveguide was milled off. The power supplies for the chips are initially blocked with 100pF single layer capacitors, and subsequently with 100nF ceramic capacitors, and are then fed out via feedthrough filters through the housing base (see **Fig 11.56**). The connections between chips and substrate (or capacitors) were created using wedge-wedge bond technology [30].

The following problem arose here: The RF connection pads of the MMICs, which were actually designed for flip-chip installation, are extremely small (see **Fig 11.57**). Directly behind these pads there are air bridges running to the chip circuit. These fragile structures are very easily caught and destroyed during the bonding by the tool that feeds the 17µm gold thread. The chip is then naturally unusable. This difficulty was avoided through the use of a still thinner needle and a correspondingly finer gold thread of 12µm. However, all other connections were created using a conventional 17µm thread.

The amplifier was initially operated at low power levels (approximately 50µW) at 76,032MHz. The gain observed was initially very disappointing (in the region of 8dB). Even after the fine adjustment of the waveguide short circuit screws the gain reading was scarcely 10dB. On the basis of experience of mis-

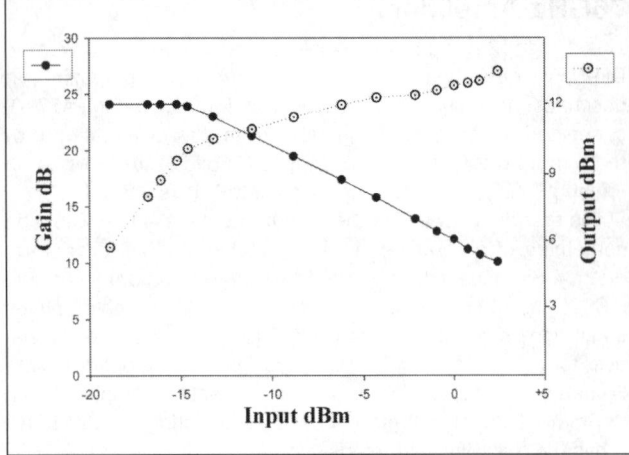

Fig 11.58: Transfer characteristics of the 76GHz amplifier at 76,032MHz.

matching of the waveguide couplings obtained during the transverter project, another series of gold threads was attached and fastened to the striplines using a UV activated adhesive (see [26]). The input power during matching amounted to -13dBm. After a laborious sequence of 9 pennants, the work was rewarded by an amplification of 24dB. That means 12dB per amplifier stage, a value which tallies well with the specifications in the data sheet [28]. A value of > 9dB (SWR < 2.1) was measured at the amplifier input, with > 25dB (SWR < 1.1) at the output. The gain and the output power for various input levels (f = 76,032 MHz) is shown in **Fig 11.58**. It can be seen that for an input power of > -15dBm the amplification is already decreasing, a behaviour to be expected. At an input power of 100µW, there is still 10mW measured at the output anyway (20dB gain). The saturation power of 12.6dBm remains unsatisfactory (approximately 18mW). Approximately 15dBm would have been expected! There could be several reasons for this behaviour. The large number of threads certainly increased the matching and thus the amplification, but at the same time a lot of energy is lost with each stub attached. Secondly, the length of the striplines at

these frequencies leads to additional losses. The power supply voltages were varied in a further attempt to attain a higher saturation power on the individual MMIC's. The optimal setting yielded the following values:

- The driver chip, gate 2 +1.4V, drains 3.0V (190mA);
- The output chip, gate 2 + 1.2V, drains 3.7V (220mA).

The result was an increase in the power of only approximately 20%. The MMIC (IAF-MPA7710) can provide good service in the manufacture of a power amplifier. Better performance could be achieved by using at least two chips operated in parallel. Possible solutions for the addition of the outputs are "magic tee" or a 3/4λ Wilkinson coupler on a quartz substrate (0.127mm!). A project showing these techniques was published in [31].

MICROWAVE SSB TRANSCEIVER DESIGN [32]

When discussing SSB transceivers, the first question to be answered is probably the following - does it make sense to develop and build new SSB radios? Today SSB transceivers are mass produced items for VHF and UHF. Most radio amateurs are therefore using a base SSB transceiver (usually a commercial product) operating on a lower frequency and suitable receive and transmit converters or transverters to operate on 1296MHz or higher frequencies. The most popular base transceiver was the IC-202. All narrow band (SSB/CW) microwave activity is therefore historically concentrated in the first 200kHz of amateur microwave segments, eg 1296.000-1296.200, 2304.000-2304.200 etc, due to the limited frequency coverage of the IC-202.

Transverters should always be considered a poor technical solution for many reasons. Receive converters usually degrade the dynamic range of the receiver while transmit converters dissipate most of the RF power generated in the base SSB transceiver. Both receive and transmit converters generate a number of spurious mixing products that are very difficult to filter out, due to the harmonic relationships among the amateur frequency bands 144/432/1296MHz. However, the worst problem of most transverters is the breakthrough of strong signals in or out of the base transceiver intermediate frequency band. This prob-

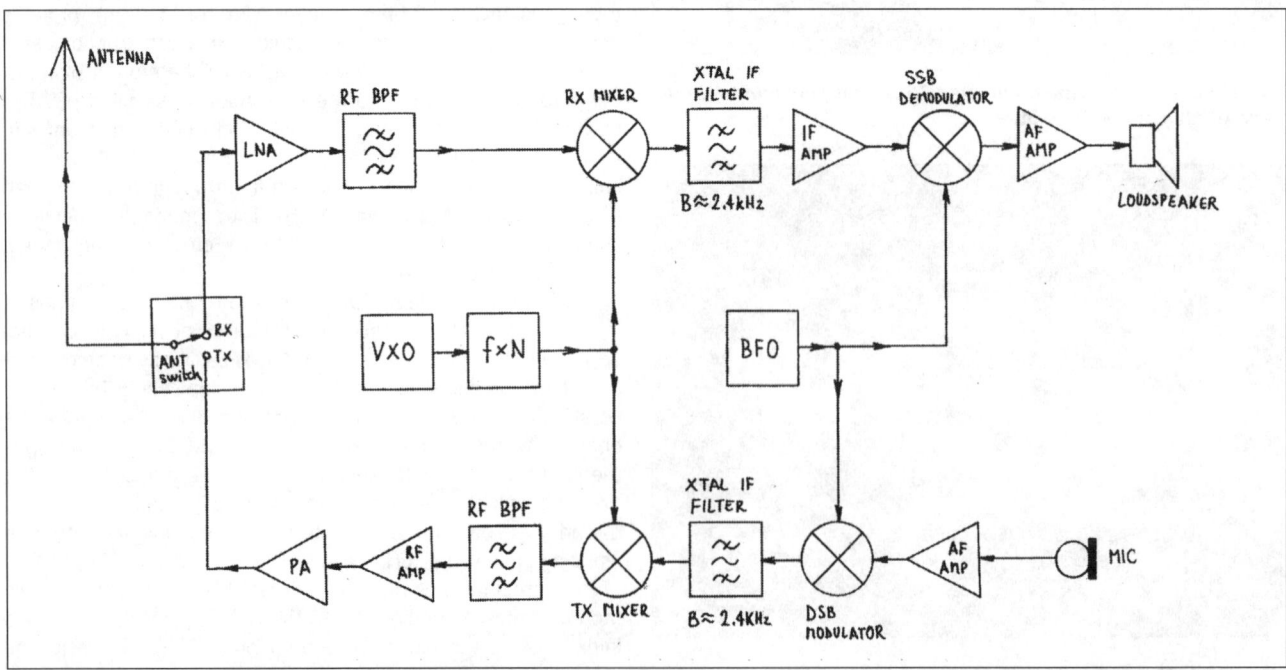

Fig 11.59: Block diagram of a conventional SSB transceiver

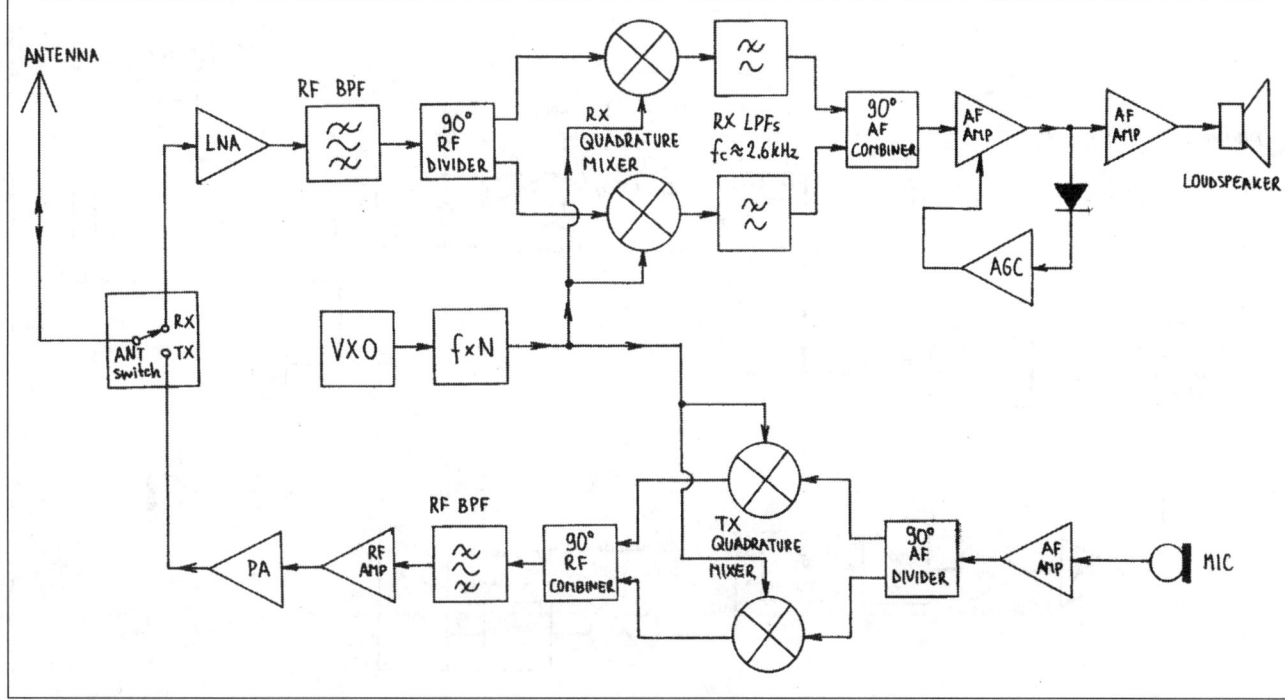

Fig 11.60: Block diagram of a direct conversion SSB transceiver

lem seems to be worst when using a 144MHz first IF. Strong 144MHz stations with big antenna arrays may break into the first IF, even at distances of 50 or 100km. Since the problem is reciprocal, a careless microwave operator may even establish two-way contacts on 144MHz although using a transverter and antenna for 1296MHz or higher frequencies.

Some microwave operators solved the above problem by installing a different crystal in the transverter; so that for example 1296.000MHz is converted to a less used segment around 144.700MHz. Serious microwave contesters use transverters with a first IF of 28MHz, 50MHz or even 70MHz to avoid this problem. Neither solution is cheap. The biggest problem is carrying a large 144MHz or HF all mode transceiver, together with a suitable power supply, on a mountaintop.

Even the IC-202 has its problems;. this radio has not been manufactured for several decades. However, there is now a growing number of lightweight transportable multimode transceivers.

As a conclusion, today it can still make sense to develop and build SSB radios for 1296MHz and higher frequencies. Since the problems of the transverters are well known and are not really new, different designers have already considered many technical solutions. Most solutions were discarded simply because they are, too complex, too expensive and too difficult to build, even when compared to the already complex combination of a base transceiver and transverter. Most commercial SSB transceivers include a modulator and a demodulator operating on a high IF, as shown in **Fig 11.59**. The resulting SSB signal is converted to the RF operating frequency in the transmitter and back to the IF in the receiver. Both the transmitter and the receiver use expensive components like crystal filters. Besides crystal filters, additional filtering is required in the RF section to attenuate image responses and spurious products of both receiving and transmitting mixers. The design of conventional (high IF) SSB transceivers dates back to the valve age, when active components (valves) were expensive and unreliable. Passive components like filters were not so critical. Complicated tuning procedures only represented a small fraction of the over-

all cost of a valve SSB transceiver. SSB crystal filters usually operate in the frequency range around 10MHz. A double or even triple up-conversion is required to reach microwave frequencies in the transmitter. On the other hand, a double or triple down-conversion is required in the receiver to get back to the crystal filter frequency. Commercial VHF/UHF SSB transceivers therefore save some expensive components by sharing some stages between the transmitter and the receiver. A conventional microwave SSB transceiver is therefore complicated and expensive. Building such a transceiver in amateur conditions is difficult at best. Lots of work and some microwave test equipment are required. The final result is certainly not cheaper and may not perform better than the familiar transverter plus base transceiver combination.

Fortunately, expensive crystal filters and complicated conversions are not essential components of a SSB transceiver. There are other SSB transceiver designs that are both cheaper and easier to build in amateur conditions. The most popular seems to be the direct conversion SSB transceiver design shown in **Fig 11.60**. A direct conversion SSB receiver achieves most of its gain in a simple audio frequency amplifier, while simple RC low pass filters achieve the selectivity.

The most important feature of a direct conversion SSB transceiver is that there are neither complicated conversions nor image frequencies to be filtered out. The RF section of a direct conversion SSB transceiver only requires simple LC filters to attenuate distant spurious responses like harmonics and sub harmonics. In a well-designed direct conversion SSB transceiver, the RF section may not require any tuning at all. The most important drawback of a direct conversion SSB transceiver is a rather poor unwanted sideband rejection. The transmitter includes two identical mixers operating at 90 degrees phase shift (quadrature mixer) to obtain only one sideband. The receiver also includes two identical mixers operating at 90 degrees phase shift to receive just one sideband and suppress the other sideband. A direct conversion SSB transceiver operates correctly only if the gain of both mixers is the same and the phase shift is exactly 90 degrees. It therefore includes some critical compo-

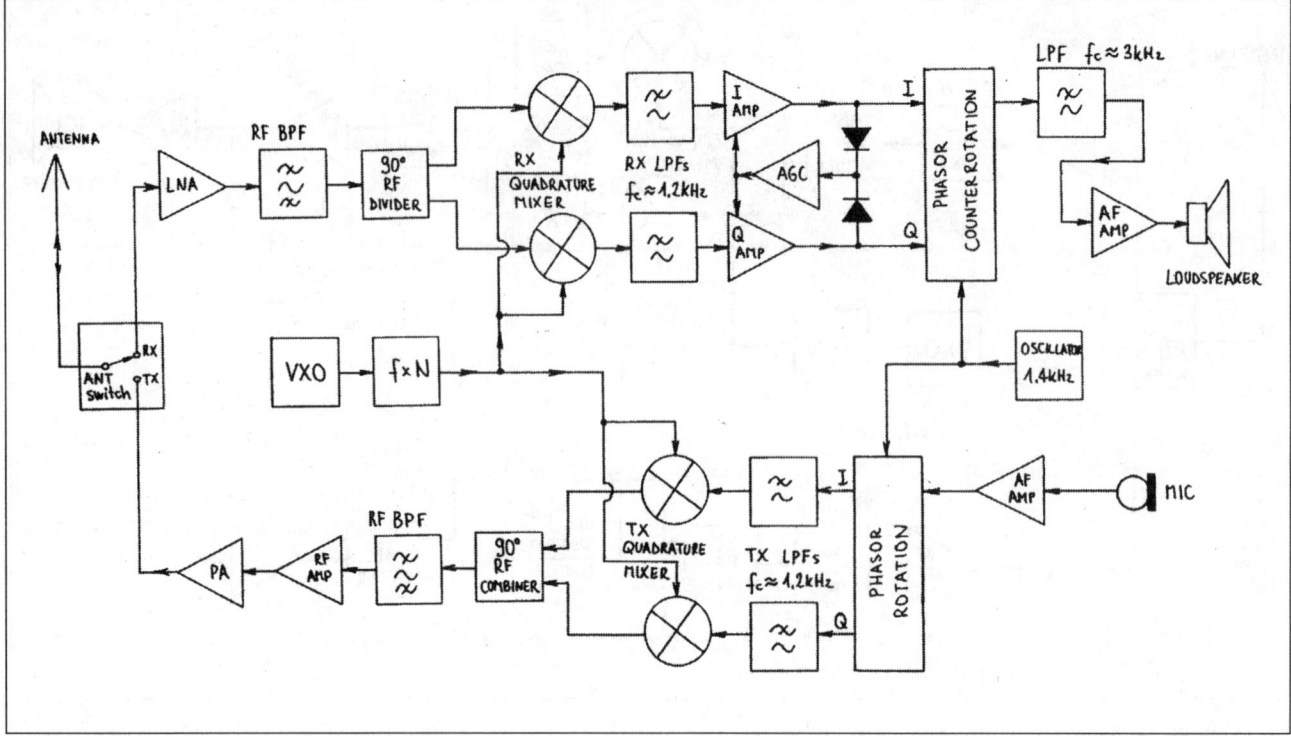

Fig 11.61: Block diagram of a Zero-IF SSB transceiver

nents like precision (1%) resistors, precision (2%) capacitors, selected or "paired" semiconductors in the mixers and complicated phase shifting networks. The most complicated part is usually the audio frequency 90 degree divider or combiner including several operational amplifiers, precision resistors and capacitors. Although using precision components, the unwanted sideband rejection will seldom be better than -40dB. This is certainly not enough for serious work on HF. In spite of these difficulties, direct conversion designs are quite popular among the builders of QRP HF transceivers. At frequencies above 30MHz it is increasingly more difficult to obtain accurate phase shifts. Due to the low natural (antenna) noise above 30MHz, a low noise RF amplifier is usually used to improve the mixer noise figure. An LNA may cause direct AM detection in the mixers. It may also corrupt the amplitude balance and phase offset of the two mixers, if the antenna picks up the local oscillator signal. A VHF direct conversion SSB transceiver is therefore not as simple as its HF counterpart.

On the other hand, a direct conversion SSB design has important advantages over conventional SSB transceivers with crystal filters, since there are no image frequencies and fewer spurious responses. Professional (military) SSB transceivers therefore use direct conversion, but the AF phase shifts are obtained by digital signal processing. The DSP uses an adaptive algorithm to measure and compensate any errors like amplitude unbalance or phase offset of the two mixers, to obtain a perfect unwanted sideband rejection. Additional AF signal processing also allows a different SSB transceiver design, for example a SSB transceiver with a zero IF as shown in **Fig 11.61**. The latter is very similar to a direct conversion transceiver except that the local oscillator is operating in the centre of the SSB signal spectrum, in other words at an offset of about 1.4kHz with respect to the SSB suppressed carrier frequency. In a zero IF SSB transceiver, the audio frequency band from 200Hz to 2600Hz is converted in two bands from 0 to 1200Hz. The low pass filters therefore have a cut-off frequency of 1200Hz, thus allowing a high rejection of the unwanted sideband. A zero IF SSB transceiver retains all of

the advantages of a direct conversion design and solves the problem of the unwanted sideband rejection.

The quadrature IF amplifier of a zero IF SSB transceiver includes two conventional AF amplifiers. Since the latter are usually AC coupled, the missing DC component will be converted in the demodulator as a hole in the AF response around 1.4kHz. Fortunately this hole is not harmful at all for voice communications, since it coincides with a hole in the spectrum of the human voice. In fact, some voice communication equipment includes notch filters to create an artificial hole around 1.4kHz to improve the signal to noise ratio and/or to add a low baud rate telemetry channel to the voice channel. Thus, for voice communications, a potential drawback of a zero IF design is actually an advantage. Like a direct conversion transceiver, a zero IF SSB transceiver also requires quadrature transmit and receive mixers. However, amplitude unbalance or phase errors are much less critical, since they only cause distortion of the recovered audio signal. Conventional components, like 5% resistors, 10% capacitors and unselected semiconductors may be used anywhere in the transceiver.

Finally, a zero IF SSB transceiver does not require complicated phase shifting networks. Both the quadrature modulator in the transmitter and the quadrature demodulator in the receiver (phasor rotation and counter rotation with 1.4kHz) are made by simple rotating switches and fixed resistor / op-amp networks. CMOS analogue switches like the 4051 are ideal for this purpose, rotated by digital signals coming from a 1.4kHz oscillator. Although the block diagram of a zero IF SSB transceiver looks complicated, such a transceiver is relatively easy to build. In particular, very little (if any) tuning is required, since there are no critical components used anywhere in the transceiver. In particular, the RF section only includes relatively wideband (10%) band pass filters that require no tuning. The IF/AF section also accepts wide component tolerances and thus requires no tuning. The only remaining circuit is the RF local oscillator. The latter may need some tuning to bring the radio to the desired operating frequency.

The Radio Communication Handbook

Microwave SSB Transceiver Implementation

The described zero IF concept should allow the design of simple and efficient SSB transceivers for an arbitrary frequency band. Four successful designs of zero IF SSB transceivers covering the amateur microwave bands of 1.3, 2.3, 5.7 and 10GHz have been published by Matjaz Vidmar, S53MV, originally in *CQ ZRS* but also in [3], the basic design is given in this chapter.

Of course several requirements and technology issues need to be considered before a theoretical concept can materialise in a real world transceiver. Fortunately, the requirements are not severe for the lower amateur microwave bands. In this frequency range no very strong signals are expected, so there are no special requirements on the dynamic range of the receiver. Only a relatively limited frequency range needs to be covered (200 to 400kHz in each band) and this can be easily achieved using a VXO and multipliers as the local oscillator. From the technology point of view it is certainly convenient to use up to date components.

High performance and inexpensive microwave semiconductors were developed, first for satellite TV receivers and then for mobile communications like GSM or DECT telephones. These new devices provide up to 25dB of gain per stage up to 2.3GHz and up to 14dB of gain per stage up to 10GHz. Many other functions, like schottky mixer diodes or antenna switching PIN diodes are also available. Using obsolete components makes designs complicated. For example, the familiar transistors BFR34A and BFR91 were introduced almost 25 years ago. At that time they were great devices providing almost 5dB of gain at 2.3GHz. Today it makes more sense to use an INA-03184 MMIC to get 25dB of gain at 2.3GHz or in other words replace a chain of 5 obsolete transistor amplifier stages.

The availability of active components also influences the selection of passive components. Many years ago, all microwave circuits were built in waveguide technology. Waveguides allow very low circuit losses and high Q resonators. Semiconductor microwave devices introduced microstrip circuits built on low loss substrates like alumina (Al2O3) ceramic or glassfibre-teflon laminates. Conventional glassfibre-epoxy laminates like FR4 were not used above 2GHz due to the high losses and poor Q of microstrip resonators. However, a zero IF SSB transceiver design does not require a very high selectivity in the RF section. If the circuit losses can be compensated for by high gain semiconductor devices, cheaper substrates like the conventional glassfibre-epoxy FR4 can be used at frequencies up to at least 10GHz. The FR4 laminate has excellent mechanical properties. Unlike soft teflon laminates, cutting, drilling and hole plating in FR4 is well known. Even more important, most SMD component packages are designed for installation on a FR4 substrate and may break or develop intermittent contacts if installed on a soft teflon board. Therefore, losses in FR4 microstrip transmission lines and filters were investigated. Surprisingly, the losses were found inversely proportional to board thickness and rather slowly increasing with frequency. This simply means that the FR4 RF losses are mainly copper losses, while dielectric losses are still rather low. FR4 RF copper losses are high since the copper surface is made very rough to ensure good mechanical bonding to the dielectric substrate. In fact, if the copper foil is peeled off a piece of FR4 laminate, the lower foil surface is rather dark. On the other hand, if the copper foil is peeled off a piece of microwave teflon laminate, the colours of both foil surfaces are similar. Since different manufacturers use different methods for bonding the copper foil, RF losses are different in different FR4 laminates. On the other hand, the dielectric constant of FR4 was found quite stable. Finally, silver or gold plating of microstrip

lines etched on FR4 laminate really makes no sense, since most of the RF losses are caused by the (inaccessible) rough foil surface bonded to the dielectric.

A practical FR4 laminate thickness for microwave circuits with SMD components is probably 0.8mm. A 50-ohm microstrip line has a width of about 1.5mm and about 0.2dB/cm of loss at 5.76GHz. Therefore microstrip lines have to be kept short if etched on FR4 laminate. For comparison, the FR4 microstrip losses are about three times larger than the microstrip losses of a glassfibre-teflon board and about ten times larger than the losses of teflon semi rigid coax cables.

Although FR4 laminate losses are high, resonators and filters can still be implemented as microstrip circuits. Considering PCB etching tolerances and especially under etching, both transmission lines and gaps in between them should not be made too narrow. A practical lower limit is 0.4mm width for the transmission lines and 0.3mm for the gaps.

As already mentioned, modern semiconductor devices are really easy to use even at microwave frequencies. Silicon MMIC amplifiers provide 25dB of gain (limited by package parasitics) up to 2.3GHz. If less gain is required, conventional silicon bipolar transistors can be used, since their input and output impedances are also close to 50 ohms. GaAs semiconductors are more practical above about 5GHz. In particular, high performance devices like HEMTs have become inexpensive since they have been mass produced for satellite TV receivers. HEMTs operate at lower voltages and higher currents than conventional GaAsFETs, so their input and output impedance are very close to 50 ohms at frequencies above 5GHz. Serious microwave engineers are afraid of using HEMTs since these devices have enough gain to oscillate at frequencies above 50GHz or even 100GHz. In this case it is actually an advantage to build the circuit on a lossy laminate like FR4,

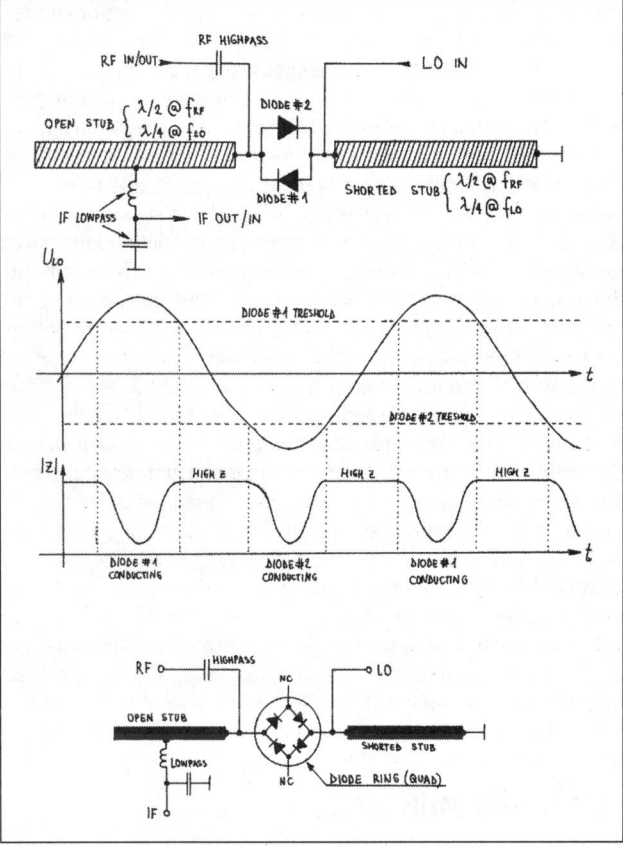

Fig 11.62: Sub harmonic mixer design

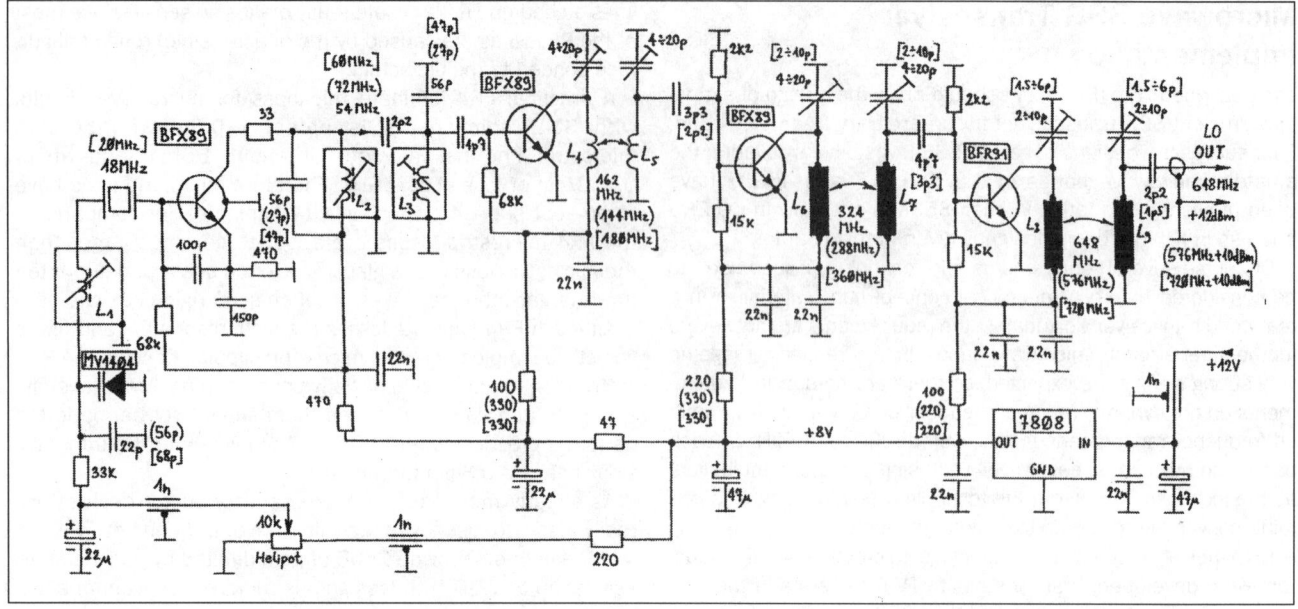

Fig 11.63: Circuit diagram of VXCO for Zero-IF transceiver

since the latter will efficiently suppress any oscillations in the millimetre frequency range. Having the ability to control the loss in a circuit therefore may represent an advantage! The availability of inexpensive power GaAsFETs greatly simplifies the construction of transmitter output stages. In particular, the high gain of power GaAsFETs in the 23cm and 13cm bands greatly reduces the number of stages when compared to silicon bipolar solutions.

Zero IF and direct conversion transceivers have some additional requirements for mixers. Mixer balancing is very important, both to suppress the unwanted residual carrier in the transmitter and to suppress the unwanted AM detection in the receiver.

At microwave frequencies, the simplest way of achieving good mixer balancing is to use a sub harmonic mixer with two anti parallel diodes as shown in **Fig 11.62**. Such a mixer requires a local oscillator at half frequency. Frequency doubling is achieved internally in the mixer circuit. A disadvantage of this mixer is a higher noise figure in the range 10 to 15dB and sensitivity to the LO signal level. Both a too low LO drive or a too high LO drive will further increase the mixer insertion loss and noise figure. On the other hand, the sub harmonic mixer only requires two non critical microstrip resonators that do not influence the balancing of the mixer. The best performances were obtained using schottky quads with the four diodes internally connected in a ring. The schottky quad BAT14-099R provides about -35dB of carrier suppression at 1296MHz and about -25dB of carrier suppression at 5760MHz with no tuning. A very important advantage of the sub harmonic mixer is that the local oscillator operates at half of the RF frequency. This reduces the RF LO crosstalk and therefore the shielding requirements in zero IF or direct conversion transceivers. A side advantage is that the half frequency LO chain requires fewer multiplier stages.

The common modules for the whole range of transceivers are described in this chapter. The modules for different frequency ranges can be found in [3]. Finally, an overview of the construction techniques is given in this chapter as well as shielding of the modules and integration of the complete transceivers.

VCXO and Multipliers

Since a relatively narrow frequency range needs to be covered, a VXO followed by multiplier stages is an efficient solution for the local oscillator. The VXO is built as a varactor tuned VCXO with a fundamental resonance crystal, since the frequency-pulling range of overtone crystals is not sufficient for this application. A fundamental resonance crystal has a lower Q and is less stable than overtone crystals, but for this application the performance is sufficient. Fundamental resonance crystals can be manufactured for frequencies up to about 25MHz. Therefore the output of the VCXO needs to be multiplied to obtain microwave frequencies.

Frequency multiplication can be obtained by a chain of conventional multipliers, including class C amplifiers and band pass filters or by a phase locked loop. Although the PLL requires almost no tuning and is easily reproducible, this solution was discarded for other reasons. A SSB transceiver requires a very clean LO signal, therefore the PLL requires buffer stages to avoid pulling the VCXO and/or the microwave VCO. Shielding and power supply regulation is also critical, making the whole PLL multiplier more complicated than a conventional multiplier chain. The circuit diagram of the VCXO and multiplier stages is shown in **Fig 11.63**. The VCXO is operating at around 18MHz in the transceivers for 1296MHz and 2304MHz, and at around 20MHz in the transceiver for 5760MHz.

All multiplier stages use silicon bipolar transistors BFX89 (BFY90) except the last stage with a BFR91. The module already supplies the required frequency of 648MHz for the 1296MHz version of the transceiver. In the 2304MHz version, the module supplies 576MHz by using different multiplication factors. The latter frequency is doubled to 1152MHz inside the transmit and receive mixer modules. In the 5760MHz version, the module supplies 720MHz and this frequency is further multiplied to 2880MHz in an additional multiplier module. Of course, the values of a few components need to be adjusted according to the exact operating frequency, shown in ( ) brackets for 2304MHz and in [ ] brackets for 5760MHz. The VCXO and multiplier chain are built on a single sided FR4 board with the dimensions of 40mm x 120mm as shown in **Fig 11.64**. (in Appendix B) The corresponding component location (for the 648MHz version) is shown in **Fig 11.65** (also in Appendix B). The exact value of L1 depends on the crystal used. Some parallel resonance crystals may even require the replacing of L1 with a capacitor. L2 and L3

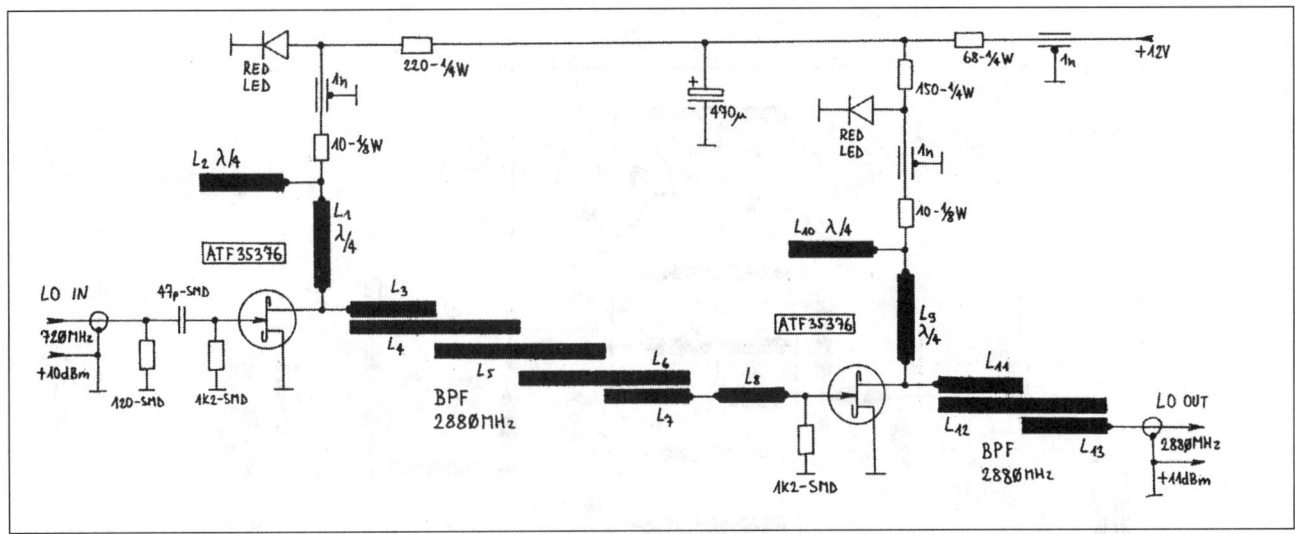

Fig 11.66: Circuit diagram of x4 multiplier to 2880MHz for Zero-IF transceiver

have about 150nH each or 4 turns of 0.25mm copper enamelled wire on a 10 x 10mm IF transformer coil former. L4 and L5 are self-supporting coils of 4 turns of 1mm copper enamelled wire each, wound on an internal diameter of 4mm. L6, L7, L8 and L9 are etched on the PCB. The VCXO module is the only part of the whole transceiver that requires tuning. L2, L3 and the capacitors in parallel with L4, L5, L6, L7, L8 and L9 should simply be tuned for the maximum output at the desired frequencies. In a multiplier chain, measuring the DC voltages over the base emitter junctions of the multiplier transistors can easily check RF signal levels.

When the multiplier chain is providing the specified output power, L1 and the capacitor in parallel with the MV1404 varactor should be set for the desired frequency coverage of the VCXO. If standard 'computer grade' 18.000MHz or 20.000MHz

crystals are used, it is recommended to select the crystal with the smallest temperature coefficient. Unfortunately not all amateurs are allowed to use the international segment around 2304MHz on 13cm. It is a little bit more difficult to find a crystal for 18.125MHz for the German segment around 2320MHz. The 5760MHz transceiver requires an additional multiplier from 720MHz to 2880MHz as shown in **Fig 11.66**. The first HEMT ATF35376 operates as a quadrupler while the second HEMT ATF35376 operates as a selective amplifier for the output frequency of 2880MHz. The additional multiplier for 2880MHz is built on a double sided microstrip FR4 board with the dimensions of 20mm x 120mm as shown in **Fig 11.67** with the corresponding component location shown in **Fig 11.68** (both in Appendix B). The 2880MHz multiplier should provide the rated output power of +11dBm without any tuning. On the other hand,

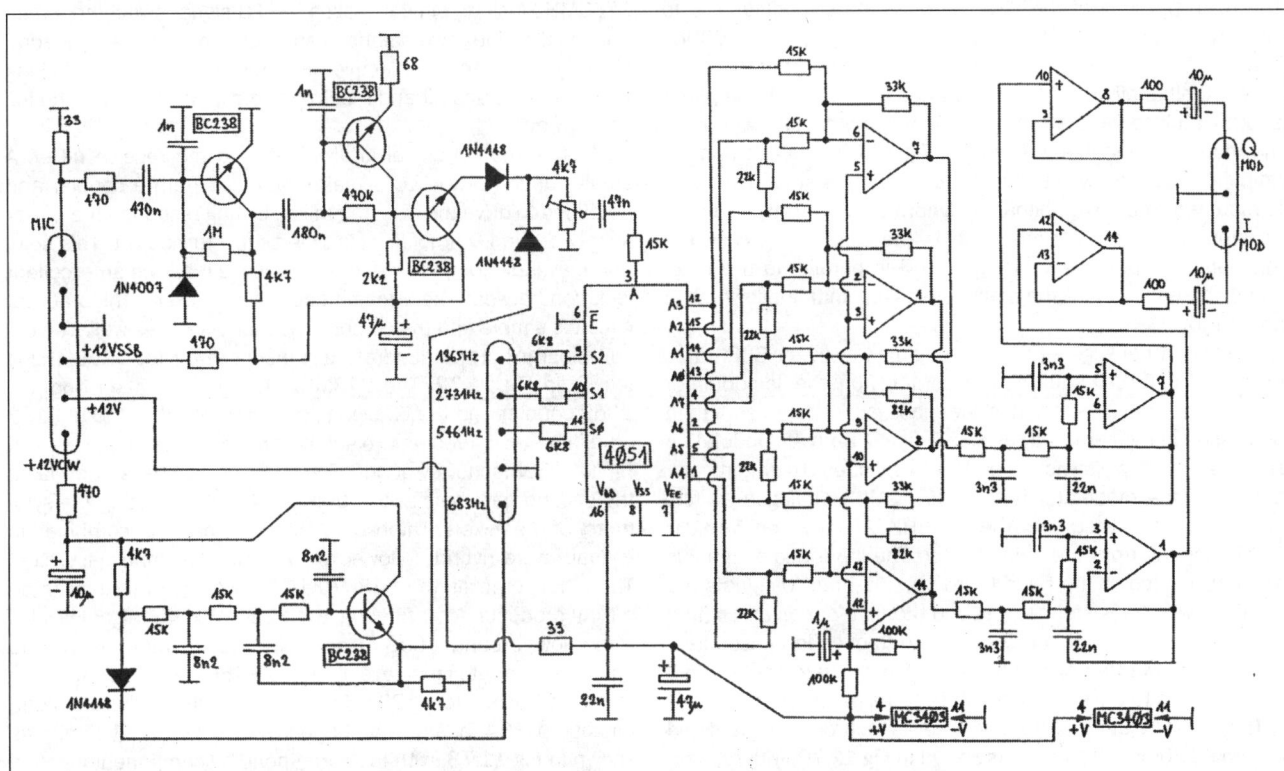

Fig 11.69: Circuit diagram of SSB/CW quadrature modulator for Zero-IF transceiver

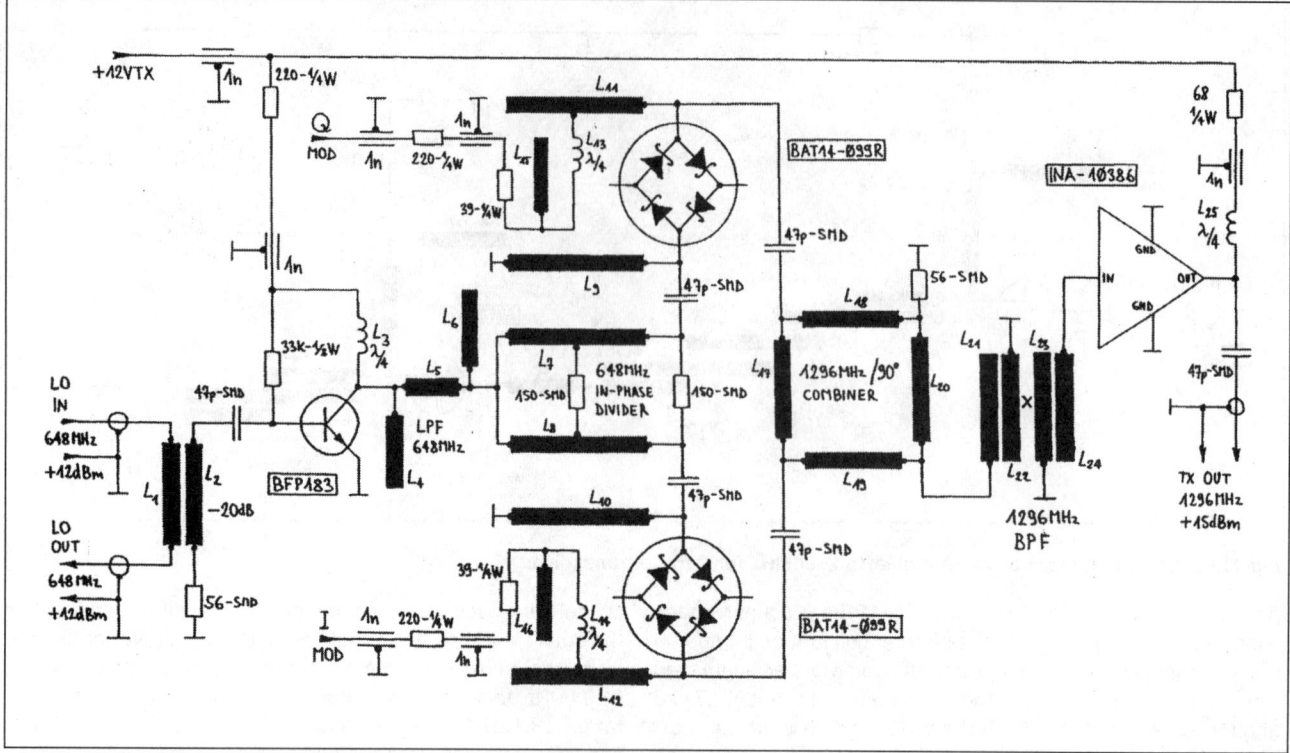

Fig 11.72: Circuit diagram of quadrature transmit modulator for 1296MHz Zero-IF transceiver

the tuning of L8 and L9 to 720MHz in the VCXO module can be optimised for the minimum DC drain current (max DC voltage) of the first HEMT. The two red LEDs are used as 2V Zeners. LEDs are in fact better than real Zeners, since they have a sharper knee and do not produce any avalanche noise.

SSB/CW Quadrature Modulator

The main purpose of the SSB/CW quadrature modulator is to convert the input audio frequency band from 200Hz to 2600Hz into two bands 0 to 1200Hz to drive the quadrature transmit mixer. Additionally the module includes a microphone amplifier and a circuit to generate the CW signal. The circuit diagram of the modulator module is shown in **Fig 11.69**. The microphone amplifier includes two stages with BC238 transistors. The input is matched to a low impedance dynamic mic with the 33-ohm resistor. The 1N4007 diode protects the input in the case the microphone input is simply connected in parallel to the loudspeaker output. Finally the output drives an emitter follower with another BC238.

The CW carrier is generated in the same way as the SSB transmission. The 683Hz square wave, coming from the demodulator module, is first cleaned in a low pass audio filter and then processed in the same way as a SSB signal. Both AF modulation sources are simply switched by 1N4148 diodes. The main component of the modulator is the 4051 CMOS analogue switch. The switch is rotated with the 1365Hz, 2731Hz and 5461Hz clocks coming from the demodulator. The input audio signal is alternatively fed to the I and Q chains. The I and Q signals are obtained with a resistor network and the first four op-amps (first MC3403). Then both I and Q signals go through low pass filters to remove unwanted mixing products. Finally there are two voltage followers to drive the quadrature transmit mixer.

The SSB/CW quadrature modulator is built on a single sided 40mm x 120mm FR4 board as shown in **Fig 11.70**, with the corresponding component location shown in **Fig 11.71** (both are in Appendix B). Most components are installed vertically to save

board space. The SSB/CW quadrature modulator does not require any alignment. The 4.7k-ohm trimmer is provided to check the overall transmitter. Full power (in CW mode) should be obtained with the trimmer in the central position.

Quadrature Transmit Mixers

All three transmit mixer modules for 1296MHz, 2304MHz and 5760MHz include similar stages: an LO signal switching, an in-phase LO divider, two balanced sub harmonic mixers, a quadrature combiner and a selective RF amplifier. LO signal switching between the transmit and receive mixers is performed in the following way:

Most of the LO signal is always fed to the receive mixer. A small fraction of the LO signal is obtained from a coupler and amplified to drive the transmit mixer. During reception the power supply of the LO amplifier stage is simply turned off. This solution may look complicated, but in practice it allows an excellent isolation between the transmit and receive mixers. The practical circuit is simple and the component count is low as well. The circuit diagram of the quadrature transmit mixer for 1296MHz is shown in **Fig 11.72**. The 648MHz LO signal is taken from a -20dB coupler and the LO signal level is restored by the BFP183 amplifier stage, feeding two sub harmonic mixers equipped with BAT14-099R schottky quads. The 648MHz low pass attenuates the second harmonic at 1296MHz to avoid corrupting the symmetry of the mixers. The two 1296MHz signals are combined in a quadrature hybrid, followed by a 1296MHz band pass filter. The latter removes the 648MHz LO as well as other unwanted mixing products. After filtering, the 1296MHz SSB signal level is rather low (around -10dBm), so an INA-10386 MMIC is used to boost the output signal level to about +15dBm. The quadrature transmit mixer for 1296MHz is built on a double-sided microstrip FR4 board with dimensions 40mm x 120mm as shown in **Fig 11.73**,.with the corresponding component location shown in **Fig 11.74** (both are in Appendix B). The circuit does not require any tuning for operation at 1296MHz or 1270MHz.

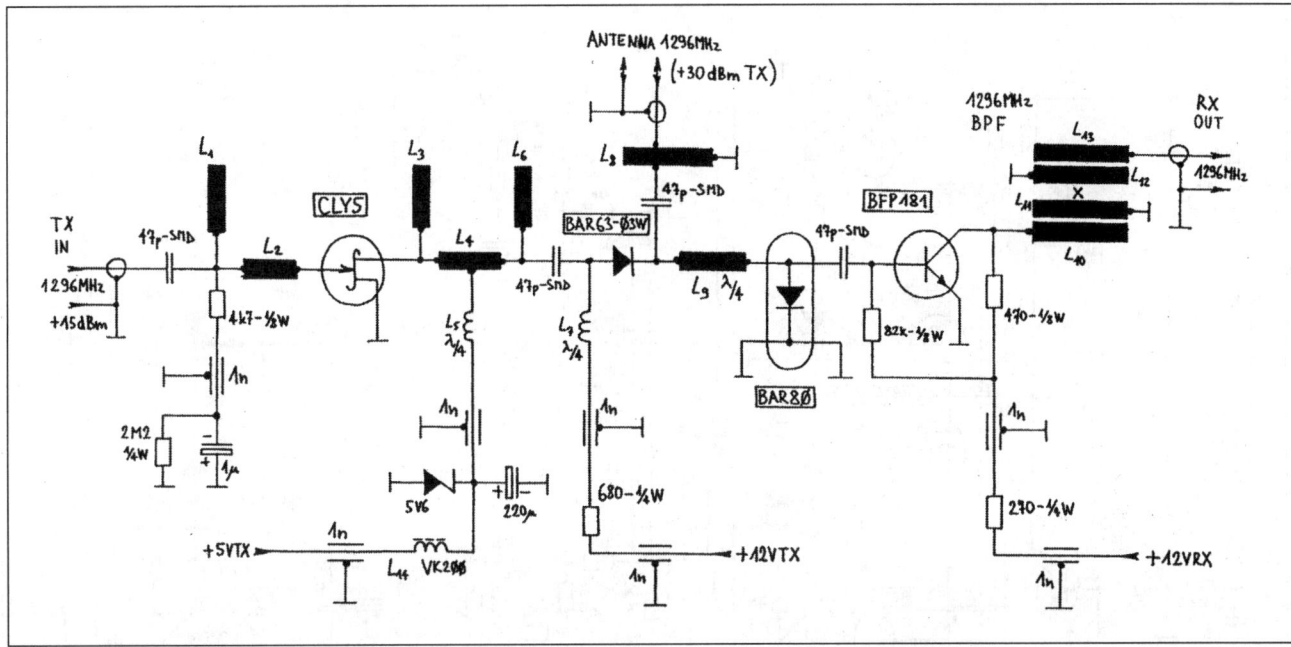

Fig 11.75: Circuit diagram of RF front end for 1296MHz Zero-IF transceiver

RF Front Ends

The RF front ends include the transmitter power amplifiers, the receiver low noise amplifiers and the antenna switching circuits. Of course there are major differences among different power amplifier designs, depending not just on the frequency, but also on the technology used and the output power desired. It no longer makes sense to use expensive coaxial relays, since PIN diodes can provide the same insertion loss and isolation at lower cost with better reliability and much shorter switching times.

The circuit diagram of the RF front-end for 1296MHz is shown in **Fig 75**. The transmitter power amplifier includes a single stage with a CLY5 power GaAsFET, providing a gain of 15dB and an output power of about 1W (+30dBm). The CLY5 is a low-voltage transistor operating at about 5V.

The negative gate bias is generated by rectification of the driving RF signal in the GS junction inside the CLY5 during modulation peaks. The gate is then held negative for a few seconds thanks to the 1μF capacitor. To prevent overheating and destruction of the CLY5, the +5VTx voltage is obtained through a current-limiting resistor. This arrangement may look strange, but it is very simple, requires no adjustments, allows a reasonably linear operation and most important of all, it proved very reliable in PSK packet-radio transceivers operating 24 hours per day in a packet-radio network.

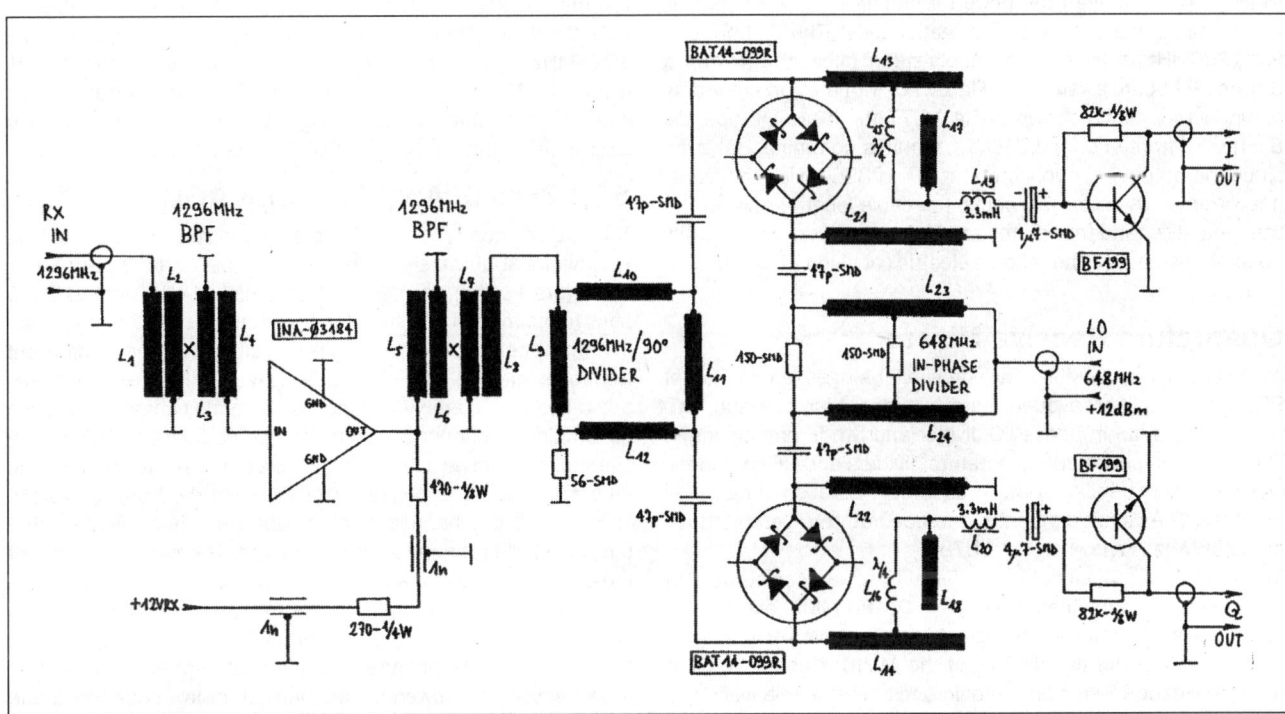

Fig 11.78: Circuit diagram of quadrature receive mixer for 1296MHz Zero-IF transceiver

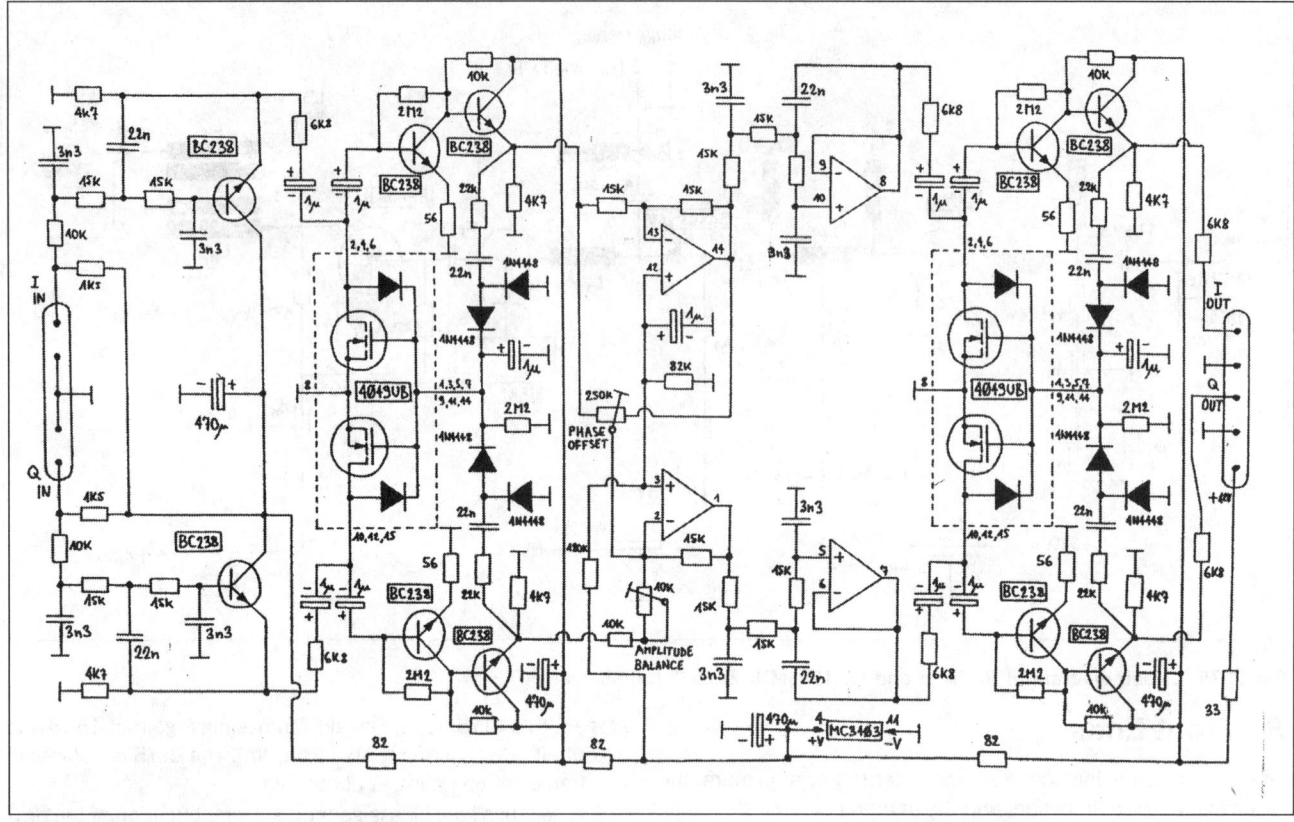

Fig 11.81: Circuit diagram of quadrature receive SSB IF amplifier for 1296MHz Zero-IF transceiver

The antenna switch includes a series diode BAR63-03W and a shunt diode BAR80. Both diodes are turned on while transmitting. L9 is a quarter wavelength line that transforms the BAR80 short circuit into an open for the transmitter. The receiving preamplifier includes a single BFP181 transistor (15dB gain) followed by a 1296MHz band pass filter (-3dB loss).

In the 1296MHz RF front-end, the LNA gain should be limited to avoid interference from powerful non-amateur users of this band (radars and other radio navigation aids). The RF front-end for 1296MHz is built on a double-sided microstrip 40mm x 80mm FR4 board as shown in **Fig 11.76**, with the corresponding component location shown in **Fig 11.77** (both are in Appendix B). The RF front end for 1296MHz requires no tuning. However, since the output impedance of the INA-10386 inside the transmit mixer is not exactly 50 ohms, the cable length between the transmit mixer and the RF front-end is critical. Therefore L1 may need adjustments if the teflon-dielectric cable length is different from 12.5cm.

Quadrature Receive Mixers

All receiver mixer modules include similar stages, an additional RF signal amplifier, a quadrature hybrid divider, two sub harmonic mixers, an in phase LO divider and two IF preamplifiers. The mixers, in phase and quadrature dividers and RF band pass filters are very similar to those used in the transmitting mixer modules. The circuit diagram of the quadrature receiving mixer for 1296MHz is shown in **Fig 11.78**.

The incoming RF signal is first fed through a microstrip band pass filter, then amplified with an INA-03184 MMIC and further filtered by another, identical microstrip band pass. The total gain of the chain of the two filters and the MMIC is about 20dB. A high gain in the RF section is required to cover the relatively high noise figure of the two sub harmonic mixers and the additional losses in the quadrature hybrid. The two receiving sub harmon-

ic mixers are also using BAT14-099R schottky quads. The mixer outputs are fed through low pass filters to the IF preamplifiers. The IF preamplifiers use BF199, HF transistors. These were found to perform better than their BCxxx counterparts in spite of the very low frequencies involved (less than 1200Hz). HF transistors have a smaller current gain, their input impedance is therefore smaller and better matches the output impedance of the mixers. Both IF preamplifiers receive their supply voltages from the IF amplifier module. The quadrature receiving mixer for 1296MHz is built on a double-sided microstrip FR4 board, 40mm x 120mm, as shown in **Fig 11.79** with the corresponding component location shown in **Fig 11.80** (both are in Appendix B). The receiving mixer for 1296MHz requires no tuning.

SSB Zero IF Amplifier with AGC

The basic feature of direct conversion and zero IF receivers is to achieve most of the signal gain with simple and inexpensive AF amplifiers. Further, the selectivity is achieved with simple RC low pass filters that require no tuning. The circuit diagram of such an IF amplifier equipped with AGC is therefore necessarily different from conventional high IF amplifiers. A zero IF receiver requires a two channel IF amplifier, since both I and Q channels need to be amplified independently before demodulation. The two IF channels should be as near identical as possible to preserve the amplitude ratio and phase offset between the I and Q signals. Therefore, both channels should have a common AGC so that the amplitude ratio remains unchanged. The circuit diagram of the quadrature SSB IF amplifier with AGC is shown in **Fig 11.81**.

The IF amplifier module includes two identical low pass filters on the input, followed by a dual amplifier stage with a common AGC. An amplitude/phase correction is performed after the first amplifier stage, followed by another pair of low pass filters and another dual amplifier stage with a common AGC. The two input low pass filters are active RC filters using BC238 emitter follow-

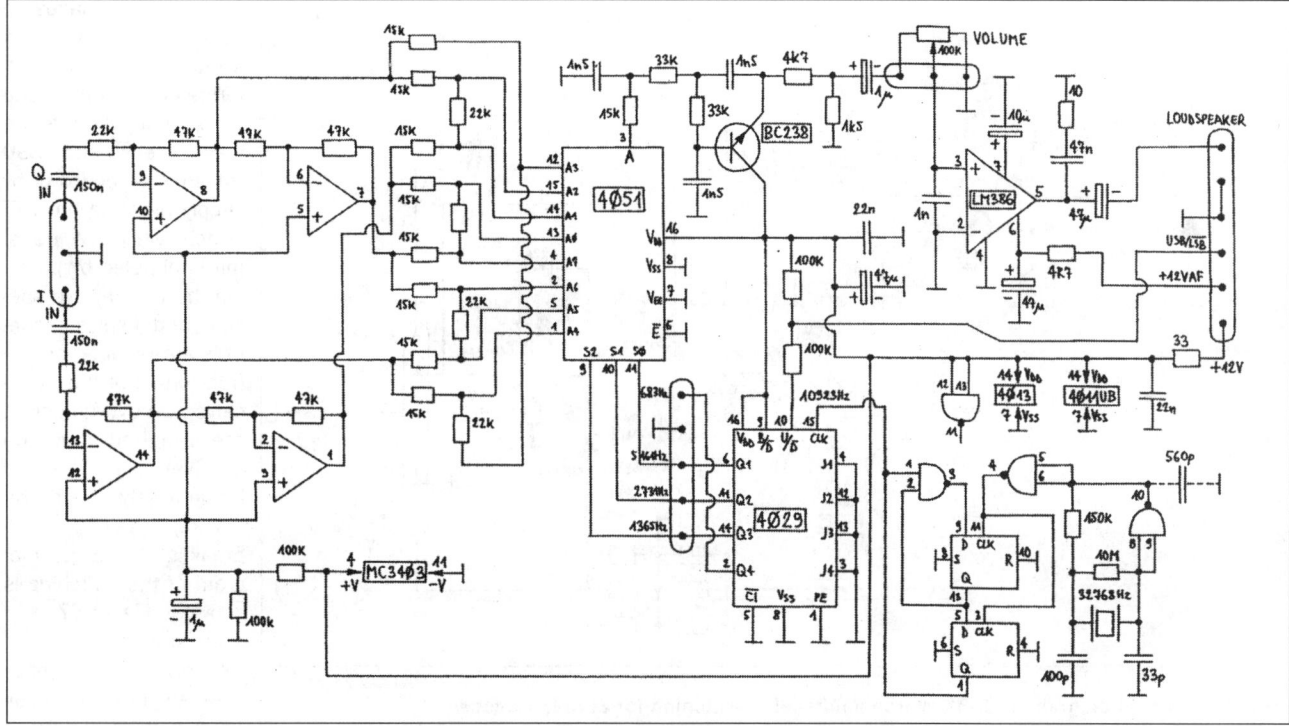

Fig 11.84: Circuit diagram of quadrature SSB demodulator and AF amplifier for 1296MHz Zero-IF transceiver

ers. Discrete bipolar transistors are used because they are much less noisy than operational amplifiers. The input circuit also provides the supply voltage to the IF preamplifiers inside the receiving mixer module through the 1.5k-ohm resistors.

The dual amplifier stages are also built with discrete BC238 bipolar transistors. Each stage includes a voltage amplifier (first BC238) followed by an emitter follower (second BC238) essentially to avoid mutual interactions when the amplifiers are chained with other circuits in the IF strip. The AGC uses MOS transistors as variable resistors on the inputs of the dual amplifier stages. To keep the gain of both I and Q channels identical, both MOS transistors are part of a single integrated circuit 4049UB. The digital CMOS integrated circuit 4049UB is used in a rather uncommon way; however the remaining components inside the 4049UB act just as diodes and do not disturb the operation of the AGC.

The IF amplifier module includes two trimmers for small corrections of the amplitude balance (10kΩ) and phase offset (250kΩ) between the two channels. The correction stage is followed by two active RC low pass filters. These use MC3403 operational amplifiers since the signals are already large enough and the op-amp noise is no longer a problem. Finally there is another, identical dual amplifier stage with its own AGC. The quadrature SSB IF amplifier is built on a single sided 50mm x 120mm FR4 board, as shown in **Fig 11.82**. with the corresponding component location shown in **Fig 11.83** (both are in Appendix B).

In order to keep the differences between the I and Q channels small, good quality components should be used in the IF amplifier. Using 5% resistors, 10% foil type capacitors and conventional BC238B transistors should keep the differences between the two channels small enough for normal operation. Most components are installed vertically to save board space. The amplitude balance (10kΩ) and phase offset (250kΩ) trimmers are initially set to their neutral (central) position. These are only used while testing the complete receiver to obtain the minimum distortion of the reproduced audio signal.

Quadrature SSB Demodulator and AF Amplifier

The main function of the quadrature SSB demodulator is the conversion of both I and Q IF signals (frequency range 0 to 1200Hz) back to the original audio frequency range: 200-2600Hz. The same module includes a power AF amplifier and a clock generator for both the phasor rotation in the transmitter and the phasor counter rotation in the receiver. The circuit diagram of the module is shown in **Fig 11.84**. The quadrature SSB demodulator includes four operational amplifiers (MC3403) to produce an 8-phase system from the I and Q signals, using a resistor network similar to that used in the modulator. The CMOS analogue switch, 4051, performs the signal demodulation or phasor counter rotation, rotating with a frequency of 1365Hz. The I and Q signals are alternatively fed to the output, or in other words the circuit performs exactly the opposite operation of the modulator. Unwanted mixing products of the phasor counter rotation are removed by an active RC low pass (BC238). The demodulated audio signal is fed to the 100kΩ volume control. A LM386 is used as the audio power amplifier due to its low current drain and small external component count.

The three clocks required to rotate both 4051 switches in the modulator and in the demodulator are supplied by a binary counter 4029. The 4029 includes an up/down input that allows the generation/demodulation of USB or LSB in this application. The up/down input has a 100kΩ pull up resistor for USB operation. LSB is obtained when the up/down input is grounded through a front panel switch. USB/LSB switching is usually not required for terrestrial microwave work; it is needed when operating through satellites or terrestrial linear transponders, or when using inverting converters or transverters for other frequency bands. Finally, USB/LSB switching may be useful to attenuate interference during CW reception.

An alternative way to switch sidebands is interchanging the I and Q channels. When assembling the transceiver modules together, it is therefore important to check the wiring so that the

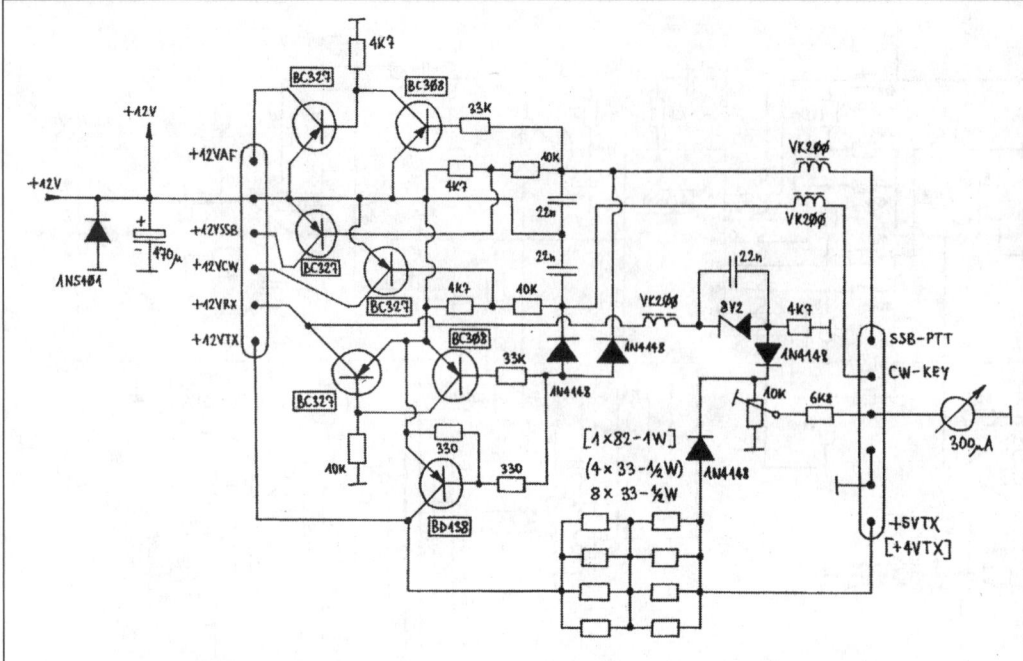

Fig 11.87: Circuit diagram of SSB/CW transmit/receive switching for Zero-IF transceiver

ic switching it makes no sense, since the Rx/Tx switching can be performed in less than one millisecond. It therefore makes sense that SSB transmit is enabled by simply pressing the PTT switch, while CW transmit is enabled by pressing the CW key. No special (and useless) controls are required on the front panel of the transceiver. On CW, no delays are required and the receiver is enabled immediately after the CW key is released (full break-in). The circuit diagram of the switching is shown in **Fig 11.87**.

In the transceivers described, most modules are enabled at all times with a continuous +12V supply to VCXO and multipliers, receiving mixer, IF amplifier, demodulator and modulator.

When enabling the transmitter, either by pressing the PTT or Morse key, the Rx LNA is turned off (+12VRX) and the Tx PA is turned on (+12VTX and +5VTX or +4VTX). During SSB transmission the receiver AF amplifier is turned off (+12VAF), to avoid disturbing the microphone amplifier (+12VSSB). On the other hand, during CW transmission the AF amplifier and most receiver stages remain on, so that the keying can be monitored in the loudspeaker or 'phones. The +12VCW supply connects the 683Hz signal to the modulator input.

The supply voltages +12VAF, +12VSSB, +12VCW and +12VRX are switched by BC327 PNP transistors. Due to the higher current drain, the +12VTX supply voltage requires a more powerful PNP transistor BD138. The transmitter PA receives its supply voltage through a current limiting resistor from the +12VTX line. Since the latter dissipates a considerable amount of power, it is built from several smaller resistors and located in the switching unit to prevent heating the PA transistor(s). The value of the current limiting resistor depends on the version of the transceiver. The 1296MHz PA with a CLY5 requires eight 33-ohm half-watt resistors for a total value of 16.5 ohms. The 2304MHz PA with a CLY2 requires four 33-ohm half-watt resistors, for a total value of 33 ohms. Finally, the 5760MHz PA with two ATF35376s requires a single 82-ohm one watt resistor.

The switching module also includes the circuits to drive the front panel meter. The latter is a moving coil type with a full scale sensitivity of about 300µA. The meter has two functions. During reception it is used to check the battery voltage. The 8V2 Zener extends the full scale of the meter to the interesting range from about 9V to about 15V. During transmission the meter is used to check the supply voltage of the PA transistor(s). Due to the self biasing operation, the PA voltage will be only 0.5 - 1V without modulation and will rise to its full value, limited by the Zener diode inside the PA, only when full drive is applied. The operation of the PA and the output RF power level can therefore be simply estimated from the PA voltage.

An S-meter is probably totally useless in small portable transceivers as those described here. If desired, the AGC voltage can

transmitter and the receiver operate on the same sideband at the same time.

The 4029 counter requires an input clock around 11kHz. This clock does not need to be particularly stable and an RC oscillator could be sufficient. In the transceivers described, a crystal source was preferred to avoid any tuning. In addition, if all transceivers use the same rotation or counter rotation frequency, the mutual interference is reduced. The oscillator uses a clock crystal, operating on a relatively low frequency of 32768Hz. The dual D flip-flop 4013 divides this frequency by 3 to obtain a 10923Hz clock for the 4029 binary counter. The resulting rotation frequency for the 4051 switches is 1365Hz. This almost perfectly matches the hole in the frequency spectrum of human voice. The same 4029 counter also supplies the CW tone, 683Hz, to reduce unwanted mixing products in the transmitter.

The quadrature SSB demodulator and AF amplifier are built on a single sided 40mm x 120mm FR4 board, as shown in **Fig 11.85** with the corresponding component location shown in **Fig 11.86**. (both are in Appendix B). Most of the components are installed vertically to save board space. The 32768Hz crystal oscillator will only operate reliably with a 4011UB (or 4001UB) integrated circuit. The commonly available "B" series CMOS integrated circuits (4011B in this case) have a too high a gain for this application. In the latter case, a 560pF capacitor may help to stabilise the oscillator. On the other hand, the oscillator circuit usually works reliably with old 4011 or 4001 circuits with an "A" suffix or no suffix letter at all.

SSB/CW Switching Rx/Tx

A SSB/CW transceiver requires different switching functions. Fortunately both SSB and CW modes of operation require the same functions in the receiver. Of course, two different operating modes are required for the transmitter: SSB Voice and CW Keying.

The Rx/Tx changeover is controlled by the PTT switch on the microphone in the SSB mode. On CW, most transceivers use an automatic delay circuit to keep the transmitter enabled during keying. This delay was perhaps required in old radios using several mechanical relays. In modern transceivers with all electron-

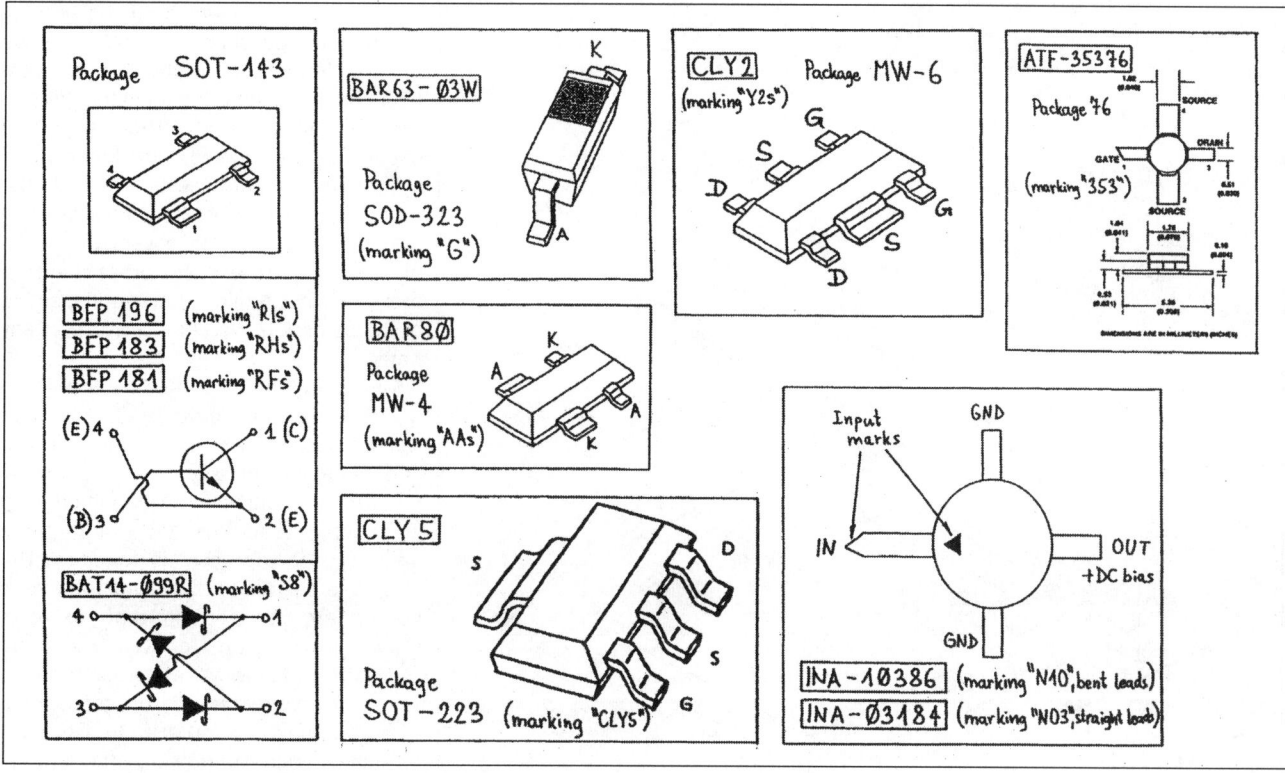

Fig 11.90: SMD semiconductor packages and pin-outs

be amplified and brought to a front panel meter. However, LED indicators are not visible in full sunshine on a mountaintop, so the choice is limited to moving-coil and LCD meters.

Most components of the SSB/CW switching are installed on a single sided FR4 board, 30mm x 80mm, as shown in **Fig 11.88** with their location (1296MHz version) shown in **Fig 11.89** (both diagrams are in Appendix B). Only the reverse polarity protection diode 1N5401 and the 470μF electrolytic capacitor are installed directly on the 12V supply connector. The 10k-ohm trimmer is used to adjust the meter sensitivity.

Construction of Zero IF SSB Transceivers

The described zero IF SSB transceivers use many SMD parts in the RF section. SMD resistors usually do not cause any problems, since they have low parasitic inductance up to at least 10GHz. On the other hand, there are big differences among SMD capacitors. For this reason, a single value (47pF) was used everywhere. The 47pF capacitors used in the prototypes are NPO type, rather large (size 1206), have a self resonance around 10GHz and introduce an insertion loss of about 0.5dB at 5.76GHz. Finally, the 4.7μF SMD tantalum capacitors can be replaced by the more popular tantalum 'drops'.

Quarter wavelength chokes are used elsewhere in the RF circuits. In the 5760MHz transceiver all quarter wavelength chokes are made as high impedance microstrips. On the other hand, to save board space in the 1296MHz and 2304MHz versions, the quarter wavelength chokes are made as small coils of 0.25mm thick copper enamelled wire of the correct length, chosen according to the frequency: 12cm for 648MHz, 9cm for 23cm mixers (648/1296MHz), 7cm for 1296MHz; and for L3, 5.5cm for 13cm mixers (1152/2304MHz) and 4cm for 2304MHz. The wire is tinned for about 5mm on each end and the remaining length is wound on an internal diameter of about

1mm. The SMD semiconductor packages and pin outs are shown in **Fig 11.90**.

Please note that due to lack of space, the SMD semiconductor markings are different from their type names. Only the relatively large CLY5 transistor in a SOT-223 package has enough space to carry the full marking "CLY5". The remaining components only carry one, two or three letter marking codes.

All of the microstrip circuits are built on double sided 0.8mm thick FR4 glassfibre-epoxy laminate. Only the top side is shown here, since the bottom side is left un-etched to act as a ground plane for the microstrips. The copper surface should not be tinned, nor silver or gold plated. The copper foil thickness should be preferably 35μm. Since the microstrip boards are not designed for plated through holes, care should be taken to ground all components properly. Microstrip lines are grounded

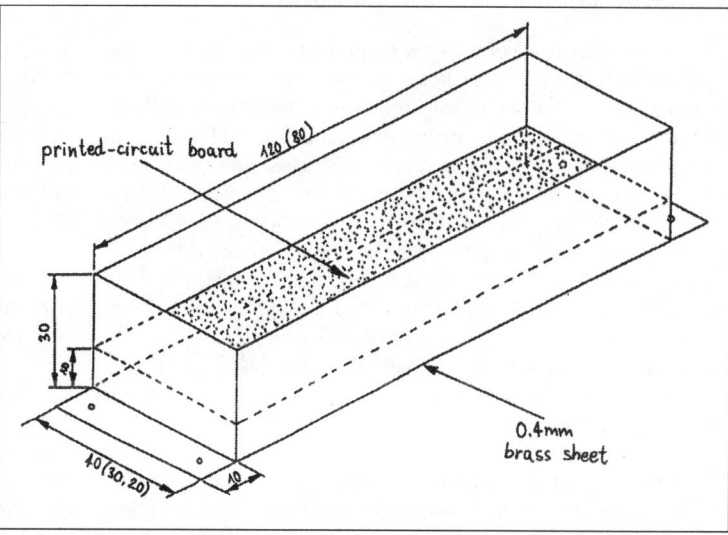

Fig 11.91: Shielded RF module enclosure for Zero-IF transceiver

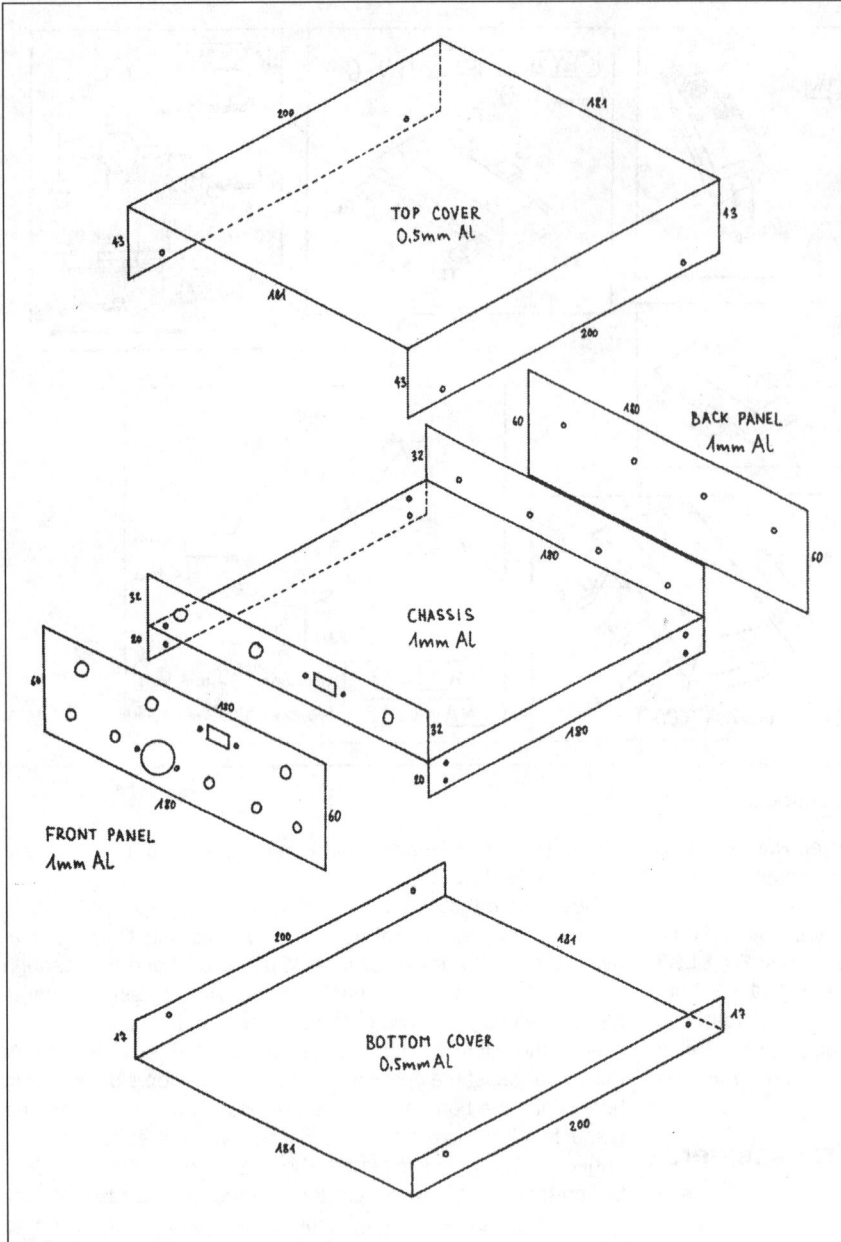

Fig 11.92: Enclosure for Zero-IF transceiver

height of about 10mm from the bottom. Additional feed through capacitors are required in the brass walls. RF signal connections are made using thin teflon dielectric coax like RG-188. The coax braid should be well soldered to the brass frame all around the entrance hole. Finally the frame is screwed on the chassis using sheet metal screws. The covers are kept in place thanks to the elastic brass sheet, so they need not be soldered. An inspection of the content is therefore possible at any time.

The VCXO/multiplier module is built on a single sided board and therefore requires both a top and a bottom cover. The remaining modules are all built as microstrip circuits, so the microstrip ground plane acts as the bottom cover and only the top cover is required. The sizes and shapes of the microstrip circuit boards are selected so that no resonances occur up to and including 2880MHz. Microwave absorber foam is therefore only required in the three modules operating at 5760MHz. To avoid disturbing the microstrip circuit, the foam is installed just below the top cover. The modules of a zero IF SSB transceiver are installed in a custom-made enclosure as shown in **Fig 11.92**. The most important component is the chassis. This must be made of a single piece of 1mm thick aluminium sheet to provide a common ground for all modules. If a common ground is not available, the receiver will probably self-oscillate in the from of 'ringing' or 'whistling' in the loudspeaker, especially at higher volume settings. The chassis carries both the front and back panels as well as the top and bottom covers.

All of the connectors and commands are available on the front panel which is screwed to the chassis using the components installed, CW pushbutton, SMA connector, meter and tuning helipot. To save weight, the top and bottom cover are made of 0.5mm thick aluminium sheet. They are screwed to the chassis using sheet metal screws. The shielded RF modules are installed on the top side of the chassis where a height up to 32mm is allowed. The audio frequency bare printed circuit boards are installed on the bottom side of the chassis. Interconnections between both sides of the chassis are made through five large diameter holes. The location of the modules as well as the location of the connectors and commands on the front panel are shown in **Fig 11.93** for both sides of the chassis.

Since the RF receiving modules are quite sensitive to vibrations, the loudspeaker should not be installed inside the transceiver. The same loudspeaker may also be used as a dynamic microphone for the transmitter. The circuit is designed so that it allows a simple parallel connection of the loudspeaker output and the microphone input. The PTT and CW keys are simple switches to ground. **Fig 11.94** shows a picture of the completed transceiver.

using 0.6mm thick silver plated wire (RG214 central conductors) inserted in 1mm diameter holes at the marked positions and soldered on both sides to the copper foil. Resistors and semiconductors are grounded through 2mm, 3.2mm and 5mm diameter holes at the marked positions. These holes are first covered on the ground plane side with pieces of thin copper foil (0.1mm). Then the holes are filled with solder and finally the SMD component is soldered into the circuit. Feed-through capacitors are also installed in 3.2mm diameter holes in the microstrip boards and soldered to the ground plane. Feed through capacitors are used for supply voltages and low frequency signals. Some components like bias resistors, Zeners and electrolytic capacitors are installed on the bottom side (as shown on the component location drawings) and connected to the feed-through capacitors.

The VCXO/multiplier module and all of the microstrip circuits are installed in shielded enclosures as shown in **Fig 11.91**. Both the frame and the cover are made of 0.4mm thick brass sheet. The printed circuit board is soldered in the frame at a

The Radio Communication Handbook

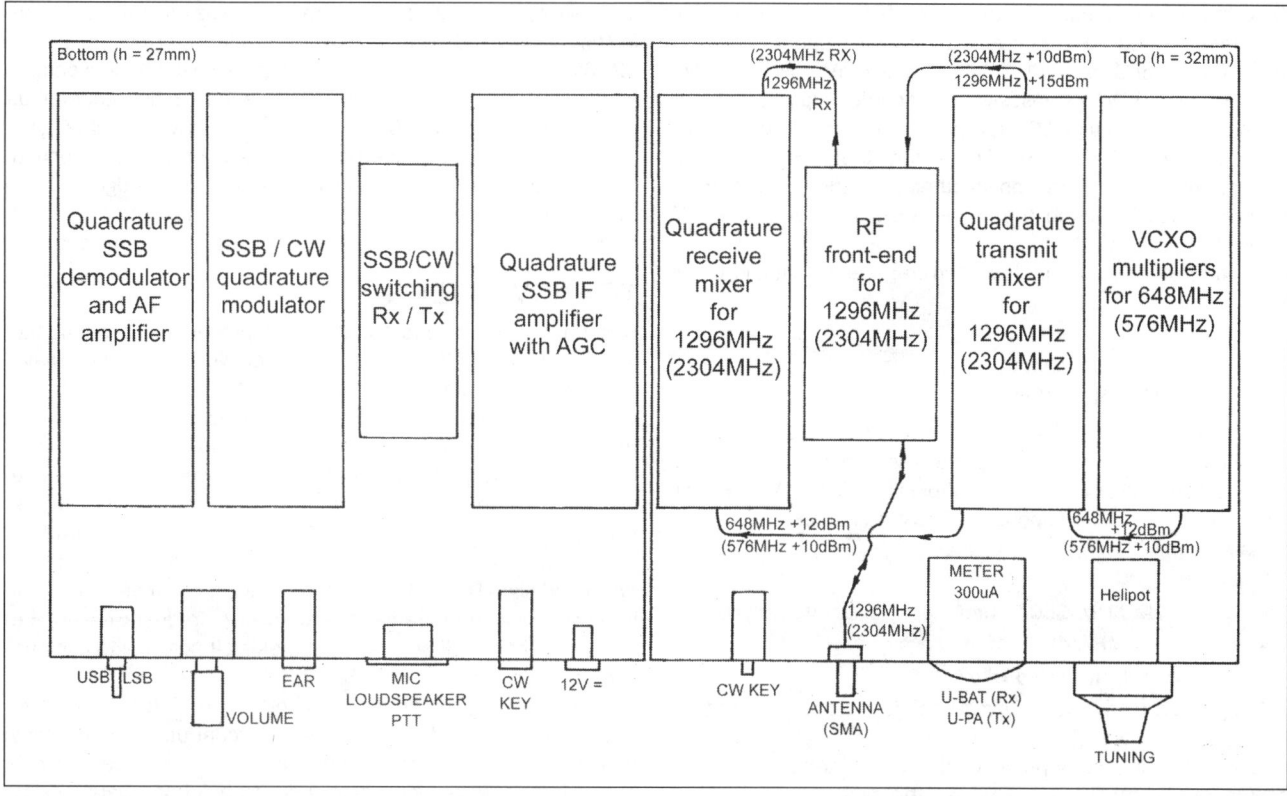

Fig 11.93: 1296MHz Zero-IF transceiver module locations

Testing the Zero IF SSB Transceivers

These SSB/CW transceivers do not require much tuning. The only module that really needs tuning is the VCXO/multiplier which is simply tuned for the maximum output on the desired frequency. Of course, the desired coverage of the VCXO has to be set with a frequency counter. After the VCXO/multiplier module is adjusted, the remaining parts of the transceiver require a check to locate defective components, soldering errors or insufficient shielding. The receiver should already work and some noise should be heard in the loudspeaker. The noise intensity should drop when the power supply to the LNA is removed. The noise should completely disappear when the receiving mixer

Fig 11.94: Picture of 1296MHz Zero-IF transceiver

module is disconnected from the IF amplifier. A similar noise should be heard if only one (I or Q) IF channel is connected.

For the next test, the receiver is connected to an outdoor antenna far away from the receiver and tuned to a weak unmodulated carrier (a beacon transmitter or another VCXO/multiplier module at a distance of a few tens of metres). Tuning the receiver around the unmodulated signal should produce both the desired tone and its much weaker mirror image changing its frequency in the opposite direction. The two trimmers in the IF amplifier should be set so that the mirror tone disappears. The correct function of the USB/LSB switch can also be checked.

Finally, the shielding of the receiver should be checked. A small, handheld antenna (10-15dBi) is connected to the receiver and the main beam of the antenna is directed into the transceiver. If the noise coming from the loudspeaker changes, the shielding of the local oscillator multiplier chain is insufficient. Next a mains operated fluorescent tube (20W or 40W) is turned on in the same room. A weak mains hum should only be heard when the handheld antenna is pointed towards the tube at 2 - 3m distance. If a clean hum without noise is heard regardless of the antenna direction, the shielding of the local oscillator multiplier chain is insufficient.

The transmitter should be checked for output power. Full power should be achieved with the modulator trimmer in the middle position whilst in CW mode. The DC voltage across the PA transistor should rise to the full value allowed by the 5.6V or 4.7V Zener diode. The output power should drop by an equal amount if only I or only Q modulation is connected to the transmit mixer. Finally the SSB modulation must be checked with another receiver for the same frequency band, or preferably in a contact with another amateur station at a distance of a few kilometres. This is the simplest way to find out the correct sideband, USB or LSB, of the transmitter, since the I and Q channels can be easily interchanged by mistake in the wiring. The residual carrier level of the transmit mixer should also be checked. Due to

the conversion principle the carrier results in a 1365Hz tone in a correctly tuned SSB receiver. The carrier suppression may range from -35dB in the 1296MHz transceiver down to only -20dB in the 5760MHz transceiver. Poor carrier suppression may be caused by a too high LO signal level or by a careless installation of the BAT14-099R mixer diodes. Note that the residual carrier cannot be monitored on another correctly tuned zero IF receiver, since it falls in the AF response hole of the zero-IF receiver.

The current drain of the described transceivers should be as follows:

Receive

- For 1296MHz: 105mA
- For 2304MHz: 175mA
- For 5760MHz: 300mA.

The current drain of the transmitters is inversely proportional to the output power due to the self biasing of the PA. The minimum current drain corresponds to SSB modulation peaks or CW transmission.

Transmitters:

- For 1296MHz: 650-870mA
- For 2304MHz: 490-640mA
- For 5760MHz: 410-440mA.

All figures are given for a typical sample at a supply voltage of 12.6V.

Finally, it should be appreciated that zero IF transceivers also have some limitations. In particular, the dynamic range of the receiver is limited by the direct AM detection in the receiving mixer. If very strong signals are expected, the LNA gain has to be reduced to avoid the above problem. This is already done in the 1296MHz receiver, since strong radar signals are quite common in the 23cm band. The sensitivity to radar interference of the 1296MHz transceiver was found comparable to the conventional transverter + 2m transceiver combination. On the other hand, the 2304MHz and 5760MHz transceivers have a higher gain LNA. If the dynamic range needs to be improved, the second LNA stage can simply be replaced with a wire bridge in both transceivers. Of course, the internal LNA gain must be reduced or the LNA completely eliminated if an external LNA is used.

TRANSVERTERS

Despite the arguments listed in the preceeding section, the most popular method of becoming active on the microwave bands today is still to use a commercial transceiver and a transverter. The transverter performs two functions, it down-converts the incoming signal, on the microwave band being used, to the chosen input frequency of the commercial transceiver (tuneable IF) and up-converts the output of the transceiver to the microwave band. The most common tuneable IF is 144MHz; using lower frequencies makes it difficult to filter out the unwanted signals. The transverter will use a common local oscillator chain so that transmit and receive frequencies on the microwave band will be the same, making it possible to use narrow band modes such as SSB.

Designs of transistorised transverters for 23cm started to appear in the mid 1970s. As semiconductor technology improved, designs for the higher amateur bands became available. Most of these were quite tricky to build and persuade to work properly. In the mid 1980s "No Tune" transverter designs started to appear. These overcame many of the construction and set-up problems. Some of the amateurs who designed

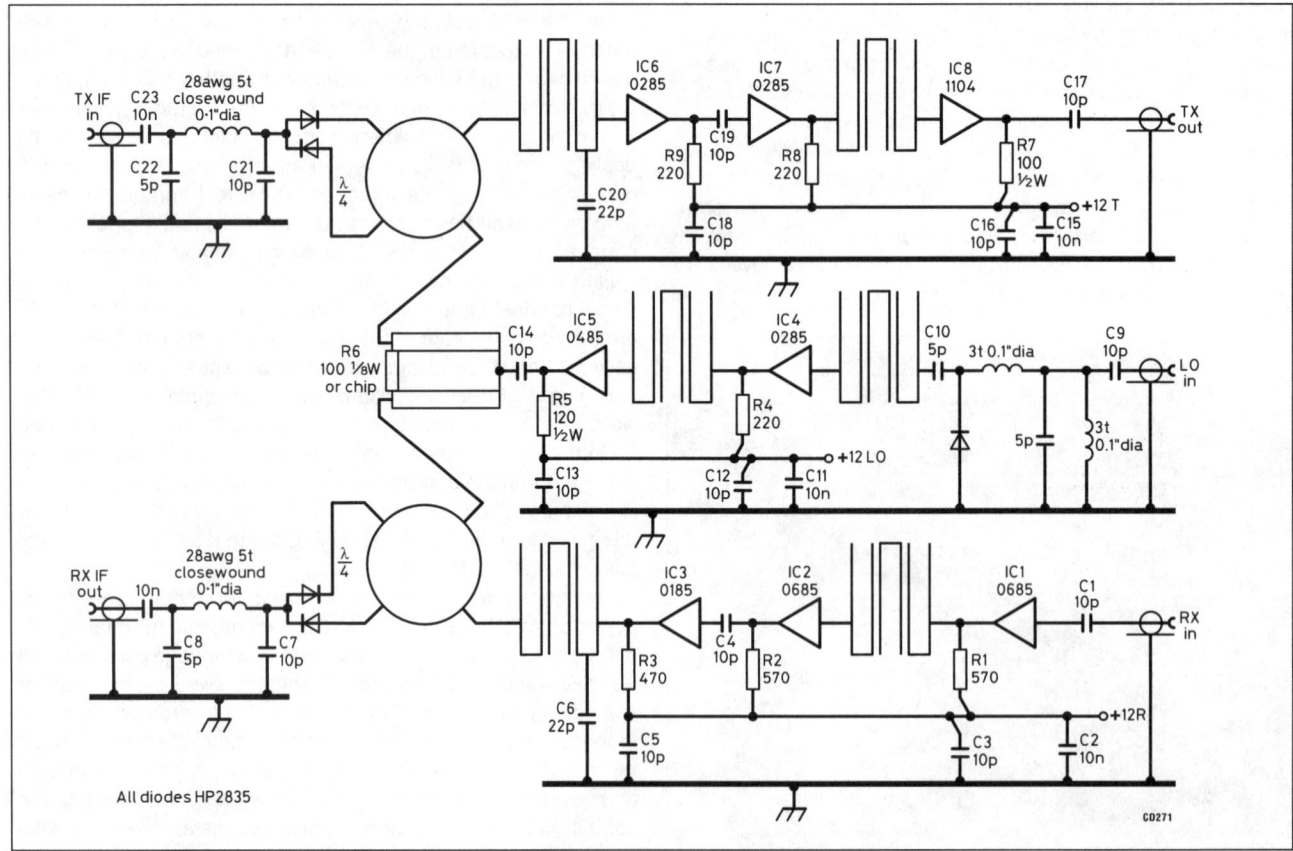

Fig 11.95: Circuit diagram of the KK7B transverter. BA481 Schottky diodes can be used instead of the HP5082-2835 diodes. Other, more modern MMICs can be substituted for the specified types, provided that the supply resistors are calculated and adjusted in value to suit the devices chosen. See reference [33]

Fig 11.96: Component layout for the KK7B transverter

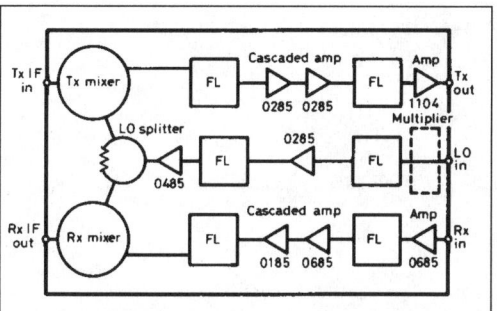

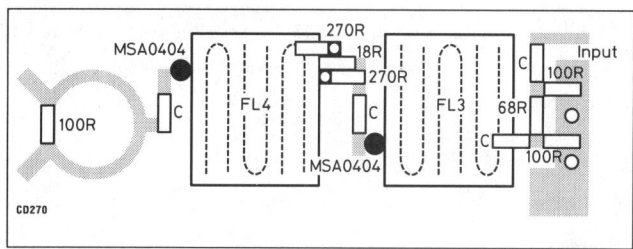

Fig 11.99: Layout of components for the modified LO chain in the KK7B transverter

these transverters now produce them as kits and ready built units, these are all listed in the bibliography.

Amateurs have always enjoyed modifying commercial equipment to work on the amateur bands, this is now possible for most of the microwave bands because there is a lot of surplus microwave link equipment on the market. The problem with this route is to recognise the potential of the equipment that we all see at rallies and equipment sales. The Internet is a wonderful source of knowledge about this, including pictures of the equipment and modification details, one such modification is shown in this chapter.

A Single-Board, No-Tune 144MHz/1296MHz Transverter

The comparatively recent development of economically priced and readily available microwave monolithic integrated circuits (MMICs) has allowed the development of a number of broadband (no-tune) low-power transverters from the 144MHz amateur band to the lower microwave bands, typically 1.3, 2.3 and 3.4GHz. Such designs use microstrip technology, including no-tune inter-stage band pass filtering in the LO, receive and transmit chains.

A 144MHz to 1296MHz transverter circuit was described by KK7B in [33]. This circuit and layout, although it does not give 'ultimate' performance in terms of either receive noise figure or transmit output power (nor is it particularly compact in terms of board size), is probably one of the simplest and most cost-effective designs available. Its simplicity also makes it suitable for novice constructors. It is also flexible enough to allow the constructor to substitute new, improved MMICs as these become available, without major re-engineering.

The receive performance can be enhanced by means of an external (possibly mast-head) low-noise amplifier (LNA), such a high-performance GaAsFET or PHEMT design. The transmit output level, at 13dBm (20mW), is ideal for driving a linear PA module such as the G4DDK-002 design [34]. This, in turn, could drive either a solid-state power block amplifier or a valve linear amplifier.

Precision printed circuit boards for this design and a similar design for the 2.3GHz band [35] have been available for some time, produced and marketed by Down East Microwave [36] in the USA. They are also available from a number of sources in the UK. A similar design concept was adopted for a transverter for the US 900MHz amateur band [37]. The Down East Microwave website [36] is worth visiting, it has a number of useful designs and application notes that can be freely downloaded

The original circuit diagram of the 1296MHz version is given in **Fig 11.95** and the physical layout of the circuit is shown, not to scale, in **Fig 11.96**. Wide use is made of hairpin-shaped, self resonant, printed microstripline filters in the LO, receive (RX) and transmit (TX) chains, together with printed microstripline transmit and receive balanced mixers and 3dB power splitter for the LO chain. An external LO source at any sub-harmonic frequency of the required injection frequency (1152MHz for the 1296-1298MHz narrow-band communications segment of the 23cm band when using an IF of 144-146MHz) was used and a simple, on-board diode multiplier produced the required injection frequency from the LO input. Although direct injection of 1152MHz was mentioned in the original description, little guidance was given as to how to achieve this. With a simple modification to the LO chain and a few changes to circuit values and devices, without need for PCB changes, it is easily possible to use the G4DDK-001 1152MHz source, already described, as the LO for this design.

Fig 11.97 gives the circuit and component values for the original LO chain, while **Fig 11.98** shows the modifications to allow the correct mixer injection levels to be attained when using the single +13dBm output option of the G4DDK-001 1152MHz source. **Fig 11.99** shows the layout of the modified circuit using the existing PCB pads and tracks.

Construction is straightforward, using surface-mount techniques, ie all components, whether SMD or conventional, are mounted on the track side of the board unlike conventional construction. All non-semiconductor components - connectors, resistors, capacitors and inductors - should be mounted first, the MMICs and mixer diodes last, taking adequate precautions to avoid both heat and static damage. Note that the

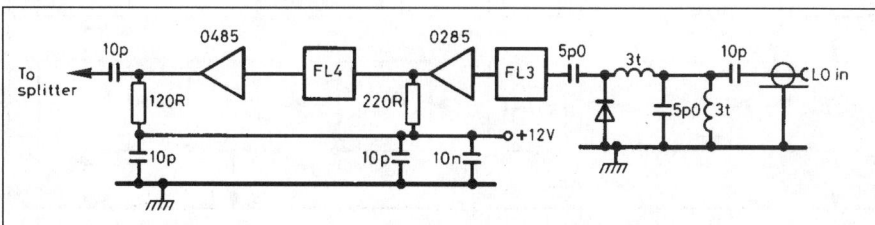

Fig 11.97: Original KK7B LO circuit used with his no-tune transverter

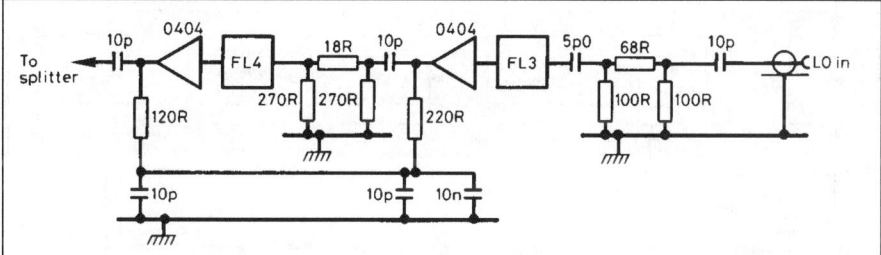

Fig 11.98: Modification to the KK7B LO circuit to allow use of the G4DDK-001 1152MHz source and the higher output level available from this circuit

3216 PLL Calculations for X Band Transverter with 144MHz 2nd IF; 1st LO=11,360MHz; 1st IF=992MHz

Ref MHz	2	Ref MHz can be 10MHz divided by any integer from 1 - 16						
VCO MHz	2,272							
PLL MHz	1,136	PLL in MHz is VCO/2 and must be an integer multiple of Ref MHz						
N	568							
		M6(Pin15)	M5(Pin14)	M4(Pin13)	M3(Pin10)	M2(Pin9)	M1(Pin8)	M0(Pin7)
M	55	0	1	1	0	1	1	1
Board as is		0	0	0	0	0	0	0
		A3(Pin21)	A2(Pin20)	A1(Pin19)	A0(Pin18)			
A	8	1	0	0	0			
Board as is		0	0	0	0			
		R2(Pin5)	R2(Pin4)	R1(Pin3)	R0(Pin2)			
R	4	0	1	0	0			
Board as is		0	0	0	0			
Lift pin22								

Reference suppression filter modifications, parallel these capacitors with the following values

Ref MHz	C1	C2,C3	Add 1pF to VCO
5	None	None	
2	1000pF	3000pF	
1	4700pF	6800pF	

Table 11.11: Synthesiser calculations for Qualcomm OmniTracks unit. 3216 PLL Calculations for X Band Transverter with 144MHz 2nd IF; 1st LO=11,360MHz; 1st IF=992MHz

values of the bias resistors, which set the working points of the MMICs, were chosen for a supply rail of +12V DC. Higher supply voltages will require recalculation of these values and the constructor should refer to either the maker's data sheets for the particular devices used or to the more general information given in reference [38]. When construction is complete and the circuit checked out for correct values and placing of components, there is no alignment as such, assuming that the LO source has already been aligned! It should simply be a matter of connecting the transverter to a suitable 144MHz (multimode) transceiver via a suitable attenuator and switching interface such as that by G3SEK [39] or the G4JNT design available from reference [40].

not provide for a two meter IF. This version uses a somewhat smaller, more recent OmniTracks unit that contains the power supply and synthesiser on the same assembly as the RF board, and utilises dual conversion high side LO to allow use of the stripline filters. The filter modification has been proven to work well by extending the filter elements to specified lengths. Some additional tuning of the transmit output stages appears to be required for maximum output.

The synthesiser VCO operates at 2,272MHz, and when multiplied by five it becomes 11,360MHz for the first LO. The first IF frequency is 992MHz which is near the original internal IF frequency of 1GHz. The second LO is derived from the synthesiser

A 10GHz Transverter from Surplus Qualcomm OmniTracks Units

These modifications were produced by Kerry Banke, N6IZW, of the San Diego Microwave Group and presented at The Microwave Update in 1999. The project offers an economical route to 10GHz, the unmodified transceiver, 10MHz TXCO and unmodified 1W PA can, at the time of writing, be ordered from Chuck Houghton for about £100 [41].

An earlier Qualcomm X-Band conversion project required considerable mechanical changes as well as electrical modifications and was based on replacing the original stripline filters with pipecap filters. These filters were required to provide sufficient LO and image rejection at 10GHz that the original stripline filters could

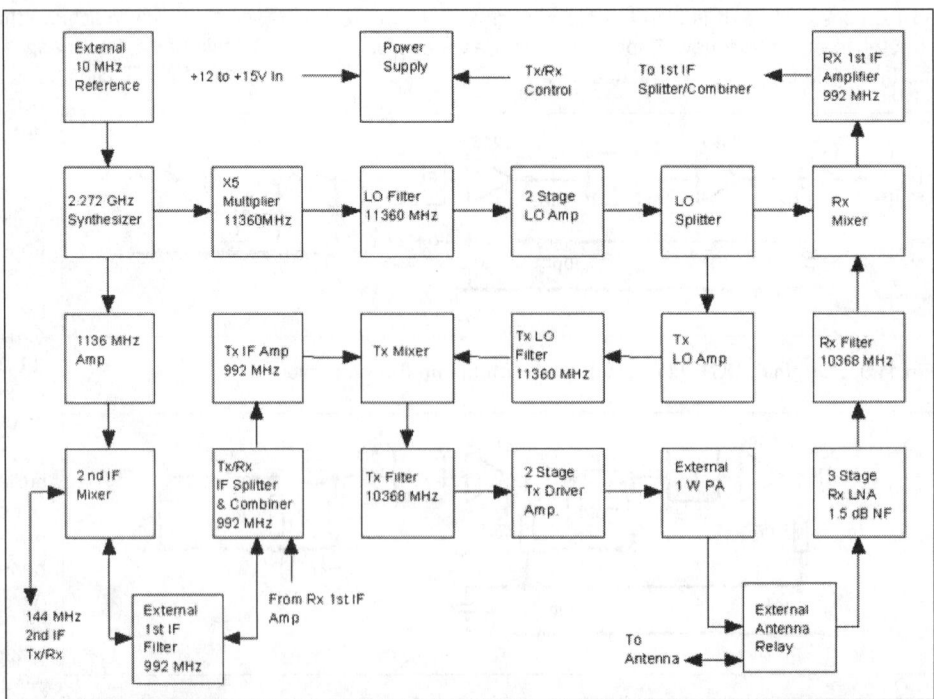

Fig 11.100: Block diagram of Qualcomm X band transverter conversion

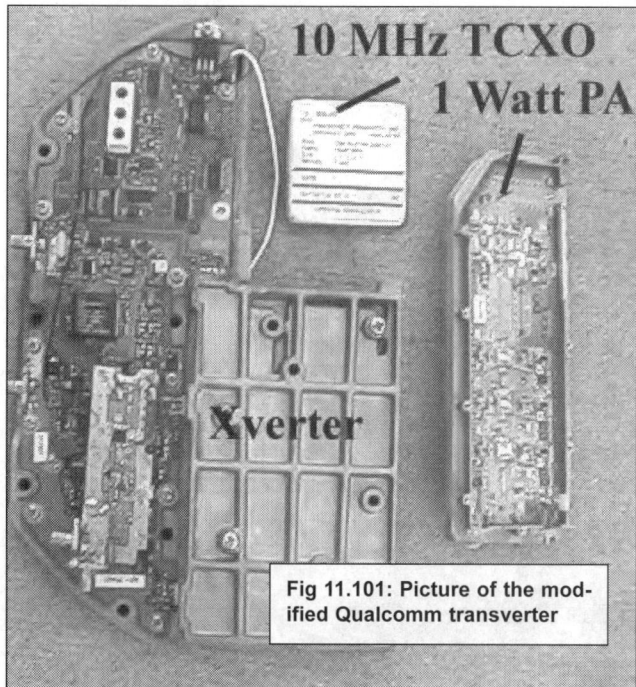

Fig 11.101: Picture of the modified Qualcomm transverter

Fig 11.103: Conversion step 1: Mark the locations of RF connectors and board cuts for coupling capacitors

pre-scaler, this divides the VCO frequency by two to produce 1,136MHz. Other second IF frequencies may be calculated using the relationship (RF-IF2)/0.9 = LO1 where RF is the 10GHz operating frequency (10,368MHz), IF2 is the second IF frequency, and LO1 is the first LO frequency. The synthesiser output frequency is then LO1 divided by five. **Table 11.11** shows the Excel spread sheet used to calculate the synthesiser programming.

The second conversion stage consists of a second LO amplifier (1,136MHz) and SRA-11 mixer converting the 992MHz 1st IF to the 144MHz 2nd IF. A 992MHz filter is required between the two conversion stages. Both Evanescent Mode and Coaxial Ceramic filters have been used. The conversion yields a reasonably high performance transverter with a noise figure of about 1.5dB and a power output of +8dBm, frequency locked to a stable 10MHz reference. Power required is +12VDC with a current consumption of about 0.5A on receive and 0.6A on transmit (about 1.5A total on transmit when including the 1W PA). **Fig 11.100** is a block diagram of the modified unit.

The unmodified circuit has a synthesiser output of 2,620MHz providing an LO of 13.1GHz. The original transmit frequency was around 14.5GHz with one watt output, and the receiver was near 12GHz. Unfortunately, the integrated PA in the original configuration provides no useful output below 12GHz and is not modifiable, so it has been removed for the 10GHz conversion. The transmit and receive IF preamplifiers make the transmit input requirement low (-10dBm) and provides high overall transverter receive gain.

Fig 11.101 shows a picture of the modified transverter, 1W amplifier and 10MHz TCXO. **Fig 11.102** shows a picture indicating the locations of the various functions. The following is an outline of the conversion procedure [42]:

1 Marking location of RF connectors and removal of circuit boards.
2 Base plate modification for mounting two SMA connectors (10GHz receive and transmit) plus four SMA connectors installed (2 RF + 1 IF and 10MHz Reference input).
3 Clearing of SMA connector pin areas in PCB ground plane.
4 Remounting of PCBs.
5 Cuts made to PCB and coupling capacitors installed.
6 Stripline filter elements extended and tuning stubs added.
7 Synthesiser reprogrammed and 4 capacitors added.
8 Add tuning stubs to the x5 Multiplier stage
9 2nd LO amplifier, mixer and 1st IF filter added.
10 Power and transmit/receive control wires added.
11 Test of all biasing.
12 Synthesiser and receiver test.
13 Transmitter test and output stage tuning.

Step 1. Mark the location of RF connectors and board cuts for coupling capacitors: Before removing the boards from the base plate, carefully drill through the board in the two places shown using a 0.050 inch diameter drill just deep enough to mark the base plate. These are the locations for receive and transmit RF SMA connectors. The upper connector hole (transmit) is located 0.5 inches to the left of the transistor case edge. The lower hole (receive) is located 0.4 inches to the left of the transistor case edge. Make the cuts as shown in **Fig 11.103** using a sharp knife.

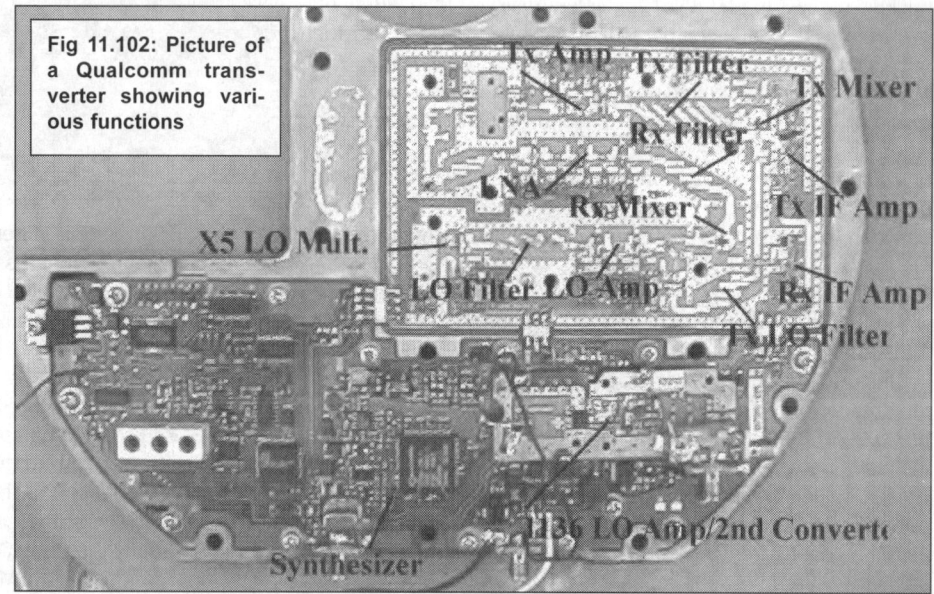

Fig 11.102: Picture of a Qualcomm transverter showing various functions

Fig 11.104: Conversion step 2: Base plate removal, modification and connector installation

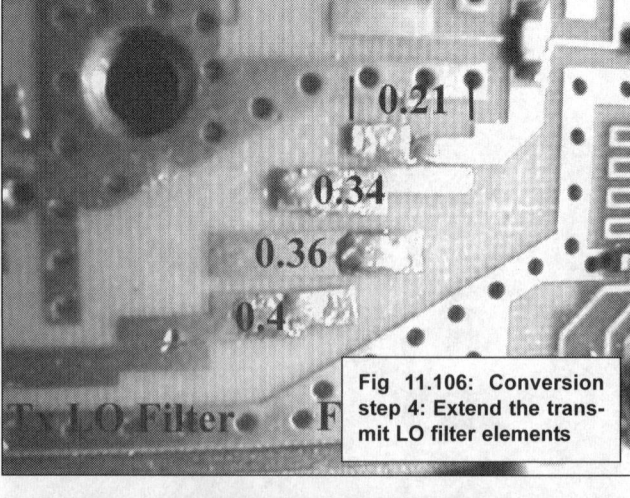

Fig 11.106: Conversion step 4: Extend the transmit LO filter elements

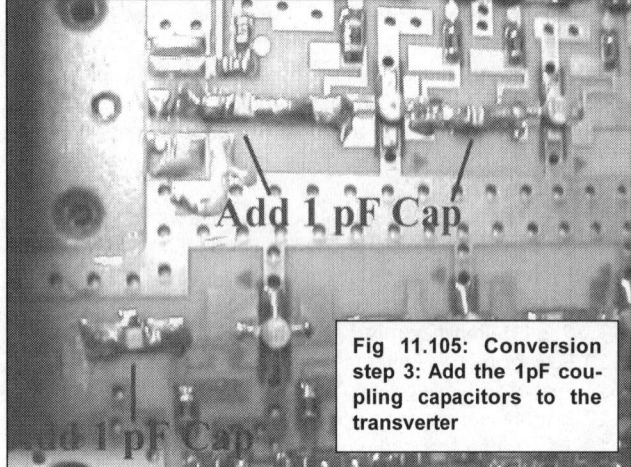

Fig 11.105: Conversion step 3: Add the 1pF coupling capacitors to the transverter

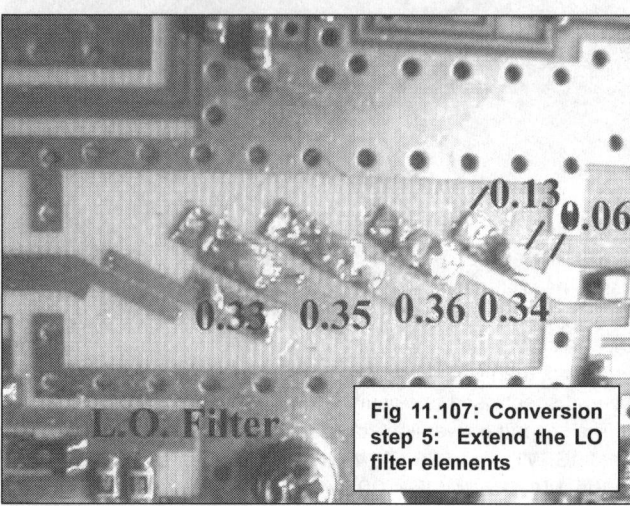

Fig 11.107: Conversion step 5: Extend the LO filter elements

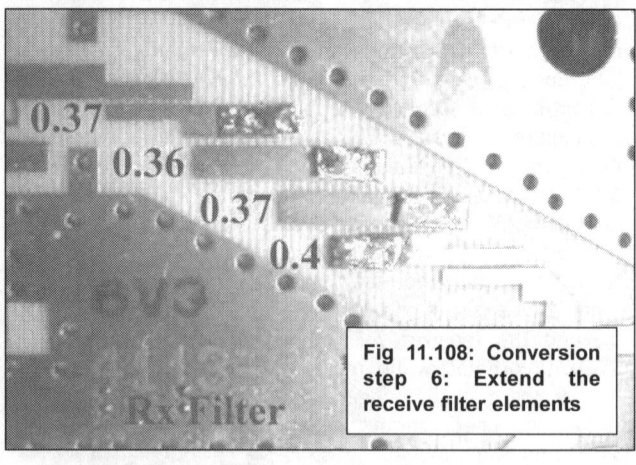

Fig 11.108: Conversion step 6: Extend the receive filter elements

Step 2. Base plate removal, modification, and connector installation: After making the holes and cuts, remove all screws and lift the boards off of the base plate. (Note: the original antenna connector pin must be de-soldered to remove the board. Once the boards are removed, drill through the plate in the 2 locations marked using a 0.161 inch drill to clear the teflon insulator of the SMA connectors. Use a milling tool to remove enough material on the back side of the base plate (see **Fig 11.104**) to clear the two SMA connector locations, taking the thickness down to about 0.125 inches (may vary depending on available SMA connector pin length). Locate, drill and tap the base plate for two 2-56 mounting screws at each connector. Mount the SMA connectors on the base plate and cut the Teflon insulator flush with the top side of the base plate (circuit board side). Carefully clear the ground plane around the two connector holes on the bottom side of the circuit board to prevent the SMA probe from being shorted (using about a 0.125 inch drill rotated between your fingers). Reinstall the circuit boards onto the base plate.

Step 3. Add coupling capacitors: Add the 3 capacitors along with the additional microstrip pieces to modify as shown in **Fig 11.105**.

Step 4. Extend the transmit LO filter elements to the total length shown in **Fig 11.106**: Filter extensions are made by cutting 0.003 - 0.005 inch copper shim stock into strips about 0.07 inches wide and tinning both sides of the strip, shaking off excess solder. No additional solder is normally needed when attaching the extensions as the tinning re-flows when touched by the soldering iron. The length of the top element (0.21 inches) is measured between the marks as shown.

Step 5. Extend the LO filter elements as shown in **Fig 11.107**: Again, total element lengths are shown except for the right-most element that has additional dimensions.

Step 6. Extend the receive filter elements as shown in **Fig 11.108**: Dimensions shown are total element length.

Step 7. Extend the transmit filter elements as shown in **Fig 11.109**: Dimensions shown are total element length.

Step 8. Add the tuning stubs to the x5 Multiplier stage: This stage is located directly to the left of the LO filter which is shown in Fig 11.107. The gate of the x5 Multiplier stage requires addition of two stripline stubs, as shown in **Fig 11.110**.

Step 9. Modify the 2nd LO amplifier board, mount onto transverter and connect 1,136MHz LO input through 1pF coupling

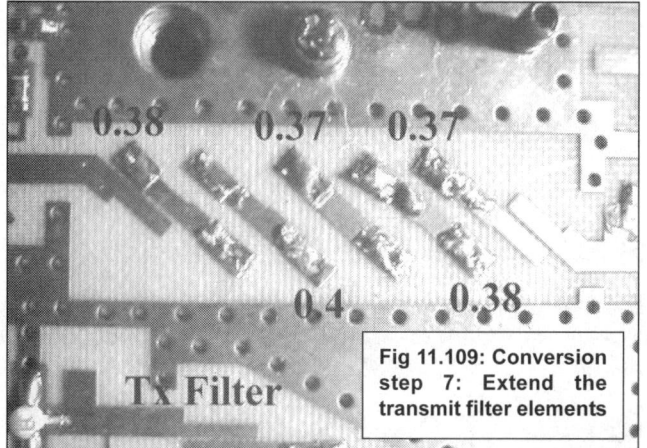

Fig 11.109: Conversion step 7: Extend the transmit filter elements

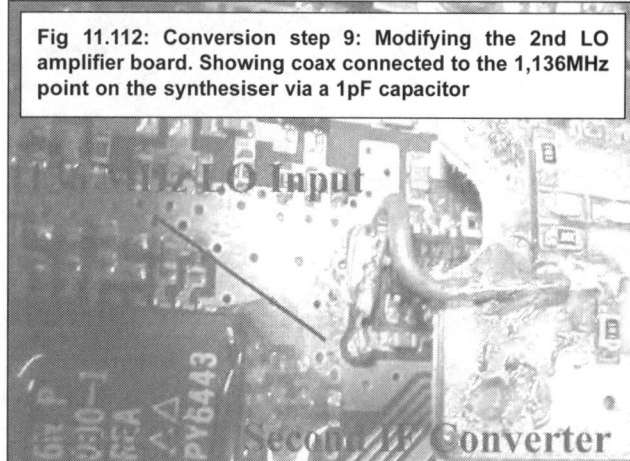

Fig 11.112: Conversion step 9: Modifying the 2nd LO amplifier board. Showing coax connected to the 1,136MHz point on the synthesiser via a 1pF capacitor

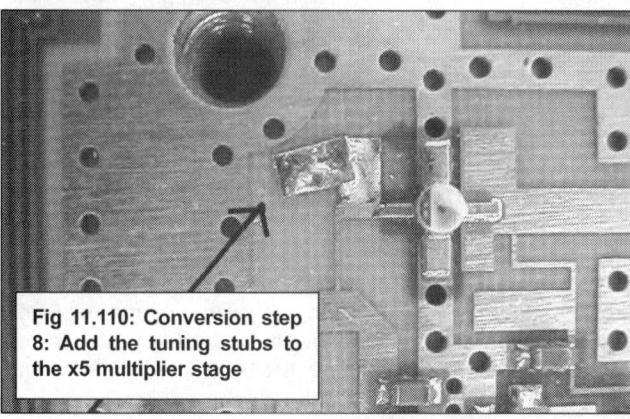

Fig 11.110: Conversion step 8: Add the tuning stubs to the x5 multiplier stage

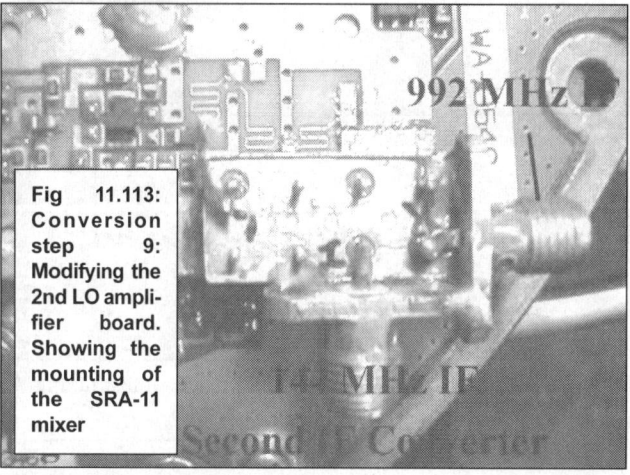

Fig 11.113: Conversion step 9: Modifying the 2nd LO amplifier board. Showing the mounting of the SRA-11 mixer

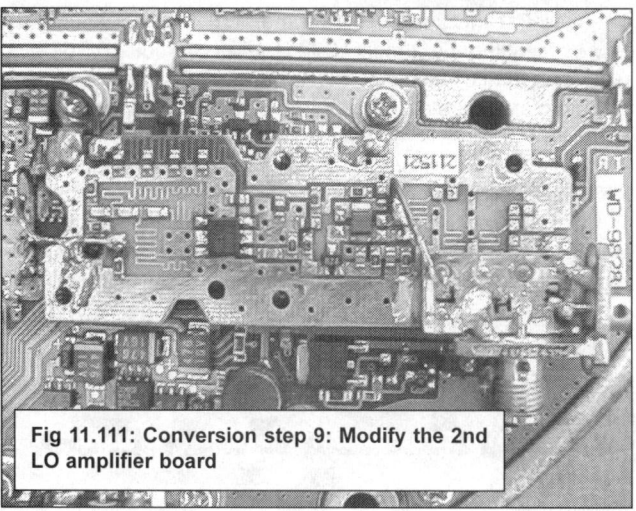

Fig 11.111: Conversion step 9: Modify the 2nd LO amplifier board

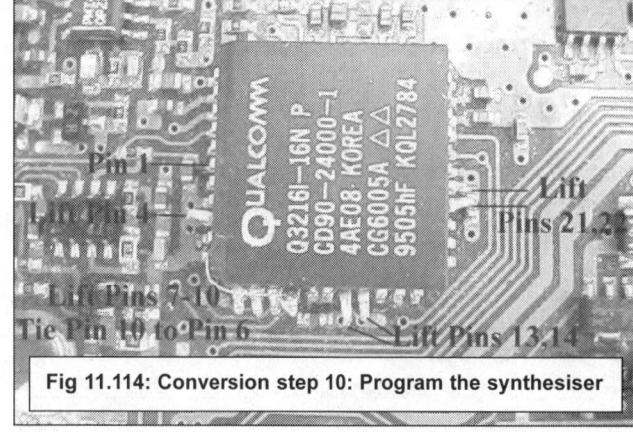

Fig 11.114: Conversion step 10: Program the synthesiser

capacitor as shown in **Figs 11.111 - 11.113**. Fig 11.111 shows the overall second IF converter which is mounted using two grounding lugs soldered to the top edge of the LO amplifier board and secured by two of the screws which mount the main transverter board. Fig 11.112 shows the coax connected to the 1,136MHz point on the synthesiser through a series 1pF capacitor. Fig 11.113 shows the mounting and wiring of the SRA-11 mixer onto the LO amplifier board. Note the cut on the original amplifier output track after the connecting point to the mixer. The mixer case is carefully soldered directly to the LO amplifier board ground plane. The IF SMA connectors are mounted by carefully soldering them directly to the top of the mixer case.

Step 10. Program the synthesiser as shown in **Fig 11.114** by carefully lifting the pins shown with a knife. Ground pin 10, connecting it to pin 6 that is ground. Add the two 3000pF and

1000pF in parallel with the existing reference filter capacitors as shown in **Fig 11.115**.

Step 11. Add a 1pF capacitor as shown in **Fig 11.116** to lower the VCO frequency

Step 12. Add three transmit mixer tuning stubs as shown in **Fig 11.117**.

Step 13. The transmit/receive control is connected as shown in **Fig 11.118**. Grounding the control line places the transverter in transmit mode. The control can be open or taken to +5V to place the transverter in receive mode.

Step 14. The +12VDC power input is connected to the point shown in **Fig 11.119**. The original air core coil, with one end connected to that point, has been removed from the board. (This choke was originally used to supply +12V to the transverter through the 1st IF port).

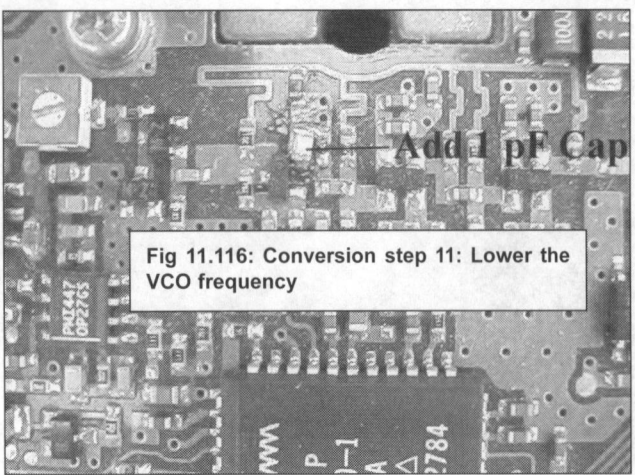

Add 1 pF Cap

Fig 11.116: Conversion step 11: Lower the VCO frequency

Add Tuning Stubs as Shown (3)

Fig 11.117: Conversion step 12: Add transmit mixer tuning stubs

Tx/RX Control Line

Fig 11.118: Conversion step 13: Transmit / receive con-

1st IF in/out   +12V Power input

Fig 11.119: Conversion step 14: Power input

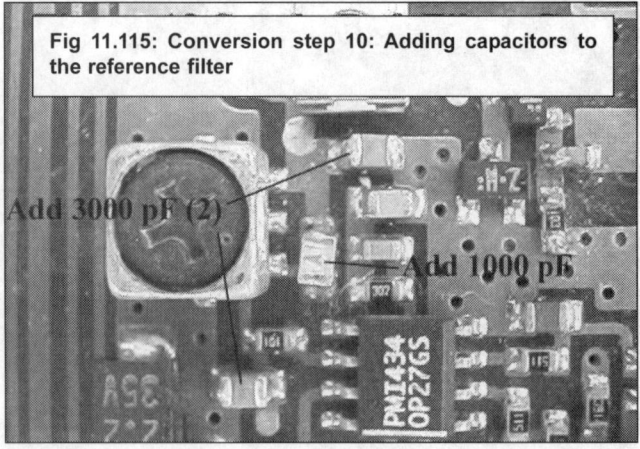

Fig 11.115: Conversion step 10: Adding capacitors to the reference filter

Add 3000 pF (2)

Add 1000 pF

Step 15. Powering up the Transverter: Apply +12V to the power connector and verify that the current drawn in receive mode is about 0.5A. Connect the 10MHz reference to the transverter board. Pin 43 of the synthesiser IC should be high when locked. If available, use a spectrum analyser to check (sniff using a short probe connected by coax) the synthesiser output frequency and spectrum. The synthesiser should be operating on 2,272MHz and no 2MHz or other spurs should be visible. Carefully probe the drain of each FET in the LO multiplier, LO amplifier, and LNA to verify biases are approximately +2 to +3VDC. A drain voltage of near 0V or 5V probably indicates a problem with that stage. Place the transverter in transmit mode and verify the biasing on the transmit LO amplifier and transmit output amp stages. Tune the 992MHz 1st IF filter (not part of the transverter board) and connect it between the 1st IF ports on the transverter board and second IF converter. The receiver noise level at the 2nd IF port on the 2nd converter should be very noticeable on a 2 meter SSB receiver. A weak 10,368MHz signal can then be connected to the receiver RF input connector and monitored on the 2m receiver. The overall gain from receiver RF input to 2nd IF output should be roughly 35 to 45 dB. Place the transverter into transmit mode and connect about –10dBm at 144MHz to the 2nd IF port. Monitor the power level at the transmit RF output port and add/move the transmit amplifier tuning stubs shown in **Fig 11.120** as required for maximum output. Typical transmit output will be about + 8dBm. This is considerably more than required to drive the one watt amp to full power.

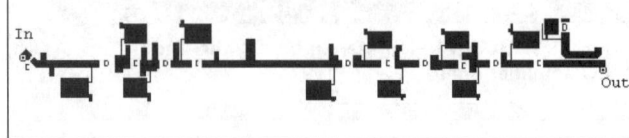

Fig 11.120: 1W PA board prior to tuning. -15dBm input gives +5 to +10dBm output with 10V at approximately 1A

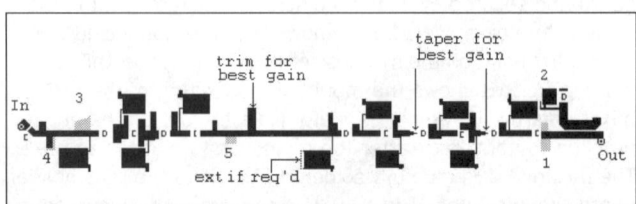

Fig 11.121: 1W PA board after tuning. The shaded tabs were added and tuned in the sequence shown. Results vary slightly from board to board. Key: [= coupling capacitor, D = devices, Input coupled with 2pF

To Convert the 1W PA

These conversion notes were produced by Ken Schofield, W1RIL [43]: Many PA boards have been successfully re-tuned for 10GHz operation. No two boards are exactly alike and each will tune a little differently from its apparent twin. The numbered steps in **Fig 11.121** will in many cases get your PA up into the gain range stated. You will find that numbered step 3 to be the most sensitive to gain increase. Unfortunately it is also one of the 'busiest' areas on the board.

Be careful! A few dos and don'ts are shown to help you bypass some of the many pitfalls that can be encountered - many are obvious and have been stated before, but bear repeating.

Do:
* Use a low voltage grounded soldering iron, and work in a static-free area.
* Check for negative bias on all stages prior to connecting Vcc voltage.
* Use good quality 50 mil chip caps - in and out approximately 1 to 2pF.
* Remove all voltages prior to soldering on board.

Don't:
* Work on board tracks when tired, shaky or after just losing an argument with your wife.
* Touch device inputs with anything that hasn't been just previously grounded.
* Apply Vcc to any stage lacking bias voltage.
* Shoot for 45dB gain - you won't get it! Be happy with 25 to 30dB

LASER DXING

Communication by light has been used for many centuries, from beacons being used to warn of advancing invaders through the use of the Aldis Lamp to transmit Morse code messages, to advanced laser communications used in today's high speed telecommunications links. For amateurs, the use of light is an extension of the frequency spectrum into the Terahertz (THz) region. Visible light is in the range 380 – 750THz and infra red from 100THz – 380THz. Use of these frequencies introduces some new challenges, not least being much more dependency on weather conditions. Amateurs in Germany and America have been actively operating in the light spectrum for many years.

In the UK there is an active group of amateurs [44] who have established the UK records for laser DX contacts. The account of this contact, by Allan Wyatt, G8LSD, (see box overleaf) emphasises the difficulties encountered in establishing a contact at

these frequencies and will hopefully encourage more amateurs to try this interesting avenue of the hobby.

OPT301 Laser Receiver

This receiver was produced by David Bowman, G0MRF [45]. The OPT301 is in a TO-99 eight lead package and has good sensitivity but a reduced bandwidth of 4kHz. The peak response is just into the Infra Red region at 750nm but its sensitivity in the visible red spectrum at 670nm is only a few percent down. The circuit is shown in **Fig 11.123**. It uses a single supply at 12 - 13.8V. The detector has a 10M-ohm feedback resistor soldered directly between pins 2 and 5. The output is followed by a NE5534 low noise op-amp (3.4nV/Hz) configured as a band pass filter. The final NE5534 is an inverting buffer amplifier with a gain of 20. The complete detector assembly is designed to be mounted in a small metal box at the focal point of a lens. **Fig 11.124** (in Appendix B) shows the PCB layout viewed from the top or component side and **Fig 11.125** shows the component layout. The detector is fitted to the other side of the board which has a ground plane in un-etched copper but has holes counter-

Fig 11.122: Picture of G0MRF's laser transmitter and receiver

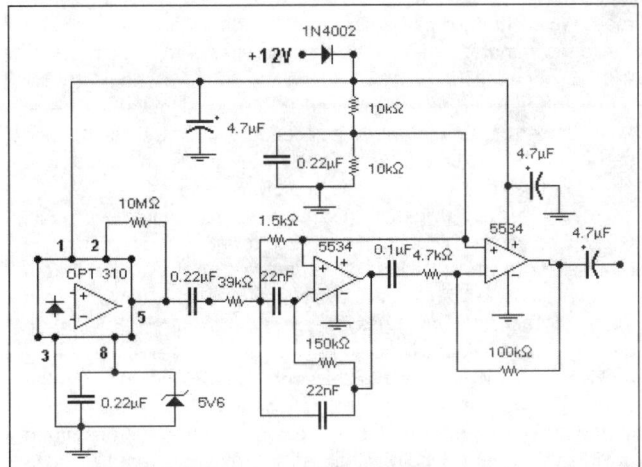

Fig 11.123: Circuit diagram of laser receiver

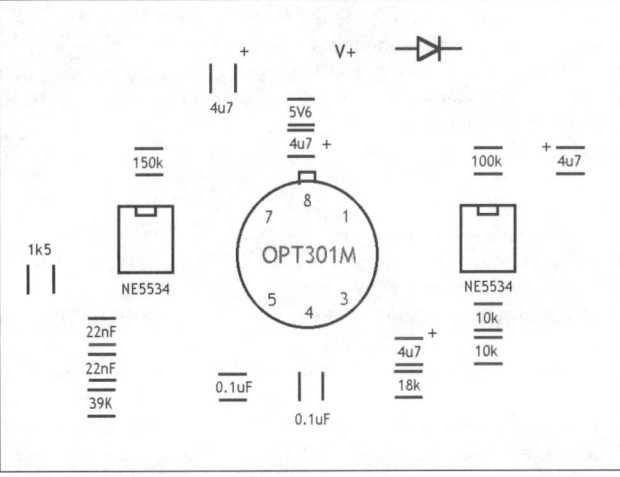

Fig 11.125: PCB layout for laser receiver

Account of setting the UK Laser DX Record – 76.1km on 8 October 2003

"It all started with a trip to Tunbridge Wells. On the return leg the air seemed very clear so I went home via the Ashdown Forest. In the binoculars I could see the towers on Truleigh Hill, the trees at Chanctonbury, then quite easily I found the twin masts at Bignor (50km+) next I could see the television mast north of Midhurst and finally I saw the stubby mast at Butser. From here I had only ever seen Butser with the aid of the telescope. Everything then moved at great speed!

"After a call to David Bowman, G0MRF [whose laser equipme nt is shown in Fig 11.122], I was back on Ashdown Forest with all the gear by 16:45 local time, but the conditions had deteriorated and to the left of the Midhurst television mast was only mid-distance hills. The cloud was very thick and black and rain looked very possible. The 70cm talkback (433.400 FM) was difficult to establish as the wind kept blowing my nine element beam round on the lighting stand that I used as a mast.

"The telescope was set-up and I confirmed that Butser was not visible. However an orange tinted light was appearing above the South Downs as a thin slit against the cloud darkened sky. The effect was to make the top of the hills glow and be exceptionally clearly defined. On Chanctonbury, the cloud covered the top of the Downs and it was less easy to make out details. Bignor Hill merged into the sky as though rain were falling. I could still see the mast at Midhurst and it was obvious that visibility was extending only just past 50km; still exceptional but less so than earlier.

"I spoke to G3JMB on 70cm to ask if he would listen-in and verify the contact that David and I hoped to achieve and then waited for G0MRF to arrive on site at Butser Hill. The weather was slowly deteriorating while I was standing by the road and I had gone from just a shirt and jumper to adding my fleece and cagoule, all in the space of 20 minutes! David arrived at Butser and had to carry all his equipment to the top of the hill wearing no wind protection. When we finally established talkback the visibility had dropped to a disappointing degree. My telescope was aimed at the dip on the horizon where Butser should have been and the horizon was a uniform grey.

"The wind was now gusting and making the image in the eyepiece jump in spite of the very heavy mounting base of the solidly made 4.5 inch Russian reflector telescope. David was unable to operate from his planned location as the local vegetation was higher than the top of his tripod. He had to move site and make a second long trek to the car for the second box of equipment. Due to the back packing required at Butser, David did not have the power for a beacon light. Any location aids had to come from my end at Ashdown. I had 250Ah of 13.8V from the batteries under the back seat of the vehicle and the associated switch mode converter, as well as 240 Volts AC from the inverter run from the starting battery. By the time David had found a suitable location the visibility had dropped even further and the wind had not abated.

"We then set about converting our usual operating plan into something to suit the conditions. David swept his laser across an angle that included Ashdown, the beam width was usually sufficient for this to work, however this time the telescope showed no sign of even the briefest of flashes of the characteristic red. Plan B time.

"As clouds were passing over the Downs to the south of me and the wind dropped. I set up the laser transmitter, the tripod being held in place by the weight of an 80Ah leisure battery, and wired up both transmitter and receiver. A second set of scans with the laser at the Butser end also failed to give any tone in the receiver. Even a strobe light operated from Ashdown found no path to Butser. Time was passing. Amid much head scratching I went back to serious scanning of the horizon with the telescope. By now a beautiful sunset was developing and the sky was a vivid orange through the broken cloud.

"For one moment I thought I could see as a very faint change in the uniform colour of the horizon, the inverted bathtub shape of Butser. I aimed the telescope at the northern end where the mast is to be found. Again no contact. Having taken note of the local terrain I had a good idea as to the aiming point and so set up my laser onto the target. David saw nothing. Light was fading fast and I expected the horizon to go black before I had a clear fix. As David had no beacon this would have ended the contact.

"Returning to the telescope, a point of light from a mercury vapour lamp at probably 45 to 50km away was revealed. This was not visible with binoculars or through the rifle sight on the transmitter. David still could see nothing. A sweep of his laser showed no flashes in the telescope. The sunset was fading into a beautiful burnt orange.

"The wind had dropped even more. Once again, I closely scanned the horizon. First I saw the faint outline of Butser Hill and then for a moment I imagined that the mast was visible against the darkening grey orange. A few minutes later after talking on the 70cm link I returned to the telescope. Several points of mercury vapour lamps acted as a horizontal grid and slightly above them was the mast on the top of Butser. The binoculars and rifle sight still showed nothing but the points of light, almost blurred into a single point. This gave just enough information to aim the laser. David still saw nothing.

"Next, as it was now quite dark, I switched on my torch. David saw it almost at once. He still could not see the laser, even though it was directly aimed at him. He swept his laser at the point where the torch had been and I saw a brief flash of red. It was momentarily bright but did not stay visible. I moved the talkback from the van closer to the receiver and was able to give a running commentary, but we could not get more than brief flashes of the laser.

"I returned to my laser and started a systematic scan of the horizon. At one point I accidentally turned the vertical control too far, and so I was seeing into the hill in the rifle sight. David saw the beam. Once again, the rifle sight prism had moved in transit and the sight had lost vertical accuracy. However, the transmitter alignment sight was still spot on in the horizontal plane. David now relayed the laser signal back over our 70cm FM link. With this feedback I could optimise the pointing and we established one way contact.

"The CW signal was clear and lacked the usual scintillation. The laser's brightness was such that David reported that I was the brightest light on the horizon. We then concentrated on receiving David's signal. This required critical adjustment but once I got a faint signal in the receiver he was able to optimise by our usual feedback route. By now I had spent 135 minutes standing by the road. We exchanged callsigns, reports and random characters in just a few minutes. David had an active filter on receive tuned to 488Hz, our tone frequency. I suggested that we should try my audio. In the past this has not worked, but as scintillation was at an all time low it seemed worth a second try. Once the microphone had made proper contact in its socket the audio came out loud and clear; a 5 by 7 report on the audio was received.

"To get further evidence of this result we contacted Jack, G3JMB, again and he pointed his beam towards Butser to receive David's 70cm transmission. While Jack set-up, David and I kept the conversation going in a cross-band full duplex contact. Jack made contact and David relayed my audio to him. He was then hearing Pulse Width Modulation to a 3mW laser diode being received on a CW optimised receiver and re-transmitted the 50km to his QTH via a hand held rig from a windy hill. It was a surreal moment. Then we reversed the process and I re-transmitted David's CW to Jack. Finally we closed the laser link that had been stable for 30 minutes and packed up."

Fig 11.126: Picture of completed laser receiver

sunk for the TO-99 package. The 1N4002 diode protects against reverse polarity. It is soldered to the un-drilled pads on the PCB.

Circuit notes:

- A 10kΩ resistor is needed between Pins 1 and 8 on the OPT301. This is shown on the PCB artwork and on the overlay.
- The capacitor coupling the signal out of the detector to the 39kΩ was changed from 0.1 to 0.22μF The capacitor across the zener diode was also changed to 0.22μF
- With a single supply line the body of the detector is held at the Zener voltage. Therefore it should be isolated from the ground plane and from any metal enclosure.

Four earth connections are required through the board:

- NE5534 pin 4 (on the left above)
- Supply decoupling 4.7μF tantalum (above and right of 5534 pin 8)
- Detector Pin 3
- Pad in lower right of PCB connected to the 10k-ohm resistor and output ground.

A picture of the completed laser receiver is shown in **Fig 11.126**. The performance of the band pass filter and buffer amplifier were evaluated by connecting a 600-ohm signal generator to the input of the band pass filter and measured the response at the output of the buffer. The nput was adjusted from 50Hz to 3kHz with a constant level of 200mV peak to peak. **Table 11.12** shows that the 6dB bandwidth is just 200Hz. This makes the detector ideal for transmitters using modulated CW on a fixed

Frequency Hz	Output mV	Relative -dB
50	20	40.80
100	80	28.80
200	215	20.20
300	430	14.20
350	630	10.86
400	980	7.02
450	1700	2.24
500	2200	0.00
550	1500	3.32
600	1050	6.42
700	620	11.00
800	445	13.90
900	360	15.70
1000	295	17.50
1500	170	22.20
2000	120	25.30
2500	95	27.30
3000	80	28.80

Table 11.12: Band pass filter / buffer amplifier characteristics for laser receiver

frequency. An OPT301, using 10MΩ feedback with a bandwidth of 200Hz gives a Noise Equivalent Power of 3×10^{-11} Watts. The transmitter uses a 4MHz crystal and a CMOS 4060 oscillator / divider to generate 488Hz. This degree of accuracy gives the option of using a laptop computer and modern DSP software (eg Argo [46] or Spectran by IK2PHD) to receive signals 20dB below normal noise level. The 40dB rejection at 50Hz gives a high tolerance to interference from street lighting. Voltage gain at the design frequency is x11 or 20.8dB. The band pass filter centre frequency is selected by two capacitors and two resistors. It can be changed to any audio frequency of your choice. Design equations are published in the *ARRL handbook*.

Laser Transmitter Circuits

Here are two circuits for laser transmitters designed by David Bowman, G0MRF. The first, shown in **Fig 11.127**, uses a simple laser pointer module. Any of the normal 3 or 5mW devices will function well. Imports from the USA or Hong Kong are frequently a higher power than the UK-approved units. The modules normally contain two or three 1.5V mercury cells.

The external modulated supply should duplicate the voltage used in your particular module.

The electronics contained in the module will include the semiconductor laser diode and a constant current source. The ability of the constant current source to be switched on and off at audio frequency will change between manufacturers, but below 1KHz, this is not normally a problem.

However, there are applications where the supply can be switched at 20 to 200kHz and voice or high-speed data can be modulated onto this subcarrier. For these high-speed applications the circuit should be tested carefully with an oscilloscope. If unsuitable, the constant current source can be modified or replaced with a circuit that has a faster response.

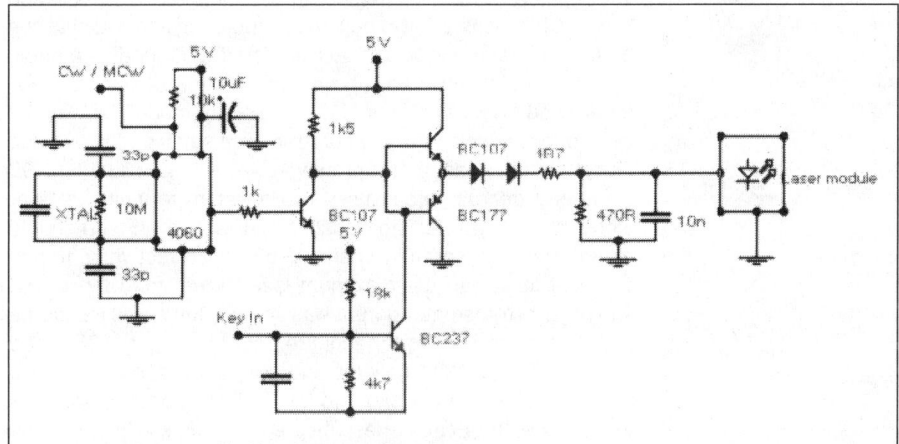

Fig 11.127: Circuit diagram of laser transmitter using a laser pointer. 4060 pins: 16; supply, 8; ground, 2; output, 10+11; crystal, 12; reset

Fig 11.128: Circuit diagram of laser transmitter using a laser diode

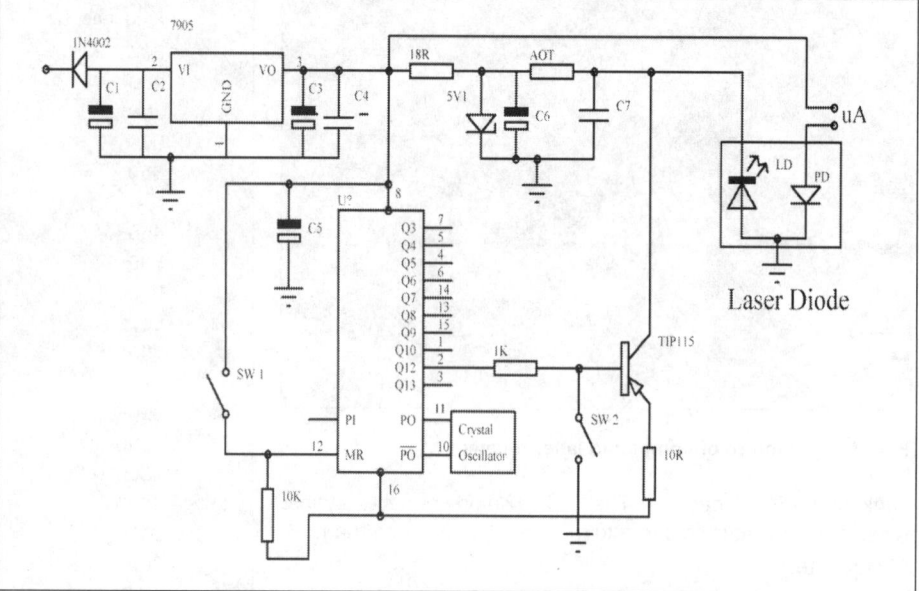

The two diodes are 1N4001 etc. The capacitor on the key input is 10nF. This circuit uses a crystal and a 4060 oscillator / divider to produce a square wave output in the audio range; in this case a 4MHz crystal giving 488Hz on pin 2. The 5Vp-p square wave is buffered by three transistors which include a totem pole driver. Two diodes have been used to reduce the voltage by 1.2V which then supplies the laser module. A switch is attached to the reset pin of the 4060. This is used to select CW (constant laser output) or Modulated CW. Another switch, wired between the key input and ground, allows the laser to be keyed via a phono socket or it can be switched permanently on.

WARNING: Lasers are dangerous

- The dangers from lasers are essentially from the amount of energy contained in a very small area. If a narrow beam is used, then all of the power can be directed into the typical 5mm diameter of the human eye causing significant damage.

- Whenever possible, ensure the beam created in home constructed or converted equipment has its beam expanded. If the beam is expanded to larger diameter then its energy is contained over a larger area and the danger of accidental eye damage is much reduced. Including a beam expander to your laser system also has the advantage of reducing the divergence of the beam, increasing the distance potential of the transmitter.

- Never leave a laser transmitter on and unattended as you are not in control of where the laser is pointing. If you move away from the transmitter always turn the laser off.

- Consider any beam, even that generated by a cheap laser pointer, as potentially dangerous until its beam has diverged to a minimum of 50mm diameter.

- Night-time eye response: At night and in dark conditions, the pupil in the human eye will dilate to increase sensitivity to light. This significantly increases the danger from lasers.

- Visible wavelength lasers can be seen by the operators and others, and their danger anticipated. However, special care should be taken when using invisible wavelengths. Infra red lasers are generally more powerful but the risks from accidentally leaving a laser source on are significantly increased. By the time you realise the source is on, it may already be too late. Consider a beam expander as mandatory with non visible wavelengths.

Further information on laser safety can be obtained from the Health and Safety Executive we site http://www.hse.gov.uk

---

This is very useful when aligning the beam onto a distant receiver. Some circuits have been published using 555 timers but it was felt that a crystal source would give greater potential for weak signal working.

Caution:

- Always use talkback to coordinate transmissions with the other station. Looking at a laser over long distances through your receive optics can be dangerous.

- Most laser modules use positive earth.

- Heating the module to solder wires can damage the optics and increase the beam divergence.

- A laser diode can fail but still appear to be working. However, what you'll have is an expensive LED with a light output that is a fraction of the laser and a beam that is not a coherent light source.

- As the laser cools down on a cold winter's night, the light output power will increase. This is not the advantage it first appears as the laser can fail because of excessive output power.

If you want to increase your laser output power, you will probably have to move on from laser pointers and use a laser diode. Then it's down to "how much you want to spend", but at least you get to choose the output power and wavelength. The circuit shown in **Fig 11.128** is adapted from an original idea suggested by K3PGP. It uses a 7905 negative voltage regulator so that the body of the laser diode can be connected to ground. The audio is generated by a 4060 as above and a PNP Darlington transistor is wired across the laser diode as a shunt modulator. You will need optics to collimate the output from a semiconductor laser. The angle of radiation of a prototype was 9 degrees high by 32 degrees wide. The circuit uses two resistors to limit the laser diode current. The resistor labelled AOT (adjust on test) should be selected *very* carefully. It should be the correct value to provide just less than the maximum laser power from the diode. The original transmitter output was set to 10mW, well inside the max ratings of the diode. Audio is generated by a 4060. The 4MHz crystal, 10MΩ resistor and two 33pF capacitors have been omitted from the diagram for clarity but are the same as used in the other circuit described earlier. C6 is 47µF, all other electrolytics are 10µF. Other capacitors are 10nF. Modulation is applied to the laser via a shunt modulator. When the Darlington transistor conducts, the current supplied from the 7905 and the

two resistors is shunted to ground through the transistor and the 10Ω resistor in its emitter. This circuit can be used with lasers up to 100mW output.

The completed PCB (**Fig 11.129** in Appendix B) is mounted on the side wall of a diecast box. The optics are contained in a small metal tube simply glued onto the PCB. A coarse adjustment of the laser can be made by carefully adjusting the position of the PCB on the four mounting bolts. The final adjustment is made with the cross hairs on the telescopic sight which is a cheap 4 x 28 (times 4 magnification and 28mm lens diameter) bought new for £19.95 from sources on the Internet.

BIBLIOGRAPHY

Books

ARRL Handbook for Radio Amateurs, ARRL, revised / re-published annually

DubusTechnik, available in the UK from G4PMK (QTHR). E-mail: dubus@marsport.demon.co.uk

International Microwave Handbook, editor Andy Barter, G8ATD, RSGB, 2002

Microwave Projects, editor Andy Barter, G8ATD, RSGB, 2003

UHF/ Microwave Experimenter's Handbook, ARRL

UHF/Microwave Projects Manual, Volumes 1 & 2, ARRL

UHF Compendium, Parts I, II, III and IV, editor K Weiner, DJ9HO, available from UKW Berichte. E-mail: info@ukwberichte.com, or from KM Publications, E-mail: andy@vhfcomm.co.uk

Magazines

CQ-TV, BATC Publications, Fern House, Church Road, Harby, Notts, NG23 7ED. E-mail: publications@batc.org.uk. Web: http://www.batc.org.uk.

Dubus magazine, available in the UK from G4PMK (QTHR). E-mail: dubus@marsport.demon.co.uk.

Proceedings of the Microwave Update, ARRL, annual publication

RadCom, RSGB members' monthly journal; it has a bi-monthly microwave column and carries occasional microwave articles. Web: http://www.rsgb.org.

Scatterpoint (formerly *Microwave Newsletter*), UK Microwave Group, 10 issues per year, contact: Martyn Kinder, G0CZD, Secretary UK Microwave Group, 12 Jessop Way, Haslington, Crewe, Cheshire, CW1 5FU. E-mail: ukug@czd.org.uk. Web: http://www.microwavers.org.

VHF Communications magazine, KM Publications, 63 Ringwood Road, Luton, Beds, LU2 7BG. E-mail: andy@vhfcomm.co.uk. Web: http://www.vhfcomm.co.uk.

Websites

A list of amateur microwave websites is given on the RSGB site at http://www.rsgb.org/books/extra/handbook.htm instead of here as URLs may change. Web users should use their preferred search engine to find the required website if the links given have changed.

Component or Kit Suppliers

Down East Microwave Inc, 954 Rt. 519 Frenchtown, NJ 08825, USA. Web: http://www.downeastmicrowave.com.

Farnell Electronic Components Limited, Canal Road, Leeds, LS12 2TU. Tel: 0870 1200 200. Fax: 0870 1200 296. Web: http://www.farnellinone.co.uk.

GH Engineering, The Forge, West End, Sherborne St. John, Hants, RG24 9LD. Tel: 01256 889295. Fax: 01256 889294. E-mail: sales@ghengineering.co.uk. Web: http://www.ghengineering.co.uk

Kuhne electronics GmbH, Scheibenacker 3, D-95180 Berg / Oberfranken, Germany. E-mail: info@kuhne-electronic.de. Web: http://www.kuhne-electronic.de or http://www.db6nt.com.

Maplin Electronics, P.O. Box 777, Wombwell, S73 0ZR. Tel: 0870 429 6000. Fax: 0870 429 6001. Web: http://www.maplin.co.uk.

Microwave Committee Component Service, P Suckling, 314A Newton Road, Rushden, Northants, NN10 0SY. Tel: 01933 411446. Fax: 01933 411446. Web: http://www.g3wdg.free-online.co.uk.

Mini-Circuits Europe, Dale House, Wharf Road, Frimley Green, Camberley, Surrey, GU16 6LF. Tel: 01252 832600. Fax: 01252 837010. Web: http://www.minicircuits.com.

R F Elettronica, Via Dante 5, 20030 Senago, MI, Italy. E-mail: info@rfmicrowave.it. Web: http://www.rfmicrowave.it.

UKW Berichte, Postfach 80, D-91081 Baiersdorf, Germany. E-mail: info@ukwberichte.com. Web: http://www.ukwberichte.com.

Some kits available in the UK from KM Publications, 63 Ringwood Road, Luton, Beds, LU2 7BG. Tel 01582 581051. Fax: 01582 581051. E-mail: andy@vhfcomm.co.uk. Web: http://www.vhfcomm.co.uk.

REFERENCES

[1] CCIR, 1937.
[2] *Admiralty Handbook of Wireless Telegraphy*, HMSO, London, 1938.
[3] *International Microwave Handbook*, editor - Andy Barter, G8ATD, RSGB 2002.
[4] *Microwave Projects*, editor - Andy Barter, G8ATD, RSGB 2003.
[5] *UHF/Microwave Experimenter's Manual*, 'Antennas, components and design', ARRL.
[6] Agilent (Hewlett Packard). Web: http://www.semiconductor.agilent.com.
[7] Mini Kits. Web: http://www.minikits.com.au.
[8] *ARRL Proceedings of Microwave Update* 1987.
[9] 'A local oscillator source for 1153MHz', S T Jewell, G4DDK, *Radio Communication*, February 1987, p218 and March 1987, pp191–201.
[10] Microwave Component Service, P Suckling, 314A Newton Road, Rushden, Northants, NN10 0SY, Tel: 01933 411446. Fax: 01933 411446. Web: http://www.g3wdg.free-online.co.uk.
[11] 'High precision frequency standard for 10MHz. Part II: Frequency control via GPS', Wolfgang Schneider, DJ8ES and Frank-Peter Richter, DL5HAT, *VHF Communications* 1/2001 pp2–8
[12] 'High precision frequency standard for 10MHz', Wolfgang Schneider, DJ8ES, *VHF Communications* 4/2000 pp177–190.
[13] 'HMET LNAs for 23cm', R Bertelsmeier, DJ9BV, *Dubus Technik IV*, pp191–197.
[14] 'No tune HEMT preamp for 13cm', R Bertelsmeier, DJ9BV, *Dubus Technik IV*, pp191–197.
[15] Franck Rousseau, F4CIB, 18 rue Colbert, Porte B AppT 31, 31400 Toulouse, France. E-mail: f4cib@ref-union.org.
[16] 'Amplificateur 4W 13cm à transistor LDMOS', Maurice Niquel, F5EFD, *Hyper* n°56, February 2001.
[17] 'Amplificateur 70W sur 1296MHz', Maurice Niquel, F5EFD, *Hyper* n°65, November 2001.
[18] 'Amplificateur 2300MHz à MRF282; 12 ou 24VDC', Jean-

Pierre Lecarpentier, F1ANH, *Hyper* n°73, July/August 2002.

[19] http://e-www.motorola.com/brdata/PDFDB/docs/MRF284.pdf

[20] '28V switched powers supply for LDMOS-PAs', Achim Volthardt, DH2VA / HB9DUN, *Dubus*, 1/2004 pp38–40.

[21] 'GaAsFET power amplifier stage up to 5W for 10GHz', Paeter Volg, DL1RQ, *VHF Communications Magazine*, 1/1995, pp.52–63.

[22] '6cm transverters in modern stripline technology', Peter Vogl, DL1RQ, *VHF – UHF* Munich 1990, pp49–66.

[23] '6cm transverters in stripline technology, part 2', Peter Vogl, DL1RQ, *VHF Communication Magazine* 2/1991, pp69–73.

[24] 'Determination of parasitic inductances on capacitors', Roland Richer, Special project at Dominicus vo Linprun Grammar School Viechtach, 1994.

[25] '76GHz amplifier', Sigurd Werner, DL9MFV, *VHF Communications Magazine*, 3/2003, pp163-169.

[26] 'A simple concept for a 76GHz transverter', Sigurd Werner, DL9MFV, *VHF Communications Magazine*, 2/2003, pp77-83.

[27] 'Medium Power Amplifier @ 77GHz', Infineon Technologies Data Sheet, T 602B_MPA_2; 05.05.99.

[28] Data Sheet IAF-MAP7710, Fraunhofer Institut for Applied Solid-State Physics, Freiburg.

[29] 'Frequency multiplier for 76GHz with an integrated amplifier', Sigurd Werner, DL9MFV, *VHF Communications Magazine*. 1/2002, pp35-41.

[30] 'Amplifier for 47GHz using chip technology', Sigurd Werner, DL9MFV, *VHF Communications Magazine*, 3/2002, pp160–164.

[31] 'Combining power at 76GHz: Three possible solutions discussed', Sigurd Werner, DL9MFV, *VHF Communications Magazine*, 1/2004, pp13–19.

[32] From article by Matjaz Vidmar, S53MV, *CQ ZRS*

[33] 'A single-board, no-tuning 23cm transverter', Richard L Campbell, KK7B, 23rd Conference of the Central States VHF Society, Rolling Meadows, Illinois, 1989, pp44–52 and 'Engineering Notes', ibid, pp53–55. Subsequently republished in *The ARRL Handbook for the Radio Amateur*, 69th edition, 1992.

[34] 'A 1W linear amplifier for 1152MHz', Sam Jewell, G4DDK, *RSGB Microwave Handbook Volume 2*, pp8.21–8.23.

[35] Richard L Campbell, KK7B, *Proceedings of the Microwave Update*, ARRL, 1988.

[36] Down East Microwave, 954 Rt. 519 Frenchtown, NJ 08825, USA. Web: http://www.downeastmicrowave.com.

[37] Richard L Campbell, KK7B, *Proceedings of the Microwave Update*, ARRL, 1989.

[38] 'VHF and microwave applications of monolithic microwave integrated circuits', Al Ward, WB5LUA, *ARRL UHF/Microwave Experimenter's Manual*, ARRL, 1990. pp7.32–7.47.

[39] *Microwave Handbook, Vol 3*, edited by M W Dixon, G3PFR, RSGB, 1992, pp18.116–18.118.

[40] Microwave Components Service, c/o Mrs P Suckling, G4KGC, 132A Newton Road, Rushden, Northants NN10 0SY, UK.

[41] For further conversion and materials availability information contact Chuck Houghton, WB6IGP, E-mail: clhough@pacbell.net, or Kerry Banke N6IZW of the San Diego Microwave Group, E-mail: kbanke@qualcomm.com.

[42] Additional conversion information articles and sources: 'Microwave GaAs FET Amps for Modification to 10GHz', C Houghton, WB6IGP & Kerry Banke, N6IZW, *Nts Feedpoint Newsletter*, December, 1993. 'UP, UP & Away to 10GHz Semi-Commercial Style', Bruce Wood, N2LIV, *Proceedings of the 20th Eastern VHF/UHF Conference*, August, 1994, p133. '10GHz Qualcomm Modification Notes' by Dale Clement, AF1T.

[43] 'Suggestions for Modifications of Qualcomm LNA Board for 10GHz', Ken Shofield, W1RIL, *Proceedings of the 21st Eastern VHF/UHF Conference*, August, 1995, p63 and 'Modification Update of Omnitrack PA Board for 10GHz'. p65. Also at http://www.uhavax.hartford.edu/disk$userdata/faculty/newsvhf/www/w1ril.html

[44] Laser communications. Web http://www.lasercomms.org.uk/basics.htm

[45] 'A simple laser communications system', David Bowman, G0MRF. Web http://www.g0mrf.freeserve.co.uk/laser.htm

[46] Argo. Web http://www.weaksignals.com

12 Propagation

IT HAS been said that, without an ionosphere, wireless telegraphy might have remained an interesting but commercially unrewarding experiment in physics. Yet, would you believe, there was no ionosphere before 1932?

That has to be a trick question, of course. For the truth is that Balfour Stewart deduced from a study of geomagnetic storm data in 1878 (which was several years before Hertz began his experiments with wireless transmission) that there had to be an electrically conducting layer high in the Earth's atmosphere.

In 1901, Marconi took a chance on it being so, and was rewarded for his faith by the achievement of the transatlantic 'first'. In the 'twenties, Appleton devised experiments to prove conclusively that such a layer existed; in fact, eventually he found more than one. And in 1932 Watson-Watt gave the region a general name - he called it the *ionosphere*.

Today, many people, including some who should know better, seem to think that we know all that we need to know about the ionosphere. But it preserves many a secret yet and there is still a place for careful experimentation by amateur operators.

This chapter can do no more than scratch the surface of a very wide-ranging topic. No attempt has been made in it to introduce divisions between LF, MF, HF, VHF and UHF because there are no such clear-cut divisions in the real world.

The story should really begin with a section about the Sun, because it is there that most of the direct influences on our signal paths have their origin. However, to appreciate that, we must first look at a few of the characteristics of our own atmosphere, and then run over some basic facts and figures.

Don't be put off by the appearance of the mathematics. There is nothing very difficult here and a simple scientific calculator or spreadsheet can handle the tasks with ease.

As you read, compare the text with your own experiences on the bands. In all probability you know already much of the 'whats', 'wheres' and 'whens' of radio propagation. With any luck, this chapter will provide you with the associated 'whys' and 'hows'. Be warned, though. This subject is strongly addictive and there is room for more up there among the 'whos'.

THE EARTH'S ATMOSPHERE

There is a certain amount of confusion surrounding any attempt to identify various portions of the Earth's atmosphere, and this has come about as a result of there being no obvious natural boundaries as there are between land and sea. Workers in different disciplines have different ideas about which functions ought to be separated and it is particularly awkward for the radio engineer that he has to deal with the *troposphere* and *stratosphere*, which are terms from one set of divisions, and the *ionosphere*, which is from another.

Fig 12.1 shows the nomenclature favoured by meteorologists and now widely accepted among physicists generally. It is based on temperature variations, as might be expected from its origin. Note, incidentally, that the height scale of kilometres is a logarithmic one, and that the atmospheric pressure has been scaled in hectopascals (hPa), which are now replacing millibars in scientific work by international agreement. However, the two units are identical in magnitude; only the name has been changed, so the numbers remain the same. See *Weather* (Royal Meteorological Society) May 1986, p172, for an explanation.

In the *troposphere*, the part of the atmosphere nearest the ground, temperature tends to fall off with height. At the *tropopause*, around 10km (although all these heights vary from day to day and from place to place), it becomes fairly uniform at first and then begins to increase again in the region known as the *stratosphere*.

This trend reverses at the *stratopause*, at an altitude of about 50km, to reach another minimum at the *mesopause* (ca 80km) after traversing the *mesosphere*. Above the mesopause temperatures begin to rise again in the *thermosphere*, soon surpassing anything encountered at lower altitudes, and levelling off at about +1200°C around 700km, where we must leave it in this survey.

The *ionosphere*, which has been defined as "the region above the Earth's surface in which ionisation takes place, with diurnal and annual variations which are regularly associated with ultraviolet radiation from the Sun, and sporadic variations arising from hydrogen bursts from sunspots" (*Chambers' Technical Dictionary*), overlaps the thermosphere, mesosphere and part of the stratosphere, but for practical purposes may be considered as lying between 60 and 700km. The name 'ionosphere' is perhaps misleading because it is the number of free electrons,

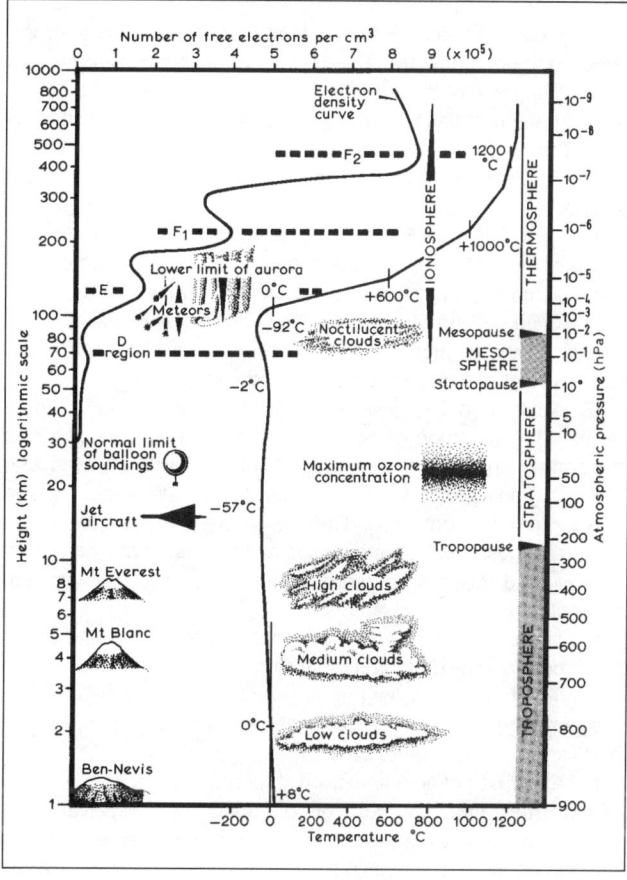

Fig 12.1: Some features of the Earth's atmosphere. The height scale is logarithmic, beginning at 1km above sea level. The equivalent pressure scale on the right is not regularly spaced because the relationship between pressure and height depends on temperature, which does not change uniformly with height

rather than the ions they have left behind, which principally determines the electrical properties of the region. The electron density curve in the diagram shows a number of 'ledges', identified by the letters D, E, F1 and F2, which are the concentrations of free electrons described as 'layers' later on, when dealing with propagation in the ionosphere.

They tend to act as mirrors to transmissions of certain wavelengths, while allowing others to pass through. The lowest layer is generally referred to as the *D-region* rather than the *D-layer* because, as we shall see, its principal role is one of absorption rather than reflection, and its presence is usually easier to infer than it is to observe, except at the lowest frequencies (LF).

It may be found helpful to refer to this diagram again when features of the ionosphere and troposphere are dealt with in later sections of this chapter.

FUNDAMENTAL CONSIDERATIONS

Radiation

The transmitted signal may be regarded as a succession of concentric spheres of ever-increasing radius, each one a unit of one wavelength apart, formed by forces moving outwards from the antenna. At great distances, these wavefronts, appear as plane surfaces as they approach the observer.

There are two inseparable fields associated with the transmitted signal, an *electric field* due to voltage changes and a *magnetic field* due to current changes, and these always remain at right-angles to one another and to the direction of propagation as the wave proceeds.

They always oscillate in phase and the ratio of their amplitudes remains constant. The lines of force in the electric field run in the plane of the transmitting antenna in the same way as would longitude lines on a globe having the antenna along its axis. The electric field is measured by the change of potential per unit distance, and this value is referred to as the *field strength*.

The two fields are constantly changing in magnitude and reverse in direction with every half-cycle of the transmitted carrier. As shown in **Fig 12.2**, successive wavefronts passing a suitably placed second antenna induce in it a received signal which follows all the changes carried by the field and therefore reproduces the character of the transmitted signal.

By convention the direction of the electric lines of force defines the direction of *polarisation* of the radio waves. Thus horizontal dipoles propagate horizontally polarised waves and vertical dipoles propagate vertically polarised waves. In free space, remote from ground effects and the influence of the Earth's atmosphere, these senses remain constant and a suitably aligned receiving antenna would respond to the whole of the incident field.

When the advancing wavefront encounters the surface of the Earth or becomes deflected by certain layers in the atmosphere, a degree of cross-polarisation may be introduced which results in signals arriving at the receiving antenna with both horizontal and vertical components present.

Circularly polarised signals, which contain equal components of both horizontal and vertical polarisation, are receivable on dipoles having any alignment in the plane of the wavefront, but the magnitude of the received signal will be only half that which would result from the use of an antenna correctly designed for such a form of polarisation (it must not be overlooked that there are two forms, differing only in their direction of rotation). Matters such as these are dealt with in detail in the chapter on VHF/UHF antennas.

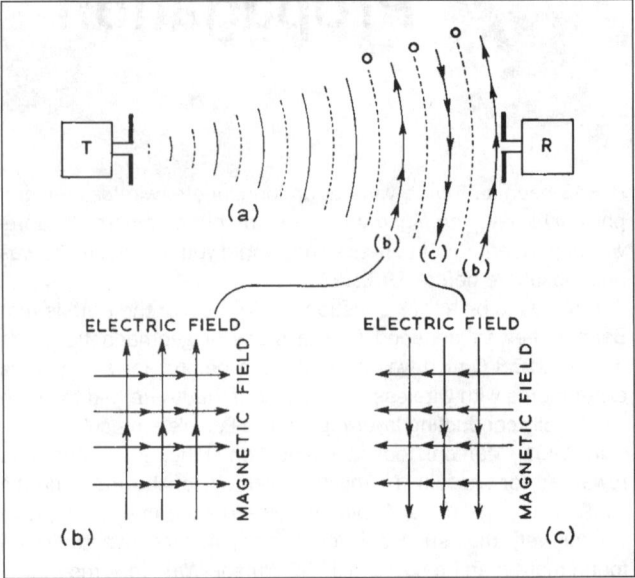

Fig 12.2: (a) A vertical section through the fields radiated from a vertically polarised transmitting antenna. The expanding spherical wavefront consists of alternate reversals of electric field, shown by the arrowed arcs; nulls are shown dotted. At right angles to the plane of the paper, but not seen here, would be simultaneous alternate reversals of magnetic field (b) and (c). These two 'snapshots' must be rotated through 90° in imagination, so that the magnetic field lines run into and out of the plane of the paper. The view is from R towards T

Field Strength

As the energy in the expanding wavefront has to cover an ever-increasing area the further it travels, the amplitude of the signal induced in the receiving antenna diminishes as a function of distance. Under free-space conditions an inverse square-law relationship applies, but in most cases the nature of the intervening medium has a profound, and often very variable, effect on the magnitude of the received signal.

The intensity of a radio wave at any point in space may be expressed in terms of the strength of its electric field at rightangles to the line of the transmission path. The units used indicate the difference of electric force between two points one metre apart, and **Fig 12.3** (which should not be used for general calculations as it relates only to free-space conditions) has been included here to give an indication of the magnitudes of fields likely to be met with in the amateur service in cases where a more-or-less direct path exists between transmitting and receiving antennas and all other considerations may be ignored. In practice signal levels very much less than the free-space values may be expected because of various losses en route, so that field strengths down to about $1\mu V/m$ need to be considered, or in the case of LF, down to the $1nV/m$ level.

Signal Input

In many calculations dealing with propagation, the parameter of interest is not the field strength at a particular place but the voltage which is induced by it across the input of a receiver. If a half-wave dipole is introduced into a field and aligned for maximum signal pick-up, the open-circuit EMF induced at its centre is given by the expression:

$$e = \frac{E\lambda}{\pi} \tag{1}$$

where e = EMF at the centre of the dipole, in volts
E = the incident field strength in volts/metre

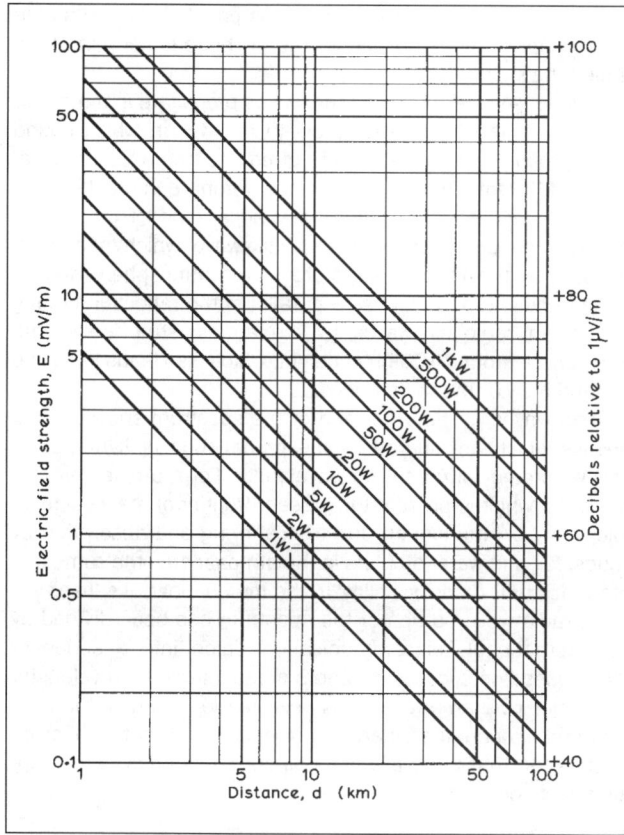

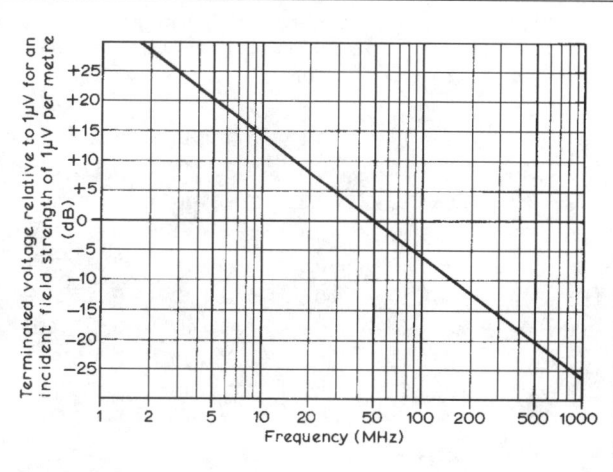

Fig 12.4: The relationship between the incident field strength on a half-wave dipole and the voltage at the receiver end of a correctly matched perfect feeder connected to it

V_i(dB) = the actual incident field strength in decibels relative to 1mV/m

G_r(dB) = the gain in decibels of the receiving antenna relative to a half-wave dipole

L_{fr}(dB) = loss in the feeder line between antenna and receiver

Example. The incident field strength of a 70MHz transmission is 100μV/m (or 40dB above 1μV/m). A three-element Yagi having a gain of 5dB over a half-wave dipole is connected to the receiver through 100ft of coaxial cable which introduces a loss of 2dB. From this, V_0 = -3.5, V_i = +40, G_r = +5 and L_{fr} = +2 (the fact that L_{fr} is a loss is allowed for in Eqn (4)), and V_r = -3.5 + 40 + 5 - 2dB = 39.5dB relative to 1μV, or a receiver input voltage of 94μV.

Table 12.1 may be found useful in converting the rather unfamiliar looking values of decibels met with in radio propagation work into their corresponding voltage or power ratios using only a cheap four-function calculator.

MODES OF PROPAGATION

Introduction

There are four principal modes by which radio waves are propagated. They are:

(a)   *Free-space waves*, which are unaffected by any consideration other than distance;

(b)   *Ionospheric waves*, which are influenced by the action of free electrons in the upper levels of the Earth's atmosphere;

(c)   *Tropospheric waves*, which are subject to deflection in the lower levels by variations in the refractive index structure of the air through which they pass; and

(d)   *Ground waves* which are modified by the nature of the terrain over which they travel.

Free-space waves propagate from point-to-point by the most direct path.

Waves in the other three categories are influenced by factors which make them tend to overcome the curvature of the Earth, either by refraction as with ionospheric waves, refraction as with tropospheric waves, or diffraction at the surface of the Earth itself as with ground waves.

Wavelength is the chief consideration which determines the mode of propagation of Earth-based transmissions.

Fig 12.3: Field strength from an omnidirectional antenna radiating from 1kW to 1W into free space. This is an application of the inverse distance law

π = the wavelength of the transmitted signal in metres. When connected to a matched feeder correctly terminated at the receiver the input voltage available will be half this, or e/2. By substituting frequency for wavelength the equation may be reduced to the more practical form:

$$V_r = \frac{47.8}{f} E \qquad (2)$$

where V_r = microvolts of signal across the receiver input

f = frequency of the transmitted signal in megahertz

E = the incident field strength in microvolts/metre.

It should be noted that for a given frequency the first term becomes a constant factor. By a further rearrangement:

$$E = \frac{f}{47.8} V_r \qquad (3)$$

which enables the field to be estimated from the magnitude of the received signal in cases where the receiver is fed from a perfectly matched half-wave dipole.

It is an advantage to work in decibels in calculations of this nature. If a standard level of 1μV/m is adopted for field strength and of 1μV for signal level it is a simple matter to take into account the various gains and losses in a practical receiving system. **Fig 12.4** shows the terminated voltage at the receiver (V_0) in terms of decibels relative to 1μV for a normalised incident field strength of 1μV/m at the transmitter frequency, derived from expression (2). Then:

$$V_r(\text{dB}) = V_0(\text{dB}) + V_i(\text{dB}) + G_r(\text{dB}) - L_{fr}(\text{dB}) \qquad (4)$$

where V_r((dB) = input to receiver in decibels relative to 1mV

V_0(dB) = input in decibels relative to 1mV, for 1mV/m incident field strength (from graph)

Combine by multiplication Voltage ratios		Combine by addition (dB)	Combine by multiplication Power ratios	
up	down		up	down
1.01×10^0	9.89×10^{-1}	0.1	1.02×10^0	9.77×10^{-1}
1.02×10^0	9.77×10^{-1}	0.2	1.05×10^0	9.55×10^{-1}
1.03×10^0	9.66×10^{-1}	0.3	1.07×10^0	9.33×10^{-1}
1.05×10^0	9.55×10^{-1}	0.4	1.10×10^0	9.12×10^{-1}
1.06×10^0	9.44×10^{-1}	0.5	1.12×10^0	8.91×10^{-1}
1.07×10^0	9.33×10^{-1}	0.6	1.15×10^0	8.71×10^{-1}
1.08×10^0	9.23×10^{-1}	0.7	1.17×10^0	8.51×10^{-1}
1.10×10^0	9.12×10^{-1}	0.8	1.20×10^0	8.32×10^{-1}
1.11×10^0	9.02×10^{-1}	0.9	1.23×10^0	8.13×10^{-1}
1.12×10^0	8.91×10^{-1}	1	1.26×10^0	7.94×10^{-1}
1.26×10^0	7.94×10^{-1}	2	1.58×10^0	6.31×10^{-1}
1.41×10^0	7.08×10^{-1}	3	2.00×10^0	5.01×10^{-1}
1.58×10^0	6.31×10^{-1}	4	2.51×10^0	3.98×10^{-1}
1.78×10^0	5.62×10^{-1}	5	3.16×10^0	3.16×10^{-1}
2.00×10^0	5.01×10^{-1}	6	3.98×10^0	2.51×10^{-1}
2.24×10^0	4.47×10^{-1}	7	5.01×10^0	1.99×10^{-1}
2.51×10^0	3.98×10^{-1}	8	6.31×10^0	1.59×10^{-1}
2.82×10^0	3.55×10^{-1}	9	7.94×10^0	1.26×10^{-1}
3.16×10^0	3.16×10^{-1}	10	1.00×10^1	1.00×10^{-1}
1.00×10^1	1.00×10^{-1}	20	1.00×10^2	1.00×10^{-2}
3.16×10^1	3.16×10^{-2}	30	1.00×10^3	1.00×10^{-3}
1.00×10^2	1.00×10^{-2}	40	1.00×10^4	1.00×10^{-4}
3.16×10^2	3.16×10^{-3}	50	1.00×10^5	1.00×10^{-5}
1.00×10^3	1.00×10^{-3}	60	1.00×10^6	1.00×10^{-6}
3.16×10^3	3.16×10^{-4}	70	1.00×10^7	1.00×10^{-7}
1.00×10^4	1.00×10^{-4}	80	1.00×10^8	1.00×10^{-8}
3.16×10^4	3.16×10^{-5}	90	1.00×10^9	1.00×10^{-9}
1.00×10^5	1.00×10^{-5}	100	1.00×10^{10}	1.00×10^{-10}
1.00×10^{10}	1.00×10^{-10}	200	1.00×10^{20}	1.00×10^{-20}

Example: 39.5dB above 1µV (voltage ratio).
39.5dB = 30 + 9 + 0.5dB
Combining equivalents by multiplication
= (3.16 x 10¹) x (2.82 x 10⁰) x (1.06 x 10⁰) = 94 times 1µV, or 94µV.

Table 12.1; Skeleton decibel table

The Spectrum of Electromagnetic Waves

The position of man-made radio waves in the electromagnetic wave spectrum is shown in **Fig 12.5**, where they can be seen to occupy an appreciable portion of a family of naturally-occurring radiations, all of which are characterised by inseparable oscillations of electric and magnetic fields and travel with the same velocity in free space. This velocity, 2.99790×10^8m/s (generally

taken as 3×10^8m/s in calculations), is popularly known as *the speed of light* although visible light forms but a minor part of the whole range.

At the long-wavelength end the waves propagate in a manner which is similar in many respects to the way in which sound waves propagate in air, although, of course, the actual mechanism is different. Thus, reports of heavy gunfire in the 1914-18 war heard at abnormal ranges beyond a zone of inaudibility revealed the presence of a sonic skywave which had been reflected by the thermal structure of the atmosphere around 30km in height, and this has a parallel in the reflection of long wavelength radio sky-waves by the atomic structure of the atmosphere around 100km in height, which also leads to a zone of inaudibility at medium ranges.

Radio waves at the other end of the spectrum show characteristics which are shared by the propagation of light waves, from which they differ only in wavelength. For example, millimetre waves, which represent the present frontier of practical technology, suffer attenuation due to scattering and absorption by clouds, fog and water droplets in the atmosphere - the same factors which determine 'visibility' in the meteorological sense.

The radio wave portion of the spectrum has been divided by the International Telecommunication Union into a series of bands based on successive orders of magnitude in wavelength.

In **Table 12.2** an attempt has been made to outline the principal propagation characteristics of each band, but it must be emphasised that there are no clear-cut boundaries to the various effects described.

Wave Propagation in Free Space

The concept of *free-space* propagation, of a transmitter radiating without restraint into an infinite empty surrounding space, has been introduced briefly in the section on field strength where it was used to illustrate, in a general way, the relationship between the strength of the field due to a transmitter and the distance over which the waves have travelled.

It is only recently, with the advent of the Space Age, that we have acquired a practical opportunity to operate long-distance circuits under true free-space conditions as, for example, between spacecraft and orbiting satellites, and it is only recently that radio amateurs have had direct access to paths of that nature. Earth-Moon-Earth contacts are becoming increasingly popular, however, and reception of satellite signals commonplace, and for these the free-space calculations apply with only relatively minor adjustments because such a large part of the transmission paths involved lies beyond the reach of terrestrial influences.

In many cases, and especially where wavelengths of less than about 10m are concerned, the free-space calculations are even applied to paths which are subject to relatively unpredictable perturbations in order to estimate a convenient (and often unobtainable) ideal - a standard for the path - against which the other losses may be compared.

The basic transmission loss in free space is given by the expressions:

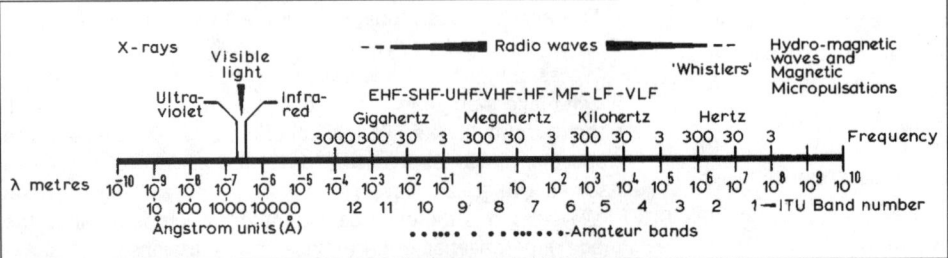

Fig 12.5: The spectrum of electromagnetic waves. This diagram shows on a logarithmic scale the relationship between X-rays, 'visible' and 'invisible' light, heat (infrared), radio waves and the very slow waves associated with geomagnetic pulsations, all of them similar in basic character

ITU Band No	Metric name of band and limits by wavelength	Alternative name of band and limits by frequency	UK amateur bands by frequency (and usual description based on wavelength)	Principal propagation modes	Principal limitations
4	Myriametric 100,000-10,000m	Very low frequency (VLF) 3-30kHz	-	Extensive surface wave Ground to ionosphere space acts as a waveguide	Very high power and very large antennas required. Few channels.
5	Kilometric 10,000-1000m	Low frequency (LF) 30-300kHz	135.7-137.8kHz	Surface wave and reflections from lower ionosphere	High power and large antennas required. Limited number of channels available. Subject to fading where surface wave and sky wave mix
6	Hectometric 1000-100m	Medium frequency (MF) 300-3000kHz	1810-2000kHz (160m band, also known as topband)	Surface wave only during daylight. At night reflection from decaying E-layer	Strong D-region absorption during day. Long ranges possible at night but signals subject to fading and considerable co-channel interference
7	Decametric 100m-10m	High frequency (HF) 3-30MHz	3.50-3.80MHz (80m) 7.00-7.20MHz (40m) 10.10-10.15MHz (30m) 14.00-14.35MHz (20m) 18.068-18.168MHz (17m) 21.00-21.45MHz (14m) 24.89-24.99MHz (12m) 28.00-29.70MHz (10m)	Short-distance working via E-layer. Nearly all long-distance working via F2-layer	Daytime attenuation by D-region, E and F1-layer absorption. Signal strength subject to diurnal, seasonal, solar-cycle and irregular changes
8	Metric 10m-1m	Very high frequency (VHF) 30-300MHz	50.00-52.00MHz (6m) 70.00-70.50MHz (4m) 144.00-146.00MHz (2m)	F2 occasionally at LF end of band around sunspot maximum. Irregularly by sporadic-E and auroral-E. Otherwise maximum range determined by temperature and humidity structure of lower troposphere	Ranges generally only just beyond the horizon but enhancements due to anomalous propagation can exceed 2000km
9	Decimetric 1m-10cm	Ultra high frequency (UHF) 300-3000MHz	430-440MHz (70cm) 1240-1325MHz (23cm) 2310-2450MHz (13cm)	Line-of-sight modified by tropospheric effects	Atmospheric absorption effects noticeable at top of band
10	Centimetric 10cm-1cm	Super high frequency (SHF) 3-30GHz	3.400-3.475GHz (9cm) 5.650-5.850GHz (6cm) 10.00-10.50GHz (3cm) 24.00-24.25GHz (12mm)	Line-of-sight	Attenuation due to oxygen, water vapour and precipitation becomes increasingly important
11	Millimetric 1cm-1mm	Extra high frequency (EHF) 30-300GHz	47.00-47.20GHz (6mm) 75.50-76.00GHz (4mm) 142.00-144.00GHz (2mm) 248.00-250.00GHz (1.2mm)	Line-of-sight	Atmospheric propagation losses create pass and stop bands. Background noise sets a threshold
12	Decimillimetric (sub-millimetric)	- 300-3000GHz	-	Line-of-sight	Present limit of technology

Table 12.2: A survey of the radio-frequency spectrum

$$L_{bf} = 32.45 + 20 \log f(\text{MHz}) + 20 \log r(\text{km}) \quad (5)$$
$$L_{br} = 36.6 + 20 \log f(\text{MHz}) + 20 \log r(\text{miles}) \quad (6)$$

where r is the straight line distance involved.

If transmitter and receiver levels are expressed in either dBW (relative to 1W) or dBm (relative to 1mW) - it does not matter which providing that the same units are used at both ends of the path - with other relevant parameters similarly given in terms of decibels, it is a relatively simple matter to determine the received power at any distance by adding to the transmitter level all the appropriate gains and subtracting all the losses. Thus:

$$P_r(\text{dBW}) = P_t(\text{dBW}) + G_t(\text{dB}) - L_{ft}(\text{dB}) - L_{bf}(\text{dB}) + G_r(\text{dB}) - L_{fr}(\text{dB}) \quad (7)$$
$$Pr(\text{dBm}) = Pt(\text{dBm}) + Gt(\text{dB}) - Lft(\text{dB}) - Lbf(\text{dB}) + Gr(\text{dB}) - Lfr(\text{dB}) \quad (8)$$

where P_r = Received power level (dBm or dBW)

P_t = Transmitted power level (dBm or dBW)

G_t = Gain of the transmitting antenna in the direction of the path, relative to an isotropic radiator

L_{ft} = Transmitting feeder loss

L_{bf} = Free-space transmission loss

G_r = Gain of the receiving antenna in the direction of the path, relative to an isotropic radiator

L_{fr} = Receiving feeder loss

The free-space transmission loss may be estimated approximately from **Fig 12.6** which perhaps conveys a better idea of its relationship to frequency (or wavelength) and distance than the nomogram generally provided. Unless a large number of calculations have to be made, it is no great hardship to use the formula for individual cases should greater accuracy be desired. It should be noted that antenna gains quoted with respect to a half-wave dipole need to be increased by 2dB to express them relative to an isotropic radiator.

Example. A 70MHz transmitter radiates 100W ERP in the direction of a receiver 550km away. The receiving antenna has

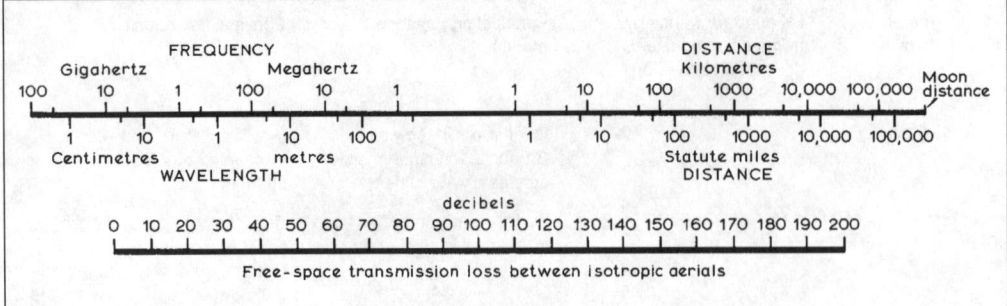

Fig 12.6: The effects of frequency (or wavelength) and distance on the free-space transmission loss between isotropic antennas. The length on the diagram between points corresponding to frequency and distance on the upper scale corresponds to the free-space transmission loss measured from the zero decibel mark on the lower scale

a gain of 5dB over a half-wave dipole (2dB more over an isotropic radiator), and there is a 2dB loss in the feeder. In this case the effective radiated power is known, which takes the place of the terms P_t, G_t and L_{ft}.

$$
\begin{array}{lll}
P_{erp} & = & 100W = 105mW = 50dBm \\
L_{bf} & = & 124 \text{ from Fig 12.6 or by use of the} \\
& & \text{expressions (5) or (6)} \\
G_r & = & 7dB \text{ over an isotropic radiator} \\
L_{fr} & = & 2dB \\
\text{Then} \quad P_t & = & P_{erp} - L_{bf} + G_r - L_{fr} \\
& = & 50 - 124 + 7 - 2 \\
& = & -69dBm, \text{ or } 69dB \text{ below } 1mW \\
& = & 12.6 \times 10^{-6} mW \\
& = & 12.6 \times 10^{-9} W
\end{array}
$$

If this power is dissipated in an input impedance of 70 ohms the voltage appearing across the receiver V_r is $\sqrt{(P_r Z_{in})}$, in this case $\sqrt{(12.6 \times 10^{-9} \times 70)}$ which is 94µV. This example deals with the same situation which was considered earlier in connection with field strength and offers an alternative method of calculation

Wave Propagation in the Ionosphere

It has been shown that a transmitted signal may be considered as consisting of a succession of spherical wavefronts, each one a wavelength apart, and they approximate to plane surfaces at great distances. At certain heights in the upper atmosphere concentrations of negatively charged free electrons occur, and these are set into oscillatory motion by the oncoming waves, which causes them to emit secondary wavelets having a phase which is 90ø in advance of the main wave. It is only in the forward direction that the original waves and their dependent wavelets combine coherently and their resultant consists of a wave in which the maxima and minima occur earlier than in the projection of the originating wave - to all intents and purposes the equivalent of the wave having travelled faster in order to arrive earlier. The amount of phase advancement is a function of the concentration of electrons and the change of speed is greatest at long wavelengths, decreasing therefore as the signal frequency increases.

The advancing wave-front, travelling, let us say, obliquely upwards from the ground, meets the layer containing the accumulation of free electrons in such a way that its upper portion passes through a greater concentration of charge than does a portion lower down. The top of the wavefront is therefore accelerated to a greater extent by the process just outlined than are the parts immediately below, which results in a gradual swing-round until the wavefront is being returned towards the ground as though it had experienced a reflection.

The nearer the wavefront is to being vertically above the transmitter the more quickly must the top accelerate relative to the bottom, and the more concentrated must be the charge of electrons. It may be that the density of electrons is sufficiently high

to turn even wavefronts propagating vertically upwards (a condition known as *vertical incidence*), although it must be appreciated that deeper penetration into the layer will occur as the propagation angle becomes steeper.

Consider the circumstances outlined in **Fig 12.7**, where T indicates the site of a transmitting station and R1, R2 and R3 three receiving sites. For a given electron concentration there is a *critical frequency* f_0 which is the highest to return from radiation directly vertically upward. Frequencies higher than this will penetrate the layer completely and be lost in space, but their reflection may still be possible at *oblique incidence* where waves have to travel a greater distance within the electron concentration. This is not the case at point A in the diagram, so that reception by sky-wave is impossible at R1 under these circumstances, but at a certain angle of incidence to the layer (as at point B) the ray bending becomes just sufficient to return signals to the ground, making R2 the nearest location relative to the transmitter at which the sky-wave could be received. The range over which no signals are possible via the ionosphere is known as the *skip distance*, and the roughly circular area described by it is called the skip zone. Lower-angle radiation results in longer ranges, for example to point R3 from a reflection at C, and a second 'hop' may result from a further reflection from the ground. The longest ranges at HF are achieved this way - and it is possible for an HF signal to travel right round the world using a succession of hops.

If the angle φ is very large and conditions are favourable, the transmission path may lead from one point on the ionosphere to another without intermediate ground reflection. This is known as *chordal hop* propagation; signal strength is usually higher than normal because there are no ground reflection points or extra transits of the absorbing D-layer to introduce losses.

From the point of view of the operator, it is φ, the angle of signal take-off relative to the horizon (**Fig 12.8**, inset diagram), that is mainly of interest in this application, not the angle of incidence at the ionosphere. The two curves in the main diagram show the distances covered for various values of θ at representative heights of 120km (E-layer) and 400km (F2-layer).

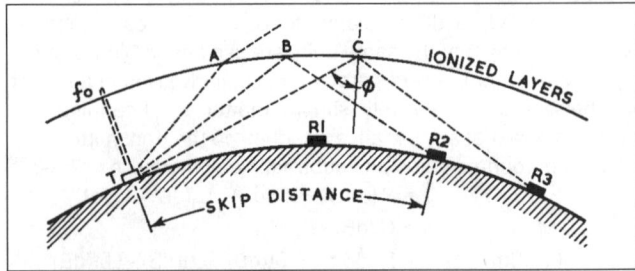

Fig 12.7: Wave propagation via the ionosphere. T is the site of a transmitter, R1, R2 and R3 are the sites of three receivers. The significance of the ray at vertical incidence (dotted) and the three oblique rays is explained in the text

The Radio Communication Handbook

For oblique incidence on a particular path (eg the ray from T to R3 via point C in the ionosphere) there is a *maximum usable frequency* (MUF), generally much higher than the critical frequency was at vertical incidence, and this is approximately equal to $f_0/\cos\varphi$, where φ is the angle of incidence of the ray to the point of reflection, as is shown in the figure. The limiting angle which defines the point at which reflections first become possible is called the *critical wave angle*, and it is this function which determines the extent of the skip zone.

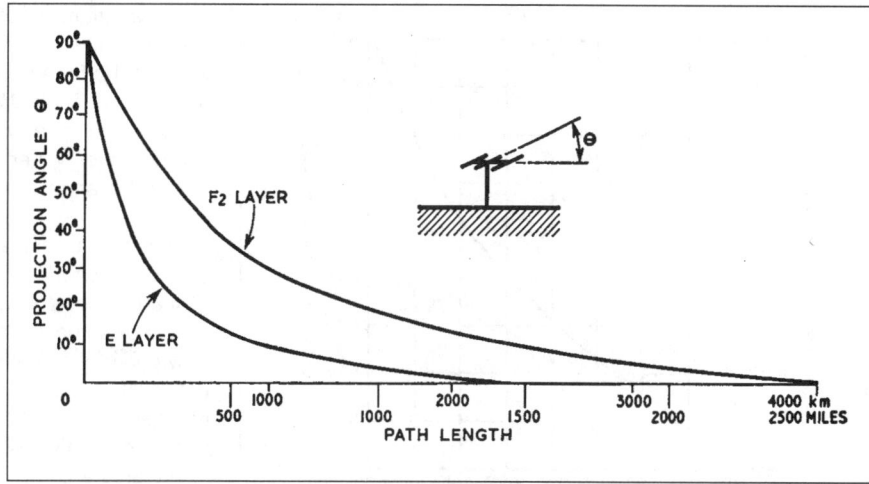

Fig 12.8: Relationship between angles of take-off and resulting path length. The two ionospheric layers are assumed to be at heights of 120km and 400km respectively

This mechanism is effective for signals in excess of about 100kHz, for which the concentration of electrons appears as a succession of layers of increasing electron density having the effect of progressively bending the rays as the region is penetrated.

Below 50kHz the change in concentration occurs within a distance which is small compared to the wavelength and which therefore appears as an almost perfect reflector. Waves are propagated in that way over great distances by virtue of being confined between two concentric spheres, one being the lower edge of the layer and the other the surface of the Earth, sometimes referred to as *waveguide mode*.

During the hours of daylight the quantity of free electrons in the lower ionosphere becomes so great that the oscillations set up by incident waves are heavily damped on account of energy lost by frequent collisions with the surrounding neutral air particles. Medium-wave broadcast band signals are so much affected by this as to have their sky-waves completely absorbed during the day, leading to the familiar rapid weakening of distant stations around dawn and their subsequent disappearance at the very time when the reflecting layers might otherwise be expected to be most effective. The shorter wavelengths are less severely affected, but suffer attenuation nevertheless. On the 136kHz band, signals can be enhanced at extreme range by 'reflection' from the D layer, the effect peaking at mid-path.

A certain amount of cross-polarisation occurs when ionospheric reflections take place so that the received signals generally contain a mixture of both horizontal and vertical components, irrespective of which predominated at the transmitting antenna.

Ionospheric scatter propagation does not make use of the regular layers of increased electron density. Instead forward scattering takes place from small irregularities in the ionosphere comparable in size with the wavelength in use (generally around 8m or about 35MHz). With high powers and very low angles of radiation, paths of some 2000km are possible and this mode has the advantage of being workable in auroral regions where conventional HF methods are often unreliable, but only a very small proportion of the transmitted power is able to find its way in the desired direction.

Wave Propagation in the Troposphere

The *troposphere* is that lower portion of the atmosphere in which the general tendency is for air temperature to decrease with height. It is separated from the *stratosphere*, the region immediately above, where the air temperature tends to remain invariant with height, by a boundary called the *tropopause* at around 10km. The troposphere contains all the well-known cloud forms and is responsible for nearly everything loosely grouped under the general heading of 'weather'.

Its effect upon radio waves is to bend them, generally in the same direction as that taken by the Earth's curvature, not by encounters with free electrons or layers of ionisation, but as a result of successive changes in the refractive index of the air through which the waves pass. In optical terms this is the mechanism responsible for the appearance of mirages, where objects beyond the horizon are brought into view by raybending, in that case resulting from temperature changes along the line-of-sight path. In the case of radio signals the distribution of water vapour also plays a part, often a major one where anomalous propagation events are concerned.

The refractive index, n, of a sample of air can be found from the expression $n = 1 + 10^{-6}N$, where N is the refractivity expressed by:

$$N = \frac{77.6P}{T} + \frac{4810e}{T^2}$$

where P = the atmospheric pressure in millibars or hectopascals

e = the water-vapour pressure, also in millibars or hectopascals

T = the temperature in kelvin or degrees absolute

Substituting successive values of P, e and T from a standard atmosphere table shows that there is a tendency for refractive index to decrease with height, from a value just above unity at the Earth's surface to unity itself in free space. This gradient is sufficient to create a condition in which rays are normally bent down towards the Earth, leading to a similar state of affairs as that which would result from the radio horizon being extended to beyond the optical line-of-sight limit by an average of about 15%.

The distance d to the horizon from an antenna of height h is approximately $\sqrt{(2ah)}$, where a is the radius of the Earth and h is small compared with it. The effect of refraction can be allowed for by increasing the true value of the Earth's radius until the ray paths, curved by the refractive index gradient, become straight again. This modified radius a' can be found from the expression:

$$\frac{1}{a'} = \frac{1}{a} + \frac{dn}{dh}$$

where dn/dh is the rate of decrease of refractive index with height. The ratio a'/a is known as the *effective Earth radius factor k*, so that the distance to the radio horizon becomes $d = \sqrt{(2kah)}$.

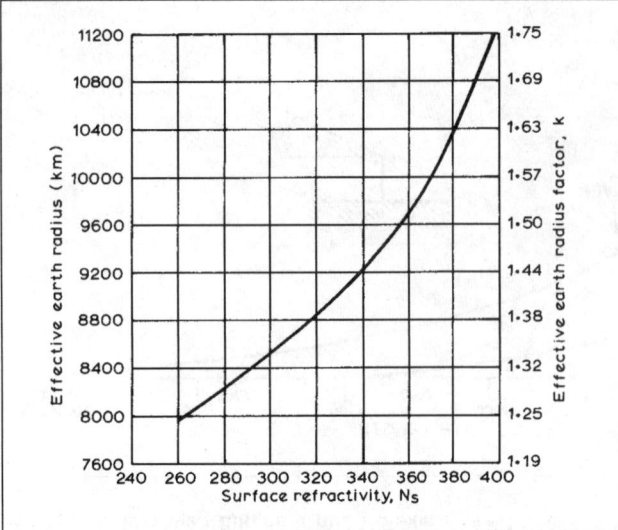

Fig 12.9: Effective Earth radius corresponding to various values of surface refractivity. A curved ray-path drawn over an Earth section between terminals may be rendered straight by exaggerating the Earth's curvature. An average value often used is equal to 1.33 times the actual radius; this is sometimes referred to as the four-thirds Earth approximation. (Based on CCIR Report 244.)

An average value for k, based on a standard atmosphere, is 1.33 or 4/3 (whence a common description of this convention, *the four-thirds Earth*). Notes relating to the construction of path profiles using this convention will be found in a later section of this chapter. When the four-thirds Earth concept is known to be inappropriate an estimate of a suitable value for k can be obtained from the surface refractivity N_s (obtained from **Fig 12.9**), using for N the value obtained from ground-level readings of pressure, temperature and vapour pressure but this method may be misleading if marked anomalies are present in the vertical refractive index structure.

Waves of widely separated length are liable to be disturbed by the troposphere in some way or other, but it is generally only those shorter than about 10m (over 30MHz) which need be considered. There are two reasons for this; one is that usually ionospheric effects are so pronounced at the longer wavelengths that attention is diverted from the comparatively minor enhancements due to the refractive index structure of the troposphere, and the other that anomalies, when they occur, are often of insufficient depth to accommodate waves as long as, or longer than, 10m. For example, it might be that the decrease of refractive index with height becomes so sharp in the lower 100m of the troposphere, that waves are trapped in an *atmospheric duct*, within which they remain confined for abnormally long distances. The maximum wavelength which can be trapped completely in a duct of 100m thickness is about 1m (corresponding to a frequency of 300MHz), for example, so that the most favourable conditions are generally found in the VHF and UHF bands or above. The relationship between maximum wavelength λ and duct thickness t is shown in the expression:

$$t = 500 \lambda^{2/3}$$

where both t and λ are expressed in centimetres.

At centimetre wavelengths signals propagated through the troposphere suffer rapid fluctuations in amplitude and phase due to irregular small-scale variations in refractive index which give rise to continuous changes known as *scintillations*, and they are also attenuated by water in the form of precipitation (rain, snow, hail, etc) or as fog or cloud. As **Table 12.3** shows, this effect increases both with radio frequency and with either the rate of rainfall or the concentration of water droplets. Precipitation causes losses by absorption and by random scattering from the liquid (or solid in the case of ice) surfaces and this scattering becomes so pronounced as to act as a 'target' for weather radars which use these *precipitation echoes* to detect rain areas.

At even shorter wavelengths resonances occur within the molecules of some of the gases which make up the atmosphere. Only one of these approximates to an amateur band, the attenuation at 22.23GHz due to water vapour. The principal oxygen resonances at around 60 and 120GHz are in the upper reaches of amateur activities, but are likely to prove to be limiting factors where amateur and professional work at millimetre wavelengths involves paths passing through the atmosphere, as opposed to outer space working.

Wave Propagation near the Ground

Diffraction is an alteration in direction of the propagation of a wave due to change in velocity over its wavefront. Radio waves meeting an obstacle tend to diffract around it, and the surface of the Earth is no exception to this. Bending comes about as a result of energy being extracted due to currents induced in the ground. These constitute an attenuation by absorption, having the effect of slowing down the lower parts of the wavefront, causing it to tilt forward in a way which follows the Earth's curvature. The amount of diffraction is dependent on the ratio of the wavelength to the radius of the Earth and so is greatest when the waves are longest. It also depends on the electrical characteristics of the surface, namely its relative permeability (generally regarded as unity for this purpose), its dielectric constant, ε (epsilon), and conductivity, σ (sigma). This diffracted wave is known as the *surface wave*.

Moisture content is probably the major factor in determining the electrical constants of the ground, which can vary considerably with the type of surface as can be seen from **Table 12.4**. The depth to which the wave penetrates is a function also of frequency, and the depth given by the δ (delta) value is that at which the wave has been attenuated to $1/e$ (or about 37%) of its surface magnitude.

At VHF and at higher frequencies the depth of penetration is relatively small and normal diffraction effects are slight. At all frequencies open to amateur use, however, the ground itself appears as a reflector, and the better the conductivity of the sur-

Frequency band (GHz)	Precipitation (mm/h)					Fog or cloud water content (g/m3) (at 0°C)		
	100	50	25	10	1	2.35	0.42	0.043
3.400-3.475	0.1	0.02	0.01	-	-	-	-	-
5.650-5.850	0.6	0.25	0.1	0.02	-	0.09	-	-
10-10.5	3.0	1.5	0.6	0.2	0.01	0.23	0.04	-
21-22	13.0	6.0	2.5	1.0	0.1	0.94	0.17	0.02
24	17.0	8.0	3.8	1.5	0.1	1.41	0.25	0.03
48-49	30.0	17.0	9.0	4.0	0.6	4.70	0.84	0.09

100mm/h = tropical downpour; 50mm/h = very heavy rain; 25mm/h = heavy rain; 10mm/h = moderate rain, 1mm/h = light rain. 2.35g/m3 visibility of 30m; 0.42g/m3 100m; 0.043g/m3 500m.

Table 12.3: The attenuation in decibels per kilometre to be expected from various rates of rainfall and for various degrees of cloud intensity

Type of surface	Dielectric constant ε	Conductivity σ (mho/m)				Depth of penetration δ (m)			
		1MHz	10MHz	100MHz	1000MHz	1MHz	10MHz	100MHz	1000MHz
Sea water (20°C)	70	5	5	5	5	0.25	0.08	0.02	0.001
Fresh water (10°C)	80	0.003	0.003	0.005	0.2	15	10	4	0.3
Very moist soil	30	0.01	0.01	0.02	0.2	5	3	1.5	0.2
Average ground	15	0.001	0.001	0.002	0.04	25	16	6	0.5
Very dry ground	3	0.0001	0.0001	0.0001	0.0002	90	90	90	40

(Top) Table 12.4: Typical values of dielectric constant, conductivity and depth of wave penetration for various types of surface at various frequencies

Distance (km)	Free-space field (dB) rel to 1µV/m for a 1kW transmitter	Sea, σ = 4			Land, σ = 3 10 2			Land, σ = 3 10 3			Land, σ = 10 3		
		1.8	3.5	7.0	1.8	3.5	7.0	1.8	3.5	7.0	1.8	3.5	7.0
3	100	1	1	1	1	3	12	8	18	30	19	28	36
10	90	1	1	1	6	8	22	18	29	42	30	38	47
30	80	1	2	2	8	18	34	28	41	51	40	48	56
100	70	2	3	7	18	33	51	44	56	68	58	63	73
300	60	10	12	23	43	61	88	68	85	-	78	-	-
1000	50	48	55	-	-	-	-	-	-	-	-	-	-

(Bottom) Table 12.5: The number of decibels to be subtracted from the calculated free-space field in order to take into account various combinations of ground conductivity and distance. The values are shown in each case for 1.8, 3.5 and 7.0MHz. Vertical polarisation is assumed

face the more effective the reflection (Table 12.5). It is this effect which makes generalisations of wave propagation near the ground difficult to make for frequencies greater than about 10MHz where the received field is, more often than not, due to the resultant of waves which have travelled by different paths.

Further aspects of propagation near the ground will be dealt with in the section on multiple-path propagation where it will be necessary to consider the consequences of operating with antennas at heights of one to several wavelengths above the ground.

MULTIPLE-PATH EFFECTS

Introduction

The preceding descriptions of the various modes of propagation do not necessarily paint a very realistic picture of the way in which the signals received at a distant location depend on the radio frequency in use and the distance from the transmitter, excluding ionospheric components. The reason for this is that the wave incident on the receiving antenna is rarely only the one which has arrived by the most direct path but is more often the resultant of two or more waves which have travelled by different routes and have covered different distances in doing so. If these waves should eventually arrive in phase they would act to reinforce one another, but should they reach the receiving antenna in antiphase they would interfere with one another and, if they happened to be equal in amplitude, would cancel one another completely.

These alternative paths may arise as a result of reflections in the horizontal plane (as in Fig 12.10(a), where a tall gasholder intercepts the oncoming waves and deflects them towards the receiving site) or in the vertical plane (as in Fig 12.10(b), where reflection occurs from a point on the ground in line-of-sight from both ends of the link). If the reflecting surface is stationary, as it ought to be in the two cases so far considered, the phase difference (whatever it may be) would be constant and a steady signal would result.

It may happen that the surface of reflection is in motion as it would be if it was part of an aeroplane flying along the transmission path. In that case the distance travelled by the reflect-

ed wave would be changing continually and the relative phases would progressively advance or retard through successive cycles (effectively an increase or decrease in frequency - the Doppler effect), leading to alternate enhancements and degradations as the two waves aid or oppose one another. This performance is one which is particularly noticeable on analogue TV receivers sited near an airport, and even non-technical viewers can instantly diagnose as aircraft flutter the fluctuations in picture brilliance which result.

Any 'ghost' image on an analogue TV picture is evidence of a second transmission path, and the amount of its horizontal displacement from the main picture is a measure of the additional transmission distance involved. So, with the reflection from the moving aircraft, the displacement of the ghost picture will change as the path length changes, and its brilliance will reach a maximum every time the difference between the direct path and the reflected path is exactly a whole number of wavelengths. An analytical treatment of the appearance of aircraft reflections on pen recordings of distant signals has appeared in

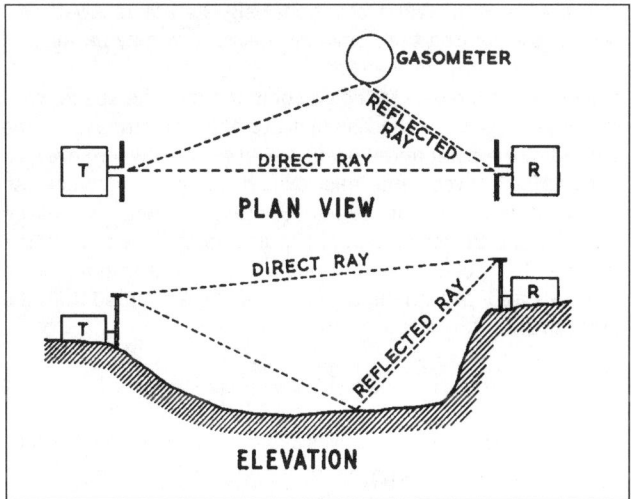

Fig 12.10: Multipath effects brought about by reflections in (a) the horizontal plane, and (b) the vertical plane

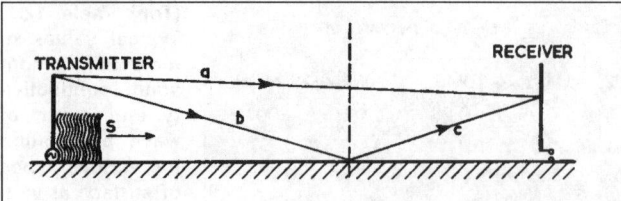

Fig 12.11: The diffracted surface wave S together with the space wave, composed in turn of a direct wave (a) and a reflected wave (b) and (c)

the pages of *Radio Communication* [1]. UHF and low microwave band operators have used reflections from aircraft to make radio contacts.

Because the waves along the reflected path repeat themselves after intervals of exactly one wavelength, it is only the portion of a wavelength 'left over' which determines the phase relationship in comparison with the direct-path wave. This suggests that relatively small changes in the position of a receiving antenna could have profound effects on the magnitude of the received signal when multiple paths are present, and this is indeed found to be the case, particularly where the point of reflection is near at hand.

Ground-wave Propagation

It should now be evident why it was not possible to generalise on the relationship between distance and received signal strength when dealing with propagation near the ground. The surface wave, influenced by the diffraction effects considered earlier, is only one of the possible paths. If the spacing of the transmitting and receiving antennas is such that they are not hidden from one another by the curvature of the Earth there will also be a space wave, made up of a direct wave and a ground-reflected wave as suggested by **Fig 12.11**.

The combination of this space wave and the diffracted surface wave form what is called the *ground wave*, and it may sometimes be difficult (and often perhaps unnecessary) to try to separate this into its three components.

Beyond the radio horizon the direct and reflected rays are blocked by the bulge of the Earth, and the range attained is then determined by the surface wave alone. This diffracted wave is strong at low frequencies (including the amateur 160m band, and more particularly at 136kHz) but becomes less so as the carrier frequency increases and may be considered negligible at VHF and beyond. When occasional signals are received well beyond the horizon the dominant mechanism may be forward scatter.

The strength of the reflected component of the space wave depends largely on the conductivity and smoothness of the ground at the point of reflection, being greatest where oversea paths are involved, and least over dry ground and rock. An extensive treatment of the various factors concerned will be found in the Society's journal [2]. If perfect reflection is presumed, the received field strength due to the interaction of one reflected ray with the direct ray can be estimated from the expression:

$$E = \frac{2E_0}{d} \sin\left(2\pi \frac{h_t h_r}{\lambda d}\right)$$

where E is the resultant received signal strength, E_0 is the direct-ray field strength, h_t and h_r are the effective antenna heights above a plane tangential to the Earth at the point of reflection, d is the distance traversed by the direct ray, and λ the wavelength, all units being consistent (eg metres).

It can be seen that the magnitude of the received signal depends on the relative heights of the two antennas, the distance between them, and, of course, the frequency.

This relationship suggests that doubling the antenna height has the same effect on the received signal strength as halving the length of the path. In view of the respective distances involved it will be appreciated that an increase in the height of one of the antennas has a greater effect than a comparable horizontal movement towards the transmitter.

The Effect of Varying Height

A few moments of experiment with two pieces of cotton representing the two alternative ray paths will provide a convincing demonstration of the effect of altering the height of one or both of the antennas. An increase in height of (say) the receiving antenna has little effect on the length of the direct path, but the ground-reflected ray has to travel further to make up the additional distance and it therefore arrives at a later point on its cycle. The consequence is even more pronounced when it is realised that at very low angles of incidence, when the two antennas are at ground level, the indirect ray may well have experienced a phase change of 180 degrees upon reflection so that the two components arrive roughly equal in magnitude but opposite in phase, so tending to cancel. As the height of one or both antennas is increased the space wave increases in magnitude and the field becomes the vector sum of the diffracted surface wave and the space wave. At even greater heights the effect of the surface wave can be neglected, while the intensity of the space wave continues to increase.

In practice these considerations apply only to antennas carrying VHF, UHF and above. This is because it is not practicable to raise antennas to the necessary heights at the longer wavelengths, and in any case the reception of ionospherically propagated waves imposes different requirements as regards angle of arrival.

As with other functions which depend on Earth constants for their effectiveness there is a marked difference between overland and over-sea conditions. There is more to be gained by raising an antenna over land than over sea, for high frequencies than for low, and for horizontally polarised waves rather than vertically polarised ones.

The ratio of the received field at any given height above ground to the field at ground level due to the surface wave alone (presuming that the two components of the space wave have cancelled one another) is known as the *height-gain factor*. This can be expressed either as a multiplier, or as the corresponding equivalent in decibels, using the voltage scale of relationships.

Over flat ground there is little to be gained from raising antennas for frequencies below about 3MHz, unless it is to clear local obstacles, but it should not be overlooked that it may be desirable to raise antennas no matter what their frequency of operation for reasons unconnected with height-gain benefits - to increase the distance to the radio horizon, for example.

The result of changing the receiving antenna height is by no means as predictable as some authorities would have us believe, and the subject is still a matter thought worthy of further investigation at some research establishments. The following figures summarise the gains to be expected after raising a receiving antenna from a height of 3m to a height of 10m above the ground, according to a current CCIR report [3], primarily concerned with television broadcasting frequencies but relevant nevertheless:

50-100MHz	Median values of height-gain 9-10dB.
180-230MHz	Median values of height-gain 7dB in flat terrain and 4-6dB in urban or hilly areas

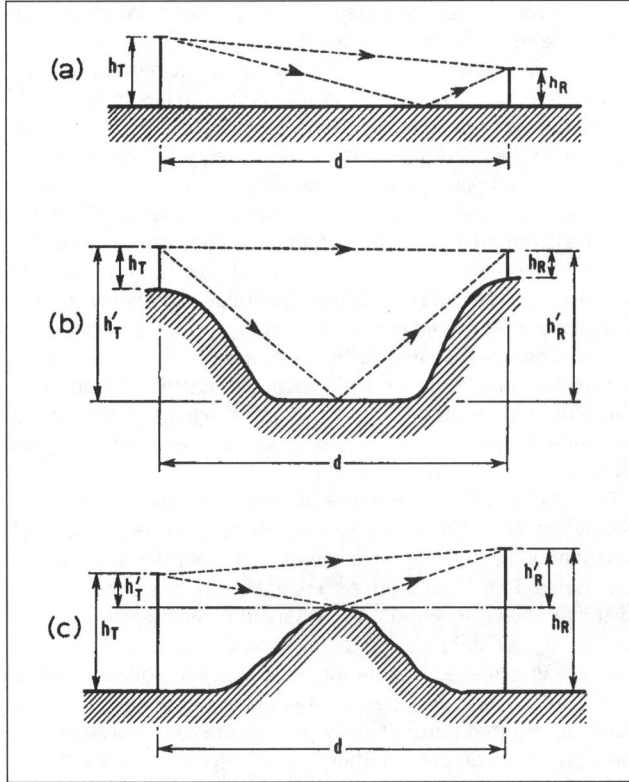

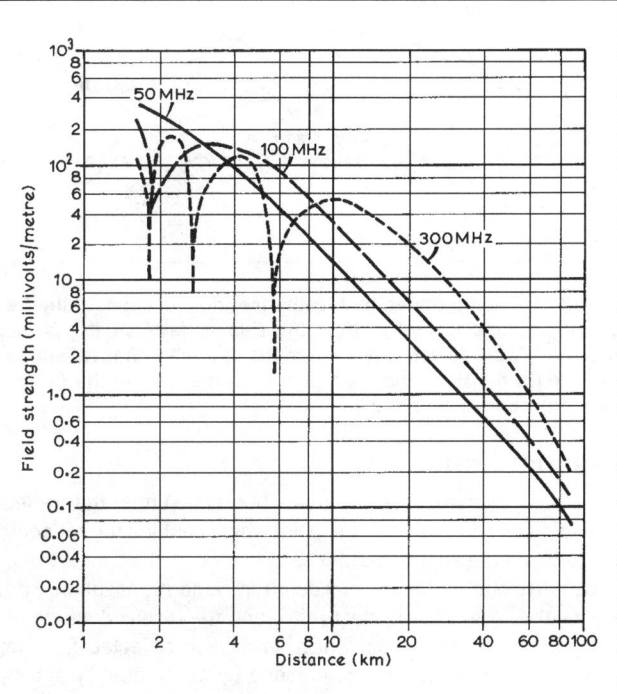

Fig 12.13: Relationship between field strength and distance at VHF and UHF

Fig 12.12: The effect of terrain on direct and indirect rays, showing the effective heights of the antennas when undertaking height-gain calculations.

450-1000MHz   Median values of height-gain very dependent on terrain irregularity. In suburban areas the median is 6-7dB, and in areas with many tall buildings 4-5dB

A simple rule-of-thumb often adopted by radio amateurs is to reckon on a height-gain of 6dB for each time that the antenna height is doubled (eg if 12dB at 3m height, then expect 18dB at 6m, 24dB at 12m, etc), but the presence of more component waves than the two considered can lead to wide departures from this relationship, particularly in urban areas.

If the terrain is not flat the result of altering the antenna height depends largely on the position in the vertical plane of the reflection points relative to the two terminals. Thus in **Fig 12.12** the situation shown in (a) corresponds to the one already considered.

Should the two antennas be sited on hills, or separated by a valley, as at (b), there will be large differences in path length between the direct and ground-reflected rays which alterations in antenna height will do little to alter, so that elevating it is unlikely to have very much effect on the received signal strength.

On the other hand, the presence of high ground between transmitter and receiver, as at (c), may make communication between them difficult at low antenna heights, and in that case there would be a great deal to be gained from raising them. In cases (b) and (c) the two antennas should be considered as having effective heights of h'_T and h'_R respectively when dealing with height-gain calculations.

If the intervening high ground has a relatively sharp and well-defined upper boundary, such as would be the case with a mountain ridge, the receiving antenna height at which signals cease might be much lower than would be expected from line-of-sight considerations, even when refractivity changes are taken into account. This is because of an effect known as *knife-edge diffraction*, which often enables 2m operators situated in the Scottish Highlands (to cite an instance) to receive signals from other stations which are apparently obscured from them by surrounding mountains.

The Effect of Varying Distance

The effect on the field strength of varying the distance between transmitter and receiver is shown in **Fig 12.13**, where the result is again due to the interference between the direct wave and the ground-reflected wave passing through successive maxima and minima as the path difference becomes an exact odd or even number of halfwavelengths. (It must be remembered that a low-angle ground-reflection itself introduces a phase change of very nearly 180 degrees).

The spacing of the maxima (which are greater in magnitude than the free-space value) is closer the higher the frequency of operation and the shorter the path for a given frequency. The most distant maximum will occur when the path difference is down to one half wavelength; beyond that the difference tends towards zero and the two waves progressively oppose one another, the field rapidly falling below the free-space estimate. If the antenna heights are raised the patterns move outwards.

As with all these matters involving ground reflection there is a difference between the behaviour of horizontally and vertically polarised waves, and the foregoing description favours the former.

The reflection coefficient and phase shift at the reflection point vary appreciably with the ground constants when vertical polarisation is employed. In practice, whichever is used, the measured field strength may vary considerably from the calculated value because of the presence of other components due to local reflections. A fairly reliable first estimate for VHF and UHF paths up to about 50km unobstructed length is just to allow for a possible increase or decrease of 10dB on the free-space figure.

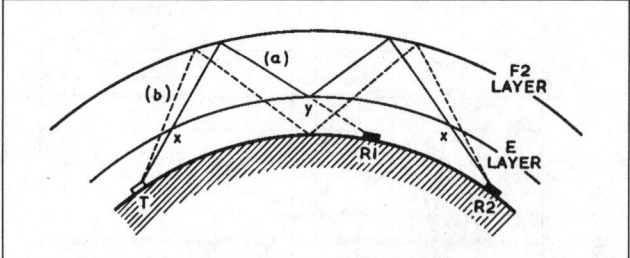

Fig 12.14: Fading due to multipath reception. In case (a) the frequency in use is higher than the E-layer MUF at the points marked 'x', but lower at 'y', where the ray will suffer reflection. In case (b) the frequency is higher than the MUF of the E-layer at all four contacts with it

Fresnel Zones

In all the explanations so far it has been presumed that reflection at a surface occurs at the point which enables the reflected ray to travel the shortest distance.

Because the surfaces considered in radio propagation work are neither plane nor perfect reflectors, the received waveform is the resultant of signals which have been reflected from an area, the size of which is determined by the frequency and by the separation of the terminal antennas, so that the individual reflected path lengths differ by no more than half a wavelength from one another. The locus of all the points surrounding the direct path which give exactly half a wavelength path difference is described by an ellipsoid of revolution having its foci at the transmitting and receiving antennas respectively. A cross-section of this volume on an intersecting plane of reflection encloses an area known as the *first Fresnel zone*.

The radius, R, of the first Fresnel zone at any point, P, is given by the expression:

$$R^2 = \frac{\lambda d_1 d_2}{d_1 + d_2}$$

where d_1 and d_2 are the two distances from P to the ends of the path, and λ is the wavelength, all quantities being in similar units. The maximum radius occurs when d_1 and d_2 are equal. Thus, on an 80km path at a wavelength of 2m, the radius of the first Fresnel zone at the midpoint is 200m, and this represents the clearance of the line-of-sight ray (corrected for refraction) necessary if the path is to be considered 'unobstructed'.

Higher orders of Fresnel zone surround the regions where similar relationships occur after separations of one wavelength, two wavelengths etc, but the conditions for reflection in the required direction rapidly become less favourable and Fresnel zones other than the first are rarely considered.

Ionospheric Multi-path Effects

The multiple-path effects so far considered have been mainly associated with VHF, UHF and SHF where very low angles of radiation and reception are generally involved, and where the wavelengths concerned are sufficiently small to enable optimum heights and favourable positions to be found for the antennas.

When ionospherically propagated signals are of interest it is generally sufficient to regard the ground wave as a single entity, without attempting to separate it into its three components. When multi-path effects occur (as they frequently do) they may be between the ground wave and an ionospheric wave, between two ionospheric waves which may have been propagated by different layers or, in the case of long range transmissions, by signals which have followed different paths entirely in different directions around the Earth's curvature, perhaps in several

'hops' between the ionosphere and the ground. Whatever the cause the result is inevitably a fading signal.

The ionosphere is not a perfect reflector; it has no definite boundaries and it is subject to frequent changes in form and intensity. These deficiencies appear on the received signal as continual small alterations in phase or frequency as the effective path length alters and, when only a single ionospheric wave is present, can pass almost unnoticed by the average HF listener, who is remarkably tolerant of imperfections on distant transmissions. However, when a second signal from the same source is present, which may be either the relatively steady ground wave or another ionospheric component, these phase or frequency changes become further emphasised by appearing as changes in amplitude as the waves alternately reinforce and interfere, and by distortion of modulated signals if the various sideband frequencies do not resolve back into their original form.

PC data recordings of signal strength generally show very clearly the period of fading which results when two modes of propagation begin to interact, continuing until the second predominates. This could occur as a result of circumstances similar to those shown in **Fig 12.14**, where the receiver at R2 may receive signals either by double-hop off the ground (as in path (b)) or double-hop off an intermediate layer, path (a), depending on the relationship between the signal frequency and the maximum usable frequency at point y. The transition between one propagation mode and another is generally accomplished within a relatively short time.

Occasionally very long-distance transmissions may be heard with a marked echo on their modulation. As the two signal components responsible have obviously travelled routes of markedly differing length they probably require very different azimuths at both ends of the path. This effect is most noticeable on omni-directional broadcast transmissions and is minimised by the use of narrow-beam antennas for transmission and reception.

Fading

Fading is generally, though not exclusively, a consequence of the presence of multiple transmission paths. For that reason it is appropriate to include a summary of its causes here, although more properly some of the comments belong elsewhere in this chapter.

Fading is a repetitive rise and fall in signal level, often described as being *deep* or *shallow* when referring to the range of amplitudes concerned, and *slow* or *rapid* when discussing the period. It is sometimes *random*, usually *periodic*, but occasionally *double periodic*, as when a signal with a rapid fade displays slow changes in mean level. Generally the fading rate increases with frequency because a particular motion in the ionosphere causes a greater phase shift at the shorter wavelengths.

At VHF and UHF the fading rate is often closely associated with the pattern of atmospheric pressure at the surface, tending to become slower during periods of high pressure. This can be particularly noticeable on a pen recording of signal strength taken while a ridge of high pressure moves along the transmission path, the slowest rate occurring as the ridge crosses the mid-point.

Interference fading, as its name implies, is caused by interference between two component waves when one or both path lengths are changing, perhaps due to fluctuations in the ionosphere or troposphere, or due to reflections from a moving surface. The period is relatively short, usually up to a few seconds. Fast interference fading is often called *flutter*. Auroral flutter comes in this category, being caused by motion of the reflecting surfaces.

Polarisation fading is brought about by continuous changes in polarisation due to the effect of the Earth's magnetic field on the ionosphere. Signals are at a maximum when they arrive with the same polarisation as the receiving antenna. Period again up to a few seconds.

Absorption fading, generally of fairly long period, is caused by inhomogeneities in the troposphere or ionosphere. Period up to an hour, or longer.

Skip fading occurs when a receiver is on the edge of a skip zone and changes in MUF cause the skip distance to shorten and lengthen. Highly irregular as regards period of fade.

Selective fading is the name given to a form of fading characterised by severe modulation distortion in which the path length in the ionosphere varies with frequency to such an extent that the various sideband frequencies are differently affected. It is most severe when ground and sky waves are of comparable intensity.

Scintillations are rapid fluctuations in amplitude, phase and angle-of-arrival of tropospheric signals, produced by irregular small-scale variations in refractive index. The term is also used to describe irregular fluctuations on HF signals transmitted through the ionosphere from satellites and other sources outside the Earth.

Diversity Reception

The effects of fading can be countered to a certain extent by the use of more than one receiving system coupled to a common network which selects at all times the strongest of the outputs available.

There are three principal versions in common use:

(a) *space diversity*, obtained by using antennas which are so positioned as to receive different combinations of components in situations where multi-path conditions exist;

(b) *frequency diversity*, realised by combining signals which have been transmitted on different frequencies; and

(c) polarisation diversity, where two receivers are fed from antennas having different planes of polarisation.

It is unlikely that any of these systems have any application in normal amateur activities, but they may be of interest in connection with research projects relating to the amateur bands.

SOLAR AND MAGNETIC INFLUENCES

The Sun

Our Sun is at the centre of a complex system consisting of nine major planets (including ours), five of which have two or more attendant moons, of several thousand minor planets (or asteroids), and of an unknown number of lesser bodies variously classed as comets or meteoroid swarms.

It is a huge sphere of incandescence of a size which is equivalent to about double our moon's orbit around the Earth, but, despite appearances to the contrary, it has no true 'edge' because nearly the whole of the Sun is gaseous and the part we see with apparently sharp boundaries is merely a layer of the solar atmosphere called the *photosphere* which has the appearance of a bright surface, preventing us from seeing anything which lies beneath.

Beyond the photosphere is a relatively cooler, transparent layer called the *chromosphere*, so named because it has a bright rose tint when visible as a bright narrow ring during total eclipses of the Sun. From the chromosphere great fiery jets of gas, known as *prominences*, extend. Some are slow changing and remain suspended for weeks, while others, called *eruptive prominences*, are like narrow jets of fire moving at high speeds and for great distances.

Outside the chromosphere is the *corona*, extending a distance of several solar diameters before it becomes lost in the general near-vacuum of interplanetary space. At the moment of totality in a solar eclipse it has the appearance of a bright halo surrounding the Sun, and at certain times photographs of it clearly show it being influenced by the lines of force of the solar magnetic field.

The visible Sun is not entirely featureless - often relatively dark *sunspots* appear and are seen to move from east to west, changing in size, number and dimensions as they go. They are of interest for two reasons: one is that they provide reference marks by which the angular rotation of the Sun can be gauged, the other that by their variations in number they reveal that the solar activity waxes and wanes in fairly regular cycles. The Sun's rotation period has been found to vary with latitude, with its maximum angular speed at the equator. The *mean synodic rotation period* - the time required for the Sun to rotate until the same part faces the Earth is 27.2753 days.

The Sun's rotations have been numbered since the year 1853 (the start of rotation 2039 is dated 18 January 2006), and observations of solar features are referred to an imaginary network of latitudes and longitudes which rotates from east to west as seen from the Earth. Remember, when looking at the mid-day Sun from the UK, north is at the top, the east limb is on the left side and the west limb on the right. A long-persistent feature which first appears on the east limb is visible for about 13.5 days before it disappears from sight over the west limb.

For many purposes it is more convenient to refer the positions of noteworthy features to a related, but stationary, set of co-ordinates, the *heliocentric latitude and longitude*, in which locations are described with respect to the centre of the visible disc. An important statistic relating to sunspots is the time of their *central meridian passage* (CMP).

Radio telescopes detect features which are usually situated in the vicinity of the solar corona, and some of them reveal disturbances beyond the limbs of the visible disc. They often travel across the face of the Sun at a faster rate than any spots beneath.

The apparent diameter of the Sun varies with the choice of radio frequency. Because the lower frequencies come from the outer parts of the corona, the width of their Sun appears to be large. At high frequencies the sources are situated below the level which provides the visible disc so that the width of the Sun then appears to be small. The optical and radio frequency diameters are similar at a frequency of 2800MHz (10.7cm wavelength), so that was chosen to provide the daily *solar flux* measurements. The daily flux figure, which can vary between about 65 at solar minimum and 300 at maximum, is now more frequently used than the sunspot number as a measure of the Sun's ionising capability. The higher the figure the better in terms of HF propagation (subject to the level of geomagnetic activity). A 90-day average of the daily figures has been found to be a good basis for home computer prediction programs.

Sunspots

Sunspots are the visible manifestation of very powerful magnetic fields; adjacent spots often have opposing polarities. These intense magnetic fields also produce *solar flares*, which are emissions of hydrogen gas. They are responsible also for the ejection of streams of charged particles and X-rays.

Sunspot numbers have been recorded for over 200 years and it has been found that their totals vary over a fairly regular cycle occupying around 22 years where magnetic polarity is the criterion, or 11 years if only the magnitude of activity is considered.

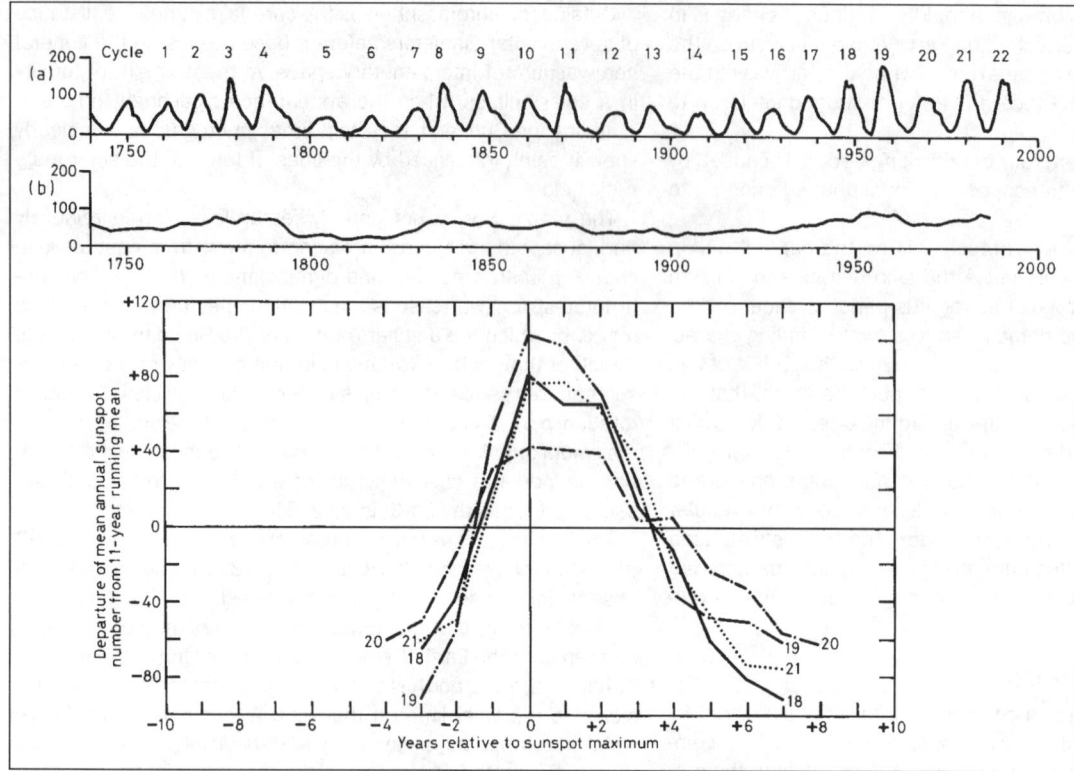

Fig 12.15: (a) Annual relative sunspot numbers, the means of monthly means of daily values. (b) The 11-year running means of the annual numbers. The succession of points reveals the underlying trend in solar activity. (c) A comparison of the four cycles most recently completed. The steep rise to maximum and the relatively slow decline thereafter are characteristic of all sunspot cycles

The 11-year peaks are known as *sunspot maxima*; the intervening troughs are *sunspot minima*.

The rise and fall times are not equal, though. Four years and seven years respectively are typical, although each cycle differs from the others in both timing and maximum value, as may be seen in **Fig 12.15**. At sunspot minimum the Sun may be completely spotless for weeks or months - or even, during the Maunder minimum in the 17th century, for years.

Tables of daily *relative sunspot numbers* are prepared monthly at the Sunspot Index Data Centre (SIDC) in Brussels, from information supplied by a network of participating observatories. It is widely supposed by radio amateurs that those figures indicate the number of visible spots but that is not so. In accordance with a formula devised by Dr Wolf in Zurich (hence the description *Wolf number*, still used professionally) the relative sunspot number, R, is found from the expression:

$$R = k (10g + t)$$

where k is a regulating factor that keeps the series to a uniform standard, g is the number of spot groups, and t is the total number of spots.

Daily figures obtained from the Ursigram messages (see later) use unity as the value of k, so, for example, Boulder figures obtained from that source record just 10 times the number of groups plus the total number of spots. Note also that those figures are provisional because they will have been prepared in haste to meet a deadline.

When the figures from the participating observatories have been combined in Brussels, a value of k which is less than unity will have been applied. That figure does not appear in the tables and it may well be subject to frequent variation, but it is currently around 0.7.

The Brussels figures are issued twice, first provisionally as soon as possible, then definitively after more careful scrutiny. From the monthly means of the definitive values a smoothed index R_{12} is obtained. This is the arithmetic mean of 12 successive monthly means, the result being ascribed to the period at the centre of the sample. In order to make that fall in the middle of the month, rather than between months, 13 months are taken but the first and the last are given only half weight in the calculations. From the nature of the 12-month running average - for that is what it is - it must be evident that R_{12} (which is the 'sunspot number' called for in ionospheric prediction programs) never reaches the peaks and troughs of the individual monthly means and it falls far short of the maxima of the daily values.

To put all these different versions of the 'sunspot number' into perspective we have only to look at a specific example, say the month which contained the peak of solar cycle 22, which was June 1989. The Boulder figures on the Ursigrams reached 401, and their monthly mean was 297. The SIDC Brussels definitive figure at maximum was 265, the monthly mean 196. However, the smoothed figure R_{12} for the month was only 158 and, remember, it is that one that you need for prediction programs, not the 401 from Boulder. You should be aware also that the latest smoothed figure available is always six months behind the current date, so a figure for this month has been a forecast six months ahead.

It is also of interest to observe that the peak Boulder figure was made up of 18 groups which, between them, contained a total of 221 spots. Put those figures in the formula and you come up with 401, the figure reported. The three largest groups accounted for 86, 53 and 26 spots respectively, and none of the others contained more than 7.

The Solar Wind

The solar corona was described in the last section as extending outwards until it becomes lost in interplanetary space. In fact it does not become lost at all, but turns into a tenuous flow of hydrogen which expands outwards through the solar system, taking with it gases evaporating from the planets, fine meteoric dust and cosmic rays. It becomes the *solar wind*.

Near the Sun the corona behaves as a static atmosphere, but once away it gradually accelerates with increasing distance to speeds of hundreds of kilometres per second. The gas particles take about nine days to travel the 150,000,000km to the Earth,

carrying with them a magnetic field (because the gas is ionised) which assumes a spiral form due to the Sun's rotation. It is the solar wind, rather than light pressure alone, which is responsible for comets' tails flowing away from the Sun, causing them to take on the appearance of celestial wind-socks.

The existence of the solar wind was first detected and measured by space vehicles such as the Luniks, Mariner II and Explorer X, which showed that its speed and turbulence are related to solar activity. Regular measurements of solar wind velocity are now routine.

There is thus a direct connection between the atmosphere of the Sun and the atmosphere of the Earth. In the circumstances it is hardly surprising that solar events, remote though they may at first seem, soon make their effects felt here on Earth.

The Earth's Magnetosphere

It is well known that the Earth possesses a magnetic field, for most of us have used a compass, at some time or another, to help us to get our 'bearings'. We know from such experiences that the field appears to be concentrated at a point somewhere near the north pole (and are prepared to believe that there is another point of opposing polarity somewhere near the south pole). Popular science articles have familiarised us with a picture of field lines surrounding the Earth like a section of a ring doughnut made up of onion-like layers.

Because the particles carried by the solar wind are charged their movement produces a magnetic field which interacts with the geomagnetic field. A blunt shock-wave is set up, called the *magnetosheath*, and the wind flows round it, rejoining behind where the field on the far side is stretched in the form of a long tail, the overall effect being reminiscent of the shape of a pear with its stalk pointing away from the Sun. The region within the magnetosheath, into which the wind does not pass, is called the *magnetosphere*.

On the Earthward side the magnetosphere merges into the ionosphere. Inside the magnetosphere there are regions where charged particles can become trapped by geomagnetic lines of force in a way which causes them to oscillate back and forth over great distances. Particles from the solar wind can enter these regions (often called *Van Allen belts* after their discoverer) in some way, as yet not perfectly understood.

The concentration of electrons in the magnetosphere can be gauged from the ground by observations on *whistlers*, naturally-occurring audio-frequency oscillations of descending pitch which are caused by waves radiated from the electric discharge in a lightning flash. These travel north and south through the ionosphere and magnetosphere from one hemisphere to another along the magnetic lines of force. The various component frequencies propagate at different speeds so that the original flash (which appears on an ordinary radio receiver as an *atmospheric*) arrives at the observer considerably spread out in time. The interval between the reception of the highest and lowest frequencies is a function of the concentration of electrons encountered along the way.

These trapped particles move backwards and forwards along the geomagnetic field lines within the Van Allen belts and some collide with atoms in the ionosphere near the poles where the belts approach the Earth most closely. Here they yield up energy either as ionisation or illumination and are said to have been dumped. These dumping regions surround the two poles, forming what are called the *auroral zones*. The radius of the circular motion in the spirals (they are of a similar form to that of a helical spring) is a function of the strength of the magnetic field, being small when the field is strong. Electrons and protons per-

form their circular motions in opposite senses, and the two kinds of spiralling columns drift sideways in opposite directions, the electrons eastward, the protons westward, around the world. Because of the different signs on the two charges these two drifts combine to give the equivalent of a current flowing in a ring around the Earth from east to west.

This ring current creates a magnetic field at the ground which combines with the more-or-less steady field produced from within the Earth. We shall see later the sort of effects that solar disturbances have on the magnetosphere, the ionosphere, and the total geomagnetic field.

The Quiet Ionosphere

With the stage set to follow the antics of the ions and electrons deposited by dumping we must pause again, this time to examine the normal day-to-day working of the ionosphere, which is dependent for its chemistry on another form of incoming solar radiation.

The gas molecules in the Earth's upper atmosphere are normally electrically neutral, that is to say the overall negative charges carried by their orbiting electrons exactly balance the overall positive charges of their nuclei. Under the influence of ultraviolet radiation from the Sun, however, some of the outer electrons can become detached from their parent atoms, leaving behind overall positive charges due to the resulting imbalance of the molecular structure. These ionised molecules are called ions, from which of course stems the word 'ionosphere'.

This process, called *disassociation*, tends to produce layers of free electrons brought about in the following manner. At the top of the atmosphere where the solar radiation is strong there are very few gas molecules and hence very few free electrons. At lower levels, as the numbers of molecules increase, more and more free electrons can be produced, but the action progressively weakens the strength of the radiation until it is unable to take full advantage of the increased availability of molecules and the electron density begins to decline. Because of this there is a tendency for a maximum (or peak) to occur in the production of electrons at the level where the increase in air density is matched by the decrease in the strength of radiation. A peak formed in this way is known as a *Chapman layer*, after the scientist who first outlined the process.

The height of the peak is determined not by the strength of the radiation but by the density/height distribution of the atmosphere and by its capability to absorb the solar radiation (which is a function of the UV wavelength), so that the layer is lower when the radiation is less readily absorbed. The strength of the radiation affects the rate of production of electrons at the peak, which is also dependent on the direction of arrival. The electron density is greatest when the radiation arrives vertically and it falls off as a function of zenith distance, being proportional to $\cos \chi$, where χ (the Greek letter chi) is the angle between the vertical and the direction of the incoming radiation. As $\cos \chi$ decreases (ie when the Sun's altitude declines) a process of recombination sets in, whereby the free electrons attach themselves to nearby ions and the gas molecules revert to their normal neutral state.

Experimental results have led to the belief that the E-layer (at about 120km) and the F1-layer (at about 200km) are formed according to Chapman's theory as a result of two different kinds of radiation with perhaps two different atmospheric constituents involved.

The uppermost layer F2 (around 400km), which normally appears only during the day, does not follow the same pattern and is thought to be formed in a different way, perhaps by the

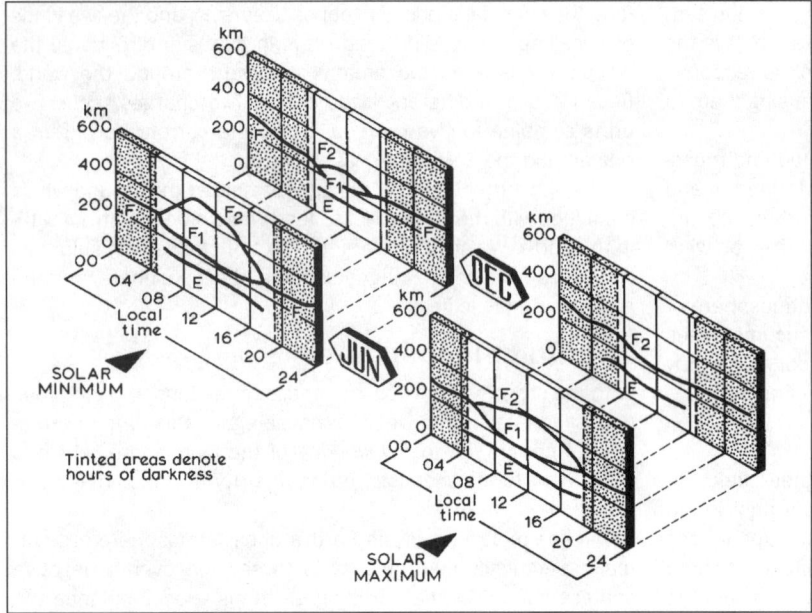

Fig 12.16; Typical diurnal variations of layer heights for summer and winter at minimum and maximum states of the solar cycle

diffusion of ions and electrons, but there are still a number of anomalies in its behaviour which are the subject of current investigation. These include the *diurnal anomaly*, when the peak occurs at an unexpected time during the day; the *night anomaly*, when the intensity of the layer increases during the hours of darkness when no radiation falls upon it; the *polar anomaly*, when peaks occur during the winter months at high latitudes, when no illumination reaches the layer at all; the *seasonal anomaly*, when magnetically quiet days in summer (with a high Sun) sometimes show lower penetration frequencies than quiet days in winter (with a low Sun); and a *geomagnetic anomaly* where, at the equinoxes, when the Sun is over the equator, the F2-layer is most intense at places to the north and south separated by a minimum along the magnetic dip equator. It is thought that topside sounding from satellites probing the ionosphere from above the active layers may help to explain some of these anomalies in F2 behaviour.

Regular Ionospheric Layers

The various regular ionospheric layers were first defined by letters by Sir Edward Appleton who gave to the one previously known as the *Kennelly-Heaviside layer* the label 'E' because he had so marked it in an earlier paper denoting the electric field reflected from it, and to the one he had discovered himself the letter 'F', rather than call it the *Appleton layer*, as some had done. To the band of absorption below thus naturally fell the choice of the letter 'D', although this was generally referred to as a 'region' rather than a 'layer' because its limits are less easy to define.

From comments already made it will be appreciated that the regular ionospheric layers which these letters define exhibit changes which are basically a function of day and night, season and solar cycle.

Most of our knowledge of the ionosphere comes from regular soundings made at vertical incidence, using a specialised form of radar called an *ionosonde* which transmits short pulses upwards using a carrier which is continuously varied in frequency from the medium-wave broadcast band through the HF bands to an upper limit of about 20MHz, but beyond if conditions warrant.

Reflections from the various layers are recorded photographically in the form of a graph called an *ionogram*, which displays *virtual height* (the apparent height from which the sounding signal is returned) as a function or signal frequency.

Critical frequencies, where the signals pass straight through the layers, are read off directly. There are many such equipments in the world; the one serving the United Kingdom is located near Slough. It is under the control of the Rutherford Appleton Laboratory, which houses one of a number of World Data Centres to which routine measurements of the ionosphere are sent from most parts of the world.

The two sets of diagrams (**Fig 12.16** and **Fig 12.17**) summarise the forms taken by the diurnal variations in height and critical frequency for two seasons of the year at both extremes of the sunspot cycle. The actual figures vary very considerably from one day to the next, but an estimate of the expected monthly median values of maximum usable frequency and optimum working frequency between two locations at any particular year, month and time of day can be obtained from predictions .

The critical frequencies of the E and F1-layers are a function of R12, the smoothed SIDC Brussels (formerly Zurich) relative sunspot number (which is predicted six months in advance for this purpose), and the cosine of the zenith distance χ, the angle

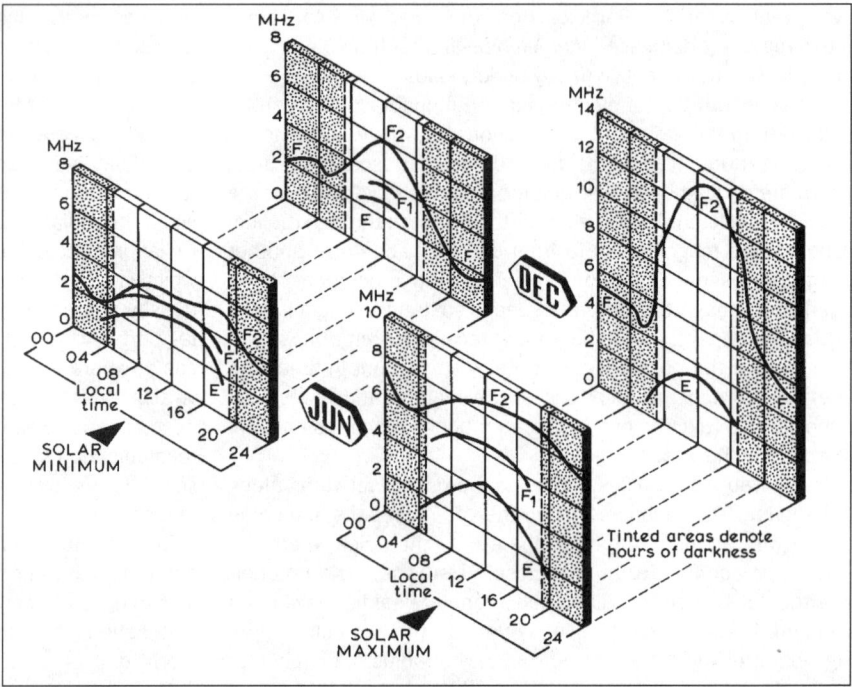

Fig 12.17: Typical diurnal variations of F-layer critical frequencies for summer and winter at the extremes of the solar cycle

	Jan	Feb	Mar	Apr	May	Jun	Jul	Aug	Sep	Oct	Nov	Dec
$R_3 = 0$	5.3	5.1	4.75	4.80	5.10	5.03	4.72	4.75	4.90	5.69	5.58	5.32
$R_3 = 150$	12.17	12.42	11.67	9.88	8.23	7.70	7.73	7.68	8.81	11.26	12.93	12.53

Table 12.6: F2-layer critical frequencies at Slough. The two rows show mean median value of f_0F2 in megahertz for three month weighted mean sunspot numbers of 0 and 150

	Distance			
Layer	1000km	2000km	3000km	4000km
Sporadic E	4.0	5.2	-	-
E	3.2	4.8	-	-
F1	2.0	3.2	3.9	-
F2 winter	1.8	3.2	3.7	4.0
F2 summer	1.5	2.4	3.0	3.3

Table 12.7: MUF (maximum usable frequency) factors for various distances assuming representative heights for the principal layers

the Sun makes with the local vertical, and thus, to a first approximation:

$$f_0E = 0.9 \, [(180 + 1.44R) \cos \chi]^{0.25}$$

(usually to within 0.2MHz of the observed values), and:

$$f_0F_1 = (4.3 + 0.01R) \cos^{0.2}\chi$$

which is less accurate because of uncertainty in the value of the exponent which varies with location and season.

The F2-layer is the most important for HF communication at a distance, but is also the most variable. It is subject to geomagnetic control which impresses a marked longitudinal effect on the overall world pattern. causing it to lag behind the sub-solar point so as to give maximum values in critical frequency during the local afternoon.

The F2 critical frequency f_0F2 varies with the solar cycle, as shown in **Table 12.6**, which shows monthly median values for Slough, applicable to sunspot numbers of 0 and 150. In recent years it has been found possible to predict the behaviour of the F2-layer by extrapolation several months ahead, using an index known as *IF2*, which is based on observations made at about 10 observatories.

The MUF which can be used on a particular circuit may be calculated from the critical frequency of the appropriate layer by applying the relationship:

$$MUF = F/\sec \varphi$$

where φ is the angle that the incident ray makes with the vertical through the point of reflection at the layer. The factor (sec φ) is called the *MUF factor*; it is a function of the path length if the height of the layer is known. **Table 12.7** shows typical figures obtained by assuming representative heights.

To an operator, the *optimum working frequency* (OWF) is the highest (of those available) which does not exceed the MUF. As will be seen later, both MUF and OWF take on different meanings in the context of ionospheric predictions.

There is a lower limit to the band of frequencies which can be selected for a particular application. This is set by the *lowest usable frequency* (LUF), below which the circuit becomes either unworkable or uneconomical due to the effects of absorption and the level of radio noise. Its calculation is quite a complicated process beyond the scope of this survey.

It is often useful to be able to estimate the radiation angle involved in one- or two-hop paths via the E and F2-layers. **Fig 12.18**, also prepared for average heights, accomplishes this. It is a useful rule-of-thumb to remember that the maximum one-

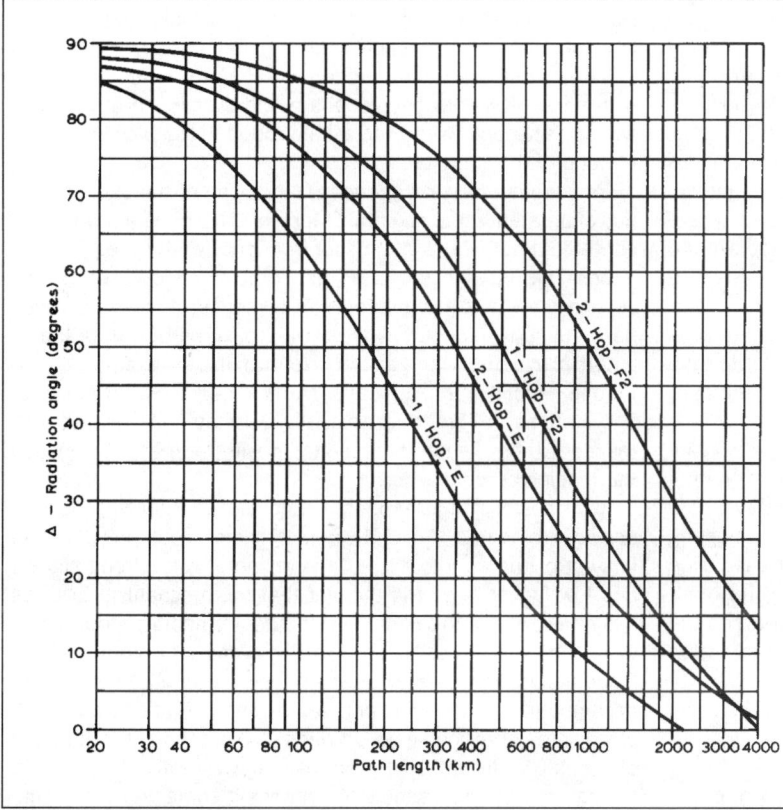

Fig 12.18: Radiation angle involved in one-hop and two-hop paths via E and F2-layers. (From NBS publication *Ionospheric Radio*

hop E range is 2000km and that the useful two-hop E range, twice that (4000km), is also the one-hop F2 range. Of course, all extreme ranges require very low angles of take-off, almost unachievable for radio amateurs.

Irregular Ionisation

Besides the regular E, F1 and F2-layers there are often more localised occurrences of ionisation which make their contribution to radio propagation. They generally occur around the heights associated with the E-layer and often the effects extend well into VHF, although the regular E-layer can never be effective at frequencies of 30MHz or more.

Sporadic E (Es) has been observed at HF on ionospheric sounding apparatus since the early 'thirties. It has been shown to take the form of clouds of high density of ionisation, forming sheets perhaps a kilometre deep and some 100km across in a typical instance, and appearing at a height of 100-120km.

However, strangely, over the years not one of the participating observatories has ever found evidence of a layer having sufficient electron density to support propagation at 144MHz - yet radio amateurs make use of something that behaves as though it was sporadic E many days of the year.

It is mainly a summertime phenomenon in May to August (the months without an 'r' in them) so far as the Northern Hemisphere is concerned, but there is also some activity during the latter part of the year. Distances worked are not often less than 500km, mainly between 1000 and 2000km, and usually with good-quality, steady, strong signals. Two-hop Es has been observed on occasion, but that is more likely to involve two small clouds rather than one big one. Also on record are cases where 2000km has been exceeded by the aid of tropospheric enhancement at one end of the path. (Generally, VHF Es and tropospheric modes are quite separate and a list of their distinguishing features will be found in **Table 12.8**.)

The numbers and durations of sporadic-E events decrease with frequency. To date, the highest recorded frequency appears to be 220MHz but that may not necessarily be the absolute limit. Due to the mystery surrounding the mechanism of the mode at VHF, many observers make a long-term specific study of Es, and there has long been an International Amateur Radio Union co-ordinator whose task is to guide national societies into setting up useful co-operative research projects.

Auroral E (Ar) is closely connected with geomagnetic disturbances. At times of high activity (popularly known as *magnetic storms*), the regions around the north and south poles where visual aurora are commonplace (the *auroral ovals*) expand towards the equator, taking with them the capability of returning VHF signals that have been directed towards them.

It used to be thought that antennas had to be turned towards the north (some even said the magnetic north) in order to take advantage of this mode but, thanks to careful observing by a group of dedicated amateurs over a period of many years, it is now known that optimum bearings can and do change considerably during an auroral event, and that from the UK a gradual swing towards the east may be expected as the activity develops. Radar measurements have shown that the reflecting regions are usually around 110km in height, but it should be noted that visual aurora extends very much beyond that. However, there is a quite good general relationship between the radio and visual forms of aurora, although attempts to match details on pen recordings of received signals against observations of changes in the structure of auroral forms seen from the transmitting site have been disappointing.

The signal paths are of necessity angled, often with one leg much longer than the other, the two antennas being directed

Tropospheric propagation	Sporadic-E propagation
May occur at any season	Mainly May, June, July and August
Associated with high pressure, or with paths parallel to fronts	No obvious connection with weather patterns
Gradual improvement and decline of signals	Quite sudden appearance and disappearance
Onset and decay times similar over a wide range of frequencies	Begins later and ends earlier as radio frequency increases
Observed at VHF, UHF, SHF	Rarely above 200MHz
Area of enhancement relatively stable for several hours at a time	Area of enhancement moves appreciably in a few hours
May last a week or more	Duration minutes or hours, never days
Wide range of distances with enhanced signals at shorter ranges	Effects mainly at 1000-2000km. No associated enhancement at short ranges

Table 12.8: Comparative characteristics of tropospheric and sporadic-E propagation

towards a common reflection point that is at a latitude higher than that of either station (forward scatter is unlikely). Stations in central Europe are able to work with beam headings considerably west of north; that sector is of little use to UK operators because there are no available contacts in the North Atlantic.

In Europe the signals have a raw, rasping tone that is readily recognised again, once identified. It is said to differ in character from the tone of auroral signals met with in North America.

A radio auroral event typically begins in the afternoon and may appear to be over by the end of the afternoon, but many events exhibit two distinct phases, and the evening phase is often the better in terms of DX worked, partly due perhaps to the greater number of stations likely to be active at that time. The event will frequently finish with dramatic suddenness, just as some of the longest paths are being achieved.

A book outlining the theory of auroral-E propagation, together with an analysis of observations made over much of the time since the second world war, will be found in the bibliography at the end of this chapter.

Trans-equatorial propagation (TEP) is one of the success stories of amateur radio research. Much of the pioneering work was carried out at 50MHz, but higher-frequency working, eg 70MHz, is possible. Instances of TEP tend to favour the years of high solar activity, but it is present, on a reduced time scale, even near sunspot minimum. Paths are typically 3000-9000km in length, usually with a north/south bias: examples are Europe/Africa, Japan/Australia, North America/South America. The stations in contact are usually symmetrically located with respect to the magnetic dip equator, with the path in between them being perpendicular to it.

Two types of TEP have been recognised. One shows a peak in activity at around 1700-1900 local time. That provides the longest of the contacts (9000km or more) with strong signals and low fading rate. It is thought that the mechanism involves two reflections at the F region without intermediate ground contact. The other tends to peak in the evening at around 2000-2300 local time. Signal strengths are high but there is an accompaniment of deep and rapid flutter. Paths are shorter than for the afternoon type, perhaps no more than 6000km, and opinions are divided as to the mechanism involved. The rapid fading characteristic seems to connect in some way with equa-

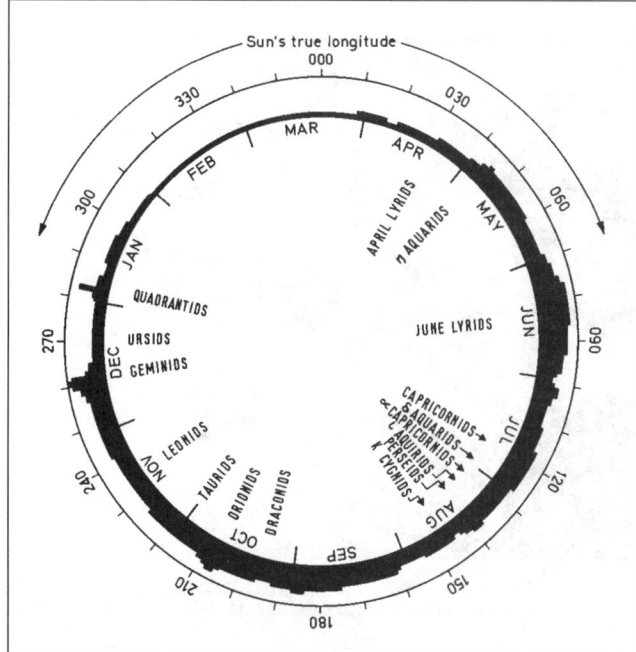

Fig 12.19: Seasonal variation of meteor activity, based on a daily relative index. Prepared from tables of 24h counts made by Dr Peter Millman, National Research Council, Ottawa. The maximum rate corresponds to an average of about 300 echoes per hour corresponding to an equivalent visual magnitude of 6 or greater

torial spread-F, which is a diffuse effect caused by irregularities in the electron density of the F region.

Before television moved from VHF to UHF and beyond, suitably placed radio amateurs in places like Greece were able to entertain visitors from less-fortunate parts of the world with an impressive display of ultra-DX TV, taking advantage of the opportunities offered by TEP.

Operators in southern Europe make use of a form of VHF propagation in which *field-aligned irregularities* (FAI) appear to play an important part. The effect has been observed at mid-latitudes both on the Continent and in North America.

The FAI 'season' runs very closely parallel to that of VHF sporadic E, May to late August, and some instances have been known to follow conventional Es openings.

The mode is characterised by signals arriving away from the expected great-circle bearing. The scattering area responsible is apparently very small and some elevation of perhaps 10-15 degrees has been required (in Italy) to find it. High-power transmitters and low-noise receivers are essential.

Italian amateurs recognise two distinct areas which appear to vary but little from one occasion to the next. One, located in the west of Switzerland, provides 2m contacts with Spain, southern France, Hungary and Yugoslavia. The other is located over Hungary itself and that provides Italian amateurs with openings to the Balkan peninsula.

Although FAI would appear to have very little direct application for UK amateurs at present, a nodding acquaintance with it might lead to fresh discoveries about its capabilities. One theory is that it is associated with anomalies in the Earth's magnetic field. Central Europe is not the only place with those.

The more-conventional *ionospheric scatter mode* is one that is developed commercially to provide communications over some 800-2000km paths, using ionisation irregularities at a height of about 85km, which is in the D-region. The frequencies

used were between 30 and 60MHz but, at the top of the range especially, very high powers were necessary.

Received signals were weak, but they had the advantage of being there during periods of severe disruption on the HF bands. The systems that were set up have fallen from favour nowadays and satellite transponders are commonly used instead for much of the traffic. Any amateur involvement would have to be limited to the 50MHz band, but the high power involved suggests that this is now one for the history books.

Meteoric ionisation is caused by the heating to incandescence by friction of small solid particles entering the Earth's atmosphere. This results in the production of a long pencil of ionisation extending over a length of 15km or more, chiefly in the height range 80 to 120km. It expands by diffusion and rapidly distorts due to vertical wind shears. Most trails detected by radio are effective for less than one second, but several last for longer periods, occasionally up to a minute and very occasionally for longer. There is a diurnal variation in activity, most trails occurring between midnight and dawn when the Earth sweeps up the particles whose motion opposes it. There is a minimum around 1800 local time, when only meteors overtaking the Earth are observed. The smaller *sporadic meteors*, most of them about the size of sand grains, are present throughout the year, but the larger *shower* meteors have definite orbits and predictable dates. **Fig 12.19** shows the daily and seasonal variation in meteor activity, based on a 24h continuous watch. Intermittent communication is possible using meteoric ionisation between stations whose antennas have been prealigned to the optimum headings of 5 to 10 degrees to one side of the great-circle path between them. Small bursts of signal, referred to as *pings*, can be received by meteor scatter from distant broadcast (or other) stations situated 1000-1200km away.

Geomagnetism

The Earth's magnetic field is the resultant of two components, a *main field* originating within the Earth, roughly equivalent to the field of a centred magnetic dipole inclined at about 11 degrees to the Earth's axis, and an *external field* produced by changes in the electric currents in the ionosphere.

The main field is strongest near the poles and exhibits slow secular changes of up to about 0.1% a year. It is believed to be due to self-exciting dynamo action in the molten metallic core of the Earth. The field originating outside the Earth is weaker and very variable, but it may amount to more than 5% of the main field in the auroral zones, where it is strongest. It fluctuates regularly in intensity according to annual, lunar and diurnal cycles, and irregularly with a complex pattern of components down to micropulsations of very short duration.

Certain observatories around the world are equipped with sensitive *magnetometers* which record changes in the field on at least three different axes, the total field being a vector quantity having both magnitude and direction. In the aspect of analysis which is of interest to us the daily records, called *magnetograms*, are read-off as eight *K-indices*, which are measures of the highest positive and negative departures from the 'normal' daily curve during successive threehourly periods, using a quasi-logarithmic scale ranging from 0 (quiet) to 9 (very disturbed). The various observatories do not all use the same scale factors in determining *K*-indices; the values are chosen so as to make the frequency distributions similar at all stations. Most large magnetic disturbances are global in nature and appear almost simultaneously all over the world. The more frequently used *planetary K index* is formed by combining the K figures for a dozen selected observatories. It is more finely graded: 0, 1-, 1o, 1+, 2-, 2o, 2+, . . . 9-, 9o, 9+. It is formed by a combination of K-

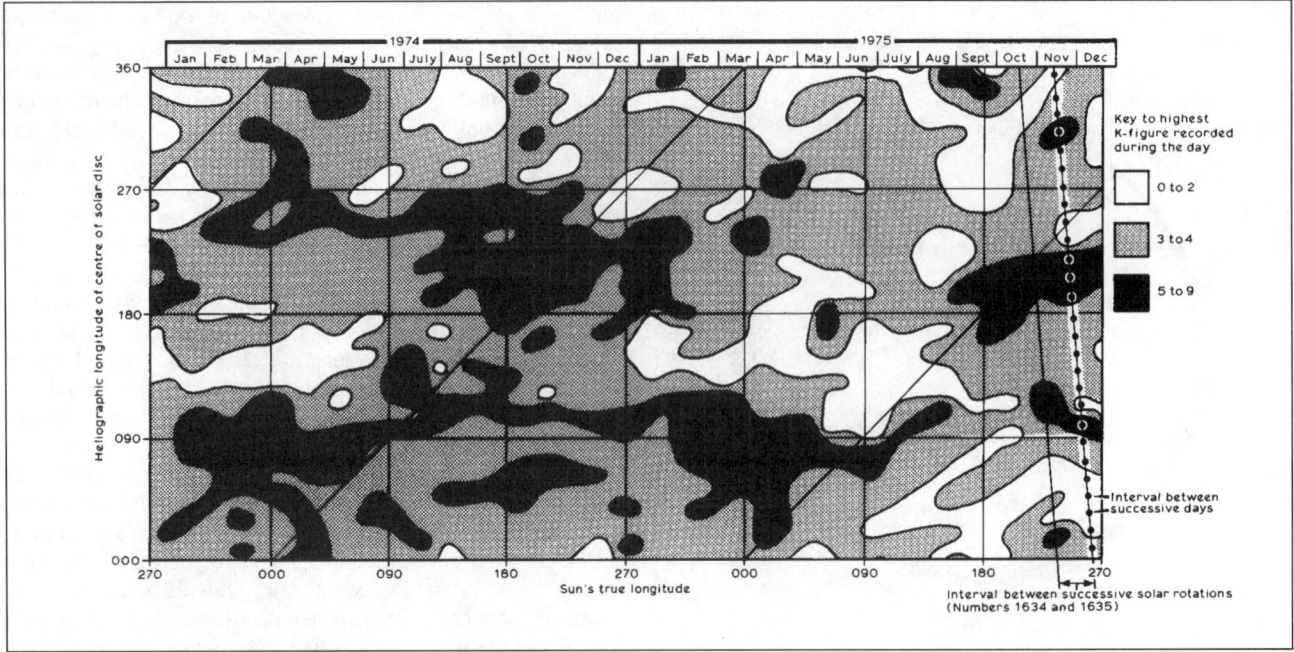

Fig 12.20: Geomagnetic activity diagram. The black areas indicate sequences where a K-figure of 5 or more was recorded at Lerwick Observatory. A horizontal line on this diagram denotes a recurrence period of 27 days, linked to the Sun's rotation period as seen from the Earth; the diagonal lines show the slope associated with a 25-day rotation period, such as the Sun has in relation to a fixed point in space

figures from 12 selected observatories. Indices of 5 or more may be regarded as being indicative of magnetic storm conditions.

K- and K_p-indices are based on quasi-logarithmic scales which place more emphasis on small changes in low activity than high. For some purposes it is more convenient to work with a linear scale, particularly if the values are to be combined to derive averages, as of the day's activity, for example. This is known as the A-index, recording the daily equivalent amplitude on a linear scale running from 0 to 400 - the maximum figure for the most severe storms. As with Kp, the daily Ap figure combines results from a selected group of observatories. Both indices are widely used as a shorthand expression of geomagnetic activity.

There is a tendency for occasions of abnormally high geomagnetic activity (and in consequence auroral activity at VHF and UHF) to recur at intervals of approximately 27 days, linked to the *solar synodic rotation period*. The chart shown in **Fig 12.20** clearly shows some long-persistent activity periods over the two-year interval 1974/75. A blank chart showing these coordinates to cover the current year with an overlap (known as a *solar rotation base map*) is published in the information section of the *RSGB Yearbook*.

The original diagram on which it is based was prepared by plotting the highest K-figure for each day on the spot determined by the longitude of the Sun's central meridian facing the Earth (thus a measure of the solar rotation), and a parameter called the *Sun's true longitude*, which indicates the position of the Earth around its orbit. Successive rotations build up a raster of daily figures in the way shown by the dots along the sloping right-hand edge, and the resulting chart should really be considered as being cylindrical, with the upper and lower edges brought together. The black areas surround the days when a K-figure of five or more was recorded - magnetic storm days - and the unshaded areas enclose relatively quiet days when the K-index was two or less. 27-day recurrences are clearly marked on this section of the record but occasionally there are periods when there is a marked tendency for storms to recur after an interval

which appears to be linked to the Sun's rotation period relative to the stars - indicated by the slope of the diagonal across the diagram.

Periods of high geomagnetic activity tend to occur somewhat more frequently around the equinoxes, while periods around the summer and winter solstice are more likely to be quieter. The geomagnetic cycle, like the solar cycle, lasts around eleven years, though tending to lag it by roughly a year or eighteen months. However, major disturbances can occur at any time of the year and any stage in the cycle - and quite often do!

Ionospheric Disturbances

Like so much that affects radio propagation ionospheric disturbances have their genesis in the Sun. Recent years have shed much light on the mechanisms involved, though they may not as yet be wholly understood. From time to time flares occur, powerful explosions that hurl vast amounts of highly charged particles out of the Sun.

This material may escape the Sun by means of holes in the Sun's outer corona and, if the hole suitably positioned in relation to Earth, the effects will reach Earth. Initially electromagnetic radiation in the form of X-rays, ultraviolet, visible light and radio waves between 3cm and 10m in length will which reach the Earth in about eight minutes. The X-rays and UV light cause immediate increases in the D-layer ionisation, leading to *short-wave* (or *Dellinger*) *fade-outs* which may persist for anything up to two hours. The effects of an SID affect the lower bands first, higher frequencies later, sometimes wiping out even high-powered transmissions for many hours - though a couple of hours is more common. Recovery works in the opposite direction, with the higher frequencies regaining propagation first. Sometimes prolonged bursts of radio noise also occur. Other effects observed are a *sudden enhancement of atmospherics* (SEA), a *sudden absorption of cosmic noise* (SCNA), and *sudden phase anomalies* (SPA) on very low-frequency transmissions.

This is followed after a few hours by the arrival of *cosmic ray particles* and perhaps the onset of *polar-cap absorption* (PCA).

The main stream of particles arrives after an interval of 20-40h and consists of protons and electrons borne by the solar wind. High-speed coronal streams, not necessarily originating in flare activity) can travel at speeds which sometimes exceed 100km/sec. What happens next greatly depends on the strength and orientation of the interplanetary magnetic field - the intensity of which is expressed in nano-Tesla (nT) units, with a southerly orientation favouring coupling with Earth's magnetic field. Where this applies, particles reaching the Earth's magnetosphere manifest themselves in visible displays of aurora and in auroral backscatter propagation at VHF, occasionally extending into the UHF range. Also, a strong polar electrojet may flow into the lower ionosphere.Changes in the make-up of the trapping regions leads to variations in the circulating ring-current which leads to violent alterations in the strength of the geomagnetic field, bringing about the sudden commencement, which is the first indication of a magnetic storm.

Associated with the magnetic storms are ionospheric storms, and both may persist for several days. The most prominent features are the reduction in F2 critical frequencies (f_0F2) and an increase in D-region absorption. During the storm period signal strengths remain very low and are subject to flutter fading. The effects of an ionospheric storm are most pronounced on paths which approach the geomagnetic poles.Conversely, LF signal levels at 136kHz are usually enhanced by up to about 10dB at the peak of a solar flare.

TROPOSPHERIC PROCESSES

Because it is all around us the troposphere is the portion of the atmosphere which we ought to know best. We are dealing here with the Sun's output of electromagnetic radiation which falls in the infrared portion of the spectrum, between 10^{-6} and 10^{-5}m wavelength, is converted to heat (by processes which need not concern us here) and is distributed about the world by radiation, conduction and convection.

At this point our link between solar actions and atmospheric reactions breaks down, because the very variable nature of the medium, and the ease by which it can be modified both by topographical features and the differing thermal conductivities of land and sea, leads to the development of air masses having such widely contrasting properties that it becomes impossible to find a direct correlation between day-to-day climatic features and solar emissions. We must accept the fact that in meteorology 'chance' plays a powerful role and look to functions of the resulting weather pattern for any relationships with signal level, without enquiring too deeply into the way in which they may be connected with events on the Sun.

Pressure Systems and Fronts

The television weatherman provides such a regular insight into the appearance and progressions of surface pressure patterns that it would be wasteful of space to repeat it all here. Suffice it to record that there are two closed systems of isobars involved, known as *anticyclones* and *depressions* (or, less-commonly nowadays, *cyclones*) within or around which appear *ridges* of high pressure, *troughs* of low pressure (whose very names betray their kinship), and *cols*, which are slack regions of even pressure, bounded by two opposing anticyclones and two opposing depressions.

The most important consideration about these pressure systems, in so far as it affects radio propagation at VHF and above, is the direction of the vertical motion associated with them.

Depressions are closed systems with low pressure at the centre. They vary considerably in size, and so also in mobility, and

frequently follow one another in quick succession across the North Atlantic. They are accompanied by circulating winds which tend to blow towards the centre of the system in an anti-clockwise direction. The air so brought in has to find an outlet, so it rises, whereby its pressure falls, the air cools and in doing so causes the relative humidity to increase. When saturation is reached, cloud forms and further rising may cause water droplets to condense out and fall as rain. Point one: depressions are associated with rising air.

Anticyclones are generally large closed systems which have high pressure in the centre. Once established they tend to persist for a relatively long time, moving but slowly and effectively blocking the path of approaching depressions which are forced to go round them. Winds circulate clockwise, spreading outwards from the centre as they do, and to replace air lost from the system in this way there is a slow downflow called *subsidence* which brings air down from aloft over a very wide area. As the subsiding air descends its pressure increases, and this produces dynamical warming by the same process which makes a bicycle pump warm when the air inside it is compressed. The amount of water vapour which can be contained in a sample of air without saturating it is a function of temperature, and in this particular case if the air was originally near saturation to begin with, by the time the subsiding air has descended from, say, 5km to 2km, it arrives considerably warmer than its surroundings and by then contains much less than a saturating charge of moisture at the new, higher, temperature. In other words, it has become warm and dry compared to the air normally found at that level. Point two: anticyclones are associated with descending air.

In addition to pressure systems the weather map is complicated by the inclusion of *fronts*, which are the boundaries between two air masses having different characteristics. They generally arrive accompanied by some form of precipitation, and they come in three varieties: *warm*, *cold* and *occluded*.

Warm fronts (indicated on a chart by a line edged with rounded 'bumps' on the forward side) are regions where warm air is meeting cold air and being forced to rise above it, precipitating on the way.

Cold fronts (indicated by triangular 'spikes' on the forward side of a line) are regions where cold air is undercutting warm. The front itself is often accompanied by towering clouds and heavy rain (sometimes thundery), followed by the sort of weather described as 'showers and bright intervals'.

Occluded fronts (shown by alternate 'bumps' and 'spikes') are really the boundary between three air masses being, in effect, a cold front which has overtaken a warm front and one or the other has been lifted up above the ground.

It is perhaps unnecessary to add that there is rather more to meteorology than it has been possible to include in this brief survey.

Vertical Motion

It is a simple matter of observation that there is some correlation between VHF signal levels and surface pressure readings, but it is generally found to be only a coarse indicator, sometimes showing little more than the fact that high signal levels accompany high pressure and low signal levels accompany low pressure. The reason that it correlates at all is due to the fact that high pressure generally indicates the presence of an anticyclone which, in turn, heralds the likelihood of descending air.

The reason that subsidence is so important stems from the fact that it causes dry air to be brought down to lower levels where it is likely to meet up with cool moist air which has been

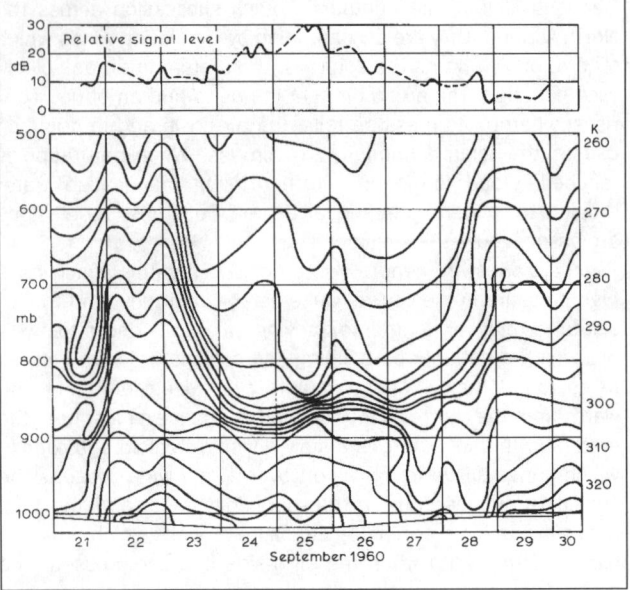

Fig 12.21: The relationship between variations in potential refractive index in the atmosphere and signal strengths over a long-distance VHF tropospheric path. (With acknowledgements to *J Atmos Terr Phys*, Pergamon Press.)

stirred up from the surface by turbulence. The result then is the appearance of a narrow boundary region in which refractive index falls off very rapidly with increasing height - the conditions needed to bring about the sharp bending of high-angle radiation which causes it to return to the ground many miles beyond the normal radio horizon. Whether you regard this in the light of being a benefit or a misfortune depends on whether you are more interested in long-range communication or in wanting to watch an interference-free television screen.

The essential part of the process is that the descending air must meet turbulent moist air before it can become effective as a boundary. If the degree of turbulence declines, the boundary descends along with the subsiding air above it, and when it reaches the ground all the abnormal conditions rapidly become subnormal - a sudden drop-out occurs. Occasionally this means that operators on a hill suffer the disappointment of hearing others below them still working DX they can no longer hear themselves. Note, however, that anticyclones are not uniformly distributed with descending air, nor is the necessary moist air always available lower down, but a situation such as a damp foggy night in the middle of an anticyclonic period is almost certain to be accompanied by a strong boundary layer. Ascending air on its own never leads to spectacular conditions. Depressions therefore result in situations in which the amount of ray bending is controlled by a fairly regular fall-off of refractive index. The passage of warm fronts is usually accompanied by declining signal strength, but occasionally cold fronts and some occlusions are preceded by a short period of enhancement.

To sum up, there is very little of value about propagation conditions which can be deduced from surface observations of atmospheric pressure. The only reliable indicator is a knowledge of the vertical refractive index structure in the neighbourhood of the transmission path.

Radio Meteorological Analysis

It remains now to consider how the vertical distribution of refractive index can be displayed in a way which gives emphasis to those features which are important in tropospheric propagation studies. Obviously the first choice would be the construction of

atmospheric cross-sections along paths of interest, at times when anomalous conditions were present, using values calculated by the normal refractive index formula. The results are often disappointing, however, because the general decrease of refractive index with height is so great compared to the magnitude of the anomalies looked for that, although they are undoubtedly there, they do not strike the eye without a search.

A closely-related function of refractive index overcomes this difficulty, with the added attraction that it can be computed graphically and easily, directly from published data obtained from upper-air meteorological soundings. It is called *potential refractive index K* and may be defined as being the refractive index which a sample of air at any level would have if brought *adiabatically* (ie without gain or loss of heat or moisture) to a standard pressure of 1000mb; see references [4] and [5].

This adiabatic process is the one which governs (among other things) the increase of temperature in air which is descending in an anticyclone, so that, besides the benefits of the normalising process (which acts in a way similar to that whereby it is easy to compare different-sized samples of statistics when they have all been converted to percentages) there is the added attraction that the subsiding air tends to retain its original value of potential refractive index all the time it is progressing on its downward journey. This means that low values of K are carried down with the subsiding air, in sharp contrast to the values normally found there. A cross-section of the atmosphere during an anticyclonic period, drawn up using potential refractive index, gives an easily-recognisable impression of this.

Fig 12.21 is not a cross-section, but a time-section, showing the way in which the vertical potential refractive index distribution over Crawley, Sussex, varied during a 10-day period in September 1960. There is no mistaking the downcoming air from the anticyclone and the establishment of the boundary layer around 850mb (about 1.5km). Note how the signal strength of the Lille television transmission on 174MHz varied on a pen recording made near Reading, Berkshire, during the period, with peak amplitudes occurring around the time when the layering was low and well-defined, and observe also the marked decline which coincided with the end of the anticyclonic period. Time-sections such as these also show very clearly the ascending air in depressions (although the K value begins to alter when saturation is reached) and the passage of any fronts which happen to be in the vicinity of the radiosonde station at ascent time.

For anyone interested in carrying out radiometeorological analysis at home - and, be assured, it is a very rewarding exercise in understanding the processes involved - the propagation chapter in the RSGB *VHF/UHF Handbook* will provide full details (see the bibliography at the end of this chapter).

PRACTICAL CONSIDERATIONS

Map Projections

Maps are very much a part of the life of a radio amateur, yet how few of us ever pause to wonder if we are using the right map for our particular purpose, or take the trouble to find out the reason why there are so many different forms of projection.

The cartographer is faced with a basic problem, namely that a piece of paper is flat and the Earth is not. For that reason his map, whatever the form it may take, can never succeed in being faithful in all respects - only a globe achieves that. The amount by which it departs from the truth depends not only on how big a portion of the globe has been displayed at one viewing, but on what quality the mapmaker has wanted to keep correct at the expense of all others.

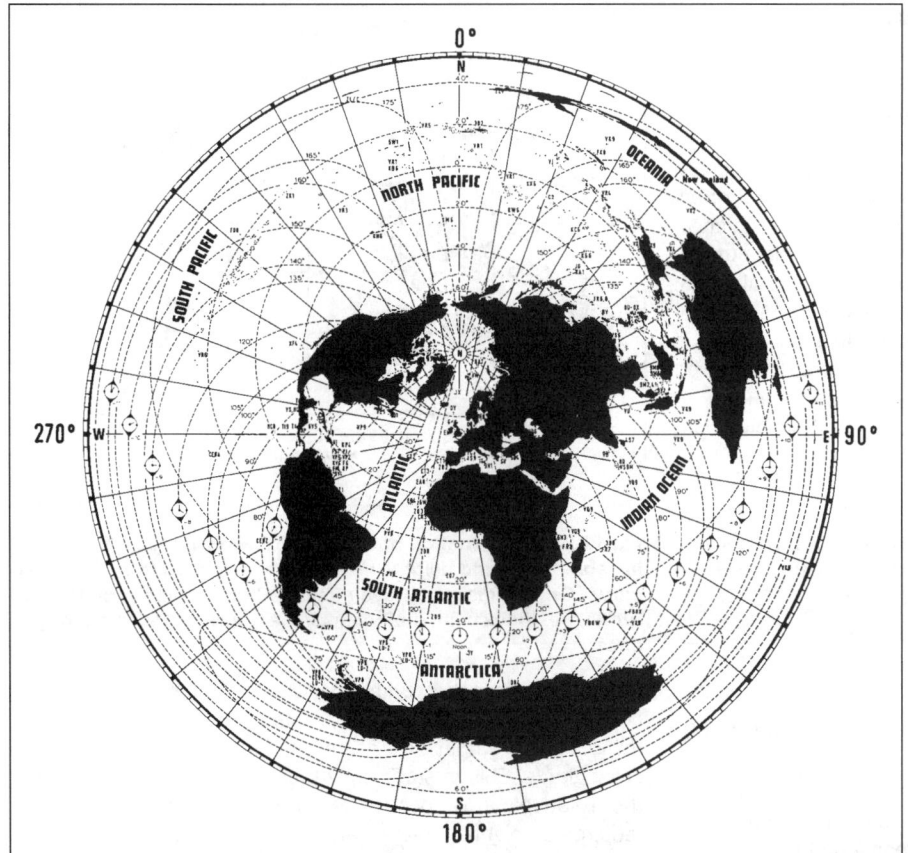

Fig 12.22: Example of an azimuthal equidistant (or great-circle) map. This map, available in a large size suitable for wall-mounting, shows the true bearing and distance from London of any place elsewhere in the world. (For magnetic bearings add 6° to the true bearing)

Projections can be divided into three groups: those which show areas correctly, described, logically enough, as *equal-area projections*; those which show the shapes of small areas correctly, known as *orthomorphic* or *conformal projections*; and those which represent neither shape nor area correctly, but which have some other property which meets a particular need.

The conformal group, useful for atlas maps generally, weather charts, satellite tracks etc includes the following:

Stereographic, where latitudes and longitudes are all either straight lines or arcs of circles, formed by projection on to a plane surface tangent either to one of the poles (*polar*), the equator (*equatorial*), or somewhere intermediate (*oblique*). Small circles on the globe remain circles on the map but the scale increases with increasing radius from the centre of projection.

Lambert's conformal conic, where all meridians are straight lines and all parallels are circles. It is formed by projection on to a cone whose axis passes through the Earth's poles.

Mercator's, where meridians and parallels are straight lines intersecting at right-angles. The meridians are equidistant, but the parallels are spaced at intervals which rapidly increase with latitude. It is formed by projection on to a cylinder which touches the globe at the equator. Any straight line is a line of constant bearing (a *rhumb line*, not the same thing as a *great circle*, which is the path a radio wave takes between two given points on the Earth's surface). There is a scale distortion which gets progressively more severe away from the equator, to such an extent that it becomes impossible to show the poles, but most people accept these distortions as being normal, because this is the best-known of all the projections.

Transverse Mercator is a modification of the 'classical' system, and is formed by projection on to a cylinder which touches the globe along selected opposing meridians. It therefore corresponds to an ordinary Mercator turned through 90 degrees, and is of value for displaying an area which is extensive in latitude but limited in longitude. A variant is the *universal transverse Mercator*, which forms the basis of a number of reference grids, including the one used on British Ordnance Survey maps.

The equal-area group is used when it is necessary to display the relative distribution of something, generally on a worldwide scale. It includes the following.

Azimuthal equal-area projection, having radial symmetry about the centre, which may be at either pole (*polar*), at the equator (*equatorial*), or intermediate (*oblique*). With this system the entire globe can be shown in a circular map, but there is severe distortion towards its periphery.

Mollweide's homolographic projection, where the central meridian is straight and the others elliptical.

Sinusoidal projection, where the central meridian is straight and the others parts of sine curves.

Homolosine projection, which is a combination of the previous two, with an irregular outline because of interruptions which are generally arranged to occur over ocean areas.

The final group includes the two following, which are of particular interest in propagation studies.

Azimuthal equidistant, centred on a particular place, from whence all straight lines are great circles at their true azimuths. The scale is constant and linear along any radius. Well known as a *great-circle map*, **Fig 12.22**.

Gnomonic, constructed by projection from a point at the Earth's centre on to a tangent plane touching the globe. Any straight line on the map is a great circle. Because the size of the map expands very rapidly with increasing distance from the centre they do not normally cover a large area. Often produced as a skeleton map on which a great circle can be drawn and used to provide a series of latitudes and longitudes by which the path can be replotted on a more detailed map based on a different projection.

Beam Heading and Locators

The shortest distance between two points on the surface of the Earth lies over the great circle that passes through them. This is easily done if you have a globe. All you need is to join the two locations with a tightly stretched thread. The shortest length of thread that can do this will be the great circle route and indicate the requisite beam heading to direct your signal the most effectively. (Irregularities in the distribution of ionization may on occasion result in signals being diverted off the great circle, but this should be understood as an exception to the general rule.). Alternatively, a great circle map along the lines of Fig 12.22 indicates optimum beam headings. For most working purposes the map works well for most UK operators. Operators far removed from London, or who simply wish to be more exact, can readily find freeware programs on the Internet that will supply beam headings and distances for any location.

Great-circle Calculations

It is sometimes useful to be able to calculate the great-circle bearing and the distance of one point from another, and the expressions which follow enable this to be done.

First label the two points A and B.

Then let L_a = latitude of point A

L_b = latitude of point B

L_0 = the difference in longitude between A and B

C = the direction of B from A, in degrees east or west from north in the northern hemisphere, or from south in the southern hemisphere.

D = the angle of arc between A and B.

It follows that:

$$\cos D = \sin L_a.\sin L_b + \cos L_a.\cos L_b.\cos L_0$$

D can be converted to distance, knowing that:

1 degree of arc = 111.2km or 69.06 miles

1 minute of arc = 1.853km or 1.151 miles

Once D is known (in angle of arc), then:

$$\cos C = \frac{\sin L_b - \sin L_a.\cos D}{\cos L_a.\sin D}$$

Note:

1. For stations in the northern hemisphere call latitudes positive.

2. For stations in the southern hemisphere call latitudes negative.

3. Cos L_a and cos L_b are always positive.

4. Cos L_0 is positive between 0 and 90°, negative between 90° and 180°.

5. Sin L_a and sin L_b are negative in the southern hemisphere.

6. The bearing for the reverse path can be found by transposing the letters on the two locations.

It is advisable to make estimates of the bearings on a globe, wherever possible, to ensure that they have been placed in the correct quadrant.

The IARU Locator

It is always nice to know where the other fellow lives and operators have been asking that question since the earliest days of operating. But all too often the answer has been something like '16k south' of some place your atlas does not deign to mention. That is particularly unsatisfactory if points in a contest or a personal distance record is at stake. Happily, the International Amateur Radio Union developed a system, originally called the Maidenhead Locator but now termed just Locator, that could be

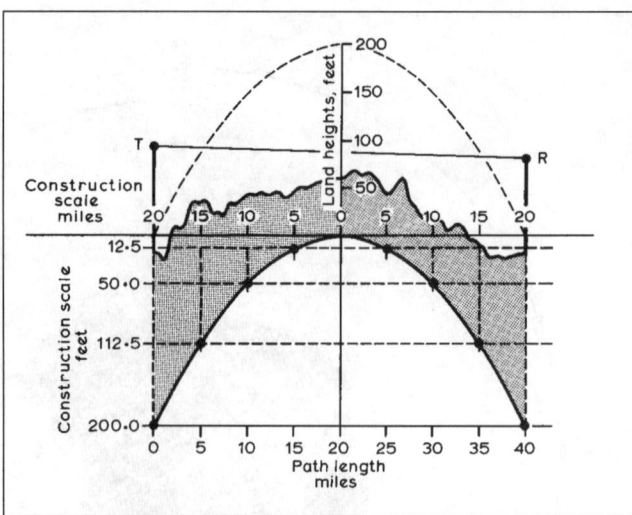

Fig 12.23: Construction of a 'four-thirds-earth' profile from data in Table 12.9. Land heights are measured upward from the lower curve. On this diagram, rays subjected to 'normal' variations of refractive index with height may be represented by straight lines

used anywhere in the world, in which information about the location is expressed in a group of six characters. The result defines the location to within 0.04 degrees of latitude and 0.08 degrees of longitude. This is sufficiently precise for most purposes; on the relatively rare occasions where greater exactitude is required an eight-character version is available.

The *RSGB Yearbook* contains a map of grid squares in Europe. Several programs for converting latitude and longitude data into 'grid locators' (and conversely) anywhere in the world, as well as the distance between any two points, are readily available on the Internet.

Plotting Path Profiles for VHF/UHF Working

For a tropospheric propagation study, it is standard practice to construct a path profile showing the curvature of the Earth as though its radius was four-thirds of its true value. This is so that, under standard conditions of refraction, the ray path may be shown as a straight line relative to the ups and downs of the intervening terrain.

It will be found convenient to construct the baseline of the chart using feet for height and miles for distance, for the reason that, by a happy coincidence, the various factors then cancel, leaving a very simple relationship that demands little more than mental arithmetic to handle.

Suppose that a profile is required for a given path 40 miles in length. First, decide on suitable scales for your type of graph paper (**Fig 12.23** was drawn originally with 1in on paper representing 100ft, height, and 10 miles, distance). Then construct a sea-level datum curve, taking the centre of the path as zero and working downwards and outwards for half the overall distance in each direction, using the expression $h = D^2/2$ to calculate points on the curve (**Table 12.9**). Remember that this works only if h is in feet and D is in miles.

Then prepare a height scale. This will be the construction scale reversed if the contours of your map are in feet, or its metric equivalent if it is one of the more recent surveys. Usually it will be sufficient to limit this scale to just the contour values encountered along the path. Plot the heights, corresponding to the contours, vertically above the datum curve (a pair of dividers may be found helpful here) and add the antenna heights at the

| | Horizontal scale | | | Vertical scale |
D (miles)	(inches)	D²	D²/2 = h	(inches)
5	0.5	25	12.5	0.125
10	1.0	100	50.0	0.500
15	1.5	225	112.5	1.125
20	2.0	400	200.0	2.000

Table 12.9: Points on the curve drawn in Fig 12.23

two terminal points. Draw a line through all the points and you are ready for business.

If you can draw a straight line from transmitting antenna to receiving antenna without meeting any obstacles along the way, then your path should be clear under normal conditions of refraction.

Ionospheric Predictions

The quality of a radio circuit is highest when it is operated at a frequency just below the maximum usable frequency (MUF) for the path. Three regions, E, F1 and F2, are considered in the determination of MUF.

For the F2-layer, which is responsible for most HF long-distance contacts, the MUF for paths less than 4000km is taken as being the MUF that applies to the mid-point. For paths longer than 4000km, the MUF for the path is the lower of the MUFs at the two ends along the path direction. For this purpose the end point locations are not those of the terminals, but of their associated *control points* where low-angle radiation from a transmitter would reach the F2-layer. Those points are taken to be 2000km from each terminal, along the great circle joining them.

To put this into perspective, a control point for a station located in the midlands of England would lie somewhere above a circle passing close by Narvik, Leningrad, Minsk, Budapest, Sicily, Algiers, Tangier, mid-Atlantic and NW Iceland, the place depending on the direction of take-off. It means that (for example) the steep rise in MUF associated with UK-end sunrise will occur something like four hours earlier on a path to the east than on a path to the west.

Taken over a month, the day-by-day path MUF at a given time can vary over a considerable range. A ratio of 2:1, as between maximum and minimum, could be considered as typical. Monthly predictions are based on median values, that is to say, on those values which have as many cases above them as below. Therefore the median MUFs in predictions represent 50% probabilities because, by definition, for half of the days of the month the operating frequency would be too high to be returned by the ionosphere. Note, though, that however certain you might be that a given circuit at a given frequency at a given time might be open 15 days in a particular month (the meaning of 50% probability in this context), there is no way of knowing which 15 days they might be. Nor can you be sure that 'good days' on one circuit will be equally rewarding on another.

If you want to ensure the most reliable communication over a particular path, say to make a daily contact at a given time, you will need to operate below the median MUF. Conventionally this optimum working frequency (OWF), at which contact should be possible on 90 per cent of days in the month, is found by multiplying the MUF figure by 0.85. Thus, for a monthly MUF of 20MHz the OWF will be around 17MHz. At the other end of the scale, there will be days when the monthly MUF figure is exceeded. This highest path frequency (HPF) is found by dividing MUF by 0.85 - in the example given earlier this would give a figure around 23.5MHz. Always remember that signals will propagate

best if you are operating on the band closest to the operational MUF on any particular day.

All this may seem to suggest that the lower the operating frequency, the greater would be the chance of success, but that is not true. The reason is that the further down from the median MUF that one operates, the greater become the losses due to absorption, and eventually these become the dominant factor. For any given path there is a *lowest useful frequency* (LUF) at which those losses become intolerable.

Unfortunately it is beyond the scope of this book to be able to provide detailed information about the preparation of monthly predictions and about how to relate the values to the amateur bands. However, each issue of *RadCom* contains a table giving current data in a form that will satisfy the needs of most operators based in the UK.

Those predictions are unique in that the information is given in the form of percentage probability for each of the HF amateur bands, taking both the HPF and the LUF into account. A figure of 1 represents 10% (or three days a month), a figure of 5 represents 50% (or 15 days a month), and so on. Multiply by three the figure given in the appropriate column and row and you will have the expected number of days in the month on which communication should be possible at a given time on a given band.

Many operators believe that if it is known that solar activity is higher (or lower) than was expected when the predictions were prepared, then they can raise (or lower) all the probabilities by a fixed amount to compensate for the changed circumstances. That is not so. The highest probability will always appear against the band closest to the OWF (as defined in the predictions sense). If the OWF is altered by fresh information, then the probabilities on one side of it will rise but on the other side will fall. The only sure way of establishing amended figures would be to enter the revised particulars into the computer program and to run off a complete set of new predictions. It is beyond the scope of this book to provide detailed descriptions of how to prepare predictions and relate them to the amateur bands. However, each issue of *RadCom* contains a table giving predictions for the month ahead for a wide range of paths, and these should suffice for most people's purposes. For those wishing to prepare your own predictions, a number of useful freeware or shareware programs are available on the Internet. These prediction programs customarily assume a quiet geomagnetic field.

Monthly prediction programs usually require an index of solar activity comparable to the 12-month smoothed relative sunspot number R_{12}, or to IF2, which is derived from ionospheric data. There is nothing to be gained by substituting the latest unsmoothed monthly sunspot figure. Also, the use of raw daily figures, such as those from Boulder or Meudon, in the expectation that they will yield meaningful daily ionospheric predictions, is a misunderstanding of the highest degree. The ionosphere does not respond to fluctuations of daily sunspot numbers. On a daily time scale signal performance is much more responsive to changes in the level of geomagnetic activity. Information about current levels of geomagnetic activity and short-term forecasts of the 'radio weather" is available from several Internet sites.

The locations of places for which predictions are given every month in *RadCom* may be seen against their relative beam headings in the great-circle outline map of Fig 12.22.

Grey-line Propagation

The grey line is the ground-based boundary around the world that separates day from night, sunlight from shadow.

Many operators believe, and with some justification, that signals beamed along the grey line near sunrise or sunset will

Fig 12.24: Grey-line propagation map

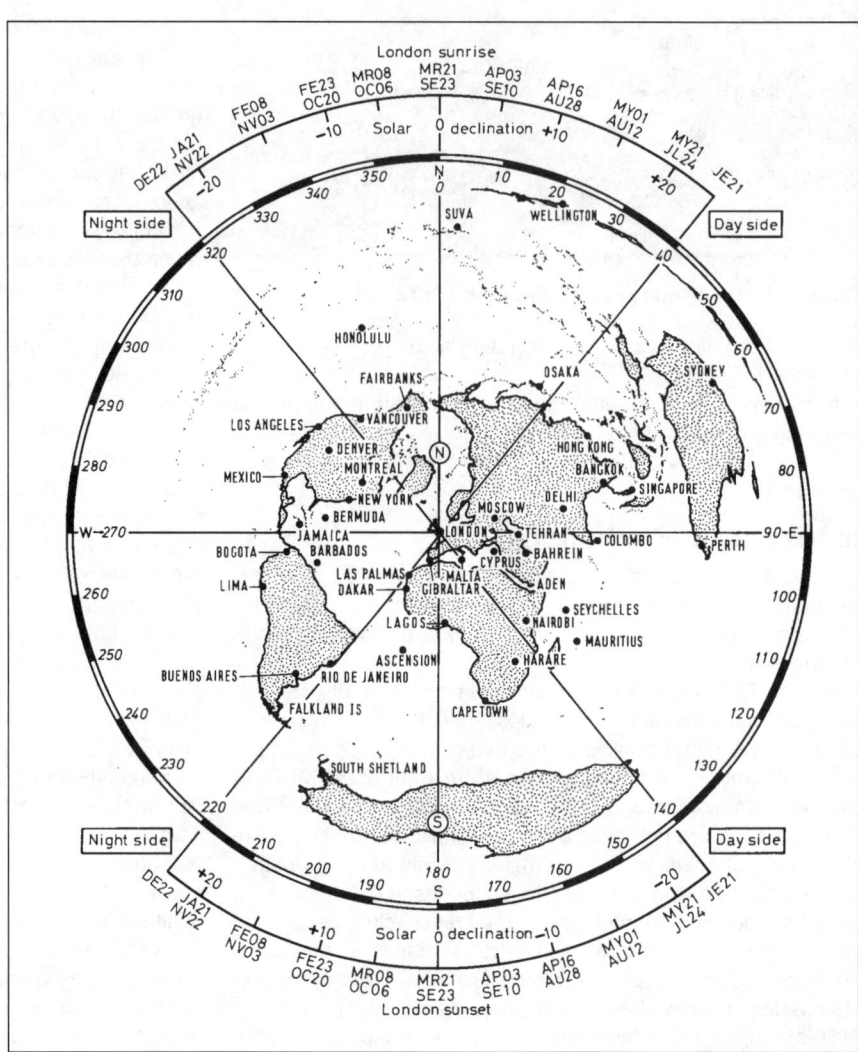

reach distant locations that are also experiencing sunrise or sunset for relatively short periods of time when conventional predictions may appear to be pessimistic.

Noon is a north-south phenomenon; all places having the same longitude encounter it at the same instant of time. But the grey line runs north and south only at the equinoxes, in March and September. At all other times it cuts across the entire range of meridians and time zones.

Sunrise and sunset are the periods when MUFs may be expected to rise or fall through their greatest range of the day.

Couple that with the fact that the one grey line represents sunrise (rising MUF) on one side of the world and sunset (falling MUF) on the other, spring or summer on one side of the equator and autumn or winter elsewhere, and you have a very strong prospect that favourable circumstances for propagation will occur somewhere along the line.

This mode is for the lower frequency bands, 1.8, 3.5 and 7MHz, because the normal house rules apply: the operating frequency has to be below the MUF for the path or the signals will pass through the ionosphere somewhere instead of being reflected.

On the ground the grey line may be considered to be a great circle, but a rather badly defined one because the Sun is not a point source of light - hence twilight time, of course.

To work the grey line you need to know three things: your latitude (atlas), sunrise/sunset times for your area (daily newspaper), and the declination of the Sun for the day in question

(*Whittaker's Almanack* or the *British Astronomical Association's Handbook*).

The ground azimuth at sunrise is given by the expression

$$\sin azimuth = \frac{\sin declination:}{\cos latitude}$$

and the ground azimuth at sunset by:

$$\sin azimuth = \frac{- \sin declination}{\cos latitude}$$

and their reciprocals. Those equations are the basis for the sunrise and sunset scales above and below the skeleton great-circle map, **Fig 12.24**. Interpolate for the appropriate date and lay a straight-edge right across the map so that it passes through the centre (London, in this case). That is your grey line and it should work in either direction, if it is to work at all. Your line will show which areas ought to be accessible if fortune smiles upon you.

That is the good news. The bad news is that it is not the ground-based shadow that determines the state of the ionosphere because the Earth's shadow is shaped like a cone; the area of darkness at F2 heights is appreciably less than has been considered in the preceding paragraph. Sunrise comes earlier than on the ground, sunset later. In fact, in mid-summer the F2-layer is in sunlight for all 24 hours of the day over the whole of the UK.

So, the ionospheric grey line (as opposed to the ground-based grey line) cannot be considered as a great circle and, therefore,

cannot be represented by the straight line on the great-circle map.

That should not stop you from trying your luck, however. But there is no point in trying to calculate the true outline of the Earth's shadow, because your signals are going to take the great-circle route no matter what you come up with on your computer. You may as well make all your plans using your ground-based data, because it is easy to come by, and make up for its likely deficiencies by being generous with your timing. At the sort of frequency you will be using, the beamwidth of the antenna will be wide enough to take care of direction.

The Beacon Network

One of the most useful aids to understanding propagation and exploiting whatever possibilities there may be effectively is offered by beacon transmissions. These also offer a useful basis for personal propagation research projects.

Beacons are found mainly on the higher HF bands upwards. For the most part they operate continuously with a simple message in Morse.

This will always include their callsign and a long dash to facilitate signal strength measurements; many also include their grid locator, town and power. Because, unlike other amateur stations, they are always there they offer a useful indication of the state of a particular path.

This is particularly true at HF of the network of beacons created by the North California DX Foundation. This consists of 18 beacons, strategically dispersed around the world, transmitting in sequence on 14.100, 18.110, 21.150, 24.930 and 28,200MHz in turn in the course of a three-minute cycle. The power is stepped down from 100W to 100mW in four stages, giving a further indication of the state of the circuit.

There are literally hundreds more beacons worldwide at HF and the VHF and UHF bands. They are far too numerous to list here but details of those likely to be heard in the UK are listed in the *RSGB Yearbook* and changes between editions can be found on the Internet.

GB2RS News

Finally, every week the weekly *GB2RS* news bulletin, posted every Friday on the RSGB website and broadcast every Sunday, contains items of propagation interest, including a summary of solar-geophysical events over the preceding week. Times and frequencies of these broadcasts vary according to location; details are printed in the *RSGB Yearbook*.

BIBLIOGRAPHY

Books

Radio Auroras, C Newton, G2FKZ, RSGB, 1991

Radio Propagation - Principles & Practice, Ian Poole, G3YWX, RSGB, 2004

Your Guide to Propagation, by Ian Poole, G3YWX, RSGB, 1998

LF Today, Mike Dennison, G3XDV, RSGB, 2004

The VHF/UHF Handbook, edited by Dick Biddulph, M0CGN, RSGB

Internet

The Radio Propagation Page: http://www.keele.ac.uk/depts/por/psc.htm. Run by the RSGB Propagation Studies Committee this carries links to a wide range of sources, ranging from the explanation of basic terms in radio propagation through the various forms of propagation to sites carrying data affecting propagation to HF prediction programs.

Amateur Radio Propagation Studies: http://www.df5ai.net

Basics of Radio Wave Propagation: http://ecjones.org/physics.html

Bouncing Radio Waves off the Sky: http://www.geocities.com/rf-man/skyrange.html

Glossary on Solar-Terrestrial Terms: www.ips.gov.au/Main.php?CatID=8

Introduction to the Ionosphere: http://www.ngdc.noaa.gov/stp/IONO/ionointro.html

Near Real-Time Global MUF Map: http://www.spacew.com/www/realtime.php

Real-Time Space Weather: http://www.spacew.com/plots.htm

SEC Radio Users' Page: http://www.sel.noaa.gov/radio.html

Solar Terrestrial Activity Report: http://www.dxlc.com/solar

Space Environment Center: http://www.sec.noaa.gov

Spaceweather.com: http://www.spaceweather.com

Search engines

A multitude of information on all aspects of radio propagation can be gleaned from a simple search request to any of the more popular Internet search engines.

REFERENCES

[1] 'Flare spot', P W Sollom, *Radio Communication* December 1970, p820; January 1971, p20; February 1971, p92

[2] 'The ground beneath us', R C Hills, *RSGB Bulletin* June 1966, p375

[3] Report 239-6 in *Propagation in non-ionised media*, Study Group 5, CCIR, XVIth Plenary Assembly, Dubrovnik, 1986, Geneva, ITU, 1986

[4] *VHF/UHF Handbook*, Dick Biddulph (ed), RSGB, 1997, Chapter 3 (Propagation)

[5] 'Patterns in propagation', R G Flavell, *Journal of the IERE*, Vol 56 No 6 (Supplement), pp175-184

ONLY
£14.99

plus p&p

Radio Propagation - Principles & Practice

Ian Poole, G3YWX

Radio propagation is a vital topic for any radio amateur or anyone with an interest in radio communications. This book provides a fascinating description of all the relevant information about radio propagation from HF to VHF, UHF and beyond.

The book includes everything you need to know including radio waves and how they travel, the atmosphere, the Sun, ionospheric propagation (with the important modes and information), ionospheric storms and aurora, how to predict and assess ionospheric propagation, tropospheric propagation, meteor scatter, and space communications. You are also guided in making the most of the available equipment choosing the right time and radio band.

This book provides the reader with a practical understanding of Radio Propagation so that they can use them to their best. Radio Propagation - Practice and Principles is an essential read for anyone associated with radio communications.

Size: 174 by 240 mm, 112 pages, ISBN: 1-872309-97-6.

RSGB Members Price: £12.74 plus p&p

E&OE

Radio Society of Great Britain
Lambda House, Cranborne Road, Potters Bar, Herts. EN6 3JE
Tel. 0870 904 7373 Fax. 0870 904 7374

ORDER 24 HOURS A DAY ON OUR WEBSITE
www.rsgb.org/shop

13 Antenna Basics and Construction

The antenna is the essential link between free space and the transmitter or receiver. As such, it plays an essential part in determining the characteristics of the complete system. The design of the antenna and its working environment will decide its effectiveness in any particular system.

In *Antennas* (2nd ed), John D Kraus defined an antenna as: "A structure that transforms electromagnetic energy contained in a guided wave to that of free-space propagation or vice versa."

THE ELECTROMAGNETIC WAVE

An electromagnetic wave in free space comprises electric (E) and magnetic (H) components perpendicular to each other. If the x co-ordinate is taken along the line in the direction of wave travel, then the E and H field vectors lie entirely in the yz plane as shown in **Fig 13.1**. The wave front is a plane surface normal to the direction of propagation and is called a transverse electromagnetic wave (TEM).

It can best be envisaged as the surface of a rapidly inflating balloon, otherwise known as a wave front, however, a small area of this expanding surface can be regarded as flat plane at a distance from the source. A TEM wave, in which the electric and magnetic vectors, while varying in magnitude and sign, remain along the same axis in space is said to be polarised, the plane of polarisation (by convention) being that contained in the electric vector, ie the xy plane in **Fig 13.1**.. Polarisation is discussed in more detail later. The wave illustrated in Fig 13.1 travelling in free space is unrestricted in its motion and is known a travelling wave.

Near and Far-Field Antenna Regions

The TEM wave described above is formed some distance away from the antenna in a region known as the far field. In this region the total electric and magnetic fields are at right-angles to each other and to the direction of propagation, and their respective maxima are phased 90° (one quarter-wavelength) apart as shown in Fig 13.1. In space, the ratio of the E/H fields yields a value of 377 ohms, which is the impedance of free space.

At distances closer to the antenna, the fields become more complex, and there are additional field components directed along the direction of propagation. This is the near field region,

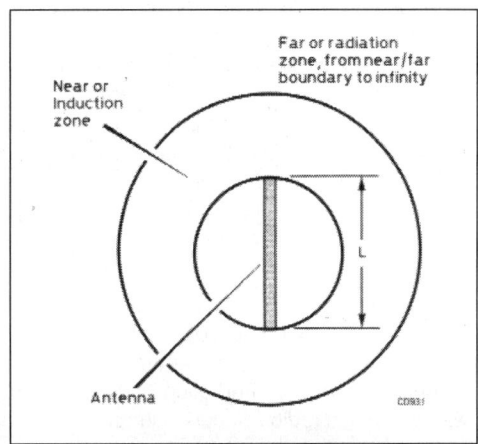

Fig 13.2: Near and far fields of an antenna

in which coupling between the antenna and adjacent conductive structures (power wires, plumbing or other antennas) becomes significantly greater than would apply from illumination by a freely propagating electromagnetic wave. Conductive objects within the near field of an antenna can seriously affect both its radiation pattern and its input impedance. The elements of antenna arrays usually lie within the near field of adjacent elements, and the mutual coupling between them must be taken into account if the best overall gain is to be achieved. Radiation pattern, input impedance and mutual coupling will described later.

The approximate near/far field boundary is defined as:

$$R = 2L^2/wavelength$$

Where R is the distance and L is the length of the antenna as illustrated in **Fig 13.2**.

RESONANCE

If an oscillatory current is passed along a wire, the electric and magnetic fields associated with it can be considered as a wave attached to the wire and travelling along it. If the wire finally terminates in an insulator the wave cannot proceed but is reflected. This reflection is an open-circuit reflection and produces standing-wave fields on the wire. How these standing waves are produced is described in the chapter on Transmission Lines.

Fig 13.3 shows two typical cases where the wire is of such a length that a number of complete cycles of the standing wave can exist along it. Since the end of the wire is an open-circuit, the current at that point must be zero and the voltage a maximum. Therefore at a point one quarter-wavelength from the end, the current must be a maximum and the voltage will be zero. At positions of current maxima, the current-to-voltage ratio is high

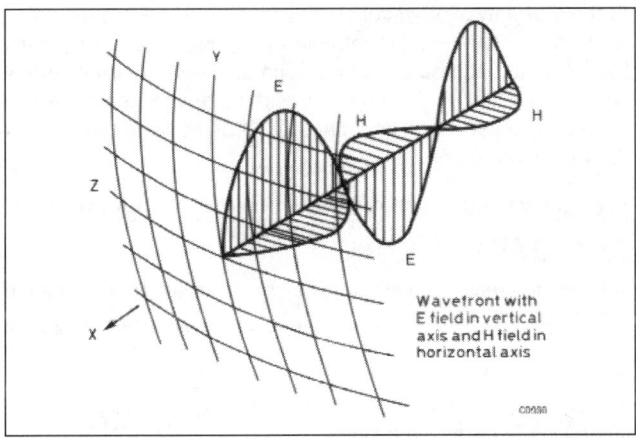

Fig 13.1: Conceptual diagram of Transverse Electromagnetic (TEM) Wave

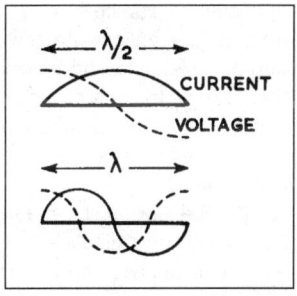

Fig 13.3: Standing waves on resonant antennas, showing voltage and current variations along the wire at its fundamental resonant frequency and at second harmonic frequency

and the wire will behave as a low-impedance circuit. At voltage maxima the condition is reversed and the wire will behave as a high-impedance circuit.

A wire carrying a standing wave as illustrated in **Fig 13.3** exhibits similar properties to a resonant circuit and is also an efficient radiator of energy. This is a resonant or standing wave antenna and the majority of the antennas met with in practice are of this general type. The length for true resonance is not quite an exact multiple of the half-wavelength because the effect of radiation causes a slight retardation of the wave on the wire and also because the supporting insulators may introduce a little extra capacitance at the ends. An approximate formula suitable for wire antennas is:

$$\text{Length (m)} = 155(n - 0.05)/f$$

or

$$\text{Length (feet)} = 485(n - 0.05)/f$$

Where n is the number of complete half-waves in the antenna and f is the frequency in megahertz.

It must be emphasised that an antenna does not have to be resonant to radiate. Radiation takes place from any elevated wire carrying a radio frequency current; if this wire is terminated in a resistor the wave will be a travelling wave rather than a resonant one. Radiation will always occur unless prevented by screening or cancelled by an opposing field of equal magnitude, as occurs in transmission lines.

RADIATION RESISTANCE

When power is delivered from the transmitter into the antenna, some small part will be lost as heat, since the material of which the antenna is made will have a finite resistance, and a current flowing in it will dissipate some power. The bulk of the power will usually be radiated and, since power can only be consumed by a resistance, it is convenient to consider the radiated power as dissipated in a fictitious resistance which is called the radiation resistance of the antenna. Using ordinary circuit relations, if a current I is flowing into the radiation resistance R, then a power of I^2R watts is being radiated.

As depicted in Fig 13.3 the RMS current distribution along a resonant antenna or indeed any standing wave antenna is not uniform but is approximately sinusoidal. It is therefore necessary to specify the point of reference for the current when formulating the value of the radiation resistance, and it is usual to assume the point of maximum current.

A halfwave dipole in free space has a radiation resistance of about 73Ω. If it is made of highly conductive material such as copper or aluminium, the loss resistance may be less than one ohm. The conductor loss is thus relatively small and the antenna provides an efficient coupling between the transmitter and free space.

FEED IMPEDANCE

When the antenna is not a resonant length, it behaves like a resistance in series with a positive (inductive) or negative (capacitive) reactance and requires the addition of an equal but opposing reactance to bring it to resonance, so that it may be effectively supplied with power by the transmitter. The combination of resistance and reactance, which would be measured at the antenna terminals with an impedance meter, is referred to in general terms as the antenna input impedance. This impedance is only a pure resistance when the antenna is at one of its resonant lengths.

Fig 13.4 shows, by means of equivalent circuits, how the impedance of a dipole varies according to the length in wavelengths. It will be seen that the components of impedance vary

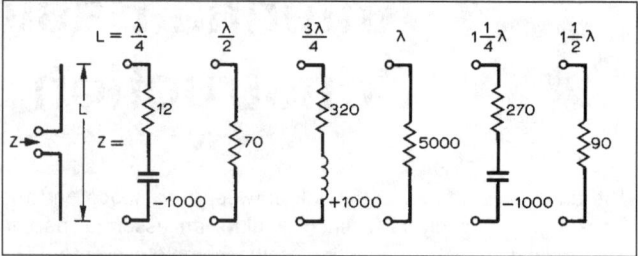

Fig 13.4: Typical input impedance (Zi) value for dipoles of various lengths

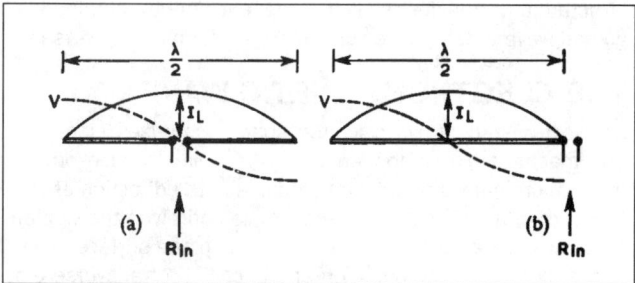

Fig 13.5: The input resistance (fixed point impedance) of a halfwave dipole is low at (a) and high at (b)

over a wide range.

The input impedance of the antenna is related specifically to the input terminals, whereas the radiation resistance is usually related to the point of current maximum. It is possible to feed power into an antenna at any point along its length so that the input impedance and the point of current maximum even of a resonant antenna may be very different in value, although in this case both are pure resistances. Only when the feed point of the antenna coincides with the position of the current maximum on a single wire will the two be approximately equal, **Fig 13.5(a)**. If the feed point occurs at a position of current minimum and voltage maximum, the input impedance will be very high, but the the point of current maximum remains unaltered **Fig 13.5(b)**. For a given power fed into the antenna, the actual feed-point current measured on an RF ammeter will be very low because the input impedance is high. Such an antenna is described as voltage fed.

Earlier it was stated that a centre fed halfwave dipole in free space has a radiation resistance of about 73Ω. However, the impedance presented at the feed point by an antenna is a complex function of the size and shape of the antenna, the frequency of operation and its environment. The impedance is affected by the proximity of other conducting objects, where the induction of RF currents alters the impedance through mutual coupling between the antenna and object. The elements of a Yagi antenna are mutually coupled together, and the driven element would present a very different impedance if measured in isolation from the rest of the structure.

RADIATION PATTERNS, DIRECTIVITY AND GAIN

The performance of an antenna can be assessed by its radiation pattern. A VHF base station or repeater usually requires antennas that distribute the signal equally in all directions, whilst a station configured for DX operation will require antennas that focus the energy in one particular direction. Methods of achieving focus of energy are described later.

Such a pattern can be made by energising the antenna with a known level of RF power and then performing a large number of

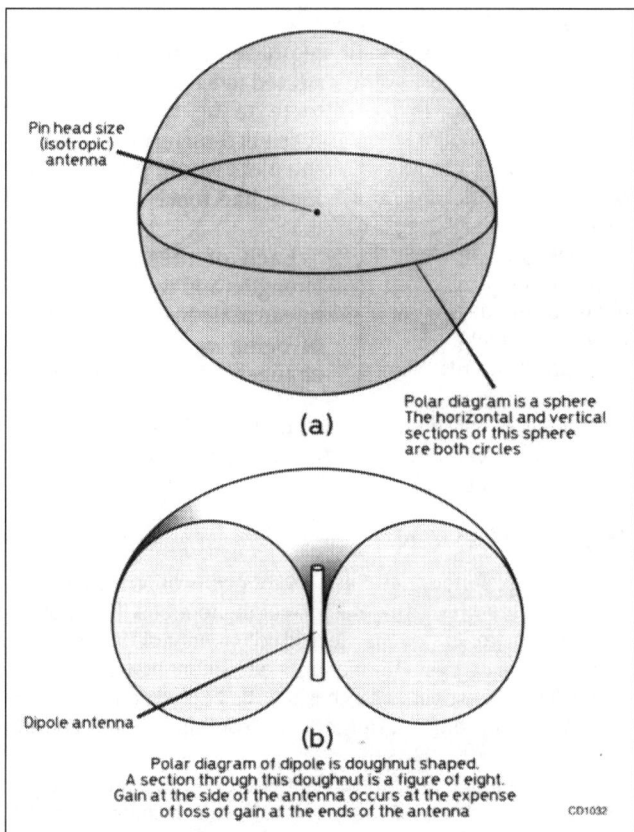

Fig 13.6: Three-dimensional free-space polar diagrams for (a) an isotropic radiator and (b) a dipole

field strength measurements at various angles, then plotting the results on a polar graph to produce an azimuth polar diagram. This diagram would then present the relative field strength or power intensity as a radial distance from the centre of the graph at the relevant angle.

The directivity of an antenna is the ratio of maximum radiation intensity to the average intensity. The isotropic antenna, see below, radiates equally in all directions and has directivity of 1, a theoretical minimum. The smaller the three-dimensional beam angle the greater the directivity.

If one antenna system can be made to concentrate more radiation in a certain direction than another antenna for the same total power supplied, it is said to exhibit gain over the other antenna in that direction. The gain of an antenna is a combination of directivity and efficiency when compared with a reference antenna.

If an antenna were minutely small and radiated equally in all three dimensions the overall radiation pattern would be a sphere. Although the construction of such an antenna is not possible it is used as a theoretical entity in antenna mathematical modelling and is known as an isotropic source; it is used as a theoretical reference for measuring antenna gain. Gain or loss relative to an isotropic radiator is stated in dBi. A radiation pattern of an isotropic source is shown in **Fig 13.6(a)**.

The simplest practical form of antenna is the dipole. Although this antenna may be of any length, the word 'dipole' usually implies a half-wavelength long resonant antenna, fed via a balanced feeder at the centre.

The dipole antenna does not radiate equally in all directions because the current along its length is not constant and it produces a three-dimensional doughnut shaped radiation pattern. This pattern and a polar diagram section is shown in **Fig 13.6(b)**.

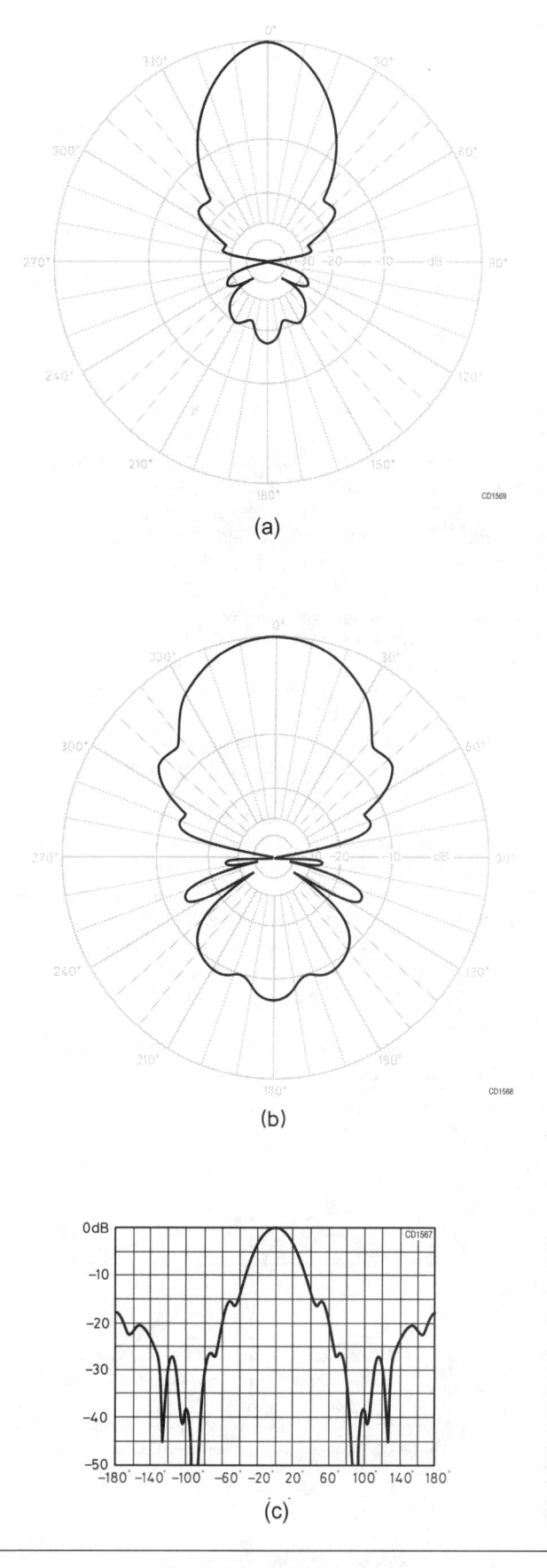

Fig 13.7: Radiation of a 12-element VHF Yagi antenna using (a) Polar format, ARRL logarithmic decibel scale. (b) Polar format, linear dB scale. (c) Rectangular format, linear dB scale

Because of its simplicity the dipole itself has become a reference standard and has a power gain of 2.15dbi. Gain figures, using the dipole radiator as a reference, are symbolised dBd.

A practical antenna may have good directivity, but low gain if the antenna has losses through poor design, the use of lossy components or poor mechanical construction. If the antenna were lossless, the gain and directivity would be the same.

Sections through a three-dimensional radiation pattern are normally either vertical (elevation) or horizontal (azimuth). The diagrams so far discussed ignore the effects of ground that could affect the diagram with reflections. Such diagrams are called free space diagrams, and like the isotropic antenna, are theoretical and only used in antenna mathematical models. Radiation patterns that include the effects of ground are described later in Computer Modelling.

Polar diagrams in early antenna literature used polar graphs plotted on a linear scale. This enabled the main lobe beamwidth to be measured but sidelobes were barely visible. The ARRL has promoted the use of a hybrid polar chart, **Fig 13.7(a)**, which combines features of both linear and logarithmic radial scaling in decibels [1], which is used in most amateur radio publications these days. The logarithmically scaled chart, **Fig 13.7(b)**, clearly shows the levels of the sidelobes at the expense of the main lobe.

There is also the more specialised rectangular format, **Fig 13.7(c)**, which uses the linear dB scale. The vertical axis of the rectangular plot represents the relative field strength or power density as a function of the angle shown on the horizontal axis. This presentation is useful for high-gain VHF/UHF antennas as a lack of symmetry can be easily seen, and is often an indication of loss of efficiency or incorrect feeding of multiple-element arrays

The radiation pattern characteristics of directional antennas are usually expressed as the beamwidth in two principal planes at right angles to each other. The beamwidth in these principal planes is usually defined as the angle including the main beam at which the radiated energy falls to one half the maximum level. This is called the half-power beamwidth, and the points on the radiation pattern are often called the 3dB or half-power points of the radiation pattern, being 3dB below the main beam as shown in **Fig 13.8**.

Key features of the radiation patterns of the antenna shown in Fig 13.8 are the main lobe or main beam, and the presence of

several sidelobes including one pointing in the opposite direction to the main lobe. The front-to-back or F/B ratio is the ratio of the energy radiated by the peak of the main lobe to that in the opposite direction, and is often used as an estimate of the 'goodness' of a beam antenna. This ratio is usually expressed in decibels. As more power is radiated in minor lobes, less power is available in the main lobe, and the gain of the antenna is reduced.

Whilst gain is usually measured by direct substitution of the antenna under test with a reference antenna, it is possible to estimate the directivity of directional antennas with fair accuracy if the half-power beamwidths can be measured in the principal (E and H) planes of the main beam. If the antenna losses can be assumed to be very small, the gain will be essentially equal to the calculated directivity. Measuring techniques with readily available amateur radio equipment are described in [2].

POLARISATION

Earlier it was stated that the plane of polarisation was, by convention, contained in the electric component of a TEM wave. A linear dipole generates the electric component of the TEM wave along its axis so this antenna, or linear antenna array, oriented vertically with respect to earth is said to be vertically polarised. The same antenna oriented horizontally is horizontally polarised.

Polarisation is important on paths that don't alter the transmitted polarisation (a line-of-sight VHF/UHF or microwave link, for example). Two such antennas must be co-polarised (polarised in the same direction) in order to communicate; totally cross-polarised antennas theoretically cannot communicate. They are also important when making antenna measurements on an antenna range.

For HF antennas, polarisation is not so important because polarisation is altered when a TEM wave is refracted by the ionosphere.

Satellite users on VHF/UHF often use circular polarisation to reduce the effects of propagation, ground reflections or the spinning motions of the satellites on the signals. The effect of circular polarisation can be visualised as a signal that would be radiated from a dipole that is spinning about its centre at the radiating frequency.

The tip of the electric vector traces out a corkscrew as it propagates away from the antenna and, like a corkscrew, the polarisation is described as right- or left-handed circular, dependent on the direction of rotation of the electric vector as seen from the transmitter. Methods for generating circular polarisation are shown in the chapter on practical VHF/UHF antennas.

A fixed linear dipole will receive an equal signal from a circularly polarised wave whether it is mounted vertically, horizontally or in an intermediate position, if there are no ground reflections.

The signal strength will be 3dB less than if a circularly polarised antenna of the same sense is used; however, a circularly polarised antenna of the opposite sense will receive no signals. Both these effects are due to polarisation mismatch between the wave and the receive antenna.

BANDWIDTH

There are no unique definitions for antenna bandwidth. Dependent upon the operational requirements of the antenna, the definitions fall into two categories: radiation pattern bandwidth and impedance bandwidth.

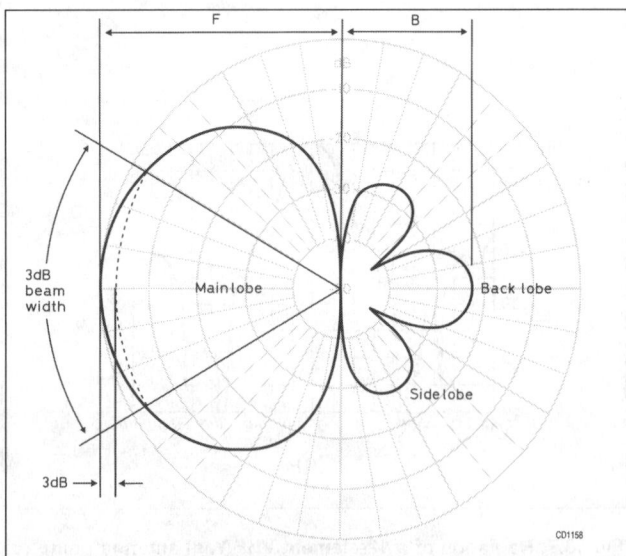

Fig 13.8: Typical polar diagram of a Yagi antenna

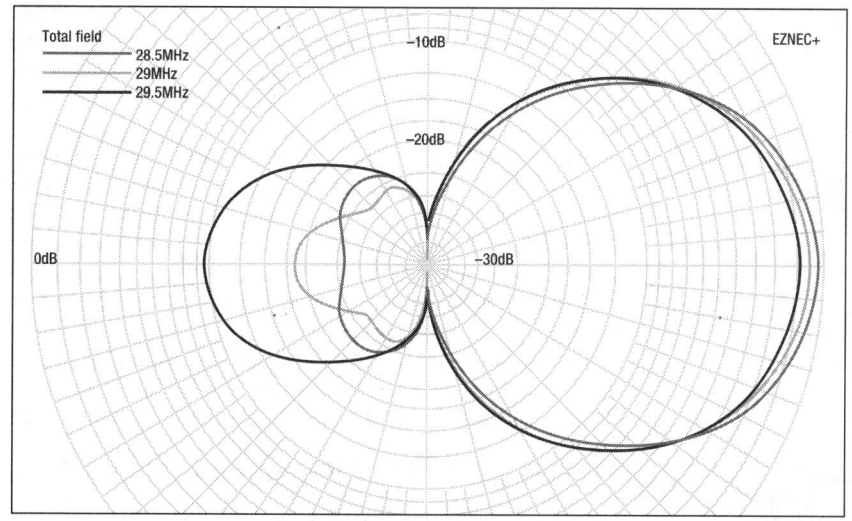

Fig 13.9: Variation in free-space radiation pattern of a three-element Yagi antenna

Radiation Pattern Bandwidth

Antenna radiation patterns are dependent upon the operating frequency. Their sensitivity to frequency changes are in turn dependent on the degree of tuning or inherent Q required to achieve the desired characteristic. Bandwidth is defined as the frequency range over which satisfactory performance can be obtained.

The criteria for defining bandwidth could be one or more of the following:

- Main lobe beamwidth
- Acceptable sidelobe level
- Minimum gain or directivity
- Polarisation qualities

With the relatively limited frequency range within the amateur bands, the gain normally does not change too radically with frequency, although this is not always the case with very-high-gain VHF/UHF Yagi antennas where the gain and the pattern shape or direction of radiation may be stable over only a very narrow frequency band. An example of this is shown in **Fig 13.9**.

For beam antennas, such as the Yagi, the radiation pattern bandwidth is often defined as the frequency range over which the main lobe gain decreases to 1dB below its maximum value. This is not to be confused with main lobe directivity beamwidth, described earlier

For electromagnetically simple, small antennas (ie when the linear dimensions are of the order of half a wavelength or less)

the limiting factor is normally the input impedance.

With circular polarisation antennas the change of the polarisation characteristic with frequency is often the limiting factor. In end-fire linear arrays, collinears and the like, the main lobe direction and shape can change considerably before the gain deteriorates significantly.

For any antenna array or multiply fed antenna, the limiting factors may be determined by the ability of the feed arrangements to maintain the correct current distribution to the antenna elements as the frequency is varied.

Such antennas bandwidths may also be limited by excursions of input impedance, as described below.

Impedance Bandwidth

The impedance bandwidth of an antenna is defined as the frequency range over which the antenna impedance results in a standing wave ratio (SWR) less than some arbitrary limit. This may be typically 1.5:1 or 2:1 for amateur operation with solid-state transmitters, or higher values for other applications. The impedance bandwidth can be very narrow on electrically small antennas such as HF mobile antennas, as shown in **Fig 13.10**.

Ideally, an antenna should be impedance matched to the feedline and thence to the transmitter or receiver. Although tuned feed arrangements are often used at HF, where a high standing wave ratio may be acceptable on the feedline, the losses in VHF feeders and tuning components usually preclude this approach at VHF and UHF.

Impedance bandwidth and radiation pattern bandwidth are independent of each other. It is quite possible for the impedance bandwidth to be greater than the radiation pattern bandwidth, especially with high-gain antennas, and to be able to feed power into an antenna that is then wasted by radiating it in other than the desired direction.

THE EFFECT OF GROUND

The ground under the antenna acts as a reflector. Electromagnetic waves from the antenna radiate in all directions and some of these waves are reflected by ground. If the reflected wave is in phase, or partially in phase, with a direct wave it enhances radiation and increases gain at a particular angle. Other combinations of reflected and direct waves, whose phases tend to cancel, reduce gain at other angles.

Waves A and C shown in **Fig 13.11** enhance gain while B and C tend to cancel and reduce the gain. This is the cause of the

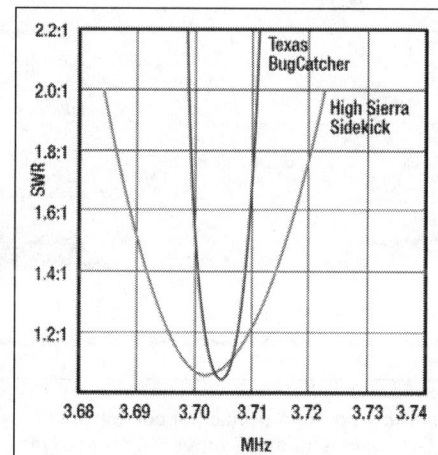

Fig 13.10: Comparative SWR curves of two commercial antennas. Assuming a SWR limit of 2:1, the Texas Bugcatcher antenna has a bandwidth of 12kHz on 80m

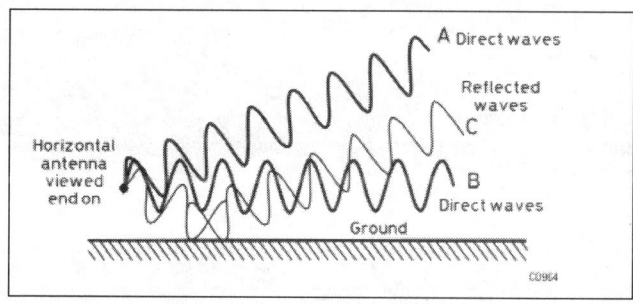

Fig 13.11: The effect of ground reflection on directly radiated waves

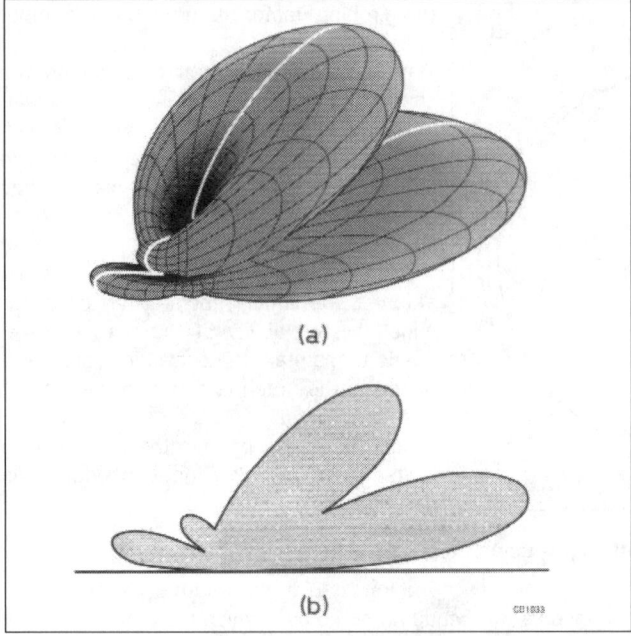

Fig 13.12: (a) Three-dimensional polar diagram of a three-element beam, showing a vertical section at the angle of maximum radiation. (b) Elevation diagram resulting from this section

familiar vertical antenna patterns. This aspect is most important and has implications viewing any horizontal polar diagram of a practical antenna. It is also important that ground effects are taken into consideration when setting up equipment on an antenna range.

Consider the three-dimensional polar diagram of a three-element beam in **Fig 13.12(a)**. If we take a vertical or elevation cross section of this diagram it produces the familiar elevation diagram shown in **Fig 13.12(b)**.

Determining the horizontal diagram is not as easy; it can not be plotted through the true horizontal because of the effect of ground (theoretically the radiation strength will be zero in the horizontal plane although this is not the case in practice). The practical solution is to plot the horizontal diagram at the angle of maximum radiation of the main lobe as shown in **Fig 13.13**.

ANTENNA MODELLING USING A COMPUTER

Modelling is the technique of evaluating the performance of one object or system by evaluating the performance of a substitute called a model. Models can be physical objects, like a VHF scale model sometimes used to evaluate a HF antenna. Models can also be purely mathematical, like the equations used in circuit analysis. The following discussion describes a mathematical model on your personal computer using readily available software.

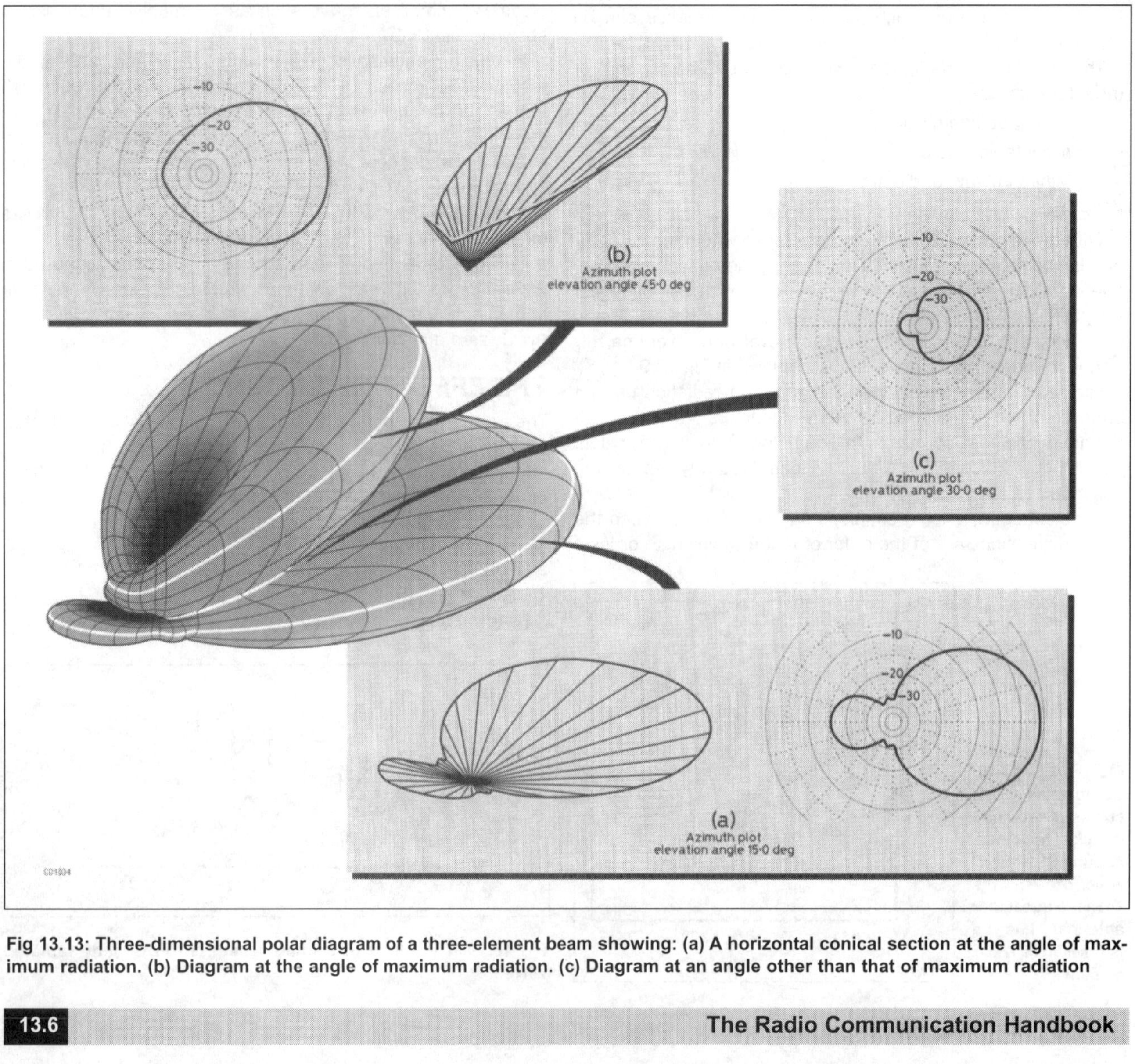

(b)
Azimuth plot
elevation angle 45·0 deg

(c)
Azimuth plot
elevation angle 30·0 deg

(a)
Azimuth plot
elevation angle 15·0 deg

Fig 13.13: Three-dimensional polar diagram of a three-element beam showing: (a) A horizontal conical section at the angle of maximum radiation. (b) Diagram at the angle of maximum radiation. (c) Diagram at an angle other than that of maximum radiation

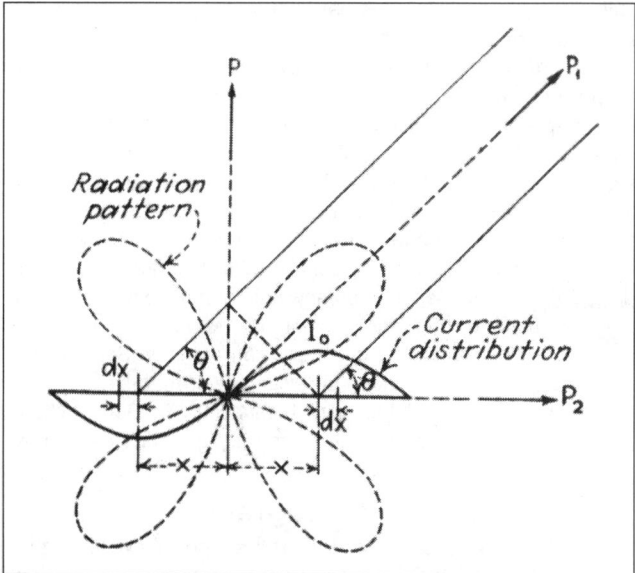

Fig 13.14: Diagram showing the factors controlling the directional characteristics of an antenna (from [3])

The essential equations describing the performance of antennas were known in the early 1900s but were complex, consequently early mathematical solutions were only used for limited conditions and special cases.

By the 1930s solutions were available for dipoles with a sinusoidal current distribution at resonance. The diagram shown in **Fig 13.14**, is from [3], which is then followed by a full page of mathematical analysis. This model is limited to the analysis of the directional characteristics of a simple antenna in free space. To model a multi-element antenna with environmental effects is a far more complex and intractable business using normal mathematical methods.

While in the past the mathematical model was impractical for all but limited conditions the situation changed with advent of the computer. It is now possible to evaluate fairly complex antennas with a relatively inexpensive computer and to even question optimistic claims made regarding some antennas. There is a further advantage of modelling an antenna using a computer; with the graphic interfaces provided with these latest programs it provides an excellent means of understanding how an antenna functions, without being bogged down with mathematical detail.

The most commonly encountered programs for antenna analysis are those derived from a program developed at govern-

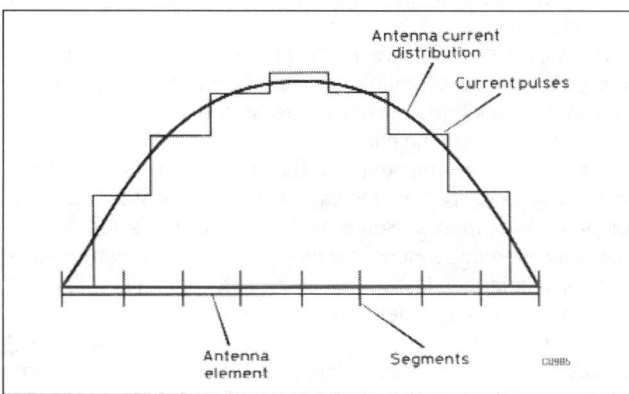

Fig 13.15: Simplified diagram of the real and modelled current distribution over a half-wave dipole

ment laboratories in the USA called NEC, short for Numerical Electromagnetics Code [4].

NEC uses a Method of Moments (MoM) algorithm, which calculates values at selected points, such as the ends of the antenna and some specified intermediate points. The accumulation of errors at the points not specified can be controlled and is known as its moment. NEC is now in its 4th implementation although most of the antenna analyser programs used by amateurs use NEC2

The first NEC program designed for use with a personal computer was MININEC [5]. Although originally visualised as a cut-down version of NEC a considerable amount of original development went into this program. The authors had to reduce some of the complex mathematical operations to a level that a PC (in the late 1980s) could handle in a reasonable amount of time. A number of compromises were necessary and most of that program's limitations were due to these consciously chosen compromises.

As the complexity of personal computer increased, so have the programs. However, this does not mean that they are more difficult to use; in fact quite the reverse, particularly in the case where adjustment to the antenna geometry can be seen simultaneously in a graphic display when using Windows.

The Analysis Program

The antenna structure is divided into a number of straight wires and each of these wires are divided into segments. The current in each of the segments of each wire is calculated by assuming a known level of RF power to the antenna.

If we consider the simplest of algorithms and the simplest of antennas, the centre-fed half-wave dipole, the actual current distribution, as modelled by earlier programs, is approximated by segments of constant current. This means that the actual current on the element (which approximates half a sine wave with the current maximum at the centre and zero at the ends as shown in **Fig 13.15**) can be modelled by a series of steps of constant current. It can be seen that the greater the number of segments, the more closely the model will represent the real current distribution. However, the more sophisticated programs now available use a sinusoidal current distribution within the segments. For example the NEC calculating engine assumes that the current has an essentially sinusoidal shape over the length of a segment, and that the currents of adjacent segments match at their junctions. This produces good results with fewer segments.

Once the magnitude and phase of the current is known then the complex impedance for any part of the element can be calculated. The total antenna electromagnetic field pattern can be built up from the magnitude and phases of the currents in the individual segments.

Conversely, current in a conductor segment, located in an electromagnetic field of known intensity, can be calculated from the current amplitudes and phases in these segments.

Calculation of magnitudes and phases of these currents should enable the model, provided that it is sophisticated enough, to represent any antenna configuration or environment. Antenna systems are often made up of more than one element. These additional elements or wires may be excited by direct connection to the wire or element energised by the source, or connected parasitically as in a Yagi.

The radiation pattern and input impedance of an antenna can be calculated, provided that the current distribution on the structure is known. The current distribution on short, thin wires, where the wire radius is a very small fraction of a wavelength approximates to a sinusoid with a minimum at the free end of

the wire. However, if the wire is thick, an appreciable fraction of a wavelength long, or close to other conductors, the current distribution deviates considerably from the sinusoidal.

There are several programs for antenna analysis available for the amateur, most of which are based on NEC2. Regardless of which program is used the antenna is modelled in three stages. In the first part the user describes the antenna using a text editor or spreadsheet. Key words and a carefully structured input enable the second part of the program to recognise both the vari-

ables and their values. This part of the program computes the antenna performance using the MoM method described earlier.

The third part of the program allows the calculated data to be displayed as an azimuth, elevation or three-dimensional plot to give a graphic representation of antenna performance. Most programs also show a graphic representation of the antenna structure to confirm the model has been constructed correctly. Most models also include sources (the point where the transmitter or feedline is connected to the antenna), loads, transmission lines, and ground media. The accuracy of the model depends on how accurately the actual antenna and its environment can be represented by the model made from these components. Some physical objects, like a physical wire or metallic tube, are easily modelled with high accuracy. Some, like a round loop or small flat metal plate, must be approximated.

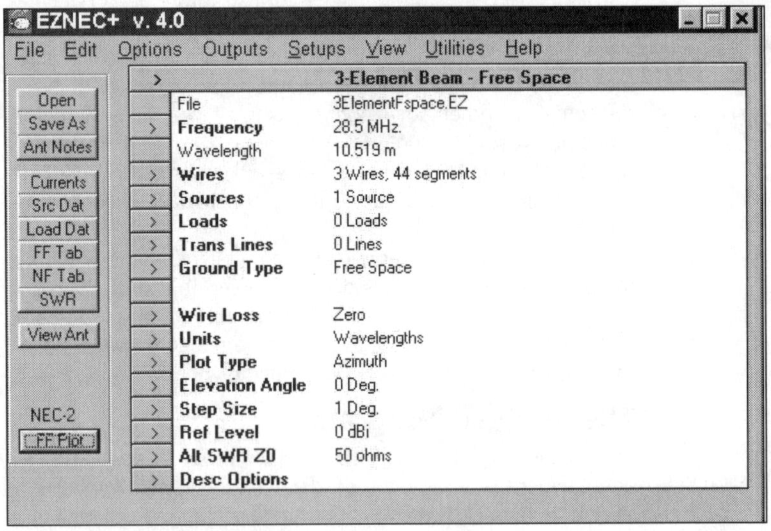

The EZNEC Program

To give some idea of the scope of available programs the antenna analysis software, EZNEC, written by Roy Lewallen, W7EL [6] is shown here as an example. It was originally available in the early 1990s as ELNEC, a DOS based program and has gone through many revisions to EZNEC+ v.4.0, now described. This description is necessarily brief - the on-disk user manual that comes with this program runs to 146 pages. In spite of its complexity the EZNEC program is one of the easiest to use if you are new to antenna modelling using a computer. This ease of use is achieved using a Control Centre Information Window, shown in **Fig 13.16**, from which all other screens are accessed.

Building the model

From this Control Centre the important WIRES spreadsheet type screen, shown in **Fig 13.17**, is used to build the model. This is the place where the basic structure of the antenna is defined.

The antenna is modelled as a set of straight conductors called 'wires', the ends of which are specified in space using X, Y and Z co-ordinates. X and Y are in the horizontal plane and Z is height. Wires are defined by specifying their end coordinates in the appropriate grid cells. The unit of measurement can be in metres, millimetres, feet, inches or wavelength. A useful feature is that the units are automatically converted when a different unit is selected.

You will notice that in the model of the 3-element Yagi, shown in Fig 13.17 the Z co-ordinate is zero. This is because this is a free-space model; the reason for using free-space models was discussed earlier in the chapter.

The number of segments is also entered using the Wires Window. As described earlier, each wire is divided into segments for analysis purposes. Some skill in modelling is required in choosing the number of segments although EZNEC has an automatic segmentation feature, which is useful particularly for newcomers to antenna modelling with a computer.

A three-dimensional view of the antenna geometry, see **Fig 13.18**, can be seen by selecting View Ant in the Control Centre Information Window. Also shown are the X, Y and Z co-ordinates and the segmentation. The antenna can also be rotated and viewed from any angle, and scaled in size.

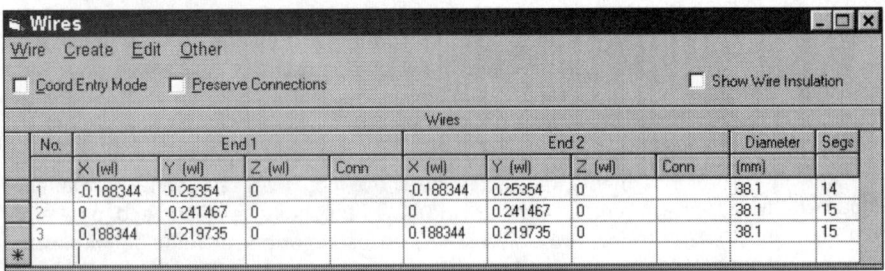

Fig 13.16: The Control Centre Information window in EZNEC, from which all other screens are accessed

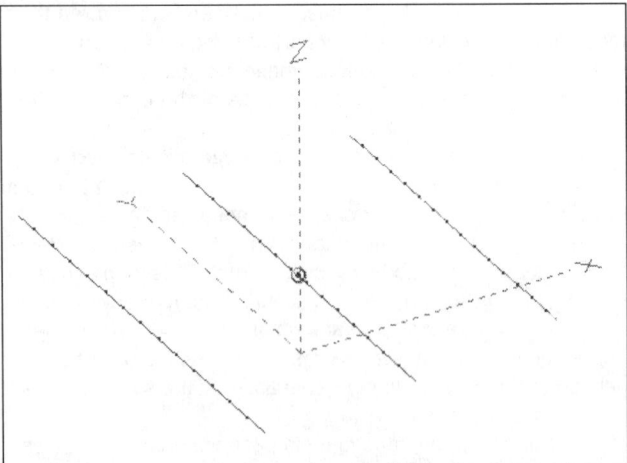

Fig 13.17: The EZNEC Wires Window Spreadsheet display showing a model of a 3-element Yagi. The Z co-ordinate is zero because this is a free-space model. The unit of measurement specified is wavelength

Fig 13.18: A three-dimensional view of the 3-element beam modelled in Fig 13.17. The feedpoint (source) is shown as a small circle

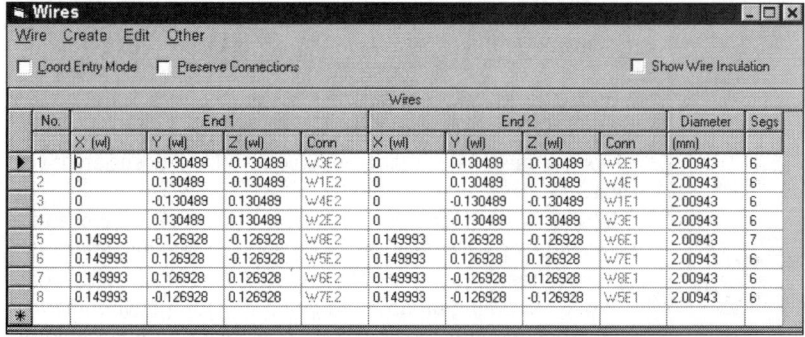

No.	End 1				End 2				Diameter	Segs
	X (wl)	Y (wl)	Z (wl)	Conn	X (wl)	Y (wl)	Z (wl)	Conn	(mm)	
1	0	-0.130489	-0.130489	W3E2	0	0.130489	-0.130489	W2E1	2.00943	6
2	0	0.130489	-0.130489	W1E2	0	0.130489	0.130489	W4E1	2.00943	6
3	0	-0.130489	0.130489	W4E2	0	-0.130489	-0.130489	W1E1	2.00943	6
4	0	0.130489	0.130489	W2E2	0	-0.130489	0.130489	W3E1	2.00943	6
5	0.149993	-0.126928	-0.126928	W8E2	0.149993	0.126928	-0.126928	W6E1	2.00943	7
6	0.149993	0.126928	-0.126928	W5E2	0.149993	0.126928	0.126928	W7E1	2.00943	6
7	0.149993	0.126928	0.126928	W6E2	0.149993	-0.126928	0.126928	W8E1	2.00943	6
8	0.149993	-0.126928	0.126928	W7E2	0.149993	-0.126928	-0.126928	W5E1	2.00943	6

Fig 13.19: The EZNEC Wires Window Spreadsheet display, showing the construction of a 2-element quad. Although this is a free-space model, Z co-ordinates are required to model the three-dimensional structure. The specified unit of measurement is wavelength

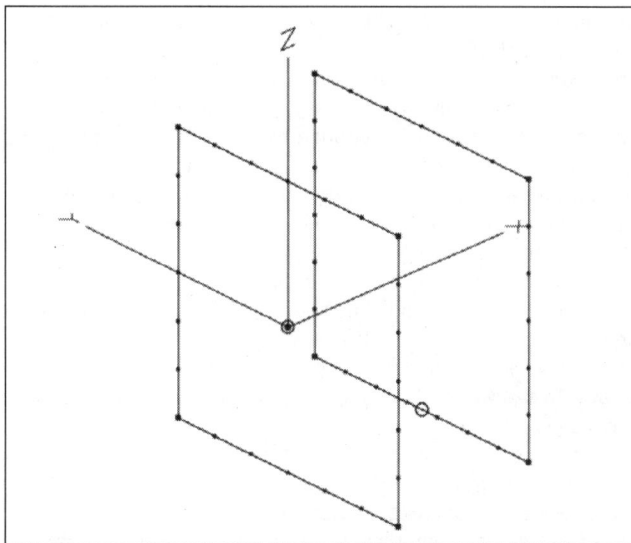

Fig 13.20: The three-dimensional view of a 2-element quad antenna showing the X, Y and Z axes. The dots at the corners of the loop are connection points. The dots along the wires are

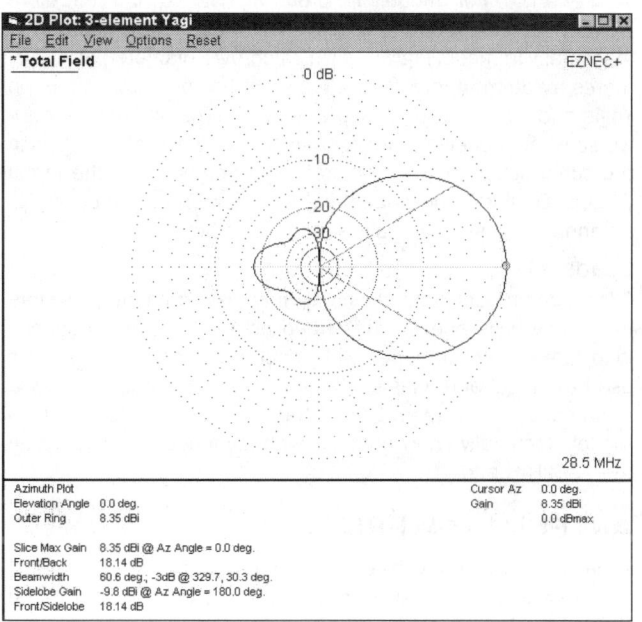

Fig 13.21: An example of an azimuth plot of the 3-element Yagi. The forward gain, front-to-back ratio, maximum sidelobe level and main lobe directivity beamwidth is shown in the data box

As mentioned earlier the antenna can only be modelled as a straight conductor. A bent wire is modelled by connecting two or more straight wires. Wires are considered to be connected when their ends share the same XYZ co-ordinates.

For example, each loop of a cubical quad antenna is described by four separate wires whose end points lie at four points as shown in **Fig 13.19**. A two-element quad, is thus modelled by eight wires, even though a real quad antenna actually has only two continuous wires strung around the spreaders. The quad structure is shown in **Fig 13.20**.

Overlapping wires aren't automatically connected by the program. For example, four wires are required to model an X-shaped structure if the conductors are connected at the centre of the X. A Yagi element composed of tapered sections of telescoping tubing may be modelled by using several connected wires having different diameters.

Plotting

Once the antenna model is built, its performance can be calculated by selecting FFPlot in the Control Centre Information Window.

The field strength diagrams can be plotted as azimuth, elevation or three-dimensional patterns and are plotted in polar co-ordinates. These are plotted in the ARRL log periodic scale, described earlier in this chapter. An example of an azimuth plot of the 3-element beam is shown in **Fig 13.21**, which shows the forward gain, front-to-back ratio, maximum sidelobe level and directivity beamwidth. The plots can be saved and viewed later without redoing the analysis.

At his stage, after the impedance values have been calculated, the View Antenna window (Fig 13.16) is modified to show the current distribution, as shown in **Fig 13.22**. Wire currents are displayed directly in relation to the wires in their true physical context, the greater the distance of the current trace from the element, the greater the relative current magnitude.

If the current phase is required then the current trace is displayed at an angle relative to the wire axis to represent phase. This aspect can be seen more easily when the antenna model is rotated to view the elements end-on so that the phase and amplitude of the currents can be seen.

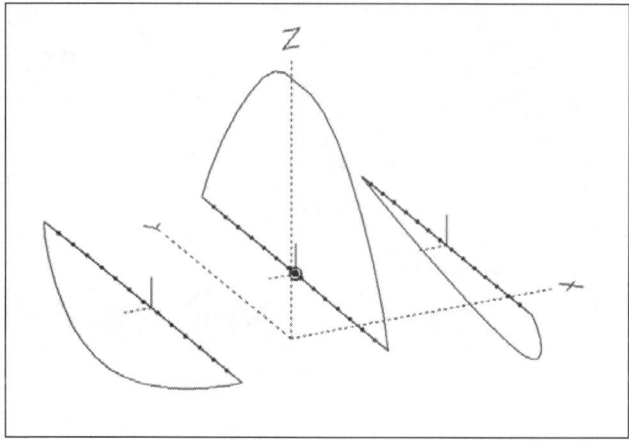

Fig 13.22: Three-dimensional view of a 3-element Yagi, showing relative current distribution and phase on the driven element and the parasitic elements. Small markers, showing 0 and 90 degrees are used as phase references

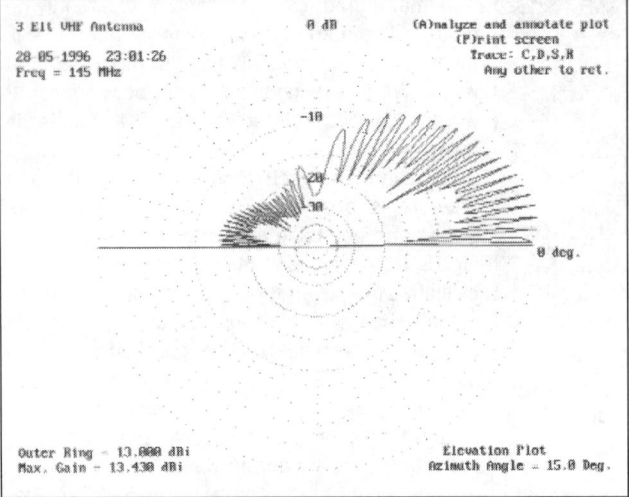

Fig 13:23: Computed elevation polar diagram of a 145MHz 3-element Yagi, 10m above ground

Modelling ground

All the polar diagrams shown so far are plotted as free space diagrams that assume no ground reflection effect. As described earlier in this chapter, a horizontal section at zero degrees relative to the X-axis through the three-dimensional diagram describes characteristics of the antenna without the complication of taking ground into consideration. This is the easiest way to model or compare antenna configurations in early stages of comparison or development.

In the real world ground affects the far field patterns of all antennas and the effect of ground on a three element beam is shown in Figs 13.12 and 13.13. Even VHF and UHF antennas located many wavelengths above ground have radiation patterns quite different to free space patterns, as shown in **Fig 13.23**.

There are various ways in which ground can be modelled and can be ideal ground, or 'real' ground environments. If either a perfect or 'real' ground is specified, EZNEC assumes a perfect ground for impedance and current calculations. The 'real' ground description is used only for determining the shape and strength of the far field (pattern). EZNEC calculates

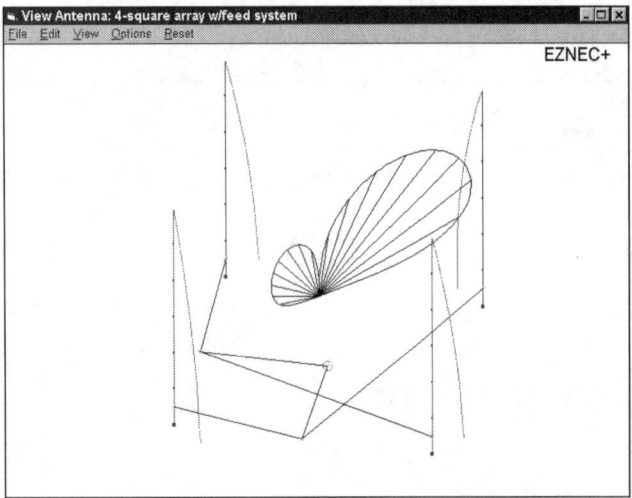

Fig 12.24: A model of the four-square array, one of the several antenna models supplied with EZNEC. Transmission lines provide the correct phasing, which is indicated by the current traces on each element. The view is overlaid with an elevation plot of the directivity pattern

the far field pattern that results from ground absorption or partial reflection due to finite ground conductivity and permittivity.

When an antenna is modelled with a ground environment using an azimuth plot, the elevation angle must also be selected for the reasons described earlier in the text associated with Fig 13.12.

EZNEC has a facility for changing the height of the entire antenna using a Change Height By function which modifies the z coordinates of both ends of all wires by the specified amount.

When modelling an antenna using a ground environment a Wire is connected to ground by specifying a zero z-co-ordinate. To give some idea of the sophistication of EZNEC, a model of a four-square array is used as an example. In **Fig 13.24** a four-square array is modelled using four verticals, each fed at the point where it contacts ground and fed via transmission lines that provide the correct phasing. This antenna model comes with a library of antenna models supplied with EZNEC.

This popular phased array has several desirable properties. Because of its symmetry, it's easy to switch in four directions. The forward lobe is broad enough that four-direction switching gives good coverage to all directions. Good rejection of signals occurs over a broad region to the rear. The small rear nulls can be eliminated and the forward gain increased slightly by increasing the element spacing.

Note that the transmission line lengths are not the same as might be expected. This is because the delay in a transmission line isn't equal to its electrical length except in special circumstances -- circumstances which don't occur in most phased arrays. To see where these feedline lengths came from, see reference [7].

As it stands, the model isn't good for testing the feed system over a range of frequencies. This is because the feedline lengths are specified in degrees, rather than length. This makes the lines magic, because they keep the same electrical length regardless of frequency. To make a realistic frequency-dependent model, the transmission line lengths would have to be specified in metres or feet.

Like any other modelling program, NEC-2 and therefore EZNEC, has limitations of its own. The most severe probably is inaccuracy in modelling wires which change diameter, as in elements made from telescoping tubing. It has a serious effect on Yagis and other sharply-tuned antennas, (The situation is even worse when modelling wire connected to the end of tubing as in the construction of the all metal quad antenna and the metal G3LDO Double-D described in the chapter on practical HF antennas, but isn't generally serious for others.

Loads

EZNEC can model lumped circuits such as terminating resistors or loading coils and are described as 'loads'. The impedance load is a resistor in series with reactance. This model can be used to model antennas containing a resistive load, such as a Rhombic or a Beverage. EZNEC has been used successfully to model electrically short antenna with loading coils, as used on the 136kHz band.

ANTENNA MATERIALS

There are two main types of material used for antenna conductors, wire and tubing. Wire antennas are generally simple and therefore easier to construct, although some arrays of wire elements can become more complex. When tubing is required, aluminium tubing is used most often because it is relatively strong and lightweight.

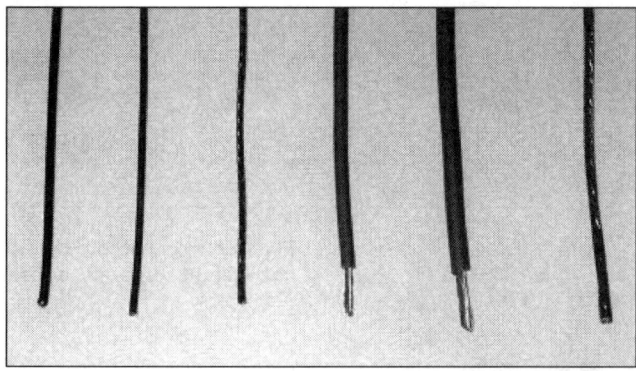

Fig 13.25: Various types of antenna wire. From the left to right (1) 16SWG hard drawn single strand, (2) 14SWG hard drawn single strand, (3) 14SWG multi-strand, (4) Plastic covered 1.5mm multi-strand (5) Plastic covered 2mm multi-strand (6) Multi-strand FLEXWEAVE antenna wire (wire samples supplied by WH Westlake Electronics)

Antenna Wire

Wire antennas can be made from any copper wire. The RF resistance of copper wire increases as the size of the wire decreases. However, in most types of antennas that are commonly constructed of wire (even quite thin wire), the radiation resistance will be much higher than the RF resistance, and the efficiency of the antenna will still be adequate. Wire sizes as small as 0.3mm have been used quite successfully in the construction of 'invisible' antennas in areas where more conventional antennas cannot be erected. In most cases, the selection of wire for an antenna will be based primarily on the physical properties of the wire. For long wire antennas the best material is 14SWG hard-drawn copper wire, which is ideal for applications where significant stretch cannot be tolerated. Care is required when handling this wire because it has a tendency to spiral when it is unrolled. Make sure that kinks do not develop; wire has a far greater tendency to break at a kink.

The most practical material for wire beams is insulated 14 to 16SWG multi-strand flexible tinned copper wire. Wire having an enamel coating is also useful and preferable to bare wire, since the coating resists oxidation and corrosion.

Wire antennas should preferably be made with unbroken lengths of wire. In instances where this is not feasible, wire sections should be spliced. The insulation should be removed for a distance of about 100mm from the end of each section by scraping with a knife or rubbing with sandpaper until the copper underneath is bright. The turns of wire should be brought up tight around the standing part of the wire by twisting with broadnose pliers.

The crevices formed by the wire should be completely filled with rosin-core solder. A large wattage soldering iron will be required to melt solder outdoors, or a propane torch can be used. The joint should be heated sufficiently so the solder flows freely into the joint when the source of heat is removed momentarily. After the joint has cooled completely, it should be wiped clean with a cloth, and then sprayed generously with acrylic to prevent corrosion.

Most antenna material dealers sell various types of antenna wire as shown in **Fig 13.25**. Antenna wire can often be obtained from scrap yards. Scrap electrical wire is usually heavily insulated and therefore too heavy for antenna elements but is fine for radials.

A cheap source of hard-drawn material is scrap outdoor telephone twin, insulated wire, used (in the UK) to distribute underground cable to subscribers via a telegraph pole.

Aluminium Tubing

Many of the beam antennas described in the antenna chapters are constructed from aluminium tubing.

Self-supporting horizontal HF beam elements require careful mechanical design to arrive at the best compromise between storm survival, sag and weight. This is done by 'tapering' the elements, ie assembling the elements from telescoping tubes, thick in the centre and thinner, in several steps, towards the tips as shown in **Fig 13.26**. G4LQI [8] has investigated the availability of aluminium tubing and what follows is the result of his work.

The most common range of American tubing comes in OD steps of 3.18mm (1/8in), and with a wall thickness of 1.47mm (0.058in). This means that each size neatly slides into the next larger size.

British amateurs who have tried to copy proven American designs have found that this could not be done with alloy tubing available in the UK. The only tubing sizes easily obtainable in the UK are Imperial, regardless whether designated in inches or millimetres. They come in outside diameters steps of 1/8in (3.18mm) but with a wall thickness of 1.63mm (0.064in), so the next smaller size does not fit into the larger one. This means that the smallest taper step is 6.35mm (1/4in), requiring the filling of the 1.55mm gap with aluminium shims. This method of construction is described later.

Metric sizes are a compromise; they fit with an easily shimmed gap. Metric size tubing (all inside and outside diameters in whole millimetres), standard in mainland Europe, are a good compromise. See **Table 13.1**,

Aluminium tubing is also available from scrap yards. However, with the exception of scaffolding poles, aluminium tubing is the scarcest material to find. This is because there are not many commodities in our society using this material, and where it is available it only comes in one-off items such as tent poles. There is a further type of material called duralumin, commonly used

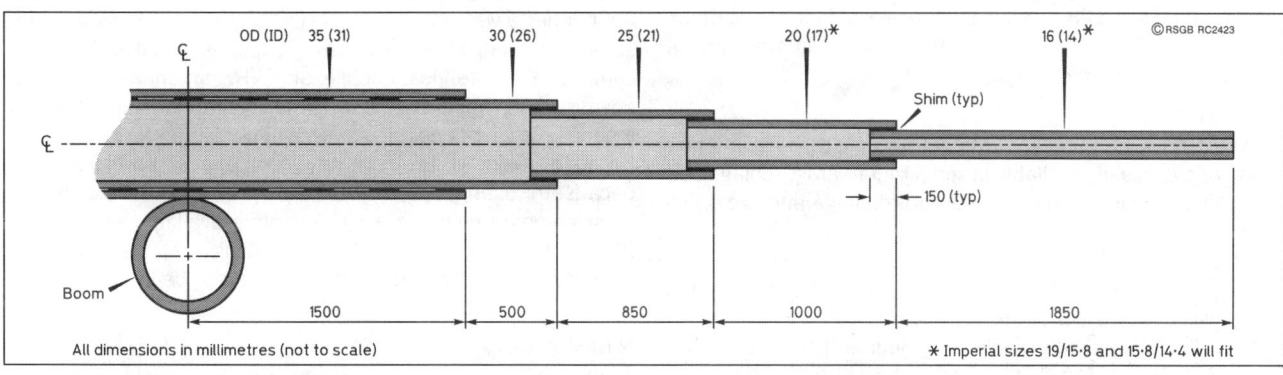

Fig 13.26: A 14MHz reflector designed to survive wind speeds of up to 159km/h, sagging 16cm and weighing 5.6kg

OD mm	Wall g/m	Weight
6	1	42
8	1	60
10	1	76
12	1	93
13	1	103
14	1	110
16	1	127
19	1.5	227
20	2	-
20	1.5	235
22	2	339
22	1.5	261
25	2.5	477
25	2	398
25	1.5	298
28	1.5	336
30	3	687
30	2	484
32	1.5	387
35	2	564
36	1.5	438
40	5	1495
40	2	644
40	1.5	489
44	1.5	541
45	2	-
48	1.5	603
50	5	1923
50	2	820

Table 13.1: Metric-size alloy tubing available in continental Europe. Material F22 (AlMgSo 0.5%). Tensile strength 22kg/mm². Standard lengths are 6m

Fig 13.27: Method of joining lightweight tube of equal diameters

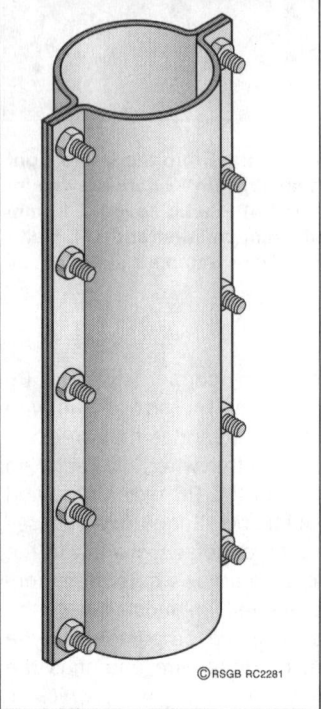

Fig 13.28: Two-piece steel sleeves of the type used to join scaffolding tubes together

for aircraft construction and boat masts. It has the advantage of being lighter and stronger than aluminium but is more brittle. In fact almost all aluminium tube is an alloy of some sort because pure aluminium is rather malleable.

Aluminium or duralumin tubing is useful for making lightweight masts. Lightweight sections of thin-wall tubing can be joined together using a short joining section, which is sliced longitudinally with a hacksaw and sprung open using a screwdriver, see **Fig 13.27**. The two sections to be joined are forced into the joining section and clamped tight using hose clamps.

Aluminium scaffolding poles are useful for masts and booms of larger HF beams. This material is thick walled and is strong and has the advantage of having clips, clamps and extension sleeves, see **Fig 13.28**, used in the process of building scaffolding platforms. Its use in the construction of a fold-over mast is described later in the chapter.

Steel tubing

Steel tubing is an excellent material for constructing antenna masts and is usually available in scrap metal yards. Tubing used for antenna masts should be free from damage and excessive corrosion.

The lower sections of a 12metre (40ft) high steel self-supporting mast should be at least 10cm (4inches) in diameter, with a wall thickness of 5mm.

Steel tubing is often available, threaded, with screw couplers. These couplers are fine for the purpose for which the tubing was designed ie piping liquid or gas. When tubing is used for antenna supports it is often under some bending stress. Couplers only have a short length of screw thread and will be a source of weakness when tubing is employed as an antenna mast, so do *not* be tempted into making a mast using these couplers.

Steel tubes should only be joined by employing lengths that telescope into each other, with at least 30cm (12inches) of overlap and secured with a nut and bolt. Do not weld the sections together, a 12-metre section of steel tubing is very heavy and difficult to manage. It is much easier to assemble a mast in sections. Details on how to construct a steel mast are described later.

When a small diameter pole is joined to larger diameter pipe, eg scaffolding pole into 8cm (3in) pipe, metal strip or angle iron shims can be used to pack any space between the differing diameters before securing with a nut and bolt.

Copper tubing

Copper has a very good conductivity but is rather heavy so is not suitable for large HF antennas. Copper is suitable for small compact HF antennas, mobile and VHF antennas. A further advantage of copper for the antenna constructor is that there is a good selection of couplings available.

Copper tubing is also relatively plentiful at scrap metal yards because of changes to central heating systems. In the UK the most common copper tubing diameters available are 16 and 22mm. However, some old scrap tubing may have imperial dimensions so check these if the tubing is to be integrated with an existing structure.

Metal plates

Aluminium plates are particularly useful for making mast-to-boom and boom-to-element fixings as shown in **Fig 13.29**.

The Radio Communication Handbook

Fig 13.29: Method of fixing mast-to-boom and boom-to-element using aluminium plate and U-clamps

Fig 13.30: (a) Mast top guy ring bearing suitable for mast diameters up to 50mm (2in) This is an upside-down view to show guy connection holes. (b) Bearing suitable for base and mast top, for mast diameters up to 9cm (3.5in). Supplied with fittings (not shown) to connect guys

Scrapyards as a Source of Material

Some antenna material can be bought from a scrap metal yard. The best yards are those located near an industrial estate. These contain a much more useful selection of material for antenna constructor. Materials obtainable from scrap metal yards, useful for antenna work construction, are listed below:

- Steel tubing (for antenna masts)
- Steel casing (for mast foundations)
- Steel angle material (for ginpoles, clamps and guy rope anchors)
- Copper and aluminium tubing (for elements and booms)
- Paxolin, Bakelite or plastic sheet (Insulators)
- Electrical wire (antenna elements)
- Electric motors and gear-boxes (for rotators)
- Aluminium angle stock (quad and Double-D spreaders)
- Aluminium plate (couplers for joining elements to booms and booms to masts)

Other Materials

The following materials are very useful for antenna construction.

Hose clamps

Jubilee clips, the antenna constructors friend. They can be used for joining different diameter sections of elements, joining sections of mast, joining wire to metal elements, joining quad spreaders to angle stock - the list is endless. They are readily available at all hardware, DIY and car (auto) part stores. When a clamp is used as part of an outdoor antenna structure, always coat it with a film of grease to prevent corrosion. Never use paint or varnish because this will make it very difficult to dismantle.

Insulators

Antenna insulators should be made of material that will not absorb moisture. The best insulators for antenna use are made of glass or glazed porcelain. Depending on the type of wire antenna, the insulator must also be capable of taking the same strain as the antenna wire and insulators.

Pulleys

Several types of pulleys are readily available at almost any hardware store. Among these are small galvanised pulleys designed for awnings and several styles and sizes of clothesline pulleys. Heavier and stronger pulleys are those used in marine work. The factors that determine how much stress a pulley will handle include the diameter of the shaft, how securely the shaft is fitted into the sheath and the size and material that the frame is made of.

Another important factor to be considered in the selection of a pulley is its ability to resist corrosion. Most good-quality clothesline pulleys are made of alloys that do not corrode readily. Since they are designed to carry at least 15m of line loaded with wet clothing in strong winds, they should be adequate for normal spans of 30 to 40m between stable supports. Choose a pulley to suit the line. The worst situation that can happen with a pulley is when a thin line gets trapped between the pulley wheel and the sheath.

Exhaust pipe clamps

In the USA these are called muffler clamps. They can be used to construct boom-to-mast and element-to-boom fittings.

Spreaders for wire beams

Cane (lightweight bamboo) or fibreglass rod. Fibreglass rods are preferred because they are lightweight and weather well. In addition they have excellent insulating properties.

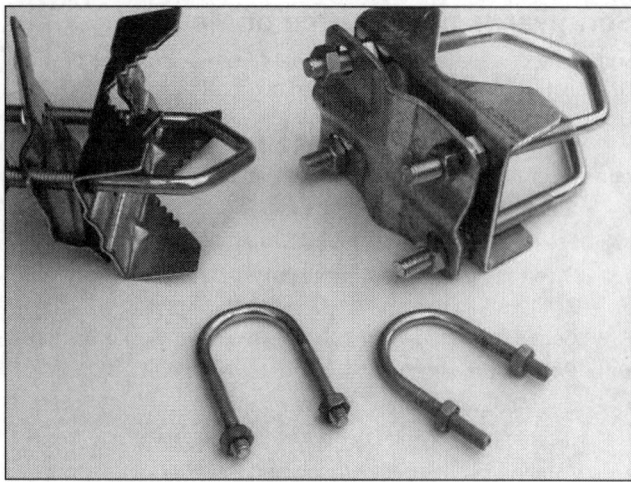

Fig 13.31: A selection of commercial antenna fittings

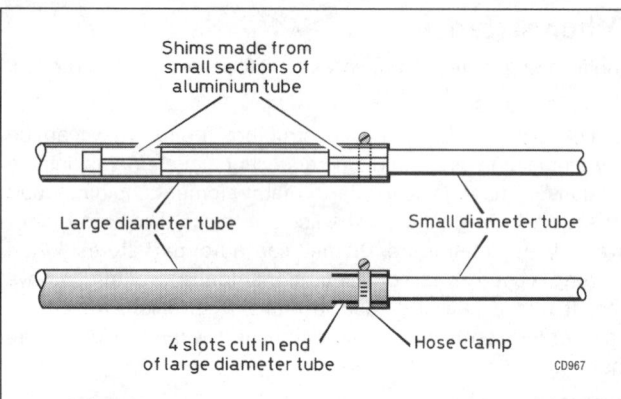

Fig 13.32: Method of joining sections of aluminium tube where the tube diameters present a poor fit (top drawing depicts a cross section of the joint)

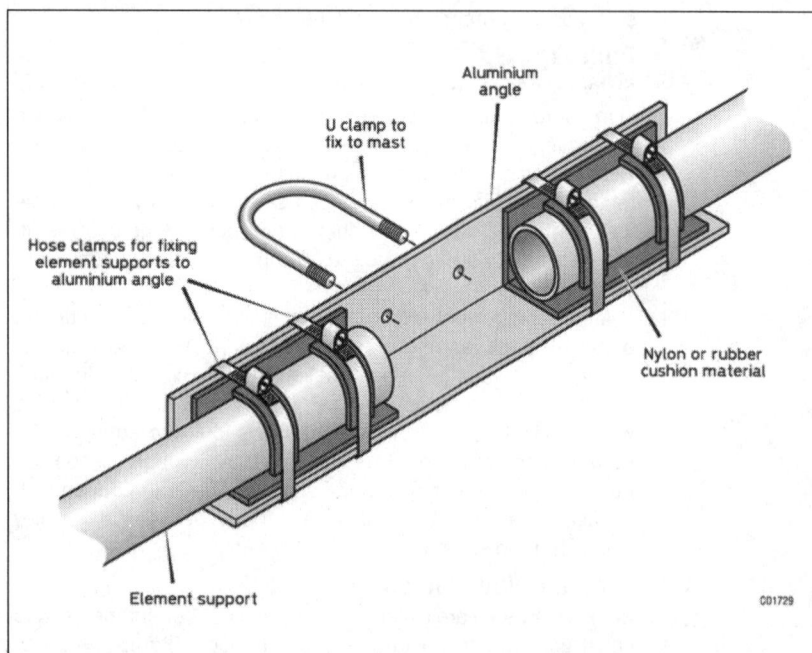

Fig 13.33: Method of fixing cane or fibreglass supports for wire beam elements to a boom or mast. The length of the aluminium angle material and the spacing between the hose clamp supports depends on the size of the antenna structure (not to scale)

Polypropylene rope

Used as halyards and guys for pulling up, and keeping up, mast and antenna structures.

Specialised antenna fittings

The importance of guying masts correctly is discussed later. The problem of how to guy a rotating mast can be overcome with appropriate fittings. There are a number of specialised commercial fittings available for antenna constructors. A mast top guy ring bearings suitable for mast diameters up to 50mm (2in) and 90mm (3.5in) are shown in **Fig 13.30**. These bearings are supplied with fittings to connect the guys.

TV type fittings

There are several fittings that are used by the TV antenna industry that can be pressed into amateur radio antenna service. These are mainly clips for fitting a small antenna to a boom, as shown in **Fig 13.31**. They can be used for fixing elements to booms of HF antennas or VHF antennas to masts

BEAM ANTENNA CONSTRUCTION

Antennas can be built using all-metal construction or with wire elements supported between insulators or on spreaders.

All-Metal Construction

The boom can be fixed to a tubular mast with a metal plate and car exhaust U-clamps as shown in Fig 13.29. Elements can be connected to booms in a similar manner.

Tapered elements can be constructed from lengths of aluminium alloy tubing with different diameters so that the lengths can be telescoped into each other. Often sections do not fit snugly at the ends of joining sections and need to be modified as shown in **Fig 13.32**.

Additionally, if there is a relatively large difference between the two joining sections, a shim can be made from a short section of tubing, slit longitudinally. Any corrosion on any of the metal surfaces that make up the join should be removed with fine sandpaper. The surfaces are then wiped with a cloth and coated with a thin film of grease to prevent corrosion. The join is clamped tight using a hose clamp. This method is far superior to using a nut and bolt where a new set of holes has to be drilled every time an adjustment to length is made. The hose clamp method also gives the joint a lower contact resistance.

If the antenna elements are constructed from tubing and insulated copper wire (such as the all metal quad, see the chapter on practical HF antennas), then a short length of the plastic insulation is stripped from the wire element extensions. These ends are then fixed with hose clamps to the end of the metal elements used to isolate the metals.

It is particularly important that these copper wire/aluminium tube joints are protected with grease to prevent corrosion. If the connections are protected with grease this should not be a problem. However, one authority [9] goes as far as to state that contact between aluminium and copper should be avoided at all costs and that a small stainless steel washer should be used to provide isolation.

Wire Beam Construction

Insulating spreaders for wire beam antennas or helically wound elements, can be con-

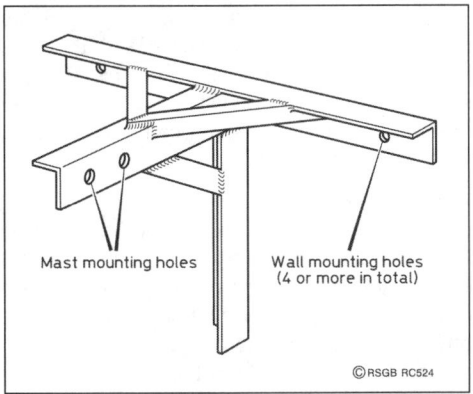

Fig 13.34: A typical well-braced wall bracket

©RSGB RC524

Mast mounting holes

Wall mounting holes (4 or more in total)

structed using cane (lightweight bamboo) or fibreglass rod. The main disadvantage of these materials is that they can easily damaged by crushing at the support point. Special support is required to avoid damaging the ends and aluminium angle stock can be used. The length of this aluminium section depends on the size and the frequency range of the antennas to be supported and for a conventional multi-band quad or Double-D a three-foot length is suitable. Two sections are required for a Double-D or four for a quad. Two holes are drilled at the centre of each section, the distance apart will depend on the size of the mast or boom and hence the size of the U-bolts.

The canes or fibreglass rods are fixed to the ends of the aluminium angle using hose clamps as shown in **Fig 13.33**. Rubber or plastic tubing cushions can be used to prevent the clamps damaging in cane or fibreglass rod supports.

ANTENNA SUPPORTS

Using the House as an Antenna Support

Placing a large antenna 10 or 20m into the air, with access for adjustment and tuning can be a minor civil engineering project. Backyard locations often do not have the space for a free-standing mast or tower. In this case the only solution is to fix the antenna to the house. The usual method of doing this is to fix it to the chimney (if you have one) using a chimney bracket or to fix it to the side of the building with a wall bracket.

Wall brackets

Useful advice for fixing wall brackets comes from G3SEK [10] as follows:

Even a small antenna installation can generate considerable wind forces on the support structure. Think about the directions in which the wind force could act. If the wind is pushing the bracket on to the wall, the force is spread over several bricks, and the fixing is as strong as the wall itself. If the wind is blow-

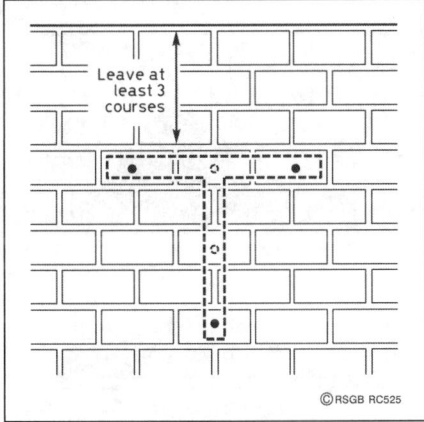

Leave at least 3 courses

©RSGB RC525

Fig 13.35: Typical drilling pattern for fixing a wall bracket. Patterns may vary, but always drill into the centre of the bricks

ing parallel to the wall, and the bracket is strong enough, most kinds of wall fixings will be extremely secure against the sideways forces. The difficult situation is when the wind is blowing away from the wall and trying to pull the bolts straight out of the bricks . . . or the bricks straight out of the wall. This latter possibility is a serious one unless the wall is well built. Older houses with mortar that has weakened over the years, and bricks made before the era of factory quality control, are simply not a good prospect for a mast bolted to the wall.

Assuming your house has reasonably sound brickwork, the aim should be to mount the top bracket as high as possible, to reduce the wind forces but always leave at least three courses of bricks between the ones you drill and the top of the wall. Also leave plenty of sideways clearance from upstairs window openings, which considerably weaken the brickwork. Obviously the best place to mount the top bracket is quite high on a gable end wall, to shorten the unsupported length of mast and reduce the wind forces.

The bracket itself is important. A cheap, poorly made wall bracket, intended for UHF TV antennas, is unsuitable. Go to an amateur radio dealer and get something substantial and well made, and preferably galvanised. A suitable bracket will look something like **Fig 13.34**, with a T-shaped piece that bolts to the wall and a well-braced arm for fixing the mast. All the component parts should be solidly double-seam welded, not just 'tacked' together. Typically there will be two or more bolt holes in the horizontal member of the T, and one or two more in the vertical member. The top row of fixings will bear almost the entire load, and **Fig 13.35** shows a typical drilling pattern.

To fix the bracket, you must use some kind of expanding wall anchor. These come in several kinds, but they all work by expanding outwards and gripping the sides of the holes. The traditional 'Rawlbolts' are best, which can give a very secure fixing. There are other anchor methods. The 'DIY' fixing using large plastic wall plugs and 'coach bolts' are not recommended.

This is a safety-critical application, so spend some money on properly engineered fixings that are designed to work together as a system, and follow the manufacturer's instructions exactly. The holes for the 'Rawlbolt' anchors should be drilled in the centre of the brick as shown in **Fig 13.35**. The optimum size for ordinary brickwork is M10, which requires a 16mm diameter hole. Three or four of these should be more than adequate to withstand the wind forces envisaged - but only if they are installed correctly.

Choose a sound set of bricks for drilling, free from any hairline cracks. If necessary, be prepared to move the mast a little from its planned location. Drill into the exact centre of each brick, not near the edges, and never into mortar. If necessary, make new mounting holes in the bracket to suit your own brickwork. The holes in the brackets should be 10mm diameter for M10 bolts, with some extra clearance to help the bolts line up. Hold the bracket to the wall, level it with a spirit level and mark the centres of the holes. Begin drilling with a small masonry bit. Before you use the electric drill, place the point of the bit exactly on your drilling mark and tap gently with a hammer to chip out a small dimple. This will prevent the point from wandering when you start drilling. If you're using a hammer-drill, start without the hammer action until you've made a deep enough hole to prevent the bit from wandering. Do the same at each change of bit as you open out the holes gradually, using progressively larger sizes. Use patience rather than brute force, and you're more likely to make good cylindrical holes, square to the wall and exactly where you want them. It's also kinder to the electric drill; and above all it's much safer for you on the ladder.

The final holes must be exactly the right diameter as specified by the manufacturer. For example, the hole for an M10 Rawlbolt

Fig 13.36: Example of a chimney lashing used to support a double-D antenna

must be 16mm diameter - not 15, not 17, but 16mm This is very important because the entire strength of any type of wall fixing comes from the contact of the anchor sleeve against the inside of the hole. The sleeve should be a gentle tap fit, so that when the bolt is tightened the anchor will immediately start to grip hard.

Tap the anchor sleeve into place just below the surface of the wall so that, when the bracket is bolted on, it contacts the wall and not sitting on the end of the sleeve. Do this without the bolt inserted, and then fit the bracket. Leave the bolts slightly loose, level the bracket, and then tighten them. The tricky part is to tighten the bolts to the correct torque - enough to expand the anchor sleeve and develop the fixing strength, but not so much that it splits the brick and ruins the whole fixing.

Although G3SEK uses Rawlbolts, he notes that they can be very prone to split the bricks if over-tightened. You might consider alternative types, such as the Fischer bolts which use a softer plastic sleeve to grip the inside of the hole. In any case, use the type of anchor with a free bolt, which screws in, and not the type with a stud that takes a nut.

If your house is built using modern bricks that have holes right through the middle, conventional expanding anchors are no use, and you'll need to investigate other systems. Wall anchors using a chemical adhesive fixing system are also available, and have the advantage of not stressing the bricks at all, while having higher claimed strengths than conventional expanding anchors. They can also be used for fixing into hollow bricks, but the strength of the bricks themselves may become a factor. As with any adhesive bonding system, success depends on careful preparation and following the instructions exactly. One suggestion when using conventional Rawlbolts in ordinary brickwork is to use epoxy resin as well, to try and obtain the best of both worlds.

If you are fixing to a gable end wall, yet another possibility is to drill right through the whole wall and into the loft space, and then use long bolts or studs to secure the bracket to a steel plate that spreads the load over the inside wall.

The lower bracket is much simpler, because it bears much less load than the upper one. Its main purpose is to steady the mast and prevent it from bowing below the upper bracket. Mark out and drill for the lower bracket after fixing the upper one, lining them up with a plumb line. The bottom of the mast should also be fixed to prevent it from moving sideways.

In the longer term, wall anchors can work loose owing to either frost or thermal expansion/contraction cycles, and then the wind will work on them further. Check the fixings every spring and autumn. If you are intending to mount a commercial mast or antenna against the wall, obtain specific advice from the manufacturer and follow it exactly.

House Chimney

The house chimney can be used for an antenna support as shown in **Fig 13.36**. The main advantage of this method is that chimney-mounting brackets are easy to obtain. Some of these mounting brackets can be seen supporting some precariously tall TV antenna structures in fringe TV signal areas.

The chimney of an older house, where the mortar that has weakened over the years, needs to be examined, and if necessary repointed before fitting a chimney mounted antenna.

The single wire lashing kits used for TV antennas are totally unsuitable for amateur radio antennas. A double TV antenna chimney lashing kit is essential and will support a large VHF array or a small sized HF beam.

ROUTING CABLES INTO THE HOUSE

While on the subject of modifying the house for amateur radio this might be a good time to consider how to route cables into the house. You might have a multi-band beam with its coax cable and rotator control cable, plus a VHF antenna and a long wire antenna for the lower frequency bands. And of course there is the earth connection.

The time-honoured way of dealing with this problem is to drill lot of holes in the window frame, however, modern houses (and a lot of older ones) use double-glazing, with its plastic and metal window frames. Using the window frame as a route for cables is not feasible and another method must be sought. G3SEK used the method of routing the cables by inserting a length of plastic drainpipe through the wall [11]. What follows is how he did it.

In a traditional British brick house with cavity outside walls, the job is well within the reach of a competent DIYer and it should make very little mess. Think of it as installing a waste pipe for the kitchen sink, because 40mm sink waste pipe is probably what you'll need - though it's always good to leave enough room for more antennas in the future! You'll also need a good electric hammer drill, at least one masonry bit of about 10mm diameter that is long enough to go right through the double wall, a shorter masonry bit of about the same size, a fairly large hammer and a long, narrow cold chisel.

Plan very carefully to find the best place to drill through. Leave at least one whole brick away from doors or window frames. Remember that the frame has a solid lintel across the top, extending outwards on both sides. Check both the inside and the outside of the wall with a live cable and metal detector to be sure that you

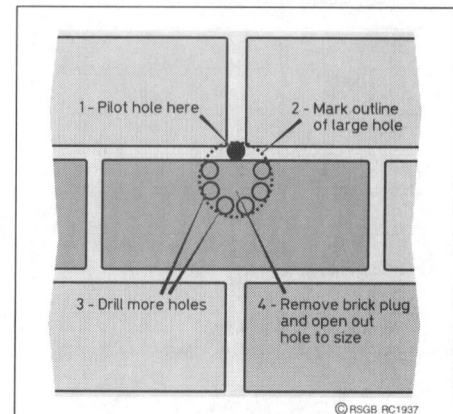

Fig 13.37: How to drill a 40mm hole through bricks. You can also use the same technique for larger holes

1 - Pilot hole here
2 - Mark outline of large hole
3 - Drill more holes
4 - Remove brick plug and open out hole to size

©RSGB RC1937

won't meet any nasty surprises. If yours is a wooden-framed house with a brick outer skin, take great care to avoid structural timbers.

First you need a pilot hole, right through the wall, and then you're going to enlarge the hole from each side by continuing around a circle. Start from the outside and drill through the mortar, halfway along a brick, as shown in **Fig 13.37**.

Use one of the shorter masonry bits to start the hole, and take care to drill it accurately at right angles to the wall - the mortar joint will guide you. It's just possible that you will hit a metal wall tie. If you do, move along to the next brick. When you break through into the wall cavity, change to the longer bit and carry on drilling. To avoid pushing off a big patch of plaster from the inside wall, stop when you're within a few centimetres of breaking through. Switch off the hammer action of the drill, and continue with very gentle pressure until you're right through both walls.

Now mark out a circle on each side of the wall, rather larger than the diameter of the pipe, as shown in Fig 13.37. The pilot hole is at the top of each circle. Work separately from each side, using the shorter masonry bit. Drill a ring of holes as close together as possible, stopping when you're through to the cavity. Because it's important to start each hole in exactly the right place, it helps to begin with a smaller bit, using the drill at slow speed with the hammer action off. When the hole is well started into solid brickwork (or breeze block on the inside wall) it's safe to change to the larger bit with the faster hammer action. Next you need to open out the hole from each side, using the cold chisel and occasionally perhaps the electric drill, until the pipe will slide right through.

Chip away carefully without too much violence so as not to crack the outside brick or do any unnecessary damage to the interior plasterwork - and try to pull the central plugs out rather than pushing them into the cavity. With care you can make almost as good a hole as a professional using a big core drill. Make sure that the pipe will slide through horizontally, or sloping a few degrees upward from the outside so that rainwater won't run in. Set the end of the pipe just proud of the inside wall, and leave any overhang outside. Fill the gaps around the pipe with mortar or exterior filler on the outside and plaster or interior filler on the inside, and let it all set solid. The next day saw off the outside end of the pipe, a few centimetres away from the wall. Now you can start to thread the cables. A 45° or right-angle pipe elbow, facing downwards, fixed to the pipe where it emerges on the outside wall, may be used to help keep the rainwater out. When all the cables are in place, stuff in plastic foam for draught proofing, or use aerosol-expanding foam.

This method of installing a pipe is quite easily reversible before you move house. The pipe will pull out from the outside with a bit of effort, and you can plug and plaster over the hole in the inside wall. On the outside, you'll only need to replace one brick if you have followed the drilling pattern in Fig 13.37.

CONSTRUCTION OF FOLD-OVER MASTS

Many radio amateurs use commercial lattice construction masts, which have a telescoping and fold-over capability. They have the advantage of having well-defined data regarding heights and wind loading, although they have a fairly high visual impact, which may be a problem in some locations.

A drawback to trying to adjust an antenna using these types of support is that they take some time and effort to raise and lower. Additionally, the winch cables are not designed for the continual raising and lowering normally encountered when a lot of antenna work is being done over a period of time. For this reason a homemade structure can be designed to be easy to raise and lower. An advantage of an easily folded-over mast is that it can be quickly lowered if severe gale force winds are forecast. Two simple designs for single-handed construction will now be described.

The G2XK Lightweight Fold-over Mast

Eric Knowles, G2XK, used the method described below to support 6-element 10 metre beam on a 11 metre (36 foot) boom at a height of 12 metres (40ft), using only 80mm (3in) diameter thin-wall duralumin tubing, see **Fig 13.38**. This large structure weathered many a gale that swept across the Vale of York.

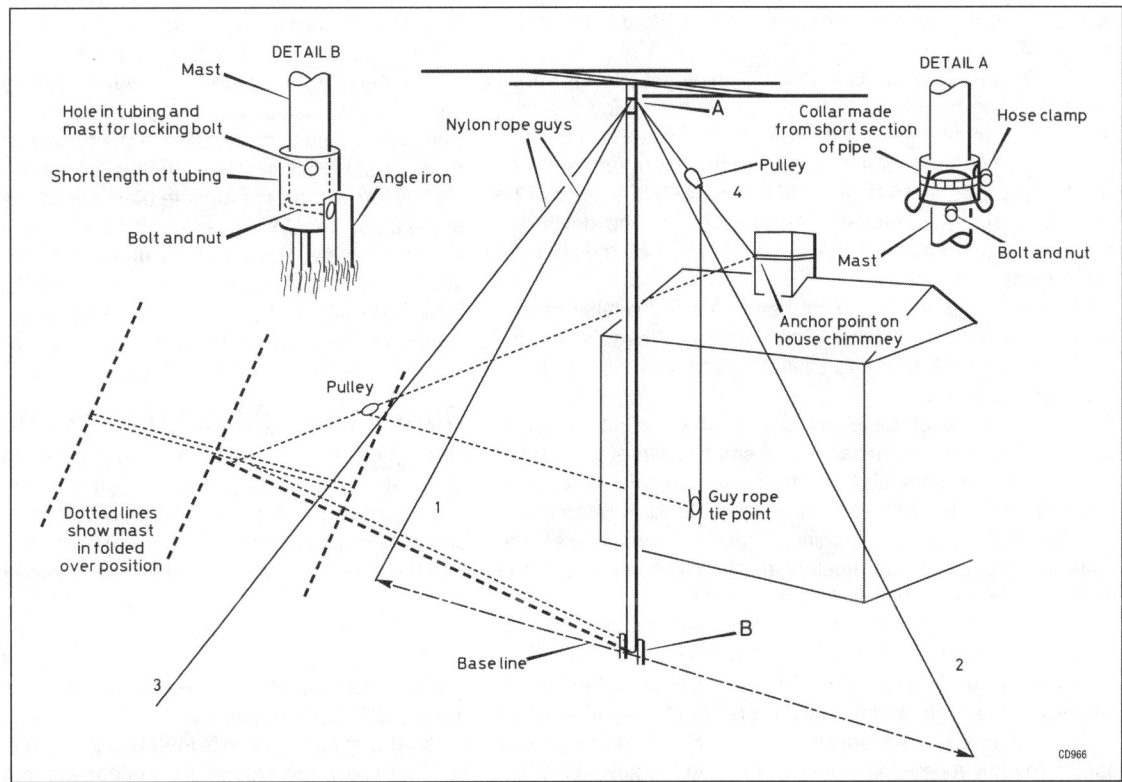

Fig 13.38: Constructional details of the G2XK type of lightweight fold-over mast

There is nothing new in this method of supporting, or raising and lowering masts using guy ropes. The military have used the method for many years for supporting wire fixed wire antennas.

The description that follows is of a similar mast; suitable for supporting a medium sized experimental beam antenna. No special tools or welding equipment are required to construct this structure and is an excellent support for experimental antennas provided the space for the guys is available. Do not use steel tubing for the mast of this design because it is too heavy. Aluminium scaffolding pole is not really suitable for this design of mast because of the weight/length ratio, although it could be used for short masts of up to 8 metre (25ft) high.

The layout is illustrated in Fig 13.38. The structure can be used with a fixed mast and rotator, or the mast can be rotated. In this case provision has to be made to allow the mast to rotate and be folded over. A minimum of four guy ropes is used.

The anchor point for guy wire 4 must be above ground level; the chimney of a nearby house is suitable with a lightweight structure. Guy ropes 1 and 2 are anchored along the baseline so that they retain the same tension when the mast is being erected or folded over. The length of guy rope 3 is adjusted so that it is under tension when the

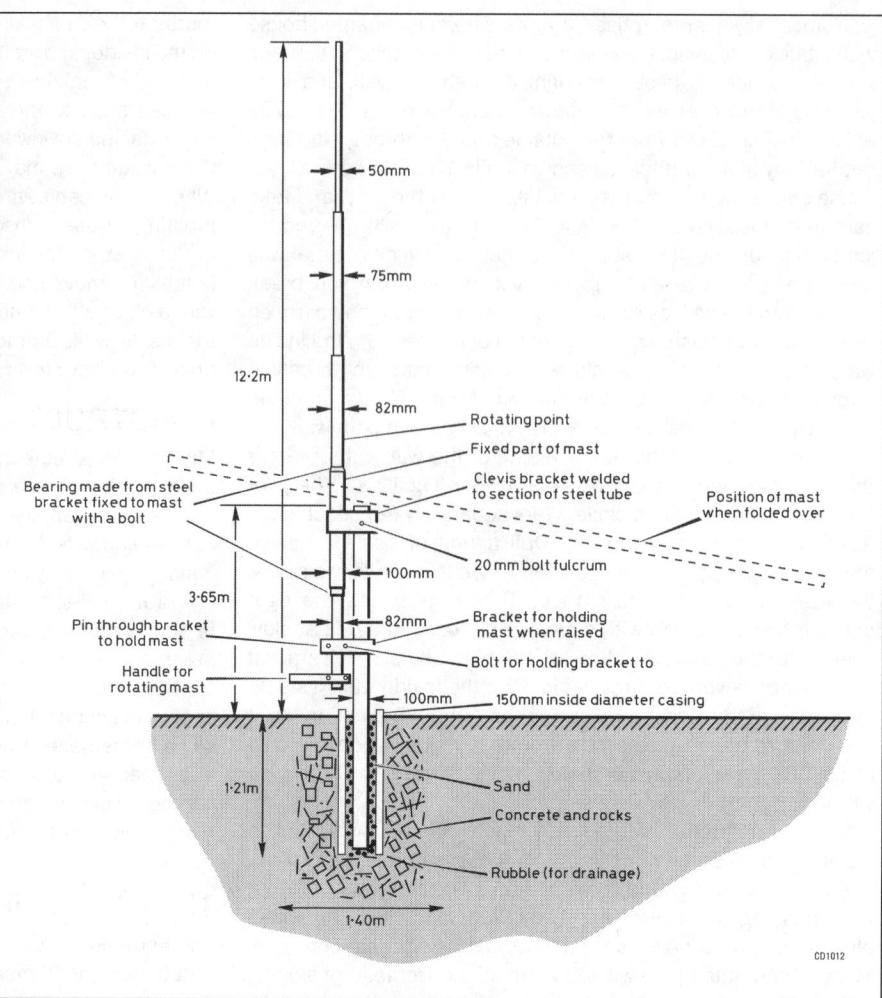

Fig 13.39: Constructional details of the counterweighted fold-over rotatable 12 metre mast

mast is in the vertical position. The original G2XK version used two sets of guy ropes but only one set is illustrated in Fig 13.38 for clarity.

This structure gains all its strength from its guys, so it is important that the guy ropes are strong and are connected securely, both at the anchorage and the top of the mast. Polypropylene rope (6mm diameter) is a suitable material for the guys, which should ideally be at 45 degrees to the mast. This angle can be reduced if space is limited but this increases the downward pressure on the mast in high winds and increases the chances of the mast buckling.

Commercial guy rope to mast bearings are available for the top of the mast, see Fig 13.30 and are recommended for this sort of application. The guys should be connected to the bearings with D-clamps.

If a commercial rotatable guy rope support bearing is unavailable, one can be constructed with a short length of steel tube, slightly larger in diameter than the mast. Very thick wire loops can be fixed to the tube using two hose clamps. The top of the mast is inserted into this tubing. A bolt and nut, through the appropriate point on the mast, holds the guy support collar in position. Detail A of Fig 13.38 illustrates this.

The guy anchorage can be constructed from a 1 metre (40in) or so length of angle iron, cut to a point one end and with a hole drilled in the other. This can be driven into the ground at 90 degrees to the angle of pull. The guy anchorage may need to be more substantial for very large masts and/or if the soil is light and sandy. The guy rope should be connected to the guy anchorage

with a D-clamp. A pulley is required for the halyard to enable the mast to be hauled up; a good quality clothesline pulley is suitable.

The base pivot point comprises two lengths of angle iron, cut and drilled in a similar way to the guy anchorages. The two angle iron pieces are driven into the ground, with the holes aligned so that the pivot bolt can be fitted. If the design calls for a rotatable mast then a small section of tubing, whose internal diameter is slightly larger than the outside diameter of the mast, is pivoted to the angle iron. The mast fits inside this section of tubing and is free to rotate. Holes can be drilled through the base tubing and the mast to enable the structure to be locked on any particular heading. Detail B of Fig 13.38 shows how this is done. Lightweight sections of thin-wall tubing can be joined together using a short joining section, see Fig 13.27.

Counter Weighted Fold-over masts

This type of support is heavier and requires more construction effort. Its main advantage is that guy ropes are not absolutely necessary. This design is based on a 18 metre (60 foot) tilt-over support designed by Alfred W.Hubbard K0OHM [12]. The original was designed to support a 3-element tri-band beam and a rotator. In this design, all sections of steel tubing of the mast were welded together. In the original design the foldover mast was partially counterweighted by filling the lower half of the tilt-over section with concrete! A pulley is used to manage the remaining 160kg (350 pound) pull.

The design of the base is interesting. It comprises a section of casing fixed in the ground with a concrete foundation. The gap

Fig 13.40: 12m (40ft) version of the fold-over mast in upright position, supporting a 14MHz metal quad antenna

Fig 13.41: 12 metre (40ft) mast and 14MHz quad in the folded over position

between the mast and the casing is filled with sand. This reduces the high-stress point at ground level that normally exists if the mast is set directly into concrete.

The sand acts as a buffer and allows the mast to flex within the base during high winds. The internal casing diameter should be around 5cm (2in) greater in diameter than the lowest section of the mast.

G3LDO built several of these masts, the largest of which were 18 metre (60 foot) high and supported an all metal quad. The mast and payload should not be fully counterweighted. A top weight imbalance of around 45kg (100 pounds), controlled with a winch will enable the momentum of the structure to be more easily managed.

Medium size 12m (40ft) fold-over steel mast

The mast, described above, is too large for most suburban sites. The following is a description of a smaller version, but even this will require a garden at least 12m long. However the design can be scaled down if required

This mast is counterweighted with approximately 15kg (30lb) of top weight so a winch is not required. It takes about 15 seconds to raise the antenna mast into the vertical position. The mast is relatively lightweight; the top third of its length is 5cm (2in) diameter scaffolding pole. The whole mast

is rotated manually using a handle fixed to the bottom of the mast.

The sections of steel tubing that make up the mast are telescoped into each other for about 30cm (12in) and secured by a bolt and nut. This allows the mast to be assembled, modified or repositioned much more easily than if the section was welded.

The detail of this mast can be seen in **Fig 13.39** and a more general views can be seen in **Figs 13.40 and 13.41**. At the time these photographs were taken, two sections of a tree trunk were used as counterweights. These weights have now been dispensed with by making the lower section of the mast out of solid 82mm steel rod.

Although these structures can be built single-handed the following are areas where some assistance would be of help.

• Inserting the lower half of the mast into the base casing. Two ropes are tied to the top of the lower section, using the holes drilled for the pivot bolt. The section can then be placed with the lower end over the base casing and the top supported on a pair of stepladders. The section can be raised using these ropes, at the same time the lower end is guided into the casing with a section of angle iron.

- Placing the clevis at the top of the mast to enable the bolt to be fitted and inserting the mast into the oversize piping used as the tiltable thrust bearing. These tasks can be eased by using a gin pole with a pulley and rope. The gin pole can be constructed from steel angle-iron and clamped to the mast with additional angle iron pieces or steel straps.

Other Fold-over Masts

Wooden masts

In the early days of amateur radio, wood was a very popular material for constructing masts and even beam antennas. These days this material is less popular because of the cost of quality seasoned timber and the lack of sensible fold-over designs that can carry the payload of a medium size beam antenna to a height of 10 to 15m. The selection of timber and weather treatment requires specialist knowledge, which is beyond the scope of this chapter.

Commercial masts

There is a range of commercial masts available (see *RadCom* or the web for sources). Most of these masts have a lattice structure, with sections of the fold over lattice mast telescoping into each other. This design enables a fairly large mast to be erected into a relatively small garden.

TREES AS ANTENNA SUPPORTS

If you have a tall tree in your garden, you may have a very good support for a wire antenna that does not require planning permission. As antenna supports, trees are unstable in windy conditions, except in the case of very large trees where the antenna support is well down from the top branches.

Tree supported antennas must be constructed much more sturdily than is necessary with stable supports. To this end, the preferred method is to use a halyard and pulley shown in **Fig 13.42**.

The use of a halyard with a mast is shown in **Fig 13.42(a)**. Here the halyard end can be lashed to a bracket. When a tree is used as the support a weight is used, see **Fig 13.42(b)**, to take up the movement of the tree. The endless loop is to allow greater control when raising and lowering the antenna.

Fixing the halyard pulley.

If the point where the pulley is to be attached can be reached using a ladder then fixing it to a branch, pole or building is the easy bit. If you cannot reach the pulley fixing point then a line has to be thrown or propelled over the anchor point.

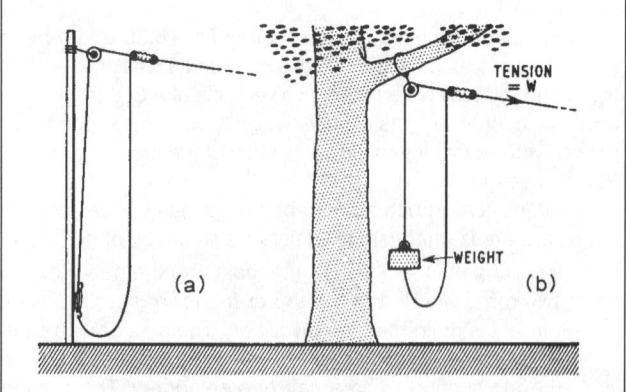

Fig 13.42: Halyard connections to (a) a pole and (b) a tree. The weight is equal to the required antenna wire tension

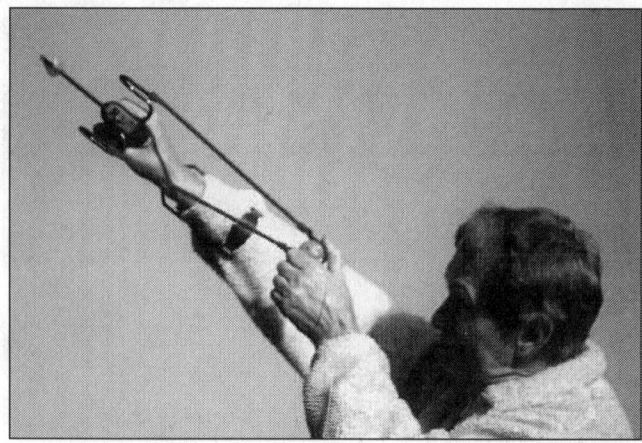

Fig 13.43: An antenna support catapult launch pad. The catapult is held nearly upside down so that the line does not get caught up with anything when fired. It is advisable to gain proficiency with few practice runs in an open space before using it to secure an antenna support.

A useful missile projector is the catapult as shown in **Fig 13.43**. A missile or weight must be used that will not cause damage or injury if things go wrong. The best object for this purpose is a squash ball or, as used in Fig 13.43, a plastic practice golf ball filled with wood filler.

The best sort of pilot line is strimmer cord. This comes in several thicknesses, the heaviest for heavy-duty petrol driven strimmers and the lightest weight for electric strimmers; the latter being the most suitable for the purpose. Strimmer line is very strong and resists kinking.

The line can be stretching out straight on the ground or better still, zigzagged left and right on the ground. If snarling of the line in a problem (due, say, to garden plants and bushes) try making a stationary reel by driving eight nails, arranged in a circle, through 20mm board. After winding the line around the circle formed by the nails, the line should reel off readily at lift-off. The board should be tilted at approximately right angles to the path of the shot.

If it is necessary to retrieve the line and start over again, the line should be drawn back very slowly; otherwise the swinging weight may wrap the line around a small branch, making retrieval impossible. This system can also be used to place an antenna support over a roof.

The pilot line can be used to pull a heavier line over the tree or roof. This line is then used to haul a pulley up into the tree after the antenna halyard has been threaded through the pulley. The line that holds the pulley must be capable of withstanding considerable chafing against the tree.

A metal ring, around 70 -100mm (3 to 4in) in diameter can be used instead of a pulley. The antenna support wire is just looped through the ring. This has more friction than a pulley but it will not jam.

Safety

Safety should be a primary consideration when erecting antenna masts.

NEVER erect an antenna and mast that could possibly come in contact with electric power lines. Never rush this sort of work.

ALWAYS stop to consider the implications of the next move, particularly when dealing with heavy sections of steel tubing.

DO NOT use an antenna support structure that requires the joint efforts of all members of the local radio club to raise and lower it, although help in the construction stages is always welcome.

Fig 13.44: A compact two-element multiband beam

WIND LOADING

The saying that "if you haven't had an antenna fall down then you don't build big enough antennas" might sound very smart but it is hardly good engineering practice. Bridge building engineers would hardly get away with it.

Most antennas are brought down by very strong winds so it is important to consider the effect of windloading of your antenna installation. While there are a couple of excellent articles on the subject [13] [14] a few suggestions are given here so that you can get some idea of the forces involved.

G3SEK [15] says that the round figures that stick in his mind are that at 100mph, every square foot of exposed area suffers a sideways pressure of 25 pounds; or that at 45-50m/s the wind force is about 150kgf/m2 (kilograms force per square metre).

So how do you know how many square metres or feet your antenna installation is? To avoid any aerodynamics the best way is to simplify the parts of the elements to a 'flat slab' area. So that a pole 3m long and 50mm (.05m) diameter is a flat slab 0.15m2.

Assessing the Exposed Areas

This is where the calculations become simple. Take, for example, a compact HF beam mounted on the roof of the house shown in **Fig 13.44**.

This beam has two elements, which are 4.49m long and 25mm in diameter. The flat slab area of the elements is 4.49 x 0.025 x 2 = 0.2245m2. The boom is 2.1m long and 0.035m diameter: 2.1 x 0.035 = 0.08m2. The loading coils and spokes at the ends of the elements are complicated so they have been modelled as four cylindrical objects 0.3m long by 0.06m in diameter. This gives an area of 0.3 x 0.06 x 4 = 0.072m2. The mast fixed to the chimney is around 1m long and 0.05mm diameter. This adds 0.05m2. For the rotator, 0.15m x 0.13m = 0.02m2 has been allocated.

This totals 0.225 + 0.072 + 0.05 + 0.02 = 0.367m2, or for rough calculation 0.4m2.

The boom has not been included because it would be end on to the wind if the beam were facing into wind. If the beam were to be rotated 90 degrees to the wind then the area facing the wind would be 0.08 + 0.06 + 0.05 + 0.02 = 0.21 or say 0.2m2.

Calculating the Wind Pressure

To work out the force acting on the antenna structure by the wind, see **Fig 13.45**.

This antenna, with a total area of 0.4m, the sideways force on this antenna with a 45m/s (100mph) wind, according to **Fig 13.45** is 128kgf/m2 x 0.4 = 51kg. By turning the antenna so that the elements are sideways to the wind the area facing the wind is reduced to 0.2m2 and the sideways force reduced to 10.3kg.

There are leverage forces that need to be taken into consideration with unsupported masts that extend above the supporting structure, such as a wall bracket, chimney bracket or guys on a mast. That is why an antenna should be fixed as close to the support or rotator as possible. If you have a Christmas tree of antennas turned by one rotator then a rotator cage is a must.

Of course, these calculations are a simplification of the real world. The wind comes in gusts and there is a lot of turbulence over the roof. Nevertheless, the method of estimating wind forces by G3SEK does give some idea of the forces that will be encountered and enable you to engineer the structure accordingly.

FURTHER READING

'Comparing MININECS, A Guide to Choosing an Antenna Optimization Program', LB Cebik, W4RNL, Communications Quarterly, Spring 1994.

'Programs for Antenna Analysis by Method of Moments', RP Haviland. W4MB. The ARRL Antenna Compendium, Vol 4

REFERENCES

[1]   The ARRL Antenna Book, All editions since 15.

[2]   The Antenna Experimenter's Guide, 2nd edition, Peter Dodd, G3LDO

[3]   Radio Engineering, 2nd edition 1937, F.E Terman

[4]   Field Computation by Moment Methods, RF Harrington, Macmillan Company, New York, 1968

[5]   The New MININEC (Version 3): A Mini-Numerical Electromagnetic Code. J C Logan and J.W. Rockway. Naval Ocean Systems Centre, San Diego, California

[6]   EZNEC+ Roy Lewallen, W7EL, PO Box 6659 Beaverton, OR 97007, USA. E mail w7el@eznec.com

[7]   'The Simplest Phased Array Feed System . . . That Works', The ARRL Antenna Compendium, Vol 2

[8]   'Eurotek', Erwin David, G4LQI, RadCom, Nov 1999

[9]   Protection Against Atmospheric Corrosion, Karel Barton, published John Wiley, 1976

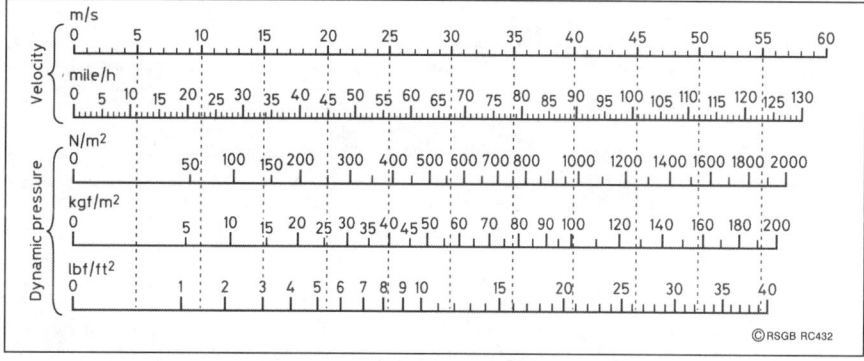

Fig 13.45: Conversion between wind speed and force per unit area

[10] 'In Practice', Ian White, G3SEK, *RadCom*, Mar 1995

[11] 'In Practice', Ian White, G3SEK, *RadCom*, Nov 1998

[12] 'The Paul Buyan Whip', Alfred W.Hubbard, K0OHM, *QST*, Mar 1963

[13] 'Wind loading' by DJ Reynolds, G3ZPF. *RadCom*, Apr and May 1988 (reprinted in *The HF Antenna Collection*, RSGB).

[14] 'Ropes and Rigging for Amateurs - A Professional Approach' by JM Gale, G3JMG, *RadCom*, March 1970 (reprinted in *HF Antenna Collection* and in the *RSGB Microwave Handbook*, Volume 1).

[15] 'In Practice' column, Ian White, G3SEK, *RadCom*, Jan,1995

14 Transmission Lines

For an antenna to function efficiently it should be installed as high and clear of buildings, telephone lines and power lines as is practically possible. On the other hand, the transmitter that generates the RF power for driving the antenna is usually located in the shack, some distance from the antenna feedpoint. The connecting link between the two is the RF transmission line or feeder. Its sole purpose is to carry RF power from the transmitter to the antenna or received signals from the antenna to the receiver as efficiently as possible.

Any conductor of appreciable length compared with the wavelength will radiate power if it is carrying RF current; in other words it becomes an antenna. The transmission line must be designed so that RF power being carried to the antenna does not radiate.

Radiation loss from transmission lines can be prevented by using two conductors so arranged and operated that the electromagnetic field from one is balanced everywhere by an equal and opposite field from the other. In such a case the resultant field is zero; ie there is no radiation. This is illustrated in **Fig 14.1**.

TRANSMISSION LINE BASICS

Characteristic Impedance

A transmission line with its two conductors in close proximity can be thought of as a series of small inductors and capacitors distributed along its whole length. Each inductance limits the rate at which each immediately following capacitor can be charged when a pulse of electrical power is fed to one end of a transmission line. The effect of the LC chain is to establish a definite relationship between current and the voltage of the pulse. Thus the line has an apparent impedance called its characteristic impedance or surge impedance, whose conventional symbol is Zo. Transmission line characteristic impedance is unaffected by the line length. A more detailed description of impedance and transmission lines is described later.

Velocity Factor

With open wire air-spaced lines the velocity of an electromagnetic wave is very close to that of light. In the presence of dielectrics other than air used in the construction of the transmission line (see below) the velocity is reduced because electromagnetic waves travel more slowly in dielectrics than they do in a vacuum. Because of this the wavelength as measured along the line will depend on the velocity factor that applies in the case of the particular type of line in use. The wavelength in a practical line is always shorter than the wavelength in free space.

Mismatch and SWR

The feedpoint impedance of an antenna may not be exactly the same as the characteristic impedance of its associated feeder. The antenna is then said to be mismatched to the feeder.

When a wave travelling along a transmission line from the transmitter to the antenna (incident wave) encounters impedance that is not the same as Zo (discontinuity) then some of the wave energy is reflected (reflected wave) back towards the transmitter. The ratio of the reflected to incident wave amplitudes is called the reflection coefficient, designated by the Greek letter ρ (Rho).

$$|\rho| = | Z_L - Zo / Z_L + Zo |$$

Where Z_L is the load impedance and Zo is the characteristic impedance of the transmission line. It follows that the magnitude of ρ lies between 0 and 1, being 0 for a perfectly matched line.

The reflectometer is an instrument for measuring ρ and comprises two power meters, one reading incident power and the other reflected power. Power detector directivity is possible because the incident wave voltage and current are in phase and in the reflected wave, 180° out of phase. The construction of a reflectometer meter is described in the test equipment chapter.

Whenever two sinusoidal waves of the same frequency propagate in opposite directions along the same transmission line, as in any system exhibiting reflections, a static interference pattern (standing wave) is formed along the line, as illustrated in **Fig 14.2**.

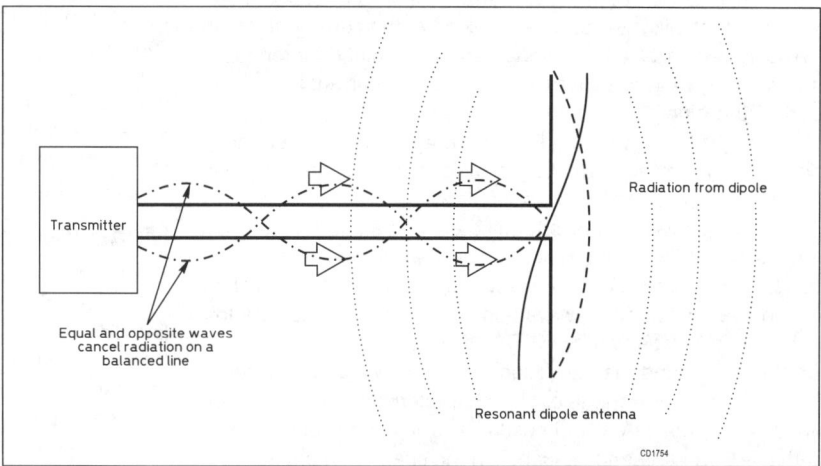

Fig 14.1: RF energy on a transmission line connected to an antenna. No radiation occurs on the line provided the RF energy on each of the lines is equal and opposite. Once the RF energy reaches the antenna there is no opposition to radiation.

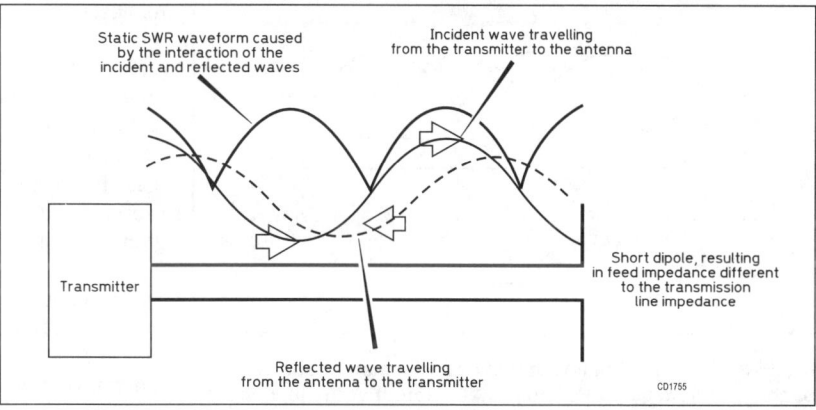

Fig 14.2: Creation of a standing wave on a transmission line.

For the purposes of quantifying reflection magnitude, however, we are interested in the amplitude of the voltage or current nodes and nulls. Standing Wave Ratio (SWR) is defined as the ratio of the voltage or current maximum to the voltage or current minimum along a transmission line, as follows:

$$SWR = V_{max}/V_{min} = I_{max}/I_{min}$$

SWR can be measured using either a current or voltage sensor. The maximum must always be greater than the minimum, thus SWR is always greater than or equal to one. If no reflections exist, no standing wave pattern exists along the line, and the voltage or current values measured at all points along the transmission line are equal. In this case impedance match is perfect, the numerator and denominator of the equation are equal, and SWR equals unity.

As can be seen from Fig 14.2 the direct measurement of SWR must be made at two positions, one quarterwave apart. However, by using the reflectometer the SWR can be measured indirectly as:

$$SWR = 1 + |\rho| / 1 - |\rho|$$

The reflectometer, calibrated in SWR, has become the standard amateur radio tool for measuring transmission line mismatch.

It is often thought that a high SWR causes the transmission line to radiate. This is not true provided the power on each line is equal and opposite as shown in Fig 14.1.

IMPEDANCE TRANSFORMATION

Impedance can be defined by the ratio of current and voltage. This will be familiar to you when looking at the current and voltage distribution of the standing wave on a dipole antenna as shown in Fig 14.1. The voltage is high and the current zero at the end of the dipole (high impedance) while at the centre of the dipole the voltage is low and the current high (low impedance). The centre is obviously the best place to feed the antenna when using low impedance transmission line.

If the transmission line is terminated with a short circuit, the impedance at that point will be very low as shown in **Fig 14.3**. It can be seen that the voltage is zero and the current very high at that point. The standing wave pattern shows that this very low impedance is repeated at every half-wave point down the line. On the other hand the impedance will be very high a quarter of a wavelength down the line. This characteristic of transmission lines is often used as an impedance transformer and is used with the G5RV antenna described in the chapter on practical HF antennas.

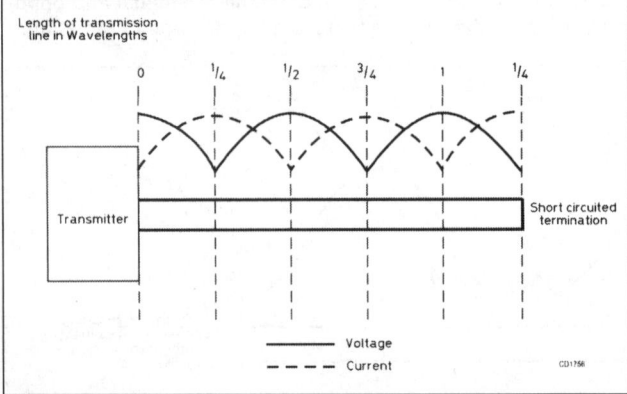

Fig 14.3: Transmission line terminated with a short showing the patterns of voltage and current SWR. Note that the low impedance caused by the short is reflected at half wavelengths from the short

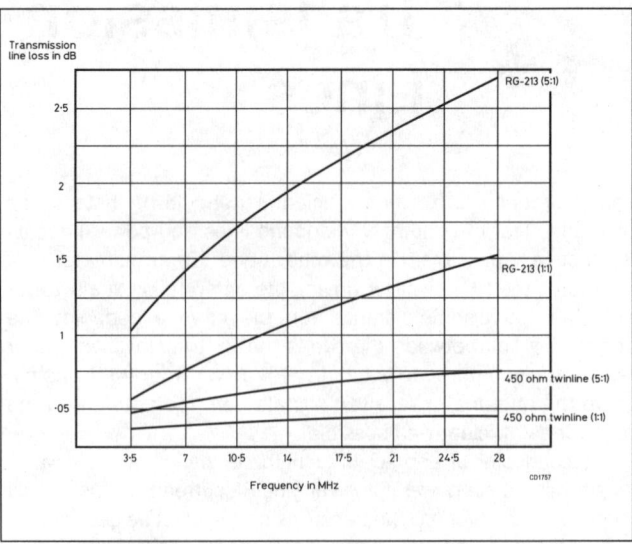

Fig 14.4: Graph showing losses on a 10m (30ft) of 450-ohm twin-line and RG-213 coaxial cable at an SWR of 1:1 and 5:1

The impedance transform affect is described later under Smith Chart.

LOSSES IN TRANSMISSION LINE

Practical transmission line has losses due to the resistance of the conductor and the dielectric between the conductors. As in the case of a two-wire line, power lost in a properly terminated coaxial line is the sum of the effective resistance loss along the length of the cable and the dielectric loss between the two conductors. Of the two losses, the resistance loss is the greater; since it is largely due to the skin effect and the loss (all other conditions remaining the same) will increase directly as the square root of the frequency.

Measurement of Coaxial Cable Loss

The classic method of measuring coaxial cable loss is to terminate the cable with a dummy load that is equal to the Zo of the line. Then use a power meter, first at the transmitter end and the load end ensuring that the transmitter power is maintained at a constant level during the test. Then calculate the loss from the difference in power readings using the formula:

$$dB\ loss = 10 \log (P1/P2)$$

where P1 is the power at the transmitter end and P2 is the power at the dummy load.

Losses Due to SWR

As described above, a transmission line has losses due to the resistance of the conductor and the dielectric between the conductors. Losses at higher frequencies can also result from a poor quality outer conductor. **Fig 14.4** shows approximate losses for 450Ω twinline and RG-213 coaxial cable. Additional losses occur due to antenna/transmission line mismatch (SWR), also shown in Fig 14.4. These losses are for a transmission line over 30m (100ft) long. SWR losses on the HF bands are not as great as is often thought, although at VHF and UHF it is a different matter. As you can see from Fig 14.4, even an SWR of 5:1 on a 30m length of RG-213 coax at 28MHz, the attenuation is only just over 1dB over the perfectly terminated loss.

A reading of SWR due to a mismatch at the transmitter end of the transmission line will be lower than if the measurement were taken at the load (antenna) end. The reason is that the losses on the line attenuate the reflected wave. This means you

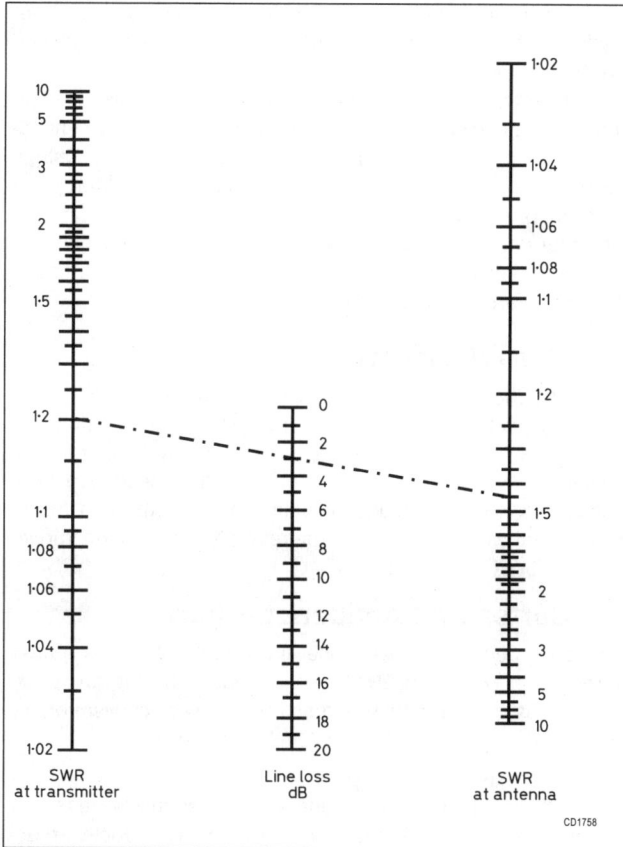

Fig 14.5: Nomograph for calculating transmission line loss using an SWR meter.

can use an SWR meter to measure transmission line loss, using the power meter method (use a load that creates a mismatch, say 100Ω) described above. Measure the SWR at the transmitter and at then at the antenna. Use the graph in **Fig 14.5** to determine the cable loss.

Low-loss Coax - is it Worth It?

The attenuation factors of various correctly terminated coax cables is shown in **Fig 14.6**. These attenuation figures are for

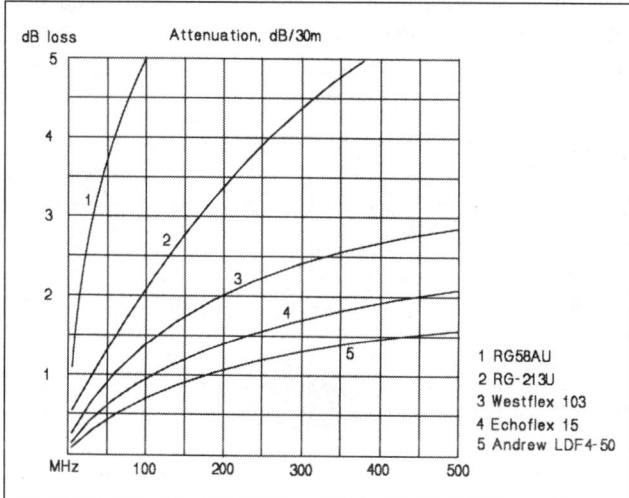

Fig 14.6: Attenuation characteristics of various 30m (100ft) lengths of correctly terminated coax cables. With the exception of the thin RG58 coax the attenuation differences of the various cables below 30MHz are not significant

30m (100ft) lengths and indicate that for frequencies below 30MHz there is not much to be gained by using expensive low-loss coax feeders. The method of construction of these cables to reduce loss is described and illustrated below.

At VHF, and particularly UHF frequencies, it is a different matter. Good quality coax can really enhance a station's performance. On a typical UHF installation at least a 3dB increase should be possible by replacing RG-213 with, say, Ecoflex. If this doesn't sound much remember that generally the size a VHF/UHF antenna array has to be doubled to get 3dB gain.

TRANSMISSION LINE CONSTRUCTION

Two types of transmission line have been used to construct antenna systems described in the antenna chapters. These are twin-line feeder and coaxial cable.

Twin-Line Feeder

Twin-line feeders can be constructed from two copper wires supported at a fixed distance apart using insulated spacers as shown in **Fig 14.7**. This type of construction is often known as 'open-wire feeder'. Spacers may be made from insulating material, such as plexiglas, polyethylene or plastic. The spacers shown in Fig 14.7 are specifically made for the job. The characteristic impedance of such a line can be calculated with the formula

$$Zo = 276 \log_{10} 2S / d$$

Where S is spacing between the wire centres and d is wire diameter.

The construction uses 1.5mm diameter copper wire, and spacers hold the wire around 75mm apart. Using the formula this gives a Zo of 550Ω. If 1mm diameter wire had been used, Zo would have been close to 600Ω.

The 300Ω twin line (the light coloured line shown in Fig 14.7) is constructed by moulding the conductors along the edges of a ribbon of polyethylene insulation and for this reason is sometimes known as ribbon line. This type of feeder is convenient to use but moisture and dirt tend to change the characteristic of the line.

A further variation of commercial twin-line feeder is 'window-line', which has windows cut in the polythene insulation at regular intervals. This reduces the weight of the line and breaks up the surface area where dirt and moisture can accumulate.

Coaxial Cable

Coaxial cable transmission line is used in most amateur installations. The two conductors of the transmission line are arranged coaxially, with the inner conductor supported within the tubular outer by means of a semisolid low-loss dielectric.

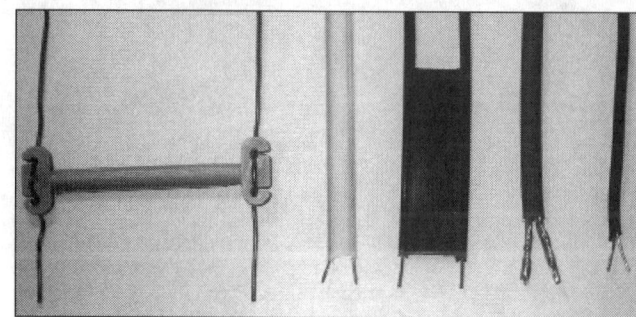

Fig 14.7: Open wire line constructed using 1mm diameter copper wire and spacers. 300Ω twin line with polyethylene insulation, 450Ω 'window' twin line, 75Ω heavy duty twin line, 75Ω light weight twin line

The characteristic impedance for concentric circular conductors is given by:

$$Z_0 = (138/\sqrt{\varepsilon}) \log_{10}(D/d)$$

Where D is the inside diameter of the outer conductor; d is the diameter of the inner conductor, and ε is the dielectric constant of the insulator.

Coaxial cable has advantages that make it very practical for efficient operation in the HF and VHF bands. It is a shielded line and has a minimum of radiation loss. Since the line has little radiation loss, nearby metallic objects have minimum effect on the line because the outer conductor serves as a shield for the inner conductor.

Electromagnetic waves tend to propagate along the surface of conductors, rather than inside, due to the phenomenon of skin effect. Coaxial cable performance depends upon the conductivity and size of the outer surface of the inside conductor and the inner surface of the outer conductor.

The centre conductor of a coaxial cable may consist of either a single wire of the desired outer diameter, or from a twisted bundle of smaller strands. Stranded centre conductors improve cable flexibility while solid centre conductors provide the greatest uniformity of outer diameter dimension, which contribute to stable electrical characteristics.

The outer conductor of coax cable ideally should be made from a solid conductive pipe but this construction makes the cable difficult to bend. The flexibility and bend radius of such cables can be improved by corrugating the outer conductor; examples are shown in **Fig 14.8**.

Nearly all of the popular flexible coaxial cables employ braided outer conductors. These are not as effective electrically as solid outer conductors because gaps in the woven outer conductor permit some signal leakage or radiation from the cable, increasing the attenuation at higher frequencies. This effect can be minimised by adding a layer of copper foil under the braid.

The dielectric material that separates the outer conductor of a coaxial cable from its center conductor determines the intensity of the electrostatic field between conductors and maintains the physical position of the inner conductor within the outer conductor. Common dielectric materials for coaxial cable include polyethylene, polystyrene and PTFE.

The least lossy dielectric material is a pure vacuum, which is totally impractical for use as a cable dielectric. However, the electromagnetic properties of air or gaseous nitrogen are very similar to a vacuum and can be used by mixing low-cost polyethylene with low-loss nitrogen. This is accomplished by bubbling nitrogen gas through molten polyethylene dielectric material before the polyethylene solidifies. This material is variously known as cellular polyethylene dielectric, foam dielectric, or poly-foam. It has half the dielectric losses of solid polyethylene at a modest increase in cost.

The characteristic impedance of most coax cable used in amateur radio installations is usually 50-ohms. Other impedance cable is used for impedance transformers and baluns (described later). The impedance of coaxial cable is often printed on the protective vinyl sheath.

In order to preserve the characteristics of the flexible, coaxial line, special coaxial fittings are available. These, and methods of fixing them, are described later.

THE SMITH CHART

The Smith Chart was invented by Phillip H Smith and described as a Transmission-Line Calculator [1]. While transmission line calculations can be done using a computer with appropriate software such as TLW [2] the Smith chart is described here because it shows very clearly the action of a transmission line as an impedance transformer and the relationship between impedance and SWR .

The Cartesian Impedance Chart

Impedance comprises resistance and reactance and is always expressed in two parts, R+jX. An impedance having an resistance of 75Ω and a inductive reactance 50Ω is conventionally written as:

$$75 + j50$$

For our consideration of impedance j can simply be regarded as a convention for reactance. The '+j' indicates inductive reactance and a '-j' indicates capacitive reactance. When the antenna is at its resonant frequency the +j and -j parts are equal and opposite so only the resistive part remains.

Impedance can be represented using a chart with Cartesian coordinates as shown in **Fig 14.9**.

This method of plotting and recording the impedance characteristics of antennas is rather like a Mercator Projection map, with the latitude and longitude of R and jX respectively plotted to define an impedance 'location'. Resonance, where the inductive and capacitive reactances in a tuned circuit or antenna element are equal and opposite, exists only on the zero reactance vertical line.

The impedance chart can be used to plot a series of measurements at various frequencies, which produces an impedance signature of the antenna or antenna system. These measure-

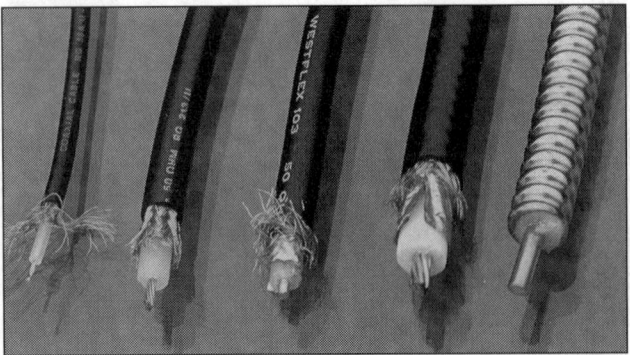

Fig 14.8: Five examples of coaxial cable. From left to right, RG58A/U, 50-ohm RG213/U, Westflex 103 and Andrew LDF4-50 with outer sheath removed

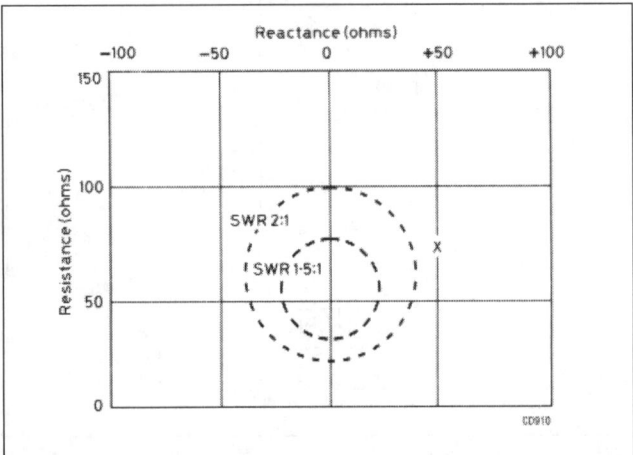

Fig 14.9: Impedance Chart, with X showing an impedance of 75R +50jX. The circles represent SWRs of 1.5:1 and 2:1 respectively for 50-ohm coaxial cable

Fig 14.10: General Radio 1606 RF impedance bridge. The indicated reactance value is valid for 1MHz and must be divided by frequency to get the true reactance. Inductive or capacitive reactance is established with the use of the switch located between the two dials

Fig 14.11: Basic simplified restricted range Smith chart

ments are done with an impedance bridge and a professional impedance bridge is shown in **Fig 14.10**. As you can see there are two calibrated controls, one for R and the other for j. Information from the calibrated dials on the instrument can be used to establish the impedance position on the chart when a measurement is made. An impedance noise bridge is described in the test equipment chapter, and other methods of measuring impedance are described in [3].

The Smith Chart

The Smith Chart is shown in **Fig 14.11** is an impedance map similar to the ones shown in Fig 14.9. It can be considered as just a different projection, just as maps have different projections, such as the Mercator Projection or the Great Circle projection. The most obvious difference with the Smith chart is that all the co-ordinate lines are sections of a circle instead of being straight. The Smith chart, by convention, has the resistance scale decreasing towards the top. With this projection the SWR circles are concentric, centred on the 50Ω point, which is known as the prime centre.

One of the advantages of the Smith impedance map projection is that it can be used for calculating impedance transforms over a length of coaxial feeder. Because the reflected impedance varies along the feeder it follows that you need to know the electrical length of your coaxial feeder to the antenna. The measured impedance, using an impedance bridge as previously described, is then modified by the impedance transform effect of the length of the feeder.

The impedance transformation Smith Chart is illustrated in **Fig 14.12**. An additional scale is added around the circumference, calibrated in electrical wave-

length. Halfway round the chart equals 0.25 or quarter wavelength, while a full rotation equals 0.5 or half wavelength.

Two lengths of 50Ω coaxial feeder are shown superimposed around the circumference of a Smith chart; one length quarter wave long and the other 3/8 wavelength). Both lengths are connected to a load having an impedance of 25 +j0. The quarter wave length of line (0.25) gives a measured impedance of 100

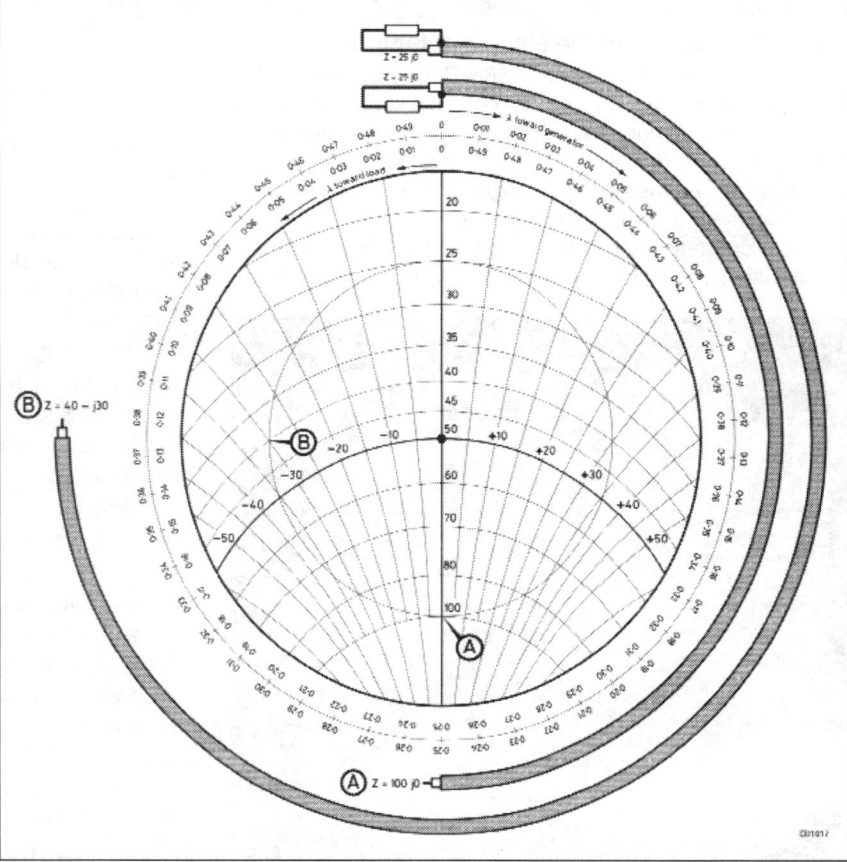

Fig 14.12: Smith chart, with transmission line electrical length scale, superimposed on two lengths of coaxial cable

+j0 at the other end while the 3/8 section (0.375) gives an impedance of 40+j30. It can also be seen from Fig 14.12 that a half wave length of coaxial would transform the impedance back to 25 j0.

A Practical Smith Chart Calculator

You can make a Smith Chart calculator using the charts shown at the end of this chapter. The first one has a prime centre of 50Ω and a restricted impedance range. This makes it easier to use but the impedance excursions are limited. The other is the standard 50Ω chart which covers impedances from (theoretically) zero to infinity.

For this exercise we will make an impedance calculator using the restricted range chart as shown in **Fig 14.13**.

Make a photocopy of the chart (at the end of this chapter), enlarging it if necessary to bring it to a usable size. The chart is then glued to a circular sheet of stiff cardboard or thin aluminium. A small hole is drilled in the chart and backing material at the 50 +j0 point.

From a piece of very thin perspex or transparent plastic or celluloid cut a circle the same size as the chart to make an overlay. A hole is then drilled exactly at the overlay centre. Identifying the centre point should be no problem if a pair of compasses are used to mark the overlay before cutting.

Make a cursor by drawing a line along the radius of the overlay, using a fine tipped marker pen. Cover the line with a strip of transparent sticky tape to prevent the line rubbing out. Trim off the excess tape.

Fix the transparent overlay to the chart with a nut and bolt, with the tape covered line against the chart. Adjust the nut and bolt so that the overlay can be easily rotated, as shown in Fig 14.13.

The uses to which this calculator can be put are numerous and just three examples are described.

Measuring coaxial cable electrical length

It is often important to know the electrical length of transmission line, either for making antenna feedpoint impedance measurements or constructing phasing lines for stacked beams or phased verticals. You can find the electrical length of coaxial cable by physically measuring its length and multiplying it by the cable velocity factor or by using a dip meter.

A more accurate method is to measure the electrical length directly using an RF impedance measuring instrument and the

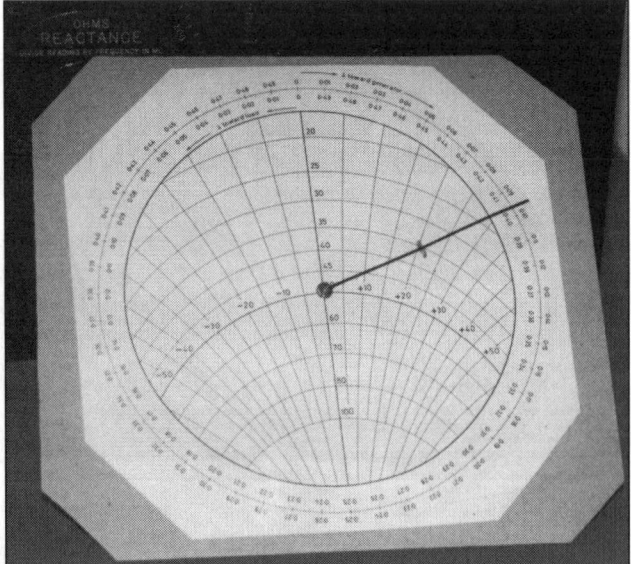

Fig 14.13: General construction of the Smith chart calculator

Smith Chart. It also assumes that the transmission line losses are low; in practice this means that the procedure will only work with relatively short lengths of fairly good quality coaxial cable..

1 Terminate the load (antenna) end of the cable with a 22Ω resistor.
2 Measure the impedance at the other end of the feeder.
3 Move the cursor so that it intersects the measured impedance point. The cursor will now point to the electrical wavelength of the feeder marked on the outer scale 'wavelengths towards generator'.

The cable may be several half wavelengths and part of a half wavelength long. The Smith chart will only register the part of a half wavelength, which is all we are interested in regarding the impedance transform effect.

Calculating antenna impedance from measured impedance

This is a method of calculating antenna impedance from a measured impedance value, using coaxial cable whose electrical length has already been determined.

1 Connect the cable to the antenna.
2 Measure the impedance at the other end of the coaxial cable.
3 Move the cursor over the measured impedance point and mark the point on the overlay with a wax pencil.
5 Follow the cursor radially outwards to the scale marked wavelengths towards load. Write this number down.
6 Add the length of cable in wavelengths to this number.
7 If the number is larger than 0.5, subtract 0.5.
8 Rotate the overlay until the cursor points to this number on the wavelengths towards load scale.
9 The antenna impedance will be found on the cursor directly under the wax pencil mark.

Example:

The measured impedance is 35+j20 ohms and the cursor points to 0.407 on the wavelengths towards load scale.

The cable electrical length was measured as 0.13 wavelengths.

Then 0.407 + 0.13 = 0.537 wavelengths. Off scale - too big! Subtract 0.5 wavelengths = 0.037 wavelengths.

Rotate the overlay until the cursor points to 0.037 on the wavelengths towards load scale.

The antenna impedance is shown as 28 -j8 ohms under the cursor at the same radius as the measured impedance.

Calculation of SWR

Calculation of SWR is very simple using the Smith chart. The result is useful for correlating impedance measurements with SWR measurements. To measure SWR:

1 Move the cursor over the measured impedance point.
2 Mark the point on the overlay with a wax pencil.
3 Move the cursor to the 0 point on the outside scales.
4 The SWR can be read off as 50 divided by the mark on the cursor. The impedance measured above gives a reading of 27 +j0. 50 divided by 27 equals 1.85; the SWR in this case is 1.85:1.

You can, of course, calibrate the cursor in SWR. Just place the cursor in the vertical zero position and place marks on the cursor at the 33.3, 25 and 20 resistance points to give SWR marks at 1.5:1, 2:1 and 2.5:1 respectively.

The Normalized Smith Chart

Most Smith charts are normalised so that they can be used at any impedance and not restricted to 50Ω, as are the ones so

ADMITTANCE

The method of calculating the total value of several resistors in series is add their individual values. And the simplest way of calculating resistors in parallel is to add their reciprocals; the answer is also a reciprocal and has to be converted back to R to be in a form which we are more familiar:

$$1/R1 + 1/R2 + 1/R3 = 1/R$$

The reciprocal of R is Conductance (symbol G) and you could work in Conductances if you were dealing with calculations involving lots of parallel circuits.

The reciprocal (or the dual) of Impedance is Admittance (symbol Y).

The reciprocal of reactance is susceptance (symbol B). The unit of conductance, susceptance and admittance is the Siemen.

It is important to know just what it is that your RF bridge is measuring. If the bridge is an Admittance bridge then the result, like the calculation of parallel resistors described above, will need to be converted into the more familiar ohms impedance.

far described. This is achieved by assigning 1 to the prime centre; other values, for example, are 0.5 for 25Ω and 2 for 100Ω in a 50Ω system. Normalization also extends the use of the chart to convert impedance to admittance (see sidebar) and vice versa. A chart for constructing a normalised Smith Chart calculator is also shown at the end of this chapter. The construction of the chart is the same as described above. To use the procedure described below, the cursor must be extended from the centre, ie the cursor line is extended from a radius to a diameter.

To convert admittance to impedance:

1   Convert the admittance to normalised admittance by multiplying the each component of the admittance value by the prime centre, usually 50.
2   Move the cursor over the measured admittance point.
3   Mark the point on the overlay with a wax pencil.
4   Move the cursor 180 degrees so that the unmarked section of the cursor lies over the measured admittance point.
5   The mark on the cursor from 3. gives the normalised impedance reading.
6   Convert to actual impedance ohms by multiplying by 50.

FITTING COAXIAL CONNECTORS

Fitting coaxial connectors to cable is something we all have to do at some time or other.

If you have had trouble in the past fitting connectors, you should find the methods described here by Roger Blackwell, G4PMK [4] helpful. Although specific styles of connector and cable are mentioned, the methods are applicable to many others.

Cables and Connectors

The main secret of success is using the right cable with the right connector. If you're buying connectors, it is important to be able to recognise good and bad types, and know what cables the good ones are for. Using the wrong connector and cable combination is sure to lead to problems. Any information you can get, such as that from old catalogues, is likely to prove useful, especially if you can get the cable cutting dimensions and equivalents lists.

Cables commonly are of one of two families, the American 'RG' (RadioGuide MIL specification) types and the English 'UR' (UniRadio) series. RG213 is equivalent to URM67, is 10.5mm in diameter and is the most common cable used with type N and PL259 connectors.

RG58 (5mm OD) is one usually used with BNC connectors, although these also fit cable URM43 since both have similar dimensions. If there is any doubt about the quality of the cable, have a look at the braid. It should cover the inner completely. If it doesn't it is unlikely to be worth buying. There are a lot of so-called 'RG8' cables about these days, intended for the cheap end of the CB market, and should be avoided - RG8 is an obsolete designation - the modern equivalent is RG213 or URM67.

The three most popular connector types are the UHF, BNC and N ranges. These will be covered in some detail, a few others mentioned later. If you can, buy connectors from a reputable manufacturer. There are some good surplus bargains about, so a trawl through the boxes at the local rally may prove worthwhile. The PL259 UHF connector is no good much beyond 200MHz, because the impedance through the plug-socket junction is not 50Ω. The suitability of N and BNC connectors for use at UHF and beyond is due to their maintaining the system impedance (50Ω) through the connector. PL259 plugs, like the RG8 cable they were intended for, have a lot of nasty imitations. Beware of any that don't have PTFE insulation. The plating should be good quality and there should be two or more solder holes in the body for soldering to the braid. There should be two small tangs on the outer mating edge of the plug, which locate in the serrated ring of the socket and stop the body rotating. If you are going to use small-diameter cable with these plugs, get the correct reducer. It is advisable to buy the reducers at the same time as buying the plugs because some manufacturers use different reducer threads.

With BNC, TNC (like the BNC but threaded) N and C (like N but bayonet) types, life can be more complicated. All these connectors are available in 50 and 75Ω versions. Be sure you get the right one! All of these connectors have evolved over the years, and consequently you will meet a number of different types. The variations are mostly to do with the cable clamping and centre pin securing method.

If you are buying new connectors, then for normal use go for the pressure-sleeve type, which is much easier to fit.

All original clamp types use a free centre pin that is held in place by its solder joint onto the inner conductor. Captive contact types have a two-part centre insulator between which fits the shoulder on the centre pin. Improved MIL clamp types may have either free or captive contacts. Pressure sleeve types have a captive centre pin. As an aid to identification, **Fig 14.14** shows these types. Pressure clamp captive pin types are easy to spot; they have a ferrule or 'top hat' that assists in terminating the braid, a two-piece insulator and a centre pin with a shoulder. Unimproved clamp types have a washer, a plain gasket, a cone-ended braid clamp and a single insulator, often fixing inside the body. Improved types have a washer, a thin ring gasket with a V-groove and usually a conical braid clamp with more of a shoulder. There are variations, so if you can get the catalogue description it helps!

Tools for the Job

To tackle this successfully, you really need a few special tools. While they may not be absolutely essential, they certainly help. Most of them you probably have anyway, so it's just a matter of sorting through the toolbox. First and foremost is a good soldering iron. If you never intend to use a PL259, a small instrument type iron is sufficient. If you use PL259s, or intend to use some

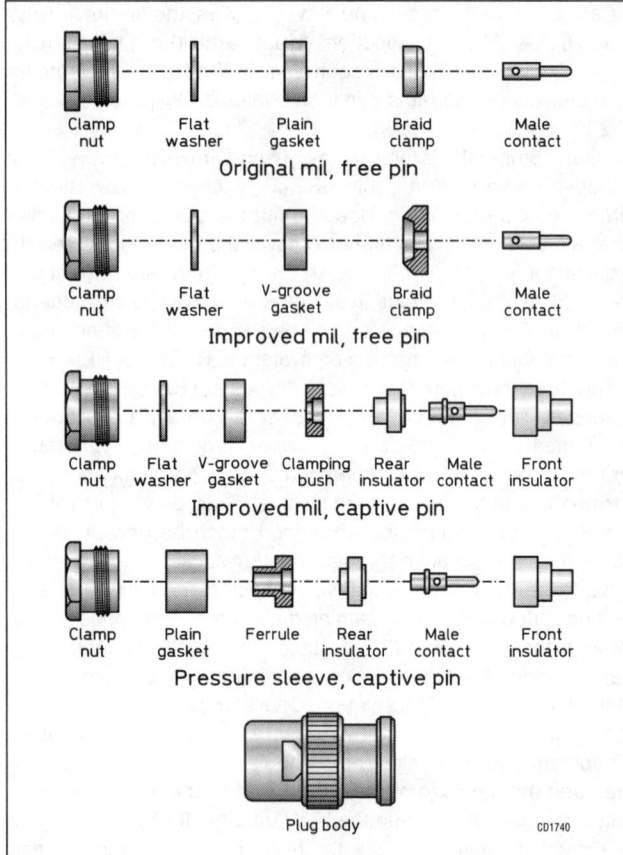

Fig 14.14: Types of BNC/N cable clamps

of the 'dirty tricks' described later, something with a lot more heat output is required. Ideally a thermostatically-controlled iron is best; as with most tools a little extra spent repays itself handsomely in the future.

A sharp knife is another must. A Stanley-type is essential for larger cables, provided that the blade is sharp. For smaller cables, you can use a craft knife or a very sharp penknife. Use sharp blades, cut away from you, and keep the object you're cutting on the bench, not in your hand. Although sharp, the steel blades are brittle and will shatter if you apply excessive force or bend them, with bits of sharp blade shooting all over the place. Dispose of used blades in a box or plastic jar. Model shops have a good range of craft knives, which will also do an excellent job. A pair of small sharp scissors is needed for cutting braids, and a blunt darning needle (mounted in a handle made from a piece of wood dowelling) is useful for unweaving the braid. A scriber is also useful for this job. You will find a small vice a great help as well. For BNC, TNC and N type connectors, some spanners are essential to tighten the gland nuts. The BNC/TNC spanners should be thin 7/16in AF. Those for type N need to be 11/16 x 5/8 AF. A junior hacksaw is needed to cut larger cables. Finally, if you intend to put heatshrink sleeves over the ends of plugs for outdoor use, some form of heat gun helps, although the shaft of a soldering iron may work. (A hot-air paint stripper can be used for this purpose - with care).

Preparing Cables

Fitting a plug requires you to remove various bits of outer sheath, braid and inner dielectric. The important knack to acquire is that of removing one at a time, without damaging what lies underneath. To remove the outer sheath, use a sharp knife or scalpel. Place the knife across the cable and rotate the

cable while applying gentle pressure. The object of doing this is to score right round the cable sheath. Now score a line from the ring you just made up to the cable end. If you have cut it just enough, it should be possible to peel away the outer sheath leaving braid intact underneath. If this is not something you've tried before, practice on a piece of cable first. For some connectors, it is important that this edge of the sheath is a smooth edge at right angles to the cable, so it really is worth getting right.

Braid removal usually just requires a bit of combing out and a pair of scissors. Removal of the inner dielectric is most difficult with large-diameter cables. Again, it is important that the end is a clean, smooth cut at right angles to the cable. This is best achieved by removing the bulk of the dielectric first, if necessary in several stages. Finally the dielectric is trimmed to length. There is a limit to how much dielectric you can remove at one go; 1-2cm is about as much as can be attempted with the larger sizes without damaging the lay of the inner. For the larger cables, it is worthwhile to pare down the bulk of the unwanted material before trying to pull the remainder off the inner. If you can, fit one plug on short cables before you cut the cable to length (or off the reel if you are so lucky). This will help to prevent the inner sliding about when you are stripping the inner dielectric.

Fitting PL259 Plugs

Without reducer, RG213 type cable

First, make a clean end. For this large cable, the only satisfactory way is to use a junior hacksaw. Chopping with cutters or a knife just spoils the whole thing. Having got a clean end, refer to **Fig 14.15** for the stripping dimensions. First, remove the sheath braid and dielectric, revealing the length of inner conductor required. Do this by cutting right through the sheath and braid, scoring the dielectric, then removing the dielectric afterwards. Next carefully remove the sheath back to the dimension indicated, without disturbing the braid. Examine the braid; it should be shiny and smooth. If you have disturbed it, or it looks tarnished, start again a little further down.

With a hot iron, tin the braid carefully. The idea is to do it with as little solder as possible; a trace of a non-corrosive flux such as Fluxite helps. Lightly tin the inner conductor also at this stage. Now slide the coupling piece onto the cable (threaded end towards the free end). Examine the plug body. If it isn't silver-

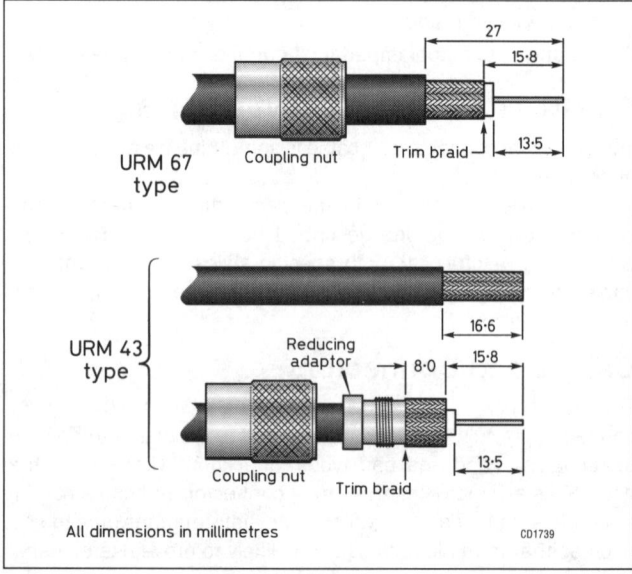

Fig 14.15: PL259 plug assembly

plated, and you think it might not solder easily, apply a file around and through the solder holes. Now screw the body onto the cable, hard. When you've finished, the sheath should have gone into the threaded end of the connector, the inner should be poking out through the hollow pin, and the end of the exposed dielectric should be hard up against the inside shoulder off the plug. Look at the braid through the solder holes. It should not have broken up into a mass of strands; that's why it was tinned. If it has, then it is best to start again.

If all is well, lightly clamp the cable in the vice, and then apply the iron to the solder holes. Heat it up and then add solder. It should flow into the holes; if it stays there as a sullen blob, the body isn't hot enough. Now leave it undisturbed to cool before soldering the inner by heating the pin and feeding solder down the inner. Finally, when its all cool, cut any excess protruding inner conductor and file flush with the pin, then screw down the coupling ring. Merely as a confidence check, of course, test for continuity on both inner and outer from one end of the cable to the other, and check that the inner isn't shortened to the braid.

With reducer, RG58 type cable

First, slide the outer coupler and the reducer on to the cable. Next, referring to Fig 14.15, remove the outer sheath without nicking the braid. Now, using a blunt needle, gently unweave the braid a bit at a time until it is all straight and sticking out like a ruff around the cable. Remove the inner dielectric, without nicking the inner conductor; so as to leave the specified amount of dielectric. Tin the inner conductor. Bring up the reducer until the end of the reducer is flush with the end of the outer sheath. Fold the braid back so it lies evenly over the shank of the reducer, then cut off the excess braid with scissors so that it is not in danger of getting trapped in the threads. Smooth it down once more, then offer up the plug body and, while holding the reducer and cable still, screw on the plug body until it is fully home. The only really good way of doing this is with two pairs of pliers. Now hold the assembly in the vice and ready the soldering iron.

There has been a spirited discussion from time to time about the advisability of soldering the braid through the holes. Professional engineers use soldered connections or compression types but see "dirty tricks" later.

Fitting BNC and Type N plugs

These are 'constant impedance' connectors; that is, when correctly made up, the system impedance of 50Ω is maintained right through the connector. It is vital that the cable fits the connector correctly, therefore check that each part fits the cable properly after you prepare it. Refer to **Fig 14.16** for BNC dimensions, and **Fig 14.17** for N types.

Original or unmodified clamp types

Slide the nut, washer and gasket onto the cable in that order. With the sharp knife, score through the outer sheath by holding the knife and rotating the cable, without nicking the braid. Run the knife along the cable from the score to the end, and then peel off the outer sheath.

Using a blunt needle, for example, start to unweave the braid enough to enable the correct length of dielectric to be removed. Now slip the braid clamp on, pushing it firmly down to the end of the outer sheath. Finish unweaving the braid, comb it smooth then trim it with scissors so that it just comes back to the end of the conical section of the clamp. Be sure that the braid wires aren't twisted.

Now fit the inner pin and make sure that the open end of the pin will fit up against the dielectric. Take the pin off and lightly tin the exposed inner conductor. Re-fit the pin and solder it in place by placing the soldering iron bit (tinned but with the solder

wiped off) on the side of the pin opposite the solder hole. Feed a small quantity of solder (22SWG or so works best) into the hole. Allow the connector to cool and then examine it. If you've been careful enough, the dielectric should not have melted. Usually it does, and swells up, so with the sharp knife trim it back to size. This is essential, as otherwise the plug will not assemble properly. Remove any excess solder from around the pin with a fine file. Now push the gasket and washer up against the clamp nut, check the braid dressing on the clamp, and then push the assembly into the plug body. Gently firm home the gasket with a small screwdriver or rod and then start the clamp nut by hand. Tighten the clamp nut by a spanner, using a second spanner to hold the plug body still; it must not rotate. Finally, check the completed job with the shack ohmmeter.

Modified or improved clamp types

In general, this is similar to the technique for unmodified clamp types described above. There are some important differences, however. The gasket has a V-shaped groove in it, which must face the cable clamp. The clamp has a corresponding V-shaped

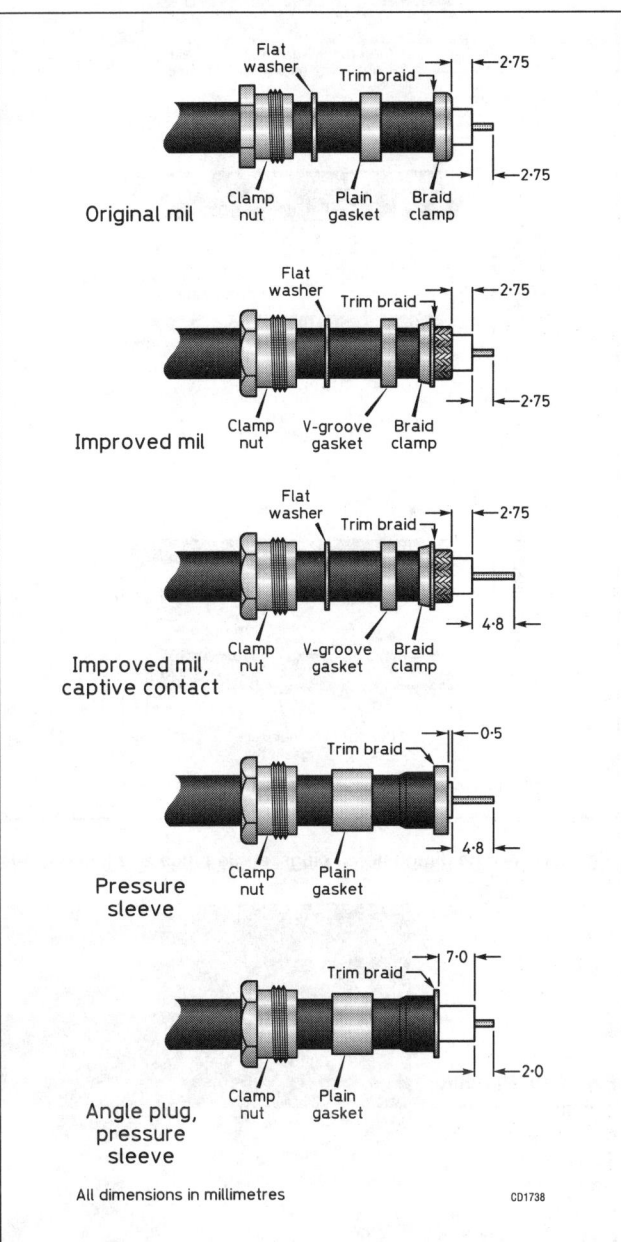

All dimensions in millimetres   CD1738

Fig 14.16: BNC dimensions, plugs and line sockets

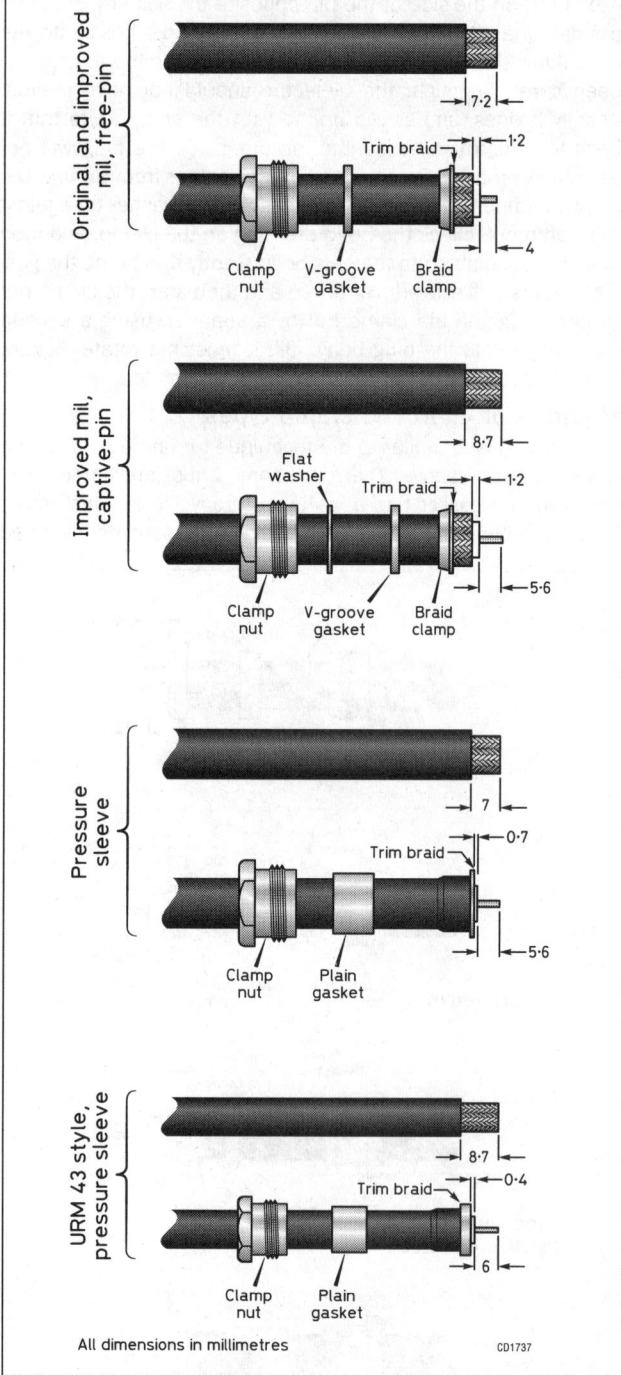

All dimensions in millimetres   CD1737

Fig 14.17: N-type dimensions, plugs, angle plugs and line sockets

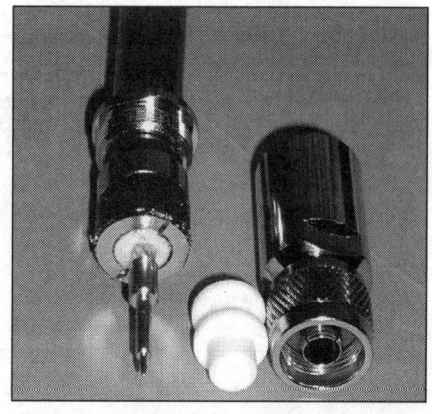

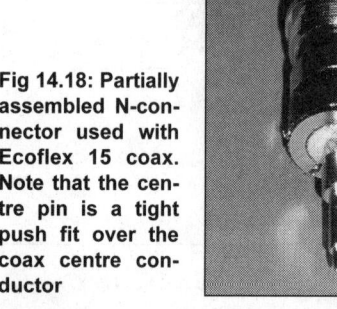

Fig 14.18: Partially assembled N-connector used with Ecoflex 15 coax. Note that the centre pin is a tight push fit over the coax centre conductor

profile on one side; the other side may be conical or straight sided, depending on the manufacturer. If the clamp end has straight sides, the braid is fanned out and cut to the edge of the clamp only, not pushed down the sides. Some types have a small PTFE insulator, which is fitted before the pin is put on (common on plugs for the small RG174 cable). You now appreciate why having the assembly instructions for your particular flavour of plug is a good idea! Still, by using these instructions as a guide, it shouldn't be too difficult to get it right, even if it does not fit the first time. One important point - if the plug has been assembled correctly and tightened up properly, the clamp will have (intentionally) cut the gasket, which is then rather difficult to re-use. This thin gasket will not stand a second attempt. The thicker gasket types will often allow careful re-use.

Captive contact types

These have a small shoulder on the pin, and a rear insulator, which fits between the pin and the cable. Most types use a thick gasket and a ferrule, although some use a V-grooved braid clamp and thin gasket. The ferrule type is described first because these are the most commonly available, and the easiest to fit.

First, slip the nut and gasket on to the cable. Refer to Figs 14.16 or 14.17 for cutting dimensions then strip off the correct amount of outer sheath by rotating the cable, producing a neat scored circle. Score back to the end of the cable and peel off the unwanted sheath. Comb out the braid, and with it fanned out evenly around the cable, slide the ferrule (small end first) on to the dielectric-covered inner conductor.

Push it home so that the narrow portion of the ferrule slides under the outer sheath, and the end of the outer sheath rests against the ferrule shoulder. Trim the braid with scissors to the edge of the ferrule.

Slide up the gasket so that it rests gently against the ferrule shoulder, which will prevent the braid from being disturbed. Using the sharp knife, trim the dielectric back to the indicated dimension, without nicking the inner conductor. Fit the rear insulator, which will have a recess on one side to accommodate the protruding dielectric. Incidentally, if you don't have the size for your particular plug, trim the dielectric until it fits; but don't overdo it!

Now trim the exposed inner conductor to length and check by fitting the pin, whose shoulder should rest on the rear insulator unless the inner has been cut too long. Tin the inner lightly, then fit the pin and solder it by applying the iron tip (cleaned of excess solder) to the side of the pin opposite from the solder hole and feed a small amount of solder into the hole.

Allow to cool, and then remove the excess solder with a fine file. Now fit the front insulator (usually separate from the body) and push the whole assembly into the body. Push down the gasket gently into the plug body with a small rod or screwdriver. Start the nut by hand, and then tighten fully with one spanner, using the other to prevent the body from rotating. Check with the ohmmeter, then start on the other end - remember to put the nut and gasket on first!

Variations

Angle plugs generally follow a similar pattern to the straight types, except that connection to the inner is via a slotted pin, accessed via a removable cap screw. Tighten the connector nut before soldering the inner. Line sockets are fitted in the same way as plugs.

Interestingly, many connectors used on high-grade low loss cables appear to be solderless. This applies to the N-type male and female connectors used with Ecoflex15, see **Fig 14.18**. This connector proved to be very easy to fit.

Dirty Tricks

We would like to use new connectors every time, but often a pressure sleeve type can be reused if the gasket is not too deformed. Get all the solder you can out of the pin and then carefully ream out the rest with a small drill, held in a pin chuck. Tarnished silver-plated connectors can be made to shine by dipping the metal parts in Goddards 'Silver Dip' silver cleaner, or a solution of photographic fixer. Rinse carefully afterwards, then bake in a slow oven.

BNC connectors for URM67 cable can be rather hard to find. A standard captive contact BNC plug can be fitted to URM67 in the following way. First, discard the nut, gasket and ferrule, and prepare the rear insulator by removing the ridge from it with a sharp knife. Now prepare the cable by cutting with a knife, right through the jacket, braid and insulator about 5mm back from the end. Cut sufficiently deep so that you notch the inner conductor strands, and remove the remains. Carefully bend the six individual outer strands of the inner so they break off flush with the end of the dielectric, leaving one straight inner strand. Now remove sufficient outer jacket (about 2cm) such that when the body is pushed on the cable, some braid is still visible. Tin the braid and inner conductor lightly, then fit the rear insulator, pin and front insulator and push home the assembly into the plug body. With the big iron, heat the plug body and feed solder down the joint with the braid. After it has cooled, put some heatshrink adhesive lined sleeving over the plug and cable join to protect it. This arrangement is almost as good as the real plug, and certainly better than an adapter; it will happily stand 100W of 1,296MHz.

An N-plug can be carefully pushed on to a BNC socket; OK for quick test equipment lash-ups, but don't do it too often or too hard as you will eventually damage the socket.

If the braid is soldered to the plug is can sometimes be difficult to judge the quality of the soldered joint. Too much heat can cause damage to the cable or the insulation of the plug while too little heat will cause a dry joint. Some amateurs avoid soldering by folding the braid down the outside of the sheath of the cable and screwing the cable entry of the plug body over it.

The most common cause of connector failure is a poor soldered joint or corrosion. If the connector is used outside the connector can be packed with any non-conducting grease, which prevents ingress of water to the connector and the coax and is very effective in eliminating corrosion.

SPLICING COAXIAL CABLE

The radio engineer's method of joining two lengths of coax together is to use coaxial connectors. However, in the description of coax cable splicing by G3SEK [5] that follows, you will see that a splice can be made entirely without connectors. A splice in coaxial cable needs to be as close as possible to an uninterrupted run of cable. In practice this requires four things:

- Constant impedance through the splice.
- As short an electrical length as possible, if it is not possible to make the impedance quite constant.
- Continuous shield coverage.
- Good mechanical properties: strong and waterproof.

At low frequencies coax can be spliced with a two-pole connector block as shown in **Fig 14.19(a)**. Tape over the joint and it's done. Even though this creates a non-constant impedance, the electrical length of the splice is so short that it's most unlikely to have any significant effect. The main drawback is that the break in the shield cover provides an opportunity for RF currents to flow out from the inside of the shield and onto the outer sur-

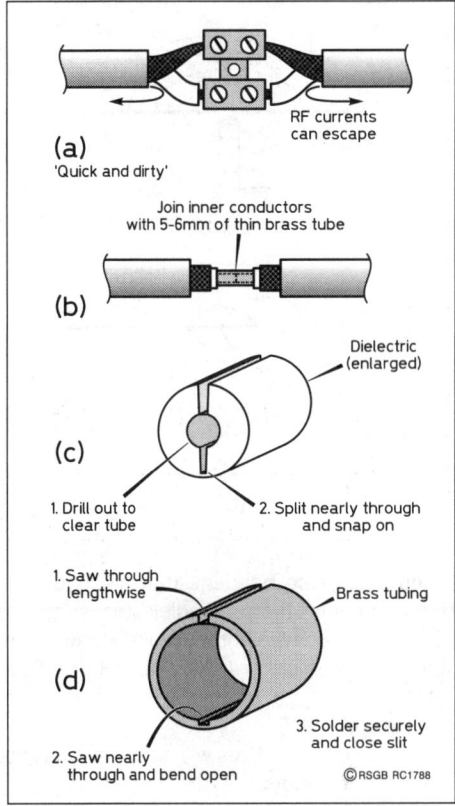

Fig 14.19: Methods of splicing coaxial cable

(a) 'Quick and dirty'

RF currents can escape

(b) Join inner conductors with 5-6mm of thin brass tube

(c) Dielectric (enlarged)
1. Drill out to clear tube
2. Split nearly through and snap on

(d)
1. Saw through lengthwise
Brass tubing
2. Saw nearly through and bend open
3. Solder securely and close slit

© RSGB RC1788

face (the skin effect makes RF currents flow only on surfaces). This may undo all your good efforts to keep RF currents off the feedline, using baluns or feedline chokes.

For a truly coaxial splice you need to join and insulate the inner conductor, and then replace the outer shield. Avoid making a big blob of twisted inner conductors and solder if you can, because that will create an impedance bump - a short section of line with a different impedance from the coax itself.

The neatest and electrically the best way to join the inner conductors is to use a 5-6mm (¼in) sleeve of thin brass tubing, see **Fig 14.19(b)**. This is available from good hobby shops in sizes from 1,6mm (1/16in) outside diameter up to 12.7mm (½in), in steps of 0.8mm (1/32in); these sizes telescope together, by the way. To replace the dielectric, take a piece of the original insulation, drill out the centre to fit over the sleeve, and split it lengthways so that it snaps over the top, see **Fig 14.19(c)**. To complete the shield on braided coax, one good way is to push the braid away from each end while you join the inner conductor, and then pull it back over the splice. Solder the braid quickly and carefully to avoid melting the dielectric underneath. For mechanical strength you can tape a rigid 'splint' alongside the joint as you waterproof it.

Alternatively the splice can be made using a very short length of air-insulated line of the same characteristic impedance. The inner conductor is joined using tubing as already described in **Fig 14.19(b)**. The outer is made from a short length of brass or copper tubing.

The outer tube is 'hinged' to fit over the joint as shown in **Fig 14.19(d)** shows how to then solder the whole thing up solidly. This method makes a very strong splice with excellent RF properties.

For 50-ohm air-spaced coax, the ratio of the inner to outer conductor diameters is 0.43, so all you need to do is to choose the right diameters of tubing for the inner and outer conductors. Remember that the relevant dimensions are the outside diameter of the inner conductor, and the inside diameter of the outer

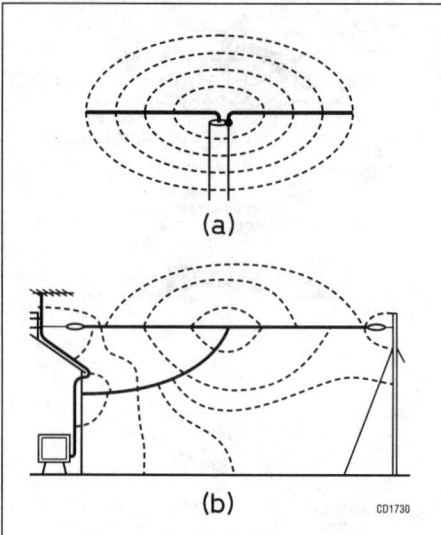

Fig 14.20: (a) Idealised picture of electric fields around a coaxial dipole. (b) The typical reality

conductor. It so happens that air-spaced line needs a larger inner diameter than solid-dielectric, semi-air spaced or foamed line, which conveniently accommodates the wall thickness of the inner sleeve. For UR67, RG213 and RG214, the best available choices are 8mm (5/16in) and 4mm (5/32in) outside diameters.

These coaxial splices will be at least as good as a splice using coaxial connectors.

A QUESTION OF BALANCE

Real-life antennas are nothing like the textbook pictures. [6] The textbooks show us a simple dipole; fed in the centre, with electric field lines neatly connecting the opposite halves, and lines of magnetic flux looping around the wires. Fig 14.20(a) is a typical version of this pretty picture, showing only the electric field lines for clarity. Everything is symmetrical, and the system is said to be 'balanced' with respect to ground.

The reality of a typical installation is very different. As Fig 14.20(b) shows, the electric field lines connect not only with the opposite half of the dipole, but also with the feedline, the ground, and any other objects nearby. The magnetic field may be less disturbed, but the overall picture is in no way symmetrical! Although the electromagnetic coupling between the opposite halves of a horizontal dipole makes the antenna 'want' to be balanced, the

coupling has to compete with the distorting effects of the asymmetrical surroundings. As a result, practical antennas can be very susceptible to the way they are installed, and are hardly ever well balanced.

Contrast the messy environment of the antenna with the tidy situation inside a coaxial cable shown in Fig 14.21. The currents on the centre core (I1) and the inside of the shield (I2) are equal and opposite, ie 180° out of phase. The two conductors are closely coupled along their entire length, so the equal and antiphase current relationship is strongly enforced. Also, what goes on inside the cable is totally independent of the situation outside. Thanks to the skin effect, which causes HF currents to flow only close to the surfaces of conductors, the inner and outer surfaces of the coaxial shield behave as two entirely independent conductors. You can hang the cable in the air, tape it to a tower or even bury it, yet the voltages and currents inside the cable remain exactly the same. About the only things you can do wrong with coax cable are to let water inside, or bend it so sharply that it kinks. That's why coax is popular - it is so easy to use.

The problems arise when you connect a coaxial cable to an antenna. If the antenna is in any way unbalanced - which it will be in any practical situation - a difference will appear between the currents flowing in the antenna at either side of the feedpoint.

This difference current is shown in Fig 14.22 as I3, and is equal to (I1-I2). The current I3 has to flow somewhere. It cannot flow down the inside of the cable because I1 and I2 must be equal, so instead it flows down the outside of the outer sheath. As a result, the feedline becomes part of the radiating antenna. This causes distortion of the radiation pattern, RF currents on metal masts and Yagi booms, and problems with 'RF in the shack'. Even worse can be RF currents flowing in the mains and on TV cables, leading to all manner of EMC problems.

Baluns

The word balun is short for 'balanced to unbalanced'. A balun is a device, which somehow connects a balanced load to an unbalanced coaxial line. It aims to prevent I3 from flowing, by placing a large series impedance on the outside of the feedline. As a result, the antenna currents can only flow on the inside of the feedline, and the properties of the coaxial cable force the antenna currents at either side of the feed point to be equal and in antiphase, ie balanced. Choke off the difference current, and the antenna currents must adjust themselves to become more symmetrical. By using an appropriate type of balun at the antenna feedpoint, you can effectively prevent stray surface currents on the feedline. As a result, the current distribution on the antenna will adjust itself to become more symmetrical, so the antenna will work better and many of your EMC problems may also disappear.

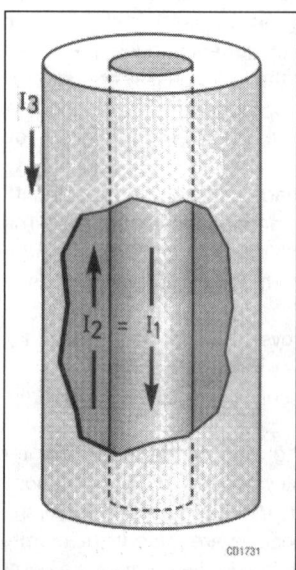

Fig 14.21: Currents on the inside of a coaxial cable (I1 and I2) are always equal and in antiphase. The skin effect allows a separate current I3 to flow on the outside

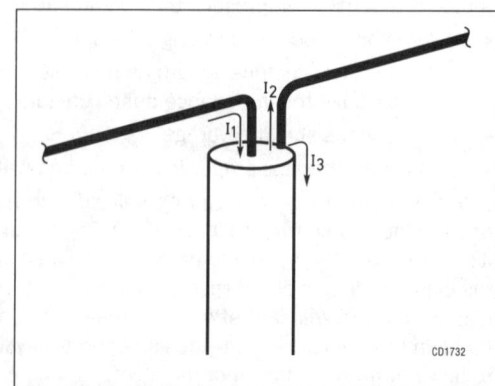

Fig 14.22: If the antenna currents on either side of the feedpoint are unequal, the difference I3 = I1 - I2 will flow down the outside of the cable

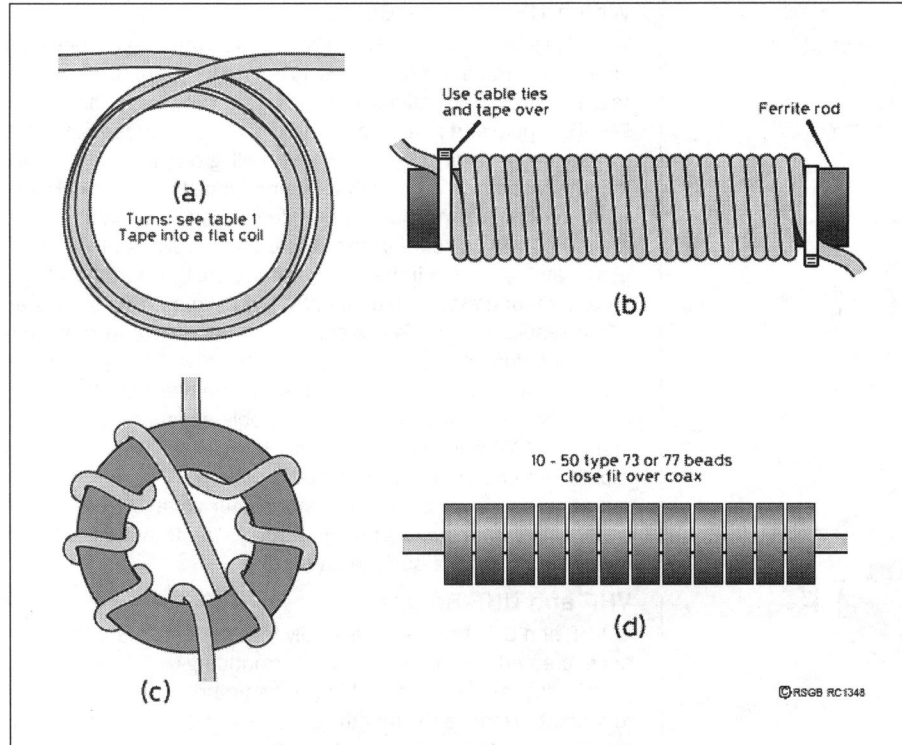

Fig 14.23: Four varieties of choke-type balun. (a) Coil of coaxial cable (typically 6-10 turns - only three turns shown). (b) Coaxial cable on toroid. (c) Coaxial cable on ferrite rod. (d) Ferrite sleeve. These chokes are designed to present a high impedance to unwantend currents on the outside of the coax cable

Choke or current baluns

The simplest baluns are the ones that prevent surface currents from flowing, by forming the coaxial cable into an RF choke at the antenna feed point (the currents flowing inside the cable are quite unaware that the cable has been coiled up.) At its very simplest, a choke balun can be just a few turns of cable in a loop of diameter 150 to 300mm (6in to12in), see **Fig 14.23(a)**, and in some circumstances this may be all you need. See **Table 14.1** for the number of turns required. Alternatives are to wind the cable around a toroid or a ferrite rod, see **Fig 14.23(b) and (c)**. Yet another alternative popularised by W2DU is to feed the cable through ferrite tubes or beads to form a sleeve as shown in **Fig 14.23(d)**; this is generally lighter and more compact than a coiled choke, but somewhat more expensive. Ideally the cable should be a close fit inside the ferrite sleeve, and this may not be easy to achieve with readily available materials.

The power rating of any choke balun is essentially that of the cable itself. Even if magnetic material is involved, the main feeder currents cancel inside the cable, leaving only the small residual surface current to magnetise the core.

Although choke baluns will prevent surface currents from flowing at the feedpoint itself, electromagnetic coupling between the antenna and points further down the feedline may induce surface currents, which the balun cannot prevent. This will happen if the installation is asymmetrical.

Induced currents become much greater if the feedline length is resonant, producing greater RF voltages on the line and on the equipment in the shack. The resonant length has to be measured from the feedpoint, down the coaxial cable, through the equipment in the shack, and away to down to RF ground via various lengths and loops of wire and metal plumbing. There can be many additional loading effects, which alter the resonant length from its free-space value, so it can be very difficult to identify the exact causes of feedline resonance. If a choke-type balun fails to suppress feedline currents, the solution is to leave the balun there, and place further RF chokes, see Figs 14.23(a), (b) and (c), at intervals down the feedline to interrupt any resonant current distributions. A suitable interval would be a quarter-wavelength, or in a multiband system you could use pairs of chokes 2.5 to 3m apart, avoiding half-wave separations at any frequency in use.

Transformer or voltage baluns

A balanced antenna feedpoint will present an equal impedance from each side to ground. The accusation levelled at choke baluns is that they treat the symptom (the surface current I3) without attempting to correct the imbalance that causes it. A transformer balun, on the other hand, does create equal and opposite RF voltages at its output terminals, relative to the grounded side of its input. This works fine in the laboratory; but up in the air where the balun is located, what exactly do we mean by 'grounded'?

Voltage baluns may or may not involve a deliberate impedance transformation. **Fig 14.24** shows a wire-wound 1:1 balun, and both wire-wound and coaxial-cable versions of the 4:1 type. These all have the common property of forcing balance at their output terminals by means of closely coupled windings within the transformer itself. The 4:1 balun in **Fig 14.24(b)** is the easiest to understand. Typically, it transforms 50Ω unbalanced to 4 x 50 = 200Ω balanced. This is achieved simply by a phase inversion. An applied voltage v at one side of the feedpoint is converted by the transformer action into a voltage - v at the other side. These two voltages 180° out of phase represent the balance we are seeking to achieve. The 4:1 impedance transformation arises as follows. If the original voltage on the feedline was v, the voltage difference between opposite sides of the feedpoint is now 2v.

MHz	RG213/UR67 feet	turns	RG58/UR76 feet	turns
	Single-band (very effective)			
3.5	22	8	20	6-8
7	22	10	15	6
10	12	10	10	7
14	10	4	8	8
21	8	6-8	6	8
28	6	6-8	4	6-8
	Multiband (good compromise)			
3.5-30	10	7	10	7
3.5-10	18	9-10	18	9-10
14-30	8	6-7	8	6-7

Table 14.1: Coiled-coax feeding chokes for HF. Wind the indicated length of coaxial cable into a flat coil and tape it together. (Source - W7EL, *ARRL Antenna Handbook*)

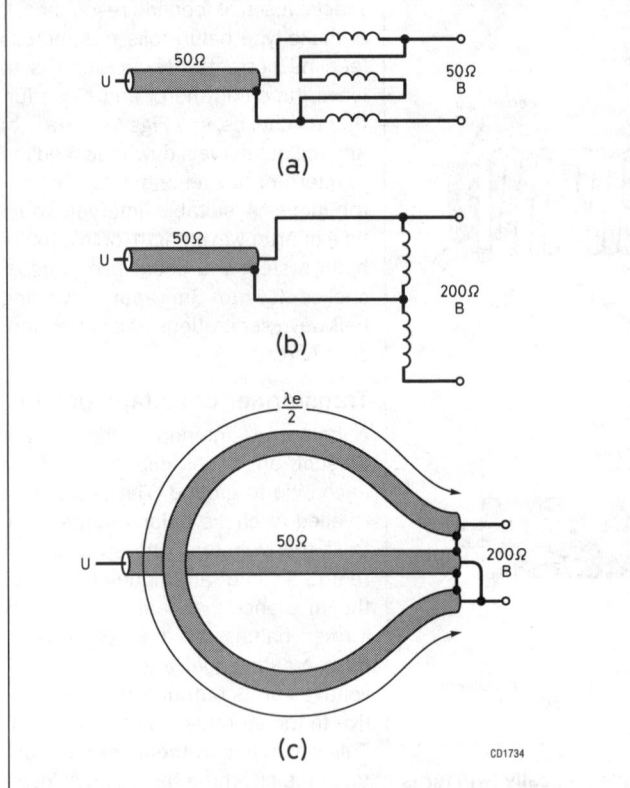

Fig 14.24: Transformer/voltage - type baluns. Windings shown separately are bifilar or trifliar wound, possibly on a ferrite rod or corn. (a) wire-wound 1:1 (trifliar). (b) Wire-wound 4:1 (bifliar). (c) Coaxial 4:1. The half-wavelength must allow for the velocity factor of the cable

Since impedance is proportional to voltage squared, and $(2v)^2 = 4v^2$, the impedance is stepped up by a factor of 4.

In the wire-wound 4:1 balun shown in Fig 14.24(b), the 180° phase inversion is achieved by the connection of the windings, while the coaxial equivalent in **Fig 14.24(c)** does it by introducing an electrical half-wavelength of cable between opposite sides of the feedpoint. Strong coupling between the windings and inside the coax forces the whole system into balance.

However, if any balun is treated as a 'black box' and we look at the currents flowing in and out, it is apparent that any current difference between the two sides of the antenna has nowhere else to flow but down the outside of the coax. But that does not mean that the balun is no better than a direct connection; when you insert the balun, it forces the currents in the antenna to readjust and become more symmetrical. To make a meaningful evaluation of the effects of a balun, you therefore need to include the entire antenna and its surroundings in your analysis.

Any unbalanced antenna/feedline system will always 'push back' against an attempt to force it into symmetry. So although the transformer balun may do a good job, it still may not achieve total equality between the currents on opposite sides of the feed point. That may leave a residual difference current I3 to flow down the outside of the coax, as noted above. Unfortunately, having made its effort to minimise that difference current, the transformer balun does nothing to prevent it from flowing onto the coax. And there is still the possibility of additional currents being induced further down the surface of the feedline. This implies that a transformer-type balun may require additional RF chokes at the balun itself and possibly further back down the line.

Which HF Balun is Best?

W7EL [7] has made some direct comparative tests between choke and transformer baluns in something approximating a real situation. The baluns were inserted between a horizontal 28MHz dipole and a half-wavelength of coaxial feeder. Since the transmitter end of the feeder was well grounded, this feeder length presents a particularly low impedance to surface currents at the feedpoint, and places great demands on the balun. W7EL measured the currents in the dipole at both sides of the feedpoint, and also the imbalance current on the surface of the coax. He found that even minor physical asymmetry in the installation resulted in marked electrical asymmetry; and that both types of baluns produced improvements.

Many amateurs have found this for themselves. Whichever type of balun cures your particular problems (such as RF in the shack or EMC), it's good enough. However, in W7EL's experiments the choke balun produced consistently better improvements in balance than the transformer or voltage type. This is consistent with the lack of direct effort in the transformer balun to suppress feedline surface currents.

VHF and UHF Baluns

At VHF and UHF there are generally fewer difficulties with imbalance created by the antenna's surroundings. The problem is usually the difficulty of making a symmetrical junction at the feedpoint, because the lengths of connecting wires and the necessary gap at the centre of a dipole become significant fractions of the wavelength.

For all its popularity the gamma match is not a balun. It does nothing to create balance between the two sides of a dipole; on the contrary, it relies entirely on electromagnetic coupling between the opposite sides of the dipole to correct the imbalance of the gamma match itself. When used with an all-metal Yagi, the direct connection of the coax shield to the centre of the dipole invites the resulting imbalance currents to travel along the boom as well as the outer surface of the feedline.

The coaxial half-wave balun is definitely the 'best buy' for all VHF/UHF bands up to at least 432MHz, see Fig 14.24(c). It strongly enforces balance, yet it does not introduce an impedance mismatch unless the cable or its length is markedly different from a true electrical half-wavelength. The problem with using this balun is that the feedpoint impedance of the antenna must be transformed up to 200Ω. Fortunately, this is often very simple. For example, the highly successful family of DL6WU long Yagis [8] have a feedpoint impedance which is close to 50Ω at the centre of the dipole driven element; this impedance can be raised to the necessary 200Ω simply by converting the driven element into a folded dipole. Other alternatives for creating a symmetrical 200Ω feedpoint impedance include the T match and the delta match.

The impedance of the coaxial cable used in a half-wave balun is not important, though characteristic impedance of one-half the load impedance (ie in most cases 100Ω) has been shown to give optimum broadband balance. Low-loss 100ohm coax is difficult to obtain, though, and it is perfectly adequate to use good-quality 50Ω cable, carefully cut to length with an allowance for the velocity factor.

The PA0SE HF Balun

This HF balun design, by Dick Rollema, PA0SE, [9] was the solution for feeding an all-band (7MHz to 29MHz) antenna. The feed impedance of this antenna ranged from 33Ω on 7MHz to 560Ω on 29MHz and highly reactive, although balanced. This ruled out direct single-coax feed and also the use of ferrite or powdered iron in the baluns. Units with a 4:1 impedance ratio promised a

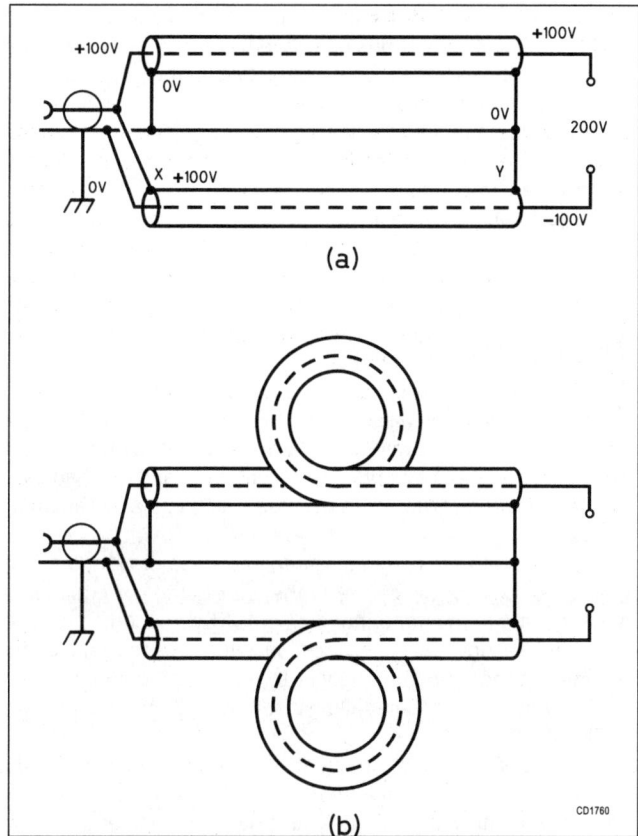

Fig 14.25: Principle of the 4:1 coax balun. Short-circuiting the voltage between X and Y in (a) is remedied by the self-inductance created by coiling the coax as in (b). The upper coax is coiled for neatness only

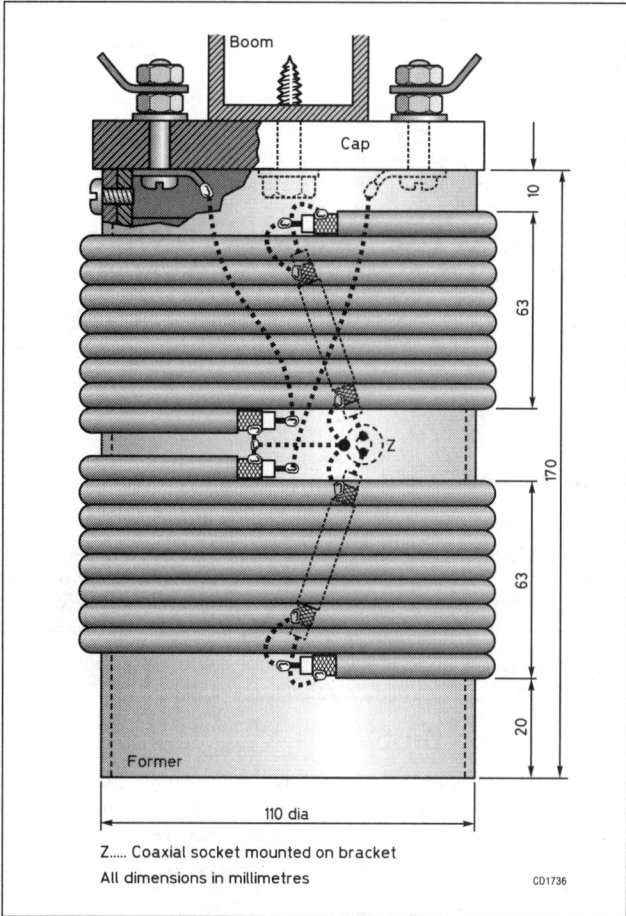

Fig 14.26: Construction of a 4:1 balun of 75ohm coax on a PVC former. The dashed connections are inside the former

Z..... Coaxial socket mounted on bracket
All dimensions in millimetres

better match to his 17m of RG21 7/U to the shack than 1:1 baluns, especially on the highest frequencies, where the coaxial cable feeder losses due to mismatch were highest.

How it works

The balun is made from coaxial cable as shown in **Fig 14.25(a)**. Two equal lengths are connected in parallel at their unbalanced left hand ends and in series at the balanced right hand ends. Let us assume that instantaneous HF voltage can be measured on the centre pin of the left hand coax connector and that this voltage is +100V with respect to earth. The result is +100V on the top balanced terminal and -100V on the bottom balanced terminal, ie 200V between them, balanced with respect to earth. However, with straight lengths of coax there would be a short-circuit between the +100V at point X and 0V at Y.

This short is eliminated by coiling the lower cable as shown in **Fig 14.25(b)**. Between X and Y there is now the reactance of that coil; if sufficiently high, point X no longer 'sees' point Y and there can be a voltage difference between them. For the top coax there is no such problem as both ends are at the same potential, but it is coiled for neatness.

This balun produces a 4:1 impedance transformation only if both cables are terminated in their characteristic impedance. PAOSE used 75Ω coax, so for 'flat' lines a purely resistive, balanced load of 2 x 75 = 150Ω would be required. Looking into the unbalanced end of the balun, an impedance of 75/2 = 37.5Ω would then be seen.

With most antennas the feed impedance is anything but 150Ω resistive. Unless the length of the cables in the balun is short with respect to the wavelength, the balun produces an additional impedance transformation which is largest when the

cable lengths are near λ/4. This can add to or subtract from the 4:1 ratio. In the extreme, the impedance at the unbalanced end can even be higher than that of the balanced load! But that does no harm; the balancing action is valid over the whole intended frequency and impedance range.

Construction

The balun assembly, with dimensions, is shown in **Fig 14.26**. None of these dimensions or the materials and cable types are critical. The former on the prototype was made of grey 110mm OD PVC waste pipe.

PAOSE used matching end caps for the balanced ends; these are not cemented on, but made removable by securing them with three self-tapping stainless steel screws. An N-socket (not shown) on a copper bracket is mounted inside the former for connection to the RG21 3/U coax feeder at point Z. N-connectors are waterproof by design and, additionally, are sheltered by the former.

On the prototype, off-cuts from a 75Ω cable TV installation, whose specification is similar to Uniradio M203, was used.

The Balun is made from two equal lengths of cable wound into the two 8-turn coils (note the winding direction). They are held in place by the ends of the bare-wire jumpers, which protrude through snug-fit holes in the former.

At the balanced end of the coils the braids are joined and connected to the N-socket outer, via a bare wire jumper. The centre conductors are connected to the balanced terminals using bare wire jumpers (shown with dotted lines). The unbalanced outer ends of the coax coils are connected by short lengths of coax inside the former to the N-socket.

Note that the inside conductor of the coax link is connected to the top coil braid and the braid to the inner conductor (on the top coil only).

All coax ends, solder connections and feed-through holes were fixed and waterproofed with epoxy cement. Three coats of clear yacht varnish protect the completed assembly against ultra-violet light. After four years use, the baluns are as good as new.

MATCHING THE ANTENNA TO THE TRANSMISSION LINE

Wire HF antennas are often used with an Antenna Tuning Unit, described in detail in the chapter on Practical HF Antennas. Another method is to use a matching arrangement at the antenna, particularly with beam antennas. Some of the more popular matching arrangements are described below.

The Direct Connection

The halfwave dipole has a theoretical centre feedpoint impedance of 75Ω at resonance. In practice this value is less, particularly at HF, because of the presence of ground. Generally the centre of a dipole can be connected directly to 50Ω coax cable as shown in the Practical HF Antennas chapter, and will almost always provide a good match. A current balun, described earlier, may be necessary at higher transmitter powers.

The Folded Dipole

A halfwave antenna that is used as the driven element will normally have a feedpoint impedance much lower that 50Ω. In this case some impedance transformation is required.

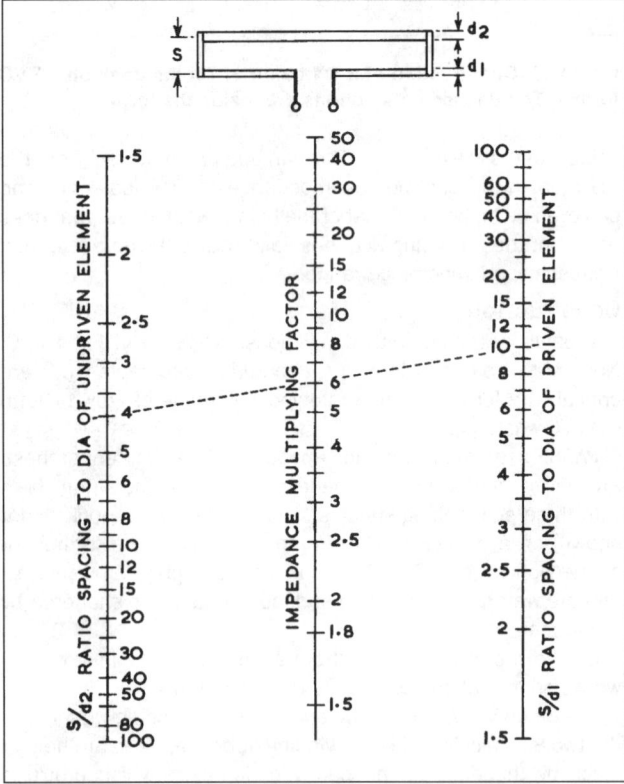

Fig 14.27: A nomogram for folded dipole impedance ratio calculations. A ruler laid across the scales will give pairs of spacing/diameter ratio for any required multiplier. In the example shown the driven element diameter is one-tenth of the spacing and the other element diameter is one-quarter of the spacing, resulting in a setup of 6:1. This shows an unlimited number of solutions for a given ratio

A transformer can be used to step the antenna impedance up to the correct value but this can have the effect of reducing the bandwidth. It has been found that by folding the antenna a 4:1 impedance step-up can often be accomplished with an increase in impedance bandwidth.

Other ratios of transformation than four can be obtained by using different conductor diameters for the elements of the radiator. When this is done, the spacing between the conductors is important and can be varied to alter the transformation ratio. The relative size and spacing can be determined with the aid of the nomogram in **Fig 14.27**.

The Gamma Match

The Gamma match is an unbalanced feed system suitable for matching coax transmission line to the driven element of a beam. Because it is well suited to plumber's delight construction, where all the metal parts are electrically and mechanically connected to the boom, it has become quite popular for amateur arrays.

A short length of conductor (often known as the gamma rod) is used to connect the centre of the coax to the correct impedance point on the antenna element. The reactance of the matching section can be cancelled either by shortening the antenna element appropriately or by using the resonant antenna element length and installing a series capacitor C, as shown in **Fig 14.28**.

Because of the many variable factors - driven-element length, gamma rod length, rod diameter, spacing between rod and driven element, and value of series capacitors - a number of combinations will provide the desired match. The task of finding a proper combination can be sometimes be tedious because the settings are interrelated.

For matching a multi-element array made of aluminum tubing to 50Ω line, the length of the gamma rod should be 0.04 to 0.05 wavelengths long and its diameter 1/3 to 1/2 that of the driven element. The centre-to-centre gamma rod / driven element (S in Fig 14.28) is approximately 0.007 wavelengths. The capacitance value should be approximately 7pF per metre of wavelength at the operating frequency. This translates to about 140pF for 20 metre operation. The exact gamma dimensions and value for the capacitor will depend on the radiation resistance of the driven element, and whether or not it is resonant. The starting-point

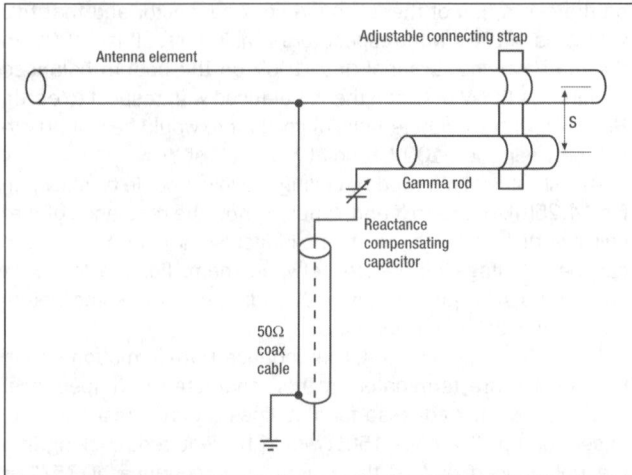

Fig.14.28: Diagram of a Gamma match. Matching is achieved by altering the position of the gamma rod adjustable connecting strap point on the antenna element. The series capacitor C also has to be adjusted to cancel the inductance of the gamma rod. See text for approximate dimensions

Fig 14.29: Simplified Gamma match by G3LDO, uses hard drawn copper wire as the gamma rod. Connection of the gamma rod to the antenna element is achieved using a hose clamp. A Philips capacitor is used as a reactance correction capacitor, which can be replaced with a fixed mica capacitor of the correct value (see text) once the adjustments are complete.

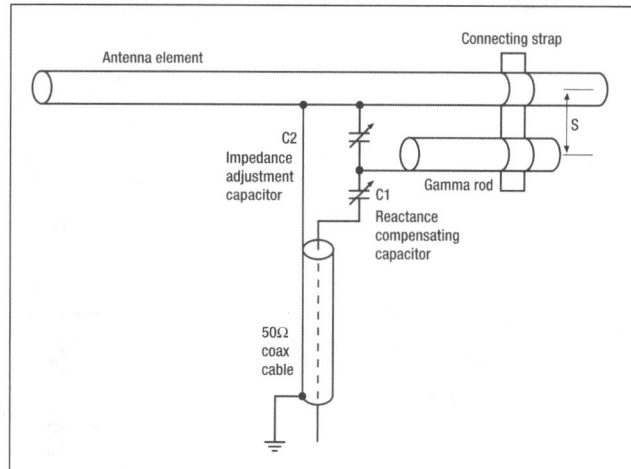

Fig.14.30: Diagram of an Omega match. Matching is achieved by adjusting the parallel capacitor C2 and the series capacitor C1

dimensions quoted are for an array having a feed-point impedance of about 25Ω, with the driven element shortened approximately 3% from resonance.

Adjustment

After installation of the antenna, the proper constants for the gamma generally must be determined experimentally. The use of the variable series capacitor, as shown in Fig 14.28, is recommended for ease of adjustment.

With a trial position of the tap or taps on the antenna, measure the SWR on the transmission line and adjust C1 for minimum SWR. If it is not close to 1:1, try another tap position and repeat.

Construction

The gamma rod is made from thin aluminium tube whose diameter recommended in most publications is 1/3 to 1/6th of the antenna element diameter. However, it is worth trying what is to hand. The Simplified Gamma match by G3LDO, uses hard drawn copper wire as the gamma rod whose connection to the antenna element is achieved using a hose clamp.

The traditional method of making a gamma match is to use an air-spaced variable capacitor and enclose it in a weatherproof metal box. Corrosion to the capacitor can still occur because of condensation.

The gamma match shown in Fig 14.29 uses a fixed capacitor whose value is determined by experiment with a variable capacitor. The value of the variable capacitor is then measured and a fixed capacitor (or several series/parallel combinations) substituted. This arrangement will handle 100W without breakdown and only requires a smear of grease to achieve weatherproofing.

The Omega Match

The Omega match is a slightly modified form of the gamma match. In addition to the series capacitor, a shunt capacitor is used to aid in cancelling a portion of the inductive reactance introduced by the gamma section. This is shown in Fig 14.30. C1 is the usual series capacitor. The addition of C2 makes it possible to use a shorter gamma rod, and makes it easier to obtain the desired match when the driven element is resonant. During adjustment, C2 will serve primarily to determine the resistive component of the load as seen by the coax line, and C1 serves to cancel any reactance. Fixed capacitors can be used to replace the variable ones once the matching procedure is complete. In general the dimensions are the same as for the Gamma match but the gamma rod can be shortened up to 50%. The maximum value of C2 is approximately 1.4pF per metre of the operating frequency.

REFERENCES

[1]   'Smith Radio Transmission-Line Calculator'. Phillip H Smith, *Electronics*, Jan 1939.

[2]   *The ARRL Antenna Book*, 20th edition, ARRL.

[3]   *The Antenna Experimenter's Guide*, 2nd Edition, Peter Dodd, G3LDO

[4]   'Fitting Coaxial Connections', Roger Blackwell, G4PMK, *RadCom* May 1988

[5]   'In Practice', Ian White, G3SEK, *Radcom,* Jun 1998

[6]   'Balanced to Unbalanced Transformers', Ian White, G3SEK, *RadCom*, Dec 1989

[7]   'Baluns: what they do and how they do it', Roy W Lewallen, W7EL, *ARRL Antenna Book*. (W7EL coined the term 'current balun' for the type described in this article as a choke balun, and 'voltage balun' for the transformer type.)

[8]   'High performance long Yagis', Ian White, G3SEK, *RadCom*, Apr 1987. See also Chapter 7

[9]   'Eurotek', Erwin David, G4LQI, *Radcom*, Aug 1992

The following three pages contain Smith charts that may be copied and enlarged to make a Smith cart calculator as described earlier in this chapter.

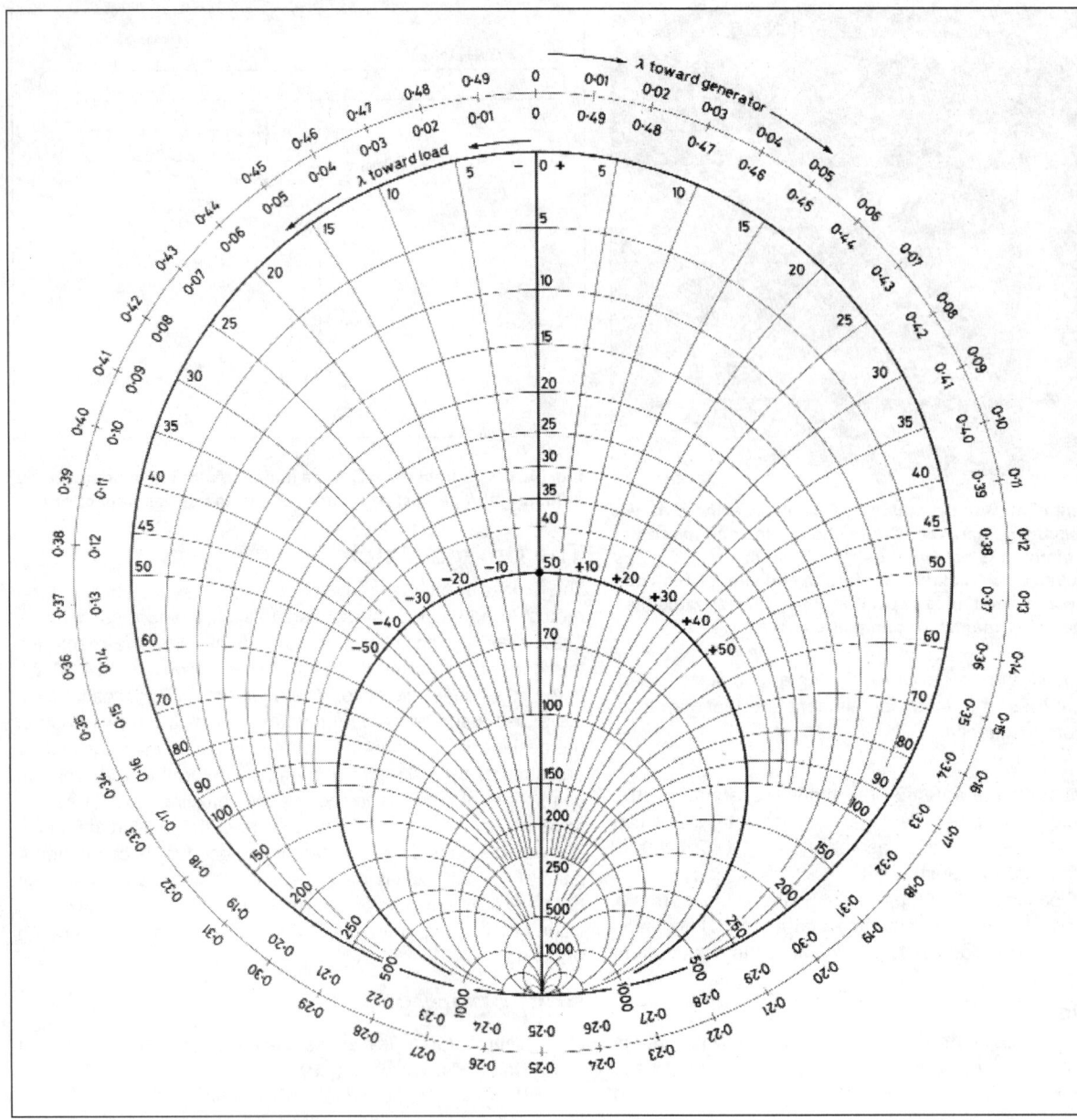

Fig.14.31: Smith chart for constructing a 50-ohm impedance calculator

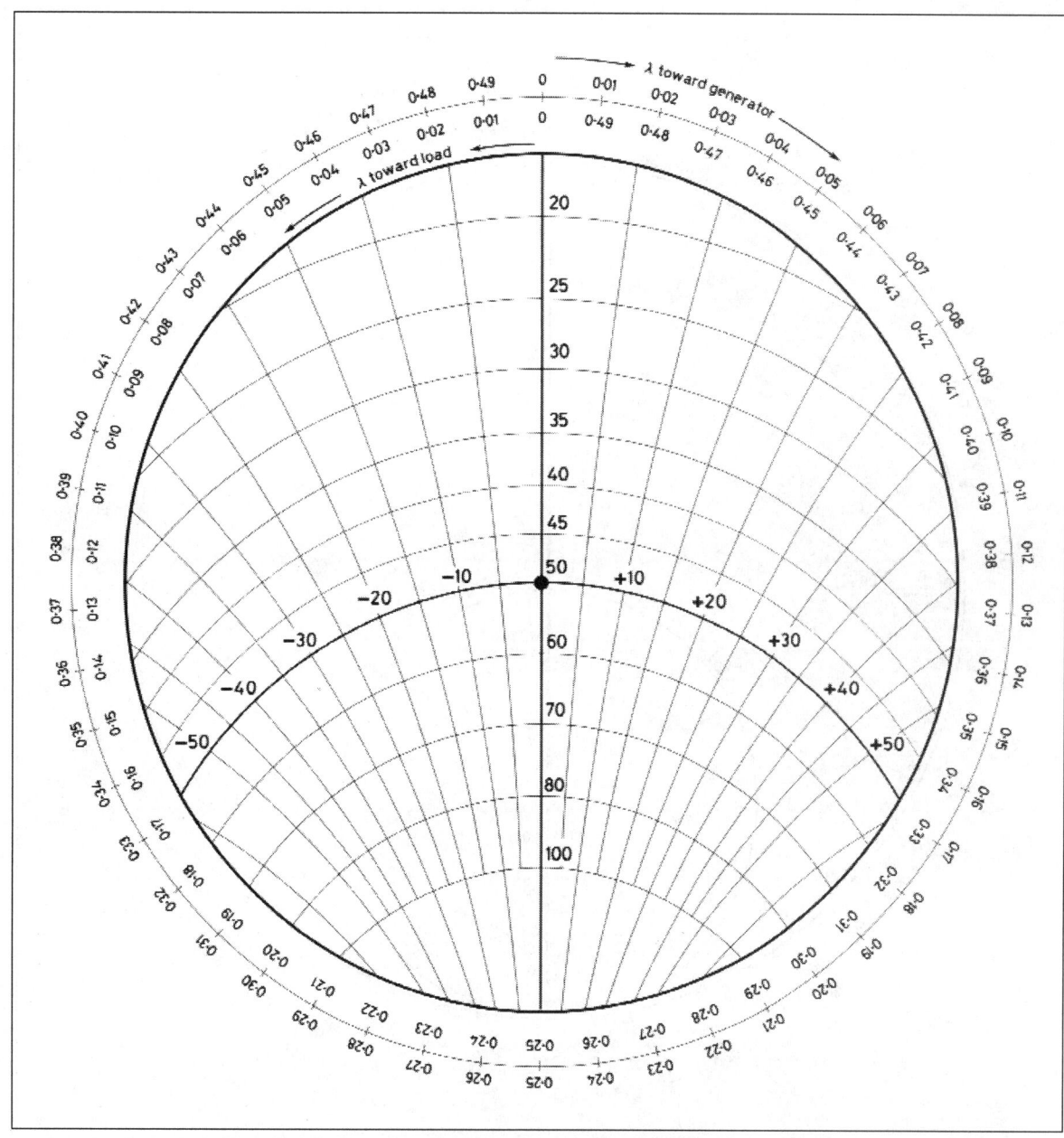

Fig.14.32: Smith chart for constructing a restricted range 50-ohm impedance calculator

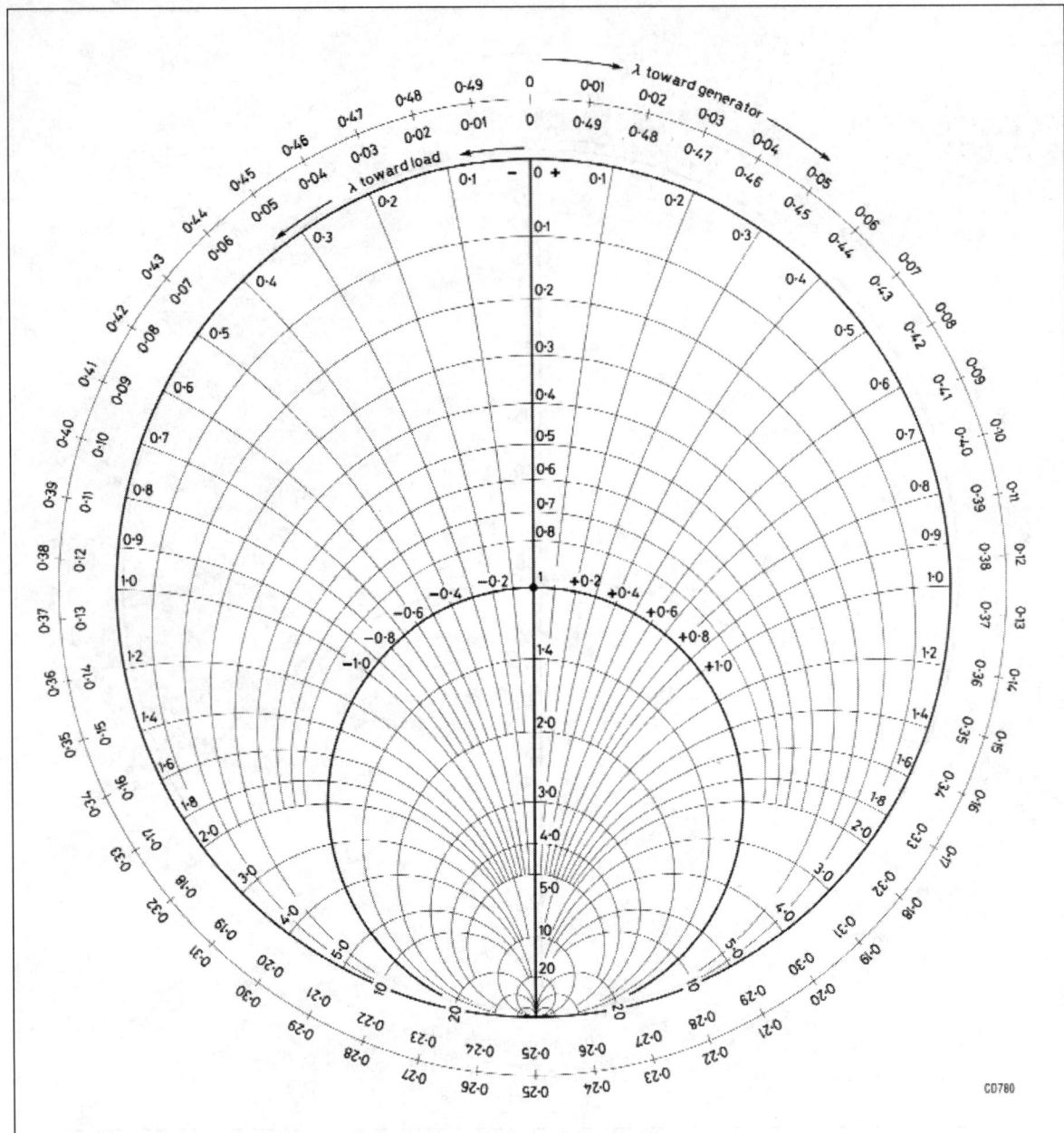

Fig.14.33: Smith chart for constructing a normalised impedance calculator

15 Practical HF Antennas

This chapter addresses the practical construction of HF antennas and ATUs. Because the scale and range of HF antennas is so extensive the chapter is confined to the description and construction of the more commonly used antennas

END-FED WIRE ANTENNAS

The Simple End-Fed Antenna

The simplest of all HF antennas is just a length of wire, one end of which is connected directly to a transmitter or antenna tuning unit (ATU). An example of such an antenna is shown in **Fig 15.1**.

Connecting an antenna directly to the transmitter is often discouraged because of the close proximity of the radiating element to house wiring and domestic equipment. This undesirable feature is aggravated by the fact that wild excursions of feed impedance can occur when changing operation from band to band. Also, good matching can sometimes be difficult to achieve. Choice of wire length may alleviate this problem and is discussed in detail later, see Matching and Tuning. An inverted L antenna, as shown in Fig 15.1 is often referred to as a Marconi antenna.

However, the end-fed antenna is simple, cheap, and easy to erect; suits many house and garden layouts and is equally amenable to base or portable operation.

Remote End-Fed Antenna

Having the antenna feedpoint remote from the shack (**Fig 15.2**) can circumvent the disadvantages of bringing the end of an antenna into the shack. Locating the long-wire antenna feedpoint away from the house minimises EMC problems on transmit on receive (electrical noise). Furthermore, it reduces the unpredictable effect on the antenna caused by possible conduit, wiring and water pipe resonances.

The disadvantage of this arrangement is that the ATU is some distance from the transceiver, and this can be rather inconvenient when it comes to making adjustments. Methods of overcoming this problem is discussed in detail later, see Matching and Tuning

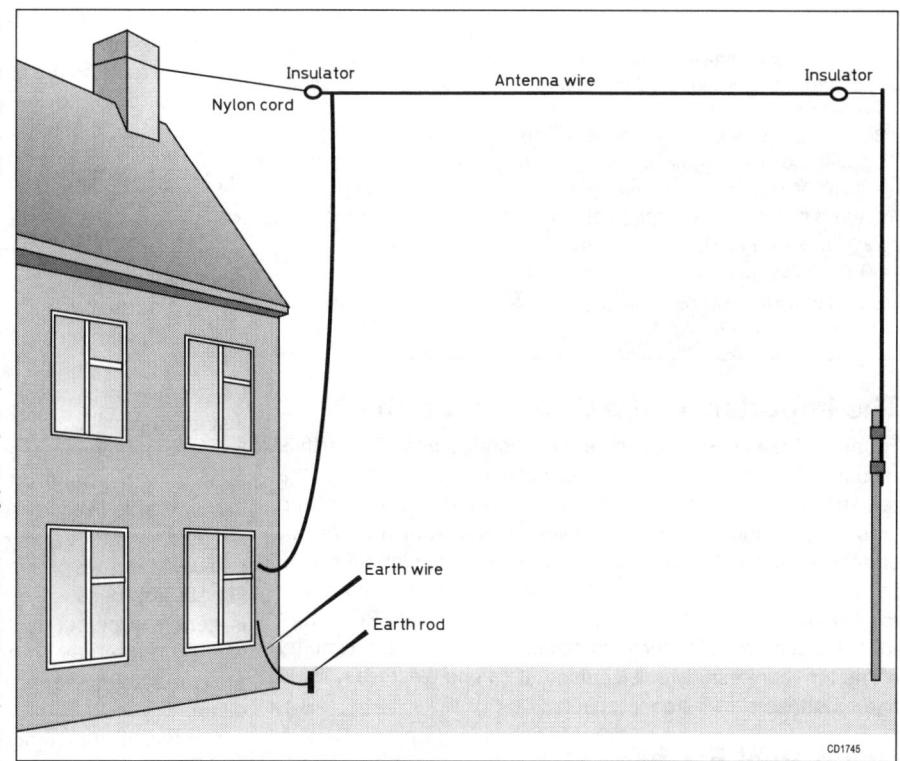

Fig 15.1: The end-fed antenna, the simplest of all multi-band antennas

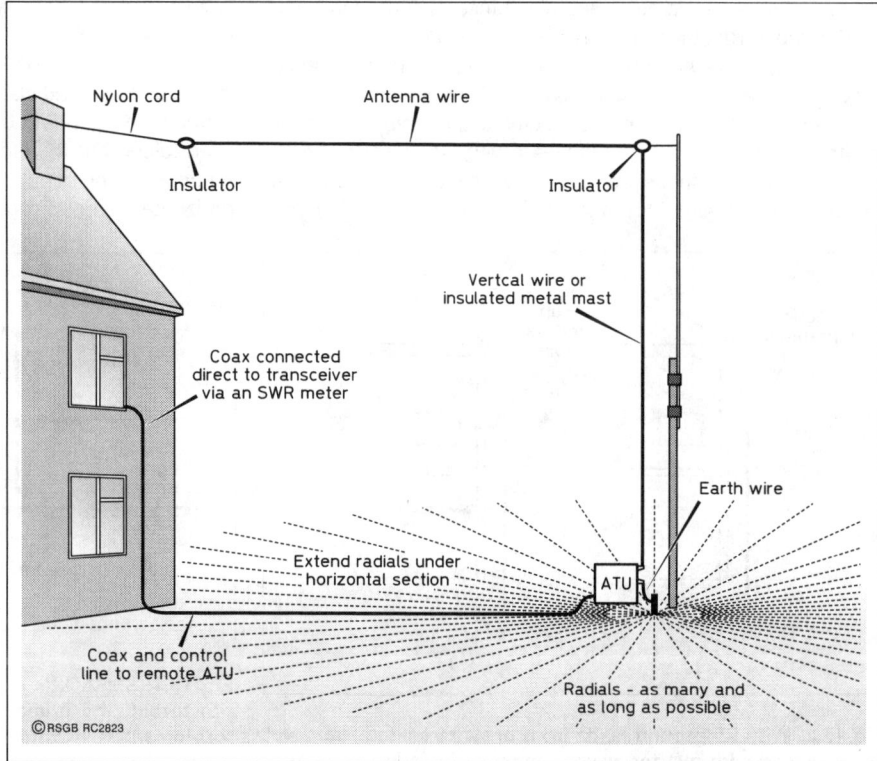

Fig 15.2: A remotely fed long wire antenna arrangement. The ATU can be either preset or automatic and may require a control cable in addition to the coaxial cable feeder

WARNING: Protective Multiple Earthing (PME)

Some houses, particularly those built or wired since the middle 'seventies, are wired on what is known as the PME system. In this system the earth conductor of the consumer's installation is bonded to the neutral close to where the supply enters the premises, and there is no separate earth conductor going back to the sub-station.

With a PME system a small voltage may exist between the consumer's earth conductor, and any metal work connected to it, and the true earth (the earth out in the garden). Under certain very rare supply system faults this voltage could rise to a dangerous level. Because of this supply companies advise certain precautions relating to the bonding of metal work inside the house, and also to the connection of external earths.

WHERE A HOUSE IS WIRED ON THE PME SYSTEM DO NOT CONNECT ANY EXTERNAL (ie radio) EARTHS TO APPARATUS INSIDE THE HOUSE unless suitable precautions are taken.

A free leaflet *EMC 07 Protective Multiple Earthing* is available on request from RSGB.

The Importance of a Good RF Earth

For an end-fed antenna to operate efficiently a good RF earth is required. The resistance of this connection is in series with the radiation resistance of the antenna so it is important to get the ground resistance as low as possible if you want an efficient end-fed antenna. A poor RF earth can result in a high RF potential on the metal cases of radio equipment. Furthermore, the microphone, key or headset leads are also 'hot' with RF so that RF feedback and BCI problems occur. Additionally, the circuitry of modern communications equipment can be electrically damaged in these circumstances.

Using Real Earth

In practice a good RF earth connection is hard to find and is only practicable from a ground floor room. The problem with the 'earth stake' is that ground has resistance and the lead connecting the earth stake to the radio has reactance.

Many ways have been tried to reduce the ground resistance. In general, the more copper you can bury in the ground the better. An old copper water tank, connected to the radio with a short length of thick copper wire, makes a very good earth. An RF earth can also be made from about 60 square metres of galvanised chicken wire. This is laid on the lawn early in the year

and pegged down with large staples made from hard-drawn copper wire. The grass will grow up through the chicken wire and as if by magic the wire netting will disappear into the ground over a period of about two months. In the early stages, the lawn has to be cut with care with the mower set so that it does not cut too close and chew up the carefully laid wire netting.

Low band DXers tend to use buried multiple radials; many wires radiating out from the earth connection. The rule seems to be the more wires the better. These types of direct connection to earth can also provide an electrical safety earth to the radio equipment in the shack.

Artificial Earths

If you operate from an upstairs shack, engineering a low-impedance earth connection at ground level using the method described above will probably be a waste of time. The reason for this is that the distance up to the shack is a significant fraction of a wavelength on the higher HF bands and above. At frequencies where this length is near one or three quarters of a wavelength, the earth connector will act as an RF insulator, which is just the opposite of what is wanted, see **Fig 15.3(a)**. This is bound to happen in one or more of our nine HF bands.

On the other hand, if the lead resonates as a half-wave, (a situation that is likely to arise on any band above 10MHz), it may act as a good RF earth. However, it also has a high-voltage point halfway down which may couple RF into the house wiring, see **Fig 15.3(b)**, because electrical wiring within the wall of a house is generally perpendicular. In other words, although an earth wire from the radio in an upstairs shack to an earth stake will provide a safety earth its usefulness as an RF earth is unpredictable.

The favoured method of obtaining a good RF earth is to connect a quarter-wave radial for each band to the transceiver and ATU earth connector, then running the free ends outside, away from the transceiver. Because the current at the end of the wire is zero and the impedance is high it follows that at a quarter wave inward, where it connects to the transceiver, the RF potential is zero (the impedance is low). The problem is where to locate all these radials; such an arrangement will require some experimenting to find the best position. Radials can be bent or even folded but the length may have to be altered to maintain resonance. The radials are best located outside the house in the horizontal plane to reduce coupling into the electrical wiring. If

Fig 15.3: Why RF ground leads from upstairs seldom work. (a) Ground lead with quarter-wave resonance (or odd multiple) is ineffective; very little current will flow into it. (b) Ground lead with half-wave resonance , (or multiples) will have high-voltage points which couple RF into house wiring

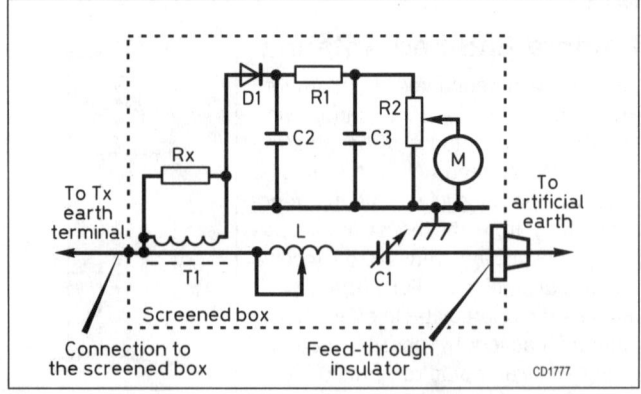

Fig 15.4: SM6AQR's earth lead tuner. T1 = Amidon T-50-43 ferrite toroid; the primary is simply the earth lead through the toroid centre; secondary = 20t small gauge enamelled wire. L = 28µH rollercoaster or multi-tapped coil with 10-position switch; see text. C1 = 200pF or more air variable, >1mm spacing, insulated from panel and case. C2 , C3 = l0nF ceramic. D1 = AA119; R1 = 1k; R2 = 10k pot, Rx see text. M = 100µA or less

The Radio Communication Handbook

the radial(s) are used indoors (say, round the skirting board) use wire with thick insulation with several additional layers of insulating tape at the ends where the RF voltage can be fairly high when the transmitter is on.

The best way to check resonance of a radial is to connect it to the radio earth, make a loop in the radial and use a dip meter to check resonance. If such an instrument is unavailable then use an RF current meter and adjust the radial length for maximum current.

Alternatively, one single length radial can be tuned to place a zero RF potential at the transceiver on any band by inserting a LC series tuning circuit between the transmitter and the radial. Such a units are commercially available, which have, in addition to the LC circuit, a through-current RF indicator which helps tuning the radial or earth lead to resonance (maximum current).

Or you can make one yourself. The unit designed by SM6AQR [1] and shown in **Fig 15.4**, uses a 200-300pF air spaced tuning capacitor with at least 1mm plate spacing; the capacitor and its shaft must be insulated from the tuner cabinet. The inductor is a 28μH roller coaster. Alternatively, a multi-tapped fixed coil plus with as many taps as possible could be used.

The tuning indicator consists of a current transformer, rectifier, smoothing filter, sensitivity potentiometer and DC microammeter. The 'primary' of the current transformer is the artificial earth lead itself; it simply passes through the centre of the T1 ferrite toroid, onto which a secondary of 20 turns of thin enamelled wire has been wound. Rx, the resistor across the T1 secondary, should be non-inductive and between 22 and 100 ohms; it is selected such that a convenient meter deflection can be set with the sensitivity control R2 on each required frequency and for the RF power used.

A separate electrical safety earth should always be used, in addition to the RF earth described above.

Using an Existing HF Wire Beam on the Lower HF Bands

A wire beam such as the quad, or any of those, described later, can be used as an end-fed antenna as shown in **Fig 15.5** provided the HF beam and mast are fairly close to the shack.

In this case, the coaxial cable is used as the antenna conductor rather than as a feeder. The inner conductor and the braid of the coax is shorted together using a PL259 socket with a shorting link and connected to the ATU. The beam itself forms a top capacitance which, provided the coaxial cable is reasonably clear of obstructions and not fixed to the tower, makes a very effective lower HF frequency antenna.

Other antennas can be used in this way. A dipole for 20 metres can be used on the lower frequency bands by connecting the coax to the ATU, as already described, so that the dipole forms a capacity top. As with all end fed antennas a good RF earth is required.

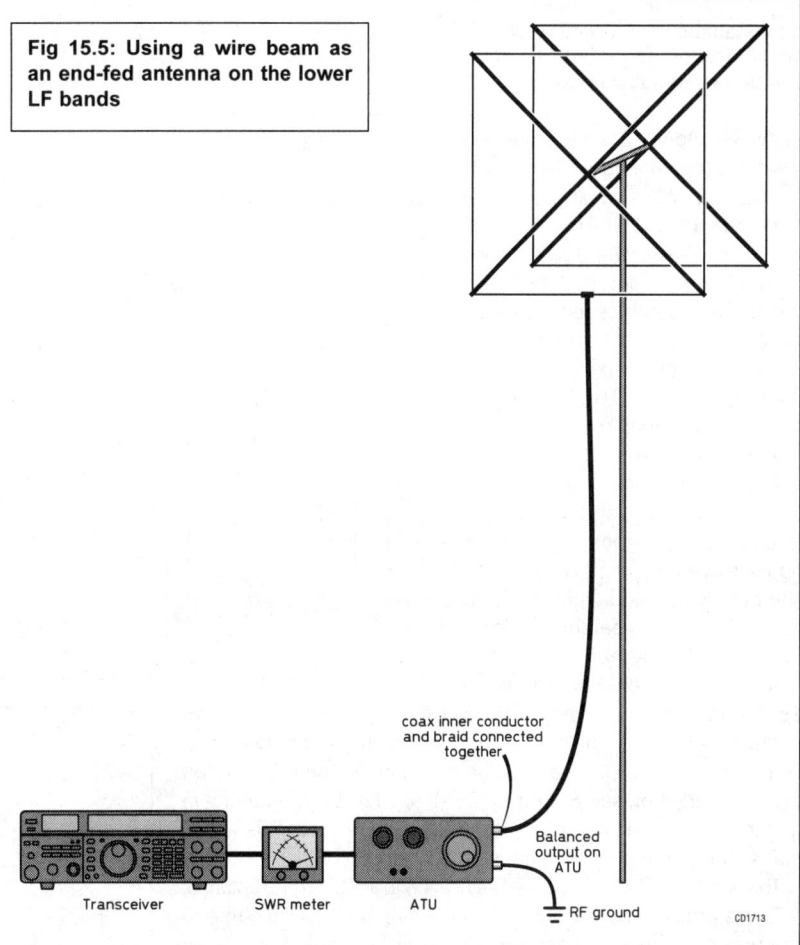

Fig 15.5: Using a wire beam as an end-fed antenna on the lower LF bands

coax inner conductor and braid connected together

Balanced output on ATU

Transceiver   SWR meter   ATU   RF ground

CD1713

CENTRE FED WIRE ANTENNAS

The Centre-Fed Dipole

Of all antennas the half-wave dipole is the most sure-fire, uncomplicated antenna that you can make and does not require an ATU. A centre fed antenna does not require connection to an earth system to function. In its basic form it is essentially a single band, half-wave balanced antenna (although normally fed in the centre with unbalanced coaxial cable). The current and voltage in one half of the dipole is matched by those values in the opposite half about the centre feed point, see **Fig 15.6**. The

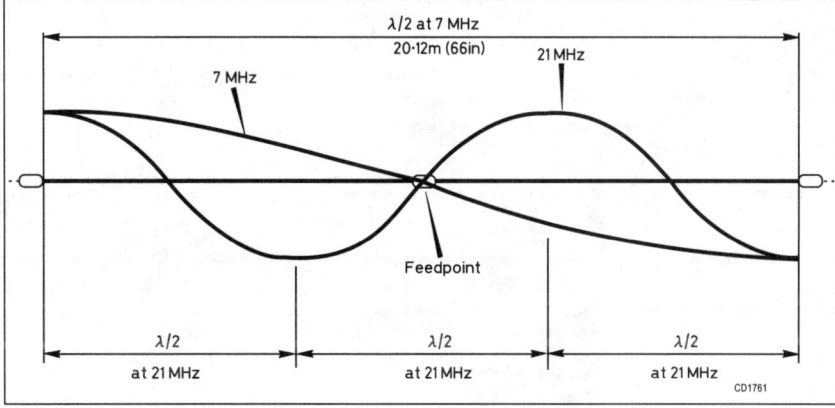

Fig 15.6: The voltage distribution on a 7MHz half-wave dipole. The coaxial cable is connected to the antenna at a point where the feed impedance is low (where the voltage is low). The 7MHz dipole will also have a low impedance at the centre on the third harmonic, at 21MHz

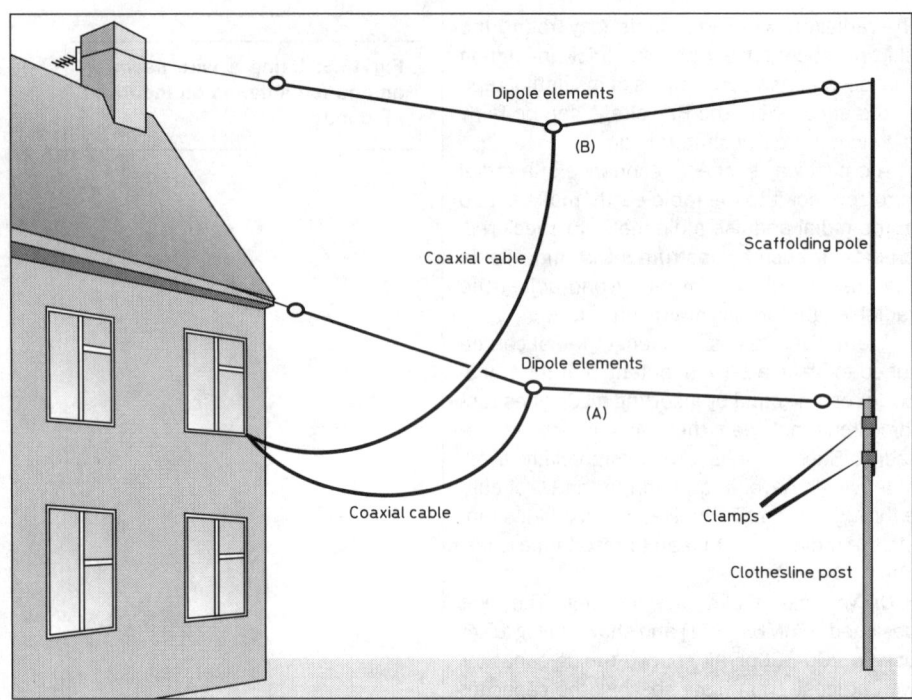

Fig 15.7: Antenna (a) shows a quick fix installation for a dipole. With the minimal effort the dipole height (b) can be raised substantially

halfwave dipole will also present a low feed impedance on its third harmonic so a 7MHz dipole will be close to resonance on 21MHz.

In most cases, the dipole is better than 95% efficient and because it has a low impedance feedpoint it can be connected the transceiver via a length of 50ohm coaxial cable without the need of an ATU. The elements can be bent, within reason, to accommodate space restrictions.

A practical dipole antenna for the higher HF bands is shown in **Fig 15.7**. It can be strung out between the house eaves and an existing clothes line post which would give the antenna an average height of around 4metres (15ft). As can be seen by the elevation radiation pattern (**see Fig 15.8**) the antenna is most effective for signals having a high angle of radiation, which would make it suitable for short skip contacts. With very little effort, the antenna can be raised to an average height of, say, 8 metres (23ft), see Fig 15.7, resulting in a much lower angle of radiation as shown in Fig 15.8. This would make the antenna much more suitable for DX contacts.

The above comments regarding height apply to any horizontal antenna.

There are various methods of connecting the coaxial cable to the antenna - two of these are shown in **Fig 15.9**. Both these arrangements take the strain of the coax from the connections. Also by having the ends of the cable facing downwards assists in preventing water entering the coax cable. It is still important to

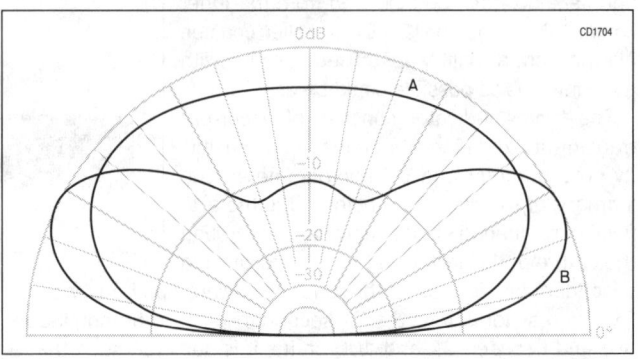

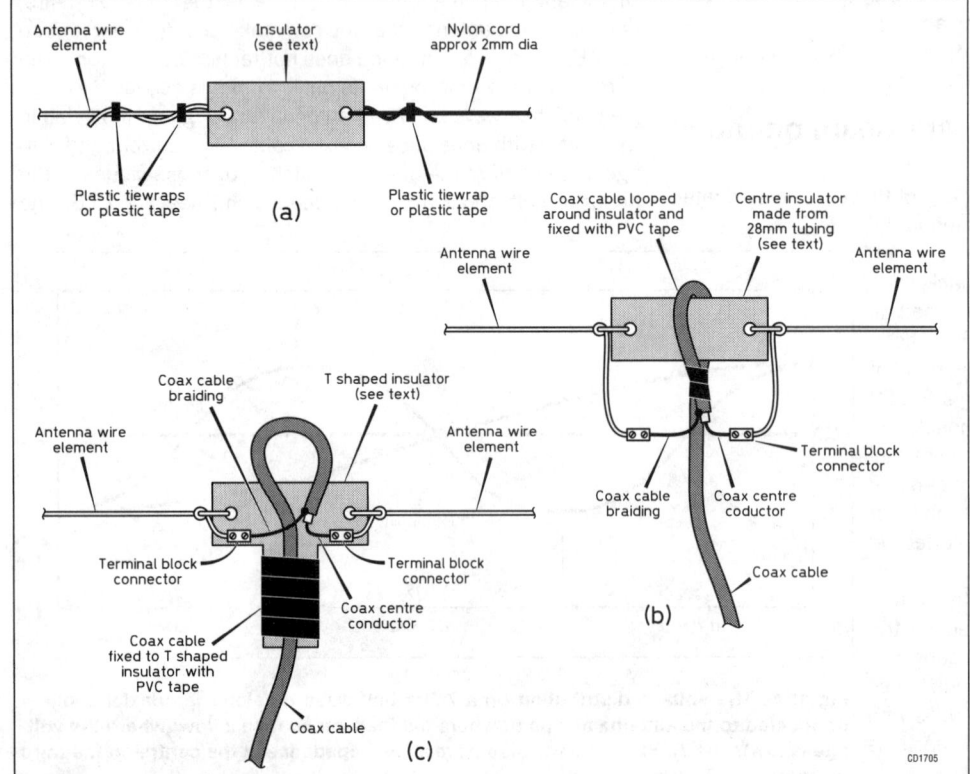

Fig 15.8: Elevation radiation pattern for a dipole at different heights above ground, see Fig 15.7. (a) is for a dipole around 4m high. (b) Is from the antenna at 8m. Path L indicates a 2dB increase in signal strength of (b) compared with (a) for low angle DX signals. Path H indicates a 2dB decrease in signal strength of (b) compared with (a) for high angle short skip non-DX signals

Fig 15.9:(a) A convenient arrangement for constructing a dipole so that the element lengths can be adjusted to make the element longer than shown in Table 15.1. The excess is taped back along the element. (b) Method of connecting coax cable to the centre of the dipole using a short length of tubing or a dog-bone insulator. (c) Method of connecting coax to the centre of a dipole using a specially constructed T insulator. A sealant should be used to prevent water ingress at the exposed coax end

Table 15.1: Wavelengths and half-wavelengths, together with resonant lengths for dipoles relative to frequency for the HF bands. (' = ft, " = in). The dipole lengths * are calculated for a wire diameter of 2mm. The dipole lengths # are calculated for a tube diameter of 25mm

Freq: MHz	λ Metres	λ Ft/in	λ/2 Metres	λ/2 Ft/in	Dipole Metres*	Dipole Ft/in*	Dipole Metres #	Dipole Ft/in #
1.83	163.82	537'6"	81.91	268'9"	80	262'	-	-
1.9	157.78	517'8"	78.89	259'10"	77	260'7"	-	-
3.52	85.17	279'5"	42.58	139'8"	41.56	135'10"	-	-
3.65	82.13	269'6"	41.07	134'9"	40.08	131.6"	-	-
7.02	42.7	140'1"	21.35	70'0"	20.76	68'2"	-	-
10.125	29.61	97'2"	14.8	48'7"	14.4	47'2"	14.2	46'7"
14.05	21.34	70'0"	10.67	35'0"	10.35	33'11"	10.20	33'5"
14.20	21.11	69'3"	10.55	34'10"	10.24	33'7"	10.09	33'1"
18.1	16.56	54'4"	8.28	27'2"	8.03	26'4"	7.88	25'10"
21.05	14.24	46'9"	7.12	23'5"	6.9	22'8"	6.78	22'3"
21.2	14.14	46'5"	7.07	23'3"	6.86	22'6"	6.73	22'1".
24.94	12.62	39'6"	6.31	19'9"	5.82	19'1"	5.70	18'8"
28.05	10.69	35'0"	5.34	17'6"	5.18	17'0"	5.05	16'7"
28.4	10.56	34'8"	5.28	17'4"	5.1	16'8"	4.99	16'4"
29.5	10.16	33'4"	5.08	16'9"	4.9	16'1"	4.80	15'9"

seal the junction against the ingress of water, using either self-amalgamating tape or a non-corrosive sealant.

The dipole can be supported using 2mm or 3mm nylon rope with 'dogbone' insulators at the ends of the elements. The method of connecting the antenna element to the insulator, shown in Fig 15.9, allows the dipole element length to be adjusted for minimum SWR.

Do not use egg insulators and wire as element end supports - the end capacity of such an arrangement can cause some very unpredictable results if the antenna is supported by wire.

The dipole is described as a half wavelength antenna. In practice the dipole length is slightly shorter than half a wavelength because of end-effect. A true wavelength on 7.02MHz is 42.7m and a halfwave 21.35m. A halfwave dipole for the same frequency will be 20.78m (68ft, 2in)

Dipole dimensions for each amateur band are shown in the **Table 15.1**, where the wire lengths have been calculated using EZNEC (see Antenna Fundamentals chapter) and assume the use of 2mm diameter wire and an antenna height of 10m (33ft).

Table 15.1 gives the equivalent wavelengths and half wavelengths for given frequencies in metric and imperial units. Half wavelengths for centre fed dipole or vertical antennas, described earlier, are also given in metric and imperial units and are calculated using EZNEC.

Most antenna books use the formula 143/f (MHz) = L (metres) or 468/f (MHz) = L (feet). This gives a close enough approximation on the higher frequency bands but may be a bit short for the lower bands. For example, the formula gives a dipole length of 40.6m for 3.52MHz while EZNEC calculates a length of 41.42m for the same frequency.

Remember, these are total lengths and the wire has to be cut in half at the centre to connect the coax and that the gap in the centre is part of the whole dipole length. You also need to be aware that around 160mm (6in) at each end of each half of the dipole elements is required to connect them to the centre insulator and the end insulator.

When a larger diameter conductor is used for the antenna element, the length has to be reduced by an amount known as the K factor (based on the length to diameter ratio). For example, the calculated length for a dipole for 21.2MHz is 6.86m, or 22'6". If the conductor diameter is increased from 2mm to 25mm (1 in), the total length should be reduced to 6.73m (22'1").

This can influence the design when making a vertical antenna with the top element of 25mm diameter tube and the lower element(s) of wire. You should use the appropriate column for

determining the length. Remember that these figures are for a half-wave antenna. For a ground plane antenna on 21.2MHz the top quarter wave 25mm diameter section should be 6.73/2 = 3.36m. The lower wire radials are 6.86/2 = 3.43m.

In practice tubular elements are best constructed using different diameter telescopic sections. This makes it easy to adjust the length on test for minimum SWR.

The 80 and 160 metre dipoles are quite long and should be made of hard drawn copper wire to reduce stretching and sagging due to the weight of the antenna and the coaxial cable.

The feed point impedance of a dipole at resonance can vary either side of the nominal 75 ohms, depending on height above ground, the proximity of buildings and any electromagnetic obstacles, together with any bends or 'dog-legs' in the wire. As a result, an SWR of 1:1 is not always possible when the antenna is fed with 50 ohm coaxial cable.

Because the dipole is a balanced symmetrical antenna, ideally it should be fed with balanced two-wire feeder. However, because almost all transmitters use a 50 ohm coaxial line antenna socket, coaxial cable is almost universally used to feed the dipole antenna. Connecting unbalanced coaxial cable to a balanced antenna does not normally affect the performance of the antenna provided the unbalanced current (antenna current) on the coaxial line is kept to a minimum. This can be done by making sure the coaxial line is not a multiple of an electrical quarter wavelength and that the coax line comes away from the antenna element at as close to 90 degrees as possible.

Antenna currents on the line, which can cause the line to radiate (and lead to EMC problems) should not be confused with SWR. A high SWR on transmission line does not cause it to radiate.

A balun can also be used to reduce these antenna currents, see Transmission Lines chapter.

The dipole antenna normally requires two supports and this may be a problem at some locations. The solution may be to mount the antenna so that it is vertical or sloping. A dipole with the centre feedpoint fixed on a single mast, with the ends sloping towards the ground, (inverted V) is a common configuration.

The Ground Plane Antenna

When a dipole is mounted vertically, it has become common practice to call the top element of the antenna a 'vertical' and the lower one a 'counterpoise'. The terminology is derived from an antenna that was once quite popular called the Ground Plane. This antenna comprises a vertical element with a counterpoise made from four wires called radials as shown

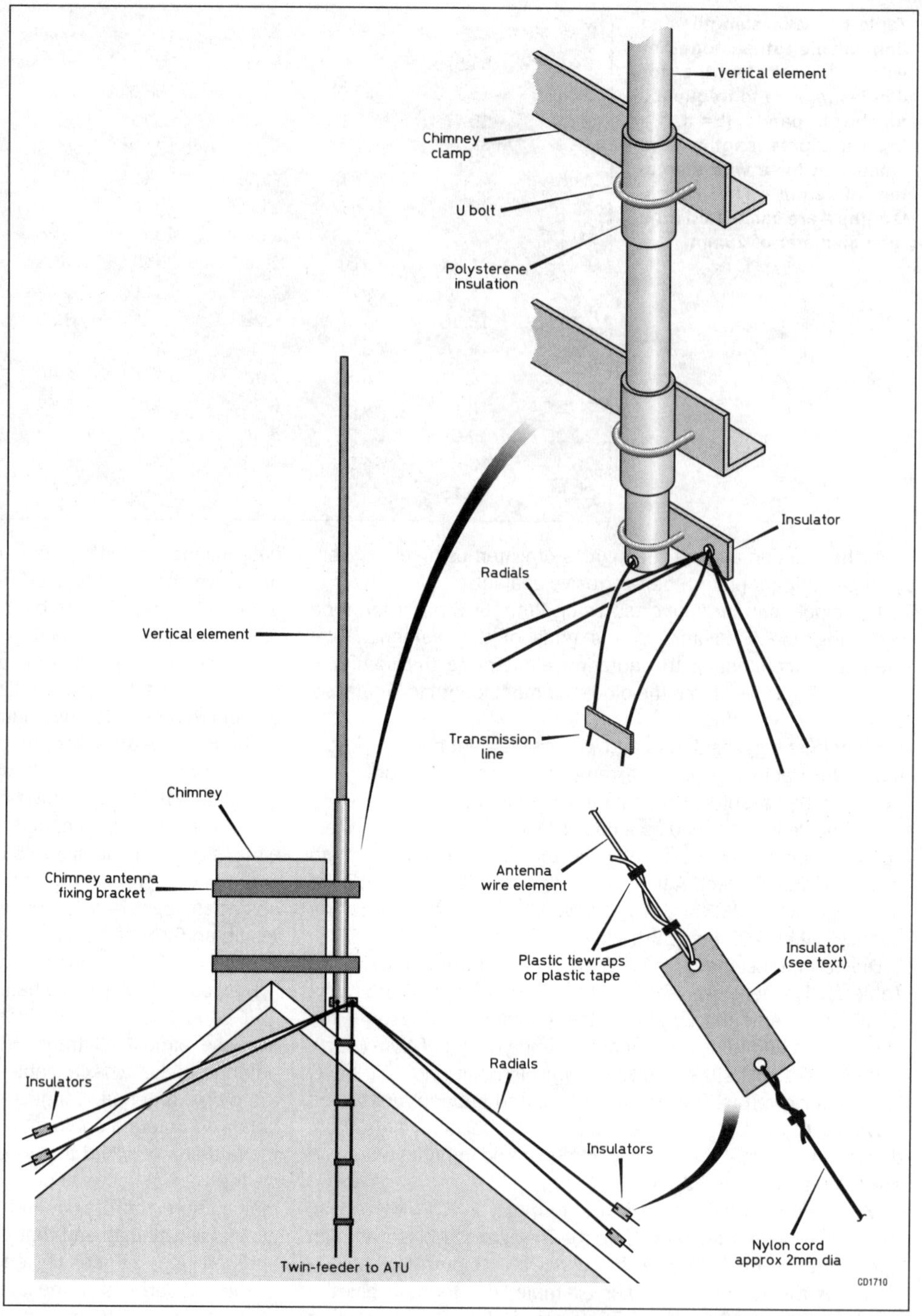

Fig 15.10: The ground plane antenna. It can be used as a single band resonant antenna, fed with coax cable; with the centre of the coax being fed to the vertical element and the braiding connected to the radials. The antenna can also be used as a multiband antenna fed with open wire line as shown. The lengths of the elements are shown in Table 15.1

in **Fig 15.10**. The radials are made to slope down from the feedpoint although the angle is not critical. If the radials are at 90 degrees to the vertical element the feed impedance is around 30 ohms; with the radials sloping down at around 45 degrees the feed impedance is around 45 to 55 ohms (depending on height), which is a good match for coax cable.

The vertically orientated antenna is often cited as having a good low angle of radiation. From the graphic data obtained from EZNEC, this appears to be true (see **Fig 15.11**). The elevation plot shows a ground plane antenna whose feedpoint is only 0.2 wavelengths above the ground, which equates to 3m (9ft) on 21MHz. It has a very deep vertical null but the maximum gain is only 0.45dBi.

If the 21MHz groundplane is raised so that the feedpoint is 10m above the ground the antenna has two elevation lobes, one at 12 degrees (1.4dBi) and the other at 38 degrees (2.6dBi), with a deep vertical null, similar to the vertical dipole shown in Fig 15.11.

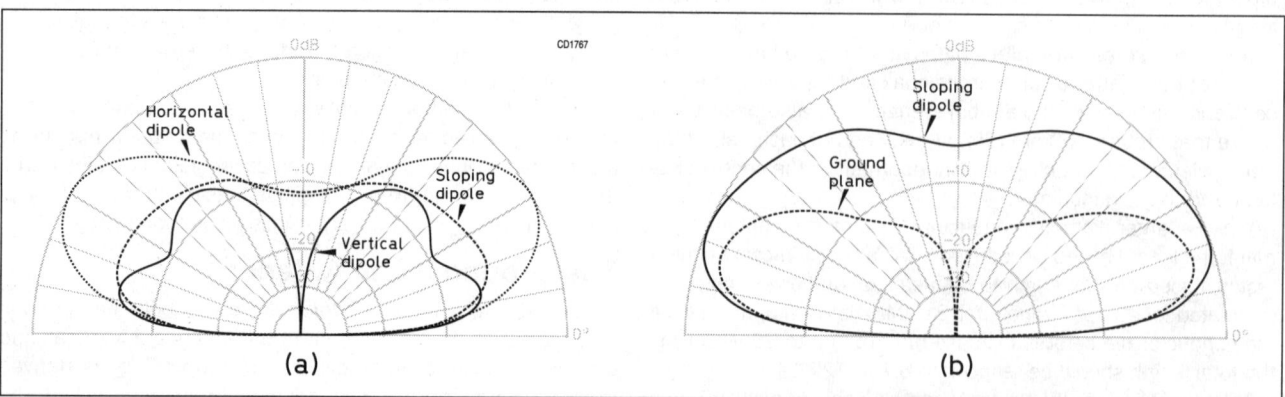

Fig 15.11: Elevation polar diagrams of different orientations of dipoles, and the ground plane 0.2 wavelengths high at the feedpoint

15: PRACTICAL HF ANTENNAS

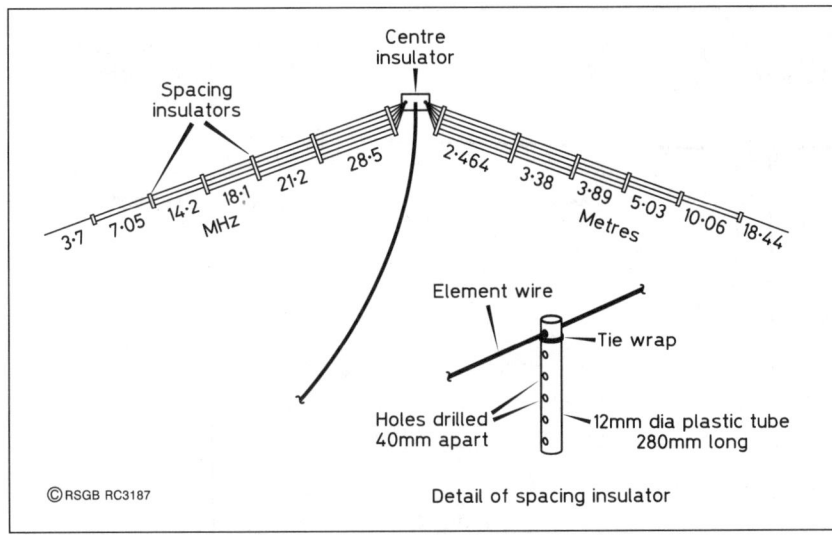

Fig 15.12: Multiband parallel dipoles. The detail shows the larger spacers to accommodate 6 wires. The outer spacers are progressively shorter with holes drilled for 5, 4, 3 and two wires respectively. The 24MHz dipole is not shown but the lengths are 2.84m (9ft 4in)

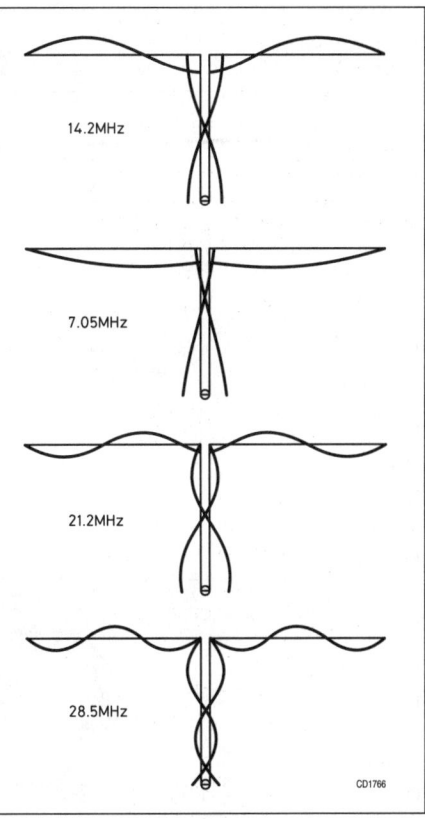

Fig 15.13: Showing the current distribution on a G5RV antenna at 14MHz; and on 7, 21 and 28MHz.

Multi-Band Centre Fed Antennas

While the resonant dipole is a very efficient antenna, which can be connected to the antenna socket of the transceiver without an ATU, using separate dipoles for each of the bands can result in a mass of wires in the back yard. Solutions to the multiband problem include using dipoles in their fundamental and harmonic modes, parallel dipoles, trap dipoles, the multiband doublet using tuned lines and the multiband doublet with an ATU.

Most multiband systems can be improved using an ATU so it is probably a good idea to invest in one, or build one. It is basic RF technology and does not date like computers, or even modern transceivers.

Parallel Dipoles

If you wish to operate on several of the HF bands and you don't have an ATU, parallel dipoles may be the answer. The design shown in **Fig 15.12**. enables all the wires of the multiple dipole to be held together in a tidy fashion. This arrangement uses the lowest frequency dipole to support the higher frequency dipoles using spacing insulators made from 11mm plastic electrical conduit. The antenna is best configured as an inverted V with the weight of the centre insulator and the 1:1 balun mounted on a suitable mast or pole. Low centre band SWRs are possible if some time is spent tuning each dipole. This can be achieved by arranging the ends of the elements so that they are clear of their support insulators by about 200mm. The dipole lengths can be reduced or increased by folding back the end and securing with plastic tape.

The resonance of these dipoles can be interactive - when you adjust one it effects the resonance of the other so be prepared to have to re-resonate elements.

The G5RV Antenna

Newcomers (and some old-timers) often regard the G5RV antenna as a panacea to the multi-band antenna problem. Louis

Varney, G5RV, designed his antenna over 40 years ago, primarily to give a clover-leaf pattern and a low feed impedance on 20 metres. The G5RV has a top of 102ft (31.27m), a total of three half wavelengths on 20 metres, which is fed in the centre.

The feed impedance on 20 metres is low because the feedpoint is at the centre of the central halfwave section. The mid-band resonant feed impedance at that point is around 90 ohms and a 34ft (10.36m) matching section of open-wire feeder is used as a 1:1 transformer, repeating the feed impedance at the other end, as shown in **Fig 15.13**.

Because of this, the lower end of the matching section can be connected to a length of 75 ohm coaxial cable as a convenient way of routing the feed to the transmitter in the shack (see **Fig 15.14**).

In addition, the antenna is presents low impedances on other bands, which were within the impedance range of earlier amateur radio transmitters with pi-output variable tuning and loading; thus the antenna could be connected directly to the transmitter without an ATU. This represented quite an advantage over routing open line feeder into the shack.

However, for the G5RV to work the top dimension must be around 31.27m (102ft) and the dimensions of the of the matching section shown in Fig 15.14 are only true for open wire feeder. If 300-ohm ribbon or slotted line is used, the length must be adjusted to take account of the velocity factor (multiply 10.36m - 34ft - by the velocity factor).

In addition, the G5RV geometry cannot be altered by, for example, converting it into an inverted-V or bending the ends to fit into a small available space, without modification to the length.

On the 10, 18 and 28MHz bands the feed impedances are likely to be fairly wild. Modern all-solid state amateur band transceivers have transmitter output stages that can be damaged when operated with high SWR on the feed cable to the antenna, or they have an ALC circuit that reduces power in some proportion to SWR. It is obvious that an ATU between the low-impedance feeder and the transceiver is required.

The Radio Communication Handbook

15.7

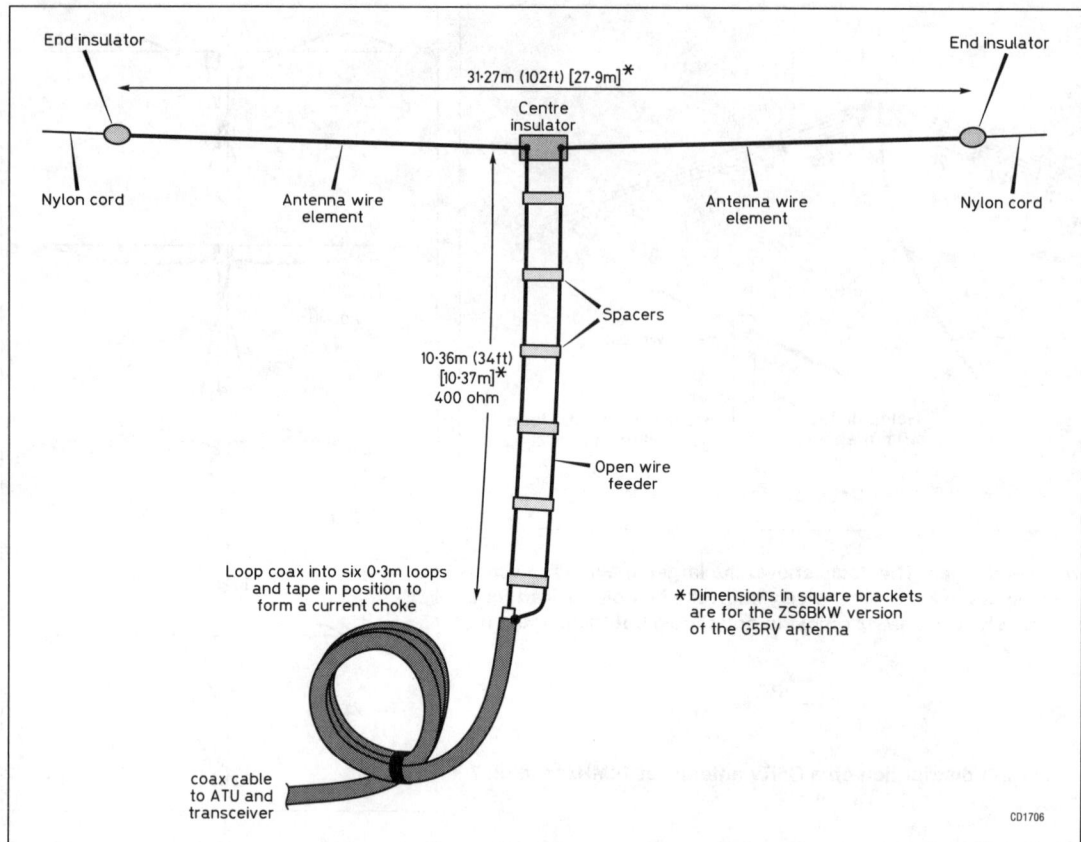

Fig 15.14: Construction of the G5RV antenna. The dimensions shown in square brackets are for the ZS6BKW version - see text

ZS6BKW developed a computer program to determine the most advantageous length and impedance of the matching section and the top length of a G5RV-type antenna. He arranged that his antenna should match as closely as possible into standard 50-ohm coaxial cable and so be more useful to the user of modern equipment. The G5RV antenna total top length of 31m was reduced to 27.9m, and the matching section was increased from 10.37m (ignoring the velocity factor). This matching section must have a characteristic impedance of 400-ohms, and it can be made up from 18SWG wires spaced at 250mm (10 in) apart.

The ZS6BKW gives improved impedance matching over the original G5RV, but still cannot be used without an ATU with modern solid state PA transmitters.

Some amateurs have reported that they get very low SWR readings on all bands. If you have consistently a low SWR using this antenna, it is possible that a test of the coaxial cable from the transmitter to the bottom of the open wire matching section might be in order, see Transmission Lines chapter.

G2BDQ notes [2] that many amateurs use the G5RV antenna with success, and that he prefers the use of either open-wire or

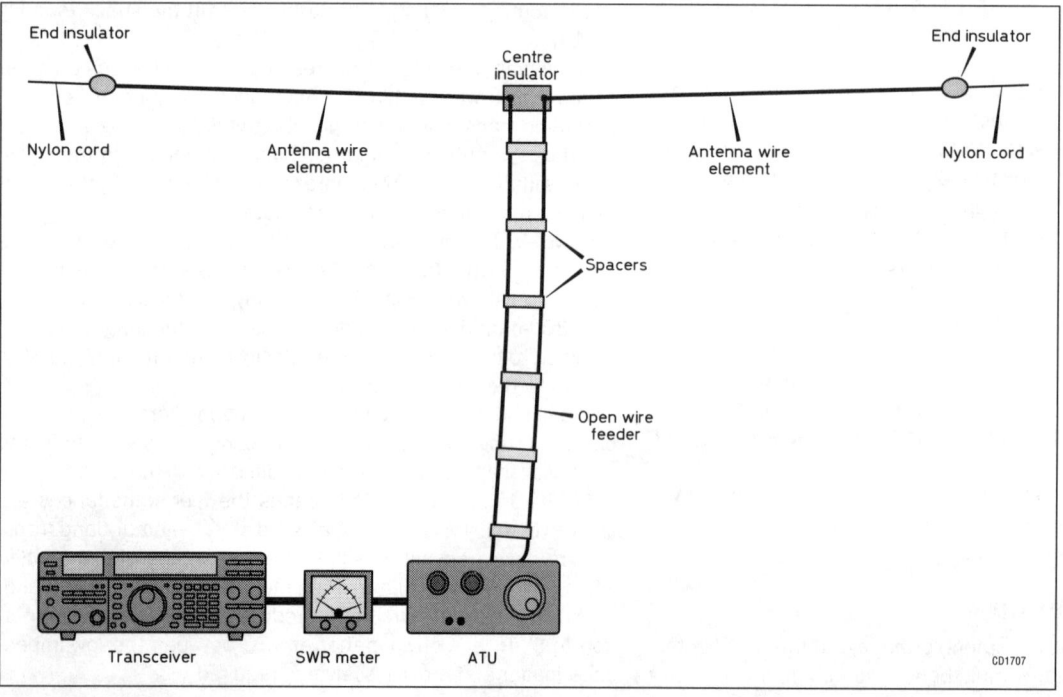

Fig 15.15: The tuned open-wire dipole using a tuned transmission line. If you are short of space the antenna could be cut for 3/8 of a wavelength on 7MHz and it will tune all bands from 7 to 28MHz. The real advantage of this antenna is that the dipole length is not critical, because the tuner provides the impedance match throughout the entire antenna system, whatever the dipole length may be

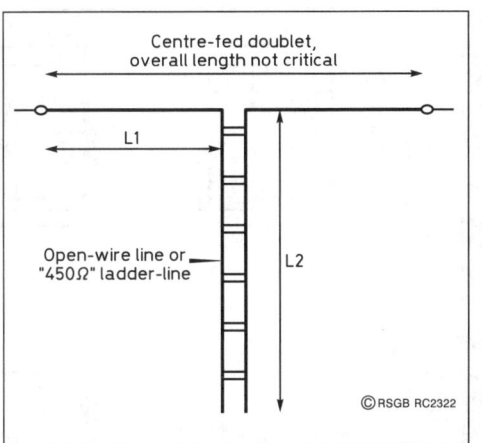

Fig 15.16: The HF Multi-band doublet fed with ladder line can have a wide range of feed impedances. To estimate the impedance, measure half the length of the doublet L, plus the electrical length of the feedline L2, (allow for velocity factor of L2)

(right) Fig 15.17: W6RCA's line-length switcher makes his 39.62m (130ft) doublet cover all eight HF bands with no ATU. Optimum dimensions will depend on local factors, but you can always change the line length to compensate

300-ohm ribbon to feed the horizontal top. With an ATU, such a feed will result in high-performance, all-band working

G5RV mentions [3] that the most efficient feeder to use is the open-wire variety, all the way down from the centre of the antenna to the equipment, in conjunction with a suitable ATU for matching. He added that by using 25.6m (84ft) of open-wire feeder the system will permit parallel tuning of the ATU on all bands; which brings us to the Open-Wire Tuned Dipole.

The Open-Wire Tuned Dipole

This antenna, also known as the Tuned Doublet or Random-Length Dipole is very simple, yet is a most effective and efficient antenna for multiband use. It is fed with open wire tuned feeders, as shown in **Fig 15.15**, and an ATU is used to take care of the wide variations of feed impedance on the different bands.

This antenna should be at least a quarter wavelength long at the lowest frequency of operation, where it radiates with an effectiveness of approximately 95% relative to a half wave dipole.

However, the feed impedance of such a short antenna results in SWR values of around 300:1 on 450-ohm line. While the antenna is quite efficient the impedances at the end of the tuned feeder will be outside the matching range of the average commercial ATU using a toroid balun to provide a balanced feed to the tuned feeders. A doublet with a length of about 3/8 wavelength on the lowest frequency would overcome this problem. This is halfway between quarter wave and half wave and will work very well if you can't erect a full half wave on 80-metres. A 3/8 wavelength dipole has an effectiveness greater than 98% relative to a half wave dipole, and the SWR values are far easier to match, being in the region of 25:1 on 600-ohm line, 24:1 on 450-ohm line, and 25:1 on 300-ohm line.

A 3/8 wavelength dipole at 3.5MHz is approximately 30m (100ft) long, which means that any length from 27m (90ft) to 30m will make an excellent radiator on all HF amateur bands, 80 through 10 metres, including the WARC bands.

If you don't have room for a 27m length of straight wire for operation on 80 metres, a 3 to 5m (10 to 16ft) portion of each end may be dropped vertically from each end support. There will be no significant change in radiation pattern on 80 and 40 metres. However, there will be a minor change in polarisation in the radiation at higher frequencies, but the effect on propagation will be negligible. Bear in mind that twin wire feeder can be affected by the close proximity of metal objects such as windows

or guttering. If this presents difficulties bringing twin feeder into the shack then the Comudipole, described later, may be a solution.

The W6RCA Multi-Band Doublet

Many antenna designs feature combinations of doublet length and feedline length resulting in a convenient impedance (one easily matched by a transceiver's internal auto-ATU) at the bottom of the feedline for a few bands, but never all of them. Hence the need for an ATU with the open wire tuned multi-band dipole described above.

The following describes a more radical approach by Cecil Moore, W6RCA, and described in [4]. His solution to the problem covers all the HF bands from 3.6 to 29.7MHz with no ATU at all. This is achieved by changing the length of the 450-ohm tuned ladder-line - and this is much more practical than it looks at first sight. The line length is adjusted for each band, so that the current maximum always coincides with the bottom of the feedline. The feed impedance at this point is then by definition low and non-reactive, and in practice the SWR is usually low enough that you can use a 1:1 choke balun, straight into coax and the transceiver. With reasonable lengths for the doublet and the permanent part of the feedline, you can always achieve an acceptable impedance match.

The requirement is that the physical half-length of the doublet L1 (see **Fig 15.16**), plus the total electrical length of the feedline L2 (allowing for the velocity factor v) must be an odd multiple of a quarter wavelength on each band:

$$L1 + L2 \times v = n \lambda/4$$
where n is 3, 5, 7 etc

From these calculations, it is obvious there are several possible solutions. The W6RCA arrangement is shown in **Fig 15.17**, with a 39.62m (130ft) centre-fed doublet and 27.5m (90ft) of 450-ohm ladder-line. The doublet is approximately a half-wave at 3.5MHz and a full-wave at 7MHz, and the 27.5m (90ft) feedline brings the current maximum to the bottom at 7.2MHZ. The big practical advantage of this combination is that all the other bands can be matched within a relatively small range of additional feedline length. The longest additional length required is 9.5m (31ft) for 3.6MHz, which extends the feedline to an electrical half-wavelength.

All other bands require a line extension somewhere between zero and 9.5m, so W6RCA built a variable-length switcher shown

Fig 15.18: Practical line switcher. The horizontal rail holds the five pairs of DPDT relays. W6RCA points out that, with hindsight, it makes more sense to start with the 2.44m (8ft) and 4.88m (16ft) loops at opposite ends of the wooden rail that holds the relays. Band changes can also be achieved manually by using plug-in lengths of line for each band

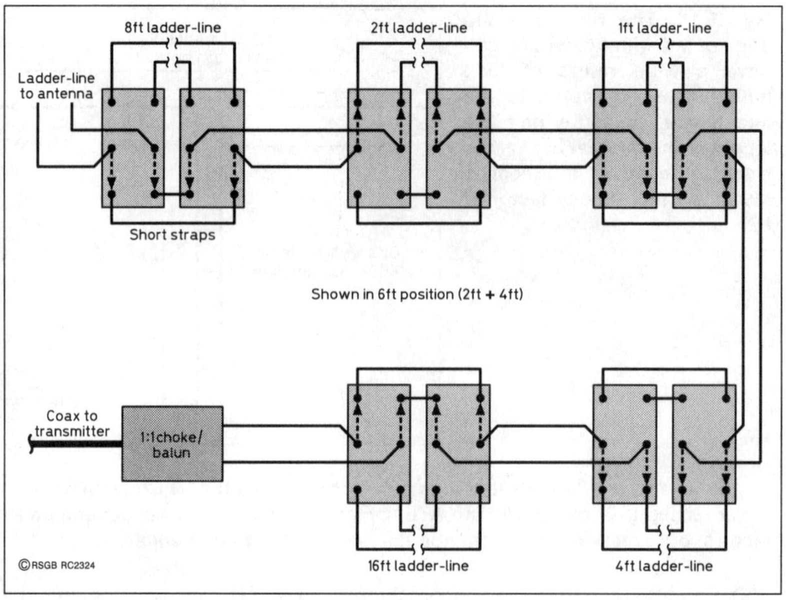

Fig 15.19: More details of W6RCA's line switcher, which uses five pairs of surplus DPDT relays. A suitable 1:1 choke balun is described in the Transmission Lines chapter

in **Figs 15.18 and 15.19**. This consists of 300mm (1ft), 600mm (2ft), 1.22m (4ft), 2.44m (8ft) and 4.88m (16ft) loops of line, which can be individually switched in or out using DPDT relays, giving any length from zero to 9.5m in 300mm steps.

W6RCA found he could cover all amateur bands from 3.6 to 29.7MHz with a SWR of better than 2:1:. The optimum dimensions will depend on a number of local factors. These include antenna height, earth properties, the use of other doublet configurations such as an inverted-V or inverted-U, and the exact type of feedline. So-called '450-ohm' ladder-line varies considerably in characteristic impedance, velocity factor and quality (conductor diameter and insulation) between different brands; hence the need to experiment.

A battery-powered tuneable SWR analyser is the perfect tool for the job of experimenting with line lengths and it should used via the 1:1 balun. You can easily make temporary splices in ladder-line using screw connectors - or just twisting the wires together. The first step would be to increase the permanent length of line by say a metre from the recommended length, then trim the feeder so that the SWR minimum occurs around 7.05MHz. You may find that the same length works well enough for 21MHz and 24.9MHz too. Next, determine the maximum extra length needed to tune all the way down to 3.5MHz with an acceptable SWR. This extra length should not be much more than 9.5m, and the optimum lengths for all the other bands will all be shorter than that.

Unfortunately, the popular 32m (102ft) G5RV-style doublet is not very well suited to this arrangement, because it requires a much wider variation in the feedline length. If you're stuck with a 102ft 'flat-top', W6RCA recommends adding a 4.6m (15ft) vertical 'drop wire' at each end, and then you're back to the much more convenient situation of Fig 15.18. For a shorter doublet covering 7 - 29.7MHz, a 20.12m (66ft) doublet and a 18.3m (60ft) feedline is a good starting-point, again with a 0 - 9.5m variable section. Note also that the system can still be used as a shortened dipole on the next band below, but you will require an ATU and there may be significant losses in the ATU and feedline due to the very low impedance.

After the initial experiments, you can think about a more permanent arrangement. You don't have to build the complete line switcher. Practical solutions range from a fully manual system to a fully automatic system linked to the transceiver's 'band data' output (ideal for HF contesting in the single-antenna section). For occasional visits to certain bands, you could insert the necessary lengths of feedline using 4mm banana plugs and sockets (the silver-plated variety can be used permanently outdoors).

It wouldn't be difficult to string something along a wooden garden fence, so long as the loops of line are suspended clear from other lines, metallic objects or the ground.

The 1:1 balun is worth a brief mention. It's important to use a balun, because any low-impedance path to ground from either side of the feedline is likely to result in very strong unbalanced radiation from the feedline itself. This is a consequence of the 'odd quarter-wavelength' principle used in selecting the feedline length. A suitable choke balun is described later.

The Comudipole
(Coaxial Cable Fed Multi-band Dipole)

In many locations, there are problems of bringing open wire feeder into the shack, particularly for apartment dwellers. One solution for a multi-band antenna was first described by Ton Verberne, PA2ABV [5] is the Comudipole.

The arrangement was used to feed an inverted-Vee dipole of about 2 x 19m mounted on the roof of a five-storey apartment building from a second floor shack. The antenna is not that much different from the tuned open wire dipole arrangement shown earlier. However, bearing in mind that twin wire feeder performance can be affected by the close proximity of metal objects such as metal structures and windows, the twin feeder is brought down to a point where it is still clear of metal objects. At this point it is connected to a 4:1 coaxial balun and there a length of RG-213 coax led to the shack where an L-network takes care of matching to the transceiver.

The comudipole overcomes the problem of running twin line feeder into the shack because the balun can be located outside the shack.

In practice the balun can be placed anywhere along the transmission line section from the antenna to the ATU as shown in **Fig 15.20**. However, the feeder system should consist of as

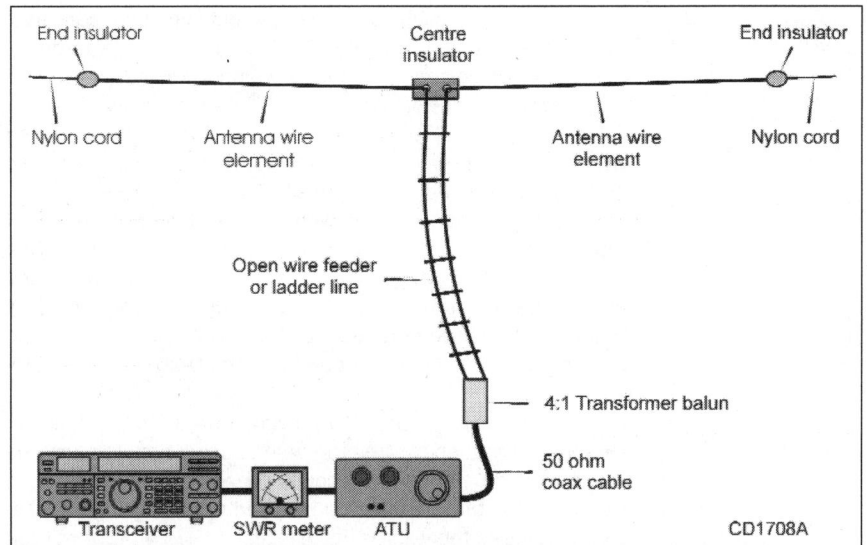

Fig 15.20: The Comudipole feed arrangement for a multiband doublet antenna

much twin wire feeder or ladder line as possible because the losses on such line with a high SWR are much lower than with coax cable. If you are restricted to a short dipole antenna, say less than 15m (45ft), then a 1:1 balun might be more appropriate, see the G3TSO ATU and balun described below.

MATCHING AND TUNING

Many of the antennas described so far may require some degree of impedance transformation before they can be connected to the station transmitter. A unit for providing this transformation is normally called an ATU (Antenna Tuning Unit) or Tuner. As the function of the unit is to match the impedance presented by the antenna system to 50 ohms, AMU (Antenna Matching Unit) might be a more accurate description, but "ATU" is much more commonly used.

There are three different antenna arrangements that may need coupling to the transmitter:
* Wire antenna fed against earth.
* Antenna fed with coaxial cable
* Antenna fed with twin-line feeder or ladder line

Matching the End-Fed Antenna

There are two aspects of the end-fed antenna, which need to be considered. The first is matching the transmitter to the range of impedances encountered at the end of wire antenna on the different bands. The other is an effective and efficient RF earth or ground, which was discussed earlier.

An end-fed antenna has traditionally been designed to resonate on one lower band in the HF spectrum, say a quarter wavelength on 80m where the feedpoint will be around 50 ohms. At a half wavelength on 40m, the input impedance will rise to a high value, presenting a voltage feed to the source. The next band, 30m, will fall in the vicinity of current feed again at three-quarter wavelength and present a fairly low impedance. The next move to 20m will once more encounter a high impedance and then through an off tune 17m to another high at 15m. The sequence continues with extra complications in that odd multiples of one wavelength will show generally increasing impedance with frequency whereas even multiples of wavelength (the halfwave points) will show decreasing impedance on the higher bands.

Fig 15.21 illustrates resistance and reactance plotted against electrical length from below λ/4 to 3λ/4 and beyond. It can be seen that dramatic changes begin to occur as the λ/2 (half-

wave) resonant point is approached. These changes are repeated at multiples of λ/2.

In spite of these wide variations of antenna feed impedance on different bands the transceiver can be matched to the antenna using a suitable ATU, which is described later.

The selection of an optimum antenna length was described in detail by Alan Chester, G3CCB [6], although this was done to meet the limitations of a wideband matching transformer system.

In **Fig 15.22**, wire length is shown against each of the nine HF bands, including 160m. The heavy lines indicate areas where impedance excursions might fall outside the matching capabilities of many ATUs. These lengths were calculated by G3CCB from the lower band edge frequency in each case and no corrections were made for the 'end effect' on a real antenna.

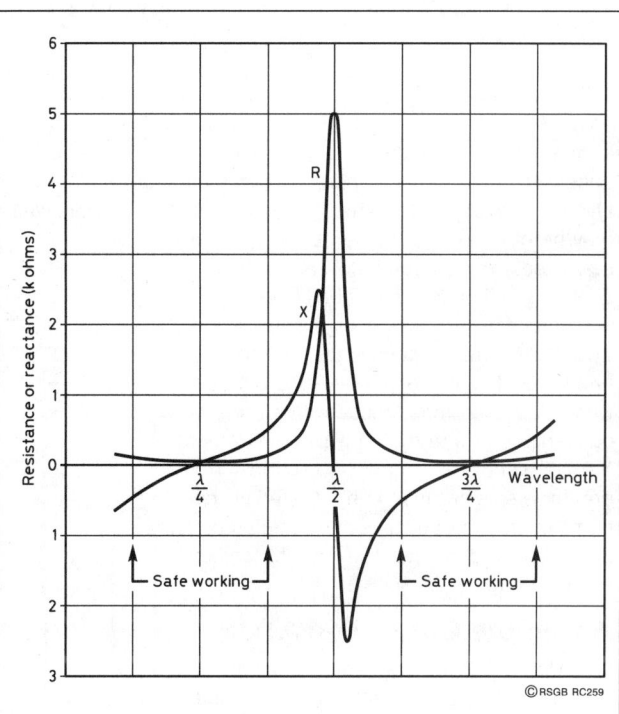

Fig 15.21: End-fed Impedance characteristics of a wire from one-quarter wavelength to three-quarter wavelengths. Values of impedance that are more easily matched using a commercial ATU are designated 'safe working'

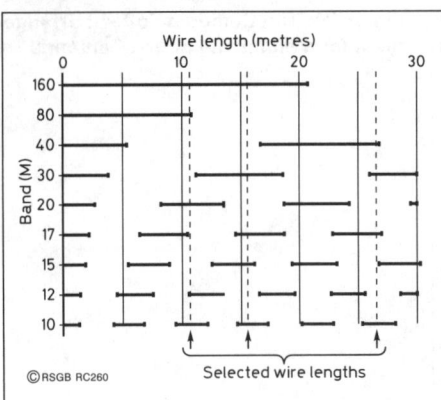

Fig 15.22: Antenna wire lengths, showing high impedance lengths for various bands

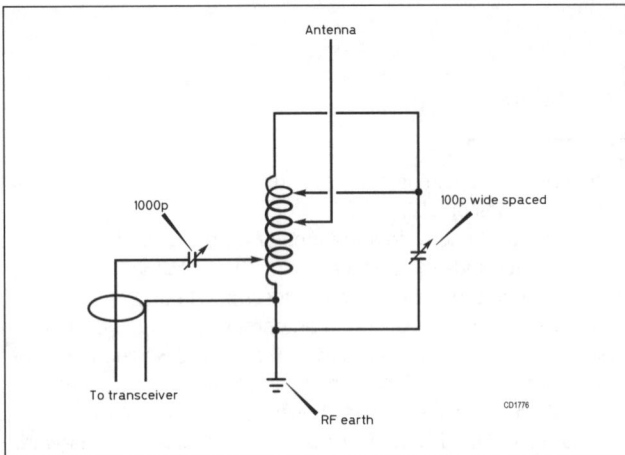

Fig 15.23: A circuit diagram of an ATU for a multiband end-fed antenna. Ideally, to match a whole range of impedances of an end-fed antenna, the coil should be tapped every turn. In practice, a limited number of coil taps can preset and selected using a clip or a switch. The three coil taps and the capacitor settings have to be changed for each band

To use the chart shown in Fig 15.22, a perpendicular straight edge is dropped from the horizontal axis and moved along until a clearest way through the gaps between the extreme impedance sectors is found. There is a minimum antenna length shown which depends on the band in use. This restriction, which may be of interest to those operating from a restricted size site, can be overcome by using a loading coil - this is described later.

In practice the ATU design shown in **Fig 15.23** gets rather complicated when multiband operation is required. Ideally, to match a whole range of impedances with all the various lengths of wire that may be encountered, the coil should be tapped every turn. There are three sets of taps to be adjusted for each band. In practice, coil taps can be adjusted on test then fixed so that they can be selected using a switch or relay.

A Remote Controlled ATU

L B Uphill, G3UCE, devised a remote controlled ATU (**Fig 15.24**) with an end-fed antenna that has proved satisfactory on all the HF bands. It sits in a rear porch and is connected to the shack by 10m of coaxial cable, plus an 8-way multicore cable for remote control of the relays. Other suitable places to house an

ATU may be used, such as the garage, outside shed, conservatory or greenhouse. Even mounting it on a post in the garden is feasible, provided the assembly is weatherproofed. A good RF earth or a counterpoise close to the ATU is necessary.

Once set up this ATU provides instant selection of preselected settings. It will, however, need several hours to set up and so should be sited in an accessible position.

The ATU has been tested with several different lengths of antenna from 18m (60ft) to 61m (200ft). Some antenna wire lengths were more difficult than others to get all six bands working, and the best turned out to be 30m (100ft) and 40m (132ft). Around 40 turns are required if 160m is the lowest band to be used; if 80m is the lowest frequency then 20-25 turns are sufficient.

A junk box coil may be used provided the turns are of a reasonably heavy copper wire. The wire spacing should allow the use of an instrument-type crocodile clip with narrow jaws to be clipped to any turn during setting-up, without shorting an adjacent turn. If a suitable coil is not available then a 40-turn coil can be wound on a 190mm (7.5in) length of 45mm diameter plastic pipe using 14 to 16SWG tinned copper wire. Fasten one end to a nut and bolt, and wind tightly using a similar thickness of string as spacing until 40 turns are wound on. Anchor the other end to a nut and bolt and carefully remove the string spacer. Apply 3 or 4 strings of adhesive, such as superglue, across the turns to hold them in place.

The capacitors used are 100pF air spaced types for the higher frequency bands and 500pF 500V working mica presets for the lower bands. A capacitor may not be required for 160m, where a direct connection is made from the coax to the coil.

The relays are 12 volt types and are not critical, provided the contacts can carry about 5A AC. A small control box with a 2-pole, 6-way Yaxley switch controls the relay switching in the shack. One pole switches the relays and the other pole switches small LED indicator lamps to show which band is selected. An 8-core miniature, screened cable is used to connect the ATU to the shack. With six bands to select, this will leave two wires spare and these are used to switch the transmitter on and off via the CW socket during adjustments from the ATU end.

The setting up procedure is as follows: Starting with the lowest frequency, tune up the transmitter on a dummy load to the centre of the band, connect the feeder, energise the appropriate

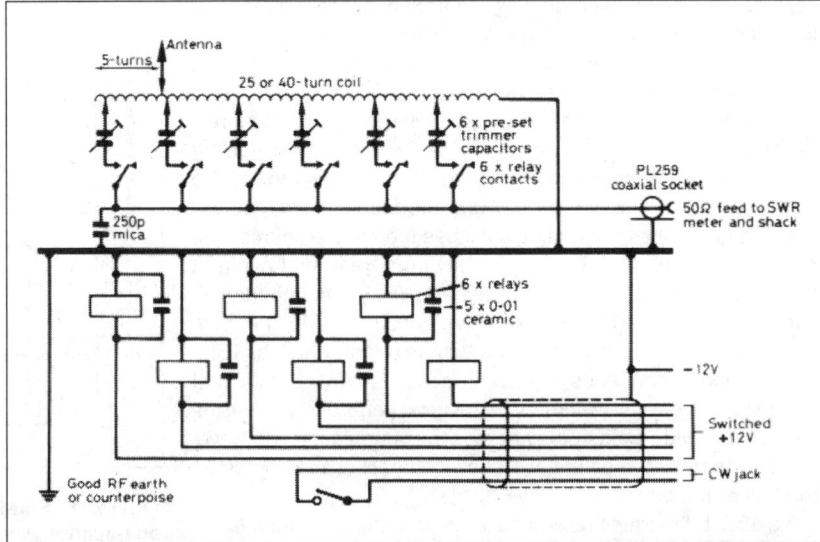

Fig 15.24: The G3UCE remote-controlled ATU

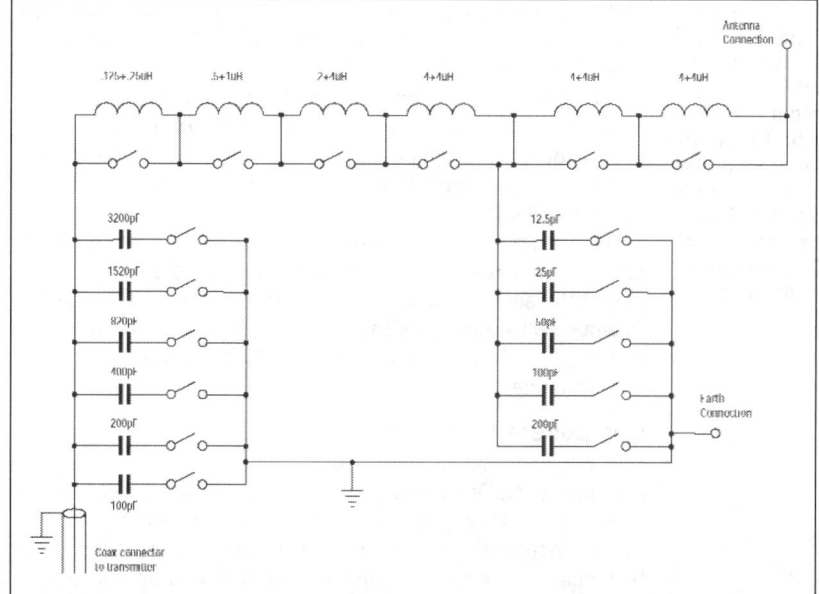

Fig 15.25: Simplified diagram of the SG-235 ATU. The Pi section inductor (the section between the two capacitor banks) is actually made up of eight inductors, while the inductors (top right) are switched in for short antennas. Switching relays are controlled by a SWR/microprocessor circuit (not shown)

SWR and impedance samples from an RF head.

The construction of an automatic tuner is beyond the scope of this chapter, however the G3XJP PicATUne automatic ATU design can be found at [7]

The SG-235 is quite a robust unit, said to be capable of handling 500W, and is obviously the big brother of the SG-230, a simplified version of which is shown in **Fig 15.25**. From this we can see that the SG-235 is a Pi-network although if one of the banks of capacitors were switched out it would be an L network.

On the left hand side of Fig 15.24 are the capacitors associated with the low impedance 50-ohm input from the transmitter. This bank of capacitors, with a total capacity of over 6000pF, can be switched in or out with relay contacts with a resolution of 100pF. On the antenna side of the ATU a total of nearly 400pF can be switched in with a resolution of 12.5pF. These capacitors are made up of groups of series/parallel capacitors to obtain a safe voltage working.

Additionally, these capacitors are switched using four sets of relay contacts in series for both switching voltage working and to reduce stray capacitance. The Pi section inductor (the section between the capacitor banks) is actually made up of eight inductors having a total inductance of 15.75μH and a switching resolution of 0.125μH. Antennas shorter that a quarter wavelength, which present low and capacitively reactive impedances are taken care of using four 4μH inductors (top right) that can be switched in or out to load a short antenna.

The microprocessor ATUs that are now on the market have made the remotely fed long wire a much more practical reality.

The T-Network Antenna Tuner

The classic pi-network, or LC/CL two-component matching networks, can be used as the basis of an ATU. These are theoretically capable of matching any transmitter to any antenna impedance (resistive or reactive). However, in practice the matching range is dependent on the component values. For the widest step-up and step-down transformations, the high-voltage variable capacitors need to have low minimum and very large maximum capacitance values - a significant disadvantage these days. The Pi-network possesses the advantage that it not only transforms impedance but also forms a low-pass filter; and so provide additional harmonic and higher frequency spurii attenuation.

Modern solid-state transceivers include built-in low-pass filtering tailored to the individual bands, with the result that there is far less requirement for the harmonic attenuation previously provided by the ATU. This has opened the way for much greater use of the T-network which can provide an acceptably wide range of impedance transformations without a requirement for large-value variable capacitors (**Fig 15.26**). The fact that they form a high-pass rather than a low-pass filter is no longer regarded as a real disadvantage.

While the T match has enjoyed considerable popularity, it does suffer losses at some transformation ratios on the higher frequencies. These losses can be mimimised by a

relay, and pass a small amount of RF to the ATU. An SWR meter must be inserted at each end of the feeder. Find a tapping on the coil, working from the aerial tap where the SWR reduces, and adjust the appropriate capacitor until a combination is found which gives the lowest SWR (ensure that the transmitter is switched off whilst manually adjusting the tapping point, to prevent the possibility of RF burn to your fingers).

Now check the shack SWR meter and if both are similar readings, the top can be soldered permanently in place on the coil. Now carry on to the next lowest frequency, remembering to switch in the appropriate relay. On the highest frequencies, (15, 12 and 10m) the tap should not need to be more than four or five turns from the antenna.

When all the bands have been satisfactorily set up, no further alterations must be made to the antenna length or the earth system or all adjustments will need to be repeated.

Commercial Remote Control ATUs

An automatic ATU provides the most satisfactory method of feeding a remote antenna.

As with all modern automatic ATUs, adjustment of the inductors and capacitors in the matching network is accomplished using relays. These relays are controlled via a microprocessor using an embedded tuning algorithm, which in turn receives

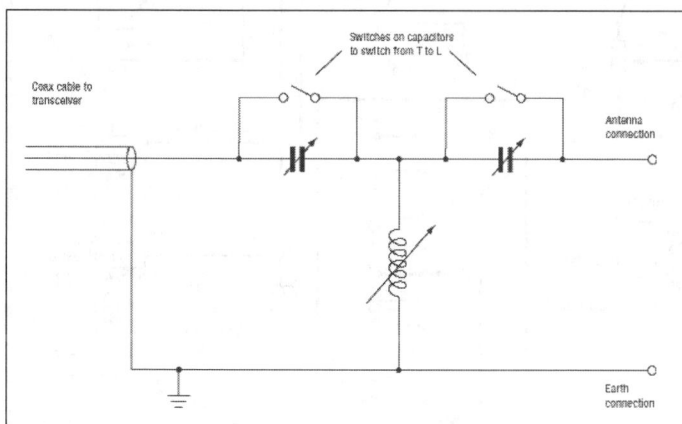

Fig 15.26: The series capacitor T network, which forms the basis of most modern ATUs. The shorting switches across the capacitors allow the unit to be switched to an L network to reduce losses

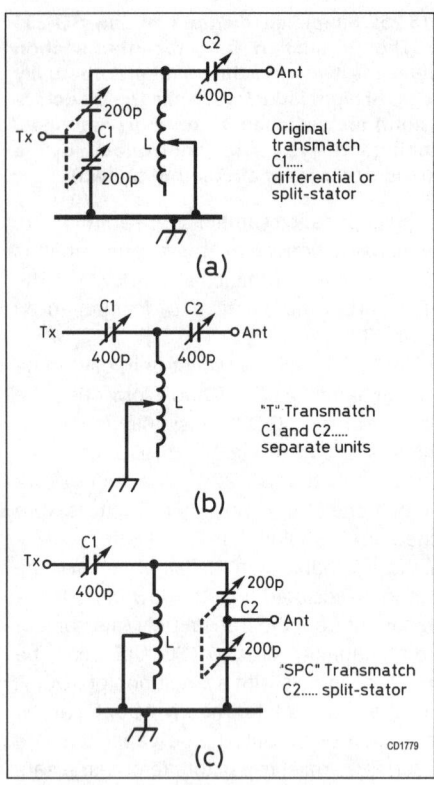

(a)

(b)

(c)

Fig 15.27: Variations of the 'Ultimate transmatch' (a) Original transmatch: CI differential or spilt stator. (b) T-transmatch: C1, 2 separate units. (c) SPC transmatch: C2 split stator

and, more recently solid state transmitters with built-in low-pass filters, harmonic suppression is not the problem it was when using Class-C AM power amplifiers.

G3TSO decided, in the interest of simplicity, to adopt the T-match variant, shown in Fig 15.27(b), of the transmatch in his general-purpose antenna tuning unit (**Fig 15.28**) This is the route taken by most ATU manufacturers at the time of writing - more of this later.

The tuning unit to be described provides operation on all bands from 1.8 to 28MHz. Other features have been added to permit the selection of different antennas as well as the facility to ground all inputs when the station is not in use. This unit also includes an SWR meter, and a balun to allow the unit to feed balanced lines.

Component selection

New components suitable for use in antenna tuners are either not readily available or very expensive, so the use of surplus components is the most economical answer. Fortunately the values of capacitors required are not too critical, and almost any high-quality wide-spaced variable capacitor can be put to use. Ideally a value of between 200pF and 400pF is suitable, and a number of surplus Johnson and Eddystone 390pF units have been seen over recent years. These units have ceramic end plates and are tested to 2,000V DC working. If in doubt, aim at a plate spacing of at least 1 .5mm between the stator and rotor plates; this is necessary to cope with the high voltages which can be developed when matching high-impedance long-wire antennas.

Inductors can be either fixed, with a number of taps selected by a rotary switch, or variable such as the roller coaster, which allows maximum flexibility in matching. Roller coasters come in a variety of different shapes and sizes, but in general are not available in other than small numbers and one-offs.

simple modification which uses a cam switch on the ends of the capacitors, and is described in [8].

The G3TSO Transmatch

The following is a description and a short history of the development of the Transmatch, plus construction details, by M J Grierson, G3TSO [9].

The original design of Transmatch, **Fig 15.27(a)**, used either a differential or a split stator input capacitor. The differential capacitor is less common than the split stator and has one section at a maximum capacitance while the other section is at minimum capacitance. This has the effect of providing a synthetic sliding tap on the inductor L, whereas the split stator capacitor tunes the inductor L, but maintains the tap centrally.

The use of a dual-type input capacitor for harmonic suppression lost all credence some years ago and the circuit was amended to the simpler T-match of **Fig 15.27(b)**. This circuit is that of a high-pass filter and provides no suppression of harmonics. More recently the 'SPC' (series-parallel capacitance) transmatch **Fig 15.27(c)** has emerged with a dual-output capacitor to providing a degree of harmonic suppression. In any event all three designs perform the task of matching a range of impedances quite successfully. As stated earlier, the advent of SSB and linear amplifiers

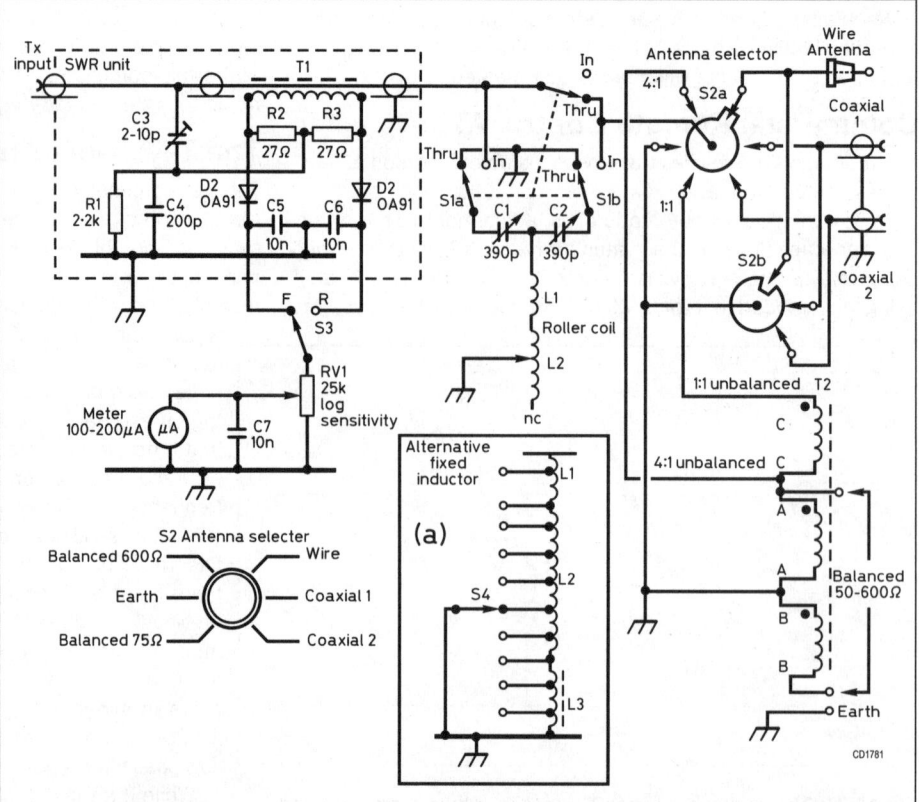

Fig 15.28: Circuit diagram of the G3TSO ATU

All switches used are of the 'Yaxley' type and use ceramic wafers; large numbers of this type of switch can often be found in junk boxes at rallies, and several switches can be broken down and reassembled to achieve the desired configuration. Paxolin wafers can be used, though they are not as good as the ceramic type.

The antenna selector switch uses a double-spaced switch unit giving six stops per revolution rather than the usual 12. The switch wafers are modified by removing alternate contacts, thus reducing the likelihood of arcing between them.

Balanced feeders

As the T-match is an unbalanced antenna tuner, some type of balun transformer must be incorporated if it is to be used successfully with balanced feeders. While a balun transformer provides a very simple solution for coupling a balanced feeder to an unbalanced tuning unit, it is not likely to be as efficient as a properly balanced ATU. Many published designs use a 4:1 balun to provide a balanced input for impedances in the range 150 to 600 ohms. However, if a low impedance feeder from either a G5RV or W3DZZ type of antenna is connected to a 4:1 balun, significant losses may occur. For this reason it was decided to use a 1:1 balun which, if fitted inside the tuning circuit, can easily be switched to 4:1 by use of the antenna selector switch. This now provides a range of balanced inputs from about 45 to 600 ohms without introducing too many losses into the system.

Balun construction

The balun transformer is wound on a single Amidon T200-2 powdered-iron core, colour coded red. For sustained high-power operation, 400W plus, two such cores can be taped together by using plumbers' PTFE tape, which can also be used to provide an added layer of insulation between the core and the windings.

Balun construction is simple, but a little cumbersome; some 14 turns of 16SWG enamelled-copper wire have to be wound trifilar fashion onto the toroidal core. That is to say, three identical windings are wound on together. Care must be taken to ensure that the windings do not overlap or cross one another and that neither the core nor enamel covering is badly scratched during construction.

Fourteen turns will require approximately 97cm (38in) of 16SWG (1.6mm) wire, so cut three equal lengths of 16SWG wire slightly longer than required and pass all three wires through the core until they have reached about halfway.

This now becomes the centre of the winding. It is easier to wind from the centre to either end rather than from one end to the other which involves passing long lengths of wire through the toroid. The T200 size core will accommodate 14 turns trifilar without any overlapping of the start and finish of the winding. Close spacing will occur at the inside of the core, and a regular spacing interval should be set up on the outside. A small gap should be left where the two ends of the winding come close together.

Connection of the balun requires care. It is necessary to identify opposite ends of the same windings, which can be done with a continuity meter, with some form of tagging or colour coding being worthwhile. On the circuit diagram a dot is used to signify the same end for separate windings. It is essential that the various windings are correctly connected if the balun is to work properly.

Details of how the balun transformer is wound and connected are shown in **Fig 15.29**. In this tuning unit, the balun is supported directly by soldering to the balanced input terminals, which are spring-loaded connectors.

A sheet of 8mm (5/16in) Perspex is then used to insulate the balun from the aluminium case. Construction of a four to one balun only is slightly simpler and only requires two (bifilar) windings.

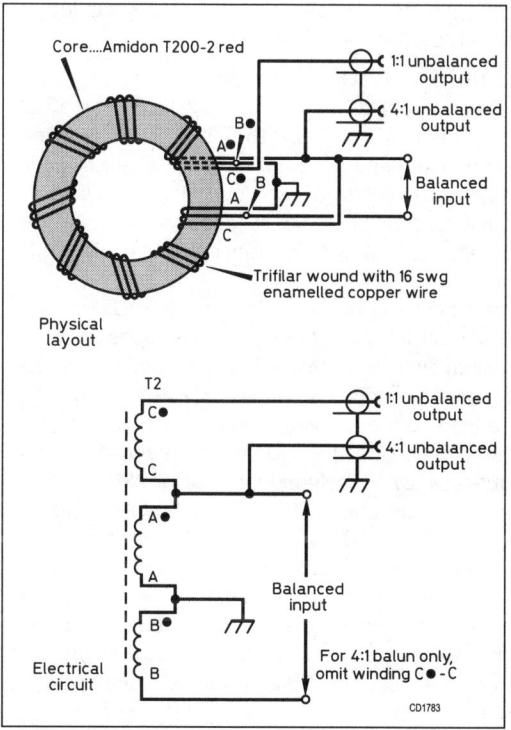

Fig 15.29: 1:1 and 4:1 balun transformer

ATU	
C1, C2	390pF 2,000VDC wkg, ceramic end-plates, eg Eddystone or Jacksons
L1	3t 10SWG, 25mm (1in) ID, 25mm (1in) long
L2	Roller coaster 36 turns, 38mm (1.5in) dia, 16SWG
T2	Amidon T200-2 (red); 14 turns trifilar 16SWG enamel
S1	Three-pole two-way ceramic Yaxley
S2	One-pole six-way double-spaced ceramic Yaxley; one-pole six-way shorting water (one pole open)

Alternative ATU circuit	
L1	2.5t 14SWG 25mm (1in) ID tapped at 1.5t
L2	14t 16SWG 1.25in ID tapped at 1, 2, 6, 9 and 14t
L3	Amidon T1 57-2; 31t 18SWG enam tapped at 6 and 27t
84	One-pole 11-way ceramic (three wafers to include S1 function)

SWR bridge	
R1	2.2kohm
C3	2-10pF trimmer
C4	200pF mica
C5, C6, C7	10nF disc ceramic
R2, R3	27ohm
RV1	25kohm log
D1, D2	Matched OA91 etc (germanium diodes)
T1	18t 22SWG 13mm (0.5in) OD ferrite ring (Amidon FT50-43, Fairite 26-43006301). Primary: 38mm (1.5in) coaxial cable, braid earthed one end only to form electrostatic shield.
Meter	100-200µA
S3	SPCO miniature toggle

Table 15.2: Components list for the G3TSO Transmatch

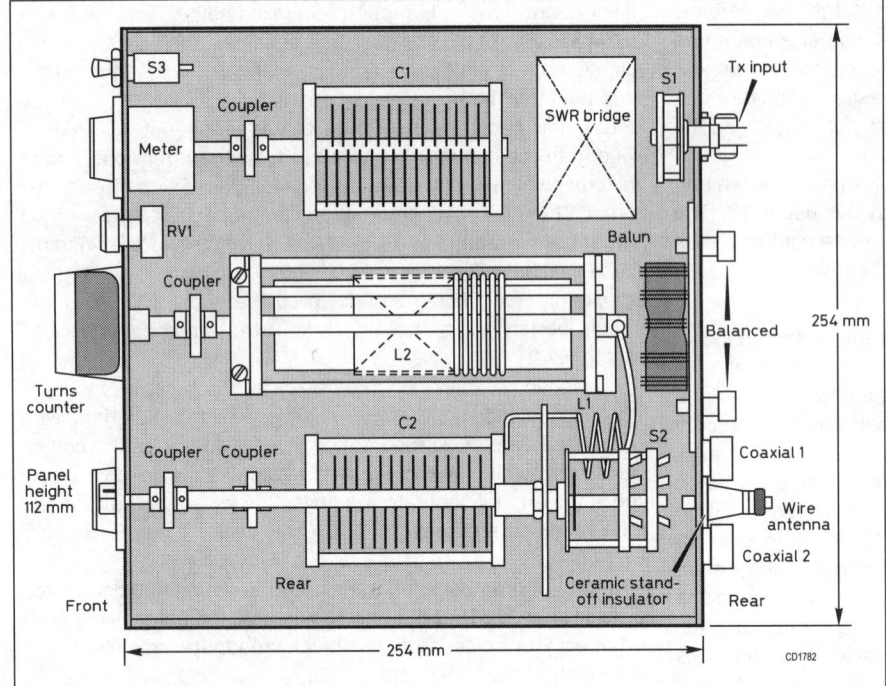

Fig 15.31: Component layout of the G3TSO ATU

A short length of coaxial cable is passed through the ferrite core to form the primary after the 18 turn secondary has been wound on. The braid of the cable can be earthed at one end to form an electrostatic screen, but on no account should both ends of the braid be earthed or it will form a shorted turn.

D1 and D2 should be a matched pair of germanium diodes, which can be selected from a number of similar-type diodes by comparing their forward and reverse resistances. Whilst this is best done with a high frequency signal, adequate matching can be achieved by using a simple multimeter.

Fig 15.30 (in Appendix B) gives a suggested layout and PCB track. The size is not at all critical, but a symmetrical layout should always be attempted.

The completed SWR bridge should be tested away from the antenna tuner by placing it in line between a suitable transmitter and a 50 ohm dummy load. The trimmer capacitor is adjusted to produce a zero-reflected reading with the forward reading at full scale. By connecting the bridge the reverse way around, some check of the diode balance can be judged by comparing the meter deflections in both directions. The forward and reverse switch selection will be reversed if the signal direction through the bridge is reversed. It is advisable to check that the bridge balances on a number of different bands, as C3 may be more sensitive at the higher frequency end of the operating range.

The sensitivity of the bridge is very dependent upon the resistance of the meter used. Comparison with a calibrated SWR bridge will enable a simple calibration of 1.5:1, 2:1 and 3:1 to be made, and in most cases a mental note of where these occur is the only calibration required, unless you wish to dismantle the meter in order to recalibrate the scale.

SWR measurement

It is often convenient to be able to connect the antenna tuner directly to the transmitter without the need for extra cables and external SWR bridges, so a built-in SWR bridge has been included in the design. The circuit, shown in Fig 15.28, is fairly conventional and is a current-sampling bridge which, unlike the voltage sampling stripline bridge, is not frequency conscious.

The current transformer T1 uses a small ferrite ring of about 12mm (0.5in) diameter, and while the size is not critical, the grade of ferrite is. Ferrite having an AL value of at least 125 should be used; the Amidon FTSO-43 ferrite core is ideally suited to this application.

Construction of the antenna tuner

The complete tuner layout is illustrated in **Figs 15.31 and 15.32** and the Components List is in **Table 15.2**. It is advisable to collect all the components and lay them out on a sheet of paper before committing yourself to a particular size. Layout is not over-critical, but a sensible approach is needed to minimise lead lengths and unnecessary stray capacitance, which could render 28MHz operation impossible.

Cases can be purchased, or prefabricated using 16 or 18SWG aluminium sheet bent into two interlocking 'U' shapes. Half-inch (12mm) aluminium angle provides stiffening as well as a means of joining the sections together. Roller coaster connections should be arranged so that minimum inductance is located at the end closest to the connections, ideally the rear of the unit. A small heavy-duty coil, L1, is included for ease of 28MHz operation and is more efficient than half a turn on the roller coil.

An alternative arrangement to the roller coaster is shown in Fig 15.28(a). Here a switched inductor is used. The switch should be ceramic with substantial contacts. A third toroidal inductor is included to permit operation on 1.8MHz. It is recommended that the bottom end of this could be shorted to ground to prevent the build-up of high voltages which could arc over.

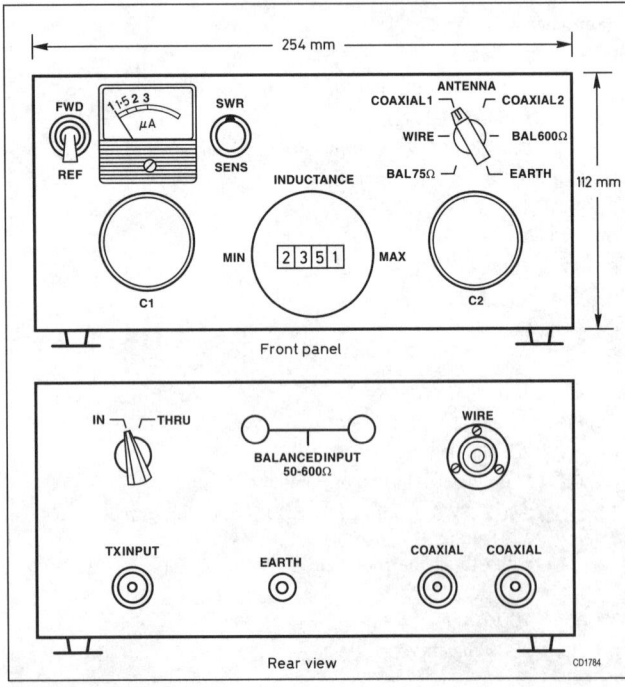

Fig 15.32: Front and rear panels of the G3TSO ATU

Fig 15.33: A variable inductance ATU coil described by Hector Cole, G3OHK. This arrangement uses two switches and just 14 taps to permit selection of from one to 50 turns of a 50-turn coil and which can be quickly reset to any number of turns previously found suitable without the turns counters required for roller coaster coils

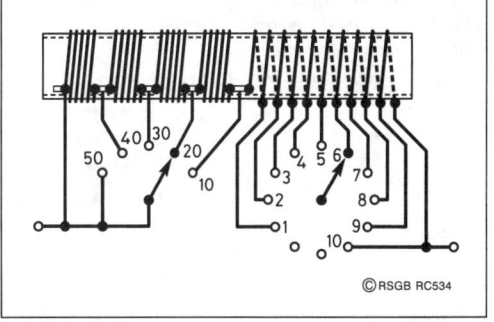

A further switched inductor was described by G3OHK, and this is shown in **Fig 15.33**.

The capacitors C1 and C2 are electrically above ground and must be mounted on insulators, a problem greatly reduced if the capacitors are constructed using ceramic end-plates. Ceramic pillars or even Perspex may be considered for mounting capacitors with metal end-plates. Additionally the shafts of the capacitors must be insulated; the use of Eddystone spindle couplers is recommended. To ease the rather sharp tuning characteristics that can be encountered on 21 and 28MHz, slow-motion drives were tried but they made tuning on the lower frequencies rather laborious and their use is not advisable. A turns counter on the roller coaster makes for much simpler operation, and may be as simple as a slot in the cabinet with a Perspex window for monitoring the position of the jockey wheel or a more sophisticated geared or direct-drive counter.

Antenna switching can introduce excessive lead lengths as well as stray capacitance, and for this reason the antenna selector switch is located on an extension shaft at the rear of the unit adjacent to the antenna inputs and the balun transformer. The wiring of the antenna switch is done strictly to achieve minimum lead lengths rather than to provide front-panel selections in any logical order. A separate IN/THROUGH switch enables the tuner to be bypassed and the antennas routed directly to the transmitter. It is located on the rear panel adjacent to the input socket to minimise lead length, and is intended only for occasional use. It is necessary to ground the tuning components in the THROUGH position to minimise capacitance effects.

Wiring of the tuner should commence after the mounting of all components, and fairly heavy wiring such as I6SWG tinned wire, coaxial cable braid or copper strip should be used. It has not been found necessary to screen the SWR bridge, but it should be located directly adjacent to the transmitter input socket and all meter leads kept away from tuning components.

The antenna selector switch has two ceramic wafers and is arranged so that every other contact is removed to give double spacing. The second wafer is used for shorting and provides a ground for all unbalanced antennas not in use. This is largely to prevent capacitive coupling to other antennas. The balanced input is grounded to DC through the balun. Balun switching is simply achieved by either taking the input from one side of the balanced input, giving a 4:1 ratio, or by selecting the third winding, giving a 1:1 ratio. An earth position enables the transceiver input to be grounded to prevent static discharge into the receiver.

Operation of the antenna tuning unit

If the SWR bridge is included in the design, it should be checked and balanced independently of the ATU, using a dummy load. Ideally it should be compared with, and calibrated against, an SWR measuring device of known accuracy.

To use the antenna tuner, select the required antenna and ensure that the THROUGH/IN switch is in the IN position. Set both C1 and C2 to halfway positions, adjust the inductance for maximum signal on receive, and one at a time adjust C1 and C2 for maximum received signal. Using low CW transmitter power, further adjust C1, C2 and the inductance to eliminate any reflected reading on the SWR meter. All tuning controls are interdependent, and settings may need to be adjusted several times before minimum SWR is achieved. In addition, more than one setting may give a matched condition, in which case the settings requiring the highest value of C1 should be used. Once the transmitter is matched on low power, increase the operating power for any final adjustments. Never attempt to tune the ATU initially on full power or with a valve power amplifier that has not been tuned up.

Generally, the higher the frequency the lower the value of inductance required, but exceptionally high impedances may require more inductance than expected. Capacitance values may vary considerably, and it is not uncommon on the higher frequencies for one capacitor to be very sharp and require a minimum value while the other is flat and unresponsive. Using the components recommended it is possible to match a wide range of impedances from 1.8 to 28MHz, but operation on 1.8MHz may become impossible if lower values of capacitance are used; however, fixed silver mica capacitors may be switched across C1 and C2 to compensate. Higher values of capacitor will almost certainly prevent operation on 28 and maybe 14MHz.

Conclusion

The antenna tuner described is not new or revolutionary in design, but probably represents the ultimate in flexibility. Performance is good and it is not inhibited by a lack of balanced input or restricted to a very narrow range of low impedances. The power handling capability of the tuner will to a large extent depend upon the impedances encountered and the spacings of the capacitors. As a rule, very high impedances should be avoided, as arcing can occur in the switches and the efficiency of the unit may well suffer. Adjustment of antenna or feeder length can remove any exceptionally high impedances that may be encountered.

G3TSO used this tuner with a 60m (180ft) doublet fed with an unknown length of 300 ohm slotted ribbon feeder, where it could be tuned to give a 1:1 SWR on all amateur bands from 1.8 to 28MHz. Using Eddystone capacitors of the type recommended, the tuning unit should be capable of handling 100W into a fairly wide range of impedances up to several thousand ohms, and the full 400W into impedances up to 600 ohms.

Two versions of the tuner have been built using the same basic circuit, one for base station operation using a roller coaster coil, and a smaller portable version using a range of switched inductors. The portable version has a slightly different layout, largely as a result of trying several other designs, and combining the IN/THROUGH facility on the inductor switch has necessitated several wafers. The balun used in this version is also the simpler 4:1 type and is connected with a flying lead.

For those who wish to adopt the 'SPC' circuit, the value of C2 should be made approximately 200pF, and an additional similar value capacitor should be ganged to C2 and connected between the antenna side of C2 and ground. Both capacitor rotors should be connected together and the stator of the new capacitor should be grounded.

The construction of the described antenna tuning unit should be well within the capabilities of most newly-licensed amateurs,

Fig 15.34: A commercial tuning unit using the T-match principle - the MFJ VersaTuner V

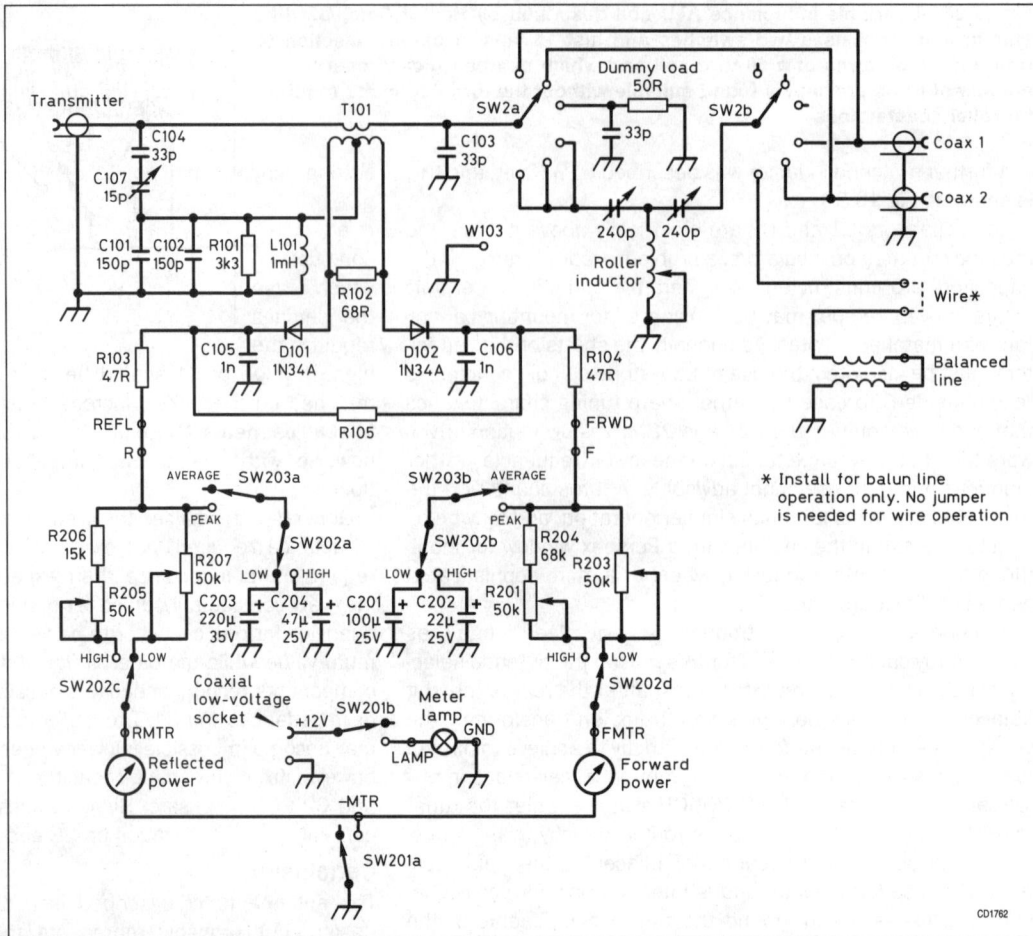

Fig 15.35: Layout of the MFJ VersaTuner V

Fig 15.36: Typical link-coupled antenna tuner circuit

and it can represent a considerable financial saving when compared to the commercial alternative.

The MFJ VersaTuner V

This commercial ATU uses the popular T-match tuning arrangement very similar to the G3TSO tuner described above. It also uses similar antenna switching and has a cross-needle power and SWR meter that is particularly convenient to use. The ability to switch in a dummy load is also a useful feature. In fact this is more than an ATU - it is an antenna management system. The circuit is shown in **Fig 15.34** and the layout in **Fig 15.35**.

The toroid balun is fixed at 4:1, with its limitations as already described.

Balanced ATUs

Many of the antennas so far described require a balanced feed. The following is material by W4RNL, who describes methods

[10] to adapt unbalanced antenna tuners (ATUs or transmatches) to service with balanced lines. Among the schemes used the following are the most common ones:

1. Float the tuner from ground and install a balun at the input end.
2. Install a balun, usually 4:1, at the antenna side of the tuner, to convert the balanced line to an unbalanced line.

Either system is subject to limitations. Floating the tuner does not guarantee freedom from common-mode currents that defeat balance. A 4:1 balun often reduces the already low impedance at the antenna terminals to a still lower one although a 1;1 balun can be used as described earlier by G3TSO.

The more classic alternative is the link-coupled or inductively coupled ATU; the basic circuitry is shown in **Fig 15.36**. The unbalanced input is inductively coupled to the main inductor. Since the mutual inductance between the coils is critical for

Resistors

R1, R2   150 ohm 2 watt metal film
RV1, 2   47k linear potentiometers with plastic shafts, Maplin FW04E

Capacitors

VC1, 3   13-250pF, 7.8kV, type TC250. Nevada
VC2   14-400pF, 1.25kV, Jackson type LAT. Cirkit
C4, C5   10nF 50V ceramic disc

Semiconductors

D1, D2   1N914

Misc

M1, M2	100µA, Maplin RX 33L
S1	4-pole 3- way rotary switch, Maplin FF76H*
Small vernier Dials	(3 off) Maplin 141 RX39N
Pointer Knob	Maplin RW75S
Spindle couplers (3 off)	
Knobs (2 off)	Maplin FK40T
GP1	Grounding Post, Maplin JL99H
TP5, 6.	Terminal Posts, Maplin HF02C
SKT1-13	4mm sockets, Maplin KC49D
SKT14	UHF chassis socket
4mm banana plugs	(54 off) Maplin JB24B (sufficient for 9 coils)
3mm panel head steel bolts (8 off). 16mm long, with nuts and washers	
Self adhesive rubber feet (4off)	
Spacer 1, 2	Terminal block (one 12-way strip, cut to suit), Maplin FE78K
TP1-4	Terminal Posts, Maplin HF02C
MDF	6mm thick, 35cm x 1.2m approx
Perspex	4mm thick, 100 x 300mm approx
Plastic tube	68mm OD, 1.25 metres
Plastic tube	36mm OD, 0.5 metres
Enamelled copper wire (159g)	1.25mm (18SWG)
Tinned copper wire (450g)	1.25mm (18SWG)
Tinned copper wire (150g)	1I.6mm (16SWG)

Two poles are spare. Use spare tags as junctions for D1, C4, VR1, and D2, CS, VR2.

Table 15.3: Components for the G0LMJ Balanced ATU

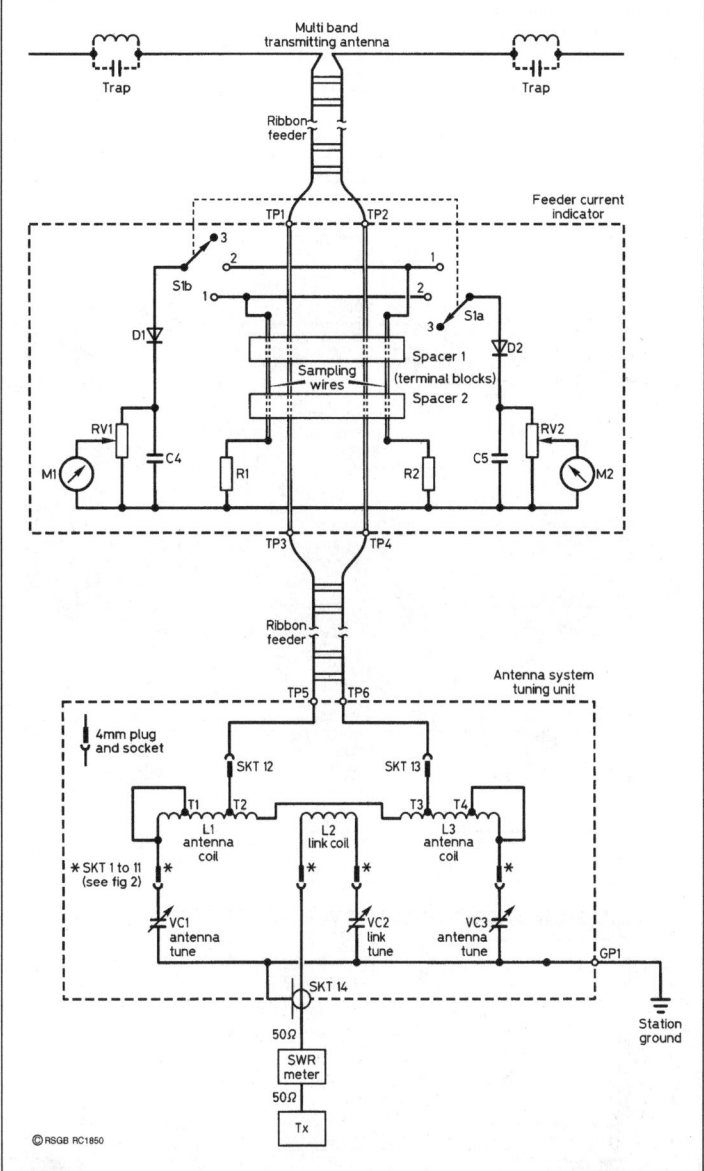

Fig 15.37: The G0LMJ balanced line ATU circuit diagram. The ATU is shown in the lower dotted box, and the feeder current indicator in the upper dotted box

maximum efficiency, the coupling is varied either by a movable link or by a series input capacitor.

Likewise, a single coil and link for all HF bands does not provide the best coupling ratios for all possible conditions. Without provision for coil tapping and series connections, the most efficient operating mode may be inaccessible, despite a 1:1 match.

For an operator who likes to change bands frequently, these inconveniences may be worse than the losses inherent in current systems pressed into balanced-line duty. However, for operators seeking the most efficient transfer of power to balanced lines, nothing beats a properly designed and constructed link-coupled ATU.

Balanced Tuner With Plug-In Coils

A properly designed and constructed link-coupled ATU is one of the most efficient methods of transferring power to balanced lines. The description of the ATU that follows is by Ted Garrott,

G0LMJ [11]. It uses plug-in coils rather than switching. Each HF band has its own coil, so nine coils are needed to cover all bands, 1.8 - 28MHz.

The ATU uses a conventional circuit, see **Fig 15.37**, except that L1 and L3 are not tuned by a split stator capacitor. Instead, two separate capacitors VCl and VC3 are used. This arrangement provides the facility of adjusting the ATU to give equal current into the two halves of the antenna. The relative values of RF current in the two feeder wires is monitored using with two meters M1 and M2.

As can be seen from **Fig 15.37**, the unit is built in two parts; the tuner and the feeder current balance indicators. An SWR meter is used between the transmitter and the ATU.

Materials and Components

Medium Density Fibreboard (MDF) is used for the chassis, and plastic drainage piping for the coil construction. All the materials and components (**Table 15.3**) have been chosen because they are relatively easy to obtain.

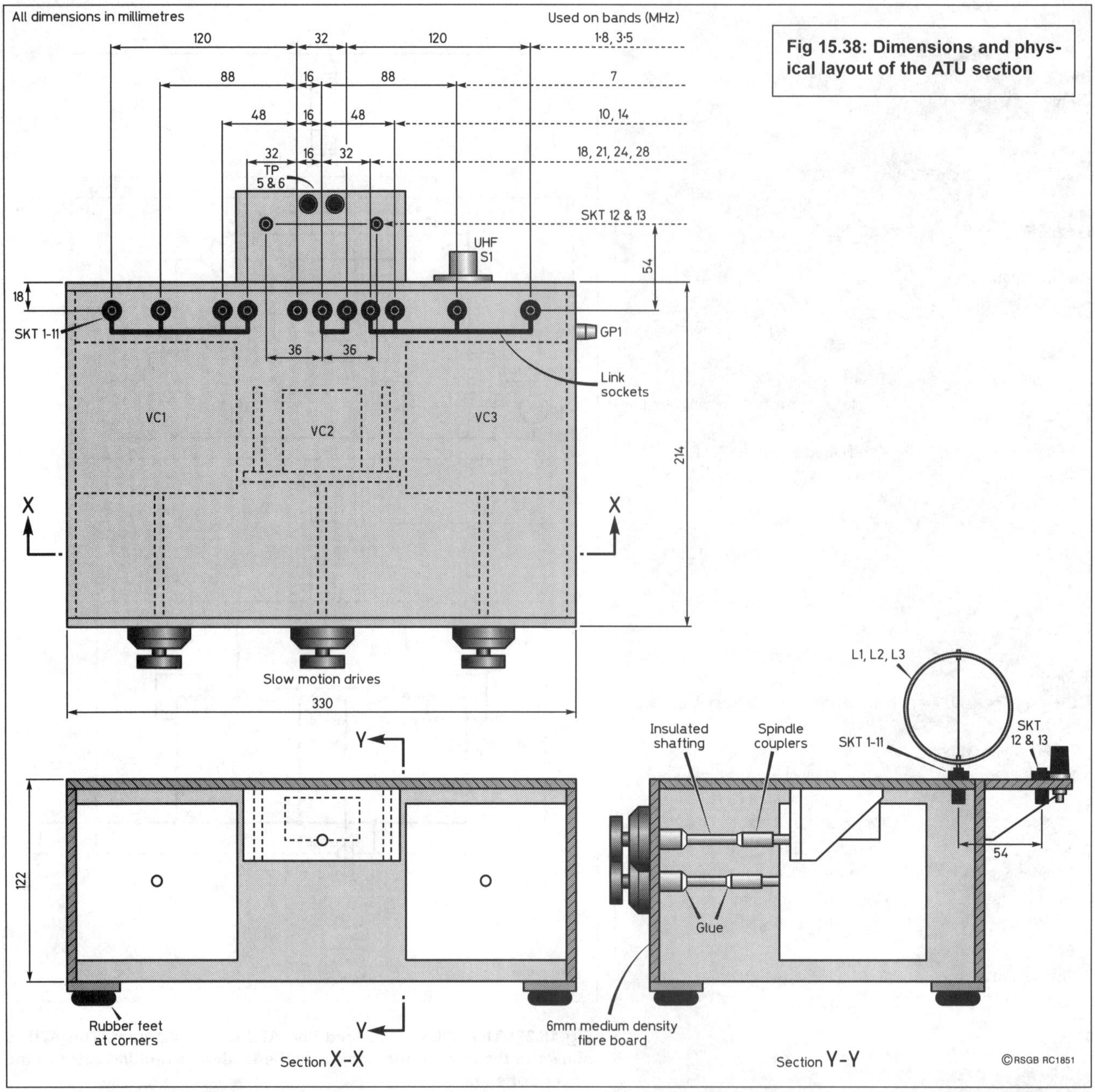

All dimensions in millimetres

Used on bands (MHz)

1·8, 3·5

7

10, 14

18, 21, 24, 28

Fig 15.38: Dimensions and physical layout of the ATU section

120   32   120

88   16   88

48   16   48

32   16   32

TP 5 & 6

SKT 12 & 13

UHF S1

54

18

SKT 1-11

GP1

36   36

Link sockets

VC1   VC2   VC3

214

X   X

Slow motion drives

330

122

Section X-X

Rubber feet at corners

L1, L2, L3

Insulated shafting   Spindle couplers   SKT 1-11

SKT 12 & 13

54

Glue

6mm medium density fibre board

Section Y-Y

©RSGB RC1851

Fig 15.39: General view of G0LMJ's Balanced Line Antenna Tuning Unit

Fig 15.40: Rear view of the ATU. Note the bracket which carries coil sockets Skt 12 and 13, and the output terminals TP 5 and 6

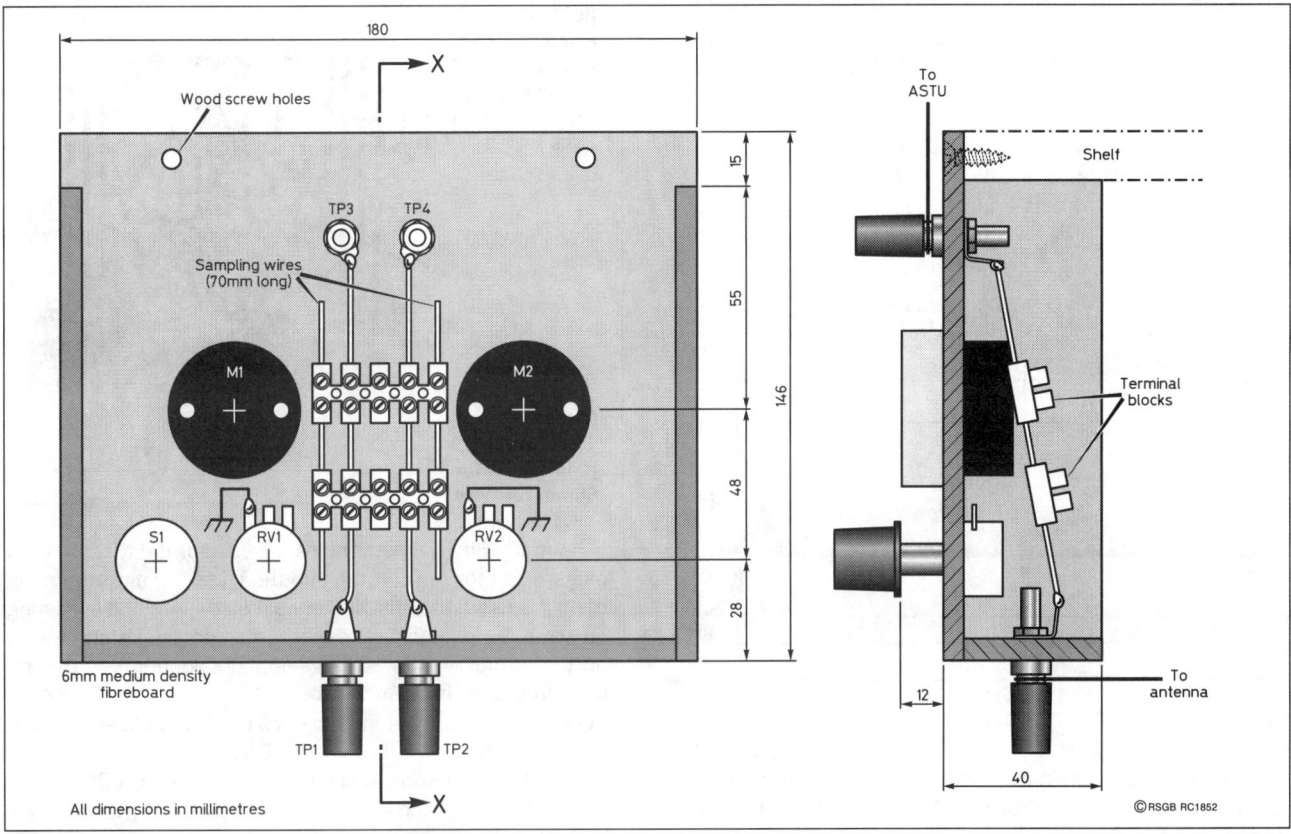

Fig 15.41: Dimensions & physical layout of feeder current indicator

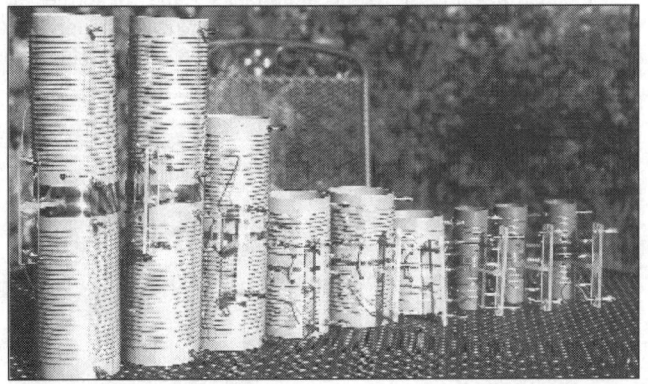

Fig 15.42: The nine coils required to cover all the HF bands

Band (MHz)	L1 (turns)	L2 (turns)	L3 (turns)	D (mm)	A (mm)	B (mm)	C (mm)
1.8	28	18	28	68	120	32	120
3.5	28	11	28	68	120	32	120
7.0	19	5	19	68	88	16	88
10.1	10	3	10	68	48	16	48
14.0	9	3	9	68	48	16	48
18.068	3	2	3	68	32	16	32
21.0	4	4	4	36	32	16	32
24.89	4	4	4	36	32	16	32
28.0	4	4	4	36	32	16	32

Table 15.4: G3LMJ tuner coil turns and dimensions. L1, L3 1.25mm tinned copper wire, L2 1.25mm enamelled copper wire, L1, L3 to fill available space and wound in same direction, L2 close wound. The coils for 21, 24 and 28MHz are identical, but the tapping points are likely to be different (see set-up procedure)

Construction

The general layout and dimensions are shown in **Fig 15.38**. The chassis is made from 6mm MDF, and the joins are fixed with wood adhesive. The outside of the chassis is smoothed with sandpaper and wiped with a damp cloth to remove the dust, and then treated with four coats of varnish. A general view of the ATU is shown in **Fig 15.39**.

VC1 and VC3 are mounted below the chassis and bolted to the side walls. VC2 is also mounted below the chassis, on its own fabricated brackets. The output terminal posts TP5 and 6, together with two 4mm coil sockets, Skt12 and 13, are mounted on a bracket glued to the rear of the chassis, as shown in **Fig 15.40**.

The coil and VCs 1, 2 and 3 are mounted to the rear of the chassis, in order to avoid hand capacity effects when tuning. VCs 1, 2 and 3 are driven by calibrated slow motion drives via lengths of insulated rod. To reduce backlash this rod is both screwed and glued to the coupler and slow motion drive.

The 4mm coil sockets, Skt 1 to 11, are arranged along the rear of the chassis to accommodate the plug-in coils. The seemingly odd spacing of these sockets is to accommodate the various lengths of coil formers and plug spacings. Note that some of the 4mm sockets are connected together, see Fig 15.38, under the chassis. Some of the sockets are too close together to use the fixing nuts, so these are glued into their holes. The sockets must be accurately set out on the chassis.

Self adhesive rubber feet are attached via pieces of 25 x 25mm MDF, glued to the corners of the chassis

The general layout of the feeder current indicator section is shown in **Fig 15.41**. This is fixed into a convenient place so that it can be seen when adjusting the ATU. In Fig 15.39 and Fig 15.40 the unit is shown fixed to a shelf.

Note the use of 5-way mains terminal blocks for spacing the sampling wires; 1.25mm wire is used for all wires through the terminal blocks.

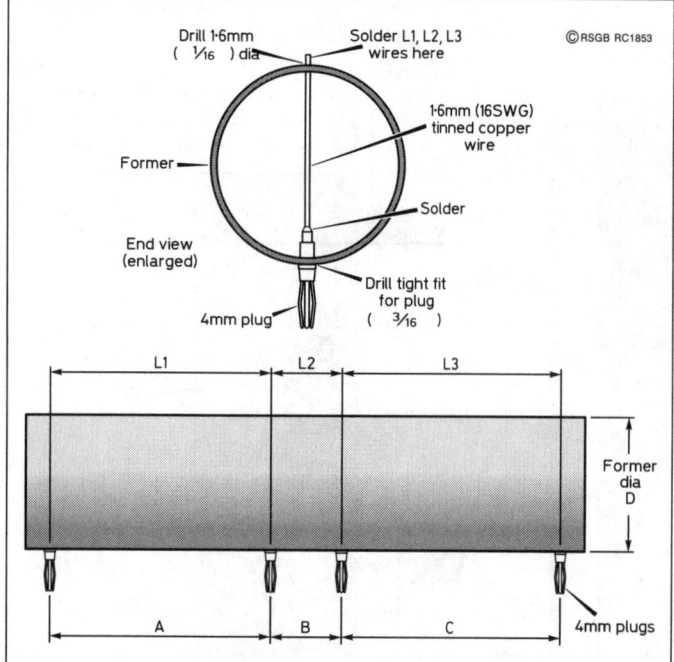

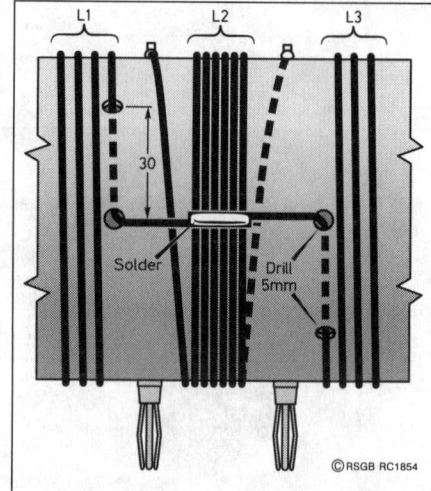

(left) Fig 15.43: Basic construction and dimensions of the plug-in coils

(right) Fig 15.44: L1 to L3 junction on the 68mm formers

Coils

The nine plug-in coils are shown in **Fig 15.42**. The formers are plastic pipes, as used for drainage. Winding details for coils to cover all nine HF bands are given in **Table 15.4**. The turns are held in place with beads of epoxy resin, four beads for 68mm and three beads for 36mm formers. The link coils L2 all resonate within the band they are wound for, using the 400pF capacitor VC2.

The method of fixing the 4mm plugs to the formers is shown in **Fig 15.43**.

5mm (3/I6in) diameter holes are fine for the Maplin 4mm plugs, and they will self tap into the plastic. If any other type of plug is used, the hole size should be determined by testing on scrap material first. The plugs must be accurately set out on the former. A 45-watt large-tipped soldering iron can be used to solder the 1.6mm wire into the plugs.

The method of terminating the ends of L1 and L3, and joining them over the top of L2, is shown in **Fig 15.44**.

This method only works well for 68mm formers. For 36mm formers the ends of L1 and L3 can be terminated using solder tags, held down to the former with self tapping screws. The coil ends are then joined with a jumper wire over the top of L2.

The construction for fixing the 4mm plugs to the coil former at Skt 12 and 13 is shown in **Fig 15.45**. The two plugs are glued to a bracket, made from 4mm perspex, which is glued to the coil former.

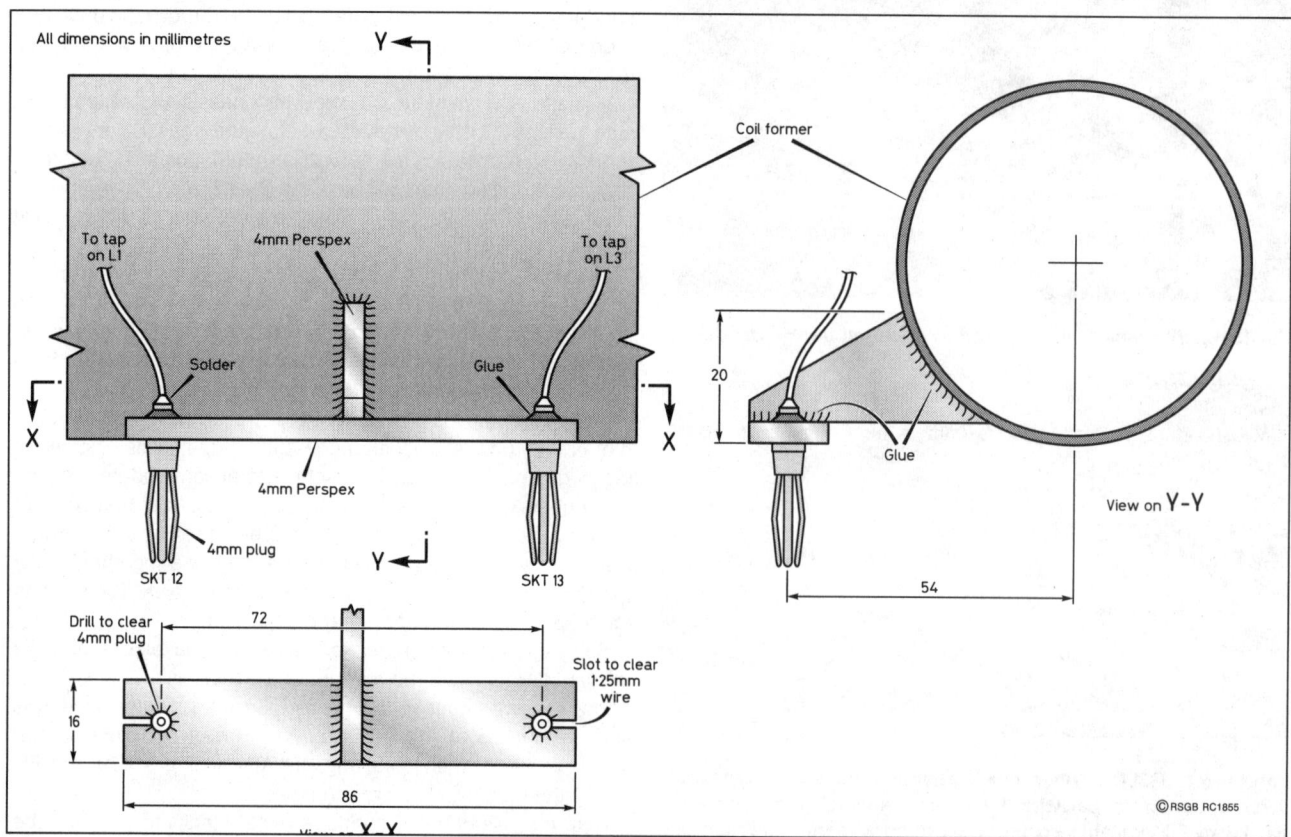

Fig 15.45: Coil connections to the 4mm plugs which connect with Skt 13 and 14

Table 15.5: Example of dial settings for the G3LMJ antenna tuner

MHz	VC1	VC2	VC3
1.81	70	50	72
2.0	32	47	38
3.5	56	55	56
3.8	39	40	37
7.0	86	50	87
7.1	79	47	79
10.1	17	34	19
10.15	16	34	20
14.0	24	27	30
14.35	22	24	28
18.068	52	60	68
18.168	51	53	66
21.0	38	27	55
21.45	37	30	48
24.89	37	15	18
24.93	37	15	18
28.0	25	26	11.0
29.7	15	43	16.5

Wire is soldered to the plugs before assembling the plugs to the perspex. Before applying glue, all components, including the coil are put into place on the chassis. The glue is then applied and allowed to set thoroughly. The result is a perfect fit of the coil plugs to the chassis sockets. In order to facilitate good bonding, the perspex is thoroughly roughened to provide a key for the epoxy resin. The bracket arms are glued to L2 on top of the 1.8, 3.5 and 7MHz coils, and direct to the former for all other coils. Coil taps T1 - T4 are determined by testing, using flexible wire and crocodile clips for attaching to the coil and sockets Skt 12 and 13. All taps are made symmetrically about the coil centre. When the correct tap positions are found, the final, permanent wiring can be made. An SWR of 1.0 is aimed for, and when achieved VCs 1, 2 and 3 should be comfortably within their working range (ie not at maximum or minimum). VC 1 and 3 should be set at about the same dial readings.

Set-up procedure

To determine the correct positions for taps T1 - T4, the following procedure is recommended:

1. Set T1 and T4 one turn from each end of the coil.
2. Transmit low power at the low frequency end of the band.
3. Try various positions for T2 and T3, tuning VCs 1, 2 and 3 for minimum SWR.
4. If an SWR of 1.0 is not achieved, carry out step No.3 again, but with T1 and T4 set nearer the centre of the coil.
5. When an SWR of 1.0 is achieved, connect the permanent taps T1, 2, 3 and 4.

Having established the correct tap positions, the settings for VCs 1, 2 and 3 must be determined for various frequencies in the band. This is done by trying various settings, until one is found that gives an SWR of 1.0. VC 1 and VC3 should be kept at about the same settings.

Current Sharing

The next stage is to ensure that the two halves of the dipole are taking the same current.

Meters M1 and M2 must be adjusted to the same sensitivity before this can be done. The adjustment is carried out as follows:

1. Set RV1 and RV2 fully anti-clockwise.
2. Transmit at low power.
3. Set S1 to position No.1.
4. Adjust M1 to mid scale, using RV1
5. Set S1 to position 2.
6. Adjust M2 to mid scale, using RV2
7. Reset S1 to position 1.

Both meters now have the same sensitivity and will show the relative currents in each half of the dipole. If these currents are not the same, they may be equalised by further adjustments to VCs 1, 2 and 3. It will be found that there are a number of VC 1,2 and 3 settings that will give an SWR of 1.0, but only one that will also give equal currents in each half of the dipole. When the correct settings are found they should be logged. Use **Table 15.5** as a guide (although these the results were obtained at G3LMJ's location with his multi-band dipole). Remember that the slow motion drives are calibrated 0 to 100. Minimum capacitance is 0, maximum is 100.

If you are using an ATU that feeds into a 450-ohm line using a voltage balun, it may be an interesting exercise to construct the feeder current indicator and check the feeder currents and see what the balance is like on your antenna feed lines.

Safety

There are two safety points which must be stressed:

1. Use a dust mask when working on MDF.
2. In thundery weather, earth both sides of the ribbon feeder to avoid a build up of high voltage on the antenna.

The Z-Match

Another link coupled ATU that has been around a long time is the Z-match. Originally it was designed as a tank circuit of a valve PA [12], the anode of which was connected to the top or 'hot' end of the multiband tuned circuit. It was fed directly from the PA valve, with its internal (source) impedance of several thousand ohms

When the circuit was adopted as an ATU [13] the tank circuit was fed directly from a source which requires a 50-ohm load via a 350pF variable coupling capacitor connected to the top (or 'hot') end of a multiband parallel-tuned LC circuit.

In spite of the great disparity between the required 50-ohm load for the transmitter and the relatively high impedance of the tank circuit the Z-match enjoyed considerable popularity, probably due to its simplicity. Z-match ATUs were produced commercially and they are described here because they are easily available and cheap. An example of such a unit is shown in **Fig 15.46**.

The design of the Z-match was improved and described by Louis Varney, G5RV [14]. All of what follows is from his article.

As you can see in **Fig 15.47(a)**, on the 3.5, 7 and 10MHz bands the main inductance, L1, is connected in parallel with the two sections of C1 which are also paralleled.

The effect of the much smaller inductance, L2, can be considered as a rather long connecting lead between the top of C1a and the top of C1b. Since the inductance of L2 is very much less

Fig: 14.46: The original KW E-ZEE MATCH, shows the general construction of a Z-match ATU

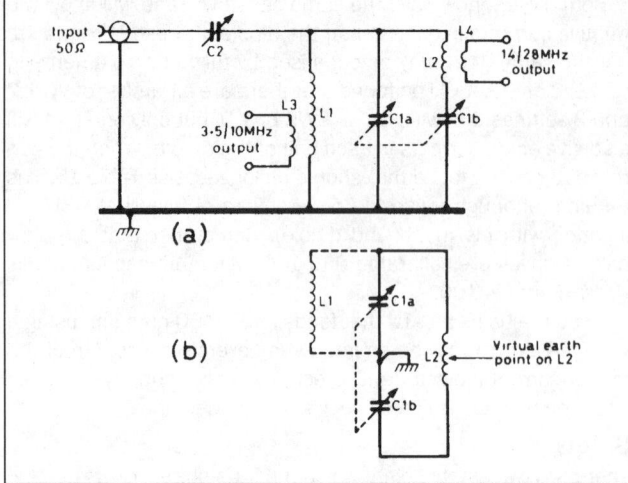

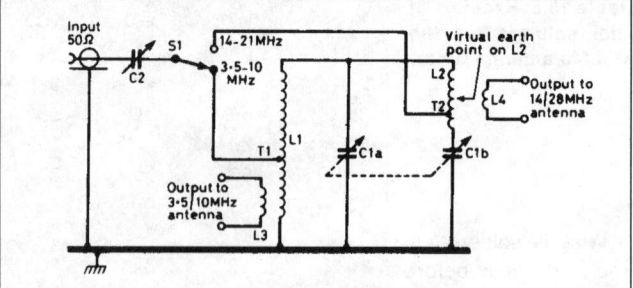

Fig 15.47: (a) The basic Z-match circuit. (b) The 14-28MHz tuned circuit shown in a more conventional form

Fig 15.48: The basic Z-match circuit showing the tapped-down feed arrangement

than that of L1, this assumption is valid for the relatively low frequencies of 3.5 and 7MHz. For these bands, therefore, L1, C1a plus C1b may be considered as a simple tuned circuit with one end earthed.

Provided the capacitance range of C1a plus C1b is sufficient, the circuit will also tune to 10MHz. It may be necessary to reduce the inductance of L1 by one or two turns to achieve resonance on that band. However, it should be noted that care must be taken to avoid the occurrence of harmonic resonance between the two circuits comprising the multiband tuned circuit; the values of the inductances. L1 and L2 must be selected with this in mind. On the 14, 18, 21, 24 and 28MHz bands the active tuned circuit consists of the two variable capacitor sections C1a, C1b as a split-stator capacitor, with the moving vanes earthed, and L2 connected between the two sets of stator vanes. Because its inductance is much greater than that of L2, L1 may be considered as an HF choke coil connected in parallel with C1a; and having no noticeable effect on the performance of the split-stator tuned circuit L2, C1a, C1b. This can be proved by first

tuning this circuit to any band from 14 to 28MHz, noting the dial-reading of C1a, C1b and then disconnecting the top of L1 and retuning for resonance. It will be found that the effect of L1 is negligible. **Fig 15.47(b)** shows the effective 14 to 28MHz tuned circuit in a more conventional manner.

The relatively high impedance LC circuits L1, C1a and C1b (paralleled for the 3.5, 7 and 10.1MHz bands) and L2, C1a, C1b (as a split-stator capacitor for the 14 to 28MHz bands) must be detuned slightly off resonance at the frequency in use; so as to present an inductive reactance component. This, in conjunction with the coupling capacitor C2, functions as a series resonant input circuit which, when correctly tuned, presents a 50ohms non-reactive load to the transmitter output.

Modifying the Z-match

Feeding the RF energy from the output of a transmitter requiring a 50Ω resistive load to the top of a parallel-tuned LC circuit cannot be the most efficient method, so G5RV felt that the circuit would benefit from modification. He performed a number of tests, which involved tapping the Coils L1 and L2 to obtain a better match. The circuit of the modified Z-match is shown in **Fig 15.48**.

The final modified Z-match

The final design is shown in **Fig 15.49,** and a list of components is in **Table 15.6**. The circuit incorporates switching for the appropriate coil coupling taps and selects the appropriate output-coupling coil (L3 or L4) to the feeder.

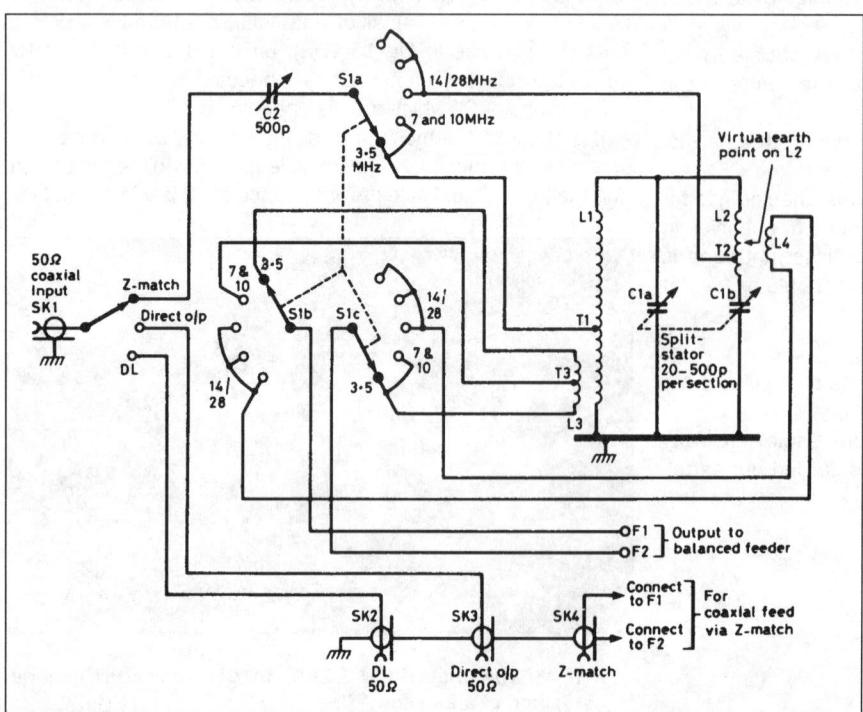

Fig 15.49: The final modified Z-match circuit

C1a-1b	Split-stator variable capacitor 20-500pF per section.
C2	500pF single-section variable capacitor (shaft insulated)
L1	10t 4cm ID C/W 14SWG enam copper wire. Tap T1 4t from earth end
L2	5t 4cm ID turns spaced wire dia 14SWG enam copper wire. T2 1.5t from centre of coil (virtual earth point
L3	8t 5cm ID C/W enam copper wire over L1. T3 at 5t from earth end.
L4	3t 5cm ID C/W over L2. 14SWG enam copper wire.
S1	Ceramic wafer switch. All sections single-pole, five positions
S2	Ceramic wafer switch. Single-pole, three positions.

Notes:

(1) A suitable 250 + 250pF (split-stator or twin-ganged) variable capacitor can be used since the capacitance required to tune L1 to 3.5MHz is approximately 420pF, and for 7.1MHz approximately 90pF. If C1a, C1b (paralleled) have a combined minimum capacitance of not more than 20pF, it should be possible also to tune L1 to 10MHz. Otherwise it may be necessary to reduce L1 to nine turns, leaving T1 at four turns from the 'earthy' end of L1. A lower minimum capacitance of C1a, C1b as a split-stator capacitor would also be an advantage for the 28-29.7MHz band.

(2) Taps on L1 and L2 soldered to inside of coil turn. Tap on L3 soldered to outside of coil turn

Table 15.6: Components list for the Z-match

For maximum output coupling efficiency on the 7MHz and 10MHz bands, a tap on L3 is selected by S1b. Provision is made for coaxial cable antennas to be fed either direct or via the Z-match.

The transmitter output can be direct to a suitable 50-ohm dummy load. The layout is not critical, but it is advisable to mount the coils L1 and L2 with their axes at right angles to prevent undesirable intercoupling. Also, all earth leads should be as short as possible and the metal front panel should, of course, be earthed.

The coupling capacitor, C2, should be mounted on an insulating sub-panel and its shaft fitted with an insulated shaft coupler to isolate it from the front panel, preventing hand-capacitance effects.

The receiving-type variable capacitors used in the experimental model Z-match have adequate plate spacing for CW and SSB (peak) output powers of up to 100W. For higher powers it would be necessary to use a transmitter-type split-stator capacitor (or two ganged single-section capacitors) for C1a, C1b. However, C2 requires only receiver-type vane spacing even for high-power operation.

Tests with additional feedpoint taps on both L1 and L2 in the modified Z-match circuit showed no practical advantage. However, the tap on the output coupling coil L3 was found to be essential on 7MHz, and 10MHz. The very tight coupling between L1/L3 and L2/L4, tends to reduce the operating Q value of the LC circuits. This renders them more 'tolerant' of the complex reactive loads presented at the input end of the feeder(s) to the antenna(s) used.

G5RV noted that the efficiency of a conventional link coupled antenna tuning unit was better than that of either form of Z-match; and that by virtue of its design, the Z-match cannot satisfy all the required circuit conditions for all bands. However, in its original form it does provide the convenience of 3.5 to 28MHz coverage without the necessity for plug-in or switched coils. Nevertheless, the inclusion of the simple switching shown in Fig 15.49 is an undoubted advantage.

LOOPS AND SLOT ANTENNAS FOR HF
Small Loop Antennas, General Comments

If space at your location is very restricted, with no place to put up a wire antenna, a small HF transmitting loop antenna may be an option. A surprising amount of information is available on these types of antennas [15].

Good efficiency can be achieved only by ensuring the loop has a very low RF resistance. Additionally its high-Q characteristic results in a narrow effective bandwidth, requiring accurate retuning for even a small change in frequency. This can be overcome by the use of complex and expensive automatic tuning systems or, more realistically for amateurs, by remote control of the tuning capacitor forming part of the loop. Another disadvantage is that even on low-power, there will be very high RF voltage across the tuning capacitor, resulting in the need for either a high-cost vacuum capacitor or a good-quality, wide-spaced transmitting capacitor.

Against these disadvantages should be set the fact that a well-constructed loop just 0.15m high can have a radiation efficiency close to that of a ground plane antenna, see Fig 15.11. Furthermore, the short loop utilises the near-field magnetic component of the electromagnetic wave, resulting in much less absorption in nearby objects. This means that a short loop can be used successfully indoors or on a balcony. For reception a 'magnetic' antenna is much less susceptible to the electric component of nearby interference sources. The reduction of man-made noise is particularly important on the lower-frequency bands, and is further enhanced by the directional properties of a loop. The loop can work effectively without any ground plane.

The high-Q characteristics of a low-loss loop also means that it forms an excellent filter in front of a receiver, reducing overload and cross-modulation from adjacent strong signals. On transmit, these properties dramatically reduce harmonic radiation and hence some forms of TVI and BCI.

Basics

To achieve good radiation efficiency in a small transmitting loop, it is essential to minimise the ratio of RF ohmic losses to radiation resistance. In a small resonant loop the RF ohmic losses are made up of the resistance of the loop and that of the tuning capacitor (which will have much lower resistive loss than a loading coil). The tendency of HF current to flow only along the surface of a conductor (skin effect) means that large diameter continuous copper tubing (or even silver-plated copper) should be used to achieve a maximum high-conductivity surface area.

Provided that the circumference of a loop is between 0.125 and 0.25 wavelengths, it can be tuned to resonance by series capacitance. If the loop is longer than 0.25λ it will lose its predominant 'magnetic' characteristic and become an 'electric' antenna of the quad or delta type but, unless approaching one wavelength in circumference, will still have relatively low radiation resistance.

The radiation resistance of a small loop is governed by the total area enclosed and is a maximum for circular loop. It is possible to build a transmitting loop antenna with a circumference less than 0.25λ, but in these circumstances the bandwidth becomes so small that it becomes practically impossible to tune the loop accurately enough. It is thus advisable to restrict the operating range of a transmitting loop to a ratio of 1:2, that is to say 3.5 to 7, 7 to 14, or 14 to 28MHz. Extending the tuning range will tend to result in a rapid falling off in efficiency. The most convenient solution for complete HF coverage is to use two loops; one for the higher frequency bands (14, 18, 21, 24 and 28MHz), the other for 3.5 and 7MHz, or 7 and 10.1MHz. For

	14MHz	21MHz	28MHz
Radiation resistance, ohms	0.09	0.46	1.68
Conductor losses, ohms	0.04	0.05	0.05
Efficiency (%)	67.3	89.5	93.3
Loop Inductance, µH	2.4	2.4	2.4
Inductive reactance, ohms	214	321	443
Q factor	789	311	127
Theoretical bandwidth, kHz	17.7	67.5	228
Voltage across tuning capacitor (100W), kV	4.1	3.1	2.3
Tuning capacitance, pF	53	23	12

Table 15.7: Calculated electrical characteristics of a one-metre diameter transmitting loop antenna using 22mm copper tubing

1.8MHz it is advisable to use a loop designed for this band, or for 1.8 and 3.5MHz.

A Small Loop for 14 to 29MHz

This design is by Roberto Craighero, I1ARZ [16]. The main physical characteristics of this antenna are:

- Circular loop: 1m diameter made from copper pipe of 22mm OD, circumference 3.14m.
- Capacitor: Split-stator or 'butterfly' type, about 120pF per section. Minimum capacitance (for 28MHz) 16pF.
- Feed: Inductive coupling with a small loop made from co-axial cable.
- Maximum power: Governed primarily by the spacing of the capacitor vanes. Suggested rating 100W maximum.
- Tuning method: Remote control of capacitor by means of electric DC motor and reduction gear. Rotation speed not faster than one turn per minute.

The electrical characteristics, calculated from the formulae given by W5QJR [17], are set out in **Table 15.7**. The overall design of the loop is shown in **Fig 15.50**.

The loop

Copper pipe of 22mm OD is generally sold in straight lengths of 3m and 6m. The only really practical way of bending 22mm pipe is with a pipe-bending tool. A suitable tool can be hired, or the pipe can be taken to a metal workshop or a plumber - this might be expensive.

Both ends of the loop must be cut longitudinally along the vertical diameter of the pipe for about 5cm, then cutting one half away. The remaining half is flattened to form a strip that can later be inserted through the insulated board used to support the tuning capacitor and connected to the stator plates. In this way only one joint will be necessary for each stator, reducing the soldering losses. At the bottom of the loop, opposite the tuning

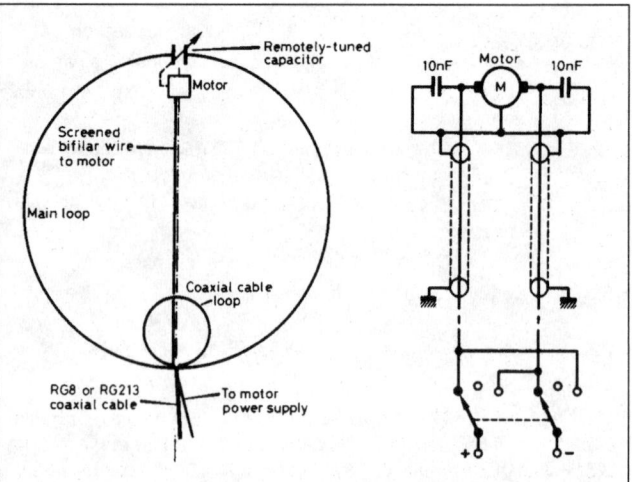

Fig 15.50: Electrical diagram of the I1ARZ loop antenna, plus the tuning motor connections

capacitor, a small copper bracket should be soldered to the loop, see **Fig 15.51**. On this bracket will be fixed the coaxial connector and the connector for the twin lead for powering the tuning motor. The bracket should be soldered to the loop using a flame- torch to ensure good electrical contact.

An external loop supporting mast

A thick PVC pipe of about 40-50mm diameter can be used for this support. Alternatively a glass fibre tube (lighter but more expensive) or a wooden mast waterproofed with plastic compound may be used. The length of this mast should be about 1.5m; with about 200mm used at the top for fixing the plastic board carrying the tuning capacitor; the remaining length is used at the base for fixing the loop to a rotor or another short mast. For obvious reasons, never use a metallic pipe across the loop.

The loop is fixed to the mast, using two U-bolts at the base, see **Fig 15.52** The ends at the top of the loop are held in place by two collar-clamps (these clamps are the cross-joints in cast aluminium, as typically used to connect the boom of television antennas to the mast.). The clamps are connected with stainless steel nuts and bolts to the back of the plastic board supporting the capacitor, see **Fig 15.53**. The bolts should be of sufficient length to act as adjustable spacers in order to have the loop completely upright. The plastic supporting mast is fixed to the back of the board by two semicircular clamps with stainless steel nuts and bolts of sufficient length to reach the front side of the plastic board. The two copper strips of the loop must be bent at 90° and inserted in suitable cuts in the board to reach the stators of the capacitor on the front side of the board (Fig 15.53 and **Fig 15.54**). The cuts should later be waterproofed with silicone compound.

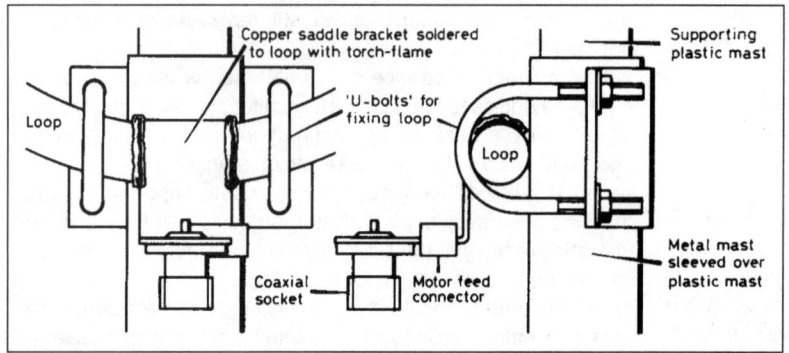

Fig 15.51: Details of the bottom of the loop - front view

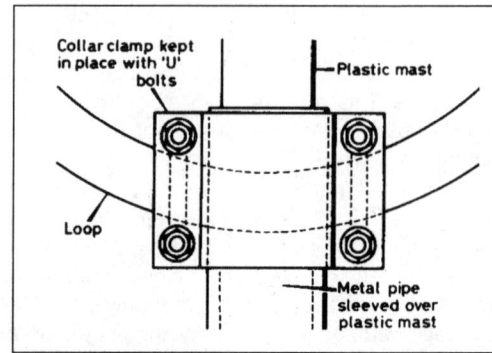

Fig 15.52: Fixing the loop to the support mast

Fig 15.53: The loop tuning board, (a) front view and (b) side view

Tuning board and cover

The size of the tuning board depends upon the dimensions of the variable capacitor and motor. The best material for high-power operation is 10mm thick Teflon; alternatively Plexiglas (Perspex) of the same thickness can be used. When calculating the size of the board, allow space for fixing the clamps of the loop and for a waterproof cover to protect the complete tuning unit. For protection, a plastic watertight box of the type used for storing food in a refrigerator is used. The original cover is cut in the centre with an opening just wide enough to permit the entry of the capacitor and motor. A layer of soft rubber is inserted between the surface of the supporting board and the cover to act as a seal. The cover is then fixed in place with several small stainless nuts and bolts, fastened tightly so that the seal is compressed between the board

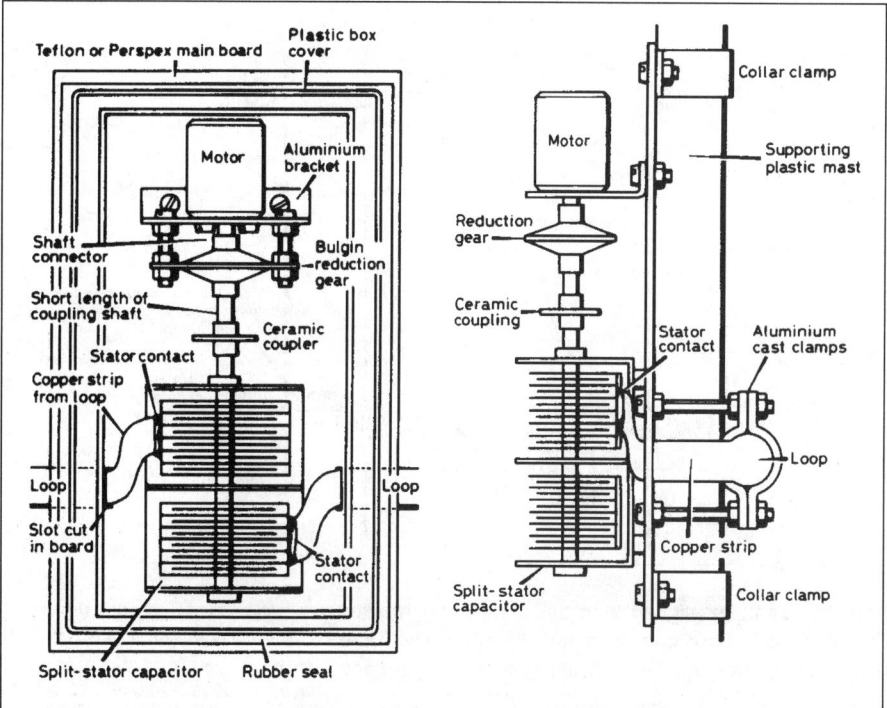

surface and cover to make it watertight. The plastic box can now be put against its cover, keeping it in place with a tight nylon lashing. Silicone compound should now be applied all round to keep out the moisture. To prevent gradual deterioration of the plastic box it is advisable to use white self-adhesive plastic sheet to protect it against ultraviolet radiation from the sun.

The tuning capacitor

It is most important to use a very good quality transmitting-type variable capacitor; otherwise the efficiency of the antenna will be reduced. Owing to the high Q of this antenna, the RF voltage across the capacitor is very high (directly proportional to the power). With 100W power, this voltage will be 4 - 5kV; with 500W it can be as high as 28kV!

It is most advisable to use a split-stator (or 'butterfly') capacitor of about 120pF per section. The advantage of this arrange-

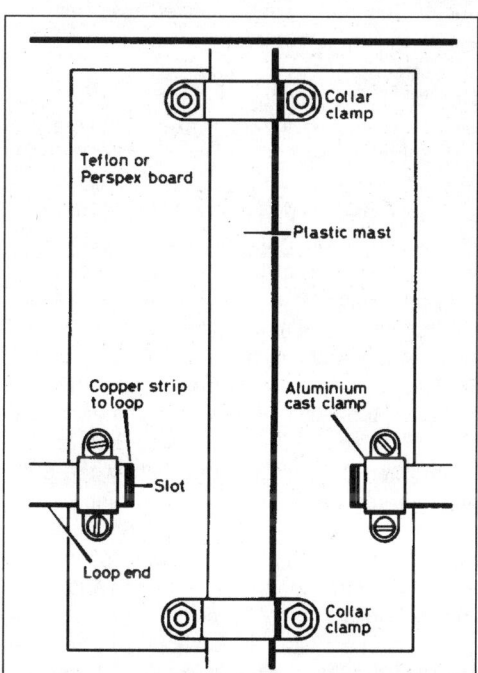

Fig 15.54: Rear view of tuning board showing loop and support mast mountings

ment is that the two sections are connected in series thereby eliminating rotor contact losses, which occur in conventional capacitors. Assuming that the loop is intended for use with a transmitter power of not more than 100W, the spacing between the vanes should be at least 1.5-2mm. A home made capacitor is described later.

A vacuum capacitor would seem a good choice. However, the high loop currents tend to heat and thereby distort thin metal in vacuum capacitors and consequently detune the loop [18]. Experiments with a vacuum capacitor tuned low-frequency loop show that with SSB (about 60W PEP) with its low power factor there is no need to retune the antenna. But when used with CW, with its greater power factor, there is a need retune the antenna from time to time.

The tuning motor

The motor forms an important part of the system; it calls for a DC motor with a reduction gear capable of providing very fine control, with the capacitor shaft rotating at only about one turn per minute or even less. Ideally, a variable speed motor is required to provide slow rotation for accurate tuning but a faster rate for changing bands.

I1ARZ used a motor with that could operate between 3 - 12V, which ran slowly at the lowest voltage and fast at 12 volts. Again, the surplus market may provide such a motor. Should you be unable to find a motor incorporating a suitable reduction gear it is possible to use a receiver-type slow-motion tuning drive; a Bulgin gear with a reduction drive ratio of 25:1 was used on the prototype.

An alternative motor control method by PA0YW is described later.

Construction of the tuning system

When estimating the dimensions of the insulated supporting board, bear in mind the following: The space required for the watertight cover, the aluminium bracket for mounting the motor, and the external reduction gearing, together with the various couplings between the capacitor spindle and the motor.

The first step is to mark the centre line of the board (ie major axis). Bolt the capacitor to the board, taking great care to ensure

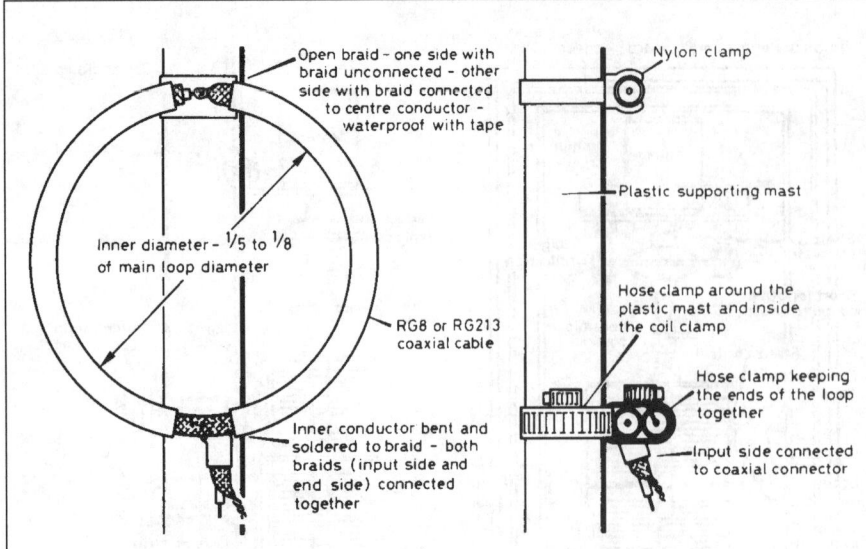

Fig 15.55: Detail of (a) the construction of the coax loop feed and (b) the plastic support mast

large wattage soldering iron, taking care that the best possible electrical contact is achieved.

Motor feedline

The feed line can be made from twin screened cable, as normally used with hi-fi audio amplifiers etc. The braid should be connected through a soldering lug to the aluminium bracket or to the motor casing. The motor must be bypassed for RF, using two 10nF ceramic capacitors connected to the braid. The cable is kept in place by means of nylon clamps along the supporting plastic mast. At the base of the loop solder a connector on the small copper bracket, with the braid soldered to the bracket. From this point to the operating position, normal electrical twin cable can be used. Some constructors have suggested inserting the feed line inside the loop pipe but this reduces the efficiency of the antenna. A small box containing the DC power supply and switch for reversing polarity of the supply is operated from the shack.

Coupling loop and matching procedure

A variety of methods for feeding the loop are shown later. I1ARZ found the most satisfactory method of coupling was a small single-turn (Faraday) coupling loop formed from a length of coaxial cable (RG8 or RG213) with a diameter one-eighth of the main loop. In practice, the optimum diameter of the coupling loop will vary slightly from this figure and it may prove worthwhile to experiment with slightly different size loops. This is done by aiming for the lowest SWR over a wide frequency range and is best achieved by constructing several coupling coils.

With the I1ARZ antenna the optimum diameter proved to be 18cm rather than the theoretical 12.5cm. The coil should have the braid open at top-centre; at this point one side is connected to the inner conductor of the coaxial cable. At the base of the loop, inner conductor and braid are connected together and jointed to the braid on the input side of the coil as shown in **Fig 15.55(a)**.

The ends and braid of the coupling loop are held together using a stainless hose clamp. This, in turn, is fixed to the mast at 90° to another hose clamp on the plastic mast, see **Fig 15.55(b)**. This provides a very simple method of adjustment by sliding the small loop up or down the mast to find the best SWR position.

The upper opening should be protected with tape and, to avoid any subsequent movement of the coil, then fixed to the mast by means of nylon clamps mounted in the same way as for the hose clamps at the base. Final matching of the antenna has to be carried out after determining the final position of the installation. An SWR bridge is connected at the base of the loop close to the input coax connector. If your transmitter covers 18MHz, make your adjustments on this band; otherwise use 21 MHz. Apply minimum power, just sufficient to deflect the SWR meter. After finding loop resonance, move the coupling coil up and down, or deform it slightly to check how the SWR varies. The coupling coil must be maintained in the same plane as the loop. After finding the lowest SWR, tighten the hose clamps and nylon clamps to keep the coil in position. The coax line and tuning motor power line must be kept vertical for about 1 m or more from their connectors at the base of the loop to avoid undesir-

that the shaft is aligned with the centre line marked on the board. A split-stator capacitor must be placed with the respective stator contacts symmetrically in the vertical plane so that the copper strips coming from the back of the board on either side of the capacitor have the same length (one being bent upwards, the other downwards). With a butterfly or conventional capacitor, the copper strips must be bent horizontally as both contacts are the same height.

Once the capacitor has been bolted to the board, measure, with callipers or dividers, the exact distance of the board from the centre of the capacitor shaft. Transfer this dimension to the centre line of the vertical side of the L-shaped aluminium bracket to be used for supporting the motor and any external reduction gear.

Drill a small pilot hole just large enough to take the motor shaft. It is important that this operation is carried out carefully since it is vital to the accurate alignment of the system.

Once it has been determined that motor and capacitor shafts are in accurate alignment, the motor may be fixed permanently to the bracket, enlarging the pilot hole and drilling holes for the motor-fixing bolts. Do not fix the bracket to the board yet .

The next step is to adapt the motor shaft to a shaft extension (as normally used for lengthening potentiometer shafts) taking care not to introduce any eccentricity. If your motor does not require external reduction-gear, you can insert the motor shaft into a ceramic coupler (circular shape with ceramic ring and flexible central bush); again making sure there is no eccentricity. The lower flange of the aluminium bracket can now be fixed to the board by means of two nuts and bolts.

If you use the Bulgin drive external reduction gear, drill two 4mm holes in the bracket; one in each side of the motor at the same distance as the mechanical connections of the gear and at the same level as the centre of the motor shaft. If the size of the motor is wider, it is necessary to join two short strips of brass or aluminium to the Bulgin gear so as to obtain an extension of the fixing points of the drive. Two long, 4mm diameter brass bolts should be inserted in the holes to hold the gear in place with the nuts. The Bulgin gear can now be fixed to the motor shaft, with the other side of the gear connected to the ceramic coupler by means of a very short piece of potentiometer-type shaft. Make a provisional check of the tuning system by temporarily connecting the power supply to ensure that everything is working smoothly. The copper strips of the loop can now be soldered to the stator capacitor vanes. This requires the use of a

Fig 15.56: Some experimental loops used by GW3JPT

able coupling with the loop itself and subsequent difficulty to achieve a proper matching. The minimum SWR should be better than 1:1.5 on all bands.

Installing and using the loop

The loop can conveniently be installed on a terrace or concrete floor or roof. One method is to use as a base or pedestal the type of plastic supports that can be filled with water or sand and often used for large sun umbrellas. Light nylon guys can be used to minimise the risk of the loop falling over in high winds. Remember that a transmitting loop operates effectively at heights of 1 to 1.5m above ground, and little will be gained by raising it any higher than say 2 or 3m at most. I1ARZ tested his loop using a telescopic mast at heights up to 9m above ground but with very little difference in performance; and that normally it was used with the mast fully retracted to about 3m high.

With a garden, the loop could be fixed directly to a short metallic mast driven into the ground. A small TV rotator could be

used but this is not essential; maximum radiation is in the plane of the loop, minimum off the sides of the loop. Large metallic masses like fences; steel plates, pipes etc reduce the efficiency of the antenna if close and in the direction of the plane of the loop. The radiation is vertically polarised at all vertical angles making the loop suitable for DX, medium and short range contacts.

There is nothing particularly complicated about operating with a small loop antenna other than the need to tune it to resonate it at the operating frequency. Tune initially for maximum signals and noise in the 'receiving' mode; this will bring the loop close to the tuning point for transmission. Using the SWR meter, tune carefully with the aid of the polarity-reversing switch to the precise point where minimum reflected power is achieved. The 'receiving-mode' procedure should always be used when changing bands.

Conclusion and final comments

I1ARZ began experimenting with small diameter transmitting loops in 1985 and he is now convinced that the loop is a thoroughly practical antenna that should not be written off as either a compromise or emergency antenna. Its performance, provided the RF ohmic losses are kept very low, is very good.

Wire Loop Antenna for the Lower HF Bands

As already mentioned, it is necessary to ensure that the resistance of the loop is as low as possible.

However, larger loops for the HF bands may be impractical due to the weight. C R Reynolds, GW3JPT, constructed many magnetic loop antennas, all of which were made from 22mm copper tubing or strip aluminium. Some of the experimental loop antennas used are shown in **Fig 15.56**.

He wanted to operate on the lower HF bands and although he found that it was possible to tune a small loop to 160m using a very large 1000pF capacitor there were two problems. On 160 metres the efficiency is rather low and on 40 metres tuning is rather critical because it only takes a few picofarads of tuning capacitor variation to tune the whole of the 40m band. This represents a very small percentage of 1000pF, requiring only a fraction of capacitor rotation to cover the band.

In an article [19] GW3JPT described a different design of a practical loop antenna for the 160, 80 and 40 metre bands. This uses a much larger square loop of a size shown in **Fig 15.57**. If this were to be made from copper tube it would be very heavy so he used a 19.5m (64ft) length of plastic covered wire. This antenna requires a 250-300pF tuning capacitor.

The Faraday coupling loop is shown in **Fig 15.58**. It is close coupled for about 0.77m (30in) each side of the centre of the tri-

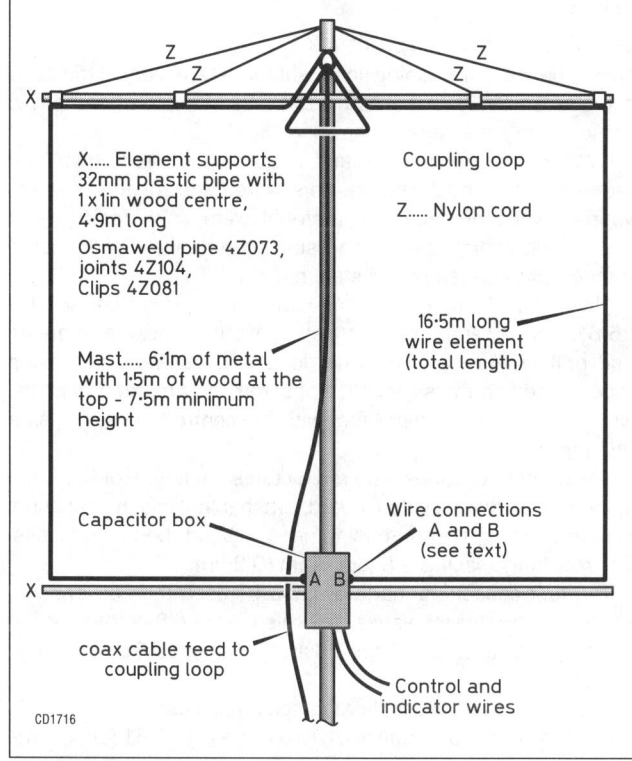

X..... Element supports 32mm plastic pipe with 1x1in wood centre, 4·9m long

Osmaweld pipe 4Z073, joints 4Z104, Clips 4Z081

Mast..... 6·1m of metal with 1·5m of wood at the top - 7·5m minimum height

Capacitor box

coax cable feed to coupling loop

Coupling loop

Z..... Nylon cord

16·5m long wire element (total length)

Wire connections A and B (see text)

A B

Control and indicator wires

CD1716

Fig 15.57: Overall view of the LF band magnetic loop

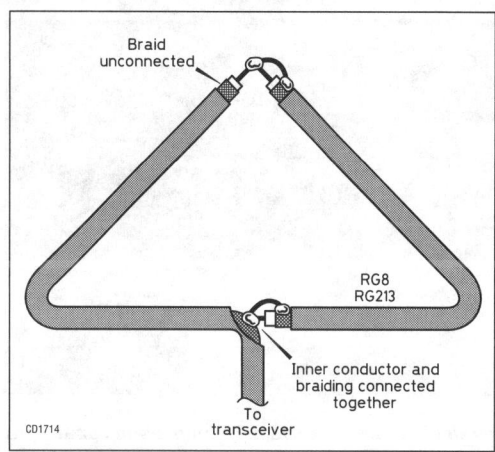

Fig 15.58: Faraday coupling loop

Braid unconnected

RG8 RG213

Inner conductor and braiding connected together

To transceiver

CD1714

Fig 15.59: Control and indicator system for the magnetic loop antenna

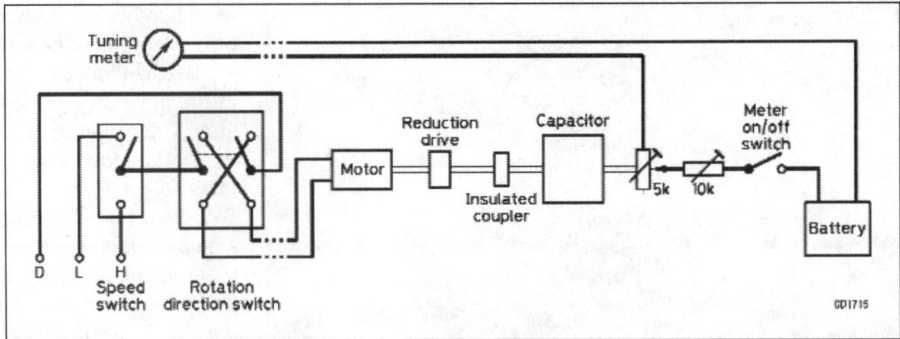

angle section of the element. This wire loop will also work on 40m. This is achieved by using a relay or a switch to disconnect the capacitor at points A and B, (Fig 15.57). The loop is then tuned by the stray capacitance of the switch or the relay. Because this stray capacity cannot be varied, the antenna element length is adjusted for correct matching using an SWR meter.

The antenna and mast can be fitted to a good ground post. It does not need any guy wire support and can be raised or lowered easily. For portable use it can be erected in a few minutes using three or four guy wires.

Capacitor drive motor

There is a reasonable range of motors available suitable for rotating the loop capacitor. The cheapest and one of the best available is a barbecue spit motor. Although this is already geared down it does require extra reduction using a 6:1 or 10:1 epicyclic drive for more precise tuning.

The motor will rotate slowly if energised by a 1.5V battery. With 3V applied the motor will run much faster. By switching from 1.5 to 3 volts a fast or slow tuning speed can be selected (**Fig 15.59**). The positive lead of the 3V battery is connected to H and the positive lead of the 1.5V battery is connected to L. The negative leads of both batteries are connected to D.

The direction of rotation is achieved using a two-pole, three-way switch. When the switch is set to the centre position the motor is disconnected from the battery (OFF position). The battery polarity to the motor is selected by the two other positions of the switch and should be labelled DOWN or UP.

The drive mechanism must be electrically isolated from the high RF voltages present at the capacitor. An insulated coupler

Fig 15.60: An example of one of GW3JPT home made capacitors

can be made from plastic petrol pipe. This pipe size should be chosen so that it is a push fit on to the drive mechanism and capacitor shafts. The pipe can then be fixed to the shafts by wrapping single strand copper wire around the ends of the pipe and tightening with a pair of pliers.

All the capacitors made by GW3JPT have the spindle extending both sides of the capacitor. One spindle is used to couple the capacitor to the drive mechanism; the other is used to connect the capacitor to a position indicator. This indicator circuitry must be electrically isolated from the capacitor as described above.

The control unit is housed in a plastic box with the fast/slow and rotation direction switches fixed to the front, together with the capacitor position meter.

Capacitor unit housing

One of the main problems of constructing any electrical circuits associated with antennas is protecting them from wind and rain. One option is to try and find some sort of suitable plastic housing and then organising the components to fit, but GW3JPT prefers to make the tuning housing from exterior plywood. The bottom and sides of the box are fixed together using 25mm square strips of timber. Glue and screws are used to make the joints waterproof. The top must, of course, be made so that it can be removed fairly easily. Paint or varnish the box as required.

Construction of capacitors

The capacitors for tuning loop antennas are very difficult to come by so GW3JPT makes his own. An example of one of his home made capacitors in shown in **Fig 15.60**.

GW3JPT used aluminium and double-sided circuit board for the vanes, and nuts and washers were used for the spacers. Various types of insulation material were used for the end plates. The centre spindle and spacing rods were constructed from 6mm-threaded plated steel rod.

Make the 76mm x 76mm (3 x 3in) end plates first, see **Fig 15.61**. These can be taped together, back-to-back, for marking and drilling. The same can be done with the vanes. Masking tape is used so the surface is not scratched around drill holes, which are drilled to clear 6mm with the centre hole acting as a bearing.

The number of vanes required dictates the length of the 6mm spindle. For double-sided board, washer/nut/washer spacers can be used so that there is no need to bond the copper sides. The resulting spacing is about 6mm (0.25in).

The first capacitors made by GW3JPT used the conventional shape for the moving vanes, but this is very difficult to cut out and fragile to use. The shape illustrated in Fig 15.61 (a) is much easier.

The fixed vane is a simple rectangle, which can be modified to reduce the minimum capacity. (Dotted line Fig 15.61 (c)). For the size shown, six pairs of vanes with 6mm (0.25in) spacing work out to about 150pf. Units using both printed circuit board and

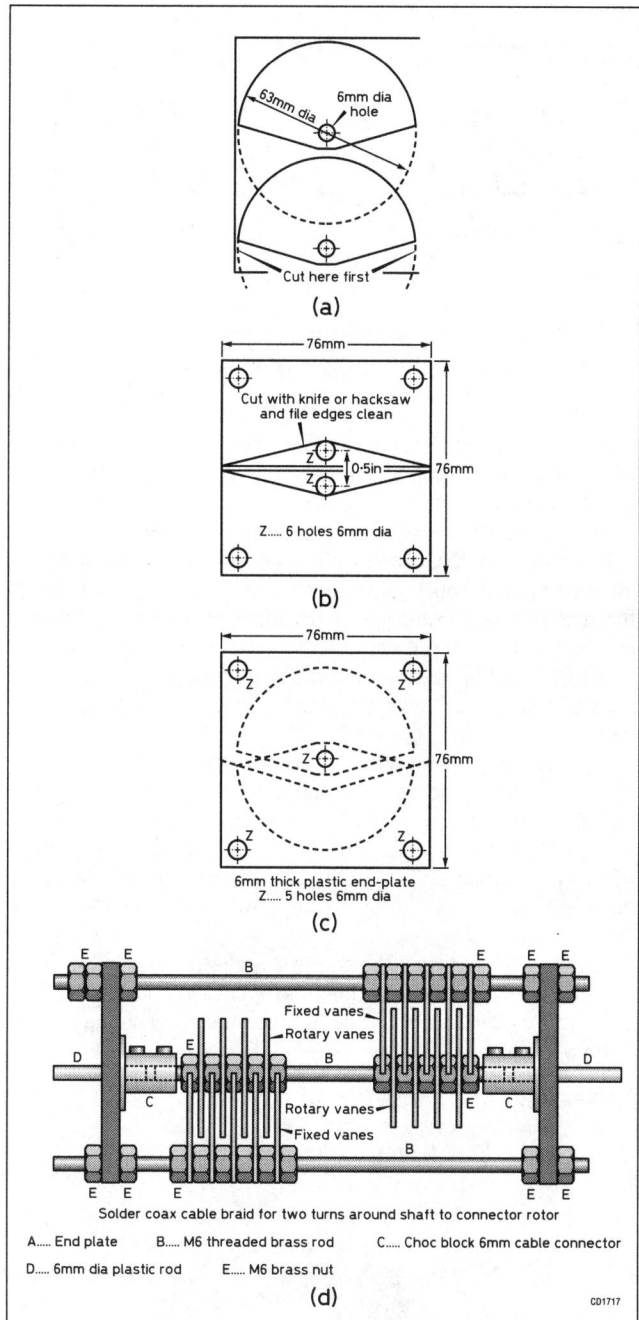

Fig 15.61: Details of homemade capacitor; (a) moving vanes; (b) fixed vanes; (c) fixed and moving vanes geometry showing minimum capacitance; (d) capacitor assembly

Table 15.8: Calculated data for the PA2JBC 80m, 2m diameter loop antenna made from 22mm tubing, using a 100W transmitter at 3.74MHz

L	5.8µH
Loaded Q	1273
Resistance	7.4mOhms
Loss resistance	46mOhms
Efficiency	14%
Bandwidth -3dB	2.94kHz
C at resonance	314pF
Capacitor Volts	8.3kV
Loop current	43.3A RMS

aluminium vanes had been in use for over two years and both were still in good working condition at the time of writing.

Operation

Loop tuning needs to be adjusted precisely for minimum SWR, which should coincide with maximum power out. This tuning is critical; a few kilohertz off tune and the SWR will rise dramatically. The best way of finding the correct position of the tuning capacitor is to listen for maximum noise, or signals, whilst tuning the loop, then fine-tune using an SWR meter.

The performance of this antenna on 80m was at least as good as a G5RV. It tuned all of 160m and gave quite good results as compared with local signals on the club nets.

A 2-Metre Diameter Loop for 80m

This loop antenna is very compact and although designed for mobile use it would be suitable for 80m operation from a very restricted location. It was designed by PA2JBC [20] to have a diameter of 2m and, for transportation purposes, be capable of being dismantled into two pieces. This feature would make it easy to get through a small loft access hatch. It also has a most interesting tuning arrangement. However, bearing in mind what has already been said about loops, an antenna this small does not have the efficiency of the larger models. The specification of the PA2JBC loop is given in Table 15.8.

The electrical characteristics were calculated for 22mm copper tubing. For practical reasons this antenna was built as an octagon. Measurements on the final antenna correlated closely with the calculations and PA2JBC has concluded that soft-soldered (rather than brazed or silver soldered) 45 degree elbows and compression couplers do not spoil the Q. This is a slightly at odds with the description of I1ARZ's loop, above. The structure of the antenna is shown in Fig 15.62.

The eight 820mm lengths of copper tubing are prepared by thoroughly cleaning the ends with fine emery paper and coating with flux. The pipe must be cut with a pipe cutter so that the ends fit snugly into the connector.

Two of the pipes to be joined are fitted into the 45° connecting elbows. The joint is heated with a blowtorch while at the same time applying multi-cored solder at the point where the pipe joins the connector. When the solder flows freely the joint is complete. Repeat for all the other joints, making sure the alignment of the loop is flat and make sure you fit the current transformer on to one of the sections of tubing (see later) before completing the loop. The completed loop is then cut in two sections as shown in Fig 15.62 and compression joints fitted at point A. If the antenna has to be frequently dismantled and reassembled, an 'olive' should be soldered to the tubing to reduce wear.

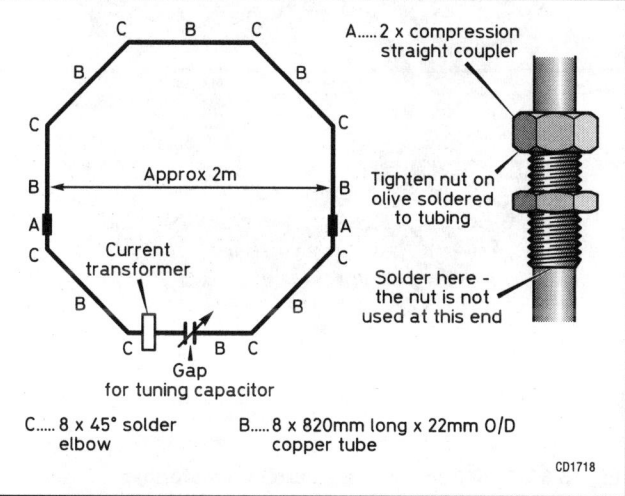

Fig 15.62: The PA2JBC 80m compact loop antenna

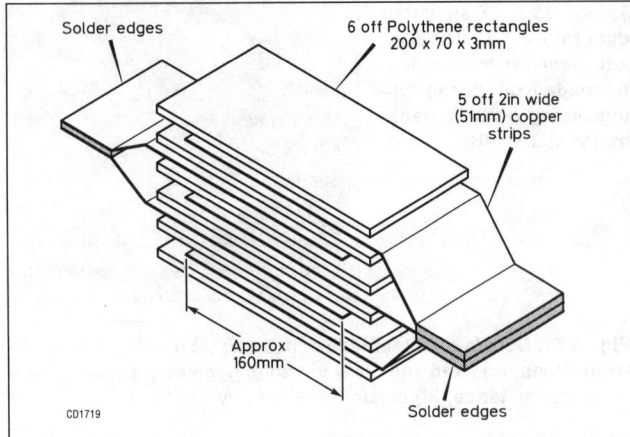

Fig 15.63: This fixed 260pF home made capacitor is good for 8kV @ 40A

The required capacity variation to cover 3.5 - 3.8MHz is 300pF to 360pF. This capacitance is made up from a 260pF fixed capacitor and a 100pF maximum variable in parallel. Using a small variable capacitor reduces the cost and improves the band-spread tuning.

The 100pF variable must be able to handle up to 9kV peak and 13A RMS when used with a 100W transmitter. A wiper connection to the rotor is unsuitable at 13A so a 2 x 200pF split-stator capacitor is used. Even then, the current path between all rotor plates must be low resistance, preferably soldered or brazed; the same goes for the stator plates and their connections to the loop tubing. At 9kV, conservative design requires 9mm spacing between the plates, or 4.5mm in a split-stator (each half takes 50% of the voltage).

The fixed capacitor is made from 51 x 0.3mm copper strips interleaved with slabs of dielectric as shown in **Fig 15.63**. Polyethylene works well as a dielectric and is inexpensive. Genuine polyethylene will not get hot! The capacity is set by adjusting the meshing of the two sets of copper 'plates' but the dielectric must extend beyond the copper by at least 6mm. After adjustment, the capacitor can be wrapped with glass-fibre-reinforced tape. Four parallel 3mm copper wires connect the fixed capacitor to the loop tubing. The 3mm-thick polyethylene is just adequate for 100W. On test the power was increased to 180W before it broke down.

With a loop of this design, PA2JBC could not get a gamma-match to work. A coupling loop also proved unsatisfactory as its shape had to be adjusted when changing frequency. The final solution was a current transformer. This transformer must match the 53 milliohm loop to a 50 ohm coax, an impedance ratio of 940:1 - a turns ratio of $\sqrt{940} \approx 30{:}1$. With this transformer the '1' is the loop tubing fixed in the centre of a toroid.

With the loop at resonance the feeder 'sees' inductance. By increasing the transformer winding to 36 turns and adding a series capacitor a 1:1 SWR can be obtained anywhere in the band. This capacitor is a receiver-type air-dielectric 250pF variable. A 1:1 balun keeps the outside of the coax 'cold'. An electrical diagram of the transformer and balun is shown in **Fig 15.64**.

Construction of the current transformer is shown in **Fig 15.65**. 6 x 6 turns of 1 mm PTFE-insulated copper wire gave the best results. Ferrite (Philips 4C6, violet) and iron powder (Amidon, red) both worked well. The transformer can be placed anywhere on the tubing, eg next to the capacitors where they and the coupler can housed be in a weatherproof box.

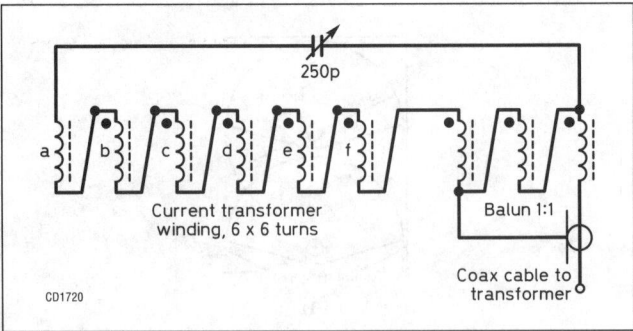

Fig 15.64: An electrical diagram of the matching transformer and balun

PA2JBC installed the antenna in his loft 3 metres above ground level and above all the electrical wiring. It should be fixed using good insulating material such as plastic pipe, as used in the I1ARZ loop.

The antenna's location close to wooden rafters and clay roofing tiles did not noticeably affect the Q of the loop, even when the roof was wet. If the loop is rotated with a TV rotator then it can be used to null out sources of interference.

With 100W PEP, only 14W is radiated from this indoor antenna, which is probably as good as the best mobile antenna for the band.

Certainly there is no problem with normal 80 metre country-wide contacts in the daytime and occasional DX at night. The high-Q characteristics of the antenna give a marked improvement in the signal to noise ratio in the presence of general electrical interference and QRN. This antenna could be used as a 'receive only' antenna in conjunction with a larger antenna for DX.

Tubing of 28mm would raise the Q and efficiency but it would also reduce the bandwidth; fine for CW, but too narrow for 80m SSB! As it is, the loop must be re-tuned for every change of frequency.

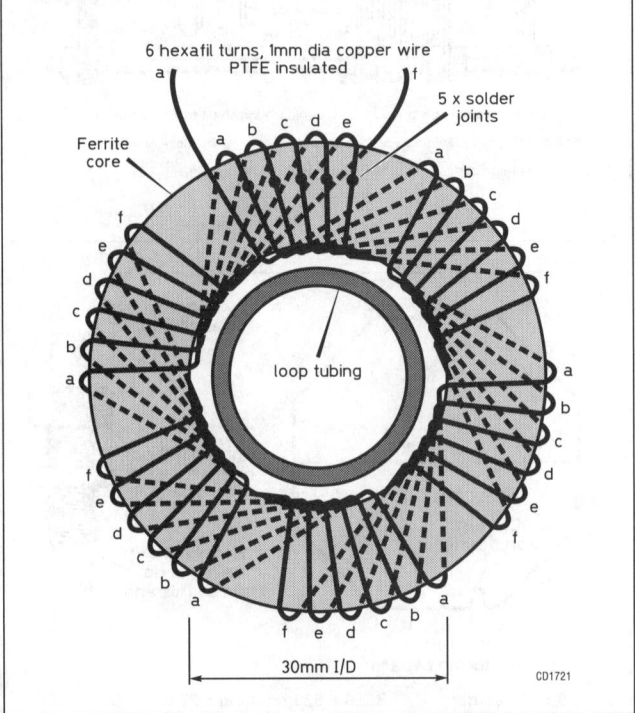

Fig 15.65: Construction of the current transformer

The Radio Communication Handbook

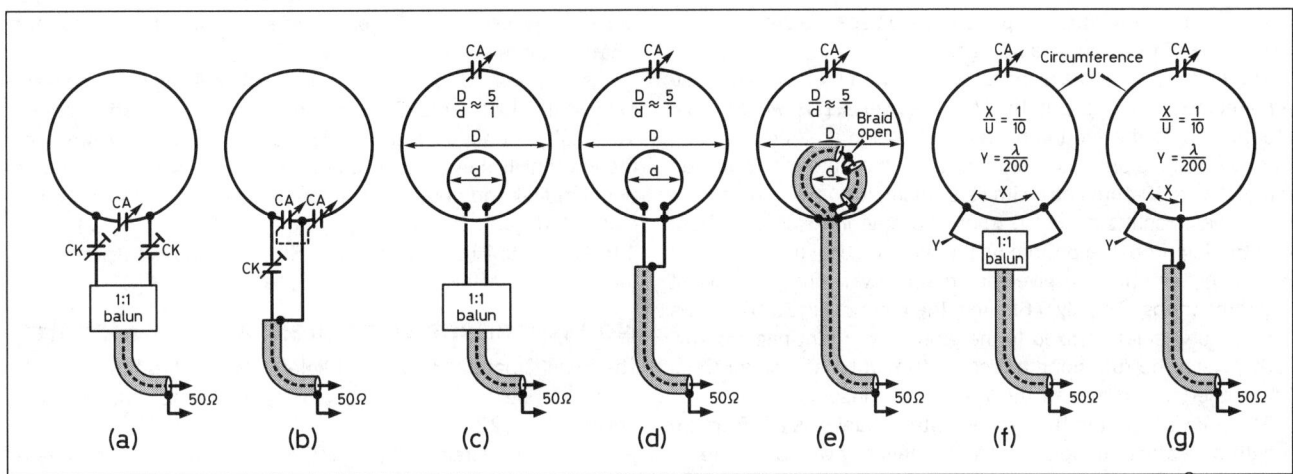

Fig 15.66: Matching a transmitting or receiving antenna to 50 ohm coaxial cable as described in 1983 by D12FA and reported in *RadCom*, **October 1996. The Faraday loop method (e) is favoured, but see text**

For outdoor use, the loop should be de-greased and painted.

Comments on Small Loop Antennas

The compact loop is a viable solution for these not able to erect a wire antenna. The compact loop by I1ARZ is, in the worst case, about 4 or 5 dB down on a dipole half a wavelength high. This is less than one S-point. It is obvious that, under conditions of normal HF fading, difficulties might be experienced when making meaningful comparisons.

The compact wire loop by GW3JPT is about 60% efficient on 160m. Making comparisons on this band is even more difficult - getting a 160m comparison dipole up half a wavelength high is

a challenge.

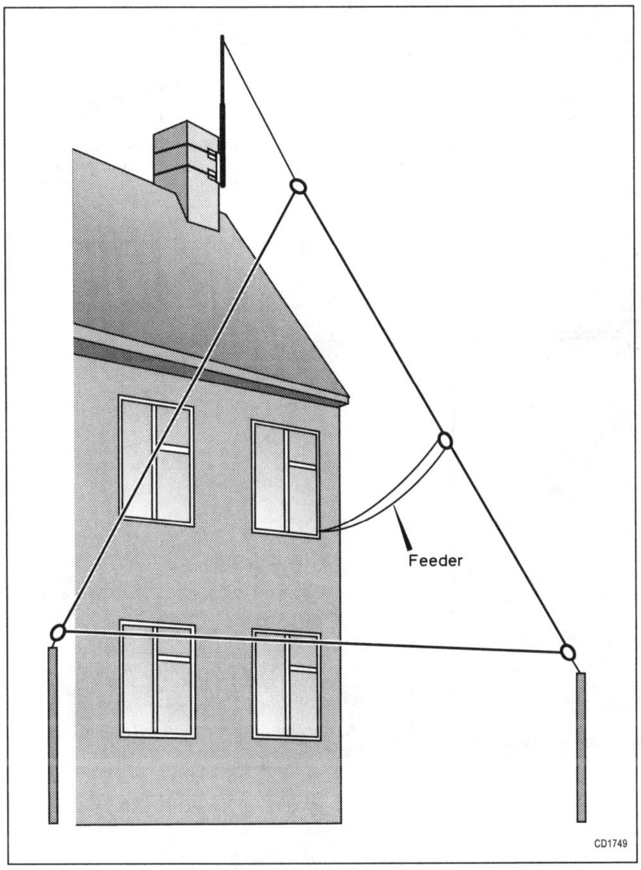

There are numerous ways of coupling a low-impedance loop to coaxial cable and most of these are shown in **Fig 15.66**. The favoured method is the Faraday loop, shown in Fig 15.66(e). There has been some comment that the Faraday loop connections, found in most descriptions of loops (including the I1ARZ and GW3PJT designs above) are incorrect. The coax inner and braid at the top or apex of the loop in Fig 15.58, for example, is shown joined, which would make a Faraday half loop. The inner to braid connection should be removed but the gap in the braid halfway round the loop should remain.

The matching methods in Fig 15.66 (a) and (b) have been used with loops of 10m or more diameter for 136kHz, for both transmit and receive. The PA2JBC compact loop for 80m, with its transformer and balun, is worthy of further development for more conventional sized loops.

Multi-Band Delta Loop Antenna

If you have the space, a larger loop is well worthwhile. If the loop is larger than 0.25λ it will lose its predominant 'magnetic' characteristic and become an 'electric' antenna of the quad or delta type. From the previous descriptions of loop antennas it can be seen that the efficiency improves with an increase in size and the resistive losses of a loop with a full wavelength circumference are very small.

A full wave loop on 7MHz can be fed with coax and will also operate on the 14, and 21MHz bands and without an ATU, provided that a transformer and balun are connected between the coax and the antenna. The shape of the loop is not too important.

If a loop antenna in the form of an equilateral triangle is used then only one support is required. If this support were a mast fixed to the chimney, it can probably circumvent planning restrictions.

The antenna is shown in **Fig 15.67**. As you can see, part of this antenna is close to the ground. This means there is a possible danger of someone receiving an RF burn if the antenna is touched when the transmitter is on. For this reason, insulated wire for the lower half of the antenna is recommended. A loop antenna of this type is not a high-Q device so very high voltages, such as those found at the tips of a dipole, do not occur.

The top half of the antenna can be constructed with bare copper wire. You could use insulated wire for all the loop, however lightweight wire for the upper half of the loop, and a lightweight support, has a low visual impact. Using lightweight thin wire

Fig 15.67: Loop antenna one wavelength circumference on 7MHz

does not affect the antenna performance because the radiation resistance of the loop is fairly high.

The first experiments were carried out with the coax connected directly to the loop, but the SWR was over 3:1. However, most literature puts the feed impedance of a loop greater than 100 ohms. A 4:1 balun was fitted, enabling the antenna to be fed directly with 50-ohm coax with little mismatch.

The best results occurred when the antenna was fed about one third up from the bottom on the most vertical of the triangle sides. This antenna will give good results even when the lowest leg of the triangle is only 0.6m from the ground. Fig 15.67 shows the corner insulators fixed to the ground with tent peg type fixtures. It can be run along a fence with shrubs and small trees being used for fixtures for the lower corner insulators.

The apex support in the experimental model was a 2.5 metre length of scaffolding pole fixed to the chimney with a double TV lashing kit. The top of the chimney was about 9m above the ground. The pole gives the antenna enough height and a reasonable clearance above the roof.

The loop proved to be a good DX transmitting antenna on 7MHz. It did tend to pick up electrical noise from the house on receive. It could be used with a smaller loop on receive if electrical noise or QRM is a problem.

Skeleton Slot Antenna

The Skeleton Slot is is a loop antenna with a difference. With the dimensions given it will operate on the 14, 18, 21, 24 and 28MHz bands using a balanced ATU.

It is very easy to construct and is a simple design with no traps or critical adjustments. This antenna has a turning radius of only 1.5m (5ft) although it is 14m (47ft) tall. However its construction means that it has a much lower visual impact than a conventional multi-band beam. The antenna is bi-directional and has a calculated gain, over average ground, of 8dBi on 14MHz and 11dBi on 28MHz. The Skeleton Slot antenna was first documented in an article by B Sykes, G2HCG, in 1953 [21].

Non-Resonant Slot Antenna for 14 - 29MHz

The main exponent of the HF Skeleton Slot, other than G2HCG, is Bill Capstick, G3JYP, whose version of this antenna was described in [22].

The G3LDO version of the slot uses wire for the vertical elements, resulting in a more simplified and rugged construction. It was first thought that this method of construction would not work because [21] and [22] gave minimum tube diameters for the elements. However, computer modelling with EZNEC4 was reassuring.

The antenna essentially comprises three aluminium tube elements fixed to the mast at 4.6m (15ft) intervals, with the lowest element only 4.6m from the ground. The mast is an integral part of the antenna, as a boom is to a Yagi. The general construction is shown in **Fig 15.68**.

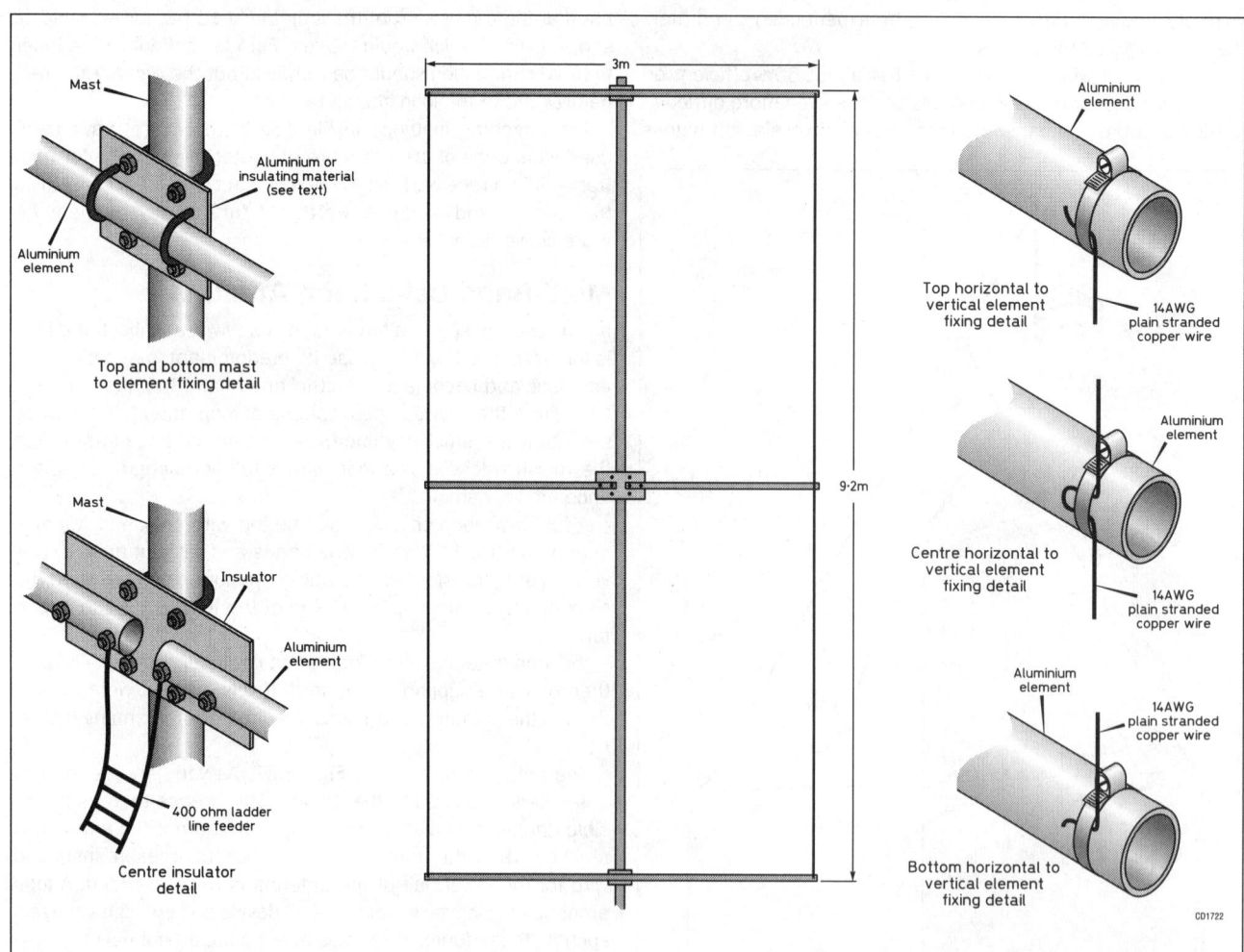

Fig 15.68: The G3LDO multiband Skeleton Slot antenna for 14 to 28MHz. The elements are fixed to the mast and the whole mast is rotated. The wire elements are fixed to the horizontal elements with hose clips. The centre insulator, as shown, is home made but a commercial one would be suitable

The centre element is fed in the centre with balanced feeder, and the upper and lower elements are fed at the ends by copper wire from the driven dipole

The aluminium tubing and copper wire are fixed using hose clips. These dissimilar metal connections have in the author's experience presented no corrosion problems, even in a location close to the sea, provided they are well coated with grease.

The centres of the upper and lower elements can be fixed directly to a metal earthed mast using an aluminium plate and U-bolts as shown in Fig 15.68. Performance on the normal HF bands is unaffected by grounding or insulating the upper and lower elements.

The diameter and length of the aluminium tube and wire are not critical.

The antenna requires a balanced feed and is fed with 450-ohm slotted line feeder, although the impedance is not critical. The feeder should be fixed on stand-off insulators about 150mm (6in) from the mast until clear of the lower element to prevent them blowing about in the wind and affecting the impedance, although this was not done in the antenna shown in **Fig 15.69**.

An ATU with a balanced output is required. A conventional Z-match was found to be adequate with the two sets of balanced outputs, one ostensibly for the higher HF frequencies and the other for the lower ones. In practice the lower frequency output worked best for all frequencies. The antenna can be fed with any of the ATUs described in earlier.

G3LDO built his skeleton slot to the size specified by Bill Capstick, and these dimensions seem nearly optimum for the

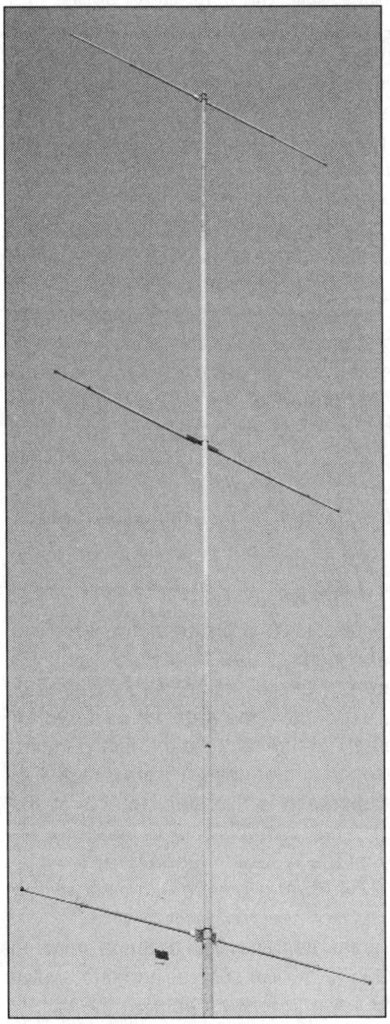

Fig 15.69: The G3LDO multiiband Skeleton slot antenna for 14 to 28MHz

Fig: 15.70: Construction of a two-element Yagi beam with dimension references for the 20, 18, 15, 12 and 10m bands

five higher HF frequency bands. While the DX performance of the Skeleton Slot is good up to 30MHz, it deteriorates at frequencies higher than this.

On the 21, 24 and 28MHz bands the antenna performed very well, particularly in marginal conditions.

ROTARY BEAM ANTENNAS

The rotary beam antenna has become standard equipment for the upper HF amateur bands, and the best known icon of amateur radio is the three-element Yagi. It offers power gain, reduction in interference from undesired directions, compactness and the ability to change the azimuth direction quickly and easily. All this has many advantages for a restricted site. All the beam antennas described in this chapter are parasitic beam antennas.

Optimum dimensioning of spacing and element lengths can only be obtained over a very narrow frequency range, and the parasitic beam will work only over a relatively restricted band of frequencies. In most cases, the bandwidth of such an array is compatible with the width of an HF amateur band.

The compactness of a parasitic beam antenna more than outweighs the disadvantage of the critical performance and no other antenna exists that can compare, size for size, with the power gain and directional characteristics of the parasitic array.

Two-Element Yagi

A two-element Yagi is shown in **Fig 15.70**. The parasitic element (Ep in the diagram) is energised by radiation from the driven element, which then re-radiates. The phase relationship between the radiated signal from the driven element and the re-radiated signal from the parasitic element causes the signal from the antenna to be 'beamed' either in the direction of Ep or away from it.

This phase relationship is effected by the length of the parasitic element. When the parasitic element is longer than the driven element it operates as a reflector and causes the power gain in a direction away from Ep. When the parasitic element is shorter it operates as a director causing the power gain in a direction towards Ep.

When the parasitic element is to be used as a director, optimum spacing between it and the driven element is around 0.1 wavelength. Optimum spacing when using the parasitic element as a reflector case is approximately 0.13 wavelength

The effect of these options can be seen in the computer simulation in **Fig 15.71**. There is very little difference in performance of a two-element beam when the parasitic element is used either as a director or reflector, with perhaps just a marginal improvement in the front-to-back ratio when the parasitic element is a director.

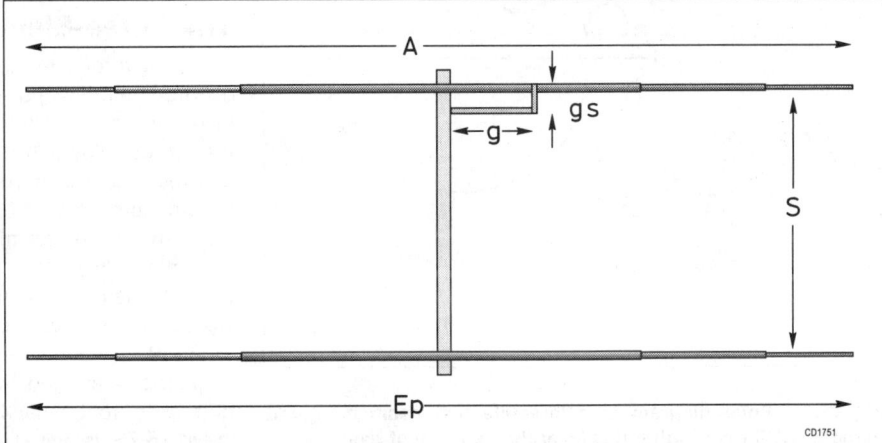

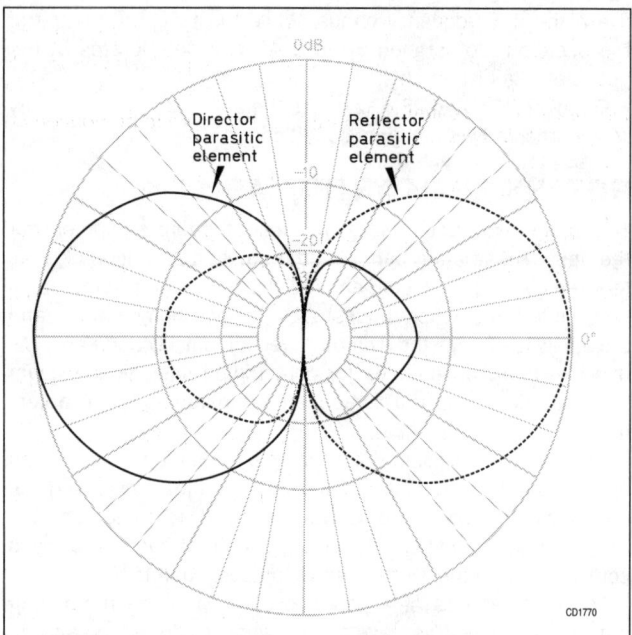

Fig: 15.71: Computer analysis of a two-element beam with the parasitic element as (a) a reflector and (b) as a director

Frequency MHz	14.1	18.1	21.2	24.9	28.5
(S) Element Spacing (m)	2.11	1.66	1.41	1.20	1.05
(Ep) Director length (m)	9.66	7.58	6.47	5.50	4.81
(A) Driven Elt (m)	10.30	8.08	6.90	5.88	5.13
(S) Element Spacing (in)	83	65	56	47	41
(Ep) Director length (in)	380	298	255	217	189
(A) Driven Elt (in)	406	318	272	231	202

Table 15.9: Dimensions for a two-element beam. Refer to Fig 15.70 for dimensions S, Ep and A. These dimensions have been calculated using EZNEC for a non-critical design to give a free-space gain better than 6dBi and a front-to-back ratio greater than

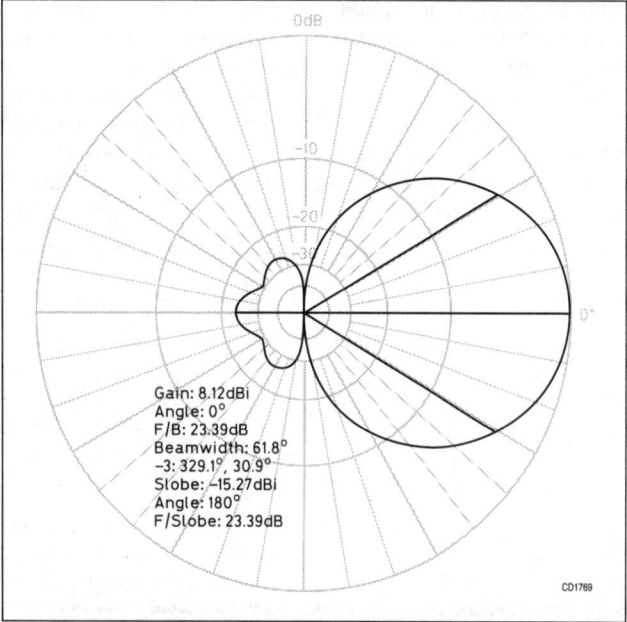

Fig 15.73: Polar diagram of a three-element beam designed using EZNEC for a high F/B ratio at the expense of gain

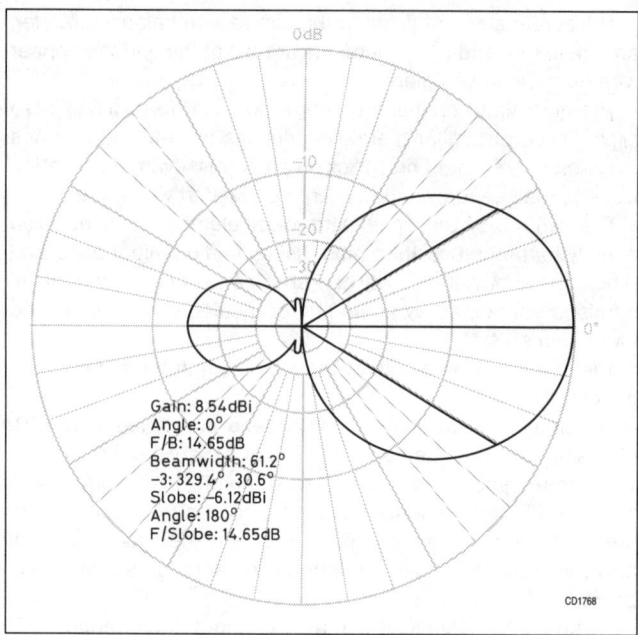

Fig 15.72: A three-element beam from the W6SAI *Beam Antenna Book* [23] shows high gain and a F/B ratio of nearly 15dB

Additionally, a two-element beam with a parasitic director will be the slightly smaller and lighter of the two options.

Practical dimensions for this option are shown in **Table 15.9**. These dimensions have been calculated using EZNEC for a non-critical design to give a free-space gain better than 6dBi and a front-to-back ratio greater than 14dB, as shown in the director parasitic element polar diagram shown in Fig 15.71. These calculations assume an average tube diameter of 20mm (0.75in) on 21MHz, which is scaled to an average of 30mm on 14MHz and 15mm on 28MHz. In practice the diameter of the tube is not critical and the diameter should be such that the antenna is mechanically stable. The elements should be made of, say, five sections that telescope into each other (Fig 15.70). Aluminium scaffolding pole 50mm (2in) in diameter is useful material for the boom for any of the bands. The construction of the elements and methods of fixing the boom to the mast are described in the antenna construction and masts section of this book.

As with all parasitic beams, the dimensions of the parasitic elements determine their performance. The length of the driven element is less critical and its length only determines the feed impedance.

The feedpoint impedance of this antenna is approximately 30 ohms. To feed it with 50-ohm coaxial cable a matching arrangement is necessary.

The Three-Element Yagi

By adding a reflector and a director to a driven element to form a three-element parasitic beam, the free-space gain is increased to over 8dBi and a front-to-back (F/B) ratio greater than 20dB, although this depends if the antenna is tuned for maximum gain or maximum front-to-back ratio. Most antenna constructors tend to tune beam antennas for maximum F/B ratio. The reason for this approach is that adjustments to the F/B ratio make a marked difference that is easy to measure. For example the polar diagram of the W3SAI antenna [24] shown in **Fig 15.72** has a gain of 8.54dBi and the front-to-back is only 14.65dB.

On the other hand, the polar diagram of an antenna selected for a good front-to-back ratio (23.4dB), such as the one shown in **Fig 15.73**, is only 0.4dB down in forward gain on the W3SAI

Frequency MHz	14.1	18.1	21.2	24.9	28.5
(S) Element Spacing (m)	2.96	2.32	1.98	1.67	1.47
(D) Director length (m)	9.66	7.60	6.47	5.52	4.83
(A) Driven Element (m)	10.30	7.92	6.76	5.76.	5.03
(R) Reflector length (m)	9.66	8.24	6.49	5.99	5.23
(S) Element Spacing (in)	116	91	77	66	58
(D) Director length (in)	382	229	254	216	190
(A) Driven Element (in)	402	314	268	229	200
(R) Reflector length (in)	414	324	276	236	206

Table 15.10: Dimensions for a three-element beam. Refer to Fig 15.74 for dimensions S, D, R and A. The dimensions are shown in metres and inches and have been calculated using EZNEC for a non-critical design to give a free-space gain better than 8dBi and a front-to-back ratio greater than 20dB

antenna. As 6dB is 1 S-point, the improvement of 1.5 'S' units on F/B is noticeable. The gain difference of 0.42dB is not going to be noticed on any 'S' meter.

The construction is similar to that described for the two-element beam. The dimensions for this antenna are given in **Table 15.10** and are read in conjunction with **Fig 15.74**.

The feedpoint impedance of this antenna is around 25 ohms. To feed it with 50-ohm coaxial cable a matching arrangement is necessary. The Gamma Match is described in the Transmission Lines chapter.

The Cubical Quad

The Cubical Quad beam is a parasitic array whose elements consist of closed loops having a circumference of one-wavelength at the design frequency. The quad construction is shown in **Fig 15.75** and the dimensions are given in **Table 15.11**.

The parasitic element is normally tuned as a reflector. It can be tuned as a director but the gain and front to back ratio is inferior. Additionally, the optimum settings are more critical.

The reflector can be constructed using the same dimensions as L, the driven element; a tuneable stub

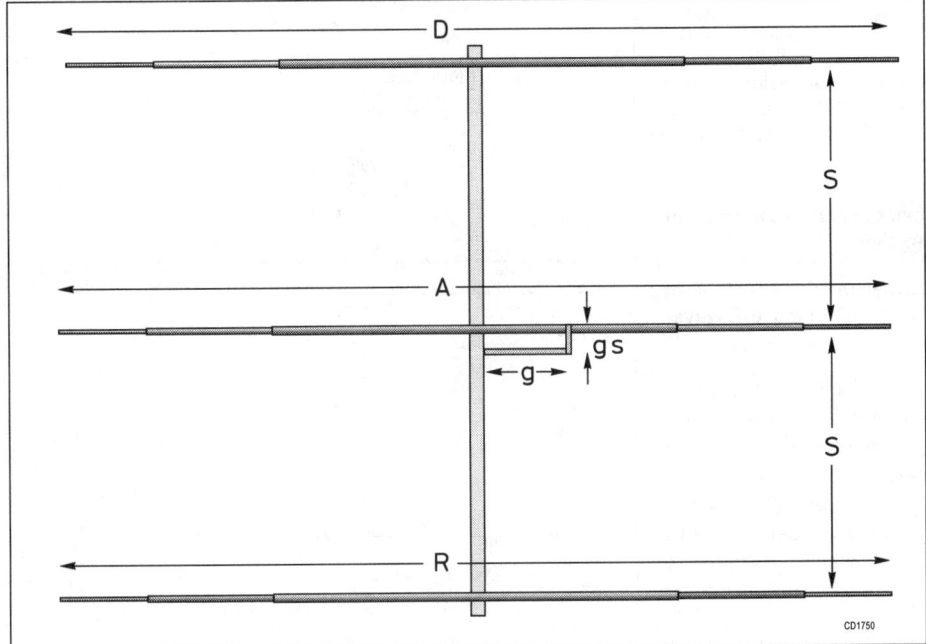

Fig 15.74: Construction of a three-element beam with dimension references for the 20, 18. 15, 12 and 10m bands

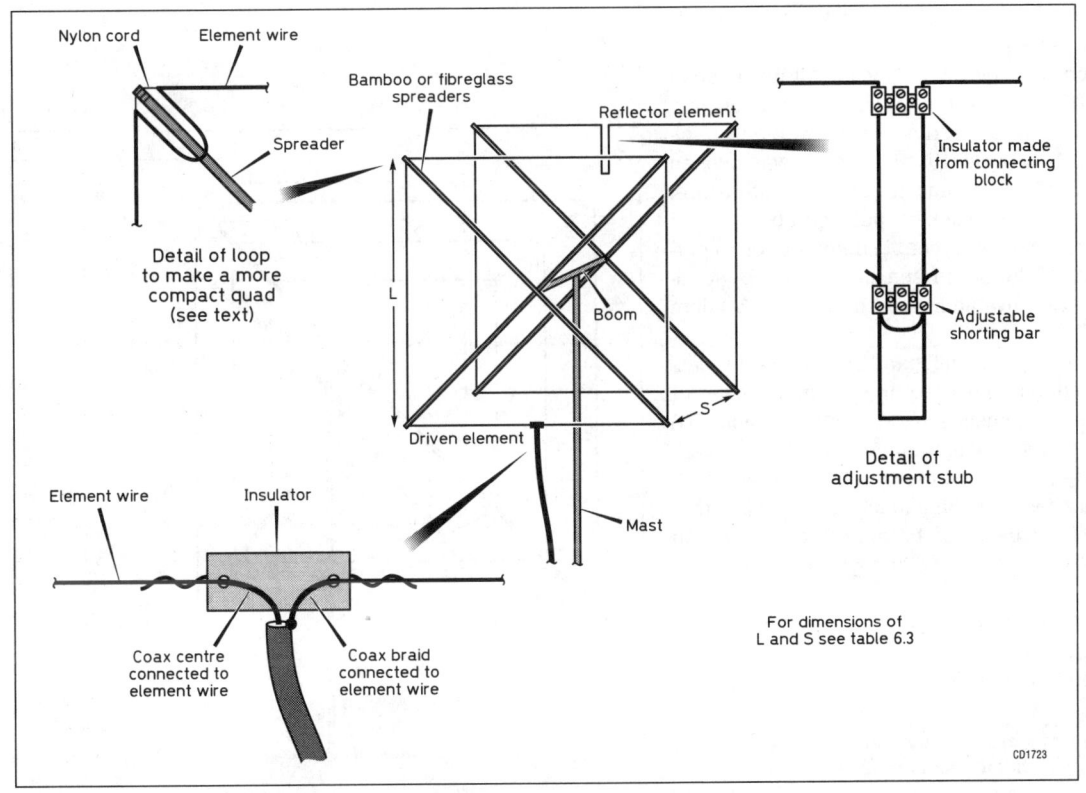

Fig 15.75: Construction of a two-element wire quad, with dimension references in Table 15.11. The reflector can be constructed using the same dimensions as L, the driven element; a tuneable stub (see detail on right) is then used to lower the frequency of the reflector. This stub can be used to tune the reflector for the greatest front-to-back ratio of the beam. The stubs on the element corner supports (detail left) are discussed in the text. A sealant should be used to prevent water ingress at the end of the coax cable

Frequency MHz	14.1	18.1	21.2	24.9	28.5
(S) Element Spacing (m)	2.98	2.34	1.99	1.70	1.49
(R) Reflector length (m)*	5.56	4.38	3.73	3.17	2.77
(ER) Element support length (m)	3.93	3.1	2.64	2.24	1.96
(A) Driven Elt (m)*	5.33	4.18	3.57	3.04.	2.65
(S) Element Spacing (in)	117	92	79	67	59
(R) Reflector length (in)*	219	172	147	125	109
(ED) Element support length (in)	155	122	104	89	77
(A) Driven Elt (in)*	210	164	140	120	105

*Note: These dimension are for one side of the quad. The total length

Table 15.11: Dimensions for a two-element quad beam. These dimensions have been calculated using EZNEC for a non-critical design to give a free-space gain around 7.5Bi and a front-to-back ratio greater than 15dB

Fig 15.76: Computer analysis of a two-element wire element quad with 0.14 wavelength element spacing

Fig 15.77: Computer analysis of a two-element wire element quad, (a) using 0.2 wavelength spacing; (b) using 0.1 wavelength spacing

is then used to lower the frequency of the reflector. This stub can be used to tune the reflector for the greatest front-to-back ratio of the beam. The stubs on the element corner supports can also be used reduce the overall size of the element and can be used with the driven element and the reflector. However the lengths will have to be determined by experiment. The easiest way to construct these stubs is to use plastic insulated wire elements and then to tape the stub along the element support, as shown in the right-hand detail of **Fig 15.75**.

The dimensions given in Table 15.11 are for a quad using an element spacing of 0.14 wavelength; the computed free-space performance is shown in **Fig 15.76**.

The lengths of the element supports could be longer than given. The length dimension is the point where the element is connected to the support.

Using the dimensions in Table 15.11, the feed impedance of the quad is around 65 ohms so the driven element can normally be connected directly to 50-ohm feedline with only minimal mismatch. The 0.14 wavelength spacing (S) was chosen because it is the most prevalent in antenna literature. However, the spacing for a two-element quad can be reduced to 0.1 wavelengths without any noticeable deterioration in performance, see **Fig 15.77**. Reduced spacing lowers the feedpoint impedance and can give an improved match to 50-ohm coaxial cable.

The quad can be made into a multi-band antenna by interlacing quad loops for the different bands on to a common support structure. In this case the element support length (ER) and (ED) should be the length for the lowest frequency band. The disadvantage of this arrangement is that the wavelength spacing (S) between the driven element and the parasitic element is different on each band. This problem can be overcome by using an element support structure with a modified geometry as shown in **Fig 15.78**

A multiband quad using this type of geometry is often referred to as a 'boomless' quad for obvious reasons. The structure, which hold the element supports in place at the correct angles

Fig 15.78: General view of a multiband variant of the quad using optimum driven element/reflector spacing for each band

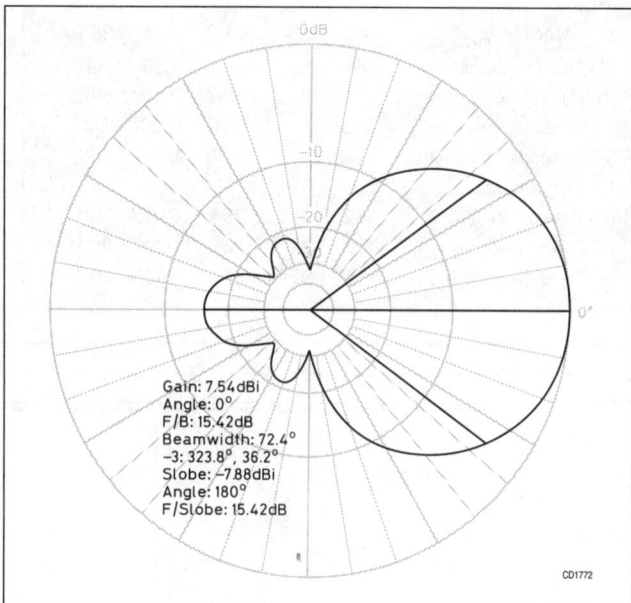

Gain: 7.54dBi
Angle: 0°
F/B: 15.42dB
Beamwidth: 72.4°
−3: 323.8°, 36.2°
Slobe: −7.88dBi
Angle: 180°
F/Slobe: 15.42dB

CD1772

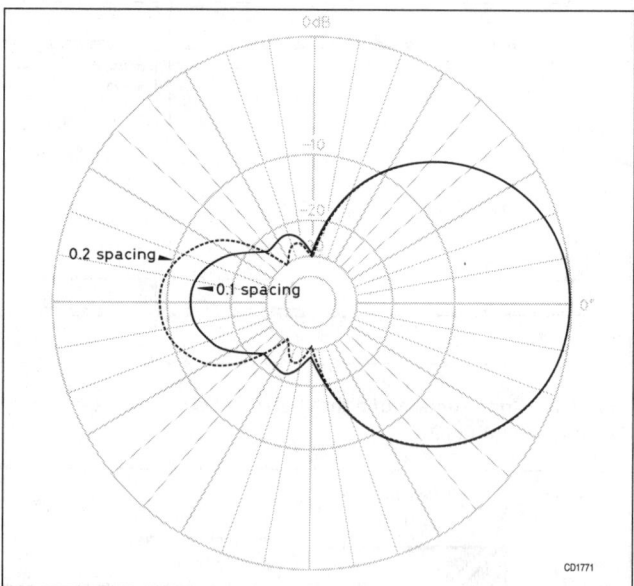

0.2 spacing
0.1 spacing

CD1771

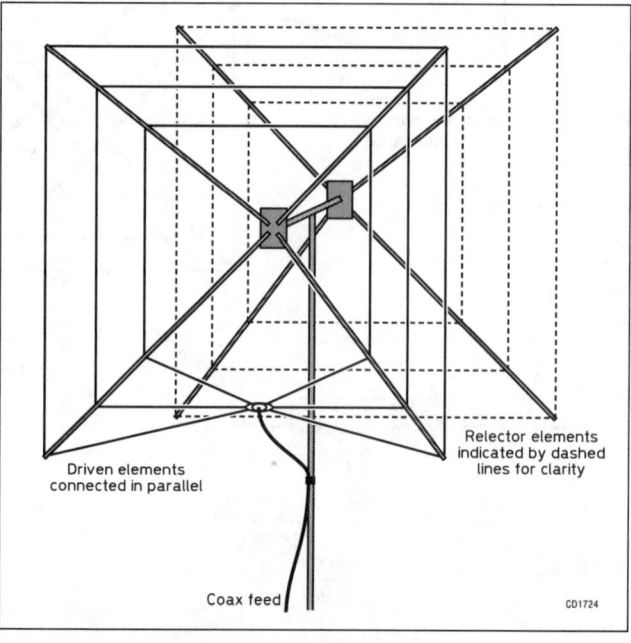

Relector elements indicated by dashed lines for clarity

Driven elements connected in parallel

Coax feed

CD1724

Fig 15.79: Construction of a commercial 'spider' for a multiband quad to allow optimum spacing on all bands

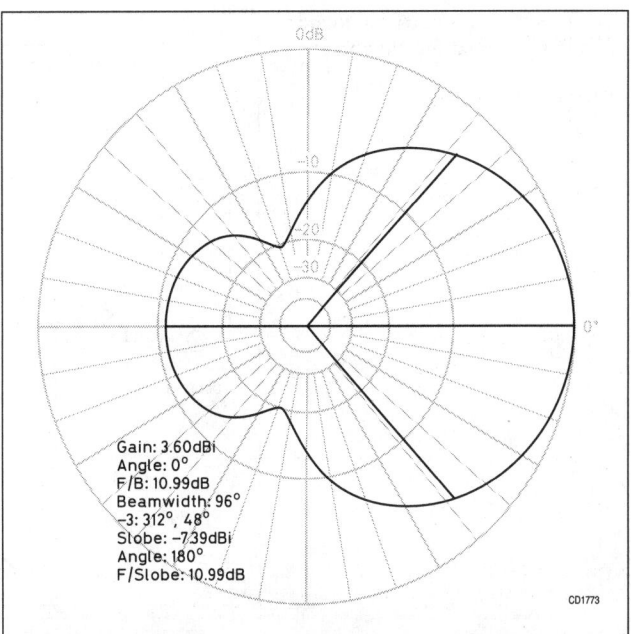

Gain: 3.60dBi
Angle: 0°
F/B: 10.99dB
Beamwidth: 96°
-3: 312°, 48°
Slobe: -7.39dBi
Angle: 180°
F/Slobe: 10.99dB

Fig 15.80: EZNEC analysis of the W1QP/W8CPC/VK2ABQ configuration

is often referred to as a 'spider'. An example of this type of structure is shown in **Fig 15.79**.

Dimensions (ER) and (ED) will have to be increased by around 5% with the boomless quad. All the driven elements can be fed in parallel, as shown in Fig 15.78, without any compromise in performance.

Methods of fixing cane or fibreglass element supports to booms are given in the Antenna construction chapter

COMPACT BEAMS WITH BENT ELEMENTS

The 10 metre 'wingspan' of a conventional Yagi for 20m can be a problem for many locations. So, can the elements be bent, as is done with a dipole when trying to fit it into a smaller space, and still retain the beam characteristics?

With antennas there is very little that is actually new. A configuration where the elements of a two-element beam were bent to halve the 'wingspan' was first suggested by John Reinartz, W1QP, in October 1937. Burton Simson, W8CPC, constructed such an antenna [24], the elements of which were supported on a wooden frame. This allowed the element ends to be folded towards each other. The 14MHz antenna was constructed from 6mm (1/4in) copper tubing with brass tuning rods that fitted snugly into the ends of the elements.

A wire edition of the W1QP/W8CPC two-element antenna was described, in 1973, by VK2ABQ [25]. In this configuration, the tips of the parasitic and driven elements support each other in the horizontal plane. The insulators are constructed so that the tips of the elements are 6mm (1/4in) apart. The gap between the tips of the elements is described as 'not critical'.

The computer model of the W1QP/W8CPC/VK2ABQ arrangement suggests that the driven element/parasitic element coupling is the same as for a wide-spaced two-element Yagi. Its performance is shown in **Fig 15.80**.

Multiband versions of this antenna were constructed without any known difficulty by nesting one antenna within the other and using a common feed for the driven element.

The G6XN antenna and the Moxon Rectangle

Les Moxon, G6XN [26], changed the structure from a square to a rectangle ,thereby reducing the centre section spacing of the elements from 0.25λ to 0.17λ, which resulted in improved gain and directivity. G6XN used loops in the elements at the element support points. This makes for a more compact antenna but increases the mechanical complexity.

C B Cebik, W4RNL [27], reduced the element spacing further to 0.14λ and obtained yet more gain and improved directivity. This antenna he called the Moxon Rectangle. The downside of this higher performance is the difficulty of making a multi-band structure due to interaction.

Tony Box, G0HAD, built a G6XN antenna whose overall size was 6.1m x 3.8m (20ft x 12ft 6in) as recommended by G6XN; this antenna out-performed his previous commercial mini beam. A computer model (EZNEC 4) was used to check these dimensions, and produced the antenna dimensions 6.92m x 3.8m (22ft 10in x 12ft 6in).

A comparison of the geometries of these antennas is shown in **Fig 15.81**. The VK2ABQ and the G6XN geometries can be multibanded.

Below are formulas to calculate dimensions for the G6XN antenna (without loops at the element support points).

Reflector 155/f = length (m)
Driven element 149.4/f = length (m)

Reflector 508/f = length (ft)
Driven element 490/f = length (ft)

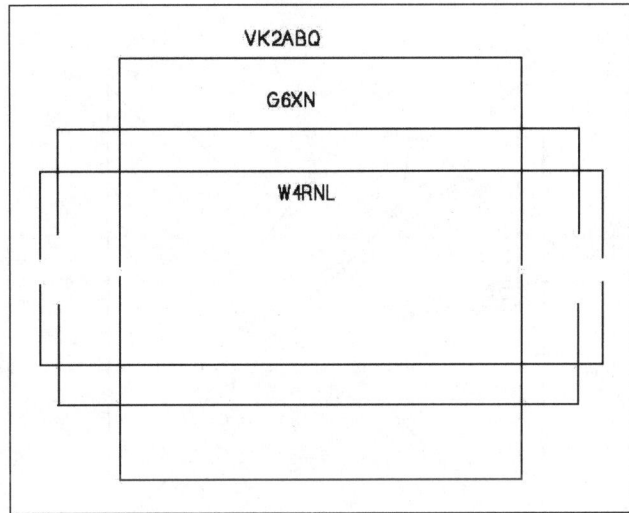

VK2ABQ
G6XN
W4RNL

Fig 15.81: The original VK2ABQ antenna structure compared with the G6XN and the W4RNL. The G6XN has a centre section spacing of the elements of around 0.17 wavelength spacing, while the W4RNL has element spacing further to 0.14 wavelength

Fig 15.82: The G6XN multiband antenna. See text for dimensions

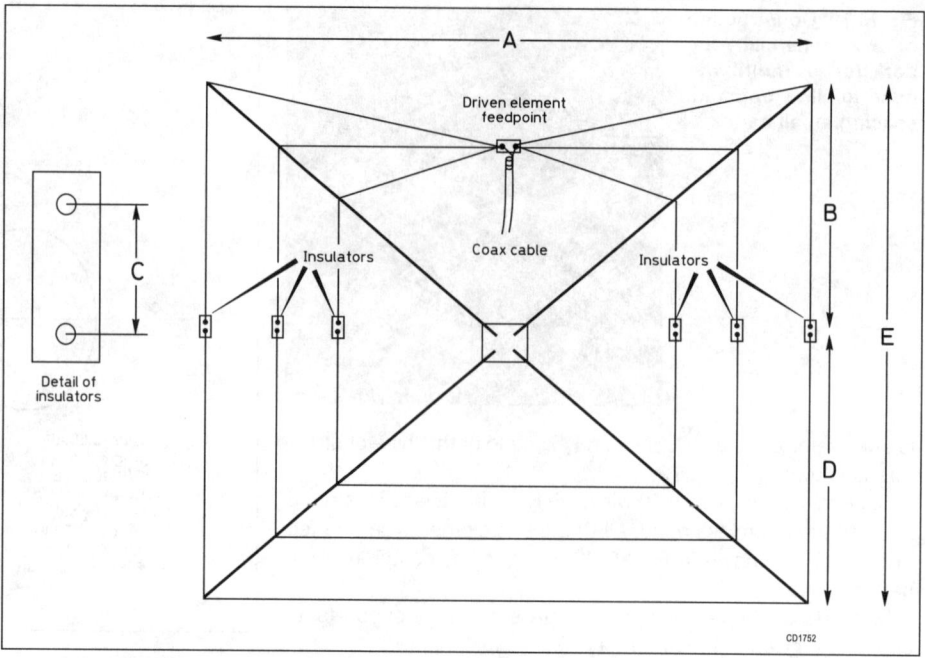

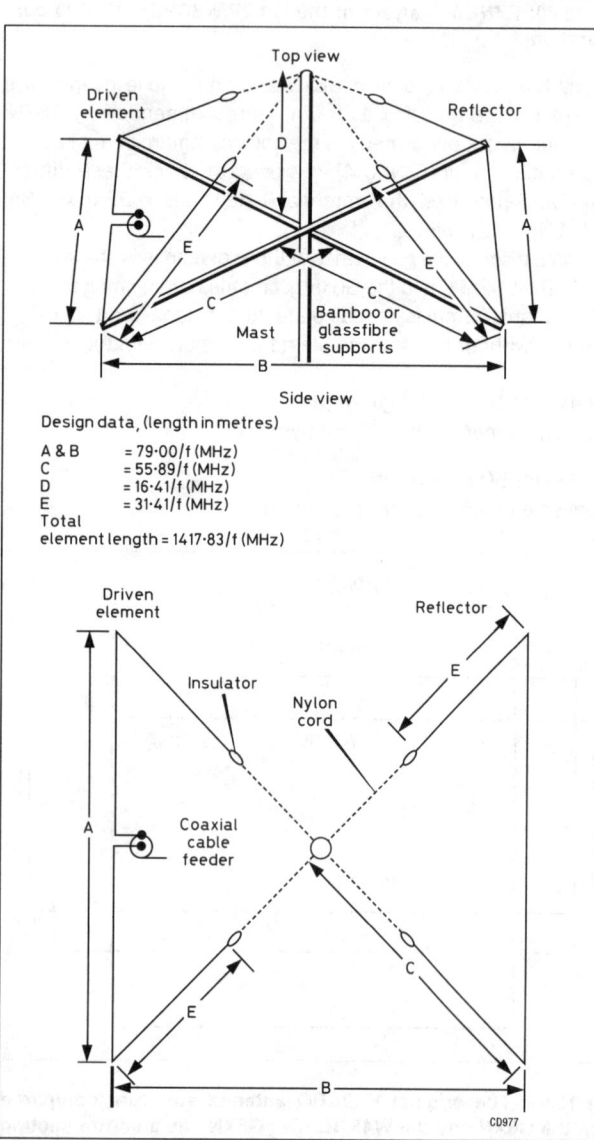

Design data, (length in metres)

A & B = 79·00/f (MHz)
C = 55·89/f (MHz)
D = 16·41/f (MHz)
E = 31·41/f (MHz)
Total
element length = 1417·83/f (MHz)

Fig 15.83: The wire Double D Antenna with approximate design data

A suggested multiband G6XN antenna is shown in **Fig 15.82**. The dimensions A and E can be found by:

A = 98.26/f = (m); 322/f = (ft)
E = 53.96/f = (m); 177/f = (ft)

C = 560mm (22in) for 14MHz, 380mm (15in) for 21MHz and 250mm (10in) for 28MHz (from experimental work by G0HAD [28]).

To work out the length of each support (cane or fibreglass rod) structure required the formula is:

56.09/f = length of diagonal in metres
184/f = diagonal length in ft.

The units of feet are indicated as a decimal number. To convert 12.5ft to feet and inches, multiply the part after the decimal by 12, eg 0.5 x 12 = 6; 12ft 6in.

The G3LDO Double D Antenna

If you want to make the bent element Yagi even smaller the ends of the elements can be folded back towards the mast in the vertical plane. This results in a pyramid configuration, and its construction was first described in [29] is shown in **Fig 15.83**. Use plastic tape to fix the wire to the canes.

This antenna will provide 3 - 4dB of gain over a dipole and a front-to-back ratio better than 14dB, which is not as good as the Moxon Rectangle but is a very compact antenna. The ends of the elements, with its 'guy' supports provide a strong lightweight structure.

Use the formula in Fig 15.83 to obtain the approximate wire lengths. In practice it is difficult to optimise the element lengths in a formula because of the geometry of the antenna. It is suggested that the ends of the elements (where they connect to the insulator) are made variable using tie wraps, as in the dipole antenna construction, shown earlier in this chapter.

Then adjust the reflector for maximum F/B and the driven element for minimum standing wave ratio. If you are in the mood to experiment you should be able to increase the gain and improve the SWR by reducing dimension B. This is achieved by altering the angles of the fixing supports relative to the mast.

The Double-D is also amenable to multibanding. A number of these antennas, for different bands, can be mounted on the same support. The simplest method of feeding turned out to be the best; paralleling the driven elements and feeding them with the one coax line as shown in Fig 15.82.

PHASED VERTICAL BEAM ANTENNAS

A phased array is a set of similar (usually identical) vertical antennas arranged in a regular geometric way and fed with a specific set of RF sources having a defined relationship to each other in terms of current magnitude and phase. Phased arrays offer a way of achieving modest gain and good reception directivity from a low profile antenna. Gains of up to 6dB, with front to back ratios of 20dB, can be achieved relative to a single vertical element with a well-designed phased vertical array.

A practical phased array system will consist of the set of radiators, earth systems, feedlines, networks to shift phase and match impedances, and a switch box. This box contains relays that switch the feedlines and allows the beam headings to be changed by changing the current distribution amongst the elements in the array. In some cases phasing and impedance matching is achieved by feeding the antennas via specific lengths of coaxial cable. Other methods use inductors or capacitors in addition to the coax feeds.

Mutual coupling between the elements in an array changes the impedances of the elements from the impedances if the elements were in isolation. These effects can be large and will change current distribution and relative phases. The performance of an array is critically dependent on errors in currents or phase relationships.

For this reason full details of phased verticals are beyond the scope of this chapter so the following are brief descriptions, with references of detailed construction and setting up. A detailed description of the phased vertical antenna was written by G3PJT [30].

One of the simplest arrays is a pair of quarter wavelength verticals, spaced at a quarter wavelength apart and fed with RF currents which are equal but 90 degrees out of phase (**Fig 15.84**). This arrangement has a gain of 3dB over a single vertical and produces a cardiod pattern, having a front-to-back ratio of 20dB with a front-to-side of about 3dB. With element one fed at 0 degrees and element two fed at 90 degrees the lobe is in line with the elements, and the arrow shows the direction of maximum radiation,

More common is the 4-element array shown in **Fig 15.85**. This arrangement uses four quarter-wave radiators, arranged in a quarter-wavelength square, the so called '4-square' array. This has a gain of up to 6dB with a front-to-back ratio of 20dB with a front-to-side of 10-15dB.

The elements are fed with equal currents in the following phase relationship.

Element 1	0 degrees
Element 2	90 degrees
Element 3,	180 degrees
Element 4	-90 degrees

The array fires diagonally across the square in the direction of element three, from element one to three as shown in Fig 15.85.

The disadvantage of the quarter wave 4-square is that it needs quite a bit of space. However there is nothing sacred about the quarter wave spacing. For example, providing that the current phase relationships are changed, satisfactory patterns can be obtained for spacings between 2.5 and 15m, corresponding to spacings of 1/16 to 3/8 wavelengths at 7MHz. The optimum phase difference lying between 160 degrees and 80 degrees respectively (for equal current amplitude).

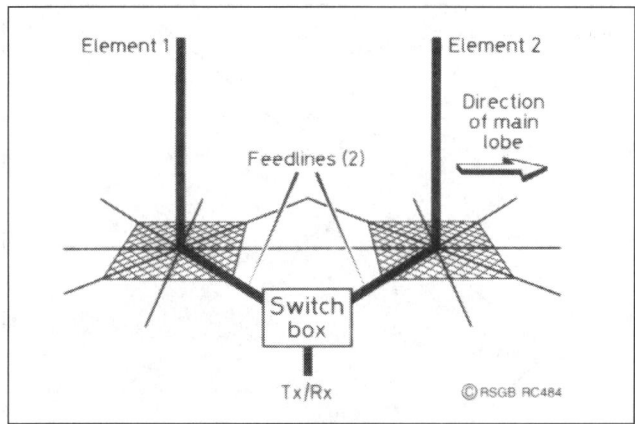

Fig 15.84 : A two-element array using quarter wavelength elements spaced quarter wavelength apart. With element 1 fed at 0 degrees and element 2 fed at 90 degrees the lobe is in line with the elements, and the arrow shows the direction of maximum radiation

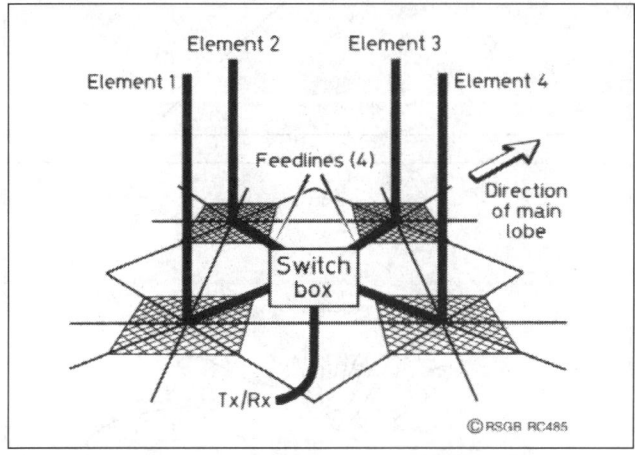

Fig 15.85: the '4-Square' array uses four quarter wave radiators, arranged in a quarter wavelength square. The elements are fed with equal currents in the following phase relationship: Element 1, 0 degrees; Element 2, 90 degrees; Element 3, 180 degrees, Element 4,-90 degrees. The array fires diagonally across the square, indicated by the arrow

G3HCT has developed an array for 7MHz using three elements, spaced only 1/8 wavelength (5.32m) apart [31]. This antenna system is capable of switching the beam in any one of six directions.

The disadvantage of closer spaced arrays is that they are more critical to set up, requiring accurate measurements of impedance. The 4-square can be set up using measurements of RF current and phase made with simple test equipment.

The performance of a phased array can more easily be achieved if all the elements in the array are identical in terms of length of element, earth system and feed impedance. Making a good earth system was described earlier in this chapter.

Further details of the phased vertical design and construction can be found at [32], [33] and [34].

FIXED LONG WIRE BEAM ANTENNAS

Generally, radio amateurs require a beam antenna that can be rotated so that it may be pointed to any position on the great circle map. Occasionally, however, a high gain fixed direction antenna can be useful for certain experimental work. An example of this was when the Narrow Band Television Association transmitted 40-line, mechanical scan, television pictures from

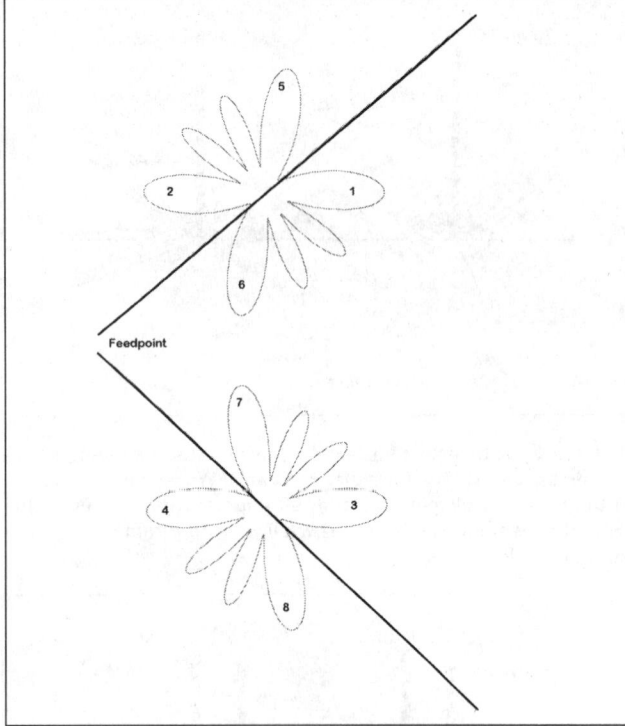

Fig 15.86: The azimuth polar diagrams of two wires, two wavelengths long, arranged in a V-beam configuration. Lobes 1, 2, 3 and 4 add in the direction of the bisector of the apex. All other lobes tend to cancel

Amberley museum in Sussex across the Atlantic to the USA in 2003. This was done to commemorate the Baird transmission on the 8th and 9th February 1928. A fairly high ERP was required to ensure the TV signals would get through so a high gain antenna was required.

The antenna used was a V-beam. This provides a combination of low cost, good gain bandwidth product, electrical and mechanical simplicity and ease of design and construction. The downside is that fixed wire beams need a great deal of space.

V-Beams

The V-beam consists of two wires made in the form of a V and fed at the apex with twin wire feeder, as shown in **Fig 15.86**. A long-wire antenna, two wavelengths long and fed at the end, has four lobes of maximum radiation at an angle of 36 degrees to the wire. If two such antennas are erected horizontally in the form of a V with an included angle of 72 degrees, and if the phasing between them is correct, lobes 1, 2, 3 and 4 add in the direction of the bisector of the apex. All other lobes tend to cancel. The result is a pronounced bi-directional beam as shown in **Fig 15.87**.

The directivity and gain of V-beams depend on the length of the legs and the angle at the apex of the V. This is likely to be the limiting factor in most amateur installations, and this is the first point to be considered in designing a V-beam. The correct angle and the gain to be expected in the most favourable direction is given in **Table 15.12**.

A practical V-beam capable of producing bi-directional high gain on the higher frequency bands (14 to 29MHz) is shown in **Fig 15.88**. V-beams are often constructed so that the apex is placed as high as possible with the ends close to the ground. This arrangement means that only one mast is required for the installation and, if space is available, several V-beams pointing in different directions could use a common support.

The V-beam so far discussed is known as resonant. The input impedance of this antenna may rise to 2000 ohms in a short V, but will be between 800 and 1000 ohms in a longer antenna. Therefore, 400 to 600ohm feed lines can be used, matched to the transceiver with a suitable balanced ATU.

The V-beam can be made unidirectional if it is terminated with resistors inserted approximately a quarter wavelength from the ends of the elements, so that final quarter-wavelengths act as artificial earths. A suitable value of resistor would be 500 ohms for each leg. Termination resistors have the effect of absorbing lobes 2, 4, 6 and 7 in Fig 15.85 and the elements become travelling wave devices.

The Narrow Band Television Association antenna was 33m high at the apex and 3m high at the ends, with the apex being supported by a tree on top of a 33m high cliff.

The Rhombic

The rhombic antenna is a V-beam with a second V added, see **Fig 15.89(a)**. The same lobe addition principle is used but there is an additional complication because the lobes from the front and rear halves must also add in phase at the required elevation angle. This introduces an extra degree of control in the design so that considerable variation of pattern can be obtained by choosing various apex angles and heights above ground.

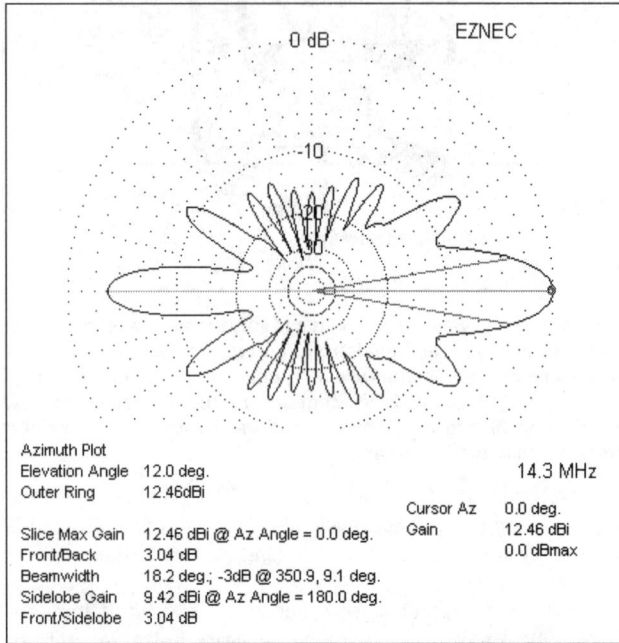

Azimuth Plot
Elevation Angle   12.0 deg.   14.3 MHz
Outer Ring   12.46dBi

                                       Cursor Az   0.0 deg.
Slice Max Gain   12.46 dBi @ Az Angle = 0.0 deg.   Gain   12.46 dBi
Front/Back   3.04 dB   0.0 dBmax
Beamwidth   18.2 deg.; -3dB @ 350.9, 9.1 deg.
Sidelobe Gain   9.42 dBi @ Az Angle = 180.0 deg.
Front/Sidelobe   3.04 dB

Fig 15.87: Prediction of the performance of the V-beam used by the Narrow Band Television Association (see text) using EZNEC4. The antenna is bi-directional because it is not terminated

Leg length Wavelength	Gain dBd	Apex Angle Degrees
1	3	108
2	4.5	70
3	5.5	57
4	6.5	47
5	7.5	43
6	8.5	37
7	9.3	34
8	10	32

Table 15.12: V-beam gain and apex angles for given lengths of element

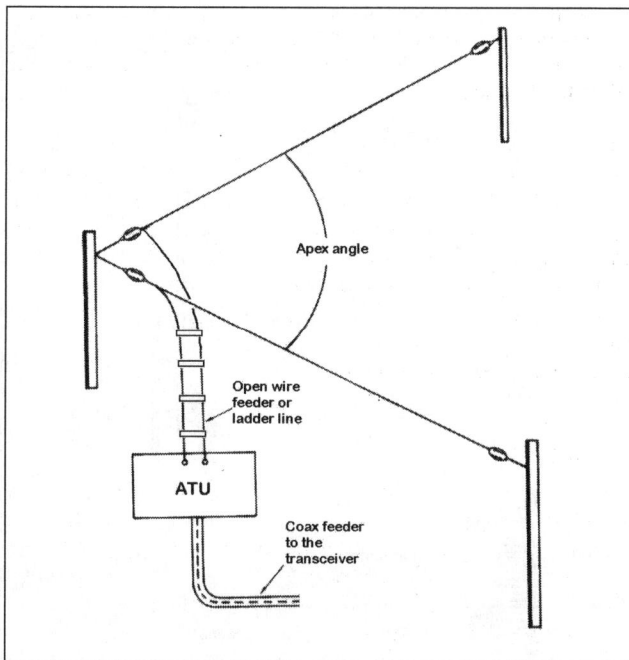

Fig 15.89: A practical resonant V-beam for the HF bands 14 to 29MHz. Ideally a balanced line ATU should be used to ensure an equal current level in each element

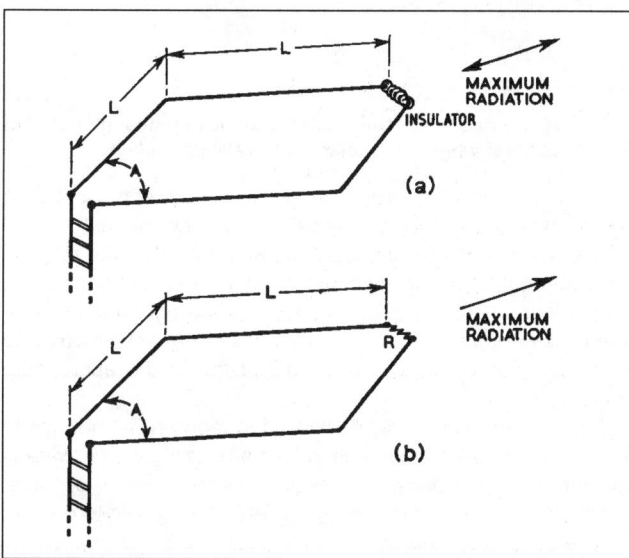

Fig 15.90: The Rhombic antenna. (a) shows the resonant unterminated version, (b) the terminated version

The rhombic gives an increased gain but takes up a lot of room and requires at least one extra support. As with the V-beam, the resonant rhombic has a bidirectional pattern, and the terminated rhombic, shown in **Fig 15.90(b)**, an unidirectional one. The terminating resistance absorbs noise and interference coming from the rear direction as well as transmitter power, which would otherwise be radiated backwards. This means that it improves signal-to-noise ratio by up to 3dB without affecting signals transmitted in the wanted direction.

The use of tuned feeders enables the rhombic, like the V-beam, to be used on several amateur bands. The non-resonant rhombic differs from the resonant type in being terminated at the far end by a non-inductive resistor comparable in value with the characteristic impedance, the optimum value being influenced by energy loss through radiation as the wave travels out-

wards. An average termination will have a value of approximately 800 ohms. It is essential that the terminating resistor be as near a pure resistance as possible, ie without inductance or capacitance - this rules out the use of wire-wound resistors. The power rating of the terminating resistor should not be less than one-third of the mean power input to the antenna. For medium powers, suitable loads can be assembled from series or parallel combinations of say, 5 ohm carbon resistors. The terminating resistor may be mounted at the extreme ends of the rhombic at the top of the supporting mast. Alternatively the resistor may be located near ground level and connected to the extreme ends of the rhombic via twin wire feeder.

The impedance at the feed point of a terminated rhombic is 700-800 ohms and a suitable feeder to match this can be made up of 16SWG wire spaced 300mm (12in) apart. The design of rhombic antennas can be based on Table 15.12, considering them to be two V-beams joined at the free ends.

The design of V and rhombic antennas is quite flexible and both types will work over a 2:1 frequency range or even more, provided the legs are at least two wavelengths at the lowest frequency. For such wide-band use the angle is chosen to suit the element length at the mid-range frequency.

Generally the beamwidth and wave angle increase at the lower frequency and decrease at the upper frequency, even though the apex angle is not quite optimum over the whole range. In general, leg lengths exceeding 10 wavelengths are impractical because the beam is then too narrow.

Advantages of the rhombic over a V-beam are that it gives about 1-2dB greater gain for the same total wire length and its directional pattern is less dependent on frequency. It also requires less space and is easier to terminate. The disadvantage is that it requires four masts.

MOBILE ANTENNAS

The antenna is the key to successful mobile operation. Because of shape of the vehicle, space limitations and the slipstream caused by the vehicle motion the vertical whip antenna is the most popular mobile antenna, regardless of the band in use. The easiest way to feed such an antenna is to make it a quarter wavelength long at the frequency in use. The resonant quarter-wavelength is a function of frequency and is 1.48m (58.5in) on 50MHz and 2.5m (8ft 2in) on 28.4MHz and progressively shorter on the higher VHF bands. Quarter wave antennas on the 28MHz bands and higher are quite practical, but on the lower HF bands it is a different matter. Even on 21MHz a quarter wavelength is 3.45m (11ft 2in) and on 14MHz is 4.99m (16ft 4in).

It follows that a practical antenna for the HF bands will be shorter than a quarter wave long. For a given antenna length, as the frequency of operation is lowered the feedpoint exhibits a decreasing resistance in series with an increasing capacitive reactance. In order to feed power to such an antenna it must be brought to resonance so that the feed point is resistive. This is achieved by adding some inductance, and is known as inductive loading.

A loading coil for a mobile antenna must be rugged to stand up to weather and the mechanical strain of a fast moving vehicle slip stream. The following represents a suitable solution.

The G3MPO Coil and Antenna

This design uses a single antenna structure with a different coil for each band. This coil was produced using workshop facilities little more than a Workmate, electric drill, and taper taps and dies. According to G3MPO, in several thousand miles of motoring this design has proved completely secure. Full use was made of a local plumber's stockist's supply of ready-made brass, stainless steel and plastic bits and pieces. 15mm plumber's brass

compression couplers were selected as both coil terminations and the means of fixing them into the antenna. The construction of the antenna and coils is shown in **Figs 15.90 and 15.91**.

Because the fittings and the antenna material are an integral part of the design these are described as well as the coil. The bottom section of the antenna is made from a length of 15mm stainless steel central heating tubing.

White polypropylene waste pipe proved a good choice for a coil former. The thread of the brass 15mm compression couplers can be screwed (with some difficulty) into the end of the 20mm (0.75in) version of this tubing to make a very strong joint. The ends of the tube can be pre-heated in hot water if necessary. Even better, a 12.5mm (0.5in) BSP taper tap can be used to cut a starting thread in the tubing. A second coupler screwed into the other end gives an excellent coil former with ready-made 15mm connections at each end, which fit and clamp directly onto the 15mm diameter lower mast section.

Varying lengths of former are used for the higher frequency coils, and where a greater diameter is needed for the lower frequencies, the same 20mm (0.75in) former is used as a spine. This runs up the middle of a larger diameter tube to which it is attached by packing the space between them at each end with postage stamp size pieces of car-repair glass mat soaked in resin. The coil assembly can then be waterproofed with a silicone rubber sealant.

The whip above the coil former comprises a short length of small diameter tube, fixed to the top of the coil with a 15mm coil coupler. A length of stainless steel whip is slid inside this tubing. Suitable tubing can be obtained from a car accessories shop in the form of copper brake tubing, which comes in outside diameter sizes from 8mm down. A nice sliding fit to the 2.5mm diameter whip was provided by a 230mm (9in) length of 8mm diameter pipe, with one end plugged by short lengths of the next two sizes down and soldered into position.

Some sort of quick release lock is required to hold the whip in position once its length is set. This is achieved by cutting a thread on the last 12mm of the end piece of tubing with a thread-cutting die and making cross cuts down its length with a mini-hacksaw.

A matching nut with a tapered thread can be made from a short length of brass rod, drilled and taper-tapped so that it would close the tube down on to the end whip as it is screwed on; thus locking it. The whip structure was completed by connecting the telescopic section onto the coil using a 15mm-to-8mm (microbore) brass reducer and a 51mm (2in) length of 15mm tubing, as shown in Fig 15..90.

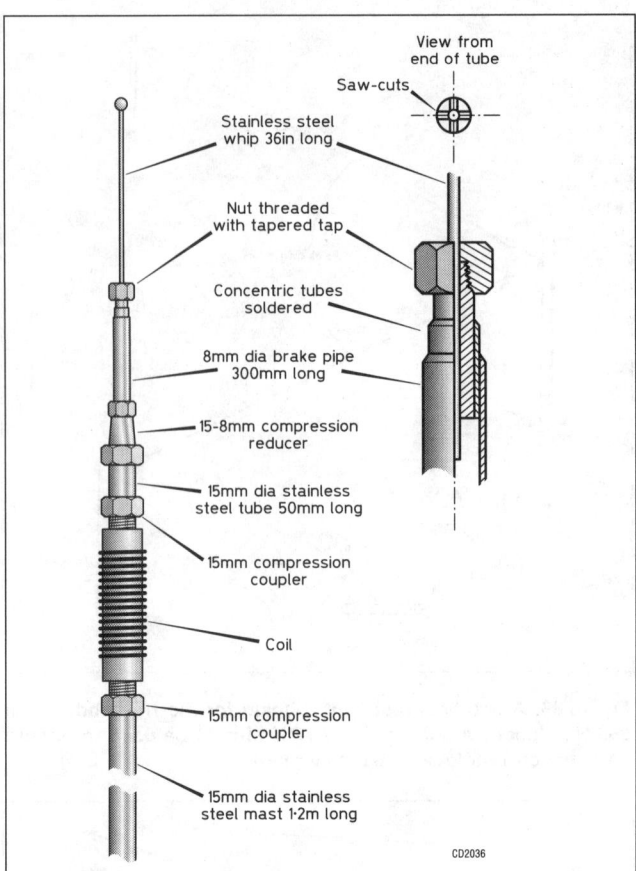

Fig 15.90: (a) Section of the G3MPO mobile antenna. (b) detail of stainless steel whip / 8mm diameter brake pipe clamp

The best method of attaching the wire to the end couplers is to drill two small holes through the polypropylene just beyond where the end of the coil would lie, and pass the wire into the tube and out through the coupler to which it is then connected. It was found best to solder a hairpin of wire onto the inside of the coupler before fitting it into the plastic former. The coil wire was then easily soldered onto this pigtail at the appropriate time.

Two or three lengths of double-sided tape are then stuck to the coil former. A sufficient length of enamelled copper wire is cut for the coil in question. Seven times the number of turns, times the diameter of former (from **Table 15.13**) allows enough

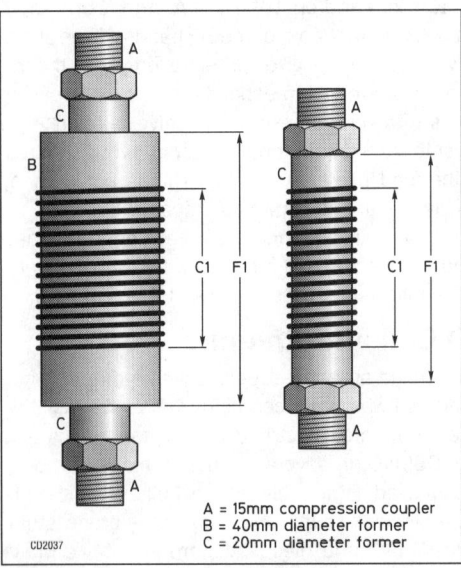

A = 15mm compression coupler
B = 40mm diameter former
C = 20mm diameter former

CD2037

Fig 15.91: Coil former construction and dimensions of G3MPO antenna. F1 is the coil former length and C1 the coil winding length, see Table 15.13

COIL DATA								
F	D	Fl	Wire	N	Cl	L	Rr	Rf
MHz	mm	mm	SWG		mm	µH	ohms	ohms
29.0	20	77	18	9	28	0.9	35	48
24.9	20	90	18	15	44	1.7	29	48
21.2	20	115	18	23	64	3.0	22	47
18.1	20	140	18	34	92	4.5	17	43
14.25	20	146	20	45	96	8.4	11	34
10.13	40	115	20	31	72	19	6	26
7.05	40	165	20	58	117	41	3	20
3.65	40	305	22	130	157	153	0.8	21
1.9	40	280	28	294	236?	558	0.2	37

Table 15.13: Read in conjunction with Fig 15.92. F = frequency (MHz), D = coil former diameter (mm), Fl = Length of coil former tube (mm), N = number of turns, Cl = Length of coil winding (mm), L = coil inductance (microhenries), Rr = Theoretical radiation resistance (ohms)

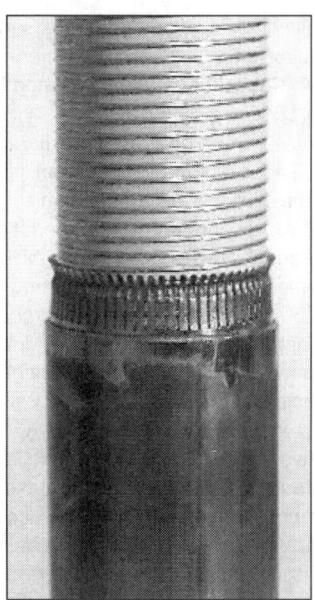

Fig 15.92: Detail of the tuning section of the 'screwdriver' antenna, which shows the coil and fingers that short the turns as it emerges. This photo, courtesy of Waters and Stanton, is of the WBB-3 derivative, see text

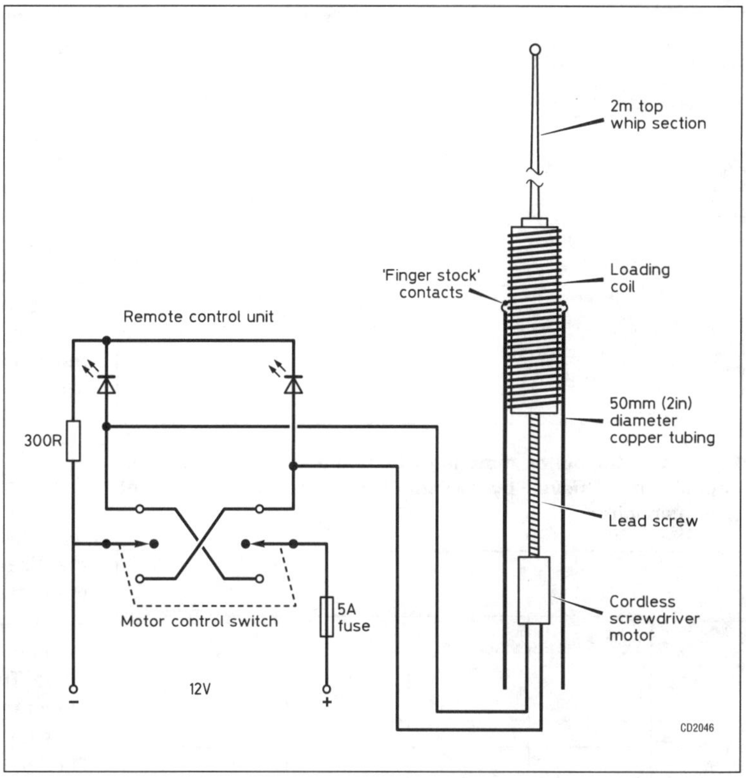

Fig 15.93: W6AAQ's DK3 mobile antenna (not to scale). The control box is located by the driver and power obtained from either the rig supply or the cigar lighter socket. The original had relay switched capacitors, selected from the drivers control box, to match the antenna on the lower frequencies

to wind the coil with some to spare. The wire spacing is achieved by winding two lengths of wire onto the former side by side and subsequently removing one of them. The double-sided tape fixed to the former holds the remaining winding in position. 10 to 12mm of wire is wound beyond the holes through which the wire endings were taken and, after removing the spacing wire, the winding is coated with polyurethane varnish. When dry, the coil is wound back at each end until the required number of turns are obtained; the ends fed through into the former, out through the end couplings and soldered to the coupling hairpins. The two small holes in the former can be sealed with varnish or mastic and the winding bound with a double layer of self-amalgamating tape.

Soldered connections are pushed well down into the coupling, out of the way, and the coil given two coats of polyurethane varnish. The self amalgamating tape can be omitted if you prefer the appearance of varnished copper coils, but it is easy to use and provides additional protection against knocks and bangs.

The W6AAQ Continuous Coverage HF Mobile Antenna

In this design the antenna resonance is adjusted from the drivers/operator's position using a cordless screwdriver electric motor.

This motor rotates a brass leadscrew via a nut fixed to the coil to cause the coil to move up or down inside in 1m (3ft) long 50mm diameter aluminium, brass or copper tube as shown in Fig 15.92. As the motor is rotated the coil is raised or lowered so that more or less of the coil is contained within the lower tube section. Finger stock connectors are used to short the coil to the tube to obtain the appropriate resonance. A circuit of the antenna and the control box is shown in Fig 15.93.

The antenna is tuned to resonance, first by listening for an increase in receiver noise, then applying transmit power and fine-tuning for the lowest SWR.

Frequency (MHz)	1.8	3.6	7.05	10	14.2	18	21.3	25	28.5
R_{rad}	0.2	0.8	3	6	12	17	21	28	36

Table 15.14: Radiation resistance (R_{rad}) of a typical mobile antenna

Matching a Mobile Antenna to the Feeder

All the antennas so far discussed are fed with 50-ohm coaxial cable; normally the centre is connected to the antenna and the braid to the vehicle body. However the radiation resistance of the antenna will generally be lower than 50 ohms and, for a given antenna size, it depends on frequency. Typical radiation resistance figures for a 2.4m (8ft) antenna are shown in **Table 15.14**:

In practice, the feed impedance will include the RF resistance of the loading coil and the resistance losses. The loss resistance, taken in total, is usually much greater than the radiation resistance, at the lower operating frequencies.

For example, the radiation resistance of an 80m antenna is around one ohm and the loading coil resistance may be around 10 ohms. The ground loss will be between 4 and 12 ohms, depending on the size of the vehicle, so the feed impedance could be in the region of 20 ohms.

This will give an SWR of 2.5:1 at resonance, which gets progressively worse very quickly as the transceiver is tuned off the antenna resonance, clearly beyond the impedance range of a modern solid state transceiver 50 ohm PA (unless it has a built-in ATU).

At the other end of the HF spectrum the radiation resistance is much higher and even though the coil losses are lower, a transceiver can be connected directly to the antenna via a length of 50-ohm coax.

There are several ways of matching the nominal 50-ohm transceiver output to the impedance encountered at the base of a resonant mobile antenna. Of these the most common are:

Figure labels (Fig 15.93):
- 2m top whip section
- 'Finger stock' contacts
- Loading coil
- 50mm (2in) diameter copper tubing
- Lead screw
- Cordless screwdriver motor
- Remote control unit
- 300R
- Motor control switch
- 5A fuse
- 12V
- CD2046

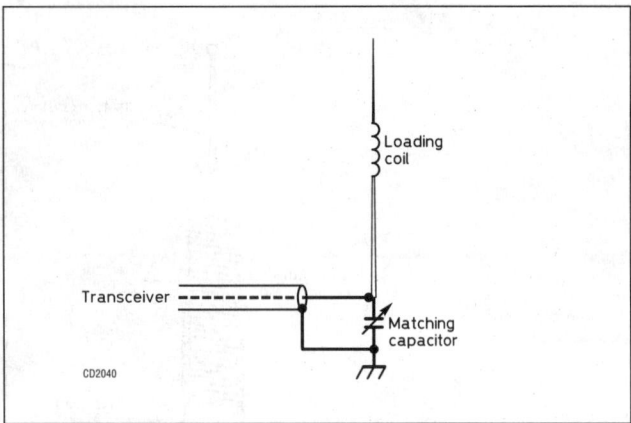

Fig 15.94: Capacitor matching. In practice the variation in capacity is achieved by switching in appropriate values of fixed capacitor

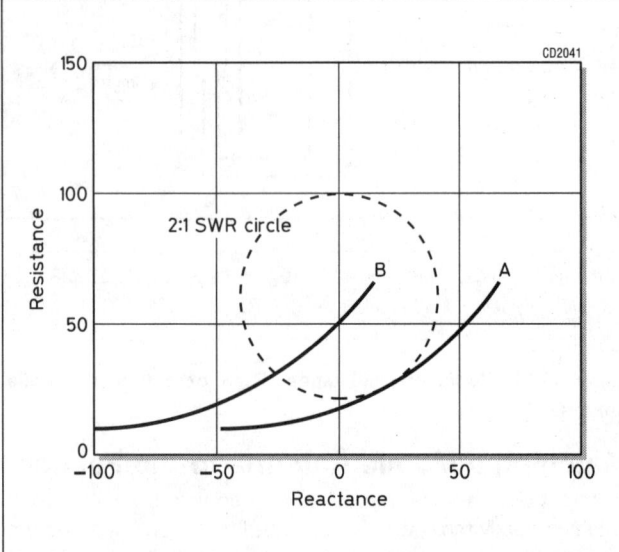

Fig 15.95: Curve A; feed impedance of an 80 metre mobile antenna in the frequency range 3.55 to 3.65MHz. On no part of the curve is the standing wave ratio better that 2:1. An improved match is achieved by increasing the inductance of the loading coil slightly, and compensating with a capacitor across the feedpoint, thereby moving the curve to B

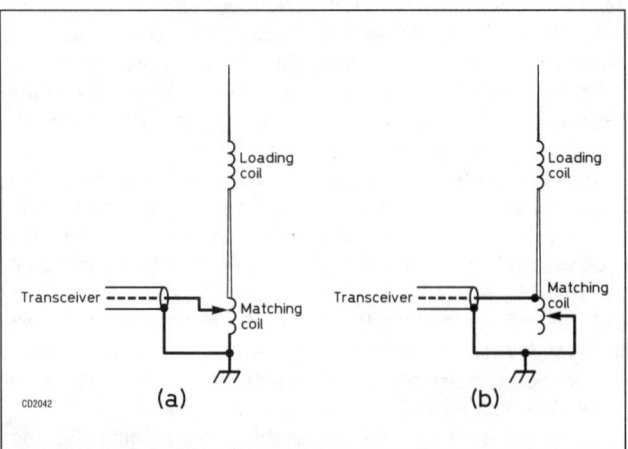

Fig 15.96: Two methods of using a tapped inductance for matching at the base of the antenna

1. Capacitive shunt feeding. This is simply the addition of a shunt capacitor directly across the antenna feedpoint as shown in **Fig 15.94**. Capacitor values calculated by G3MPO are shown in **Table 15.15**:

Exact values can be determined experimentally and will need to be switched for multiband operation.

The way that this works can be seen by referring to **Fig 15.95**. The curve A represents the feed impedance of a Pro Am antenna in the frequency range 3.55 to 3.65MHz, measured using the 3-m impedance box [7]. At the lower frequency the impedance is about R10-50jX, while at the higher frequency it is R70+70jX. On no part of the curve is the SWR better that 2:1. By increasing the inductance of the loading coil slightly and compensating with a capacitor across the feedpoint, the curve can be shifted to B to achieve an improved match

2. Inductive shunt feeding. This is achieved with the addition of a small tapped inductance at the base of the antenna. With the loading coil adjusted to take into account the effect of the base coil the antenna base impedance is raised in proportion to the size of the base inductance. There are two ways that this method of feeding can be implemented.

Selecting an inductor that results in a value greater than 50 ohms at the lowest point of the antenna impedance/frequency curve. The transceiver connection is then tapped down the base inductance to obtain the best match as shown in **Fig 15.96(a)**.

Using a variable inductance across the feedpoint as shown in **Fig 15.96(b)**. The inductance value or tapping point must be changed when the frequency band is changed.

3. Transformer matching. This arrangement uses a conventional RF transformer wound on a toroid core as shown in **Fig 15.97**. A commercial or home-brew matching transformer can be used. The one described is designed by G3TSO and uses toroid cores such as the Amidon T 157-2. It is wound with 20 bifilar turns using 18SWG (1 2mm) wire. Both windings are connected in series, in phase, and the second winding is tapped every other turn as shown in **Fig 15.98**. With the loading coil adjusted to take into account the inductance of the transformer windings, antenna impedances from 50 ohms down to 12 ohms can be matched.

The matching arrangements shown can be located part way between the transceiver and the antenna. As stated earlier, the

F, MHz	29	24.9	21.2	18.1	14.25	10.13	7.05	3.65	1.9
pF	18	27	37	74	150	300	544	1000	1000

Table 15.15: Values for capacitive shunt feeding of a mobile antenna

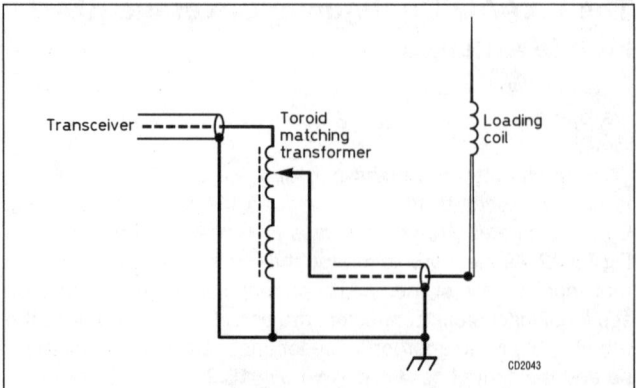

Fig 15.97: Matching arrangement for a mobile antenna using a variable ratio RF transformer

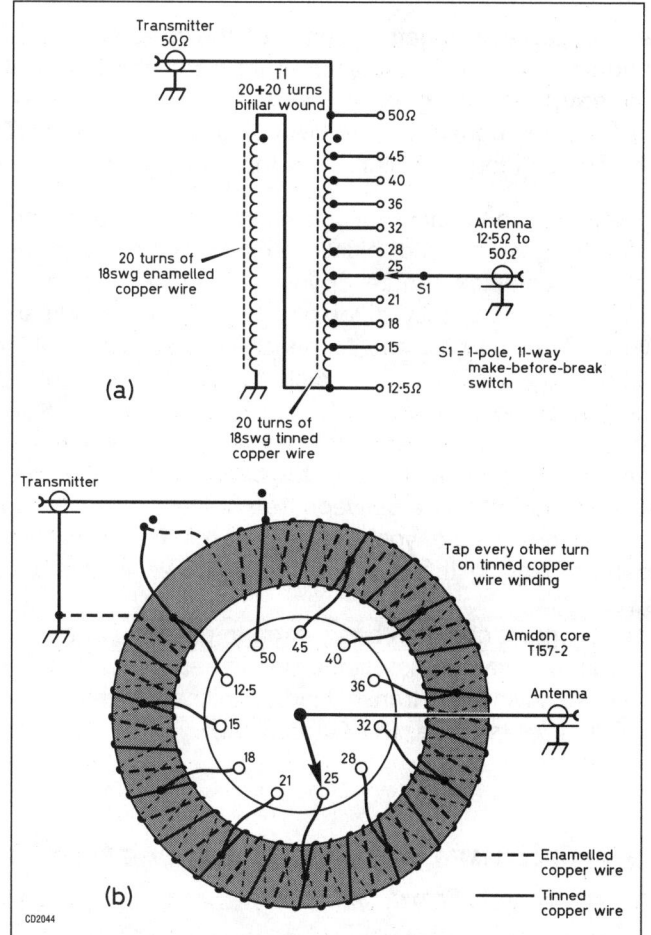

Fig 15.98: Details of the G3TSO RF transformer. (a) The circuit and impedance matching range. (b) Constructional details

feed impedance comprises the radiation resistance, earth resistance and coil resistance in series, which means that, in practice, the transceiver can be connected directly to the antenna without any matching at frequencies 14MHz and above. On the lower frequency bands the coax feeder is so electrically short in a mobile installation that losses caused by a higher SWR are minimal. If the matching arrangement can be adjusted from the driver's seat then this represents a greater degree of operator convenience that if it were located at the base of the antenna.

Further Mobile Antenna Information

Details of mobile antenna construction, mounting and matching can be found in *The Amateur Radio Mobile Handbook*, RSGB [37].

REFERENCES

[1]   'Eurotek', *Radcom*, Sep 1993
[2]   *Practical Wire Antennas*, John D Heys, G3BDQ, RSGB
[3]   'The G5RV Multiband Antenna', *Radcom*, Jul 1984
[4]   'In Practice', *Radcom*, Aug 1999
[5]   *Electron* (December 1992) and reported in 'Technical Topics', *RadCom*, May 1993
[6]   'Taming the End-Fed Antenna', Alan Chester, G3CCB, *Radcom* Sep 1994
[7]   'PicATUne - the Intelligent ATU', Peter Rhodes, G3XJP, *RadCom*, Sep 2001 - Jan 2002
[8]   'Save your tuner for two pence', Tony Preedy, G3LNP, *Radcom*, May 2000
[9]   'A General-Purpose Antenna Tuning Unit', M J Grierson, G3TSO, *RadCom*, Jan 1987
[10]   'Link Coupled Antenna Tuners: A Tutorial', L B Cebik, W4RNL, http//www.cebik.com/
[11]   'A balanced Line ASTU', Ted Garrott, G0LMJ, *Radcom* Jul/Aug 1998
[12]   'To Turrets - Just Tune', King, W1CJL, *QST*, Mar 1948
[13]   'The Z-match Antenna Coupler', King, W1CJL, *QST*, May 1955
[14]   An improved Z-match ATU, Louis Varney CEng MIEE AIL, G5RV, *Radcom* Oct 1998
[15]   'Loop Antennas, Fact not Fiction', A J Henk, G4XVF, *Radcom* Sep/Oct 1991
[16]   'Electrically Tunable Loop', Roberto Craighero, I1ARZ, *Radcom* Feb 1989
[17]   'Small high-efficiency antennas - alias the loop', 100-page booklet published by Ted Hart, W5QJR
[18]   DK5CZ's comments on the use of vacuum cap, 'Eurotek', *Radcom* Oct 1991,p39
[19]   'Experimental Magnetic Loop Antenna' C R Reynolds, GW3JPT, *Radcom* Feb, 1994
[20]   'A Magnetic loop for 80m mobile', Loek d'Hunt, PA2JBC, 'Eurotek', *Radcom* May 1993
[21]   'Skeleton Slot Aerials', B Sykes, *RSGB Bulletin* (forerunner of the RSGBs Radcom) Jan 1953
[22]   'The HF Skeleton Slot Antenna', Bill Capstick, *Radcom* Jun 1996
[23]   *Beam Antenna Handbook*, Bill Orr, W6SAI, Radio Publications Inc.
[24]   'Concentrated Directional Antennas for Transmission and Reception', John Reinartz, W1QP and Burton Simson W8CPC, *QST* Oct 1937
[25]   'VK2ABQ Antenna', Fred Caton VK2ABQ, *Electronics Australia*, Oct 1973,
[26]   *HF Antennas for all Locations*, 2nd Edn, Les Moxon, G6XN, RSGB
[27]   L B Cebik, W4RNL, http//www.cebik.com/
[28]   'Antennas', *Radcom*, Mar 2002
[29]   'Wire Beam Antennas and the Evolution of the Double-D', Peter Dodd, G3LDO, *QST* Oct 1984, also *Radio Communication*, Jun/Jul 1980 (RSGB)
[30]   'Phased Vertical LF Band Antennas', Bob Whelan, G3PJT, *RadCom*, May/Jun, 1995
[31]   'Phased Arrays for 7MHz', John Bazley, G3HCT, *RadCom*, Mar/Apr 1998
[32]   *Low Band DXing*, John Devoldere, ON4UN
[33]   *ARRL Antenna Book* (1988). pages 8-8 to 8-31, Lewallen, W7EL, (also described in later editions of the *ARRL Antenna Book*
[34]   'Vertical Phased Arrays', Gehrke. *Ham Radio*, May, Jun, Jul, Oct, Dec 1983
[35]   'An All-Band Antenna for Mobile or Home', John Robinson, G3MPO. *Radcom*, Dec 1992 and Jan 1993. Also reproduced in [36]
[36]   *Backyard Antennas*. Peter Dodd, G3LDO
[37]   *The Amateur Radio Mobile Handbook*, Peter Dodd, G3LDO

International Antenna Collection 2
Edited by George Brown, M5ACN

This book collects together some of the best articles from around the world on the subject of antennas. It will appeal to radio amateurs in general, whether they be antenna enthusiasts or not. It is a follow-up to the successful The International Antenna Collection, compiled by the same editor.

You will find antennas for most of the amateur bands. Traditional designs and highly original designs are here, simple and complex. Whatever your requirement, you will find something that is directly suited or that sets you thinking about how to solve your problem. Amongst the practical and highly erudite contents is an invited article by one of America's most respected authors on the subject of aerials. He is Kurt N Sterba, the regular 'Aerials' columnist of WorldRadio magazine. He considers one of his pet subjects - the much - misunderstood interface between transceiver and aerial. Does your ATU really tune your aerial? All the facts are clearly presented, leaving the author in no doubt as to the correct answer.

As before, great care has been taken to ensure that there are antennas to cover almost all the bands between 136kHz and 2.4GHz, receiving and transmitting, mobile and fixed.
RSGB Members Price: £11.04 plus p&p

ONLY £12.99 plus p&p

International Antenna Collection
Edited by George Brown, M5ACN

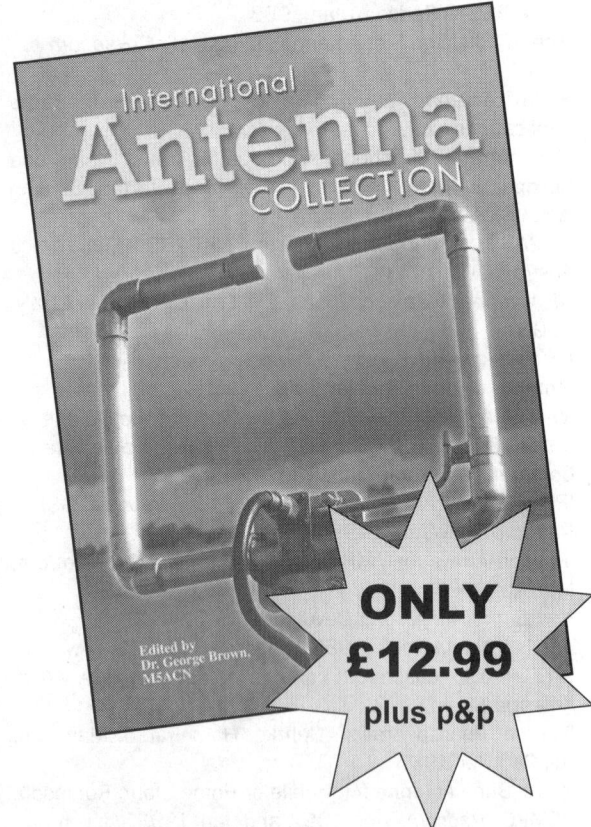

This book is a collection of over 50 of the very best articles published on antennas from around the world. The book is wide ranging and offers solutions to many problems experienced by the antenna enthusiast. Amongst the articles are antenna designs for most amateur bands. Stealthy and invisible antennas are covered alongside many interesting traditional designs. The book also benefits from two articles specially commissioned for inclusion here. The first, by Professor Mike Underhill, G3LHZ, of the University of Surrey at Guildford, UK, entitled 'The Truth About Loops', gives an exhaustive account of the performance of the much-maligned small loop, which also takes into account the loop's situation. approaching the same problem from the computer modelling angle is the subject of the second invited article, 'A Brief Overview of the Performance of Wire Aerials in their Operating Environments', from Jack Belrose, VE2CV. Great care has been taken to ensure that there are antennas to cover the range from 136kHz to 1.3GHz, receiving and transmitting, fixed and mobile. Everyone interested in antenna design and construction will find something in this book.
Size: 272 x 200mm, 256 pages, ISBN: 1-872309-93-3
RSGB Members Price: £11.04 plus p&p

ONLY £12.99 plus p&p

E&OE

Radio Society of Great Britain
Lambda House, Cranborne Road, Potters Bar, Herts. EN6 3JE
Tel. 0870 904 7373 Fax. 0870 904 7374

ORDER 24 HOURS A DAY ON OUR WEBSITE
www.rsgb.org/shop

16 Practical VHF/UHF Antennas

VHF and UHF antennas differ from their HF counterparts in that the diameter of their elements are relatively thick in relationship to their length and the operating wavelength, and transmission line feeding and matching arrangements are used in place of lumped elements and ATUs.

THE (VHF) DIPOLE ANTENNA

At VHF and UHF, most antenna systems are derived from the dipole or its complement, the slot antenna. Many antennas are based on half-wave dipoles fabricated from wire or tubing. The feed point is usually placed at the centre of the dipole, for although this is not absolutely necessary, it can help prevent asymmetry in the presence of other conducting structures.

The input impedance is a function of both the dipole length and diameter. A radiator measuring exactly one half wavelength from end to end will be resonant (ie will present a purely resistive imped-ance) at a frequency somewhat lower than would be expected from its dimensions. Curves of 'end correction' such as **Fig 16.1** show by how much a dipole should be shortened from the expected half wavelength to be resonant at the desired frequency.

The change of reactance close to half-wavelength resonance as a function of the dipole diameter is shown in **Fig 16.2**.

In its simplest form, dipole antennas for 2m and 70cm can be constructed from 2mm diameter enamelled copper wire and fed directly by a coaxial cable as shown in **Fig 16.3**. The total ele-ment length (tip to tip) should be 992mm for 145MHz operation and 326mm to cover the band 432 to 438MHz. The impedance will be around 70 ohms for most installations, so that a 50-ohm coaxial cable would present a VSWR of around 1.4:1 at the transceiver end.

A more robust construction can be achieved using tubing for the elements and moulded dipole centre boxes, available from a number of amateur radio antenna manufacturers and at radio rallies. The dipole length should be shortened in accordance with Fig 16.1 to compensate for the larger element diameters. Construction ideas and UK sources of materials can be found at [1].

Note that this simple feed may result in currents on the out-side of the cable, and consequently a potential to cause inter-ference to other electronic equipment when the antenna is used for transmitting. This can be reduced or eliminated by using a balun at the feed point.

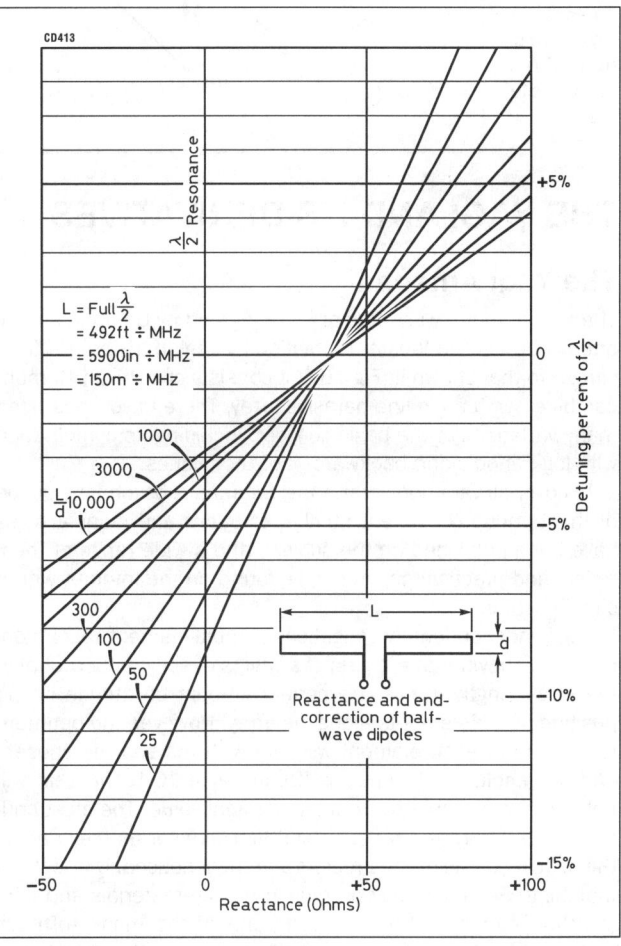

Fig 16.2: Tuning and reactance chart for half-wave dipoles as a function of diameter

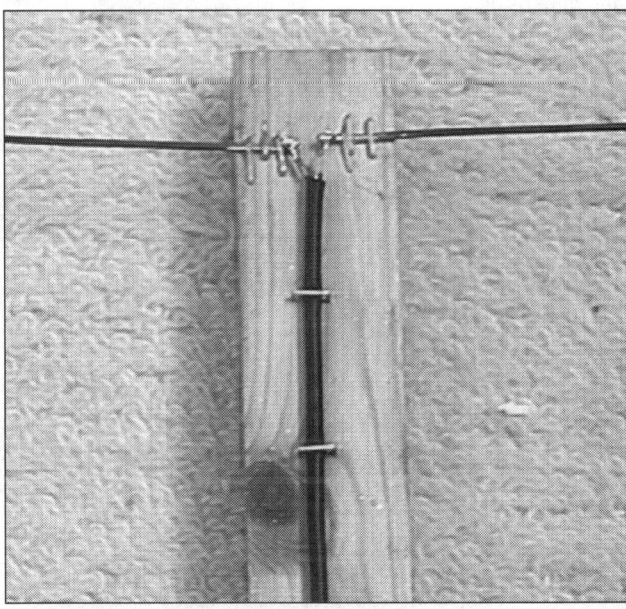

Fig 16.3: Simple dipole construction for 2m and 70cm

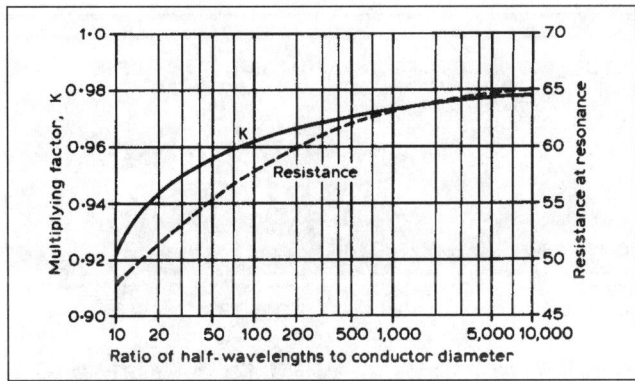

Fig 16.1: Length correction factor for half-wave dipole as a func-tion of diameter

Fig 16.4: Simple Yagi antenna structure, using two directors and one reflector in conjunction with a driven element

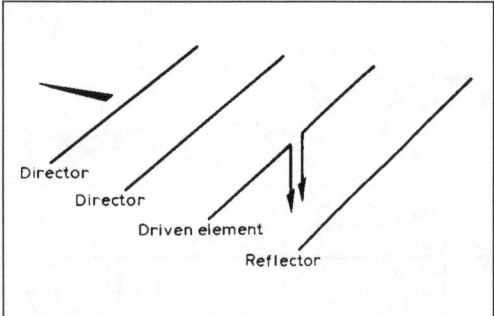

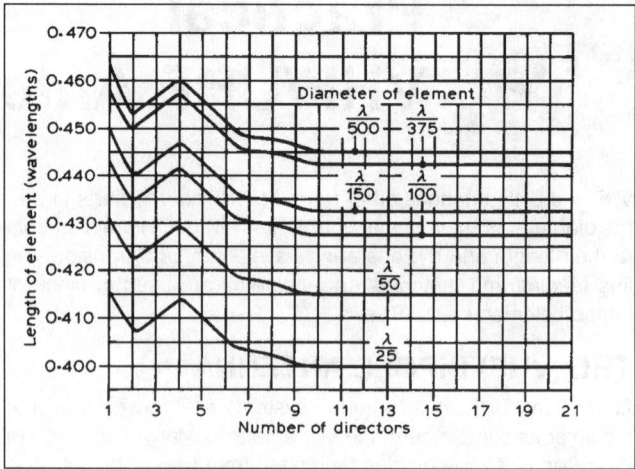

Fig 16.5: Length of director position in the array for various element thicknesses (*ARRL Antenna Book*)

THE YAGI AND ITS DERIVATIVES

The Yagi Antenna

The Yagi antenna was originally investigated by Uda and subsequently brought to Western attention by Yagi in 1928 in a form similar to that shown in **Fig 16.4**. It consists of a driven element combined with an in-line parasitic array. There have since been many variations of the basic concept, including its combination with log-periodic and backward-wave techniques.

To cover all variations of the Yagi antenna is beyond the scope of this handbook. A great number of books and many articles have been published on the subject, and a wide range of theoretical and practical pages can be found on the Internet with a simple search.

Many independent investigations of multi-element Yagi antennas have shown that the gain of a Yagi is directly proportional to the array length. There is a certain amount of latitude in the position of the elements along the array. However, the optimum resonance of each element will vary with the spacing chosen. With Greenblum's dimensions [2], in **Table 16.1**, the gain will not vary more than 1dB from the nominal value. The most critical elements are the reflector and first director as they decide the spacing for all other directors and most noticeably affect the matching. Solutions may be refined for the materials and construction methods available using one of the many software tools now freely available from the Internet, and discussed elsewhere in this handbook. These tools can be used to assess the sensitivity of a given design to alternative diameter elements and dimensions.

The optimum director lengths are normally greater the closer the particular director is to the driven element. (The increase of capacitance between elements is balanced by an increase of inductance, ie length through mutual coupling.) However, the length does not decrease uniformly with increasing distance from the driven element.

Fig 16.5 shows experimentally derived element lengths for various material diameters. Elements are mounted through a cylindrical metal boom that is two or three diameters larger than the elements.

Some variation in element lengths will occur using different materials or sizes for the support booms. This will be increasingly critical as frequency increases. The water absorbency of insulating materials will also affect the element lengths, particularly when in use, although plastics other than nylon are usually satisfactory.

Fig 16.6 shows the expected gain for various numbers of elements if the array length complies with **Fig 16.7**.

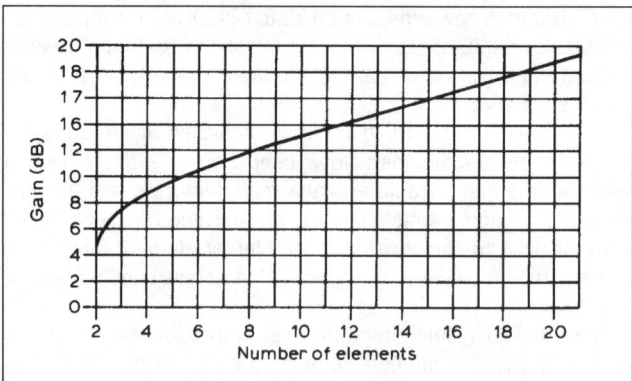

Fig 16.6: Gain over a half-wave dipole (dBd) versus the number of elements of the Yagi array (*ARRL Antenna Book*)

Number of elements	R-DE	DE-D1	D1-D2	D2-D3	D3-D4	D4-D5	D5-D6
2	0.15 -0.20						
2		0.07 -0.11					
3	0.16 -0.23	0.16 -0.19					
4	0.18 -0.22	0.13 -0.17	0.14 -0.18				
5	0.18 -0.22	0.14 -0.17	0.14 -0.20	0.17 -0.23			
6	0.16 -0.20	0.14 -0.17	0.16 -0.25	0.22 -0.30	0.25 -0.32		
8	0.16 -0.20	0.14 -0.16	0.18 -0.25	0.25 -0.35	0.27 -0.32	0.27 -0.33	0.30 -0.40
8 to N	0.16 -0.20	0.14 -0.16	0.18 -0.25	0.25 -0.35	0.27 -0.32	0.27 -0.32	0.35 -0.42

DE = driven element, R = reflector and D = director. N = any number. Director spacing beyond D6 should be 0.35-0.42

Table 16.1: Greenblum's optimisation for multielement Yagis

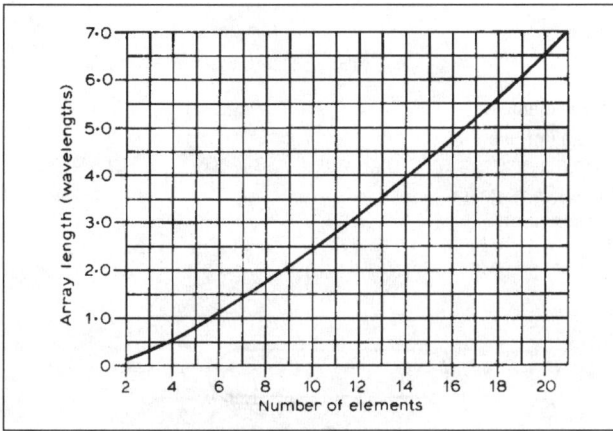

Fig 16.7: Optimum length of Yagi antenna as a function of number of elements (*ARRL Antenna Book*)

The results obtained by G8CKN using the 'centre spacing' of Greenblum's optimum dimensions shown in Table 16.1 produced identical gains to those shown in **Fig 16.8**. Almost identical radiation patterns were obtained for both the E and H planes (V or H polarisation). Sidelobes were at a minimum and a fair front-to-back ratio was obtained.

Considerable work has been carried out by Chen and Cheng on the optimising of Yagis by varying both the spacing and resonant lengths of the elements [3].

Table 16.2 and **Table 16.3** show some of their results obtained in 1974, by optimising both spacing and resonant lengths of elements in a six element array.

Table 16.3 shows comparative gain of a six element array with conventional shortening of the elements or varying the element lengths alone. The gain figure produced using conventional shortening formulas was 8.77dB relative to a λ/2 dipole (dBd). Optimising the element lengths produced a forward gain of 10dBd. Returning to the original element lengths and optimising the element spacing produced a forward gain of 10.68dBd. This is identical to the gain shown for a six-element Yagi in Fig 16.6. Using a combination of spacing and element length adjustment obtained a further 0.57dBd, giving 11.25dBd as the final forward gain as shown in **Table 16.**3.

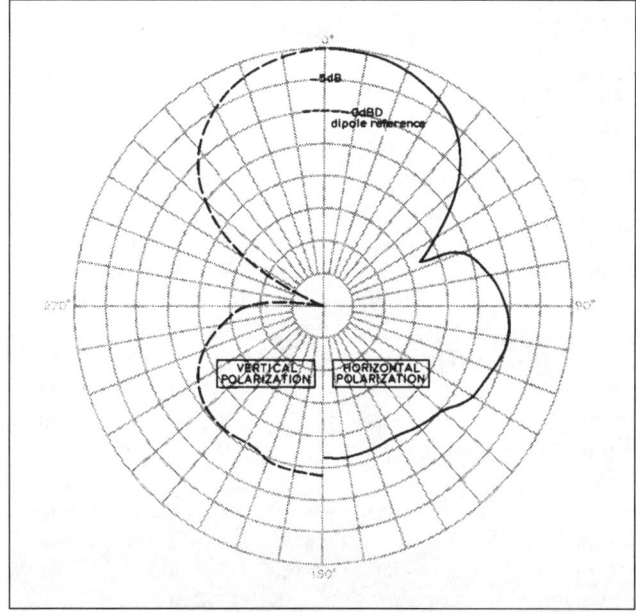

Fig 16.8: Radiation pattern for a four element Yagi using Greenblum's dimensions

A publication of the US Department of Commerce and National Bureau of Standards [4], [5] provides very detailed experimental information on Yagi dimensions. Results were obtained from measurements to optimise designs at 400MHz using a model antenna range.

The information, presented largely in graphical form, shows very clearly the effect of different antenna parameters on realisable gain. For example, it shows the extra gain that can be achieved by optimising the lengths of the different directors, rather than making them all of uniform length. It also shows just what extra gain can be achieved by stacking two elements, or from a 'two-over-two' array.

The paper presents:

(a) The effect of reflector spacing on the gain of a dipole.

(b) Effect of different equal-length directors, their spacing and number on realisable gain.

	h_1/λ	h_2/λ	h_3/λ	h_4/λ	h_5/λ	h_6/λ	Directivity (referring to λ/2 dipole)	Gain (dBD)
Initial array	0.255	0.245	0.215	0.215	0.215	0.215	7.544	8.78
Length-perturbed array	0.236	0.228	0.219	0.222	0.216	0.202	10.012	10.00
$b_i1 = 0.250\lambda$, $b_i2 = 0.310\lambda$ (i = 3, 4, 5, 6), $a = 0.003369\lambda$								

Table 16.2: Directivity optimisation of six element Yagi-Uda array (perturbation of element lengths)

	h1/λ	h2/λ	h3/λ	h4/λ	h5/λ	h6/λ	b21/λ	b22/λ	b43/λ	b34/λ	b35/λ	Directivity (referring to λ/2 dipole)	Gain (dBD)
Initial array	0.255	0.245	0.215	0.215	0.215	0.215	0.250	0.310	0.310	0.310	0.310	7.544	8.78
Array after spacing perturbation	0.255	0.245	0.215	0.215	0.215	0.215	0.250	0.289	0.406	0.323	0.422	11.687	10.68
Optimum array after spacing and length perturbations	0.238	0.226	0.218	0.215	0.217	0.215	0.250	0.289	0.406	0.323	0.422	13.356	11.26

Table 16.3: Directivity optimisation for six-element Yagi-Uda array (perturbation of element spacings and element lengths)

Length of Yagi (λ)	0.4	0.8	1.20	2.2	3.2	4.2
Length of reflector (λ)	0.482	0.482	0.482	0.482	0.482	0.475
Length of directors (λ):						
1st	0.424	0.428	0.428	0.432	0.428	0.424
2nd	-	0.424	0.420	0.415	0.420	0.424
3rd	-	0.428	0.420	0.407	0.407	0.420
4th	-	-	0.428	0.398	0.398	0.407
5th	-	-	-	0.390	0.394	0.403
6th	-	-	-	0.390	0.390	0.398
7th	-	-	-	0.390	0.386	0.394
8th	-	-	-	0.390	0.386	0.390
9th	-	-	-	0.398	0.386	0.390
10th	-	-	-	0.407	0.386	0.390
11th	-	-	-	-	0.386	0.390
12th	-	-	-	-	0.386	0.390
13th	-	-	-	-	0.386	0.390
14th	-	-	-	-	0.386	-
15th	-	-	-	-	0.386	-
Director spacing (λ)	0.20	0.20	0.25	0.20	0.20	0.308
Gain (dBD)	7.1	9.2	10.2	12.25	13.4	14.2

Element diameter 0.0085λ. Reflector spaced 0.2λ behind driven element. Measurements are for 400MHz by P P Viezbicke.

Table 16.4: Optimised lengths of parasitic elements for Yagi antennas of six different boom lengths

(c)   Effect of different diameters and lengths of directors on realisable gain.

(d)   Effect of the size of a supporting boom on the optimum length of parasitic elements.

(e)   Effect of spacing and stacking of antennas on gain.

(f)   The difference in measured radiation patterns for various Yagi configurations.

The highest gain reported for a single boom structure is 14.2dBd for a 15-element array (4.2λ long reflector spaced at 0.2λ, 13 graduated directors). See **Table 16.4**.

It has been found that array length is of greater importance than the number of elements, within the limit of a maximum element spacing of just over 0.4λ. Reflector spacing and, to a lesser degree, the first director position affects the matching of the Yagi. Optimum tuning of the elements, and therefore gain and pattern shape, varies with different element spacing.

Near-optimum patterns and gain can be obtained using Greenblum's dimensions for up to six elements. Good results for a Yagi in excess of six elements can still be obtained where ground reflections need to be minimised.

Chen and Cheng employed what is commonly called the long Yagi technique. Yagis with more than six elements start to show an improvement in gain with fewer elements for a given boom length when this technique is employed.

As greater computing power has become available, it has been possible to investigate the optimisation of Yagi antenna gain more extensively, taking into account the effects of mounting the elements on both dielectric and metallic booms, and the effects of tapering the elements at lower frequencies. Dr J Lawson, W2PV, carried out an extensive series of calculations and parametric analyses, collated in reference [6], which

Table 16.5: Typical dimensions of Yagi antenna components. Dimensions are in inches with metric equivalents in brackets

	Length		
	70.3MHz	145MHz	433MHz
Driven elements			
Dipole (for use with gamma match)	79 (2000)	38 (960)	12 3/4 (320)
Diameter range for length given	1/2 - 3/4		1/4 - 3/8
	1/8 - 1/4		
	(12.7 - 19.0)	(6.35 - 9.5)	(3.17 - 6.35)

Folded dipole 70-ohm feed			
l length centre-centre	77 1/2 (1970)	38 1/2 (980)	12 1/2 (318)
d spacing centre-centre	2 1/2 (64)	7/8 (22)	1/2 (13)
Diameter of element	1/2 (12.7)	1/4 (6.35)	1/8 (3.17)

a centre/centre	32 (810)	15 (390)	5 1/8 (132)
b centre/centre	96 (2440)	46 (1180)	152 (395)
Delta feed sections (length for 70Ω feed)	22½ (570)	12 (300)	42 (110)
Diameter of slot and delta feed material	1/4 (6.35)	3/8 (9.5)	3/8 (9.5)
Parasitic elements			
Element			
Reflector	85 1/2 (2170)	40 (1010)	13 1/4 (337)
Director D1	74 (1880)	35 1/2 (902)	11 1/4 (286)
Director D2	73 (1854)	35 1/4 (895)	11 1/8 (282)
Director D3	72 (1830)	35 (890)	11 (279)
Succeeding directors	1in less (25)	1/2in less (13)	1/8in less
Final director	2in less (50)	1in less (25)	3/4in less
One wavelength (for reference)	168 3/4 (4286)	81 1/2 (2069)	27 1/4 (693)
Diameter range for length given	1/2 - 3/4	1/4 - 3/8	1/8 - ¾
	(12.7 - 19.0)	(6.35 - 9.5)	(3.17 - 6.35)
Spacing between elements			
Reflector to radiator	22 1/2 (572)	17 1/2 (445)	5 1/2 (140)
Radiator to director 1	29 (737)	17 1/2 (445)	5 1/2 (140)
Director 1 to director 2	29 (737)	17 1/2 (445)	7 (178)
Director 2 to director 3, etc	29 (737)	17 1/2 (445)	7 (178)

Dimensions are in inches with millimetre equivalents in brackets.

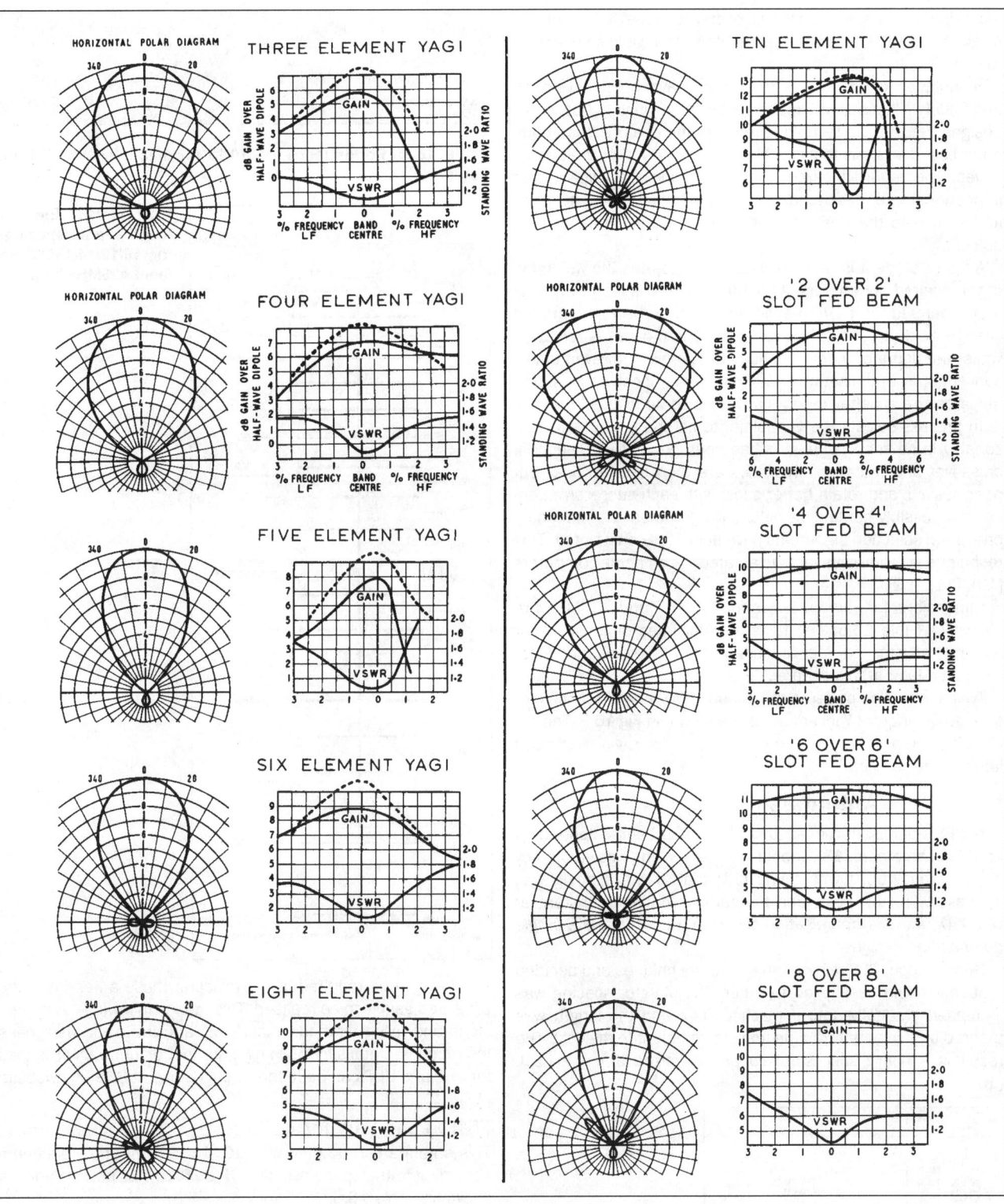

Fig 16.9: Charts showing voltage polar diagram and gain against VSWR of Yagi and skeleton-slot antennas. In the case of the six Yagi antennas the solid line is for conventional dimensions and the dotted lines for optimised results discussed in the text.

although specifically addressing HF Yagi design, explain many of the disappointing results achieved by constructors at VHF and above. In particular, the extreme sensitivity of some designs to minor variations of element length or position are revealed in a series of graphs which enable the interested constructor to select designs that will be readily realisable.

The keen constructor with a personal computer may now also take advantage of modelling tools specifically designed for optimisation of Yagi antennas and arrays, eg [7], although some care is needed in their use if meaningful results are to be assured. The Internet is a good source for Yagi antenna design and optimisation programmes, many of which can be obtained free of charge, or for a nominal sum.

From the foregoing, it can be seen that several techniques can be used to optimise the gain of Yagi antennas. In some circumstances, minimisation of sidelobes is more important than maximum gain, and a different set of element spacings and lengths would be required to achieve this. Optimisation with so many

independent variables is difficult, even with powerful computing methods, as there may be many solutions that yield comparable results.

Techniques of 'genetic optimisation' have been developed and widely adopted, which can result in surprising, but viable designs [8], [9]. The technique requires the use of proven computer-based analysis tools such as NEC, MININEC or their derivatives. The required parameters (gain, sidelobe levels, input impedance) are described and weighted according to their importance to the designer, together with the permitted variables.

A figure of merit is defined, which incorporates the weighting of the desired parameters. An initial structure is input, which is then analysed, its performance recorded, and an incremental change made to one of the variables. The process is repeated whilst the figure of merit continues to improve. However, unlike conventional optimisation methods, where local optimisation may obscure a better result that may also be available, a random process selects the variable(s) to be changed until a reasonably large seed population has been generated. Selection, crossover and mutation processes are then used to filter out poor designs and retain better ones, with each successive generation possibly containing better designs than the preceding one, if the selection algorithms have been well constructed. This technique is readily available to amateurs with home computers [10], [11], [12]

Dimensions for Yagi antennas for 70, 145 and 433MHz are shown in **Table 16.5**. The table also includes dimensions for feeding two stacked Yagi antennas with a *skeleton slot feed*, described later in this chapter.

Typical radiation patterns, gains and VSWR characteristics for a range of different Yagi antennas are shown in **Fig 16.9**. The figure also contains information on skeleton slot Yagis, discussed later in this chapter.

Long Yagi Antennas

The NBS optimisation described above has been extended by American amateurs [13]. Tapering of the spacing was studied by W2NLY and W6QKI who found [14] that, if the spacing was increased up to a point and thereafter remained constant at 0.3-0.4λ, another optimisation occurred. Both these are *single optimisation* designs.

Günter Hoch, DL6WU, looked at both techniques and decided that they could be applied together. The director spacing was increased gradually until it reached 0.4λ and the length was tapered by a constant *fraction* from one element to the next. The result is a highly successful *doubly optimised* antenna [15], [16].

Number of elements	10	13	14	19	23
Gain (dBd)	11.7	13	13.3	15	16
Horizontal beamwidth	37°	30.5°	30°	26.5°	24°
Vertical beamwidth	41°	33°	32°	28°	24.5°

Table 16.6: Performance of 10/13/14/19/23 element 435MHz Yagis

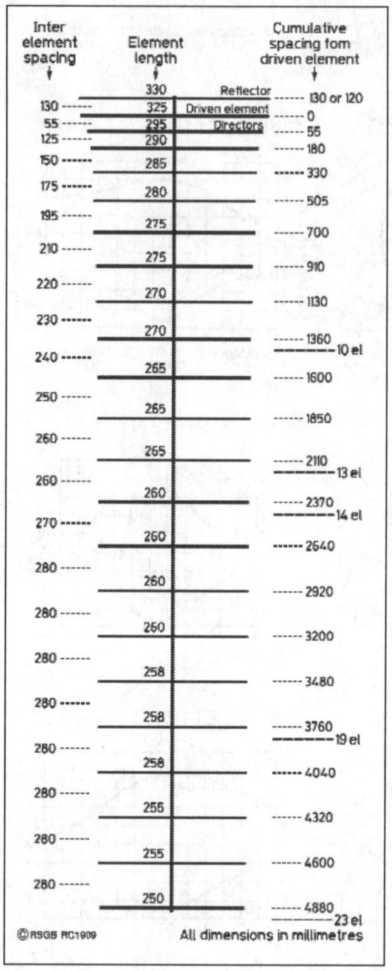

Fig 16.10: Element lengths and spacings for 10/13/14/19/23 element 435MHz Yagi

Great care is required in constructing these antennas if the predicted gain is to be realised. This means following the dimensions and fixing methods *exactly* as laid out in the designer's instructions. Details for building a number of long Yagi antennas for VHF and UHF can be found through links at G3SEK's website [17].

F5JIO Long Yagi for 435MHz

This antenna can be built with 10, 13, 14, 19 or 23 elements according to the space available (**Fig 16.10**). Its performance is shown in **Table 16.6**.

An extra 0.2dB gain and some reduction of backlobes can be obtained by fitting twin reflectors (**Fig 16.11**), but note that the spacing between the driven element and the reflector is reduced from 130mm to 120mm.

The 23 element version requires a boom length in excess of 5010mm, and must be solidly constructed and supported. The boom is made from 20 x 20mm square aluminium tubing, and all elements from 8mm diameter (round) tubing. All elements except the driven element must be *insulated* from the boom and mounted so that their centres are 8mm above its upper surface. Dimensions for the driven element and cable balun construction to provide a feed impedance of 50 ohms

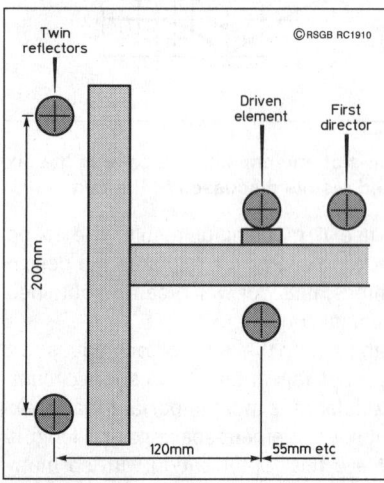

Fig 16.11: Twin reflector details

Fig 16.12: 435MHz long Yagi driven element and balun

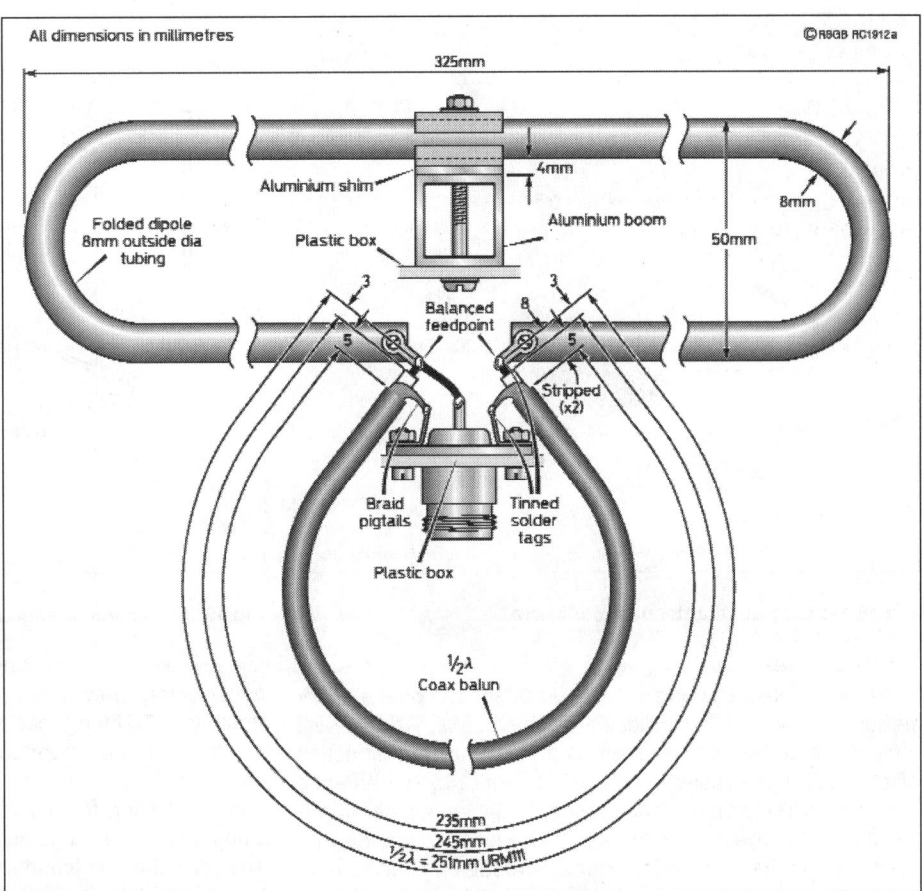

Table 16.7: Balun cable length calculation

λ/2@435MHz = 300,000/435 x 2 = 345mm (in air)
In URM111: 16mm of stripped end (@v=0.9) =18mm (electrical)

Cutting length = 345mm-18mm (@v=0.72 (PTFE insulation) = 235mm (unstripped)

Note: Use v=0.66 for Polyethylene insulation

are shown in **Fig 16.12**. Balun cable lengths and calculations are shown in **Table 16.7**. The cable should be a 75-ohm miniature PTFE insulated type, such as URM111 or equivalent. A small weatherproof box should be fitted over the ends of the element, inside which the balun cable may also be coiled. The driven element may be made from 9.5 x 1.6mm flat aluminium bar which is easier to bend and drill.

Quad Antenna

The Quad antenna can be thought of as a Yagi antenna comprising pairs of vertically stacked, horizontal dipoles with their ends bent towards each other and joined, **Fig 16.13**. The antenna produces horizontally polarised signals, and in spite of its relatively small physical size a forward gain of 5.5 to 6dB can be obtained with good front-to-back ratio. Additional quad or single element directors can be added to the basic two element array in the same manner as the Yagi.

Typical dimensions for lightweight wire 51, 71 and 145MHz Quad antennas are given in **Table 16.8**, and a photograph of the 145MHz version is shown in **Fig 16.14**. This variant has equal

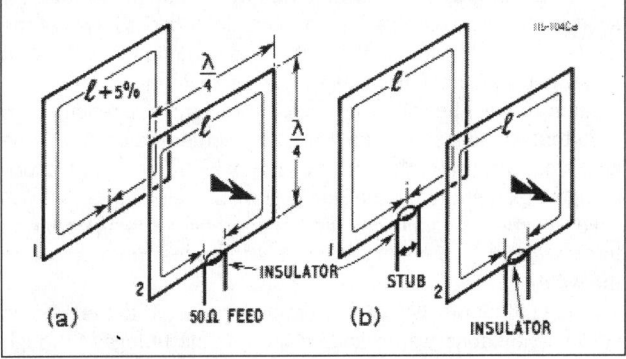

Fig 16.13: Quad antenna structure and electrical dimensions

size loops and uses a stub to tune the reflector. The boom is made from 15mm copper tubing with a T-piece in the centre for fixing to the mast or rotator. The element supports are made from 10 or 12mm square wooden dowelling fixed to square pieces of plywood using nuts and bolts. The plywood centres are fixed to the boom using L-brackets and hose clamps. A 50-ohm coaxial cable can be connected directly to the driven element. A 1:1 balun will minimise currents on the outer of the cable, preventing distortion of the radiation pattern and potential EMC problems.

Quads may be stacked or built into a four square assembly in the same way as the basic Yagi (see below).

Band	Element spacing, mm	Reflector sides, mm	Driven element sides, mm
51MHz	840	1560	1500
71.5MHz	600	1210	1080
145MHz	294	548	524

Table 16.8: Design dimensions for 51, 70 and 144MHz quad antennas

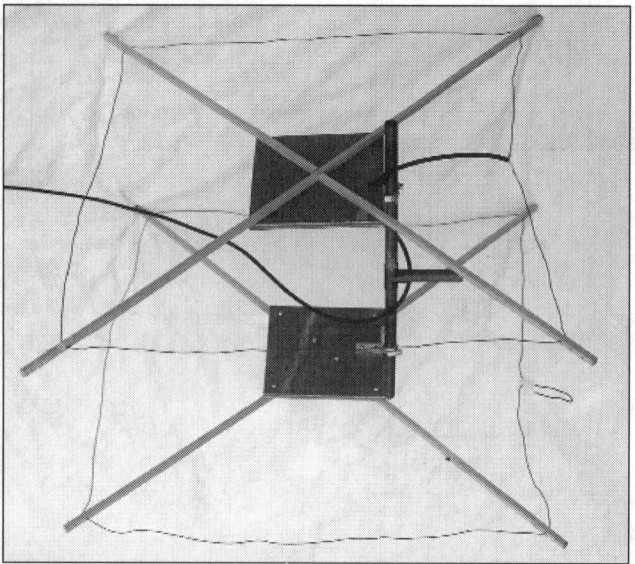

Fig 16.14: Wire quad antenna for 145MHz

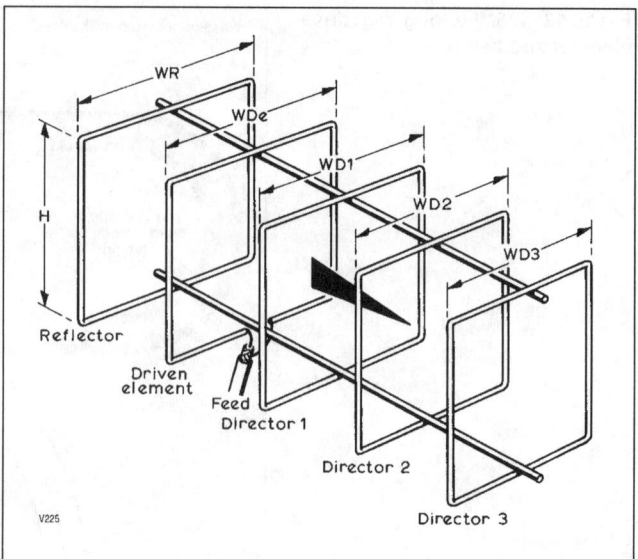

Fig 16.15: General arrangement for a multi-element quad antenna

Multi-element quad (Quagi)

The multiple element quad antenna or "Quagi" can offer a better performance with reduced sidelobes compared with the average simple Yagi, whilst retaining a simple robust form of construction (**Fig 16.15**). Dimensions for a four element, 145MHz antenna are given in **Table 16.9**. Generally the maximum number of elements used is five. Where more gain is needed, a pair may be stacked vertically or horizontally, although for maximum mechanical strength the vertical arrangement is to be preferred.

The whole structure may be made up of aluminium tube (or solid rod for the elements). The only insulator necessary is at the feed point of the driven element. In construction, it is best to make each element from one piece of material. A 3/8in aluminium rod will bend to form corners much more readily than tube that would also need a 'filler'. The corner radius should be kept small, and allowance must be made for the resultant 'shortening' of element length, ie side of the quad element.

For mechanical simplicity (and appearance) it is a good idea to arrange for all the element heights to be the same, and vary the width.

Fixing the elements to the boom and the boom to the mast is conveniently done with standard TV antenna fittings. Although

suitable blocks or clamping arrangements can be made by the constructor, they often tend to be unnecessarily heavy. Purchased TV fittings can be more cost-effective than obtaining raw materials and there is also much less effort involved in construction. There are also several antenna manufacturing companies catering for the radio amateur who sell tubing, mast clamps and small components for securing elements to booms. They can often be found at rallies and amateur radio events, or advertise in the pages of *RadCom*.

If preferred, the reflector may be made the same size as the driven element, and tuned with a suitable stub. If vertical polarisation is required, instead of horizontal, then the feeder can be attached to the centre of one of the vertical sides of the driven element. (The same 'side' must always be used for correct phase relationship within stacked arrays.)

The relative performance of multi-element quad and Yagi antennas is shown in **Fig 16.16**, demonstrating that the shorter quad structures can provide gains comparable with longer Yagi antennas. This may be of benefit if turning space is limited (eg inside a loft). However, there is no such thing as a free lunch, and in general, the weight and wind loading of the multi-element quad antenna will be slightly greater than its Yagi counterpart.

Height H	21 (533)	21	21	21
Width reflector WR	24½ (622)	24½	24½	24½
Driven WDe	20½ (520)	20½	20½	20½
Director 1 WD1	-	18 (457)	18	18
Director 2 WD2	-	-	16 (406)	16
Director 3 WD3	-	-	-	14 (356)
Spacing				
Reflector to Driven	7 (178)	19 (483)	20 (508)	20
Driven to Director 1	-	12 (305)	14½ (368)	14½
Director 1 to Director 2	-	-	14½	14½
Director 2 to Director 3	-	-	-	14½
Approx gain (dBd)	5	7	10.5	12.5

Element diameters all 3/8in (9.35mm). Feed impedance in all cases is 75 . Dimensions are in inches with millimetre equivalents in brackets.

Table 16.9: Dimensions for a multi-element quad antenna for 144MHz

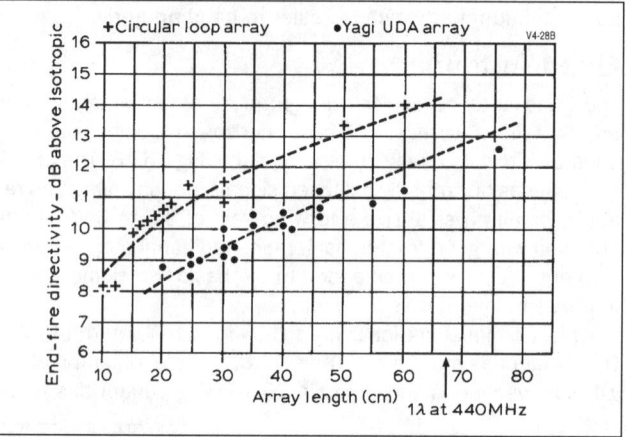

Fig 16.16: Comparative directivity of the Yagi and Quad as a function of overall array length. Although measured with circular loops, performance with square loops is comparable (*ARRL Antenna Book*). Note the gains are in dBi, not dBd

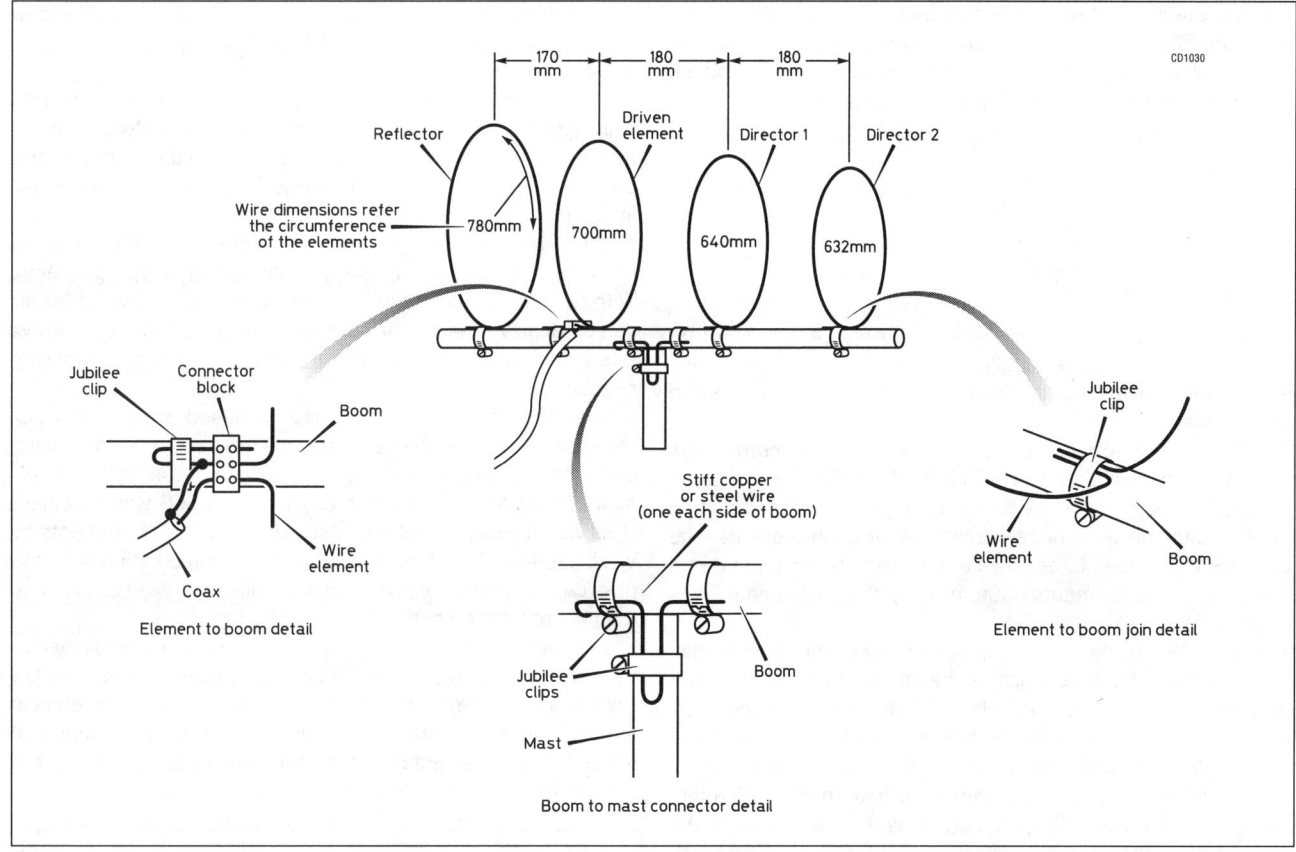

Fig 16.17: Four element quad loop Yagi for 435MHz

The Loop Yagi

At frequencies above 433MHz, the construction of multiple quad antennas can be considerably simplified by bending the elements into circular loops. High gains can be achieved by using large numbers of elements, and the relatively simple construction allows gains up to around 20dBi to be realised with manageable boom lengths [18].

A practical horizontally polarised four element loop Yagi antenna for 435MHz is shown in **Fig 16.17**. 2mm diameter enamelled copper wire elements are fixed to a tubular metal boom using hose clamps. A three terminal, plastic mains power connector block is used to connect the coaxial cable and provide the method for fastening the driven element to the boom. the enamel insulation is removed from the ends of the driven element to a distance of 20mm at one end and 50mm at the other. The 50mm end is folded into a loop and passed back into the connector block. The parasitic elements should be made 40mm longer than the dimensions shown, and the enamel removed from the last 20mm at each end. These ends should be bent at right angles and the remaining wire formed into a loop. The bent ends should be soldered together to simplify assembly. The boom and mast can be connected together using thick wire loops as shown, or the boom could be made from copper water pipe with a T-piece to connect to the support mast if preferred. A gain of around 9dBi should be achieved.

ANTENNA ARRAYS

Array Principles

The gain achievable with any antenna structure is ultimately limited by the fundamentals of its operation. However, higher gains can be achieved by using several antenna elements in an *array*.

The array can comprise antennas *stacked* vertically above each other, or arranged side by side in *bays*, or a combination of both. These are *broadside* arrays, where most of the radiated power is projected at right angles to the plane in which the array elements lie. An array can also be formed where the main beam is projected along the array of elements; these are *endfire* arrays, of which the HB9CV and Yagi antennas are examples.

An array of elements has a narrower beamwidth, and hence a higher gain than the individual antennas. The maximum achievable gain could be N times greater than one element fed with the same power ($10\log_{10}$ N decibels) if there are N elements in the array. However, more complex feed arrangements can reduce the VSWR bandwidth and introduce losses, reducing the array gain. Arrays need care in construction and attention to detail, especially at UHF and above, but the results reward the effort expended.

Antenna array theory can be found in almost any book devoted to antennas. However, a good treatment with many radiation pattern examples can be found in Refs [19] and [20].

Disadvantages of Multi-element Arrays

High gain cannot be achieved by simply stacking many elements close together. If we consider a dipole collecting power from an incident field for delivery to a load (receiver), it can be thought of as having a collecting area or effective aperture that is somewhat larger than the dipole itself. The higher the directivity of the antenna, the larger the effective aperture, as given by the relationship:

$$A_{eff} = \frac{\lambda^2}{4\pi} D$$

where   D is the directivity of the antenna
           λ is the working wavelength

If the effective apertures of adjacent antennas overlap, the incoming RF energy is shared between them, and the maximum possible directivity (or gain) of the elements cannot be attained.

The generalised optimum stacking distance is a function of the half power beamwidth of the elements in the array, and is given by:

$$S_{opt} = \frac{\lambda}{\left[2\sin\left(\frac{\phi}{2}\right)\right]}$$

where ϕ is the half power beamwidth and S_{opt} is in wavelengths. Note that this is usually different for the E and H planes, so that the spacing of the elements is also usually different in each plane.

Also, when antennas are placed close together, *mutual coupling* between elements occurs. This leads to changes in the current distribution on the elements, changing both the radiation pattern and the feed point impedance of each element. The changes to the feed impedance often result in unequal powers being fed to the elements of the array, with consequential loss of gain.

Optimum stacking rules are based on the assumption of minimum mutual influence, which can be difficult to predict for complex antennas such as Yagis. However, antennas with low sidelobe levels are less susceptible than those with high sidelobes, as might be expected intuitively.

The coupling and effective aperture overlap problems cannot simply be solved by arbitrarily increasing the separation of the elements. As the element spacing increases beyond one half wavelength, *grating sidelobes* appear, which can reduce the forward gain. The grating lobes are due solely to the array dimensions, and can be seen by plotting the array factor for the chosen configuration.

Arrays of Identical Antennas

A parasitic array such as the Yagi can be stacked either vertically or horizontally to obtain additional directivity and gain. This is often called *collinear and broadside* stacking.

In stacking it is assumed that the antennas are identical in pattern and gain and will be matched to each other with the correct phase relationship, that is, 'fed in phase'. It is also assumed that for broadside stacking the corresponding elements are parallel and in planes perpendicular to the axis of the individual arrays. With vertical stacking it is assumed the corresponding elements are collinear and all elements of the individual arrays are in the same plane.

The combination of the radiation patterns can add but can also cancel. The phase relationships, particularly from the side of the Yagi, are very complex. Because of this complexity the spacing to obtain maximum forward gain does not coincide with the best sidelobe structure. Usually maximum gain is less important than reducing signals to the sides or behind the array.

If this is the case, 'optimum spacing' is one that gives as much forward gain as possible as long as the sidelobe structure does not exceed a specific amplitude compared with the main lobe. There will be different 'optimum' spacings according to the acceptable sidelobe levels.

Fig 16.18 gives typical optimum spacing for two arrays under three conditions:

(a) optimum forward gain with sidelobe down 10dB,

(b) sidelobe 20dB down and

(c) virtually no sidelobe.

The no-sidelobe case can correspond to no additional forward gain over a single antenna. **Fig 16.19** shows the optimum stacking spacing for four-unit arrays.

The maximum forward gain of two stacked arrays is theoretically +3dB, and +6dB for four stacked arrays. More complex arrays could produce higher gain but losses in the matching and phasing links between the individual arrays can outweigh this improvement.

When stacking two arrays, the extra achievable gain is reduced at close spacing due to high mutual coupling effects. With two seven-element arrays a maximum gain of about 2.5dB can be achieved with 1.6λ spacing; with two 15-element arrays it is also possible to achieve the extra 2.5dB but the spacing needs to be 2λ.

The use of four arrays, in correctly phased two-over-two systems, can increase the realisable gain by about 5.2dB. Using seven-element Yagis produced a total gain of 14.2dB. With 15-element optimised Yagis a total gain of 19.6dB was obtained. (This was the highest gain measured during the experiments by Viezbicke [4].) The effects of stacking in combination with the physical and electrical phase relationship can be used to reduce directional interference.

An improvement in front-to-back ratio can be accomplished in vertical stacking by placing the top Yagi a quarter-wavelength in front of the lower Yagi as shown in **Fig 16.20**. The top antenna is fed 90° later than the bottom antenna by placing additional cable in the upper antenna feed run. The velocity factor of the cable must be taken into account.

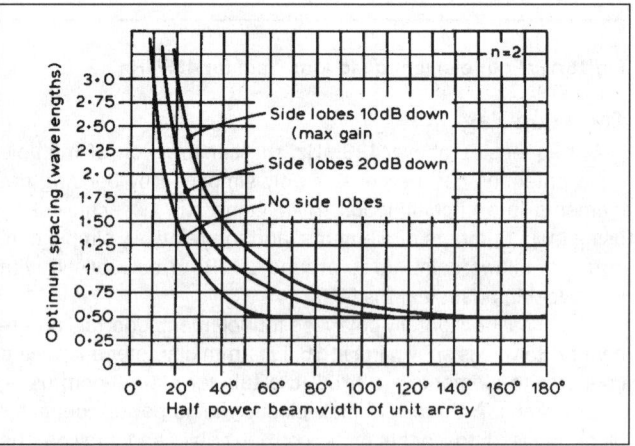

Fig 16.18: Optimum stacking spacing for two-unit arrays. The spacing for no sidelobes, especially for small beamwidths, may result in no gain improvement over a single array element (*ARRL Antenna Handbook*)

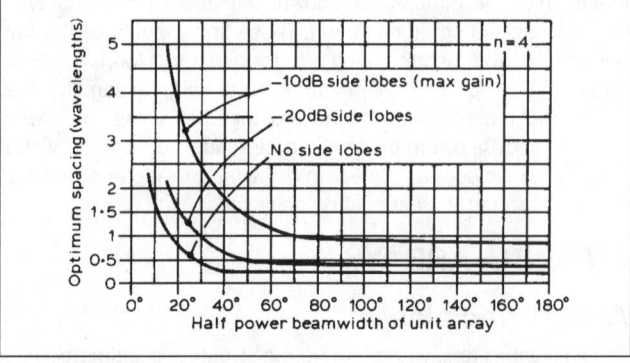

Fig 16.19: Optimum stacking spacing for four-unit arrays (ARRL *Antenna Handbook*)

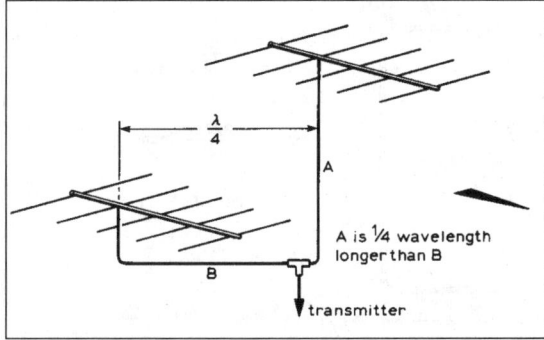

Fig 16.20: Improving front to back ratio of stacked Yagi antennas with offset vertical mounting

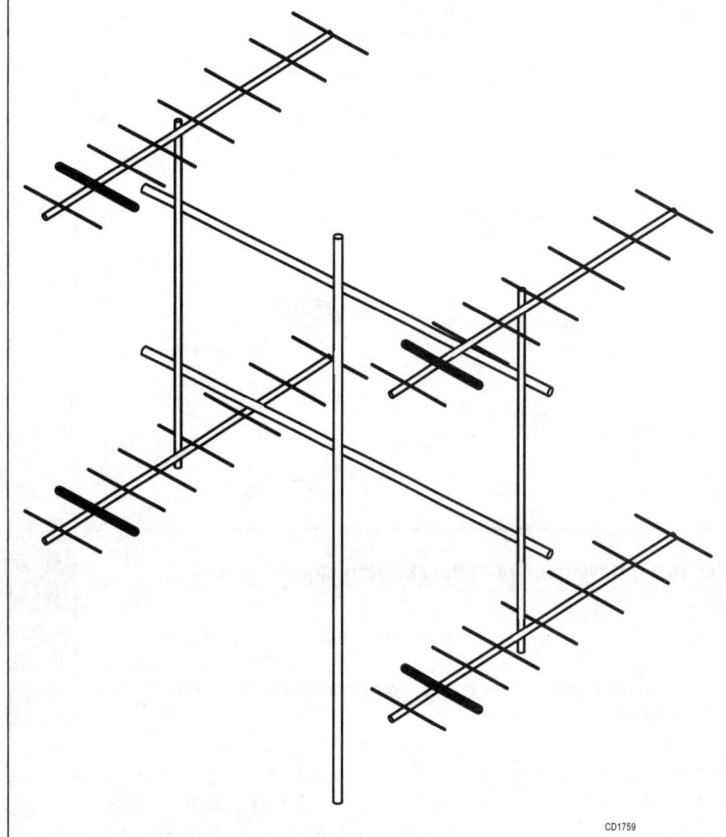

Fig 16.21: Four Yagi antennas stacked and bayed

A Coaxial Cable Harness for Feeding Four Antennas

Four identical antennas such as Yagis can be mounted at the corners of a rectangle as a stacked and bayed array as shown in **Fig 16.21**, with separations determined by their beamwidths as described above. Whilst feed harnesses can be purchased with the antennas, they can also be constructed using standard coaxial cables and connectors, as shown in **Fig 16.22**. Each antenna and all cables must have an impedance of 50 ohms. The two feeders L1 and L3 connected in parallel result in 25 ohms at Point A. This is transformed to 100 ohms by the cable between A and B, which must be an odd number of quarter wavelengths long. The two 100-ohm impedances connected in parallel at point B result in a 50-ohm impedance presented to the transceiver feeder. Feeders L1-L4 may be any convenient length, provided that they are all identical.

Skeleton Slot Feed for Two Stacked Yagis

A serious disadvantage of the Yagi array is that variation of the element lengths and spacing causes interrelated changes in the feed impedance. To obtain the maximum possible forward gain experimentally is extremely difficult. For each change of element length it is necessary to readjust the matching either by moving the reflector or by resetting a matching device.

However, a method has been devised for overcoming these practical disadvantages. It involves the use of a radiating element in the form of a skeleton slot. This is far less susceptible to the changes in impedance caused by changes in the length of the parasitic elements. A true slot would be a slot cut in an infinite sheet of metal. Such a slot, when approximately $\lambda/2$ long, would behave in a similar way to a dipole radiator. In contrast with a dipole, however, the electric field produced by a vertical slot is horizontally polarised.

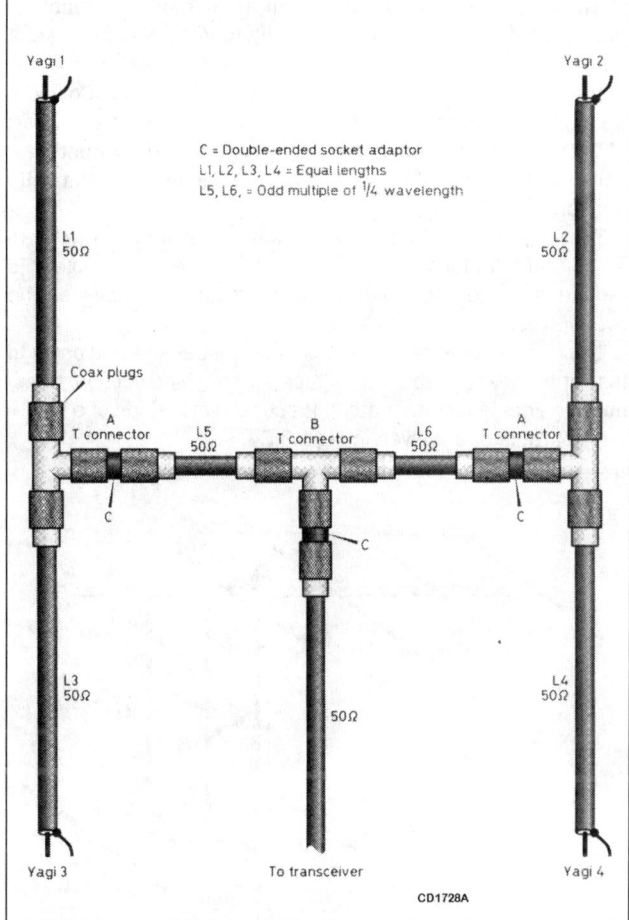

Fig 16.22: Coaxial feed harness for four antennas

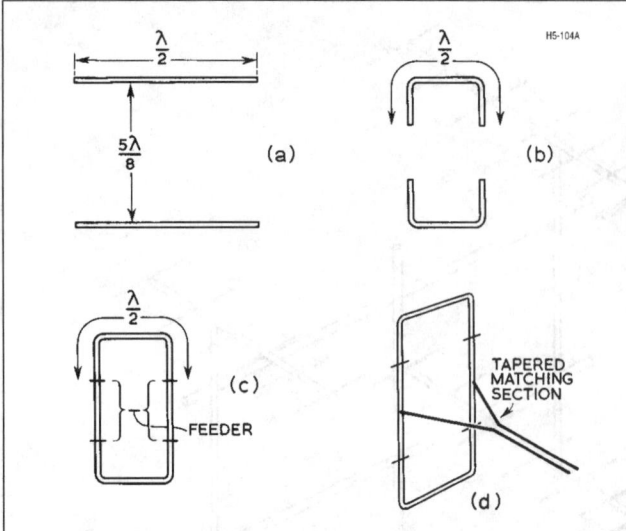

Fig 16.23: Development of the skeleton-slot radiator

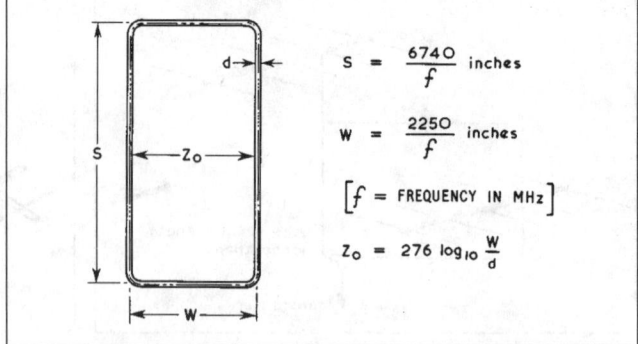

Fig 16.24: Dimensional relationships of a skeleton-slot radiator. Both S and W may be varied experimentally, and will not change the radiation characteristics of the slot greatly. See text

$$S = \frac{6740}{f} \text{ inches}$$

$$W = \frac{2250}{f} \text{ inches}$$

$$[f = \text{FREQUENCY IN MHz}]$$

$$Z_0 = 276 \log_{10} \frac{W}{d}$$

Element	Length, in (mm)	Element spacing, in (mm)
A	34 (864)	A - B, 14.25 (362)
B	34 (864)	B - C, 14.25 (362)
C	34 (864)	C - D, 14.25 (362)
D	34 (864)	D - E, 34.25 (870)
E	40 (1016)	D - Slot, 14.5 (368)
Boom	70 (1720)	

Table 16.10: Dimensions for six-over-six slot fed Yagi antenna for 145MHz

The skeleton slot was developed during experiments to find how much the 'infinite' sheet of metal could be reduced without the slot antenna losing its radiating property. The limit was found to occur when there remained approximately $\lambda/2$ of metal beyond the slot edges. However, further experiments showed that a thin rod bent to form a 'skeleton slot' (approximately $5\lambda/8$ by $5\lambda/24$) exhibited similar properties to those of a true slot.

The way a skeleton slot works is shown in **Fig 16.23**. Consider two $\lambda/2$ dipoles spaced vertically by 5 /8. Since the greater part of the radiation from each dipole takes place at the current maximum (ie the centre) the ends of the dipoles may be bent without serious effect.

These 'ends' are joined together with a high-impedance feeder, so that 'end feeding' can be applied to the bent dipoles. To radiate in phase, the power should be fed midway between the two dipoles.

The high impedance at this point may be transformed down to that of the feeder cable with a tapered matching section/transmission line (ie a delta match). Practical dimensions of a skeleton-slot radiator are given in **Fig 1.5.24**.

These dimensions are not critical, and may be varied over a modest range without affecting the radiating characteristics of the slot. However, the feed impedance is very sensitive to dimensional changes, and must be properly matched by altering the length and shape of the delta section after completing all other adjustments.

It is important to note that two sets of parasitic elements are required with a skeleton-slot radiator and not one set as with a true slot. One further property of the skeleton slot is that its bandwidth is greater than a pair of stacked dipoles.

Radiation patterns and VSWR data for some typical slot-fed Yagi antennas are shown in Fig 16.9 earlier in this chapter.

Details of a practical 'six-over-six' skeleton slot Yagi antenna are shown in **Fig 16.25** with essential dimensions listed in **Table 16.10**.

Skeleton Slot Yagi Arrays

Skeleton-slot Yagi arrays may be stacked to increase the gain but the same considerations of optimum stacking distance as previously discussed apply. The centre-to-centre spacing of a pair of skeleton-slot Yagi arrays should typically vary between

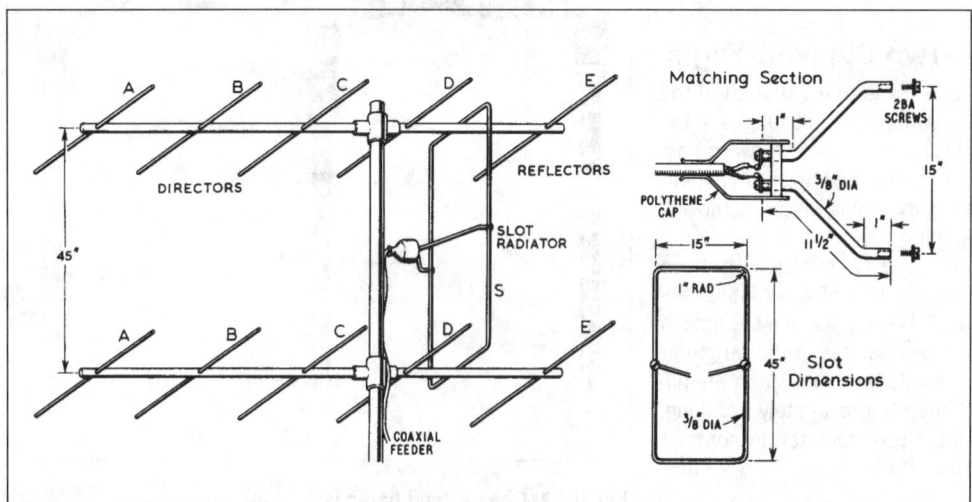

Fig 16.25: A six-over-six skeleton slot fed Yagi antenna for 145MHz

Fig 16.26: A high gain 432MHz antenna consisting of four 8-over-8 slot-fed Yagi antennas arranged in a square formation

1λ and 3λ depending on the number of elements in each Yagi array. A typical 4 x 4 array of 8-over-8 slot-fed Yagis for 432MHz is shown in **Fig 16.26**.

Quadruple Quad for 144MHz

This collapsible antenna was designed for portable use [21], but is equally useable as a fixed antenna for use indoors or in a loft, and can achieve gains of between 10 and 11dBi (8 - 9dBd) on 2 metres. It is effectively a stacked quad using mutual coupling instead of a phasing harness to excite the outer elements. Constructional details are shown in **Fig 16.27**.

Each section has a circumference of around 1.04 wavelengths, which is not as would be expected for conventional quads. The dimensions are the result of experiments to obtain the best front to back ratio and least sensitivity to adjacent objects, which can be important for portable or loft operation, ensuring that the antenna will work without extensive adjustment.

Note that the antenna was designed for low-power (1 watt) operation; the ferrite bead must not be allowed to saturate magnetically, or harmonic generation may occur. The bead may also become hot and shatter. For higher power operation, ferrite rings could be considered for the balun transformer, or a sleeve balun constructed as appropriate.

Yagi Antenna Mounting Arrangements

The performance of Yagi antennas can be greatly degraded if they are not correctly installed and the feeder routed to minimise unwanted interaction with the antenna. Many commercial designs have fixed clamping positions which have been optimised to minimise coupling into the support mast. However, there are a number of precautions that can be taken when installing any Yagi antennas, whether operating in the same band as an array, or operating in several different bands.

Antenna performance may be completely destroyed if the mast is installed parallel, and through the Yagi antenna as in **Fig 16.28a**. The mast should be mounted at right angles to the antenna elements to minimise coupling between the elements and the mast, **Fig 16.28c**. If the antenna is to provide vertical polarisation, it should be offset from the main support mast with a stub mast if possible, **Fig 16.28b**. Mechanical balance can be restored (and the Yagi-mast separation increased) by using a symmetrical stub mast and a second antenna for the same band (bayed array), or for another frequency.

Fig 16.27: Quadruple Quad. The match point xx should be found experimentally and will be approximately 200mm from the open end (VHF Communications)

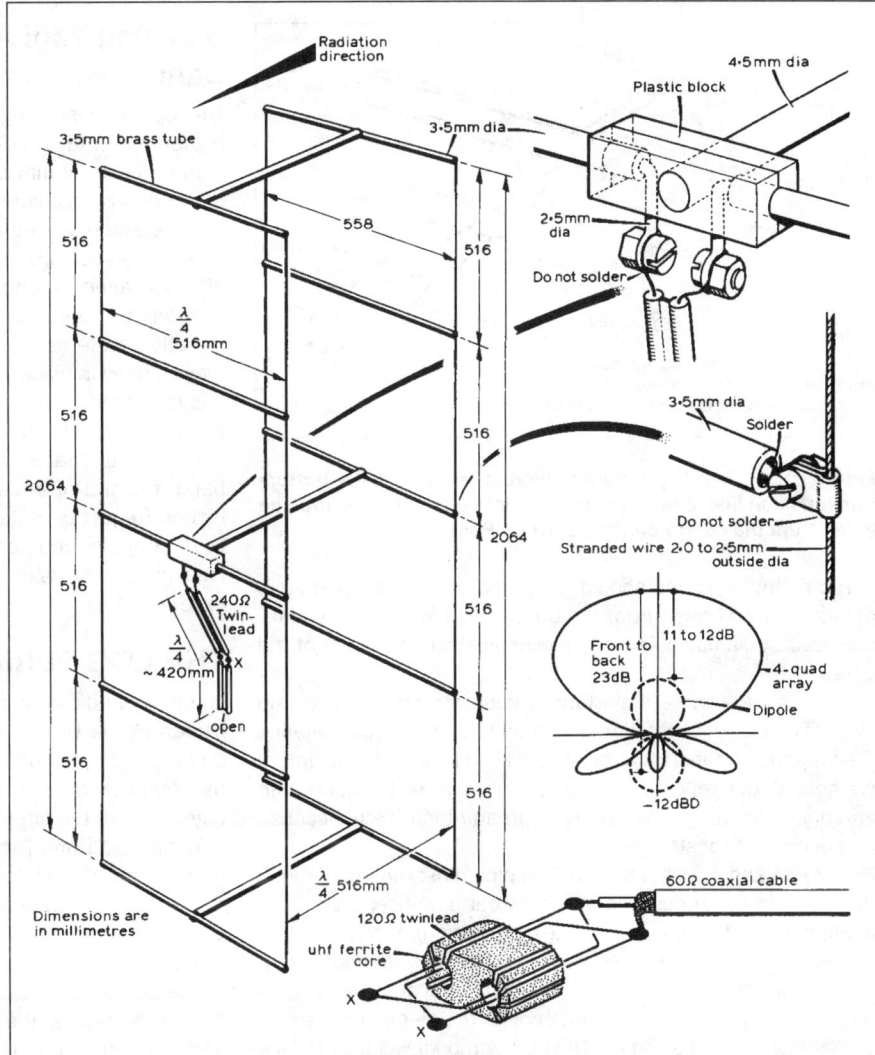

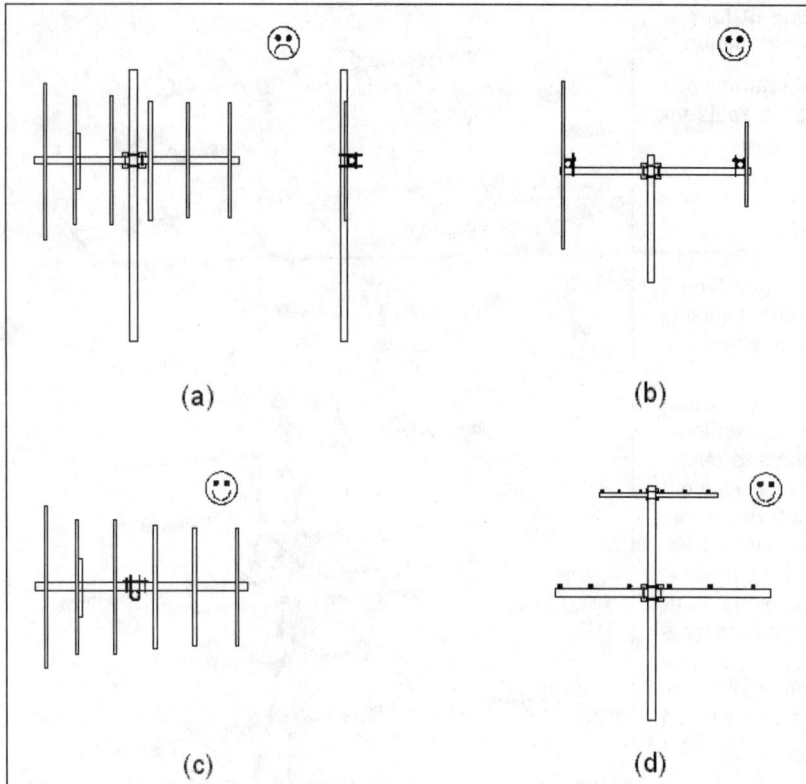

(a)

(b)

(c)

(d)

Fig 16.28: Yagi mounting methods (a) the wrong way, antenna couples strongly with the mast, destroying performance (b) best arrangement for vertically polarised antennas, offset from the mast at least λ/4 with a horizontal tube (c) mast should be fastened mid way between elements, but not next to the driven element (d) put highest frequency antenna at top of a multi-band stack

these effects, and is generally mechanically adequate to support higher frequency antennas.

The feed cable should be arranged to lie in a plane at right angles to the Yagi elements, or be taped below and along the boom until it can be run down the support mast.

In the case of circularly polarised antennas, for example, crossed Yagi antennas fed in phase quadrature, the elements should be arranged at 45 degrees to the support mast when the antenna is viewed along its boom. There will be some degradation of circularity, but it can be minimised if the support mast is not an odd multiple of quarter wavelengths long. The feed cable should be taped to the boom and dressed on to the support mast with the minimum bend radius for which the cable is designed.

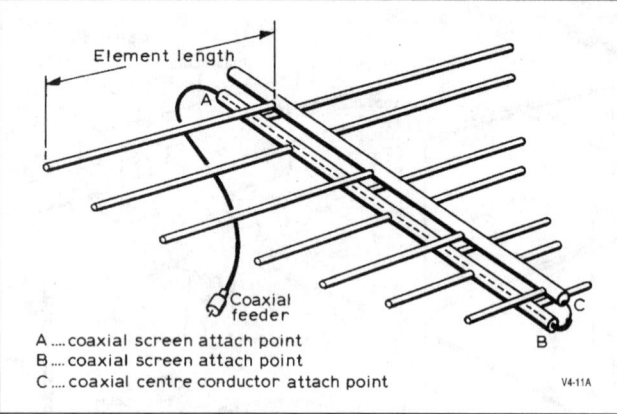

A....coaxial screen attach point
B....coaxial screen attach point
C....coaxial centre conductor attach point

V4-11A

Fig 16.29: Typical log-periodic antenna. Note that the bottom transmission line is fed from the coaxial outer while the top line is fed from the centre conductor (*Ham Radio*)

The mounting clamp should be placed mid-way between elements, and well away from the driven element. This is usually achieved by clamping near the mechanical balance point of the antenna.

However, it is more important to keep the mast and clamp away from the adjacent elements than to mechanically balance the antenna. In theory, the mast could be clamped to the antenna behind the reflector element(s). This is rarely done with antennas operating at wavelengths greater than 23cm because of mechanical constraints.

At 23cm and above, a 50mm (2in) pipe mast running through the antenna will seriously degrade its performance, even if the elements are at right angles to the pipe. Performance will not be so badly affected if the (horizontally polarised) antenna is right at the top of the mast, with the minimum amount of pipe required for clamping projecting through the elements. A smaller diameter pipe, eg 25mm (1in) for the mount will also reduce

Stacking Yagi Antennas for Different Bands

The optimum spacing between identical antennas to create higher gain arrays is discussed earlier in the chapter. However, in many cases, it may be desired to put several antennas for different bands on a common rotating mast.

If the antennas are all pointing in the same direction on the mast, they should be separated sufficiently to ensure that their effective apertures do not overlap (see formula earlier in this chapter) to avoid interaction and mutual degradation of their radiation patterns. To a first approximation, the antenna gain may be used in place of the directivity in the formula. Yagi antennas may be stacked more closely together if alternate antennas point in directions at 90 degrees to each other. The separation may then be reduced so that the effective aperture of the lowest band antenna of any pair is not physically encroached by the higher frequency antenna. Closer spacings may be possible without excessive interaction, but need to be investigated on a case by case basis using antenna modelling software or careful experiment.

THE LOG PERIODIC ANTENNA

The log-periodic antenna **Fig 16.29** was originally designed and proven at the University of Illinois in the USA in 1955 [22].Its properties are an almost infinite bandwidth, governed only by the number of elements used and mechanical limitations, together with the directive qualities of a Yagi antenna [23].

Table 16.11 and **Table 16.12** show typical dimensions for element spacing and length for log-periodic arrays. These are derived from a computer-aided design produced by W3DUQ [24]. Other frequency bands can be produced by scaling all dimensions.

The tabulated parameters have a 5% overshoot of the working frequency range at the low end and a 45% overshoot at the high-frequency end. This is done to maintain logarithmic response

Element	21-55MHz array Length (ft)	(mm)	Diameter (in)	(mm)	Spacing (ft)	(mm)	50-150MHz array Length (ft)	(mm)	Diameter (in)	(mm)	Spacing (ft)	(mm)	140-450MHz array Length (ft)	(mm)	Diameter (in)	(mm)	Spacing (ft)	(mm)
1	12.240	3731	1.50	38.1	3.444	1050	5.256	1602	1.00	2.54	2.066	630	1.755	535	0.25	6.7	0.738	225
2	11.190	3411	1.25	31.8	3.099	945	4.739	1444	1.00	2.54	1.860	567	1.570	479	0.25	6.7	0.664	202
3	10.083	3073	1.25	31.8	2.789	850	4.274	1303	1.00	2.54	1.674	510	1.304	397	0.25	6.7	0.598	182
4	9.087	2770	1.25	31.8	2.510	765	3.856	1175	0.75	19.1	1.506	459	1.255	383	0.25	6.7	0.538	164
5	8.190	2496	1.25	31.8	2.259	689	3.479	1060	0.75	19.1	1.356	413	1.120	341	0.25	6.7	0.484	148
6	7.383	2250	1.00	25.4	2.033	620	3.140	957	0.75	19.1	1.220	372	0.999	304	0.25	6.7	0.436	133
7	6.657	2029	1.00	25.4	1.830	558	2.835	864	0.75	19.1	1.098	335	0.890	271	0.25	6.7	0.392	119
8	6.003	1830	0.75	19.1	1.647	500	2.561	781	0.50	12.7	0.988	301	0.792	241	0.25	6.7	0.353	108
9	5.414	1650	0.75	19.1	1.482	452	2.313	705	0.50	12.7	0.889	271	0.704	215	0.25	6.7	0.318	97
10	4.885	1489	0.75	19.1	1.334	407	2.091	637	0.50	12.7	0.800	244	0.624	190	0.25	6.7	0.286	87
11	4.409	1344	0.75	19.1	1.200	366	1.891	576	0.50	12.7	0.720	219	0.553	169	0.25	6.7	0.257	78
12	3.980	1213	0.50	12.7	1.080	329	1.711	522	0.375	9.5	0.648	198	0.489	149	0.25	6.7	0.231	70
13	3.593	1095	0.50	12.7	0.000		1.549	472	0.375	9.5	0.584	178	0.431	131	0.25	6.7	0.208	63
14							1.403	428	0.375	9.5	0.525		0.378	115	0.25	6.7	0.187	57
15							1.272	388	0.375	9.5	0.000		0.332	101	0.25	6.7	0.169	52
16													0.290	88	0.25	6.7	0.000	
Boom	25.0	7620	2.0	50.8	0.5	12.7	16.17	5090	1.5	38.1	0.5	152	5.98	1823	1.5	38.1	0.5	152

Table 16.11: Spacing and dimensions for log-periodic VHF antennas

Element	Length (ft)	(mm)	Diameter (ft)	(mm)	Spacing (ft)	(mm)
1	0.585	178	0.083	2.1	0.246	75
2	0.523	159	0.083	2.1	0.221	67
3	0.435	133	0.083	2.1	0.199	61
4	0.418	127	0.083	2.1	0.179	55
5	0.373	114	0.083	2.1	0.161	49
6	0.333	101	0.083	2.1	0.145	44
7	0.297	91	0.083	2.1	0.131	40
8	0.264	80	0.083	2.1	0.118	36
9	0.235	72	0.083	2.1	0.106	32
10	0.208	63	0.083	2.1	0.095	29
11	0.184	56	0.083	2.1	0.086	26
12	0.163	50	0.083	2.1	0.077	23
13	0.144	44	0.083	2.1	0.069	21
14	0.126	38	0.083	2.1	0.062	19
15	0.111	34	0.083	2.1	0.056	17
16	0.097	30	0.083	2.1	0.000	0
Boom	1.99	607	0.5	12.7		

Table 16.12: Spacing and dimensions for log-periodic UHF antenna (420-1350MHz)

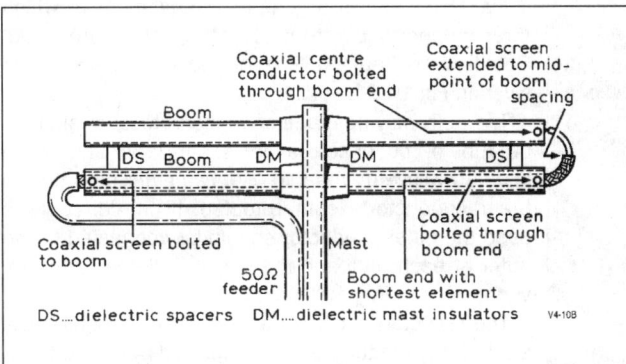

Fig 16.30: Log-periodic antenna mast mounting and feeder arrangements (Ham Radio)

over the complete frequency range specified as the log-periodic cell is active over approximately four elements at any one specific frequency. The logarithmic element taper (α) is 28° for all three antennas. They have a forward gain of 6.55dBd, with a front-to-back ratio of typically 15dB and a VSWR better than 1.8:1 over the specified frequency range.

Construction is straightforward. The element lengths for the highest-frequency antenna allow for the elements to be inserted completely through the boom, ie flush with the far wall. The two lower-frequency antennas have element lengths calculated to butt flush against the element side of the boom, and a length correction factor must be added to each element if through-boom mounting is used.

The supporting booms are also the transmission line between the elements for a log-periodic antenna. They must be supported with a dielectric spacer from the mast of at least twice the boom-to-boom spacing. Feed-line connection and the arrangement to produce an 'infinite balun' is shown in **Fig 16.30**. Any change in the boom diameters will require a change in the boom-to-boom spacing to maintain the transmission line impedance. The formula to achieve this is:

$$Z_0 = 273 \log_{10} D/d$$

where D is the distance between boom centres and d the diameter of the booms. Mounting arrangements are shown in Fig 16.30. The antenna can be oriented for either horizontal or vertical polarisation if a non-conductive mast section is used. The horizontal half-power beamwidth will be typically 60° with a vertical half-power beamwidth of typically 100°.

THE AXIAL MODE HELIX

The axial mode helix antenna provides a simple means of obtaining high gain and a wide-band frequency characteristic. When the circumference of the helix is of the order of one wavelength, axial radiation occurs, ie the maximum field strength is found to lie along the axis of the helix. This radiation is circularly polarised, the sense of the polarisation depending on whether the helix has a right or left-hand thread. The polarisation can be determined by standing behind the antenna. If a clockwise motion would be required to travel along the helix to its far end, the helix will generate and receive Right Hand Circularly Polarised (RHCP) waves.

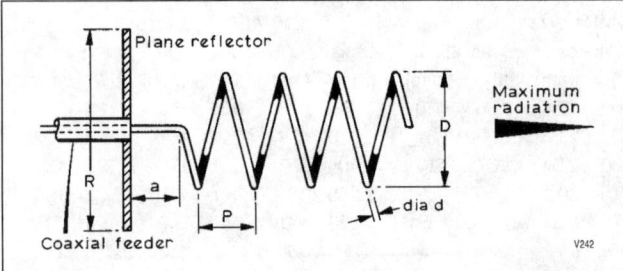

Fig 16.31: The axial mode helix antenna. The plane reflector may take the form of a dartboard type wire grid or mesh. The dimensions given in Table 13 are based on a pitch angle of 12 degrees. The helix tube or wire diameter is not critical, but it must be supported by low loss insulators

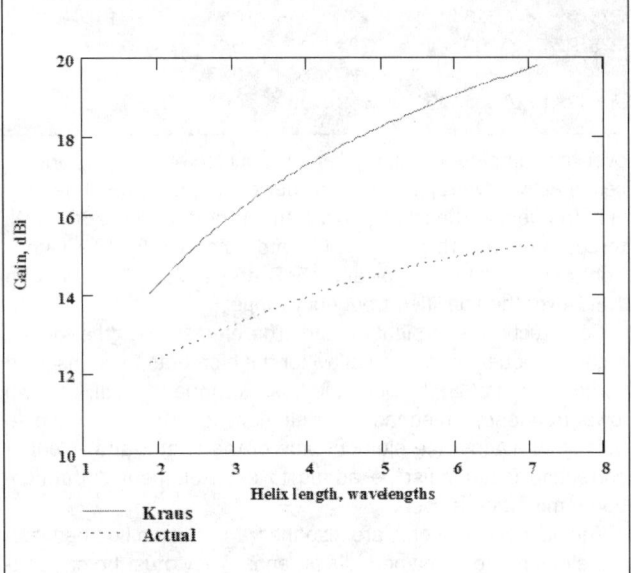

Fig 16.32: Kraus's theoretical gain and realisable gain for a helix antenna of different lengths. The antenna has a circumference of 1.06λ and α = 13°

Band	Dimensions				
	D	R	P	a	d
General	0.32	0.8	0.22	0.12	
144MHz	25 1/2 (648)	64 (1626)	17 3/4 (450)	8 3/4 (222)	1/2 (12.7)
433MHz	8 3/4 (222)	22 (559)	6 (152)	3 (76)	3/16-1/2 (4.8-12.7)
1296MHz	3 (76)	7 (178)	2 (50)	1 1/8 (28)	1/4 - 1/8 (3.2-6.4)
Turns	6	8	10	12	20
Gain	12dBi	14dBi	15dBi	16dBi	17dBi
Beamwidth	47°	41°	36°	31°	24°

Dimensions in inches, millimetres are given in brackets. The gain and beamwidth of the helical antenna are dependent upon the total number of turns as shown above.

Bandwidth = 0.75 to 1.3λ

Feed impedance = 140 x $\frac{circumference}{\lambda}$ ohms

(Note: λ and circumference must be in the same units.)

Beamwidth (degrees) = $\sqrt{\frac{12,300}{No\ of\ turns}}$

Table 16.13: General dimensions for 144, 433 and 1296MHz helix antennas

A helix may be used to receive plane or circularly polarised waves. When signals are received from a transmitting helix care must be taken to ensure that the receiving helix has a 'thread' with the same hand of rotation as the radiator, or significant signal will be lost due to *polarisation mismatch*.

The properties of the helical antenna are determined by the diameter of the helix D and the pitch P (see **Fig 16.31**). It is also dependent on radiation taking place all along the helical conductor. The gain of the antenna depends on the number of turns in the helix.

The diameter of the reflector R should be at least λ/2, with the diameter of the helix D about λ/3 and the pitch P about 0.24λ. A detailed description of the way in which the antenna radiates, and the relationships between pitch and diameter for different antenna characteristics are described by its inventor, J D Kraus in [25].

A helix of this design will have a termination / feed impedance of about 140 ohms. A 50-ohm impedance can be obtained by shaping the last quarter turn from the feedpoint to lie close to the reflector by reducing the pitch of the helix over the last turn. Gain of the antenna is proportional to the number of turns in the helix, and may be enhanced slightly by tapering the open end towards the centre.

At higher frequencies an additional 1dB can be obtained by replacing the flat reflector with a cup that encloses the first turn. However, the theoretical gains published by Kraus and others are optimistic. (**Fig 16.32**) Maximum realisable gains are given by following formula for helix lengths between 2 and 7 wavelengths.

$$G_{max} = 10.25 + 1.22L - 0.0726L^2\ dBi$$

where L is the length of the antenna in wavelengths.

A typical antenna with a seven turn helix has a gain of approximately 12dBi over a 2:1 frequency range. To fully utilise this gain it is necessary to use a circularly polarised antenna (eg a helix of the same sense) for both transmission and reception. If a plane-polarised antenna, such as a dipole, is used there will be an effective loss of 3dB due to polarisation mismatch.

General dimensions for helix antennas are shown in **Table 16.13**.

A Practical Helix Antenna for 144MHz

The greatest problem to be overcome with this type of antenna for 144MHz operation, with a helix diameter of 24½in, is the provision of a suitable support structure.

Fig 16.33 shows a general arrangement in which three supports per turn (120° spacing) are used. Details of suitable drilling of the centre boom are given in **Fig 16.34**.

The helix may be made of copper, brass, or aluminium tube or rod, or coaxial cable. This latter alternative is an attractive material to use, being flexible with the braid 'conductor' weatherproofed. If coaxial cable is used the inner conductor should be connected to the outer at each end, and the jacket well sealed to prevent moisture ingress and corrosion.

The reflector is located at a distance a behind the start of the first turn, and is supported by crossed supports from the central boom. Material for the reflector can be any kind of metal mesh such as chicken netting. Radial spokes alone are not sufficient and will

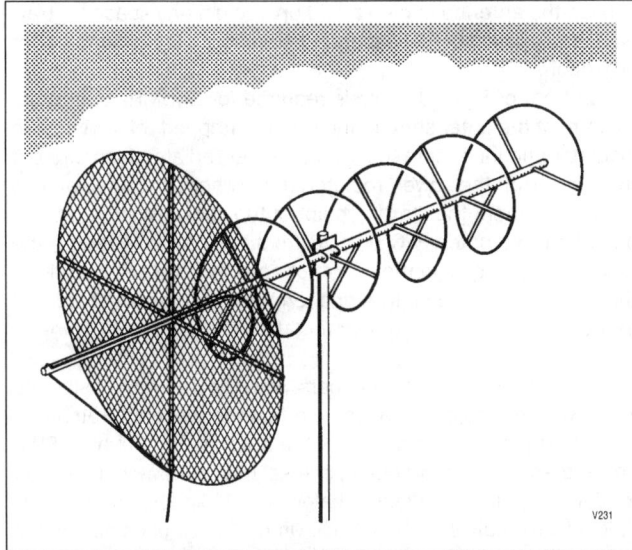

Fig 16.33: General arrangements of support structure for a five-turn helical antenna for 144MHz. The antenna is right hand circularly polarised

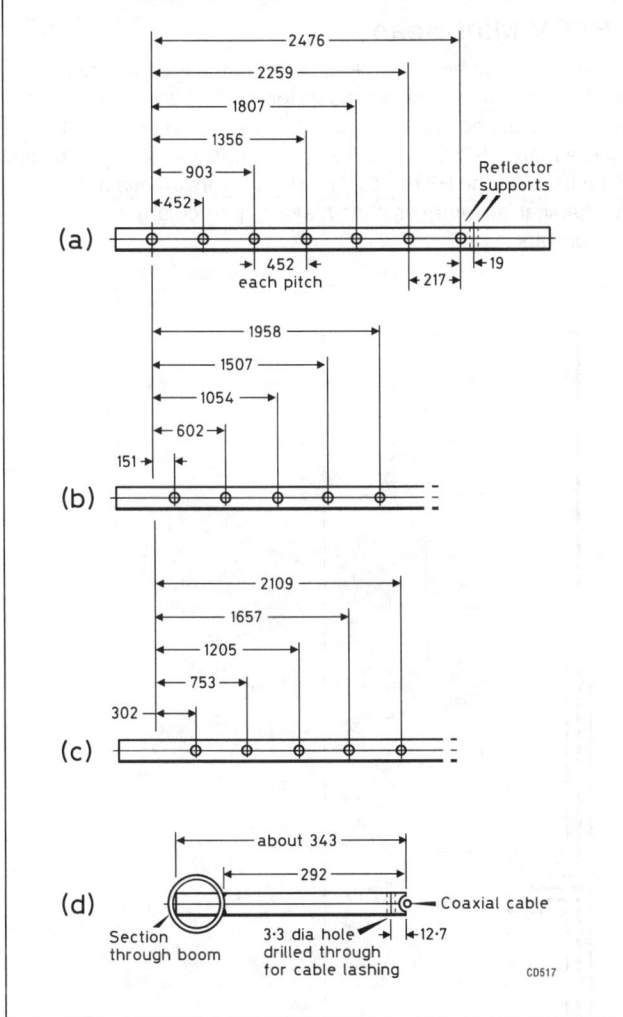

Fig 16.34: 144MHz helix first side drilling dimensions (a), reflector holes are drilled at right angles; (b) and (c) are drilled at intervals of 120 and 240 degrees respectively from (a). (d) cutting and filing dimensions for the element stand-offs

reduce the gain by 1-2dB, unless connected together with wires in dartboard fashion.

It is not essential that the central boom should be constructed of non conductive material. Metal booms may be used provided that they are centrally placed along the axis of the helix. This can lead to a simple construction using square aluminium tubing, as sold for self-construct shelving in some DiY centres. Corners and end fixtures can also be used to fasten the boom rigidly to the reflector without having to resort to machining or fabricating brackets. A square section also simplifies the mounting of the insulators, which can be made from Delrin or other plastic rod, and secured through the boom by a long screw or bolt with a single hole at the far end through which to thread the helix. The number of insulators will depend upon the rigidity of the helix material. At 433MHz, supports every 1.25 turns are adequate for a helix made from copper tubing.

Although probably too heavy for 144MHz designs, copper tubing for small-bore central heating is suitable for 433MHz helices. It is readily obtainable in DiY centres in malleable coils that can be easily shaped over a suitable former. Draw a line with a wax pencil or paint along adjacent turns whilst they are still on the former, which if sized correctly, will allow the turns to be drawn out to the correct positions whilst the marks remain in a straight line. This helps considerably when ensuring the turns diameter and pitch are maintained along the length of the helix. The ends of the tube should be hammered flat and soldered up to prevent the ingress of water.

The last fractional turn of the helix closest to the reflector should be brought very close to the reflector as it approaches the connector, to bring the impedance of the helix to 50 ohms. Helix antennas for higher frequencies are easier to construct than Yagi antennas of comparable gain and require little adjustment. Detailed instructions for building 435MHz and 1296MHz helix antennas for satellite communications have been published in [26] and [27].

HAND-HELD AND PORTABLE ANTENNAS

Normal Mode Helix

The normal mode helix antenna comprises a length of spring wire wound such that the diameter of the spring is less than 0.1λ, and typically of order 0.01λ. Such antennas become resonant when their axial length is around 0.1λ, and can be designed to offer manageable impedances at the base. The resonance occurs over a relatively narrow band, and is heavily influenced by any jacket or sleeve fitted over the helix, and by the nature of the groundplane (if any) against which it is fed. The current distribution along the length of the helix is similar to that of a whip antenna, but compressed into the much shorter length of the helix.

For hand-held radios, the length of the helix is dimensioned such that the current distribution is similar to that expected on a 5/8λ whip, ie the current maximum occurs about one third of the overall height above the feed point. This helps to maximise the radiation efficiency of the antenna, whilst also minimising effects of the variability of the ground plane (hand held radio) and body proximity on both the input impedance and radiating efficiency.

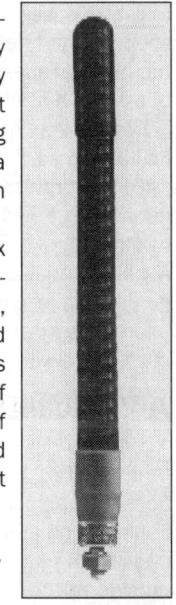

Fig 16.35: A typical commercial helical antenna with screw mounting facility

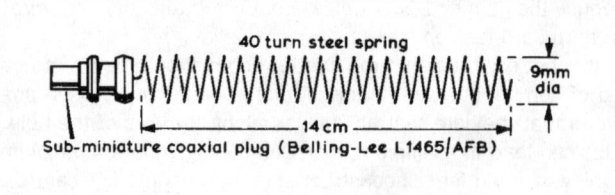

Fig 16.36: Details of a home-made helical whip for 145MHz. A BNC plug could also be used

A 3/4λ whip over a moderate ground plane has a resistive match very close to 50 ohms. If this whip is coiled into a helical spring it will match to approximately 50 ohms and resonate at a lower frequency, partly due to capacitance between the coil turns.

If the spring is trimmed to the original frequency the result will be an antenna of about 0.1λ overall height. The actual length of wire is between 1/2λ and 5/8λ at the working frequency. Electrically it is still a resonant 3/4λ antenna. Near-base capacitance also modifies the matching under certain frequency and ground plane conditions.

If the turns are very close together, the helical antenna will resonate at a frequency approaching the axial length, because of strong coupling between the turns. There is an optimum 'spacing' between turns for best performance. A 145MHz helical antenna typically has a spacing between turns equal to twice the diameter of the wire used.

The helical whip is very reactive off-resonance. It is very important that it is resonated for the specific conditions that prevail in its working environment.

Fortunately, it is often only necessary to change the number of turns to resonate the spring over such diverse conditions, ie a large ground plane or no ground plane at all. The resistive part of the impedance can vary between 30 and 150 ohms at the extremities.

Under typical 'hand-held' conditions (**Fig 16.35**), although to a small extent depending on the frequency of operation, the spring can offer something close to a 50-ohm impedance match. **Fig 16.36** shows the number of turns required for a typical 9mm diameter helix for 3/4λ resonance.

As the helical is reduced in length two effects occur. First, the radiation resistance is lower than the equivalent linear whip so the choice of a good conducting material is important to reduce resistive losses. A plain steel spring compared with a brass or copper-plated helix can waste 3dB of power as heat. Secondly, the physical aperture of the helical whip is around one third that of a λ/4 whip, which would imply a loss of 4.77dB.

Results obtained from copper-plated, Neoprene-sheathed helical antennas, correctly matched to a hand-held transmitter at 145MHz, provided signals at worst 3dB and at best +1dB compared with a λ/4 whip. A λ/4 whip with minimal ground plane would offer signals about 6dB compared to a λ/2 dipole. A helical antenna, resonant and matched, on a λ/2 square ground plane can give results 2-3dB below a λ/2 dipole. An alternative arrangement using a bifilar-wound helix gives identical results (within 0.2dB) to a λ/2 dipole.

A Vertical Dipole for Portable Operation

A practical dipole for portable operation on in either the 2m or 6m band [28] is shown in **Fig 16.37**. The upper and lower sections together form a centre-fed half wavelength dipole. The feed cable is wound into a resonant choke to present a high impedance to the lower end of the dipole to reduce currents on the outside of the feed cable. Constructed from RG58CU cable or

similar, the antenna can be rolled up into a small space for travel, then unrolled and suspended from a suitable support for operation.

3870mm of RG8CU cable is required for 145MHz operation. 470mm of the outer sheath and braid is stripped off, leaving the insulator and inner core to form the upper radiator. Measure out the length of the lower radiator from where the insulator is exposed to mark the starting point of the choke. Wind 4.6 turns on 32mm diameter PVC pipe to from the choke. Feeding the cable through holes in the centre of end caps on the pipe allows the antenna to hang tidily. A ring terminal or solder tag soldered to the tip may be used for hoisting the antenna on nylon line or similar.

Tuning should be done outdoors, with the antenna positioned well away from objects that could affect the resonant frequency. Trim short pieces from the tip of the antenna to obtain a VSWR better than 1.3:1 (in 50 ohms) across 144 - 146MHz. If a longer feeder is required, the length below the choke should be a multiple of one half wavelength (680mm to compensate for the velocity factor of the dielectric) to minimise de-tuning.

A 6m variant can be constructed using 7280mm of RG858CU cable. 11.8 turns of cable should be wound on a 50mm diameter PVC tube to form the choke. Any additional feeder should be a multiple of 1980mm.

HB9CV Mini Beam

The HB9CV mini-beam, because of its compact and straightforward construction, is suitable for both base station and portable use, and can be particularly useful in confined spaces such as lofts. Similar antennas are the lazy-H and ZL Special often used on the HF bands. The HB9CV version has one or two mechanical advantages that make it particularly suitable for VHF portable use.

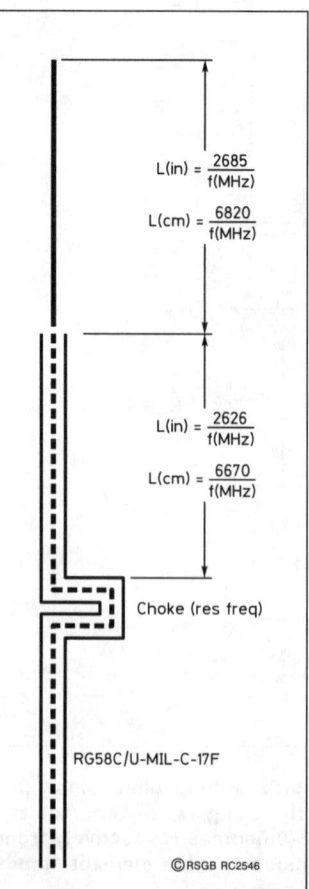

$$L(in) = \frac{2685}{f(MHz)}$$

$$L(cm) = \frac{6820}{f(MHz)}$$

$$L(in) = \frac{2626}{f(MHz)}$$

$$L(cm) = \frac{6670}{f(MHz)}$$

Choke (res freq)

RG58C/U-MIL-C-17F

© RSGB RC2546

Fig 16.37: The "feedline vertical" antenna

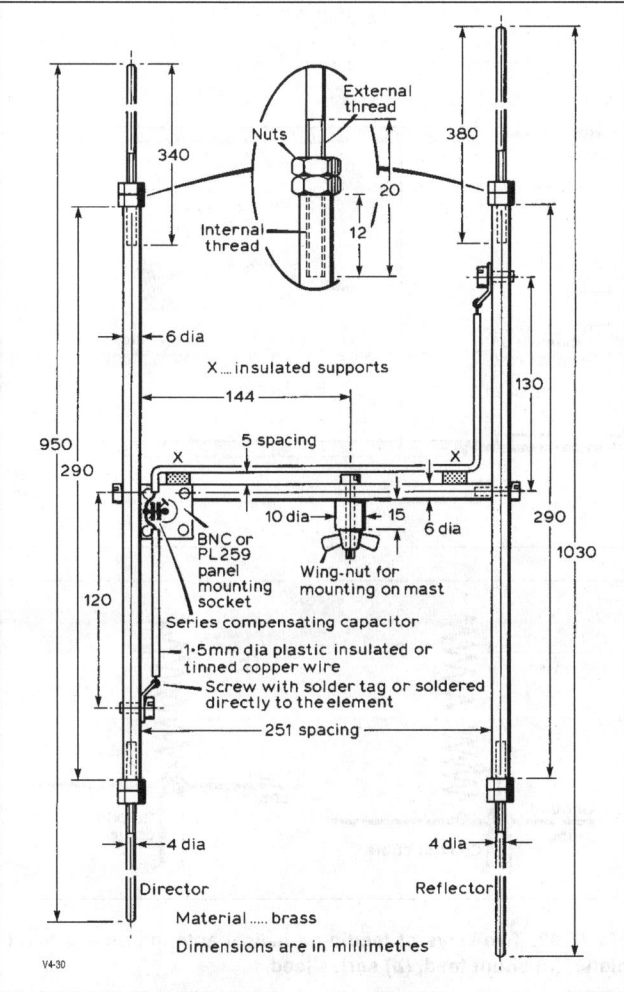

Fig 16.38 A collapsible HB9CV antenna for the 144MHz band (*VHF Communications*)

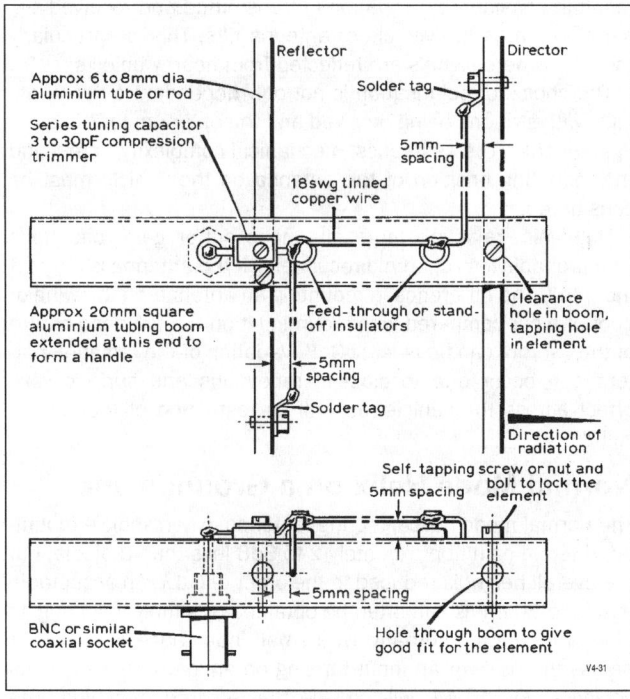

Fig 16.39: Alternative boom and feed arrangement for the 144MHz HB9CV antenna

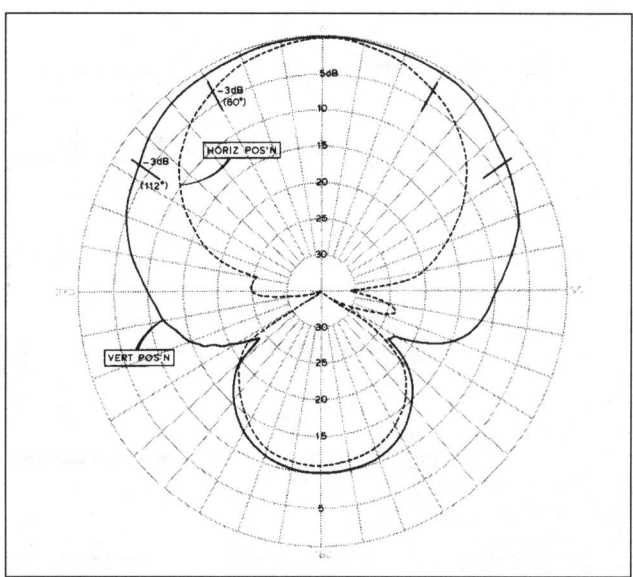

Fig 16.40: HB9CV antenna radiation patterns. Antenna 10m above ground

Fig 16.38 [29] and Fig 16.39 show two methods of construction for the HB9CV antenna. Note that a series capacitor of 3-15pF is required to adjust the gamma match/phasing combination to a VSWR of about 1.3:1 referred to 50Ω. The element spacing, and in particularly the transmission line spacing (5mm in this case), is critical for optimum impedance matching and phasing, and therefore gain and front-to-back ratio.

The principle of operation is as follows. If two dipoles are close spaced (typically 0.1 - 0.2λ) and fed with equal currents with a phase difference corresponding to the separation of the dipoles, 'end-fire' radiation will occur along the line between the dipoles in one direction, and almost no radiation will occur in the reverse direction as explained earlier in this chapter in the section on arrays.

The different element lengths found on most HB9CV antennas improve the VSWR bandwidth, not the directivity as might at first be thought by comparison with a two element Yagi antenna.

The end at which the beam is fed defines the direction of radiation. A theoretical gain in excess of 6dBd should be possi-

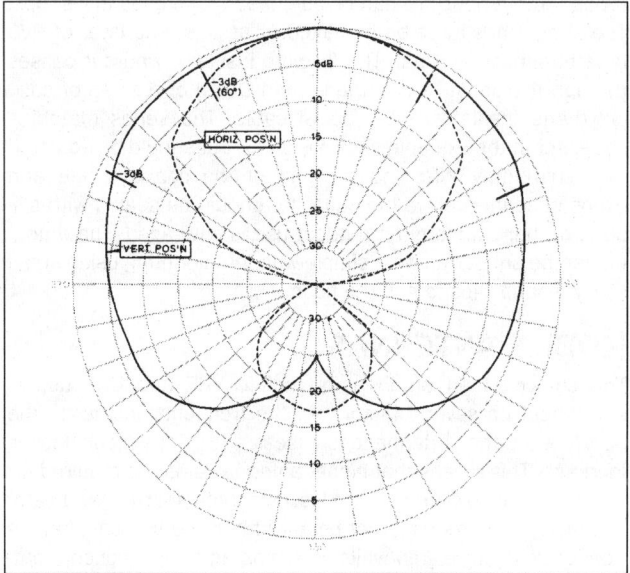

Fig 41: HB9CV antenna radiation patterns. Antenna hand-held, 1-2 metres above ground

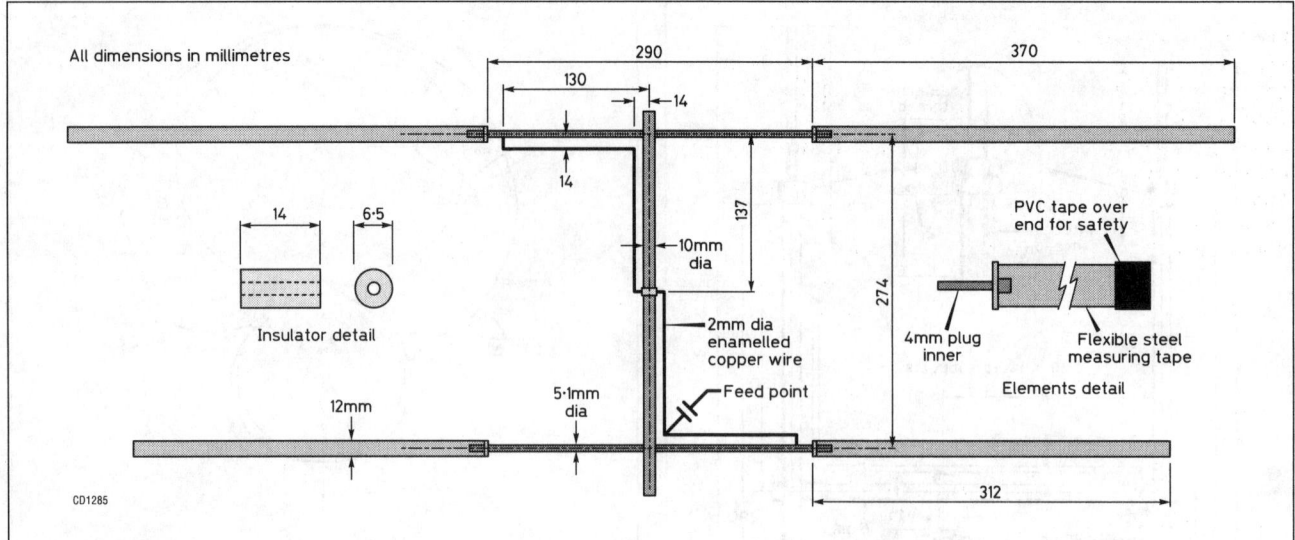

Fig 16.42: Construction details of lightweight HB9CV antenna for 144MHz

ble. Depending on construction techniques, gains of 4 to 5dBd with front-to-back ratios of 10 to 20dB tend to be obtained in practice. The radiation patterns shown in **Fig 16.40** and **Fig 16.41** are for the antenna of Fig 16.38. This antenna has a typical gain of 5dBd. Note the difference obtained when mounted at 10m (30ft) above the ground compared with hand-held measurements 1-2m above the ground. The latter height is typical for the antenna being used for direction finding.

Lightweight HB9CV for 144MHz

The compact size of the HB9CV design makes it eminently suitable for direction finding contests, EMC or interference probing, and portable work. The need for a very lightweight directional antenna for EMC investigations led to the design shown in **Fig 16.42**.

The boom and stub elements are made from thin walled brass tubing, soft soldered or brazed together. The removable elements are made from an old 12mm wide spring steel measuring tape soldered on to 4mm 'banana' plugs, although replacement tapes without housings can be purchased from good tool shops. The sharp ends must be protected by at least one layer of PVC tape or similar material. The feedline insulator where it passes through the boom can be made from Delrin or a scrap of solid polythene insulator from coaxial cable. The series matching capacitor in the example shown is 13pF, but should be adjusted for minimum VSWR, and the end of the coaxial cable and exposed connection to the capacitor should be sealed with silicone rubber compound if outdoor use is envisaged. The antenna can be supported on a simple wooden mounting using small 'Terry' spring clips to grip the boom.

MOBILE ANTENNAS

The choice of an antenna for mobile VHF and UHF use is dependent on several factors. As the frequency increases the aperture of the antenna decreases, and propagation losses increase. This means that higher antenna gains are required for UHF than VHF to overcome the losses of both aperture and path.

Considerable reduction of beamwidth in the vertical plane is needed to achieve gain whilst retaining an omnidirectional pattern in the horizontal plane. A compromise has to be made to obtain maximum gain in the best direction that gives minimum disruption of signals when mobile.

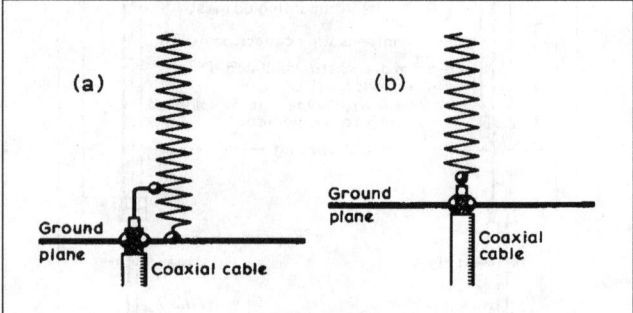

Fig 16.43: Two ways of feeding a helical antenna on a groundplane: (a) shunt feed, (b) series feed

For example an omnidirectional antenna of 6dBd gain will have a typical half-power beamwidth in the vertical plane of under 30°. The narrow disc shaped beam that is produced can result in considerable variation in transmitted and received signal strength as the vehicle or antenna tilts. This is particularly the case where signals are reflected from nearby objects.

The choice of polarisation is not only dependent on compatibility with stations being received and the optimum for the propagation path. The aesthetics, mechanical complexity, safety and the mounting position of the antenna on the vehicle must be considered.

High-gain, relatively large, antennas suffer gain reductions with probable loss of omnidirectionality if the antenna is not roof mounted. The difference in mounting an antenna on the wing or boot of a car compared with mounting it on the top dead centre of the car roof can be at least 3dB. Variation of the radiation pattern can occur due to close-in reflections and surface-wave effect across the vehicle, as well as restriction of the 'line of sight'.

Normal Mode Helix on a Groundplane

The normal-mode helical (spring) antenna, when vehicle mounted, offers a gain approximately 2 to 3dB less than a dipole, but the overall height is reduced to the order of 0.1λ. An acceptable match to 50 ohms can often be obtained by simply adjusting its resonant length. Alternatively, a small inductance or capacitor across the base or an input tapping on an 'earthed' helical, as shown in **Fig 16.43**, will provide the required matching. The design and limitations of the normal mode helix were discussed earlier in this chapter under the heading of hand-held antennas.

Fig 16.44: The λ/2 antenna and its grounded λ/4 counterpart. The missing λ/4 can be considered to be supplied by the image in ground of good conductivity

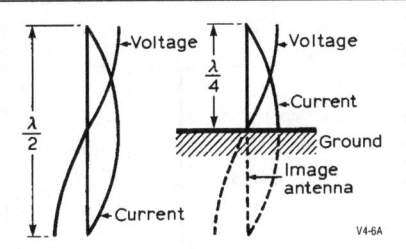

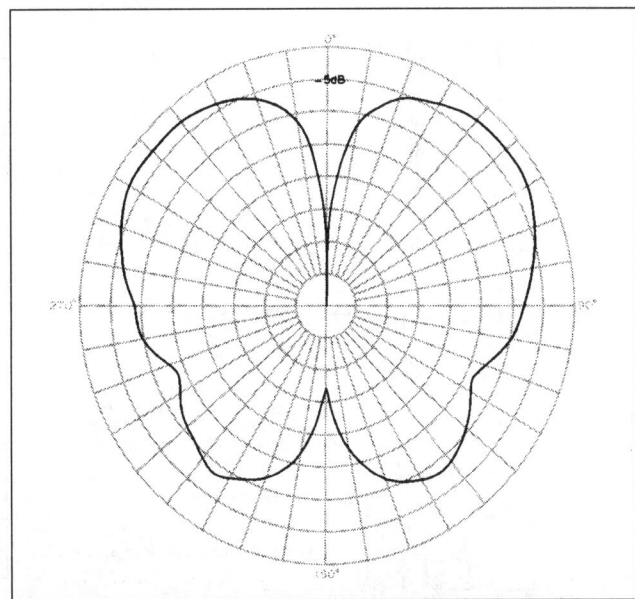

Fig 16.46: Decibel radiation pattern of a λ/4 monopole over a 1λ square groundplane at 145MHz

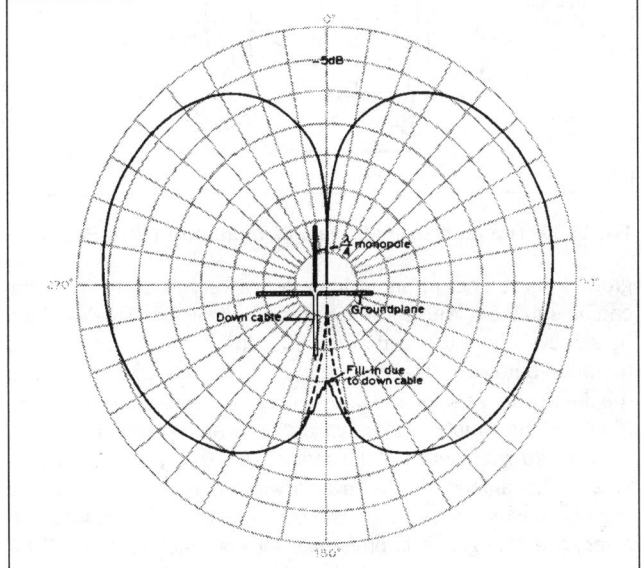

Fig 16.45: Decibel radiation pattern of a λ/4 monopole over a λ/2 square groundplane at 145MHz

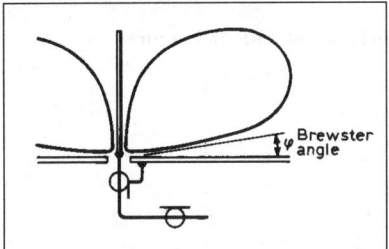

Fig 16.47: Radiation pattern of whip on large groundplane showing elevation of the main lobe

Quarter-wave Whip

This is the simplest and most basic mobile antenna. The image of the vertical λ/4 section is 'reflected' in the ground plane, producing an antenna that is substantially the same as a dipole, provided that the ground plane is infinitely large and made of a perfectly conducting material (**Fig 16.44**). In this case, all of the radiation associated with the lower half of the dipole is radiated by the top half, resulting in a 3dB improvement in signal strength in a given direction for the same power input to the antenna.

In practice the size of the ground plane and its resistive losses modify the pattern. The full 3dB is never realised. Measurement of a 5GHz monopole on an aluminium ground plane of 40 wavelengths diameter showed a gain of 2.63dBd. **Fig 16.45** and **Fig 16.46** show optimum patterns of a λ/4 whip measured on a ground plane of λ/2 sides and 1λ sides. Although the pattern is raised from the horizontal, on a medium sized ground plane the loss of horizontal gain is relatively small (20° and 1dB at 0° in Figure 45, but 40° and 6dB at 0° in Fig 16.46). However, as the groundplane size increases, the main lobe continues to rise until the situation of **Fig 16.47** pertains.

When a vertical radiator is mounted over a ground plane as described, the input impedance is typically halved. For the λ/4 whip or monopole, the input impedance is typically 36Ω - jX, that is to say approximately half the resistance of the dipole but with an additional reactive component. With 50 ohm cable impedance this would produce a standing wave ratio at the antenna base of about 1.5:1.

The simplest way to overcome this mismatch is to increase the length of the whip to produce an inductive reactance to cancel the capacitive reactance normally obtained. In practice an increase in length also raises the resistive value of the whip and a close match can usually be obtained to a 50-ohm cable.

At VHF (145MHz) the λ/4 whip's simplicity and limited height (about 49cm/19in) is often an accepted compromise. At 70MHz the physical dimensions (about 102cm/40in) are such that size is the usual limit, making a 1/4λ whip preferable to a 'gain' antenna. The effective aperture of the antenna at this frequency is compatible with path loss conditions, and the ground-plane size, when roof-mounted on a vehicle, is such that the radiation angle is fairly low. However, the shape of the radiation pattern can result in a gain reduction of 3dB to each side of the vehicle.

Half-wave and Five-eighths-wave Antennas

Ground-plane techniques described for the 1/4λ whip can be used for vertical gain antennas. If the 1/2λ dipole is extended in length, maximum forward gain (before the pattern divides into several lobes) is obtained when the dipole is about 1.2λ. This corresponds to a maximum length of 5/8λ for a ground-plane antenna.

A natural extension to the 1/4λ whip is the 1/2λ whip. However, such a radiator fed against a ground plane has a high input impedance. On the other hand, a 3/4λ radiator fed against a ground plane has a resistive input close to 50 ohms. Unfortunately, the resultant radiation pattern in the elevation plane is less than optimum.

If the 1/2λ whip could be made to look like a 3/4λ radiator then it would be possible to obtain a 50-ohm resistive input. A series coil at the ground-plane end of a 1/2λ radiator can be used to resonate at 3/4λ, but the input is still of fairly high impedance and reactive. If, however, the coil is shorted to the ground plane, tapping up the coil will provide the required

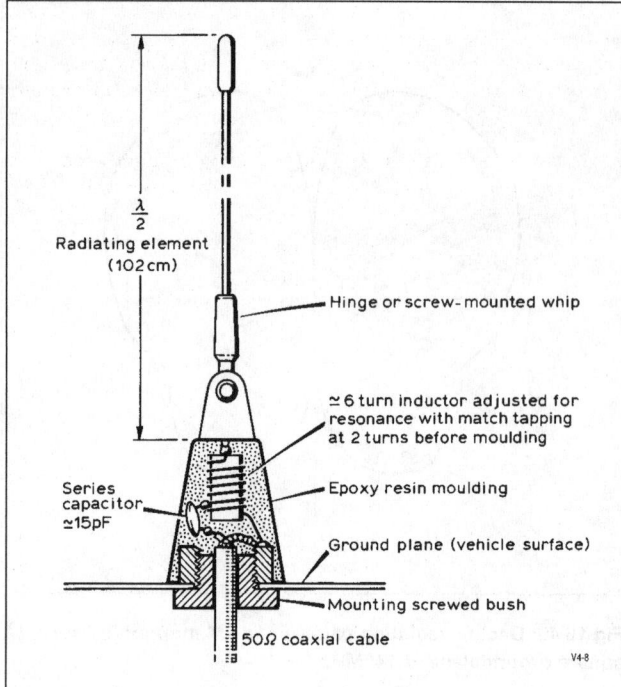

Fig 16.48: A home-built mobile antenna and mount

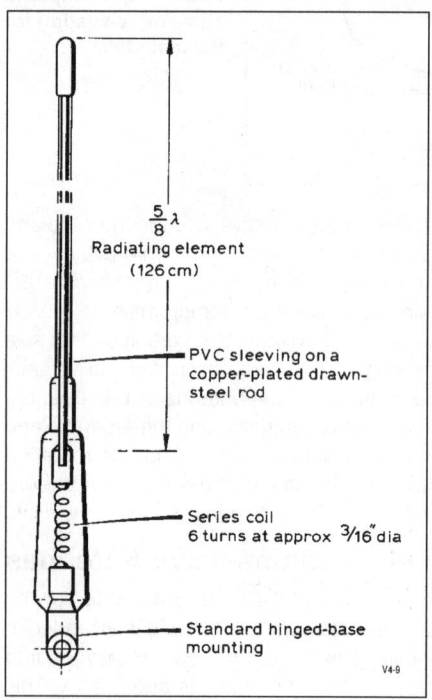

Fig 16.49: A typical commercial 5/8λ mobile antenna and mount

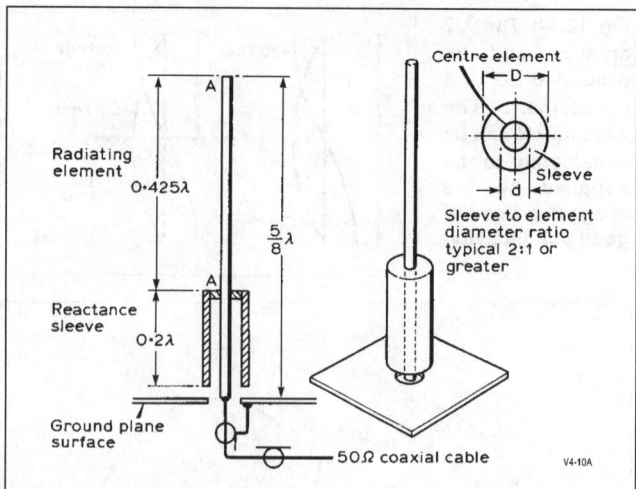

Fig 16.50: The reactance-sleeve 5/8λ monopole (*Ham Radio*)

ground plane end, an input impedance very close to 50 ohms can be obtained. With correct materials a gain close to 4dBd can be achieved from the further increase in effective aperture. The radiation pattern is raised more than that of a 1/2λ antenna, so the improved gain of the 5/8λ may not always be realised. However, the simplicity of construction is an advantage.

Fig 16.49 gives details of the series 5/8λ whip. One other advantage of this antenna is that over a wide range of mounting and ground-plane conditions it will self-compensate for impedance and resonance changes. It is preferable for both the 1/2λ and 5/8λ antennas to be 'hinged', particularly if roof-mounted, to enable folding or 'knock down' by obstructions, eg trees and garages.

Various gain figures have been reported for the 5/8λ whip antenna. Unfortunately not all antennas use optimum materials. Resistive steel wires or rods produce heating loss, and the use of a glass fibre-covered wire changes the resonant length by as much as 20%. The radiator therefore has to be cut shorter than 5/8λ, with an accompanying loss of aperture.

The construction of the series coil is important. Movement of the coil turns will change the antenna's resonance, giving apparent flutter. Some transceivers with VSWR-activated transmitter close-down will be affected by change of resonance of the antenna. This can make the power output of the transmitter continually turn down or be switched off, producing what appears as extremely severe 'flutter' on the transmission.

Several of the '5/8λ ground-plane antennas' discussed in various articles are in fact not truly antennas of this nature.

One of these devices worth considering for its own merits is that shown in Fig 16.50. It consists of a 5/8λ vertical element with a reactive sleeve of 0.2λ at the ground-plane end. The gain obtained from this antenna is typically 1.8dBd. As can be seen, the actual radiating element A-A is shorter than a 1/2λ antenna.

Another antenna family, with similar properties but different in construction, includes the 'J' and Slim Jim. These are described later in this chapter.

Seven-eighths-wave Whip

This mobile antenna is derived from the Franklin collinear shown later in this chapter. It consists of two 1/2λ elements coupled by a series 'phasing' capacitor. One effect of the capacitor is to resonate the combined elements at a lower frequency than that of a single 1/2λ element. However, reducing the length of the top element tunes the arrangement back to the original frequency.

The base impedance above a perfect ground plane is 300-400 ohms with some capacitive reactance. A series loading coil

match/input point. The addition of a capacitor in series with the input will compensate for the remaining reactive component. Fig 16.48 shows details of such an antenna.

As the aperture of the antenna has been doubled compared with the 1/4λ whip, the gain over the whip approaches 3dB. Achievement of this figure requires minimum losses in the radiating element, ie it must be copper-plated or made from a good conducting material.

The maximum radiator size of 5/8λ for a single-lobe pattern can also make use of the impedance characteristics of the 3/4λ radiator.

Construction is simpler than for a 1/2λ antenna. If the radiating element is made 5/8λ long, and a series coil is placed at the

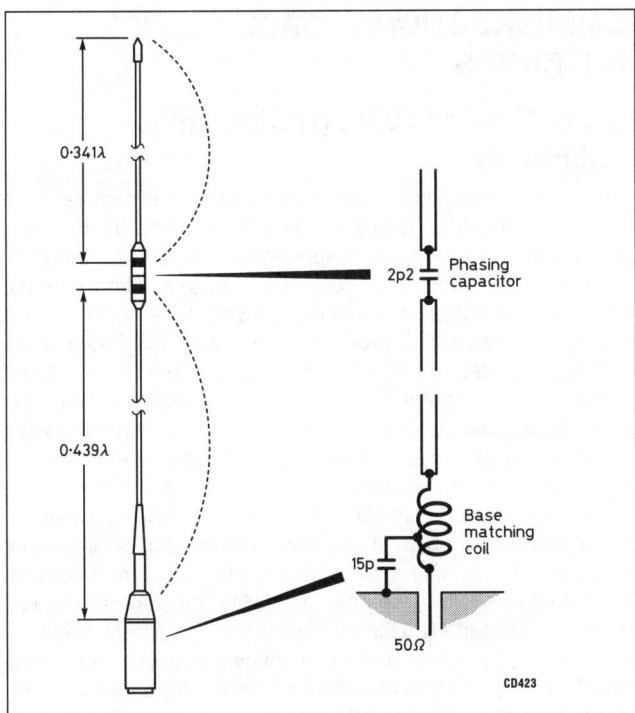

Fig 16.51: The 7/8λ whip antenna. This is effectively two short-ened half-wave elements in series with a series phasing capac-itor between them. The assembly shown is that of a commercial form of the antenna. The dotted lines show the approximate current distribution

in combination with an L-matching section gives a good match to 50-ohm coaxial feeder. The match is maintained with quite modest ground plane size (1/4λ radials or 1/2λ diameter metal surface). This makes the 7/8λ whip suitable for vehicle mount-ing or for use as a base-station antenna.

The final length of the two radiator elements is somewhat dependent on their diameters and the design of the series capacitor and matching unit. **Fig 16.51** shows the general appearance and dimensions of a commercial version of the 7λ/8 whip, together with typical circuit components and current distribution in each element.

The theoretical gain of this antenna is 4.95dBi (2.8dBd) over a perfect ground plane. The professionally measured gain, with the whip on a 1m ground plane, was slightly over 4.7dBi for the full 144MHz band. The radiation pattern in the E (vertical) plane was predominantly a single lobe (torus/doughnut) peaking at 4° above the horizon and with a 3dB beamwidth of 38.5°.

Fig 16.52: The mechanical construction of the Omni-V (dimen-sions are in millimetres)

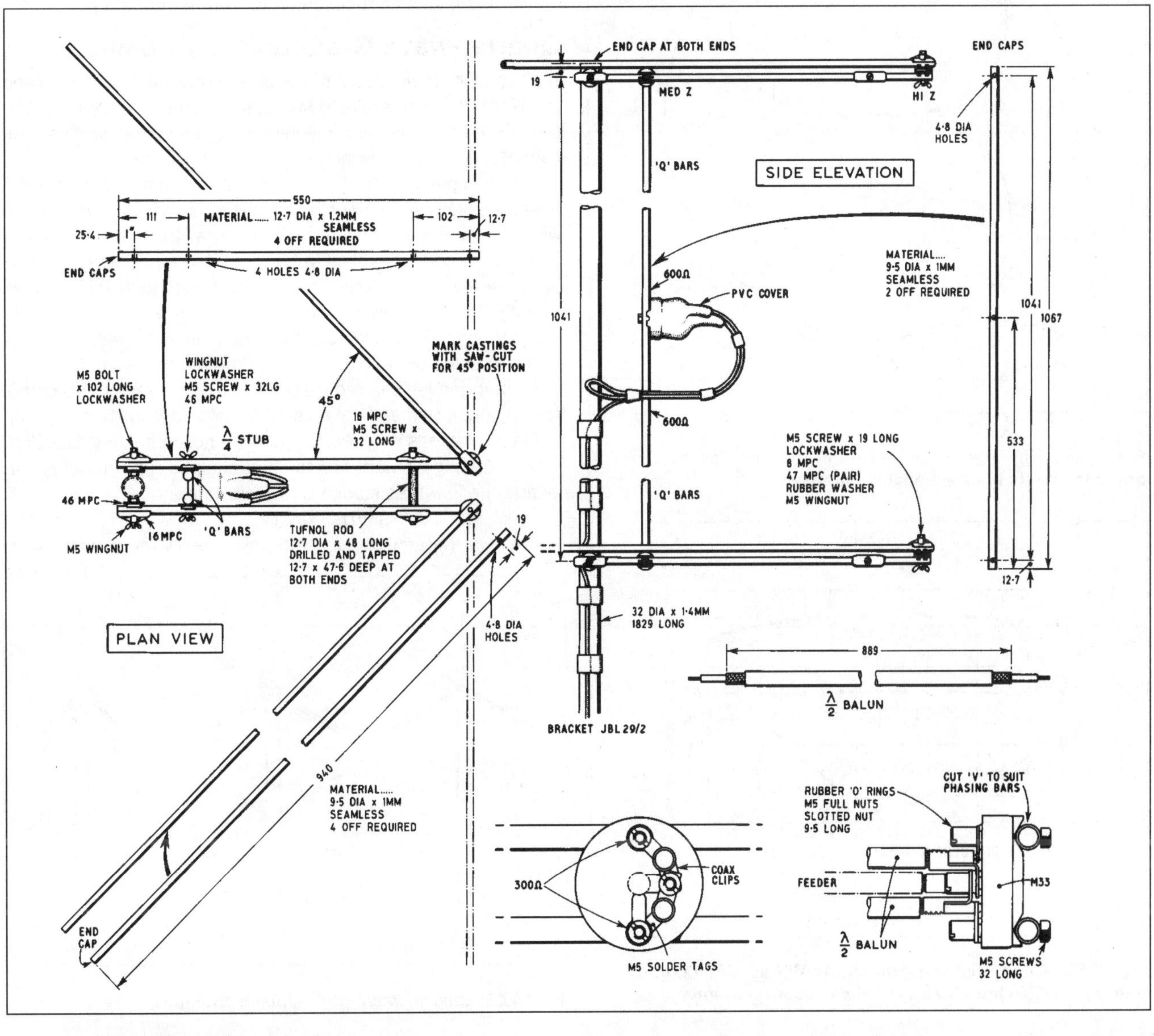

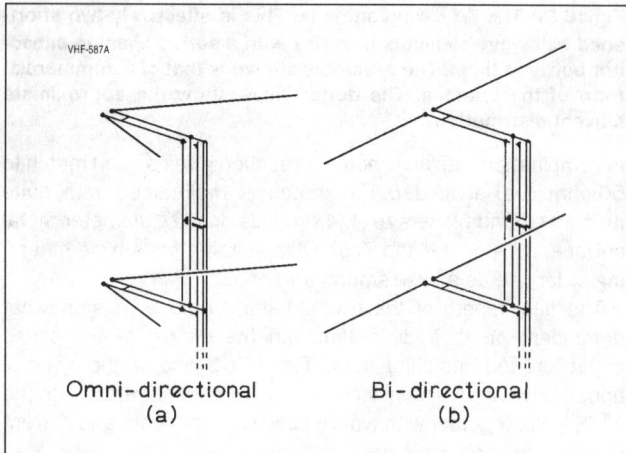

Fig 16.53: Formation of the Omni-V antenna

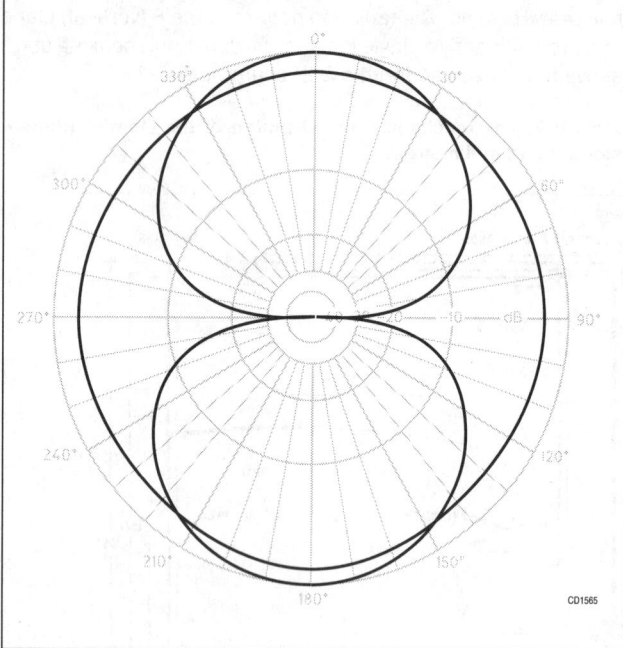

Fig 16.54: The horizontal radiation patterns for an Omni-V antenna and its bi-directional counterpart

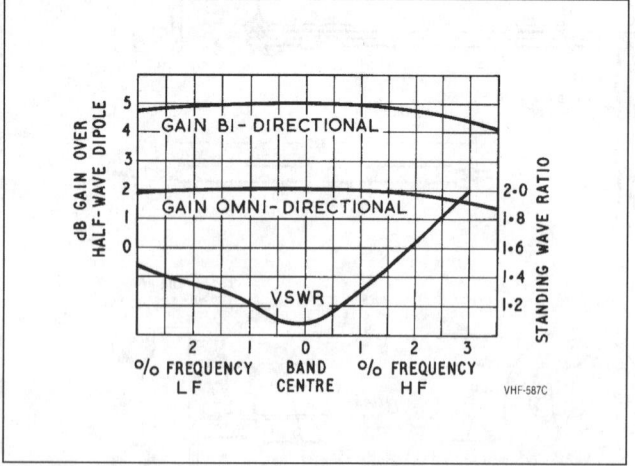

Fig 16.55: Chart showing gain and VSWR as a function of frequency for the Omni-V and the bi-directional antenna

OMNIDIRECTIONAL BASE STATION ANTENNAS

Omni-V for 144MHz (Horizontally Polarised)

This antenna consists of a pair of λ/2 dipoles. The centres of the dipoles are physically displaced to produce quadrature radiation with the ends of each dipole supported on a λ/4 shorted stub. A pair of Q-bars are tapped down the stubs to a point where the impedance is 600Ω as shown in **Fig 16.52**. When the two units are fed in parallel, they produce an impedance of 300Ω at the centre. A 4:1 balance-to-unbalance coaxial transformer is fitted to the centre point of the Q-bars to enable a 75Ω coaxial feeder cable to be used. A 50Ω feed can be arranged by repositioning the Q bars on the antenna stubs to provide a tap at 400Ω on the stubs. This can be achieved by monitoring the VSWR on the coaxial feeder whilst adjusting the Q bar position by small but equal amounts on both stubs. The balun should, of course, be constructed from 50Ω coaxial cable (transforming from 200Ω in the balanced section) for a match to 50Ω. The general arrangement is shown in **Fig 16.53** showing how the antenna may be arranged to give either an omnidirectional or bi-directional radiation pattern, and typical radiation patterns for either case are shown in **Fig 16.54**. **Fig 16.55** shows the gain and VSWR of these antennas as a function of the centre frequency.

Quarter-wave Groundplane Antenna

This is one of the simplest omnidirectional antennas to construct and usually yields good results. However, some unexpected effects may occur when the antenna is mounted on a conductive mast, or if RF current is allowed to flow on the outside of the feeder.

In its simplest form, the ground plane antenna comprises a quarter-wavelength extension to the inner of a coaxial cable, with several wires extending radially away from the end of the outer of the coaxial cable, **Fig 16.56(a)**. The input resistance will be quite low, in the order of 20 ohms, although this may be transformed to a higher impedance by using a folded monopole radiator as shown in **Fig 16.56(b)**. Equal diameter elements provide a 4:1 step-up ratio to around 80 ohms, and a smaller diameter grounded leg can reduce the input impedance to 50 ohms. The feedpoint impedance can be modified by bending the groundplane rods downwards from the horizontal, **Fig 16.56(c)**. If the radiating element and the groundplane rods are all λ/4 in length, the input resistance is approximately:

$$R = 18(1+\sin\theta)^2 \quad \text{ohms}$$

where θ is the groundplane rod angle below the horizontal, in

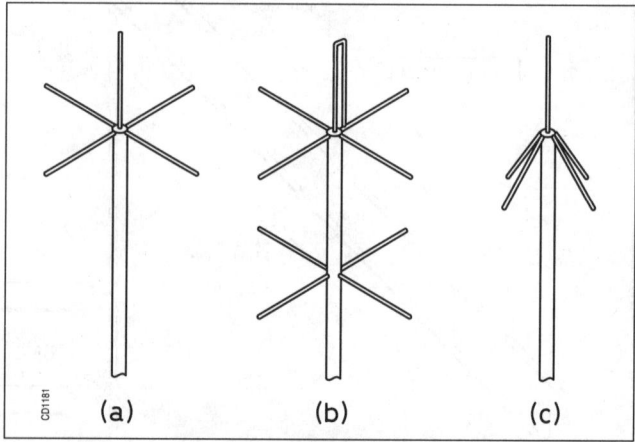

Fig 16.56: Quarter wave groundplane antennas

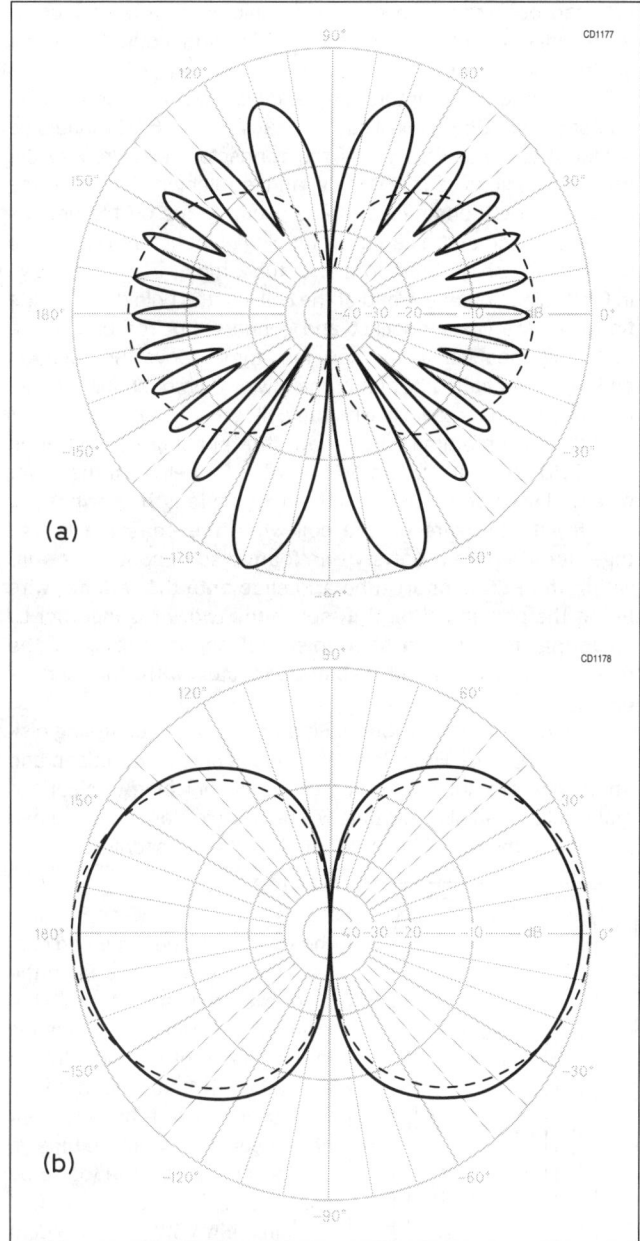

Fig 16.57: Radiation patterns of quarter-wave groundplane antenna, (a) with and (b) without a metallic mast

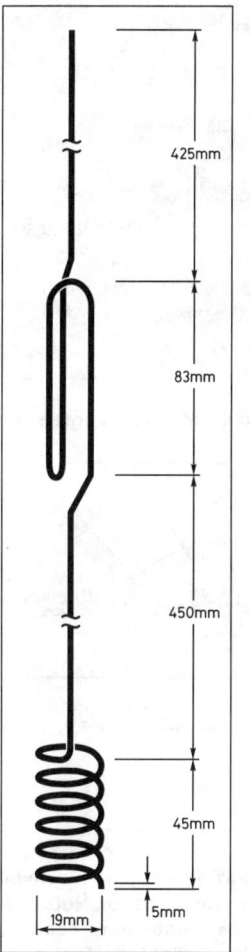

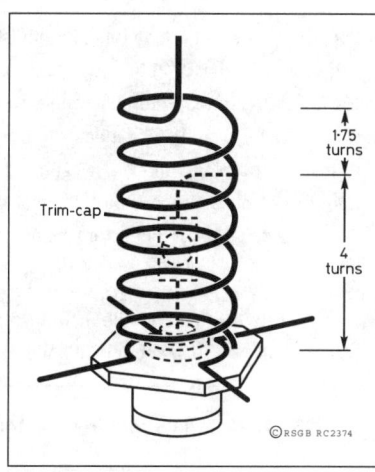

Fig 16.58: Shape and dimensions of the radiator wire

Fig 16.59: Matching section and resonant groundplane radials

degrees. A 50 ohm resistance is achieved when θ is 42 degrees.

The ends of the groundplane rods are sometimes joined together with a conductive ring to provide additional mechanical stability. The ring increases the electrical size of the groundplane, and the length of the radials can be reduced by about 5%.

The few rods forming the groundplane usually do not prevent current flowing on any conductive supporting mast, or on the outside of a coaxial feeder. The mast or feeder can become a long radiating element which may enhance or destroy the radiation pattern of the antenna, dependent upon the magnitude and phase of the mast currents relative to that on the antenna. An example of this is shown in Fig 16.57(a), where the monopole and groundplane is mounted on a 5λ mast (about 10 metres). The corresponding radiation patterns without mast or cable influences are shown in Fig 16.57(b). The effects of ground reflections have been ignored in both cases.

Some antenna designs make use of these currents to enhance the gain of the monopole, and sometimes have a sec-

ond set of groundplane rods further down the mast, tuned to present a high impedance to reduce currents flowing below that point. The mast currents can be reduced a little by using more radials in the groundplane or extending their length to around 0.3λ.

An open circuited choke sleeve can be more effective than radial wires for mast current control. This technique is used in the skirted antenna described later in this Chapter.

A Dual-band Whip for 145/435MHz

This base station antenna, devised by Bert Veuskens, PA0HMV acts as an end fed antenna on 145MHz with a gain of 0dBd, and two stacked 5λ/8 radiators with a gain of about 5dBd at 435MHz [30]. The radiator is made from 2mm copper wire, formed as shown in Fig 16.58. The coil at the base and a series capacitor provides the matching network at 145MHz. The folded stub section at the centre ensures the upper and lower 5/8λ sections radiate in phase at 435MHz. A groundplane comprising four short radials resonant at 435MHz is fixed to the input connector, see Fig 16.59. No groundplane is required for operation at 145MHz, although the coaxial feed can be coiled close to the connector to choke off any current on the outside of the cable. The whole assembly is fitted inside a section of PVC tube for support and protection from the weather.

Assembly

A list of materials is given in Table 16.14. The corners of the coaxial socket are filed off to make it a snug fit in the 28mm copper pipe, and a hole is required for the earthy end of the matching coil as shown in Fig 16.60. The radials should be cut slightly too long to allow for trimming after assembly. A jig to hold the radials and connector in place during soldering is essential,

Copper or brass mounting tube, 28mm OD x 220mm

PVC tube, 32mm OD, 28mm ID, x 1.2m

Jubilee clip, 32mm, preferably stainless steel

N-type 50Ω socket, PTFE dielectric, with square flange

Brass rod or tubing, 3mm OD x 720mm for radials

Bare or enamelled copper wire, 2mm (14SWG) 1.6m

Copper/brass tube or inner from connector block to butt-joint radiating elements (discard screws)

Tubular trimmer capacitor, 10pF or RG58 cable per Figure 61. Ensure that the capacitor can handle the transmitter power safely.

Silicone rubber sealing compound or bathroom sealant

Table 16.14: Materials required for 145/435MHz omni antenna

Fig 16.60: Details of the coaxial socket and attachment of radials

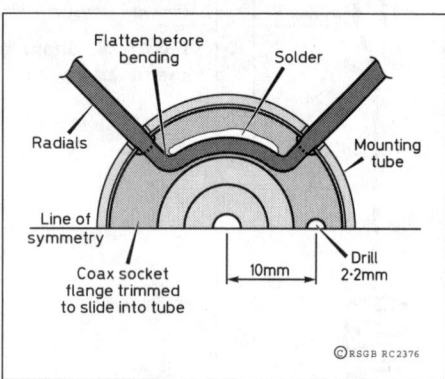

Fig 16.61: RG58 coaxial cable capacitor. Note that the length may need to be adjusted for best match

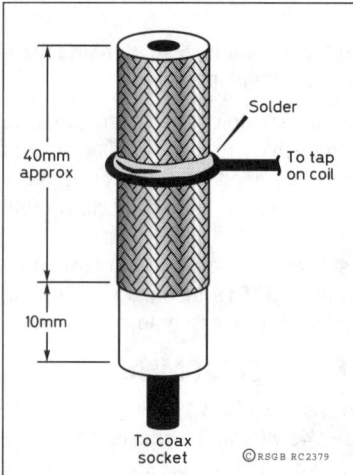

Fig 16.62: Phasing stub bending detail

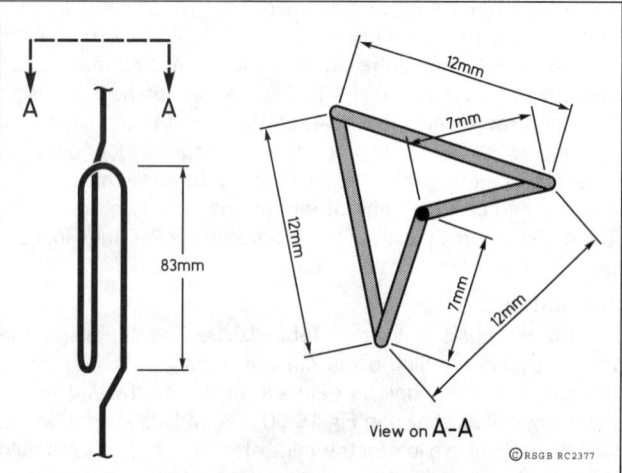

and can be made from a piece of chipboard. After soldering, each radial should be trimmed to 173mm measured from the centre of the socket.

The rotor terminal of the trimmer capacitor, or the centre conductor of the RG58 coaxial cable capacitor (**Fig 16.61**) should be soldered to the centre pin of the connector next. Prepare the matching coil by straightening and stretching 60cm of the antenna wire. If enamelled wire is used, scrape off the enamel at one end and at the capacitor tapping point before winding up and shaping on a 19mm former of tube, rod or dowel as shown in Fig 16.59. Solder the end of the coil into the hole drilled in the flange of the coaxial socket and connect the free end of the matching capacitor to the coil, four turns above the soldered end. Approximately 200mm of the lower radiator should be projecting upwards, coaxially with the coil.

With the remaining wire, shape the phasing section as in **Fig 16.62** using a 9.5mm drill bit as a former. Trim the lower wire end so that it makes up the 450mm length shown in Fig 16.58 with the wire on the coil when the ends are butted together. Prepare a polystyrene foam disc to centre the wire within the PVC support tube and slide onto the radiator wire below the phasing stub. Butt-splice the two wires together by soldering them into a short piece of copper tubing or the inner part of a small cable connector with the screws removed.

Cut the top wire to 460mm. Fit a second foam centreing disk on to the top wire. Four slots, 90 degrees apart, 7mm deep and 4mm wide are cut at one end of the copper pipe to clear the groundplane radials. This pipe will be used to clamp the finished antenna to the top of its mast with U-bolts and saddles.

Tuning

Pass a short 50-ohm cable through the copper pipe and connect it to the N socket. Push the pipe over the socket until the groundplane radials are resting in the bottom of the four slots in the tube. Set up the antenna without its cover tube well clear of objects that could de-tune it, but low enough for the top to be accessible. Trim the top element for minimum VSWR in the 70cm band at a frequency 3MHz higher than required, eg 438MHz for 435MHz operation. The PVC cover tube will reduce the frequency by this amount when it is installed. Fix the two centreing discs in place with a drop of epoxy glue, at 170mm below the end of the top wire and halfway between the top of the coil and the bottom of the phasing stub for the lower wire.

Slide the PVC cover tube (**Fig 16.63**) over the antenna until the radials are fully seated in the slots, check that the minimum VSWR is below 1.5:1 and at the desired frequency in the 70cm

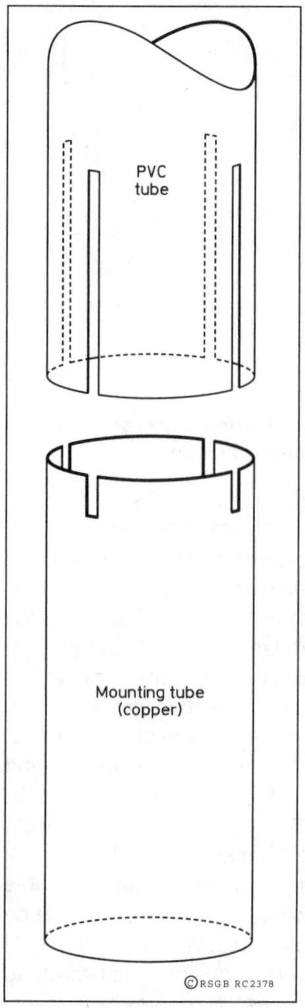

Fig 16.63: Mounting tube and PVC cover details showing slots to clear radials

Figure 15.64: Photograph of finished antenna installed on mast

band. Raise the PVC tube just enough to allow access to the trimmer/cable capacitor and adjust for minimum VSWR at 145MHz. Push the PVC tube down again and secure with the Jubilee clip below the radials. Cap and seal the top and weatherproof the slots with sealant to keep rain out, whilst leaving a hole to prevent condensation from being trapped. The finished amntenna is shown in **Fig 16.64**.

The Skirted Dipole Antenna

The skirted dipole antenna (**Fig 16.65**) does not require ground-plane radials, and can be mounted in a cylindrical radome for better appearance and lower wind induced noise. The skirt forms the lower part of a half wave dipole, and being one quarter wavelength long, presents a high impedance at its lower end, reducing unwanted currents on the mast. The current is further reduced by a second choke, with its open, high impedance end placed one quarter wavelength below the dipole skirt for best effect. The radiation pattern of this antenna closely resembles that of a half-wave dipole in free space.

Gain sleeve dipole

The *gain sleeve dipole* (**Fig 16.66**) is derived from the 1.8dBd shunt-fed 5/8λ antenna described in the section on mobile antennas.

The radiating element B-B is a 1 element fed part way along its length with a 0.2λ series short circuited stub to provide the transformation to 50 ohms and phasing. The impedance of asymmetrical antennas was investigated by R W P King [31], and has a number of applications in the design of groundplane antennas with elevated feedpoints. Having approximately twice the aperture of the λ/2 dipole, a gain of typically 2.5-3dBd is achieved.

Mechanical construction is open to interpretation but 'beer can' or plastic water pipe formats offer two solutions. Note that the mounting point should be at A-A and not on the 0.25λ sleeve.

The Discone

The discone is often used where a single omnidirectional antenna covering several VHF/UHF bands is required. A single antenna is capable of covering the 70, 144 and 432MHz bands or 144, 432 and 1296MHz. However, as the antenna can operate over roughly a 10:1 frequency range, it will more readily radiate harmonics present in the transmitter output. It is therefore important to use a suitable filter to provide adequate attenuation. The radiation angle tends to rise after the first octave of frequency and this is the normal acceptable working range. If correctly constructed, a VSWR of less than 2:1 can be obtained over the octave range. One characteristic of the basic discone is a very sharp deterioration of the VSWR at the lowest frequency of operation.

The discone consists of a disc mounted above a cone, and ideally should be constructed from sheet material. However, with only a little loss of performance the components may be made of rods or tubes as illustrated in **Fig 16.67**. At least eight or preferably 16 rods are required for the 'disk' and 'cone' for reasonable results. Open mesh may be used as an alternative to sheet metal or rods.

The important dimensions are the end diameter of the cone and the spacing of this from the centre of the disc. These are instrumental in obtaining the best termination impedance, ie 50 ohms [32].

Fig 16.68 shows the key dimensions, which must satisfy the following requirements:

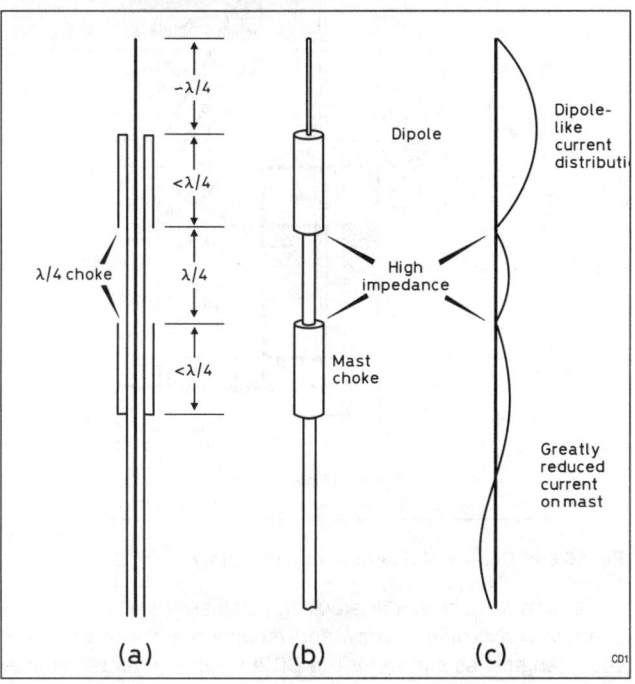

Figure 15.65: Skirted dipole antenna with mast choke

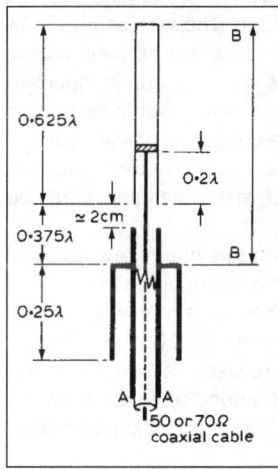

Fig 16.66: Gain sleeve dipole

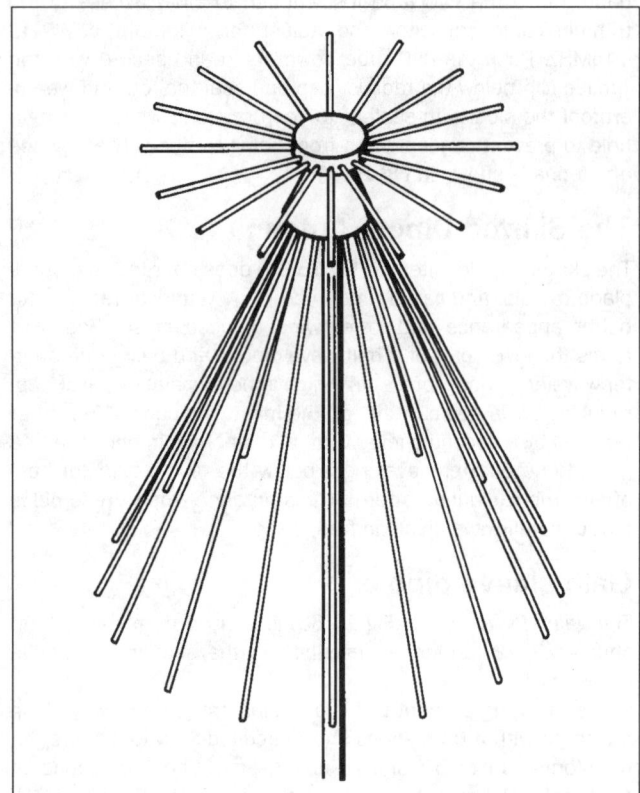

Fig 16.67: General arrangement of skeleton form of discone

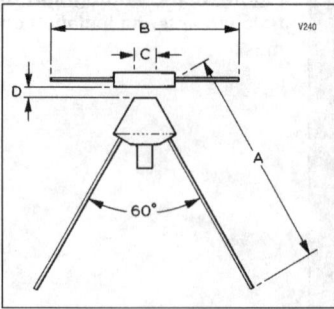

Fig 16.68: Key dimensions of discone antenna

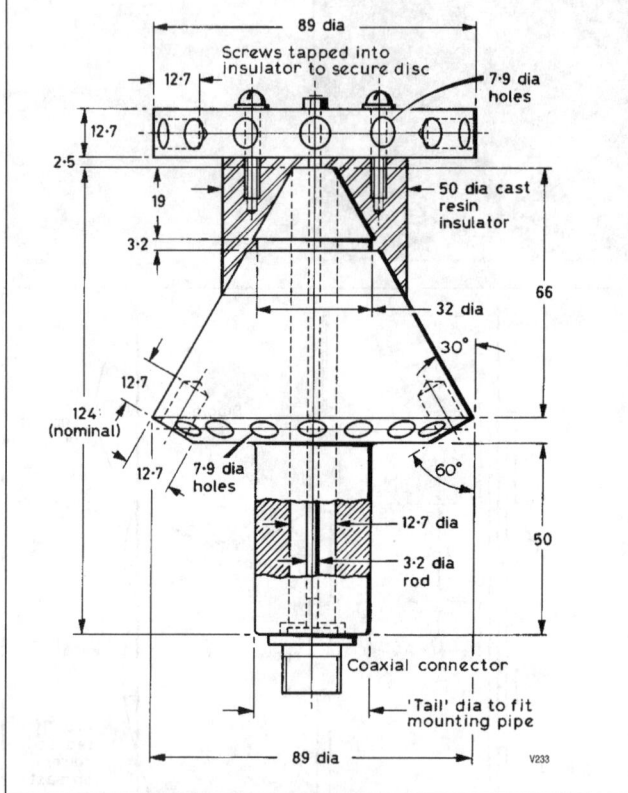

Fig 16.69: Details of discone hub assembly

A   The length of the cone elements are λ/4 at the *lowest* operating frequency (2952/*f* MHz inches).
B   The overall disc diameter should be 70% of λ/4.
C   The diameter of the top of the cone is mainly decided by the diameter of the coaxial feeder cable. For most purposes 0.5in will be suitable.
D   Spacing between top disc and the cone top is 20% of *C* or 0.1in for 50 ohms.

The detail given in **Fig 16.69** of the 'hub' construction will be suitable for any design using a 50-ohm feeder cable and may be taken as an example. A suitable insulator can be made with a potting resin or turned from nylon, PTFE or other stable low loss material.

The low frequency coverage can be extended by fitting a whip antenna to the centre of the disk. The antenna then operates like a quarter wave whip on a skeleton groundplane (the cone providing the groundplane). The VSWR can be optimised at a particular frequency below discone cut-off by adjusting the length of the whip, or several whips can be fitted, resonating at different frequencies, in the style of multi-band dipoles. This can be useful if the antenna is used for other purposes in addition to amateur band coverage. **Fig 16.70** shows the calculated elevation radiation patterns for a conventional discone at 145, 435 and 1296MHz, together with those of a discone of the same dimensions fitted with a 850mm whip to provide 70MHz coverage.

Collinear Dipole Arrays

Communication with mobile stations is best achieved with a vertically polarised omnidirectional antenna, as there is no need to point the antenna in the direction of the mobile. However, a fixed station is not as constrained by mechanical considerations as a mobile, and can thus be fitted with longer, higher gain antennas.

This can be achieved by stacking dipoles vertically above one another in a collinear array, and feeding them with cables of equal lengths, as shown for the GB2ER repeater antenna later in this chapter. Another method of achieving gain with simpler feed arrangements is discussed below.

The current on a length of wire several wavelengths long will be distributed as shown in **Fig 16.71(a)**. The wire shown is 2λ long. Radiation at right angles to the wire will be poor, as the successive half wavelength current maxima are in opposite phases, and if the currents were equal, there would be perfect cancellation of the radiation from the oppositely phased pairs of current maxima. However, if all the current maxima were in phase, the radiated fields would add, and a high gain could be achieved. **Fig 16.71(b)**.

There are several ways of achieving this phase reversal. The simplest is to insert an anti-resonant network or a non-radiating half-wavelength of transmission line as a phasing section between the half-wave radiating elements, **Fig 16.71(c)**. The half wavelength transmission line can be realised as a quarter wavelength of ribbon cable, which can be wound around the insulator between the radiating elements (see the section on mobile antennas).

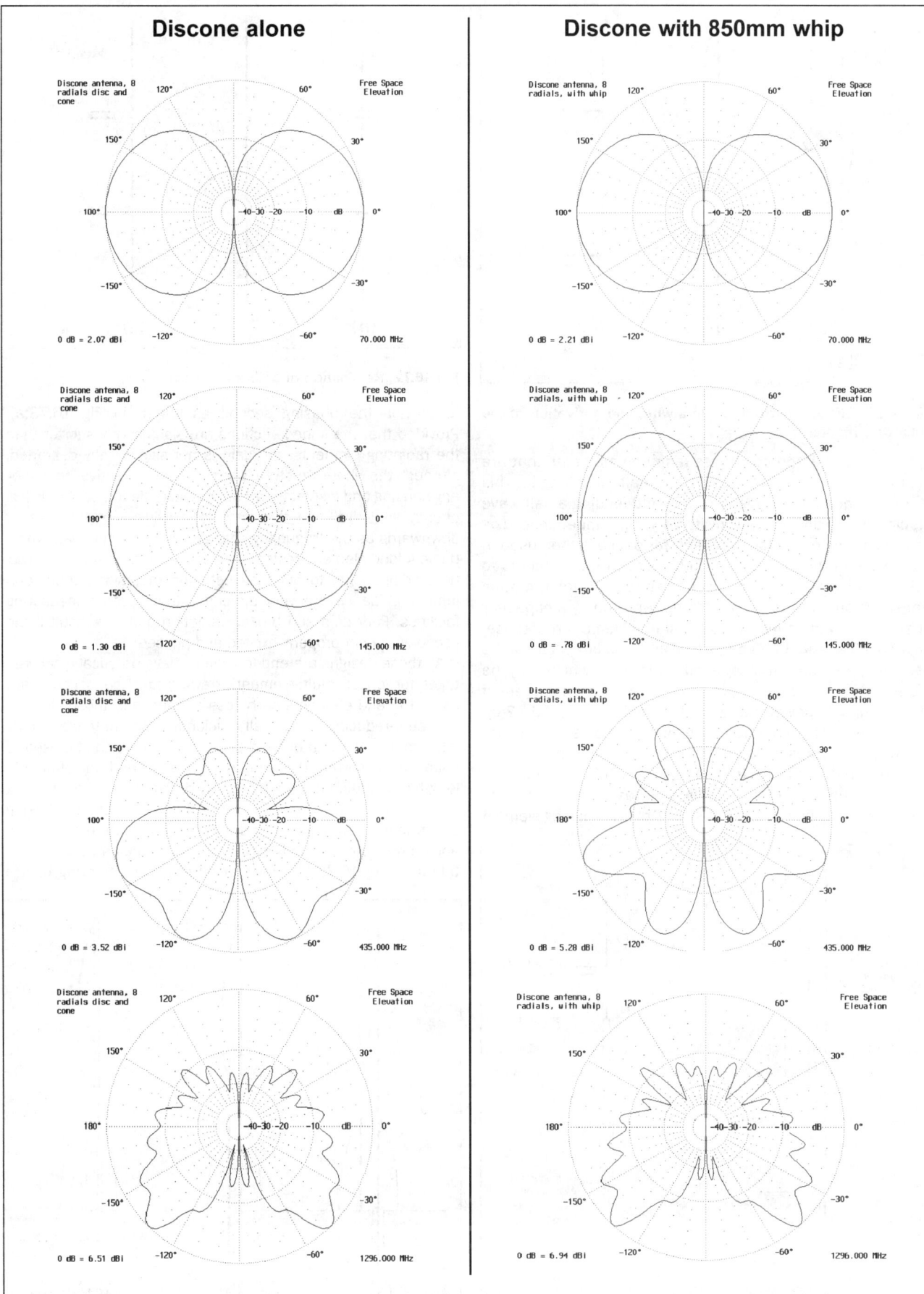

Fig 16.70: Discone elevation radiation patterns, with and without whip for low-frequency operation. Top radials are 305mm, skirt radials 915mm and whip 850mm in length. The whip length was not optimised for 70MHz.

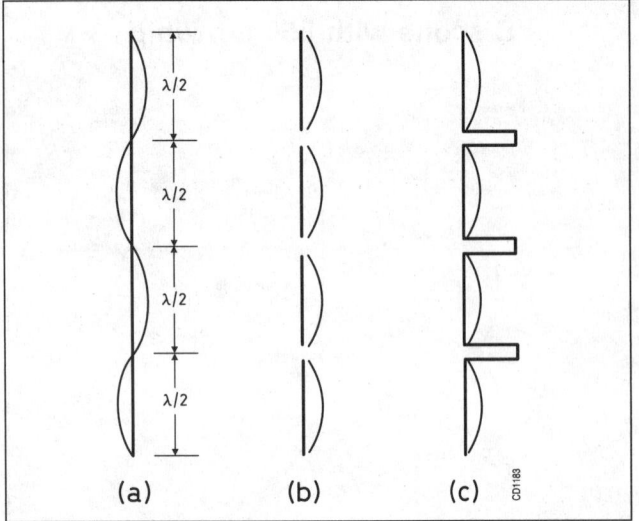

Fig 16.71: Current distribution on a wire, and derivation of the collinear antenna

A more subtle approach uses radiating elements that are a little longer or shorter than one half wavelength. This helps the feeding arrangements, as end feeding a half wave dipole is difficult because of its very high impedance. The self reactance of the longer or shorter dipole is then used in the design of the phasing network between the elements to achieve the desired overall phase shift. The non-radiating transmission line can then often be replaced by a capacitor or an inductor in series with the residual element reactance, **Figs 15.72(a) and (b)**. Again, a transmission line stub can be used to synthesise the required reactance, which may be more convenient or cheaper than a lumped component if significant RF power handling is required, **Fig 16.72(c)**. Sometimes a parallel tuned circuit is realised as an inductor resonated by the self-capacitance of the insulator separating the radiating elements, and upon which it is wound.

A technique devised by Franklin that has been attractive to VHF antenna manufacturers folds parts of the radiating element

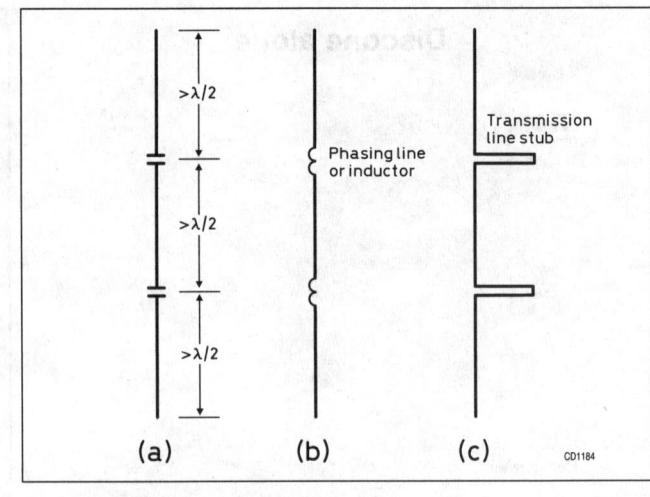

Fig 16.72: Realisation of collinear antennas

to provide the phasing section as shown in **Fig 16.73(a)**. Provided that the folded sections are significantly shorter than the radiating elements, the gain is not significantly degraded, although the whole structure is sensitive to capacitive loading by any housing and insulators required. The radiation pattern is frequency sensitive, and the main lobe will squint upwards or downwards as the frequency changes from the nominal. Whilst these folded element designs look attractive for home construction, adjustments to optimise both the radiation pattern and input impedance are very difficult without proper measuring facilities. Poor gain and broken radiation patterns result if the sections are not properly excited and phased.

All these designs are end-fed, which have practical disadvantages for longer, multi-element arrays. If identical sections are used, the end elements carry less current than those close to the feed, reducing the overall efficiency of the antenna. Whilst different length radiators and phasing elements can be used to equalise the current distribution, the design and adjustment is lengthy, and definitely requires good radiation measurement facilities. If the array can be centre fed, any residual phasing errors tend to cancel out, and for a given length, the performance tends to be better because of a more uniform current distribution. **Fig 16.73(b)** shows one means of achieving centre

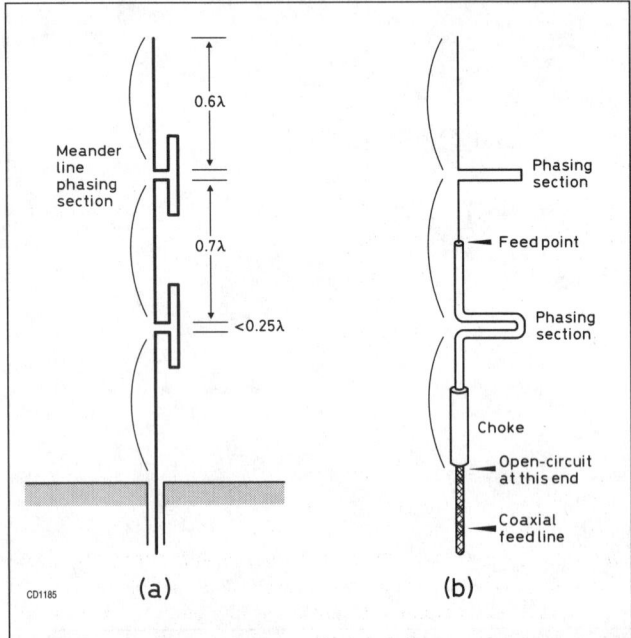

Fig 16.73: Franklin collinear antennas, end-fed and centre fed

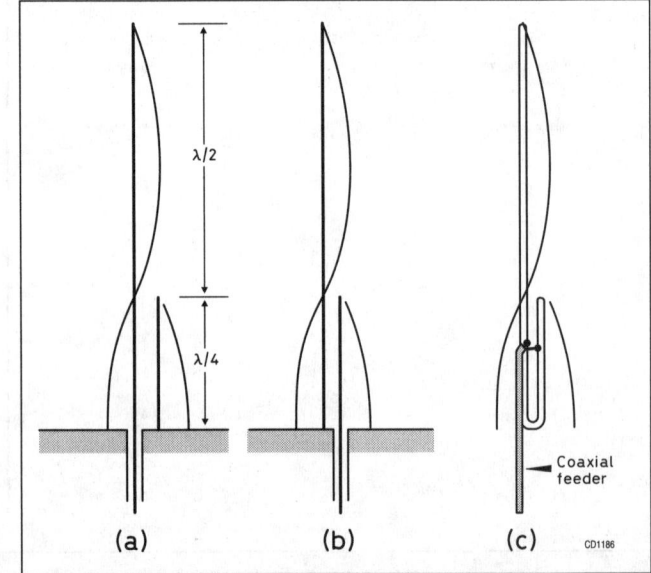

Fig 16.74: The J antenna

feeding with a Franklin array. Note the use of the quarter wave choke section at the base of the array, which is essential to prevent current flowing down the outer of the coaxial cable and destroying the performance of the antenna. The practical gain limit of the singly-fed collinear antenna is around 10dBi.

Practical collinear antennas in radio amateur use tend to use variations on Fig 16.72. The radiating elements may comprise combinations of lengths up to 5/8λ, with or without ground planes. The presence of a good groundplane increases the gain as the image or reflection effectively doubles the length of the array (see also the section on mobile antennas). However, good results can be achieved with collinear antennas directly mounted on pipe masts, especially if care is taken to minimise unwanted currents flowing on the mast.

The J Antenna

Collinear antennas, by virtue of their operation as end-fed structures, have high feed point impedances. A good feed arrangement, valid for both ground-plane and mast mounted antennas, is the use of a quarter-wave short-circuited transmission line, as described in the chapter on antenna fundamentals.

Fig 16.74 shows such an arrangement to end feed a half wave dipole mounted over a groundplane. The matching section should not radiate, and the overall effect is that of a half wave radiator raised λ/4 above the groundplane. Either leg of the quarter wave section can be fed, leading to the structure in Fig 16.74(b), which is identical to Fig 16.74(a) in terms of current distribution, and hence radiation performance. The evolution can be taken a stage further by removing the groundplane and feeding either leg of the quarter wave section as in Fig 16.74(c); this is the 'J' or 'J-pole' antenna, which may use different diameters of tubing for the radiator and stub.

The 'Slim Jim' antenna provides an elegant solution for a simple, mechanically robust antenna made from a single piece of tubing as shown in Fig 16.75. This antenna, described in [33] comprises a folded, open circuit 1/2λ radiator above a 1/4λ transformer section, and is a derivative of the 'J' antenna. The folded stub characteristics of the radiator provide some control over the reactive element of the input impedance. The two ends of the tube can be joined by an insulator, eg a piece of stiff plastic tubing, to provide weather proofing and enhanced mechanical rigidity. Either balanced or unbalanced feeds can be used, tapped on to the 1/4λ transformer section at the point that provides the best match to the feeder. Coaxial feeders should be strapped or bonded to the quar-

ter wave section to reduce unwanted currents on the outer of the cable. The antenna has a maximum gain of around 2.8dBi (0.6dBd) in free space, although the main lobe is tilted up about 10 degrees. The main lobe can be brought to the horizontal by reducing the length of the upper section to about 0.4 wavelengths. This reduces the peak gain to around 2.5dBi (0.3dBd), and can make the feed impedance capacitive.

Phasing sections and additional elements can be combined to produce a collinear form for the 'J' as shown in Fig 16.76(a). This antenna and that of Fig 16.76(c) have been used successfully to produce low-angle radiation for the GB3SN 144MHz repeater.

A variation of the techniques described, using coils as with the original Marconi concept, is shown in Fig 16.77(a) for 432MHz and Fig 16.77(b) for 144MHz. The expected gain is between 6 and 7dBd.

Materials required for Fig 16.77(a) are as follows:
- One 2.5cm diameter 10cm long glassfibre tube.
- One 4.0mm diameter 1.2m long glassfibre rod.
- Four 2.0mm diameter 20cm long glassfibre rods.
- Length of braiding from 'junk' large coaxial or multicore cable.
- Length of 1.2mm wire for matching coils.
- Approximately 5cm square of singled sided PCB.

First, adjust the bottom 5/8λ element to give minimum VSWR by moving the tapping point on the bottom coil (approximately four turns). A fine adjustment can be made by altering the length of the first 5/8λ element.

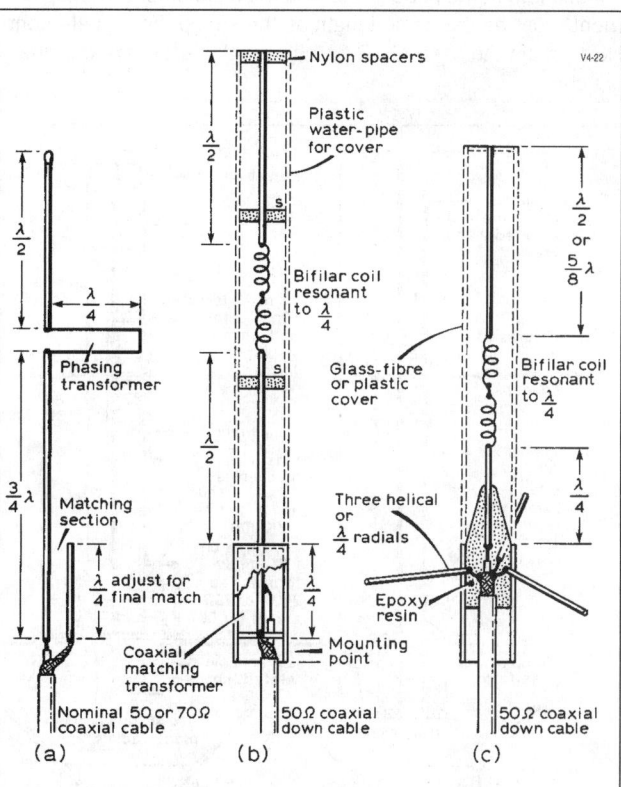

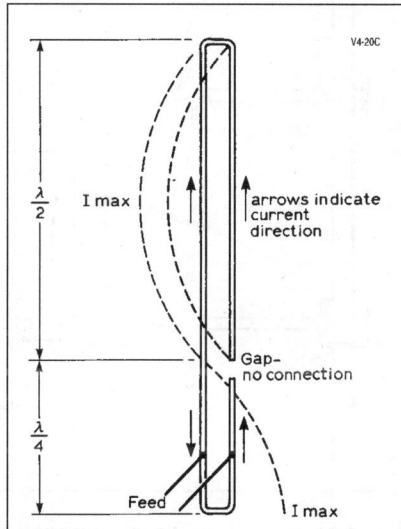

Fig 16.75: The basic Slim Jim, showing direction of current flow and phase reversal (*Practical Wireless*)

Fig 16.76: A collinear form of 'J' antenna. (a) the addition of λ/4 sections as suggested by Franklin. (b) Use of a coaxial short-circuit λ/4 transformer to give an unbalanced input. The tapping point of the matching transformer is approximately 0.15λ from the 'earthy' end. (c) A variant of (b) with radials, With both (b) and (c) the λ/4 phasing transformer has been 'wound up' as a bifilar coil (each coil being wound in the opposite hand). While the inductive component is cancelled, the mutual capacitance of the windings makes them physically shorter than λ/4

Next fit the centre matching coil and the top element. To obtain the best results, both elements should be the same length and approximately 5/8λ. Further improvement in VSWR is obtained by adjusting the centre matching coil (the coil is spread over 1/4λ).

The matching coil provides the phase change necessary to feed the top element and so adjustment is quite critical. If the matching coil has to be 'squeezed up' to obtain a good VSWR, the coil has too many turns. The opposite is true if the coil has to be greater than 1/2λ for a good VSWR.

To prevent the collinear going 'off tune' once set up, the elements are secured to the centre glassfibre rod and the matching coil taped with self amalgamating tape. Provided care is taken in setting up, a VSWR close to 1.1:1 can be obtained.

Materials required for Fig 16.77(b) are as follows:
- Two 12.7mm diameter by 1206 ±12mm, 5/8λ elements (adjustable).
- Four 495mm rods for the ground plane.
- One 6.4mm diameter by 762mm insulated rod.
- One 25mm diameter insulated tube (a cotton reel can be used).
- 1.6mm wire for matching and phasing coils.

The diagram shows extra insulated tubing over the matching and phasing coils to give more mechanical strength and weatherproofing.

Setting up is carried out as follows. First, adjust the length of the bottom 5/8 element to give minimum VSWR.

Secondly, fit the phasing coil and the top element. The top element must be the same length as the set-up, bottom element. Next obtain the best VSWR by 'adjusting' the turns of the phasing coil.

The coil provides the phase change necessary to 'feed' the top element. It consists of a length of 1.6mm wire, (about 1λ), coiled up to give 70-72 turns on a 6.4mm diameter former. The λ/4 spacing between the two elements is more critical than the number of turns. 68 turns gave a satisfactory VSWR with the prototype.

Some difficulty may occur in setting up the phasing coil. If more than seven turns have to be removed, go back to the first adjustment stage to ensure the bottom 5/8λ element is correctly matched. If the bottom element is not correctly set up the collinear will not tune up. Careful adjustment should produce a VSWR of 1.1:1 at the chosen operating frequency.

A technique, widely used for commercial systems, combines λ/2 dipoles fed in phase from a single source, or alternatively with an appropriate variation of feeder cable length between dipoles to provide phasing.

The disadvantage with this form of antenna array is that some interaction occurs between cables and radiating elements. However, the disadvantage is balanced by the ability to modify the radiation pattern shape by simple adjustment of dipole spacing or phasing cable length.

The example given in **Figs 15.78, 15.79 and 15.80** is probably the simplest to set up and was devised for the GB3ER 432MHz band repeater. If the cables are made to be an odd number of quarter wavelengths long, an equal current feed to each dipole is assured.

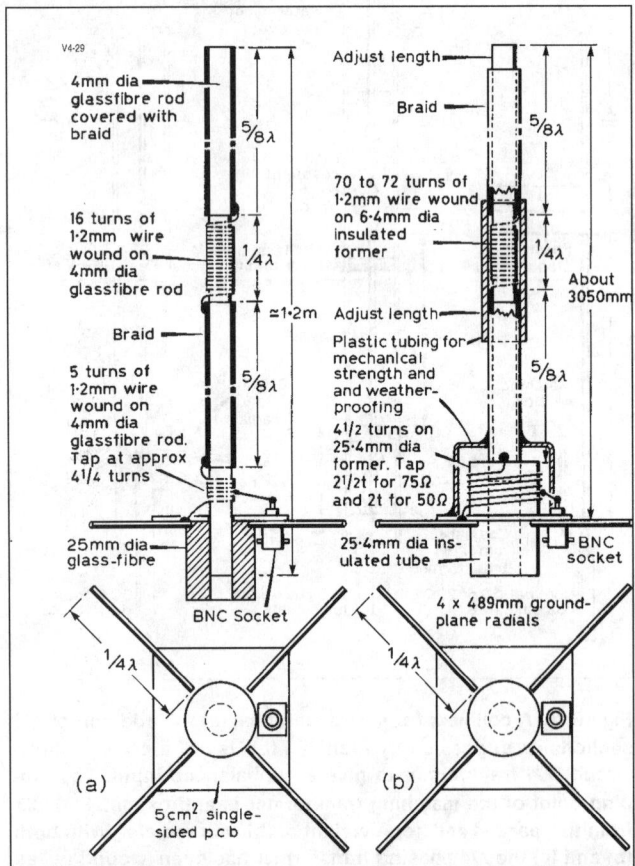

Fig 16.77: (a) A 432MHz collinear antenna (b) A 144MHz collinear antenna (*UK Southern FM Journal*)

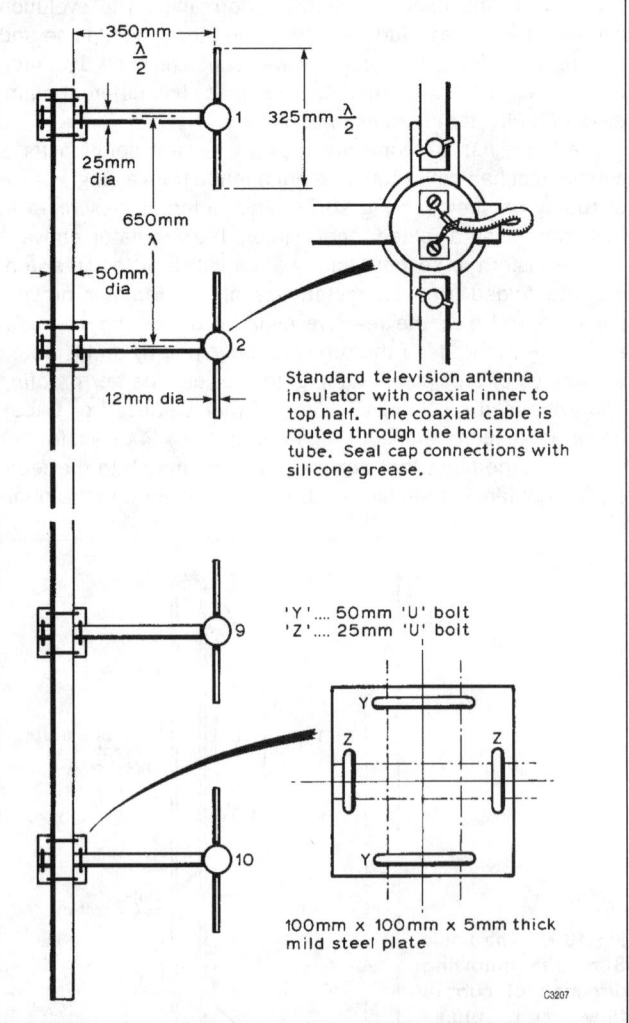

Fig 16.78: Mechanical details of GB3ER collinear

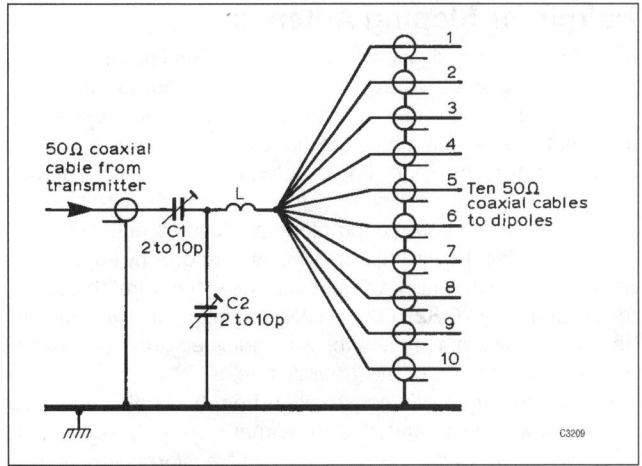

Fig 16.79: Matching Unit of GB3ER collinear

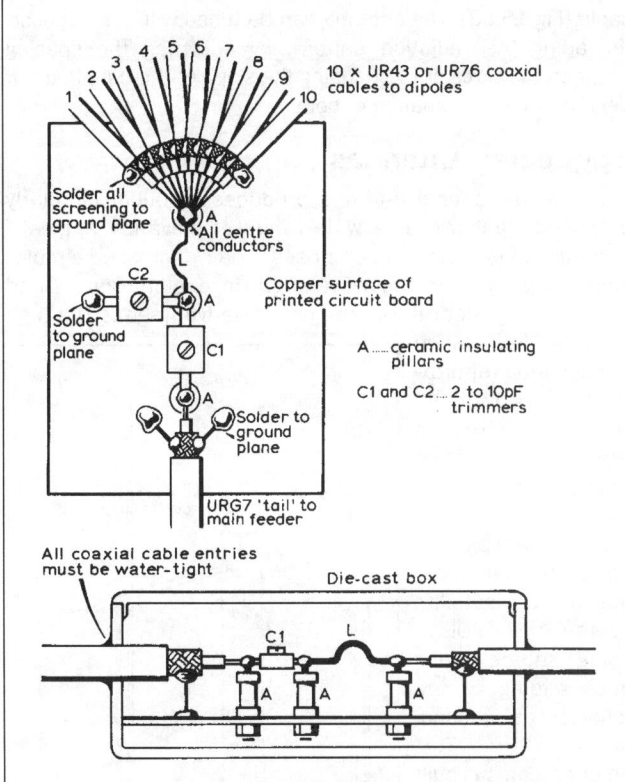

Fig 16.80: Matching unit layout for GB3ER collinear

ANTENNAS FOR SATELLITE COMMUNI-CATION

For the radio amateur, satellite ground station antennas fall into two groups. The first group comprises *steerable antennas*, which enable the passage of the satellite to be tracked across the sky. The second group consists of *fixed antennas*, which have essentially hemispherical radiation patterns to receive the satellite signals equally from any direction. These antennas do not need to be steered to receive signals during the satellite's passage. The tracking antennas are usually of high gain, while the fixed antennas are usually of low gain, due to their hemispherical coverage. Fortunately, signal losses between ground and the satelite in line-of-sight are relatively low. With no obstructions, low-gain antennas of the fixed variety are often acceptable for reception of amateur or

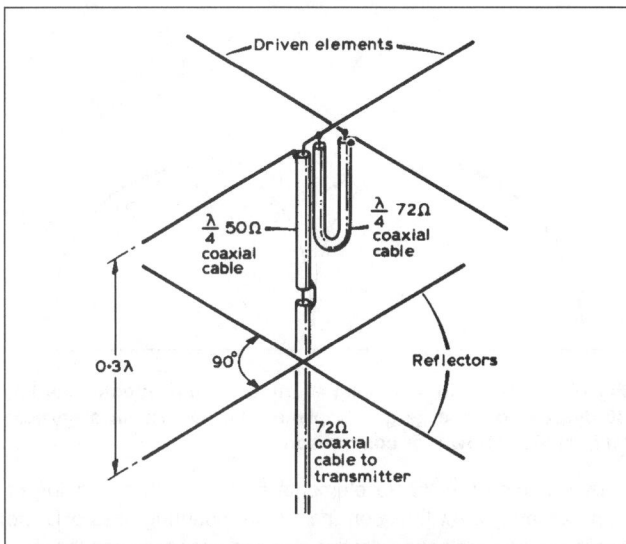

Fig 16.81: A crossed dipole antenna for 145MHz

weather satellites, helped by the higher radiated powers available from current low-earth orbiting amateur satellites.

As many satellites rotate or change their orientation with respect to the ground station, both groups of antennas are designed to provide circular polarisation. Amateur convention calls for the use of Right-Hand circular polarisation for earth/space communications. However, the downlink may have either Left or Right handed polarisation according to the satellite and frequency in use.

Of the higher-gain tracking antennas, the crossed Yagi and the helix antenna are the main ones used. The crossed Yagi is probably the easiest to construct and most readily available commercially. Construction details for these antennas were described earlier, and polarisation switching schemes are discussed at the end of this chapter.

Crossed Dipole or Turnstile Antenna

Fig 16.81 shows a simple arrangement of crossed dipoles above a ground plane for 145MHz. This type of antenna can be scaled for use at 29, 145 or 432MHz. Mechanical problems may make the reflectors inadvisable in a 29MHz version. The height above ground can be about 2m for 145MHz and 3m for 29MHz.

Typical dimensions are:

29MHz	driven elements ($\lambda/2$)	188in	4775mm
145MHz	driven elements ($\lambda/2$)	38in	965mm
	reflectors	40.5in	1030mm
	spacing (0.3λ)	24.5in	622mm

The phasing line comprises $\lambda/4$ of 72-ohm coaxial cable. The matching section for a 72-ohm feed is $\lambda/4$ of 50-ohm cable. When calculating the length of the $\lambda/4$ sections, the velocity factor of the cable must be taken into account. Typically this is 0.8 for cellular and semi-airspaced cables, and 0.66 for solid dielectric cables, but verification of the figure for the particular cable used should be obtained. As an example, a matching section of RG59/U would be 13in (330mm) in length. Omit the transformer section for 50-ohm operation. This will result in an input VSWR of less than 1.4:1.

For a centre-fed crossed dipole, it is advisable to have a 1:1 balun to ensure a consistent pattern through 360° of azimuth. Dependent on the spacing between the dipoles and ground plane, the radiation pattern can be directed predominantly to the side, for satellites low on the horizon, or upwards for overhead passes.

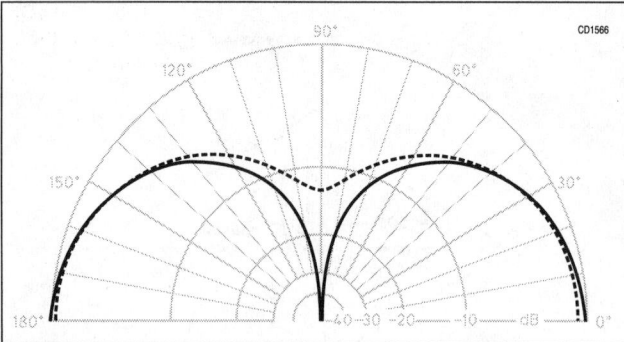

Fig 16.82: Elevation radiation pattern for a whip antenna sloping 30 degrees on a large groundplane. The pattern for a vertical whip is also shown for comparison.

By drooping the dipole elements at 45°, with a spacing of approximately 0.4λ between the dipole mounting boss and the ground plane/reflectors, better coverage towards the horizon can be obtained. As ground reflections affect horizontal and vertical polarisation differently, low-to-horizon flight paths will not produce circular polarisation. This is due to 'ground scatter' of the satellite signal when it is low on the horizon, and ground reflections locally at the ground-based antenna.

Circular polarisation is normally produced by feeding one dipole 90° out of phase to the second dipole, and can be achieved by having an extra λ/4 of cable on one side of a combining harness.

An alternate approach to this method of phasing is to use the phasing properties of a capacitive or inductive reactance.

Suppose, for example, that the length and diameter of the dipoles are set to give a terminal impedance of 70Ω - j70Ω (capacitive). If a second, crossed dipole is set to be 70Ω + j70Ω (inductive) the combined terminal impedance of the arrangement becomes 35Ω ± j0, ie 35 ohms resistive. As the two dipoles are connected in parallel, the current in each dipole is equal in magnitude. However, due to the opposite phase differences of 45° produced by the capacitive and inductive reactances, the radiated fields are in phase quadrature (a 90° phase difference) which results in circularly polarised radiation.

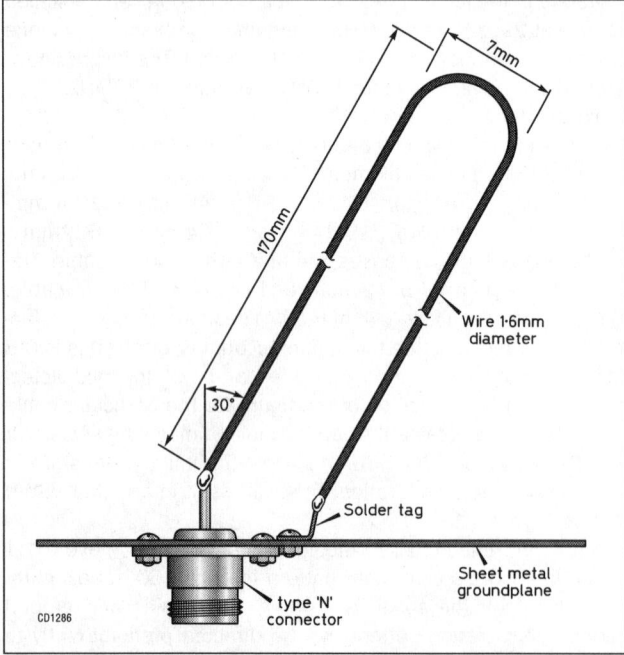

Fig 16.83: Construction of 435MHz hairpin antenna and groundplane

Hairpin or Sloping Antennas

As technology has advanced, so has the radiated power from low earth orbiting amateur radio satellites increased, to the extent that communications can be established using handheld radios if the satellite is well above the horizon. Accordingly, it can be worth constructing very simple antennas to get started with satellites.

A vertical whip on a ground plane produces a null along the axis of the whip. However, if the whip is bent over by about 30°, the overhead null is filled without degrading the azimuth pattern too severely (**Fig 16.82**). For 145MHz operation, the whip should be around 480mm in length, and adjusted for best match according to the size of the groundplane.

At 435MHz, a *hairpin* construction from 2mm diameter wire (or even a wire coat hanger) is somewhat more rugged, and can offer wider bandwidths because of the transforming action of the folded dipole. The connector/feed point joint should be potted or sealed with silicon rubber or flexible epoxy resin if the antenna is to be used outdoors, to prevent moisture entering the cable (**Fig 16.83**). The antenna can be tuned, without seriously degrading the radiation pattern, by adjusting the spacing between the wires or by changing the slope of the hairpin a few degrees, or a combination of both.

Eggbeater Antennas

This omnidirectional antenna produces excellent circularly polarised signals over a wide range of elevation angles if carefully constructed. It comprises a pair of crossed circular loops, slightly over one wavelength in circumference, and fed in phase quadrature. The only disadvantage is the high terminal impedance of each loop (approximately 140 ohms, which requires a transformer section to match to 50 ohms [34].

Performance has been further optimised for satellite operation by K9OE by using square loops and adding tuned reflectors as a groundplane [35]. This antenna can be built with an excellent match to 50 ohms without recourse to special impedance cables.

The general arrangement is shown in **Fig 16.84**. The loops are constructed of rod, tubing, or 2mm enamelled copper wire for 70cm, bolted to a 22mm plastic water pipe coupler. Critical dimensions

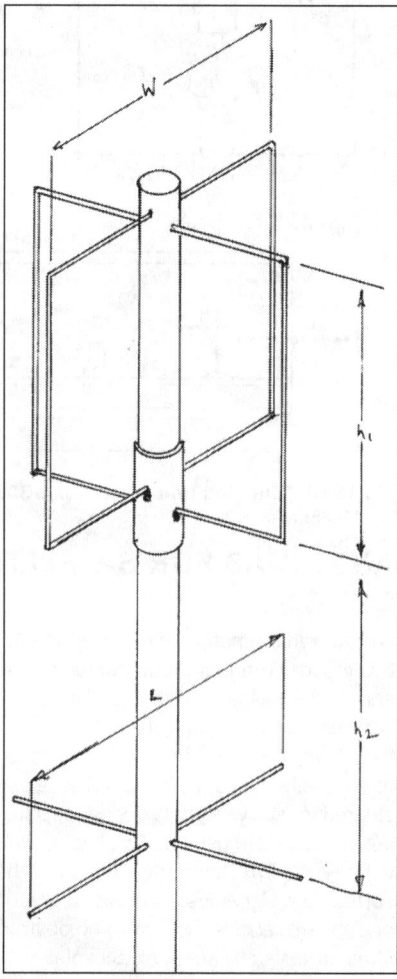

Fig 16.84: General construction arrangements for K5OE Eggbeater II antenna

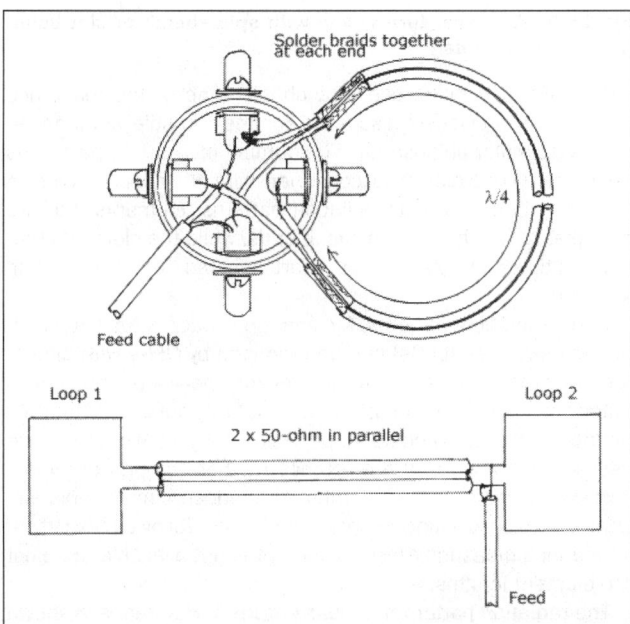

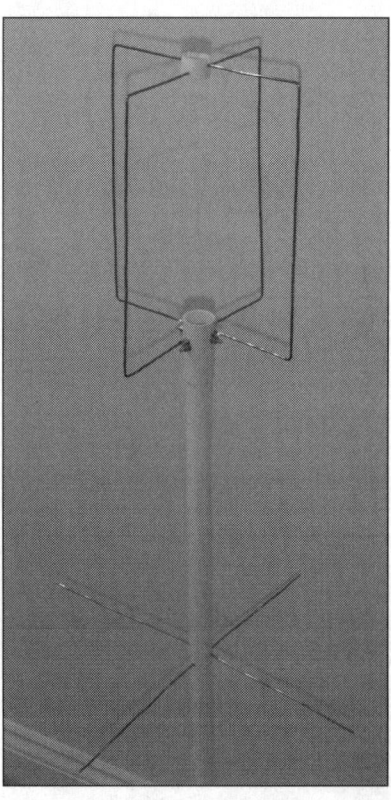

Fig 16.87: K5OE Eggbeater II without weatherproofing

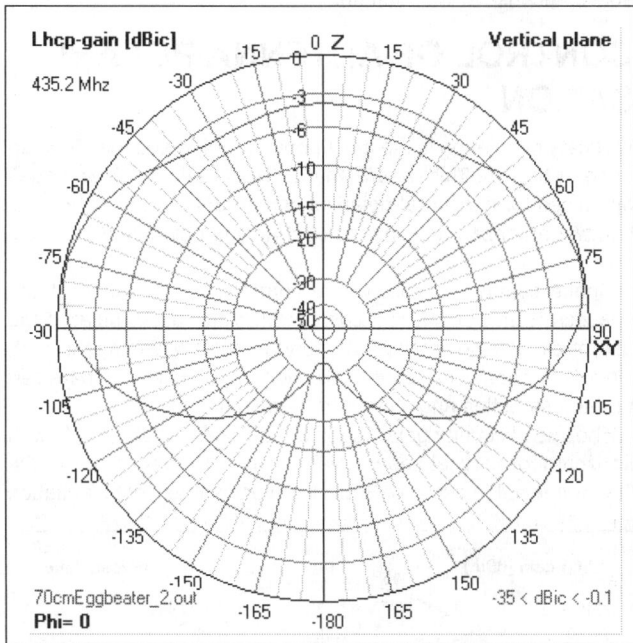

Fig 16.85: Feed point construction detail and phasing line connections for Eggbeater antenna

Fig 16.86: Elevation radiation pattern for 70cm Eggbeater antenna. The azimuth pattern is essentially omnidirectional. Note the figures are in dBi

Design frequency	w	h1	L	h2	Loop wire dia	Phasing cable length
137MHz	540	665	1065	1080	5	390
145.5MHz	510	625	1000	1010	5	370
435MHz	170	210	335	330	2	124

Table 16.15: K5OE Eggbeater antenna critical dimensions in mm. Phasing cable lengths are for solid PTFE dielectric, and should be measured between where the inner dielectric clears the braided coaxial outer

are shown in **Table 16.15**. The tops of the loops are supported by a section of (plastic) water pipe that fits into the coupler, and capped to prevent water ingress. One loop should be set 5 - 10mm higher than the other so that the top of the loops can cross without touching. Instead of the 95-ohm RG62 cable described by K5OE, the phasing section can be made from two pieces of 50-ohm cable, each one quarter-wavelength long, with the outers connected together at each end. This makes a 100-ohm balanced wire transmission line, which can be coiled or folded up if required. All four braids can be connected together if the connections are conveniently close together (e.g. the 70cm variant). The antenna is Left Hand Circularly Polarised (LHCP) if the connections are as shown in **Fig 16.85** when the antenna is viewed from below. To change the polarisation to RHCP, transpose the two phasing cable connections at *one* end only.

For many purposes, a 50-ohm coaxial feeder can be connected directly as shown. Five or six turns of feeder closely wound can provide a balun and choke against current on the outside of the cable if required. A ferrite bead fitted on the feed cable close to the connection can be effective, and there is space for a low noise amplifier inside the water pipe between the coupler and the tuned reflectors if the antenna is to be used only for reception.

Calculated radiation patterns for the 435MHz variant are shown at **Fig 16.86**. **Fig 16.87** shows the antenna construction.

Quadrifilar Helix or Volute Antenna

The quadrifilar helix (QFH) or volute is a four-element helical antenna which can be used to give either directional gain or hemispherical circular polarised coverage as originally described by Kilgus [36][37]. The general form of a QFH is shown in **Fig 16.88**.

Radiation patterns produced for several combinations of turns and resonant lengths are shown in **Fig 16.89(a) to (d)**, and generally produce better circularly polarised signals at low elevation angles than a turnstile with drooping arms. A range of dimensions and resulting radiation patterns is discussed in [38].

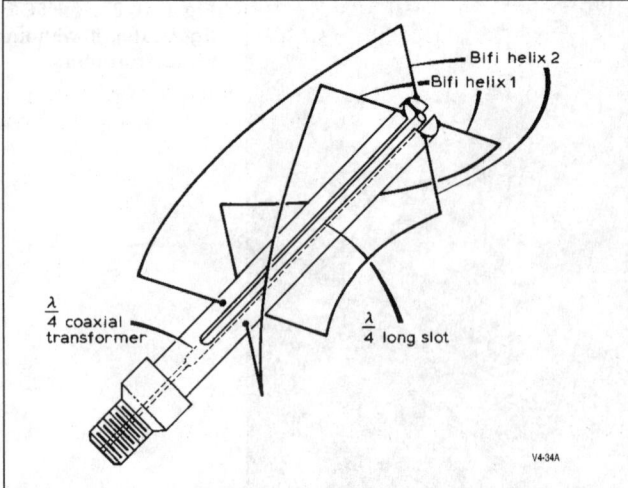

Fig 16.88: A quarter turn volute with split sheath or slot balun (*Microwave Journal*)

The QFH can make use of a phasing line or the reactance method (as previously discussed for the turnstile antenna) to produce circular polarisation. The number of 'turns' or part turns of the radiating elements combined with their length can be used to produce various radiation patterns. Elements that are multiples of $\lambda/4$ have open-circuit ends, while the elements that are multiples of $\lambda/2$ can be short-circuited to the mounting structure.

Good radiation patterns for communication with low earth orbiting satellites (LEOs) can be produced by QFHs with diameters less than 0.1λ and one quarter turn per loop. This makes antennas for 145MHz possible without large support structures or incurring high wind load penalties [39] [40]. Both these designs make use of the reactance method (as previously discussed for the turnstile antenna) to produce circular polarisation, and constructional details must be followed exactly if impedance measuring test equipment is not available to adjust the element lengths.

The radiation pattern of a quarter turn QFH antenna is shown in **Fig 16.90**.

'Long' QFH with multiple turns can produce even better low angle coverage, but are generally only suitable for 435MHz or above because of their length.

CONTROL OF ANTENNA POLARISATION

Vertical polarisation is the most popular for FM and mobile operation in the UK. This means that a fixed station with antennas optimised for horizontal polarisation can only operate effectively if two antennas or a means of changing polarisation is available.

Space communication, where control of polarisation at the spacecraft end of the link can be difficult, has stimulated the use of circular polarisation. Its fundamental advantage is that, since reflections change the direction of polarisation, there can be far less fading and 'flutter' from reflections.

The use of circular polarisation at only one end of the link, with horizontal or vertical linear polarisation at the other end of the link, will result in a loss of 3dB. However, changes of polarisation

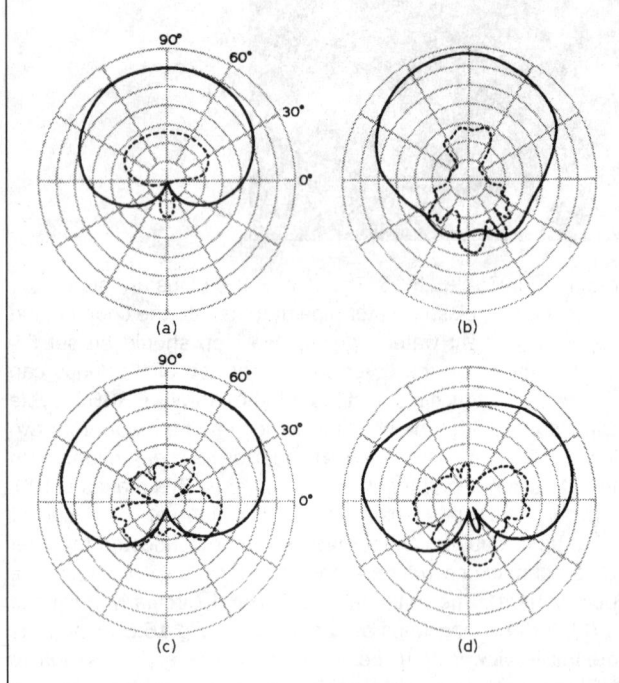

Fig 16.89: Volute radiation patterns, co- and cross- circular polarisation (a) 3/4 turn λ/4 volute. (b) 3/4 turn λ/2 volute. (c) 3/4 turn 3/4λ volute. 3/4 turn 1λ volute (*Microwave Journal*)

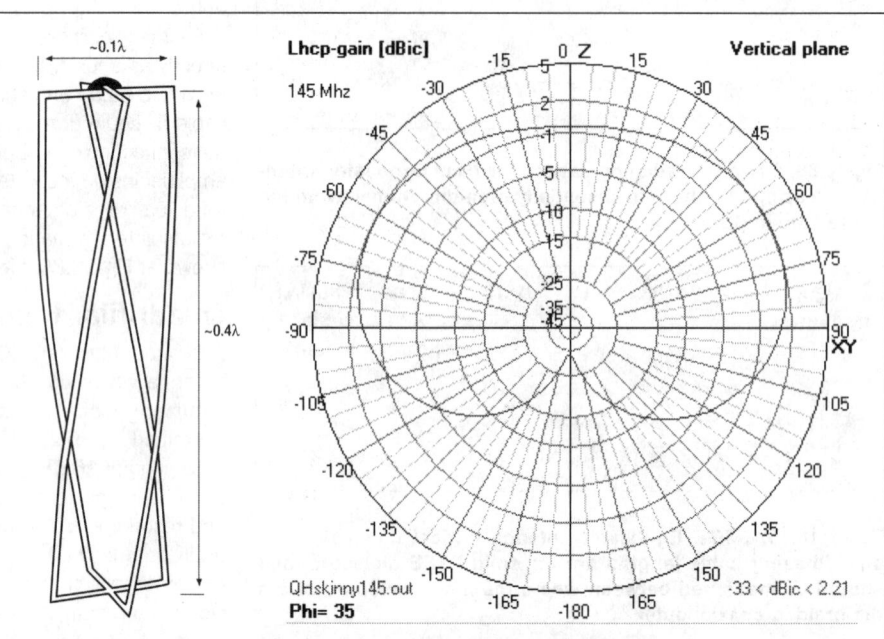

Fig 16.90: Quarter turn slim quadrifilar helix antenna and elevation radiation pattern. Each loop is approximately 1λ in circumference. Support pole not shown

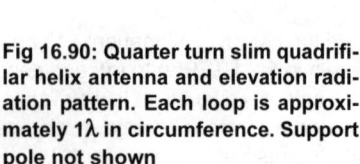

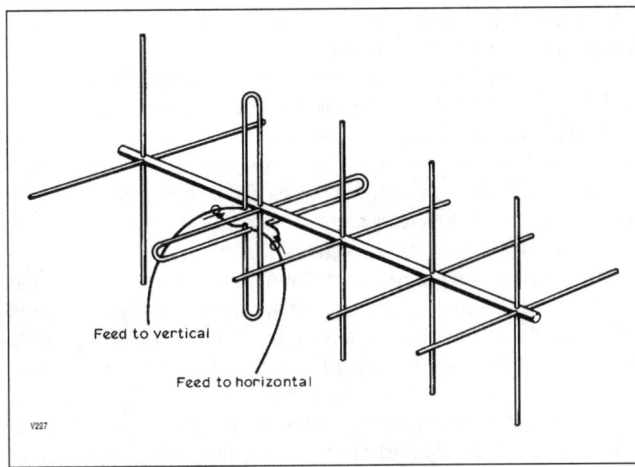

Fig 16.91: General arrangement of a crossed Yagi antenna

caused by propagation often result in better reception of linearly polarised signals with a circularly polarised antenna. Conversely, if circular polarisation is used at both ends of the link, but the two stations use oppositely polarised antennas, the loss of signal due to polarisation mismatch can be 10dB or more.

It has been usual to standardise on Right-Hand Circular polarisation (RHCP) in the northern hemisphere for fixed communications. Space communications may use either Right or Left Hand Circular polarisation according to the frequency used and the satellite of interest. The handedness of circular polarisation is defined by the direction in which the electric vector rotates as the wave travels away from the observer. Thus, if in following the conductor of a helix antenna along its length away from the observer, the motion is clockwise, the antenna polarisation is said to be Right Handed.

The helix antenna produces Right or Left-Handed circular polarisation dependent upon whether the antenna element is wound clockwise or anti clockwise. Horizontal or vertical linear polarisation is also possible from helix antennas by using two helices and suitable phasing arrangements.

A compromise arrangement for receiving circular-polarisation signals is to use slant polarisation. To obtain this, a single Yagi is set at an angle of 45°. This enables horizontal and vertical signals to be received almost equally. At first sight one would expect a loss of 3dB for H and V polarised signals compared with an appropriately aligned Yagi. However, long-term practical measurements have shown when averaged that this arrangement gives a 6dB improvement with typical mixed polarised signals. In addition this arrangement it is only a little affected by the mounting mast, unlike a vertical polarised Yagi.

The simplest way of being able to select polarisation is to mount a horizontal Yagi and a vertical Yagi on the same boom, giving the well-known *crossed Yagi* antenna configuration (**Fig 16.91**). Separate feeds to each section of the Yagi brought down to the operating position enable the user to switch to either horizontal or vertical. It is perhaps not generally realised that it is quite simple to alter the phasing of the two Yagis in the shack and obtain six polarisation options. These are two slant positions (45° and 135°), two circular positions (clockwise and anticlockwise) and the original horizontal and vertical polarisation.

The presence of the mast in the same plane as the vertical elements on a Yagi considerably detracts from performance, but a simple solution is to mount the antennas at 45° relative to the vertical mast. With appropriate phasing, vertical and horizontal polarised radiation patterns can be obtained that are unchanged by the presence of the mast.

If a crossed Yagi is mounted at 45°, with individual feeders to the operating position, the polarisation available and the phasing required is as follows:

(a)   Slant position 45° and 135°. Antennas fed individually.

(b)   Circular positions clockwise and anti-clockwise. Both antennas fed 90° + or 90° - phase relationship respectively.

(c)   Horizontal and vertical polarisation. Both antennas fed with 0° or 180° phase relationship respectively.

Phasing is simply the alteration of the length of the feeders of each crossed Yagi to change the polarisation. Where a 90° phase shift is required, λ/4 of cable is inserted and, where a 180° phase shift is required, λ/2 cable is inserted. The polarisation switch must switch in the appropriate λ/4 'impedance transformer' and correct phasing by connecting the appropriate length(s) of cable.

Mast-head Multiple Polarisation Switch for 145MHz

There are several disadvantages in locating the polarisation switch in the shack. Two feeders are required from the antenna to the equipment, and RF losses may be significant if the antennas are positioned on a mast. Whilst it is technically feasible to improve signal to noise performance by installing twin preamplifiers at the antennas, the problems of equalising gain and ensuring relative phase stability are formidable. An alternative solution is to place the polarisation switch close to the antennas, which minimises the feeder losses, and to utilise a single preamplifier (with transmitter bypass arrangements if required) immediately after the switch to improve the system signal to noise ratio before despatching the signals to the receiver via a single feeder.

The polarisation switching circuit (**Fig 16.92**) offers a fair compromise between cost, complexity and performance. A pair of half crystal-can double pole changeover relays (**Table 16.16**) insert extra lengths of line into the feeds of a 50-ohm crossed Yagi anten-

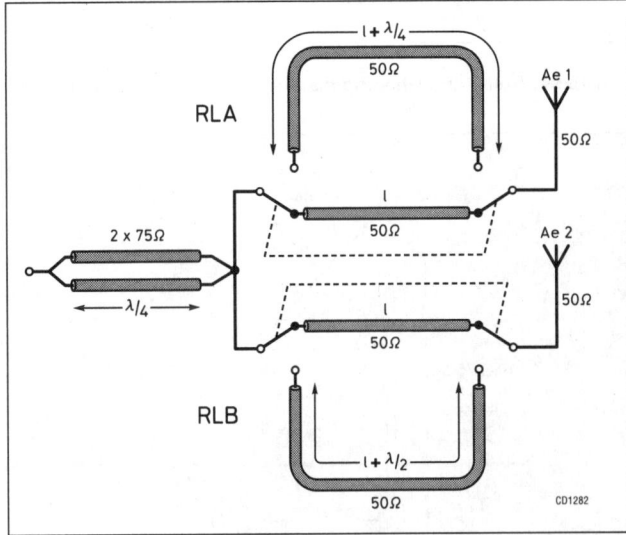

Fig 16.92: Masthead antenna polarisation switch

Characteristic	145MHz	435MHz
Insertion loss	<0.2dB	<0.3dB
Mismatch	<1.03:1	<1.2:1
Isolation between normally open and moving contact	42dB	30dB
Isolation between moving contacts	39dB	28dB

Table 16.16: Half crystal can relay measured characteristics

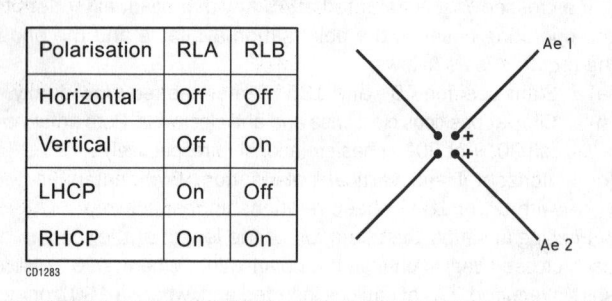

Polarisation	RLA	RLB
Horizontal	Off	Off
Vertical	Off	On
LHCP	On	Off
RHCP	On	On

CD1283

Fig 16.93: Rear view of crossed antenna with polarisation table. The radiated wave propagates into the paper.

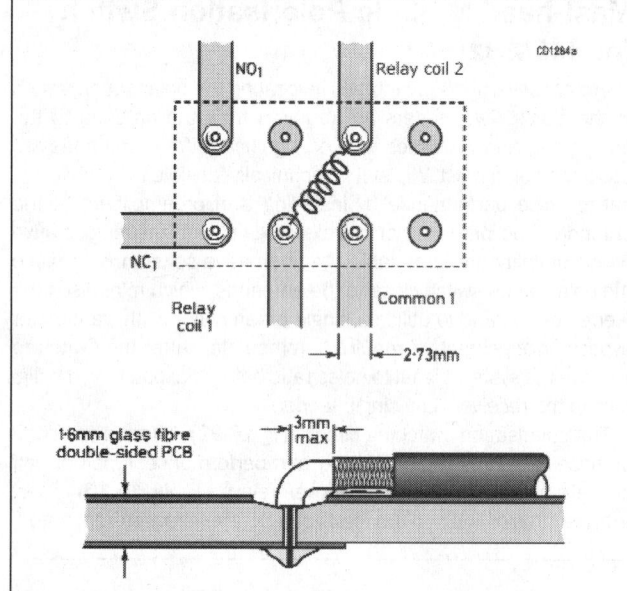

Fig 16.94: Relay and phasing/matching cable mounting details

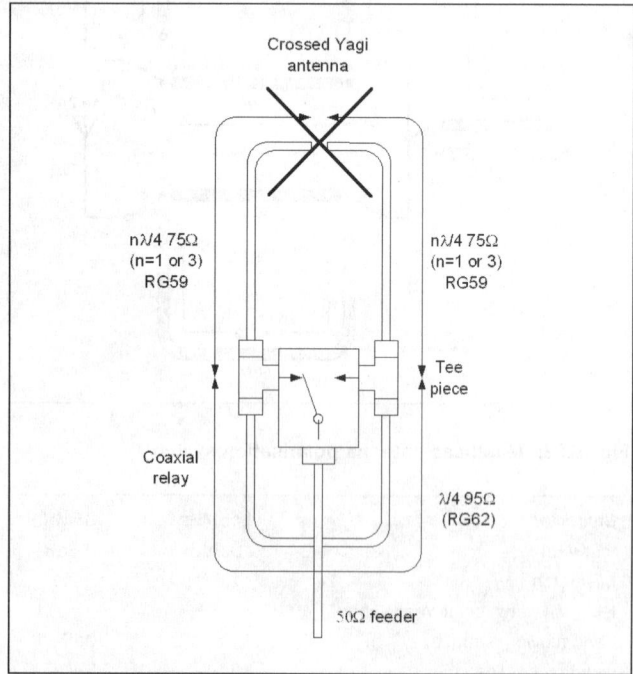

Fig 16.95: Simple switch and phasing line to produce reversible circular polarisation

na mounted at 45 degrees to its support mast, to provide four polarisations (**Fig 16.93**). The antenna cable lengths must be cut to compensate for any phase differences between the antennas due to physical offset in the direction of propagation. Under ideal conditions with perfectly matched antennas, the four main polarisations will be achieved. If the antennas are not well matched to 50-ohm, the power division will not be equal, and mixed polarisation will result for each relay setting. If the VSWR is less than 1.2:1, the depolarisation will not be serious.

The relays are mounted on 1.6mm double sided copper clad glass fibre PCB with 50-ohm microstrip tracks for the RF circuits (**Fig 16.94**). RF interconnection and phasing lines are made from 3mm PTFE cable (URM 111 and T3264) which provides a good trade-off between losses and size for coiling up within a diecast weatherproof box. Polythene dielectric cables could possibly be used if soldered very quickly, but this is not recommended. The cable ends must be kept as short as possible as shown in Fig 16.94. The polarisation unit will handle 100 watts of CW or FM, provided that the antenna system VSWR at the relay is better than about 1.5:1.

If a transient suppression diode is to be fitted across the relay coil, it should be bypassed with a 1nF capacitor, or rectified RF pickup during transmission may cause the relay to change state. Experience has shown that it is better to dispense with the diode at the relay altogether if high transmitter powers are involved.

Circular Polarisation Feed and Reversing Switch

The following simple arrangement generates remotely selectable Right or Left hand circular polarisation from a crossed Yagi antenna, or two separate, identical Yagi antennas with their elements set at right angles to each other. It will work well provided that both antennas are closely matched to 50 ohms, and it uses readily obtainable cables, adaptors and connectors throughout (**Fig 16.95**). It should be mounted close to the antenna(s).

The electrical length of the two antenna cables must be an *odd* number of quarter wavelengths. $\lambda/4$ should be sufficiently long to make the connections for 2m operation, and $3\lambda/4$ for 70cm. Higher multiples of $\lambda/4$ will reduce the VSWR and polarisation bandwidths of the antenna. If longer antenna cables are necessary, equal lengths of 50-ohm cable should be used to connect the antennas to the $\lambda/4$ 75-ohm sections at the relay.

The 50-ohm impedance of the antenna is transformed by the quarter wave sections to 100 ohms at the T-piece. A quarter wavelength of RG62 (95-ohm) cable provides the phasing needed to produce circular polarisation, and the common relay contact, in either position, 'sees' two 100-ohm loads in parallel, ie 50 ohms, for presentation to the downlead.

Suitably weatherproofed PL239 connectors will work for 2m operation, but type N connectors should be used for 70cm. The electrical length of the connectors and tee-pieces must be taken into account when cutting the quarter wavelength cables.

For 1.3GHz and above, use separate circularly polarised antennas and a change-over relay if selectable polarisations are required.

REFERENCES

[1] http://www.vhfman.freeuk.com/radio/antennas/html. General construction notes and sources of materials (UK)

[2] 'Notes on the Development of Yagi arrays', C Greenblum, *QST* and other ARRL publications, edited by Ed Tilton, W1HDQ

[3] 'Yagi-Uda Arrays', Chen and Cheng, *Proc IEEE*, 1975

[4] 'Yagi Antenna Design', P Viezbicke, *National Bureau of*

Standards Technical Note 688, 1976

[5]   http://www.boulder.nist.gov/timefreq/general/pdf/451.pdf. Copy of P Viezbicke's *Technical Note 688* in PDF format

[6]   *Yagi Antenna Design*, Dr J L Lawson, W2PV, ARRL, 1986

[7]   Yagi Optimiser YO, B Beezley, 3532 Linda Vista Drive, San Marcos, Ca 92069. Tel (619) 599-4962

[8]   'How good are genetically designed Yagis?', R A Formato, WW1RF, *VHF Communications* Vol. 2/98 pp87-93

[9]   'A genetically designed Yagi'" R A Formato, WW1RF, *VHF Communications* Vol, 2/97 pp 116-123

[10] http://www.si-list.org/swindex2.html. YGO2, YGO3 Yagi Genetic Optimiser software

[11] http://www.g8wrb.org/yagi/optimise.html. A particular Yagi optimisation study

[12] http://ic.arc.nasa.gov/people/jlohn/Papers/ices2001/pdf

[13] 'How to Design Yagis', W1JR, *Ham Radio*, Aug 1977, pp22-30

[14] 'Long Yagis', W2NLY and W6KQI, *QST* Jan 1956

[15] 'Extremely long Yagi Antennas', G Hoch, DL6WU, *VHF Communications*, Mar 1982, pp131-138; Apr 1977, pp204-21

[16] 'DL6WU Yagi Designs', *VHF/UHF DX Book,* ed I White, G3SEK, DIR Publishing / RSGB 1993

[17] http://www.ifwtech.co.uk/g3sek/diy-yagi. Various long Yagi designs with construction details and other information

[18] 'The JVL LoopQuad', *Microwave Handbook,* Vol 1, M W Dixon, G3PFR, (ed) p4.7ff, RSGB

[19] *Antennas*, J D Kraus, 2nd edn, McGraw-Hill, 1988, Chap 4

[20] *Antenna Theory, Analysis and Design*, Constantine A Balanis, Harper and Row, 1928 pp204-243

[21] 'A Quadruple Quad Antenna - an Efficient Portable Antenna for 2 Metres', M Ragaller, DL6DW, *VHF Communications* 2/1971, pp82-84

[22] 'Analysis and Design of the Log-Periodic Dipole Antenna' *Antenna Laboratory Report No.52*, University of Illinois

[23] 'Log-Periodic Dipole Arrays', D E Isbell, *IRE Transactions on Antennas and Propagation* Vol AP-8, May 1960, pp260-267

[24] W3DUQ, *Ham Radio*, Aug 1970

[25] *Antennas*, J D Kraus, 2nd edn, McGraw-Hill, 1988 pp265-319

[26] 'Helical Antennas for 435MHz', J Miller, G3RUH, *Electronics and Wireless World* Jun 1985, pp43-46

[27] 'A Helical antenna for the 23cm Band', Hans-J Griem, DJ1SL, *VHF Communications* 3/83 pp184-189

[28] 'Feedline Verticals for 2m & 6m', Rolf Brevig LA1IC, *RadCom* Mar 2000 p36

[29] 'The HB9CV Antenna for VHF and UHF', H J Franke, DK1PN, *VHF Communications* Feb 1969

[30] 'Eurotek', E David, G4LQI, *RadCom,* Sep 1999 pp31-32

[31] 'Asymmetrically Driven Antennas and the Sleeve Dipole', R W P King, *Proc IRE,* Oct 1955 pp1154-1164

[32] 'Designing Discone Antennas', J J Nail, *Electronics* Vol 26, Aug 1953, pp167-169

[33] 'Slim Jim Antenna', F C Judd, G2BCX, *Practical Wireless*, Apr 1978, pp899-901

[34] 'Eurotek' *RadCom* May 1996 p59 & 61, Circular loop Eggbeater antenna

[35] http://members.aol.com/k5oejerry/eggbeater2.htm. 'Eggbeater II omni LEO antennas'

[36] 'Resonant Quadrifilar Helix Design' C C Kilgus, *Microwave Journal*, Vol 13-12, Dec 1970, pp49-54

[37] 'Shaped-Conical Radiation Pattern Performance of the Backfire Quadrifilar Helix', C C Kilgus, *IEEE Trans Antennas and Propagation* May 1975, pp392-397

[38] http://www.ee.surrey.ac.uk/Personal/A.Agius/qha_perf.html

[39] http://www.pilotltd.net/qha.htm. Construction details for a very slim quadrifilar helix antenna for 2m or 137.5MHz. Note that the elements are twisted just a quarter of a turn.

[40] http://web.ukonline.co.uk/phqfh/qfh.pdf. Conventional half-turn QFH for 137MHz with very detailed assembly instructions

VHF/UHF Antennas

Ian Poole, G3YWX

ONLY £13.99 plus p&p

This great new book from the popular author Ian Poole investigates the exciting area of VHF and UHF antennas. The VHF and UHF bands provide an exciting opportunity for those wishing to experiment, whilst the antenna sizes at these frequencies mean that they do not occupy great amounts of space.

VHF / UHF Antennas contains both the basic theory and constructional details for many antenna designs, taking the reader through the essentials in an easy to understand fashion. Chapters are devoted to different types of antenna from dipoles to Yagis, and verticals to log periodic antennas, giving details of the way in which they work and constructional information for a variety of designs. Simple antennas for broadcast reception supplement the amateur radio projects. The final chapters in the book cover measurements and installation techniques.

The book is a mine of information for anyone wanting to understand more about antennas for the VHF and UHF bands, or wants to construct them. It is a valuable resource for anyone interested in antennas, whether newcomer or experienced hand.

Size: 240 x 174mm, 128 pages, ISBN 1-872309-76-3
RSGB Members Price £11.89 plus p&p

E&OE

Radio Society of Great Britain
Lambda House, Cranborne Road, Potters Bar, Herts. EN6 3JE
Tel. 0870 904 7373 Fax. 0870 904 7374

ORDER 24 HOURS A DAY ON OUR WEBSITE
www.rsgb.org/shop

RSGB SHOP

17 Practical Microwave Antennas

The big advantage that microwave antennas have over those used for other amateur bands is that they are relatively small. We can let our imagination loose and have antennas with gains that are not achievable elsewhere and mount them at heights, in wavelengths, that users of the lower frequency bands can only dream of.

This chapter has been organised by type of antenna rather than by band usage, the types being:

- **Patch antennas:** These are easily constructed because they are etched onto printed circuit board. They are small and have found lots of applications in commercial electronics such as mobile phones, cordless phones and local area networks. If you need a small antenna, the patch may suit your needs.
- **Slot antennas:** These are omni directional and can be used for repeaters or mobile use.
- **Helical antennas:** The big advantage of the helical antenna is that it is circularly polarised. This has made it widely used for satellite operation.
- **Yagi antennas:** The Yagi is probably the best known antenna. With the reduced element size on the microwave bands some impressive gains can be achieved.
- **Horn antennas:** As the frequency increases, the horn antenna comes into its own at 10GHz and above.

PATCH ANTENNA

Antenna for 5.8GHz

This design was produced by Gunthard Kraus, DG8GB as a practical project to design a patch antenna for 5.8GHz [1].

The techniques described here can be used to design patch antennas for any of the lower amateur microwave bands, ie 23cm to 3cm. The main design work uses PUFF CAD software, this is will run on a wide range of PCs running DOS to Windows XP, the use of the software is described in much more detail in the reference material [1], [2].

The patch antenna was developed for the 5.8GHz ISM (Industrial, Scientific and Medical) band for monitoring video feeds two floors away in the new Tettnanger Electronic Museum in Germany. Since no holes could be bored and no slots could be cut out for cables in the historic museum building, the designers simply used the wall of the house opposite as the reflector for the 5.8GHz signal; the patch antenna was aligned with it. The 5.8GHz ISM band contains 16 channels at 9MHz intervals. between 5732MHz and 5867MHz.

The antenna thus has to display a self-resonant frequency of 5800MHz and a bandwidth of approximately 140MHz (2.4%). The input resist-

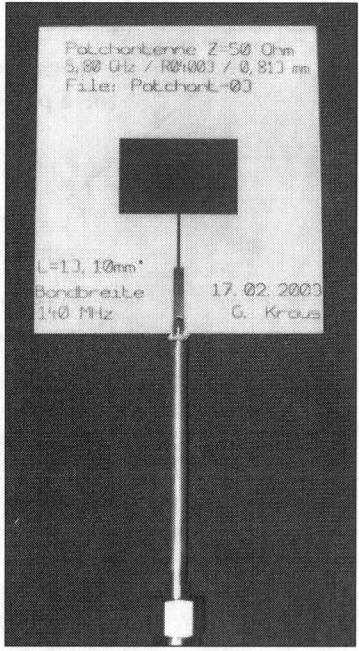

Fig 17.1: The finished patch antenna for 5.8GHz

ance should be 50 ohms, with a semi-rigid cable having a soldered SMA plug. The circuit board is made from Rogers R04003, which is a very stable material from the mechanical point of view and is easy to machine. It should have the following specification:

- ε_r = 3.38
- Printed circuit board thickness = 32mil   (0.813mm)
- Dielectric loss factor = 0.001
- Copper coating = 35μm

A $\lambda/4$ line transforms the antenna radiation resistance to the required 50 ohms. The circuit board size should be 50mm x 50mm and a short 50-ohm microstrip is needed from the transformation line to the cable connection on the circuit board. This can be seen clearly in the photograph of the finished product in **Fig 17.1**.

Design Procedure

The basic procedure can be found in an article on the subject from *VHF Communications* [2], [3].

After a little experimenting with the *patch16* program (available from the Internet [4]), an initial design was produced with a

```
These are the design parameters:

Length (L) = .5271  inches
Width (W) = .8  inches
Height (H) =  .032  inches
Dielectric Constant (D) =  3.38
Loss Tangent (T) =  .001
Feedpoint Distance (F) =  0  inches

Do you wish to edit any value? (Y/N):
```

Fig 17.2: The input parameters for *Patch 16*

```
The Resonant Frequency is   5.800 GHz
Qo is  26.0

The Edge Radiation Resistance is  142.42 ohms
Zc of Quarter-wave transformer is  84.4 ohms
Approx. width of the Quarter-wave transformer is 0.028 inches
Length of Quarter-wave transformer is 0.321 inches at the Resonant Freq.

Input Resistance at probe location is 142.42 ohms

The 2:1 VSWR Bandwidth is  2.9%
Upper Frequency Limit = 5.883 GHz
Lower Frequency Limit = 5.716 GHz

Press 'ENTER' to continue: ▮
```

Fig 17.3: Simulation results for the patch antenna

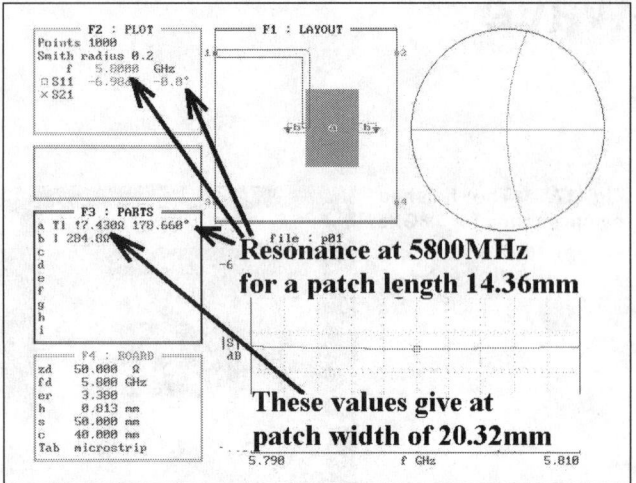

Fig 17.4: Simulation of the patch antenna using *PUFF*

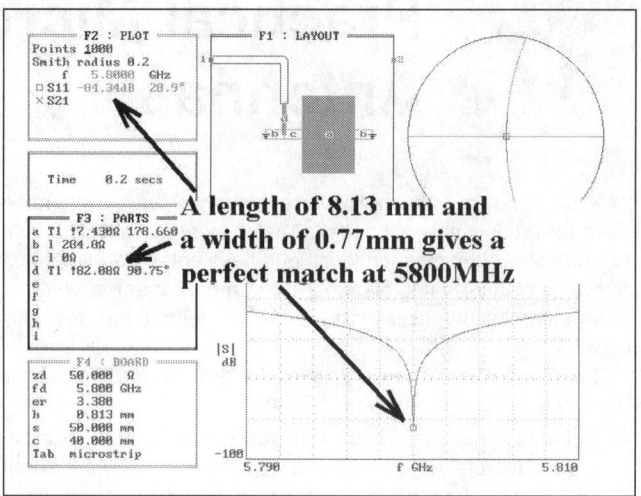

Fig 17.5: Using a quarter-wave transmission line, the input resistance is matched to 50Ω

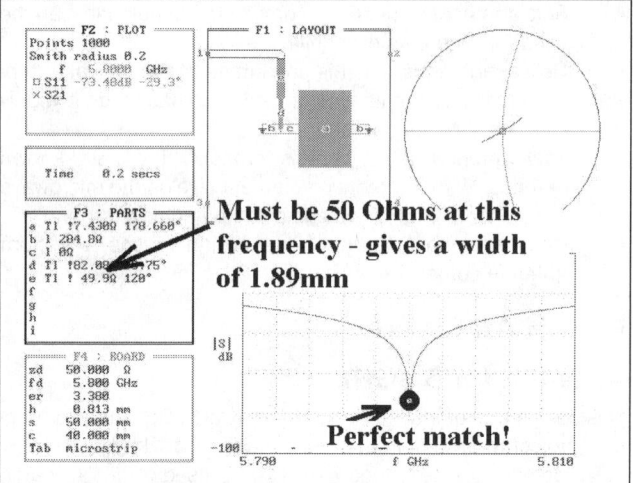

Fig 17.6: Matching patch antenna to 50Ω

centre frequency of exactly 5800MHz, while the bandwidth was deliberately increased to 2.9%.

The input and calculated characteristics of the antenna can be seen in **Figs 17.2 and 17.3**. The values required for the subsequent work using *PUFF* [5] must first be converted from inches into millimetres:

- Patch width = 20.32mm
- Patch length = 13.39mm
- Total radiation resistance = 142.4Ω, which gives 284.8Ω on each patch edge.

A text editor is needed to open the setup file of *PUFF21*, and to enter the R04003 material and circuit board data. *PUFF* is started up (work with the protected mode version by loading *puffp.exe*) and the patch is modelled as a large width, lossy transmission line. The two radiation resistances are positioned on the two patch edges and the centre frequency is set at 5.8GHz.

First, delete the exclamation mark after the entry "TL" in field F3, and experiment with the characteristic impedance of the line until (once the equals sign has been entered) a width of w = 20.32mm is set.

Now replace the exclamation mark, and vary the electrical length of the line until the configuration is as near as possible to resonance. This situation can easily be recognised, since then the phase angle of S11 is exactly zero degrees. Remember to select a swept frequency range as small as possible, and also try to obtain the highest possible resolution for the amplitude range for |S11|. The process is clarified in **Fig 17.4**, which immediately supplies the required data. For a patch width of 20.32mm, a microstrip line is needed with a characteristic impedance of 7.43Ω at low frequencies. A mechanical length of 14.36mm has an electrical length of λ/2 at 5.8GHz.

Note: If you're surprised by the big difference between this length value of 14.36mm and the *patch16* suggestion of L = 13.39mm, the following will clear up the mystery. First the patch has to be shortened by the open-end extension on both sides. More on this subject later, but the result can be given in advance: it is 0.41mm on each side. But this only reduces the length to 14.36mm - 2 x 0.41mm = 13.53mm. The difference is provided by something discovered by chance in the specialist literature: ". . . . the difference between the patch resonance and the electrical length for the corresponding λ/2 microstrip is about 1 % . . .". Thus 0.99 x 13.53mm = 13.40mm.

Using a λ/4 transformation line, the input resistance of the configuration must be brought to precisely 50Ω. In **Fig 17.5**, this

has already happened, and the required data are:

- Line length = 8.18mm
- Line width = 0.765mm

Providing the configuration with a 50-ohm feed (required length about 11mm up to the edge of the board), is shown in **Fig 17.6**. From this, the width of the feed can be determined (once again, after removing the exclamation mark after "TL" and pressing the equals sign) at w = 1.89mm. First, the open-end extensions for the microstrip line must be determined. The quickest way is still to use the appropriate diagram from the *PUFF* manual. **Fig 17.7** shows the necessary procedure and also supplies the raw data required:

- For a patch with Z = 7.4Ω use 51% of the board thickness of 0.813mm = 0.42mm
- For the 50Ω feed, use 45% of the board thickness of 0.813mm = 0.37mm

Since the transformation line displays the highest characteristic impedance and thus the smallest width, it must be extended by the following amounts at both ends (the correction formulas can be found on the same page in the *PUFF* manual):
For the patch side, the result is:

$$\Delta L = \left(1 - \frac{w2}{w1}\right) \cdot 0.42 = \left(1 - \frac{0.77}{20.32}\right) \cdot 0.42 = 0.41mm$$

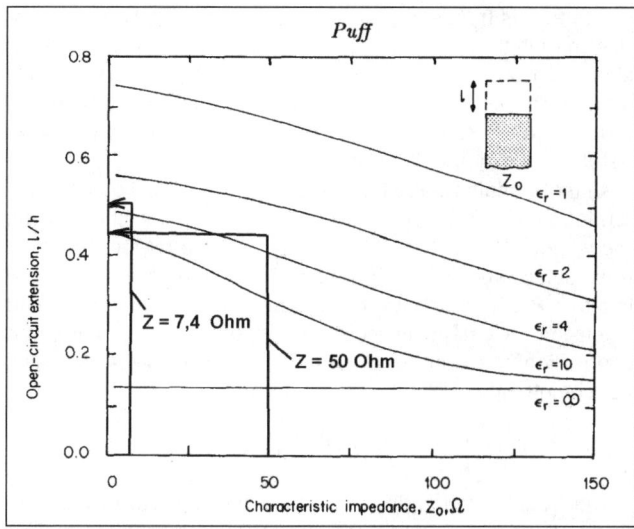

Fig 17.7: Determining the open end extension using the graph published in the *PUFF* manual

For the feed side, the result is:

$$\Delta L = \left(1 - \frac{w2}{w1}\right) \cdot 0.37 = \left(1 - \frac{0.77}{1.89}\right) \cdot 0.37 = 0.22\text{mm}$$

So for the transformation line, the requirements are a length of 8.18mm + 0.41mm + 0.22mm = 8.81mm, and a width of 0.77mm. The best way to work on the patch is to use the length supplied by the *Patch16* program, ie 13.39mm, with a patch width of 20.32mm.

Using a printed circuit board CAD program, centre the patch on the selected circuit board (dimensions 50mm x 50mm), add the transformation line, and finally connect the feed line, with a width of 1.89mm, up to the board edge as in Fig 17.1.

Now the semi-rigid cable, with an SMA plug, must be connected to the circuit board with the minimum of electrical irregularities as shown in **Fig 17.8**: The method is as follows:

- First, saw off the cable and carefully file the end flat at an angle. Do not forget to trim it.

- Using a fine saw (eg a jig saw), make a cut parallel to the cable and a few millimetres long, precisely following the inner conductor.

- Carefully cut perpendicular to the cable to expose the inner conductor completely and remove the internal Teflon insulation.

- Push the circuit board into this cut solder the inner conductor to the 50Ω microstrip feed line. Carefully solder the remainder of the cable sheathing to the underside.

An investigation of the antenna using an HP8410 network analyser, HP5245L microwave counter and an HP5257 transfer oscillator, gave the resonance at 5690MHz, with a value of |S11| = -16dB.

A more precise examination, using a polar display, showed that the input resistance here exceeds the system resistance of 50 ohms and consequently the radiation resistance must have a higher value. These findings were immediately converted into a *PUFF* simulation, and it became clear that in reality a resistance of 407.5 ohms should be assumed on each patch edge (**Fig 17.9**).

The patch length required for 5890MHz should be displayed immediately in the F3 parts list - it amounts to L = 14.65mm. Now, in the F4 field, simply change the design frequency to the required 5800MHz, thus once again simulating the patch reso-

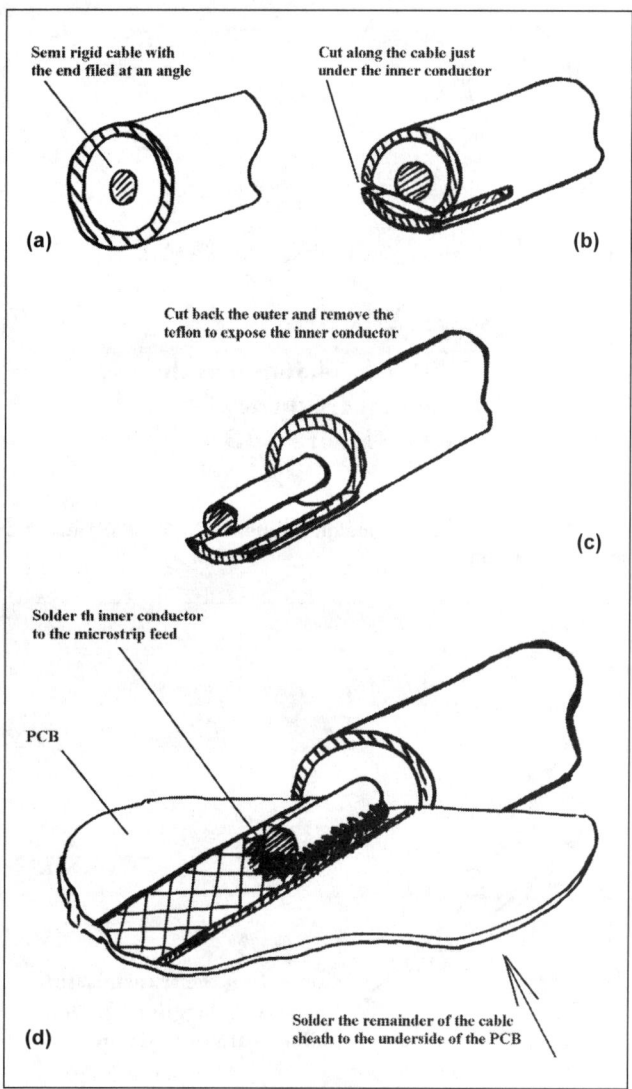

Fig 17.8: Details of how to fit the semi-rigid feeder. (8a) Semi rigid cable with the end filed at an angle. (8b) Cut along the cable just under the inner conductor. (8c) Cut back the outer and remove the teflon to expose the inner conductor. (8d) Solder the inner conductor to the microstrip feed and the cable sheath to the underside of the PCB

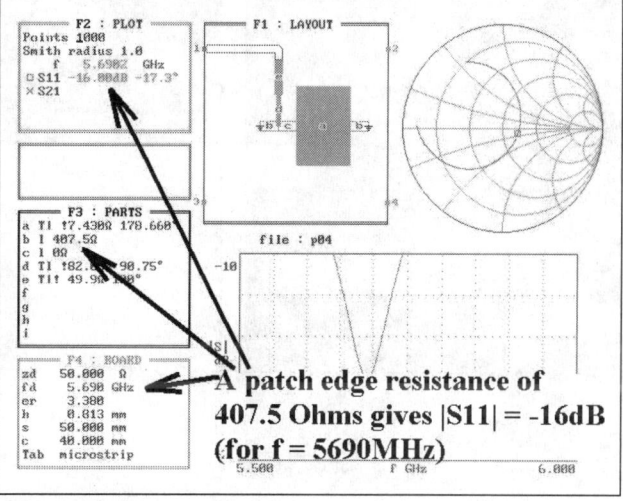

Fig 17.9: *PUFF* simulation showing the patch edge resistance of 407.5Ω and |S11| of -16dB

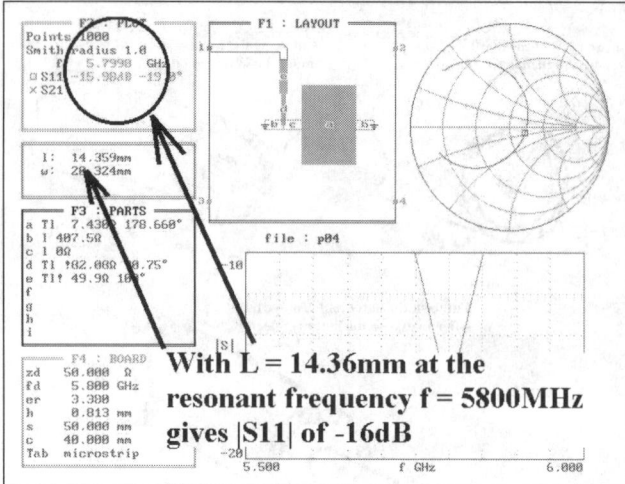

With L = 14.36mm at the resonant frequency f = 5800MHz gives |S11| of -16dB

Fig 17.10: Changing the design frequency to 5800MHz gives a length of 14.36mm

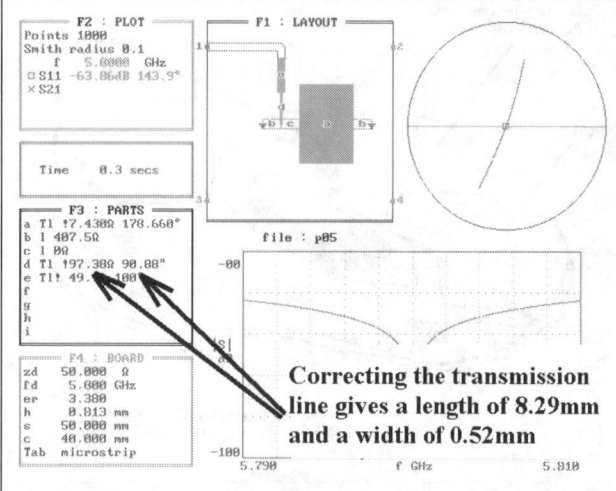

Correcting the transmission line gives a length of 8.29mm and a width of 0.52mm

Fig 17.11: Correcting the matching transformer

nance at this frequency. It can be seen from **Fig 17.10** that for this the length must be reduced to 14.36mm - consequently, shortened by 0.29mm! Thus, in the printed circuit board CAD system, the length used for the initial design of 13.39mm is reduced to 13.39mm - 0.29mm = 13.1mm. The patch width, naturally, remains at 20.32mm.

All that remains is the final change - correcting the λ/4 transformation line to obtain better values for the matching of the measured - 16dB. This is also a very easy matter for *PUFF*, and the result can be seen in **Fig 17.11**. The new values required are: length = 8.29mm and width = 0.52mm. The required open-end extension for each side must be added to this. For the patch connection, the value of 0.41mm used previously is still valid, but something is altered on the feed side:

$$\Delta L = \left(1 - \frac{0.52}{1.89}\right) \times 0.37 = 0.27mm$$

The transformation line, therefore, has a width of 0.52mm and a length of 8.29mm + 0.41mm + 0.27mm = 8.97mm. Test readings on a second prototype are shown in **Fig 17.12**.

The design was also evaluated with two other simulation programs, *Mstrip40* and *Sonnet Lite*, descriptions of these and/or projects using them can be found in *VHF Communications* [6] and [7].

Sonnet was used on the radiating patch, in order to re-check the resonance frequency and also to investigate the matter of the much higher radiation resistance. Using an online menu made it easy to use this program, but it is important to pay heed to the rules of simulating such antenna structures in the *Sonnet* manual. It is essential to obtain (free) the additional license for extending the maximum usable PC working memory to 16 megabytes, in order to carry out the simulation successfully. **Fig 13** shows the *Sonnet* editor screen with the selected box and cell dimensions, together with the new 'Quick Start Guide'. Owing to the restrictions on the Lite version, the transformation line is simply omitted and replaced by a 50-ohm power feed taken right up to the box wall. The actual radiation resistance of the antenna can then be determined easily and directly from the reflection factor determined in this way. This is how to do it:

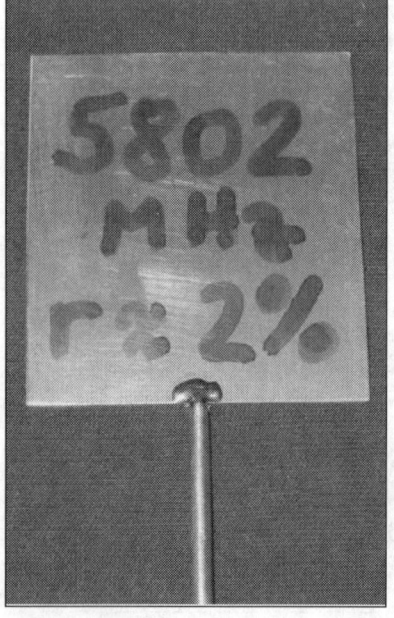

Fig 17.12: A good test result with this antenna

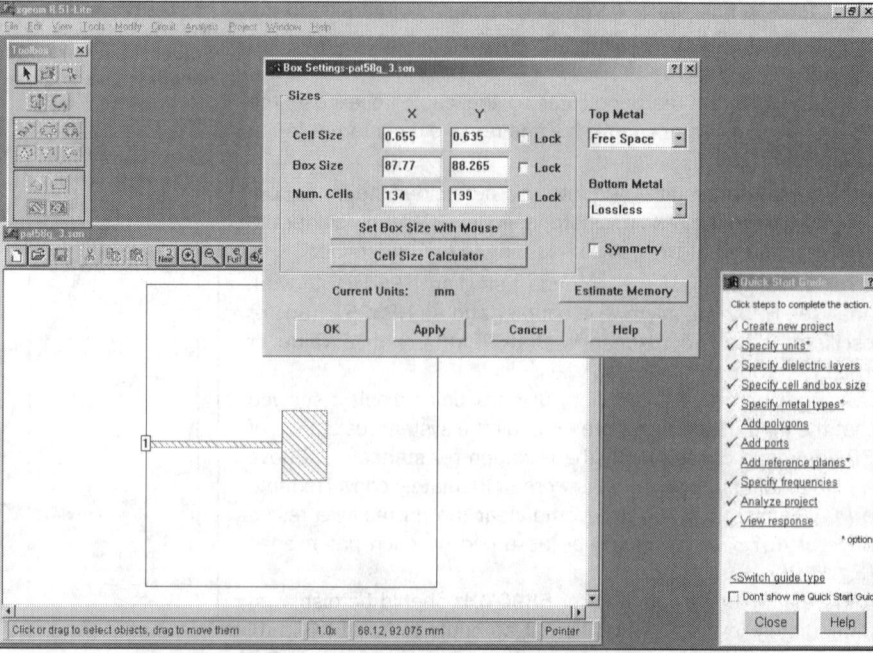

Fig 17.13: Using the *Sonnet-Lite* simulator for the patch antenna

The simulation result can be seen in **Fig 17.14**, with a resonance of 6.03GHz and |S11| = -4.4dB. This gives a reflection factor of:

$$r = \log\frac{-4.4\text{dB}}{20\text{dB}} = 0.60$$

Because a 50Ω feed is being used, it is necessary to reduce the circuit length in the Smith diagram until resistances exceeding 50Ω are arrived at on the real axis. Then, in accordance with the following relationship, the total resistance on the patch edge is obtained:

$$R = \frac{1+r}{1-r}\cdot Z = \frac{1+0.6}{1-0.6}\cdot 50\Omega = 200\Omega$$

Consequently, each radiating edge is affected by a radiation resistance of 400Ω. Compared with the measured value of 407.5Ω, this is an extremely satisfactory result, and thus confirms the validity of the measurement. The resonance frequency was simply predicted too high, with an error of:

$$\frac{6030\text{ MHz} - 5800\text{ MHz}}{5800\text{ MHz}}\cdot 100\% = 3.96\%$$

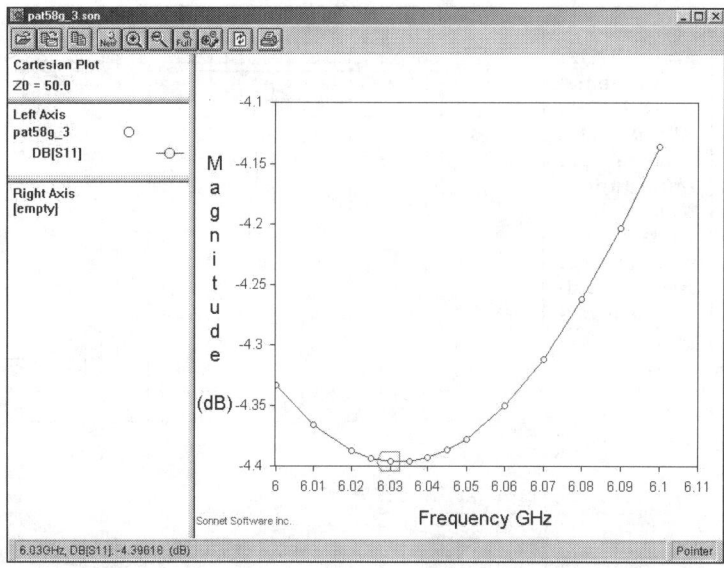

Fig 17.14: Simulation results from *Sonnet-Lite*

For a comparison simulation, entering the cell data used for *Sonnet* into the *mstrip40* program gives precisely this resonant frequency, but with rather larger discrepancies of approximately 10 to 15% in the radiation resistance.

Nowadays, the most modern design aids are available, even to private developers without an expensive industrial scale test rig, and at reasonable cost.

SLOT ANTENNA

The slot antenna that has become known in the amateur world as the *Alford slot* actually derives from work by Alan D Blumicin of London, and is detailed in his patent number 515684 dated 7 March 1938. The work by Andrew Alford was carried out during the mid 1940s and 50s, and not applied to microwave bands but to VHF/UHF broadcasting transmitters in the USA.

Development by M Walters, G3JVL, was carried out during 1978 when designs for the GB3IOW 1.3GHz beacon were being investigated. The initial experiment was carried out at 10GHz as the testing was found to be much easier, especially when conducted in a relatively confined space. A rolled copper foil cylinder produced results close to those suggested in the original work. Initially, it was thought that the skeleton version would be best used at the lower frequencies only. However, several models for use on 144MHz have been constructed and they performed very well, but the design was, at the time, not regarded by G3JVL as being of interest.

Further developments have resulted in working models being constructed for the 50MHz, 432MHz, 900MHz and 1.3GHz bands. However, some aspects were not easy to explain and valuable assistance was provided by G3YGF. This assistance provided more than just an explanation for the fact that early skeleton versions worked at a lower than designed frequency. A working operational theory was developed as a direct result, along with a better understanding of the strange effects that were observed.

The 2.3GHz version developed by G3JVL fulfils the need for an omni directional horizontally polarises antenna. This makes it particularly useful for beacons, fixed station monitoring purposes and mobile operation. Mechanical details of this antenna are shown in **Fig 17.15**. The prototype was made from 22mm outside diameter copper water pipe. Material is removed from one part of the tubing to produce a slotted tube with an outside diameter of 18.5mm and a slot width of 2.6mm. To ensure cir-

culariity the tube is best formed around a suitable diameter mandrel. Small tabs are soldered at the top and bottom of the tube to define the slot length of 229mm. A plate is soldered across the bottom of the tube to strengthen the structure.

The RF is fed via a length of 0.141in (3.6mm) RG141 semi rigid coaxial cable up the centre of the tube to the centre of the slot via a 4:1 balun constructed at the end of the cable. The detailed construction of the balun is shown in Fig 17.15. The two diametrically opposite slots are cut carefully using a small hacksaw with a new blade. The inner and outer of the cable are shorted using the shortest possible connection and the balun is attached to the slot using two thin copper foil tabs. If suitable test gear is available, the match of the antenna can be optimised by carefully adjusting the width of the slot by squeezing

Fig 17.15: Construction of the 2.3GHz Alford slot antenna using a dual slotted cylinder. The feed point impedance is 200Ω. (a) Dimensions for 2,320MHz are: slot length 280mm, slot width 3mm, tube diameter 19mm by 18SWG. (b) Construction of a suitable balun. The balun slots are 1mm wide and 26mm long

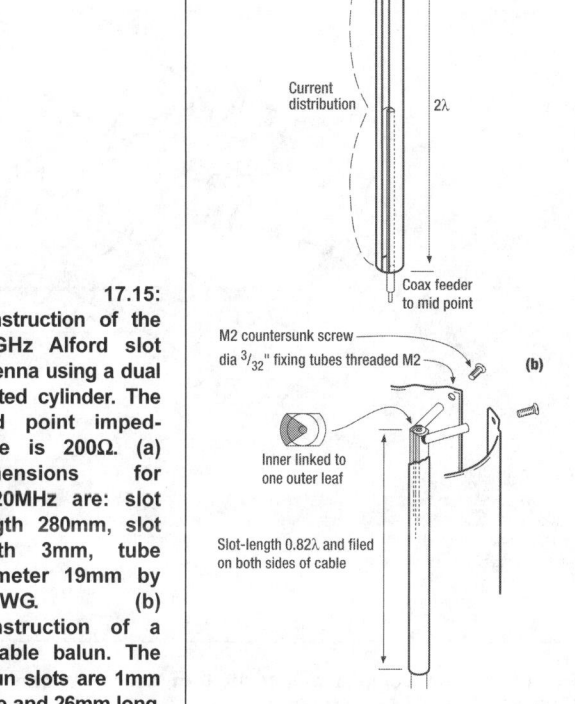

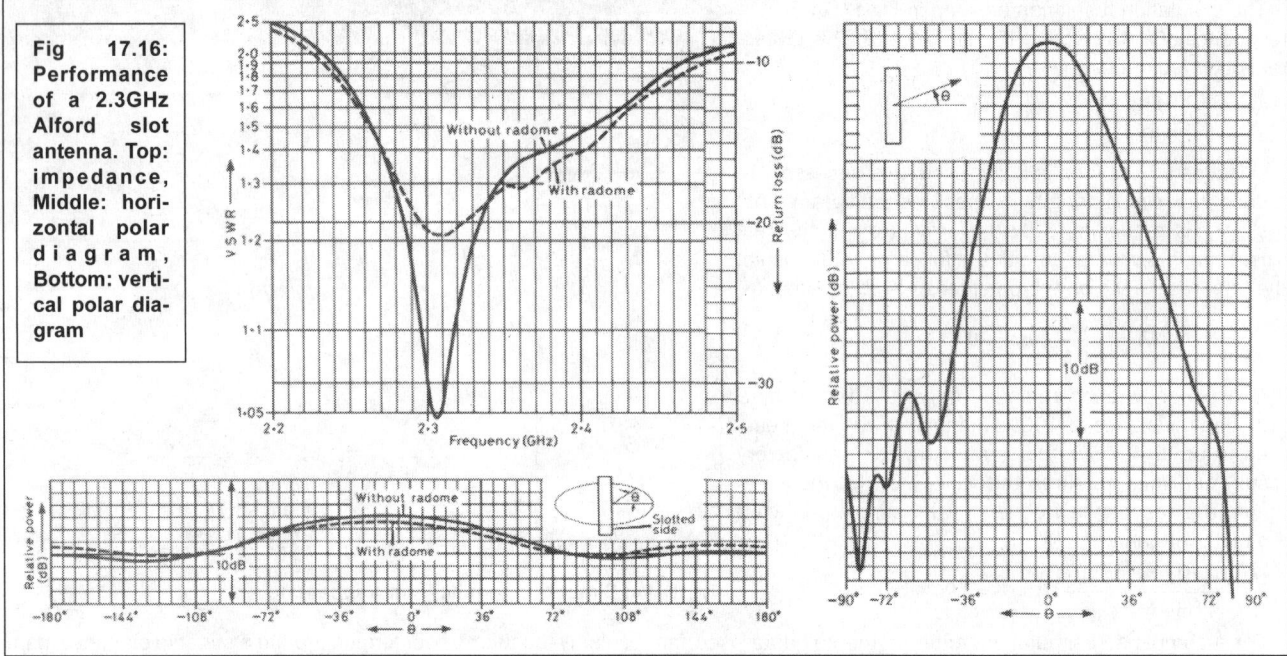

Fig 17.16: Performance of a 2.3GHz Alford slot antenna. Top: impedance, Middle: horizontal polar diagram, Bottom: vertical polar diagram

the tube in a vice, or by prising the slot apart with a small screwdriver. Typical antenna characteristics are shown in **Fig 17.16**. The gain of the antenna has been measured to 6.4dBi.

HELICAL ANTENNA

Paolo Pitacco, IW3QBN, designed this array of 4 x 16 turn helix antennas for 2402MHz [8]. Such a project is an interesting challenge for design, manufacture and testing. The designer had no space for a parabolic reflector and required easy mounting and dismounting in case of strong wind. The idea was to design a system capable of operating as a satellite and terrestrial antenna with discrete gain, good capture area and rapid installation. As a starting point a simple 16-turn helix antenna [9] was used to hear signals from the DOVE satellite (2401MHz), and to work

Fig 17.17: The completed 4 x 16 turn helical antenna for 2402MHz

with home made ATV systems. The endpoint is an array of four of these helices, shown in **Fig 17.17**.

Radio amateurs today have many 'ready to use' computer programs available for helical antenna design, from simple DOS to complex Windows or Linux applications. It is possible to choose from any of the following techniques:

• Reproduce the design by G3RUH [10]
• Make a new design using KA1GT [11] program
• Use Peter Ward's *Excel* worksheet, 'RF2'
• Follow design formulas from antenna engineering literature [12 & 13]

This excludes all commercial and costly programs. A mix of these ways was followed to check the results from outputs of several programs, and validate and refine using formulas [12]. The same input data (central frequency, number of turns and mechanical dimension) were used in each case.

The design was centred on 2402MHz, to obtain sufficient bandwidth to cover AO-40 and ATV frequencies. Using this centre frequency, a support diameter of 38mm was chosen, together with 31mm spacing between turns. Copper wire of 3mm diameter was used because it is self supporting without the need for any type of plastic tube which would cause up to 15% detuning of the helix due to dielectric action.

Matching a single helix is not difficult. It is possible to achieve a broadband match without use of a network analyser or other measuring instruments, but multiple helix arrays are more difficult. Multiple antennas require a coupling system and matching to the transmission line, and the first is related to the second. The *aperture area* of each antenna must be evaluated, and a method found to couple the array to a single cable.

There are four methods:

1) Each antenna matched to 50Ω with a 4:1 coupler to the feeder
2) Unmatched (Z~140Ω) with a coaxial quarter-wavelength transformer and a 4:1 coupler to the feeder
3) Unmatched (Z~140Ω) with a quarter-wavelength wireline transformer to the feeder
4) Unmatched (Z~140Ω) with a coaxial quarter-wavelength transformer to the feeder

Fig 17.18: Quarter wavelength transformer for 2402MHz 4 x 16 turn helical antenna

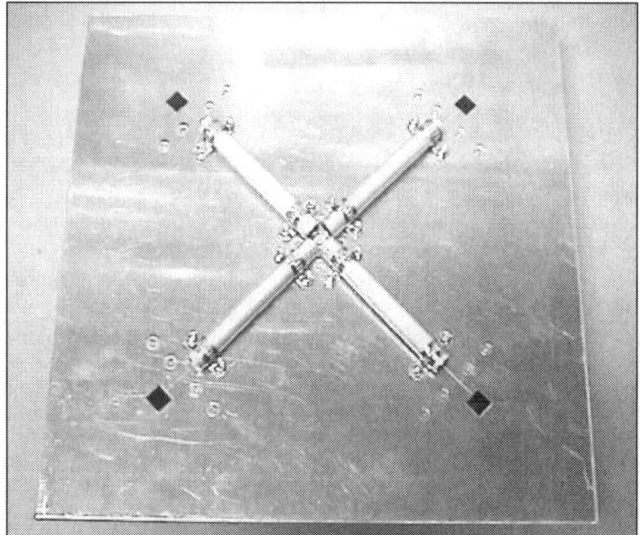

Fig 17.19: Mounting of the four quarter wavelength transformers for 2402MHz 4 x 16 turn helical antenna

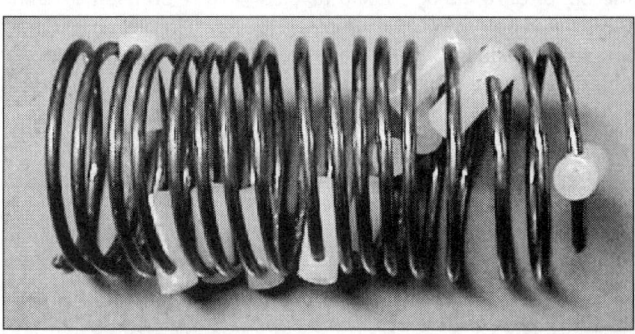

Fig 17.20: Fitting spacers onto wound helix for 2402MHz 4 x 16 turn helical antenna

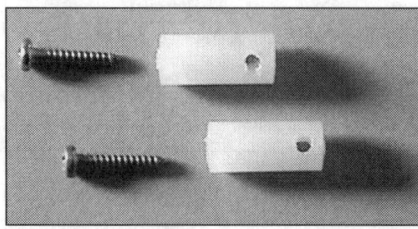

Fig 17.21: Spacers for helix of 2402MHz 4 x 16 turn helical antenna

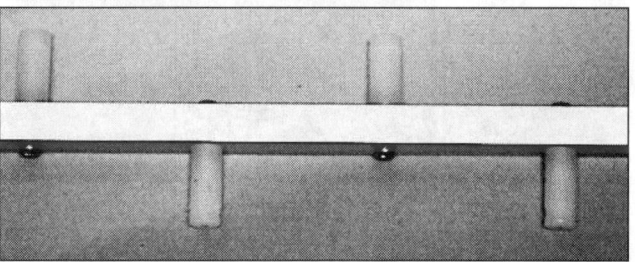

Fig 17.22: Aluminium support tube for 2402MHz 4 x 16 turn helical antenna

Solution 1 is easy and is used by G6LVB. Good connectors and cable, and great precision is needed to make four equal helix antennas and cables.

Solution 2 is difficult and no longer used. Mechanical tools and a lot of connectors and cable are needed to make a good transformer.

By using Solution 3, the helix antennas and feeder can matched in a single pass as a quarter-wavelength transformer with an output impedance equal to 4 x 50Ω is easy, but a wire-line is a potential antenna.

Solution 4 has the advantages of 3, but by using coaxial line you avoid any problem of radiation and the matching line has a more stable impedance. It is a mechanically complex solution, but requires only one connector.

For this project, Solution 4 was chosen,and this drove subsequent mechanical and electrical issues. The helices must be placed at the correct spacing, and a quarter wavelength transformer. must be constructed. For 2402MHz, lambda (λ) is 12.48cm and λ/4 is 3.12cm, but calculations indicate an aperture area of 1.4λ or 17.5cm. A mechanical solution had to be found, using an odd multiple for quarter wavelength, exactly 3λ/4 (93mm) to maintain transformation properties and distance between helices.

Impedance matching with coaxial transformer is straightforward when mechanical machining isn't a problem, using the equation:

$$Z_m = \sqrt{Z_a \times (Z_l \times N_a)}$$
(1)

Where Z_a is the antenna impedance, N_a is the number of antennas, Z_l is the impedance of main feeder and Z_m the required impedance for a match.

Using equation below, it is possible to determine tube and conductor diameter to build the matching section.

$$Z_{coax} = \frac{138}{\sqrt{\varepsilon_r}} \cdot \log\left(\frac{D_g}{D_p}\right)$$
(2)

Where D_g stands for inner diameter of tube, D_p is diameter of wire, ε_r is 1.001 for air).

Because the computed helix impedance is 153Ω, a characteristic impedance of 173Ω is needed for this line. The quarter wavelength transformer was made from a commercially available aluminium tube with an external diameter of 12mm (10mm inside), and a 0.6mm diameter silvered copper wire. This represented a good compromise (about 170Ω).

The wire is held in position with a couple of thin (3mm) centre drilled nylon plugs, this means the transformer's dielectric will be near that of air (see **Fig 17.18**).

All four transformers are locked in position (on the upper side of the reflector) by means of two U-shaped brackets also used as a ground connection. One side is soldered on to an N connector (see **Fig 17.19**).

The reflector plate is a made of a 3mm thick square of aluminium sheet with 30cm sides.

Each helix is made with 3mm diameter copper wire, wound on a 35mm diameter support, then gently relaxed and loaded using 11 spacers (every 1.5 turns), as shown in **Fig 17.20**. The spacers are small cylinders of nylon, 20mm long and 8mm diameter, each pre-drilled with a 3.2mm hole 15.6mm from the bottom (**Fig 17.21**). The first spacer is located at the end of the helix and last, one turn before the feed point. The spacers are subsequently locked on a square aluminium tube (10mm side, 1mm depth) pre-drilled with 3mm holes at distances of 47mm (one each 1.5 turns) as shown in **Fig 17.22**. This tube, which is 57cm long, has no effect on performance, and is used to hold the helix

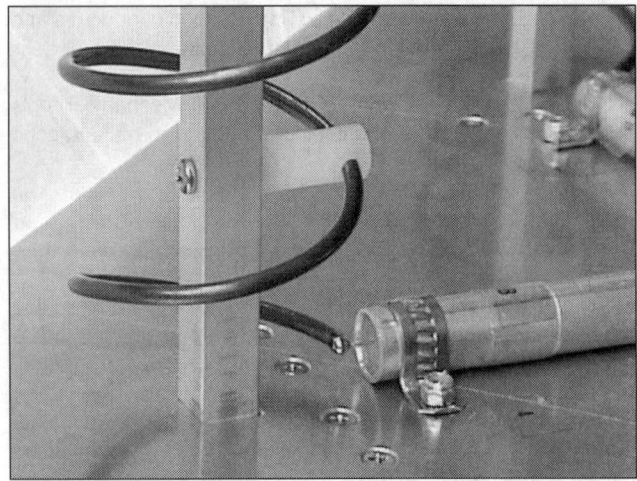

Fig 17.23: Close up view of connection to quarter wavelength transformer for 2402MHz 4 x 16 turn helical antenna

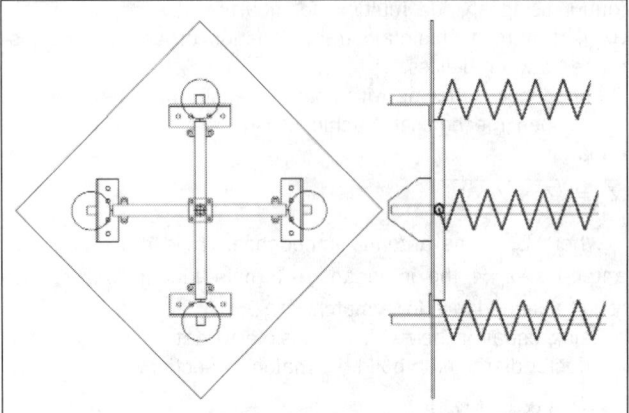

Fig 17.24: Complete mechanical drawing for 2402MHz 4 x 16 turn helical antenna

in position by means of L-shaped aluminium brackets (measuring 25 x 50mm).

Another U-shaped bracket is used to hold converters, preamplifiers or other devices, and as a mounting point (a tripod for example).

All helices are wound in the same direction (counter clockwise as seen from rear) in order to maintain the phase of the feed point. **Fig 17.23** shows the electrical and mechanical connection between the helix and the coaxial transformer. Attention should be given to soldering the inner transformer conductor; pre-solder the 3mm copper wire beforehand. **Fig 24** is a complete mechanical drawing of this array.

The first measurement was made with the aid of an Agilent E8714ES Network Analyser, and the results were close to what was expected as shown in **Fig 17.25**. A low SWR was obtained on 2430MHz (near the centre of S band), but the antenna gives good results over the entire band. Other lengths of transformer (92 and 94 mm respectively) have been tried, but 93mm represents the best compromise for amateur use. The calculated gain is 18dB (including losses), and tests on AO-40 signals demonstrate this figure. A comparison was made with a 80cm dish with 3 turn helix, and this gave 6-8 dB higher gain.

YAGI ANTENNAS

Wide Band Yagi for 23cm

Noel Hunkeler, F5JIO, designed an array of six half-wave dipoles with a reflector plane for a 23cm packet radio link [14]. It provides 10dBd gain and an SWR not exceeding 1.1:1 throughout the band. Curtains of phased dipoles have been used by amateurs since before WWII, examples are shown in the 1937 Jones antenna handbook. F5JIO consulted *Rothammel*, the German antenna bible, which gives the following guidelines for the reflector plane:

> *For best F/B ratio, it should extend at least a half wave beyond the perimeter of the curtain on all sides. If made of wire mesh instead of solid sheet to reduce*

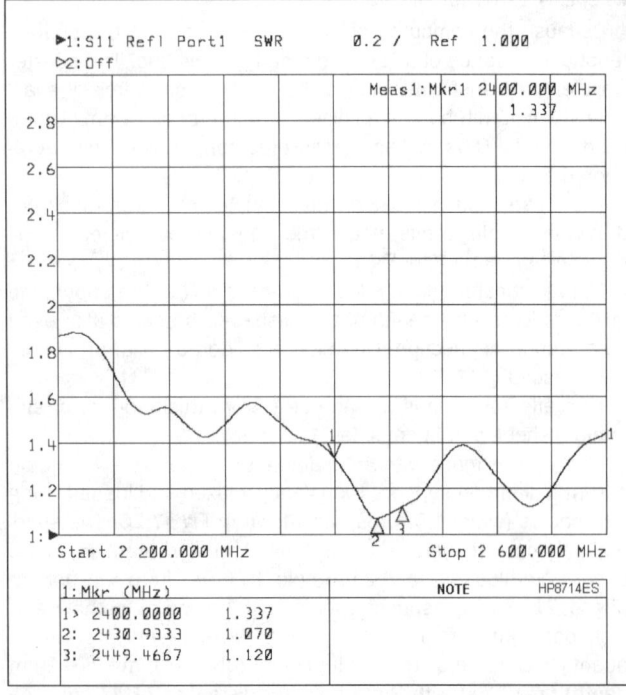

Fig 17.25: Results for 2402MHz 4 x 16 turn helical antenna measured on a network analyser

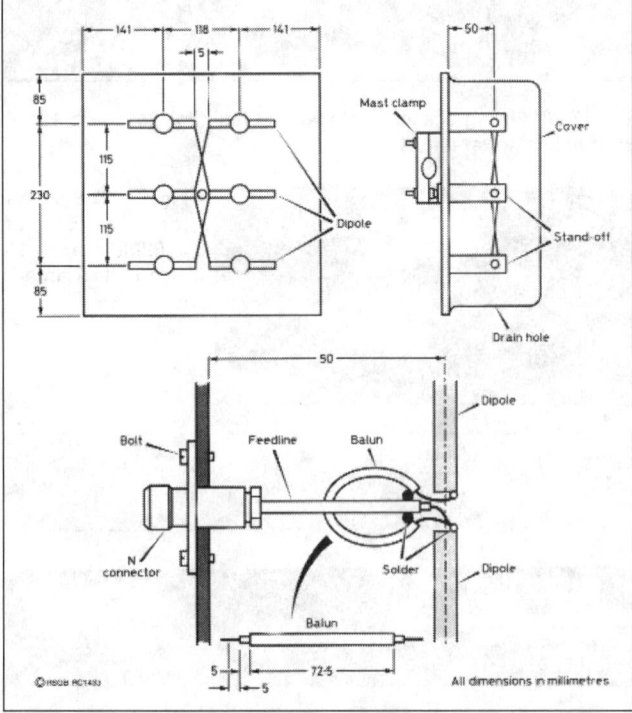

Fig 17.26: Details of wide band yagi for 23cm

Reflector	400 x 400 (340 min) 2.5 thick aluminium sheet (Qty 1)
Stand-off	Teflon (or PVC) 60L x 20D (Qty 6)
Dipole	Brass, silvered, 108L x 6D (Qty 6)
Rod, phasing	Wire, silvered, 2D (Qty 4)
N connector	(Qty 1)
Feedline	Semi rigid coax, 50 , approximately 4D (Qty 1)
Balun	as above, 92.5L (Qty 1)
Bolt	M3 x 8, SS (Qty 4)
Cover	Plastic food container (Qty 1)
Mast clamp	From TV antenna (Qty 1)

Table 17.1: Component list for wideband 23cm beam. All dimensions are in mm

Fig 17.27: Picture of the 22 element Yagi antenna for the 13cm band

windage, the wire pitch should be λ/20 or less. A reflector plane spaced 5λ/8 behind the radiator adds maximum gain, up to 7dB, but a spacing of 0.1 - 0.3λ gives better F/B ratio. If spaced at least 0.2λ behind the curtain, the reflector plane does not affect the feed point impedance of the array.

With the dimensions given in **Fig 17.26** and the parts shown in **Table 17.1**, the feed point impedance of each dipole pair is approximately 600Ω balanced. Three pairs in parallel give 200Ω balanced. The 4:1 re-entrant line balun transforms this to 50Ω, unbalanced. Note that each dipole is supported at its voltage node; hence the standoffs need not be high quality insulators.

For 23cm this antenna is small, so a solid aluminium reflector is practical; this plate supports all other components. Slightly bend the phasing rods so they do not touch at the crossovers. For weather protection, a plastic food container serves as a radome. Its RF absorption seems negligible for our application and it is much cheaper than Teflon. Although precision is required, construction of this antenna is not difficult.

DL Design Yagi for 13cm

Leo Lorentzen, OZ3TZ, designed these 10 and 22 element Yagi antennas for 13cm [15]. The design comes from a popular antenna designer, DL6WU, who has been much copied. While very popular, his writings are also courtesy of Mr H Yagi from Japan, dating right back to 1928 (see the Antenna Fundamentals chapters for more on Yagis).

The Danish version of this Yagi (**Fig 17.27**) is made of standard dimension brass materials, worth mentioning, as you will often see measurements for the materials used that are not in stock at your local metal goods dealer. Common standard dimensions like 2.0, 2.5, 3.0mm etc are off the peg goods, so once the shopping has been done and the brass has been delivered, you're well on your way towards your 13cm antenna project.

The customary practice for brass work is to use good tools. Mark out the antenna boom and check the marking an extra time before positioning the twenty centre punch marks and then drilling twenty 2mm diameter holes for the directors, which first

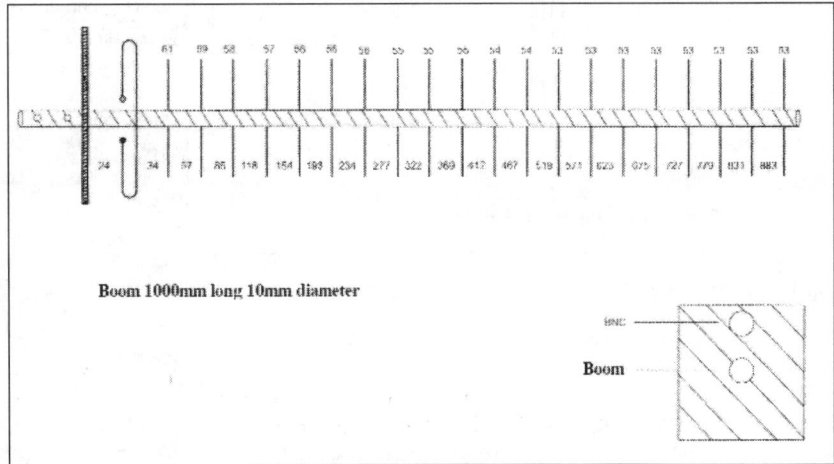

Fig 17.28: Element dimensions for the 22 element 13cm yagi

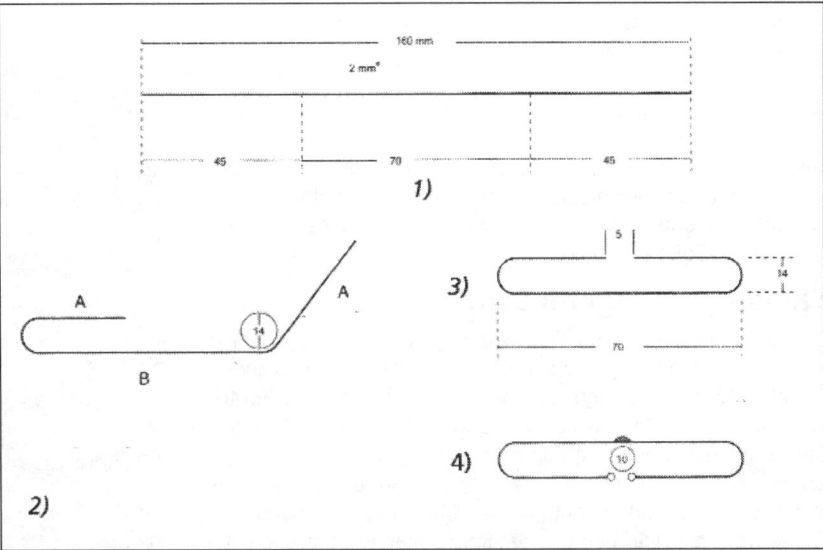

Fig 17.29: Details of the dipole for the 22 element 13cm yagi

need to be soldered into the boom. Solder the dipole on top of the boom. **Fig 17.28** shows the dimensions of the elements and **Fig 17.29** shows the detail of the dipole. Drill two 10mm diameter holes in the 80 x 80mm, 1mm brass plate reflector plate, soldering it firmly to the antenna boom as well.

Behind the reflector, drill two 4mm diameter holes, spaced 75mm apart. On top of the holes two 4mm diameter nuts are soldered so that the antenna can be secured to a mast fixture or something else. Mount a BNC socket in the reflector plate to

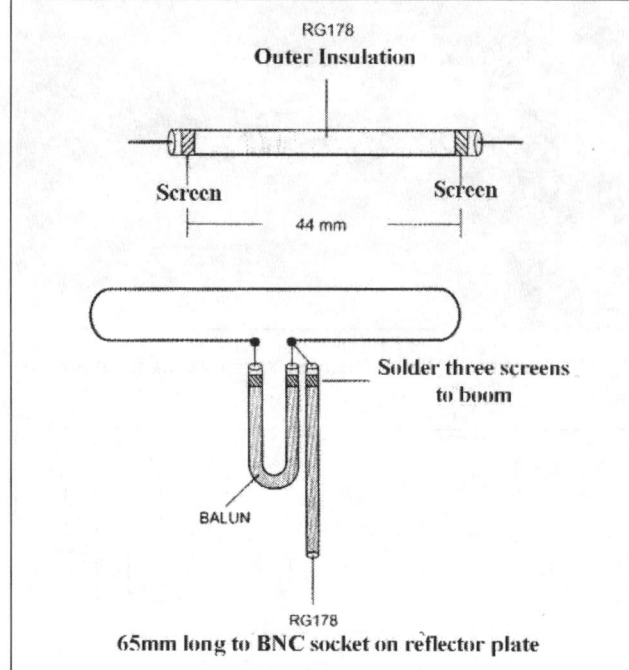

Fig 17.30: Details of the balun for the 22 element 13cm yagi

accept the 65mm RG178s from the dipole/balun, see **Fig 17.30**.

The really important thing is to solder the screen directly onto the antenna socket and allow the screen to go right up to the inside pin on the BNC socket, where the inner conductor in the cable will be soldered. If the antenna is to be sited outdoors, Araldite can be used to make 50Ω cabling etc watertight.

The finished antenna fares well in practical use. To measure the gain, an absolutely stable test signal was received on a half-wave dipole connected to a 13cm converter. With the dipole antenna, an approximately S1 signal was received. A 10 element yagi was then substituted for the dipole and an approximately S3 signal was received. Then the 22 element antenna was connected, and an impressive S7 signal was received. According to the original designer of the antenna, a boom length of seven wavelengths and approximately 20 elements makes for something over 15dBd.

Quad Loop Yagi for 9cm

This antenna was originally designed by M Walters, G3JVL. The design is shown in **Fig 17.31** with the critical dimensions shown in **Table 17.2**. The design is a scaled version of the 1,296MHz design, adapted to the narrow band segment at 3,456MHz. There are 61 directors giving a boom length of 2m. The construction of the antenna is quite straightforward providing care is taken in the marking out process. Measurements should be made from a single point or datum. In marking the boom for instance, measurements of the position of the elements should be made from a single point rather than marking out individual spacings.

Boom:

Boom diameter	12.5mm od
Boom length	2.0m
Boom material	aluminium alloy

Elements:

Driven element	1.6mm diameter welding rod
All other elements	1.6mm diameter welding rod

Reflector size:

Reflector plate	52.4 x 42.9mm

Element lengths:

Length 1 (mm) (see Fig 17.32a):		Directors 21-30	78.2
Reflector loop	99.2	Directors 31-40	76.1
Driven element	92.7	Directors 41-50	75.1
Directors 1-12	83.9	Directors 51-60	74.6
Directors 13-20	81.2		

Cumulative element spacings (mm):

RP		0.0	RL	29.5
DE	38.6		D1	49.2
D2	57.2		D3	74.1
D4	91.1		D5	103.0
D6	125.0		D7	158.9
D8	192.8		D9	226.7
D10	260.6		D11	294.5
D12	328.4		D13	362.3
D14	396.2		D15	430.1
D16	464.1		D17	498.0
D18	531.9		D19	565.8
D20	599.7		D21	633.6
D22	667.5		D23	701.4
D24	735.3		D25	769.2
D26	803.1		D27	837.0
D28	871.0		D29	904.9
D30	938.8		D31	972.7
D32	1006.6		D33	1040.5
D34	1074.4		D35	1108.3
D36	1141.2		D37	1176.1
D38	1210.0		D39	1244.0
D40	1277.9		D41	1311.8
D42	1345.7		D43	1379.6
D44	1413.5		D45	1447.4
D46	1481.3		D47	1515.2
D48	1549.1		D49	1583.1
D50	1617.0		D51	1650.9
D52	1684.8		D53	1718.7
D54	1752.6		D55	1786.5
D56	1820.4		D57	1854.3
D58	1888.2		D59	1922.1
D60	1956.1		D61	1990.0

Table 17.2: Dimensions of 3.4GHz loop quad

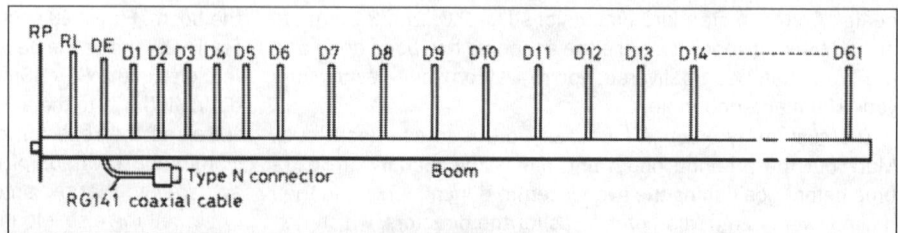

Fig 17.31: The 9cm quad loop yagi. Dimensions are shown in Table 17.2

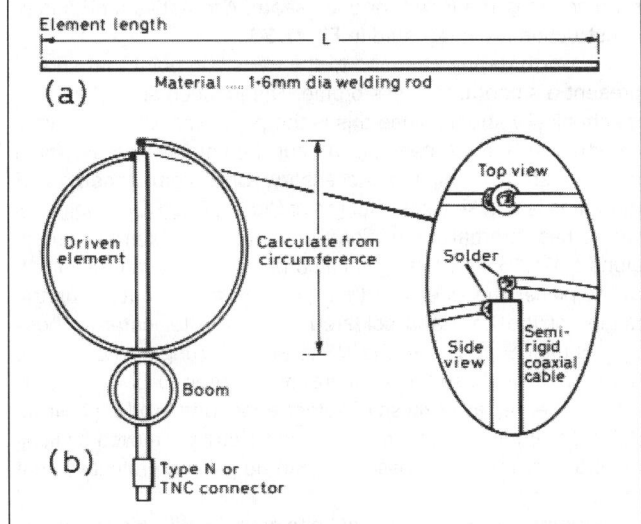

Fig 17.32: (a) Detail of element construction. (b) Assembly of the driven element for the 9cm quad loop yagi

All elements are made from 1.6mm diameter welding rod cut to the lengths shown in the table, then formed into a loop as shown in **Fig 17.32a**. The driven element is brazed to a M6 x 25 countersunk screw drilled 3.6mm to accept the semi rigid coaxial cable. All other elements are brazed onto the heads of M4 x 25 countersunk screws. All elements, screws and joints should be protected with a coat of polyurethane varnish after assembly. If inadequate attention is paid to weatherproofing the antenna, its performance will gradually deteriorate as a result of corrosion.

Provided the antenna is carefully constructed, its feed impedance will be close to 50Ω. If a suitably rated power meter or impedance bridge is available, the match may be optimised by carefully bending the reflector loop toward or away from the driven element.

The antenna can be mounted using a suitable antenna clamp. It is essential that the antenna be mounted on a vertical support, as horizontal metalwork in its vicinity can cause severe degradation in its performance.

HORN ANTENNAS

3cm Horn

Large pyramidal horns can be an attractive form of antenna for use at 10GHz and above. They are fundamentally broadband devices showing virtually perfect match over a wide range of frequencies, certainly over an amateur band. They are simple to design, tolerant of dimensional inaccuracies during construction and need no adjustment. Their gain can be predicted within a dB

Fig 17.33: Picture of a large 10GHz horn

or so (by simple measurement of the size of the aperture and length) which makes them useful for both the initial checking of the performance of systems and as references against which other antennas can be judged. Their main disadvantage is that they are bulky compared with other antennas having the same gain.

Large (long) horns, such as that illustrated in **Fig 17.33**, result in an emerging wave which is nearly planar and the gain of the horn is close to the theoretical value of $2\pi AB/\lambda^2$, where A and B are the dimensions of the aperture. For horns which are shorter than optimum for a given aperture, the field near the edge lags in relation to the field along the centre line of the horn and causes a loss in gain.

For very short horns, this leads to the production of large minor lobes in the radiation pattern. Such short horns can, however, be used quite effectively as feeds for a dish. The dimensions for an optimum horn for 10GHz can be calculated from the information given in **Fig 17.34** and, for a 20dB horn, are typically:

- A = 5.19in (132mm)
- B = 4.25in (108mm)
- L = 7.67in (195mm)

There is, inevitably, a trade off between gain and physical size of the horn. At 10GHz this is in the region of 20dB or perhaps slightly higher. Beyond this point it is better to use a small dish. For instance a 27dB horn at 10GHz would have an aperture of 11.8in (300mm) by 8.3in (210mm) and a length of 40.1in (1,019mm) compared to a focal plane dish that would be 12in (305mm) in diameter and have a length of 3in (76mm) for the same gain.

Horns are usually fabricated from solid sheet metal such as brass, copper or tinplate. There is no reason why they should not be made from perforated or expanded metal mesh, provided that the size and spacing of the holes is kept below about λ/10. Construction is simplified if the thickness of the sheet metal is close to the wall thickness of WG16, ie 0.05in or approximately 1.3mm. This simplifies construction of the transition from the waveguide into the horn. The geometry of the horn is not quite as simple as appears at first sight since it involves a taper from an aspect ratio of about 1:0.8 at the aperture to approximately

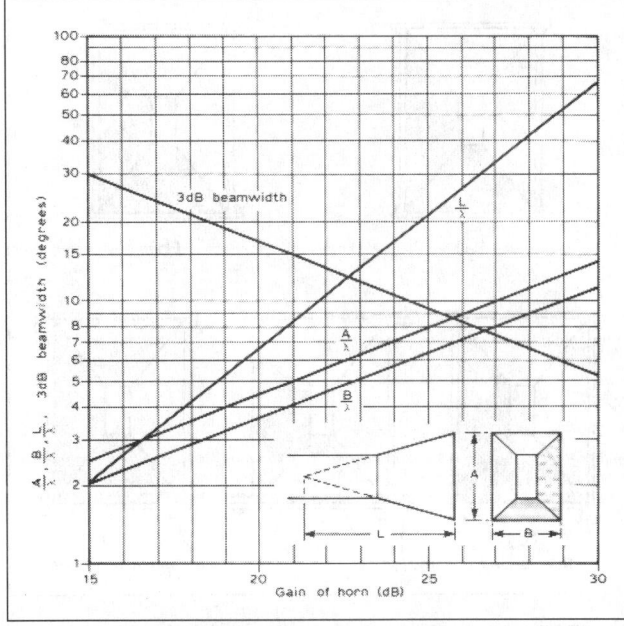

Fig 17.34: Horn antenna design chart

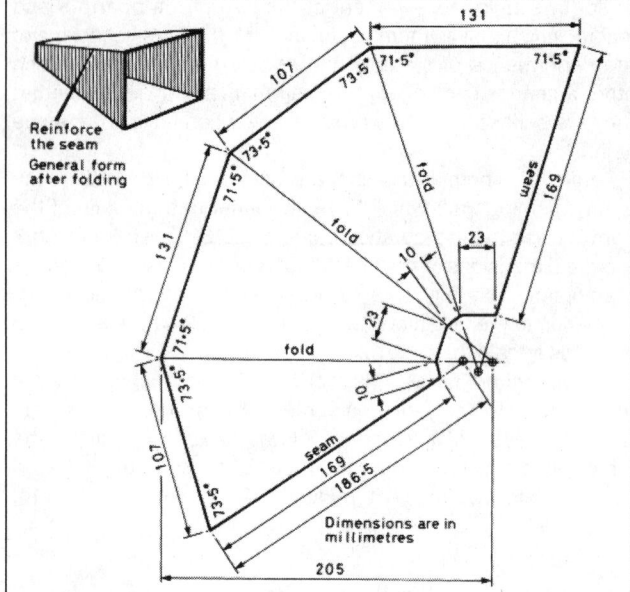

Fig 17.35: Dimensioned template for single piece construction of a 20dB horn

2:1 at the waveguide transition. For a superficially rectangular object, a horn contains few right angles, as shown in **Fig 17.35** an approximately quarter scale template for a nominal 20dB horn at 10.4GHz. If the constructor opts to use the one piece cut and fold method suggested by this figure, it is strongly recommended that a full sized template be drafted on stiff card that can be lightly scored to facilitate bending to final form. This will give the opportunity to correct errors in measurement before transfer onto sheet metal and to prove to the constructor that, on folding, a pyramidal horn is formed!

The sheet is best sawn (or guillotined) rather than cut with tin snips, so that the metal remains flat and undistorted. If the constructor has difficulty in folding sheet metal, then the horn can be made in two or more pieces, although this will introduce more soldered seams which may need jigging during assembly and also strengthening by means of externally soldered angle pieces

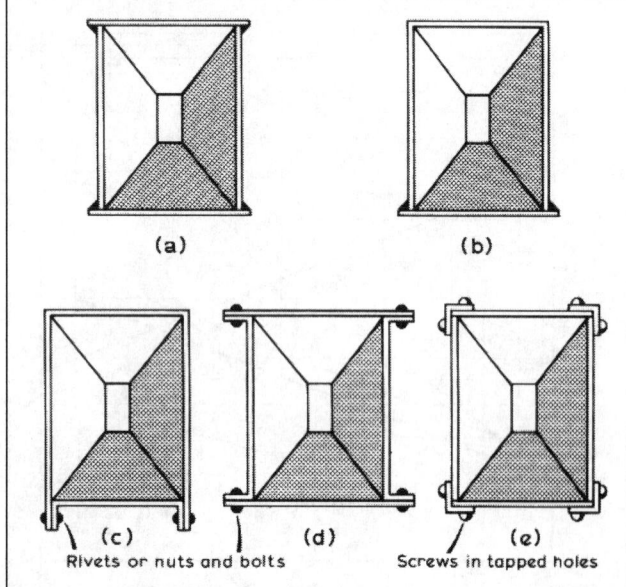

Fig 17.36: Alternative construction methods for constructing horn antennas

running along the length of each seam. Alternative methods of construction are suggested in **Fig 17.36**.

It is worth paying attention to the transition point that should present a smooth, stepless profile. The junction should also be mechanically strong, since this is the point where the mechanical stresses are greatest. For all but the smallest horns, some form of strengthening is necessary. One simple method of mounting is to take a short length of Old English (OE) waveguide, which has internal dimensions matching the external dimensions of WG16, and slitting each corner for about half the length of the piece. The sides can then be bent (flared) out to suit the angles of the horn and soldered in place after carefully positioning the OE guide over the WG16 and inserting the horn in the flares. One single soldering operation will then fix both in place. After soldering, any excess of solder appearing inside the waveguide or throat of the horn should be carefully removed by filing or scraping. The whole assembly can be given a protective coat of paint.

An alternative method would be to omit the WG16 section and to mount the horn directly into a modified WG16 flange. In this case the thickness of the horn material should be a close match with that of WG16 wall thickness and the flange modified by filing a taper of suitable profile into the flange.

Whichever method of fabrication and assembly is used, good metallic contact at the corners is essential. Soldered joints are very satisfactory provided that the amount of solder in the horn is minimised. If sections of the horn are bolted or riveted together, it is essential that many, close spaced bolts or rivets are used to ensure good contact. Spacing between adjacent fixing points should be less than a wavelength, ie less than 30mm.

24GHz Horn Antenna

A 24GHz horn is a very easy antenna to construct, as its dimensions are not critical and it will provide a good match without any tuning. The dimensions, as per **Fig 17.37**, for optimum gain horns of various gains (at 24GHz) are shown in **Table 17.3**. Above 25dB the horn becomes very long and unwieldy and it becomes more practical to use a dish.

Suitable materials are PCB material, brass or copper sheet, or tin plate. If PCB material is used, the doubled-sided type will enable the joins to be soldered both inside and out for extra strength. An ideal source of tin plate is an empty oil can.

The transition from the guide to the horn should be smooth. Use either a butt joint, or file the waveguide walls to a sharp

Fig 17.37: Dimensions of horn antenna for 24GHz

Gain (dB)	Beamwidth (degrees)	Length (mm)	A (mm)	B (mm)
15	30	26	31	25
20	17	81	55	45
25	9	270	98	79
30	5	819	174	141

Table 17.3: Gain of 24GHz horn antennas

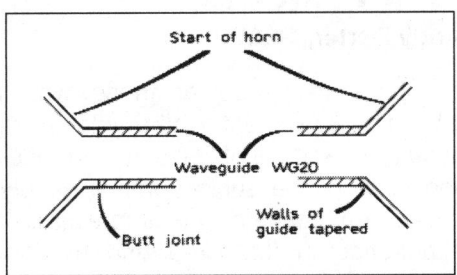

Fig 17.38: Joining a 24GHz horn directly to a waveguide

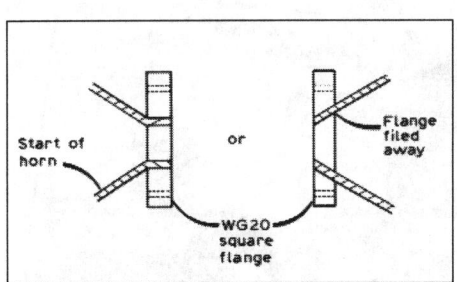

Fig 17.39: Joining a 24GHz horn directly to a flange

edge, see **Fig 17.38**. The material should be cut to size and then soldered together and onto the end of a piece of waveguide. Alternatively the horn may be coupled directly to the inside of a flange. In this case the material should be the same thickness as the waveguide would be and it is bent as it enters the flange, or the flange may be filed to be part of the taper, see **Fig 17.39**.

REFERENCES

[1] 'Practical project: A patch antenna for 5.8GHz', Gunthard Kraus, DG8GB, *VHF Communications Magazine* 1/2004, pp 20 - 29

[2] 'Modern patch antenna design', Part 1, Gunthard Kraus, DG8GB, *VHF Communications Magazine* 1/2001, pp 49 - 63

[3] 'Modern patch antenna design', Part 2, Gunthard Kraus, DG8GB, *VHF Communications Magazine* 2/2001, pp. 66 - 86

[4] Search for *patch16.zip*, by WB0DGF, on the Internet. One source is: virtual.xs4all.nl/nl3asd/antenna3.html

[5] 'PUFF - a CAD program for microwave stripline circuits', Robert E Lentz, DL3WR, *VHF Communications Magazine* 2/1991, pp 66 - 69. 'PUFF 2.1 - improved and expanded version', Dipl Ing A Gerstlauer, DG5SEB, *VHF Communications Magazine* 2/1998, pp 97 - 101

[6] 'An interesting program MSTRIP', Gunthard Kraus, DG8GB, *VHF Communications Magazine* 2/2002, pp 69 - 85

[7] 'An interesting program. SonnetLite 9.51', Gunthard Kraus, DG8GB, *VHF Communications Magazine* 3/2004, pp 156 - 178

[8] 'An array of 4 x 16 turn helix antennas for 2402MHz', Paolo Pitacco, IW3QBN, *VHF Communications Magazine* 1/2004, pp 2 - 6

[9] 'Antenne a elica, esperienze utili per tutti', Paolo Pitacco, IW3QBN, *AMSAT-I News* Vol8, N3, p13

[10] 'Un'elica di 16 spire per la banda S', G3RUH (translated IØQIT), *AMSAT-I News* Vol5, N1, p19

[11] 'HELIX.BAS Basic program for Helix design', KA1GT, *UHF/Microwave Experimenters Manual* (ARRL), pp 9-49

[12] *Antennas*, J D Kraus, R J Marhefka, McGraw Hill - 3th ed.

[13] *Antenna Engineering*, Jasik

[14] Contribution by Noel Hunkeler, F5JIO, to 'Technical topics', *RadCom* July 1997, p 72

[15] From an article by Leo Lorentzen, OZ3TZ, *OZ* 1/2001, pp 18 -20 and *OZ* 11/2000, pp. 633 - 635

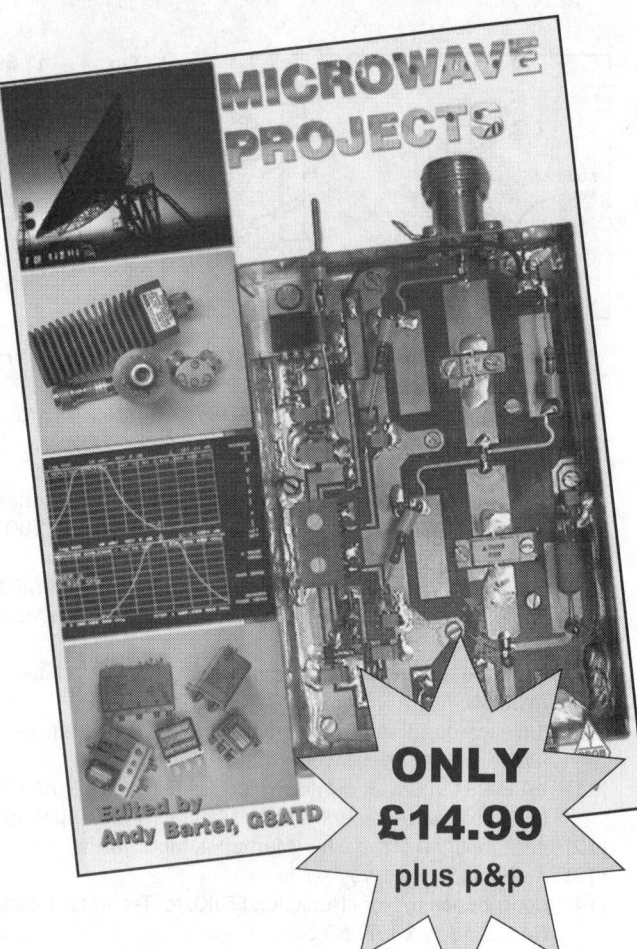

Microwave Projects

Edited by: Andy Barter, G8ATD

Microwave Projects is aimed at those who are interested in building equipment for the amateur radio microwave bands. Packed full of ideas from around the world this book covers the subject with a variety of projects. The book has many contributors who have a wealth of experience in this area and they have produced many projects, design ideas, complete designs and modifications of commercial equipment, for the book. This title provides much useful information as to what can be achieved effectively and economically. Aimed at both the relative novice and the "old hand" the book also covers useful theory of designing microwave circuits and test equipment for the projects. The book includes chapters covering:

· Signal Sources
· Transverters
· Power Amplifiers
· Test Equipment
· Design

Microwave Projects is a must have book for all those who are already active on the microwave bands and those looking for interesting projects to embark on.
Size:173 x 240mm. 200 pages. ISBN: 1-872309-90-9,
RSGB Members Price: £12.74 plus p&p

ONLY £14.99 plus p&p

COMING SOON !

Microwave Projects 2

More great projects from Andy Barter, G8ATD

Available late 2005

RSGB Members Price: £12.74 plus p&p

ONLY £14.99 plus p&p

E&OE

Radio Society of Great Britain
Lambda House, Cranborne Road, Potters Bar, Herts. EN6 3JE
Tel. 0870 904 7373 Fax. 0870 904 7374

ORDER 24 HOURS A DAY ON OUR WEBSITE
www.rsgb.org/shop

18 Morse Code

In the UK and an increasing number of countries, it is no longer necessary to pass a Morse test in order to gain access to the bands below 30MHz. This might be thought to lead to an overall decline in Morse activity, but there has been a welcome influx of new operators trying their hand at Morse, at their own speed, and without the daunting hurdle of a formal test to overcome.

The IARU band plans reserve the bottom portion of most bands for Morse, with the faster ('QRQ') signals generally occupying the bottom 30kHz of each HF band, and slower stations higher up. By agreement within IARU, the 10MHz band is not to be used for SSB or other wide bandwidth modes, and is almost exclusively given over to Morse code.

Radio amateurs use the terms 'Morse' and 'CW' interchangeably. This may seem strange, since CW stands for Continuous Wave, whereas Morse code is anything but continuous. However, almost all amateur Morse is sent by on-off keying of a continuous carrier so CW is widely used as a synonym for Morse.

Operators come to use Morse for a variety of reasons, and compulsion is no longer one of them. **Table 18.1** lists ten arguments in favour of the use of Morse:

History

Morse code isn't a code, and it wasn't invented by Morse. Strictly speaking, a scheme in which each letter is represented by a symbol (combination of dots and dashes in this case) is referred to as a cypher, not a code, and the basic idea of using dots and dashes was invented by Samuel Morse's assistant, Alfred Vail.

Starting in 1837 and building on the work of Joseph Henry, the artist Samuel Finley Breese Morse worked to develop a practical electric telegraph. Morse's idea was that common words in a message would be represented by numbers, and these numbers would be transmitted as a series of dots. At the receiving end, the dots would be marked by the equipment on paper tape. The operator would then have to decode the dots to work out which numbers were being sent, and look up the number combinations in a dictionary to find the word. Less common words would be spelled out, with numbers again representing letters. At about the same time in England, Cooke and Wheatstone were also working on the development of an electric telegraph.

While Morse was working on his system in New York City, his partner Alfred Vail thirty miles away in Morristown, New Jersey made numerous improvements to Morse's equipment. It was Vail's idea that instead of using a series of dots for numbers, the code could be made up of combinations of dots, dashes and spaces to represent letters directly. In this way a message could be spelled out without the need to look up code numbers in a book.

Samuel Morse was an incurable self-publicist and failed to give Vail credit in his lifetime as the true inventor of the code that bears Morse's name. When developing his code, Vail visited a local printer in Morristown to find out which letters occurred most frequently in English text, and he assigned the shortest combinations of dots and dashes to these letters. At the time of the first public demonstrations of Morse code, the code was made up of four elements in addition to the standard space: dot, dash, long dash and long space. The code underwent further changes but remained in use as American Morse code until well into the 20th century.

Advantages of using Morse code

1. **Simple equipment.** It is possible to build a simple CW transmitter using fewer than a dozen parts, and a direct conversion receiver can give acceptable performance on CW. The art of 'homebrewing' is alive and well, and building a CW transmitter and/or receiver can get an amateur onto the air on CW very easily and cheaply.

2. **International.** It is easy to get around the language barrier by the use of abbreviations such as *QTH* and *73*, so that two amateurs can have an elementary QSO without knowing each other's language, and without accent or phonetic problems that may arise on phone.

3. **Silent operation.** Wearing headphones and using a straight key or bug, it is possible to operate CW silently, at night time without disturbing others sleeping in the house. Similarly, holiday operation on CW from a hotel can be done in 'stealth' mode.

4. **Morse gets through.** Cross-mode contacts aren't very common, perhaps because they're not valid for most awards, but when struggling to copy a weak phone station on a crowded band it's surprising how often a switch to CW will enable the contact to be completed.

5. **Spectrum efficiency.** The minimum bandwidth needed to copy an SSB signal is about 1.8kHz, and to sound natural the requirement is more like 2.5kHz. On the other hand IF filters for CW operation are typically only 500Hz wide, and if necessary CW can be copied through filters of 100Hz or less. Put simply, at least five times as many CW contacts can fit in the bandwidth required for SSB.

6. **Less breakthrough.** Those who operate both modes know that breakthrough problems are worse on SSB than CW. It is always better to try and resolve any TVI or RFI problem but if this is not possible it may be that switching to CW enables operation to continue, when it is impossible on phone. And if power has to be reduced, CW comes into its own.

7. **More competitive.** The difference between 'big gun' and 'little pistol' seems to be accentuated on phone. The low power or antenna-limited operator may struggle for contacts on phone, whereas CW is a great leveller.

8. **Morse is a skill.** It's wrong to say that phone operation doesn't require skill, but the basic skill required is that of speech, which just about everyone has. On the other hand, a new skill has to be learned in order to be able to communicate using CW, and the sense of achievement can be considerable.

9. **Automation.** There are computer programs which make it possible to automate the transmission and reception of Morse code. This makes it possible under some conditions to engage in CW contacts without knowing the code, but this would be to miss the whole point: it would be much better to use a more efficient automated mode such as PSK31. In any case, a good human operator can easily outperform most computer techniques when copying a CW signal on a channel with even a moderate amount of interference or fading.

10. **Morse is easy.** The general public thinks that Morse is a language, and at special event stations Morse operation always proves fascinating to visitors. All that is needed to acquire Morse skills is to learn the symbols for 26 letters, ten numbers and a few special characters. This is a very great deal easier than learning a foreign language complete with all its grammar and vocabulary.

Table 18.1: Why operating using Morse is still popular, despite learning the code no longer being compulsory

Alphabet and numerals		Accented letters		Punctuation and other codes	
A	di-dah	à, á, â	di-dah-dah-di-dah	Full stop (.)	di-dah-di-dah-di-dah
B	dah-di-di-dit	ä	di-dah-di-dah	Comma (,)	dah-dah-di-di-dah-dah
C	dah-di-dah-dit	ç	dah-di-dah-di-dit	Colon (:)	dah-dah-dah-di-di-dit
D	dah-di-dit	ch	dah-dah-dah-dah	Question mark (?)	di-di-dah-dah-di-dit
E	dit	è, é	di-di-dah-di-dit	Apostrophe (')	di-dah-dah-dah-dah-dit
F	di-di-dah-dit	ê	dah-di-di-dah-dit	Hyphen or dash (-)	dah-di-di-di-di-dah
G	dah-dah-dit	ñ	dah-dah-di-dah-dah	Fraction or slash (/)	dah-di-di-dah-dit
H	di-di-di-dit	ö, ó, ô	dah-dah-dah-dit	Brackets - open [(]	dah-di-dah-dah-dit
I	di-dit	ü, û	di-di-dah-dah	- close [)]	dah-di-dah-dah-di-dah
J	di-dah-dah-dah			Double hyphen (=)	dah-di-di-di-dah
K	dah-di-dah	**Abbreviated numerals**		Quotation marks (")	di-dah-di-dah-di-dit
L	di-dah-di-dit	1	di-dah 6 dah-di-di-di-dit	Error	di-di-di-di-di-di-di-dit
M	dah-dah	2	di-di-dah 7 dah-di-di-dit	Message starts (CT)	dah-di-dah-di-dah
N	dah-dit	3	di-di-di-dah 8 dah-di-dit	Message ends (AR)	di-dah-di-dah-dit
O	dah-dah-dah	4	di-di-di-di-dah 9 dah-di-dit	End of work (VA)	di-di-di-dah-di-dah
P	di-dah-dah-dit	5	di-di-di-di-dit 0 daaah (long dash)	Wait (AS)	di-dah-di-di-dit
Q	dah-dah-di-dah	1	di-dah-dah-dah-dah	Understood (SN)	di-di-di-dah-dit
R	di-dah-dit	2	di-di-dah-dah-dah	The @ in e-mails	di-dah-dah-di-dah-dit
S	di-di-dit	3	di-di-di-dah-dah		
T	dah	4	di-di-di-di-dah	**Spacing and length of signals**	
U	di-di-dah	5	di-di-di-di-di	1 A dash is equal to three dots.	
V	di-di-di-dah	6	dah-di-di-di-dit	2 The space between the signals which form the same letter is equal to one dot.	
W	di-dah-dah	7	dah-dah-di-di-dit	3 The space between two letters is equal to three dots.	
X	dah-di-di-dah	8	dah-dah-dah-di-dit	4 The space between two words is equal to seven dots.	
Y	dah-di-dah-dah	9	dah-dah-dah-dah-dit		
Z	dah-dah-di-dit	0	dah-dah-dah-dah-dah		

Table 18.2: The Morse code and its sound equivalents

The electric telegraph and Morse code started to be used in Europe in the mid to late 1840s. A German, Frederick Gerke, addressed the problem of distinguishing between standard and long dashes, and standard and long spaces, by reducing the code to just the elements of dot and dash that we are familiar with today. More changes were made and in 1865 the 'continental' or 'international' Morse code came into being, in very much the form that exists today. Small changes have been made to the code since then, the most recent being the adoption in 2004 of a new Morse code symbol for the '@' used in Internet addresses.

Both Vail and Morse intended that Morse characters should be printed onto paper tape and read as dots and dashes from there, but it soon became apparent that operators were learning to decode from the clicks of the machinery marking the tape. This led to the development of the sounder and now whenever Morse is written down it is always written down directly as letters, never as individual dots and dashes. A modern exception to this is QRSS, or extremely slow Morse decoded using a FFT program (as used on 136kHz and for very low power experiments) which is displayed as dots and dashes on a computer screen and decoded by eye.

THE CODE

Although Morse is composed of dots and dashes it is better to think of the elements as 'dits' and 'dahs'. These are in the correct proportions so that saying 'di-dah-di-dit' out loud is the correct representation of the letter 'L'. Spacing and rhythm are essential to good Morse code. The inter-element space is the same duration as a dit length, while the length of a dah is three times the dit length. The space between letters of a word is three dit lengths and the space between words is seven dits.

As well as the letters and numbers, **Table 18.2** lists commonly used punctuation and procedural symbols ('prosigns') in common use on the amateur bands.

SPEED

For many years it was necessary in order to gain an amateur licence, to pass a Morse test at 12 words per. minute (WPM). In practice this is the minimum speed that is generally heard on the amateur bands. A word is defined as 50 dit lengths, and this is referred to as the Paris standard since the word 'PARIS', including the seven dit lengths at the end of the word, is exactly the right length. By coincidence the word 'MORSE' is also exactly the right length.

A simple method of measuring speed is repeatedly to send PARIS, including the correct inter-word gap, and count how many are sent in a minute. At a speed of 12WPM the word can be sent just once in five seconds. Another method of measurement which is simple to do but overstates the speed by 4 percent is to count the number of dashes in a five second period.

LEARNING MORSE CODE

A later chapter in this manual shows the extent to which the personal computer has become an integral part of the radio amateur's shack. There is a method of learning Morse code which goes back to the 1930s but has come into its own in the age of the computer. That is the Koch method.

Ludwig Koch was a German psychologist who carried out a systematic study of skilled Morse operators, and then conducted a series of trials to find the most efficient way of teaching Morse to new students. Existing methods at that time were based on learning the letters either visually, or audibly at a slow speed and then increasing speed by constant practice. This approach to learning Morse is fundamentally flawed. Learning all the letters at slow speed requires a conscious mental translation from the combination of dits and dahs to the corresponding letter. After all letters have been memorised, the student hears a combination of

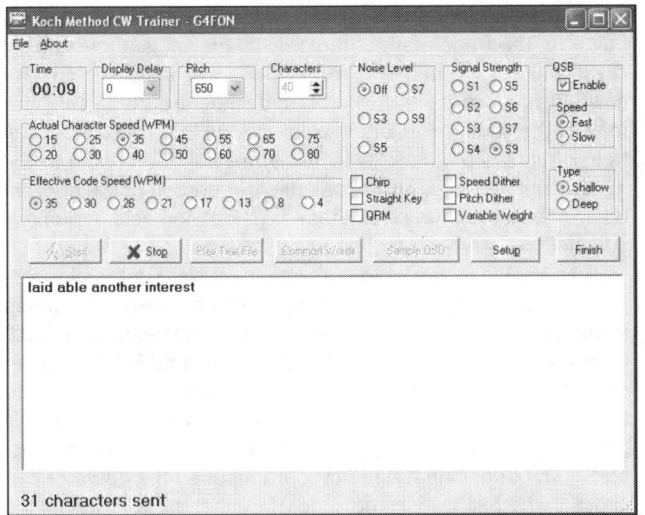

Fig 18.1: Part of the screen of the G4FON Koch Morse trainer program

dits and dahs and searches in a mental lookup-table for the right one. With a great deal of practice, speed increases but everyone who learns by this method experiences a 'plateau' at 8-10WPM where no more progress seems possible. Many aspiring CW operators give up in frustration at this point.

What is happening is that the process of responding to individual dits and dahs and looking them up in a mental table is hitting a natural upper speed limit. The student who can move beyond that limit has made the unconscious transition from the old 'lookup' method to a stage where the whole Morse character is recognised by reflex. Koch's insight was to see that it was essential that each character retain its acoustic wholeness and never be broken down by the student into dits and dahs. The essence of the Koch method is that the letters are learned one at a time, but with the correct rhythm and at full speed. There is no process of decoding the dits and dahs, but from the beginning a reflex is built, so that when 'dah-di-di-dah' is heard, it is automatically recognised as X without any conscious decoding process. It's the same way that a touch typist knows where the letters are on a keyboard without having to think about it.

Another key feature of the Koch method is that it provides positive reinforcement. Once a letter has been learned, it has been learned at full speed and it is possible to measure one's progress towards mastery of the full alphabet.

Koch's technique was difficult to implement for classes of students in the 1930s but has come into its own in the day of the personal computer, when each student can learn at his or her own pace. A program written by Ray Goff, G4FON, implements the Koch method and can be downloaded from [1]. Much of the credit for rediscovery and promotion of the Koch method must go to N1IRZ, and the G4FON program follows his suggested implementation.

The student starts by learning just two letters, at a speed of 15WPM. The program sends random combinations of these letters for a fixed period, typically three to five minutes, and the student must copy them onto paper. At the end of the run the copy is compared with what the computer sent, as displayed on the screen. If 90 percent or greater correct copy is received, a third character is added to the set and another run started. It is up to the student how many runs to do in a day. When Ludwig Koch carried out his investigations in the thirties he was seeking to train commercial operators but found that too many sessions in a day gave diminishing returns and the amateur student is more likely to have to fit in training sessions around other activities.

As the training programme progresses and more and more Morse characters are added, the student gains a real sense of achievement, and proof that it is possible to master the copying of Morse characters at full speed. G4FON's program introduces new characters in the order suggested by N1IRZ which mixes easy and difficult letters, numbers and punctuation in an apparently random order.

Once all 43 characters have been learned the program lets the student practice with text files (**Fig 18.1**), or with simulated contacts. At this stage, most people would also have started to listen to real contacts on the amateur bands, though the quality of 'real' Morse code can be rather variable.

The ubiquitous computer can have its dangers, as well. The Koch method is the quickest means of learning the code but some students in their impatience have resorted to use of Morse decoding software as an aid to getting on the air on CW more quickly. This approach leads to reliance on the computer and the temptation must be resisted.

Sending

Although the PC makes it possible for a student to learn Morse code without assistance, when it comes to sending by far the best approach is to get an experienced operator to listen and correct any bad habits at the earliest opportunity. The student who has learned using the Koch method may have fewer bad habits than most. If no-one is available to listen in person then one method that can work is to make tape recordings and play them back to try and discover weaknesses. Only when the student's sending sounds perfect should it be tried on the air.

MORSE KEYS

It is important to start by getting a good quality straight key ('pump handle') such as the one in **Fig 18.2**, and position it correctly on the table. Adjust the height of the seat so that when not sending, the arm is resting horizontally, and it requires only a small lift of the forearm to hold the knob of the Morse key. This should be held with the forefinger on top of the knob, the thumb to the left and slightly underneath the knob, and the second finger either on top or to the right of the knob (assuming a right-handed operator). Dots are made by a small wrist movement whereas dashes require a more pronounced downward movement of the wrist.

The forearm and the upper arm should make an angle of approximately 90 degrees. When sending, movement of the key should come from a combination of the elbow and the wrist. Often the key is mounted at the edge of the table but if it is further back, the forearm should be slightly above the table, not resting on it.

Bugs and Elbugs

With practice it is possible to send at speeds over 20WPM on a straight key, but the operator who aspires to high-speed (QRQ) working will sooner or later wish to learn how to use a bug.

Fig 18.2: A modern straight key (photo: RA Kent (Engineers))

Fig 18.3: A mechanical bug key

Fig 18.4: A paddle for use with an electronic keyer

The semi-automatic key (**Fig 18.3**), generally known as the bug after the trade mark of the original maker, is a mechanical key in which the arm moves from side to side instead of up and down. It has two pairs of contacts; dashes are made singly by moving the knob to the left, thereby closing the front contacts. A train of dots is produced by similarly moving the paddle to the right against a stop. This causes the rear portion of the horizontal arm to vibrate and close the rear pair of contacts. A properly adjusted bug key will produce at least 25 dots.

Bugs were in widespread use until the advent of the electronic keyer, or 'elbug'. The first elbugs were cumbersome affairs using combinations of relays and valves. Transistor types followed, giving way to IC-based keyers and now PIC and other microprocessor keyers. Most modern transceivers incorporate an electronic keyer, and station logging and computer logging programs provide a CW keyer as standard. Many programs support both paddle input and keyboard input so the operator can choose whether to use the program either as a keyer taking input from a paddle, or as a keyboard sender.

The key or 'paddle' (**Fig 18.4**) used with an electronic keyer is a derivative of the mechanical bug key, with movement from side to side instead of up and down. For the right-handed operator, moving the paddle to the right with the thumb produces a train of dots. Moving the paddle to the left with the side of the index finger produces a train of dashes. An iambic or 'squeeze' keyer requires a twin paddle key, and as well as movement of each paddle separately they can be squeezed together, hence the name. According to which contact is closed first, a train of either di-dah-di-dah or dah-di-dah-dit is produced.

As with the straight key, the newly acquired bug or electronic key should not be tried on the air until the operator has gained confidence through practice. Some electronic keys provide auto-spacing in an attempt to enforce the three- or seven-dit gaps between characters and words, but it is still just as easy to send bad Morse on an elbug as on a straight key.

A Simple Electronic Keyer

This keyer designed by E. Chicken G3BIK, is of basic design in that it uses only four integrated circuits and does not include the iambic facility. According to which side the paddle is moved, a

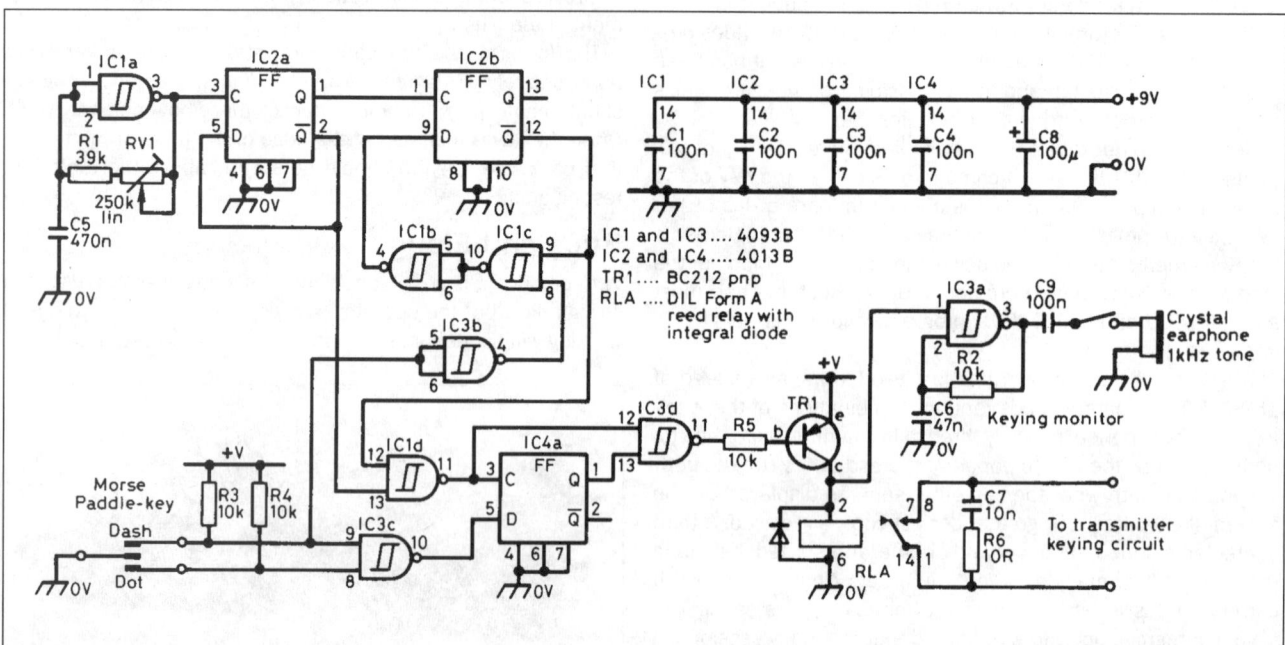

Fig 18.5: The G3BIK simple keyer uses four ICs for precision Morse

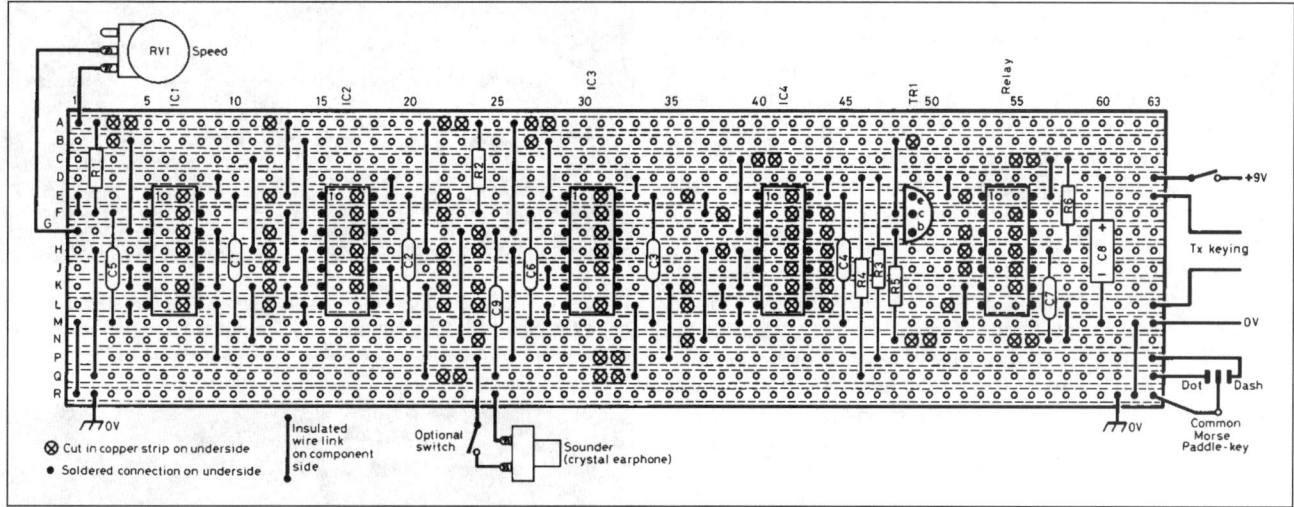

Fig 18.6: Construction of the G3BIK keyer is easy using strip board. The component side is shown

train of dots or dashes is produced. The speed is adjustable in the range 5-35wpm. A small sounder is included as a side tone (keying monitor). The circuit diagram is shown in **Fig 18.5**, and the stripboard layout is in **Fig 18.6** . Details of the paddle key are contained in the full article [2].

CW TRANSMISSION AND RECEPTION

A crystal-controlled QRP transmitter for CW can be made with fewer than a dozen components, and give many contacts under favourable conditions. There are many designs for simple CW-only homebrew transmitters or transceivers and several kits are available that offer CW-only transmission.

Almost all commercial transceivers are designed with SSB in mind, and sometimes give the impression that CW has been added as an afterthought. Even the top of the range Yaesu transceiver, the FT1000MP, is known for its key clicks and several modifications have been developed - by individuals and not the manufacturer - to address this shortcoming in the design of an otherwise excellent piece of equipment.

In many transceivers a dedicated IF filter for CW is an optional extra. Some rigs incorporate passband tuning which can be used to narrow the receive bandwidth without the expense of buying additional internal filters. There are occasions under crowded band conditions when it is extremely difficult to separate the many CW signals that can pass together through an SSB filter, which may have a bandwidth of up to 2.5kHz. IF filters most commonly available to fit in a transceiver for CW reception have a bandwidth of 500Hz though 250Hz filters can sometimes be obtained.

In cases where IF filtering is not available, CW filtering can be done at audio frequencies, and combinations of IF and AF filtering can be extremely effective. The problem with audio filtering on its own is that any filtering done after the AGC detector will result in 'pumping' whereby strong signals which are not heard by the operator because of the audio filter nevertheless are let through by the IF filter and cause desensitisation of the receiv-

er. More recent designs using IF DSP (digital signal processing) can help to address this problem by implementing the filter before detection, though the dynamic range of the DSP system may then become an issue.

Different operators have different styles, and some of the best contest and DXpedition operators like to use relatively wide filters and do most of the separation of CW signals in their head: it is sometimes said that the best CW filter is between the ears. Nevertheless there are times when a narrow IF filter is essential for pileup and weak signal CW work.

A very simple add-on audio filter which provides some selectivity as well as helping to clean up the hiss, clicks, hum and thumps which spoil the audio of some rigs, has been described by Fraser Robertson, G4BJM [3]. In its simplest form it consists of a simple series-resonant circuit and with the inductor and capacitor values shown in **Fig 18.7** it resonates at around 730Hz. The components can simply be wired inline as a head-phone extension lead, and covered with self amalgamating tape or heat-shrink tubing to provide some mechanical stability. Alternatively, it can housed in a small box as in **Fig 18.8**.

Most transceivers incorporate semi break-in or 'VOX' keying on CW. This means that the user does not have to operate a separate transmit/receive switch, but transmission starts automatically almost at the instant that the key is pressed. The equipment returns to receive a short period after the key is opened, and this period can usually be varied. The intention is to ensure the rig stays in transmit between Morse characters, but returns again to receive without the operator having to throw a switch.

Full break-in ('QSK') is also frequently implemented, with a varying degree of success. Full break-in means that the operator

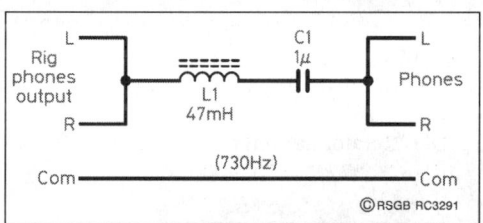

Fig 18.7: Circuit of the two-component CW filter

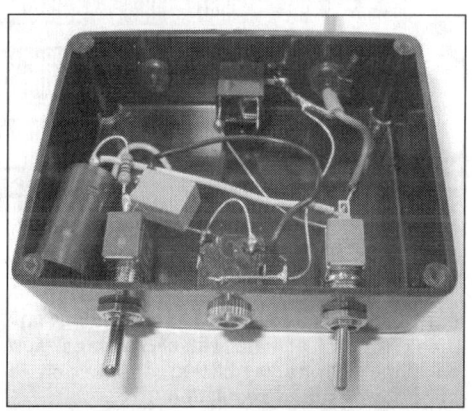

Fig 18.8: The deluxe version of G4BJM's two-component filter includes switches to connect a loudspeaker, and to bring the filter in and out of circuit

can listen between the individual dits and dahs of a Morse character, right up to 40WPM and beyond. With careful transceiver design this goal can be achieved, but design compromises and shortfalls generally mean that a number of manufacturer's implementations are far from perfect. Ten-Tec and Elecraft have, however, really understood the design principles involved.

When examining the keying waveform of many amateur transceivers, the correct 1:1:3 ratio between spaces, dits and dahs is not always maintained or if it is, this is at the expense of reduced receive time when using full break-in. A common feature of poorly designed rigs when using semi break-in is a short first dit, because of time taken in the transceiver to switch between transmit and receive. If this is used to drive an amplifier which itself has a slow T/R relay, the overall effect in extreme cases can be to lose the first dit of a transmission altogether.

If an oscilloscope is available it can be used to display the outline of the pattern made by the signal, this is called the keying envelope. The keyed RF signal is fed to the Y-plates or vertical amplifier of a slow-scan oscilloscope, with the timebase set in synchronism with the keying speed. The square shape in **Fig 18.9(a)** is a very 'hard' signal and will radiate key clicks over a wide range of frequencies. The transceiver design should ensure that the rise and fall times are lengthened such that interference is no longer objectionable, but without impairing intelligibility at high speed. The goal should be to achieve rise and fall times of about 10% of the dit length **(Fig 18.9(c))**.

The characteristics of the power supply may contribute to the envelope shape, as the voltage from a power unit with poor regulation will drop quickly each time the key is closed, and rise when it is released. This can lead to the shape in **Fig 18.9(d)**.

QRSS AND DFCW

Not all Morse used on the amateur bands is intended for reception by ear. When a signal is transmitted over fairly stable paths, such as on 136kHz, it can be received well below the threshold of audibility by using a computer's sound card and software [4] that displays the signal as dots and dashes on the screen.

QRSS is extremely slow speed CW, with dot lengths from three to 120 seconds. The name is derived from the Q-code QRS (reduce your speed). To take advantage of the very narrow bandwidth of the transmitted signal an appropriate filter at the receiver end is needed. Making a 'software filter' using Fast Fourier Transform (FFT) [5] has some advantages over the old fashioned hardware filter. One of the main advantages, when using it for reception of slow CW signals, is that FFT does not provide a single filter but a series of filters with which it is possible to monitor a complete spectrum at once. This means that it is not necessary to tune exactly into the signal, what can be very delicate

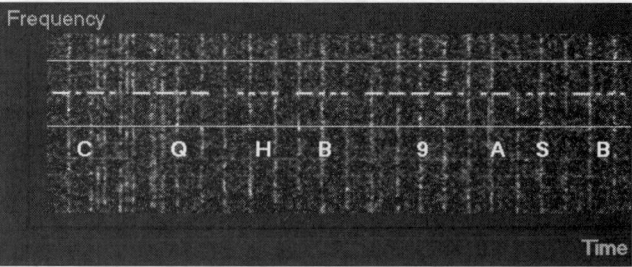
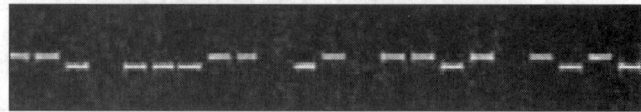

Fig 18.10: A QRSS signal on a curtain display. The translated letters have been added to the picture; they are not present on the original display. The vertical lines are static crashes [pic: ON7YD]

Fig 18.11: DFCW uses different frequencies instead of dots and dashes. This says "G3AQC" [pic: ON7YD]

at sub-Hertz bandwidths. Also it is possible to monitor more than one QRSS signal at the same time.

At first glance it looks as if it is complicated to do this, even if FFT presents you this nice multi-channel filter it might be difficult to monitor all these channels. Further the long duration of the dots and dashes is unfavourable for aural monitoring.

A solution to the above problems is to show the outcome of the FFT on screen rather than making it audible. The result is a graphic where one axis represents time, the other axis represents frequency and the colour represents the signal strength. If the vertical axis represents time it is called a waterfall display, while a curtain display is where the horizontal axis represents time **(Fig 18.10)**.

A variant of QRSS is Dual Frequency CW (DFCW) where the dots are transmitted on one frequency, and the dashes on a slightly higher frequency **(Fig 18..11)**. This saves time as the dashes can be the same length as the dots. In fact, a time reduction of better than 50% can be achieved, which can be very important at these extremely slow speeds, especially when trying to fit a contact into a relatively small ionospheric propagation window.

SUMMARY

In summary, Morse should not be seen as daunting. Thousands - millions - of people have mastered the use of Morse code over the years and many amateurs continue to use it on a daily basis. A 5WPM Morse test is counterproductive, because it encourages people to start slow and try to build up speed gradually. Ludwig Koch showed in the 1930s that there was a better way and by using G4FON's program it's possible to acquire CW receiving skills which can be of practical use on the amateur bands.

REFERENCES

[1]   Computer program for learning Morse code. http://www.g4fon.net
[2]   'The BIK Simple Electronic Key', Ed Chicken, G3BIK, *Radio Communication* Aug 1993
[3]   'A Two-Component CW Filter' Fraser Robertson, G4BJM, *RadCom*, Aug 2002
[4]   Viewers for QRSS and DFCW can be downloaded from http://www.weaksignals.com
[5]   QRSS and DFCW information at ON7YD's web site http://www.qsl.net/on7yd/136narro.htm

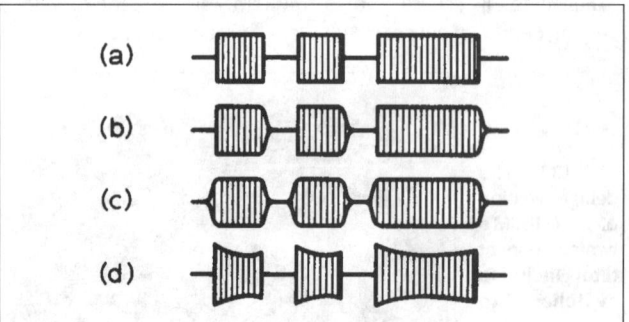

Fig 18.9: Keying envelope characteristics. (a) Click at make and break; (b) click at make with click at break suppressed; (c) ideal envelope with no key clicks; (d) affect on keying envelope of poor power supply regulation

19 Data Communications

It is generally assumed that data communication is a modern aspect of amateur radio, but of course Morse code transmission can also be regarded as a data mode. Recent deregulation of international and local Morse requirements has done nothing to weaken the interest in Morse, and the mode remains one of the most popular on the DX bands.

The mode of operation first associated with data communications by most amateurs is radio teletype (RTTY). Amateurs in the UK first began using RTTY on the air in the late 1950s with surplus machines such as the Creed 7B. Despite inferior performance in comparison to that of more recently developed modes, RTTY still has its enthusiasts. There are still many RTTY contests, although now computers are used in place of mechanical machines.

DATA MODE DEVELOPMENTS

Digital electronics and then computers allowed technically improved data modes to be developed. The first, developed by Peter Martinez, G3PLX, was called AmTOR (from 'amateur teletype over radio'). Based on the commercial SiTOR system, this was the first amateur data communications mode to make use of error-detection and correction techniques, albeit in a very simple form. This was to be followed by many other specially designed data modes. Commercially developed modems offering protocols such as PacTOR, G-TOR and CLOVER were introduced, and these continue to provide error-free (but expensive) HF communications for bulletin-board and automated mail systems; these modes correct errors using two-way transactions. **Fig 19.1** shows how modern modes have developed from RTTY.

There has also been strong interest in modes that (like RTTY) offered the amateur operator real-time keyboard to keyboard chatting. From the mid 1990s, a serious effort was made to replace RTTY with modes designed specifically to take into account the requirements of amateur chat modes, and the characteristics of HF propagation. The first of these new 'designer' modes was PSK31, developed by G3PLX. An amazingly fruitful period of new mode development followed, based on the PC sound card, and as one experienced operator commented:"More modes have been developed in the last few years than in the previous century!" [1]. During this period, Hellschreiber was revived, MT63 was developed, and MFSK16 was introduced. Other less successful modes have come and gone, or remain as curiosities. Some newer developments are

used for special applications, for example slow PBSK modes and the MFSK mode JASON (weak signal modes for LF); and WSJT and high speed Hellschreiber (for meteor scatter use).

The late 1990s and early 2000s saw the ascendancy of the high-speed personal computer with sound card as the preferred method of signal processing. In addition to the development of new modes that would not have been possible without digital signal processing, the sound card technique has made digital mode operation easy and inexpensive for beginners. It has also paved the way for a range of new tools, such as the spectrogram, used to detect weak signals and monitor propagation.

Although not exactly a data mode, Slow Scan Television (SSTV) technology also advanced significantly during this period, to become a popular addition to the SSB QSO. Through the use of the sound card and digital signal processing, inexpensive SSTV operation is now widely enjoyed. See the Image Techniques chapter for more on SSTV.

The trend in HF data modes is continuing to move away from specialised hardware toward general-purpose computers using digital signal processing. Performance never dreamed possible is becoming a reality, and indications are that before long the receiver and transmitter as we know them will be completely absorbed into the computer - the so-called software defined radio [2].

Towards the end of the 1970s, amateurs in North America began experimenting with microprocessor and digital techniques to generate and process data. A standard based on the CCITT X.25 public packet switched network protocol was finally agreed on, and this Amateur X.25 (AX.25) protocol became the standard for amateur packet radio. During the 1980s and 1990s the technique grew into a global amateur integrated network offering error corrected communications between any two stations in a network, file transfer, automatic message storage and forwarding, and bulletin dissemination. While packet radio no longer has the following it once had, largely due to the incredible popularity of the Internet, useful work is still continuing in the area of packet data handling. There is especially strong interest in international message handling for remote users, computer networking, and in technology related to telemetry and the tracking of vehicles and other assets. The leaders in this area have been Bob Bruninga, WB4APR, (APRS) [3]; Phil Karn, KA9Q [4]; Ian Wade , G3NRW (radio network operating systems) [5], and the late Roger Barker, G4IDE (UIVIEW and WINPACK) [6]. Packet radio has also been widely used to disseminate hints to DX operators ("DX spotting").

GETTING STARTED

There is no need to be daunted by the prospect of operating HF data modes. While some systems require skill and experience, there are also some very effective computer modes that are easy to set up, and a pleasure to use. The simplest are Hellschreiber and PSK31.

HF keyboard mode operation is now a very cost-effective addition to the ham shack. Not only is the equipment inexpensive, if an HF rig and computer are available, but also the more effective modes do not require high power or large antennas for good DX. All the keyboard to keyboard modes now use sound card technology, and there is software for the most popular modes

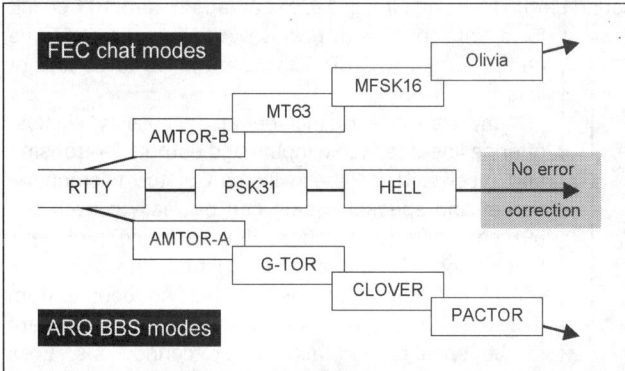

Fig 19.1: Development of HF Digital Modes

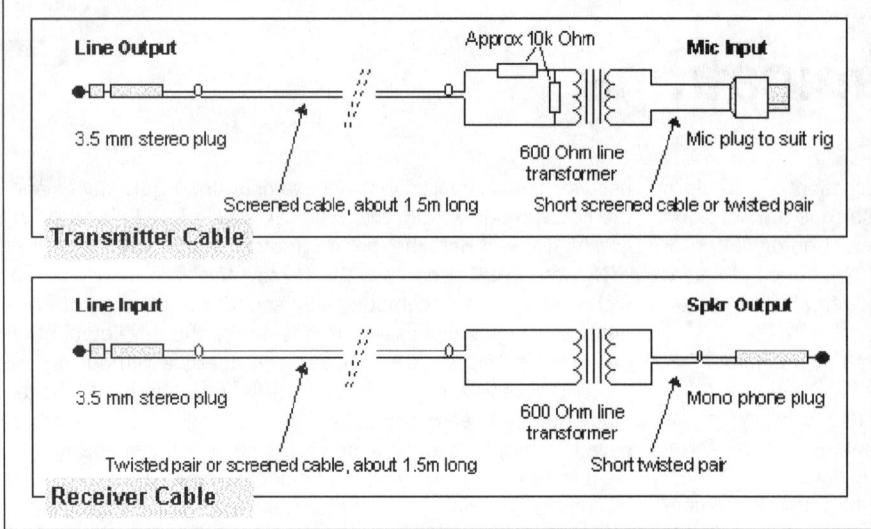

Fig 19.2: The simple cables used to connect PC to radio

available for several different platforms - the most popular being the ubiquitous PC with some version of Windows™ or LINUX™ operating system. Some excellent commercial software provides the widest range of features and operating modes. There is certainly no performance compromise involved in operating using a computer rather than a specialised hardware solution.

To operate these digital modes, a modern computer with sound card is required (computer specification depends on software, but even a 233MHz Pentium will provide a lot of pleasure). Some simple cables, easy to make, are also required, or a commercial interface such as the RIGBLASTER™ can be purchased. A conventional HF SSB transceiver is used, connected via the receiver audio output and microphone audio input. It is best if this is a modern solid-state unit, with good filters and low drift, but many operators use older rigs for RTTY, MT63, Hellschreiber and SSTV with no particular problems. These are the modes least affected by drift and poor frequency netting.

Some of the newer modes require very high stability, and very low frequency offset is necessary between transmit and receive. Most synthesised transceivers will suffice. A transceiver that drifts less than 5Hz per over will operate the newer modes very successfully. Unfortunately offset cannot be accurately corrected by using the Receiver Incremental Tuning (RIT).

The connections between computer and transceiver are quite straightforward. Most amateurs should be able to build the required cables. **See Fig 19.2**.

The resistors in the transmit cable are used to attenuate the sound card signal so that it does not overload the transceiver. If the microphone socket is used, a lower value of resistor may be

necessary across the transformer to further attenuate the audio. If an accessory socket is used, the values shown may suffice.

While the receiver cable is shown with a connector to be directly plugged into the external speaker socket on the receiver, with many rigs this will disconnect the speaker, which isn't helpful. It is best in this case to use an adaptor allowing both the PC cable and an external speaker to be connected.

There is a very good reason for not using the computer speakers instead of an external speaker on the transceiver. Computer speakers receive their audio from the sound card output, and by connecting the LINE IN signal from the radio to the LINE OUT or SPEAKER OUT and the speakers, the receiver output signal will also be sent to the microphone input of the transceiver. This causes feedback problems, especially if VOX is used.

The Importance of Isolation

The transformers shown in Fig 19.2 provide complete DC isolation between the computer and the radio transceiver. The most compelling reason to do this is to prevent serious damage to the radio and computer. Most power supplies are grounded for safety reasons. If the power supply cable to the transmitter becomes loose, the full 20A transmitter current can pass through the microphone circuit, down the cable and through the computer sound card to ground via the PC power cable. Even if the transmitter power cable is considered reliable, significant current could still flow through the sound card cable, causing instability, hum and RF feedback. The simple expedient of isolating the connections also reduces the risk of RF in the computer, and computer noises in the radio.

VOX and PTT Control

Most operators find VOX operation of digital modes quite appropriate and reliable, although the delay may need to be set longer than for Morse or SSB. If for some reason direct control of the rig is necessary, the transmit control must also use an isolated circuit. An opto-coupler does this nicely, driving the Press - to - Talk (PTT) directly without requiring a relay or any further power supplies.

The digital mode software usually controls the transceiver via a serial port, by driving RTS or DTR (often both) positive on transmit, with an appropriate delay before sending tones out from the sound card. The design in **Fig 19.3** is an appropriate PTT circuit for a transceiver with positive voltage on the PTT line and a current when PTT is closed of up to 100mA or less.

Many transceivers include an 'accessory socket', offering line-level audio inputs and outputs for transmit and receive. Using these instead of the microphone socket and speaker socket can be really convenient, but can lead to a range of unexpected problems. Sometimes PTT is not available from the accessory socket, and sometimes the VOX does not operate from this socket. The signal levels can also be quite different to the speaker and microphone connections. Even more troublesome, some transceivers leave the microphone operating while sending data through the acces-

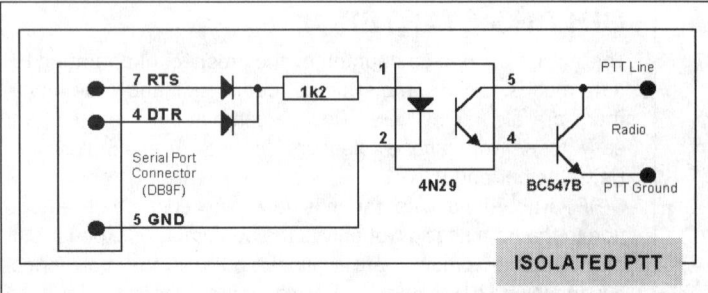

Fig 19.3: An opto-isolated PTT circuit

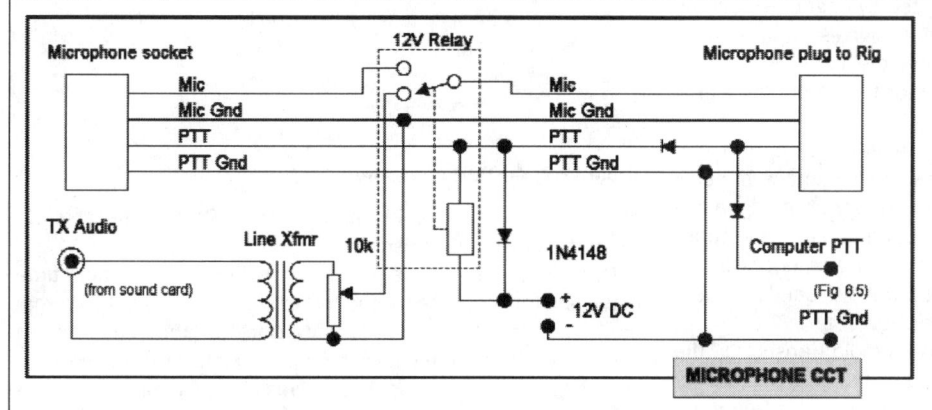

Fig 19.4: A simple way to connect microphone and computer

sory socket, so coughs, mutterings and keyboard clatter go out over the air!

A simple home-made adaptor (see **Fig 19.4**) provides a way to operate voice and data modes interchangeably without disconnecting anything. Using this design, the data transmit cable is connected by default, but when the microphone PTT switch is depressed, the relay switches over the audio input and normal microphone use occurs. Operation is simple, and feels natural. The isolated PTT circuit of Fig 19.3 can be built into the same box, and the whole assembly replaces the transmit cable in Fig 19.2. There are several similar designs offered as kits [7].

There are several commercial interface designs available for users not disposed to building a home-made or a kitset interface, but not all provide full isolation. There are also USB interfaces suitable for laptop computers and others with no serial port or sound card.

An area that causes confusion among beginners is the business of setting up and adjusting the sound card. The adjustments are all performed in software, mostly using an application provided with the operating system, and once set for one mode or program, the settings should be correct for all the rest. There are two main software adjustments, for transmit and for receive, and it is not very obvious where to find these, especially the receiver adjustments. The better applications provide direct access to the adjustments. In addition to the gain settings, you need to select the correct inputs and outputs, and disable those not being used. The procedure and these adjustments are described in detail in the RSGB publication *Digital Modes for All Occasions* [8], a reference work recommended for both novice and experienced operators.

RTTY

RTTY is now almost exclusively operated using computers, but the actual data signalling remains the same as it was in the 1950s, when mechanical machines were used.

RTTY uses five sequential pulses to represent each of the letters, figures, symbols and machine functions. Start and stop pulses are added to facilitate serial transmission. This code is now recognised as the International Telegraph Alphabet No 2 (ITA2) [9], an international standard with national variations. A

Symbol Rate	45.45 or 50 baud
Typing Speed	60 or 66WPM
Bandwidth	270Hz
ITU-R Description	270HF1B

Table 19.1: RTTY summary

summary of the specification of the RTTY mode is shown in **Table 19.1**.

The RTTY technique as we know it took many decades to become what it is today. Engineers Emile Baudot (who developed the system of multiplexing the data), Donald Murray (who developed the actual alphabet), Frederick Creed (who sold telegram printers to the GPO), Howard Krum (who developed the start-stop technique), and finally Edwin Armstrong (who developed the FSK keying technique) all played significant parts. Perversely, the alphabet developed by Murray is also known as the Baudot code, although it wasn't developed by Baudot, and is an alphabet, rather than a code!

The ITA2 five-pulse alphabet has only two conditions for each of the five data pulses (binary 0 or 1, space or mark), allowing for 32 different combinations. Because it was necessary to provide 26 letters, 10 figures and punctuation marks, the 32 combinations were not enough, and the problem was resolved by using most combinations twice; once in letters (LTRS) case, and again in figures (FIGS) case. Two special characters were assigned, LTRS and FIGS, to indicate which case was in use. The receiving station continued to use the case indicated by the last received case command character until it received a different one. Control functions such as LTRS, FIGS, CR, LF, space and blank were made available in either case. The remaining 26 have different meanings depending on whether the LTRS or FIGS case is selected.

Each mode in this chapter is accompanied by a spectrogram that illustrates the bandwidth properties and appearance of the signal. **Fig 19.5** is the first of these. The spectrogram is a three-dimensional record of the radio signal - frequency vertically, time

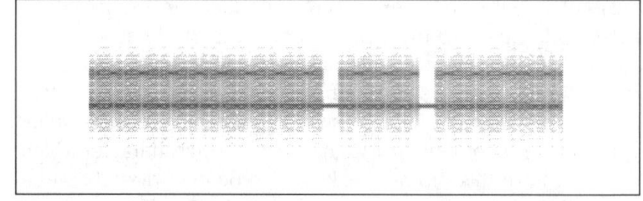

Fig 19.5: The RTTY Spectrogram

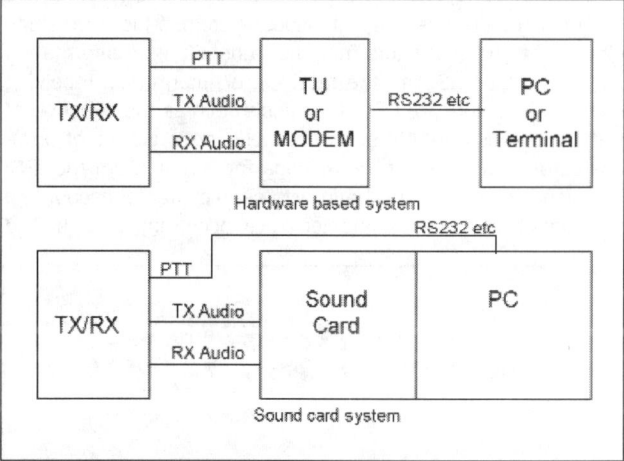

Fig 19.6: Block Diagrams of hardware and sound card systems

horizontally, and signal strength indicated by brightness. All the spectrograms were recorded from live signals at the same bandwidth and time settings, allowing the bandwidth to be assessed and should assist the operator to identify the signal from its appearance.

Compatibility was a problem in the early days of RTTY. Different types of surplus mechanical equipment was being used in different countries, many with different signalling speeds, and the American 45.45 baud speed eventually became the standard. At 45.45 baud each element is 22ms long, and each character takes about 165ms to send, so there are about six characters per second, or 60WPM.

So that the signals can be carried on a radio transmitter, the low frequency data pulses must be modulated onto a carrier or tone (sub-carrier). Similarly, at the receiver, the audio frequency sounds must be demodulated back into pulses. Equipment to do this is called a terminal unit (TU) or modem, although in computer systems a sound card and software inside the computer now perform these functions (**Fig 19.6**). Most operators now use audio frequency shift keying (AFSK) and an SSB transmitter.

AMTOR

RTTY suffers from problems such as multi-path reception (fading) and noise, which make successful decoding difficult or impossible on many occasions. In order to overcome these problems, it is useful to be able to compare multiple versions of the same transmission. This can be achieved by using frequency, polarisation, space or time diversity. AmTOR uses time diversity, sending groups of characters twice, spaced by a small time interval. AmTOR has two modes, Mode A (ARQ - automatic repeat request) and Mode B (FEC - forward error correction), which use this time diversity in different ways. In Mode A, a repeat is only sent when requested by the receiving station, while in Mode B each character is always sent twice.

AmTOR was developed from the commercial SiTOR system, which was devised to improve the communication between teleprinters using the ITA2 alphabet. The system uses a seven-pulse alphabet with an exact correspondence to the ITA2 five-pulse code - the two extra pulses provide error detection information [10]. This Moore code was designed to have a constant ratio of four binary 1s to three binary 0s in all valid combinations, so only 35 out of the possible 128 combinations are valid. This provides a form of error detection since any character which does not have this 4:3 ratio is known to be in error and can be rejected. In addition to accommodating the 32 ITA2 combinations, a further three are available for link information signals (Idle Signal Alpha, Idle Signal Beta and Repeat request (RQ)).

AMTOR Mode A is a synchronous system, which transmits blocks of three characters from the transmitting or information sending station (ISS) to the receiving or information receiving station (IRS). During a QSO, the roles switch as the direction of traffic changes. The ISS sends its message in groups of three characters, pausing between groups for a reply from the IRS. The signalling rate is 100 baud, with each character accounting for 70ms and a three-character block occupying 210ms. The

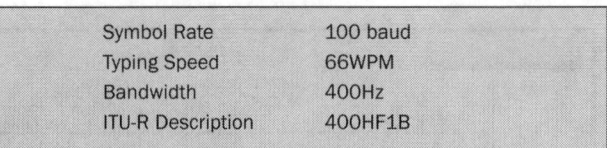

Symbol Rate	100 baud
Typing Speed	66WPM
Bandwidth	400Hz
ITU-R Description	400HF1B

Table 19.2: AmTOR summary

block repetition time is 450ms, so there is 240ms in each cycle when the ISS is not transmitting.

This 240ms period is taken up by the propagation time between stations, time for the IRS to send back its link information, time for the return journey back to the ISS, and an allowance for switching delay from transmit to receive. This time should be less than 20ms. The 450ms block repetition cycle limits the distance over which a Mode A QSO can take place.

A spectrogram of an AmTOR Mode A transmission is shown in **Fig 19.7**, and one showing AmTor Mode B is in **Fig 19.8**. **Table 19.2** shows a summary of the AmTOR specification.

When it is required to transmit to no particular station (for example when calling CQ, operating in a net, or transmitting a bulletin) there is no one station to act as IRS. Similarly, Mode A isn't helpful to others listening in, as they do not get corrections, but do receive repeats they may not need. Mode B is designed for these applications, and achieves a simple forward error correction (FEC) technique by sending each character twice. In order to provide time diversity, each character is repeated after four other characters have been transmitted, thus avoiding errors associated with bursts of noise. The receiving station tests for the constant four to three ratio, and prints only correct characters. If neither version is correct then an error symbol is displayed.

AMTOR Mode A is little used these days, although commercial traffic is still widely heard. Unfortunately timing restrictions make PC sound card programs for AMTOR Mode A impractical. AMTOR Mode B has enjoyed a longer life since it can be successfully operated using a computer with sound card. It is a useful mode for bulletin broadcast.

PSK31

Also developed by G3PLX, and based on an idea by Pawel Jalocha, SP9VRC, PSK31 was intended to replace RTTY as a simple to use and easy to tune keyboard chat mode. The PSK31 mode first used a low-cost DSP starter kit as modem, but is now firmly in the realm of PC sound cards and public-domain software, using modern DSP techniques [11]. The bandwidth of PSK31 is much lower than most other data modes, and has high sensitivity, which means it can work at lower signal levels in today's crowded bands.

Keying is achieved by phase-shifting the carrier by 180°, rather than frequency-shifting it, resulting in a very narrow-band signal. The technique is called differential binary phase shift keying, or BPSK. Data is encoded in the phase difference, rather than absolute phase, since phase is not constant due to ionospheric effects. With the chosen baud-rate of 31.25, the bandwidth is down from the 300-500Hz of other modes to only about

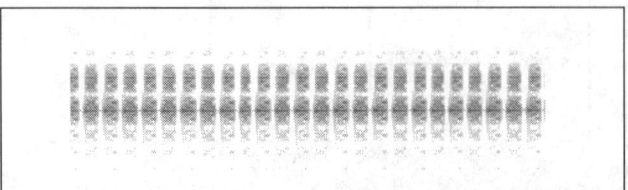

Fig 19.7: The AmTOR-A Spectrogram

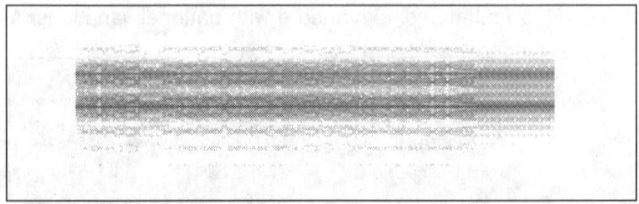

Fig 19.8: The AmTOR-B Spectrogram

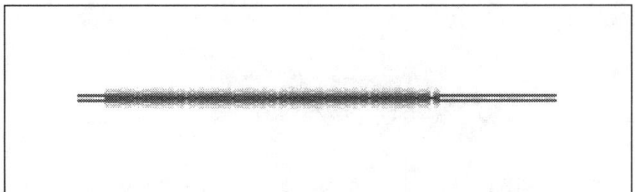
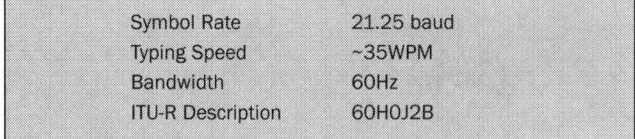

Fig 19.9: The PSK31 Spectrogram

Symbol Rate	21.25 baud
Typing Speed	~35WPM
Bandwidth	60Hz
ITU-R Description	60H0J2B

Table 19.3: PSK31 summary

62.5Hz. By using an alphabet with properties similar to Morse, ie with short codes for common letters, the text speed of PSK31 is about 35WPM. By using narrow filters in the receiver, the performance of PSK31, even without error correction, is certainly better than RTTY and AMTOR. In addition, the transmitted signal is carefully shaped to minimise bandwidth as the phase is switched [12].

A spectrogram of a PSK31 transmission is shown in **Fig 19.9**, and **Table 19.3** gives a summary of the specification.

PSK31 has no error correction, but on average gives much better reception than RTTY. Radio paths with fading and especially high phase shift caused by ionospheric Doppler effects can prove very difficult, as the incidental phase shift can easily exceed the differential phase shift of the intended modulation. A version of PSK31 with convolutional error correction (QPSK31) works well on VHF and paths with burst noise, but is even more adversely affected by Doppler on HF. This is because the phase shift is reduced to 90° in order to accommodate twice as much data (Quadrature PSK). Several other successful variants exist - for example FSK31, developed by UT2UZ, which uses MSK modulation, and PSK63F, developed by IZ8BLY, which uses convolutional FEC on a single 62.5 baud PSK bitstream (rather than QPSK).

PSK31 is popular on the HF bands, and is probably the most widely used digital mode. It is also effective on VHF. The latest software is easy to use, and little transmitter power is required for good DX. Perhaps the most popular spot is around 14.070MHz on the 20m band.

HELLSCHREIBER

There is a grey area (pun intended) between digital data modes (such as RTTY and PSK31) and analogue data modes (such as SSTV or HFFAX), of modes which cannot adequately be described as totally analogue or totally digital.

Arguably, Morse is one of these modes, because while it is apparently a digital transmission, reception occurs at an analogue human-readable level. For example, experienced Morse operators can identify the sender by his 'fist', can read the signal better than any electronic means, and can tell much about propagation and the transmitter from the sound of the signal.

Of the other modes in this category, Hellschreiber is the favourite. The term Fuzzy Modes has been coined to describe these modes with both analogue and digital features. Fuzzy modes use the human brain to assist in interpretation of an analogue presentation of the received signal, rather than electronic decisions made by hardware or computer.

While Hellschreiber is a relatively recent arrival on the amateur scene, its origins are old. It was developed as a means of sending press messages by telephone line. Hellschreiber, even

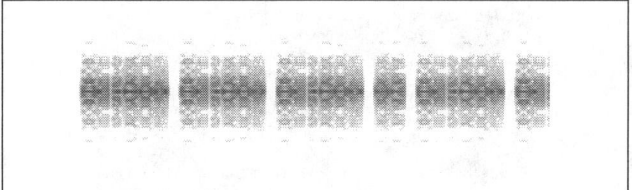

Fig 19.10: The Feld-Hell Spectrogram

Symbol Rate	112.5 baud
Typing Speed	25WPM
Bandwidth	350Hz
ITU-R Description	350HA1C

Table 19.4: Hellschreiber summary

in the 1930s, was an audio sub-carrier mode, and was soon sent by radio, predating RTTY for this purpose by at least 15 years. The mode was used to send press traffic right up to the 1960s.

Developed by Rudolf Hell [13], the technique involves sending each character as a pattern of timed dots, rather like a dot matrix printer [14]. Black dots are sent (key down) and white spaces are not sent (key up) by scanning each character vertically upwards, then moving along from left to right at a constant rate. The column data can be analogue, but typically consists of 14 black or white dot positions per column, and seven columns per character, including the space between characters. Each character takes 400ms to send, so the typing speed is about 25WPM. At the receiver, the incoming dots are presented to the reader as dots of varying greyness according to signal strength, allowing the reader to discern the transmitted text by eye, and so enabling text to be recognised in the presence of considerable noise.

A number of clever techniques were developed to improve reception and minimise transmission bandwidth. For example, the received dots are displayed twice, spaced vertically, making synchronisation unnecessary. The font developed by Hell did not permit individual pixels to be sent, rather at least two were sent consecutively, so the characters could have 14 x 7 resolution with the bandwidth of a 7 x 7 font. These features are retained today in PC sound card software for Hellschreiber. Modern techniques include rendered characters (grey pixels on corners), raised cosine dot shaping for minimum bandwidth, and proportional fonts, which are faster to send.

The most popular Hell mode is the original one used over military radio links from 1944 by portable mechanical machines such as the Siemens A2, and for that reason is called Feld-Hell. The signal is on/off keyed at 122.5 baud with carefully shaped dots, and has a bandwidth of about 350Hz. Feld-Hell is especially useful on noisy bands, and because the transmitter duty cycle is only about 20%, is ideally suited to QRP and portable operation. It is badly affected by multi-path, which causes interesting ghosting effects. These can often be minimised by careful adjustment of receiver gain. Another popular mode is FM-Hell, developed by Nino Porcino IZ8BLY, which uses minimum shift keying (MSK), has similar bandwidth, but is more robust and sensitive. FM-Hell is not so affected by multi-path, but operates the transmitter at 100% duty cycle.

A spectrogram of a Feld-Hell signal is shown in **Fig 19.10**, and **Table 19.4** shows a summary of the specification.

There are many free software packages for Hellschreiber modes, the most popular being IZ8BLY Hellschreiber, written by Nino Porcino IZ8BLY. This software includes several other interesting Hell-related modes. Hell signals can be found on most

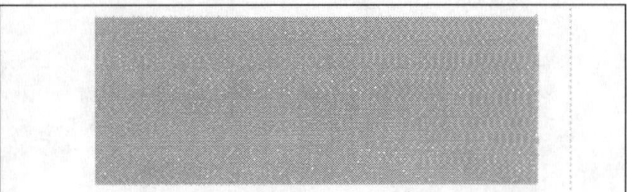

Fig 19.11: The MT63 Spectrogram

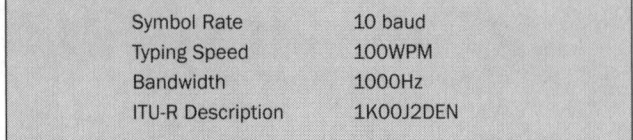

Symbol Rate	10 baud
Typing Speed	100WPM
Bandwidth	1000Hz
ITU-R Description	1K00J2DEN

Table 19.5: MT63 summary

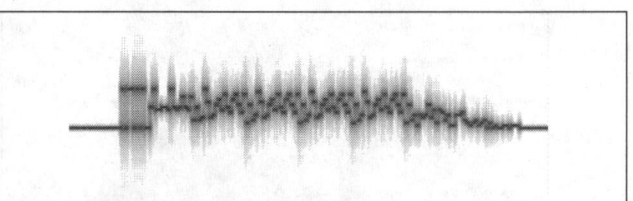

Fig 19.12: The MFSK16 Spectrogram

Symbol Rate	15.625 baud
Typing Speed	40WPM
Bandwidth	316Hz
ITU-R Description	316HF1B

Table 19.6: MSK16 summary

bands, and is most popular on 80m and 20m. Check around 14.075MHz.

MT63

This remarkable mode has been likened to a juggernaut driving down the high street at rush hour - nobody dares get in its way! Developed by Pawel Jalocha, SP9VRC, this is definitely *not* a mode to be used on a crowded band, but it has some very special properties. Not unlike many PSK transmissions at the same time, MT63 uses 64 carriers, spaced 15.625Hz apart, each one phase modulated at 10 baud. The resulting signal bandwidth is 1kHz, and the signal sounds just like noise [15].

What makes the mode particularly unusual is the way the data is coded. The raw data rate is 640 bits per second, but a very strong FEC system is used, and the resulting text rate is 100WPM. The FEC system uses a Walsh-Hadamard transform, where the seven ASCII text data bits index a table of carefully selected 64-bit words to be transmitted. These 64 data bits are then spread across the 64 tones, and also spread over six seconds of transmission. The result is a sensitive mode of incredible robustness, reasonably immune to burst noise and interference, and also able to operate under conditions of ionospheric instability that would stop most other modes.

A spectrogram of MT63 is shown in **Fig 19.11**, and **Table 19.5** shows a summary of the specification.

Because of its bandwidth and slow turnaround (12 seconds between overs!), MT63 is little used for DXing. However, little transmitter power is needed and tolerance to drift and mistuning is about 50Hz. MT63 is a good choice for maintaining regular contact with friends over trans-polar and long path routes. SP9VRC has recently developed an MFSK mode of similar bandwidth which uses the same FEC system. This new mode, named OLIVIA, is one of the most sensitive modes designed yet, and has a typing speed of about 17.5WPM. Both modes have wider and narrower variants, and both can be found regularly just above 14.1MHz.

MFSK16

Multi-frequency shift keying (MFSK) operates like RTTY, but instead of just two tones, four or more are used, allowing more data to be sent at a lower keying rate. MFSK has been used commercially since the 1950s when the Coquelet electro-mechanical system was developed in Belgium. A more sophisticated electronic system called Piccolo was developed for the British Foreign and Commonwealth office and publicly demonstrated in 1963. These two systems were initially designed for the ITA2 alphabet, and converted the signals to and from standard teleprinters. Over the years a number of versions of Piccolo with differing numbers of tones and different speeds were developed, and all exhibited good sensitivity and considerable superiority to teletype on long distance circuits. These systems operated without FEC. It made sense therefore to develop an improved MFSK mode for amateur use.

The benefits of MFSK are:
- Sensitivity improves with the number of tones used.
- More data can be sent as the number of tones increases.
- Since the same data can be sent at a lower keying rate, immunity to multi-path effects is improved.
- Immunity to interference depends on the signalling rate, not the overall signal bandwidth, improving sensitivity and robustness.

These benefits mean that a very robust typing-speed mode can be developed that has high sensitivity. Multi-path reception causes timing errors which cause individual keying elements to run into each other. Because the keying rate of MFSK is much lower for the same data rate, the technique is useful for avoiding the timing problems that cause such difficulty to other modes. One disadvantage of MFSK is that it requires rather accurate tuning and stable equipment.

The first and most successful amateur MFSK mode is MFSK16 [16], designed by Murray Greenman ZL1BPU, specifically for long-path keyboard to keyboard operation. It has also proved to be excellent on 80m where NVIS multi-path problems are extreme. The first MFSK16 computer program was STREAM by Nino Porcino, IZ8BLY, and it is still the most popular. MFSK16 uses 16 tones spaced 15.625Hz apart, and a signalling rate of 15.625 baud. Because each tone represents four bits of data, the data rate is 62.5BPS. MFSK16 uses a very powerful convolutional code FEC system, with an interleaver to provide time diversity. Both the FEC code and interleaver are tied to the data bit weighting, and so no synchronism is required. MFSK16 also uses a variable length character set like Morse and PSK31, which results in a typing speed of over 40WPM.

Fig 19.12 shows the spectrum and **Table 19.6** gives a summary of the specification.

MFSK16 is one of the best DX modes, requiring little power, and capable of operating under rather poor conditions. Users report reception with no errors even when the signal cannot be heard! The FEC system ensures that copy is virtually perfect until printing simply stops as the signal is lost, which cannot be said for most other modes. There are numerous computer programs offering MFSK16. Most operation is close to where RTTY is found, for example just below 14.080MHz on 20m.

The sound of MFSK16 is distinctive, and although tuning the signal is tricky, users soon learn to align the receiver with the

lowest (idle) tone. This tone appears at the start of each over and during pauses in the transmission (see **Fig 19.12**).

Other amateur-developed MFSK modes include: MFSK8, also by IZ8BLY; THROB by Lionel Sear, G3PPT; OLIVIA by SP9VRC; FSK441 by Joe Taylor, K1JT; Domino by Con Wassilief, ZL2AFP; and JASON, a narrow band LF mode by Alberto deBene, I2PHD. FSK441 is a high speed four-tone mode for meteor scatter use. JASON and Domino use Incremental Frequency Keying (IFK), encoding the data as differences in frequency, rather than absolute frequency.

MFSK8 is the same bandwidth as MFSK16, but uses 32 tones spaced 8Hz apart at 8 baud. It is extremely difficult to tune accurately. THROB uses an unusual combination of single and dual tones to encode a restricted character set, and operates at 1, 2 or 4 baud. Despite the very low signalling rate, the typing speed is reasonable, since each signal is a complete character. There is no FEC.

Domino is designed for HF band chatting, and encodes each character of a limited (6-bit) character set into two successive tones. The tones are in two interleaved sets of eight, odd and even, and the data is recovered by measuring the distance between successive tones. As you can imagine, if one measurement is in error, the next will be in error in the opposite direction - this is the main flaw with IFK. The receiver synchronises easily because of the odd-even tone sets, and the order of the tone sets is determined by analysis of the received data. Domino has no error correction in its experimental form, and yet it is remarkably robust and forgiving. The ZL2AFP software is very easy to use.

The particular advantage of the IFK technique is much reduced sensitivity to drift and poor tuning. For example, Domino can be received while the receiver is slowly tuned across the signal! Other similar modes with IFK coding and FEC are likely to be developed in the future.

HF ARQ MODES

These modes were developed to provide improved automatic operation on HF. When forwarding mail or in communication with a bulletin-board system, it is important for communications to be letter perfect, or the commands could be misinterpreted or data corrupted. Since these operations are invariably station to station and automated (rather than nets or broadcasts), an ARQ mode is more appropriate.

The first automated systems used AMTOR, which maintained links well, but data rate was poor. Some also used HF (300 baud) packet, which performed very poorly unless propagation was perfect. Most systems now use commercial (and not inex-

Fig 19.13: The PACTOR Spectrogram

Symbol Rate	100 or 200 baud
Typing Speed	66 wpm (300WPM PACTOR 2)
Bandwidth	500-600Hz
ITU-R Description	600HF1B (500HG1B PACTOR2)

Table 19.7: PACTOR summary

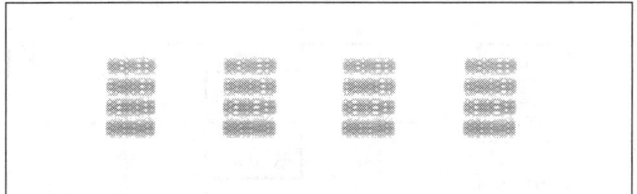

Fig 19.14: The CLOVERII Spectrogram

Symbol Rate	31.25 baud
Typing Speed	30 - 500WPM
Bandwidth	500Hz
ITU-R Description	500HJ2DEN

Table 19.8: CLOVERII summary

pensive) modems, which operate the specialised modes PACTOR, PACTOR2, PACTOR3, CLOVER II and G-TOR.

The original PACTOR mode is FSK, not unlike AmTOR, except that the data is ASCII, transmitted in longer blocks (1.25s period) and much better error detection is used. In addition, a scheme known as Memory ARQ allows data to be corrected by processing multiple corrupted versions of the same data. Compression techniques are used to reduce the number of bits transmitted.

A summary of the PACTOR specification is in **Table 19.13** and a spectrogram is in Fig 19.13.

Later versions PACTOR2 and PACTOR3 use PSK modulation on multiple carriers, and are considerably faster and more robust. The calling and linking functions retain the original PACTOR FSK modulation mode for compatibility. It is important to appreciate that these are commercial and proprietary modes (not public domain) and therefore their amateur use may be prohibited or restricted in some countries. A special hardware modem is required, and it is also not possible to "listen in" to a transmission in these modes.

CLOVER II has a wide range of different modulation schemes, but is best described as an orthogonal frequency division multiplex (OFDM) system. There are four tone frequencies, each amplitude and phase modulated. Special hardware is required, and the equipment can automatically switch between the available modes in an attempt to provide best throughput. Clover uses Reed-Solomon FEC in addition to its ARQ system. The protocol is proprietary. It is not now used very widely.

Fig 19.14 shows a spectrogram of CLOVERII, and a summary of the specification is in **Table 19.8**.

Of the ARQ modes, G-TOR is the most similar to AMTOR. It has the same FSK modulation, but differs in using the ASCII character set, and in the use of a very strong Golay FEC error correction system, which transmits two differently coded versions of the data. Requests for repeat are reduced because the system is often able to reconstruct the data from the first transmission, and if the second is required the ability to reconstruct the data accurately is enhanced further. G-TOR is proprietary and only available using suitably equipped hardware. Although a good system, unfortunately G-TOR has never enjoyed wide popularity.

PACKET RADIO

Packet radio was the first true amateur digital, as opposed to analogue, transmission system. This makes the relaying of signals much more efficient since the data is reconstituted at each stage of the link and any end-to-end noise and distortion is simply that of the digitising process and not the transmission of the

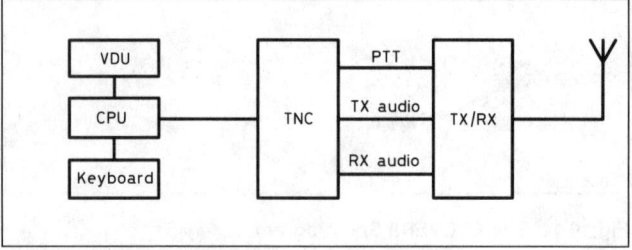

Fig 19.15: Block diagram of a typical packet radio station

digital information. One of the other main benefits of this mode of operation was always assumed to be that the channel could be shared by many users. Unfortunately the radio-based systems are different to computer networks in that not all stations can receive each other, thus making it more difficult for channel sharing.

As with other methods of data communications, packet radio commonly makes use of a terminal unit (Terminal Node Controller, or TNC), either a stand-alone unit or as part of a PC-based system using the sound card as an interface.

Very simply, the function of the TNC is to take the arriving data and assemble it into packets which are then passed to the on-board modem (or PC sound card under PC control) for conversion into audio tones. The receive side of the TNC performs the reverse of the tasks outlined. On VHF the transmission speed for most end-user access is 1200 baud with tone frequencies of 1200Hz (mark) and 2200Hz (space), with 300 baud and 200Hz shift being employed for HF applications. These standards coincide with Bell 202 and 103 modems for VHF and HF respectively. 9600 baud is commonly used for inter-site linking and satellite communications on VHF and UHF.

A block diagram of a typical packet station is shown in **Fig 19.15**. Although the drawing shows a computer, a simple dumb terminal can be used; however, to make use of the full facilities for file transfer etc a computer is essential.

Channel Access

The basis of a packet radio contact is that each station transmits some information and receives an acknowledgement. If no acknowledgement is received then the information is retransmitted. One of the main causes of non-receipt of acknowledgement is collision with another transmission of either the main transmission or the acknowledgement.

Early packet radio experiments made use of a channel access system in which a station transmitted without checking if the channel was free. If the transmission was not acknowledged within the correct time slot, the TNC waited a random length of time before retrying. Current packet systems make use of data carrier detect (DCD) - they listen for an empty channel before transmitting. This is not a guarantee against collisions, because two stations may 'decide' to transmit at the same time, but it is an improvement.

AX.25 Level 2 Link Layer Protocol

Version 2 of the AX.25 Level 2 protocol was adopted by the ARRL back in October 1984. This protocol follows that of CCITT Recommendation X.25 except that the address field has been extended to accommodate amateur callsigns, and an Unnumbered Information (UI) frame has been added. This protocol formally specifies the format of a packet radio frame and the action a station must take when it transmits or receives such a frame.

At this link layer, data is sent in blocks called frames. As well as carrying data, each frame carries addressing, error checking and control information. The addressing information carries details of the station which sent the frame, who it is intended for and which station should relay it. This forms the basis of many stations sharing the channel since any station can be set up to monitor all frames on the channel, through various stages to monitor only those intended for it and ignore any others. The error-checking information allows the intended recipient to determine if the frame has been received free of errors. If this is the case and the two stations have previously established a connection, an acknowledgement is generated by the receiving station. If errors are detected the frame is ignored and some time later the sending station resends the frame.

AX.25 Format

Packet radio transmissions are sent in frames with each frame divided into fields. Each frame consists of a start flag, address field, control field, network protocol identifier, information field, frame check sum (FCS), and an end flag. **Fig 19.16** shows the format of a frame and **Fig 19.17** shows a typical address field.

Flag field

Each frame starts and ends with a flag which has a particular bit pattern: 01111110. This pattern appears only at the beginning and end of frames. If five 1 bits show up elsewhere in the frame, a procedure called zero insertion (more commonly called bit stuffing) takes place and a 0 is inserted by the sending station

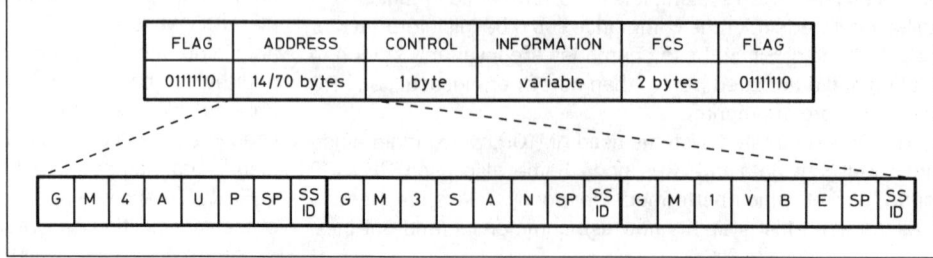

Fig 19.16: Format of a frame

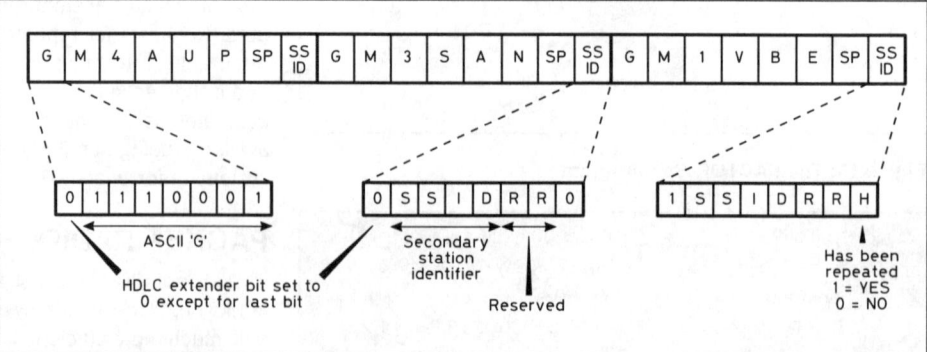

Fig 19.17: A typical address field

and deleted by the receiving station. The receiver will therefore delete any 0 bit which follows five consecutive 1 bits that occur between the flag fields.

Address field

The address field consists of the destination field, source field and up to eight optional relay or digipeat stations. These fields usually contain callsigns and space is available for up to six characters per callsign with a seventh available as a secondary station identifier (SSID). This allows up to 16 different packet radio stations to operate with one callsign. The default is an SSID of 0. For example, GM4AUP-0 could be the real-time station, GM4AUP-2 could be a personal message system (PMS) and GM4AUP-4 could be a node station. The SSID byte in the digipeater address also contains information as to whether it is repeating a frame or not.

Control field

The control field is used to identify the type of frame being transmitted and the frame number.

Protocol identifier field

This field is contained within the information field and identifies what, if any, network layer protocol is being used.

Information field

The information field contains the data to be transmitted and can contain any number of bytes, up to a maximum of 256, of information.

Frame checksum field

The FCS is a 16-bit number calculated by the sender. On receipt of a frame the receiving station calculates a FCS and compares it with that received in the FCS field. If the two match then the receiving station acknowledges the frame.

AX.25 operation

As previously described, the TNC is the device which assembles the data into frames as above. When first powered up, the TNC is in a disconnected state and is monitoring traffic on the appropriate radio channel.

In order to communicate with another station it is necessary to enter the connected state. This is done by issuing a connect frame which contains the callsign it is requesting connect status with as the addressee. If the other station is on the air it responds with an acknowledgement frame and the stations become connected. If no acknowledge frame is received the requesting station re-issues the command a pre-determined time later and continues to do so until a preset number of tries has taken place. If no connection is established the requesting TNC issues a failure notification.

Once a link is established the TNCs enter the connected or information transfer state and exchange information and supervisory frames. The control field contains information about the number of the frame being sent and the number of the last one received (0 to 7). This allows both TNCs to know the current link status and which to repeat if necessary.

When in the connected state either station may request a disconnection which occurs after an acknowledgement is received or if no response is received after several attempts.

Packet Operation

Packet operation currently makes use of the HF, VHF, UHF and SHF parts of the spectrum with both terrestrial and satellite links being utilised. In the early days, much packet operation was real-time person-to-person operation, either direct or through a digipeater.

Most TNCs are capable of digipeat operation and this enables stations who cannot contact each other direct to do so by the on-frequency retransmission of the digipeater. As packet became more popular the real-time operation tended to be replaced terrestrially by store-and-forward systems such as nodes, although there are Earth-orbiting digipeaters placed into operation periodically from amateur-radio equipped space stations.

Digipeaters

Most TNCs can be used as a digipeater as this function is usually contained within the AX.25 Level 2 firmware.

Fig 19.18 shows how two stations A and D can connect to each other using digipeaters B and C. In order for information to be passed from station A to station D via the digipeaters B and C, the information frame must be received by station D and the acknowledgement frame received by station A before a frame can be said to be successfully sent. Digipeaters B and C play no part in the acknowledgement process; they merely retransmit any frames that contain their callsigns in the digipeat portion of the address field. If the acknowledgement is not received by station A then the frame is retried over the whole path. The use of digipeaters has reduced dramatically in recent years with the advent of the network nodes.

Network Nodes

The network node significantly improved the packet radio system as a means of communicating between packet-equipped stations in both real time and by the use of mailboxes. The major advantage of a network node over a digipeater is that any frame which is being transmitted is separately acknowledged between each individual element rather along the whole chain.

Fig 19.19 shows a system with station A trying to communicate with station D via the nodes B and C. In trying to communicate with each other the information is sent from station A to node B and acknowledged back to station A. Node B then passes the frame on to node C and receives an acknowledgement back. Node C then passes the frame to station D who acknowledges it back to Node C. If anywhere in the path no acknowledgement is received then the frame is retried only over the part of the path for which no acknowledgement has been received.

There are two types of network protocol in use - virtual-circuit and datagram. In the virtual-circuit protocol the appearance of a

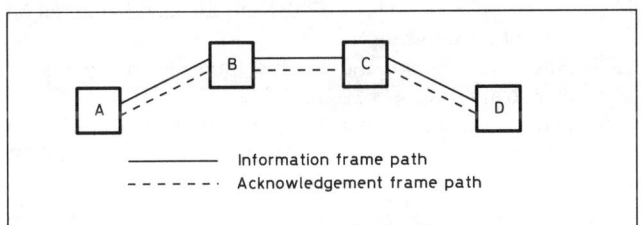

Fig 19.18: How two stations can connect to each other using digipeaters

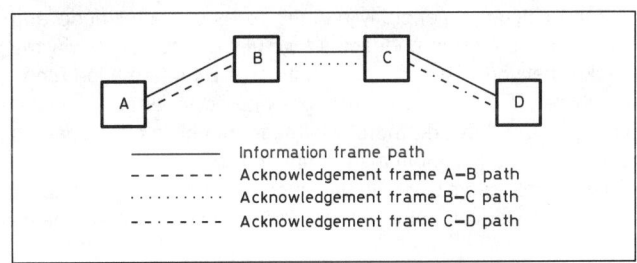

Fig 19.19: How two stations can connect to each other using network nodes

direct connection between the two stations is provided. In order to establish communications a 'call set-up' packet is sent through the network to make a path to the other station. Once this path is established information is sent through the circuit. Any packets sent do not have the full address of the required path because the network attempts to maintain this path for the duration of the contact. After the contact is completed the virtual circuit is cleared by removing the information on the path along the network. An example of a virtual-circuit protocol is the RATS Open System Environment (ROSE) developed by the Radio Amateur Telecommunications Society (RATS) of New Jersey. ROSE is a firmware replacement for TNC2 clones. The virtual-circuit protocol is not very common in the UK and most networking is done using the datagram protocol.

In the datagram protocol each packet contains full network addressing and routing information. This enables a packet to reach its destination via any route still open, regardless of how reliable the network may be. The network overhead is greater in this protocol but it has much greater flexibility and the end user does not need to know the route, only the node nearest him and the node nearest the station with which he desires to connect. Datagram protocols used in the UK are NET/ROM (and clones such as TheNET), TheNODE and Internet.

TCP/IP

The Internet protocol software was written by Phil Karn, KA9Q, and is more commonly known as TCP/IP which is an acronym for two protocols, the Internet Protocol (IP) and the Transmission Control Protocol (TCP).

In reality KA9Q's TCP/IP consists of a suite of individual protocols, Address Resolution Protocol (ARP), File Transfer Protocol (FTP), Serial Line Transfer Protocol (SLIP), Simple Mail Transfer Protocol (SMTP), Telnet Protocol, User Datagram Protocol (UDP) as well as TCP and IP.

Each station using TCP/IP is a network node with a unique IP address that has been assigned by the local IP address co-ordinator. The amateur TCP/IP network has been assigned the network name AMPRNET and all amateur addresses commence with the two digits 44, followed by three digits indicating the country code (as an example of a full address '44.131.5.2' is assigned to G3NRW). TCP/IP is becoming very popular in the UK and is said to offer many advantages over 'ordinary' AX.25.

DX Clusters

A DX Cluster provides information on DX stations being worked/heard along with information on QSL managers, WWV propagation and prefixes for example. The operation is not dissimilar to mailbox operation but users stay connected to their local DX Cluster for as long as they wish to receive announcements. The type of announcement the user receives is customised to suit his own needs and can be used to select prefix information, band information, mode information or a combination of all those.

Each cluster is generally referred to as a cluster node and these nodes can be connected together to each other via the packet network. This enables an item of DX information (commonly referred to as a spot) to propagated to all other cluster nodes in the network, thereby in theory enabling all connected users to see this spot in a short timeframe.

DX Cluster spots are quite common on virtually all contesting software, in lots of major SSB and CW contests, as well as being

in use by all major DXpeditions. Some stations remain connected to the Cluster all the time and have an audio warning from their PC to tell them about any DX that might be available. Internet-based DX Clusters are commonly used as an alternative to radio-based DX Cluster connections.

Satellite Communications

Using satellites for communications can be a very satisfying achievement. There are several data satellites orbiting the Earth. Using a dedicated set-up, it is possible to run an automatic station to track, send and receive mail to several different satellites. Full duplex mode is used, and 9600 baud is the standard used. Packet signals have even been bounced off the Moon, although the distortion on that path prevents regular communication.

More information can be found in the chapter on satellite communication.

Packet Radio Bibilography

Further information can found from the following books and periodicals:

Your First Packet Station, Steve Jelly, G0WSJ, RSGB, 1996.

Packet Radio Primer, Dave Coomber, G8UYZ, and Martyn Croft, G8NZU, RSGB, 2nd edition 1995.

NOSIntro, Ian Wade, G3NRW.

AX.25 Link Layer Protocol, ARRL.

ARRL Handbook, ARRL.

RadCom, the RSGB members' magazine.

Information on licensing and policy matters with regard to data communications is available from the chairman of the Data Communications Committee c/o RSGB, Lambda House, Potters Bar, Hertfordshire EN6 3JE.

REFERENCES

[1]   Comment made in conversation by Dr Gary Bold ZL1AN
[2]   See http://www.sdradio.org/
[3]   See http://web.usna.navy.mil/~bruninga/aprs.html
[4]   See http://www.ka9q.net/
[5]   *NOSIntro: TCP/IP over Packet Radio*, Ian Wade, G3NRW, TAPR
[6]   See http://www.ui-view.org/ and http://www.winpack.org.uk/
[7]   The BARTG publish designs from time to time. See http://www.bartg.demon.co.uk/
[8]   Appendix D, *Digital Modes for All Occasions*, Murray Greenman ZL1BPU, RSGB
[9]   A table of the ITA2 character set is to be found in Digital Modes for All Occasions, Murray Greenman ZL1BPU, RSGB
[10] A table of this character set is to be found in Appendix B, Digital Modes for All Occasions, Murray Greenman ZL1BPU, RSGB .
[11] See http://www.aintel.bi.ehu.es/psk31.html
[12] The technical details of PSK31 are explained in detail on the PSK31 web site.
[13] Patent "Device for the electric transmission of written characters" received 1929.
[14] See www.qsl.net/zl1bpu/FUZZY/Contents.html
[15] More details can be found at http://www.qsl.net/zl1bpu/MT63
[16] See www.qsl.net/zl1bpu/MFSK

20 Imaging Techniques

Although amateur radio has a long history of audible communication, whether by voice or Morse tones, it has also had a following of enthusiasts exploring the visual facets of the hobby. Imaging, whether used for information such a weather maps or used for live television pictures, is simply the name for viewing a picture at one location that has been produced at another, remote location. The imaging side of the hobby has many aspects, ranging from the reception of pictures from weather satellites to the transmission and reception of very high quality digital 'live' television signals.

It is a common misconception that it is expensive and complicated to get involved in amateur television, in fact the opposite is true. The spin-off of domestic analogue and digital broadcasts has been a surplus of very cheap receiving equipment which is ideal for conversion to amateur use. The transmitting equipment is less accessible and is one of the few remaining areas where real experimenting and 'home-brewing' is still common place. Putting technicalities aside, it is still quite feasible to build a complete television station for less than the cost of a new hand-held VHF rig.

Broadly speaking, there are two categories of amateur television, the narrow bandwidth modes which typically convert images to sound so they can be sent over voice channels and wide bandwidth modes which may occupy many megahertz and are therefore only allowed on UHF frequencies and above. From a picture perspective, the narrow band modes can only transfer still images whereas the wide band modes can carry enough information to convey live colour pictures and even stereo sound in some cases.

Each of these categories can be further divided into transmissions using analogue techniques and those using digital modulation. 'Digital' can be a scary word to people familiar with things that can be measured with a meter and dazed by 'new technology'. It is worthwhile reminding them that the first digital long distance transmissions were made by Samuel Morse in 1843, more than 30 years before the telephone was invented and 50 years before the first man-made RF hit the airwaves!

SLOW SCAN TV AND FACSIMILE

These are also known as SSTV and Fax respectively; they are technically very similar. The bandwidth required to send information is proportional to the rate it is sent and both these modes slow that rate to a pace that allows the information to be encoded as low frequency tones. By reducing the rate, the ability to send pictures quickly enough to be seen as motion video is lost, some modes can take more than a minute to send a single picture (a frame). In contrast, normal broadcast TV and fast scan modes send at 25 or 30 frames *per second*. However, for sending still pictures, these modes are perfectly adequate. Because they convert the picture to tones within the voice frequency range, it is possible to transmit them by simply feeding the tones to a transmitter microphone socket. The transmission is frequently FM on the VHF/UHF bands and SSB on the HF bands. It should be noted that throughout transmission the audio level will remain high and therefore an SSB transmitter will likely run at peak power for several continuous seconds (high duty cycle), it is advisable to 'back off' the output power to prevent over stressing the PA stages.

SSTV and Fax both work in the same way, only the scan frequency and number of scan lines distinguishes them. The picture being sent is divided into horizontal lines and each line is preceded by a synchronising (sync) pulse. The sync pulse is sent as a particular audio tone then the line of picture is scanned from left to right. As the scan point traverses the picture, the brightness at that point is converted to another tone. A low tone represents a dark part of the line and high tone a bright part. The shades of grey between them produces a tone proportional to the brightness.

For colour modes, each line is scanned three times and the brightness of the three primary colours, red, green and blue are sent in the same way as grey. When the end of the scan line is reached, the next one starts, again with a sync pulse but this time the line is a little further down the picture. When all the lines have been scanned, the bottom of the picture has been reached and (normally), the transmission ends.

At the receiving station, the tones are taken from the receiver, either through the headphone socket or another audio outlet. The sound is filtered to extract the sync pulse tones and these provides a reference point to align the start of each line. The following tones are converted back to an image, the higher tones producing a brighter trace than the low ones.

In the same way as the sending station scanned the picture to produce tones, the receiving station converts them back again and if all goes well, the same picture is reproduced. For colour pictures, tones conveying the red, green and blue content of each line are sent in sequence and then overlaid to reproduce a composite of the three before moving to the next scan line.

The sequence of colours, the scan rates and tone range have to be the same at the transmitting and receiving stations and for this reason, standards have been defined. Unfortunately, there are around ten different standards in use but two, the 'Scottie1' and 'Martin1' seem most popular. The majority of SSTV and Fax stations use computers both to generate tones and to decode them back to pictures again and there are many software applications around to do this. The screen shot in **Fig 20.1** was taken from MMSSTV [1] written by JE3HHT.

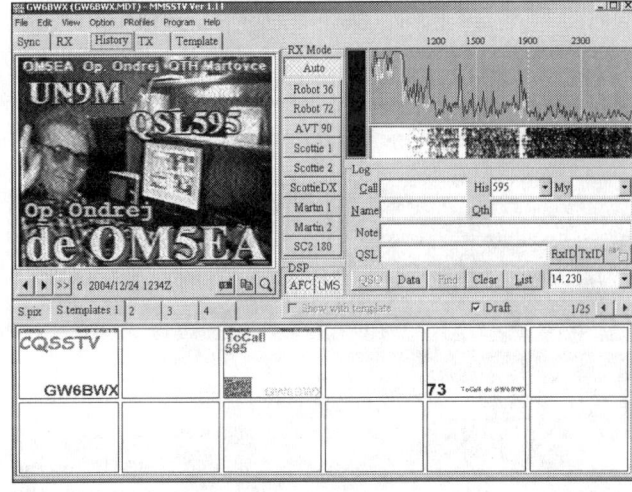

Fig 20.1: The slow scan window of JE3HHT's MMSTV program

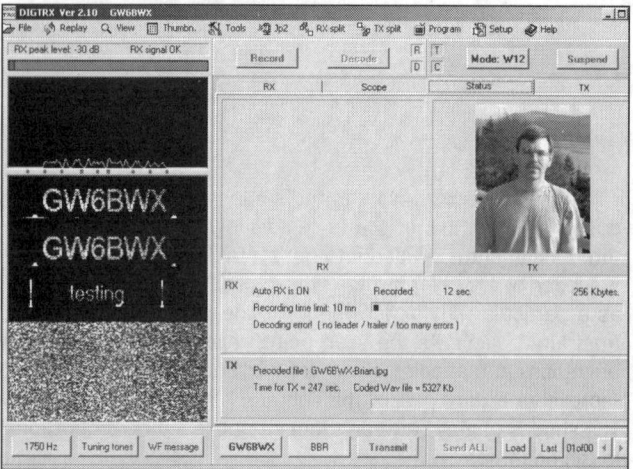

Fig 20.2: The DIGTRX control window

In most cases, all that is needed to get on the air with these modes is a cable to link the computer's sound system to the rig microphone and headphone sockets.

In days of old, it was common practice to decode SSTV with quite complicated audio filtering circuits. Typically, a narrow band filter would separate the sync signals from the rest of the tones and a filter with a linear gradient across the voice spectrum would convert the picture tones to different voltage levels. The voltage samples then had to be stored in order that the picture could be built up over its 15 seconds transmission duration. Originally this was done with long-persistence phosphor CRTs and later with semiconductor memories.

The advent of the personal computer brought about many changes to SSTV, making the process of creating, decoding and storing the pictures much easier. Images are now usually pre-prepared using scanned photographs or digital art programmes. They are converted to tones using synthesis techniques, utilising the capabilities of the computer's sound card. Decoding is achieved by performing a Fourier analysis of sound samples, a method which helps to isolate the dominant frequencies from the background noise to give a better signal to noise ratio. Storage while viewing is in the computer's video memory, and optionally the images can be permanently stored on disc or printed on paper.

A recent development, similar in speed to the method described earlier is 'DIGTRX' by PY4ZBZ [2]. It is a system of sending data files or pictures by using a different tone for each of the eight bits in each data byte. The bytes are sent one after another and the pattern of bits in each byte produces up to eight tones which are spaced equally in the voice frequency range. It can therefore be sent in a narrow bandwidth in the same way as conventional SSTV. A degree of pre-processing has to be done to the bit pattern before the bits are converted to tones to lessen

Fig 20.3: An SSTV image received by DIGTRX

the effects of errors and to produce a more even spread of tones. Take the case for example of a long string of zeroes being sent, which happens frequently in data transmission, without any bits being set, no tones would be produced and no transmission would take place. The pre-processing ensures that a signal is always present no matter what the content of the data is. An additional DIGTRX feature is being able to send short messages or callsigns as tone patterns, rather like very low resolution analogue slow scan but the DIGTRX software uses it for station identification. **Figs 20.2 and 20.3** show screen shots from the DIGTRX program.

The best place to hear SSTV is around 14.230MHz and for DIGTRX around 14.240MHz. Both modes are easy to identify by their distinctive sounds. SSTV produces a repetitive 'blip' against a background of sweeping tones while DIGTRX makes eerie, shrill sounds.

FAST SCAN TELEVISION

As with slow scan modes, there are analogue and digital versions of fast scan amateur television. At the time of writing this, it is still quite expensive to buy or build digital transmitters but the very low cost of digital receivers is persuading more and more people to try this mode. The digital standards being adopted by amateur stations are very similar to those used by commercial broadcasters and this has resulted in mass manufactured domestic receivers becoming easily obtainable at very reasonable prices. As time goes by and home users 'upgrade' or replace their receivers, they will become disposable items and easy to get hold of for repair or modification.

Analogue and digital fast scan modes can only be used on the 430MHz band and above because of the wide bandwidths they occupy. Indeed, on the 430MHz band itself, only limited quality can be sent, otherwise the edges of the modulation envelope would spread outside the band allocation. Using such a wide portion of band does not necessarily imply that interference will be caused to other users because spreading the power over a wider space means that less lies on any particular frequency. The power distribution changes with picture content and, coupled with the universally adopted horizontal antenna polarisation, the presence of properly engineered television transmissions hardly affects other stations.

On 1.3GHz and above, the bands are sufficiently wide, and FM is also normally used as it makes the transmitting equipment much easier to build. In the transmitter, it is much easier to apply the modulation signal to a varactor (varicap) diode in the oscillator tuning circuit than to control the carrier amplitude linearly. The power amplifying stages are also simplified as they too no longer have to be linear. In fact, most FM television transmitters are run flat out to achieve the best efficiency. Harmonic suppression, if necessary, is achieved in the normal way with LC or cavity filters.

The bands used for fast scan TV run from 430MHz up to the high microwave bands, and with a spread as wide as this, the techniques used to transmit and receive vary widely. The most popular bands are 1.3GHz, 2.3GHz and 10GHz so the detail here is concentrated on equipment suitable for them. These are not the only bands used by any means but they provide a useful starting point for looking at the methods used.

1.3GHz Equipment

This is the most popular ATV band and has wide repeater coverage throughout Europe and beyond. Almost, but not all, transmissions use FM because producing appreciable power from linear devices is quite expensive. FM does not need linear amplifying stages and it is considerably more efficient to run with a con-

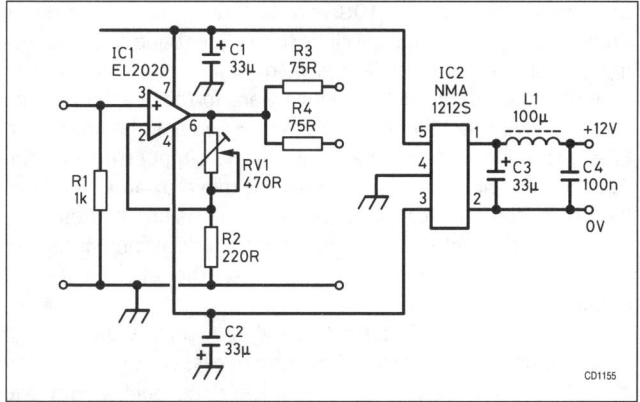

Fig 20.4: Amplifier suitable for raising the video level from a satellite receiver

R1	1K	C1	33µF
R2	220R	C2	33µF
R3	75R	C3	33µF
R4	75R	C4	100nF
RV1	470R	L1	100µH
IC1	EL2020	IC2	NMA1212S

Table 20.1: Parts list for the video amplifier

stant hard-driven PA stage. The popularity comes primarily from the ease of getting hold of suitable equipment. Almost all the domestic satellite receivers used in the world are capable of receiving the whole of the 23cm band and as more users switch to digital satellite television, their old analogue receivers are given away or discarded. With nothing more than adding a suitable antenna, a complete receiving system is ready made.

Although an unmodified satellite receiver can be used, a few simple modifications can dramatically improve their performance for amateur use. To see why the modifications help, it is first necessary to understand how the receivers were intended to work when picking up satellite transmissions. A domestic receiving system has two parts, one mounted at the focal point of the satellite dish, the other somewhere indoors, usually near the home TV. The outdoor part is a dish that is either a parabolic or 'prime focus' reflector but more likely an offset-feed dish which is actually a slanted section of a parabola. A prime focus dish has its receiving antenna on the centre axis of the bowl, equally spaced from the edge of the dish and usually mounted on three support arms. A line extended from the middle of the bowl, through the antenna would head straight toward the satellite signal source.

This type of construction has two disadvantages, firstly the antenna and support arms are casting a shadow on the dish, obstructing some of the signal path and secondly, they are difficult to mount as their centre of gravity is inconveniently under the dish where attaching a pole is difficult. In an offset design, the curve of the bowl is modified so the focal point is no longer directly in front of the dish, this slightly reduces its ability to concentrate as much signal on the antenna but without the shadowing the loss is more than overcome. The weight distribution is also improved, making it possible to attach to a pole behind the dish. Whichever type is used, the intention is to capture and concentrate as much RF as possible on to the antenna. A typical domestic satellite dish will give an effective gain of about 35dBd.

The antenna is connected to what is similar to the front end of a superhet receiver. It firstly connects to several amplifier

stages, then to a mixer where a local oscillator at about 9.75GHz is added, then to several IF amplifying stages. A filter between the mixer and first IF stage makes sure that only the RF minus LO product is selected. Satellite signals in the 10 to 12 GHz region are down-converted to around 750 - 1750MHz so they can be fed indoors through conventional co-axial cable. Importantly, this encompasses the 23cm band. Collectively, the electronics at the dish end is called an LNB or 'Low Noise Block converter' and it will amplify by about 60dB. We will revisit the LNB and dish later when we move on to the 10GHz band.

As you will have seen, the indoor satellite receiver box is expecting around 100dB of gain in line with its input socket and consequently isn't designed to be particularly sensitive or have a low noise figure. A DC feed to the LNB is also present on the input socket, this would normally be the supply to dish so a separate cable is not needed. Signal and power are split at the receiver input and the LNB output using an LC network.

An antenna connected directly to the input of the receiver will give good results when the signal strength is fairly high but it is absolutely essential to insert a capacitor in line with the co-axial cable centre conductor. A conventional folded dipole has a continuous conduction path up the centre wire, around the dipole and back down the co-axial braid, this will short out the DC feed and could damage the receiver circuitry. A better option is to utilise the DC feed to power a pre-amplifier stage. This not only overcomes the danger of shorting out the supply, it gives the benefit of increasing the signal sensitivity as well. The power is already present, it may as well be put to good use.

Another drawback to satellite receivers is their wide IF bandwidth. Domestic broadcasts use a relatively high modulation index and consequently occupy a chunk of spectrum about 25MHz wide. On the satellite downlink band this is not a problem, there is plenty of empty band to use but when used for amateur purposes, the bandwidth is far in excess of that needed. A wider than necessary bandwidth causes a decrease in signal to noise ratio and makes the system prone to interference from adjacent frequencies. There is no easy solution to this problem as the bandwidth is normally constrained by a SAW (Surface Acoustic Wave) filter which is not 'tweakable'. In practice, directional antennas and tuned pre-amplifiers go a long way toward keeping unwanted signals out of the bandwith and the problem is not as serious as it may first seem. Amateur broadcasts, being relatively narrow band, do however, result in less recovered video from the detector (frequency discriminator) stages. Domestic receivers expect to recover quite a lot of video voltage from the wide satellite broadcast deviation and when presented with a signal of smaller deviation, may not recover enough signal to be useable. The fix for this is simple though, add an extra video amplifier stage. A design for such an amplifier is shown in **Fig 20.4**; a components list is in **Table 20.1**. The PCB is small enough to fit upright behind the rear panels of most

Fig 20.7: The completed video amplifier

receiver boxes (see **Figs 20.5 and 20.6** in the Appendix for PCB and component layouts). **Fig 20.7** is a photograph of the completed amplifier project.

A bonus to using a domestic satellite receiver is that almost all models have an on-screen display of the receiver frequency. In most cases this takes into account that the LNB is downshifting the band by 9.75GHz but to cater for older LNB designs which used 10GHz local oscillators, there is usually an option to select either 9.75 or 10.0GHz offsets. If the latter is chosen, the frequency selected at the input socket on the receiver is exactly 10.0GHz below the frequency indicated so ignoring the first displayed digit gives the effective frequency being received. For example if the receiver indicated 11.296GHz it would actually be tuned to 1296MHz. The same applies to using an LO of 9.75GHz but a little calculation is required to add an extra 250MHz to the frequency shown.

In recent months a new source of receivers has hit the market. These are actually intended to be used with domestic 'video senders', devices that use low RF power to distribute television signals around the home. For example to allow a VCR in one room to be watched in another without running cables between them. They generally come in pairs, one receiver and one transmitter module and unfortunately rarely utilise pre and de-emphasis. This makes them work well together but incompatible with other equipment unless additional filter circuits are added to the input of the transmitter and output of the receiver. They normally use synthesized tuning which gives the option of the user selecting one of four pre-set channels. The synthesizer normally comprises a small microcontroller chip and by careful re-programming, the frequency can be shifted so it works within the permitted amateur bands.

The receiver is generally more sensitive than a satellite box but the additional work needed to ensure compatibility with other equipment can be a problem. The transmitter modules normally only produce a few milliwatts of RF power and will require several stages of amplification to be usable for longer distances. Despite their drawbacks, these modules can make the basis of a very inexpensive television transceiver and they are readily available in ranges covering both the 23cm and 13cm bands.

10GHz Equipment

Despite 10GHz being well into the microwave region, it is remarkably easy to send and receive images on this band. We return to the domestic satellite equipment mentioned earlier but utilise it in a slightly different way. This time, instead of only using the indoor unit, we also use the original dish and LNB. Satellite downlink frequencies are just above the 3cm band and with a little modification it is possible to modify an LNB so it tunes lower than originally intended and sends 3cm signals to an indoor unit instead. Although retuning is quite simple, a screw is turned to move the LO to a lower frequency, there is another obstacle to be overcome. Internally, most LNBs consist of two input stages, one with a horizontal antenna and one with a vertical antenna.

These input stages are selectively powered on and off so only the desired polarisation is picked up. The input stages are combined and fed through several more amplifiers and then through a filter to the mixer. The mixer subtracts the LO, shifting the received frequency down low enough to be carried by co-axial cable to the indoor unit. The filter is the obstacle since it is designed to block stray LO signals from reaching the amplification stages. It is designed to cut off somewhere between the bottom of the satellite broadcast band, around 10.7GHz

and the LO frequency of 10GHz or 9.75GHz. That 'somewhere' unfortunately lies perilously close to the 3cm band and depending on manufacturing tolerances, may well block or at least seriously attenuate the signals we want to receive. To use the LNB on 3cm, the filter cut-off frequency needs to be lowered a little. Most filters are fabricated as parallel copper tracks on the PCB and it is not too difficult to extend these by about one millimetre by soldering copper or tin-plate extensions to them. The new filter characteristic is not too important and it is not unknown for the filter to be removed altogether and replaced by a link.

Sometimes, the LO tuning screw may not allow enough adjustment. In these cases it may be necessary to replace the dielectric resonator element (known as "the puck") with one designed for a lower frequency. As these can be difficult to get hold of in small quantities, it is worthwhile removing the old one and raising it away from the PCB by placing a thin plastic washer underneath it. Mounting it on a pedestal like this will often shift it's frequency enough but be careful to secure it with the minimum amount of glue. A single dot of Cyanoacrylate 'Superglue' will suffice; too much glue may reduce its Q factor enough to stop the oscillator running.

Going back to the dish, offset-feed dishes are very inexpensive and already have suitable mounting arrangement to hold an LNB. They do however, look upwards rather than toward the horizon as would be required for point-to-point communication. If a line is drawn along the axis of the LNB feed horn it will touch the dish close to its centre point but approaching it from an angle well below horizontal. Now imagine a horizontal line drawn across from rim to rim of the dish, immediately in front of its centre, the direction the dish is actually looking is at the same angle as the LNB axis follows but above the second line instead of below it. For example, if the LNB axis approached the centre from 20 degrees below the centre line, it would be focussed on a signal 20 degrees above it. This would be a problem normally as the dish would have to be angled downward to look at the horizon but a simple trick can be used to overcome it: rotating the dish by 90 degrees so its LNB arm is horizontal will change the upward offset to a sideways one. Now the dish simply has to be rotated around its mounting pole so it looks as though it is aiming about 40 degrees off target. The exact angle depends upon the dish design and is best learned by experimentation. It may look strange when the dish is apparently pointing off-course but you get used to it.

Various methods of producing TV signals on 3cm have been tried with great success. Most popular is the modified Gunn Diode module transmitter. These are the units used to detect motion by reporting Doppler shift and are commonly used in automatic door opening units and security alarms. They are actually a simple transceiver, producing an LF output in which the frequency is proportional to the rate of movement nearby. They are simple in operation: a Gunn effect diode is mounted in a cavity tuned to about 10.7GHz. Inside this high-Q tuned circuit, the diode produces a carrier signal. Most of the signal leaves the cavity and exits the unit through a short waveguide, a small amount is picked up by a mixer diode which sits inside that waveguide.

Ordinarily, the diode would produce a small DC voltage by rectifying the carrier signal and indeed this is what it does if there is nothing in front of the waveguide's exit. If an object reflects any of the signal back, it enters the waveguide and also reaches the mixer diode. The direct carrier and reflected one may or may not be in phase, depending on the distance the waves have travelled to and from the source of the reflection and the mixer out-

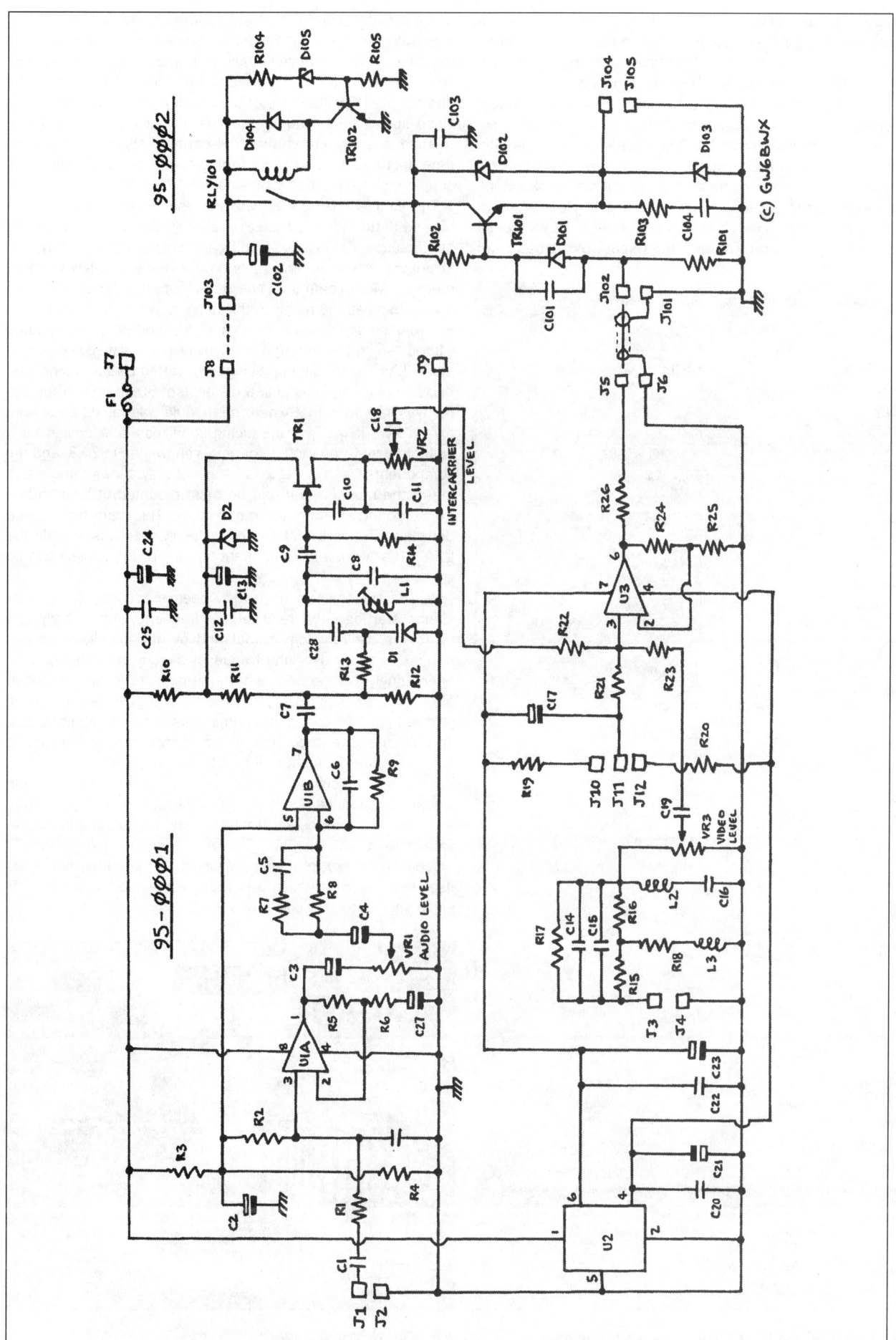

Fig 20.8: Modulator for a 3m Gunn diode. Remote tuning and a sound channel are included

put may vary from close to zero if the waves cancel up to close to twice the voltage if they are in phase and add to each other. Movement in the object causes the phase of the reflected wave to change and an alternating voltage is produced.

For TV purposes we can use another property of Gunn effect modules, that the exact frequency of oscillation is proportional to the voltage applied across the Gunn diode. A word of caution here; these are negative resistance devices, they defy Ohms law, over part of their operating range, the current they draw actually increases as the applied voltage falls. It is very easy to burn them out with excess current by applying insufficient voltage so ensure they are not run with less than about 5V across them

Part / Value	Quantity	Schematic Reference
18R	1	R18
75R	3	R15, 16, 26 (see text)
82R	1	R101 (see text)
300R	1	R17
390R	1	R10
470R	1	R103
1K	5	R23, 24, 25, 102
2K2	3	R6, 7, 22
10K	5	R3, 4, 5, 13, 21
47K	1	R2
68K	1	R8
100K	3	R11, 12, 105
330K	2	R9, 14
1K variable	3	VR1, 2, 3
22p	3	C6, 8, 18
33p	2	C10, 11
68p	1	C16 (if required)
100p	1	C28
680p	2	C15, 16
1n	4	C5, 9, 14, 101
0.1µ (100n)	9	C1, 7, 12, 19, 20, 22, 25, 103,104
2µ2	2	C4,27
10µ	3	C2, 3, 17
47µ	5	C13, 21,23, 24,102
10µH	2	L2, 3 (L2 if required)
15µH	1	L1
6V8	1	D102
8V2	2	D2, 101
9V1	2	D103, 105
1N4148	1	D104
MV1208	1	D1
BF244	1	TR1
BD131	1	TR101
BC337	1	TR102
TL072	1	U1
NMA1212S	1	U2
EL2020	1	U3
Relay	1	RLY1
Veropins	17	J1-12, J101-105

Tuning control either 10 turn or single turn but MUST be 10k value and preferably linear track. This part is not mounted on the PCBs. Use a type that suits your preferred box or enclosure. A heatsink can be fitted to TR101 as it runs quite warm. Fold a 30mm x 15mm aluminium strip at 90 degrees, half way along its length. Then drill a 3mm mounting hole in the centre of one of the 'wings', 5mm from one side.

Table 20.2: Parts list for the 3cm Gunn diode modulator

and if possible, current limit their power supply as well. Typically they work at about 7-8V and a current of around 100mA. Because the frequency changes with applied voltage, they are very easy to frequency modulate. All you need to do is apply the appropriate DC voltage to set the frequency and superimpose a video signal on it. Usually there is also a mechanical tuning method, a screw penetrating the cavity to slightly alter its volume, this acts like a coarse tuning control while changing the supply works like a fine tuning control.

The simplest TV link is two of these modules, facing each other with the video applied to one module and the mixer on the other, suitably amplified, feeding a monitor. Unfortunately, this arrangement will probably only have a range of a few hundred metres. More commonly, the mixer output is ignored and an LNB is used as the receiver, the range now increases manifoldly. From personal experience, a Gunn module sending video without any other antenna or dish, just the waveguide output alone, has been picked up using a converted LNB and produced perfect colour pictures over a distance of 30km. In fact the signal was so strong that when the module was facing away from my receiver it was still producing good results. A design for a suitable 3cm Gunn modulator unit is shown in **Fig 20.8**, and the components list is in **Fig 20.2**. This not only allows remote tuning of the module so it can be mast mounted, it also adds a 6MHz sound channel to the picture. This can be retuned between 5.5 and 6.5MHz to cater for TV receivers outside the UK. PCB artwork and component layouts for this project can be found at [3].

10GHz transmitters have also been made using multiplier chains, starting with a relatively low frequency and multiplying it up to the desired 3cm output and by utilising dielectric resonators to set an oscillator frequency directly in the band. Most interestingly, several people have managed to reverse the operation of LNBs by starting with the LO and physically reversing part of their PCB so the input amplifiers become output stages. This has the advantage the LNB is already optimally designed to fit in a standard dish and of course, no cost is involved. Typical output power from a reversed LNB is 50mW where a Gunn would only produce about one tenth of that. Options are retuning the LO to the desired frequency and modulating its supply voltage to achieve the FM signal, or utilising the original mixer to up-convert a lower frequency FM signal. The latter method lends itself to using the LNB completely in reverse, feeding a 23cm FM signal in with the DC power and getting 3cm out.

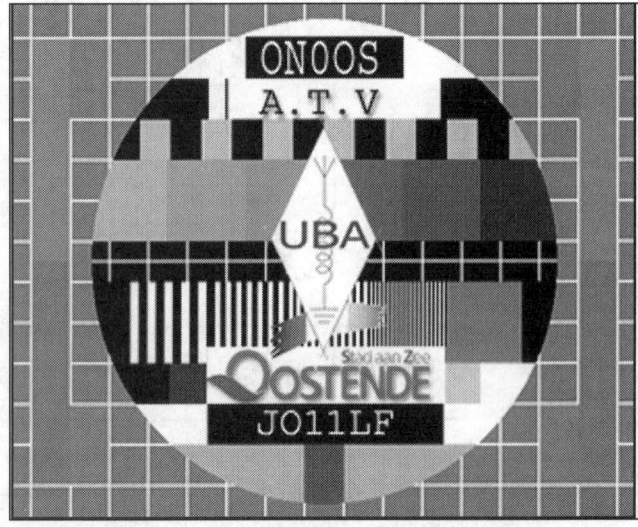

Fig 20.9: An ATV test card

The Radio Communication Handbook

OPERATING PRACTICE

Television is a far more 'friendly' mode of operation, no longer is the other person just a voice, they can now show facial expression. How many times have you passed right by someone you have spoken to on the air without realising who they were? It also allows items to be demonstrated rather than just described; that's very useful, particularly if a talk-back channel is used so the demonstration becomes interactive. When using ATV, the principles are the same as for voice communication. There are 'centre of activity' frequencies, calling frequencies and a wide coverage of repeaters. CQ calls are either made by voice or by transmitting a suitably worded caption as a picture. When live pictures are sent, be careful about their content; remember that the camera can also see what is going on in the background. It is also useful to monitor your own pictures and if you are showing a face shot, place the camera close to the monitor screen, preferably right above it. This makes it look as though you are looking straight at the person receiving the picture while you are actually monitoring your own video. It is considered impolite to be looking away from the camera while talking to someone, just as it would be in close company.

A club, affiliated to the RSGB, exists to support enthusiasts of television and imaging. It is the British Amateur Television Club (BATC), membership is open to anyone around the world. A colour magazine, *CQ-TV*, is distributed to members quarterly and the club can supply some ATV items from its Members Services Department. More details are available from [4].

Full construction details of the equipment mentioned in this chapter, including PCB designs are available at [3] which also carries other video and imaging projects.

REFERENCES

[1] MMSSTV slow scan television software can be downloaded from http://mmhamsoft.ham-radio.ch/mmsstv/index.htm

[2] DIDTRX digital slow scan software can be downloaded from PY4ZBZ's web page at: http://planeta.terra.com.br/lazer/py4zbz/

[4] Construction details of the equipment mentioned in this chapter, including PCB designs. http://www.atv-projects.com.

[3] British Amateur Television Club. http://www.batc.org.uk. Membership Secretary, The Villa, Plas Panteidal, Aberdyfi, LL35 0RF, UK.

Practical Projects

Edited by George Bown M5ACN

Packed with fifty "weekend projects" Practical Projects is a book of simple construction projects for the radio amateur and those just interested electronics. A wide variety of radio ideas are covered with everything from an 80m Transceiver, Antennas, ATUs and simple electronic keyers all included. Other simple electronic designs are such as dry battery testers, mobile microphones and various meters and monitors are also added. The book also contains a handy section on "now I've built it what shall I do with it?" questions answered. This book is excellent those just looking for interesting ideas to construct and for the newcomers to the hobby looking to expand their knowledge.
Size: 240 x 174mm, 224 pages, ISBN: 1-872309-88-7
RSGB Member's Price £11.04 plus p&p

E&OE

Radio Society of Great Britain
Lambda House, Cranborne Road, Potters Bar, Herts. EN6 3JE
Tel. 0870 904 7373 Fax. 0870 904 7374

ORDER 24 HOURS A DAY ON OUR WEBSITE
www.rsgb.org/shop

21 Satellites and space

Radio Amateurs have always been keen to exploit new means of communication and within only three years of the launch of Sputnik 1 a group of radio amateurs in the USA had designed and built the first Orbiting Satellite Carrying Amateur Radio (*Oscar*). The first non-military, non-governmental payload to go into space, Oscar 1 was launched on 12 December, 1961 from Vandenberg air force base in California. Carrying a small CW beacon transmitter (**Fig 21.1**), it sent "HI HI" and was copied by radio amateur around the world.

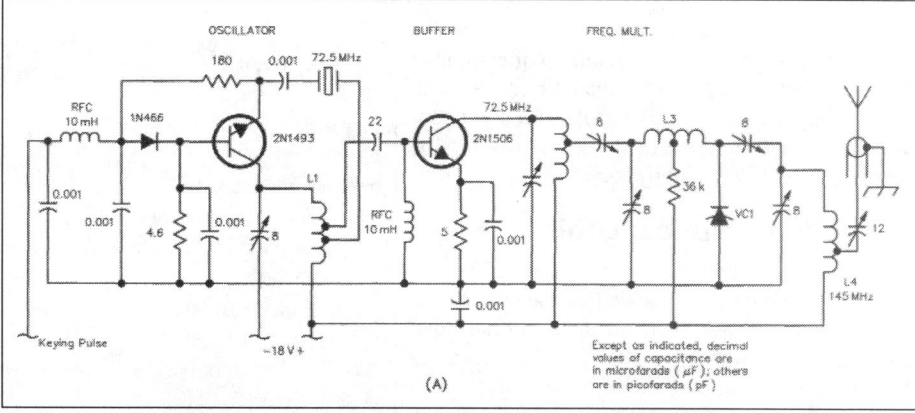

Fig 21.1: Oscar 1 beacon transmitter that delivered 140mW at 145MHz

With over 40 years of experience in designing and building satellites the amateur radio community has pioneered many new techniques and contributed greatly to the science and engineering of spacecraft and space communications. Today we have a whole range of very capable satellites for CW, voice and data communications. The manned space missions, the Shuttle, and the International Space Station (ISS) also play their part. Amateur radio satellite groups are active in many countries including AMSAT-UK (United Kingdom) [1], AMSAT-NA (USA) [2] and AMSAT-DL (Germany) [3]. AMSAT groups provide a structure for planning, designing and building satellites and the programs are always ambitious. AMSAT-DL is looking beyond Earth orbit with their plans for a satellite to go to Mars.

THE SATELLITE SERVICE

The Amateur Satellite Service is a separate user service (from *the Amateur Service*) under the terms of international licensing regulations. Fortunately this does not mean that a separate license is required. All licensed radio amateurs are welcome to use the satellites according their own licence conditions.

Amateur radio satellites are costly to build, launch and operate. Just about all of this funding comes from the amateur radio community, so if you do use the satellites, please consider joining AMSAT-UK or one of the other AMSAT groups and help to fund new satellites for us all to enjoy.

SATELLITE ORBITS

Fig 21.2 shows some of the fundamental concepts. Footprint is the term used to describe the area of the Earth which is in radio reception range of the satellite.

In amateur radio terms any stations within the footprint can communicate via the satellite. The area of the footprint is determined by the height of the satellite with the best DX contacts being available between stations on the extreme edges of the footprint.

Geostationary

Geostationary, or more correctly geosynchronous, satellites are located at around 40,000km into space. Orbiting at the same rate as the Earth, they appear to the Earth-bound observer to be stationary. At such a great distance they have a large footprint.

They are familiar as TV and communications satellites. It takes a lot of launch energy, and therefore cost, to get a satellite into this orbit and to date there are no geosynchronous amateur radio satellites. GEO does offer the possibility of fixed antennas and communications 24 hours a day, seven days a week.

LEO, Low Earth Orbit

This is where we find most of the amateur radio satellites. There is no fixed definition of LEO, but typically the satellites are in orbits of about 800km altitude. LEO is very useful for weather satellites, science satellites, environmental monitoring satellites, and amateur radio. If the orbit is chosen to go over the poles then the satellite will pass over all parts of the globe several times each day. Messages can be collected in one part of

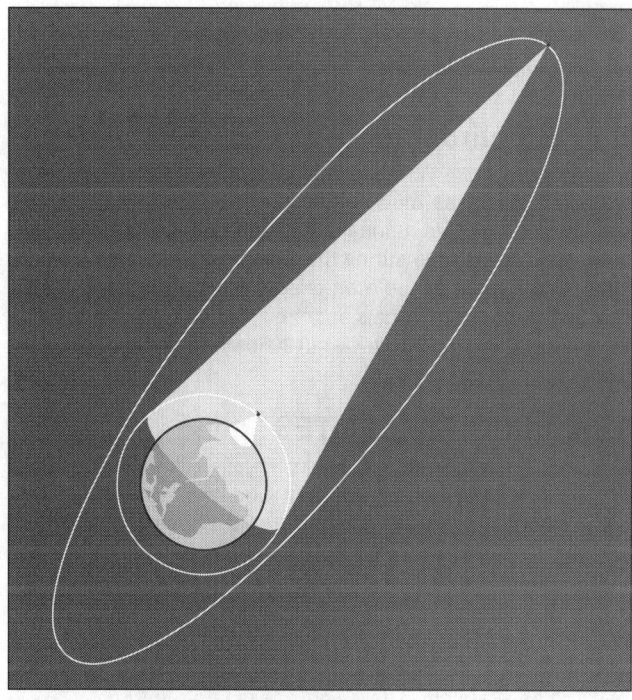

Fig 21.2: Footprint for high and low altitude orbits: LEO satellite at 1,000km; HEO satellite at 60,000km

the world and then re broadcast over another, this is known as *store and forward*. Similarly, the satellite can collect scientific data throughout an orbit then broadcast it as it passes over the control ground station; described as *Whole Orbit Data* it can be of considerable scientific value.

LEO orbits are the easiest to achieve in terms of launch energy. From a communications standpoint they are limited. At these low altitudes the satellite will be in range of a ground station for only a maximum of 15 minutes or so. This makes for short contacts, but a lot of fun with the possibility for UK based stations to work many countries.

HEO, High Elliptical Orbit

This is typically around 300km at the point of closest approach to the Earth, a position on the orbit called *perigee*, to 20,000km or more at its furthest distance, *apogee*. It is easy to remember: furthest **a**way at its **a**pogee.

This orbit is very useful for communications as it has a large footprint for most of its orbit (**Fig 21.2**). From the ground station's point of view, as the satellite approaches and leaves Apogee it appears to 'hang in the sky', simplifying antenna tracking.

A specific variant of the HEO is the Molynia orbit. By positioning the ellipse of the orbit in a particular way, the ground track repeats each day. First used by the Russians in April 1965 for a communication satellite, the orbit had a perigee of 309km and an apogee of 24470km.

This gave the benefit of long periods of reliable communications over a large part of the Northern Hemisphere. The special characteristics of the Molnya orbit have been exploited for amateur radio HEOs, OSCARs 10, 13 and 40. Quite large payloads can be put into HEO with far less energy than needed to reach GEO.

The International Space Station

Orbiting at around 350km altitude the ISS is a particularly low LEO. The orbit is frequently boosted using the thrust from the regular supply vessels. Without this intervention, the ISS would lose altitude and re-enter the atmosphere.

Sun Synchronous

A particular type of LEO, this orbit is chosen to keep the spacecraft in sunlight as much as possible. Space is an extreme environment; power management and thermal stability are both vital factors in ensuring the long life of the satellites payload. A sun synchronous orbit ensures that batteries are kept charged by the solar panels and the spacecraft warmed by the Sun. Imparting a spin to the craft helps with temperature stability.

TRACKING SATELLITES

If you don't know where the satellite is you can't work through it. In the pioneer days, radio amateurs produced ingenious graphical methods and charts such as the 'Oscarlocator' to find a satellite's position. Today we use computers.

The satellite travels around its orbit independently of the earth. Given the co-ordinates of our ground station, some facts about the geometry of the orbit, and a reference point from which to start the calculation, we can predict the satellite's position several hours, days and weeks into the future.

The basic data needed for tracking comes in two parts. The first is a set of parameters that define the shape and size of

```
FO-29
1 24278U 96046B   04364.77938403 -.00000040  00000-0 -18814-5 0  8617
2 24278  98.5724  71.0280 0349969 255.1477 101.0648 13.52906415413283
```

Table 21.1: A set of keplerian Elements for Fuji Oscar 29, downloaded from the Internet

Satellite: FO-29	Satellite name
Catalog number: 24278	Satellite number
Epoch time: 04364.77938403	Year day and time of the observation
Element set: 861	Sequential number
Inclination: 98.5724	0 = equatorial, 90 = polar
RA of node: 71.0280	Astronomical term
Eccentricity: 0.0349969	0 = perfect circle 1= infinite ellipse
Argument of perigee: 255.1477	Astronomical term
Mean anomaly: 101.0648	Satellite's position around its orbit 0-256
Mean motion: 13.52906415	Number of orbits per day
Decay rate: 0.00000040	Drag factor
Epoch rev: 41328	Orbit number

Table 21.2: Keplerian Elements in the AMSAT format

the orbit, and its location in space relative to the Earth. The second is a recent observation which tells us where the satellite was on this orbital track at a precise moment in time.

This set of numbers used for satellite tracking are known as *Keplerian Elements* and are named after the 17th century astronomer Johannes Kepler, who worked out the equations that described planetary motion. Popularly called 'Keps', they originate from NORAD (North American Aerospace Defense Command) in Cheyenne Mountain. The satellite positions are determined by radar and optical methods and a set of Keps are constructed to give the best approximations to the observations. Not all sets of observations are equally good so the careful satelliter will take care to keep the previous element set, just in case the new set proves to be less reliable.

A set of Keps downloaded from the internet will typically be a text file made up of entries like the one shown in **Table 21.1**.

The last digit in each row, 7 and 3 in this case, are not part of the satellite data. Each is the modulo 10 check digit for that line. The check digit ensures that if there have been errors in transmission of any of data your tracking software will report it.

Known as NASA two line elements they are ideal for computers but not very human friendly The AMSAT format, shown in **Table 21.2**, is easier to work with, especially if you want to read them over the air. Some of the figures relate to astronomers' terms which describe the orientation of the orbital plane in space.

There is no room here to deal with orbital mechanics, so here are a few practical tips when using elements and tracking software.

• Use recent element sets, forward predictions become less valid the further you go from the reference data. Satellites in LEO are affected by atmospheric density; even though the atmosphere is incredibly thin the speed of the spacecraft makes atmospheric drag a factor.

• Take care with the ISS. Often the orbit is boosted, so always use the most up to date element set you can get. With the exception of the ISS you don't need to renew your element sets daily; weekly is OK.

- Keep accurate time in the shack. Radio synchronized clocks are cheap now and ideal. Also there are plenty of time services on the net. For most applications, accuracy to within a couple of seconds is fine.

- Make sure your computer's internal clock is set to match the shack clock. It is not unknown to be waiting for a predicted pass of a satellite at the wrong time, through not checking the computer clock.

- If you edit any downloaded element sets make sure your word processor can output a text file, otherwise it may add control characters which your tracking software won't recognise. If in doubt, use something simple like *Notepad*.

Tracking Software

There is plenty to choose from, running under Windows, DOS and Linux. Some is free to download from the Internet. If you purchase software from AMSAT groups, part of the proceeds go towards providing future satellites.

One of my favourites is *Instantrak* donated to AMSAT by Franklin Antonio, N6NKF. Running under DOS, the code was written to be very fast because old PCs were much slower. Consequently it will run perfectly on old laptop computers and discarded office PCs, leaving the main shack computer free for other duties. *Instantrak* will run under Windows, including XP Home and Pro.

When choosing software, look for the ability to feed output to rotator interfaces. You may want to automate your station and have the computer tracking software provide the signals to drive your rotators. The other feature to look for is Doppler Tracking. Modern transceivers have CAT ports so that they can be controlled from computers. Linking your radio to your tracking software lets the computer tune the radio for you, adjusting the frequencies to compensate for Doppler shift.

Here are a few tips in setting up any tracking software.

- Time: Check that your PC has accurate time. Look for any time zone setting requirements in the setup notes. Most trackers will default to an off set from UTC.

  Decide how to deal with time zones in your shack. I set my tracking computer clock to UTC since this is the time reference used in all satellite work. In DOS the command line to put into your AUTOEXEC.BAT file is SET TZ=0. Make a backup copy of the file first.

- Location: You will need to specify your location using Latitude and Longitude. The software may require this in degrees and minutes or decimal degrees. Carefully check to see if a plus or minus sign is needed to designate east or West of Greenwich, there is no agreed convention for this.

- Keplers: Don't expect the software to be shipped with up to date elements pre loaded.

- For useable predictions you will need to update the Keps file. In modern trackers this can often be done automatically by the software over the Internet. There should also be a facility to read a text file from a disk. Recommended sources of Keplerian elements are Space-Track.com, Celestrak.com and of course AMSAT.

The web site www.heavens-above.com is very useful if you need some satellite pass times and don't have tracking software. For example, If you are keen to listen for a few satellites before getting involved with your own software, or want predictions for visible satellites or the ISS on a clear evening. It also gives the opportunity to get some comparison pass times.

It can be very frustrating listening for a satellite that does not turn up because your tracker is not properly set up. If you do make comparisons with heavens-above, or with other tracking software don't expect a 100% match with predictions. The predicted time for the satellite depends on the the algorithms used in the calculations and the reference orbit data.

USING SATELLITES FOR COMMUNICATIONS

Frequencies

Transmitting to the satellite is known as the *Uplink*, and the signal we receive from the satellite is known as the *Downlink*. In satellite work, the uplink and downlink are in two different bands. There are several reasons for this, but one of the simplest is that if using the same band a very large cavity filter would be needed on the spacecraft to give the required isolation between its transmitter and receiver. The most used amateur bands are 2m and 70cm with 2.4GHz (S band) coming along rapidly as a good choice for the downlink. This is part of a general migration to higher bands for satellite work, moving away from crowded frequencies, and to much smaller antenna arrays, both on the spacecraft and at ground stations. Uplinks on 1.2GHz are also in use, particularly on SO-51.

The frequency allocations for satellite working need to be taken into consideration. The L band allocation for satellite working is 1260-1270MHz, not the same as the terrestrial amateur allocation. At the ground station, extensive use is made of up-converters and down-converters, and with the availability of new PA designs and RF efficient devices it's often practical to mount up-converters near to the antenna to reduce feeder losses.

Doppler Shift

Doppler shift is the observed change in frequency due to the relative motion of the object and the observer. The usual example given is the change in note of the siren as a speeding police car approaches and recedes. The amount of Doppler shift is related to the speeds involved and the frequency.

An observer that is at rest with respect to a transmitter will measure a frequency f_0 while an observer who is moving with respect to the transmitter will measure a different frequency $f*$. The relation is given by:

$$f* = f_0 - \frac{v_r}{c} f_0$$

where

f_0 = frequency as measured by an observer at rest with respect to the source (source frequency)

$f*$ = frequency as measured by an observer who is moving with respect to the source (apparent frequency)

v_r = relatively velocity of observer with respect to source

c = speed of light = 3×10^8 m/s

Beacon Frequency	* Earth Rotation	ISS 370 km	AO-16 800 km	HEO sat. Perigee 2545 km	HEO Sat. Apogee 36365 km
29.5MHz	0.045	0.76	0.7	0.52	0.09
146MHz	0.226	3.76	3.45	2.56	0.45
435MHz	0.67	11.2	10.3	7.6	1.33
1.27GHz	1.97	30.7	30.1	22.2	3.9
2.4GHz	3.72	61.8	56.7	41.9	7.4
10.5GHz	15.5	269.5	247.4	182.8	31.5
* Contribution due to the Earth's rotation					

Table 21.3: Calculated maximum doppler shift in kHz for a variety of frequencies and satellite altitudes

This equation is often written:

$$\text{Doppler shift: } \Delta f = f^* - f_0 = -\frac{v_r}{c} f_0$$

Note that v_r is negative when a spacecraft is approaching. The apparent frequency will therefore be higher than the source frequency. When a spacecraft is receding, v_r is positive and the apparent frequency is lower than the source frequency.

The practical consequences during communications via a rapidly moving satellite are that the satellite receives our uplink signal on a shifting frequency; we receive the satellites downlink on a shifting frequency. It was expected that SSB communications would be very difficult under these conditions but in practice, small compensation adjustments during the QSO keeps both stations on frequency. Using computer control of the transceiver automates the process as the software calculates the Doppler shift and updates the radio every second or so.

An interesting fact is that the satellite's speed and direction will be different for each station in the footprint. Consequently, the Doppler shift experienced by each of the partners in the QSO will be different.

The amount of Doppler shift encountered by a satellite user is a function of the frequencies involved, and the velocity of the satellite. **Table 21.3** is based on a mathematical calculation, and gives an indication of the maximum shift for a variety of frequencies and satellite heights. The height or altitude of the satellite is a factor simply because to stay aloft satellites at low altitudes must travel faster than satellites at high altitudes.

Antennas

The choice of antennas will be determined by the type of operation envisaged; fixed station, portable or hand held. It would be very easy to get bogged down here in the science and engineering aspects of the perfect antenna array for a satellite ground station. However, as radio amateurs we generally have to take a practical approach and work within budgetary and other constraints. Fortunately, its quite possible to 'home brew' effective antennas for space communications. It is also possible to purchase good equipment commercially.

The newer generation of FM satellites can be worked with a hand-held radio with 2m and 70cm with 5 watts. Some skill is needed though, and results are much better with a small dual band Yagi which can be hand held. Using such equipment, contacts from the UK into North America happen regularly. The

Fig 21.3: Andy Thomas, operating as YL/G0SFJ/P, and working through SO-50 from Riga. He used an Arrow antenna and 5W from a Kenwood THD-7E

Arrow antenna, available from AMSAT-UK [1] is ideal for this sort of work as can be seen in **Fig 21.3**. Designed to dismantle easily and pack into a small bag, it is very light and can easily be used on holiday.

For fixed stations, probably an early consideration is what can you can do with existing equipment. A cross boom can easily be fixed up to accommodate the 2m and 70cm antennas. If using crossed Yagis for circular polarisation, mount them in the X configuration to reduce interaction with the cross boom (see the VHF/UHF Antennas chapter for more on circular polarisation and crossed Yagis). This can be either aluminum or fibreglass. Fibreglass is often recommended, but experience has shown that when the antennas are mounted as suggested there is no discernable interaction with a metal boom. The professional satellite engineer puts circular polarisation antennas at both ends of the link, satellite and ground station. This reduces the effects of fading, and attenuates reflections which tend to be of linear polarisation. In practice, the signals arriving from space are subject to large polarisation changes as they pass through the ionized layers of the atmosphere. There is a wide variation in antenna configurations used by satelliters, including many with simple linear polarisation. Don't let the additional complexity of producing circular polarisation on 2m and 70cm put you off; give linear polarisation a try first.

At 2.4GHz and 1.2 GHz you will probably be using a small dish antenna. When the signal is reflected off the dish the polarisation is reversed. A Right Hand Circular Signal arriving at your dish needs to see a Left Hand Circular Polarised feed. There are plenty of designs around for dish feeds, with full constructional details. Those from G6LVB and G3RUH for example.

Rotators

Antenna rotators for *azimuth* and *elevation* are costly. That level of purchase is not recommend until you are sure of a long term interest in satellite working. Instead, use an azimuth rotator and angle your antennas up by about 20 degrees. Unless you are a very serious DXer, the effect on terrestrial work will not be noticeable and you will be able to work through plenty of satellites when at low elevation. (If you recall the footprint discussion, these are the times for the best DX).

Rotator Controllers

Most proprietary rotators can be interfaced to the PC to give automatic satellite tracking. Howard Long, G6LVB, has developed a low cost design which is available in kit form from Amsat-UK [1].

Pre-amplifiers, Feeders and Connectors

Use the best cable you can afford, and the best connectors. A well constructed set up will last 10 years or more so it's worth the investment. Cable such as RG58 is pretty useless at the frequencies in question and even RG213 is quite lossy at 70cm. Study the cable specifications at various frequencies, and the cost of the cable before deciding what to purchase. Westflex 103 is very popular with satelliters being a good compromise between cost and performance.

The centre core is too big for standard N connectors, buy the special Ns or file down the centre core. If you decide to file, don't allow the filings to fall inside the cellular structure of the cable. Use good quality N connectors, they will last for years. For receive only applications, satellite TV cable gives good results at low cost.

Mast head pre-amplifiers may be needed if you have a long cable run. High gain pre-amps designed with the terrestrial DXer in mind sometimes give problems with desensitisation ("de-

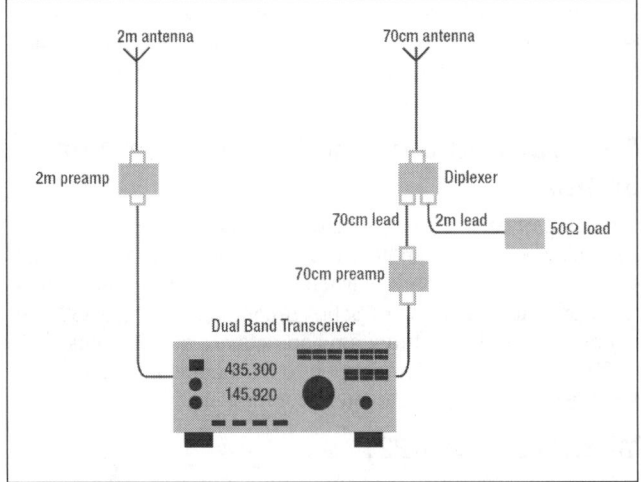

Fig 21.4: Using a commercial diplexer to reduce receiver desensitisation on mode V/U

sense") which is caused by the third harmonic from the 2m uplink signal. Changing the layout of the cables and antennas can sometime reduce the problem but a better solution is to use a stub filter as shown in **Fig 21.4**.

TRANSCEIVERS FOR SATELLITE WORKING

Hand Helds

A Dual band, 2m/70cm radio with about 5W output will be suitable for working the FM LEO satellites. If you are interested in digipeating APRS (Amateur Position Reporting System) via satellite, a built in TNC is useful.

Full duplex is recommended so that you can hear your own downlink from the satellite whilst transmitting. Doppler compensation is dealt with by using the radio's programmable memories to set the transmit/receive frequency pairs. The operating technique is to receive about 10kHz HF of the published frequency at the start of the pass, and click through the pre-set memories as the pass progresses (**Table 21.4**). Always have the squelch fully open; not doing so is often the cause of frustration for new operators. A recording device is a good idea to help with logging.

Fixed and Portable Stations

It's not essential to buy the latest fully featured multi band transceiver for satellite working. Many stations use two separate radios; one for transmitting and one for receiving, this gives full duplex working and in some respects can be easier than learning to operate a multi band rig with lots of features. A secondhand transmitter or receiver might be all you need to get you going.

Modern multi-band transceivers come with many operating features for the satelliter. Radios with a VHF/UHF heritage seem

	Rx	Tx
AOS	435.310	145.910
	435.305	145.915
TCA	435.300	145.920
	435.295	145.925
LOS	435.290	145.930

AOS =*Acquisition of signal, the start of the pass*
TCA = *Time of closest Approach*
LOS = *Loss of signal, the end of the pass*

Table 21.4: Frequency memory settings for working SO-51 on a hand-held radio

to offer the best range of added satellite features but study the specifications and see what other people are using successfully on satellites.

Whichever route you choose here are a few pointers:

- Transmit power should be variable, ideally from full to a few watts. With modern high power radios the full output is seldom required for satellite working.
- If buying a combined transceiver, the 2m and 70cm sections should each be separately controllable during a contact.
- Computer control or a CAT port will be useful, if not when you first start then a bit later on. Most modern radios only need an RS232 cable to link the radio to the PC. Older radios may need an interface. Check that these are still available or that you can home brew one.
- Packet ports are a useful feature for 9600 and 1200 baud data applications, and are essential if your main interest is in digital satellites. They are also useful if you want to capture telemetry, as 9600 is becoming the preferred data rate on newer satellites. If you have an older radio you may need to make some internal modifications to successfully use data modes.
- Built in doppler tracking is a useful feature if not using external computer control. When using SSB, this synchronizes your uplink and downlink frequencies. Then, as you move up and down the band the rig adjusts the transmit frequency. Only minor re-tuning will be needed to be on frequency.

SATELLITE SUMMARY

The information for this section came from a variety of sources, but principally from AMSAT [2]. It was accurate at the time of writing, but satellites have a finite life, so for the most up to date information check the amsat.org web site.

AMSAT Oscar 7 (AO-7)

Launched in 1974, AO-7 suffered battery failure in 1981. In 2002 its shorted battery cells became open circuit enabling the spacecraft to run directly off its solar panels. Semi-operational, this satellite can still give some good contacts when it is in sunlight.

UoSAT Oscar 11 (UO-11)

A University of Surrey (SSTL) built satellite, launched in 1984, Oscar 11 is now operating in default mode. It continues to send ASCII telemetry data for 10 days with a 10 day break. Operation can be terminated at any time by the ground controllers. The 'mode S' beacon is on continuously transmitting on 2401.5MHz with an unmodulated carrier. Beacon reception reports are welcomed by Clive Wallis, G3CWV [4].

Radio Sport (RS-15)

One of the last of the successful RS series of Russian amateur radio satellites, RS-15 is only semi-operational with its CW beacon occasionally heard. Some contacts have been made via RS-15.

AMSAT Oscar 16

Launched in July 1991, Oscar 16 was a major store and forward digital satellite. It now operates only in digipeater mode.

LUSAT Oscar 19 (LO-19)

Launched in 1991, the BBS is non operational. The CW telemetry beacon is fully operational producing strong signals.

UoSAT 5 (UO-22)

Launched in July 1991, and built by the University of Surrey, UO-22 was a major digital satellite handling large volumes of traffic at 9600 baud. It is currently semi operational whilst SSTL are trouble shooting. Operating only when in full sunlight, operations could be terminated by the ground controllers at any time.

AMRAD Oscar 27 (AO-27)

AO-27 was launched September 1993, as part of a commercial satellite. With limited battery capability, AO-27 switches to analogue operations six minutes after coming into sunlight. After six minutes of analogue there is one minute of digital telemetry. The operating schedule changes frequently. Latest information from the control operator, N3UC, can be found at [5].

Fuji Oscar 29 (FO-29)

Launched in August 1996 on a Japanese H2 launcher from Tanegashima Space Centre Japan, FO-29 failed briefly in June 2003, probably due to a major solar flare. Now operational again, FO-29 is a popular SSB/CW satellite. From time to time the transponder is switched off and an automated voice message is broadcast so that school children in Japan can hear the message from space as part of their science course. The message generally starts "This is Jas two.......", followed by a bird call.

Gurwin TechSat1b (GO-32)

TechSat1b was launched in July 1998. Users may be able to use the BBS. Check the GO-32 web site and satellite beacons for updates [6].

AMSAT Oscar 40 (AO-40)

After providing several months of excellent contacts and experiments, since its launch in November 2000, AO-40 suffered a catastrophic failure of its onboard systems. Recovery efforts continue but it is unlikely to return to service. See Amsat bulletins for the latest information [7].

Saudi Oscar SaudiSat 1a (SO-41)

Launched in September 2000, SO-41 is semi operational, turned on by an internal timer programmed to activate the satellite over land masses. Sporadic operation.

PCSAT Navy Oscar 44 (NO-44)

Launched September 2001, NO-44 is currently semi operational because the battery is not maintaining sufficient charge for operations in eclipse. Used extensively for 1200 baud APRS digipeating from handhelds and mobile stations, NO-44 could return to full service when the eclipse period decreases.

Tiung SAT (MO-46)

Malaya's first micro-satellite was built in co-operation with SSTL and launched in September 2000. It carries commercial land and water imaging as well as FM, FSK amateur radio data communications using a 38k4 baud downlink.

Saudi Oscar SaudiSat 1c (SO-50)

Launched in December 2002, SO-50 requires a 74.4Hz CTCSS tone to turn on the transponder. This activates a 10 minute timer. All transmissions then need a 67.0Hz CTCSS tone to access the transponder. Another 74.4Hz tone will reset the timer for another 10 minute period.

AMSAT Echo (AO-51)

Launched June 2004, AO-51 has FM voice, PSK 31 and data modes. This is a new generation microsat with many advanced systems on board. Wednesday is experimenter's day when different configurations of uplink and downlink frequencies are enabled. Visit the web site for up to date operating information [8].

Amateur Radio on the International Space Station

The ISS carries a variety of voice and packet equipment and SSTV may be added. Amateur radio is an approved recreational activity for the crew, many of whom become licensed radio amateurs prior to the mission. Schools contacts with the ISS are arranged via ARISS but random contacts are always a possibility. Check out the crew's work and rest schedule at [9]. Visit the ARISS web site [10] for the latest information.

Mozhayets (RS-22)

This is a training satellite, built by students at a military academy in St Petersburg. The CW beacon sends telemetry at around 5WPM with a 10 second break between transmissions. Data blocks start with "rs22" This satellite is sometimes wrongly attributed in sets of Keplerian elements. The NASA catalogue number corresponding with RS-22 is 27939.

SSETI EXPRESS

A European Space Agency (ESA) satellite built by engineering students, it is expected to be launched in 2005. Several AMSAT groups, including AMSAT-UK have assisted students with the design and construction of the amateur radio payload. It will carry an FM transponder and packet radio. There will be downlinks on 2.4GHz and 70cm with data rates of 38k4 and 9k6 respectively. ESA will be offering a prize for the station that captures and e-mails them most telemetry.

Hamsat VUSAT (VO-52)

Hamsat, built by Amsat-India and ISRO, was launched in early 2005. It carries beacons in the 2m band and two 1W analogue transponders for CW and SSB.

Amsat Eagle

A future launch, this is a major International AMSAT project led by AMSAT-NA. In a high elliptical orbit it will offer long operating times and a large footprint. Visit [2] for latest information.

ESA Sputnik 50

In 2007, to celebrate the 50th anniversary of the launch of Sputnik 1, ESA plans to launch 50 student/university built satellites.

Cubesat

An initiative for Universities, this project provides launch opportunities for tiny science and engineering satellites, measuring 100mm cube and weighing up to a kilo. Several universities have flown cubesats and their telemetry and data can be copied on amateur frequencies. Put "cubesats" into an internet search engine or visit [2] for up to date information.

P3E

This is an AMSAT-DL led project to build an HEO satellite similar to the very successful AO-13. For more details, visit the AMSAT-DL web site [3] which is available in English.

P5A

An ambitious project from AMSAT-DL [3] to take amateur radio to Mars.

Weather Satellites

The NOAA polar orbiting weather satellites transmit in the 137MHz band. They are easily copied on amateur radio equipment. Visit the GEO and RIG web sites [11, 12] for more information.

RADIO ASTRONOMY

The well equipped amateur radio station already has many of the facilities needed for radio astronomy. The shack PC, soundcard, and software such as *Sky-Pipe* gives access to DSP and data recording techniques that at one time only existed in University departments. Visit the JOVE website [13] for more information.

RECOMMENDED READING

A Guide to Oscar Operating, edited Richard Limebear, G3RWL

Available from AMSAT-UK [1], this comprehensive guide is kept at a low price to assist beginners. It contains much valuable information based on years of practical satellite operating.

The Radio Amateur's Satellite Handbook, Martin Davidoff, K2UBC (ARRL)

Essential material for beginners and for the experienced operator. The author, an experienced satelliter, covers the whole subject in a practical and accessible way from antennas to orbital mechanics. The mathematics are well explained with worked examples. A book to dip into on a regular basis. Available in the UK from the RSGB [14].

WEB RESOURCES

A major source of information is the AMSAT-UK web site [1], and also the AMSAT NA site [2]. All of the topics covered briefly in this chapter are fully explained. The AMSAT-NA web site has a section entitled: 'New to Satellites' which is a good place to start.

Also, try putting a satellite's name into an Internet search engine. This will in many cases take you to radio amateurs' own web sites where you can find operating hints and tips from someone who regularly uses a particular satellite.

MOONBOUNCE COMMUNICATIONS

Moonbounce or EME (Earth-Moon-Earth) communications presents one of the most significant challenges in amateur radio. Stations who wish to communicate simply point their antennas at the Moon which is then used as a passive reflector. After a number of tests by the US military and others in the 'forties it was fairly soon realised that EME propagation was within the reach of amateur stations. The first two-way amateur contact took place in 1960 between W6HB and W1BU on 1296MHz and has now been followed by others on every amateur band from 28MHz to 10GHz.

From the basic radar equations and the knowledge of the reflectivity of the Moon (about 6%) it is quite easy to calculate the propagation loss which ranges from 242dB at 50MHz to 276dB at 2320MHz. This on its own would call for exceptional station performance but, with the addition of somewhat unpredictable propagation through the Earth's ionosphere, EME communications really does require the ultimate from an amateur station. It not only places strenuous demands upon the station but also requires excellence in weak-signal operating. Although SSB operation is sometimes possible, nearly all EME operation

is on CW so be prepared for long periods of 'rushing' and 'ringing' in the headphones.

In line with terrestrial activity the majority of EME operation is on the 144MHz band closely followed by 432MHz. 1296MHz is popular among those already active on one of the lower bands, leaving the other bands with very sparse activity - QSOs on the microwave bands occur only as a result of prearranged schedules. Despite all of these difficulties there are currently several hundred amateurs across the world who are regularly active on this mode with a handful or so at the forefront having attained DXCC via EME.

Propagation

The principle of EME communications is simple - both stations must be able to see the Moon, they point their antennas in the right direction and communications should be possible. In practice it is not quite that simple as there are a large number of other factors which need to be taken into consideration.

Celestial objects and noise

Although the Moon is about a quarter of a million miles away from the Earth it subtends an arc of about one-half degree to a terrestrial observer. On all but the higher microwave bands this is significantly less than the beamwidth of most amateur antennas used for EME. This means that when the antenna is directed towards the Moon it will be significantly illuminated by the cosmic background. This background is not quiet - it contributes to the overall received noise power and indeed there are a number of strong sources especially towards the centre of our own galaxy. These sources are so strong that for several days each month EME operation is ruled out by high galactic noise. Likewise there are days each month when the Sun is in a similar direction to the Moon which again raises the received noise floor and drowns out all amateur signals.

The Moon as a reflector

Another source of loss comes as a result of the Moon being an imperfect reflector and the relative motions between the Earth and the Moon called *librations*. In simple terms, because the surface of the Moon is rough the reflected wave will consist of a large number of small reflections with differing phases. The signal observed back on Earth will be the sum of these reflections which is somewhat less than if the Moon were a perfect reflector. As the Moon and the Earth are moving relative to each other so the incident wavefront 'moves' across the surface of the Moon. The result is that the reflected signal becomes the sum of a large number of varying multiple reflections changing in amplitude and phase from moment to moment. *Libration fading* is the term used to describe this complex effect which manifests itself as rapid fluttering with deep fades and occasional peaks.

The EME path

First let's consider how much signal gets lost on the trip to the Moon and back. The basic path loss can be calculated from the radar equation and the average lunar reflectivity, approximately 7%. A small complication is that the Moon has a slightly elliptical orbit but this only results in a 2dB additional loss at apogee. **Table 21.5** shows the EME path loss for the most popular bands when the Moon is at its at closest approach, ie at perigee.

Spatial offset

Most amateur VHF and UHF stations use linearly polarised antennas, normally horizontal for DX working. Even if all operators were

Frequency (MHz)	50	144	432	1296	2320
Loss (dB)	244	252	261	271	276

Table 21.5: EME path loss when Moon is at perigee

using horizontal polarisation the curvature of the Earth's surface introduces a problem. When viewed from the Moon, signals from stations from across the Earth's globe will arrive with differing linear polarisations. Therefore the signal from one part of the Earth's surface reflected from the Moon will arrive at another spot on the Earth with a different linear polarisation. This would suggest that for successful EME communications antenna systems would need fully rotatable polarisation. Fortunately ionospheric effects cause rotation of polarisation which makes communications between amateurs with fixed polarisation arrays possible.

Ionospheric effects

The biggest source of signal variation on the EME path at VHF comes from the ionosphere. Ionospheric absorption does occur but its effects are only significant (>0.5dB) at low antenna elevations during daylight hours at 50MHz and below. Although strong absorption at frequencies up to 432MHz and polarisation mixing is often reported during ionospheric storms, there are many reports of strange signal enhancements. What is sure is that when the ionosphere is disturbed EME conditions can never be described as 'normal'.

Faraday rotation

By far the greatest problem and also the most unpredictable effect on the whole EME path is that of *Faraday rotation*. At radio wavelengths the ionosphere is *birefringent*, that is to say it has multiple refractive indices. These occur as a result of the effect of the Earth's magnetic field upon ionised electrons in the ionosphere.

These refractive indices are normally calculated for propagation directions longitudinal and transverse to the Earth's magnetic field. Faraday rotation occurs as a result of radio waves passing through this birefringent layer in the ionosphere (remember for EME you get two passes, one in each direction). The refractive indices are not fixed but vary directly with the solar input to the ionosphere which gives rise to the great variability. Hence during the daytime the Faraday rotation is far greater than at night - typical values of Faraday rotation for a single pass through the ionosphere are given in **Fig 21.5**.

One of the important things to remember is that when the signals reflected back from the Moon pass back through the ionosphere (even through the same volume such as when listening to one's own echoes) the Faraday rotation doesn't 'unwind' the rotation. Instead, it adds onto the twist imparted upon the signal on the way out. It can be readily visualised that it only takes small variations in ionospheric conditions, especially when one ionospheric path is in daylight, to cause significant variation in the reception of EME signals.

With linearly polarised antennas a 90° shift will cause complete extinction of an otherwise strong signal. The overall effect is a near continuous 'rolling' of signal polarisation on 144MHz, slow variations on 432MHz and near stability on 1296MHz and above. To compensate for the unpredictable nature of Faraday rotation many operators are tending to use antennas with switchable and rotatable receive polarisation.

EME Station Equipment

Frequency stability & accuracy

Although contacts have been made with relatively simple equipment the basic requirement is for a transceiver or a separate receiver and transmitter with excellent frequency stability and readout.

The main receiver should have good CW capabilities, preferably with a range of narrow IF filters and a well-calibrated IRT. As the majority of EME operators use a quality HF transceiver with VHF/UHF external transverters these too must have good frequency stability. It can be really annoying if the station which you are listening to fades down in QSB only to reappear 15 minutes later outside your receiver passband. Likewise you might not be able to complete a contact if you drift by only a few hundred hertz and your QSO partner fails to re-tune his receiver.

Sorting out drift in a transverter can be a lot more difficult especially with kits and small commercial units - some can take as long as an hour before they settle down and often the exact frequency changes with external temperature. Whilst not a universal cure the addition of one of the cheap Murata clip-on crystal temperature regulators can help a lot. These regulators heat the crystal casing up to about 35°C and then maintain it at this temperature. The result is greatly increased drift for the first few minutes after switch-on but this is followed by excellent medium-term stability even on the microwave bands.

Another very desirable and essential feature for those looking for EME schedules with other stations is accurate frequency readout. Although you might be able to make do with 1kHz accuracy on 144 and 432MHz, being able to place yourself within 100Hz will save a lot of time and heartache when looking for expedition stations or for those contemplating running schedules. With complex multi-digit transceiver readouts it is important to understand the difference between what the display says and the actual transmitted/received signal frequency. This will need to be carried out or repeated via any transverter used in order to determine the true transmitter frequency.

Receivers, audio filters and headphones

The vast majority of modern HF and many all-mode VHF transceivers will meet the frequency stability and readout criteria - apart

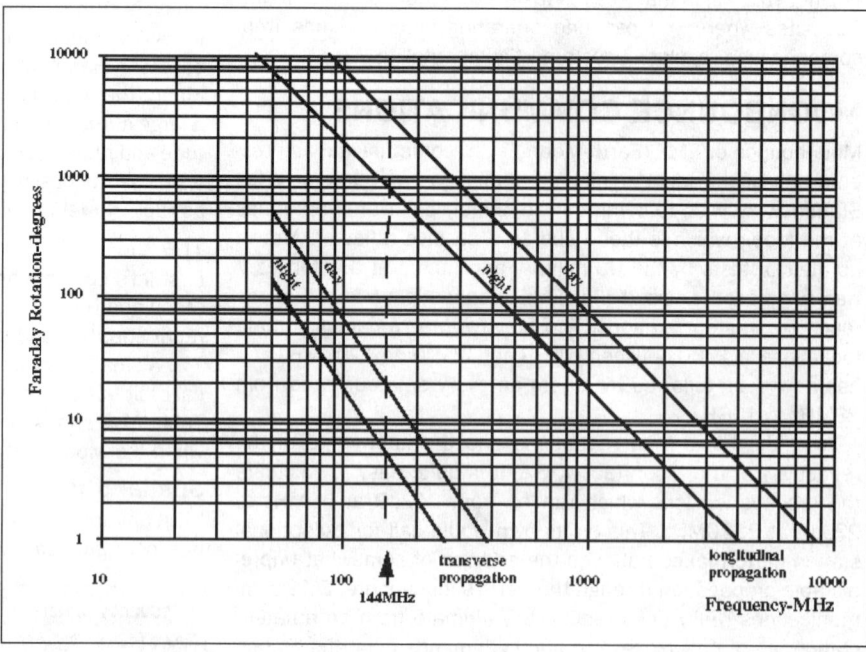

Fig 21.5: Typical values of Faraday rotation for one pass through the ionosphere constructed from data obtained from [15]

from a good CW IF filter the other features come down to user choice. An important issue with any rig used for EME operation is operator comfort and this means choosing your filters carefully.

An IF filter is almost essential, with 300-500Hz bandwidth being commonly used - filters which are too narrow or peaky can result in undesirable ringing which can be a real pain. In addition to a good IF filter most operators use an external audio filter, usually tuneable and often with variable bandwidth. As computer technology has advanced so have DSP audio filters, and as they have improved the majority of operators have changed from using more traditional analogue audio peak filters. When calculating EME station performance an audio filter bandwidth of 100Hz is often assumed, but it can be argued that 'natural' filtering by the ear and the brain results in an effective receiver bandwidth of about 50Hz. A part of the receiver that is very often forgotten, but very important for EME operation, is the operator interface - the headphones. As you will be dealing with weak and noisy signals for long periods of time the issues here are comfort and keeping out background noise. While there are many 'communications' types on sale, many of the hi-fi types can often offer superior performance at lower cost.

Preamplifiers

A low-noise masthead preamplifier is near essential for any successful EME operation. Although shack-mounted preamplifiers may well produce acceptable results for terrestrial and amateur satellite operation, the addition of the feeder loss in front of the preamplifier will always result in an inferior, sub-optimal noise figure.

If you are considering serious EME operation you should also consider mounting your main transmit/receive changeover relay

Fig 21.6: The 5.4m dish antenna built by G4CCH for 23cm moon-bounce

at the masthead. High-performance, high-power coaxial relays generally have lower insertion loss than preamplifiers with on-board relays. The choice of preamp is relatively easy - go for the lowest noise figure that you can afford. If you can't afford to buy a commercial unit there are designs available for all of the VHF, UHF and microwave amateur bands which can realise the very low noise figures (preferably <0.5dB) that are needed for EME. Often you may be unsure as to the actual noise figure of your preamp and it is well worthwhile taking it to one of the many rallies or conventions where noise figure measurement facilities are provided.

Although GaAsFET preamplifiers are not renowned for exceptional dynamic range, the majority of strong-signal problems are generated in the later RF stages of the receiver. If you do suffer from local strong out-of-band signals causing intermodulation or even blocking problems then place a bandpass filter between the preamplifier and your main receiver. Placing filters in front of the preamplifier will almost certainly degrade your hard-won receiver noise figure.

Transmitters and amplifiers

As you will have already realised, the very high path losses mean that high output power is essential for all EME operation. This means that whenever possible you will need to be running close to the maximum permitted licensed power at the antenna. On the microwave bands, where high-power generation is difficult, most EME stations increase their antenna size to compensate. There are also trade-offs which can be made between feedline loss and transmitter power output but lower-loss feeder usually turns out to be the most economical answer. There are suitable designs for high-power amplifiers in this handbook and in elsewhere. If you are going to build an amplifier and are contemplating EME operation make sure that it has adequate cooling to cope with the heavyweight duty cycles required when running schedules.

When running high power there is no excuse for radiating a poor signal - hum and chirp can detract from your readability and key-clicks are likely to alienate all of your locals. It is also important to ensure that you remain within your licence conditions especially with regard to harmonic output, and a low-pass filter is a 'must'. A number of high-performance filters capable of high-power operation have been published.

Keyers

Although a few stations still prefer to use a straight key the majority of EME operation is made using some form of memory keyer. There is no doubt that precise, well-formed characters and regular rhythm can help improve readability at very low signal strengths. Information is usually sent without punctuation and regularly repeated for the duration of a transmit period, making keyer programming an easy job. Although there are a number of commercial models available, and you can even use your home computer to perform this function, many EME operators choose to build one of the many designs available.

Computer

EME operation is made easier by using customised computer software (eg WSJT) to make the most of this mode. Furher information is available from [16]. A computer is invaluable to arrange skeds by e-mail and through special interest groups. It can also be used to help align an antenna system with the moon.

Antennas

The antenna system is probably the most important part of an amateur EME station - it is often very large and requires significant effort in its construction. Most texts usually include calcu-

lations based on the path loss (**Table 21.5**), maximum legal power at the antenna and system noise figure to arrive at the minimum antenna gain to detect echoes. These calculations result in a gain requirement of approximately 20dBd at 144MHz and 23dBd at 432MHz. On 144MHz four stacked long Yagis will yield the required gain whilst on 432MHz eight long Yagis will be needed. In practice stations operating using such antenna systems will quite often hear their own echoes. However, many stations have had very good results with 3dB less antenna gain - this is mainly because the lunar reflectivity of 7% is only an average and libration fading will produce signal peaks several decibels above 'normal' signal levels. Operation with sub-optimal antenna systems will usually require narrower receiver bandwidths and well-honed operating skills. On 1296MHz and above the majority of EME stations use dish antennas ranging from 3m C band TVRO to 12m ex-commercial systems. These antenna sizes are made on the assumption that the station is capable of a contact with a station of similar size - there are a number of amateurs that have systems up to eight times larger than those quoted above. As a result even the most simple one-Yagi stations are capable of EME QSOs with such 'big-guns'.

As for amateur satellite operation, EME antenna systems also require elevation control in addition to azimuth rotation. Because of their size and weight EME arrays usually require the largest heavy-duty rotators. Lightweight elevation rotators, such as are used for amateur satellite antennas, are rarely adequate for EME arrays.

Although there are specialised chain-drive elevations systems available, many amateurs build their own elevation systems, often using 'screw-jacks' used for pointing TVRO dishes. One feature that can ease construction of an EME elevation system is that in the UK the Moon only reaches a maximum elevation of 60°. Pointing accuracy depends upon the antenna beamwidth; simple direct-reading meters are usually OK for Yagi arrays but greater accuracy, preferably with digital readout, is required for dish antennas.

Ground gain

As with HF antennas, reflections from the ground can give VHF and UHF antennas additional gain with a small upwards tilt in the main lobe. This additional gain can often make EME QSOs possible between stations which would otherwise not be able to hear each other. It is by using such 'ground gain' that many smaller 144MHz stations gain their first experience of EME. It is very difficult to calculate the exact angle at which the maximum ground gain will occur - a common practice is to arrange tests for half an hour from moonrise or before moonset. On 432MHz and above the effect of the ground will also increase the background noise floor which may well negate the effect of the ground gain on the received S/N.

Operating

The majority of those who operate on EME started off by listening - although not essential, such an apprenticeship will give you a good idea of the callsigns of the strongest active stations and the procedures in use on that particular band.

More information on EME operation and procedures can be found in [16].

REFERENCES

[1]   AMSAT-UK: www.uk.amsat.org
[2]   AMSAT - the Radio Amateur Satellite Corporation (or AMSAT-NA): www.amsat.org
[3]   AMSAT-DL: www.amsat-dl.org. Pages in German and English
[4]   Beacon reception reports for UO-11 should go to Clive Wallis, G3CWV, e-mail g3cwv@amsat.org
[5]   Web site of N3UC, control operator of AO-27, www.ao27.org
[6]   Gurwin TechSat1b (GO-32) details: www.iarc.org/tech-sat/techsat.html
[7]   AMSAT News Bulletins: www.amsat.org/amsat/news.html
[8]   AMSAT Echo web page: www.amsat.org/amsat-new/echo/
[9]   International Space Station: spaceflight.nasa.gov/station/timelines/
[10] Amateur Radio on the International Space Station: www.ariss-eu.org/
[11] GEO weather satellites: www.geo-web.org.uk
[12] Remote Imaging Group (RIG): www.rig.org.uk/
[13] JOVE, radio astronomy: radiojove.gsfc.nasa.gov/
[14] Radio Society of Great Britain, Lambda House, Cranborne Road, Potters Bar, Herts EN6 3JE. www.rsgb.org
[15] Faraday rotation - *Radio Wave Propagation*, Lucien Boithias, North Oxford Academic Publishers Ltd, 1987. ISBN 0 946536 06 6.
[16] *Amateur Radio Operating Manual*, Don Field, G3XTT, RSGB

22 Computers in the Shack

Computers have been part of a well-equipped radio shack for some years, mostly for data communications and logging/contesting.

More recently, cheap online connections have brought Internet resources into many amateur stations. This chapter aims to show that there are many other uses for the shack computer, making it an essential tool for the constructor.

INSIDE A COMPUTER

Before discussing what a computer can do for you as a radio amateur, it is useful to take a brief look at how it works.

For the purposes of this book, it will be assumed that the computer is a 'PC' operating under Windows. Other computers such as the Apple Mac and older 'hobby' machines (eg Sinclair Spectrum, BBC 'B') could be pressed into service with the appropriate software. Other operating systems such as DOS or Unix can also be used.

The essential components of a computer (shown in **Fig 22.1**) are a processor to do the work, memory to store information, inputs and outputs for communication betweeen the computer and the operator (eg keyboard and screen), software to give it instructions and a power supply.

Central Processing Unit

The CPU is the engine of a computer and is measured by the number of cycles of work it can carry out in one second. Thus a '800MHz' computer carries out 800 million instructions per second. Modern CPU chips get very hot and have heat sinks or even their own cooling fans.

Memory

Just like the human brain, a computer cannot work without being able to store information. There are several types of memory, divided into their function, the storage medium and the amount of time that information can be stored.

Bootstrap

This is a tiny, permanent, memory on a chip. It is known as Read Only Memory (ROM) as it cannot be over-written with new data. Its function is to give the CPU the very basic information it needs to start up and function as a computer.

RAM

In contrast to ROM, Random Access Memory (RAM) is designed to be continuously re-used. It is the temporary storage used to hold all of the data required during processing. RAM is located on chips so it can be 'written to' and 'read' very rapidly. It is commonly described in 'Megs', though this is Megabytes, not MHz.

Hard disk

Most of a computer's storage is done on a magnetic disk. Although reading and writing is nowhere near as fast as RAM, it has the advantage that it keeps its information indefinitely, even when the computer is not powered up.

Removable memory

This refers to the disks that can be taken out of a computer for future reading by the same of another computer at a later date. Older machines have 3.5in so called 'floppy disks', or 'diskettes' which can store about 1 megabyte of data, whereas all modern computers use CDs capable of storing up to 800Mb or DVDs which can store several gigabytes - that's thousands of megabytes. Additionally it is now possible to plug in an external memory chip capable of storing up to 1 gigabyte in a small space and with rapid read and write.

Input / Output

Abbreviated to I/O, these devices are what is needed for human beings to interact with the computer. They include the keyboard, screen, mouse and sound card.

Operating System

Usually stored on the hard disk, this is the permanently installed software, that makes the CPU into a usable computer. The most commonly used operating system is Microsoft's Windows, although there are amateur radio programs that run under DOS or Unix. The operating system defines how the various parts of the computer work together and how it connects to real people.

Software

Although the operating system is an essential piece of software, it cannot do anything other than make a computer. To perform any useful task, such as word processing or sending e-mail, additional software known as programs must be installed.

A new computer will usually come with some programs, usually an Internet browser and some office functions, but there is an almost infinite number of additional programs that can be added to perform specialist functions. Although some programs, especially those for commercial applications such as producing this book, are very expensive, many are quite cheap or even free. Fortunately, many amateur radio programs are in the latter category.

Power Supply Unit

Like any piece of electronic equipment, the computer has a PSU, to run from the mains. Additionally lap-top computers have hefty batteries capable of running the unit for an hour or two.

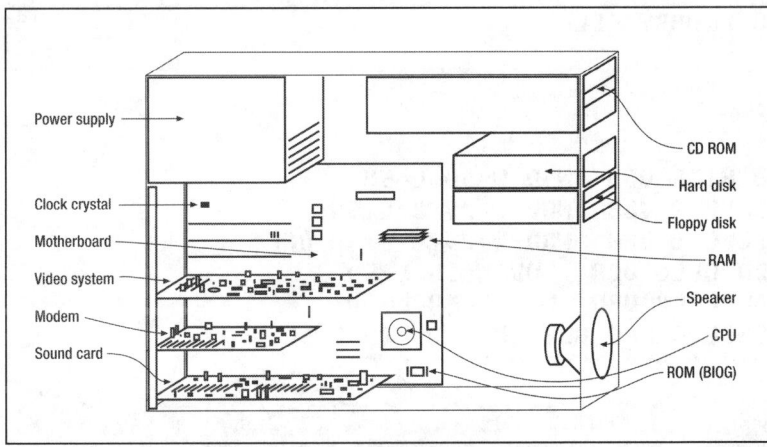

Fig 21.1: Hardware in a typical desk-top computer

OPERATING AIDS

Computers are used to enhance the shacks of many keen DXers. Facilities available to the operator include:

- Logging
- Contest aids
- Rig control
- Maps
- Data communications
- DXCluster
- Propagation information
- Maps and locators

Most of these are outside the scope of this *Handbook*, but detailed descriptions of all of the above can be found in [1].

CIRCUIT DESIGN

Drawing

If you are not good at drawing circuits, good quality illustrations, such as many of those in this book, may be drawn using 'computer aided design' (CAD) software. There are many generic drawing packages, from professional quality software such as CorelDraw to inexpensive or even free programs available for download on the Internet These save the work of producing neat straight lines, boxes and circles but components must be individually drawn. An alternative is to use a CAD program tailored for electronic circuit design. These have the facilities of a generic program, but also have a library of component symbols. Most of the simulator progams described below include schematic drawing facilities.

Circuit Simulators

These allow circuits to be drawn and then analysed, and most are based on the industry standard SPICE. Several are available to try out as free demo versions, usually with restrictions. A useful list can be found at [2].

Some simulators incorporate printed circuit board (PCB) design, whilst others can export data to a dedicated design program. The result can be printed and used in producing the PCB itself.

It is possible to simulate both analogue and digital circuits, or even a mixture of the two. Several of the projects in this book have been initially designed by this type of program, most notably PUFF.

The user starts by drawing the circuit diagram, made from graphical elements provided with the software. The result can be analysed by the program's 'virtual' test equipment, to check how well (or whether) it works. The information displayed can include amplitude vs frequency response, phase vs frequency, group delay vs frequency, gain or loss, and the impedance at input and output. A typical display is shown in **Fig 22.2**.

Changes can be made and the results tested, all before any real construction takes place. As with all software simulations, it is no substitute for knowledge and errors can be reduced by having a basic understanding of how real electronic circuits work.

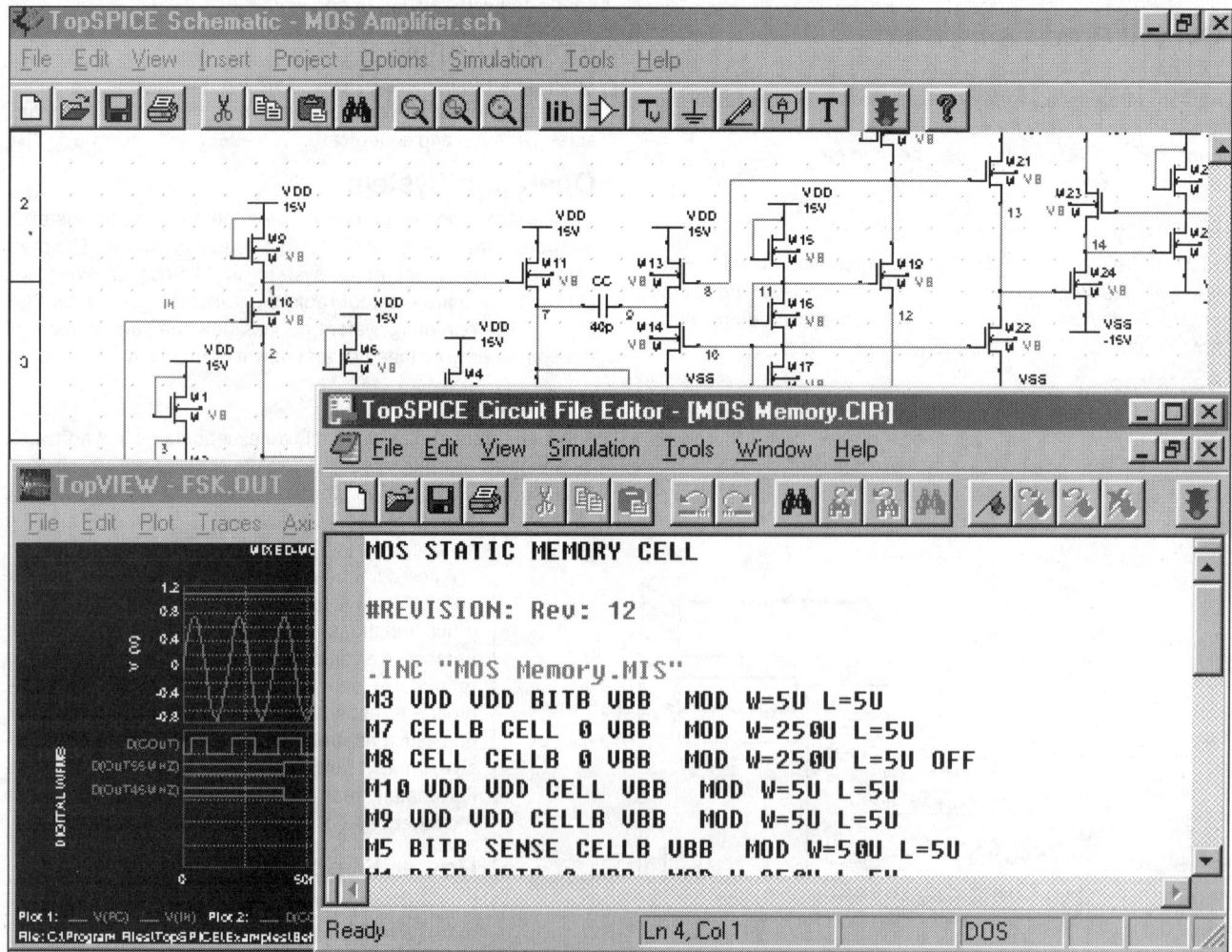

Fig 22.2: A typical circuit simulator / analysis screen. [Source http://penzar.com/topspice/topspice.htm]

Fig 22.3: A Smith Chart produced by ELSIE, the filter design program from Tonne Software.

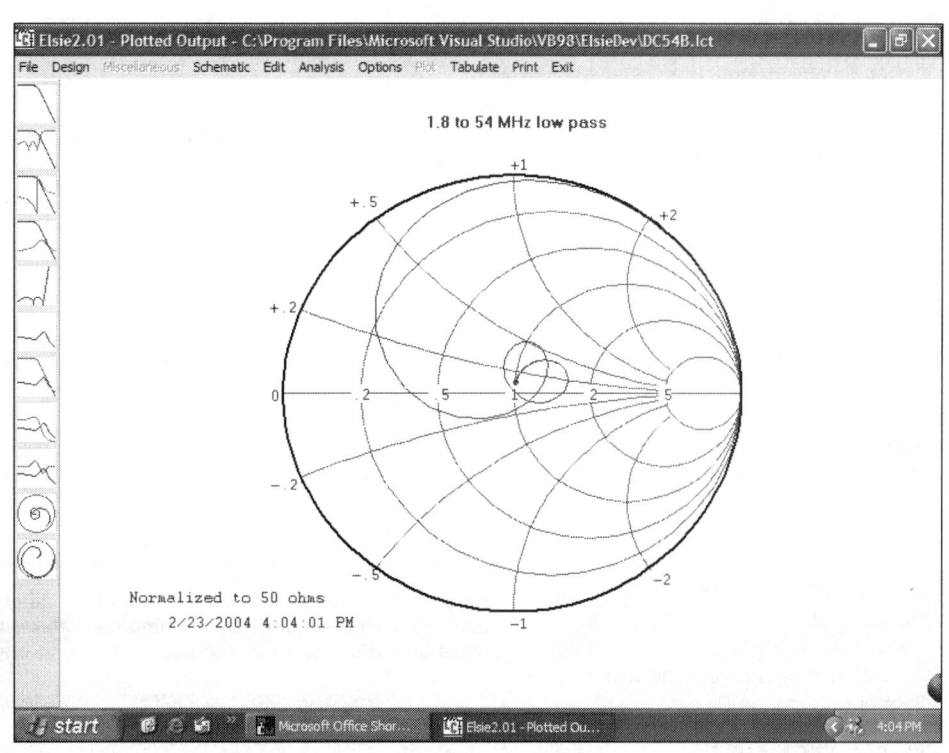

Simulators can usually predict performance quite accurately (mostly for analogue circuits), although they do require substantial effort to learn to use effectively.

A circuit simulator won't do the designing for you, but it does help a lot in finding out what is wrong with a design, and predicting what will happen before you actually build the circuit. Note that the cheaper programs will not operate with time-varying inputs, eg intermodulation analysis.

Spreadsheets

Many aspects of circuit design can be calculated, so a spreadsheet program is useful for making the most complex and interactive calculations. Examples are the optimisation of inductors and filters, Yagi dimensions and linear amplifiers.

Alan Melia, G3NYK, has published Loopcalc [3], an Excel spreadsheet which calculates the radiation resistance and efficiency of a 'square' loop and compares it with the radiation resistance and efficiency of a inverted L aerial of a similar size.

Word Processors

In the past, constructors would use rub-down lettering or adhesive characters to produce professional-looking legends on a front-panel. Nowadays a word-processor can generate text in many fonts, and with a little practice it can also be used to produce arrows, meter scales, etc.

Other Software

A useful tool is AADE Filter Design which can be downloaded free from [4]. It will calculate gains, impedances, group delay, phase, return loss and more. Filter types handled include Butterworth, Chebyshev, elliptic (Caur), Bessel, Legendre and linear phase low-pass, high-pass, band-pass, and band-reject filters, as well as coupled resonator band-pass and crystal Ladder band-pass filters using identical crystals. A tutorial is included for those wishing to know more about filters.

A large range of programs can also be downloaded from the website of Reg Edwards, G4FGQ [5]. These include calculators for the design of antennas, transmission lines, inductors, filters and amplifiers. Also included are propagation path-loss calculators.

Resonance of an inductor and a capacitor can be calculated using Resonate from G3NYK's website [3] which has several pieces of software.

Another useful source is Tonne Software [6] which has programs for filter design, meter scales, customised maps, antennas and matching. Most are totally free, or available free in a cut-down version. The filter design program, ELSIE (**Fig 22.3**), was used in the PIC-A-STAR project described in an earlier chapter.

TEST AND MEASUREMENT

A computer can be used to make all sorts of measurements with the appropriate software, and either a sound card or additional hardware (eg an analogue to digital converter).

Audio analysis software such as Spectrum Lab [7] uses the computer's sound card and can be used for such things as measuring distortion, intermodulation, noise, oscillator drift and modulation. It has a built-in tone generator.

Also using the sound card are programs that can measure and display audio signals (in fact up to several tens of kHz), including those at the output of a radio, and may also be used as a data logger for monitoring beacons. One of these is shown in **Fig 22.4**.

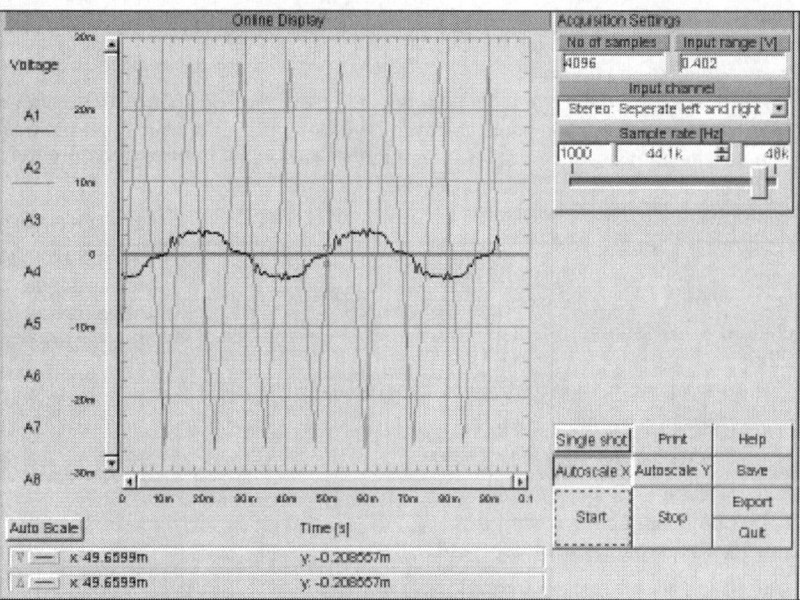

Fig 22.4: A sound-card-based signal analyser with data logging facilities [From http://www.hacker-technology.com/4361/30004.html]

ANTENNA SIMULATION

It is no longer essential to dismantle your antenna in order to experiment with ideas for a new one. Thanks to software originally developed for the US military, and made available in a usable form by amateurs, it is possible to make a 'virtual' antenna on the computer. You can then measure its performance, alter it and measure it again until something practical is found - or the project is abandoned as a failure. Work on a practical antenna need only take place when the 'virtual' one has been optimised.

As with all computer programs, it is possible to get the wrong answer, but this is usually a case of 'garbage in, garbage out'. A good knowledge of antenna theory and practice will help you avoid the major pitfalls. Information on how to obtain antenna simulation software can be found at [8].

Much more about the use of antenna simulation programs can be found in the chapter on Antenna Basics.

THE INTERNET

For very little money, you can have facilities at your fingertips that only a few years ago would have been the envy of the best equipped library in the world, together with communications facilities that have almost totally eclipsed the postal service.

Although the Internet is often used to describe the vast array of web sites, it is actually the network on which sits the web, e-mail and many other facilities. The most common are:

World Wide Web

The 'web' is a repository for many millions of documents, including text, pictures, sound and video. Amateur radio is well represented with information on just about every aspect of the hobby.

There are two main ways to find the information you want. You could start with a large site that covers most aspects of amateur radio, such as those run by the Radio Society of Great Britain [9] or the American Radio Relay League [10], and follow the links to sites carrying additional information. Alternatively, use a 'search engine' such as Google [11] or Yahoo [12], type in one or more keywords (eg circuit simulation) to get a large list of sites that may be useful (see **Fig 22.5**).

The sort of information that can be found on the web includes:

- News and events diaries
- Calculators for component values
- Equipment modifications
- Component catalogues
- Advice from experienced amateurs
- DXpedition information and logs
- Circuits and construction projects
- Propagation and solar data

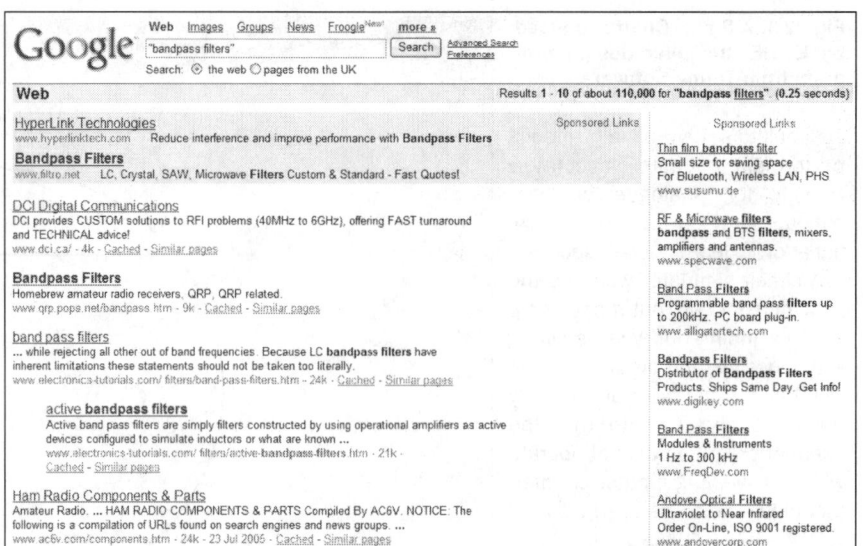

Fig 22.5: An Internet search for "bandpass filters" produces over 100,000 results, including advertisements for commercial filters, technical papers, descriptions and advice by radio amateurs, and lists of where to buy components

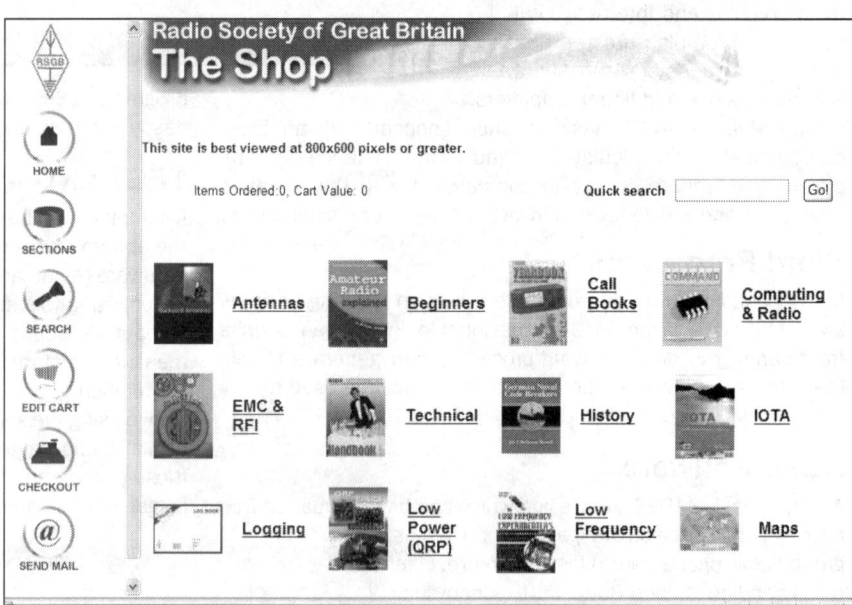

Fig 22.6: Part of the RSGB's e-commerce site where books, CD-ROMs, maps etc can be bought by credit card. It includes a search facility and detailed descriptions for those who like to browse before they buy

E-commerce

Most shops have a presence on the world wide web, and some are available only via the 'web'. This includes major component suppliers as well as much smaller specialist outlets. All display their wares, and most encourage electronic sales by credit card. This has two advantages, you can browse without leaving your home and you can buy from overseas shops - note, though that VAT, import duty and other charges may be payable on entry to the UK.

The on-line catalogues of Farnell [13], Maplin Electronics [14}, and RS Components [15] and many others are not only places where components can be purchased, but also a really good source of information such as data sheets.

Auction sites such as eBay [16] are where second-hand (and new) bargains may be found. Simply enter a search term (eg "ATU") and choose from an array of items. As with all 'blind' pur-

Fig 22.7: Input and output connectors on a laptop computer. From left: Two USB ports, network cable, video out, parallel port (25-way), serial port (9-way) and external monitor.

chases, such as a newspaper advertisement, the buyer should take precautions to prevent fraud. If possible visit the seller and check before you buy.

One of the first examples of e-commerce was book selling, and the Internet makes it possible to search for what you want from millions of books, including many specialist publications that you will never see in a high street shop. Amateur radio books can be bought direct from the ARRL [10] and the RSGB [9] (**Fig 22.6**).

All sorts of radio and electronics books - even technical papers -can be searched for (by keyword, author or title) on big sites such as Amazon [17] which often has second-hand books listed alongside the new ones. Again, auction sites are a good source of second-hand and antique books.

E-mail

Although your main contact with other amateurs may well be on the air, or at the local radio club, e-mail can still be useful. It can be used to maintain a dialogue with like-minded amateurs all over the world, and allows the exchange not only of text but also pictures (including circuit diagrams), sound files and programs.

Groups

Similar to e-mail are newsgroups and reflectors. These allow groups of amateurs who have something in common, for instance an interest in VHF contesting or the city they live in, to share news and information with all of the other members. if you are new to a particular aspect of amateur radio, this is often the place to get advice from those with much more experience than you.

Some groups, such as those hosted by Yahoo or Google require you to enter a password (available free) before gaining access.

A good example of a special interest group is http://uk.groups.yahoo.com/group/picastar/ which is used by those building the PIC-A-STAR project (see the earlier chapter on this transceiver) to compare notes and get help.

CDs and DVDs

An alternative to the Internet is the compact disk. Many computer programs are available, both free and paid-for, on disk. Advantages over downloading include no phone costs and the ability to re-install the software on a replacement computer with ease.

Publications such as the ARRL's *QST*, RSGB;s *RadCom* (including archives going back to the first ever edition) and books such as this one are also available in full on disk. In addition to saving shelf space, digital books and magazines are usually fully-searchable (ie any word or phrase can be searched for - much more useful than a conventional Index). Visually impaired people can considerably magnify each page on screen if required, or even use a program that reads the publication out loud.

9-pin connector:		25-pin connector:	
1	Carrier Detect	2	Transmit Data
2	Receive Data	3	Receive Data
3	Transmit Data	4	Request To Send
4	Data Terminal Ready	5	Clear To Send
5	Signal Ground	6	Data Set Ready
6	Data Set Ready	7	Signal Ground
7	Request To Send	8	Received Line Signal Detector
8	Clear To Send	20	Data Terminal Ready
9	Ring Indicator	22	Ring Indicator
		Pins 1, 9-19, 21, 23-25 are not used	

Table 22.1: Serial port connections for two types of socket

CONNECTING TO THE REAL WORLD

Much can be done with your computer with just an Internet connection. However, it can be connected to other items such as your radio or equipment under test.

It can also be used to develop software that will eventually be copied to a chip to drive stand-alone equipment. This is the technique used in the PIC-A-STAR transceiver project described in detail in an earlier chapter.

As described earlier, the computer's sound card is an input/output (I/O) device that can be used in amateur radio for digital communications, audio filters, making measurements and so on.

Serial and Parallel

Other I/O sockets on a standard computer include serial and parallel ports. Often just a couple of pins from these multi-way connectors is used for our purposes. For example, the chapter on digital communications shows how transmit/receive switching can be controlled through a serial port.

Serial means that the eight bits in a byte of data are sent one at a time down a single wire, whereas a Parallel port sends the bits simultaneously but needs eight wires. Voltage on the pins switches between minus 3-25 volts and plus 3-25 volts.

Serial ports, sometimes called 'comm' or RS-232, are used for external modems (including packet radio TNCs). Parallel connectors are often called printer ports as they were commonly used for connecting printers before USB ports became universal.

Table 22.1 gives the pin connections for 9-pin and 25-pin serial connectors.

Nowadays, the USB port reigns supreme on modern computers. It has solved many of the compatibility problems experienced with the older I/O systems, but is not easy to use in amateur radio projects.

Fig 22.7 shows the range of input and output ports available on a typical laptop computer.

Controlling your Radio

Many modern base-station transceivers incorporate sockets so that they can be controlled by a computer using special software.

Station control programs, such as those used by DXpeditions and contesters, have facilities to control at least transmit and receive frequency. This can be extended to being able to control your station remotely over a modem (with the appropriate approval, if necessary), as described in [18].

COMMUNICATIONS

Amateurs have been using home computers for making contacts ever since they became available - RTTY programs were available for the 'BBC-B' computer, for example. The chapter on Digital Communications gives much more information.

Towards the end of the 20th century, ways were developed to link amateur radio and the Internet, so that radio amateurs can communicate even though one of them has no radio, or to inter-link repeaters via the Internet. Three systems are in use at the time of writing: Echolink [19], eQSO and IRLP [20, 21].

EMC

Sometimes the introduction of a computer into a radio shack causes unforeseen EMC problems. Either the computer (and/or its peripherals) radiates noise on one or more amateur bands, or the operation of the station transmitter causes the computer system to fail or operate erratically.

It is quite possible to make these two equipments compatible by following the measures recommended in the chapter on Electromagnetic Compatibility.

THERE'S MORE . . .

This chapter has barely scratched the surface of how computers can be of benefit radio amateurs and , for that matter, anyone interested in building electronic equipment. No doubt each reader will have his or her own favourite program or Internet site that could have been added.

It is hoped that this brief overview will have sparked some ideas and will set readers searching the Internet, or asking their friends, for new ideas. Computers and amateur radio are a marriage that has only just begun.

REFERENCES

[1]   The Amateur Radio Operating Manual, 6th ed, Don Field, G3XTT, RSGB 2004
[2]   http://www.terrypin.dial.pipex.com/ECADList.html
[3]   http://www.alan.melia.btinternet.co.uk/programs.htm
[4]   http://www.aade.com/filter.htm
[5]   http://www.btinternet.com/~g4fgq.regp/
[6]   http://www.tonnesoftware.com/
[7]   http://www.qsl.net/dl4yhf/spectra1.html
[8]   http://www.cebik.com/model/nec.html
[9]   RSGB web site, http://www.rsgb.org
[10] ARRL web site, http://www.arrl.org
[11] Google search engine, http://www.google.co.uk, or http://www.google.com
[12] Yahoo search engine, http://www.yahoo.co.uk, or http://www.yahoo.com
[13] http://uk.farnell.com
[14] http://www.maplin.co.uk/
[15] http://rswww.com/
[16] http://www.ebay.co.uk/
[17] http://www.amazon.co.uk
[18] 'There's a remote possibility . . .', David Gould, G3UEG, *RadCom*, Aug/Sep 2005
[19] http://www.echolink.org/
[20] http://www.irlp.net/
[21] http://www.ukirlp.co.uk/

23 Electromagnetic Compatibility

Electromagnetic compatibility (always abbreviated to 'EMC') is the ability of various pieces of electrical and electronic equipment to operate without mutual interference. So far as amateur radio is concerned, the object is to achieve good EMC performance: that is, not to suffer from received interference or to cause interference to others.

In practice, the amateur must endeavour to minimise interference caused by his (or her) station and, where appropriate, increasing the immunity of susceptible local domestic radio and electronic equipment. Complementary to this is interference to amateur reception. In recent years this has become increasingly important. The number of potential interference generators in the typical home is increasing all the time and in addition, the transmission of high speed data on cables which are less than optimum from the EMC point of view, may become a serious problem.

Most national administrations have enacted legislation defining minimum EMC standards which products on sale to the public must meet. These standards lay down maximum permitted emitted interference, and also the minimum immunity which equipment must have to unwanted signals. In the UK the standards are issued by the British Standards Institute, and are harmonised to the common standards of the European Community. In general, EMC standards are framed round a normal domestic or industrial radio environment, and fall short of what would be ideal from the radio amateur's point of view. Amateurs tend to generate high field strengths and attempt to receive smaller signals than other radio users in a typical residential area.

So, for the radio amateur, interference breaks down into two major categories:

- Interference caused by operation of the transmitter.
- Interference to reception, usually simply called radio frequency interference or RFI.

INTERFERENCE CAUSED BY THE TRANSMITTER

This is a common occurrence and most amateurs have suffered from it at some time or other. In all too many cases it has been the cause of bad feeling with neighbours. Fortunately, in recent years, the number of cases of serious disagreements with neighbours has declined markedly. This is due to a number of factors one of which is the coming of the EMC regulations which have made EMC a major factor in the design of all types of domestic electrical and electronic equipment. An even more important factor, however, is the appreciation by amateurs themselves of the need to design and operate their stations to minimise all types of interference.

Interference from the transmitter also falls into two categories.

- Interference caused by the legitimate amateur signal, on the normal operating frequency, breaking through into some piece of susceptible

equipment. This is usually called breakthrough to emphasise that it is not really a transmitter fault, but rather a defect in the equipment which is being interfered with.
- Interference due to unwanted emissions from the amateur station. The general name for unwanted emissions is spurious emissions.

Nowadays, breakthrough is much more likely to be a cause of interference than spurious emissions, but before jumping to any conclusions carry out a few simple checks.

First consider what sort of equipment is affected. If it does not use radio in any way, then the cause must be breakthrough of some sort. If the susceptible equipment makes use of radio in some way (usually broadcast radio or TV but it could be some form of communication or control device) then the cause could be either breakthrough or a spurious emission from the transmitter. The next step is to find out if the interference only occurs when the equipment being interfered with is tuned to specific frequencies. If so, check whether similar equipment in your own house (or in neighbours' houses) is affected when tuned to the same frequencies. In effect, if your transmitter is radiating a

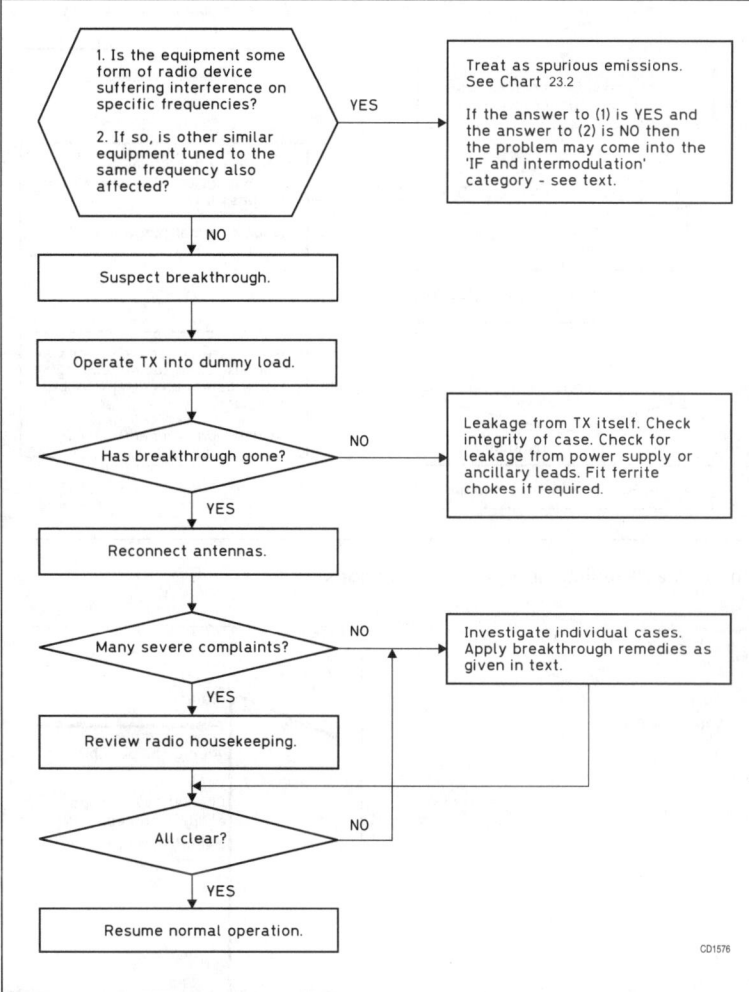

Chart 23.1: Breakthrough

spurious signal, then anything tuned to that frequency is likely to be affected. **Charts 23.1 and 23.2** show the general procedure.

BREAKTHROUGH

All electronic equipment is to some extent vulnerable to strong radio frequency fields and is a potential EMC threat. Any amateur who lives in close proximity to neighbours will have to give some thought to the avoidance of breakthrough. This can be tackled in two ways:

- By installing and operating the station so as to reduce the amount of radio frequency energy reaching neighbouring

domestic equipment. The term good radio housekeeping has been coined to cover all aspects of this activity.

- By increasing the immunity of the affected installation to the amateur signal.

Good Radio Housekeeping

The essence of good radio housekeeping is keeping your RF under reasonable control, putting as much as possible where it is wanted (in the direction of the distant station) and as little as possible into the local environment.

It is fortunate that installations designed to achieve this are also likely to minimise the pick-up of locally generated interference - adding a bonus to good neighbourliness.

Where a station is to be installed in a typical urban environment, in close proximity to neighbours, it is essential to plan with EMC in mind

Antennas

It is always good practice to erect any antenna as far from houses as possible, and as high as practical, but for HF operation the relatively long wavelengths in use give rise to special problems. In locations where breakthrough is likely to be a problem, HF antennas should be:

- Horizontally polarised. House wiring and other leads tend to look like antennas working against ground, and hence are more susceptible to vertically polarised signals.

- Balanced, to minimise out-of-balance currents, which can be injected into the house wiring, particularly in situations where a good earth is not practical. These currents will also give rise to unwanted radiation.

- Compact, so that the whole of the radiating part of the antenna can be kept as far from house wiring as possible. Try to avoid antennas where one end is close to the house while the other end is relatively far away. This encourages direct coupling between the near end and the house wiring, inducing RF currents which will be greater than would otherwise be the case.

The sort of thing to aim for is a dipole, or small beam, located at least 15m from the house (and neighbouring houses), and fed with coaxial cable via a balun (**Fig 23.1**). This is not too difficult to achieve above 10MHz, but at lower frequencies

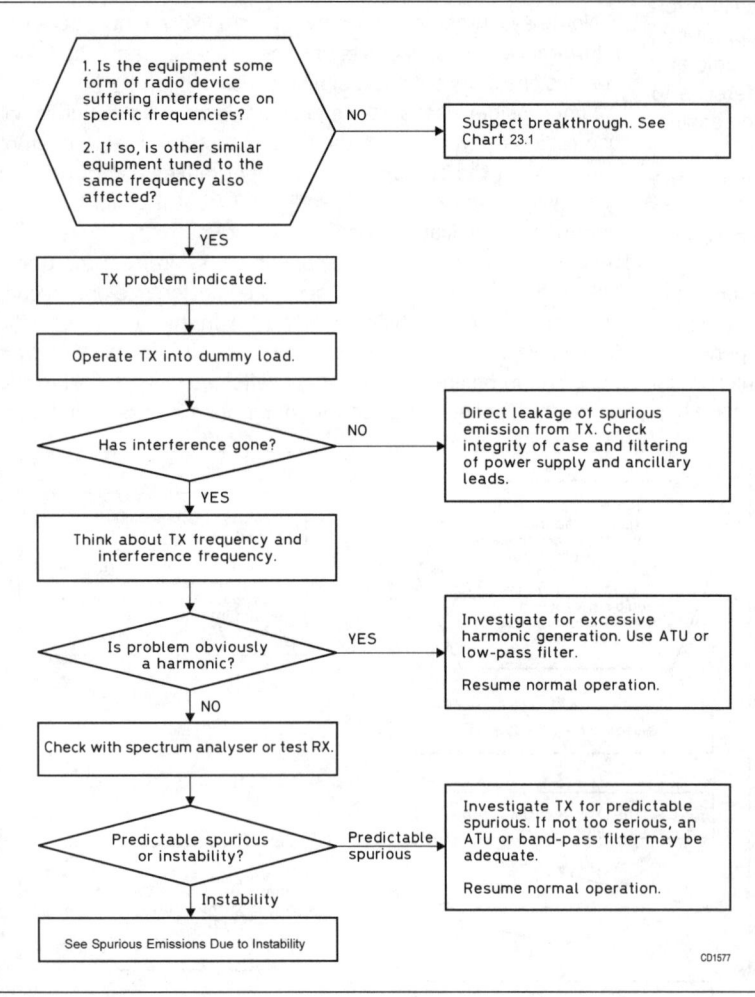

Chart 23.2: 'Predictable' spurious emissions

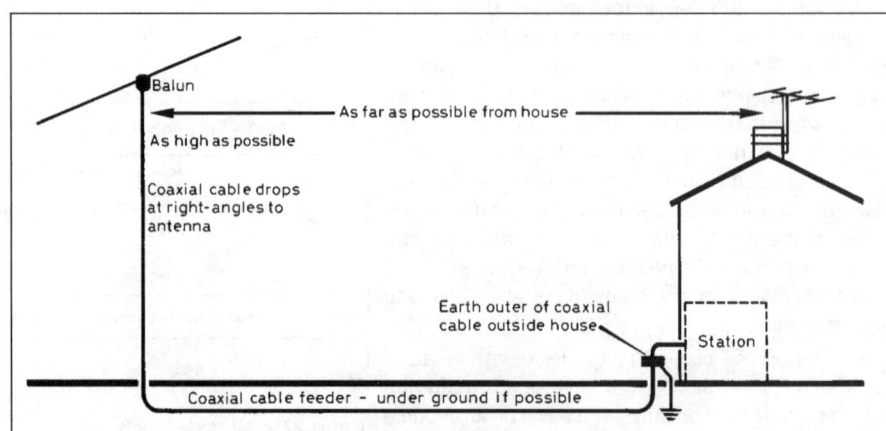

Fig 23.1: Antenna and feeder system with EMC in mind

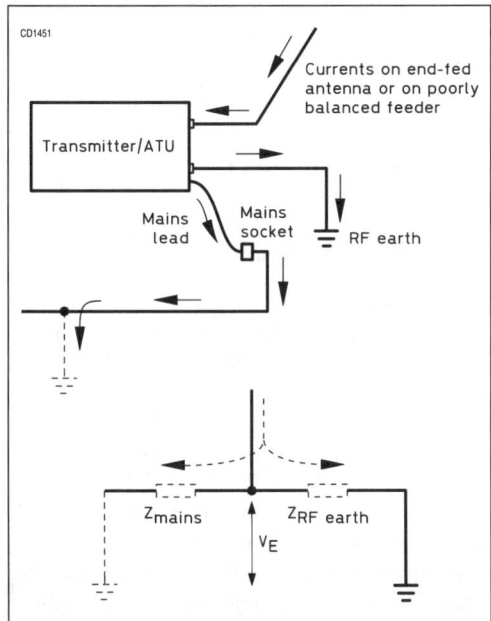

Fig 23.2: Earth current divides between RF earth and mains. The current down each path will depend on the impedances. The transmitter earth terminal will be at V_E relative to 'true' earth potential

a compromise is usually necessary. A balanced feeder can be used, but coaxial feeder is convenient and allows the use of ferrite chokes to reduce any unwanted currents which may find their way on to the braid.

Earths

From the EMC point of view, the purpose of an earth is to provide a low-impedance path for RF currents which would otherwise find their way into household wiring, and hence into susceptible electronic equipment in the vicinity. In effect, the RF earth is in parallel with the mains earth path as in **Fig 23.2**. Good EMC practice dictates that any earth currents should be reduced to a minimum by making sure that antennas are balanced as well as possible. An inductively coupled ATU can be used to improve the isolation between the antenna/RF earth system and the mains earth [1]. The impedance of the mains earth path can be increased by winding the mains lead supplying the transceiver and its ancillaries onto a stack of ferrite cores as described below for breakthrough reduction.

Antennas which use the earth as part of the radiating system - ie antennas tuned against earth - should be avoided since these inevitably involve large RF currents flowing in the earth system. If this type of antenna must be used, arrange for it to be

WARNING: Protective Multiple Earthing (PME)

Some houses, particularly those built or wired since the middle 'seventies, are wired on what is known as the PME system. In this system the earth conductor of the consumer's installation is bonded to the neutral close to where the supply enters the premises, and there is no separate earth conductor going back to the sub-station.

With a PME system a small voltage may exist between the consumer's earth conductor, and any metal work connected to it, and the true earth (the earth out in the garden). Under certain very rare supply system faults this voltage could rise to a dangerous level. Because of this supply companies advise certain precautions relating to the bonding of metal work inside the house, and also to the connection of external earths.

WHERE A HOUSE IS WIRED ON THE PME SYSTEM DO NOT CONNECT ANY EXTERNAL (ie radio) EARTHS TO APPARATUS INSIDE THE HOUSE unless suitable precautions are taken.

A free leaflet *EMC 07 Protective Multiple Earthing* is available on request from RSGB, or from the.RSGB EMC Committee web site [7]

fed through coaxial cable so that the earth, or better some form of counterpoise, can be arranged at some distance from the house.

The minimum requirement for an RF earth is several copper pipes 1.5m long or more, driven into the ground at least 1m apart and connected together by thick cable. The connection to the station should be as short as possible using thick cable or alternatively flat copper strip or braid.

Where the equipment is installed in an upstairs room, the provision of a satisfactory RF earth is a difficult problem, and sometimes it may found that connecting an RF earth makes interference problems worse. In such cases it is probably best to avoid the need for an RF earth by using a well-balanced antenna system.

For more on choosing antennas and earths, see the antenna chapters in this book.

Operating from Difficult Locations

If there is no choice but to have antennas very close to the house or even in the loft, then it will almost certainly be necessary to restrict the transmitted power. Some modes are more likely to give breakthrough problems than others, and it is worth looking at some of the more frequently used modes from this point of view.

SSB is a popular mode but also the most likely to give breakthrough problems, particularly where audio breakthrough is concerned.

FM is a very EMC friendly mode, mainly because in most cases the susceptible equipment sees only a constant carrier turned on and off every so often.

CW has two big advantages. First, providing the keying waveform is well shaped, the rectified carrier is not such a problem to audio equipment as SSB. The slow rise and fall gives relatively soft clicks which cause less annoyance than SSB. The second advantage is that it is possible to use lower power for a given contact than with SSB.

Data modes used by amateurs are generally based on frequency-shift keying (FSK), and should be EMC friendly. All data systems involve the carrier being keyed on and off - when going from receive to transmit, and vice versa - and consideration should be given the carrier rise and fall times just as in CW.

Passive Intermodulation Products (PIPs)

This phenomenon is a fairly unusual aspect of good radio housekeeping but one which has turned up from time to time since the early days of radio. Traditionally it is known as the rusty bolt effect. It occasionally causes harmonic interference to be generated by amateur transmissions, but far more often it simply degrades receiver performance without being identified as a problem.

All mixing and harmonic generating circuits use non-linear elements such as diodes to distort the current waveform and hence to generate the required frequency components. A similar effect will be produced whenever the naturally produced semiconductor layer in a corroded metal joint forms an unwanted diode. These unwanted diodes are usually most troublesome in the antenna system itself, particularly in corroded connectors. In the case of a single transmitter the effect simply causes excessive harmonic radiation, but where two or more transmitters are operating in close proximity the result can be spectacular intermodulation product generation. On receive, the result is much-reduced receiver intermodulation performance, and in severe cases there will be a noticeably high background noise level which consists of a mishmash of unwanted signals. The best way to avoid troubles of this sort is to keep the antenna system

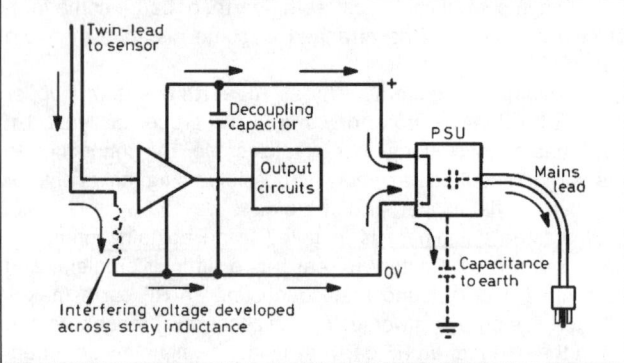

Fig 23.3: Path of RF signal in a typical sensor/alarm device

in good repair and to examine all connections every few months. PIPs can be generated in corroded metal gutters and similar structures not directly associated with the antenna system. The solution is to clean up or remove the corroded metalwork or, where this is not possible, to short-circuit the corroded junctions with a conductive path. It is obviously good practice to keep antennas away from doubtful metalwork.

Dealing with Breakthrough at the Receiving End

Breakthrough is simply unwanted reception, and the basic mechanism by which signals are picked up is the same as for any other reception. Breakthrough can occur to such a wide range of equipment that for the sake of general discussion it is simpler to assume some non-specific device, in other words a black box.

For signals to get into our black box, they must be picked up on a wire which is a significant fraction of a wavelength long, which means that on HF the external leads are the most common mode of entry and direct pick-up by the equipment itself is unlikely unless the transmitter field strength is very high. At frequencies above about 50MHz pick-up by wiring inside the black box becomes more likely but only if the box is made of non-conducting material.

Fig 23.3 shows how unwanted RF signals might get into the black box - in this case, some sort of alarm circuit is assumed with a sensor connected by twin cable, and an amplifier and power supply inside the box. The sensor lead acts as a crude earthed antenna so that electromagnetic energy from the transmitter causes currents to flow in the antenna formed by the sensor lead, through any stray impedance between the input connection and the 0V rail, and through the power supply to earth. These currents are called common-mode currents because they flow on both conductors in the cable in phase - in effect they act

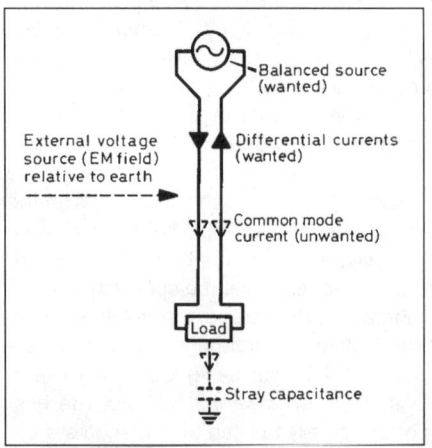

Fig 23.4: Differential-mode and common-mode currents

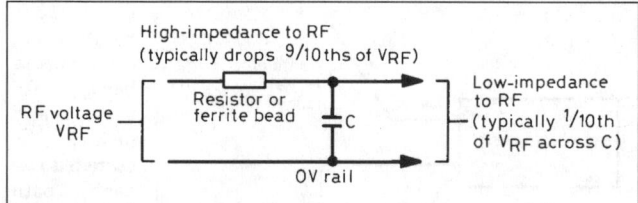

Fig 23.5: The principle of by-passing or decoupling

as if the pair of wires were one conductor. The wanted signals are differential, flowing in one direction on one wire, and the reverse direction in the other. Common-mode and differential currents are illustrated in **Fig 23.4**.

The key to avoiding breakthrough is to prevent the unwanted signals picked up on the external lead from getting into the circuits inside the box. There are two ways of doing this:

• Bypass the unwanted RF by providing a low-impedance path across the vulnerable input circuit.

• Use a ferrite choke to increase the impedance of the unwanted antenna where it enters the black box, effectively reducing the currents getting into the sensitive internal circuitry.

Bypassing the unwanted RF

The principle of bypassing is to arrange for a potential divider to be formed in which the majority of the unwanted signal is dropped across a series impedance, as shown in **Fig 23.5**. In some instances the series element may be a ferrite bead or, where circuit conditions permit, a low-value resistor, but in many cases no series element is used and the stray series impedance of the lead provides the series element. A ceramic capacitor with a value in the region of 1 to 10nF would be typical. It is important to keep the leads as short as possible, and to connect the 'earthy' end of the capacitor to the correct 0V point - usually the point to which the amplifier 0V is connected, and to which the inputs are referred. ('0V' in this context has much the same meaning as 'ground', but avoids any confusion with other meanings of 'ground' used in radio discussions such as in 'capacitance to ground' etc.)

Generally, bypassing is not very practical in domestic situations and is included as food for thought for anyone building their own equipment. It is inadvisable for the amateur to attempt to modify commercial equipment unless he (or she) has expert knowledge - in particular, never attempt to modify equipment belonging to someone else.

Ferrite chokes

Ferrite chokes are also used to form a potential divider to the unwanted RF currents, though in this case the series impedance is increased outside the black box, and the stray capacitance or resistance inside the box forms the shunt element (**Fig 23.6**).

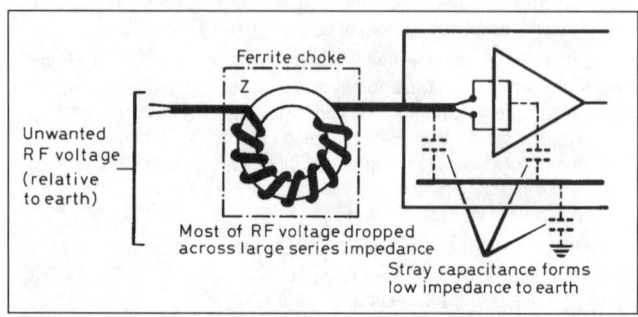

Fig 23.6: The series choke. The impedance of Z forms a potential divider with stray capacitance

The great benefit of this technique is that it does not involve any modification inside the box. All that is required is to wind the susceptible lead on to a suitable ferrite core. Ferrite cores which are designed for EMC purposes have a reasonably high permeability, which is combined with a relatively high loss at the frequencies of interest. This enables a high impedance to be achieved without resonance effects becoming dominant. Further information on core types will be found in Appendix 3 of [1] and in [2].

A very important feature of this type of ferrite choke is that it acts only on the common-mode interfering currents - differential-mode wanted signals will not be affected. The go and return currents of the differential signal are equal and opposite at any instant, and so their magnetic fields cancel out (except in the space between the wires) and there is no external field to interact with the ferrite.

The most popular core for EMC use is the toroid or ring, because the ring shape means that the magnetic field is confined inside the core, giving a relatively high inductance for a given material. Where a toroid cannot be used, either because the lead is unusually thick, or because large connectors cannot be removed, it may be possible to use the ferrite yoke deflection assembly from a scrap TV set. (Take great care when removing a scrap yoke assembly. Imploding CRTs can cause serious injury. In addition the EHT connection can be at high potential due to stored charge long after the power has been switched off.) In some circumstances split ferrite cores can be used. Another option is to wind the lead on a length of ferrite rod such as that used for medium-wave radio antennas, though in this case large number of turns will be required. Further information can be found in Appendix 3 of [1] .

How many turns?

The inductance of a ferrite ring choke is proportional to the length of ferrite through which each turn passes (ie the thickness or 'depth' of the ring) and also to the square of the number of turns. Traditionally cores about 6.5mm thick have been used, two of these being stacked to give an effective thickness of 13mm or so. To make a choke for HF frequencies, 12 to 14 turns on two such rings should be satisfactory. For VHF, fewer turns should be used - about seven turns on a single ring will be effective at 144MHz. At the time of writing the ferrite cores available from the RSGB are about 12.7mm thick so that one of these rings is equivalent to two of the thinner rings. It is most important to use the correct type of ferrite ring. There is a very wide

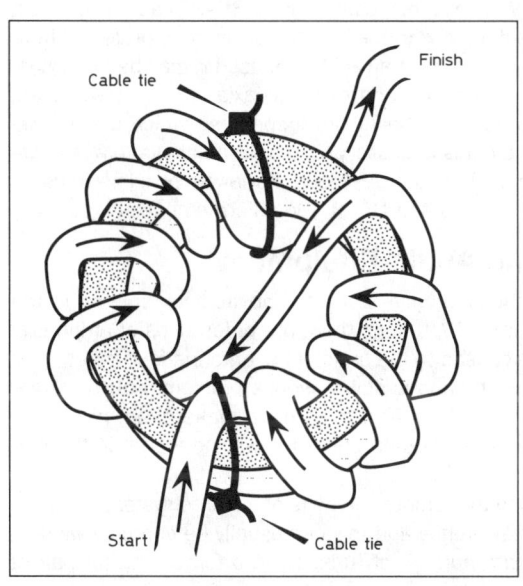

Fig 23.7: Winding a ferrite choke for minimum shunt capacitance

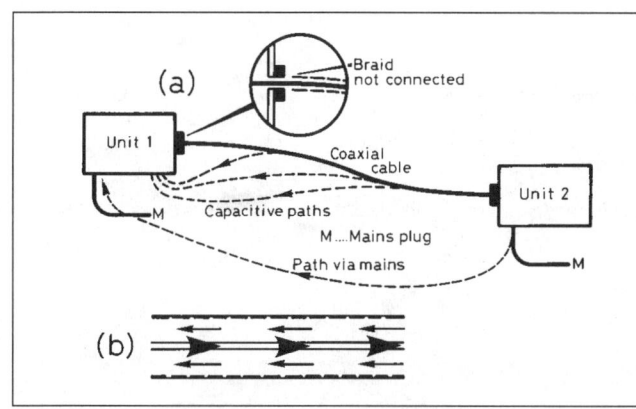

Fig 23.8: (a) Broken braid connection causes the current which would normally flow on the inner surface of the braid to return by random paths. (b) In a correctly connected coaxial cable the currents balance, giving no external field

range of ferrite materials, and some rings are unsuitable for EMC purposes. The simplest way to ensure that the correct rings are obtained is to purchase them from a reliable supplier, advertising rings specifically for EMC purposes. Ferrite rods require more turns (20 or more) to make an effective choke, and are generally only effective above 20MHz.- see Appendix 3 of [1].

The shunt capacitance across a ferrite ring choke can be minimised by winding the core as in Fig 23.7.

Breakthrough to TV and Video Equipment

TV installations started to become more complicated in the early eighties, with the coming of the domestic video cassette recorder (VCR). Since then we have seen satellite TV and digital terrestrial TV. This means more interconnections between the various units and "set-top" boxes. It is no longer realistic to talk of a typical TV installation and the following discussion assumes a simple installation with TV set and a video recorder to illustrate the way in which breakthrough occurs. In general the mechanism in more complex installations is basically the same.

In any TV installation, the most vulnerable point is the antenna input, and in the majority of cases eliminating unwanted signals at this point will clear up the problem. The most common route for the unwanted signals to get in to the set is via the braid of the coaxial feeder. The small size of UHF TV antennas makes it unlikely that large interfering signals will be picked up on the antenna itself, unless the frequency is relatively high - above about 100MHz or so.

First have a look at the general installation. Are the coaxial connectors correctly fitted with the braid firmly gripped and making good electrical contact with the outer connection of the plug? Ideally the centre pin of antenna connectors should be soldered, to avoid poor contact, but in most installations it will not be. Corrosion of the connector at the end of the cable coming down from the antenna may indicate that water has got into the cable; the commonest cause is chaffing or splitting of the sheath, though it could be a failure of the seal at the antenna itself.

In a complicated installation comprising several units, make sure that all connectors are correctly mated, and that coaxial leads are correctly 'made-off' with the braid properly connected. In many cases, equipment will work satisfactorily with the braid of an interconnecting cable disconnected. In this case the return path for the signals in the cable is via random earth paths through other parts of the system. This reduces the immunity of the installation to breakthrough, and allows the leakage of radio frequency interference (Fig 23.8).

Fig 23.9: A ferrite ring choke suitable for TV coaxial down lead

TV antenna amplifiers

A well-designed masthead amplifier which is correctly installed, close to the TV antenna, is unlikely to give problems at HF because the length of cable, before the amplifier, is short compared to the wavelength of the amateur signal, but at VHF the conditions could be very different.

Problems arise if the unwanted signal reaching the input to the amplifier is large enough to drive it into non-linearity and cause cross-modulation. Masthead amplifiers get their DC power supplies through the coaxial feeder, so that any filters or braid-breakers must pass (and not short-circuit) the required DC supply. Unless specific information is available, it is best to use only ferrite-ring chokes in the down leads from such amplifiers.

Indoor amplifiers are much more likely to cause problems, especially where HF is concerned, but fortunately they can be easily removed from the circuit for test purposes. Many amplifiers have several outputs, allowing more than one set to be operated from the same TV antenna. If the amplifier has to be retained in the installation it should be considered as part of the TV set (or sets), and the standard techniques, using chokes and filters, on the antenna side of the amplifier will still apply.

TV amplifiers vary in input bandwidth. Some will accept all signals from about 40MHz and below up to the top of the UHF TV band. Others are designed for UHF only. In countries such as the UK where all terrestrial TV is on the UHF bands it is obviously better to choose a UHF only amplifier.

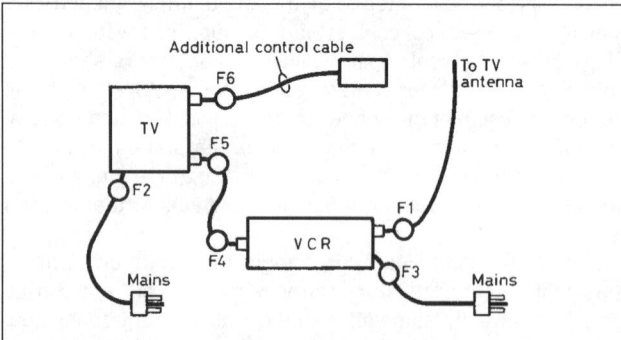

Fig 23.10: Typical domestic TV installation, showing positions of ferrite chokes. F1: antenna coaxial feeder choke (or filter). F2/F3: mains chokes. F4/F5: coaxial chokes on RF lead (VCR to TV). F6: Choke on additional leads to controls or external speakers or displays

Chokes and filters for TV installations

The simplest way of choking-off signals travelling down the braid is to use a ferrite ring choke; used in this way it is often called a ferrite-ring braid-breaker. The low-loss cable commonly used for UHF TV is not suitable for winding on to the core. It is best to make the choke separately, using a length of smaller-diameter coaxial cable, with suitable TV connectors at each end. **Fig 23.9** shows a choke for HF breakthrough. The coaxial cable used for the choke should, of course, be 75 ohms, but there is no need to use low-loss cable since the length involved is quite short.

Braid-breakers and filters are available from a number of commercial manufacturers, and these are usually effective provided that the correct type is used. Some devices are simply braid-breakers, with no filtering of signals travelling down the inside of the coaxial cable, while others include high-pass or band-stop filters optimised for specific bands, so it is important to study the specification before making a decision. In some instances it may be necessary to use two devices (for instance, a braid-breaker in series with a high-pass filter) to achieve the desired rejection. Further information on braid-breakers and filters can be found in Appendix 3 of [1] and in [2].

The RSGB stocks suitable braid-breakers and filters. Information can be found on the Society's web site, http://www.rsgb.org.

Fitting the chokes and filters

Try the chokes and filters in order, starting with the antenna choke (F1 in **Fig 23.10**). Do not remove a device if it does not seem to work - in many cases of breakthrough, the unwanted signals enter the installation by more than one route, and the observed interference is the result of the signals adding or subtracting. If a ferrite choke is only partially successful, another should be tried in series to increase the rejection, but it is usually not worthwhile to go beyond two. When the sources have been identified, chokes which are not needed can be removed.

The chokes fitted to the mains leads (F2 and F3 in Fig 23.10) are made by winding the cable onto ferrite rings in the same way as for a braid-breaker. It goes without saying that in this case it is essential to avoid damaging the cable in any way. If the cable is unusually thick it may be necessary to use fewer turns and more rings.

Due to the closed magnetic circuit in a toroidal core, it is not necessary to pull the cable tight onto the ferrite material - all that is required is for the winding to be secure and the ends not too close together.

Modern TV installations usually use a SCART cable to connect the VCR or set-top box to the TV set. This is to be preferred from the EMC point of view. If an RF link is used it may be necessary to fit chokes on either end of the coaxial cable between the VCR/set top box and the TV set, depending which unit is susceptible. Where it is necessary to fit a ferrite choke to a relatively thick lead such as a SCART lead, it may be possible to use a ferrite yoke from a scrap TV set. (See "Ferrite chokes" above).

Radio and Audio Equipment

In general the approach to radio and audio problems is the same as that for TV, though there are differences of emphasis. Interference caused by the braid of the antenna feeder acting as an unwanted antenna is still a serious problem, but the larger elements on a VHF broadcast antenna make direct pick-up of interference (which is then passed down the feeder in the normal way) more likely.

The leads to the remote speakers of a stereo system frequently pick up interference and this can usually be cured by winding the appropriate number of turns on to a ferrite ring (or pair of

rings) to form a choke at the amplifier end of the lead. Where this is not possible, a split core or a ferrite rod can be used.

Where breakthrough to a portable radio is experienced, there is not very much that can be done from the outside of the set since there are no external leads to choke. However, it is often possible to move the set to a position where the interference is negligible. Frequently portables only give trouble when operated from the mains, and in this case a ferrite choke in the mains lead is likely to be effective.

IF and Intermodulation Breakthrough

In many cases of interference to radio and TV, breakthrough is caused by overloading RF circuits or by rectified RF affecting audio or control circuits. In addition, radio and TV sets can suffer from more subtle forms of breakthrough. These fall into four categories:

- IF breakthrough, where the IF of some piece of radio equipment falls in or near an amateur band. The most common example is a transmission on the 10MHz band getting into the 10.7MHz IF of a VHF receiver. This is breakthrough caused by insufficient IF rejection in the receiver.
- IF breakthrough caused by harmonics of the amateur transmitter being picked up in the receiver IF, for example, the second harmonic of the 18MHz band entering the IF of a TV set. (The standard TV IF band in the UK includes 36MHz.) A less likely possibility is the third harmonic of 3.5MHz entering the 10.7MHz IF of a VHF receiver. In cases of this sort the fault lies with the transmitter which is radiating too much harmonic energy - it is no excuse to say that the susceptible receiver should be better designed.
- Image interference. This where a signal on the 'wrong side' of the local oscillator beats with it to give the IF. It is fairly common on the 1.8MHz band, where amateur signals give image responses on medium-wave receivers. For instance, a receiver with a 455kHz IF, tuned to 990kHz (303m) would have a local oscillator of 1445kHz. This would beat with a strong amateur signal on 1.9MHz, which would be tuned in on the medium-wave band like any other signal. This is a case of breakthrough caused by poor image rejection of the susceptible receiver.
- Amateur signals intermodulating with the harmonics of the local oscillator or other oscillators in the susceptible equipment, causing spurious responses. These give rise to interference may be tuneable at the receiver, but is nevertheless a case of breakthrough.

Security Systems

Breakthrough to security systems has become a problem in recent years. Experience has shown that the standard breakthrough measures such as ferrite chokes and by-pass capacitors, are rarely effective. The most usual cause of breakthrough to security systems is poor immunity in the passive infra red sensors (PIRs). The solution is to replace the offending units with ones which are more immune to RF signals. Occasionally the control panel has been found to have insufficient immunity, but this is unusual. More information will be found in the RSGB's leaflet EMC 03 Dealing with alarm EMC problems [3]. The best place to find up to date information on security system problems is the 'EMC' column of RadCom.

Most security systems are professionally installed, and it is up to the installer to ensure that it is sufficiently immune to operate in the environment in which it is installed. A well installed system with suitably immune PIRs is unlikely to be affected by an amateur station practising reasonable radio housekeeping.

If a complaint of breakthrough to a security system is received, it is advisable to be open with the complainant and to carry out checks to see exactly what the problem is. As in all breakthrough problems check your station and your radio housekeeping, before going any further. Once you are sure of the problem, advise the complainant that the installer should be approached, and give them a copy of the RSGB's Leaflet EMC 02 Radio Transmitters and Home Security Systems [4].

All new installations should be using CE marked units, including the PIRs, though the mark may not always be on the unit itself. It could be on the instruction or on the packing.

Breakthrough to Telephones

The main problem with modern telephones is breakthrough, caused by rectification and amplification of RF currents, and the solution is to avoid the RF getting into the equipment in the first place. There are two courses of action:

- First check your radio housekeeping.
- Prevent common-mode interfering currents getting into vulnerable circuits by fitting ferrite chokes as close to the susceptible units as possible.

Most households have extension telephones connected by plug and socket to the 'line jack' provided by the telephone company and these may involve quite long lengths of interconnecting cable. At the start of an investigation, unplug any extension units and their cables, leaving only the instrument near the line jack connected. Clear any breakthrough on this, and then reconnect the extension leads and telephones one by one, dealing with breakthrough problems as they arise. In some cases it may be necessary to re-route vulnerable extension leads, but in most cases liberal use of ferrite chokes will prove effective. Some telephones may have unusually poor immunity. If this is suspected check by substituting a known good unit. In such cases replacing the telephone may be the most practical option.

Breakthrough may be caused by signals being picked up on the lines before they come into the house. Commercial filters are available which fit into the master line jack, but at the time of writing the only one available in the UK is really intended to deal with medium wave broadcast signals and is only effective on the lower HF bands. However the 'microfilters' used for self-install ADSL have characteristics which might make them useful in cases of HF breakthrough [5]. Where it is suspected that the problem is due to a fault on the lines themselves, contact the authority responsible for the lines, but before doing this it is important to ensure that your radio housekeeping is in order and that you are operating reasonably with regard to all the circumstances.

Breakthrough to Internet Access Systems

This is a very quickly developing field. At the time of writing there are three types of access system:

- Fibre and cable systems
- Radio systems using frequencies above 1GHz.
- Systems which use the telephone lines.

The first two of these do not appear to give rise to many problems. It remains to be seen whether there will be problems should the use of these systems become more widespread.

The dial-up modem which was once the most popular means of internet access is being replaced by DSL systems. DSL stands for Digital Subscriber Line and there is a whole family of DSL systems. The common one in the UK is ADSL. There have been few cases of breakthrough to dial-up modems reported and even fewer to ADSL systems, even though there are now millions of ADSL installations in the UK. It is possible that, in future,

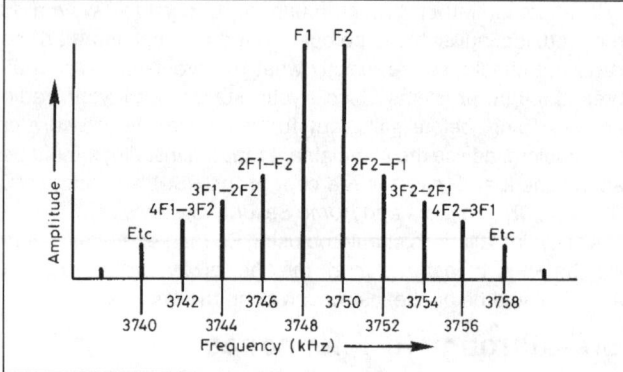

Fig 23.11: Odd-order 'intermods' from a tow-tone SSB signal on the 3.5MHz band

enhanced DSL systems using a wider RF bandwidth might give rise to breakthrough problems. The best place to look for the latest information is in *RadCom* and on the RSGB web site.

INTERFERENCE CAUSED BY UNWANTED EMISSIONS FROM THE TRANSMITTER

As mentioned above, where interference is caused by spurious emissions from a transmitter, the interference can only be to equipment which uses radio in some way, and will be evident on similar equipment in the vicinity which is tuned to the same frequency. This is in contrast to breakthrough which is dependent on the susceptibility of the 'victim' equipment rather than any defect in the transmitter.

Spurious emissions fall into two major categories:

- Predictable spurious emissions which are generated as part of the process of carrier generation, and hence are (at least in principle) predictable.
- Oscillations caused by faults in the design or construction of a transmitter. These give rise to unexpected emissions which can occur on almost any frequency. This is known as instability.

Predictable Spurious Emissions

The commonest predictable spurious emissions are harmonics of the carrier. These harmonics are simply a measure of the distortion of the sine wave which constitutes the carrier; an absolutely pure sine wave would have no harmonic content. In practice the object is to generate a carrier which is as undistorted as possible, and where necessary to filter it so that harmonics are kept to a negligible level.

The other major class of spurious emissions is the products caused by multiplication and mixing processes in the transmitter. In almost all modern transmitters the carrier is generated by mixing the outputs of oscillators (fixed and variable) to produce the final frequency. In most cases there are several mixer stages and at each stage unwanted products will be generated as well as the wanted signals.

The secret of good design is to make sure that unwanted products are at frequencies widely separated from the wanted

signals so that they can be readily filtered out. As with harmonics, the object is to ensure that everything except the wanted output is attenuated to a negligible level.

How small 'negligible' is depends on the frequency bands in use and the power of the transmitter. As a general rule on HF, harmonics should be at least 50dB, and other spurii at least 60dB, below the carrier. On VHF the figures should be 60dB and 80dB respectively.

There is another class of predictable spurious emissions, and this is splatter - the generation of unwanted intermodulation products by the modulation process in an SSB transmitter. The important intermodulation products are the odd-order products because they appear close to the carrier and hence cannot be filtered out. (See also the Building Blocks chapters.) There will always be some intermodulation products generated in any SSB transmitter, and the object is to design and operate the transmitter in such a way that they are kept within reasonable bounds. About 35 or 40dB down on the carrier is a reasonable target to aim for (see **Fig 23.11**).

Identifying and Rectifying Predictable Spurious Emissions

Interference caused by harmonics of a transmitter can usually be identified by consideration of the frequencies on which interference is evident and the frequency on which the transmitter is operating. For instance, if the problem is interference to a VHF broadcast signal on 100.5MHz when the transmitter is operating on 50.25MHz, then it is very likely that second-harmonic radiation is the culprit. Where excessive harmonic radiation is suspected, first check the transmitter and its adjustment and, if the problem persists, fit a low-pass filter as in **Fig 23.12**. Note that the filter comes after the VSWR meter and, where an ATU is used, it should come after the filter.

With non-harmonic spurious emissions (which are often just called spurious, spurii or spurs) the situation is more difficult. The best procedure is to set up some way of monitoring the output from the transmitter to see if any excessive spurious signals are present. If a spectrum analyser or measuring receiver is available the task will be simple, but in most cases resort will have to be made to a general-coverage communications receiver. This can be used to tune round the frequencies on which interference is experienced while the transmitter is operated.

All receivers have spurious responses which can be difficult to differentiate from transmitted spurious emissions, and these will become much worse if the receiver is overloaded. Make sure that the signal from the transmitter, as received on the test receiver, is not greater than about S9 plus 20dB; this will avoid overloading and at the same time enable fairly weak spurious signals to be detected. Once the spurious signal has been identified, the transmitter should be investigated; the frequency of the spurious signal will often indicate the stage which is likely to be at fault. Where the problem is not too severe, using an ATU on HF or a suitable band-pass filter on VHF may be all that is required.

If a spectrum analyser or a measuring receiver is available it should be connected up as in **Fig 23.13**. It is good practice to start with more attenuation than is expected to be required and

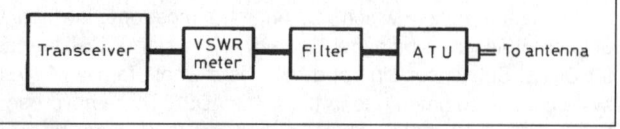

Fig 23.12: Position of the filter. The ATU ensures that the filter sees a 50-ohm load

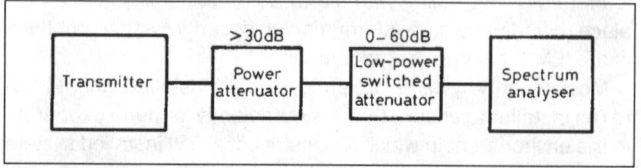

Fig 23.13: Connecting a spectrum analyser to a transmitter

to remove it to establish the desired level. Remember that the attenuator connected to the transmitter must be capable of handling the full transmitter power. More information on suitable high-power attenuators will be found in Chapter 6 of [1].

Where a communication receiver has to be pressed into service, it will probably be simpler to set up the receiver with a short antenna some distance from the transmitter. In such an arrangement the magnitude of any spurious signals detected will be dictated by the antenna characteristics at the receiver, but the technique will none the less be adequate to show up serious spurious emissions.

Spurious Emissions Due to Instability

This is caused by unintentional feedback in some stage of a transmitter causing oscillation. Instability also falls into two categories:

- Direct instability where feedback in one of the amplifying stages of the transmitter causes oscillation on or near the normal operating frequency.

- Parasitic oscillation, where incidental inductance and capacitance cause oscillation at a frequency far removed from that at which the circuit would normally operate.

Both these effects are serious but direct instability is perhaps the most dangerous because it can result in very large spurious signals being generated at a frequency where the antenna can be expected to be a good radiator. In very bad cases the spurious signal can be comparable with the normal output power, causing widespread interference. Fortunately both types of instability are rare in modern, well-designed, transceivers but care should be exercised when repairs have been carried out. This is particularly true where power transistors have been replaced with 'near-equivalents'. The way to avoid instability is by employing good design and construction, with particular reference to layout, screening and decoupling. Where appropriate, parasitic stoppers should be included. See the chapters on transmitters.

Evidence of instability

The two main indicators of instability are bad signal reports and erratic operation of the transmitter. If instability is suspected and suitable test gear is not available, the best procedure is to arrange for a report from someone who can receive you at good strength, and who knows what they are looking for. Ask them to look as far either side of your signal as possible, looking for unstable signals coming and going in sympathy with your main signal. Vary the drive and loading conditions of the transmitter - though in some modern transceiver the ability to do this may be limited. It is important to remember that a transmitter may be stable when operating into a good 50 ohm load but unstable when connected to an antenna.

Erratic operation usually means that the RF output and power supply current varies unpredictably as the drive is varied or as the ATU is tuned but, before jumping to the conclusion that the transmitter is faulty, make sure that the problem is not caused by poor station lay-out. It is not uncommon for RF from the transmitter to find its way into the power supply, or into ancillaries. This can cause very erratic operation. In a modern commercial transceiver, which is normally well behaved, this is more likely to be the cause of erratic operation than instability in the transceiver itself. Check the antenna/earth arrangements, and review your radio housekeeping - see Chapter 3 of [1].

Frequency halving

A spurious emission which doesn't come into either of the above categories is frequency halving. This occurs in bipolar transistor power amplifiers, when the input and output conditions are such

that the parameters of the device are modified between one cycle and the next. The result is a significant output at half the input frequency. The practical effect is that an output at half frequency appears at certain tuning conditions. Divisions other than two are possible but less common. Harmonics are also present in the output and often the most obvious symptom of this effect, is a 'harmonic' at one and a half times the nominal frequency. When tuning a VHF power amplifier into a dummy load, it is advisable to make sure that it is tuned well away from any frequency halving condition otherwise a change of load, such as connecting an antenna, may cause it to change to the halving condition.

INTERFERENCE TO RECEPTION (RFI)

Background Interference Level

On HF there will always be noise picked up by the antenna, however well it is sited - this is usually called the ambient noise. This noise may be either man made or natural, and in most cases is much greater than the thermal noise generated in the front-end of the receiver. The 'bottom line' of this noise is the galactic noise plus the natural static from atmospheric discharges which are always taking place somewhere in the world and are propagated by the ionosphere. Added to this is the man made noise from electrical and electronic devices of various sorts. As frequencies rise into the VHF region, the external noise reduces and becomes 'whiter', until above about 100MHz the thermal noise in the receiver front-end predominates.

There is a great deal of misunderstanding about the ambient noise level on the HF bands. A casual measurement using the 9kHz bandwidth, usually used for EMC measurements, could give the impression that the noise level is so high that relatively high levels of broadband emissions would not cause a significant problem. Arguments such as this have been used in attempts to justify relaxations of the EMC standards. However this is a complete misunderstanding of the HF electromagnetic environment.

Sources of noise in typical residential location tend to fall into two broad categories. Firstly there are the traditional sources such as motors and other sparking devices. Sometimes these can cause quite high levels of interference but in most cases they are in use for only a limited time. The other category is interference from electronic equipment such as computers. In this case the interference may be present for a large part of the day,

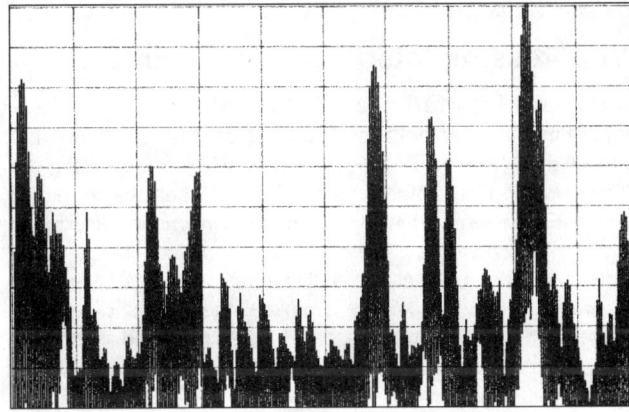

Fig 23.14: Plot taken in the 9kHz IF bandwidth of a measuring receiver tuned to 7.025MHz. Resolution bandwidth 100Hz; vertical scale 5dB/division; horizontal scale 1kHz/division. The bottom line of the graticule represents -15dBμV (-122dBm) at the receiver input. The antenna was an inverted-V dipole

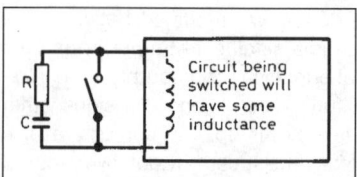

Fig 23.15: A resistor and capacitor used to absorb the energy released when contact is broken

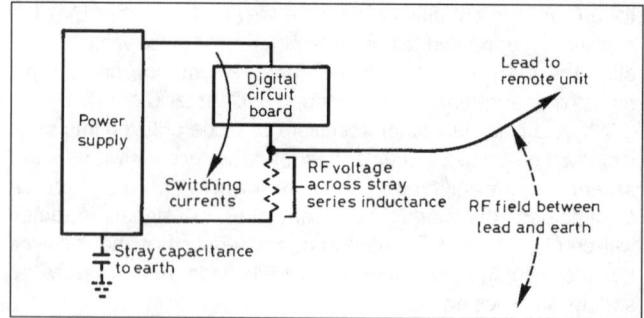

Fig 23.17: External lead, nominally at 0V, acting as an unwanted antenna

but usually the interference approaches the maximum level permitted by the EMC standards only on a few narrow frequency bands. The upshot is that, if the ambient noise is observed on a reasonably sited antenna such as that illustrated in Fig 23.1, a "floor" will be seen consisting of the natural noise, mentioned above, plus an aggregate of distant man made sources. This is fairly constant in any location and has been called the ambient noise floor. To this will be added occasional bursts of higher level broadband interference from local sources, and discrete emissions from computer-like sources, which may or may not fall into the frequency band of interest.

A typical example might be the 7MHz amateur band in a reasonably quiet suburban location around about midday. In an audio bandwidth, the floor of the noise might be in the region of -110dBm. **Fig 23.14** illustrates this. The plot is taken from a spectrum analyser connected to the IF output of a measuring receiver tuned to 7.025MHz. The IF bandwidth of the receiver is the 9kHz "CISPR" bandwidth normally used for HF EMC measurements. This is displayed at 1kHz per division, with a resolution bandwidth of 100Hz. The plot was taken on a Sunday morning, in a typical suburban location.

HF communication services such as amateur radio has grown up coping with localised, intermittent, wideband noise bursts or with discrete emissions, but widely distributed continuous high levels of broadband interference such as might be generated by inappropriate broadband data transmission technologies would be a very much more serious problem.

It is a general rule of radio that good transmitting antennas are also good receiving antennas but, so far as HF is concerned, this is modified by the fact that the ambient noise sets a limit to small-signal reception. For HF reception it may be better to mount a relatively poor antenna in a good location than to have an efficient antenna mounted where the man made noise is high. In difficult situations a relatively small active antenna mounted high up and well away from noise sources may out-perform a much larger antenna which cannot be so well sited - see Chapter 7 of [1]. Tips on identifying and tracking down sources of interference can be found in the leaflet *Interference to amateur Radio Reception* EMC 04 [6]

Suppressing Interference at Source

Impulsive interference
Where mechanical contacts are concerned, there are a number of well-tried remedies based on the principle of absorbing the energy which would otherwise be released when the contact is broken. The energy is initially stored in the magnetic field, due to the normal operating current flowing in any inductance which may be present in the circuit, and in many cases this will be considerable. When the contact is broken, the magnetic field collapses and a large voltage appears for a short period as the con-

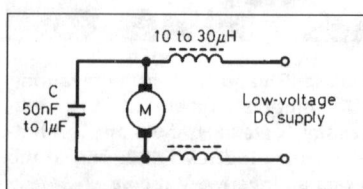

Fig 23.16: Suppressing a small low-voltage DC motor

tacts open, causing a spark. Radio frequency currents are exchanged between the inductance and capacitance in the vicinity of the contact, using the ionised air of the spark as a bridge.

The traditional way of absorbing the energy is to connect a resistor and capacitor across the contacts as in **Fig 23.15**. This effectively quenches the spark, by dissipating the unwanted energy in the resistor. The capacitor should be between 0.01 and 0.1µF and the voltage rating must be several times the voltage being switched. Special capacitors rated for use on AC mains are available, and these must always be used where mains voltages are involved. Units containing a resistor and capacitor in one encapsulated unit can be purchased from component suppliers.

Small, low-voltage DC motors can be suppressed by using a shunt capacitor of between 0.05 and 1µF, and series ferrite cored chokes of 10 to 30µH, as in **Fig 23.16**. The chokes are more effective at higher frequencies, and may not be required if only low frequencies are involved. Mains motors are best dealt with by purchasing a suitable mains filter. This should be installed as close to the machine as practical.

Interference from digital equipment
It is relatively easy to reduce the leakage of interference from digital equipment at the design stage - it is really a matter of good engineering practice. Good decoupling and the provision of a substantial ground plane for the common 0V rail is a good start. It is important to prevent external leads having energy coupled into them from shared return paths, and so acting as antennas (**Fig 23.17**). Ideally, interference-generating circuits should be completely screened. The screen should be connected to the common 0V point through a path which has the lowest possible impedance. Leads should be decoupled where they pass out through the screen.

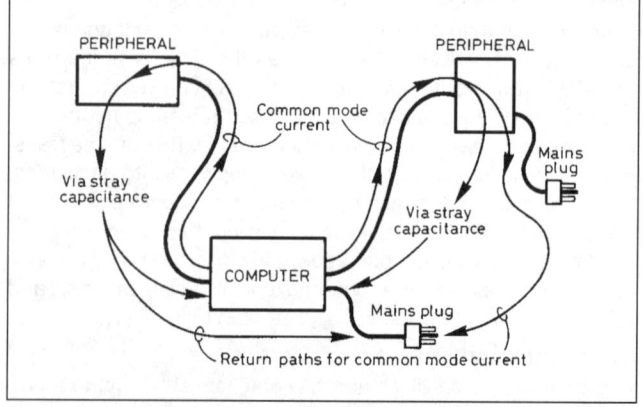

Fig 23.18: Common-mode current paths

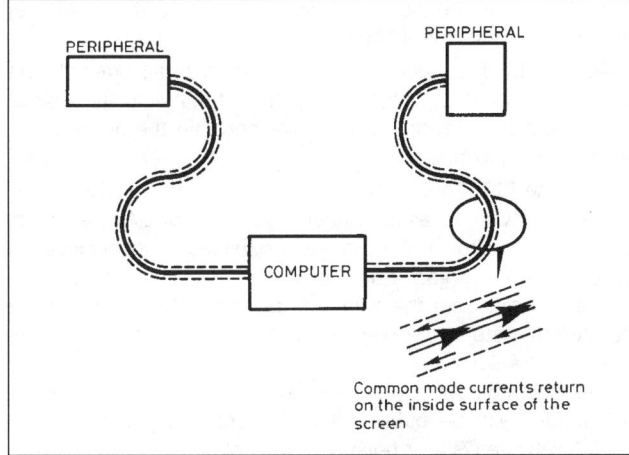

Fig 23.19: Screening confines common mode currents

In situations where interference reduction is a major factor, for instance where digital control circuits are actually part of the receiver, special attention should be paid to screening and feedthrough capacitors should be used on all leads except those carrying RF. It is important to choose the correct value - too large a capacitance will distort fast digital signals. So far as possible, the screen should be continuous and, where there are any joints, there must be good electrical contact along the mating surfaces.

Interference from computer installations

Most modern computer installations are put together from units complying with the requirements of the relevant EMC standards. If the various units are connected together using good quality screened leads, then the EMC performance should at least be reasonable. At HF frequencies most RFI problems are caused by common-mode currents flowing on the interconnecting leads, as illustrated in **Fig 23 18**. Where a screened lead is used to connect two units together, the screen has two functions. As might be expected, it does reduce radiation by forming an electrostatic screen around the conductors, but more importantly it provides a low-impedance path back to the computer for the common-mode currents which would otherwise leak back by devious routes, as illustrated in **Fig 23.19**.

As with all electronic equipment the 0V arrangements on the computer are a compromise, and this sometimes leads to emissions along the lines of Fig 23.17, even if the lead is screened. In such cases a ferrite choke on the lead can be beneficial in reducing RFI. Information on reducing RFI from home computers will be found in [1].

Mains filters

There are a large number of proprietary devices available, mainly advertised as preventing disturbances on the mains from entering computers and similar equipment, but they can be effective against RF interference entering or leaving radio equip-

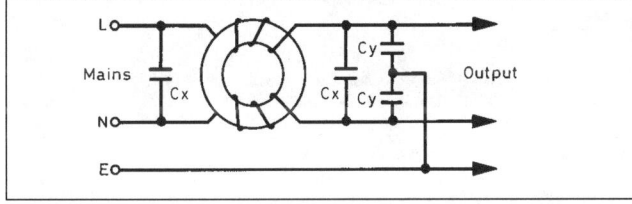

Fig 23.20: Mains filter with balanced inductor (the action of the coil is similar to a very large number of turns of two-core lead on a ferrite ring)

ment. Simple filters consist only of capacitors but more complicated devices include inductors, usually in a balanced arrangement as in **Fig 23.20**. Most filters do not include any inductance in the mains earth line, but there are some more expensive ones which do. Some devices also include semiconductor surge suppressers which serve to limit mains spikes.

When considering a mains filter, make sure that it contains components to suppress both differential and common-mode currents. Some mains filters are sold with a simplified circuit diagram, and if this indicates that it contains an arrangement similar to **Fig 23.20**, then the filter will probably give a reasonable degree of attenuation to both modes. If a mains filter is being considered as an interference reduction measure, it is worth trying a ferrite choke made by winding the mains lead on a ferrite ring, before buying a mains filter. A ferrite choke of this sort has the advantage of reducing common-mode currents on all three conductors (live, neutral and earth). It may be more effective, particularly at higher frequencies and will certainly be much cheaper.

Reducing the Effects of Interference at the Receiver

Where it is not possible to suppress the interference at source, the only solution is to attempt to mitigate its effects at the receiver. There are two traditional techniques available: the first is some form of noise limiter or blanker and the second is by cancellation.

The noise limiter has been around for a long time, but in recent years the it has been displaced by the blanker, which does a similar job in a rather more complex way, and generally much more effectively. More information on this topic will be found in the chapter on HF receivers.

Cancellation is a very powerful tool for dealing with received interference and deserves to be much more widely known. The principle is to receive both the wanted signal and the interference on two antennas and to adjust conditions so that the interference is cancelled, leaving the wanted signal largely unaffected. This is not as difficult as it sounds, and is well worth considering in cases of severe interference. Further information on cancelling will be found in Chapter 7 of [1]

Digital signal processing (DSP) techniques are now being used in receivers to enhance the extraction of wanted signals from interference. These techniques can be very powerful, but determining what can be achieved in any particular situation is a matter of practical testing. If you are attempting to combat specific interference, study all the information, and if possible arrange a test on the interference in question..

INTERNET ACCESS USING TELEPHONE LINES AND MAINS WIRING

Systems Using Telephone Lines - the DSL Systems

ADSL (Asymmetric Digital Subscriber Line)

At the time of writing many households are changing to broadband internet access, and many of these are using systems in which data signals are transmitted to the subscriber's house over the telephone lines. The basic principle is to spread the data signals over relatively broad frequency bands so that transmission rates greatly exceed those obtainable with a dial-up modem. Almost all of these are ADSL (Asymmetric Digital Subscriber Line) which may use frequencies up to about 1.1MHz, but many lower cost systems do not use the whole fre-

quency range, so that energy is concentrated at lower frequencies. It is likely that bandwidths will increase to facilitate services such as home entertainment, pushing the maximum frequency up to about 2.2MHz. ADSL systems have a range of a few kilometres and usually signals are fed into the telephone line at the exchange. At the beginning of 2005 there were over four million subscribers in the UK. So far there have been very few cases of interference, either to or from amateur radio stations reported, so it is not possible to give useful advise on identifying and fixing EMC problems. Information will be published in *RadCom* and on the RSGB EMCC web site [7] as it becomes available.

VDSL (Very high frequency Digital Subscriber Line)

This version of DSL uses frequencies in the HF band and could be more of a problem to amateur radio. At the time of writing it is not deployed in the UK to any significant extent. Because of the high frequencies involved the signals have only a short range - about a kilometre - so its deployment is likely to be more specialised. It is interesting to note that the ETSI VDSL Standard calls for systems to have the potential to be "notched" for the amateur bands should the need arise.

Since all DSL systems use the existing telecommunications cables the powers which can be injected are limited by mutual interference considerations in the cables (cross talk). This has the effect of putting a ceiling on potential radio interference, independent of the EMC Standards.

Systems Using the Electricity Supply Cables

PLT (PLC) (Powerline Telecommunication or Communication)

This is the use of electricity wiring to carry data signals. Low frequency signalling on electricity supplies has quite a long history, but this involves very low frequencies, and is completely different from the internet access systems which use frequencies in the HF band.

PLT is not really a single identifiable technology but a name for systems which use the electricity cabling to carry high speed data signals. It is worth noting that while information on frequencies and launch powers on DSL systems are in the public domain, similar information on PLT is not readily available.

There are two categories of PLT, Access PLT which provides internet access via the electricity supply cables and In-House PLT which uses the electricity wiring to link together computer equipment in the same house.

Access PLT: The proposed systems vary in detail, but in the UK they involve injecting the data signals at the electricity sub-station. The signal travels up the cable and into the houses supplied by the sub-station.

In-House PLT: This links computers and other devices using the electricity wiring as the transmission medium. At the time of writing most in-house PLT systems are based on a specification agreed by an organisation of interested companies. This includes notches for the amateur bands. It is known that wider bandwidth systems are being considered, but, to date, no information is available.

The situation on all aspects of PLT is changing rapidly. Information will be published in *RadCom* and on the RSGB EMCC web site [7] as it becomes available.

FURTHER ASSISTANCE

Where an EMC problem fails to yield to the procedures that have been discussed in this chapter, it is time to call for some help. In most cases this means contacting your National Society. In the UK, the RSGB's EMC Committee has set up a countrywide network of EMC co-ordinators to advise members on EMC problems. Contact information is published in the *RSGB Yearbook* [2] and on the EMC Committee web site [7] which also has general EMC information and links to other sites. The EMCC web site can be accessed via the main RSGB site.

REFERENCES

[1]   *The RSGB Guide to EMC*, Robin Page-Jones, G3JWI, RSGB, 1998.

[2]   *The RSGB Yearbook*, RSGB, published annually

[3]   'Dealing with Alarm EMC problems', leaflet EMC 03. Available from RSGB*

[4]   'Radio Transmitters and Home Security Systems', leaflet EMC 02. Available from RSGB*

[5]   'EMC' column, *RadCom*, Feb 2005

[6]   'Interference to Amateur Radio Reception', leaflet EMC 04. Available from RSGB*

[7]   RSGB EMC Committee web site. Follow the link from http://www.rsgb.org

*   *The EMC leaflets can be downloaded from the RSGB EMC Committee web site, (see above)*

24 Power Supplies

Amateur radio equipment normally derives its power from one of four sources:

- The public AC mains which is nominally 230 volts at 50Hz in the EU, though 240 volts still exists in the UK.
- Batteries, which are either primary (non-rechargeable) or secondary (re-chargeable)
- Engine driven DC generators or alternators, whether separate or incorporated in a vehicle.
- Renewable sources such as wind, possibly water turbine or pedal driven generators, solar cells or rarely, thermocouples. As these are intermittent in nature, a secondary battery with a regulator must be used in conjunction.

For fixed stations the AC mains is readily available, is relatively cheap and is almost always used. It can be converted by transformers, rectifiers, smoothing circuits or switch mode (high frequency) circuits to a wide range of direct voltages and currents necessary for amateur equipment use.

Batteries, both primary and secondary (accumulators) have always provided a convenient though relatively expensive source of power, especially for low powered or portable rigs, or test equipment.

Discrete engine driven generators can give an output of DC or AC, but the most popular give an output of 240/230 V AC at a nominal 50Hz, matching the domestic mains. Car alternators at present provide charging for a 12V accumulator, though 42V may soon be common.

Renewable sources provide AC or DC according to type, and are discussed in detail later.

SAFETY

The operation of all power supplies (except perhaps low-voltage, low current primary batteries) can be dangerous if proper precautions are not taken. The domestic mains can be lethal. High voltage power supplies for valve equipment can also be lethal and great care must be taken with them. There is a case on record of 12 volts proving fatal, for it is not voltage which kills you but current, and the law of the late German doctor (Georg Siemon Ohm) applies. Having wet hands is asking for trouble; if you must handle high voltages, do so with one hand in your pocket.

Where petrol or liquefied gas is used for your engine's fuel, there is a fire hazard, particularly when re-fuelling. If you persist in re-fuelling while the engine is running, and have a fire, you will not get much sympathy from either your insurance company or relatives.

SUPPLIES FROM THE PUBLIC MAINS

It is assumed that the supply will be a nominal 230 volts at 50Hz in accordance with EU regulations.

A rectifier (or rectifiers) will be needed to convert AC to DC. Rectifiers are nearly always silicon PN junction or Schottky diodes but Silicon Carbide diodes will be met occasionally (see the chapter on semi-conductors for more on rectifier diodes).

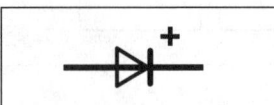

Fig 24.1: Rectifier symbol

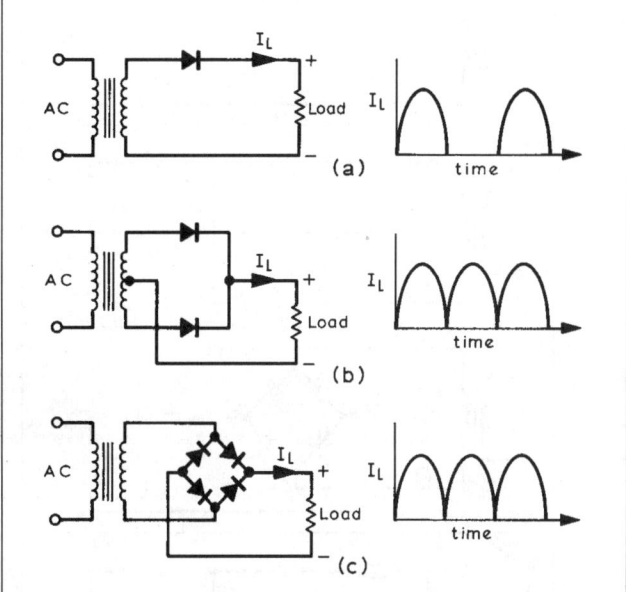

Fig 24.2: Rectifier circuits showing the output current waveforms with resistive loads. (a) half-wave, (b) full-wave or bi-phase half-wave, (c) bridge

These are all very efficient in that they have a very low forward voltage drop and a very high reverse resistance, within their rating. Copper oxide, selenium and germanium diodes are obsolete for this purpose. The diode has a conventional symbol (**Fig. 24.1**) in which the arrow points in the direction of current flow, not electron flow - so the arrow head is the anode, and the plate the cathode. By convention it is the cathode, which is marked +, banded or coloured red because it is positive when rectifying AC.

Fig. 24.2 shows three types of rectifier circuits which are widely used in amateur Power Supply Units (PSUs), together with waveforms of the current supplied by the rectifier(s). In all cases this can be resolved into a DC and an AC component. The latter is called ripple, and requires removal. It can be seen that the half-wave circuit of **Fig 24.2(a)** has a worse ripple than either of the full wave circuits, and has the lowest frequency component (50Hz). Also notice that DC flows through the transformer, which may cause saturation and over-heating if the transformer is not designed for this purpose. Consequently it is not much used.

The full-wave circuit of **Fig 24.2(b)** needs a centre-tapped transformer while the bridge circuit of **Fig 24.2(c)** does not. The bridge incurs two diode (voltage) drops but uses the transformer winding more efficiently. Microwave oven transformers are not designed to use bridge rectifiers, and the insulation (if any) of the low potential end is inadequate, but the circuit of Fig 24.2(a) may be used if the high leakage reactance of the transformer can be tolerated. The oven power supply cannot be used without modification, as it is neither smooth enough nor of the right polarity to feed a valve amplifier.

In most cases the rectifier diodes feed a large capacitor (the smoothing or reservoir capacitor), which stores energy during the part of the cycle when the diodes are not conducting. In a few cases the diodes feed an inductor (choke) which is then followed by a capacitor. The choke is heavy and expensive, but

Circuit	DC output voltage	PIV across diode	Diode DC current	Diode peak current	Secondary RMS current
	$0.45V_{ac}$	$1.4V_{ac}$	I_L	$3.14I_L$	$1.57I_L$
	$0.9V_{ac}$	$2.8V_{ac}$	$0.5I_L$	$1.57I_L$	$0.785I_L$
	$0.9V_{ac}$	$1.4V_{ac}$	$0.5I_L$	$1.57I_L$	$1.11I_L$
	$1.4V_{ac}$ (no load)	$2.8V_{ac}$ maximum	I_L	See Fig 15.8	= Diode RMS current See Fig 15.7
	$1.41V_{ac}$	$2.82V_{ac}$	$0.5I_L$	See Fig 15.6	I_L
	$1.4V_{ac}$ (no load) See Fig 15.6	$1.4V_{ac}$ maximum	$0.5I_L$	See Fig 15.8	= Diode RMS current x 1.4 See Fig 15.7
	$0.9V_{ac}$	$1.41V_{ac}$	$0.5I_L$	$2I_L$ when $L = L_C$	$0.65I_L$
	$0.9V_{ac}$	$1.4V_{ac}$	$0.5I_L$	$2I_L$ when $L = L_C$	$1.22I_L$ when $L = L_C$

CD1665

Table 24.1: Operating conditions of single-phase rectifier circuits

The Radio Communication Handbook

Type	VRRM	Iav	IFRM	IFSM
Diodes				
1N4001*	50	1.0	10	20
1N4007*	1000	1.0	10	20
1N5401†	100	3.0	20	60
1N5408†	1000	3.0	20	60
BY98-300	300	10	50	100
BY98-1200	1200	10	50	100
BY96-300	300	30	100	200
BY96-1200	1200	30	100	200
Bridge rectifiers				
1KAB10E	100	1.2	25	50
1KAB100E	1000	1.2	25	50
MB151	100	15	150	300
MB156	600	15	150	300
GBPC3502	200	35	200	400
GBPC3506	600	35	200	400

Note: The diodes marked * and † are wire ended - the rest are mounted on screwed studs. V_{RRM} is the maximum reverse voltage or peak inverse voltage, Iav is the average output current in amps, I_{FRM} is the maximum repetitive peak current in amps, I_{FSM} is the maximum non-repetitive peak forward current with a maximum duration of 5ms.

Table 24.2: Electrical characteristics of some common diodes and bridge rectifiers

does lower the peak current in both diodes and capacitor.

A rectifier diode for mains frequency has three important parameters:

- Maximum mean forward current
- Maximum peak forward current
- The peak inverse voltage, which is encountered by the diode when it is not conducting. This is made up of the instantaneous voltage of the transformer, ie when it is at its negative peak, added to the voltage of the smoothing capacitor. Placing diodes in series is a means of increasing peak inverse voltage, but shunt resistors may be needed to ensure correct voltage distribution

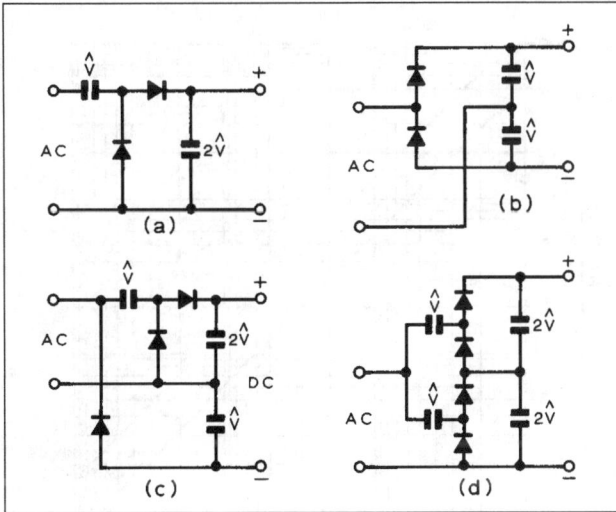

Fig 24.3: Voltage multiplier circuits. (a) Half-wave voltage doubler; (b) full-wave voltage doubler; (c) voltage tripler; (d) voltage quadrupler. V^ = peak value of the AC input voltage. The working voltages of the capacitors should not be less than the values shown

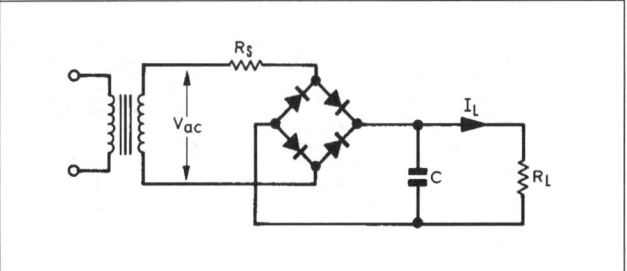

Fig 24.4: Bridge rectifier with capacitor input filter

Table 24.1 shows the voltages and currents associated with various configurations. Table 24.2 gives the parameters for some of the more common of the many diodes now available.

Other categories are fast recovery diodes which are used along with Schottky (low forward voltage and fast recovery time) diodes in switch mode power supplies and soft recovery diodes, which because they switch relatively slowly, cause less RF interference.

Avalanche diodes break down on over-voltage throughout the silicon chip and not a localised spot.

There are many packages of four diodes as a bridge, ready to mount on a heat-sink. These are relatively cheap, but watch the manufacturer's peak inverse voltage rating; does it apply to the individual diodes or to the bridge? Table 24.2 also lists some common bridge packages.

Voltage Multipliers

When a DC voltage greater than the peak of the available AC is needed, a voltage multiplier circuit can be used. These can give a large voltage multiplication, but with poor regulation (decrease of output voltage with increase of load current). The operation may be visualised roughly by thinking of the diodes as 'ratchets'. The mechanical ratchet passes motion in one direction only, and in the multiplier, each stage 'jacks up' the voltage on the following stage. Fig 24.3 shows some of the possible circuits.

Smoothing Circuits

These are low pass filters which follow the rectifier diodes, and the behaviour of the circuit depends on the input element of the filter. This is usually a capacitor in mains frequency circuits and a choke in switch mode PSUs. There may be further components where a greater ripple reduction is needed (see later in this chapter).

Capacitor input

An example is the bridge rectifier of Fig 24.4 in which R_s is the effective resistance of the transformer (resistance of the secondary, plus the turns ratio times the primary resistance).

Fig 24.5 shows voltage and current waveforms of the capacitor as it charges up during part of the cycle and discharges during the rest. The ratio of output voltage to peak input voltage depends on the size of the capacitor, the load and effective resistance in series with the rectifier. Fig 24.6 shows the relationship graphically being 2π times the input frequency.

The charging and discharging of the reservoir capacitor constitutes the ripple current and all electrolytic capacitors have a maximum ripple current rating (see the chapter on passive components). Exceeding the maximum will overheat the capacitor and shorten its life. Ripple current is difficult to calculate, but can be measured by using a true RMS ammeter, or estimated at three times the DC rating of the PSU.

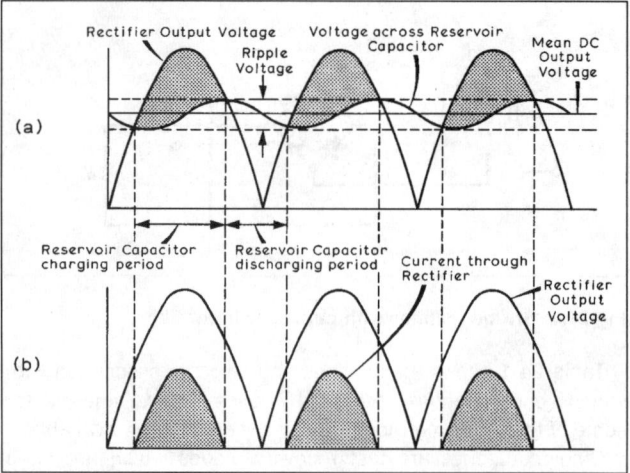

Fig 24.5: Curves illustrating the output voltage and current waveforms from a full-wave rectifier with capacitor input filter. The shaded portions in (a) represent periods during which the rectifier input voltage exceeds the voltage across the reservoir capacitor, causing charging current to flow into it from the rectifier

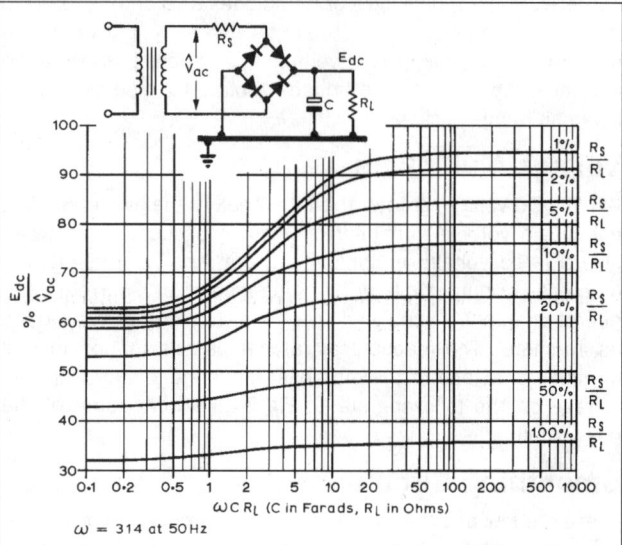

Fig 24.6: Output DC voltage as a percentage of peak AC input voltage for a bridge rectifier with capacitor input filter

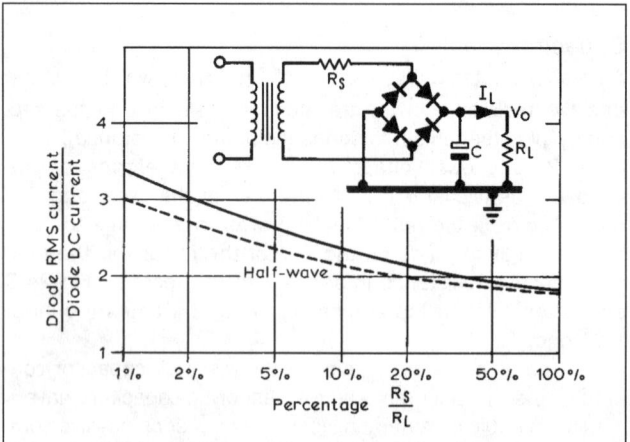

Fig 24.7: Relationship between diode RMS current and percentage R_S/R_L for values of ωCR_L greater than 10. (ω = 314 for 50Hz mains). The dotted line applies to half-wave rectifiers

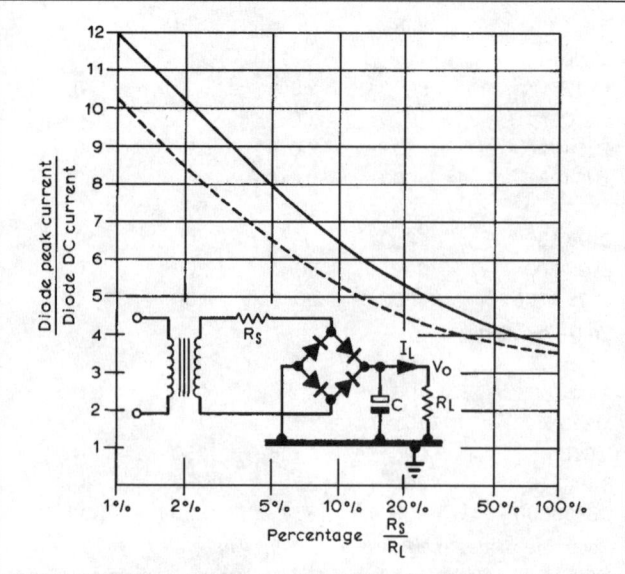

Fig 24.8: Diode peak current as a ratio of diode DC current for values of CRL greater than 10. Note: in a bridge rectifier circuit, diode DC current is half the load current. The dotted line applies to half-wave rectifiers; in this case the diode DC current is equal to the load current

As the effective resistance of the transformer becomes a smaller and smaller fraction of the load resistance, R_L, the peak rectifier current increases (see **Figs 24.7 and 24.8**). With increasing C (Fig 24.4) the peak rectifier current increases and the ripple current decreases (**Fig 24.9**). The efficiency also decreases, meaning more transformer and diode heating. A simple idea to avoid destroying the diodes by excessive current, is to limit the maximum peak current by adding a series resistor to augment the R_s of Fig 24.4. The minimum value of R_s is given by

$$R_s = V/I_{FRM} \qquad (1)$$

where V is the output voltage of the transformer and I_{FRM} the maximum peak current for the diode.

Then calculate the effective resistance of the transformer as already explained. If it is more than the value calculated above,

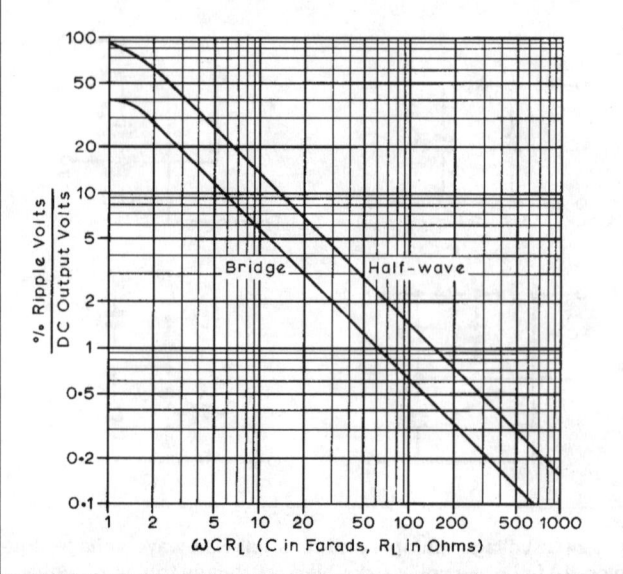

Fig 24.9: Percentage ripple voltage (RMS) against values of ωCR_L ($\omega = 2\pi f$ where f is the mains supply frequency)

24.4

no augmentation is necessary. If it is less, add a series resistor to make up the difference, bearing in mind the necessary power rating.

Soft starting

When first switched on, the capacitor's inrush current may be excessive and there are means of limiting it.

A simple way is to connect an NTC thermistor (see the chapter on passive components) of correct current rating in series with the primary of the transformer, as in **Fig 24.10(a)**. Another way is to have a resistor in series with the primary, which is shorted by a relay whose coil is in parallel with the reservoir capacitor (**Fig 24.10(b)**).

This relay must be chosen so that it closes at about 75% of the normal output voltage. By putting the limiter in the primary, the doubling of the magnetising current and possible magnetic saturation when the transformer is switched on at a zero crossing of the mains, is avoided.

A refinement of this circuit is shown in **Fig 24.10(c)**, where over-voltage protection is also provided.

Inductor (choke) input

Here the situation is different (see **Fig 24.11**) - if the inductor's value is above a certain limit (see below), current flows during the whole time (**Fig 24.12**), much reducing the peak value The critical value, (L_c) for the inductance in a full wave circuit is:

$$L_c = \frac{R_S + R_L}{6\pi f} \qquad (2)$$

Where L_c is in Henrys, f is the supply frequency in Hertz, and resistances are in ohms. If R_S is much less than R_L and the frequency of the supply is 50Hz, for a full wave rectifier, this reduces to:

$$L_c = \frac{R_L}{940} \qquad (3)$$

It will be seen that the inductance required increases as the load current decreases, (the load resistance R_L increases), so it may be necessary to provide a minimum current by means of a bleeder resistor if the inductor input is to remain effective (the output voltage will rise if it is not).

The value of this resistor in ohms is 940 times the maximum value of the inductor in henrys. The inductance of an iron cored coil depends on the current through it.

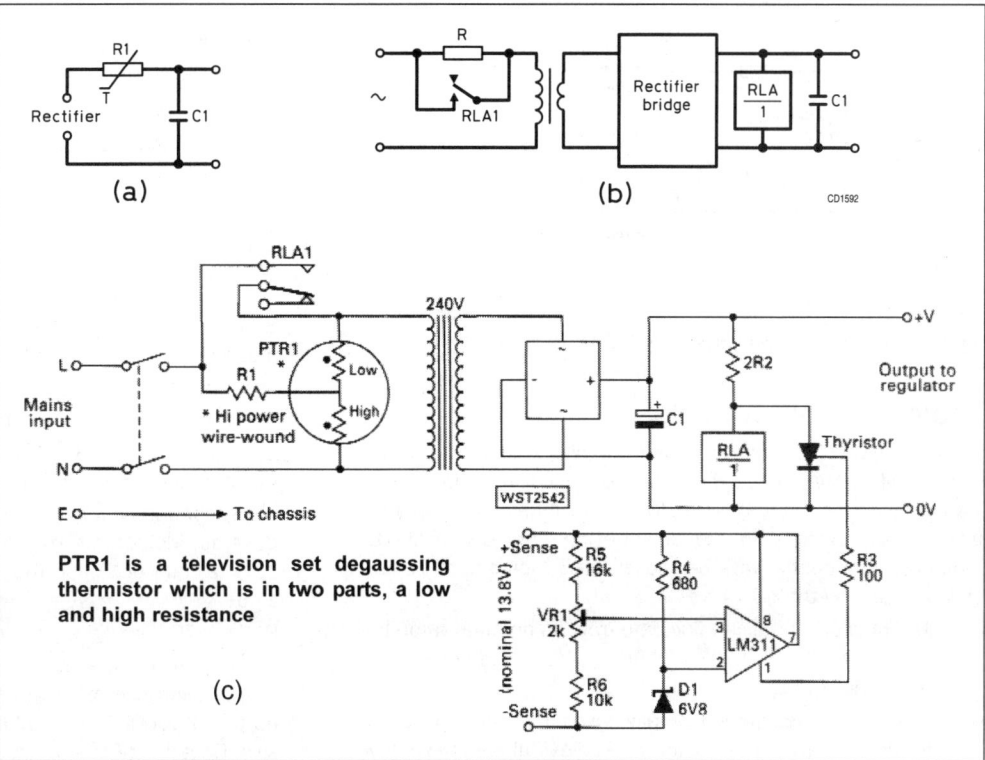

(a)

(b)

(c)

PTR1 is a television set degaussing thermistor which is in two parts, a low and high resistance

Fig 24.10: Soft starting circuits. C1 is the smoothing capacitor. (a) Using R1, an NTC thermistor. (b) Using a relay to short out a series resistor. (c) A more sophisticated circuit. When power is applied, C1 charges up slowly because R1 and the low thermistor limit the inrush current. When C1 is sufficiently charged to operate RLA, the relay closes and RLA1 puts full mains voltage across the transformer primary. The high resistance part of the thermistor remains hot and keeps the 'low' part high. The circuit also provides over-voltage protection by 'crow-barring' C1 through the 2R2 resistor. An enhancement is to place a neon lamp in series with a 150Ω resistor across the relay contact; this flashes briefly at switch-on and stays on if overvoltage occurs or an attempt is made to switch on with a load connected. [Fig 24.10(c) is reproduced by permisson from *Practical Wireless*]

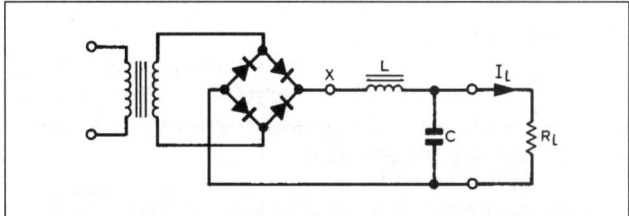

Fig 24.11: Bridge rectifier with choke input filter

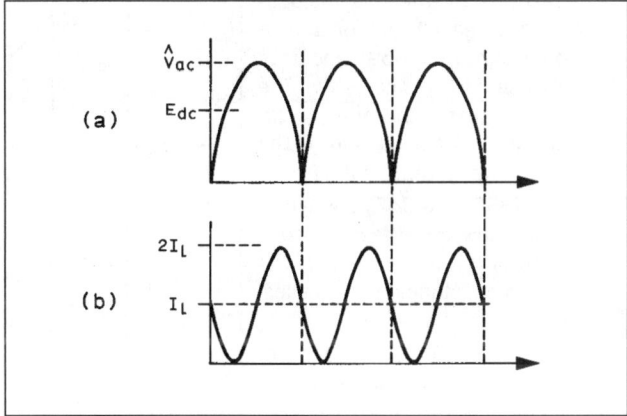

Fig 24.12: Waveforms at rectifier output (point X in Fig 24.11) in a choke input circuit. (a) Voltage waveform. (b) Current waveform (L = L_C)

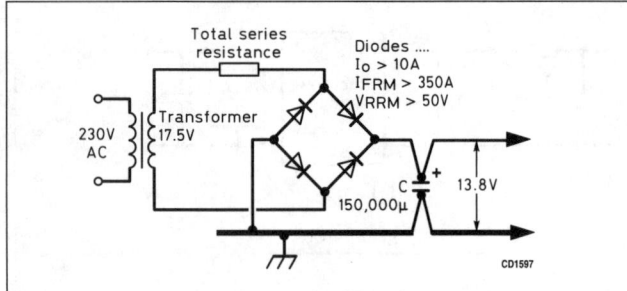

Fig 24.13: Circuit of a power supply for 13.8V at 5A

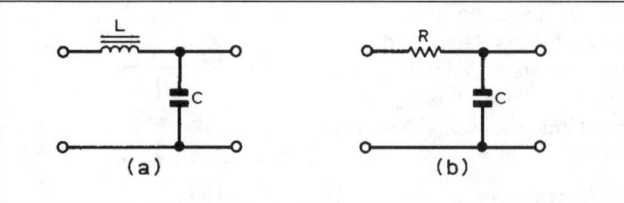

Fig 24.14: Additional smoothing sections to follow the circuits of Fig 24.4 or 24.11

Choice of Components

The ideas given here are for power supplies which will always work but which may not be the most economical in components. The reason for this is that generally components are cheap, but time is important and troubleshooting can be difficult. Components should always be chosen on a 'worst case' basis. That is assuming that:

- The mains voltage can fluctuate from its nominal value by ±10%. This does not allow for drastic load shedding during a very hard winter!

- Electrolytic capacitors generally have a tolerance of +50% to -20%, so that a capacitor marked 100µF can have any value between 150µF and 80µF

- Rectifier diodes should have a peak inverse voltage rating of at least three times the output voltage of the transformer (if one is used), when the mains voltage is 10% higher than nominal. This does not allow for spikes on the mains, and either an even higher rating should be adopted or some means of spike reduction installed.

- Choose a diode or diodes with an average current rating of at least twice the required value.

Capacitor input

1   The secondary resistance of the transformer is assumed to be 0.1Ω and the primary resistance 5Ω. The turns ratio (see below) is 0.0729 to 1 so the effective resistance of the primary transferred to the secondary will be 0.027Ω and the total effective resistance of the transformer is 0.127Ω.

2   The bridge rectifier is a type SKB25/02, which is rated at an average current of 10A, a peak current of 359A and peak inverse voltage of 200V. The RMS transformer output voltage is 17.7V (see below). The minimum value of R_s is 17.7/350 = 0.047Ω, which is well below the effective resistance of the transformer, so no added resistance will be necessary.

3   R_s/R_L = 0.047 /2.76 = 0.017 or 1.7%. The average diode current is 2.5A (two diodes share the 5A).

4   Referring to Fig 24.8, the peak diode current is 12 times this ie

30A and the RMS current from Fig 24.7 is 3.2 times the mean ie 8A.

5   Assume ωCR_L = 100 (this is an arbitrary choice based on the need for low ripple voltage - see later), so E_{DC}/E_{peak} = 0.95 (from Fig 24.6 which also shows that there is not much to be gained by increasing ωCR_L further). Therefore the secondary voltage = 13.8/0.95 = 14.5. However this does not include the voltage drop in each diode of about 1.5V (a total of 3V) so the secondary voltage required is 17.5V.

6   R_L is 2.76 so C is 100/(2π50 x 2.76) which equals 115,000µF, the mains frequency being 50Hz. Electrolytics have a tolerance of -20 to +50% so this would be scaled up to 150,000µF, an available value if somewhat expensive. Because of this, a higher voltage is often rectified, a smaller capacitor used, and the ripple removed by a voltage regulator circuit - see below.

7   Fig 24.9 gives the ripple as about 0.6%.

The input and output leads to the capacitor should have as little in common as possible to avoid the introduction of ripple. Take them independently to the capacitor terminals as shown in **Fig 24.13**.

The design of all other power supplies working from the mains, no matter what voltage or current, follows the same rules except the following:

- When an input inductor (choke) is used (see below).

- Where a 12 or 24V secondary battery (usually a vehicle battery) is 'floated' across a DC supply and takes the place of the reservoir capacitor.

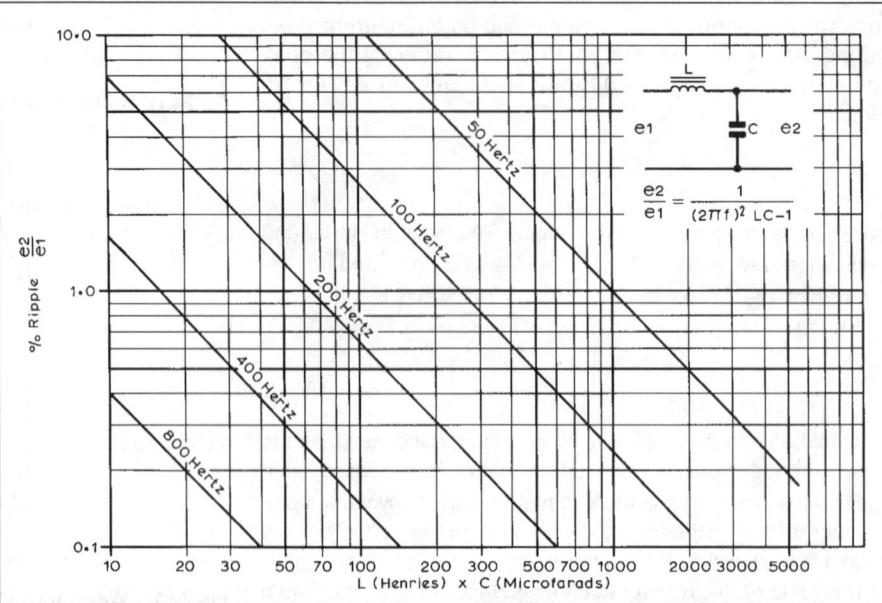

Fig 24.15: Relationship between large ripple and product of LC

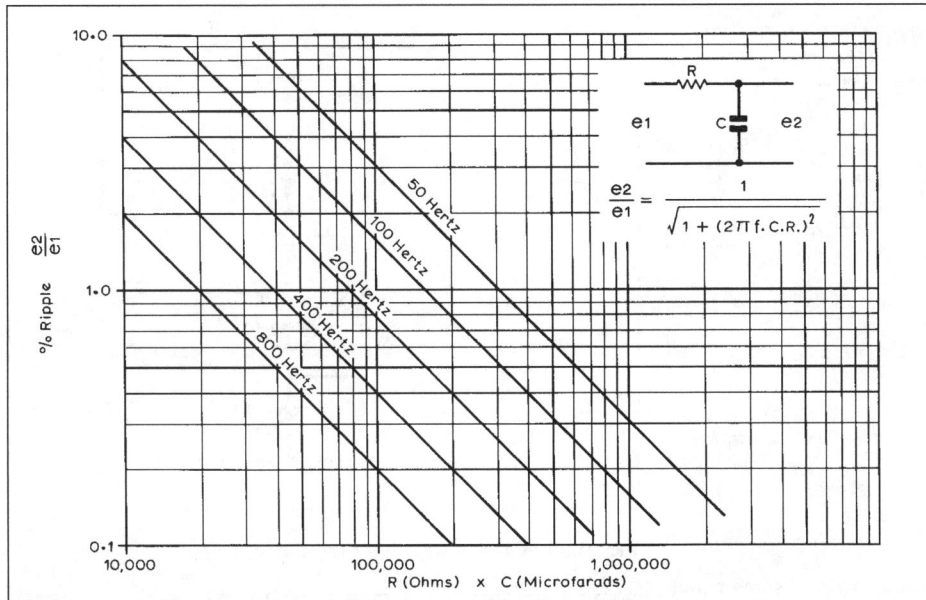

Inductor (Choke) input

Here the inductor is connected directly to the rectifier diodes (Fig 24.11) and is followed by a smoothing capacitor. Note that it is equally possible to use a centre-tapped transformer and just two diodes to make a full wave rectifier.

As mentioned above in equations (2) and (3), the inductor must have a certain minimum value for a given load. This must be calculated for the smallest current expected, which may only be that of a bleeder, if fitted.

The voltage and current values and waveforms for a circuit using at least this value of inductance are shown in Fig 24.12. It is clear that the peak rectifier current is only double the mean current. Under these conditions the transformer RMS current is 1.22 times the load current and the average current per diode is half the load current.

The output voltage is 0.9 of the transformer RMS output minus voltage drops in winding resistance of both transformer and choke less the diode(s) drop. The diode drop can be neglected for power supplies above say 100V and estimated from manufacturer's data where significant. (Power diodes do not just drop 0.6V!)

The value of the filter capacitor (the first if more than one) is arranged to give the wanted ripple voltage ER from the equation:

$$ER = \frac{\text{output voltage}}{0.8LC} \quad (4)$$

where L is in henrys, C in microfarads and the supply frequency 50Hz (full wave rectification).

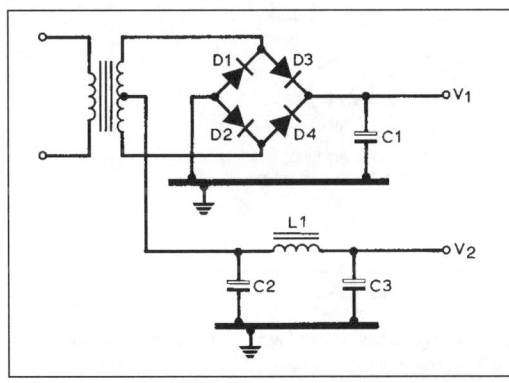

Fig 24.17: Circuit of a dual-voltage power supply

Further smoothing can be added to either capacitor or inductor input by means of LC or RC circuits as in **Figs 24.14, 24.15 and 24.16**. The RC circuit is unsuitable for high currents because it drops the voltage.

Care must be taken that L and C do not resonate at the ripple frequency or any harmonics thereof. For a full wave circuit on 50Hz, the lowest ripple component is at 100Hz and the LC product is 2.53 for resonance. Normally LC is very much higher than this (see Figs 24.15 and 24.16). Note that the choke should be designed so that it does not become magnetically saturated at full DC output. An air gap in the core will help in this respect. The 'swinging choke' is so designed that it does approach saturation at full output and has a higher inductance at low currents where it is needed to satisfy equation (3).

Dual Power Supply

The circuit of **Fig 24.17** shows a dual voltage supply from a centre-tapped transformer. On analysis, although it uses a bridge rectifier, the two halves of the bridge feed the two supplies separately. The earth point may be chosen to give V and V/2 or +V and -V as required.

Voltage Regulators or Stabilisers

These are circuits that give a virtually constant voltage output regardless of load or input, within certain limits. They are necessary for supplying variable frequency oscillators (VFOs), DC amplifiers and some logic circuits. At higher voltages they are necessary for supplying screen and/or grid bias for valve amplifiers.

Shunt regulators - zener diodes

These are diodes with a sharp breakdown voltage, (see the chapter on semi-conductors). If fed from an un-regulated source through a resistor (**Fig 24.18**) it forms a simple regulator for more or less constant loads, with the advantage of being able to source or sink current.

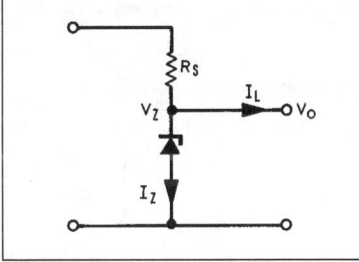

Fig 24.18: A simple zener diode voltage regulator

Type	Zener voltage (V)	Normal operating current (A)	Zener slope resistance (Ω)	Maximum dissipation (W)	Temp coeff (mV/°C)
BZX79C2V4	2.4	0.005	100	0.4	3.5
BZX79C6V2	6.2	0.005	10	0.4	+2.0
BZX79C75	75	0.002	255	0.4	+80
BZT03C7V5	7.5	0.1	2	3.25	+2.2
BZT03C270	270	0.002	1000	3.25	+300
BZY91C10	10	5.0	NA	75*	NA
BZY91C75	75	0.5	NA	75*	NA

* On a heatsink.

'NA' means 'not available'.

While these are all Philips devices, all semiconductor manufacturers make zener diodes in one or more sizes. The data on Philips devices were used for this table because they were to hand and not because they are recommended above other makes. This table represents the extremes of size, power dissipation and voltage.

Table 24.3: Electrical and thermal characteristics of some zener diodes

There are reference diodes whose breakdown voltage is nearly independent of temperature, usually available in the 8-10 volt region, at a specified current. Below 5 to 6 volts the zener diode has a negative temperature coefficient and above, a positive one. **Table 24.3** gives some figures for typical zener diodes made by Philips.

The series resistor value (R_s) is calculated so that the diode provides regulation when the input voltage (V_s) is at its minimum when the load current (I_L) is a maximum. It is important to check that the diode's maximum dissipation is not exceeded when these conditions are reversed, ie when V_s is at its maximum and I_L a minimum. The expression for the series resistor is:

$$R_s = \frac{V_{s(min)} - V_{zener}}{I_{L(max)} + I_{zener(min)}}$$

The resistor must be rated for

$$\frac{(V_{s(max)} - V_{zener})^2}{R_s} \text{ watts.}$$

Shunt regulators - integrated circuits

There are many voltage reference devices made for shunt regulator purposes, ranging from 1.22 up to about 36V. Some have three terminals to allow fine adjustment. They are used in the same way as the zener diodes previously described

Shunt regulators - high voltage

Beam tetrodes used for linear power amplifiers need a screen supply, and the 4CX250 series in particular can sink or source screen current. A series string of zener diodes could be used to cope with this, but would use expensive high wattage diodes. The circuit of **Fig 24.19** transfers the problem to a cheaper

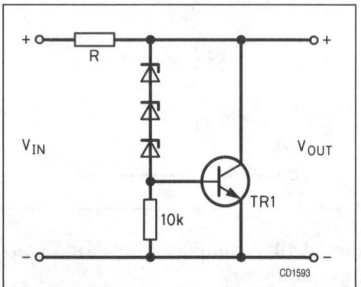

Fig 24.19: Stabiliser for screen of beam tetrode

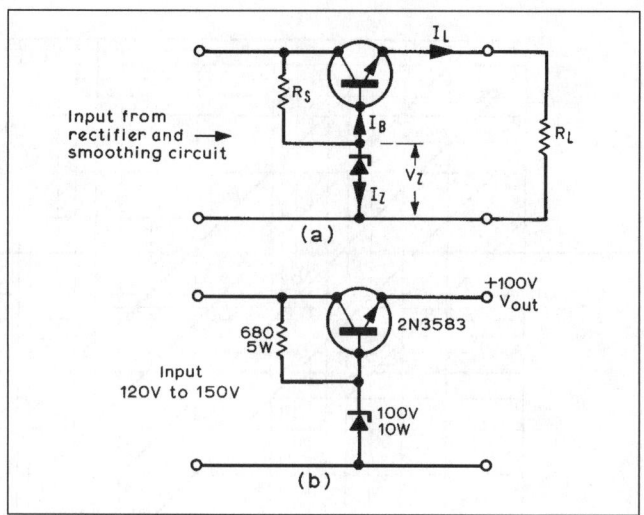

Fig 24.20: A series transistor regulator

power transistor, but does not remove the need for a series string of diodes. These can now be of low wattage and so cheaper. The higher the h_{FE} of the transistor or Darlington pair, the cheaper will be the diodes. Formerly gas filled voltage regulators were used, and are still available. As long as you allow for the difference between the striking and running voltages, the procedure is the same as that for zener diodes.

Series regulators

In these an active device is placed in series with the supply, and negative feedback applied in such a way that the output voltage remains constant in spite of varying load and input voltage. The output voltage, or a definite fraction of it, is compared with a reference voltage and the difference amplified to control the series pass element in such a way as to minimise the difference. The greater the gain of the amplifier, the better will be the final result, provided the circuit is stable. A single transistor may be good enough in less demanding situations.

The pass element can be either a bipolar or field effect transistor. Protection from failure of the pass transistor is advised to avoid damage from over-voltage (see below).

Figs 24.20 and 24.21 show the simplest type of regulator using only two semiconductors. As previously said, they are only suitable in less demanding situations. There is not much excuse for building voltage regulators out of discrete components, as integrated circuits are cheap and often better.

IC voltage regulators

There are many IC voltage regulators, of which only linear types will be considered here. They all have the elements so far

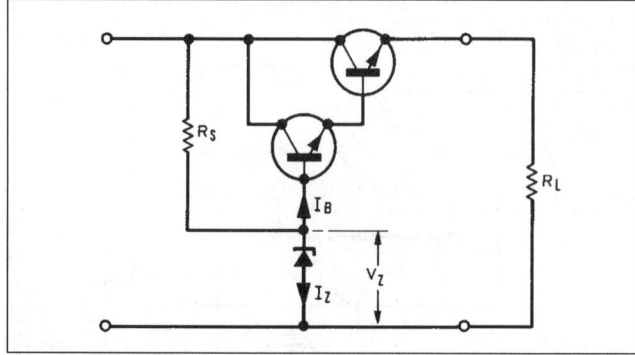

Fig 24.21: A series regulator using two compounded transistors as the series element

The Radio Communication Handbook

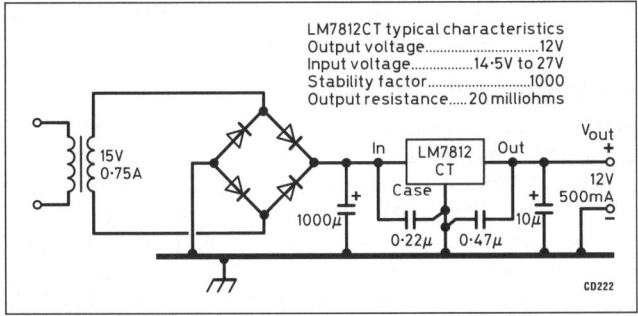

Fig 24.22: A 12V 500mA power unit using a regulator type LM7812CT

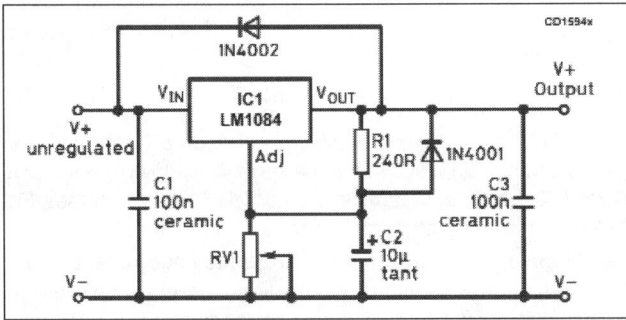

Fig 24.23: Adjustable voltage regulator using an IC. The value of RV1 is $(V_{out} - 1.25)$R1/1.25 ohms

Type	Voltage	Current	Polarity	Vin(min)	Vin(max)
Fixed voltage					
MC78L05APC	5.0	0.1	+	6.9	30
MC79L05APC	5.0	0.1		6.9	30
LM78 12CT	12.0	1.0	+	14.6	35
LM79 12CT	12.0	1.0		14.5	35
LT1086CT12	12.0	1.5	+	13.5	25
MC78T 15CT	15.0	3.0	+	17.8	40
78P 05SC	5.0	10.0	+	8.25	40
LM2931A	5.0	0.4	+	5.65	25*
Adjustable voltage					
LM317LZ	1.2-37	0.1	+	NA	40
TL783C	1.25-120	0.7	+	NA	125
LM317T	1.2-37	1.5	+	Vo + 3	Vo + 40
LT1086CT	1.2-29	1.5	+	Vo + 1.5	Vo + 30
79HGSC	2.1-24	5.0		NA	35

* Low 'drop-out' type, ie it has a low voltage drop across the series transistor. More of these are now available.

Notes. There are many other different voltage and current rated stabilisers and they are made by many different manufacturers. All the high-powered devices must be fixed to a suitable heatsink. All 'fixed voltage' devices can have their voltage adjusted upwards by adding a resistor, a diode or a zener diode in series with their 'common' lead. The value of resistor varies with the device and the manufacturer, and the latter's literature should be consulted.

Table 24.4: IC voltage regulators

described incorporated into a single chip, and often include various over-load and over-temperature protection. There are four main types:

1   Positive fixed voltage eg 5, 12, 15 and 24 V
2   Positive adjustable voltage, adjustable by external resistors
3   Negative fixed voltage
4   Negative adjustable voltage, adjustable by external resistors

All need capacitors connected close to their input and output pins to prevent oscillation, see **Fig 24.22**, and it is advisable to fit diodes to prevent capacitors in the load from discharging back through the IC.

This is particularly important for LM317 and LM1084 type adjustable types (**Fig 24.23**) - take the maker's advice! There is no excuse for attempting to add components to a fixed regulator to get an increased voltage; the adjustable ones cost little more.

Table 24.4 gives some typical examples. The circuits for adjustable types do vary from manufacturer to manufacturer (Fig 24.23 again).

Most IC voltage regulators have internal current limiting and some have 'fold back' current limiting in which the voltage falls sharply if you attempt to exceed the current limit, see **Fig 24.24**.

The input voltage must exceed the output voltage by a stated amount; this is called the drop-out voltage.

An external pass transistor, connected as shown in **Fig 24.25** for a positive regulator, may increase the output current. The resistor is chosen so that the transistor is turned on (by 0.6V on its base) when just before the IC's current maximum is reached. This applies particularly to the 723 type of IC where its ability can be extended almost indefinitely (note that the correct compensation capacitor must be connected according to the data sheet) [1]

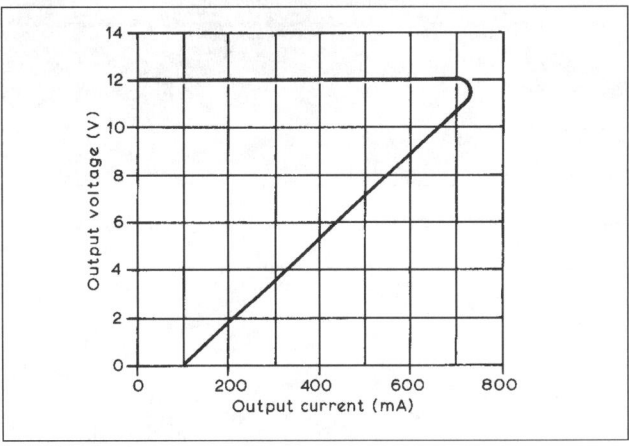

Fig 24.24: Voltage current characteristic showing 'fold-back'

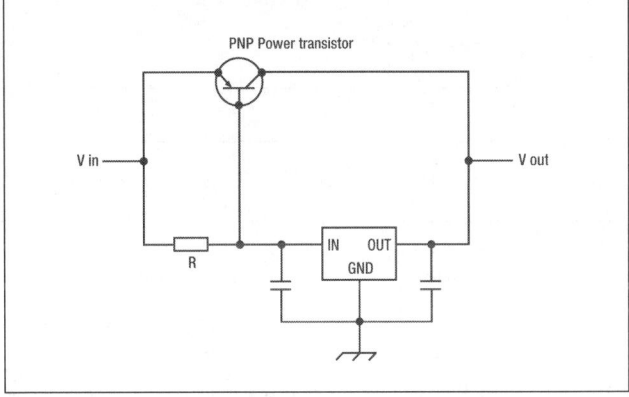

Fig 24.25: Connecting an external pass transistor to increase the output current

Field effect transistors (FETs)

Instead of using bipolar transistors for the pass element, a MOS-FET (see the chapter on semiconductors) can be used with advantage. They are now cheap, and have the advantages that they are not subject to thermal run-away, have a reasonably constant gain with drain current, and a very low on-resistance. A disadvantage is that as they are only made in enhancement versions, the gate requires a voltage somewhere between 2 and 8V (according to type) above the source. For efficiency this should be derived from a low power auxiliary rectifier and transformer or winding. As MOSFETs have an integral reverse diode, protection from damage due to charge stored in capacitors after the regulator is not required.

Over-voltage protection

If the pass transistor fails to a short circuit condition, the whole input voltage (perhaps 18-20V for a 13.8V supply) will reach the output. Many rigs do not like this very much, and to avoid expense, some form of over-voltage protection ought to be included. The most common is called a crow-bar, so called because it short circuits the supply and either blows a fuse or operates a relay to open circuit the rectifier (see Fig 24.10(c)).

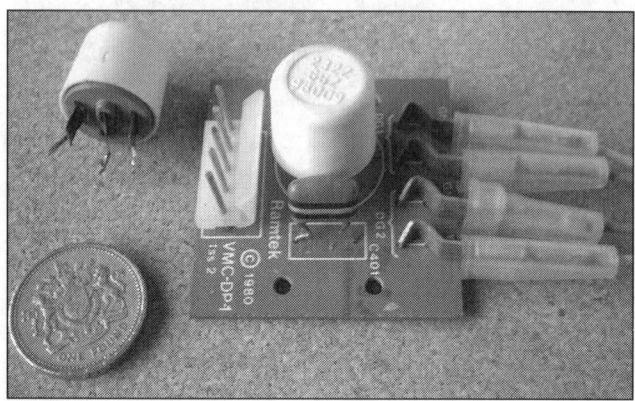

Fig 24.26: The TV de-gaussing thermistor used in Fig 24.10(c)

The type of thermistor used here comes from the de-gaussing circuit of a TV set, and is shown in **Fig 24.26**.

Over-current Protection

Most ICs have internal current limiting, but where augmentation of the ICs current is used, often with the 723, the internal limit may not work or may need too many millivolts in a current sensing resistor. The Maxim MAX 4373 is a very useful IC that needs very few millivolts from the current sensing resistor, and provides a latched over-current flag. **Fig 24.27** shows it being used to sense the voltage across an ammeter, and shut down a 723 IC by crowbarring an internal amplifier transistor. If you use this circuit, you must put a 1k resistor between pin 1 (Current Sense) of the 723 IC and ground, to prevent destruction of the transistor by excess current. You restore the latch by grounding pin 5 of the MAX 4373. Full details of the Max 4373 series are available from Maxim.

Constant Current Circuits

From time to time, a constant current source is needed, for example to charge a secondary battery (see below). A depletion mode FET with a source bias resistor will do the trick (see **Fig 24.28**), but the I_{DSS} of a junction transistor is only loosely specified. Corresponding to drop out voltage, the knee voltage of the FET has to be overcome before constant current is achieved. To save trouble, FETs with built in source resistors can be bought.

A similar arrangement with a bipolar transistor, whose base is held at a fixed potential, can provide a constant current up to several amps if required.

The LM317 adjustable regulator can be used in the circuit of **Fig 24.29**. The LM317 passes enough current to maintain 1.25V between the output and Adj pins, so the current in amps will be 1.25/R where R is in Ohms. As the circuit only has two connections, it can be used for either positive or negative supply, having due regard to polarity.

For a really high current supply, put a suitable choke in series with the primary of the transformer of a normal full wave circuit. Achieving the correct choke is easier done by trial and error!

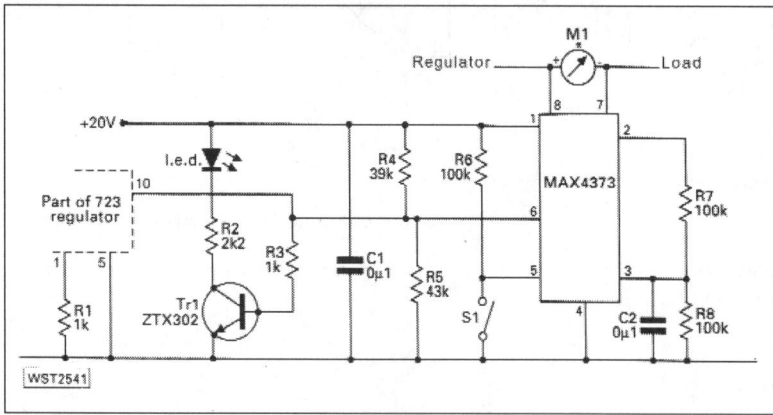

24.27: Current limiting and protection [*Practical Wireless*]

SWITCH MODE POWER SUPPLIES

All the previously mentioned DC supplies transformed the mains, rectified it and perhaps regulated it. At 50Hz, the transformer is bulky and heavy. If a higher frequency could be used, the transformer and smoothing arrangements could be made much smaller. The switch mode device does just this, and was used in the thirties as a means of effectively transforming DC. With the advent of transistors, much higher frequencies, up to the MHz region could be used, also making efficient regulation possible, if needed.

The subject would fill a book not much smaller than this, so you should either buy a cheap switch mode power unit, or read a suitable maker's application note covering your need. In either case you

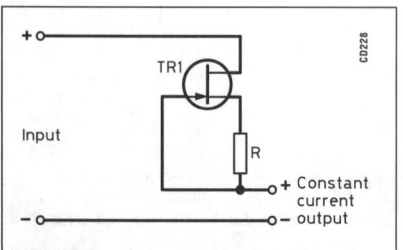

Fig 24.28: Principle of the constant-current circuit. TR1 small JFET; R to set current - depends on G_m of FET

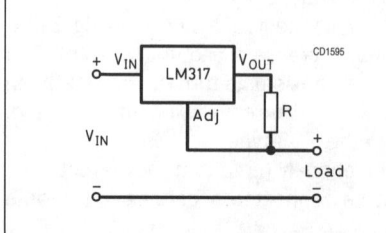

Fig 24.29: LM317 used as a constant-current regulator I_{LOAD} = 1.25/R

Type	Voltage (V)	Weight (g)	Maximum size (mm)	Current range (mA)
AAA	1.5	9	45 10.5	0-25
AA	1.5	18	50.5 14.5	0-40
C	1.5	48	50 26.2	20-60
D	1.5	110	61.5 34.2	25-100
PP3	9	38	48.5 17.5 26.5	1-10
PP9	9	410	81 52 66	5-50
C (HP)	1.5	as 'C' above		0-1000
D (HP)	1.5	as 'D' above		0-2000

Note. Where there are two dimensions, the first is the length and the second the diameter. The current range is that which gives a reasonable life. Manufacturers do not often give capacities in ampere-hours. The shelf life of either type of cell is about a year although it can be improved by keeping it cold. 'HP' is the high-powered type.

Table 24.5: Characteristics of typical zinc-carbon cells and batteries

have to consider the EMC problem, for SMPSs are notorious for the interference they can create.

BATTERIES

The basic types of batteries have been described in the chapter on Principles.

There are two types, primary or one shot, and secondary or rechargeable. Strictly speaking, a 'battery' is an assembly of two or more 'cells', but a single cell is commonly also called a battery.

Primary Batteries

At present there are two main varieties, that are suitable for amateur use, based on zinc or lithium. The battery derives its energy from the metal used as the negative electrode. The positive electrode has an effect as well and has to be able to dispose of the hydrogen, which would otherwise be liberated there. A depolariser surrounds the electrode if it is unable to do this. For watches and similar purposes zinc-mercury oxide and zinc-silver oxide cells are available at some cost.

Zinc-carbon

These form the oldest and cheapest primary cells and are called dry cells as the electrolyte, although not dry, is immobilised so that it cannot spill.

Three different electrolytes are used, an aqueous solution of ammonium chloride (sal ammoniac) in the cheapest, zinc chloride in 'high power' cells, and sodium hydroxide (caustic soda) in manganese-alkaline cells. **Table 24.5** gives some parameters of typical types.

Manganese-alkaline cells are made in the same sizes and **Table 24.6** gives details.

Type	Voltage	Weight (g)	Capacity (Ah)
AAA	1.5	11	1.2
AA	1.5	22	2.7
C	1.5	67	7.8
D	1.5	141	18.4
PP3	9	45	0.55

Note. The dimensions are as above and the capacity is in ampere-hours. They have a shelf life of several years.

Table 24.6: Characteristics of typical manganese-alkaline batteries

Zinc-air

These are similar to zinc-carbon, but use the oxygen of the air as the depolariser. You buy them sealed and they only 'come to life' when the seal is removed. Potassium hydroxide (caustic potash) is used as the electrolyte, and slowly absorbs carbon dioxide from the air, ending the life of the cell. They must be used in a well-ventilated housing. They have a higher energy density than zinc-carbon cells ie they pack more energy into a given weight or space.

Lithium

The negative electrode is the highly reactive metal lithium, so the electrolyte contains no water. The positive electrode is either iron disulphide (1.5V) or manganese dioxide (3V on load). The electrolyte is either an organic liquid or thionyl chloride (2.9V).

These cells have a long shelf life making them good for battery back up, and have a low internal resistance. They also present a fire hazard if broken or an attempt to charge is made.

Secondary Batteries

Lead-acid, Nickel-Cadmium (Ni-Cad), Nickel-Metal Hydride (Ni-MH) and Lithium are the only types in amateur use which will be described.

Lead-acid

This the earliest cell, with lead plates and dilute sulphuric acid electrolyte. Vehicle batteries nearly always use these, as the alternatives (Ni-Cad or Ni-MH, see below) are too expensive. They are heavy, but have very low resistance. One feature is that if left discharged for some time, the plates sulphate irreversibly and the battery is almost always ruined. The usual charging is at a constant voltage of 13.8V, with some form of current limiting to prevent too large a current flowing initially. Do not try to charge a 12V battery from a 13.8V PSU. If the mains fails or is disconnected, the battery will discharge into the PSU, possibly damaging it. Any charger should therefore contain a diode to prevent reverse current flow. Over-charging a sealed battery will result in explosive gases being generated and may burst the vent.

Nickel-Cadmium

Ni-Cads (1.2V) have an electrolyte of potassium hydroxide in water, which is attacked by carbon dioxide in the air. Today the cells are sealed to prevent this, and as such are much used by amateurs. **Table 24.7** gives data on a selection. Cadmium compounds are toxic and Ni-Cads should be disposed of with care; many local authorities make provision for this and should be consulted.

Charging can be done at a constant current of C/10 amps where C is the capacity in ampere-hours. This will be complete in about 14 hours (allowing for inefficiency) and moderate overcharging at this rate does not result in harm. This long charging time is a nuisance, so fast charge circuits have been developed. When fully charged, the Ni-Cad cell voltage actually decreases

Size	Voltage (V)	Capacity (Ah)	Weight (g)
AAA	1.2	0.18	10
AA	1.2	0.5	22
C	1.2	2.2	70
D	1.2	4.0	135
PP3	8.4	0.11	46
PP9	8.4	1.2	377

Note. The dimensions are the same as those for the zinc-carbon cells/batteries above.

Table 24.7: Characteristics of typical nicad (NiCd) cells and batteries

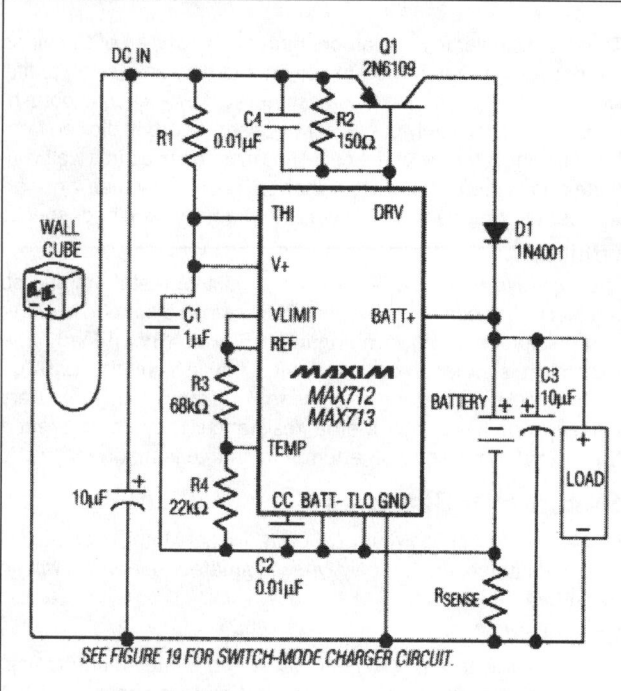

Fig 24.30: The Maxim MAX 712/713 Ni-Cad and Ni-MH fast charger ICs

with time. Maxim make a charger IC for Ni-Cads, the MAX 713. This detects the decrease and stops fast charge, **Fig 24.30**.

There is a memory effect with Ni-Cads if re-charged before being completely discharged; the cell 'remembers' that it was not fully discharged and will not be able to discharge fully after subsequent charge. Some authorities dispute this, believing that it only occurs after repeated discharge to less than complete. If your cell suffers from this, short out the cell only when discharged as much as it will. Beware of reversing the current through the cells of a battery; this damages them. Connecting a resistor across each cell of a battery is recommended, but not usually possible.

Opinion also differs with the procedure for storing cells that are not required. Unlike the lead-acid cell, Ni-Cads may be left discharged, and some have found this better than leaving them fully charged. Maker's advice is not readily available.

Nickel-metal hydride cells
They have advantages over nicads in that they have a higher energy density and they do not contain cadmium. A simple comparison between one of these and the same size in nicad is given in **Table 24.8**.

	Metal-hydride	Nicad
Capacity (Ah)	3.5	2.0
Voltage (V)	1.2	1.2
'Memory'*	None	Severe
Toxics	None	Cadmium
Discharge rate (A)	<12-15	<15
Overcharge capability	Cont. at C/5	Cont. at C/5

'Memory' is an alleged effect [3] which shows up if a cell is only partially discharged before recharging. It is said to reduce the capacity of the cell. This has been disputed [4].

Table 24.8: Comparison of metal hydride and nicad 'C' size cells [2]

Nickel-Metal Hydride
Ni-MH cells are very similar to Ni-Cads, the voltage is the same, but capacity for the same size is higher. There is no memory effect, but the self-discharge rate is higher. Ni-MH cells do not exhibit the decrease of voltage with time at full charge (like the Ni-Cads do); the voltage remains constant. The MAX 712 (Fig 24.30) detects this and stops fast charge.

Lithium-ion
Note that these are not lithium-iron! The negative electrode is carbon, and the positive lithium-cobalt oxide in most cells. The voltage starts at about 4V after charge and drops to 3V when discharged. Charging is best done at a constant voltage of 4.2V (4.1V if the positive electrode is lithium-nickel oxide). Analogue Devices market an ADP 3820 in versions suitable for charging both types (**Fig 24.31**).

Safety
As mentioned above, the electrolyte of lead-acid cells is sulphuric acid, and it should be treated with the greatest respect, and not allowed to touch the skin. If it does, it should be immediately washed off with running water. In particular the eyes should be protected from it. If it gets on clothes, if left, it will slowly make a hole. The explosive nature of the gas evolved by unsealed cells has already been noted. Sparks from the terminals on connection or disconnection can ignite the gas, and the entire cell could explode.

The electrolytes of the other types are undesirable also, and should be washed off if they contact the skin. Lithium is an extremely reactive metal, and if a cell containing it bursts, a fire may start. Do not attempt to destroy such a cell by burning.

REVERSE BATTERY PROTECTION
Applying power with the wrong polarity can damage equipment. Here are four simple ways to prevent this:

1 Put a power diode in series with the load. This has the disadvantage of wasting a volt or so across the diode, but a Schottky diode would be better (0.31V at 8A for the 95SQ015). **See Fig 24.32(a)**.

2 Put a power diode in parallel with the load and a fuse in series. If the power is incorrectly applied, the diode conducts and the fuse will blow, see **Fig 24.32(b)**.

3 Use a relay to switch the power with a diode in series with the relay, which will then only operate if the power is correctly applied, **Fig 24.32(c)**.

4 Use a power MOSFET in the circuit reported by G4CLF. The MOSFET is only turned on by the correct power supply polarity. The 'on' resistance of available MOSFETs is so low that it may be neglected. See **Fig 24.32(d)**.

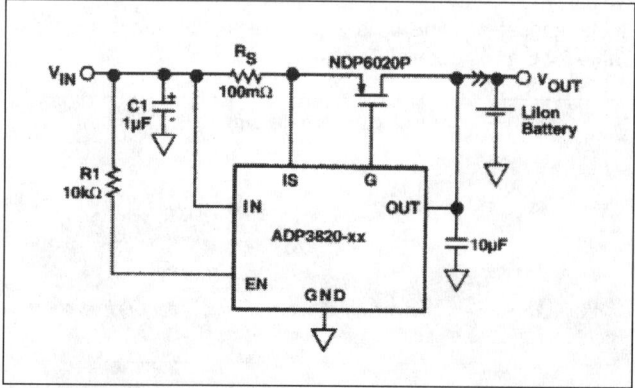

Fig 24.31: The Analog Devices Lithium-ion fast charger IC

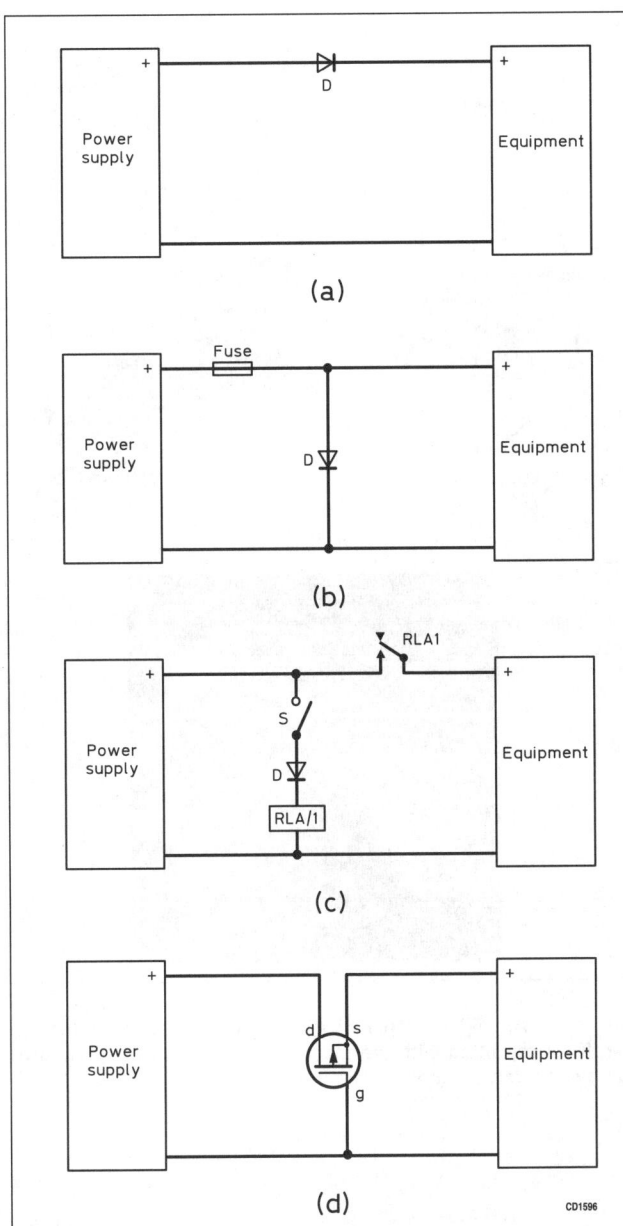

(a)

(b)

(c)

(d)

CD1596

Fig 24.32: Four methods of reverse-polarity protection. D1 is a silicon diode; in (a) it must carry the whole current. In (c), S is the on-off switch, A/1 is the relay operating coil and A1 is the normally open relay contact. (d) Using a p-channel MOSFET

Fig 24.33: Wind generator *[Marlec]*

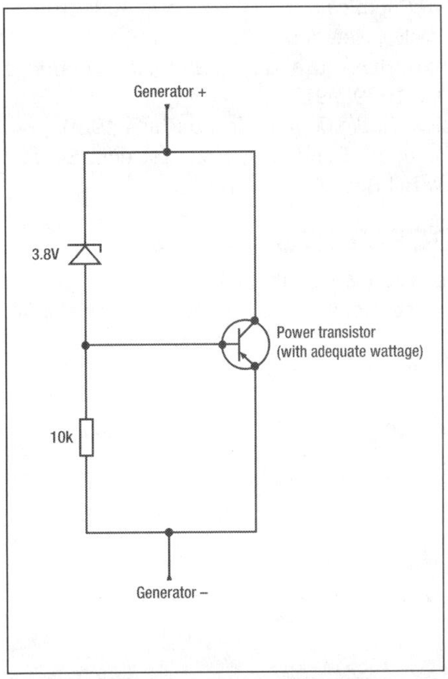

Fig 24.34: A shunt regulator suitable for intermittent generators

RENEWABLE ENERGY SOURCES

Wind, Water and Pedal Generators

These may not find much application at home, but could be useful in portable operation. As all of them supply power intermittently, a rechargeable battery will be needed. This implies a regulator, and a simple shunt regulator is all that is necessary, plus a means of preventing reverse current. A wind generator is shown in **Fig 24.33**, and one type of regulator is shown in **Fig 24.34**.

Solar Cells

A photograph of a solar cell is shown in **Fig 24.35**; they come in various sizes. You cannot estimate the available power by exposing it to bright sunlight, then measuring the no-load voltage and the short circuit current. Thévenin's theorem does not

Fig 24.35: One type of solar cell

apply, but many makers will provide you with performance curves.

Thermocouples

In about 1820, Seebeck noticed that if one junction between two metals was heated, and the other cooled, an EMF was generated. Modern materials [5] make possible a power generator based on this effect, but the efficiency is not high. **Fig 24.36** shows one made to use the waste heat from a kerosene lamp to power a valve radio. 90V HT, 9V grid bias and 1.4V filament supply was available. Home construction might be attempted, but there may be a difficulty in getting the materials.

REFERENCES

[1] LM723 data sheet. Available on the Internet at http://www1.jaycar.com.au/images_uploaded/ LM723.PDF

[2] *Encyclopedia of Chemical Technology*, Vol 3, Kirk-Othmer, 4th edn, John Wiley, New York, 1992, p1070.

[3] *Art of Electronics*, Horowitz and Hall, 2nd edn, Cambridge University Press, 1989, p927.

[4] 'Technical Topics', *Radio Communication* June 1989, p34.

[5] *Electronics Engineers Handbook*, Fink, 1st edn, Section 27-14. McGraw Hill Book Co

ACKNOWLEDGEMENTS

Grateful thanks are extended to the following who provided information and/or illustrations during the revision of this chapter:

Amberley Working Museum
Analog Devices
Chloride Power Protection
Dallas Maxim
Future Electronics
Marlec Engineering
National Semiconductor
Practical Wireless

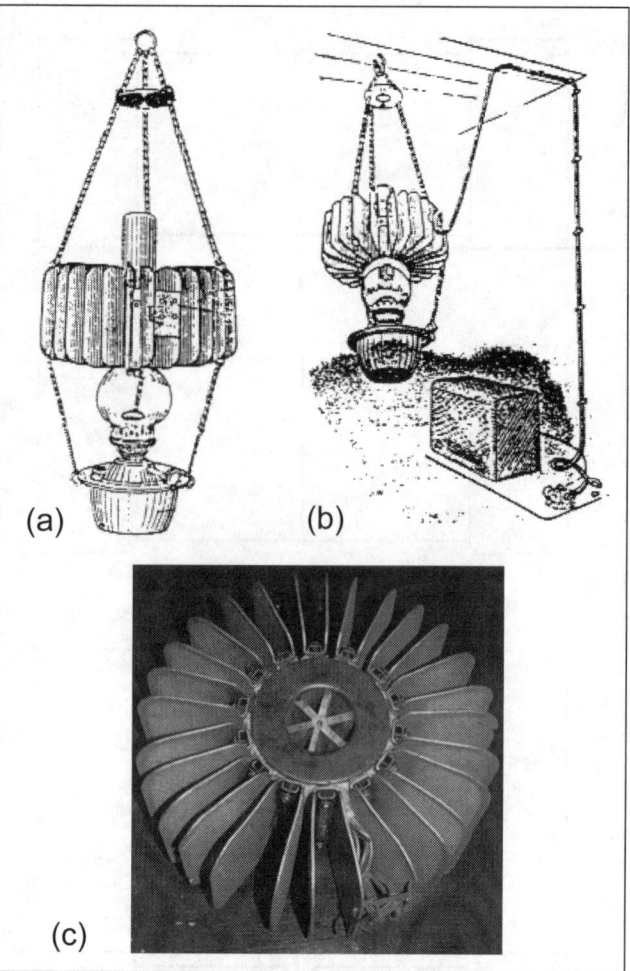

Fig 24.36: An old example of a thermo-electric generator. (a) Paraffin lamp fitted with thermocouple. (b) Using the generator to power a valve radio. (c) The thermal head

25 Measurement and Test Equipment

Correct operation of amateur radio equipment involves measurements to ensure optimum performance, in order to comply with the licence conditions and to avoid interference to other users. This will involve the use of test equipment, as will the repair and maintenance of equipment.

Some professional test gear is very expensive but it is the intent of this chapter to show how some of the cheaper (and perhaps home-built) equipment can be used to good effect.

For further information, equipment and a more detailed discussion of some of the topics the reader should consult [1]. Useful material is also contained in [2] and [3].

One word of caution - whilst the components used are presently available it is a fast changing situation and some components may become obsolete during the life of this book. If a component appears to be unavailable it is always worth putting the part number into a search engine on the Internet.

CURRENT AND VOLTAGE MEASUREMENTS

Most electrical tests rely on the measurement of voltage and current. To this end many types of instrument have been developed, such as meters, oscilloscopes, spectrum analysers etc. These are all examined in this chapter.

The ubiquitous multimeter (**Fig 25.1**) tends to be used for many voltage and current measurements nowadays. The units are either analogue or digital, they are relatively cheap and usually provide resistance measurement as well. Because they are so cheap it is usually not worth making one, except for the experience.

However, when making power supplies, amplifiers etc, it is important to have meters dedicated to a single function, or a group of functions. The following sections deal with this, and how the meters can be adapted to the ranges that need to be used.

One problem with all measuring instruments is how they affect the circuit they are measuring, due to the power they require to provide the input signal.

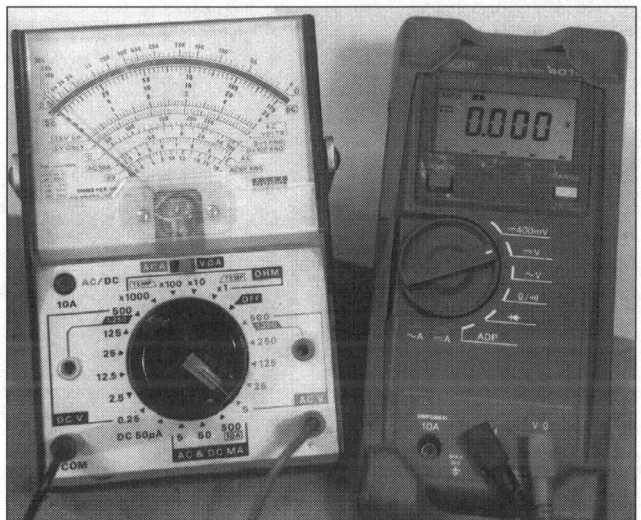

Fig 24.1: Typical analogue and digital multimeters

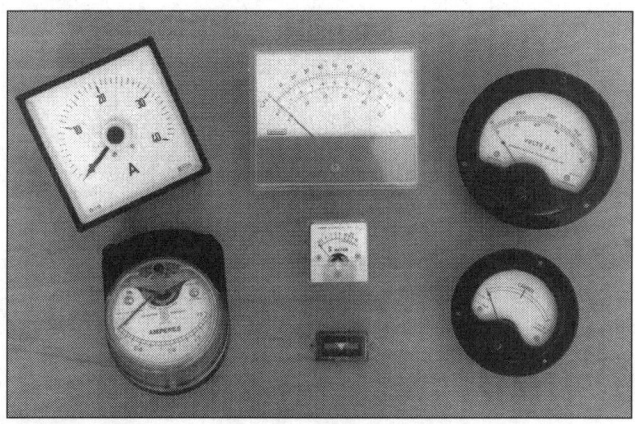

Fig 25.2: Various moving-coil and moving-iron meters

ANALOGUE METERS

These are of electromechanical design and consist (amongst others) of the moving-coil and moving-iron type meters (**Fig 25.2**). The moving coil meter has a linear scale while the moving iron meter is non-linear, the scale being very cramped at the lower end.

The moving-coil meter is the most sensitive and the most accurate, but will respond to DC only, while the moving-iron instrument is AC/DC with a response up to about 60Hz. The modern analogue meter tends to be rectangular, older types usually being round.

The sensitivity of analogue meters is defined by the current that must flow through them in order to provide *full-scale deflection* (FSD). The moving-coil range starts at about 50µA FSD while the moving iron range works from about 100mA FSD.

These analogue meters do draw current from the circuit under test to operate and the coil has resistance (Rm) because it is made from wire. When used for measuring current, **Fig 25.3(a)**, the meter is placed in series with the circuit and so there is a voltage drop - typically 100 to 200mV. When used as a voltmeter, **Fig 25.3(b)**, the current drawn depends on the basic meter movement and, if this takes more than 10% of what is flowing in the circuit, then the circuit conditions are being progressively affected.

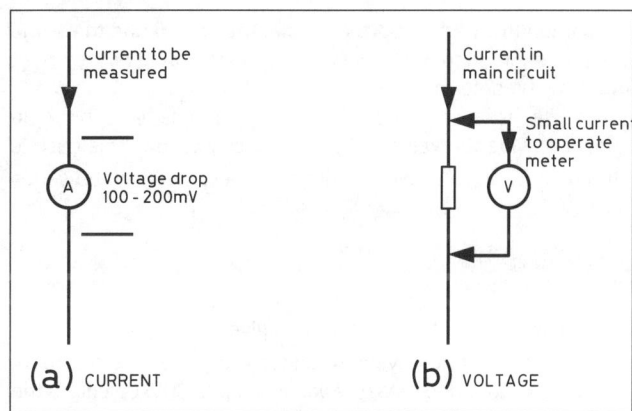

Fig 25.3: The use of meters for measurement

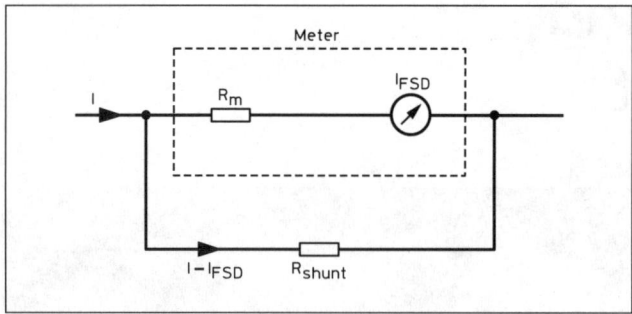

Fig 25.4: Arrangement for current shunts

Extending the Range of Analogue Meters

The meters referred to above come in various fixed arrangements and may not suit the ranges it is desired to measure. It is possible, by the addition of resistors, to extend the range of meters, possibly still using the original scaling.

There is no reason why the scale should not be redrawn by hand or by using transfers. The scale plate can often be removed.

An analogue meter requires current to operate; consider the measurement of current initially. The FSD of the meter cannot be changed so it is necessary to shunt some of the current to be measured around the meter, the typical circuit being shown on **Fig 25.4**.

Here, assuming the maximum current to be measured is I, the shunt resistance is given by:

$$R_{shunt} = \frac{R_m I_{FSD}}{I - I_{FSD}}$$

where I_{FSD} is the current for full-scale deflection of the meter and R_m the resistance of the meter. It is normal to choose I so that only a multiplying factor is required of the scale reading. The power rating of the shunt can be calculated and is:

$$(I - I_{FSD})^2 R_{shunt}.$$

Example: It is desired to use a 100μA FSD meter to measure a maximum current of 500μA. The resistance of the basic movement is 2000Ω. Substituting these values in the above formula gives:

$$R_{shunt} = 500Ω \text{ with a power rating of 80μW}$$

An alternative way of considering this problem is to consider what the multiplying factor (n) of the scale must be. Using the previous definitions of resistors, the formula for the shunt becomes

$$R_{shunt} = \frac{R_m}{n - 1}$$

Applying this to the above example, then n = 5 and the same value of shunt is found. However, the power rating of the shunt must still be determined.

For use as a voltmeter, the maximum voltage to be read should provide the value of I_{FSD}. The circuit used in this case is shown on **Fig 25.5**. The equation for the resistance of the series resistor R_{mult} is given by

$$R_{mult} = \frac{V}{I_{FSD}} - R_m$$

The power rating for the resistor is given by $I^2_{FSD} R_{mult}.$

Example: A 50μA movement meter with a coil resistance of 3000Ω is required to measure voltages up to 30V. Calculate the multiplier resistor.

$$R_{mult} = 597kΩ \text{ with a power rating of 1.5mW}$$

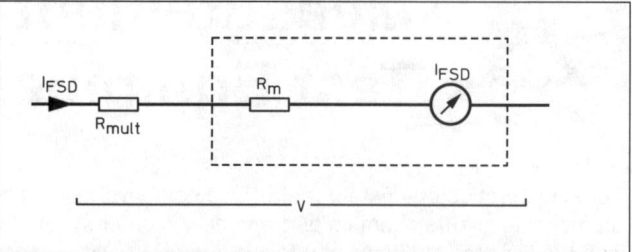

Fig 25. 5: Arrangement for voltage multipliers

These simple calculations show the basis on which the familiar multimeter is based and how they are designed. The switch on the multimeter merely switches in different shunt and multiplier resistors. Remember, these calculations only apply to DC for the moving-coil meter.

Meter Sensitivity

The sensitivity of a voltmeter is usually expressed in ohms/volt. This is merely the reciprocal of the full-scale current sensitivity I_{FSD} of the basic meter. Hence, a 1mA meter used as a voltmeter would be described as 1000Ω/V and a 50μA meter as 20,000Ω/V.

Effect on Circuit Readings

Putting a voltmeter across a resistor may upset the circuit conditions, and the loading effect of a meter has to be considered. For example, putting a meter which requires 50μA across a resistor through which only 100μA flows will disturb the circuit significantly. Putting the same meter across a resistor through which 10mA flows will have little effect. How can one gauge this or guard against it?

Consider a 20,000Ω/V meter. Set on the 10V range this will have a resistance of 10 x 20,000Ω = 200kΩ. It is suggested that any resistance across which this voltmeter is placed should have a maximum value of one-tenth of this, eg 20kΩ. Hence, for any range one can use this rule-of-thumb method. The smaller the percentage, the more accurate will be the reading.

For ammeters the point that must be considered is the voltage drop across the ammeter in relatively low voltage circuits (ie I_{FSD} x R_m). For example, a 0.5V drop across an ammeter is unacceptable in a 12V circuit but it is immaterial in a 100V circuit. One must therefore choose a meter that has as low a coil resistance as possible. This reduces the in-circuit voltage drop and keeps any shunt resistance value as high as possible. If possible, use an ammeter of I_{FSD} equal or just greater than the range required.

For mains circuits of 100V or above the moving-iron meter represents a more viable alternative and tends to be cheaper.

Meter Switching

In order to save cost (and sometimes panel space), it may be worthwhile for a meter to serve several functions. This is more likely to be used in valve circuits for measuring grid and anode voltages and currents. These normally require different ranges for the various parameters being measured. For convenience, two meters would be used - a voltmeter and an ammeter.

In all instances a break-before-make switch should be used. Care should also be exercised in selecting the switch when used in high-voltage circuits.

When measuring current, the resistance of wire and switch contacts may affect the value of low-value shunts.

Voltage measurements are normally made with respect to 0V or earth. This means that one end of the voltmeter is fixed - see

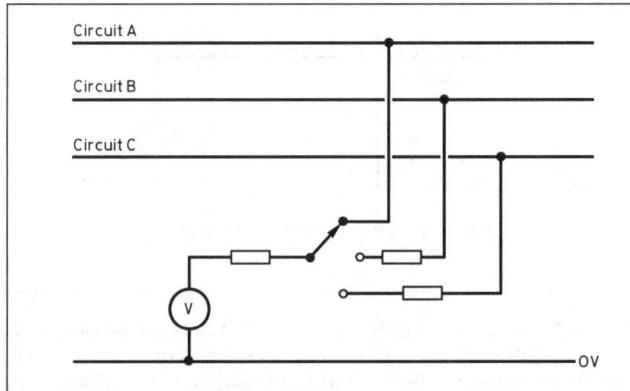

Fig 25.6: Switched voltage measurements

Fig 25.6. Knowing the characteristics of the meter, the various values of series resistance can be calculated. It is suggested that the lowest value is usually wired directly in series with the meter and then the other values chosen such that this value plus the additional one equals the value calculated. Assuming that circuit A in Fig 25.6 has the lowest voltage to be measured, then some current limiting always exists in series with the meter. For current measurements the problem is overcoming contact and wire resistance when low-value shunts are used (ie less than 0.1Ω).

For the purposes of this discussion, a meter is assumed to have 1mA FSD and coil resistance of 100Ω. **Fig 25.7** shows how switching could be arranged for the measurement of current on three ranges. Switching/conductor resistance is unlikely to be a problem with circuit A, but it may be a problem on circuit B and certainly will be on circuit C.

One solution is to use a non-switchable meter for any current range which requires a low shunt value, typically less than about 0.5Ω.

A different approach is to consider the meter as measuring volts across a resistor.

The problems of measuring a voltage and the current taken must then be considered as previously discussed. If a 50µA meter was used, then it must be possible to develop a minimum voltage drop of about 150mV; for a 1mA movement it should be about 100mV.

The voltage drop should be equal to or greater than I_{FSD} x R_m. The typical circuit used for this arrangement is given in **Fig 25.8**.

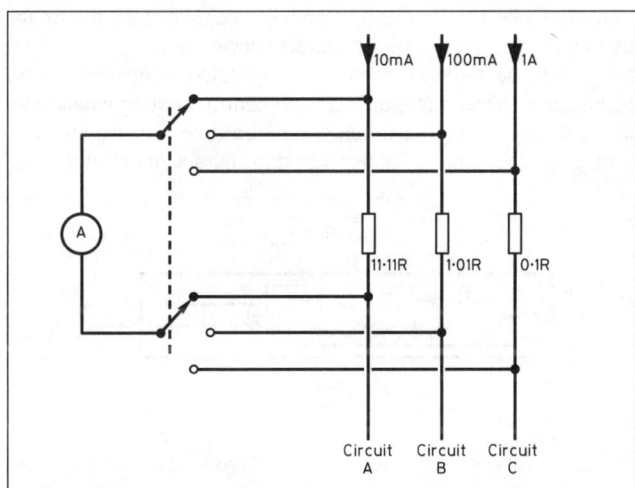

Fig 25.7 Switched current measurements

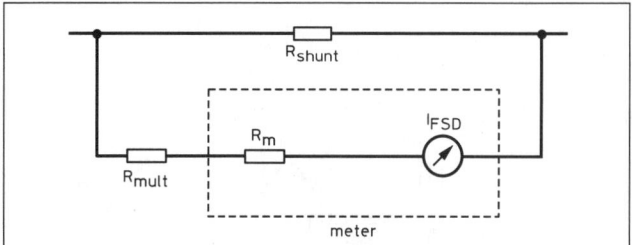

Fig 25.8: Current measurement by volt-drop method

Meter Protection

Meters are relatively expensive and easily damaged if subjected to excessive current. Damage can be prevented simply and cheaply by connecting two silicon diodes in parallel (anode to cathode) across the meter terminals as in **Fig 25.9**, and this should be regarded as standard practice. No perceptible change of sensitivity or scale shape need occur.

A characteristic of silicon diodes is that they remain very high resistance until the anode is some 400mV above the cathode, at which point they start to conduct and the resistance falls to a low value. Since the voltage drop across the average meter is around 200mV, it follows that a silicon diode connected across the meter will have no effect even when the meter shows full-scale deflection. If, however, the meter is overloaded to twice the FSD and the voltage across the meter rises to 400mV, the diodes will begin to conduct and shunt the meter against further increase of fault current.

Most meters will stand an overload of at least twice the FSD without damage but it is wise to include a series resistor as shown on Fig 25.9 to ensure the protection afforded by parallel diodes without affecting the meter. The series resistance ensures that the voltage drop across the meter/resistor combination is 200mV minimum. Parallel diodes are used because excessive current in either direction can damage the meter.

Example: What series resistor should be added to a 1mA FSD meter with a coil resistance of 100Ω?

At 1mA FSD the voltage drop across the meter is 1mA x 100Ω = 100mV. Thus the drop across the series resistor should also be 100mV, and this requires a resistance of 100Ω. This then means that the meter is protected for currents in excess of 400mV/200Ω = 2mA.

If an additional series resistor is to be included then any shunts to be included to increase the current range should be placed across this combination and the series resistance taken into account when making the calculations.

For most cases, small-signal silicon diodes such as the OA202, 1N914 or 1N4148 are satisfactory - they have the advantage of having an inherently high reverse resistance - ie a low reverse leakage current is required as this shunts the meter circuit. However, it is important that under the worst fault conditions the diode will not fail and go open-circuit, thus affording no protection. An example of this with small-signal diodes would be

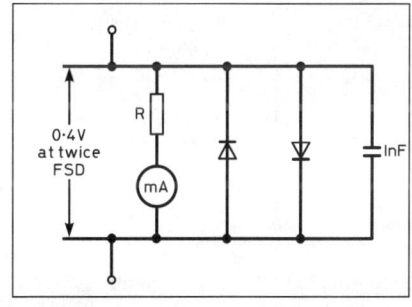

Fig 25.9: Meter protection using diodes

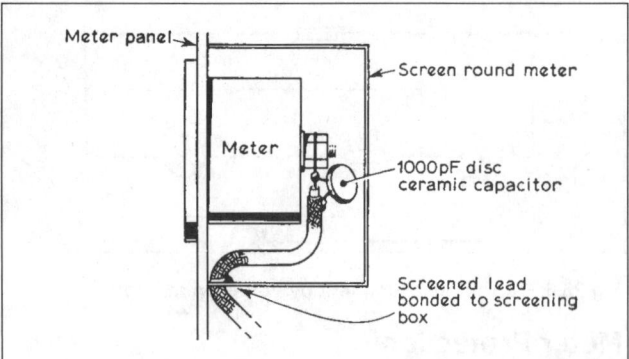

Fig 25.10: Screening and by-passing a meter in a transmitter

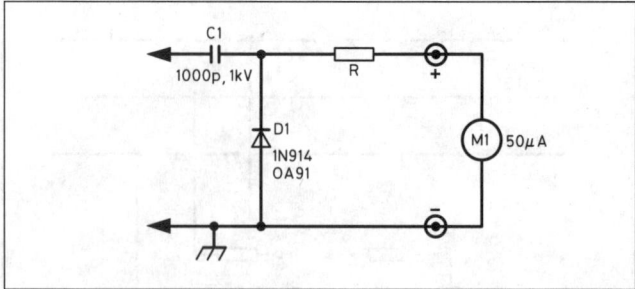

Fig 25.12: RF probe. For R = 270k + 12k, the meter scaling is 0-10V, and full-scale, power in 50 ohms is 2W. For R = 820 + 27k, the meter scaling is 0-30V, and full-scale, power in 50 ohms is 18W

in a high-voltage supply where a large current could flow in the event of a short-circuit of the power supply. In these cases a rectifier diode should be used, such as the 1N400X or 1N540X series. The reverse current of these diodes may be a few microamps and, depending on the current to be measured, may have a slight effect on the sensitivity of the meter circuit.

Although diode protection should be applied as routine in order to safeguard instruments, it can cause some unusual effects if measurements are made with an AC signal imposed on a DC signal. This AC component, providing it is symmetrical, should not normally introduce any error but, if the AC is large enough to bring the diodes into conduction at the peak of the cycle, it introduces a dynamic shunt on the meter. This can be partly confusing when back-to-back diodes are used as the meter sensitivity will drop without any offset reading to warn what is happening. These effects are most likely to occur when measuring rectified mains or when RF is present.

Whenever a meter is to be used when RF may be present (this includes even a power supply output voltmeter) it is wise to shunt the meter with a capacitor, typically a 1000pF ceramic type - see **Fig 25.10**. In addition, if strong RF fields are likely to be present, eg in a transmitter, it would also be wise to shield the meter and possibly feed it via screened cable.

AC Measurements

If an alternating current is passed through a moving-coil meter there will normally be no deflection since the meter will indicate the mean value and, in the case of a waveform symmetrical about zero, this is zero. If, however, the AC is rectified so that the meter sees a series of half-sine pulses (full-wave rectification) it will indicate the mean value ($2/\pi$ or 0.637 of the peak value). Commercial instruments using moving-coil instruments for AC sine-wave measurements therefore incorporate a rectifier (see **Fig 25.11** for a typical arrangement) and the scale is adjusted to read RMS values (0.707 of the peak value). They will read incorrectly on any waveform that does not have these relationships. The moral is: *do not use the meter on any waveform other than a sine wave*. This arrangement is normally only used for voltage

measurements - AC current measurements pose additional problems and are not considered further. The typical frequency range extends to between 10Hz and 20kHz.

Moving-iron instruments, as previously mentioned, do respond to an alternating current and can be used for measurements without rectifiers. This type of meter unfortunately has a square-law characteristic and so the scale tends to be cramped at the lower end. Moving-iron meters normally have a full-scale reading of about 20% more than the normal value to be displayed. They are not used for multimeters.

Other AC measurements can be accomplished by means of electronic voltmeters or oscilloscopes.

ICs do exist (eg AD536, 636, 736, 737, SSM2110) which will provide the RMS of any waveform but their frequency range is limited.

RF Measurements

These probably pose the biggest problem: the circuit under test should not be loaded, capacitance has an increasing effect as frequency rises and the diodes used for rectification must handle the frequencies concerned. The diode characteristics required mean they have a relatively low reverse-voltage rating (1N914 is 100V, OA202 is 150V with slightly poorer RF capabilities) and the forward diode voltage drop. The approach in measuring RF voltages is to rectify as soon as possible and then use DC measuring circuits.

Fig 25.12 shows a typical probe for measuring RF voltages. Capacitor C1 provides DC isolation, D1 rectifies the signal and the resistor is used to convert what is essentially a peak reading to an RMS reading on the meter. For the 50μA meter it is possible to use an individual meter or the most sensitive range on many multimeters. If possible use the precautions for the meter as depicted on Fig 25.10. **Fig 25.13** shows the typical construction of a probe, the exact method being left to the ingenuity of the constructor. A scrap length of 15mm central heating piping may make a good tube. The probe should be useful for frequencies from 50kHz to about 150MHz with an accuracy of about ±10%.

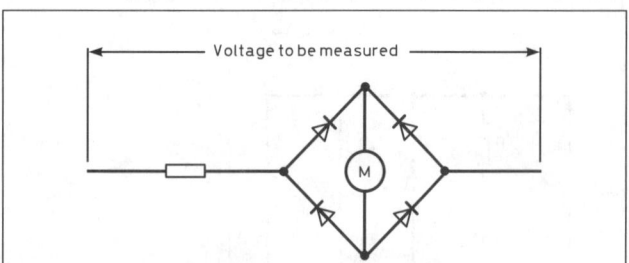

Fig 25.11: Typical arrangements for AC measurements

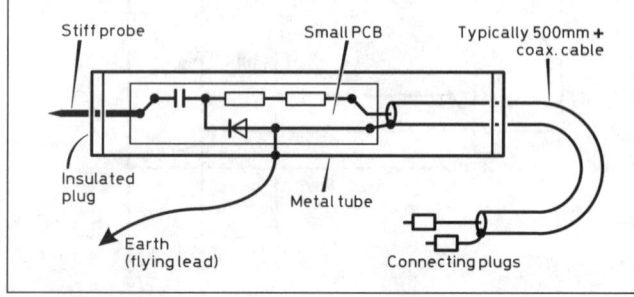

Fig 25.13: Typical construction of an RF probe

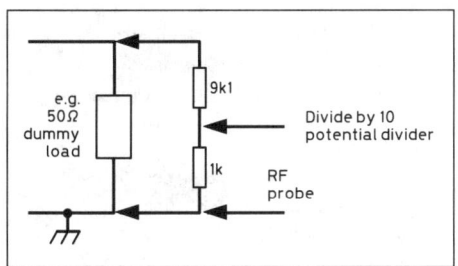

Fig 25.14: Suggested method for higher voltages

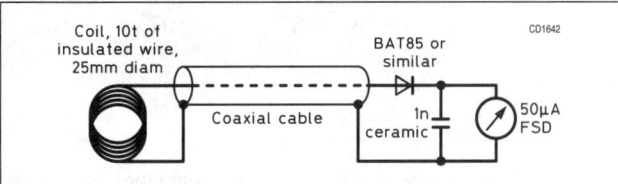

Fig 23.15: Pick-up loop with a diode

Because of reverse-voltage limitations of the diodes, it is necessary to make modifications to take higher voltage readings. **Fig 25.14** shows how a resistive potential divider can be used to effect a ten-fold reduction in voltage to be measured. The resistors should of course be suitable for RF and of adequate power rating. An alternative approach is to use several diodes in series but they will need equalising resistors across them.

An alternative to a probe that makes physical contact with a circuit is the use of a pick-up loop with a diode - see **Fig 25.15**. This cannot give a direct reading of voltage but is capable of indicating the presence of RF energy and may be useful for tuning purposes, ie looking for a maximum or minimum reading. The diode used should have a low forward drop - a germanium or Schottky type would be suitable. To minimise disturbing the RF circuit the pick-up coil should be placed for minimum coupling but give an adequate deflection on the meter.

DIGITAL METERS

The digital meter is fast becoming more common than analogue types and its price is now comparable in most instances. It provides a very accurate meter at reasonable price. Its disadvantages are that the smallest digit can only jump in discrete steps (hence digital) and that it requires a battery.

The digital meter (**Fig 25.16**) works by converting an input analogue voltage to a digital signal that can be used to drive either an LED (light-emitting diode) or LCD (liquid crystal display).

The conversion technique used is either an analogue-to-digital (A-D) converter or the dual ramp technique. A digital meter is often quoted as having, for example, a 3½ digit display. This means that it will display three digits 0 - 9, with the most significant being only a 0 (normally suppressed) or a 1, ie a maximum display showing 1999 as well as + or - signs.

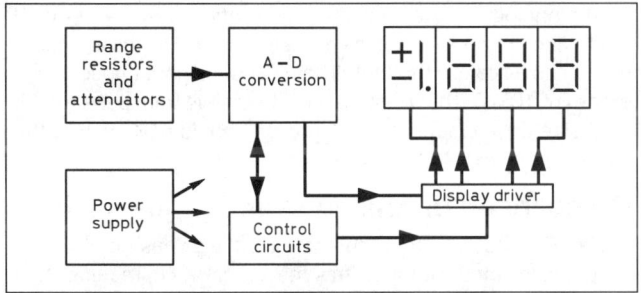

Fig 25.16: Block diagram of a digital meter (3½-digit display)

Fig 25.17: A typical digital panel meter

There are quite a few ICs made by various manufacturers that provide a basic digital voltmeter, external components being required for extending the range, over-voltage protection and displays. These ICs have outputs suitable for driving LEDs, LCDs or provide BCD outputs for further processing.

The digital meter is essentially a DC voltage measuring device (as opposed to the moving-coil meter which is current controlled). Hence all measurements to be made must be converted to a voltage.

The digital form of the multimeter is readily available at reasonable cost and it is not worth the exercise of making one of these meters. The approach here, as for the analogue meter, is to understand the basic principles and how to apply them to specific situations.

The Digital Panel Meter

The best approach for a digital display is to use a panel meter module (which includes the above ICs) and comes with a 3½ or 4½ digit display. These are relatively cheap and provide a good basis for making various types of metering system.

They are normally modules based on LCDs and either plug into a DIL socket or are on a small PCB. They have (typically) a full scale reading for a 199.9mV DC input, work over different supply ranges, from 5V to 14V (depends on model), consume very low current (eg 150-300µA on a 9V supply) and have an input resistance of at least 100MΩ. Because of this high input resistance they present virtually no loading on the circuit under test.

The panel meter itself will provide an accuracy of 0.1% or better but this does not take into account any external signal conditioning circuits such as amplifiers or attenuators. In addition to these parameters, some of the displays will also show units or prefixes such as µ, m, V, A, Ω, Hz etc (referred to as annunciators).

They can be purchased with and without backlighting; a typical meter is shown in **Fig 25.17**. The main design consideration in using these units is to get the parameter to be measured to a DC voltage in the range 0-199.9mV. This can include amplifiers, attenuators and rectifiers.

The following designs are based on the Anders OEM22 module which is readily available. The panel meter consists of a liquid crystal display driven by a 7136 IC which contains an A-D converter and LCD drivers. The unit can be driven from 5V (typically 5mA) providing two links are made on the board or direct from 9V (typically 500µA). It comes with a leaflet containing technical details [4] and the pin designations are marked on the board. The principles explained can, however, be applied to modules available from other manufacturers.

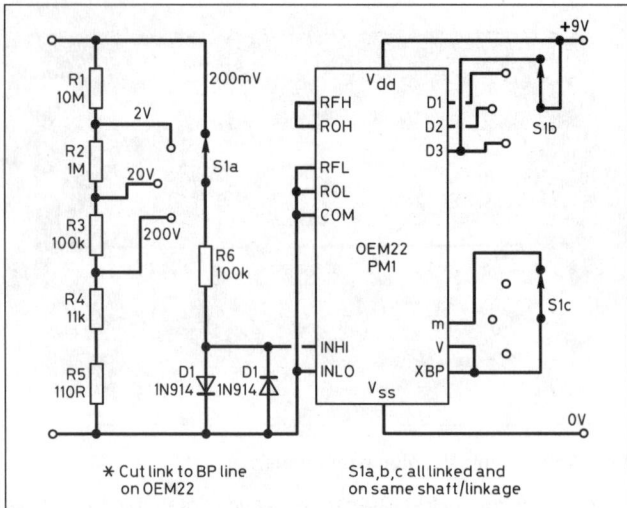

Fig 25.18: A practical digital voltmeter

PM1	Anders Panel Meter	R2	1M, 0.5W, 1%
	type OEM22	R3	100k, 0.5W, 1%
SW1	Rotary switch 3p, 4w	R4	11k, 0.5W, 1%
D1,D2	1N914 or similar	R5	110R, 0.5W, 1%
R1	10M, 0.5W, 1%	R6	100k, 0.5W, 5%

Table 25.1: Components for the practical digital voltmeter

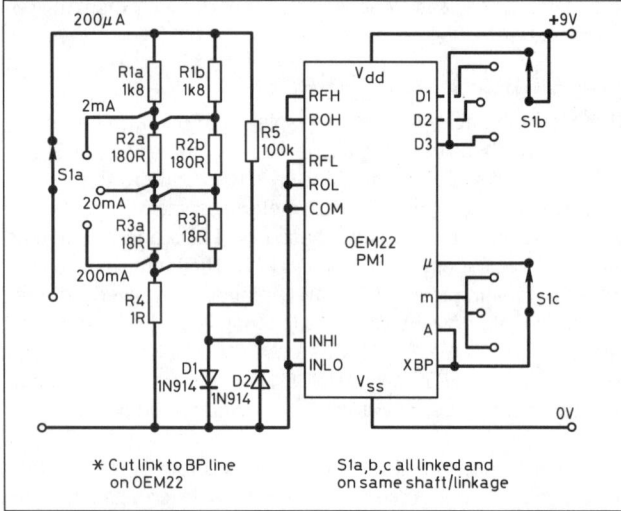

Fig 25.19: A practical digital ammeter

R1a, R1b	1k8, 0.5W, 1%	R5	100k, 0.5W, 5%
R2a, R2b	180R, 0.5W, 1%	PM1	Anders panel meter,
R3a, R3b	18R, 0.5W, 1%		type OEM22
R4	1R, 0.5W, 1%	SW1	Rotary switch, 3p, 4w
D1,D2	1N914 or similar		

Table 25.2: Components for the practical digital ammeter

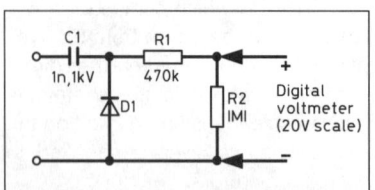

Fig 25.20: RF probe for digital voltmeter

Feature	Analogue Meter	Digital Meter
Operation	Current	Voltage
AC or DC	DC Moving Coil (AC with rectifiers) AC/DC Moving Iron	DC (AC with rectifiers or converters)
Display	Electro-mechanical	Semiconductor
Power supply required	None (taken from circuit under test)	DC supply
Best Sensitivity	50µA FSD typical	199.9mV typical
Circuit Loading	Depends on circuit and sensitivity of meter	Input >10MΩ, may affect high impedance circuit
RF interference	None	Possible due to internal oscillator

Table 25.3: Comparison of Analogue and digital meters

A Practical Digital Voltmeter

Fig 25.18 shows the arrangement for a digital voltmeter for DC voltage ranges of 200mV, 2V, 20V and 200V. The unit requires a 9V DC supply. Components are listed in **Table 25.1**.

Resistors R1 to R5 form a potential divider network with switch S1a selecting the correct input, ie the maximum voltage to the panel meter is to be 199.9mV. Resistor R6 and diodes D1/D2 provide protection for the panel meter should the wrong range (S1a) be inadvertently selected and introduce an error of less than 0.1%. S1b selects the position of the decimal point while S1c selects the annotation to be shown. The link *must* be cut to the BP line on the panel meter. Because the input resistance of the meter module is of the order of 100MΩ it represents negligible loading on the potential divider chain. The overall input resistance of the meter is about 10MΩ.

Sufficient information is provided for the reader to adapt this design to cope with other ranges. For inputs lower than 200mV, then an amplifier is required ahead of the meter input.

A Practical Digital Ammeter

This relies on measuring the voltage drop developed by the current to be measured passing through the measurement resistor, and it must be 200mV for full scale. Hence the circuit of **Fig 25.19** results in a meter measuring 200µA to 200mA in decade ranges.

Resistors R1 to R4 form the load across which the voltage is developed from the current being measured. Resistors R1 to R3 involve resistors in parallel to make up the correct value required. Switch S1a selects the input, S1b selects the decimal point positions and S1c selects the correct annotation for the range being used. The combination R5/D1/D2 provides protection for the panel meter input. The unit can be powered from a PP3 battery or equivalent. **Table 25.2** lists the components used.

RF Measurements

Similar problems arise for the digital meter as explained earlier for the analogue meter. A slight modification is made to the RF probe circuit and this is shown in **Fig 25.20**. This assumes the meter has a scale with 20V full scale and an input impedance in excess of 10MΩ. The resistors provide scaling from peak to RMS for a sine-wave input. The construction should be similar to that shown in Fig 25.13.

Comparison of Analogue and Digital Meters

Table 25.3 assumes that the analogue meter has no electronic circuit associated with it as this may alter its characteristics. It should also be borne in mind that the input to a digital meter might be affected by input amplifiers and attenuators.

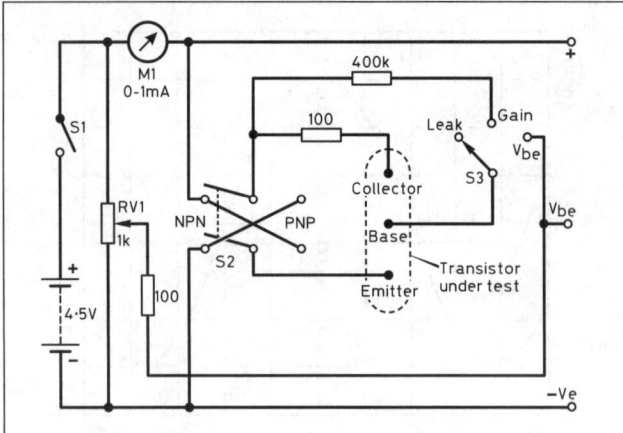

Fig 25.21: A dual trace oscilloscope

Fig 25.23: A typical displayed waveform

THE OSCILLOSCOPE

The oscilloscope (**Fig 25.21**) is a general-purpose instrument for examining electrical waveforms. It can be used for various sets of measurements depending on how it is has been set up. It is the intention of this section to explain briefly how an oscilloscope works and how it can be used for taking various measurements.

The Basic Oscilloscope

Some oscilloscopes can display a single trace whilst others can display two traces, or even more with adapters. The single-trace oscilloscope has a cathode-ray tube with a single electron gun firing at the phosphor.

The two-trace oscilloscopes fall into two categories - the dual-trace and the dual-beam. In the dual-trace oscilloscope there is a single electron gun but the control of it is split between the two traces to be shown - first one and then the other etc, but using the same timebase. In the dual-beam type there are two electron guns in the same cathode-ray tube which are independent of each other, hence the two beams can use different timebase settings.

Fig 25.22 shows the absolute basics for an oscilloscope. It consists of a display (usually a cathode-ray tube) which shows an electrical waveform. These signals have been processed in some way (eg amplified) for them to be suitable for display. The oscilloscope also contains an oscillator (or timebase) which causes the display beam to traverse the display face in the horizontal plane.

Fig 25.22: Basic block diagram of an oscilloscope

In addition a power supply is required to provide the amplifier and timebase with low-voltage supplies, and the tube with a high voltage (usually in the kilovolt region). The screen is split into two directions, the X (or horizontal) and Y (vertical direction).

Voltage Measurements

Consider the oscilloscope screen display as depicted in **Fig 25.23**. This is obviously a sine wave but what is its voltage? What voltage are we talking about? The easiest voltage measurement to take is the peak-to-peak value. The vertical displacement (Y) is 6 divisions. The setting of the Y-controls must also be taken into account; say these are at 0.5V/div. The peak-to-peak voltage is therefore 6 x 0.5 = 3V. The peak value is half of this, ie 1.5V; the RMS value is 0.7071 times this value, ie 1.06V.

Frequency Measurements

The method for making a frequency measurement is similar to the above, except that now the horizontal (X) axis is used with its setting. One problem with something like a sine wave is estimating a point on a curve and so it is an estimate only. Measure from like point to like point, eg the two negative peaks. The distance between the two negative peaks is 8 units. If the X or timebase setting is 0.5ms/div, then this represents a period of 8 x 0.5 = 4ms. The frequency is the reciprocal of this, ie 250Hz. It should be noted from this exercise that the period of a rectangular waveform is easier to estimate than that of a sine wave.

Equipment Limitations

The Y-amplifiers (plus the tube) limit the frequency response of the oscilloscope. This means that after a certain point the oscilloscope calibration is not valid, but comparative measurements can still be made above this point providing the frequency is not changed.

The capacitance of the oscilloscope and/or its probe may affect the circuit under test if the capacitance in the latter is of the same order as the oscilloscope input (20-40pF), eg in a tuned circuit.

The input voltage on an oscilloscope is normally quoted as x volts DC plus peak AC. Typical of these figures are 400V DC plus the peak AC signal that can be displayed. Exceeding this will damage internal components of the oscilloscope. Although a divide-by-10 probe can be used to extend the voltage range, these have a voltage limit (typically 600V DC) but may have to be derated as frequency rises - check the specification. The divide-by-10 probe will also reduce any loading effects on the circuit.

Fig 25.24: Simple transistor and diode tester

For most semiconductor applications the voltage limit never causes a problem, but with high-voltage valve circuits due regard must be paid to the limitations.

COMPONENT MEASUREMENTS

The cost of test equipment has decreased and the complexity of it has increased. It is now possible to purchase relatively cheap instruments to measure resistance, capacitance and inductance as well as testing transistors and diodes.

There are the standard type of analogue LCR meters which normally use a bridge technique for measuring impedance. It is possible to buy these but you may also find ex-commercial units at rallies.

A typical digital LCR tester will cost between £35 and £100 (2005 prices). However, you get what you pay for and the typical resolution for capacitance is 1pF and for inductance is 1µH. The measurement frequency depends on model and varies between about 1kHz and 200kHz. For resistance the minimum resolution is of the order of 1Ω.

The diode and transistor testers again vary in price and similar to the LCR meters. They will certainly test the basic operation of the device but they may not, for example, test the high frequency response.

Below are two circuits that can be built for diode/transistor testing and capacitance measurement.

A DIODE AND TRANSISTOR TESTER

The circuit of **Fig 25.24** shows a simple tester which will identify the polarity and measure the leakage and small-signal gain of transistors plus the forward resistance of diodes.

Testing Transistors

To check the DC current gain h_{FE} (which approximates to the small signal current gain h_{fe} or ß), the transistor is connected to the collector, base and emitter terminals and S2 switched for the transistor type. Moving switch S3 to the GAIN position applies 10mA of base current and meter M1 will show the emitter current. With S3 at the LEAK position, any common-emitter leakage current is shown, which for silicon transistors should be barely perceptible. The difference between the two values of current divided by 10mA gives the approximate value of $h_{FE} + 1$ which is close to h_{fe} for most practical purposes.

A high value of leakage current probably indicates a short-circuited transistor, while absence of current in the GAIN position indicates either an open-circuited transistor or one of reversed polarity. No damage is done by reverse connection, and PNP and

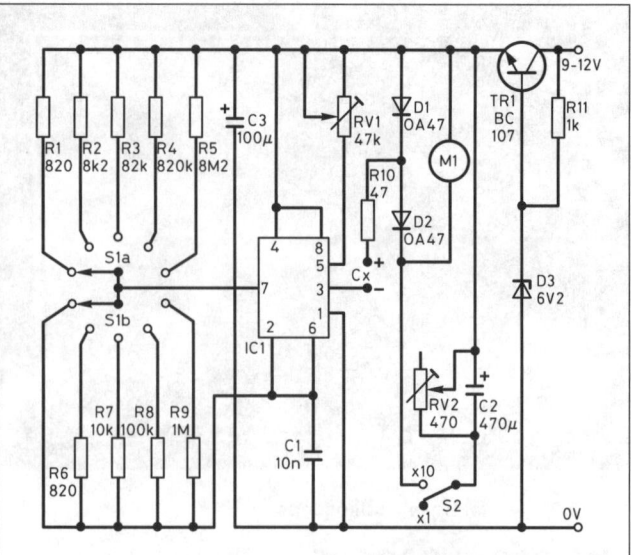

Fig 25.25: Circuit of the linear-scale capacitance meter

NPN transistors may be identified by finding the polarity which gives normal gain.

With S3 in the V_{be} position, the base-emitter voltage is controlled by RV1 which should be near the negative end for NPN and near the positive end for PNP. V_{be} may be measured by a voltmeter connected between the terminal marked 'V_{be}' and either the positive or negative rail depending on the polarity of the device. This test position may be used for FETs but only positive or zero bias is possible.

Testing Diodes

The forward voltage drop across a diode may be measured by connecting it across the terminals marked '+' and 'V_{be}' with a voltmeter in parallel. The forward current is set by RV1.

Diodes may be matched for forward resistance and, by reversing the diode, the reverse leakage can be seen (which for silicon diodes should be barely perceptible). The value of forward voltage drop can be used to differentiate between germanium and silicon diodes.

The unit can also be used to check the polarity of LEDs as the maximum reverse voltage of 4.5V is hardly likely to damage the device (note: the reverse voltage applied to an LED should not exceed 5V). For this test RV1 should be set to about mid-position.

A LINEAR-SCALE CAPACITANCE METER

This instrument is based on the familiar 555 timer and the circuit is shown in **Fig 25.25**. It has five basic ranges with a x10 multiplier. This gives the equivalent of six ranges of full scale values 100pF, 1nF, 10nF, 100nF, 1µF and 10µF.

The meter works by charging the unknown capacitor Cx to a fixed voltage and then discharging it into a meter circuit. The average current is proportional to the capacitance and hence a direct reading on the meter. If measuring small electrolytic capacitors please observe the polarity. The unit requires a low current 9V DC supply.

Construction

A components list is given in **Table 25.4**. The layout of the components is not critical (**Fig 25.26**). A PCB pattern (**Fig 25.27**) is given in Appendix B. The builder can either make a box or, as is more usual, purchase one of the cheaper plastic types.

R1, R6	820R	RV2	Single turn trimmer, 470R, 0.5W
R2	8k2	C1	Polystyrene, 10nF, ±1%
R3	82k	C2	Electrolytic, 470µF, 16V
R4	820k	D1,D2	OA47 or BAT85
R5	8M2	D3	6V2, 400mW zener
R7	10k	TR1	BC107 or similar NPN
R8	100k	IC1	555 Timer
R9	1M	M1	Moving Coil, 50µA FSD
R10	47R	S1	2p, 6way, Rotary, PCB mounting
R11	1k	S2	PCB Mount SPCO switch
RV1	Single turn trimmer, 47k, 0.5W		

Resistors are metal film type, MRS25, 1% unless specified otherwise.

Table 25.4: Components for the linear-scale capacitance meter

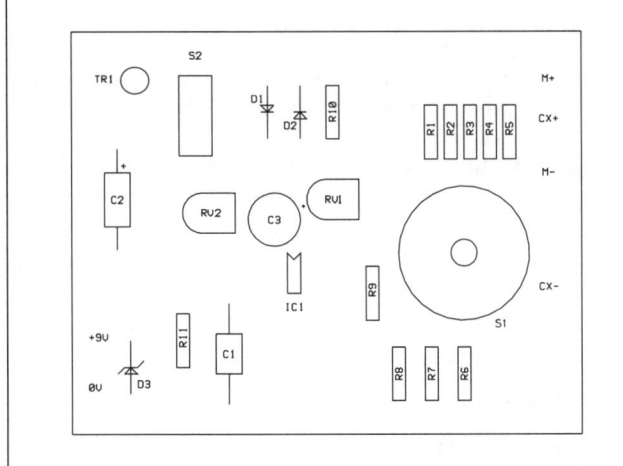

Fig 25.26: Component layout for the linear-scale capacitance meter (not to scale). The PCB layout can be found in Appendix B

Calibration

Calibration may be carried out on any range; if possible obtain 100pF, 1nF and 10nF capacitors with ±1% tolerance. With the range switch set to position 2 and multiplier switch S2 in the x1 position, connect the 1nF capacitor. Adjust RV1 for full-scale deflection. Switch to the x10 position of S2 and adjust RV2 for a meter reading of 0.1. Use the other capacitors to check the other ranges. Calibration is now complete.

Warning: If a large-value capacitor is to be measured, the meter will be overloaded.

IMPEDANCE MEASUREMENTS

Impedance measurement for the radio amateur probably means antenna or feed point impedance. The following two items of home-built equipment allow impedance of various circuits to be analysed. The first one merely provides the impedance value. It does not give the resistive and reactance components separately, the second one will provide this in series form. A previous design in [1] (page 70) will give the equivalent parallel components of an unknown impedance.

A good design was produced by G3BIK and is described in *RadCom*, Dec 1999 with full constructional details; this has an internal VFO and does not use the noise generator/communications receiver as described here.

Proprietary equipment can be purchased [2] which is designed to cover the amateur bands up to 470MHz and would

be useful for other frequencies as well. However they are likely to set you back about £350. The alternative is to look for second-hand commercial gear or build your own.

Further information can be obtained from more specialised books on amateur antennas and experimentation.

An RF Impedance Bridge

The need for an instrument which will measure impedance is felt at some time or other by every experimenting amateur. The instrument normally used is the full RF bridge, but commercial RF bridges are elaborate and expensive. On the other hand it is possible to build a simple RF bridge which, provided the limitations are appreciated, can be inexpensive and a most useful adjunct in the amateur workshop. In fact it is essential if experiments with antennas are undertaken.

The instrument described here will measure impedances from 0 to 400Ω at frequencies up to 30MHz. It does not measure reactance or indicate if the impedance is capacitive or inductive. A good indication of the reactance present can be obtained from the fact that any reactance will mean a higher minimum meter reading.

Circuit description

There are many possible circuits, some using potentiometers as the variable arm and others variable capacitors, but a typical circuit is shown on **Fig 25.28**. The capacitors have to be differential in action, mounted in such a way that as the capacitance of one decreases the capacitance of the other increases. The capacitors should be of the type which have a spindle protruding at either end so that they can be connected together by a shaft coupler. To avoid hand-capacitance effects, the control knob on the outside of the instrument should be connected to the nearest capacitor by a short length of plastic coupling rod. These capacitors form two arms of the bridge, the third arm being the 100Ω non-inductive resistor and the fourth arm the impedance to be measured. Balance of the bridge is indicated by a minimum reading on the meter M1.

Construction

Construction is straightforward, but keep all leads as short as possible. The unit should be built into a metal box and screening provided as shown in Fig 25.28.

Signal source and calibration

The instrument can be calibrated by placing across the load terminals various non-reactive resistors (ie not wire-wound) of known value. The calibration should preferably be made at a low

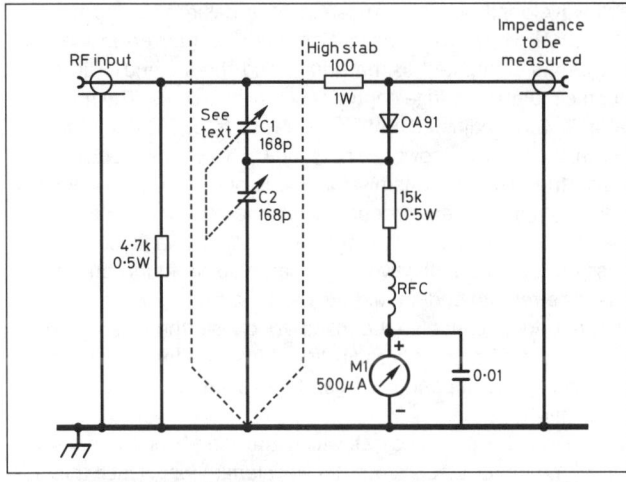

Fig 25.28: Simple RF bridge. Note that a BAT85 diode may be used instead of the OA91

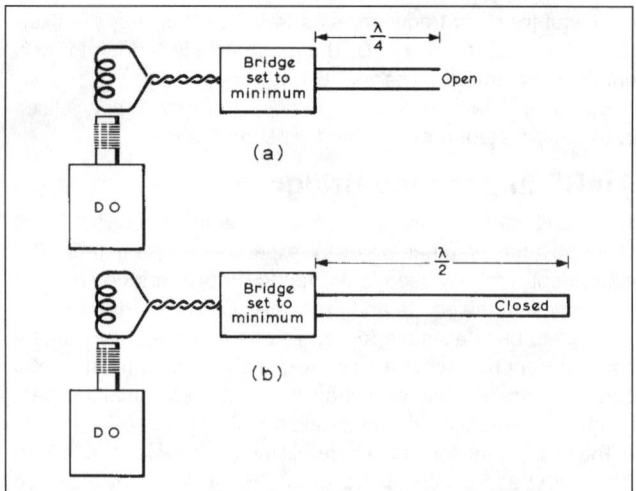

Fig 25.29: Use of the RF bridge with a dip oscillator

frequency where stray capacitance effects are at a minimum, but calibration holds good throughout the frequency range. In using the instrument, it should be remembered that an exact null will only be obtained on the meter when the instrument has a purely resistive load. When reactance is present, however, it becomes obvious from the behaviour of the meter; adjusting the control knob will give a minimum reading but a complete null cannot be obtained.

The RF input to drive the bridge can be obtained from a dip oscillator, signal generator or low-power transmitter capable of giving up to about 1W of signal power. The signal source can be coupled to the bridge by a short length of coaxial cable directly or via a link coil of about four turns as shown on **Fig 25.29**.

If using the dip oscillator, care should be exercised in order to not over-couple with it as it may pull the frequency or, in the worst circumstances, stop oscillating. As the coupling is increased it will be seen that the meter reading of the bridge increases up to a certain point, after which further increase in coupling causes the meter reading to fall. A little less coupling than that which gives the maximum bridge meter reading is the best to use. The bridge can be used to find antenna impedance and also used for many other purposes, eg to find the input impedance of a receiver on a particular frequency.

Some practical uses

One useful application of this type of simple bridge is to find the frequency at which a length of transmission line is a quarter- or half-wavelength long electrically. If it is desired to find the frequency at which the transmission line is a quarter-wavelength, the line is connected to the bridge and the far end of it is left open-circuit. The bridge control is set to zero ohms. The dip oscillator is then adjusted until the lowest frequency is found at which the bridge shows a sharp null. This is the frequency at which the piece of transmission line is one quarter-wavelength. Odd multiples of this frequency can be checked in the same manner. In a similar way the frequency at which a piece of transmission line is a half-wavelength can also be found but in this case the remote end should be a short-circuit.

The bridge can also be used to check the characteristic impedance of a transmission line. This is often a worthwhile exercise, since appearances can be misleading. The procedure is as follows.

1. Find the frequency at which the length of transmission line under test is a quarter-wavelength long. Once this has been found, leave the oscillator set to this frequency.
2. Select a carbon resistor of approximately the same value

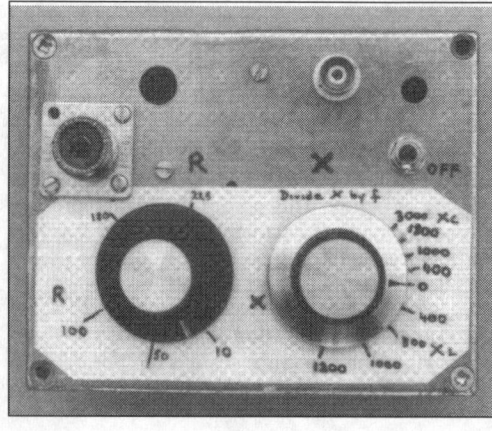

Fig 25.30: The G3ZOM noise bridge

as the probable characteristic impedance of the transmission line. Replace the transmission line by this resistor and measure the value of this resistor at the preset frequency. (Note: this will not necessarily be identical with its DC value).
3. Disconnect the resistor and reconnect the transmission line. Connect the resistor across the remote end of the transmission line.
4. Measure the impedance now presented by the transmission line at the preset frequency. The characteristic impedance (Z_0) is then given by:

$$Z_0 = \sqrt{Z_s \times Z_r}$$

where Z_s is the impedance presented by the line plus load and Z_r is the resistor value.

An RF Noise Bridge

The noise bridge uses the null method. Wide-band RF noise is used as a source, and a receiver is used as frequency-selective null detector. Noise bridges do not have a reputation for accuracy but they are small and convenient to use. The accuracy and the depth of the null depends mostly on the layout of the bridge network and the care taken in balancing out the bridge.

The description of a noise bridge that follows is by G3ZOM. The front panel is illustrated in **Fig 25.30**, the circuit diagram is shown in **Fig 25.31** and parts list in **Table 24.5**. **Fig 25.32** shows the layout of the bridge components. It can be used to measure impedance in terms of series resistance and reactance, within the frequency limits 1 to 30MHz. The useful range is approximately 0 to 200Ω. The reactance range is dependent on frequency and the capacitance swing of the variable capacitor used in the variable arm of the bridge. As a rough guide, using the suggested 250pF variable:

R1	220R	ZD1	5V6 Zener, 400mW
R2	100k	C1, C2	10n
R3	1k2	C3,C4	10n
R4	68k	C5	82p
R5	680R	C6	10μ, 16V electrolytic
RV1	220R pot, carbon	VC1	250p variable
TR1,TR2	BC182, 2N3903 or similar	S1	SPCO switch
T1	Dust iron core FT50-43, FT50-5 or similar. See Fig 25.31 for winding details		
Resistors are 0.25W/0.5W, 5% unless specified otherwise.			

Table 25.5: RF Noise bridge components list

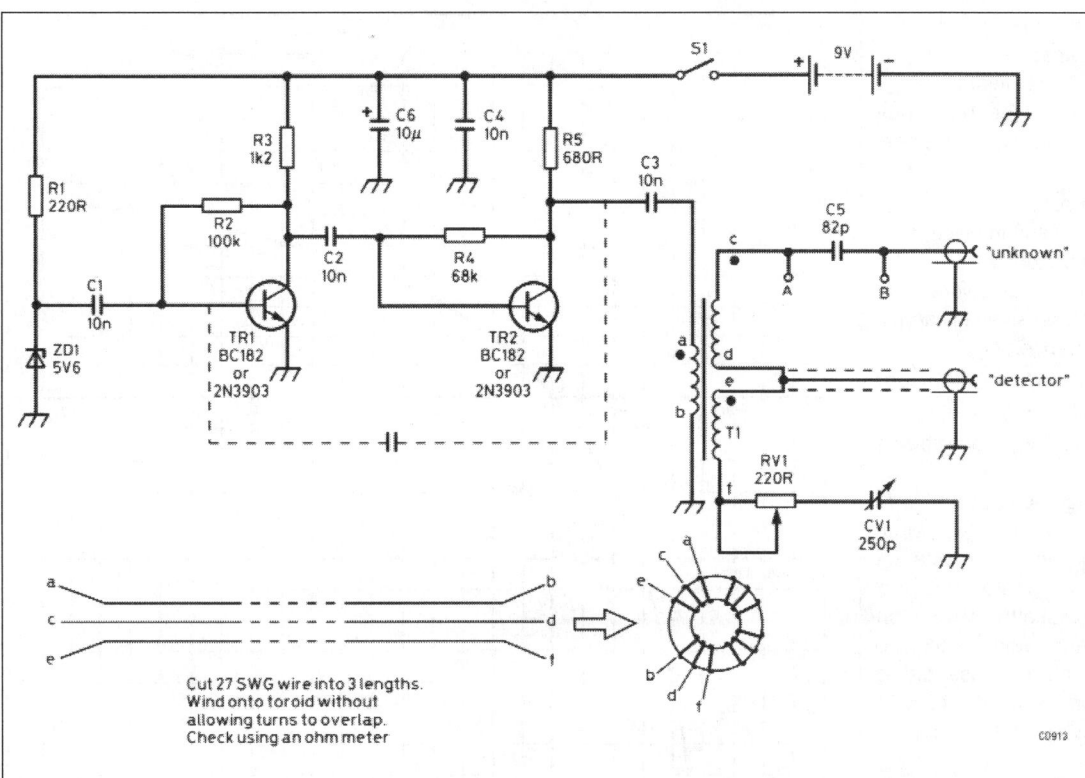

Fig 25.31: Circuit of the G3ZOM noise bridge

Cut 27 SWG wire into 3 lengths.
Wind onto toroid without
allowing turns to overlap.
Check using an ohmmeter

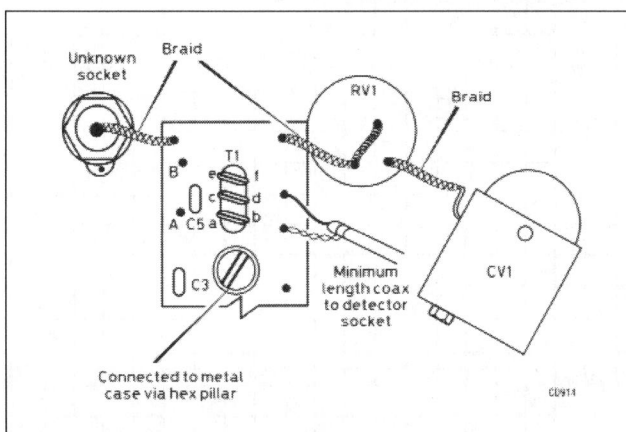

Fig 25.32: Layout of bridge components

At 1MHz: 5000Ω capacitive to 1200Ω inductive
At 30MHz: 170Ω capacitive to 40Ω inductive

Balancing the bridge

The bridge has to be balanced to obtain a reasonable calibration over the intended range. Connect a suitable receiver to the detector socket and a non-reactive (carbon or metal film) across the UNKNOWN socket. The resistor leads must be kept as short as possible to reduce the unwanted reactance to a minimum (important at the high-frequency end of the range). Set RV1 to maximum resistance and CV1 to maximum capacitance (fully meshed). Tune the receiver to around 14MHz and switch the noise bridge on. A loud 'rushing' noise should be heard from the receiver, and the S-meter (if fitted) should show a good signal strength.

By listening to the noise level, and observing the S-meter (if fitted), adjust CV1 to obtain a decrease in volume (a null). Then adjust RV1 for a deeper null. Repeat these two adjustments until the deepest null has been reached. Temporarily mark the null positions of RV1 200Ω and CV1 to zero.

Set the receiver within the 1 to 2MHz range and repeat the nulling procedure. This time the null will be much sharper so careful adjustment is needed. The positions of RV1 and CV1 should be the same or close to those obtained previously. If not, the wiring around the bridge components is probably too long. Short wiring lengths are essential.

Repeat the procedure again with the receiver set to around 30MHz. This time the null will he much wider. The position of RV1 should again be close to that obtained previously, but it will probably be found that the position of CV1 is somewhat different to before. If this is the case, the situation can be remedied by adding a small-value balancing capacitor between pin A and chassis in Fig 25.31.

Both the value and the position of this balancing capacitor will need to be determined by trial and error. Try, say, 10pF to pin A and repeat the nulling procedure at 2 and 30MHz. If the situation is worse than before, try a 10pF capacitor between the chassis and RV1 (where it connects to T1). One or the other position will result in an improvement which is worth the effort to obtain reliable measurements. Even greater accuracy can be obtained by adding compensating inductance to the bridge but this has not been found necessary to date.

The instrument shown in Fig 25.31 has a better than 50dB null at 10MHz.

Calibration

Calibration can now be carried out. Tune the receiver to around 14MHz.

Resistance scale

Connect suitable resistors, one at a time across the UNKNOWN socket, nulling the bridge and marking the resistance values of the test resistors on the RV1 scale. CV1 should remain at its zero position. The resistance scale should be fairly linear, allowing simple interpolation of unmarked values.

Reactance scale

Connect a 51Ω resistor across the UNKNOWN socket, using short leads. Null the bridge and mark the position CV1 as zero '0'.

Leave the 51Ω resistor in place and connect a selection of capacitors across C5 (use terminals A and B), nulling the bridge each time and marking the reactance scale with the capacitance value. This part of the scale represents series inductance (positive reactance or X_L).

With the 51Ω resistor still in place, connect a series of capacitors across CV1, again nulling and marking as before. This part of the scale represents series capacitance (negative reactance, or X_C). Note that the scale will only be linear if a linear capacitance law variable is used for CV1.

Reactance scale calibration - capacitance or ohms?

At this stage, the reactance scale is temporarily calibrated in capacitance. Most published designs leave the reactance scale calibrated this way and use either a graph or a formula to make the conversion to the required reactance value in positive or negative ohms. You can use either method of calibration, using the conversion graph of **Fig 25.33** or the formula -

$$X = \frac{10^6}{2\pi f} \times \frac{S}{C_5(S+C_5)}$$

where X is the reactance in ohms, f is the frequency in MHz, S is the scale reading (+ or -) and C5 = 82 (the value of C5 in pF).

Using the noise bridge

This bridge, in common with all other impedance measurement bridges, measures the impedance presented to the UNKNOWN socket. This may not be the same as the antenna impedance because of the impedance transformation effect of the coaxial cable connecting the antenna to the bridge.

RV1 and CV1 are then varied alternately to obtain the best null. The equivalent series resistance is obtained directly from the RV1 scale.

Impedance at the UNKNOWN socket is measured by connecting the noise bridge as shown in **Fig 25.34**. The receiver or transceiver is tuned to the measurement frequency and the R and X controls adjusted for minimum noise. These controls interact and the sharpest dip must be found by trial and error. Antenna impedance measurements can be accomplished in one of two ways:

1. At the transmitter end of the feeder, **Fig 25.34(a)**. By using a multiple of λ/2 at the frequency of measurement, the antenna feed impedance is reflected back to the transmitter end of the feeder. The disadvantage of this method is that the antenna matching network (eg gamma match) is at the antenna, remote from the impedance measurement, making the method rather cumbersome.

2. At the antenna end of the feeder, **Fig 25.34(b)**. The adjustment, and the measurement of the

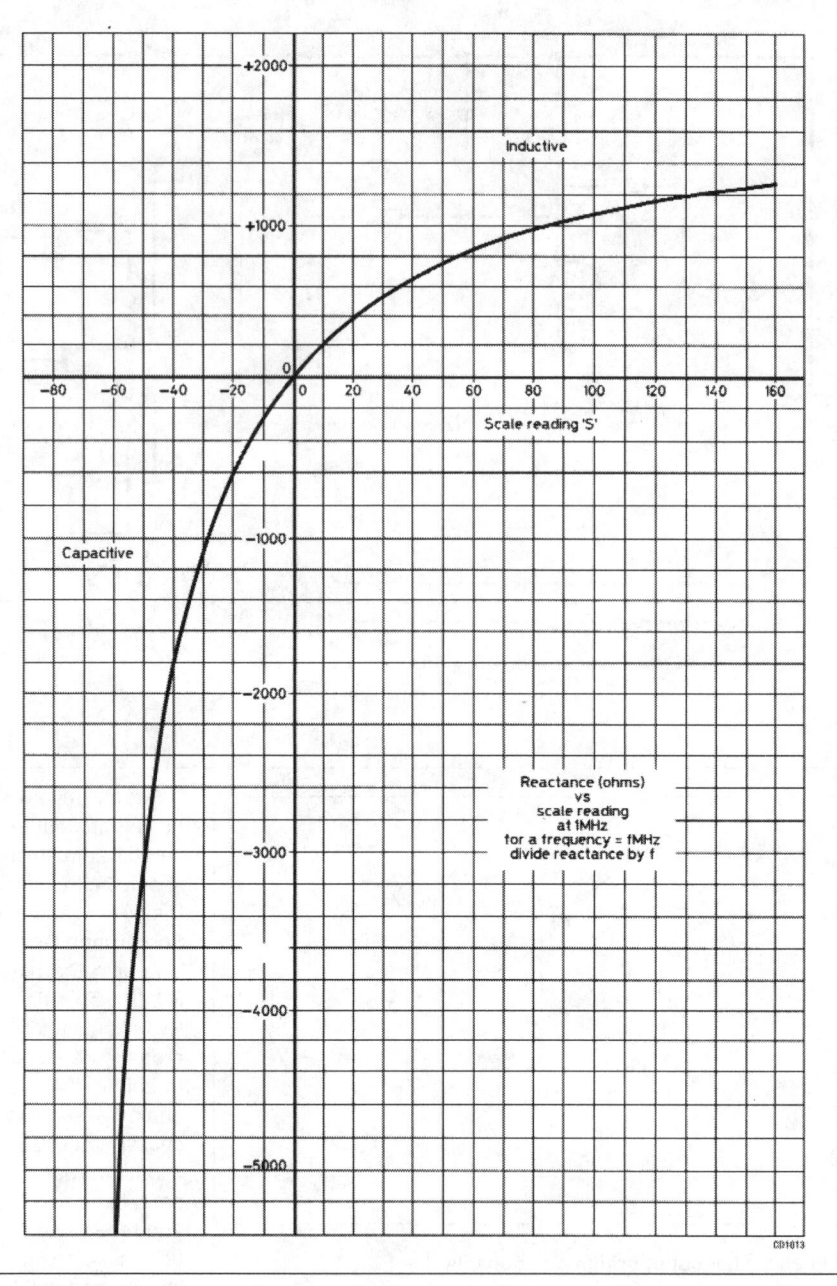

Fig 25.33: Calibration graph for converting capacitance value s to reactance (+ or - j)

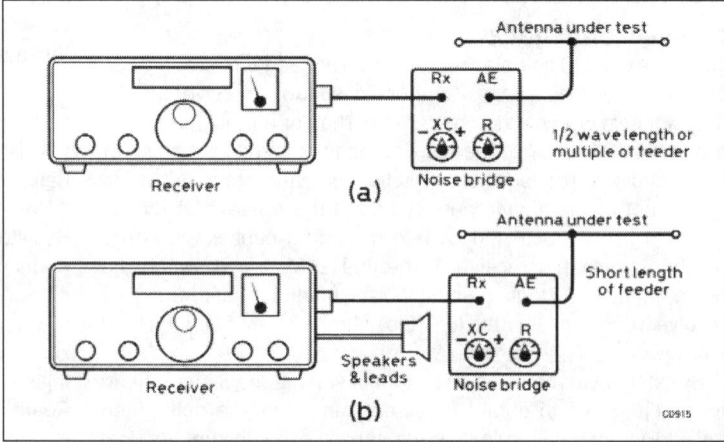

Fig 25.34: Noise bridge and receiver connections for antenna impedance measurements

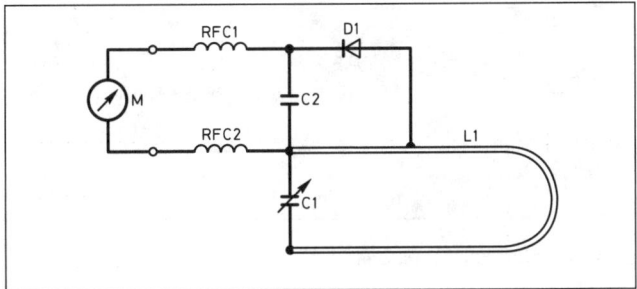

Fig 25.35: A simple absorption wavemeter for 65-230MHz

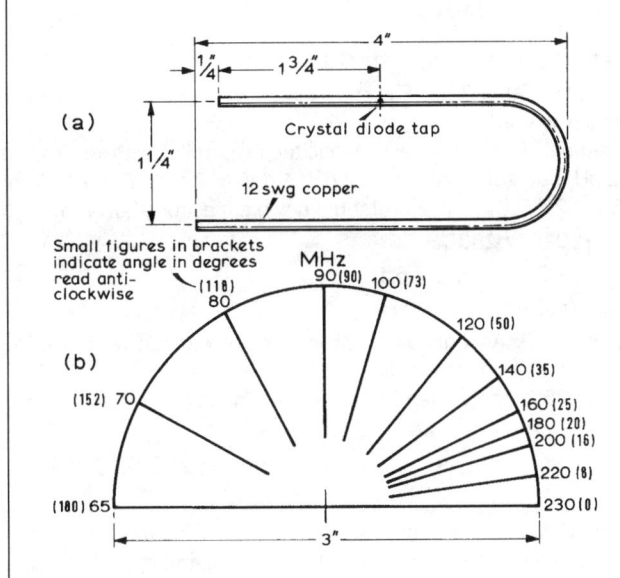

**Fig 25.36: Constructional details of simple absorption waveme-
ter: (a) inductor; (b) dial plate.** ¼in = 6.3mm, 1¼in = 31.8mm,
1¾in = 44.5mm, 3in = 76.2mm, 4in = 101.6mm

RFC1,2	80t of 40SWG ECW wound on 10k, 0.5W resistor	M1	1mA FSD or better
		C1	4-50p, Jackson C804 or equivalent
L1	See Fig 25.36(a)	C2	470p ceramic
D1	OA91, BAT85 or similar		

Table 25.6: Simple absorption wavemeter components list

results of the adjustment, is far more convenient.
However, the method is limited to situations where there
is access to the antenna in situ. A further disadvantage is
that the noise null detector, the receiver, also has to be
close at hand, which may be rather inconvenient 20m up
a mast or on the roof of a house. The problem can be overcome by leaving the receiver in the shack. A small speaker or a pair of headphones can be connected to the output
of the receiver via another feeder or a couple of wires from
the rotator cable. The feeder length is immaterial. *Make
sure that the receiver/headphone arrangement is earthed to prevent an electric shock hazard.* Alternatively, a pair
of low-cost PMR446 handheld transceivers may be used
to relay the audio.

A SIMPLE ABSORPTION WAVEMETER FOR 65-230MHz

The absorption wavemeter shown in **Fig 25.35** is an easily built
unit covering 65-230MHz. For a lower-range unit the dip oscillator described in the next section can be used.

Construction is straightforward and all the components, apart
from the meter, are mounted on a Perspex plate of thickness 3
or 4mm and measuring 190 x 75mm. Details of the tuned circuit are shown on **Fig 25.36(a)** and should be closely followed.
The layout of the other components is not critical provided they
are kept away from the inductor. A components list is given in
Table 25.6.

For accurate calibration, a signal generator should be used
but, provided the inductance loop is carefully constructed and
the knob and scale are non-metallic, the dial markings can be
determined from **Fig 25.36(b)**.

In operation the unit should be loosely coupled to the circuit
under test and the capacitor tuned until the meter indicates resonance (a maximum). For low-power oscillators etc a more sensitive meter should be used (eg 50µA or 100µA).

The wavemeter can also be used as a field strength indicator
when making adjustments to VHF antennas. A single-turn coil
should be loosely coupled to the wavemeter loop and connected
via a low-impedance feeder to a dipole directed towards the
antenna under test.

DIP OSCILLATORS

It is possible to buy these units and radio rallies would be a good
starting point. Alternatively one can build them - see later and
reference [1] and [5].

Although the dip
oscillator has a wide
range of uses for
measurements on
both complete equip-
ment and individual
components, these all
rely on its ability to
measure the frequen-
cy of a tuned circuit. In
use, the coil of the dip
oscillator is coupled
indirectly to the circuit
under test, with maxi-
mum coupling being
obtained with the axis
of the oscillator coil at
right-angles to the
direction of current

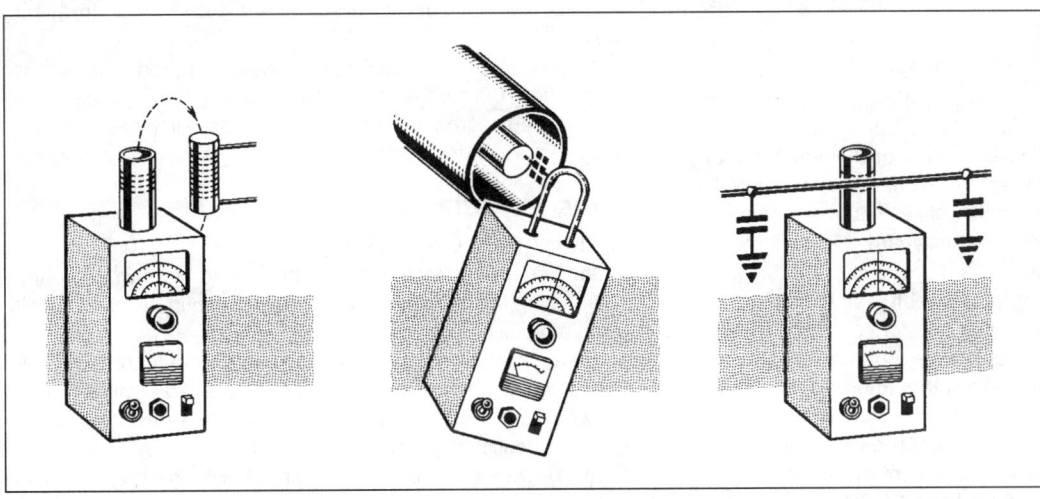

Fig 25.37: Using a dip oscillator

flow. Coupling should be no greater than that necessary to give a moderate change on the dip oscillator meter. These are shown diagrammatically in **Fig 25.37**.

If the tuned circuit being investigated is well shielded magnetically (eg a coaxial line) it may be difficult to use inductive coupling. In such cases it may be possible to use capacitive coupling by placing the open end of the line near to one end of the dip oscillator coil.

A completely enclosed cavity is likely to have some form of coupling loop and the dip meter coil can usually be coupled inductively by means of a low-impedance transmission line such as a twisted pair with a coupling loop.

When used as a wavemeter, the oscillator is not energised and the tuned circuit acts as a pick-up loop. This arrangement is useful when looking for harmonic output of a multiplier or transmitter or for spurious oscillations.

Determination of the Resonant Frequency of a Tuned Circuit

The resonant frequency of a tuned circuit can be found by placing the dip oscillator close to that of the circuit and tuning for resonance.

No power should be applied to the circuit under test and the coupling should be as loose as possible consistent with a reasonable dip being produced on the indicating meter.

The size of the dip is dependent on the Q of the circuit under test, a circuit having a high Q producing a more pronounced dip than one only having low or moderate Q.

Measurement of L and C

The following is by G3BIK [6] and describes a method of measuring L (inductance) and C (capacitance).

The dip oscillator provides a quick and easy means of checking (to a degree of accuracy acceptable for experimental purposes) the inductance value of coils in the microhenry (µH) range and capacitors in the picofarad (pF) range, such as are commonly used in radio circuits. This can be very useful, for example, when constructing an ATU, a crystal set, a short-wave receiver, a VFO or a band-pass filter for a direct conversion receiver.

For this purpose the following are kept with the dip oscillator:

- two fixed value RF coils of known inductance - 4.7µH and 10µH
- one capacitor each of 47pF and 100pF

The choice is yours and you can decide to keep several of each to be selected from **Table 25.7**. A personal choice of coil-type is the moulded RF choke (Maplin) or RF inductor (Mainline or RS). These are axial-leaded, ferrite based, encapsulated, easy to handle, and readily available at low cost in a range of fixed-value microhenries. The capacitors are 5% tolerance polystyrene, also axial-leaded.

To determine or verify the value of either an RF coil or a capacitor, simply connect the unknown component in parallel with the appropriate known component to form a parallel LC tuned circuit, ie an unknown L in parallel with a known C (or vice versa), then use the dip oscillator to determine the resonant frequency of the parallel LC circuit. The value of the unknown component can then be obtained easily to an acceptable approximation, by using the relevant formula from **Table 25.7** and a pocket calculator. The formulas were derived from the accepted formula for

To determine unknown capacitor						
Known L µH	1.0	2.0	4.7	6.8	10	22
C pF is	25330÷F²	11513÷F²	5389÷F²	3725÷F²	2533÷F²	1151÷F²
To determine unknown inductor						
Known C pF	10	22	33	47	68	100
L µH is	2533÷F²	1151÷F²	768÷F²	539÷F²	373÷F²	253÷F²

Table 25.7: Determination of L and C using known values

the resonant frequency of a parallel tuned circuit-

$$f = \frac{1}{2\pi\sqrt{LC}}$$

(f in Hz, L in Henries and C in Farads)

NOTE: in **Table 25.7**, F is the frequency in MHz as given by the dip oscillator.

Example 1: An unknown capacitor in parallel with an known 10µH inductor, produces a dip at 6.1MHz, hence F = 6.1. From **Table 25.7**, the value of the unknown capacitor is given by:

C pF = 2533 ÷ F²

= 2533 ÷ 6.1 ÷ 6.1

= 68pF

Example 2: An unknown coil in parallel with an known 47pF capacitor, produces a dip at 12.8MHz, hence F = 12.8. From **Table 25.7**, the value of the unknown inductance is given by:

L µH= 539 ÷ F²

= 539 ÷ 12.8 ÷ 12.8

= 3.3µH

Bear in mind that because the accuracy of results relies upon the frequency as derived from the dip oscillator, it would be sensible to keep the coupling between the LC circuit and the dip oscillator as loose as possible, consistent with an observable dip. This minimizes pulling of the dip oscillator frequency. Also, rather than relying upon the frequency calibration of the dip oscillator itself, it might be useful to monitor the frequency on an HF receiver or a digital frequency meter.

A final point worth considering is that each fixed-value inductor of the type mentioned might have its own self-resonant frequency, but these would typically lie above the HF range so should not be a problem. For example, the self-resonance of the selected 10µH inductor is about 50MHz and that of 4.7µH is about 70MHz. You could quickly and simply find out the self-resonant frequency of an inductor, by taping it to each of the dip oscillator coils in turn and tuning across the full frequency span.

It is best to make L and C measurements at frequencies much lower than the self-resonant frequency of your chosen test-inductor, but perhaps better be safe than sorry and stick with the lower µH values if your interest lies between 1.8 and 30MHz.

Tone Modulation

The following is also by G3BIK [6].

Sometimes it is also useful to be able to hear an audio tone when using the GDO as an RF signal source in association with a radio receiver.

If your dip oscillator does not have tone modulation, you might like to construct the simple add-on 1kHz audio oscillator circuit shown in **Fig 25.38**. It uses a unijunction transistor, the frequency of oscillation being given approximately by 1/(R1 x C1).

The 1kHz tone output connects via C2 to the positive supply line of the dip oscillator, which it modulates. R2 acts as the modulator load and its value helps to determine the level of modulation. This

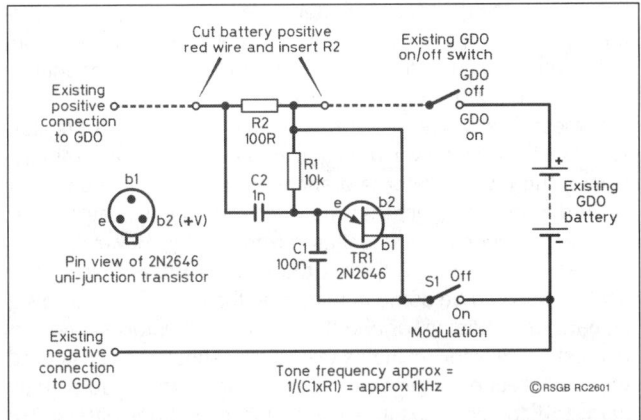

Fig 25.38: Add-on audio 1kHz oscillator for a dip oscillator (sometimes referred to as a GDO)

produces a simple but effective tone modulation of the dip oscillator's RF signal, which can be heard on an AM or an FM receiver.

The method of construction is really left to the builder but it can be fitted on a small piece of strip board without having to cut copper tracks. The finished board might conveniently mount onto one of the dip oscillator meter terminals, provided care is taken to isolate the copper-tracks from the terminal.

Use as an Absorption Wavemeter

A dip oscillator may be used as an absorption wavemeter by switching off the oscillator's power supply and then using in the normal way. In this case, power has to be applied to the circuit under test. Resonance is detected by a deflection on the meter due to rectified RF received. It should be noted that an absorption wavemeter will respond to a harmonic if the wavemeter is tuned to its frequency.

A SIMPLE DIP OSCILLATOR [2]

The best of the simple solid-state dip oscillators so far discovered were two very similar designs published in the 'Technical Correspondence' column of *QST* [7, 8]. The circuit of **Fig 25.39** is a variation of these two designs, plus improvements incorporated by G3ZOM.

This circuit does not measure gate current directly; instead it measures the total current through the FET. This is large compared with that flowing in a base or gate of a solid-state oscillator. However, the variation of current through resonance is only a small part of the total current through the FET. The dip is enhanced by offsetting the meter reading using a potentiometer in a bleeder network. This is set so that the meter reads about 75% FSD) when the instrument is not coupled to a load.

This design does not perform very well on the VHF bands with the circuit values shown and the dip tends to reverse if the coupling is too tight. This suggests that reducing the 100pF capacitors in the tuned circuit to a smaller value would improve the performance at VHF. This instrument met all the criteria of a good dip oscillator on all the HF and lower VHF bands.

The simple dip oscillator is easy to construct and, provided the necessary components are available, can be constructed in an evening. It is possible to extend the range up to 60MHz plus by designing another plug in coil L1.

This is not a complete description of how to make the instrument, but rather a few notes to emphasize the important aspects of construction.

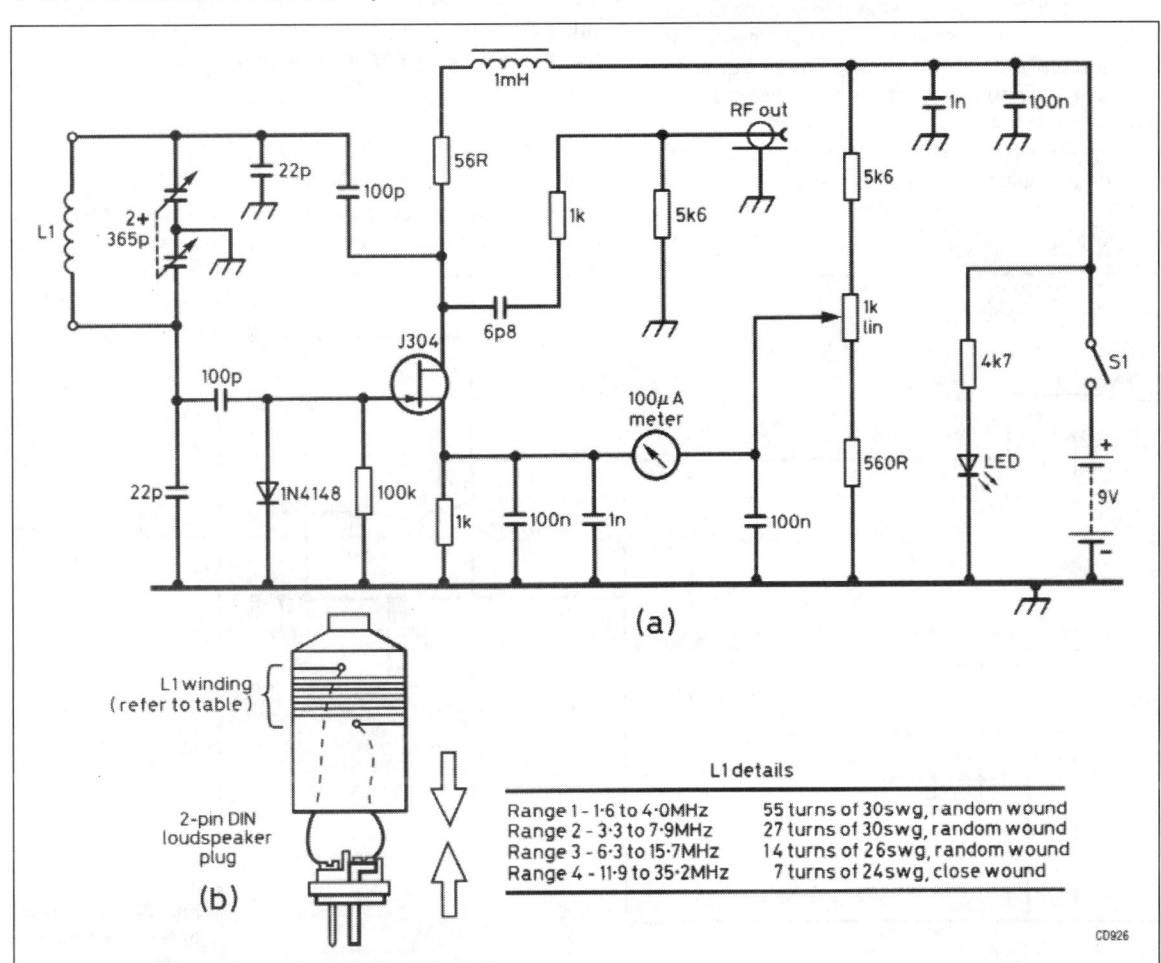

L1 details

Range 1 - 1·6 to 4·0MHz	55 turns of 30swg, random wound
Range 2 - 3·3 to 7·9MHz	27 turns of 30swg, random wound
Range 3 - 6·3 to 15·7MHz	14 turns of 26swg, random wound
Range 4 - 11·9 to 35·2MHz	7 turns of 24swg, close wound

Fig 25.39: A simple dip oscillator

The most important part of a dip oscillator is the tuning capacitor and frequency read-out dial. Sometimes a whole assembly can be obtained from an old transistor radio. The coil socket should be to-as close to the tuning capacitor as possible so that the coil leads can be kept short. The rest of the circuit can be wired around these main components. Choose a coil plug and socket arrangement that is practical. All the circuits so far discussed use two-pin coil plugs and sockets. This means that simple arrangements using crystal holders or phono plugs sockets can be used. The arrangement shown in Fig 25.39 uses two-pin DIN loudspeaker plugs for the coils.

CALIBRATION OF DIP OSCILLATORS

A frequency counter is the most convenient instrument for calibrating the dial although it is possible to check the calibration by listening for the output on a general-coverage receiver, an amateur receiver or scanner. This probably allows a good check on the calibration into the VHF range. Additional points can be found by using the second-channel response provided that the IF is known (the second-channel response is 2 x IF removed from the normal response).

Another method is to use the resonances of lengths of feeder cables, providing that the velocity factor for the particular cable is known so that the physical length corresponding to the wanted electrical half-waves and quarter-waves can be found.

SIGNAL SOURCES

Signal sources of controlled frequency and amplitude are necessary for setting up both transmitters and receivers. Ideally, for receiver adjustment, it is desirable to have an rf source covering from a few hundred kilohertz up to the highest frequency used at the station. The amplitude should be known from a fraction of a microvolt up to tens of millivolts. In a good signal generator both frequency and amplitude are accurately known but such instruments are costly and certainly difficult to make and calibrate in an amateur workshop.

Fortunately many good instruments appear on the surplus market although the frequency calibration is sometimes not too accurate. This is not too important as the amateur almost always has means of checking frequency. Therefore, in selecting an instrument, the quality of the attenuator and the effectiveness of the screening are all-important. At very low levels, a poorly screened oscillator will emit sufficient to by-pass the attenuator and prevent low microvolt output levels being attained.

For less onerous requirements, simple oscillators can be constructed for tuning over a limited range and several examples follow. The dip oscillator is a simple form of signal source but suffers from the defect that the frequency is easily pulled with changes in coupling and it has no attenuator. However, a dip oscillator placed remotely from the receiver under test is often useful. It should be borne in mind that the output may have significant harmonic power and the possibility of interference with domestic receivers, including televisions, should be considered. An audio frequency generator is useful for testing audio amplifiers and for checking the performance of transmitters. The design given in this chapter provides a sinewave output and a frequency range well in excess of the audio range.

LOW-FREQUENCY SINEWAVE OSCILLATOR FOR 10Hz-100kHz

The circuit diagram for this oscillator is shown on **Fig 25.40**. It is based on a Wien bridge oscillator formed around IC1a and buffered by IC1b. The main frequency-determining components are R1/R2 and RV2 with capacitors C1 to C8. In the configuration shown, stable oscillation can occur only if the loop gain remains at unity at the oscillation frequency. The circuit achieves this control by using the positive temperature coefficient of a small lamp to regulate the gain as the oscillator varies its output. Potentiometer RV3 forms the output level control, with R4 giving a defined output resistance of approximately 600Ω and C11 providing DC isolation. Capacitors C9 and C10 provide power supply line decoupling.

The approximate ranges provided are:

Range 1	10Hz-100Hz
Range 2	100Hz-1kHz
Range 3	1kHz-10kHz
Range 4	10kHz-100kHz

The exact range is dependent on the tolerance of the components used and ambient temperature variations.

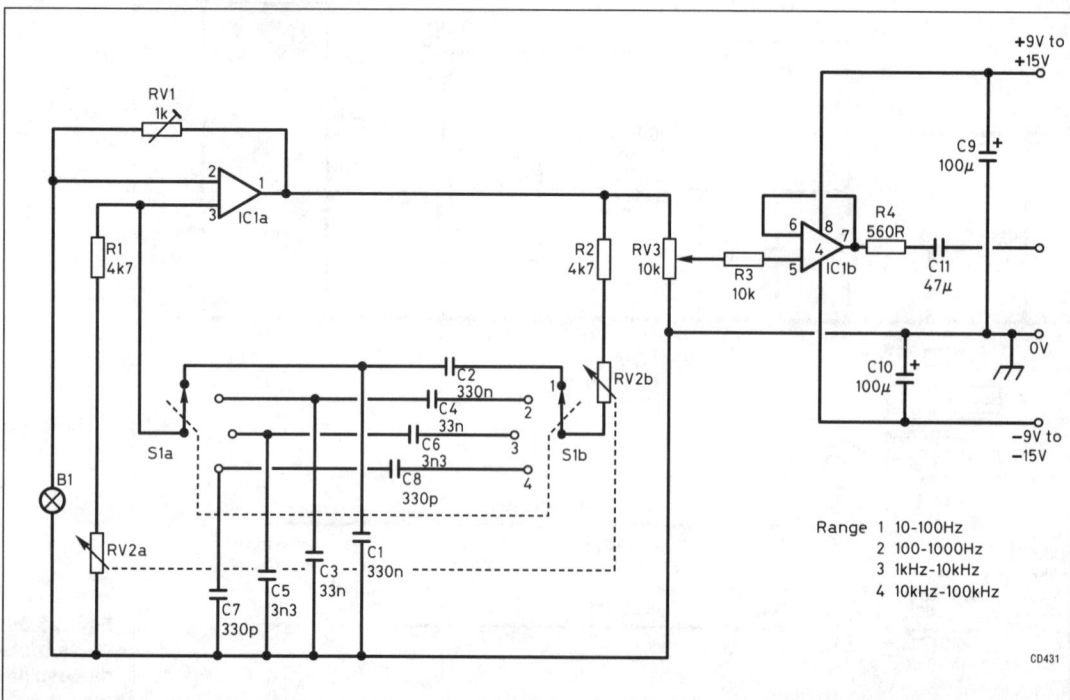

Fig 25.40: Circuit of the low frequency oscillator

R1,R2	4k7	C1,C2	330n
R3	10k	C3,C4	33n
R4	560R	C5,C6	3n3
RV1	1k trimmer	C7,C8	330p
RV2	47k dual gang pot.	C9,C10	100µ, 25V
RV3	10k lin. pot.	C11	47µ, bipolar
B1	28V, 40mA bulb	IC1	LM358

Table 25.8: Low Frequency Oscillator, 10Hz to 100kHz

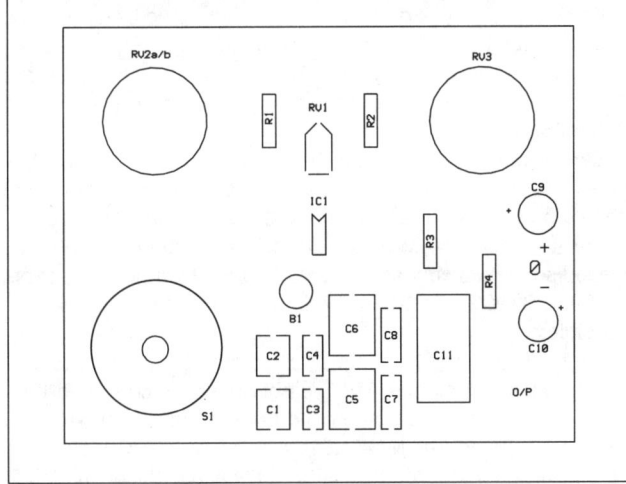

Fig 25.42: Component layout of the low frequency oscillator (not to scale)

The circuit requires a symmetrical plus and minus supply between 9 and 15V.

Construction

A components list is given in **Table 25.8**. The layout of the circuit is not critical but a PCB pattern (**Fig 25.41**) is given in Appendix B and a component layout in **Fig 25.42**. If some ranges, or the output level control, are not required then the layout can be tailored accordingly. The feedback resistor RV1 should be adjusted so that the output on all ranges is just below the clipping level.

Testing

No frequency calibration is required of this circuit but it would be wise to check with a frequency counter that the ranges are as suggested. An oscilloscope is required for setting up the adjustment of RV1.

A CRYSTAL-BASED FREQUENCY MARKER

The purpose of this unit is to produce a 'comb' of output frequencies which are all based on a crystal. The unit described here gives outputs at harmonics of 1MHz, 100kHz, 25kHz, 12.5kHz and 10kHz with an additional output of a sine wave at 1kHz which may be useful as an accurate modulation signal. The sine wave output has an output resistance of approximately 600Ω and maximum amplitude of approximately 2.5V peak to peak.

Circuit Description

The circuit diagram is shown in **Fig 25.43**. The signal is derived from a 1MHz crystal-controlled oscillator formed by XL1 and IC1 plus various components. Capacitor VC1 allows a slight varia-

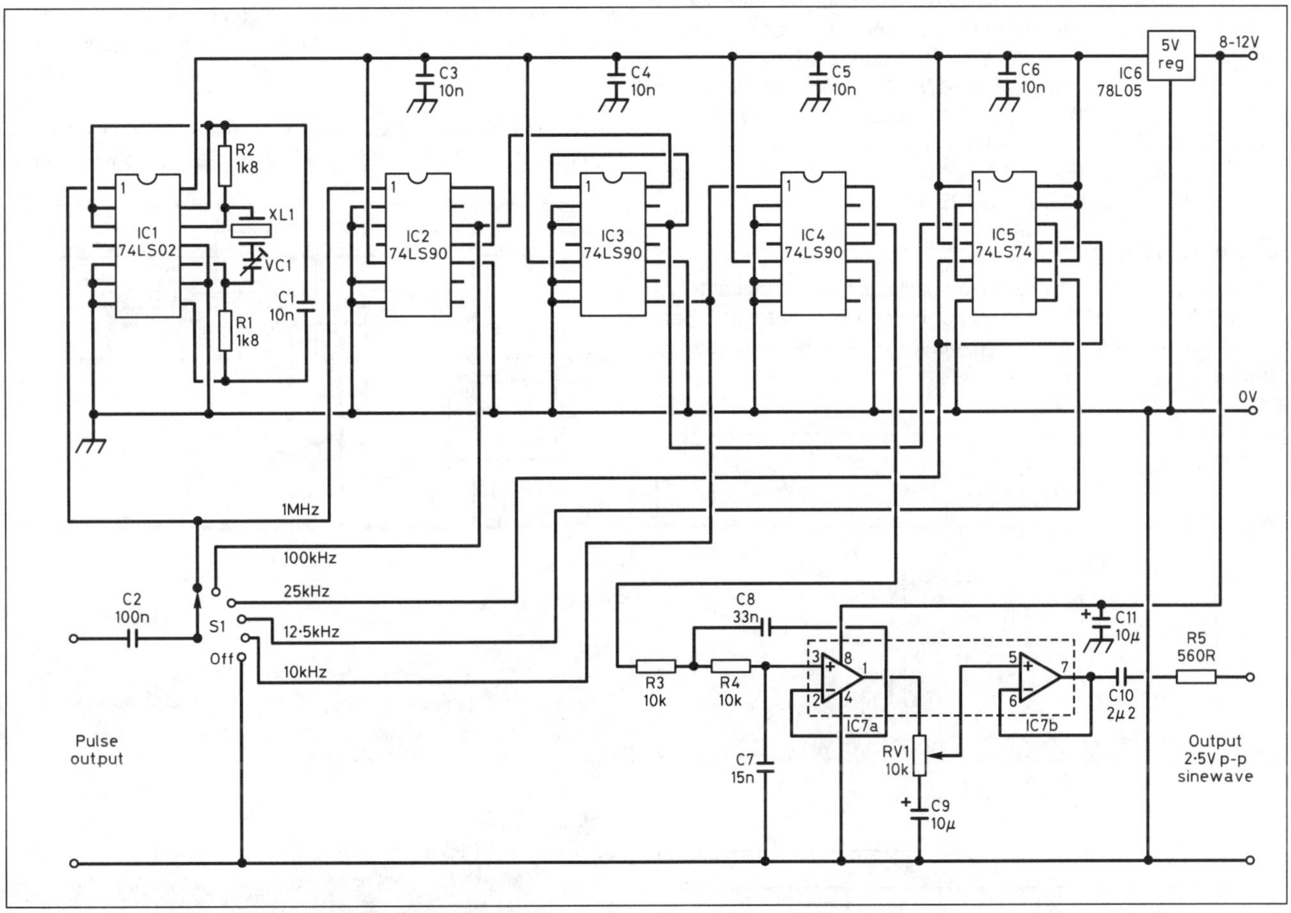

Fig 25.43: Circuit of the crystal-based frequency marker

R1,R2	1k8	C10	2µ2
R3,R4	10k	IC1	74LS02
R5	560R	IC2,IC3,IC4	74LS90
VR1	10k lin pot.	IC5	74LS74
C1	10n, ceramic	IC6	78L05
C2	100n, ceramic	IC7	LM358
C3,C4	10n, ceramic	SW1	2p, 6w rotary switch
C5,C6	10n, ceramic	XL1	1MHz crystal, HC6U
C7	15n	VC1	30p trimmer
C8	33n	IC Sockets	14p, 5 off; 8p, 1 off
C9,C11	10µ, 25V tant bead		

Resistors are 0.25W/0.5W, 5% unless specified otherwise.

Table 25.9: Crystal-based frequency marker components list

tion of the crystal frequency for calibration as described later. This 1MHz signal is divided by 10 by IC2 to give a 100kHz signal. This signal is then passed to IC3 which has a 50kHz output and also a 10kHz output. The 50kHz output is divided by dual flip-flop IC5 to give a 25kHz and 12.5kHz output. The 10kHz signal from IC3 is divided by 10 by IC4 to give a 1kHz square-wave output.

The 1kHz square wave is then filtered by an active low-pass filter formed by IC7a. The variable-amplitude sine-wave output is then buffered by IC7b. R5 forms the output resistance of the buffer.

Construction

A components list is given in **Table 25.9**. The layout for this circuit is not critical but the completed circuit should be housed in a metal box to prevent unwanted radiations. The output should be via a coaxial socket to a small antenna when in use. It requires a power supply of 8 to 12V DC at about 50mA. If the voltage regulator IC6 is omitted the circuit can be fed straight from a 5V supply but ensure there is a supply to the 1kHz filter IC7. A PCB pattern (**Fig 25.44**) is given in Appendix B and component placement details in **Fig 25.45**.

Calibration

The frequency of the 1MHz crystal oscillator can be adjusted by a small amount by VC1. The output from the oscillator or a harmonic should be checked against an accurate frequency source.

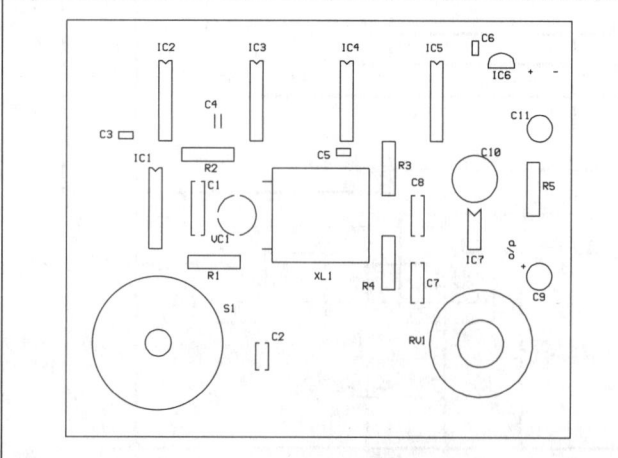

Fig 25.45: Component placement details for the crystal-based frequency marker (not to scale)

AN HF SIGNAL SOURCE

This project uses a Maxim MAX038 IC which is available from several sources. This IC is a high-frequency precision function generator and can produce triangular, sawtooth, square and pulse waveforms as well as a sine wave. In this application it is configured for a sine-wave output of 2V p-p from the IC before any load or filtering. Further details can be obtained from the data sheet [9]. To make this into a useful HF source it should be followed by an attenuator - see end of this chapter.

Fig 25.46 shows the circuit diagram for this signal source which should provide an output frequency from about 2MHz to 20MHz. Frequency of oscillation fo is determined by R1, R2 and C3. The circuit can also be adapted so that the output frequency can be controlled using a voltage applied to pin 8 - this is further explained in the data sheet. A component list is given in **Table 25.10**.

Care should be taken with the circuit layout - use a double-sided circuit board with the top side as an earth plane. C3 should be chosen for low temperature coefficient - NPO ceramics or silvered mica types should be satisfactory. It should be placed so that its connection to the ground plane is close to pin 6 of IC1; C1, C2 and C4 should be placed as close to the IC as possible.

Keep *all* capacitor leads as short as possible in order to minimise series inductance. The variable resistor R2 should be a multi-turn cermet type potentiometer or a single-turn type with slow-motion drive. The MAX038 is followed by a 25MHz low-pass filter. The inductors for this filter can be bought or made. To make L1/L2; for each use a 10kΩ, 0.25W resistor and wind on it 18 turns of 30SWG (0.315mm dia).

The circuit must be supplied from a well-regulated ±5V supply with adequate decoupling at RF. The complete circuit should be placed in a well-screened metal box and the capacitance to earth held constant by fixing the circuit board on suitable spacers.

A suitable double-sided board for experimentation is shown in Appendix B (**Fig 25.47**) and **Fig 25.48** is the component overlay;

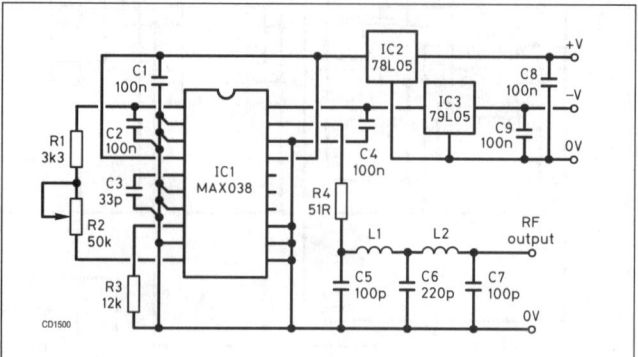

Fig 25.46: Circuit of the HF signal source

R1	3k3	C5,C7	100p low K, ±5%
R2	50k multi-turn	C6	220 low K, ±5%
	(see text)	C8,C9	100n ceramic
R3	12k		±10%
R4	51R	L1, L2	390nH (see text)
C1,C2,C4	100n ceramic ±10%	IC1	MAX038
C3	33p NPO ceramic or	IC2	78L05
	silver mica, ±2% or better	IC3	79L05

Table 25.10. HF signal source component list

The Radio Communication Handbook

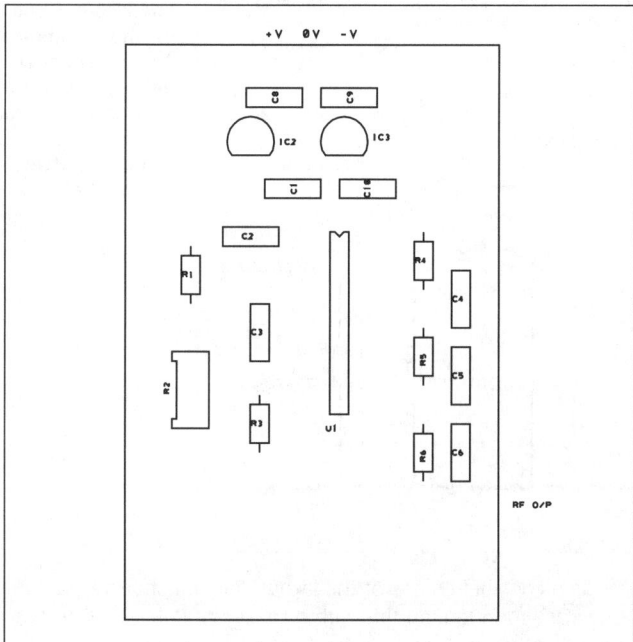

Fig 25.48: Component placement for the HF signal source not to scale)

R1	3k8
R2	1k
R3	68R
R4, 5, 6, 7	1k SMD type 1206
R8	See text. For MSA0304 use 270R, 0.5W
R9, R11	292R (270R + 22R) SMD type 1206
R10	18R
C1	4µ7 tantalum
C2	100n
C3, C4	1n SMD type 0805
C5 - C9	100p type 0805
RFC1, RFC2	1mH
L1	44nH (3T of 0.56mm ecw, 65mm long, inside 47.5mm, pulled out until there is 10mm between the ends
L2	7.5nH (loop of 0.56mm ecw 15mm long, ends 10mm apart
IC1	48MHz crystal oscillator module
TR1	BFR96
TR2	MMIC. See text. Original used MSA0304
VR1	78L05
D1 - D4	1N914 or any suitable switching diode
F1	Toko CBT3
F2	Toko 7HW
S1	2-way single pole switch
Double sided glass-fibre PCB type FR4, 16mm thick	
Tin-plate box, 74 x 148 x 30 mm	
Track pins	
All resistors are 0.25W, 5% tolerance unless specified otherwise	

Table 25.11. Component list for combined 2m and 70cm signal source

tor R8 must be changed - see *Biasing MMC Amplifiers* at www.minicircuits.com/application.html.

Preview

The component list is given in **Table 25.11**. The circuit, shown in **Fig 25.50**, starts with a TTL crystal oscillator module. The 2.5V rectangular-wave output it provides is fed into a BFR96 amplifier/multiplier, the output of which is selected by diode switches between the filters. The output from the filters, 144 or 432MHz, is then amplified by a standard MMIC (monolithic microwave integrated circuit); sometimes called a modamp. In spite of the fearsome name, MMICs are easy to handle and will work over a wide range of frequencies without fuss or the need for tuned circuits. After amplification the output is passed through a 3dB attenuator, connecting finally to a BNC socket.

The power supply is external. VR1 provides the necessary +5V for the crystal oscillator module. As the unit consumes a total of 130mA, an internal battery is not used. The crystal oscillator module takes 80mA and the MMIC will vary according to device used.

The Circuit in Detail

The output from the crystal oscillator module (IC1) is rich in harmonics. This brew is fed into a UHF-type transistor (TR1), working as a frequency multiplier/amplifier. A wide range of harmonics is produced at the collector, up to and including the 32nd. The output from the collector is diode switched between two filters. The filters are standard types by TOKO.

The switch selects the filter to be used by connecting a positive voltage to one pair of diodes or the other, causing them to conduct and thus complete one of the circuits.

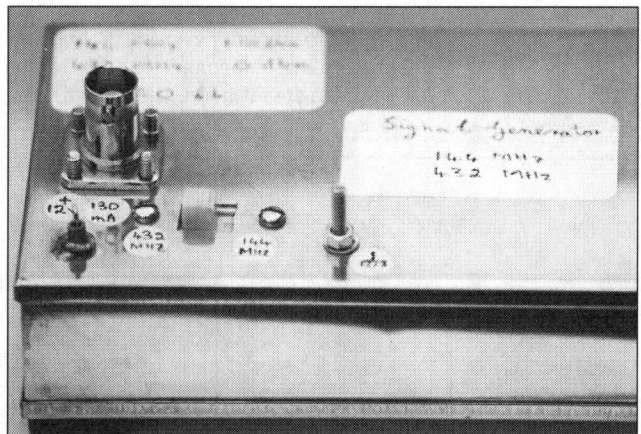

Fig 25.49: The combined 2m and 70cm signal source showing the BNC output socket, power supply terminals and frequency selector circuit

this includes the regulated power supply. *Do not* use an IC socket for the MAX038 as this will affect capacitance to earth.

A COMBINED 2M AND 70CM SIGNAL SOURCE

This signal source was developed by John Brown, G3DVV [10]. The project (**Fig 25.49**) uses one of the low-cost crystal oscillator modules that are available - the one that is really useful is the module for 48MHz. To a VHF/UHF enthusiast 48 x 3 = 144MHz and 48 x 9 = 432MHz, ie 2m and 70cm. These modules have a lot going for them. Their cost is less than £10 - compare that with purchasing a separate crystal, semiconductors, capacitors and resistors, and then putting them all together on a PCB. What's more, the modules work first time, which makes a pleasant change.

NOTE: The original device for TR2 (MSA0304) appears to be obsolete. If you can find one then all well and good. An alternative is one of the MAR monolithic amplifiers from Mini-Circuits. Several of the devices match the specification but the bias resis-

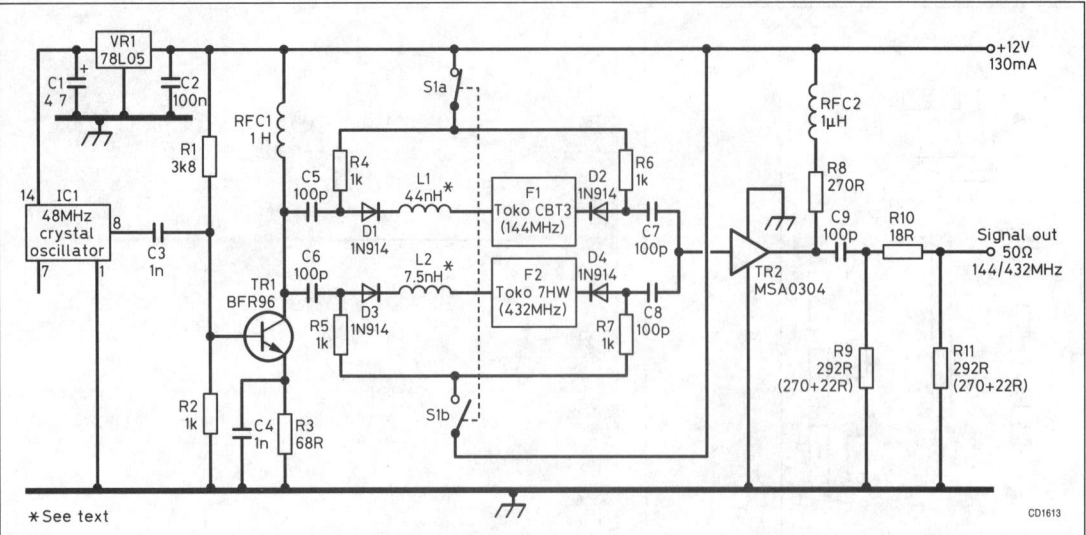

Fig 25.50: Circuit of the combined 2m and 70cm signal source. Steering diodes are used to switch between filters F1 and F2

The two coils (LI and L2), one for each filter, give an approximate match from the collector of the BFR96 into the 50Ω input of the filters. Their value is not critical. During development, a number of diodes were tried as switches, and the 1N914 was found to be as good as any.

The output from the filters, both at 50Ω, is fed into the 50Ω input of the MMIC. After being amplified by the MMIC, the output is still 50Ω, as is the 3dB attenuator and the BNC socket. The attenuator acts as a buffer against open-circuits and short-circuits.

Construction

The PCB pattern is given in **Fig 25.51** (in Appendix B) and the layout in **Fig 25.52(a)**. It is important to use the right type of printed circuit board, as the tracks form a transmission line with the ground plane, thus cutting down radiation and providing the correct matching. The PCB material (FR4) and track widths are critical in maintaining transmission line impedance; components are mounted on both sides of the board. The choice of etching method is yours but it must be a double-sided board, one side forming a ground plane. After cutting the PCB to size, make sure that it fits under the lid of the box.

On the component side of the board, use 0.1in (2.5mm) tracks. The reason for doing this is that this size of stripline forms a 50Ω transmission line with the ground plane. It is known as microstrip. For power supply lines, the microstrip size makes no difference, but the same width is used for convenience. While you are working try to keep your fingers off the copper, as finger marks do not etch very well. Wash the board well and dry after etching, then drill.

The only difficult items to mount are the two filters (**Fig 25.52(b)**) because the lugs as well as the pins have to be accommodated. Holes are required for TR1 and TR2, and pin through holes for IC1 (note that there is no connection to pin 1 - clear the copper away from pins 1, 8 and 14). TR1 and TR2 should fit snugly into their holes, with their leads lying flat on the copper strips.

This just leaves the holes to be drilled for the circuit pins, otherwise known as half-track pins,

Veropins or vias. Wherever the layout diagram shows a pad or a track as grounded, do this with a track pin. Push a pin through from the component side and, after soldering to track and ground plane, cut it off at board level. Note that three pins are not cut, namely those marked as 'S1a', 'S1b' and '+12V'. Clear the copper away from the base of these three pins; these are hard wired.

Next, mount the SMDs (surface mount devices). Tin one side of the strip or pad, then place the component in position and hold down firmly. You will need a steady hand. The tip of a small screwdriver or a toothpick is placed on the centre of the compo-

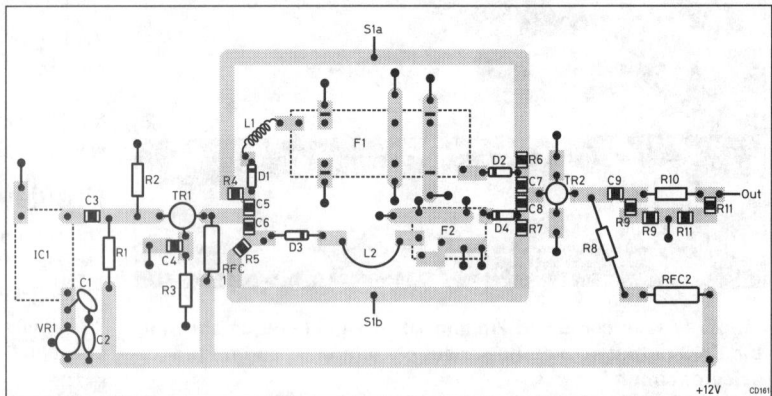

Fig 25.52 (a): Component placement for the signal source

Fig 25.52 (b): The signal source showing the two filters and the crystal oscillator module

nent while a soldering iron with a small tip is applied to one end. It is very easy to play 'tiddlywinks' with these devices, and once in orbit, that's it. When one side has been soldered satisfactorily, solder the other side. Then re-solder the first connection if necessary. Finally, solder in the rest of the components, connecting the semiconductors last.

Take the lid of the box and fit the switch, the connections for the power supply and the BNC socket. The position of the BNC socket has to be exact, as the inner terminal of the socket must touch the pad marked 'out'. The wires from the switch and power supply are brought out around the sides of the board, which is held in position by screws.

When all appears to be in order, solder the tip of the BNC socket to the output pad, which is also connected to R10 and

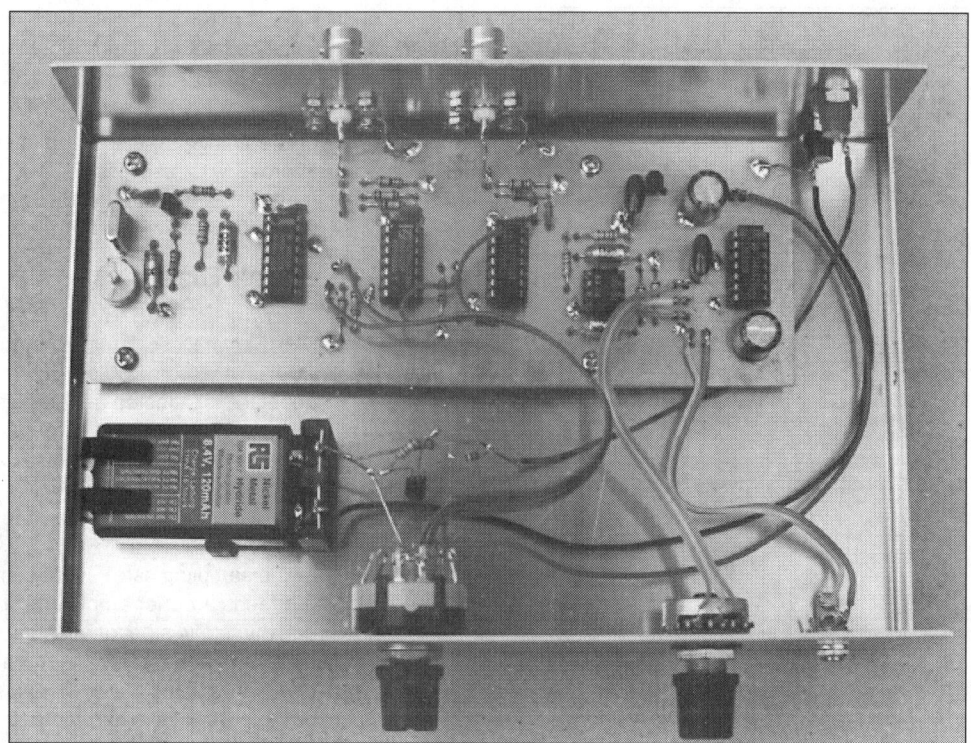

Fig 25.53: The receiver calibrator and transmitter monitor

R11. Complete the wiring to the three uncut circuit pins.

You may find that the bottom lid of the box fits without trouble, but in G3DVV's case the protruding screws from the 2m Toko filter push against the lid, requiring two accommodation holes to be drilled.

Final Test

The components face the underside of the lid and are difficult to check in their final position. Therefore, check that the unit is working before finally fixing it in position. There is nothing in the circuit which requires adjustment. *Do not* try altering the filter settings.

The unit should work from switch-on, but if it doesn't try the following:

- Check that all the track pins are properly soldered.
- Check all the voltages.
- Use a probe to check the output from IC1, TR1 and TR2. The probe can be a diode, a link to the coil of a dip meter, or (probably) a 2m and 70cm receiver with a piece of coaxial cable and a wire loop as an RF sniffer.

Results

The output on 2m is 10mW, and on 70cm 1mW. On 2m the nearest harmonic is on 288MHz and 16dB down. On 70cm the nearest harmonic is on 366MHz and 24dB down.

Although there is a whole range of harmonics, the temptation to use the 27th for 1296MHz was not followed because of the difficulty of cramming the components onto the printed board.

RECEIVER CALIBRATOR & TRANSMITTER MONITOR

The receiver calibrator and transmitter monitor (**Fig 25.53**) described is by G4COL [11]. It offers the following facilities:

- Frequency calibration of receivers and transmitters from

50kHz to over 150MHz.

- Receiver sensitivity measurement from 50kHz to over 50MHz.
- Calibration of attenuators and receiver S meters.
- Monitoring transmissions for stability, speech and keying quality.

Frequency intervals of 5MHz, 500kHz and 50kHz are provided, and frequency can be set accurately against broadcast standards. Receiver measurements need an external attenuator. The unit is battery portable, and was designed to be reproducible with components readily available and minimal test equipment.

Principle of Operation

The heart of the unit is a harmonic comb generator, which produces very short duration pulses, quite commonly used for frequency calibration. Its use for providing a reasonably accurate signal level is less common. The main properties of relevance to this project are below. Some of them are not readily found and had to be calculated:

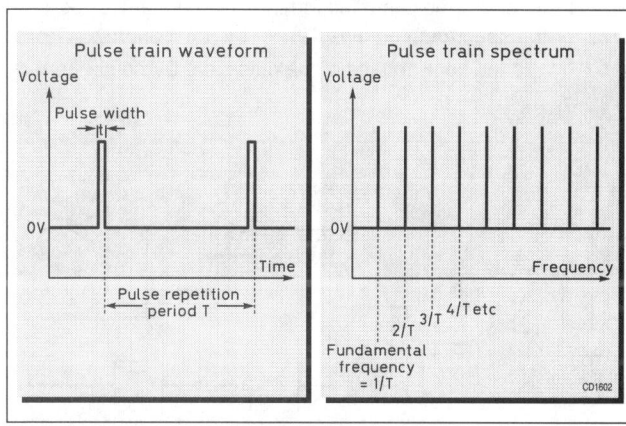

Fig 25.54: Waveform and spectrum of a train of narrow pulses

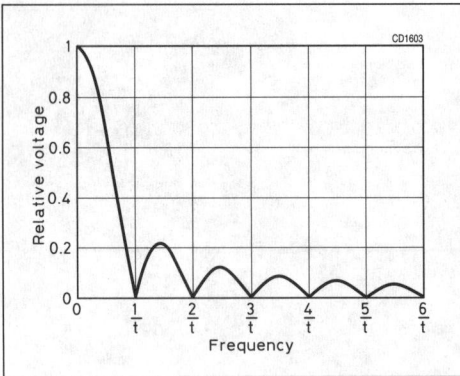

Fig 25.55: Enveolpe of the harmonic comb's spectrum, viewing a greater frequency than Fig 25.54

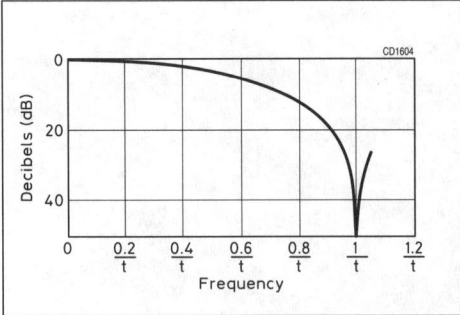

Fig 25.56: Lowest frequency lobe of the harmonic comb's spectrum. Note the logarithmic Y axis

- The RMS voltage of the low-frequency harmonics is the DC voltage x 1.414 (the square root of 2).
- The DC component can be measured relatively easily.

Don't worry if quite a bit of this seems hard to take in. Some of the points are discussed later, and understanding is likely to improve with use of the test gear for making real measurements. If you want to delve further, consult an engineering maths textbook.

Block Diagram

Referring to **Fig 25.57**, the reference frequency is provided by a 5MHz crystal oscillator, the output of which is divided by 10 and 100 to produce 500kHz and 50kHz. One of these three frequencies is selected as the pulse repetition frequency, to drive a pulse generator which produces very narrow pulses, giving the spectral comb. The pulses are fed to the output socket via an attenuator and also drive the sampling gate, which is essentially a switch that acts as the input stage of the transmitter monitor.

The sampling gate's input comes from a panel socket. Its output feeds a buffer amplifier, volume control and output amplifier (which drives headphones or a small loudspeaker). The sampling gate behaves as the mixer of a direct-conversion receiver, in which an audible beat signal can be produced by a signal close to any of the harmonics of the pulse repetition frequency. Suppose we are switching the sampling gate at 500kHz; a 1kHz beat note will be produced by an input signal present at any of 499, 501, 999, 1001, 1499, 1501, 1999, 2001, 2499, 2501kHz and so on.

Circuit Description

Most of the circuit, shown in **Fig 25.58**, operates from a 5V supply derived from the 9V battery by three-terminal regulator IC1. The 5MHz crystal reference frequency oscillator is a Colpitts, based around TR1. IC2 divides this continuously by 10 and 100 to produce 500kHz and 50kHz. The pulse repetition frequency is selected by three sections of analogue switch IC3, controlled by wafer switch S1b. IC4 generates the narrow pulses, using AC series TTL.

The pulse width is set by the delay section R7 with IC4b. Negative-going pulses from IC3c feed an output attenuator consisting of R11 and R12. An inverter stage IC4d drives the sampling gate IC3d with positive-going pulses. The attenuator gives the unit an output resistance very close to 50Ω. That is the extent of the calibrator, the rest of the circuitry being the monitor.

A plot of the negative-going output pulse is shown in **Fig 25.59**. The original of this was captured using a fast digital storage oscilloscope and shows the pulse width to be within the 5 nanosecond per division horizontal graticule. The corresponding spectrum was found to have its first null at 270MHz, indicating a pulse width of 3.7 nanoseconds.

- The term 'harmonic comb' describes the spectrum produced by very narrow pulses. As **Fig 25.54** shows, energy has shifted into harmonics to such an extent that the fundamental frequency component and many harmonics have similar amplitude. Note that the two voltage axes are not to scale. The individual 'teeth' of the harmonic comb have a much lower voltage than the peak voltage of the pulse - the energy of the pulse has been spread among its many components.

- The pulse width determines how the amplitudes of the different harmonics vary with frequency. **Fig 25.55** shows the 'envelope' of the spectrum, looking much higher in frequency than Fig 25.54. The tip of each spectrum component lies on this curve.

- The lowest-frequency lobe is shown in more detail in **Fig 25.56**. It contains 90% of the total energy and is 3dB down with respect to low frequencies, at 44% of the first null frequency. 72% of the total power of the comb falls below this '3dB down' frequency.

- The lobe is 1dB down at 26% of the first null frequency. This means that if we are to use the harmonic comb for calibration only up to a frequency of 26% of the first null, the teeth of the comb fall within a 1dB range. This portion of the spectrum contains 49% of the total energy.

- If V_p is the peak voltage of the pulse, the DC component is $V_p \times t/T$

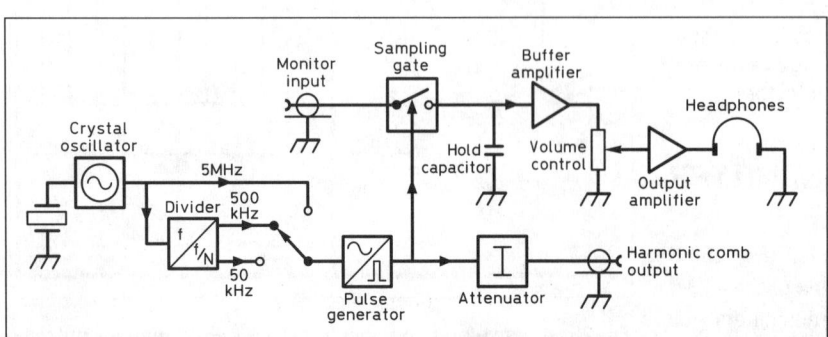

Fig 25.57: Block diagram of the receiver calibrator and transmitter monitor

The Radio Communication Handbook

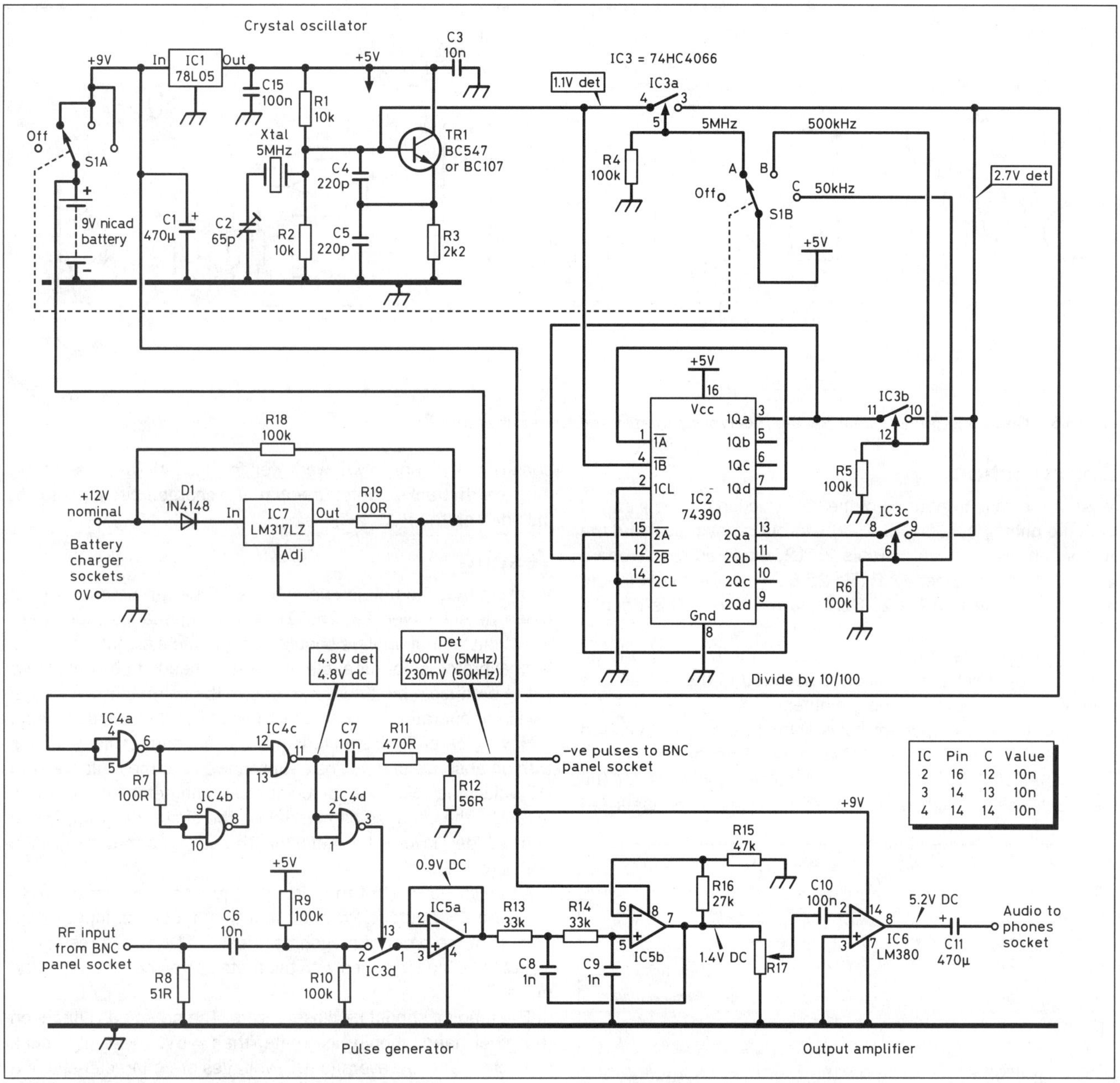

Crystal oscillator

+9V — In IC1 78L05 Out — +5V — C3 10n — IC3 = 74HC4066

Off — S1A — 9V nicad battery — C1 470µ — C2 65p — C15 100n — R1 10k — Xtal 5MHz — C4 220p — R2 10k — C5 220p — TR1 BC547 or BC107 — R3 2k2

1.1V det — IC3a — R4 100k — 5MHz — 500kHz — A B — C 50kHz — Off — S1B — 50kHz — +5V — 2.7V det

R18 100k — R19 100R — +12V nominal — D1 1N4148 — In IC7 LM317LZ Out — Adj — Battery charger sockets — 0V

+5V — Vcc 16 — 1 1A — 4 1B — 2 1CL — 15 2A — 12 2B — 14 2CL — IC2 74390 — 3 1Qa — 5 1Qb — 7 1Qc — 1Qd — 13 2Qa — 11 2Qb — 2Qc — 9 2Qd — Gnd 8 — Divide by 10/100

IC3b — 11 10 — 12 — R5 100k — IC3c — 8 9 — 6 — R6 100k

4.8V det 4.8V dc — Det 400mV (5MHz) 230mV (50kHz)

IC4a — 4 6 5 — R7 100R — IC4b — 9 8 10 — IC4c — 12 11 13 — IC4d — 2 3 1 — C7 10n — R11 470R — R12 56R — -ve pulses to BNC panel socket

IC	Pin	C	Value
2	16	12	10n
3	14	13	10n
4	14	14	10n

+5V — R9 100k — 0.9V DC — IC5a — R13 33k — R14 33k — C8 1n — C9 1n — IC5b — R15 47k — R16 27k — 1.4V DC — R17 — C10 100n — +9V — IC6 LM380 — 5.2V DC — C11 470µ — Audio to phones socket

RF input from BNC panel socket — C6 10n — R8 51R — R10 100k — IC3d

Pulse generator — Output amplifier

Fig 25.58: Circuit diagram of the receiver calibrator and transmitter monitor. S1 is used to control the 'comb' of frequencies at 5MHz, 50kHz or 50kHz intervals

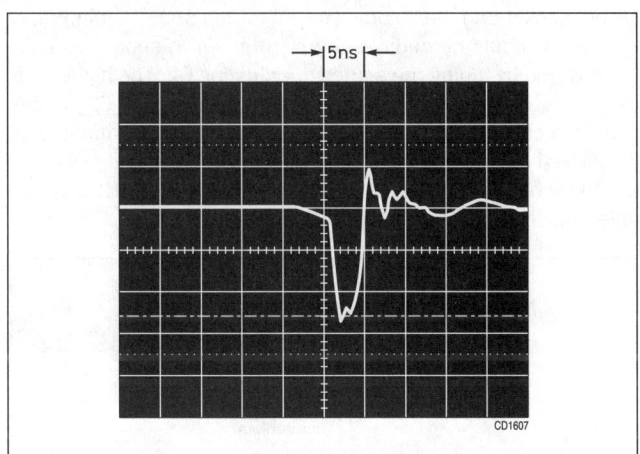

|5ns|

CD1607

Fig 25.59: Oscilloscope trace of negative-going calibrator output pulse

R8 sets the monitor input resistance. R9 and R10 bias the sampling gate switch IC3d into its operating range, between 0 and 5V. The sampling gate's 'hold' capacitance is not shown on the circuit diagram.

It consists of stray capacitance plus the input capacitance of IC5a, a high-impedance, unity-gain buffer. The buffer is followed by a low-pass filter to attenuate the sampling pulses which inevitably feed through the gate. After the volume control, IC6 provides audio power output.

Part of Fig 25.58 is a battery charger circuit, for use with nickel cadmium (NiCd or 'nicad') or nickel metal hydride (NiMH) batteries. It is based on IC7, which is configured as a current source of about 12mA (set by R19). The charger allows the internal battery to be recharged by connecting a power source of 12 to 24V to the terminals.

D1 prevents damage from a reversed supply and R18 allows the battery voltage (when not being charged) to be monitored by a high-impedance voltmeter (eg digital multimeter) at the charger terminals.

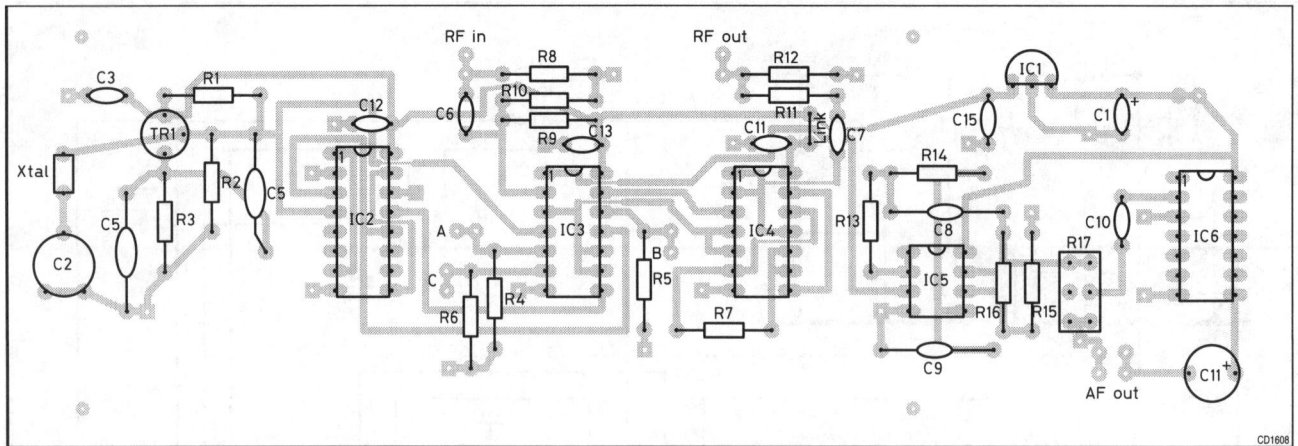

Fig 25.61: Component placement for the receiver calibrator and transmitter monitor

Construction

Most of the salient features of the construction can be gleaned from the photograph, Fig 25.53. The layout is not critical and the component list is given in **Table 25.12**. A printed circuit board layout is given in Appendix B (**Fig 25.60**) whilst the component placement is given in **Fig 25.61**. Note that the PCB is double-sided.

The square pads denote connection to the copper plane on the component side; wire links are soldered to the pad on one side and the copper plane on the other.

The remaining component holes should be carefully cleared of copper from the component side using a 4mm drill bit to clear a diameter of 2mm or so. The four isolated pads are for the mounting holes, which were opened up to 4mm. Single-sided construction is unlikely to work well for this unit, because of the long earth tracks. Note: the battery charger circuitry is not included on the PCB.

Testing

Verifying that the unit is operating correctly can be done with a diode detector (see **Fig 25.62**), a high-impedance multimeter (most digital multimeters should be fine) and a signal source. As the calibrator is not tuneable, the source needs to be able to be set to produce a beat note with one of the comb teeth. A crystal oscillator operating close to a multiple of 500kHz could be used.

Having carefully inspected the unit for correct connections with an absence of dry joints, apply power. Set the multimeter to DC volts and, with the detector-plus-multimeter combination, measure the voltages as marked on the circuit diagram, Those marked 'det' involve the detector. The voltages may differ by a few percent.

Although not part of the testing sequence, it is instructive to measure IC4d output. Provided that the multimeter input resistance is of the order of megohms, measure DC with a resistor of $10k\Omega$ to $100k\Omega$ in series with the meter, so as not to disturb the pulse.

The reading should be close to zero. The detected voltage, on the other hand, should be greater than 4.5V. These high peak (detected) and low average (DC) voltages are characteristic of a very narrow pulse.

Connect the signal source (level around 100mV RMS) to the BNC input connector, with headphones plugged into the audio output socket and the volume control turned above minimum. A beat note should be audible, the pitch of which should be able to be varied by tuning the source or adjusting C2. The higher the frequency of the source, the greater the effect of C2. If the source is close to an exact number of megahertz or multiple of 500kHz, a beat note should be produced with the unit switched to 50kHz or 500kHz. If a source at a multiple of 5MHz is available, this too should give a beat.

R1, R2	10k	C3	10n, 63V ceramic
R3	2k2	C4, C5	220p, 160V polystyrene
R4, R5,		C6, C7	10n, 63V ceramic
R6	100k	C8, C9	1n, 160V polystyrene
R7	100R	C10	100n, 100V mylar
R8	51R	C11	470µ, 16V electrolytic
R9, R10	100k	C12,C13,	
R11	470R	C14	10n, 63V ceramic
R12	56R	C15	100n, 100V mylar
R13, R14	33k	IC1	78L05ACZ
R15	47k	IC2	74HC390N
R16	27k	IC3	74HC4066P
R17	100k log pot,	IC4	74AC00
	carbon or similar	IC5	CA3240E
R18	100k	IC6	LM380N
R19	100R	IC7	LM317LZ
C1	470µ, 16V	TR1	BC547, BC107
	electrolytic	D1	1N4148
C2	5p5 to 65p trimmer	X1	5MHz crystal, HC-49/U

S1   Rotary switch, 3pole 4 way
3.5mm jack socket
BNC square flange chassis socket, 50Ω
Case, knobs, PCB battery holder

All resistors 0.6W metal film, 1% tolerance unless specified otherwise

Table 25.12: Components list for receiver calibrator and transmitter monitor

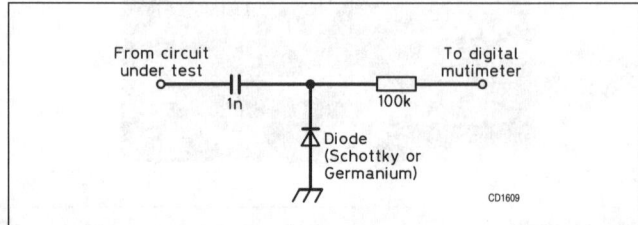

Fig 25.62: The diode detector used for testing

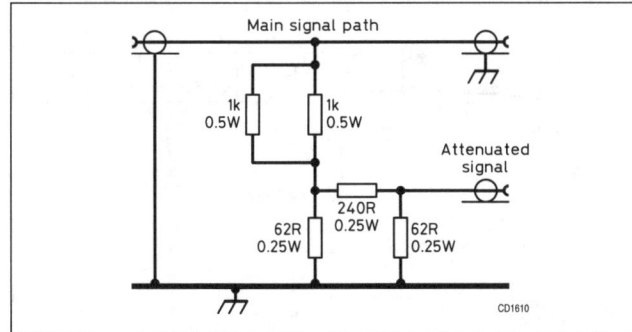

Fig 25.63: Signal pick-off

Operation

Some additional items are needed:

• For monitoring transmitters a high-power attenuator or signal pick-off plus dummy load;

• For receiver measurements a low-power attenuator will suffice and the signal pick-off can be used. A suitable signal pick-off is shown in **Fig 25.63**.

Fig 25.64 shows the measurement set-ups. Where 'calibrator' is shown, connect the coaxial cable to the pulse output, and for 'monitor', to the monitor input. Similarly 'receiver' and 'transmitter' can be a transceiver in receive and transmit modes respectively.

Frequency Calibration

See **Figs 25.64(a) and (b)**. If a receiver is available covering 10MHz, the calibrator crystal oscillator frequency can be adjusted for zero beat with the WWV frequency and time standard signal.

Measuring Receiver Sensitivity

See Figs 25.64(a) and (b). The harmonic comb's principal advantage for this task is that the signal power in each 'tooth' is already low. One of the reasons professional-quality signal generators are expensive is that it is difficult to produce a pure signal with an accurate low level. Much careful screening and filtering is involved, as well as an accurate switched output attenuator of up to 130dB total attenuation, which itself demands extreme isolation from input to output.

The output level is given by:

pulse height x pulse width ' repetition frequency x
attenuation factor x √2

The pulse width is nominally 4 nanoseconds. If the pulse height is 5V, the repetition frequency 5MHz, and the output attenuation factor into a 50Ω load is 0.05, then the level of the lowest frequency teeth is:

$$5 \times 4 \times 10^{-9} \times 5 \times 10^{6} \times 0.05 \times \sqrt{2}$$
$$= 7.07\text{mV or } -30\text{dBm}$$

Similarly, for 500kHz and 50kHz repetition frequency, the corresponding outputs are 707mV and 70.7mV respectively.

At 50kHz repetition frequency, a 40dB attenuator will present the receiver with 0.7mV signal. A 40dB attenuator is not unduly difficult to make, or may be purchased, sometimes in the form of a switched attenuator or perhaps a pair of 20dB coaxial attenuators used in series. The theoretical -1dB frequency of the comb is 65MHz. A spectrum analyser plot of the comb at 5MHz repetition frequency is shown in **Fig 25.65**. The agreement with the predicted level is really quite good (the spectrum analyser used has much better amplitude accuracy than most). The reason for the difference between teeth is not clear but nevertheless the plot shows that there is relatively little variation up to 50MHz. The main sources of uncertainty will be the height and

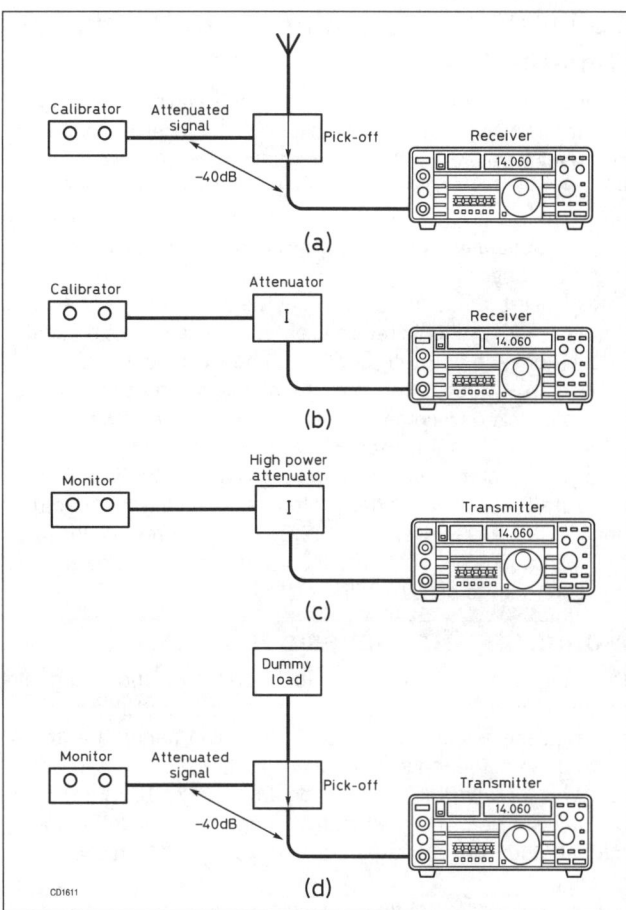

24.64: Measurement set-ups

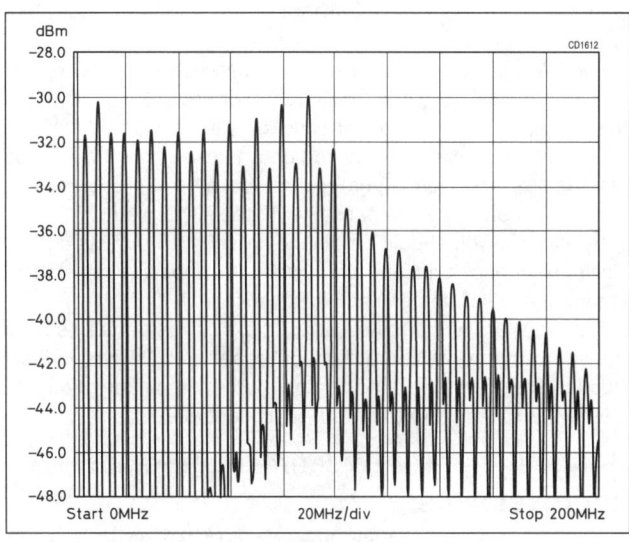

Fig 25.65: Spectrum analyser plot of 5MHz pulse output

width of the pulse. A very rough calculation indicates that the signal level should be within ±3dB of nominal.

Beware: when testing receivers with broad-band front-ends (no preselector), it is advisable to use at least 20dB attenuation between calibrator and receiver. This is because the receiver mixer is subjected to many teeth of the harmonic comb. The peak pulse voltage, which includes all the teeth, is 250mV (5V x the output attenuation factor of 0.05). Although this is not likely to damage the mixer, most will be driven outside their normal operating range.

Calibrating Attenuators and Checking S-meters

Referring to Fig 25.64(b), 20dB attenuators can be checked very accurately using the receiver's S-meter. The receiver itself does not require calibration. Couple the calibrator to the receiver. Set the calibrator to 500kHz and tune the receiver to a harmonic of 5MHz (this may not be possible on an amateur-bands-only receiver), obtaining a beat from the calibrator signal. Note the S-meter reading.

Now insert the attenuator and set the calibrator to 5MHz. The S-meter reading should return to exactly the same position if the attenuator is an accurate 20dB, any change indicating an error.

Once an accurate attenuator is available, the receiver S-meter can be checked by comparing its readings with and without the attenuator in the signal path.

To give an idea of the accuracy which can be obtained, G4COL measured a nominal 20dB attenuator (using a spectrum analyser as the receiver) to be 19.3dB. The figures produced by the calibration company were then consulted and showed their measurement to be 19.22dB!

Monitoring Transmissions

Referring to Fig 25.64(c) and (d), make sure that adequate attenuation is used between the transmitter and the monitor, and that the attenuator or dummy load can handle the transmitter power. The input to the monitor should be kept to 0dBm (1mW or 224mV RMS), but can be driven up to 10dBm without undue distortion. This means that for up to 10W (40dBm) transmitter output, a 40dB attenuator or the signal pick-off shown are suitable.

Tune the transmitter for an audible beat note for CW and intelligible speech for SSB. With headphones which are fairly acoustically opaque, it is quite feasible to assess one's own speech quality speaking into the microphone, though if another source is available so much the better.

For CW signals, if you have an oscilloscope, monitoring the audio output with the beat note around or below 1kHz will show the shape of the actual radio frequency waveform and the effect of adding filters. Harmonic distortion was evident on the output of G4COL's HP606A signal generator at 25MHz and was removed by adding a low-pass filter to the signal path.

By listening you should be able to detect such defects as key clicks, frequency drift and chirp, speech clipping and splatter. Frequency calibration can also be checked.

Note that the audio bandwidth of the monitor, actually set by the sampling process, is around 5kHz on the 50kHz repetition frequency setting. Since most transceivers contain steep cut-off filters, this is not likely to have a significant effect.

Suggestions for Experimentation and Improvements

Here are a few suggestions that can be used as a basis for improvement and experiment:

(a) The calibrator could be built on its own without the monitor receiver.

(b) The monitor noise could be reduced. The aim here has been simplicity. An earlier, more complex, version was quieter.

FREQUENCY COUNTERS

The cost of frequency counters has decreased over the last few years and handheld types operating in excess of 1GHz cost less than £100, with some units going up to 3GHz not much more

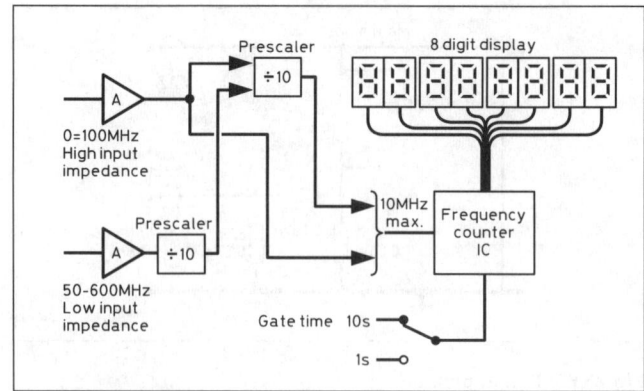

Fig 25.66: Block diagram of a frequency counter

expensive. It is the number of digits shown which is important. It has come to a point where the cost of construction and availability of components makes construction at home not worthwhile.

The purpose of this section is to explain briefly how a frequency counter works and gives tips for making measurements. **Fig 25.66** shows the typical block diagram of a frequency counter but there will be variations on the input arrangements. There is usually a high (1MΩ) and low input impedance input - the low input is typically 50Ω for the higher frequencies. The problem with a high input impedance is the effect of the shunt capacitance, eg 10pF at 144MHz has a reactance of 110Ω. The actual inputs can either be from a test probe or direct pick-up off-air using an antenna. The input to a frequency counter is fairly sensitive and it may only require 50mV of RF in order to obtain a digital readout. Consider also the maximum signal the frequency counter can cope with and any DC voltage present - above this you are likely to damage an expensive piece of test equipment.

When using a probe with physical connection to a circuit, care should be taken that the probe does not affect circuit conditions. The main problem is the probe impedance (especially the shunt capacitance) changing the frequency of operation of the circuit, eg 'pulling' of the oscillator. Also the probe may affect the DC operating conditions. An oscilloscope probe is a good unit to use. If trying to measure frequency in a 50Ω system, make sure the probe does not cause a mismatch or overload the circuit. These points are summarised as follows:

• Do not attach a probe to frequency-determining elements.
• Measure oscillator frequency after the buffer amplifier.
• Use AC coupling, especially in higher-voltage valve circuits.
• Put a 1kΩ resistor in series with the probe.
• Do not allow a 50Ω system to become mismatched.

An alternative is to use a pick-up loop as shown in **Fig 25.67**. Ensuring only loose coupling, this overcomes some of the inherent problems in trying to minimise influence on the circuit and damage to the counter. Another similar way is to use an antenna as the input device which is coupled straight to the frequency counter input.

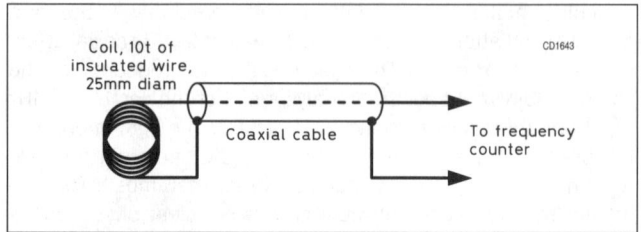

Fig 25.67: Pick-up loop

When taking frequency measurements, especially off-air, only an unmodulated carrier should be used. The following guidelines should be used:

- With AM and FM radios, *Do not* speak into the microphone or provide modulation.
- With SSB sets, the CW position should be used with key down.
- With digital transmissions, no superimposed data should be transmitted, just a carrier.

More expensive frequency counters will allow the period to be measured in usually microseconds or milliseconds. This is useful for frequencies less than about 100Hz and will generally give a more accurate result but you will need to use a calculator to revert to frequency.

SPECTRUM ANALYSERS

What is the purpose of a spectrum analyser? It is a piece of equipment that can receive a signal (or group of signals) and give a display of the frequency components present and the relative amplitudes.

It is possible to purchase a brand new one or perhaps a second-hand can be found, but the cost will still be high! There are add-on units for oscilloscopes which will function to 1GHz and cost about £600 at 2005 prices.

A Simple Spectrum Analyser (SSA), designed by Roger Blackwell, G4PMK (see [12] for the original article) can be found at http://www.qsl.net/g3pho/ssa.html. It offers reasonable performance over the approximate range 1-90MHz, is fairly cheap to build and utilises almost any oscilloscope for its display.

POWER OUTPUT MEASUREMENTS

The UK Amateur Licence requires that you should be able to measure transmitter output power in order to comply with the licence conditions.

The following information is taken from the Amateur Radio Licence Terms and Limitations Booklet BR68. Only the relevant paragraphs have been included.

Notes to the Schedule

(a) Maximum power *refers to the RF power supplied to the antenna. Maximum power levels will be specified by the peak envelope power (pep).*

(e) *Interpretation*
   (i) Effective Radiated Power (erp): *The product of the power supplied to the antenna and its gain in the direction of maximum radiation.*
   (ii) Gain of an antenna: *The ratio, usually expressed in decibels, of the power required at the input of a loss free reference antenna to the power supplied to the input of the antenna to produce, in a given direction, the same field strength or the same power flux-density at the same distance. When not otherwise specified, the gain refers to the direction of maximum radiation. The gain may be considered for a specified polarisation. The reference antenna is usually a half-wave dipole. The gain may be referred to as decibels relative to a half-wave dipole (dBd).*
   (iii) Peak Envelope Power (pep): *The average power supplied to the antenna by a transmitter during one radio frequency cycle at the crest of the modulation envelope taken under normal operating conditions.*

The oscilloscope can be used up to about 30MHz to monitor modulated waveforms and measure output power, but above this it becomes an expensive item and may provide unwanted loading effects on the equipment being monitored. The familiar

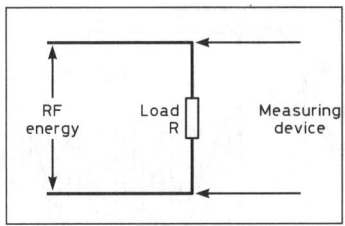

Fig 25.68: Power measurement arrangement

VSWR meter monitors forward and reflected signals and the scale can be made to represent power in a 50Ω line. It is possible to use an RF voltmeter across a given load to measure power. The higher you go in frequency, the more difficult it becomes to measure the modulation and power with relatively cheap equipment. Yet it is a condition of the licence that these parameters can be monitored. It is more difficult to measure PEP than average carrier power.

Constant-amplitude Signals

In a carrier-wave situation (CW, FM or unmodulated AM), the output is of constant amplitude and so it is relatively easy to measure the output power. To measure these signals using the circuit as shown on **Fig 25.68**, the power output is given by:

$$P_{out} = V^2 / R$$

where V must be the RMS value of the voltage.

This voltage measurement can be carried out using an oscilloscope or RF voltage probe. The SWR meter described later can also provide this value.

Amplitude-modulated Signals

These pose more of a problem and two cases are dealt with below.

Amplitude modulation (A3E)

With no modulation, the problem reverts to the measurement of power of a constant carrier as described above. If the carrier is amplitude modulated (A3E) then the overall output power increases. The power is divided between the sidebands and the carrier component. With 100% modulation the output power increases to 1.5 times the unmodulated condition - the power contained in each of the two sidebands being one-quarter that in the carrier. It is suggested that for this form of modulation the carrier power is measured (ie no modulation) and multiplied by 1.5 to give the maximum output power.

If an exact value for the output power is required it is necessary to determine the modulation index. This can be carried out using an oscilloscope of adequate frequency response. Set the oscilloscope as shown in **Fig 25.69** and calculate the modula-

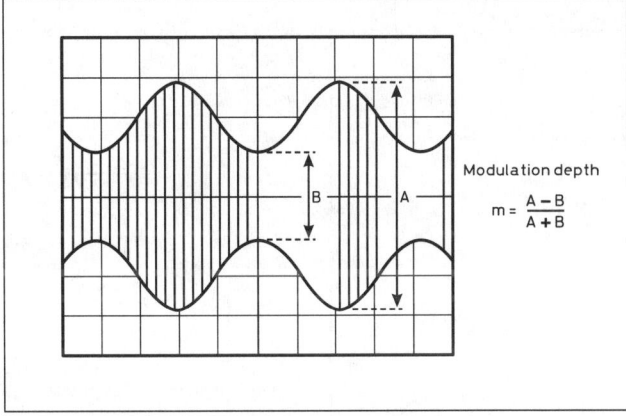

Modulation depth

$$m = \frac{A - B}{A + B}$$

Fig 25.69: Modulation depth measurement

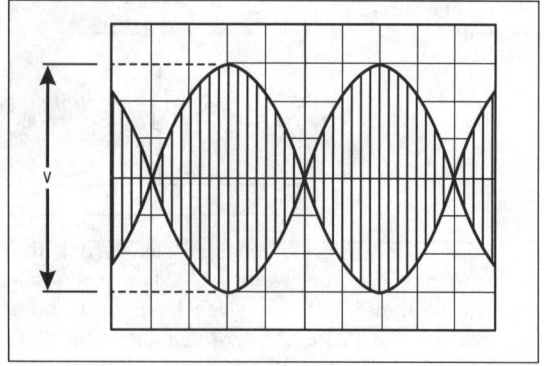

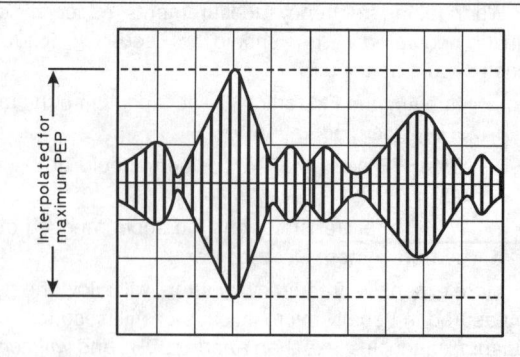

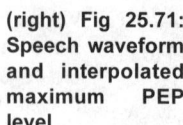

(left) Fig 25.70: Two-tone test display

(right) Fig 25.71: Speech waveform and interpolated maximum PEP level

tion depth m. The output power is then given by:

$$P_{out} = \frac{V^2}{R}\left(1 + \frac{1}{2}m^2\right)$$

where V is the RMS value of the unmodulated carrier and R the load.

For test purposes an audio signal can be fed in at the microphone socket of the transmitter using the low-frequency oscillator described earlier.

Single sideband (J3E)

With single sideband, no power is output until modulation is applied. The output envelope is non-sinusoidal in appearance. The normal method for measuring output power is by observation of the modulation envelope and determination of the peak envelope power - this is the parameter defined by the UK licensing authority. This can be accomplished using an oscilloscope of suitable frequency response as described below or using a SWR meter that will respond to peak envelope power - see later in this chapter or reference [1].

Fig 25.70 shows the display when a two-tone test signal is fed in via the microphone socket. If the peak-to-peak voltage V at the crest of the envelope is measured across a load of value R, then the PEP is given by:

$$P_{out} = \frac{V^2}{8R}$$

The equivalent peak-to-peak voltage reading can then be interpolated for the maximum allowable PEP and the position noted on the display. **Fig 25.71** shows a typical display for a speech waveform and the interpolated maximum PEP level.
VSWR

Every station should have a VSWR meter somewhere in its line-up. These can be bought commercially or made. When building, the higher the frequency range required, the more careful the constructor should be in placing the components so that the forward and reverse measuring circuits are symmetrical.

In theory a 1:1 VSWR is desirable but this is a condition that is often impossible to achieve. Looking at the problem from a practical viewpoint, it is worth trying to get a VSWR of better than 2:1 (equivalent to 11% reflected power). The guidelines shown in **Table 25.13** are suggested practical conditions and the actions that should be taken.

A VSWR METER

Reflectometers designed as VSWR indicators have normally used sampling loops capacitively coupled to a length of transmission line. This results in a meter deflection that is roughly proportional to frequency and they are therefore unsuitable for power measurement unless calibrated for use over a narrow band.

By the use of lumped components this shortcoming can be largely eliminated and the following design may be regarded as independent of frequency up to about 70MHz.

Fig 25.72: Circuit of the frequency-independent VSWR meter

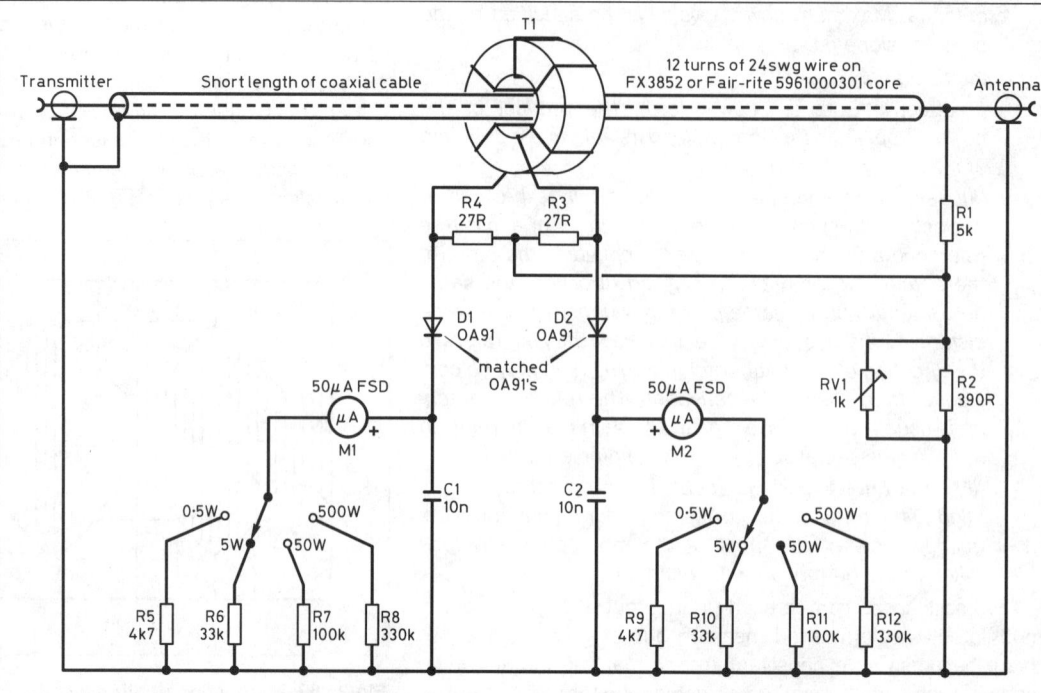

VSWR	% reflected power	Comment
1 - 2.5	0 - 18	Solid-state transmitter SWR protection starts to operate, try looking for an improvement at the higher SWR value
2.5 - 5	18 - 45	Valve equipment probably OK but start looking for a problem or improve the SWR to get closer to 2:1
5 - ∞	45 - 100	Check the feed/antenna system; there is a problem!

Table 25.13: Guideline for various VSWRs

R1	5k carbon (see text)	C1, C2	10n ceramic
R2	390R carbon	T1	Philips FX3852,
R3, R4	27R, 2W carbon		4332202097180 or
R5,9	4k7		Fair-rite 596 1000301
R6, R10	33k	D1, D2	OA91 (matched) - see
R7, R11	100k		text
R8, R12	330k	M1, M2	50µA FSD meters
RV1	1k skeleton pot.,	Switches	2 off, 1 p, 4w but
	0.5W		good quality
All resistors are 0.25W, 5% tolerance unless specified otherwise			

Table 25.14: VSWR meter components list

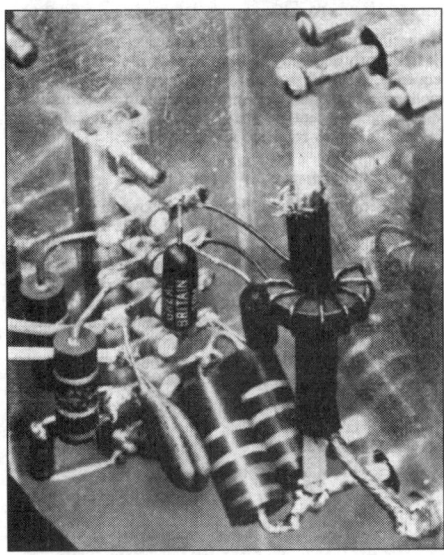

Fig 25.73: Construction of the frequency-independent VSWR meter

The ratio of the sampling resistors R1 and R2 is determined by the sensitivity of the current sensing circuit. As the two sampling voltages must be equal in magnitude under matched conditions, RV1 provides a fine adjustment of the ratio.

Germanium diodes as specified are essential if an instrument is to be used at low power levels, otherwise silicon diodes such as 1N914 or Schottky diodes such as the BAT85 may be substituted. To increase the sensitivity at low power levels, eg 1W, then the feed line could be looped through the toroid. It may then be necessary to use a large toroid or smaller coaxial cable (but this will not cope with high powers!).

A components list for this project is given in **Table 25.14**.

Calibration

Accurate calibration requires a transmitter and an RF voltmeter or possibly an oscilloscope. The wattmeter is calibrated by feeding power through the meter into a dummy load of 50wΩ. RV1 is adjusted for minimum reflected power indication and the power scale calibrated according to the RF voltage appearing across the load. The reflected power meter is calibrated by reversing the connections to the coaxial line.

The instrument has full-scale deflections of 0.5, 5, 50 and 500 watts, selected by the range switch. These should not normally be ganged since the reverse power will normally be much less than the forward power.

A SENSITIVE ANTENNA BRIDGE

The bridge (**Fig 25.74**) to be described (by G4COL [13]) enables antenna systems to be adjusted for minimum VSWR using single-frequency 'CW' signal powers as low as a few microwatts over a frequency range of 1.8 to greater than 60MHz (it should be

Circuit Description

The circuit is shown in **Fig 25.72** and uses a current transformer in which the low resistance at the secondary is split into two equal parts, R3 and R4. The centre section is taken to the voltage-sampling network (R1, R2, RV1) so that the sum and difference voltages are available at the ends of the transformer secondary winding.

Layout of the sampling circuit is fairly critical. The input and output sockets should be a few inches apart and connected together with a short length of coaxial cable. The coaxial cable outer must be earthed at one end only so that it acts as an electrostatic screen between the primary and secondary of the toroidal transformer. The layout of the sensing circuits in a similar instrument is shown in **Fig 25.73**.

The primary of the toroidal transformer is formed by threading a ferrite ring on to the coaxial cable. Twelve turns of 24SWG (0.56mm) enamelled copper wire are equally spaced around the entire circumference of the ring to form the secondary winding. The ferrite material should maintain a high permeability over the frequency range to be used: the original used a Mullard FX1596 which is no longer available but suggested alternatives are Philips FX3852 or 432202097180 and Fair-rite 5961000301, other types may also be suitable.

The remaining components in the sampling circuits should have the shortest possible leads. R1 and R2 should be non-inductive carbon types. For powers above about 100W, R1 can consist of several 2W carbon resistors in parallel. RV1 should be a miniature skeleton potentiometer in order to keep stray reactance to a minimum. The detector diodes D1 and D2 should be matched point-contact germanium types with a PIV rating of about 50V; OA91 diodes are suitable. The resistors R3 and R4 should be matched to 5% or better.

Fig 25.74: The frequency-independent VSWR meter

useable on 70MHz). It is most useful during extended periods of experimentation, where very-low-power transmissions should be used to avoid inconvenience to other band users and possible exposure of the experimenter to high radio frequency (RF) voltages or powers. This instrument is not intended to supplant the inline VSWR meter used for monitoring during normal transmissions. Its function is essentially the same as a 'resistance only' noise bridge but works with a low-power external signal source. It uses a built-in moving-coil meter to monitor the signal and does not require a receiver.

The bridge is so simple that it can be built in a few hours. Suitable signal sources include a signal generator, crystal oscillator, dip meter and transmitter plus attenuator. An outline specification is as follows:

- Operating signal power: 25 to +3dBm (3mW to 2mW) at 10MHz, for full-scale meter deflection

- Frequency response: less than 2dB variation from 1.8 to 65MHz

- Power supply: 9V PP3 battery with current drain of 18mA

Achieving Sensitivity

The bridge uses conventional diode detection. The most common method of driving a meter is to follow the diode detector by a DC amplifier as shown in **Fig 25.75(a)** which shows a 'system gain' G, which is the meter current produced by a given radio frequency input voltage (or power). In **Fig 25.75(b)**, the same sys-

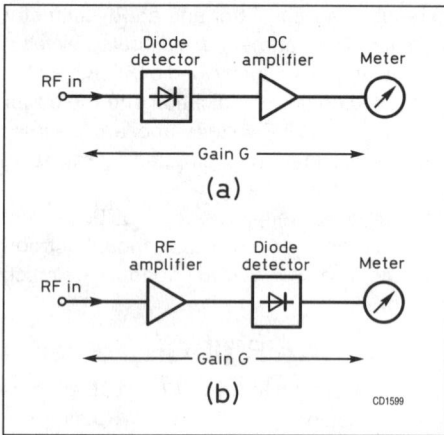

Fig 25.75: (a) Diode detector followed by DC amplifier. (b) RF amplifier followed by diode detector

tem gain has been obtained by amplifying the RF signal prior to detection.

Given enough RF input, the two methods produce the same result in principle. The first approach has a number of advantages: the DC amplifier does not affect the frequency response, is simpler and less dependent on layout.

However, at low RF levels, such as the operating range of this bridge, the diode exhibits a 'threshold effect' whereby its sensitivity (and so also the system gain) falls with signal level and the two approaches cease to be equivalent. The effect of this on an antenna bridge is to produce an erroneously wide null that masks the point of optimum match.

The bridge uses the method of Fig 25.75(b) and takes advantage of one of the low-cost, readily available RF (or video) integrated circuit amplifiers currently on the market. The amplifier boosts the RF signal level to the point where efficient detection can take place.

Circuit Description

The circuit is shown in **Fig 25.76**. Resistors R1 and R3, together with the impedance connected to SK2, form a bridge that will be perfectly balanced by a 51Ω resistive load. The bridge voltage is amplified by differential RF amplifier IC1 and detected by a full-wave rectifier using diodes D1 and D2. The differential amplifier obviates the need for the usual balun transformer. The rectified voltage is buffered at DC by IC2a/IC2b and at RF by R9/R10. Potentiometer R11 allows the meter sensitivity to be set.

Overall sensitivity can be further altered by changing the RF amplifier gain-setting resistor R8. Raising its value lowers the gain but increases the bandwidth, and conversely for a lower value. If working in a different system impedance, such as 75Ω, R1 to R3 should be made equal to this value.

Constructional Notes

The component list is given in **Table 25.15**. First prepare the case by drilling holes for the BNC connectors, potentiometer and panel meter (see photograph, Fig 25.74). The meter cut-out can be made by drilling a pattern of small holes and filing to final size, or by opening up a circular hole using small files. Next cut a piece of copper-clad board approximately 65mm square and hold it inside the case at the socket end. Drill two 3mm holes through the case and the board close to the BNC connectors. Once drilling and de-burring are complete, apply dry transfer let-

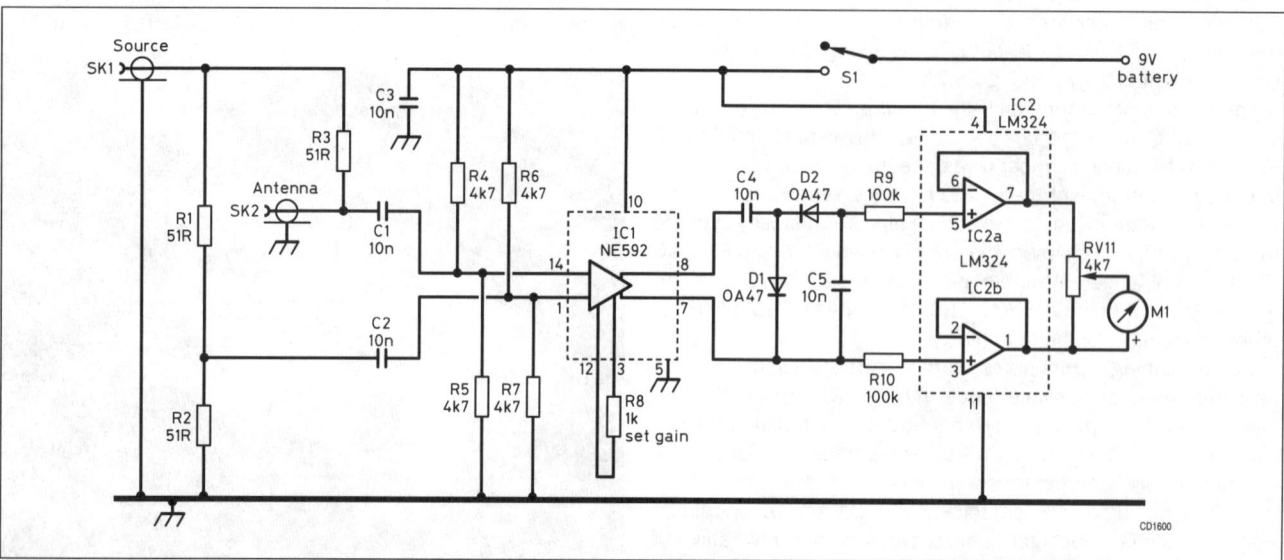

Fig 25.76: Circuit diagram of the bridge

R1, R2	51R	C1-C5	10n ceramic disc, 12V
R3, R12	51R		or greater
R4, R5	4k7	D1, D2	OA47, OA91 or similar
R6, R7	4k7		germanium or BAR21,
R8	1k		BAT42 Schottky diode
R9, R10	100k	IC1	NE592
R11	4k7 linear pot	IC2	LM324

M1   250µA FSD meter
SK1, SK2 BNC chassis sockets or other preferred type such as SO239
Aluminium box, suggested size 133xx70x38
PP3 battery and connector
Insulated wire
Knob
Dymo tape or dry transfer lettering or similar
Velcro for securing battery
Single-sided copper-clad circuit board
M2.5 x 6mm screws (10 off)
M2.5 nut (2 off)
Double-sided tape for securing meter
All resistors are 0.125W or greater, metal film

Table 25.15: Components list for sensitive antenna bridge

tering if desired, and spray the case lightly with lacquer to protect the surface. Clean the copper board and fix it in place with two M2.5 screws and nuts using the 3mm holes. Mount the potentiometer and BNC connectors. The meter can be fixed to the case with double-sided adhesive tape. Self-adhesive Velcro strips hold the 9V PP3 battery in place.

As can be seen in the photo, the circuitry can be built up quickly and easily on the copper board using 'ugly' style construction. Start by positioning the ICs with their ground pins bent down to touch the copper and solder them down - place IC1 quite close to the BNC sockets. The other components can then be added at will, finishing off with the few connecting wires.

Operation

Other than the sensitivity panel potentiometer, there are no adjustments. After checking the wiring, connect the battery and switch on. Applying a CW signal within the specified range (see introductory paragraphs) to the source socket with nothing connected to the load socket should give a meter deflection which can be adjusted to full scale using the potentiometer. *Do not* connect a transmitter output directly to the bridge as this will damage the unit. If a 51Ω resistor is placed across the load socket, the meter reading should drop to zero.

The meter reading is proportional to reflection coefficient: for instance, half-scale corresponds to a reflection coefficient of 0.5. This is related to VSWR by:

$$VSWR = \frac{1+\rho}{1-\rho}$$

where ρ is the reflection coefficient.

This means that the half-scale reading corresponds to a VSWR of 3. As the RF signal level falls, accuracy at low VSWRs will decline due to the fall-off in diode detector sensitivity. Calibration, if required, can be established with a set of resistors. For example, a 75Ω resistor presents a VSWR of 1.5, corresponding to a reflection coefficient of 0.2 which would ideally give a meter reading of 20% of full scale. Use carbon composition (check the resistance carefully), metal film or carbon film

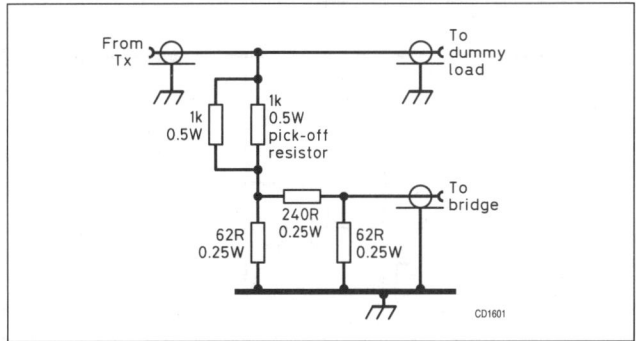

Fig 25.77: Signal pick-off

resistors for calibration and avoid wire-wound type since these have significant inductance.

Obtaining a sufficiently low power using a transmitter can be a problem and for this a signal pick-off, as shown in **Fig 25.77**, can be used. The values shown are suitable for up to 10W (40dBm), and attenuate the signals reaching the bridge by about 40dB. Make sure that the dummy load can dissipate the transmitter power. Below 100mW (20dBm) a low-power attenuator can be used directly without a dummy load and signal pick-off.

When making measurements, as opposed to simply using the bridge to indicate a dip for best match, remember to set the meter reading to full scale with an open-circuit or short-circuit at the load socket.

POWER METERS

Described below is a Precision Peak-following Power Meter but the reader may also like to consider units in chapter 6 of reference [1], and the 'Crawley Power Meter' by G3GRO and G3YSX [14].

A Precision Peak-following Power Meter

The following is an abridged description of a power meter by G3GKG [15] that requires no setting up and no adjustment during use. The meter covers the HF bands and copes with powers up to about 450W. A finished unit is shown in **Fig 25.78**.

The heart of this instrument is the type of coupler known variously as a Tandem Match, a bi-directional coupler and a 4-port hybrid transformer.

(NOTE: To quote from a reference by G4ZNQ:[16]:

"A hybrid is a very simple circuit - just two transformers and four connectors - with some amazing properties. The connectors or ports are best thought of as two pairs. If a signal is fed into one connector and out of the other

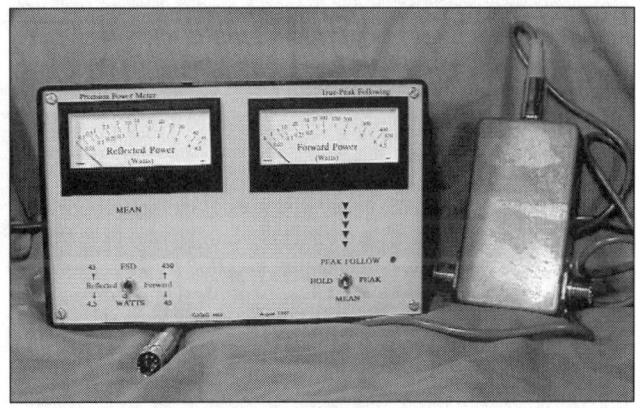

Fig 25.78: Precision peak-following power meter

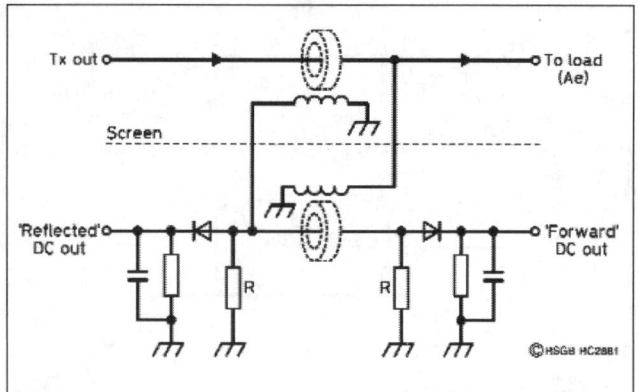

Fig 25.79: The Tandem Match - symmetrical between input and output

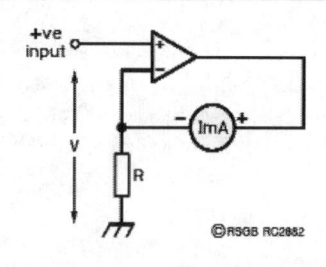

Fig 25.80: The basic meter circuit, showing the meter in the feedback loop of the op-amp driver

of a pair, into some unknown impedance load (say an antenna) then, if both the other connectors are terminated in the intended system impedance (say 50Ω), the hybrid feeds a fraction of the power passing forwards through the first pair of connectors into one of the terminations."

"It feeds an equal fraction of the reverse power flow into the other termination. Hybrids can be designed to have different sampling fractions, usually quoted in decibels, so that a 20dB hybrid diverts 1% of the flowing power to the appropriate terminated port".)

Various authors have suggested different coupling factors according to application and by juggling the figures around, G3GKG derived some simple formulas to produce an essentially standard output voltage which can be defined for any particular system. (Calculations made during the course of the development of the power meter are provided as spreadsheets on the RSGB Members Only web-site [17]. The spreadsheets may be freely downloaded and used to assist readers to develop their own circuits for use at different powers and with different meter sensitivities.)

Using the correct type of toroid core (which must be of high permeability ferrite), this is a precision circuit (**Fig 25.79**) which produces voltages at both the Forward and Reflected output ports which are strictly and predictably defined by the RF power, the designed load resistance and the number of turns on the secondary winding of the toroid. Used with the amplifier and display units to be described, the calibration is constant throughout (at least) the HF range of the amateur frequency bands and is accomplished completely and accurately just by using the calculated design parameters.

It is readily apparent from Fig 25.79 that the circuit is completely symmetrical and this is indeed borne out by its performance. Reversing the transmitter and aerial connections merely causes the Forward and Reflected output ports to interchange positions, as does reversing the connections to one or other of the toroidal windings.

RF to DC

The display unit (Fig 25.78) uses individual meters for Forward and Reflected power, so the outputs are brought out from the RF section separately, after rectification and buffering.

The Forward metering circuit has been designed to accept a DC voltage range close to an optimum of 10V FSD. To set the power range, it is therefore arranged for the coupler to produce this voltage from the designated maximum forward power (peak), by first finding the required number of turns on the toroids to produce 10V as closely as possible from that power,

and then calculating the actual precise voltage for that number of turns. (Fractional turns cannot be wound on a toroid!)

With a 10V range and the employment of Schottky diodes for both RF rectification and for an op-amp linearising circuit, the errors are reduced to negligible proportions, the DC output tracking the RF voltage accurately down to about 30mV (representing a power level of 18µW with a 50Ω load) with little deviation well below that. It is important for this tracking that the pair of diodes in each of the detector/op-amp circuits is initially matched regarding forward voltage drop and also that they remain at the same ambient temperature during use.

The same operational amplifier also provides a convenient low impedance DC output from this part of the circuit to the main Display Unit, allowing the RF unit to be constructed in a separate housing which also caters for the requirement regarding ambient temperature. The RF Head Unit can then be installed in the direct coaxial line between transmitter and aerial matching unit, well away from the main measuring and display instrument, which can therefore be located in the optimum position for viewing.

Metering

Each meter circuit incorporates the meter itself in the feedback loop of an op-amp driver, **Fig 25.80**. The voltage range is determined only by the current range of the individual meter (irrespective of its inherent resistance) and the scaling resistor, R, the value of which is given simply by dividing the actual full scale voltage required, V, by the nominal full-scale deflection (FSD) current sensitivity, I, of the meter. For supreme accuracy, the FSD current can be individually measured and used in the calculation.

As the whole of the Head Unit circuitry is completely symmetrical (as Fig 25.79 shows), the reflected output voltage of the coupler for the same full-scale power would, of course, be the same, at 10 volts, but we can select a range for that metering circuit of something less - ie an FSD which is more commensurate with the maximum reflected power likely to be encountered - bearing in mind the square law relationship which dictates that half the full scale voltage represents a quarter of the power. If the Forward meter is calibrated so that 10V FSD represents 400W, a companion Reflected meter calibrated for 5V FSD will read up to 100W, which would be equivalent to an SWR of 3:1.

It is also convenient and easy with this degree of sensitivity to provide accurate, alternative, very low power ranges, so that the transmitter and aerial system can be tuned and matched with minimum chance of causing interference. This entails incorporating a two-way toggle switch to select different values of scaling resistors which set the scaling of both meters appropriately.

For a station using the full legal limit, convenient ranges might be -

Forward Power: 450W or 45W;
Reflected Power: 45W or 4.5W

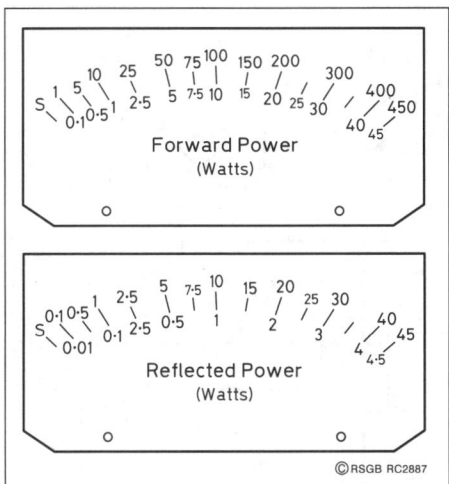

Fig 25.81: Meter scales showing the common calibration marks for the two power ranges

Peak Reading

Many commercially-designed power meters include a function labelled 'Peak' or 'PEP'. Occasionally, such a meter will indicate something more or less close to the true peak power, at the expense of a very long decay time. Most of those which are manufactured as separate items, or included in an aerial matching unit, do not and cannot measure or indicate the instantaneous peaks of an SSB speech waveform because they do not include any active circuitry. It is frequently maintained that if the unit requires an external power supply it will most likely contain such a refinement and be capable of capturing these peaks, but this can be a snare and a delusion. The power supply frequently serves only to provide the illumination for the meter(s)!

Let's be sure what we are talking about. With a continuous carrier, which is of course just an RF sine wave, the peak power we want to measure is, in fact, what we know as the peak envelope power or PEP. The picture on a monitor scope will show a solid band (the envelope), the vertical width of which varies with the output power of the transmitter and clearly illustrates that, in this sense, the mean and peak powers are the same. Any of the instruments on the market should produce the same reading in either measuring mode. An SSB speech waveform, on the other hand, shows a band of power that is constantly varying at a syllabic rate and further illustrates that, without compression or other processing, the power only reaches local maxima for very brief periods of time. It is these brief peaks that we wish to measure when we refer to 'peak' power - more properly called the instantaneous peak power. When we switch to read 'mean' power, we want to know the average power output over a period of time, and we need the meters to read in this mode when tuning or adjusting the rig. With a properly adjusted transmitter in the SSB mode (ie no compression), the mean power will be quite small compared with the peak power.

thus enabling a single set of scale markings to be calculated and used for all ranges, with different figures on the two meters such as those shown on **Fig 25.81**.

Consider now the actual scaling resistors. Because the voltage ranges are all calculable, these can be fixed components which, although unlikely to be easily available in the exact values, can be made up from suitable series or parallel combinations of 1% tolerance resistors. Alternatively, by using pre-set variable resistors, the completed instrument could be calibrated using precise values of DC voltage injected at the input socket to the main unit in place of the output from the head unit.

All the calculations are available on spreadsheets - see comment earlier. These are for determining the required number of turns on the toroids to suit a given power range, the actual output full-scale voltage, values of scaling resistors (R1 - R4) for all four chosen ranges with particular meter movement sensitivities (with the facility of calculating series combinations of preferred value resistors to produce the value required) and the meter needle deflections in degrees for the required calibration points.

The peak-reading circuits normally encountered all employ the same sort of 'diode-pump-charging-a-capacitor' circuitry, with varying degrees of sophistication designed to overcome the inherent drawbacks of the circuit. In order to capture brief peaks accurately, the diode non-linearity and knee voltage must be

Fig 25.82: Inside the RF head unit

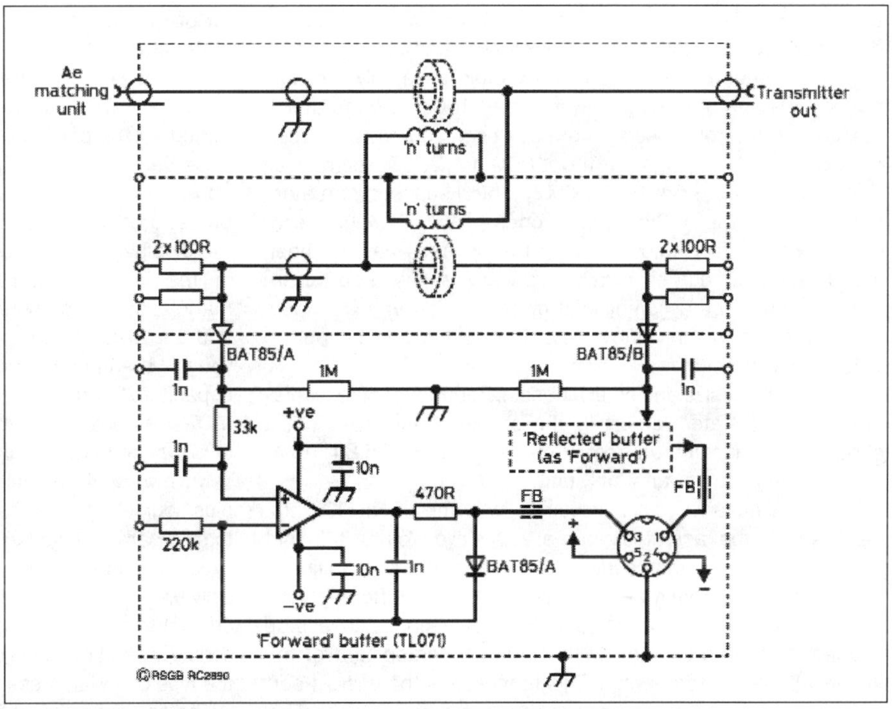

Fig 25.83: Circuit of the head unit

overcome, the charging circuit must have a fast attack (implying a very low source impedance) and it must have a decay time long enough to enable the peak to be read - often accomplished by incorporating a 'peak-hold' feature which removes the normal discharge resistor. In normal SSB use, most of them are inevitably very sluggish in their response.

In order to follow the peaks of an SSB signal, either at a syllabic rate or by capturing the peak amplitude in each short phrase of speech and still provide time to read the meter, it is necessary to provide a fast attack, a preset 'hang' period and a rapid decay, in order to capture the next peak. I have tried several approaches to this idea and the most successful is one using a sample-and-hold chip, type LF398 (of which the N version is the best in this application).

Construction

It was not the intention that this description should provide full constructional details - everyone will have their own requirements and preferences. As can be seen in the photograph, the Head Unit (Fig 25.82) is built into a die-cast aluminium box measuring about 115 x 62 x 29mm and includes the tandem match and detection circuitry (Fig 25.83). It is connected to the main display unit (Fig 25.78) by a 5-pin DIN to 5-pin DIN with screened lead. Thin double-sided copper-clad fibreboard is used in the construction of the screened compartments and for the TL071 linearising buffer amplifiers (with the rectifier diodes passing through holes in the final screen).

The two ends of the second coax-toroid assembly are supported by a pair of orthogonally-mounted 100Ω, 2W metal film resistors. The toroids used so far have been generously proportioned with Al ratings in the 1800 to 2000 region (eg Electrovalue type B 64290K 632X27), but I suspect from more recent testing that rather smaller ones (...45X27) would serve equally well. Offset adjustment for the op-amps has not been found necessary but, especially if very low ranges of power are required, could be accomplished by connecting an experimentally determined, high-value resistor from the negative supply to either pin 1 or pin 5.

Obviously, the size and type of housing required for the display unit will be dictated largely by the choice of meters used. It should ideally include the rest of the circuitry (Fig 25.84) apart from possibly the power supply unit.

(NOTE: A negative rail is required by some of the ICs but, because all the actual signals are positive going, it is not required to be more than a few volts and need not be stabilised.)

Author's Note: From my experience with several different types of meter, most are over-damped (ie the response to a step change in input is too slow, with the needle creeping over the last few percent of its swing). This response is the easiest sort to improve, but the compensation must be done very carefully and precisely so as to prevent over-swing with consequent false

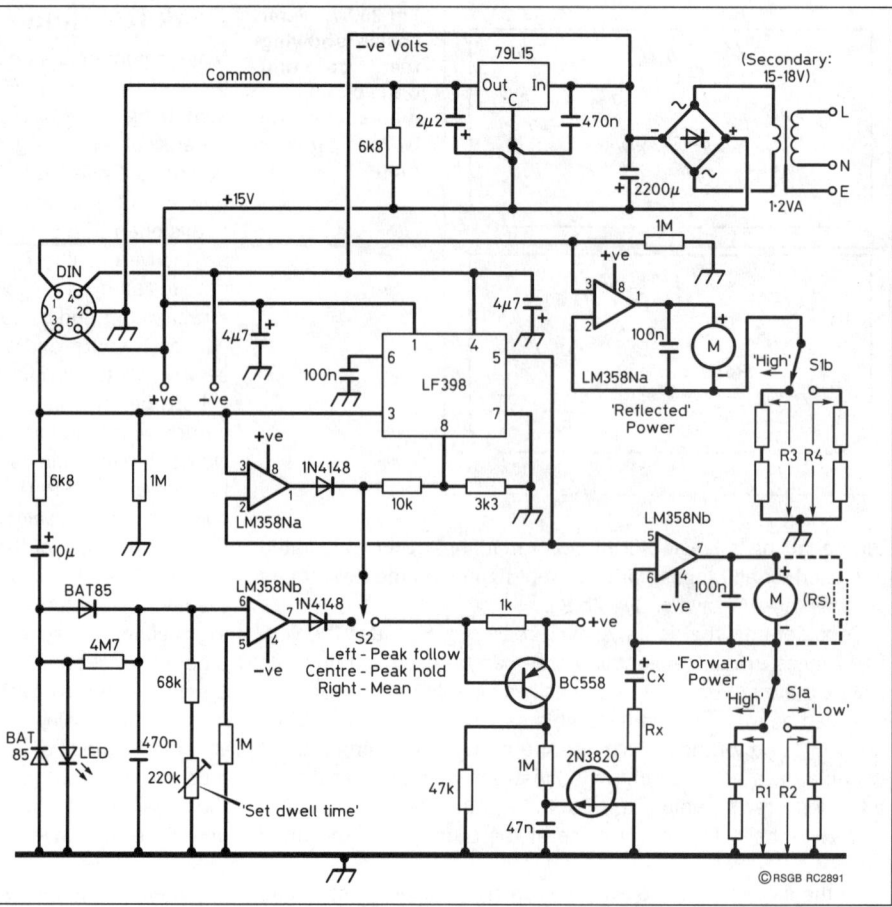

Fig 25.84: Complete circuit of the main display

readings. It is only required in the peak-following mode and only on the Forward power meter, as the Reflected meter always reads mean power. The required components form a series combination of resistance (Rx) and capacitance (Cx), switched into circuit in the appropriate mode, across the calibrating resistor of the op-amp (see Fig 25.84). Part of the resistive component is the 'on' resistance of the FET (which, with the associated transistor, performs the necessary part of the switching function from 'peak' to 'mean' reading), whilst the other values are determined for the particular meter, as follows.

Leave out these components until the instrument is completed and working. Then, remove the plug coming from the head unit and connect a signal from a rectangular-wave generator to the display unit, (between the forward voltage pin and common of the input socket). Use about 8V positive-going with a mark/space ratio about 1:1 and a repetition rate of about 1Hz. With the switches set to the High power range and Peak (follow), you will then need to adjust the values of both resistor (Rx) and capacitor (Cx) until the meter follows the amplitude excursions as fast as possible without over-swinging on either the upward or downward swings - it is easier than it sounds!

My own Mk4 instrument uses 1mA Sifam meters (from a surplus source) and the compensation consists of two 6.8μF capacitors in parallel together with just the 'on' resistance of the FET. Exceptions to my earlier statement, these meters are in fact basically under-damped so that the Forward one also had to be shunted (Rs) to achieve the required critical damping, thus altering its sensitivity and requiring changes to the scaling resistors. Whatever the type of meter, these values must be critically determined.

The original article [15] also had a variant on this design using a digital bargraph instead of meters.

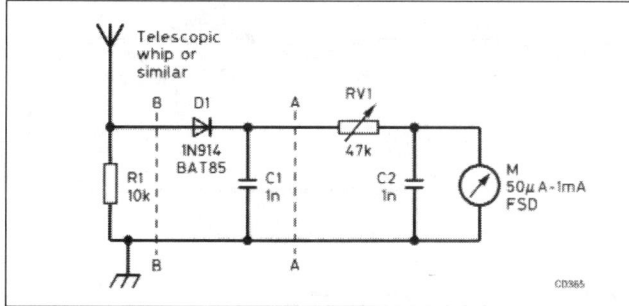

Fig 25.85: Simple untuned field strength meter

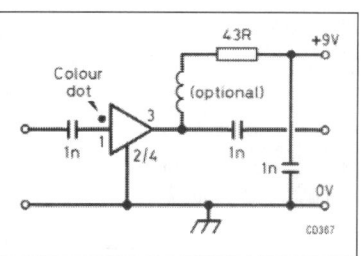

Fig 25.86: Broad-band amplifier based around a MAR8

FIELD STRENGTH METERS

Field strength meters used by amateurs are normally used as indicators to maximise the radiated power and not to make an actual field strength measurement at a receiving site. The absorption wavemeter or dip oscillator (in absorption mode) can be used for this purpose. These units may need some form of external telescopic antenna to be fitted. However, the use of a tuned circuit is sometimes inconvenient as no attempt is being made to differentiate between wanted and unwanted transmitted signals (this should already have been dealt with at the transmitter!).

The alternative is to use a simple type of system such as that shown on **Fig 25.85**. A signal is picked up by the antenna, rectified and smoothed by D1/C1, the resulting DC signal is then indicated on the meter M1, with RV1 acting as a sensitivity control. Capacitor C2 provides an AC short across the meter for any unwanted RF signals.

Construct the unit in a box, using either a telescopic whip or a loop of wire. The unit can be used for relative field strength measurements at a given frequency. It should not be used for relative measurements between different frequencies as the efficiency of the antenna and rectifier will affect readings. It should be a useful device for tuning a transmitter to obtain maximum radiated power. It could also be used for adjusting an antenna for maximum radiated power, provided the unit is far enough away from the antenna. By splitting the circuit at AA, the antenna/rectifier combination could be used as a remote reading head, with the meter/sensitivity control being in the shack.

Higher Sensitivity Broadband Field Strength Meter

The concept here is to amplify the received signal first and then to detect it to drive a meter. This can be accomplished by using one of the relatively inexpensive broadband RF amplifiers such as the MAR series from Mini circuits; there may be alternatives on the surplus market.

These amplifiers would be placed in the circuit of Fig 25.85 at position BB. They should be constructed on circuit board with a good ground plane. All components should have short leads to minimise lead inductance and the capacitors carefully chosen for the frequency range envisaged. These devices generally have outputs of the order of +10dBm, it is therefore imperative that diode D1 is of a type with low forward volt drop such as Schottky type BAT85. If the field strengths being measured are very low, it may be possible to cascade two or more such amplifiers.

Fig 25.86 shows a circuit based around a MAR8 monolithic amplifier produced by Mini Circuits. The circuit has a response from DC to 1GHz with a quoted gain at 100MHz of 33dB and 23dB at 1GHz; the maximum output is about +10dBm. The device requires 7.5V at 36mA, the circuit shows a series resistor for operation from a 9v DC supply.

It is also worth consulting reference [2] where other circuits are suggested as well as the use of a communications receiver.

ATTENUATORS

Attenuators are useful for receiver measurements, especially when testing from signal generators that have minimal or no attenuators within them. The greatest problem is the radiation and leakage of signals from within the unit. Because of this the attenuator should consist of a good RF-tight metal box with high quality connectors. An attenuator might also be useful between a transmitter and transverter.

Attenuators are normally made from Pi or T networks. For this exercise it is assumed that load and source impedances are equal. See **Table 25.16** for component values for 50 and 75-ohm source/loads.

The circuit shown on **Fig 25.87** is a low power switched attenuator using a combination of Pi and T networks, this enables preferred value resistors to be used. A 1-2-4-8-...dB switching sequence is used so that the maximum attenuation range is obtained for a given number of sections.

The switches used are a standard wafer type panel mounting slide switches, for which the effective transfer capacitance in the circuit used is only 0.8pF. With 5% carbon film resistors the attenuator accuracy is ±0.5dB on the 1,2,4 and 8dB positions

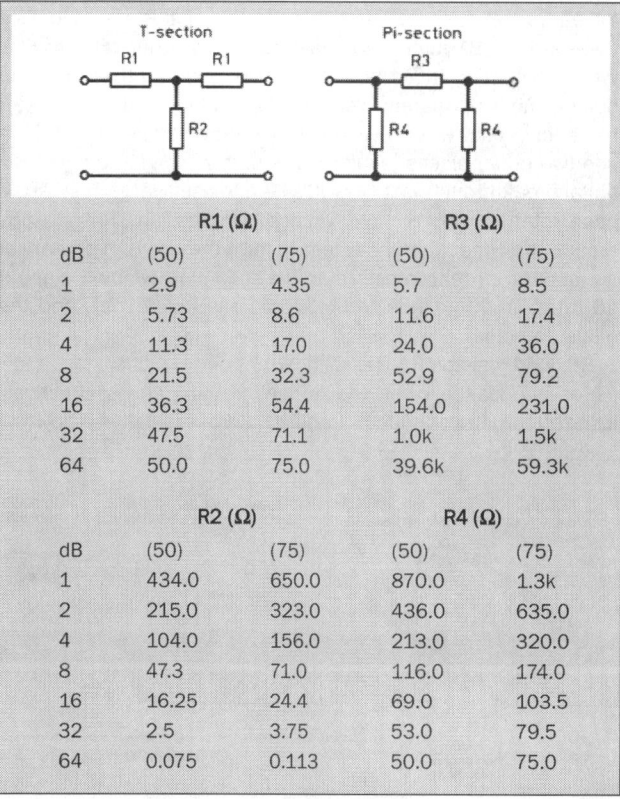

	T-section			Pi-section		
	R1	R1		R3		
		R2		R4	R4	

	R1 (Ω)		R3 (Ω)	
dB	(50)	(75)	(50)	(75)
1	2.9	4.35	5.7	8.5
2	5.73	8.6	11.6	17.4
4	11.3	17.0	24.0	36.0
8	21.5	32.3	52.9	79.2
16	36.3	54.4	154.0	231.0
32	47.5	71.1	1.0k	1.5k
64	50.0	75.0	39.6k	59.3k

	R2 (Ω)		R4 (Ω)	
dB	(50)	(75)	(50)	(75)
1	434.0	650.0	870.0	1.3k
2	215.0	323.0	436.0	635.0
4	104.0	156.0	213.0	320.0
8	47.3	71.0	116.0	174.0
16	16.25	24.4	69.0	103.5
32	2.5	3.75	53.0	79.5
64	0.075	0.113	50.0	75.0

Table 25.16: Resistors for 50-ohm and 75-ohm attenuators based on pi and T sections

Fig 25.87: Circuit of a 50Ω attenuator using preferred values of resistor

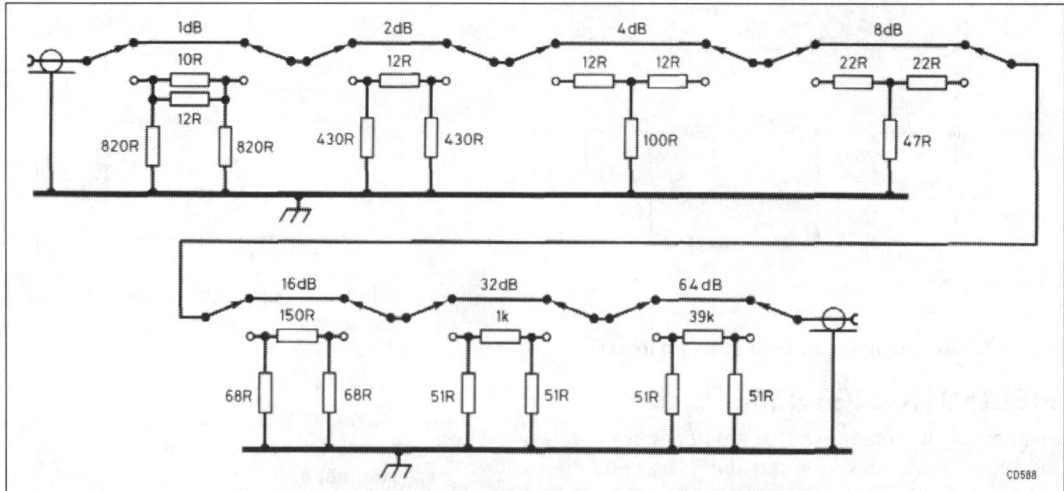

Fig 25.88: Construction of the 50Ω attenuator

up to 500MHz. The 16dB position is 1dB low at 500MHz, the 32dB position 1dB low at 30MHz and the 64dB position 1dB low at 750kHz. The photograph on **Fig 25.88** shows the construction of the unit.

An alternative design is given on page 51 of reference [2].

A CLIP-ON RF CURRENT METER

This article is from 'In Practice' in *RadCom* [18], see this article for earlier references. The basic version of this handy device takes about 10 minutes to tack-solder together (**Fig 25.89**). When you're convinced how useful it is, you can then go on to build a more permanent version. The clip-on RF current meter has a long history, early versions involved breaking a ferrite ring into two equal pieces - which takes some doing! The constructional breakthrough was G0SNO's idea to use a large split ferrite bead intended for HF interference suppression. This clamps around the conductor under test, to form the one-turn primary of a wideband current transformer. The secondary winding is about 10 turns, and is connected to a load resistor, R1-R2, and the diode detector.

The load resistor, R1-R2, is important because it creates a low series impedance when the current transformer is effectively inserted into the conductor under test. For the values shown in

Fig 25.89. (10-turn secondary, 2 x 100Ω) this is $50/10^2 = 0.5\Omega$. Some circuits omit this resistor, but that creates a high insertion impedance - exactly the opposite of what is needed. Also, more secondary turns create a lower insertion impedance, but at the expense of HF bandwidth.

The other components in Fig 25.89. are discussed in G0SNO's article which is reproduced on the 'In Practice' website [19]. Component types and values are critical only if you want to make a fully calibrated meter with switchable current ranges. However, for a first try, and for most general RFI investigations, the meter is almost as useful without any need for calibration. Make R4 about 4.7-10kΩ, and omit R3 and S1. If the meter is either too sensitive or not sensitive enough, either change R4 or change the HF power level.

Just about any split ferrite core intended for RFI suppression will do the job, but there are a few practical points. Choose a large core, typically with a 13mm diameter hole. This allows you to clip the core onto large coax, mains and other multi-core cables while still leaving enough space for the secondary winding (which should be made using very thin enamelled or other insulated wire). It is important that the core closes with no air gap - and that can be a problem. A major disadvantage of the basic split ferrite core in its plastic housing is that the housing is not meant to be repeatedly opened and closed, so the hinge will soon break. By all means try out this gadget in the basic form but it is likely that you will soon be thinking about something more permanent. The classic way to do this is using a clothes-peg **Fig 25.90** but there are now better alternatives.

For example, the first photograph (Fig 25.90) shows the rather heavy-duty version using two strong clothes-pegs, fibreglass sheet and epoxy glue (more details at [19]). The second photograph shows G1OXAC's neat and simple version using a giant plastic paper-clip, with a small plastic-cased meter stuck on the side. The only requirement of the clip is that it must be basi-

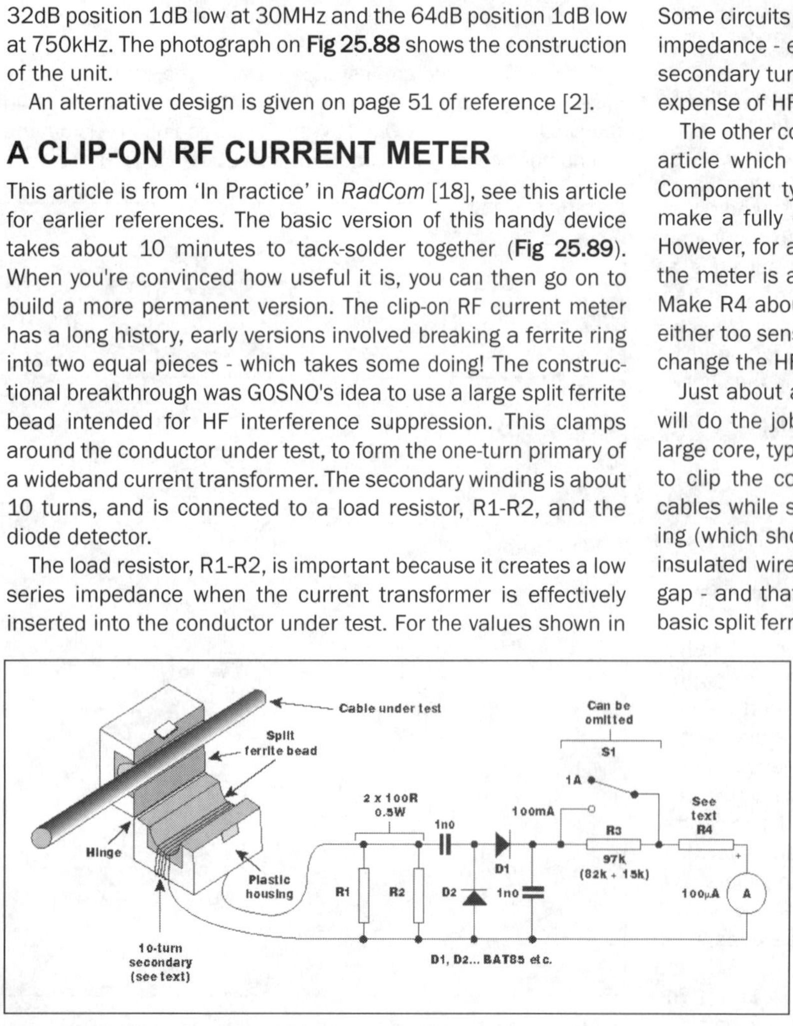

Fig 25.89: G0SNO's clipon RF current meter

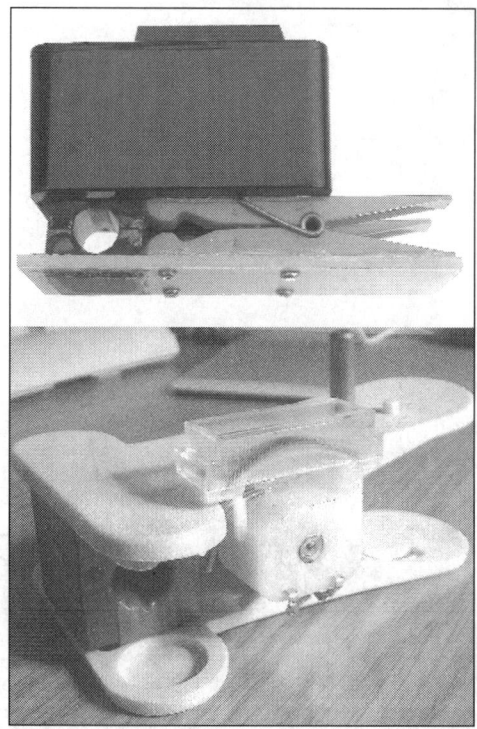

ATA	India	10,000kHz
DCF77	Germany	77.5kHz
CHU	Canada	3330, 7335, 14,670kHz
HLA	Korea	5000kHz
MSF	UK	60kHz
LOL2	Argentina	5000, 10,000, 15,000kHz
RWM	Russia	4996, 9996, 14,996kHz
TDF	France	162kHz
WWV	USA	2500, 5000, 10,000, 15,000, 20,000, 25,000kHz
WWVB	USA	60kHz
WWVH	USA	2500, 5000, 10,000, 15,000, 20,000, 25,000kHz

Table 25.17: A selection of standard frequency transmissions

may well be present way up into the HF bands - the 64th harmonic is 1MHz and this would be accurate to ±0.01%. However, newer generations of TV sets may not provide such a good source if additional screening has been added to help minimise extraneous radiations.

REFERENCES

[1] *Test Equipment for the Radio Amateur*, 3rd edition, Clive Smith, G4FZH, RSGB, 1994.

[2] *The Antenna Experimenter's Guide*, 2nd edition, Peter Dodd, G3LDO, RSGB 1996.

[3] *VHF/UHF Handbook*, Chapter 11,. Edited by Dick Biddulph, G8DPS. RSGB, 1997

[4] 'Anders OEM22 Panel Meter Data Sheet'. Website http://www.anders.co.uk.

[5] The G3WPO FET dip oscillator Mk2, Tony Bailey, G3WPO, Radio Communication Apr 1987.

[6] 'More from your dip oscillator', Ed Chicken, G3BIK, *RadCom* May 2000

[7] 'Technical Correspondence', Peter Lumb, G3IRM, *QST* Jun 1972

[8] 'Technical Correspondence', W1CER, *QST* Nov 1971

[9] 'Maxim Data Sheet for MAX038'. Website http://www.maxim.com

[10] 'A combined 2m and 70cm signal source', J Brown, G3DVV, *RadCom* Aug 1998.

[11] 'A receiver calibrator and transmitter monitor', Ian Braithwaite, G4COL, *RadCom* Jun 1998.

[12] 'Simple Spectrum Analyser', Roger Blackwell, G4PMK, *Radio Communication* Nov 1989

[13] 'A sensitive antenna bridge', Ian Braithwaite, G4COL, *RadCom* Jul 1997.

[14] 'The Crawley Power Meter', Derek Atter G3GRO and Stewart Bryant G3YSX, *RadCom* Jan/Feb 2000

[15] 'A Precision Peak-following Power Meter', Brian Horsfall, G3GKG, *RadCom* Mar 2001

[16] 'Bi-directional In-line Wattmeter', G4ZNQ, *Antenna Handbook*, G-QRP Club, pp29-32

[17] RSGB Members Only web site: http://www.rsgb.org.uk/ membersonly/

[18] 'In Practice', Ian White G3SEK, *RadCom* Dec 2003

[19] 'In Practice' website: http://www.ifwtech.co.uk/g3sek

cally non-metallic, and that it can hold the two halves of the core accurately together while the whole weight of the meter is dangling from the cable. Another option worth investigating would be the pliers-style plastic work clamps that are sold in a range of sizes by hobby shops. Whatever you use, it is vital that you glue the two halves of the core to the clip in such a way that they always close tightly together with no air gap. Hint: glue one half of the core to one side of the clip first, and let that side set; don't try to glue the second half until the first is good and solid.

A clip-on RF current meter could hardly be simpler to build. It's an ideal project for beginners and clubs. Once upon a time, every amateur station was required to have an absorption wavemeter, which achieved almost nothing; if every amateur station today had a clip-on RF current meter, we'd see a lot less RFI and a lot more confidence about going on the air!

STANDARD FREQUENCY SERVICES

There are various standard frequencies transmitted throughout the world and these can be harnessed in order to check other equipment against them. Typical of these transmissions are those shown in **Table 25.17**.

These standard frequencies are maintained to an accuracy of typically one part in 10^{11}. However, if the sky-wave is used there could be a large error in reception due to Doppler shift and there will be fading of the signal. These problems can be avoided by using a low-frequency transmission such as those from MSF or WWVB. Timing information is also impressed on the signals in either GMT or UTC.

In addition in the UK the BBC maintains the accuracy of the Droitwich 198kHz (formerly 200kHz) carrier to high accuracy - on a long-term basis being 2 parts in 10^{11}.

An additional method that may be available (providing the TV radiates well!) is the line timebase of a TV and the associated harmonics. The line timebase is at 15.625kHz and harmonics

ONLY £18.99
plus p&p

Backyard Antennas

Peter Dodd G3LDO

Radio amateurs and short-wave listeners all want to achieve the very best from their HF and VHF equipment. Receivers and transmitters are available to professional standards, but very few people have the real estate to erect the sort of antenna used by a commercial radio station.

Antenna guru Peter Dodd explains how, by using a variety of simple techniques, it is possible to achieve very high performance from a compact antenna. Also detailed is how to make an antenna efficient on several bands at once. The book covers end-fed and centre-fed antennas, rotary beams, loops, tuning units, VHF/UHF antennas, antenna and mast construction, transmission lines, and how to estimate and measure the performance of your antenna. Whether you have a house, bungalow or apartment, Backyard Antennas will help you find the solution to radiating a good signal on your favourite band.

Size: 244 x 183mm, 208 pages, ISBN: 1-872309-59-3.
RSGB Member's Price £16.14

E&OE

Radio Society of Great Britain
Lambda House, Cranborne Road, Potters Bar, Herts. EN6 3JE
Tel. 0870 904 7373 Fax. 0870 904 7374

ORDER 24 HOURS A DAY ON OUR WEBSITE
www.rsgb.org/shop

26 Construction and Workshop Practice

In spite of the now worldwide commercial market of amateur radio there are still many items of equipment that are worth making for yourself. Cost is an important factor and often things may be made very cheaply by utilising ex-equipment components. For the newcomer it is worthwhile starting the 'junk box' (usually many boxes) of a variety of useful bits and pieces collected from junk sales, rallies and the like. If you belong to a radio club, many of its members may have items to exchange or give away which will just 'do nicely' for your project. The creatively satisfying work of home construction requires many skills, all of which make this an interesting and rewarding facet of the amateur radio hobby. The skills learnt and the pleasures gained are well worth the effort. It all takes time, and for the newcomer it is better to start on some simple project, just to get the 'feel' of things, rather than to dive in on some marathon project which usually finishes up as a complete deterrent to construction or as a collection of bits in someone else's junk box!

Most amateurs deciding to build some piece of equipment start collecting components, information on circuits, source of components and other technical data, but often forget to enquire about the manufacturing techniques required. Materials, tools and how to use them; methods of component assembly; and making the finished job look good are just some of these techniques.

The first requirement is an elementary knowledge of the materials useful and normally available to constructors.

METALS

Aluminium and aluminium alloys

Typical uses: cabinets, boxes, panels, masts, beam antennas, heatsinks.

These are good electrical conductors and are the lightest in weight of the normal metals available. They are non-magnetic, of medium to high cost, and available in sheet, rod, tube and other forms. The annealed quality usually bends and machines easily but soldering requires special fluxes and solders, and often the soldered joint is not good electrically. With the unknown quality of aluminium that most amateurs encounter, soldering is best left alone or experimented with before committing the project to this method of fastening. Long-term corrosion is also a problem with soldered joints in aluminium unless specialist methods are applied to prevent this. Adhesive bonding is very good but such joints are not usually reliable electrically. Aluminium is non-corroding in normal use, but direct contact with brass or copper should be avoided as these react and encourage corrosion with the resultant troubles of poor or non-existent electrical contact (see the section on corrosion.) Nickel-plated or stainless-steel screws, nuts and washers, rather than brass or steel, should be used to reduce the chances of corrosion around this type of fastening.

A metal that looks similar to aluminium, and which may come into the hands of the unwitting, is magnesium alloy. Filings and chippings from this metal are highly flammable. They can burst into a glaring flame with the heat generated by filing and drilling, and trying to put out such a fire with water only makes matters worse. (Factories where this alloy is machined use a chemical fire extinguisher, one type of which goes under the title of DX pow-

der!) It is difficult to tell the difference by just looking, but size-for-size magnesium alloy sheet weighs less than aluminium and usually on bending there is a 'granulating-cum-scrunching' feel.

For reliable bending of sheet aluminium up to 1.5mm thick, it should be possible to bend a sample strip back on itself and hammer the fold flat without breakage. Annealing aluminium is possible, but under specialised conditions.

Before any heating is attempted it is essential to degrease the part thoroughly.

A workshop 'dodge' which sometimes works, and is certainly worth a try if bending or forming proves difficult, is to heat the aluminium slowly (the slower the better) to such a temperature that a small pine stick, when rubbed on the heated surface, just chars, but does not ignite (equivalent to approximately 300°C) Care needs to be taken to avoid overheating, for this will destroy the properties of the aluminium. This method of localised annealing has proved effective when bending tube or rod for the folded dipole elements of beams. In this case the area to be annealed is small and the 'dodge' can be applied easily. Another method, suitable for larger areas, is to rub soap on the surface and warm the aluminium until the soap turns black. However, it is essential that it is the aluminium which is warmed by the blow torch or hot air gun and not the soap. This is ensured by applying the heat to the non-soaped face. Any bending or forming should be carried out as soon as the annealed piece has cooled, because most grades of aluminium alloys start to re-harden and may actually improve in strength after such heating. Do not attempt to speed up the cooling process by quenching in water. Allow the part to air cool. Note that during the heating of the component there must be no visible change in the surface appearance of the aluminium, otherwise the aluminium will lose its properties and disintegrate. This fact can be verified, if you are so minded, by attempting to bring a small piece of aluminium sheet to red heat.

Heat treatment of aluminium and its alloys is a specialised process used to improve a particular characteristic of the metal, and is carried out in controlled conditions not usually available in the home workshop. Technically the above techniques are more of a part softening process rather than true annealing.

For interest, the usual method of commercially annealing most aluminium alloys is that the part is 'soaked' for about two hours at 420°C, air cooled, and then soaked for a further two to four hours at 225°C. All this is performed in a furnace or salt bath which has its temperature closely controlled. Most, if not all, of the aluminium used in extruded tube or sections has undergone this type of treatment immediately prior to extrusion, ie in the billet form.

Brass

Typical uses: Morse keys, terminal posts, weatherproof boxes, extension spindles, waveguides and other microwave components.

This is an expensive but good non-magnetic electrical conductor, and is available in sheet, rod, tube and other forms. It can be soldered easily but adhesive bonding can prove difficult. For work involving bending or forming, the most suitable grade is ductile brass. For panels and non-formed parts, the half-hard and engraving brasses are adequate.

It can be annealed like copper, though care is necessary as brass is nearing its melting point when heated to bright red. Brass is non-corroding in normal use but reacts with aluminium and zinc.

Copper

Typical uses: heatsinks, coils, antenna wires, tuned lines, waveguides, earthing stakes and straps.

This is a very good, though expensive, electrical conductor. It is non-magnetic and is available in sheet, rod, tube and other forms. Before work of a forming or bending nature is attempted, this metal should be annealed by heating as uniformly as possible to a bright red heat and air or water cooled. If considerable bending is required, this annealing should be repeated as soon as the metal begins to resist the bending action. In the annealed state, copper bends very easily. Soft or hard soldering present no problems but adhesive bonding can be troublesome. It is non-corroding in normal use but does react with aluminium and zinc, especially in an out-of-doors environment. Cuts on the hands etc caused by copper should be cleansed and treated immediately they occur, for such cuts can turn septic very quickly.

Steels

Typical uses: Masts and tabernacles, screw fixings, guy wire stakes and parts subject to high wear or heavy loads.

These are electrical conductors, which are magnetic except for the expensive stainless types. They are cheap and available in numerous forms and qualities. The common grades are called mild steel or GCQ (good commercial quality) steel. The black quality steels are usually cheaper and used for such things as the stakes for mast guys, or similar 'rough' work. Silver steel is a special grade suitable for making tools, pivot pins and other items which require the parts to be tough or hardened and tempered. Most steel sheet forms commonly available will bend, solder and machine easily, and can be annealed by heating to bright red and allowing to cool slowly in air. Do not quench in water or in any way cool rapidly, for this may cause some steels to harden. Corrosion is a problem unless plated or well painted. For outdoor use the commercial process of galvanising is perhaps the best form of protection. The next best thing is a few coats of paint.

Tin Plate

Typical uses: Boxes, screening cans and plates, light-duty brackets, retaining clips, spacers.

This is a good electrical conductor. It is very thin steel coated with an even thinner layer of tin on each side, and is magnetic, cheap and available in sheet form up to 0.5mm thick. It can be easily soldered, bonded, bent and machined, and is non-corroding in normal use. Cut edges should be re-tinned with solder if the full benefits of the non-corroding properties are required. It can be annealed but this will destroy the tin coating. This is a 'friendly' metal to use and is normally readily available in the form of biscuit tins and similar containers, hence the old timers' expression 'an OXO tin special'.

General Comments on Metals

All of the above metals work-harden and will break if repeatedly flexed at the same point. Annealing removes the effects of work-hardening, providing it is carried out before the part is over-stressed. There are professional standards which classify the above metals and each is given a specific identifier code. Fortunately, amateur constructors do not normally need to enter this maze of professional standards, and metal suppliers usually understand that to most of us steel is steel and brass is brass!

PLASTICS

Plastics are electrical and thermal insulators, and are not suitable as RF screens unless the plastic is specially metal coated or impregnated. The insulation can normally be considered as excellent for most amateur purposes. The following is a brief description of the more commonly available and useful plastics, and is a very small selection of the many plastics in use today.

Laminates

Typical uses: PCBs, matrix boards, coil formers, insulating spacers.

Various base fabrics such as paper, cotton, glass, asbestos etc, are bonded together by selected resins, and usually compacted and cured under pressure. The combination of the resin and the base fabrics produce laminates which may be used for many applications. Most are available in sheet, rod and tube form. The cured laminate cannot be formed easily. Normal machining is possible, particularly if attention is paid to the lay of the base material. Drilled or tapped holes should be arranged so that they go through at right-angles to, and not in the same plane as, the laminations.

Where components made from this material are exposed to the elements and expected to insulate, the glass-based laminates should be preferred. This also applies where the dielectric properties are important (VHF converters, RF amplifiers etc). The normal heat generated by valves and similar electronic components will not harm these laminates but the glass or asbestos fabric-based laminates should be used for higher temperatures (100-140°C). Costs range from expensive for the paper and cotton bases to very expensive for the nylon and glass bases. Glassfibre repair kits for cars are a useful laminate for weatherproofing antenna loading coils and making special covers or insulators. The filler putty supplied with these kits may contain metal.

ABS (acrylonitrile butadiene styrene)

Typical uses: antenna insulators, coil formers, handles, equipment enclosures.

This is expensive and is available in natural white coloured sheet. It machines easily and can be formed by heating, similar to Perspex®. Bonding requires proprietary adhesives. A tough plastic and a good insulator.

Acetal Copolymer

Typical uses: bearing bushes for rotating beams, Morse key paddles or electrical parts such as insulators or feeder spacers.

This is a medium- to low-cost plastic and is available in white or black rod and sheet. It machines very easily without specialist tools and is a useful plastic to have available in the workshop. It cannot be formed easily. Bonding requires proprietary adhesives and such joints are usually the weakest part of any assembly.

Acrylics (Perspex®)

Typical uses: decorative and protective panels, dials, Morse key paddles, insulated fabricated boxes and covers.

A medium-cost plastic which is available in clear or coloured sheet, rod or tube. It is non-flexible and can shatter or crack under shock or excessive loads, although it is often used for see-through machine guards. The clear sheet is ideal for covering and protecting the front panels of equipment.

Perspex® may be formed by heating but not with a flame as this plastic is combustible and gives off unpleasant fumes if burnt. If placed in a pan of water and simmered (or in an oven)

at around 95°C, the plastic softens and can be formed or bent very easily. Forming should be stopped and the work re-heated the moment hardening or resistance to bending is felt, otherwise breakage will occur.

Bonding requires proprietary adhesives. A properly made bond is structurally sound and can be transparent. Normal drilling, sawing and filing are straightforward, providing the work is adequately supported. Most sheet forms are supplied covered with protective paper and it should be worked with this left in place. It is not very heat resistant and should not be placed in direct contact with any heat source such as lamps, heatsinks and valves. Where transparency is not required a better plastic to use would be Acetal or, for particularly tough applications, ABS.

Nylon 66

This is a cheap- to medium-cost plastic, and is available in sheet, rod or tube. It is usually supplied in its natural creamy white colour. It can be machined but it does tend to spring away from any cutting edge, making tapping or threading difficult. It is not easily formed. Proprietary adhesives are available which claim to bond nylon successfully. There are other types of nylon but most of these are expensive and intended for special applications, such as bearings, gears etc. For most amateur purposes the Acetal copolymer, mentioned previously, is generally an easier material to use.

Polyethylene (Ultra High Molecular Weight)

A medium- to high-cost plastic, normally available in sheet and rod. It is usually supplied in its natural white colour and it can be machined easily but not formed. Proprietary adhesives are available for bonding. This is an ideal plastic for outdoor components such as insulators, feeder separators etc as it is virtually rot-proof.

Polycarbonate

This is an expensive material, normally available in transparent sheet form. It can usually be bought at builders' merchants where it is sold as vandal-proof glazing. A very tough plastic, virtually unbreakable (bullet proof!) and, though it can be machined, it will wear out normal tools very quickly. It is ideal for making an insulated base for a vertical mast/antenna and in other areas where impact, high loads and temperature changes would rule out other less-durable plastics.

Polypropylene

This is an expensive plastic, normally available in opaque-coloured rod form. It can be machined and formed but not bonded. It can be sensitive to prolonged frictional contact with metals, particularly copper, and disintegration can occur in these circumstances. Because of its strength and resistance to atmospheric attack, it is usually used by amateurs in its rope form for halyards, mast guys etc.

The twisted strands are normally melt-welded together to prevent fraying, using a soldering iron. A naked flame should not be used as a substitute for the soldering iron as this plastic burns and melts, and burning droplets can go anywhere, even on the hands!

Polystyrene

This is a relative cheap plastic and is available in a variety of types, shapes and colours from black to transparent. It can be formed, machined, painted and bonded very easily - a model maker's delight! Usually used by radio amateurs for coil formers, insulated extender spindles and in other areas requiring insulation. It is a particularly tough flexible material, although some transparent types can be brittle. Some forms are also heat sensitive. A paper/card laminate of this plastic is available and this is very useful for making mock-ups of cabinets, boxes etc. Model-making suppliers usually stock extruded polystyrene sections, some shapes of which can be utilised in making bezels and other cabinet embellishments.

PTFE

This is an expensive material, noted for its excellent dielectric performance and low frictional properties. It is available in sheet, tube and rod, and is normally supplied in its natural off-white colour. Extremely difficult to machine and cannot normally be formed or bonded. Usually used for low-friction bearings, insulators (up to UHF), capacitor dielectric and the nozzles of de-soldering guns. The fumes from overheated PTFE are very toxic.

PVC

This cheap material is available in many forms including rod, tube and sheet. It is usually grey or black in colour, and can be easily machined, formed and bonded. Proprietary adhesives should be used (although hot-air welding with a filler rod is also possible) but skill is required to produce structurally sound joints. Certain of the building types of PVC encountered seem to have some conducting capabilities which can lead to problems if used in electrical or RF applications. A suggested test for this is to try to cook a small sample in a microwave oven (alongside a cup half full with water), and if no metal is present the PVC should stay cool to just warm.

PVC insulating tapes are strong, cheap and normally self-adhesive, and are supplied in a variety of colours and widths intended for wrapped insulation. Some of the poorer-quality tapes do not weather very well and suffer adhesion failure with the first frost.

ADHESIVES

Many modern adhesives are hazardous and it is essential to follow meticulously the manufacturer's instructions when mixing, using and curing them. Most are insulators and unsuitable for electrical joints.

Five general rules should be applied for bonding:

- Degrease the parts thoroughly; even finger marks impair results.
- Roughen the joint faces unless a transparent joint is required.
- Do not place bonded joints under a peeling type of load.
- Ensure that the work is dry and warm.
- Wear protective glasses and gloves and have the necessary first aid chemicals readily to hand.

Epoxides

A group of medium to expensively priced, cold- or heat-setting resins (usually self-generated heat) that can be used for bonding, surface coating, laminating or encapsulation. Air and gas bubbles are the biggest problem with encapsulation (this work is carried out professionally under vacuum). The problem can be minimised by warming the work and the resin to around 40°C and providing a generous shrinkage allowance with a large pouring area which can be cut off from the cured encapsulation. Careful thought should be given to the necessity of encapsulation, for once completed, the encapsulated module cannot be

altered or repaired. Encapsulation is usually used when circuits are subjected to harsh vibrational and environmental conditions. (See also 'silicone sealants').

These are usually two-part adhesives and require careful mixing just prior to use. A structural joint should not be over-clamped during bonding, and a bonding gap of typically around 0.05mm is required for the joint to be made properly. In other words, don't squeeze out all of the glue!

Surface coatings can be applied by dipping, spraying or brushing. Flexible resins are usually used for this type of work and are ideal for protecting beams, traps etc.

Cyanoacrylates ('Superglue')

These expensive adhesives are available in various grades, each intended for bonding a particular set of materials. The low- to medium-viscosity grades are suitable for most amateur work. They are scientific marvels of bonding and as such require correct and proper application to ensure success. Releasing or debonding agents are available and it is a wise precaution to keep some of this handy in case of accidental bonding of fingers etc. These adhesives should be used and cured in well-ventilated conditions. It is advisable to wear protective glasses when using this adhesive.

Toughened Acrylics

These are expensive, fast-curing adhesives intended for structural joints. Various types are available and are usually supplied in two parts - the 'glue' and the primer/activator. The glue is applied to one side and the primer to the other. They are suitable for use on most of the materials already mentioned, but some may not be used with certain plastics as they dissolve the material and eventually the joint fails. They are usually easier to use than the cyanoacrylates.

OTHER MATERIALS

Silicone Rubber Compounds

These are medium-priced materials, available as paste in squeeze tubes and as a liquid in tins. When cured, they normally set to give a white or translucent silicone rubber finish. They are ideal for encapsulation and the sealing or weatherproofing of antenna connecting boxes and similar out-of-door items. The electrical insulating characteristics are excellent and can be used to prevent parts from vibrating in equipment used for mobile or portable work. One type of this compound emits acetic acid during curing and this may damage some insulators and component connections.

Though not normally sold as such, one type of this compound has been used successfully as a resilient adhesive for structural and pressure-sealing joints on metal, plastic and glass.

Self-amalgamating Tapes

These are a form of insulating tape which, when stretched and overlap-wrapped around cables, coaxial plugs etc, will amalgamate or flow together as one. They are reasonably priced and available in widths up to 50mm either with or without a self-adhesive face. Excellent weather-resistant properties. The self-adhesive form of this tape is ideal for waterproofing antenna traps, joints and connectors.

Nickel Silver

Typical uses: Boxes, screens, nuts, bolts and washers etc where the non-corroding properties are required.

An expensive non-magnetic electrical conductor which is corrosion resistant. Available in sheet and rod. It is often used by railway modeller's for it is a very 'friendly' metal to use, soldering and bending easily. It has some resistivity.

Wood

Typical uses: Aerial masts, booms, spreaders, storage boxes, and other parts which do not require RFI protection.

A common material, not usually used today for radio work as it has no ability to conduct electricity or to screen against RFI. Its insulating properties are marginal unless dried and treated to prevent ingress of water/moisture. The radios of the 1920s and 1930s were cabinetmaker's delights but this was before the 'enlightened' days of electromagnetic compatibility. Other joinery-type materials such as chip board, MDF, plywood etc are equally unsuitable for RF applications but may be utilised to make mock-ups of proposed designs, especially for the panels and cabinets.

CORROSION

There are two main processes of metal corrosion. The first relies on environmental conditions such as rain or condensation which results in an acidic electrolytic liquid being formed on the surface of the corroding metal. As the metal corrodes, the acidity of the liquid increases until the electrolytic process of corrosion becomes almost self-sustaining. The second occurs due to the electrolytic action occurring between dissimilar metals in contact, and is referred to as galvanic action. Both processes change the metal into a different form, which in the case of steel or iron we know as 'rust', and often refer to the process as oxidisation.

Table 26.1 shows the galvanic relationship between metals. The numbers are item numbers only to show the position of each metal in the galvanic series. The actual values depend on several factors such as temperature, radiation and acidity etc. This list enables metals to be selected which will have the minimum galvanic corrosion effect on one another. The greater the list separation, the greater will be the possibility of corrosion.

For example, brass and copper are adjacent and would therefore not cause problems if in direct contact with each other. Brass and aluminium alloy are widely separated and corrosion occurs if these metals are in direct contact with each other. This state can be reduced by tinning the brass with lead-tin solder, which falls about halfway between the other two metals (galvanic interleaving). The higher item-numbered metal will normally promote the corrosion of the lower item-numbered one.

Dissimilar-metal galvanic action corrosion is avoided by ensuring that the galvanic series separation is minimal. If widely separated metals do need to contact each other, then a suitable interleaving material should be used to reduce the galvanic separation level.

Corrosion by moisture or rain is more difficult to combat effectively. The commonly accepted anti-corrosion treatment consists of protection by paint and, providing the paint coat remains

1.	Magnesium alloys	8.	Lead
2.	Zinc	9.	Tin
3.	Aluminium alloys	10.	Nickel
4.	Cadmium	11.	Brass
5.	Mild steel	12.	Copper
6.	Stainless steel 18/8	13.	Silver
7.	Lead-tin solders	14.	Gold

Note: the numbers are item numbers only and are not values!

Table 26.1: Galvanic series

intact or is renewed regularly, this is a very effective treatment. Steel and iron can also be coated by protective metals such as zinc (galvanising) or nickel (plating), both methods being normally outside the range of the home workshop. Aluminium is slightly different, for this metal forms its own protective oxide barrier which, providing it is not disturbed, will prevent further corrosion. It is this oxide barrier which makes soldering aluminium difficult. Unfortunately, this natural protective oxide layer can be disturbed by stress or galvanic action, and the corrosion process bites deeper into the metal. Anodising is a process on aluminium which forms a controlled layer of oxide on the surface and presents a toughened surface finish which can be coloured by dyeing. Brass and copper can be considered as corrosion resistant for most amateur purposes. However, their surfaces do oxidise and this can impair good electrical contact. Certain platings, such as silver, nickel or chromium, can reduce this. The platings themselves can also corrode but usually at a much slower rate than the parent metal.

If not adequately protected, corroding metal will gradually lose strength and the device from which it is made will fall apart. This is usually seen as collapsed masts, broken antennas and similar expensive disasters.

Corrosion also affects RF and electrical connectors, particularly feeder-to-antenna connections, and causes a gradual decline in the overall performance of the system. Signals become weaker and calls to DX stations which used to be answered are ignored. Most observed lowering of performance in this area is usually (or eventually), discovered to be due to corrosion. This can be practically eliminated by first ensuring that no dissimilar metals are in contact to cause galvanic corrosion, and second that water is excluded from all connections. Copper coaxial-cable inners should never be connected directly to the aluminium elements of a beam but should be tinned first as previously stated. Coaxial plugs and sockets should be fitted with heat-shrink sleeving and wrapped with amalgamating tape to prevent the ingress of water. Connector boxes can be filled with silicone-rubber compound for a like purpose. Stainless steel antenna fittings are the least affected by corrosion but even these would benefit from a coating of protective lacquer, particularly on screwed fittings. The position of stainless steel in the above list is interesting, for sometimes the metal in contact with the stainless steel will corrode due to galvanic action, particularly in wet, smoggy or salt-laden conditions, and some stainless steel fittings are supplied with plastic interleaving spacers and washers to prevent this.

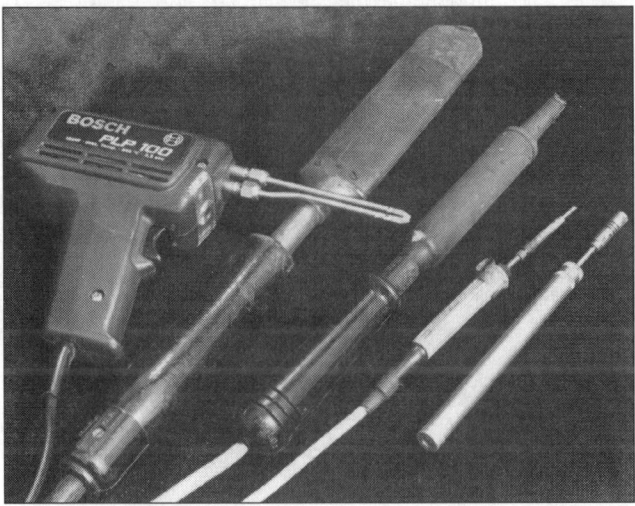

Fig 26.1: Soldering iron

Condensation is also a problem with outdoor enclosed or boxed-in items such as rotators, remote tuners, antenna traps, coaxial cable and the like - where there is condensation, there will be corrosion. The simplest, but not necessarily the easiest, solution is to allow the enclosed area to 'breathe' by introducing suitable weather-resistant holes as drainage vents, while ensuring that these are at the lowest point of the enclosed area and cannot be inlets for rain or the run-off water from the outside of the enclosure. Complete sealing usually makes matters worse, for a fully water and air-tight enclosure still produces internal moisture by condensation. Full hermetic sealing is difficult to apply for it normally requires the ability to pressurise the enclosure with an inert gas, as with some military or maritime equipment. If the items within the enclosure can be protected as if they were exposed to the elements, much of the corrosive effect of condensation is reduced.

In the case of rotators, attempts at filling the voids with grease does not help, for the grease forms small pockets which can hold water. It is better to lacquer or wax oil the moving parts and the inside of the housing, and to use grease for its intended purpose of lubrication. Any electrical items within the rotator should be sprayed with a commercial, non-insulating, waterproofing liquid.

The threads, screws, nuts, bolts etc should be given a light coating of anti-seize compound on assembly and, after assembly, sprayed over with lacquer or wax oil, or coated with a water-repellent grease such as lanolin. This makes for easier maintenance and reduces the possibilities of corrosion.

Corrosion is not limited to outdoor items. Corrosion of connector pins of microphones, plug-in PCBs, computers etc is not uncommon. Careful selection of mating materials to avoid galvanic action, combined with appropriate painting, plating etc and regular maintenance, will reduce the effects of corrosion.

TOOLS

Any tool bought with reliable use in mind should be the best you can afford, for cheap tools usually lead to frustration, like a cheap pair of wire-cutters that has its cutting edges notched by copper wire the first time used, and from then on will not cut but only fold the wire! Retiring toolmakers and fitters often sell their 'kit', and some very good but used tools can be obtained this way at reasonable prices. Tools also make very good presents!

Most amateurs have a shack, room or some place with a bench, hacksaw and a vice of one type or another. Accepting these, the range and type of useful tools available is virtually limitless, unlike most pockets. Basic tools and a few extras - some home-made - are listed here.

- Soldering irons (**Fig 26.1**). 15W instrument, 50W electrician's (and a useful extra, a 200W heavy-duty iron) all mains powered. The choice of type of iron in the above selection is a personal one, bearing in mind the cost. Temperature-controlled irons are excellent but expensive. The main factors to consider are the availability of the replacement parts (bits, heating elements, handles) and the ease with which these may be fitted. The anti-static properties of an iron should also be considered. Battery and gas-powered irons are worth considering if on-site or outdoor work is likely. Similarly, the rapid heat soldering gun is a worthwhile addition.
- De-soldering 'gun' (**Fig 26.2 bottom**). The miniature anti-static, hand-operated type is an effective and value-for-money device .
- Electrician's pliers (**Fig 26.2 top centre**). Also known as combination pliers (180mm).

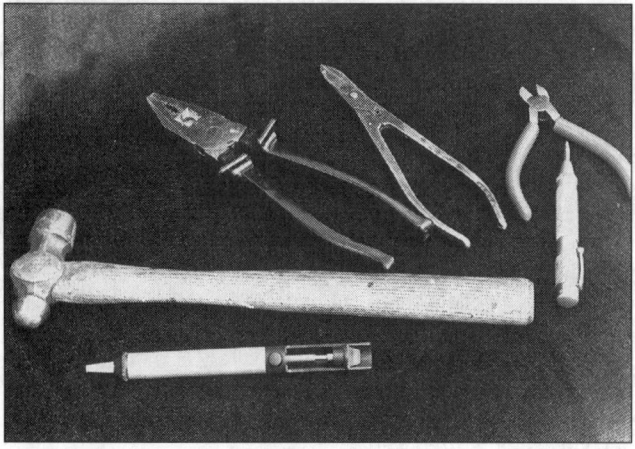

Fig 26.2: A selection of hand tools

- Side cutters (**Fig 26.2 top right**). Also referred to as diagonal cutters (120-180mm). There is a variety of wire-cutters and the type specified here is suggested as the most suitable for general use. They will cut most wires up to 2mm diameter (barbed or fencing wire excepted!)
- Watch-maker's shears (**Fig 26.2 top, half right**).
- Long-nosed pliers (also known as Snipe nose) (120-180mm).
- 8oz ball pein hammer (**Fig 26.2**).
- 8oz soft-faced hammer with replaceable heads.
- Twist drills. These should be of high-speed steel in at least 1.0-12mm diameter (fractional sizes from one-sixteenth to half inch diameter). Carbon-steel drills are cheaper but require sharpening much more often.
- A centre drill (BS1 size) is a worthwhile addition (**Fig 26.3**). Unfortunately, it is very easy to break the tips of these centre drills and it is worth considering keeping a few 'in stock'!

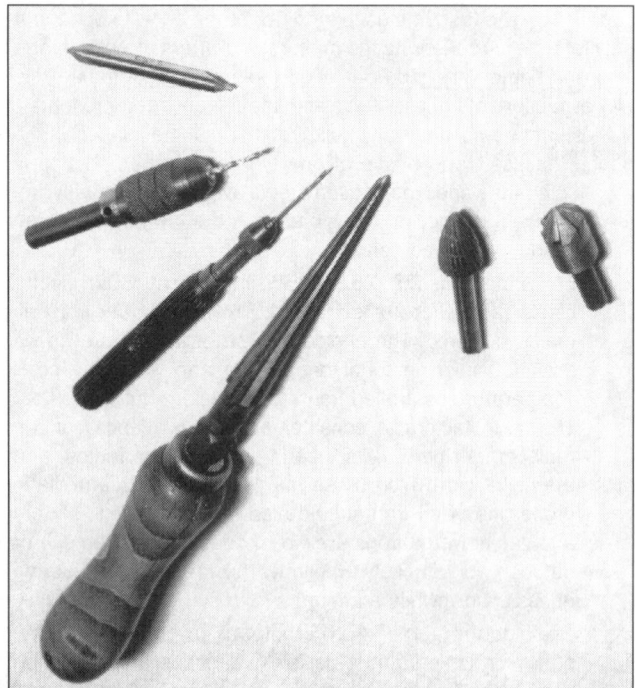

Fig 26.3: Anti-clockwise from the top left: centre drill, pin chuck, pin vice, taper reamer, file burr, countersink

- Drills need regular sharpening and some drill-sharpening device should be considered. There are many types available and the main points to consider when buying are that:
  - (a) it will sharpen the range of drills in use;
  - (b) spare grinding stones can be purchased;
  - (c) little or no skill is required to obtain the correct drill point; and
  - (d) the speed is not so high that it will soften instead of grind the drills.
- Electric drill. An electric drill should preferably be of the continuously variable speed type and should have a drill chuck capacity to at least match the range of drills on-hand. A drill stand is a valuable addition. The model-makers' variable speed, hand-held or stand-mounted drilling and grinding unit is a very useful extra, especially for the finer work of PCB making.
- Pin chuck (Fig 26.3). This device enables small-diameter drills (1mm downwards) to be held in the normal drill chuck. Most pin chucks come with interchangeable collets to cover a range of small-diameter drills. This pin chuck should not be confused with the pin vice which is intended to hold small components during filing or fitting, and not drills.
- Screwdrivers (**Fig 26.4**). Minimum requirement: parallel flat blade 3.2 x 100mm, 5 x 150mm and similar sizes for the cross-head types of screw (Phillips™, Pozidriv™ etc). A set of watch or instrument makers' screwdrivers are very useful.
- The interchangeable-bit screwdriver is also worth considering but it can become annoying to use - the right bit never seems to be in place when it's wanted!
- Spanners. Box and open-ended types in the BA and ISO metric sizes, plus a small adjustable spanner as the minimum starter requirement. Many older components require imperial/BA spanners but the newer European parts are usually ISO metric. Pliers are a poor substitute for spanners, leading to mangled nuts and scratched panels!
- Hexagon socket keys. BA and metric sizes up to 8mm. These keys are available mounted in screwdriver handles. The interchangeable-bit screwdriver mentioned previously often has hexagon key bits.
- Files with handles (**Fig 26.5**). 150-200mm second cut. Hand, half-round, round and three-square, also the same shapes in round-handled needle files. In the interests of

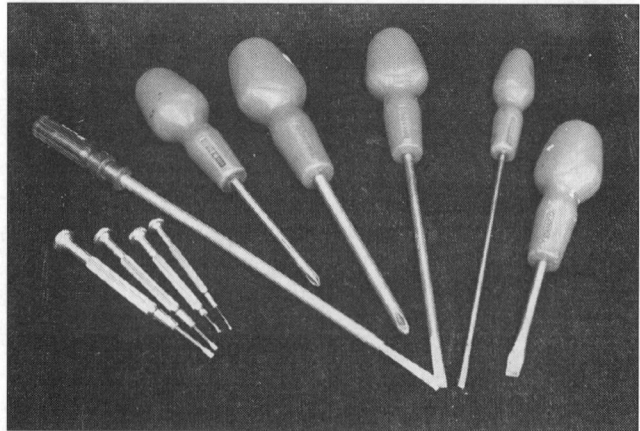

Fig 26.4: A selection of screwdrivers, including screw-holding type

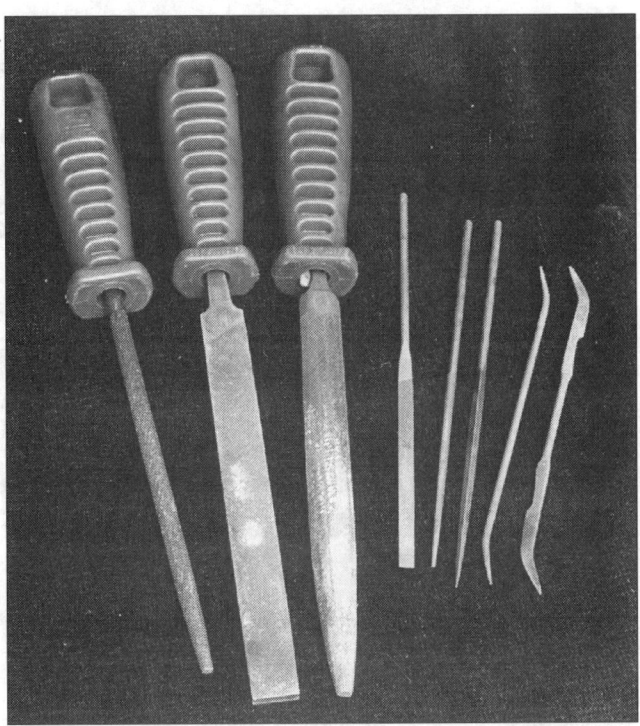

Fig 26.5: A selection of files. The two on the right are Reifler files which are useful for filing awkward shapes

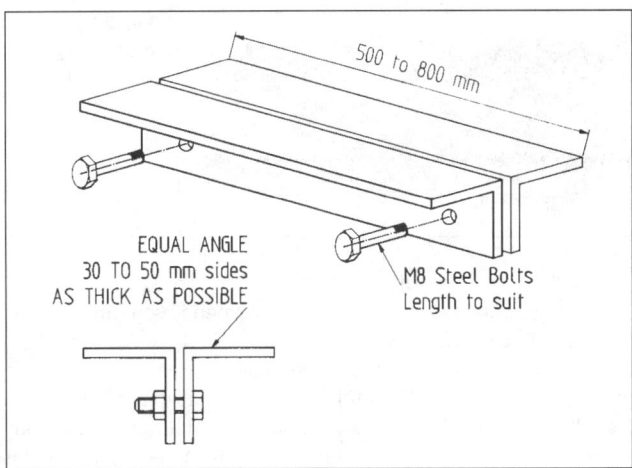

Fig 26.6: Bending bars

personal safety, files should always be fitted and used with handles, for file tangs produce nasty cuts. There are sets of files available in the above selection, which are supplied complete with fitted plastic handles. The wood-worker's shaper plane/file (Surform™) is very useful for the rapid trimming of aluminium sheet. Files should be stored where they cannot contact and damage other tools.

- Tapered reamer (see Fig 26.3). Sometimes known as repairman's reamer. The 3-12mm size with handle will suit most requirements.
- Hand countersink (Fig 26.3). The 12mm size with five or more cutting edges will cover most work and, mounted in a plastic handle, it is very useful for deburring holes.

The above list of tools can be considered as a starter kit for the newcomer to construction. With these few tools it is possible to attempt many of the jobs encountered in making or repairing something but, as construction skills grow, so will the tool chest! A tap and die set is extremely useful and should be placed high on the 'tools required next' list.

Useful Extras

One of the problems most of us have in metalwork and the like is cutting or filing things square and straight. Also, marking out to ensure that parts will fit together correctly is sometimes not as easy as it appears. The few extras described here should sim-plify things.

Bending bars

See **Fig 26.6**. These are normally a home-made item, and the things to check when obtaining the steel angle are its straightness and squareness. If these are not 'true' the corrective actions needed will require skills which may be beyond those so far

acquired. An old bed-frame angle, provided it has not rusted too badly, usually makes very good bending bars. One bed frame can be used to make several bending bars of different lengths. The choice of length and distance between the clamping bolts of the bars is related to the maximum width of metal or other sheet purchased and the maximum size of panel worked. Aluminium sheets, for example, usually come in 1.8 x 1.2m size and trying to cut this is rather difficult. It would be better, if buying in these sort of quantities, to have the sheets guillotined by the supplier to widths which will fit between the bending bar clamp bolts. (Suggest 482mm.) The length of the sheet will not matter.

Hole embossing tool

See **Fig 26.7 and 26.8**. This very simple home-made tool forms holes along the edges of metal panels to accept self-tapping screws of about 2.5mm diameter (depending on metal thick-ness), and with an engagement length of about 12mm. It facili-

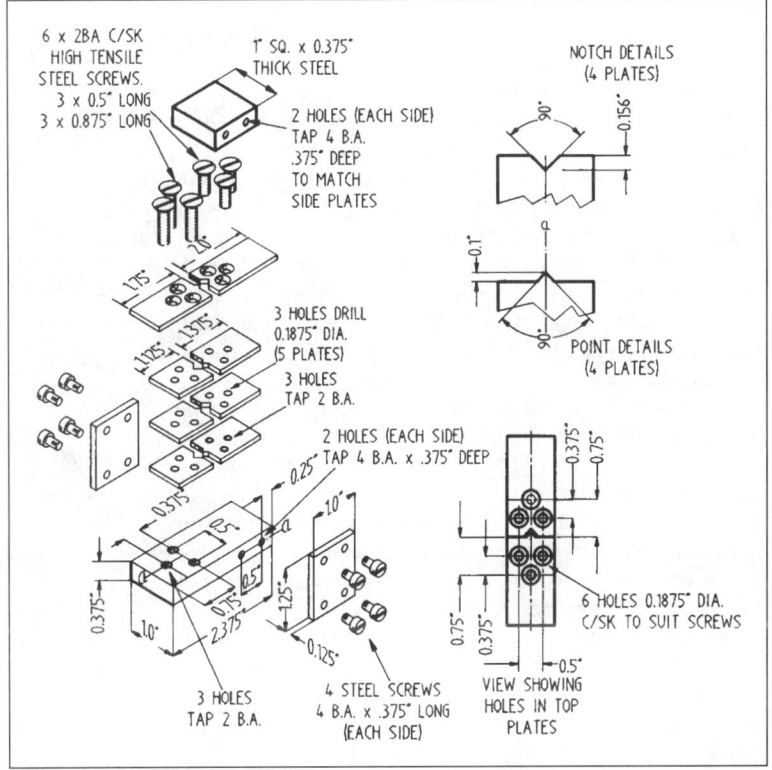

Fig 26.7: Hole embossing tool

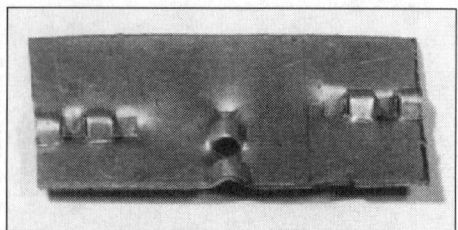

Fig 26.8: Holes made by the embossing tool

Fig 26.9: Toolmaker's clamp and joiner's G-cramp

tates the rapid manufacture of cabinets, boxes and screens, eliminating much of the accurate metal bending normally associated with this type of work.

The working parts and the two side plates of the tool are made from 1in wide by 0.125in thick-ground gauge plate, which is a tough but workable steel capable of being hardened and tempered. It is not necessary to harden the finished tool, unless it is intended to form mild steel but, if facilities are available to do this, it will improve the durability. The base and top plates are made from 1in wide by 0.375in thick mild steel.

The vee points must be on the centre-line of the tool. The vee notches are not so critical for they play no part in the shape of the hole produced, but they should allow sufficient clearance for the formed metal to flow into. The point and notch sizes shown are suitable for most metal thicknesses up to 1.6mm (16SWG). The vee points can be made as a set, with the four plates clamped together. The fully shaped points should be polished with an oilstone to obtain square, burr-free, sharp edges along each face of the point. The edges of each point are the working parts of the tool and should be made with care. It is essential that the flat faces on both sides of the notch and point ends of each set of plates are square, parallel and level with each other. Tapping-size holes should be marked out and drilled on each top plate only, and these used as a template to drill the remaining holes on each plate. The holes in each bottom plate should be tapped 2BA and the holes in the remaining plates opened up to 0.1875in diameter - a good fit for the high tensile steel 2BA screws. ISO M5 may be used in place of the 2BA screws and tapped holes. The outside faces of the holes in the top plates should be countersunk deep enough to ensure that the screw heads are slightly under-flush. It is good practice to lightly countersink both sides of every hole to remove burrs and facilitate assembly.

The back edges should also be square and level to ensure that the vice pressure is applied evenly during forming. The 0.125in thick side plates and the 0.375in thick top and bottom plates form the bearing for the sliding part of the tool, which must slide easily with the minimum of play in any direction. Failure to get this right will ruin the action of the tool, for each point must just slide over the opposing one or two points with the minimum of clearance.

To operate, the tool is placed between the jaws of a 4in vice and supported by the extended top plates. The metal to be holed is rested on the base of the tool and the marked out hole position is aligned with the tip of the visible point. The vice is then closed with the minimum of force until each face of the working ends of the tool just contacts the metal sheet. It is pointless trying to go beyond this, for squeezing the tool and the work by excessive pressure from the vice will only ruin the tool and the work. The vice is then opened and the metal sheet gently prised away from the points. This method of releasing the formed sheet could be improved upon, but the tool would become more complex to make and, as a spot of oil on the working faces eases the problem, this extra complexity is not worthwhile.

The tool performs very well on aluminium, tin plate and annealed brass or copper in thicknesses up to 0.0625in

(16SWG). It is not recommended for use on sheet steel unless the points have been hardened and tempered.

Tool-maker's clamps

See **Fig 26.9**. These are available in several sizes. Two 100mm clamps are suitable for most amateur purposes. These clamps should not be confused with the joiner's C- or G-cramps, which are also useful and certainly better than no clamping device at all.

Measuring and marking-out tools

After the soldering iron, these will probably be the most used tools and buying the best possible will pay in the long term.

- Engineers' combination square (**Fig 26.10**). The 300mm/12in ruler with square head only is preferred from the usefulness-to-cost viewpoint. Most of these combination squares have a small scriber, housed in a hole on the square head, and this saves buying a scriber separately. A complete quality combination square is expensive and, as well as the square head, it has a centre square and a clinometer head, neither of which is essential for normal amateur work. Beware of the cheap bazaar-type combination squares, for they are a waste of money and lead to corners which are not square and inaccurate measurements.
- Spring dividers (100mm size) (Fig 26.10).
- Jenny callipers (100-200mm size) (Fig 26.10). These are also known as odd-legs.

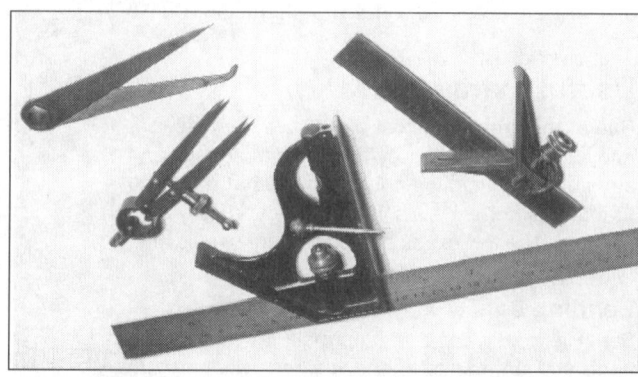

Fig 26.10: Anti-clockwise from the top left. Jenny callipers and spring dividers. 12in combination square with with square head fitted. 6in combination square with centre square fitted

Fig 26.11: Finishing and polishing tools

Fig 26.12: Dip oscillator using home-made box described in the text

- Centre punch (Fig 26.2). An automatic centre punch is preferred, but a simple and cost-effective punch can be made from a round 100mm long steel nail sharpened to a suitably tapered point.

Finishing Tools and Aids

Most construction work will require some form of finishing or pre-finishing such as deburring, emery dressing and polishing. The tools and other aids for this work are numerous, and a brief description of some of them is given here (**Fig 26.11**).

Abrasives

Emery paper or cloth is the most common for metal work. Grades are referred to as 'very coarse' (80 grit) to 'very fine' (800 grit). The 240 (medium) to 400 (fine) grades are suitable for most amateur uses.

Flap wheels for mounting in the electric drill come in various grit sizes and can save a great deal of work. The suggested grades are 60 grit to 120 grit, or finer.

Wet and dry paper is useful for the final smoothing before painting, or between coats of paint. Fine wire wool or plastic pan scouring pads are good substitutes for wet and dry paper.

Blocks of abrasive mounted in a rubber-like material (rather like erasers for paper) are also extremely useful and are available in fine to coarse grades. One of each grade is suggested. The super-fine grade block containing non-metallic abrasive is ideal for the cleaning and polishing of PCBs. (It cleans without removing too much copper.)

A glassfibre brush pencil is useful for cleaning the smaller areas prior to soldering. A brass wire brush insert is also available for this pencil.

An electric drill fitted with a hard felt disc or a calico polishing wheel used with polishing compound will save considerable effort when trying to polish a component to a bright and scratch-free finish. Polishing aluminium by this method is not recommended, for aluminium tends to 'pick-up' on the wheel and produce deep score marks on the panel being polished. The fine-grit flap wheel or the abrasive eraser are more suitable for aluminium, and the final polish can be made using a rotary bristle brush and thinned-down polishing compound.

Steel or brass wire brushes, either hand or machine types of various grades (coarse to fine), are very useful where paint, rust or other corrosion needs to be removed. These brushes can also be used (in skilled hands) to produce decorative effects on the surface of aluminium, brass, or stainless steel.

Warning

The dust and fumes created whilst polishing or grinding can be harmful and a face mask and protective glasses should be worn. Always ensure that the work place is well ventilated. Some polishing compounds can cause skin irritation and the use of barrier cream on the hands is advised. The wearing of protective gloves can save the hands getting filthy, but take care when working with powered tools to ensure that the gloves do not become entangled with the rotating parts. Even a DiY electric drill can break a finger!

USING THE TOOLS

Marking out and measuring

This is the important bit of construction. If not done carefully and accurately, failure is certain, hence the saying: "cut once, measure twice". All measurements and squareness checks should be made from one or two datum edges.

Consider marking out the box used for the dip oscillator shown in **Fig 26.12**. The first operation is to obtain two edges straight and at right-angles to each other. The method of achieving this should be clear from the illustrations (**Fig 26.13**). These two

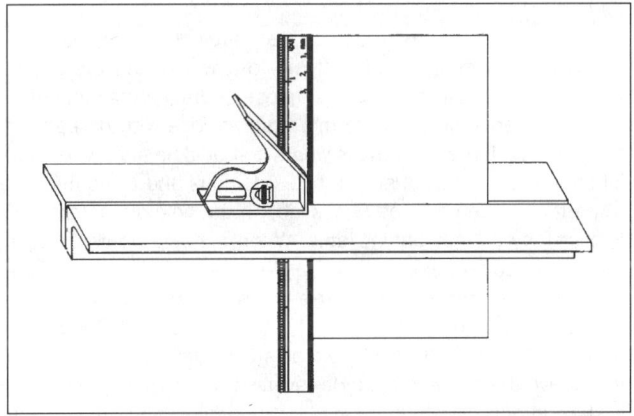

Fig 26.13: Squaring up the work prior to bending or cutting

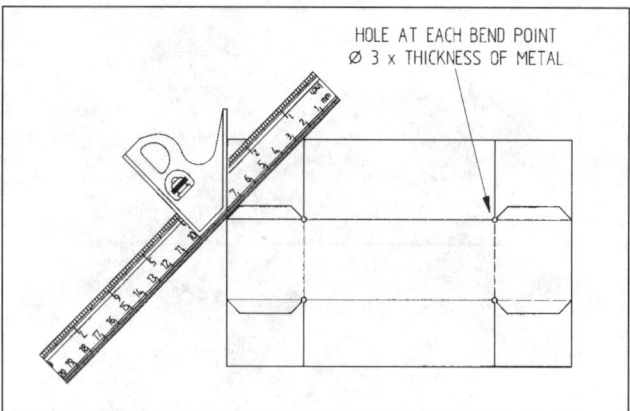

Fig 26.14: Marking out the box

edges are now the datum edges for vertical and horizontal measurements, and should be clearly marked. The overall size of sheet is marked out and cut to size. This results in a sheet with all corners square and all edges straight. With the bending bars as a guide it is really very simple! In this case the sides of the box are the same depth. Adjust the Jenny callipers to the required measurement and, locating from each edge in turn, scribe round the sheet. Check with the combination square from the datum edges that these lines are parallel and square with their respective edges. If Jenny callipers are not available, the combination square may be used by setting the ruler to the required measurement and scribing round the sheet with the scriber held against the end of the ruler. These lines show where the bending is to occur and are used to align the sheet in the bending bars (**Fig 26.14**). It is good practice to lightly centre-pop along the lines at about 20mm intervals to make 'sighting-up' in the bending bars easier - half the centre pop should be visible and the other half invisible inside the bending bars.

Existing boxes, panels etc, can be similarly marked out using two datum edges that are straight and at right-angles. If a truly square corner is not available, then the longest and straightest side should be used as a datum and the end of the combination ruler used to mark out the lines at right-angles. A suitably placed line at right-angles to this single datum can be used to mark off the measurements in this plane.

Lines marking the centres of holes and cut-outs should be scribed in, for the centre punch can then be located by feel into the notch formed by their respective intersections. It is worthwhile using the dividers to scribe-in each hole diameter around its centre-popped centre. This helps to ensure that the correct size hole is drilled and in checking that holes do not foul. If screw-up chassis punches are used, these scribed outlines can be used to locate the punch accurately.

When a gleaming or ready-painted cabinet, box or panel is purchased, marking out directly on to it would ruin the finish. Instead mark out in pencil on to a piece of draughting film cut to suit and then secure this to the required face with draughting tape. Check that everything is where it should be and, when satisfied, centre-pop through the hole positions and the outline of any cut-outs. Remove the film, scribe in the hole diameters and join up any cut-out outline dots. When scribing holes on to a ready painted surface, it is worthwhile gently scoring through the paint to reveal the metal, for this helps prevent the paint chipping off around the holes during drilling and reaming. Some constructors drill through masking tape stuck over the hole position to achieve the same results. However, some masking tapes really stick, especially when warmed by the drilling, and are very difficult to remove without lifting the paint.

Fixing holes for meters, sockets, plugs etc should be carefully measured from the component and these measurements transferred to the work. Where holes and cut-outs are related to each other, and not to some other edge of the chassis or panel, vertical and horizontal datum lines for these holes alone should be used. For example, a meter may require a hole for the meter body and four holes related to this for the fixing screws or studs. The centre lines of the meter body hole should be used as datum lines to mark out the fixing hole centres. The centres of the hole for the meter body will of course be related to the two main datum edges.

The centres of holes or studs of the same diameter can be found by simply measuring from the inside edge of one to the outside edge of the other.

It is tempting to use, and easier to see, markings made with a fine-point marker pen. Beware - the marking ink used is normally very difficult to remove, and if not removed, will usually bleed slowly through any subsequent paint work. A soft lead pencil, sharpened to chisel point, is also useful for marking out, particularly when 'trying things for size'. The marks can be wiped off easily using a dampened cloth or a very soft eraser.

Hole Making

Drilling is a straightforward operation providing a few key guidelines are observed. The drill speed should be adjusted to match the size of drill and the material being drilled. As a general rule, the smaller the drill, the higher the speed. A correctly sharpened drill should not require heavy pressure to cut. If it does, either the material is too hard or the drill needs to run slower. There is a tendency for normally sharpened drills not to cut some brass properly. This is due to the incorrect rake angle of the cutting edges of the drill. The problem can be solved by sharpening the helix edge on the face of the flute to give a rake angle of about 15 degrees, ie leaning backwards away from the normal cutting edge (**Fig 26.15**). Drilling some plastics can also present problems, such as chipping of the edges and breakaway of the material as the drill breaks through. Some improvement is possible by sharpening the drill to produce an included point angle of around 80 degrees. The standard included point angle is 118 degrees.

When drilling steel, it is advisable to use a coolant such as soluble cutting oil to keep the drill and the work cool. It also saves having to re-sharpen the drill so often. Paraffin is a good coolant for aluminium and copper. Brass and most plastics do not normally need a coolant, providing the drilling speed is correct, but treat as steel if necessary. These coolants may be applied either by an oilcan, or a brush. (Old liquid soap squeeze bottles make good coolant dispensers and any left-in detergent is not detrimental.) Soapy water is also a good coolant and certainly better than no coolant at all - it is best for most plastics.

Holes should always be centre-popped before drilling and, if the BS1 centre drill referred to previously is available, this should be used next to provide an accurate location for the drill. The holes drilled in thin sheet are often anything but round because the drill does not have enough depth of metal to round

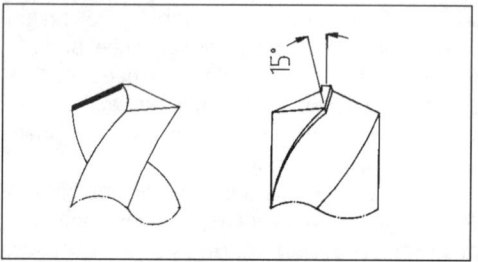

Fig 26.15: Modified drill rake angle to improve the drilling of brass

Fig 26.16: Large round hole making tools. Clockwise: deluxe home-made tank cutter, commercial tank cutter, screw-up punch, hole saw

the hole properly before it breaks through. There are at least two ways round this. The easiest is to drill the hole undersize and bring to size using the taper reamer, which must be allowed to cut without forcing, otherwise another fancy-shaped hole will be produced (another reason for scribing-in the holes beforehand.) Another way is to alter the cutting angle of the drill to an included angle of about 140 degrees and thin the chisel edge of the drill to a point. It is worthwhile keeping a set of drills sharpened in this manner especially for thin sheet drilling. These sheet metal drills normally require a much slower speed and the holes should also be pre-drilled using the centre drill.

Whatever is being drilled should be well supported and clamped to prevent rotation and lifting. Failure in this direction can lead at best to a broken drill and at worst to serious personal injury, for drills have a habit of picking-up just as they are about to break through the hole. (Plastic and copper are particularly susceptible to this.) A panel whirling round on the end of a 10mm diameter drill is a frightening sight!

All drilling will produce burrs around the hole edges and it is good engineering practice to remove these using the hand-held countersink bit referred to previously. A file will only scratch the rest of the surrounding surface and bend the burrs into the hole. The use of a large drill for hole de-burring is not recommended, particularly on the softer metals, unless the 'touch' for this method has been acquired.

It is usual to step drill holes larger than 10mm diameter, that is a hole of smaller diameter (3mm) is drilled first, then another slightly larger (+1mm), and so on until the finished size is reached. Step drilling is unsuitable for sheet material. There are special, though expensive, stepped drills available for drilling holes up to 40mm diameter in sheet material.

Making large round holes can be tackled in at least two ways. The first requires a washer or tank cutter (**Fig 26.16**) and an extra slow speed drill (a joiners' brace is effective). The biggest snag with this method is trying to obtain an even cut around the full circle. By clamping the work to a block of wood and drilling through into the wood for the centre pilot of the tank cutter, a guide is provided which improves things a little. The main thing with this method is not to be in a hurry. The second method can be applied equally well to non-round holes. Contiguous holes of about 5mm diameter are drilled on the waste side of the hole or cut-out and 1 to 2mm away from the finished size markings. The waste is removed, using tin snips. The hole is then carefully filed to size, using the bending bars for support, and as a guide for any straight portions.

To de-burr large holes, a small half-round needle file can be used in the 'draw' fashion. The file is held at both ends and drawn round the edges to be de-burred in a manner similar to

using a spoke shave. The file should be held at an angle to produce a small 45 degree chamfer around each edge of the hole.

Machine countersinking of holes requires a very slow speed drill (60RPM). They can be produced using the handle-mounted countersink bit referred to previously. Even the multi-toothed rose-countersink bit will chatter and leave a very unsightly surface if too high a speed is used. Countersunk screws will not sit properly on such holes. There are countersink file burrs available which do a similar job and can be used in the DIY drill, providing the work is held securely with the burr square to the work. They do have a tendency to skid over the surface, leaving a trail of deep scratches, if not located properly.

Drilling accurately positioned holes can be a problem and templates can be used with success. Keeping the small holes for multi-pin DIL ICs in-line and at the correct spacing is difficult, but the job is made easy by using a piece of the correct pitch matrix board as a drilling template. One of the holes (pin 1 is suggested) is drilled first in the required position. The template is located from this by passing a drill through the hole and the matrix board. The matrix board is then aligned with the rest of the marked-out hole positions and clamped using the toolmakers' clamps. The remaining holes are drilled through using the matrix board as a template. It helps to mark each row of holes on the matrix board template. If a considerable amount of such accurate drilling is to be done, it is worth making a set of metal templates for each size of IC using this method.

Sometimes the component itself can be used as a template to reduce the risk of error, eg slide switches, coaxial sockets, dial fixing holes etc.

Filing

Accurate filing is a skill which can only be acquired by practice. General rules are:

- Always use a file handle. This eliminates the risk of running a file tang into the wrist and enables the file to be guided properly.

- Use a sharp file. It is normal practice to use new files for brass and, as their sharpness wears off, use them for filing steel or aluminium. New files should be kept separately or otherwise identified.

- Do not force the file to cut. Only a light relaxed pressure is required, which also aids the accuracy of filing.

- Keep the file clean by brushing with a file card or by rubbing a piece of soft brass or copper along the teeth grooves.

- Support the work properly. Trying to file with the work held in one hand and the file in the other is a guarantee of failure.

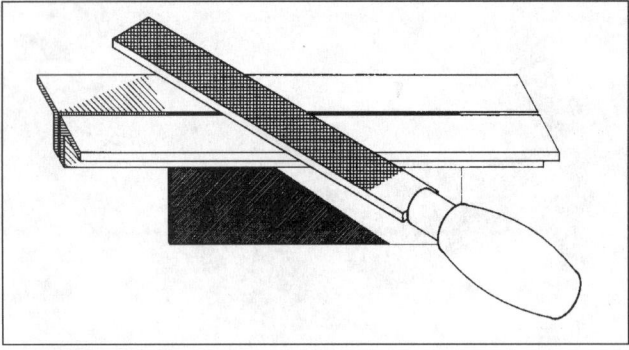

Fig 26.17: Bending bars as a file guide

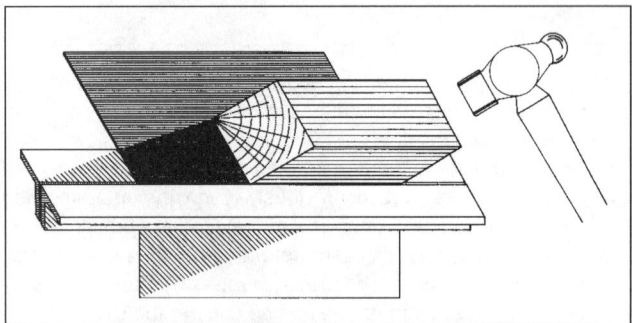

Fig 26.18: Method of bending in the bending bars. Note that the cutter block is held into the point of the bend

The height of the vice should be such that when the filer's arm is bent to place the fisted hand under the chin, the point of the elbow should just rest on the top of the vice jaws. An old but effective form of ergonomics!

A strip of emery cloth wrapped along the file can be used to obtain a better surface finish. This method was frowned upon in apprentice schools but it works and saves time. The draw filing technique referred to previously imparts the final touches to the filed edges and removes the filing burrs. The bending bars can be used as a filing guide (**Fig 26.17**) and a set of bars can be made especially for this purpose, checking them regularly to see that the guiding surfaces are not bowed by too much one-spot filing.

Bending and forming

The first essential is to ensure that the material can be bent. This sounds obvious but it is better to check first than to find out after all but the bending work has been done. Annealing can be applied as mentioned previously under the section on materials. A metal hammer should never be used directly on the metal when bending or forming. Either use a soft-faced hammer or a block of wood or plastic as a buffer for the hammer blows (**Fig 26.18**). This prevents all the humps and hollows which would otherwise occur. Do not use any metal as a buffer block as there is a danger of the metal chipping and flying into the face or eyes. The block should be kept into and near the point of bending. It seems easier and quicker to try to use the sheet as a lever and hammer as far away from the point of bending as possible. This will only produce a bend which curves up and back again. Where three sides are to be formed by bending, the point of intersection of the marked-out bend lines should be drilled before bending, with a hole diameter three times the thickness of the metal. This prevents corner bulge (Fig 26.14).

Fig 26.19 shows how to ensure that the end lugs of the box shown in Fig 26.12 fit snugly inside the box. **Fig 26.20** shows one method of bending the ends of a similar box.

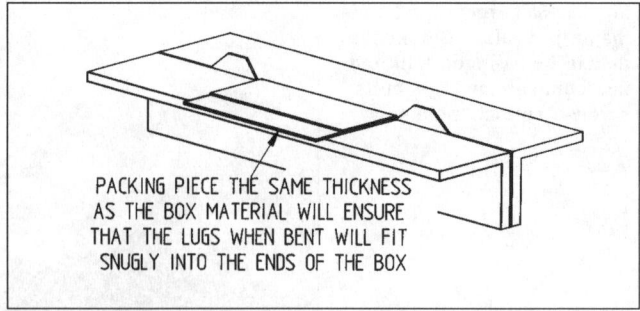

PACKING PIECE THE SAME THICKNESS AS THE BOX MATERIAL WILL ENSURE THAT THE LUGS WHEN BENT WILL FIT SNUGLY INTO THE ENDS OF THE BOX

Fig 17.19: Making allowance for bending the end lugs of the box

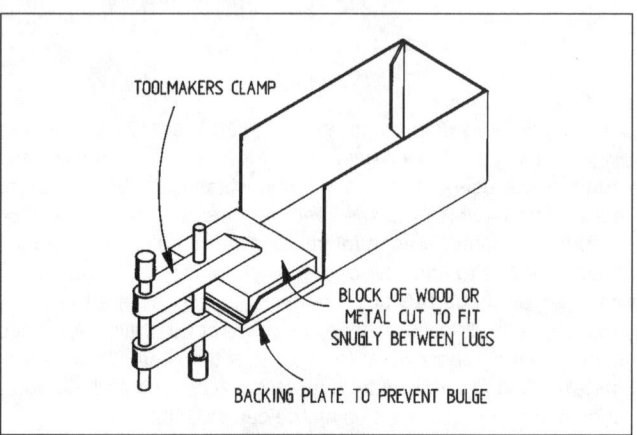

TOOLMAKERS CLAMP

BLOCK OF WOOD OR METAL CUT TO FIT SNUGLY BETWEEN LUGS

BACKING PLATE TO PREVENT BULGE

Fig 26.20: Bending the ends of the box using a toolmaker's clamp to secure the forming block

Tube bending for beams, tuned lines etc is not difficult and flattening or kinking can be reduced by observing the following:

- Ensure that the tube is suitable for bending. Anneal as necessary and re-anneal as soon as resistance to bending is felt.

- Unless skilled or equipped with specialised tooling, do not attempt to bend to a radius of less than three times the outside diameter of the tube, eg 12mm OD tube should have a 36mm minimum bend radius.

- Always bend round a former shaped to the required radius (**Fig 26.21**).

- Pack the tube tightly with fine sand (clean birdcage or builders' sand is ideal). Wet the sand and cork both ends. This will minimise the risk of kinking during bending, and the sand can be washed out afterwards. The tube can be re-annealed if necessary with the sand left in place but the corks should be removed to let the hot gases escape.

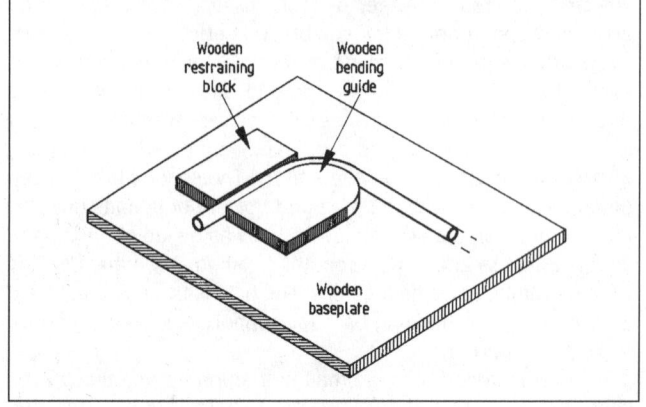

Wooden restraining block

Wooden bending guide

Wooden baseplate

Fig 26.21: Tube or rod bending set-up

A low-melting-point, lead-like material is available which can be used in place of sand, and is known as bending metal. It is poured molten into the blocked, ended tube, allowed to set and, when the bending has been completed, is melted out to be used again. This process should be carried out with the tube at an angle to allow the hot air to escape. The molten bending metal should be poured very slowly for it has a tendency to 'blow-back'. Normal lead is not a substitute for bending metal and should not be used, particularly with aluminium tubes, for the melting temperature required for lead may destroy the aluminium's properties.

Modelling wax (also known as American wax) can be used in place of sand or the bending metal. The wax is heated, not boiled, until molten and then poured into the tube in the same manner as the bending metal. The wax should be allowed to set hard before bending. Unfortunately it is not possible to re-anneal with this wax left in place.

Another tube-bending tool is the bending spring. This is often used by plumbers and pipe fitters. Each tube size should have a matching bending spring, for it is all too easy to use the wrongly matched spring and have it permanently trapped inside the bent tube!

Cutting

Cutting long strips of metal with tin snips or shears is an expert's job. The cut edges usually produced by non-experts are anything but straight and they require flattening to remove the cutting curl. Tin snips are best used where a one-snip cut will remove the required amount of metal, such as 45-degree corners, or the trimming to length of narrow strips. Snips should not be used on plastics, laminates or copper-clad board (PCBs), for the cutting action can cause de-lamination or shattering of the edges.

A guillotine is by far the quickest way of cutting most sheet material but it is not a normal home-workshop tool. For the home workshop the cutting of nearly all materials is best done by sawing, using a hacksaw and the bending bars as a guide and support (**Fig 26.22**). The hacksaw should be fitted with a 24 to 32 teeth per inch blade for most sheet work, and coarser blades for cutting blocks of material. The blades known as ding-dong style, ie with their cutting teeth arranged along a wavy edge, are very good for most work. When sawing copper-clad laminates such as PCBs, it is advisable to score through the copper beforehand using a sharp chisel. This will bevel the edge of the copper and prevent delamination due to sawing. The coolants recommended for drilling can be used if required. A fine-toothed roofers' saw is very useful to cut sheets which are too big to pass inside the hacksaw frame. This is similar to a joiners' saw but shorter with a stiffer fine-toothed blade. A pad saw or fine-toothed machine hacksaw blade which is handle mounted is equally effective in these circumstances. Some constructors use a fretsaw with either a metal-cutting blade or an Abrafiler to overcome the problem, particularly when making large holes.

Hand-operated nibblers are available for cutting most thin sheet materials, including laminates, although they are sometimes difficult to guide properly. The power-operated nibblers are not normally a home-workshop tool. The DiY jig-saw fitted with a fine-toothed, metal-cutting blade is very good but it is essential to support the work very well and even then it is a noisy process. Similarly the DiY router-cutter is extremely useful, especially for large holes, but it can be a dangerous tool to use.

Cutting and sawing is one of those areas where gadgets are forever appearing, each claiming to save time and produce a better job! The hacksaw has not yet been replaced as a good general-purpose cutting tool.

Soldering

Learning to solder properly is not difficult and it is an essential skill for the home-constructor.

There are two main types of soldering: soft and hard. For most amateur purposes soft soldering is the norm. The type of soft solder most commonly used in electronic work is that known as 60/40, ie 60% tin and 40% lead. A flux is required to enable the solder to 'wet' the surfaces to be joined, and in the case of electronic work it must be non-corrosive. To make things easier this type of solder is usually supplied with a built-in flux, in the form of core or cores surrounded by the solder alloy, and is known as 60/40 multi-cored solder. This multi-cored solder comes in various diameters and the 22SWG size is suitable for most electronic work. The melting point of this solder is around 185°C, which is low enough to reduce the risk of heat damage to most of the items being soldered. There are numerous other soft-solder alloys with differing characteristics, such as low-melting-point, low-residue, high-melting-point and others. The low-melting-point and low-residue solders are normally used with surface-mounted components. The high-melting-point solders are used for joints requiring high mechanical strength. Another type of solder is known as universal or all-purpose, and this can be used for soldering aluminium or stainless steel. Its main claim to fame is that it will solder most metals, but it requires too much heat for normal electronic work and is therefore best used for structural applications or work which is less sensitive to heat.

Structural parts may be joined by Tinman's solder. This has a high melting point, around 200°C, and will therefore require a hotter iron or even a gas torch. It is normally supplied in round or square bars, about 300mm long, and not usually in multi-cored form, to enable suitable external fluxes to be used for the metals being soldered. A suitable flux for this type of solder is a liquid known as Baker's fluid, which comes in various grades; the number 3 grade is suitable for most jobs. It is sometimes referred to as killed spirits, from the method used to make it - a hazardous process. This flux is corrosive and should not be used for electronic or PCB work. Soldered joints made using this flux may be neutralised by a good washing and brushing with hot soapy water. Commercially this neutralisation is carried out chemically but these chemicals are not too friendly to have around the home workshop. Paste fluxes are available which can be wiped along the faces before joining. Most paste fluxes are corrosive and some are susceptible to overheating which can nullify their fluxing action during structural soldering. Flux is used to enable the solder to wet or bond with the required surfaces to be joined. Its primary function is to prevent these surfaces oxidising during soldering.

Various grades/melting point solders with flux are available in paste/cream form. Some of these are particularly suitable

Fig 26.22: Bending bars as a sawing guide

for soldering surface-mounted components, where the exact amount of cream can be accurately applied before soldering. This is usually done using syringe-like dispensers. These paste solders are also helpful when it is required to join several parts together at one go. The paste can be applied and the parts positioned as required, then the complete assembly is brought up to soldering heat either in an oven or by a blow torch or hot-air gun.

Hard or silver soldering produces very strong structural joints, which are also sound electrically, on brass, copper and most steels. The solder is available in various grades, shapes and sizes to suit the type of work to be joined. Typical applications are self-supporting waveguide assemblies and in areas where lower-temperature soft-soldering is required at a later stage of assembly. A silver-soldered component can be heated to accept soft solder without falling apart. The parts to be silver soldered need to be heated to around 600 to 700°C, and a gas torch is essential.

The process is thus unsuitable for PCBs. The fluxes used are related to the melting temperature of the silver solder and the type of work - brass, copper, stainless steel etc. The standard Easy-floTM flux powder and No 2 silver solder in 22SWG wire form should cover most requirements. It is difficult to silver solder joints which were previously made with soft solder. No amount of cleaning of the previously soft soldered faces will alter this because the lead/tin in the former soft-soldered joint will not allow the silver solder to wet the surface properly. Similarly some metals with a high lead content will not silver solder.

Another process known as brazing is about the nearest most home workshops can come to welding. A workshop equipped for brazing is more for model making than amateur radio construction, although it is a useful process to have available.

Rules for soldering

- All surfaces must be scrupulously clean. This applies equally well to the bit of the soldering iron. These bits are often iron plated to prevent rotting, and this plating should not be removed but wiped clean using a damp cloth or sponge. The bit should be kept 'tinned' with the solder and flux in use. Note that some fluxes will corrode rapidly the iron plated soldering bits. In the case of PCBs, the copper cladding should be cleaned using the fine abrasive eraser block referred to previously. Component wires and pins etc should be similarly cleaned for, although many brand-new parts come with the wires dipped in a solder-through or flux coating, it is still worthwhile just making certain. There must be no dirty or greasy patches on any of the surfaces to be soldered and corrosion spots must be removed. Even finger marks can spoil the soldering, so "clean it and when you think it's clean, clean it again!"
- Ensure that there is just sufficient heat to make the joint. Too much heat is harmful to electronic work and, in the case of structural work, it causes distortion. Not enough and the solder will not bond correctly.
- Heat and apply solder to the greater conductor of heat first, which in the case of PCBs is usually the copper pad and in the case of switches etc. is the solder tag, and then finally the wire, (the lesser conductor of heat) in one continuous process.

Soldering technique

After cleaning the parts, the following technique should be used to make a sound mechanical and electrical joint on PCBs.

Apply the clean, tinned hot iron to the pad adjacent to the component wire to be soldered, and tin the copper surface by

applying just sufficient multi-cored solder to the point where the tip of the iron is in contact with the copper. When the solder and flux starts to wet the copper surface, slide the iron into contact with the wire and apply a little more solder. As soon as both the wire and the copper surface are blended together in a small pool of molten solder, draw the iron up the wire away from the work.

This whole process should take seconds, not minutes. The less time any component is under this form of heat stress the better. In the case of particularly heat-sensitive components, it is advisable to heatsink the component's leads during soldering. A good soldering heatsink can be made from a crocodile clip, on to each jaw of which is soldered a small strip of copper. A small set of metal forceps will also serve as a clamp-on heatsink. The clamp-on action is required so that both hands are left free to do the soldering. Commercial soldering heatsink clips are available.

This method of soldering applies equally well to the joining of wires to components such as switches, potentiometers etc. In this instance, good practice demands that the wires should be wrapped and secured mechanically before soldering, but bear in mind that the joint may need to be unsoldered later.

Good soft-soldered joints should look neat and smooth, with no draw-off points, and be continuous around or along the joint. Large blobs of solder do not ensure that a joint is well-made. The cored solder should always be applied to the work and the iron at the same time, not carried on the iron to the work.

The results of incorrect soldering procedures are 'dry joints' which are either open-circuit or high-resistance points and will prevent the circuit from functioning correctly. Often the effects of such soldering will appear after several hours of operation, and this really leads to frustration especially to the novice constructor. Commercial equipment has exactly the same problems, if this is any consolation! In most instances it is difficult to locate a dry joint visually. Sometimes a perfectly sound looking joint turns out to be 'dry' immediately at the contact point, which is hidden under the solder. As a rough guide most 'good' joints appear shiny and devoid of any 'craters' or occlusions, whereas some 'dry' joints appear dull and pitted. There is no easy way to identify 'dry' joints but there is an easy way of avoiding them - follow the rules and guidelines set out above.

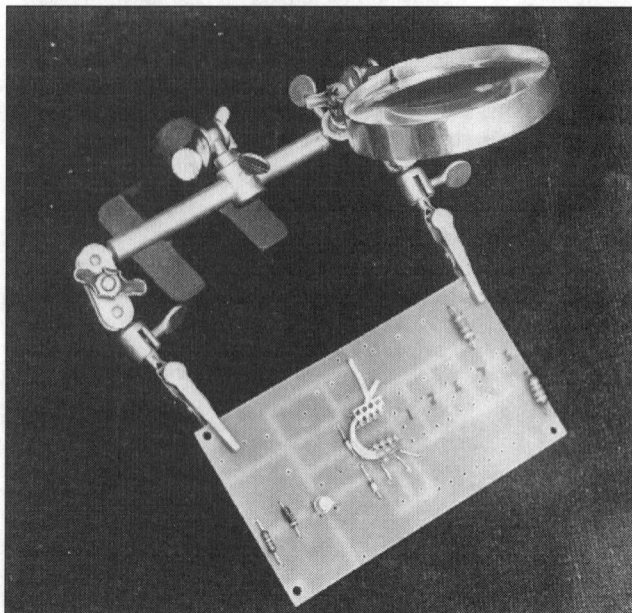

Fig 26.23: Work-holding fixture (the 'third hand')

To join structural parts together by soft soldering, each mating face should be pre-tinned, ie given a thin coat of solder using the appropriate flux to ensure proper 'wetting'. The tinned faces must be clean and shiny with no burnt or blackened flux spots. The two tinned faces are then painted with flux, brought together, heated to just melt the solder, and clamped together until the solder has cooled. There is no need to apply any more solder. If the job has been done properly, a neat, clean solder fillet will be seen along the mating edges. The flux residue must be washed away immediately after soldering to prevent long-term corrosion.

It is not good practice to have long soldered joints on PCB laminates. The expansion caused by the necessary heat will de-laminate the copper. In these circumstances it is better to tack solder along the required structural joint, placing 2mm long solder-tacks at about 10mm intervals. Any such joints on copper laminates will be as mechanically strong as the laminate-to-copper bond.

It is better practice to secure mechanical parts such as angle brackets, shaft housings etc by screws or clips which pass right through the laminate, and to use solder only to make a sound electrical connection.

General comments on soldering

All soldering should be carried out in a well-ventilated room. The fumes from flux and molten solder are irritants, especially to people with asthma. In the interests of safety, it is advisable to wear safety glasses when soldering. Molten solder and flux have a nasty habit of spitting and minute spots can fly anywhere, even into the eyes!

All work being soldered should be supported to leave the hands free to hold the soldering iron and apply the solder (**Fig 26.23**).

The greater the area to be soldered, the larger the soldering iron bit needs to be to transfer the heat to the work. A hotter iron is not a substitute for the correct bit size. The quantity of heat is related to the wattage of an electric soldering iron, and the larger the area to be soldered, the greater the wattage required. The soldering irons of various wattage specified previously should enable most soft-soldering jobs to be carried out.

Figs 26.24 and 26.25 show one type of soldering iron bit-saver. When the iron is placed on the holder arm, the diode is switched in series with the heater of the iron, approximately halving the voltage. This considerably reduces the 'bit rot' referred to previously, but at the same time allows the iron to remain hot - only a small delay occurs on picking it up again before soldering temperature is regained.

HEATSINKS

Heat levels for semiconductors are usually defined as 'not to be exceeded' junction temperatures. The power dissipation at these junction temperatures is specified in watts, and can range from milliwatts to several hundred watts. Whenever the device is operating, heat is being generated. Unless some means of heat dissipation is used, the recommended operational limits will be exceeded very rapidly and the device will fail. A semiconductor junction takes microseconds to reach its operating temperature and only a few more microseconds to destroy itself if heat is not dissipated. Electronic protection circuits can help safeguard the device but will not eliminate the requirement for heat dissipation. Semiconductor devices generate heat very rapidly and it is essential to ensure proper cooling.

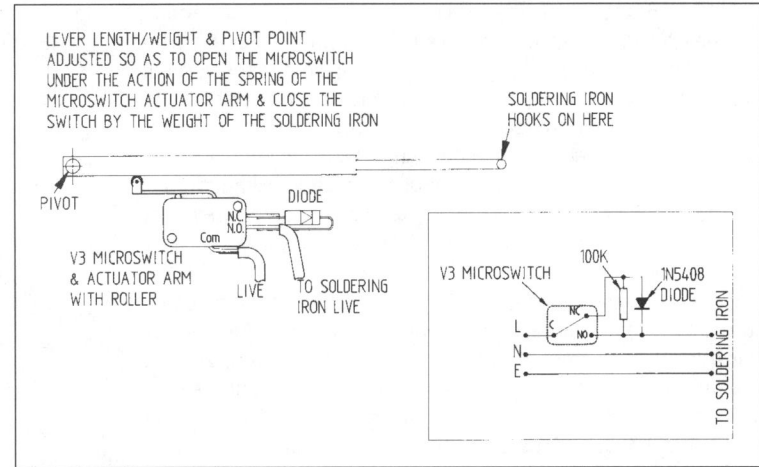

Fig 26.24: Circuit and mechanical details of soldering iron saver

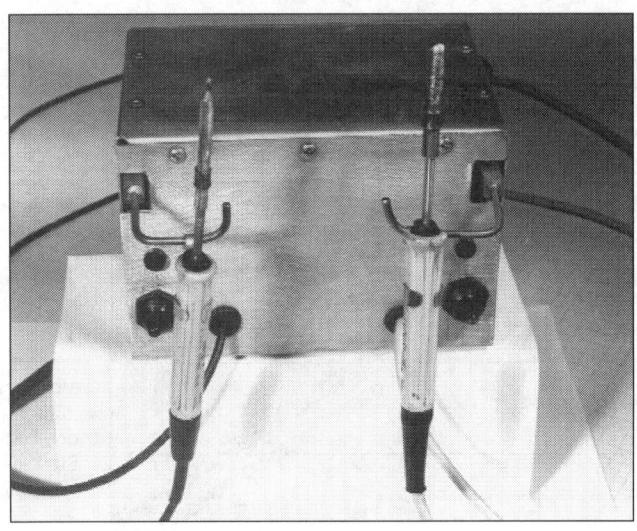

Fig 26.25: Dual soldering iron saver

The heat must be dissipated at the same rate as it is being generated in order to ensure that the device remains within its specified temperature operating limits. The usual system consists of a heat-conducting material which takes the heat from the device and transfers it to the atmosphere. The materials used for heatsinks should have high thermal conductivity to conduct the heat away from the heat source, and high emissivity to radiate this heat to the atmosphere. Emissivity depends on the surface finish and texture of the radiating elements, rather than the type of material, but thermal conductivity depends on the type of material. Most commercial heatsinks are made from aluminium, for it is lighter in weight, cheaper and can be extruded or cast into the complex shapes required for compact, efficient heatsinks. Aluminium can also be anodised and dyed black to increase its emissivity.

Heatsinks should be made and positioned such that the fins radiate to atmosphere. Fan cooling enables heatsinks to be smaller for the same thermal resistance, and as a guide, a substantial airflow over the fins can reduce the existing thermal resistance by up to 40%, depending on the number of fins receiving air. Painting a natural metallic finished heatsink with a very thin coating of matt black paint will also produce a reduction in overall thermal resistance by increasing the emissivity. The actual amount of reduction depends on the mattness of the black paint and the thickness of the coating but a 10 to 25% improvement could be expected.

Fig 26.26 shows the thermal resistance possible at various volumes for different metals, finishes and fin positioning. The bands indicate the effects of a matt black finish and the vertical or horizontal positioning of the fins. The top edge of each band represents the thermal resistance of natural-finished horizontal fins. The bottom edge represents matt-black-finished vertical fins. The band for aluminium or brass is an approximation, for the thermal conductivity of either metal varies widely depending on the alloying metals, but it can be considered reasonably accurate for most amateur purposes.

The heat transfer path is considered as flowing from the semiconductor (heat source) to ambient air, with each junction or transitional point treated as a thermal resistance (**Fig 26.27**). The thermal resistances of fins and heatsink to ambient are considered in parallel to give the total heatsink value. Similarly the thermal resistance of the device junction to air is considered to be in parallel with the total thermal resistance.

Heatsink formulas:

$$P_D = \frac{T_j - T_a}{\theta_{jc} + \theta_{cs} + \theta_{sa}} \qquad (1)$$

and

Total thermal resistance x Power dissipation
= Temperature rise in °C above ambient of the transistor junction (2)

from which:

$$\theta_{sa} = \frac{T_j - T_a}{P_D} - (\theta_{jc} + \theta_{cs}) \qquad (3)$$

also:

$$T_j - T_a = P_D \times \theta_{ja} \qquad (4)$$

where:

T_j = Maximum allowable junction temperature (°C)
T_a = Ambient temperature (°C)
θ_{ja} = Thermal resistance, junction to air (°C/W)
θ_{jc} = Thermal resistance, junction to case (°C/W)
θ_{cs} = Thermal resistance, case to heatsink, plus any insulating washer and heat-conducting compound (°C/W). (Can be assumed to be between 0.05 and 0.2°C/W).
θ_{sa} = Thermal resistance, heatsink to ambient air (°C/W).
P_D = Power dissipation (W)

Readily available semiconductor data do not always specify all of the above details but approximations may be used to design a heatsink suitable for most amateur applications. The usual data includes P_{tot} in watts (P_D) at a case temperature of usually 25°C. This can be used to derive the junction temperature (T_j) for substitution in formula (1). Because $\theta_{ja} = 25°C/P_{tot}$ and $T_j = 25°C/\theta_{ja}$ then $T_j = P_{tot}°C$.

Example: A 2N3055 is to be used as the pass transistor in a 13.8V DC regulated PSU. The available data shows P_{tot} as 115W at 25°C, thus $T_j = 115°C$ and $\theta_{jc} = 0.2°C/W$ (25°C/P_{tot}). It is estimated that, for most applications, $\theta_{cs} = 0.1°C/W$, assuming thermally conducting compound and a mica insulating washer are used. (If no insulating washer is used, $\theta_{cs} = 0.05°C/W$ but the heatsink may be 'live'.) Substituting in formula (3):

$$\theta_{sa} = \frac{115 - 25}{115} - (0.2 + 0.1)$$
$$= 0.3°C/W$$

which is the heatsink thermal resistance required to ensure correct thermal operation.

The above example assumes that the 2N3055 is working at its limit, and this should seldom be the case. In this example it would be usual to supply the pass transistor with 18V DC to allow adequate regulation, and the transistor would be run at 5A, or half the rated amperage. The difference between the supply voltage and the output voltage is 18 - 13.8 = 4.2V DC, hence the power to be dissipated will be 4.2 x 5 = 21W. Substituting this in formula (3) gives:

$$\theta_{sa} = \frac{115 - 25}{21} - (0.2 + 0.1)$$
$$= 4.0°C/W$$

This is the thermal resistance required for this power supply. If necessary a further allowance could be made for the inefficiency of the semiconductor and the calculated thermal resistance decreased to allow for this. In this example this factor has been ignored.

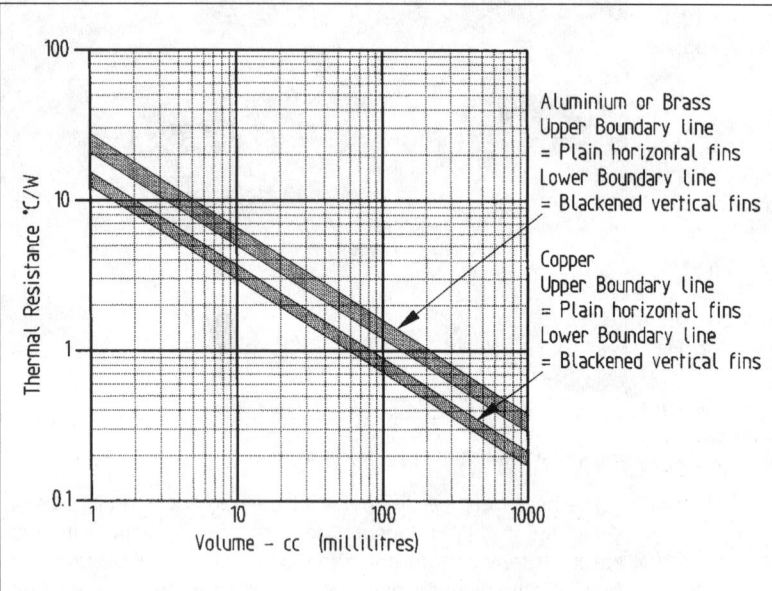

Fig 26.26: Heatsink thermal resistance / size chart

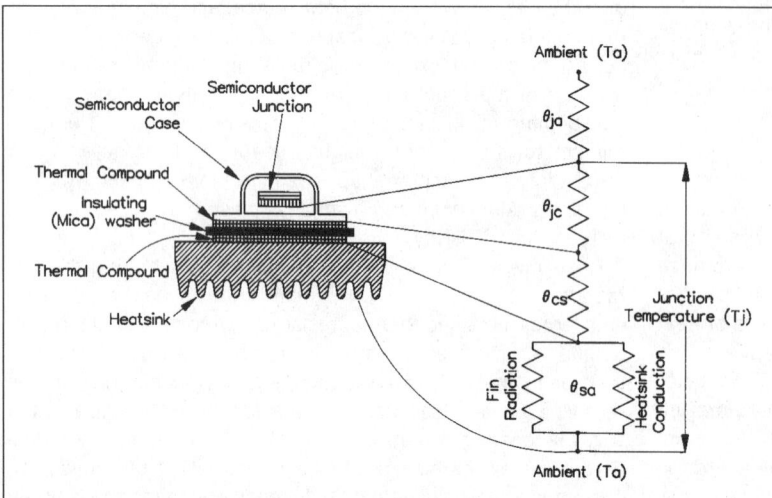

Fig 26.27: Heat transfer path and thermal resistance zones

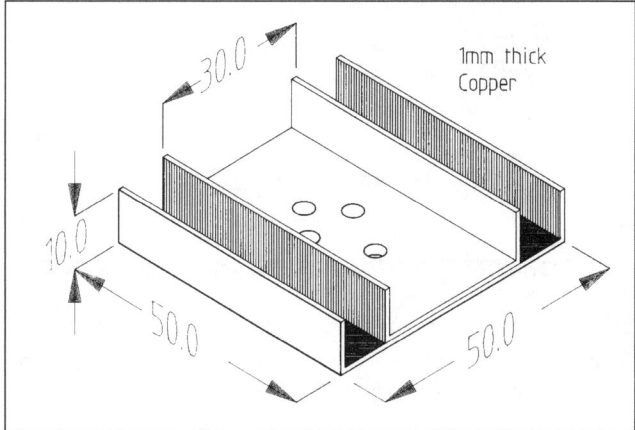

Fig 26.28: Heatsink based on the calculations of the example in the text

From Fig 26.26 it can be seen that about 6cm³ of blackened or 9cm³ of natural finished copper (vertical fins) will be a suitable minimum size. **Fig 26.28** shows a typical home-made heatsink based on the above calculations. Ideally the two bent plates should be soldered together to minimise any thermal resistance at this junction. Thermal conducting compound should be used if this joint is not soldered. The copper should be annealed before bending.

Using a heatsink with a lower-than-calculated thermal resistance will not affect the operation of the cooled semiconductor. It would be unwise to use a heatsink with a higher-than-calculated thermal resistance. In other words, a larger heatsink is better than one too small !

In use the heatsink will become warm to the touch but it should not become untouchable. If it does, then the semiconductor providing the heat is probably being overworked. It would be worthwhile checking the calculations and ensuring that each mating joint is made properly. Some semiconductor devices are designed to run at high temperatures and a hot heatsink would be expected. This heat represents waste energy, and the necessity of large cooling systems, though seeming to be the norm, indicate inefficiency.

Professional designers use very sophisticated computer programs to design their heatsinks. Even then they rely on trials and measurements to tune-up their results! The above formulas and chart are accurate for most amateur purposes, eliminating the need for complex equations or computers.

Heat can also be dissipated from the connecting pins, together with the case, of semiconductor devices, and this can be useful if trying to keep the finished unit compact. A heatsink can be attached to the pin side of the board by a layer of thermally conducting, electrically insulating, elastomer and mechanically fastened to the board with screws. This method can provide about a 5 to 10% reduction in the required overall size of the main heatsink. Large soldering pads for the pin connections also act as heatsinks, and these can be used in conjunction with a suitable heatsink for such devices as audio amplifiers, power regulators, rectifiers etc.

The duty cycle of the device also affects the size of heatsink required. A 50% duty cycle can allow about a 20% reduction in the size of heatsink, depending on the heatsink design. During the OFF cycle, the heatsink must be able to dissipate all of the heat generated during the ON cycle, and it is usual in these circumstance to provide forced-air (or water) cooling to ensure that this is achieved. Many solid-state transceivers specify ON and OFF times for continuous full-power carrier (FSK, AM etc) opera-tion. The small handheld transceivers start to get very warm if they are transmitting for longer than they are receiving. In the case of handheld transceivers, the batteries usually limit the transmit time but, if such transceivers were operated from a mains PSU, it would be essential to ensure that the heatsink arrangements were adequate.

MAKING CIRCUITS

Translation from diagram to working circuit is another accomplishment. The systems available are numerous and varied. Before making any circuit, experimental or permanent, it is worth considering what is expected of the finished project and relate this to the type of circuit construction available. All of the components should be collected together, tested and identified A well-organised storage system, good test equipment and a magnifying glass are invaluable. It is worth trying to understand how the proposed circuit is supposed to work and what function each component performs.

If it is intended for personal use only, then the circuit board needs to be functional rather than reproducible. If it is to be portable in use, as distinct from transportable, then weight, power consumption and compactness are important. If more than a few of the same circuits are needed, then it will be worth considering setting up to design and make printed circuit boards.

The following list suggests some of the factors worth considering before starting on any circuit design and making:

1. Permanent or experimental?
2. How will the circuit/s be housed and mounted?
3. What power supplies are required (battery, mains, internal, external etc)?
4. What types of inputs/outputs are required?
5. What controls are required?
6. What safeguards are required (eg accidental switch-on, wrong polarity, over-voltage etc)?
7. Methods of construction. (Can you make it?)

In the rush to make the circuit, it is all too easy to discover afterwards that some of the above factors were important!

Many solid-state devices are susceptible to damage by static discharge. Handling and soldering such components requires care. It is safer to assume that all semiconductor devices are prone to static discharge damage and treat accordingly. Assembly and repair should be carried out using the normal anti-static precaution of connecting the soldering iron, PCB and operator to a common point. This equalises the static level and is further improved if the common point is properly grounded.

Commercial equipment is available, consisting of an anti-static mat, wrist strap and connecting leads for the other tools. Semiconductors which are very sensitive to static are usually supplied mounted in a metal strip or wire clip (or a carbon-conductive foam in the case of ICs) which shorts together every pin of the device. This anti-static protection should be left in place until the device is plugged or soldered in position, and then carefully removed. A suitable size 'Bulldog' clip connected to the ground mat by a length of flexible wire makes an excellent anti-static connector for ICs.

Methods for Experimental or Temporary Circuits

The main factor with this sort of work is adaptability. It should be possible to change components, and even the whole circuit, with the minimum of effort. The variety of methods devised over the years by fellow amateurs are ingenious and effective.

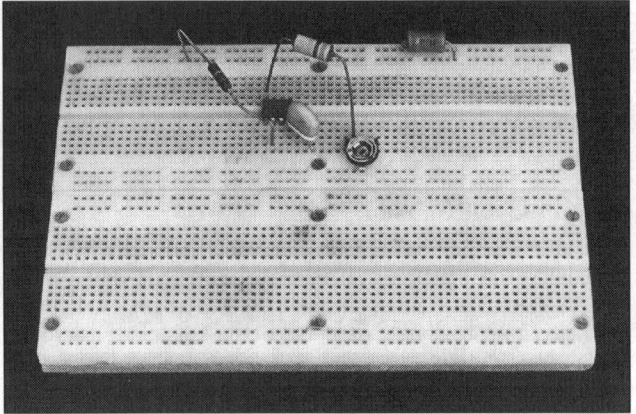

Fig 26.29: Example of a plug-in prototype board

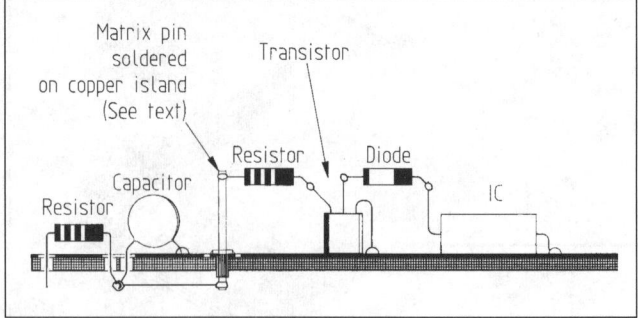

Fig 26.31: Various component mounting and connnecting methods for experimental circuit construction

An old-time experimental circuit construction method was to use pins nailed into a wooden board and wrap or solder the wires and components to these. This was further improved upon by using drawing pins in place of pins and trapping the wires under the heads. If brass-headed drawing pins were used the wires could be soldered in place. Several of the subsequent methods of experimental circuit construction can trace their origins to these early techniques.

Many experimental circuits may be made using plug-in matrix boards of the type shown in **Fig 26.29** which are also known as prototyping boards.) The RF characteristics of these boards are not of the best, but for digital, AF and other work they are ideal. This board can also be used to do trial component layouts during the designing of a printed circuit or a strip board layout.

Copper-clad laminates have become the accepted base for most circuit assembly. The types of laminates available have been mentioned previously in the section on materials. The glassfibre boards are preferred, whether single or double-sided, for they are kinder to repeated soldering and de-soldering than other laminates, reducing the chances of copper de-lamination. For some experimental work, PCBs are made consisting of small copper pads in a grid pattern on the board (**Fig 26.30**). Components are soldered directly on to these pads without drilling holes (**Fig 26.31**). Commercial boards of this type are available for the experimenter who does not wish to go to the trouble and hazards of etching.

An allied technique is that of sticking the main components such as transistors and ICs etc on to the board with their wires uppermost and interconnecting as required with wire, resistors or capacitors, using the copper or one of the supply rails as ground. Tag strips or insulated connection pillars can be fastened to the board to provide more support for components and connections. This is a very quick way of making a circuit and has the advantage that changes may be made with the minimum of trouble. However, it's not a 'pretty sight', and it has been referred to as the ugly system, although it works very well, particularly for RF circuits.

Some of the simpler circuits can be made using only tag strips, eliminating the need for any copper laminate board.

Fig 26.32 shows a commercial system which provides a PCB with the same conductor layout as its matching plug-in prototype board.

Drilled matrix board without copper cladding, using special pins which are press-fitted into the appropriate holes for component mounting, is yet another method.

A combination of the above methods may be used to achieve a working circuit. A set of counter-boring drills are very useful, for they can be used to remove circles of copper around holes or to form small annular islands of copper to secure items which need to be insulated from the rest of the board. A counterboring drill is similar to an ordinary drill but with the cutting faces square to the axis of the drill.

Fig 26.30: Grid pattern PCB for experimental circuits

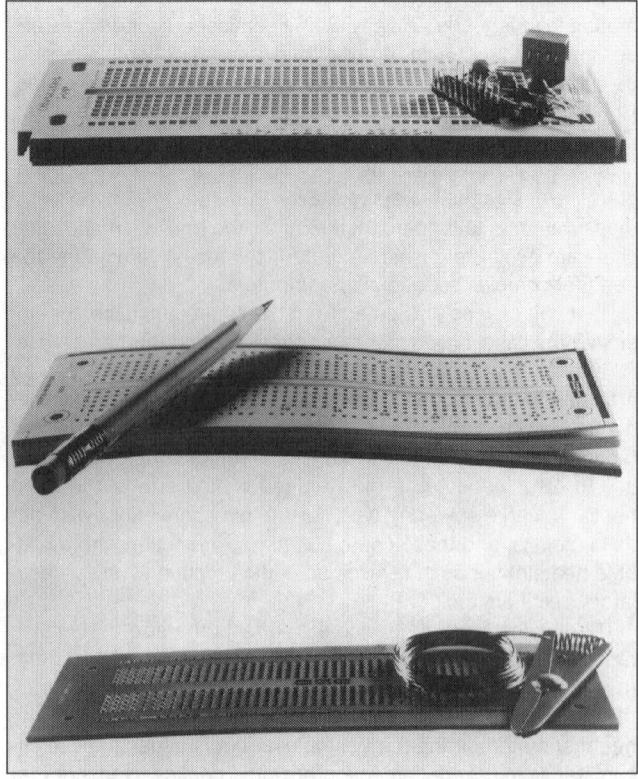

Fig 26.32: Another prototype-to-PCB system: plug-in board, sketch pad, PCB

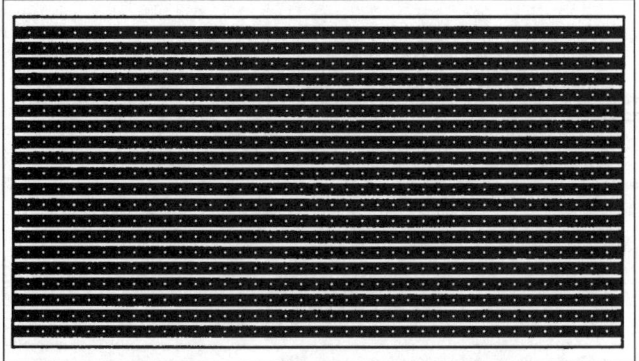

Fig 26.33: Strip board PCB suitable for experimental or permanent circuits

Permanent Circuits

Many of the above-mentioned experimental circuit assembly systems can be used for permanent circuits, but for repeatability and more robust construction it is usual to turn to other methods.

One form of permanent circuit assembly which requires no home etching of boards uses the 0.1in pitch matrix of holes and copper strips, commonly known as strip board (**Fig 26.33**). The techniques of using this sort of board are relatively straightforward and it is very easy to translate from the experimental plug-in board assembly mentioned earlier. The main consideration is the number of holes required per connection and this is derived by careful scrutiny of the circuit diagram. For the more complicated circuits, it is possible to use a grid reference system; the copper strips lettered 'A' to 'Z' and the holes numbered '1' to 'n'. The connections per component may then be identified on the circuit diagram by its appropriate grid reference. A drawing of the layout beforehand can often show up better ways of placement.

The unwanted strips of copper may be cut away using either a sharp knife or a special strip board cutter. The finished assembly can be made to look neat and tidy, particularly if any track interconnecting wires are dressed against the board. These track interconnecting wires do sometimes present the problem of the insulation melting during soldering. To avoid this it is possible to use either bare wire or sleeve the wire with heat-resisting sleeving. Connecting wires for power supplies or controls should be wrapped to pins soldered to the board. This eliminates the problem of wire fracture that often occurs when these wires are passed through the holes and soldered to the tracks. Similar boards are available for digital circuits, with holes and copper pads suitable for dual-in-line ICs. Many of these digital boards are available for wire wrap, or point-to-point wire and solder methods. These normally require expensive tools, components and wires but are favoured by digital equipment experimenters. They can save considerable time for things like memory boards, microprocessors etc and save the effort of designing and making special one-off PCBs for such circuits.

Etching

The next set of techniques involves etching. A brief outline of the equipment, materials and methods required for this are given as a guide.

The usual etchant is ferric chloride which is corrosive. It should not be allowed to contact any part of the body, particularly the mouth and eyes. Work should be carried out in a well-ventilated room, and protective glasses, gloves and overalls should be worn. No child or pet should be allowed near when using or preparing etchants. The initial action in the event of accidental skin contact is to wash immediately with running water. Splashes to the eyes should be first-aided using a proprietary clinical eye wash and medical advice sought immediately. This etchant is made by dissolving ferric chloride (hexahydrate) crystals in water. The proportions are around 1kg of crystals to 2 litres of boiled water. It is difficult to mix ferric chloride crystals and warm, not boiling, water helps. The solution should be mixed in a non-metallic vessel using a plastic or glass stirring rod. Nothing metallic should be used in the preparation and storage of this substance; plastic containers with plastic screw-top caps are ideal. Child-proof caps are recommended where there is the slightest chance of exploring little fingers. Most chemists will be pleased to advise and possibly supply suitably safe containers.

The container should be clearly labelled 'POISON - FERRIC CHLORIDE'. A permanent marker pen can be used to write on the plastic container, eliminating the risk of a sticky label dropping-off or becoming obliterated.

A small quantity (10-20ml for the above) of hydrochloric acid may be added to improve the solution.

Fig 26.34 shows a suggested arrangement for an etching bath. The essential requirements are that the etchant can be warmed to around 50°C and agitated continuously during etching. The etching bath should be large enough for the work to be fully immersed. The aquarium aerator block is glued to the bottom of the bath and air from an aquarium electric aerator pump is fed through a small bore plastic pipe passed over the lip of the container. The pump should be placed clear of the bath to prevent splashes of the etchant ruining the pump.

The PCB to be etched should be clamped in a plastic carrier and placed in the warm (60-80°C) foaming etchant. It usually takes 10min or more with this set-up to completely etch a PCB. Check from time to time. The shorter the etch time, the less the risk of under-cut and of the etch-resistant materials being washed or dissolved away. The board should be washed and scrubbed in water immediately after etching to remove all traces of the etchant. Failure to do this results in continued erosion even after the board has been assembled. Some of the faults of early commercial PCBs were found to be caused by inadequate cleaning after etching.

The cooled etchant can be poured back into its container and saved for future use. The partially used etchant can be kept for many months, particularly if hydrochloric acid has been added. It does not deteriorate but becomes saturated with copper and it is this which prevents further etching. It is interesting to place a small iron bar or plate into the seemingly spent etchant. The copper attaches itself to the bar and forms artistic surface patterns depending on the cleanliness of the bar - it might even be a method of reclamation!

Check with the local authority before disposing of ferric chloride for it is unlikely to be acceptable to just pour it down the drain.

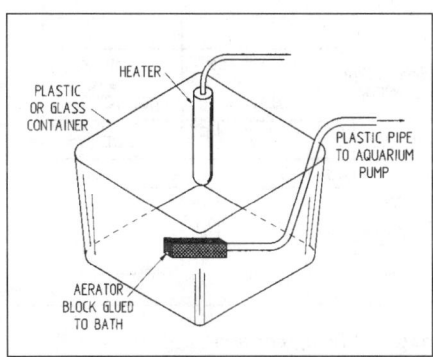

Fig 26.34: Etching bath using aquarium devices

Fig 26.35: Component wire bending jigs

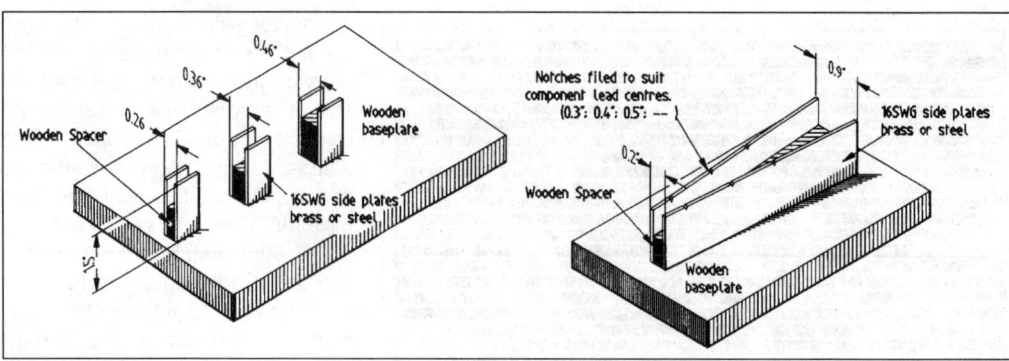

Designing Printed Circuit Boards

Designing a PCB requires skill, time and a great deal of patience. The process is intended for quantity reproduction of circuits and its value to the amateur constructor is worth careful thought. For the individual constructor, as distinct from a club project or a published design, it is better to consider 'modularisation' rather than an 'all-on-one-board' approach to printed circuits. This is the exact opposite of some commercial philosophy but it gives the home constructor much more flexibility, as well as being less prone to error.

Standardisation of board size helps during the design and making stages, and later when fitting the assemblies into a cabinet.

The standardising of the fixing centres of resistors is also helpful. **Fig 26.35** shows two types of component wire-bending jigs which can be used to bend the leads to accurate centres for resistors of various wattage. It is particularly necessary to consider all of the previously mentioned design criteria, for PCBs are very restricting, especially in the size of components used and in the ability to make changes. For example, it is difficult to use, say, 0.5W resistors in place of the specified 0.125W ones or to change from a discrete transistor to an integrated circuit. With this in mind, the first essential before designing a PCB is to have either the components on-hand or to have accurate details of their sizes and fixings: it is pointless guessing (see **Fig 26.36**). It

is possible to design the circuit such that more than one size of component may be used (**Fig 26.37**), but this may detract from the intended compactness and repeatability of the PCB.

Commercial PCB designers use computer software to speed up the very tedious task of track routing and component layout. The software even produces data for controlling such things as hole drilling, board shaping and the mechanised placing of components on the board.

There are some very good programs available for the home computer which facilitate the designing of printed and schematic circuits (see the chapter on computers).

Some home constructors use less expensive methods, the most popular being a sheet of 0.1in pitch graph paper as a guide, placed under tracing paper.

The approach to designing the printed circuit, by any method, consists of four main areas, as follows.

- Positioning the active devices (semiconductors, ICs etc).
- Positioning the passive components such as resistors, capacitors etc.
- Positioning inputs and outputs (connections).
- Linking together the tracks.

A different coloured pen or pencil and a separate sheet of tracing paper may be used for each of the above areas. Draw the active devices, say, in red on sheet one and resistors etc in blue on sheet two and so on. This way it is possible to try various combinations without having to redraw everything. The more compli-

Size / Wattage	A mm	B inches	TYPE
0.125	4.1	0.5	CARBON
0.25	8	0.6	HIGH STABILITY
0.5	11	0.7	HIGH STABILITY
1.0	16	1.0	HIGH STABILITY
2.0	24	1.3	HIGH STABILITY

Mini Polyester Dip Coated 250 VDC Wkg		Silvered Mica 350 VDC Wkg		Mini Polyester Layer 100 VDC Wkg		Monolithic Ceramic 100 VDC Wkg	
VALUE	D"	VALUE	E"	VALUE	F"	VALUE	G"
0.01–0.1μF	0.4	2–70pF	0.3	0.1–0.47μF	0.3	10pF – 0.047μF	0.1
0.22μF	0.6	80–220pF	0.5	1μF	0.4	0.1–1μF	0.2
0.47μF	0.8	220–10000pF	0.8	2.2μF	0.6		
1–2.2μF	1.1						

Fig 26.36: Some component fixing centres

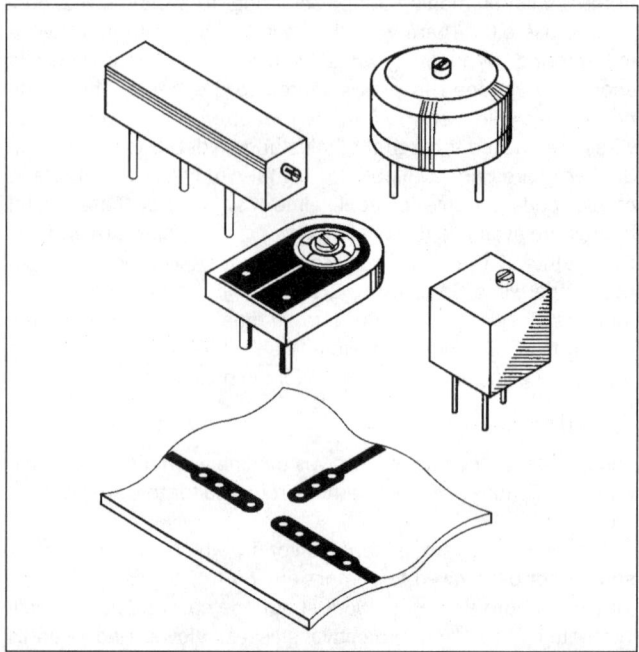

Fig 26.37: Accommodating different sizes of components

cated the circuit, the greater the likelihood of many, many trials. A multiplicity of circuit jumper links usually indicates either a poor layout design or that the designer has had enough!

It is also essential to adopt some standard form of viewing, to avoid the all-too-easy pitfall of producing a beautiful but mirror-imaged board. Most designers seem to adopt the 'as viewed from the component side' standard even for double-sided boards.

Eventually, an acceptable circuit layout will be produced. At this stage a thorough check should be made to ensure that there are no crossed tracks, mirror-imaged components or other mistakes which would be annoying to discover when the board has been made. If you don't find any errors, look again! Some designers tick off each item on a copy of the schematic diagram as they verify that it is on the PCB layout. As with the previously mentioned strip board construction, it is worthwhile marking the number of components connected to each interconnecting track. It is all too easy to either miss one or to find that connections are made to the wrong track.

Guidelines for RF circuits, including high-speed digital

1   Use double-sided board with one face, usually the component side, as a ground plane.

2   Keep the input and output of each stage as separate as possible consistent with (3).

3   Keep tracks short.

4   Decouple all supply leads at the point of entry to the board as well as at the points specified by the circuit. Small ferrite beads placed over each supply lead usually eliminate any possibility of circuit interaction occurring through these leads. These ferrite beads can sometimes be used on signal leads in digital circuits to reduce the possibility of interaction or external RFI, and if these sort of problems occur it is worth trying.

General PCB guidelines

1   Keep mains and high-voltage circuits well away from other circuits, preferably by using separate boards. In the interests of safety, any tracks etc carrying high voltages should be covered by an insulating material to prevent accidental contact.

2   Avoid high current density points. Rapid changes of line width should be avoided, particularly in circuits carrying several amperes. The recommended current density I in amps is:

$I = 3(^3\sqrt{w^2})$

where w is the width of the copper track in millimetres. Circuit board tracks make excellent fuses if incorrectly designed!

3   The capacitance C in picofarads between each side of a double-sided board is:

$C = 0.0885 \times K \times (A/h)$.

where A is the area (sq cm) of the smallest side; h is the thickness (cm) of the laminate excluding copper, and K is the dielectric constant of the laminate material, which for glassfibre is about 4.5.

4   The recommended line or track width varies with wattage, but for most home production a 1mm width is about the smallest possible, with a minimum gap between tracks of not less than 1mm unless you are good at drawing. Commercial artwork for PCBs is often drawn four times full size and photographically reduced.

5   An 8mm minimum margin should be allowed around the board for handling during working and etching. This margin can be removed later.

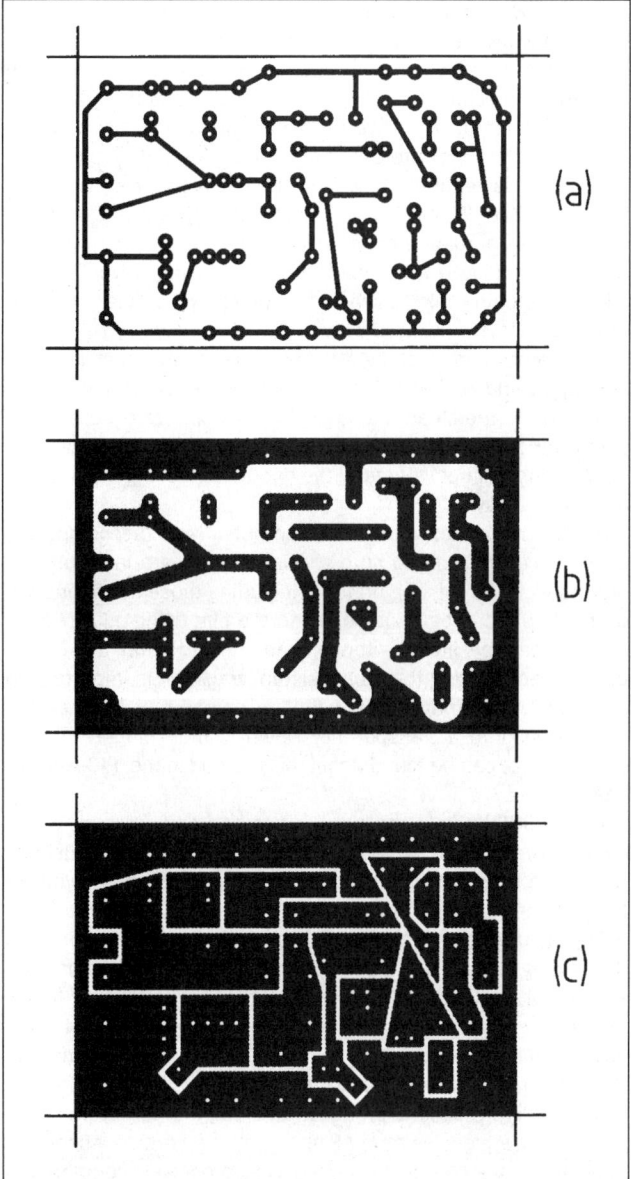

Fig 26.38: Three layout modes for PCB. An alternative is to use a computer PCB design program, a laser printer and special iron-on film (see text)

Making Printed Circuits

There are several methods of transferring a home-designed or a published layout to the copper laminate. Which one to use depends on the number of boards required and the facilities available.

Suggested methods for a one-off PCB

The traced PCB layout is attached to the appropriate side of the laminate; which side depends as previously mentioned on the drawing system used. All the hole positions are carefully marked through using either an automatic centre punch set at its lowest level, or an ordinary centre punch which should be very lightly pressed to mark the surface. Heavy thumping centre popping will cause de-lamination and should be avoided. The tracing paper is removed and all the holes drilled using a 0.8-1.0mm diameter drill. In the case of dual-in-line IC holes, the matrix template mentioned previously can be used to assist accuracy. The holes are deburred and the board cleaned. Three different styles are shown in **Fig 26.38**

Fig 26.38(a) is the usual type of line track and pads. This can be hand drawn using a special fibre-tipped pen or a tubular draughting pen. Fibre pens with a special etch-resistant ink are available especially for drawing directly onto copper. The fibre pens used for drawing overhead transparencies may also be used, providing they are of the etching or permanent ink types. Special ink is required for use in tubular drawing pens and is known as 'K' or 'P' ink. These inks etch into plastic and must be used in special tubular pens, which should be cleaned out thoroughly immediately after use with the matching solvent for the ink. The etching ink dissolves the plastic of ordinary tubular drawing pens.

An alternative to drawing with ink is to use special etch-resistant tapes and rub-on transfers. These must be very well burnished on to the clean and grease-free copper to ensure adhesion during the subsequent etching process. Most normal rub-on lettering will not withstand the rigours of etching and is easily washed away.

A PCB that is designed on a computer (see the computers chapter) can be printed onto special film using a laser printer. Alternatively, an existing printout (including those in Appendix B of this book) can be photocopied to the film using a laser copier. This can be literally ironed onto the PCB and the filmed peeled back leaving the etch-resistant tracks. It is important to ensure that the image is the correct way round - it may need to be printed onto a transparency which is then flipped and re-copied. More can be found about this method in the 'PIC-A-STAR' chapter.

Fig 26.38(b) shows the simple lines and pads system, which can be drawn with either the pens mentioned previously or which can be painted using an artist's fine brush and cellulose paint or the special 'P' ink.

Fig 26.38(c) shows the scratching system, which requires no draughting skills. The copper side of the drilled board is painted all over with either cellulose paint or the 'P' ink; allowed to dry thoroughly, and the gaps between the tracks scratched away using a 0.5mm wide chisel-pointed piece of metal. It may be helpful to first mark out the scratching lines using a soft lead or wax pencil. The circuits produced by this method perform very well at all frequencies from AF up to UHF, and are very useful in providing extra heatsinking when heat-producing components are used. As a result, soldering requires more heat due to the larger areas of copper.

The circuit may also be transferred to the copper by using carbon paper and going over the tracing paper circuit with the carbon paper in contact with the copper. However, the transferred image needs to be inked-in and the transferred carbon often prevents this being done successfully.

Suggested method for multiple board production
This method uses photographic techniques to copy the circuit layout on to the copper. It is an expensive process and, as well as the etching bath, a further similar bath and agitator is required for developing the exposed board.

A source of ultra-violet light is also necessary, preferably with some form of accurate automatic timer. As UV light is harmful, this light source needs to be totally enclosed and incapable of being switched on until fully covered. Commercial high-intensity UV light exposure boxes are available which meet all of these requirements and, unless fully conversant with the hazards of UV light, it is safer to use one of these commercial units rather than attempt to make your own. High-intensity UV light produces serious permanent eye damage with effects similar to that experienced by the accidental viewing of an electric welding arc. Lower-intensity UV light boxes are available but exposure times

usually become very long and the safety factor is not much improved.

Full darkroom facilities are not required, only protection from direct and reflected sunlight. A normally darkened room is usually adequate with subdued tungsten or non-UV lighting.

The copper-clad board is coated with a photosensitive etch-resistant coating, the thickness and smoothness of which controls the quality of the finished PCB. The positive type of photo-resist is the easiest to use for it does not require the making of a negative 'master'. With positive photo-resists, the unexposed parts remain in place during developing and the exposed parts are washed away. All photo-resists are expensive and usually have a limited shelf life.

Ready-photosensitised copper laminates are available and are usually supplied with an opaque plastic covering which is peeled away immediately prior to exposure. The board is usually cut with this plastic left in place to protect the sensitised coating and to ensure that no fogging occurs. The margins already referred to should be used, for sometimes light does fog the edges of these sensitised boards. However, if required, it is possible to 'roll your own' using either a spray or a paint-on photo-resist. Full instructions are usually supplied with either and these should be followed. It is essential that during and after coating no UV light strikes the coated surface, for this will cause 'fogging' and attendant poor quality of the finished circuit.

The coating must be of even thickness and free from dust. It is this which makes DiY coating very difficult, especially if repeatable results are required. Spin coating is about the best and an old variable-speed record player turntable set at about 100RPM makes an ideal spinner. The immaculately clean, grease-free and smooth board to be coated is secured as centrally as possible, copper side up, to the turntable. Allow the turntable to spin for a few minutes just to ensure that the board is securely fastened and, using a glass squeeze-dropper, gently apply drops of the photo-resist to the centre of the spinning board until it is seen that the whole board is covered. Photo-resist flies everywhere and it is preferable to place the turntable in a high-sided box. Usually by the time the turntable has come to rest the coating is dry enough to touch but only by the margins.

Normally the board is then dried by heat at the temperature specified by the manufacturer or else it is just air dried. Baking the coating at the correct temperature hardens it and improves the potential quality of the finished board. A clean and light-proof metal box (biscuit tin) can be used to house the freshly coated board, reducing the risk of dust settling on the surface and preventing UV light from pre-exposing the board. The same spin method can be used with the spray-on photo-resists for it does ensure fairly repeatable even coatings. If the coatings vary in thickness, the exposure times will vary, which makes repeat production difficult. It is similar to having a camera loaded with film with each frame having a different and unknown exposure rating.

Double-sided boards may be coated in a similar manner, coating and drying each side in turn. The double-sided board should be attached to the spinner using the 'margins' to ensure that the already-coated side is not scratched or marred by the fastenings. Obviously, UV light must not be allowed to fog the existing coating.

Using the previously generated circuit board design, a further drawing is required which must be on fully UV-transparent material (usually clear plastic film), and the tracks and pads etc must be completely opaque to UV light or failure is certain. The previously mentioned pens, inks, tapes and rub-on transfers can be

Fig 26.39: Exposure test strip for photo-sensitised PCB

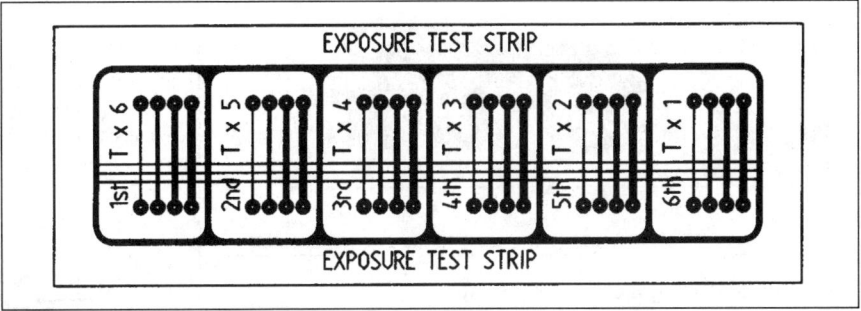

used to make this 'master' transparency. Ordinary rub-on transfers can be used, for there is no etchant to resist. If the 'as viewed from the components side' system of drawing was used, this should be continued for single-sided boards when applying the ink or the transfers etc. These ought to come into direct contact with the photosensitised copper surface of the PCB during the exposure stage to ensure the best possible clarity of lines etc. The master for the component side of a double-sided board should also be produced to ensure this.

A strip of the photosensitised board should be used to assess the exposure time required. **Fig 26.39** shows a suggested exposure testing strip, which should be made of the same materials as those used for the master transparency. Each section of the test strip is progressively uncovered until the complete strip has been exposed. For example, if the timer is set to, say, 20s then each step would be 20s greater than the previous one. This is normal photographic practice. The fully exposed strip is now developed. The image of the test strip should be visible, at least in sections, and at this stage the board is washed thoroughly in running water, dried and then etched. It is possible to assess the correct exposure without etching but etching is more reliable and much easier to see. The developing is usually carried out by using sodium hydroxide dissolved in water in the ratio of between 7 and 16g per litre, depending on the resist type. Cooled boiled water is preferred. During mixing, the sodium hydroxide tends to produce a boiling effect and great care should be taken to prevent contact with the eyes etc. The usual safety precautions should be observed: glasses, gloves, no children or pets etc. The mixed solution has a limited life and a fresh solution should be made for each batch of developing.

Some types of photosensitive resists use developers other than sodium hydroxide and most need careful handling, particularly in the home workshop.

Having established an exposure time, the set of PCBs can now be made. The master transparency is attached to the photo-sensitised board by one margin edge only, using either masking or double-sided tape. The assembly is placed face down towards the light source in the UV light box and the lid closed, ensuring that the master is in full contact with the photosensitised face of the board. Some exposure devices use a vacuum hold-down system to ensure this but, providing the board is not too big and the master transparency is allowed to roll into contact with the board, all should be well, with no air bubbles trapped to distort the image. Expose for the time determined by the test strip, develop, wash thoroughly and then allow the board to dry thoroughly before etching. If the coating has been hardened by baking, it is possible to go straight into the etching process, for the coating should not have softened too much during development. To give some idea of the time involved, providing the boards are correctly coated, it is possible to produce a board ready for etching about every five minutes using the set-up and methods described. In the case of double-sided boards, each side is exposed before developing, and great care is needed to ensure

that the previously exposed side is not fogged during the second exposure.

Registration is a problem with double-sided boards. The simplest method is to drill a few of the holes beforehand in a manner similar to that described in the one-off method and locate each 'master' to these holes. The holes are best drilled after the board is coated, otherwise the holes will cause variations in coating thickness. In the case of the ready-coated boards, the plastic protection is left in place during drilling. This is a technique commonly used in commercial prototype work and is referred to as spotting through. In this case all of the holes are drilled using computer control. Spotting through is made simple by holding the work against a non-UV light source.

Some constructors make a master transparency from either a published circuit or their own layout, using a photocopier. Unfortunately, some of these copiers do not produce opaque enough images for use with UV light. Also, a slight, non-uniform, scale change usually occurs, which makes hole centres inaccurate. This is particularly noticeable on the multi-pin dual-in-line ICs. A transparent photocopy of the exposure test strip shown in Fig 26.39 may be used to find out the quality of transparency produced, and this could save a considerable amount of wasted effort.

The completed boards made by either of the above methods should be washed thoroughly to remove all traces of etchant, and the resist, ink, transfers or paint should be removed using the appropriate solvent. Proprietary brush-cleaning solvent usually works. Some photosensitive resists need not be removed for they can be soldered through. Final drilling should be carried out and the last stage should be the removal of the handling margins. Commercial boards are usually tinned using either chemical plating or hot dip tinning. Tinning by running the soldering iron around the tracks is unnecessary and can cause de-lamination.

CONSTRUCTION USING SURFACE MOUNT TECHNOLOGY

Although most people have seen surface mount technology (SMT), relatively few without a professional involvement in electronics have used it in construction. SMT dominates commercial technology where its high-density capability and automated PCB manufacture compatibility are of great value. The arrival of SMT combined with the increasing dominance of monolithic IC parts over discrete components means that amateurs will have to move with the technology or rapidly lose the ability to build and modify equipment. SMT has an image of being difficult but this is not necessarily the case and it is quite practical for amateurs to use the technology. An incentive is that traditional leaded components are rapidly disappearing with many recently common parts no longer being manufactured and new parts not being offered in through-hole format. The cost of SMT parts in small quantities has fallen dramatically and through-hole construction will become increasingly expensive in comparison when it is possible at all.

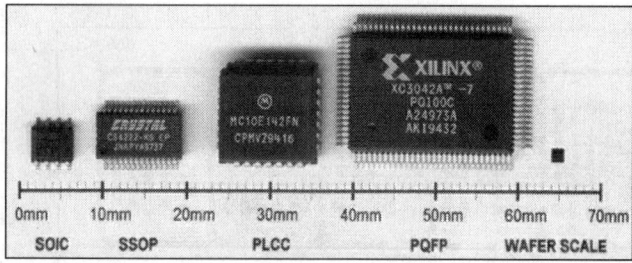

Fig 26.40: SMT integrated circuit packages

The variety of available SM components is enormous and growing rapidly. While the fundamental component types are familiar, the packaging, benefits and limitations of SM components are significantly different to their through-hole equivalents and a brief review is worthwhile.

Active Components

There is a wide range of packages in use at present and the competitive drives for miniaturisation, higher densities and lower costs mean that packaging designs are changing quite rapidly. It is not possible to discuss all types but there are a number of dominant styles that are worth gaining familiarity with. Some SMT packages are illustrated in the chapter on Passive Components.

The small outline diode (SOD) package is a simple two-leaded plastic package for diodes. There is a wide range of competing styles and sizes (SMA, B, C, D etc) and care must be taken to check the actual package style of individual parts.

The small outline transistor (SOT) package family is dominant for parts with three to five leads. The name is totally misleading as the package is used for diodes, ICs and transistors. The commonest types are the small SOT23 series used initially for discrete transistors and diodes in three-lead form and now typically in five-lead form for small monolithic parts such as op-amps and individual logic gates. A number of larger thermally efficient packages has emerged such as SOT223 for higher-power parts such as voltage regulators which are replacing the familiar TO220 through-hole parts.

Metal electrode leadless face (MELF) parts are essentially surface-mount versions of existing axial-leaded packages that have the leads removed and replaced by end metallisation. They are becoming less common as SOD and SOT packages dominate for devices of less than eight leads.

Modpacks are small four-leaded parts that are frequently used to house monolithic gain blocks. They are quite suitable for amateur construction and are very similar to the early RF transistor packages. They are normally a plastic package but sometimes ceramic package parts will be encountered. These must be treated with caution if the part cannot be identified, as a few higher-power parts use beryllium in their construction and a health hazard may exist if the parts are cracked or damaged.

The plastic leaded chip carrier (PLCC) package (**Fig 26.40**) is intended to be either socket mounted or soldered directly to a PCB. These are sometimes referred to as J-lead parts due to the way that the leads come down the body side before bending under around the body. They will probably become less common as they are relatively bulky but are fairly easy to work with.

The small outline package (SOP) or small outline integrated circuit (SOIC) housing is very common with a 1.27mm lead pitch and between eight and 28 leads but with various body widths. They are quite suitable for hand construction but are being displaced by the shrunken small outline package (SSOP) (**Fig 26.40**) and thin small outline package (TSOP) packages with 0.635mm

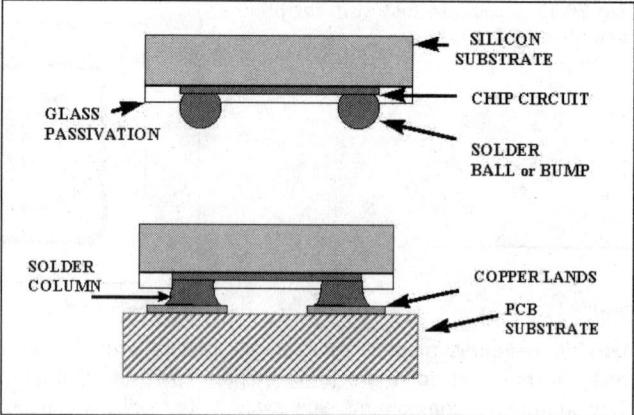

Fig 26.41: Wafer-scale package

and 0.5mm lead pitches. Although care is required to solder or lift individual pins on these denser parts it is relatively easy to solder or remove entire parts with the correct techniques. These parts are sometimes called gull wing due to the way the leads come out of the package when viewed from the end.

The plastic quad flat pack (PQFP) (Fig 26.40) is common for parts with between 40 and 208 leads which are arranged on four sides of a rectangular package. The lead pitches vary with package size from 0.5 to 0.8mm. Care is required to solder to or lift individual pins on these parts but it is possible to solder or remove entire parts with the correct techniques.

Amateurs are less likely to use ball grid array (BGA) packages that are becoming common on devices with high pin counts (typically 225-560 connections). The connections are underneath the device in the form of an array of small solder balls formed on top of connecting pads on the device base. The part is carefully placed on top of the PCB land pattern. During subsequent reflow soldering the balls are melted and surface tension draws the package slightly closer to the PCB, leaving small columns of solder linking the pads on the device to pads on the PCB. These devices require specialist tools and techniques to mount and X-ray inspection to assess the solder bonds which are almost entirely hidden under the device. BGA parts are not well suited to amateur use.

The wafer-scale packages (Fig 26.40) can be regarded for amateur purposes as a smaller version of the BGA with solder balls typically on a 0.5mm pitch. The reduction in size is possible since there is no actual package, simply a 'bare' IC with glass coating for protection and solder balls for connections (**Fig 26.41**). The solder balls are sitting directly on metallised areas on the chip surface which have been left clear of the outer glass passivation. The resulting device is little larger than the minimum size of the chip which enables parts with 100 connections to be fabricated at sizes of under 1cm square. This interconnect density hidden under the chip causes even more difficulties than the BGA. A recent development from National Semiconductors is to use this technology for low-lead-count ICs such as the LMC60351BP dual CMOS op-amp, sized 1.45mm square and 0.9mm high. This is a 95% PCB area reduction in comparison to an equivalent SOIC part. Wafer-scale packages may largely replace other packages due to the size and possible long-term cost advantage. There are problems, however, such as the need to match thermal expansion coefficients of chip and substrate as well as PCB design and manufacturing difficulties. When using surface-mount active devices additional care must often be taken over PCB design as many parts rely on conduction cooling via the PCB copper tracks to stay within design limits. Moderate power devices such as SOT223 regulators require

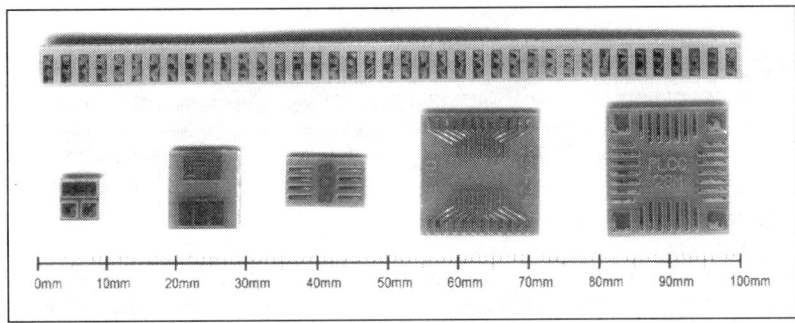

Fig 26.42: Mini Mounts

particular attention to the manufacturer's PCB specifications to ensure long-term reliability.

Passive Components

Surface mount resistor, capacitors, inductors and connectors, together with their component markings, are described in the chapter on Passive Components.

Construction Techniques

PCB

This is the ideal construction method and can make surface mount construction easier than through-hole construction due to the reduced number of holes that need to be accurately drilled. The limit of PCB etching resolution for the amateur may make finer-pitch ICs difficult to use but it is feasible to use SSOP and TSOP parts with care. An additional limit is imposed by the difficulty of making multi-layer PCBs and this makes larger parts more tricky to use. An alternative method to etching PCBs is the use of small PC-controlled milling systems that mill the non-required copper off a plain copper-clad PCB to leave the required track pattern. These are commercially available (at the price of a second-hand car!) but have been built by some amateurs using pen plotter mechanisms.

Veroboard

Strip board still has uses in surface mount work as resistors and capacitors can be soldered directly between tracks and across cuts in tracks. Transistors and diodes are also simple as SOT23 and SOT143 can also be mounted directly on the track side. The big problem is how to handle ICs but SOIC parts can just be managed since the pin pitch is 0.05in, enabling every second pin to be soldered to a 0.1in strip with the unsoldered ones bent up 90 degrees with flying leads soldered directly to them.

Mini Mounts

The German company Wainwright manufacture small pieces of self-adhesive PCB called Mini Mounts with land patterns for a wide variety of components (**Fig 26.42**). These can be placed onto either plain copper board or a PCB and inter-wired to rap-

idly build prototypes or one-offs. The pads are available in the UK via Wessex Electronics [1] but are normally subject to a minimum order.

Ugly construction

There are numerous styles of 'dirty' construction and most SMT parts can be used at least as easily as leaded components. One trick that can make things easier is to use small pieces of Veroboard to make 0.1in pitch pads or busses when glued onto plain copper board.

The method shown in **Fig 26.43** can be extended to make mounts for devices such as SOT23 which can then be glued down where required.

An even simpler method is to use self-adhesive copper tape intended for EMC purposes stuck on top of self-adhesive KaptonTM (polyimide) tape to form self-adhesive insulated tracks of pads (**Fig 26.44**). Kapton tape is often used as masking tape in PCB soldering operations and is very resistant to heat. The tapes can be cut to any desired shape or size then be stuck where required, eg on a copper ground plane. The only real problem is that the adhesive used on copper tapes is very soft when heated so care must be taken when soldering. It is best to tin the copper tape before cutting it to size as smaller parts will be more difficult to solder as the adhesive melts. Extending the method slightly, multiple layers of tape and foil can be used to run tracks past each other without wire links. A final variation would be to photo-etch the tape composite to obtain a self-adhesive flexible PCB structure but a better adhesive is really required for the copper tape. If 50-ohm transmission lines are required then the Kapton tape is problematic as track widths will be impracticably narrow due to very thin dielectric. Single-sided glassfibre PCB should then be used with the copper side used as a ground plane and the copper tape applied to the glassfibre side to form microstrip. The copper plane side can still be used with tape to add more circuitry.

When soldering this type of work ensure proper ventilation to avoid breathing fumes from flux, adhesives or plastics.

Soldering and desoldering

For hand soldering there is no requirement to use solder paste as standard soldering tools and materials are adequate. Similarly there is no need to glue (stake) components down before soldering and this is only done on certain automated soldering systems. When desoldering components, great care must be taken to avoid damaging the PCB as it is very easy to lift pads when removing components. Similarly, care must be taken to avoid excessive temperatures or heating for excessive times as components are heated much more directly than their through-hole equivalents and are thus more vulnerable. Working on valuable equipment should be left to the experienced as it is easy to damage assemblies when learning.

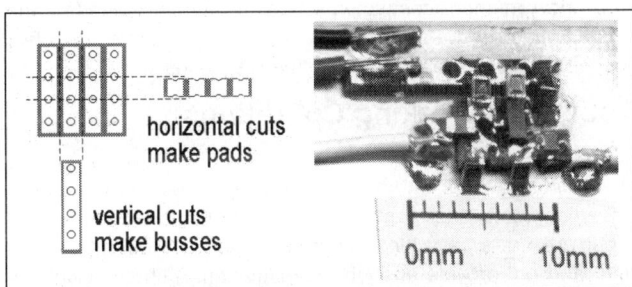

Fig 26.43: 'Ugly' construction using Veroboard strips

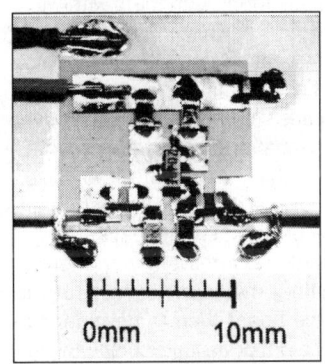

Fig 26.44: Kapton/copper tape construction

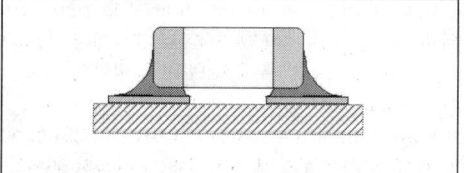

Fig 26.45: Ideal solder meniscus

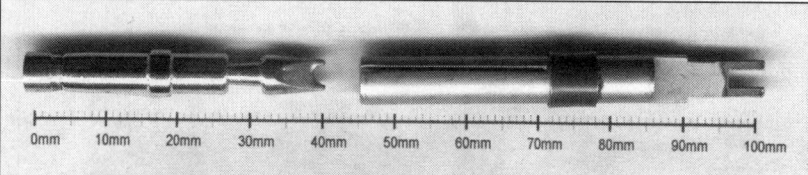

Fig 26.46: Desoldering bits

Resistors and capacitors are soldered quite easily and the key is to prepare the pads or lands prior to presenting the component. Tin the pads first and use solder wick to get a flat surface on one of the pads. Hold the component on the pads using tweezers and use the iron to heat the junction of the component and the unlevelled solder pad until the solder melts and the component is pressed gently so that it is parallel to the board. Apply a little solder to the opposite pad/component end and then a little more to the first end again. There should be a meniscus of solder at both ends of the component with a bright shiny surface (**Fig 26.45**).

Desoldering passive components is normally done with a pair of desoldering tongs or a special forked bit in a standard iron (**Fig 26.46**). If a spare iron is to hand then it is possible to desolder both ends simultaneously with two standard irons but with a little desperation and practice a single iron can be used to desolder these parts.

Soldering active devices requires more care but discrete semiconductors and SOIC ICs are quite easy. For these parts prepare the lands by tinning and flattening with solder wick. Hold the part in place with tweezers and without using solder press one of the leads down onto the pad for a couple of seconds using the iron to form a weak joint. Go to another lead and repeat the process so it should then be possible to release the part and it will not move. Then individually solder down the other leads and finally solder down the 'tacked' leads.

If specialist desoldering tools are not available desoldering SOIC ICs is easier if a slightly aggressive approach is taken. Using a large-sized solder bit, apply a large amount of solder quickly down one side if the IC; this will probably short out quite a few leads but this does not matter. Then get a thin pointed object such as a thin knife blade or a probe and, while pushing it in from under one end of the IC body so as to gently lift the soldered side, run the iron back over the newly applied solder so that it is simultaneously melted. This will allow one side of the IC to be lifted off the PCB. When it has cooled press the IC back to its original position and repeat the process on the other side so the IC becomes free. Done quickly, this method will not harm the IC and it can be reused if necessary.

Soldering fine pitch parts is less easy with hand tools but with care it is possible to attach wires to or lift individual pins on 0.5mm pitch parts. Soldering down or removing a whole part is actually slightly easier with a slight variation of the above techniques. To fit a part, first tin the pads then clean them with wick. Then present the part and gently push down some of the corner leads with a soldering iron to form weak joints. Using a large soldering bit and a good quantity of solder run the iron and solder along the outer end of a line of leads so that there is solder over every pin and possibly many of them shorted out. Repeat on the leads on the other side then use wick to remove the surplus solder - use an eyeglass to check that there are no shorts remaining. It is best to avoid getting solder between the leads and the component body as it is much harder to remove it from there.

If a part is being removed another method is to first cut the leads to remove the part from the board then to desolder the leads. This can be done with a scalpel or a similar blade by cut-

ting directly at the junction of the IC body and the emerging lead. With any of the techniques for working on surface mount boards it is best to practice them first, and scrap SMT consumer electronic assemblies are readily obtained for this purpose. It is very easy to accidentally remove PCB lands when removing components and some practice is essential.

Some final points must be made about cleanliness as it can cause significant problems with high-density boards. Flux splashes often cause mating problems on high-density connectors so ensure that these are kept protected when soldering and clean them afterwards. A general point with flux is that some very-high-impedance amplifiers must be de-fluxed after soldering as the flux can represent a significant shunt resistance. Tiny solder flecks can also cause shorts on connectors or between IC pins so again keep all work and work surfaces clean.

Commercial Manufacturing Methods

It is worth mentioning commercial methods as these give some idea as to what conditions components are designed to survive. The first consideration is should the board be surface mount or not? If it is necessary to use many through-hole parts it may be best to avoid mixing technologies and only use through-hole parts to arrive at the minimal cost solution.

The commercial boards use solder paste (tiny solder spheres within a liquid flux) which is applied by a silk screen process to the solder pads. To help avoid shorts it is normal to apply a varnish known as the solder resist over areas that are not to be soldered. The parts are then loaded onto the boards using automated 'pick and place' machines. Depending on the subsequent solder process, dots of glue may have been applied to the PCB prior to placing the components to help retention during soldering.

There are two basic types of process - reflow and wave soldering. Since boards may have components on both sides and pass through several processes, occasionally both processes may be used on a single assembly.

The wave soldering process originated with through-hole boards where the board is passed just above a bath of liquid solder. The solder has waves on its surface which brush the underside of the board and so solder the components - surface mount components will be totally immersed by this solder wave and will have been glued to the board. The reflow processes are simply means of melting the solder paste either by infra-red heating or by passing the assembly through a bath of high-temperature inert gas. With the reflow method, components are not normally glued to the board. Whatever the process components are completely subject to molten solder temperatures for many seconds.

Tools and Working Conditions

Lighting

Good lighting is essential and can be provided with relatively cheap domestic desk lights. The lamps that also include a magnifier are not always ideal since they may provide insufficient light and not provide enough magnification. A cheap jeweller's eyeglass is very useful - a magnification of about x9 is about optimum as very high magnification can be difficult to use.

Component storage and control

Components must be stored properly since the parts are small and fragile. Ideally they should be stored in whatever packaging they are delivered in and only removed when they are going to be used. Components should not be handled directly as oils, acids and salts from the skin will affect the solder quality. It is particularly important always to use tweezers to take things out of storage containers as fingers touching other components which are than going to remain in storage will result in significant corrosion before the stored components are finally used. Unless care is taken it is very easy to confuse small components - only have one container open at a time and seal the container as soon as the required part has been removed. It is very easy to drop and lose these parts - always keep the work surface clean and clear of extraneous items.

Tweezers

For handling small parts needle-pointed tweezers are essential and generally the types which are curved or angled are more convenient than the straight types.

Desoldering braid

Solder wick is invaluable for cleaning tinned solder lands prior to soldering components and removing excessive solder. Even on very small parts braid with widths of 2 to 3mm is generally most useful as the very fine braid has only a limited ability to remove solder.

Solder

Getting hold of the finest available is to be recommended but there do not appear to be any sources of small quantities for amateurs. Unfortunately very fine (26SWG) solder only appears to be available in 0.25kg reels and can cost two to three times as much as 18SWG solder. At present 22SWG is probably the finest readily and economically available size. The solder should be flux cored and of medium or low melting temperature. Remember that the iron temperature and the solder melting point must be compatible.

Flux

Flux is useful for soldering and desoldering larger parts but only use electronic fluxes and not the plumber's types.

Flux cleaner

Ideally the use of solvents and cleaners should be minimised for environmental and health reasons but even with 'no clean' solders it is sometimes necessary to remove flux from critical areas such as connector surfaces.

Soldering and desoldering bits

Surface mount parts are soldered directly with much shorter leads than through-hole components and it is therefore much easier to damage parts in the process. A temperature-controlled iron is ideal but small low-power irons can be used. A large copper area such as a ground plane is difficult to solder with low-power irons but a good PCB design can usually avoid having large copper areas directly contacting components.

It is useful to have a variety of bits with conical bits for fine work and chisel bits for larger components. Special desoldering bits (Fig 26.46) can be useful but if you only have a single iron it can be inefficient to be constantly changing the bits as you add and remove parts.

Soldering and fumes

When soldering, great care must be taken to avoid breathing fumes and it is advisable to wear safety glasses. The use of tiny components means that the eyes tend to be close to the point of soldering and are vulnerable to solder/flux splashes and fumes from adhesives (see also 'Safety' below).

Static control

Static control must be exercised to avoid damage to semiconductors. This damage is not always immediate in nature and can result in mysterious failures at a later date. Static control involves grounds and the amateur must make personal safety the main concern. Wrist straps and ground cords must contain appropriate current limiting resistors and should not be worn if high voltages are potentially exposed.

Safety

- When working with live equipment or soldering, eye protection is advised and most DiY stores have suitable items available.
- Soldering should take place in a well-ventilated place and the fumes should not be breathed.
- Caution must be exercised with any chemicals used such as solvents or PCB cleaners as some people are particularly vulnerable to these materials.
- Some SMT devices may contain beryllium in their construction - this represents a serious health hazard if the package is broken or damaged.

CABINETS

Housing the finished project is another important area. The cabinet is always on view and therefore should reflect the care and time devoted to the project. Commercially made cabinets appear to be the easy way out but often these cabinets do not meet the requirements exactly. They are either too big or too small. Also, they do not always allow for easy changes in layout. These restrictions, combined with the expense of commercial cabinets, are a great encouragement to make your own. This is

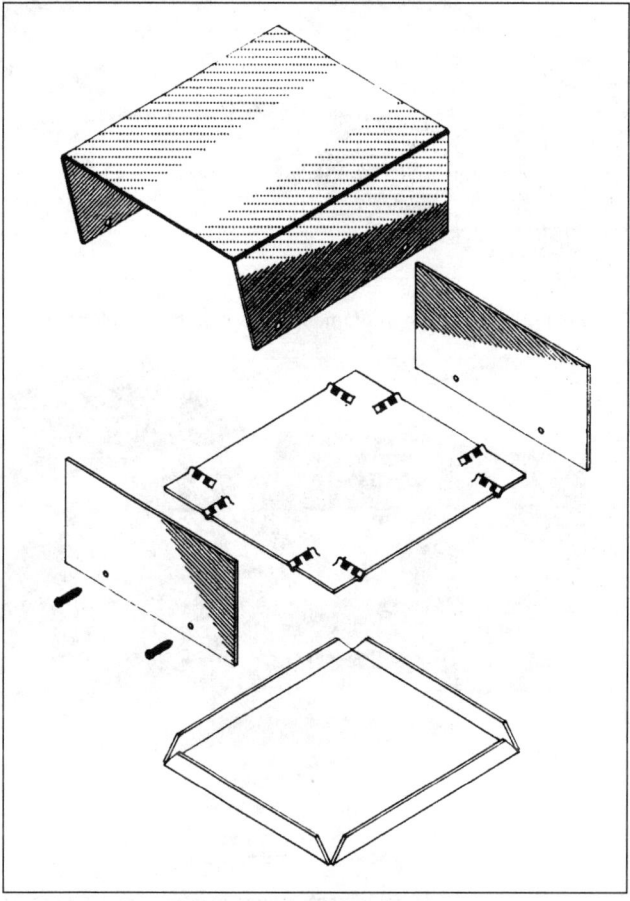

Fig 26.47: Small cabinet design with alternative base plates

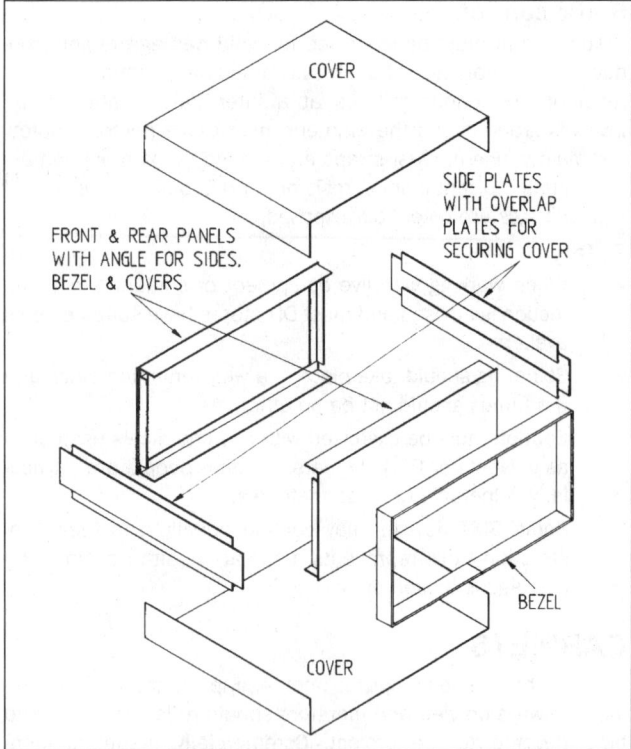

Fig 26.48: Large cabinet design

Fig 26.49: Home-made equipment housed in the small cabinet design

Fig 26.50: Home-made equipment housed in commercial boxes. The items in the foreground and on the right use the paper/Perspex panel system described in the text

a relatively easy thing to do, especially if the techniques and tools previously mentioned are acquired. It is also possible to re-use old cabinets from obsolete equipment, modifying these to suit your requirements.

Ideally any box or cabinet should be able to be altered with the minimum of effort and waste. To this end, the base and the front and rear panels should be separate items. The two cabinets shown in **Figs 26.47 and 26.48** are very easy to make and meet this requirement. That shown in Fig 26.47 is intended for the smaller projects, such as keyers, QRP ATUs and active filters etc. Typical sizes are 130mm wide, 80mm high and 150mm deep. It can be sized for larger work but the cover becomes difficult to make. The cover of the larger version really needs making from heavy-gauge material to reduce flexing, and bending becomes difficult. The cabinet shown in Fig 26.48 overcomes this problem and is suitable for such things as high-wattage ATUs and linears etc. The cabinets can be made from steel or aluminium, and a small version of Fig 26.47 can be made from tin plate.

Fig 26.47 shows two possible base plates. The flat base plate utilises the hole-embossing tool, and the edges of the cover should extend beyond the base plate to form feet. These feet can be edged with plastic strip of the type that is used to bind papers together. This plastic binding is available in various colours and, if used correctly, the finished job is very professional looking. The alternative bent base plate can be made using a former. The finished base plate and the former can then be used as a former to bend the cover, thus ensuring a good fit. The sides of the cover do not need extending if using the bent base plate.

In the transmitting radio shack, it is advisable to use metal boxes or cabinets to house electronic devices. This gives some measure of screening, which helps to reduce mutual RF interference. Plastic, wood and other non-conducting materials, though often making things easier to construct, do not provide such protection unless lined with metal foil which is grounded. These materials are best used to provide cabinet embellishments, such as bezels, feet, handles etc.

Fig 26.49 shows home-constructed equipment using the small cabinet or box. **Fig 26.50** shows home-constructed equipment fitted into commercially made boxes. Also, the front and right-hand side pieces of equipment shown in this diagram utilise the paper panel methods referred to later.

FINISHING

Having built the equipment, the urge to use it may be so great that there is no time to finish it off. This is a shame, because the finish affects the appearance, which in turn adds to the pleasure of using home-made equipment. The time spent on construction may be considerable but that spent on finishing is undoubtedly the most rewarding.

Painting

This is probably the most common form of finish used. The combination of two or more colours can enhance the equipment. The choice is limitless and it is usual to try to adopt some standard colour scheme. If commercial equipment is already in the shack, it is possible to try to match or blend with its colour

scheme. Brush painting very rarely gives a good surface over large areas and a considerable amount of work is required to burnish this form of painting into anything like a smooth finish. Brush-applied hammer or crinkle finish paints are useful for covering the outsides of cabinets but front panels look far superior and workmanlike with a monotone finish.

Spray painting has been available for sometime in the form of aerosol sprays and these are a reasonably cost-effective form of painting. A durable and pleasing appearance can be obtained, once the simple technique of using them has been learnt. The directions, usually given with product, for the method of application should be followed. The main area of difficulty usually arises when spraying the three-sided corners of a box. It often happens that too much paint goes on to one surface and runs; the other surfaces are left unpainted or just lightly coated. The trick is to spray each corner face in turn, masking-off the other faces, which may or may not be freshly painted, by holding a piece of cardboard in the path of the unwanted spray. A few practice runs on an unwanted piece of metal enables the technique of aerosol spray painting to be acquired quickly.

Before any painting is done, the surfaces should be suitably prepared, for the resultant ease of obtaining a smooth finish depends on the quality of the unpainted surface. The surface must be smooth and burr-free. This can be achieved quickly by using the flapwheel and drill mentioned previously or by rubbing down with emery cloth. In the case of aluminium, ensure that the emery cloth does not pick up the aluminium dust and form an abrasive lump which will score the surface. Any scratches or score marks should be smoothed out or, after priming, filled with one of the proprietary fillers which should be blended-in completely to leave a smooth flat surface. The abrasive block should be used for the final rub-down. The surfaces should then be washed and degreased thoroughly with scouring powder and the scouring pad. The now clean and grease-free surface should not be handled, for any finger marks will affect the finish.

Surfaces can also be prepared and sometimes finished by etching but, as most of the substances used for this technique are hazardous, this method is not recommended for the home workshop. A solution of ferric chloride will etch to produce a reasonable ready-to-paint surface on copper, brass, mild steel and aluminium. The surfaces must be clean and grease-free otherwise uneven etching will occur.

The next stage in painting is priming. Most aerosol paints have details of the recommended primer to use printed on the can, and these instructions should be followed because the manufacturer is as anxious as you that the painting works properly. If no instructions are given then the following types of primer should be used:

- For steels use the oxide primers to suit the type of paint that will be used for the final finish.
- For aluminium use the etch-type primers.
- If it really is necessary to paint brass or copper then the oil-bound undercoat paints should be used. Zinc or aluminium-based primers do not always work well on brass or copper and tend to lift after a short period of time.

After priming, and when the work has dried thoroughly, carry out any filler work necessary. When this has dried a light rubdown with very fine wet and dry paper will smooth off ready for the finishing coat or coats in the colour of your choice.

Many of the modern enamel spray paints require neither primers nor undercoats and still produce a durable pleasing finish, providing the surface of application has been well prepared. There are so many types of paint to choose from that a little personal experimenting is advisable. Most paint manufacturers will supply literature on their products, giving full details of the methods of application, durability etc. A few types are listed here.

Acrylic paints
A good gloss finish is possible with these reasonably priced paints but they are susceptible to some cleaning solvents. They are suitable for use on most metals or plastics and are available in aerosol cans. These paints are probably the best all-round paint for the finishing of home-constructed projects. Some of the hammer finish paints are of this type.

Bitumen paint
A useful and cheap paint. It is protective rather than decorative and is suitable for the protection of masts and other outdoor metal work. Usually it gives a black semi-gloss finish and is normally brush applied. White spirit can be used to clean brushes etc.

Cellulose dope
This is a fast-drying liquid which, like shellac, can be used as an adhesive or a coating to secure and waterproof. Unlike shellac, the dope may dissolve some plastics. It is usually available from model aircraft shops in clear or coloured form and is reasonably priced. The coloured dopes contain pigments and in some cases metal powder. These may provide an electrical conductive coating. The non-shrinking clear dope weathers very well. It is a good protective coating for beam elements and does not appear to upset the beam's performance. Normally it is brush applied, although it can be thinned with cellulose thinners for use in DiY-type paint sprayers.

Cellulose lacquer
A reasonably priced, quick-drying lacquer, giving a high and reasonably durable gloss which can be improved by polishing. It can be affected by some cleaning solvents. The rapid drying properties considerably reduce the risk of dust-marred work. It is available in aerosol cans and as normal paint. Special thinners are required for cleaning brushes etc.

Stoving paints
These are usually of the acrylic or alkyd types, which rely on stoving to complete the chemical changes of drying. They are usually applied by spray and the resultant finishes are hard, durable and can be very glossy. This is about the best paint finish possible but the stoving and spraying requirements usually present a problem in the home workshop. However, some commercial concerns do offer a small-quantity painting and stoving service, which is worth considering for that extra-special job. Some car body repair shops have these facilities also.

The number of coats of paint and the method of application is entirely a matter of choice. All painting should be carried out in an atmosphere as warm, dry and dust-free as possible, and in very well-ventilated conditions. If the painted work can be placed in an electrically heated oven at around 70°C, drying will be greatly assisted.

Putting the work in a tin with a lid and placing this on top of a stove or central heating boiler is the next best thing, for it will keep dust from marring the work during drying. Rubbing down and cutting-in compounds supplied for car paint retouching are also very useful in obtaining that final smooth finish.

Other Finishing Materials

Shellac lacquer
A lacquer which is made by dissolving shellac flakes in methylated spirits or similar solvents. This is a well-tried, inexpensive waterproofing, insulating and light-duty adhesive-cum-lacquer. It can be used to protect antenna elements and connectors, and

to secure wires, coils and components. The lacquer dries to a hard brown translucent finish. The drying process can be speeded up by the application of heat.

The shellac flakes and methylated spirits are normally available from dispensing chemists, or it can be bought ready made as French/Button polish or knotting compound. The ready-made lacquers are usually very 'watery' and may contain solvents other than methylated spirits.

Self-adhesive plastic sheets

These are an alternative finish that can be used to good effect on the outsides of cabinets and boxes, saving a great deal of painting and rubbing down. Leather-effect cloth provides a very professional touch, particularly on the larger surfaces such as covers of ATUs, PSUs etc. This cloth, if not self-adhesive, may be glued down using contact adhesive very evenly applied to both surfaces. The previously mentioned panel cleaning should be carried out before applying either of these types of sheet finishings. The main problem with using these sheets is that of shrinkage and the edges curling with age. This can be eliminated or reduced by heating the panel to around 60°C (just too hot to hold) immediately before applying the sheet. Also, wrapping the sheet around the edges can sometimes improve the problem. With wrapping, edge bulge can become obtrusive. Trapped air is the bane of this process (a bit like wall-papering), and the knack is to peel-on the sheet using a straight edge or a rod to peel against from the non-adhesive side.

Self-adhesive coloured or metallic finish tapes can be used to line-in features of the panel or cabinet. There is a strong tendency, due to the ease of application, to overdo this and finish up with a cross between a juke box and a 1960s car radiator grill.

Lettering Panels

This is now very easy and, though machine engraving is still considered the best, it is little used because of cost. Most modern commercial equipment has the lettering screen printed on to panels, knobs and push-buttons etc. Screen printing can be another home workshop process.

The wet slide-on transfers are effective and can be complementary to the more popular rub-on lettering such as Letraset™, Chartpak™ and others. These are expensive and it is worth considering carefully what style and sizes suit your tastes. The Helvetica Medium style seems to match the lettering on most commercial equipment, and is available in black or white in various point sizes. The range 10, 12 and 14 point should cover most requirements. A sheet of numbers only in the same range is also worthwhile. Variations of importance can be made using a combination of upper and lower case, with different point sizes of the same style. Mixing styles of lettering is not usually aesthetically effective.

Sometimes these rub-on letters can be difficult to apply, usually because the surface is uneven, cold, damp, greasy or too glossy. A very light rub-down with wet and dry paper should be adequate preparation for the surface. The paint will have a matt finish but this is taken care of at a later stage after lettering. The pre-release method will usually ensure success, particularly on the smaller characters. To pre-release the required character, the lettering sheet should be placed on to its non-stick backing paper and, with the two sheets resting on a hard surface, the area of the character should be rubbed over as normal. A change of texture can be seen through the carrier sheet when the character has released. In this state, the character, still lightly attached to its carrier sheet, may be applied where required, and usually a very light pressure will allow the carrier sheet to be

The earlier chapter on the PIC-A-STAR transceiver project contains much additional information on construction strategies and techniques

withdrawn, leaving the character in place. The newly placed character must be burnished immediately into place through the non-stick paper, to ensure that it will not be lifted off during the next character application.

Alignment of the lettering during application is facilitated by using the guide markings on the lettering sheet. Some have the guides under each letter and some at the ends of each row of lettering. A strip of card can be arranged as a ruler guide and its alignment checked using the combination square. Lining-up the bottom of each letter as the lettering progresses is difficult and will not produce correctly aligned work. Also, for some reason, there is a tendency to gradually curve downwards, particularly if referencing from the immediately previous character. Spacing the characters equally, say around the centre line of a switch or volume control, is best carried out by working outwards from the centre letter or space. Typesetting is an art and a few experiments with the chosen style can help in getting things right on the finished work. The guide-bar under each letter on some lettering sheets is also a guide to spacing. Each bar should be positioned such that a continuous straight line is produced, and then the lettering will be at the correct spacing for that style. Rubbing on characters is a slow process, but this serves to ensure that only essentials are given, which in turn produces a good workmanlike labelled panel, not an instruction book! For interest, using the smallest letters available, it is worthwhile to date the work, say, in the bottom right-hand corner - this way you can see how your workmanship has improved over the years!

Unfortunately, transfers are easily rubbed off and require some form of protective covering. The quickest way is to spray the completed panel with clear lacquer but some lacquers react with the surface of application and this should be checked beforehand. Another effective way is to face-up the work using a thin sheet of transparent Perspex®, which can be secured to the panel using screws and/or the existing component fixings. Self-adhesive clear sheets can also be used but this has the problem that they must be applied right first time, for any attempt to peel off and start again will remove all of the lettering. This technique can be used to remove unwanted or misplaced letters at the time of lettering and is superior to scraping. A quick dab with the sticky side of masking tape on the unwanted letter usually does the trick.

If clear plastic sheet is used to face the panel, other possibilities occur, which can eliminate painting the latter. Paper of the coloured art sort can be cut to size with all holes cut to match the panel, and the lettering etc applied to this. The lettered paper or card can be secured to the panel by the Perspex® sheet. It should not be glued to the panel, for this will cause the paper or card to buckle and not sit flat. If a computer with a desktop publishing (DTP) package and printer is available, this can be used to produce paper/card panels very easily. Coloured card (which will pass through the printer) has been found very effective with this system. When properly applied, this paper or card system of panel marking and covering is very effective. A very pleasing effect can be produced by placing the coloured or metal finish narrow, self-adhesive tape, as previously mentioned, along the edges of the card/paper. Do not try to stick the card to the panel in this manner, as the differential expansion and contraction will cause the card to buckle even under the Perspex®.

Etch engraving is also very effective and the same techniques used in the production of printed circuits are applied. It is a system worthwhile experimenting with in those lulls between projects.

Self-adhesive strip labels made on hand-operated lettering machines are a useful way of marking controls but do not add much to the appearance of the equipment. However, this type of strip is very useful in making Braille labels for equipment used by a blind operator. Matrix board (0.1in pitch) can be used as a guide for the Braille characters, for it is about the correct spacing for the Braille dots. These are formed through the guide and from the peel-off backing side of the tape, using a suitably thinned centre-punch, with the tape resting on a strip of rubber or soft plastic.

Home-made dials for dip oscillators and field strength meters etc can be made by attaching card on to a metal or plastic supporting disc and then calibrating and lettering as required. A final spray with clear lacquer and a lasting job is produced. Using these lettering techniques, it is also possible to recalibrate meter dials and so utilise functional meters which others would have thrown away.

Other Finishes

Plating, anodising and colouring etc which have to be carried out by an appropriate firm are well worthwhile but a few words of advice may be helpful. Most finishing firms do not pre-polish or dress the work prior to finishing. With plating, the article is usually plated 'as received', scratch marks and all. The techniques described in the sections on materials and tools should be applied therefore as required to obtain a smooth and scratch-free surface before any item is sent for plating. Aluminium is usually anodised and coloured and does not require a highly polished surface, only a scratch-free one. The contractor may, on request, carry out some form of etch treatment prior to anodising. It is possible for lettering and other characters to be colour dyed on to aluminium during anodising and a master 'see-through' layout of this, to size, should be supplied to the finisher if this is required. These finishes are expensive but do provide corrosion protection as well as enhancing appearance.

If none of the aforementioned methods of finishing appeal, an old-fashioned method can be used to produce whirl patterns on the surface of the work. This is done by sticking a disc of emery cloth on the end of a piece of 12mm diameter dowel rod mounted in the electric drill. The patterns can be generated, either randomly or sequentially to suit, by bringing the emery/dowel straight down on to the work, holding it there briefly, and repeating as required. The same problem of metal pick-up with aluminium can occur. This process, derived from the days of hand scraping, is a quick way of visually achieving what was then considered a quality finish. Nowadays, it is considered wasteful and uneconomical! Yet another alternative is to 'grain' the work, using either fine emery cloth or an abrasive eraser, retaining and enlivening this surface by a spray of clear lacquer.

The polishing of cut edges or the removal of scratch marks on Perspex® and most plastics can be done by sequentially filing, emerying (using first the medium grade and working down to the Crocus paper grade) and finishing off with a mixture of metal polish and powder cleaner, the proportions of which are reduced until the final polish is achieved with metal polish alone. Commercial compounds are available to do the same job. A hard felt pad mounted on the end of a drill can also be used with the normal buffing compounds to polish and remove the sharp edges. However, care should be taken with this method of polishing to avoid overheating which produces an 'ocean wave' effect on the surface being polished!

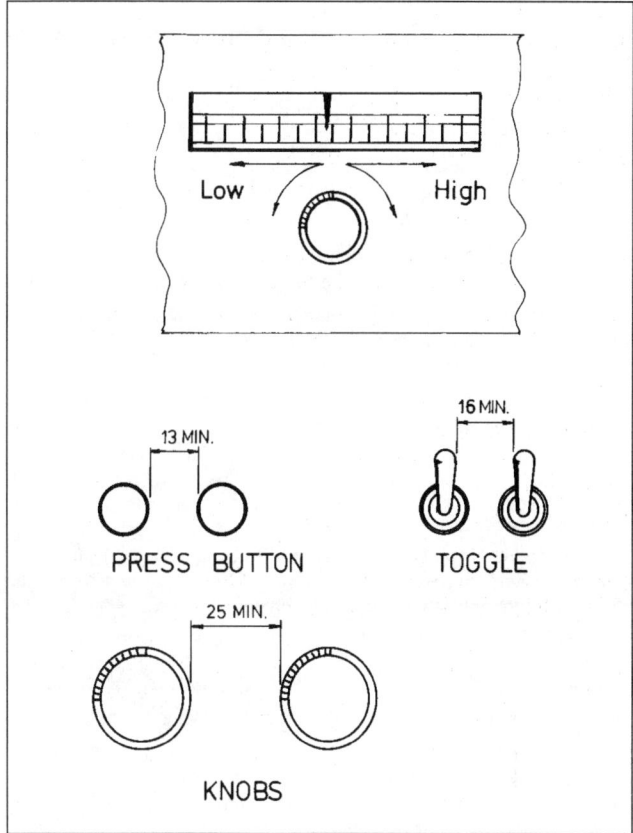

Fig 26.51: Suggested ergonomic strategy for some controls

DESIGN THOUGHTS

Home construction allows individual ideas to be designed into the project being made. With commercial equipment 'you pays your money' but you don't always get exactly what you would like. Some controls you find unnecessary and others which you want are not there or are hidden inside the case. These factors are often overlooked until you make something for yourself and then aspects of commercial designs take on a different meaning. The quality or lack of it becomes apparent and it is very difficult to acquire this appraisal ability without having made, or having tried to make, something for yourself.

Over the years there has become an awareness of the value of ergonomics (see **Fig 17.51**). Most of us now know that controls should be positioned according to their purpose and that this purpose should be self-evident from either the type of control or the area in which it is placed. For example, most transceiver tuning knobs appear adjacent to the tuning display and, when rotated clockwise, the frequency increases. Usually, it is evident by looking at most transceivers which is the tuning knob. Unfortunately the same cannot be said about press buttons and other controls. The use of press buttons needs great care, especially where accidental operation could cause damage. If correctly designed, it should be impossible to accidentally press two or more buttons at the same time.

Also, if such an action is done accidentally or deliberately, it should not destroy the device, or send it into an unusual operating or locked-up state. Ideally, any button-type control should be accompanied by an indicator which shows what mode the button is in. Power supply switches are particularly important in this respect, especially with portable equipment, where it is all too easy to switch it on and not be aware of this fact.

Safety recommendations for the amateur radio workshop

1. All equipment should be controlled by one master switch, the position of which should be well known to others in the house or club.

2. All equipment should be properly connected to a good and permanent earth (but see the box about PME in the chapter on Practical HF Antennas, and Note A).

3. Wiring should be adequately insulated, especially where voltages greater than 500V are used. Terminals should be suitably protected.

4. Transformers operating at more than 100V RMS should be fitted with an earthed screen between the primary and secondary windings or have them in separate slots in the bobbin.

5. Capacitors of more than 0.01µF capacitance operating in power packs etc (other than for RF bypass or coupling) should have a bleeder resistor connected directly across their terminals. The value of the bleeder resistor should be low enough to ensure rapid discharge. A value of 1/C megohms (where C is in microfarads) is recommended. The use of earthed probe leads for discharging capacitors in case the bleeder resistor is defective is also recommended. (Note B). Low-leakage capacitors, such as paper and oil-filled types, should be stored with their terminals short-circuited to prevent static charging.

6. Indicator lamps should be installed showing that the equipment is live. These should be clearly visible at the operating and test position. Faulty indicator lamps should be replaced immediately. Gas-filled (neon) lamps and LEDs are more reliable than filament types.

7. Double-pole switches should be used for breaking mains circuits on equipment. Fuses of correct rating should be connected to the equipment side of each switch in the live lead only. (Note C.) Always switch off before changing a fuse.

8. In metal-enclosed equipment install primary circuit breakers, such as micro-switches, which operate when the door or lid is opened. Check their operation frequently.

9. Test prods and test lamps should be of the insulated pattern.

10. A rubber mat should be used when the equipment is installed on a floor that is likely to become damp.

11. Switch off before making any adjustments. If adjustments must be made while the equipment is live, use one hand only and keep the other in your pocket. Never attempt two-handed work without switching off first. Use good-quality insulated tools for adjustments.

12. Do not wear headphones while making internal adjustments on live equipment.

13. Ensure that the metal cases of microphones, Morse keys etc are properly connected to the chassis.

14. Do not use meters with metal zero-adjusting screws in high-voltage circuits. Beware of live shafts projecting through panels, particularly when metal grub screws are used in control knobs.

15. Certain chemicals occur in electronic devices which are harmful. Notable amongst these are the polychlorinated biphenyls (PCBs) which have been used in the past to fill transformers and high-voltage capacitors and beryllium oxide (BeO) which is used as an insulator inside the case of some high-power semiconductors. In the case of PCBs, the names to look out for on capacitors are: ARACLOR, PYROCHLOR, PYRANOL, ASBESTOL, NO-FLAMOL, SAF-T-KUL and others [3]. If one of these is present in a device, it must be disposed of carefully. The local Health and Safety Authority will advise. In the case of beryllium oxide, the simple rule is DON'T OPEN ANY DEVICE THAT MAY CONTAIN IT.

Note A. - Owing to the common use of plastic water main and sections of plastic pipe in effecting repairs, it is no longer safe to assume that a mains water pipe is effectively connected to earth. Steps must be taken, therefore, to ensure that the earth connection is of sufficiently low resistance to provide safety in the event of a fault. Checks should be made whenever repairs are made to the mains water system in the building.

Note B. - A 'wandering earth lead' or an 'insulated earthed probe lead' is an insulated lead permanently connected via a high-power 1kΩ resistor or a 15W 250V lamp at one end to the chassis of the equipment; at the other end a suitable length of bare wire with an insulated handle is provided for touch contacting the high-potential terminals to be discharged.

Note C. - Where necessary, surge-proof fuses can be used.

Rotary switches are less susceptible to accidental switching but they must be positioned and have suitable style knobs to facilitate switching. Most rotary switches occupy more space than push buttons and it is this which usually restricts their use on the compact rigs of today. Rotary switches are ideal for such controls as mode and band selection. One look at the switch position shows immediately what mode or band is selected. It is impossible to attempt to select two modes or two bands and there is no need for any other form of indicator: a power-saving factor also.

Toggle and slide switches are an equally effective form of self-indicating switches but toggle switches are prone to accidental operation.

Potentiometric rotary controls, such as AF volume, Morse speed and RF gain etc should all be self-indicating. Also, clockwise rotation should increase the function. Rotary concentric controls should give each control a related and easily identified function.

Knob styles can also affect the quality of presentation of the finished work. Unfortunately, knobs are expensive but ex-equipment knobs can often be used as a cheaper alternative. The size of knob should relate to the accuracy required from the function it controls. For example, a tuning knob would not feel right if it was less than say 40mm diameter, but the same diameter for a volume control would be unsuitable and usually unnecessary. Slider controls are very good, especially when simultaneous operation is required on such as faders and mixers, but this facility is not usually required on radio equipment. These controls also take up space if they are to be accessed easily. Some commercial amateur radio equipment use the smaller type of this control for the less-utilised or preset controls.

Many more factors can be found, and it is discovering these which adds to the pleasure of home construction.

REFERENCES

[1]   http://www.wessexelectronics.co.uk/

Appendix A
General Data

Capacitance

The capacitance of a parallel-plate capacitor is:

$$C = \frac{0.224\ KA}{d}\quad \text{picofarads}$$

where K is the dielectric constant (air = 1.0), A is the area of dielectric (sq in), and d is the thickness of dielectric (in).

If A is expressed in centimetres squared and d in centimetres, then:

$$C = \frac{0.0885\ KA}{d}\quad \text{picofarads}$$

For multi-plate capacitors, multiply by the number of dielectric thicknesses.

The capacitance of a coaxial cylinder is:

$$C = \frac{0.242}{\log_{10}(D/d)}\quad \text{picofarads per centimetre length}$$

where D is the inside diameter of the outer and d is the outside diameter of the inner.

Capacitors in series or parallel

The effective capacitance of a number of capacitors in *series* is:

$$C = \frac{1}{\dfrac{1}{C_1} + \dfrac{1}{C_2} + \dfrac{1}{C_3} + \text{etc}}$$

The effective capacitance of a number of capacitors in *parallel* is:

$$C = C_1 + C_2 + C_3 + \text{etc}$$

Characteristic impedance

The characteristic impedance Z_0 of a feeder or transmission line depends on its cross-sectional dimensions.

(i) *Open-wire line*:

$$Z_0 = 276\ \log_{10}\frac{2D}{d}\quad \text{ohms}$$

where D is the centre-to-centre spacing of wires (mm) and d is the wire diameter (mm).

(ii) *Coaxial line*:

$$Z_0 = \frac{138}{\sqrt{K}}\ \log_{10}\frac{d_o}{d_i}\quad \text{ohms}$$

where K is the dielectric constant of insulation between the conductors (eg 2.3 for polythene, 1.0 for air), d_o is the inside diameter of the outer conductor and d_i is the diameter of the inner conductor.

Decibel

The *decibel* is the unit commonly used for expressing the relationship between two power levels (or between two voltages or two currents). A decibel (dB) is one-tenth of a *bel* (B). The number of decibels N representing the ratio of two power levels P_1 and P_2 is 10 times the common logarithm of the power ratio, thus:

$$\text{The } ratio\ N = 10\ \log_{10}\frac{P_2}{P_1}\quad \text{decibels}$$

If it is required to express *voltage* (or *current*) ratios in this way, they must relate to identical impedance values, ie the two different voltages must appear across equal impedances (or the two different currents must flow through equal impedances). Under such conditions the *power* ratio is proportional to the square of the *voltage* (or the *current*) ratio, and hence:

$$N = 20\ \log_{10}\frac{V_2}{V_1}\quad \text{decibels}$$

$$N = 20\ \log_{10}\frac{I_2}{I_1}\quad \text{decibels}$$

Dynamic resistance

In a parallel-tuned circuit at resonance the dynamic resistance is:

$$R_D = \frac{L}{Cr} = Q\omega L = \frac{Q}{\omega C}\quad \text{ohms}$$

where L is the inductance (henrys), C is the capacitance (farads), r is the effective series resistance (ohms), Q is the Q-value of the coil and $\omega = 2\pi \times$ frequency (hertz).

Frequency – wavelength – velocity

The velocity of propagation of a wave is:

$$v = f\lambda\quad \text{metres per second}$$

where f is the frequency (hertz) and λ is the wavelength (metres).

For electromagnetic waves in free space the velocity of propagation v is approximately 3×10^8 m/s and, if f is expressed in kilohertz and λ in metres:

$$f = \frac{300{,}000}{\lambda}\quad \text{kilohertz}$$

$$\lambda = \frac{300{,}000}{f}\quad \text{metres}$$

$$\text{Free space } \frac{\lambda}{2} = \frac{492}{\text{MHz}}\quad \text{feet}$$

$$\text{Free space } \frac{\lambda}{4} = \frac{246}{\text{MHz}}\quad \text{feet}$$

Note that the true value of v is 2.99776×10^8 m/s.

Impedance

The impedance of a circuit comprising inductance, capacitance and resistance in series is:

$$Z = \sqrt{R^2 + \left(\omega L - \frac{1}{\omega C}\right)^2}$$

where R is the resistance (ohms), L is the inductance (henrys), C is the capacitance (farads) and $\omega = 2\pi \times$ frequency (hertz).

Inductors in series or parallel

The total effective value of a number of inductors connected in *series* (assuming that there is no mutual coupling) is given by:

$$L = L_1 + L_2 + L_3 + \text{etc}$$

If they are connected in *parallel*, the total effective value is:

$$L = \frac{1}{\dfrac{1}{L_1} + \dfrac{1}{L_2} + \dfrac{1}{L_3} + \text{etc}}$$

When there is mutual coupling M, the total effective value of two inductors connected in series is:

$$L = L_1 + L_2 + 2M \text{ (windings aiding)}$$

$$\text{or } L = L_1 + L_2 - 2M \text{ (windings opposing)}$$

Ohm's Law

For a unidirectional current of constant magnitude flowing in a metallic conductor:

$$I = \frac{E}{R} \qquad E = I R \qquad R = \frac{E}{I}$$

where I is the current (amperes), E is the voltage (volts) and R is the resistance (ohms).

Power

In a DC circuit, the power developed is given by:

$$W = E I = \frac{E^2}{R} = I^2 R \text{ watts}$$

where E is the voltage (volts), I is the current (amperes) and R is the resistance (ohms).

Q

The Q-value of an inductance is given by:

$$Q = \frac{\omega L}{R}$$

where L is the inductance (henrys), R is the effective resistance (ohms) and $\omega = 2\pi \times$ frequency (hertz).

Reactance

The reactance of an inductance and a capacitance respectively is given by:

$$X_L = \omega L \text{ ohms}$$

$$X_C = \frac{1}{\omega C} \text{ ohms}$$

where L is the inductance in henrys, C is the capacitance in farads and $\omega = 2\pi \times$ frequency (hertz).

The total reactance of an inductance and a capacitance in series is $X_L - X_C$.

Resistors in series or parallel

The effective value of several resistors connected in *series* is:

$$R = R_1 + R_2 + R_3 + \text{etc}$$

When several resistors are connected in *parallel* the effective total resistance is:

$$R = \frac{1}{\dfrac{1}{R_1} + \dfrac{1}{R_2} + \dfrac{1}{R_3} + \text{etc}}$$

Resonance

The resonant frequency of a tuned circuit is given by:

$$f = \frac{1}{2\pi\sqrt{LC}} \text{ hertz}$$

where L is the inductance (henrys) and C is the capacitance (farads).

If L is in microhenrys (μH) and C is picofarads (pF), this formula becomes:

$$f = \frac{10^3}{2\pi\sqrt{LC}} \text{ megahertz}$$

The basic formula can be rearranged thus:

$$L = \frac{1}{4\pi^2 f^2 C} \text{ henrys}$$

$$C = \frac{1}{4\pi^2 f^2 L} \text{ farads}$$

Since $2\pi f$ is commonly represented by ω, these expressions can be written as:

$$L = \frac{1}{\omega^2 C} \text{ henrys}$$

$$C = \frac{1}{\omega^2 L} \text{ farads}$$

See Figs 23.1 and 23.2.

Time constant

For a combination of inductance and resistance in series the time constant (ie the time required for the current to reach $1/\varepsilon$ or 63% of its final value) is given by:

$$t = \frac{L}{R} \text{ seconds}$$

where L is the inductance (henrys) and R is the resistance (ohms).

For a combination of capacitance and resistance in series, the time constant (ie the time required for the voltage across the capacitance to reach $1/\varepsilon$ or 63% of its final value) is given by:

$$t = C R \text{ seconds}$$

where C is the capacitance (farads) and R is the resistance (ohms).

Transformer ratios

The ratio of a transformer refers to the ratio of the number of turns in one winding to the number of turns in the other winding. To avoid confusion it is always desirable to state in which sense the ratio is being expressed, eg the 'primary-to-secondary' ratio n_p/n_s. The turns ratio is related to the impedance ratio thus:

$$\frac{n_p}{n_s} = \sqrt{\frac{Z_p}{Z_s}}$$

where n_p is the number of primary turns, n_s is the number of secondary turns, Z_p is the impedance of the primary circuit (ohms) and Z_s is the impedance of the secondary circuit (ohms).

COIL WINDING

Most inductors for tuning in the HF bands are single-layer coils and they are designed as follows. Multilayer coils will not be dealt with here.

The inductance of a single-layer coil is given by:

$$L\ (\mu H) = \frac{D^2 \times T^2}{457.2 \times D + 1016 \times L}$$

where D is the diameter of the coil (millimetres), T is the number of turns and L is the length (millimetres). Alternatively:

$$L\ (\mu H) = \frac{R^2 \times T^2}{9 \times R + 10 \times L}$$

where R is the radius of the coil (inches), T is the number of turns and L is the length (inches).

Note that when a ferrite or iron dust core is used, the inductance will be increased by up to twice the value without the core. The choice of which to use depends on frequency. Generally, ferrite cores are used at the lower HF bands and iron dust cores at the higher. At VHF, the iron dust cores are usually coloured purple. Cores need to be moveable for tuning but fixed thereafter and this can be done with a variety of fixatives. A strip of flexible polyurethane foam will do.

Designing inductors with ferrite pot cores

This is a simple matter of taking the *factor* given by the makers and multiplying it by the square of the number of turns.

Example

A RM6-S pot core in 3H1 grade ferrite has a 'factor' of 1900 nanohenrys for one turn. Therefore 100 turns will give an inductance of:

$$100^2 \times 1900nH = 10000 \times 1900nH = 19mH$$

There are a large number of different grades of ferrite; for example, the same pot as above is also available in grade 3E4 with a 'factor' of 3300. Manufacturers' literature should be consulted to find these 'factors'.

Table A.1

Diameter (mm)	Approx SWG	Turns/cm	Turns/in
1.5	16–17	6.6	16.8
1.25	18	7.9	20.7
1.0	19	9.9	25
0.8	21	12.3	31
0.71	22	13.9	35
0.56	24	17.5	45
0.50	25	19.6	50
0.40	27	24.4	62
0.315	30	30.8	78
0.25	33	38.5	97
0.224	34–35	42.7	108
0.20	35–36	47.6	121

Note: SWG is Imperial standard wire gauge. The diameters listed are those which appear to be most popular; ie they are listed in distributor's catalogues. The 'turns/cm' and 'turns/in' are for enamelled wire.

Table A.3. Wire table

Diameter (mm)	Approx SWG	Max current (A)	Fusing current (A)	Resistance at 20°C (Ω/km)
2.5	12	7.6	325	3.5
2.0	14	4.9	225	5.4
1.5	16–17	2.7	147	9.7
1.0	19	1.2	81	22
0.71	22	0.61	46	43
0.5	26	0.30	28	87
0.25	32	0.076	10	351
0.20	36	0.049	7.1	541

'Max current' is the carrying capacity at 1.55A/mm². This is a very conservative figure and can usually be doubled. The 'fusing current' is approximate since it depends also on thermal conditions, ie if the wire is thermally insulated, it will fuse at a lower current.

Table A.2. Coaxial cables

Type	Nominal impedance (ohms)	Outside diameter (mm)	Velocity factor	Capacitance (pF/m)	Attenuation per 30m (100ft) of cable 10MHz (dB)	100MHz (dB)	1000MHz (dB)
RG58-U/UR43	52	5.0	0.66	100	1.0	3.3	10.6
RG213/UR67	50	10.3	0.66	100	0.68	2.26	8.0
Westflex 103	50	10.3	0.85	78	0.27	0.85	2.7
EchoFlex 15	50	14.6	0.86	77	0.26	0.28	2.9
LDF4-50	50	16	0.88	88	0.21	0.68	2.5
UR70	75	5.8	0.66	67	0.5	1.5	5.2
RG59BU	75	6.15	0.66	68	0.5	1.5	4.6
RG62AU	95	6.15	0.84	44	0.3	0.9	2.9

This short list of coaxial cables represents what is available in distributors' advertisements.
The data in this table has been obtained from these catalogues and from the Internet.
Note that there are various standards of measuring coax loss, such as dB/100m, dB/10m and dB/100ft and at various frequencies.
The data above uses dB/30m (approximately 100ft) representing the most common length from the shack to the antenna.

Table A.16. Chebyshev low-pass filter (Pi configuration)

	Ripple (dB)	C_1	C_2	C_3	C_4	C_5	L_1	L_2	L_3	L_4
Single section (3-pole)	1	6441.3	6441.3	—	—	—	7.911	—	—	—
	0.1	3283.6	3283.6	—	—	—	9.131	—	—	—
	0.01	2007.7	2007.7	—	—	—	7.721	—	—	—
	0.001	1301.2	1301.2	—	—	—	5.781	—	—	—
Two-section (5-pole)	1	6795.5	9552.2	6795.5	—	—	8.683	8.683	—	—
	0.1	3650.4	6286.6	3650.4	—	—	10.91	10.91	—	—
	0.01	2407.5	5020.7	2407.5	—	—	10.38	10.38	—	—
	0.001	1727.3	4170.5	1727.3	—	—	8.928	8.928	—	—
Three-section (7-pole)	1	3538	5052	5052	3538	—	17.24	18.20	17.24	—
	0.1	3759.8	6673.9	6673.9	3759.8	—	11.32	12.52	11.32	—
	0.01	2536.8	5564.5	5564.5	2536.8	—	11.08	13.00	11.08	—
	0.001	1875.7	4875.9	4875.9	1875.7	—	9.879	12.31	9.879	—
Four-section (9-pole)	1	6938.3	9935.8	10,105	9935.8	6938.3	8.906	9.467	9.467	8.906
	0.1	3805.9	6794.5	7019.9	6794.5	3805.9	11.48	12.87	12.87	11.48
	0.01	2592.5	5743.5	6066.3	5743.5	2592.5	11.36	13.63	13.63	11.36
	0.001	1941.7	5124.6	5553.2	5124.6	1941.7	10.27	13.25	13.25	10.27

Inductance in microhenrys, capacitance in picofarads. Component values normalised to 1MHz and 50Ω.

Table A.17. Chebyshev high-pass filter ('T' configuration)

	Ripple (dB)	C_1	C_2	C_3	C_4	C_5	L_1	L_2	L_3	L_4
Single section (3-pole)	1	1573	1573	—	—	—	8.005	—	—	—
	0.1	3085.7	3085.7	—	—	—	6.935	—	—	—
	0.01	5059.1	5059.1	—	—	—	8.201	—	—	—
	0.001	7786.9	7786.9	—	—	—	10.95	—	—	—
Two-section (5-pole)	1	1491	1060.7	1491	—	—	7.293	7.293	—	—
	0.1	2775.6	1611.7	2775.6	—	—	5.803	5.803	—	—
	0.01	4208.6	2018.6	4208.6	—	—	6.098	6.098	—	—
	0.001	5865.7	2429.5	5865.7	—	—	7.093	7.093	—	—
Three-section (7-pole)	1	1469.2	1028.9	1028.9	1469.2	—	7.160	6.781	7.160	—
	0.1	2694.9	1518.2	1518.2	2694.9	—	5.593	5.058	5.593	—
	0.01	3994.1	1820.9	1820.9	3994.1	—	5.715	4.873	5.715	—
	0.001	5401.7	2078	2078	5401.7	—	6.410	5.144	6.410	—
Four-section (9-pole)	1	1460.3	1019.8	1002.7	1019.8	1460.3	7.110	6.689	6.689	7.110
	0.1	2662.2	1491.2	1443.3	1491.2	2662.2	5.516	4.922	4.922	5.516
	0.01	3908.2	1764.1	1670.2	1764.1	3908.2	5.578	4.647	4.647	5.578
	0.001	5216.3	1977.1	1824.6	1977.1	5216.3	6.657	4.780	4.780	6.657

Inductance in microhenrys, capacitance in picofarads. Component values normalised to 1MHz and 50Ω.

Table A.18. Chebyshev high-pass filter (Pi configuration)

	Ripple (dB)	L_1	L_2	L_3	L_4	L_5	C_1	C_2	C_3	C_4
Single section (3-pole)	1	3.932	3.932	—	—	—	3201.7	—	—	—
	0.1	7.714	7.714	—	—	—	2774.2	—	—	—
	0.01	12.65	12.65	—	—	—	3280.5	—	—	—
	0.001	19.47	19.47	—	—	—	4381.4	—	—	—
Two-section (5-pole)	1	3.727	2.652	3.727	—	—	2917.3	2917.3	—	—
	0.1	6.939	4.029	6.939	—	—	2321.4	2321.4	—	—
	0.01	10.52	5.045	10.52	—	—	2439.3	2439.3	—	—
	0.001	14.66	6.074	14.66	—	—	2837.3	2837.3	—	—
Three-section (7-pole)	1	7.159	5.014	5.014	7.159	—	1469.2	1391.6	1469.2	—
	0.1	8.737	3.795	3.795	8.737	—	2237.2	2023.1	2237.2	—
	0.01	9.985	4.552	4.552	9.985	—	2286.0	1949.1	2286.0	—
	0.001	13.50	5.195	5.195	13.50	—	2584.1	2057.7	2584.1	—
Four-section (9-pole)	1	3.651	2.549	2.507	2.549	3.651	2844.1	2675.6	2675.6	2844.1
	0.1	6.656	3.728	3.608	3.728	6.656	2206.5	1968.9	1968.9	2206.5
	0.01	9.772	4.410	4.176	4.410	9.772	2230.5	1858.7	1858.7	2230.5
	0.001	13.05	4.943	4.561	4.943	13.05	2466.3	1911.8	1911.8	2466.3

Inductance in microhenrys, capacitance in picofarads. Component values normalised to 1MHz and 50Ω.

would use two inductances of a value equal to $L_k/2$, while the balanced constant-k π-section high-pass filter would use two capacitors of a value equal to $2C_k$.

If several low- (or high-) pass sections are to be used, it is advisable to use m-derived end sections on either side of a constant-k section, although an m-derived centre section can be used.

Table A.11. Component colour codes

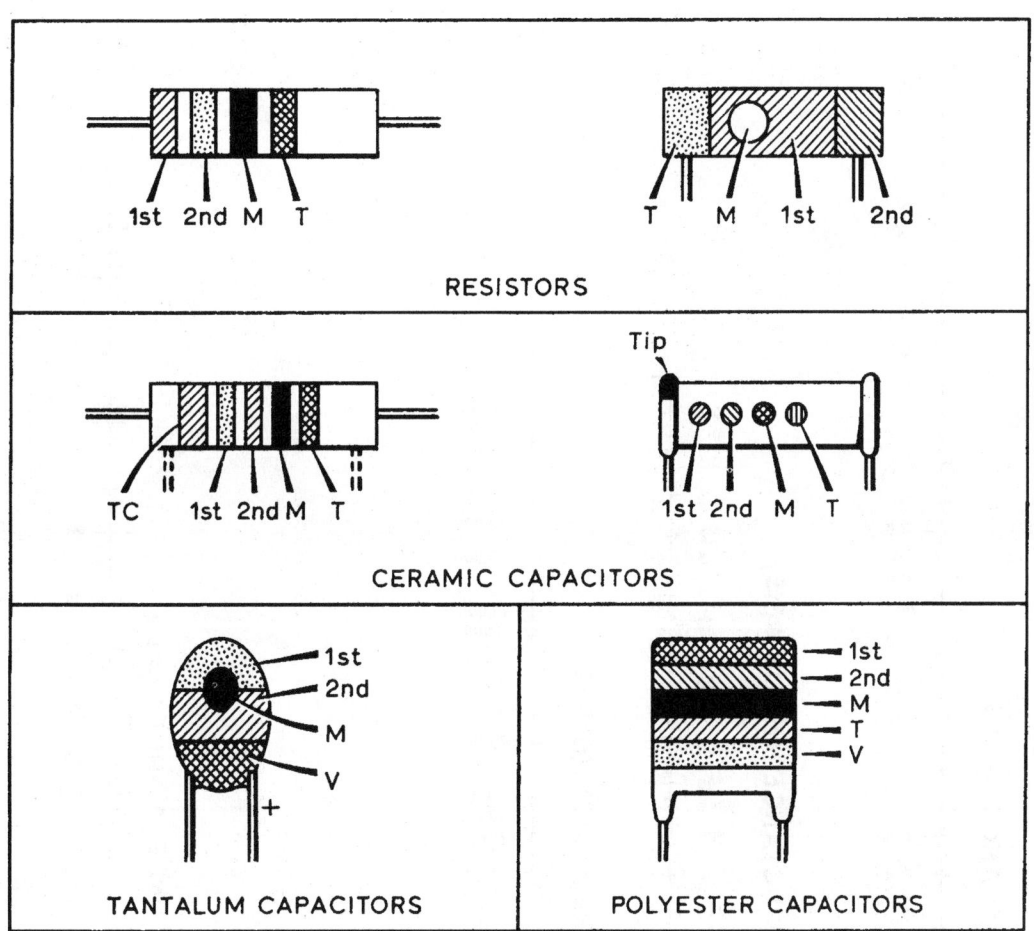

Colour (cap)	Significant figure (1st, 2nd)	Decimal multiplier (M)	Tolerance (T) (per cent)	Temp coeff (TC) (parts/10⁶/°C)	Voltage (V) (tantalum cap)	Voltage (V) (polyester
Black	0	1	±20	0	10	—
Brown	1	10	±1	−30	—	100
Red	2	100	±2	−80	—	250
Orange	3	1000	±3	−150	—	—
Yellow	4	10,000	+100, −0	−220	6.3	400
Green	5	100,000	±5	−330	16	—
Blue	6	1,000,000	±6	−470	20	—
Violet	7	10,000,000	—	−750	—	—
Grey	8	100,000,000	—	+30	25	—
White	9	1,000,000,000	±10	+100 to −750	3	—
Gold	—	0.1	±5	—	—	—
Silver	—	0.01	±10	—	—	—
Pink	—	—	—	—	35	—
No colour	—	—	±20	—	—	—

Units used are ohms for resistors, picofarads for ceramic and polyester capacitors, and microfarads for tantalum capacitors.

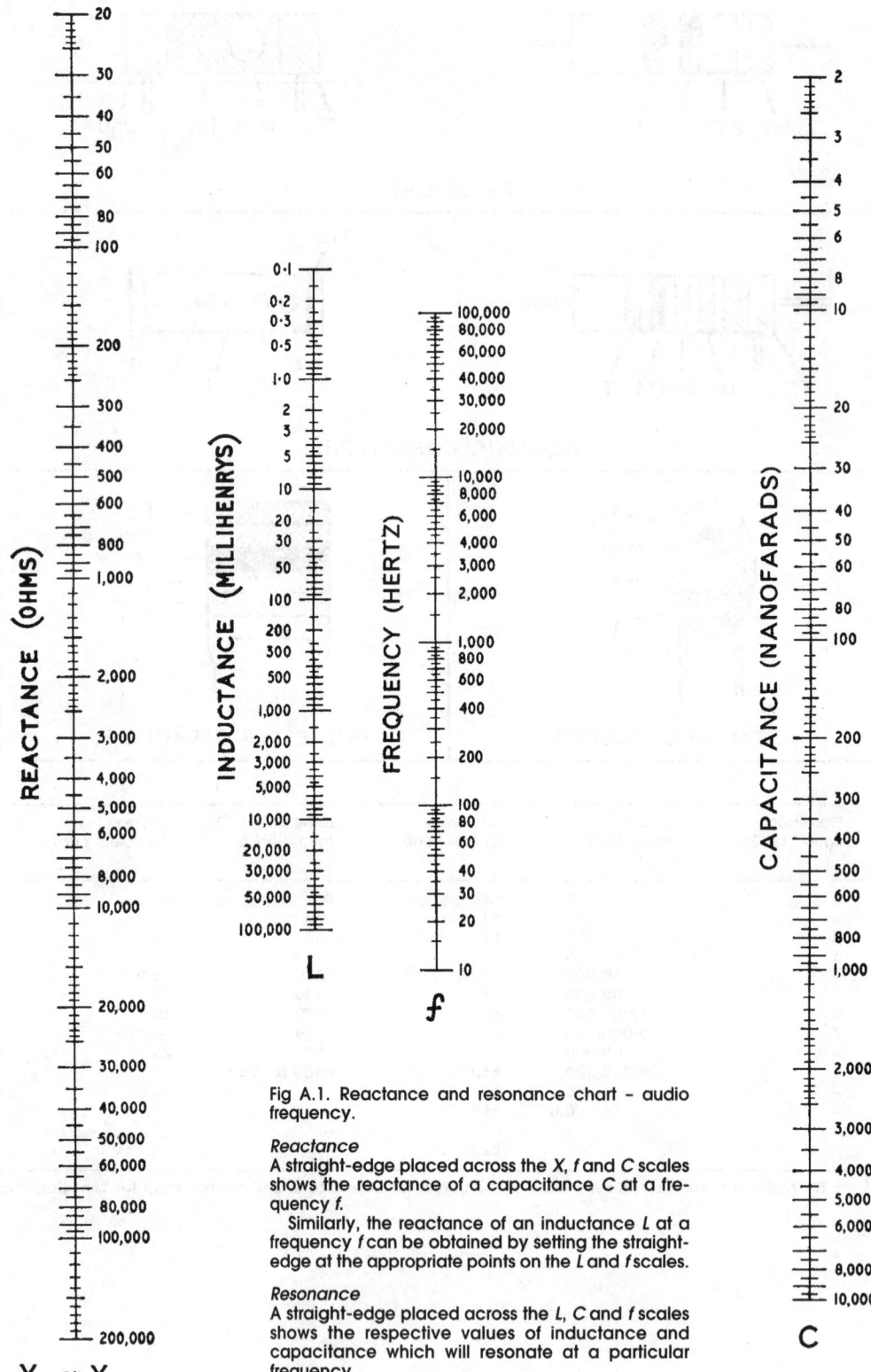

Fig A.1. Reactance and resonance chart – audio frequency.

Reactance
A straight-edge placed across the X, f and C scales shows the reactance of a capacitance C at a frequency f.
Similarly, the reactance of an inductance L at a frequency f can be obtained by setting the straight-edge at the appropriate points on the L and f scales.

Resonance
A straight-edge placed across the L, C and f scales shows the respective values of inductance and capacitance which will resonate at a particular frequency.

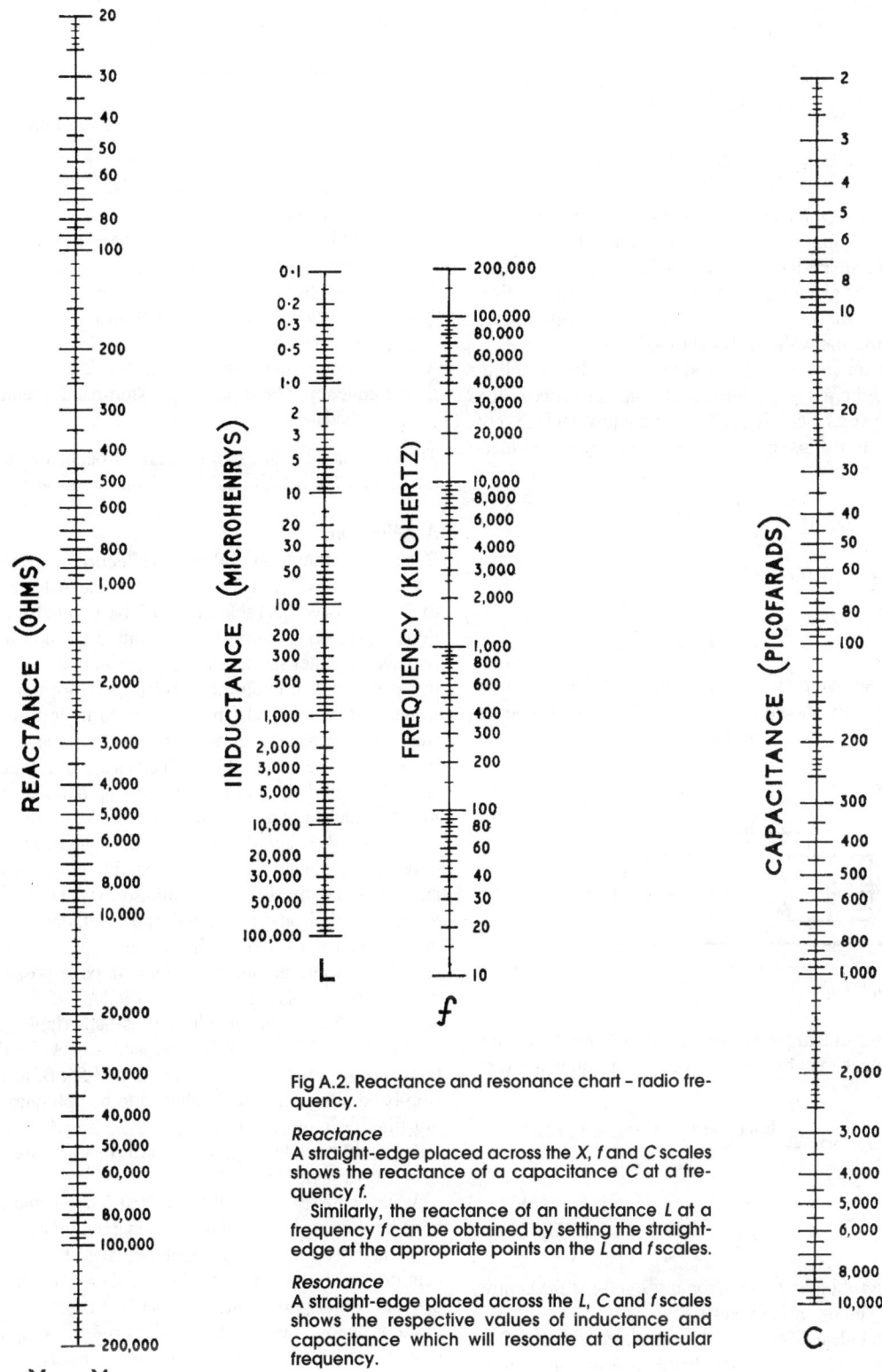

Fig A.2. Reactance and resonance chart – radio frequency.

Reactance
A straight-edge placed across the X, f and C scales shows the reactance of a capacitance C at a frequency f.
  Similarly, the reactance of an inductance L at a frequency f can be obtained by setting the straight-edge at the appropriate points on the L and f scales.

Resonance
A straight-edge placed across the L, C and f scales shows the respective values of inductance and capacitance which will resonate at a particular frequency.

FILTER DESIGN CALCULATIONS
Coupling between two resonant circuits tuned to the same frequency [1]

The coupling coefficient is the ratio of the mutual inductance between windings to the inductance of one winding. This is true where the primary and secondary are identical; for simplicity, this is taken to be the case.

When the peak of the response is flat and on the point of splitting, the coupling is at its critical value, which is given by:

$$k_c = \frac{1}{Q} \quad (Q_p = Q_s)$$

Hence, the higher the Q, the lower the coupling required. In an IF transformer, the coupling is set at the critical value; however, for use in wide-band couplers it is convenient to have it slightly higher. The design formulae given below are based on a coupling/critical coupling ratio of 1.86, corresponding to a peak-to-trough ratio of 1.2:1, or a response flat within 2dB over the band.

The most convenient way of introducing variable coupling between two tuned circuits is with a small trimmer between the 'hot' ends of the coils (see Fig A.3). This is equivalent, except where phase relationships are concerned, to a mutual inductance of the value:

$$M = \frac{C_1}{C_1 + C} L$$

Hence the coupling coefficient is:

$$k = \frac{C_1}{C_1 + C}$$

The purpose of the damping resistors R in Fig A.3 is to obtain correct circuit Q; they should not be omitted unless the source or load provides the proper termination.

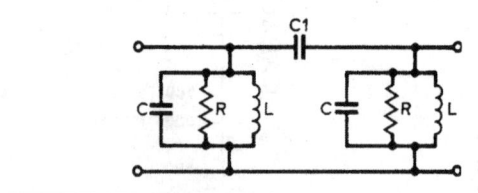

Fig A.3. Basic coupler circuit

Given set values of damping resistance, pass-band and centre frequency, all values may be calculated from the following formulae:

$$k = 0.84 \frac{\text{Bandwidth (kHz)}}{\text{Centre frequency (kHz)}}$$

$$Q = \frac{1.86}{k} \qquad L = \frac{R}{2\pi f Q} \qquad C = \frac{1}{L}\left(\frac{1}{2\pi f}\right)^2$$

where C is in microfarads, L in microhenrys and f is the centre frequency in megahertz. R is in ohms.

Note that C includes all strays: if the calculated value of C is less than the estmated strays on any band, a lower value of R should be used. The bandswitch can increase the strays to 20pF or more.

Coupling capacitance C_1 is given by:

$$C_1 = \frac{k}{1-k} C$$

Elliptic filters [1]

Using modern design procedure, a 'normalised' filter having the desired performance is chosen from a series of precalculated designs. The following presentation, originally due to W3NQN, uses normalisation to a cut-off frequency of 1Hz and termination resistance of 1Ω, and all that is required to ascertain the constants of a practical filter is to specify the actual cut-off frequency and termination resistance required and to scale the normalised filter data to those parameters.

The following abbreviations are used in these curves:

A = attenuation (dB),
A_p = maximum attenuation in pass-band,
f_4 = first attenuation peak,
f_2 = second attenuation peak with two-section filter or third attenuation peak with three-section filter,
f_6 = second attenuation peak with three-section filter,
f_{co} = frequency where the attenuation first exceeds that in the pass-band,
A_s = minimum attenuation in stop-band,
f_s = frequency where minimum stop-band attenuation is first reached.

The attenuation peaks f_4, f_6 or f_2 are associated with the resonant circuits L4/C4, L6,C6 and L2/C2 on the respective diagrams.

Applications

Because of their low value of reflection coefficient (P) and VSWR, Tables 1-1, 1-2 and 1-3 of Table A.12 and Tables 2-1 to 2-6 inclusive of Table A.13 are best suited for RF applications where power must be transmitted through the filter. The two-section filter has a relatively gradual attenuation slope and the stop-band attenuation level (A_s) is not achieved until a frquency f_s is reached which is two to three times the cut-off frequency. If a more abrupt attenuation slope is desired, then one of the three-section filters (Tables 2-1 to 2-6 in Table A.13) should be used. In these cases the stop-band attenuation level may be reached at a frequency only 1.25 to 2 times f_{co}.

Tables 1-4 to 1-6 of Table A.12 are intended for AF applications where transmission of appreciable power is not required, and consequently the filter response may have a much higher value of VSWR and pass-band ripple without adversely affecting the filter performance. If the higher pass-band ripple is acceptable, a more abrupt attenuation slope is possible. This can be seen by comparing the different values of f_s at 50dB in Tables 1-4, 1-5 and 1-6 which have pass-band ripple peaks of 0.28, 0.50 and 1.0dB respectively. The values of A_s for the audio filters were selected to be between 35 and 55dB, as this range of stop-band attenuation was believed to be optimum for most audio filtering requirements.

It should be noted that *all* C and L tabular data *must be multiplied by a factor of 10^{-3}*.

With one exception, all the C and L tabulated data of each table have a consistent but unequal increase or decrease in value, a characteristic of most computer-derived filter tables. An exception will be noted in Table 1-5, $A_s = 50$, column C1. The original author points out that this is not an error but arose from a minor change necessitated in the original computer program to eliminate unrealisable component values.

How to use the filter tables

After the desired cut-off frequency has been chosen, the frequencies of f_s and the attenuation peaks may be calculated by multiplying their corresponding tabular values by the required cut-off frequency (f_{co}). The component values of the desired

Table A.12. Two-section elliptic-function filters normalised for a cut-off frequency of 1Hz and terminations of 1Ω

REFLECTION COEFFICIENT, VSWR & Ap	As dB	fs Hz	f4 Hz	f2 Hz	C1 Farad	C3 Farad	C5 Farad	C2 Farad	L2 Henry	C4 Farad	L4 Henry
Table 1−1	70	3·24	3·39	5·42	110·4	235	103·5	4·34	199·0	11·72	187·5
p = 4%	65	2·92	3·07	4·88	109·6	233	101·0	5·39	197·9	14·67	183·7
VSWR = 1·08	60	2·56	2·68	4·24	108·2	229	96·9	7·20	195·8	19·88	177·3
	55	2·37	2·48	3·90	107·2	227	93·8	8·57	194·3	23·9	172·7
Ap = 0·0069 dB	50	2·13	2·23	3·48	105·5	223	88·6	10·88	192·0	31·0	164·7
Table 1−2	70	3·07	3·22	5·13	118·3	243	110·8	4·73	203	12·78	191·0
p = 5%	65	2·79	2·92	4·64	117·4	241	108·3	5·82	202	15·82	187·2
VSWR = 1·11	60	2·46	2·57	4·06	116·0	237	104·0	7·67	200	21·2	180·7
	55	2·28	2·39	3·75	115·0	234	100·8	9·07	198·5	25·3	175·9
Ap = 0·011 dB	50	2·06	2·16	3·36	113·2	230	95·6	11·43	196·0	32·4	168·1
Table 1−3	70	2·79	2·92	4·64	138·4	262	129·6	5·59	210	15·09	196·4
p = 8%	65	2·56	2·68	4·24	137·4	259	126·9	6·75	208	18·32	192·4
VSWR = 1·17	60	2·28	2·39	3·75	135·9	255	122·4	8·72	206	23·9	185·7
	55	2·06	2·16	3·36	134·2	251	117·4	10·98	204	30·6	178·4
Ap = 0·028 dB	50	1·887	1·970	3·05	132·2	245	111·8	13·55	201	38·4	170·3
Table 1−4	55	1·701	1·773	2·71	217	317	190·8	18·03	191·5	49·7	162·3
p = 25%	50	1·556	1·617	2·44	213	306	181·3	22·8	187·3	63·8	151·9
VSWR = 1·67	45	1·440	1·493	2·22	209	295	170·6	28·3	182·7	80·9	140·5
	40	1·325	1·369	1·988	203	279	155·8	36·4	176·0	108·0	125·1
Ap = 0·28 dB	35	1·236	1·273	1·802	195·9	262	139·2	46·4	168·2	144·3	108·3
Table 1−5	55	1·618	1·690	2·56	248	348	214	21·3	181·4	58·7	151·0
p = 33%	50	1·481	1·540	2·30	249	336	210	27·4	174·9	76·7	139·3
VSWR = 2·00	45	1·369	1·416	2·08	244	318	197·5	34·7	169·2	99·8	126·5
	40	1·270	1·308	1·878	238	299	177·3	44·4	161·7	133·7	110·8
Ap = 0·50 dB	35	1·186	1·222	1·700	229	280	163·3	57·0	153·9	177·6	95·5
Table 1−6	55	1·528	1·591	2·39	314	401	276	28·3	156·9	77·5	129·1
p = 45%	50	1·407	1·459	2·16	308	381	260	35·5	153·3	99·6	119·4
VSWR = 2·67	45	1·245	1·313	1·898	306	365	247	46·6	150·7	135·0	108·9
	40	1·217	1·250	1·755	296	341	227	59·2	138·9	176·2	92·0
Ap = 1·00 dB	35	1·145	1·174	1·597	284	315	203	75·4	131·6	237	77·7
	As dB	fs Hz	f4 Hz	f2 Hz	L1 Henry	L3 Henry	L5 Henry	L2 Henry	C2 Farad	L4 Henry	C4 Farad

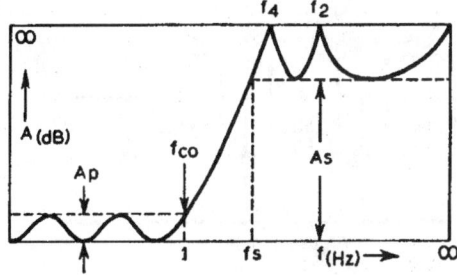

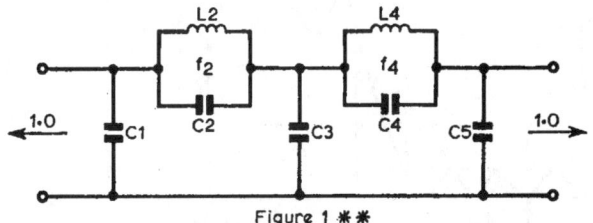

Figure 1 ✳✳

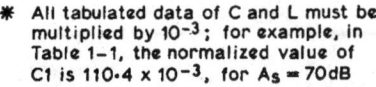

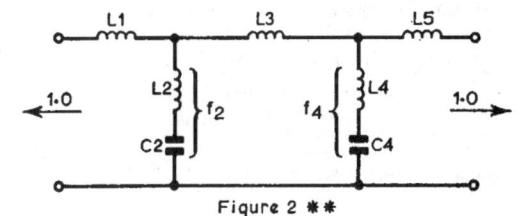

Figure 2 ✳✳

✳ All tabulated data of C and L must be multiplied by 10⁻³; for example, in Table 1−1, the normalized value of C1 is 110·4 × 10⁻³, for As = 70dB

✳✳ In the above tabulation, the top column headings pertain to Figure 1 while the bottom column headings pertain to Figure 2

filter are then found by multiplying C and L values in the tables by $1/Rf_{co}$ and R/f_{co} respectively.

Example 1

A low-pass audio filter to attenuate speech frequencies above 3kHz with a minimum attenuation of 40dB for all frequencies above 3.8kHz, and to be terminated in resistive loads of 1.63kΩ. (This odd value has been chosen merely for convenience in demonstrating the design procedure.)

The circuit of Fig 1 in the Tables is chosen because this has the minimum number of inductors, which are both more expensive and have higher losses than do capacitors. The parameters are:

$$A_s = 40\text{dB} \qquad f_{co} = 3\text{kHz} \qquad R = 1.63\text{k}\Omega$$

From Table 1-5 of Table 23.12, $A_s = 40$dB, calculate f'_s, f'_4 and f'_2. (Numbers with the prime (′) are the frequency and component values of the final design: numbers without the prime are from the filter catalogue.)

(1) $f'_s = f_s(f_{co}) = 1.270 \times 3 = 3.81$kHz.
$f'_4 = f_4(f_{co}) = 1.308 \times 3 = 3.92$kHz
$f'_2 = f_2(f_{co}) = 1.878 \times 3 = 5.63$kHz

Table A.13. Three-section elliptic-function filters normalised for a cut-off frequency of 1Hz and terminations of 1Ω

REFLECTION COEFFICIENT, VSWR & Ap	As dB	fs Hz	f4 Hz	f6 Hz	f2 Hz	C1 Farad	C3 Farad	C5 Farad	C7 Farad	C2 Farad	L2 Henry	C4 Farad	L4 Henry	C6 Farad	L6 Henry
Table 2-1 $p=1\%$ VSWR $=1.02$ Ap $=0.43 \times 10^{-3}$dB	70	2.00	2.04	2.49	4.35	79.6	209	201	63.1	7.42	180.2	30.9	196.4	26.3	155.0
	64	1.836	1.876	2.27	3.95	78.3	204	194.8	58.2	9.10	178.4	38.4	187.6	33.0	148.3
	60	1.743	1.780	2.15	3.72	77.3	200	190.3	54.5	10.35	177.1	44.1	181.4	38.2	143.5
	55	1.624	1.657	1.990	3.41	75.8	194.2	183.5	48.5	12.42	175.2	53.8	171.4	47.2	135.6
	50	1.524	1.554	1.854	3.15	74.1	187.8	176.3	41.8	14.75	172.8	65.3	160.7	58.0	127.1
Table 2-2 $p=2\%$ VSWR $=1.04$ Ap $=1.7 \times 10^{-3}$dB	70	1.836	1.876	2.27	3.95	93.8	222	212	75.7	8.34	194.8	35.8	201	29.4	167.0
	64	1.701	1.737	2.09	3.61	92.5	216	205	70.7	10.08	193.1	43.8	191.6	36.2	160.0
	60	1.624	1.657	1.990	3.41	91.5	212	200	67.1	11.35	191.6	49.8	185.1	41.3	154.8
	55	1.524	1.554	1.854	3.15	89.9	206	192.7	61.1	13.47	189.4	60.0	174.8	50.2	146.7
	50	1.414	1.440	1.702	2.86	87.5	196.9	182.1	52.2	16.70	186.1	76.4	160.0	64.8	135.0
Table 2-3 $p=3\%$ VSWR $=1.06$ Ap $=3.9 \times 10^{-3}$dB	70	1.743	1.780	2.15	3.72	104.2	230	219	84.7	9.06	203	39.7	201	31.8	172.5
	65	1.624	1.657	1.990	3.41	102.8	224	211	79.7	10.84	201	48.1	191.8	38.7	165.4
	60	1.524	1.554	1.854	3.15	101.2	217	203	74.1	12.86	198.3	57.8	181.6	46.8	157.5
	55	1.440	1.466	1.737	2.92	99.5	211	194.8	67.9	15.12	195.9	69.0	170.8	56.3	149.1
	50	1.367	1.391	1.636	2.73	97.6	203	186.2	61.2	17.65	193.1	82.2	159.2	67.5	140.1
Table 2-4 $p=4\%$ VSWR $=1.08$ Ap $=6.9 \times 10^{-3}$dB	70	1.701	1.737	2.09	3.61	113.0	236	224	93.0	9.37	208	41.6	202	32.7	177.0
	65	1.589	1.621	1.942	3.32	111.6	230	217	88.0	11.18	205	50.2	192.3	39.6	170.0
	60	1.494	1.523	1.813	3.07	110.0	224	208	82.4	13.20	203	60.0	181.9	47.6	161.9
	55	1.414	1.440	1.702	2.86	108.3	217	199.6	76.3	15.47	201	71.4	171.1	57.0	153.4
	50	1.325	1.347	1.576	2.61	105.6	206	187.5	67.3	18.94	196.9	89.7	155.6	72.2	141.3
Table 2-5 $p=5\%$ VSWR $=1.11$ Ap $=11 \times 10^{-3}$dB	70	1.662	1.696	2.04	3.51	120.6	242	229	99.9	9.77	211	43.9	201	33.9	179.4
	65	1.556	1.586	1.897	3.23	119.2	235	221	94.9	11.61	209	52.7	191.1	40.9	172.0
	60	1.466	1.494	1.774	3.00	117.6	228	212	89.3	13.67	206	62.8	180.8	49.0	164.1
	55	1.367	1.391	1.636	2.73	115.2	219	199.7	81.0	16.81	203	78.8	166.2	61.9	152.7
	51.5	1.325	1.347	1.576	2.61	113.8	213	193.4	76.5	18.57	201	88.2	158.3	69.5	146.6
	50	1.305	1.327	1.548	2.55	113.1	211	190.2	74.1	19.51	199.7	93.2	154.4	73.7	143.5
Table 2-6 $p=8\%$ VSWR $=1.17$ Ap $=28 \times 10^{-3}$dB	70	1.556	1.586	1.897	3.23	139.7	252	237	116.2	11.30	214	52.0	193.4	39.1	180.0
	65	1.466	1.494	1.774	3.00	138.1	245	228	110.9	13.30	212	61.9	183.5	46.6	172.5
	60	1.390	1.415	1.668	2.79	136.3	238	218	105.0	15.54	210	73.2	173.0	55.3	164.4
	55	1.325	1.347	1.576	2.61	134.4	230	208	98.6	18.05	207	86.3	161.9	65.4	155.8
	50	1.252	1.271	1.471	2.39	131.4	218	193.9	89.2	21.9	202	107.3	146.1	81.6	143.4
	As dB	fs Hz	f4 Hz	f6 Hz	f2 Hz	L1 Henry	L3 Henry	L5 Henry	L7 Henry	L2 Henry	C2 Farad	L4 Henry	C4 Farad	L6 Henry	C6 Farad

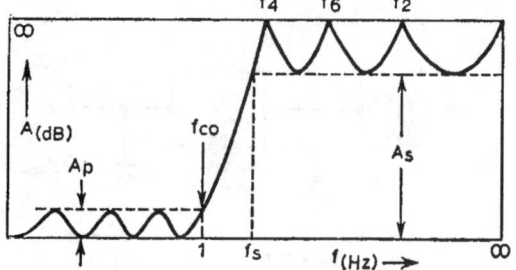

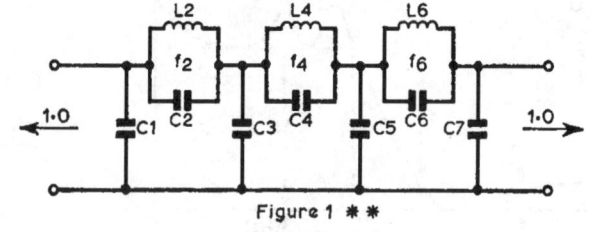

Figure 1 ＊＊

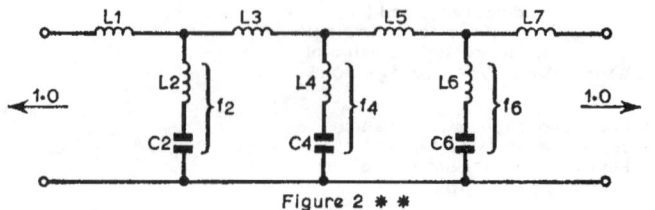

Figure 2 ＊＊

＊ All tabulated data of C and L must be multiplied by 10^{-3}; for example, in Table 2-1, the normalized value of C1 is 79.6×10^{-3}, for As $=70$dB

＊＊ In the above tabulation, the top column headings pertain to Figure 1 while the bottom column headings pertain to Figure 2

(2) Calculate factors $1/Rf_{co}$ and R/f_{co} to determine the capacitor and inductor values.

$1/Rf_{co} = 1/(1.63 \times 10^3)(3 \times 10^3)$
$= 1/(4.89 \times 10^6)$
$= 0.2045 \times 10^{-6}$

$R/f_{co} = (1.63 \times 10^3)/(3 \times 10^3) = 0.543$

(3) Calculate the component values of the desired filter by multiplying all the catalogue tabular values of C by $1/Rf_{co}$ and L by R/f_{co} as shown below:

$C'1 = C1(1/Rf_{co}) = (238 \times 10^{-3})(0.2045)10^{-6} = 0.0487\mu F$
$C'3 = C3(1/Rf_{co}) = (299 \times 10^{-3})(0.2045)10^{-6} = 0.0612\mu F$
$C'5 = C5(1/Rf_{co}) = (177.3 \times 10^{-3})(0.2045)10^{-6} = 0.0363\mu F$
$C'2 = C2(1/Rf_{co}) = (44.4 \times 10^{-3})(0.2045)10^{-6} = 0.00908\mu F$
$C'4 = C4(1/Rf_{co}) = (133.7 \times 10^{-3})(0.2045)10^{-6} = 0.00273\mu F$
$L'2 = L2(R/f_{co}) = (161.7 \times 10^{-3})(0.543) = 87.8mH$
$L'4 = L4(R/f_{co}) = (110.8 \times 10^{-3})(0.543) = 60.1mH$

These calculations, which may conveniently be performed with a pocket calculator, complete the design of the filter.

It should be noted that all the elliptic-function data is based

Table A.14. Butterworth filters

K	C_1 L_1	C_2 L_2	C_3 L_3	C_4 L_4	C_5 L_5	C_6 L_6	C_7 L_7	C_8 L_8	C_9 L_9	C_{10} L_{10}
1	2.000	—	—	—	—	—	—	—	—	—
2	1.4142	1.4142	—	—	—	—	—	—	—	—
3	1.000	2.000	1.000	—	—	—	—	—	—	—
4	0.7654	1.8478	1.8478	0.7654	—	—	—	—	—	—
5	0.6180	1.6180	2.000	1.6180	0.6180	—	—	—	—	—
6	0.5176	1.4142	1.9319	1.9319	1.4142	0.5176	—	—	—	—
7	0.4450	1.2470	1.8019	2.000	1.8019	1.2470	0.4450	—	—	—
8	0.3902	1.1111	1.6629	1.9616	1.9616	1.6629	1.1111	0.3902	—	—
9	0.3473	1.000	1.5321	1.8794	2.000	1.8794	1.5321	1.000	0.3473	—
10	0.3129	0.9080	1.4142	1.7820	1.9754	1.9754	1.7820	1.4142	0.9080	0.3129

on the use of lossless components and purely resistive terminations. Therefore components of the highest possible Q should be used and precautions taken to ensure that the filter is properly terminated.

It will be noticed that some rather curious values of both capacitance and inductance may emerge from the calculations but these may be rationalised to the extent that the tolerance on the values of components need not be closer than some ±3%.

Example 2
A three-section low-pass filter to suppress harmonics at the output of a transmitter covering the HF bands up to a frequency of 30MHz with a matching impedance of 50Ω and a minimum attenuation in the stop-band of 50dB.

The parameters are, from Table 2-2 (circuit Fig 2) of Table A.13:

$$A_s = 50\text{dB} \qquad f_{co} = 30\text{MHz} \qquad R = 50\Omega$$

From Table 2-2 (bottom line) of Table A.13, calculate f'_s, f'_4, f'_6 and f'_2.

(1) $f'_s = f_s(f_{co}) = 1.414 \times 30 = 42.4\text{MHz}.$
$f'_4 = f_4(f_{co}) = 1.440 \times 30 = 43.2\text{MHz}$
$f'_6 = f_6(f_{co}) = 1.702 \times 30 = 51\text{MHz}$
$f'_2 = f_2(f_{co}) = 2.860 \times 30 = 85.8\text{MHz}$

(2) Calculate factors $1/Rf_{co}$ and R/f_{co} to determine the capacitor and inductor values respectively.

$1/Rf_{co} = 1/50(30 \times 10^{-6}) = 66 \times 10^{-11}$
$R/f_{co} = 50/(30 \times 10^6) = 1.67 \times 10^{-6}$

(3) Calculate component values of the desired filter by multiplying all tabular values of C by $1/Rf_{co}$ and L by R/f_{co}, remembering to multiply *all* values in the tables by 10^{-3}.

$\begin{aligned}
C'2 &= C2(66 \times 10^{-11}) \\
&= (186.1 \times 10^{-3})(66 \times 10^{-11}) \\
&= 12{,}286.6 \times 10^{-14}\text{F} \\
&= 12{,}282.6 \times 10^{-2}\text{pF} \\
&= 122.8\text{pF}
\end{aligned}$
$\begin{aligned}
L'1 &= L1(1.67 \times 10^{-6}) \\
&= (87.5 \times 10^{-3})(1.67 \times 10^{-6}) \\
&= 146.1 \times 10^{-9}\text{H} \\
&= 0.15\mu\text{H}
\end{aligned}$

$C'4 = (160 \times 10^{-3})(66 \times 10^{-11})$ $L'2 = 0.03\mu\text{H}$ $L'3 = 0.33\mu\text{H}$
$ = 105.6\text{pF}$ $L'4 = 0.13\mu\text{H}$ $L'5 = 0.30\mu\text{H}$
$C'6 = 89.1\text{pF}$ $L'6 = 0.11\mu\text{H}$ $L'7 = 0.09\mu\text{H}$

As a check, it will be found that the combination C4, L4 tunes to 43.2MHz and that the other two series-tuned circuits tune to the other two points of maximum attenuation previously specified.

In order to convert the values in the filter just designed to match an impedance of 75Ω it is only necessary to multiply all values of capacitance by 2/3 and all values of inductance by 3/2. Thus C6 and L6 in a 75Ω filter become approximately 59.4pF and 0.17μH respectively.

Butterworth filters [1]
Frequency response curve:

$$A = 10 \log_{10}\left[1 + \left(\frac{f}{f_c}\right)^{2K}\right]$$

where A is the attenuation, f is the frequency for an insertion loss of 3.01dB, and K is the number of circuit elements.

Low- and high-pass filters
Table A.14 is for normalised element values of K from 1 to 10 (number of sections) reduced to 1Ω source and load resistance (zero reactance) and a 3.01dB cut-off frequency of 1 radian/s (0.1592Hz). In both low-pass and high-pass filters:

$$L = \frac{R}{2\pi f_c} = L \ (1\Omega/\text{radian}) \qquad C = \frac{1}{2\pi f_c R} = C \ (1\Omega/\text{radian})$$

where R is the load resistance in ohms and f_c is the desired 3.01dB frequency (Hz).

An example of a Butterworth low-pass filter is given in Fig A.4 (see Table A.14 for element values). In these examples of five-element filters (a) has a shunt element next to the load and (b) has a series element next to the load. Either filter will have the same response. In the examples of five-element filters given in Fig A.5, (a) has a series element next to the load and (b) has a shunt element next to the load. Either filter will have the same response.

Butterworth band-pass filters
Centre frequency $f_0 = \sqrt{f_1 f_2}$
Bandwidth $BW = f_2 - f_1$

If the bandwidth specified is not the 3.01dB bandwidth (BW_c), the latter can be determined from:

$$BW_c = \frac{BW}{(10^{0.1A} - 1)/2K}$$

where A is the required attenuation at cut-off frequencies.

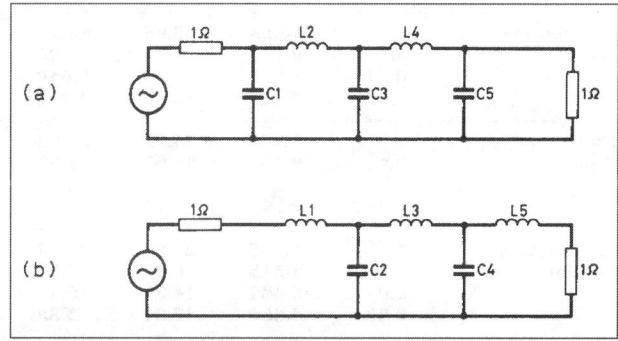

Fig A.4. Butterworth low-pass filter

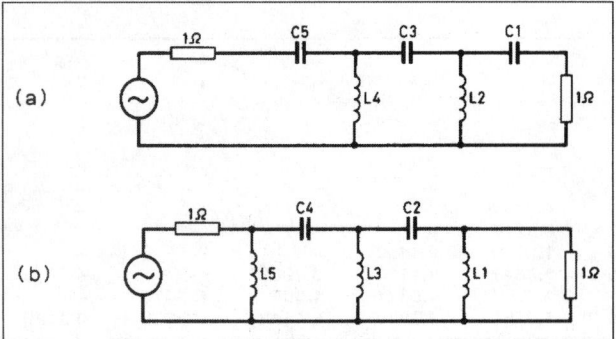

Fig A.5. Butterworth high-pass filter

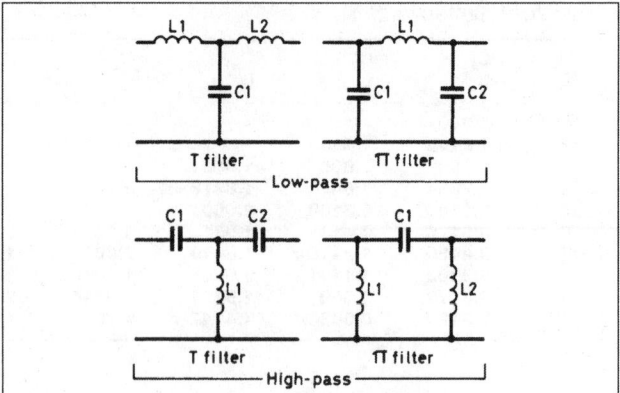

Fig A.6. Single-section three-pole filter elements

Lower cut-off frequency:

$$f_{cl} = \frac{-BW_c + \sqrt{(BW_c)^2 + 4f_0^2}}{2}$$

Upper cut-off frequency: $f_{cu} = f_{cl} + BW_c$

An alternative, more-convenient method, is to choose a 3.01dB bandwidth (as wide as possible) around the desired centre frequency and compute the attenuation at other frequencies of interest by using the transformation:

$$\frac{f}{f_c} = \left[\left(\frac{f}{f_0} - \frac{f_0}{f} \right) \frac{f_0}{BW_c} \right]$$

Chebyshev filters [1]

Tables A.15 to A.18 provide the essential information for both high-pass and low-pass filters of T and π form. Figures are given for pass-band ripples of 1, 0.1, 0.01, and 0.001dB which respectively correspond to VSWR of 2.66, 1.36, 1.10 and 1.03.

The filters in this case are normalised to a frequency of 1MHz and an input and output impedance of 50Ω. This means that for any particular desired frequency the component values simply have to be divided by the required frequency in megahertz.

The 1MHz is the cut-off frequency; attenuation increases rapidly above the frequency for a low-pass filter and correspondly below for a high-pass type.

The filter data is also dependent on the impedance which as given is for 50Ω. For other impedances the component values need to be modified by the following:

$$\frac{Z_n}{50} \text{ for inductors} \qquad \frac{50}{Z_n} \text{ for capacitors}$$

where Z_n is the required impedance.

There is an advantage in using toroidal-form inductors due to their self-screening (confined-field) properties. Mica or silver mica capacitors are superior to other types for filter applications.

Practical filters for the amateur HF bands are given in Table A.19.

Constant-*k* and *m*-derived filters [2]

The filter sections shown in Fig A.7 can be used alone or, if greater attenuation and sharper cut-off is required, several sections can be connected in series. In the low-pass and high-pass filters, f_c represents the cut-off frequency, the highest (for the low-pass) or the lowest (for the high-pass) frequency transmitted without attenuation. In the band-pass filter designs, f_1 is the low-frequency cut-off and f_2 the high-frequency cut-off. The units for *L*, *C*, *R* and *f* are henrys, farads, ohms and hertz respectively.

All the types shown are for use in an unbalanced line (one side grounded), and thus they are suitable for use in coaxial line or any other unbalanced circuit. To transform them for balanced lines (eg 300Ω transmission line or push-pull audio circuits), the series reactances should be equally divided between the two legs. Thus the balanced constant-*k* π-section low-pass filter

Table A.15. Chebyshev low-pass filter ('T' configuration)

	Ripple (dB)	L_1	L_2	L_3	L_4	L_5	C_1	C_2	C_3	C_4
Single section (3-pole)	1	16.10	16.10	—	—	—	3164.3	—	—	—
	0.1	8.209	8.209	—	—	—	3652.3	—	—	—
	0.01	5.007	5.007	—	—	—	3088.5	—	—	—
	0.001	3.253	3.253	—	—	—	2312.6	—	—	—
Two-section (5-pole)	1	16.99	23.88	16.99	—	—	3473.1	3473.1	—	—
	0.1	9.126	15.72	9.126	—	—	4364.7	4364.7	—	—
	0.01	6.019	12.55	6.019	—	—	4153.7	4153.7	—	—
	0.001	4.318	10.43	4.318	—	—	3571.1	3571.1	—	—
Three-section (7-pole)	1	17.24	24.62	24.62	17.24	—	3538.0	3735.4	3538.0	—
	0.1	9.40	16.68	16.68	9.40	—	4528.9	5008.3	4528.9	—
	0.01	6.342	13.91	13.91	6.342	—	4432.2	5198.4	4432.2	—
	0.001	4.69	12.19	12.19	4.69	—	3951.5	4924.1	3981.5	—
Four-section (9-pole)	1	17.35	24.84	25.26	24.84	17.35	3562.5	3786.9	3786.9	3562.5
	0.1	9.515	16.99	17.55	16.99	9.515	4591.9	5146.2	5146.2	4591.9
	0.01	6.481	14.36	15.17	14.36	6.481	4542.5	5451.2	5451.2	4542.5
	0.001	4.854	12.81	13.88	12.81	4.854	4108.2	5299.0	5299.0	4108.2

Inductance in microhenrys, capacitance in picofarads. Component values normalised to 1MHz and 50Ω.

Table A.16. Chebyshev low-pass filter (Pi configuration)

	Ripple (dB)	C_1	C_2	C_3	C_4	C_5	L_1	L_2	L_3	L_4
Single section (3-pole)	1	6441.3	6441.3	—	—	—	7.911	—	—	—
	0.1	3283.6	3283.6	—	—	—	9.131	—	—	—
	0.01	2007.7	2007.7	—	—	—	7.721	—	—	—
	0.001	1301.2	1301.2	—	—	—	5.781	—	—	—
Two-section (5-pole)	1	6795.5	9552.2	6795.5	—	—	8.683	8.683	—	—
	0.1	3650.4	6286.6	3650.4	—	—	10.91	10.91	—	—
	0.01	2407.5	5020.7	2407.5	—	—	10.38	10.38	—	—
	0.001	1727.3	4170.5	1727.3	—	—	8.928	8.928	—	—
Three-section (7-pole)	1	3538	5052	5052	3538	—	17.24	18.20	17.24	—
	0.1	3759.8	6673.9	6673.9	3759.8	—	11.32	12.52	11.32	—
	0.01	2536.8	5564.5	5564.5	2536.8	—	11.08	13.00	11.08	—
	0.001	1875.7	4875.9	4875.9	1875.7	—	9.879	12.31	9.879	—
Four-section (9-pole)	1	6938.3	9935.8	10,105	9935.8	6938.3	8.906	9.467	9.467	8.906
	0.1	3805.9	6794.5	7019.9	6794.5	3805.9	11.48	12.87	12.87	11.48
	0.01	2592.5	5743.5	6066.3	5743.5	2592.5	11.36	13.63	13.63	11.36
	0.001	1941.7	5124.6	5553.2	5124.6	1941.7	10.27	13.25	13.25	10.27

Inductance in microhenrys, capacitance in picofarads. Component values normalised to 1MHz and 50Ω.

Table A.17. Chebyshev high-pass filter ('T' configuration)

	Ripple (dB)	C_1	C_2	C_3	C_4	C_5	L_1	L_2	L_3	L_4
Single section (3-pole)	1	1573	1573	—	—	—	8.005	—	—	—
	0.1	3085.7	3085.7	—	—	—	6.935	—	—	—
	0.01	5059.1	5059.1	—	—	—	8.201	—	—	—
	0.001	7786.9	7786.9	—	—	—	10.95	—	—	—
Two-section (5-pole)	1	1491	1060.7	1491	—	—	7.293	7.293	—	—
	0.1	2775.6	1611.7	2775.6	—	—	5.803	5.803	—	—
	0.01	4208.6	2018.6	4208.6	—	—	6.098	6.098	—	—
	0.001	5865.7	2429.5	5865.7	—	—	7.093	7.093	—	—
Three-section (7-pole)	1	1469.2	1028.9	1028.9	1469.2	—	7.160	6.781	7.160	—
	0.1	2694.9	1518.2	1518.2	2694.9	—	5.593	5.058	5.593	—
	0.01	3994.1	1820.9	1820.9	3994.1	—	5.715	4.873	5.715	—
	0.001	5401.7	2078	2078	5401.7	—	6.410	5.144	6.410	—
Four-section (9-pole)	1	1460.3	1019.8	1002.7	1019.8	1460.3	7.110	6.689	6.689	7.110
	0.1	2662.2	1491.2	1443.3	1491.2	2662.2	5.516	4.922	4.922	5.516
	0.01	3908.2	1764.1	1670.2	1764.1	3908.2	5.578	4.647	4.647	5.578
	0.001	5216.3	1977.1	1824.6	1977.1	5216.3	6.657	4.780	4.780	6.657

Inductance in microhenrys, capacitance in picofarads. Component values normalised to 1MHz and 50Ω.

Table A.18. Chebyshev high-pass filter (Pi configuration)

	Ripple (dB)	L_1	L_2	L_3	L_4	L_5	C_1	C_2	C_3	C_4
Single section (3-pole)	1	3.932	3.932	—	—	—	3201.7	—	—	—
	0.1	7.714	7.714	—	—	—	2774.2	—	—	—
	0.01	12.65	12.65	—	—	—	3280.5	—	—	—
	0.001	19.47	19.47	—	—	—	4381.4	—	—	—
Two-section (5-pole)	1	3.727	2.652	3.727	—	—	2917.3	2917.3	—	—
	0.1	6.939	4.029	6.939	—	—	2321.4	2321.4	—	—
	0.01	10.52	5.045	10.52	—	—	2439.3	2439.3	—	—
	0.001	14.66	6.074	14.66	—	—	2837.3	2837.3	—	—
Three-section (7-pole)	1	7.159	5.014	5.014	7.159	—	1469.2	1391.6	1469.2	—
	0.1	8.737	3.795	3.795	8.737	—	2237.2	2023.1	2237.2	—
	0.01	9.985	4.552	4.552	9.985	—	2286.0	1949.1	2286.0	—
	0.001	13.50	5.195	5.195	13.50	—	2584.1	2057.7	2584.1	—
Four-section (9-pole)	1	3.651	2.549	2.507	2.549	3.651	2844.1	2675.6	2675.6	2844.1
	0.1	6.656	3.728	3.608	3.728	6.656	2206.5	1968.9	1968.9	2206.5
	0.01	9.772	4.410	4.176	4.410	9.772	2230.5	1858.7	1858.7	2230.5
	0.001	13.05	4.943	4.561	4.943	13.05	2466.3	1911.8	1911.8	2466.3

Inductance in microhenrys, capacitance in picofarads. Component values normalised to 1MHz and 50Ω.

would use two inductances of a value equal to $L_k/2$, while the balanced constant-k π-section high-pass filter would use two capacitors of a value equal to $2C_k$.

If several low- (or high-) pass sections are to be used, it is advisable to use m-derived end sections on either side of a constant-k section, although an m-derived centre section can be used.

Table A.19. Practical Chebyshev low-pass filters (3-section, 7-pole)

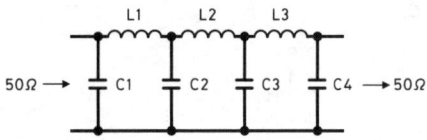

Amateur band	28	21	14	7	3.5	1.8	MHz
F_c	30.9	21.69	15.16	7.98	4.11	2.05	MHz
VSWR	1.10	1.06	1.09	1.08	1.07	1.09	
C1, C4	82	100	160	300	560	1200	pF
C2, C3	180	240	360	680	1300	2700	pF
L1, L3	0.36	0.49	0.72	1.37	2.62	5.42	µH
L2	0.42	0.59	0.85	1.62	3.13	6.41	µH

Ripple is 0.01dB.

The factor m relates the ratio of the cut-off frequency and f_∞, a frequency of high attenuation. Where only one m-derived section is used, a value of 0.6 is generally used for m, although a deviation of 10 or 15% from this value is not too serious in amateur work. For a value of $m = 0.6$, f will be $1.25f_c$ for the low-pass filter and $0.8f_c$ for the high-pass filter. Other values can be found from:

$$m = \sqrt{1 - \left(\frac{f_c}{f_\infty}\right)^2}$$

for the low-pass filter and:

$$m = \sqrt{1 - \left(\frac{f_\infty}{f_c}\right)^2}$$

for the high-pass filter.

The filters shown should be terminated in a resistance R, and there should be little or no reactive component in the termination.

Microstrip circuit elements [3, 4]

In the calculation of microstrip circuit elements it is necessary to establish the dielectric constant for the material. This can be done by measuring the capacitance of a typical sample.

$$\text{Dielectric constant } e = \frac{113 \times C \times h}{a}$$

where C is in picofarads, h is the thickness in millimetres and a is the area in square millimetres. This should be done with a sample about 25mm square to minimise the effects of the edges. Having found the dielectric constant, it is now necessary to calculate the characteristic impedance (Z_0) of the microstrip. There are many approximations for this but the following is simple and accurate enough (±5%) for amateur use since microstrip is fairly forgiving of small errors:

$$Z_0 = \frac{131}{\sqrt{(e + 0.47)}} \times \log_{10}\left(\frac{13.5h}{w}\right)$$

where e is the dielectric constant, h is the dielectric thickness (see Fig 23.8), and w is the conductor width. Note that h and w *must* be in the same units, eg both in centimetres. The formula assumes that the conductor is thin relative to the dielectric.

An accurate plot of Z_0 against w/h for dielectric constants between 2 (approximately that of PTFE) through 4 (approx that of epoxy-glassfibre) to 6 is given in Fig A.9. The next operation is to determine the *velocity factor*, the ratio of the velocity of electromagnetic waves in the dielectric to that in free space. Here, too, the equations are complex but the factor only changes slowly as w/h changes. Fig 23.10 gives figures for the above range of dielectric constants.

A microstrip resonator has the form of a strip on one side of a double-sided PCB with the other side as a ground plane. It is usually a quarter-wavelength long. The length is calculated from the free space length multiplied by the velocity factor as estimated above. The line width depends on the required Z_0 and a starting point for experiment would be 50–100Ω. If it is necessary to tune the resonator accurately, it should be made shorter

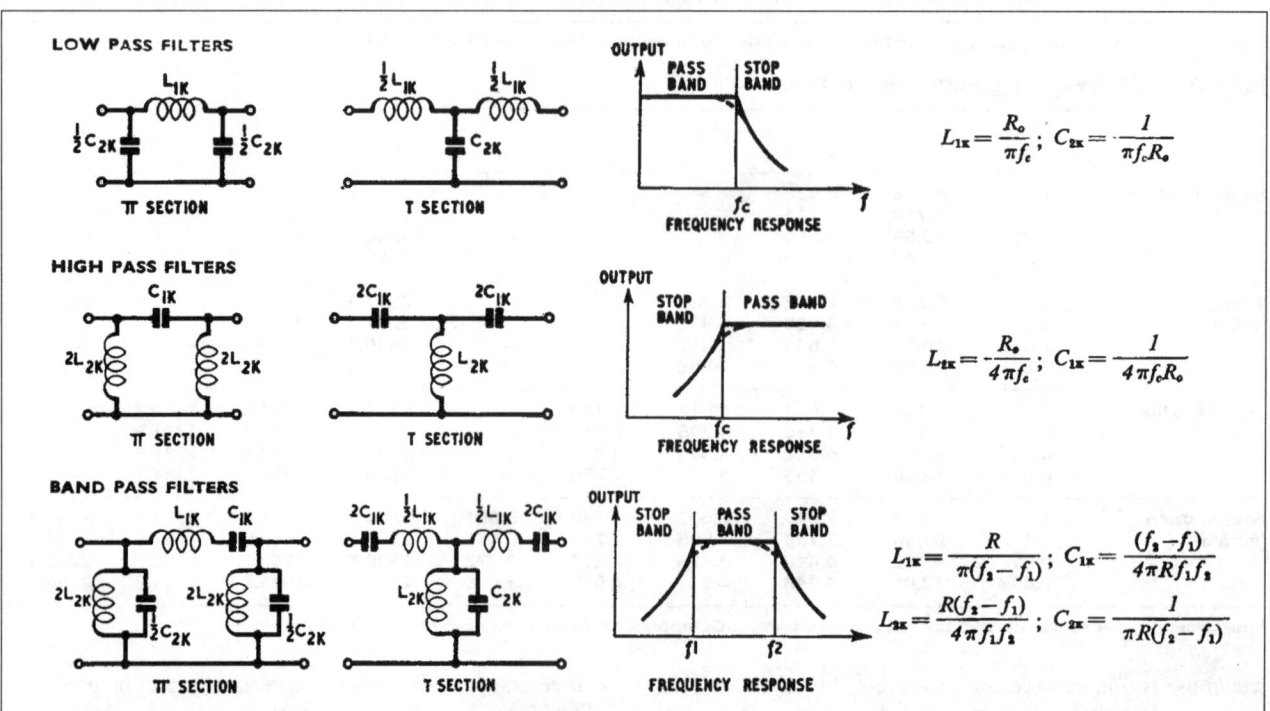

Fig A.7(a). Constant-k filters

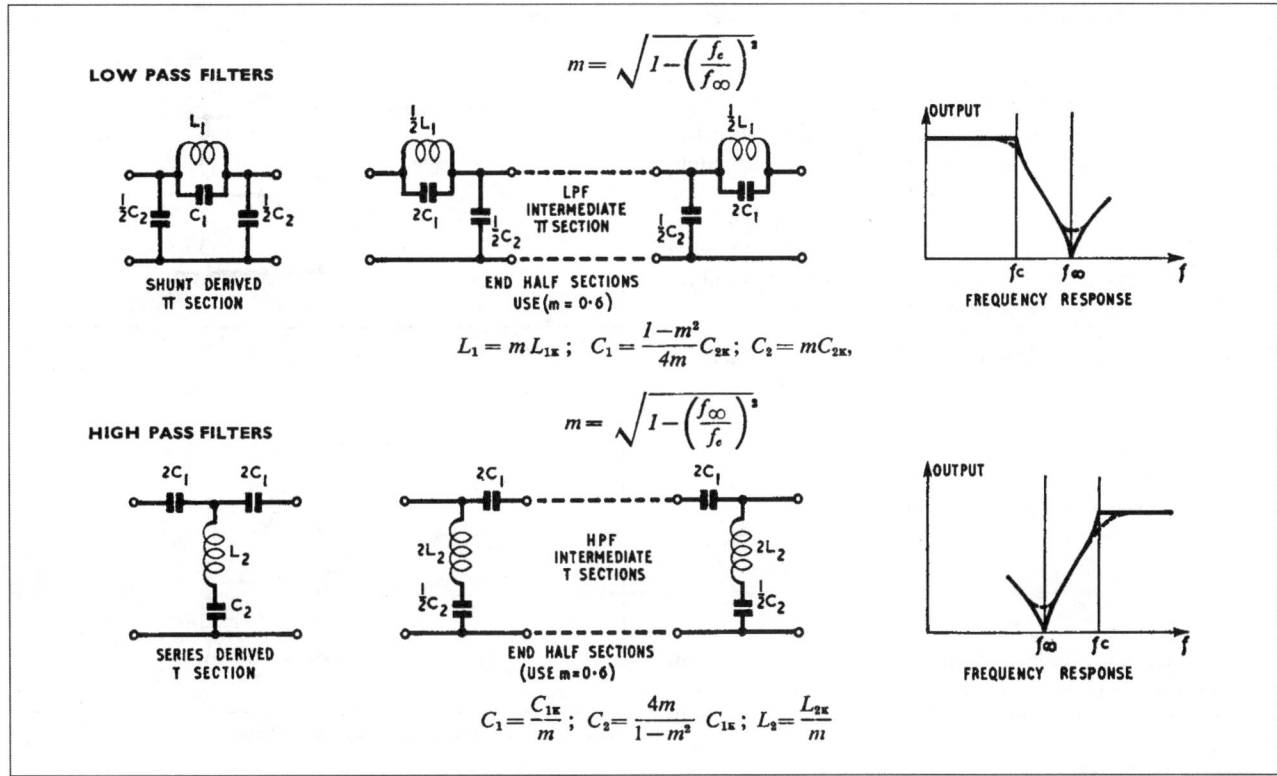

Fig A.7(b). *m*-derived filters

than calculated above and a trimmer capacitor connected between the 'hot' end and the ground plane. The new length can be calculated from:

$$l = 0.0028\lambda \times \tan^{-1}\left(\frac{\lambda}{0.188CZ_0}\right)$$

where l is the length in centimetres, λ is the wavelength in centimetres, C is the capacitance (say at half maximum) in picofarads, Z_0 is the characteristic impedance in ohms and $\tan^{-1}(*)$ is the angle in degrees of which * is the tangent. * represents the figures in the bracket.

Coupling into and out of the line may be directly via tapping(s) or by additional line(s) placed close to the tuned line. A spacing of one line width and a length of 10–20% of the tuned line would be a starting point for experiment.

Materials

The most-used material for amateur purposes is glassfibre reinforced epoxy double-sided PCB. It has a dielectric constant of 4.0–4.5, depending on the resin used. Accurate lines may be made by scoring through the copper carefully with a scalpel or modelling knife and lifting the unwanted copper foil after heating it with a soldering iron to weaken the bond to the plastic.

For microwave use, glassfibre-reinforced PTFE with a dielectric constant close to 2.5 is the preferred material. Further information is given in the *Microwave Handbook* [4].

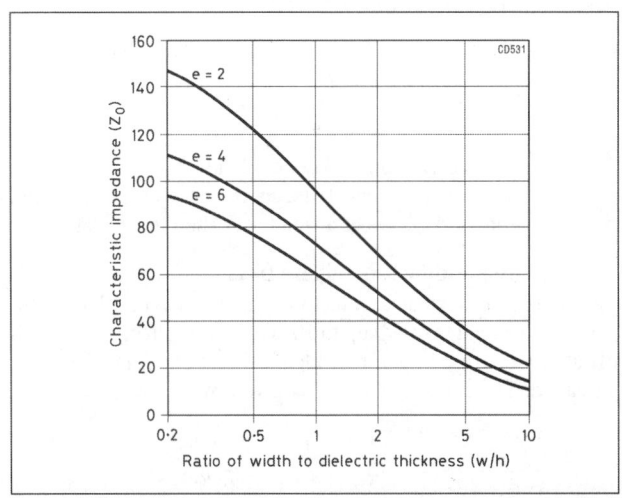

Fig A.9. Characteristic impedance versus *w/h* for dielectric constants between 2 and 6

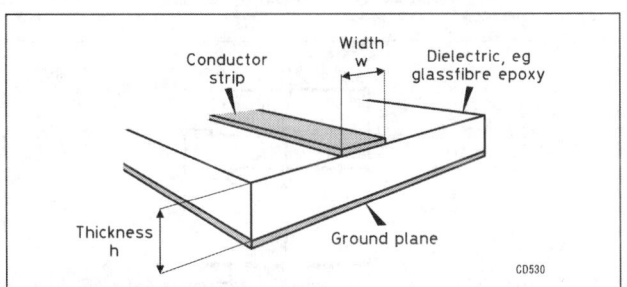

Fig A.8. Dimensions involed in calculating the characteristic impedance of microstrip

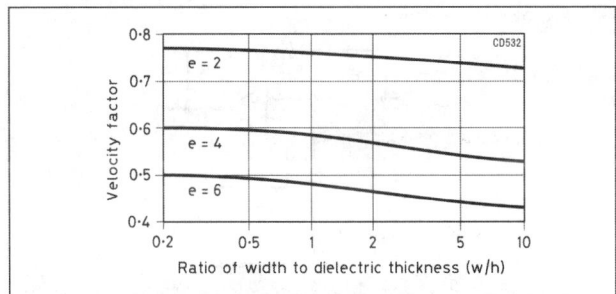

Fig A.10. Velocity factor versus *w/h* for dielectric constants between 2 and 6

Op-amp-based active filters [5]

Design information (taken, by permission, from reference [5]) will be given for four common filter configurations. All are based on inexpensive op-amps such as the 741 and 301A (or their duals or quads in one package) which are adequate when frequencies are in the voice range, insertion gain is between unity and two (0–6dB), signal input and output voltages are in the range between a few millivolts and a few volts, and signal (input, feedback and output) currents between a microamp and a milliamp. This covers the bulk of common amateur applications. No DC supplies to the op-amps are shown.

2nd order 'Sallen and Key' Butterworth low-pass filter

Referring to Fig 23.11, the cut-off (−3dB) frequency f_c is:

$$f_c = \frac{1}{2\pi\sqrt{(R_1 R_2 C_1 C_2)}}$$

Choosing 'equal components', meaning $R_1 = R_2 = R$ and $C_1 = C_2 = C$, then:

$$f_c = \frac{1}{2\pi RC}$$

For a second-order Butterworth response, the pass-band gain *must* be 4dB or ×1.586. This is achieved by making $(R_A + R_B)/R_A = 1.586$. This is implemented with sufficient accuracy with 5% standard-value resistors of $R_A = 47k\Omega$ and $R_B = 27k\Omega$. This means that a 1V input generates an output of 1.586V at a frequency in the pass-band and $0.707 \times 1.586 = 1.12V$ at the −3dB cut-off frequency; the roll-off above f_c is 12dB/octave or 20dB/decade.

Example. Design a two-pole 'equal component' Butterworth low-pass filter (Fig 23.12) with $f_c = 2700$Hz.

Choosing for C a convenient value of 1nF and solving:

$$R = \frac{1}{2\pi f_c C} = 59k\Omega$$

This can be made up from 56kΩ and 2.7kΩ in series.

Should R come out below 10kΩ, choose a larger C; if R would be larger than 100kΩ, select a smaller C; then recalculate R.

The multiple-feedback bandpass filter

Providing two feedback paths to a single op-amp, a band-pass filter can be made with Q up to 10. To get reasonably steep roll-off at low Q, from two to four identical sections (Fig 23.14) are cascaded. The centre frequency is given by:

$$f_0 = \frac{1}{2\pi C} \sqrt{\frac{1}{R_3} \cdot \frac{R_1 + R_2}{R_1 R_2}}$$

for which the three resistors can be calculated from:

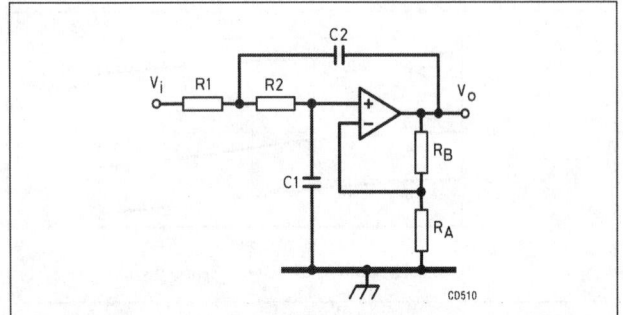

Fig A.11. 'Sallen & Key' Butterworth low-pass filter

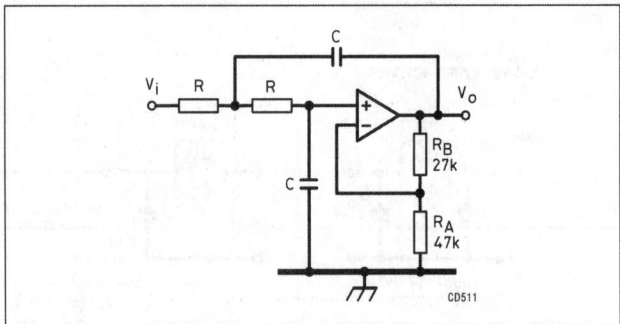

Fig A.12. Equal-component low-pass filter

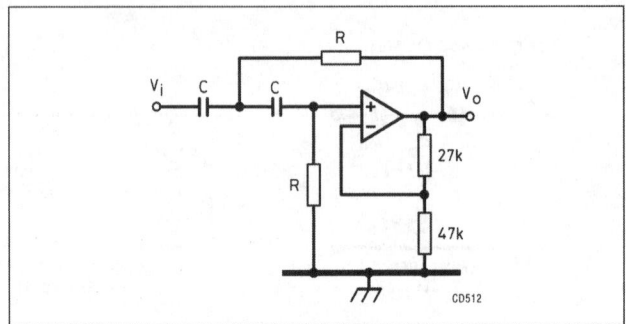

Fig A.13. Equal component high-pass filter

$$R_1 = \frac{Q}{2\pi f_0 G_0 C}$$

$$R_2 = \frac{Q}{2\pi f_0 C(2Q^2 - G_0)}$$

$$R_3 = \frac{Q}{\pi f_0 C}$$

The equations for R_1 and R_3 combine into:

$$G_0 = R_3/2R_1$$

Also, the denominator in the formula for R_2 yields:

$$Q > \sqrt{(G_0/2)}$$

Example. Design a band-pass filter with centre frequency 800Hz, −6dB bandwidth of 200Hz and centre-frequency gain of 2.

A two-section filter, with each section having a 200Hz −3dB bandwidth, is indicated.

$$Q = 800/200 = 4$$

$$G_0 = \sqrt{2} = 1.4$$

Select a convenient C, say 10nF.

$$R_1 = \frac{4}{6.28 \times 800 \times 1.4 \times 10^{-8}}$$

$$= 56.9k\Omega, \text{ (use 56k}\Omega\text{)}$$

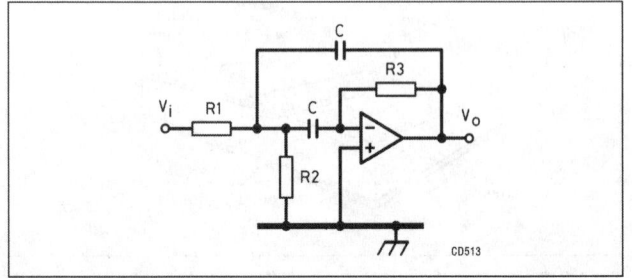

Fig A.14. Multiple-feedback band-pass filter

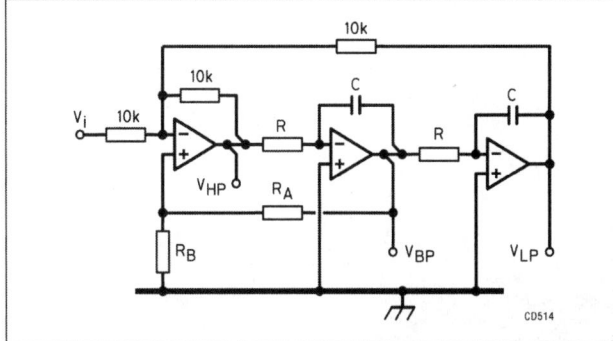

Fig A.15. State-variable filter

$$R_3 = 2 \times 56.9 \times 1.4$$
$$= 159\text{k}\Omega \text{ (use } 100\text{k}\Omega + 56\text{k}\Omega)$$

$$R_2 = \frac{4}{6.28 \times 800 \times 10^{-8} \times (2 \times 4^2 - 1.4)}$$
$$= 2.60\text{k}\Omega \text{ (use } 5.1\text{k}\Omega \text{ in parallel with } 5.1\text{k}\Omega)$$

Note that the centre frequency can be shifted up or down at constant bandwidth and centre frequency gain by changing R_2 only, using ganged variable resistors for cascaded sections:

$$R_2' = R_2 \left(\frac{f_0}{f_0'}\right)^2$$

The state-variable or 'universal' filter

Three op-amps, connected as shown in Fig 23.15, can simultaneously provide second-order high-pass, low-pass and band-pass responses. The filter is composed of a difference amplifier and two integrators. The common cut-off/centre frequency is given by:

$$f_{cL} = f_{cH} = f_0 = \frac{1}{2\pi RC}$$

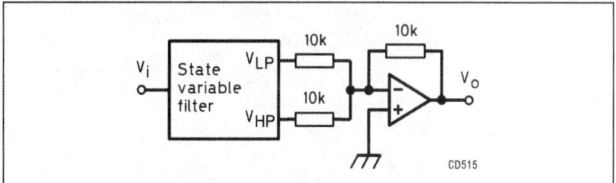

Fig A.16. Using the state-variable filter to obtain a notch response

The filter's Q depends only on R_A and R_B:

$$R_A = (3Q - 1)R_B$$

There is no way to simultaneously optimise the performance of high-/low-pass and band-pass performance. For a Butterworth response, Q must be 0.7 and even for a second-order 3dB-ripple Chebyshev response the Q is no more than 1.3, obviously too low for good band-pass response. No DC voltage should be applied to the input of this filter and there should be no significant DC load on its outputs.

By adding the low-pass and high-pass outputs from a variable-state filter in a summing amplifier, a notch response is obtained. See Fig 23.16.

For an application of the variable-state filter refer to the active filters section in Chapter 5 and Fig 5.108 in particular.

REFERENCES

[1] *Radio Data Reference Book*, 5th edn, G R Jessop, G6JP, RSGB, 1985.

[2] *ARRL Radio Amateur's Handbook*, 1953 edn, ARRL, p542.

[3] *VHF/UHF Manual*, 4th edn, ed G R Jessop, G6JP, RSGB, 1983, p3.10.

[4] *Microwave Handbook*, Vol 1, ed M W Dixon, G3PFR, RSGB, 1989, p5.14*ff*.

[5] *The Design of Operational Amplifier Circuits, with Experiments*, Howard M Berlin, W3HB, E & L Instruments Inc, Derby, Conn, USA, 1977.

The RSGB Guide to EMC

Robin Page Jones, G3JWI

Achieving electromagnetic compatibility (EMC) with the increasing number of electronic devices in surrounding buildings can be a major problem for anyone operating radio equipment.

This timely guide will help you avoid EMC problems by practising good radio housekeeping, and assist you in the diagnosis and cure of any ones which do occur. The underlying causes, as well as the remedies, are given so that you should be well prepared to tackle any problems which turn up now or in the future. There is also a considerable amount of reference data presented concerning suitable filters and braid-breakers. The social dimension is not forgotten, and a whole chapter is devoted to dealing with neighbours.

Considerable revisions of the text have been made for this edition, including coverage of the important new EU EMC regulations which came into effect in 1996, and the impact of computers on radio reception. This book continues to be an invaluable reference for all users of the radio spectrum.
Size: 244 x 172 mm, 208 pages, ISBN: 1-872309-48-8
RSGB Member's Price £16.99 plus p&p

E&OE

ONLY £19.99 plus p&p

Radio Society of Great Britain
Lambda House, Cranborne Road, Potters Bar, Herts. EN6 3JE
Tel. 0870 904 7373 Fax. 0870 904 7374

ORDER 24 HOURS A DAY ON OUR WEBSITE
www.rsgb.org/shop

RSGB SHOP

Appendix B
Printed Circuit Board Artwork

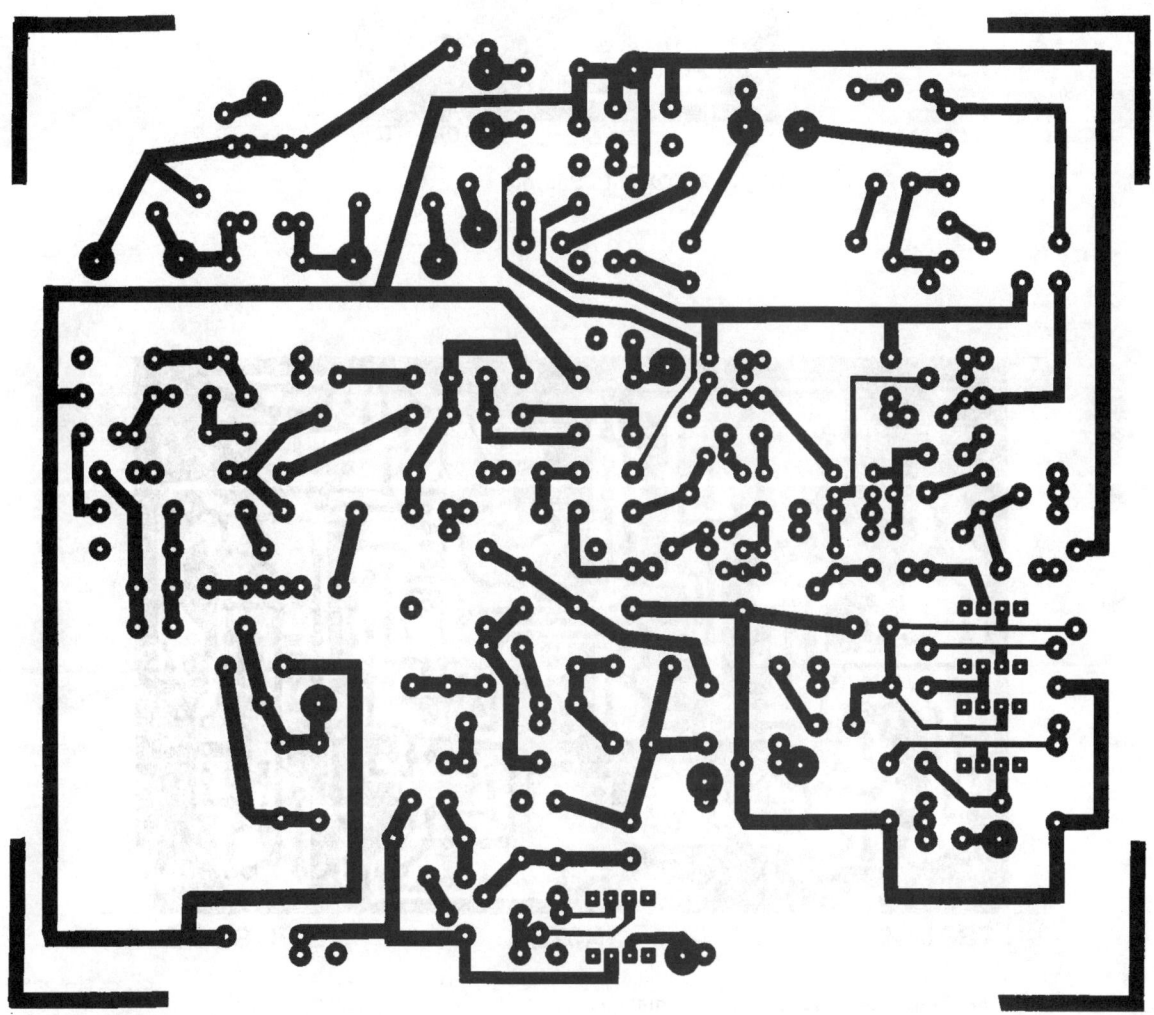

Fig 7.45: 1.8MHz QRP transceiver PCB layout

IMPORTANT NOTICE

Whilst every effort has been made to ensure that the artwork in this appendix is displayed accurately, it is the responsibilty of the reader to check that he/she is using the correct drawing, that it is accurate and the correct size (some may need re-scaling on a photocopier) and whether the artwork is 'mirror image'.It is important to read the associated text in the relevant chapter.

The RSGB cannot be held responsible for any errors or consequential losses incurred by the use of this artwork.

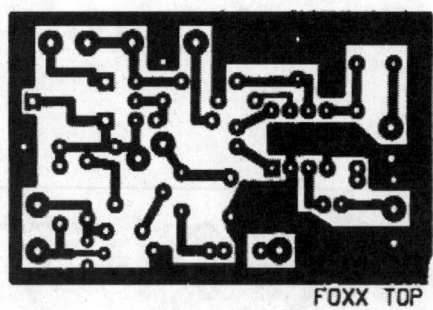

Fig 7.52: FOXX2 PCB layout

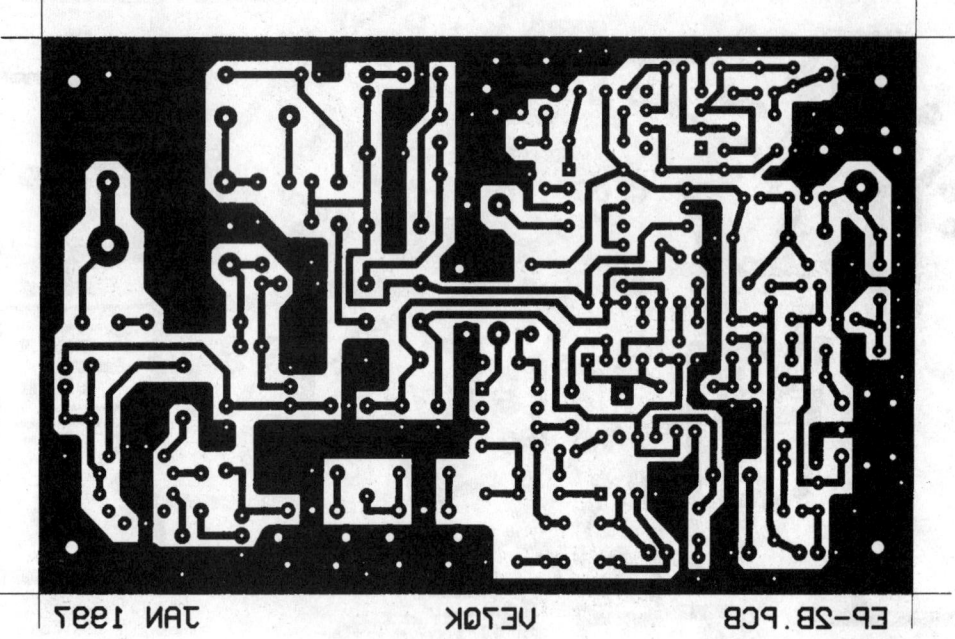

Fig 7.58: Epiphyte-2 PCB layout. This image is reversed

Before using this artwork, please read the IMPORTANT NOTICE on page B.1 of this Appendix

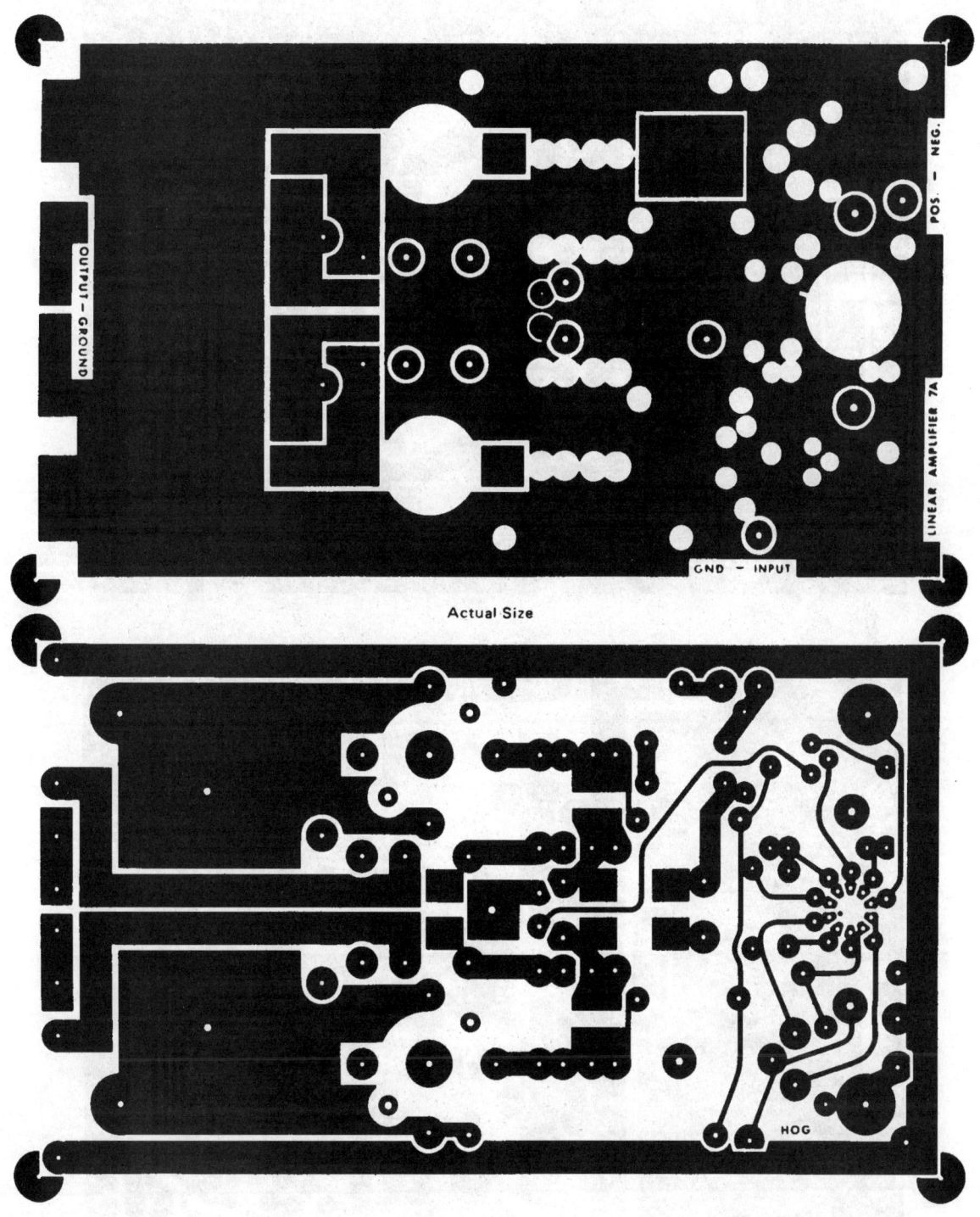

Actual Size

Fig 7.66: The PCB layouts for the 140-300W amplifiers (Motorola)

Before using this artwork, please read the IMPORTANT NOTICE on page B.1 of this Appendix

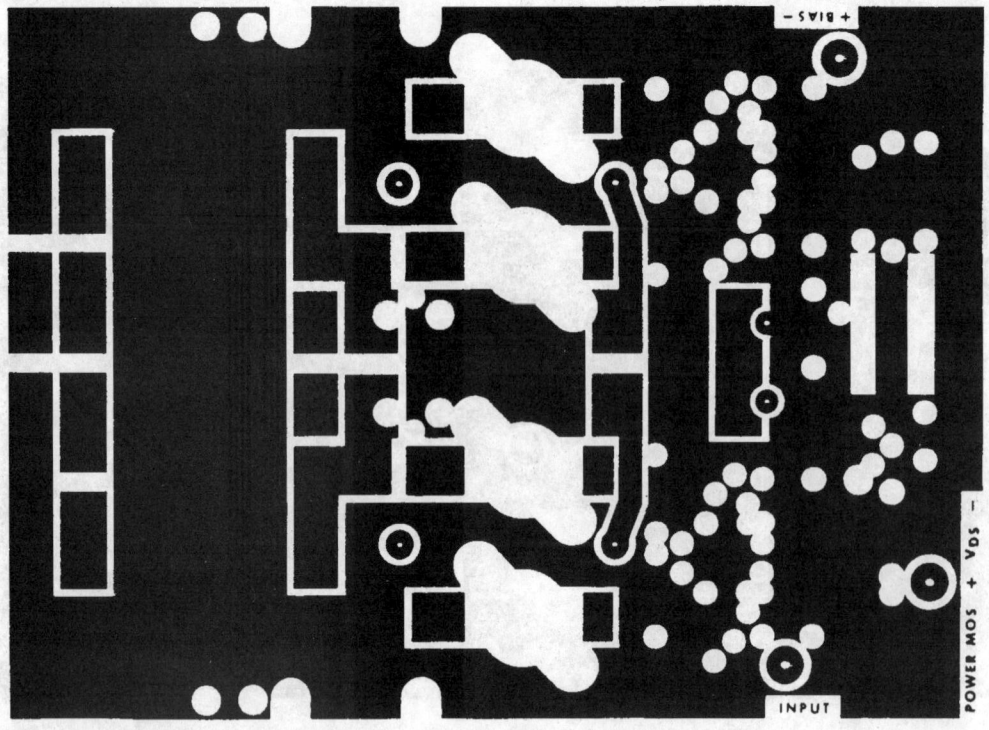

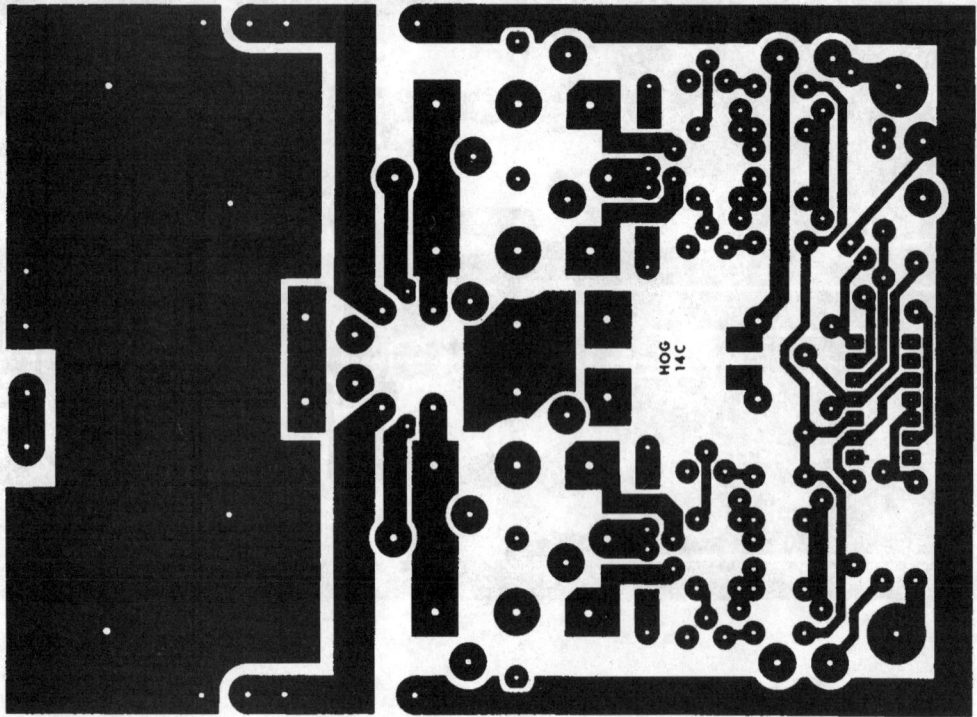

Fig 7.73: PCB layouts for the 600W amplifier (Motorola)

Before using this artwork, please read the IMPORTANT NOTICE on page B.1 of this Appendix

NOTE: Fig 8.8 is shown overleaf >>>>>>>>>>>>>>

(a) Mother board 90 x 139.7mm

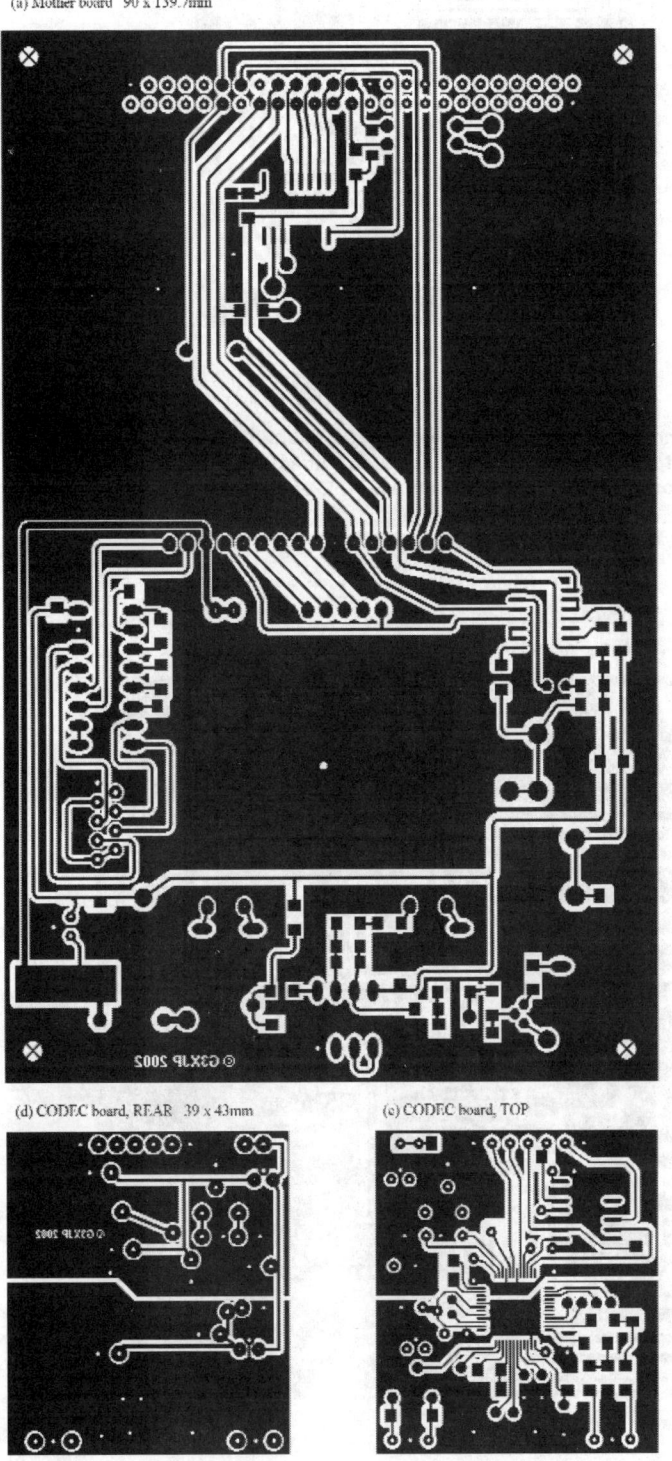

(b) Processor board, TOP 88.9 x 66.4mm

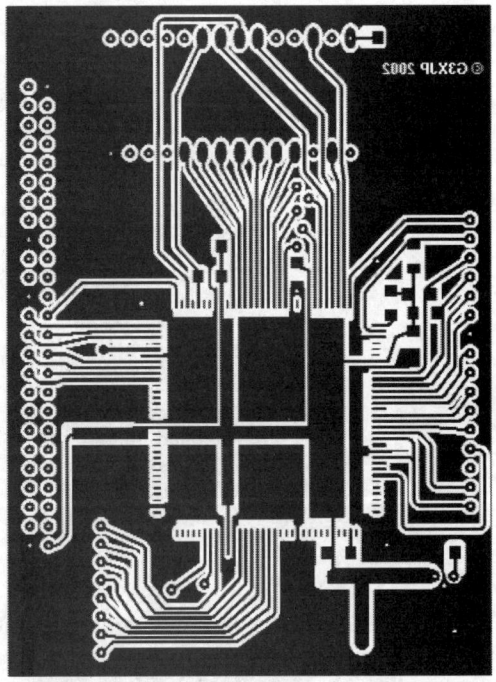

(c) Processor board, REAR

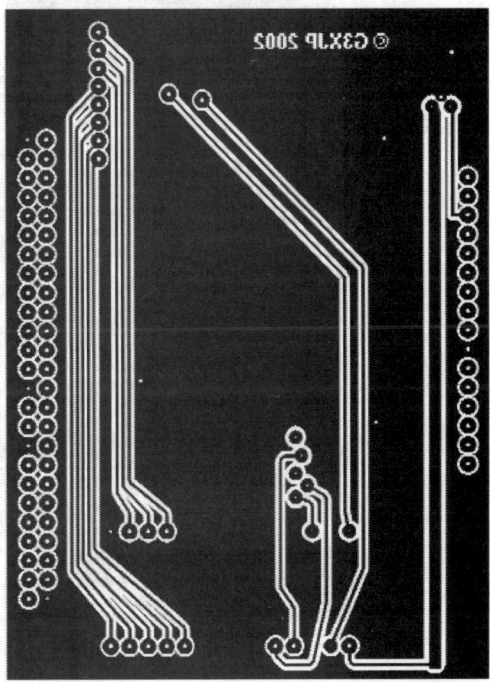

(d) CODEC board, REAR 39 x 43mm (e) CODEC board, TOP

Fig 8.20: DSP boards PCB artwork. NB all these images are mirrored for direct copying to laser film. Should you wish to use an indelible pen to apply the artwork to the back of the daughter boards, the image needs to be flipped (or simply viewed in a mirror). All holes are 0.7mm

Before using this artwork, please read the IMPORTANT NOTICE on page B.1 of this Appendix

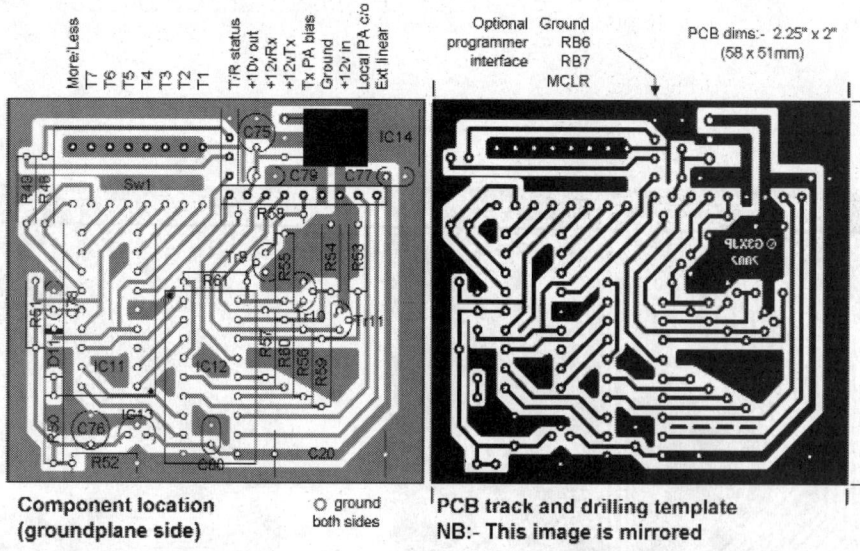

Fig 8.8: Timer board PCB - with the component side unetched. Countersink the ungrounded holes on the component side. The input/output connector (if any) is not specified, but is 0.1in pitch. The author used SIL plug/socket strip for this and all arbitrary connectors

PCB tracking and hole drilling template. NB:- This image is mirrored

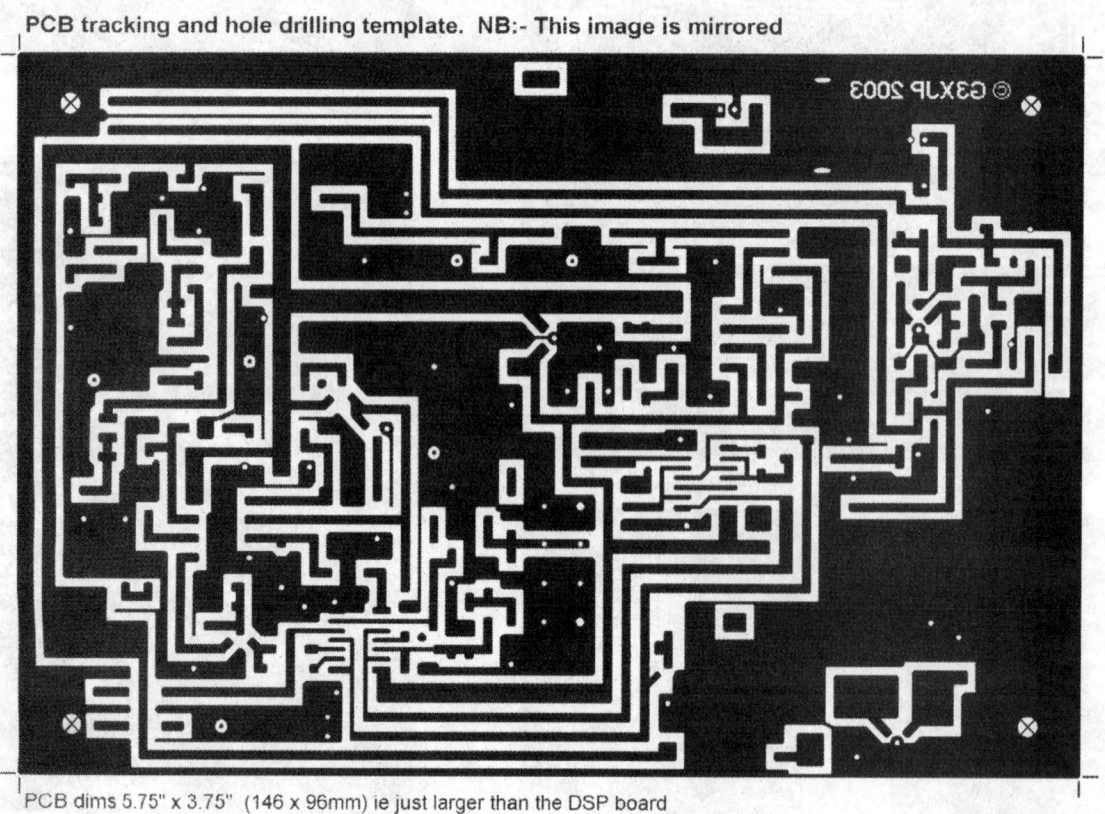

PCB dims 5.75" x 3.75" (146 x 96mm) ie just larger than the DSP board

Fig 8.24(b): The IF board

Before using this artwork, please read the IMPORTANT NOTICE on page B.1 of this Appendix

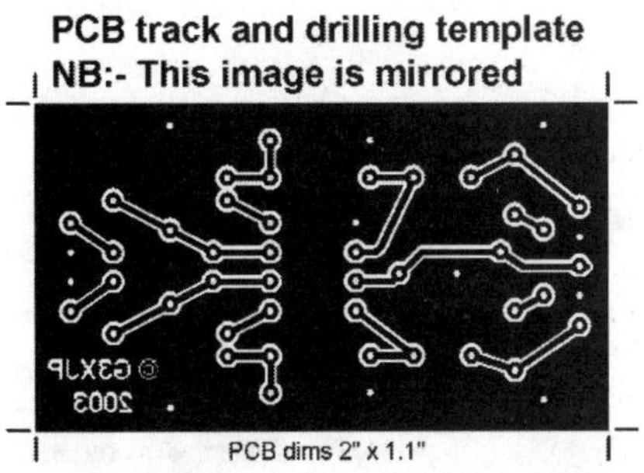

PCB track and drilling template
NB:- This image is mirrored

PCB dims 2" x 1.1"

Fig 8.35: Stereo amplifier PCB track and drilling template. NB: This image is reversed

PCB track and drilling template
NB:- This image is mirrored

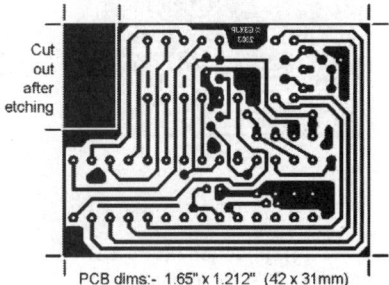

Cut out after etching

PCB dims:- 1.65" x 1.212" (42 x 31mm)

Fig 8.37: PicAdapter board PCB track and drilling template. NB: This image is reversed

PCB track and drilling template
NB:- This image is mirrored

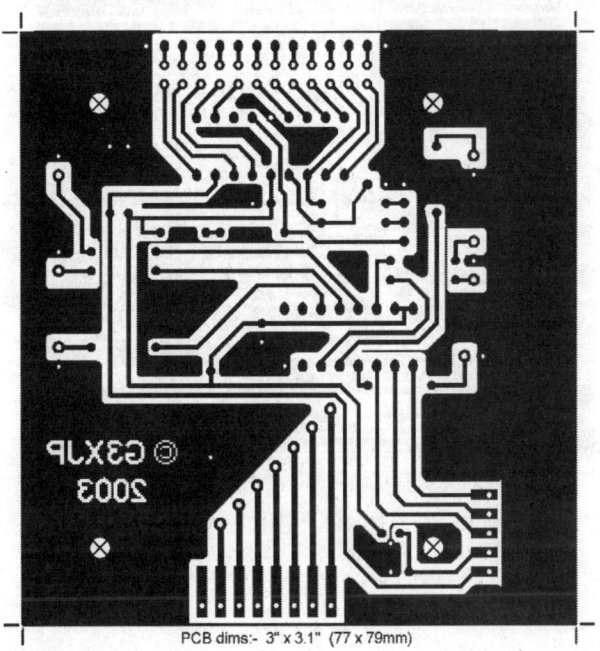

PCB dims:- 3" x 3.1" (77 x 79mm)

Fig 8.39: Status Board PCB track and drilling template. NB: This image is reversed

Before using this artwork, please read the IMPORTANT NOTICE on page B.1 of this Appendix

PCB track and drilling template

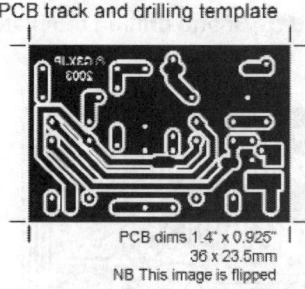

PCB dims 1.4" x 0.925"
36 x 23.5mm
NB This image is flipped

**Fig 8.41(b): Injection filter PCB
track and drilling template**

PCB tracking

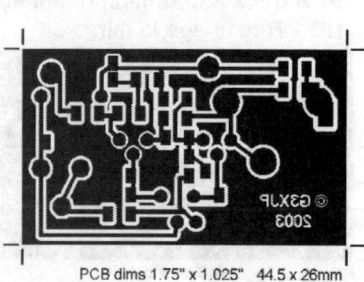

PCB dims 1.75" x 1.025"   44.5 x 26mm
May be made from 1- or 2-sided material
NB This image is flipped

Fig 8.43(b): Reference oscillator PCB

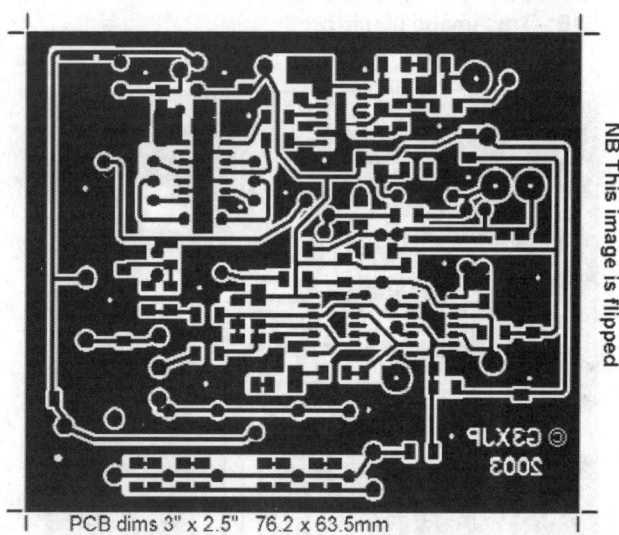

PCB dims 3" x 2.5"   76.2 x 63.5mm

Fig 8.55: 'Magic Roundabout' PCB artwork

Before using this artwork, please read the IMPORTANT NOTICE on page B.1 of this Appendix

Fig 8.61: Band-pass filters. PCB layout for nine filter blocks

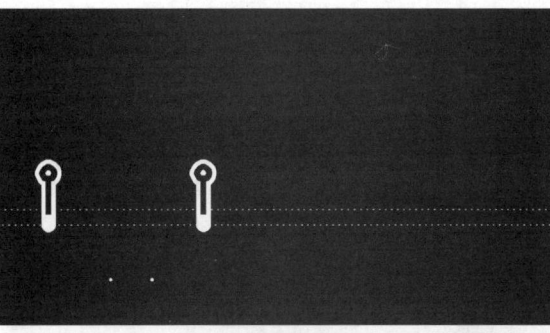

<u>PCB dims</u>:-
Filter board 2.90" x 5.60"
Side plates 1.65" x 5.60"
End plates 3.00" x 1.65"
<u>Complete assembly</u>:-
3" x 1.65" x 5.757"
assuming 2mm thick PCB

NB This image is mirrored

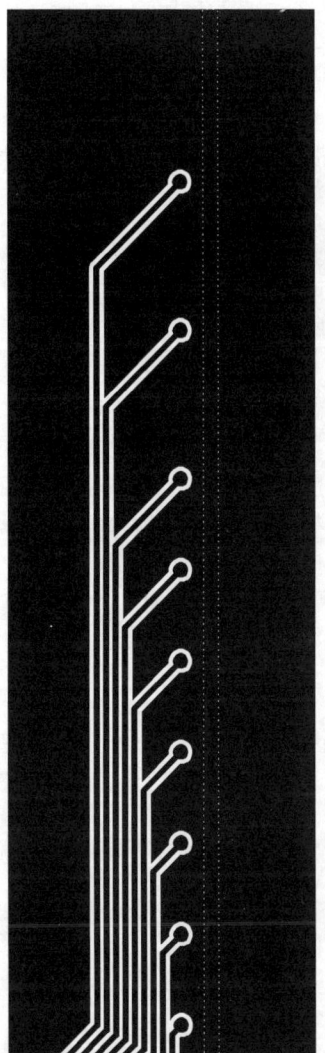

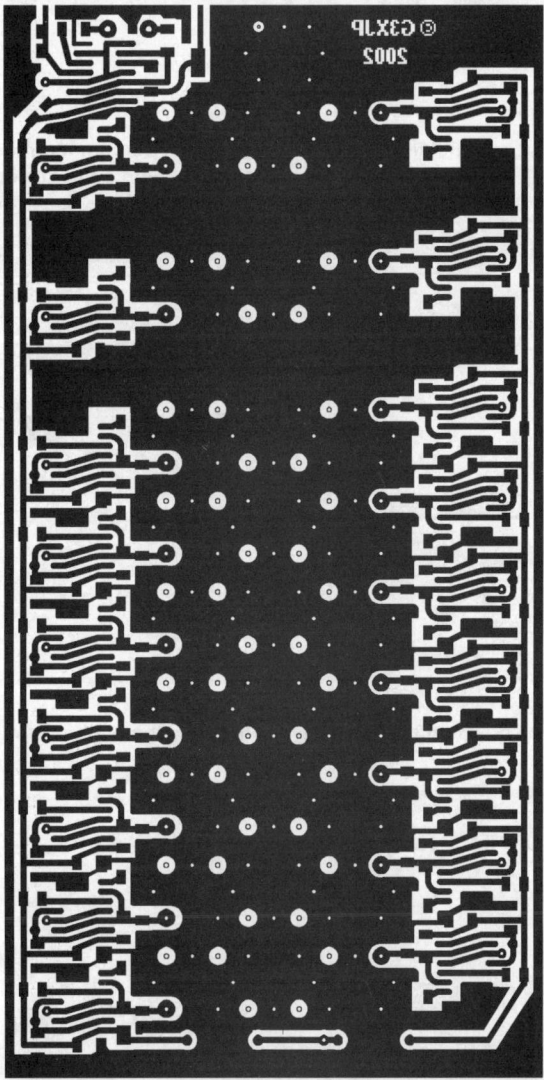

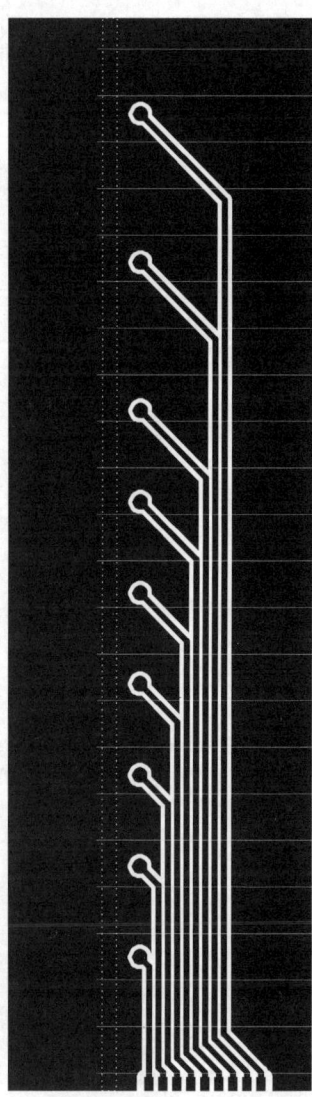

Drill these holes for tight fit on 1206 resistors

Drill several holes (not shown) through each side- and end-plate - and intimately connect the inner and outer grounded faces.

Before using this artwork, please read the IMPORTANT NOTICE on page B.1 of this Appendix

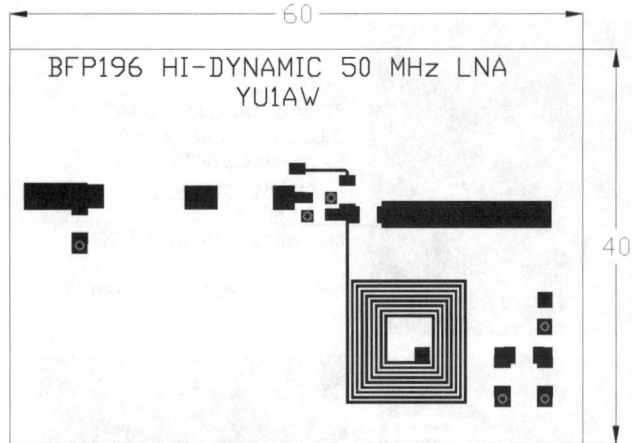

Fig 9.36: PCB layout for the 6m low noise preamplifier. The illustration should be re-scaled as shown

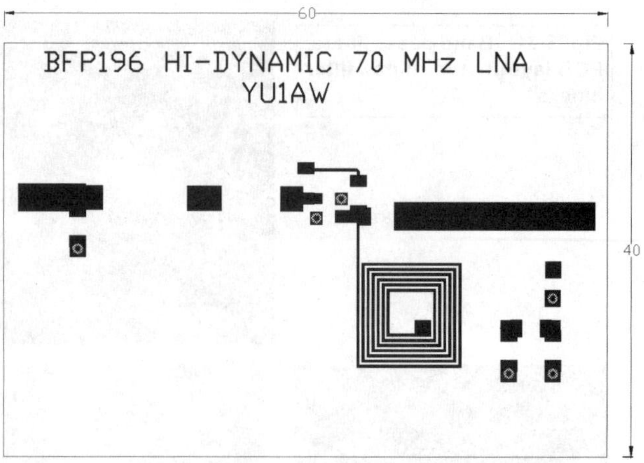

Fig 9.37: PCB layout for the 4m low noise preamplifier. The illustration should be re-scaled as shown

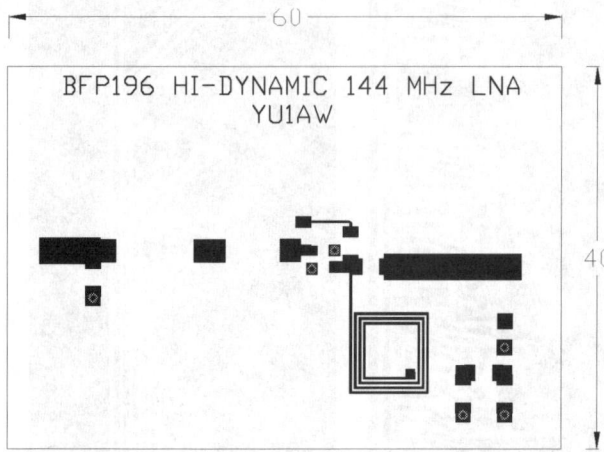

Fig 9.38: PCB layout for the 2m low noise preamplifier. The illustration should be re-scaled as shown

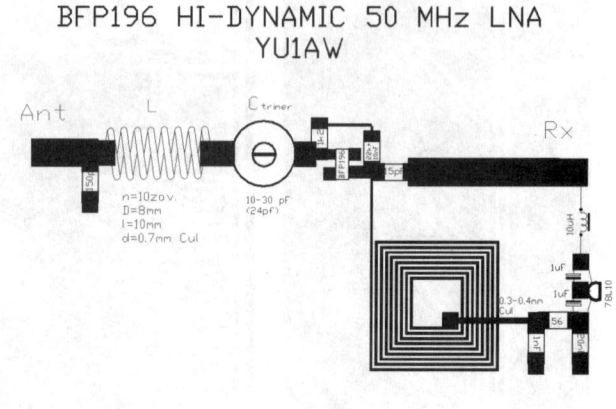

Fig 9.46: Component layout for the 6m low noise preamplifier

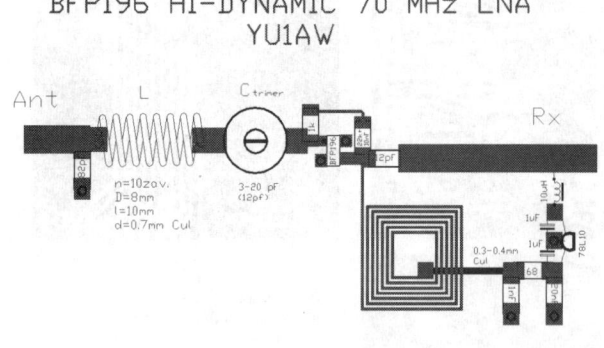

Fig 9.47: Component layout for the 4m low noise preamplifier

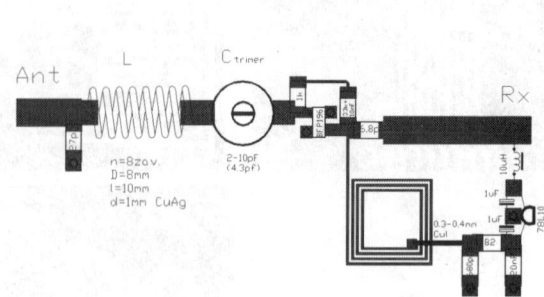

Fig 9.48: Component layout for the 2m low noise preamplifier

Before using this artwork, please read the IMPORTANT NOTICE on page B.1 of this Appendix

The Radio Communication Handbook

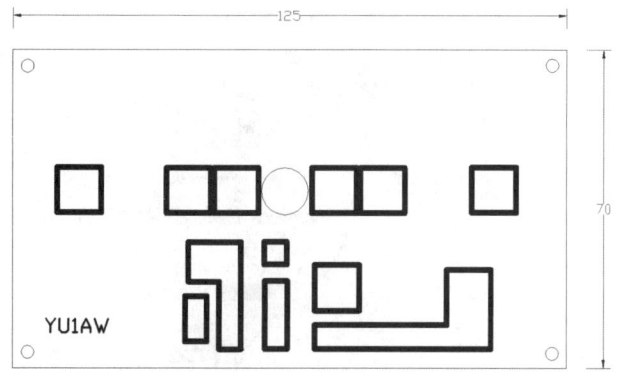

Fig 9. 72: PCB layout for the one transistor power amplifier. The illustration should be re-scaled as shown

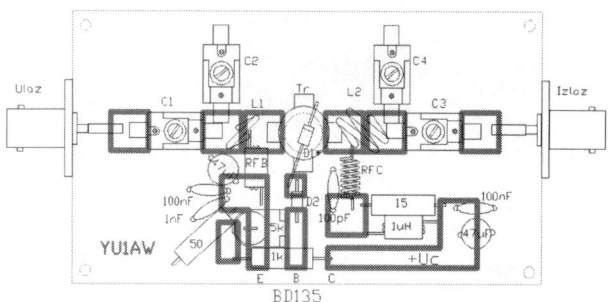

Fig 9.73: Component layout for the one transistor power amplifier

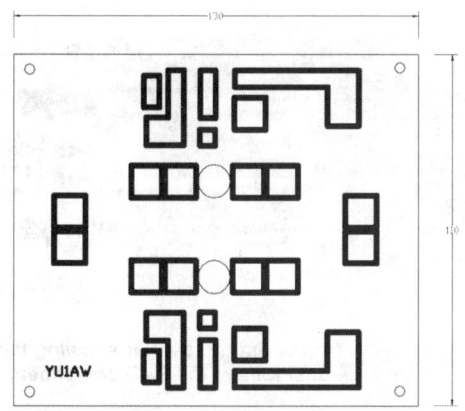

Fig 9.76: PCB layout for the two transistor power amplifier. The illustration should be re-scaled as shown

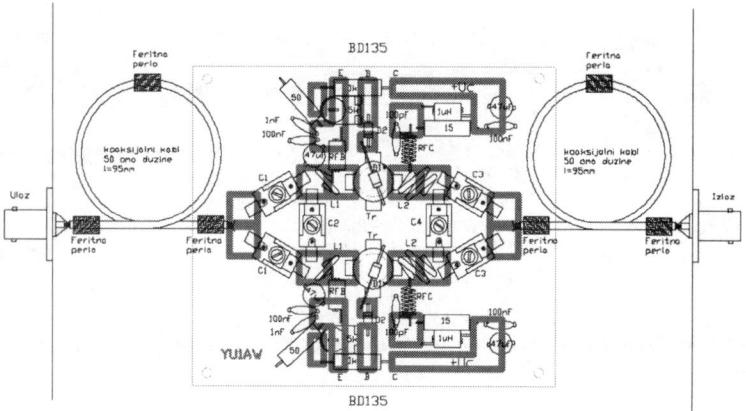

Fig 9.77: Component layout for the two transistor power amplifier

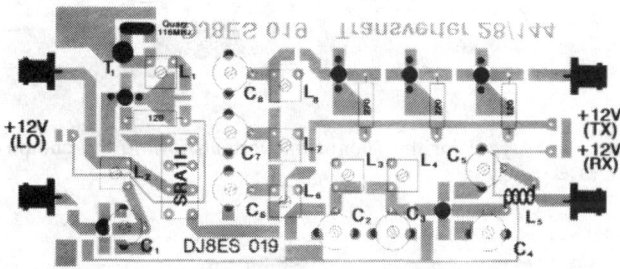

Fig 9.88: Component layout for the 2m transverter showing the component side of the PCB. The board size is 54 x 108mm

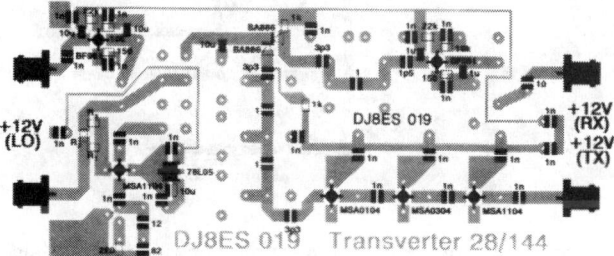

Fig 9.89: Component layout for the 2m transverter showing the track side of the PCB and the positions of the SMD components

Before using this artwork, please read the IMPORTANT NOTICE on page B.1 of this Appendix

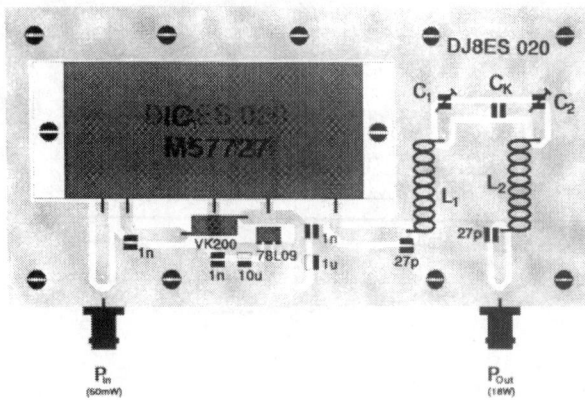

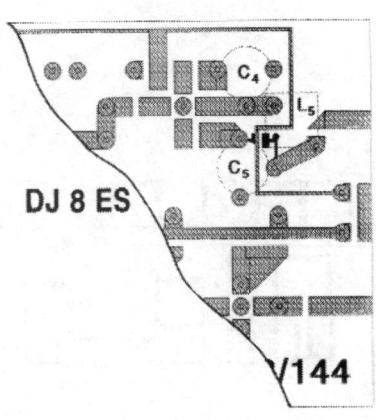

Fig 9.95: Component layout of the 2m power amplifier to be used with the 2m transverter. The board size is 54 x 108mm

Fig 9.97: Details of modifications to the 2m transverter PCB for use as 6m transverter

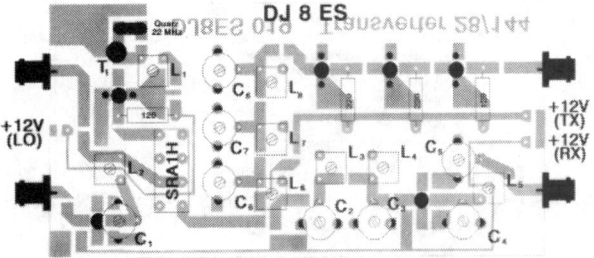

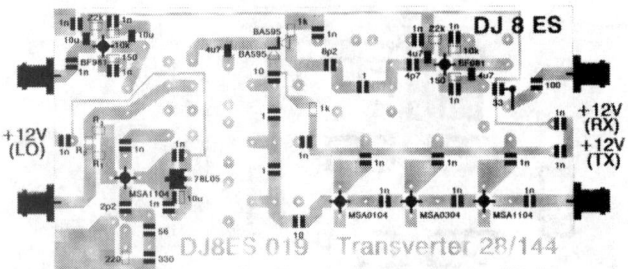

Fig 9.98: Component layout for the 6m transverter showing the component side of the PCB

Fig 9.99: Component layout for the 6m transverter showing the track side of the PCB and the positions of the SMD components

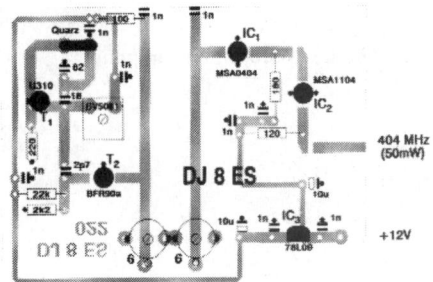

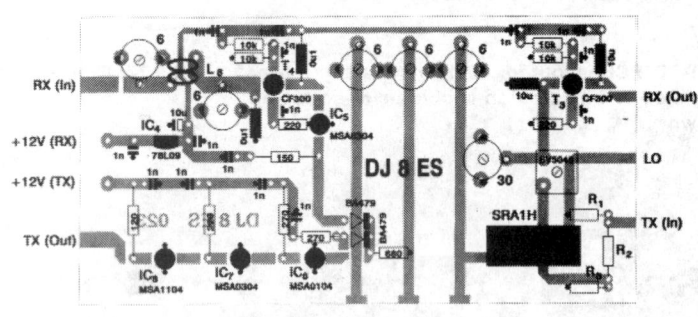

Fig 9.105: Component layout of the local oscillator used on the 70cm transverter showing the component side of the PCB.

Fig 9.107: Component layout for the 70cm transverter showing the component side of the PCB

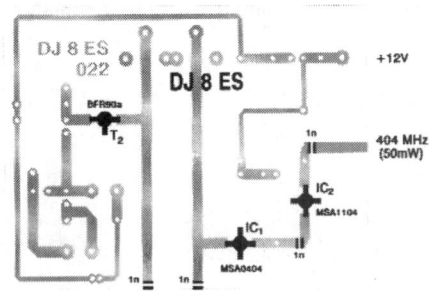

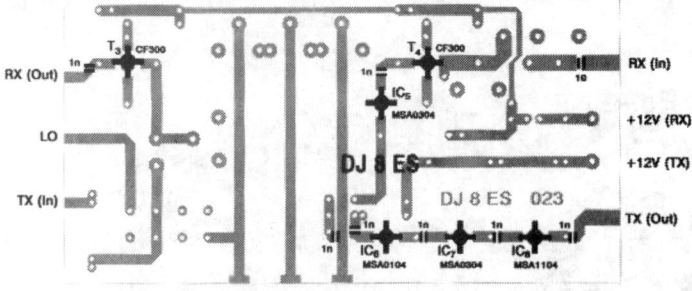

Fig 9.106: Component layout of the LO on the 70cm transverter showing the track side of the PCB and the positions of the SMD components

Fig 9.108: Component layout for the 70cm transverter showing the track side of the PCB and the positions of the SMD components

Before using this artwork, please read the IMPORTANT NOTICE on page B.1 of this Appendix

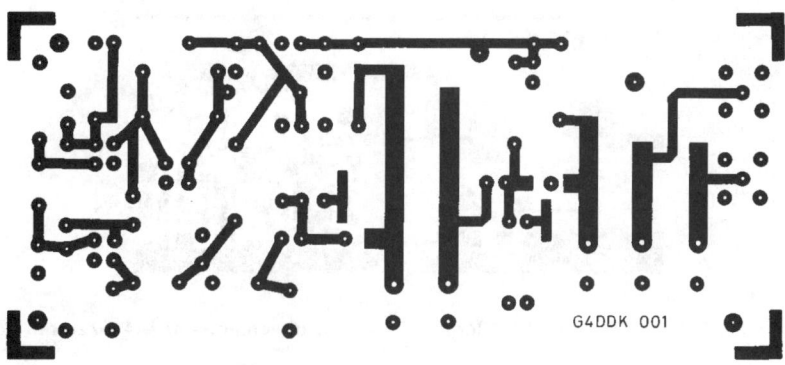

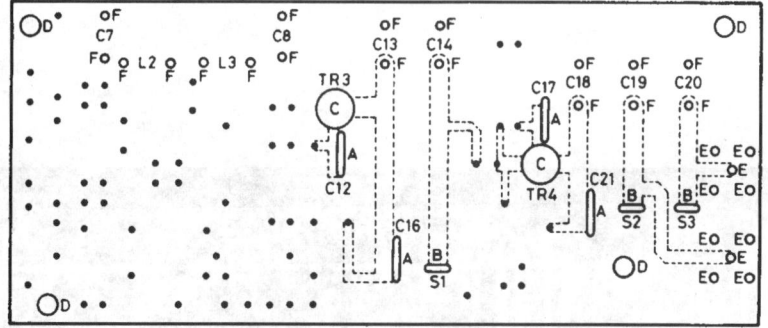

Slots 'A'.... 6·4mm long x 1·2mm wide   Slots 'B'.... 3mm long x 0·8mm wide
Holes 'C'.... 5mm dia   Holes 'D'.... 2·5mm dia   Holes 'E'.... 1·2mm dia   Holes 'F'.... 1mm dia
Holes marked ● are 0·8mm dia although 1mm dia is permissible if more convenient

Fig 11.17: Printed circuit board artwork and drilling pattern for the microwave source G4DDK-001

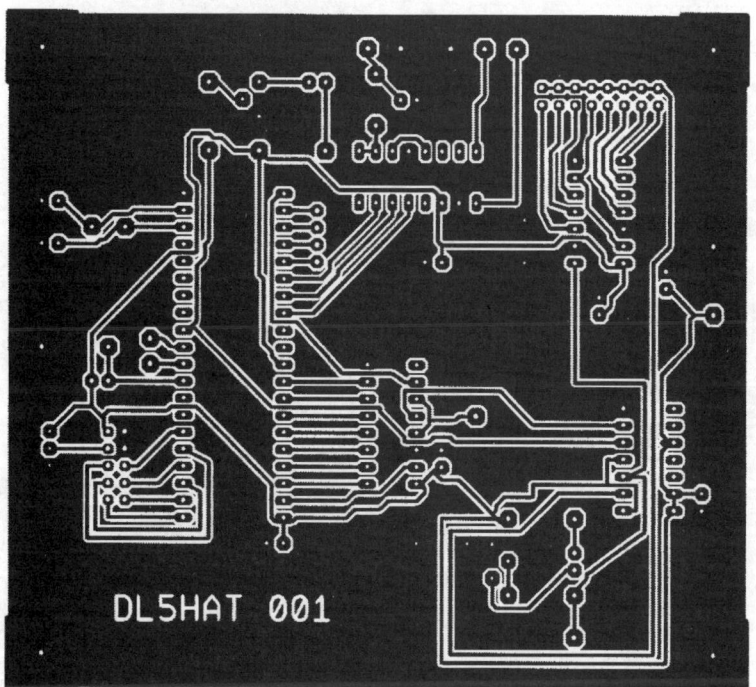

Fig 11.24: PCB layout (DL5HAT-001) for GPs control stage of the high precision frequency standard for 10MHz. The finished PCB should be 100mm x 100m

Before using this artwork, please read the IMPORTANT NOTICE on page B.1 of this Appendix

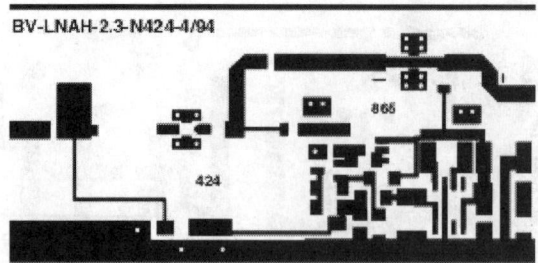

BV-LNAH-2.3-N424-4/94

Fig 11.30: PCB layout for 13cm PHEMT. PCB dimensions are 34 x 72mm

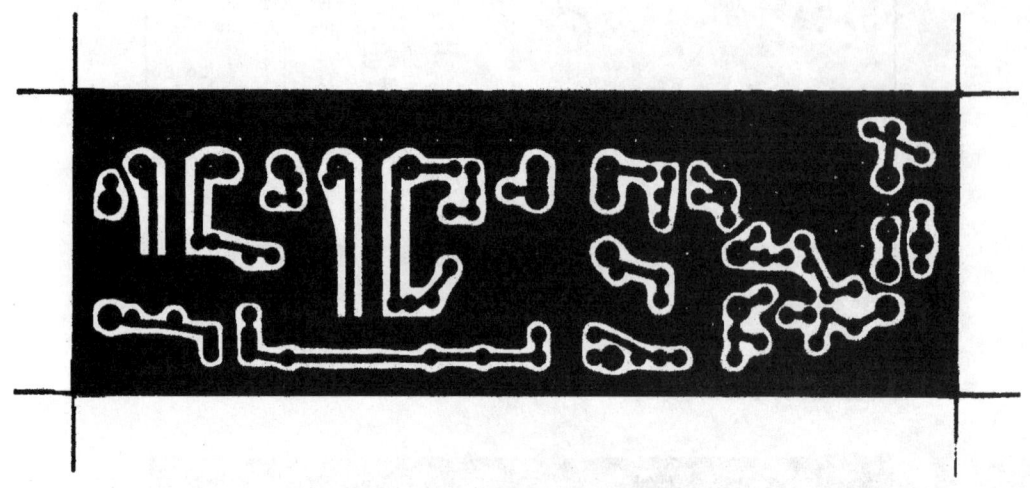

Fig 11.64: VCXO PCB for Zero-IF transceiver

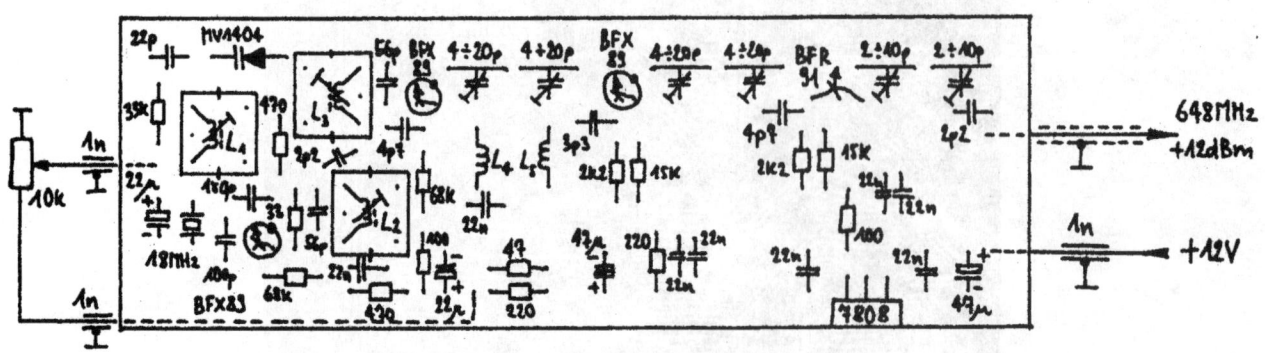

Fig 11.65: VCXO component layout for Zero-IF transceiver

Before using this artwork, please read the IMPORTANT NOTICE on page B.1 of this Appendix

Fig 11.67: PCB for x4 multiplier to 2880MHz for Zero-IF transceiver

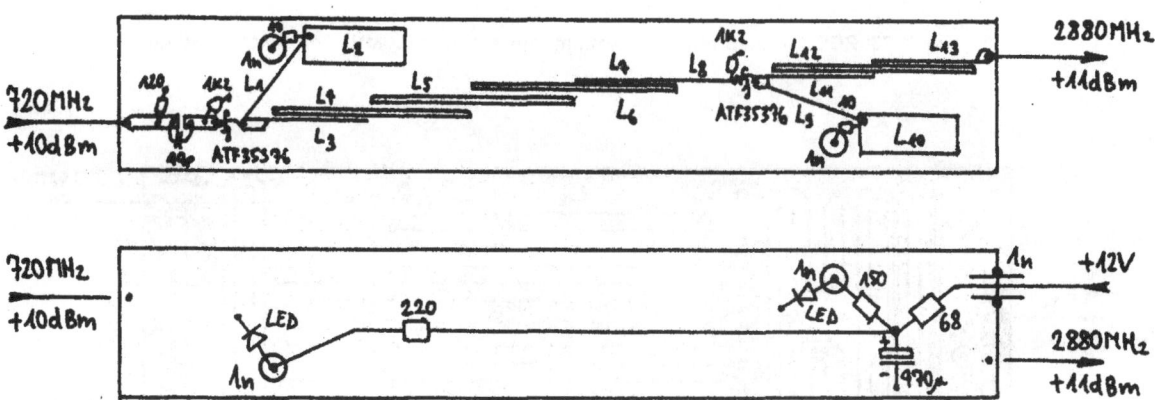

Fig 11.68: Component layout for x4 multiplier to 2880MHz for Zero-IF transceiver

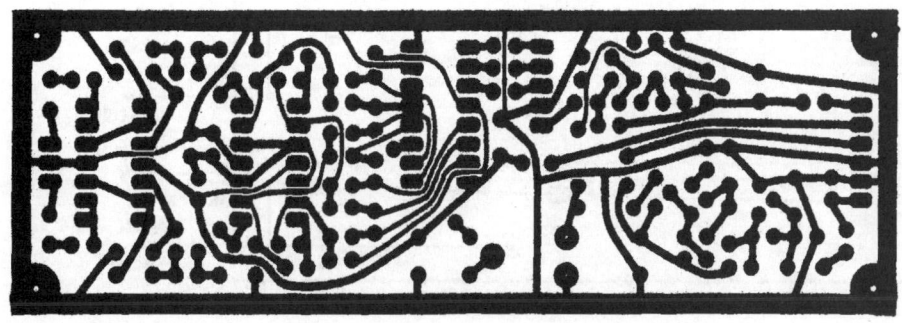

Fig 11.70: PCB for SSB/CW quadrature modulator for Zero-IF transceiver

Fig 11.71: Component layout for SSB/CW quadrature modulator for Zero-IF transceiver

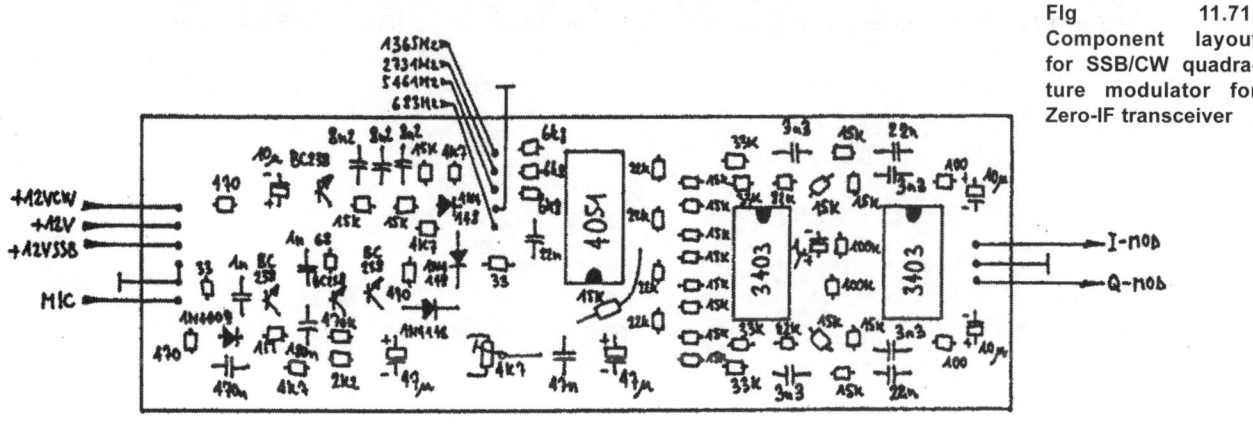

Before using this artwork, please read the IMPORTANT NOTICE on page B.1 of this Appendix

Fig 11.73: PCB for quadrature transmit modulator for 1296MHz Zero-IF transceiver

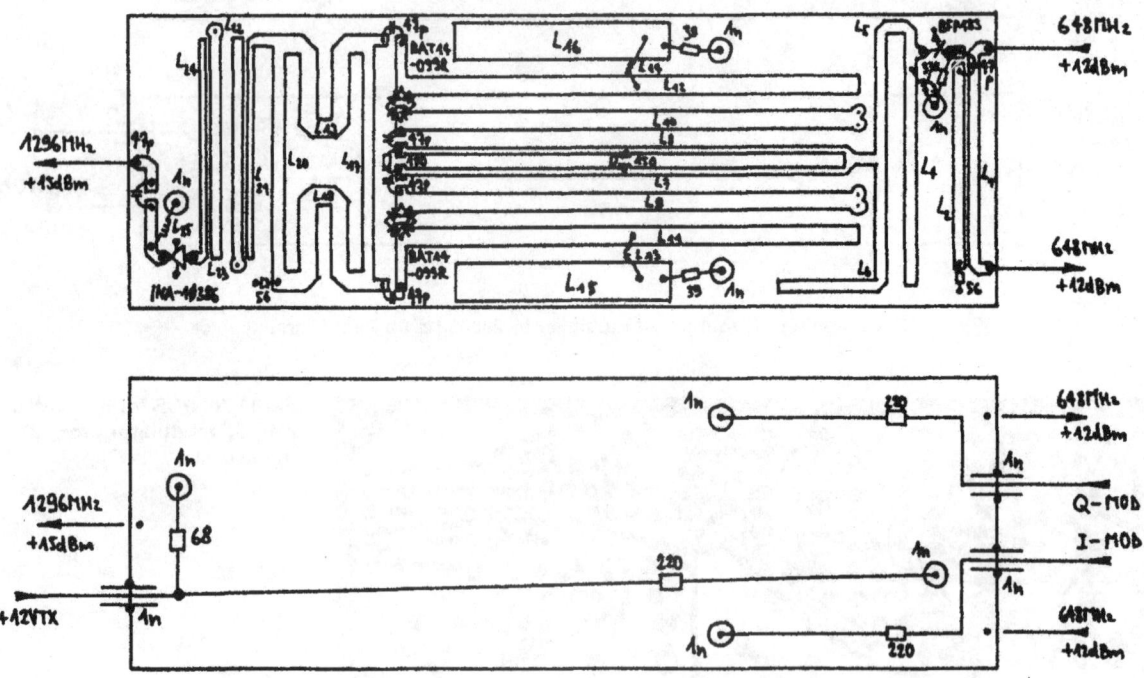

Fig 11.74: Component layout for quadrature transmit modulator for 1296MHz Zero-IF transceiver

Before using this artwork, please read the IMPORTANT NOTICE on page B.1 of this Appendix

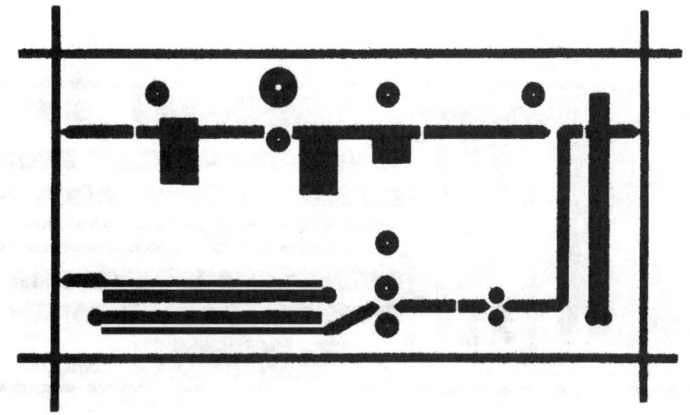

Fig 11.76: PCB for RF front-end for 1296MHz Zero-IF transceiver

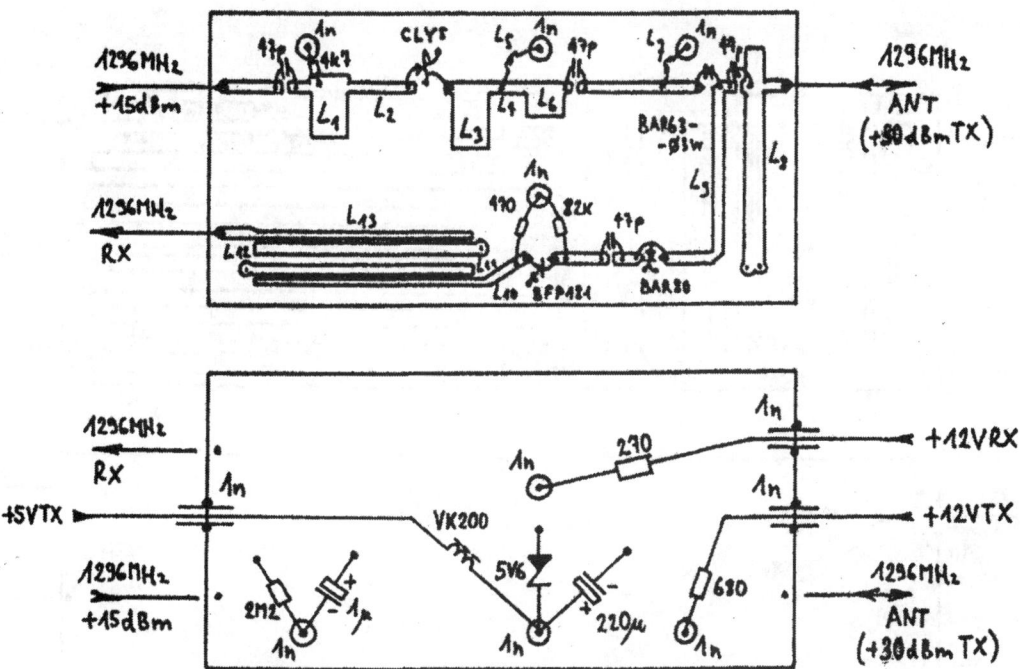

Fig 11.77: Component layout for RF front-end for 1296MHz Zero-IF transceiver

Before using this artwork, please read the IMPORTANT NOTICE on page B.1 of this Appendix

Fig 11.79: PCB for quadrature receive mixer for 1296MHz Zero-IF transceiver

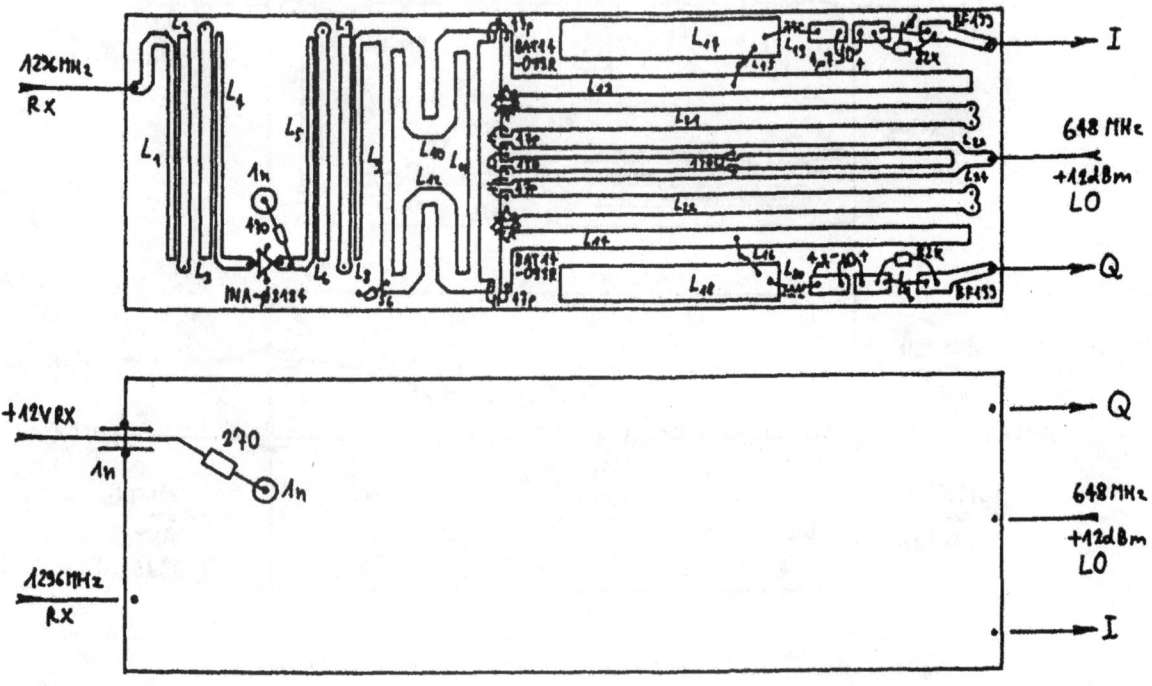

Fig 11.80: Component layout for quadrature receive mixer for 1296MHz Zero-IF transceiver

Before using this artwork, please read the IMPORTANT NOTICE on page B.1 of this Appendix

Fig 11.82: PCB for quadrature receive SSB IF amplifier for 1296MHz Zero-IF transceiver

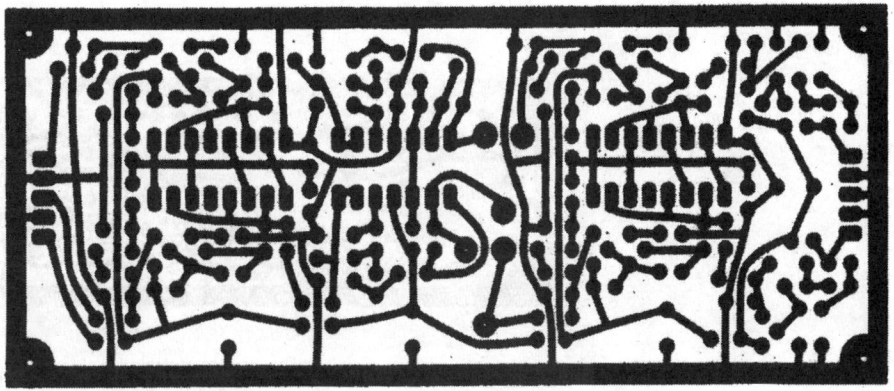

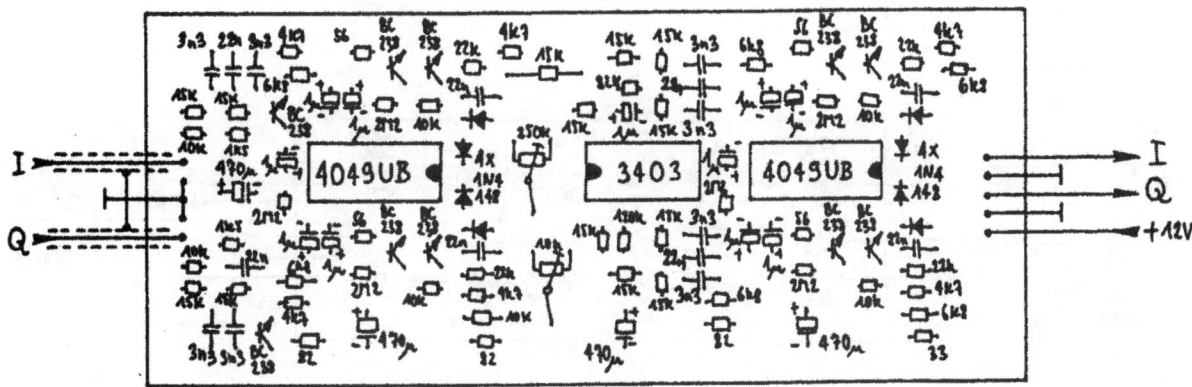

Fig 11.83: Component layout for quadrature receive SSB IF amplifier for 1296MHz Zero-IF transceiver

Fig 11.85: PCB for quadrature SSB demodulator and AF amplifier for 1296MHz Zero-IF transceiver

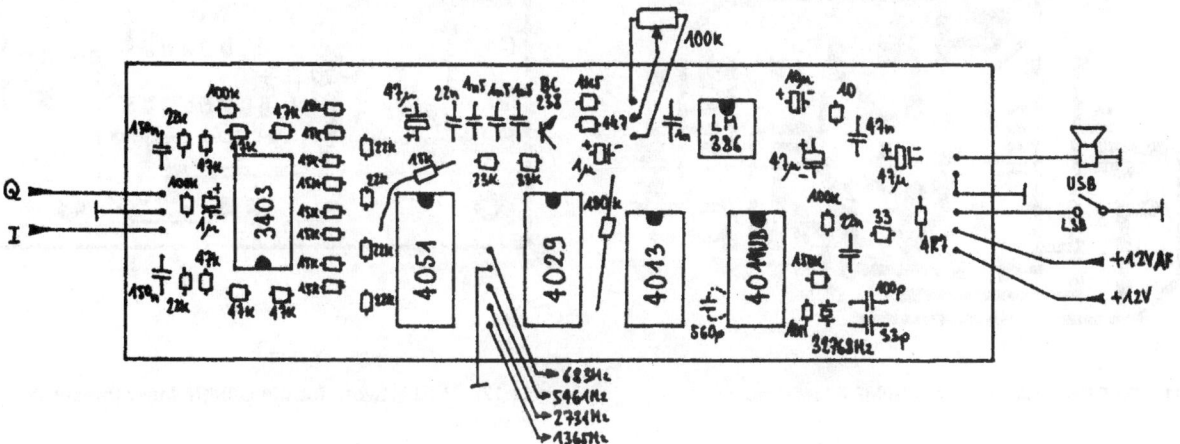

Fig 11.86: Component layout for quadrature SSB demodulator and AF amplifier for 1296MHz Zero-IF transceiver

Before using this artwork, please read the IMPORTANT NOTICE on page B.1 of this Appendix

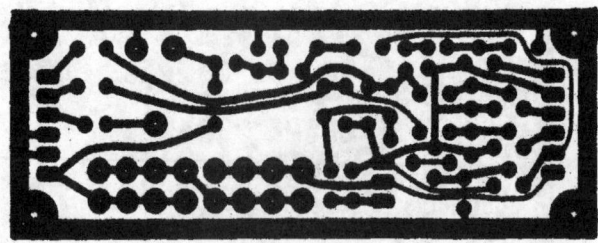

Fig 11.88: PCB for SSB/CW transmit/receive switching for Zero-IF transceiver

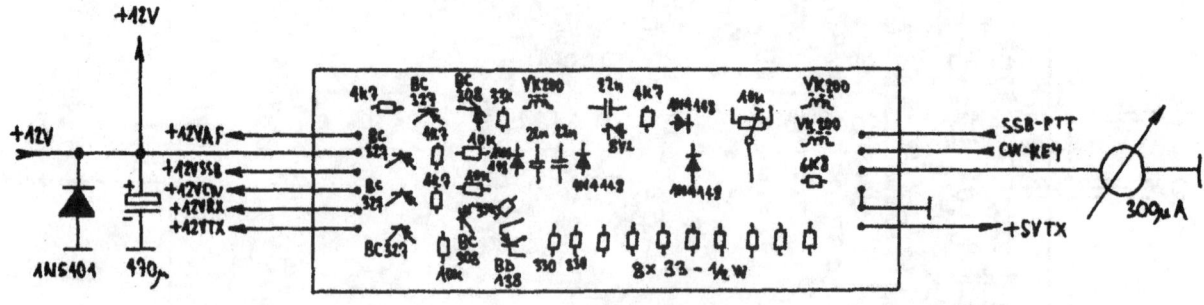

Fig 11.89: Component layout for SSB/CW transmit/receive switching for Zero-IF transceiver

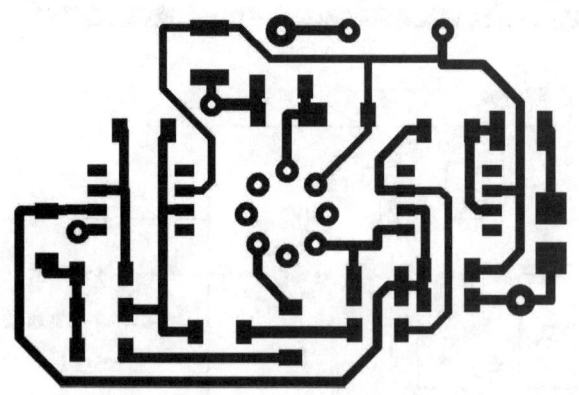

Fig 11.123: PCB artwork for the G0MRF laser receiver

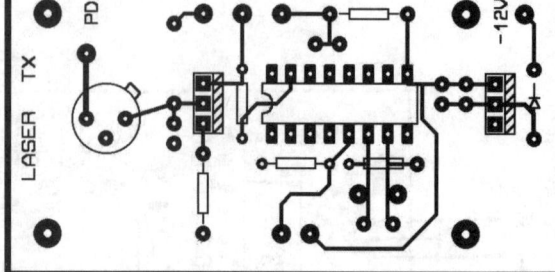

Fig 11.128: PCB artwork for the G0MRF laser transmitter

Before using this artwork, please read the IMPORTANT NOTICE on page B.1 of this Appendix

Fig 15.30: PCB and layout for the SWR bridge

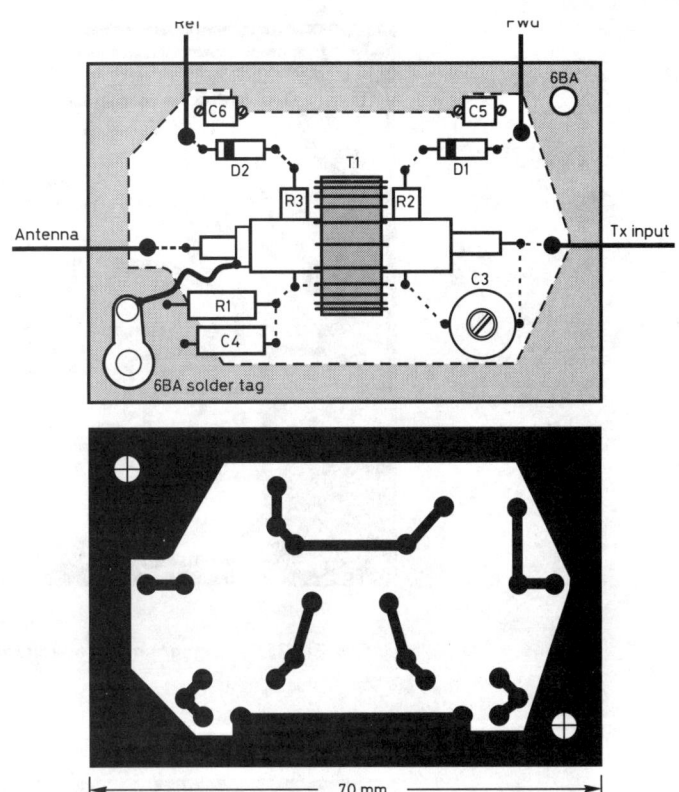

(left) Fig 20.5: PCB for video amplifier

(right) Fig 20.6: Component layout for video amplifier

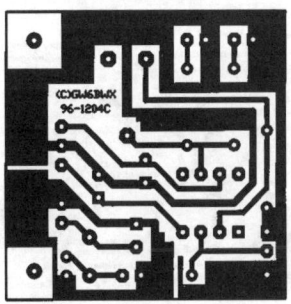

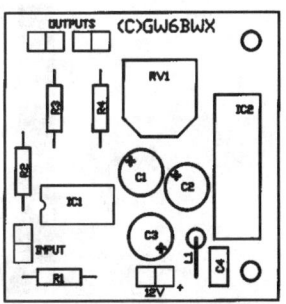

Before using this artwork, please read the IMPORTANT NOTICE on page B.1 of this Appendix

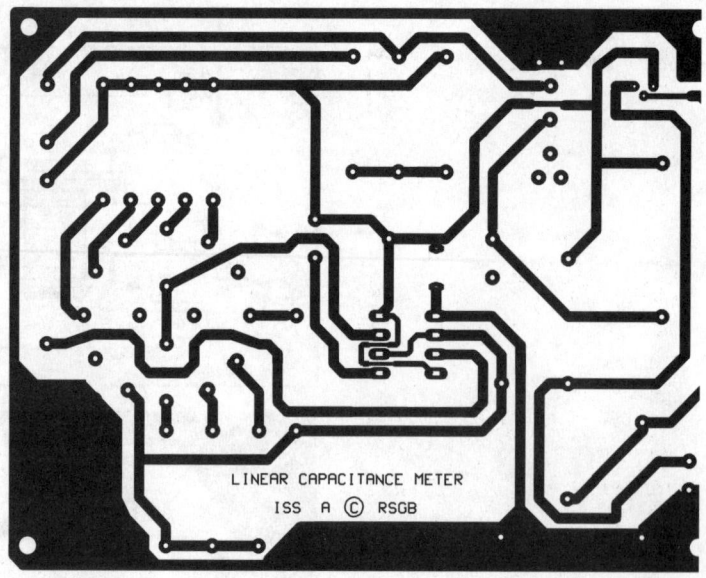

Fig 25.27: Linear scale capacitance meter PCB layout

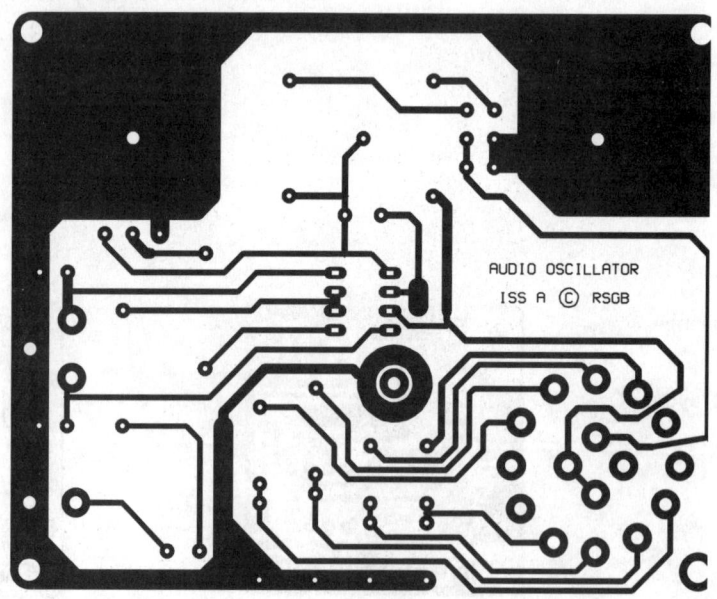

Fig 25.41: Low frequency oscillator PCB layout

Before using this artwork, please read the IMPORTANT NOTICE on page B.1 of this Appendix

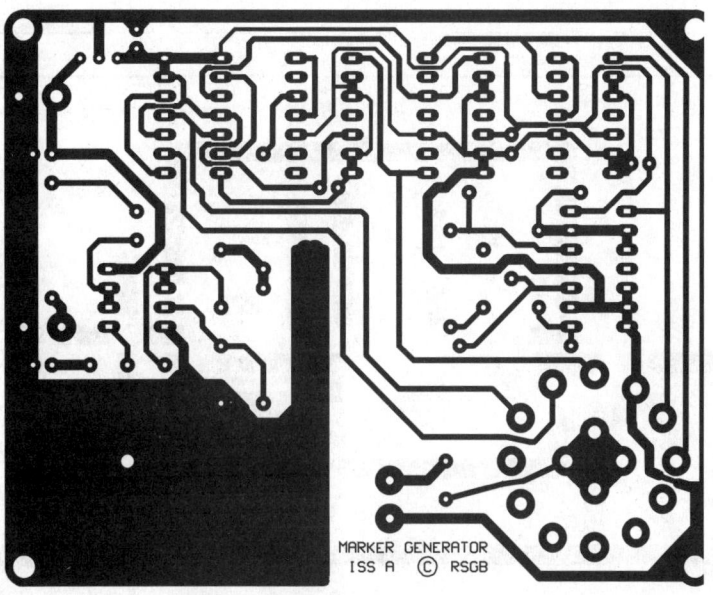

Fig 25.24: Frequency marker PCB layout

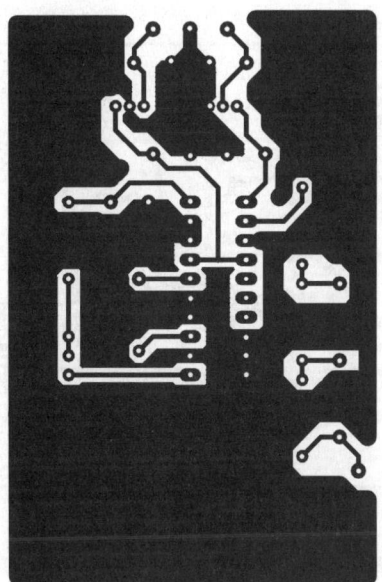

Fig 25.47: HF signal source PCB layout

Before using this artwork, please read the IMPORTANT NOTICE on page B.1 of this Appendix

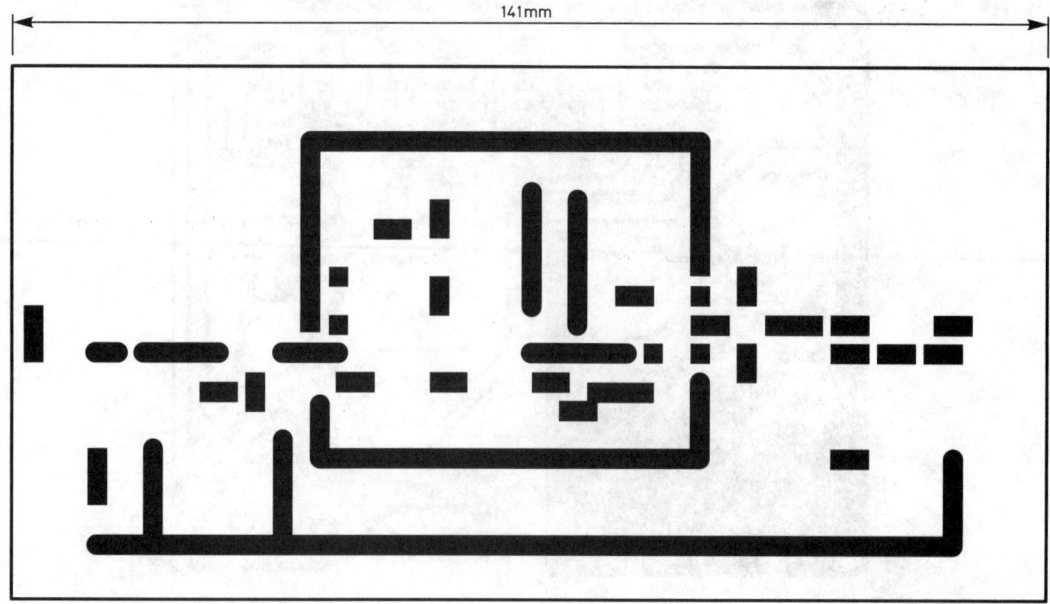

141mm

Fig 25.51: PCB layout for the combined 2m and 70cm signal source

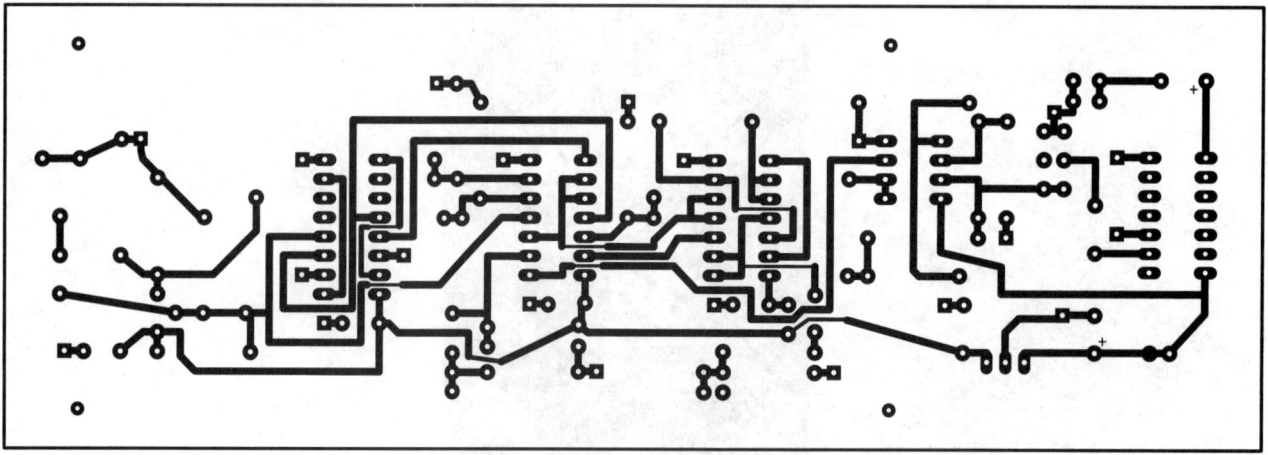

Fig 25.60: PCB layout for the receiver calibrator and transmitter monitor

Before using this artwork, please read the IMPORTANT NOTICE on page B.1 of this Appendix

Index

INDEX

INDEX